# Gear Cutting Tool



This new edition of *Gear Cutting Tools* has been updated with revised chapters and illustrations as well as additional, new material with the aim to provide a systematic and comprehensive discussion on modern designs, kinematics, and cutting geometry of gear-cutting tools.

This book presents the *DG/K*-based method of surface generation—a practical mathematical method for designing gear-cutting tools with optimal parameters. The text addresses the evolution of gear-cutting tools and scientific classification for all types of gear machining meshes before discussing optimal cutting tool designs. Designs currently used and those being planned are covered, and the approach allows for the development of scientific predictions and optimal designs. Solutions appear in analytical form and/or graphical form, with a wealth of new figures added, and new appendices offer additional data for readers.

This is an essential reading for engineers who work in the field of gear-cutting, especially those involved in the manufacturing of autos, aircraft, agriculture machines, and metal-cutting machines. This book would also be relevant to various mechanical engineering courses for graduate students, such as machine elements, machine and mechanism science, and theory of machines and mechanisms.

# Gear Cutting Tools

## Science and Engineering

Third Edition

Stephen P. Radzevich

CRC Press
Taylor & Francis Group
Boca Raton London New York

CRC Press is an imprint of the
Taylor & Francis Group, an **informa** business

Designed cover image: Stephen P. Radzevich

Third edition published 2025
by CRC Press
2385 NW Executive Center Drive, Suite 320, Boca Raton FL 33431

and by CRC Press
4 Park Square, Milton Park, Abingdon, Oxon, OX14 4RN

*CRC Press is an imprint of Taylor & Francis Group, LLC*



[First edition published by CRC Press 2010]

[Second edition published by CRC Press 2017]



**Library of Congress Cataloging-in-Publication Data**
Names: Radzevich, S. P. (Stepan Pavlovich), author.
Title: Gear cutting tools : science and engineering / Stephen P. Radzevich.
Description: Third edition. | Boca Raton : CRC Press, 2025. | Includes bibliographical references and index. | Summary: "This new edition of Gear Cutting Tools has been updated with revised chapters and illustrations as well as additional, new material with the aim to provide a systematic and comprehensive discussion on modern designs, kinematics, and cutting geometry of gear cutting tools. This book presents the DG/K based method of surface generation, a practical mathematical method for designing gear cutting tools with optimal parameters. The text addresses the evolution of gear cutting tool and scientific classification for all types of gear machining meshes before discussing optimal cutting tool designs. Designs currently used and those being planned are covered, and the approach allows for development of scientific predictions and optimal designs. Solutions appear in analytical form and or graphical form, with a wealth of new figures added, and new appendices offer additional data for readers. This is an essential reading for engineers who work in the field of gear cutting, especially those involved in the manufacturing of autos, aircraft, agriculture machines, metal-cutting machines, etc. This book would also be relevant to various mechanical engineering courses for graduate students, such as machine elements, machine and mechanism science, and theory of machines and mechanisms"-- Provided by publisher.
Identifiers: LCCN 2024026656 (print) | LCCN 2024026657 (ebook) | ISBN 9781032856438 (hardback) | ISBN 9781032859262 (paperback) | ISBN 9781003520528 (ebook)
Subjects: LCSH: Gear-cutting machines. | Gearing--Design and construction. | Metal-cutting tools--Design and construction--Mathematical models. | Machinery, Kinematics of--Mathematical models.
Classification: LCC TJ187 .R34 2025 (print) | LCC TJ187 (ebook) | DDC 621.9/44--dc23/eng/20241030
LC record available at https://lccn.loc.gov/2024026656
LC ebook record available at https://lccn.loc.gov/2024026657

ISBN: 978-1-032-85643-8 (hbk)
ISBN: 978-1-032-85926-2 (pbk)
ISBN: 978-1-003-52052-8 (ebk)

DOI: 10.1201/9781003520528

Typeset in Times LT Std
by KnowledgeWorks Global Ltd.

# Contents

## *APPENDICES*

# Preface

*Gear Cutting Tools: Science and Engineering* is intended for mechanical engineers and manufacturing engineers who are interested in the scientific basis and the practical aspects of the advances in gear machining.

This book is neither a textbook nor a manual. It combines science and engineering. The reader will find enough material to understand the geometry of gear-cutting tools, as well as the kinematics of a gear machining processes.

This book deals with the science of surface generation. The investigation of surface generation process offers an excellent context and motivation for exploring the pervasive ties between *kinematics* and *geometry*.

The discussion in the book is based on the "*DG/K*-based approach" of part surface generation. The term *DG/K-based approach* is derived from two terms—namely, **D**ifferential **G**eometry (*DG*) and **K**inematics (*K*) of multi-parametric motion in Euclidean space $E_3$. The author is the key developer of the *DG/K*-based approach of part surface generation. The approach is disclosed in more detail in some of his earlier published books [133, 144, 148, 149, 151, 152, 155, 168, 171], as well as in other sources.

So far, when designing gear-cutting tools, and when developing novel methods of cutting gears, gear-cutting tool designers are governed mostly by common sense, by accumulated experience, and by empiricism. The lack of the use of the scientific approach is a significant obstacle to attaining the most efficient designs of gear-cutting tools.

To the author's knowledge, this book represents the first attempt to disclose the details of designing the most favorable gear-cutting tools based on the fundamental scientific theory. Most, but not all, important topics in the field of gear-cutting tool design are covered in this book. However, this area of gear engineering is broad, and there remains plenty of room for further improvements in this field.

Special-purpose gear-cutting tools for machining noncircular gears are not considering in this monograph. However, the approach that is disclosed in this work can be extended to the area of mechanical engineering as well.

# Acknowledgments

The author would like to share credit for any success of the *Gear Cutting Tools: Science and Engineering* to plenty discussions on the subject with numerous representatives of the gear community, both domestic and international.

The contribution of many friends and colleagues in overwhelming numbers cannot be acknowledged individually, and as much as our benefactors have contributed, even though their kindness and help must go unrecorded.

Special thanks to Ms. Katya Porter, commissioning editor for *Mechanical Engineering*, who strongly supported this project.

# Author's Biography

**Dr. Stephen P. Radzevich** is a Professor of Mechanical Engineering and a Professor of Manufacturing Engineering. He received his M.Sc. in 1976, Ph.D. in 1982, and Dr. (Eng)Sc. in 1991, all in mechanical engineering. Dr. Radzevich has extensive industrial experience in gear design and manufacture. He has developed numerous software packages dealing with computer-aided design (CAD) and computer-aided machining (CAM) of precise gear finishing for a variety of industrial sponsors. His primary research interest is in kinematic geometry of part surface generation, with a particular focus on precision gear design, high-power-density gear trains, torque share in multi-flow gear trains, the design of special purpose gear-cutting and finishing tools, and the design and machining (finish) of precision gears for low-noise and noiseless transmissions of cars, light trucks, and the like.

Dr. Radzevich has spent over 40 years developing software, hardware, and other processes for gear design and optimization. In addition to his work in the industry, he trains engineering students at universities and gear engineers in corporate settings.

He has authored and coauthored over 30 monographs, handbooks, and textbooks. The monographs *Generation of Surfaces*, *Kinematic Geometry of Surface Machining* (*CRC Press*, 2nd Edition 2014), *CAD/CAM of Sculptured Surfaces on Multi-Axis NC Machine: The DG/K-Based Approach*, *Gear Cutting Tools: Fundamentals of Design and Computation* (*CRC Press*, 2010), *Precision Gear Shaving*, *Dudley's Handbook of Practical Gear Design and Manufacture* (*CRC Press*, 2012), *Geometry of Surfaces: A Practical Guide for Mechanical Engineers* are among his recently published volumes. He has also authored and coauthored over 300 scientific papers and holds over 250 patents on inventions in the field (US, Japan, Russia, Europe, Canada, Soviet Union, South Korea, Mexico, and others).

# Introduction

"Simplex sigillum veri" (Simplicity is the seal of truth)

**Latin proverb**

This book deals with cutting tools used for machining gears. Gears are produced in enormous amount—billions of gears are produced every year. While the automotive industry ranks as the primary consumer of gears, numerous other industries also require huge amounts of gears: construction machinery, agricultural machinery, and the aerospace industry, to name a few.

The gear-cutting process is costly, with the cost of gear-cutting tools exceeding 50% of the total cost of the gear machining operation. If we place cost savings in the production of every gear in the range of just 10 cents, the total cost savings could reach hundreds of millions dollars. This is an appealing incentive for engineers and scientists to turn their attention to the gear-machining process and the design of gear-cutting tools used in gear machining.

In writing this book, the author has tried to expand the theory and fill various gaps.

The approach used in the text is primarily analytical. However, geometric interpretations are given, and, in places, synthetic reasoning is also applied. Furthermore, the author has largely limited himself to the mathematics of algebraic geometry, vector and matrix algebra, and elementary calculus. Special mathematical methods are avoided, even in the final chapters.

The use of powerful computers makes it reasonable not to derive equations in their final form (as they are often bulky) but just outline the major steps with which a problem can be solved.

At this point, it is fitting to recall the old Chinese proverb: "The beginning of wisdom is to call things with their right names". Unfortunately, even among gear specialists, there is some ambiguity in the terms used to describe gears and gear-related parameters. In this work, to the extent possible, we follow the conventional terminology. When necessary, definitions of new terms are provided.

The development of the scientific classification of kinematic schemes of the gear-machining process and illustration of the classification with practical designs of gear-cutting tools and advanced methods of gear machining are two of the biggest reasons, among various others, that the author was highly motivated to write this text.

The development of the scientific classification of kinematic schemes of gear machining is among the major objectives of this book. The identification of the proper place of all known gear-cutting tool designs in the developed classification is another important objective. The identification of areas where there is plenty of room for additional developments in the field of designing favorable gear-cutting tools is the third major goal of this book. However, the primary aims of this book are not just limited to these primary objectives.

## HISTORICAL BACKGROUND

This study surveys and assesses the considerable body of research on gears and gear-machining processes that has accumulated to date. It is not the aim of the author, who has been actively involved in this field of research for more than 40 years, to develop all possible gear-cutting tool designs. That massive undertaking is not within the scope of this text. However, the potential disclosure of all possible conceptual designs of gear-cutting tools is another major objective of this book.

## UNIQUENESS OF THIS PUBLICATION

This book is unique for many reasons.

Most of the material used in this book is new and cannot be found elsewhere.

This is the second edition of the first work in English dedicated solely to the favorable design of gear-cutting tools. The treatment is rigorous and elegant, focusing on the mathematical development of the subject apart from any particular applications. This book is also unique in the sense that most of the material was developed by the author.

The discussion in this text starts with general concepts and problems, and then the scope gradually narrows to more specialized simple cases. The kinematics of the gear machining mesh is a key point for proper understanding of the disclosed approach. Eventually it can be understood that by just playing with two rotation vectors,[1] one can develop a novel method of gear machining, as well as a novel gear-cutting tool design for this purpose (playing with two rotation vectors $\boldsymbol{\omega}_g$ and $\boldsymbol{\omega}_{ct}$, along with one or two additional vectors of prime motion, feed motion, etc.).

Numerous examples and an extensive bibliography further enhance the usefulness of the book.

Because of its generality and lucid style, this classic work will be invaluable not only to specialists in gear-cutting tool design, but also to those working in general areas of mechanical/manufacturing engineering.

## INTENDED AUDIENCE

Since perusal of the table of contents may leave the reader wondering whether this volume was intended as a textbook, a research monograph, or a historical treatise, some

[1] Angular velocity is considered in this book as a vector along the axis of rotation that is pointed in a direction defined by the right-hand screw rule. It is understood here and below that rotations $\omega_g$ and $\omega_p$ are not vectors in nature. Therefore, special care is required when treating rotations as vectors. In particular, when coordinate system transformations are analytically described using matrices, the order of the matrices in the matrix product is predetermined by the order of the elementary coordinate system transformations—that is, the order of the matrices in the matrix product cannot be altered.

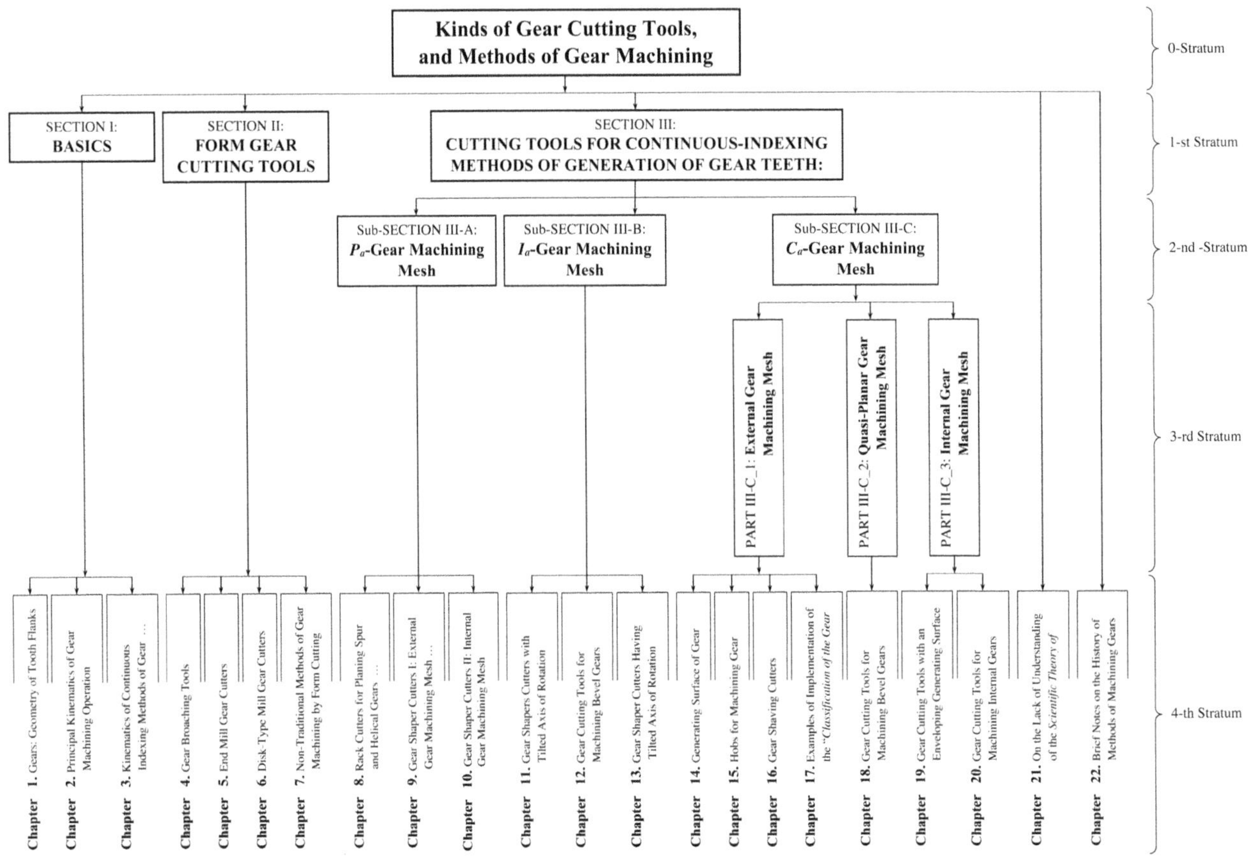


**FIGURE 0.1** Organizational chart of the book.

explanatory remarks are perhaps in order. There was, in fact, no conscious intent to aim for any of these, and what has transpired seems to bear the fact that it became a combination of all three. Gear experts working in various industries and in academia and university students (senior graduate, graduate, and postgraduate students) are the intended audience of the book.

## ORGANIZATION OF THIS BOOK

The book that the reader holds in his/her hands, is subdivided on sections, sub-sections, parts, and chapters.

As this book is lengthy [646 pages in total], for the readers' convenience, an organization chart has been provided in Figure 0.1.

This book is comprised of three sections:

Section I: Basics
Section II: Form Gear-Cutting Tools
Section III: Cutting Tools for Continuous-Indexing Methods of Generation of Gear Teeth

**Section I** consists of three chapters:
- **Chapter 1:** Gears: Geometry of Tooth Flanks
- **Chapter 2:** Principal Kinematics of Gear-Machining Operation
- **Chapter 3:** Kinematics of Continuous Indexing Methods of Gear-Machining Operations

**Section II** consists of four chapters:
- **Chapter 4:** Gear-Broaching Tools
- **Chapter 5:** End Mill Gear Cutters
- **Chapter 6:** Disk-Type Mill Gear Cutters
- **Chapter 7:** Nontraditional Methods of Gear Machining with Form Cutting Tools

**Section III** consists of three sub-sections:
- **Sub-section III-A:** Parallel-Axes Gear Machining Mesh
- **Sub-section III-B:** Intersected-Axes Gear Machining Mesh
- **Sub-section III-C:** Crossed-Axes Gear Machining Mesh

**Sub-section III-A** consists of three chapters:
- **Chapter 8:** Rack Cutters for Planing Spur and Helical Gears
- **Chapter 9:** Gear Shaper Cutters I: External Gear Machining Mesh
- **Chapter 10:** Gear Shaper Cutters II: Internal Gear Machining Mesh

**Sub-section III-B** consists of three chapters:
- **Chapter 11:** Gear Shapers Cutters with Tilted Axis of Rotation
- **Chapter 12:** Gear-Cutting Tools for Machining Bevel Gears
- **Chapter 13:** Gear Shaper Cutters Having Tilted Axis of Rotation: *Internal Gear Machining Mesh*

**Sub-section III-C** consists of three parts:

- **Part III-C_1:** Design of Gear-Cutting Tools: *External Gear Machining Mesh*
- **Part III-C_2:** Design of Gear-Cutting Tools: *Quasi-Planar Gear Machining Mesh*
- **Part III-C_3:** Design of Gear-Cutting Tools: *Internal Gear Machining Mesh*

**Part III-C_1** consists of four chapters:

- **Chapter 14:** Generating Surface of Gear-Cutting Tool
- **Chapter 15:** Hobs for Machining Gears
- **Chapter 16:** Gear Shaving Cutters
- **Chapter 17:** Examples of Implementation of the *Classification of Gear-Machining Meshes*

**Part III-C_2** consists of just one chapter:

- **Chapter 18:** Gear-Cutting Tools for Machining Bevel Gears

**Part III-C_3** consists of two chapters:

- **Chapter 19:** Gear-Cutting Tools with an Enveloping Generating Surface
- **Chapter 20:** Gear-Cutting Tools for Machining Internal Gears

Two additional chapters that are not included the earlier listed section/sub-section/part finalize the contents of this book. These chapters are titled:

- **Chapter 21:** On the Lack of Understanding of the Scientific Theory of Gearing by the Majority of Gear Scientists and Engineers
- **Chapter 22:** Brief Notes on the History of Methods of Machining Gears and of Design of Gear-Cutting Tools

The following sections discuss the organization of the book in more detail.

## Section I. Basics

Fundamental issues of gear-cutting tools are covered in this section of the book. The discussion encompasses types of gears to be machined, the analytical representation of the gear tooth flanks, types of relative motions of the work gear and the gear-cutting tool in gear machining process. The section ends with a brief consideration of the linear transformation methods that are practical in designing gear-cutting tools.

The first part of the book consists of Chapters 1 through 3.

**Chapter 1:** The geometry of tooth flanks of common types of gears is discussed in the Chapter 1. Along with numerous practical examples of gear designs, equations for the tooth flanks of spur, helical, straight bevel, and helical bevel gears are derived there. The derived equations illustrate the powerful method that can be used to derive an equation for the tooth flank of any given design. An analytical description of the desired gear tooth flank is a good starting point for the gear-cutting tool designer.

**Chapter 2:** General aspects of the kinematics of gear machining processes are considered in Chapter 2. The analysis is primarily focused on two issues: (a) possible types of relative motions in the gear-machining process, including but not limited to elementary relative motions of the work gear and the gear-cutting tool and (b) rolling of the conjugate surfaces over each other. In particular, the reader's attention is drawn to the practical applications of special types of motions under which a surface allows sliding *over itself*.

**Chapter 3:** This chapter focuses on the kinematics of special methods of gear-machining processes. The methods are referred to as the *continuous indexing methods* of gear-machining processes. Vector representation of the gear machining mesh and the kinematic relationships for the gear machining mesh are covered in this chapter. Based on the in-depth analysis of the principal features of configuration of the rotation vectors that specify the gear machining mesh, a scientific classification of types of gear machining meshes is developed. The classification is the key for the development of novel designs of gear-cutting tools and new methods of machining gears. Major steps for transformation of kinematics of the gear machining mesh into the corresponding kinematics of the gear-machining process are considered.

## Section II. Form Gear-Cutting Tools

Gear-cutting tools featuring the generating surface that is either congruent to tooth surfaces of the gear to be machined or makes line contact with the tooth flank of the work gear are discussed in this part of the book. The kinematics of the gear-machining processes used for this purpose of form gear-cutting tools is the simplest possible. However, simple kinematics of surface machining often entails the necessity in the cutting tools of very complex designs.

The second part of the book consists of Chapters 4 through 7.

**Chapter 4:** Design of cutting tools for broaching gears is the subject of this chapter. The consideration begins with the analysis of the kinematics of the gear-broaching process and is followed by generation of the generating surface of the gear-cutting tool, transformation of the generating surface into a workable gear-cutting tool, cutting edge geometry, and chip removal diagrams. Both analytical methods and *DG*-based methods of analysis (that is, descriptive geometry–based methods) are widely used in this chapter. Sharpening of gear broaches is an issue of particular consideration. The discussion of gear-broaching tools ends with a detailed disclosure of the concept of precision gear-broaching tools for machining involute gears and with application issues of gear-broaching tools.

The concept of broaching is the cornerstone of other methods of gear-cutting. These methods include, but are not limited to, (a) shear-speed cutting of gears, (b) rotary broaching with slater tools, and (c) the "Revacycle" method of gear cutting. The principles of gear-cutting tool designs for these methods of gear machining are also covered in this chapter.

**Chapter 5:** Design methods of end-type mill cutters for machining spur and helical gears are discussed in this chapter. At the outset, the kinematics of the gear-cutting process with end-type mill cutter is discussed. This is followed by an analytical description of the secondary generating surface of the end-type mill cutter, and by an investigation of the cutting edge geometry. The analysis makes it possible to calculate the cutting wedge angles at any point of interest within the cutting edge. Accuracy of the milled gears is discussed from the perspective of the geometry of interacting surfaces and the kinematics of the gear-cutting process. Application of the end-type mill cutters is considered at the end of the chapter.

**Chapter 6:** Disk-type mill cutters and their design and analysis are discussed in this chapter. This is followed by an analytical description of the generating surfaces of mill cutters designed for machining of spur and helical gears. Next, the intrinsic geometry of the generating surfaces of the mill cutters is investigated. Considered together, an equation of the secondary generating surface of the mill cutter, an equation of the rake surface, and an equation of the clearance surface of the mill-cutter tooth allow for the analytical representation of the cutting edge of the cutting tool. Profiling of mill cutters is considered from the perspective of implementing both the *DG*-based method and analytical methods in profiling of form cutting tools. The cutting edge geometry of the disk-type mill cutter is explored using analytical elements of vector algebra and calculus. The peculiarities of mill cutter designs for the roughing of gears are considered. Accuracy issues are investigated from the perspective of fulfillment/violation of the conditions of proper part surface generation. The chapter ends with a discussion of the principal features of the practical application of disk-type mill cutters for cutting spur and helical gears.

**Chapter 7:** Investigation of the kinematics of gear-machining operations makes it possible to develop novel gear-machining methods as well as novel gear-cutting tool designs. All possible methods of the gear cutting are referred to as nontraditional methods of gear machining with form cutting tools.

The consideration begins with the systematic investigation of all possible types of single-parametric motions. Analysis reveals that novel schematics of gear cutting with (a) end-type mill cutters, (b) disk-type mill cutters, (c) face gear mill cutters, (d) with rotary broaches, and so forth, can be developed. In particular, various methods and cutting tools of various designs intended for machining worms are considered. In closing, the possibility of classifying form gear-cutting tools is discussed.

## Section III. Cutting Tools for Continuous-Indexing Methods of Generating Gear Teeth

### Sub-Section III-A. Parallel-Axes Gear Machining Mesh

Gear-cutting tool designs and gear-cutting methods featuring parallel axes of the work gear and of the gear-cutting tool are investigated in this part of the book. This section begins with a discussion of the kinematics of the parallel-axes *gear machining mesh*, which is the key issue for proper understanding of design and operation of many gear-cutting tools designs.

The third part of the book consists of Chapters 8 through 10.

**Chapter 8:** Rack cutters for planing gears is the simplest example of generating gear-cutting tools featuring parallel-axes gear machining mesh. The geometry and kinematics of machining of gears with rack cutters are analyzed. For profiling rack cutters, use of the graphical method based on wide implementation of *DG*-based methods and analytical methods is recommended. Methods of vector algebra are used to investigate the cutting edge geometry of rack cutters. Based on the results of the analysis, improvements are made in the geometry of the lateral cutting edges. The chip thickness cut by cutting edges of the rack cutter tooth is analyzed. Accuracy of the tooth flanks of machined gears is analyzed from the perspective of using the approach that is commonly referred to as the *kinematic geometry of surface machining*. Application of rack cutters as well as potential gear-cutting methods and rack-type gear-cutting tool designs are discussed in the final two sections.

**Chapter 9:** The design of shaper cutters and methods of shaping of external gears are investigated in Chapter 9. The external parallel-axes gear machining mesh is complemented with the cutting motion. In this way, the kinematics of the gear shaping process is derived. The analytical description of the generating surface of the shaper cutter, types of the rake surface, and geometry of the clearance surface are discussed. Methods of vector algebra are used to investigate the cutting edge geometry of shaper cutters. Valuable improvements in cutting edge geometry are discovered based on the results of the research. Elements of the kinematics of gear meshing are used for the analytical description of thickness of chip cut by the shaper cutter tooth, as well as for the analytical description of the deviations of the machined gear tooth surface from the desired geometry of the gear tooth flanks. The analysis is focused on verifying whether the fifth and sixth conditions of proper part surface generation are satisfied. Issues relating to the application of gear shaper cutters cover both conventional and special-purpose designs of gear shaper cutters for machining spur and helical gears. As an example, typical operations of gear shaping are briefly described. Methods of grinding the rake and clearance surfaces of gear shaper cutters are discussed at the end of the chapter.

**Chapter 10:** The design of gear-cutting tools that are used in the parallel-axes internal gear machining mesh is discussed in this chapter of the book. The gear machining mesh is complemented with the primary motion. The design procedure encompasses (1) generation of the generating surface of the cutting tool, (2) profiling of the gear shaper cutters, and (3) exploration of the cutting edge geometry of the gear shaper cutter. Calculation of the thickness of chip cut by the gear shaper cutter tooth and accuracy of the shaped internal gears are also covered in this chapter. Because the same parallel-axes internal gear machining mesh is not limited to the shaping of internal gears but can also be applied for the design purposes of enveloping shaper cutters for machining external gears, the related issues are briefly discussed in this chapter. Some important issues relating to the practical application of the gear shaper cutters are discussed at the end of the chapter.

### Sub-Section III-B. Intersected-Axes Gear Machining Mesh

Gear-cutting tool designs and gear-cutting methods featuring intersecting axes of the work gear and the gear-cutting tool are investigated in this part of the book. The discussion starts with the investigation of the kinematics of the intersecting-axes *gear machining mesh*, which is the key issue for proper understanding of the design and operation of gear-cutting tools of many designs.

The fourth part of the book consists of Chapters 11 through 13.

**Chapter 11:** Gear shapers with a tilted axis of rotation are examples of gear-cutting tools designed based on the principle of the external intersecting axis gear machining mesh. The kinematics of the intersecting-axes gear machining mesh is complemented with an additional primary motion (motion of cut) of the cutting tool to be designed. In this way, the kinematics of the gear-machining process using gear shaper cutters with a tilted axis of rotation is derived. For the particular kinematics of gear cutting, the generating surface of the gear shaper cutter with a tilted axis of rotation is derived. The capabilities of the external gear machining mesh are illustrated with several cases of its applications in gear-cutting tool design.

**Chapter 12:** The design of gear-cutting tools for machining bevel gears is discussed in this chapter of the book. Gear-cutting tools are designed using a round rack that properly meshes with the gear to be machined. The discussion begins with the analysis of the principal elements of the kinematics of bevel gear generation. The geometry of the interacting surfaces—namely, the involute straight bevel gear flank and the generating surface of the gear-cutting tool—along with the geometry of tooth flanks of the generated gears is analyzed. Generation of straight bevel gears with offset teeth is also covered in this chapter. The discussed geometrical and kinematical aspects are used for the comprehensive investigation of methods of planing and milling of straight bevel gears and milling of bevel gears with curved teeth. The design of gear-cutting tools used for this purposes is discussed, along with the corresponding methods of gear machining. It is stressed here that most bevel gears are a type of practical approximation to the desired geometrically accurate tooth geometry.

**Chapter 13:** The possibility of designing gear-cutting tools based on an internal intersected-axes gear machining mesh is discussed in this chapter. The shaping of internal gears using the gear shaper cutter with a tilted axis of rotation, as well as cutting of external gears with enveloping gear shaper cutters with a tilted axis of rotation, are good examples in this regard. The shaping of external recessed tooth forms using enveloping shaper cutter is one more opportunity to utilize an internal intersecting-axes gear machining mesh for the purposes of designing gear-cutting tools. However, gear machining meshes of this type have not been subjected to a comprehensive investigation as of yet.

### Sub-Section III-C. Crossed-Axes Gear Machining Mesh

Most gear-cutting tool designs are based on the continuously generating principle. Spatial gear machining mesh is simulated when machining a gear using these gear-cutting tools. The huge numbers of known and possible designs of gear-cutting tools results in discussion of considering them as three sections, each of which is devoted to: (a) external gear machining mesh, (b) quasi-planar gear machining mesh, and (c) internal gear machining mesh.

The rest of the book consists of by Part III-C_1 through Part III-C_3.

**Part III-C_1:** Design of Gear-Cutting Tools: *External Gear Machining Mesh*

Gear-cutting tool designs based on the principle of external spatial gear machining mesh are considered in this section. The section starts with the investigation of the possible types of generating surfaces of the gear-cutting tools. Designs of gear hobs, shaver cutters, and the like are considered in subsequent chapters.

This section of the book consists of Chapters 15 through 18.

**Chapter 14:** The generation principles of the generating surfaces of gear-cutting tools are investigated in this chapter. The discussion begins with an analysis of the kinematics of the external spatial gear machining mesh. Possible types of auxiliary generating surfaces, together with practical methods of their generation, are considered. Generation of generating surface of the gear-cutting tool is based on the possible types of auxiliary generating surfaces. Equations of generating surfaces are derived and design parameters of the surfaces are calculated. For this analysis, descriptive-geometry (*DG*)-based methods and analytical methods are used. Next, various types of generating surfaces are constructed. These include, but are not limited to, conventional cylindrical generating surfaces, cylindrical generating surfaces with a zero profile angle, conical generating surfaces, generating surfaces featuring an asymmetrical tooth profile, and surfaces with a torus-shaped pitch surface. The chapter ends with a brief discussion of the geometrical and kinematical constraints on the design parameters of the generating surface of gear-cutting tools.

**Chapter 15:** Hobs are a perfect example of gear-cutting tool design based on the external spatial gear machining mesh. In this chapter, the geometry of the generating surface of gear hobs along with geometry of the rake and clearance surfaces are analyzed. Generation methods for the rake and clearance surfaces of hob teeth are also considered. The derived equations for the working surfaces of the gear hob enable an in-depth analysis of the accuracy of hobs for machining of involute gears. Numerous advanced designs of hobs for machining gears are discussed as examples of implementation. The cutting edge geometry of gear hobs is analytically described in the tool-in-use reference system. To satisfy the necessary conditions of proper part surface generation in a gear-machining process, constraints on the parameters of modification of the hob tooth profile are investigated. This discussion is followed by several numerical examples of calculation of the design parameters of the hobs. The obtained results can be enhanced for application in designing hobs for machining non-involute profiles. The applications of hobs for machining gears are discussed from the perspective of the analytical research on the geometry and kinematics of the gear-hobbing process.

**Chapter 16:** Gear-shaving cutters represent another type of gear-cutting tool designed on the premises of spatial external gear machining mesh. The procedure of designing a shaving cutter begins with the transformation of the generating surface into the workable gear-shaving cutter. Peculiarities of the geometry of the rake surface, the clearance surface, and the cutting edge geometry are discussed. Issues on the design parameters of a shaving cutter, design features of the serrations, and re-sharpening of shaving cutters are considered. Four basic methods for gear shaving – (1) axial shaving, (2) diagonal shaving, (3) tangential shaving, and (4) plunge shaving – are discussed from the perspective of the kinematic geometry of surface machining. The discussion ends with an analysis of the advance designs of gear shaving cutters and practical issues relating to the implementation of gear shaving processes for shaving precision gears.

**Chapter 17:** Examples illustrating the capabilities of the external crossed-axes gear machining mesh are discussed in this chapter. These examples can be divided into two sections. The design of gear-cutting tools and methods for machining cylindrical gears are considered in the first section. Hobs for tangential hobbing, hobs for plunge hobbing, hobs for machining face gears, continuous generating cutting of worms, gear-cutting tools for scudding process, and gear shaper cutters with a tilted axis of rotation for machining cylindrical gears along with rack-type gear shaper cutters are also discussed in the first section. The design of gear-cutting tools and methods for machining conical gears are discussed in the second section. Discussion in this section is limited to special-purpose tools for burnishing gears through surface plastic deformation and the conical hobs of the *Paloid* method of gear cutting.

**Part III-C_2:** Design of Gear-Cutting Tools: "*Quasi-Planar Gear Machining Mesh*"

Design of gear-cutting tools based on the principle of the kinematics of *quasi-planar* gear machining mesh is discussed in this section. The section begins with a detailed analysis of the vector representation of the kinematics of *quasi-planar* gear machining mesh.

This section of the book consists only of Chapter 18.

**Chapter 18:** This chapter is devoted to cutting tools for machining bevel gears. Gear-cutting tool designs for novel gear-cutting methods are discussed. Topics include cutting tools for plunge cutting of bevel gears, face hobs for continuously indexing method of gear cutting, and possible directions for further development.

**Part III-C_3:** Design of Gear-Cutting Tools: *Internal Gear Machining Mesh*

The final section of the book deals with designs of gear-cutting tools that use internal spatial gear machining mesh. Again, the consideration begins with a detailed analysis of the kinematics of the internal spatial gear machining mesh. For this purpose, vector representation of internal spatial gear machining mesh is widely used.

This section of the book consists of Chapters 19 and 20.

**Chapter 19:** Designs of gear-cutting tools with an enveloping generating surface are discussed in this chapter. Gear-cutting tools with cylindrical, conical, and toroidal generating surfaces are also discussed, including the analytical determination of the generating surface itself, as well as some aspects relating to the accuracy of the machined gears. This is followed by examples of gear-cutting tool designs relating to this group of gear-cutting tools.

**Chapter 20:** The kinematics and design of gear-cutting tools is briefly discussed in this last chapter. The principal design parameters of gear-cutting tools are determined, followed by examples of novel designs of gear-cutting tools for machining internal gears.

Two chapters—Chapters 21 and 22—are not covered by the earlier considered sections/sub-sections/parts.

**Chapter 21:** The discussion in this chapter is focused on the key mistakes in the understanding and application of the scientific theory of gearing by the majority of gear scientists and engineers.

**Chapter 22:** This chapter provides a brief (but still systematical!!) analysis of the key accomplishments that formed the foundation of the scientific theory of gearing. The main feature of the organization of this chapter of the book is as follows.

The advances and achievements in the scientific theory of gearing and the gear researchers who they are attributed to deserve to be named. As such, the known fundamental accomplishments in the scientific theory of gearing are listed in chronological order. Where possible, the name of the gear researcher who each accomplishment is attributed to is provided.

The completion of this book was an endeavor that the author devoted years to.

It is inevitable that a study of this nature will lean toward greater emphasis on the author's own contributions, if only because they reflect his perspective on the subject matter. Nevertheless, an effort has been made to summarize the key ideas (if not the technical details) of the most significant developments in the field, and otherwise give pointers to many others.

A book of this size is likely to contain omissions and errors. If you have any constructive suggestions, please convey these to Dr. Stephen Radzevich (radzevich@usa.com).

**Stephen P. Radzevich**
*Sterling Heights, Michigan*
*April 22, 2024*

# Section I

## Basics

Gear-cutting tools are used for machining gear teeth. When machining a gear, the gear and the gear-cutting tool are performing a motion relative to each other. Under such motions, the work gear and the gear-cutting tool often comprise a virtual gear pair. This means that the following elements are interdependent:

- The gear tooth shape.
- The kinematic parameters of the relative motion of the gear and the tool.
- The tool's design parameters.

Therefore, any change in the gear tooth shape necessitates corresponding changes to the shape of the gear-cutting tool teeth, the kinematics of the gear-machining process, or both.

Numerous methods for machining gears are used in the industry. As an example, Figure I.1 illustrates power skiving of an internal involute gear.

**FIGURE I.1** Power skiving of an internal involute gear.

DOI: 10.1201/9781003520528-1

When machining a gear, the interaction of the work gear and of the gear-cutting tool closely resembles the interaction of gear tooth flanks in a gear pair. A gear pair serves to transform the given input motion into the required output motion. Gear pairs have three principal components:

- Tooth shape of the input gear.
- Tooth shape of the conjugate output gear.
- Motion of the output gear relative to the input gear.

An infinite number of tooth shape combinations are capable of ensuring the desired motion transformation.

A similar situation can be observed in gear-machining operations. Generally, a gear-machining operation also features three principal components:

- Tooth shape of the gear-cutting tool to be applied.
- Tooth shape of the gear to be machined.
- Motion of the tool and the gear in relation to each other.

Machining the desired gear tooth shape is the primary goal of the gear-machining process. An infinite number of combinations of (a) the shape of the gear-cutting tool teeth and (b) kinematics of the relative motion of the work gear and the gear-cutting tool can ensure this goal is met.

Before beginning to design a gear-cutting tool, major design parameters of the gear need to be determined. Certain kinematics of the gear-machining process can be chosen among feasible relative motions of the work gear and the gear-cutting tool. Ultimately, a gear-cutting tool can only be designed for machining a given gear if the kinematics of the gear-machining process is pre-specified.

Two issues can be drawn up from these brief considerations.

***First***, before designing a gear-cutting tool, it is necessary to specify the geometry of the tooth flanks of the gear to be machined. Answering the question, "What are the specific features of the geometry of the gear tooth surface to be machined?" is a must.

***Second***, it is necessary to determine exactly how the work gear and the gear-cutting tool move with respect to each other when machining the gear teeth. Investigation and analysis of all the feasible relative motions of the work gear and the gear-cutting tool are of critical importance. Parameters of the most favorable kinematics of a gear-machining process should be established prior to the design stage. In most practical cases, not the most favorable kinematics, but suitable and reasonable kinematics of the gear-machining process is often implemented instead.

Both issues are critically important to the gear-cutting tool designer.

The so-called *direct problem of gear-cutting tool design* encompasses determining the parameters of the geometry of the gear-cutting tool's generating surface on the premises of (a) the known geometry of the tooth flank of a gear to be machined and (b) the predetermined kinematics of the gear-machining process. Another problem relates to determining the geometry of the tooth flank of the gear that is machined by means of a given gear-cutting tool. Often, problems of this type are referred to as *inverse problems of a gear-cutting tool design*.

The inverse problem of a gear-cutting tool design arises in two cases. First, when an approximation of the tooth profile of the designed gear-cutting tool has been implemented, then for the purpose of calculating the actual accuracy of the gear-cutting tool, the inverse problem of gear-cutting tool design is analyzed. The second case arises when a gear-cutting tool is given, and it is necessary to determine the geometry of the tooth flank of the gear that is machined with the given gear-cutting tool.

The following discussion begins with an investigation of the possible options to solve the direct problem of gear-cutting tool design.

The first part of the book is comprised by Chapter 1 through Chapter 3.

# 1 Gears
## *Geometry of Tooth Flanks*

Gears are used as elements of gear trains for transmission and transformation of a motion from a driving shaft to a driven shaft. A large number of gears are used in the automotive industry (Figure 1.1). It is commonly understood that a gear is a machine element having gear teeth. Usually, though not necessarily, gears are round.[1] There is a large variety of gear forms, as well tooth forms in the lengthwise direction. Gear teeth are also available in a wide variety of profiles.

The performance of a gear transmission strongly depends on the accuracy of the gears that comprise it. Precision gears are mainly produced by cutting with gear-cutting tools. For the purposes of gear machining and of gear-cutting tool design, the gear teeth flanks to be machined are specified. Precise analytical specification of the gear tooth flank is required.

**FIGURE 1.1** Gears of various types are used in the design of the automotive transmission.

## 1.1 BASIC TYPES OF GEARS

Although there is a large number of gear forms, as well as numerous gear teeth shapes and profiles, only a small number of basic types of gear design are commonly recognized. For illustrative purposes, several basic types of gears are considered in this chapter. The following discussion aims to provide the reader with an overview of the geometry of gear tooth flanks, a both qualitative and quantitative understanding of how complex the geometry of gear tooth flanks can be, and the types of problems that often arise when machining gear teeth.

A *spur involute gear* is the simplest type of involute gear. Straight in the lengthwise direction teeth of a spur gear are uniformly distributed around the cylindrical body of the gear. This design includes both external spur involute gears (see Figure 1.2) and internal spur involute gears (see Figure 1.3).

A gear sector can be cut off from either an external or an internal spur helical gear. In a particular case, when the pitch diameter of a spur gear sector approaches infinity, the gear sector transforms into a straight-sided *spur rack* (see Figure 1.4).

A *helical involute gear* features helical teeth (see Figure 1.5). This refers to gear teeth shaped in the form of a helix that wraps around a cylinder. Helix angle ranging from only a few degrees up to about 45° are practical. Helical gears may be made with either external or internal teeth.

**FIGURE 1.2** External spur involute gear.

[1] This book is focused not on the design features and production methods and means for non-circular gears, scroll gears, and special-purpose elements of mechanical transmissions.

DOI: 10.1201/9781003520528-2

FIGURE 1.3 Internal spur involute gear.

FIGURE 1.5 A helical involute gear.

A straight-sided *helical rack* has a plane pitch surface and straight teeth that are oblique to the face surfaces. A straight-sided helical rack resembles a corresponding straight-sided spur rack, with the exception of the direction of the teeth. In a helical rack, the teeth are at an angle $(90° - \psi_b)$ in relation to the direction of the straight motion of the rack, where $\psi_b$ is the base helix angle of the rack.

*Herringbone involute gear* [see Figure 1.6(a)] and *double-helical involute gear* (see Figure 1.6(b)) can both be described as the gears composed of two helical gears facing in opposite direction of the helix. When two helical gears are separated

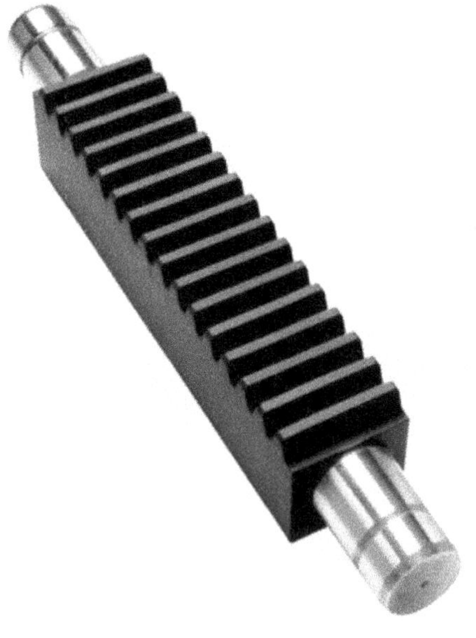

FIGURE 1.4 Straight-sided spur rack.

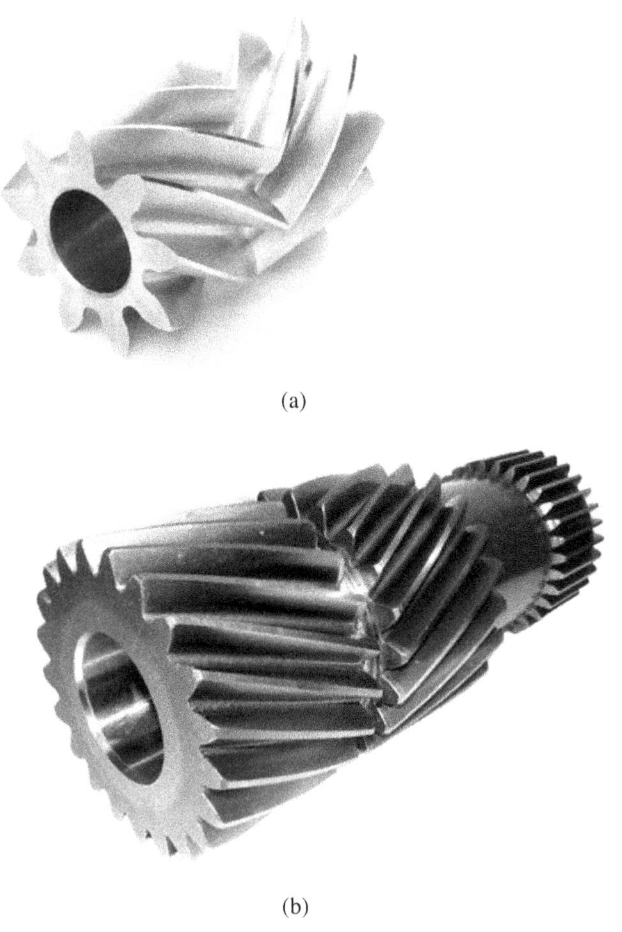

FIGURE 1.6 Herring-bone involute gear (a) and double-helical involute gear (b).

**FIGURE 1.7** A cluster gear.

from each other at a certain distance, they are often referred to as a double-helical involute gear. The separation distance for herringbone involute gears is equal to zero. As with spur involute gears, this includes both external and internal herringbone and double-helical involute gears. In a particular case, a herringbone or double-helical straight-sided rack can be formed.

A *cluster gear* (see Figure 1.7) can be thought of as a gear that is composed of two or more gears with different design parameters, such as different helices—that is, differences in the handedness of the helices, the magnitudes of the helices, or both.

When the helix angle of a helical gear is large—for example, when it differs from 90° by a few degrees and up to about 45°—the helical gear transforms into a worm (see Figure 1.8). Worms can make proper mesh with ordinary spur and helical gears, as well as with appropriately

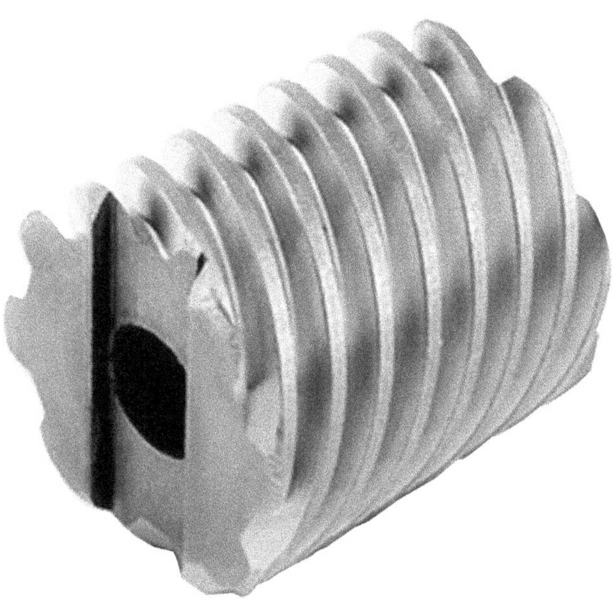

**FIGURE 1.8** A cylindrical worm.

**FIGURE 1.9** A worm gear.

designed worm gears (see Figure 1.9). In certain cases, double-enveloping worms are used. The elementary gear drive that is composed of a worm gear and a double-enveloping worm is commonly referred to as a *double-enveloping worm gear drive* or, simply, a *Cone drive.*

Straight bevel gears (see Figure 1.10) are used as machine elements of gear drives whose main purpose is the transmission of motion between two shafts that intersect each other. Spiral bevel gears (see Figure 1.11) are widely used for this purpose as well.

Face gears (see Figure 1.12) are capable of transmitting rotation between either the intersecting axes or the crossing axes of rotation. The capacity to make an approximate mesh with cylindrical pinion is an important advantage of face gearing.

**FIGURE 1.10** A straight bevel gear.

**FIGURE 1.11** A spiral bevel gear.

While it is not the goal of this chapter to cover a large number of gear designs widely used in the industry, this brief overview covered some of the most commonly used gears. The interested reader may wish to seek out the numerous special texts on gearing, where various gear designs are discussed in more detail. However, this introductory consideration more serves to establish the importance of the development of analytical representation of gear tooth flanks to be machined.

**FIGURE 1.12** A face gear.

## 1.2 ANALYTICAL DESCRIPTION OF GEAR TOOTH FLANKS

When designing gear-cutting tools, accurate specifications for the gear tooth flanks to be machined are of critical importance. For this purpose, the analytical description of gear tooth flanks is preferable from many standpoints.

The total number of tooth flank geometries of *geometrically accurate gears* (or *perfect gears*) is limited just to three types of the tooth flanks geometries [171]. Geometrically accurate gears are those meet all three fundamental laws of gearing [172]. The said, tooth flank geometries are limited to:

- Involute tooth flank geometry (for gears that operate on parallel axes).
- Spherical involute tooth flank geometry (for gears that operate on intersected axes).
- Generalized spherical involute tooth flank geometry (for gears that operate on crossing axes).

No other types of gear tooth flank geometries for perfect gear pairs are feasible.

The total number of tooth flank geometries that *approximately* fulfill the condition of meshing is large. However, the tooth flank geometries commonly used in the industry are limited just to a few, as follows:

- Cycloid cylinder and helix.
- Helix with a circular-arc or double-circular-arc profile (or *Novikov tooth profile*).
- Rectangular splines.
- Modified and special tooth geometries.

Similarly, there is a large variety of gear tooth shapes, but commonly used gear tooth forms with gear teeth in a lengthwise direction are usually limited to just a few forms, including:

1. Straight teeth.
2. Helical teeth.
3. Herringbone and double helical teeth.
4. Circular-arc teeth.
5. Cycloid teeth.

This is not an exhaustive list.

All of the listed tooth flank geometries, as well as most other practical geometries of gear tooth flanks, can be described analytically.

A comprehensive analysis of the geometry of gear tooth forms to be machined is not within the scope of this text, but this is discussed in more detail in advanced sources [171, 172]. Only a few illustrative examples of derivation of equation for gear tooth flanks are presented in this book. The examples demonstrate the capabilities of one of the approaches—that is, the vector-based approach—for derivation of such an equation.

A set of the design parameters of a gear is usually indicated on the gear blueprint. The indicated data could be sufficient for the specification of the geometry of the gear tooth flank and are usually used for the design procedure. This approach is referred to the *engineering approach* for gear tooth flank specification.

The analytical representation of a gear tooth flank is another means of the specifying the geometry of the gear tooth flank. This method of specification is referred to as the *scientific approach* for gear tooth flank specification and is preferred for analysis, development, and research.

The use of engineering or scientific approaches are equivalent as far as their results of the calculations are the same. Moreover, both approaches can be converted to the other. For this purpose, basic gear formulae can be used [13, 217]. For convenience, these equations are summarized in Appendix F.

The analytical description of gear tooth flanks can be easily obtained using the *scientific approach*.

### 1.2.1 Tooth Flank of Spur Involute Gear

Consider a spur gear with an involute tooth profile (see Figure 1.2). The generation of the involute tooth flank is illustrated in Figure 1.13. In the coordinate system $X_gY_g$, (see Figure 1.13,a) position vector $\mathbf{r}_{inv}(V_g)$ of the point of an involute tooth profile allows matrix representation in the form:

$$\mathbf{r}_{inv}(V_g) = \begin{bmatrix} r_{b.g} \cdot (\sin V_g - V_g \cdot \cos V_g) \\ r_{b.g} \cdot (\cos V_g + V_g \cdot \sin V_g) \\ 0 \\ 1 \end{bmatrix}, \quad V_g^{(l)} \le V_g \le V_g^{(a)} \tag{1.1}$$

Here, $V_g^{(l)}$ and $V_g^{(a)}$ are the values of parameter $V_g$, which correspond to the start of active profile (*SAP*) point of the tooth profile and to the end of active profile (*EAP*) point of the tooth profile—that is, the point in the tooth profile that is commonly located on the major diameter of the gear.

The tooth flank of a spur involute gear can be represented as the locus of consecutive positions of the involute tooth profile $\mathbf{r}_{inv}(V_g)$, which is moving directly forward in the direction of the gear axis $Z_g$. Let us denote the parameter of the motion of the tooth profile as $U_g$. Equation (1.1) immediately yields the expression for the position vector $\mathbf{r}_g(U_g, V_g)$ of the point of tooth flank $\mathscr{G}$ of the spur involute gear (see Figure 1.13,b):

$$\mathbf{r}_g(U_g, V_g) = \begin{bmatrix} r_{b.g} \cdot (\sin V_g - V_g \cdot \cos V_g) \\ r_{b.g} \cdot (\cos V_g + V_g \cdot \sin V_g) \\ U_g \\ 1 \end{bmatrix}, \quad \begin{matrix} V_g^{(l)} \le V_g \le V_g^{(a)} \\ 0 \le U_g \le B_g \end{matrix} \tag{1.2}$$

The current value of parameter $U_g$ is within the gear face width $B_g$.

One can see that the identity $U_g \equiv Z_g$ is observed for the selected type of parameterization of gear tooth flank $\mathscr{G}$.

### 1.2.2 Tooth Flank of Helical Involute Gear

The tooth flanks of helical gears are shaped in the form of a screw involute surface. Generation of a screw involute surface by a straight line that rolls with no sliding over the base cylinder of the gear is illustrated in Figure 1.14.

When performing a screw motion, the straight line generates a screw involute surface, $\mathscr{G}$, which is represented as a locus of consecutive positions along the line. This generating line is commonly referred to the *characteristic line* $\mathscr{E}$. Line $\mathscr{E}$ is tangent to the helix on the base cylinder of a radius $r_{b.g}$. The helix in question is traced out on the base cylinder by point of tangency of the generating line $\mathscr{E}$ with the cylinder. The generating line $\mathscr{E}$ forms base lead angle $\lambda_{b.g}$, with the plane perpendicular to the $Z_g$ – axis of the *Cartesian coordinate system* $X_gY_gZ_g$.

The position vector $\mathbf{r}_g$ of the point of the screw involute surface $\mathscr{G}$ can be represented as a sum of three vectors:

$$\mathbf{r}_g = \mathbf{A} + \mathbf{B} + \mathbf{C} \tag{1.3}$$

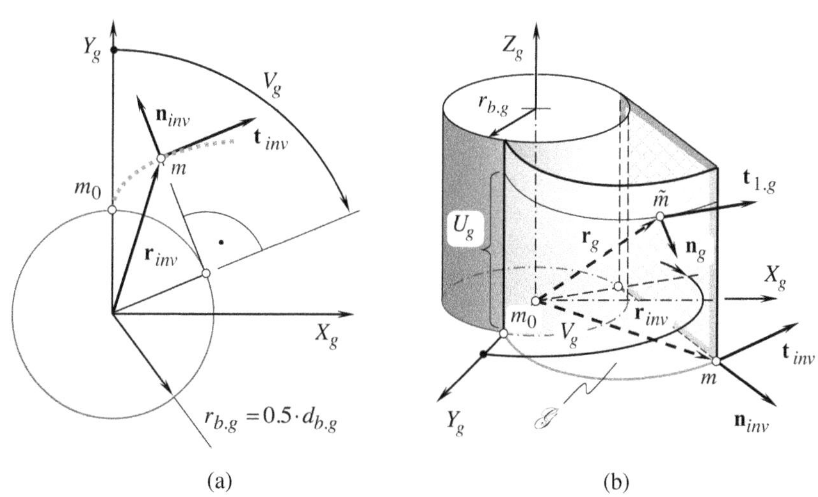


**FIGURE 1.13** Schematic for derivation of equation of tooth flank $\mathscr{G}$ of a spur involute gear: (a) section of a gear tooth flank by a plane perpendicular to the gear axis of rotation, and (b) close-up of the gear tooth flank.

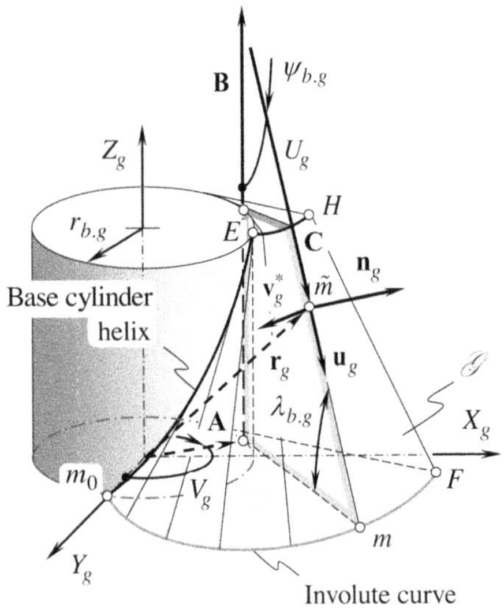


**FIGURE 1.14** Schematic for derivation of equation of tooth flank $\mathscr{G}$ of a helical involute gear.

Here, $|\mathbf{A}|$ is the base cylinder radius (this is due to the equality $|\mathbf{A}| = r_{b.g}$ is valid). Vector **A** makes roll angle $V_g$ with the $Y_g$ – axis. $|\mathbf{B}| = p_g \cdot V_g$ is the axial displacement in screw motion corresponding to the rotation angle $V_g$, and $p_g$ is the screw parameter (or the *reduced pitch*) of the flank surface $\mathscr{G}$. Ultimately, $|\mathbf{C}| = U_g$ is the segment of the generating straight line $\mathscr{E}$ measured from the tangency point on the base cylinder to the current point on the screw involute surface $\mathscr{G}$.

Projecting the vectors **A**, **B**, and **C** onto the coordinate axes $X_g$, $Y_g$, and $Z_g$ can produce an equation for the position vector of point $\mathbf{r}_g$ on the screw involute tooth flank, $\mathscr{G}$. After the necessary formulae transformations are performed, the equation for the position vector of point $\mathbf{r}_g$ of the screw involute surface $\mathscr{G}$ can be represented in matrix form:

$$\mathbf{r}_g(U_g, V_g) = \begin{bmatrix} r_{b.g}\cos V_g + U_g\cos\lambda_{b.g}\sin V_g \\ r_{b.g}\sin V_g - U_g\sin\lambda_{b.g}\sin V_g \\ r_{b.g}\tan\lambda_{b.g} - U_g\sin\lambda_{b.g} \\ 1 \end{bmatrix} \quad \begin{matrix} V_g^{(l)} \le V_g \le V_g^{(a)} \\ 0 \le U_g \le [U_g] \end{matrix} \tag{1.4}$$

Here, $[U_g]$ denotes the maximum allowed value of the parameter $U_g$. The value of the parameter $[U_g]$ can be expressed in terms of the base diameter $d_{b.g} = 2r_{b.g}$ of the gear, base lead angle $\lambda_{b.g}$, and the gear face width, $B_g$. The interested reader is referred to [124, 139, 148, 171, 172] for more detail on derivation of Eq. (1.4).

When a pair of perfect parallel-axes gears are engaged in mesh, the rolling plane—that is the plane through the vectors **B** and **C**—is congruent to the plane of action *PA* of the gear pair.

A brief analysis of the local geometry of the screw involute tooth surface $\mathscr{G}$ follows.

Equation (1.4) yields a calculation of tangent vectors $\mathbf{U}_g(U_g, V_g)$ and $\mathbf{V}_g(U_g, V_g)$, which are correspondingly equal:

$$\mathbf{U}_g(U_g, V_g) = \frac{\partial \mathbf{r}_g}{\partial U_g}(U_g, V_g) = \begin{bmatrix} \cos\lambda_{b.g}\sin V_g \\ -\cos\lambda_{b.g}\cos V_g \\ -\sin\lambda_{b.g} \\ 1 \end{bmatrix} \tag{1.5}$$

$$\mathbf{V}_g(U_g, V_g) = \frac{\partial \mathbf{r}_g}{\partial V_g}(U_g, V_g) = \begin{bmatrix} -r_{b.g}\sin V_g + U_g\cos\lambda_{b.g}\cos V_g \\ r_{b.g}\cos V_g + U_g\cos\lambda_{b.g}\sin V_g \\ r_{b.g}\tan\lambda_{b.g} \\ 1 \end{bmatrix} \tag{1.6}$$

Accordingly, the corresponding unit tangent vectors $\mathbf{u}_g$ and $\mathbf{v}_g$ are equal:

$$\mathbf{u}_g(U_g, V_g) = \frac{\mathbf{U}_g}{|\mathbf{U}_g|} \tag{1.7}$$

$$\mathbf{v}_g(U_g, V_g) = \frac{\mathbf{V}_g}{|\mathbf{V}_g|} \tag{1.8}$$

The direction of the tangent to the $U_g$ – coordinate curve through the given point on the gear tooth flank $\mathscr{G}$ is specified by the unit vector $\mathbf{u}_g$. Similarly, the direction of the tangent to the $V_g$ – coordinate curve through the same point on the surface $\mathscr{G}$ is specified by the unit vector $\mathbf{v}_g$.

The derived expressions for the vectors $\mathbf{U}_g$ and $\mathbf{V}_g$ can be used for the calculation of the fundamental magnitudes of the first order:

$$E_g = \mathbf{U}_g \cdot \mathbf{U}_g, \tag{1.9}$$

$$F_g = \mathbf{U}_g \cdot \mathbf{V}_g \tag{1.10}$$

$$G_g = \mathbf{V}_g \cdot \mathbf{V}_g \tag{1.11}$$

For screw involute surface $\mathscr{G}$, Eqs. (1.9) through (1.11) return:

$$E_g = 1, \tag{1.12}$$

$$F_g = -\frac{r_{b.g}}{\cos\lambda_{b.g}} \tag{1.13}$$

$$G_g = \frac{U_g^2\cos^4\lambda_{b.g} + r_{b.g}^2}{\cos^2\lambda_{b.g}} \tag{1.14}$$

These equations yield an expression for the first fundamental form of the gear tooth flank $\mathscr{G}$:

$$\Phi_{1.g} \Rightarrow dU_g^2 - 2\frac{r_{b.g}}{\cos\lambda_{b.g}}dU_g dV_g + \frac{U_g^2\cos^4\lambda_{b.g} + r_{b.g}^2}{\cos^2\lambda_{b.g}}dV_g^2 \tag{1.15}$$

The discriminant $H_g = \sqrt{E_g G_g - F_g^2}$ of the first fundamental form $\Phi_{1.g}$ of the tooth flank $\mathscr{G}$ is calculated from the formula:

$$H_g = U_g \cos\lambda_{b.g} \tag{1.16}$$

To derive an equation for the second fundamental form $\Phi_{2.g}$ of gear tooth surface $\mathscr{G}$, the second derivatives of $\mathbf{r}_g(U_g, V_g)$ with respect to the $U_g$ – and $V_g$ – parameters are necessary. Equations (1.5) and (1.6) for the vectors $\mathbf{U}_g$ and $\mathbf{V}_g$ yield the expressions:

$$\frac{\partial\mathbf{U}_g}{\partial U_P}(U_g, V_g) = \begin{bmatrix} 0 \\ 0 \\ 0 \\ 1 \end{bmatrix} \tag{1.17}$$

$$\frac{\partial\mathbf{U}_g}{\partial V_g}(U_g, V_g) \equiv \frac{\partial\mathbf{V}_g}{\partial U_g}(U_g, V_g) = \begin{bmatrix} \cos\lambda_{b.g}\cos V_g \\ \cos\lambda_{b.g}\sin V_g \\ 0 \\ 1 \end{bmatrix} \tag{1.18}$$

$$\frac{\partial\mathbf{V}_g}{\partial V_g}(U_g, V_g) = \begin{bmatrix} -r_{b.g}\cos V_g - U_g\cos\lambda_{b.g}\sin V_g \\ -r_{b.g}\sin V_g + U_g\cos\lambda_{b.g}\cos V_g \\ 0 \\ 1 \end{bmatrix} \tag{1.19}$$

for the derivatives of the vectors $\mathbf{U}_g$ and $\mathbf{V}_g$ with respect to $U_g$ – and $V_g$ – parameters.

By definition, the fundamental magnitudes of the second order are represented in the form:

$$L_g = \frac{\frac{\partial\mathbf{U}_g}{\partial U_g} \times \mathbf{U}_g \cdot \mathbf{V}_g}{H_g} \tag{1.20}$$

$$M_g = \frac{\frac{\partial\mathbf{U}_g}{\partial V_g} \times \mathbf{U}_g \cdot \mathbf{V}_g}{H_g} \tag{1.21}$$

$$N_g = \frac{\frac{\partial\mathbf{V}_g}{\partial V_g} \times \mathbf{U}_g \cdot \mathbf{V}_g}{H_g} \tag{1.22}$$

Equations (1.20) through (1.22) allow for the calculation of the second fundamental magnitudes of the helical gear tooth flank surface $\mathscr{G}$:

$$L_g = 0 \tag{1.23}$$

$$M_g = 0 \tag{1.24}$$

$$N_g = -U_g \sin\lambda_{b.g}\cos\lambda_{b.g} \tag{1.25}$$

Ultimately, an equation can be derived to calculate the second fundamental form of helical involute gear tooth flank $\mathscr{G}$:

$$\Phi_{2.g} \quad \Rightarrow \quad -d\mathbf{r}_g \cdot d\mathbf{N}_g = -U_g \sin\lambda_{b.g}\cos\tau\lambda_{b.g} dV_g^2 \tag{1.26}$$

The discriminant, $T_g$, of the second fundamental form $\Phi_{2.g}$ of the tooth flank, $\mathscr{G}$, is determined as:

$$T_g = \sqrt{L_g M_g - N_g^2} = 0 \tag{1.27}$$

Equations (1.15) and (1.26) are helpful for solving geometrical problems pertaining to the design of gear-cutting tools. For example, a simple expression can be used to calculate the actual value of radius $R_g$ for normal curvature at the point of gear tooth flank $\mathscr{G}$: $R_g = \Phi_{1.g}/\Phi_{2.g}$.

Many other parameters of the geometry of the gear tooth flank can be expressed in terms of the first $\Phi_{1.g}$ and the second $\Phi_{2.g}$ fundamental forms of the gear tooth flank $\mathscr{G}$.

The set of six equations for the calculation of fundamental magnitudes of the first $\Phi_{1.g}$ and the second $\Phi_{2.g}$ fundamental forms is known as the *natural parameterization* of the gear tooth flank surface $\mathscr{G}$. These equations are summarized in Table 1.1.

A corresponding equation of the tooth flank of a gear of any design can be derived in a similar manner.

### 1.2.3 Tooth Flank of Straight Bevel Gear

Generation of the gear tooth surface $\mathscr{G}$ of a straight bevel gear can be presented as rolling with no sliding of a plane over the gear base cone (see Figure 1.15). Position vector $\mathbf{r}_g$ of a point of the tooth surface $\mathscr{G}$ allows for the following representation:

**TABLE 1.1**
**Fundamental Magnitudes of Screw Involute Surface $\mathscr{G}$.**

| First Order, $\Phi_{1.g}$ | Second Order, $\Phi_{2.g}$ |
|---|---|
| $E_g = 1$ | $L_g = 0$ |
| $F_g = -\frac{r_{b.g}}{\cos\lambda_{b.g}}$ | $M_g = 0$ |
| $G_g = \frac{U_g^2\cos^4\lambda_{b.g} + r_{b.g}^2}{\cos^2\lambda_{b.g}}$ | $N_g = -U_g\sin\lambda_{b.g}\cos\lambda_{b.g}$ |

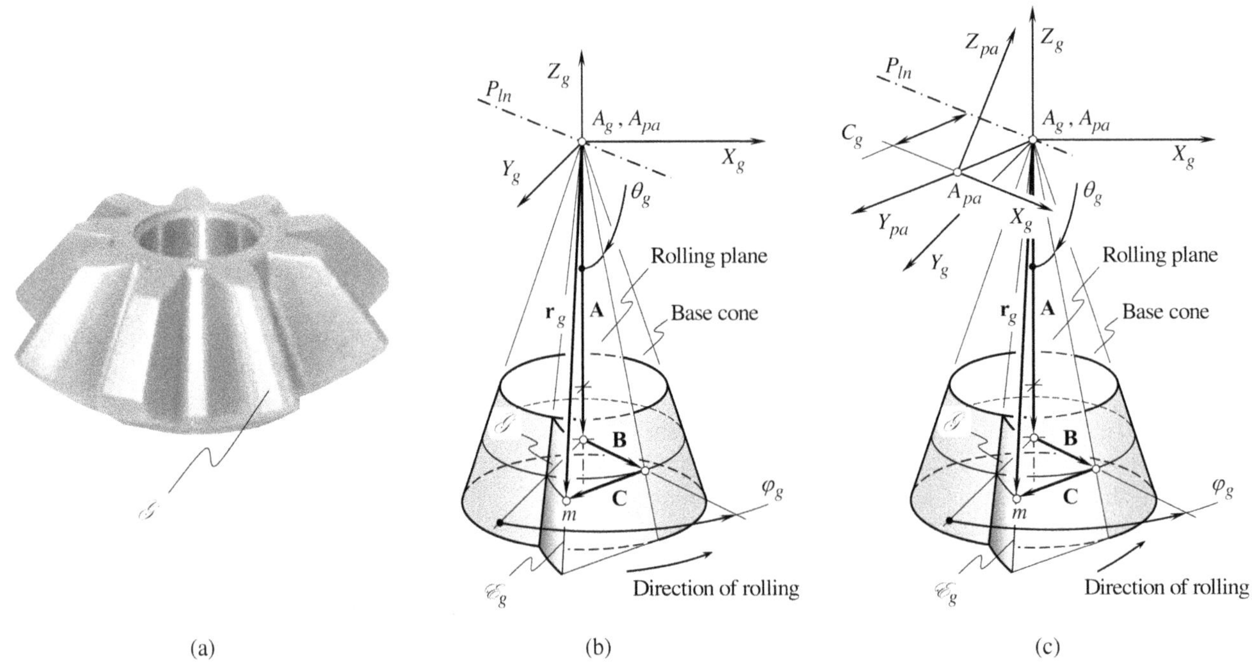


**FIGURE 1.15** Schematic of derivation of the equation of tooth flank $\mathscr{G}$ of a bevel gear: (a) straight bevel gear; (b) schematic of derivation of the equation of the tooth flank of a bevel gear for an intersected-axes gear pair; (c) schematic of derivation of the equation of the tooth flank of a bevel gear for a crossed-axes gear pair.

A straight bevel gear is shown in Figure 1.15(a). In Figure 1.15(b), a schematic for derivation of the equation of the tooth flank of a bevel gear for an intersected-axes gear pair is shown. Here, the position vector of point $\mathbf{r}_g$ of the gear tooth flank is represented as the sum of three vectors:

$$\mathbf{r}_g = \mathbf{A} + \mathbf{B} + \mathbf{C}, \tag{1.28}$$

where:

$$\mathbf{A} = -\mathbf{k} \cdot U_g \tag{1.29}$$

$$\mathbf{B} = \mathbf{i} \cdot U_g \tan\theta_g \sin\varphi_g + \mathbf{j} \cdot U_g \tan\theta_g \cos\varphi_g \tag{1.30}$$

$$\mathbf{C} = -\mathbf{i} \cdot \varphi_g U_g \tan\theta_g \cos\varphi_g + \mathbf{j} \cdot \varphi_g U_g \tan\theta_g \sin\varphi_g \tag{1.31}$$

By substituting vectors **A**, **B**, and **C** into $\mathbf{r}_g = \mathbf{A} + \mathbf{B} + \mathbf{C}$, an equation can be created for the position vector of point $\mathbf{r}_g$ of the surface $\mathscr{G}$ for a straight bevel gear, as represented in the matrix in Eq. (1.32):

$$\mathbf{r}_g(U_g, \varphi_g) = \begin{bmatrix} U_g \tan\theta_g \sin\varphi_g - \varphi_g \cdot U_g \cdot \tan\theta_g \cos\varphi_g \\ U_g \tan\theta_g \cos\varphi_g + \varphi_g \cdot U_g \cdot \tan\theta_g \sin\varphi_g \\ -U_g \\ 1 \end{bmatrix} \tag{1.32}$$

For gears that operate on intersected axes of rotation (see Figure 1.15(b)), the gear tooth flank $\mathscr{G}$ is represented in the reference system $X_g Y_g Z_g$, which originates at the gear base-cone-apex $A_g$. The gear base-cone-apex $A_g$ is located within the axis of instant rotation $P_{ln}$ and is coincident with the *plane-of- action-apex* $A_{pa}$. Reference system $X_g Y_g Z_g$ for the gear tooth flank $\mathscr{G}$ that operates on crossing axes of rotation (see Figure 1.15(c)) is also located at the gear base-cone-apex $A_g$. However, this apex is displaced from the axis of instant rotation $P_{ln}$ at a *gear-center-distance* $C_g$ and does not coincide with the *plane-of- action-apex* $A_{pa}$.

A comparison of Figure 1.15(b) and Figure 1.15(c) reveals that bevel gears for two applications—namely, for intersected-axis and for crossed-axis gearing—feature tooth flanks of different geometry, and thus, these gears are not interchangeable [172].

Evidently, this approach to deriving the equation for a bevel gear tooth flank $\mathscr{G}$ is not the only one possible. There is plenty of room to explore other approaches [171].

### 1.2.4 Tooth Flank of Skew Bevel Gear

For a bevel gear having skew teeth (see Figure 1.16), the position vector $\mathbf{r}_g$ of point $m$ can be expressed in terms of vectors **A**, **B**, **D**, and **E**, which comprise $\mathbf{r}_g$:

$$\mathbf{r}_g = \mathbf{A} + \mathbf{B} + \mathbf{D} + \mathbf{E} \tag{1.33}$$

Here, the vectors **A** and **B** are comprise the vector $\mathbf{r}_g$ for a spur bevel gear. The vectors **D** and **E** are those to be expressed in terms of the geometrical parameters of the gear tooth surface $\mathscr{G}$.

Let us denote the magnitudes of the vectors **D** and **E** as:

$$d = |\mathbf{D}| = |\mathbf{C}| \cdot \tan\psi_{b.g} \tag{1.34}$$

$$e = |\mathbf{E}| = |\mathbf{C}| / \cos\psi_{b.g} \tag{1.35}$$

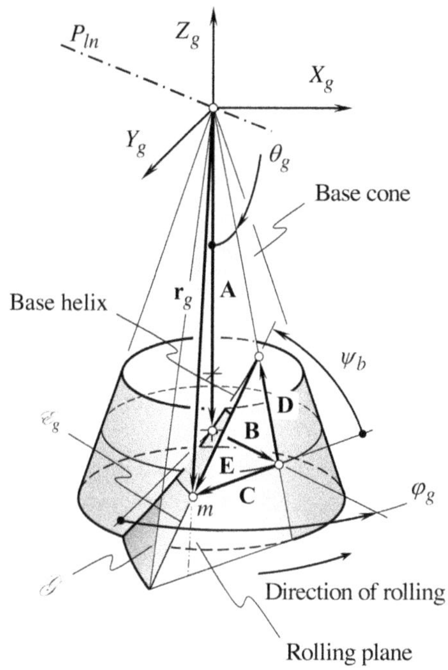


**FIGURE 1.16** Schematic of derivation of the equation of the tooth flank $\mathscr{G}$ of a skew bevel gear.

Here, $|\mathbf{C}| = \varphi_g U_g \tan\theta_g$. Then, the following expression can be produced for vector **D**:

$$\mathbf{D} = -\mathbf{i}\, d \sin\theta_g \sin\varphi_g - \mathbf{j}\, d \sin\theta_g \cos\varphi_g + \mathbf{k}\, d \cos\theta_g \quad (1.36)$$

The vector **E** yields representation in the form of $\mathbf{E} = \mathbf{C} - \mathbf{D}$. Therefore,

$$\begin{aligned}\mathbf{E} = \mathbf{i}\,&(\varphi_g U_g \tan\theta_g \cos\varphi_g - d \sin\theta_g \sin\varphi_g)\\ &+\mathbf{j}(\varphi_g U_g \tan\theta_g \sin\varphi_g + d \sin\theta_g \cos\varphi_g)\\ &-\mathbf{k}\, d \cos\theta_g\end{aligned} \quad (1.37)$$

Substituting $d = \varphi_g U_g \tan\theta_g \tan\psi_{b.g}$ into the equations for vectors **D** and for **E** results in the following expressions:

$$\begin{aligned}\mathbf{D} = &-\mathbf{i}\,\varphi_g U_g \tan\theta_g \tan\psi_{b.g} \sin\theta_g \sin\varphi_g\\ &-\mathbf{j}\,\varphi_g U_g \tan\theta_g \tan\psi_{b.g} \sin\theta_g \cos\varphi_g\\ &+\mathbf{k}\,\varphi_g U_g \tan\theta_g \tan\psi_{b.g} \cos\theta_g\end{aligned} \quad (1.38)$$

$$\begin{aligned}\mathbf{E} = \mathbf{i}\,&(\varphi_g U_g \tan\theta_g \cos\varphi_g - \varphi_g U_g \tan\theta_g \tan\psi_{b.g} \sin\theta_g \sin\varphi_g)\\ &+\mathbf{j}(\varphi_g U_g \tan\theta_g \sin\varphi_g + \varphi_g U_g \tan\theta_g \tan\psi_{b.g} \sin\theta_g \cos\varphi_g)\\ &-\mathbf{k}\,\varphi_g U_g \tan\theta_g \tan\psi_{b.g} \cos\theta_g\end{aligned} \quad (1.39)$$

Ultimately, an expression for the position vector $\mathbf{r}_g$ of the point of the tooth flank $\mathscr{G}$ of a skew bevel gear is represented in matrix form:

$$\mathbf{r}_g = \begin{bmatrix} (\sin\varphi_g - 2\varphi_g \tan\psi_{b.g} \sin\theta_g \sin\varphi_g)\cdot U_g \tan\theta_g \\ [\cos\varphi_g + 2\varphi_g \sin\varphi_g - \tan\psi_{b.g} \sin\theta_g \cos\varphi_g (\tan\theta_g - \varphi_g)]\cdot U_g \tan\theta_g \\ -U_g \\ 1 \end{bmatrix} \quad (1.40)$$

In a similar manner to that described in Section 1.2.2, an analysis of the local geometry at the point of a tooth flank for both a bevel gear and a skew bevel gear can be performed.

When a pair of geometrically accurate intersected-axes gears are engaged in mesh, in both cases (see Figures 1.15 and 1.16), the rolling plane—that is, the plane through the vectors **C** and $\mathbf{r}_g$ is congruent to the plane of action *PA* of the gear pair.

## 1.3 GEAR TOOTH FLANKS THAT *ALLOW FOR SLIDING OVER THEM*

Another cluster of surfaces used for the design of gear teeth comprises those surfaces that *allow for sliding on themselves*. As will be discussed later in this text, this type of surface is also widely used in the design of gear-cutting tools.

There is a certain property of a special type of surface $\mathscr{G}$ that allows for sliding over the surface. When this motion occurs, the surface enveloping consecutive positions of the moving surface $\mathscr{G}$ is congruent to surface $\mathscr{G}$ itself. Generally speaking, this type of motion can be:

- Single-parametric,
- Bi-parametric, or
- Tri-parametric.

The screw surface of constant axial pitch[2] ($p_x = Const$) is the most general type of surface that allows for sliding over itself. When performing a screw motion of that same axial pitch $p_x$, surface $\mathscr{G}$ is sliding over itself, similar to a bolt and a nut. Surfaces of this type are found in the design of helical gears (see Figure 1.5). If the involute gear performs a screw motion about its axis, and the parameter of the screw motion is equal to the parameter of the involute tooth surface, then the surface enveloping consecutive positions of the moving tooth flank $\mathscr{G}$ is congruent to the original moving surface. This means that during such motion, the gear tooth flank *slides over itself.* The same is true with respect to the surface of a cylindrical worm (see Figure 1.8).

For the general type of cylinders,[3] the axial pitch of the screw surface inflates to an infinitely large value. Surfaces of that type allow straight motion along the straight generating lines of the surface. The tooth flanks of spur gears (see Figures 1.2 and 1.3) are shaped in the form of general-type cylinders (the so-called *surfaces of translation*). When a spur gear performs a straight motion along its axis, the surface enveloping consecutive positions of the moving tooth flank $\mathscr{G}$ is congruent to itself. This means that as a motion of this type occurs, the teeth flanks of the spur gear slide over themselves.

Finally, the axial pitch of the screw surface could be equal to zero (that is, $p_x = 0$), in which case the screw surface

[2] Also called *general helicoid.*
[3] Also called *surface of translation.*

FIGURE 1.17 A spur gear with modified (*barreled*) teeth.

reduces to a surface of revolution. Every surface of revolution slides over itself when it is rotated about its axis. For example, the tooth flank of the modified (*barreled* or *crowned*) spur gear (see Figure 1.17) could be shaped in the form of a surface of revolution. A schematic layout of how the tooth flank of a modified spur gear can be shaped in the form of a surface of revolution is shown in Figure 1.18.

Screw surfaces of constant axial pitch $p_x = Const$ cylinders of the general type and surfaces of revolution are the only surfaces that allow a mono-parametric motion that causes the surface to slide over it.

Surfaces like a round cylinder allow rotation as well as straight motion along the axis of the cylinder. In this case, the surface motion is bi-parametric (rotation and translation can be performed independently).

A sphere allows rotations of about three axes independently. A plane surface allows straight motion in two different directions, as well as a rotation about an axis that is orthogonal to the plane. The surface motion in these two cases (for a sphere and for plane) is tri-parametric.

Ultimately, surfaces *allowing for sliding over themselves* can be summarized as being limited to:

1. Screw surfaces of constant axial pitch.
2. Cylinders of the general type (surfaces of translation).
3. Surfaces of revolution.
4. Round cylinders.
5. Spheres.
6. Plane.

Research has found [144, 148, 168, 189] no other types of surfaces that share this particular property.

A brief look at the gears shown in Figure 1.9 through Figure 1.12 reveals that none of these tooth flanks *allow for sliding on themselves*.

Surfaces that *allow for sliding on themselves* have very convenient uses in manufacturing, as well as in other engineering applications. Most of the surfaces that are machined in various industries are surfaces of this nature, and many gear-machining processes are utilizing surfaces with this property. Almost all designs of gear-cutting tools are also developed on the premises of surfaces that allow sliding.

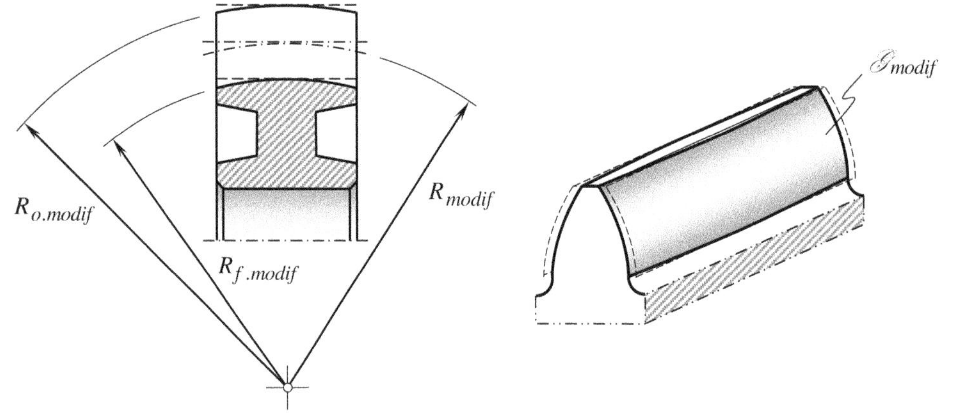


FIGURE 1.18 A spur gear with modified (*crowned*) teeth (tooth flanks of the gear are shaped in the form of a surface of revolution).

# 2 Principal Kinematics of a Gear-Machining Operation

In a gear-machining operation, a gear and a gear-cutting tool are viewed as a virtual mechanism with two functional surfaces, $\mathscr{G}$ and $T$, respectively, that interact with each other. When solving problems of gear-cutting tool design, two types of problems are distinguished: *direct* and *inverse*.

The parameters of the generating surface $T$ of a gear-cutting tool are determined as solutions to direct problems of gear-cutting tool design. An actual gear-cutting tool is designed on the premise of this generating surface $T$.

Parameters of the generated tooth flank $\mathscr{G}_{act}$ of the gear are determined as solutions to *inverse* problems of gear-cutting tool design. The accuracy of the generation of the gear tooth flank is dependent on the derived generating surface $T$ of the gear-cutting tool.

The geometry of the generating surface is the main output when solving the direct problem of the theory of part surface generation [33, 148, 151, 158]. In this chapter, the generating surface $T$ of the gear-cutting tool is viewed in the context of being an enveloping (in the case of a form gear-cutting tool) or a conjugate[1] (in the case of a generating gear-cutting tool) surface to consecutive positions of the gear tooth flank $\mathscr{G}$ in the motion of the work gear, relative to a reference system at which the generating surface $T$ will be analytically described. When continuous gear-machining indexing methods of are used, the generating surface of the geometrically accurate gear-cutting tool is conjugate to the tooth flank of the gear to be machined. In this case, surface $T$ can be congruent to tooth surface $\mathscr{G}$ of the gear.

Most of the major elements of the geometry of gear-cutting tools can be calculated as a solution to the direct problem of the theory of part surface generation.

The geometry of the gear tooth flank is the first necessary major component for solving the direct problem of part surface generation. Gear tooth flank $\mathscr{G}$ can be uniquely specified based on the data available from the gear blueprint. Examples of the gear tooth flank to be machined are briefly discussed in Chapter 1.

The kinematics of the gear-machining operation is the second necessary major component for solving the problem under consideration.

[1] To clarify the difference between enveloping and conjugate surfaces: Conjugate surfaces are always enveloping to each other, but the inverse is not true—that is, not all enveloping surfaces are conjugate to one another. At every point of the line of contact of two conjugate surfaces, the common perpendicular intersects the axis of instant rotation of the surfaces. *Reversibly enveloping surfaces* or just $R_e$ – *surfaces* (for simplicity) is another term for the conjugate surfaces [152].

The development of the generalized theory of part surface generation on machine tools can be traced back to the middle of twentieth century [25, 56, 192, 195]. The body of research on part surface generation was later significantly enhanced by Prof. S.P. Radzevich [133, 144, 148, 151, 158, 168], who is the principal developer of the DG/K-based approach of part surface generation. In the following discussion, the kinematics of the gear-machining process, with a focus on gear-cutting tool design, is largely based on the research results obtained by the author.

## 2.1 RELATIVE MOTIONS IN THE GEAR-MACHINING PROCESS

Every gear-machining operation features certain kinematics of part surface machining. For example, when grinding a helical involute gear with an abrasive worm (see Figure 2.1), the work gear rotates about its axis with a rotation $\omega_g$. The grinding worm rotates about its axis with an angular velocity $\omega_c$. Rotations $\omega_g$ and $\omega_c$ are synchronized. When spur gears are machined, the required ratio $\omega_g \cdot N_g = \omega_c \cdot N_c$ in the gear-machining operation must be fulfilled. Here, $N_g$ denotes the work gear tooth number, and $N_c$ is the number of starts of the abrasive worm.

In addition to synchronization of rotations $\omega_g$ and $\omega_c$ with the tooth counts $N_g$ and $N_c$, the work gear and the gear-cutting tool travel relative to each other, with a feed $F_c$ in the axial direction of the gear. No kinematic restrictions are imposed on feed $F_c$ when spur gears are machined. If a helical gear is machined, then, in addition to being synchronized with one another, the parameters of the kinematics of the gear-machining operation must be synchronized, so as to account for the pitch helix angle $\psi_g$ of the work gear and the axial feed $F_c$.

The example (see Figure 2.1) illustrates that relative motions of the work gear and the gear-cutting tool are critical elements in every gear-machining operation. It also reveals that the actual variety of feasible relative motions of the work gear blank and the gear-cutting tool is not limited to just a few motions. It is strongly recommended to take into account all feasible motions of the gear-cutting tool with respect to the work gear if one aims to develop an optimal design for the gear-cutting tool. Only on the basis of a profound analysis of the kinematics of the gear-machining operation can a favorable design for the gear-cutting tool can be developed.

DOI: 10.1201/9781003520528-3

FIGURE 2.1 Key elements of kinematics of a gear-machining operation.

### 2.1.1 Elementary Relative Motions of the Work Gear and the Gear-cutting Tool

The geometry of the work gear tooth flank is the only information available to the gear-cutting tool designer when embarking on designing a gear-cutting tool.

Consider a Cartesian coordinate system $X_gY_gZ_g$ that is associated with the work gear, as shown in Figure 2.2. In the coordinate system $X_gY_gZ_g$, the gear tooth flanks $\mathscr{G}$ can be analytically described—for example, by an equation in either a vector or a matrix representation.

The geometry of the generating surface of the gear-cutting tool for machining a given work gear is not known at this point. Therefore, it is not a gear-cutting tool, but the coordinate system $X_{ct}Y_{ct}Z_{ct}$ can be considered instead. After it has been designed, the gear-cutting tool will be described in coordinate system $X_{ct}Y_{ct}Z_{ct}$.

Let us assume that any desirable motion of the gear-cutting tool with respect to the work gear can be performed on a specified machine tool. Any motion of the gear-cutting tool relative to the work gear can be substituted with the corresponding motion of coordinate system $X_{ct}Y_{ct}Z_{ct}$ with respect to coordinate system $X_gY_gZ_g$. This is because coordinate system $X_gY_gZ_g$ is rigidly associated with the work gear, and coordinate system $X_{ct}Y_{ct}Z_{ct}$ is rigidly connected to the gear-cutting tool to be designed.

Without loss of generality, coordinate system $X_gY_gZ_g$ can be assumed to be stationary. In this scenario, the principle of inversion of relative motions can be implemented. Following this principle, in addition to its original motions, coordinate system $X_{ct}Y_{ct}Z_{ct}$ performs all the motions originally performed by coordinate system $X_gY_gZ_g$. However,

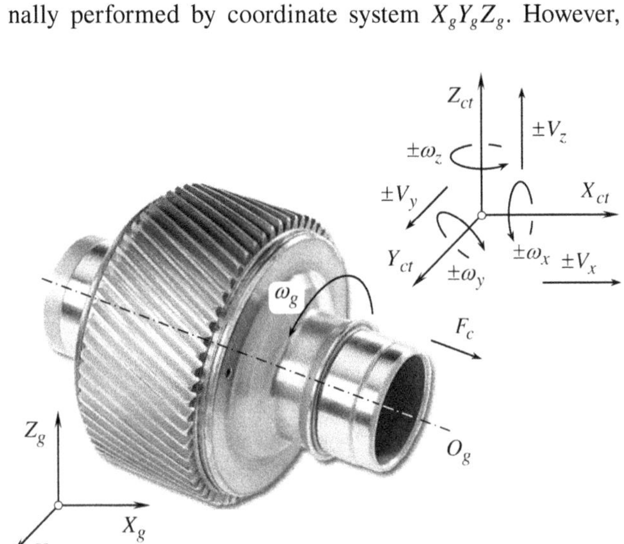


FIGURE 2.2 Permissible elementary relative motions of the work gear and the gear-cutting tool.

the additional motions of coordinate system $X_{ct}Y_{ct}Z_{ct}$ are performed opposite of their original directions.

Any resultant relative motion of the gear-cutting tool relative to the work gear can be resolved into translations along and rotations about the axes of coordinate system $X_{ct}Y_{ct}Z_{ct}$. The total number of elementary translations is limited to three translations: $\pm V_x$, $\pm V_y$, and $\pm V_z$. Similarly, the total number of elementary rotations is limited to three rotations: $\pm \omega_x$, $\pm \omega_y$, and $\pm \omega_z$.

A brief look at the schematic shown in Figure 2.2 makes it clear that not all elementary translations, elementary rotations, or their combinations are permissible when machining a given work gear. Motions of the gear-cutting tool relative to the work gear that were not permissible were eliminated from further consideration. To do this, it is necessary to distinguish the relative motions of the gear-cutting tool that are permissible in machining a given gear from those that are not. The relative motions of the gear-cutting tool relative to the work have to be considered in close connection with the geometry of the work gear tooth flanks [134, 144, 148]. Peculiarities of the geometry of tooth flanks to be machined entail the corresponding peculiarities in the kinematics of the gear-machining operation.

### 2.1.2 Permissible Relative Motions of the Work Gear and the Gear-cutting Tool

Generally speaking, the relative motion of the work gear and the gear-cutting tool is a type of spatial motion. When machining a gear, the gear-cutting tool travels with respect to the work gear. While traveling, the gear-cutting tool removes the stock from the gear blank. Machining the gear is accomplished when the whole stock is removed from the work gear teeth.

To distinguish the permissible relative motions of the gear-cutting tool relative to the work gear from those that are not permissible, it is necessary to conduct a comprehensive analysis of the gear-cutting tool's spatial motion with respect to the machined gear. However, when just the qualitative analysis of relative motions is sufficient, for simplicity, a 2D rather than a 3D analysis can be performed. With no principal restrictions, the results of a 2D analysis could be easily extended to a 3D case.

Generally, three phases are distinguished in the gear-cutting tool's relative motion with respect to the work gear:

1. The *roughing phase* is observed when a tooth of the gear-cutting tool approaches the work gear tooth surface.
2. The *generating phase* is observed when the motion of the tooth of the gear-cutting tool moves tangentially to the work gear tooth flank
3. The *departure phase* is observed when a tooth of the gear-cutting tool breaks contact with the machined part surface of the work gear; neither roughing nor surface generation occurs.

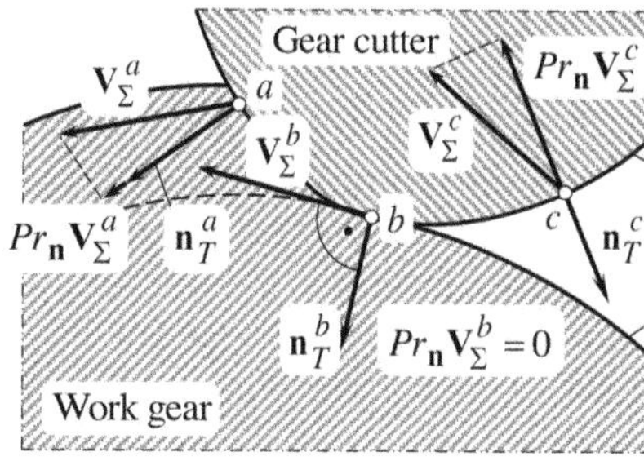


**FIGURE 2.3** Permissible and impermissible motions of gear-cutting tool $T$ relative to the work gear.

The kinematic analysis in this book is based on particular locations of vector $\mathbf{V}_{c/w} \equiv \mathbf{V}_\Sigma$ of the motion of the generating surface $T$ of the gear-cutting tool relative to the work gear.

Let's take a more in-depth look at the generation of the tooth flank $\mathcal{G}$ being machined.

The gear-cutting tool performs a certain motion relative to the work gear. The gear tooth flank $\mathcal{G}$ is generated as an envelope surface to consecutive positions of generating surface $T$ of the gear-cutting tool. There is an instant when three different types of points can be distinguished on generating surface $T$.

At point $a$ (see Figure 2.3), the vector of the resultant motion of the gear-cutting tool with respect to the work gear is denoted by $\mathbf{V}_\Sigma^a$. Projection $\mathrm{Pr}_\mathbf{n}\, \mathbf{V}_\Sigma^a$ of vector $\mathbf{V}_\Sigma^a$ onto the unit's normal vector $\mathbf{n}_T^a$ to generating surface $T$ is pointed to the interior of the body of the work gear. Therefore, in the vicinity of point $a$, the gear-cutting tool penetrates the body of the work gear. In this manner, the roughing cutting edges of the cutting tool cut out the stock. No geometrical or kinematic constraints are imposed on the direction of vector $\mathbf{V}_\Sigma^a$. Points such as point $a$ are observed in the *roughing phase* of the relative motion of the gear-cutting tool.

Relative motions such as $\mathbf{V}_\Sigma^a$ are of critical importance when designing the roughing teeth of gear-cutting tools.

At contact point $b$ (see Figure 2.3), the vector of the resultant motion of the gear-cutting tool with respect the work gear is denoted by $\mathbf{V}_\Sigma^b$. The projection $\mathrm{Pr}_\mathbf{n}\, \mathbf{V}_\Sigma^b$ of the vector $\mathbf{V}_\Sigma^b$ onto the unit's normal vector $\mathbf{n}_T^b$ to the generating surface $T$ at the point $b$ is perpendicular to the normal vector—that is, it is tangent to the work gear tooth flank, $\mathcal{G}$—and, therefore, the equality $\mathrm{Pr}_\mathbf{n}\, \mathbf{V}_\Sigma^b = 0$ is valid. Given this, when in the differential vicinity of point $b$, the cutting tool does not penetrate the body of the work gear. The cutting edges of the gear-cutting tool do not cut out the stock. Generating surface $T$ of the gear-cutting tool generates the work gear tooth surface $\mathcal{G}$ in the vicinity of the point $b$. At point $b$, the motion of the gear-cutting tool toward the machined tooth surface $\mathcal{G}$ is not permissible. The vector $\mathbf{V}_{c/w} \equiv \mathbf{V}_\Sigma$ is located within the common tangent plane to surfaces $\mathcal{G}$ and $T$ at point $b$. This means that only two types of motion are permissible for moving surface $T$ and the envelope tooth flank $\mathscr{G}$. The surfaces $\mathscr{G}$ and $T$ can roll and slide over one another. The component of the resultant relative motion $\mathbf{V}_\Sigma^b$ in the direction of the common perpendicular to the surfaces is always equal to zero (see Figure 2.3). Points such as point $b$ are observed in the *generating phase* of the relative motion of the gear-cutting tool.

Relative motions such as $\mathbf{V}_\Sigma^b$ are investigated in this text in detail. They are of critical importance when designing finishing teeth of gear-cutting tools.

At point $c$ (see Figure 2.3), the vector of the resultant motion of the gear-cutting tool with respect the work gear is denoted by $\mathbf{V}_\Sigma^c$. Projection $\text{Pr}_{\mathbf{n}}\,\mathbf{V}_\Sigma^c$ of vector $\mathbf{V}_\Sigma^c$ onto the unit's normal vector $\mathbf{n}_T^c$ to generating surface $T$ is pointed outside of the gear body. Therefore, in the vicinity of point $c$, the cutting edges of the gear-cutting tool leave the machined gear tooth flank $\mathscr{G}$. Points such as point $c$ are observed in the *departing phase* of the relative motion of the gear-cutting tool.

Evidently, in the vicinity of the third type of points, the tool cutting edges do not cut out the stock, and the generating surface $T$ of the gear-cutting tool does not generate the work gear tooth flank $\mathscr{G}$.

The type of relative motions represented by $\mathbf{V}_\Sigma^c$ are not covered in detail in this discussion.

When analyzing possible motions of the gear-cutting tool relative to the work gear (see Figure 2.2), the results derived from the analysis of Figure 2.3 are useful.

## 2.2 ROLLING OF CONJUGATE SURFACES IN RELATION TO ONE ANOTHER

Two types of relative motions are of particular interest to the gear-cutting tool designer. The first is the sliding motion of surfaces $\mathscr{G}$ and $T$ over themselves. Very few types of surfaces allow for this sort of motion. This type of surface is briefly discussed in Section 1.3. The interested reader is referred to [166, 189] and other advanced sources for more details on this subject.

The second type of relative motion of interest is the rolling of the generating surface $T$ of the gear-cutting tool over the work gear tooth flank $\mathscr{G}$. Fulfillment of the *Shishkov equation of contact*, $\mathbf{n}\cdot\mathbf{V}_\Sigma = 0$, is a necessary analytical requirement for rolling motion.[2] Conjugacy of interacting surfaces $\mathscr{G}$ and $T$ is also necessary in the rolling motion of the surfaces.

In most cases of rolling, sliding of interacting surfaces $\mathscr{G}$ and $T$ also occurs. Because of this, pure rolling is distinguished from rolling with sliding. In the first case, no sliding of surfaces $\mathscr{G}$ and $T$ is observed ($\mathbf{V}_\Sigma = 0$). In the second ($\mathbf{V}_\Sigma \neq 0$), surfaces $\mathscr{G}$ and $T$ move with respect to each other in a direction that is within the common tangent plane.

When tooth flank $\mathscr{G}$ of the work gear and generating surface $T$ of the gear-cutting tool roll over each other—either with or without sliding—two auxiliary surfaces can be rigidly associated with them. In certain cases, auxiliary surfaces have a property that allows the surfaces to roll over each other with no sliding between them. Auxiliary surfaces of this geometry are referred to as the *axodes*. The axode of the gear tooth surface is rigidly connected to tooth flank $\mathscr{G}$. The axode of the generating surface $T$ of the gear-cutting tool is rigidly connected to the gear-cutting tool itself. Thus, the relative motion of the work gear and the gear-cutting tool can be viewed as the rolling of one axode over another.

Depending on the actual configuration of the axes of rotation of the axodes, the axodes' instant relative motion can be interpreted as a type of instant screw motion. This instant screw motion features a certain axis, as well as a screw parameter of the motion. Relative motion of the general type of axodes is not relevant for the gear-cutting tool designer; however, specific types of relative motion of the general type could be.

The relative screw motion of the axodes can occur about the screw motion axis, the location and orientation of which do not alter in space. This type of relative motion of axodes is referred to as an *ordinary screw motion*. The ordinary screw motion can be broken down into two elementary motions: a translation and a rotation. Furthermore, the speed of one of these two elementary motions can be equal to zero. If the speed of the translation motion becomes zero, then the ordinary screw motion reduces, corresponding to the rotation. Conversely, when the speed of the rotation is equal to zero, the elementary screw motion reduces corresponding to the translation. In this manner, the rolling of axodes in the form of ordinary screw surfaces reduces to the rolling of axodes of a simpler shape. The rolling of two cylinders of revolution with parallel axes is a perfect example in this setup.

Particular kinds of relative motions of the work gear and the gear-cutting tool are of critical importance for the gear-cutting tool designer. No doubt, this issue warrants more detailed consideration.

## 2.3 PRINCIPAL PLANES AND MAIN REFERENCE SYSTEMS ASSOCIATED WITH GEAR-MACHINING OPERATION

For a specified gear and gear-cutting tool with a known set of design parameters, a corresponding gear vector diagram for the rotation vectors can be constructed. Several principal directions are associated with a *gear-to-gear cutting-tool pair* (also called a *gear-machining mesh*). It can be said that the principal directions are associated with the gear-machining mesh in natural manner. These directions are defined by:

a. The rotation vector of gear $\boldsymbol{\omega}_g$.
b. The rotation vector of the gear-cutting tool $\boldsymbol{\omega}_{ct}$.
c. The vector of instant rotation $\boldsymbol{\omega}_{pl}$.
d. The centerline, $\mathbb{C}$.

[2] Prof. V.A. Shishkov was the first to propose the equation $\mathbf{n}\cdot\mathbf{V}_\Sigma = 0$ (as early as 1948 [205, 207]) for the analytical description of the condition of contact, which was verbally formulated earlier by F. Reuleaux [194].

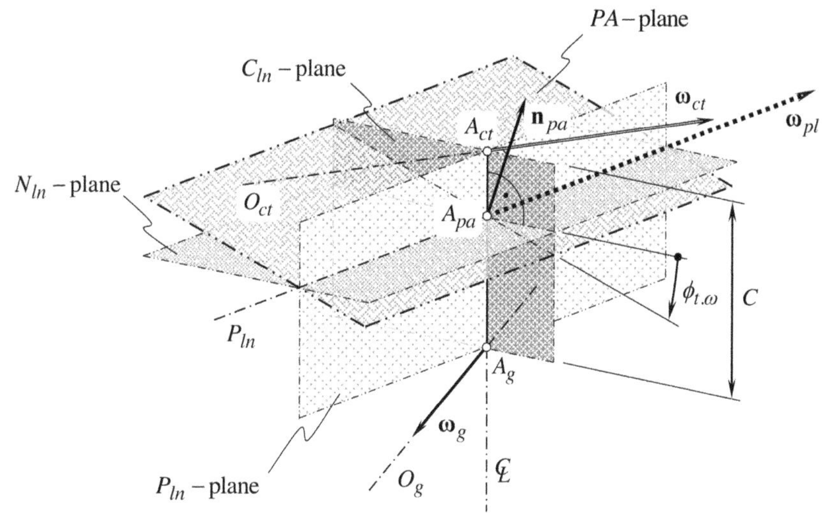


**FIGURE 2.4** Principal planes associated with gear machining mesh.

The use of principal directions allows for the construction of a set of principal planes and a set of main reference systems associated with a gear machining mesh. An analysis of any and all kinds of gear-machining operations is significantly simplified when the principal planes and the main reference systems are used for this purpose.

Let us begin the analysis with consideration of the principal planes associated with a gear machining mesh.

### 2.3.1 Principal Planes Associated with Gear-Machining Mesh

Consider an arbitrary gear machining mesh, and construct the axis of instant rotation $P_{ln}$ and the centerline ℄ for this mesh.

A set of three principal planes can be associated with the gear machining mesh.

The *pitch-line plane* ($P_{ln}$ – *plane*) is the first of three principal planes associated with gear machining mesh (see Figure 2.4).

> **Definition 2.1:** *The pitch-line plane is the plane through the axis of instant rotation $P_{ln}$ and the centerline ℄ in the gear machining mesh.*

At the *plane-of-action-apex* $A_{pa}$, the principal unit normal vector $\mathbf{n}_{p.ln}$ is constructed to the $P_{ln}$ – plane. The parameters of the normal vector $\mathbf{n}_{p.ln}$ can be expressed in terms of the parameters of the $P_{ln}$ – plane, and vice versa—the $P_{ln}$ – plane can be expressed in terms of the unit normal vector $\mathbf{n}_{p.ln}$. The principal unit normal vector $\mathbf{n}_{p.ln}$ is used in calculations of an angle formed by the pitch-line-plane and other planes, as well as in other applications.

The *centerline plane* ($C_{ln}$ – *plane*) is the second principal plane associated with gear machining mesh (see Figure 2.4).

> **Definition 2.2:** *The centerline plane is the plane through the centerline ℄ in the gear pair, perpendicular to the axis of instant rotation $P_{ln}$.*

At the plane-of-action-apex $A_{pa}$, the principal unit normal vector $\mathbf{n}_{c.ln}$ is constructed to the $C_{ln}$ – plane. The parameters of the normal vector $\mathbf{n}_{c.ln}$ can be expressed in terms of the parameters of the $C_{ln}$ – *plane*, and vice versa. The principal unit vector $\mathbf{n}_{c.ln}$ is used in calculations of an angle formed by the centerline plane and other planes, as well as in other applications.

Finally, the normal plane ($N_{ln}$ – *plane*) is the third of three principal planes associated with gear machining mesh (see Figure 2.4).

> **Definition 2.3:** *The normal plane is the plane through the plane-of-action-apex $A_{pa}$, perpendicular to the centerline ℄ in the gear machining mesh.*

At the plane-of-action-apex $A_{pa}$, the principal unit normal vector $\mathbf{n}_{n.ln}$ is constructed to the $N_{ln}$ – plane. The parameters of the normal vector $\mathbf{n}_{n.ln}$ can be expressed in terms of the parameters of the $N_{ln}$ – plane, and vice versa—the $N_{ln}$ – *plane* can be expressed in terms of the unit normal vector $\mathbf{n}_{n.ln}$. The unit vector $\mathbf{n}_{n.ln}$ is used in calculations of an angle formed by the normal plane and other planes, as well as in other applications.

Altogether, these three planes—$P_{ln}$ – *plane*, $C_{ln}$ – *plane*, and $N_{ln}$ – *plane*—form a set of principal planes associated with gear machining mesh (see Figure 2.4).

Once a set of principal planes in a gear machining mesh is determined, it is possible to introduce the plane of action *PA* in the gear machining mesh (see Figure 2.4) that complements the set.

> **Definition 2.4:** *The plane of action in a gear machining mesh is the plane through the axis of instant rotation $P_{ln}$ at a transverse pressure angle $\phi_{t.\omega}$, in relation to the perpendicular to the centerline ℄ of the gear machining mesh that is constructed within the centerline plane.*

The plane of action $PA$ forms a transverse pressure angle $\phi_{t.\omega}$ with the $N_{ln}$ – *plane*.

A schematic configuration of the plane of action $PA$ in relation to the set of principal planes in a gear machining mesh is shown in Figure 2.4.

At the plane-of-action-apex $A_{pa}$, a principal unit normal vector $\mathbf{n}_{pa}$ is constructed to the plane of action $PA$. The parameters of the normal vector $\mathbf{n}_{pa}$ can be expressed in terms of the parameters of the plane $PA$, and vice versa—the plane of action $PA$ can be expressed in terms of the unit normal vector $\mathbf{n}_{p.ln}$. The unit vector $\mathbf{n}_{pa}$ is used in calculations of an angle formed by the plane of action and other planes, as well as in other applications.

The provided definition of the term *plane of action PA* is also valid for the reduced cases of gear machining meshes—namely, for intersected axes—as well as for gear machining meshes.

### 2.3.2 Main Reference Systems Associated with Gear Machining Mesh

For the analytical description and investigation of gear machining mesh kinematics, as well as the geometry of the interacting tooth flanks of a gear and that of the gear-cutting tool, corresponding reference systems are associated with the gear, with the gear-cutting tool, and with the gear machining mesh. Orthogonal Cartesian coordinate systems are used extensively for these purposes.

Both the kinematics of machining and the geometry of the tooth flanks of a gear and of the gear-cutting tool can be analytically described in an arbitrary reference system. However, it is convenient to choose reference systems so as to simplify analytical expressions for all the elements of the gear machining mesh.

The simplest possible analytical representation of an element of a gear machining mesh can be derived from specific reference systems. These reference systems are associated with the gear machining mesh in a specific manner. Reference systems of this type are referred to as the gear machining mesh's *main reference systems*.

To determine a set of main reference systems, consider a *gear vector diagram* for an arbitrary gear machining mesh (see Figure 2.5). The rotation vectors $\boldsymbol{\omega}_g$ of the gear and $\boldsymbol{\omega}_{ct}$ of the gear-cutting tool are positioned at a center-distance $C$, and they cross each other. Points $A_g$ and $A_{ct}$ are the points of intersection of the centerline ℄ by the gear axis of rotation $O_g$ and the gear-cutting tool axis of rotation $O_{ct}$, respectively. Point $A_g$ is referred to as the *gear base cone apex*, and point $A_{ct}$ is referred to as the *gear-cutting tool base cone apex*.

The vector of instant rotation $\boldsymbol{\omega}_{pl}$ of the gear-cutting tool in relation to the gear is a vector through the point $A_{pa}$. This point is located within the centerline ℄. Point $A_{pa}$ is referred to as the *plane-of-action-apex*.

The axis of instant rotation $P_{ln}$ is the straight line through the plane-of-action-apex $A_{pa}$ that is pointed along the vector of instant rotation $\boldsymbol{\omega}_{pl}$.

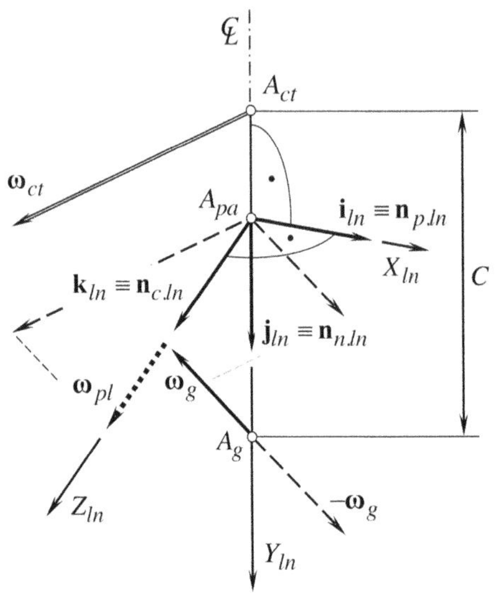


**FIGURE 2.5** Main reference systems associated with gear machining mesh.

Two straight lines through a common point uniquely specify a plane through this point. In the case under consideration, this is the plane through both the axis of instant rotation $P_{ln}$ and the centerline ℄.

The main reference system $X_{ln}Y_{ln}Z_{ln}$ is the first principal reference system associated with gear machining mesh. The origin of the main reference system $X_{ln}Y_{ln}Z_{ln}$ is coincident with the plane-of-action-apex $A_{pa}$. The axes $X_{ln}$, $Y_{ln}$, and $Z_{ln}$ are pointed along the lines of intersection of the principal planes, as illustrated in Figure 2.5: The axis $Z_{ln}$ is along the vector of instant rotation $\boldsymbol{\omega}_{pl}$; the axis $Y_{ln}$ is along the centerline ℄; and the $X_{ln}$ – axis complements the axes $Y_{ln}$ and $Z_{ln}$ to a left-hand-oriented Cartesian coordinate system $X_{ln}Y_{ln}Z_{ln}$. The unit vectors along the axes $X_{ln}$, $Y_{ln}$, and $Z_{ln}$ are labeled $\mathbf{i}_{ln}$, $\mathbf{j}_{ln}$, and $\mathbf{k}_{ln}$, respectively. It is evident that the identities $\mathbf{i}_{ln} \equiv \mathbf{n}_{p.ln}$, $\mathbf{j}_{ln} \equiv \mathbf{n}_{n.ln}$, and $\mathbf{k}_{ln} \equiv \mathbf{n}_{c.ln}$ are valid.

The stationary gear reference system $X_{g.s}Y_{g.s}Z_{g.s}$ is the second principal reference system associated with gear machining mesh. The origin of the reference system $X_{g.s}Y_{g.s}Z_{g.s}$ is coincident with the gear base cone apex $A_g$, as shown in Figure 2.6. The axis $Z_{g.s}$ is pointed along the rotation vector of the gear $\boldsymbol{\omega}_g$; the axis $Y_{g.s}$ is along the centerline ℄; and the $X_{g.s}$ – axis complements the axes $Y_{g.s}$ and $Z_{g.s}$ to a left-hand-oriented Cartesian coordinate system $X_{g.s}Y_{g.s}Z_{g.s}$. The stationary gear reference system $X_{g.s}Y_{g.s}Z_{g.s}$ is turned through the gear cone angle $\Sigma_g$ about $Z_{ln}$ – axis of the reference system $X_{ln}Y_{ln}Z_{ln}$ associated with a gear pair.

The gear reference system $X_gY_gZ_g$ is the third principal reference system associated with the gear machining mesh. The reference system $X_gY_gZ_g$ is rigidly associated with the gear and is rotated together with the gear about its axis of rotation $O_g$. The reference systems $X_gY_gZ_g$ and $X_{g.s}Y_{g.s}Z_{g.s}$ share the common axis $Z_g \equiv Z_{g.s}$. The reference system $X_gY_gZ_g$ is turned about the $Z_{g.s}$ – axis through a gear rotation angle $\varphi_g$.

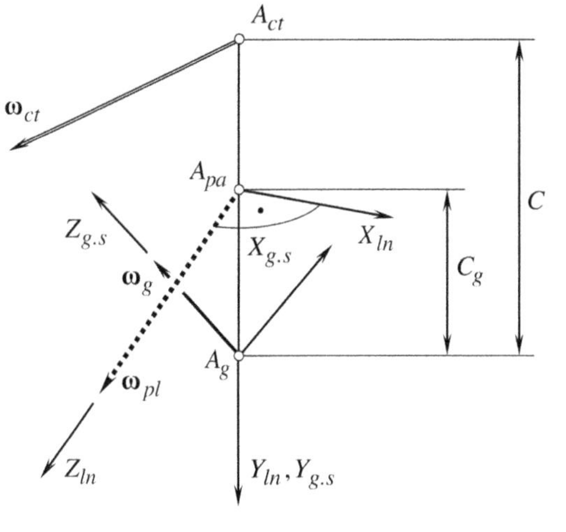


**FIGURE 2.6** Configuration of the motionless gear principal reference system $X_{g.s}Y_{g.s}Z_{g.s}$ in relation to the main principal reference system $X_{ln}Y_{ln}Z_{ln}$ associated with the gear machining mesh.

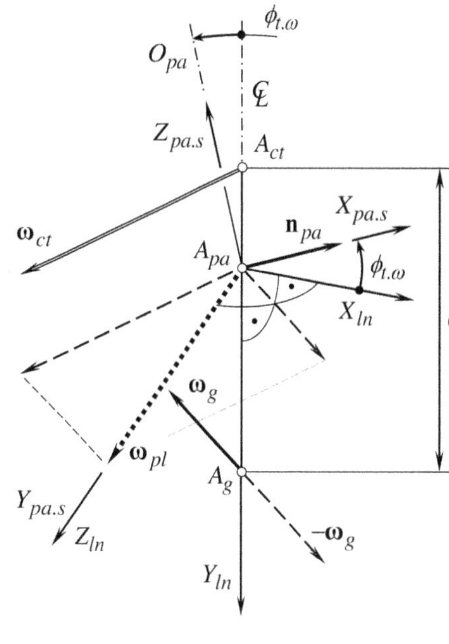


**FIGURE 2.7** Configuration of the stationary plane-of-action principal reference system $X_{pa.s}Y_{pa.s}Z_{pa.s}$ in relation to the main principal reference system $X_{ln}Y_{ln}Z_{ln}$ associated with a *gear machining mesh.*

The stationary gear-cutting tool reference system $X_{ct.s}Y_{ct.s}Z_{ct.s}$ is the fourth principal reference system associated with the gear machining mesh. The origin of the reference system $X_{ct.s}Y_{ct.s}Z_{ct.s}$ is coincident with the gear-cutting-tool apex $A_{ct}$. The axis $Z_{ct.s}$ is pointed along the rotation vector of the gear-cutting tool $\omega_{ct}$; the axis $Y_{ct.s}$ is along the centerline ℄; and the $X_{ct.s}$ – axis complements the axes $Y_{ct.s}$ and $Z_{ct.s}$ of the left-hand-oriented Cartesian coordinate system $X_{ct.s}Y_{ct.s}Z_{ct.s}$. The stationary gear-cutting tool reference system $X_{ct.s}Y_{ct.s}Z_{ct.s}$ is turned through the pinion cone angle $\Sigma_{ct}$ about the $Z_{ln}$ – axis of the principal reference system $X_{ln}Y_{ln}Z_{ln}$ associated with a gear machining mesh.

The gear-cutting tool reference system $X_{ct}Y_{ct}Z_{ct}$ is the fifth principal reference system associated with the gear machining mesh. Reference system $X_{ct}Y_{ct}Z_{ct}$ is rigidly associated with the gear-cutting tool and is rotated together with the gear-cutting tool about its axis of rotation $O_{ct}$. The reference systems $X_{ct}Y_{ct}Z_{ct}$ and $X_{ct.s}Y_{ct.s}Z_{ct.s}$ share the common axis $Z_{ct} \equiv Z_{ct.s}$. The reference system $X_{ct}Y_{ct}Z_{ct}$ is turned in relation to the reference system $X_{ct.s}Y_{ct.s}Z_{ct.s}$ about the $Z_{p.s}$ – axis through a gear-cutting tool rotation angle $\varphi_{ct}$.

A gear rotation angle $\varphi_g$ and a *gear-cutting tool rotation angle* $\varphi_{ct}$ correlate to each other in the following manner: $\varphi_{ct} = u\,\varphi_g$, where $u$ is the gear ratio in the *gear machining mesh.*

The five *principal reference systems* introduced are as follows:

- The *main principal* reference system $X_{ln}Y_{ln}Z_{ln}$.
- The *stationary gear principal* reference system $X_{g.s}Y_{g.s}Z_{g.s}$.
- The *gear principal* reference system $X_gY_gZ_g$.
- The *stationary gear-cutting tool principal* reference system $X_{ct.s}Y_{ct.s}Z_{ct.s}$.
- The *gear-cutting tool principal* reference system $X_{ct}Y_{ct}Z_{ct}$.

These form a set of main reference systems associated with a gear pair.

Two more reference systems have to be considered here:

- The *stationary plane-of-action principal* reference system $X_{pa.s}Y_{pa.s}Z_{pa.s}$.
- The *plane-of-action principal* reference system $X_{pa}Y_{pa}Z_{pa}$.

A current configuration of the *stationary plane-of-action principal* reference system $X_{pa.s}Y_{pa.s}Z_{pa.s}$ in relation to the *main principal* reference system $X_{ln}Y_{ln}Z_{ln}$ associated with a gear pair, is illustrated in Figure 2.7. The $Z_{pa.s}$ – axis is pointed along the vector of instant rotation $\boldsymbol{\omega}_{pl}$; the $X_{pa.s}$ – axis forms the transverse pressure angle $\phi_{t.\omega}$ with the axis $X_{ln}$ of the main principal reference system $X_{ln}Y_{ln}Z_{ln}$; and the $Y_{pa.s}$ – axis complements the axes $X_{pa.s}$ and $Z_{pa.s}$ to the left-hand-oriented Cartesian reference system $X_{pa.s}Y_{pa.s}Z_{pa.s}$.

The unit normal vector $\mathbf{n}_{pa}$ to the plane of action $PA$ can be expressed in terms of the unit vectors $\mathbf{i}_{ln}$ and $\mathbf{j}_{ln}$ along the axes $X_{ln}$ and $Y_{ln}$ and the transverse pressure angle $\phi_{t.\omega}$:

$$\mathbf{n}_{pa} = \mathbf{i}_{ln}\cos\phi_{t.\omega} + \mathbf{j}_{ln}\sin\phi_{t.\omega} \tag{2.1}$$

The reference system $X_{pa}Y_{pa}Z_{pa}$ is turned about the coordinate axis $Z_{pa.s}$ in relation to the references system $X_{pa.s}Y_{pa.s}Z_{pa.s}$ through an angle of rotation $\varphi_{pa}$ of the plane of action $PA$.

A few more auxiliary reference systems are commonly used when a gear machining mesh is investigated.

### 2.3.3 Transformation of the Main Coordinate Systems

Multiple reference systems are commonly applied when investigating a gear machining mesh. It is common practice that one of the coordinate systems is associated with the gear, another is associated with the gear-cutting tool, and the third is associated with the housing for the gear machining mesh. Numerous auxiliary (intermediate) reference systems are also extensively used when necessary.

To investigate the kinematics and the geometry of a gear machining mesh, all the components have to be analytically described in a common reference system. As every component of a gear pair can effectively be represented in a specific reference system, coordinate system transformations are used extensively to represent the entire gear machining mesh in a common coordinate system. The operators of elementary coordinate transformations are used for this purpose (see Appendix D for details).

#### 2.3.3.1 Transition from the Gear Reference System to the Main Reference System

The transition from the gear coordinate system $X_g Y_g Z_g$ to the main reference system $X_{ln} Y_{ln} Z_{ln}$ can be executed in three steps.

First, the gear coordinate system $X_g Y_g Z_g$ is turned counterclockwise about its axis $Z_g$ through a gear rotation angle $\varphi_g$. This elementary coordinate system transformation is analytically described by the operator of rotation $\mathbf{Rt}(\varphi_g, Z_g)$:

$$\mathbf{Rt}(\varphi_g, Z_g) = \begin{bmatrix} \cos\varphi_g & \sin\varphi_g & 0 & 0 \\ -\sin\varphi_g & \cos\varphi_g & 0 & 0 \\ 0 & 0 & 1 & 0 \\ 0 & 0 & 0 & 1 \end{bmatrix} \tag{2.2}$$

Second, the stationary gear coordinate system $X_{g.s} Y_{g.s} Z_{g.s}$ is translated through the gear center-distance $C_g$ along the axis $Y_g$. In this intermediate position, the coordinate system $X_{g.s} Y_{g.s} Z_{g.s}$ is labeled as $X^*_{g.s} Y^*_{g.s} Z^*_{g.s}$. For the analytical description of the transition from the stationary gear coordinate system $X_{g.s} Y_{g.s} Z_{g.s}$ to the intermediate auxiliary coordinate system $X^*_{g.s} Y^*_{g.s} Z^*_{g.s}$, the operator of translation $\mathbf{Tr}(C_g, Y^*_{g.s})$ is used:

$$\mathbf{Tr}(C_g, Y^*_{g.s}) = \begin{bmatrix} 1 & 0 & 0 & 0 \\ 0 & 1 & 0 & C_g \\ 0 & 0 & 1 & 0 \\ 0 & 0 & 0 & 1 \end{bmatrix} \tag{2.3}$$

Then, the transition from the auxiliary reference system $X^*_{g.s} Y^*_{g.s} Z^*_{g.s}$ to the main reference system $X_{ln} Y_{ln} Z_{ln}$ is analytically described by the operator of rotation $\mathbf{Rt}(\Sigma_g, Y_{g.s})$ of coordinate system $X_{g.s} Y_{g.s} Z_{g.s}$ counterclockwise about its axis $Y_{g.s}$ through the gear cone angle $\Sigma_g$:

$$\mathbf{Rt}(\Sigma_g, Y_{g.s}) = \begin{bmatrix} \cos\Sigma_g & 0 & \sin\Sigma_g & 0 \\ 0 & 1 & 0 & 0 \\ -\sin\Sigma_g & 0 & \cos\Sigma_g & 0 \\ 0 & 0 & 0 & 1 \end{bmatrix} \tag{2.4}$$

The resultant transition from the gear coordinate system $X_g Y_g Z_g$ to the main reference system $X_{ln} Y_{ln} Z_{ln}$ is analytically described by the operator of the resultant coordinate system transformation $\mathbf{Rs}(g \mapsto ln)$:

$$\mathbf{Rs}(g \mapsto ln) = \mathbf{Rt}(\Sigma_g, Y_{g.s}) \cdot \mathbf{Tr}(C_g, Y_{ln}) \cdot \mathbf{Rt}(\varphi_g, Z_g) \tag{2.5}$$

For the inverse coordinate system transformation—that is, for transition from the main reference system $X_{ln} Y_{ln} Z_{ln}$ to the gear coordinate system $X_g Y_g Z_g$—the operator of the resultant coordinate system transformation $\mathbf{Rs}(ln \mapsto g)$ can be used:

$$\mathbf{Rs}(ln \mapsto g) = \mathbf{Rs}^{-1}(g \mapsto ln) \tag{2.6}$$

#### 2.3.3.2 Coupled Linear Transformations

Coupled linear transformations [150, 152] can be effectively used for the derivation of the operators $\mathbf{Rs}(g \mapsto ln)$ and $\mathbf{Rs}(ln \mapsto g)$ of the resultant coordinate system transformations. Coupled linear transformations combine translation along and rotation about a coordinate axis.[3]

Only rotation is observed in the first step of the coordinate system transformation—that is, the translation is zero. Therefore, in the case under consideration, the coupled operator of the linear transformation $\mathbf{Cp}_z(\varphi_g, 0)$ is identical to the operator of rotation $\mathbf{Rt}(\varphi_g, Z_g)$:

$$\mathbf{Cp}_z(\varphi_g, 0) = \begin{bmatrix} \cos\varphi_g & \sin\varphi_g & 0 & 0 \\ -\sin\varphi_g & \cos\varphi_g & 0 & 0 \\ 0 & 0 & 1 & 0 \\ 0 & 0 & 0 & 1 \end{bmatrix} \tag{2.7}$$

The next two separate linear transformations—that is, $\mathbf{Tr}(C_g, Y_{ln})$ and $\mathbf{Rt}(\Sigma_g, Y_{g.s})$—are analytically described

[3] It is instructive to note here that only the initial and the final configurations of elementary reference system are accounted by the coupled linear transformation operators. No intermediate configurations of the reference system are described by the coupled linear transformation operators. Aiming an analytical description of all intermediate configurations of the reference system, an operator of screw motion $\mathbf{Sc}_x[a_x(t), \varphi_x(t)]$ is used instead of the coupled operator of the linear transformation $\mathbf{Cp}_x(a_x, \varphi_x)$ (see Appendix D).

by the coupled operator of the linear transformation $\mathbf{Cp}_y(\Sigma_g, C_g)$. This operator equals:

$$\mathbf{Cp}_y(\Sigma_g, C_g) = \begin{bmatrix} \cos\Sigma_g & 0 & \sin\Sigma_g & 0 \\ 0 & 1 & 0 & C_g \\ -\sin\Sigma_g & 0 & \cos\Sigma_g & 0 \\ 0 & 0 & 0 & 1 \end{bmatrix} \quad (2.8)$$

Ultimately, the resultant transition from the gear coordinate system $X_gY_gZ_g$ to the main reference system $X_{ln}Y_{ln}Z_{ln}$ is analytically described by the operator of the resultant coordinate system transformation $\mathbf{Rs}(g \mapsto ln)$:

$$\mathbf{Rs}(g \mapsto ln) = \mathbf{Cp}_y(\Sigma_g, C_g) \cdot \mathbf{Cp}_z(\varphi_g, 0) \quad (2.9)$$

The use of the operators of coupled linear transformations allows a reduction in the total number of the operators of elementary coordinate system transformations that makes sense when long chains of consequent coordinate system transformations are necessary.

#### 2.3.3.3 Transition from the Gear-Cutting Tool Reference System to the Main Reference System

The transition from the gear-cutting tool coordinate system $X_{ct}Y_{ct}Z_{ct}$ to the main reference system $X_{ln}Y_{ln}Z_{ln}$ can be performed as described in this section.

First, the gear-cutting tool coordinate system $X_{ct}Y_{ct}Z_{ct}$ is turned clockwise about its axis $Z_{ct}$ through a gear-cutting tool rotation angle $\varphi_{ct}$ to the position of the stationary gear-cutting tool coordinate system $X_{ct.s}Y_{ct.s}Z_{ct.s}$. This elementary coordinate system transformation is analytically described by the operator of rotation $\mathbf{Rt}(\varphi_{ct}, Z_{ct})$:

$$\mathbf{Rt}(\varphi_{ct}, Z_{ct}) = \begin{bmatrix} \cos\varphi_{ct} & -\sin\varphi_{ct} & 0 & 0 \\ \sin\varphi_{ct} & \cos\varphi_{ct} & 0 & 0 \\ 0 & 0 & 1 & 0 \\ 0 & 0 & 0 & 1 \end{bmatrix} \quad (2.10)$$

Next, the stationary pinion coordinate system $X_{ct.s}Y_{ct.s}Z_{ct.s}$ is turned about the coordinate axis $X_{ct.s}$ through an angle of 180° to a position of the reference system $X^*_{ct.s}Y^*_{ct.s}Z^*_{ct.s}$. This linear transformation is analytically described by means of the operator $\mathbf{Rt}(180°, X_{ct.s})$:

$$\mathbf{Rt}(180°, X_{ct.s}) = \begin{bmatrix} 1 & 0 & 0 & 0 \\ 0 & -1 & 0 & 0 \\ 0 & 0 & -1 & 0 \\ 0 & 0 & 0 & 1 \end{bmatrix} \quad (2.11)$$

Then, the intermediate pinion coordinate system $X^*_{ct.s}Y^*_{ct.s}Z^*_{ct.s}$ is translated at the pinion center-distance $-C_{ct}$ along the axis $Y^*_{ct.s}$. In this intermediate position, the coordinate system $X^*_{ct.s}Y^*_{ct.s}Z^*_{ct.s}$ is labeled $X^{**}_{ct.s}Y^{**}_{ct.s}Z^{**}_{ct.s}$. For analytical description of the transition from the stationary pinion coordinate system $X^*_{ct.s}Y^*_{ct.s}Z^*_{ct.s}$ to the intermediate auxiliary coordinate system $X^{**}_{ct.s}Y^{**}_{ct.s}Z^{**}_{ct.s}$, the operator of translation $\mathbf{Tr}(r_p, Y^*_{p.s})$ is used:

$$\mathbf{Tr}(C_{ct}, Y^*_{ct.s}) = \begin{bmatrix} 1 & 0 & 0 & 0 \\ 0 & 1 & 0 & -C_{ct} \\ 0 & 0 & 1 & 0 \\ 0 & 0 & 0 & 1 \end{bmatrix} \quad (2.12)$$

Finally, the transition from the auxiliary reference system $X^{**}_{ct.s}Y^{**}_{ct.s}Z^{**}_{ct.s}$ to the main reference system $X_{ln}Y_{ln}Z_{ln}$ is analytically described by the operator of rotation $\mathbf{Rt}(\Sigma_{ct}, Y^{**}_{ct.s})$ of the coordinate system $X_{ct.s}Y_{ct.s}Z_{ct.s}$ counterclockwise about its axis $Y_{ct.s}$ through the angle that is equal to $(180° - \Sigma_{ct})$:

$$\mathbf{Rt}(\Sigma_{ct}, Y^{**}_{ct.s}) = \begin{bmatrix} \cos(180° - \Sigma_{ct}) & 0 & \sin(180° - \Sigma_{ct}) & 0 \\ 0 & 1 & 0 & 0 \\ -\sin(180° - \Sigma_{ct}) & 0 & \cos(180° - \Sigma_{ct}) & 0 \\ 0 & 0 & 0 & 1 \end{bmatrix} \quad (2.13)$$

With that said, the resultant transition from the pinion coordinate system $X_{ct}Y_{ct}Z_{ct}$ to the main reference system $X_{ln}Y_{ln}Z_{ln}$ is analytically described by the operator of the resultant coordinate system transformation $\mathbf{Rs}(ct \mapsto ln)$:

$$\mathbf{Rs}(ct \mapsto ln) = \mathbf{Rt}(\Sigma_{ct}, Y^{**}_{ct.s}) \cdot \mathbf{Tr}(C_{ct}, Y^*_{ct.s}) \cdot \mathbf{Rt}(180°, X_{ct.s}) \cdot \mathbf{Rt}(\varphi_{ct}, Z_{ct}) \quad (2.14)$$

For the inverse coordinate system transformation—that is, for the transition from the main reference system $X_{ln}Y_{ln}Z_{ln}$ to the pinion coordinate system $X_{ct}Y_{ct}Z_{ct}$, the operator of the resultant coordinate system transformation $\mathbf{Rs}(ln \mapsto ct)$ is used:

$$\mathbf{Rs}(ln \mapsto ct) = \mathbf{Rs}^{-1}(ct \mapsto ln) \quad (2.15)$$

Again, coupled linear transformations [150, 152] can be used to derive the operators $\mathbf{Rs}(ct \mapsto ln)$ and $\mathbf{Rs}(ln \mapsto ct)$ of resultant coordinate system transformations.

Only rotation is observed in the first step of the coordinate system transformation, meaning the translation is zero. Therefore, in the case under consideration, the coupled operator of the linear transformation $\mathbf{Cp}_z(\varphi_{ct}, 0)$ is identical to the operator of rotation $\mathbf{Rt}(\varphi_{ct}, Z_{ct})$:

$$\mathbf{Cp}_z(\varphi_{ct}, 0) = \begin{bmatrix} \cos\varphi_{ct} & -\sin\varphi_{ct} & 0 & 0 \\ \sin\varphi_{ct} & \cos\varphi_{ct} & 0 & 0 \\ 0 & 0 & 1 & 0 \\ 0 & 0 & 0 & 1 \end{bmatrix} \quad (2.16)$$

The transition from the stationary pinion coordinate system $X_{ct.s}Y_{ct.s}Z_{ct.s}$ to a position of the reference system $X^*_{ct.s}Y^*_{ct.s}Z^*_{ct.s}$ is analytically described by means of the operator $\mathbf{Cp}_x\,(180°, 0)$:

$$\mathbf{Cp}_x\,(180°, 0) = \begin{bmatrix} 1 & 0 & 0 & 0 \\ 0 & -1 & 0 & 0 \\ 0 & 0 & -1 & 0 \\ 0 & 0 & 0 & 1 \end{bmatrix} \tag{2.17}$$

The next two separate linear transformations (which are analytically described by the operators $\mathbf{Tr}\,(r_{ct}, Y^*_{ct.s})$ and $\mathbf{Rt}\,(\Sigma_{ct}, Y^{**}_{ct.s}))$ can be analytically described by the coupled operator of the linear transformation $\mathbf{Cp}_y\,[(180° - \Sigma_{ct}), -C_{ct}]$. This operator is equal:

$$\mathbf{Cp}_y\,[(180° - \Sigma_{ct}), - C_{ct}] = \begin{bmatrix} \cos(180° - \Sigma_{ct}) & 0 & -\sin(180° - \Sigma_{ct}) & 0 \\ 0 & 1 & 0 & -C_{ct} \\ \sin(180° - \Sigma_{ct}) & 0 & \cos(180° - \Sigma_{ct}) & 0 \\ 0 & 0 & 0 & 1 \end{bmatrix} \tag{2.18}$$

Ultimately, the resultant transition from the gear-cutting coordinate system $X_{ct}Y_{ct}Z_{ct}$ to the main reference system $X_{ln}Y_{ln}Z_{ln}$ is analytically described by the operator of the resultant coordinate system transformation $\mathbf{Rs}\,(ct \mapsto ln)$:

$$\mathbf{Rs}\,(ct \mapsto ln) = \mathbf{Cp}_y\,[(180° - \Sigma_{ct}), - C_p] \cdot \mathbf{Cp}_x\,(180°, 0) \cdot \mathbf{Cp}_z\,(\varphi_{ct}, 0) \tag{2.19}$$

The use of the operators of coupled linear transformations allows for a considerable reduction in the total number of the operators of elementary coordinate system transformations. This is reasonable, especially in cases when long chains of consequent coordinate system transformations need to be described analytically.

The derived operators of coordinate system transformations are also valid for the reduced cases of gear machining mesh.

In a similar manner, any and all coordinate system transformations can be analytically described by means either of the operators of elementary coordinate system transformations—that is, by means of operators of translations $\mathbf{Tr}\,(x_i, X_i)$ and operators of rotations $\mathbf{Rt}\,(\varphi_{x_i}, X_i)$ or by means just of operators of coupled coordinate system transformations $\mathbf{Cp}_{x_i}\,(\varphi_{x_i}, x_i)$. The latter is more practical.

### 2.3.4 Transition from the Plane-of-Action Reference System to the Main Reference System

The plane-of-action reference system $X_{pa}Y_{pa}Z_{pa}$ is rigidly connected to the plane of action *PA* and is rotated with the plane *PA*. At a current instant of time, the plane-of-action reference system $X_{pa}Y_{pa}Z_{pa}$ is turned about the $Z_{pa.s}$ – axis in relation to the stationary plane-of-action reference system $X_{pa.s}Y_{pa.s}Z_{pa.s}$ associated with a gear pair through an angle $\varphi_{pa}$. For the analytical description of transition from the reference system $X_{pa}Y_{pa}Z_{pa}$ to the reference system $X_{pa.s}Y_{pa.s}Z_{pa.s}$, the following equation is used:

$$\mathbf{Rt}\,(\varphi_{pa}, Z_{pa.s}) = \begin{bmatrix} \cos\varphi_{pa} & -\sin\varphi_{pa} & 0 & 0 \\ \sin\varphi_{pa} & \cos\varphi_{pa} & 0 & 0 \\ 0 & 0 & 1 & 0 \\ 0 & 0 & 0 & 1 \end{bmatrix} \tag{2.20}$$

The transition from the stationary plane-of-action reference system $X_{pa.s}Y_{pa.s}Z_{pa.s}$ to the main reference system $X_{ln}Y_{ln}Z_{ln}$ associated with a gear pair can be performed in two steps.

First, the reference system $X_{pa.s}Y_{pa.s}Z_{pa.s}$ has to be turned counterclockwise about the axis $X_{pa.s}$ through a right angle to a position $X^*_{pa.s}Y^*_{pa.s}Z^*_{pa.s}$. For analytical description of this kind of linear transition, an operator of rotation is commonly used:

$$\mathbf{Rt}\,(90°, X_{pa.s}) = \begin{bmatrix} 1 & 0 & 0 & 0 \\ 0 & 0 & -1 & 0 \\ 0 & 1 & 0 & 0 \\ 0 & 0 & 0 & 1 \end{bmatrix} \tag{2.21}$$

Second, the reference system $X^*_{pa.s}Y^*_{pa.s}Z^*_{pa.s}$ has to be turned counterclockwise about the axis $Z^*_{pa.s}$ through a transverse pressure angle $\phi_{t.\omega}$. For analytical description of this kind of transition, an operator of rotation is used:

$$\mathbf{Rt}\,(\phi_{t.\omega}, Z^*_{pa.s}) = \begin{bmatrix} \cos\phi_{t.\omega} & -\sin\phi_{t.\omega} & 0 & 0 \\ \sin\phi_{t.\omega} & \cos\phi_{t.\omega} & 0 & 0 \\ 0 & 0 & 1 & 0 \\ 0 & 0 & 0 & 1 \end{bmatrix} \tag{2.22}$$

The resultant coordinate system transformation is analytically described by the operator of the resultant transformation $\mathbf{Rs}\,(pa.s \mapsto ln)$:

$$\mathbf{Rs}\,(pa.s \mapsto ln) = \mathbf{Rt}\,(\phi_{t.\omega}, Z^*_{pa.s}) \cdot \mathbf{Rt}\,(90°, X_{pa.s}) \tag{2.23}$$

For the inverse coordinate system transformation, the operator of the coordinate system transformation is used.:

$$\mathbf{Rs}\,(ln \mapsto pa.s) = \mathbf{Rs}^{-1}\,(pa.s \mapsto ln) \tag{2.24}$$

Additional reference systems associated with a gear pair can be introduced when necessary.

# 3 Kinematics of Continuous-Indexing Methods of Gear-Machining Operations

Many commercially produced gears are manufactured using methods based on continuous indexing operations, wherein the gear being machined and the gear-cutting tool are engaged in a gear-cutting mesh. Gear hobbing, gear shaping, and gear shaving are perfect examples of gear-machining operations of this type.

There has been a concerted effort in recent decades to investigate gear tooth flank generation process using continuous indexing methods of surface machining. The results of this research can be found in [4, 23, 148, 151, 158, 198, 199, 206], as well as other sources.

Meshing of two-toothed bodies is imitated when machining a gear using a continuous indexing method of tooth flank machining. Under such a scenario, the work gear and the gear-cutting tool perform motions in a properly timed manner. Interaction of the imaginary gear to be machined and the generating surface of the gear-cutting tool (which is also a phantom surface) can be construed as a kind of virtual gear mesh. The virtual mesh of the imaginary gear to be machined and the generating surface of the gear-cutting tool is referred to as the *gear machining mesh*. The gear machining mesh allows for interpretation in the form of a *work-gear-to-generating-surface mesh*, or, more concisely, *$\mathcal{G}$ – to – T mesh*.

The tooth ratio of the gear machining mesh is a constant value when ordinary gears are being machined (see Section 1.1). Tooth ratio can be expressed in terms of the angle of rotation of the work gear when machining a noncircular gear. This issue, however, is out of the scope of this book.

The gear machining mesh plays an important role in the investigation and design of gear-cutting tools. Any kind of gear-cutting tool is designed based on a certain gear machining mesh. Because of this, the investigation of gear machining meshes having tooth ratio as a constant value is of critical importance to the researchers and gear-cutting tool designers.

An analytical study of the gear machining mesh undertaken in the following sections of this book is based on the vector representation of the gear machining mesh.

## 3.1 VECTOR REPRESENTATION OF THE "GEAR-MACHINING PROCESS"

Relative motions of the work gear and the generating surface of the gear-cutting tool, which take place when the gear is being machined, are a type of uniform motion. This means that the rotation of the work gear $\omega_g$ and the rotation of the generating surface of the gear-cutting tool $\omega_{ct}$ are synchronized so as to keep the ratio $\omega_g/\omega_{ct}$ constant in time.

A uniform translation and a uniform rotation are perfect examples of uniform motions. Under certain circumstances, a uniform translation and a uniform rotation can be combined to form a kind of uniform screw motion. Screw motion is the most general type of uniform motion.

A uniform screw motion allows for representation in the form of summa of two motions: a translation and a rotation. The translation vector is designated $\mathbf{V}$, whereas rotation is denoted by $\boldsymbol{\omega}$.

The translation vector $\mathbf{V}$ and the rotation vector $\boldsymbol{\omega}$ are sliding vectors. This means that a gear-cutting tool designer is free to choose an appropriate point at which the vectors $\mathbf{V}$ and $\boldsymbol{\omega}$ are applied.

Both vectors $\boldsymbol{\omega}$ and $\mathbf{V}$, which comprise the uniform screw motion, are collinear vectors. They act along the axis $O_{sc}$ of the screw motion. Axis $O_{sc}$ of the screw motion is aligned with axis $O$ of rotation $\boldsymbol{\omega}$.

In reality, the vectors $\boldsymbol{\omega}$ and $\mathbf{V}$ are pointed either in the same direction or in opposite directions. Depending on the actual configuration of vectors $\boldsymbol{\omega}$ and $\mathbf{V}$, the resultant uniform screw motion is recognized as either *left-hand-oriented* or *right-hand-oriented*. When the vectors $\boldsymbol{\omega}$ and $\mathbf{V}$ are pointed in the same direction, the resultant screw motion is known as a left-hand-oriented screw motion. Conversely, when vectors $\boldsymbol{\omega}$ and $\mathbf{V}$ are in opposite directions, the resultant screw motion is a right-hand-oriented screw motion.

Theoretically, a screw motion can be simultaneously performed by the work gear and the gear-cutting tool. The kinematics of this type of gear-machining process are physically permissible; however, this case has no practical importance. Machine tools of complex, and thus of impractical, kinematics are required to achieve this. Further analysis is limited to the kinematics of the gear machining mesh composed just of two rotations about a skew axis. The results of the analysis obtained in this manner will cover all known gear-machining operation methods. They will also cover all known possible shapes of generating surfaces of gear-cutting tools. The theory derived this way becomes complete and consistent.

Consider a gear machining mesh that is specified in a gear-machining-mesh reference system $X_{gmm}Y_{gmm}Z_{gmm}$. In the case under consideration, the reference system is an orthogonal Cartesian coordinate system $X_{gmm}Y_{gmm}Z_{gmm}$.

DOI: 10.1201/9781003520528-4

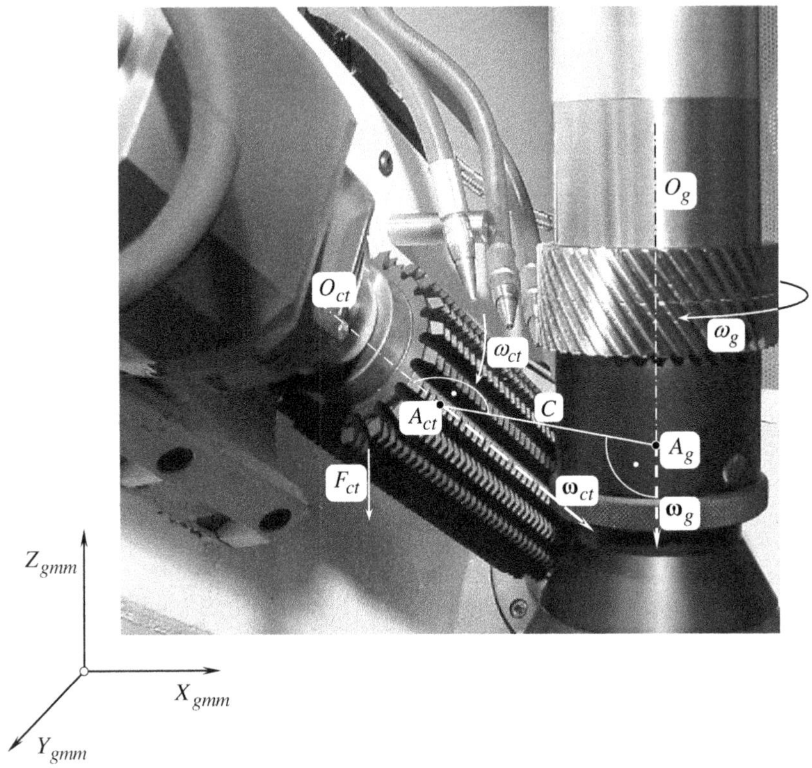


**FIGURE 3.1** Configuration of the rotation vectors $\boldsymbol{\omega}_g$ of the work gear and of the gear cutting tool $\boldsymbol{\omega}_{ct}$ in the gear machining mesh.

In a gear machining mesh, the work gear and the gear-cutting tool rotate about their axes, $O_g$ and $O_{ct}$, respectively. The rotation vector of the work gear is denoted as $\boldsymbol{\omega}_g$, and $\boldsymbol{\omega}_{ct}$ is the rotation vector of the gear-cutting tool. The inequalities $\boldsymbol{\omega}_g \neq 0$ and $\boldsymbol{\omega}_{ct} \neq 0$ are commonly observed. Only in particular cases, for example, when machining a gear by the rack cutter and so forth, the cutting tool rotation is equal to zero $\boldsymbol{\omega}_{ct} = 0$.

The rotation vector $\boldsymbol{\omega}_g$ of the gear is aligned with axis of rotation $O_g$ of the gear. The vector $\boldsymbol{\omega}_g$ is pointed in the direction that corresponds to the right-hand screw (see Figure 3.1). The magnitude $\omega_g$ of the rotation vector $\boldsymbol{\omega}_g$ is proportional to the RPMs of the gear. Discussion of a point at which the rotation vector $\boldsymbol{\omega}_g$ is applied will be further discussed in this section.

In Figure 3.1, the rotation vector $\boldsymbol{\omega}_{ct}$ is assigned to the hob (the gear-cutting tool). The vector $\boldsymbol{\omega}_{ct}$ is very similar to rotation vector $\boldsymbol{\omega}_g$. The vector $\boldsymbol{\omega}_{ct}$ is aligned with the axis of rotation, $O_{ct}$, of the hob. The direction of the rotation vector $\boldsymbol{\omega}_{ct}$ corresponds to the hand of the screw of the hob's generating surface. The magnitude $\omega_{ct}$ of the rotation vector $\boldsymbol{\omega}_{ct}$ is proportional to the hob rotation. A point at which the vector $\boldsymbol{\omega}_{ct}$ is applied is further discussed in this section.

The rotations $\omega_g$ and $\omega_{ct}$ relate to each other in compliance with the proportion.

$$\omega_g \cdot N_g = \omega_{ct} \cdot N_{ct} \tag{3.1}$$

Here, $N_g$ denotes the number of teeth of the work gear, $N_{ct}$ is the number of teeth/threads of the gear-cutting tool, $\omega_g$ and $\omega_{ct}$ are the magnitudes of the rotation vectors $\boldsymbol{\omega}_g$ and $\boldsymbol{\omega}_{ct}$ (the equalities $\omega_g = |\boldsymbol{\omega}_g|$ and $\omega_{ct} = |\boldsymbol{\omega}_{ct}|$ are valid).

The axes $O_g$ and $O_{ct}$ of the rotations $\boldsymbol{\omega}_g$ and $\boldsymbol{\omega}_{ct}$ are positioned at a certain distance $C$ from each other. The distance $C$ is measured along the centerline ℄ and is referred to as the *center distance*. The length of the center distance is equal to the closest distance of approach of the crossing axes $O_g$ and $O_{ct}$. In Figure 3.1, the length of the straight-line segment $A_g A_{ct}$ is equal to the center distance $C$.

The angle between the crossing axes $O_g$ and $O_{ct}$ is an important consideration.

As shown in Figure 3.2, the angle between two skew straight lines $L_1$ and $L_2$ is usually measured between one of the lines—say the line $L_1$—and some other straight line that intersects the line $L_1$ and is parallel to the line $L_2$. In this case, either the acute angle $\Sigma$ or the obtuse angle $\Sigma^*$ can be used for the purposes of specifying the crossed angle between the two skew lines $L_1$ and $L_2$ [Figure 3.2(a)]. These cases are equivalent.

When unit vectors $\mathbf{s}_1$ and $\mathbf{s}_2$ are associated with the straight lines $L_1$ and $L_2$, the crossed-axis angle between the lines can be uniquely specified. Depending on the actual directions of unit vectors $\mathbf{s}_1$ and $\mathbf{s}_2$, the crossed-axis angle $\Sigma$ is either acute [see Figure 3.2(b) or obtuse [see Figure 3.2(c)]. No ambiguities are observed with the crossed-axis angle $\Sigma$ if the vectors are assigned to crossing straight lines.

The crossing angle between the axes $O_g$ and $O_{ct}$ in the gear machining mesh is designated in the following

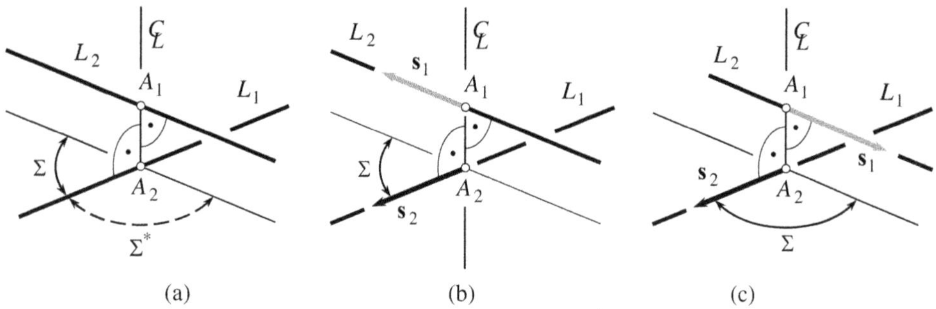


**FIGURE 3.2** (a–c) Crossed-axis angle Σ between two skew straight lines $L_1$ and $L_2$.

equations as Σ. The crossed-axis angle Σ is the angle that the rotation vectors $\boldsymbol{\omega}_g$ and $\boldsymbol{\omega}_{ct}$ form with each other, that is:

$$\Sigma = \angle(\boldsymbol{\omega}_g, \boldsymbol{\omega}_{ct}) \quad (3.2)$$

For an acute crossed-axis angle ($\Sigma < 90°$), the actual value of the angle Σ is equal to that which is commonly understood when using the term *crossed-axes angle*. For an obtuse crossed-axis angle ($\Sigma > 90°$), the equality $\bar{\Sigma} = 180° - \Sigma$ is observed. Here, $\bar{\Sigma}$ denotes the crossed-axis angle in the common meaning of the term.

The kinematics of most gear-machining methods is not limited just to the motions required for gear machining meshes. In the example shown in Figure 3.1, in addition to rotations $\boldsymbol{\omega}_g$ and $\boldsymbol{\omega}_{ct}$, one more motion is taking place during gear machining. This motion is the axial feed $F_{ct}$ of the cutter. The feed motion $F_{ct}$ occurs in the axial direction of the work gear. Depending on the design peculiarities of the hobber, either the hob moves with respect to the work gear, or the hobber's worktable travels relative to the rotating hob.

The feed motion $F_{ct}$ of the gear cutter causes sliding of the gear tooth flank $\mathscr{G}$ over itself. When analyzing kinematics of a gear-machining operation, feed motions of this type can be omitted from consideration as the geometry of the gear tooth flank $\mathscr{G}$ is not affected by the feed motion $F_{ct}$.

The kinematics of gear-machining operations shown in Figure 3.1 is extensively used in conventional hobbers for machining cylindrical gears, both spur gears and helical gears. It is easy to see (see Figure 3.1) that by nature, the kinematics of gear-machining operations is an invertible one. The invertibility of the kinematics of the gear-machining operation means that the work gear can be replaced with the gear-cutting tool, and vice versa. For example, in Figure 3.1, the work gear can be replaced with the gear shaper, whereas the hob can be replaced with the blank of a worm. In this scenario, the worm can be machined when performing the same rotations $\boldsymbol{\omega}_g$ and $\boldsymbol{\omega}_{ct}$ and moving the cutter in the axial direction of the work gear axis of rotation $O_g$ with the feed motion $F_{ct}$. In this case, the feed motion $F_{ct}$ causes sliding of the worm thread over itself. Machining operations of this type are physically permissible; however, it is not yet used in practice. Conversions similar to the aforementioned are valid in all kinds of gear-machining kinematics. When converting principal kinematics of the gear-machining process into it a practical version, it is recommended to keep in mind the possibility of both principal kinematics of gear machining—that is, the original and the converted kinematics of the gear-machining operation.

Relative motion of the work gear and the gear-cutting tool can be specified by the vector $\boldsymbol{\omega}_{pl}$ of instant rotation. For the determination of the vector of instant rotation $\boldsymbol{\omega}_{pl}$, implementing the principle of inversion of a motion is convenient.[1] This principle is illustrated in Figure 3.3.

In the reality, the work gear and the gear-cutting tool rotate about their axes, $O_g$ and $O_{ct}$, respectively. The rotations are specified by rotation vectors[2] $\boldsymbol{\omega}_g$ and $\boldsymbol{\omega}_{ct}$, as shown in Figure 3.3(a). To make the work gear stationary, the work gear must undergo one more rotation $-\boldsymbol{\omega}_g$. Summa of the initial rotation $\boldsymbol{\omega}_g$ and of the additional rotation $-\boldsymbol{\omega}_g$ in the opposite direction is equal to zero, $\boldsymbol{\omega}_g + (-\boldsymbol{\omega}_g) \equiv 0$. In this manner, the work gear is rendered stationary. To avoid making any changes to the rest of the elements of the gear machining mesh, the gear-cutting tool must also perform an additional rotation $-\boldsymbol{\omega}_g$ [see Figure 3.3(b)]. In this scenario, the resultant rotation of the gear-cutting tool relative to the motionless work gear is specified by the vector of instant rotation:

$$\boldsymbol{\omega}_{ct/g} = \boldsymbol{\omega}_{ct} + (-\boldsymbol{\omega}_g) \quad (3.3)$$

The vector $\boldsymbol{\omega}_{ct/g}$ of the instant rotation of the gear-cutting tool in relation to the work gear is applied at the plane-of-action-apex[3] $A_{pa}$ [see Figure 3.3(b)].

In a manner similar to that just discussed, the vector $\boldsymbol{\omega}_{g/ct}$ of instant rotation of the work gear with respect to

[1] The process of holding different links of a kinematic chain stationary is known as *inversion*.

[2] Angular velocity is considered in this monograph as a vector along the axis of rotation that is pointed in a direction defined by the right-hand screw rule. It is understood here that rotations $\boldsymbol{\omega}_g$ and $\boldsymbol{\omega}_{ct}$ are not vectors by nature. Therefore, special care is required when treating rotations as vectors. In particular, when a coordinate systems transformation is analytically described through matrices, the order of the matrices in the matrix product is predetermined by the order of the elementary coordinate system transformations, and this order is not allowed to be altered.

[3] The designation $A_{pa}$ corresponds to the term *plane-of-action apex, $A_{pa}$* extensively used in the scientific theory of gearing (Radzevich, S.P., *Theory of Gearing: Kinematics, Geometry, and Synthesis*, *CRC* Press, Boca Raton, Florida, 2012, 743 pages) when discussing the cases of intersected-axes gearing, as well as crossed-axes gearing [171, 172].

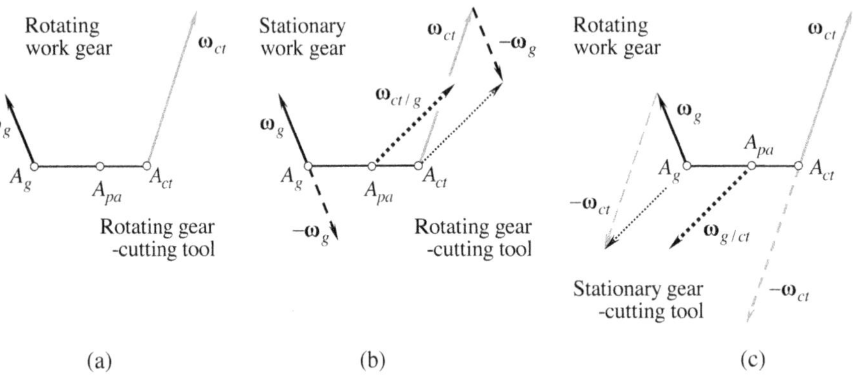


**FIGURE 3.3** (a–c) The principle of inversion of two rotations.

the gear-cutting tool can be constructed [see Figure 3.3(c)]. This rotation vector $\boldsymbol{\omega}_{g/ct}$ is calculated as:

$$\boldsymbol{\omega}_{g/c} = \boldsymbol{\omega}_g + (-\boldsymbol{\omega}_{ct}) \tag{3.4}$$

Evidently, the rotation vectors $\boldsymbol{\omega}_{ct/g}$ and $\boldsymbol{\omega}_{g/ct}$ are of the same magnitude ($|\boldsymbol{\omega}_{ct/g}| = |\boldsymbol{\omega}_{g/ct}|$) and are pointed opposite of each other—that is, $\boldsymbol{\omega}_{ct/g} = -\boldsymbol{\omega}_{g/ct}$.

Assume that the work gear and the gear-cutting tool together are rotated about the work gear axis $O_g$. This rotation is specified by the rotation vector $-\boldsymbol{\omega}_g$. As the rotation $-\boldsymbol{\omega}_g$ is applied, the work gear gets stationary:

$$\boldsymbol{\omega}_g + (-\boldsymbol{\omega}_g) \equiv 0 \tag{3.5}$$

The resultant rotation of the gear-cutting tool can be specified by the vector $\boldsymbol{\omega}_{ct} + (-\boldsymbol{\omega}_g)$, which is equal to the vector $\boldsymbol{\omega}_{pl}$ of instant rotation of the gear-cutting tool relative to the work gear:

$$\boldsymbol{\omega}_{pl} = \boldsymbol{\omega}_{ct} + (-\boldsymbol{\omega}_g) \tag{3.6}$$

Similarly, the rotation vector of the work gear relative to the gear-cutting tool can be determined. Evidently, this rotation vector equals to $-\boldsymbol{\omega}_{pl}$.

The vector of instant rotation $\boldsymbol{\omega}_{pl}$ is aligned with the axis of instant rotation $P_{ln}$. The axis of instant rotation is sometimes called the *pitch line*, $P_{ln}$. Once the configuration of rotation vectors $\boldsymbol{\omega}_g$ and $\boldsymbol{\omega}_{ct}$ is known, say rotation vectors $\boldsymbol{\omega}_g$ and $\boldsymbol{\omega}_{ct}$ are specified in a certain reference system, the vector $\boldsymbol{\omega}_{pl}$ of the instant rotation and the pitch line[4] $P_{ln}$ can be obtained following the common rules known from vector algebra.

As an example, consider two rotation vectors, $\boldsymbol{\omega}_g$ and $\boldsymbol{\omega}_{ct}$, as shown in Figure 3.4. The configuration of the rotation vectors is initially given in the system of two planes of projections, $\pi_1$ and $\pi_2$. The rotation vectors $\boldsymbol{\omega}_g$ and $\boldsymbol{\omega}_{ct}$ are the key elements of the gear machining mesh. The rest of the important elements of the kinematics of the gear-machining operation can be drawn up from the *vector diagram* of the gear machining mesh. The diagram is constructed based on the rotation vectors $\boldsymbol{\omega}_g$ and $\boldsymbol{\omega}_{ct}$ of the work gear and the gear-cutting tool, respectively.

Generally, two rotation vectors $\boldsymbol{\omega}_g$ and $\boldsymbol{\omega}_{ct}$ can occupy an arbitrary location and orientation in the reference system $\pi_1\pi_2$. For convenience of the analysis, the rotation vectors $\boldsymbol{\omega}_g$ and $\boldsymbol{\omega}_{ct}$ are shown parallel to the plane of projections $\pi_1$. The crossed-axis angle $\Sigma$ is projected onto the horizontal plane of projections $\pi_1$ with no distortion.

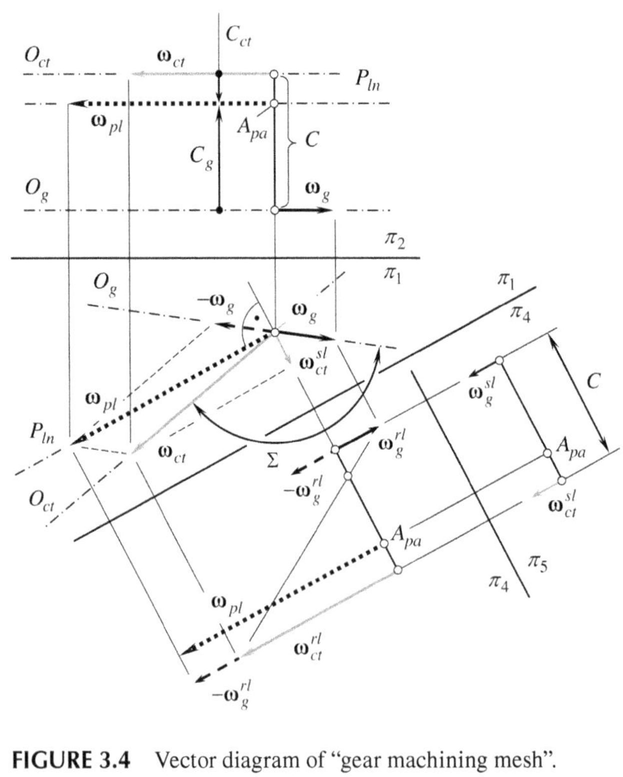


**FIGURE 3.4** Vector diagram of "gear machining mesh".

[4] It is worth noting here the reason that the axis of instant rotation of a gear-cutting tool in relation to a gear to be machined is designated $P_{ln}$. In the simplest cases of spur and helical gearing, it is a commonly adopted practice to consider instant motion of a pinion in relation to its mating gear as an instant rotation about the *pitch point P*. Actually, the instant rotation of the pinion occurs not about the pitch point, *P*, but, instead, it occurs about a straight line through the pitch point *P*. Therefore, it is natural to refer to the axis of instant rotation as to the *pitch line* and to designate such a straight line as $P_{ln}$, for convenience. However, in this scenario, the term *pitch line*, $P_{ln}$ that is used in the gear kinematics needs to be distinguished from a similar term *pitch line* that is used, for example, with respect to a gear rack. Here and through the rest of the book, the term the *pitch line* $P_{ln}$ is understood in the sense of the axis of instant rotation of a gear-cutting tool in relation to a gear to be machined.

Similarly, the center distance $C$ is projected onto the vertical plane of projections $\pi_2$ with no distortion.

In the plane of projections $\pi_1$, the vector $\omega_{pl}$ of instant rotation is determined in terms of the rotation vectors $\omega_g$ and $\omega_{ct}$ [see Eq. (3.6)]. The vector $\omega_{pl}$ is applied at the apex point $A_{pa}$, the location of which is not specified yet. It is only known that point $A_{pa}$ is located within the centerline ℄.

An additional plane of projections $\pi_4$ is constructed for the purposes of determination of location of the apex point $A_{pa}$. Within the plane of projections $\pi_1$, the axis of projections $\pi_1/\pi_4$ is passing parallel to the vector of instant rotation $\omega_{pl}$. Because of this, the components $\omega_g^{rl}$ and $\omega_{ct}^{rl}$ of the rotation vectors $\omega_g$ and $\omega_{ct}$ are projected with no distortions onto the plane of projections $\pi_4$. Both components $\omega_g^{rl}$ and $\omega_{ct}^{rl}$ are parallel to the vector of instant rotation $\omega_{pl}$:

$$\omega_g^{rl} \| \omega_{ct}^{rl} \| \omega_{pl} \tag{3.7}$$

The magnitudes $\omega_g^{rl}$ and $\omega_{ct}^{rl}$ of the rotation vectors $\omega_g^{rl}$ and $\omega_{ct}^{rl}$ relate to each other in the way that is observed in the equality:

$$\omega_g^{rl} \cdot C_g = \omega_{ct}^{rl} \cdot C_{ct} \tag{3.8}$$

Here, $C_g$ and $C_{ct}$ denote distances of point $A_{pa}$ from gear axis of rotation $O_g$ and from gear-cutting tool axis of rotation $O_{ct}$, respectively (these are so-called the *gear-center-distance* $C_g$ and the gear-cutting tool *gear-center-distance* $C_{ct}$).

The center-distances $C_g$ and $C_{ct}$ are signed values. Their algebraic summa is equal to the center-distance $C$:

$$C_g + C_{ct} = C \tag{3.9}$$

Ultimately, the components $\omega_g^{rl}$ and $\omega_{ct}^{rl}$ of the rotation vectors result in pure rolling of axodes of the work gear and of the gear-cutting tool over each other.

It is important to point out that the vectors $\omega_g^{rl}$ and $\omega_{ct}^{rl}$ represent instant rotations, whereas vectors $\omega_g$ and $\omega_{ct}$ describe continuous rotations.

Within the plane of projections $\pi_4$, the rotation vectors $\omega_g$ and $\omega_{ct}$ are constructed, following the conventional rules developed in descriptive geometry. The location of the apex point $A_{pa}$ within the centerline distance ℄ becomes evident immediately after the projections of the rotation vectors $\omega_g$ and $\omega_{ct}$ are constructed within the plane of projections $\pi_4$. Furthermore, the projection of the apex point $A_{pa}$ can be constructed within the plane of projections $\pi_2$.

Two other components $\omega_g^{sl}$ and $\omega_{ct}^{sl}$ of the rotation vectors $\omega_g$ and $\omega_{ct}$ are perpendicular to the vector of instant rotation $\omega_{pl}$:

$$\omega_g^{sl} \perp \omega_{pl} \quad \text{and} \quad \omega_{ct}^{sl} \perp \omega_{pl} \tag{3.10}$$

With no distortions, the components $\omega_g^{sl}$ and $\omega_{ct}^{sl}$ are projected onto the plane of projections $\pi_5$. The axis of projections $\pi_4/\pi_5$ is erected perpendicular to the axis of projections $\pi_1/\pi_4$. The vectors $\omega_g^{sl}$ and $\omega_{ct}^{sl}$ are equal to each other; therefore, the equality $\omega_g^{sl} = \omega_{ct}^{sl}$ is valid. The components $\omega_g^{sl}$ and $\omega_{ct}^{sl}$ cause pure sliding of the axodes of the work gear and the gear-cutting tool relative to each other.

It is important to point out here that, similar to the vectors $\omega_g^{rl}$ and $\omega_{ct}^{sl}$, the vectors $\omega_g^{sl}$ and $\omega_{ct}^{sl}$ also represent instant rotations, whereas the vectors $\omega_g$ and $\omega_{ct}$ describe continuous rotations.

The equality $\omega_g^{rl} = \omega_{ct}^{rl}$ should not be confusing. Aside from the fact that the components $\omega_g^{sl}$ and $\omega_{ct}^{sl}$ are pointed in the same direction, the corresponding linear sliding velocities $\mathbf{V}_g^{sl}$ and $\mathbf{V}_{ct}^{sl}$ possess the following properties:

- The linear velocity vectors $\mathbf{V}_g^{sl}$ and $\mathbf{V}_{ct}^{sl}$ are pointed in the opposite direction.
- These vectors are of different magnitudes.

The first property is because the apex point $A_{pa}$ is located within the centerline ℄ and not outside of this straight-line segment. The second property is attributed to the fact that, usually, the center-distances $C_g$ and $C_{ct}$ are not equal to each other (the inequality $C_g \neq C_{ct}$ is valid).

Occasionally, axodes of the rotating work gear and the gear-cutting tool are used for the analysis of the gear machining mesh. The axodes are two surfaces, one of which is associated with the work gear and the other with the gear-cutting tool. When a continuous indexing method of machining is used for machining a work gear, the axodes roll over one another. It is commonly understood that axodes roll with no sliding between them. It is noteworthy that the components $\omega_g^{rl}$ and $\omega_{ct}^{rl}$ of the rotations $\omega_g$ and $\omega_{ct}$ result in pure rolling (with no sliding) of the axodes. However, the rotation vectors $\omega_g^{sl}$ and $\omega_{ct}^{sl}$ always cause sliding of the axodes when the rotation vectors $\omega_g$ and $\omega_{ct}$ are crossed. This type of sliding is inevitable by nature when the rotations take place about crossing axes of a work-gear and the gear-cutting tool.

The components $\omega_g^{sl}$ and $\omega_{ct}^{sl}$ are usually strongly undesirable when designing an elementary gear drive (gear-to-gear mesh). It is common practice for the gear designers to reduce the sliding velocity to the lowest possible range. The components $\omega_g^{sl}$ and $\omega_{ct}^{sl}$ result in sliding of the axodes of the work gear and of the gear-cutting tool. Sliding of the axodes inevitably entails the corresponding sliding of the tooth flanks in the lengthwise direction of the gear tooth. In this way, the sliding components $\omega_g^{sl}$ and $\omega_{ct}^{sl}$ reduce the performance of the elementary gear drive.

On the other hand, the components $\omega_g^{sl}$ and $\omega_{ct}^{sl}$ of the rotation vectors $\omega_g$ and $\omega_{ct}$ can also be useful when designing a gear-cutting tool. Relative sliding of the axodes can be used as the primary (cutting) motion. In this scenario, it is highly desirable to increase the sliding velocity to the highest possible range. In this way, the efficiency of the gear-machining operation can be significantly improved.

The analysis shown in Figure 3.4 makes it possible to conclude that pure rolling of the axodes of the work gear and the gear-cutting tool occurs if and only if rotation vectors $\omega_g$ and $\omega_{ct}$ are located within a common plane. Otherwise, sliding of the axodes is inevitable by nature. For a gear machining mesh that features either parallel or intersecting axes of rotation, the axes $O_g$ and $O_{ct}$ are by definition located within a common plane. Therefore, no sliding occurs in these cases. Sliding of axodes occurs if and only if the work gear and the gear-cutting tool are rotating about skew axis $O_g$ and $O_{ct}$.

Sliding of the gear tooth flanks is proportional to the sliding rotation $\omega^{sl} = \omega_g^{sl} \equiv \omega_{ct}^{sl}$. For a given point of interest within the line of contact of the axodes, gear teeth sliding is also proportional to the distance of the point from the axis of instant rotation $P_{ln}$. Keep in mind that configuration of the axis of instant rotation $P_{ln}$ is a function of the center-distance $C$.

The following conclusions can be drawn from the preceding analysis:

1. No sliding of axodes is observed when axes of rotation of the work gear and the gear-cutting tool are located within a common plane.
2. Only rotations about skew axes feature relative sliding of axodes.
3. For rotations about skew axes, sliding of axodes of the gear machining mesh becomes more significant as the crossed-axis angle $\Sigma$ increases. Thus, the condition $\Sigma \to 180°$ can be interpreted as the condition for maximal sliding of axodes in the gear machining mesh.
4. For rotations about skew axes, sliding of axodes in the gear machining mesh become smaller as the crossed-axis angle $\Sigma$ decreases. Thus, the condition $\Sigma \to 0°$ can be interpreted as the condition for minimum sliding of axodes of the gear machining mesh.

In reality, a gear machining mesh can be specified by just two rotation vectors, $\omega_g$ and $\omega_{ct}$, and by center-distance $C$. The rest of the vectors used for the analysis of the kinematics of the gear machining mesh can be expressed in terms of the vectors $\omega_g$ and $\omega_{ct}$ and the center-distance $C$.

## 3.2 KINEMATIC RELATIONSHIPS FOR THE "GEAR MACHINING MESH"

Consider two rotations about skew axes, as schematically illustrated in Figure 3.5. From a kinematic and geometric perspective, two rotations about skew axes $O_g$ and $O_{ct}$ can be construed as rolling of a hyperboloid of one sheet over a hyperboloid of another sheet. These two one-sheet hyperboloids of revolution serve as axodes of the gear machining mesh. Axes of rotation of the axodes align with the rotation vectors $\omega_g$ and $\omega_c$. The straight generating line of the axodes aligns with the vector of instant rotation $\omega_{pl}$. The vector $\omega_{pl}$ describes the instant rotation of the axodes about the axis of instant rotation $P_{ln}$.

Rotation vectors $\omega_g$ and $\omega_c$ cross at a certain crossed-axis angle $\Sigma$. The axes $O_g$ and $O_{ct}$ of the rotations $\omega_g$ and $\omega_{ct}$ are remote from each other at a center-distance $C$, as shown in Figure 3.5. For the given configuration of the rotation vectors $\omega_g$ and $\omega_{ct}$, the corresponding vector of instant rotation $\omega_{pl}$ is constructed in Figure 3.5(a).

Consider the projections of rotation vectors $\omega_g$, $\omega_{ct}$, and $\omega_{pl}$ onto a plane that is orthogonal to the centerline ℄. Projections of rotation vectors $\omega_g$ and $\omega_{ct}$ onto the axis of instant rotation $P_{ln}$ are designated as $\omega_g^{rl}$ and $\omega_{ct}^{rl}$, respectively. These components cause pure rotation of the axodes. Projections of the rotation vectors $\omega_g$ and $\omega_{ct}$ onto a direction that is perpendicular to the axis of instant rotation $P_{ln}$ are designated $\omega_g^{sl}$ and $\omega_{ct}^{sl}$. These components cause sliding of the axodes. It can be proven that the rotation vectors $\omega_g^{sl}$ and $\omega_{ct}^{sl}$ always have the same magnitude and direction. Therefore, the equality $\omega_g^{sl} = \omega_{ct}^{sl}$ is observed.

The rotation vectors $\omega_g^{rl}$ and $\omega_{ct}^{rl}$ are situated within a common plane through the centerline ℄. The location of apex point $A_{pa}$ within the centerline ℄ is specified by the expression $\omega_g^{rl} \cdot C_g = \omega_{ct}^{rl} \cdot C_{ct}$ [see Eq. (3.6)], and is presented in Figure 3.4.

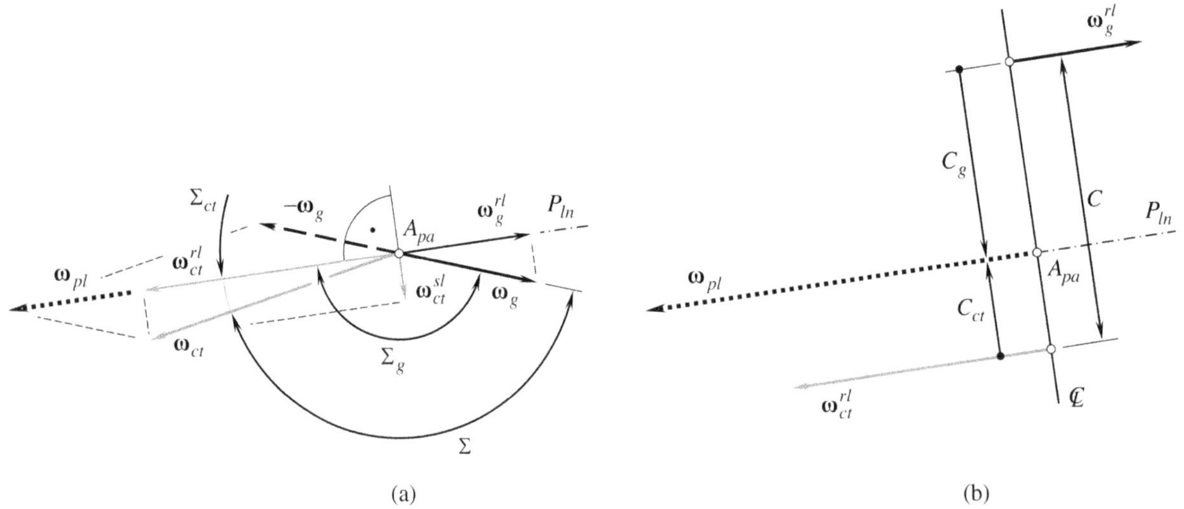


**FIGURE 3.5** Derivation of the principal kinematic relationships for the gear machining mesh: projection onto the plane perpendicular to the center-line (a), and onto the plane through the axis of instant rotation and the center-line (b).

It is noteworthy that in certain applications, it is convenient to represent Eq. (3.6) in the form:

$$\frac{C_g}{(\omega_g^{rl})^{-1}} = \frac{C_{ct}}{(\omega_{ct}^{rl})^{-1}} \quad (3.11)$$

where it is not the rotations $\omega_g^{rl}$ and $\omega_{ct}^{rl}$ themselves, but the inverse values $(\omega_g^{rl})^{-1}$ and $(\omega_{ct}^{rl})^{-1}$ are used instead. The inverse values $(\omega_g^{rl})^{-1}$ and $(\omega_{ct}^{rl})^{-1}$ are often referred to as the *rotation frequencies*.

As the equality $C_g + C_{ct} = C$ is valid [see Eq. (3.9)] for the center-distance $C_{ct}$, the expression:

$$C_{ct} = C - C_g \quad (3.12)$$

is valid. After substituting of the later expression for the center-distance $C_{ct}$, Eq. (3.8) takes on the following form:

$$\omega_g^{rl} \cdot C_g = \omega_{ct}^{rl} \cdot (C - C_g) \quad (3.13)$$

Elementary formulae transformations yield the expression:

$$r_g = \frac{1 + \omega_{ct} - \omega_g}{1 + \omega_{ct}} \cdot C \quad (3.14)$$

for the calculation of center-distance $C_g$ [Figure 3.5(b)].

After the actual value of center-distance $C_g$ is calculated, the following formula is used for the calculation of center-distance $C_{ct}$:

$$C_{ct} = C - C_g \quad (3.15)$$

When solving a particular problem that relates to the generation of gear tooth flanks, in addition to the vectors $\boldsymbol{\omega}_g^{rl}$ and $\boldsymbol{\omega}_{ct}^{rl}$, it is often convenient to use the corresponding normalized vectors $\bar{\boldsymbol{\omega}}_g^{rl}$ and $\bar{\boldsymbol{\omega}}_{ct}^{rl}$. The initially given vectors $\boldsymbol{\omega}_g^{rl}$ and $\boldsymbol{\omega}_{ct}^{rl}$ are normalized to $\bar{\boldsymbol{\omega}}_g^{rl}$ and $\bar{\boldsymbol{\omega}}_{ct}^{rl}$ by the value of $\omega_{ct}^{rl}$. In this case, the normalized angular velocity vector $\bar{\boldsymbol{\omega}}_{ct}^{rl}$ of the gear-cutting tool is equal to the unit vector:

$$\bar{\boldsymbol{\omega}}_{ct}^{rl} = \frac{\boldsymbol{\omega}_{ct}^{rl}}{\omega_{ct}^{rl}}, \quad \left(\left|\bar{\boldsymbol{\omega}}_{ct}^{rl}\right| \equiv 1\right) \quad (3.16)$$

Here, magnitude of the vector $\boldsymbol{\omega}_{ct}^{rl}$ is designated as $\omega_{ct}^{rl}$. The normalized angular velocity $\bar{\boldsymbol{\omega}}_g^{rl}$ of the gear is equal to a certain value:

$$\bar{\boldsymbol{\omega}}_g^{rl} = \frac{\boldsymbol{\omega}_g^{rl}}{\omega_{ct}^{rl}}, \quad \left(0 \le \bar{\omega}_g^{rl} < 1\right) \quad (3.17)$$

The normalized value $\bar{\omega}_{ct}^{rl}$ always exceeds or is equal to the normalized value $\bar{\omega}_g^{rl} = \left|\bar{\boldsymbol{\omega}}_g^{rl}\right|$, and the inequality $\bar{\omega}_{ct}^{rl} \ge \bar{\omega}_g^{rl}$ is valid.

Center-distances $C_g$ and $C_{ct}$ of the axodes can also be normalized by $C_{ct}$. In this case, the normalized gear-cutting tool center-distance $\bar{C}_{ct}$ is equal to a unit:

$$\bar{C}_{ct} = \frac{C_{ct}}{C_{ct}} \equiv 1 \quad (3.18)$$

Accordingly, the normalized gear-center-distance $\bar{C}_g$ is equal to a certain value:

$$\bar{C}_g = \frac{C_g}{C_{ct}} \ge 1 \quad (3.19)$$

The normalized gear-cutting tool center-distance $\bar{C}_{ct}$ is always smaller than or equal to the normalized gear-center-distance $\bar{C}_g$. Thus, the inequality $\bar{C}_{ct} \le \bar{C}_g$ is observed.

A few more formulae are useful for the calculations.

To calculate for the magnitude $\omega_{pl}$ of the vector of instant rotation $\boldsymbol{\omega}_{pl}$, the formula:

$$\omega_{pl} = \sqrt{(\omega_g^{rl})^2 + (\omega_{ct}^{rl})^2 - 2 \cdot \omega_g^{rl} \cdot \omega_{ct}^{rl} \cdot \cos\Sigma} \quad (3.20)$$

can be used.

Gear cone angle $\Sigma_g$ is the angle that the rotation vector $\boldsymbol{\omega}_g$ forms with the vector of instant rotation $\boldsymbol{\omega}_{pl}$—that is, gear cone angle $\Sigma_g$ is defined as:

$$\Sigma_g = \angle(\boldsymbol{\omega}_g, \boldsymbol{\omega}_{pl}) \quad (3.21)$$

To calculate for $\Sigma_g$, the formula:

$$\Sigma_g = \frac{1 + \omega_{ct} - \omega_g}{1 + \omega_{ct}} \cdot \Sigma \quad (3.22)$$

is used. This formula is composed similar to Eq. (3.14). Equation (3.22) can be derived based on the relationship:

$$\frac{\Sigma_g}{\Sigma_{ct}} = \frac{\omega_{ct}}{\omega_g} \quad (3.23)$$

which is evident.

Gear-cutting tool cone angle $\Sigma_{ct}$ is the angle that the rotation vector $\boldsymbol{\omega}_{ct}$ forms with the vector $\boldsymbol{\omega}_{pl}$ of instant rotation—namely, gear-cutting tool cone angle $\Sigma_{ct}$ is defined as:

$$\Sigma_{ct} = \angle(\boldsymbol{\omega}_{ct}, \boldsymbol{\omega}_{pl}) \quad (3.24)$$

Gear-cutting tool cone angle $\Sigma_{ct}$ is calculated from the equation:

$$\Sigma_{ct} = \frac{1 + \omega_g - \omega_{ct}}{1 + \omega_g} \cdot \Sigma \quad (3.25)$$

Following the analysis presented in Figure 3.5 for cone angels $\Sigma_g$ and $\Sigma_{ct}$, the equality:

$$\Sigma = \Sigma_g - \Sigma_{ct} \quad (3.26)$$

is valid.

In certain cases, gear cone angle $\Sigma_g$ can be equal to a right angle. The orthogonality of the rotation vectors

$\boldsymbol{\omega}_g^{rl} \perp \boldsymbol{\omega}_{pl}$ occurs when the equality $\omega_{pl} = \sqrt{(\omega_{ct}^{rl})^2 - (\omega_g^{rl})^2}$ is fulfilled.

The condition under which rotation vectors $\boldsymbol{\omega}_g$ and $\boldsymbol{\omega}_{pl}$ are orthogonal to each other can be represented in the form of the dot product of the following vectors:

$$\boldsymbol{\omega}_g \cdot \boldsymbol{\omega}_{pl} = 0 \tag{3.27}$$

After the gear cone angle $\Sigma_g$ and the gear-cutting tool cone angle $\Sigma_{ct}$ have been calculated, it can be shown that the following formulae for the projections $\boldsymbol{\omega}_g^{sl}$ and $\boldsymbol{\omega}_{ct}^{sl}$:

$$\omega_g^{sl} = \omega_g \cdot \cos(\Sigma_g - 90°) = \omega_g \cdot \sin \Sigma_g \tag{3.28}$$

$$\omega_{ct}^{sl} = \omega_{ct} \cdot \cos(\Sigma_{ct} - 90°) = \omega_{ct} \cdot \sin \Sigma_{ct} \equiv \omega_g^{sl} \tag{3.29}$$

are valid.

Similarly, to calculate for magnitudes $\omega_g^{rl}$ and $\omega_{ct}^{rl}$, the formulae:

$$\omega_g^{rl} = \omega_g \cdot \cos \Sigma_g \tag{3.30}$$

$$\omega_{ct}^{rl} = \omega_{ct} \cdot \cos \Sigma_{ct} \tag{3.31}$$

$$\frac{\omega_g}{\omega_{ct}} = \frac{\omega_g^{rl}}{\omega_{ct}^{rl}} \cdot \frac{\cos \Sigma_{ct}}{\cos \Sigma_g} = \frac{N_{ct}}{N_g} \tag{3.32}$$

are used.

Magnitudes $\omega_g^{rl} = \left|\boldsymbol{\omega}_g^{rl}\right|$ and $\omega_{ct}^{rl} = \left|\boldsymbol{\omega}_{ct}^{rl}\right|$ of the components $\boldsymbol{\omega}_g^{rl}$ and $\boldsymbol{\omega}_{ct}^{rl}$ are equal [(Figure 3.5(a)] to:

$$\omega_g^{rl} = \omega_g \cdot \cos \Sigma_g \tag{3.33}$$

$$\omega_{ct}^{rl} = \omega_{ct} \cdot \cos \Sigma_{ct} \tag{3.34}$$

Magnitudes $\omega_g^{sl} = \left|\boldsymbol{\omega}_g^{sl}\right|$ and $\omega_{ct}^{sl} = \left|\boldsymbol{\omega}_{ct}^{sl}\right|$ of the components $\boldsymbol{\omega}_g^{sl}$ and $\boldsymbol{\omega}_{ct}^{sl}$ are calculated from:

$$\omega_g^{sl} = \omega_g \cdot \sin \Sigma_g \tag{3.35}$$

$$\omega_{ct}^{sl} = \omega_{ct} \cdot \sin \Sigma_{ct} \tag{3.36}$$

Rotation vector $\boldsymbol{\omega}_{pl}$ is applied at apex point $A_{pa}$, which is situated within the centerline ℄. The following conditions:

$$\frac{C_g}{\omega_{ct} \cdot \cos \Sigma_{ct}} = \frac{C_{ct}}{\omega_g \cdot \cos \Sigma_g} \tag{3.37}$$

$$\frac{C_g}{\omega_{pl}} = \frac{C}{\omega_g \cdot \cos \Sigma_g} \tag{3.38}$$

$$\frac{C_{ct}}{\omega_{pl}} = \frac{C}{\omega_{ct} \cdot \cos \Sigma_{ct}} \tag{3.39}$$

are required at the apex point $A_{pa}$. The latter is used when calculating the actual location of the apex point $A_{pa}$.

When the principle of inversion is applied to the rotations of the work gear $\boldsymbol{\omega}_g$ and the gear-cutting tool $\boldsymbol{\omega}_{ct}$, the work gear becomes stationary, and the gear-cutting tool performs the resultant instant rotation $\boldsymbol{\omega}_{pl}$ in relation to the work gear. The instant rotation vector $\boldsymbol{\omega}_{pl}$ is calculated from $\boldsymbol{\omega}_{pl} = \boldsymbol{\omega}_{ct} + (-\boldsymbol{\omega}_g)$ [see Eq. (3.6)]. The sliding components $\boldsymbol{\omega}_g^{sl}$ and $(-\boldsymbol{\omega}_{ct}^{sl})$ of the rotation vectors $\boldsymbol{\omega}_{ct}$ and $(-\boldsymbol{\omega}_g)$ have the same magnitude but opposite directions. They comprise a rotation pair. This rotation pair is equivalent to a translation motion. Vector $\mathbf{V}^{sl}$ of the translation motion is parallel to the instant rotation vector $\boldsymbol{\omega}_{pl}$. Magnitude $V^{sl}$ of the speed of the translation vector $\mathbf{V}^{sl}$ is:

$$V^{sl} = \left|\mathbf{V}^{sl}\right| = C \cdot \omega_g \cdot \sin \Sigma_g = C \cdot \omega_{ct} \cdot \sin \Sigma_{ct} \tag{3.40}$$

The translation vector $\mathbf{V}^{sl}$ is parallel to the rotation vector $\boldsymbol{\omega}_{pl}$. Superposition of two motions—namely, the instant rotation $\boldsymbol{\omega}_{pl}$ and translation $\mathbf{V}^{sl}$—allows for interpretation in the form of a screw motion. Parameter (reduced pitch) $p_{pl}$ of the screw motion is calculated from the formula:

$$p_{pl} = \frac{V^{sl}}{\omega_{pl}} = \frac{C \cdot \omega_g \cdot \sin \Sigma_g}{\omega_{pl}} = \frac{C \cdot \omega_{ct} \cdot \sin \Sigma_{ct}}{\omega_{pl}} \tag{3.41}$$

Magnitude $\omega_{pl}$ of the instant rotation vector $\boldsymbol{\omega}_{pl}$ equals:

$$\omega_{pl} = \frac{C \cdot \omega_{ct} \cdot \cos \Sigma_{ct}}{C_g} = \frac{C \cdot \omega_g \cdot \cos \Sigma_g}{C_{ct}} \tag{3.42}$$

The calculated magnitude $\omega_{pl}$ from Eq. (3.42) is then substituted into Eq. (3.41). Equation (3.41) recasts into the formula:

$$p_{pl} = C_g \cdot \tan \Sigma_{ct} = C_{ct} \cdot \tan \Sigma_g \tag{3.43}$$

for calculation of the parameter $p_{pl}$ of the resultant instant screw motion.

Ultimately, an expression for the ratio:

$$\frac{C_g}{C_{ct}} = \frac{\tan \Sigma_g}{\tan \Sigma_{ct}} \tag{3.44}$$

can be derived.

The preceding analysis reveals that summa of two rotations about skew axes allows for interpretation in the form of an instant screw motion.

## 3.3 CONFIGURATION OF VECTORS OF RELATIVE MOTION

The variety of all possible types of gear machining mesh strongly depends on the permissible configurations of rotation vectors $\boldsymbol{\omega}_g$ and $\boldsymbol{\omega}_{ct}$. The rotations $\boldsymbol{\omega}_g$ and $\boldsymbol{\omega}_{ct}$ can be implemented to develop a scientific classification of all possible types of gear machining mesh.

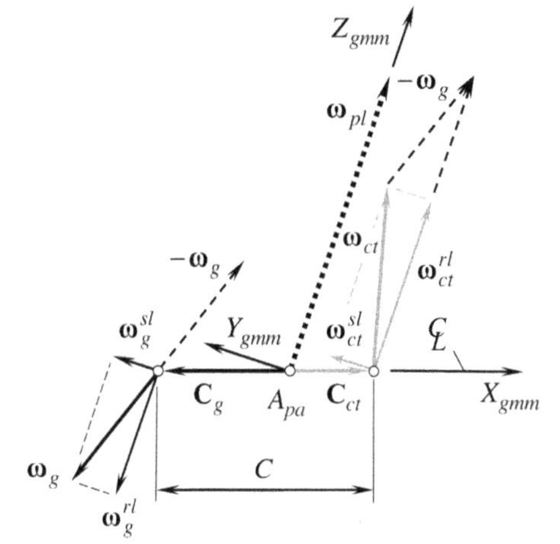


**FIGURE 3.6** Specification of principal parameters of "gear machining mesh".

### 3.3.1 Principal Features of Configuration of Two Rotation Vectors

Two arbitrary rotation vectors $\omega_g$ and $\omega_{ct}$ are shown in Figure 3.6. The rotation vectors $\omega_g$ and $\omega_{ct}$ are positioned at a certain centerline ℄. A Cartesian coordinate system $X_{gmm}Y_{gmm}Z_{gmm}$ is associated with the rotation vectors $\omega_g$ and $\omega_{ct}$ and the straight-line segment $C$. The origin of the reference system $X_{gmm}Y_{gmm}Z_{gmm}$ is coincident with the apex point $A_{pa}$.[5] The axis $X_{gmm}$ is aligned with the centerline ℄; the axis $Z_{gmm}$ is aligned with the vector of instant rotation $\omega_{pl}$; and the axis $Y_{gmm}$ complements the axes $X_{gmm}$ and $Z_{gmm}$ of the left-hand-oriented Cartesian coordinate system $X_{gmm}Y_{gmm}Z_{gmm}$.

For further analysis, it is convenient to introduce a unit vector $\mathbf{c}$. The vector $\mathbf{c}$ is along the $X_{gmm}$ – axis and is pointed in the positive direction of the $X_{gmm}$ – axis.

The newly introduced unit vector $\mathbf{c}$ allows for the simple analytical representation of two vectors $\mathbf{C}_g$ and $\mathbf{C}_{ct}$ (see Figure 3.6). The closest distance of approach $C_g$ of the axes $O_g$ and $P_{ln}$ of the rotation vectors $\omega_g$ and $\omega_{pl}$ is specified by the vector $\mathbf{C}_g$. The closest distance of approach $C_{ct}$ of the axes $O_c$ and $P_{ln}$ of the rotation vectors $\omega_{ct}$ and $\omega_{pl}$ is specified by vector $\mathbf{C}_{ct}$.

The vector $\mathbf{C}_{ct}$ is equal to $\mathbf{C}_{ct} = C_{ct} \cdot \mathbf{c}$. This vector is applied at the apex point $A_{pa}$. The normalized vector $\bar{\mathbf{C}}_{ct}$ is calculated from the formula $\bar{\mathbf{C}}_{ct} = C_{ct} \cdot \mathbf{c}$.

The vector $\mathbf{C}_g$ is equal to $\mathbf{C}_g = -C_g \cdot \mathbf{c}$. This vector is also applied at the apex point $A_{pa}$. The normalized vector $\bar{\mathbf{C}}_g$ is calculated from the formula $\bar{\mathbf{C}}_g = -\bar{C}_g \cdot \mathbf{c}$.

[5] In a reduced case of parallel-axes gearing, the apex point $A_{pa}$ is remote to infinity. The pitch point $P$ is used to construct the vector diagram in a case of $P_a$ – gearing. Cutting gears using form tools (e.g., gear broaching, milling gears by end-mill, and disk-mill cutters) represents more reduced cases compared to the machining processes based on the kinematics of $P_a$ – gearing. In these later cases, even the pitch point $P$ does not exist. Therefore, a point $P_0$ within the work gear axis of rotation $O_g$ is used to construct the vector diagram in a completely reduced case of cutting gears with form tools.

The magnitude of the vector $\mathbf{C}_g$ always exceeds or is equal to the length of the vector $\mathbf{C}_{ct}$, and thus, the inequality $|\mathbf{C}_g| \geq |\mathbf{C}_{ct}|$ is always observed. The same is valid with respect to the normalized vectors $\bar{\mathbf{C}}_g$ and $\bar{\mathbf{C}}_{ct}$.

The center-distance $C$ can be expressed in terms of the vectors $\mathbf{C}_g$ and $\mathbf{C}_{ct}$. For this purpose, the expression:

$$C = |\mathbf{C}| = |\mathbf{C}_{ct} - \mathbf{C}_g| \tag{3.45}$$

is used. The similar equation:

$$\bar{C} = |\bar{\mathbf{C}}| = |\bar{\mathbf{C}}_{ct} - \bar{\mathbf{C}}_g| \tag{3.46}$$

is valid for the normalized parameters $\bar{\mathbf{C}}_g$, $\bar{\mathbf{C}}_{ct}$, and $\bar{C}$.

By definition, the normalized center-distance is $\bar{C}_{ct} = 1$. Because an inequality $\bar{C}_g \geq \bar{C}_{ct}$ is always observed, the expression:

$$\bar{C} = \bar{C}_g + \bar{C}_{ct} \geq 2 \tag{3.47}$$

is valid for the normalized center distance $\bar{C}$.

Equations (3.46) and (3.47) impose strong constraints onto configuration of axodes in the gear machining mesh.

In general cases, the rotation vectors $\omega_g$ and $\omega_{ct}$ cross at a certain crossed-axis angle $\Sigma$. For all permissible combinations of the rotation vectors $\omega_g$ and $\omega_{ct}$, the crossed-axis angle $\Sigma$ falls into the interval $0° \leq \Sigma \leq 180°$.

Depending on the design parameters of the actual gear machining mesh, three locations of apex point $A_{pa}$ are recognized:

- The apex point $A_{pa}$ is located within the straight-line segment $C$.
- The apex point $A_{pa}$ is located outside the straight-line segment $C$.
- The apex point $A_{pa}$ is located at one of the two endpoints of the straight-line segment $C$.

No other locations of the apex point $A_{pa}$ are permissible.

The actual location of the apex point $A_{pa}$ within the straight-line segment $C$ indicates that for a given configuration of the rotation vectors $\omega_g$ and $\omega_{ct}$, the corresponding axodes are in external tangency to each other. When apex point $A_{pa}$ is located outside the straight-line segment $C$, the axodes are in internal tangency to each other.

The third case, wherein apex point $A_{pa}$ is located at one of the two endpoints of the straight-line segment $C$ requires more in-depth consideration. Because the configuration of the rotation vectors in this case relates exactly to what is between the first case (external tangency of the axodes) and the second case (internal tangency of the axodes), the third case is viewed as having *critical configuration* of the rotation vectors $\omega_g$ and $\omega_{ct}$.

The term *critical configuration* reflects a particular configuration of the rotation vectors $\omega_g$ and $\omega_{ct}$, for which apex point $A_{pa}$ is located at one of the two ends of the straight-line segment $C$.

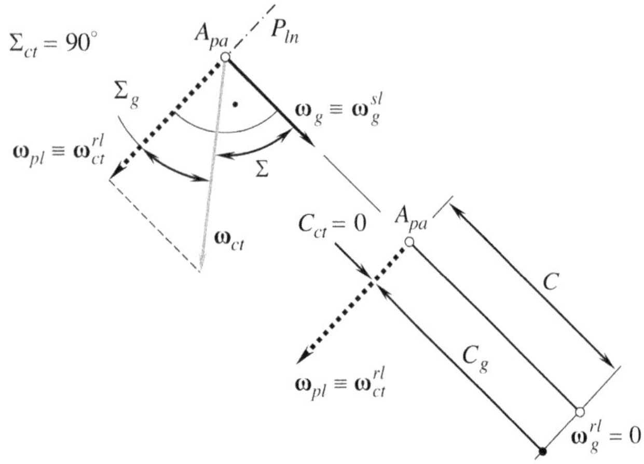


**FIGURE 3.7** An example of "critical configuration" of rotation vectors $\omega_g$ and $\omega_{ct}$.

For a certain configuration of the rotation vectors $\boldsymbol{\omega}_g$ and $\boldsymbol{\omega}_{ct}$, the rotation vector $\boldsymbol{\omega}_g$ becomes orthogonal to the vector of instant rotation $\boldsymbol{\omega}_{pl}$ (see Figure 3.7). In this case, the apex point $A_{pa}$ is at zero center-distance $C_{ct} = 0$ from the axis of rotation $O_{ct}$ of the gear-cutting tool. For orthogonal vectors $\boldsymbol{\omega}_g$ and $\boldsymbol{\omega}_{pl}$, the equality $\boldsymbol{\omega}_g \cdot \boldsymbol{\omega}_{pl} = 0$ is always valid.

Based on the formula in Eq. (3.25) $\Sigma_{ct} = 90°$, one can come up with an expression:

$$\Sigma_{cr}^{(1)} = \frac{1+\omega_g}{1+\omega_g-\omega_{ct}} \cdot \frac{\pi}{2} \tag{3.48}$$

for calculation the first critical value $\Sigma_{cr}^{(1)}$ of the crossed-axis angle $\Sigma$.

This case [Eq. (3.48)] pertains to the scenario under which the rotation vector $\boldsymbol{\omega}_{ct}$ of the gear-cutting tool is orthogonal to the vector of instant rotation $\boldsymbol{\omega}_{pl}$.

Apex point $A_{pa}$ can be located at the opposite end of the straight-line segment $C$. In this second case, plugging $\Sigma_g = 90°$ into Eq. (3.22), the expression:

$$\Sigma_{cr}^{(2)} = \frac{1+\omega_{ct}}{1+\omega_{ct}-\omega_g} \cdot \frac{\pi}{2} \tag{3.49}$$

can be derived for the second critical value $\Sigma_{cr}^{(2)}$ of the crossed-axis angle $\Sigma$.

### 3.3.2 Classification of the "Gear Machining Meshes"

A scientific classification[6] of all permissible kinds of gear machining meshes was proposed by Prof. S.P. Radzevich (~2008). This classification is based on the developed vector representation of gear machining meshes. Despite the axodes [that are associated (a) with the gear to be machined and (b) with the gear-cutting tool to be used for this purpose] occasionally being mentioned in this section of the book, the classification is developed based on vector diagrams and not on axodes, as vector diagrams represent more advanced means than axodes for the gear researcher.

All possible kinds of the gear machining meshes are classified as shown in Figure 3.8.

All gear machining meshes are categorized into three groups:

a. *External gear machining meshes* (also referred to as *u – negative gear machining meshes*) are those meshes whose crossed-axis angle $\Sigma$ exceeds its critical value $\Sigma_{cr}$ (the inequality $\Sigma > \Sigma_{cr}$ is valid). The apex point $A_{pa}$ is located within the center-distance $C$.
b. *Planar gear machining meshes* (also referred to *u – zero gear machining meshes*) are those meshes whose crossed-axis angle $\Sigma$ is equal to its critical value $\Sigma_{cr}$ (the equality $\Sigma = \Sigma_{cr}$ is valid). The apex point $A_{pa}$ is located at one of two ends of the center-distance $C$
c. *Internal gear machining meshes* (also referred to *u – positive gear machining meshes*) are those whose crossed-axis angle $\Sigma$ is smaller than its critical value $\Sigma_{cr}$ (the inequality $\Sigma < \Sigma_{cr}$ is valid). The apex point $A_{pa}$ is located within the center-distance $C$ beyond the end point of the center distance $C$.

There are five different types of gear machining meshes in total (see Figure 3.8).

First, no difference is observed in the configuration of the rotation vectors $\boldsymbol{\omega}_g$ and $\boldsymbol{\omega}_{ct}$, which represent an external gear machining mesh—that is, when the inequality $\Sigma_{cr} < \Sigma < 180°$ is valid. Any of two rotation vectors can be associated with either the work gear or the gear-cutting tool. Therefore, only one type of vector diagram can be associated with an external gear machining mesh.

Second, for planar gear machining meshes—that is, when the inequality $\Sigma = \Sigma_{cr}$ is valid—*two different kinds of configurations* of the rotation vectors $\boldsymbol{\omega}_g$ and $\boldsymbol{\omega}_{ct}$ are distinguished. For one of these configurations, the rotation vector $\boldsymbol{\omega}_{ct}$ of the gear-cutting tool is perpendicular to the vector of instant rotation $\boldsymbol{\omega}_{pl}$. For the other configuration, the work gear rotation vector $\boldsymbol{\omega}_g$ is perpendicular to the vector of instant rotation $\boldsymbol{\omega}_{pl}$.

The gear machining meshes those featuring the configuration ($\Sigma = \Sigma_{cr}$) of the rotation vectors is of particular importance to designers of gear-cutting tools. The values of the design parameters of the gear-cutting tool, which entail an increase of components $\boldsymbol{\omega}_g^{sl}$ and $\boldsymbol{\omega}_{ct}^{sl}$ of rotation vectors $\boldsymbol{\omega}_g$ and $\boldsymbol{\omega}_{ct}$, are often preferred when designing a gear-cutting tool.

Third, internal gear machining meshes (that is, when the inequality $0° < \Sigma < \Sigma_{cr}$ is valid) also feature *two different kinds of configurations* of the rotation vectors $\boldsymbol{\omega}_g$ and $\boldsymbol{\omega}_{ct}$. In one configuration, magnitude $\omega_g$ of the work gear's rotation vector $\boldsymbol{\omega}_g$ is greater than magnitude $\omega_{ct}$ of the gear-cutting tool's rotation vector $\boldsymbol{\omega}_{ct}$ ($\omega_g > \omega_{ct}$). Gear machining meshes of this type correspond to the machining processes of an external work gear with the enveloping gear-cutting

[6] A similar classification of vector diagrams is a cornerstone in the *Theory of Gearing* [171, 172].

tool. For the other configurations, magnitude $\omega_g$ of the work gear's rotation vector $\boldsymbol{\omega}_g$ is smaller than magnitude $\omega_{ct}$ of the gear-cutting tool's rotation vector $\boldsymbol{\omega}_c$ ($\omega_g < \omega_{ct}$). Gear meshes of this type correspond to the machining process of an internal work gear with the external gear-cutting tool.

Three design parameters are critical for these five different kinds of gear machining meshes:

- Center distance $C$.
- Crossed-axes angle $\Sigma$.
- Magnitudes $\omega_g$ and $\omega_{ct}$ of the rotation vectors $\boldsymbol{\omega}_g$ and $\boldsymbol{\omega}_{ct}$.

If one of these design parameters reaches an extremal value, the corresponding gear machining mesh transforms into one of the particular cases of gear machining meshes.

Zero value is critical to the center-distance $C$: When the center-distance vanishes ($C = 0$), a spatial gear machining mesh reduces to a mesh of bevel gears of a corresponding type. An external spatial gear machining mesh reduces to an external mesh of bevel gears. A planar gear machining mesh reduces into two corresponding gear machining meshes, which can be simulated by axodes in the form of a cone of revolution and plane. An internal spatial gear machining mesh reduces into two corresponding gear machining

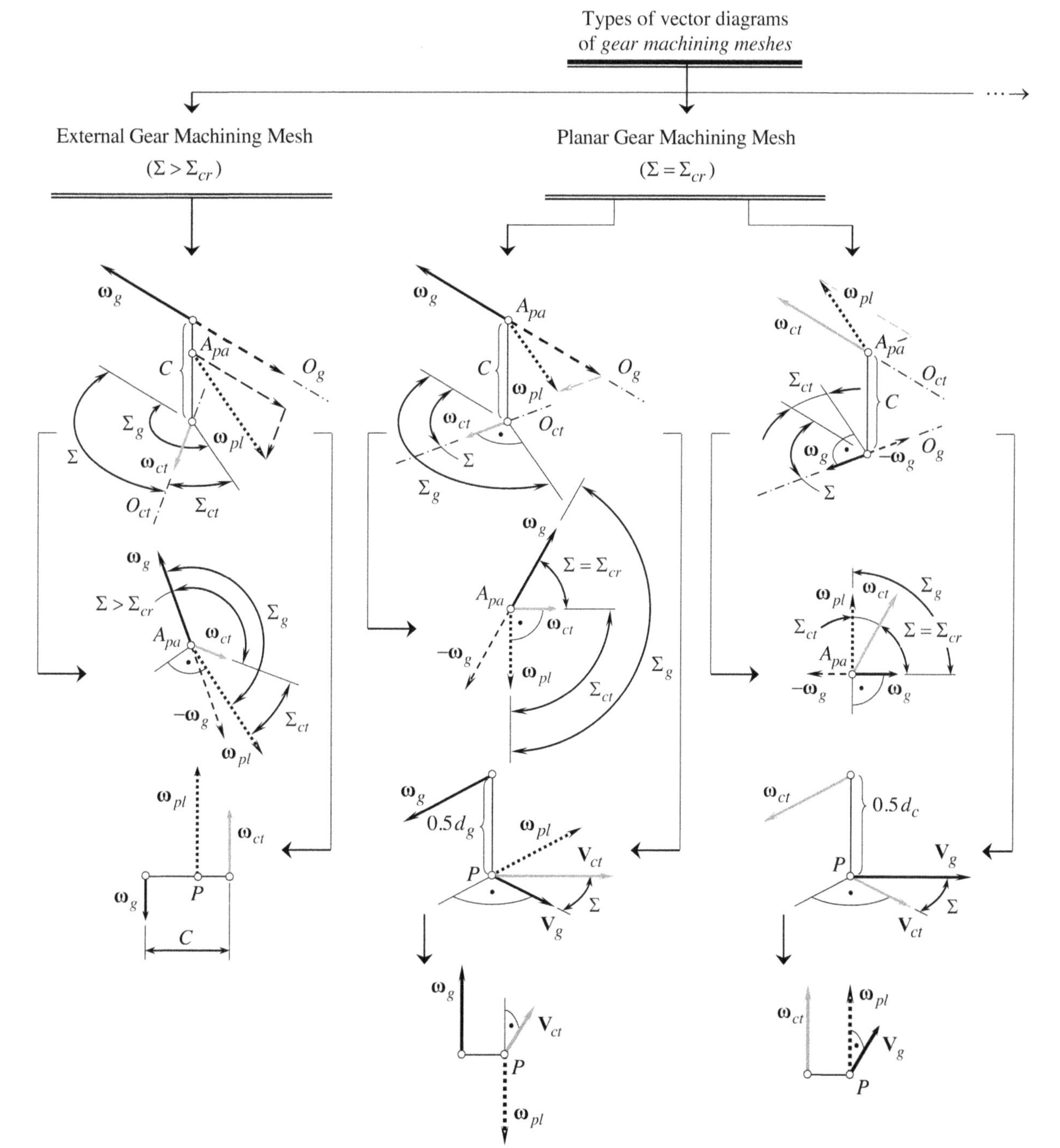


**FIGURE 3.8** Classification of kinds of the gear machining meshes. *(Continued)*

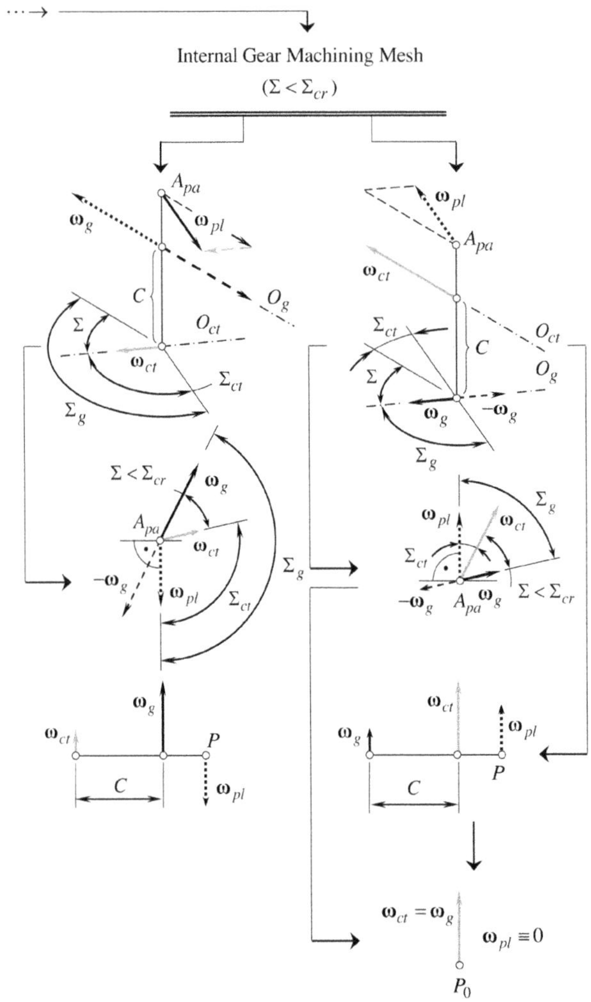


**FIGURE 3.8** *(Continued)*

meshes, which can be simulated by axodes in the form of two cones of revolution. Ultimately, if the center-distance vanishes, five spatial gear machining meshes reduce into five corresponding planar gear machining meshes.

Zero value is also critical to the crossed-axis angle $\Sigma$: When the crossed-axis angle reaches zero value ($\Sigma = 0°$), a spatial gear machining mesh is reduced to a mesh of gears with parallel axes of the corresponding type. Three possible types of gear machining mesh with parallel rotation vectors $\boldsymbol{\omega}_g$ and $\boldsymbol{\omega}_{ct}$ are shown in Figure 3.8.

When the magnitude of one of the rotation vectors $\boldsymbol{\omega}_g$ and $\boldsymbol{\omega}_{ct}$ reaches a zero value, the corresponding spatial gear machining mesh reduces into two gear machining meshes, which are specified by either the vector pair $\boldsymbol{\omega}_g$ and $\mathbf{V}_{ct}$ or the vector pair $\mathbf{V}_g$ and $\boldsymbol{\omega}_{ct}$. These two types of gear machining mesh are shown in Figure 3.8. In certain cases, both the vector pairs $\boldsymbol{\omega}_g$ and $\mathbf{V}_{ct}$ and the vector pairs $\mathbf{V}_g$ and $\boldsymbol{\omega}_{ct}$ can be perpendicular to each other (that is, $\boldsymbol{\omega}_g \perp \mathbf{V}_{ct}$ and $\mathbf{V}_g \perp \boldsymbol{\omega}_{ct}$). These two types of gear machining mesh are also illustrated in Figure 3.8.

In a reduced case, not just one but two design parameters (that is, both parameters $C$ and $\Sigma$) can simultaneously have a zero value ($C = 0$ and $\Sigma = 0°$). An example of this type of gear machining mesh is illustrated in Figure 3.8. In the interest of thoroughness, a few more critical combinations of the design parameters of the gear machining mesh should be noted:

- $C = 0$ and either $\omega_g = 0$ or $\omega_{ct} = 0$.
- $\Sigma = 0°$ and either $\omega_g = 0$ or $\omega_{ct} = 0$.
- $C = 0$, $\Sigma = 0°$ and either $\omega_g = 0$ or $\omega_{ct} = 0$.

Ultimately, the total number of possible types of gear machining meshes is limited to just 21. Eighteen gear machining meshes are depicted in Figure 3.8, and three more are specified by the critical values of the design parameters of the gear machining mesh.

The discussion of classification of possible types of vector diagrams of gear machining meshes (see Figure 3.8) is the key (or a *road map*) to the development of novel designs of gear-cutting tools and new methods of machining gears.[7]

[7] It could happen that future research in the field will result in certain alterations to the proposed scientific classification of possible types of vector diagrams of the *gear-machining operations*. These alterations, should any occur, are unlikely to change the core of the classification and its concept, and would, rather, be local in nature.

## 3.4 KINEMATICS OF GEAR-MACHINING OPERATIONS

The kinematics of gear machining mesh is the core of the kinematics of the entire gear-machining operation. However, the kinematics of the gear machining mesh is not always sufficient for determining of the generating surface of the gear-cutting tool, nor for cutting the work gear. For these purposes, in addition to the kinematics of the gear machining mesh, the kinematics of the gear-machining operation includes more relative motions of the work gear and the gear-cutting tool.

Additional motions are usually those that allow for sliding of the generating surface of the gear-cutting tool over itself. Through motions of this type, the gear machining mesh can be complete in most cases to the kinematics of the gear-machining process.

Ultimately, designing a gear-cutting tool begins with the analysis of kinematics:

Kinematics of the gear machining mesh is analyzed from the standpoint of whether a given work gear can be machined using only a certain gear machining mesh. If the kinematics of the gear machining mesh is sufficient for the generation of generating surface $T$ of the gear-cutting tool, as well as for material removal, then this surface $T$ is used for the design of the gear-cutting tool.

If no kind of kinematics of the gear machining mesh is feasible for a particular gear-machining operation, those motions that allow for the sliding of the surface $T$ over itself are added to the kinematics of the gear machining mesh. In this manner, the kinematics of the gear machining mesh evolves to the corresponding kinematics of the gear-machining operation. Motions that allow for the sliding of the work gear tooth flank $\mathscr{G}$ over itself are also often incorporated into the kinematics of the gear-machining operation.

There are cases when the generating surface of the gear-cutting tool does not allow for sliding motion over itself. This type of case is the most inconvenient challenge for both the gear-cutting tool designer and the gear manufacturer. However, even problems of this kind are solvable if the scientific theory of gearing, and the theory of gear tooth flank generation are used correctly.

# Section II

# Form Gear-Cutting Tools

Form gear-cutting tools represent the oldest known design of special purpose cutting tools: those used for cutting gears by discretely generating methods of gear machining. The kinematics of discretely generating methods of gear machining is the simplest possible. It is often represented by one of those motions that cause a surface (that is, either a gear tooth flank $\mathscr{G}$ or generating surface of the gear-cutting tool $T$) to slide on themselves. A screw motion of a constant pitch is the most general type of motion that causes a surface to slide on itself. Consider a screw surface performing a screw motion. Let us assume that the axis of the screw motion is aligned with the axis of the screw surface, and screw parameters of the surface and of the screw motion are identical to each other. In such a scenario, the enveloping surface of consecutive positions of the screw surface in the screw motion is congruent to that same screw surface. This means that the screw motion causes the screw surface to slide on itself. Physically, this can be observed in a bolt-and-nut pair, the screw surfaces of which allow for sliding over each other.

When the parameter of a screw motion increases and approaches an infinite value, the screw motion reduces to a translation. When a cylindrical surface (that is, a surface of translation) performs a straight motion, consequently generating a straight line, the enveloping surface of consecutive positions of the cylindrical surface in the straight motion is congruent to the cylinder itself. This property of translation allows for the interpretation that a motion of this kind causes cylinders to slide over them.

When the parameter of a screw motion decreases and approaches a zero value, the screw motion reduces to a rotation. When a surface of revolution performs a rotation about its axis of rotation, the enveloping surface to consecutive positions of the surface of revolution in its rotational motion is congruent to the surface of revolution itself. This property of rotational motion allows for the interpretation that motions of this type cause surfaces of revolution to slide on themselves. As shown in Figure II.1, a gear broaching operation is a perfect example of machining operations that utilize the property of part surfaces to allow sliding on themselves.

**FIGURE II.1** Gear broaching operation.

DOI: 10.1201/9781003520528-5

There are three different kinds of motions that cause surfaces to slide on themselves:

- Screw motion of a constant pitch
- Translational motion
- Rotational motion

The various designs of gear cutting tools and the corresponding methods of gear machining discussed in Part II—gear broaching, gear milling, gear grinding, and so forth—are based on the fundamental property of gear tooth flanks that allow for sliding on themselves. Relative motions of this type are extensively used when machining tooth flanks of work gears. Tooth flanks of gears of many geometries allow for sliding over themselves [see Chapter 1]. Part surfaces of this particular kind are easy for broaching, milling, grinding, and so forth.

The second part of the book consists of Chapter 4 through Chapter 7.

# 4 Gear Broaching Tools

A gear broach is a cutting tool for rapid machining of a desired contour in a workpiece surface by moving a cutter entirely past the workpiece. Broaches are often used to cut internal and external gear teeth, racks, and gear segments on small gears. They are usually designed to cut all teeth at the same time.

The broach has a long series of cutting teeth (see Figure 4.1). Broach teeth are divided into two groups: roughing cutting teeth and finishing cutting teeth. Roughing cutting teeth of the broach gradually increase in height. What would be the feed in other types of machining is designed into the broach and is called chip per tooth. The last few teeth of the broach are designed to finish the cut rather than remove a considerable amount of stock. These cutting teeth are referred to as the finishing teeth. Normally, finishing teeth have uniform dimensions to take off the final increment of stock, completing the form to the major tooth shape. The cutting edges of the finishing teeth of the gear broach are located within the generating surface of the gear broach.

## 4.1 KINEMATICS OF GEAR BROACHING OPERATION

In the classification of vector diagrams of gear machining meshes, the vector diagram for the gear broaching operation is reduced to the simplest kind. The kinematics of gear tooth flank generation consists of no relative motions of part surface generation.

To transform the vector diagram of the gear machining mesh into the corresponding kinematics of the gear broaching operation, a cutting motion is added to the vector diagram. In Figure 4.2, the cutting motion is represented as a superposition of the straight motion $\mathbf{V}_{cut}$ and the rotation $\boldsymbol{\omega}_{cut}$. Superposition of two motions $\mathbf{V}_{cut}$ and $\boldsymbol{\omega}_{cut}$ enables broaching of a helical work gear.

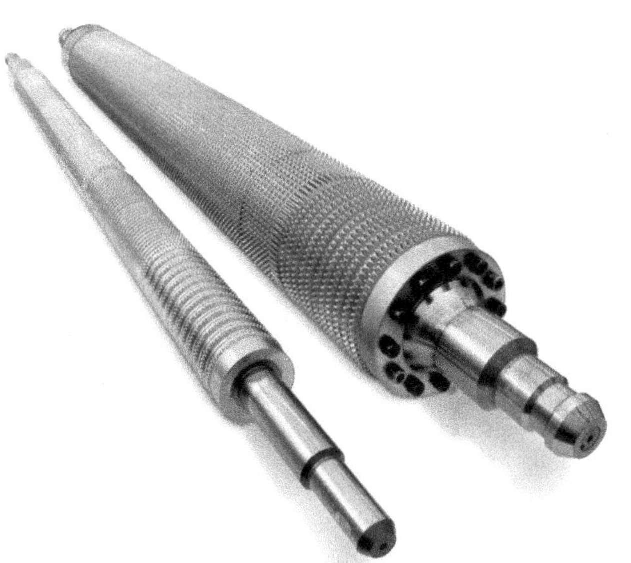

FIGURE 4.1 Involute gear broaches.

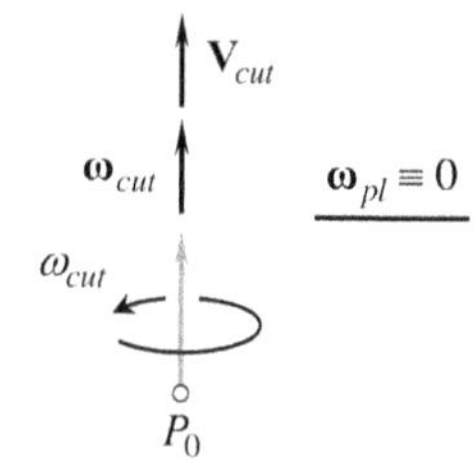


FIGURE 4.2 Kinematics of gear broaching operation.

The motions $\mathbf{V}_{cut}$ and $\boldsymbol{\omega}_{cut}$ do not affect the geometry of the work gear tooth flank $\mathscr{G}$, as these motions cause sliding of the gear tooth flank $\mathscr{G}$ over itself.

The ratio of the translation $V_{cut}$ and the rotation $\omega_{cut}$ is specified by the pitch helix angle $\psi_g$ of the work gear. Broaches for machining spur gears ($\psi_g = 0°$) are designed to keep the relative velocities $V_{cut} \neq 0$ and $\omega_{cut} = 0$. If the relative velocities are set up so that $V_{cut} = 0$ and $\omega_{cut} \neq 0$, a surface of revolution $\mathscr{G}$ is generated in the broaching operation. Methods of gear machining and corresponding designs of the gear-broaching tools based on the reduced kinematics of the gear machining operation have not yet been thoroughly investigated.

## 4.2 GENERATING SURFACE OF A GEAR BROACH

The generating surface of the gear broach is congruent to the tooth surfaces of a gear to be machined. As such, the generating surface $T$ of the broach for machining spur gears is congruent to the spur gear being machined. Similarly, the generating surface of the broach for machining helical gears is congruent to the helical gear being machined. This promptly yields an equation for the generating surface $T$ of the gear broach:

$$\mathbf{r}_c(U_c, V_c) = \begin{bmatrix} r_{b.c} \cos V_c + U_c \sin \psi_{b.c} \sin V_c \\ r_{b.c} \sin V_c - U_c \cos \psi_{b.c} \sin V_c \\ r_{b.c} \cot \psi_{b.c} - U_c \cos \psi_{b.c} \\ 1 \end{bmatrix} \tag{4.1}$$

where:

- $U_c$, $V_c$ are the curvilinear coordinates of the generating surface $T$ of the gear broach
- $r_{b.c}$ is the radius of the base cylinder of the generating surface $T$ (Figure 4.3)
- $\psi_{b.c}$ is the base helix angle of the surface $T$

DOI: 10.1201/9781003520528-6

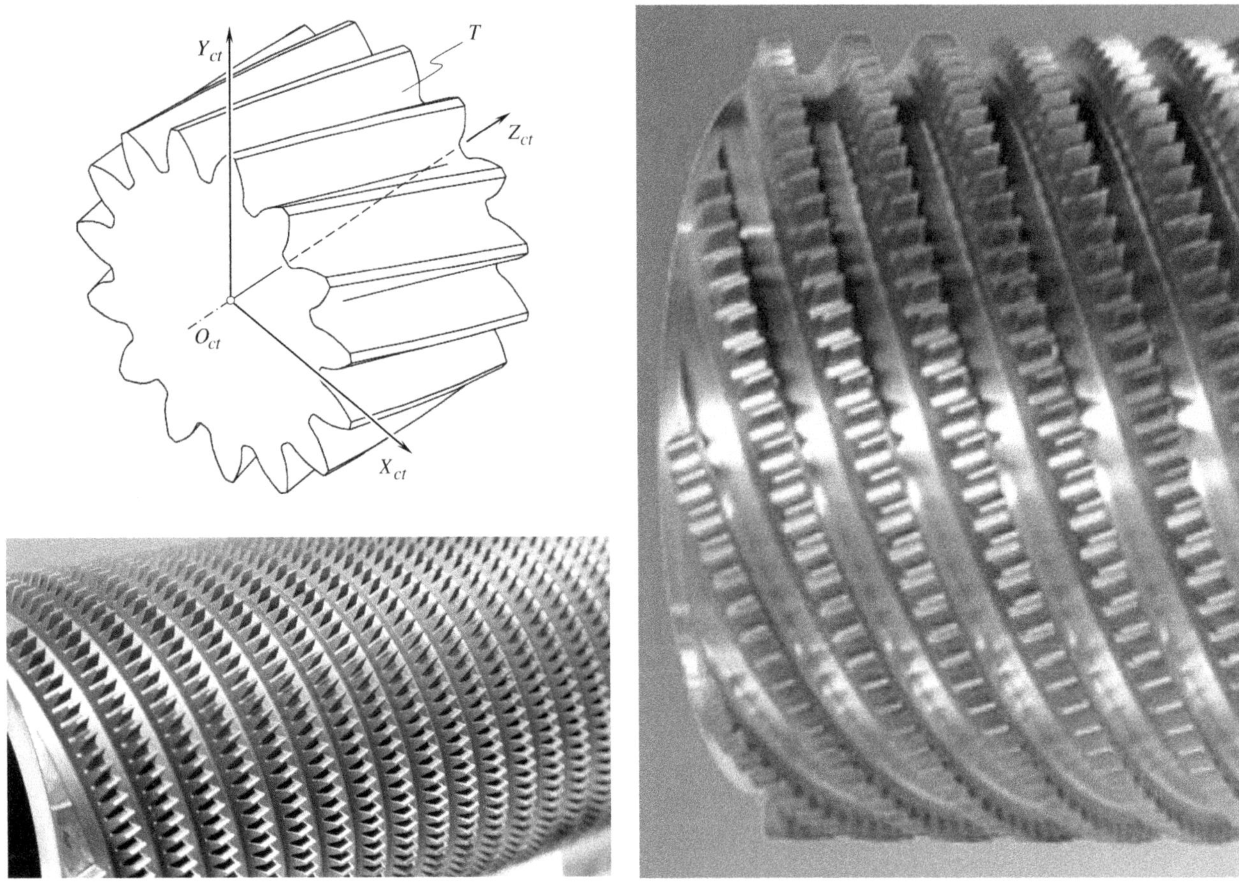


**FIGURE 4.3** Geometry of the generating surface $T$ of an involute gear broach.

All roughing cutting edges of the gear broach are shifted inward, the body bounded by the generating surface $T$. Finishing cutting edges are located within the generating surface $T$ of the gear broach. Theoretically, any line within the generating surface $T$ can be used as the cutting edge $CE$ of finishing teeth of the gear broach. Deviations of the finishing cutting edges from the generating surface $T$ are the root cause of the corresponding deviation of the actually machined gear from its desirable geometry.

Equation (4.1) yields an expression:

$$\mathbf{G}_T = \begin{bmatrix} 1 & -\dfrac{r_{b.c}}{\sin\psi_{b.c}} \\ -\dfrac{r_{b.c}}{\sin\psi_{b.c}} & \dfrac{U_c^2 \sin^4\psi_{b.c} + r_{bc}^2}{\sin^2\psi_{b.c}} \end{bmatrix} \tag{4.2}$$

for the first fundamental matrix of the generating surface $T$ of the gear broach.

The second fundamental matrix of the generating surface $T$ of the gear broach can be expressed in the form:

$$\mathbf{D}_T = \begin{bmatrix} 0 & 0 \\ 0 & U_c \sin\psi_{b.c} \cos\psi_{b.c} \end{bmatrix} \tag{4.3}$$

Use of the first $\mathbf{G}_T$ and second $\mathbf{D}_T$ fundamental matrices of the generating surface $T$ of the gear broach simplifies the derivation of equations relating to the geometry and design parameters of the gear broach.

In a manner largely similar to that previously described, an equation for the generating surface $T$ of the gear broach for machining non-involute tooth profiles can be derived. Ultimately, the derived equation for the generating surface $T$ allows for simple formulae for the calculation of many of the design parameters of the gear broach.

## 4.3 CUTTING EDGES OF GEAR BROACHING TOOL

The cutting edges $CE$ of a gear-cutting tool are viewed in the sense of lines of intersection of the rake surface $R_s$ and the clearance surface $C_s$. The cutting edges of roughing and semi-roughing teeth are situated beneath the generating surface $T$ of the gear broach. Cutting edges of finishing teeth of the gear broach are located within the generating surface $T$.

There is greater freedom in designing roughing and semi-roughing teeth of a gear broach. Design of finishing cutting edges is more constrained by the necessity to constrain them within the generating surface $T$.

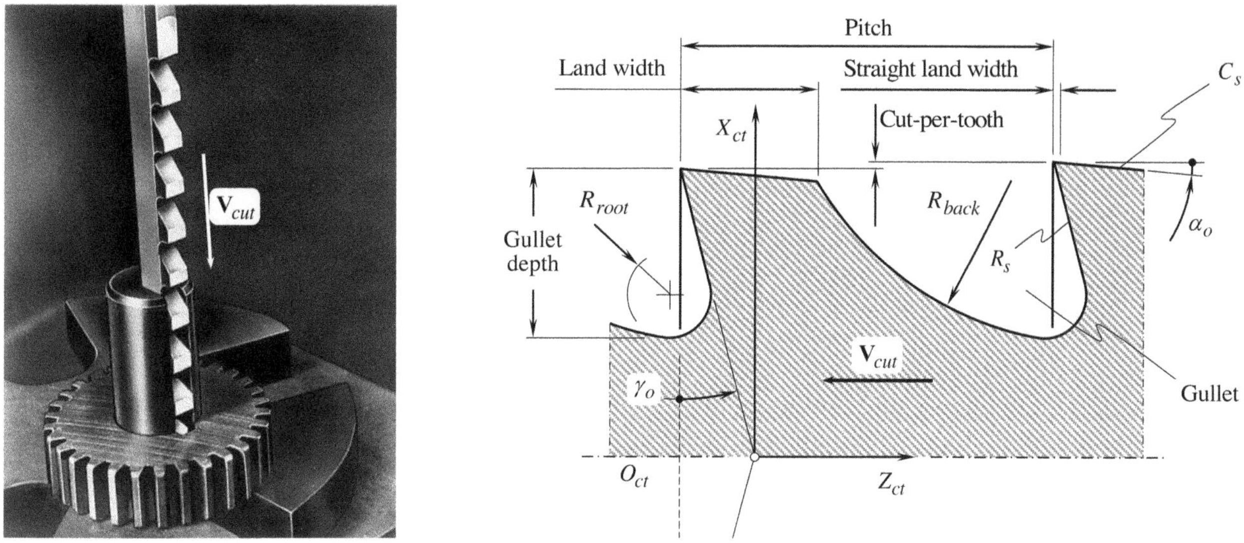


**FIGURE 4.4** Principal design parameters of the cutting tooth of a gear broach.

The desirable cutting edges of the finishing cutting teeth of a gear broach can be specified in terms of the generating surface $T$ of the gear broach and of the rake surface $R_s$.

### 4.3.1 Rake Surface of the Finishing Teeth of the Gear Broach

Designing of a gear broach can be viewed as the transformation of a rigid body bounded by the generating surface $T$ into the corresponding cutting tool capable of cutting off the stock. Following this concept, gullets need to be machined. The term *gullet* is often applied to *chip space*.

The axial profile of a gear broach gullet is shown in Figure 4.4. This profile is shaped by:

- The axial profile of the rake surface $R_s$.
- The root radius $R_{root}$, the radius just below the cutting edge $CE$ that blends into the back of the tooth radius (and sometimes referred to as face angle radius).
- The back radius $R_{back}$, the radius on the back of the tooth in the chip space.
- The axial profile of the clearance surface $C_s$ which is often referred to as land.

For finishing teeth of a gear broach, the cutting edge can follow by straight land, which is used for finishing teeth to retain broach size after a series of re-sharpenings. For the gullet depth, the term tooth depth is also used. The distance between two adjacent broach teeth is referred to as the pitch of the broach teeth. The pitch is the distance from the cutting edge $CE$ of one tooth to the corresponding point on the next tooth.

Chips from a broach are necessarily confined in the space between the teeth. Therefore, it is absolutely necessary to provide an adequate chip space when designing the broach. If this is not done, there is a chance of dulling or even breaking the tool and roughing or tearing the surface of the work being broached. Chip space is limited by the pitch of the teeth and, in small broaches, by the diameter of the broach.

Broaching in general requires the use of chip breakers.[1] Otherwise, an overly wide chip may be produced. Such chips are difficult to control and are usually handled by providing staggered chip breakers on successive broach teeth.

Two major types of surfaces are practically used to form rake surfaces of gear broaches:

- The surface of a cone of revolution.
- The screw Archimedean surface.

Conical rake surface. Rake surface $R_s$ in the form of a cone of revolution is coaxial with the gear broach itself. Equation of the conical rake surface is necessary for further analysis.

A Cartesian coordinate system $X_{ct}Y_{ct}Z_{ct}$ is associated with the gear broach, as shown in Figure 4.4. In this coordinate system, every point of the generating straight line of the rake surface $R_s$ fulfills the equality:

$$Z_{ct} = -X_{ct} \cdot \tan\gamma_o \tag{4.4}$$

where $\gamma_o$ denotes the rake angle at the top cutting edge of the gear broach tooth.[2]

In the case being considered, position vector of a point $\mathbf{r}_{rs}$ of the rake surface $R_s$ can be expressed in terms of the

[1] Chip-breakers are notches in the teeth of broaches that divide the width of chips, facilitating their removal.

[2] The rake angle $\gamma_o$ at the top cutting edge of gear broach tooth is often referred to as the *face angle*, or sometimes the *hook angle*.

current diameter $d_{y.c}$ of the broach, the rake angle $\gamma_o$ and the angular parameter $\vartheta_{rs}$:

$$\mathbf{r}_{rs}(d_{y.c}, \vartheta_{rs}) = \begin{bmatrix} \frac{d_{y.c}}{2} \cdot \sin \vartheta_{rs} \\ \frac{d_{y.c}}{2} \cdot \cos \vartheta_{rs} \\ -\frac{d_{y.c}}{2} \cdot \tan \gamma_o \\ 1 \end{bmatrix} \quad (4.5)$$

where the diameter $d_{y.c}$ and the angular parameter $\vartheta_{rs}$ are viewed in the sense of Gaussian coordinates of the rake surface $R_s$.

Rake surface in the form of screw Archimedean surface. Gear broaches with the rake surface $R_s$ in the form of a screw Archimedean surface are convenient in re-sharpening the worn broach. The screw rake surface eliminates the necessity of indexing after sharpening each tooth of the gear broach.

A screw rake surface $R_s$ is generated by the straight line:

$$Z_{ct} = -X_{ct} \cdot \tan \gamma_o \quad (4.6)$$

that is performing a screw motion with a certain screw parameter $p_{rs}$ of the screw motion. If the axial pitch of the gear broach is designated as $p_c$, the screw parameter (that is, the reduced pitch) is equal to $p_{rs} = \frac{p_c}{2 \cdot \pi}$. Ultimately, one can come up with the matrix representation:

$$\mathbf{r}_{rs}(d_{y.c}, \vartheta_{rs}) = \begin{bmatrix} \frac{d_{y.c}}{2} \cdot \sin \vartheta_{rs} \\ \frac{d_{y.c}}{2} \cdot \cos \vartheta_{rs} \\ -\frac{d_{y.c}}{2} \cdot \tan \gamma_o + p_{rs} \cdot \vartheta_{rs} \\ 1 \end{bmatrix} \quad (4.7)$$

of the screw Archimedean rake surface $R_s$.

In Eq. (4.7), the screw parameter $p_{rs}$ is a signed value. It is of a positive value for the right-hand screw Archimedean rake surface and of a negative value for the left-hand rake surface $R_s$.

Rake angle $\gamma_o$ for gear broaches does not exceed $\gamma_o \leq 10°$. When the equality $\gamma_o = 0°$ is observed, the Eq. (4.7) reduces to:

$$\mathbf{r}_{rs}(d_{y.c}, \vartheta_{rs}) = \begin{bmatrix} \frac{d_{y.c}}{2} \cdot \sin \vartheta_{rs} \\ \frac{d_{y.c}}{2} \cdot \cos \vartheta_{rs} \\ p_{rs} \cdot \vartheta_{rs} \\ 1 \end{bmatrix} \quad (4.8)$$

Equations (4.5) and (4.7) make derivation of the first $\mathbf{G}_{rs}$ and second $\mathbf{D}_{rs}$ fundamental matrices of the rake surface $R_s$ of a gear broach possible. By means of the fundamental matrices $\mathbf{G}_{rs}$ and $\mathbf{D}_{rs}$, all the parameters of the rake surface $R_s$ can be readily calculated.

### 4.3.2 Clearance Surface of Gear Broach Tooth

The purpose of the clearance surface is twofold. First, the clearance surface $C_s$ provides clearance between the broach tooth and the machined surface. The clearance is necessary to reduce friction that causes heat generation and increases the tooth wear of the broach. Second, the clearance surface must be of a shape that is capable of maintaining the tooth profile after every re-sharpening of the worn broach.

Gear broaches feature very small clearance angles.[3] The clearance angle $\alpha_o$ at the top cutting edges of roughing teeth is at the range of $\alpha_o \cong 4°$. For semi-finishing cutting edges, the clearance angle is reduced to $\alpha_o \cong 2°$. Finally, the clearance angle at finishing cutting edges is in the range of $\alpha_o \cong 0.5°$.

The desirable clearance surface $C_s$ of the lateral profile of the gear broach tooth can be generated by the cutting edge that is performing a screw motion.

To derive an equation of the cutting edge $CE$ of the gear broach tooth, the equation for the generating surface $T$ of the broach [see Eq. (4.1)] is considered together with equation for the rake surface $R_s$ [see Eqs. (4.5) through (4.8)]. Without going into details of derivation, let us assume that equation of the cutting edge $CE$ is derived and is represented in vector form:

$$\mathbf{r}_{ce} = \mathbf{r}_{ce}(\xi_{ce}) \quad (4.9)$$

where $\xi_{ce}$ denotes the parameter of the cutting edge of the gear broach tooth.

Furthermore, in order to generate the clearance surface $C_s$ of a desired shape, the cutting edge $\mathbf{r}_{ce} = \mathbf{r}_{ce}(\xi_{ce})$ undergoes a screw motion with a certain screw parameter $p_{cs}$. The actual value of the screw parameter $p_{cs}$ depends on the required value of the normal clearance angle $\alpha_N$ at the lateral cutting edge of the gear broach tooth and on the design parameters of the gear broach. The screw motion of the cutting edge $CE$ can be analytically described using corresponding matrices of translation along and rotation about the axis of the gear broach, or the corresponding operator of the screw motion can be implemented for this purpose. Ultimately, the use of the operator $\mathbf{Sc}_z(\vartheta_{cs}, p_{cs})$ of the screw motion (see Appendix D for details) yields an expression:

$$\mathbf{r}_{cs}(\xi_{ce}, \vartheta_{cs}) = \mathbf{Sc}_z(\vartheta_{cs}, p_{cs}) \cdot \mathbf{r}_{ce}(\xi_{ce}) \quad (4.10)$$

for the position vector $\mathbf{r}_{cs}$ of point of the clearance surface $C_s$ of the gear broach tooth.

[3] The clearance angle of gear broach is also referred to as the *back-off angle*. This is a relief angle back of the cutting edge.

The roughing cutting teeth of a gear broach do not require an accurate profiling. This is because the final accuracy of the machined gears depends on the accuracy of the finishing teeth of the broach and not on the accuracy of its roughing teeth. The rake angle $\gamma_o$ and the clearance angle $\alpha_o$ of small values ($\gamma_o = 0°$ and $\alpha_o \cong 0.5°$, respectively) allow for simplification of the problem of profiling of the finishing cutting edges.

The generating surface $T$ of a gear broach is congruent to the tooth flank surface of the gear being broached. Therefore, no cusps occur on tooth flanks of the machined gear.

## 4.4 CHIP REMOVAL DIAGRAMS

When broaching gears, almost the entire material found in the space between two adjacent gear teeth is removed by roughing broach teeth. Broach performance largely depends on the type of chip removal diagram that has been used in the design of the broach. Chip removal diagrams specify the shape and parameters of the cross section of the stock removed by the broach teeth. Numerous types of chip removal diagrams are known.

Without going into details of the analysis, consider how the stock between two adjacent teeth of the gear is removed by the gear broach using a conventional cutting diagram (Figure 4.5).

A gear broach features a progressive increase in tooth height from tooth to tooth of the broach. The progressive increase in tooth height is often referred to as cut per tooth. Cut per tooth is also referred to as chip per tooth. The action of the step height produces the equivalent of the feed in other kinds of machining.

Broaches are designed with the aim of simple production and easy maintenance of gear-cutting tools.

The production of gear broaches can be simplified if the roughing teeth of the broach are designed in the same way when they roughly generate the teeth flanks of the gear being machined. The cut per tooth for roughing cutting teeth of the gear broach is of a certain value $t_r$. This means that every roughing cutting tooth of the gear broach is $t_r$ higher the previous one. Cross sections of the removed chips in this case have an almost rectangular form. The removed material is easily deformed into the chip. No interference in chip flows is observed in this case, and thus it is easy to accommodate the chip in the gullet.

Cutting teeth of the semi-finishing section of the gear broach are designed in the same way when they are copying the tooth flank of the gear being machined. Cut-per-tooth for semi-roughing cutting teeth of the gear broach is of a certain value $t_{sr}$. Every cutting tooth of the semi-roughing section of the gear broach is $2t_{sr}$ wider than the previous one. Cross sections of the removed chips are slightly curved following the involute profile of the tooth flanks. However, the removed material in this case is also easily deformed into the chip. No interference of chip flows is observed in this case, and it is easy to accommodate the chip in the gullet.

It has been proven to be practical to set cut-per-tooth $t_r$ within the roughing section of the gear broach greater than $t_{sr}$ for the semi-roughing teeth of the broach.

The teeth of the work gear are finally generated by the finishing teeth of the gear broach. Cut-per-tooth for finishing cutting teeth of the gear broach has a zero value.

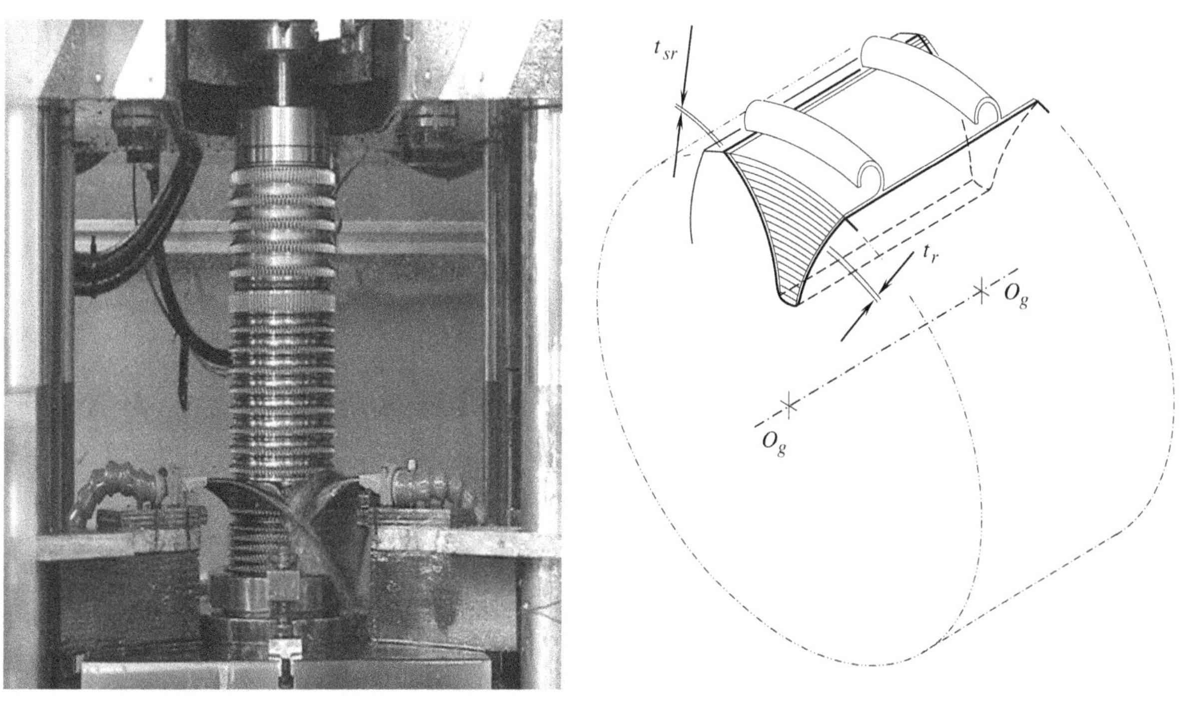


**FIGURE 4.5** Chip removal diagram of an ordinary gear broach.

## 4.5 SHARPENING OF GEAR BROACHES

Sharpening as well as re-sharpening of a gear broach is a kind of grinding operation. A close-up of this grinding operation is depicted in Figure 4.6. When sharpening, the broach is installed in the centers. When sharpening long broaches, steady rests are used. The longer the broach, and the higher the requirements for the accuracy of its sharpening, the more steady rests are required to sharpen it.

Gear broach re-sharpening is performed on the rake surface. The rake surface of the roughing broach teeth is shaped in the form of an internal cone of revolution. This is done through grinding by the external cone surface of the grinding wheel (Figure 4.7). Because of this, grinding the rake surface of gear broaches must deal with a geometrical problem that strictly relates to the fulfillment or violation of the third necessary condition of proper part surface generation [71, 72, 131, 136, 144, 148, 151, 158] (see Appendix E).

The rake surface $R_s$ of a gear broach tooth is shaped in the form of an internal cone of revolution (see Figure 4.6). The cone surface $R_s$ can be specified in terms of the rake angle[4] $\gamma_o$ of the gear broach and the broach diameter $d_{ct}^{\min}$. The closest point $a$ of the straight profile of the rake surface $R_s$ is remote from the centerline $O_{ct}$ of the broach at the distance $0.5 \cdot d_{ct}^{\min}$.

The generating surface $T_{gw}$ of the grinding wheel is shaped in the form of an external cone of revolution. The cone surface $T_{gw}$ can be specified in terms of its diameter $d_{gw}$ and the cone angle $\varphi_{gw}$. The actual location of point $a$ on the generating surface $T_{gw}$ of the grinding wheel is specified by the diameter $d_{gw}$.

Angle $\beta$ is the angle that the axis of rotation $O_{gw}$ of the grinding wheel forms with the centerline $O_{ct}$ of the gear broach. Two of the three angles $\gamma_o$, $\varphi_{gw}$, and $\beta$ are given. The third can be expressed in terms of the two given angles. Usually, the angles $\gamma_o$ and $\beta$ are known. In this case, the expression:

$$\varphi_{gw} = \beta - \gamma_o \tag{4.11}$$

is used to calculate for the actual value of the grinding wheel angle $\varphi_{gw}$.

When re-sharpening, the gear broach is supported at the centers. The grinding wheel rotates about its axis $O_{gw}$ with a certain angular velocity $\omega_{gw}$. The rotation $\omega_{gw}$ provides the necessary cut speed. The broach rotates about its axis of rotation $O_{ct}$ with an angular velocity $\omega_{ct}$. The rotation $\omega_{ct}$ serves as the feed motion.

For a gear broach, it is necessary to choose the appropriate grinding wheel size so that it does not interfere with the broach tooth—no interference of the surfaces $R_s$ and $T_{gw}$ is permissible at all.

The rake surface $R_s$ and the generating surface $T_{gw}$ of the grinding wheel make contact along a straight line. This line

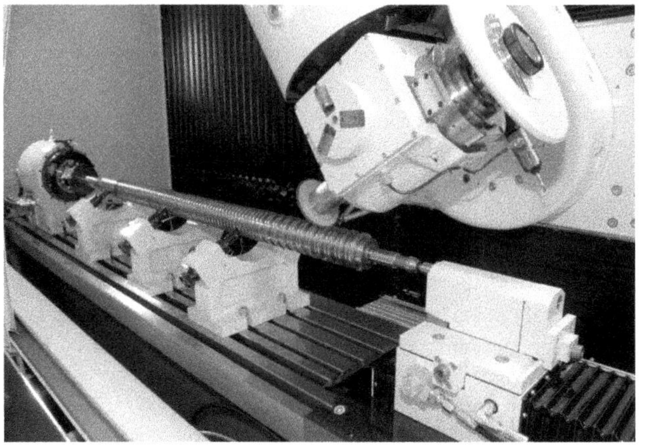

**FIGURE 4.6** Close-Up of gear broach sharpening operation.

is the common straight generating line for the cone surfaces $R_s$ and $T_{gw}$. For the surface of internal cone $R_s$, all radii of normal curvature $\rho_{rs}$ are of a negative value ($\rho_{rs} < 0$). For the generating surface $T_{gw}$ of the grinding wheel, all radii of normal curvature are of a positive value ($\rho_{gw} > 0$). Magnitudes of the radii of normal curvature $\rho_{rs}$ and $\rho_{gw}$ are not of a constant value within the line of contact of the surfaces $R_s$ and $T_{gw}$. Interference of the generating surface $T_{gw}$ into the rake surface $R_s$ of the gear broach may occur if the inequality $\rho_{gw} \geq |\rho_{rs}|$ is observed. Within the line of contact of the surfaces $R_s$ and $T_{gw}$, the point $a$ is candidate number one, at which the third necessary condition of proper part surface generation is violated. If the third condition is fulfilled at point $a$, it is fulfilled at all other points of the line of contact of the surfaces $R_s$ and $T_{gw}$.

The inequality $\rho_{gw} \geq |\rho_{rs}|$ casts into the formula for the calculation of the maximal permissible outer diameter $d_{gw}$ of the grinding wheel:

$$d_{gw} \leq d_{ct}^{\min} \cdot \frac{\sin(\beta - \gamma_o)}{\sin \gamma_o} \tag{4.12}$$

Details on the derivation of Eq. (4.12) are available from [151, 158] and other sources.

For machining gears, gear broaches with helical gullets[5] are widely used. The term helical gullet means chip space that wraps around a broach tool helical line (a thread[6]).

In a manner similar to that just discussed, a formula for the calculation of the maximal permissible outer diameter $d_{gw}^{\max}$ of the grinding wheel for sharpening gear broaches with helical gullets can be derived. Definitely, in the latter case,

---

[4] Another term is also applied for the rake angle $\gamma_o$. Sometimes, it is referred to either as the *face angle* or as the *hook angle*.

[5] The *helical gullet* also referred to as *spiral gullet*.

[6] Gear broaches with helical gullets also feature the *shear cutting tooth*, which is specifically positioned so that it does not make a right angle with the direction of broach motion. The shear tooth cuts with a shear action with maximum tooth overlap. The top cutting edge of the shear cutting tooth is at a *shear angle* with respect to the broach axis. The shear angle is the angle between the cutting edge of a shear tooth and a line perpendicular to the broach axis, or a line of travel on surface broaches. In reality, the term shear angle is equivalent to the conventional term *angle of inclination* of the cutting edge.

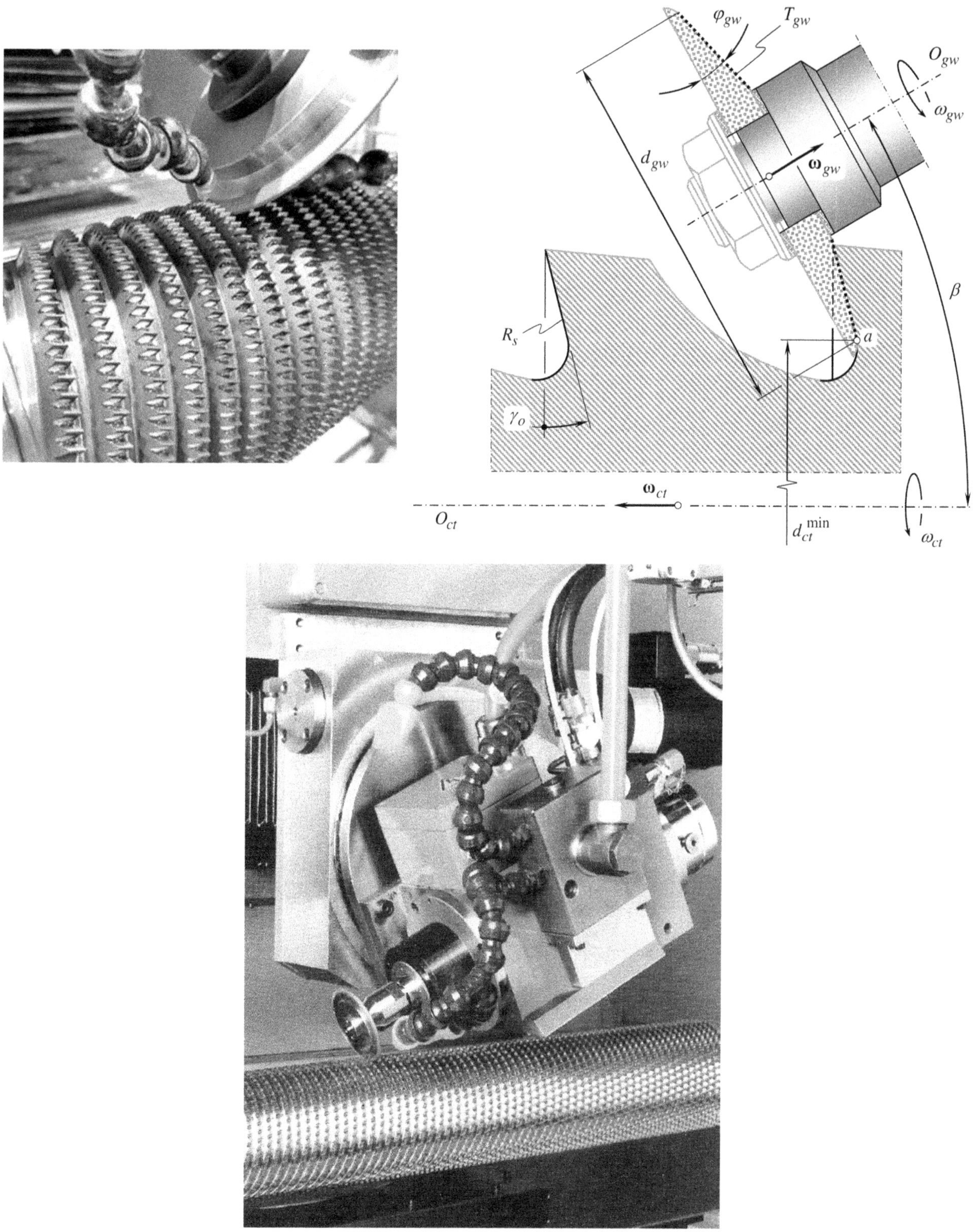


**FIGURE 4.7** Geometrical constraints on design parameters and setup parameters of the grinding wheel for sharpening a gear broach.

the maximal diameter $d_{gw}^{max}$ of the grinding wheel would depend on the helix angle of the grooves of the gear broach.

More detailed analysis of the contact geometry of the interacting rake surface $R_s$ and the generating surface of the grinding wheel $T_{gw}$ can be found out immediately below in Figure 4.8.

In most practical cases, a broach is reground over the rake surface $R_s$. The rake surface $R_s$ of a broach is shaped in the form of an internal cone of revolution (see Figure 4.8). The working surface $T_{gw}$ of the grinding wheel is shaped in the form of a surface of an external cone of revolution. The grinding wheel rotates about its axis $O_{gw}$ with a certain

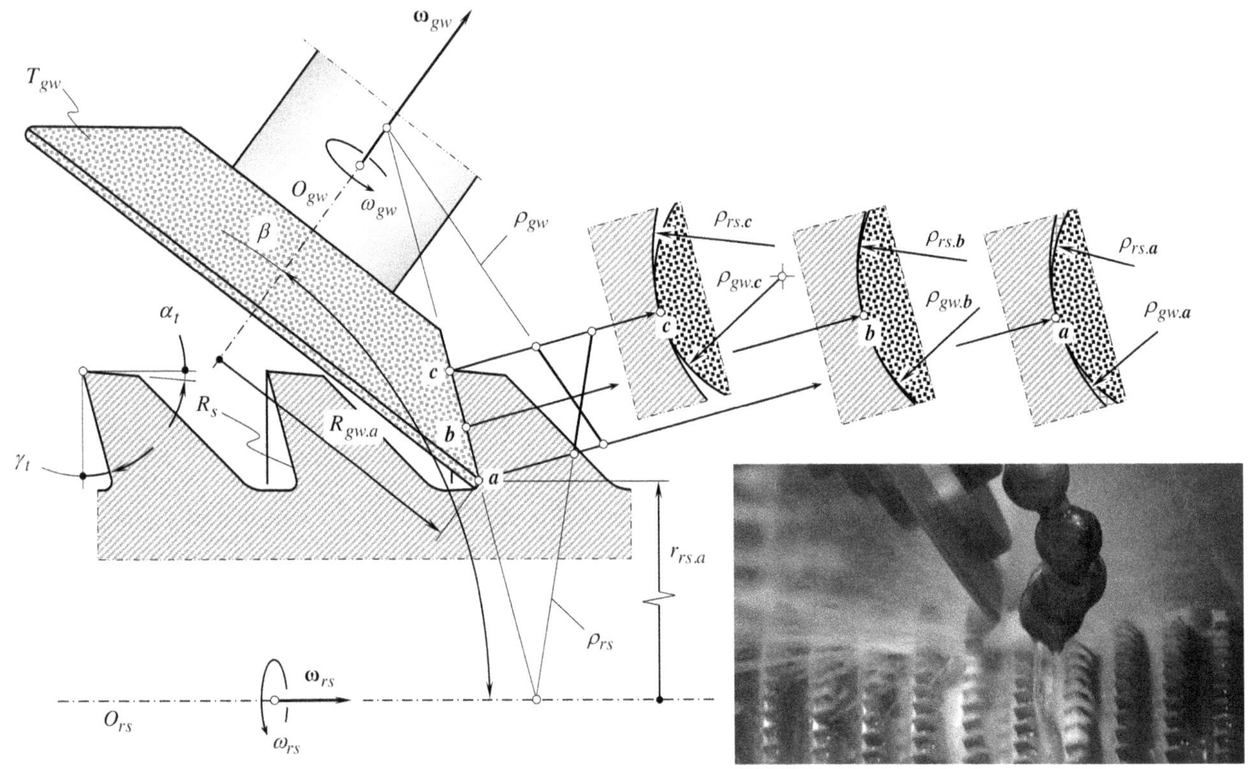


**FIGURE 4.8** Example of violation of third necessary condition of proper part surface generation when regrinding gear broach.

angular velocity $\omega_{gw}$. The rotation $\omega_{gw}$ provides the necessary speed of cut. The broach is rotating about its axis of rotation $O_{rs}$ with an angular velocity $\omega_{rs}$. The rotation $\omega_{rs}$ can be viewed as the feed motion. The inequality $\omega_{rs} << \omega_{gw}$ is always observed.

The rake surface $R_s$ and the generating surface $T_{gw}$ of the grinding wheel make contact along a straight line. Points $a$, $b$, and $c$ are within the line of contact of the surfaces $R_s$ and $T_{gw}$. At different points of the line of contact, the radii of normal curvature $\rho_{rs.i}$ and $\rho_{gw.i}$ are not of the same value within the line of contact of the surfaces $R_s$ and $T_{gw}$. For the surface of internal cone $R_s$, the radii of normal curvature have a negative value. For this surface, the inequality $|\rho_{rs.a}| < |\rho_{rs.b}| < |\rho_{rs.c}|$ is valid. For the generating surface $T_{gw}$ of the grinding wheel, all the radii of normal curvature are of a positive value. For the surface $T_{gw}$ the inequality $\rho_{gw.a} > \rho_{gw.b} > \rho_{gw.c}$ is observed. Without going into detail, it is evident from the analysis in Figure 4.8 that the point $a$ is the first point at which the third necessary condition of proper part surface generation can be violated [151]. If the third condition of proper part surface generation is fulfilled at point $a$, it is fulfilled for all the rest points of the line of contact of the surfaces $R_s$ and $T_{gw}$.

Radius of normal curvature $\rho_{rs.a}$ of the rake surface $R_s$ can be calculated using the Meusnier formula [152]:

$$\rho_{rs.a} = \frac{r_{rs.a}}{\sin \gamma_t} \tag{4.13}$$

Here, $r_{rs.a}$ designates the distance of the point $a$ from the axis $O_{rs}$ of the broach, and $\gamma_t$ designates the rake angle of the broach.

The radius of normal curvature $\rho_{gw.a}$ of the generating surface $T_{gw}$ of the grinding wheel can also be calculated on the premises of the Meusnier formula (see Figure 4.8):

$$\rho_{gw.a} = \frac{R_{gw.a}}{\sin(\beta - \gamma_t)} \tag{4.14}$$

Here, $\rho_{gw.a}$ designates the distance of the point $c$ from the axis $O_{gw}$ of the grinding wheel, and $\beta$ designates the angle that forms the axes $O_{rs}$ and $O_{gw}$.

In order to fulfil the third necessary condition of proper part surface generation [151], fulfillment of the inequality $\rho_{gw.a} \le |\rho_{rs.a}|$ is the must. Therefore, the inequality:

$$\frac{r_{rs.a}}{\sin \gamma_t} \ge \frac{R_{gw.a}}{\sin(\beta - \gamma_t)} \tag{4.15}$$

must be fulfilled when re-grinding the rake surface $R_s$ of the broach.

The Eq. (4.15) casts into the formula for the calculation of the maximal permissible outer radius $R_{gw.a}$ (diameter) of the grinding wheel:

$$R_{gw.a} \le \frac{r_{rs.a} \sin(\beta - \gamma_t)}{\sin \gamma_t} \tag{4.16}$$

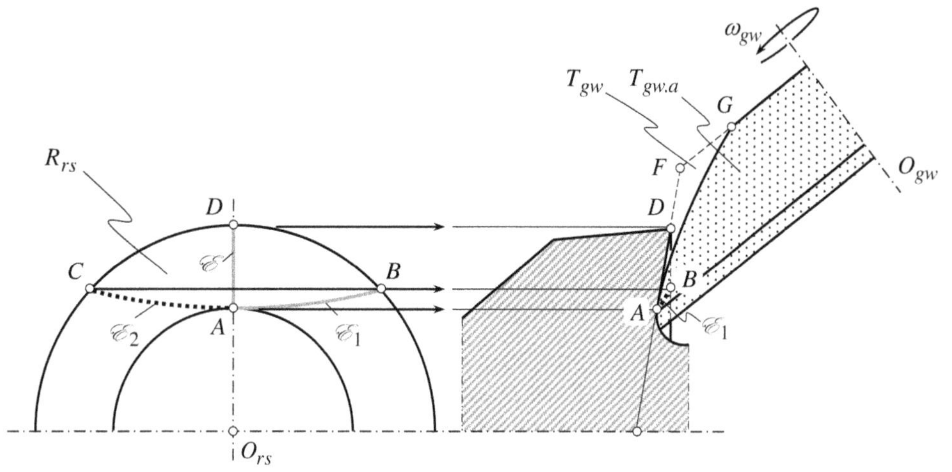


**FIGURE 4.9** A particular case of broach grinding using a form grinding wheel (an example of part surface generation when a surface characteristic splits on two portions).

In a similar manner, a formula for the calculation of the maximal permissible outer radius $R_{gw.a}$ (diameter) of the grinding wheel for grinding broaches that have helical grooves can be derived.

The maximum permissible outside diameter of the grinding wheel can be increased, for example, by means of alteration of the schematic of generation of the rake surface $R_s$ of the broach, as will be further discussed.

Consider the generation of the surface of internal circular cone, say, the generation of the rake surface of the broach for machining a gear, as schematically illustrated in Figure 4.9. An approximate generating surface $T_{gw.a}$ of the grinding wheel is used for this purpose. Usually, grinding wheels having the generating surface $T_{gw}$ that is shaped in the form of an external cone of revolution are used. The design parameters of the grinding wheel, as well as its setup parameters, are determined in accordance with accurate generation of the rake surface $R_s$ of the gear broach being ground.

The generating straight line $AF$ of the grinding wheel $T_{gw}$ is aligned with the generating straight line $AD$ of the rake face $R_s$ of the broach. In this case, the straight-line segment $AD$ is the surfaces' contact line $\mathscr{E}$. Strict constraints are imposed onto the maximal diameter of the grinding wheel by the necessity of fulfilling the conditions of proper part surface generation.

A method of re-sharpening a gear broach on the rake surface [151, 158] features line contact of the rake face $R_s$ with the grinding wheel surface $T_{gw}$. The line of contact splits into two branches. Generation of the rake surface $R_s$ is performed either by the line (by the circle) or by narrow generating surface $T_{gw}$ of the grinding wheel. The rake surface $R_s$ is generated approximately in the first case and accurately in the second case.

The broaches can be re-ground with the form grinding wheels of an appropriate design. The form grinding wheel is rotated about its axis $O_{gw}$ with a certain angular velocity $\omega_{gw}$. The curve $AG$ is used as the generating curve of the approximate generating surface $T_{gw.a}$ of the grinding wheel (see Figure 4.9). The grinding wheel surface $T_{gw.a}$ and the rake surface $R_s$ of the gear broach touch each other along the line of contact $BAC$. The portion $BA$ of this curve is the surface characteristic line $\mathscr{E}_1$, and the portion $AC$ serves as the surface characteristic line $\mathscr{E}_2$. The approximate generating surface $T_{gw.a}$ of the form grinding wheel is represented by a surface of revolution that is tangent to the rake surface $R_s$ of the gear broach along two symmetrically located characteristic lines $\mathscr{E}_1$ and $\mathscr{E}_2$. The parameters of the curvilinear generating line $AG$ meets the conditions under which the deviations of the actually generated rake surface $R_s$ from the desired shape do not exceed the tolerance for the accuracy of the rake surface $R_s$.

The discussed alteration to the schematic of generation of the rake surface $R_s$ of a gear broach is useful, as it gives an opportunity to increase the maximum permissible outer diameter of the grinding wheel $T_{gw.a}$.

Because the rake angle $\gamma_o$ of the gear broach is of a positive value ($\gamma_o > 0°$), the rake surface $R_s$ of the gear broach tooth is shaped in the form of an inner cone of revolution. Constraints on the maximal permissible outer diameter of the grinding wheel are imposed when grinding the concave rake surface $R_s$ with a convex generating surface of the grinding wheel. In order not to violate the third necessary condition of proper part surface generation [71, 72, 131, 136, 144, 148, 151, 158] (see Appendix E), the outer diameter of the grinding wheel must meet the condition specified by the Eq. (4.12). It is definitely more practical to sharpen the gear broach with a grinding wheel of the largest possible outer diameter. However, no possibilities to increase the grinding wheel diameter have been observed in the discussed method (see Figure 4.7) of sharpening of gear broaches.

The issue is resolved by a novel method of re-sharpening gear broaches, as illustrated in Figure 4.10. The generating surface $T_{gw}$ of the grinding wheel is shaped in the form of an external cone of revolution. Sharpening of the broach

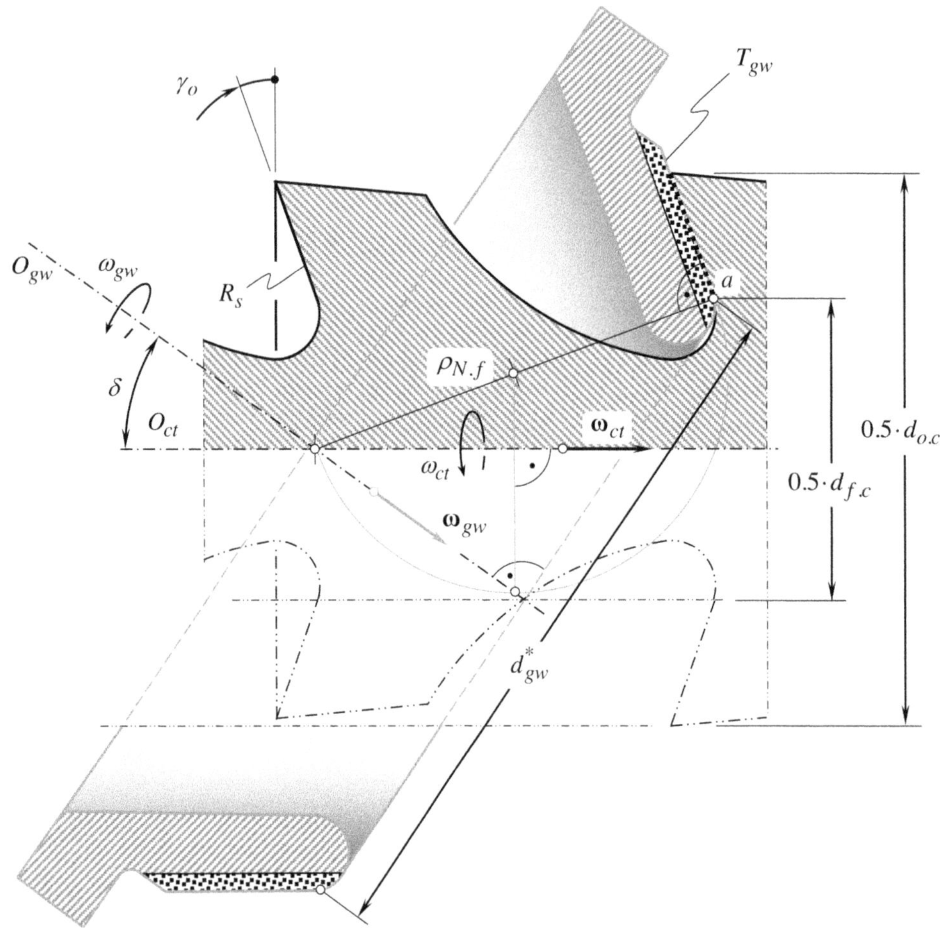


**FIGURE 4.10** Schematic diagram of a novel method of re-sharpening a gear broach.

on the rake surface $R_s$ is performed with the grinding wheel that features a hole in the center. The gear broach being sharpening passes through the hole as shown in Figure 4.10.

When sharpening a gear broach of an outer diameter $d_{o.c}$, the broach is rotating about its axis of rotation $O_{ct}$. In Figure 4.10, the rotation of the broach is designated as $\omega_{ct}$ (or as $\boldsymbol{\omega}_{ct}$, in vector notation). The grinding wheel rotates about its axis of rotation $O_{gw}$. The rotation of the grinding wheel is designated as $\omega_{gw}$. The axis of rotation of the grinding wheel $O_{gw}$ intersects the axis of rotation of the gear broach $O_{ct}$ at a certain angle $\delta$.

To avoid violating the third condition of proper part surface generation, both the design parameters of the grinding wheel and its setup parameters should be determined in compliance with the design parameters of the gear broach to be re-sharpened.

Point $a$ of the axial profile of the rake surface $R_s$ is located at the distance of $0.5 \cdot d_{f.c}$ from the axis $O_{ct}$ of the gear broach. This point is the critical point from the standpoint of fulfillment or violation of the third necessary condition of proper part surface generation. It is the closest point of the axial profile of the rake surface $R_s$ to the axis $O_{ct}$ of the gear broach. If the condition is fulfilled at point $a$, it is fulfilled at all the rest points of the axial profile of the rake surface $R_s$.

Consider a normal section of the rake surface $R_s$ and the generating surface $T_{gw}$ of the grinding wheel. The normal section is through point $a$. The normal cross section is perpendicular to the line of contact of the surfaces $R_s$ and $T_{gw}$. The radius of curvature of the rake surface $R_s$ at this normal section is designated as $\rho_{N.f}$. The radius of normal curvature $\rho_{N.f}$ is of a negative value ($\rho_{N.f} < 0$). This is because the rake surface $R_s$ is a kind of concave surfaces. The corresponding radius of the normal curvature of the generating surface $T_{gw}$ of the grinding wheel is designated as $\rho_{N.c}$. The radius $\rho_{N.c}$ is not shown in Figure 4.10. Because the generating surface $T_{gw}$ is a convex surface, the radius of normal curvature $\rho_{N.c}$ has a positive value (that is, $\rho_{N.c} > 0$).

The third necessary condition of proper part surface generation is fulfilled if and only if the inequality:

$$\rho_{N.c} \leq |\rho_{N.f}| \tag{4.17}$$

is fulfilled.

The radii of normal curvature $\rho_{N.f}$ and $\rho_{N.c}$ can be expressed in terms of the design parameters of the gear broach and the grinding wheel and then substituted into

Eq. (4.17). In this manner, the maximum permissible outer diameter $d^*_{gw}$ of the grinding wheel and the setup angle $\delta$ can be determined. The diameter $d^*_{gw}$ of the grinding wheel significantly exceeds that for the grinding wheel that is applicable in the method illustrated in Figure 4.7.

The presented method of calculation for the design parameters and setup parameters of the grinding wheel can be enhanced to determination the maximal permissible outer diameter $d^*_{gw}$ of the grinding wheel. This is a particular example of the problem of optimization that can be solved using conventional methods of part surface machining [144, 148, 151, 158, 168].

The method illustrated in Figure 4.10 can also be applied to grind gear broaches that have helical gullets. For this case, the axes $O_{ct}$ and $O_{gw}$ of the gear broach and the grinding wheel are skewed and positioned at a certain center-distance $C_{ct/gw}$ from each other. The same condition [see Eq. (4.17)] should be fulfilled when helical gullets are ground. Fulfillment of the third necessary condition of proper part surface generation is the key to solving problems of this sort.

## 4.6 NOVEL CONCEPT OF PRECISION GEAR BROACHING TOOL FOR MACHINING HELICAL INVOLUTE GEARS

The finishing cutting edges of a gear broach are located within the generating surface $T$ of the broach—this is required for precision gear broaches. A considerable freedom is given to the gear broach designer due to this requirement. Any line of a reasonable geometry that is within the generating surface $T$ of the gear broach can be used as the finishing cutting edge. This setup is illustrated in Figure 4.11.

Consider a curved segment $ab$ that is entirely located within the generating surface $T$ of the gear broach. The finishing cutting edge $CE_{ab}$ can be configured so as to incline at a certain positive rake angle $\gamma_o > 0°$ at the top cutting edge. No physical constraints are imposed on such a configuration of the finishing cutting edge $CE_{ab}$. As long as the finishing cutting edge $CE_{ab}$ is located within the generating surface $T$, the accuracy of the gear broach is assured.

Consider a curved segment $cd$ that is entirely located within the generating surface $T$ of the gear broach. Because of chip, the cut by the finishing cutting edges of the gear broach is thin; the rake angle $\gamma_o$ doesn't affect the material removal process. The roundness of the cutting edge $\rho_c$ is a more critical design parameter of the finishing cutting edges. Because of this, the finishing cutting edge $CE_{cd}$ of the gear broach is configured so as to be at the rake angle $\gamma_o = 0°$. This configuration of the finishing cutting edges has proved practical.

As discussed earlier, the generating surface $T$ of the gear broach is a screw involute surface, which can be generated by the corresponding screw motion of a straight line. This means that a straight line can be drawn up on any screw involute surface. As an example, a straight-line segment $ef$ is drawn up in Figure 4.11. All points of the straight-line segment $ef$ are located within the generating surface $T$ of the gear broach. Only a particular configuration of the straight line results in this straight line being entirely located within the screw involute surface $T$: the straight line segment $ef$ is tangent to the base cylinder of the involute gear broach and also tangent to the base helix on it. [The straight line $ef$

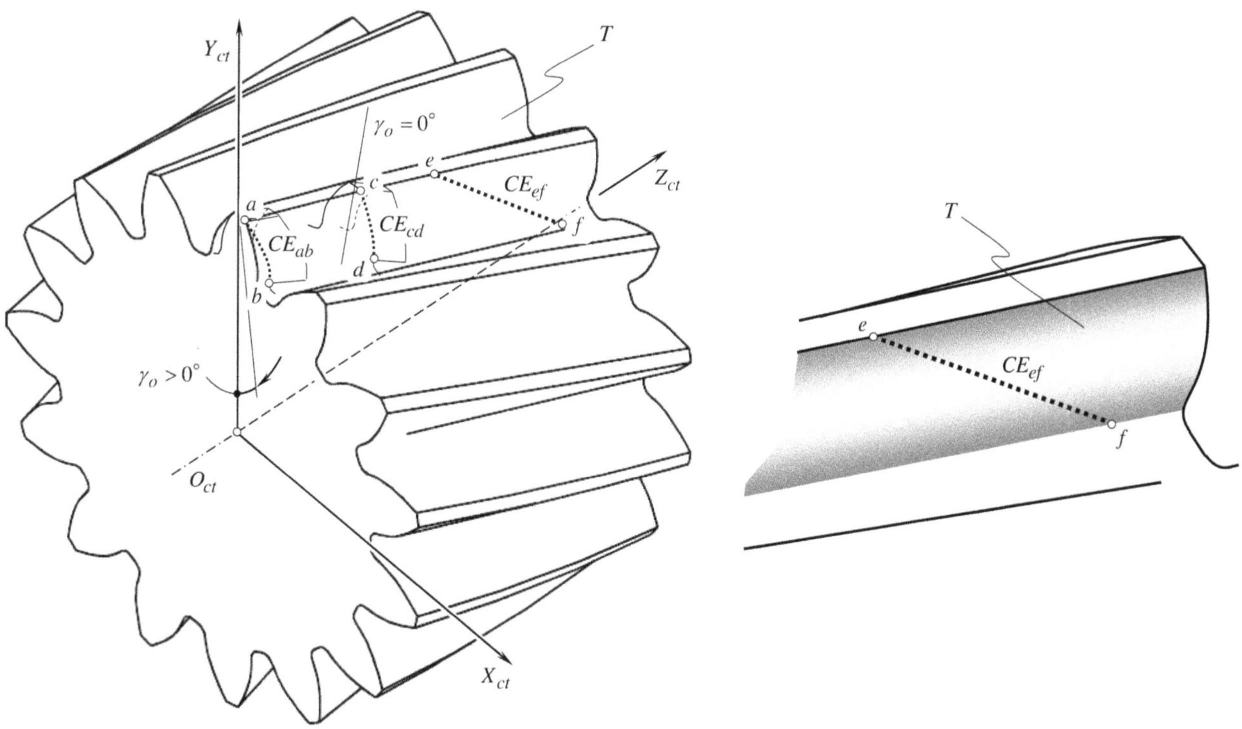


**FIGURE 4.11** Permissible shapes and configurations of the finishing cutting edges of a helical involute gear broach.

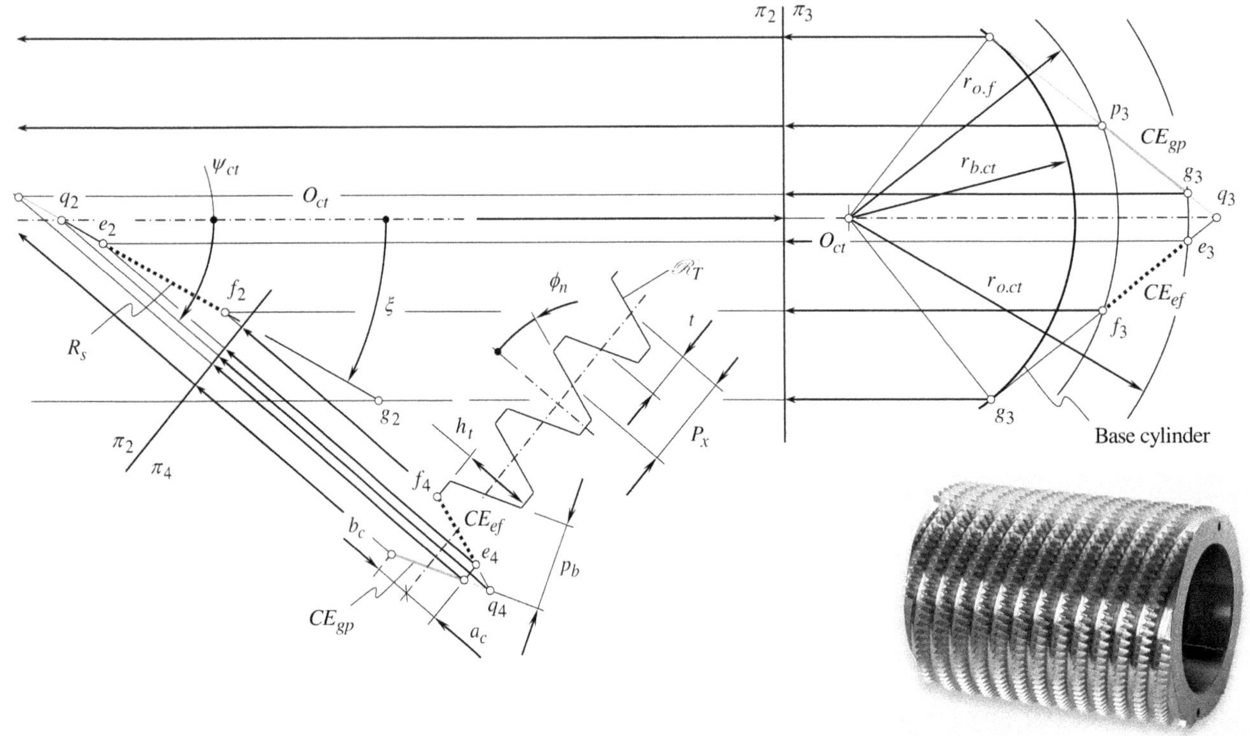


**FIGURE 4.12** Determination of the configuration of straight lateral cutting edges within the rake plane $R_s$ of the finishing tooth of an involute gear broach.

considered in Figure 4.11 can be compared with the straight line along the vector **C** in Figure 1.24]. This unique configuration of the lateral cutting edge $CE_{ef}$ of the involute gear broach can be expressed in terms of its design parameters.

Various options are available to the involute gear broach designer when designing a gear broach with straight cutting edges. Numerous configurations of straight lateral cutting edges at opposite sides of the tooth profile can be of practical importance. As an example, consider a configuration of two lateral straight cutting edges that are within a common plane. First, this particular configuration of the cutting edges of the gear broach tooth is helpful in understanding the principal features of the geometry of the involute broach. Second, analysis of the location of straight lateral cutting edges of the involute broach can be used for the development of other practical configurations.

Without loss of generality, consider the determination of the desirable configuration of just two straight lateral cutting edges of the finishing section of the precision involute broach (see Figure 4.12). The descriptive geometry-based method is used for this purpose.

The involute gear broach is shown in the system of projective planes so that the centerline $O_{ct}$ is perpendicular to the plane of projections $\pi_3$. The outer diameter of the broach is designated as $d_{o.c} = 2 \cdot r_{o.c}$, and its base diameter is denoted as $d_{b.ct} = 2 \cdot r_{b.ct}$. The diameter $d_{f.ct} = 2 \cdot r_{f.ct}$ is the smallest diameter of the circle, at which the point of the lateral cutting edge (that still generated the gear tooth profile) is located.

The normal section of the auxiliary rack $\mathscr{R}_T$ of the involute broach is shown in the plane of projections $\pi_4$. The location of this plane of projections is specified by the pitch helix angle $\psi_{ct}$ of the broach. The tooth flanks of the opposite sides of the tooth profile of the generating surface $T$ of the broach intersect each other along a screw line. This line intersects the plane of projections $\pi_4$ at point $q_4$. The projections $q_2$ and $q_3$ of point $q$ are constructed using conventional methods developed in descriptive geometry.

The straight lateral cutting edges $CE$ of the involute broach align with two straight lines that are tangent from the opposite sides to the base helices of the generating surface $T$ of the gear broach. In the plane of projections $\pi_3$, these straight lines are tangent from the opposite sides to the base cylinder of the broach. The projection $e_3 f_3$ of the lateral cutting edge $CE_{ef}$ is used for the construction of other projections $e_2 f_2$ and $e_4 f_4$ of the cutting edge $CE_{ef}$. The cutting edge $ef$ in Figure 4.12 corresponds to the cutting edge $ef$ in Figure 4.11. The opposite straight lateral cutting edge $CE_{gp}$ is located within the rake plane $R_s$ and is symmetrical to the cutting edge $CE_{ef}$. The constructed rake plane $R_s$ forms a certain angle $\xi$ with the broach centerline $O_c$. The required value of the angle $\xi$ can be expressed in terms of the design parameters of the involute broach:

$$\xi = \tan^{-1} \frac{\cos \psi_c}{\sin \psi_c + \sqrt{\dfrac{(d_c + t_c \cdot \cot \phi_n)^2}{d_{b.c}^2} - 1} \cdot \tan \phi_n} \tag{4.18}$$

where:

$d_{ct}$ is the pitch diameter of the involute broach
$t_{ct}$ is the tooth thickness
$\phi_n$ is the normal pressure angle

For the practical purposes, the straight lateral cutting edges $CE_{ef}$ and $CE_{gp}$ of the opposite sides of the tooth profile can be shifted in the axial direction with respect to each other at a certain distance. The decision that the involute gear broach designer makes in this regard depends on the particular requirements that the gear broach needs to meet.

The disclosed design concept of precision gear-cutting tool with straight lateral cutting edges for machining involute gears was proposed by Prof. S.P. Radzevich in the late 1970s [138, 169].

## 4.7 APPLICATION OF GEAR BROACHING TOOLS

The forming of gear teeth has traditionally been a time-consuming, heavy stock-removal operation for which close tooth size, shape, run-out, and spacing accuracy are required.

A gear-machining operation in which material from a work gear is removed by pulling or pushing a broach along the axis of the work gear is called broaching. The actual cutting action of any single broach tooth is very much similar to that of a single form tool. Each section of the broach has as many teeth as there are tooth spaces on the gear.

Gear broaching is primarily applicable to high-volume jobs that require high production rates. Gears can be broached accurately since a number of the elements are controlled by the accuracy of the broach. In addition, broaching can produce a fine finish on the gear tooth profile. Occasionally, it may be used in fairly small lot production to obtain an accuracy level on a critical part that cannot readily be made with a high degree of accuracy by other methods. However, a gear broach will cut only parts that have identical gear specifications.

### 4.7.1 Broaching Internal Gears

A broach can (see Figure 4.13) be used to machine almost any internal form. The part teeth can be either equally or unequally spaced, and tooth forms can be either symmetrical or asymmetrical. The tooth forms must be uniform in the direction of the broach axis, and the wall of the part must be strong enough to withstand the broaching pressure.

Small internal gears can be cut in one pass of a broach (see Figure 4.14). Large internal gears can be made by using

**FIGURE 4.14** Broaching internal gears.

a surface type of broaches to make several teeth at a pass. Indexing the gear and repeated passes of the broach can produce a complete gear. Gears as large as 1.5 meters (60 in.) in diameter are made through this process.

Today, the broached internal helical running gears in automatic transmissions in automobiles are exclusively produced using full-form finishing high-speed steel broaches. An example of a broach for helical internal gears is illustrated in Figure 4.15. Full-form finish broaching provides fine surface finishes, precision involute form, accurate tooth thickness and precision tooth spacing, and lead.

In some cases, broaches can be designed with a removable finishing section that finishes the complete tooth form (see Figure 4.16). This side-shaving section of the broach finishes the entire profile of each gear tooth. A full form-finish section solves the profile problems caused by

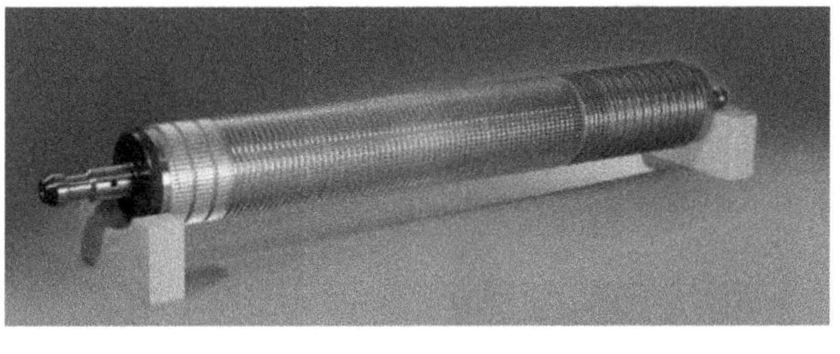

**FIGURE 4.13** Broach for machining internal spur involute gears.

**FIGURE 4.15** A broach for machining helical internal gears.

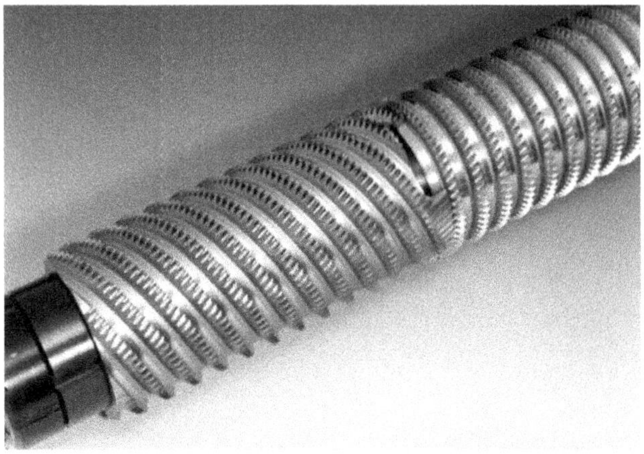

**FIGURE 4.16** Removable finishing section of broach.

off-center drift of the broach, attributed to poor maintenance of machine alignment and perpendicularity or nonuniform tooth dulling.

In certain cases, gear broaches are designed with a so-called burnishing button. A burnishing button is a broach tooth without a cutting edge. A series of buttons is sometimes placed after the cutting teeth of a broach to produce a smooth surface through material compression. A burnishing button can also be design as a separate broaching tool, as shown in Figure 4.17.

Pull broaches and push broaches are used in the gear production.

Push broaches are normally short and therefore less expensive than pull broaches, which are longer. However, one pull broach can sometimes remove as much stock as several push broaches. Most internal broaching is carried out with pull broaches. Not only do they remove more stock than push broaches, they are also capable of making much longer cuts.

Gear broaching usually requires a part with a through hole. However, blind holes can be broached if there is sufficient clearance beyond the broached section and a series of punch-type broaches are used.

Broaching can produce close tolerances on profile, spacing, lead, and size.

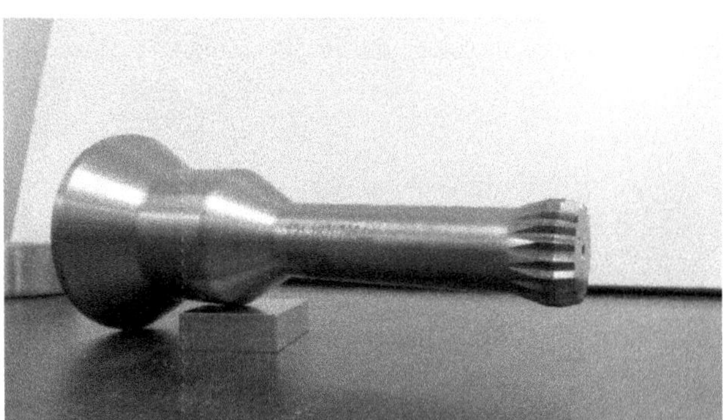

**FIGURE 4.17** A broach for burnishing internal gears.

### 4.7.2 Broaching External Gears

The fastest way to produce medium and high production external gears is by pot broaching. The basic process consists of a tool with internal teeth (see Figure 4.18) that are held in a pot being passed over a round part, producing external shapes in the form of splines and involute gear teeth as illustrated in Figure 4.19.

In the past, external gears were broached by either pushing the workpiece down through a stationary pot broach or mounting the workpiece on a stationary post and pushing the pot broach down over it. This resulted in problems related to chips packing in the chip gullets when using the push-down stationary pot broach method. Slow workpiece loading, attributed to the use of a plug-type approximate locator for each part, is a problem when broaching parts with the push-down stationary workpiece method.

**FIGURE 4.18** Pot-broach for machining external gears.

**FIGURE 4.19** Broaching external gears.

Two more methods used are push-up pot broach and pull-up broaching.

Push-up pot broaching consists of pushing workpiece upward through a fixed pot broaching tool to produce external teeth under ideal conditions, assuring quick and complete chip removal from the broach teeth. Chips fall by gravity away from the tool and work gear. The coolant is flushed into the tool area through a quick-disconnect coupling. Automatic loading is simplified because the part is elevated by the process, and gravity is utilized to unload parts onto a conveyor [13].

Pull-up broaching is used when the part diameter and broach length are such that the required post diameter and length do not provide sufficient strength to rigidly support the workpiece. In this method, the work gear is pulled up through a stationary pot broaching tool with a pull rod. This process permits broaching parts with deeper teeth and wider faces by using longer tools.

Pot broaching tools are classified into two different types:

- The ring-type, for parts where tooth form and spacing are critical.
- The lower-cost stick-type, which is used where accuracy permits.

The ring-type pot broaching tool, shown in Figure 4.20(a), is made up of a holder with a series of individual precision ground keyed high-speed steel rings or wafers, each of which has internal cutting teeth that are individually backed off. These rings can be sharpened via either the conventional face-sharpening method or the progressive method, in which several face sharpenings are performed, followed by inner diameter grinding of all rings. Ring-type pot broaches are used for precision running gears and splines where close tooth tolerances are required.

The stick-type pot broaching tool [see Figure 4.20(b)], which has lower-cost individual high-speed steel slab broach inserts, includes a holder that supports a series of keyed ground rings that locate the broach sticks in internal ground slots. On thin parts, the teeth in the stick inserts can have wide, long-life tooth lands and also be axially staggered to provide a helical pattern to balance the cutting forces.

These tools are particularly adapted to short face-width splines and gears, and they are often used in combination.

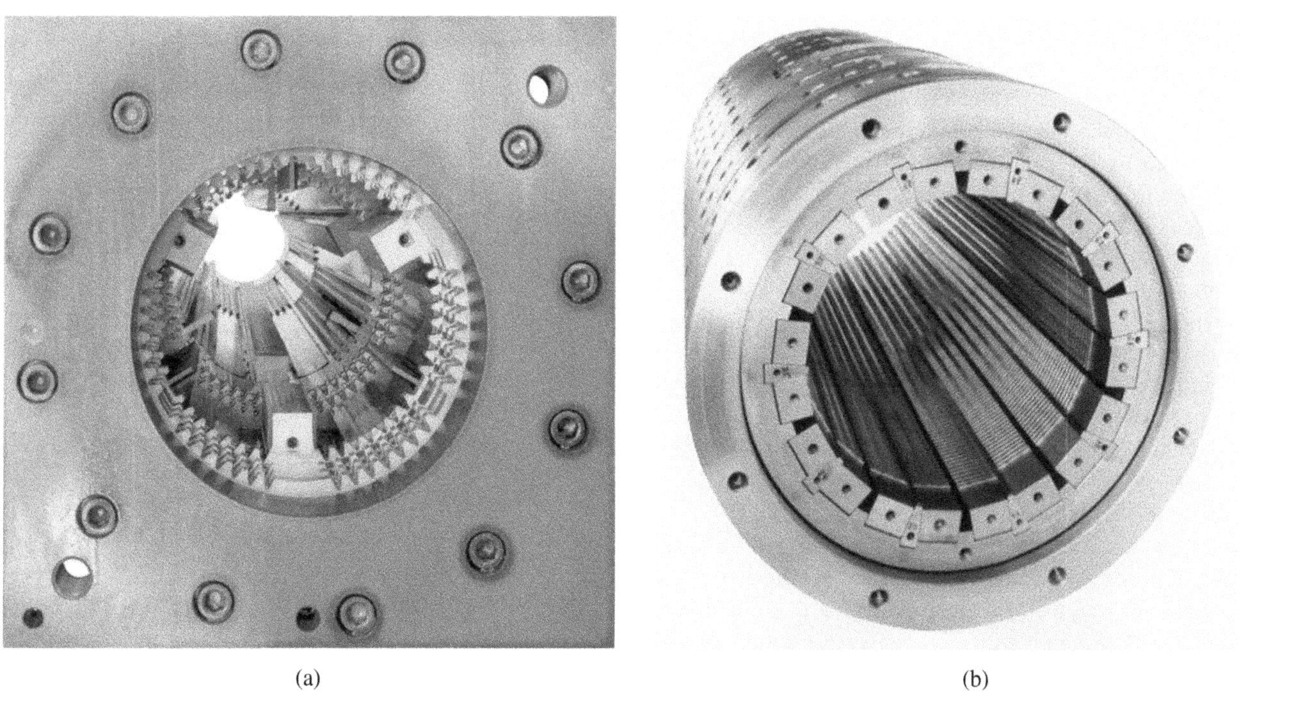

(a) (b)

**FIGURE 4.20** Types of pot broaches: (a) ring-type pot broach (b) lower-cost stick-type pot broach.

**FIGURE 4.21** Generating broaching of an internal gear by means of the broach-shaper cutter.

## 4.8 ON THE POSSIBILITY OF FURTHER DEVELOPMENT OF GEAR BROACHING METHODS

Various aspects of the gear broaching process can be improved. This includes the kinematics of the processing process, the design of the tool used, etc. As an illustrative example, consider a method of broaching internal gears that features continuous rolling relative motion of the gear and the broaching tool (see Figure 4.21).

In conventional gear broaching operation [see Figure 4.21(a)], the relative straight motion of the gear and the broaching tool along the common axis is the only motion that comprise the gear machining kinematics. The finishing cutting edges of the gear cutting tool are situated on the gear machining surface $T$ of the broach. A specially designed broach is necessary for machining gears of each particular design, as the gear tooth profile is copied from the finishing teeth of the gear broach.

In the novel method of machining of gears [see Figure 4.21(b)], the broach of conventional design is replaced with the generating gear broach. A gear cutting tool of such a design resembles a broach of conventional design, and, in addition, the gear broach teeth are shaped similar to the teeth of a shaper-cutter.

When machining a gear, the axis of the broach-shaper cutter is positioned at a center-distance $C$ parallel to the work gear axis of rotation. The gear cutter travels up and down while the work gear and the broach-shaper cutter roll relative one another. Gears of various designs can be broached by a broach-shaper cutter of that same design, which is not possible in conventional gear broaching operations.

## 4.9 *SHEAR-SPEED* CUTTING

With no or very limited changes, the principles of profiling gear broaching tools can be implemented for the purpose of profiling others gear-cutting tools with a similar vector diagram of the gear machining operation. Cutting tools for Shear-Speed[7] cutting of gears is a perfect example in this regard.

### 4.9.1 Principle of Shear-Speed Cutting of Gears

The nature of the operation of shear cutting is quite similar to broaching. Form tool blades are used to shear-cut each space between the gear teeth. The same number of gear-cutting tools as there are teeth in the work gear (see Figure 4.22) is assembled radially in the cutting tool holder mounted at the head of the machine. The cutting tools can be compared with the teeth of the finishing section on a broach. All teeth in a gear are cut simultaneously in the Shear-Speed gear-cutting process.

The principle of Shear-Speed cutting of gears is illustrated in Figure 4.23. The tool head assembly consists of three main parts:

- A retained housing.
- A radially slotted member that controls tool alignment.
- A movable, double cone-shaped guiding unit that controls the feed.

[7] Shear-Speed is a registered trademark of the Michigan Tool Co. of Detroit, Michigan.

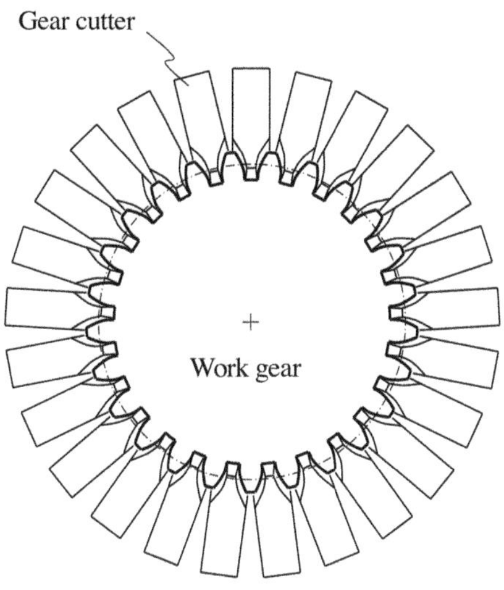


**FIGURE 4.22** A set of form tools for machining spur gear using the *Shear-Speed* cutting principle.

With the tool head being stationary, the part is reciprocated past the cutting tools. Each tool is fed radially at a predetermined distance with each stroke until the full tooth depth is cut. All tools are retracted on each return stroke of the gear to avoid drag and are returned to their initial position at the end of each cycle.

### 4.9.2 Profiling of Form Tools for Shear-Speed Cutting of Gears

Cutting edges *CE* of the form tools reproduce the generating surface when the Shear-Speed gear shaper is operating. The generating surface *T* is congruent to the gear tooth surfaces including surfaces of the bottom lands between the gear teeth. Because form tools are designed with a positive rake angle $\gamma_o > 0°$ and the clearance angle $\alpha_o$ is always positive ($\alpha_o > 0°$), the transverse section of the generating surface *T* is not identical either to the tooth profile within the rake surface $R_s$ of the form tool 6 or to the normal section of the clearance surface $C_s$.

For the manufacture of the form tools and the re-sharpening of worn form tools, it is necessary to express the design parameters of the form tool in terms of the design parameters of the work gear. This particular problem in the design of form cutting tools can be solved either analytically or using descriptive geometry-based methods.

#### 4.9.2.1 Descriptive Geometry-Based Method for Profiling of Form Tool

The tool heads for Shear-Speed cutting of gears (see Figure 4.24) are normally designed to produce a specific part. In Figure 4.25, the generating surface *T* of the Shear-Speed gear cutter is shown is the system of projective planes $\pi_2\pi_3$. For profiling of the gear-cutting tool, multiple points are chosen on the transverse profile of the generating surface *T*. As an example, points *a*, *b*, *c*, ..., *k* are shown. A subscript 3 is assigned to the projections of all points onto the frontal plane of projections $\pi_3$. Corresponding projections of those same points onto the vertical plane of projections $\pi_2$ are labeled with a subscript 2.

Rake surfaces of two types are practical for the Shear-Speed cutting tool. Usually, the rake surface is shaped either in the form of a plane or in the form of a cone of revolution. As an example, the rake plane $R_s$ is depicted in Figure 4.24. The rake plane $R_s$ is a projective plane onto the frontal plane of projections $\pi_3$ and is inclined at the rake angle $\gamma_o$. The cutting edge *CE* is viewed as the line of intersection of the generating surface *T* by the rake plane $R_s$. With no distortions, the cutting edge *CE* is constructed in the lower portion of Figure 4.25. The subscript *rs*, indicating rake surface, is assigned to all the points *a*, *b*, *c*, ..., *k* of the cutting edge *CE*.

All tools are form-relived to maintain the correct form after re-sharpening. The clearance surface $C_s$ with a cylindrical shape (that is, the clearance surface $C_s$ is a surface of translation) can be used in designing the Shear-Speed cutting tool. The clearance surface is a cylinder that passes at the clearance angle $\alpha_o$ through the cutting edge *CE*. The normal section $N_{cs} - N_{cs}$ is specified by points *a*, *b*, *c*, ..., *k*, to which the subscript *cs*, indicating clearance surface, is assigned.

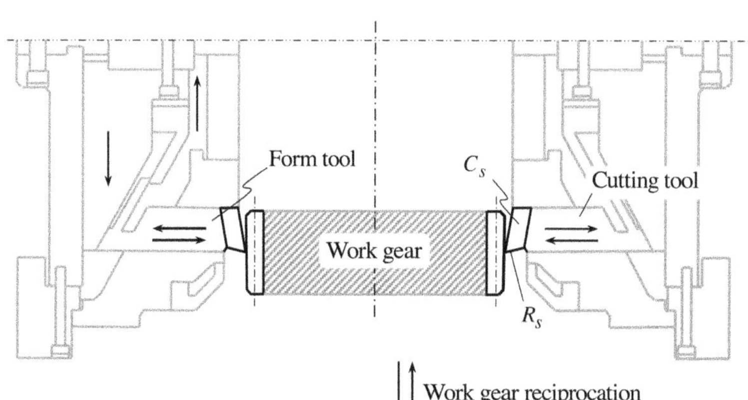


**FIGURE 4.23** Kinematics of *Shear-Speed* cutting of spur gears.

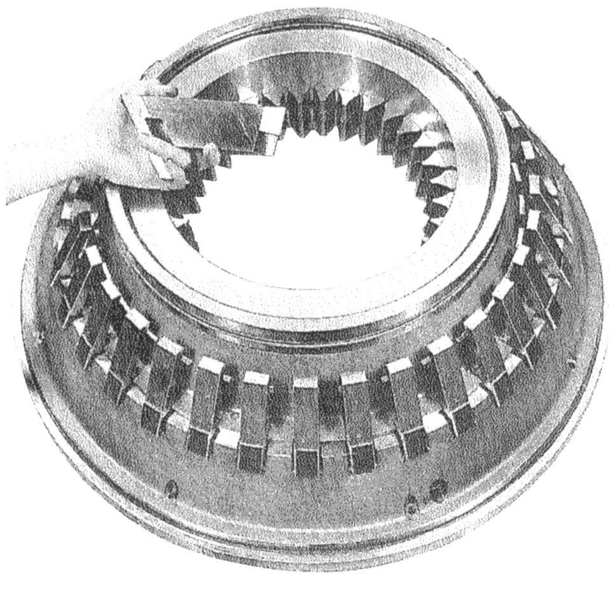

FIGURE 4.24 The tool head for *Shear-Speed* cutting of gears.

The problem of profiling the Shear-Speed cutting tool is solved immediately after two profiles of the gear-cutting tool are constructed. The tooth profile $a_{rs}$, $b_{rs}$, $c_{rs}$, …, $k_{rs}$ within the rake plane $R_s$ is one, and the normal section of the clearance surface $a_{cs}$, $b_{cs}$, $c_{cs}$, …, $k_{cs}$ is the second. Both profiles of the tooth are used for manufacturing, re-sharpening, and inspection purposes of the gear-cutting tool.

Similarly, the form cutting tools can be designed with the rake surface and the clearance surface having any other geometry that is desired in a particular case of gear machining.

### 4.9.2.2 Analytical Profiling of Form Tool

The analytical solution to the problem of profiling of form tools for Shear-Speed cutting of gears is conceptually similar to the graphical solution to that same problem.

Consider an analytical solution to the problem of profiling of the form tool for machining a spur gear with an involute tooth profile. First of all, it is necessary to analytically describe the generating surface of the form tool. In the Cartesian coordinate system $X_cY_cZ_c$ (see Figure 4.26), the equation of the generating surface $T$ can be represented in the form:

$$\mathbf{r}_c(U_c,\phi_c)=\begin{bmatrix} \dfrac{r_{b.c}}{\cos\phi_c}\cdot\sin(\alpha+\text{inv}\phi_c) \\ -\dfrac{r_{b.c}}{\cos\phi_c}\cdot\cos(\alpha-\text{inv}\phi_c) \\ U_c \\ 1 \end{bmatrix} \quad (4.19)$$

where:

- $\mathbf{r}_c$ is the position vector of a point of the generating surface $T$ of the form tool
- $U_c$, $\phi_c$ are the Gaussian (curvilinear) coordinates on surface $T$
- $r_{b.c}$ is the base cylinder radius of involute profile of the form tool
- $\alpha$ is the angle that specifies the location of the start point of the involute curve on the base cylinder of the generating surface $T$ of the form tool (see Figure 4.26)

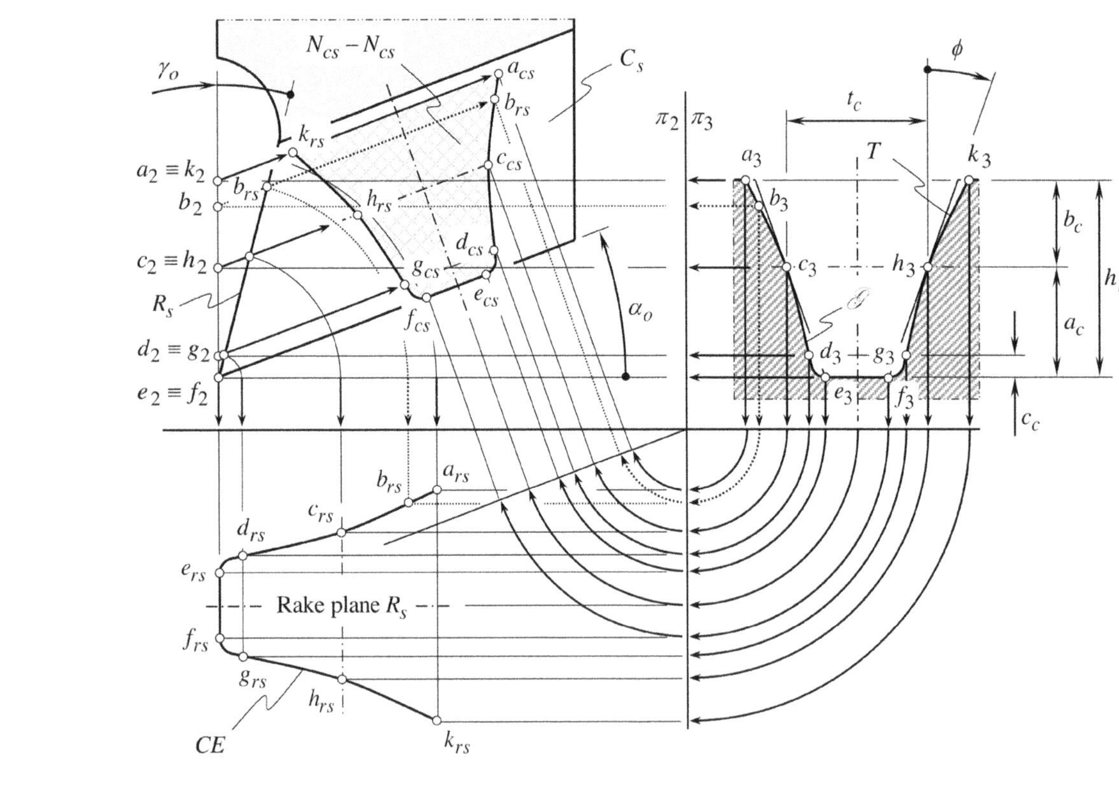


FIGURE 4.25 Descriptive geometry-based solution to the problem of profiling of *Shear-Speed* cutting tool.

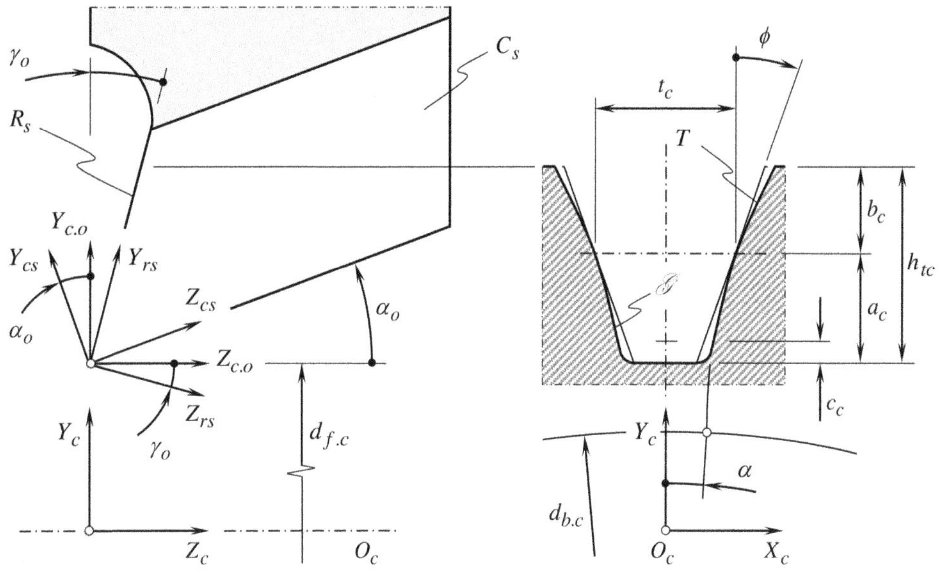


**FIGURE 4.26** Cartesian coordinate systems used for analytical profiling of form tool.

Equation (4.19) is based on the premise presented in Eq. (1.2).

Since the generating surface $T$ has already been determined, an equation of the cutting edge $CE$ of the form tool can be derived in the form of an equation of the line of intersection of the surface $T$ of the form tool by the rake plane $R_s$. Various opportunities are available to the form tool designer in this regard.

For example, the simplest form of analytical representation of the rake plane $R_s$ can be derived in a particular coordinate system. A Cartesian coordinate system $X_{rs}Y_{rs}Z_{rs}$ is associated with the rake plane $R_s$, as shown in Figure 4.25. The coordinate plane $X_{rs}Y_{rs}$ is congruent with the plane $R_s$. Therefore, in the reference system $X_{rs}Y_{rs}Z_{rs}$ equation of the rake plane fulfills the equality $Z_{rs} = 0$. To derive the equation for the cutting edge $CE$, the generating surface $T$ of the form tool can be represented in the coordinate system $X_{rs}Y_{rs}Z_{rs}$. Then, the value of $Z_{rs} = 0$ is substituted into the equation for the surface $T$. This returns an equation for the cutting edge $CE$.

In the reference system $X_{rs}Y_{rs}Z_{rs}$, the position vector of a point $\mathbf{r}_c^{rs}(U_c, \phi_c)$ of the generating surface $T$ of the form tool can be calculated from the expression:

$$\mathbf{r}_c^{rs}(U_c, \phi_c) = \mathbf{Rt}(\gamma_o, X_{c.o}) \cdot \mathbf{Tr}(0.5 \cdot d_{f.c}, Y_c) \cdot \mathbf{r}_c(U_c, \phi_c) \quad (4.20)$$

where $d_{f.c}$ denotes the inner diameter of the gear-cutting head.

Next, an expression:

$$\mathbf{r}_{ce}^{rs}(U_c, \phi_c) = \begin{bmatrix} X_{ce}(U_c, \phi_c) \\ Y_{ce}(U_c, \phi_c) \\ 0 \\ 1 \end{bmatrix} \quad (4.21)$$

is valid for the position vector of a point of the cutting edge $CE$ of the form tool.

It is important to stress here that the parameters $U_c$ and $\phi_c$ in Eq. (4.21) are not independent, and one can be expressed in terms of the other, either in the form $U_c = U_c(\phi_c)$ or in the form $\phi_c = \phi_c(U_c)$. This means that the position vector of point $\mathbf{r}_{ce}^{rs}(U_c, \phi_c)$ of the cutting edge $CE$ depends on just one, rather than two, of these parameters.

Several opportunities are also available for the form tool designer when deriving equation of the normal cross section of the clearance surface $C_s$ of the form tool.

The simplest analytical representation of the normal section of the clearance surface can be obtained in a specific reference system. A Cartesian coordinate system $X_{cs}Y_{cs}Z_{cs}$ is associated with the clearance surface $C_s$, as shown in Figure 4.26. The coordinate plane $X_{cs}Y_{cs}$ is congruent to the plane of normal section $N_{cs} - N_{cs}$ (see Figure 4.25). Therefore, in the reference system $X_{cs}Y_{cs}Z_{cs}$, the equation of the normal plane meets the requirement $Z_{cs} = 0$. To derive an equation of the normal section of the clearance surface $C_s$, the equation of the cutting edge $CE$ of the form tool can be represented in the coordinate system $X_{cs}Y_{cs}Z_{cs}$. Next, the cutting edge $CE$ travels in the direction specified by the clearance angle $\alpha_o$. In this manner, an equation for the clearance $C_s$ is derived. An equation for the normal section of the clearance surface can be derived when the value of $Z_{cs} = 0$ is substituted into the derived equation of the clearance surface $C_s$.

In the reference system $X_{cs}Y_{cs}Z_{cs}$, the position vector of a point $\mathbf{r}_{ce}^{cs}(U_c)$ of the cutting edge $CE$ of the form tool is calculated from the expression:

$$\mathbf{r}_{ce}^{cs}(U_c, V_{cs}) = \mathbf{Tr}(V_{cs}, Z_{cs}) \cdot \mathbf{Rt}[(\alpha_o + \gamma_o), X_{c.o}] \cdot \mathbf{r}_{ce}^{rs}[U_c, \phi_c(U_c)] \quad (4.22)$$

Equation (4.22) yields the expression:

$$\mathbf{r}_{nn}^{cs}(U_c, V_{cs}) = \begin{bmatrix} X_{nn}(U_c, V_{cs}) \\ Y_{nn}(U_c, V_{cs}) \\ 0 \\ 1 \end{bmatrix} \quad (4.23)$$

for the profile of the normal section of the clearance surface $C_s$ of the form tool.

Again, the parameters $U_c$ and $V_{cs}$ in Eq. (4.23) are not independent, and one can be expressed in terms of the other one, either in the form $U_c = U_c(V_{cs})$ or in the form $V_{cs} = V_{cs}(U_c)$. This means that position vector of a point $\mathbf{r}_{nn}^{cs}(U_c, V_{cs})$ of the normal section of the clearance surface $C_s$ of the form tool depends on just one parameter rather than two.

The derived analytical descriptions of the cutting edge *CE* and the normal section of the clearance surface are the solution to the problem of profiling the form tool for Shear-Speed gear-cutting head.

### 4.9.3 Application of Shear-Speed Cutting

The Shear-Speed process is a high-production method of producing gear teeth. The Shear-Speed cuts all teeth of external spur forms simultaneously. The production advantage is greater for gears with large numbers of teeth.

In the Shear-Speed process, it is possible to form cut involute spur gear teeth with practically any required tooth modification. This includes pre-shave root relief, semi-topping, full fillet modifications, and modifications in the involute form.

Although some clearance on both sides of the part is required for overtravel of the form cutting tool, parts that are integral with closely spaced flanges or shoulders, such as cluster gears, can be readily cut by the Shear-Speed process.

The number of tools depends on the number of spaces to be cut in the periphery of the work gear. All tools may have the same form, or the forms may vary depending on the part specifications. The tools may be equally or unequally spaced. The cutting tools are easily removed for sharpening and all are sharpened at one time on a surface grinder.

Although the majority of forms shaped by the Shear-Speed are external, there are cases wherein this method is most advantageous in the shaping of internal forms.

The process is limited to coarse-pitch gears with spur teeth. In addition, each model is limited to a small range of diameters and pitches.

Helical forms are not applicable to this method.

All three cutting edges of every form tool cut simultaneously. Chip flows cut by the adjusting cutting edges interfere with each other. Because of this, tool life is relatively short and this method is normally applied to semi-finishing or finishing of gear teeth.

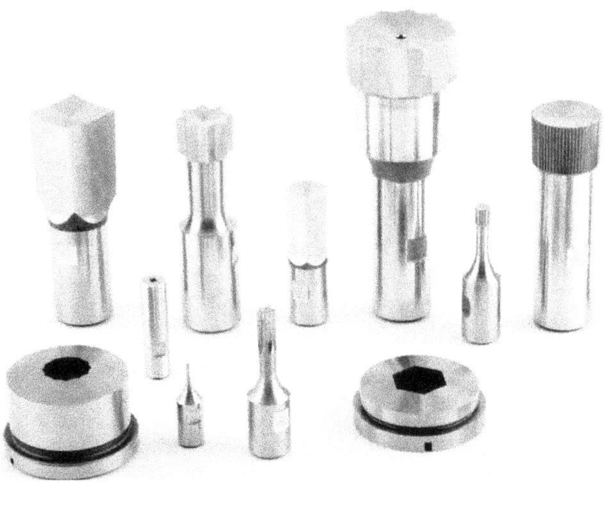

**FIGURE 4.27** Slater tools.

## 4.10 ROTARY BROACHES: SLATER TOOLS

External and internal tooth forms can be produced via rotary broaches.[8] Cutting tools of this type are often called slater tools.

The slater tool seems to be similar to one finishing section of a gear broach that is attached to the shank. The clearance angle $\alpha_o$ at the outer diameter of the tool is greater than that for a regular gear broach.

Examples of slater tools are depicted in Figure 4.27.

The kinematics of the rotary broaching process is illustrated in Figure 4.28.

When broaching, the work gear rotates $\boldsymbol{\omega}_g$ about its axis. Because the generating surface of the slater tool is congruent to the work gear tooth surface, the generating surface *T* rotates $\boldsymbol{\omega}_{ct}$ about its axis, together with the rotation of the work gear.

The shank axis is at a certain angle $\vartheta_{cut}$ in relation to the rotation vectors $\boldsymbol{\omega}_g$ and $\boldsymbol{\omega}_c$. The angle $\vartheta_{cut}$ has a positive value; however, it does not exceed the clearance angle $\alpha_o$. Ultimately, the active value of the clearance angle in rotary broaching is equal to the difference $(\alpha_o - \vartheta_{cut})$.

When machining the work gear, the slater tool comes into contact with and is driven by the work gear. The rotation $\boldsymbol{\omega}_{cut}$ of the slater tool is the primary motion in rotary broaching process. While rotating, the slater tool travels under the force $\mathbf{F}_c$ in the axial direction of the work gear. The gear is machined in only one path of the slater tool. Because of this, rotary broaching is a very fast process.

Most rotary broach tools are made of M-2 high speed steel, a material that provides the required edge toughness in an operation that, by nature, does not generate much heat. Powder metal PM4 is also used in the production of rotary broach tools. Coated with TiN or TiCN, gear slater tools have longer tool life and better cutting performance.

For rotary broaching of a gear, turning, milling, drilling, or screw machines can be used. The slater tool is installed

[8] Produced by Slater Tools Inc., USA.

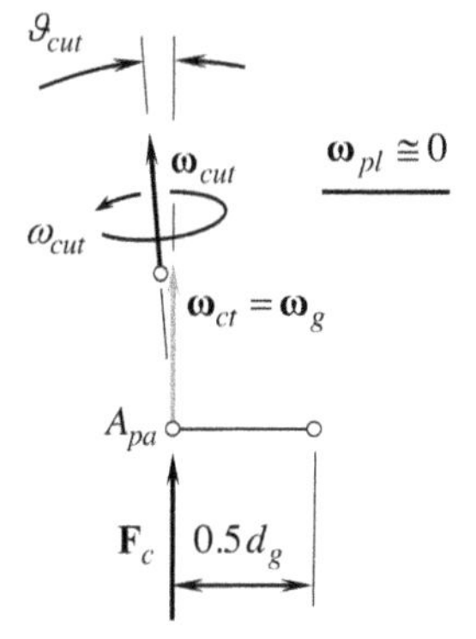


**FIGURE 4.28** Kinematics of rotary broaching process.

in a rotary broach tool holder, which has an internal live spindle that holds an end cutting broach tool. The holder along with the end cutting broach tool is designed for producing polygon forms while the machine spindle rotates in a forward or reverse direction.

In a turning or screw machine, the holder is mounted stationary, whereas its internal live spindle along with the end cutting broach tool rotates with and is driven by the work gear rotation. While rotating with the work gear, the end cutting broach tool's pressure and contacting points are continually changing on the work gear (wobbling type action), making broached forms easily produced.

In a milling or drilling machine, the holder is mounted into and rotates with the machine spindle, whereas its internal live spindle along with the end cutting broach tool remains stationary upon contact with the work gear.

Form sizes can be broached up to 2 inches for aluminum, 1.5 inches for brass, and 1 inch for steel. The practical forming length of rotary broaching is up to 1.5 times the distance across flats or the diameter of the inscribed circle of the profile to be broached.

It is necessary to align the end of the broach tool to the centerline of the work gear. Both internal and external forms require a pilot hole and a lead chamfer. For internal broaching, the diameter of the pilot hole should be, at minimum, equal in size to the distance across the flats or inscribed circle diameter of the form to be broached. The lead chamfer diameter must be slightly larger in diameter than the form being broached. For external broaching, the turned diameter on the work gear should be smaller than the major diameter of the form being broached. If full form is required, turn the diameter .0008″ to .0015″ larger than the final diameter. For starting the external broach tool on the work gear, a 45° chamfer should be smaller than the minor diameter and larger than the major diameter.

## 4.11 BROACHING BEVEL GEAR TEETH

For broaching bevel gear teeth, the Revacycle process is used. The Revacycle process is the fastest method for cutting straight bevel gears of commercial quality. It is primarily

**FIGURE 4.29** Revacycle cutter.

used for large-volume production of gears for such applications as automotive-type differentials and agricultural implements [211].

### 4.11.1 Principle of the Revacycle Process of Cutting Gear Teeth

Most Revacycle-cut gears are produced with one completing operation using a cutter that continuously rotates at a uniform rate. An example of the cutter that is used in the Revacycle process is depicted in Figure 4.29. The cutter blades, which extend radially outward from the cutter head, have concave edges that produce convex profiles on the gear teeth (see Figure 4.30). During the cutting operation (see Figure 4.31), the workpiece is held motionless while the cutter is moved by means of a cam in a straight line across the face of the gear and substantially parallel to its root line. This motion makes it possible to produce a straight tooth bottom, whereas the desired tooth shape is produced by the combined effect of the motion of the cutter and the shapes of the cutter blades.

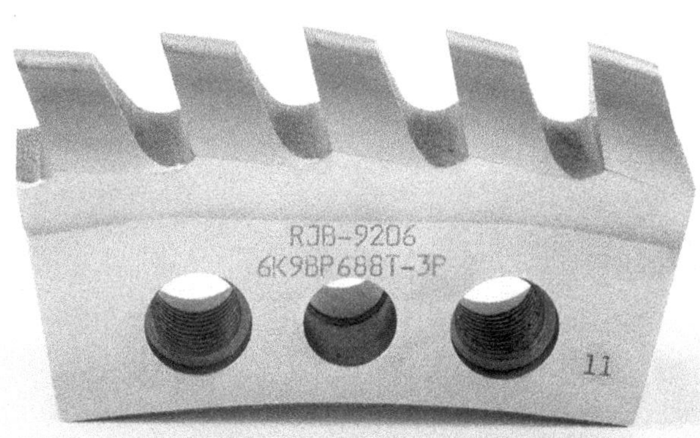


**FIGURE 4.30** Cutting sector of the Revacycle cutter shown in Figure 4.29.

FIGURE 4.31 Set-up for Revacycle-cut straight bevel gears.

There is no depth-wise feed of the cutter into the work; the effective feed being obtained by making each cutter blade progressively longer than the one before it.

The completing cutter contains three kinds of blades: roughing, semi-finishing, and finishing. One revolution of the cutter completes each tooth space, and the work is indexed in the gap between the last finishing blade and first roughing blade.

The Revacycle process of broaching bevel gear teeth is illustrated in Figure 4.32.

Figure 4.32 illustrates the Revacycle cutter in position at the beginning of the cut. As the cutter rotates counterclockwise, blades of gradually increasing length contact the work gear until the root line of the tooth is reached. Relative configuration of the work gear and of the gear cutting tool is illustrated in Figure 4.33(a). Figure 4.339(b) shows a transverse view of the tooth space as roughed to full depth. Each

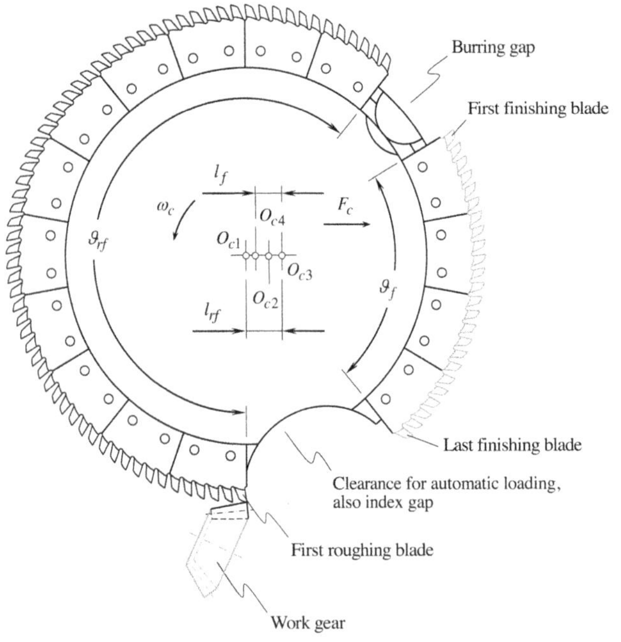


FIGURE 4.32 Kinematics of the Revacycle process of machining bevel gear teeth.

chip extends the full width of the slot, except for the amount of stock left for finishing. Figure 4.33(c) is an axial section of the same tooth, showing that the roughing chips extend the full length of the tooth. During the first part of the roughing, the cutter is fed $F_c$ from $O_{c1}$ to $O_{c2}$. It then remains stationary at $O_{c2}$ until full depth is reached.

The roughed tooth does not have the proper taper. It is substantially correct along the diagonal line in Figure 4.33(b), but the portions of the tooth surface lying to the right of this line toward the heel of the tooth still have considerable stock that must be removed before finishing. This is accomplished by the semi-finishing blades, which cut while the cutter center travels from $O_{c2}$ to $O_{c3}$ (see Figure 4.32).

Finishing is carried out while the cutter center returns from $O_{c3}$ to $O_{c4}$ at a uniform rate. The finishing blades are given the proper shape to produce the correct tooth taper and proper profile at every point along the tooth. The blades cutting at either end of the tooth space are made slightly wider than necessary for correct taper in order to ease off the ends of the tooth and produce a crowning, which results in localized tooth bearing [211].

The resulting finished tooth surface is generated, being made up of a series of inclined cuts similar to those shown in Figure 4.33(d). In general, a different cutter is required for each gear specification.

For Revacycle work that is too deep to complete in one cut, separate roughing and finishing operations are used. Separate cutters and separate machine setups are required for each operation. The cutters and machine cycle are similar to those previously described, with some slight differences. The roughing cutter has no semi-finishing or finishing blades, and there is no translation of the cutter in the roughing operation. The finishing cutter, however, translates as in the completing operation, with semi-finishing blades cutting on the first translation and finishing blades cutting during the return stroke.

### 4.11.2 The Revacycle Cutting Tools

The Revacycle cutter blades are relief ground when manufactured and thus require sharpening on the front face only. Blade-to-blade spacing, the angle of the plane of the front face, and surface finish of the front face must be all closely controlled during sharpening. In addition, when new segments are assembled in heads, assembly cleanliness, accuracy of the position of segment-locating keys, and close control of segment-holding bolt tension are all necessities for proper cutting results.

The lateral cutting edges of the finishing teeth of the Revacycle cutting tool are located within the generating surface $T$. The generating surface $T$ of the Revacycle cutting tool is tangent to the bevel gear tooth flanks $\mathscr{G}$ at every instant of cutting of the gear tooth. After it has been determined, the generating surface $T$ is used for the analytical description of the finishing cutting edges and the clearance surfaces of the finishing teeth of the Revacycle cutting tool.

The generating surface of the Revacycle cutting tool is also important for profiling roughing and semi-finishing cutting blades.

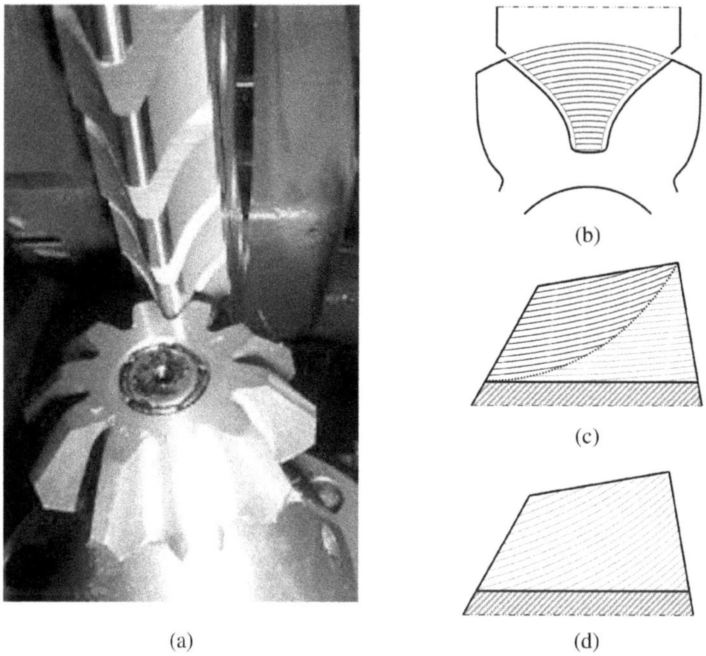


**FIGURE 4.33** Tooth space generation using the Revacycle process of machining straight bevel gear teeth: schematic of the gear machining process (a), and the gear cutting diagrams (b) though (d).

Cutting edges of the roughing blades are at a certain distance $t_{sr}$ inward of the generating surface $T$. The distance $t_{sr}$ is equal to the stock thickness to be removed by the semi-finishing blades. Therefore, cutting edges of the roughing blades are within a surface that is offset at the distance $t_{sr}$ with respect to the generating surface $T$ of the Revacycle cutting tool.

Similarly, cutting edges of the semi-roughing blades are located within a surface that is offset at distance $t_f$ with respect to the generating surface $T$ of the Revacycle cutting tool, where $t_f$ denotes the portion of stock to be removed by finishing blades of the cutting tool.

This reveals the importance of the generating surface $T$ for the purpose of profiling the Revacycle cutting tool.

In the case under consideration, the generating surface $T$ of the cutting tool can be determined in the following manner.

Consider the tooth flank of a straight bevel gear to be machined using the Revacycle process. For the position vector of point $\mathbf{r}_g$ of the straight bevel gear tooth flank $\mathscr{G}$, Eq. (1.32) can be used:

$$\mathbf{r}_g(U_g, \varphi_g) = \begin{bmatrix} U_g \cdot \tan\theta_{b.g} \cdot \sin\varphi_g - \varphi_g \cdot U_g \cdot \tan\theta_{b.g} \cdot \cos\varphi_g \\ U_g \cdot \tan\theta_{b.g} \cdot \cos\varphi_g + \varphi_g \cdot U_g \cdot \tan\theta_{b.g} \cdot \sin\varphi_g \\ -U_g \\ 1 \end{bmatrix} \tag{4.24}$$

where:

$U_g$ and $\varphi_g$ are the curvilinear (Gaussian) coordinates of the bevel gear tooth flank $\mathscr{G}$
$\theta_{b.g}$ is the half of the base cone angle of the bevel gear

When machining bevel gear teeth $\mathscr{G}$, the Revacycle cutting tool rotates and travels in the lengthwise direction of the gear tooth. The rotation $\omega_{ct}$ of the cutter about its axis of rotation and the translation of the cutter with a feed $F_{ct}$ are synchronized. Within the roughing section, the rotation $\omega_c$ and translation $F_{ct}$ are timed so that when the cutting tool turns through the angle $\vartheta_{rf}$, it travels at a distance $l_{rf}$. Here, $\vartheta_{rf}$ denotes the angle that spans over the roughing section of the cutting tool, and $l_{rf}$ denotes the distance between the two consequent positions $O_{c1}$ and $O_{c3}$ of the axis of rotation of the Revacycle cutting tool.

Within the finishing section of the cutting tool, the rotation $\omega_{ct}$ and the translation $F_{ct}$ are timed with each other in a manner that is largely similar to that just discussed. When the cutting tool turns through the angle $\vartheta_f$, it travels at a distance $l_f$. Here, $\vartheta_f$ denotes the angle that spans over the finishing section of the cutting tool and $l_f$ denotes the distance between the two consequent positions $O_{c3}$ and $O_{c4}$ of the axis of rotation of the Revacycle cutting tool.

The analytical representation of the bevel gear tooth flank to be machined [see Eq. (4.24)], along with the just outlined kinematics of the Revacycle process, makes possible derivation of an equation for the both portions of the generating surface of the gear-cutting tool—that is, for the roughing portion, as well as for finishing portion of the Revacycle cutting tool.

### 4.11.3 Profiling Cutter for Machining Straight Bevel Gears Using Revacycle Process

A cutter for machining bevel gear teeth using the Revacycle process consists of a large diameter disc 550 or 650 mm (21 or 25 in.) with the cutting blades projecting radially from the periphery, commonly arranged in groups and each blade progressively longer than the preceding one.

Most of the material from the tooth space is removed by teeth of the first group. Roughing and semi-roughing teeth of the cutter are referred to as the teeth of the first group. The main purpose of the first group of teeth is to remove material from the tooth space in the gear. The teeth of the first group do not generate the bevel gear tooth flanks. Therefore, requirements to the accuracy of the gear tooth profile are not close and can vary significantly.

Teeth of the second group are the finishing teeth. Only a small portion of material is removed by the finishing teeth. The main purpose of the second group of teeth is to generate the bevel gear tooth flanks in compliance with the blueprint requirements. Because of this, the requirements to the accuracy of the gear tooth profile are close and can vary just within very narrow intervals.

Solution to the problem of profiling of the Revacycle gear cutter can be broken down into two steps.

First, profiling of the finishing teeth is considered. Next, profiling of the roughing and semi-roughing teeth is performed based on the solution obtained in the first step.

#### 4.11.3.1 Profiling the Finishing Teeth of the Revacycle Gear Cutter

A left-hand-oriented Cartesian coordinate system $X_{bg}Y_{bg}Z_{bg}$ is associated with a straight bevel gear to be machined, as shown in Figure 4.34. In the coordinate system $X_{bg}Y_{bg}Z_{bg}$, the bevel gear tooth flank $\mathscr{G}$ is analytically described by Eq. (4.24). See Chapter 1 for more details on the analytical description of the tooth flank of the bevel gear (see Figure 1.15).

A left-hand-oriented Cartesian coordinate system $X_{ct}Y_{ct}Z_{ct}$ is associated with the cutter. This coordinate system rotates $\omega_{ct}$, together with the cutter when machining the gear. In the initial position of the cutter, the coordinate system $X_{ct}Y_{ct}Z_{ct}$ is aligned with the coordinate system $X_{c0}Y_{c0}Z_{c0}$, as shown in Figure 4.34.

While rotating $\omega_{ct}$ through a certain angle $\vartheta_{f.x}$ (here, the angle $\vartheta_{f.x}$ falls into the range of $0° \leq \vartheta_{f.x} \leq \vartheta_f$), the cutter travels along the $X_{c0}-$ axis at a certain distance $a_x$. The angular displacement $\vartheta_{f.x}$ and the linear displacement $a_x$ correlate to each other in the known manner—that is, $a_x = a_x(\vartheta_{f.x})$. For practicality, this relationship is reduced to the linear expression $a_x = k_{rc} \cdot \vartheta_{f.x}$, where $k_{rc}$ is constant.

When deriving an equation for the generating surface $T$ of the Revacycle gear cutter, it is necessary to represent the equation of the straight bevel gear tooth flank $\mathscr{G}$ in the coordinate system $X_{ct}Y_{ct}Z_{ct}$ associated with the gear cutter. The corresponding operators of the coordinate system transformations (see Appendix D) can be used for this purpose.

At the beginning of the machining process, configuration of the work gear and of the Revacycle gear cutter is known. This statement implies that the operator $\mathbf{Rs}(bg \mapsto c_0)$ of the resultant coordinate system transformation, say from the bevel gear coordinate system $X_{bg}Y_{bg}Z_{bg}$ to the gear cutter coordinate system $X_{c0}Y_{c0}Z_{c0}$, is either given from the very beginning or can be composed using the actual design parameters of the bevel gear and the *Revacycle* gear cutter. Consequently, the operator $\mathbf{Rs}(c_0 \mapsto bg)$ of the inverse coordinate system transformation can be determined as well. Recall that the relationship $\mathbf{Rs}(c_0 \mapsto bg) = \mathbf{Rs}^{-1}(bg \mapsto c_0)$ for operators of the coordinate system transformations is

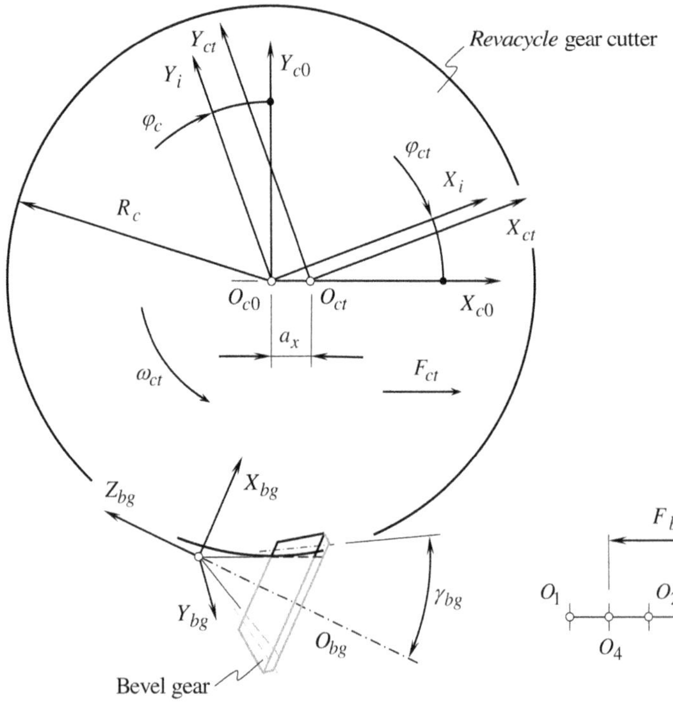


**FIGURE 4.34** Reference systems used for profiling the Revacycle cutter.

valid. The operator $\mathbf{Rs}(bg \mapsto c_0)$ of the coordinate system transformation is used for representation of the position vector $\mathbf{r}_g$ [see Eq. (4.24)] of a point of the gear tooth flank $\mathcal{G}$ in the coordinate system $X_{c0}Y_{c0}Z_{c0}$ that corresponds to the initial configuration of the work gear and the gear cutter. In the reference system $X_{c0}Y_{c0}Z_{c0}$, the position vector of point $\mathbf{r}_g^{c0}$ of the bevel gear tooth flank $\mathcal{G}$ is calculated from the expression:

$$\mathbf{r}_g^{c0}(U_g, \varphi_g) = \mathbf{Rs}(bg \mapsto c_0) \cdot \mathbf{r}_g(U_g, \varphi_g) \qquad (4.25)$$

Then, in the current instant of time, the coordinate system $X_{ct}Y_{ct}Z_{ct}$ of the Revacycle gear cutter is turned through an angle $\varphi_c$ about the $Z_{ct}$ – axis and is shifted along the $X_{c0}$ – axis at a certain distance $a_x$. Recall that the displacements $a_x$ and $\varphi_c$ correlate to each other. The correlation is expressed in the form of a function $a_x = a_x(\vartheta_{f.x})$ [or in the form $\vartheta_{f.x} = \vartheta_{f.x}(a_x)$, which is equivalent to $a_x = a_x(\vartheta_{f.x})$]. In a particular case of bevel gear machining, the equality $a_x = k_{rc} \cdot \vartheta_{f.x}$ is observed.

The rotation of the coordinate system $X_{c0}Y_{c0}Z_{c0}$ through the angle $\varphi_c$ to the position of the intermediate reference system $X_iY_iZ_i$ can be analytically described by the operator of rotation $\mathbf{Rt}(\varphi_c, Z_{c0})$. Then, for the analytical description of translation of the intermediate coordinate system $X_iY_iZ_i$ at the distance $a_x$ to the position of the gear cutter coordinate system $X_{ct}Y_{ct}Z_{ct}$, the operator of translation $\mathbf{Tr}(a_x, X_i)$ is implemented. Ultimately, the expression:

$$\mathbf{r}_g^{c}(U_g, \varphi_g, \varphi_c) = \mathbf{Tr}(a_x, X_i) \cdot \mathbf{Rt}(\varphi_c, Z_{c0}) \cdot \mathbf{Rs}(bg \mapsto c_0) \cdot \mathbf{r}_g(U_g, \varphi_g) \qquad (4.26)$$

can be used for the analytical description of the straight bevel gear tooth flank $\mathcal{G}$ in the coordinate system $X_{ct}Y_{ct}Z_{ct}$ associated with the gear cutter.

It is instructive to note here that taking into account the direction of the translation of the Revacycle gear cutter from the point $O_1$ through the point $O_2$ to the point $O_3$, and its rotation through the angle $\vartheta_f$ (see Figure 4.34), both displacements $a_x$ and $\varphi_c$ in the operators $\mathbf{Tr}(a_x, X_i)$ and $\mathbf{Rt}(\varphi_c, Z_{c0})$ of the linear transformations have negative values ($a_x < 0$ and $\varphi_c < 0°$).

At a current location of the gear cutter, the position vector of point $\mathbf{r}_g^c(U_g, \varphi_g, \varphi_c)$ in the coordinate system $X_{ct}Y_{ct}Z_{ct}$ also depends on the current value of the $\varphi_{ct}$ – parameter. The angle $\varphi_{ct}$ is a time-dependent parameter. It can be expressed in terms of the rotation of the cutter $\omega_{ct}$ and of the time $t$. This relation is described by the function:

$$\varphi_{ct} = \omega_{ct} \cdot t \qquad (4.27)$$

With Eq. (4.26), which analytically describes the current configuration of the straight bevel gear tooth flank $\mathcal{G}$ in the reference system $X_{ct}Y_{ct}Z_{ct}$, position vector of a point $\mathbf{r}_{ct}$ of the generating surface $T$ of the Revacycle gear cutter can be derived using the Shishkov equation of contact[9] (1948) [172, 206]:

$$\mathbf{n}_{bg} \cdot \mathbf{V}_\Sigma = 0 \qquad (4.28)$$

Here, $\mathbf{n}_{bg}$ denotes the unit normal vector to the bevel gear tooth flank $\mathcal{G}$ and $\mathbf{V}_\Sigma$ is the vector of the resultant motion of the tooth flank $\mathcal{G}$ in the coordinate system $X_{ct}Y_{ct}Z_{ct}$. The equation for the position vector of a point of the generating surface $T$ of the Revacycle gear cutter can be obtained as the solution to a set of two equations:

$$\mathbf{r}_{ct}(U_g, \varphi_g) \Rightarrow \begin{cases} \mathbf{r}_g^c = \mathbf{r}_g^c(U_g, \varphi_g, \varphi_{ct}) \\ \begin{vmatrix} \dfrac{\partial \mathbf{r}_g^c}{\partial U_g}(U_g, \varphi_g, \varphi_{ct}) \\ \dfrac{\partial \mathbf{r}_g^c}{\partial \varphi_g}(U_g, \varphi_g, \varphi_{ct}) \\ \dfrac{\partial \mathbf{r}_g^c}{\partial \varphi_{ct}}(U_g, \varphi_g, \varphi_{ct}) \end{vmatrix} = 0 \end{cases} \qquad (4.29)$$

After the second equation in Eq. (4.29) is solved with respect to the parameter of relative motion $\varphi_{ct}$, the calculated solution $\varphi_{ct}$ is substituted into the first. Ultimately, the equation of the generating surface $T$ of the Revacycle gear cutter can be represented in the form:

$$T \Rightarrow \mathbf{r}_{ct} = \mathbf{r}_{ct}(U_{ct}, V_{ct}) \qquad (4.30)$$

It is important to note here that, for the finishing group of teeth, the rotation $\omega_{ct}$ of the reference system $X_{ct}Y_{ct}Z_{ct}$ is performed through the angle $\vartheta_f$ (see Figure 4.34), whereas the translation is performed at the distance $O_3O_4$. Theoretically, the distance $O_3O_4$ should span over the entire face width of the tooth flank in the direction of the bottom land of the work gear. However, in practice, reasonable deviations of the actual distance traveled by the coordinate system $X_{ct}Y_{ct}Z_{ct}$ from the $O_3O_4$ could be permissible.

It is of critical importance to point out that the generating surface $T$ [see Eq. (4.30)] does not allow for sliding motion on itself.

Equation (4.30) of the generating surface $T$ makes designing an accurate Revacycle gear cutter possible. The accuracy of the Revacycle cut gears can even exceed the precision of generated bevel gears.

Cutting edges of the finishing teeth of the Revacycle gear cutter are within the generating surface $T$ [see Eq. (4.30)].

---

[9] Shishkov, V.A., "Elements of Kinematics of Generating and Conjugating in Gearing", In: *Theory and Calculation of Gears*, Vol. 6, Leningrad: LONITOMASH, 1948.

Shishkov, V.A., *Generation of Surfaces Using Continuous Indexing Methods of Machining*, Moscow, Mashgiz, 1951, 152 pages; also see the earlier publications on the subject by the same author.

The total number of the finishing teeth is denoted by $n_c$. The finishing teeth are within the angle $\vartheta_f$ that spans over the finishing section. The finishing teeth are spaced at the angular increment $\Delta\varphi_c$, which is equal to $\Delta\varphi_c = \vartheta_f / n_c$. The rake angle $\gamma_o$ at the outer diameter of the gear cutter is equal to zero ($\gamma_o = 0°$). Therefore, position vector of a point $\mathbf{r}_{rs}$ of the rake plane for each finishing tooth of the gear cutter can be represented in the form:

$$\mathbf{r}_{rs}(U_{rs}, V_{rs}) = \begin{bmatrix} U_{rs} \\ U_{rs} \cdot \tan\left[\vartheta_f^0 + \Delta\varphi_c \cdot (\text{No} - 1)\right] \\ V_{rs} \\ 1 \end{bmatrix} \quad (4.31)$$

where $U_{rs}$ and $V_{rs}$ are the curvilinear coordinates of point of the rake plane $R_s$ [tangent vectors $\mathbf{t}_{u.rs}$ and $\mathbf{t}_{v.rs}$ to the coordinate lines $U_{rs}$ and $V_{rs}$ are parallel to the coordinate axes of the reference system $X_{ct}Y_{ct}Z_{ct}$ (the unit vector $\mathbf{t}_{u.rs}$, $X_{ct}$ – axis, and the unit vector $\mathbf{t}_{v.rs}$ are parallel to the axis $Z_{ct}$)]; $\vartheta_f^0$ is the angle at which the finishing section begins; and the actual number of a current finishing tooth is denoted by No.

The cutting edge $CE$ of every finishing tooth fulfills both Eq. (4.30) and (4.31). Therefore, an analytical description of the cutting edge $CE$ can be derived as a solution to the set of two Eqs. (4.30) and (4.31). The solution yields representation in matrix form:

$$\mathbf{r}_{ce}(V_{rs}, No) = \begin{bmatrix} X(V_{rs}, \text{No}) \\ Y(V_{rs}, \text{No}) \\ Z(V_{rs}, \text{No}) \\ 1 \end{bmatrix} \quad (4.32)$$

The actual shape of every finishing tooth depends on the parameter No. This means that the particular shape of every finishing tooth has to have is dependent on the location of the tooth within the finishing sector. In practice, reasonable approximation is permissible for the desired shape of the finishing teeth of the Revacycle gear cutter.

A clearance surface is a type of relived surface. Clearance surface $C_s$ is a surface through the cutting edge $CE$ [see Eq. (4.32)]. It forms the desirable clearance angle $\alpha_o$ with the generating surface $T$ of the gear cutter.

#### 4.11.3.2 Shape of Roughing Teeth of Revacycle Gear Cutter

The cutting edges of the roughing teeth are situated beneath the generating surface $T$ of the gear cutter. They are within the interior of the generating body of the Revacycle gear cutter. The purpose of the roughing teeth is twofold:

- Most of the material removal from the tooth space is performed by the roughing teeth.
- After roughing, a little stock for finishing is left.

It is desired that the stock for finishing be uniformly distributed over the bevel gear tooth flank.

Because roughing teeth do not directly affect the final tooth flank geometry, the tolerance for the accuracy of roughing teeth is not as close as it is for finishing teeth.

The desirable shape of roughing teeth can be determined in the following manner.

First, let us determine a surface that is enveloping to the bevel gear tooth flank when roughing of the bevel gear teeth is performed. When roughing, the Revacycle gear cutter rotates through an angle $\vartheta_f$ and traveling from $O_1$ through $O_2$ to $O_3$. The kinematics of roughing is similar to that for the finishing teeth. Therefore, for the surface $T_{rf}$ that is enveloping the bevel gear tooth flank $\mathscr{G}$, an analytical description can be derived using the same process for deriving the generating surface $T$ of the Revacycle gear cutter [see Eq. (4.32)]. Following this process, the following expression:

$$T_{rf} \Rightarrow \mathbf{r}_{cr} = \mathbf{r}_{cr}(U_{cr}, V_{cr}) \quad (4.33)$$

can be derived for the position vector of point $\mathbf{r}_{cr}$ of the generating surface $T_{rf}$. If one assumes that the straight bevel gear tooth flanks are profiled by the roughing teeth, then their cutting edges would be within the surface $T_{rf}$. In reality, the roughing teeth do not contact with the bevel gear tooth flanks. The closest distance of approach of the roughing teeth to the bevel gear tooth flank $\mathscr{G}$ is equal to the stock thickness $t_{rf}$ left for the finishing teeth. The stock thickness is measured along the perpendicular to the gear tooth flank $\mathscr{G}$. This immediately yields an equation for the position vector $\mathbf{r}_{cr}^a$ of the surface within which location of the cutting edges of the roughing teeth is desired:

$$T_{rf}^a \Rightarrow \mathbf{r}_{cr}^a(U_{cr}, V_{cr}) = \mathbf{r}_{cr}(U_{cr}, V_{cr}) - \mathbf{n}_{bg} \cdot t_{rf} \quad (4.34)$$

where $\mathbf{n}_{bg}$ denotes the unit normal vector to the gear tooth flank $\mathscr{G}$. After the necessary formula transformation is accomplished, the surface $T_{rf}^a$ for the roughing teeth can be analytically described by the expression:

$$T_{rf}^a \Rightarrow \mathbf{r}_{cr}^a = \mathbf{r}_{cr}^a(U_{cr}, V_{cr}) \quad (4.35)$$

Reasonable deviations in the actual location of the roughing cutting edges from those specified by Eq. (4.35) are permissible.

Similar to the finishing group of teeth, for the roughing group of teeth, the rotation $\omega_{ct}$ of the reference system $X_{ct}Y_{ct}Z_{ct}$ is performed through an angle $\vartheta_{rf}$ (see Figure 4.34), whereas translation is performed at the distance $O_1O_3$. The distance $O_1O_3$ correlates with the face width of the straight bevel gear to be machined.

The total number of roughing teeth is denoted by $n_{rc}$.

The first roughing tooth is the shortest one, and the last roughing tooth is of full tooth height $h_{rc} = h_{bg}$ of the tallest bevel gear tooth. The roughing tooth increment $\Delta h_{bg}$ in this case is calculated from the formula:

$$\Delta h_{rc} = \frac{h_{rc}}{n_{rc} - 1} \tag{4.36}$$

To calculate the tooth height of a current roughing tooth of the Revacycle gear cutter, the following expression is used:

$$h_{rc}^{i} = \Delta h_{rc} \cdot \text{No}_r \tag{4.37}$$

where $\text{No}_r$ denotes the number of the current roughing cutting tooth.

The roughing teeth are located within the angle $\vartheta_{rf}$, which spans over the roughing section. The finishing teeth are located at the angular increment $\Delta\varphi_{rc}$. The increment equals to $\Delta\varphi_{rc} = \vartheta_{rf}/n_{rc}$. In practical terms, the increment $\Delta\varphi_{ct}$ and the increment $\Delta\varphi_{rc}$ are equal to each other (that is, $\Delta\varphi_{rc} = \Delta\varphi_{ct}$); however, this is not a required condition. Rake angle $\gamma_{ro}$ at the outer cutting edges of the gear cutter is of a positive value ($\gamma_{ro} > 0°$). Therefore, position vector of a point $\mathbf{r}_{rrf}$ of the rake plane for each finishing tooth of the gear cutter can take the form:

$$\mathbf{r}_{rsr}(U_{rsr}, V_{rsr}) = \begin{bmatrix} U_{rsr} \\ U_{rsr} \cdot \tan\left[\vartheta_{rf}^{0} + \Delta\varphi_{rc} \cdot (\text{No}_r - 1)\right] + 0.5 \cdot d_{o.c} - \left(h_{rc} - h_{rc}^{i}\right) \\ V_{rsr} \\ 1 \end{bmatrix} \tag{4.38}$$

where $U_{rsr}$ and $V_{rsr}$ denote the curvilinear coordinates of the rake plane $R_{rs}$ and are parallel to the coordinate axes of the reference system $X_{ct}Y_{ct}Z_{ct}$ (that is, $U_{rsr} = X_{ct}$ and $V_{rsr} = Z_{ct}$); $\vartheta_{rf}^{0}$ denotes the angle at which the roughing section begins.

The cutting edge of every roughing cutting tooth should meet both Eqs. (4.35) and (4.38). Therefore, an analytical description of the roughing cutting edge $CE_r$ can be derived as a solution to Eqs. (4.35) and (4.38). The solution yields representation in matrix form:

$$\mathbf{r}_{rce}(V_{rsr}, \text{No}_r) = \begin{bmatrix} X(V_{rsr}, \text{No}_r) \\ Y(V_{rsr}, \text{No}_r) \\ Z(V_{rsr}, \text{No}_r) \\ 1 \end{bmatrix} \tag{4.39}$$

The actual shape of every roughing tooth depends on the parameter $\text{No}_r$. This means that the shape of every specific roughing tooth is dependent on the location of the roughing tooth within the roughing sector of the Revacycle gear cutter.

Clearance surface $C_{rs}$ is a type of relived surface. The clearance surface is a surface through the cutting edge $CE_r$ [see Eq. (4.39)] and forms the desirable clearance angle $\alpha_{ro}$ with the surface $T_{rf}^{a}$ of the Revacycle gear cutter.

### 4.11.4 Application of the Revacycle Process of Cutting Gear Teeth

The Revacycle process provides the quickest method of machining tooth flanks in straight bevel gears generally intended for automotive differential units. For broaching of bevel gear teeth, large-diameter [53- or 64 cm (20- or 25-in.)] cutters are used. In one revolution of the cutter, utilizing roughing, semi-finishing, and finishing cutters, a tooth space is completely cut. The center of rotation of the cutter changes as different operations are performed. This provides a constant root clearance in the work gear, and the concave finishing cutters give the correct tooth profile. Between the finishing and the starting cutting blades, a space is provided that permits the work gear to be indexed to the next tooth space while the tool continues to rotate at a uniform rate.

In common with all high-production-rate machines, the cutting cycle is entirely automatic and ceases after the last tooth space in a blank has been completed. It is often combined with automatic loading and unloading devices for continuous production. The quality of the Revacycle cut gears is inferior to that of the generated types, but the pinions produced are suitable for specific purposes. Naturally, the standard of maintenance of the machine and cutting tools is of considerable importance when the quality of the finished gear is being considered. Tooth proportions differ from the usual standards and special blanks are required.

Although the initial tooling cost is greater than for other types of straight bevel gear machines, the extremely fast production rate results in the lowest overall cost for large-scale production. Revacycle-cut gears are not interchangeable with gears produced by other methods. Revacycle tooth proportions differ from those of other straight bevel gears [191].

# 5 End Mill Gear Cutters

End mill gear cutters are extensively used in roughing and finishing external spur gears and in roughing external helical gears. Worms and racks may be cut by end mill gear cutters. The use of end mill cutters does not require special gear-cutting machines. Cutting spur and helical gears with end mill gear cutters can be done in universal milling machines, as illustrated in Figure 5.1.

Before designing an end mill gear cutter, the geometry of the generating surface of the gear-cutting tool must be predetermined. Two critical issues should be considered when determining the generating surface of the end mill gear cutter. The first is the geometry of the tooth flank of the work gear. The geometry of spur and helical gear tooth flanks allows for analytical representation of the tooth flanks these gears [see Eqs. (1.2) and (1.4)]. The second issue is the kinematics of the gear milling process with end mill gear cutters.

## 5.1 KINEMATICS OF GEAR CUTTING WITH END MILL CUTTERS

The kinematics of the gear-machining process when cutting a gear with an end mill gear cutter is the simplest among those shown in the classification of types of vector diagrams of gear-machining processes (see Figure 3.8). To increase the cutting speed, the sliding motion of the generating surface of the mill cutter is not used as the primary motion, as it be would in gear broaching; instead, an additional rotation of the mill cutter is used for this purpose. This motion enables the work gear tooth flank $\mathcal{G}$ to allow sliding on itself is used as the feed motion. Generally, the feed motion in this case is a kind of screw motion, with the screw motion and the work gear having identical screw parameters.

The kinematics of the gear-machining process with end mill gear cutters is composed of the vector diagram of the gear-machining processes and the additional rotation and feed motion. The rotations $\boldsymbol{\omega}_g$ and $\boldsymbol{\omega}_{ct}$ are not utilized when cutting a gear with an end mill gear cutter. For such cases, the equality $\omega_g = \omega_{ct} = 0$ is observed.

The kinematics of machining a work gear with an end mill gear cutter is illustrated in Figure 5.2.

The vector of the additional rotation $\boldsymbol{\omega}_{cut}$ is perpendicular to the centerline $O_g$ of the work gear. The screw feed motion is shown in Figure 5.2 as a superposition of the rotation $\boldsymbol{\omega}_{fc}$ about the work gear centerline $O_g$ and the translation $\mathbf{F}_{ct}$ along the centerline $O_g$. The parameter of the screw motion $p_{fc}$ is equal to:

$$p_{fc} = \frac{|\mathbf{F}_{ct}|}{|\boldsymbol{\omega}_{fc}|} \tag{5.1}$$

and is identical to the reduced pitch $p_g$ of the tooth flank of the helical work gear ($p_{fc} \equiv p_g$). For spur gears, the screw motion reduces to straight motion $\mathbf{F}_{ct}$.

Once the geometry of the tooth flank $\mathcal{G}$ of the gear to be machined is known [see Eq. (1.2) for spur gear and Eq. (1.4) for helical gear tooth flank] and the kinematics of the

**FIGURE 5.1** Milling helical gear by end mill cutter in universal milling machine.


DOI: 10.1201/9781003520528-7

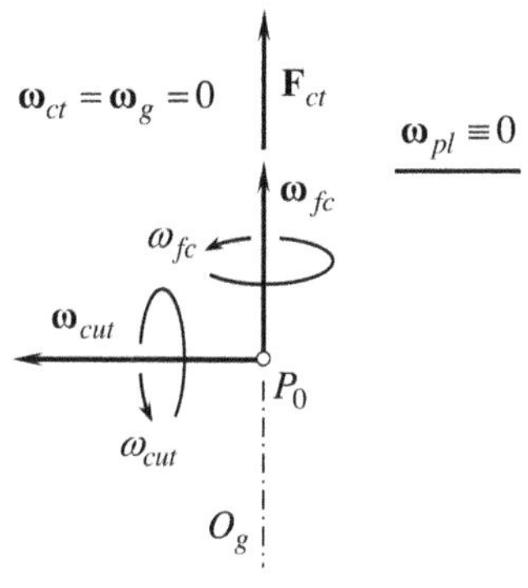


**FIGURE 5.2** Kinematics of the gear-machining process with an end mill gear cutter.

gear-machining process is predetermined (see Figure 5.2), the generating surface of the gear-cutting tool $T$ can be determined on the premise of this input information.

## 5.2 GENERATING SURFACE OF END MILL GEAR CUTTER

When a gear-cutting tool is specifically designed so as to implement the kinematics of the gear-machining processes alone, then in the case under consideration, the generating surface of the gear-cutting tool $T$ is congruent to the tooth flank $\mathcal{G}$ of the gear being machined. Aside from the fact that this type of kinematics is practical and used for designing gear-cutting tools, it is important to consider other possibilities, especially from the standpoint of the potentially increasing speed of the primary motion.

The rotation $\omega_{cut}$ allows for an increase of speed of the primary motion. Moreover, there is no kinematic correlation between the newly introduced rotation $\omega_{cut}$ and the other components of the kinematics of the gear-machining process using an end mill gear cutter.

Because of the additional rotation $\omega_{cut}$, the generating surface of the end mill gear cutter $T$ differs from that initially derived solely on the basis of the vector diagram of the gear-machining process [see Equation (4.1)].

### 5.2.1 Equation of Generating Surface of End Mill Gear Cutter for Machining Spur Involute Gears

Gear tooth profiles are commonly symmetrical. For material removal from tooth spaces between two neighboring teeth with symmetrical profiles, end mill gear cutters are used.[1]

Spur gears can be formed by milling a slot at a time and indexing to the next slot. The end mill gear cutter is specifically designed so that its axial section has an involute curve that matches that of the gear tooth. As shown in Figure 5.3, the axial cross section of the mill cutter is the same as that through the space between two adjacent gear teeth. This observation is

[1] Asymmetrical tooth profiles cannot be machined with end mill gear cutters.

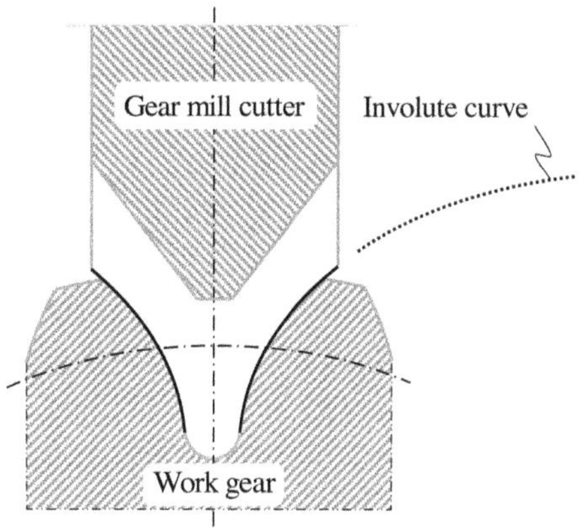


**FIGURE 5.3** Correspondence between the spur gear tooth profile and the mill cutter tooth profile.

helpful when deriving an equation for the generating surface $T$ of the end mill gear cutter.

The equation for the generating surface $T$ of an end mill for machining spur involute gears can be derived in three steps. First, it is necessary to derive an equation for the tooth profile of the spur work gear. Second, it is required to represent the equation of the spur gear tooth profile in a reference system associated with the mill cutter. In the third step, the generated surface of the end mill gear cutter $T$ is derived as the locus of consecutive positions of the involute gear tooth profile when it rotates about the mill cutter's axis of rotation.

A Cartesian coordinate system $X_{inv}Y_{inv}$ is associated with the work gear, as shown in Figure 5.4. In this reference system, the involute tooth profile of the work gear is analytically described by the vector equation:

$$\mathbf{r}_{inv}(\phi_g) = \frac{d_{b.g}}{2 \cdot \cos\phi_g} \cdot [\mathbf{i} \cdot \sin(\text{inv}\phi_g) + \mathbf{j} \cdot \cos(\text{inv}\phi_g)] \tag{5.2}$$

$$\text{inv}\phi_g = \tan\phi_g - \phi_g \tag{5.3}$$

where $d_{b.g}$ denotes the base diameter of the work gear, and $\phi_g$ (expressed in radians) is the profile angle at a point of interest of the involute tooth profile.

Another form of analytical presentation of the involute tooth profile is convenient for particular applications:

$$\begin{aligned}\mathbf{r}_{inv}(\psi_g) = r_{b.g} \cdot [&\mathbf{i} \cdot (\sin\psi_g - \psi_g \cdot \cos\psi_g) \\ &+ \mathbf{j} \cdot (\cos\psi_g + \psi_g \cdot \sin\psi_g)]\end{aligned} \tag{5.4}$$

Here, the equality $d_{b.g} = 2 \cdot r_{b.g}$ is observed.

The angle $\psi_g$ is correlated to the tooth profile angle $\phi_g$ as (see Figure 5.4):

$$\psi_g = \text{inv}\phi_g + \phi_g \tag{5.5}$$

Next, the involute tooth profile [either in the form of Eqs. (5.2) and (5.3), or in the form of Eq. (5.4)] should be represented in a reference system associated with the mill cutter. For this purpose, a Cartesian coordinate system $X_gY_g$

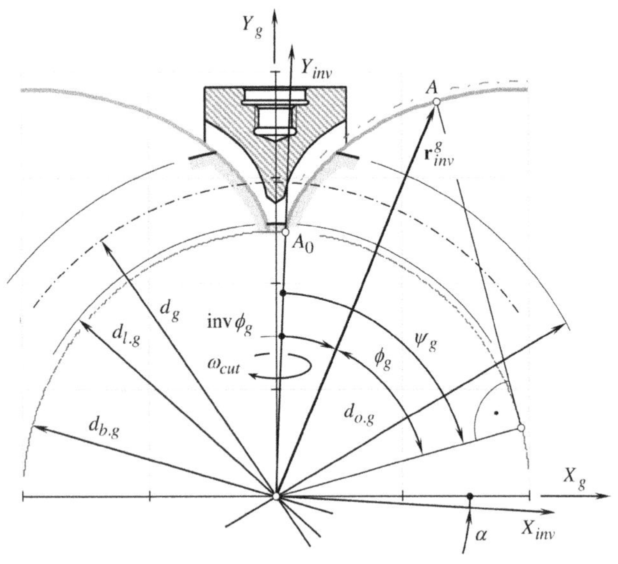


**FIGURE 5.4** Determination of the shape of the generating surface $T$ of the end mill gear cutter.

is introduced. The axis $Y_g$ of the reference system $X_gY_g$ is aligned with the line of symmetry of the gear tooth space. The coordinate system $X_gY_g$ is turned about the axis of the gear rotation with respect to the coordinate system $X_{inv}Y_{inv}$ through a certain angle $\alpha$. The angle $\alpha$ is the angle between the corresponding axes of the coordinate systems $X_gY_g$ and $X_{inv}Y_{inv}$. To calculate for the actual value of the angle $\alpha$, the following formula is used:

$$\alpha = \frac{2\pi}{N_g} \cdot \frac{\breve{s}_g}{\breve{t}_{t.g}} - \text{inv}\phi_g \tag{5.6}$$

where $N_g$ denotes the tooth number, $\breve{s}_g$ is the space width, and $\breve{t}_{t.g}$ is the arc tooth thickness. For gears having standard tooth profile, the equality $\breve{t}_{t.g} = \breve{s}_g$ applies. In this particular case, Eq. (5.6) can be reduced to:

$$\alpha = \frac{\pi}{N_g} - \text{inv}\phi_g \tag{5.7}$$

The required coordinate system transformation can be implemented by using the operator of rotation $\mathbf{Rt}(\alpha, Z_g)$ (see Appendix D).

The following equation:

$$\mathbf{r}^g_{inv}(\phi_g) = \mathbf{Rt}(\alpha, Z_g) \cdot \mathbf{r}_{inv}(\phi_g) \tag{5.8}$$

is valid for the analytical description of the involute gear tooth profile in the coordinate system $X_gY_g$.

Equation (5.2) allows for representation in matrix form:

$$\mathbf{r}_{inv}(\phi_g) = \begin{bmatrix} \dfrac{r_{b.g}}{\cos\phi_g} \cdot \sin(\text{inv}\phi_g) \\ \dfrac{r_{b.g}}{\cos\phi_g} \cdot \cos(\text{inv}\phi_g) \\ 0 \\ 1 \end{bmatrix} \tag{5.9}$$

After the vector $\mathbf{r}_{inv}(\phi_g)$ has been pre-multiplied by:

$$\mathbf{Rt}(\alpha, Z_g) = \begin{bmatrix} \cos\alpha & \sin\alpha & 0 & 0 \\ -\sin\alpha & \cos\alpha & 0 & 0 \\ 0 & 0 & 1 & 0 \\ 0 & 0 & 0 & 1 \end{bmatrix} \tag{5.10}$$

Equation (5.8) casts into:

$$\mathbf{r}^g_{inv}(\phi_g) = \begin{bmatrix} \dfrac{r_{b.g}}{\cos\phi_g} \cdot \sin(\alpha + \text{inv}\phi_g) \\ -\dfrac{r_{b.g}}{\cos\phi_g} \cdot \cos(\alpha - \text{inv}\phi_g) \\ 0 \\ 1 \end{bmatrix} \tag{5.11}$$

of the involute gear tooth profile that is specified in the coordinate system $X_gY_gZ_g$ associated with the gear.

Ultimately, generating surface $T$ of the end mill gear cutter for machining spur involute gears can be represented as a locus of consecutive positions of the involute gear tooth profile [see Eq. (5.11)] when the profile rotates $\omega_{cut}$ about the mill cutter axis of rotation $O_{ct}$. When deriving the equation for the generating surface $T$, it is convenient to use the operator of rotation $\mathbf{Rt}(\theta, Y_g)$ about the $Y_g$ – axis: This coordinate axis is aligned with the axis of rotation of the mill cutter. In this manner, the equation for the generating surface $T$ can be presented in a generalized form as the product:

$$\mathbf{r}^g_T(\phi_g, \theta) = \mathbf{Rt}(\theta, Y_g) \cdot \mathbf{r}^g_{inv}(\phi_g) \tag{5.12}$$

Multiplying the vector $\mathbf{r}^g_{inv}(\phi_g)$ [see Eq. (5.11)] at left by:

$$\mathbf{Rt}(\theta, Y_g) = \begin{bmatrix} \cos\theta & 0 & \sin\theta & 0 \\ 0 & 1 & 0 & 0 \\ -\sin\theta & 0 & \cos\theta & 0 \\ 0 & 0 & 0 & 1 \end{bmatrix} \tag{5.13}$$

returns the equation:

$$\mathbf{r}^g_T(\phi_g, \theta) = \begin{bmatrix} \dfrac{r_{b.g}}{\cos\phi_g} \cdot \sin(\alpha + \text{inv}\phi_g) \cdot \cos\theta \\ -\dfrac{r_{b.g}}{\cos\phi_g} \cdot \cos(\alpha - \text{inv}\phi_g) \\ \dfrac{r_{b.g}}{\cos\phi_g} \cdot \sin(\alpha + \text{inv}\phi_g) \cdot \sin\theta \\ 1 \end{bmatrix} \tag{5.14}$$

of the generating surface $T$ of the end mill gear cutter for machining spur gears.

The generating surface $T$ of the end mill gear cutter [Eq. (5.14)] is a kind of surface of revolution. The axis of rotation $O_{ct}$ of the cutter is aligned with the $Y_{ct} \equiv Y_g$ – axis of the reference system $X_{ct}Y_{ct}Z_{ct}$. Here, $Y_{ct}$ denotes the axis of the coordinate system $X_{ct}Y_{ct}Z_{ct}$ associated with the end mill gear cutter.

The generating surface of the end mill gear cutter for machining spur gears is expressed in terms of two parameters—that is, of the profile angle $\phi_g$ of the involute gear tooth profile at a point of interest, and the angle $\theta$ of rotation of the involute tooth profile about the centerline of the mill cutter ($-\infty < \theta < +\infty$).

### 5.2.2 Equation of Generating Surface of End Mill Gear Cutter for Machining Helical Involute Gears

When milling helical gears, the line of contact between the helical gear tooth flank $\mathscr{G}$ and the generating surface $T$ of the end mill gear cutter is a 3D spatial curve. No analytical description of the spatial line of contact is known at this point. Because of this, the earlier implemented approach [see section 5.2.1] cannot be used in deriving an equation for the generating surface $T$ of the end mill gear cutter for machining helical involute gears.

It is helpful to use the methods developed in the theory of enveloping surfaces to solve the problem of profiling with an end mill gear cutter. To solve this problem, two main reference systems are applied. The first coordinate system is associated with the helical gear to be machined and the second with the mill cutter. It should be stressed that, at this point, the mill cutter does not yet exist. Therefore, we are operating only with a coordinate system that the mill cutter will be associated with and not with the mill cutter itself.

When milling a helical gear, the mill cutter performs two motions: (1) a rotation $\omega_{cut}$ of the mill cutter about its axis of rotation $O_{ct}$ and (2) a screw motion of the mill cutter in relation to the work gear. The screw relative motion is composed of the translation $\mathbf{F}_c$ and the rotation $\omega_{fc}$ (see Figure 5.2).

The screw relative motion causes sliding of the helical gear tooth flank $\mathscr{G}$ on itself. Under the screw relative motion, the helical gear tooth flank $\mathscr{G}$ slides on it. This relative motion does not affect the shape or parameters of the generating surface $T$ of the mill cutter. Therefore, this motion can be eliminated from the list of considerations when deriving an equation for the generating surface $T$ of the end mill gear cutter. Ultimately, the generating surface of the end mill gear cutter can be viewed as the envelope to consecutive positions of the helical gear tooth flank $\mathscr{G}$ that is rotating about to the mill cutter axis of rotation.

An equation for the generating surface of the end mill gear cutter can be derived in three steps.

- First, analytically describe the helical gear tooth flank $\mathscr{G}$ in an appropriate coordinate system. For this purpose, the equation derived earlier for the screw involute surface [see Eq. (1.4)] can be used.
- Second, derive an equation for the gear tooth flank $\mathscr{G}$ in its current location with respect to the coordinate system associated with the mill cutter. In this step, as well as in the previous step, implementation of the operators of coordinate system transformations is convenient (see Appendix D).
- Third, compose a set of two equations that specify the enveloping surface. The solution to the set of two equations returns an equation of the generating surface $T$ of the end mill gear cutter for machining helical gears.

Similar to the way the equation for the generating surface $T$ of the end mill gear cutter for machining spur gears (see Figure 5.4) was derived earlier in this section of the book, the screw involute tooth flank surface $\mathscr{G}$ should be analytically described [see Eq. (1.4)]:

$$\mathbf{r}_g(U_g, V_g) = \begin{bmatrix} r_{b.g}\cos V_g + U_g \cos\lambda_{b.g}\sin V_g \\ r_{b.g}\sin V_g - U_g \sin\lambda_{b.g}\sin V_g \\ r_{b.g}\tan\lambda_{b.g} - U_g \sin\lambda_{b.g} \\ 1 \end{bmatrix} \quad \begin{matrix} V_g^{(l)} \le V_g \le V_g^{(a)} \\ 0 \le U_g \le [U_g] \end{matrix} \tag{5.15}$$

in a certain coordinate system $X_gY_gZ_g$. The $Y_g$ – axis of the coordinate system $X_gY_gZ_g$ is aligned with the line of symmetry of the transverse profile of the gear space width.

Use of the operator of rotation $\mathbf{Rt}(\alpha_t, Z_g)$ makes it possible to derive an equation for the tooth flank $\mathscr{G}$ with respect to the reference system associated with the work gear. The angle $\alpha_t$ that is used in the operator of rotation $\mathbf{Rt}(\alpha_t, Z_g)$ is measured in the transverse section of the helical gear. This angle is analogous to the angle $\alpha$ in the linear transformation operator $\mathbf{Rt}(\alpha, Z_g)$ [see Eq. (5.10)].

The following equation:

$$\mathbf{r}_g^*(U_g, V_g) = \mathbf{Rt}(\alpha_t, Z_g)\cdot \mathbf{r}_g(U_g, V_g) \tag{5.16}$$

describes the gear tooth flank $\mathscr{G}$ in the required reference system $X_gY_gZ_g$.

Equation (5.16) is the solution to the first step of the routing of derivation of the equation for the generating surface $T$ of the end mill gear cutter.

To derive an expression that describes the gear tooth flank $\mathscr{G}$ in its current location with respect to the coordinate system associated with the mill cutter, a Cartesian reference system $X_{ct}Y_{ct}Z_{ct}$ is introduced (see Figure 5.5). The coordinate system $X_{ct}Y_{ct}Z_{ct}$ is turned through a certain angle $\theta$ about the $Y_g$ – axis of the coordinate system $X_gY_gZ_g$. This angle $\theta$ allows for interpretation in the form $\theta = \omega_{cut}\cdot t$, where $\omega_{cut}$ denotes the rotation of the mill cutter about its axis when machining the work gear, and time is denoted by $t$.

After it has been designed, the end mill gear cutter will be associated with the coordinate system $X_{ct}Y_{ct}Z_{ct}$.

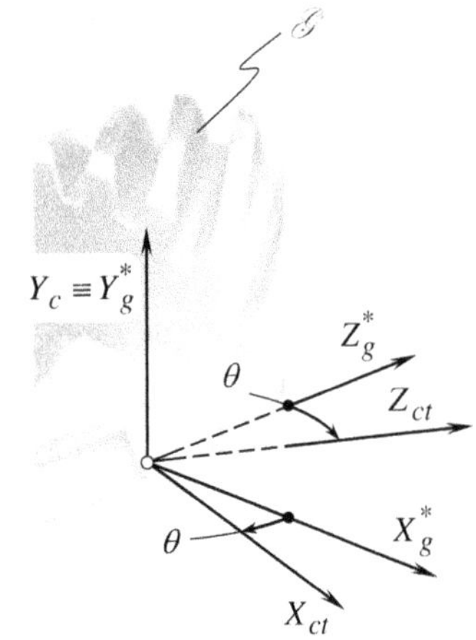


**FIGURE 5.5** Configuration of the reference system $X_{ct}Y_{ct}Z_{ct}$ associated with the end mill gear cutter relative to the helical work gear.

When designing the mill cutter, only the relative motion of the coordinate systems is important. It makes no difference whether the mill cutter coordinate system $X_{ct}Y_{ct}Z_{ct}$ is rotating about its axis $O_{ct} \equiv Y_g$ or the tooth flank $\mathscr{G}$ of the helical gear is rotated about the $Y_g$ – axis.

The position vector of a point $\mathbf{r}_g^c$ of the work gear tooth flank $\mathscr{G}$ in its current location is analytically described by:

$$\mathbf{r}_g^c(U_g, V_g, \theta) = \mathbf{Rt}(\theta, Y_g) \cdot \mathbf{r}_g^*(U_g, V_g) \tag{5.17}$$

Equation (5.16) presents the solution to the second step of routing the derivation of equation of the generating surface $T$ of the end mill gear cutter.

The third step in deriving the equation of the generating surface of the mill cutter can be carried out in two ways. One way is based on elements of the theory of enveloping surfaces, while the other approach is based on the kinematic properties of a moving surface $\mathscr{G}$.

The first (traditional) method is developed in the differential geometry of surfaces. Following this method, the equation of the generating surface $T$ of the end mill gear cutter can be derived as the solution to the set of two equations:

$$\mathbf{r}_g^c = \mathbf{r}_g^c(U_g, V_g, \theta) \tag{5.18}$$

$$\begin{vmatrix} \dfrac{\partial \mathbf{r}_g^c}{\partial U_g} & \dfrac{\partial \mathbf{r}_g^c}{\partial U_g} & \dfrac{\partial \mathbf{r}_g^c}{\partial U_g} \\ \dfrac{\partial \mathbf{r}_g^c}{\partial V_g} & \dfrac{\partial \mathbf{r}_g^c}{\partial V_g} & \dfrac{\partial \mathbf{r}_g^c}{\partial V_g} \\ \dfrac{\partial \mathbf{r}_g^c}{\partial \theta} & \dfrac{\partial \mathbf{r}_g^c}{\partial \theta} & \dfrac{\partial \mathbf{r}_g^c}{\partial \theta} \end{vmatrix} = 0 \tag{5.19}$$

There are three unknowns in the set of Eqs. (5.18) and (5.19): $U_g$, $V_g$, and $\theta$. Two parameters alone are sufficient for an analytical description of a surface. To eliminate the parameter $\theta$, solve Eq. (5.19) with respect to the angle $\theta$. Next, substitute the derived expression for the parameter $\theta$ into Eq. (5.18). After all necessary formula transformations are executed, the equation for the generating surface $T$ of the end mill gear cutter for machining involute gears casts into the equation:

$$\mathbf{r}_{ct} = \mathbf{r}_{ct}(U_{ct}, V_{ct}) \tag{5.20}$$

The second method of deriving an equation for the generating surface $T$ of the mill cutter is based on implementation of the Shishkov equation of contact[2] for moving surfaces [203, 204]:

$$\mathbf{n}_g \cdot \mathbf{V}_\Sigma = 0 \tag{5.21}$$

where $\mathbf{n}_g$ is the common perpendicular to the moving surface and its enveloping surface, and $\mathbf{V}_\Sigma$ is the vector of the resultant motion of the moving surface in relation to the enveloping surface at point(s) of the surfaces contact.

The Shishkov equation of contact is the basic equation of the so-called kinematic method of surface generation. In compliance with the kinematic method, at the points of contact of two surfaces, which move in relation to each other (that is, at the points of contact of the moving tooth flank $\mathscr{G}$ and the enveloping surface that is generated by the moving tooth flank $\mathscr{G}$), the vector of the common perpendicular $\mathbf{n}_g$ and the vector of the relative motion $\mathbf{V}_\Sigma$ of the surfaces must be orthogonal to each other.

When the tooth flank of a helical gear rotates about the mill cutter axis, only those points of tooth flank $\mathscr{G}$ generate the surface $T$, at which point the equation of contact is fulfilled.

Equation (5.17) allows for the calculation of the unit normal vector $\mathbf{n}_g$ at a point of interest of the helical gear tooth flank $\mathscr{G}$:

$$\mathbf{n}_g(U_g, V_g, \theta) = \frac{\mathbf{U}_g \times \mathbf{V}_g}{|\mathbf{U}_g \times \mathbf{V}_g|} = \mathbf{u}_g \times \mathbf{v}_g \tag{5.22}$$

For the given parameterization of Eq. (5.17), the vector $\mathbf{V}_\Sigma$ of the resultant motion of a point of the tooth flank $\mathscr{G}$ is calculated from the formula $\mathbf{r}_g^c(U_g, V_g, \theta)$:

$$\mathbf{V}_\Sigma(U_g, V_g, \theta) = \frac{\partial \mathbf{r}_g^c}{\partial \theta} \tag{5.23}$$

Equations (5.22) and (5.23) yield the equation of contact ($\mathbf{n}_g \cdot \mathbf{V}_\Sigma = 0$) in the form:

$$\mathbf{u}_g \times \mathbf{v}_g \cdot \frac{\partial \mathbf{r}_g^c}{\partial \theta} = 0 \tag{5.24}$$

[2] (a) Shishkov, V.A., "Elements of Kinematics of Generating and Conjugating in Gearing", in: *Theory and Calculation of Gears*, Vol. 6, Leningrad: LONITOMASH, 1948.
(b) Shishkov, V.A., *Generation of Surfaces Using Continuous Indexing Methods of Part Surface Machining*, Moscow, Mashgiz, 1951, 152 pages, as well as earlier publications on the subject by that same Author.

The solution to Eq. (5.18) and (5.24):

$$\mathbf{r}_g^c = \mathbf{r}_g^c(U_g, V_g, \theta) \tag{5.25}$$

$$\mathbf{u}_g \times \mathbf{v}_g \cdot \frac{\partial \mathbf{r}_g^c}{\partial \theta} = 0 \tag{5.26}$$

returns an equation for the generating surface $T$ of the end mill gear cutter for machining helical involute gears.

After all the necessary formula transformations have been performed, the equation for the surface $T$ can be presented in the form:

$$\mathbf{r}_{ct} = \mathbf{r}_{ct}(U_{ct}, V_{ct}) \tag{5.27}$$

which is identical to the earlier derived Eq. (5.20).

This method is referred to as the kinematic method of determination of enveloping surfaces.

The generating surface of end mill gear cutter [see Eqs. (5.20) and (5.27)] is a kind of surface of revolution. The axis of rotation $O_{ct}$ of the mill cutter is aligned with the $Y_{ct} \equiv Y_g$ – axis of the reference system $X_{ct}Y_{ct}Z_{ct}$. Here, $Y_{ct}$ denotes the axis of the coordinate system $X_{ct}Y_{ct}Z_{ct}$ associated with the mill cutter.

Very similar to what has just been discussed, corresponding equations for generating surfaces of end mill gear cutters for machining spur and helical gears with non-involute tooth profiles can be derived. Various tooth profiles can be machined with end mill gear cutters, except those with asymmetrical tooth profiles.

### 5.2.3 Elements of Intrinsic Geometry of Generating Surface of End Mill Gear Cutter

Formulae for calculation of the major elements of the local geometry of the generating surface of the gear-cutting tool can be significantly simplified if they are expressed in terms of fundamental magnitudes of the first and the second order of the generating surface $T$.

Position vector of a point of the generating surface $T$ of a gear-cutting tool can be initially specified by a vector equation $\mathbf{r}_{ct} = \mathbf{r}_{ct}(U_{ct}, V_{ct})$. Two tangent vectors $\mathbf{U}_{ct} = \partial \mathbf{r}_{ct} / \partial U_{ct}$ and $\mathbf{V}_{ct} = \partial \mathbf{r}_{ct} / \partial V_{ct}$ can be calculated at a point of the generating surface $T$. The vectors $\mathbf{U}_{ct}$ and $\mathbf{V}_{ct}$ are used to calculate for the fundamental magnitudes of the first order:

$$E_{ct} = \mathbf{U}_{ct} \cdot \mathbf{U}_{ct} \tag{5.28}$$

$$F_{ct} = \mathbf{U}_{ct} \cdot \mathbf{V}_{ct} = \mathbf{V}_{ct} \cdot \mathbf{U}_{ct} \tag{5.29}$$

$$G_{ct} = \mathbf{V}_{ct} \cdot \mathbf{V}_{ct} \tag{5.30}$$

The vectors $\mathbf{U}_{ct}$ and $\mathbf{V}_{ct}$ allow the users to compose a matrix $\mathbf{A}_{ct}$ of the first derivatives:

$$\mathbf{A}_{ct} = [\mathbf{U}_{ct} | \mathbf{V}_{ct}] \tag{5.31}$$

The use of this matrix leads to the following expression for the first fundamental matrix $\mathbf{G}_{ct}$:

$$\mathbf{G}_{ct} = \mathbf{A}_{ct}^T \cdot \mathbf{A}_{ct} = \begin{bmatrix} E_{ct} & F_{ct} \\ F_{ct} & G_{ct} \end{bmatrix} \tag{5.32}$$

The importance of this matrix is explained in the following discussion.

By definition, fundamental magnitudes of the second order can be presented in the form:

$$L_{ct} = \frac{\frac{\partial \mathbf{U}_{ct}}{\partial U_{ct}} \times \mathbf{U}_{ct} \cdot \mathbf{V}_{ct}}{H_{ct}} \tag{5.33}$$

$$M_{ct} = \frac{\frac{\partial \mathbf{U}_{ct}}{\partial V_{ct}} \times \mathbf{U}_{ct} \cdot \mathbf{V}_{ct}}{H_{ct}} = \frac{\frac{\partial \mathbf{V}_{ct}}{\partial U_{ct}} \times \mathbf{U}_{ct} \cdot \mathbf{V}_{ct}}{H_{ct}} \tag{5.34}$$

$$N_{ct} = \frac{\frac{\partial \mathbf{V}_{ct}}{\partial V_{ct}} \times \mathbf{U}_{ct} \cdot \mathbf{V}_{ct}}{H_{ct}} \tag{5.35}$$

Similar to Eq. (5.32), an expression for the second fundamental matrix can be composed:

$$\mathbf{D}_{ct} = \begin{bmatrix} \mathbf{n}_{ct} \cdot \frac{\partial \mathbf{U}_{ct}}{\partial U_{ct}} & \mathbf{n}_{ct} \cdot \frac{\partial \mathbf{U}_{ct}}{\partial V_{ct}} \\ \mathbf{n}_{ct} \cdot \frac{\partial \mathbf{V}_{ct}}{\partial U_{ct}} & \mathbf{n}_{ct} \cdot \frac{\partial \mathbf{V}_{ct}}{\partial V_{ct}} \end{bmatrix} \tag{5.36}$$

Since $\partial \mathbf{U}_{ct} / \partial V_{ct} = \partial \mathbf{V}_{ct} / \partial U_{ct}$ for all of the surfaces we are concerned with, matrix $\mathbf{D}_{ct}$ will be symmetric:

$$\mathbf{D}_{ct} = \begin{bmatrix} L_{ct} & M_{ct} \\ M_{ct} & N_{ct} \end{bmatrix} \tag{5.37}$$

The calculated values of the fundamental magnitudes of the first and the second order allow for significant simplification of the analysis of generating surface of the gear-cutting tool.

## 5.3 CUTTING EDGE OF END MILL GEAR CUTTER

End mill gear cutters can be considered a kind of form mill cutters. The determination of the generating surface of the mill cutter is just a prerequisite for designing an end mill gear cutter for machining either an involute gear or a gear with a non-involute tooth profile. Determination of the cutting edge of the end mill gear cutter is the next step in designing the gear-cutting tool.

Cutting edges of a precision mill cutter are located within the generating surface $T$ of the mill cutter. Any deviation of the cutting edge from the generating surface $T$ entails the corresponding deviation of the machined gear tooth flank $\mathscr{G}$.

On the other hand, cutting edge $CE$ is an element of the cutting wedge that is formed by the rake surface and the clearance surface. The cutting edge $CE$ can be viewed as the line of intersection of the rake surface $R_s$ of the mill cutter tooth, and the clearance surface $C_s$ of the mill cutter tooth. In addition, this line of intersection has to be located within the generating surface $T$. Ultimately, three surfaces—the rake surface $R_s$, the clearance surface $C_s$, and the generating surface $T$—pass through the cutting edge $CE$ of the end mill gear cutter. Physically, the cutting edge aligns with the line of intersection of the surfaces $R_s$ and $C_s$, as the generating surface of the mill cutter does not exist in reality (this surface is a virtual surface). However, virtually the cutting edge $CE$ must be located within the surface $T$ of the end mill gear cutter.

This brief discussion reveals the importance of the geometry of the rake surface $R_s$ and the clearance surface $C_s$ to the cutting tool designer.

### 5.3.1 Rake Surface of Mill Cutter for Machining Involute Gears

The location of the cutting edges of the mill cutter within the generating surface $T$ is one of the major requirements that a precision mill cutter must comply with. Any curve of a reasonable geometry that is entirely located within the generating surface $T$ is capable of precisely generating the involute gear tooth flank $\mathscr{G}$. However, the shape of the cutting edge must be convenient for manufacturing purposes. The rake surface and the clearance surface that physically form the cutting edge have to be machined somehow. The accuracy of the machining of the surfaces $R_s$ and $C_s$ needs to be inspected, and the regrinding of the worn mill cutter must be simple. For these reasons, as well as other factors, it is not practical to include all cutting edges of every shape and geometry. In the following, preference is given only to the shapes of the cutting edges, formed by the surfaces $R_s$ and $C_s$, that are practical and simple to machine and inspect.

There are two kinds of surfaces of are practical to use as the rake surfaces of the end mill gear cutters: a plane and an Archimedean screw surface. Implementation of both surfaces provides the mill cutter with a zero rake angle. When regrinding worn cutting tools, mill cutters with a zero rake angle are simpler to machine. The cutting edge of the mill cutter for machining spur gear is identical to the involute tooth profile of the gear to be machined, which significantly simplifies inspection of the mill cutter. However, in terms of the length of the tool's life, mill cutters with a zero rake angle do not offer maximum durability.

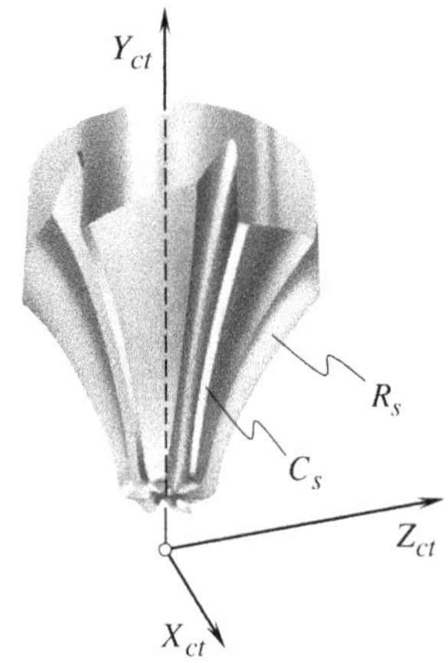


**FIGURE 5.6** Rake surface $R_s$ and clearance surface $C_s$ of precision mill cutter for finishing involute gears.

#### 5.3.1.1 Rake Surface in Form of Plane

The rake surface $R_s$ of a precision mill cutter for finishing an involute gear can be shaped in a form of plane. For the mill cutter with a zero rake angle, the rake plane is a plane through the centerline $O_{ct}$ of the cutter (Figure 5.6). In Figure 5.6, the mill cutter axis $O_{ct}$ is aligned with the $Y_{ct}$ – axis of the coordinate system $X_{ct}Y_{ct}Z_{ct}$ associated with the mill cutter. Any plane through the axis $O_{ct}$ could serve as the rake plane of the mill cutter. For example, the coordinate plane $X_cY_c$ of the coordinate system $X_{ct}Y_{ct}Z_{ct}$—that is, the plane given by equation $Z_{ct} = 0$—can be employed as the rake plane. In such a scenario, the cutting edge $CE$ of the mill cutter is determined as the line of intersection of the generating surface $T$ of the mill cutter [see either Eq. (5.14) for mill cutters for machining spur gears or Eq. (5.20) for mill cutters for machining helical gears] and the rake plane $R_s$ ($Z_{ct} = 0$).

In deriving the equation for the cutting edge $CE$ of the gear mill cutter, the format of the analytical representation of the rake plane $R_s$ must be the same as that for the generating surface $T$ of the mill cutter. For this reason, the equation $Z_{ct} = 0$ is presented in matrix form:

$$R_s \quad \Rightarrow \quad \mathbf{r}_{rs}(X_{rs}, Y_{rs}) = \begin{bmatrix} X_{rs} \\ Y_{rs} \\ 0 \\ 1 \end{bmatrix} \tag{5.38}$$

where coordinates of the points of rake plane $R_s$ are expressed in terms of $X_{rs}$ and $Y_{rs}$.

Parameterizations of the generating surface $T$ and of the rake plane $R_s$ have to be compatible with each other when aiming for mutual consideration of the generating surface $T$ [see Eqs. (5.14) and (5.20)], together with the rake plane $R_s$ [see Eq. (5.38)]. Usually, this issue can be resolved by changing the kind of parameterization (by the so-called re-parametrization) either of the generating surface $T$ or of the rake surface $R_s$ (see Appendix C). For example, the parameters $X_{rs}$ and $Y_{rs}$ can be expressed in terms of $U_{ct}$ − and $V_{ct}$ − parameters as functions $X_{rs} = X_{rs}(U_{c2}, V_{c2})$ and $Y_{rs} = Y_{rs}(U_{c2}, V_{c2})$.

#### 5.3.1.2 Rake Surface in Form of *Screw* Archimedean Surface

The rake surface of a roughing mill cutter for machining involute gear is often shaped in the form of a screw Archimedean surface (see Figure 5.7). In the coordinate system $X_{ct}Y_{ct}Z_{ct}$ associated with the mill cutter, an equation of the position vector of a point $\mathbf{r}_{rs}$ for the rake surface $R_s$ allows matrix representation in the form:

$$R_s \quad \Rightarrow \quad \mathbf{r}_{rs}(\rho_{rs}, \theta) = \begin{bmatrix} \rho_{rs} \cdot \cos\theta \\ p_{rs} \cdot \theta \\ \rho_{rs} \cdot \sin\theta \\ 1 \end{bmatrix} \tag{5.39}$$

where $\rho_{rs}$ denotes the distance of a point of the rake surface $R_s$ from the axis of rotation $O_{ct}$ of the mill cutter $\theta$ is the polar angle, and $p_{rs}$ is the helical parameter of the screw Archimedean surface. In Figure 5.7, the axis $O_{ct}$ is aligned with the $Y_{ct}$ − axis of the coordinate system $X_{ct}Y_{ct}Z_{ct}$ associated with the mill cutter.

For the calculation of the fundamental magnitudes of the first order of the rake surface, $R_s$, Eq. (5.39) and Eqs. (5.28) through (5.30) can be used. In compliance with Eq. (5.32), one can derive the first fundamental matrix:

$$\mathbf{G}_{ct} = \begin{bmatrix} 1 & p_{rs} \\ p_{rs} & \rho_{rs} + p_{rs} \end{bmatrix} \tag{5.40}$$

Similarly, for the calculation of the fundamental magnitudes of the second order of the rake surface, $R_s$, Eq. (5.39) and Eqs. (5.33) through (5.35) can be used. Using Eq. (5.37), one can derive the second fundamental matrix

$$\mathbf{D}_{ct} = \begin{bmatrix} 0 & \dfrac{p_{rs}}{\sqrt{p_{rs}^2 + \rho_{rs}^2}} \\ \dfrac{p_{rs}}{\sqrt{p_{rs}^2 + \rho_{rs}^2}} & 0 \end{bmatrix} \tag{5.41}$$

for the rake surface $R_s$.

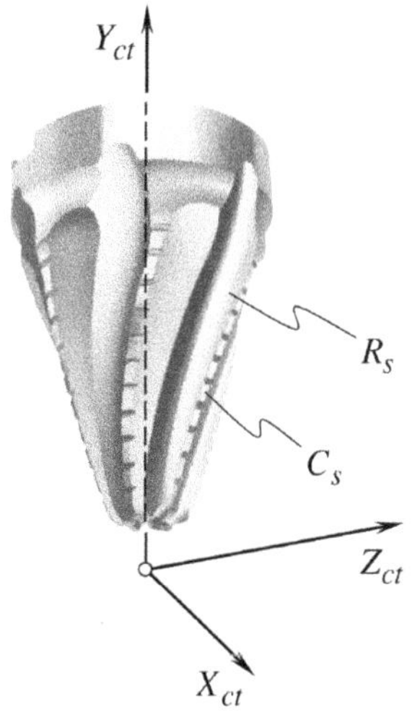


**FIGURE 5.7** Rake surface $R_s$ and clearance surface $C_s$ of a roughing mill cutter for machining of involute gears.

Parameterizations of the generating surface $T$ and the rake surface $R_s$ have to be compatible with each other when aiming mutual consideration of the generating surface $T$ [see Eqs. (5.14) and (5.20)] and the screw Archimedean rake surface $R_s$ [see Eq. (5.39)]. Because of this, for a point of interest of the rake surface $R_s$ the polar angle $\theta$ in Eq. (5.39) is of the same value as for the generating surface $T$ of the mill cutter [see Eq. (5.14)]. The polar distance $\rho_{rs}$ can be expressed in terms of the parameters $\phi_g$ and $\theta$ of the generating surface $T$ of the mill cutter by the expression:

$$\rho_{rs}(\phi_g, \theta) = \frac{r_{b.g}}{\cos\phi_g} \tag{5.42}$$

After substituting the expression for $\rho_{rs}$, Eq. (5.39) casts into:

$$R_s \quad \Rightarrow \quad \mathbf{r}_{rs}(\phi_g, \theta) = \begin{bmatrix} \dfrac{r_{b.g}}{\cos\phi_g} \cdot \cos\theta \\ p_{rs} \cdot \theta \\ \dfrac{r_{b.g}}{\cos\phi_g} \cdot \sin\theta \\ 1 \end{bmatrix} \tag{5.43}$$

Equations (5.14) and (5.43) can be solved together. The solution to Eq. (5.14) and (5.43) analytically describes the 3D spatial cutting edge $CE$ of the mill cutter. The equation for the cutting edge $CE$ of the end mill cutter can be presented in matrix form:

$$\mathbf{r}_{ce}(\vartheta_{ce}) = \begin{bmatrix} X(\vartheta_{ce}) \\ Y(\vartheta_{ce}) \\ Z(\vartheta_{ce}) \\ 1 \end{bmatrix} \tag{5.44}$$

where $\vartheta_{ce}$ is a parameter of the cutting edge $CE$ of the mill cutter.

When the generating surface $T$ of the mill cutter and the rake surface $R_s$ are expressed in terms of common parameters, one of the parameters can be expressed in terms of another one. In this manner, the parameter $\vartheta_{ce}$ of the cutting edge $CE$ of the mill cutter can be identified.

### 5.3.2 Clearance Surface of Mill Cutter for Machining Involute Gears

Clearance surface of a precision mill cutter is a surface through the cutting edge $CE$ that forms a clearance angle with the surface of cut $S_{cut}$. Unfortunately, no equation of the surface of cut $S_{cut}$ is known. Moreover, derivation of such an equation is bulky and thus inconvenient.

For the end mill cutters, the generating surface $T$ of the mill cutter is a perfect local approximation of the surface of cut. Therefore, two clearance angles are distinguished:

- The actual clearance angle $\alpha^{ac}$, which is the angle between the clearance surface $C_s$ and the surface of cut $S_{cut}$
- The calculated clearance angle $\alpha$, which is the angle between the clearance surface and the generating surface $T$ of the mill cutter

In most practical cases of milling involute gears, the difference between the clearance angles $\alpha^{ac}$ and $\alpha$ is negligibly small ($\Delta\alpha = \alpha^{ac} - \alpha \cong 0$). Thus, the generating surface $T$ of the mill cutter—and not the surface of cut $S_{cut}$—is used as the reference surface for the calculation of the rake angle of the end mill cutter.

As noted, the cutting edge $CE$ of the mill cutter can be interpreted as the line of intersection of threes surfaces—that is, of the:

- Generating surface $T$.
- Rake surface $R_s$.
- Clearance surface $C_s$.

Using the equation for cutting edge $CE$ [see Eq. (5.44)] allows for the construction of the clearance surface $C_s$ of the end mill cutter. In this method, the clearance surface $C_s$ is presented as a locus of consecutive positions of the cutting edge $CE$ that performs a certain motion.

In deriving the equation of clearance surface $C_s$, it is convenient to resolve the resultant instant motion $\mathbf{v}_{ce}$ of the cutting edge $CE$ onto three components [see Figure 5.8(a)]:

$$\mathbf{v}_{ce} = \mathbf{v}_t + \mathbf{v}_n + \mathbf{v}_{ax} \tag{5.45}$$

The component $\mathbf{v}_t$ is tangent to the generating surface $T$ of the mill cutter. This component is the linear velocity of the rotation $\boldsymbol{\omega}_{ce}$ of the cutting edge about the axis of rotation $O_{ct}$ of the mill cutter. For an arbitrary point $m$ within the cutting edge $AB$, the component $\mathbf{v}_t$ can be expressed in terms of two parameters: of the distance $r_m$, which represents the distance of point $m$ from the axis of rotation of the mill cutter $O_{ct}$ and of the angle $\theta_{ce}$—that is, of the angle through which the cutting edge turns about the axis of rotation $O_{ct}$ during certain time $t$. Angle $\theta_{ce}$ is equal to $\theta_{ce} = \omega_{ce} \cdot t$. Ultimately, the expression:

$$\mathbf{v}_t(\theta_{ce}) = \begin{bmatrix} \omega_{ce} \cdot r_m \cdot \cos\theta_{ce} \\ 0 \\ \omega_{ce} \cdot r_m \cdot \sin\theta_{ce} \\ 1 \end{bmatrix} \tag{5.46}$$

can be derived for the tangential component $\mathbf{v}_t$.

The normal component $\mathbf{v}_n$ is in the radial direction of the generating surface $T$ of the end mill gear cutter. At point $m$, the magnitude $|\mathbf{v}_n|$ of the component correlates to the magnitude $|\mathbf{v}_t|$ of the component $\mathbf{v}_t$ as:

$$|\mathbf{v}_n| = |\mathbf{v}_t| \cdot \tan\alpha_{xz} \tag{5.47}$$

where $\alpha_{xz}$ denotes the rake angle of the end mill gear cutter in the transverse cross section through $m$.

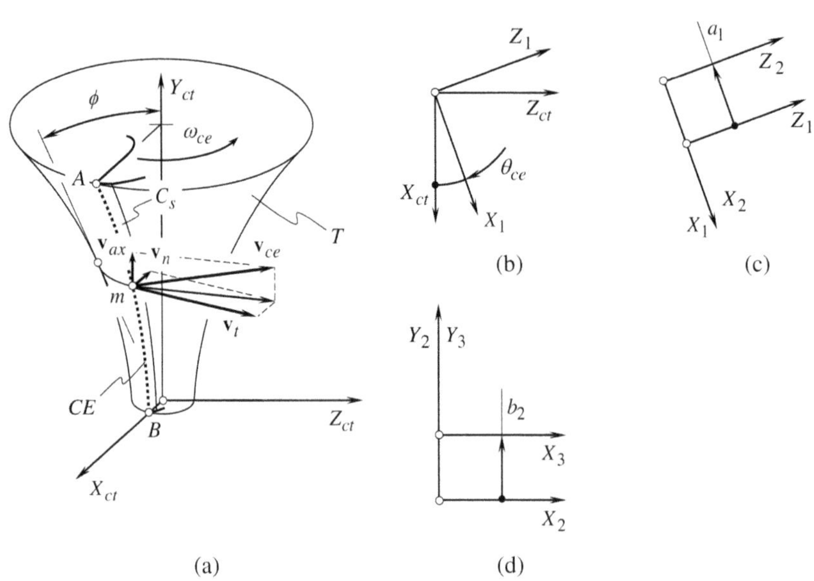


**FIGURE 5.8** Construction of the clearance surface $C_s$ of the end mill cutter: kinematics of generation of the clearance surface (a), and a set of Cartesian reference systems applied for the consequent coordinate system transformations (b) though (d).

For point $m$, an expression for component $\mathbf{v}_n$ yields representation in matrix form:

$$\mathbf{v}_n(\theta_{ce}) = \begin{bmatrix} -\omega_{ce} \cdot r_m \cdot \sin\theta_{ce} \cdot \tan\alpha_{xz} \\ 0 \\ -\omega_{ce} \cdot r_m \cdot \cos\theta_{ce} \cdot \tan\alpha_{xz} \\ 1 \end{bmatrix} \quad (5.48)$$

The component $\mathbf{v}_{ax}$ is pointed in the axial direction of the end mill gear cutter. The magnitude $|\mathbf{v}_{ax}|$ of the axial component $\mathbf{v}_{ax}$ somehow correlates to the magnitude $|\mathbf{v}_n|$ of the normal component $\mathbf{v}_n$. It has been proved practical to establish this correlation in the form:

$$|\mathbf{v}_{ax}| = |\mathbf{v}_n| \cdot \tan\phi \quad (5.49)$$

where the pressure angle on the pitch diameter of the work gear to be machined is denoted by $\phi$. If so, the point $m$ has to be coincident with the cutting edge point at which tangent to the axial profile of the surface $T$ forms the pressure angle $\phi$ with the cutter axis of rotation $O_{ct}$.

The axial component $\mathbf{v}_{ax}$ of the resultant motion $\mathbf{v}_{ce}$ of the cutting edge $CE$ is analytically described by the Eq. (5.49):

$$\mathbf{v}_{ax} = \begin{bmatrix} 0 \\ \omega_{ce} \cdot r_m \cdot \tan\alpha_{xz} \cdot \tan\phi \\ 0 \\ 1 \end{bmatrix} \quad (5.50)$$

Consider the resultant motion of the cutting edge $AB$ with respect to the coordinate system $X_{ct}Y_{ct}Z_{ct}$ associated with the body of the end mill gear cutter. Assume that the cutting edge is analytically described in the coordinate system $X_{ce}Y_{ce}Z_{ce}$. Initially, the coordinate system $X_{ce}Y_{ce}Z_{ce}$ is aligned to the coordinates system $X_{ct}Y_{ct}Z_{ct}$.

In compliance with Eq. (5.45), the resultant motion of the coordinate system $X_{ce}Y_{ce}Z_{ce}$ in relation to the coordinate system $X_{ct}Y_{ct}Z_{ct}$ can be resolved into three components—that is, onto $\mathbf{v}_t$, $\mathbf{v}_n$, and $\mathbf{v}_{ax}$.

The tangential component $\mathbf{v}_t$ of the resultant motion is caused by the rotation $\omega_{ce}$ of the cutting edge $CE$ about the axis of rotation $O_{ct}$ of the end mill cutter. After it has been turned through the angle $\theta_{ce}$ about the $Y_{ce}-$ axis, the coordinate system $X_{ce}Y_{ce}Z_{ce}$ occupies the first intermediate configuration $X_1Y_1Z_1$ [see Figure 5.8(b)]. Coordinate transformation of this type is analytically described by the operator of rotation $\mathbf{Rt}(\theta_{ce}, Y_{ct})$. In the case under consideration, the operator $\mathbf{Rt}(\theta_{ce}, Y_{ct})$ equals:

$$\mathbf{Rt}(\theta_{ce}, Y_{ct}) = \begin{bmatrix} \cos\theta_{ce} & 0 & -\sin\theta_{ce} & 0 \\ 0 & 1 & 0 & 0 \\ \sin\theta_{ce} & 0 & \cos\theta_{ce} & 0 \\ 0 & 0 & 0 & 1 \end{bmatrix} \quad (5.51)$$

The normal component $\mathbf{v}_n$ of the resultant motion causes the coordinate system $X_1Y_1Z_1$ to shift along its $X_1-$ axis at a certain distance $a_1$ to the position of the second intermediate configuration $X_2Y_2Z_2$ [see Figure 5.8(c)]. The current value of the distance $a_1$ can be calculated as:

$$a_1 = \theta_{ce} \cdot r_m \cdot \tan\alpha_{xz} \quad (5.52)$$

Therefore, the translation of the coordinates system $X_1Y_1Z_1$ to the position of the coordinate system $X_2Y_2Z_2$ is analytically described by the operator of translation $\mathbf{Tr}(a_1, X_1)$. For the end mill gear cutter, the operator of translation $\mathbf{Tr}(a_1, X_1)$ is derived:

$$\mathbf{Tr}(a_1, X_1) = \begin{bmatrix} 1 & 0 & 0 & \theta_{ce} \cdot r_m \cdot \tan\alpha_{xz} \\ 0 & 1 & 0 & 0 \\ 0 & 0 & 1 & 0 \\ 0 & 0 & 0 & 1 \end{bmatrix} \quad (5.53)$$

Ultimately, the axial component $\mathbf{v}_{ax}$ of the resultant motion results in the shift of the coordinate system $X_2Y_2Z_2$ along its $Y_2-$ axis at a certain distance $b_2$ to the position of the third coordinate system $X_3Y_3Z_3$ [see Figure 5.8(d)]. To calculate for the current value of the distance $b_2$, the expression:

$$b_2 = \theta_{ce} \cdot r_m \cdot \tan\alpha_{xz} \cdot \tan\phi \quad (5.54)$$

applies.

For the analytical description of translation of the coordinate system $X_2Y_2Z_2$ to the position of the coordinate system $X_3Y_3Z_3$, the operator of translation $\mathbf{Tr}(b_2, Y_2)$ can be composed. For the case under consideration, this operator of the linear transformation can be presented as:

$$\mathbf{Tr}(b_2, Y_2) = \begin{bmatrix} 1 & 0 & 0 & 0 \\ 0 & 1 & 0 & -\theta_{ce} \cdot r_m \cdot \tan\alpha_{xz} \cdot \tan\phi \\ 0 & 0 & 1 & 0 \\ 0 & 0 & 0 & 1 \end{bmatrix} \quad (5.55)$$

The equation of the clearance surface $C_s$ of the end mill gear cutter yields representation in the form of dot product of two expressions. To come up with an equation for the position vector $\mathbf{r}_{cs}(\vartheta_{ce}, \theta_{ce})$ of point of interest of the clearance surface $C_s$, it is necessary to pre-multiply the position vector $\mathbf{r}_{ce}(\vartheta_{ce})$ of point of the cutting edge $CE$ [see Eq. (5.44)] by the operator $\mathbf{Rs}(3 \mapsto T)$ of the resultant coordinate system transformation:

$$\mathbf{r}_{cs}(\vartheta_{ce}, \theta_{ce}) = \mathbf{Rs}(3 \mapsto T) \cdot \mathbf{r}_{ce}(\vartheta_{ce}) \quad (5.56)$$

Here, the operator of the resultant linear transformation $\mathbf{Rs}(3 \mapsto T)$ can be presented as the product of the operators of elementary coordinate systems transformations:

$$\mathbf{Rs}(3 \mapsto T_2) = \mathbf{Tr}(b_2, Y_2) \cdot \mathbf{Tr}(a_1, X_1) \cdot \mathbf{Rt}(\theta_{ce}, Y_c) \quad (5.57)$$

Finally, the equation for the clearance surface $C_s$ [see Eq. (5.56)] is expressed in terms of two independent parameters $\vartheta_{ce}$ and $\theta_{ce}$. The location of point of interest within the cutting edge $AB$ is specified by the parameter $\vartheta_{ce}$, whereas the current location of the cutting edge within the clearance surface is specified by the parameter $\theta_{ce}$.

### 5.3.3 Cutting Edge Geometry of the End Mill Gear Cutter

The cutting wedge of an end mill gear cutter is comprised by the rake surface $R_s$ and of the clearance surface $C_s$. For the determination of the cutting edge geometry, the configuration of the cutting wedge in relation to the surface of cut $S_{cut}$, as well as with respect to the speed of cut $\mathbf{V}_{cut}$, needs to be specified. When determining the cutting edge geometry, it is assumed that the speed of cut $\mathbf{V}_{cut}$ is equal to the speed $\mathbf{V}_\Sigma$ of the relative motion of a point of interest within the cutting edge $CE$ relative to the surface of cut. The actual configuration of the cutting wedge relative to the surface of cut, and to the vector $\mathbf{V}_{cut}$ is specified by the geometrical parameters of the cutting edge. Rake angle $\gamma$, clearance angle $\alpha$, and angle of inclination $\lambda$ of the cutting edge $CE$ are of prime importance to end mill gear cutter for machining involute gears.

To achieve the best possible performance of the end mill gear cutter, all the geometrical parameters of the cutting edge are necessary to be equal to their optimal values. Performance of the gear-cutting tool, for example, tool life $TL$ of the end mill gear cutter, strongly depends on the actual values of the geometrical parameters of the cutting edge. A function of the gear-cutting tool performance versus the actual value of the geometrical parameters of the cutting edge features a maximum, is shown in Figure 5.9. The graph of the function $TL = TL(\alpha)$ has a maximum $TL = TL_{\max}$, which is observed when the corresponding geometrical parameter $\alpha$ of the cutting edge is equal to its optimal value $\alpha = \alpha_{opt}$. This means that there always exists an optimal value of each geometrical parameter (namely, of $\alpha$, $\gamma$, $\lambda$, and so forth) under which tool performance of the gear-cutting tool reaches its maximal possible value. For example, if the clearance angle of the end mill gear cutter is equal to its optimal value $\alpha_{opt}$, this corresponds to a maximum of the function $TL = TL(\alpha)$ in Figure 5.9. In designing an end mill gear cutter—or any type of cutting tool, for that matter—it is highly desirable to ensure that all the geometrical parameters are equal to their optimal values.

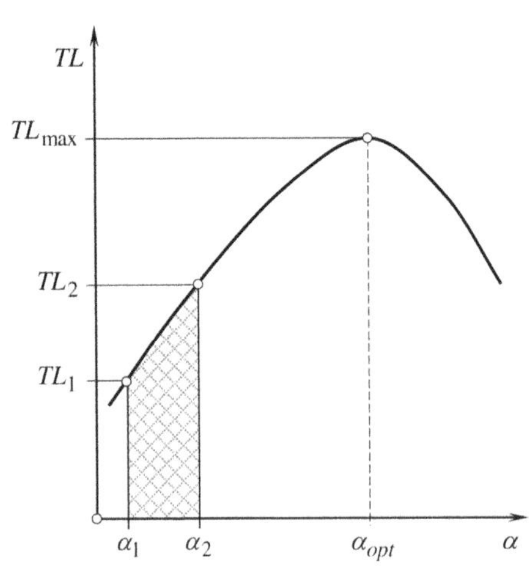


**FIGURE 5.9** Qualitative relation of cutting performances (the tool life $TL$) vs. clearance angle $TL = TL(\alpha)$ in the end mill gear cutter.

Unfortunately, because of numerous constraints, actual values of the geometrical parameters of cutting tools are too far from their optimal values. Therefore, gear-cutting tools are designed not with optimal values of the clearance angle $\alpha$, but with a clearance angle $\alpha$ that falls into a certain interval $\alpha_1 < \alpha < \alpha_2$, where the inequality $\alpha_2 \ll \alpha_{opt}$ is often observed. As a result, the actual tool life of the end mill gear cutter falls into a certain range $TL_1 < TL < TL_2$. The commonly achieved tool life $TL_2$ is often much shorter compared to the maximal possible tool life $TL_{\max}$. The shortened tool life reduces the efficiency of the gear-machining process, as gear-cutting tools of all kinds are costly and tool longevity is strongly desired.

Without going into details of optimality of the geometrical parameters of gear-cutting tools, it is necessary to stress on that:

- The optimal combination of geometrical parameters of the cutting edge can be determined for gear-cutting tool of any design.
- In designing the most economical gear-cutting tool, actual values of geometrical parameters of cutting edges must either be equal to their optimal values, or their deviation from optimal values must be the smallest possible.

For this purpose, it is necessary to calculate the actual values of the geometrical parameters at every point of the cutting edge of the gear-cutting tool.

In calculating the geometrical parameters at point of interest of the cutting edge of an end mill gear cutter, an infinitesimally short portion $dl_{ce}$ of the cutting edge is considered. This assumption allows us to neglect the curvature within the differential $dl_{ce}$ of the actually curved cutting edge $CE$.

#### 5.3.3.1 Rake Surface Is a Plane

The geometry of the end mill gear cutter, having rake surface $R_s$ in the form of a plane through the axis of rotation of the mill cutter, can be specified by three angles:

- Rake angle $\gamma = 0°$
- Angle of inclination of the cutting edge $\lambda = 0°$
- Clearance angle $\alpha > 0°$

Because both the rake angle $\gamma$ and angle of inclination $\lambda$ are of zero values ($\gamma = 0°$ and $\lambda = 0°$), no analysis of the geometry of the rake surface of the end mill gear cutter is necessary. To predict the performance of the end mill

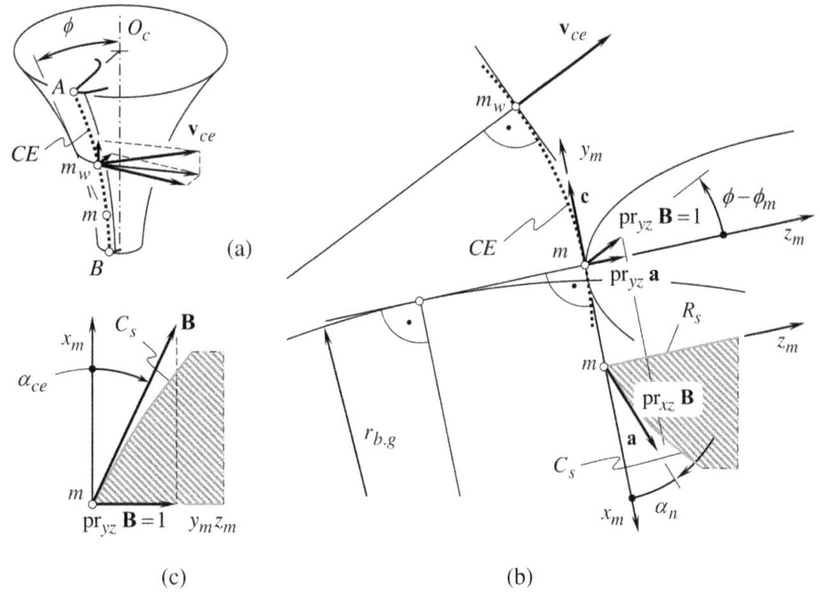


**FIGURE 5.10** Analysis of the geometry of clearance surface $C_s$ of a gear finishing end mill cutter: the geometry of clearance surface (a), and sections of the cutting wedge by the planes through a point of interest (b) and (c).

cutter, only the geometry of the clearance surface needs to be investigated.

It has been proven convenient to use a method of investigation of the cutting tool geometry based on the implementation of elements of vector algebra. To the best of the author's knowledge, Dr. S.S. Mozhayev [49] was the first (1948) to use elements of vector calculus for solving problems that relate to the geometry of the active part of cutting tools. The interested reader may wish to check other sources for details on the method [148, 158, 168].

Consider a cutting edge $CE$ of an end mill gear cutter shown in Figure 5.10(a). The actual value of the normal rake angle $\alpha_{ce}$ at a point $m_w$ within the cutting edge $AB$ of the end mill gear cutter is given.[3] It is required to derive an equation for the calculation of the normal rake angle $\alpha_n$ at a point of interest $m$ within the cutting edge of the end mill gear cutter.

A local Cartesian coordinate system $x_m y_m z_m$ is associated with the cutting edge $AB$. The origin of the reference system $x_m y_m z_m$ coincides with a point of interest $m$ within the cutting edge $AB$. The axis $x_m$ is pointed along the assumed direction of the speed of cut; the axis $y_m$ is tangent to the cutting edge $CE$ and the axis $z_m$ is perpendicular to the surface of cut $S_{cut}$. The surface of cut $S_{cut}$ is not determined yet. However, the generating surface $T$ of the end mill gear cutter is a perfect local approximation to the surface of cut $S_{cut}$. Therefore, instead, the $z_m$ – axis can be constructed perpendicular to the surface $T$ of the end mill gear cutter.

To derive an equation for the calculation of the normal rake angle $\alpha_n$, three vectors—namely, the vectors **a**, **B**, and **c**—are constructed. The vectors **a**, **B**, and **c**, have to be expressed in terms of the clearance angle $\alpha_n$ to be determined, and in terms of the known parameters of the geometry of the cutting wedge of the cutting tool.

Here and in the following, unit vectors are denoted by lowercase characters, whereas vectors of other magnitudes are denoted by uppercase characters. In reality, only the directions of the vectors are of critical importance. The length of the vectors can be arbitrary. Solely for the convenience of derivation of equations, magnitudes of some vectors were chosen equal to unit vectors, whereas magnitudes of other vectors of different lengths were chosen for specific cases.

Unit vector **a** is tangent to the line of intersection of the clearance surface $C_s$ by the plane perpendicular to the cutting edge $CE$. In the local coordinate system $x_m y_m z_m$, the vector **a** can be expressed by the equation [see Figure 5.10(b)]:

$$\mathbf{a} = \mathbf{i}\cos\alpha_n + \mathbf{k}\sin\alpha_n \tag{5.58}$$

The magnitude of the vector **B** is chosen as the length, and the projection of the vector **B** onto the coordinate plane $y_m z_m$ is equal to unit ($\text{pr}_{yz}\,\mathbf{B} = 1$). Hence, the following expression:

$$\mathbf{B} = \mathbf{i}\cot\alpha_{ce} + \mathbf{j}\sin(\phi - \phi_m) + \mathbf{k}\cos(\phi - \phi_m) \tag{5.59}$$

where $\alpha_{ce}$ denotes the clearance angle in the direction of $\mathbf{v}_{ce}$ [see Eq. (5.45)] ($\alpha_{ce}$ is equal to the normal clearance angle at point $m_w$ of the cutting edge); $\phi$ is the pressure angle of the gear to be machined and $\phi_m$ is the pressure angle at point $m$. The angle $\phi_m$ can be readily expressed in terms of the design parameters of the gear to be machined.

Unit vector **c** is aligned with the $y_m$ – axis of the coordinate system $x_m y_m z_m$ and simply expressed as:

$$\mathbf{c} = \mathbf{i} \tag{5.60}$$

[3] The subscript $w$ here is due to point $m_w$ of the cutting edge $CE$ is commonly corresponded to a point of the involute profile on the pitch diameter $d_w$ of the gear to be machined.

By construction, the vectors **a**, **B**, and **c** are situated within a common plane. Therefore, their triple scalar product is identical to zero (that is, $\mathbf{a} \times \mathbf{B} \cdot \mathbf{c} \equiv 0$). The triple scalar product of the vectors can be expressed in the form of a determinant. This immediately yields the following expression:

$$\mathbf{a} \times \mathbf{B} \cdot \mathbf{c} = \begin{vmatrix} \cos\alpha_n & 0 & \sin\alpha_n \\ \cot\alpha_{ce} & \sin(\phi - \phi_m) & \cos(\phi - \phi_m) \\ 0 & 1 & 0 \end{vmatrix} = 0 \quad (5.61)$$

Expression (5.61) casts into the formula:

$$\alpha_n(\phi_m) = \tan^{-1}[\tan\alpha_{ce} \cdot \cos(\phi - \phi_m)] \quad (5.62)$$

where the normal clearance angle $\alpha_n$ is expressed in terms of the pressure angle $\phi_m$ at a point of interest $m$, within the cutting edge $CE$. The actual value of the normal angle $\alpha_n$ can be expressed in terms of the design parameter of the gear to be machined. For this purpose, the equations listed in Appendix F apply. For convenience, pressure angle $\phi_m$ can be expressed in terms of the design parameters of the gear to be milled. For this purpose, the following formula applies:

$$\phi_m = \cos^{-1}\frac{r_{b.g}}{r_{m.g}} \quad (5.63)$$

where $r_{b.g}$ is the radius of base cylinder of the gear to be machined, and $r_{m.g}$ is the radius of the circle through point $m$ of the gear tooth profile.

Figure 5.11 illustrates an example of variation of the normal clearance angle $\alpha_n$ within the cutting edge $CE$ of the end mill gear cutter. The computations are performed for the end mill gear cutter having a clearance angle of $\alpha_{ce} = 12°$. The shape of the curves $\alpha_n = \alpha_n(m)$ depends on the number of teeth in the gear. Involute gears with a smaller number of teeth are subject to more severe changes in normal clearance angle $\alpha_n$. In the plot, "normal clearance angle vs. height of work gear tooth" does not cover the portion of the gear tooth within the transient curve at the bottom of the gear tooth space. This is because the shape of the fillet significantly differs from the true involute form of the gear tooth.

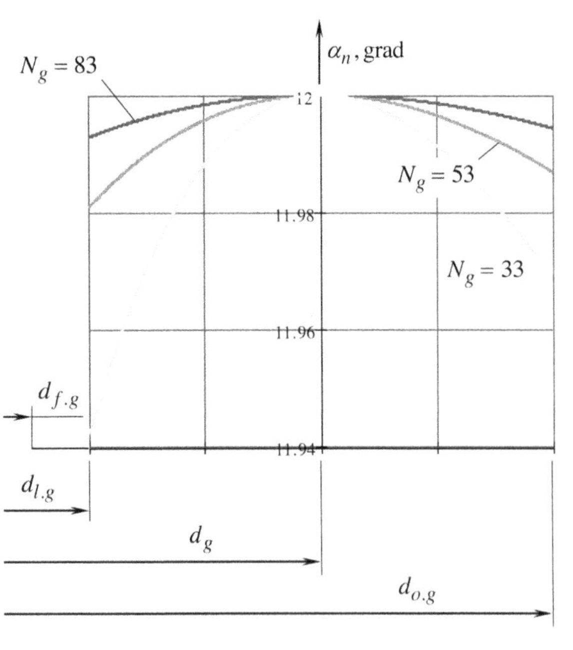


**FIGURE 5.11** Normal clearance angel $\alpha_n$ at point of interest $m$ of cutting edge $CE$ of the end mill gear cutter (normal clearance angle $\alpha_{ce}$ at point $m_w$ equals $\alpha_{ce} = 12°$).

### 5.3.3.2 Rake Surface Is a Screw Surface

Consider a roughing end mill gear cutter with a rake surface that is shaped in the form of a screw surface (see Figure 5.12). At point $m_w$, the helix angle $\psi$ of the rake face $R_s$ is given. It is necessary to derive an equation to calculate for the angle of inclination $\lambda$ of the cutting edge $CE$ at a point of interest $m$ within the cutting edge $AB$ of the end mill gear cutter.

By definition [148, 158, 168], the angle of inclination $\lambda$ of the cutting edge:

- is measured within the plane of cut $P_{cut}$ (the plane of cut is a plane that is tangent to the surface of cut $S_{cut}$ at point of interest $m$ within the cutting edge $CE$ of the end mill cutter).
- is equal to the angle that the vector of speed of cut $\mathbf{V}_{cut}$ forms with the perpendicular $\mathbf{n}_{ce}$ to the cutting edge $CE$ at point of interest $[\lambda = \angle(\mathbf{V}_{cut}, \mathbf{n}_{ce})]$.

As viewed from the end of the unit normal vector $\mathbf{n}_{se}$ to the surface of cut $S_{cut}$, the angle of inclination $\lambda$ has a positive

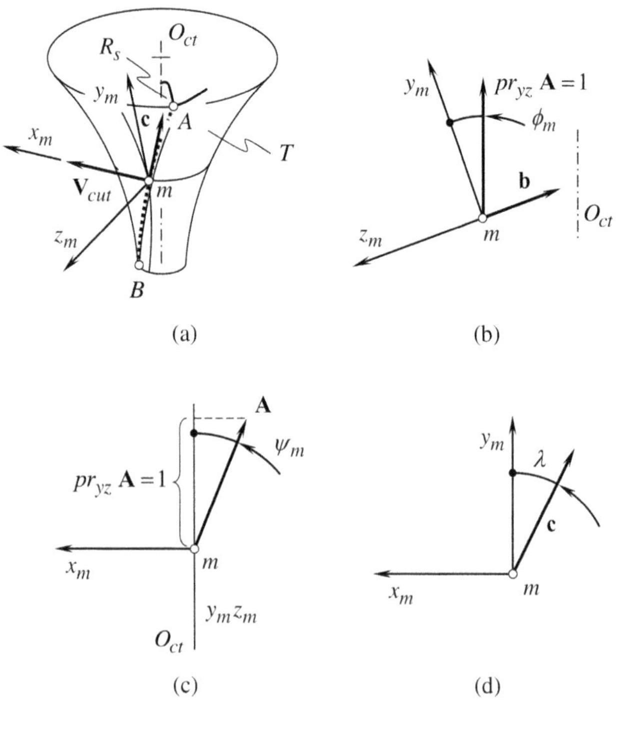


**FIGURE 5.12** Analysis of the geometry of the clearance surface $C_s$ of the roughing end mill gear cutter: the geometry of clearance surface (a), and sections of the cutting wedge by the planes of a local Cartesian coordinate system (b) through (d).

value when the motion from $\mathbf{V}_{cut}$ toward $\mathbf{n}_{ce}$ is counterclockwise. Otherwise, the angle of inclination is of a negative value. (See Appendix G ).

To derive a formula for calculation of the actual value of the angle of inclination $\lambda$ at a point of interest $m$ of the cutting edge $CE$ of the roughing end mill cutter, a set of three coplanar vectors $\mathbf{A}$, $\mathbf{b}$, and $\mathbf{c}$ through point $m$ is constructed. The vectors $\mathbf{A}$, $\mathbf{b}$, and $\mathbf{c}$ need to be expressed in terms of the angle of inclination $\lambda$ to be determined in terms of the angle of known parameters of the geometry of the cutting wedge of the cutting tool.

The vectors $\mathbf{A}$, $\mathbf{b}$, and $\mathbf{c}$ are analytically described in the local Cartesian reference system $x_m y_m z_m$ [see Figure 5.12(a)]. The coordinate system $x_m y_m z_m$ is identical to the reference system $x_m y_m z_m$ that was used earlier (see Figure 5.10) in this chapter of the book.

The family of helices can be drawn up on the screw rake surface $R_s$ of the end mill gear cutter. One of the helices is through point of interest $m$. Vector $\mathbf{A}$ is tangent to the screw line through point of interest $m$. The vector $\mathbf{A}$ is of a magnitude under which the projection of $\mathbf{A}$ onto the coordinate plane $y_m z_m$ is equal to unit, say $\mathrm{pr}_{yz}\,\mathbf{A} = 1$ [see Figure 5.12(b)]. Under such an assumption, an expression for the vector $\mathbf{A}$ [see Figure 5.12(c)] can be expressed in the form:

$$\mathbf{A} = \mathbf{i}\tan\psi_m + \mathbf{j}\cos\phi_m - \mathbf{k}\sin\phi_m \tag{5.64}$$

where $\psi_m$ is the helix angle of the screw rake surface $R_s$, and $\phi_m$ is the pressure angle at the point of interest $m$ within the cutting edge $CE$ of the end mill gear cutter.

The unit vector $\mathbf{b}$ is tangent to the rake surface $R_s$ of the end mill gear cutter and is pointed along the $z_m$ – axis toward axis of rotation $O_{ct}$ of the end mill cutter [see Figure 5.12(b)]. Therefore,

$$\mathbf{b} = -\mathbf{k} \tag{5.65}$$

Finally, the unit vector $\mathbf{c}$ is tangent to the cutting edge $CE$, and it is therefore also tangent to the surface of cut $S_{cut}$. For the vector $\mathbf{c}$, the following expression is composed [see Figure 5.12(d)]:

$$\mathbf{c} = -\mathbf{i}\sin\lambda + \mathbf{j}\cos\lambda \tag{5.66}$$

By construction, three vectors $\mathbf{A}$, $\mathbf{b}$, and $\mathbf{c}$ form a set of three coplanar vectors. Therefore, their triple scalar product is identical to zero (that is, $\mathbf{A}\times\mathbf{b}\cdot\mathbf{c} \equiv 0$). The triple scalar product of the vectors yields interpretation in the form of the determinant:

$$\mathbf{A}\times\mathbf{b}\cdot\mathbf{c} \equiv \begin{vmatrix} \tan\psi_m & \cos\phi_m & \sin\phi_m \\ 0 & 0 & -1 \\ -\sin\lambda & \cos\lambda & 0 \end{vmatrix} = 0 \tag{5.67}$$

After the determinant has been expanded, Eq. (5.67) casts into the formula:

$$\lambda = \tan^{-1}\left(-\frac{\tan\psi_m}{\cos\phi_m}\right) \tag{5.68}$$

for the calculation of the angle of inclination $\lambda$ of the cutting edge of the roughing end mill gear cutter.

In Eq. (5.68), both the parameters $\psi_m$ and $\phi_m$ can be expressed in terms of the design parameters of the end mill gear cutter (see Appendix G).

Helix angle $\psi_m$ at a point of interest $m$ within the cutting edge $CE$ of the end mill gear cutter can be expressed in terms of its known value $\psi$ at the pre-specified point $m_w$. It is common practice that point $m_w$ corresponds to the pitch diameter $d_{w.ct}$ of the end mill cutter. The end mill gear cutter is sliced by plane sections. All of the plane sections are perpendicular to the axis $O_{ct}$ of the end mill gear cutter. Outer diameter of the section through the point $m$ of the cutting edge is denoted by $d_{w.ct}$. For the case under consideration, the following expression:

$$\psi_m = \tan^{-1}\left(\frac{d_{ct}}{d_{w.ct}}\cdot\tan\psi\right) \tag{5.69}$$

is derived for the calculation of angle $\psi_m$.

To calculate for the end mill gear cutter geometry, it can be assumed that the outer diameter of the plane section through the point $m_w$ of the end mill gear cutter is equal to $d_{w.ct} = 0.5\pi m$ (where $m$ is the module of the end mill gear cutter).

To calculate for the diameter $d_{w.ct}$, the following equality:

$$\pi\cdot m - \frac{d_{w.ct}}{d_{g.m}} + \mathrm{inv}\phi_m = \frac{d_{ct}}{d_g} + \mathrm{inv}\phi \tag{5.70}$$

applies.

In the case of end mill cutters for machining internal involute gears, the equality:

$$\frac{d_{w.ct}}{d_{g.m}} + \mathrm{inv}\phi_m = \frac{d_{ct}}{d_g} + \mathrm{inv}\phi \tag{5.71}$$

is valid.

Pressure angle $\phi_m$ at a point of interest $m$ within the cutting edge $CE$ can be determined from Eq. (5.63).

We can calculate for the inclination angle $\lambda$ at a point of interest $m$ within the cutting edge of the roughing end mill gear cutter using the equations derived. An example of the computations is shown in Figure 5.13.

It is common practice to designate the lead angle in a helical gear as $\lambda$. It is also common practice to designate the angle of inclination of the cutting edge in a cutting tool as $\lambda$. To avoid confusion, it is preferred to specify the helix angle in a gear-cutting tool rather than its lead angle $\lambda$. Lead

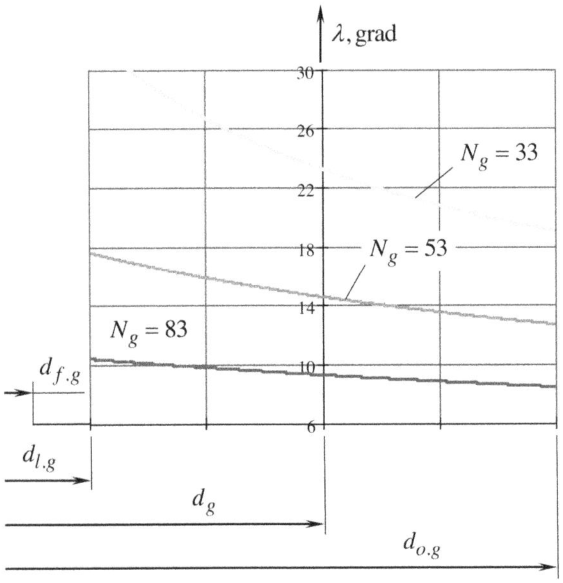


**FIGURE 5.13** Angle of inclination λ at a current point $m$ of the cutting edge of the end mill gear cutter (lead angle at point $M$ is 20°).

angle λ and helix angle ψ in a gear-cutting tool complement each other to 90°—that is, the equality ($\lambda + \psi = 90°$) is valid. Therefore, the screw rake surface $R_s$ of the roughing end mill gear cutter can be specified not by the lead angle λ, but by the helix angle ψ instead.

It is common practice to grind rake surfaces, both plane rake surfaces, as well as screw rake surfaces, for sharpening and re-sharpening of worn end mill gear cutters. Clearance surfaces are ground only on relieving operation. Commonly, clearance surfaces of the worn mill cutters are not reground by the user.

### 5.3.4 Design Features of End Mill Gear Cutters with Cutting Inserts

For machining involute gears of coarse-pitch, end mill gear cutters with cutting inserts are used. The cutting inserts are made up either of high-speed-steel, carbide alloys, cutting ceramics, or polycrystals of superhard materials. The cutting inserts feature cutting edges with short length—shorter than cutting edges of the end mill gear cutter. The cutting edges of end mill gear cutter of this kind (see Figure 5.14) are comprised of a plurality of the relatively short cutting edges of the cutting inserts that are installed on the body of the cutting tool.

The cutting inserts are shifted in relation to one another in order to overlap distinct cutting edges so as to be capable of providing smooth finish of the tooth flanks of the gear to be machined. A corresponding configuration of the cutting insert in relation to the cutting tool body is required in order to ensure the required values of cutting angles (the rake angle γ, the clearance angle $\alpha_{ce}$, and the angle of inclination of the cutting edge λ).

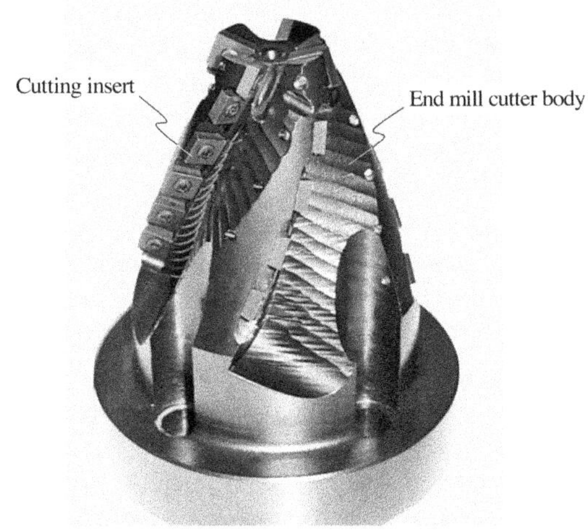


**FIGURE 5.14** A coarse-pitch end mill gear cutter with the cutting inserts made up of carbide alloy.

Many designs of end mill gear cutters with cutting inserts are developed in the industry. The use of the cutting inserts in the design of coarse diametral pitch end mill cutter make the gear production more economical, as the worn cutting inserts can be replaced in the body of the end mill cutter—there is no need to replace the cutting tool body. The coarser the diametral pitch of the gear to be cut, the more efficient the application of end mills with cutting inserts, and vice versa.

## 5.4 ACCURACY OF GEAR TOOTH FLANKS MACHINED BY END MILL CUTTERS

The accuracy of machined gears is a very challenging problem. This problem definitely deserves to be investigated and has been discussed in a separate volume. In this text, only particular problems relating to the accuracy of machined gears are discussed. These particular problems of gear machining could be referred to as the accuracy of generation of the gear tooth flanks. The major difference between the accuracy of gear machining and the accuracy of gear generation is as follows.

The deviation of a gear tooth flank from its desirable geometry and configuration is mostly due to the impact of physical, geometric, kinematic, and other factors of the machining operation. The gear generation encompasses deviations of the machined gear tooth flank that are caused only by geometric and kinematic factors of the gear-machining process. In contrast, the gear machining itself causes additional deviations of the machined gears tooth flank. This includes but is not limited to the impact of:

- Dynamics of the gear-machining operation.
- Tool wear.
- Heat generation and thermal extension of the elements of the machine tool.
- Elastic deformation of the elements of the machine tool.

Thus, the process of gear generation can be interpreted as a simplified model of the actual gear-machining operation. As shown in the next section, numerous useful results can be drawn up using the gear-generation process as the simplified model of the gear-machining process.

### 5.4.1 Background of Approach

Line contact between the gear tooth flank $\mathscr{G}$ and the generating surface $T$ of the mill cutter is observed when machining a gear using an end mill gear cutter. The line of contact of the surfaces $\mathscr{G}$ and $T$ is referred to as the characteristic $\mathscr{E}$ (or as the characteristic line $\mathscr{E}$). When the end mill gear cutter is traveling in the axial direction of the gear from one face of the gear toward the opposite face, the characteristic $\mathscr{E}$ travels together with the end mill cutter. The gear tooth flank $\mathscr{G}$ is generated as the family of consecutive positions of the characteristic lines $\mathscr{E}$. No deviations between the (actual) machined tooth flank from the desired tooth flank is observed if the generating surface $T$ is a continuous surface—namely, when the cutting tool reproduces the entire generating surface $T$. One can imagine a grinding wheel with a generating surface $T$ that is identical to that of the end mill gear cutter. Finishing of the gear with the grinding wheel of such design produces a precision tooth flank $\mathscr{G}$ with no deviations from the desired shape of the actually machined tooth flank.

In designing an end mill gear cutter, the entire generating surface $T$ cannot be reproduced. In reality, the generating surface $T$ is represented not continuously, but discretely by a certain number of distinct curves—the cutting edges $CE$. Therefore, despite the fact that the interacting surfaces $\mathscr{G}$ and $T$ are in line contact, cusps on the milled tooth flanks $\mathscr{G}$ of the gear become inevitable.

The cusps' height must fall into the tolerance for the accuracy of the gear tooth flank $\mathscr{G}$. Otherwise, the sixth necessary condition of proper part surface generation is violated (see Appendix E). For the calculation of the cusp's height use of the DG/K-based method of part surface generation [133, 144, 148, 158, 168] is helpful.

### 5.4.2 Cusps on Tooth Flanks of a Spur Gear

The end mill gear cutter shown in Figure 5.15 is spun about its axis of rotation $O_{ct}$ while simultaneously being fed in the direction of the feed $\mathbf{F}_{ct}$ parallel to the axis $O_g$ of the gear blank in a gear-milling process. After each pass of the end mill cutter, the path cut by the mill cutter is the resulting shape of the tooth profile. The gear blank is indexed one transverse pitch ($2\pi/N_g$ radians), and the cutting process is repeated. Accurate indexing of the gear blank *exactly* one transverse pitch is required when using this method for fabricating gears.

The number of teeth in the end mill gear cutter is limited. When machining a spur gear, residual cusps are generated on the gear tooth flanks (see Figure 5.16) because the end mill cutter has a limited number of teeth. Cusps of this type

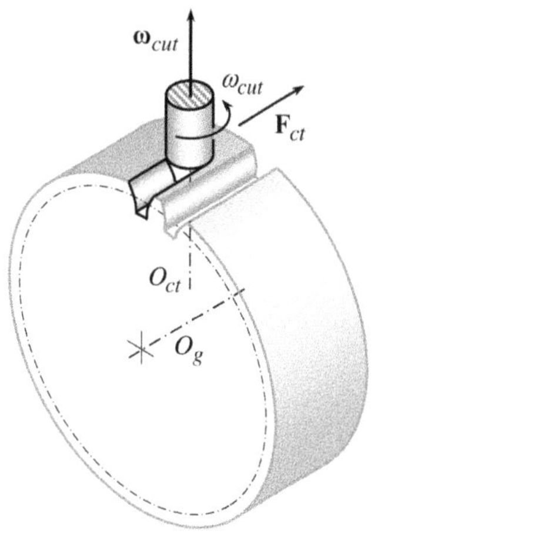


**FIGURE 5.15** Schematic of the machining process of a spur gear with an end mill gear cutter.

are often referred to as waviness of the machined tooth flank $\mathscr{G}$. The length $ab$ of the cusps is equal to the feed per tooth $F_{ct}$ of the end mill cutter. The straight-line segment $ab$ is a portion of the straight generating line (see Figure 1.13) that is parallel to the gear axis of rotation. The waviness height $cd$ is designated $h_w$. The actual profile of the waviness is shaped in the form of epitrochoid. The length $F_{ct}$ of the waviness is much smaller compared to the radius of curvature of the epitrochoid at any of its points within the length $F_{ct}$, which allows for a reasonable approximation of the epitrochoid by a circular arc. The radius of the circular arc $R_w$ is equal to the radius of curvature of the epitrochoid at its apex—at point $a$. Under this assumption, the maximal deviation $h_w$ between the actually machined tooth flank from its desirable surface $\mathscr{G}$ is calculated from the equation:

$$h_w = R_w - \sqrt{R_w^2 - 0.25 \cdot F_{ct}^2} \tag{5.72}$$

In this equation, the radius of the approximating circular arc has not yet been determined. To calculate for the radius $R_w$, Meusnier's formula [144, 148, 158, 168] applies. Implementation of Meusnier's formula allows an expression of the radius $R_w$ in terms of the design parameters of the end mill gear cutter:

$$R_w = \frac{d_{c.m}}{2 \cdot \cos\phi_m} \tag{5.73}$$

In this equation, the diameter of the mill cutter $d_{c.m}$ at point of interest $m$ within the cutting edge $CE$ is calculated on the premises of the equality [see Eq. (5.70)] for machining external involute gears, or on the premises of the expression [see Eq. (5.71)] for machining internal gears.

The derived equations allow for the calculation of the deviation $h_w$ within the tooth height of the spur involute gear machined with an end mill cutter. Figure 5.17 illustrates these computations. The deviation $h_w$ varies within the gear tooth height. The value of the deviation $h_w$ is larger

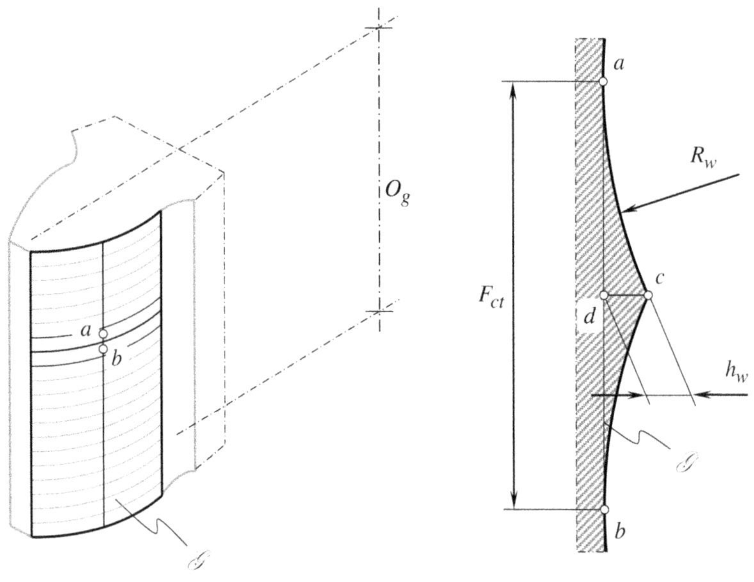


**FIGURE 5.16** Deviation (waviness) of an actual tooth flank from the desired tooth flank $\mathscr{G}$ in a spur gear machined with an end mill cutter.

at smaller diameters of the end mill cutter and becomes smaller for those points of the gear tooth profile closer to the outer diameter of the machined gear. Variation in the mill cutter diameter is the root cause for the variation in the deviation $h_w$.

### 5.4.3 Cusps on Tooth Flanks of a Helical Gear

The curved shape of the tooth flank of a helical gear is the main reason for the deviation from the desired gear tooth flank geometry of the helical involute gear and the actual gear. In the direction of the feed $\mathbf{F}_{ct}$, the tooth flank of a spur gear is straight and the helical gear tooth flank is curved. The curvedness of the tooth flank of the helical gear should be calculated and accounted for when calculating the deviations between the actual and desired gear tooth flank.

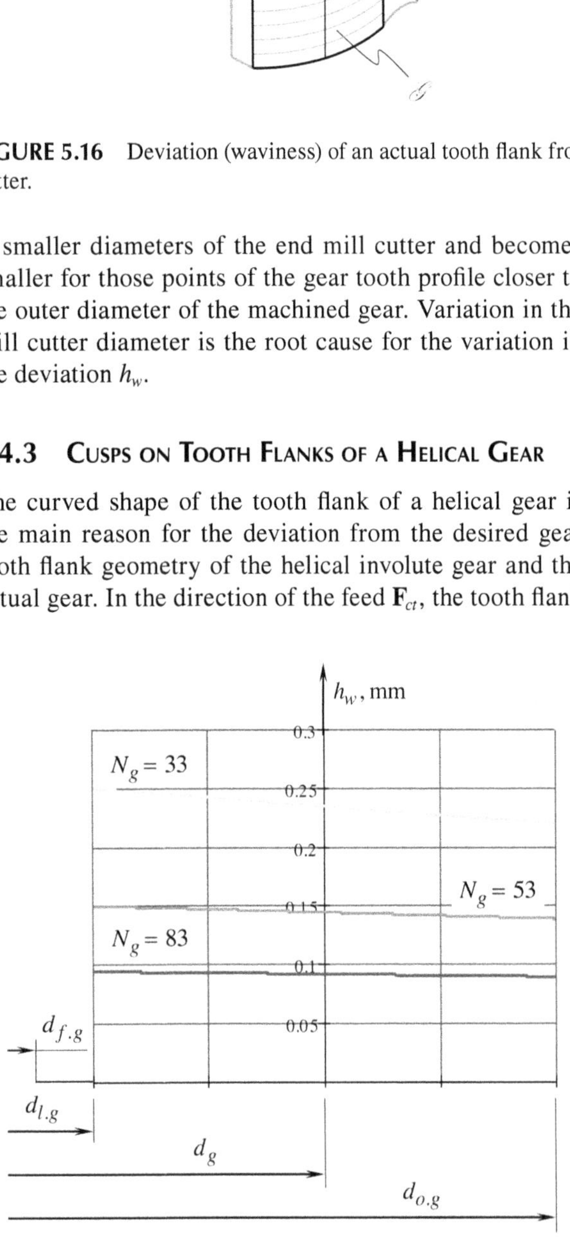


**FIGURE 5.17** Deviation of the actual tooth flank from the desired tooth flank $\mathscr{G}$ within the tooth height of a spur gear machined by the end mill cutter.

#### 5.4.3.1 Normal Curvature of a Helical Gear Tooth Flank in the Direction of Feed Motion, $\mathbf{F}_{ct}$

Similar to generation of the tooth flank of spur gears, the tooth flank of helical gears can also be generated by a straight line (see Figure 1.14). However, in this case, the straight generating line is not parallel to the axis of rotation of the gear, but it forms a base helix angle $\psi_{b.g}$ with the gear axis $O_g$. In the direction of the feed of the end mill cutter, the normal section is curved rather than straight. The radius of curvature $R_g$ of this section can be calculated from Euler's formula:

$$R_g = \frac{1}{\dfrac{R_{1.g}}{\cos^2 \psi_{b.g}} + \dfrac{R_{2.g}}{\sin^2 \psi_{b.g}}} \tag{5.74}$$

where $R_{1.g}$ and $R_{2.g}$ denote the principal radii of curvature at a point of the gear tooth flank $\mathscr{G}$. To calculate for the actual values of the principal radii of curvature $R_{1.g}$ and $R_{2.g}$ the following two formulae:

$$R_{1.g} = \frac{\sqrt{d_{g.m}^2 - d_{b.g}^2}\,\cos\psi_{b.g}}{2\sin^2\psi_{b.g}} \quad \text{and} \quad R_{2.g} \to \infty \tag{5.75}$$

are derived. For details regarding the derivation of Eqs. (5.74) and (5.75), the interested reader is referred to [98, 148, 158], and [175].

Finally, one can come up with the equation:

$$R_g = \frac{R_{1.g}}{\cos^2 \psi_{b.g}} \tag{5.76}$$

for the calculation of the actual value of the required normal radius of curvature $R_g$ of the gear tooth flank $\mathcal{G}$.

#### 5.4.3.2 Calculation of Deviation of a Machined Tooth Flank of a Helical Gear from Its Desired Shape

The calculated value of the normal radius of curvature $R_g$ can be substituted into the approximate expression:

$$h_w \cong \frac{R_g \cdot (R_g + R_w) \cdot \left(1 - \cos \frac{\breve{F}_w}{2 \cdot R_g}\right)}{R_g - (R_g + R_w) \cdot \cos \frac{\breve{F}_w}{2 \cdot R_g}} \tag{5.77}$$

for the calculation of the cusp height $h_w$ [144, 148, 158, 168].

In Eq. (5.77), the radius of normal curvature $R_g$ is calculated via the formula in Eq. (5.76), and Eq. (5.73) is used to calculate for the mill cutter radius $R_w$. The feed per tooth $\breve{F}_{ct}$ of the mill cutter is equal to the length of the arc segment *ab* in Figure 5.18.

Another expression can be implemented for the calculation of the cusp height $h_w$.

Consider two straight lines. One is the straight line through points *c* and *d* (see Figure 5.18). The second (not shown in Figure 5.18) is the straight line through the centers of curvature of the circular arcs of the radii $R_g$ and $R_w$. The angle between these two straight lines is denoted by $\varphi$. The value of the angle $\varphi$ (in radians) is known or can be calculated; next, for the calculation of the cusp's height $h_w$, the approximate equation [144, 148, 158, 168]:

$$h_w \cong \frac{\varphi^2}{2 \cdot R_w} \cdot R_g \cdot (R_g + R_w) \tag{5.78}$$

can be used.

Consider a cylinder of an arbitrary diameter $d_{y.g}$. The axis of the cylinder is aligned with the gear axis $O_g$. The diameter $d_{y.g}$ is between the outer diameter of the gear $d_{o.g}$ and the limit diameter $d_{l.g}$—that is, the inequality

$$d_{l.g} \le d_{y.g} \le d_{o.g} \tag{5.79}$$

is observed. The gear tooth flank deviation $h_w$ is smaller for the larger diameter $d_{y.g}$ (see Figure 5.17), and vice versa.

The deviation $h_w$ depends on the number of teeth of the machined gear, and its value is larger for gears with a fewer tooth.

The deviation of tooth flanks for a helical gear machined by an end mill gear cutter differs from that of tooth flank for spur gears. The deviation $h_w$ of tooth flanks for external helical gears is larger compared to that of spur gears, with smaller deviation $h_w$ of tooth flanks for internal helical gear compared to that of spur gears.

## 5.5 APPLICATION OF END MILL GEAR CUTTERS

End mill gear cutters are universal in application. Gear milling finds it's the most extensive application in roughing and finishing external spur gears, and in roughing external helical gears. Although it is usually applied to external tooth forms, internal tooth forms can also be produced by milling.

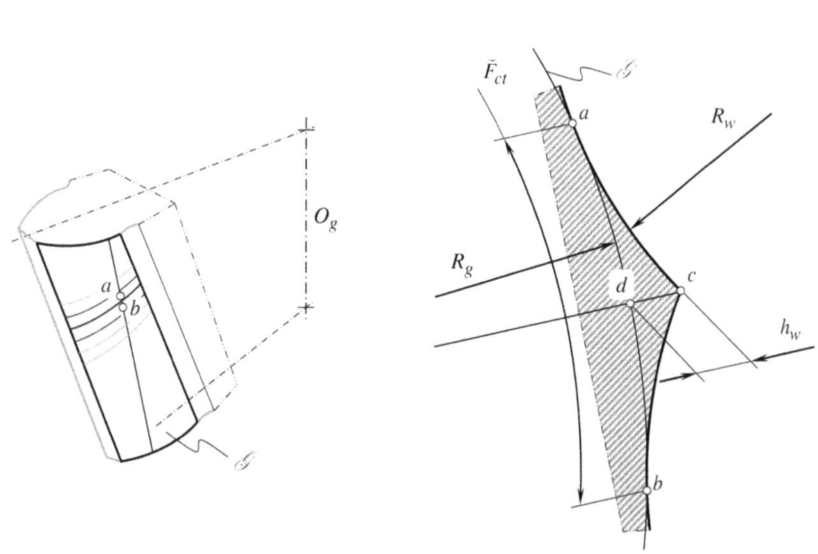


**FIGURE 5.18** Deviation $h_w$ of the actual tooth flank from the desired tooth flank $\mathcal{G}$ in helical gear machined by end mill cutter.

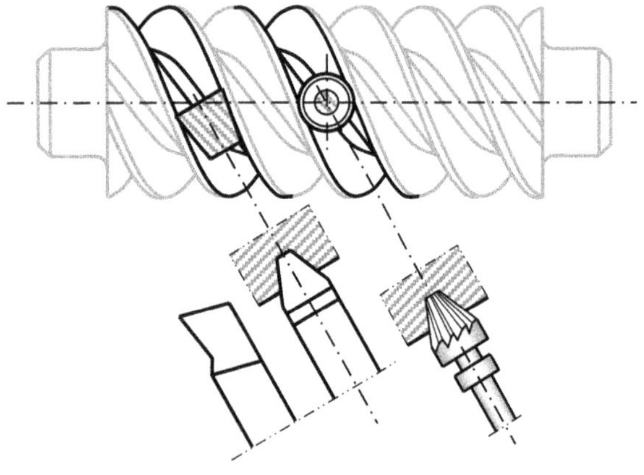

FIGURE 5.19 Machining of an involute worm by an end mill cutter on a lathe.

The milling process is practically suitable for making racks, especially large racks. Worms may also be produced by the milling process (see Figure 5.19). Straight bevel gears are often roughed out by milling prior to generating, and in some cases, these gears are finished by milling.

Most present-day users of gear mill cutters are in coarse-pitch applications. A common application in coarse-pitch gears is the roughing operation, leaving stock for finish hobbing or shaving.

In practice, however, gear mill cutters are usually confined to producing replacement gears or small-lot gears with special tooth forms.

Gear teeth may be milled in one cut, or they may be given a rough cut almost to size and then finish-milled.

Parts to be milled must allow room for run-out of the cutter at each end of the tooth. Wide face widths can be milled, and it is usually possible to have fairly long shaft extension on the gear.

The end mill gear cutter is a type of form mill cutter. In the case of spur gears, the form of the mill cutter teeth is reproduced on the gear. When helical gears are milled, the gear mill cutter tooth form is not reproduced on the gear.

Gear cutters are not universal for a complete range of teeth. To produce theoretically correct gear teeth, the cuter tooth must be designed for a specific number of teeth. However, where a small deviation in tooth form is acceptable, cutters that are designed for a range of teeth can be used.

For spur gears, a gear cutter will produce a tooth space that has the same form as the cutter teeth. Since the cutter axis is at 90° from the gear axis, the axis of the cutter is contained in the transverse plane of the gear. Under these conditions, the gear tooth profiles are finished when the cutter teeth pass through this plane. Therefore, the form on the cutter teeth is reproduced in the gear.

In practice, standard cutters are sometimes used to cut helical gears. The range of the cutter to be used can be determined by finding the equivalent number of spur teeth. For small helix angles, a standard cutter for the proper range of teeth will produce a close approximation to the desired gear tooth form. However, gear cutters for gears with large helix angles are usually designed specifically to produce the desired form.

**TABLE 5.1**
**Standard Set of Mill Cutter for Machining Involute Gears**

| Cutter No. | Will Cut Gears From |
|---|---|
| 1 | 135 teeth to rack |
| 2 | 55–134 teeth |
| 3 | 35–54 teeth |
| 4 | 26–34 teeth |
| 5 | 21–25 teeth |
| 6 | 17–20 teeth |
| 7 | 14–16 teeth |
| 8 | 12–13 teeth |

Standard involute gear cutters are designed to cut a range of gear-tooth numbers. For each pitch, eight cutters are made as follows (see Table 5.1).

The cutters are made in either 14.5° or 20° profile angles, and a backed-off unground form.

The cutter is designed for the lowest number of teeth in the range.

As higher accuracy of tooth form is frequently required, the half-number gear mill cutters are used (see Table 5.2). These provide a better tooth form for the high end of each tooth range.

The form on the cutter is made correct for the lowest number of teeth in that particular range. Thus, all teeth within the range are provided with sufficient tip relief. The same form is produced on all tooth spaces within that range. If close accuracy is desired in form-milled teeth, it is necessary to use a cutter with a curvature that is appropriate for the exact number of teeth. This can be done by purchasing a special single-purpose cutter.

**TABLE 5.2**
**Standard Set of Mill Cutter for Machining Precision Involute Gears**

| Cutter No. | Will Cut Gears From |
|---|---|
| $1\frac{1}{2}$ | 80–114 teeth |
| $2\frac{1}{2}$ | 42–54 teeth |
| $3\frac{1}{2}$ | 30–34 teeth |
| $4\frac{1}{2}$ | 23–25 teeth |
| $5\frac{1}{2}$ | 19–20 teeth |
| $6\frac{1}{2}$ | 15–16 teeth |
| $7\frac{1}{2}$ | 13 teeth |

Theoretically, a special cutter is required to mill a helical gear of a specific tooth form and helix angle. Determination of the equivalent number of spur teeth provides an approximation that is often used to find the standard-range cutter required. The equivalent number of spur teeth is found by dividing the actual number of gear teeth by the cosine of the helix angle cubed.

Milling gear teeth is primarily a form-milling operation to which the general techniques of milling apply. Except for the addition of an indexing mechanism, and the determination of lead gears and setup for helical teeth, machine operation is comparable with conventional milling practice.

Conventional milling machines equipped with a dividing head may be used to mill gear teeth. The cutter tooth form is centered correctly with the gear axis so that a symmetrical tooth space is produced. After each tooth space is milled, the gear blank is indexed to the next cutting position.

Depending on the size of the gear and the capacity of the machine, standard milling machines are used with either manual or automatic indexing mechanisms. On a production basis, where milling is used to produce large coarse-pitch gears, special automatic indexing gear-cutting machines are used. Small automatic gear-cutting machines are applied to the volume production of fine-pitch gears having special tooth forms.

Whether the machine is hand-operated or automatic, the accuracy of tooth spacing depends on the indexing mechanism. The impact of the indexing on the accuracy of the milled gear can be eliminated when the multiple end mill gear cutters are used (see Figure 5.20).

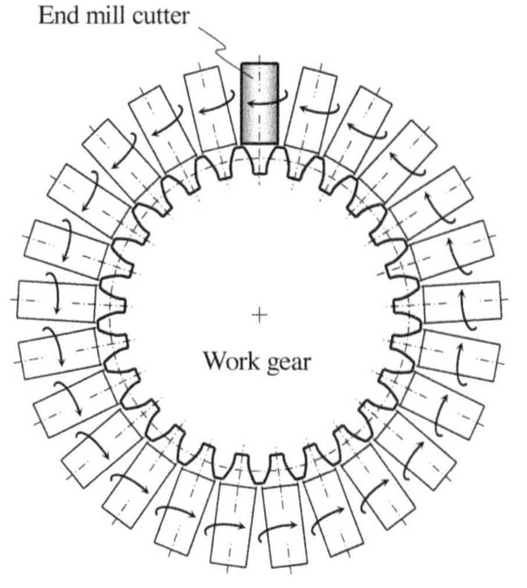


**FIGURE 5.20** Machining of involute gear by multiple end mill cutters.

Since milling gear teeth is a form-cutting operation in which the cutter tooth produces the complete gear tooth profile, a fine surface finish can be obtained.

The number of teeth in an end mill gear cutter is usually even, and it is assigned in the range of 2–8 teeth for the mill cutter with a diameter in the range of $40 \div 220$ mm.

Relieving of the end mill gear cutter is performing at the angle of 10...15° to the cutter's axis of rotation.

# 6 Disk-Type Mill Gear Cutters

Disk-type mill gear cutters are extensively used in roughing and finishing external spur and helical gears. Milling a spur gear with a disk mill cutter in a universal milling machine is illustrated in Figure 6.1. Worms and racks may also be cut by disk mill gear cutters. The use of disk-type mill cutters does not require the use of special gear-cutting machines.

The geometry of the generating surface $T$ of the gear-cutting tool must be predetermined before a disk-type mill gear cutter is designed. Two issues are of critical importance when determining the generating surface of the disk-type mill gear cutter:

- The geometry of the tooth flank of the gear being machining is one of the issues to be considered.
- The geometry of the tooth flanks of spur and helical gears allows for analytical representation [see Eq. (1.2) for spur gear tooth flanks and Eq. (1.4) for helical gear tooth flanks].
- The kinematics of the gear-milling process using a disk-type mill cutter is another issue to be considered.

These two issues are the starting point when designing disk-type mill gear cutters.

## 6.1 KINEMATICS OF GEAR CUTTING BY DISK-TYPE MILL CUTTER

The kinematics of the gear-machining process when cutting a gear using a disk mill gear cutter is simple. In fact, it is one of the simplest one among those shown for various types of gear-machining processes (see Figure 3.8).

The kinematics of the gear-machining process using disk-type mill gear cutters resembles that of milling gears with end mill gear cutters (see Figure 6.2). The rotation of the disk-type mill gear cutter is applied to create the speed of cut. The motion that enables the gear tooth flank to allow for sliding on itself is used as the feed motion. Generally speaking, this feed motion is a type of screw motion. The parameter of the screw motion is identical to the screw parameter of the gear being machining.

The kinematics of the gear-machining process with a disk-type mill cutter is comprised of the kinematics of the gear-machining process and of the introduced feed motion as illustrated in Figure 6.2.

The rotation vector $\boldsymbol{\omega}_{cut}$ is positioned at a certain center distance $C$ apart from the axis $O_g$ of the gear to be machined [Figure 6.2(a)]. The vector of the additional rotation $\boldsymbol{\omega}_{cut}$ crosses at a right angle to the gear axis $O_g$. The straight feed motion $\mathbf{F}_{ct}$ is pointed along the gear axis $O_g$. The kinematics of the gear-machining process of this kind is used when machining a spur gear.

When a helical gear is machined by a disk-type mill cutter [see Figure 6.2(b)], the rotation vector $\boldsymbol{\omega}_{cut}$ is crossing with the axis $O_g$ of the gear to be machined. The crossed-axis angle is equal to the pitch helix angle $\psi_g$ of the gear. The feed motion is a kind of screw motion. The screw feed motion is represented in Figure 6.2(b) as a superposition of the rotation $\boldsymbol{\omega}_{fc}$ about the gear axis $O_g$ and the translation $\mathbf{F}_{ct}$ along the gear axis $O_g$. The parameter $p_{fc}$ of the screw feed motion equal:

$$p_{fc} = \frac{|\mathbf{F}_{ct}|}{|\boldsymbol{\omega}_{fc}|} = \frac{F_{ct}}{\omega_{fc}} \tag{6.1}$$

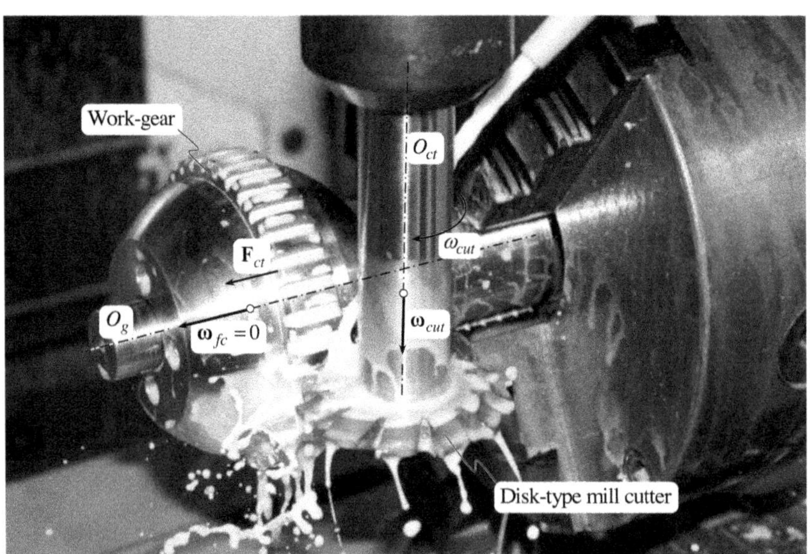


**FIGURE 6.1** Milling spur gear by disk-type mill cutter in universal milling machine.


DOI: 10.1201/9781003520528-8

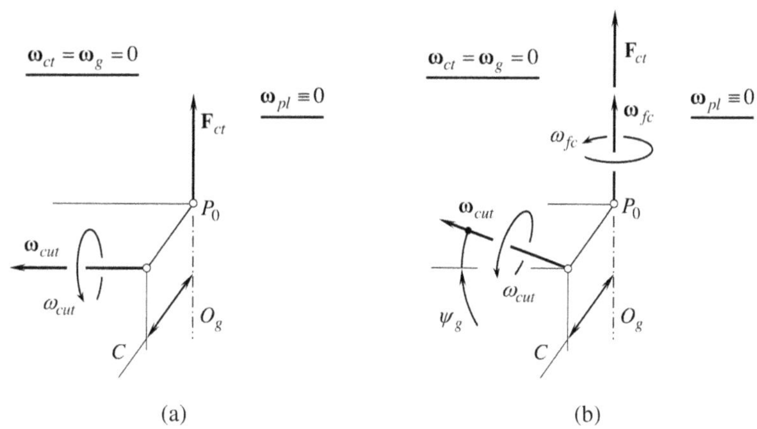


**FIGURE 6.2** Kinematics of the gear-machining process with a disk-type mill gear cutter when milling (a) a spur gear and (b) a helical gear.

and is identical to the reduced pitch $p_g$ of the tooth flank $\mathscr{G}$ of the helical gear to be machined—that is, the identity $p_{fc} \equiv p_g$ is observed. For a specified helical gear, the reduced pitch $p_g$ is given as:

$$p_g = \frac{L_g}{2\pi} \tag{6.2}$$

where $L_g$ denotes the tooth lead of the gear to be machined.

The reduced pitch can be expressed in terms of the pitch diameter $d_g$ and the pitch helix angle $\psi_g$ of the gear being machining:

$$p_g = \frac{d_g}{2}\cot\psi_g \tag{6.3}$$

Once the geometry of the tooth flank $\mathscr{G}$ of the gear to be machined is known [see Eq. (1.2) for spur gear tooth flanks and Eq. (1.4) for helical gear tooth flanks] and the kinematics of the gear-machining process has been predetermined (see Figure 6.2), the generating surface of the gear-cutting tool $T$ can be determined.

## 6.2 GENERATING SURFACE OF DISK-TYPE MILL GEAR CUTTER

It is assumed in the following that when designing a disk-type mill gear cutter, the generating surface of the gear-cutting tool is already determined.

### 6.2.1 Equation for Generating Surface of a Disk-Type Mill Gear Cutter for Machining Spur Involute Gears

In the case of spur gears, the translation motion $\mathbf{F}_{ct}$ causes the gear tooth flank $\mathscr{G}$ to slide on itself. This motion does not affect the shape of the generating surface $T$ of the disk-type mill gear cutter. This motion could be omitted when determining the generating surface $T$ of the disk-type mill cutter.

Disk-type mill cutters can be used for machining of both symmetrical [see Figure 6.3(a)] and asymmetrical [see Figure 6.3(b)] tooth profiles.

Derivation of an equation for the generating surface of the disk-type mill cutter for machining spur involute gears can be executed in three steps:

- First, an equation for the tooth profile of the spur gear being machining is derived.
- Second, the derived equation of the spur gear tooth profile is represented in a reference system associated with the disk-type mill gear cutter.
- Third, the generating surface of the disk-type mill cutter is represented as the locus of consecutive positions of the involute gear tooth profile when this profile is rotating about the axis of rotation of the disk-type mill cutter.

To derive an equation of the generating surface $T$ of the disk gear mill cutter, the earlier derived analytical

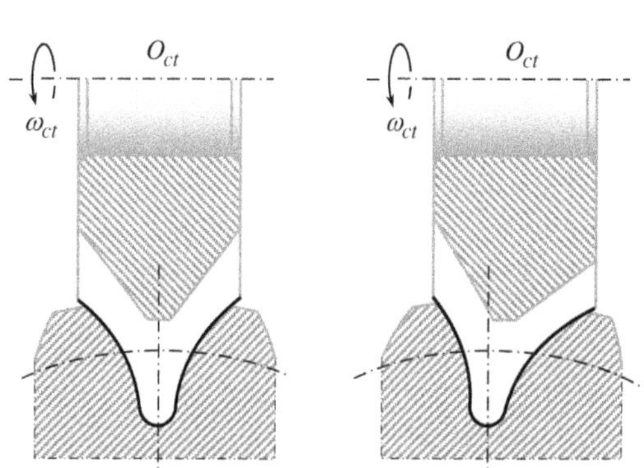


**FIGURE 6.3** Tooth profile of disk-type mill gear cutters for machining (a) a symmetrical gear tooth profile and (b) an asymmetrical gear tooth profile.

description of the involute gear tooth profile [see Eqs. (5.2) and (5.3)]:

$$\mathbf{r}_{inv}(\phi_g) = \frac{d_{b.g}}{2 \cdot \cos\phi_g} \cdot [\mathbf{i} \cdot \sin(\text{inv}\phi_g) + \mathbf{j} \cdot \cos(\text{inv}\phi_g)] \quad (6.4)$$

$$\text{inv}\phi_g = \tan\phi_g - \phi_g \quad (6.5)$$

applies.

Equations (6.4) and (6.5) return the solution to the first step of the problem under consideration.

In the second step, the operator of rotation $\mathbf{Rt}(a, Z_g)$ about the gear axis $Z_g$ is composed. The operator $\mathbf{Rt}(a, Z_g)$ is implemented for transferring of the gear tooth profile [see Eqs. (6.4) and (6.5)] into the properly oriented Cartesian coordinate system $X_g Y_g Z_g$ (see Figure 6.4). This allows us to account for the space width of the gear being machining. Next, the operator of translation $\mathbf{Tr}(C, Y_g)$ is used for the purpose of representation of the tooth profile in the appropriate reference system that the disk-type mill cutter will be associated with.

Ultimately, the equation for the involute tooth profile in the coordinate system $X_{ct} Y_{ct} Z_{ct}$ allows for matrix representation:

$$\mathbf{r}_{inv}^{c}(\phi_g) = \mathbf{Tr}(C, Y_g) \cdot \mathbf{Rt}(\alpha, Z_g) \cdot \mathbf{r}_{inv}(\phi_g) \quad (6.6)$$

In the third step, the generating surface $T$ of the disk-type mill cutter is constructed as the locus of consecutive positions of the involute tooth profile [see Eq. (6.6)] with a rotation $\omega_{cut}$ about the disk-type mill cutter axis $O_{ct}$ (see Figure 6.4). Implementation of the operator of rotation $\mathbf{Rt}(\theta, X_{ct})$ of the involute tooth profile $\mathbf{r}_{inv}^{c}(\phi_g)$ about the $X_{ct}$ – axis is convenient for this purpose. The $X_{ct}$ coordinate axis is aligned with the axis of rotation $O_{ct}$ of the disk-type mill cutter:

$$\mathbf{r}_T^{c}(\phi_g, \theta) = \mathbf{Rt}(\theta, X_{ct}) \cdot \mathbf{r}_{inv}^{c}(\phi_g) \quad (6.7)$$

In this equation, the operator of rotation $\mathbf{Rt}(\theta, X_c)$ allows for matrix representation:

$$\mathbf{Rt}(\theta, X_{ct}) = \begin{bmatrix} 1 & 0 & 0 & 0 \\ 0 & \cos\theta & \sin\theta & 0 \\ 0 & -\sin\theta & \cos\theta & 0 \\ 0 & 0 & 0 & 1 \end{bmatrix} \quad (6.8)$$

After the position vector $\mathbf{r}_{inv}^{c}(\phi_g)$ is pre-multiplied by the operator of rotation $\mathbf{Rt}(\theta, X_{ct})$, the equation:

$$\mathbf{r}_T^{c}(\phi_g, \theta) = \begin{bmatrix} \dfrac{r_{b.g}}{\cos\phi_g} \cdot \sin(\alpha + \text{inv}\phi_g) \\ \left(C - \dfrac{r_{b.g}}{\cos\phi_g} \cdot \cos(\alpha - \text{inv}\phi_g)\right) \cdot \cos\theta \\ -\left(C - \dfrac{r_{b.g}}{\cos\phi_g} \cdot \cos(\alpha - \text{inv}\phi_g)\right) \cdot \sin\theta \\ 1 \end{bmatrix} \quad (6.9)$$

can be derived for the generating surface $T$ of the disk-type mill cutter for machining spur involute gears.

The generating surface $T$ [see Eq. (6.9)] is a surface of revolution. It is expressed in terms of two parameters: the profile angle $\phi_g$ of the involute at a point of interest of the gear tooth profile and the angle $\theta$ of rotation of the involute profile about the axis of rotation of the disk mill gear cutter $(-\infty < \theta < +\infty)$.

Equation (6.9) is the solution to the problem of profiling a disk mill cutter for machining spur gears.

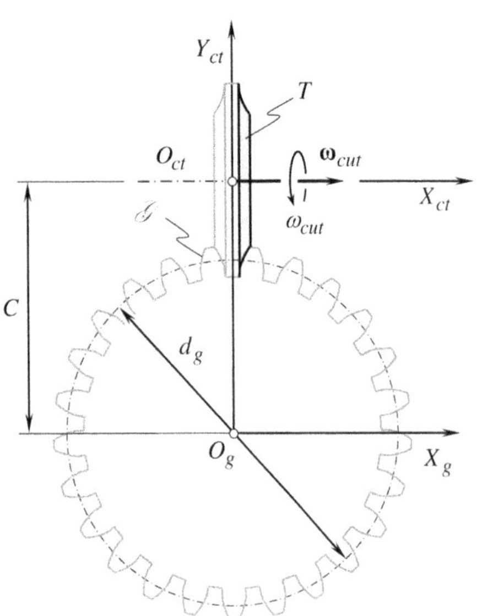


FIGURE 6.4 Reference systems used in deriving the equation for the generating surface $T$ of disk-type mill cutter for machining spur gears.

### 6.2.2 Equation for the Generating Surface of a Disk-Type Mill Cutter for Machining Helical Involute Gears

When a helical gear is machined, the line of contact between the helical gear tooth flank $\mathscr{G}$ and of the generating surface of the disk-type mill cutter $T$ is a kind of spatial curve rather than a planar curve. This is the major reason that the implementation of elements of the theory of enveloping surfaces is helpful for solving this particular problem in the design of disk-type mill gear cutter. According to this approach, two main reference systems are used. The first coordinate system is associated with the helical gear being machined, and the second is the coordinate system that the designed disk-type mill cutter will be associated with. It should be stressed that the disk-type mill cutter doesn't yet exist. Therefore, we are operating with a coordinate system that the disk mill cutter will be associated with.

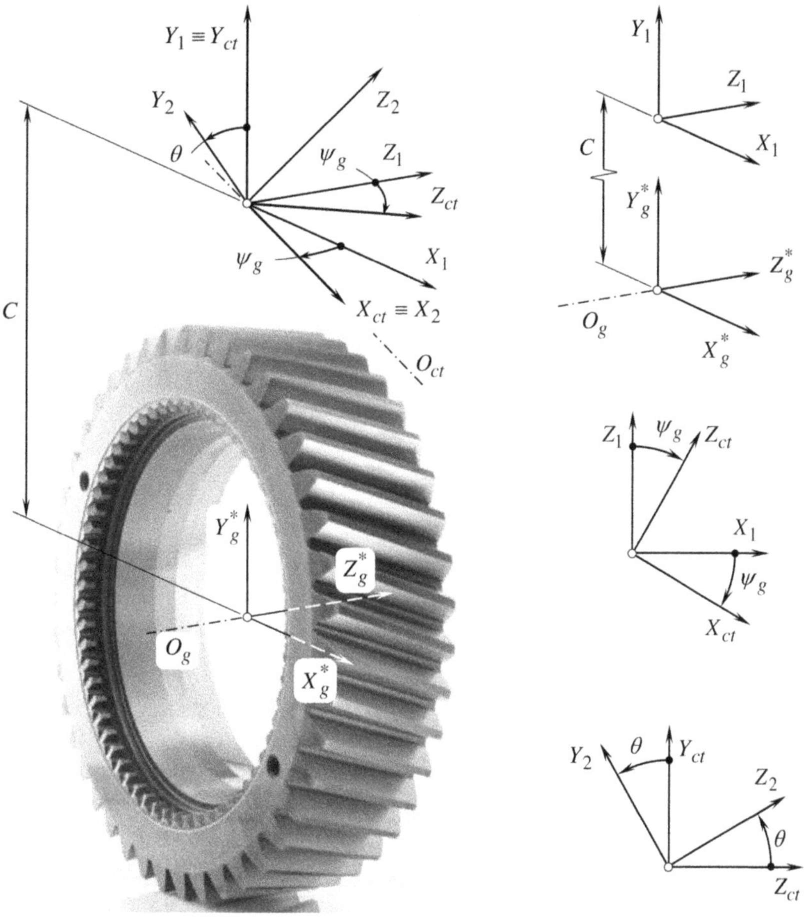


**FIGURE 6.5** Reference systems used in deriving the equation for the generating surface $T$ of disk-type mill cutter for machining helical gears.

When milling a helical gear, the disk-type mill cutter performs two motions:

- The rotation of the mill cutter about its axis of rotation.
- The screw feed motion of the mill cutter with respect to the gear to be machined.

When the screw relative motion of the helical gear is performed, the gear tooth flank $\mathcal{G}$ slides on itself. This relative motion does not affect the shape and parameters of the generating surface of the disk-type mill cutter. Therefore, this motion can be eliminated from the list of considerations when deriving an equation of generating surface $T$ of the disk-type mill cutter. Ultimately, the generating surface of the disk-type mill cutter can be interpreted as the envelope to consecutive positions of the helical gear tooth flank $\mathcal{G}$ that is rotated relative to the mill cutter axis of rotation.

An equation for the generating surface of the disk-type mill cutter can be derived in three steps.

First, it is necessary to analytically describe the helical gear tooth flank $\mathcal{G}$ in an appropriate coordinate system. For this purpose, the earlier derived equation for a screw involute surface [see Eq. (1.4)]:

$$\mathbf{r}_g^* = \mathbf{r}_g^*(U_g, V_g) \tag{6.10}$$

can be utilized.

Second, it is necessary to derive an equation for the gear tooth flank $\mathcal{G}$ in its current location with respect to the coordinate system associated with the disk-type mill cutter.

For this purpose, consider the helical involute gear shown in Figure 6.5. A Cartesian coordinate system $X_g^* Y_g^* Z_g^*$ is associated with the gear. After it has been designed, the disk-type mill cutter will be associated with the reference system $X_{ct}Y_{ct}Z_{ct}$. The coordinate system $X_{ct}Y_{ct}Z_{ct}$ is shifted in relation to the coordinate system $X_g^* Y_g^* Z_g^*$ at the center distance $C$ and turned about the $Y_g^*$ – axis through the gear pitch helix angle $\psi_g$.

Translation of the coordinate system $X_g^* Y_g^* Z_g^*$ along the $Y_g^*$ – axis at the center distance $C$ can be analytically described by the operator of translation $\mathbf{Tr}(C, Y_g^*)$. After it has been translated, the coordinate system $X_g^* Y_g^* Z_g^*$ occupies the position of the intermediate coordinate system $X_1Y_1Z_1$.

The rotation of the coordinate system $X_1Y_1Z_1$ about the $Y_1-$ axis through the pitch helix angle $\psi_g$ is analytically described by the operator of rotation $\mathbf{Rt}(\psi_g, Y_1)$. After it has been rotated, the coordinate system $X_1Y_1Z_1$ occupies the position of the disk-type mill cutter coordinate system $X_{ct}Y_{ct}Z_{ct}$.

The resultant coordinate system transformation—that is, of the transformation from the gear to be machined coordinate system $X_g^*Y_g^*Z_g^*$ to the disk-type mill cutter coordinate system $X_{ct}Y_{ct}Z_{ct}$—is performed by the operator $\mathbf{Rs}(g^* \mapsto ct)$ of the resultant coordinate system transformation. In this case under consideration, the operator $\mathbf{Rs}(g^* \mapsto ct)$ is calculated from the formula:

$$\mathbf{Rs}(g^* \mapsto ct) = \mathbf{Rt}(\psi_g, Y_1) \cdot \mathbf{Tr}(C, Y_g^*) \quad (6.11)$$

To derive an equation for the gear tooth flank $\mathscr{G}$ in its current location in relation to the coordinate system $X_{ct}Y_{ct}Z_{ct}$, it is convenient to introduce one more reference system. As shown in Figure 6.5, the Cartesian coordinate system $X_2Y_2Z_2$ is turned about the $X_{ct}-$ axis through a certain angle $\theta$. The angle $\theta$ allows for interpretation in the form:

$$\theta = \omega_{cut} \cdot t \quad (6.12)$$

where $\omega_{cut}$ is the rotation of the disk-type mill cutter about its axis of rotation $O_c$ and $t$ denotes time.

The rotation of the coordinate system $X_{ct}Y_{ct}Z_{ct}$ about the $X_c-$ axis through the angle $\theta$ can be analytically described by the operator of rotation $\mathbf{Rt}(\theta, X_{ct})$. After it has rotated, the coordinate system $X_{ct}Y_{ct}Z_{ct}$ occupies the position of the disk-type mill cutter coordinate system $X_2Y_2Z_2$.

When designing a gear disk-type mill cutter, only relative motion of the coordinate systems is important. It makes no difference whether the coordinate system $X_{ct}Y_{ct}Z_{ct}$ of the disk mill cutter is rotating about its axis $O_{ct} \equiv X_{ct}$ or the gear tooth flank $\mathscr{G}$ is rotating in the opposite direction about that same axis of rotation.

At a certain instant in time, the gear tooth flank $\mathscr{G}$ in its current location in relation to the coordinate system $X_{ct}Y_{ct}Z_{ct}$ can be analytically described by the following equation:

$$\mathbf{r}_g^c(U_g, V_g, \theta) = \mathbf{Rt}(\theta, X_{ct}) \cdot \mathbf{Rs}(g^* \mapsto ct) \cdot \mathbf{r}_g^*(U_g, V_g) \quad (6.13)$$

Equation (6.13) represents the solution to the second step of deriving the equation for the generating surface $T$ of the disk-type mill cutter.

Third, the kinematic method[1] of determining the enveloping surface can be applied in the derivation of an equation of the generating surface $T$ of the disk-type mill cutter. It is well known that this method is based on the application of the Shishkov equation of contact:

$$\mathbf{n} \cdot \mathbf{V}_\Sigma = 0 \quad (6.14)$$

The unit normal vector $\mathbf{n}_g^c$ at a point of the gear tooth flank $\mathscr{G}$ in its current position with respect to the coordinate system $X_{ct}Y_{ct}Z_{ct}$ can be calculated from the expression:

$$\mathbf{n}_g^c(U_g, V_g, \theta) = \mathbf{u}_g^c \times \mathbf{v}_g^c \quad (6.15)$$

Equation (6.13) yields calculation for the unit tangent vectors $\mathbf{u}_g^c$ and $\mathbf{v}_g^c$.

To calculate for the velocity vector $\mathbf{V}_\Sigma$ of the resultant motion of a point of the tooth flank $\mathscr{G}$, Eq. (6.13) is used. Therefore,

$$\mathbf{V}_\Sigma(U_g, V_g, \theta) = \frac{\partial \mathbf{r}_g^c(U_g, V_g, \theta)}{\partial \theta} \quad (6.16)$$

A solution to the set of two equations:

$$\mathbf{r}_g^c = \mathbf{r}_g^c(U_g, V_g, \theta) \quad (6.17)$$

$$\mathbf{n}_g^c(U_g, V_g, \theta) \cdot \mathbf{V}_\Sigma(U_g, V_g, \theta) = 0 \quad (6.18)$$

returns an equation of the generating surface $T$ of the disk-type mill cutter for machining helical involute gears. After all necessary formula transformations are performed, the equation of the surface $T$ can be represented in vector notation:

$$\mathbf{r}_T = \mathbf{r}_T(U_T, V_T) \quad (6.19)$$

When pitch helix angle of a helical gear is equal to zero ($\psi_g = 0°$), the helical gear reduces to a spur gear. Under such a scenario, Eq. (6.19), the generating surface $T$ of the disk-type mill cutter for machining helical gear reduces to Eq. (6.9) for the generating surface $T$ of the disk-type mill cutter for machining spur gears.

Very similar to what has just been discussed, corresponding equations for the generating surfaces of disk-type mill cutters for machining spur and helical gears with non-involute tooth profiles can also be derived. Various tooth profiles can be machined by disk-type mill cutters, including but not limited to ratchets and gears with asymmetrical involute tooth profiles, especially those that cannot be machined by conventionally designed end mill gear cutters.

### 6.2.3 Elements of the Intrinsic Geometry of the Generating Surface of a Disk-Type Mill Cutter

The derived equation of the generating surface of the disk-type mill cutter [see Equation (6.19)] allows us to calculate for the first $\mathbf{G}_{ct}$ and the second $\mathbf{D}_{ct}$ fundamental matrices of the generating surface $T$ of the mill cutter. To calculate for the fundamental matrices $\mathbf{G}_{ct}$ and $\mathbf{D}_{ct}$, the equations:

$$\mathbf{G}_{ct} = \begin{bmatrix} E_{ct} & F_{ct} \\ F_{ct} & G_{ct} \end{bmatrix} \quad (6.20)$$

[1] Shishkov, V.A., *Generation of Surfaces in Continuous Indexing Methods of Part Surface Machining*, Mashgiz, Moscow, 1951, 152 pages.

$$\mathbf{D}_{ct} = \begin{bmatrix} L_{ct} & M_{ct} \\ M_{ct} & N_{ct} \end{bmatrix} \tag{6.21}$$

are used.

Equations (5.28) through (5.30) can be used for the calculation of the fundamental magnitudes of the first order $E_{ct}$, $F_{ct}$, and $G_{ct}$ of the generating surface of the disk-type mill cutter. To calculate for the fundamental magnitudes of the second order—that is, $L_{ct}$, $M_{ct}$, and $N_{ct}$—of the generating cutter surface $T$ of the disk-type mill cutter, Eqs (5.33) through (5.35) are used.

As an example of the implementation of the elements of intrinsic geometry of surfaces, consider the following practical gear-machining problem. An involute helical gear being machined is given, and it is necessary to determine a particular number of the mill cutter from the standard set of mill cutters that is appropriate for machining the gear.

Standard mill cutters (see Table 5.1) are designed for machining spur gears. When machining a helical gear with a mill gear cutter, an appropriate approximation of the helical gear tooth profile with the axial profile of the mill cutter is required. In other words, the problem under consideration can be formulated as follows: What is the radius of the normal curvature of the mill gear cutter having a number $N_{eq}$ that is the closest to the radius of the normal curvature of the helical gear?

Principal radii of curvature $R_{1.g}$ and $R_{2.g}$ at a point of the tooth flank $\mathscr{G}$ of the helical gear being machining are the roots of the square equation:

$$\begin{vmatrix} L_g - \dfrac{E_g}{R_g} & M_g - \dfrac{F_g}{R_g} \\ M_g - \dfrac{F_g}{R_g} & N_g - \dfrac{G_g}{R_g} \end{vmatrix} = 0 \tag{6.22}$$

This equation casts into:

$$R_g = \frac{(E_g N_g - 2F_g M_g + G_g L_g) \pm \sqrt{(E_g N_g - 2F_g M_g + G_g L_g)^2 - 4(E_g G_g - F_g^2)(L_g N_g - M_g^2)}}{2(L_g N_g - M_g^2)} \tag{6.23}$$

Omitting intermediate formulae transformations, the expressions [98, 138, 144, 151, 158, 172]:

$$R_{1.g} = \frac{\sqrt{d_g^2 - d_{b.g}^2} \cdot \cos\psi_{b.g}}{2 \cdot \sin^2 \psi_{b.g}} \tag{6.24}$$

$$R_{2.g} \to \infty \tag{6.25}$$

for the calculation of the principal radii of curvature $R_{1.g}$ and $R_{2.g}$ at a point of the helical gear tooth flank $\mathscr{G}$ are derived. The second principal radius of curvature $R_{2.g}$ of the gear tooth flank $\mathscr{G}$ to be machined is not of importance in solving the problem under consideration.

In Eq. (6.24), the following designations apply:

$d_g$ is the pitch diameter of the helical gear
$d_{b.g}$ is the base diameter of the gear being machined
$\psi_{b.g}$ is the base helix angle

The parameters $d_{b.g}$ and $\psi_{b.g}$ can be expressed in terms of the design parameters of the helical gear to be machined (see Appendix F).

Furthermore, to calculate for the first principal radius of curvature $R_{1.ct}$ at a point of the generating surface $T$ of a standard disk mill gear cutter, the following expression can be used:

$$R_{1.ct} = \frac{m \cdot N_{eq}}{2} \cdot \sin\phi \tag{6.26}$$

where:

$m$ is the module of the gear being machined
$\phi$ is the profile angle of the helical gear
$N_{eq}$ is the equivalent tooth number to be determined

In reality, the equivalent tooth number $N_{eq}$ corresponds to a phantom gear, which does not physically exist. Therefore, there should be no confusion when the calculated value of $N_{eq}$ is represented with a number with fractions.

Ideally, the error of approximation of the principal radius of curvature $R_{1.g}$ of the tooth flank $\mathscr{G}$ of the helical gear by the principal radius of curvature $R_{1.ct}$ at a point of the generating surface $T$ of the disk-type mill cutter is the minimal possible. Under such a scenario, the equality $R_{1.ct} = R_{1.g}$ is valid. This equality, together with Eqs. (6.24) and (6.26), yields the expression:

$$N_{eq} = \frac{\sqrt{d_g^2 - d_{b.g}^2} \cdot \cos\psi_{b.g}}{m \cdot \sin\phi \cdot \sin^2 \psi_{b.g}} \tag{6.27}$$

for the calculation of the equivalent tooth number $N_{eq}$ for the identification of the required disk mill gear cutter.

Next, the calculated value of $N_{eq}$ is compared to the data shown in Table 5.1. In this manner, the appropriate number of the disk-type mill cutter is determined. For example, if the calculated number of $N_{eq}$ equals 67.375, then the disk mill gear cutter #2 is appropriate for machining the given helical gear.

There are numerous practical examples of implementation of the elements of intrinsic geometry of the generating surface $T$ of the disk-type mill gear cutters.

## 6.3 CUTTING EDGE OF THE DISK-TYPE MILL GEAR CUTTER

The cutting edge of a disk-type mill cutter is physically formed by the rake surface $R_s$ and the clearance surface $C_s$. For the geometrical and kinematical analysis, the cutting edge $CE$ is construed as the line of intersection of the surfaces $R_s$ and $C_s$ (see Figure 6.6).

The cutting edges of a finishing disk-type mill cutter are located within the generating surface $T$ of the cutting tool. Therefore, three surfaces—the rake surface $R_s$, the clearance surface $C_s$, and the generating surface $T$—of the disk-type mill gear cutter are the surfaces through the cutting edge $CE$. For roughing, as well as for semi-roughing, disk-type mill cutters, reasonable deviations of the cutting edge from the generating surface $T$ are permissible.

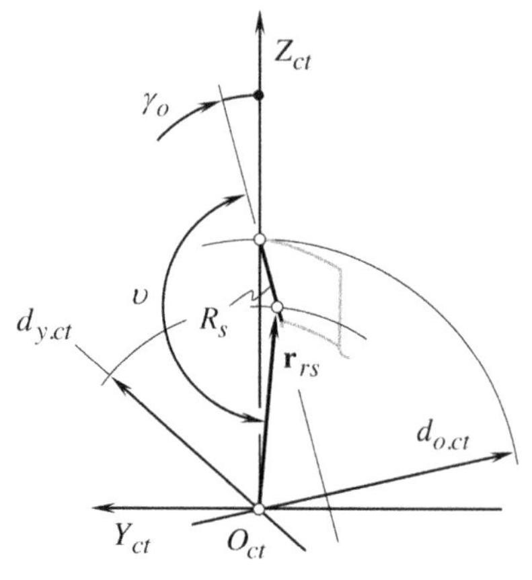


**FIGURE 6.7** Rational parameterization of the rake plane $R_s$ of the disk-type mill cutter having nonzero rake angle $\gamma_o \neq 0°$.

### 6.3.1 Rake Surface of Disk-Type Mill Cutter for Machining Involute Gears

The rake surface $R_s$ of a finishing disk-type mill cutter of most practical designs is a plane through the axis of rotation of the cutting tool. Because of this, the rake angle is equal to zero ($\gamma_o = 0°$). After regrinding, the accuracy of the worn finishing disk mill cutters is less vulnerable to diameter alteration of the reground disk mill cutters if the rake angle is zero ($\gamma_o = 0°$).

Equation for the rake surface $R_s$ of the finishing disk-type mill cutter can be presented in the form $Y_c = 0$. This equation casts into its matrix representation:

$$\mathbf{r}_{rs}(X_{ct}, Z_{ct}) = \begin{bmatrix} X_{ct} \\ 0 \\ Z_{ct} \\ 1 \end{bmatrix} \tag{6.28}$$

where the position vector of a point of the rake surface $R_s$ is denoted by $\mathbf{r}_{rs}$.

Disk-type mill cutters for roughing and semi-finishing gears are manufactured with a positive rake angle $\gamma_o > 0°$. For semi-finishing and finishing hardened gears, disk-type mill cutters with a negative rake angle at the top cutting edge ($\gamma_o << 0°$) can be applied.

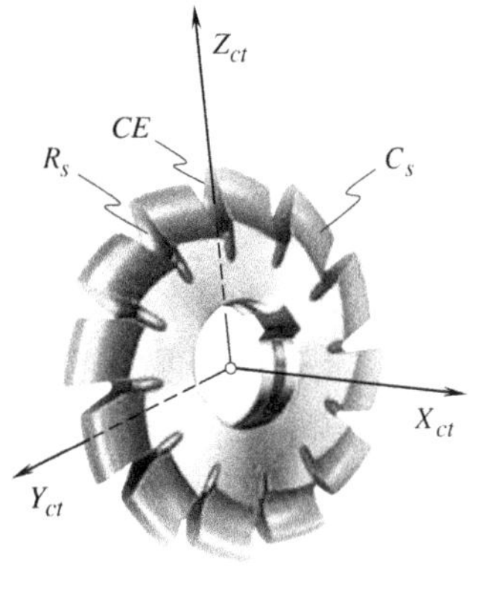


**FIGURE 6.6** Rake surface $R_s$, clearance surface $C_s$, and the cutting edge $CE$ of a disk-type mill gear cutter.

The equation for the rake plane $R_s$ of disk-type mill cutters with either a positive or a negative rake angle $\gamma_o$ allows for matrix representation (see Figure 6.7):

$$\mathbf{r}_{rs}(X_{ct}, d_{y.ct}) = \begin{bmatrix} X_{ct} \\ \dfrac{d_{y.ct}}{2} \cdot \cos(\upsilon - \gamma_o) \\ \dfrac{d_{y.ct}}{2} \cdot \sin(\upsilon - \gamma_o) \\ 1 \end{bmatrix} \tag{6.29}$$

where the angle $\upsilon$ is calculated from the formula:

$$\upsilon = \sin^{-1}\left( \frac{d_{o.ct}}{d_{y.ct}} \cdot \sin \gamma_o \right) \tag{6.30}$$

where:

$d_{o.ct}$ the outer diameter of the disk-type mill cutter
$d_{y.ct}$ is the diameter through a point of interest of the surface, $R_s$
$\gamma_o$ is the rake angle at the outer diameter of the disk-type mill cutter

Equation (6.19) for the generating surface $T$ and Eq. (6.29) for the rake plane $R_s$ can be solved together. The solution to this set of two equations analytically describes the planar cutting edge $CE$ of the disk-type mill cutter. Equation for the cutting edge of the disk-type mill cutter can be presented in matrix form:

$$\mathbf{r}_{ce}(\vartheta_{ce}) = \begin{bmatrix} X(\vartheta_{ce}) \\ Y(\vartheta_{ce}) \\ Z(\vartheta_{ce}) \\ 1 \end{bmatrix} \tag{6.31}$$

where $\vartheta_{ce}$ denotes the parameter of the cutting edge $CE$ of the disk mill gear cutter.

When the generating surface $T$ of the disk-type mill cutter and the rake surface $R_s$ are expressed in terms of common parameters, one of the parameters can be expressed in terms of another parameter. In this manner, the angular parameter $\vartheta_{ce}$ of the cutting edge $CE$ of the disk-type mill cutter can be identified.

The possible shapes of the rake surface of the disk-type mill cutter are not limited to just a plane surface. In certain cases, rake surfaces of special geometry are used as well.

### 6.3.2 Clearance Surface of a Disk-Type Mill Cutter for Machining Involute Gears

The purpose of the clearance surface is twofold:

- First, the shape of the clearance surface $C_s$ provides the required clearance between the cutting wedge and the machined surface of the gear, reducing friction and wear of the disk mill gear cutter.
- Second, the clearance surface $C_s$ is designed so as to allow for avoiding any alterations in the shape of the cutting edge $CE$ after each regrinding of the disk-type mill cutter.

In most practical cases, the Archimedean spiral curve is used in the design of the clearance surface of the disk-type mill gear cutter. In this case, the equation for the clearance surface can be derived in the following way.

A Cartesian coordinate system $X_{ct}Y_{ct}Z_{ct}$ is associated with the disk-type mill cutter, as shown in Figure 6.8. This reference system is identical to that used in Figure 6.6. In the coordinate system $X_{ct}Y_{ct}Z_{ct}$, the cutting edge $CE$ is analytically described by Eq. (6.31). The clearance surface $C_s$ of the disk-type mill cutter yields interpretation in the form of consecutive positions of the cutting edge $CE$ that is performing a certain motion in the reference system $X_{ct}Y_{ct}Z_{ct}$. When the Archimedean spiral curve is applied, the motion of the cutting edge is a superposition of two motions. A uniform rotation of the cutting edge through a certain angle about the axis of rotation $O_{ct}$ is one of the motions. In Figure 6.8, the current value of the angle of rotation is designated as $\theta_{cs}$. The straight motion toward the axis of rotation $O_{ct}$ through a certain distance is the second motion of the cutting edge $CE$. The current value of the distance traveled due to this translation is designated as $a_z$. Because of the Archimedean spiral curve, both motions are uniform. Therefore, the current value of the angle of rotation $\theta_{cs}$ and the current value of the displacement $a_z$ correlate to each other linearly—that is, the equality:

$$a_z = k \cdot \theta_{cs} \tag{6.32}$$

is observed. Here $k$ is a coefficient of proportionality.

The coefficient of proportionality $k$ can be expressed in terms of the desired clearance angle $\alpha_o$ and the design parameters of the gear disk-type mill gear cutter. For a disk-type mill cutter of an outer diameter $d_{o.c}$ and of $N_{ct}$ teeth, the angle $\theta_{cs}$ between two adjacent teeth equals:

$$\theta_{cs} = \frac{2 \cdot \pi}{N_{ct}} \tag{6.33}$$

For this specific value of the central angle $\theta_{cs}$, the coefficient of proportionality $k$ is designated as $K$. To calculate for the coefficient $K$, the expression:

$$K = \frac{\pi \cdot d_{o.ct}}{N_{ct}} \cdot \tan \alpha_o \tag{6.34}$$

is commonly used.

The coefficient $K$ is an important design parameter of the disk-type mill cutter, particularly for manufacturers of the gear-cutting tool.

Under the resultant motion, an Archimedean spiral curve is the trajectory of every point of the cutting edge $CE$. In Figure 6.8, the arc segment $ab$ represents an example of such a trajectory. The trajectory of every point of the cutting edge $CE$ is equidistant to the arc segment $ab$.

We introduce a coordinate system $X_{ce}Y_{ce}Z_{ce}$ that the cutting edge $CE$ is rigidly connected to. When generating the clearance surface $C_s$, the coordinate system $X_{ce}Y_{ce}Z_{ce}$ travels together with the cutting edge itself.

In the initial position, axes of the coordinate system $X_{ce}Y_{ce}Z_{ce}$ are aligned with corresponding axes of the disk-type mill cutter coordinate system $X_{ct}Y_{ct}Z_{ct}$. Because of this, the coordinate system $X_{ce}Y_{ce}Z_{ce}$ that is associated with the cutting edge is not shown in its initial position in Figure 6.8. When the cutting edge $CE$ turns through a certain angle $\theta_{cs}$, the coordinate system $X_{ce}Y_{ce}Z_{ce}$ turns to the position of the coordinate system $X_1Y_1Z_1$. Furthermore, when the cutting edge is translated at a certain distance $a_z$, the coordinate system $X_{ce}Y_{ce}Z_{ce}$ travels to the position of the coordinate system $X_2Y_2Z_2$. In the last position of

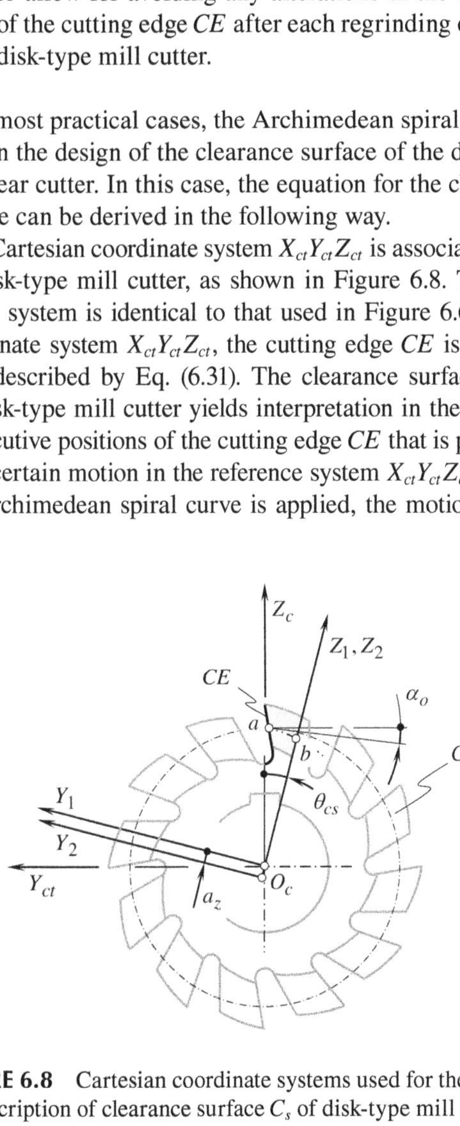


FIGURE 6.8 Cartesian coordinate systems used for the analytical description of clearance surface $C_s$ of disk-type mill cutter.

the coordinate system $X_2Y_2Z_2$, the equation for the cutting edge $CE$ remains the same [see Eq. (6.31)]. Therefore, in order to derive an equation for the clearance surface $C_s$, it is necessary to analytically describe the final position of the cutting edge $CE$ in the coordinate system $X_{ct}Y_{ct}Z_{ct}$ associated with the disk-type mill cutter.

The operator of translation $\mathbf{Tr}(a_z, Z_2)$ can be used for the analytical description of the transition from the coordinate system $X_2Y_2Z_2$ to the coordinate system $X_1Y_1Z_1$. Furthermore, rotation from the coordinate system $X_2Y_2Z_2$ to the coordinate system $X_{ct}Y_{ct}Z_{ct}$ associated with the disk-type mill cutter can be analytically described by the operator of rotation $\mathbf{Rt}(\theta_{cs}, X_2)$. Ultimately, use of the operators of linear transformations $\mathbf{Tr}(a_z, Z_2)$ and $\mathbf{Rt}(\theta_{cs}, X_2)$ yields an equation for the clearance surface $C_s$ of the disk-type mill gear cutter:

$$\mathbf{r}_{cs}(\theta_{cs}, \vartheta_{ce}) = \mathbf{Rt}(\theta_{cs}, X_2) \cdot \mathbf{Tr}(a_z, Z_2) \cdot \mathbf{r}_{ce}(\vartheta_{ce}) \tag{6.35}$$

where the clearance surface $C_s$ is described by Eq. (6.35) in terms of two parameters: the angle $\theta_{cs}$ and the parameter $\vartheta_{ce}$ of the cutting edge [see Eq. (6.31)].

Similarly, corresponding equations of the clearance surface $C_s$, which is shaped with the help either of a logarithmic spiral curve or of a circular arc, as well as by means of another appropriate curve, can be derived. For this purpose, it is sufficient to just enter an appropriate function $a_z = a_z(\theta_{cs})$ into the operator of translation $\mathbf{Tr}(a_z, Z_2)$. An appropriate correspondence between the displacement $a_z$ and the angle $\theta_{cs}$ is established by the function $a_z = a_z(\theta_{cs})$. Finally, the operator of translation could be represented in the form $\mathbf{Tr}[a_z(\theta_{cs}), Z_2]$. No other alterations to Eq. (6.35) need to be introduced at this point.

## 6.4 PROFILING DISK-TYPE MILL GEAR CUTTERS

When machining a spur gear, the axial profile of the generating surface of the disk-type mill cutter is identical to the normal section of the space width between the gear teeth. Neither the face surface nor the clearance surface of the disk-type mill cutter coincides with the normal section of the space between two neighboring gear teeth. This causes distortions in the tooth profile within the face surface of the disk-type mill cutter and the tooth profile in the normal section of the clearance surface. The tooth profile within the face surface of the disk-type mill cutter is used for the inspection purposes, whereas the tooth profile in the normal section of the clearance surface of the disk-type mill cutter tooth is used for the purpose of manufacturing the gear-cutting tool.

### 6.4.1 Use of Descriptive-Geometry-Based Method of Profiling

When tooth profile of a spur gear is specified, the disk-type mill cutter tooth profile within the face surface, as well as the normal profile of the clearance surface can be determined on the premise of descriptive-geometry-based method of profiling form cutting tools. In the remainder of this chapter, these method of profiling of gear-cutting tools are referred to as the *DG*-based method. The high efficiency of implementation of *DG*-based method is practically proven and widely recognized [98, 144, 148, 151, 158].

Consider an example of profiling of a disk-type mill gear cutter according to *DG*-based method.

In Figure 6.9, a spur gear being machining is depicted in the system of the projective planes $\pi_2$ and $\pi_3$. The axis of

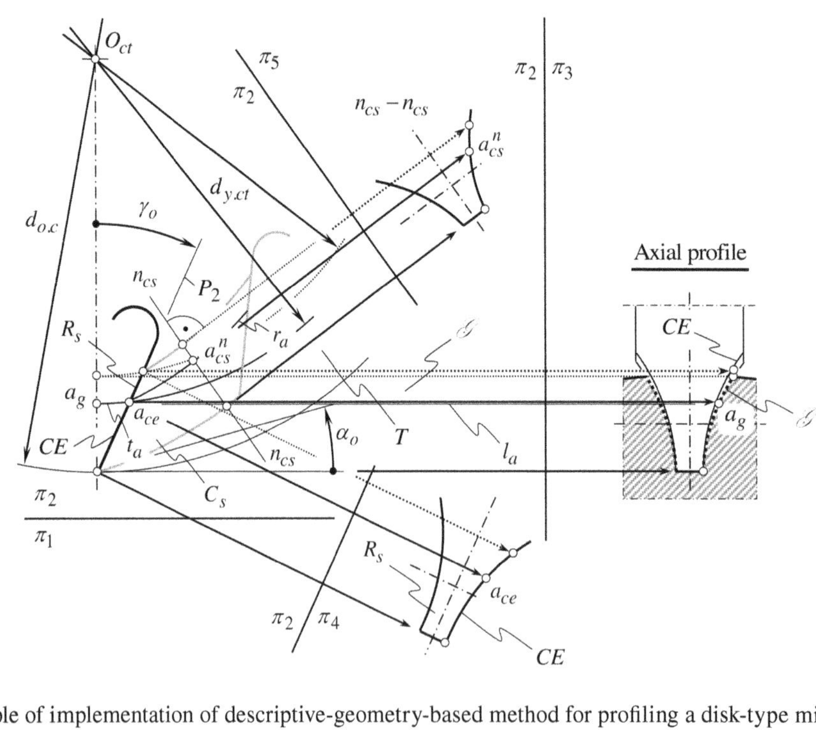


**FIGURE 6.9** Example of implementation of descriptive-geometry-based method for profiling a disk-type milling cutter.

rotation $O_g$ of the gear is parallel to the axis of projections $\pi_1/\pi_2$. Tooth space of the gear is projected onto the frontal plane of projections $\pi_3$ with no distortions.

For profiling a disk-type mill cutter, the gear tooth profile is subdivided by points $a, b, c, \ldots$ into a number of reasonably short segments. It is assumed below that the linear approximation of every short segment of the gear tooth profile in $\pi_2$ is permissible. For illustrative purposes, only point $a$ of the gear tooth profile is depicted in Figure 6.9. Within the plane of projections $\pi_2$, the point $a$ of the gear tooth profile is designated as $a_g$. The rest of the points of the cutting edge are treated similarly to that the point $a$ is treated.

Within the vertical plane of projections $\pi_2$, point $a$ of the gear tooth flank $\mathscr{G}$ is defined by the set of straight lines $l_a$, $l_b$, $l_c$, .... These lines are through the points $a, b, c, \ldots$ are parallel to the axis of projections $\pi_1/\pi_2$.

Generating surface $T$ of the disk-type mill cutter can be viewed as an enveloping surface to consecutive positions of the gear tooth flank $\mathscr{G}$ if one assumes that the surface $\mathscr{G}$ rotates about the disk-type mill cutter axis $O_{ct}$. In Figure 6.9, the surface $T$ is defined by a set of circular arcs $t_a, t_b, t_c, \ldots$.

For profiling the disk-type mill cutter, an auxiliary projective plane is constructed. This plane is a projective plane onto the vertical plane of projections $\pi_2$. Within the vertical plane of projections $\pi_2$, the auxiliary projective plane is depicted as the trace $P_2$. By construction, the auxiliary projective plane is a plane through the top cutting edge of the disk-type mill cutter and is inclined at the rake angle $\gamma_o$. Therefore, the auxiliary plane is congruent to the face surface $R_s$ of the disk-type mill cutter. This plane of projections is designated as $\pi_4$. Cutting edge $CE$ of the disk-type mill cutter is aligned with the line of intersection of the generating surface $T$ by the auxiliary projective plane $\pi_4$. Point $a_{ce}$ is located within the cutting edge $CE$. With no distortions, the tooth profile of the disk-type mill cutter within the rake plane $R_s$ is shown in the plane of projections $\pi_4$ (see Figure 6.9). The axis of projections $\pi_2/\pi_4$ is parallel to the trace $P_2$. Projection of the cutting edge $CE$ onto the frontal plane of projections $\pi_3$ is also shown.

A relieving curve is a curve $r_a$ through point $a_{ce}$ that forms a desired clearance angle $\alpha_o$ in relation to the generating surface $T$ of the disk-type mill cutter. This could be either the Archimedean spiral curve or a circular arc or another planar curve of an appropriate geometry.

The clearance surface, $C_s$, of the disk-type mill cutter tooth is defined in Figure 6.9 as the set of the relieving curves $r_a, r_b, r_c, \ldots$.

Ultimately, the clearance surface $C_s$ is intersected by the plane $n_s - n_s$ that is orthogonal to the surface $C_s$. In this way, the normal section of the clearance surface $C_s$ is constructed. In Figure 6.9, the point $a$ of the normal section of the clearance surface is denoted by $a_{cs}^n$. This projection of the point $a$ is located within the plane of projections $\pi_5$. The axis of projections $\pi_2/\pi_5$ is parallel to the to the normal section $n_s - n_s$.

The use of the considered $DG$-based approach of profiling can be expanded to profiling disk-type mill cutters having angle of inclination at the top cutting edge.

Based on the analysis illustrated in Figure 6.9, the configuration of an arbitrary point $a$ remains the same in the direction of the tooth width, but it is altered in the direction of the tooth height.

Graphical methods of profiling of gear-cutting tools lead to a clear understanding of the problem under consideration. They also return a solution to the problem that enables elimination of rough errors in the determination of the disk-type mill cutter tooth profile in various sections. An accurate solution to the problem of profiling of disk-type mill gear cutter can be derived on the premise of analytical methods.

### 6.4.2 Analytical Profiling of Disk-Type Mill Cutters

Numerous approaches can be used aiming for an analytical solution to the problem of profiling of disk-type mill cutters. A convenient method for solving this particular problem uses elements of linear transformations.

Consider the analytical profiling of the disk-type mill cutter for machining a spur gear.

In a coordinate system $X_gY_gZ_g$ associated with the gear, the tooth profile of the spur involute gear being machining is given by Eq. (5.11):

$$\mathbf{r}_{inv}^{g}(\phi_g) = \begin{bmatrix} \dfrac{r_{b.g}}{\cos\phi_g}\cdot\sin(\alpha+\mathrm{inv}\phi_g) \\ -\dfrac{r_{b.g}}{\cos\phi_g}\cdot\cos(\alpha-\mathrm{inv}\phi_g) \\ 0 \\ 1 \end{bmatrix} \tag{6.36}$$

The axial profile of the disk-type mill cutter is congruent with the tooth profile of the gear to be machined [see Eq. (6.36)].

The coordinate system $X_{ct}Y_{ct}Z_{ct}$ is associated with the disk-type mill cutter, as shown in Figure 6.10. The operator of translation $\mathbf{Tr}(C, Y_g)$ is used for the analytical description of the transition from the coordinate system $X_gY_gZ_g$ to the coordinate system $X_{ct}Y_{ct}Z_{ct}$. Here, the center distance is designated as $C$.

In the reference system $X_{ct}Y_{ct}Z_{ct}$, the axial profile of the generating surface $T$ of the disk-type mill cutter is analytically described by:

$$\mathbf{r}_{inv}^{c}(\phi_g) = \mathbf{Tr}(C, Y_g)\cdot\mathbf{r}_{inv}^{g}(\phi_g) \tag{6.37}$$

To generate the surface $T$, it is necessary to rotate the axial profile $\mathbf{r}_{inv}^{c}(\phi_g)$ about the axis $O_{ct}$ of the disk-type mill cutter. For an analytical description of the rotation, the Cartesian coordinate system $X_1Y_1Z_1$ is introduced. The coordinate system $X_1Y_1Z_1$ shares the axis $X_1 \equiv X_{ct}$ with the coordinate system $X_{ct}Y_{ct}Z_{ct}$. The reference system $X_1Y_1Z_1$ is turned about the $X_{ct}$ − axis through a certain angle $\varphi$. The transition

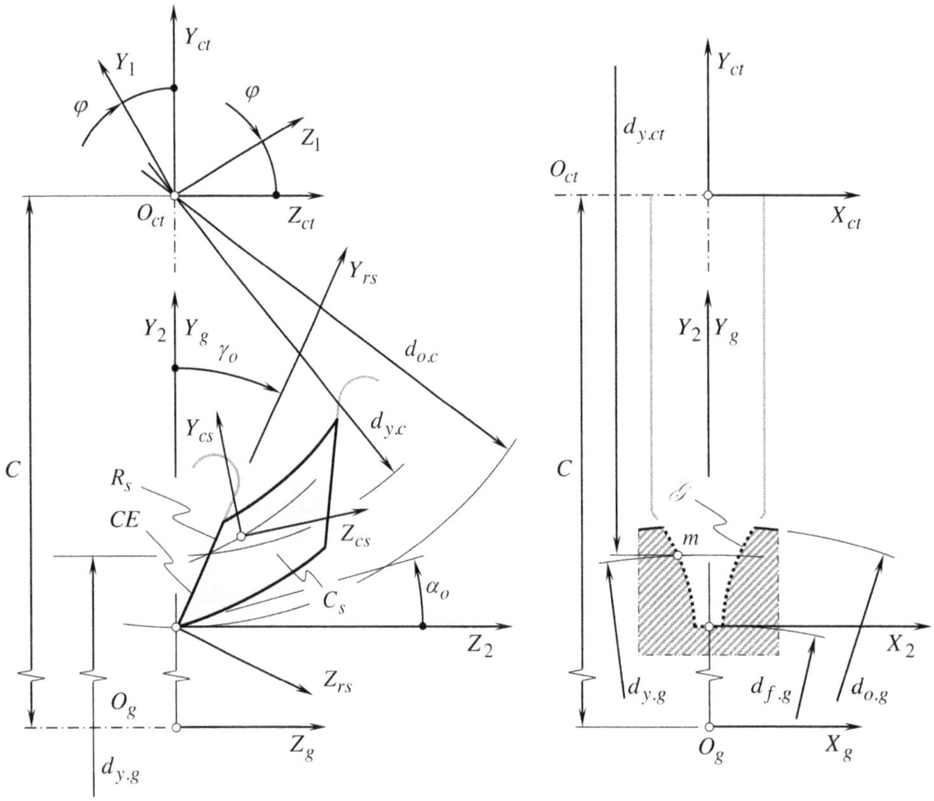


**FIGURE 6.10** Analytical profiling of a disk-type mill cutter.

from the coordinate system $X_1Y_1Z_1$ to the coordinate system $X_{ct}Y_{ct}Z_{ct}$ is analytically described by the operator of rotation $\mathbf{Rt}(\varphi, X_{c2})$. The position vector $\mathbf{r}_{ct}$ of a point of the generating surface $T$ of the disk-type mill cutter can be obtained by left multiplying the position vector of a point $\mathbf{r}_{inv}^{c}(\phi_g)$ of the axial profile of the generating surface $T$ [see Eq. (6.37)] by the operator of rotation $\mathbf{Rt}(\varphi, X_{c2})$:

$$\mathbf{r}_c(\phi_g, \varphi) = \mathbf{Rt}(\varphi, X_{ct}) \cdot \mathbf{r}_{inv}^{c}(\phi_g) \tag{6.38}$$

Omitting bulky formula transformation, the following equation:

$$\mathbf{r}_{ct}(\phi_{y.g}, \varphi) = \begin{bmatrix} \dfrac{r_{b.g}}{\cos\phi_{y.g}} \cdot \sin(\alpha + \mathrm{inv}\phi_{y.g}) \\ \dfrac{d_{o.ct} - d_{y.ct}(\phi_{y.g})}{2} \sin\phi_{y.g} \\ \dfrac{d_{o.ct} - d_{y.ct}(\phi_{y.g})}{2} \cos\phi_{y.g} \\ 1 \end{bmatrix} \tag{6.39}$$

can be derived for position vector $\mathbf{r}_{ct}$ of a point of the generating surface $T$ of the gear disk-type mill cutter.

The use of Eq. (6.39) for the generating surface $T$ corresponds to the grinding operation of a gear with a form grinding wheel for an external [see Figure 6.11(a)] and an internal [see Figure 6.11(b)] spur involute gear. Gears with non-involute tooth profiles can be ground in a similar manner [see Figure 6.11(c)]. In this case, the axial profile of the generating surface $T$ [see Eq. (6.37)] can be used for dressing the form grinding wheel. The approach can be enhanced to grinding helical involute gears with a form grinding wheel (see Figure 6.12).

The formula:

$$C = 0.5 \cdot (d_{y.g} + d_{y.ct}) \tag{6.40}$$

is used for the calculation of the center distance $C$.

In Eq. (6.40), $d_{y.g}$ and $d_{y.ct}$ are the diameters of the gear being machining and the gear-cutting tool through that same point of interest $m$ of the gear tooth profile (see Figure 6.10). The diameter $d_{y.g}$ can be expressed in terms of the gear pressure angle $\phi_{y.g}$ at the point of interest of the tooth profile:

$$d_{y.g} = d_{y.g}(\phi_{y.g}) \tag{6.41}$$

For this purpose, the engineering formulae for the specification of the gear tooth flank can be used (see Appendix F). Ultimately, this yields an expression:

$$d_{y.ct} = d_{y.ct}(\phi_{y.g}) \tag{6.42}$$

for the calculation of the diameter $d_{y.ct}$.

Equation (6.42) is used for the reparameterization in Eq. (6.39).

(a) (b) (c)

**FIGURE 6.11** Grinding (a) an external and (b) an internal, and (c) a non-involute spur gear with a form grinding wheel.

Furthermore, in calculating for the coordinates of points of the disk-type mill cutter tooth profile those located within the rake plane $R_s$, an appropriate equation is necessary. The required equation can be derived from the following consideration.

The coordinate system $X_{rs}Y_{rs}Z_{rs}$ is associated with the rake surface $R_s$, as shown in Figure 6.10. Transition from the coordinate system $X_{ct}Y_{ct}Z_{ct}$ that is associated with the disk-type mill cutter, to the rake surface coordinate system $X_{rs}Y_{rs}Z_{rs}$ can be executed in two steps. First, it is necessary to translate the coordinate system $X_{ct}Y_{ct}Z_{ct}$ along the $Y_{ct}$ – axis at the distance $0.5 \cdot d_{oc}$ (see Figure 6.10) to the position of the coordinate system $X_2Y_2Z_2$. A linear transformation of this type is analytically described by the operator of translation $\mathbf{Tr}(0.5 \cdot d_{oc}, Y_{ct})$.

Next, the coordinate system $X_2Y_2Z_2$ is turned about the axis $X_2$ through the rake angle $\gamma_o$ to the position of the coordinate system $X_{rs}Y_{rs}Z_{rs}$ of the rake surface. A linear transformation of this type is analytically described by the operator of rotation $\mathbf{Rt}(\gamma_o, X_2)$.

In the coordinate system $X_{rs}Y_{rs}Z_{rs}$ the equation of the generating surface $T$ of the disk-type mill cutter [see Eq. (6.35)] casts into:

$$\mathbf{r}_{ct}^{rs}(\phi_{y.g}, \varphi) = \mathbf{Rt}(\gamma_o, X_2) \cdot \mathbf{Tr}(0.5 \cdot d_{o.ct}, Y_{ct}) \cdot \mathbf{r}_{ct}(\phi_{y.g}, \varphi) \quad (6.43)$$

In the coordinate system, $X_{rs}Y_{rs}Z_{rs}$ the equation of the rake plane $R_s$ of the disk-type mill cutter (see Figure 6.10) allows for representation in the form $Z_{rs} = 0$. Substituting this value of $Z_{rs}$ into Eq. (6.43), one can derive an expression:

$$\mathbf{r}_{ce}^{rs}(\phi_{y.g}, \varphi) = \begin{bmatrix} \dfrac{r_{b.g}}{\cos\phi_{y.g}} \cdot \sin(\alpha + \operatorname{inv}\phi_{y.g}) \\ \dfrac{d_{y.ct}\sin\varphi}{\sin\gamma_o} \\ 0 \\ 1 \end{bmatrix} \quad (6.44)$$

for the calculation of the coordinates of points of the disk-type mill cutter tooth profile located within the rake plane $R_s$—that is, calculation of coordinates of points of the cutting edge $CE$ of the disk mill gear cutter.

It is important to note that the following equality:

$$\sin(\gamma_o + \varphi) = \frac{d_{o.ct} \cdot \sin\gamma_o}{d_{y.ct}} \quad (6.45)$$

is valid for the parameters in Eq. (6.44).

Inverse linear transformation can be used to represent the expression for the position vector of a point $\mathbf{r}_{ce}^{rs}$ of the cutting edge $CE$ in the coordinate system $X_{ct}Y_{ct}Z_{ct}$ associated with the disk-type mill cutter. The operator of the resultant inverse coordinate system transformation $\mathbf{Rs}(rs \mapsto ct)$ is equal to the dot product:

$$\mathbf{Rs}(rs \mapsto ct) = \mathbf{Tr}^{-1}(0.5 \cdot d_{o.ct}, Y_{ct}) \cdot \mathbf{Rt}^{-1}(\gamma_o, X_2) \quad (6.46)$$

By left multiplying the position vector $\mathbf{r}_{ce}^{rs}$ by the operator $\mathbf{Rs}(rs \mapsto ct)$:

$$\mathbf{r}_{ce}^{c} = \mathbf{Rs}(rs \mapsto ct) \cdot \mathbf{r}_{ce}^{rs} \quad (6.47)$$

**FIGURE 6.12** Grinding a helical gear with a form grinding wheel.

The equation for the position vector of a point $\mathbf{r}_{ce}^{c}$ of the cutting edge $CE$ that is represented in the coordinate system $X_{ct}Y_{ct}Z_{ct}$ can be derived.

To calculate for the coordinates of points of the axial profile of the clearance surface $C_s$ of the disk-type mill cutter, it is necessary:

- To present the formula [Eq. (6.35)] for the surface $C_s$ in an expended form.
- To express the equation of the surface $C_s$ in terms of the parameters that are convenient for the analysis of this particular type.

For this purpose, a Cartesian coordinate system $X_{cs}Y_{cs}Z_{cs}$ is introduced. The coordinate system $X_{cs}Y_{cs}Z_{cs}$ is associated with the moving cutting edge $CE$ that serves as the generating curve for the generation of the clearance surface $C_s$ of the disk-type mill cutter tooth.

When the clearance surface $C_s$ is generating by the cutting edge, the cutting edge $CE$ and the coordinate system $X_{cs}Y_{cs}Z_{cs}$ perform a motion, which is commonly referred to as the relieving motion. The relieving motion can be viewed as a superposition of two motions, that is, of:

- The uniform rotation of the disk-type mill gear cutter about its axis $O_{ct}$.
- The straight motion toward the axis $O_{ct}$

Depending on the parameters of the straight motion—say, depending on the type of the function $k = k(\theta_{cs})$ [see Section 6.2.2]—clearance surfaces of various geometries can be generated (see Figure 6.13). An Archimedean clearance surface $C_s$ is considered as an example. For the Archimedean clearance surface, the equality:

$$a_y = k \cdot \theta_{cs} \tag{6.48}$$

is valid.

When the disk-type mill cutter turns through an angle $\theta_{cs}$, the cutting edge $CE$, together with the coordinate system $X_{cs}Y_{cs}Z_{cs}$, approaches the axis $O_{ct}$ at the distance $a_y = a_y(\theta_{cs})$, as shown in Figure 6.10. For an analytical description of the linear transformation of this type, the operator $\mathbf{Rs}(ct \mapsto cs)$ of the resultant coordinate system transformation is composed. The operator $\mathbf{Rs}(ct \mapsto cs)$ is equal to the product:

$$\mathbf{Rs}(ct \mapsto cs) = \mathbf{Tr}(a_y, Y_{cs}) \cdot \mathbf{Rt}(\theta_{cs}, X_{cs}) \tag{6.49}$$

The operator $\mathbf{Rs}(ct \mapsto cs)$ of the linear transformation is used to obtain an expanded form of the equation for position vector of a point $\mathbf{r}_{cs}^{cs}$ of the clearance surface $C_s$ of the disk-type mill cutter tooth. By left multiplying the position vector $\mathbf{r}_{ce}^{c}$ of a point of interest of the cutting edge $CE$ by the operator $\mathbf{Rs}(c \mapsto cs)$ of the resultant coordinate system transformation, the equation:

$$\mathbf{r}_{cs}^{cs} = \mathbf{Rs}(ct \mapsto cs) \cdot \mathbf{r}_{ce}^{c} \tag{6.50}$$

can be derived for the position vector of a point $\mathbf{r}_{cs}^{cs}$ of the clearance surface $C_s$ that is given in the local reference system $X_{cs}Y_{cs}Z_{cs}$.

Substituting $Z_{cs} = 0$ into Eq. (6.50) returns an equation for the axial profile of the disk-type mill cutter tooth. The position vector of a point $\mathbf{r}_{ax}^{cs}$ of the axial section of the clearance surface $C_s$ yields matrix representation in the form:

$$\mathbf{r}_{ax}^{cs}(\phi_{y.g}, \theta_{cs}) = \begin{bmatrix} \dfrac{r_{b.g}}{\cos\phi_{y.g}} \cdot \sin(\alpha + \mathrm{inv}\phi_{y.g}) \\ d_{o.ct} - d_{y.ct} - a_y(\theta_{cs}) \\ 0 \\ 1 \end{bmatrix} \tag{6.51}$$

**FIGURE 6.13** Relieving teeth of disk-type mill cutter.

where the condition:

$$\sin(\gamma_o + \varphi) = \frac{d_{o.ct} \cdot \sin \gamma_o}{d_{y.ct}} \tag{6.52}$$

is fulfilled.

Similarly, the equation for the position vector of a point $\mathbf{r}_{ax}^{cs}$ of the axial profile is derived [see Eq. (6.51)], and the equation of the normal section of the clearance surface $C_s$ can be derived as well. To derive the equation for the position vector of a point $\mathbf{r}_{nn}^{cs}$ of the clearance surface $C_s$, it is convenient to introduce an auxiliary coordinate system $X_{nn}Y_{nn}Z_{nn}$. The coordinate plane $Z_{nn} = 0$ of the auxiliary coordinate system $X_{nn}Y_{nn}Z_{nn}$ is orthogonal to the clearance surface $C_s$. The interested reader may wish to work out this partial problem independently.

The results of the analysis can be generalized for the case of disk-type mill cutters with a certain inclination angle $\lambda_o$ at the top cutting edge. Disk-type mill cutters of such a geometry have limited use in the industry. Because of this, the generalized case is not disclosed in detail here.

An important intermediate conclusion: As follows from Eqs. (6.44), (6.45), (6.51), and (6.52), the parameters of the tooth profile of the disk-type mill cutter depend on the outer diameter $d_{o.ct}$ of the disk-type mill cutter. The outer diameter $d_{o.ct}$ becomes smaller after every regrinding of the worn disk mill gear cutter. A decrease in the outer diameter unavoidably entails a corresponding decrease in accuracy of the disk mill gear cutter. A reduction in the accuracy is more significant for disk-type mill cutters with a larger tooth height ($h$)/ outer diameter ($d_{o.ct}$) ratio, where the tooth height is given by $h = (d_{o.ct} - d_{f.ct})/2$. Therefore, coarse pitch disk-type mill cutters with a small outer diameter are more subject to the impact of regrinding onto the accuracy of the cutting edges.

This means that the parameters of the tooth profile of the disk-type mill cutter have to be changed after every regrinding. Unfortunately, conventional methods of relieving disk-type mill cutter (see Figure 6.13) do not allow for alteration of the shape of the disk-type mill cutter clearance surface. Ultimately, after every regrind of the gear-cutting tool reduces the accuracy of disk-type mill cutters. To accommodate the deviations in shape of the clearance surface $C_s$, appropriate improvements in the relieving operation of precision disk-type mill cutters are of practical importance. Perfect results can be achieved when a special method[2] of relieving of disk-type mill cutters is used. Implementation of this method [89] is efficient for relieving green disk-type mill cutters before heat treatment and, in particular, for finish relieving hardened disk-type mill cutters.

[2] SU Pat. No. 848.290, *A Method of Relieving of Gear Hob*/S.P. Radzevich, Int. Cl. B 24 b 3/12, Filed: August 06, 1979.

## 6.5 CUTTING EDGE GEOMETRY OF DISK-TYPE MILL GEAR CUTTERS

The cutting-edge geometry of the disk-type mill cutter can be specified in terms of:

- The rake angle $\gamma$
- The angle of inclination $\lambda$ of the cutting edge
- The clearance angle $\alpha$
- The radius of curvature $R_{ce}$ of the cutting edge
- The roundness $\rho$ of the cutting edge

and a few other parameters.

Under certain gear machining conditions, some of these parameters have a more severe impact on the disk-type mill cutter performance, whereas other geometrical parameters are more important under different of conditions.

Analytical methods can be implemented in the investigation of the cutting edge geometry of the disk-type mill cutter. Analysis of the geometry of the clearance surface is carried out as an example.

A disk-type mill cutter is depicted in Figure 6.14.

The references system and the principal vectors for the determination of the normal clearance angle $\alpha_{N.y}$ at a point of interest of the cutting edge $CE$ of a disk mill gear cutter are shown in Figure 6.15.

The clearance surface $C_s$ of a disk-type mill cutter is a type of relieved surface. It is shaped using an Archimedean spiral curve as the relieving curve. The outer diameter $d_{o.ct}$ of the disk-type mill cutter and the clearance angle $\alpha_o$ at the top cutting edge are given. The rake angle $\gamma_o$ at the top cutting edge of the disk-type mill cutter tooth is equal to zero. The parameters of the involute tooth profile being machined are also known. It is necessary to determine the normal clearance angle $\alpha_{N.y}$ at a point of interest $m$ of the cutting edge $CE$. The cutting-edge point $m$ is located on a circle of an arbitrary diameter $d_{y.ct}$, the value of which is specified.

FIGURE 6.14 Disk-type mill cutter.

With the aim of expressing the normal clearance angle $\alpha_{N.y}$ at point $m$ of the cutting edge in terms of the design parameters of the disk-type mill cutter, a local Cartesian coordinate system $x_m y_m z_m$ is associated with the cutting edge $CE$, as shown in Figure 6.15. A set of three vectors **a**, **b**, and **C** is constructed within the plane that is tangent at point $m$ to the clearance surface $C_s$. The vectors **a**, **b**, and **C** need to be expressed in terms of the normal clearance angle $\alpha_{N.y}$ to be determined and known parameters of the geometry of the cutting wedge of the cutting tool.

The unit vector **a** is tangent to the relieving curve through point $m$. In terms of projections onto the coordinate axes, the unit vector **a** can be expressed as:

$$\mathbf{a} = \mathbf{i} \cdot \sin\alpha_{o.y} + \mathbf{k} \cdot \cos\alpha_{o.y} \tag{6.53}$$

where the clearance angle $\alpha_{o.y}$ in transverse section of the clearance surface $C_s$ is calculated from the well-known formula:

$$\alpha_{o.yt} = \tan^{-1}\left(\frac{d_{o.ct}}{d_{y.ct}} \cdot \tan\alpha_o\right) \tag{6.54}$$

The unit vector **b** is tangent to the lateral cutting edge $CE$ at the point of interest $m$. To calculate for this vector, the following expression:

$$\mathbf{b} = -\mathbf{i} \cdot \cos\phi_y + \mathbf{j} \cdot \sin\phi_y \tag{6.55}$$

is valid.

To calculate for the profile angle $\phi_y$ at a point of interest of the cutting edge of the disk-type mill cutter, the following equation can be used:

$$\phi_y = \cos^{-1}\left(\frac{d_{b.g}}{d_{y.g}}\right) \tag{6.56}$$

where $d_{b.g}$ denotes the base diameter of the gear to be machined, and $d_{y.g}$ denotes the gear diameter through the point of interest $m$.

Ultimately, magnitude of the vector **C** is chosen so that the length of the projection of the vector **C** onto the coordinate plane $z_m = 0$ is equal to unit. For the calculation of the vector **C**, the formula:

$$\mathbf{C} = \mathbf{i} \cdot \sin\phi_y + \mathbf{j} \cdot \cos\phi_y + \mathbf{k} \cdot \cot\alpha_{N.y} \tag{6.57}$$

is composed (see Figure 6.15).

By construction, three vectors **a**, **b**, and **C** are situated within the plane that is tangent at $m$ to the clearance surface $C_s$. Once these vectors are within the common plane, their triple scalar product is identical to zero ($\mathbf{a} \times \mathbf{b} \cdot \mathbf{C} \equiv 0$).

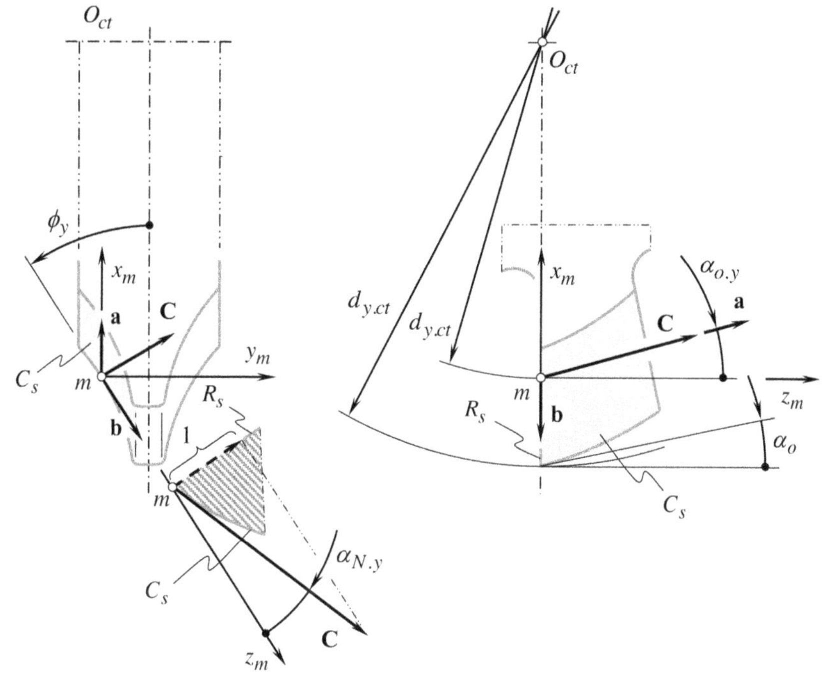


**FIGURE 6.15** References system and the principal vectors for the determination of the clearance angle $\alpha_{N.y}$ at a point of interest of the cutting edge $CE$ of a disk-type mill cutter.

The triple scalar product of the vectors **a**, **b**, and **C** can be expressed in the form of the determinant:

$$\begin{vmatrix} \sin\alpha_{o.y} & 0 & \cos\alpha_{o.y} \\ -\cos\phi_y & \sin\phi_y & 0 \\ \sin\phi_y & \cos\phi_y & \cot\alpha_{N.y} \end{vmatrix} = 0 \tag{6.58}$$

Equation (6.54), which is used to calculate the clearance angle $\alpha_{o.y}$, and Eq. (6.56), used to calculate for the pressure angle $\phi_y$, can be substituted into Eq. (6.58). After the necessary formula transformations are executed, Eq. (6.58) casts into:

$$\alpha_{N.y} = \tan^{-1}\left(\frac{d_{o.c}}{d_{y.c}} \cdot \sqrt{d_{y.c}^2 - d_{b.c}^2} \cdot \tan\alpha_o\right) \tag{6.59}$$

for the calculation of the normal clearance angle $\alpha_{N.y}$ at point of interest $m$ of the cutting edge $CE$ of the disk-type mill cutter.

An example of a graphical interpretation of Eq. (6.59) is illustrated in Figure 6.16. The current value of the normal clearance angle $\alpha_{N.y}$ is larger at points closer to the outer diameter $d_{o.ct}$ of the disk-type mill cutter and becomes progressively smaller as the cutting edge point approaches the axis of rotation $O_{ct}$ of the disk mill gear cutter. For portions of the cutting edge that are closer to the axis $O_{ct}$ values of the normal clearance angle can reach as low of a value as about $\alpha_{N.y} \leq 2°$. The cutting performance of a gear-cutting tool with an unsuitable normal clearance angle $\alpha_{N.y}$ is very often insufficient from a practical standpoint. This means that appropriate improvements in the design of the disk-type mill cutter are needed.

Disk-type mill cutters for machining involute gears with fewer teeth are more vulnerable to change in the normal clearance angle $\alpha_{N.y}$ within the cutting edge $CE$ compared to disk-type mill cutters used for machining of gears with a larger number of teeth.

Analysis of Equation (6.59) also reveals that a particular type of relieving curve does not affect the actual value of

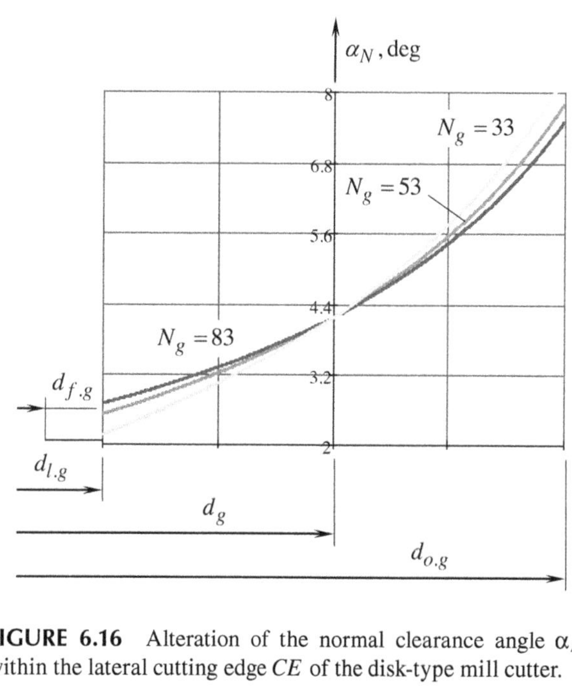


**FIGURE 6.16** Alteration of the normal clearance angle $\alpha_{N.y}$ within the lateral cutting edge $CE$ of the disk-type mill cutter.

the normal clearance angle $\alpha_{N.y}$ at a point of the cutting edge of the disk-type mill cutter.

The same analytical method used for the investigation of the geometry of the clearance surface $C_s$ can be implemented to analyze the geometry of the rake surface $R_s$ of the disk-type mill cutter.

Disk-type mill cutters are used for roughing and finishing of gears. Finishing disk-type mill cutters are produced with relieved teeth. The clearance surface of the finishing disk-type mill cutters is generated during the relieving operation. The clearance angle at the top cutting edges (at the outer diameter of the disk-type mill cutter) falls in the range of $\alpha_o = 12\ldots15°$, which means the normal clearance angle at lateral cutting edges is at the range of $\alpha_{N.y} = 1.5\ldots3.0°$.

Finishing disk-type mill cutters have zero rake angle $\gamma_o$ at the top cutting edge ($\gamma_o = 0°$). Therefore, both the normal rake angle $\gamma_{N.y}$ and the angle of inclination $\lambda_y$ of the cutting edge have zero values at every point of the lateral cutting edge $CE$ of the cutter tooth ($\gamma_{N.y} = 0°$ and $\lambda_y = 0°$).

Shown in Figure 6.17, a disk mill cutter for roughing gears is designed to have rake angle of either zero value ($\gamma_o = 0°$) or equal to $\gamma_o = +10°$. For these rake angle $\gamma_o$ values, changes in the normal rake angle $\gamma_{N.y}$ and the angle of inclination $\lambda_y$ of the cutting edge within the lateral cutting edge are negligible and do not affect the cutting performance of the disk mill gear cutter.

Disk-type mill cutters for semi-finishing of hardened gears (gear to be machined tooth surface hardness at the range of up to $HRC$ 60...62) can be designed with a negative

FIGURE 6.17 Disk-type mill cutter for roughing gears.

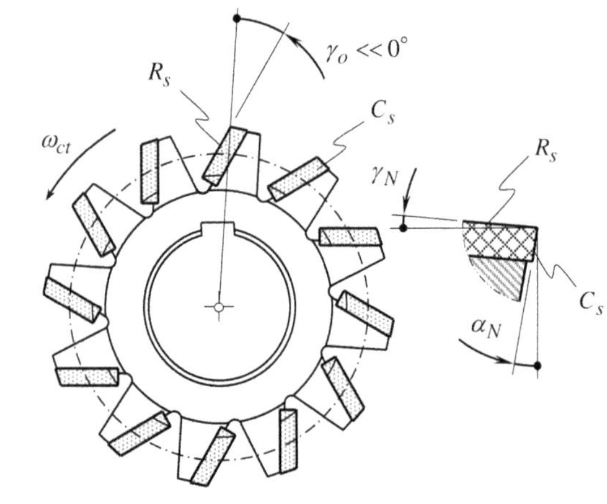


FIGURE 6.18 Disk-type mill cutter for finishing and/or semi-finishing hardened gears.

rake angle $\gamma_o$ value. For this purpose, the rake angle is increased up to $\gamma_o = -30°\ldots-40°$ (see Figure 6.18). The rake angle $\gamma_{o.y}$ varies within the lateral cutting edges, and variation in the angle $\gamma_{o.y}$ can reach up to $\Delta\gamma_{o.y} \cong 20°$.

Because of an increased negative rake angle ($\gamma_o = -30°\ldots-40°$), the angle of inclination $\lambda$ of the cutting edge $CE$ is also approximately of that same range. Similarly, the variation in the angle of inclination $\lambda_y$ of the cutting edge within the lateral cutting edges can reach up to $\Delta\lambda_y \cong 20°$.

Because of an increased negative rake angle $\gamma_o$, the normal rake angles $\gamma_N$ at the lateral cutting edges are also negative ($\gamma_N \cong -1 \div -5°$). It is practical to maintain the normal clearance angle at the lateral cutting edges $\alpha_N$ at the range of $\alpha_N = 8 \div 10°$. For this purpose, worn disk-type mill cutters are reground on the clearance surfaces and not on the rake surfaces.

Top edges of the disk-type mill cutter of this design do not cut the stock. Therefore, the geometry of the top cutting edges is not a focus of interest for cutting tool designers.

The analytical method for the calculating cutting edge geometry can also be used to analyze the geometry of the lateral cutting edges of disk-type mill cutters for semi-finishing hardened gears.

## 6.6 DISK-TYPE MILL CUTTERS FOR ROUGHING GEARS

Roughing disk-type mill cutters are mostly used for machining of coarse pitch gears and are also used for machining medium height gear teeth.

Design of roughing disk-type mill cutters is similar to that of finishing gear disk-type mill cutters (see Figure 6.19). However, the accuracy requirements for roughed gears are not as strict. This allows for improvements in the design of disk-type mill cutters used only for roughing gears.

Rake angle $\gamma_o$ of roughing disk-type mill cutters is either zero ($\gamma_o = 0°$) or can be increased to $\gamma_o = 10°$ [see Figure 6.19(a)]. Clearance angle can be increased up to

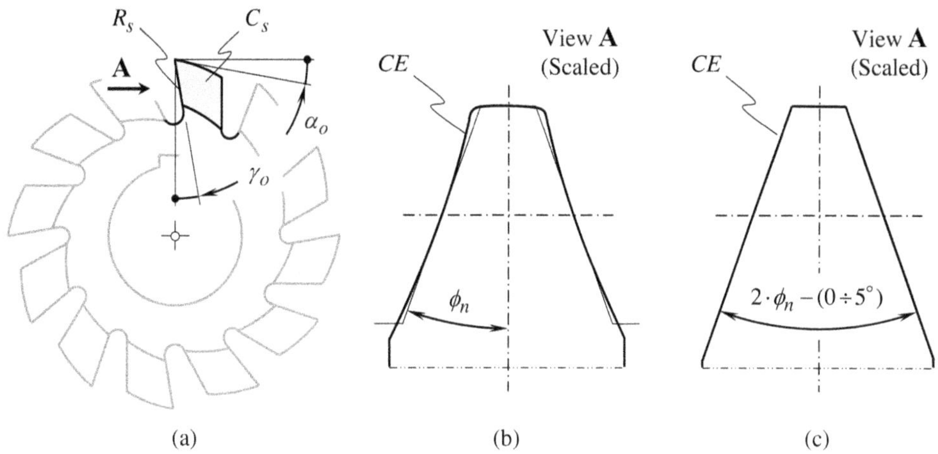


**FIGURE 6.19** Design parameters of the disk-type mill cutters for roughing involute gears.

$\alpha_o = 15 \div 18°$. Both the increased rake angle $\gamma_o$ and the increased clearance angle $\alpha_o$ entail an improved cutting performance of disk mill gear cutters.

The geometry of the tooth profile of the disk-type mill cutter depends on the tooth number of the gear being machined. When the tooth number is larger, the radius of curvature of the disk-type mill cutter tooth profile also becomes larger. When the tooth number approaches infinity, the gear transforms into a rack with a straight tooth profile. This property of the disk-type mill gear cutter is used for improvement in the design of roughing disk-type mill cutters. For machining gears with a reasonably small tooth number, disk-type mill cutters with an involute tooth profile are used [see Figure 6.19(b)]. For milling gears with a large tooth number—and consequently, with an almost straight tooth profile—disk-type mill cutters with straight lateral cutting edges [see Figure 6.19(c)] are extensively used.

When machining a coarse pitch gear, the most intensive material removal is observed during roughing of the gear. For this purpose, it is recommended to use roughing disk-type mill cutters with a profile angle at the range of $2\phi = 35\ldots40°$. Use of disk-type mill cutters with a slightly reduced profile angle ($2\phi = 35°$) causes good stock to be removed when finishing the gear.

Roughing disk-type mill cutters whose clearance surface is used for regrinding have an increased normal clearance angle on the lateral cutting edges (up to $\alpha_N = 8\ldots10°$).

For grinding worn disk-type mill cutters (see Figure 6.19), conventional and special[3] methods of grinding are used. Using the methods discussed elsewhere [86, 190], the disk-type mill cutters can be reground on both the rake surfaces $R_s$ and the clearance surfaces $C_s$.

For machining coarse pitch gears, implementation of assembled disk-type mill cutters (see Figure 6.20) is recommended. Use of carbide inserts is efficient, especially for roughing gears made of materials that are difficult to machine.

One more issue is of critical importance with respect to the design of disk-type mill cutters for roughing of gears.

Consider a roughing design of disk-type mill cutter of conventional design. Every cutting tooth of the roughing design of disk-type mill cutter is trapezoidal and has thus three cutting edges: top cutting edge and two lateral cutting edges. All three remove the stock simultaneously.

**FIGURE 6.20** An assembled disk-type mill cutter for machining coarse pitch gears.

[3] SU Pat. No. 975.339, *A Method of Grinding of Gear Disk mill gear cutter*/S.P. Radzevich, Int. Cl. B 24 b 3/12, Filed: October 10, 1980.
SU Pat. No. 1.627.375, *A Method of Grinding of Gear Disk mill gear cutter having Straight Slots*/S.P. Radzevich, and I.V. Vinokourvov, Int. Cl. B 24 b 3/12, Filed: October 17, 1988.

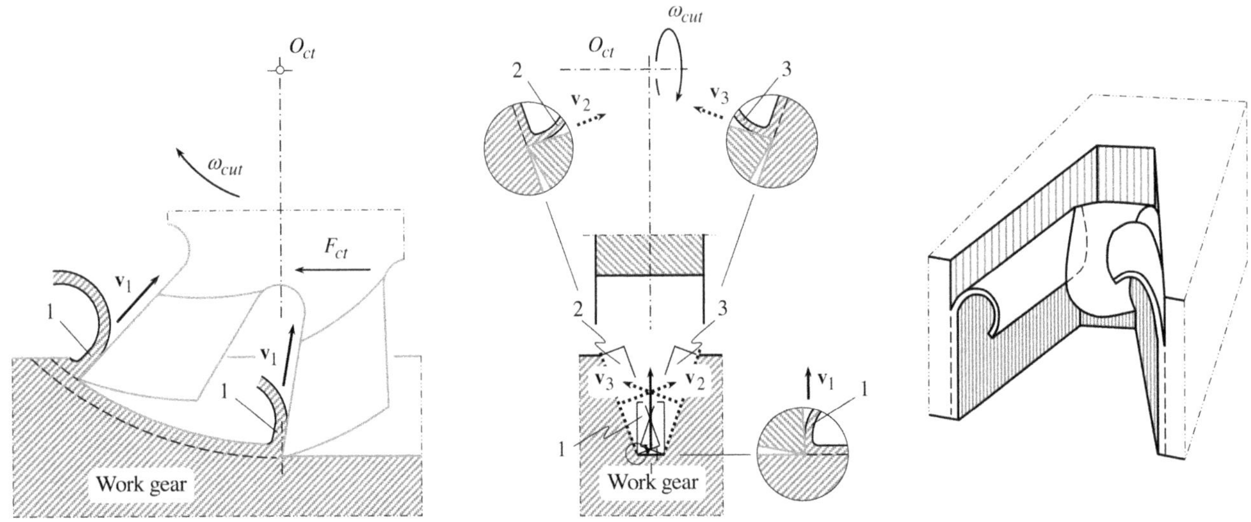


**FIGURE 6.21** Interference of the chip flows from neighboring cutting edges of the disk-type mill cutter tooth.

When machining a gear, the design of disk-type mill cutter rotates about its axis of rotation $O_{ct}$ (see Figure 6.21). The rotation of the disk mill gear cutter is designated as $\omega_{cut}$. At the same time, the design of disk-type mill cutter travels along the gear axis of rotation $O_g$ with a feed motion $F_{ct}$. The major portion of the stock is removed by the top cutting edges. Chip 1 from the top cutting edge flows in the direction $\mathbf{v}_1$. This direction of the chip flow is orthogonal to the top cutting edge. The rest portions of the stock are removed by the lateral cutting edges of the disk mill gear cutter. The desired direction of chip flow of the chip 2 from the lateral cutting edge is designated as $\mathbf{v}_2$. The direction of $\mathbf{v}_2$ is perpendicular to the lateral cutting edge of the disk mill gear cutter. The opposite lateral cutting edge removes chip 3. The desired direction of flow of the chip 3 from the lateral cutting edge is designated as $\mathbf{v}_3$. The direction of $\mathbf{v}_3$ is perpendicular to the lateral cutting edge of the disk mill gear cutter.

The vectors $\mathbf{v}_1$, $\mathbf{v}_2$, and $\mathbf{v}_3$ are located within the face surface $R_s$ of the of disk-type mill cutter. This means that the chips 1, 2, and 3, are constrained to flow in the desired directions $\mathbf{v}_1$, $\mathbf{v}_2$, and $\mathbf{v}_3$. The direction $\mathbf{v}_2$ of the flow of chip 2 interferes with the direction $\mathbf{v}_1$ of the flow of chip 1. The direction $\mathbf{v}_3$ of the flow of chip 3 interferes with the direction $\mathbf{v}_1$ of the flow of chip 1. Ultimately, chips from lateral cutting edges cannot move in the directions $\mathbf{v}_2$ and $\mathbf{v}_3$. The thickest chip 1 that flows from the top cutting edge, significantly affects the directions of the resultant chip flow.

As a result of the chip flow interference, an intensive deformation of the material being removing is observed in the corners made by the top cutting edge and by the lateral cutting edges.

The interference of chip flows from neighboring cutting edges of the disk-type mill cutter results in elevated cutting temperature, increased cutting forces, and intensive cutting tool wear. The interested reader is referred to [37] for details on this matter.

To improve the cutting performance of disk-type mill, it is highly desirable to eliminate the interference of chip flows from the neighboring cutting edges. Modification of the disk-type mill cutter tooth profile is an efficient means for solving problems of this sort.

For example, consider the design of the disk-type mill cutter that removes the stock in compliance with the so-called progressive cutting diagram.[4]

Cutting teeth of disk-type mill cutters with an even number of teeth [see Figure 6.22(a)] can be divided onto two groups: those that are even-numbered and those that are odd-numbered. For this purpose, all the teeth are numerated.

Even-numbered disk-type mill cutter teeth form the first group of teeth. Disk-type mill cutter teeth in this group have regular tooth thickness $t_{ct}$. However, they are shorter at a certain value $e_1$ with respect to disk-type mill cutter teeth of regular height [see Figure 6.22(b)]. The parameter $e_1$ is the smallest permissible value that exceeds the thickness of the chip being cutting by the top cutting edge. Therefore, only the lateral cutting edges of the disk-type mill cutter teeth of the first group remove the stock. Top cutting edges don't cut the gear to be machined.

Odd-numbered milling-cutter teeth from the second group. Disk-type mill teeth in this group are of regular height. However, the tooth thickness of these teeth is smaller compared to the space-width of the gear being machining at a certain value $2 \cdot e_2 / \cos\phi$ [see Figure 6.22(b)]. The parameter $e_2$ is the smallest permissible value that exceeds the thickness of the chip to be cut by the lateral cutting edges. Therefore, only the top cutting edges of the disk-type mill cutter teeth of the second group remove the stock. Lateral cutting edges do not cut the gear.

[4] Pat. No.167.118, USSR as quoted in Eq. [41].

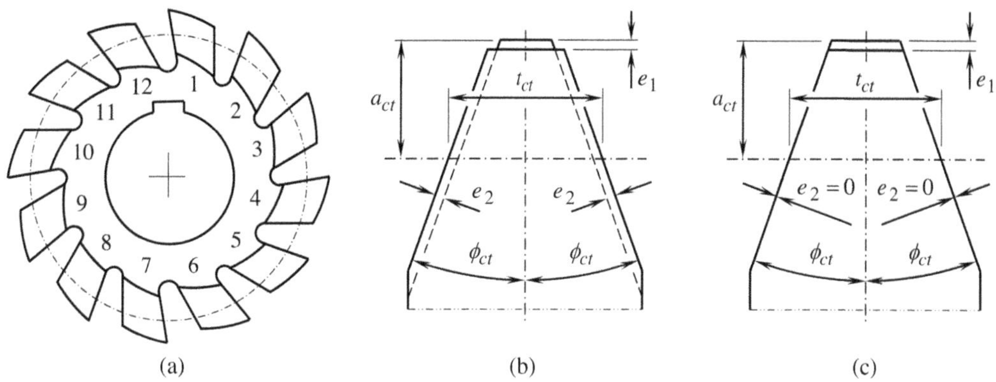


**FIGURE 6.22** Cutting diagrams for disk-type mill cutters for roughing gears.

Practical values of the parameters of modification of the disk-type mill cutter teeth $e_1$ and $e_2$ are calculated using the empirical formulae [41]:

$$e_1 = a_1 + 0.1 \text{ mm} \tag{6.60}$$

$$e_2 = a_{2(3)} + 0.05 \text{ mm} \tag{6.61}$$

where $a_1$ denotes the stock thickness to be removed by the top cutting edge and $a_{2(3)}$ denotes the stock thickness to be removed by the lateral cutting edges.

In the design of disk-type mill cutter, the possibility of chip flow interference is eliminated. In this way, the tool life of the disk-type mill cutter can be increased by a factor of 2, or even higher [41].

A simplified version of the design of the disk-type mill cutter [see Figure 6.22(c)] is also practical. The disk-type mill cutter of this design removes the stock following the so-called top-loaded cutting diagram.[5] The top-loaded cutting diagram features zero value of the parameter $e_2$.

Only lateral cutting edges of the disk-type mill cutter teeth of the first group remove the stock. Top cutting edges do not cut the gear to be machined. Disk-type mill cutter teeth in the second group cut the stock using all three cutting edges. However, the top cutting edge cuts the stock at double thickness. The chip flow from the top cutting edge dominates over chip flows from the lateral cutting edges. In this way, the cutting performance of the disk-type mill cutter can be increased by a factor of $1.5 \div 1.8$ [41].

Standard disk-type mill cutters can be reground in order to obtain the top-loaded disk mill gear cutters shown in Figure 6.22(c). For this purpose, a special method of grinding[6] clearance surfaces of the disk-type mill cutter was developed by Prof. S.P. Radzevich [176].

Numerous cutting diagrams have been developed for various applications. The interested reader may wish to refer to [42, 159] and other advanced sources for more details on this topic.

For roughing coarse pitch gears, use of stocking disk mill cutters with wavy lateral cutting edges (see Figure 6.23) is preferred. These are roughing cutters designed for high-speed cutting. They are used singly as well as in duplex or triplex setups. Several styles of stocking disk mill gear cutters are available on the market. Wavy cutting edges subdivide wide chip into numerous narrow chips. This results in improved cutting performance of stocking disk mill gear cutters.

Similarly, wide chips can be subdivided into multiple narrow chips using the assembled disk-type mill cutter with a special disposition of the cutting inserts (see Figure 6.24) made up of carbide alloys. Shown in Figure 6.24, disk-type mill cutters are used mostly for roughing and for gashing (see Figure 6.25) accurate spur and helical gears, or similar purposes.

Disk-type mill cutters of various designs for roughing gears are currently available on the market.

## 6.7 ACCURACY OF GEAR TOOTH FLANKS MACHINED BY DISK-TYPE MILL CUTTERS

Line contact between the tooth flank $\mathscr{G}$ of the gear being machined and of the generating surface $T$ of the disk-type mill cutter is observed when machining the work gear

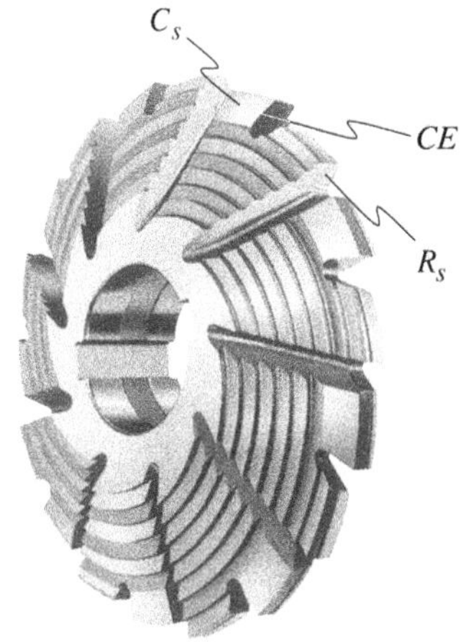


**FIGURE 6.23** Stocking mill cutter for roughing coarse pitch gears.

[5] SU Pat. No. 181.953 as quoted in Eq. [41].

[6] SU Pat. No. 1.335.425, *A Method of Reliving of Gear Hobs*/S.P. Radzevich, and V.I. Lempert, Int. Cl. B 24b, 3/12, Filed: April 24, 1986.

**FIGURE 6.24** An assembled mill cutter for roughing coarse pitch gears.

with a disk-type mill cutter. The accuracy of the gear tooth flank $\mathcal{G}$ machined with the disk-type mill cutter is viewed here from the standpoint of fulfillment or violation of the sixth necessary condition of proper part surface generation (see [151, 158, 162, 168] for details on conditions of proper part surface generation). This means that deviations of the machined tooth surface of the gear from the desirable geometry of the tooth flank $\mathcal{G}$ are expressed in terms of:

- The design parameters of the disk-type mill cutter.
- The parameters of the relative motions of the disk-type mill cutter and the work gear.

When the disk-type mill cutter travels in the axial direction of the gear to be machined from one of its faces toward the opposite face, the line of contact $\mathcal{E}$ of epy surfaces $\mathcal{G}$ and $T$ travels relative to the work together with the disk-type mill cutter. The machined tooth адфтл of the gear is generated as the family of consecutive positions of the characteristic $\mathcal{E}$.

In reality, the entire generating surface $T$ is not reproduced by the disk-type mill cutter. In fact, the surface $T$ is represented discretely by a certain number of distinct curves—the cutting edges $CE$ of the disk-type mill cutter—rather than continuously. As the result, cusps are observed on the machined gear tooth flank.

When machining a spur gear, the disk-type mill cutter (see Figure 6.26) rotates about its axis of rotation $O_{ct}$. Rotation of the disk-type mill cutter is denoted as $\omega_{cut}$. Simultaneously with the rotation $\omega_{cut}$, the disk-type mill cutter is fed $\mathbf{F}_{ct}$ in the axial direction of the gear. After each pass of the disk-type mill cutter, the path cut by the disk-type mill cutter is the resulting shape of the gear being machining tooth flank.

**FIGURE 6.25** Gashing a helical gear by the assembled milling cutter shown in Figure 6.24.

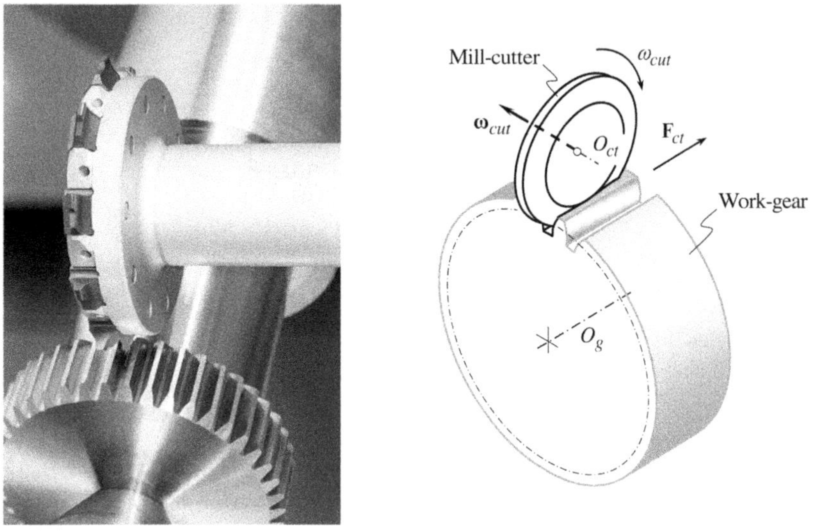


**FIGURE 6.26** Schematic of machining a spur gear with a disk-type mill cutter.

The number of teeth in a disk-type mill cutter is limited. When machining a gear, residual waviness is generated on tooth flanks of the machined gear.

For milled spur gears (see Figures 5.16 and 5.18), the maximal cusps height $h_w$ on the machined gear tooth flank can be calculated from the formula (5.72):

$$h_w = R_w - \sqrt{R_w^2 - 0.25 \cdot F_{ct}^2} \quad (6.62)$$

where $R_w$ is the radius of the circular arc cusp profile. This parameter of the machined gear tooth surface can be calculated using the Meusnier formula [144, 148, 151, 158, 168]. For gears machined by disk-type mill cutters, implementation of Meusnier formula yields an expression for the calculation of the radius $R_w$:

$$R_w = \frac{d_{c.m}}{2 \cdot \sin \phi_m} \quad (6.63)$$

The diameter $d_{c.m}$ of the disk-type mill cutter at a point of interest $m$ within the cutting edge $CE$ is calculated using either Eq. (5.70) for machining of external involute gears or Eq. (5.71) when internal gears are machined.

The derived equations allow for calculation of the deviation $h_w$, as well as the variation in the deviation $h_w$ within the tooth height of a spur involute gear machined by the disk-type mill cutter. Some results of the calculations are shown in Figure 6.27. The deviation $h_w$ varies within the tooth height. Its value is smaller for smaller diameter of the gear to be machined and becomes larger for points of the gear tooth profile closer to the outer diameter $d_{o.g}$. Variation in the disk-type mill cutter diameter is the major reason for the variation in the deviation values $h_w$ within the tooth height of the spur involute gear.

It is of interest to compare the waviness of the tooth flank that is machined by an end mill gear cutter with that machined by the Figure 6.27 cutter. To make the comparison easier, the results of the calculations in Figure 6.27 are plotted in the coordinates that are identical to that in Figure 5.17. Aside from the results of calculations for both cases, which are more qualitative than quantitative, they give a clear understanding of how the deviations $h_w$ are distributed within the machined gear teeth for both cases of gear milling.

When machining a helical gear, the curvedness of the helical gear tooth flank contributes to the resultant deviation $h_w$ of the machined tooth flank from its desired shape.

The expression:

$$R_g = \frac{R_{1.g}}{\cos^2 \psi_{b.g}} \quad (6.64)$$

can be used for the calculation of the radius of normal curvature $R_g$ of the gear tooth flank $\mathscr{G}$ in the direction of the feed motion $\mathbf{F}_{ct}$. Equation (6.64) is similar to the earlier derived Eq. (5.76).

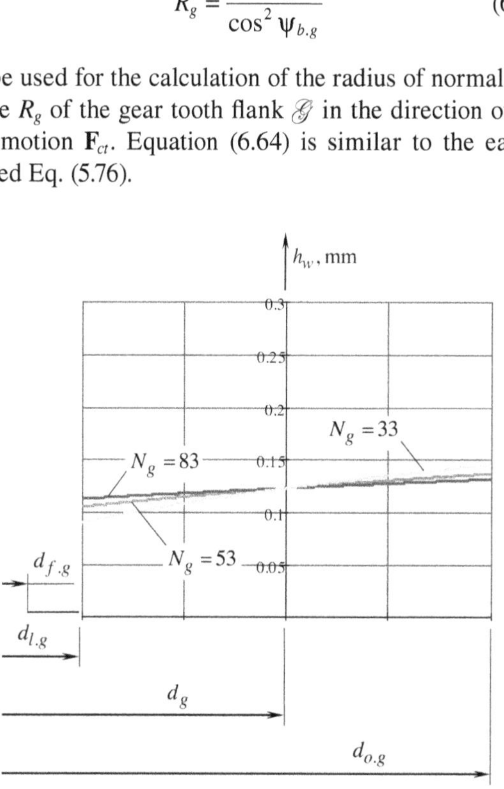


**FIGURE 6.27** Deviation of the actual tooth flank from the desired tooth flank $\mathscr{G}$ within the tooth height of a spur gear machined by a disk-type mill cutter.

To calculate for the gear tooth waviness $h_w$ an approximate formula [see Eq. (5.77)]:

$$h_w = \frac{R_g \cdot (R_g + R_w) \cdot \left(1 - \cos\frac{F_w}{2 \cdot R_g}\right)}{R_g - (R_g + R_w) \cdot \cos\frac{F_w}{2 \cdot R_g}} \tag{6.65}$$

is used.

The performed analysis reveals a possibility of calculation for the deviation of the machined tooth flank of the machined gear from its desired shape.

## 6.8 APPLICATION OF DISK-TYPE MILL GEAR CUTTERS

Most of the present-day users of disk-type mill cutters are on coarse-pitch applications. A common application on coarse pitch gears is the roughing operation, leaving stock for finish-hobbing or shaving. In practice, however, disk-type mill cutters are usually confined to producing replacement gears or small-lot gears with special tooth forms. The milling process is practically suitable for making racks, especially large racks.

The cutters have either 14.5° or 20° pressure angles, radial face sharpening, and a "backed-off" unground form.

Gear cutters are not universal for a complete range of teeth. To produce theoretically correct gear teeth, the cuter tooth form must be designed for the specific number of teeth. However, where a small error in tooth form is acceptable, cutters that are designed for a range of teeth can be used. Standard cutters are available for established ranges of teeth. Thus, all teeth within the range are provided with sufficient tip relief. The same form is produced on all tooth spaces within that range.

Similar to end mill gear cutters, two types of sets of disk-type mill cutters are available on the market. One is of eight mill cutters (see Table 5.1). The tooth profile of the disk-type mill cutter of each number differs from that of the disk-type mill cutter of another number, as shown in Figure 6.28. The form on the cutter is made to be correct for the lowest number of teeth in that particular range.

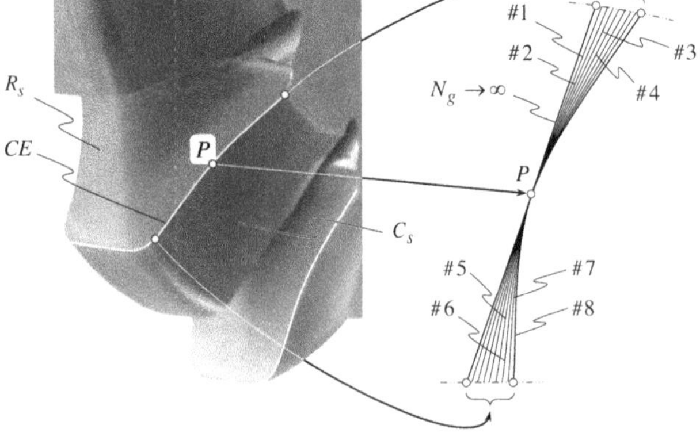


**FIGURE 6.28** Tooth profiles of the standard set of eight disk-type mill cutters.

**FIGURE 6.29** The use of two form disk-type mill cutters for cutting gears in both directions.

Another set is of 15 gear disk-type mill cutters (see Table 5.2). If close accuracy is desired in form-milled teeth, it is necessary to obtain a cutter with a curvature that is right for the exact number of teeth. This can be done by purchasing a special single-purpose cutter.

Gear teeth may be milled in one cut or given a rough cut almost to size and then finish-milled.

To increase the productivity rate of the gear-milling process, two form milling cutters can be ganged on the machine tool spindle, as illustrated in Figure 6.29. One gash in the gear is machined when the disk-type mill cutter travels down past the work gear, then the next gash is machined when the disk-type mill cutter travels up past the work gear.

With a single cutter, the cutter tooth form is centered in line with the gear axis so that a symmetrical tooth space is produced. More than one tooth space can be cut by using a gang of single cutters. Normally, a roughing and a finishing cutter are ganged, with the finishing cutter being centered with respect to the gear. Gang-type disk-type mill gear cutters and multiple-tooth cutters are usually single-purpose cutters used only for cutting the number of teeth for which they are designed. They are made in sets to cut two or three teeth at one pass. Figure 6.30 shows disk-type mill cutters ganged together. Similarly, grinding wheels for grinding spur gears can also be ganged together, as illustrated in Figure 6.31.

Gang-type disk mill cutters for machining gears with a relatively small number of teeth, which can be composed of roughing disk-type mill cutters either of regular design [see Figure 6.32(a)] or with an oblique tooth profile milling [see Figure 6.32(b)], can be implemented for this purpose. In the latter case, the centerline of the disk-type mill cutter tooth profile is at the angle of 10° in relation to the centerline of the disk-type mill tooth of conventional design.

In the method of roughing coarse pitch gears,[7] the concept of the inclined tooth profile is extended to its limit. The method [193] is illustrated in Figure 6.33. Two disk

[7] SU Pat. No. 1.815.024, *A Method of Roughing of Coarse Pitch Gear*/ S.P. Radzevich, I.V. Vinokourvov, A.P. Radzevich, G.A. Androshchuk, N.Ya. Smol'nikov, Int. Cl. B 23 f 5/20, 17/00, Filed: November 11, 1990.

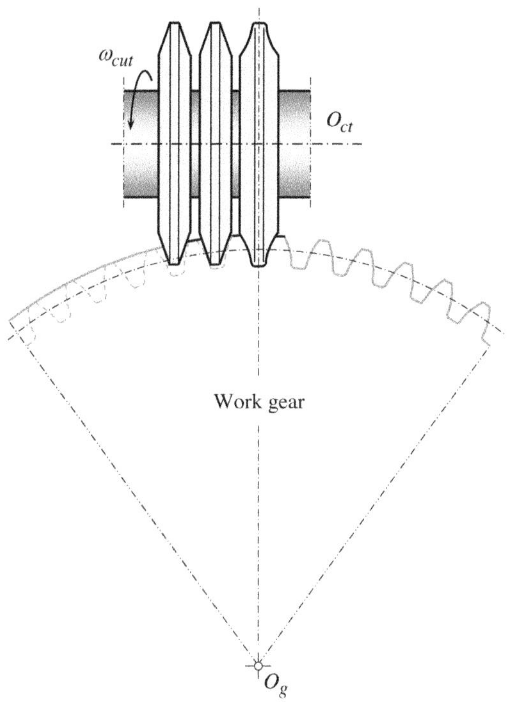


FIGURE 6.30 Three disk-type mill cutters ganged together.

mill gear cutters are used for the machining the work gear. The tooth profile of the disk-type mill cutters is of profile angle $2\phi_n$. One working surface of the disk-type mill cutter is flat. The opposite working surface is conical. A transient working surface is located between the flat portion of the generating surface $T$ of the disk-type mill cutter and its conical portion. The disk-type mill cutters are installed on a common arbor, allowing the flat working surface to be parallel to each other. The design and configuration of the disk-type mill cutters create a favorable clearance angle within the cutting edges that improves cutting performance of the gear-cutting tool.

The distance between the disk-type mill cutter faces is $(W+2\cdot t)$, where $W$ denotes the span measurement of the machined gear and $t$ denotes the stock for finishing of the gear tooth profile. The geometry of the implemented

FIGURE 6.31 Three grinding wheels ganged together.

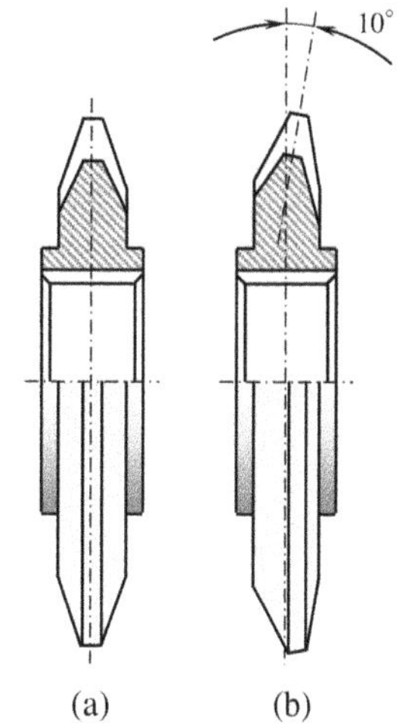


FIGURE 6.32 Roughing disk mill gear cutters (a) of ordinary design and (b) with an inclined tooth profile.

configuration of the disk-type mill cutters (see Figure 6.33) resembles that in the span measurement of an involute gear. To calculate for the span measurement $W$, the formula is derived:

$$W = d_{b.g} \frac{\tan \phi_t}{\sin \psi_g} \tag{6.66}$$

where:

$d_{b.g}$ is the base diameter of the gear being machined
$\phi_t$ is the transverse profile angle at the pitch diameter of the gear to be machined
$\psi_g$ is the pitch helix angle of the gear

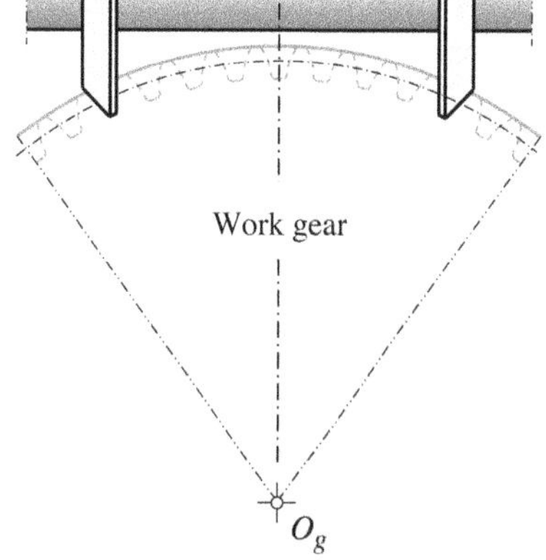


FIGURE 6.33 A method of roughing of a coarse pitch spur gear by two disk mill gear cutters with flat portions of the generating surface $T$ (SU pat. No. 1815024).

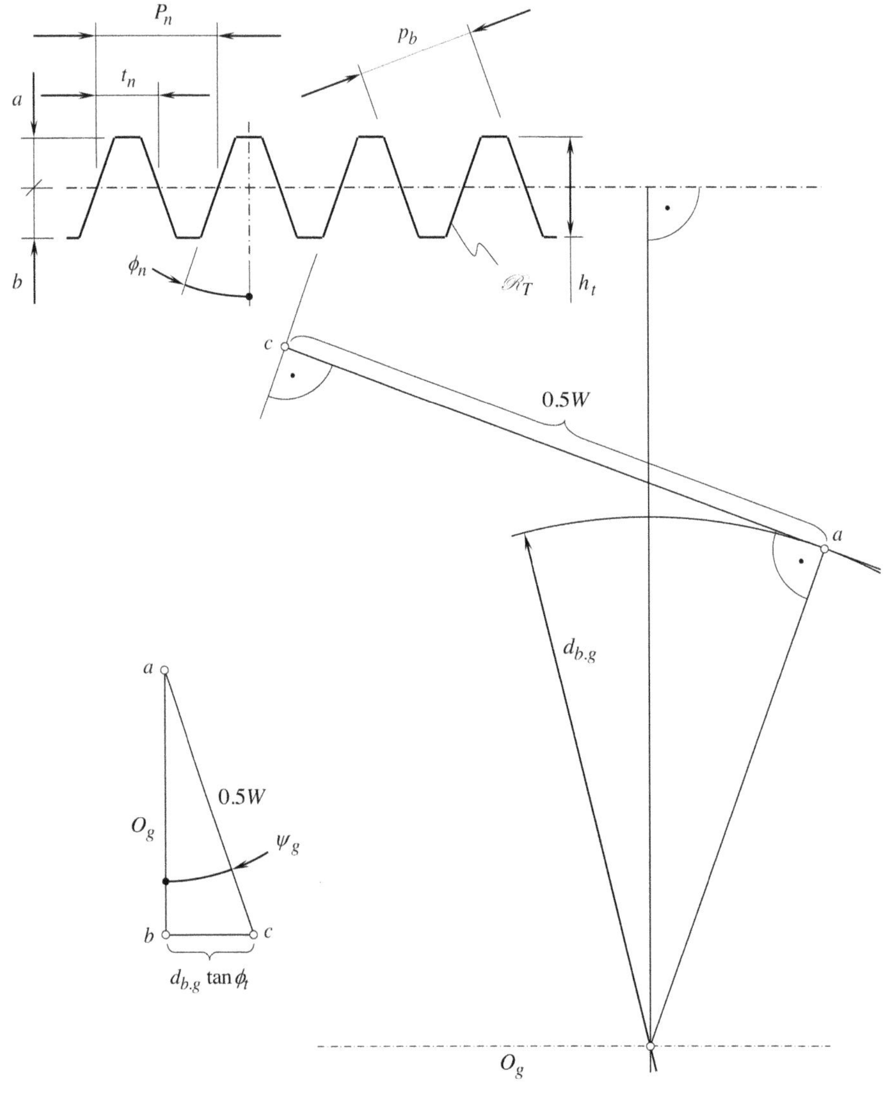


**FIGURE 6.34** Calculation of span measurement $W$ in helical gear.

Figure 6.34 illustrates the concept that is used for the derivation of Eq. (6.66). In the normal sectional view of the auxiliary generating rack $\mathscr{R}_T$ is depicted in relation to the helical gear with an involute tooth profile. An elementary trigonometrical analysis immediately returns Eq. (6.66) for the span measurement $W$.

Indexing after machining a space between the gear teeth can be executed into a certain integer number of the gear teeth $N_{ind}$.

When possible, it is recommended to synchronize the gear tooth number $N_g$, the number of the gear teeth within the span $N_{span}$, and indexing tooth number $N_{ind}$ in such a way that all gear teeth are machined without changing the $N_{ind}$.

As shown in Figure 6.33, the disk-type mill cutter configuration is not limited to just roughing of coarse pitch gears. Finishing gears can be done under a same configuration (see Figure 6.35).

The profiling of the gear disk-type mill cutter (see Figure 6.35) is not identical to that of ordinary disk-type

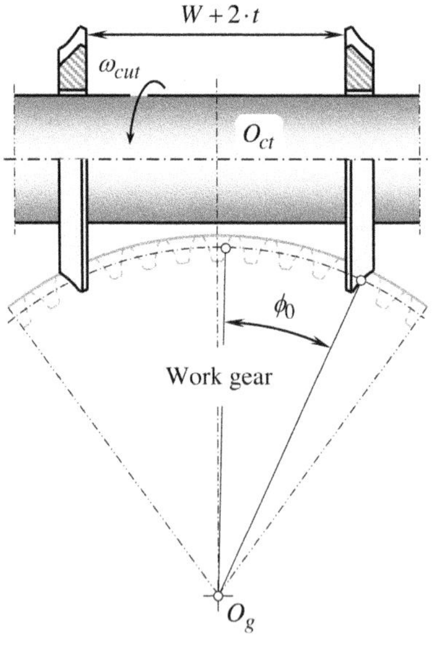


**FIGURE 6.35** Finishing of coarse pitch spur gear by two disk-type mill cutters with flat portions of the generating surface $T$.

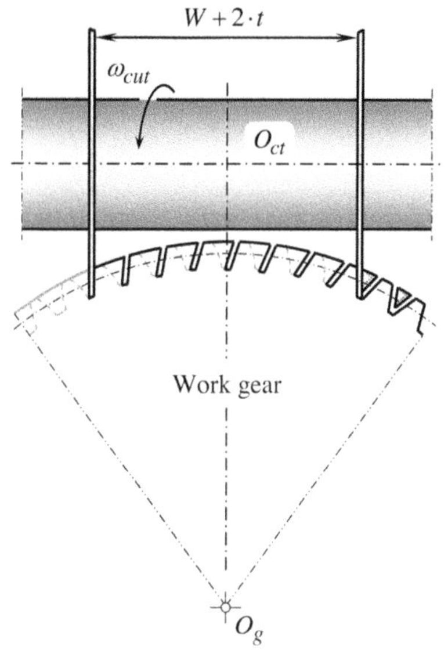


**FIGURE 6.36** Implementation of two circular saws for the machining of extremely coarse pitch spur gears.

mill cutters. The gear tooth profile machining in the position shown in Figure 6.33 is turned through an angle $\phi_0$ about the gear axis of rotation. This necessitates the use of the operator of rotation $\mathbf{Rt}(\phi_0, Z_g)$, in addition to those used in Eq. (5.4), for the position vector $\mathbf{r}_{inv}^{c}(\phi_g)$ of a point of interest of the gear tooth involute profile (see Figure 5.3).

The design of disk-type mill cutters (Figure 6.35) features favorable clearance angles at every point within the cutting edge. The tool life, the accuracy of the machined gear, and other parameters of the gear-machining operation correlate with the geometry of the cutting edge of the used disk-type mill cutter.

For machining extremely coarse spur gears, two circular saws can be implemented. Following this method (see Figure 6.36), two reasonably narrow circular saws are ganged at the distance $(W + 2 \cdot t)$ on the arbor. While traveling along the gear axis $O_g$, each saw cuts the narrow slot. Lateral surfaces of the slots are tangent the gear tooth profiles on the pitch diameter. When the tooth number of the gear to be machined is relatively small, the two saws can be used for roughing of the gear. When machining a gear with a large number of teeth of almost straight profiles, the method shown in Figure 6.36 is applicable for finishing the gear.

An important feature of the method (see Figure 6.36) is that not the whole material from the space between the gear teeth is cut in form of chips. Large size portions of material from the space remain solid and can be used for other purposes. This makes the method of gear machining more economical.

In producing replacement gears or small-lot gears with special tooth forms, fly-cutters are used. A fly-cutter can be viewed as a disk-type mill cutter with just one tooth. A schematic of the machining process of a gear by the fly-cutter is illustrated in Figure 6.37. Fly-cutters are easier in production rather than the disk-type mill cutters.

Internal gears can also be machined by disk-type mill cutters. A schematic of the machining process of an internal gear is illustrated in Figure 6.38.

It is widespread practice to cut the gear rack on universal milling machine, as illustrated in Figure 6.39. This process is common in low volume production of gear racks.

Rack teeth are often cut by the set of disk-type mill cutters (see Figure 6.40). Trapezoidal tooth profile is used in the design of the disk-type mill cutter for this purpose. In cases of zero rake angle ($\gamma_o = 0°$), the lateral cutting edges of the disk-type mill cutter form an angle with the tooth centerline. This angle equals to profile angle $\phi$ of the gear rack to be cut.

**FIGURE 6.37** Machining of a gear by the fly-cutter.

**FIGURE 6.38** Cutting an internal gear by the disk-type mill cutters.

Multiple-tooth cutters are often used to mill rack teeth (see Figure 6.41). Rack tooth gang cutters for the simultaneous machining of several tooth spaces are made up of several single-profile circular-type disk-type mill cutters. Through the continual offsetting of the gashes in relation to the drive slot, the individual cutters successively come (helically) to the point of cutting addition. This feature particularly promotes quiet running of the milling machine under heavy chip loads.

To allow for regrinding combined in a gang, the individual cutters are provided with an additional closely tolerated

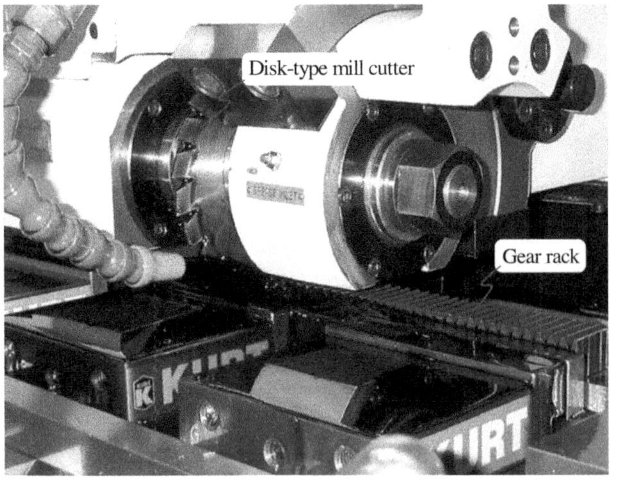


**FIGURE 6.39** Cutting a rack by a disk-type mill cutter.

**FIGURE 6.40** Cutting a rack by the disk-type mill cutters.

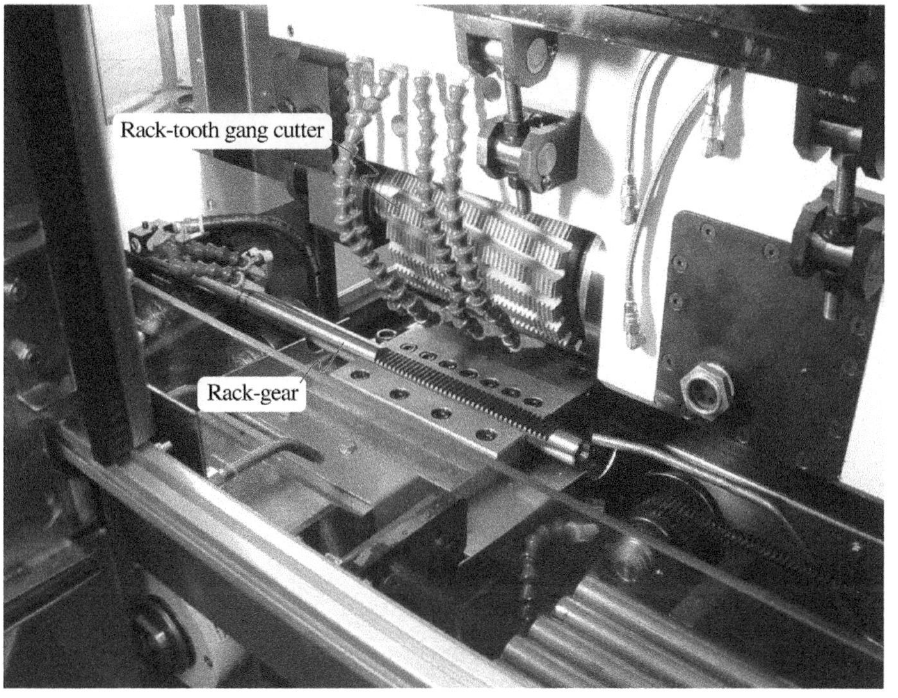


**FIGURE 6.41** Milling of a rack by the rack tooth gang cutters. (Courtesy of Mr. J. Vosmik, Drake Manufacturing, Warren, OH, USA.)

keyway (grinding slot), so that an axially parallel position of the cutting faces on the arbor is guaranteed.

Rack tooth gang cutters are profile relief ground with parallel flat lapping of the contact faces.

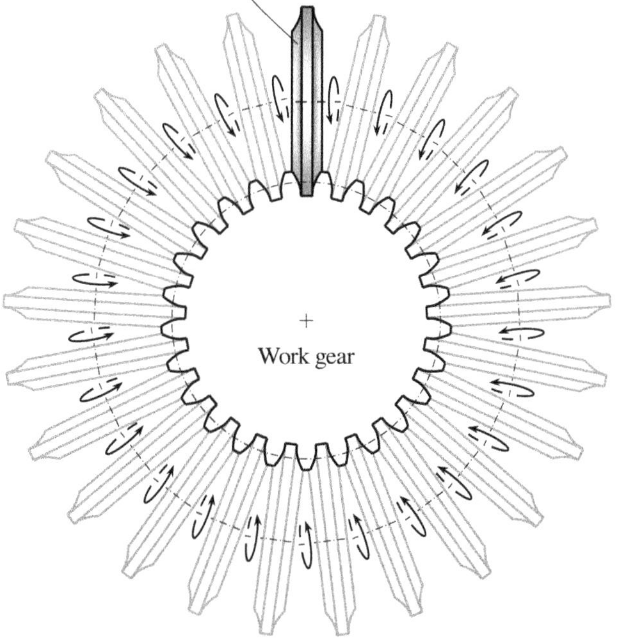


**FIGURE 6.42** Machining an involute gear by the multiple disk-type mill cutters.

Whether the machine is hand-operated or automatic, the accuracy of tooth spacing depends on the indexing mechanism. Impact of the indexing on the accuracy of the milled gear can be eliminated when the multiple disk-type mill cutters are used (see Figure 6.42). When the gear to be machined is passing through the set of gear disk-type mill cutters, all the teeth of the gear are cut in a single path.

Indexable mill cutters are used to cut spur and helical gears using CNC technology (see Figure 6.43).

The form of helical gear tooth spaces will not be the same as the cutter tooth form. In milling helical teeth,

**FIGURE 6.43** Cutting a helical gear with the indexable mill cutter.

**FIGURE 6.44** Grinding a worm by the disk-type grinding wheel.

the cutter axis is usually set at the helix angle of the gear. At this setting, the axis of the cutter is contained in the normal plane through the center of the gear tooth space. All other points are on the finished profile are reproduced in different planes. Therefore, the form on the cutter teeth is not reproduced in the gear. In addition to the setting angle, the diameter of the cutter affects the gear tooth form and must be considered in designing the proper cutter.

Grinding of an involute worm (see Figure 6.44) features numerous similarities to the grinding of a helical involute gear.

# 7 Non-Traditional Methods of Gear Machining by Form Cutting Tools

The kinematics of gear-machining processes with form cutting tools is simple. Only one motion is required for generating gear tooth flanks. This generating motion could be either a straight motion or a rotation. A combination of a translational motion with a rotational motion can also be implemented. In such a case, a translation and a rotation are timed with each other so as to comprise a screw motion. A screw motion and a reduced kind of screw motion—that is, either a translation or a rotation—both represent a single-parametric relative motion. Only single-parametric motions are used for profiling form tools for machining gears.

Once the kinematics of machining of a work gear with the form cutting tool is limited to a single-parametric relative motion, it becomes possible to carry out a comprehensive search for novel methods of machining a given gear, as well as for novel designs of form gear-cutting tools for these purposes.

## 7.1 PLURALITY OF SINGLE-PARAMETRIC MOTIONS

Consider a work gear to be stationary. A left-hand oriented Cartesian coordinate system $X_gY_gZ_g$ is associated with the gear as shown in Figure 7.1. A form gear-cutting tool can perform a variety of single-parametric motions relative to the reference system $X_gY_gZ_g$ when it is considered as a rigid body. Generally speaking, all permissible single-parametric relative motions of the form cutting tool relative to the reference system $X_gY_gZ_g$ are either a kind of a screw motion or a reduced kind of screw motion—namely, either a translation or a rotation. In such a scenario, the generating surface $T$ of the form cutting tool makes line contact with the gear tooth flank $\mathcal{G}$.

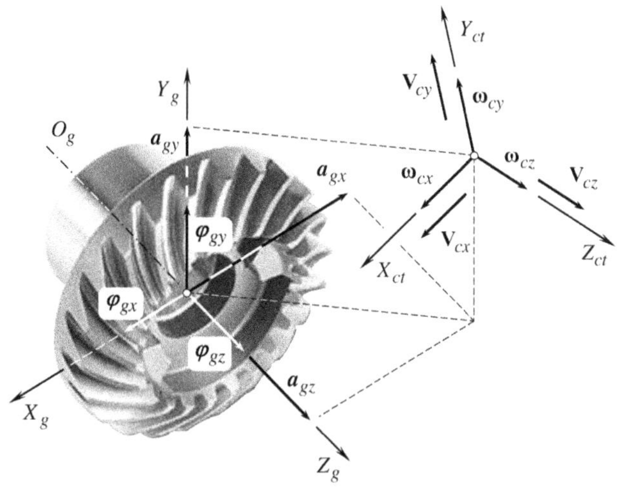


**FIGURE 7.1** An arbitrary configuration of the reference system $X_{ct}Y_{ct}Z_{ct}$ associated with the form gear-cutting tool relative to the reference system $X_gY_gZ_g$ associated with the gear.

A Cartesian coordinate system $X_{ct}Y_{ct}Z_{ct}$ is the cutting tool reference system. After the form gear-cutting tool has been designed, it will be represented in this reference system.

In the most general case, the location and orientation of the coordinate system $X_{ct}Y_{ct}Z_{ct}$ in relation to the coordinate system $X_gY_gZ_g$ can be specified by the following parameters.

Coordinates of the origin of the coordinate system $X_{ct}Y_{ct}Z_{ct}$ are completely specified by three vectors: $\boldsymbol{a}_{gx}$, $\boldsymbol{a}_{gy}$, and $\boldsymbol{a}_{gz}$. These vectors are pointed along the corresponding coordinate axes $X_g$, $Y_g$, and $Z_g$. Therefore, the position vector $\boldsymbol{a}_{c2}$ of the origin of the cutting tool coordinate system $X_{ct}Y_{ct}Z_{ct}$ can be expressed in terms of the vectors $\boldsymbol{a}_{gx}$, $\boldsymbol{a}_{gy}$, and $\boldsymbol{a}_{gz}$ of the elementary linear displacements:

$$\boldsymbol{a}_{c2} = \boldsymbol{a}_{gx} + \boldsymbol{a}_{gy} + \boldsymbol{a}_{gz} \tag{7.1}$$

In reality, every elementary linear displacement can be performed either along a coordinate axis or along two, or even three, coordinate axes simultaneously. Therefore, the total number $n_a$ of the permissible displacements is limited to the following seven linear displacements: $\boldsymbol{a}_{gx}$, $\boldsymbol{a}_{gy}$, $\boldsymbol{a}_{gz}$, $(\boldsymbol{a}_{gx} + \boldsymbol{a}_{gy})$, $(\boldsymbol{a}_{gy} + \boldsymbol{a}_{gz})$, $(\boldsymbol{a}_{gx} + \boldsymbol{a}_{gz})$, and $(\boldsymbol{a}_{gx} + \boldsymbol{a}_{gy} + \boldsymbol{a}_{gz})$.

The orientation of the coordinate system $X_{ct}Y_{ct}Z_{ct}$ with respect to the gear reference system $X_gY_gZ_g$ is completely specified by three rotation vectors[1] $\varphi_{gx}$, $\varphi_{gy}$, and $\varphi_{gz}$, which are a kind of angular displacement. These vectors are about the corresponding coordinate axes $X_g$, $Y_g$, and $Z_g$. Therefore, the resultant orientation vector $\varphi_c$ of the cutting tool coordinate system $X_{ct}Y_{ct}Z_{ct}$ can be expressed in terms of the vectors $\varphi_{gx}$, $\varphi_{gy}$, and $\varphi_{gz}$ of the elementary angular displacements:

$$\varphi_c = \varphi_{gx} + \varphi_{gy} + \varphi_{gz} \tag{7.2}$$

Each elementary angular displacement can be performed either about a coordinate axis or about two or three coordinate axes simultaneously. Therefore, the total number

[1] It should be noted here that the angular displacements $\varphi_{gx}$, $\varphi_{gy}$, and $\varphi_{gz}$ are not vectors in nature. Therefore, special care is required when treating the displacements $\varphi_{gx}$, $\varphi_{gy}$, and $\varphi_{gz}$ as vectors.

DOI: 10.1201/9781003520528-9

$n_\varphi$ of the feasible displacements of this type is limited to just seven angular displacements: $\varphi_{gx}$, $\varphi_{gy}$, $\varphi_{gz}$, $(\varphi_{gx}+\varphi_{gy})$, $(\varphi_{gy}+\varphi_{gz})$, $(\varphi_{gx}+\varphi_{gz})$, and $(\varphi_{gx}+\varphi_{gy}+\varphi_{gz})$.

The total number of possible displacements $n_{a\varphi}$ of the coordinate system $X_{ct}Y_{ct}Z_{ct}$ in relation to the coordinate system $X_gY_gZ_g$ is equal to:

$$n_{a\varphi}=(n_a+1)\cdot(n_\varphi+1)-1=63 \tag{7.3}$$

The cutting tool is allowed to travel along any of the coordinate axes $X_{ct}$, $Y_{ct}$, or $Z_{ct}$. Velocities of these translational motions are denoted by $\mathbf{V}_{cx}$, $\mathbf{V}_{cy}$, and $\mathbf{V}_{cz}$, respectively. Any two or even all three translations can be performed simultaneously. Without going into detailed analysis, it is easy to see that in addition to the elementary translations $\mathbf{V}_{cx}$, $\mathbf{V}_{cy}$, and $\mathbf{V}_{cz}$, three pairs of translations—$(\mathbf{V}_{cx},\mathbf{V}_{cy})$, $(\mathbf{V}_{cy},\mathbf{V}_{cz})$, and $(\mathbf{V}_{cz},\mathbf{V}_{cx})$—and one triple translation $(\mathbf{V}_{cx},\mathbf{V}_{cy},\mathbf{V}_{cz})$ are permissible. Therefore, seven combinations of the elementary translations are allowed. All permissible combinations of the translations $\mathbf{V}_{cx}$, $\mathbf{V}_{cy}$, and $\mathbf{V}_{cz}$ cause the resultant motion, which is also a kind of translational motion:

$$\mathbf{V}_{c\Sigma}=\mathbf{V}_{cx}+\mathbf{V}_{cy}+\mathbf{V}_{cz} \tag{7.4}$$

Ultimately, the total number $n_{\mathbf{V}}$ of permissible translations of the cutting tool coordinate system $X_{ct}Y_{ct}Z_{ct}$ relative to the coordinate system $X_gY_gZ_g$ associated with the gear is limited to $n_{\mathbf{V}}=7$ translational motions.

The cutting tool can rotate about any of the coordinate axes $X_{ct}$, $Y_{ct}$, or $Z_{ct}$. The angular velocities of these rotations are respectively denoted by $\boldsymbol{\omega}_{cx}$, $\boldsymbol{\omega}_{cy}$, and $\boldsymbol{\omega}_{cz}$. Any two or three rotations can be performed simultaneously. Without going into detailed analysis, it is easy to see that in addition to the elementary rotations $\boldsymbol{\omega}_{cx}$, $\boldsymbol{\omega}_{cy}$, and $\boldsymbol{\omega}_{cz}$, three pairs of rotations—$(\boldsymbol{\omega}_{cx},\boldsymbol{\omega}_{cy})$, $(\boldsymbol{\omega}_{cy},\boldsymbol{\omega}_{cz})$, and $(\boldsymbol{\omega}_{cz},\boldsymbol{\omega}_{cx})$—and one triple rotation $(\boldsymbol{\omega}_{cx},\boldsymbol{\omega}_{cy},\boldsymbol{\omega}_{cz})$ are permissible. Therefore, seven combinations of elementary rotations are possible. All possible combinations of the rotations $\boldsymbol{\omega}_{cx}$, $\boldsymbol{\omega}_{cy}$, and $\boldsymbol{\omega}_{cz}$ cause the resultant rotation $\boldsymbol{\omega}_{c\Sigma}$. The rotation $\boldsymbol{\omega}_{c\Sigma}$ can be expressed in terms of the elementary rotations as:

$$\boldsymbol{\omega}_{c\Sigma}=\boldsymbol{\omega}_{cx}+\boldsymbol{\omega}_{cy}+\boldsymbol{\omega}_{cz} \tag{7.5}$$

Similarly, the total number $n_\omega$ of permissible rotations of the cutting tool coordinate system $X_{ct}Y_{ct}Z_{ct}$ relative to the coordinate system $X_gY_gZ_g$ associated with the gear is limited to $n_\omega=7$ rotational motions.

The translations $\mathbf{V}_{cx}$, $\mathbf{V}_{cy}$, and $\mathbf{V}_{cz}$ and the rotations $\boldsymbol{\omega}_{cx}$, $\boldsymbol{\omega}_{cy}$, and $\boldsymbol{\omega}_{cz}$ represent the components of single-parametric motions.

The translations and the rotations can be performed simultaneously—that is, they can be combined with one another. The total number of possible motions $n_{\mathbf{V}\omega}$ of the coordinate system $X_{ct}Y_{ct}Z_{ct}$ with respect to the coordinate system $X_gY_gZ_g$ is equal to:

$$n_{\mathbf{V}\omega}=(n_{\mathbf{V}}+1)\cdot(n_\omega+1)-1=63 \tag{7.6}$$

Neither the linear nor the angular displacements of the cutting tool reference system relative to the gear reference system are strictly required. They can be observed or some of them can be of a zero value. It is essential that a relative motion of the coordinate system $X_{ct}Y_{ct}Z_{ct}$ relative to the coordinate system $X_gY_gZ_g$ is observed. At least one of the permissible relative motions must be observed. This means that the total number $n_{cg}$ of all possible combinations of the displacements and the relative motions is limited to:

$$n_{cg}=(n_{a\varphi}+1)\cdot n_{\mathbf{V}\omega}=4032 \tag{7.7}$$

Design of a cutting tool for machining a given gear cannot be performed in compliance with all possible combinations [see Eq. (7.7)], as many of the combinations are not physically feasible.

In the most general case, the resultant translation $\mathbf{V}_{c\Sigma}$ and the resultant rotation $\boldsymbol{\omega}_{c\Sigma}$ can be performed simultaneously. When this happens, the motions $\mathbf{V}_{c\Sigma}$ and $\boldsymbol{\omega}_{c\Sigma}$ are in time with each other in a certain manner, and they comprise a screw motion. The actual configuration of the resultant screw motion can be expressed in terms of the parameters of the elementary translations and elementary rotations.

## 7.2 IMPLEMENTATION OF SINGLE-PARAMETRIC MOTIONS FOR DESIGNING FORM GEAR-CUTTING TOOLS

Investigation of the single-parametric motions (or, the same, investigation of all possible configurations of the resultant screw motion relative to the work gear) reveals that not all single-parametric motions are permissible from the standpoint of their implementation in the design of gear-cutting tools. However, the potential capabilities of some of these motions sound promising.

Consider a few examples where the results of the analysis of single-parametric motions are implemented for the design of gear-machining methods and gear-cutting tools. The examples discussed in the following sections are aimed at illustrating the importance of the kinematical approach in designing gear-cutting tools.

### 7.2.1 End Mill Gear Cutter

When designing a conventional end mill gear cutter (see Figure 5.2), it is assumed from the beginning that the axis of rotation of the mill cutter $O_{ct}$ intersects the gear axis of rotation $O_g$ at a right angle. This means that the mill cutter rotation vector $\boldsymbol{\omega}_{ct}$ is aligned with the rotation vector $\boldsymbol{\omega}_{cy}$. Let us assume that, in addition to the rotation $\boldsymbol{\omega}_{cy}$, the end mill gear cutter is also rotating with the angular velocity $\boldsymbol{\omega}_{cz}$ (see Figure 7.2). Two rotations $\boldsymbol{\omega}_{cy}$ and $\boldsymbol{\omega}_{cz}$, which are performed simultaneously, cause the resultant rotation $\boldsymbol{\omega}_{cut}$ about the axis $O_{ct}$ (here, $\boldsymbol{\omega}_{cut}=\boldsymbol{\omega}_{cy}+\boldsymbol{\omega}_{cz}$). The axis $O_{ct}$ of the resultant rotation $\boldsymbol{\omega}_{cut}$ is inclined within the coordinate plane $Y_gZ_g$ and forms a certain angle $\sigma_c$ with the $Y_g$ – axis (see Figure 7.2).

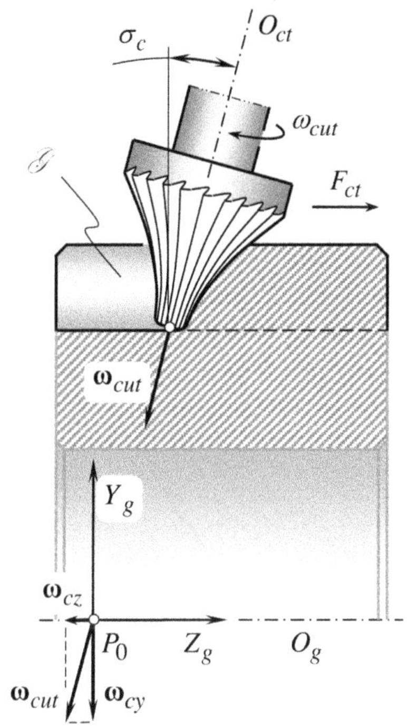


**FIGURE 7.2** End mill gear cutter with an inclined axis of rotation.

In determining of the generating surface $T$ of the mill cutter, the principle of inversion of motions can be implemented. In this case, the mill cutter surface $T$ can be generated as an enveloping surface to consecutive positions of the gear tooth flank $\mathscr{G}$ when it rotates about the $O_{ct}$ – axis.

Once the generating surface $T$ of the end mill gear cutter is determined, all the subsequent steps in the process of designing a gear-cutting tool are quite similar to those of end mill gear cutters with a conventional design.

Inclination of the end mill cutter's axis of rotation $O_{ct}$ at angle $\sigma_c$ provides further control over the gear-machining operation. In this manner, an area at which cutting speed $\mathbf{V}_{cut}$ is of zero value (that is, $\mathbf{V}_{cut} = 0$) is eliminated from contact with the work gear.

### 7.2.2 Disk-Type Mill Gear Cutter

A mill cutter for machining gears can be designed in a way that the work gear is located inside the form gear-cutting tool. An example of a disk-type mill gear cutter of this kind is illustrated in Figure 7.3.

The work gear is stationary when machining a space between the gear teeth takes place. The gear mill cutter rotates about its axis $O_{ct}$. Rotation of the mill gear cutter is denoted by $\omega_{cut}$. The rotation vector $\boldsymbol{\omega}_{cut}$ is parallel to the $X_g$ – axis of the Cartesian coordinate system $X_gY_gZ_g$ associated with the gear.

The work gear is fed outward from the mill cutter's axis of rotation $O_{ct}$. The feed motion is denoted by $F_{ct}$. This motion is terminated when the full depth of the tooth space is machined. After machining of the tooth space is accomplished, the work gear is moved toward the mill cutter's axis of rotation $O_{ct}$ for indexing. This is followed by machining of the next tooth space. Machining of the gear is finished when all the tooth spaces are cut.

Because the work gear is located within the interior of the gear mill-cutter, crowning of the machined gear teeth flanks is achieved in a natural way. For correct crowning of the gear tooth flanks, a proper correlation between the gear face width and the inner diameter of the cutting tool is the required.[2]

Multiple gears can be cut simultaneously using the enveloping disk-type mill form cutter, wherein the work gears are located within the interior of the cutting tool. The productivity rate of gear machining in the last case is significantly higher.

Profiling of the mill-cutter teeth (see Figure 7.3) is similar to that in conventional disk-type mill cutters (see Chapter 6). The equation of generating surface $T$ of the disk-type mill cutter is of similar structure. Differences are mostly attributed to two factors:

- Different design parameters of the actual mill cutter.
- Specific location of the work gear relative to the gear-cutting tool.

It is convenient to implement plane face surfaces of the cutting tooth of the mill cutter. The rake plane $R_s$ can be either a plane through the axis of rotation of the mill cutter (the rake angle at the top cutting edge is zero $\gamma_o = 0°$), or the rake plane can be offset at a certain distance from the axis $O_{ct}$ (the rake angle at the top cutting edge has a positive value $\gamma_o > 0°$).

The clearance surfaces of the mill gear cutter teeth are relief ground for accuracy.

### 7.2.3 Face Mill Gear Cutter

For machining of crowned gears, face mill gear cutters can be used. Gear mill cutters of such a design feature cutting teeth located circumferentially within the face of the cutting tool, as illustrated in Figure 7.4. When machining a work gear, the workpiece is stationary, whereas two face form mill gear cutters rotate about the common axis of rotation $O_{ct}$. In Figure 7.4, the rotations of the cutters are denoted by $\omega_{cut}$. The rotation vector $\boldsymbol{\omega}_{cut}$ is parallel to the $Y_g$ – axis of the Cartesian coordinate system $X_gY_gZ_g$ associated with the gear.

The mill gear cutters are fed along the cutter axis $O_{ct}$ toward the gear being machined. The feed motion $F_{ct}$ is terminated when the full depth of the tooth space is machined. After machining the tooth space, the mill gear cutters are

[2] For details on this particular method for crowning of gears, see: [172] Radzevich, S.P., *Theory of Gearing: Kinematics, Geometry, and Synthesis*, 3rd Edition, revised and expanded, CRC Press, Boca Raton, FL, 2022, 1154 pages; [2nd Edition, revised and expanded, 2018, 934 pages [171]; 1st Edition, 2012, 760 pages].

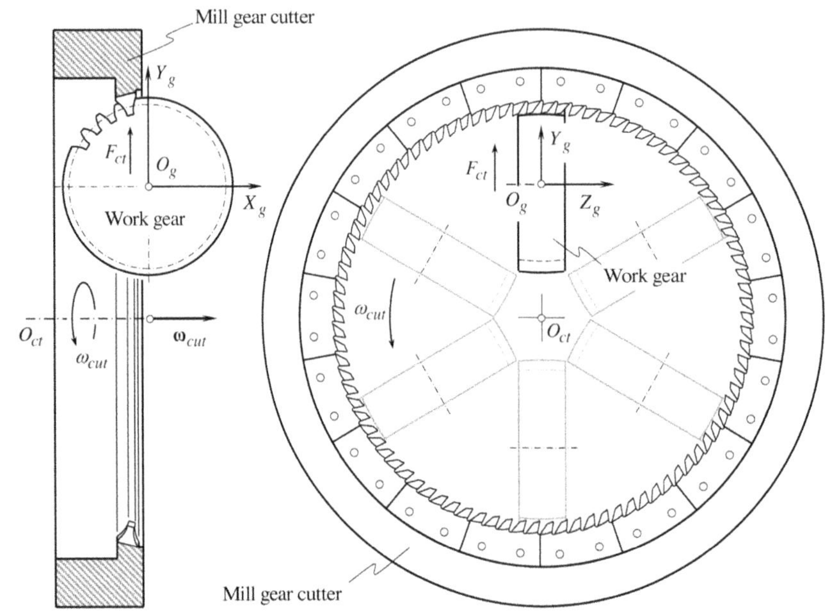


**FIGURE 7.3** An enveloping form disk-type mill gear cutter (the work gear is located inside the cutting tool).

moved out of the work gear for indexing. This is followed by machining the next tooth space. The gear is done being machined when all the tooth spaces are cut.

Because the work gear is located between the form mill gear cutters, crowning of the gear tooth flanks is achieved in a natural way. One side of the work gear tooth flank is generated by one of two face form mill cutters. In this way, the tooth flank becomes convex in the longitudinal direction of the gear tooth. In a similar manner, the opposite side of the gear tooth flank is generated by another face mill gear cutter. It also becomes convex in the longitudinal direction of the gear tooth.

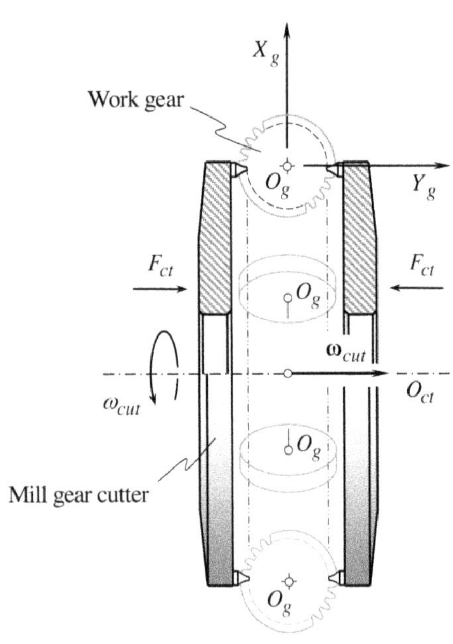


**FIGURE 7.4** Face mill gear cutters for machining crowned gears.

For correct crowning of the gear teeth, a proper correlation between the gear face width and the diameter of the cutting tool at which the cutting teeth are located is the required.

Multiple work gears can be simultaneously cut using face mill gear cutters. Under such a scenario, the work gears are located between two face mill cutters. The productivity rate of the gear-machining process in the last case is significantly higher.

Conventional methods can be used for profiling the form mill gear cutter is shown in Figure 7.4.

It is convenient to implement plane face surfaces $R_s$ of the cutting tooth of the mill cutter. The rake planes can be either through the axis of rotation of the form mill cutter (the rake angle at the top cutting edge is zero $\gamma_o = 0°$), or can intersect the axis $O_{ct}$ (the rake angle at the top cutting edge has a positive value $\gamma_o > 0°$).

Clearance surfaces of the mill gear cutter teeth are relief ground for accuracy.

### 7.2.4 Enveloping Round Broach for Cutting Spur and Helical Gears

For machining gears using the internal disk-type mill cutter discussed earlier (see Figure 7.3), a feed motion $F_{ct}$ of the work gear toward the gear-cutting tool is required. The feed motion can be eliminated from the kinematics of the gear-machining process. For this purpose, the enveloping round broach can be used (see Figure 7.5).

The work gear remains stationary when broaching is performed. The enveloping round gear broach rotates $\omega_{cut}$ about its axis of rotation, designated as $O_{ct}$ in Figure 7.5. All cutting teeth of the gear broach are divided into several sections. The roughing teeth of the enveloping broach

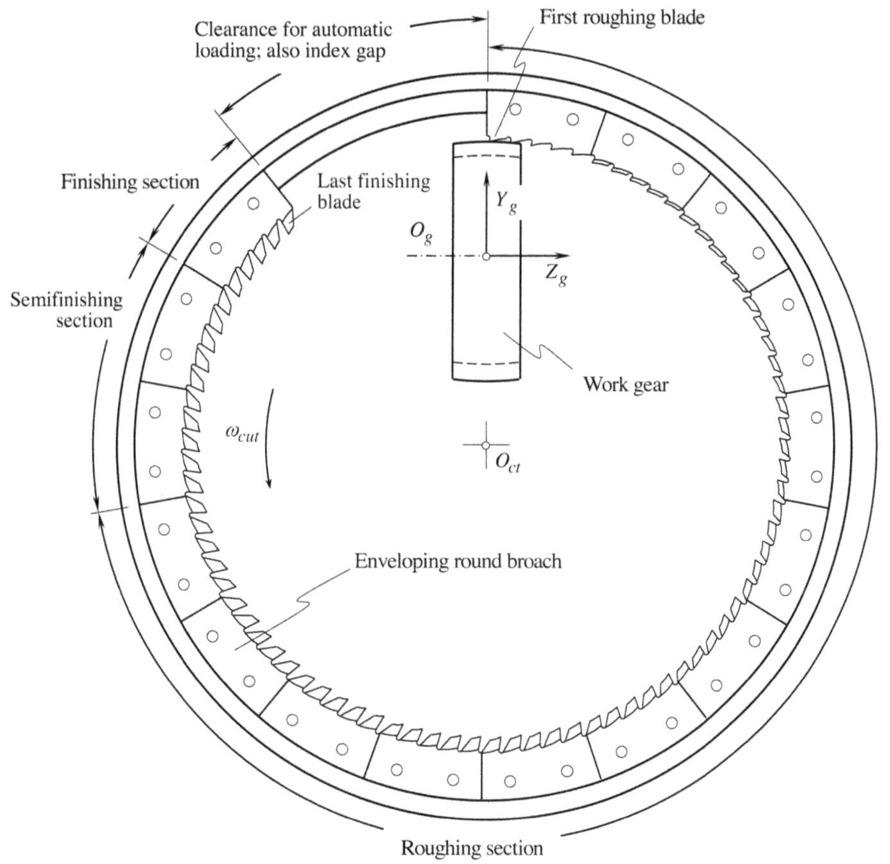


**FIGURE 7.5** Enveloping round broach for cutting spur and helical gears.

remove most of the material from the tooth space of the work gear. The tooth height of the cutting teeth incrementally increases from the first roughing blade to the last cutting blade within the roughing section of the broach teeth. The roughing teeth cut the chips of approximately rectangular cross-sections, thereby making chip curling easy. No chip interference coming from adjacent cutting edges is observed in roughing teeth. These allow for predicting higher tool longevity for the enveloping gear broach.

The roughing teeth are followed by the semi-finishing teeth, which remove a limited portion of the stock from the space width of the gear blank. The main purpose of the semi-finishing teeth is to leave thin and uniformly distributed portions of stock for the finishing teeth.

The tooth flanks of the gear are generated by finishing teeth. The accuracy of the finishing teeth directly affects the accuracy of the broached gears.

For both automatic loading and indexing, the enveloping round gear broach features a clearance between the last finishing cutting tooth and the first roughing cutting tooth.

Profiling finishing teeth of the enveloping round broach is identical to that for the enveloping round disk mill gear cutter. The cutting edges of both the roughing and the semi-finishing cutting teeth of the enveloping round gear broach are shifted inward on the generating body of the gear-cutting tool.

Plane surfaces are convenient for use as rake surfaces $R_s$ of the enveloping round gear broach. The rake planes can be either through the axis of rotation of the gear broach (the rake angle at the top cutting edge is zero, $\gamma_o = 0°$) or offset at a certain distance from the axis $O_{ct}$ (the rake angle at the top cutting edge is of a positive value $\gamma_o > 0°$).

Clearance surfaces $C_s$ of the enveloping round broach teeth are relief grounds for accuracy.

### 7.2.5 Enveloping Round Broach for Machining Straight Bevel Gears

For cutting straight bevel gears, the enveloping round broaches (see Figure 7.6) can be used.

When machining bevel gear teeth, the enveloping round broach rotates about its axis of rotation $O_{ct}$. The work gear is turning about an axis $O_{fr}$. The $O_{fr}$ – axis crosses at right angle with the axis of rotation $O_{ct}$ of the enveloping round broach. The rotation $\omega_{cut}$ of the broach about its axis and the rotation of the cutter with a feed rate $\omega_{fr}$ are timed with each other (see Figure 7.6). The required values of the feed rate rotation $\omega_{fr}$ of the work gear and the required location of the center $O_{fr}$ of the rotation $\omega_{fr}$ can be expressed in terms of the design parameters of the enveloping round broach and of the straight bevel gear being machined.

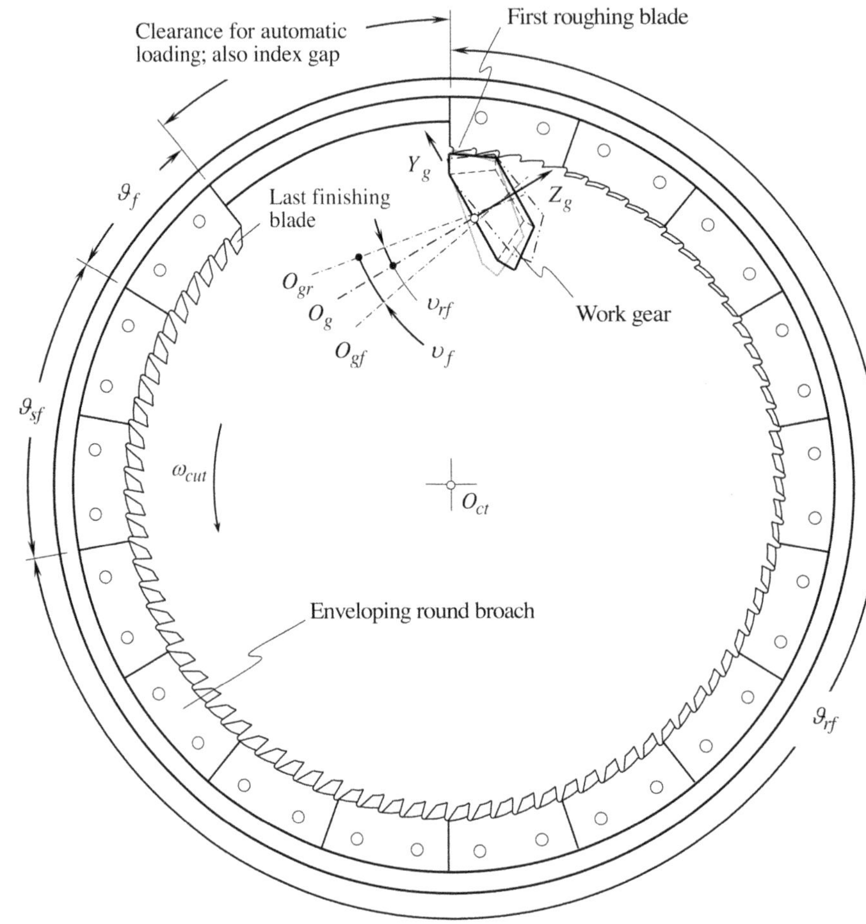


**FIGURE 7.6** Enveloping round broach for machining straight bevel gears.

Three sections of cutting teeth are recognized in the design of the enveloping round broach:

- Roughing cutting teeth, which are located within the angle $\vartheta_{rf}$.
- Semi-finishing cutting teeth, which are located within the angle $\vartheta_{sf}$.
- Finishing cutting teeth, which are located within the angle $\vartheta_f$.

The roughing teeth of the enveloping broach remove most of the stock from the space between two neighboring teeth of the work gear. The tooth height of the cutting teeth incrementally increases from the first roughing blade to the last cutting blade within the roughing section of the broach teeth. The roughing teeth cut the chips of approximately rectangular cross-sections, which makes chip curling easy. No chip interference coming from adjacent cutting edges is observed in roughing teeth. These allow predicting high tool longevity for the gear broach.

The roughing teeth are followed by the semi-finishing teeth of the enveloping broach. The semi-finishing teeth remove a limited portion of the stock from the tooth space of the bevel gear blank. The main purpose of the semi-finishing teeth is to leave thin and uniformly distributed portions of stock for the finishing teeth.

The finishing teeth of the round broach generate flank surfaces of the straight bevel gear teeth. The accuracy of the finishing teeth directly affects the accuracy of the broached gears.

For both automatic loading and indexing, the enveloping round gear broach features a clearance between the last finishing cutting tooth and the first roughing cutting tooth.

Within the roughing section and the semi-finishing section, the rotation $\omega_{cut}$ of the cutter and the rotation $\omega_g$ of the gear blank are appropriately timed so that when the cutting tool turns through the angle $\vartheta_{rf}$, the gear blank simultaneously turns through the angle $(\upsilon_{rf} + \upsilon_{sf})$. Here, $\vartheta_{rf}$ denotes the angle spanning over the roughing section of the cutting tool $\upsilon_{rf}$, and $\upsilon_{sf}$ denotes the angles between the two consequent positions $O_g$ and $O_{gr}$ of the axis of rotation of the work gear. The angles of the work gear rotation $\upsilon_{rf}$ and $\upsilon_{sf}$ correspond to the roughing and semi-finishing cycles of the bevel gear-machining operation.

Within the finishing section of the cutting teeth, the rotation $\omega_{cut}$ and the rotation through the angle $\upsilon_f$ are timed in a manner that is very similar to that just discussed. When the cutting tool turns through the angle $\vartheta_f$, the work gear turns through the angle $\upsilon_f$. Here, $\vartheta_f$ denotes the angle spanning over the finishing section of the cutting tool, and $\upsilon_f$ denotes the angular displacement between the two consequent positions $O_{gr}$ and $O_{gf}$ of the axis of rotation of the work gear.

Profiling of the finishing teeth of the enveloping round broach can be performed on the premise of the known geometry of the straight bevel gear tooth flank and the kinematics of the gear-machining process we have just discussed. The use of elements of the theory of enveloping surfaces can be decisive in this concern. The cutting edges of the roughing and of the semi-roughing cutting teeth of the enveloping round gear broach are shifted inward to the generating body of the gear-cutting tool.

The enveloping round broach blades are relief grounds when manufactured, and thus require sharpening on the rake face only. Blade-to-blade spacing, the angle of the plane of the front face, and the surface finish of the front face must all be closely controlled in sharpening. In addition, when new segments are assembled in heads, the cleanliness of assembly, accuracy of position of segment-locating keys, and close control of segment-holding bolt tension are all necessary for satisfactory cutting results.

## 7.3 DIVERSITY OF FORM TOOLS FOR MACHINING A GIVEN GEAR

The diversity of form tools for cutting the screw involute surface of a worm is used to illustrate the diversity of possible designs of form tools for cutting a given gear using the various kinematics of the gear-machining process.

### 7.3.1 Machining a Worm on a Lathe

Methods of machining a worm on a lathe represent the simplest examples of worm-cutting operations. When machining a worm (see Figure 7.7), the work gear rotates about its axis and the form tool travels along the axis of the work gear $O_g$. The rotation of the work gear $\omega_g$ and the translation of the form cutting tool $F_{ct}$ are timed with each other so that when the work gear makes one revolution about its axis $O_g$, the form tool travels at a distance that is equal to the axial pitch $P_{x.g}$ of the worm times the number of starts of the worm. This distance is equal to the lead of the screw involute surface to be machined.

Form cutters of several designs can be used for this purpose.

For turning an Archimedean worm, a form cutter with straight lateral cutting edges is used (see Figure 7.7). The rake plane $R_s$ of the form cutter is the plane through the axis of the worm being machined. The trapezoidal profile of the form cutter within the rake plane is identical to the axial profile of the Archimedean worm.

Two designs of form cutters are used for machining the worm. The first features a cutter profile that is coincident with the profile of the space between two adjacent threads of the worm [see Figure 7.7(a)]. The second design features a cutter profile that is congruent to the profile of the threads of the worm [see Figure 7.7(b)].

The form tool design shown in Figure 7.7(a) is simpler in production, as well as in re-sharpening after being worn. However, when cutting the worm, chip flows from all three cutting edges interfere with one another. This causes intensive wear of the form tool. The form tool design shown in Figure 7.7(b) is not as convenient to produce or re-sharpen compared to the first design. However, only two cutting edges cut the stock, rather than three. Because of this, interference of chip flows does not result in intensive wear on the form tool.

Indexing is executed when a multi-start worm is machined. In this case, after one thread (or after one space between two adjacent threads) is machined, indexing is performed for machining the next thread (or the next space between two adjacent threads).

Such a schematic layout of machining a worm can also be implemented for turning involute worms. In this case,

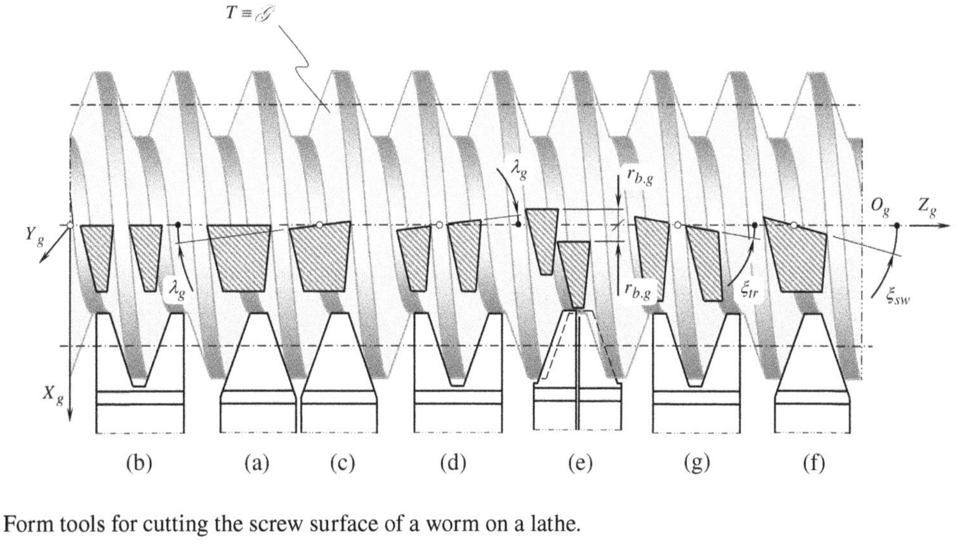


**FIGURE 7.7** Form tools for cutting the screw surface of a worm on a lathe.

the lateral cutting edges of the form tool should be slightly concave rather than straight. The equation of the lateral cutting edges of the form tool for turning involute worm immediately follows from the equation of the screw involute surface $\mathcal{G}$ [see Eq. (1.4)]. After substituting of $Y_g = 0$, Eq. (1.4) casts into equation:

$$\mathbf{r}_g(U_g, V_g) = \begin{bmatrix} r_{b.g} \cos V_g + U_g \cos \lambda_{b.g} \sin V_g \\ 0 \\ r_{b.g} \tan \lambda_{b.g} - U_g \sin \lambda_{b.g} \\ 1 \end{bmatrix} \begin{matrix} V_g^{(l)} \le V_g \le V_g^{(a)} \\ 0 \le U_g \le [U_g] \end{matrix}, \quad (7.8)$$

of the lateral cutting edge $CE$ of the form tool.

Equation (7.8) is valid for both designs of the form tool, namely:

- When the form tool profile follows the profile of the space between two adjacent threads of the worm [see Figure 7.7(a)]
- When the form tool profile is congruent to the profile of threads of the worm [see Figure 7.7(b)].

Form tools of similar design are used for turning a convolute worm. The rake plane of the form tool forms a certain angle with the axis $O_g$ of the worm. This angle is equal to lead angle $\lambda_g$ on the pitch cylinder of the worm. Again, two types of form tools are used for machining convolute worms. The profile of one of these form tools is identical to the normal profile of the space between two adjacent threads of the worm [see Figure 7.7(c)]. The profile of the other is identical to the normal profile of the thread of the worm [see Figure 7.7(d)].

The schematic of machining of convolute worms can be adjusted for the turning of involute worms. In this case, form tools with curved lateral cutting edges should be implemented. To derive the equation for the required geometry of lateral cutting edges of both types of form tools, implementation of Eq. (1.4) would be helpful.

Two separate form tools with straight lateral cutting edges can be used to machine a precision involute worm [see Figure 8.7(e)]. In this case, the rake plane of one of the cutters is shifted upward from the axis of rotation of the worm $O_g$, whereas the rake plane of the other cutter is shifted downward from the axis of rotation of the worm. With this configuration of the cutters, the straight cutting edges are tangent to the base helices of the worm threads. The use of cutters with straight profiling cutting edges offers higher accuracy of machining of involute worms.

Investigation into the geometry of involute worms has been conducted by Prof. S.P. Radzevich [170]. The performed analysis allows for the development of two designs of form tools for machining involute worms.

A screw involute surface of one side of the thread profile of a worm can be generated by a straight line that is rolling with no sliding over the base helix within the base cylinder. This is similarly true with respect to the screw involute surface of the opposite side of thread profile of the worm. For a particular configuration of the cutters, two straight generating lines are located within a common plane, which is the concept for designing cutters for machining screw involute surfaces of a worm. To take advantage of the concept, the parameters of this particular configuration of two straight generating lines should be determined.

Cutters of both designs feature straight lateral cutting edges that are located in a common plane. This plane serves as the rake plane $R_s$ of the cutter.

To cut the worm properly, the angle that the rake plane $R_s$ forms with the axis of the worm $O_g$ and the profile angle of the cutter $\phi_{rf}$ should be predetermined.

The angle $\xi_{tr}$ that the rake plane $R_s$ of the cutter forms with the axis of the worm $O_g$ [see Figure 7.7(g)] is calculated from the equation:

$$\xi_{tr} = \tan^{-1} \frac{\cos \psi_g \cdot \tan \nu_g}{\tan \phi_n + \sin \psi_g \cdot \tan \nu_g} \quad (7.9)$$

To calculate for the form tool profile angle $\phi_{rf}$, the following expression:

$$\phi_{rf} = \tan^{-1} \frac{\tan \nu_g}{\sin \xi_{tr}} \quad (7.10)$$

is used.

The following expression is used to calculate for the angle $\nu_g$ which will be entered into Eqs. (7.9) and (7.10):

$$\nu_g = \tan^{-1} \frac{d_{b.g}}{\sqrt{(d_g + t_{n.g} \cot \phi_n)^2 - d_{b.g}^2}} \quad (7.11)$$

where [for Eqs. (7.9) through (7.11)]:

$\psi_g$ is the pitch helix angle of the worm
$\phi_n$ is the normal profile angle of the involute worm
$d_{b.g}$ is the base diameter of the involute worm
$d_g$ is the pitch diameter of the worm
$t_{n.g}$ is the normal thickness of the worm thread

Equations (7.9) and (7.10) are also valid for the calculation of the design parameters of the cutter for the second type of machining of the involute worm [see Figure 7.7(f)]. In this case, the formula:

$$\nu_g = \tan^{-1} \frac{d_{b.g}}{\sqrt{(d_g - w_{n.g} \cot \phi_n)^2 - d_{b.g}^2}} \quad (7.12)$$

is used to calculate for angle $\nu_g$.

In Eq. (7.12), $w_{n.g}$ denotes the normal width of the space between two adjacent threads of the worm.

The form tools shown in Figure 7.7(g) and Figure 7.7(f) are capable of precise machining of involute worms, and both are easy to produce and resharpen.

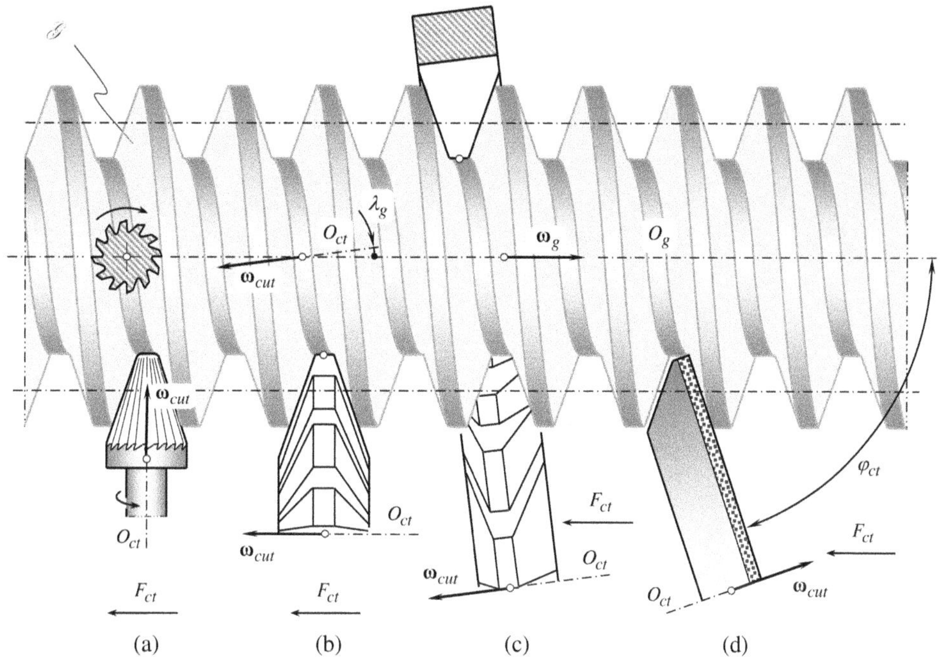


**FIGURE 7.8** Form mill cutters for machining involute worm.

The interested reader is referred to [138] and [170] for details on how Eqs. (7.9) through (7.12) are derived.

The given examples illustrate some of the possibilities of the kinematic approach to the design of form-cutting tools for machining gears.

### 7.3.2 Milling of Involute Worm

A high productivity rate can be achieved when an involute worm is machined by a form mill cutter. When machining, the worm rotates about its axis of rotation $O_g$. The rotation vector of the worm $\boldsymbol{\omega}_g$ is pointed along the $O_g$ – axis (see Figure 7.8). The mill gear cutter rotates about its axis of rotation $O_{ct}$. The rotation vector of the mill cutter $\boldsymbol{\omega}_{cut}$ is aligned with the $O_{ct}$ – axis. In addition to the rotation $\boldsymbol{\omega}_{cut}$, the cutting tool travels along the worm's axis of rotation. This motion represents the feed and is denoted by $F_{ct}$. The rotation of the worm $\omega_g$ and the feed motion $F_{ct}$ are timed with each other. Timing of the motions $\omega_g$ and $F_{ct}$ is performed in such a way that the mill gear cutter travels at the distance of the axial pitch times the number of starts per one revolution of the worm.

Various configurations of the rotation vectors $\boldsymbol{\omega}_g$ and $\boldsymbol{\omega}_{cut}$ are permissible when machining a worm with a mill gear cutter.

Machining an involute worm with an end mill gear cutter [see Figure 7.8(a)] is a conventional means of machining a coarse pitch worm. The axis of rotation $O_{ct}$ of the mill gear cutter intersects at a right angle with the axis of rotation $O_g$ of the worm. In particular cases, the axes $O_{ct}$ and $O_g$ can be at a certain angle in relation to one another, as shown in Figure 7.2. End mill gear cutters can be designed for machining of the worm when the axes $O_{ct}$ and $O_g$ cross each other either at a right angle or at a certain acute angle. No physical constraints are imposed on such a configuration of the axes of rotation of the worm and the cutting tool.

Profiling the end mill gear cutter for machining a worm is similar to that for machining of involute gear (see Chapter 5).

Disk mill gear cutters can also be implemented for machining worms [see Figure 7.8(b)]. Usually, the axis of rotation $O_{ct}$ of the mill gear cutter crosses the axis of rotation of the worm $O_g$ at an angle that is equal to the worm pitch lead angle $\lambda_g$ (see Figure 7.9). Profiling disk mill gear cutters of this design is similar to that considered in Chapter 6. In reality, in cases of necessity, a special-purpose mill gear cutter can be designed for machining the worm when the axes of rotation $O_g$ and $O_{ct}$ form an angle that differs from the pitch lead angle $\lambda_g$ of the worm. In particular, the angle that the axis of rotation $O_{ct}$ forms with the axis of rotation $O_g$ can be equal to zero. In this scenario, the axes $O_g$ and $O_{ct}$ are parallel to each other. In this case, the gang cutter can be used for machining the worm. There is an important advantage of the considered configuration of the axes of rotation of the worm and the mill gear cutter.

When milling of an involute worm, the enveloping round mill gear cutter can be used [see Figure 7.8(c)]. Either the angle that the rotation vector of the worm $\boldsymbol{\omega}_g$ forms with the rotation vector of the mill gear cutter $\boldsymbol{\omega}_{cut}$ can be of a certain value or the rotation vectors $\boldsymbol{\omega}_g$ and $\boldsymbol{\omega}_{cut}$ can be parallel to each other. In a particular case, the angle between the rotation vectors $\boldsymbol{\omega}_g$ and $\boldsymbol{\omega}_{cut}$ can be equal to the lead angle $\lambda_g$ on the pitch cylinder of the worm.

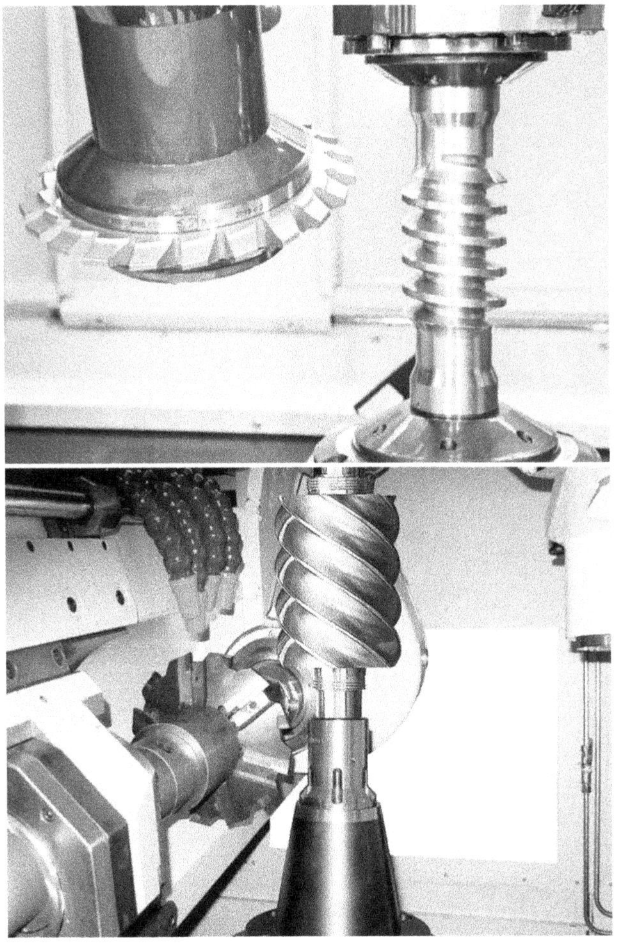

**FIGURE 7.9** Machining an involute worm by disk-type form mill cutters.

Profiling the enveloping round mill gear cutter for machining involute worms can be performed following the procedure for profiling disk mill gear cutters of conventional design.

Diversity of designs of mill gear cutters for machining involute worms is not limited to those considered in this discussion. There is much room for further improvement in this regard.

One more particular configuration of the rotation vectors $\omega_g$ and $\omega_{cut}$ deserves to be mentioned here. This configuration can be implemented for the purpose of designing a mill gear cutter. However, for practical reasons, it is used for grinding involute worms.

The configuration of the rotation vectors $\omega_g$ and $\omega_{cut}$ under consideration relates to the case of tangency of a plane to the screw involute surface. The approach is based on the well-known property of the screw involute surface, according to which the surface of this geometry can be generated as the envelope surface to consecutive positions of a plane that is performing a screw motion. This property of screw involute surfaces can be found in many sources (e.g., [138] and [170]).

For this configuration, the rotation vectors $\omega_g$ and $\omega_{cut}$ form the base lead angle $\lambda_{b.g}$. Therefore, the plane that intersects the worm axis of rotation $O_g$ at the angle:

$$\varphi_c = 90° - \lambda_{b.g} \tag{7.13}$$

is tangent to the surface $\mathscr{G}$ of the worm thread. This plane is used as the generating surface $T$ of the grinding wheel for finishing a precision involute worm [see Figure 7.8(d)]. To calculate for the required angle $\varphi_c$, the following expression:

$$\varphi_c = \sin^{-1}(\cos\phi_n \cdot \cos\psi_g) \tag{7.14}$$

was derived by Prof. S.P. Radzevich [138]. The angle $\varphi_c$ can also be calculated from an equivalent formula:

$$\varphi_c = \tan^{-1}\frac{\cos\phi_n}{\sqrt{\sin^2\phi_n + \tan^2\psi_g}} \tag{7.15}$$

which is known from many advanced sources.

In Eqs. (7.14) and (7.15), $\phi_n$ denotes the involute worm normal pressure angle and $\psi_g$ is the worm helix angle in the pitch cylinder.

### 7.3.3 Thread Whirling

A version of the profile milling process used for cutting worms is provided by thread whirling [219]. Whirling shows impressive returns in producing worms for gear worm reducers. Difficult threading operations, such as very long shafts or very wide profiles, can easily be machined by whirling.

The cutter head encircles the work and is driven by a separate motor while the work is rotated at a much slower speed (see Figure 7.10). During cutting, the head is swung over to the required helix angle $\psi_g$, and the cutters are ground to suit the space profile necessary to impart the desired thread shape.

Material removal occurs as a result of the whirling ring rotating $\omega_{cut}$ at high speed around a slowly rotating $\omega_g$ workpiece. The rotating workpiece combined with the advancement of the longitudinal slide produces the required pitch of the worm. The amount that the worm's centerline eccentric to the whirling ring determines the over-pin dimension. The tangential cutting action of the whirling process leaves the tool in contact with the workpiece for a much longer time along the arc of the cutting circle of the insert. Each subsequent pass of the insert removes four to six times the amount of material. The tangential cutting action also permits much higher cutting speeds, thus shortening cycle times again. The low approach angle of the insert when it first makes contact with the workpiece ensures that the whirling tools achieve a good tool life.

There are limitations to the size of worm thread that can be cut with a particular whirling head, but these are made to

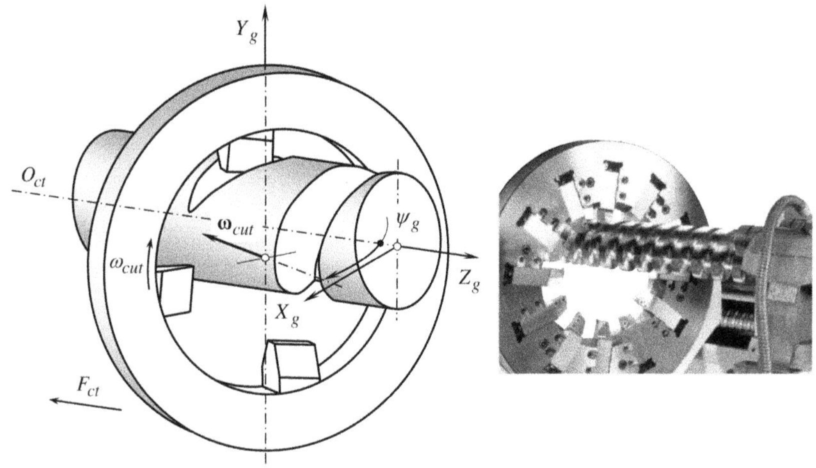


**FIGURE 7.10** Concept of thread whirling.

cut a range of sizes. Normally, single and two start worms are suitable since the maximum permissible helix angle is $\psi_g = 30°$, owing to the interference between the head and the work, but a useful range of worms can be successfully cut by the process.

Production rates of whirling are at least four times faster than thread milling or hobbing.

Whirling of internal worms is possible as well.

### 7.3.4 Grinding of Involute Worm

Profiling the form grinding wheel for finishing an involute worm can be done following the procedure used for mill gear cutters.

High cutting speed is a prerequisite for grinding operations. The required cutting speed can easily be achieved when the outer diameter of the grinding tools is bigger. For this purpose, implementation of face-type form grinding tools sounds promising.

A machining operation of a worm by a face gear grinding wheel is illustrated in Figure 7.11. The worm rotates about its axis of rotation $O_g$ and the rotation vector of the worm $\omega_g$ is along the $O_g$ – axis (see Figure 7.12). The face gear grinding wheel rotates about its axis of rotation $O_{ct}$. The rotation vector of the face gear grinding wheel $\boldsymbol{\omega}_{cut}$ is aligned with the $O_{ct}$ – axis. In addition to the rotation $\omega_g$, the grinding wheel travels along the worm axis of rotation with feed motion $F_c$. The rotation of the worm $\omega_g$ and the feed motion $F_{ct}$ are timed with one another. Timing of the motions $\boldsymbol{\omega}_g$ and $F_{ct}$ is performed in such a way that the face gear grinding wheel travels at a distance of the axial pitch times the worm number of starts per one revolution of the worm.

**FIGURE 7.11** Involute worm grinding operation (see Figure 7.12).

When planning to use form tools for grinding involute worms, the following types of kinematics of the worm threads can be implemented.

The rotation vector $\boldsymbol{\omega}_{cut}$ of the grinding wheel intersects the worm axis of rotation $O_g$ at a right angle. No physical constraints are imposed on the configuration on the rotation vectors $\boldsymbol{\omega}_g$ and $\boldsymbol{\omega}_{cut}$ when they are along crossing straight lines and, thus, the closest distance of approach of the straight lines is not equal to zero. A special axial profile is necessary to ensure the accuracy of machining of the worm in both cases. The generating surface $T$ of the face gear grinding wheel in this case can be determined as the enveloping surface to consecutive positions of the thread surface of the worm when it rotates about the axis of rotation of the grinding wheel $O_{ct}$.

The necessity in profiling of the grinding wheel following the special shape can be eliminated if limited changes are introduced to the kinematics of the worm grinding process. In a certain cases, when the closest distance of approach between the axes of rotation $O_g$ and $O_{ct}$ is equal to the radius $r_{b.g} = 0.5 \cdot d_{b.g}$ of the base cylinder of the worm, the grinding wheel having a straight axial profile can be used for finishing the worm. Under such a configuration of the rotation vectors $\boldsymbol{\omega}_g$ and $\boldsymbol{\omega}_{cut}$, the cone angle $\theta_c$ of the generating surface $T$ of the grinding wheel must be equal to the base helix angle $\psi_{b.g}$ of the worm. This means that fulfillment of the equality $\theta_c = \psi_{b.g}$ is required for precision machining of the worm when the axes $O_g$ and $O_{ct}$ are remote from each other at the pre-specified distance $r_{b.g}$. Offsetting of the rotation axes $O_g$ and $O_{ct}$ at the pre-specified distance $r_{b.g}$ is of critical importance, especially when grinding precise multi-start involute worms.

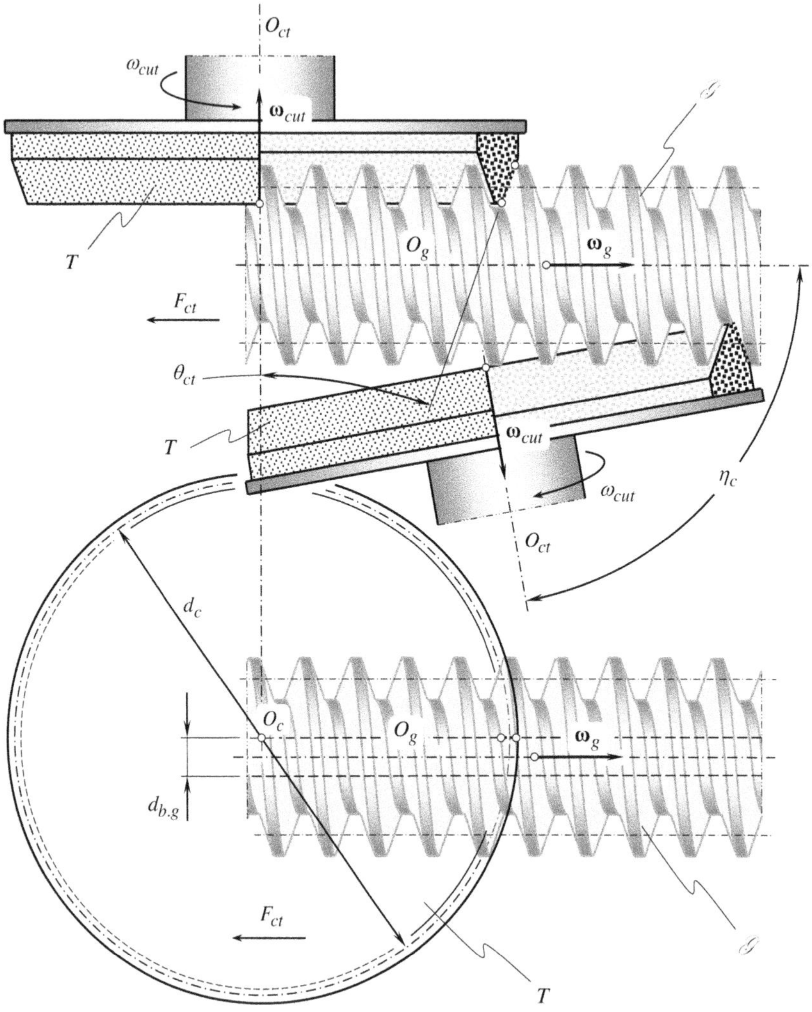


**FIGURE 7.12** Grinding wheels for finishing an involute worm.

Only short-length worms can be ground with a face gear grinding wheel when the rotation axes $O_g$ and $O_{ct}$ are at a right angle to each other. The maximal feasible length of the ground worm is limited to a length for which there is enough room within the interior of the grinding wheel. For grinding lengthy involute worms—say, for grinding worms theoretically of infinite length—a schematic of machining of worms when the rotation vectors $\omega_g$ and $\omega_{cut}$ are at a certain angle $\eta_c$ (see Figure 7.12) can be implemented. If the angle $\eta_c$ is calculated properly, worms of an arbitrary length can be ground by the face gear grinding wheel.

Again, the axes of rotation $O_g$ and $O_{ct}$ can either intersect each other or they can cross each other at a certain crossed axes angle. In both cases, special profiling of the face gear grinding wheel is required for accuracy purposes. Only at a particular configuration of the rotation vectors $\omega_g$ and $\omega_{cut}$ does the grinding wheel axial profile becomes straight. For this particular configuration, the axes of rotation $O_g$ and $O_{ct}$ are remote at distance $r_{b.g}$ of the radius of base cylinder of the involute worm. To calculate for the required value of cone angle $\theta_c$ of the generating surface $T$ of the grinding wheel, for which the rotation axis $O_{ct}$ is inclined at an angle $\eta_c$ with respect to the axis of rotation $O_g$ of the worm, the following expression:

$$\theta_c = \psi_{b.g} + \eta_c - 90^\circ \tag{7.16}$$

can be used.

A scenario wherein the rotation vectors $\omega_g$ and $\omega_{cut}$ form a certain acute angle enables machining of internal involute worm, as shown in Figure 7.13. In this case, the angle $\eta_c$ and the design parameters of the grinding wheel must be calculated so as to meet the set of conditions of proper part surface generation [136, 144, 148, 151, 158] and so forth (see Appendix E). The possibility of grinding of internal involute worms using face gear grinding wheels having an inclined axis of rotation is an important advantage of the discussed design of the gear grinding wheel.

Similar to the concept of the involute worm being ground via the outer side of the conical grinding wheel, as shown in Figure 7.12, it can also be ground by the inner side

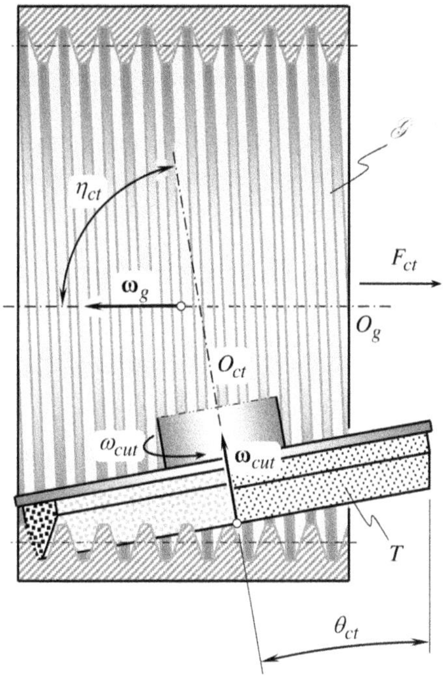


**FIGURE 7.13** An example of grinding of internal involute worm.

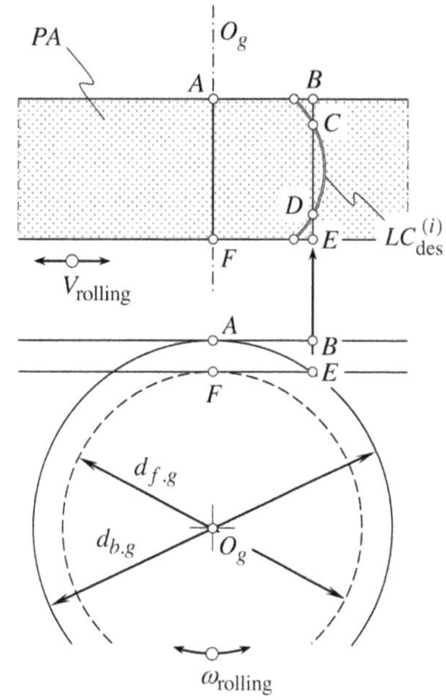


**FIGURE 7.15** A schematic of the method of CNC machining (cutting/grinding) gear tooth flanks.

of the conical grinding wheel. Two possible schematics for grinding involute worms using the grinding wheel with the generating surface $T$ in the form of the inner cone of revolution are shown in Figure 7.14.

Involute worms are not the only type of worms that can be machined with face gear grinding wheels. Worms with a thread profile of other geometries can also be machined with face gear grinding wheels. Moreover, a cutting tool of this type can be used for machining conical worms.

Tooth flanks of gears of all kinds—namely, gears for parallel-axes gearing, for intersected-axes gearing, and for crossed-axes gearing—can be machined on Computer Numerically Controlled (CNC) machines. As shown in Figure 7.15, the key feature of the method of machining is that the tool paths next to the root circle are represented by a corresponding portion of a desirable line of contact $LC_{des}$, terminated by points of intersection of the line of contact $LC_{des}$ by the base cone/cylinder.

Evidently, there is much room for improvement in finishing involute worms by alternating the parameters of the schematic of worm machining.

## 7.4 ON CLASSIFICATION OF FORM GEAR-CUTTING TOOLS

The classification of every possible kind of form gear-cutting tools for machining gears has not been developed yet. However, the need for classification is evident. Such a classification could be helpful in selecting a

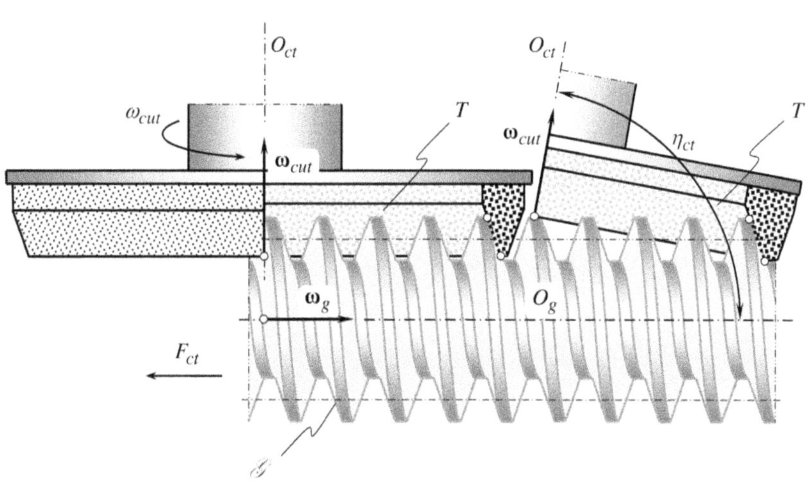


**FIGURE 7.14** Grinding of an involute worm by the inner side of the conical grinding wheel.

cutting tool design that best fits the needs of a given gear manufacturer.

It is important to bear in mind that the total number of possible configurations of the coordinate system associated with the gear-cutting tool in relation to the coordinate system associated with the gear to be machined is not infinite, but is limited to $n_{cg} = 4032$ combinations [see Eq. (7.7)]. The limited number of configurations makes it possible to carefully analyze each of the configurations individually and, in this way, make a decision on whether a particular combination is permissible for the requirements of gear machining. Computer methods can be used for the purpose of such an analysis. After all permissible configurations have been selected from the total number of the combinations (that is, from $n_{cg} = 4032$ combinations), the formal classification of every kind of form tool for machining gears can be developed. No principal constraints are observed in this regard.

The development of the classification of types of form tools for machining gears is not the goal of this text. For further consideration, it is sufficient to know that:

- The total number of configurations of the cutting tool relative the work is limited, not infinite
- The total number of the configurations ($n_{cg}$) is limited to 4032.
- All permissible configurations can be analyzed from the standpoint of their implementation in the design of gear-cutting tools.
- Computers can be used to analyze the feasibility of all the configurations of the coordinate systems $X_gY_gZ_g$ and $X_{ct}Y_{ct}Z_{ct}$.

The interesting reader may wish to exercise this challenging scientific problem independently.

# Section III

## Cutting Tools for Continuous Indexing Methods of the Generation of Gear Teeth

Section III is the largest section of this book. For the reader's convenience, this section is divided onto three subsections, beginning with Subsection III-A: Parallel-Axes Gear Machining Mesh.

# Subsection III-A

## Parallel-Axes Gear Machining Mesh

Continuous indexing methods of generating gear teeth are extensively used in the industry. High productivity rates and high accuracy of the machined gears are the main advantages of these methods of gear manufacturing.

Gear-cutting tools that utilize continuous indexing generating principle of gear tooth flank generation are the most sophisticated kinds of gear-cutting tools. To succeed in designing gear-cutting tools of this sort, an extensive use of accomplishments attained using the scientific theory of gearing [171] is a must. Otherwise, there is a limited chance to produce high quality gears.

When machining a work gear in a continuous indexing generating gear machining process, a rotation of the work gear and a rotation of the gear-cutting tool are synchronized with one another in a timely manner (see Figure III-A.1). It is commonly said that when machining a gear, rolling with no sliding of the axode associated with the work gear over the axode associated with the gear-cutting tool is always observed. It should be stressed here that the concept of axodes is an obsolete one. It is critical for the further discussion that the rotations $\omega_g$ and $\omega_{ct}$ of the work gear and of the gear-cutting tool, respectively, are somehow timed with one another, useless of axodes that will become evident in the following.

In this section, we review only those designs of gear-cutting tools that are engaged into a parallel-axes mesh with the work gear. The latter means that in all of the continuous generating methods of gear machining discussed in this section, the rotation vectors $\omega_g$ and $\omega_{ct}$ are always parallel to one another.

For the reader's convenience, this review of the designs of gear-cutting tools begins with the analysis of the kinematics of the parallel-axes gear machining mesh, or the $P_a$ – gear machining mesh.

DOI: 10.1201/9781003520528-10

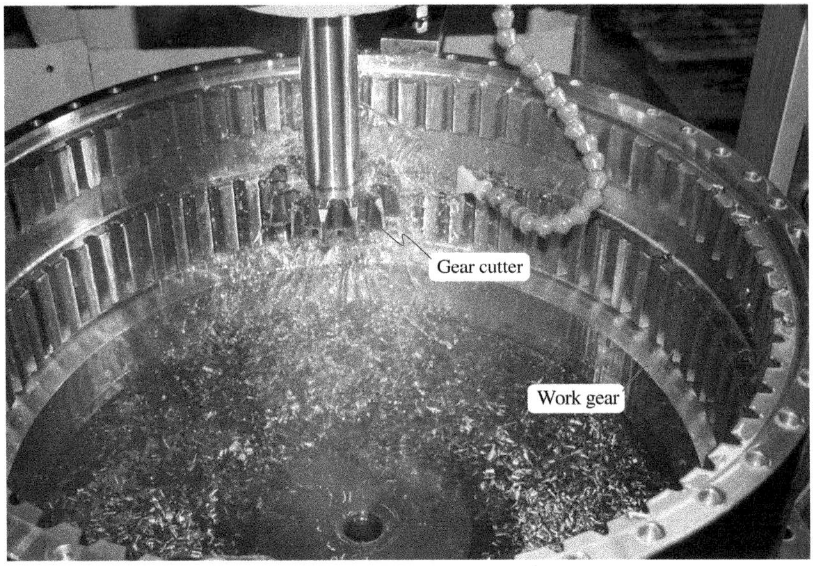


**FIGURE III-A.1** An example of parallel-axes gear-machining mesh.

## III-A.1 KINEMATICS OF PARALLEL-AXES GEAR MACHINING MESH

Parallel-axes gear machining mesh can be diagrammatically depicted by two rotation vectors that are parallel to each other—the rotation vector of the work gear $\boldsymbol{\omega}_g$ and the rotation vector of the gear-cutting tool $\boldsymbol{\omega}_{ct}$.

The rotation vectors $\boldsymbol{\omega}_g$ and $\boldsymbol{\omega}_{ct}$ are positioned at a certain center distance $C$ apart from each other. They are pointed along the axes $O_g$ and $O_{ct}$ of rotation of the work gear and of the gear-cutting tool, respectively. Commonly, the vectors $\boldsymbol{\omega}_g$ and $\boldsymbol{\omega}_{ct}$, are applied at the ends of the straight-line segment $C$, as shown in Figure III-A.2. However, because the vectors $\boldsymbol{\omega}_g$ and $\boldsymbol{\omega}_{ct}$ are a type of sliding vectors, in particular cases and for convenience, they can be applied at other points within the corresponding axis of rotation.

Because the rotation vectors $\boldsymbol{\omega}_g$ and $\boldsymbol{\omega}_{ct}$ are parallel, the vector of instant rotation $\boldsymbol{\omega}_{pl} = \boldsymbol{\omega}_{ct} - \boldsymbol{\omega}_g$ is also parallel to these vectors. The vector $\boldsymbol{\omega}_{pl}$ of instant rotation is applied at pitch point $P$. The pitch point $P$ is located within the centerline ℄. The ratio of the distance $r_g$ from the apex point to the work gear axis $O_g$ to the distance $r_{ct}$ from the apex point to the cutting tool axis $O_{ct}$ is inverse to the ratio of the magnitudes of the rotation vectors $\boldsymbol{\omega}_g$ and $\boldsymbol{\omega}_{ct}$. This relationship is analytically described by the expression:

$$r_g \cdot \omega_g = r_{ct} \cdot \omega_{ct} \tag{III.1}$$

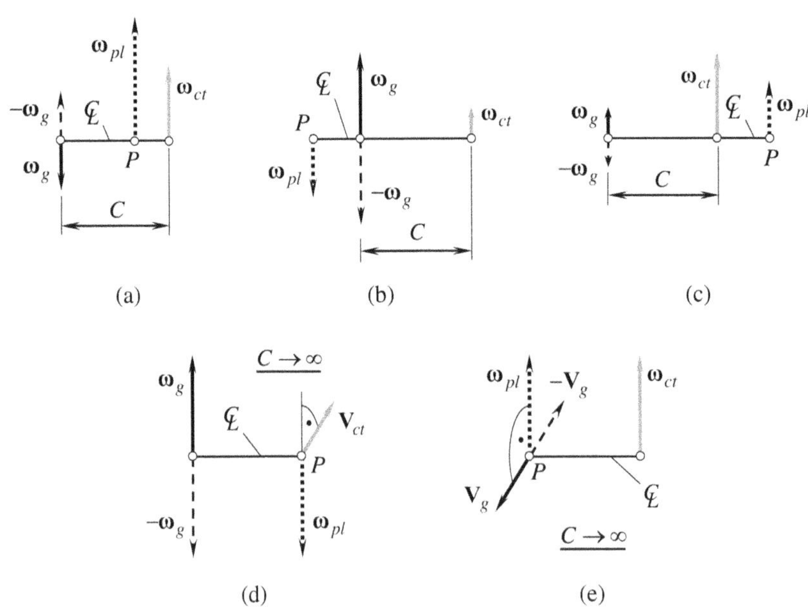


**FIGURE III-A.2** A variety of possible kinds of parallel-axes gear machining meshes.

Evidently, the actual location of pitch point $P$ within the centerline ℄, depends upon the ratio, $\omega_g/\omega_{ct}$, of the angular velocities. This ratio is commonly referred to as the tooth ratio in the gear machining mesh. Tooth ratio is designated as $u$. To calculate for $u$, the expression:

$$u = \frac{\omega_g}{\omega_{ct}} \equiv \frac{r_{ct}}{r_g} \tag{III.2}$$

is valid.

Shown in Figure III-A.2(a), the vector diagram corresponds to an external gear machining mesh. For an external gear machining mesh, the tooth ratio, $u$, is of a positive value (that is, $u > 0$). In this particular case, the vector of instant rotation $\boldsymbol{\omega}_{pl}$ is located between the rotation vectors $\boldsymbol{\omega}_g$ and $\boldsymbol{\omega}_{ct}$ of the gear and of the gear-cutting tool. Any alteration in the $u = \omega_g/\omega_{ct}$ ratio immediately entails a corresponding change in the location of the pitch point $P$ within the straight centerline ℄. However, as long as the ratio $u$ is positive, the vector of instant rotation $\boldsymbol{\omega}_{pl}$ would be located in between the rotation vectors $\boldsymbol{\omega}_g$ and $\boldsymbol{\omega}_{ct}$ occupying various positions within the centerline ℄. Under such a scenario, the pitch point $P$ is migrating within the centerline ℄ and moving closer either to the axis $O_g$ of the work gear or to the cutting tool axis $O_{ct}$ of rotation.

If the pitch point $P$ is located beyond one of the ends of the center distance $C$, the corresponding vector diagram corresponds to an internal gear machining mesh. Internal gear machining meshes of two different kinds are distinguished.

First, an external gear can be machined by the enveloping gear-cutting tool. A vector diagram for this particular case of gear machining is shown in Figure III-A.2(b).

Second, an internal gear can be machined by an external gear-cutting tool (see Figure III-A.2(c)].

In both cases, the tooth ratio $u$ is of a negative value ($u < 0$). In the first case [see Figure III-A.2(b)], the negative sign of $u$ is attributed to $r_{ct} < 0$ (whereas $r_g > 0$). In the second case [see Figure III-A.2(c)], the negative sign of $u$ is attributed to $r_g < 0$ (whereas $r_{ct} > 0$).

For an external gear machining mesh, the vector $\boldsymbol{\omega}_{pl}$ of instant rotation is applied at a point within the straight-line segment $C$. In contrast, for an internal gear machining mesh, the vector $\boldsymbol{\omega}_{pl}$ of instant rotation is applied at a point outside the straight-line segment $C$. What is in between these two cases? Or, in other words, which type of gear machining mesh corresponds to the case(s) for which the vector $\boldsymbol{\omega}_{pl}$ of instant rotation is applied at one of two ends of the straight-line segment $C$?

The vector diagram for this case, which corresponds to the rotating work gear ($\boldsymbol{\omega}_g$) and translating gear-cutting tool ($\mathbf{V}_c$), is schematically illustrated in Figure III-A.2(d). The pitch point $P$ is positioned at a distance $r_g$ from the work gear axis $O_g$. Because the rotation vector $\boldsymbol{\omega}_{pl}$ passes through the pitch point $P$, this means that the cutting tool axis $O_{ct}$ is remote to an infinity (that is, $r_{ct} \to \infty$). Since $r_{ct} \to \infty$, this results in the equality $\omega_{ct} = 0$ being valid, and the relative motion can be expressed in terms of the linear velocity $\mathbf{V}_{ct}$.

The vector diagram for the case that is inverse to that just discussed, which corresponds to the translating work gear ($\mathbf{V}_g$) and rotating gear-cutting tool ($\boldsymbol{\omega}_{ct}$), is schematically illustrated in Figure III-A.2(e). The apex point $A_{pa}$ is positioned at a distance $r_{ct}$ from the cutting tool axis $O_{ct}$. Because the rotation vector $\boldsymbol{\omega}_{pl}$ passes through the pitch point $P$, this means that the work gear axis of rotation $O_g$ is remote to infinity (that is, $r_g \to \infty$). Because $r_g \to \infty$, this results in the equality $\omega_{ct} = 0$ being valid, and the relative motion of the gear and the gear-cutting tool can be expressed in terms of the linear velocity $\mathbf{V}_g$.

A gear-cutting tool can be designed based on each of the vector diagrams shown in Figure III-A.2. Various designs of gear-cutting tools featuring the parallel-axes work gear-to-cutting tool mesh can be developed. It is convenient to begin this consideration by focusing on the design of a gear-cutting tool featuring the parallel-axis gear machining mesh shown in Figure III-A.2(d).

The third part of the book consists of Chapters 8 through 10.

# 8 Rack Cutters for Planing Spur and Helical Gears

Designing a rack cutter for planing spur and helical gears, as well as herring-bone gears and double-helical gears, begins with a comprehensive analysis of the kinematics of the gear machining mesh in the gear planing operation. The kinematics of gear-machining operation in all possible parallel-axes gear machining meshes, including but not limited to the kinematics of gear planing operation by rack cutters, is briefly outlined in the following.

## 8.1 GENERATING SURFACE OF RACK CUTTER

Spur and helical gear may be planned by a rack cutter. The planing process for a helical involute gear is illustrated in Figure 8.1. Rack cutters are also suitable for cutting sector gears. The generating action in the gear planing operation is similar to rolling of a gear along a mating rack. For profiling rack cutters, as well as for the design purposes, the generation surface $T$ of the rack cutter must be determined.

Analysis of the kinematics of machining gears by rack cutters reveals that generation of the gear tooth flank $\mathscr{G}$ can be interpreted as the enveloping surface to consecutive positions of the rack when the pitch surface of the rack rolls with no sliding over the pitch cylinder of the gear being machined. The kinematics of the gear-machining operation in this case corresponds to the gear machining mesh shown in Figure III.2(d).

The generating surface $T$ of the rack cutter can be viewed as the enveloping surface to consecutive positions of the gear tooth flank when the pitch cylinder of the gear being machined rolls with no sliding over the pitch plane of the rack cutter. In practice, a rack cutter can be designed based on the corresponding virtual rack that is engaged in proper mesh with the gear.

However, in order to enhance the capabilities of the tool designer, representation of the generating surface $T$ of the rack cutter in terms of the design parameters of the gear to be machined and the kinematics of the gear-machining operation is strongly desired.

### 8.1.1 Equation for the Generating Surface of the Rack Cutter

To derive an equation for the generating surface $T$ of the rack cutter, consider a left-hand-oriented Cartesian coordinate system $X_gY_gZ_g$ associated with the work gear, as shown in Figure 8.2. The axis $Z_g$ of the coordinate system $X_gY_gZ_g$ is aligned with axis of rotation $O_g$ of the work gear. In Figure 8.2, the pitch diameter of the gear is denoted by $d_g$. In the gear machining mesh, the pitch diameter exceeds the base diameter of the gear ($d_g \geq d_{b.g}$), where $d_{b.g}$ denotes the base diameter of the gear. On the other hand, the pitch diameter $d_g$ does not exceed a certain limit value $[d_{o.g}]$, under which either pointing of the rack cutter teeth is observed or the actual length of the top cutting

**FIGURE 8.1** Planing helical involute gear teeth.


DOI: 10.1201/9781003520528-11

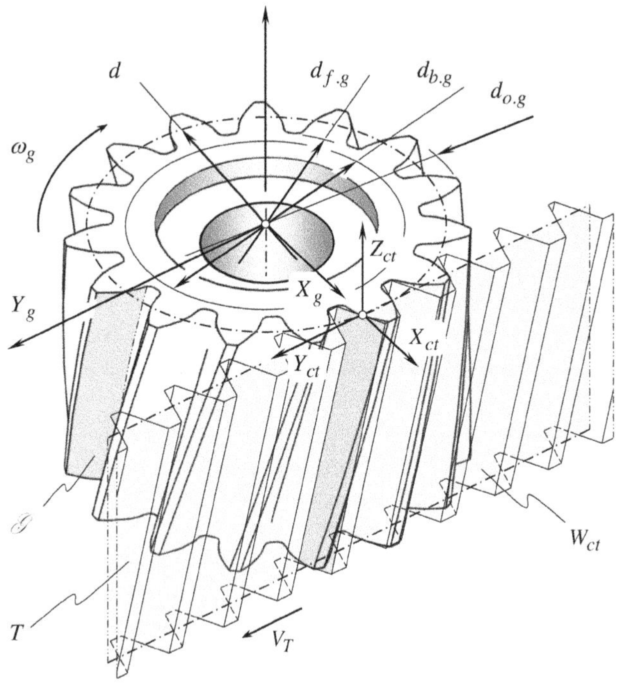


**FIGURE 8.2** Determination of the generating surface $T$ of the rack cutter for machining a helical involute gear.

edge is insufficiently short ($d_g < [d_{o.g}]$). To summarize, one can conclude that the pitch diameter of the gear in the gear machining mesh falls into the interval:

$$d_{b.g} \le d_g < [d_{o.g}] \tag{8.1}$$

In Figure 8.2, the pitch plane of the rack cutter is designated as $W_{ct}$. The pitch plane $W_{ct}$ is positioned at a distance $r_g = 0.5 \cdot d_g$ from the gear axis $O_g$. When designing a rack cutter, the tool designer is free to vary the distance $r_g$ in the range of:

$$0.5 \cdot d_{b.g} \le r_g < 0.5 \cdot [d_{o.g}] \tag{8.2}$$

At this point, the generating surface $T$ is not yet determined. Therefore, the motion of a certain reference system relative to the coordinate system $X_gY_gZ_g$—and not the motion of the rack cutter relative to the work gear—is considered. For this purpose, a left-hand-oriented Cartesian coordinate system $X_{ct}Y_{ct}Z_{ct}$ is implemented. The reference system $X_{ct}Y_{ct}Z_{ct}$ is somehow associated with the pitch plane $W_{ct}$. The gear-cutting tool designer is free to make the decision on how the pitch plane $W_{ct}$ is represented in the reference system $X_{ct}Y_{ct}Z_{ct}$. Convenience of analytical representation of the pitch plane $W_{ct}$ in the reference system $X_{ct}Y_{ct}Z_{ct}$ is the only criterion for selection of the desired configuration of the coordinate system $X_{ct}Y_{ct}Z_{ct}$ relative to the pitch plane $W_{ct}$. For convenience, at the beginning, the reference system $X_{ct}Y_{ct}Z_{ct}$ is oriented so that it is just shifted along the $X_g$ – axis at the distance $r_g$, as shown in Figure 8.2.

After it has been derived, the generating surface $T$ of the rack cutter will be represented in the reference system $X_{ct}Y_{ct}Z_{ct}$.

When machining, the work gear rotates $\omega_g$ about its axis $O_g$. The rack cutter tangentially reciprocates relative to the work gear tooth flank $\mathscr{G}$, and it travels forward with a certain velocity $V_T$.

The reciprocation of the rack cutter causes the generating surface $T$ to slide on itself. Because of this, the reciprocation does not affect the shape of the generating surface $T$. Therefore, this motion can be omitted from further analysis when determining the design parameters of the rack surface $T$.

The straight motion $V_T$ of the rack cutter $T$ is synchronized with the rotation $\omega_g$ of the work gear. When the reference system associated with the gear $X_gY_gZ_g$ turns about the $Z_g$ – axis through a certain angle $\varphi_g = \omega_g \cdot t$, the coordinate system $X_{ct}Y_{ct}Z_{ct}$ associated with rack cutter to be designed translates along the $Y_g$ – axis at a distance $l_{yc} = V_T \cdot t$, where $t$ denotes time. Because the equality $V_T = \omega_g \cdot r_g$ is valid, for the calculation of the displacement $l_{yc}$ the expression:

$$l_{yc} = \varphi_g \cdot r_g \tag{8.3}$$

is used.

To derive an equation of the generating surface $T$ of the rack cutter, an expression that analytically describes the tooth flank of the gear $\mathscr{G}$ in a current location of the work gear with respect to the reference system $X_{ct}Y_{ct}Z_{ct}$ is necessary. The equation for the involute gear tooth flank $\mathscr{G}$ [Eq. (1.4)]:

$$\mathbf{r}_g(U_g, V_g) = \begin{bmatrix} r_{b.g}\cos V_g + U_g \cos\lambda_{b.g} \sin V_g \\ r_{b.g}\sin V_g - U_g \sin\lambda_{b.g} \sin V_g \\ r_{b.g}\tan\lambda_{b.g} - U_g \sin\lambda_{b.g} \\ 1 \end{bmatrix} \quad \begin{matrix} V_g^{(l)} \le V_g \le V_g^{(a)} \\ 0 \le U_g \le [U_g] \end{matrix} \tag{8.4}$$

is implemented for this purpose.

Equation (8.4) describes the gear tooth flank $\mathscr{G}$ in the reference system $X_gY_gZ_g$ associated with the work gear itself. Implementation of operators of linear transformations is helpful for the transition from the coordinate system $X_gY_gZ_g$ to the coordinate system $X_{ct}Y_{ct}Z_{ct}$.

During the time $t$, the work gear, together with the reference system $X_gY_gZ_g$, turns about the $Z_g$ – axis through a certain angle $\varphi_g$. Transition from the coordinate system $X_gY_gZ_g$ to the reference system $X_g^*Y_g^*Z_g^*$ in the initial position of the work gear is described by the operator of rotation $\mathbf{Rt}(-\varphi_g, Z_g)$. For the analytical description of the transition from the coordinate system $X_g^*Y_g^*Z_g^*$ to the reference system $X_{ct}^*Y_{ct}^*Z_{ct}^*$, the operator of translation $\mathbf{Tr}(-r_g, X_g^*)$ is used. Here, the coordinate system corresponding to the initial position of the rack cutter is designated as $X^*_{ct}Y^*_{ct}Z^*_{ct}$. Ultimately, the operator of translation $\mathbf{Tr}(-l_{yc}, Y^*_c)$ can be used for the transition from the coordinate system $X^*_{ct}Y^*_{ct}Z^*_{ct}$ to the rack cutter coordinate system $X_{ct}Y_{ct}Z_{ct}$. Taking into account the equality $l_{yc} = \varphi_g \cdot r_g$, the operator of translation $\mathbf{Tr}(-l_{yc}, Y^*_c)$ can be represented in the form $\mathbf{Tr}(-\varphi_g \cdot r_g, Y^*_c)$.

Gear tooth flank $\mathscr{G}$ is described in the form of a function of two parameters: $U_g$ and $V_g$ [see Eq. (8.4)]. The operators of linear transformations are functions of the angle $\varphi_g$, through which the gear turns about its axis of rotation $O_g$. Therefore, an expression for the gear tooth flank $\mathscr{G}$ in the coordinate system $X_{ct}Y_{ct}Z_{ct}$ will be presented in the form of a function of three independent parameters—that is, of the Gaussian coordinates $U_g$ and $V_g$, and the angle $\varphi_g$.

Finally, an expression:

$$\begin{aligned}\mathbf{r}_g(U_g, V_g, \varphi_g) = \mathbf{Tr}(-\varphi_g \cdot r_g, Y^*_c) \cdot \mathbf{Tr}(-r_g, X^*_g)\\ \cdot \mathbf{Rt}(-\varphi_g, Z_g) \cdot \mathbf{r}_g(U_g, V_g)\end{aligned} \tag{8.5}$$

can be derived for the position vector $\mathbf{r}_g$ of a point of the gear tooth flank $G$. By means of this expression, the position vector $\mathbf{r}_g$ is represented in the reference system $X_{ct}Y_{ct}Z_{ct}$ associated with the rack cutter.

The equation for the generating surface $T$ of the rack cutter can be derived as the solution to the set of two equations:

$$\begin{cases}\mathbf{r}_g = \mathbf{r}_g(U_g, V_g, \varphi_g)\\ \dfrac{\partial}{\partial \varphi_g}\mathbf{r}_g(U_g, V_g, \varphi_g) = 0\end{cases} \tag{8.6}$$

After the second equation is solved with respect to the angular parameter $\varphi_g$ and the solution is substituted into the first equation in Eq. (8.6), an expression for position vector $\mathbf{r}_{ct}$ of a point of the generating surface $T$ can be presented in vector form:

$$\mathbf{r}_{ct} = \mathbf{r}_{ct}(U_{ct}, V_{ct}) \tag{8.7}$$

Without going into details of the derivation, the generating surface $T$ of the rack cutter is expressed below in terms of the design parameters of the gear to be machined:

$$\mathbf{r}_{ct}(U_{ct}, V_{ct}) = \begin{bmatrix} U_{ct} \\ \left(U_{ct} \cdot \cot\phi_g + r_g \cdot \tan\phi_g \pm \dfrac{n_{ct} \cdot p_{b.g}}{\cos\phi_g}\right)\cos\psi_g - V_{ct} \cdot \sin\psi_g \\ \left(U_{ct} \cdot \cot\phi_g + r_g \cdot \tan\phi_g \pm \dfrac{n_{ct} \cdot p_{b.g}}{\cos\phi_g}\right)\sin\psi_g + V_{ct} \cdot \cos\psi_g \\ 1 \end{bmatrix} \tag{8.8}$$

Tooth flanks of the generating surface $T$ are shaped in the form of two families of planes. These families of planes form the teeth of the phantom rack. The number of the current tooth of the rack is denoted by $n_{ct}$. This virtual rack is often referred to as the basic rack.

The planes of one family are parallel to one another. The distance between every two neighboring planes is equal to the normal base pitch $p_{b.g}$ of the work gear teeth.

The lines of intersection of the family of planes by the $X_gY_g$ coordinate plane form a certain profile angle $\phi_g$ with the $X_g-$ axis. The angle $\phi_g$ is equal to the transverse profile angle of the work gear tooth at the pitch cylinder of radius $r_g$.

The lines of intersection of the planes by the $Y_gZ_g$ coordinate plane form a certain helix angle $\psi_g$ with the $Z_g-$ axis. The angle $\psi_g$ is equal to the helix angle at the pitch cylinder of the radius $r_g$ of the gear.

The planes of another family of planes are also parallel to one another. The only difference between the planes of the two families is in the sign of the profile angle $\phi_g$. For the second family of parallel planes, this angle is not equal to $\phi_g$, but it is equal to $-\phi_g$ instead.[1]

The configuration of these two families of parallel planes relative each other is an important consideration. The tooth thickness and the space width at the pitch cylinder of a standard gear are equal to each other. When a standard gear is machined by a standard rack cutter, the configuration of one family of planes relative to the other family is uniquely specified by that. When a standard gear is machined by a nonstandard rack cutter, the tooth thickness and the space width are not equal to each other. Moreover, these two parameters become dependent on the number of teeth of the gear to be machined. Therefore, in this particular case, for the purposes of specifying the configuration of two families of parallel planes, the rolling motion of the pitch surfaces should be considered. The tooth thickness of the generating surface $T$ at the pitch plane must be equal to the corresponding space width of the gear teeth at the pitch cylinder of the radius $r_g$. Similarly, the space width of the generating surface $T$ at the pitch plane must be equal to the corresponding tooth thickness on the work gear pitch cylinder of the radius $r_g$. The same approach is applicable when nonstandard gears are required to be machined by a rack cutter.

The tooth flank surface of a rack cutter for machining a helical gear is described by Eq. (8.8). Substitution of the helix angle $\psi_g = 0°$ reduces Eq. (8.8) to a simpler expression, which corresponds to the tooth flank surface of a rack cutter for machining a spur gear.

Because the tooth flanks of the generating surface of the rack cutter are planes, this eliminates the necessity of analyzing the intrinsic geometry of the generating surface $T$.

In a particular case of gear machining, the variables that are to be entered into in Eq. (8.8)—base pitch $p_{b.g}$, profile angle $\phi_g$, and helix angle $\psi_g$ on the pitch cylinder

[1] Gears with asymmetric tooth profiles are not considered here.

of radius $r_g$—can be expressed in terms of tooth count $N_g$, normal pitch $P_{n.g}$, normal pressure angle $\phi_n$, and so forth. For this purpose, the engineering formulae listed in Appendix F are helpful.

A clear understanding of the geometry and the kinematics of the gear machining mesh is of critical importance to a gear-cutting tool designer.

## 8.2 ON THE VARIETY OF PERMISSIBLE TOOTH PROFILES OF RACK CUTTERS

The actual tooth profile of the generating surface $T$ of the rack cutter strongly depends on the type of timing of the work gear rotation and the translation of a reference system associated with the rack cutter. Because the pitch plane of the rack cutter rolls over the pitch cylinder of the work gear with no sliding, the linear velocity of the rotational motion of points on the pitch cylinder $\omega_g \cdot r_g$ is equal to the translational velocity of the rack cutter $V_T$—that is, the equality:

$$\omega_g \cdot r_g = V_T \tag{8.9}$$

is observed.

This equality immediately yields an expression for the radius $r_g$ of the pitch cylinder of the work gear when it is cutting by the rack cutter:

$$r_g = \frac{V_T}{\omega_g} \tag{8.10}$$

Equation (8.9) shows that variation of the ratio $V_T/\omega_g$ causes corresponding alterations in the radius of the pitch cylinder of the work gear in the gear-machining process. Ultimately, the alterations result in corresponding changes in the parameters of the tooth profile of rack $T$.

The impact of the radius of the pitch cylinder $r_g$ of the work gear (in other words, impact of the ratio $V_T/\omega_g$) on the shape of the tooth profile of the generating surface $T$ of the rack cutter is illustrated in Figure 8.3.

For the minimum permissible radius $r_g$ of the pitch cylinder, the equality $r_g = r_{b.g}$ is valid [see Figure 8.3(a)]. The profile angle of the rack cutter tooth is equal to the pressure angle of the work gear at a point on the pitch cylinder ($\phi_{ct} = \phi_g$). Because the equality $r_g = r_{b.g}$ is observed, the profile angle $\phi_{ct}$ of the rack cutter is equal to zero ($\phi_{ct} = 0°$).

The helix angle $\psi_{ct}$ of the generating surface $T$ of the rack cutter is equal to the helix angle $\psi_g$ on the pitch cylinder of the gear being machined ($\psi_{ct} = \psi_g$). Again, because the equality $r_g = r_{b.g}$ is valid, the helix angle $\psi_{ct}$ of the generating surface $T$ of the rack cutter in the case under consideration is equal to base helix angle of the work gear ($\psi_{ct} = \psi_{b.g}$).

The width of the top land $t_o$ of rack $T$ with a profile angle with a zero value $\phi_c = 0°$ is the maximal possible. The base pitch of the tooth flank surfaces of the rack cutter is designated as $p_b$. The base pitch $p_b$ must be identical to the base pitch of the gear to be machined (that is, $p_{b.g} \equiv p_{b.ct} \equiv p_b$). Because of this, no additional subscripts are used in the designation of base pitch $p_b$.

When the tooth profile angle is equal to zero, the tooth addendum vanishes ($a = 0$), and the tooth dedendum $b$ becomes equal to the whole tooth height ($b = h_t$).

For a larger ratio $V_T/\omega_g$ [see Figure 8.3(b)], which corresponds to a larger radius $r_g$ of the pitch cylinder of the work gear, the profile angle of the rack cutter $\phi_{ct}$ is also larger, and the width of the top land $t_o$ is also smaller. The base pitch $p_b$, however, remains of the same value and is equal to the base pitch of the gear to be machined—this is the law. For a larger ratio $V_T/\omega_g$, the expression $\phi_{ct} = \phi_g$ for the cutter tooth profile angle and the expression $\psi_{ct} = \psi_g$ for the cutter helix angle are still valid. The tooth addendum $a$ of the rack $T$ and the tooth dedendum $b$ are of certain values. The whole tooth height is equal to $h_t = a + b$.

The tooth thickness of the generating surface $T$ measured within the pitch line is equal to the arc width of the tooth space of the work gear measured on the pitch circle. In both cases described above, the tooth thickness of the rack and width of space between adjacent teeth are not equal to each other.

In a particular case of gear machining, the radius of the pitch cylinder of the gear can be of a certain value when the tooth thickness of rack $T$ is equal to the space width of the rack [see Figure 8.3(c)].

Although the major relationships for the generating surface $T$ of the rack cutter remain valid ($p_{b.g} \equiv p_{b.ct} \equiv p_b$, $\phi_{ct} = \phi_g$, $\psi_{ct} = \psi_g$, and so forth), this particular case features an important advantage for practical use of rack cutters. The equality between the tooth thickness of the rack cutter and the space width makes the rack cutter versatile—the same rack cutter can be used for machining gears with different tooth counts.

There are no physical constraints for further increase of the ratio $V_T/\omega_g$, as shown in Figure 8.3(d). However, at some point, tooth pointing of the generating surface $T$ of the rack cutter is observed. Physically, a rack cutter can have teeth with pointing, but this is completely impractical. Geometrically, the ratio $V_T/\omega_g$ can be increased beyond the value of the radius $r_g$ when the tooth pointing occurs [see Figure 8.3(e)]. In such a scenario, the top land of the rack cutter vanishes. Analytically, it can be expressed in negative values; however, the practicality of the rack cutter is at best doubtful.

The following conclusions can be draw up from the preceding discussion:

- Variation in the kinematic parameters of the gear-machining operation provides the ability to control the main design parameters of the rack cutter.
- The actual tooth profile angle of the rack cutter and the actual helix angle of the rack cutter are equal to the corresponding design parameters of

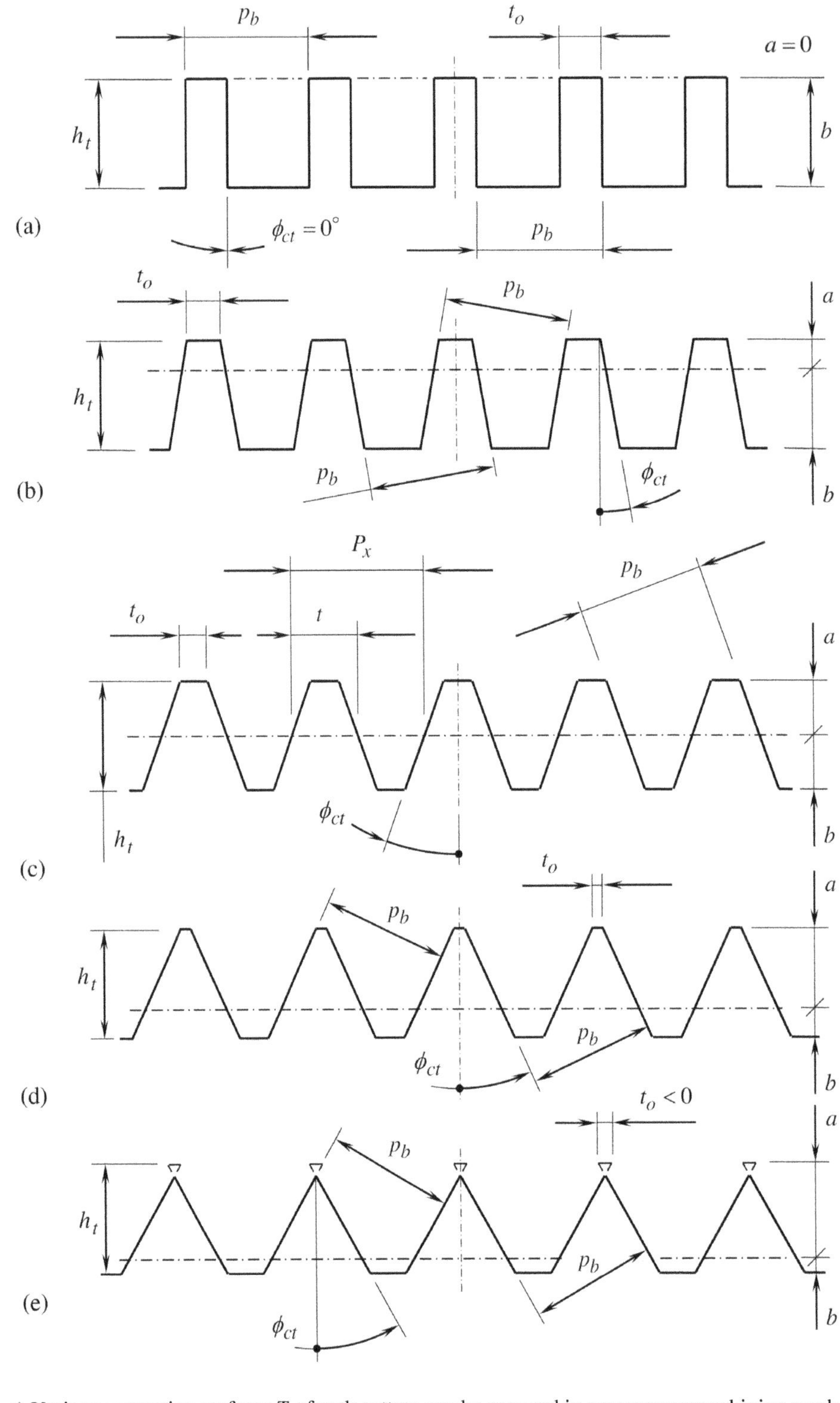


**FIGURE 8.3** (a–e) Various generating surfaces $T$ of rack cutters can be engaged in proper gear machining mesh with the work gear.

the gear to be machined, as measured on its pitch cylinder.

- The particular value of the ratio $V_T/\omega_g$ under which tooth thickness of the rack cutter is equal to tooth space width is of practical importance. In this particular case, the same rack cutter can be used for machining gears with a different number of teeth.
- The width of the rack cutter top land is under control and reaches its maximum value when the tooth profile angle of the rack is equal to zero.
- The width of the rack cutter top land can be reduced to undesirably small values or tooth pointing could be observed when the rack cutter tooth profile angle exceeds a certain allowable value.

These features are among the important issues that a cutting tool designer has to keep in mind when aiming to design a high-performance gear-cutting tool.

## 8.3 CUTTING EDGE OF THE RACK CUTTER

Generating surface is a key element in designing a rack cutter for machining a gear.

Every tooth of the rack cutter has three cutting edges. The top cutting edge is shaped in the form of a straight-line segment. The bottom land of tooth space of the work gear is machined by this cutting edge. The gear tooth profile is generated by two lateral cutting edges. Lateral cutting edges of the rack cutter are straight. The last is a solid advantage of the gear-cutting tool of this design.

Cutting edges of a rack cutter for machining gears are commonly viewed as the line of intersection of the generating surface $T$ of the gear-cutting tool by the rake surface $R_s$.

### 8.3.1 Rake Surface of the Rack Cutter

Initially, rack cutters were designed with the rake surface shaped in the form of a plane. The rake plane $R_s$ forms certain rake angle $\gamma_o$ with the perpendicular to the top land of the generating surface $T$ of the rack cutter. This angle is referred to as rake angle $\gamma_o$, whose value could be positive ($\gamma_o > 0°$), negative ($\gamma_o < 0°$), or equal to zero.

Two different configurations of the rake plane $R_s$ are used in the design of rack cutters. One is used in designing rack cutters for machining both spur and helical gears. The other is used in designing rack cutters for machining only helical gears. The latter features the rake plane that forms the pitch helix angle $\psi_g$ with the generating surface $T$ of the rack cutter. Here, $\psi_g$ denotes the pitch helix angle of the helical gear to be machined.

#### 8.3.1.1 The Rake Plane Is Perpendicular to the Generating Straight Lines of Surface $T$

Consider the cutting edges of the rack cutter teeth when the rake plane is perpendicular to the generating surface (see Figure 8.4).

The generating surface $T$ of the rack cutter is depicted in the system of planes of projections $\pi_1\pi_2\pi_3$, as shown in Figure 8.4. The shape of the generating surface $T$ is specified by a set of the design parameters. For this purpose, conventional design parameters—base pitch $p_b$, tooth profile angle $\phi_n$, transverse diametral pitch $P_x$, tooth thickness $t$, tooth addendum $a$, tooth dedendum $b$, and whole tooth height $h_t$—are used.

For the generating surface $T$ of the given orientation and location in the system of planes of projections $\pi_1\pi_2\pi_3$, the rake plane $R_s$ is a projective plane onto the frontal plane of projections $\pi_3$. Because of this, the rake plane $R_s$ in $\pi_3$ is shown as the trace of this plane. The trace forms the rake angle $\gamma_o$ with the perpendicular to the top land of the generating surface $T$. The actual value of the rake angle $\gamma_o$ has to be known.

Once the rake plane is constructed in $\pi_3$, projections of the rake plane $R_s$ onto the planes of projections $\pi_1$ and $\pi_2$

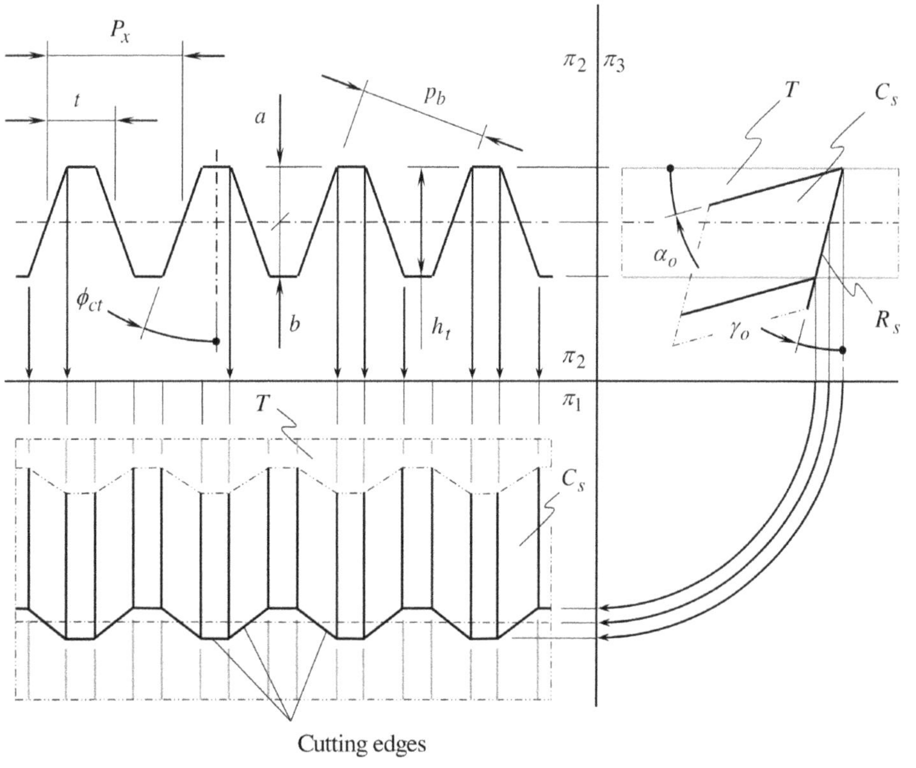


**FIGURE 8.4** Configuration of working surfaces of a rack cutter tooth.

can be constructed following well-known rules, which are developed in descriptive geometry (*DG*) (see Figure 8.4).

#### 8.3.1.2 The Rake Plane Is at the Helix Angle to the Generating Straight Lines of Surface *T*

Rack cutters can be used for machining gears with shoulders, for example, when machining a cluster gear. For this particular case, rack cutters are designed with the rake plane $R_s$ at the helix angle to the generating lines of surface $T$ of the rack cutter.

In Figure 8.5, the generating surface $T$ of the rack cutter is depicted in the system of planes of projections $\pi_1\pi_2\pi_3$. Here, the surface $T$ is identical to that shown in Figure 8.4. By definition, the rake plane $R_s$ is at the helix angle $\psi_g$ relative to the generating surface $T$. At the same time, the rake plane forms rake angle $\gamma_o$ with the perpendicular to the top land of the generating surface $T$.

An additional plane of projections $\pi_4$ is constructed so that the rake plane $R_s$ occupies the position of a projective plane onto this plane of projections. Therefore, in $\pi_4$, the rake plane $R_s$ is shown as the trace, and the rake angle $\gamma_o$ can be readily constructed. The actual value of the rake angle $\gamma_o$ has to be known.

Once the rake plane is constructed in $\pi_4$, then the projections of the rake plane $R_s$ onto the planes of projections $\pi_1$ and $\pi_2$ can be constructed following well-known rules, which are developed in descriptive geometry (see Figure 8.5).

Designs of rack cutters often feature one of the various modifications of the rake plane $R_s$. These modifications aim to improve the cutting performance of the rack cutter. Several types of modifications of the rake surface are discussed in this chapter.

Cutting edges of the rack cutter are considered in this text from the standpoint of the kinematic geometry of surface machining [133, 144, 148, 151, 158, 168]. This means that the cutting edges *CE* are viewed as the line of intersection of the generating surface $T$ by the rake plane $R_s$.

### 8.3.2 Clearance Surface of the Rack Cutter Tooth

Rack cutters are designed with a clearance surface $C_s$ that is shaped in the form of a plane. The clearance plane through the top cutting edge is at the clearance angle $\alpha_o$ relative to the generating surface $T$ of the rack cutter (see Figures 8.4 and 8.5).

The clearance planes $C_s$ through the lateral cutting edges *CE* of the rack cutter tooth form a certain clearance angle $\alpha_{cl}$ with the generating surface $T$. The clearance angle $\alpha_{cl}$ correlates to the clearance angle $\alpha_o$ at the top cutting edge of the rack cutter. To clarify the correlation between the clearance angles $\alpha_{cl}$ and $\alpha_o$, consider the section of the clearance surface $C_s$ through the top cutting edge of the rack cutter tooth by a plane that is perpendicular to the top land of the generating surface $T$ of the rack cutter. This plane is perpendicular to the top cutting edge *CE* in the case shown in Figure 8.4. For the case shown in Figure 8.5, the section forms the helix angle $\psi_g$ with the top cutting edge *CE*. In both cases, the line of intersection of the clearance plane $C_s$ is at the clearance angle $\alpha_o$ with respect to the top land of the generating surface $T$ of the rack cutter.

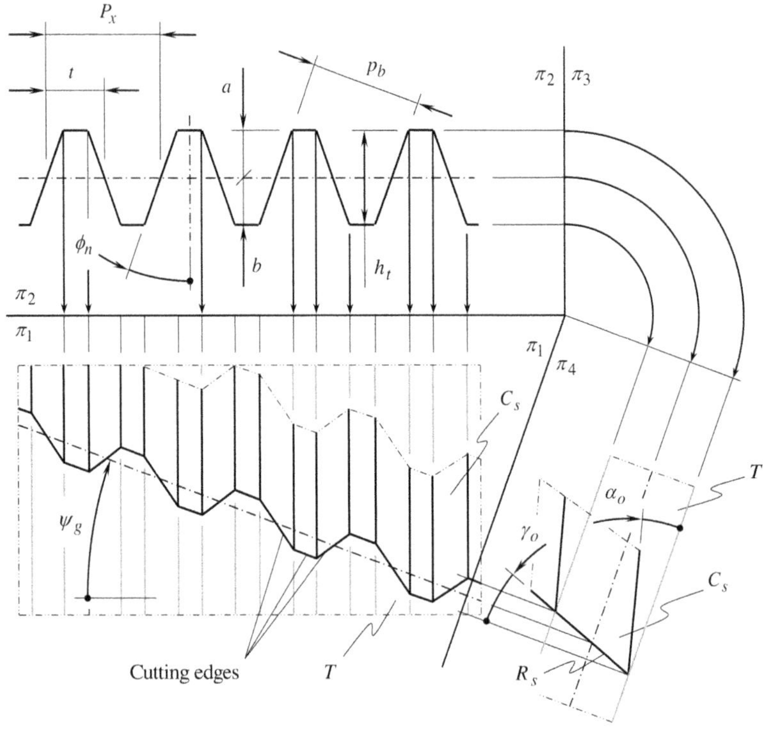


**FIGURE 8.5** Configuration of the working tooth surfaces of a helical rack cutter.

Furthermore, consider the section of the clearance surface $C_s$ through the lateral cutting edge of the rack cutter tooth. The section plane is parallel to that just considered. The line of intersection of the clearance surface $C_s$ by the plane forms an angle with respect to the top land of the generating surface $T$. This angle is equal to the clearance angle $\alpha_o$.

For such a configuration of the clearance surfaces through the top cutting edge and through the lateral cutting edges of the rack cutter, the normal clearance angles at the cutting edges correlate with each other. The considered configuration of the clearance surfaces resembles the configuration of the clearance surfaces of a cutting tool with relieved clearance surfaces.

Clearance surfaces are the surfaces through the cutting edges *CE* of the rack cutter teeth.

## 8.4 PROFILING OF RACK CUTTERS TEETH

For profiling of the rack cutters for machining gears, both descriptive-geometry-based (*DG* – based) methods and analytical methods of profiling are used. The purpose of profiling the rack cutters is twofold. It is used to determine the rack cutter tooth profile:

- Within the rake plane $R_s$.
- In the normal section of the clearance surface $C_s$ of the rack cutter teeth.

Implementation of both *DG* – based methods and analytical methods of profiling rack cutters is discussed in the following sections.

### 8.4.1 Profiling Rack Cutter Teeth Using *DG*-based Methods

It is convenient to begin this discussion by focusing on profiling a rack cutter for machining spur gears. For the purposes of profiling, the generating surface $T$ of the rack cutter is depicted in the system of planes of projections $\pi_2\pi_3$ so that all the planes of the generating surface $T$ are projective planes onto the plane $\pi_2$ (see Figure 8.6). In such a scenario, the rake plane $R_s$ and the normal section of the clearance surface $C_s$ are the projective planes onto the plane of projections $\pi_3$. The use of conventional methods developed in descriptive geometry immediately yields the rake plane profile of the rack cutter tooth, as well as the normal section of the clearance surface $C_s$ (see Figure 8.6).

For profiling a helical rack cutter, the generating surface $T$ is depicted in the system of planes of projections $\pi_1\pi_2\pi_3$ so that all planes of the generating surface $T$ are projective planes onto the plane $\pi_2$ (see Figure 8.5). The process of profiling a rack tooth for machining a helical gear with a shoulder additionally features the determination of the generating surface $T$ in a particular plane of projections. This plane of projections is labeled as $\pi_4$. The axis of projections $\pi_1/\pi_4$ is perpendicular to the rake plane $R_s$. Consequently, the rake plane $R_s$ and the normal section are the projective planes onto the plane of projections $\pi_4$. After the rake plane $R_s$ and the normal section are constructed in $\pi_4$, the rake face profile of the rack cutter tooth, as well as the normal section of the clearance surface $C_s$, can be readily constructed within the plane of projections $\pi_1$ (see Figure 8.7).

It is recommended to perform profiling of a rack cutter using the *DG* – based methods before profiling them analytically. This results in the elimination of the rough errors of profiling.

### 8.4.2 Analytical Profiling of Rack Cutter Teeth

Analytical methods of profiling rack cutter teeth are used to attain high accuracy in profiling. The required equations for the calculation of coordinates of points of the tooth profile can be drawn up from the graphical solution derived on the basis of the implementation of the *DG* – based methods of profiling. Without going into details of elementary

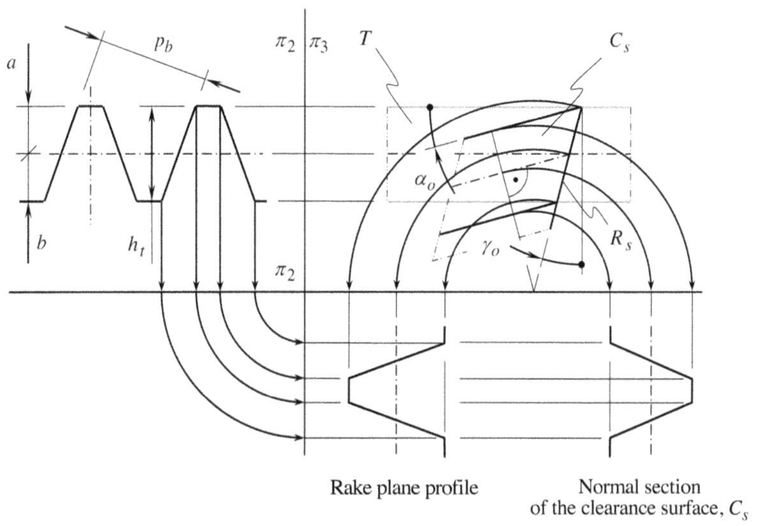


**FIGURE 8.6** Profiling a spur rack cutter using the descriptive-geometry-based method of profiling cutting tools.

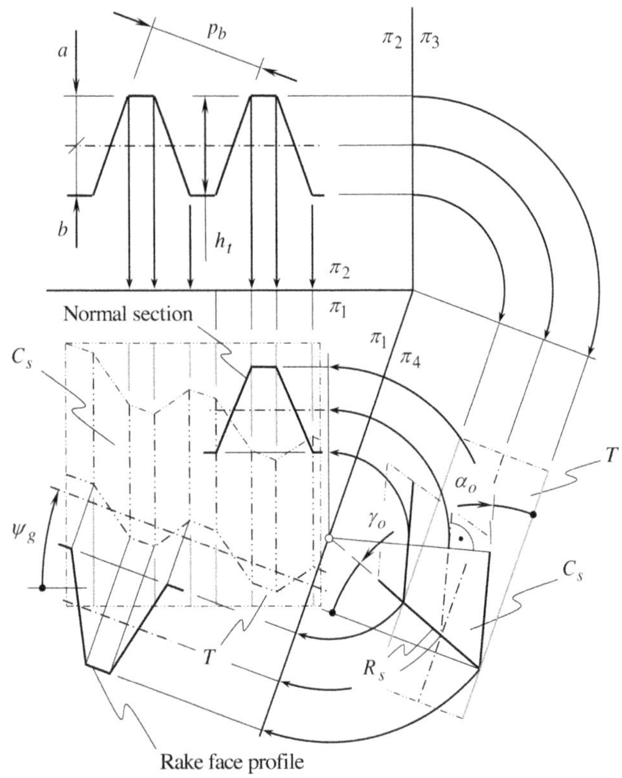


**FIGURE 8.7** Profiling a helical rack cutter tooth using the *DG*–based method.

derivation, the resultant formulae for the calculation are given in the following.

Consider a rack cutter for machining a spur gear (see Figure 8.8). The current point of the tooth profile of the generating surface $T$ is at a certain distance $\tau_y$ from the tooth centerline. The parameter $\tau_y$ is referred to as the width coordinate of the rack tooth profile. For a given value of the width coordinate $\tau_y$, the corresponding height coordinate $h_y$ is either known or can be calculated.

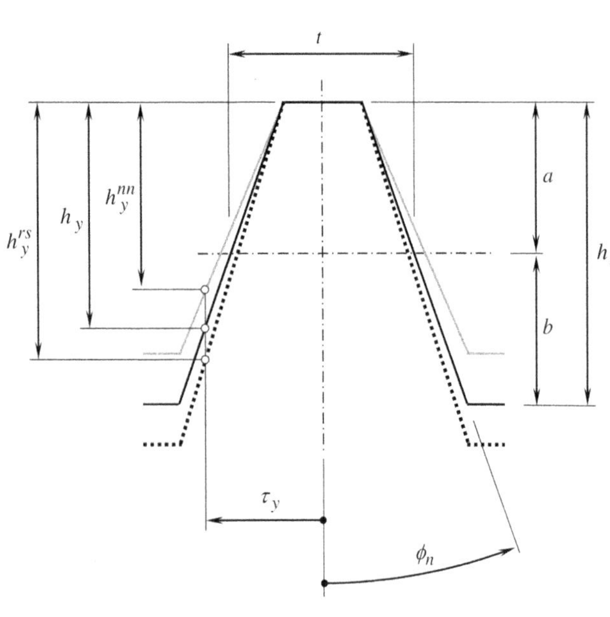


**FIGURE 8.8** Correspondence between the design parameters of rack cutter tooth profile measured in different sections of the generating surface $T$ of the cutting tool.

For that same width coordinate $\tau_y$, the corresponding height coordinate $h_y^{rs}$ within the rake plane $R_s$ of the rack cutter tooth is calculated from the equation:

$$h_y^{rs} = \frac{h_y}{\cos\gamma_o} \tag{8.11}$$

The rack cutter tooth profile angle is designated as $\phi_n$. The actual value of the angle $\phi_n$ is commonly known. The corresponding tooth profile angle $\phi_{rs}$ within the rake plane $R_s$ is equal to:

$$\phi_{rs} = \tan^{-1}(\tan\phi_n \cdot \cos\gamma_o) \tag{8.12}$$

Similarly, equations for the height coordinate $h_y^{nn}$ and the profile angle $\phi_{nn}$ in the normal section of the clearance surface $C_s$:

$$h_y^{nn} = h_y \cdot \frac{\cos(\alpha_o + \gamma_o)}{\cos\gamma_o} \tag{8.13}$$

$$\phi_{nn} = \tan^{-1}\left[\frac{\cos\gamma_o}{\cos(\alpha_o + \gamma_o)} \cdot \tan\phi_n\right] \tag{8.14}$$

can be derived. The last two formulae are valid for both spur rack cutters and helical rack cutters.

The preceding discussion reveals that the profiling of rack cutter teeth for machining involute gears is easy to perform. These simple examples are helpful for a better understanding of the complex problems of profiling gear-cutting tools considered in the following chapters.

It is of critical importance to note here that only involute gears can be precisely generated in the gear-planing process. Gears with no other tooth flank geometry can be perfectly generated by the rack cutter—only approximate non-involute gears can be cut this way. This is attributed to the kinematics of the gear-planing process and to violation of the condition of conjugacy of the tooth flanks $\mathcal{G}$ of a gear being machined and of the generating surface $T$ of the gear-cutting tool. In compliance with the condition of conjugacy of tooth flanks, at every point of contact of the surfaces $\mathcal{G}$ and $T$, the common perpendicular **n** must intersect the axis of instant rotation $P_{ln}$ of the gear and the gear-cutting tool in the gear machining mesh [171].

## 8.5 CUTTING EDGE GEOMETRY OF RACK CUTTER

Rack cutters for roughing gears are often designed with rake angle $\gamma_o$ at the top cutting edge, which is equal to $\gamma_o = 6°30'$. The clearance angle $\alpha_o$ at the top cutting edge of the rack cutters of this design is equal to $\alpha_o = 5°30'$. The geometry of the top cutting edges of rack cutters for finishing gears is slightly different. For the finishing rack cutters, the rake angle $\gamma_o$ is equal to $\gamma_o = 4°$, and the clearance angle $\alpha_o$ is equal to $\alpha_o = 6°52'$.

Consider a rack cutter with standard normal profile angle $\phi_n = 20°$. When the geometry at the top cutting edge $CE$ of the rack cutter is specified by the angles $\gamma_o = 6°30'$ and $\alpha_o = 5°30'$, the profile angle within the rake plane $R_s$ of the rack cutter tooth is equal to $\phi_{rs} = 19°25'54''$. The profile angle of the normal section of the clearance surface $C_s$ in this case is equal to $\phi_{nn} = 20°17'25''$.

### 8.5.1 Calculation of Cutting Tool Geometry for Lateral Cutting Edges

When the angles $\gamma_o$, $\alpha_o$, and $\phi_n$ are given, the geometry of the lateral cutting edges of the rack cutter tooth can be determined analytically.

First, let us determine the actual value of the inclination angle $\lambda$ of the lateral cutting edge $CE$ of the rack cutter tooth.

Consider the rack cutter tooth depicted in Figure 8.9. A Cartesian coordinate system $XYZ$ is associated with the rack cutter tooth, as shown in Figure 8.9. To derive an equation for the inclination angle $\lambda$ of the cutting edge, it is convenient to apply the vector method of analysis.[2] For this purpose, three vectors **a**, **B**, and **C**, which are associated with the lateral cutting edge $CE$, are constructed. The vectors **a**, **B**, and **C** need to be expressed in terms of the inclination angle $\lambda$ to be determined and the known parameters of the geometry of the cutting wedge of the cutting tool.

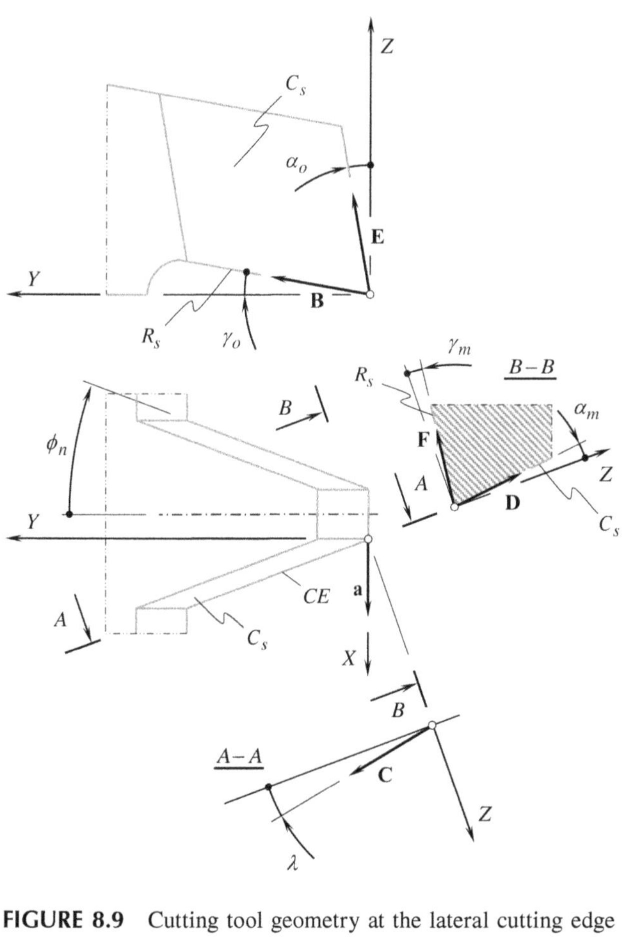


**FIGURE 8.9** Cutting tool geometry at the lateral cutting edge $CE$ of a rack cutter tooth.

The inclination angle $\lambda$ is measured within the plane of cut [144, 148, 151, 158]. The plane of cut is tangent to the surface of cut $S_c$ at a point of interest within the cutting edge $CE$. The plane of cut is designated in Figure 8.9 as $A-A$.

The vector **a** is a unit vector. This vector is pointed along the top cutting edge. In the reference system $XYZ$, the following expression:

$$\mathbf{a} = \mathbf{i} \tag{8.15}$$

is valid for the unit vector **a**.

The vector **B** is aligned with the line of intersection of the rake plane $R_s$ by the coordinate plane $YZ$. To solve the problem under consideration, only the direction of the vector **B** is important. The length of the vector **B** is out of current interests. Because of this, the vector **B** of the length for which the projection of the vector onto the $Y-$ axis is equal to unit ($\text{pr}_y\mathbf{B} = 1$) can be applied. In such a scenario, the vector **B** can be analytically described as:

$$\mathbf{B} = \mathbf{j} + \mathbf{k}\tan\gamma_o \tag{8.16}$$

The vector **C** is pointed along the lateral cutting edge $CE$ of the rack cutter. The following expression:

$$\mathbf{C} = -\mathbf{i}\sin\phi_n + \mathbf{j}\cos\phi_n + \mathbf{k}\tan\lambda \tag{8.17}$$

is valid for the analytical representation of the vector **C** in terms of its projections onto the axes of the reference system $XYZ$.

As follows from the analysis in Eq. (8.17), in the case under consideration, the vector **C** is chosen for which length of the projection of the vector **C** onto the coordinate plane $XY$ is equal to the unit—that is, $\text{pr}_{xy}\,\mathbf{C} = 1$.

By construction, three vectors **a**, **B**, and **C** comprise a set of coplanar vectors. All three are within the rake plane $R_s$ of the rack cutter. The triple scalar product of coplanar vectors is identical to zero: $\mathbf{a} \times \mathbf{B} \cdot \mathbf{C} \equiv 0$. Therefore,

$$\mathbf{a} \times \mathbf{B} \cdot \mathbf{C} \equiv \begin{vmatrix} 1 & 0 & 0 \\ 0 & 1 & \tan\gamma_o \\ -\sin\phi_n & \cos\phi_n & \tan\lambda \end{vmatrix} = 0 \tag{8.18}$$

[2] Implementation of the vector method for the calculation of the cutting edge geometry can be traced back to the mid-1940s and should be attributed to Dr. S.S. Mozhayev [49].

This expression [see Eq. (8.18)] casts into the formula:

$$\lambda = \tan^{-1}(\tan\gamma_o \cdot \cos\phi_n) \quad (8.19)$$

for the calculation of the angle of inclination $\lambda$ of the lateral cutting edge $CE$ of the rack cutter tooth.

Second, an equation can be derived to calculate for the actual value of the rake angle $\gamma_m$ [144, 148, 151, 158]. The angle $\gamma_m$ is measured within the major section plane of the lateral cutting wedge of the rack cutter. For this purpose, a method similar to that just implemented for the derivation of Eq. (8.19) can be used.

Consider the straight line of intersection of the rake plane $R_s$ by the major section plane (see Figure G.3 in Appendix G for details on the major section plane $P_{ms}$). In Figure 8.9, the major section plane $P_{ms}$ is designated as $B-B$. The vector $\mathbf{D}$ is aligned with this straight line (see Figure 8.9).

The length of the vector $\mathbf{D}$ is chosen so for which the projection of $\mathbf{D}$ onto the coordinate plane $XY$ is equal to the unit—that is, $\text{pr}_{xy}\mathbf{D} = 1$. This yields an expression:

$$\mathbf{D} = \mathbf{i}\cos\phi_n + \mathbf{j}\sin\phi_n + \mathbf{k}\tan\gamma_m \quad (8.20)$$

for the vector $\mathbf{D}$.

By construction, the vectors $\mathbf{a}$, $\mathbf{B}$, and $\mathbf{D}$ comprise a set of coplanar vectors. Therefore, the scalar triple product of the vectors is equal to the unit. Because of this, the expression $\mathbf{a}\times\mathbf{B}\cdot\mathbf{D} \equiv 0$ is valid. This expression yields representation in the form:

$$\mathbf{a}\times\mathbf{B}\cdot\mathbf{D} \equiv \begin{vmatrix} 1 & 0 & 0 \\ 0 & 1 & \tan\gamma_o \\ \cos\phi_n & \sin\phi_n & \tan\gamma_m \end{vmatrix} = 0 \quad (8.21)$$

Equation (8.21) casts into the formula:

$$\gamma_m = \tan^{-1}(\tan\gamma_o \cdot \sin\phi_n) \quad (8.22)$$

for the calculation of the rake angle $\gamma_m$ in the section of the lateral cutting edge $CE$ of the rack cutter tooth by the major section plane $P_{ms}$.

Third, consider three vectors $\mathbf{C}$, $\mathbf{E}$, and $\mathbf{F}$, all of which are within the clearance surface $C_s$ of the lateral cutting edge $CE$ of the rack cutter tooth (see Figure 8.9).

The vector $\mathbf{E}$ is pointed along the line of intersection of the clearance plane $C_s$ by the coordinate plane $YZ$. The vector $\mathbf{E}$ is of a length for which the projection of $\mathbf{E}$ onto the $Z-$ axis is equal to unit ($\text{pr}_z\mathbf{E} = 1$). Because of this, in the coordinate system $XYZ$, the following expression:

$$\mathbf{E} = \mathbf{j}\tan\alpha_m + \mathbf{k} \quad (8.23)$$

is valid for the vector $\mathbf{E}$. Here, $\alpha_m$ denotes the clearance angle that is measured within the major section plane $P_{ms}$ of the lateral cutting wedge of the rack cutter tooth [144, 148, 151, 158].

The vector $\mathbf{F}$ is pointed along the line of intersection of the clearance surface $C_s$ by the major section plane $P_{ms}$. The magnitude of the vector $\mathbf{F}$ is chosen for which projection of the vector $\mathbf{F}$ onto the coordinate plane $XY$ is equal to the unit—that is, $pr_{xy}\,\mathbf{F} = 1$. For the vector $\mathbf{F}$, the expression:

$$\mathbf{F} = \mathbf{i}\cos\phi_n + \mathbf{j}\sin\phi_n + \mathbf{k}\cot\alpha_m \quad (8.24)$$

can be readily composed.

By construction, three vectors $\mathbf{C}$, $\mathbf{E}$, and $\mathbf{F}$ are located within a common plane. Because of this, the identity $\mathbf{C}\times\mathbf{E}\cdot\mathbf{F} \equiv 0$ takes place. The identity $\mathbf{C}\times\mathbf{E}\cdot\mathbf{F} \equiv 0$ yields representation in the form:

$$\mathbf{C}\times\mathbf{E}\cdot\mathbf{F} \equiv \begin{vmatrix} -\sin\phi_n & \cos\phi_n & \tan\lambda \\ 0 & \tan\alpha_o & 1 \\ \cos\phi_n & \sin\phi_n & \cot\alpha_m \end{vmatrix} = 0 \quad (8.25)$$

Ultimately, to calculate for the clearance angle $\alpha_m$, the following formula:

$$\alpha_m = \tan^{-1}\left[\frac{\sin\phi_n}{\cot\alpha_o - \tan\gamma_o\cos^2\phi_n}\right] \quad (8.26)$$

can be derived from Eq. (8.25).

### 8.5.2 Possible Improvements to the Geometry of Lateral Cutting Edges of Rack Cutters

The derived equations make it possible to calculate the geometrical parameters of the cutting edges of the rack cutters. The calculations reveal that, for example, when the rake angle $\gamma_o$ and the clearance angle $\alpha_o$ fall into the range of 5...6°, the corresponding geometrical parameters $\gamma_m$ and $\alpha_m$ measured in the major section plane $P_{ms}$ are in the range of 2° or so. These values of the angles $\gamma_m$ and $\alpha_m$ are insufficient for high-performance gear-cutting tools. Because of this, numerous modifications in the working surfaces of the rack cutters are proposed to improve the geometry of lateral cutting edges.

#### 8.5.2.1 Modifications to the Rake Surface

Improvement in the cutting tool geometry is the major purpose of rake surface modifications. To create favorable conditions for cutting, it is necessary to increase the rake angle $\gamma_m$ at the lateral cutting edge of the rack cutter teeth. For this purpose, round groves are ground along the lateral cutting edges $CE$ of the coarse pitch rack cutters (for the rack cutters of module $m \geq 10$ mm), as shown in Figure 8.10(a). In this manner, an increased rake angle $\gamma^*$ is created in the plane section of the lateral cutting edge $CE$ by a plane that is perpendicular to the grinding wheel axis of rotation, as shown in Figure 8.10(b). The actual value of the rake angle $\gamma^*$ depends on the diameter $d_{gw}$ and the setup parameters of the grinding wheel. When the width of the ground grove is

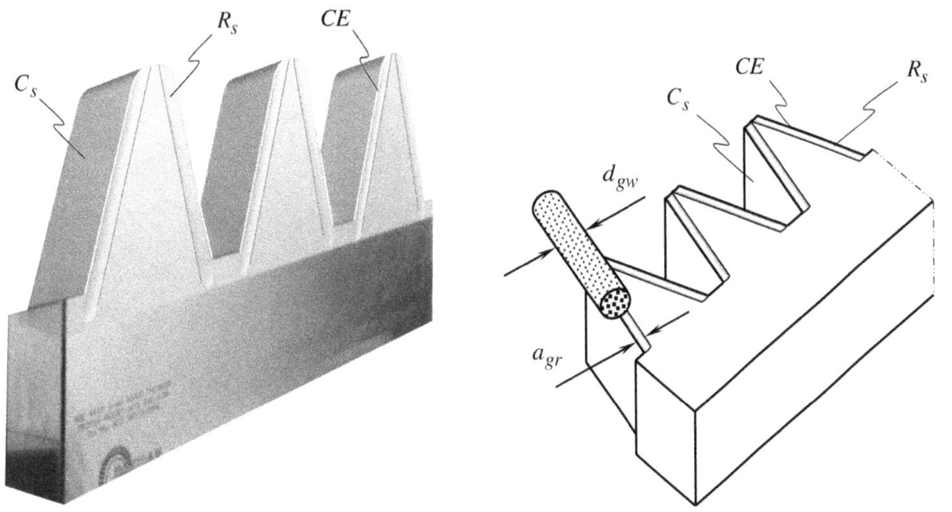


**FIGURE 8.10** Modified rake surface $R_s$ of the coarse pitch rack cutter for planing spur gears: (a) close-up view, (b) parameters of the modification in the rake surface $R_s$.

equal to a certain value $a_{gr}$, the rake angle $\gamma^*$ of the value [see Figure 8.10(b)]:

$$\gamma^* = \sin^{-1}\left(\frac{a_{gr}}{d_{gw}}\right) \tag{8.27}$$

is created at the lateral cutting edge *CE* of the rack cutter.

For the modification of the rake surface $R_s$ of rack cutters of module in the range of $m = 10...24$ mm, grinding wheels of diameters $d_{gw} = 35...70$ mm are usually suited. Width $a_{gr}$ of the round grove is in the range of $a_{gr} = 7...15$ mm. In such a scenario, a rake angle $\gamma^*$ in the range of $\gamma^* = 11°30'...12°30'$ is created.

To eliminate distortions of the rack cutter tooth profile, a narrow land with a width of around 0.05 mm is retained along the lateral cutting edge *CE* of the rack cutter.

Fine pitch rack cutters could have entirely modified the rake surface $R_s$ [see Figure 8.11(a)]. For a rake surface modification of this type, a cylindrical grinding wheel is used [see Figure 8.11(b)]. The axis of rotation of the grinding wheel is positioned at a certain angle with respect to the rake plane $R_s$. For this purpose, it is common practice to use grinding wheels of diameter $d_{gw} = 8m$ mm. The axis of rotation of the grinding wheel is inclined at the angle 4°10′ relative to the rake plane $R_s$.

Because it is formed in this manner, the rake angle $\gamma^*$ is not constant within the lateral cutting edge *CE* of the rack cutter tooth. The rake angle is of the minimal value in the vicinity of the top cutting edge and of the maximal value at the bottom of the rack cutter tooth. The width of the narrow land is also not constant along the lateral cutting edge.

Rack cutters for machining of helical gears with shoulders feature an increased positive rake angle $\gamma_m > 0°$ at one side. For high-performance rack cutters of this type, modification of the rake surface is a must. The required modification is performed by grinding additional portions of the rake surface $R_s$, as illustrated in Figure 8.13. These portions of the rake surface are located close to the lateral cutting edges *CE* and have a favorable value of the rake angle $\gamma_m$.

### 8.5.2.2 Modifications to the Clearance Surface

Insufficient geometry of the clearance surface $C_s$ is among the major reasons cited for the limited cutting performance in rack cutters. Clearance angle $\alpha_o$ at the top cutting edge is usually in the range of 5...6°. Clearance angles $\alpha_m$ at the lateral cutting edge are commonly in the range of 2°. These rages for the clearance angles are insufficient if high-performance gear-cutting tools are required. Considerable improvements in the cutting performance can be achieved with an appropriate modification of the clearance surfaces $C_s$ of the rack cutter teeth.

In the design of a standard rack cutter, clearance angle $\alpha_o$ at the top cutting edge *CE* and clearance angle $\alpha_m$ at the lateral cutting edge correlate to each other [see Eq. (8.26)]. The desired increase in the clearance angle $\alpha_m$ requires a corresponding increase in the clearance angle $\alpha_o$. The latter is often undesired.

The rack cutter shown in Figure 8.14 features the combined clearance surface through the top cutting edge. Clearance surfaces through the lateral cutting edges of the rack cutter tooth are designed so as to have an increased clearance angle $\alpha_m$. For this purpose, the clearance angle $\alpha_o$ is increased in accordance to Eq. (8.26). To eliminate the negative impact of the increased clearance angle $\alpha_o$, an auxiliary clearance surface through the top cutting edge is designed. Clearance angle $\alpha_o^{eff}$ at this portion of the clearance surface $C_s$ is smaller than the clearance angle $\alpha_o$, and the inequality $\alpha_o^{eff} < \alpha_o$ is observed. In this way, the clearance angles $\alpha_o$ and $\alpha_m$ are independent of each other. This makes it possible to assign the actual values of the clearance angles $\alpha_o$ and $\alpha_m$ close to their optimal values. Ultimately, the cutting performance of

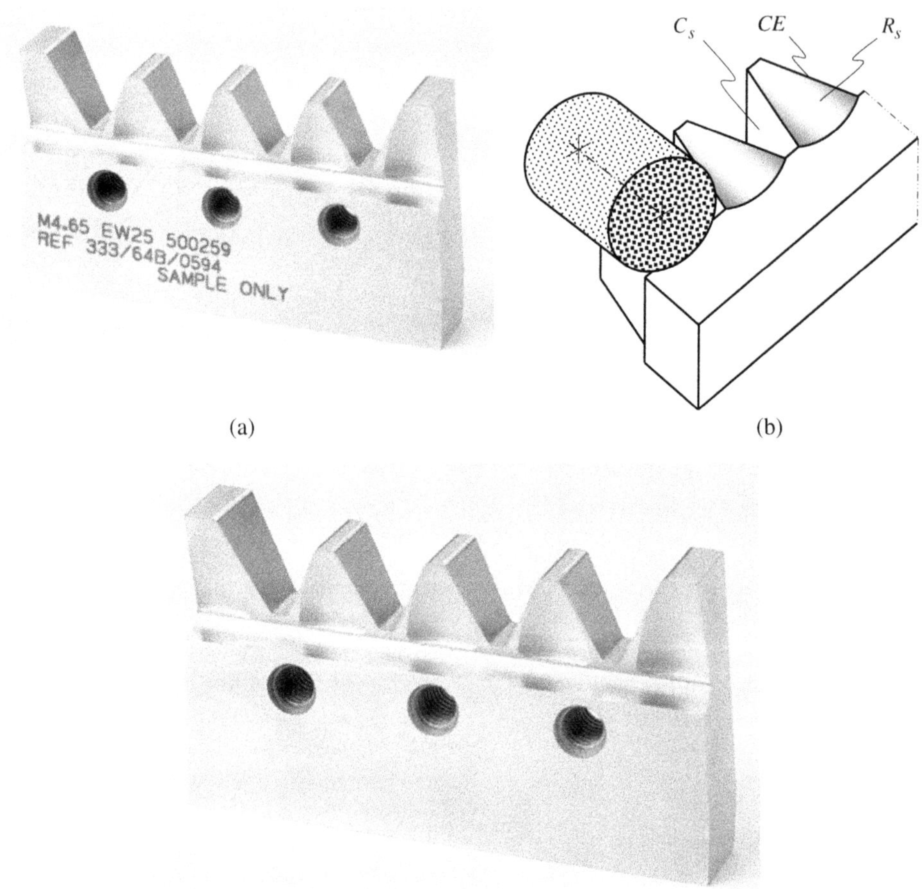


**FIGURE 8.11** Modified rake surface $R_s$ of the fine pitch rack cutter for planing spur gears: (a) close-up view, (b) parameters of the modification in the rake surface $R_s$.

the rack cutter (see Figure 8.14) is higher compared to that of rack cutters of the standard design.

An increased clearance angle $\alpha_m$ at the lateral cutting edges of the rack cutters makes it possible to cut gears with a modified tooth profile. For this purpose, the rack cutter tooth profile also requires modification. It is common practice to make the corresponding modification to the clearance surface $C_s$ of the rack cutter tooth. However, that same problem can be resolved by a corresponding alteration to the shape of the rake surface $C_s$. In certain cases of gear planing, this option is preferred.

The tooth design of a rack cutter for machining gears with a modified tooth profile[3] is illustrated in Figure 8.15. Originally, the active gear tooth profile is shaped in the form of an involute curve with a certain normal profile angle $\phi_n$ on the pitch diameter. A relatively short portion of the gear tooth profile next to the top land is modified. This portion is shaped in the form of another involute curve with a slightly increased profile angle $\phi_n^m > \phi_n$. For planing a work gear with this type of tooth profile modification, it is possible to apply a rack cutter with the combined rake

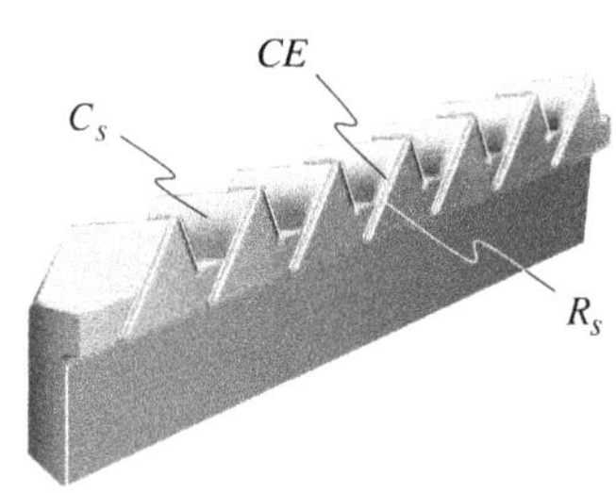


**FIGURE 8.12** Rack cutter for planing helical gears with a shoulder.

**FIGURE 8.13** Grinding of the modified rake surface $R_s$ of the rack cutter shown in Figure 8.12.

[3] SU Pat. No. 1017444, *A Gear Cutting Tool.*/S.P. Radzevich, Int. Cl. B 23 f 21/16, Filed: April 26, 1982.

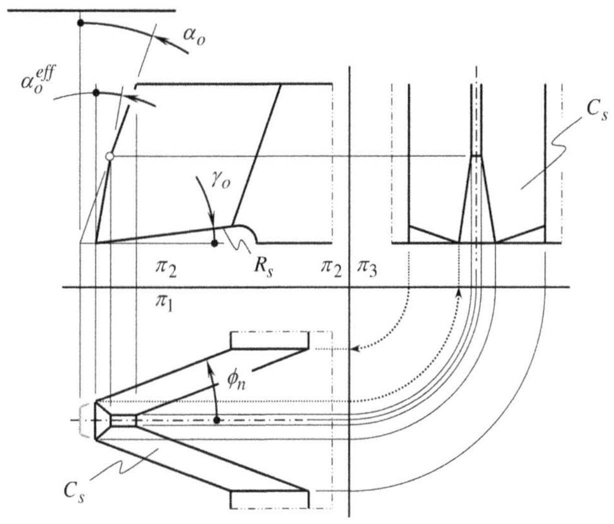


**FIGURE 8.14** Rack cutter with an increased clearance angle at the lateral cutting edges.

surface $R_s$ [68]. For this purpose, it is necessary to determine a corresponding value of the rake angle $\gamma_o^m$. When the clearance angle $\alpha_m$ is increased, use of the rack cutter with the combined rake surface makes planing the modified gear tooth profile possible.

Consider the determination of the desired rake angle $\gamma_o^m$ of the additional portion of the rake surface of the rack cutter tooth. It is assumed in the following that the nominal profile angle $\phi_n$, the modified profile angle $\phi_n^m$, the nominal rake angle $\gamma_o$, and the nominal clearance angle $\alpha_o$ are known. It is necessary to express the desired rake angle $\gamma_o^m$ of the auxiliary portion of the rake surface in terms of these design parameters of the rack cutter tooth.

In order to solve the problem under consideration, it is convenient to construct three vectors **A**, **b**, and **C**, as shown in Figure 8.15.

The vector **A** is pointed along the major portion of the lateral cutting edge *CE* of the rack cutter tooth. It can be assumed that projection of the vector **A** onto the coordinate plane *XZ* is equal to unit ($\text{pr}_{xz}\,\mathbf{A} = 1$). If so, the vector **A** can

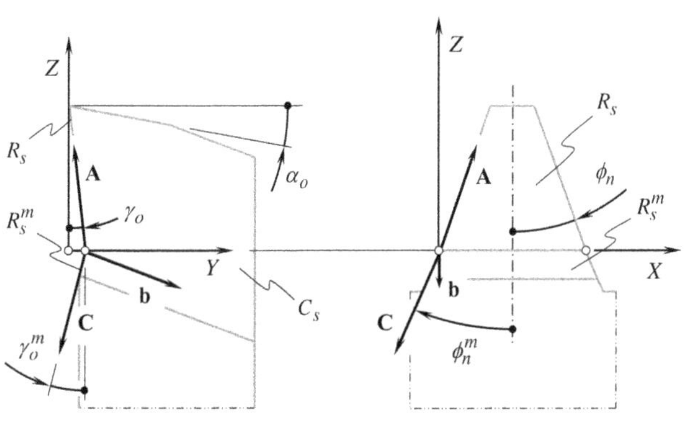


**FIGURE 8.15** A rack cutter for planing gears with a modified tooth profile.

be represented in the coordinate system *XYZ* associated with the rack cutter tooth in the form of vector equation:

$$\mathbf{A} = \mathbf{i}\tan\phi_n - \mathbf{j}\tan\gamma_o + \mathbf{k} \tag{8.28}$$

The unit vector **b** is pointed along the line of intersection of the clearance surface $C_s$ of the rack cutter tooth by the major section plane $P_{ms}$ (see Appendix G). An expression:

$$\mathbf{b} = \mathbf{j}\cos\alpha_o - \mathbf{k}\sin\alpha_o \tag{8.29}$$

is used to express the vector **b** in terms of the clearance angle $\alpha_o$.

The vector **C** is pointed along the auxiliary portion of the lateral cutting edge of the rack cutter tooth. The auxiliary lateral cutting edge generates the modified portion of the work gear tooth. It is assumed below that projection of the vector **C** onto the coordinate plane *XZ* is equal to the unit ($\text{pr}_{xz}\,\mathbf{C} = 1$). For the analytical representation of the vector **C**, the following expression:

$$\mathbf{C} = -\mathbf{i}\tan\phi_n^m - \mathbf{j}\tan\gamma_o^m - \mathbf{k} \tag{8.30}$$

is composed.

By construction, the vectors **A**, **b**, and **C** comprise a set of coplanar vectors situated within the clearance plane of the lateral side profile of the rack cutter tooth. Consequently, the scalar triple product of the vectors is identical to zero ($\mathbf{A}\times\mathbf{b}\cdot\mathbf{C} \equiv 0$). The identity yields the expression:

$$\mathbf{A}\times\mathbf{b}\cdot\mathbf{C} \equiv \begin{vmatrix} \tan\phi_n & -\tan\gamma_o & 1 \\ 0 & \cos\alpha_o & -\sin\alpha_o \\ -\tan\phi_n^m & -\tan\gamma_o^m & -1 \end{vmatrix} = 0 \tag{8.31}$$

Equation (8.31) casts into the formula of the calculation of the desired value of the auxiliary rake angle $\gamma_o^m$:

$$\gamma_o^m = \tan^{-1}\left[\frac{(1-\tan\gamma_o\tan\alpha_o)\cdot k_m - 1}{\tan\alpha_o}\right] \tag{8.32}$$

Here, in Eq. (8.32), the coefficient of the tooth profile modification $k_m$ is equal to:

$$k_m = \frac{\tan\phi_n^m}{\tan\phi_n} \tag{8.33}$$

Equation (8.32) reveals that the greater the angles $\gamma_o$ and $\alpha_o$, the more readily the required modification of the rack cutter tooth profile can be attained. This is the reason it is recommended to use the modified rake surface (see Figure 8.15) in the design of rack cutters with an increased clearance angle $\alpha_o$.

## 8.6 CHIP THICKNESS CUT BY CUTTING EDGES OF A RACK CUTTER TOOTH

Cutting a work gear with a rack cutter is a type of a continuous indexing method of gear machining. When planing a gear with a rack cutter (Figure 8.16), the work gear rotates about its axis of rotation $O_g$. The rotation of the work gear is denoted in Figure 8.16 as $\omega_g$.

The rack cutter reciprocates in the direction that is tangent to the flank surface $\mathcal{G}$ of the work gear tooth. The direction of the reciprocation $\mathbf{V}_{cut}$ of the rack cutter is parallel to the axis $O_g$ when planing spur gears. When planing helical gears, the direction of the reciprocation $\mathbf{V}_{cut}$ of the rack cutter is at the pitch helix angle $\psi_g$ relative to the axis of rotation $O_g$ of the work gear.

In addition to the reciprocation $\mathbf{V}_{cut}$, the rack cutter travels tangential to the work gear. This translational motion $\mathbf{V}_{ct}$ of the rack cutter, together with the rotation $\omega_g$ of the work gear, yields interpretation in the form of rolling with no sliding of the pitch plane of the rack cutter over the pitch cylinder of the work gear. The rolling motion of the pitch surfaces of the rack cutter and the work gear affects the actual value of the chip thickness cut by the cutting edges of the rack cutter.

For machining teeth of a work gear, two motions must be performed in the gear-machining process: (a) the cut motion and (b) the feed motion.

Reciprocation of the rack cutter serves as the cutting motion $\mathbf{V}_{cut}$ in the gear-machining operation (see Figure 8.16). It provides the necessary cut speed. The cut speed is equal to the speed $\mathbf{V}_{cut}$ of the reciprocation of the rack cutter.

One of the surprising developments in gear cutting is the G-TRAC generator. This machine uses rack-type cutting tools mounted on an endless chain. The gear being cut rolls in mesh with the passing rack teeth in the chain. Figure 8.17 shows a stack of parts being cut to have helical gear teeth.

It has been claimed that the G-TRAC machine is intended for high production, low cost, and high accuracy. It is possible, though, to use the machine effectively for small production as well.

**FIGURE 8.17** Generating helical gear teeth on a stack of gear blanks with the G-TRAC machine.

It was anticipated that simple rack tools can be made by gear shops using the G-TRAC. This feature should make it easy to change tooth module (pitch), tooth pressure angle, or tooth depth when a gear unit is under development. (Normally, gear-cutting tools have to be ordered from a toolmaking company, and some months' time is lost when new, nonstandard gear tools are needed). It was also anticipated that the concept of G-TRAC machine would open several

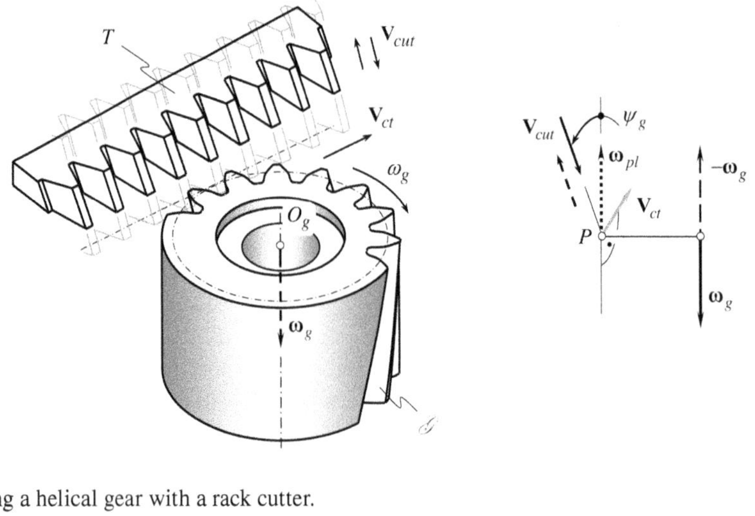


**FIGURE 8.16** Machining a helical gear with a rack cutter.

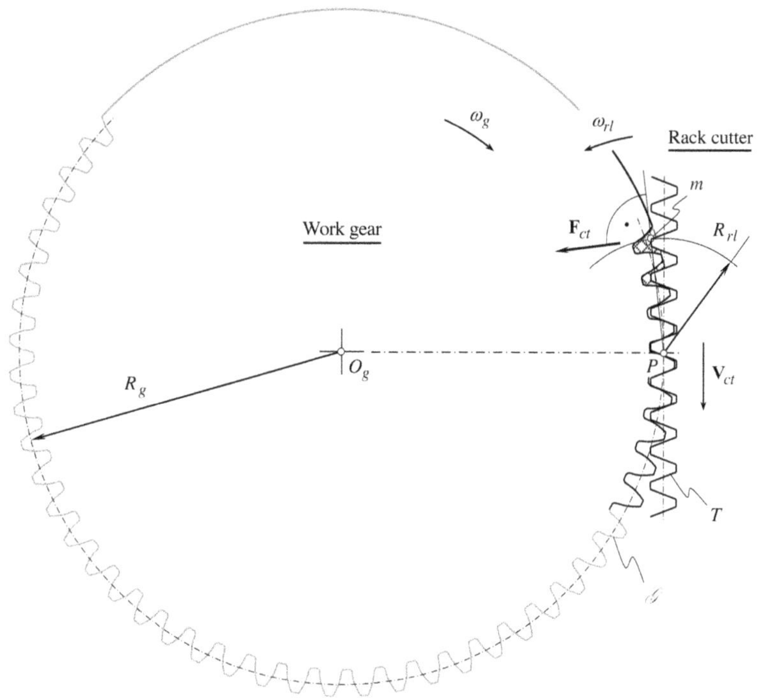


**FIGURE 8.18** Feed motion $F_{ct}$ when machining a spur gear with a rack cutter.

new possibilities in gear machine tools. Unfortunately, the just discussed approach to gear generating is nothing more than an example of the situation when unlimited funds spent on the project are not supported by a corresponding proper design of the gear generator—in other words, this is a good illustration of what results can be anticipated from those who are not sufficiently familiar with the gear theory.

Feed motion in gear planing is attributed to rolling of the pitch plane of the rack cutter over the pitch cylinder of the work gear. When the work gear rotates about its axis $O_g$ with the rotation $\omega_g$, the rack cutter travels with a linear velocity $\mathbf{V}_{ct}$ (see Figure 8.18). No sliding is observed in the rolling motion of the pitch surfaces.

Rolling of the pitch surfaces can be viewed as the instant rotation of the generating surface $T$ of the rack cutter about the pitch point $P$. The speed of the instant rotation $\omega_{rl}$ can be expressed in terms of the speed of translational motion $V_c = |\mathbf{V}_c|$ and the radius of the pitch cylinder $R_g$ of the work gear:

$$\omega_{rl} = \frac{V_c}{R_g} \tag{8.34}$$

Because the equality $V_c = \omega_g \cdot R_g$ is observed, the speed of instant rotation is equal to the speed of work gear rotation ($\omega_{rl} = \omega_g$).

The speed of the feed motion $F_{ct} = |\mathbf{F}_{ct}|$ is equal to the speed of linear motion of the cutting edge of the rack cutter tooth when it is performing the instant rotation $\omega_{rl}$:

$$F_{ct} = \omega_{rl} \cdot R_{rl} \tag{8.35}$$

where $R_{rl}$ denotes the distance of a point of interest $m$ within the cutting edge $CE$ of the rack cutter tooth from pitch point $P$. The point of interest can be located within any of the three cutting edges of the rack cutter tooth.

When machining a work gear with the rack cutter, instant rotation occurs with a constant angular velocity $\omega_{rl}$. This means that the equality $\omega_{rl} = const$ takes place.

Different points of the cutting edge of the rack cutter tooth are remote at different distances from the pitch point $P$—that is, the equality $R_{rl} = var$ is valid. This means that the magnitude of the instant feed motion $F_{ct}$ depends on the coordinates of the point of interest $m$ within the cutting edge $CE$ of the rack cutter tooth. Moreover, the orientation of the position vector $\mathbf{r}_m$ of the point of interest is a function of the coordinates of the point $m$. Because the direction of the instant feed motion $F_{ct}$ is perpendicular to the position vector $\mathbf{r}_m$, the direction of the motion $\mathbf{F}_{ct}$ at different points of the cutting edge $CE$ is also different. To determine the direction of the feed motion $\mathbf{F}_{ct}$, it is helpful to use the equality:

$$\mathbf{F}_{ct} = \omega_{rl} \times \mathbf{r}_m \tag{8.36}$$

Next, the parameters of a portion of the stock cut out by the rack cutter tooth per one stroke can be calculated. For this purpose, the implementation of the approach proposed by Prof. V.A. Shishkov [206] is important.

To calculate the thickness of cut, it is convenient to consider two consequent configurations of the rack cutter. The location of the rack cutter in its initial position can be specified by the configuration of a Cartesian coordinate system $X_gY_gZ_g$ associated with the work gear (see Figure 8.19). During one of the reciprocation strokes of the rack cutter, the work gear turns about its axis of rotation $O_g$ through a certain angle $\varepsilon_{st}$.

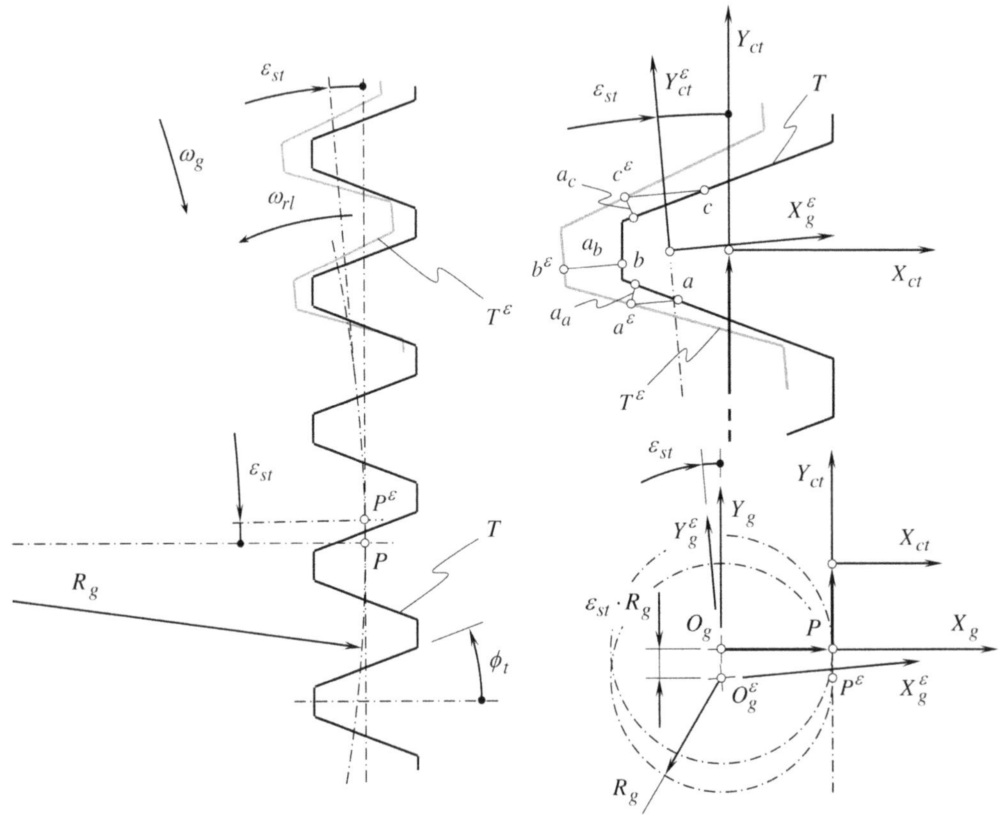


**FIGURE 8.19** Thickness of cut when machining a spur work gear with the rack cutter.

The angle $\varepsilon_{st}$ is a portion of the roll angle $\varepsilon$. In other words, the angle $\varepsilon_{st}$ can be interpreted as a roll angle per stroke of the rack cutter. The actual value of the angle $\varepsilon_{st}$ depends on the difference $(\varepsilon_o - \varepsilon_l)$ between the roll angle $\varepsilon_o$ at the outer diameter of the gear tooth profile and the roll angle $\varepsilon_l$ at the inner diameter of the gear tooth. It also depends on the number of strokes of the rack cutter $n_{st}$ that is required for machining the entire tooth flank surface:

$$\varepsilon_{st} = \frac{\varepsilon_o - \varepsilon_l}{n_{st}} \tag{8.37}$$

The angle $\varepsilon_{st}$ is small, and its value is in the range of $\varepsilon_{st} \cong 1°$ or so. Its value can vary depending on the requirements of the particular gear-planing operation.

When rotating, the work gear travels along the pitch line. Because rolling with no sliding of the pitch surfaces is observed, the travel distance of the gear axis of rotation from the initial position $O_g$ to its current position $O^{\varepsilon}_g$ is equal to $\varepsilon_{st} \cdot R_g$, where $R_g$ denotes the pitch radius of the work gear in the gear planing operation.

For the analytical description of the translation of the coordinate system at the distance $\varepsilon_{st} \cdot R_g$, the operator of translation $\mathbf{Tr}(\varepsilon_{st} \cdot R_g, Y_g)$ is used. Accordingly, the operator of the rotation $\mathbf{Rt}(\varepsilon_{st}, Z_g)$ is implemented for the analytical description of the rotation of the coordinate system through the angle $\varepsilon_{st}$. The resultant linear transformation in the case under consideration is analytically described by two operators of the linear transformations $\mathbf{Tr}(\varepsilon_{st} \cdot R_g, Y_g)$ and $\mathbf{Rt}(\varepsilon_{st}, Z_g)$. For simplification, one operator[4] $\mathbf{Rl}_y(\varepsilon_{st}, Z_g)$ alone—rather than two operators, $\mathbf{Tr}(\varepsilon_{st} \cdot R_g, Y_g)$ and $\mathbf{Rt}(\varepsilon_{st}, Z_g)$—is used (see Figure 8.19). Ultimately, use of the operator of rolling $\mathbf{Rl}_y(\varepsilon_{st}, Z_g)$ allows for transition from the coordinate system $X_g Y_g Z_g$ in its initial location to the same coordinate system in its current location. The latter is designated as $X^{\varepsilon}_g Y^{\varepsilon}_g Z^{\varepsilon}_g$.

Furthermore, consider a point of interest within the cutting edge $CE$ of the rack cutter tooth. This might be either point $a$ within the lateral cutting edge of the rack cutter tooth, point $b$ within the top cutting edge, or point $c$ within the lateral cutting edge of the opposite side of the tooth profile (see Figure 8.19). To specify the coordinates of the point of interest, use of the local coordinate system $X_{ct} Y_{ct} Z_{ct}$ is commonly helpful.

The position vector $\mathbf{r}_a$ of the point of interest $a$ for a certain configuration of the generating surface $T$ is known. (The same is valid with respect to any other point of interest). Equation (8.8) allows for the calculation of the unit normal vector $\mathbf{n}_{T.a}$ to the generating surface $T$ at point $a$. When cutting a work gear with the rack cutter, the pitch plane of the rack $T$ rolls with no sliding over the pitch cylinder of the gear. For the consequent cutting stroke, the rack cutter turns through the roll angle $\varepsilon_{st}$ and travels through the distance $\varepsilon_{st} \cdot R_g$. In the new position, the generating surface

[4] See Appendix D for details on the operators of rolling.

of the rack cutter is designated as $T^{\varepsilon}$. During the rolling motion, the point of interest $a$ travels to the position $a^{\varepsilon}$. The position vector $\mathbf{r}_a^{\varepsilon}$ of the point of interest in the new location of the rack cutter can be expressed in terms of the position vector $\mathbf{r}_a$ and the operator of rolling $\mathbf{Rl}_y(\varepsilon_{st}, Z_g)$ in the following manner:

$$\mathbf{r}_a^{\varepsilon} = \mathbf{Rl}_y(\varepsilon_{st}, Z_g) \cdot \mathbf{r}_a \tag{8.38}$$

Similarly, the following expression:

$$\mathbf{n}_{T.a}^{\varepsilon} = \mathbf{Rl}_y(\varepsilon_{st}, Z_g) \cdot \mathbf{n}_{T.a} \tag{8.39}$$

is valid for the unit normal vector $\mathbf{n}_{T.a}^{\varepsilon}$ to the generating surface $T^{\varepsilon}$ through the point $a^{\varepsilon}$.

The direction along which cut thickness is measured is perpendicular to the machined surface of the work gear. The direction varies within the cutting edge of the rack cutter. At point $a^{\varepsilon}$, cut thickness is denoted as $a_a$. Corresponding designations $a_b$ and $a_c$ are used to denote the cut thickness at the points $b^{\varepsilon}$ and $c^{\varepsilon}$, respectively.

The cut thickness $a_a$ is equal to the length of the straight-line segment between point $a^{\varepsilon}$ and generating surface $T$ of the rack cutter. The point of interception of the straight line $\mathbf{a}_a(t)$ through point $a^{\varepsilon}$ along the unit normal vector $\mathbf{n}_{T.a}^{\varepsilon}$:

$$\mathbf{a}_a(t) = \mathbf{r}_a^{\varepsilon} + t \cdot \mathbf{n}_{T.a}^{\varepsilon} \tag{8.40}$$

is shown (but not labeled) in Figure 8.19. Here, $t$ denotes the parameter of the straight line.

The equation for the generating surface $T$ of the rack cutter [Eq. (8.8) describes only one lateral flank of the surface $T$; a similar equation is valid for the opposite flank of the tooth of the surface $T$; and the equation for the top land of the surface $T$ is a trivial one] together with Eq. (8.40) returns the position vector $\hat{\mathbf{r}}_a^{\varepsilon}$ of the point of interception of the straight line $\mathbf{a}_a(t)$ with the generating surface $T$ of the rack cutter. This immediately yields a formula:

$$a_a = \left| \mathbf{r}_a^{\varepsilon} - \hat{\mathbf{r}}_a^{\varepsilon} \right| \tag{8.41}$$

for the calculation of cut thickness at the point of interest $m$ within the cutting edge $CE$ of the rack cutter tooth.

Equation (8.41), used for the calculation of the thickness of cut at a point of interest of the cutting edge of the rack cutter, is derived with no simplifications, and it is an accurate one. It can be used when the point of interest is a far remote point from the pitch point in a gear-planing operation and the chip to be cut is thick. It is also valid when the point of interest is close to the pitch point and the cutting edge—cut thin chip—when the chip thickness is in the range close to the roundness of the cutting edge.

For approximate calculations, simpler equations are known. However, the equations are applicable for remote portions of the cutting edges, and, therefore, they are not applicable for remote portions of the cutting edges of the rack cutter.

Equation (8.41) can be used to calculate all the other parameters of the portion of the stock that has been removed by a cutting edge of the rack cutter—that is, the shape of the section of the portion of the stock, its area, and so forth.

Cut thickness is important because it significantly affects tooth wear and the longevity of the rack cutter.

Cutting edges of the rack cutter teeth cut chips of different thickness.

First, cut thickness depends on the distance between the point of interest and the pitch point. The greater the distance, the larger the cut thickness becomes.

Second, different cutting edges of the rack cutter tooth cut the chips of varying thickness. The top cutting edge cuts the thickest chip. The thickness of the chip that is cut by the lateral cutting edge of the entering side of the tooth profile is larger, rather than that of the recessing side of the rack cutter tooth profile.

Lateral cutting edges of the rack cutter teeth cut the stock mostly within the distance between the pitch point and the point where the cutting begins. A small amount of the stock is also cut beyond the pitch point. The top land cutting edge does not cut the stock beyond the pitch point.

A detailed analysis of chip thickness is beyond the scope of the book.

## 8.7 ACCURACY OF A PLANED GEAR

The accuracy of the planed gear is discussed here from the standpoint of the kinematic geometry of surface generation [151, 158].

When cutting a work gear with the rack cutter of conventional design, most of the conditions of proper part surface generation (PSG) can be readily fulfilled. Usually, it is not necessary to verify whether they are all fulfilled or some of them are violated. In particular cases of gear machining with a rack cutter, only the fifth and sixth conditions of proper PSG must be met in more careful analysis (see Appendix E for details on conditions of proper PSG).

### 8.7.1 Fulfillment of the Fifth Condition of Proper Part Surface Generation

To meet the fifth condition of proper PSG, no intersection of the adjacent portions of the tooth profile of the generating surface $T$ of the rack cutter has to be observed. When machining a work gear with the rack cutter, the pitch plane of the rack cutter rolls with no sliding over the pitch cylinder of the work gear. Violation of the fifth condition of proper PSG occurs mostly because the pitch surfaces roll over each other.

The teeth of the gear being machined can be considered as a certain number $N_g$ of portions of the circular cylinder surface (root lands, as well as top lands), and that same number $N_g$ of the right-hand-oriented and left-hand-oriented screw involute flank surfaces $\mathscr{G}$. Therefore, the generating surface $T$ of the rack cutter is composed of $N_{ct}$ portions of plane surface (root lands and top lands) and that same number $N_{ct}$ of pairs of lateral planes $T$. When planing a work gear, the lateral surface make contact with different portions of the gear tooth flanks $\mathcal{G}$.

Various relative configurations of adjacent portions of tooth profile of the generating surface $T$ of the rack cutter can be observed. The adjacent portions of tooth profile of the surface $T$ can:

1. Share no common points—that is, they can be apart from each other
2. Be connected to each other at the endpoints.
3. Intersect each other.

In conditions (1) and (2), no violation of the fifth condition of proper PSG occurs. In condition (3), intersection of the adjacent portions of the rack cutter tooth profile is observed. Because of the intersection, the fifth necessary condition of proper PSG is violated. Ultimately, violation of the fifth condition of proper PSG results in the occurrence of transient curves at the bottom land of the gear tooth profile.

For example, for the generation of an involute gear tooth profile, the generating surface $T$ of the rack cutter with a transverse profile angle $\phi_t$ is used. The desired tooth profile of the surface $T$ is composed of three straight-line segments (see Figure 8.20):

- Left-side lateral straight-line segment $a_l^* b_l^*$.
- Top land straight-line segment $a_l^{**} a_r^{**}$.
- Right-side lateral straight-line segment $a_r^* b_r^*$.

The straight-line segments intersect each other at points $f_l^*$ and $f_r^*$, as shown in Figure 8.20. Therefore, the portions $a_l^* b_l^*$ and $a_l^{**} f_l^*$, as well as the portions $a_r^* f_r^*$ and $a_r^{**} f_r^*$, of the rack cutter tooth profile do not exist physically. This results in the portions $a_l c_l$ and $a_r c_r$ of the gear tooth profiles $a_l b_l$ and $a_r b_r$ (see Figure 8.20) being substituted with the transient curves $c_l d_l$ and $c_r d_r$ that form the fillets.[5] For the same reason, the portions $a_l d_l$ and $a_r d_r$ of the cylinder root land are also replaced with the transition curves $c_l d_l$ and $c_r d_r$.

To fulfil the fifth condition of proper PSG, it is necessary to keep the transient curve between the circle of root diameter $d_{f.g}$ and the circle of limit diameter $d_{l.g}$. This criterion can be stated analytically as:

$$[\Delta] \leq 0.5(d_{l.g} - d_{f.g}) \tag{8.42}$$

When the inequality in Eq. (8.42) is fulfilled, the fifth condition of proper PSG is fulfilled as well, and, thus, the work gear can be machined by the rack cutter in compliance with the blueprint specification. Therefore, Eq. (8.42) is vital for the calculation of the desired design parameters of the rack cutter.

[5] Reminder: The start of active profile (SAP) is the intersection of the limit diameter and the involute profile (Figure 8.20). Points $c_l$ and $c_r$ within the circle of limit diameter $d_{l.g}$ are the points of SAP.

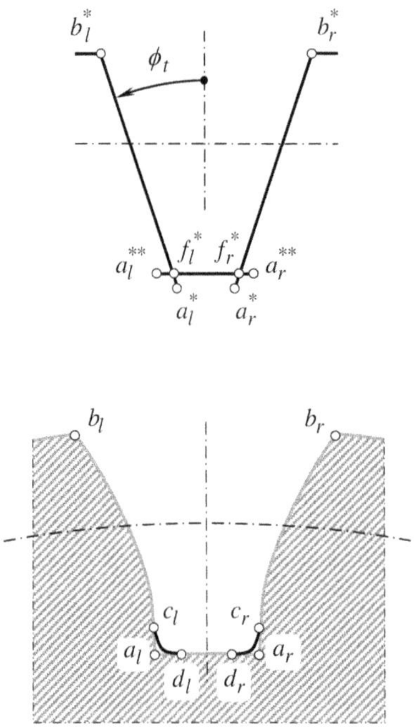


FIGURE 8.20 Intersection of the adjacent cutting edges of the rack cutter tooth results in the occurrence of transient curves.

The inequality in Eq. (8.42) can also be used for the determination and graphical interpretation of a permissible zone of variation (a) of the rack cutter profile angle $\phi_t$ and (b) of the work gear tooth count $N_g$, within which the fifth condition of proper PSG is fulfilled.

Consider generation of a gear tooth profile with a rack cutter (see Figure 8.21). The left-hand side of the work gear tooth profile is generated along the active portion of the left-hand side line of action $LA_l$. Similarly, the right-hand side of the gear tooth profile is generated within the active potion of the right-hand side line of action $LA_r$. The tooth profile of the rack cutter does not project beyond the work gear root circle of diameter $d_{f.g}$. Because of this, portions of the active lines of action $LA_l$ and $LA_r$ cannot be represented in the gear-machining operation. Therefore, within these portions of the lines of action, the generation of the transient curves of a certain height $\Delta$ is inevitable.

An equation can be derived for the calculation of the height $\Delta$ of the transient curve.

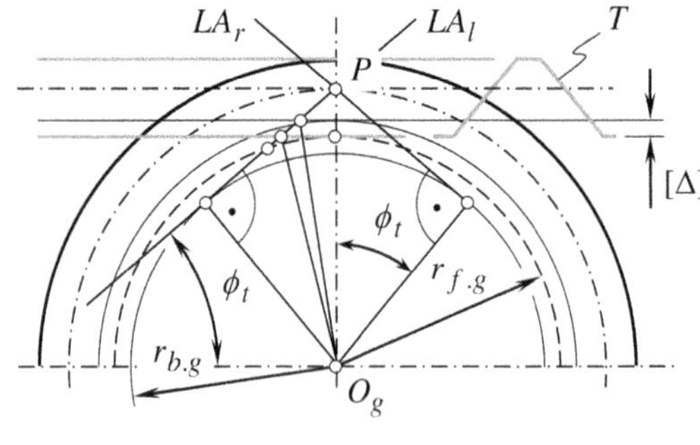


FIGURE 8.21 Gear-to-rack-cutter meshing diagram.

Analysis of the gear-to-rack cutter meshing diagram (see Figure 8.21) allows us to compose two equations for the lines of action $LA_l$ and $LA_r$, as well as an equation for the root circle of diameter $d_{f.g}$. These equations yield the formula:

$$\Delta = 0.5 \cdot \left( \sqrt{\left( d_{f.g} - \frac{d_{b.g}}{\cos \phi_t} \right)^2 \cot^2 \phi_t + d_{f.g}^2} - d_{f.g} \right) \quad (8.43)$$

for the calculation of the actual value of the deviation $\Delta$.

The performed analysis of the gear-to-rack-cutter meshing diagram also allows for a convenient graphical interpretation of the zone within which the fifth condition of proper PSG in gear planing operation is fulfilled.

The maximum value of the deviation $\Delta$ is limited to its maximal permissible value $[\Delta]$. Therefore, the interval $0 \leq \Delta \leq [\Delta]$ is the permissible interval for the deviation $\Delta$. The latter inequality can be expressed in terms of the design parameters of the rack cutter and of the gear to be machined.

To keep the deviation $\Delta$ within the permissible range $0 \leq \Delta \leq [\Delta]$, the transversal profile angle of the rack cutter teeth has to fulfil the inequality (see Figure 8.21):

$$\frac{d_{f.g} \cdot d_{b.g} - A_c}{(d_{f.g} + 2 \cdot [\Delta])^2} \leq \cos \phi_t \leq \frac{d_{f.g} \cdot d_{b.g} + A_c}{(d_{f.g} + 2 \cdot [\Delta])^2} \quad (8.44)$$

where $A_c = \sqrt{d_{f.g}^2 \cdot d_{b.g}^2 - (d_{f.g} + 2 \cdot [\Delta])^2 \cdot (d_{b.g}^2 - 4 \cdot d_{f.g} \cdot [\Delta] - 4 \cdot [\Delta]^2)}$.

As an example, consider planing a spur involute gear of modulus $m = 10$ mm with normal pressure angle $\phi_g = 20°$, tooth count $N_g = 75$, and radial clearance $c = 0.25\,m$. The rest of the design parameters of the gear can be calculated: the pitch diameter $d_g = m \cdot N_g = 750$ mm, the outer diameter $d_{a.g} = d_g + 2 \cdot m = 770$ mm, the root diameter $d_{f.g} = d_g - 2 \cdot 1.25 \cdot m = 725$ mm, and the base diameter $d_{b.g} = d_g \cdot \cos \phi_g = 704.769$ mm. For planing the spur involute gear, use of Eq. (8.44) allows for calculation of the following permissible interval for the rack cutter transverse pressure angle: $8.398° \leq \phi_t \leq 21.817°$. Machining the work gear using the rack cutter with a transverse pressure angle within the calculated interval $8.398° \leq \phi_t \leq 21.817°$ results in a deviation $\Delta$ that does not exceed the radial clearance $\Delta < c$.

Similar calculations can be performed for the case of planing a gear with a different tooth number $N_g$. The results of the calculations can be interpreted graphically in the form of the permissible zone within which the fifth condition of proper PSG is met. Figure 8.22 illustrates an example of the permissible zone. This permissible zone is calculated for the case of planing a spur involute gear with a profile angle $\phi_g = 20°$ and cut using a rack cutter with a tooth profile angle $\phi_t$. The deviation $\Delta$ of the gear tooth is within the tolerance $0 \leq \Delta \leq [\Delta] = 0.25 \cdot m$. A similar allowed zone can be constructed for the case of planing a spur or helical gear of any design.

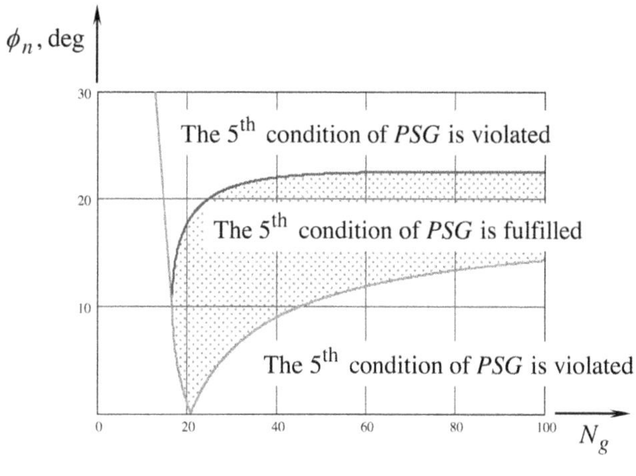


**FIGURE 8.22** Zones of fulfillment and of violation of the fifth condition of proper PSG in gear planing.

Application of the rack cutter, design parameters of which correspond with the design parameters of the gear to be machined [see Eq. (8.44)], results in an increased accuracy of the machined gear.

### 8.7.2 Fulfillment of the Sixth Condition of Proper Part Surface Generation

When machining a work gear with the rack cutter, a discrete type of generation of the gear tooth flank is observed. The machined gear tooth flank $\mathcal{G}$ is shaped in the form of numerous straight cuts, each of which is tangent to the desired involute tooth form. Cusps on the machined gear tooth flanks are inevitable when a gear is cut with the rack cutter.

For the fulfillment of the sixth condition of proper PSG (see Appendix E), the cusp height on the machined gear tooth flank must be within the tolerance for accuracy on the tooth flank. The cusp height can also be interpreted as the deviation of the actual gear tooth surface from the desired tooth surface. From this standpoint, cusps result in low accuracy of the machined gear.

To verify whether the sixth condition of proper PSG is fulfilled, determination of the actual cusp height is required.

Consider the tooth of the gear being machining (see Figure 8.23). At an arbitrary point of the gear tooth profile, say at point $i$, the cutting edge of the rack cutter is tangent to the involute profile of the gear tooth. On the next stroke, the cutting edge of the rack cutter is tangent to the gear tooth profile at the adjacent point $(i+1)$. The perpendiculars to the involute tooth profile at points $i$ and $(i+1)$ form the angle that is equal to the roll angle per stroke $\varepsilon_{st}$ of the rack cutter. Ultimately, when the machining process is over, the desired smooth involute gear tooth profile is substituted with a polygonal shape. The deviation of the actual gear tooth profile in relation to the desired tooth profile is denoted by $h_{ss}$. The deviation $h_{ss}$ is measured along the perpendicular to the desired gear tooth surface that passes through the point of intersection of the $i$-th and of the $(i+1)$-th cuts.

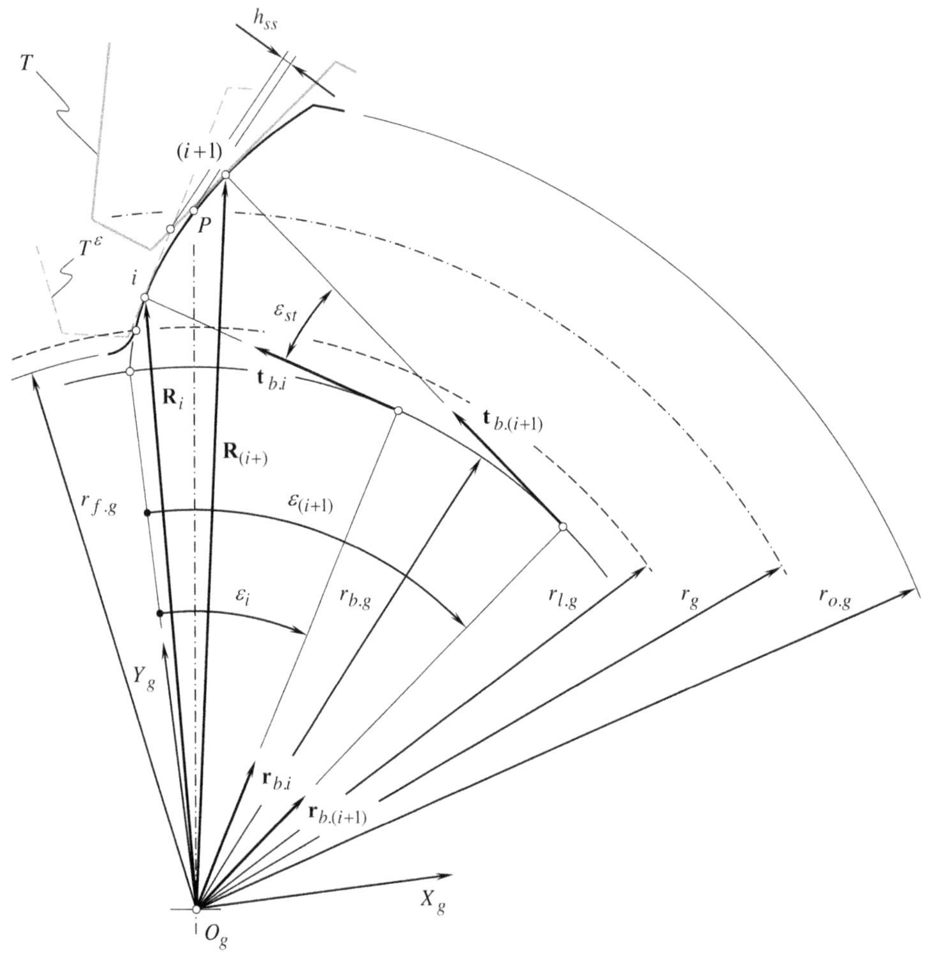


FIGURE 8.23 Major parameters of the discrete generation of the gear tooth profile by the rack cutter.

For the calculation of the deviation $h_{ss}$, it is handy to implement the straight-line segment through the points $i$ and $(i+1)$. Only the length of the straight-line segment is of importance in the problem under consideration.

The position vector of point $i$ is designated as $\mathbf{R}_i$. Analytically, the vector $\mathbf{R}_i$ can be expressed as:

$$\mathbf{R}_i = \mathbf{r}_{b.i} \cdot r_{b.g} + \mathbf{t}_{b.i} \cdot r_{b.g} \cdot \varepsilon_i \tag{8.45}$$

where $\mathbf{r}_{b.i}$ and $\mathbf{t}_{b.i}$ denote unit vectors in the corresponding directions.

In projections onto coordinate axes of the reference system $X_g Y_g Z_g$, Eq. (8.45) casts into:

$$\mathbf{R}_i = \mathbf{i} \cdot r_{b.g} \cdot (\sin \varepsilon_i - \cos \varepsilon_i) + \mathbf{j} \cdot r_{b.g} \cdot (\sin \varepsilon_i + \cos \varepsilon_i) \tag{8.46}$$

Similarly, for the position vector $\mathbf{R}_{(i+1)}$ of the point $(i+1)$, the following expression:

$$\begin{aligned} \mathbf{R}_{(i+1)} = \mathbf{i} \cdot r_{b.g} \cdot [\sin \varepsilon_{(i+1)} - \cos \varepsilon_{(i+1)}] \\ + \mathbf{j} \cdot r_{b.g} \cdot [\sin \varepsilon_{(i+1)} + \cos \varepsilon_{(i+1)}] \end{aligned} \tag{8.47}$$

can be composed.

The distance between the points $i$ and $(i+1)$ is equal to $|\Delta \mathbf{R}_i|$ (see Figure 8.24). Here, the equality:

$$|\Delta \mathbf{R}_i| = |\mathbf{R}_i - \mathbf{R}_{(i+1)}| \tag{8.48}$$

is valid.

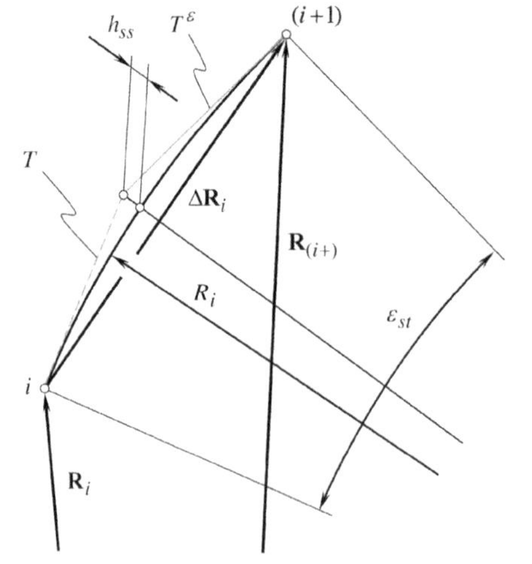


FIGURE 8.24 Geometry of an elementary cusp on the planed gear tooth profile.

Once the distance $|\Delta \mathbf{R}_i|$ is calculated, an expression:

$$h_{ss} = 0.5\sqrt{r_{b.g}^2 \cdot [\varepsilon_i + \varepsilon_{(i+1)}]^2 \cdot \sin^2(0.5 \cdot \varepsilon_{st}) - |\Delta \mathbf{R}_i|^2} \quad (8.49)$$

for the calculation of the cusp height $h_{ss}$ immediately follows from the analysis in Figure 8.24.

The sixth condition of proper PSG is fulfilled if, and only if, the inequality:

$$h_{ss} \le [h_{ss}] \quad (8.50)$$

is valid.

## 8.8 APPLICATION OF RACK CUTTERS

The rack cutter, mounted in a clapper box, reciprocates while the gear rolls past its cutting field. Cutting generally takes place during the down-stroke, and the clapper box clears the tool from the work on the up-stroke (see Figure 8.16).

Three types of rack cutters are used:

- Rough rack cutters
- Finish rack cutters
- Pre-grind rack cutters

To calculate for the design parameters of rack cutters, the following expressions (see Figure 8.25):

$$\frac{t_{finish} - t_{rough}}{2} = 0.2 \cdot \sqrt{m} \quad (8.51)$$

$$\frac{t_{finish} - t_{grind}}{2} = 0.1 \cdot \sqrt{m} \quad (8.52)$$

commonly apply.

An increase in tooth height for roughing rack cutters is equal to $\varepsilon_{rough} = 0.2 \cdot \sqrt{m}$, mm, and for grind rack cutters it is $\varepsilon_{grind} = 0.1 \cdot \sqrt{m}$, mm.

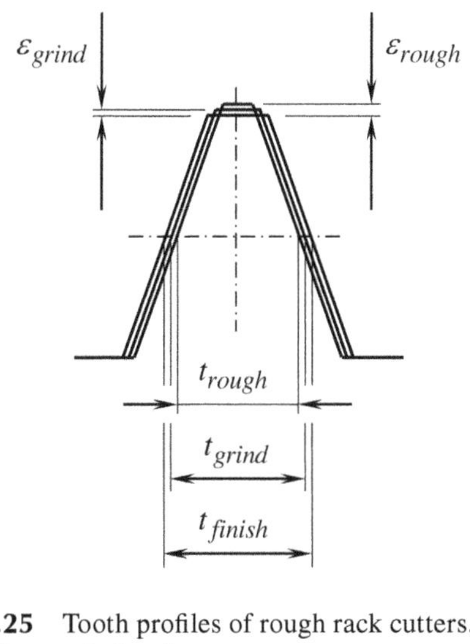


**FIGURE 8.25** Tooth profiles of rough rack cutters, finish rack cutters, and pre-grind rack cutters.

The use of rack cutters makes it possible to machine any gear with continuous indexing and with no interruption in the machining process. For this purpose, at the end of every stroke of the rack cutter, it is necessary to turn the work gear about its axis through an angle that corresponds to one tooth of the work gear. This additional discrete indexing is superposed with the original continuous indexing.

A rack tool cuts only a few teeth in one generating cycle, then indexes to pick up the next few teeth. The number of teeth per generation, the number of strokes per tooth (feed), and the stroking speed are all individually variable.

The most common use of rack cutters is for coarse pitches, high-hardness material, and narrow-gap double helical gears. It is also ideal for cutting segments, since only the toothed portion must be rolled past the tool. The rack cutter can also machine two- and three-lobe rotors as a result of the large tools size and the large number of strokes per tooth.

The indexing time varies from about 0.08 minute on small machines to around 1.0 minute on large machines. When finishing high-precision gears, the machines are indexed once per tooth so that the same tool finishes all the gear teeth, thereby preventing tool pitch errors or mounting errors being transferred to the gear. In roughing, more teeth may be cut per index, depending on the gear diameter and tooth size.

The fewer the strokes, the larger the generating marks on the given tooth. Gears of 100 to 200 teeth have almost straight-line profiles, whereas smaller numbers of teeth have more profile curvature. For this reason, it is necessary to use more strokes per tooth when finishing gears with a small number of teeth. In fine pitches, the roughing cuts do not remove too much metal. This makes it possible to use fewer strokes per tooth in roughing than in finishing. In coarse pitches, however, a substantial amount of metal has to be removed in roughing. This makes it necessary to use more strokes per tooth in roughing than in finishing. With heavy-duty machines, coarse pitch "gashing" is usually carried out with step-type plunge cutters, which can remove a large volume of material with fewer strokes per tooth.

## 8.9 POTENTIAL IMPROVEMENTS IN METHODS OF GEAR CUTTING AND DESIGNS OF RACK GEAR CUTTERS

Investigation of the kinematics of the parallel-axis gear machining mesh is the key for the development of novel methods and designs for gear-cutting tools for gear machining. A particular case of gear machining mesh that is schematically depicted in Figure III.2 was discussed. By adding a cutting motion, this simple gear machining mesh is enhanced to a corresponding kinematics of a gear-machining process. More opportunities appear there for a gear engineer, especially when considering the cutting motion that is represented by the translation vector, the rotation vector, or combinations of these two vectors. Depending on the actual configuration of the vector of cutting motion, various methods and designs for gear-cutting tools can be developed.

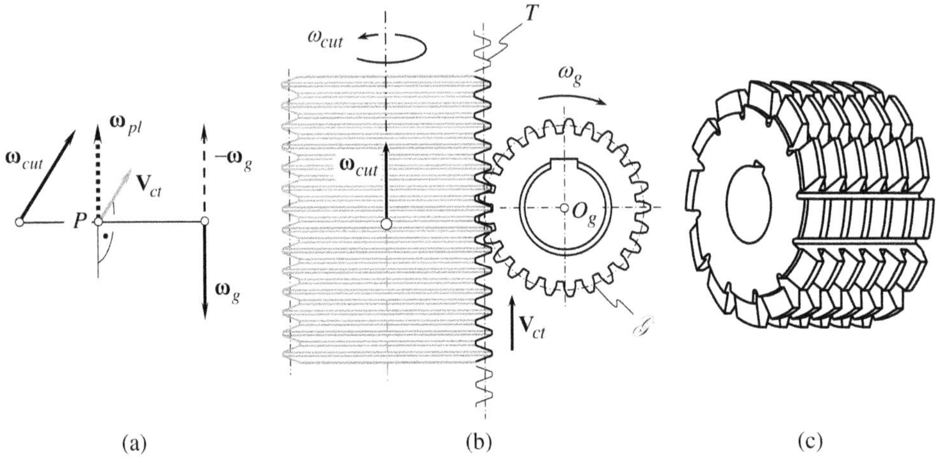


**FIGURE 8.26** (a–c) Kinematics of milling/grinding of a narrow face work gear.

As shown in Figure 8.15, even the simplest case of cutting a gear with the rack cutter allows for significant improvements. A method of gear cutting on the G-TRAC generator is a perfect example in this concern (see Figure 16.26 on page 913 in [146]). Rack cutters mounted on an endless chain are used in this method of gear machining. The gear being cut rolls in mesh with the passing rack teeth assembled in the chain. This method is intended for high volume production, low tool cost, and high machining accuracy. It is possible, though, to use the method effectively for small production. In this last case, a single row of cutting tools is used, and one tooth slot is formed at a time.

In a conventional gear planing operation, the reciprocation of the rack cutter is utilized as the cutting motion. Work gears with a narrow face width can be generated with round rack cutters or with gang milling cutter, as schematically shown in Figure 8.26. In this case, it is not the reciprocation of the cutting tool, but the rotation $\boldsymbol{\omega}_{cut}$ that is used as the primary motion. Vector representation of the kinematics of the gear-machining operation [see Figure 8.26(a)], the work gear in mesh with the gear-cutting tool [see Figure 8.26(b)], and the gear-cutting tool itself [see Figure 8.26(c)] are illustrated in this figure. Gear-cutting tools of this type have limited application for machining of gears under specific manufacturing conditions. However, an operation featuring the kinematics of the gear-machining operation that is an inverse one to that shown in Figure 8.26 is used for cutting of a round rack with the gear shaper cutter.

Another configuration of the rotation $\boldsymbol{\omega}_{cut}$ of the primary motion (see Figure 8.27) in relation to that same gear

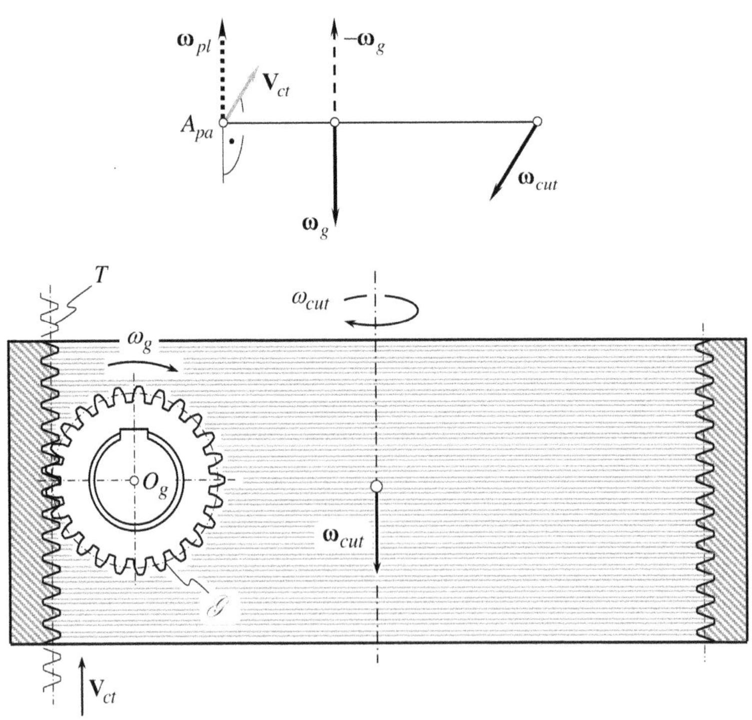


**FIGURE 8.27** Kinematics of milling/grinding of a work gear with an enveloping gear-cutting tool.

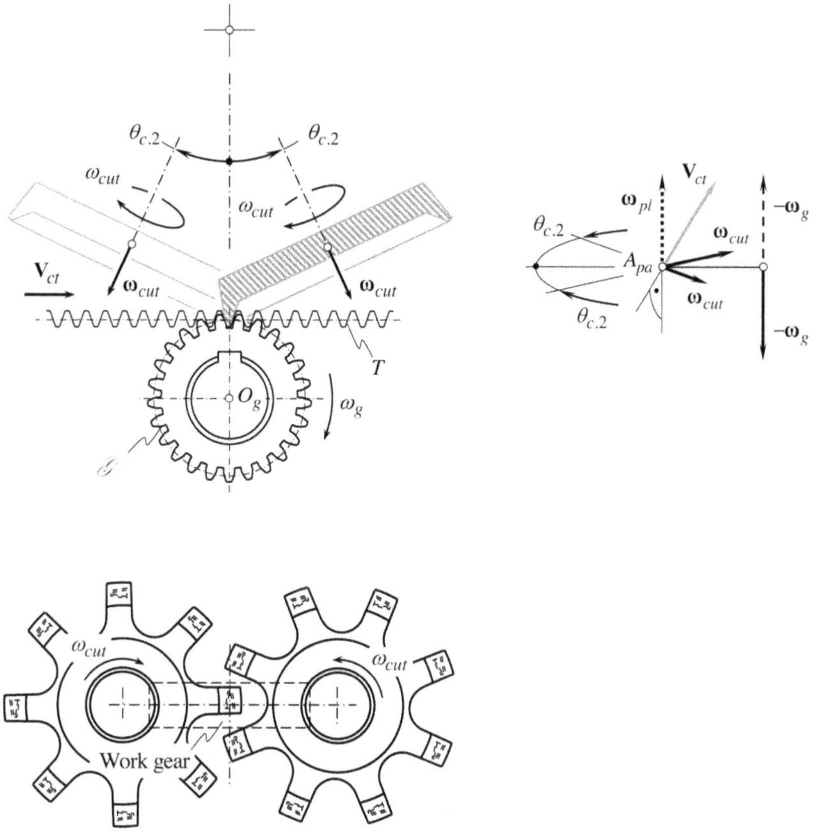


**FIGURE 8.28** Kinematics of generation of a work gear by two tilted face mill gear cutters.

machining mesh corresponds to generation of gears by means of the enveloping round rack cutter. The alteration of configuration of the vectors $\boldsymbol{\omega}_g$, $\mathbf{V}_{ct}$, and $\boldsymbol{\omega}_{cut}$ in the vector representation of the kinematics of the gear-machining operation makes the novel method for gear cutting and a novel design of the gear-cutting tool for this purpose possible.

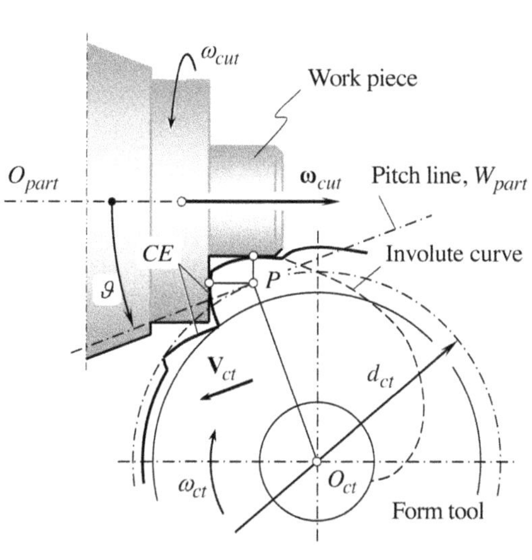


**FIGURE 8.29** Machining of a surface of revolution with the generating rotary form tool.

More examples in this regard can be developed using two cutting tools simultaneously, rather than just one. Modification of vector representation of the kinematics of the gear-machining process allows for a method of the generation of the cylindrical gear with two milling cutters (or with two grinding wheels instead). In this method, the axes of rotation of the gear-cutting tools intersect each other at a certain angle $2\cdot\theta_{c.2}$ (see Figure 8.28). The angle $\theta_{c.2}$ is within the interval $0° \leq \theta_{c.2} \leq \phi_n$ (In a particular case of gear machining, axes of the cutting tools' rotation can be parallel to each other).

Vector representation of the kinematics of the gear-machining operation can be designed so that the vectors $\mathbf{V}_{ct}$ and $\boldsymbol{\omega}_{cut}$ are at a certain angle $(180° - \vartheta)$ to each other. An example is illustrated in Figure 8.29. Despite the fact that the kinematics of this type is not of great interest for the gear manufacturers, it is used for other purposes, such as for turning surfaces of revolution on a lathe.

The considered examples clearly illustrate the extensive opportunities that the kinematics of the gear-machining processes can offer to a gear-cutting tool designer.

# 9 Gear Shaper Cutters I
## *External Gear Machining Mesh*

A gear shaper cutter is one of numerous designs of cutting tools that are extensively used in the production of spur, helical, herring-bone, and double-helical gears. A close-up of a gear-shaping process is illustrated in Figure 9.1. Various designs of gear shaper cutters that are commonly used in the industry are illustrated in Figure 9.2.

Gear shaper cutters, which are often referred to as pinion cutters, are especially efficient when cutting gears with a large tooth count and a narrow face width, with internal teeth, and with shoulders. In addition, gear shaper cutters may be used for machining face gears and involute worms, as well as for rough relieving hobs.

Designing of shaper cutter begins with the determination of the generating surface of the cutting tool. For this purpose, the kinematics of the gear-shaping operation is utilized.

### 9.1 KINEMATICS OF THE GEAR-SHAPING OPERATION

The generating surface of a gear shaper cutter is conjugate to the involute gear tooth flank. When machining the work gear, the generating surface $T$ of the shaper cutter and the gear being machined are in proper mesh with one another.

Kinematics of gear generation is a critical factor in the determination of the generating surface $T$ of the gear shaper cutter. When shaping a work gear by the shaper cutter, the work gear and the shaper cutter rotate $\boldsymbol{\omega}_g$ and $\boldsymbol{\omega}_{ct}$ about their axes of rotation $O_g$ and $O_{ct}$ (see Figure 9.3). The axes $O_g$ and $O_{ct}$ of rotation are parallel to each other ($O_g \parallel O_{ct}$). The magnitudes $\omega_g$ and $\omega_{ct}$ of the rotation vectors $\boldsymbol{\omega}_g$ and $\boldsymbol{\omega}_{ct}$ are timed with each other in compliance with the formula[1]:

$$\frac{\omega_g}{\omega_{ct}} = \frac{N_{ct}}{N_g} \tag{9.1}$$

where $N_g$ denotes the tooth number of the work gear and $N_{ct}$ denotes the tooth number of the gear shaper cutter.

[1] It is instructive to point out here that that same kinematics of the gear machining mesh is featured in another gear-machining process—parallel-axes shaving. In this application [147, 219], the shaving cutter and the work gear are rotating in mesh with the axes parallel. Without the benefit of the crossed-axis angle, there is no helical sliding of the tooth flanks of the shaving cutter and the work gear. Therefore, a reciprocating motion must be induced at a comparatively high frequency. This requires a special machine. The tool cuts only on its face as it rotates in one direction and then the other while the reciprocation takes place. It is generally applied to internal shaving. This method is not commonly used in the industry.

The gear shaper cutter reciprocates in its axial direction simultaneously with the rotations $\boldsymbol{\omega}_g$ and $\boldsymbol{\omega}_{ct}$. The speed of the reciprocation is designated as $\mathbf{V}_{cut}$. The reciprocation of the shaper cutter is required for stock removal. In the case of a helical work gear, the rotation of the shaper cutter $\boldsymbol{\omega}_{cut}$ is performed in addition to the reciprocation $\mathbf{V}_{cut}$. Timing of the motions $\mathbf{V}_{cut}$ and $\boldsymbol{\omega}_{cut}$ depends on the hand of the tooth helix of the work gear. The resultant screw motion that is composed by the motions $\mathbf{V}_{cut}$ and $\boldsymbol{\omega}_{cut}$ is a screw cutting motion of the gear shaper cutter. This screw motion does not correlate to the rotations $\boldsymbol{\omega}_g$ and $\boldsymbol{\omega}_{ct}$ and is independent of the rotations $\boldsymbol{\omega}_g$ and $\boldsymbol{\omega}_{ct}$. The parameter of the screw motion is identical to the screw parameter of the work gear helix. The vector of the screw cutting motion can be pointed either in the same direction as or in the opposite direction of the vector $\boldsymbol{\omega}_{ct}$.

The distance $C$ between the axes $O_g$ and $O_{ct}$ of the rotations changes when the shaper cutter reciprocates back and forth (see Figure 9.4). When the shaper cutter moves down and when it cuts the stock, the distance $C$ is equal to the nominal center distance in the virtual gear machining mesh—it is equal to the distance between the axis of rotation of the work gear and the axis of rotation of the generating surface $T$ of the gear shaper cutter. When the gear shaper cutter moves back, the distance between the axes $O_g$ and $O_{ct}$ increases to a certain value. For this purpose, in most of gear-shaping machine designs, the work-table with the gear reciprocates in the direction that is perpendicular to the axis of rotation $O_g$ of the work gear. The speed of this reciprocation is designated as $V_g$. The reciprocation of the work gear is necessary to eliminate the contact of the clearance surface of the gear shaper cutter teeth and the machined surface of the work gear. In this manner, the tool-life of the gear shaper cutter is increased.

Several different types of in-feed motion of the gear shaper cutter are used in practice. The conventional method of gear-shaping features either an intermittent radial feed of the shaper cutter with a continuous rotary feed or a radial feed without a rotary feed. Continuous radial and rotary feeds are available in modern gear-shaping machines. Numerous other types of in-feed motion of the shaper cutter are also known.

### 9.2 GENERATING SURFACE OF A GEAR SHAPER CUTTER

Not all motions of the work gear and the gear shaper cutter already discussed are necessarily considered when determining the generating surface of the gear shaper cutter.


DOI: 10.1201/9781003520528-12

**FIGURE 9.1** Shaping of a spur gear.

Only those motions that actually affect the meshing of the gear being machined, and the virtual gear represented by the generating surface of the gear shaper cutter must be considered.

Once the kinematics of the gear-machining process is identified (see Figure 9.3), an equation for the gear tooth surface being machined is the only necessary input for determining the generating surface of the gear shaper cutter. As an example, consider machining a helical gear. The case of machining a spur gear can be drawn up from the given case as the simplification under the assumption that the pitch helix angle of the gear is equal to zero ($\psi_g = 0°$). The set of reference systems in both cases is similar.

Consider shaping of a helical gear using the gear shaper cutter (see Figure 9.5). In the Cartesian coordinate system $X_g Y_g Z_g$ associated with the gear, the gear tooth flank surface $\mathcal{G}$ can be analytically described as [see Eq. (1.4)]:

$$\mathbf{r}_g(U_g, V_g) = \begin{bmatrix} r_{b.g} \cos V_g + U_g \cos \lambda_{b.g} \sin V_g \\ r_{b.g} \sin V_g - U_g \sin \lambda_{b.g} \sin V_g \\ r_{b.g} \tan \lambda_{b.g} - U_g \sin \lambda_{b.g} \\ 1 \end{bmatrix} \qquad (9.2)$$

where the position vector $\mathbf{r}_g$ of a point of the gear tooth flank $\mathcal{G}$ is expressed in terms of Gaussian coordinates $U_g$ and $V_g$ and the design parameters $r_{b.g}$ and $\lambda_{b.g}$ of the gear.

The kinematic method of surface generation can be implemented to derive an equation for the tooth flank of the generating surface $T$ of the gear shaper cutter. Following

**FIGURE 9.2** Designs of gear shaper cutters that are commonly used in the industry.

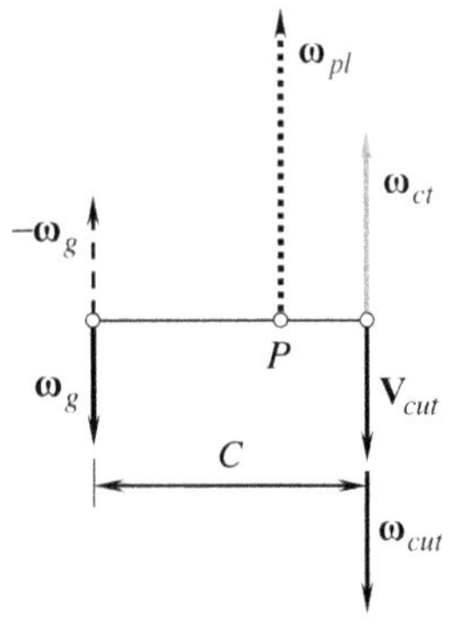


**FIGURE 9.3** Vector diagram of the kinematics of machining an external gear with a shaper cutter.

this method, it is necessary to represent the gear tooth flank $\mathcal{G}$ at its current configuration in the coordinate system $X_{ct}Y_{ct}Z_{ct}$ associated with the gear shaper cutter.

The position vector of a point $\mathbf{r}_g^c$ of the tooth flank $\mathcal{G}$ of the gear allows for representation in the form:

$$\mathbf{r}_g^c(U_g, V_g, \varphi_g, \varphi_{ct}) = \mathbf{Rt}(\varphi_{ct}, Z_{ct}) \cdot \mathbf{Tr}(C, Y_g^*) \cdot \mathbf{Rt}(\varphi_g, Z_g) \cdot \mathbf{r}_g(U_g, V_g) \quad (9.3)$$

where:

$\varphi_g$ is the angle through which the coordinate system $X_gY_gZ_g$ turns relative to the stationary coordinate system $X_g^*Y_g^*Z_g^*$; initially, before the rotation starts, axes of the coordinate system $X_gY_gZ_g$ align to the corresponding axes of the coordinate system $X_g^*Y_g^*Z_g^*$

$\varphi_{ct}$ is the angle through which the coordinate system $X_{ct}Y_{ct}Z_{ct}$ turns relative to the stationary coordinate system $X_{ct}^*Y_{ct}^*Z_{ct}^*$; initially, before the rotation starts, axes of the coordinate system $X_{ct}Y_{ct}Z_{ct}$ align to the corresponding axes of the coordinate system $X_{ct}^*Y_{ct}^*Z_{ct}^*$

$C$ is the center distance between the axis of rotation $O_g$ of the work gear and the axis of rotation $O_{ct}$ of the gear shaper cutter.

The angle $\varphi_g$ may be measured between the coordinate axes $X_g$ and $X_g^*$ (see Figure 9.5). Similarly, the angle $\varphi_{ct}$ may be measured between the coordinate axes $X_{ct}$ and $X_{ct}^*$.

Of the two angles $\varphi_g$ and $\varphi_{ct}$, one may be considered an independent parameter, whereas the other is dependent on the value of the first one. The correlation between the angles $\varphi_g$ and $\varphi_{ct}$ can be expressed by the ratio:

$$\frac{\varphi_g}{\varphi_{ct}} = \frac{N_{ct}}{N_g} \quad (9.4)$$

Ultimately, Eq. (9.3) reduces to the form:

$$\mathbf{r}_g^c = \mathbf{r}_g^c(U_g, V_g, \varphi_g) \quad (9.5)$$

Equation (9.5) describes the position vector of a point $\mathbf{r}_g^c$ of the gear tooth flank $\mathcal{G}$ in its current position with respect to the gear shaper cutter coordinate system $X_{ct}Y_{ct}Z_{ct}$. The parameter $\varphi_g$ serves as the parameter of motion.

To derive the equation of the tooth surface $T$, it is necessary to eliminate the parameter $\varphi_g$ from Eq. (9.5).

At a certain instant of time, not all points of the tooth flank $\mathcal{G}$ make contact with the tooth flank of the generating

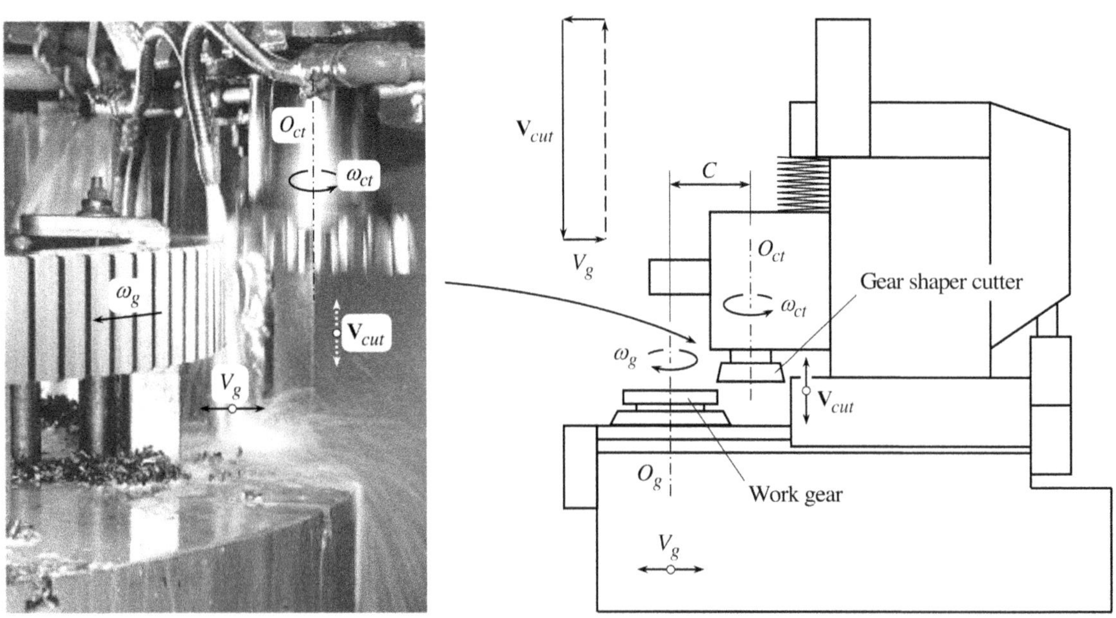


**FIGURE 9.4** Principal elements of the kinematics of gear-shaping operation.

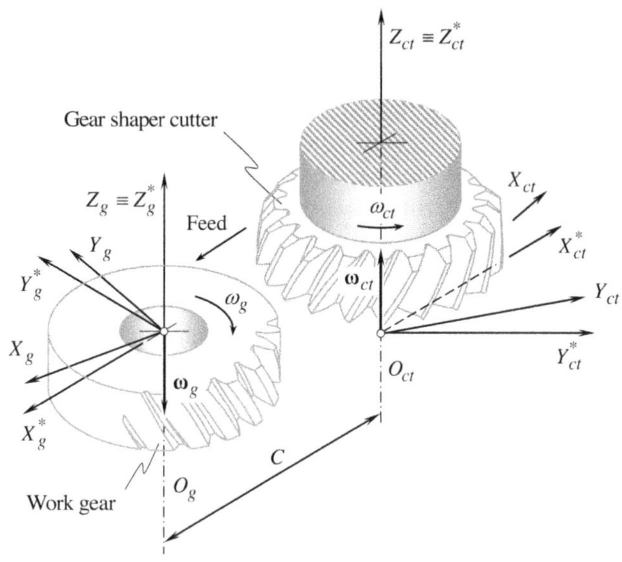


**FIGURE 9.5** The applied coordinate systems.

surface $T$. Only certain points of the tooth flank $\mathcal{G}$ make contact at which the condition of contact:

$$\mathbf{n}_g^c \cdot \mathbf{V}_{g-ct} = 0 \tag{9.6}$$

is fulfilled. In this equation, $\mathbf{n}_g^c$ denotes the unit normal vector to the surface $\mathbf{r}_g^c$, and $\mathbf{V}_{g-ct}$ is a vector of speed of the relative motion of the gear tooth flank $\mathcal{G}$ relative to the coordinate system $X_{ct}Y_{ct}Z_{ct}$.

It should be stressed here that Eq. (9.6) is commonly referred to as the Shishkov equation of contact [204, 206].

The vector $\mathbf{V}_{g-ct}$ of speed of the relative motion may be represented in the form of sum of two vectors, namely:

- The vector $\mathbf{V}_g$, which is the vector of linear speed of the rotational motion of a point of the work gear tooth flank $\mathcal{G}$ about the axis of rotation of the work gear.
- The vector $\mathbf{V}_{pl}$, which is the vector of linear speed of rotation of the gear axis of rotation $O_g$ about the axis of rotation $O_{ct}$ of the coordinate system $X_{ct}Y_{ct}Z_{ct}$.

Ultimately, the equality:

$$\mathbf{V}_{g-ct} = \mathbf{V}_g + \mathbf{V}_{pl} \tag{9.7}$$

is valid.

Note that only the direction of the velocity vector $\mathbf{V}_{g-ct}$ is of importance for further consideration. The magnitude of the vector $\mathbf{V}_{g-ct}$ is not important when the equation of contact is used. This is because the dot product in Eq. (9.6) is equal to zero.

Both the vectors $\mathbf{n}_g^c$ and $\mathbf{V}_{g-ct}$ are functions of coordinates of a point of the moving surface $\mathbf{r}_g^c$. In other words, the dot product of the vectors $\mathbf{n}_g^c$ and $\mathbf{V}_{g-ct}$ allows for interpretation in the form of a function of the parameters $U_g$, $V_g$, and $\varphi_g$. After solving the function with respect to the parameter of motion $\varphi_g$, the derived expression for $\varphi_g$ is substituted into Eq. (9.5). This yields for an equation of the tooth flank $T$ of the generating surface of the gear shaper cutter. Ultimately, in the coordinate system $X_{ct}Y_{ct}Z_{ct}$, the derived equation for the generating surface $T$ can be represented in vector form:

$$\mathbf{r}_{ct} = \mathbf{r}_{ct}(U_{ct}, V_{ct}) \tag{9.8}$$

In a particular case of shaping a helical involute gear, Eq. (9.8) casts into the expression in matrix form:

$$\mathbf{r}_{ct}(U_{ct}, V_{ct}) = \begin{bmatrix} r_{b.ct}\cos V_{ct} + U_{ct}\cos\lambda_{b.ct}\sin V_{ct} \\ r_{b.ct}\sin V_{ct} - U_{ct}\sin\lambda_{b.ct}\sin V_{ct} \\ r_{b.ct}\tan\lambda_{b.ct} - U_{ct}\sin\lambda_{b.ct} \\ 1 \end{bmatrix} \tag{9.9}$$

where:

$r_{b.ct}$ is the radius of the base cylinder of the generating surface $T$ of the gear shaper cutter.
$\lambda_{b.ct}$ is the base lead angle of the generating surface $T$ of the gear shaper cutter.

Equation (9.9) is very similar to Eq. (9.2). This is a common occurrence because a screw involute surface can be properly meshed only with another screw involute surface.

Equation (9.9) can be expressed in terms of the design parameters of the gear shaper cutter. For this purpose, engineering formulae listed in Appendix F can be used.

The normal profile angle $\phi_{n.c}$ on the pitch cylinder of the gear shaper cutter is equal to the normal pressure angle $\phi_{n.g}$ in a gear-shaping operation ($\phi_{n.c} = \phi_{n.g} = \phi_n$). Therefore, this angle is designated as $\phi_n$.

Similarly, the pitch helix angle $\psi_{ct}$ of the gear shaper cutter is equal to $-\psi_g$, where $\psi_g$ is the pitch helix angle of the work gear.

The aforementioned correlation between the normal profile angles of the gear shaper cutter and of the work gear, as well as between the pitch helix angles $\psi_{ct}$ and $\psi_g$, gives the cutting tool designer the opportunity to alter the design parameters of the shaper cutter. The possibility of alteration the design parameters can be used to improve the cutting performance of the gear shaper cutter. Equality of the base pitch of the gear shaper cutter $p_{b.c}$ to the base pitch of the work gear $p_{b.g}$ is a must ($p_{b.ct} \equiv p_{b.g}$).

## 9.3 CUTTING EDGES OF SHAPER CUTTER TEETH

When designing a shaper cutter, the cutting edges $CE$ are commonly viewed as the lines of intersection of the generating surface $T$ of the shaper cutter by the rake surface $R_s$. Furthermore, the clearance surface $C_s$ is designed so as to pass through the cutting edge $CE$ at a favorable clearance angle with respect to the generating surface $T$ of the shaper cutter.

Another way of creating the cutting edges of gear shaper cutter is also possible. In compliance with this second approach, the cutting edges $CE$ are created as lines of intersection of the generating surface $T$ by the clearance surface $C_s$ of the shaper cutter. After that, the rake surface $R_s$ is designed so as to pass through the cutting edge $CE$ at a desired rake angle $\gamma_o$ perpendicular to the generating surface $T$.

In both cases, there are three surfaces through the cutting edge $CE$ of the gear shaper cutter. They are:

- The generating surface $T$.
- The rake surface $R_s$.
- The clearance surface $C_s$.

Only the rake surface $R_s$ and the clearance surface $C_s$ exist physically; the generating surface $T$ of the gear shaper cutter is, instead, a virtual surface. However, the surface $T$ imposes constraints onto the required location of the cutting edges $CE$.

### 9.3.1 Rake Surface of a Gear Shaper Cutter

It is reasonable to distinguish the difference between the geometry of the rake surfaces of shaper cutters for machining spur gears from those for machining helical gears.

#### 9.3.1.1 Geometry of the Rake Surface of a Spur Gear Shaper Cutter

The rake surface of a spur gear shaper cutter is shaped in the form of one of three surfaces:

- A plane that is perpendicular to the axis of rotation of the shaper cutter.
- An internal cone of revolution.
- A surface of revolution of more complex geometry (in special cases).

Spur gear shaper cutters with a rake surface that is shaped in the form of a plane perpendicular to the axis of rotation of the shaper cutter feature zero rake angle ($\gamma_o = 0°$). The rake surface $R_s$ of this geometry meets the equation $Z_{ct} = 0$ [see Figure 9.6(a)]. For profiling of the gear shaper cutter teeth, it is preferrable to use the matrix representation of the equation $Z_{ct} = 0$:

$$\mathbf{r}_{rs}(U_{ct}, V_{ct}) = \begin{bmatrix} U_{ct} \\ V_{ct} \\ 0 \\ 1 \end{bmatrix} \tag{9.10}$$

where:

$\mathbf{r}_{rs}$ is the position vector of a point of the rake plane $R_s$

$U_{ct}$, $V_{ct}$ are the Gaussian coordinates of the rake plane $R_s$

Convenience of application is the only criterion for selecting Gaussian coordinates $U_{ct}$ and $V_{ct}$ of the rake plane $R_s$ for a particular design of gear shaper cutters. For example, Gaussian coordinates $U_{ct}$ and $V_{ct}$ could be equal to the corresponding Cartesian coordinates $X_{ct}$ and $Y_{ct}$ of the rack plane $R_s$—that is, $U_{ct} = X_{ct}$ and $V_{ct} = Y_{ct}$.

The cutting performance of the gear shaper cutter can be significantly enhanced when the rake surface $R_s$ forms a certain favorable rake angle $\gamma_o$ with the perpendicular to the generating surface $T$ of the shaper cutter. For this purpose, the conical rake surface $R_s$ is commonly used in the design of gear shaper cutters [see Figure 9.6(b)]. Position vector of a point $\mathbf{r}_{rs}$ of the conical rake surface $R_s$ can be analytically described by the matrix equation:

$$\mathbf{r}_{rs}(Z_{ct}, V_{ct}) = \begin{bmatrix} Z_{ct} \cdot \tan\gamma_o \cdot \cos V_{ct} \\ Z_{ct} \cdot \tan\gamma_o \cdot \sin V_{ct} \\ Z_{ct} \\ 1 \end{bmatrix} \tag{9.11}$$

In special cases, such as for precision gear shaper cutters, implementation of rake surface $R_s$ that is shaped in the form of a surface of revolution of a more complex geometry is known [see Figure 9.6(c)]. In this case, for the analytical description of the position vector of a point $\mathbf{r}_{rs}$ of the rake surface $R_s$ use of matrix equation:

$$\mathbf{r}_{rs}(U_{ct}, V_{ct}) = \begin{bmatrix} r_{rs}(U_{ct}) \cdot \cos U_{ct} \cdot \cos V_{ct} \\ r_{rs}(U_{ct}) \cdot \cos U_{ct} \cdot \sin V_{ct} \\ r_{rs}(U_{ct}) \cdot \sin U_{ct} \\ 1 \end{bmatrix} \tag{9.12}$$

has proven to be convenient. Here, the equality $r_{rs} = |\mathbf{r}_{rs}|$ is valid.

In all three cases those considered above in this section of the book, a common surface (either a plane, an internal cone of revolution, or a surface of revolution) serves as the rake surface for all the gear shaper cutter teeth. Local modifications of the rake surface within the narrow land along the cutting edge are not considered here.

#### 9.3.1.2 Geometry of the Rake Surface of Helical Gear Shaper Cutter

Helical gear shaper cutters with a reasonably small pitch helix angle $\psi_{ct}$ are manufactured with the rake plane $R_s$, similar to that shown in Figure 9.6(a). Helical gear shaper cutters of this design have a common rake plane $R_s$ for all the gear shaper cutter teeth.

Helical gear shaper cutters with a larger helix angle are designed so each tooth has an individual rake plane $R_s$ (see Figure 9.7). This is mostly attributed to the impact of the pitch helix angle $\psi_{cts}$ onto the rake angle of the lateral cutting edges of the shaper cutter tooth. This impact cannot be neglected in the case under consideration.

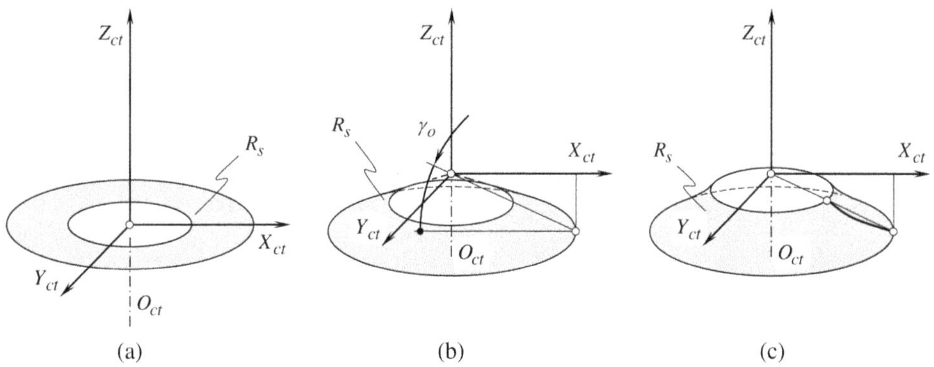


**FIGURE 9.6** Practical kinds of the rake surface $R_s$ of gear shaper cutters for machining spur gears: (a) a plane perpendicular to the shaper cutter axis of rotation, (b) a cone of revolution, and (c) a surface of revolution.

Figure 9.7 provides insight on the derivation of equation of the rake plane $R_s$ of the helical gear shaper cutter:

$$\mathbf{r}_{rs}(X_{ct}, Y_{ct}) = \begin{bmatrix} X_{ct} \\ Y_{ct} \\ Y_{ct} \cdot \tan \psi_{ct} \\ 1 \end{bmatrix} \tag{9.13}$$

The gear shaper cutter teeth are turned through a certain angle relative to each other about the axis of rotation of the cutting tool. The actual value of this angle $\theta_{cn}$ equals $\theta_{cn} = 2\pi n / N_{ct}$. Here, $n$ denotes the number of the tooth counting from the tooth for which Eq. (9.13) is derived. Use of the operator of rotation $\mathbf{Rt}(\theta_{cn}, Z_{ct})$ yields an expression:

$$\mathbf{r}_{rs}^{(n)}(X_{ct}, Y_{ct}) = \mathbf{Rt}(\theta_{cn}, Z_{ct}) \cdot \mathbf{r}_{rs}(X_{ct}, Y_{ct}) \tag{9.14}$$

for the rake plane $R_s$ of the gear shaper cutter tooth that is numbered with $n$.

### 9.3.2 Clearance Surface of the Gear Shaper Cutter Tooth

A screw surface through the cutting edge can be utilized as the desired clearance surface of a shaper cutter tooth. For this purpose, the screw surface $C_s$ has to have an appropriate pitch.

When the equation for the generating surface $T$ of the gear shaper cutter [Eq. (9.9)] is considered together with the equation for the rake surface $R_s$ [see Eqs. (9.10) through (9.14)], points that fulfil both equations simultaneously specify the cutting edge $CE$ of the gear shaper cutter tooth. Generally, the cutting edge of the shaper cutter tooth can be viewed as a spatial curve of intersection of the generating surface $T$ by the rake surface $R_s$.

To derive an equation for the cutting edge $CE$ of the gear shaper cutter tooth, the surfaces $T$ and $R_s$ need to be properly parameterized. Often, a reparameterization of one of the surfaces needs to be carried out for this purpose (see Appendix C on details of re-parameterization of surfaces). After the re-parameterization is complete, both the surfaces $T$ and $R_s$ are expressed in terms of the common parameters $U$ and $V$. In other words, the equation for the generating surface $T$ [Eq. (9.9)] and a corresponding equation of the rake surface $R_s$ [see Eq. (9.10) through (9.14)] are represented in the form:

$$\mathbf{r}_{ct} = \mathbf{r}_{ct}(U, V) \tag{9.15}$$

$$\mathbf{r}_{rs} = \mathbf{r}_{rs}(U, V) \tag{9.16}$$

respectively.

One of the parameters (either the $U-$ or $V-$ parameter) can be eliminated from the set of two equations $\mathbf{r}_{ct} = \mathbf{r}_{ct}(U, V)$ and $\mathbf{r}_{rs} = \mathbf{r}_{rs}(U, V)$. Ultimately, this yields an equation for the cutting edge $CE$ of the gear shaper cutter tooth in the form

$$\mathbf{r}_{ce} = \mathbf{r}_{ce}(t_{ce}) \tag{9.17}$$

where $t_{ce}$ is a parameter of the cutting edge $CE$.

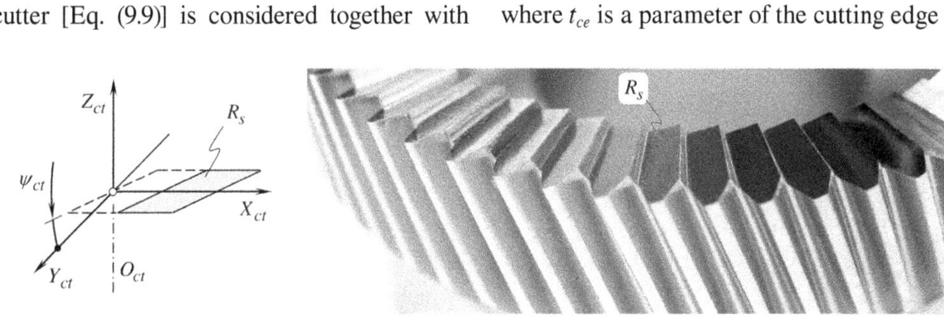


**FIGURE 9.7** Rake plane $R_s$ in the design of a helical gear shaper cutter tooth.

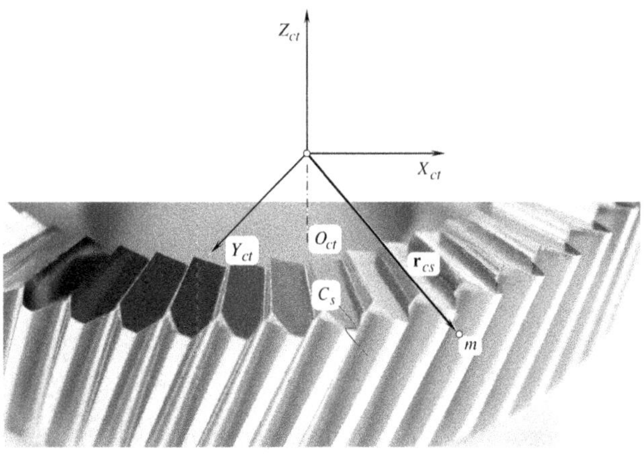


**FIGURE 9.8** Clearance surface $C_s$ in the design of a helical gear shaper cutter tooth.

For the construction of the clearance surface $C_s$ of the gear shaper cutter tooth, a screw motion of the cutting edge *CE* is considered. When the cutting edge *CE* performs an appropriate screw motion, the clearance surface $C_s$ is generated as the family of consecutive positions of the cutting edge *CE*. To construct the desired clearance surface $C_s$ of the gear shaper cutter tooth, it is sufficient to derive an equation for the cutting edge *CE* in its current location and orientation while performing the screw motion. Use of the operator of screw motion $\mathbf{Sc}_z(\varphi_z, p_z)$ is helpful when deriving the equation for the desired clearance surface $C_s$ (see Appendix D for details on the operator of screw motion). Ultimately, position vector $\mathbf{r}_{cs}$ of a point of the desired clearance surface $C_s$ of the gear shaper cutter tooth yields vector interpretation in the form (see Figure 9.8):

$$\mathbf{r}_{cs}(t_{ce}, \varphi_z) = \mathbf{Sc}_z(\varphi_z, p_z) \cdot \mathbf{r}_{ce}(t_{ce}) \tag{9.18}$$

In this equation, position vector $\mathbf{r}_{cs}$ of a point of the clearance surface $C_s$ is expressed in terms of two parameters, namely, $t_{ce}$ and $\varphi_z$.

The reduced pitch $p_z$ of the clearance surface $C_s$ can be expressed in terms of two parameters: the clearance angle $\alpha_m$ in the main section of the cutting edge and the pitch diameter $d_{w.ct}$ of the gear shaper cutter. Finally, the following expression:

$$p_z = 2\pi \cot \alpha_m \tag{9.19}$$

is derived for the calculation of the reduced pitch $p_z$ of the clearance surface $C_s$ of the gear shaper cutter tooth.

Unfortunately, the desired clearance surface $C_s$ of the gear shaper cutter tooth [see Eq. (9.18)] is not convenient for grinding. This is the major reason that, in practice, the desired clearance surface $C_s$ of the gear shaper cutter tooth is approximated by a screw surface that is more convenient for grinding.

## 9.4 PROFILING OF GEAR SHAPER CUTTERS TEETH

Gear shaper cutters are round cutting tools. The pitch diameter of a gear shaper cutter becomes progressively smaller after each resharpening of the worn tool. Because of the change in the pitch diameter, the shape of the cutting edges of a new gear shaper cutter and its shape after resharpening are not identical. This factor must be taken into account when designing a gear shaper cutter.

The following discussion is largely based on the analysis of meshing of a partially ground gear shaper cutter with the associated auxiliary virtual rack $\mathscr{R}_T$.

In the case under consideration, the generating surface *T* of the gear shaper cutter is formed as the envelope surface to consecutive positions of a virtual rack surface. The virtual rack surface $\mathscr{R}_T$ serves as the auxiliary generating surface of the gear shaper cutter. The rack surface $\mathscr{R}_T$ properly meshes with the gear being machined, as well as with the generating surface *T* of the gear shaper cutter. Application of the second approach of forming the generating surface *T* of the gear shaper cutter is convenient from the standpoint of the reduction of diameters of the gear cutting after every resharpening of the shaper cutter.

Consider an example of meshing of the auxiliary rack surface $\mathscr{R}_T$ with the generating surface *T* of the gear shaper cutter (see Figure 9.9). This is used to illustrate how different portions of the generating surface *T* have to change their relative orientation when the pitch diameter $d_{w.ct}$ of the gear shaper cutter is gets increasingly smaller after every resharpening of the gear-cutting tool. The geometry of the portions of the generating surface remains the same as it is for the new gear shaper cutter. It has been proven convenient to use the auxiliary rack surface $\mathscr{R}_T$ for illustrative purposes.

Consider the sectional view of the rack by a plane that is perpendicular to the axis of rotation $O_{ct}$ of the gear shaper cutter. When the surface *T* is generating, the pitch line *w* of the auxiliary rack rolls without sliding over the pitch circle of the diameter $d_{w.ct}$. The involute tooth profile of the generating surface *T* is shaped in the form of the enveloping surface to consecutive positions of the auxiliary rack $\mathscr{R}_T$ when the pitch line *w* rolls over the pitch circle.

To be able to cut the stock, a positive clearance angle is necessary at all cutting edges of the tooth of the gear shaper cutter. To design the gear shaper cutter in such a way that the clearance angle $\alpha_o$ at the top cutting edge corresponds to the clearance angles at the lateral cutting edges is the practical way to create the required clearance angles at all cutting edges of the gear shaper cutter tooth. To do this, the configuration of portions of the generating surface *T* that are through the lateral sides of the tooth profile have to be changed as the outer diameter of the gear shaper cutter continues to get smaller. In this way, the clearance angles of a certain value can be obtained on the lateral cutting edges of the gear shaper cutter.

Let us assume that the nominal section of the generating surface *T* is aligned with the section by the plane through

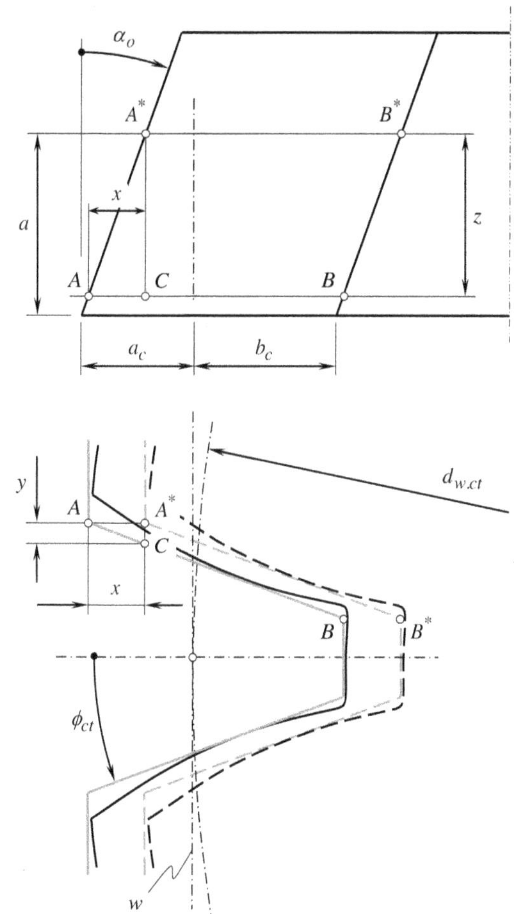


**FIGURE 9.9** Diagram of the meshing of the auxiliary rack surface $\mathscr{R}_T$ with the generating surface $T$ of the gear shaper cutter.

points $A$ and $B$. The plane of the nominal section is perpendicular to the axis of rotation $O_{ct}$. The ordinary involute tooth profile of the generating surface $T$ is formed in the nominal section of the gear shaper cutter. Calculations of most of the design parameters of the gear shaper cutter are usually performed for the nominal section.

In the nominal section, the tooth profile of the gear shaper cutter can be specified in terms of the following design parameters: addendum $a_c = 1.25 \cdot m$ [mm] (the gear shaper cutter module is denoted by $m$), dedendum $b_c = 1.25 \cdot m$ [mm], pitch diameter $d_{w.ct} = m \cdot N_{ct}$ (where the gear shaper cutter tooth number is denoted by $N_{ct}$), and nominal circular tooth thickness $t_n = \pi \cdot m/2$ [mm]; the circular space width is equal to the circular tooth thickness. The actual value of the circular tooth thickness exceeds its nominal value:

$$t = t_n + 0.5 \cdot B, \text{ mm} \tag{9.20}$$

where $B$ designates backlash in the gear transmission that is composed by two gears in mesh.

The section of the plane through points $A^*$ and $B^*$ is an arbitrary normal section of the generating surface $T$ (see Figure 9.9). An arbitrary normal section is at a certain distance $z$ from the nominal normal section. Within the arbitrary normal section, the tooth profile of the generating surface $T$ of the gear shaper cutter can be generated by the auxiliary rack that is identical to the auxiliary rack $\mathscr{R}_T$. However, in this section, the auxiliary rack $\mathscr{R}_T$ is closer to the axis of rotation $O_{ct}$ of the gear shaper cutter. The distance $x$ at which the auxiliary rack $\mathscr{R}_T$ within the arbitrary normal section is shifted toward the axis $O_{ct}$ relative to its initial position can be calculated from the formula:

$$x = z \cdot \tan \alpha_o \tag{9.21}$$

where $\alpha_o$ is the desired clearance angle at the top cutting edge of the gear shaper cutter tooth.

After it has been shifted, the auxiliary rack $\mathscr{R}_T$ generates the corresponding tooth profile of the generating surface $T$. The profile shift correction coefficient $\xi_c$ in this case is:

$$\xi_c = \frac{x}{m} = \frac{z \cdot \tan \alpha_o}{m} \tag{9.22}$$

It is critically important to point out that the coefficient of the correction $\xi_c$ is a function of the $z$ – coordinate—that is, $\xi_c = \xi_c(z)$.

The parameters of the rolling motion of the auxiliary rack $\mathscr{R}_T$ remain the same for both the initial normal section and any arbitrary normal section. Therefore, the pitch diameter $d_{w.c}$ of the gear shaper cutter when it meshes with the auxiliary rack $\mathscr{R}_T$ also retains its original value. The pitch circle of a diameter $d_{w.c}$ is the circle at which the circular pitch $P_c$ of the gear shaper cutter teeth is equal to the pitch of the auxiliary rack $\mathscr{R}_T$ teeth—that is, $P_c = \pi \cdot m$.

The base diameter $d_{b.c}$ of the gear shaper cutter can be calculated from the formula:

$$d_{b.c} = d_{w.ct} \cdot \cos \phi_{ct} \tag{9.23}$$

where $\phi_{ct}$ is the profile angle of the auxiliary rack $\mathscr{R}_T$.

As follows from Eq. (9.23), the base diameter at any arbitrary normal section has the same value as that in the nominal normal section.

The rest of the design parameters of the gear shaper cutter tooth in an arbitrary normal section can be expressed in terms of the corresponding design parameters those measured in the nominal normal section of the gear shaper cutter. For example, the gear cutter tooth addendum $a_c^*$ is equal to:

$$a_c^* = a_c + z \cdot \tan \alpha_o = 1.25 \cdot m + z \cdot \tan \alpha_o, \text{ mm} \tag{9.24}$$

To calculate for the gear shaper cutter tooth dedendum $b_c^*$, the formula:

$$b_c^* = b_c - z \cdot \tan \alpha_o = 1.25 \cdot m - z \cdot \tan \alpha_o, \text{ mm} \tag{9.25}$$

can be used.

The value of the whole depth $h_{t.c}$ remains the same:

$$h_{t.c} = a_c + b_c = a_c^* + b_c^* = 2.25 \cdot m, \text{ mm} \tag{9.26}$$

The tooth thickness when using the gear shaper cutter requires a special treatment. In an arbitrary normal section, tooth thickness $t^*$ can be expressed in terms of the design parameters of the gear-cutting tool. For this purpose, the expression:

$$t^* = \frac{\pi \cdot m}{2} + B + 2 \cdot z \cdot \tan\alpha_o, \text{ mm} \tag{9.27}$$

is derived.

The involute tooth profile within an arbitrary normal section of the generating surface $T$ of the gear shaper cutter is constructed from the same base circle of diameter $d_{b.c}$ as the involute profile within which the nominal normal section is constructed. This means that the same involute curve is used for the tooth profile, both in the nominal and in the arbitrary normal section of the gear shaper cutter. However, the two tooth profiles differ from each other. This is attributed to different segments of the same involute curve used in these two cases.

An angular displacement of the involute tooth profile in the arbitrary normal section of the gear shaper tooth with respect to the tooth profile within the nominal section is observed. To calculate for the current value of the angular displacement $\varepsilon$ of the tooth profiles, the following formula:

$$\varepsilon = \frac{t^* - t}{d_{w.ct}} = \frac{\tan\alpha_o \cdot \tan\phi_{ct}}{d_{w.ct}} \cdot z \tag{9.28}$$

can be used.

As follows from Eq. (9.28), the angular displacement $\varepsilon$ is a linear function of the $z-$ coordinate. Consequently, the locus of consecutive positions of the involute tooth profile in various locations of the arbitrary normal section when it moves along the axis of rotation $O_{ct}$ of the gear shaper cutter is within a screw surface of a constant axial pitch. The axial pitch $P_x$ of this surface is equal to:

$$P_x = \frac{\pi \cdot d_{w.ct}}{\tan\alpha_o \cdot \tan\phi_{ct}} \tag{9.29}$$

Equation (9.29) immediately follows from Eq. (9.28) when the angle of $2\pi$ is substituted instead of an arbitrary value of the angular displacement $\varepsilon$. In such a scenario, the equality $z = P_x$ is observed.

### 9.4.1 Features of Profiling Shaper Cutters for Machining Non-involute Shapes

Shaper cutters are also used for machining non-involute profiles. For profiling shaper cutters intended for machining non-involute profiles, the kinematic method of profiling cutting tools is commonly used.

When shaping parts with a non-involute tooth profile, the kinematics of part surface generation (PSG) is similar to that of machining involute tooth profiles. The pitch cylinder of a radius $R_{wg}$ of the gear to be machined rolls without sliding over the pitch cylinder of radius $R_{wc}$ of the shaper cutter, as shown in Figure 9.10. An orthogonal Cartesian coordinate system $X_gY_gZ_g$ is associated with the work gear and rotates with the gear with a certain angular velocity $\omega_g$. The initial position of the coordinate system $X_gY_gZ_g$ is designated as $X_{gs}Y_{gs}Z_{gs}$. Another Cartesian coordinate system $X_{ct}Y_{ct}Z_{ct}$ is the cutting tool reference system with which the shaper cutter will be associated after it has been designed. This reference system is rotating with the angular velocity $\omega_{ct}$. The rotations $\omega_g$ and $\omega_{ct}$ are synchronized with each other so as to keep the ratio $\omega_g \cdot R_g = \omega_{ct} \cdot R_{ct}$ valid. When the work gear turns about its axis of rotation $O_g$ through a certain angle $\varphi_g$, the shaper cutter turns about its axis of rotation $O_{ct}$ through the angle $\varphi_{ct} = \varphi_g \cdot (R_g/R_{ct})$.

Shishkov equation of contact $\mathbf{n}_g^m \cdot \mathbf{V}_\Sigma^m = 0$ is used for the purpose of profiling the shaper cutter for machining the work gear that features non-involute tooth profile. To compose the equation of contact, an analytical representation

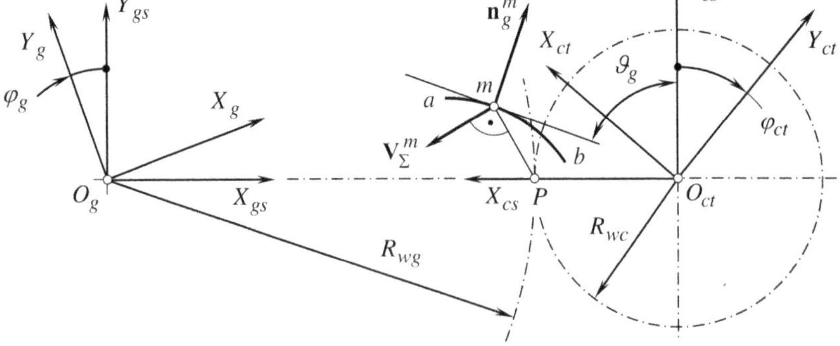


**FIGURE 9.10** Schematic of implementation of kinematic method of profiling cutting tools on shaper cutters for machining non-involute tooth profiles.

of the vector of relative motion $\mathbf{V}_\Sigma^m$ at a point of interest within the gear tooth profile is required. It is also necessary to calculate a unit normal vector $\mathbf{n}_g^m$ to the gear tooth profile at that same point of interest.

A point of interest $m$ is chosen within the work gear tooth profile $ab$. In the motionless coordinate system $X_{gs}Y_{gs}Z_{gs}$, the unit normal vector $\mathbf{n}_g^m$ to the gear tooth profile at $m$ is analytically described by an equation:

$$\mathbf{n}_g^m = -\mathbf{i}\cdot\tan(\vartheta_g + \varphi_g) + \mathbf{j} \tag{9.30}$$

Here and in the following, only the direction (and not the magnitude) of the relative motion $\mathbf{V}_\Sigma^m$, as well as the direction of the normal vector $\mathbf{n}_g^m$, are of importance.

In Eq. (9.30), the angle $\vartheta_g$ designates the angle that the tangent to the tooth profile, $ab$ at the point of interest $m$ forms with the $Y_g$ – axis of the reference system $X_gY_gZ_g$.

The vector of linear motion of instant rotation of point $m$ about pitch point $P$ is designated as $\mathbf{V}_\Sigma^m$. This allows for representation of the vector $\mathbf{V}_\Sigma^m$ in the form of vector equation:

$$\mathbf{V}_\Sigma^m = \mathbf{i}\cdot Y_{gs} - \mathbf{j}\cdot X_{gs} \tag{9.31}$$

Substituting Eqs. (9.30) and (9.31) into the equation of contact $\mathbf{n}_g^m \cdot \mathbf{V}_\Sigma^m = 0$, the particular form of the equation of contact:

$$\sin(\vartheta_g + \varphi_g) = \frac{X_g \cos\vartheta_g + Y_g \sin\vartheta_g}{R_{wg}} \tag{9.32}$$

is derived.

Equation (9.32) yields for calculation of the angle $\varphi_g$ at which a given point of the gear tooth profile makes contact with the corresponding point of the shaper cutter tooth profile point.[2] Furthermore, coordinates of the point of the shaper cutter tooth profile are calculated using operators of linear transformations.

It is right point to stress here that only involute gears can be precisely generated in this gear-shaping process. No gears with other tooth flank geometry can be perfectly generated by the gear shaper cutter—only approximate non-involute gears can be cut this way. This is attributed to the kinematics of the gear-shaping process and to violation of the condition of conjugacy of the tooth flanks $\mathscr{G}$ of both a gear being machined and the generating surface $T$ of the gear shaper cutter. In compliance with the condition of conjugacy of the interacting tooth flanks, at every point of contact of the surfaces $\mathscr{G}$ and $T$, the common perpendicular $\mathbf{n}$ must intersect the axis of instant rotation $P_{ln}$ of the gear and the gear-cutting tool in the gear machining mesh [171].

[2] Fulfillment of the condition of contact is necessary, but this is not a sufficient condition for profiling perfect—that is, geometrically accurate—shaper cutter teeth. Another condition to be fulfilled is the condition of conjugacy of both the tooth flanks of the gear being cut and the gear-cutting tool. To comply with the condition of conjugacy, at every point of contact of the interacting tooth flanks of the gear and the gear-cutting tool, a common perpendicular to these surfaces must intersect the axis of instant rotation $P_{ln}$. Considered together, the condition of contact and the condition of conjugacy are sufficient for profiling perfect shaper cutter teeth. If the condition of conjugacy is not fulfilled, only approximate cutting tools can be designed. On top of that, the base pitch of the gear shaper cutter and the operating base pitch in the gear machining mesh must be equal to the base pitch of the gear being machined [172]. In parallel-axes gearing, the base pitch for a non-involute tooth profile does not exist, and, thus, it cannot be defined.

## 9.5 CRITICAL DISTANCE TO *NOMINAL* PLANE SECTION OF SHAPER CUTTER TOOTH

A plane section of the generating surface $T$ of the gear shaper cutter at which the design parameters of the cutting tool are calculated is commonly referred to as the nominal plane section of the gear shaper cutter.

The parameters of the tooth profile of the gear shaper cutter after being resharpened differ from that of the new gear shaper cutter for many reasons. It is practical to design a gear shaper cutter so the deviations of tooth profile of the new gear shaper cutter are of maximal permissible values. After every re-sharpening of the gear shaper cutter, the tooth profile deviations become smaller. At a certain section, the tooth profile deviations of the gear shaper cutter are equal to zero. Further re-sharpening cause progressively larger deviations of the gear shaper cutter tooth profile. However, the sign of the deviations is opposite to the sign of the deviations in the new shaper cutter. Ultimately, at a certain section, the deviations reach values of corresponding tolerances. The distance in the axial direction of the gear shaper cutter between the initial plane section of the generating surface $T$ and the plane section that corresponds to the completely worn-out gear shaper cutter is of critical importance to the cutting tool designer. The location of the nominal plane section of the gear shaper cutter is specified by this distance.

The nominal plane section of the gear shaper cutter is at a distance $a$ in the axial direction from the corresponding plane section of the new gear shaper cutter. The distance $a$ is half a distance between the sections of the new and the worn-out gear shaper cutters.

It is desirable to design a gear shaper cutter with the greatest possible distance $a$. The greater the distance $a$, the more re-sharpenings the gear shaper cutter can withstand. This means that a larger number of gears can be machined during the lifespan of the given gear shaper cutter. In this way, the cost of the gear shaper cutter per machined gear can be significantly reduced.

The maximal length of the distance $a$ is limited by two major factors.

First, the distance $a$ affects the length of the top cutting edge of the gear shaper cutter. The greater the distance $a$, the shorter the top cutting edge becomes. If the gear shaper cutter is inappropriately designed, either the length of the top cutting edge is insufficiently short or gear shaper cutter tooth pointing may occur. The lifespan of a gear shaper cutter with a short top cutting edge is shorter compared to lifespan of a properly designed gear shaper cutter.

Second, the distance $a$ also affects the parameters of the transient profile that connects the bottom land and the involute tooth profile of the cut gear. When the distance $a$ exceeds a certain permissible value, the transient profile can spread out beyond the allowed limits.

It has been found that, in practice, the distance $a$ for which the length of the top cutting edge is sufficient also fulfils the requirements imposed by the parameters of the transient profile. Therefore, the favorable value of the distance $a$ can be calculated from the condition of sufficient length of the top cutting edge of the gear shaper cutter.

The length $t_{o.ct}$ of the top cutting edge must not be shorter than a specific allowed value $t_{o.ct} \le [t_{o.ct}]$. The shortest allowed length $[t_{o.ct}]$ of the top cutting edge depends on module $m$ of the gear shaper cutter. This can be calculated from the empirical formula:

$$[t_{o.ct}] = \sqrt{0.2594 \cdot m - 0.0375} \tag{9.33}$$

To calculate for the actual length of the top cutting edge $t_{o.ct}$ of the gear shaper cutter, the following equation can be used (see Figure 9.11):

$$t_{o.ct} = 0.5 \cdot d_{o.ct} \cdot \upsilon \tag{9.34}$$

where $\upsilon$ denotes the central angle that the top cutting edge spans over and $d_{o.ct}$ is the outer diameter of the gear shaper cutter.

Figure 9.11 shows that for calculation of the angle $\upsilon$ the formula is valid:

$$\upsilon = \frac{2 \cdot t_{w.ct}}{d_{w.ct}} - 2 \cdot (\operatorname{inv}\phi_o - \operatorname{inv}\phi_w) \tag{9.35}$$

where $\phi_o$ denotes the profile angle at the outer diameter $d_{o.ct}$ of the generating surface $T$ of the gear shaper cutter and $\phi_w$ is the profile angle at the pitch diameter $d_{w.ct}$ of the generating surface $T$.

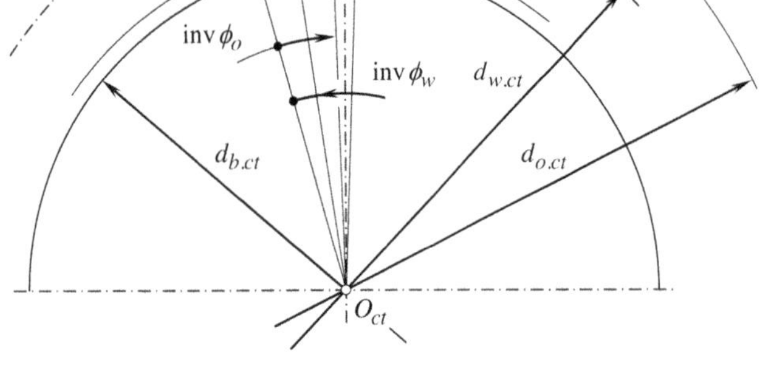


**FIGURE 9.11** Design parameters of the gear shaper cutter that affect the actual length $t_{o.ct}$ of the top cutting edge.

Equation (9.35) can be substituted into Eq. (9.30). This yields an expression:

$$t_{o.ct} = d_{o.ct}\left(\frac{t_{w.ct}}{d_{w.ct}} - \operatorname{inv}\phi_o + \operatorname{inv}\phi_w\right) \tag{9.36}$$

for the calculation of the length of the top cutting edge of the gear shaper cutter.

The parameters that will be entered in Eq. (9.36) can be expressed in terms of the design parameters of the gear shaper cutter:

$$d_{o.ct} = d_{w.ct} + 2.5 \cdot m + 2 \cdot a \cdot \tan\alpha_o \tag{9.37}$$

$$t_{w.ct} = \frac{\pi \cdot m}{2} + B + 2 \cdot a \cdot \tan\alpha_o \cdot \tan\phi_w \tag{9.38}$$

$$\phi_o = \cos^{-1}\frac{d_{w.ct} \cdot \cos\phi_w}{d_{o.ct}} \tag{9.39}$$

To compute the length of the top cutting edge $t_{o.ct}$ of the gear shaper cutter, use of commercial software (e.g., MathCAD) has proven to be handy.

Another approach in determining the length of the top cutting edge $t_{o.ct}$ of the gear shaper cutter is used in practice. The shortest allowed length $[t_{o.ct}]$ of the top cutting edge can be calculated from Eq. (9.33). Moreover, the desired distance $a$ can be determined using the corresponding value of the profile shift correction coefficient $\xi_c$.

In the nominal plane section, the addendum of standard shaper cutters is equal to $1.25 \cdot m$. Figure 9.12 illustrates the function of the reduced length of the top cutting edge $t_{o.ct}/m$ of the new gear shaper cutter versus the profile shift correction coefficient $\xi_c$ for standard gear shaper cutters of different a tooth count $N_{ct}$. When computing for the coordinates

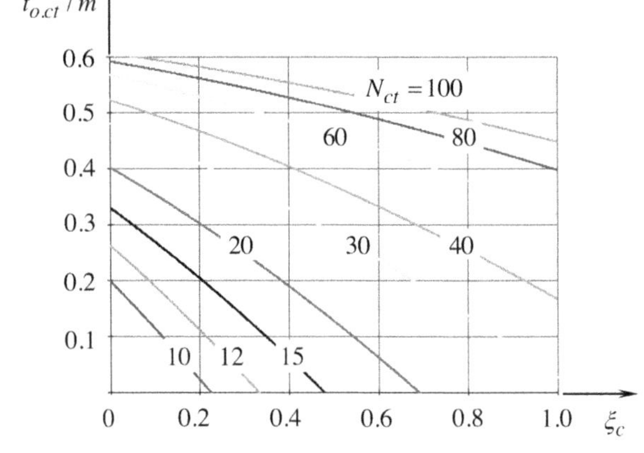


**FIGURE 9.12** Graphical interpretation of the function of the reduced length of the top cutting edge $t_{o.ct}/m$ of the new gear shaper cutter vs. the profile shift correction coefficient $\xi_c$ for standard gear shaper cutters with a different tooth count $N_{ct}$.

of points of the curves in Figure 9.12, it is assumed that the backlash is equal to $B = 0.01 \cdot m$.

The graph, shown in Figure 9.12, allows for a reasonable determination of an approximate value of the profile shift correction coefficient $\xi_c$. Thus, the required distance $a$ can be calculated from the expression:

$$a = \frac{\xi_c}{\tan \alpha_o} \cdot m \tag{9.40}$$

After the desired distance $a$ is calculated, the parameters of the plane section of the generating surface $T$ of the gear shaper cutter are as follows:

$$\text{Tooth addendum} \Rightarrow a_c = 1.25 \cdot m + a \cdot \tan \alpha_o \tag{9.41}$$

$$\text{Tooth dedendum} \Rightarrow b_c = 1.25 \cdot m - a \cdot \tan \alpha_o \tag{9.42}$$

To calculate for the circular tooth thickness $t_{w.ct}$, Eq. (9.38) can be used.

Ultimately, verification of the absence of interference of the designed gear shaper cutter with the gear teeth must be performed.

## 9.6 CUTTING EDGE GEOMETRY OF GEAR SHAPER CUTTER TOOTH

Gear shaper cutters for machining involute gears feature small values of the major geometrical parameters of the cutting edges. Because of this, even small deviations in the geometrical parameters of the cutting edges from their calculated values can significantly affect the gear shaper cutter tool life, as well as the accuracy of the machined gears. Therefore, rigorous control over the geometry of the cutting edges of gear shaper cutters tooth is advantageous.

Standard gear shaper cutters are designed with the rake angle $\gamma_o = 5°$ at the top cutting edge and the clearance angle $\alpha_o = 6°$. The inclination angle of the top cutting edge is zero $\lambda_o = 0°$. The values of angles $\gamma_o$, $\alpha_o$, and $\lambda_o$ do not alter within the top cutting edge. Thus, they do not require further investigation.

Special care should be given to the lateral cutting edges of the gear shaper cutter tooth.

### 9.6.1 ANGLE OF INCLINATION OF THE LATERAL CUTTING EDGE

For the calculation of the actual value of the angle of inclination at a current point of interest within the cutting edge of a gear shaper cutter tooth, elements of vector algebra are helpful.

The local Cartesian coordinate system $x_m y_m z_m$ is associated with the cutting edge $CE$ of the gear shaper cutter tooth, as shown in Figure 9.13. The origin of the coordinate system $x_m y_m z_m$ is at the current point of interest $m$ of the lateral cutting edge $CE$. In reality, a Cartesian coordinate system with arbitrary orientation of the coordinate axes can be implemented. However, solely for the reader's convenience, the coordinate axes are pointed, as shown in Figure 9.13.

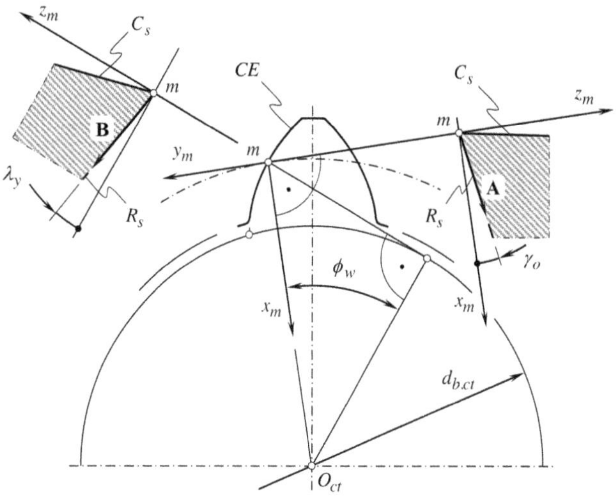


**FIGURE 9.13** Cutting edge geometry of involute gear shaper cutter tooth: The angle of inclination $\lambda$ of the lateral cutting edge $CE$.

To derive an equation for the angle of inclination $\lambda_y$ at point $m$, three vectors **A**, **B**, and **j**, are constructed. The vectors **A**, **B**, and **j** need to be expressed in terms of the clearance angle $\alpha_f$ to be determined and of the known parameters of the geometry of the cutting wedge of the cutting tool.

The vector **A** is the vector pointed along the line of intersection of the rake surface $R_s$ of the gear shaper cutter by the axial section through point $m$. The magnitude of the vector **A** is chosen such that the projection of the vector **A** onto the coordinate plane $x_m y_m$ is equal to the unit (that is, the equality $\mathrm{pr}_{xy}\mathbf{A} = 1$ takes place). The following expression:

$$\mathbf{A} = \mathbf{i} + \mathbf{k} \tan \gamma_o \tag{9.43}$$

is valid for the analytical description of the vector **A**.

The vector **B** is tangent to the lateral cutting edge $CE$ of the gear shaper cutter tooth. The magnitude of the vector **B** is chosen such that the projection of the vector **B** onto the coordinate plane $x_m y_m$ is equal to unit (that is, the equality $\mathrm{pr}_{xy}\mathbf{B} = 1$ is valid). For the analytical representation of the vector **B**, the following expression:

$$\mathbf{B} = \mathbf{i} \cos \phi_{y.c} - \mathbf{j} \sin \phi_{y.c} + \mathbf{k} \tan \lambda_y \tag{9.44}$$

can be composed.

The unit vector **j** is used to solve the problem under consideration.

By construction, the vectors **A**, **B**, and **j** are located within a common plane. At point $m$ within the lateral cutting edge $CE$, this plane is tangent to the rake surface $R_s$ of the gear shaper cutter tooth. Because the vectors **A**, **B**, and **j** comprise a set of coplanar vectors, their scalar triple product is identical to zero ($\mathbf{A} \times \mathbf{B} \cdot \mathbf{j} \equiv 0$). An expanded expression for the scalar triple product can be represented in the form of a determinant:

$$\mathbf{A}\times\mathbf{B}\cdot\mathbf{j}\equiv\begin{vmatrix}0 & 1 & 0\\ 1 & 0 & \tan\gamma_o\\ \cos\phi_{y.c} & -\sin\phi_{y.c} & \tan\lambda_y\end{vmatrix}=0 \qquad (9.45)$$

A formula:

$$\lambda_y=\tan^{-1}(\tan\gamma_o\cdot\cos\phi_{y.c}) \qquad (9.46)$$

for the calculation of the inclination angle $\lambda_y$ immediately follows from Eq. (9.45).

Eq. (9.46) reveals that a very limited variation in the angle of inclination $\lambda_y$ within the lateral cutting edge of the gear shaper cutter is observed. This variation can be neglected when designing a gear shaper cutter.

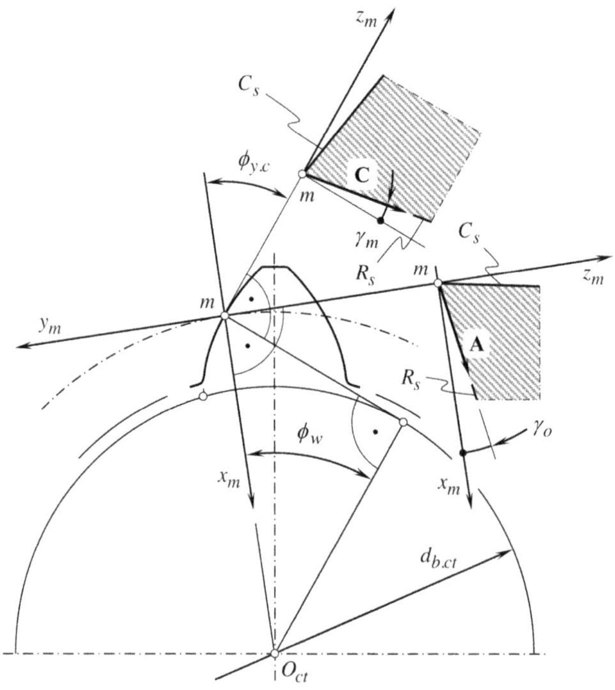


**FIGURE 9.14** Cutting edge geometry of the involute gear shaper cutter tooth: normal rake angle $\gamma_N$ at a point of the lateral cutting edge $CE$.

### 9.6.2 Rake Angle of the Lateral Cutting Edge

The normal rake angle at the lateral cutting edge of the gear shaper cutter tooth is measured within a section of the cutting wedge by a plane that is perpendicular to the cutting edge at the point of interest.

The rake surface $R_s$ in a standard gear shaper cutter is shaped in the form of a cone of revolution. The axis of rotation of the cone of revolution is aligned with the axis of the gear shaper cutter.

To determine the normal rake angle $\gamma_N$, a local Cartesian coordinate system $x_m y_m z_m$ is used. The origin of the coordinate system $x_m y_m z_m$ is at the point of interest within the lateral cutting edge $CE$. The coordinate axes are oriented as shown in Figure 9.14.

Three vectors are constructed to derive an equation for the normal rake angle $\gamma_N$ of the involute gear shaper cutter.

The vector **B** derived earlier [see Eq. (9.44)] is the first vector used to derive the equation for the rake angle $\gamma_N$.

The vector **C** is the second vector in the set of vectors to be used. The vector **C** is pointed along the line of intersection of the rake surface $R_s$ by the plane that is tangent to the base cylinder of the gear shaper cutter. The magnitude of the vector **C** is chosen so that the projection of the vector **C** onto the coordinate plane $x_m y_m$ is equal to the unit (that is, the equality $\text{pr}_{xy}\mathbf{C}=1$ is valid). This yields the following expression for the analytical representation of the vector **C** in the local coordinate system $x_m y_m z_m$:

$$\mathbf{C}=\mathbf{i}\sin\phi_{y.c}+\mathbf{j}\cos\phi_{y.c}+\mathbf{k}\tan\gamma_m \qquad (9.47)$$

where $\gamma_m$ denotes the rake angle in the major section plane.

Finally, the unit vector **j** along the $y_m$ – axis is the third vector used.

By construction, the vectors **A**, **C**, and **j** are situated within a common plane. At point of interest $m$ of the cutting edge $CE$, this plane is tangent to the rake surface $R_s$ of the gear shaper cutter tooth. Because the vectors **A**, **C**, and **j** comprise a set of coplanar vectors, their scalar triple product is identical to zero ($\mathbf{A}\times\mathbf{C}\cdot\mathbf{j}\equiv 0$). The scalar triple product $\mathbf{A}\times\mathbf{C}\cdot\mathbf{j}\equiv 0$ can be expressed in the form of determinant:

$$\mathbf{A}\times\mathbf{C}\cdot\mathbf{j}\equiv\begin{vmatrix}0 & 1 & 0\\ 1 & 0 & \tan\gamma_o\\ \sin\phi_{y.c} & \cos\phi_{y.c} & \tan\gamma_m\end{vmatrix}=0 \qquad (9.48)$$

Equation (9.44) casts into the formula:

$$\gamma_m=\tan^{-1}(\tan\gamma_o\cdot\sin\phi_{y.c}) \qquad (9.49)$$

for the calculation of the rake angle $\gamma_m$ at the point of interest within the lateral cutting edge of the shaper cutter tooth.

The normal rake angle $\gamma_N$ can be expressed in terms of:

- The rake angle $\gamma_m$.
- The angle of inclination $\lambda_y$.

For this purpose, the formula:

$$\gamma_N=\tan^{-1}(\tan\gamma_m\cdot\cos\lambda_y) \qquad (9.50)$$

can be used.

Equation (9.50) can be found in numerous advanced sources (e.g., [144, 148, 151, 158]).

Ultimately, after Eqs. (9.46) and (9.49) are substituted into Eq. (9.50), the following formula:

$$\gamma_N = \tan^{-1}\left(\frac{\tan\gamma_o \cdot \sin\phi_{y.c}}{\sqrt{1+\tan^2\gamma_o \cdot \cos^2\phi_{y.c}}}\right) \tag{9.51}$$

can be obtained for the calculation of the normal rake angle $\gamma_N$ at a point within the lateral cutting edge of the gear shaper cutter tooth.

### 9.6.3 Clearance Angle of the Lateral Cutting Edge

A screw involute surface through the lateral cutting edge serves as the clearance surface of the involute gear shaper cutter tooth. In the section of the generating surface $T$, with a cylinder of revolution that is coaxial with the gear shaper cutter, the surface $T$ and the clearance surface $C_s$ form the clearance angle $\alpha_f$. The value of the clearance angle $\alpha_f$ is equal to the lead angle of the helix within the intersecting cylinder. Thus, the following formula can be used to calculate for the angle $\alpha_f$:

$$\alpha_f = \tan^{-1}\left(\frac{d_{y.c}}{d_{w.c}} \cdot \tan\alpha_o \cdot \tan\phi_w\right) \tag{9.52}$$

where $d_{y.c}$ denotes the diameter of the intersecting cylinder through the point of interest $m$ within the lateral cutting edge $CE$ of the gear shaper cutter tooth.

For the calculation of the clearance angle $\alpha_f$ at the pitch diameter of the gear shaper cutter, Eq. (9.52) simplifies to:

$$\alpha_f = \tan^{-1}(\tan\alpha_o \cdot \tan\phi_w) \tag{9.53}$$

When the clearance angle $\alpha_f$ is known, the normal clearance angle $\alpha_N$ can also be calculated. The clearance angle $\alpha_N$ is measured within the section of the cutting wedge by a plane that is perpendicular to the cutting edge $CE$ at the point of interest $m$.

To derive an equation for the calculation of the normal clearance angle $\alpha_N$ of the gear shaper cutter for machining involute gears, it is convenient to use the method based on elements of vector algebra.

The origin of a local Cartesian coordinate system $x_m y_m z_m$ is located at the point of interest $m$ within the lateral cutting edge $CE$ of the gear shaper cutter tooth. The axes of the coordinate system $x_m y_m z_m$ are pointed, as shown in Figure 9.15.

The vector **D** in Figure 9.15 is constructed so as to be tangent to the line of intersection of the clearance surface $C_s$ by a plane that is tangent to the base cylinder of the gear shaper cutter. In the coordinate system $x_m y_m z_m$, the vector **D** is analytically expressed by vector equation:

$$\mathbf{D} = \mathbf{i}\tan\alpha_m + \mathbf{k} \tag{9.54}$$

where $\alpha_m$ denotes the clearance angle that is measured in the major section of the cutting wedge.

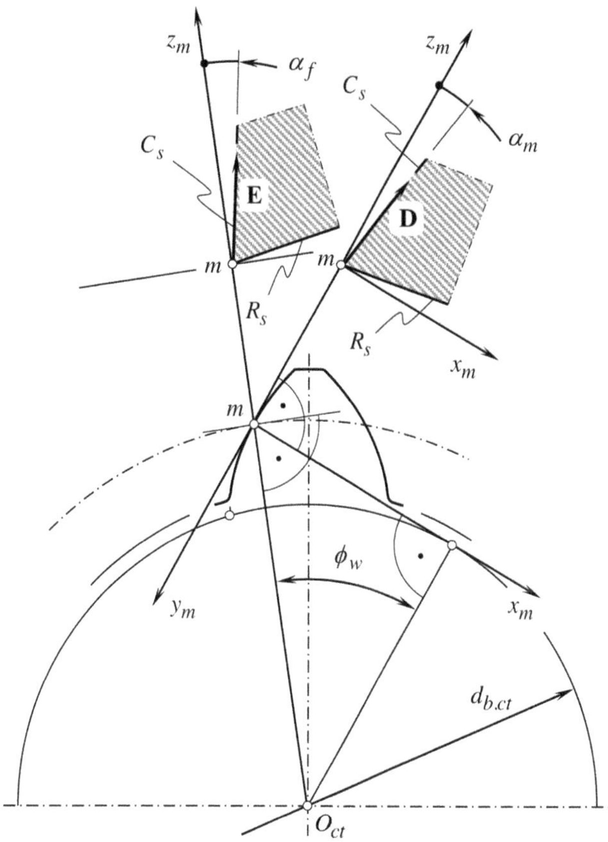


**FIGURE 9.15** Cutting edge geometry of involute gear shaper cutter tooth: the normal clearance angle $\alpha_N$ at a point of the lateral cutting edge $CE$.

The vector **E** through the point of interest $m$ is constructed so as to be tangent to the helix within the intersecting cylinder. This allows the following expression:

$$\mathbf{E} = \mathbf{i}\tan\alpha_f\cos\phi_{y.c} - \mathbf{j}\tan\alpha_f\sin\phi_{y.c} + \mathbf{k} \tag{9.55}$$

for the vector **E**.

The vectors **D**, **E**, and **j** are tangent to the clearance surface $C_s$ of the gear shaper cutter tooth. They comprise a set of coplanar vectors for which the identity $\mathbf{D}\times\mathbf{E}\cdot\mathbf{j}\equiv 0$ is valid. The identity can be expressed in the form of a determinant:

$$\mathbf{D}\times\mathbf{E}\cdot\mathbf{j}\equiv\begin{vmatrix} 0 & 1 & 0 \\ \tan\alpha_m & 0 & 1 \\ \tan\alpha_f\cos\phi_{y.c} & -\tan\alpha_f\sin\phi_{y.c} & 1 \end{vmatrix} = 0 \tag{9.56}$$

Taking Eq. (9.53) into account allows us to cast Eq. (9.56) into the formula:

$$\alpha_m = \tan^{-1}(\tan\alpha_o \cdot \sin\phi_w) \tag{9.57}$$

for the calculation of the clearance angle $\alpha_m$.

The performed analysis of Eq. (9.57) shows that the value of clearance angle $\alpha_m$ does not alter within the lateral cutting edge $CE$ of the gear shaper cutter tooth, as no distance of the point of interest $m$ to the axis of rotation $O_{ct}$ of the gear shaper cutter is entered into Eq. (9.57).

The normal clearance angle $\alpha_N$ can be expressed in terms of the clearance angle $\alpha_m$ and the angle of inclination $\lambda_y$ of the cutting edge $CE$. For this purpose, the formula:

$$\alpha_N = \cot^{-1}(\cot\alpha_m \cdot \cos\lambda_y) \tag{9.58}$$

can be used. This formula is known from many advanced sources (e.g., [144, 148, 151, 158]).

Ultimately, after Eqs. (9.46) and (9.57) are substituted into Eq. (9.58), the following formula:

$$\alpha_N = \tan^{-1}\left(\tan\alpha_o \cdot \sin\phi_w \cdot \sqrt{1+\tan^2\gamma_o \cdot \cos^2\phi_w}\right) \tag{9.59}$$

is obtained for the calculation of the normal clearance angle $\alpha_N$ at a point within the lateral cutting edge of the gear shaper cutter tooth.

As follows from Eq. (9.59), the value of the normal clearance angle $\alpha_N$ at a point of interest within the lateral cutting edge of the gear shaper cutter tooth depends on the rake angle $\gamma_o$.

The following values of the design parameters are common for standard gear shaper cutters: clearance angle $\alpha_o = 6°$ and profile angle $\phi_w = 20°$. In such a scenario, the normal clearance angle is $\alpha_N \cong 2°10'$. This value of the normal clearance angle is too far from its favorable value ($\alpha_N \cong 10 \div 15°$).

### 9.6.4 Improvement in the Geometry of the Lateral Cutting Edge

The actual values of the geometrical parameters of the cutting edge of the gear shaper cutter are insufficient and are significantly far from their optimum values.

The profile angle of the gear shaper cutter tooth $\phi_{y.ct}$ vary within the lateral cutting edge. Its value is the smallest at the start of active profile diameter $d_{l.ct}$ and largest at the outer diameter $d_{o.ct}$ (that is, at the end of active profile). The variation of the profile angle $\phi_{y.ct}$ depends on the tooth count $N_{ct}$ of the gear shaper cutter. The cutting edge geometry of the gear shaper cutter is significantly affected by the variation of the tooth profile angle.

#### 9.6.4.1 Enhancement of the Geometry of the Rake Surface of the Gear Shaper Cutter Tooth

For the standard gear shaper cutter with the rake angle $\gamma_o = 5°$, the smallest value of the normal rake angle $\gamma_N$ can be as small as $\gamma_N = 0°$. This occurs later, when the base diameter $d_{b.ct}$ is equal to the start of active profile diameter $d_{l.ct}$ of the gear shaper cutter. The largest normal rake angle $\gamma_N = 2°...3°$ is observed at the outer diameter of the gear shaper cutter. Evidently, the normal rake angle in the range of $\gamma_N = 0°...3°$ is too small. The cutting performance of the gear shaper cutter can be significantly enhanced in a case when the rake angle $\gamma_N$ of the cutting tool is increased.

The normal rake angle $\gamma_N$ can be increased when the rake angle $\gamma_o$ at the outer diameter of the gear shaper cutter is larger [see Eq. (9.51)]. Improving the cutting performance of the gear shaper cutter in this manner, however, is not promising for two reasons.

First, a significant increase of the rake angle $\gamma_o$ is required for a reasonable increase in the normal rake angle $\gamma_N$. Second, the accuracy of the gear shaper cutter is significantly affected by an increased rake angle $\gamma_o$. The larger the rake angle $\gamma_o$, the larger the deviations of the actual tooth profile of the machined gears from the desired tooth profile. For this reasons, improvement of the cutting performance of the gear shaper cutter using this approach is impractical.

For particular cases of gear shaping, shaper cutters with modified rake surfaces can be implemented.

An example of the modified rake surface $R_s$ of a coarse pitch shaper cutter is depicted in Figure 9.16. The portion of the rake surface $R_s^o$ at the outer cylinder of the gear shaper cutter is formed by a cylindrical grinding wheel. The grinding wheel of a diameter $d_{gw}$ is used for this purpose. Portions of the rake surface $R_s^f$ close to the lateral cutting edges are generated by a conical grinding wheel. The design parameters of the grinding wheel, for example, the outer diameter $d_{gw}$ and the profile angle $\theta_{gw}$, as well as the setup parameters—(a) the angle $\varphi_{gw}$ that the grinding wheel axis of rotation $O_{gw}$ forms with the centerline of the gear shaper cutter tooth, (b) the location of the grinding wheel apex $A_{gw}$ in radial direction $a_{gw}^r$ and (c) the location of the grinding wheel apex $A_{gw}$ in the tangential direction $a_{gw}^t$—are calculated for the gear shaper cutter of a given design. In Figure 9.16, the modification of the rake surface of the gear shaper cutter makes it possible to design the gear-cutting tool with the optimal values of the normal rake angle $\gamma_N$. In this way, the cutting performance of the gear shaper cutter can be significantly enhanced. Use of modification of this type (see Figure 9.16) is practical for gear shaper cutters with a large outer diameter ($d_{o.ct} \geq 360$ mm).

The rake surfaces of each tooth of gear shaper cutters of smaller outer diameter can be ground using either double-conical grinding wheel or the form cutting tool, as shown in Figure 9.17. The shape of the generating surface $T_{gw}$ of the form grinding wheel is designed so as to enable the lateral cutting edges of the ground gear shaper cutter to be identical to that of the new one. The outer diameter of the grinding wheel $d_{gw}$ and the distance $a_{gw}$ between the axis of rotation $O_{gw}$ of the grinding wheel and the axis of rotation $O_{ct}$ of the gear shaper cutter and the length $b_{gw}$ of the grinding wheel top land are calculated so as to enable a design of the gear shaper cutter with an optimal normal rake angle $\gamma_N$ value.

Shaping the coarse pitch spur gears with a large tooth count $N_g$ is an appropriate area of the implementation of a gear shaper cutter with a modified rake surface. The modification of the rake surface of the gear shaper cutter tooth becomes more practical when the tool life of the gear shaper cutter has to be sufficient for machining at least one gear of a large diameter.

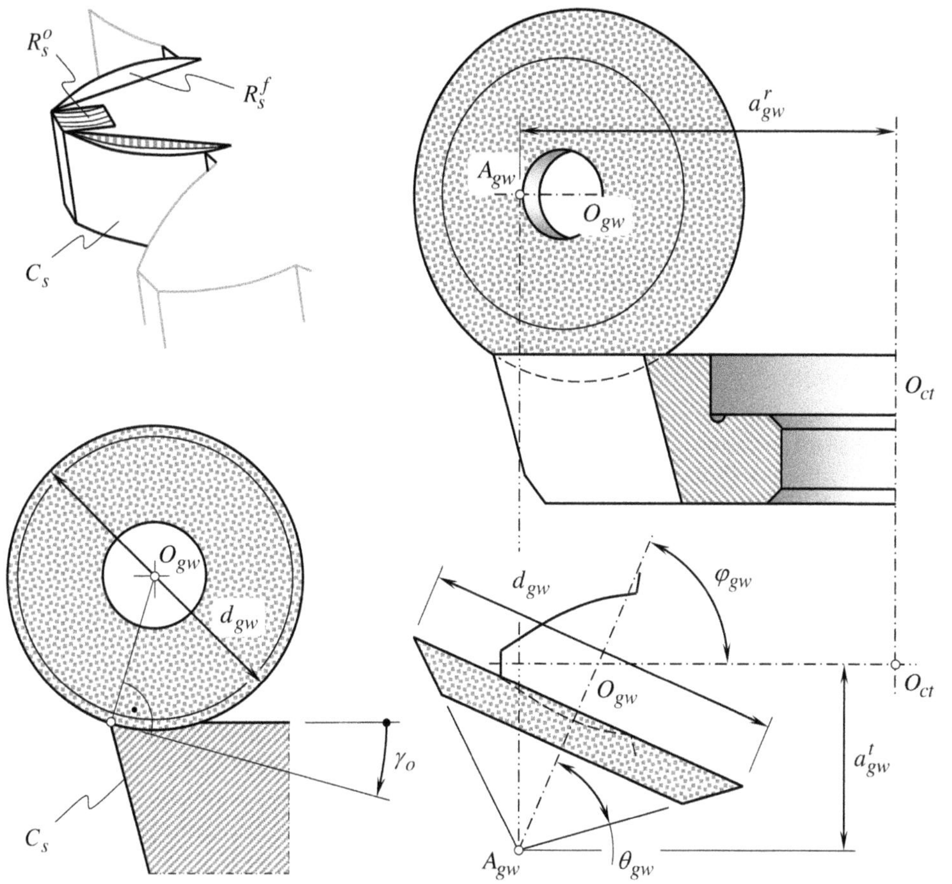


**FIGURE 9.16** Kinds of modification of the rake surface $R_s$ of the coarse pitch involute gear shaper cutter.

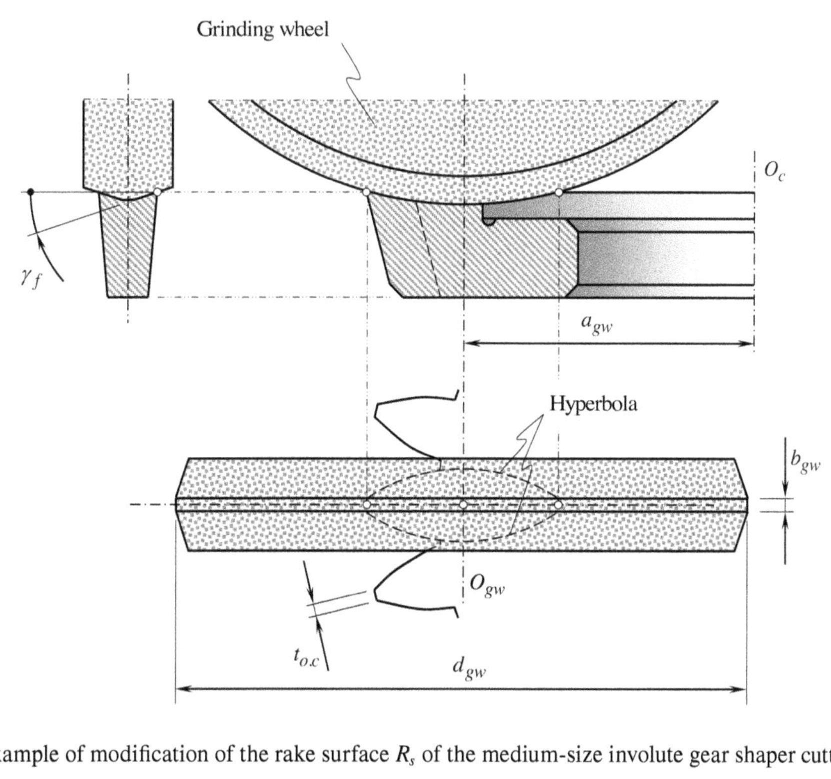


**FIGURE 9.17** An example of modification of the rake surface $R_s$ of the medium-size involute gear shaper cutter.

#### 9.6.4.2 Enhancement of the Geometry of the Clearance Surface of the Gear Shaper Cutter Tooth

The normal clearance angle of the gear shaper cutter is also very far from the optimal value for this parameter of the cutting edge geometry. Both spur and helical gear shaper cutters have undesirably small normal clearance angle $\alpha_N$. It is much more difficult to improve the geometry of the clearance surface of the gear shaper cutter. However, such possibilities exist, and they can be determined. As an example, consider the helical gear shaper cutter[3] for machining involute gears (see Figure 9.18) [73].

The generating surface of the helical gear shaper cutter is shaped in the form of a screw involute surface $T$. Theoretically, any line within the surface $T$ can serve as a cutting edge of the gear shaper cutter. The screw involute surface can be generated by a straight line performing the screw motion. There are no principal constraints on using the straight generating line as the lateral cutting edge $CE$ of the gear shaper cutter.

A plane that is tangent to the generating surface $T$ is a plane $T_{pl}$ through the straight generating line. In the lower portion of Figure 9.18, the particular cutting edge $CE$ that is aligned with the straight generating line of the surface $T$ is projected into the point $CE$. In this illustration, the tangent plane is projected into a straight line $T_{pl}$.

Because the lateral cutting edge $CE$ of the involute gear shaper cutter tooth is straight, a plane can be chosen as the clearance plane $C_s$ (another plane can also be chosen as a rake plane $R_s$) of the gear shaper cutter tooth. The plane $C_s$ is a plane through the cutting edge $CE$. This plane can be constructed so as to pass at the optimal normal clearance angle $\alpha_N^{\text{opt}}$ relative to the tangent plane $T_{pl}$. This means that the normal clearance angle of the gear shaper cutter is of a constant value within the lateral cutting edge, and it could be of optimal value at every point of the lateral cutting edge $CE$. Use of the gear shaper cutters with an optimal value of the normal clearance angle extends the tool life of the gear shaper cutter and enhances the accuracy of the machined gears.

The same is true with respect to the rake plane. The rake surface $R_s$ is a plane through the cutting edge $CE$. It can be designed so as to form an optimal normal rake angle $\gamma_N^{\text{opt}}$ with respect to a perpendicular to the tangent plane $T_{pl}$. Again, the normal rake angle is of a constant value within the lateral cutting edge and the value of the rake angle $\gamma_N^{\text{opt}}$ can be optimal at every point of the lateral cutting edge $CE$. The use of gear shaper cutters with an optimal value of the normal rake angle extends the tool life of the gear shaper cutter and enhances the accuracy of the machined gears.

Similar improvements are feasible for the lateral cutting edge of the opposite side of the gear shaper cutter tooth profile. Prof. S.P. Radzevich was the first to propose this design for helical gear shaper cutter [73]. The design has since been investigated in another work [169]. There is a plenty of room for improvements in design of involute gear-cutting tools using this concept.

[3] Pat. No. 956187 (USSR), *A Helical Gear Shaper Cutter.*/S.P. Radzevich, Int. Cl. B23F 21/10, Filed: September 26, 1980.

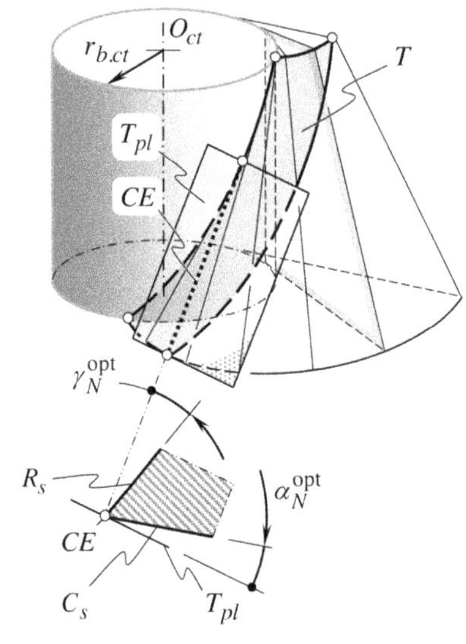


**FIGURE 9.18** Possibility of optimization of geometry of the cutting edge of a helical gear shaper cutter.

## 9.7 DESIRED CORRECTIONS TO THE GEAR SHAPER CUTTER TOOTH PROFILE

Working surfaces formed by the cutting wedge of the gear shaper cutter tooth are shaped in the form of:

- A cone of revolution for the rake surface $R_s$.
- A cone of revolution for the clearance surface $C_s$ at the top cutting edges.
- A screw involute surface for the clearance surface $C_s$ at the lateral cutting edges of the gear shaper cutter tooth.

The lines of intersection of the clearance surfaces $C_s$ by the rake surface $R_s$ are within the corresponding portions of the generating surface of the gear shaper cutter $T$. Otherwise, the deviations of the actual cutting edges of the gear shaper cutter from its generating surface are all but inevitable.

Resharpening of standard gear shaper cutters is performed over a conical rake face. The addendum modification coefficient changes after each resharpening of the gear shaper cutter. However, it is known that the same involute gear can be properly machined by the gear shaper cutters with a different addendum modification coefficient value. This means that use of a screw involute surface as the clearance surface of the lateral tooth profiles of the gear shaper cutter tooth is reasonable.

Because the rake angle exceeds a zero value ($\gamma_o > 0°$), the projection of the actual lateral cutting edge onto the plane perpendicular to the axis of rotation of the gear shaper cutter deviates from the true involute curve. The deviations of the gear shaper cutter tooth are transferred to the machined gear. The actual values of the deviations strongly depend on the value of the rake angle $\gamma_o$ and the value of the clearance angle $\alpha_o$. This is the major reason that standard gear shaper cutters are designed with very small values of these geometrical parameters of the cutting wedge ($\gamma_o = 5°$ and $\alpha_o = 6°$). No practical possibilities are available that would result in an increase in the values of angles $\gamma_o$ and $\alpha_o$ for gear shaper cutters of standard design.

The impact of the geometry of the cutting edge onto the accuracy of the involute gear shaper cutter tooth profile can be evaluated. For this purpose, it is convenient to determine the value of the tooth profile angle of the rack, which can be generated by the projection of the lateral cutting edge, onto the plane perpendicular to the gear shaper cutter axis of rotation.

Consider a point $m$ at the intersection of the cutting edge and the circle of pitch diameter of the gear shaper cutter. A local Cartesian coordinate system $x_m y_m z_m$ is associated with the gear shaper cutter. The origin of the coordinate system $x_m y_m z_m$ coincides with the point of interest $m$. The $z_m$ – axis is aligned with the axis of rotation $O_{ct}$ of the gear shaper cutter. The rests of the coordinate axes are pointed as shown in Figure 9.19.

Three vectors **A**, **B**, and **C** through point $m$ are constructed so that they are situated within a common plane tangent to the clearance surface $C_s$.

The vector **A** is located within the plane that is perpendicular to the axis of rotation $O_{ct}$ of the gear shaper cutter. This vector is tangent to the clearance surface $C_s$ of the shaper cutter tooth. In the local coordinate system $x_m y_m z_m$, the vector **A** is analytically described by the formula:

$$\mathbf{A} = \mathbf{i} + \mathbf{j} \cdot \tan\phi_w \tag{9.60}$$

The vector **B** is tangent to the helix of intersection of the clearance surface $C_s$ by the pitch cylinder of the gear shaper cutter. This vector forms the clearance angle $\alpha_f$ with the axis of rotation $O_{ct}$ of the gear shaper cutter. For the vector **B**, the following expression:

$$\mathbf{B} = -\mathbf{j} \cdot \tan\alpha_f + \mathbf{k} \tag{9.61}$$

is valid.

Taking into account Eq. (9.52) for the calculation of the clearance angle $\alpha_f$, this expression for the vector **B** can be represented in the form:

$$\mathbf{B} = -\mathbf{j} \cdot \tan\alpha_o \cdot \tan\phi_w + \mathbf{k} \tag{9.62}$$

The vector **C** is tangent to the cutting edge $CE$. Therefore, in the local coordinate system $x_m y_m z_m$, it can be analytically represented by the expression:

$$\mathbf{C} = \mathbf{i} + \mathbf{j} \cdot \tan\phi_{ce} + \mathbf{k} \cdot \tan\gamma_o \tag{9.63}$$

where $\phi_{ce}$ denotes tooth profile angle of the auxiliary rack for the projection of the lateral cutting edge.

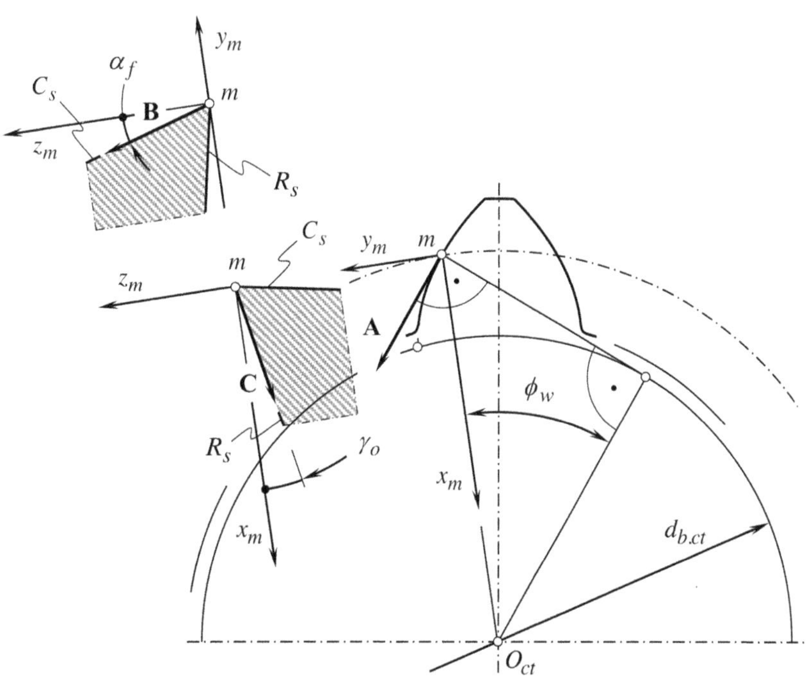


**FIGURE 9.19** Calculation of the desired correction to the profile angle $\phi_w$ of the gear shaper cutter tooth.

As by construction the vectors **A**, **B**, and **C** comprise a set of coplanar vectors, the identity:

$$\mathbf{A}\times\mathbf{B}\cdot\mathbf{C}\equiv 0 \tag{9.64}$$

is observed.

Equation (9.64) can be represented in the form:

$$\mathbf{A}\times\mathbf{B}\cdot\mathbf{C}\equiv\begin{vmatrix}1 & \tan\phi_w & 0\\ 0 & -\tan\alpha_o\cdot\tan\phi_w & 1\\ 1 & \tan\phi_{ce} & \tan\gamma_o\end{vmatrix}=0 \tag{9.65}$$

Ultimately, Eq. (9.65) casts into a formula:

$$\phi_{ce}=\tan^{-1}\left[\tan\phi_w(1-\tan\alpha_o\cdot\tan\gamma_o)\right] \tag{9.66}$$

for the calculation of the corrected tooth profile angle $\phi_{ce}$ of the involute gear shaper cutter.

Calculations in compliance with Eq. (9.66) indicate that, for a gear shaper cutter with a profile angle $\phi_c = 20°$, a rake angle $\gamma_o = 5°$, and a clearance angle $\alpha_o = 6°$, the profile angle $\phi_{ce}$ is equal to $\phi_{ce} = 20°10'$.

Generating the clearance surface of the lateral sides of tooth profile of standard gear shaper cutters with the auxiliary rack with the modified tooth profile is recommended. Then, the pitch diameter of the base cylinder of the clearance surfaces can be calculated as:

$$d^*_{w.ct}=d_{w.ct}\cdot\cos\phi_{ce}=m\cdot N_{ct}\cdot\cos\phi_{ce} \tag{9.67}$$

Special care is required to avoid confusion between the different base cylinders of the involute gear shaper cutter—that is, between the base cylinder of the generating surface $T$ of the gear shaper cutter and the base cylinder of the clearance surface $C_s$ of the lateral cutting edge of the gear shaper cutter tooth. Because the generating surface $T$ and the clearance surface $C_s$ are two different screw involute surfaces, it is evident that they are constructed from different base cylinders.

## 9.8 THICKNESS OF A CHIP CUT BY THE GEAR SHAPER CUTTER TOOTH

The gear-shaping operation is a type of continuous indexing method of gear machining. The kinematics of the machining process of a work gear with the gear shaper cutter is illustrated in Figure 9.3. When shaping a gear using the gear shaper cutter, the work gear rotates about its axis of rotation $O_g$. The gear shaper cutter rotates about its axis of rotation $O_{ct}$. Rotation of the gear $\omega_g$ is synchronized with the rotation of the gear shaper cutter $\omega_c$ so that the ratio $\omega_g/\omega_c = N_c/N_g$ is valid.

Two motions are required for machining a work gear with the gear shaper cutter: the primary motion (the motion of cut) and the feed motion.

Reciprocation of the gear shaper cutter is used here as the primary motion. The reciprocation of the gear shaper cutter provides the necessary speed of cut $\mathbf{V}_{cut}$. The speed of cut is equal to the speed of the shaper cutter reciprocation $V_{rc}$.

The rotations of the work gear $\omega_g$ and of the gear shaper cutter $\omega_{ct}$ cause the feed motion $F_c$ (see Figure 9.20). The rolling of the pitch surfaces of the work gear and the gear shaper cutter about their axes of rotation allows for interpretation in the form of instant rotation of the gear shaper cutter pitch cylinder about the pitch point. The speed of the instant rotation $\omega_{rl}$ can be expressed in terms of the rotations $\omega_g$ and $\omega_{ct}$ and the design parameters of the gear and the gear shaper cutter. Ultimately, the speed of the feed motion $F_c$ is equal to the speed of the linear motion of the cutting edge of the gear shaper cutter tooth in its instant rotation (see Figure 9.21):

$$F_c=\omega_{rl}\cdot R_{rl} \tag{9.68}$$

where $R_{rl}$ is the distance of a point of interest $m$ of the gear shaper cutter cutting edge from the pitch point $P$. The point of interest $m$ can be chosen within any of three cutting edges of the gear shaper cutter tooth.

For a particular gear-shaping operation, the instant rotation $\omega_{rl}$ has a constant value ($\omega_{rl} = const$). The distance $R_{rl}$

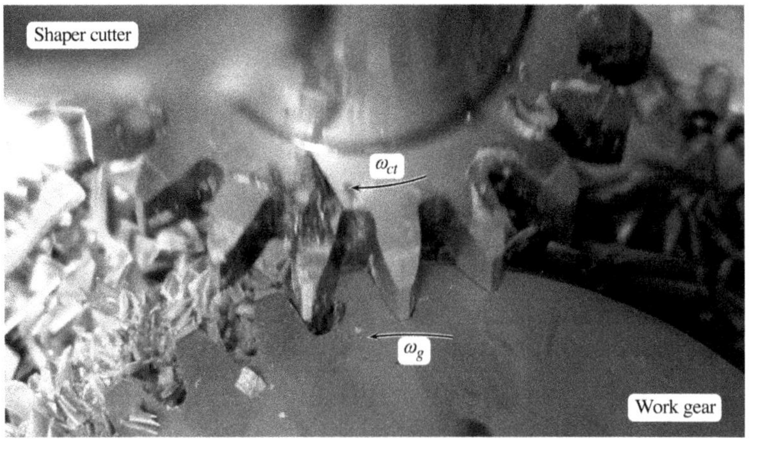


**FIGURE 9.20** Generation of the involute gear tooth flanks $\mathscr{G}$ in the gear-shaping process.

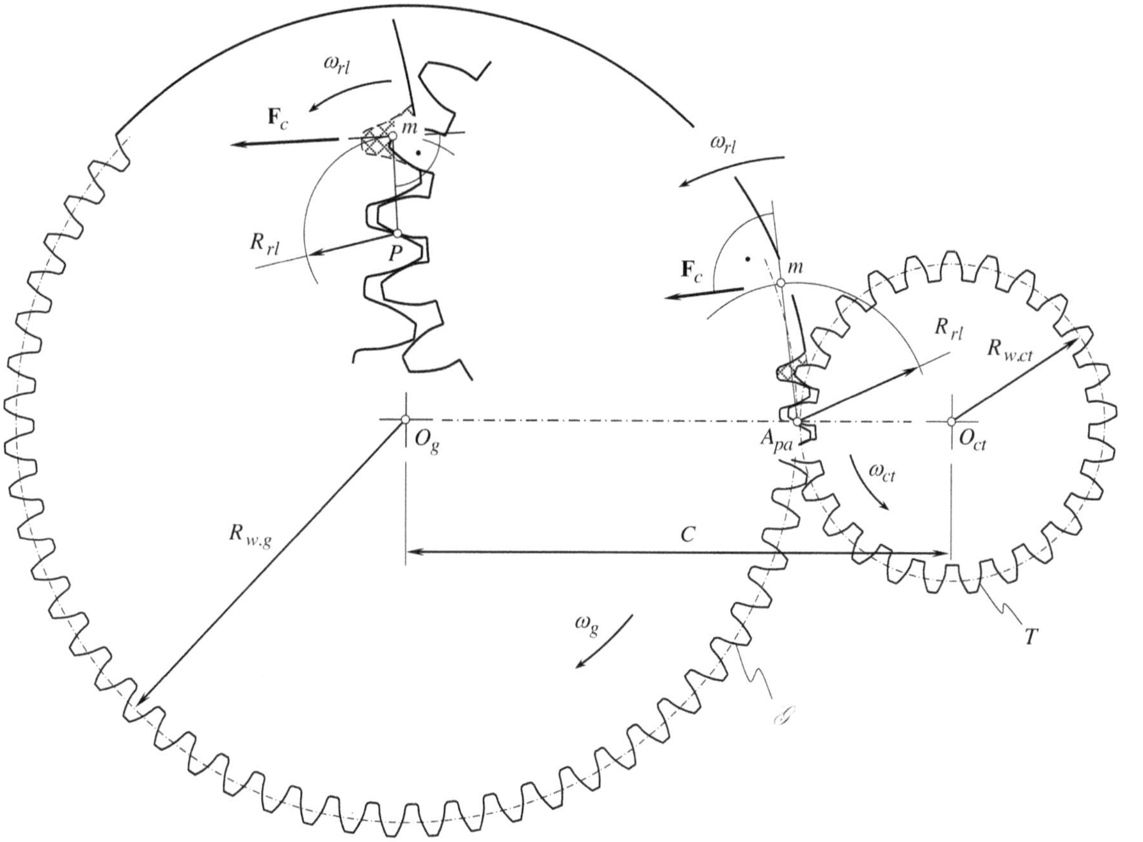


**FIGURE 9.21** Feed motion $F_c$ in shaping operation of a spur gear by the gear shaper cutter.

depends on coordinates of the point of interest $m$ within the cutting edge $CE$ of the gear shaper cutter. This can be specified by the position vector $\mathbf{r}_m$. Two vectors $\boldsymbol{\omega}_{rl}$ and $\mathbf{r}_m$ allow for an expression:

$$\mathbf{F}_c = \boldsymbol{\omega}_{rl} \times \mathbf{r}_m \tag{9.69}$$

for the calculation of the vector $\mathbf{F}_c$ of the feed motion.

To calculate the parameters of a portion of stock cut out by the gear shaper cutter tooth per stroke, the approach proposed by Prof. V.A. Shishkov [204, 206] can be used.

Two consequent locations of the gear shaper cutter tooth are considered for the calculation of the cut thickness. The initial position of the gear shaper cutter is defined by a Cartesian coordinate system $X_gY_gZ_g$ that is embedded to the work gear as shown in Figure 9.21. When machining, the work gear turns about its axis of rotation $O_g$ through a certain portion $\varepsilon_{st}$ of the roll angle $\varepsilon$. The angle $\varepsilon_{st}$ is viewed as a roll angle per stroke of the gear shaper cutter. The actual value of the angle $\varepsilon_{st}$ can be expressed in terms of:

- The roll angle $\varepsilon_o$ for the tooth profile point at the outer diameter of the gear tooth profile.
- The roll angle $\varepsilon_l$ for the tooth profile point at the inner diameter of the gear tooth profile.
- The number of strokes $n_{st}$ of the gear shaper cutter necessary to shape the whole tooth flank surface.

With that said, the following equation can be used for the calculation of the angle $\varepsilon_{st}$:

$$\varepsilon_{st} = \frac{\varepsilon_o - \varepsilon_l}{n_{st}} \tag{9.70}$$

Usually, the value of the angle $\varepsilon_{st}$ is in the range of $\varepsilon_{st} \cong 1°$ or so. The angle $\varepsilon_{st}$ can vary depending on the requirements to a particular gear-shaping operation.

The rotation of the gear shaper cutter $\omega_{ct}$ is timed with the rotation of the work gear $\omega_g$. The rotations $\omega_g$ and $\omega_{ct}$ are timed so that the ratio:

$$\frac{\omega_g}{\omega_{ct}} = \frac{N_{ct}}{N_g} \tag{9.71}$$

is observed.

The tooth ratio $N_g/N_{ct}$ is designated as $u$. The angle of rotation of the cutter $\varepsilon_{st}^c$ can be expressed in terms of the angle of rotation of the work gear $\varepsilon_{st}$ and the tooth ratio $u$:

$$\varepsilon_{st}^c = \frac{\varepsilon_{st}}{u} \tag{9.72}$$

When the work gear turns about its axis $O_g$ through the angle $\varepsilon_{st}$, the gear shaper cutter turns about its axis of rotation $O_{ct}$ through the corresponding angle $\varepsilon_{st}^c$. Moreover, in the relative motion of the gear shaper cutter with respect the

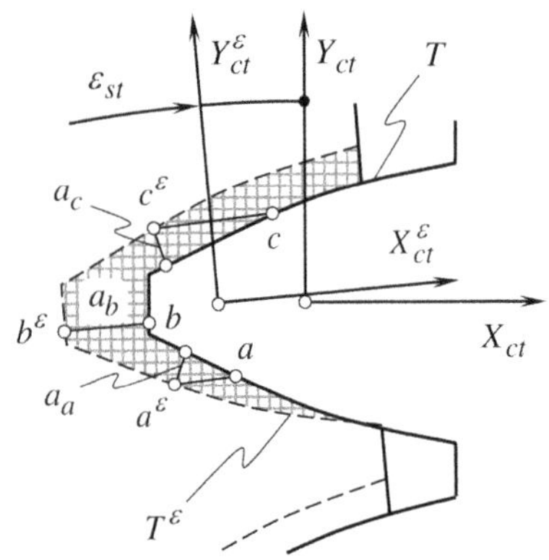


**FIGURE 9.22** Cut thickness in the process of shaping a spur gear by the gear shaper cutter.

stationary work gear, the axis of rotation $O_{ct}$ turns about the axis of rotation of the gear $O_g$ through the angle $\varepsilon_{st}$.

The operator of rotation $\mathbf{Rt}[\varepsilon_{st}^c, Z_c]$ is used for the analytical description of the rotation of the gear shaper cutter about the axis $O_{ct}$. Then, for the analytical description of the rotation of the gear shaper cutter about the axis $O_g$ through the angle $\varepsilon_{st}$, the operator of rotation $\mathbf{Rt}(\varepsilon_{st}, Z_g)$ is used. Ultimately, translation at the center distance $C$ along the $X_g$ – axis is analytically described by the operator of translation $\mathbf{Tr}(C, X_g)$.

In the case under consideration, the resultant linear transformation is analytically described by three operators of elementary transformations, namely, $\mathbf{Rt}[\varepsilon_{st}^c, Z_c]$, $\mathbf{Rt}(\varepsilon_{st}, Z_g)$, and $\mathbf{Tr}(C, X_g)$. It is not necessary to use three operators of linear transformations—that is, the operators $\mathbf{Rt}[\varepsilon_{st}^c, Z_c]$, $\mathbf{Rt}(\varepsilon_{st}, Z_g)$, and $\mathbf{Tr}(C, X_g)$—to simplify the derivation of the required equations; instead, only the operator of rolling[4] $\mathbf{Rr}_u(\varepsilon_{st}, Z_g)$ is used (see Figure 9.22). Ultimately, use of the operator of rolling $\mathbf{Rr}_u(\varepsilon_{st}, Z_g)$ enables the transition from the coordinate system $X_{ct}Y_{ct}Z_{ct}$ in its initial location to that same coordinate system in its current location, which is designated as $X_c^{\varepsilon}Y_c^{\varepsilon}Z_c^{\varepsilon}$.

Consider a point of interest within the cutting edge of the gear shaper cutter tooth. This might be a point $a$ within the lateral cutting edge of the gear shaper cutter tooth, a point $b$ within the top cutting edge, or a point $c$ within the lateral cutting edge of the opposite side of the tooth profile (see Figure 9.22). To specify the coordinates of the point of interest, use of the local Cartesian coordinate system $X_{ct}Y_{ct}Z_{ct}$ is often helpful.

For a given configuration of the generating surface $T$ of the gear shaper cutter, the position vector $\mathbf{r}_a$ of the point of interest $a$ is known (The same is valid with respect to any other point of interest within the cutting edge). Equation (9.8) allows for calculation of the unit normal vector $\mathbf{n}_{T.a}$ to the surface $T$ at point $a$. When shaping a gear by the gear shaper cutter, the pitch cylinder of the gear shaper cutter rolls without sliding over the pitch cylinder of the work gear. For the consequent stroke of cutting, the work gear turns about its axis of rotation through the roll angle $\varepsilon_{st}$ and the gear shaper cutter turns about its axis of rotation through the corresponding roll angle $\varepsilon_{st}^c$.

The new position of the gear shaper cutter is designated in Figure 9.22 as $T^{\varepsilon}$. Under the rolling motion of the pitch cylinders, the point of interest $a$ moves to the position $a^{\varepsilon}$. The position vector $\mathbf{r}_a^{\varepsilon}$ of the point of interest in the new configuration of the gear shaper cutter can be expressed in terms of the position vector $\mathbf{r}_a$ and the operator of rolling $\mathbf{Rr}_u(\varepsilon_{st}, Z_g)$, as follows:

$$\mathbf{r}_a^{\varepsilon} = \mathbf{Rr}_u(\varepsilon_{st}, Z_g) \cdot \mathbf{r}_a \tag{9.73}$$

Similarly, the following expression:

$$\mathbf{n}_{T.a}^{\varepsilon} = \mathbf{Rr}_u(\varepsilon_{st}, Z_g) \cdot \mathbf{n}_{T.a} \tag{9.74}$$

is valid for the unit normal vector $\mathbf{n}_{T.a}^{\varepsilon}$ to the generating surface $T^{\varepsilon}$ through the point $a^{\varepsilon}$.

Cut thickness is measured in the direction perpendicular to the machined surface of the work gear. Its value varies within the cutting edge of the gear shaper cutter. At point $a^{\varepsilon}$, the thickness of cut is denoted by $a_a$. Corresponding designations $a_b$ and $a_c$ are used to denote thickness of cut at points $b^{\varepsilon}$ and $c^{\varepsilon}$, respectively. The cut thickness of x $a_a$ is equal to the length of the straight-line segment between point $a^{\varepsilon}$ and the surface $T$. The point of interception of the straight line $\mathbf{a}_a(t)$ through the point $a^{\varepsilon}$ along the unit normal vector $\mathbf{n}_{T.a}^{\varepsilon}$ (shown but not labeled in Figure 9.22) is calculated from the expression:

$$\mathbf{a}_a(t) = \mathbf{r}_a^{\varepsilon} + t \cdot \mathbf{n}_{T.a}^{\varepsilon} \tag{9.75}$$

where $t$ is a parameter of the straight line $\mathbf{a}_a(t)$.

Solving the equation for the generating surface $T$ of the gear shaper cutter (note that Eq. (9.9) describes only one lateral flank of the surface $T$; a similar equation is valid for the opposite flank of the tooth of the surface $T$; and the equation for the top land of the surface $T$ is a trivial one), together with Eq. (9.75), returns the position vector $\hat{\mathbf{r}}_a^{\varepsilon}$ of the point of interception of the straight line $\mathbf{a}_a(t)$ with the generating surface $T$. This immediately yields a formula:

$$a_a = \left| \mathbf{r}_a^{\varepsilon} - \hat{\mathbf{r}}_a^{\varepsilon} \right| \tag{9.76}$$

for the calculation of the cut thickness at the point of interest within the cutting edge $CE$ of the gear shaper cutter tooth.

Equation (9.76), used to calculate for the cut thickness at a point of interest within the cutting edge $CE$, is derived with no simplifications, and, thus, it is accurate. This equation can be used when the point of interest is a far remote point from the pitch point $P$ in the gear-shaping operation. In this case, the chip thickness is greater. The equation is

[4] See Appendix D for details on the operator of rolling $\mathbf{Rr}_u(\varepsilon_{st}, Z_g)$, as well as on operators of rolling along and about other axes of a reference system.

also valid when the point of interest is close to the pitch point $P$, in which case the cutting edge cuts a very thin chip. The chip thickness can be in the range of or close to the roundness of the cutting edge.

For approximate calculations, simpler equations are known. However, either the simplified equations are applicable for remote portions of the cutting edges and, thus, are not applicable for the cutting edges close to the pitch point, or, conversely, they are applicable for the points of the cutting edge close to the pitch point and therefore are not applicable for the remote portions of the cutting edges of the gear shaper cutter.

Equation (9.76) can be used to calculate all the other parameters of the portion of the stock that has been removed by a cutting edge of the gear shaper cutter—that is, the shape of the section of the portion of the stock, its area, and so forth.

Cut thickness is an important issue for a gear manufacturer because it significantly affects the wear of the gear shaper cutter teeth.

The cutting edges of the gear shaper cutter teeth cut chips of different thickness.

First, cut thickness depends on the distance between the point of interest and the pitch point. The greater the distance between these two points, the larger the thickness of cut, and vice versa.

Second, different cutting edges of the gear shaper cutter tooth cut chips of varying thickness. The top cutting edge cuts the thickest chips. The thickness of the chip that is cut by the lateral cutting edge of the entering side of the tooth profile is larger than that of the recessing side of the gear shaper cutter tooth profile.

The lateral cutting edges of the gear shaper cutter teeth cut the stock mostly within the distance between the pitch point in the gear machining mesh and the point where the cutting begins. A small amount of the stock is also cut beyond the pitch point. The top cutting edge does not cut the stock beyond the pitch point.

A detailed analysis of cut thickness is beyond the scope of the book.

## 9.9 ACCURACY OF GEARS CUT BY A GEAR SHAPER CUTTER

The accuracy of gears cut by a gear shaper cutter is discussed here from the standpoint of fulfillment or violation of the necessary conditions of proper PSG. Fulfillment of the set of necessary conditions of proper PSG is a must when machining gears with the gear shaper cutter (See Appendix E for details on conditions of proper PSG).

When shaping a gear with a conventionally designed gear shaper cutter, most of the conditions of proper PSG are met, and it is usually not necessary to verify whether they have been fulfilled. In particular cases of gear machining, only the fifth and the sixth conditions of proper PSG are required in a more in detail analysis.

### 9.9.1 Fulfillment of the Fifth Condition of Proper Part Surface Generation

To fulfil the fifth condition of proper PSG, no intersection of the adjacent portions of the tooth profile of the generating surface $T$ of the gear shaper cutter must be observed. When shaping a gear with the gear shaper cutter, the pitch cylinder of the gear shaper cutter is rolling without sliding over the pitch cylinder of the work gear. In such kinematics of gear machining, violation of the fifth conditions of proper PSG may occur.

The teeth of the gear being machined can be considered as a certain number $N_g$ of portions of round cylinder surface (root lands, as well as top lands), and the corresponding number of pairs $N_g$ of screw involute flank surfaces $\mathscr{G}$ for each of the gear teeth. Therefore, the generating surface $T$ of the gear shaper cutter is composed of portions of the cylinders of revolution ($N_{ct}$ root lands and $N_{ct}$ top lands) and the corresponding number of the screw involute surfaces. When shaping the gear, the generating surface $T$ makes contact with different portions of the gear tooth surface $\mathscr{G}$.

Various configurations of adjacent portions of the tooth profile of the generating surface of the gear shaper cutter are observed. The adjacent portions of the generating surface $T$ of the gear shaper cutter tooth profile can:

- Share no common points with each other—that is, they can be apart from each other, or
- Be connected to each other at the endpoints, or
- Intersect each other.

The first two conditions are out of interest from the perspective of fulfillment or violation of the fifth condition of proper PSG. No violation of the fifth condition of proper PSG occurs in the first two cases. In the third condition, adjacent portions of the gear shaper cutter tooth profile intersect each other. Because of the intersection, the fifth condition of proper PSG is violated. Ultimately, this results in transient curves on the gear tooth profile.

For example, when generating the involute gear tooth profile, the generating surface $T$ of the gear shaper cutter is used. The desired tooth profile of the generating surface $T$ is composed of three arc segments (see Figure 9.22):

- The left-side lateral arc segment $a_l^* b_l^*$.
- The top land circular arc segment $a_l^{**} a_r^{**}$.
- The right-side lateral arc segment $a_r^* b_r^*$.

The arc segments intersect each other, as shown in Figure 9.23, and the points $f_l^*$ and $f_r^*$ are the points of the intersection. Therefore, the portions $a_l^* f_l^*$ and $a_l^{**} f_l^*$, as well as the portions $a_r^* f_r^*$ and $a_r^{**} f_r^*$ of the gear shaper cutter tooth profile, do not exist physically. As a consequence, the portions $a_l c_l$ and $a_r c_r$ of the gear tooth profiles $a_l b_l$ and $a_r b_r$ (see Figure 9.23) are substituted with the transient curves

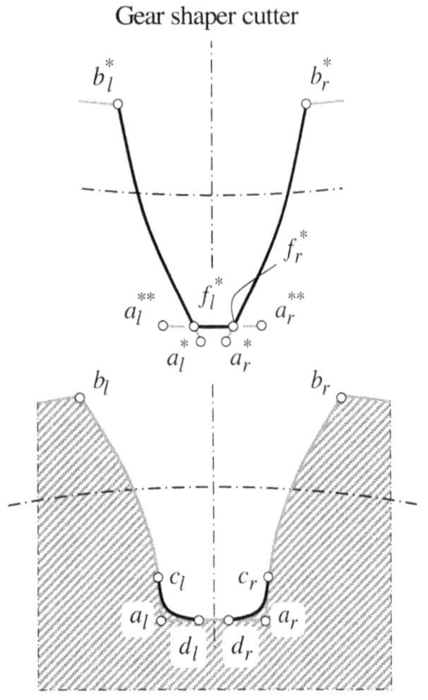


FIGURE 9.23 Violation of the fifth necessary condition of proper PSG in gear machining with the gear shaper cutter (transient curves are observed).

$c_l d_l$ and $c_r d_r$, which form the fillets.[5] For the same reason, the portions $a_l d_l$ and $a_r d_r$ of the cylinder root land are also replaced with the transient curves $c_l d_l$ and $c_r d_r$.

To fulfil the fifth condition of proper PSG, it is necessary to keep the transient curve within the space between the circle of root diameter $d_{f.g}$ and the circle of limit diameter $d_{l.g}$. This criterion can be stated analytically as:

$$\Delta \le 0.5(d_{l.g} - d_{f.g}) = [\Delta] \tag{9.77}$$

When Eq. (9.77) is fulfilled, the fifth condition of proper PSG is fulfilled as well, and thus the gear can be machined with the gear shaper cutter in compliance with the blueprint specification. Therefore, fulfillment of Eq. (9.77) is vital for the calculation of the desired design parameters of the gear shaper cutter.

Eq. (9.77) can also be used for the determination and graphical interpretation of a zone of variation of the gear shaper cutter profile angle $\phi_t$—that is, the zone within which fulfillment of the fifth condition of proper PSG is guaranteed.

Consider the generation of the tooth profile of the work gear by the gear shaper cutter (see Figure 9.24). The left-hand side of the work gear tooth profile is generated along the active portion of the left-hand side line of action $LA_l$. Similarly, the right-hand side of the gear tooth profile is generating within the active potion of the right-hand side line of action $LA_r$. The tooth profile of the gear shaper cutter does not project beyond the root circle of diameter $d_{f.g}$ of the work gear. Because of this, certain portions of the active lines of action $LA_l$ and $LA_r$ cannot be represented physically in the gear-machining operation. Therefore, within these portions of the lines of action, transient curves of a certain height $\Delta$ are generated.

[5] Reminder: The start of active profile (*SAP*) is the intersection of the limit diameter and the involute profile (see Figure 9.24). The points $c_l$ and $c_r$ within the circle of limit diameter $d_{l.g}$ are the points of SAP.

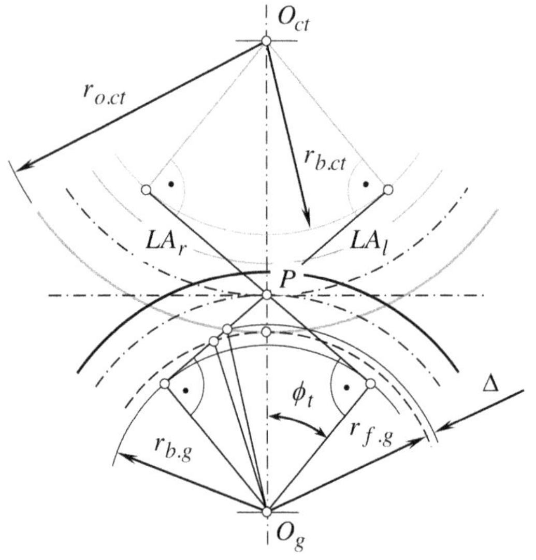


FIGURE 9.24 Gear-to-gear-shaper cutter meshing diagram.

For the calculation of the height $\Delta$ of the transient curve, a corresponding equation can be derived.

The performed analysis of the gear-to-gear-shaper cutter meshing diagram (see Figure 9.24) allows us to compose equations for the lines of action $LA_l$ and $LA_r$, as well as an equation for the root circle of a diameter $d_{f.g}$. To calculate for the actual value of deviation $\Delta$, these equations yield the formula:

$$\Delta = 0.5 \cdot \left( \sqrt{d_{o.c}^2 + 4 \cdot C^2 - 8 \cdot d_{o.c} \cdot C \cdot \cos(\text{inv}\,\phi_{t.o})} - d_{f.g} \right) \tag{9.78}$$

where:

- $d_{o.ct}$ is the outer diameter of the gear shaper cutter
- $C$ is the center distance in the work gear-to-gear-shaper cutter mesh
- $\phi_{t.o}$ is the transverse tooth profile angle at the outer diameter ($d_{o.ct}$) of the gear shape cutter $[\phi_{t.o} = \cos^{-1}(d_{b.c}/d_{o.c})]$
- $d_{b.ct}$ is the base diameter of the gear shaper cutter
- $d_{f.g}$ is the root diameter of the work gear

Informative graphical interpretation of fulfillment or violation of the fifth condition of proper PSG in gear-shaping operation can be drawn up from the analysis of the work gear-to-gear-shaper cutter meshing diagram.

The maximum value of the deviation $\Delta$ is limited to its maximal permissible value $[\Delta]$. This means that the interval $0 \le \Delta \le [\Delta]$ for the permissible value of the deviation $\Delta$ can be established. This inequality can be expressed in terms of the design parameters of the gear shaper cutter and the design parameters of the work gear.

To keep the deviation $\Delta$ within the allowed interval $0 \le \Delta \le [\Delta]$, the transvers profile angle $\phi_{t.c}$ of the gear shaper cutter teeth has to meet the inequality (see Figure 9.24):

$$\phi_{t.c}^{\min} \le \phi_{t.c} \le \phi_{t.c}^{\max} \tag{9.79}$$

Considered together, the inequality $0 \le \Delta \le [\Delta]$ and Eqs. (9.78) and (9.79) allow for a graphical interpretation of the allowed zone, within which the fifth condition of proper PSG is fulfilled (similar to the allowed zone shown in Figure 8.21 that is constructed for the case of planing gears with the rack cutter). The allowed zone can be constructed for the shaping of spur or helical gear of any design.

The fulfillment of the fifth condition of proper PSG must be verified for three cases:

- New gear shaper cutter (when the distance $a$ to the nominal plane section of the gear shaper cutter is of the maximal value $a = a_{\max} > 0$).
- Partially worn gear shaper cutter (when the distance $a$ to the nominal plane section of the gear shaper cutter is zero $a \cong 0$).
- Completely worn gear shaper cutter (when the distance $a$ to the nominal plane section of the gear shaper cutter is the minimal $a = a_{\min} < 0$).

The accuracy of the shaped gear can be increased if the design parameters of the gear shaper cutter correlate with the corresponding design parameters of the work gear. This is a way to meet the fifth condition of proper PSG.

### 9.9.2 Fulfillment of the Sixth Condition of Proper Part Surface Generation

When machining a gear by a gear shaper cutter, the discrete type of generation of the gear tooth flank rather than a smooth profile is observed. The machined gear tooth flank is shaped in the form of numerous curved cuts, each of which is tangent to the desired involute tooth form. Cusps are inevitable on the gear tooth surface that is machined with the gear shaper cutter.

The cusps height on the machined gear tooth flank must be kept within the tolerance for accuracy of the tooth flank in order to satisfy the sixth condition of proper PSG (see Appendix E). The cusp height can also be interpreted as the deviation of the actual gear tooth flank from the desired tooth surface. From this perspective, the cusps result in low accuracy of the machined gear.

To conclude whether the sixth condition of proper PSG is fulfilled, determination of the actual cusp height is required.

Consider a tooth of the work gear being machined by the gear shaper cutter (see Figure 9.25). At an arbitrary point of the gear tooth profile, say at point $i$, the cutting edge of the gear shaper cutter is tangent to the involute profile of the gear tooth. On the consequent stroke, the cutting edge of the gear shaper cutter is tangent to the gear tooth profile at the adjacent point $(i+1)$. The perpendiculars to the involute tooth profile at points $i$ and $(i+1)$ form the angle that is equal to the roll angle per stroke $\varepsilon_{st}$ of the gear shaper cutter. Ultimately, the desired smooth involute gear tooth profile is substituted with a polygonal shape with curved cusps on its profile. The deviation of the actual gear tooth profile from the desired tooth profile is denoted as $h_{ss}$. The deviation $h_{ss}$ is measured along the perpendicular to the desired gear tooth flank that passes through the point of intersection of the $i-$ th and of the $(i+1)$-th cuts.

As the roll angle per stroke $\varepsilon_{st}$ of the gear shaper cutter is of a small value, the corresponding portions of the contacting tooth profiles can be approximated by circular arc segments through points $i$ and $(i+1)$. To calculate for the deviation $h_{ss}$, use of the circular arc segments through points $i$ and $(i+1)$ is convenient. Only the length of the circular arc segment is of importance from the standpoint of solving the problem under consideration.

To calculate for deviation $h_{ss}$, Eq. (8.14) from [151, 158] (see Figure 9.26):

$$h_{ss} \cong \frac{\varepsilon_{st}}{r_i^c} \cdot R_i \cdot [R_i - r_i^c] \tag{9.80}$$

can be applied.

In Eq. (9.80), a current value of the radius of curvature $R_i$ of the gear tooth profile is calculated from the equation:

$$R_i = \frac{\varepsilon_i + \varepsilon_{i+1}}{2} \cdot r_{b.g} \tag{9.81}$$

To calculate for the corresponding radius of curvature $r_i^c$ of the gear shaper cutter tooth profile, the equation (see Figure 9.26):

$$r_i^c \cong (r_{b.g} + r_{b.c}) \cdot \tan \phi_{t.c} - R_i \tag{9.82}$$

applies.

The sixth condition of proper PSG is fulfilled if and only if the inequality:

$$h_{ss} \le [h_{ss}] \tag{9.83}$$

is valid.

## 9.10 APPLICATION OF GEAR SHAPER CUTTERS

Cylindrical gears of any design can be cut by the gear shaper cutter. In particular, shaper cutters are used for machining of shoulder gears, cluster gears, gear rings for internal gear pairs, and so forth. Machining cylindrical gears of small modules with a large tooth count, a narrow face width, and a big helix angle is especially efficient when using gear shaper cutters.

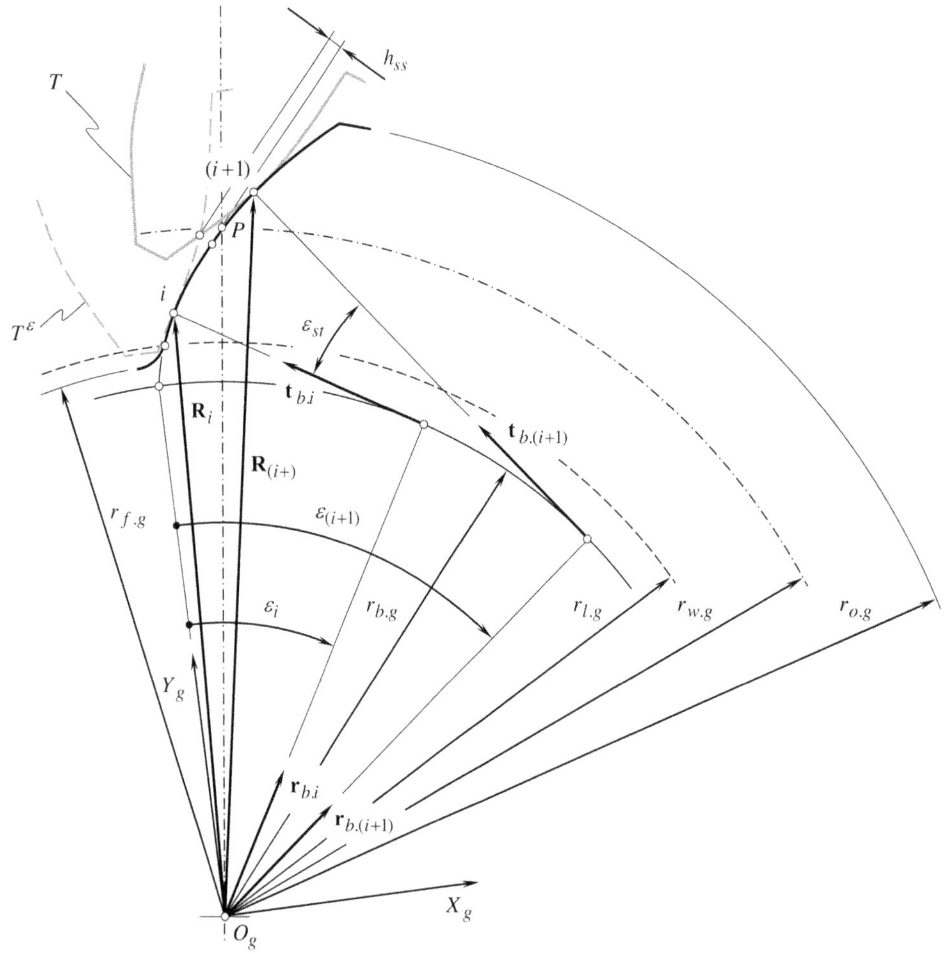


**FIGURE 9.25** Principal elements of the discrete generation of the gear tooth profile by the gear shaper cutter.

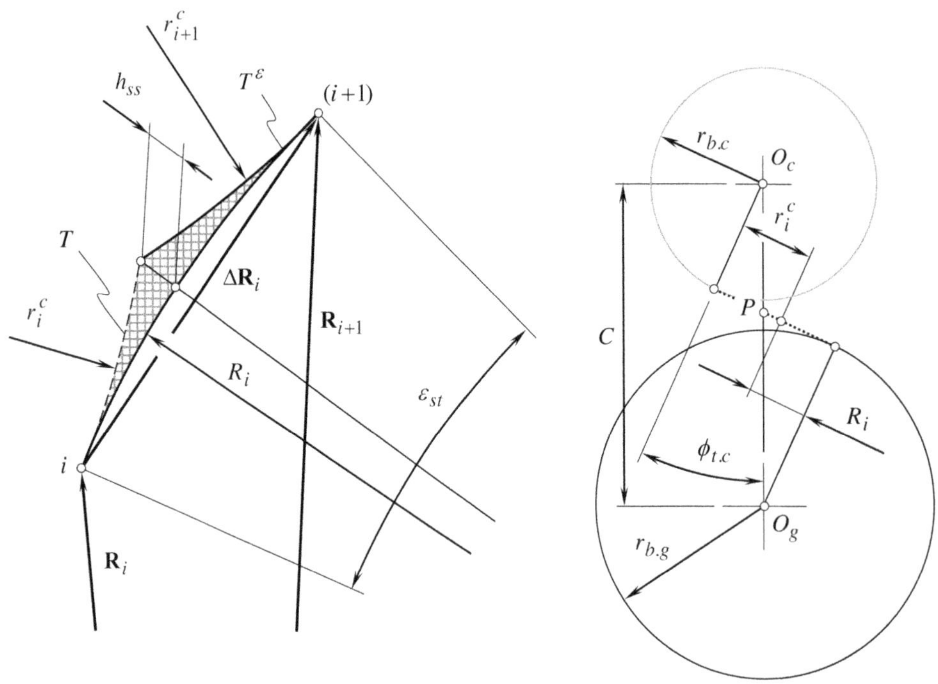


**FIGURE 9.26** Geometry of an elementary cusp on the machined gear tooth profile.

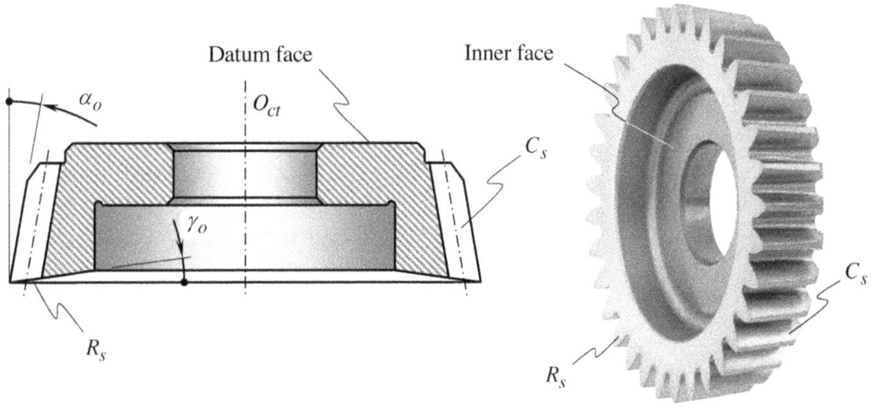


**FIGURE 9.27** A typical disk-type gear shaper cutter.

## 9.10.1 Design of Gear Shaper Cutters

Various types and sizes of gear shaper cutters are available in the market. Commonly, two basic designs of gear shaper cutters are used for machining external gears: the disk and the deep counter-bore types of gear shaper cutters.

Disk gear shaper cutters are the most commonly used cutters for machining of external gears and splines. Figure 9.27 shows a typical gear shaper cutter of this design. (The shaper cutters shown in Figure 9.27 [as well as in Figure 9.29] are referred to as Fellows gear shaper cutters for machining spur gears). This is the common type of gear shaper cutters that is normally installed directly on the cutter spindle (see Figure 9.28). Disk-type gear shapers are also used on adapters, especially when the necessary size is required between a disk and taper shank cutter.




**FIGURE 9.28** Shaping a spur gear with a disk-type gear shaper cutter.

Deep counter-bore type gear shaper cutters are similar to disk-type ones, except that the blank thickness is increased to make the room for holding the nut or screw above the cutter's lifeline (see Figure 9.29). Deep counterbore-type cutters are normally used for cutting shoulder gears and cluster gears (see Figure 9.30), as well as for machining of internal gears.

Gear shaper cutters are made of high-speed-steels containing molybdenum and tungsten. They are heat treated to a Rockwell C hardness of 64 to 65. A nitriding surface treatment further improves their wear resistance.

Titanium nitride coating is often applied to increase permissible feeds and speeds on modern shaping machines. Gear shaper cutters coated with titanium nitride increase productivity by as much as 30% without decrease tool life.

It is common to assign the rake angle at the top cutting edge of the gear shaper cutter equal to $\gamma_o = 5°$. Roughing gear shaper cutters sometimes have a rake angle as high as $\gamma_o = 10°$.

The clearance angle at the top cutting edge of the gear shaper cutter is commonly equal to $\alpha_o = 6°$. The sides of the cutter have a side clearance of about $\alpha_f = 2°$.

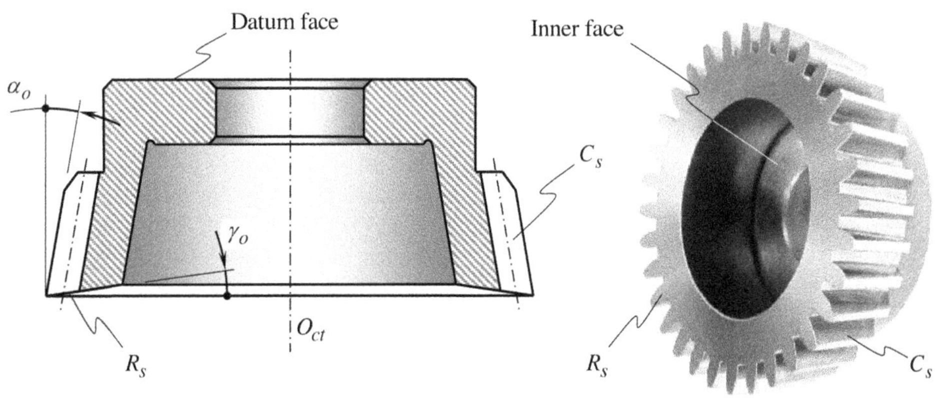


**FIGURE 9.29** A typical counter-bore type gear shaper cutter.

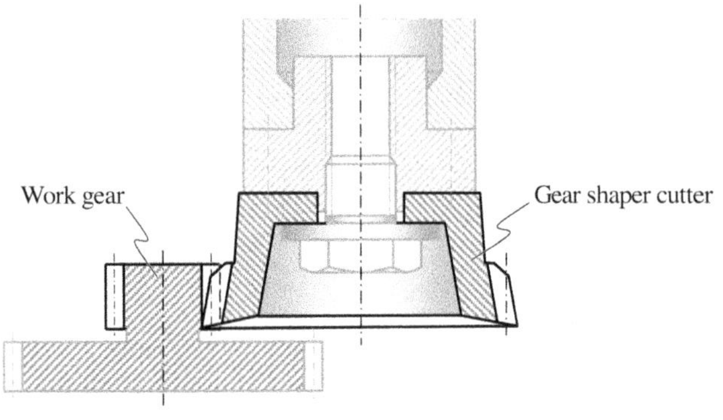


**FIGURE 9.30** Shaping a spur gear with a counter-bore type gear shaper cutter.

The rake angle $\gamma_o$ and the clearance angles $\alpha_o$ and $\alpha_f$ are all necessary to make the gear shaper cutter an efficient metal-cutting tool.

Commercial-grade gear shaper cutters for spur gears with $\phi = 20°$ profile angle are usually available from retailers. High-accuracy precision ground cutters for finishing operations and ground cutters for pre-shaving, pre-grinding, and roughing operations are special-order items. Gear shaper cutters are made in two quality classes: class A cutters are for finishing and pre-shaving and class B cutters are for roughing and pre-grinding. Gear shaper cutters for finishing work are usually made to a very high degree of precision.

In many high-production jobs, it is desirable to design the gear shaper cutter for a particular gear so that the cutter will do exactly what is wanted throughout its life. This is particularly true when cutters are used to pre-shave cut a gear and leave an undercut. The position of the undercut must remain constant within close limits to tie in with the gear shaving cutter design.

### 9.10.2 Special Features of a Gear Shaper Cutter Tooth Profile

Standard gear shaper cutters are designed to produce a true involute without modifications to the gear tooth profile. Often, gear design mandates the use of modified profiles to improve performance. Gear shaper cutters can be ordered with various profile modifications.

There is a variety of special features that can be provided—and are frequently needed—in gear shaper cutter teeth. Figure 9.31 illustrates six different special features.

a. **Root fillet:** Root fillet radius modification consists of blending the corners of the tool tip with a radius that produces a controlled fillet in the root corners of the generated teeth [see Figure 9.31(a)]. A root fillet increases bending strength of the gear teeth and improves fatigue life. Tool wear is also improved, as a sharp corner would chip off very quickly.
b. **Tip relief:** Tip relief cutters remove slightly more material near the top of the gear tooth [see Figure 9.31(b)]. Tip relief compensates for interference between mating profiles and usually results in reduced gear noise. In high-speed gears, it is often desirable to minimize noise and tip bearing resulting from tooth deflection under heavy loads. For the tooth profile modification of this type, implementation of the method [68] proposed by Prof. S.P. Radzevich can be helpful.[6]
c. **Semi-topping** (or chamfer): The root area of semi-topping gear shaper cutters is modified at an angle to create a chamfer at the tip of the generated gear teeth [see Figure 9.31(c)]. The root of the cutter is filled in to generate a sharp corner break of chamfer on the tips of the gear, which minimizes tip buildup during heat treatment due to nicks incurred during handling. This feature is very helpful in minimizing damage due to nicks and burrs, which could result in gear noise, during part handling.
d. **Full topping:** The gear shaper cutter tooth is ground equal to the whole depth (*WD*) of the gear tooth [see Figure 9.31(d)]. The outside diameter of the gear is topped to size when the teeth are cut. Modification of this type is used more frequently for fine pitch gearing.
e. **Protuberance:** The gear shaper cutter tooth profile is built up on the cutter tip to provide an undercut near the root of the gear being generated [see Figure 9.31(e)]. The undercut in the gear root area acts as a clearance. The undercut is intended to facilitate subsequent finishing operations. Protuberance cutters are frequently applied for gears that are finished by shaving or grinding.
f. **Modified pressure angle:** The gear shaper cutter tooth profile is ground to a slightly lower pressure angle to provide for a constantly increasing amount of stock from root to tip of the gear generated [see Figure 9.31(f)]. The tooth profile modification provides relief for subsequent finishing operations.

In particular cases of gear machining, customized tooth profile modifications can also be used.

### 9.10.3 Gear Shaper Cutters for Machining Helical and Herring-Bone Gears

Helical and double-helical gears can be cut with helical gear shaper cutters (see Figure 9.32), provided that the cutter has an appropriate helix angle and the shaping machine has a helical guide of appropriate lead to twist the cutter as it strokes back and forth. Each helix angle requires its own special helical guide. A helical gear shaper cutter and its guide form a set of tooling required to cut a specific helical gear. Two helical cutters and two guides are required to cut a pair of meshing helical gears (left- and right-hand gears).

[6] Pat. No. 1017444 (USSR), *A Gear Cutting Tool*/S.P. Radzevich, Int. Cl. B23F 21/16, Filed: April 26, 1982.

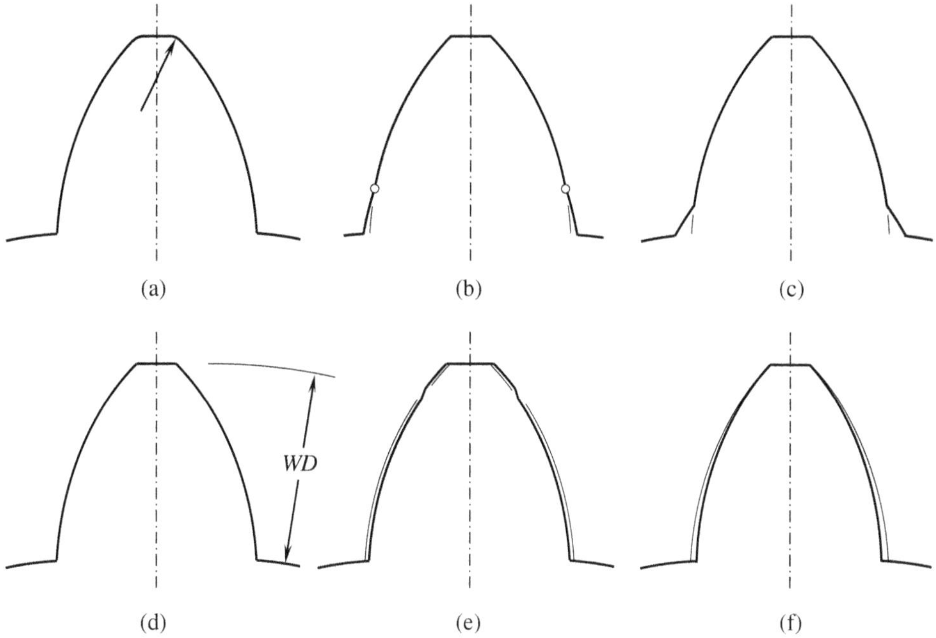


**FIGURE 9.31** (a–f) Special features of the gear shaper cutter tooth profile.

The helix angle that a gear shaper cutter produces depends on both the lead of the guide and the number of cutter teeth. The helix of the cutter must also agree with the helix angle being cut, or serious cutter rub will occur. The formula for the relation of cutter teeth to lead of guide is:

$$\frac{\text{No. of teeth in cutter}}{\text{Lead of guide}} = \frac{\text{no. of teeth in gear}}{\text{lead of gear}} \tag{9.84}$$

Ordinarily, helical gears are cut with gear shaper cutters with a rake face normal to the helix of the cutter tooth. If the helix angle is low, the transverse section could serve as the rake face of the gear shaper tooth. Otherwise, gear shaper cutters of the design shown in Figure 9.33 are used. In this design, the rake surface of each tooth is perpendicular to the pitch helix of the cutter.

Helical shaper cutters are used to machine helical involute gears. A few designs of helical gear shaper cutters are illustrated in Figure 9.34. The shaper cutters designs shown in Figure 9.34 are referred to as Fellows helical gear shaper cutters. The configuration of the rake plane $R_s$ (see Figure 9.35) is the key feature of Fellows helical gear shaper cutters. (An alternative design of helical shaper cutter, the so-called Sykes helical gear shaper cutter, is briefly discussed in Chapter 22).

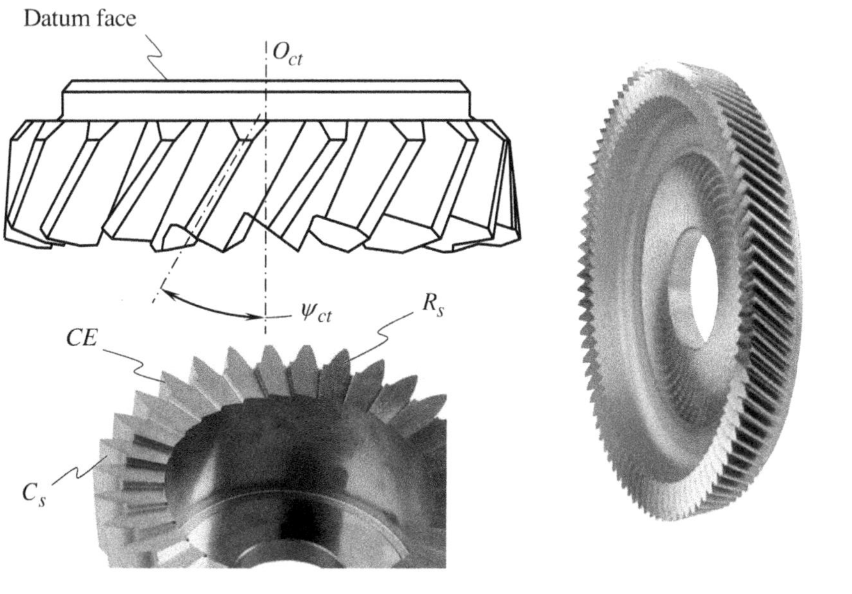


**FIGURE 9.32** Design features of a helical gear shaper cutter.

FIGURE 9.33 Gear shaper cutter for machining helical and herring-bone gears.

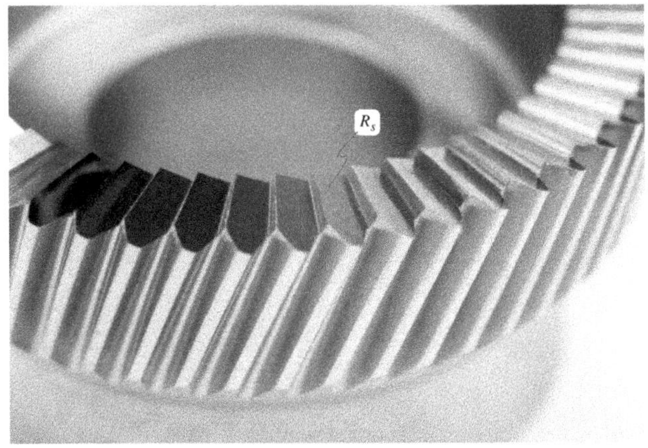


FIGURE 9.35 Configuration of the rake plane $R_s$ in the design of fellows helical gear shaper cutters.

FIGURE 9.34 Numerous examples of design of helical shaper cutters for machining helical involute gears.

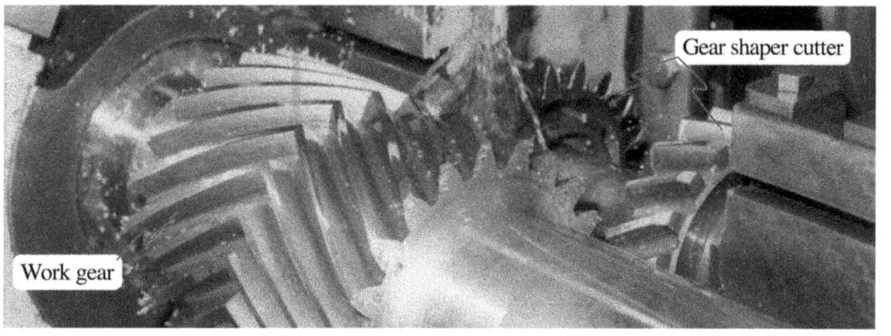


**FIGURE 9.36** Shaping of a herring-bone gear by gear shaper cutters working together.

Helical gear shaper cutters generally require a wider recess at the bottom of the cut for full cutter runout.

Gear shaper cutters do not usually cut the same whole depth throughout their life. A gear shaper cutter can be designed with a front clearance angle that has such a relation to the side clearance angle that a certain number of teeth may be cut to an exact depth and thickness for the life of the cutter. However, if this cutter is used to cut gear of substantially larger or smaller numbers of teeth, the whole depth cut will vary slightly from the design value. If a gear shaper cutter designed to cut an external gear is used to cut an internal gear of the same tooth thickness, the discrepancy in whole depth may be quite appreciable. In many cases, a designer of a gear shaper cutter does not know all the numbers of teeth that the cutter may have to cut during its life. This leads the designer to make the front clearance angle large enough that the cutter usually cuts a little extra on the whole depth. It is commonly reasoned that the problem of a little extra depth is less than the problem of having the depth too shallow.

Herring-bone gears of the continuous-tooth type must be cut by a pair of gear shaper cutters working together, as illustrated in Figure 9.36. To make the cutting match from both sides, it is necessary to use a cutter with the top face ground normal to the cutter axis. This makes the top face angle $\gamma_o = 0°$, and it gives one side of the cutter tooth an acute angle and the other side an obtuse angle. These features do not aid the cutting action of the tool, but they are necessary to produce the continuous tooth. A special sharpening technique has produced a good cutting edge, even on the obtuse-angle side.

### 9.10.4 Gear Shaper Cutters of Special Designs

Possible designs of gear shaper cutters are not limited to the disk-type shaper cutters and deep counterbore-type gear shaper cutters just discussed. Many other designs of gear shaper cutters are developed for industrial applications. In this section, only a few designs of gear shaper cutters are briefly discussed. Moreover, they could be of importance for practical application—those chosen to be considered for designs of gear shaper cutters are interesting from a scientific perspective.

#### 9.10.4.1 Gear Shaper Cutters with an Improved Geometry of Cutting Edge

An investigation into the geometry of the rake surface of gear shaper cutters was conducted by Dr. Kau [30] and Dr. Ngok [50]. Their research resulted in a novel design of gear shaper cutter with an optimal value of the rake angle $\gamma_o^{opt}$ (see Figure 9.37). The gear shaper cutter features the rake surface $R_s^{cmb}$ of a combined geometry.

The proposed gear shaper cutter design is developed on the premise of conventional its design. The rake surface $R_s$ and the clearance surface $C_s$ form the cutting wedge of the conventionally designed gear shaper cutter. The rake angle of the gear shaper cutter is designated as $\gamma_o$. The actual value of the rake angle $\gamma_o$ is small and is too far from to be equal to its optimal value, $\gamma_o^{opt}$. Because the inequality $\gamma_o << \gamma_o^{opt}$ is observed, the tool life of conventionally designed gear shaper cutters is far from optimal.

The proposed improvements in the design of the gear shaper cutter incorporate the tolerance [δ] on the accuracy

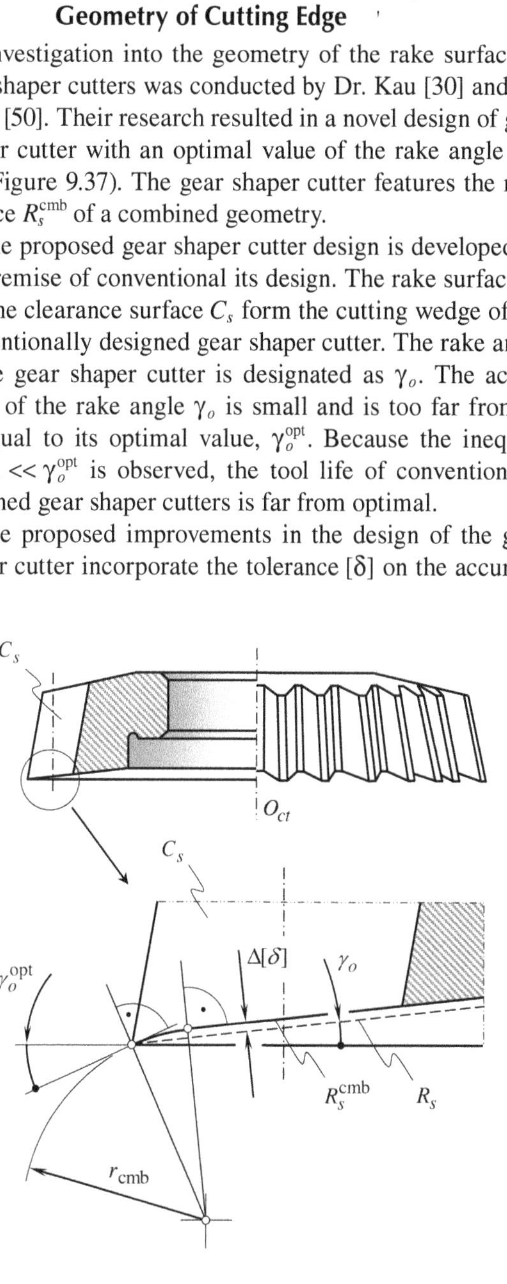


**FIGURE 9.37** Gear shaper cutter with the rake surface $R_s$ of a combined geometry.

of the rake face $R_s$ of the gear shaper cutter. Any and all deviations in the shape of the surface $R_s$ within the tolerance [δ] are permissible.

In particular cases, the tolerance [δ] can be tightened to a certain value $[\delta^*] < [\delta]$. This enables a considerable improvement in the geometry of the rake surface. The available portion $\Delta[\delta] = [\delta] - [\delta^*]$ of the tolerance [δ] can be used for controllable deviation of the actual rake surface from its initial form $R_s$. The modified rake surface is referred to as the combined rake surface of the gear shaper cutter and is designated as $R_s^{\text{cmb}}$.

The rake surface $R_s^{\text{cmb}}$ is composed of the conical portion blended with the portion of torus surface. The torus surface is generated by a circle of a certain radius $r_{\text{cmb}}$. If the available portion Δ[δ] of the tolerance [δ] is known, the value of the radius $r_{\text{cmb}}$ can be expressed in terms of the optimal rake angle $\gamma_o^{\text{opt}}$ and the portion Δ[δ] of the tolerance [δ]:

$$r_{\text{cmb}} = \frac{\Delta[\delta]}{1 - \cos \gamma_o^{\text{opt}}} \tag{9.85}$$

Reasonable corrections to the calculated value of the radius $r_{\text{cmb}}$ are permissible.

The cutting performance and the tool life of the gear shaper cutter with an optimal value of the rake angle are significantly higher compared to that of conventionally designed gear shaper cutters.

Improvements to the geometry of the clearance surface of the gear shaper cutter are constricted more severe.

The insufficiently small clearance angle $\alpha_f$ is among a few major factors that significantly limit the cutting performance of the gear shaper cutters. The analysis of the kinematics of a gear-shaping operation [177] reveals that an opportunity for the improvement of the geometry of the clearance surface is available. First of all, it is desirable to increase the clearance angle $\alpha_f$ at the entering side of the gear shaper cutter tooth profile.[7]. For this purpose, the clearance surfaces of the gear shaper cutter teeth are constructed from different base cylinders. In this way, the clearance angle $\alpha_f$ at the entering cutting edge is slightly increased; however, the clearance angle $\alpha_f$ at the opposite cutting edge is slightly reduced. Ultimately, reasonable changes (that is, the redistribution) in clearance angles at the lateral cutting edges result in an improved cutting performance of the gear shaper cutter. It should be pointed out here that no changes occur in the base cylinder of the generating surface of the gear shaper cutter. It retains its original value before the recommended changes are implemented in the geometry of the clearance surfaces.

The cutting performance of the gear shaper cutter can be significantly increased if the design allows for the elimination of interference of chip flows from the adjacent cutting edges of the gear shaper cutter tooth. Use of sets of two gear shaper cutters set up on a common spindle is a perfect example in this regard.[8] The chip removal is performed by the top cutting edges of one of two gear shaper cutters (lateral cutting edges of this gear shaper cutter do not cut the material), while the second one generates the tooth profile of the work gear with the lateral cutting edges (top cutting edges of this second gear shaper cutter do not cut the material).

Extremely small values of the clearance angle $\alpha_f$ at the lateral cutting edges is the root cause of the insufficient cutting performance and low tool life of conventionally designed gear shaper cutters. Clearance surfaces of the set of two gear shaper cutters shown in Figure 9.38 are designed so as to make it possible to increase the clearance angle $\alpha_f$ at one side of the tooth profile of one of the gear shaper cutters and an increased clearance angle $\alpha_f$ at the opposite side of the tooth profile of the other gear shaper cutter. Ultimately, the working cutting edges of both lateral sides of the tooth profile of the gear shaper cutter set have an increased clearance angle $\alpha_f$ [107].

The top cutting edges and lateral cutting edges of one side of the tooth profile of the lower gear shaper cutter cut the stock. Similarly, the top cutting edges and lateral cutting edge of the opposite side of the tooth profile of the upper gear shaper cutter cut the rest of the stock. The greater clearance angle $\alpha_f$ at the lateral cutting edges results in a significant increase in the productivity rate of machining gears by the set of two gear shaper cutters, compared to cutting gears with conventionally designed gear shaper cutters.

To perform the regrinding, the worn set of two gear shaper cutters must be disassembled. Next, each of the gear shaper cutters can be ground separately. Either the rake faces $R_s$ (a conventional approach) or the clearance surfaces $C_s$ (an advanced approach) can be ground up. In regrinding gear shaper cutters, grinding clearance surfaces requires less material to be removed. This process causes less reduction of the tooth width compared to grinding over the rake face. In certain cases, grinding on clearance surfaces of the gear shaper cutters is preferred.

After being reground, the gear shaper cutters should be re-assembled on the common spindle. Because the teeth of reground gear shaper cutters get narrower, this in turn requires the shaper cutters to be at a certain angle in relation to each other. In the design of a set of two gear shaper cutters, shown in Figure 9.38, the required rotation of the gear shaper cutters with respect to each other is performed using specially designed methods.

Key slots on the upper gear shaper cutter are made at a certain angle φ to one another. The key slots can have an even circumferential distribution, but this is not required. Key slots on the lower gear shaper cutter are at a certain angle θ to one another. The angular pitches φ and θ are

[7] Pat. No. 1641535 (USSR), *A Gear Shaper Cutter.*/S.P. Radzevich and S.N. Medveditskov, Int. Cl. B23F 21/10, Filed: June 26, 1988.

[8] Pat. No. 1039657 (USSR), *A Gear Shaper Cutter.*/S.P. Radzevich, Int. Cl. B23F 21/10, Filed: March 16, 1982; and: Pat. No. 1172660, (USSR), *A Gear Shaper Cutter.*/S.P. Radzevich, Int. Cl. B23F 21/10, Filed: February 13, 1984.

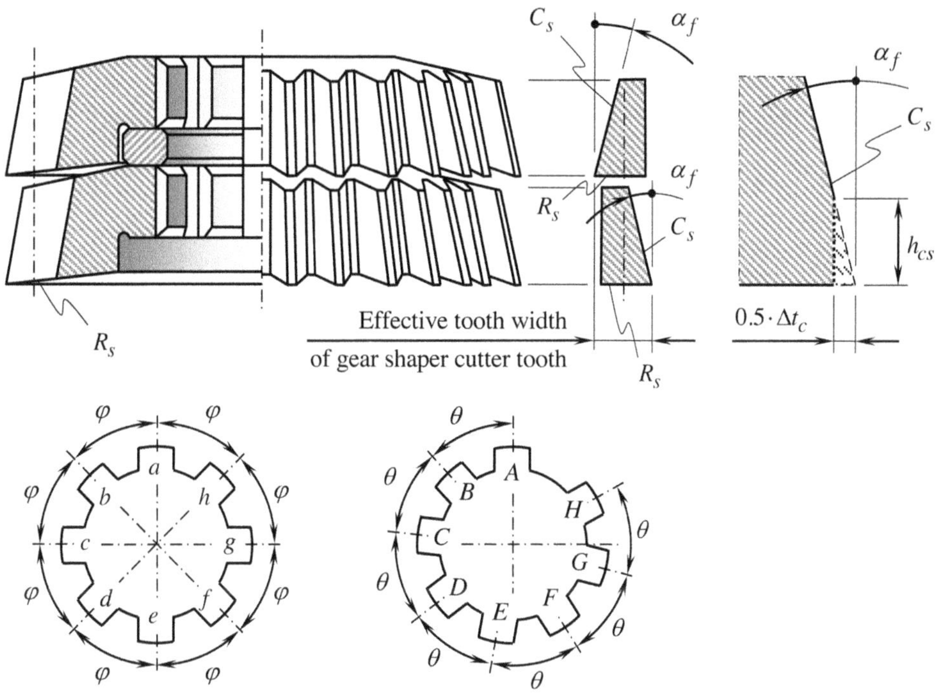


**FIGURE 9.38** A set of two gear shaper cutters: Design i. (From Radzevich, S.P., USSR pat. 1039657, int. Cl. B23F 21/10, March 16, 1982.)

slightly different from each other. The difference between the angular pitches is calculated from the following formula:

$$\varphi - \theta = 2 \cdot \sin^{-1}\left(\frac{\Delta t_{ct}}{d_{ct}}\right) \tag{9.86}$$

where $\Delta t_{ct}$ denotes the reduction of tooth thickness of the gear shaper cutter (see Figure 9.38), and $d_{ct}$ is the pitch diameter of the gear shaper cutter.

First, the key slots labeled *a* and *A* are used for assembling a new gear shaper cutter. After the first regrinding of the worn gear-cutting tool, the slots *b* and *B* are used for assembling the gear shaper cutter. After the second regrinding of the worn gear-cutting tool, the slots *c* and *C* are used for assembling the gear shaper cutter, and so forth. Because of the difference $(\varphi - \theta)$ between the angular pitches of the key slots, the upper and lower portions of the two-cutter set turn relative to each other through the angle $\varphi - \theta$.

A result similar to that just discussed can be achieved by implementing the set of two gear shaper cutters shown in Figure 9.39. The upper portion of the gear-cutting tool has only one key slot [109], while the lower portion of the

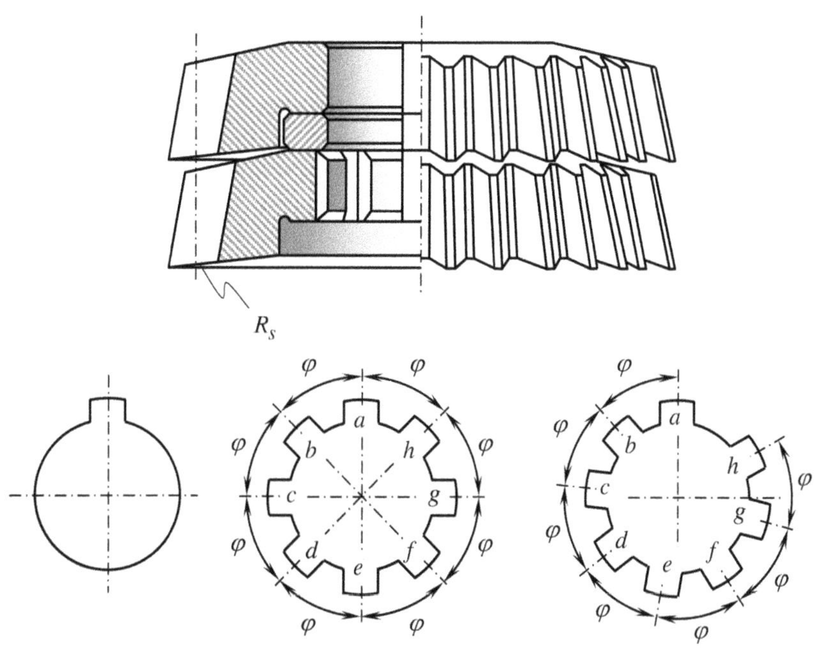


**FIGURE 9.39** A set of two gear shaper cutters: Design II. (From Radzevich, S.P., USSR pat. 1172660, int. Cl. B23F 21/10, Feb. 13, 1984.)

gear-cutting tool is designed with multiple key slots. The key slots of the lower portion of the gear-cutting tool are at a certain angular pitch φ. The angle φ is a function of the angular pitch $2 \cdot \pi / N_{ct}$ of the gear shaper cutter teeth:

$$\varphi - \frac{2 \cdot \pi \cdot n}{N_{ct}} = 4 \cdot \cos\frac{\varphi}{2} \tag{9.87}$$

where $n$ denotes a reasonable integer number.

The gear-cutting tool determined in this manner enables precise turns through an angle σ of the upper portion of the gear shaper cutter relative to its lower portion through the angle:

$$\sigma = \frac{4 \cdot h_{cs}}{d_{ct}} \cdot \tan\alpha_f \tag{9.88}$$

Sufficiency in just one key slot at the upper portion provides an advantage over the gear-cutting tool shown in Figure 9.38. However, timing of the angular pitches φ and $2 \cdot \pi / N_{ct}$ could be a challenge.

A correlation between the angular pitch of the gear shaper cutter teeth and the space between the angular pitch of the slots is advantageous, as it provides the possibility of increasing the accuracy of the gear-cutting tool, as well as increasing the total number of permissible re-grindings of the gear shaper cutter.

Implementation of the gear-cutting tools (see Figure 9.38 and Figure 9.39), which are composed of two gear shaper cutters, make it possible to apply the modified gear-cutting diagram that eliminates the interference of the chip flows from the adjacent cutting edges of conventionally designed gear shaper cutters.

#### 9.10.4.2 Precision Gear Shaper Cutters

The possibility of increasing accuracy is strongly restricted by the permissible values of the rest of the design parameters of the gear shaper cutter. As a result, any attempt to remedy this problem is especially valuable.

A design of precision gear shaper cutter can be developed on the premise of the implementation of the rake surface $R_s$ that is generated via rotation of the lateral cutting edge $CE$ of a corresponding geometry. This is constructed as the line of intersection of the generating surface $T$ of the gear shaper cutter by the actual clearance surface $C_s$ rather than as the line of intersection of the generating surface $T$ of the gear shaper cutter by the rake surface $R_s$. The cutting edge $CE$ that is constructed as the line of intersection of the surfaces $T$ and $C_s$ is rotated about the gear shaper cutter axis $O_{ct}$. Ultimately, the rake surface $R_s$ is shaped in the form of a surface of revolution, which is generated by means of the previously constructed cutting edge $CE$.

In compliance with this concept, a design for a precision gear shaper cutter is developed [51]. The gear shaper cutter is illustrated in Figure 9.40. The generating profile of the rake surface $R_s$ is convex for gear shaper cutters with a positive rake angle $\gamma_o$. When the rake angle has a negative value, the generating profile of the rake surface $R_s$ becomes concave.

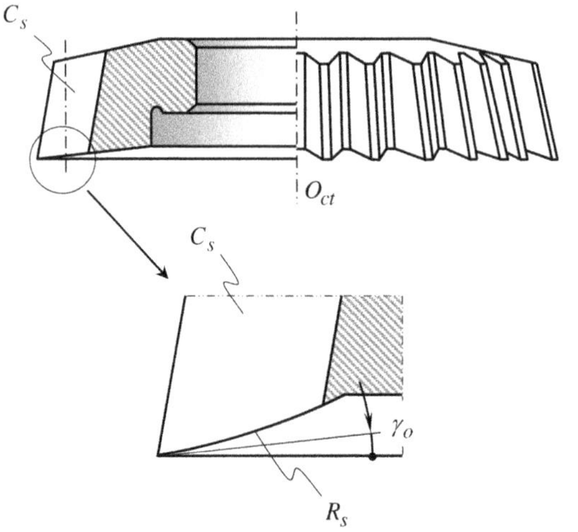


**FIGURE 9.40** Design of a geometrically accurate (precision) gear shaper cutter with a modified rake surface $R_s$.

The disclosed concept can be implemented to increase the accuracy of spur gear shaper cutters alone and is not applicable with respect to helical gear shaper cutters. For helical gear shaper cutters, another approach [73] can be implemented to improve accuracy.

The helical gear shaper cutter[9] features straight lateral cutting edges, as shown in Figure 9.41. The cutting edges $CE$ are aligned with the straight lines that are tangent to the base helices of the generating surface $T$ of the helical gear shaper cutter. Therefore, the lateral cutting edges cross the axis of rotation $O_{ct}$ of the gear shaper cutter at the base helix angle $\psi_{b.ct}$. Tangency of the lateral cutting edges to the base helices of the gear shaper cutter is the essence of the helical shaper cutter design under consideration. For the configuration of the lateral cutting edges that is shown in Figure 9.41, the cutting edges are situated within a common plane that serves as the rake plane $R_s$ of the gear shaper cutter tooth. However, this is not essential—other configurations of the lateral cutting edges are permissible as well. The lateral cutting edges can align with a pair of skew straight lines if this happen to be of importance for a particular application. Numerous designs of precision gear shaper cutters can be developed on the premise of this concept.

The concept of the precision helical gear shaper cutter with straight lateral cutting edges can be enhanced to the area of verification of accuracy of the clearance surfaces[10] of the gear shaper cutter [95].

[9] Pat. No. 956187 (USSR), *A Helical Gear Shaper Cutter*/S.P. Radzevich, Int. Cl. B23F 21/10, Filed: September 9, 1980.

[10] Pat. No. 1710231 (USSR), *A Method to Control the Accuracy of a Helical Gear Shaper Cutter*/S.P. Radzevich, Int. Cl. B23F 21/10, G01B 5/20, Filed: September 13, 1989.

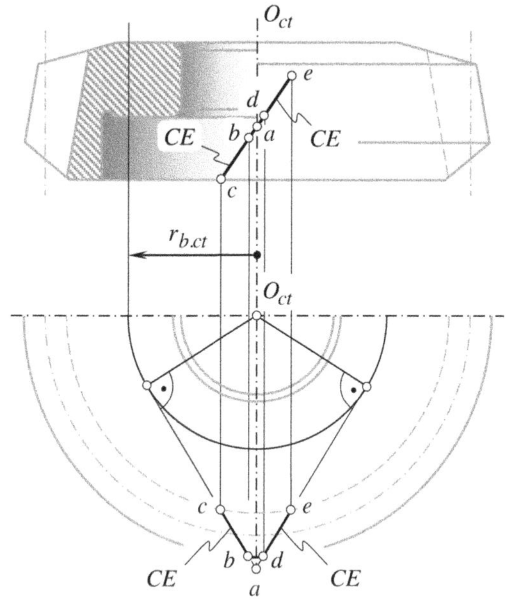


**FIGURE 9.41** The concept of precision helical gear shaper cutter with straight lateral cutting edges. (From Radzevich, S.P., USSR pat. 956187, int. Cl. B24B 3/34, Jan. 12, 1988.)

### 9.10.4.3 Shaper Cutters for Machining Hardened Gears

Shaping hardened gears is a challenging problem for gear manufacturers. The high level of hardness of the tooth flanks of these gears is the major reason that gear manufacturers are faced with significant difficulties in sharpening hardened gears. This largely explains why shaping hardened gears has limited application in the industry to date.

Carbide gear shaper cutters are rarely used in normal gear production. The advantages of carbide tools cannot be fully exploited on gear shapers because machines are limited in terms of the maximum reciprocating speed of the cutter spindle. As opposed to hobbing, where the hob material limits the permissible cutting speed, the limiting factor in shapers is the maximum speed of the reciprocating mechanism of the cutter spindle.

Carbide gear shaper cutters are used when it is necessary to machine gears made of hard steel. The design parameters and the geometry of the cutting edge of the carbide shaper cutters are very similar to those used for shaper cutters made of high-speed steel.

For particular cases of semi-finishing or finishing hardened gears, specially designed carbide gear shaper cutters are developed. An example of carbide gear shaper cutters for semi-finishing of hardened gears is depicted in Figure 9.42. Except for the geometry of the rake face $R_s$, the carbide gear shaper cutter is very similar to that of conventionally designed gear shaper cutters. The rake surface $R_s$ of the semi-finishing gear shaper cutter is shaped either in the form of a cone of revolution or an individual plane $R_s$ (facet) for each tooth of the gear shaper cutter.

The rake surface $R_s$ of the semi-finishing gear shaper cutter is at a negative rake angle $\gamma_o = -10°$ to $-25°$. Due to

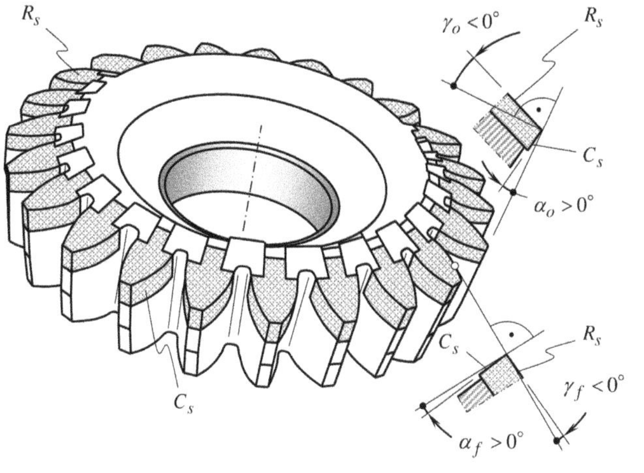


**FIGURE 9.42** Carbide alloy gear shaper cutter of a special design for semi-finishing or finishing hardened gears.

the negative rake angle $\gamma_o < 0°$, the negative rake angles $\gamma_f < 0°$ of corresponding values are created at the lateral cutting edges. These enables the gear shaper cutter teeth to withhold high loads when shaping hardened gears.

The use of cutting elements made of super-hard materials (borazon, for example) proved to be useful for machining hardened surfaces. It is natural to assume that the use of cutting elements made of super-hard materials can solve numerous problems that pertain to machining gears with gear shaper cutters. Unfortunately, the dimensions of the cutting elements are often too small to span over the entire lateral cutting edge of the gear shaper cutter with a standard design. Small dimensions of the cutting elements impose strong constraints in the implementation of the cutting elements made of super-hard materials in the design of gear shaper cutter. This particular problem can be resolved by using gear shaper cutters[11] [108] specifically designed so that the pitch diameter and the outer diameter of the gear shaper cutter are equal to each other (see Figure 9.43). When the equality $r_{w.ct} = r_{o.ct}$ is observed, the lateral cutting edge shrinks to a point [108]. In reality, the generating surface of the gear shaper cutter is reduced to the set of straight lines through the corners of the top cutting edge. The straight lines are parallel to the axis of rotation $O_{ct}$ of the gear shaper cutter. This makes it possible to place a small sized cutting element into that point. In such a case, the gear shaper cutter (see Figure 9.44) can be used for machining a gear of any coarse pitch.

In reality, the gear tooth profile that is machined with the gear shaper cutter (see Figure 9.44) is shaped in the form of an epicycloid when machining external gears and in the form of a hypocycloid when machining internal gears. However, reasonable approximation of the desired involute tooth profile with a cycloidal curve is often possible.

One more advantage of the gear shaper cutter of the design depicted in Figure 9.44 is attributed to the fact that

[11] Pat. No. 1174187 (USSR), *A Gear Shaper Cutter*/S.P. Radzevich, Int. Cl. B23F 21/10, Filed: April 4, 1981.

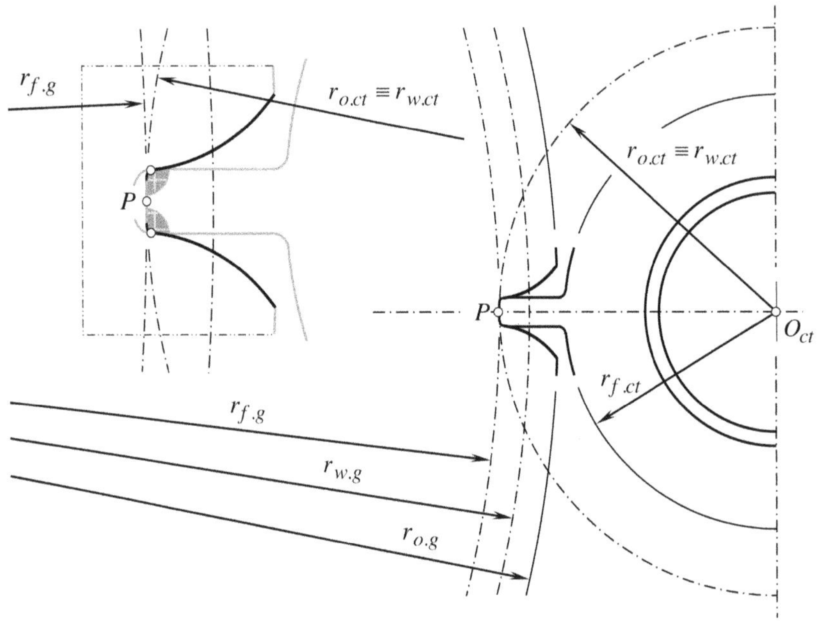


**FIGURE 9.43** Concept of the finishing gear shaper cutter with shortened cutting edges (for superhard material).

its cutting edge geometry does not depend on the gear tooth profile to be machined. Because of this, the gear shaper cutter can be designed with the optimal values for the rake angle and the clearance angle. The design of the gear shaper cutter (see Figure 9.44) does not entail problems with the kinematics of grinding of the rake surface $R_s$ and the clearance surface $C_s$.

### 9.10.5 Typical Gear-Shaping Operations

Spur and helical gears can be cut by a pinion-type gear shaper cutter. Either internal or external gears can be cut. Parts running from less than 1 mm to over 3000 mm (1/16 to over 120 in.) may be shaped. Relatively wide face widths may be cut, but in certain cases, shaping will not handle as much face width as the hobbing process.

Furthermore, with the advent of the modern cutter spindle back-off type machines, it is not uncommon to see stroking rates of 1000 strokes/minute for face width 25 mm (1 in.) or smaller. In fact, for narrow face widths of 6 mm (1/4 in.), high-speed shapers capable of stroking rates of more than 2000 strokes/minute are in use.

Gear shaping is quite advantageous for parts with narrow face width. In hobbing, it takes time for the hob to travel into and out of the cut. For helical gears, the speed the hob travels must be increased proportionally. In shaping, there is a minimum of overtravel for spur gears, and this overtravel does not increase for helical gears.

The stroke length in the gear-shaping process can be calculated as the sum of:

- The gear face width.
- The length of approach.
- The overrun length.

The gear face width is a value that is read directly from the part blueprint. The approach and the overrun are the only requirements to ensure safe cutting conditions. As a general rule, 15% should be added to the face width to account for approach and overrun distance (see Figure 9.45).

In designing gears to be shaped, it is necessary to machine a groove as deep as the gear tooth at the end of the face width for runout of the cutter. Gear shaper cutters need only a small amount of cutter-runout clearance at the

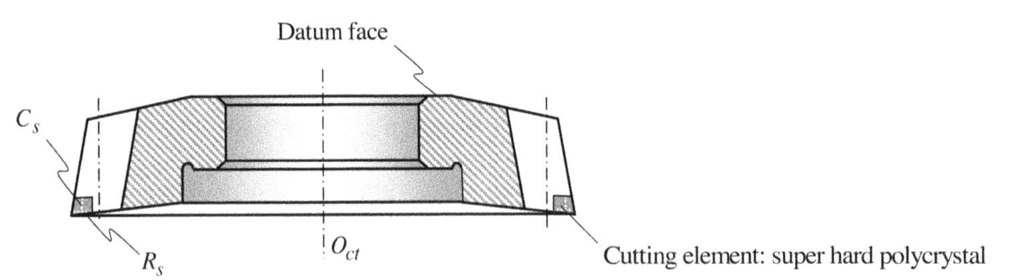


**FIGURE 9.44** Schematic of the finishing gear shaper cutter with cutting elements made out of a super-hard polycrystal.

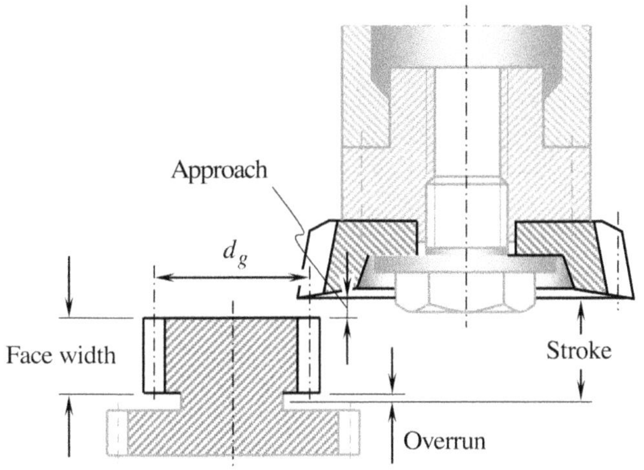

**FIGURE 9.45** Stroke length in the gear-shaping operation.

end of the cut or stroke. Shaped teeth may be located close to shoulders.

Where it may be impossible to hob a cluster gear because of insufficient hob-runout clearance, the same gear can be readily shaper-cut. One *NC* shaper could conceivably cut a cluster gear in one setup, depending on gear data. Double helical gears can be cut with very narrow gaps between helices. In fact, one gear shaper cutter design can actually produce a continuous double helical tooth.

Several types of gear-shaping processes are practically used in the industry for machining cluster gears [21].

It is likely that shaping a spur gear is an example of one of the simplest gear-shaping processes. However, even in a simple gear-shaping operation, special care should be taken when cutting low-tooth-count gears (see Figure 9.46) of coarse pitch, as in this case the torques acting against the work gear and against the gear shaper cutter are not always balanced. Thus, reliable clamping of these components must be ensured.

**FIGURE 9.46** Shaping a low-tooth-count spur gear.

*Combination Shaping*: Two gears with the same number of teeth can be cut simultaneously by two gear shaper cutters mounted on the same arbor [see Figure 9.47(a)]. The tooth forms may differ either in pressure angle or tooth thickness, but they must have the same number of teeth. This setup requires only half the stroke length of a single cutter operation. Machining time is, therefore, reduced by 50%.

*Cutting Two Gears with the Same Diametral Pitch*: This a typical *NC* gear-shaping operation, where two gears with same diametral pitch and same pressure angle are cut in sequence during the same cycle [see Figure 9.47(b)]. The machining parameters are automatically adjusted for each gear by *NC* control so that both gears are produced in optimum cutting conditions. It is essential for this application that one gear shaper cutter design is suitable for gears.

*Cutting Two Gears with Different Diametral Pitches*: When the gears have different diametral pitches, the same gear cutter obviously cannot be used. The solution is to mount two gear cutters on the same arbor, as shown in Figure 9.47(c).

*Cutting Two Gears with Up and Down Shaping:* The cluster gear shown in Figure 9.47(d) can be cut in one cycle by mounting two gear shaper cutters back-to-back on the same arbor. The upper gear is cut with the downstroke, whereas the lower gear is cut with the upstroke.

The special drives resulted in the development of a new cutting method, sometimes referred to as spiral infeed. This cutting method tends to improve chip loading conditions and more evenly distributes tool wear around the cutter.

*Cutting Three Gears with Different Diametral Pitches*: When three gears have different diametral pitches, two gear shaper cutters mounted back-to-back on the same arbor can be used (see Figure 9.48). The middle and the upper gears are cut by the lower gear shaper cutter. Then, the lower gear is cut by the upper gear shaper cutter.

One possibility for shaping an external gear and an internal gear in single setup on CN gear shaper is illustrated in Figure 9.49.

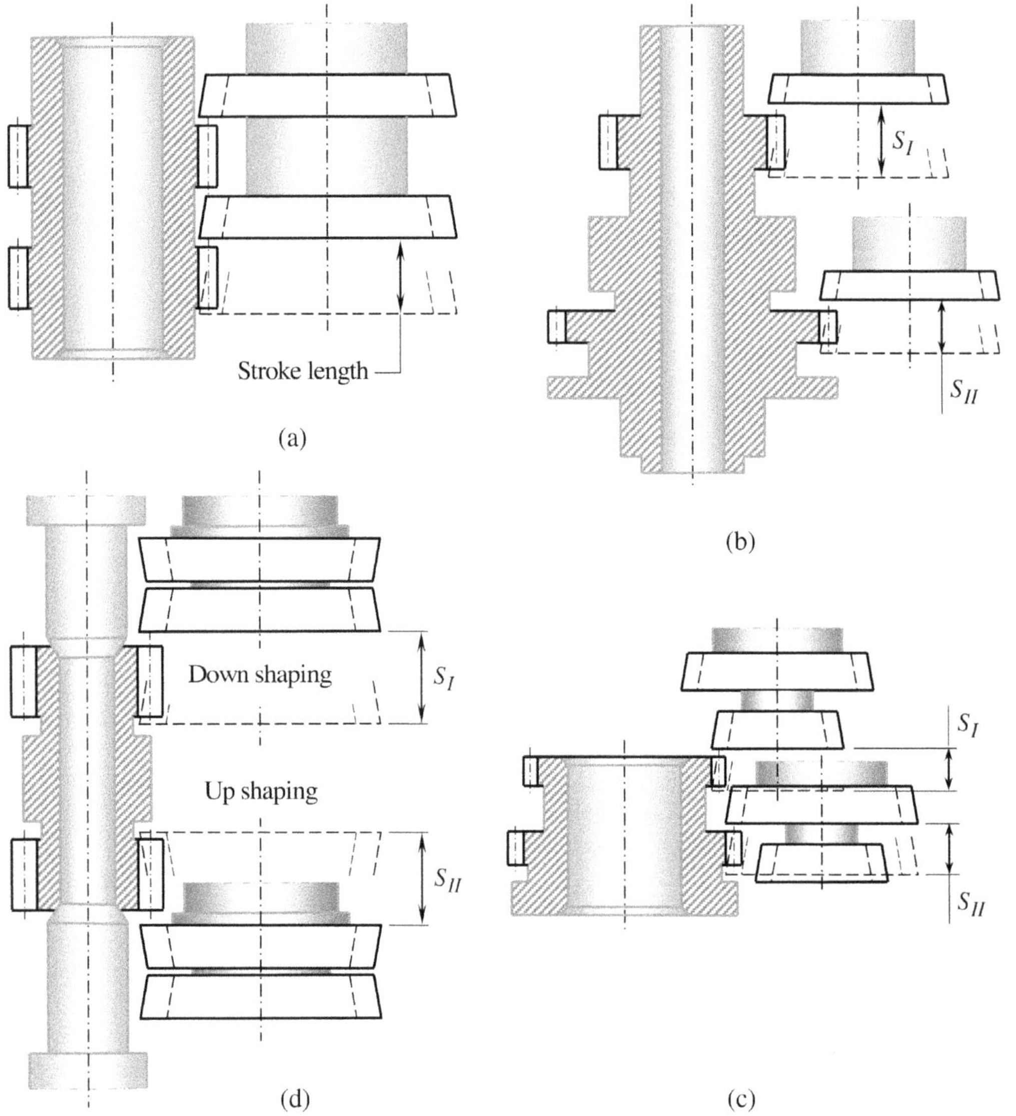


**FIGURE 9.47** Consecutive cutting of two gears [parts (a) through (d) are discussed in detail within the text body of the chapter].

**FIGURE 9.48** Shaping a cluster gear in single setup.

**FIGURE 9.49** Shaping an external and an internal gear in single setup.

### 9.10.6 Grinding of Worn Gear Shaper Cutters

Gear shaper cutter teeth can be ground either on rake surfaces or on clearance surfaces.

#### 9.10.6.1 Grinding Shaper Cutter Teeth on Clearance Surfaces

Clearance surface of involute gear shaper cutter for machining both, spur gears as well as helical gears, is a kind of screw involute surface.

Clearance surfaces $C_s$ are commonly ground in the production of gear shaper cutters. For grinding, the gear shaper cutter is set up on the grinder, as schematically depicted in Figure 9.50. The grinding wheel rotates $\omega_{gw}$ about its axis $O_{gw}$. When grinding the clearance surfaces $C_s$, both the gear shaper cutter and the grinding wheel roll over each other. The relative rolling motion is a superposition of the reciprocation $\mathbf{F}_{rl}$ of the grinding wheel and the rotation $\omega_{rl}$ of the gear shaper cutter. In addition, either the grinding wheel or the gear shaper cutter reciprocates $\mathbf{F}_c$ in the lengthwise direction of the gear shaper cutter tooth. In this way, the entire clearance surface $C_s$ of the gear shaper cutter tooth is ground. The relative rolling motion ($\mathbf{F}_{rl}$ and $\omega_{rl}$) and the reciprocation $\mathbf{F}_c$ should be properly synchronized with each other in order to avoid nonground patches on the surface $C_s$.

The grinding wheel is gradually fed $\mathbf{F}_{gw}$ toward the gear shaper cutter axis of rotation $O_{ct}$.

#### 9.10.6.2 Grinding Shaper Cutter Teeth on Rake Surfaces

Wear to the cutting wedge is usually observed over the clearance surface of the gear shaper cutter teeth. When the extent of the worn or rubbed areas on the cutter tooth tips or flanks reaches about 0.25–0.40 mm (0.010–0/015 in.), they should be resharpened to remove the worn surface by grinding material off the front face. It is preferable to grind the worn gear shaper cutter over clearances surfaces of their teeth. Because of the absence of grinders with appropriate capabilities, regrinding worn gear shaper cutters on face surfaces is implemented extensively.

Grinding spur gear shaper cutters is performed either on specially designed grinders or on conventional grinders.

A standard gear shaper cutter is an example of the Fellows gear shaper cutter for machining spur gears.

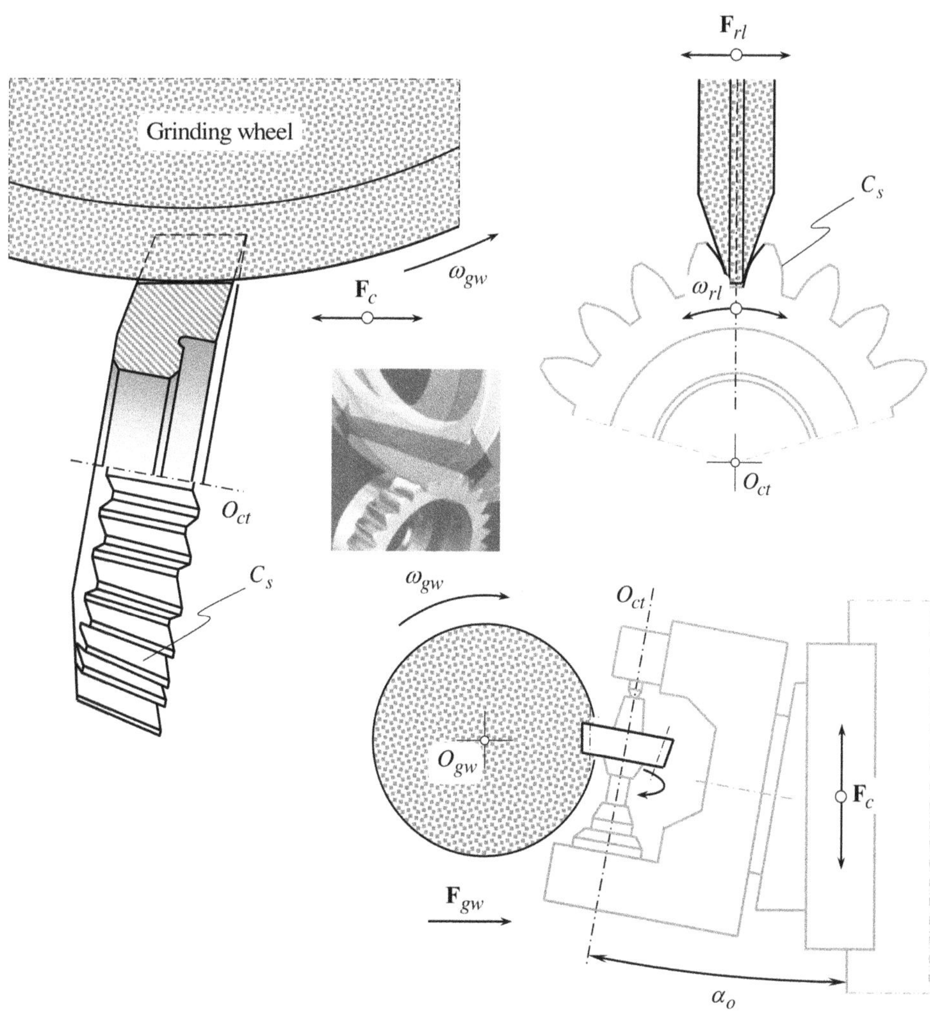


FIGURE 9.50 Grinding a gear shape cutter on the clearance surface $C_s$.

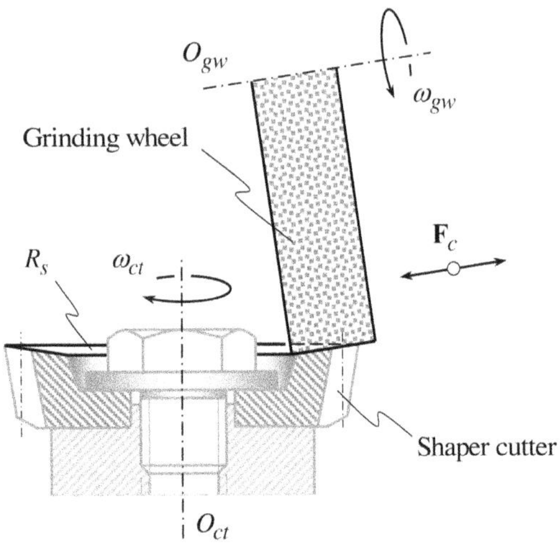


**FIGURE 9.51** Kinematics of sharpening of standard gear shaper cutter for machining spur gears.

When grinding the rake surface $R_s$ (see Figure 9.51), the gear shaper cutter rotates about its axis $O_{ct}$. The rotation $\omega_{ct}$ of the gear shaper cutter serves the feed motion. The grinding wheel axis of rotation $O_{gw}$ forms an obtuse angle $(90° - y_o)$ with the gear shaper cutter axis of rotation $O_{ct}$. The rotation $\omega_{gw}$ of the grinding wheel creates the primary motion of cutting in the gear shaper cutter grinding operation. When capabilities of the actual grinder allow, it is advisable to reciprocate $\mathbf{F}_c$ the grinding wheel along the straight generating line of the conical rake surface $R_s$.

Because the rake angle $y_o$ at the top cutting edges of standard gear shaper cutters is small ($y_o \cong 6°$), no violation of the third necessary condition of proper PSG is observed when grinding gear shaper cutters on the rake surface. However, this problem could become critical when grinding gear shaper cutters of special design, such as the precision gear shaper cutters shown in Figure 9.40.

Depending on features of the design, two approaches are used for sharpening helical gear shaper cutters. The first approach is applicable for grinding gear shaper cutters for machining helical gears, and the second is used for grinding gear shaper cutters for machining herringbone gears.

The gear shaper cutter shown in Figure 9.52 is an example of a Fellows gear shaper cutter for machining helical gears.

Each rake plane $R_s$ of the helical gear shaper cutter (see Figure 9.52) can be ground separately. Indexing of the gear

**FIGURE 9.52** Schematic of sharpening a helical involute gear shaper cutter.

FIGURE 9.53 An alternative arrangement of re-sharpening process of helical involute fellows gear shaper cutter for machining helical gears.

shaper cutter after grinding the rake plane is required in this method. Accumulated wear of the grinding wheel reduces the accuracy of the ground helical gear shaper cutter (Figure 9.53).

To eliminate the negative impact of the accumulated wear of the grinding wheel onto the accuracy of the gear shaper cutter, a continuous-indexing method of resharpening of helical gear shaper cutters is used.[12] When sharpening, the helical gear shaper cutter rotates about its axis of rotation $O_c$ (see Figure 9.54). Continuous rotation of the gear shaper cutter $\omega_{ct}$ makes it possible to transfer from sharpening the rake plane of the gear shaper cutter tooth to sharpening the rake plane of the next tooth of the gear shaper cutter [62].

For sharpening, the grinding wheel with a plane working surface is used. At the instant of generation of the rake plane $R_s$, the working surface of the grinding wheel is congruent to the rake plane $R_s$. The grinding wheel rotates about its axis of rotation $O_{gw}$. The rotation of the grinding wheel serves as the primary (cutting) motion in the gear shaper cutter grinding operation.

Simultaneous to the rotation of the gear shaper cutter, the grinding wheel reciprocates toward the rake surface being ground in a backward direction. It is desirable to keep the reciprocation distance of the shortest possible length; however, it must be sufficient for free rotation of the gear shaper cutter when the grinding wheel travels backward.

The grinding wheel slowly approaches the gear shaper cutter. Then, at a very short instant in time, it makes contact with the rake plane $R_s$. After that, it quickly (instantly) moves backward. The speed of this latter motion is much greater than the speed the gear shaper cutter approaches the grinding wheel.

Between every two consequent contacts of the gear shaper cutter tooth and the grinding wheel, the gear shaper cutter turns about its axis $O_{ct}$ through a whole number of

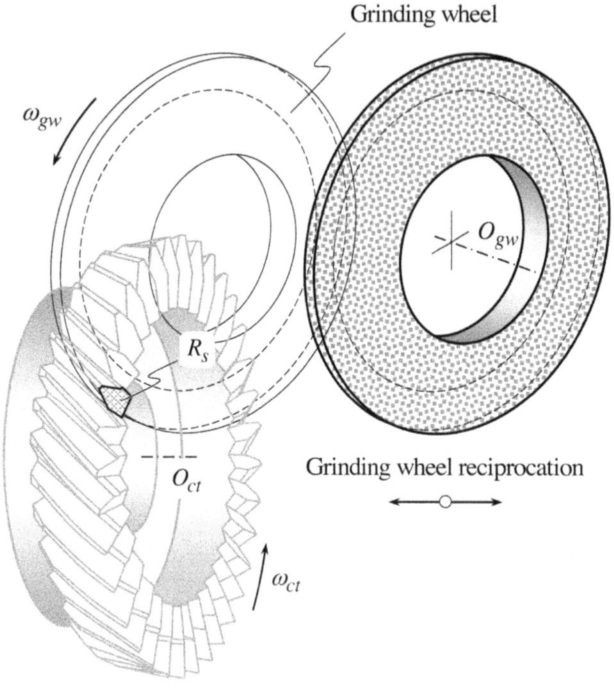


FIGURE 9.54 Schematic of the continuous-indexing method of sharpening helical involute gear shaper cutters. (From Radzevich, S.P., USSR pat. 1604567, int. Cl. B23F 21/10, Sept. 9, 1980.)

teeth for indexing. There must be no common multipliers between the whole number of teeth and the gear shaper cutter tooth number.

The setup parameters of the grinding wheel (see Figure 9.54) are calculated in terms of the design parameters of the helical involute gear shaper cutter to be ground and the design features of the grinder.

When grinding helical gear shaper cutters in compliance to the method shown in Figure 9.54, the impact of the grinding wheel wear onto the accuracy of the ground gear shaper cutter is much smaller. Grinding productivity is significantly higher, primarily because the indexing is continuous.

Herring-bone gears of the continuous tooth type are usually cut by a pair of gear shaper cutters working together. To make the cutting match from both sides, it is necessary to use a cutter with the top face ground normal to the cutter axis. This makes the top rake angle 0°, and it gives one side of the cutter tooth an acute flank rake angle ($\gamma_f < 90°$) and the other side an obtuse angle ($\gamma_f > 90°$). These features do not aid the cutting action of the tool, but they are necessary to produce the continuous tooth.

When sharpening gear shaper cutters for machining herring-bone gears, special care is required. Sharpening is performed with grinding wheels with a small outer diameter [see Figure 9.55(a)]. Making a groove from one side of the gear shaper cutter tooth profile and a corresponding chamfer from the opposite side of the gear shaper cutter tooth profile produces optimal values of the rake angles $\gamma_f$ at both sides of the cutter tooth profile [see Figure 9.55(b)].

[12] Pat. No. 1604567 (USSR), *A Continuous indexing Method of Sharpening of Gear Shaper Cutters*/S.P. Radzevich, Int. Cl. B24B 3/34, Filed: January 12, 1988.

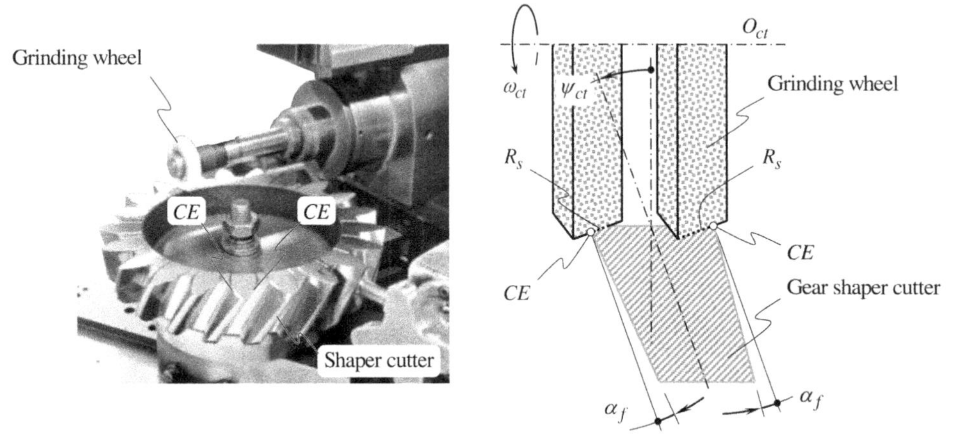


(a) (b)

**FIGURE 9.55** Sharpening a helical gear shaper cutter for machining herring-bone gears—sykes gear shaper cutter: (a) close-up of the sharpening process, and (b) interaction of the grinding wheels with a tooth of the helical gear shaper cutter.

In this way, the cutting performance of the gear shaper cutter can be significantly improved.

As discussed earlier (see Figure 9.16), the cutting performance of the gear shaper cutter can be improved if additional rake surfaces $R_s^f$ are ground next to lateral cutting edges of the shaper cutter teeth. An alternative arrangement for grinding of the additional rake surfaces $R_s^f$ is illustrated in Figure 9.56. The setup parameters of the grinding wheel for these purposes and its design parameters are calculated in terms of the design parameters of the gear shaper cutter to be ground. It is also advisable to leave a narrow land (a portion of the rake surface $R_s$) of even width along the lateral cutting edges of the gear shaper cutter.

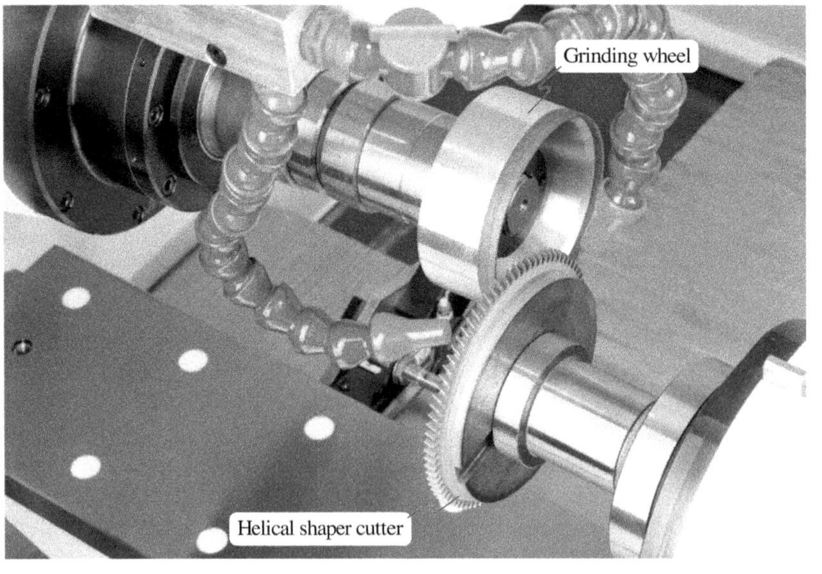


**FIGURE 9.56** An alternative arrangement of the re-sharpening process for helical involute sykes gear shaper cutter for machining herring-bone gears (see Figure 9.16).

# 10 Gear Shaper Cutters II
## *Internal Gear Machining Mesh*

The gear shaper cutter is the primary type of gear-cutting tool for machining internal spur, helical, and herring-bone gears. A close-up of the process of shaping an internal spur involute gear is illustrated in Figure 10.1.

The generating surface of the gear-cutting tool is required to be determined prior to designing a gear shaper cutter for machining a given internal gear. To determine the generating surface of the gear shaper cutter, the internal gear machining mesh applies. The internal gear machining mesh is one of the many possible types of gear machining meshes classified in Figure 3.8.

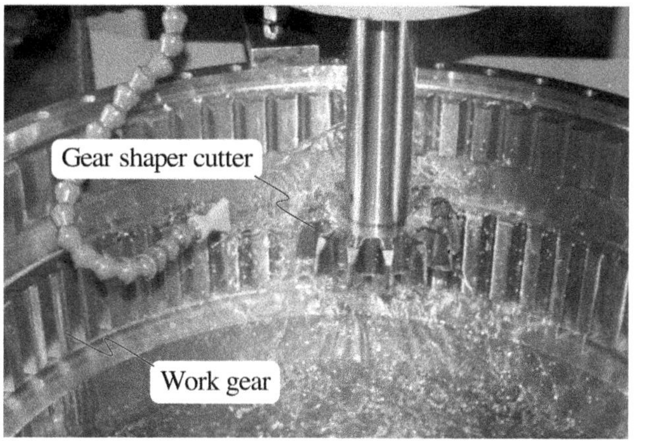


**FIGURE 10.1** Shaping an internal spur involute gear.

### 10.1 KINEMATICS OF SHAPING OF INTERNAL GEAR

The internal gear machining mesh, together with a primary motion (cutting motion) of the gear-cutting tool, comprises the kinematics of shaping operation of an internal gear.

In the internal gear machining mesh, the rotation vector of the work gear $\boldsymbol{\omega}_g$ and that of the gear-cutting tool $\boldsymbol{\omega}_{ct}$ are parallel and positioned at a center distance $C$ from each other. In this type of gear machining mesh, the vectors $\boldsymbol{\omega}_g$ and $\boldsymbol{\omega}_{ct}$ are pointed in the same direction. Under such a scenario, the pitch point $P$ of the gear machining mesh is located not within the center distance $C$, but, instead, it is located outside of it.[1] The vector of instant rotation $\boldsymbol{\omega}_{pl}$ is a vector through the pitch point $P$ and is parallel to the vectors $\boldsymbol{\omega}_g$ and $\boldsymbol{\omega}_{ct}$. An example of configuration of the rotation vectors $\boldsymbol{\omega}_g$, $\boldsymbol{\omega}_{ct}$, and $\boldsymbol{\omega}_{pl}$ is illustrated in Figure 10.2.

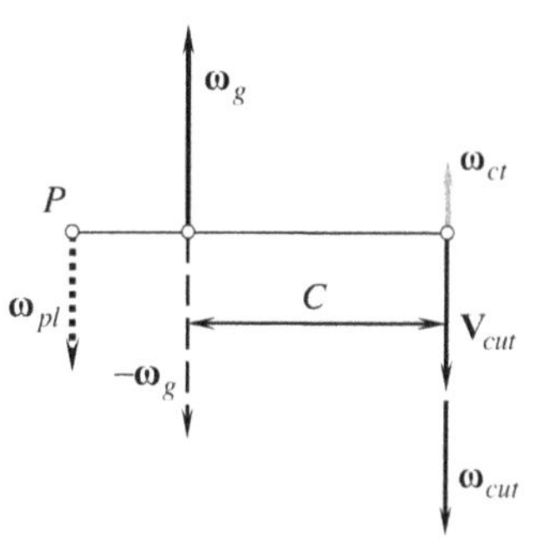


**FIGURE 10.2** Vector diagram of the kinematics of the machining process of internal gear by the gear shaper cutter.

Because in the gear shaping operation (see Figure 10.2) the pitch point is located outside the center distance $C$, the rotation vectors $\boldsymbol{\omega}_g$ and $\boldsymbol{\omega}_{ct}$ are not allowed to be of the same magnitude. Therefore, fulfillment of the inequality $\omega_g \neq \omega_{ct}$ is a must for gear shaping operations of this type. Here, $\omega_g$ and $\omega_{ct}$ denote the magnitudes of the rotation vectors $\boldsymbol{\omega}_g$ and $\boldsymbol{\omega}_{ct}$—that is, the equalities $\omega_g = |\boldsymbol{\omega}_g|$ and $\omega_{ct} = |\boldsymbol{\omega}_{ct}|$ are valid.

To fulfil the inequality $\omega_g \neq \omega_{ct}$, either the rotation $\omega_{ct}$ must exceed the rotation $\omega_g$ or the rotation $\omega_g$ must exceed the rotation $\omega_{ct}$. In the first case, the inequality $\omega_g < \omega_{ct}$ is valid, whereas in the second case, the inequality $\omega_g > \omega_{ct}$ is observed. The first case ($\omega_g < \omega_{ct}$) corresponds to the operation of machining of an internal gear by the shaper cutter. The shaping of an internal work gear with the gear shaper cutter is schematically illustrated in Figure 10.3.

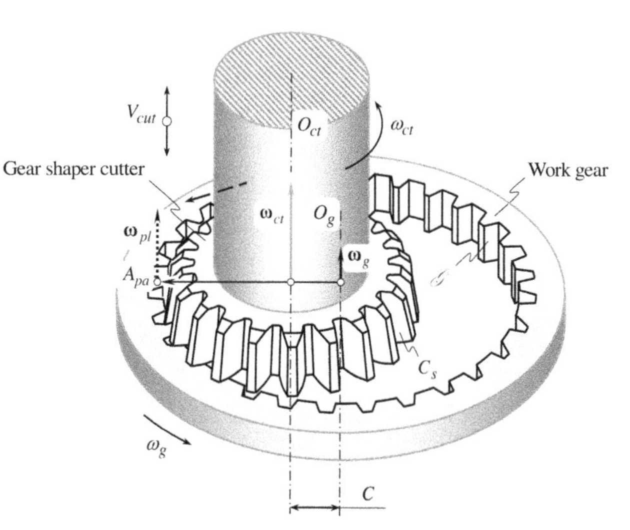


**FIGURE 10.3** Shaping an internal work gear with the gear shaper cutter.

[1] It should be noted here that the same kinematics of the gear machining mesh is featured by another gear-machining process—parallel axes shaving [147], [219].

DOI: 10.1201/9781003520528-13

The inequality $\omega_g \neq \omega_{ct}$ is not the only constraint onto the ratio between the rotations $\omega_g$ and $\omega_{ct}$. One more constraint is attributed to the numbers of tooth of the work gear $N_g$ and the gear shaper cutter $N_{ct}$, under which no interference of the gear teeth and the gear shaper cutter teeth is observed before the teeth are engaged in the mesh and after they are disengaged from of the mesh. This issue needs to be discussed more extensively as a separate topic.

The generating surface of a gear shaper cutter is conjugate to the work gear tooth flank. When shaping the work gear teeth, the generating surface $T$ of the gear shaper cutter and the gear to be machined are in proper tight (that is, with zero backlash) mesh with one another.

The kinematics of gear generation is of critical importance for the determination of the generating surface $T$ of the gear shaper cutter. When shaping an internal gear by a gear shaper cutter, the work gear and the gear shaper cutter rotate $\boldsymbol{\omega}_g$ and $\boldsymbol{\omega}_{ct}$ about their axes of rotation $O_g$ and $O_{ct}$ (see Figure 10.2). The axes $O_g$ and $O_{ct}$ of the rotations are parallel to each other ($O_g \parallel O_{ct}$). The rotations $\omega_g$ and $\omega_{ct}$ are synchronized with one another in such a way that the ratio:

$$\frac{\omega_g}{\omega_{ct}} = \frac{N_{ct}}{N_g} \tag{10.1}$$

is valid.

The gear shaper cutter reciprocates in its axial direction simultaneously with the rotations $\omega_g$ and $\omega_{ct}$. The speed of the reciprocation is designated $\mathbf{V}_{cut}$. The reciprocation of the gear shaper cutter is necessary for the stock removal.

When shaping a helical work gear, the rotation of the gear shaper cutter $\boldsymbol{\omega}_{cut}$ is performed in addition to the reciprocation $\mathbf{V}_{cut}$. Timing of the motions $\mathbf{V}_{cut}$ and $\omega_{cut}$ depends on the hand of the tooth helix of the work gear. The resultant screw motion that is comprised by the translation $\mathbf{V}_{cut}$ and the rotation $\omega_{cut}$ is a screw cutting motion of the gear shaper cutter along and about its axis $O_{ct}$. This screw motion of the shaper cutter does not correlate to the rotations $\omega_g$ and $\omega_{ct}$ and is independent of the rotations $\boldsymbol{\omega}_g$ and $\omega_{ct}$. The parameter of the screw motion is identical to the screw parameter of the work gear helix. The vector of the screw cutting motion can be pointed either in the same direction or in the opposite direction of the rotation vector $\boldsymbol{\omega}_{ct}$.

The distance between the axes $O_g$ and $O_{ct}$ of the rotations alters when the gear shaper cutter reciprocates back and forth, similar to that shown in Figure 10.3.

## 10.2 DESIGN OF GEAR SHAPER CUTTERS

An in-detail discussion of the design of gear shaper cutters reproducing an external gear machining mesh (see Chapter 9 for details) makes it easier to form a concise analysis of the design of gear shaper cutters engaged in an internal gear machining mesh when shaping gears.

### 10.2.1 Generating Surface of the Gear Shaper Cutter

Most of internal gears machined by gear shaper cutters are spur gears. However, internal gears with helical teeth can also be machined by gear shaper cutters.

The generating surface of a gear shaper cutter for machining internal gears is similar to that used for machining external gears. Therefore, without going into details of the analysis, Equation (9.9):

$$\mathbf{r}_{ct}(U_{ct}, V_{ct}) = \begin{bmatrix} r_{b.ct} \cos V_{ct} + U_{ct} \cos \lambda_{b.ct} \sin V_{ct} \\ r_{b.ct} \sin V_{ct} - U_{ct} \cos \lambda_{b.ct} \sin V_{ct} \\ r_{b.ct} \tan \lambda_{b.ct} - U_{ct} \sin \lambda_{b.ct} \\ 1 \end{bmatrix} \tag{10.2}$$

can be used for the analytical description of the generating surface $T$ of the gear shaper cutter for machining internal gears.

Equation (10.2) is applicable for both helical and spur gear shaper cutters. For the latter, it is necessary to make the base lead angle $\lambda_{b.ct}$ of the generating surface $T$ equal to the right angle ($\lambda_{b.ct} = 90°$).

### 10.2.2 Profiling of the Gear Shaper Cutter Tooth

When profiling teeth of a gear shaper cutter, it is necessary to meet the condition under which three surfaces pass through a common line. These surfaces are listed immediately below:

- The generating surface of the gear shaper cutter $T$.
- The rake surface $R_s$.
- The clearance surface $C_s$.

These surfaces have to pass through the common line—that is, through the cutting edge $CE$ of the gear shaper cutter tooth.

There are three methods of transforming the generating body of the cutting tool[2] into a workable edge cutting tool [46]:

1. Intersect the generating surface $T$ with the rake surface $R_s$. Then, the clearance surface $C_s$ is a surface at a clearance angle $\alpha$ through the cutting edge $CE$.
2. Intersect the generating surface $T$ with the clearance surface $C_s$. Then, the rake surface $R_s$ is a surface at a rake angle $\gamma$ through the cutting edge $CE$.
3. Construct on the generating surface $T$ the cutting edge $CE$ with the optimal inclination angle $\lambda$. Then, the rake surface $R_s$ and the clearance surface $C_s$ are the surfaces through the cutting edge

[2] A body that is made of cutting material and is bounded by the generating surface $T$ of the cutting tool is referred to as the generating body of the cutting tool.

*CE* (the surface $R_s$ forms rake angle $\gamma$ with the surface of cut *SC* and the surface $C_s$ forms clearance angle $\alpha$ with the surface of cut *SC*).

In the first method, considered the most practical option for this particular case, the generating surface *T* of the gear shaper cutter is intersected by the rake surface $R_s$. The cutting edge *CE* of the gear shaper cutter is viewed here as the line of intersection of the surfaces *T* and $R_s$.

The generating surface of the gear shaper cutter was previously defined [see Equation (10.2)]. Either an internal cone of revolution or a plane (orthogonal to the gear shaper cutter axis of rotation) is employed as the rake surface $R_s$ of the gear shaper cutter tooth (see Figure 9.6 in Chapter 9). For helical gear shaper cutters, the plane that is orthogonal to the pitch helix of the gear shaper cutter tooth (see Figure 9.7 in Chapter 9) can be used as the rake surface as well. Coordinates of points of the cutting edge *CE* meet both the equation for the generating surface *T* of the gear shaper cutter and the equation for the rake surface $R_s$. Next, the clearance surface $C_s$ is specifically constructed so as to pass through the cutting edge *CE*. For the convenience of the manufacturer of gear shaper cutters, the desired clearance surface is approximated by a screw involute surface. The clearance surface $C_s$ of the chosen geometry passes close to but not exactly through the cutting edge *CE*, as an actual clearance surface is a kind of approximation of the desirable clearance surface. Because of the approximation, deviations of the actual location of the cutting edges from their desired geometry, location, and orientation become inevitable. To minimize the deviations, certain corrections can be introduced to the tooth profile of the gear shaper cutter. Determination of the critical distance to the nominal plane section is also a critical issue when designing gear shaper cutters.

Only involute gears can be precisely generated in internal gear shaping process. No internal gears with other tooth flank geometry can be perfectly generated by the shaper cutter—only approximate non-involute gears can be cut this way. This is attributed to the kinematics of the gear shaping process and to violation of the condition of conjugacy for the tooth flanks $\mathscr{G}$ of an internal gear to be machined and of the generating surface *T* of the gear-cutting tool. In compliance with the condition of conjugacy of the tooth flanks, at every point of contact of the surfaces $\mathscr{G}$ and *T*, the common perpendicular **n** must intersect the axis of instant rotation $P_{\text{ln}}$ of the gear and the gear-cutting tool in the gear machining mesh [171].

### 10.2.3 Cutting Edge Geometry of the Gear Shaper Cutters Tooth

The rake angles $\gamma_o$ and $\gamma_f$ at the top and the lateral cutting edges, respectively, as well as the clearance angles $\alpha_o$ and $\alpha_f$ are all necessary to make the gear shaper cutter an efficient metal-cutting tool.

The geometry of three major surfaces, namely, *T*, $R_s$, and $C_s$, in shaper cutters for machining gears in the internal gear machining mesh is similar to the geometry of the same major surfaces in the shaper cutters for machining gears in external gear machining mesh. Thus, for the calculation of the angle of inclination $\lambda$, the rake angles $\gamma_o$ and $\gamma_f$, and the clearance angles $\alpha_o$ and $\alpha_f$, the earlier developed formulae are applicable (see Chapter 9).

Typically, the rake angle on the top cutting edge of a gear shaper cutter is set equal to $\gamma_o = 5°$. Occasionally, roughing shaper cutters have a rake angle as high as $\gamma_o = 10°$. The clearance angle at the top cutting edge of the gear shaper cutter is usually set to $\alpha_o = 6°$. The lateral cutting edges of the cutter have a side clearance of about $\alpha_f = 2°$.

Because of a smaller tooth count, the actual values of rake angle $\gamma_f$ and the clearance angle $\alpha_f$ alters along the lateral cutting edges of shaper cutters of some designs. Shank-type gear shaper cutters and small-diameter disk-type gear shaper cutter with deep counterbore are good examples in this regard: the alterations in the corresponding angles $\gamma_f$ and $\alpha_f$ exceed that for gear shaper cutters for cutting external gears.

## 10.3 THICKNESS OF CHIPS CUT BY GEAR SHAPER CUTTER TOOTH

Cut thickness is an important consideration because it significantly affects the wear of the gear shaper cutter teeth. The parameters of the cut thickness in the gear-shaping operation can be calculated.

The kinematics of the gear shaping operation features continuous indexing of rotations of the work gear and the gear shaper cutter. When shaping an internal gear with the gear shaper cutter (see Figure 10.2), the work gear rotates about its axis of rotation $O_g$. The gear shaper cutter rotates about its axis of rotation $O_{ct}$. In the case under consideration, the vectors of both rotations $\omega_g$ and $\omega_{ct}$, as well as the vector $\boldsymbol{\omega}_{pl}$ of instant relative rotation, are pointed in the same direction. Thus, the equality $\omega_{pl} = \omega_{ct} - \omega_g$ is observed when shaping an internal gear. The rotation of the gear $\omega_g$ is synchronized with the rotation of the gear shaper cutter $\omega_{ct}$ so that the ratio $\omega_g / \omega_{ct} = N_{ct} / N_g$ is valid.

Two motions are necessary for machining a work gear by the gear shaper cutter: the cut motion and the feed motion.

Reciprocation of the gear shaper cutter serves as the primary motion (cut motion). Reciprocation of the gear shaper cutter provides the necessary cutting speed. Cutting speed equals to the speed $V_{rc}$ of the gear shaper cutter reciprocation.

The feed motion $F_c$ is attributed to the instant relative rotation $\omega_{pl}$ of the gear shaper cutter relative to the work gear—that is, to the rotation $\boldsymbol{\omega}_g$ of the work gear and the rotation $\boldsymbol{\omega}_{ct}$ of the gear shaper cutter. The rolling of the pitch surfaces of the work gear and the gear shaper cutter about their axes of rotation allows for interpretation in the form of instant rotation of the gear shaper cutter pitch cylinder about the pitch line. The speed of the instant rotation $\omega_{pl}$ can be expressed in terms of the rotations $\omega_g$ and $\omega_{ct}$ and in terms of the design parameters of the gear being machined and the gear shaper cutter. Ultimately, the speed of the feed motion $F_c$ allows for interpretation in the form of product of the speed of rotary motion of the elementary portion of the cutting edge of the gear shaper cutter tooth in its instant relative rotation, by the distance of point of interest *m* from the axis of the rotation:

$$F_c = \omega_{pl} \cdot R_{rl} \tag{10.3}$$

where:

$\omega_{pl}$ is the instant relative rotation of the gear shaper cutter

$R_{rl}$ is the distance of point of interest *m* within the gear shaper cutter cutting edge (the point of interest *m* can be chosen within any of three cutting edges of the shaper cutter tooth)

For a given gear shaping operation, the instant rotation $\omega_{pl}$ has a constant value ($\omega_{pl} = const$). The distance $R_{rl}$ depends on the coordinates of point of interest *m* of the gear shaper cutter cutting edge. This can be specified by the position vector $\mathbf{r}_m$ of the point of interest *m*. Thus, a formula:

$$\mathbf{F}_c = \boldsymbol{\omega}_{pl} \times \mathbf{r}_m \tag{10.4}$$

can be used for the calculation of the vector $\mathbf{F}_c$ of the feed motion. This formula reveals that both the magnitude and direction of the feed motion depend on location of the chosen point of interest within the corresponding cutting edge of the gear cutter tooth.

The approach proposed by Prof. V.A. Shishkov [204, 206] can be used to calculate the parameters of a portion of stock cut out per stroke by the gear shaper cutter tooth.

Two consequent locations of the gear shaper cutter tooth are considered for the calculation of cut thickness. The initial position of the gear shaper cutter tooth is defined by a Cartesian coordinate system $X_gY_gZ_g$ associated with the work gear, as shown in Figure 10.4. Similar to Equation (10.70), the angle per stroke $\varepsilon_{st}$ of the gear shaper cutter can be expressed in terms of:

- The roll angle $\varepsilon_o$ at the outer diameter of the gear tooth profile.
- The roll angle $\varepsilon_l$ at the inner diameter of the gear tooth profile.
- The number of strokes $n_{st}$ of the gear shaper cutter necessary to shape the whole tooth flank of the gear.

This yields a formula for the calculation of the angle $\varepsilon_{st}$:

$$\varepsilon_{st} = \frac{\varepsilon_o - \varepsilon_l}{n_{st}} \tag{10.5}$$

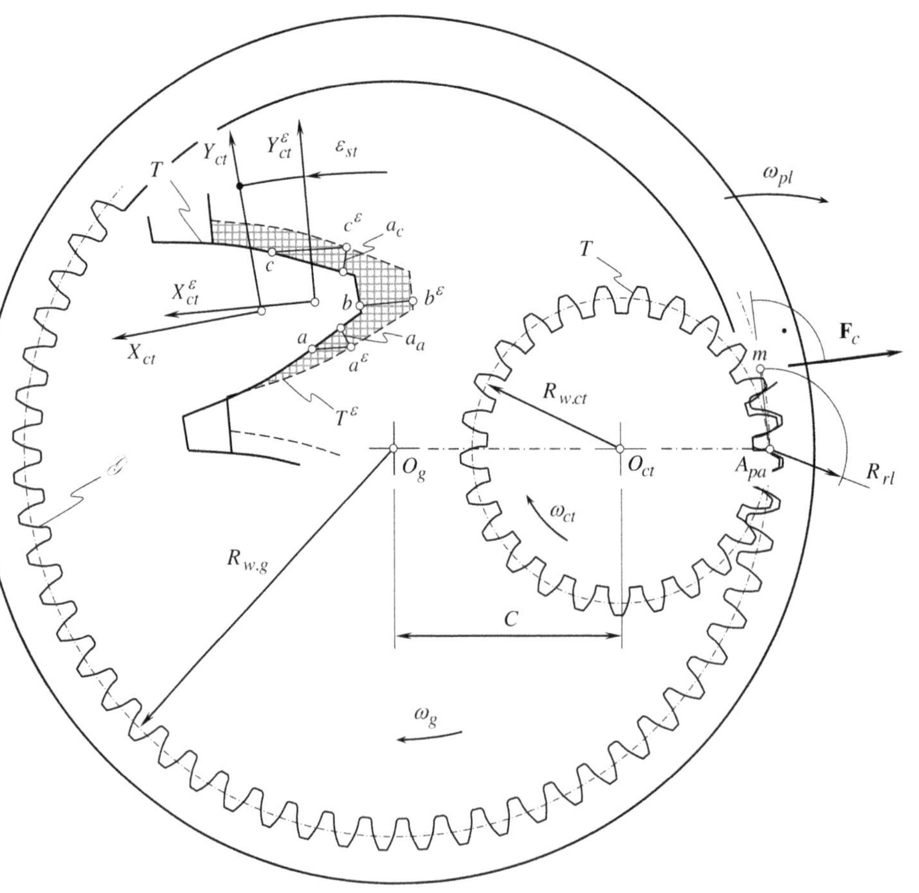


**FIGURE 10.4** Determination of the cut thickness in shaping internal spur gear with the gear shaper cutter.

Usually, the actual value of the angle $\varepsilon_{st}$ is in the range of $\varepsilon_{st} \cong 1°$. However, when shaping internal gears, it can exceed the corresponding value calculated for the case of shaping of external gears. This is mostly due to the convex tooth profile of the gear shaper cutter making contact with the concave tooth profile of the gear being machined, rather than in the case when two convex tooth profiles make contact when machining external gears with a standard gear shaper cutter design.

The rotation of the gear shaper cutter $\omega_{ct}$ is synchronized with the rotation of the work gear $\omega_g$ to fulfil the ratio:

$$\frac{\omega_g}{\omega_{ct}} = \frac{N_g}{N_{ct}} \tag{10.6}$$

The gear tooth ratio $N_g/N_{ct}$ is commonly denoted by $u$. The angle of rotation of the cutter $\varepsilon_{st}^c$ can be expressed in terms of the angle of rotation of the work gear $\varepsilon_{st}$ and of the tooth ratio $u$. For this purpose, the expression:

$$\varepsilon_{st}^c = \frac{\varepsilon_{st}}{u} \tag{10.7}$$

applies.

When the work gear turns about its axis $O_g$ through an angle $\varepsilon_{st}$, the gear shaper cutter turns about its axis of rotation $O_{ct}$ through the corresponding angle $\varepsilon_{st}^c$. In addition, as this relative motion takes place, the axis of rotation $O_{ct}$ turns about the axis of rotation $O_g$ through the angle $\varepsilon_{st}$. Under the rotations $\omega_g$ and $\omega_{ct}$, the pitch cylinder of the gear shaper cutter rolls with no sliding over the pitch cylinder of the work gear. The relative motion of the coordinate system of the gear shaper cutter relative to the coordinate system associated with the work gear is the result of rolling of pitch cylinders. The Cartesian coordinates system associated with the gear shaper cutter in its initial position is designated as $X_{ct}Y_{ct}Z_{ct}$. In the current configuration of the gear shaper cutter relative to the work gear, that same coordinate system is designated as $X_{ct}^{\varepsilon}Y_{ct}^{\varepsilon}Z_{ct}^{\varepsilon}$.

The resultant linear transformation that describes the transition from the coordinate system $X_{ct}Y_{ct}Z_{ct}$ to the coordinate system $X_{ct}^{\varepsilon}Y_{ct}^{\varepsilon}Z_{ct}^{\varepsilon}$ is analytically expressed by the operator of rolling $\mathbf{Rr}_u(\varepsilon_{st}, Z_g)$ (see Figure 10.4). The operator of rolling $\mathbf{Rr}_u(\varepsilon_{st}, Z_g)$ can be viewed as a product of three consequent operators of elementary transformations, namely, the operator of rotation $\mathbf{Rt}[\varepsilon_{st}^c, Z_{ct}]$, the operator of rotation $\mathbf{Rt}(\varepsilon_{st}, Z_g)$, and the operator of translation $\mathbf{Tr}(C, X_g)$. See Appendix D for details on the operator of rolling $\mathbf{Rr}_u(\varepsilon_{st}, Z_g)$.

Consider a point of interest within the cutting edge of the gear shaper cutter tooth. This may be point $a$ within the lateral cutting edge of the gear shaper cutter tooth, point $b$ within the top cutting edge, or point $c$ within the lateral cutting edge of the opposite side of the tooth profile (see Figure 10.4). In the case under consideration, use of the local coordinate system $X_{ct}Y_{ct}Z_{ct}$ is helpful.

For a given configuration of the generating surface $T$ of the gear shaper cutter, the position vector $\mathbf{r}_a$ of the point of interest $a$ is specified (the same is valid with respect to any other point of interest within the cutting edge $CE$ of the gear shaper cutter tooth). The unit normal vector $\mathbf{n}_{T.a}$ to the generating surface $T$ at point $a$ can be calculated using Equation (10.2).

After one stroke of the gear shaper cutter is complete, its generating surface in the new position is designated as $T^{\varepsilon}$ (see Figure 10.4). Under the rolling of the pitch cylinders, the point of interest $a$ travels to a new position $a^{\varepsilon}$. The position vector $\mathbf{r}_a^{\varepsilon}$ of the point of interest in the new location of the gear shaper cutter can be expressed in terms of the position vector $\mathbf{r}_a$ and the operator of rolling $\mathbf{Rr}_u(\varepsilon_{st}, Z_g)$ in the following manner:

$$\mathbf{r}_a^{\varepsilon} = \mathbf{Rr}_u(\varepsilon_{st}, Z_g) \cdot \mathbf{r}_a \tag{10.8}$$

Similarly, the following expression:

$$\mathbf{n}_{T.a}^{\varepsilon} = \mathbf{Rr}_u(\varepsilon_{st}, Z_g) \cdot \mathbf{n}_{T.a} \tag{10.9}$$

is valid for the unit normal vector $\mathbf{n}_{T.a}^{\varepsilon}$ to the generating surface $T^{\varepsilon}$ through the point $a^{\varepsilon}$.

Cut thickness is measured perpendicular to the machined tooth flank of the gear. Its value varies along the cutting edge of the gear shaper cutter. At point $a^{\varepsilon}$, the cut thickness is denoted by $a_a$. The parameters $a_b$ and $a_c$ are implemented for the cut thickness at points $b^{\varepsilon}$ and $c^{\varepsilon}$, respectively.

Cut thickness $a_a$ is equal to the length of straight-line segment between the point $a^{\varepsilon}$ and the generating surface $T$. The point of interception of the straight line $\mathbf{a}_a(t)$ through the point $a^{\varepsilon}$ along the unit normal vector $\mathbf{n}_{T.a}^{\varepsilon}$ (shown but not labeled in Figure 10.4) is calculated as follows:

$$\mathbf{a}_a(t) = \mathbf{r}_a^{\varepsilon} + t \cdot \mathbf{n}_{T.a}^{\varepsilon} \tag{10.10}$$

where $t$ is a parameter of the straight line.

Solving the equation for the generating surface $T$ of the gear shaper cutter [note that (a) Equation (10.2) describes only one lateral flank of the surface $T$; a similar equation is valid for the opposite flank of the tooth of the generating surface $T$; and (b) equation for the top land of the generating surface $T$ is a trivial one], together with Equation (10.10), returns the position vector $\hat{\mathbf{r}}_a^{\varepsilon}$ of the point of interception of the straight line $\mathbf{a}_a(t)$ with the generating surface $T$. This immediately yields a formula:

$$a_a = \left| \mathbf{r}_a^{\varepsilon} - \hat{\mathbf{r}}_a^{\varepsilon} \right| \tag{10.11}$$

for the calculation of the cut thickness at a point of interest within the cutting edge of the gear shaper cutter tooth.

Equation (10.11), used to calculate for thickness of cut at a point of interest within the cutting edge of the gear shaper cutter, is derived with no assumptions and simplifications and is an accurate equation. It can be used for the cases when point of interest is a far remote point from the pitch point of the gear machining mesh in a gear shaping operation, and the cutting edge consequently cuts a thick chip. It is also valid when the point of interest is close to the pitch point, and the cutting edge consequently cuts a thin chip, whose thickness is in the range comparable with the roundness of the cutting edge.

For approximate calculations, simpler equations are known. However, either the approximate equations are applicable for remote portions of the cutting edges and are therefore not applicable for portions of the cutting edges close to the pitch point, or they are applicable for the points of the cutting edge close to the pitch point and are therefore not applicable for the remote portions of the cutting edges of the gear shaper cutter.

Equation (10.11) can be used to calculate the other parameters of the portion of stock that has been cut out by the cutting edge of a gear shaper cutter: the shape of the cross section of the portion of stock, its area, and so forth.

The performed analysis reveals that cutting edges of the gear shaper cutter teeth cut chips of varying thickness.

First, cut thickness depends on the distance at which the point of interest is remote from the pitch point in the gear machining mesh. The greater this distance, the thicker the cut.

Second, the different cutting edges of the gear shaper cutter tooth cut chips of varying thickness. The chip that is cut by the top cutting edge is the thickest. The thickness of the chip that is cut by the lateral cutting edge of the entering side of the tooth profile is greater than that of the recessing side of the gear shaper cutter tooth profile.

The lateral cutting edges of the gear shaper cutter teeth cut the stock mostly within the distance between the pitch point and the point where the cutting begins. A small amount of the stock is also cut beyond the pitch point. The top land cutting edge does not cut the stock beyond the pitch point.

A detailed analysis of cut thickness is beyond the scope of the book.

## 10.4 ACCURACY OF THE SHAPED INTERNAL GEARS

The accuracy of the internal gears machined by the gear shaper cutters is discussed from the perspective of fulfillment of a set of necessary conditions for proper part surface generation.

Fulfillment of the conditions of proper part surface generation is a must when machining gears with the gear shaper cutter (see Appendix E for details on conditions of proper part surface generation). When shaping an internal gear with a conventionally designed gear shaper cutter, the pitch surface of the work gear is shaped in the form of an inner cylinder of revolution, whereas the pitch surface of the gear shaper cutter is shaped in the form of an external cylinder of revolution.

Because the pitch cylinders make inner tangency with each other, the importance of the fifth condition of proper part surface generation is not as critical as when shaping external gears. Only in very special cases of gear machining does it become necessary to verify whether the fifth condition of proper part surface generation is fulfilled.

Similarly, when machining an internal gear with a conventionally designed gear shaper cutter, the convex tooth profile of the gear shaper cutter interacts with the concave tooth profile of the work gear. The cusp height is significantly reduced due to this type of tangency. Therefore, the sixth condition of proper part surface generation is also not essential for the practical needs of the gear manufacturer.

When machining an internal gear with a conventional gear shaper cutter, most of the conditions of proper part surface generation are fulfilled, and there is usually no need for verification. Violation of the conditions of proper part surface generation may occur when the difference between the number of teeth of the internal work gear and the gear shaper cutter is small. In such a scenario, interference of the work gear teeth and the gear shaper cutter teeth is likely. The latter should not be allowed to occur.

Elimination of interference of internal work gear and the gear shaper cutter teeth. Interference of the internal work gear and the gear shaper cutter teeth can be caused by contact with non-involute portions of the tooth flank. As internal gears are usually machined with pinion-type cutters, special attention has to be paid to the length of the involute tooth flanks.

Where the difference in the number teeth of the work gear and the gear shaper cutter is very small, tip interference could occur because of the collision between the tooth tips of the gear and the gear shaper cutter beyond the line of action [38]. This is caused by the special geometry of engagement of internal gears and is not connected with the choice of gear shaper cutter. To remedy this type of interference, it is necessary to change the geometrical layout of the pair of gears in the gear machining mesh. In particular, an adequately large working pressure angle is recommended.

The following gear shaper cutter data are required to check for interference:

- $N_{ct}$ tooth number of the gear shaper cutter
- $d_{b.ct}$ gear shaper cutter base diameter
- $x_{ct}$ addendum modification coefficient in reground condition of the gear shaper cutter (where applicable)
- $d_{o.ct}$ outer diameter of the gear shaper cutter in reground condition (where applicable)

The pressure angle $\phi_{o.c}$ at the gear shaper cutter tooth tip is given by:

$$\phi_{o.c} = \cos^{-1}\left(\frac{d_{b.ct}}{d_{o.ct}}\right) \tag{10.12}$$

The nominal pressure angle usually differs from the generating pressure angle when an internal gear is cut by a gear shaper cutter. When the gear shaper cutter reaches the full theoretical tooth depth on the internal gear, the transverse generating pressure angle is given by [37]:

$$\text{inv}\phi_{gt} = \text{inv}\phi_t + 2\cdot\frac{(x_g - x_{ct})}{N_g - N_{ct}}\cdot\tan\phi_n \quad (10.13)$$

The center distance between the gear and the gear shaper cutter becomes:

$$C = \frac{d_{b.g} - d_{b.ct}}{2\cdot\cos\phi_{gt}} \quad (10.14)$$

The interference between an internal gear and the gear shaper cutter occurs if the number of teeth of the internal gear and the gear shaper cutter differ only by a few teeth. Whether such interference occurs can be analytically verified. Note that results of the calculations are very sensitive to small changes in center distance. Therefore, it is advisable to verify the upper and lower limit of the center distance $C$.

To prevent interference between an internal gear and the gear shaper cutter with a small difference in the number of teeth, the following inequality [37]:

$$\frac{d_{b.ct}}{2}\cdot(\text{inv}\phi_{o.ct} - \text{inv}\phi^* + \text{arc}\delta_{ct}) - \frac{d_{b.g}}{2}\cdot(\text{inv}\phi_{o.g} - \text{inv}\phi^* + \text{arc}\delta_g) > 0 \quad (10.15)$$

has to be met.

The design parameters of the internal gear and the gear shaper cutter to be entered in Equation (10.14) can be calculated from the following formulae:

$$\phi^* = \cos^{-1}\left(\frac{d_{b.g} - d_{b.ct}}{2\cdot C}\right) \quad (10.16)$$

$$\phi_{o.ct} = \cos^{-1}\left(\frac{d_{b.ct}}{d_{o.ct}}\right) \quad (10.17)$$

$$\delta_c = \cos^{-1}\left[\left(1+\frac{d_{o.g}}{d_{o.ct}}\right)\cdot\frac{d_{o.g} - d_{o.ct}}{4\cdot C} - \frac{C}{d_{o.ct}}\right] \quad (10.18)$$

$$\phi_{o.g} = \cos^{-1}\left(\frac{d_{b.g}}{d_{o.g}}\right) \quad (10.19)$$

$$\delta_g = \cos^{-1}\left[\left(1+\frac{d_{o.ct}}{d_{o.g}}\right)\cdot\frac{d_{o.g} - d_{o.ct}}{4\cdot C} - \frac{C}{d_{o.g}}\right] \quad (10.20)$$

When the gear shaper cutter tooth number $N_{ct}$ is too small, cutter interference at the tips of the teeth of the machined internal gear may occur. This results in the loss of a portion of the involute profile. This trimming of the involute profile results in a smaller transverse contact ratio. Provided the transverse contact ratio is still greater than 1, it will not affect the transmission of motion but will reduce the load carrying capacity of the mating gears.

To prevent this from happening, the gear shaper cutter tooth number $N_{ct}$ should be greater than the required minimum number of teeth, as given by following formula [37]:

$$N_{ct} \geq N_g\cdot\left(1 - \frac{\tan\phi_{o.g}}{\tan\phi_{gt}}\right) \quad (10.21)$$

In this expression, the value of the profile angle $\phi_{o.g}$ is calculated from Equation (10.19), and the value of the profile angle $\phi_{gt}$ is calculated from Equation (10.13).

## 10.5 ENVELOPING GEAR SHAPER CUTTERS

Shaping external gears with an enveloping gear shaper cutter is another example of the implementation of the internal gear machining mesh. This kinematics corresponds to the case when rotation of the work gear $\omega_g$ exceeds the rotation of the gear shaper cutter ($\omega_g > \omega_{ct}$). A vector diagram of the kinematics of the machining process of an external gear with the enveloping gear shaper cutter is illustrated in Figure 10.5. Again, the rotation vectors $\omega_g$ and $\omega_{ct}$ are parallel to each other and are pointed in the same direction. The vector of instant rotation $\boldsymbol{\omega}_{pl}$ is parallel to the rotation vectors $\omega_g$ and $\omega_{ct}$ and pointed in the same direction.

Enveloping gear shaper cutters are used for machining spur gears. The generating surface of an enveloping gear shaper cutter is shaped in the form of the corresponding internal spur gear. Enveloping gear shaper cutters can be also designed and manufactured for machining helical gears. No physical constraints are imposed in this regard. Helical enveloping gear shaper cutters are not used in the industry, which is attributed to difficulties in grinding of internal screw involute clearance surfaces.

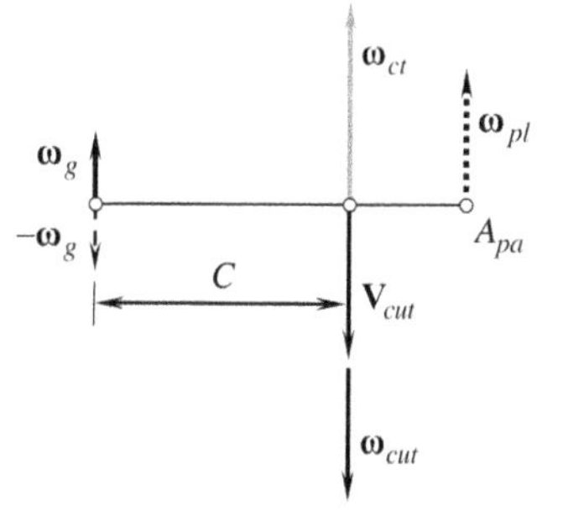


**FIGURE 10.5** Vector diagram of the kinematics of the machining process of cutting an external gear with the enveloping gear shaper cutter.

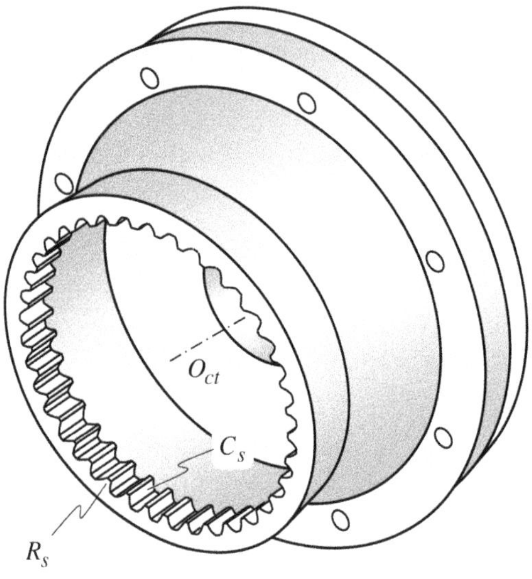


**FIGURE 10.6** Enveloping gear shaper cutter: Close-up.

The principal steps of designing an enveloping gear shaper cutter for machining external gears are the same as those for the external gear shaper cutter for machining internal gears. A close-up view of the enveloping gear shaper cutter is shown in Figure 10.6. This type of gear shaper cutter may be used in the production of external gears where it is impracticable to use a normal external disc cutter—for example, because of lack of space between the external teeth of the component and some protrusion in the component. Often, a shank-type cutter is used to perform the task, although this may have to be of a very small diameter and contain very few teeth. The internal cutter is fully generated to provide excellent tool life and much higher rates of productivity compared to a small shank.

## 10.6 APPLICATION OF GEAR SHAPER CUTTERS

Spur and helical internal involute gears are commonly shaped by gear shaper cutters. A close-up of the shaping process of an internal spur involute gear is shown in Figure 10.7, and the shaping process of an internal helical involute gear is similarly depicted in Figure 10.8.

For internal gears, it is frequently necessary to use small cutters when the disc-type construction cannot be used. The smallest cutters are usually shank-type gear shaper cutters. When the cutters are small in diameter for the tooth size, they are made integral with a shank. If the face width to be cut is wide, the shank has to be rather long and sturdy.

Two basic designs of gear shaper cutters for machining internal gears have been developed:

- The shank-type gear shaper cutter.
- The deep counter-bore gear shaper cutter that is assembled on a special adapter.

Figure 10.9 shows a shank-type gear shaper cutter and its nomenclature.

Gear shaper cutters of this type are generally used in cutting small pitch diameter internal parts. The cutter length below the taper must be adequate to accommodate the face width of the gear to be cut, plus the required over-travel at the bottom of the machine stroke and available life in the cutter. The work-holding fixture, or a recess of the gear teeth in the blank may require extra length. Taper shank-type cutters are made in four taper sizes (as measured at the large end of the taper; see Table 10.1).

The pitch diameter of the cutter should approximate the diameter of the taper. Flutes can be added to long

**FIGURE 10.7** Cutting of internal spur involute gear with the shaper cutter.

**FIGURE 10.8** Cutting of internal helical involute gear with the shaper cutter.

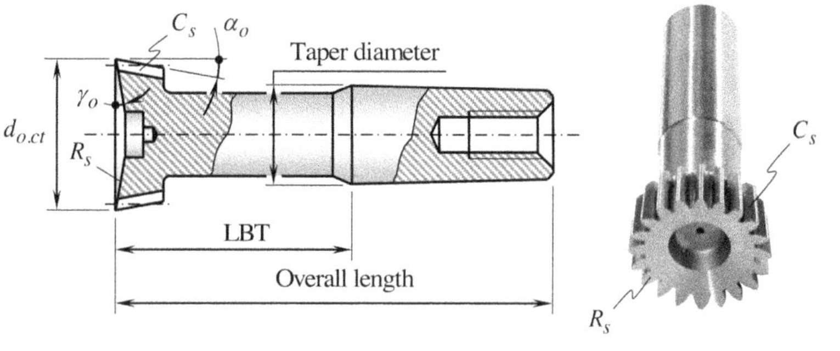


**FIGURE 10.9** Close-up of shank-type gear shaper cutters.

cutters of small pitch diameter to minimize deflection when cutting.

Taper shank cutters, which come in various sizes, normally cut small internal gears or splines. The taper fits directly into the machine spindle or indirectly with a reduction taper bushing.

The cutter has external teeth and must be small enough to avoid damaging the corners of the internal teeth.

Gear shaper cutters frequently have small enough numbers of teeth to cause the base circle to come high on the tooth flank. The region below the base circle may be left as a simple radial flank or it may be filled in. The filled-in design can be used to break the top corner of the gear tooth being cut.

**TABLE 10.1**
**Taper Sizes for Shank Type Gear Shaper Cutters**

| Taper Diameter (in.) | Taper Type |
|---|---|
| 1.0625 in. | 0.6235 in tpf |
| 0.700 in. | #2 Morse |
| 0.475 in. | #1 Morse |
| 0.250 in. | #2 Jarno |

Gear shaper cutters can be made with a protuberance at the tip. The protuberance cuts an undercut at the root of the gear tooth. This provides a desirable relief for a shaving tool. The protuberance design is also used in some cases to permit the sides of gear teeth to be ground without having to grind the root fillet.

A small disk-type gear shaper cutter with a deep counterbore mounted on a special adapter is shown in Figure 10.10. Gear shaper cutters of this design are also often used for cutting the internal gear teeth.

Gear shaper cutters are made of high-speed steels containing molybdenum and tungsten. They are heat-treated to a Rockwell C hardness of 64 to 65. A nitriding surface treatment further improves their wear resistance.

Titanium nitride coating is often applied to increase permissible feeds and speeds on modern gear shaping machines. Gear shaper cutters coated with titanium nitride increase productivity by as much as 30% without decreasing tool life.

For shaping internal gears, kinematic schemes of machining gear teeth featuring parallel rotation vectors of the same direction are used. An additional advantage of using this schematic of gear machining can be attained when shaping an internal gear with a shoulder, as depicted in Figure 10.11.

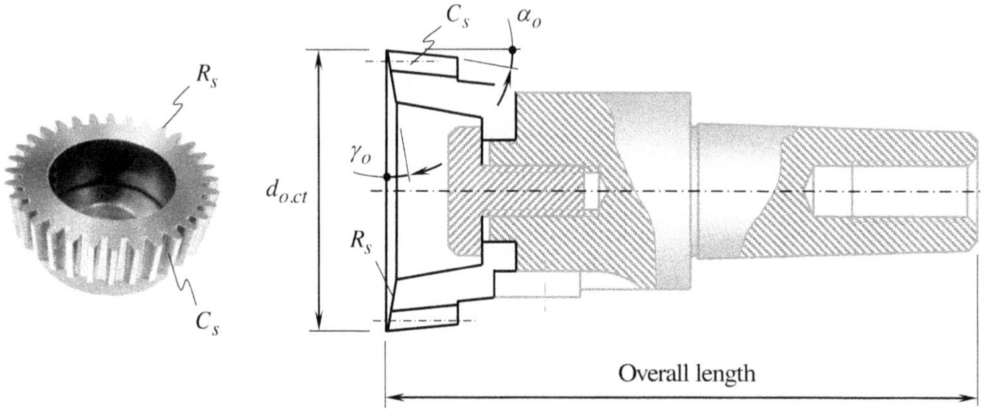


**FIGURE 10.10** Small disk-type gear shaper cutter with deep counterbore mounted on a special adapter.

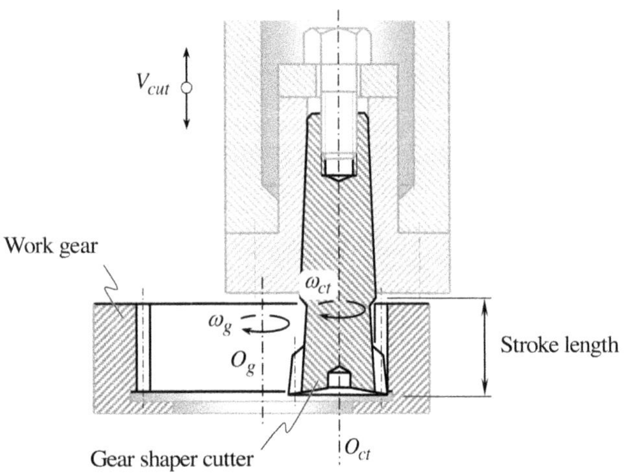


**FIGURE 10.11** Shank-type gear shaper cutter for machining of small-diameter internal gears with a shoulder.

An internal gear and an external gear can be cut in the same cycle by mounting an internal and an external gear shaper cutter on the same arbor (see Figure 10.12). Machining parameters for each cut are performed in the numerical control. Machining the cluster gear in a single setup is economical because it saves time in remounting of the gear shaper cutters.

In the gear shaping operation, cutting is accomplished by feeding the gear shaper cutter radially into the annulus, and the number of cuts required is controlled by the material, the size of the work gear, and the quality expected. It is common practice to finish with a light finishing cut, and with a depth range of about 0.05–0.25 mm (0.002–0.010 in.). Internal gears tend to distort easily unless their thickness is substantial. For epicyclic (planetary) gear units, some flexibility in the annulus is beneficial, but during cutting, it is

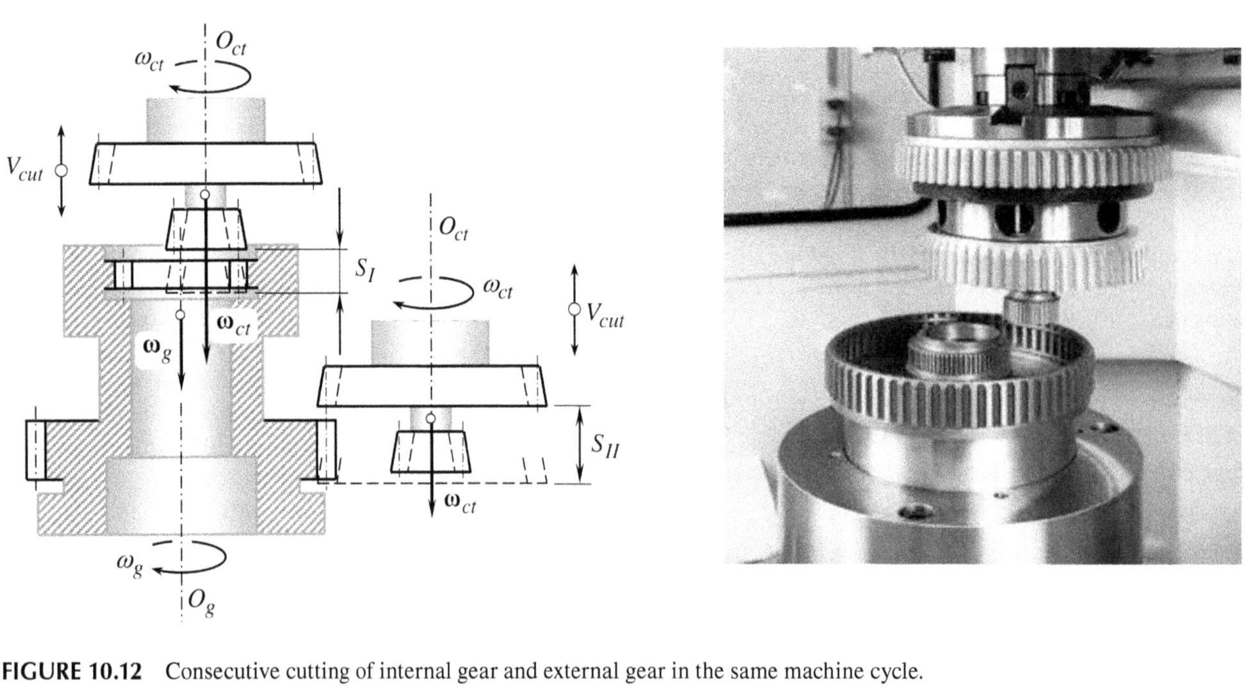


**FIGURE 10.12** Consecutive cutting of internal gear and external gear in the same machine cycle.

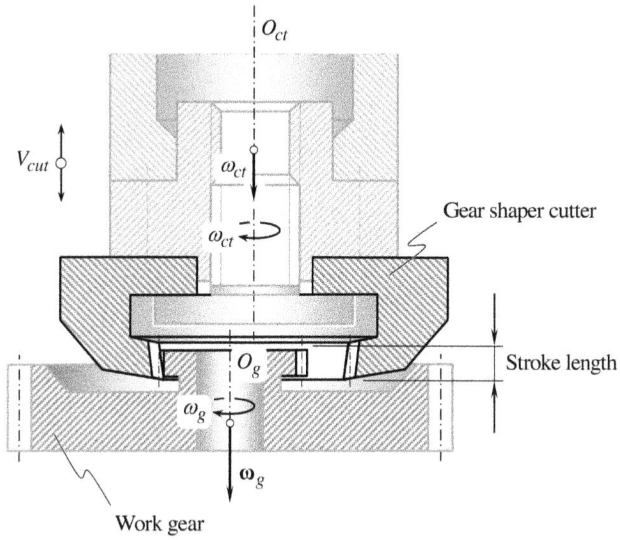


**FIGURE 10.13** Enveloping gear shaper cutter for machining external gear.

necessary to prevent radial deflection by whatever means is most suitable.

Shaping external gears with enveloping gear shaper cutters is not often done (see Figure 10.13). However, in some cases, the enveloping gear shaper cutters can be substituted with no gear-cutting tool of any other design instead. High cost is the primary reason for limited use of enveloping gear shaper cutters.

# Subsection III B

# Intersected-Axes Gear Machining Mesh

*Intersected-axes gear machining mesh* is another example of gear machining meshes. Gear-cutting tools for continuously indexing methods of gear teeth generation can be designed on the premise of the intersected-axes gear machining mesh. When machining a work gear with a gear-cutting tool, a synchronized rotation of the work gear and of the gear-cutting tool is always observed. In the past, such a relative motion of the work gear and of the gear-cutting tool were commonly viewed as rolling with no sliding of one surface of revolution (associated with the work gear) over another surface of revolution (associated with the gear-cutting tool).

Solely for convenience, this review of gear-cutting tool designs begins with the analysis of the kinematics of the intersected-axes gear machining mesh.

## III B.1 KINEMATICS OF CONTINUOUS-INDEXING INTERSECTED-AXES GEAR MACHINING MESH

The intersected-axes gear machining mesh can be schematically depicted by two rotation vectors[1] $\boldsymbol{\omega}_g$ and $\boldsymbol{\omega}_{ct}$, lines of action which intersect each other. These vectors are the rotation vector $\boldsymbol{\omega}_g$ of the work gear and the rotation vector $\boldsymbol{\omega}_{ct}$ of the gear-cutting tool.

The rotation vectors $\boldsymbol{\omega}_g$ and $\boldsymbol{\omega}_{ct}$ form a certain intersected-axes angle, $\Sigma$. The vectors $\boldsymbol{\omega}_g$ and $\boldsymbol{\omega}_{ct}$ are along the axes of rotation, $O_g$ and $O_{ct}$, of the work gear and the gear-cutting tool, respectively. Commonly, the rotation vectors $\boldsymbol{\omega}_g$ and $\boldsymbol{\omega}_{ct}$ are applied at the point of intersection of the axes $O_g$ and $O_{ct}$. As shown in Figure III B.1, this point is referred to as the *apex,* $A_{pa}$. However, because the vectors $\boldsymbol{\omega}_g$ and $\boldsymbol{\omega}_{ct}$ are sliding vectors, in particular cases for convenience they can be applied at another point within the corresponding axis of rotation, either within the axis $O_g$ (for the rotation vector $\boldsymbol{\omega}_g$), or within the axis $O_{ct}$ (for the rotation vector $\boldsymbol{\omega}_{ct}$).

[1] ***Reminder:*** Rotations, by nature, are not vectors. Therefore, special care is required to be undertaken when treating rotations as vectors.

The gear cone apex, $A_g$, and the gear-cutting tool cone apex, $A_{ct}$, always coincide with the apex $A_{pa}$ in the intersected-axes gear machining mesh. (As it is shown below in this chapter of the book, the apex $A_{pa}$ in nature is the *plane-of-action apex* in the intersected-axes gear machining mesh.)

Because the rotation vectors $\boldsymbol{\omega}_g$ and $\boldsymbol{\omega}_{ct}$ intersect one another, the vector of instant rotation, $\boldsymbol{\omega}_{pl} = \boldsymbol{\omega}_{ct} - \boldsymbol{\omega}_g$, is also located within the plane through these vectors. The vector $\boldsymbol{\omega}_{pl}$ of instant rotation is applied at that same point of intersection, $A_{pa}$, of the axes $O_g$ and $O_{ct}$.

The rotation vector of the work gear, $\boldsymbol{\omega}_g$, and the vector of instant rotation, $\boldsymbol{\omega}_{pl}$, form an angle that is denoted by $\Sigma_g$. The angle $\Sigma_g$ is referred to as the *gear cone angle.* Similarly, the angle between the rotation vector of the gear-cutting tool, $\boldsymbol{\omega}_{ct}$, and the vector of instant rotation $\boldsymbol{\omega}_{pl}$ is denoted as $\Sigma_c$. The angle $\Sigma_{ct}$ is referred to as the *gear-cutting tool cone angle.* A simple correlation is observed between the magnitudes $\omega_g$ and $\omega_{ct}$ of the rotation vectors, $\boldsymbol{\omega}_g$ and $\boldsymbol{\omega}_{ct}$, and between the angles $\Sigma_g$ and $\Sigma_{ct}$:

$$\frac{\sin \Sigma_g}{\sin \Sigma_{ct}} = \frac{\omega_{ct}}{\omega_g} \tag{III B.1}$$

The actual configuration of the vector of instant rotation, $\boldsymbol{\omega}_{pl}$, within the plane through the rotation vectors $\boldsymbol{\omega}_g$ and $\boldsymbol{\omega}_{ct}$ depends on the ratio $\omega_{ct}/\omega_g$ of the rotations of the gear and that of the mating pinion. This ratio is commonly referred to as the *tooth ratio* in the gear machining mesh. Tooth ratio is designated as $u$. For the calculation of the tooth ratio $u$, the following expression is applied:

$$u = \frac{\omega_{ct}}{\omega_g} \equiv \frac{\sin \Sigma_{ct}}{\sin \Sigma_g} \tag{III B.2}$$

In Figure III B.1(a), the vector diagram corresponds to an external *work gear-to-cutting tool mesh.* For an external work gear-to-cutting tool mesh, the tooth ratio $u$ has a negative value ($u < 0$). In this particular case, the rotation

DOI: 10.1201/9781003520528-14

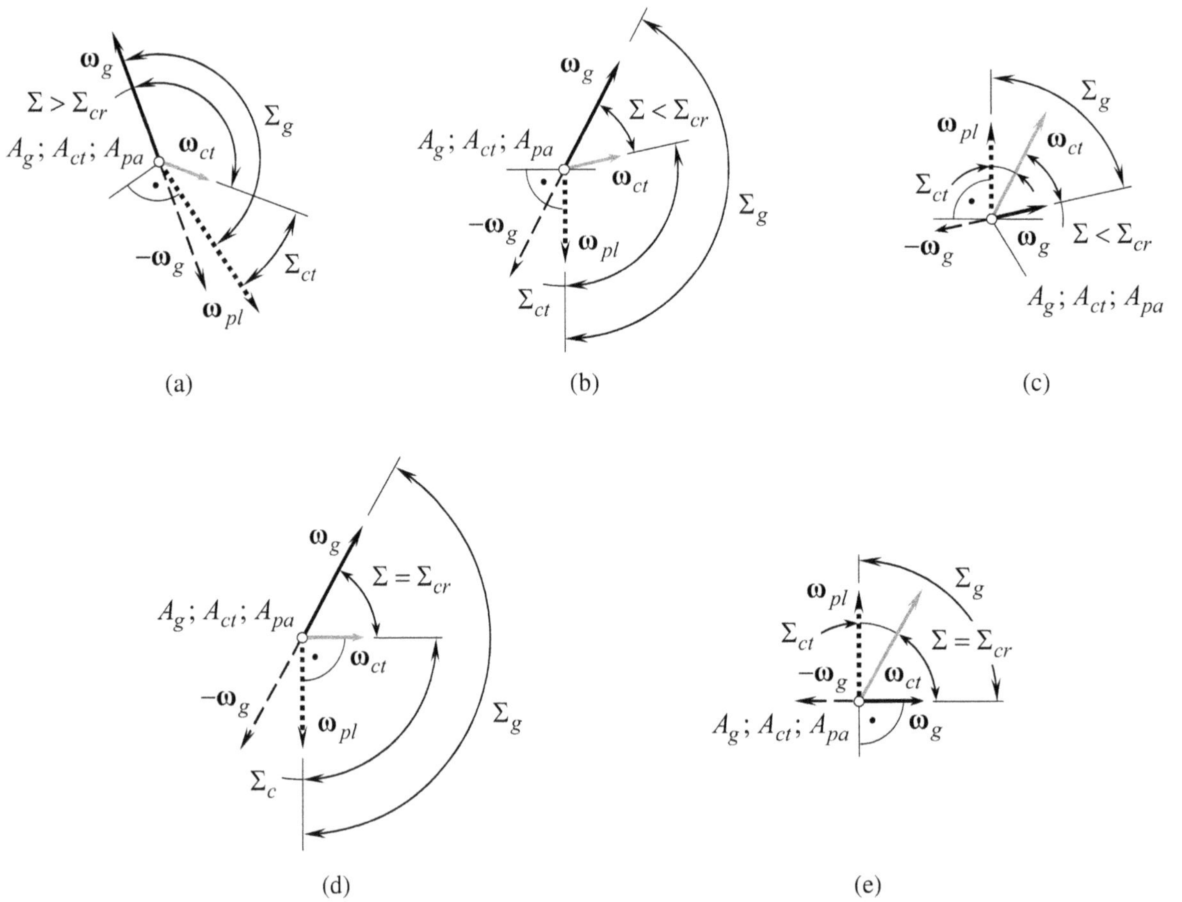


**FIGURE III B.1** Types of vector diagrams of intersected-axes gear machining meshes [parts (a) through (e) are discussed in detail within the text body of the chapter].

vectors $\boldsymbol{\omega}_g$ and $\boldsymbol{\omega}_{ct}$ form an obtuse intersected-axes angle $\Sigma$. Any alteration to the tooth ratio, $u = \omega_g / \omega_{ct}$, immediately entails a corresponding change in the configuration of the vector of instant rotation, $\boldsymbol{\omega}_{pl}$.

If the intersected-axes angle $\Sigma$ is an acute angle, the corresponding vector diagram corresponds to an *internal gear machining mesh*. There are two types of meshes for machining internal gears.

***First***, an external gear can be machined with an enveloping gear-cutting tool. A vector diagram for this particular case of gear machining is shown in Figure III B.1(b).

***Second***, an internal gear can be machined with an external gear-cutting tool. A vector diagram for this case of gear machining is shown in Figure III B.1(c). In both cases, the tooth ratio $u$ has a positive value ($u > 0$, as the input and output rotations are of the same sign).

In the first case, the positive sign of $u$ is attributed to $\Sigma_g > 0$ (whereas $\Sigma_{ct} < 0$). In the second case, the negative sign of $u$ is attributed to $\Sigma_{ct} > 0$ (whereas $\Sigma_g < 0$).

For an *external gear machining mesh*, the intersected-axes angle $\Sigma$ is obtuse. In contrast, for an internal gear machining mesh, the angle $\Sigma$ is acute. What is between these two cases? In other words, what type of gear machining mesh corresponds to the case or cases which separate an external intersected-axes gear machining mesh from an internal intersected-axes gear machining mesh?

A vector diagram for this case shows that either the rotation vector of the gear-cutting tool $\boldsymbol{\omega}_{ct}$ is perpendicular to the vector of instant rotation $\boldsymbol{\omega}_{pl}$ [see Figure III B.1(d)], or the rotation vector of the work gear $\boldsymbol{\omega}_g$ is perpendicular to the vector of instant rotation $\boldsymbol{\omega}_{pl}$ [see Figure III B.1(e)].

In the first case [see Figure III B.1(d)], the cutting tool transforms into a round rack. Because the vector of instant rotation $\boldsymbol{\omega}_{pl}$ is specified as the difference, ($\boldsymbol{\omega}_{pl} = \boldsymbol{\omega}_{ct} - \boldsymbol{\omega}_g$), between the rotation vectors $\boldsymbol{\omega}_g$ and $\boldsymbol{\omega}_{ct}$, this immediately yields an analytical expression

$$(\boldsymbol{\omega}_{ct} - \boldsymbol{\omega}_g) \cdot \boldsymbol{\omega}_{ct} = 0 \tag{III B.3}$$

that makes it possible to identify a gear machining mesh of this type.

In the second case [see Figure III B.1(e)], the work gear is reduced to a round rack. The expression

$$(\boldsymbol{\omega}_{ct} - \boldsymbol{\omega}_g) \cdot \boldsymbol{\omega}_g = 0 \tag{III B.4}$$

can be used to identify a gear machining mesh of this type.

The derived analytical criteria [Eq. (III-B.2) and Eq. (III-B.3)] allow for the corresponding expressions

$$(\omega_{ct} - \omega_g) \cdot \omega_g < 0 \quad \text{(III B.5)}$$

$$(\omega_{ct} - \omega_g) \cdot \omega_c < 0 \quad \text{(III B.6)}$$

which are valid for an external gear machining mesh, as well as the expressions

$$(\omega_{ct} - \omega_g) \cdot \omega_g > 0 \quad \text{(III B.7)}$$

$$(\omega_{ct} - \omega_g) \cdot \omega_{ct} > 0 \quad \text{(III B.8)}$$

which are valid for an internal gear machining mesh.

A gear-cutting tool can be designed based on each of the vector diagrams shown in Figure III B.1. Various designs of gear-cutting tools featuring intersected-axes work gear-to-cutting tool mesh can be developed. It is convenient to begin this discussion by focusing on the design of a gear-cutting tool featuring the intersected-axes gear machining mesh shown in Figure III B.1(b).

The fourth part of this book comprises three chapters, Chapter 11 through Chapter 13.

# 11 Gear Shaper Cutters with Tilted Axis of Rotation

The classification of all possible types of gear machining meshes (see Figure 3.8) offers several advantages:

- Systematization of all known designs of gear-cutting tools as well as of all known methods of gear cutting.
- Development of novel designs of gear-cutting tools, and novel methods of gear machining.

Many gear-cutting tool designs in industry nowadays take advantage of the planar intersected-axes gear machining mesh. The gear shaper with a tilted axis of rotation is a good example in this regard.

## 11.1 KINEMATICS OF GEAR SHAPING OPERATION WITH SHAPER CUTTER HAVING TILTED AXIS OF ROTATION

Consider the two rotations about intersecting axes, $O_g$ and $O_{ct}$. In Figure 11.1, the vector of one of the rotations is designated $\omega_g$. This vector is pointed along the axis of rotation $O_g$ of the gear being machined. The vector of another rotation is designated as $\boldsymbol{\omega}_{ct}$. In Figure 11.1, this vector is pointed along the axis $O_{ct}$ of the gear shaper cutter to be designed. The rotation vectors, $\boldsymbol{\omega}_g$ and $\omega_{ct}$, are positioned at a certain intersected-axes angle, $\Sigma$. The vector of instant rotation $\omega_{pl}$ of the gear-cutting tool relative to the work gear, ($\boldsymbol{\omega}_{pl} = \omega_{ct} - \omega_g$), is situated within the plane through the rotation vectors $\boldsymbol{\omega}_g$ and $\omega_{ct}$.

The rotation vector of the work gear $\boldsymbol{\omega}_g$ forms a certain gear cone angle $\boldsymbol{\Sigma}_g$ with the vector of instant rotation $\omega_{pl}$. For an external gear machining mesh, the gear angle equals $\Sigma_g = 180° - \gamma_g$, where $\gamma_g$ denotes the pitch cone angle of the work gear. The rotation vector of the gear-cutting tool, $\boldsymbol{\omega}_{ct}$, forms a certain gear-cutting tool cone angle, $\Sigma_{ct}$, with the vector of instant rotation, $\boldsymbol{\omega}_{pl}$. This angle is equal to the pitch cone angle of the gear-cutting tool ($\Sigma_{ct} = \gamma_c$).

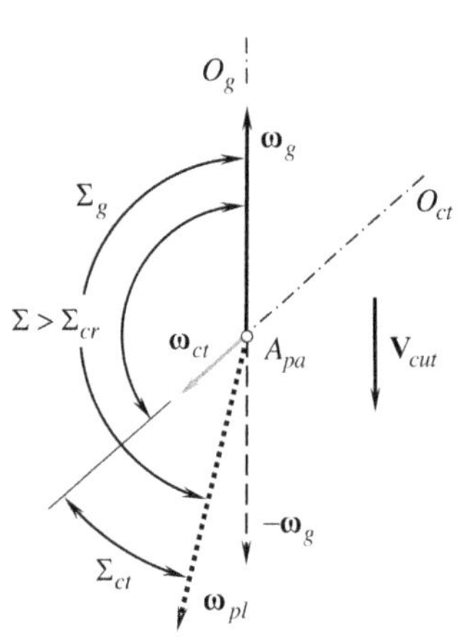


**FIGURE 11.1** Kinematics of gear shaping operation using the shaper cutter with tilted axis of rotation.

In the case under consideration (see Figure 11.1), the following relationship:

$$\gamma_g + \gamma_{ct} = \pi - \Sigma \tag{11.1}$$

is valid.

When the equality [see Eq. (11.1)] is known, the following ratio

$$\frac{\sin \gamma_g}{\sin \gamma_{ct}} = \frac{\omega_{ct}}{\omega_g} \tag{11.2}$$

can be used for the calculation of the cone angles $\gamma_g$ and $\gamma_c$ of the gear being machined and of the gear-cutting tool.

The ratio can be expressed in terms of the tooth numbers of the gear, $N_g$, and of the gear-cutting tool, $N_{ct}$:

$$\frac{\sin \gamma_g}{\sin \gamma_{ct}} = \frac{N_g}{N_{ct}} \tag{11.3}$$

The rotations $\omega_g$ and $\omega_{ct}$, which are synchronized with one another in a properly timed manner, result in the generating action of the gear-cutting tool. To make the material removal possible when machining a work gear, a corresponding primary motion, $\mathbf{V}_{cut}$ (motion of cut), must be performed. Usually, several solutions are available to the problem of selecting the primary motion. For this purpose, it is convenient to investigate any and all possible motions of the reference system $X_{ct}Y_{ct}Z_{ct}$ associated with a gear-cutting tool, relative to the coordinate system $X_gY_gZ_g$ associated with the work gear. When analyzing a possible candidate for the primary motion $\mathbf{V}_{cut}$ one should consider the constraints imposed by the rolling motion in the gear machining mesh on the feasible additional relative motion $\mathbf{V}_{cut}$ of the work gear and the gear-cutting tool.

When no constraints are imposed on the additional relative motion, the coordinate system $X_{ct}Y_{ct}Z_{ct}$ of the gear-cutting tool can perform three elementary translations ($\mathbf{V}_x$, $\mathbf{V}_y$, and $\mathbf{V}_z$) along the coordinate axes, and three elementary rotations ($\omega_x$, $\omega_y$, and $\omega_z$) about the coordinate axis of the reference system $X_gY_gZ_g$. Consider all six elementary motions ($\mathbf{V}_x$, $\mathbf{V}_y$, $\mathbf{V}_z$, and $\omega_x$, $\omega_y$, $\omega_z$) and all their possible combinations by 1, 2, ..., 6. Various values of the speed ratio of elementary motions must be considered. Not all


DOI: 10.1201/9781003520528-15

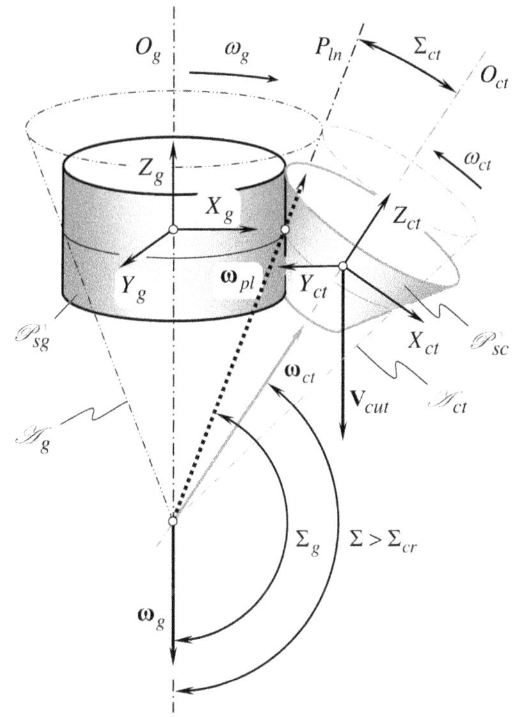


**FIGURE 11.2** Straight primary motion $\mathbf{V}_{cut}$ in the kinematics of machining a spur gear by the gear shaper with tilted axis of rotation.

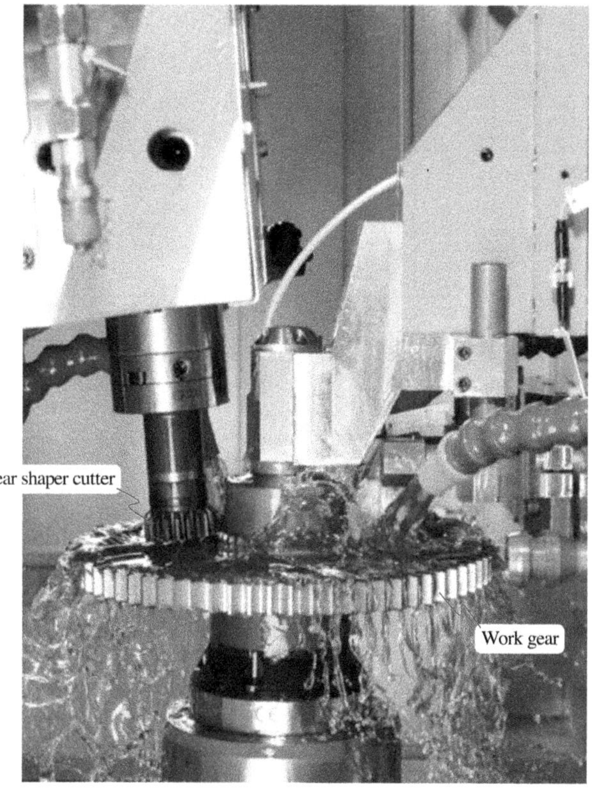


**FIGURE 11.3** Shaping of a spur gear by the shaper cutter with a tilted axis of rotation.

relative motions or combinations of motions are permissible for machining a given gear. The motions that are not permissible should be omitted from further analysis. In this way, all suitable primary motion(s) can be determined.

As an example, let us assume that associated with the gear-cutting tool reference system, $X_{ct}Y_{ct}Z_{ct}$, a straight motion of cut $\mathbf{V}_{cut}$ is performed. The direction of this motion is parallel to the work gear axis of rotation, $O_g$. Such a configuration of the primary motion is illustrated in Figure 11.2. Solely for the readers' convenience, in addition to the rotation vectors $\boldsymbol{\omega}_g$, $\boldsymbol{\omega}_{ct}$, and $\boldsymbol{\omega}_{pl}$, the axodes $\mathscr{A}_g$ and $\mathscr{A}_{ct}$, as well as the pitch surfaces $\mathscr{P}_{sg}$ and $\mathscr{P}_{sc}$, of the work gear and of the gear-cutting tool, correspondingly, are shown in Figure 11.2.

A close-up of the method of shaping a spur gear by the gear shaper cutter with a tilted axis of rotation is shown in Figure 11.3. The method of gear cutting is based on the kinematics of machining shown in Figure 11.2. In this method,[2] the work gear is rotating about its axis of rotation $O_g$ (see Figure 11.4). The vector of rotation $\boldsymbol{\omega}_g$ of the work gear is aligned with the gear axis of rotation $O_g$. The gear shaper cutter is rotating about its axis of rotation $O_{ct}$. The rotation vector $\boldsymbol{\omega}_{ct}$ is aligned with the shaper axis $O_{ct}$. The gear shaper cutter axis of rotation $O_{ct}$ is at a certain angle relative to the axis of rotation $O_g$ of the work gear. This angle is equal to the clearance angle $\alpha_o$ at the top cutting edge $CE$ of the gear shaper cutter.

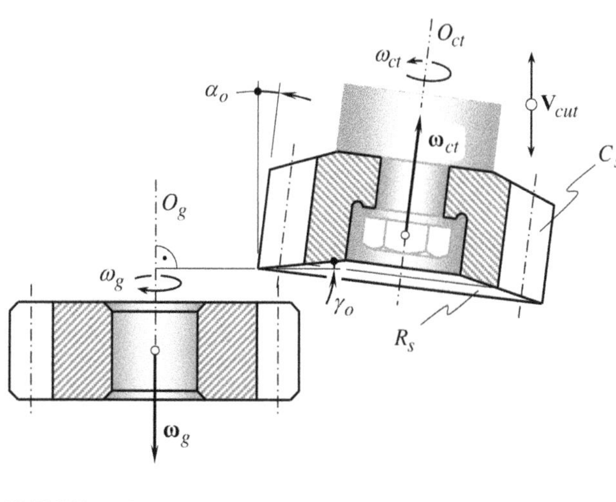


**FIGURE 11.4** A method of shaping spur gears by the shaper with a tilted axis of rotation.

[2] The concept of shaping spur gears by the gear shaper cutter with a tilted axis of rotation can be traced back to 1930, when E. Wildhaber filed his invention [US Pat. No. 1.913.865, *Cutting Tool, Ernst Wildhaber*, Filed: July 14, 1933, Published: June 13, 1930]. Later on, in the late 1930s, Prof. Ye. M. Khaimovich of the Kiev Polytechnic Institute (Ukraine) was granted a USSR patent on an invention titled "Gear Shaping Machine." In accordance with the description of this invention, the gear shaping machine featured the spindle of the gear shaper cutter that is tilted at an angle to the work gear arbor. Reciprocation of the gear shaper cutter performs in the direction that is parallel to the axis of rotation of the work gear. Since the time the patent was issued, much research has been carried out on the method of gear machining as well as on the design of the gear shaper cutters.

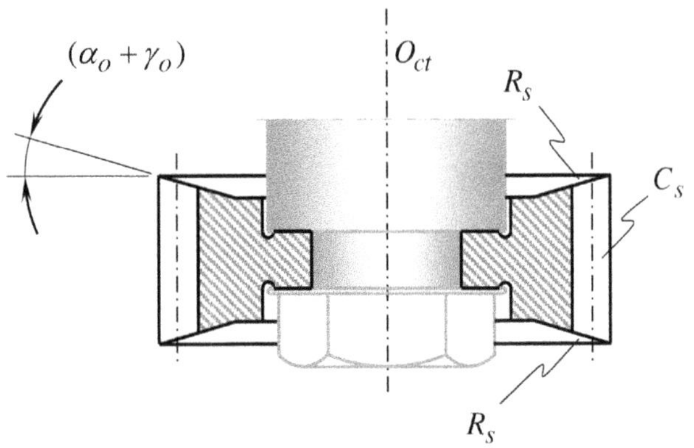


**FIGURE 11.5** Design of gear shaper cutter having the rake surfaces at opposite ends.

In addition to the rotations $\omega_g$ and $\omega_{ct}$, either the gear shaper cutter or the work gear reciprocates along the axis of rotation of the work gear. The speed of the reciprocation is equal to the speed of the primary motion $\mathbf{V}_{cut}$.

The rake surface $R_s$ of the gear shaper cutter teeth is shaped in the form of a cone of revolution. The kinematics of gear machining used in the method of shaping spur gears (see Figure 11.4) makes it possible to implement clearance surfaces $C_s$ that are shaped in the form of involute cylinders. Because of this feature, this gear shaper cutter has three strong advantages over the gear shaper cutter of conventional designs:

- The accuracy of the gear shaper cutter with the tilted axis of rotation does not depend on the number of its regrinding.
- The gear shaper cutter can be designed to have two rake surfaces $R_s$ which are located on opposite sides of the gear shaper cutter (see Figure 11.5). This makes it possible to use the rake surface at one yTB, and after it is worn out, to use the rake surface of the opposite yTB of the gear shaper cutter. This doubles the time between two consequent regrindings of the gear shaper cutter.
- Grinding of the rake surfaces of the gear shaper cutter teeth is easier because they have equal helix angles.

Much research has been undertaken on gear shaper cutters with a tilted axis of rotation since the time this concept was proposed by Dr. Wildhaber [215].

## 11.2 DETERMINATION OF GENERATING SURFACE OF GEAR SHAPER CUTTER WITH TILTED AXIS OF ROTATION

To determine the generating surface of a gear shaper cutter with tilted axis of rotation, the kinematic method for the determination of enveloping surfaces is extensively used.[3] Three Cartesian reference systems are used for the purpose of derivation of an equation of the generating surface $T$ of the gear shaper cutter (see Figure 11.6).

A left-hand-oriented Cartesian coordinate system $X_g Y_g Z_g$ that is associated with the gear is the first reference system. The $Z_g$ – axis of this coordinate system is aligned with the rotation vector $\boldsymbol{\omega}_g$ of the work gear. The coordinate system $X_g Y_g Z_g$ is rotating together with the work gear.

A left-hand-oriented coordinate system $X_{ct} Y_{ct} Z_{ct}$ that is associated with the gear shaper cutter is the second reference system. The $Z_{ct}$ – axis of this coordinate system is aligned with the rotation vector $\boldsymbol{\omega}_{ct}$ of the gear shaper cutter. The coordinate system $X_{ct} Y_{ct} Z_{ct}$ is rotating together with the gear shaper cutter.

A stationary left-hand-oriented coordinate system $X_{pl} Y_{pl} Z_{pl}$ is the third reference system to be used. The $Z_{pl}$ – axis of coordinate system $X_{pl} Y_{pl} Z_{pl}$ is aligned with the vector of the instant rotation $\boldsymbol{\omega}_{pl}$.

Origins of the coordinate systems $X_g Y_g Z_g$, $X_{ct} Y_{ct} Z_{ct}$, and $X_{pl} Y_{pl} Z_{pl}$ are snapped together at the point of intersection of the axes of rotation $O_g$, $O_{ct}$, and $P_{\text{ln}}$.

The coordinate axes $Z_g$, $Z_{ct}$, and $Z_{pl}$ are situated within a common plane. Let us assume that initially the axes $Y_g$, $Y_{ct}$, and $Y_{pl}$ align with each other as shown in Figure 11.6.

When the coordinate system $X_g Y_g Z_g$ is turned about the gear axis $O_g$ through a certain angle $\varphi_g$, the coordinate system $X_{ct} Y_{ct} Z_{ct}$ is turned about the gear shaper cutter axis $O_{ct}$ through the corresponding angle $\varphi_{ct}$. The ratio of the angles of the rotations $\varphi_g$ and $\varphi_{ct}$ fulfils the proportion

$$\frac{\varphi_g}{\varphi_{ct}} = \frac{\sin \gamma_{ct}}{\sin \gamma_g} \tag{11.4}$$

The generating surface of the gear shaper cutter $T$ allows for interpretation in the form of an enveloping surface to consecutive positions of the gear tooth flank surface $\mathscr{G}$ in the relative motion of the gear and the gear shaper cutter. To

[3] Prof. A. V. Shishkov is one of the key developers of the so-called kinematic method for the determination of enveloping surfaces. In the late 1940s, he summarized the concept of the method in a monograph [206]. The method uses the so-called equation of contact, $\mathbf{n}_g \cdot \mathbf{V}_\Sigma = 0$, where $\mathbf{n}_g$ is the (unit) normal vector to the gear tooth flank, and $\mathbf{V}_\Sigma$ is the vector of the relative motion of the cutting tool relative to the gear to be machined at the point of their contact. The Shishkov equation of contact, $\mathbf{n}_g \cdot \mathbf{V}_\Sigma = 0$, is the name that is commonly used in the current scientific literature on gearing (as well as in other areas of mechanical engineering) when referring to the equation of contact in this form. Often, the gear tooth flank has a relatively simple geometry that allows for the determination of the vector $\mathbf{n}_g$ without the necessity of using the derivatives of the position vector of a point with respect to the surface parameters. The kinematics of the relative motion of the gear-cutting tool is also often simple, which makes it possible to determine the vector $\mathbf{V}_\Sigma$ without the necessity of using the derivatives of the position vector of a point with respect to the parameter of the relative motion. Ultimately, equations for vectors $\mathbf{n}_g$ and $\mathbf{V}_\Sigma$ can be derived directly from the geometry of the gear tooth flank, and from the kinematics of meshing. Because no derivatives are required, this results in significant simplification when determining enveloping surfaces. No advantages can be obtained from the use of the kinematic method if the derivatives are used for the calculation of the vectors $\mathbf{n}_g$ and $\mathbf{V}_\Sigma$. Under such a scenario, the kinematic method reduces to the conventional method known from the theory of enveloping surfaces.

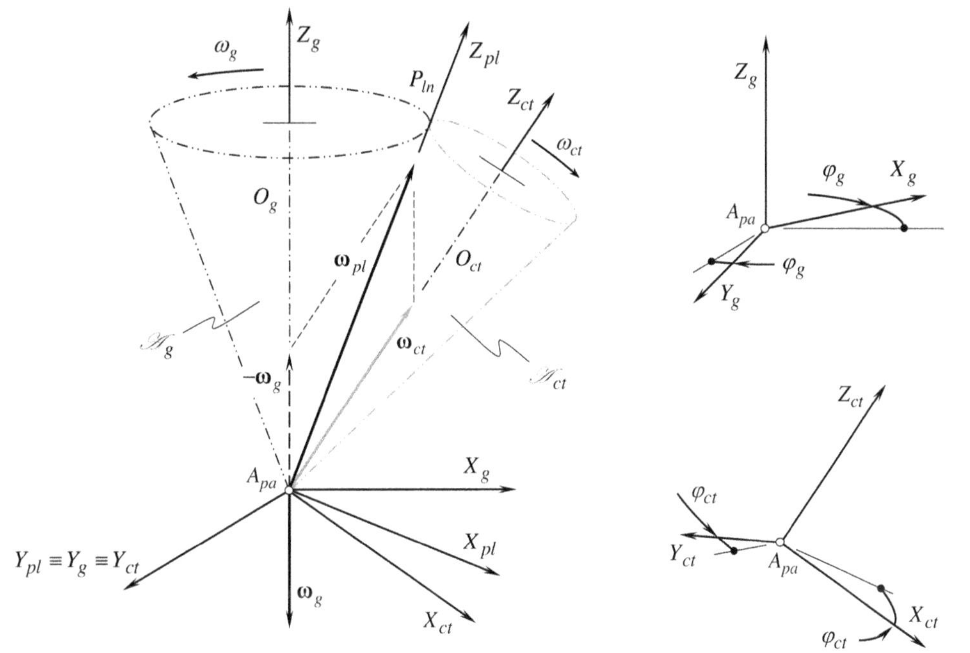


**FIGURE 11.6** Reference systems used for the determination of generating surface of gear shaper cutter with tilted axis of rotation.

use this concept, the following operators of the coordinate system transformations must be composed for the derivation of the equation of the generating surface $T$.

***First***, the operator $\mathbf{Rs}(g \mapsto pl)$ of the transition from the coordinate system $X_g Y_g Z_g$ that is rotating together with the gear to the motionless coordinate system $X_{pl} Y_{pl} Z_{pl}$ must be composed. Without going into details of the derivation, an expression for the operator $\mathbf{Rs}(g \mapsto pl)$ can be represented in matrix form:

$$\mathbf{Rs}(g \mapsto pl) = \begin{bmatrix} \sin\varphi_g \cos\gamma_g & \cos\varphi_g \cos\gamma_g & -\sin\gamma_g & 0 \\ \cos\varphi_g & -\sin\varphi_g & 0 & 0 \\ \sin\varphi_g \sin\gamma_g & \cos\varphi_g \sin\gamma_g & \cos\gamma_g & 0 \\ 0 & 0 & 0 & 1 \end{bmatrix} \quad (11.5)$$

For the inverse coordinate system transformation, the operator $\mathbf{Rs}(pl \mapsto g) = \mathbf{Rs}^{-1}(g \mapsto pl)$ is used.

***Second***, the operator $\mathbf{Rs}(ct \mapsto pl)$ of the transition from the coordinate system $X_{ct} Y_{ct} Z_{ct}$ that is rotating together with the gear shaper cutter to the motionless coordinate system $X_{pl} Y_{pl} Z_{pl}$ must also be composed. Without going into details of the derivation, an expression for the operator $\mathbf{Rs}(c \mapsto pl)$ can be represented in matrix form:

$$\mathbf{Rs}(ct \mapsto pl) = \begin{bmatrix} -\sin\varphi_{ct} \cos\gamma_{ct} & \cos\varphi_{ct} \cos\gamma_{ct} & \sin\gamma_{ct} & 0 \\ \cos\varphi_{ct} & \sin\varphi_{ct} & 0 & 0 \\ \sin\varphi_{ct} \sin\gamma_{ct} & -\cos\varphi_{ct} \sin\gamma_{ct} & \cos\gamma_{ct} & 0 \\ 0 & 0 & 0 & 1 \end{bmatrix} \quad (11.6)$$

For the inverse coordinate system transformation, the operator $\mathbf{Rs}(pl \mapsto ct) = \mathbf{Rs}^{-1}(ct \mapsto pl)$ can be used.

The operators of the coordinate system transformations $\mathbf{Rs}(g \mapsto pl)$ and $\mathbf{Rs}(pl \mapsto ct)$ allow for the derivation of an expression

$$\mathbf{Rs}(g \mapsto ct) = \mathbf{Rs}(pl \mapsto ct) \cdot \mathbf{Rs}(g \mapsto pl) \quad (11.7)$$

for the operator $\mathbf{Rs}(g \mapsto ct)$ of the direct resultant transition from the work gear coordinate system $X_g Y_g Z_g$ to the gear shaper cutter coordinate system $X_{ct} Y_{ct} Z_{ct}$. The operator $\mathbf{Rs}(g \mapsto ct)$ yields matrix representation as

$$\mathbf{Rs}(g \mapsto ct) = \begin{bmatrix} \begin{matrix} \cos\varphi_g \sin\varphi_{ct} + \\ +\sin\varphi_g \cos\varphi_{ct} \cos(\gamma_g + \gamma_{ct}) \end{matrix} & \begin{matrix} \cos\varphi_g \cos\varphi_{ct} \cos(\gamma_g + \gamma_{ct}) - \\ -\sin\varphi_g \sin\varphi_{ct} \end{matrix} & -\cos\varphi_{ct} \sin(\gamma_g + \gamma_{ct}) & 0 \\ \begin{matrix} \cos\varphi_g \cos\varphi_{ct} - \\ -\sin\varphi_g \sin\varphi_{ct} \cos(\gamma_g + \gamma_{ct}) \end{matrix} & \begin{matrix} -\sin\varphi_g \cos\varphi_{ct} - \\ -\cos\varphi_g \sin\varphi_{ct} \cos(\gamma_g + \gamma_{ct}) \end{matrix} & \sin\varphi_{ct} \sin(\gamma_g + \gamma_{ct}) & 0 \\ \sin\varphi_g \sin(\gamma_g + \gamma_{ct}) & \cos\varphi_g \sin(\gamma_g + \gamma_{ct}) & \cos(\gamma_g + \gamma_{ct}) & 0 \\ 0 & 0 & 0 & 1 \end{bmatrix} \quad (11.8)$$

Suppose that the gear tooth profile is given analytically in the coordinate plane $X_g Y_g$ by the vector equation

$$\mathbf{r}_{g.pr} = \mathbf{r}_{g.pr}(t) \tag{11.9}$$

where $t$ is a parameter of the gear tooth profile $\mathbf{r}_{g.pr}$.

Equation (11.9) allows for the calculation of the unit tangent vector $\mathbf{t}_{g.pr}(t)$

$$\mathbf{T}_{g.pr}(t) = \frac{\partial \mathbf{r}_{g.pr}(t)}{\partial t} \tag{11.10}$$

$$\mathbf{t}_{g.pr}(t) = \frac{\mathbf{T}_{g.pr}(t)}{|\mathbf{T}_{g.pr}(t)|} \tag{11.11}$$

as well as the unit normal vector $\mathbf{n}_{g.pr}(t)$

$$\mathbf{N}_{g.pr}(t) = \frac{\partial \mathbf{r}^2_{g.pr}(t)}{\partial t^2} \tag{11.12}$$

$$\mathbf{n}_{g.pr}(t) = \frac{\mathbf{N}_{g.pr}(t)}{|\mathbf{N}_{g.pr}(t)|} \tag{11.13}$$

at a point of interest of the gear tooth profile.

An equation for the unit normal vector $\mathbf{n}_{g.pr}$ can be represented in the form

$$\mathbf{n}_{g.pr} = \mathbf{i} \cdot \sin\vartheta + \mathbf{j} \cdot \cos\vartheta \tag{11.14}$$

where $\vartheta$ denotes the angle that the unit tangent vector $\mathbf{t}_{g.pr}$ forms with the $X_g -$ axis of the work gear coordinate system $X_g Y_g Z_g$.

The unit normal vector $\mathbf{n}_{g.pr}$ can be represented in the stationary reference system $X_{pl} Y_{pl} Z_{pl}$. For this purpose, the operator $\mathbf{Rs}(g \mapsto pl)$ of the coordinate system transformation is used:

$$\mathbf{n}^{pl}_{g.pr} = \mathbf{Rs}(g \mapsto pl) \cdot \mathbf{n}_{g.pr} \tag{11.15}$$

In the instant rotation of the work gear relative to the gear shaper cutter, the linear speed of the rotation can be expressed as

$$\mathbf{V}_{g/ct} = \boldsymbol{\omega}_{pl} \times \mathbf{r}_{g.pr} \tag{11.16}$$

Substituting Eq. (11.15) and Eq. (11.16) into the equation of contact [206] ($\mathbf{n}^{pl}_{g.pr} \cdot \mathbf{V}_{g/ct} = 0$), the following equation can be derived for the calculation of the angle $\vartheta$:

$$\cos(\vartheta + \varphi_g) = 2 \cdot \frac{X \cdot \cos\vartheta - Y \cdot \sin\vartheta}{d_g} \tag{11.17}$$

Equation (11.17) allows for the selection of points of the gear tooth profile $\mathbf{r}_{g.pr}$ which make contact with the gear shaper cutter tooth profile at the given angle of rotation $\varphi_g$.

Coordinates of points within the tooth profile of the generating surface $T$ of the gear shaper cutter for machining a given gear are determined in the following order:

1. At a point within the tooth profile of the gear being machined, the unit tangent vector $\mathbf{t}_{g.pr}$ is calculated. This immediately yields the current value of the angle of inclination $\vartheta$ of the unit vector $\mathbf{t}_{g.pr}$.
2. From the equation of contact [see Eq. (11.17)], a corresponding rotation angle $\varphi_g$ of the gear is determined. This angle corresponds to the instant of time when the chosen point within the gear tooth profile makes contact with the gear shaper cutter tooth profile.
3. Coordinates of the chosen point within the gear tooth profile along with the calculated value of the rotation angle $\varphi_g$ can be used to compute the coordinates of the corresponding point in the gear shaper cutter tooth profile.

It is useful to approximate a desired tooth profile of the gear shaper cutter with a smooth regular curve. Such an approximation is advantageous because the involute tooth profile is easier to machine. Under such a scenario, the gear shaper cutter tooth flanks would be identical to the tooth flanks of a corresponding spur gear.

Let us determine an involute profile that is tangent to the gear shaper cutter tooth profile on the pitch cylinder.

As shown in Figure 11.7, the point of tangency $m$ of the tooth profiles is located on the axode of the work gear. The actual location of point $m$ within the face width of the work gear affects the actual value of the profile shift correction coefficient $\xi_{ct} = x/m$. It is recommended to design the gear shaper cutter with a zero profile shift coefficient ($\xi_{ct} = 0$). Consequently, point $m$ is located in the middle of the gear face width $F_g$.

Point $m$ is remote from the axis of rotation of the work gear $O_g$ at a distance $0.5 \cdot d_g$. The corresponding diameter of the gear shaper cutter is designated as $d_{ct}$.

A local reference system $xyz$ is associated with cutting edge $CE$ of the gear shaper cutter. The origin of the coordinate system $xyz$ is at point $m$. Axes of the coordinate system $xyz$ are parallel to the corresponding axes of the Cartesian coordinate system $X_g Y_g Z_g$ associated with the work gear.

Consider a plane through point $m$ that is tangent to the gear tooth profile. This plane is a projecting plane onto the transverse plane of the work gear and forms a profile angle $\phi_g$ with the $x -$ axis [see Figure 11.7(b)]. The plane is also tangent to the generating surface $T$ of the gear shaper cutter.

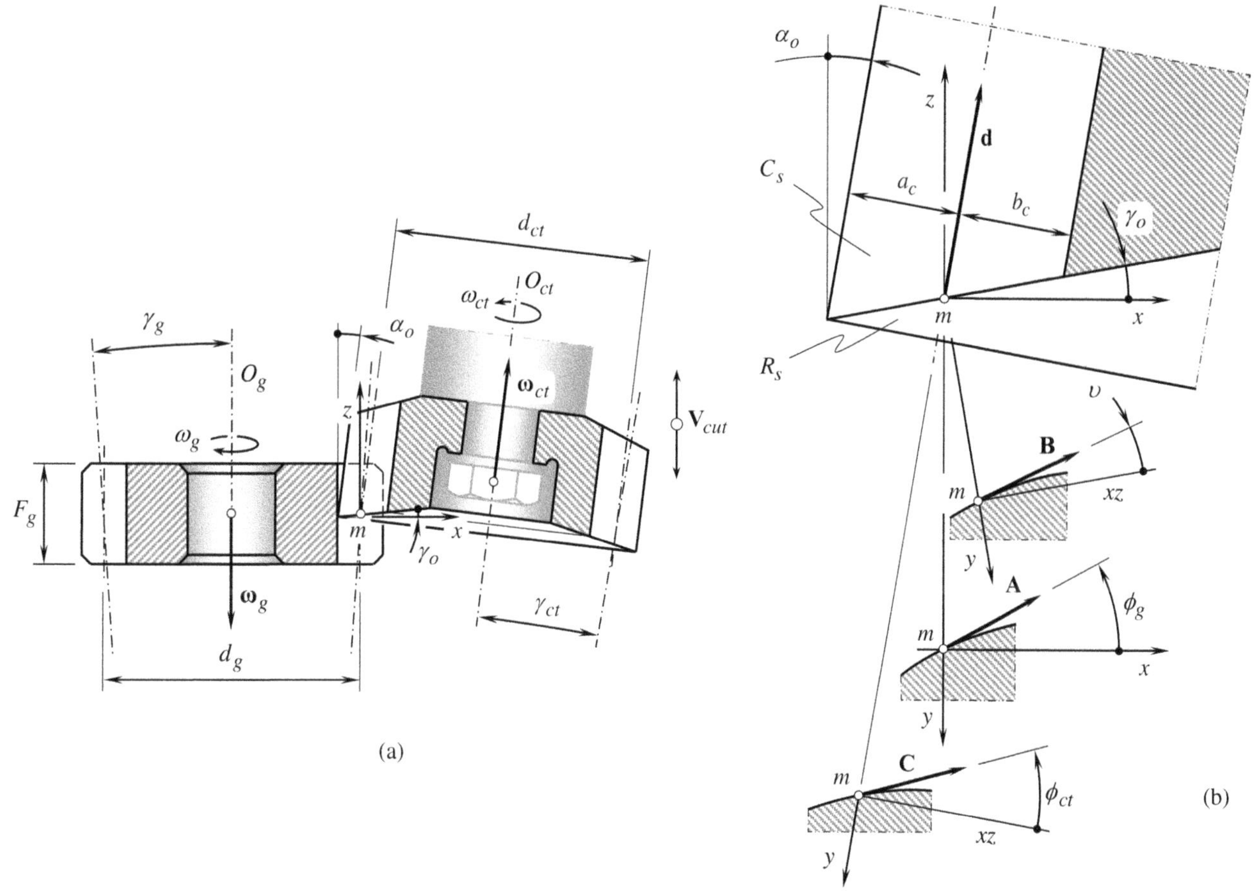


**FIGURE 11.7** Determination of involute profile of the gear shaper cutter for machining involute gear: (a) kinematics of gear shaping operation by the shaper cutter with a tilted axis of rotation, and (b) elements of geometry of lateral cutting edge of the shaper cutter in different plane sections.

Three vectors **A**, **B**, and **k** through point $m$ within the cutting edge $CE$ are constructed. The vectors **A**, **B**, and **k** need to be expressed in terms of the angle $\upsilon$ to be determined and the known parameters of the geometry of the cutting wedge of the cutting tool.

The vector **A** is constructed so that it is tangent to the generating surface $T$ and is located within the $xy-$ plane. The following expression

$$\mathbf{A} = \mathbf{i} - \mathbf{j} \cdot \tan\phi_g \tag{11.18}$$

is valid for the vector **A**.

The vector **B** is tangent to the cutting edge $CE$ of the gear shaper cutter. It can be analytically represented as

$$\mathbf{B} = \mathbf{i} \cdot \cos\gamma_o - \mathbf{j} \cdot \tan\upsilon + \mathbf{k} \cdot \sin\gamma_o \tag{11.19}$$

where $\upsilon$ denotes the angle that the tangent line to the cutting edge makes with the $xz-$ plane, and **k** is the unit vector along the $z-$ axis.

The vectors **A**, **B**, and **k** comprise a set of three coplanar vectors: they are within a plane that is tangent to the gear tooth flank surface. Therefore, the triple scalar product of these vectors is identical to zero ($\mathbf{A} \times \mathbf{B} \cdot \mathbf{k} \equiv 0$). The triple scalar product of the vectors **A**, **B**, and **k** can be represented in the form of a determinant:

$$\mathbf{A} \times \mathbf{B} \cdot \mathbf{k} \equiv \begin{vmatrix} 1 & -\tan\phi_g & 0 \\ \cos\gamma_o & -\tan\upsilon & \sin\gamma_o \\ 0 & 0 & 1 \end{vmatrix} = 0 \tag{11.20}$$

A formula for the calculation of the angle $\upsilon$:

$$\upsilon = \tan^{-1}(\tan\phi_g \cdot \cos\gamma_o) \tag{11.21}$$

immediately follows from Eq. (11.19).

The vector **C** is constructed so that it is located within the plane that is perpendicular to the gear shaper cutter axis $O_{ct}$. For the vector **C**, the expression:

$$\begin{aligned} \mathbf{C} = \mathbf{i} \cdot \cos(\gamma_g + \gamma_{ct}) - \mathbf{j} \cdot \tan\phi_{ct} \\ -\mathbf{k} \cdot \sin(\gamma_g + \gamma_{ct}) \end{aligned} \tag{11.22}$$

is valid.

The unit vector **d** is pointed along the generating line of the clearance surface $C_s$ of the gear shaper cutter tooth. The vector **d** can be expressed in terms of its projections onto the axes of the reference system $xyz$:

$$\mathbf{d} = \mathbf{i} \cdot \sin(\gamma_g + \gamma_{ct}) + \mathbf{k} \cdot \cos(\gamma_g + \gamma_{ct}) \qquad (11.23)$$

Again, the vectors **B**, **C**, and **d** comprise a set of coplanar vectors, and their triple scalar product is identical to zero ($\mathbf{B} \times \mathbf{C} \cdot \mathbf{d} \equiv 0$). This identity allows for the expression

$$\mathbf{B} \times \mathbf{C} \cdot \mathbf{d} \equiv \begin{vmatrix} \cos\gamma_o & -\tan\upsilon & \sin\gamma_o \\ \cos(\gamma_g + \gamma_{ct}) & -\tan\phi_{ct} & -\sin(\gamma_g + \gamma_{ct}) \\ \sin(\gamma_g + \gamma_{ct}) & 0 & \cos(\gamma_g + \gamma_{ct}) \end{vmatrix} = 0 \qquad (11.24)$$

from which the formula for the profile angle $\phi_{ct}$

$$\phi_{ct} = \tan^{-1}\left(\frac{\tan\upsilon}{\cos(\gamma_o + \gamma_g + \gamma_{ct})}\right) \qquad (11.25)$$

is derived.

Substituting Eq. (11.28) into Eq. (11.32), the formula

$$\phi_{ct} = \tan^{-1}\left(\frac{\tan\phi_g \cdot \cos\gamma_o}{\cos(\gamma_o + \gamma_g + \gamma_{ct})}\right) \qquad (11.26)$$

can be derived for the calculation of the gear shaper cutter profile angle $\phi_{ct}$.

The base diameter of the gear shaper cutter can be calculated from the equation

$$d_{b.ct} = d_{ct} \cdot \cos\phi_{ct} \qquad (11.27)$$

The tooth thickness on the pitch diameter $d_{ct}$ of the gear shaper cutter is equal to the space width on the pitch diameter $d_g$ of the gear being machined. The addendum $a_c$ and the dedendum $b_c$ of the gear shaper cutter tooth are assumed to be equal to each other. They are calculated by using the formula

$$a_c = b_c = 1.25 \cdot m \cdot \frac{\cos(\gamma_o + \gamma_g + \gamma_c)}{\cos\gamma_o} \qquad (11.28)$$

The rest of the design parameters of the gear shaper cutter with a tilted axis of rotation can be determined in the same way that the corresponding design parameters of a conventional gear shaper cutter were determined.

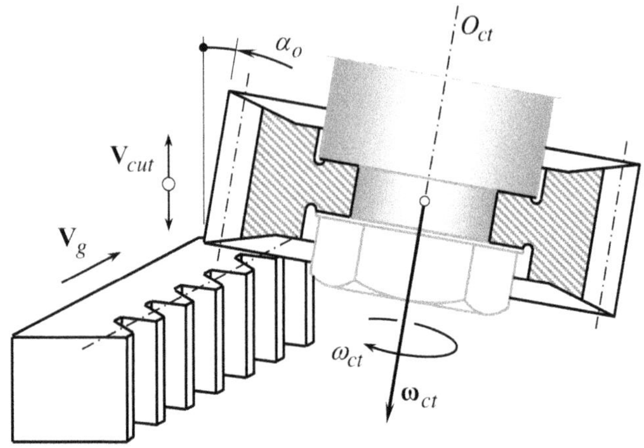


**FIGURE 11.8** Shaping a rack by the gear shaper cutter with a tilted axis of its rotation.

Gear shaper cutters having a tilted axis of rotation can also be used for machining racks as schematically illustrated in Figure 11.8. In this case, the rack is viewed as a work gear with an infinite tooth count. The axis of rotation $O_{ct}$ of the gear shaper cutter is tilted at the clearance angle $\alpha_o$ relative to the pitch plane of the rack. The direction of the reciprocation $\mathbf{V}_{cut}$ is parallel to the pitch plane of the rack.

Gear shaper cutters with a tilted axis of rotation are easier in production. This is mostly because the straight generating lines of the cylindrical clearance surface are parallel to the axis of rotation of the gear shaper cutter. The design of the gear shaper cutters makes it possible to increase the rake angle $\gamma_o$ and—more importantly—to increase the clearance angle $\alpha_o$. In this way, the geometry of the cutting edges results in a better cutting performance by the gear shaper cutter. The accuracy of the gear shaper cutter does not depend on the number of resharpenings, and it remains the same throughout the tool's life span. If accuracy is a critical issue, it can be improved by implementing a specially designed curved generating profile of the gear shaper cutter rake surface $R_s$. However, the use of gear shaper cutters with a tilted axis of rotation requires specially designed gear shaping machines that are capable of handling the tilted axis of the gear shaper cutter spindle.

## 11.3 ILLUSTRATION OF CAPABILITIES OF THE EXTERNAL INTERSECTING-AXIS GEAR MACHINING MESH

The potential capabilities of the external intersecting-axis gear machining mesh [see Figure IV.1(a)] are not limited to the gear shaping operation shown in Figure 11.4. Other combinations of translations and rotations of the reference system $X_{ct}Y_{ct}Z_{ct}$ associated with the gear-cutting tool relative to the work gear are also permissible. As an example,

**FIGURE 11.9** Shaping a straight bevel gear.

two more practical methods of gear machining are discussed next.

### 11.3.1 Shaping Straight Bevel Gears

Straight bevel gears are used in helicopter transmissions as well as in the design of other gear units. More examples of industrial applications of straight bevel gears are reported elsewhere (e.g., [12, 45]). Tapered splines are often used in synchronizer and coupler designs to prevent mechanisms from jumping out of gear. A tapered tooth form also compensates for heat treatment distortions and unequal shrinkage [7].

Straight bevel gears can be cut by gear shaper cutters with a tilted axis of rotation. A close-up of the process is illustrated in Figure 11.9.

When the tilt angle is zero, the equations derived for the gear shaper cutter with the tilted axis of rotation reduce to the case of a gear shaper cutter of conventional design. Evidently, the cut gears are far from being geometrically accurate, however, even gears cut this way are applied in industry (in the design of gear units that spin at low velocities and transmit low power density).

The kinematic scheme that is shown in Figure 11.10 can be used in designing a gear-cutting tool for machining conical involute gears. The kinematic scheme is composed of two rotations $\boldsymbol{\omega}_g$ and $\boldsymbol{\omega}_{ct}$. The shaft angle is denoted by $\Sigma$. The primary motion (the motion of cut $\mathbf{V}_{cut}$) is along the axis of rotation $O_{ct}$ of the gear shaper cutter.

Conical involute gears (or a tapered tooth form) can be produced by tilting the axis of the gear shaper cutter as shown in Figure 11.11, or by tilting the axis of rotation $O_g$ of the work gear. The work gear and the gear shaper cutter are rotating about their axes of rotation $O_g$ and $O_{ct}$ in a properly timed manner. The direction of the primary motion $\mathbf{V}_{cut}$ is at a certain angle with respect to the gear axis of rotation $O_g$. This angle is equal to the pitch cone angle of the conical involute gear.

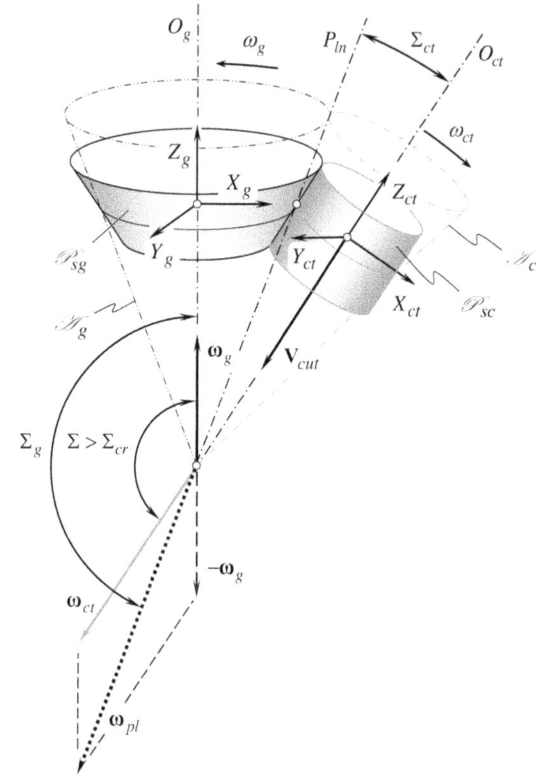


**FIGURE 11.10** Kinematics of shaping a conical involute gear (a tapered tooth form of the gear) using a gear shaper cutter of conventional design. (Straight primary motion $\mathbf{V}_{cut}$ is pointed along the axis of the gear shaper cutter.)

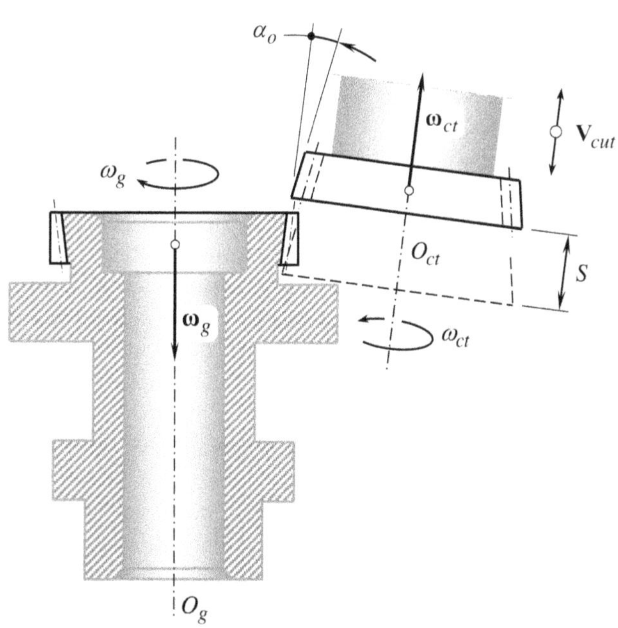


**FIGURE 11.11** Shaping of a conical involute gear by a gear shaper cutter of conventional design.

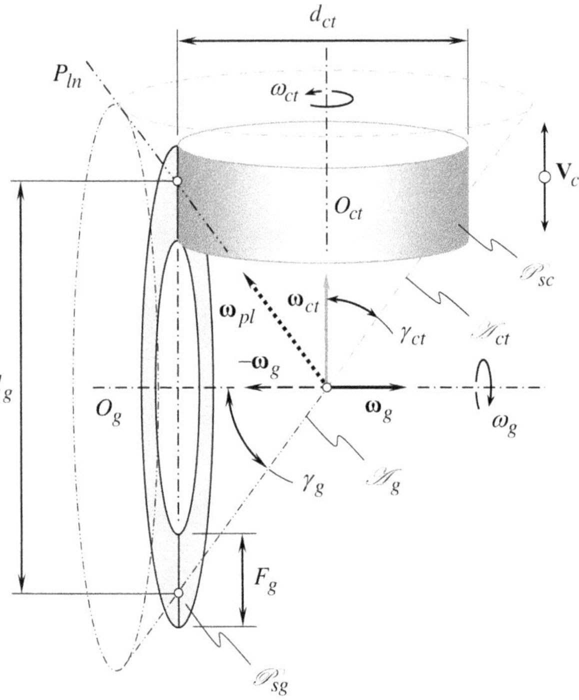


**FIGURE 11.12** Straight primary motion (the motion of cut $\mathbf{V}_{cut}$) in shaping operation of face gear by the standard gear shaper cutter.

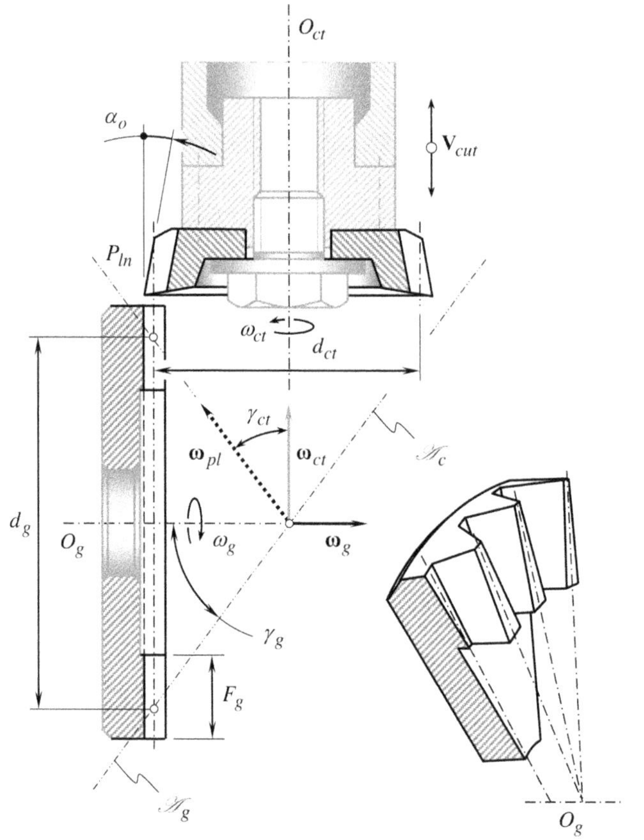


**FIGURE 11.13** Shaping face gear (a crown gear) by a gear shaper cutter of conventional design.

It is likely that the concept of gear machining using a gear shaper cutter with a tilted axis of rotation (see Figure 11.4) can be applied to machining of conical involute gears.

### 11.3.2 Shaping Face Gears

Face gear drives are widely used in industry [3]. Examples can be easily found in the aerospace industry (helicopter transmissions), in the design of automobile differentials, and so forth. Shaping is the easiest way to machine the teeth of face (crown) gears.

The kinematics of machining a face gear by the gear shaper cutter can be composed based on the gear machining mesh shown in Figure IV.1(a). Standard gear shaper cutters can be used in this operation. Either conventional or special gear shaper cutter machines can be used for machining of crown gears. A special attachment is required in the first case to enable the rotations of the work gear and of the gear shaper cutter about the axes that intersect at a right angle.

Figure 11.12 illustrates the kinematics of shaping a conventional face gear by the standard gear shaper cutter. When shaping the face gear, the rotation vectors of the work gear $\boldsymbol{\omega}_g$ and that of the gear shaper cutter $\boldsymbol{\omega}_{ct}$ are at a right angle. In the relative motion, the gear shaper cutter is performing an instant rotation $\boldsymbol{\omega}_{pl}$ about the axis $P_{\text{ln}}$ of the instant rotation.

The direction of the primary motion (that is, of the reciprocation of the gear shaper cutter $\mathbf{V}_{cut}$) is orthogonal to the face gear axis of rotation $O_g$. The diameter of the operating pitch circle of the gear is designated as $d_g$, and the corresponding diameter of the operating pitch cylinder of the gear shaper cutter is denoted by $d_{ct}$.

When shaping the face gear (see Figure 11.13), the work gear and the gear shaper cutter are rotating about their axes of rotation $O_g$ and $O_{ct}$ correspondingly. The rotations $\omega_g$ and $\omega_{ct}$ are synchronized with one another so that the ratio $\omega_g \cdot N_g = \omega_{ct} \cdot N_{ct}$ is observed. Here, $N_g$ is the tooth number of the face gear, and $N_{ct}$ is the tooth number of the gear shaper cutter. The gear shaper cutter is reciprocating in the direction toward the axis $O_g$ of the work gear and back. Proper face gear tooth flank generation is feasible within a reasonable face width $F_g$. The feasibility of proper tooth generation strongly depends on tooth pointing at the outer diameter of the face gear and on tooth undercutting at the inner diameter of the gear shaper cutter as shown in Figure 11.14.

The concept of gear shaping by a gear shaper cutter having a tilted axis of rotation (see Figure 11.4) can be extended to machining face gears as well. A schematic diagram of this concept is illustrated in Figure 11.14.

To date, the kinematic scheme of gear generation (see Figure 11.14) has not been investigated thoroughly. Analysis of the feasibility of other combinations of

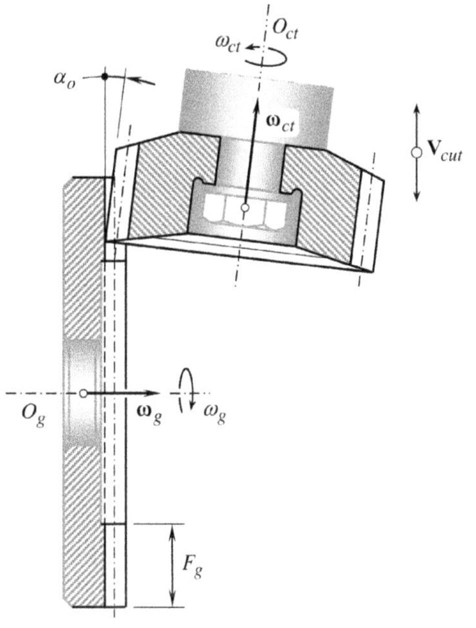


**FIGURE 11.14** Shaping of face gear (a crown gear) by a gear shaper cutter that features a tilted axis of rotation.

translations and rotations of the reference system associated with the gear-cutting tool relative to the work gear reference system could reveal novel methods of gear generating, as well as new design concepts for gear-cutting tools. There remains a considerable amount of research to be conducted in this field.

Only approximate cylindrical gears, bevel gears, face gears, and rack gears can be cut by gear-cutting tools when the axes of rotation of the work gear and the gear-cutting tool intersect one another. Geometrically accurate gears of these types cannot be cut by gear shaper cutters when the axes of rotation $O_g$ and $O_{ct}$ intersect. This is attributed to the kinematics of the gear shaping process, and to the violation of the condition of conjugacy of the tooth flanks $\mathscr{G}$ of a gear being machined and of the generating surface $T$ of the gear-cutting tool. In compliance with the condition of conjugacy of the tooth flanks, at every point of contact of the surfaces $\mathscr{G}$ and $T$ the common perpendicular $\mathbf{n}$ must intersect the axis of instant rotation $P_{\text{ln}}$ of the gear and the gear-cutting tool in the gear machining mesh [171]. This is a must if we are about machining precision (i.e., geometrically accurate) gears.

# 12 Gear-Cutting Tools for Machining Bevel Gears

Two rotation vectors—one of which is associated with the work gear and the other one with the gear-cutting tool—uniquely determine the vector of instant rotation of the gear-cutting tool relative to the work gear. The latter makes possible the classification of various types of gear machining meshes (see Figure 3.8 in Chapter 3). The actual configuration of the vector of instant rotation $\boldsymbol{\omega}_{pl}$ depends on the rotation vectors $\boldsymbol{\omega}_g$ and $\boldsymbol{\omega}_{ct}$ of the work gear and of the gear-cutting tool. The angle $\Sigma = \angle(\boldsymbol{\omega}_g, \boldsymbol{\omega}_{ct})$ that is measured between the rotation vectors $\boldsymbol{\omega}_g$ and $\boldsymbol{\omega}_{ct}$, along with magnitudes $\omega_g$ and $\omega_{ct}$ of the vectors $\boldsymbol{\omega}_g$ and $\boldsymbol{\omega}_{ct}$, uniquely determine the angles $\angle(\boldsymbol{\omega}_{pl}, \boldsymbol{\omega}_g)$ and $\angle(\boldsymbol{\omega}_{pl}, \boldsymbol{\omega}_{ct})$ which the rotation vectors $\boldsymbol{\omega}_g$ and $\boldsymbol{\omega}_{ct}$ form with the vector of instant rotation $\boldsymbol{\omega}_{pl}$.

Depending on the actual value of the magnitude and orientation of the rotation vectors $\boldsymbol{\omega}_g$ and $\boldsymbol{\omega}_{ct}$, each of the angles $\Sigma_g$ and $\Sigma_{ct}$ can either be obtuse or acute. In a particular case, one of these angles is equal to a right angle [either $\Sigma_g = \angle(\boldsymbol{\omega}_{pl}, \boldsymbol{\omega}_g) = 90°$ or $\Sigma_{ct} = \angle(\boldsymbol{\omega}_{pl}, \boldsymbol{\omega}_{ct}) = 90°$].

The first case (that is, $\boldsymbol{\omega}_{pl} \perp \boldsymbol{\omega}_g$) relates to the intersecting-axes gear machining mesh whose vector diagram is shown in Figure III-B.1(e). Additional details of this gear machining mesh are shown in Figure 12.1. Here, in addition to the vector diagram [see Figure 12.1(a)], axodes of the work gear $\mathscr{A}_g$ and the gear-cutting tool $\mathscr{A}_{ct}$ are also depicted [see Figure 12.1(b)]. To date, this type of gear machining mesh has not been thoroughly investigated. To fully utilize its potential in industry, an extensive investigation of its capabilities must be carried out. Because of the limited amount of available research results, it will not be discussed further in this work.

The second case (that is, $\boldsymbol{\omega}_{pl} \perp \boldsymbol{\omega}_{ct}$) relates to the intersecting-axes gear machining mesh whose vector diagram is shown in Figure III-B.1(d). Additional details of this gear machining mesh are illustrated in Figure 12.2. Here, in addition to the vector diagram [see Figure 12.2(a)], axodes of the work gear $\mathscr{A}_g$ and of the gear-cutting tool $\mathscr{A}_{ct}$ are also depicted [see Figure 12.2(b)]. This type of gear machining mesh is extensively used in industry. Numerous designs of gear-cutting tools, as well as gear machining methods, are developed based on the gear machining mesh schematically shown in Figure 12.2. The analysis performed in this chapter is focused on the application of the intersecting-axes gear machining mesh for which the condition $(\boldsymbol{\omega}_{pl} \perp \boldsymbol{\omega}_{ct})$ is observed. Gear-cutting tools of various designs for machining bevel gears can be designed based on a gear machining mesh of this type. However, the capabilities of the gear machining mesh (see Figure 12.2) are not limited to the design of gear-cutting tools for cutting bevel gears; gear-cutting tools for machining gears of other types can also be designed on the premise of this particular gear machining mesh.

## 12.1 PRINCIPAL ELEMENTS OF KINEMATICS OF BEVEL GEAR GENERATION

Spiral bevel gears, when manufactured in a metal-cutting process, can either be produced in single indexing or continuous-indexing operations, which govern the form of the tooth trace.

For generation of a bevel gear, an intersecting-axes gear machining mesh is used (see Figure 12.2). The rotation vectors $\boldsymbol{\omega}_g$ and $\boldsymbol{\omega}_{ct}$ are along the corresponding axes of rotation $O_g$ and $O_{ct}$. The rotation vectors $\boldsymbol{\omega}_g$ and $\boldsymbol{\omega}_{ct}$ are at a

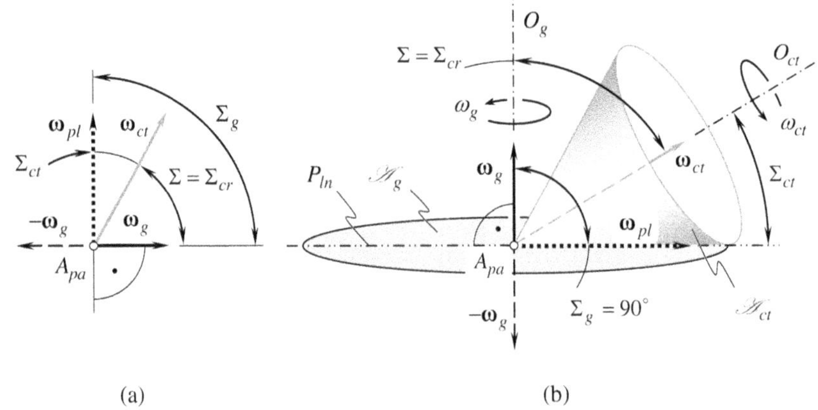


**FIGURE 12.1** Vector representation of the intersecting-axes gear machining mesh for machining round rack: (a) the gear vector diagram, and (b) the corresponding axodes for the gear machining mesh.



DOI: 10.1201/9781003520528-16

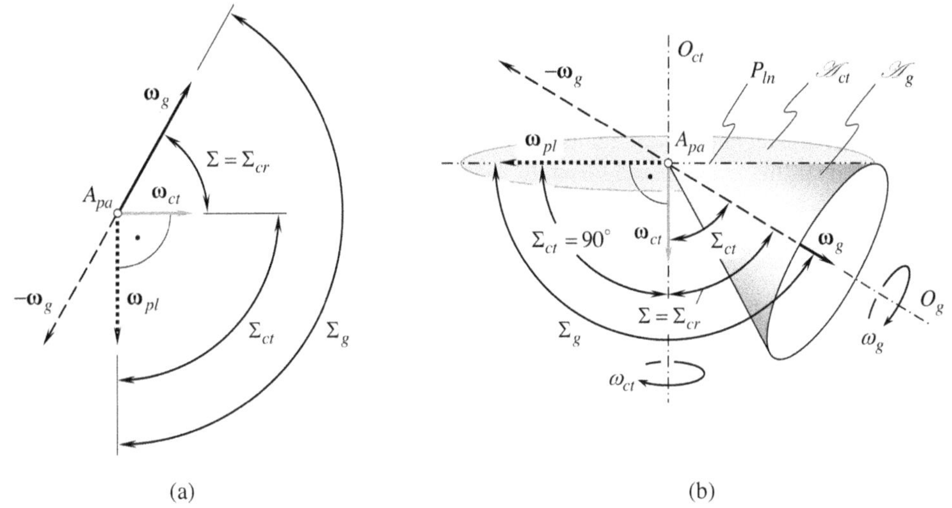


**FIGURE 12.2** Vector representation of the round rack–type intersected-axes gear machining mesh: (a) the gear vector diagram, and (b) the corresponding axodes for the gear machining mesh.

certain intersected-axes angle $\Sigma$ relative to each other. For this case, the intersected-axes angle $\Sigma$ is equal to its critical value $\Sigma_{cr}$, that is, the equality $\Sigma = \Sigma_{cr}$ is observed. This occurs because the condition $\boldsymbol{\omega}_{pl} \perp \boldsymbol{\omega}_{ct}$ is observed. By definition, the critical value of the intersected-axes angle $\Sigma_{cr}$ is that for which the rotation vector of the gear-cutting tool $\boldsymbol{\omega}_{ct}$ is perpendicular to the vector of instant rotation $\boldsymbol{\omega}_{pl}$.

For the readers who prefer to visualize the gear machining mesh, two axodes[1] can be associated with the rotation vectors $\boldsymbol{\omega}_g$ and $\boldsymbol{\omega}_{ct}$. The gear axode $\mathscr{A}_g$ is associated with the rotation vector $\boldsymbol{\omega}_g$, whereas the cutting tool axode $\mathscr{A}_{ct}$ is associated with the rotation vector $\boldsymbol{\omega}_{ct}$. Generally speaking, each of the axodes is shaped in the form of two cones of revolution.

The parameters of the axodes $\mathscr{A}_g$ and $\mathscr{A}_{ct}$ can be specified in terms of the pitch angles $\gamma_g$ and $\gamma_c$. In the case under consideration, the pitch angle $\gamma_c = \pi/2$. This means that the axode of the gear-cutting tool is transformed into a plane $\mathscr{A}_{ct}$ that is rotating about the axis $O_{ct}$. The axis of rotation $O_{ct}$ is perpendicular to the plane $\mathscr{A}_{ct}$. It is advantageous to have the axode $\mathscr{A}_{ct}$ in the form of a plane when designing a gear-cutting tool for machining bevel gears, that is, those bevel gears having either straight teeth, or skew teeth, or circular arc teeth, and so forth.

When the intersected-axes angle $\Sigma$ is given, for the calculation of the pitch angle $\gamma_g$ of the axode of the work gear, the formula $\gamma_g = \pi/2 - \Sigma$ applies.

To calculate the pitch angle $\gamma_g$, the formula applies:

$$\gamma_g = \sin^{-1}\left(\frac{\omega_{ct}}{\omega_g}\right) \tag{12.1}$$

[1] The axodes themselves, those associated with the gear and with the gear-cutting tool, are not the main focus in this book. Any and all results here can be derived without applying the concept of axodes. In some illustrations, the axodes are shown only for the readers who have special preferences.

The formula for the pitch cone angle $\gamma_g$ can be expressed in terms of the tooth numbers of the work gear $N_g$ and of the generating surface of the gear-cutting tool $N_{ct}$:

$$\gamma_g = \sin^{-1}\left(\frac{N_g}{N_{ct}}\right) \tag{12.2}$$

The axis of instant rotation $P_{\text{ln}}$ is aligned with the vector $\boldsymbol{\omega}_{pl}$ of the instant rotation of the axodes $\mathscr{A}_g$ and $\mathscr{A}_{ct}$.

When machining a work gear with a gear-cutting tool that is designed in compliance with the gear machining mesh (see Figure 12.2), the cone $\mathscr{A}_g$ rolls over the plane $\mathscr{A}_{ct}$ with no sliding between them. The relative motion of the axode $\mathscr{A}_{ct}$ of the gear cutter with respect to the axode $\mathscr{A}_g$ of the work gear allows for interpretation in the form of instant rotation $\boldsymbol{\omega}_{pl}$ about the axis of instant rotation $P_{\text{ln}}$.

The generating action of the gear-cutting tool is the result of the two rotations $\boldsymbol{\omega}_g$ and $\boldsymbol{\omega}_{ct}$. The rotations are synchronized with one another in a timely manner. To make material removal possible, the corresponding primary motion (the motion of cut) is required when machining the work gear. Several solutions are available when determining the primary motion. Any and all permissible motions of a reference system $X_{ct}Y_{ct}Z_{ct}$ associated with the gear-cutting tool, relative to the coordinate system $X_gY_gZ_g$ associated with the work gear, are considered when selecting an appropriate motion of cut. When performing the analysis, one should bear in mind the constraints imposed by the rolling motion of the axodes onto the permissible relative motion of the work gear and the gear-cutting tool.

When no constraints are imposed, the relative motion of the coordinate system $X_{ct}Y_{ct}Z_{ct}$ of the gear-cutting tool can be reduced to not more than three elementary translations ($\mathbf{V}_x$, $\mathbf{V}_y$, and $\mathbf{V}_z$) along the coordinate axes, and not more than three elementary rotations ($\boldsymbol{\omega}_x$, $\boldsymbol{\omega}_y$, and $\boldsymbol{\omega}_z$) about the coordinate axis of the reference system $X_{ct}Y_{ct}Z_{ct}$.

Consider all six possible elementary motions, and all possible combinations as 1, 2, …, 6. Various ratios of speed of the elementary motions need to be accounted for. Not all relative motions and their combinations are permissible when machining a given gear. Motions that are not permissible for implementation in cutting gears should be omitted from further analysis. In this way, all permissible motion(s) of cut can be determined.

## 12.2 GEOMETRY OF INTERACTING TOOTH SURFACES

The gear machining mesh that is schematically depicted in Figure 12.2 is used in industry for the design of gear-cutting tools used for machining straight bevel gears as well as bevel gears having skew teeth. Gear-cutting tools used in these machining processes and methods of gear machining are good illustrations of the classification of possible types of gear machining meshes (see Fig. 3.8 in Chapter 3).

To design a cutting tool for machining a given bevel gear, a comprehensive analysis of the geometry of the tooth flank of the work gear should be done.

### 12.2.1 Principal Elements of the Geometry of Straight Bevel Gear Tooth Flank

The shape of the straight tooth of a bevel gear is illustrated in Figure 12.3. Major design parameters of the bevel gear are also indicated in the figure. In addition to the conventional design parameters, the addendum angle $\gamma_a$ and the dedendum angle $\gamma_d$ in the design of a bevel gear tooth are shown in Figure 12.3. Generally speaking, the tooth flank $\mathscr{G}$ of a straight bevel gear is shaped in the form of a conical surface. The apex of the cone is coincident with the gear base cone apex $A_g$. In the design of perfect bevel gears, the tooth profile is viewed as a spherical involute. The tooth profile serves as the director line of the conical surface $\mathscr{G}$ (see Figure 7.5 on page 160 in [172]). This means that the tooth flank can be interpreted as a locus of straight lines through a point of the tooth profile and the base cone apex $A_g$. Therefore, the straight bevel gear tooth flank $\mathscr{G}$ is a type of developable surface.

The straight bevel gear tooth flank $\mathscr{G}$ is analytically described by Eq. (2.32):

$$\mathbf{r}_g(U_g,\varphi_g) = \begin{bmatrix} U_g \tan\theta_g \sin\varphi_g - \varphi_g \cdot U_g \cdot \tan\theta_g \cos\varphi_g \\ U_g \tan\theta_g \cos\varphi_g + \varphi_g \cdot U_g \cdot \tan\theta_g \sin\varphi_g \\ -U_g \\ 1 \end{bmatrix} \quad (12.3)$$

The generating surface $T$ of a gear-cutting tool must be conjugate to the bevel gear tooth flank given by Eq. (12.3). This means that at every point of contact of the surfaces $\mathscr{G}$ and $T$, the common perpendicular $\mathbf{n}_g$ must intersect the axis of instant rotation $P_{ln}$ in the gear machining mesh.

The geometry and the generation of a straight bevel gear tooth flank were thoroughly investigated by Prof. N. I. Kolchin [32]. It has been shown that the tooth flank $\mathscr{G}$ of a straight bevel gear is a surface of complex geometry. This surface significantly differs from that of a spur gear. The profile of a straight bevel gear tooth can be constructed on a sphere having its center coincident with the base cone apex $A_g$ of the bevel gear. The actual tooth profile[2] of a straight bevel gear tooth is commonly referred to as a *spherical involute*.

### 12.2.2 Generating Surface of the Gear-Cutting Tool

In the gear machining mesh (see Figure 12.2), the axis of rotation $O_{ct}$ of the generating surface $T$ of the cutter is perpendicular to the vector of instant rotation $\boldsymbol{\omega}_{pl}$. The generating straight line of the cutting tool axode $\mathscr{A}_{ct}$ is aligned to the rotation vector $\boldsymbol{\omega}_{pl}$. In the case under consideration, the generating surface $T$ of a gear-cutting tool can be viewed as a surface of straight bevel gear having a pitch cone angle $\gamma_{ct} = 90°$.

The geometry of involute straight bevel gear tooth flank $\mathscr{G}$, having a right base cone angle $\gamma_{ct} = 90°$, has not been thoroughly investigated. A bevel gear of this geometry is often referred to either as a *round rack* or a *crown gear*. Tooth flanks of the round rack (of the crown gear, in other terminology) have a spherical involute profile (see Figure 12.4). A tooth profile of this type is often viewed as an involute curve constructed on a sphere of the corresponding diameter. The spherical involute features a cusp (at the points $A$ and $C$ in Figure 12.4) that is located on the spherical surface, and a point of inflection (at the points $B$ and $D$ in Figure 12.4) that is located within the axode $\mathscr{A}_{ct}$.

A bevel round involute rack $T$ having the tooth form shown in Figure 12.4 can be engaged in correct mesh with the perfect straight bevel involute gear being machined. Potentially, a round rack of such design could be used as the generating surface of the gear-cutting tool for machining bevel involute gears. However, the use of the bevel round involute rack for this purpose is inconvenient in practice for several reasons.

*First*, although proportions remain the same, dimensions of the tooth profile of the round rack reduce toward the base cone apex $A_{ct} \equiv A_{pa}$ of the round rack. The cutting edge $CE$ of the gear-cutting tool is rigid, and it is not capable of

[2] The commonly used term *profile* is incorrect with respect to intersected-axes gears. By definition, the term *profile* means a planar curve. The plane within which the profile is located is perpendicular to the direction of the profile motion. A surface is generated when the profile is traveling. This can be either a surface of translation (that is, a cylinder), or a surface of revolution (and not a helical surface). No straight bevel gear tooth flank $\mathscr{G}$ can be generated by a moving spherical involute. Therefore, the term *spherical involute profile* has to be eliminated from the list of terms that pertains to intersected-axes gear machining mesh.

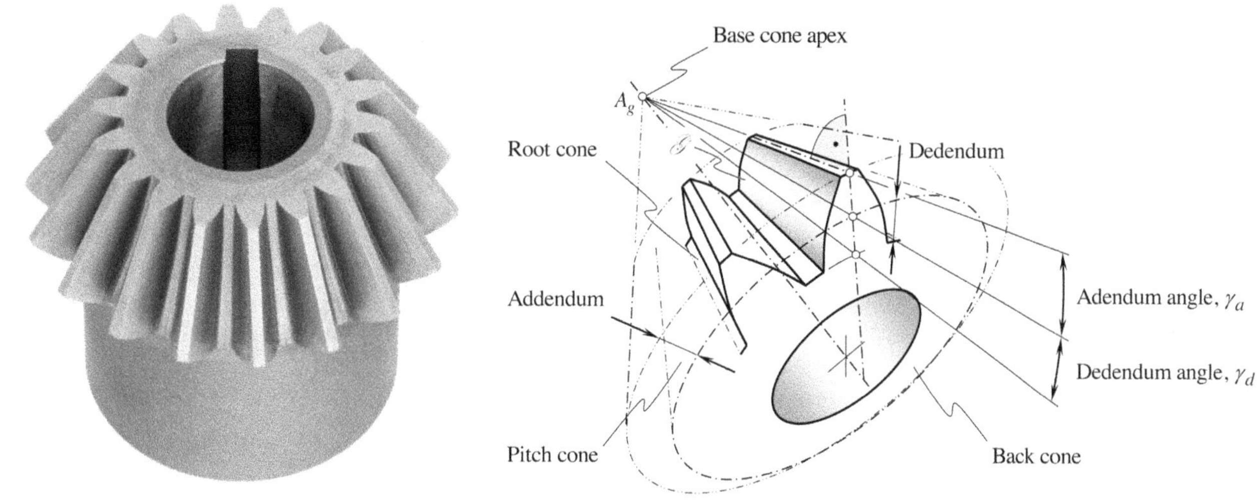


**FIGURE 12.3** Geometry of tooth flank $\mathscr{G}$ and the key design parameters of a straight bevel gear.

changing its shape in response to the change in the dimensions of the round rack tooth profile.

*Second*, in order to avoid interference of the cutting edge beneath the bottom land of the bevel gear, the direction of the primary motion $\mathbf{V}_{cut}$ of the cutting tool has to be pointed at the dedendum angle $\gamma_d$ relative to the vector of instant rotation $\boldsymbol{\omega}_{pl}$ (see Figure 12.5). The corresponding mesh of the work-gear teeth and the round rack teeth is illustrated in Figure 12.6. Under such a scenario, the rotation vector of the cutting tool $\boldsymbol{\omega}_{ct}$ is perpendicular to the vector of instant rotation $\boldsymbol{\omega}_{pl}$, whereas the vector of the primary motion $\mathbf{V}_{cut}$ is not orthogonal to the rotation vector of the gear-cutting tool $\boldsymbol{\omega}_{ct}$.

Fortunately, the curvature of the tooth spherical profile above and below the axode $\mathscr{A}_{ct}$ is small enough. Figure 12.7 provides an insight into the actual shape of the spherical mapping of the round rack having a mean cone distance $A_m = 200$ mm. This makes it possible to substitute the curved cutting edge with a straight one. The approximation of the curved cutting edge with the straight one could be reasonable in many practical cases of bevel gear design and manufacture. (Be careful when machining bevel gears with a low tooth count; the curvature of cutting edges in *LTC*− gears can be significant and cannot be ignored!) The straight cutting edge does not need to change its shape when the cutting tool moves toward the axis of rotation of the cutting tool. This is an important advantage of the approximation.

Moreover, in most practical cases of bevel gear design and manufacture, the actual value of the dedendum angle $\gamma_d$ is small compared to the base cone angle. Because of that, the direction of the primary motion $\mathbf{V}_{cut}$ is practically

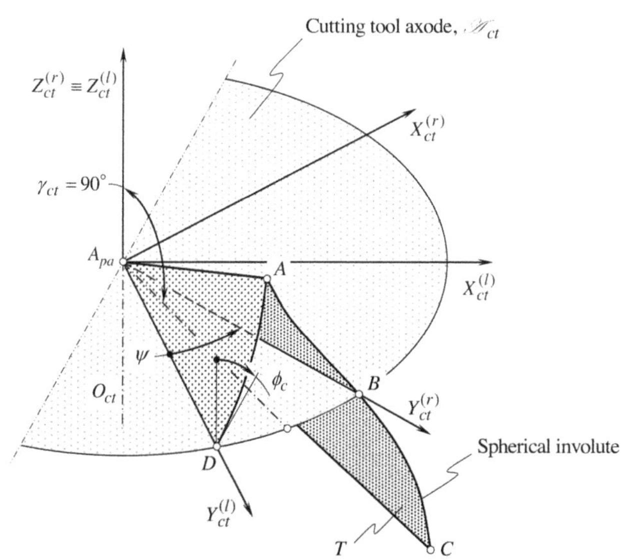


**FIGURE 12.4** Generating surface $T$ of the gear-cutting tool for machining a perfect straight bevel gear tooth form.

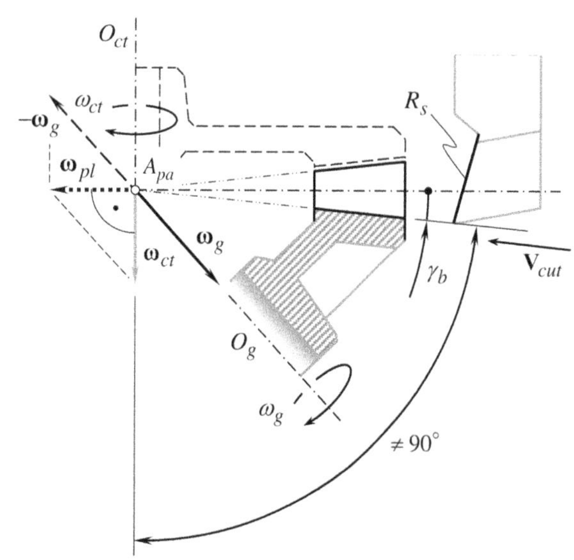


**FIGURE 12.5** The desired direction of the primary motion $\mathbf{V}_{cut}$ is not perpendicular to the rotation vector $\boldsymbol{\omega}_{ct}$.

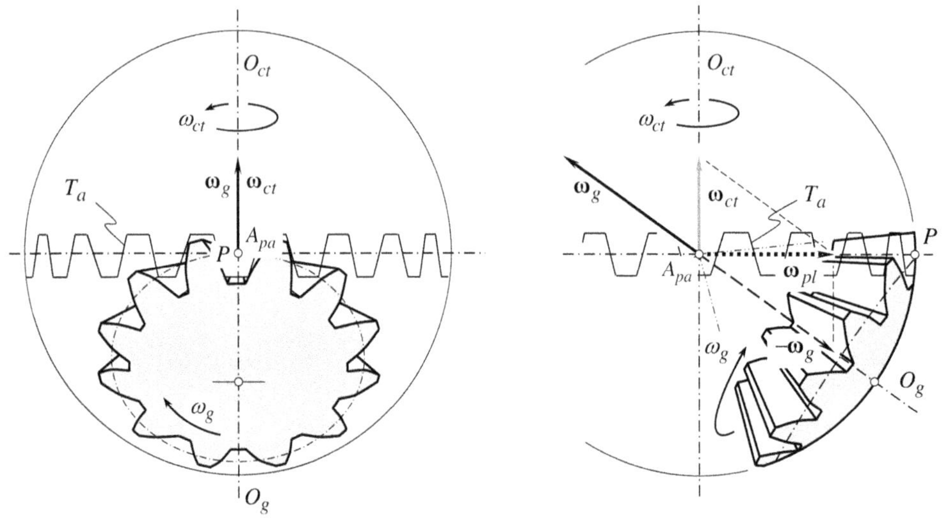


**FIGURE 12.6** A bevel work gear in mesh with the straight-sided bevel crown rack.

pointed along the vector of instant rotation $\boldsymbol{\omega}_{pl}$ as shown in Figure 12.8. As a result, the rotation vector $\boldsymbol{\omega}_{ct}$ of the gear-cutting tool is not perpendicular to the vector of instant rotation $\boldsymbol{\omega}_{pl}$. In such a condition, the pitch surface of the round rack is no longer a plane. It reverts to a cone having the cone angle $\gamma_c$ whose magnitude is close to a right

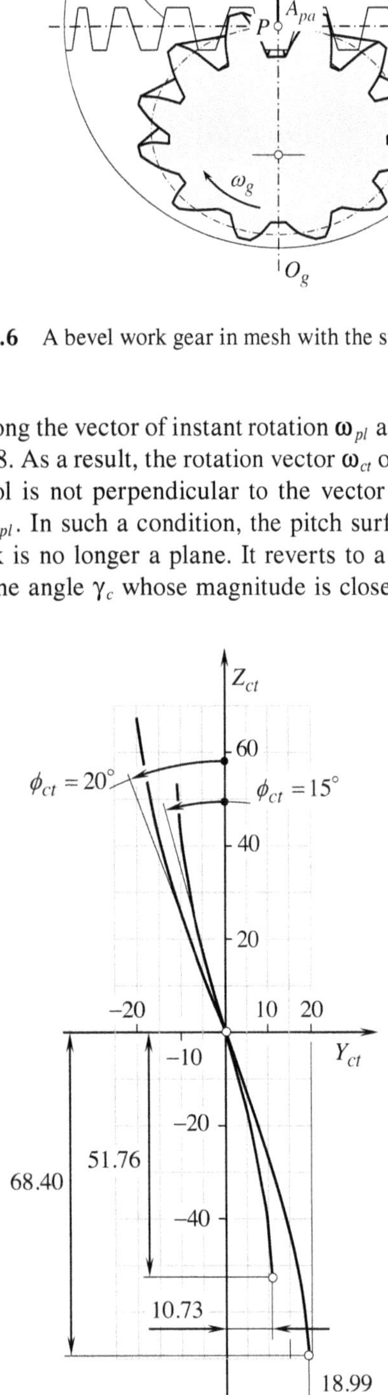


**FIGURE 12.7** Spherical mapping of tooth spherical profiles of the bevel round involute rack (after Kolchin, N. I., *Analytical calculation of planar and spatial gearing: with application to profiling of the gear cutting tools and to calculation of the deviations in gearing*, Moscow, Mashgiz, 1949, 210 pages).

angle ($\gamma_c \approx 90°$). This particular design of the round rack is referred to as the *planar-top-land round rack* ($PTL-$ *rack*). $PTL-$ rack is a type of practical approximation to the desired bevel round involute rack.

Two simplifications—(1) approximation of the curved cutting edge $CE$ of the gear-cutting tool with a straight cutting edge, and (2) substitution of the bevel round involute rack with the $PTL-$ rack—affect the accuracy of the machined bevel gears. However, in most practical cases of bevel gear machining, both simplifications are reasonable and allowable.

Tooth flanks of the $PTL-$ rack are not conjugate to the bevel involute gear tooth flanks $\mathscr{G}$. Therefore, the $PTL-$ rack can be considered a kind of practical approximation $T_a$ of the desired generating surface $T$ of the gear-cutting tool. Thus, straight bevel gears—and not the desired involute bevel gears—can be generated with the approximated surface $T_a$. In reality, the generated straight bevel gear is a type of approximation of the desired bevel involute gear.

The generation of straight bevel gear tooth flanks is a perfect example of when the generating surface $T$ is not the envelope surface of consecutive positions of the desired bevel involute gear tooth flanks, which is given by the blueprint. Instead, the approximate generating surface $T_a$ is considered the given surface. For such a case, the straight bevel gear tooth flanks $\mathscr{G}_a$ can be interpreted as an approximation of the desired tooth flanks $\mathscr{G}$.

In such a scenario, an equation of the approximate generating surface of the gear-cutting tool is derived differently from the way it has often been done before. Tooth flanks of the approximate generating surface $T_a$ (namely, of the $PTL-$ rack) are shaped in the form of planes through the apex of the base cone of the rack, and they form the profile angle $\phi_{ct}$ with the rotation vector $\boldsymbol{\omega}_{ct}$ of the cutting tool. This makes it possible to derive an equation for the approximate surface $T_a$ of the desired bevel involute

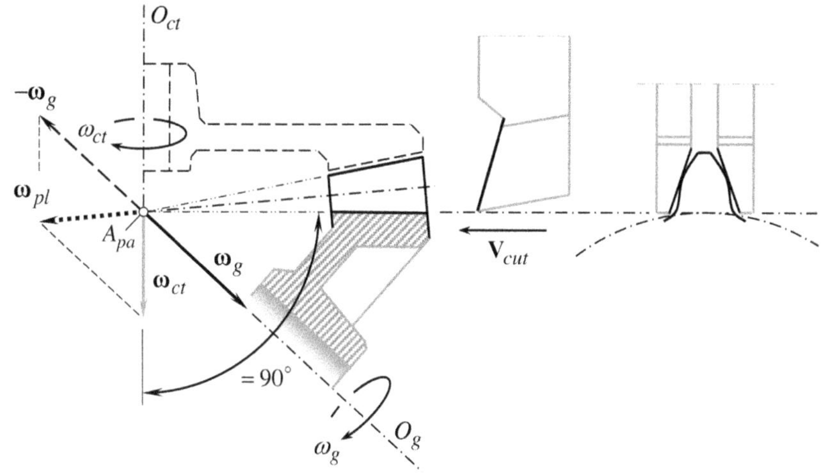


**FIGURE 12.8** The actual direction of the primary motion $\mathbf{V}_{cut}$ is perpendicular to the rotation vector $\omega_{ct}$.

gear $\mathscr{G}$ independently of Eq. (12.3). To derive the equation of the approximate generating surface $T_a$, only the design parameters of the bevel gear being machined are used as input.

Figure 12.9 shows a Cartesian coordinate system $X_{ct}Y_{ct}Z_{ct}$ with which the gear-cutting tool will be associated after it has been designed. In the coordinate system $X_{ct}Y_{ct}Z_{ct}$, a plane $T_a^*$ is the plane through the $X_{ct}-$ axis. The plane $T_a^*$ is at the profile angle $\phi_{ct}$ relative to the $Z_{ct}-$ axis. This immediately yields an equation

$$\mathbf{r}_{ta}^* = \begin{bmatrix} X_{ct} \\ -Y_{ct} \\ Y_{ct}\cdot\cot\phi_{ct} \\ 1 \end{bmatrix} \quad (12.4)$$

for the position vector $\mathbf{r}_{ta}^*$ of a point of the plane $T_a^*$.

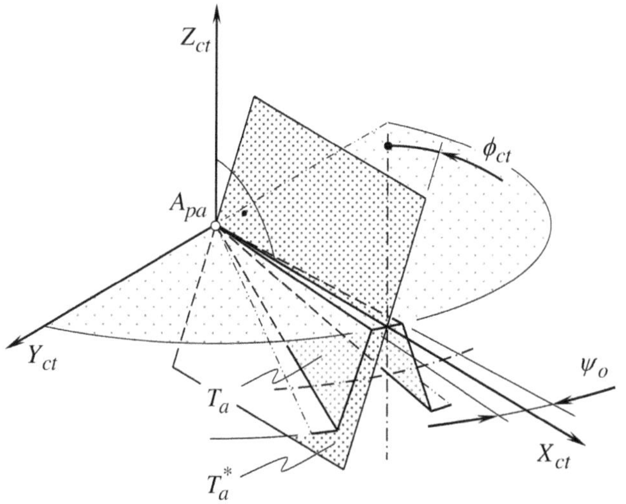


**FIGURE 12.9** Determination of the configuration of the lateral tooth flank in the $PTL-$ rack.

Let us turn the plane $T_a^*$ about the $Z_{ct}-$ axis through the angle $0.5\cdot\psi_o$. Here, $\psi_o$ denotes the angle over which the top land of the generating surface $T_a$ spans (see Figure 12.9). After it has been turned, the plane $T_a^*$ occupies the position of the lateral plane of the generating surface $T_a$ of the gear-cutting tool. Therefore, the equation for the position vector $\mathbf{r}_{ta}$ of a point of the lateral plane $T_a$ can take the form

$$\mathbf{r}_{ta} = \mathbf{Rt}(0.5\cdot\psi_o, Z_{ct})\cdot\mathbf{r}_{ta}^* \quad (12.5)$$

It is important to stress here that the plane $T_a$ is a plane through the apex of the $PTL-$ rack.

Similarly, an equation for the opposite tooth flank of the $PTL-$ rack can be derived. The opposite tooth flank is also a plane through the apex of the $PTL-$ rack.

### 12.2.3 Geometry of Approximated Tooth Flank of Generated Bevel Gear

An analytical description for the generated tooth flank $\mathscr{G}_a$ of the straight bevel gear can be derived using the kinematic method of generation of enveloping surfaces. For this purpose, it is necessary

- to represent the lateral plane $T_a$ in its current location and orientation in the reference system $X_gY_gZ_g$ associated with the work gear, and
- to find out in the coordinate system $X_gY_gZ_g$ the loci of lines of contact of the known plane $T_a$ with the straight bevel gear tooth flank $\mathscr{G}_a$ to be determined.

The coordinate systems $X_gY_gZ_g$ and $X_{ct}Y_{ct}Z_{ct}$ are associated with the work gear and with the gear-cutting tool, respectively, as illustrated in Figure 12.10. For convenience, the origins of booth reference systems are snapped together at the apex $A_{pa}$ of the base cone of the work gear and of the gear-cutting tool. The selected configuration of the

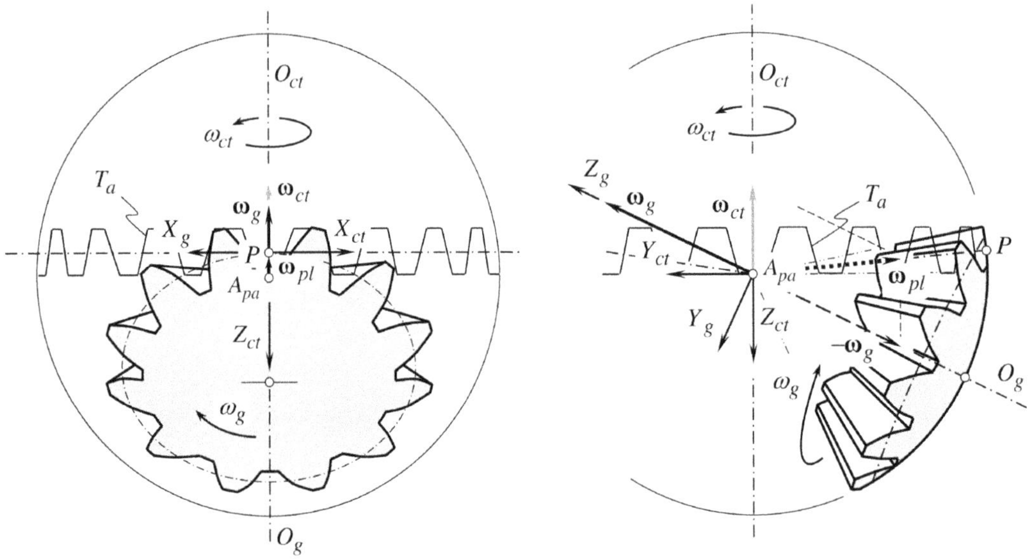


**FIGURE 12.10** Determination of the generated tooth flank $\mathcal{G}_a$ of the straight bevel gear.

reference systems makes it easier to compose the operator $\mathbf{Rs}(ct \mapsto g)$ of the resultant coordinate system transformation, namely, the operator of the transition from the cutting tool coordinate system $X_{ct}Y_{ct}Z_{ct}$ to the coordinate system $X_gY_gZ_g$ associated with the straight bevel gear. The operator $\mathbf{Rs}(ct \mapsto g)$ is equal to the product of the corresponding operators of elementary transformations (see Appendix D for details).

The current location and orientation of the lateral plane $T_a$ in the reference system $X_gY_gZ_g$ can be analytically expressed in terms of the position vector $\mathbf{r}_{ta}$ of a point of the lateral plane $T_a$ [see Eq. (12.5)] and the operator of coordinate system transformation $\mathbf{Rs}(ct \mapsto g)$. For the position vector of a point $\mathbf{r}_{ta}^g$ of the lateral plane $T_a$ in the reference system $X_gY_gZ_g$, the expression

$$\mathbf{r}_{ta}^g(\varphi_g) = \mathbf{Rs}(ct \mapsto g)\cdot \mathbf{r}_{ta} \tag{12.6}$$

is valid.

It should be stressed here that the position vector $\mathbf{r}_{ta}^g$ is a function of the angle of rotation $\varphi_g$ of the gear. Because of this, the operator of the resultant coordinate system transformation $\mathbf{Rs}(ct \mapsto g)$ depends on the current value of the angle $\varphi_g$.

Next, it is necessary to compose the equation of contact for the plane $T_a$ in its motion relative to the reference system $X_gY_gZ_g$. The equation can be represented in the form of a dot product [203, 204]

$$\mathbf{n}_{ta}^g \cdot \mathbf{v}_{ct/g}^g = 0 \tag{12.7}$$

where $\mathbf{n}_{ta}^g$ designates the unit normal vector to the lateral plane $T_a$ and $\mathbf{v}_{c/g}^g$ is the vector of the relative motion of the plane $T_a$. Note that both the vectors have to be represented in the common coordinate system $X_gY_gZ_g$.

Equation (12.4) yields an expression

$$\mathbf{n}_{ta} = \frac{Y_{ct}}{\cos\phi_{ct}}\cdot \mathbf{Rt}(0.5\cdot\psi_o, Z_{ct})\cdot \begin{bmatrix} X_{ct} \\ Y_{ct} \\ Y_{ct}\cdot\tan\phi_{ct} \\ 1 \end{bmatrix} \tag{12.8}$$

for the unit normal vector $\mathbf{n}_{ta}$ to the lateral plane $T_a$ in the reference system $X_{ct}Y_{ct}Z_{ct}$. Next, the unit normal vector $\mathbf{n}_{ta}^g$ can be calculated as

$$\mathbf{n}_{ta}^g(\varphi_g) = \mathbf{Rs}(ct \mapsto g)\cdot \mathbf{n}_{ta} \tag{12.9}$$

It is convenient to compose an expression for the vector $\mathbf{v}_{c/g}^g$ of relative motion of the plane $T_a$ using, for this purpose, the vector $\mathbf{v}_{ct/g}$ of the linear speed of the plane $T_a$ in the reference system $X_{ct}Y_{ct}Z_{ct}$.

In the case under consideration, the vector $\boldsymbol{\omega}_{ct}$ is the rotation vector of the plane $T_a$. This immediately yields the formula

$$\mathbf{v}_{ct/g} = \boldsymbol{\omega}_{ct} \times \mathbf{r}_{ta} \tag{12.10}$$

for the calculation of the vector $\mathbf{v}_{ct/g}$.

In this equation, position vector $\mathbf{r}_{ta}$ is given by Eq. (12.5). The rotation vector $\boldsymbol{\omega}_{ct}$ is calculated as the product

$$\boldsymbol{\omega}_{ct} = \omega_{ct}\cdot \mathbf{k}_{ct} \tag{12.11}$$

where $\omega_{ct}$ is the magnitude of the rotation vector $\boldsymbol{\omega}_{ct}$, and $\mathbf{k}_{ct}$ is the unit vector along the $Z_{ct}$ – axis.

The vector $\mathbf{v}^{g}_{ct/g}$ of the relative motion of the plane $T_a$ can be expressed in terms of the vector $\mathbf{v}_{ct/g}$ and the operator $\mathbf{Rs}(ct \mapsto g)$ of the resultant coordinate system transformation:

$$\mathbf{v}^{g}_{ct/g}(\varphi_g) = \mathbf{Rs}(ct \mapsto g) \cdot \mathbf{v}_{ct/g} \tag{12.12}$$

Substitution of the vectors $\mathbf{n}^{g}_{ta}$ and $\mathbf{v}^{g}_{ct/g}$ into Eq. (12.7) yields the equation of contact in the form

$$\mathbf{n}^{g}_{ta}(\varphi_g) \cdot \mathbf{v}^{g}_{ct/g}(\varphi_g) = 0 \tag{12.13}$$

Furthermore, we are required to solve Eq. (12.13) with respect to the angle of rotation $\varphi_g$. The expression for $\varphi_g$ derived in this manner is then substituted into Eq. (12.6), which makes it possible to eliminate the angle $\varphi_g$ from the expression for the position vector $\mathbf{r}^{g}_{ta}(\varphi_g)$. The position vector $\mathbf{r}^{g}_{ta}$ is no longer dependent on the angle $\varphi_g$, and in its current form it describes points of the tooth flank $\mathscr{G}_a$ of the generated straight bevel gear tooth.

The generated tooth flank $\mathscr{G}_a$ of the straight bevel gear deviates from the corresponding tooth flank $\mathscr{G}$ of the perfect involute straight bevel gear. At current point $i$ of the tooth flank $\mathscr{G}$, the deviation $\delta_i$ of the approximate tooth flank $\mathscr{G}_a$ from the perfect tooth flank $\mathscr{G}$ is measured along the perpendicular to the tooth flank $\mathscr{G}$. The deviation $\delta_i$ can be expressed in terms of the parameters of the geometry of the tooth flanks $\mathscr{G}$ and $\mathscr{G}_a$.

The position vector of a point within the tooth flank $\mathscr{G}$ of the gear is designated as $\mathbf{r}^{i}_{g}$ [see Eq. (12.3)]. The unit normal vector $\mathbf{n}^{i}_{g}$ to the tooth flank $\mathscr{G}$ at point $i$ can be calculated on the premise of Eq. (12.3):

$$\mathbf{N}_g(U_g, \varphi_g)\Big|_i = \frac{\partial \mathbf{r}_g}{\partial U_g} \times \frac{\partial \mathbf{r}_g}{\partial \varphi_g}\Bigg|_i \tag{12.14}$$

$$\mathbf{n}_g(U_g, \varphi_g)\Big|_i = \frac{\mathbf{N}_g(U_g, \varphi_g)}{\left|\mathbf{N}_g(U_g, \varphi_g)\right|}\Bigg|_i \tag{12.15}$$

The straight line along the unit normal vector $\mathbf{n}^{i}_{g}$ intersects the approximate tooth flank $\mathscr{G}_a$ of the straight bevel gear at a point whose position vector is denoted by $\mathbf{r}^{i}_{ga}$. The deviation $\delta_i$ can be expressed in terms of two vectors, $\mathbf{r}^{i}_{g}$ and $\mathbf{r}^{i}_{ga}$. For this purpose, the formula

$$\delta_i = \left|\mathbf{r}^{i}_{ga} - \mathbf{r}^{i}_{g}\right| \tag{12.16}$$

applies.

At some point $i$ within the tooth flank $\mathscr{G}$, the deviation $\delta_i$ reaches its maximum value $\delta_{\max}$. The deviation $\delta_{\max}$ determined in this manner, equals the maximal deviation of the approximate tooth flank $\mathscr{G}_a$ from the desired tooth flank $\mathscr{G}$.

## 12.3 PECULIARITIES OF GENERATION OF STRAIGHT BEVEL GEARS WITH OFFSET TEETH

Bevel gears can be designed not just with straight teeth, but with helical teeth as well (see Figure 12.11). Bevel gears of this type are also referred to as *bevel gear with offset teeth* (and also as *bevel gears with skew teeth*).

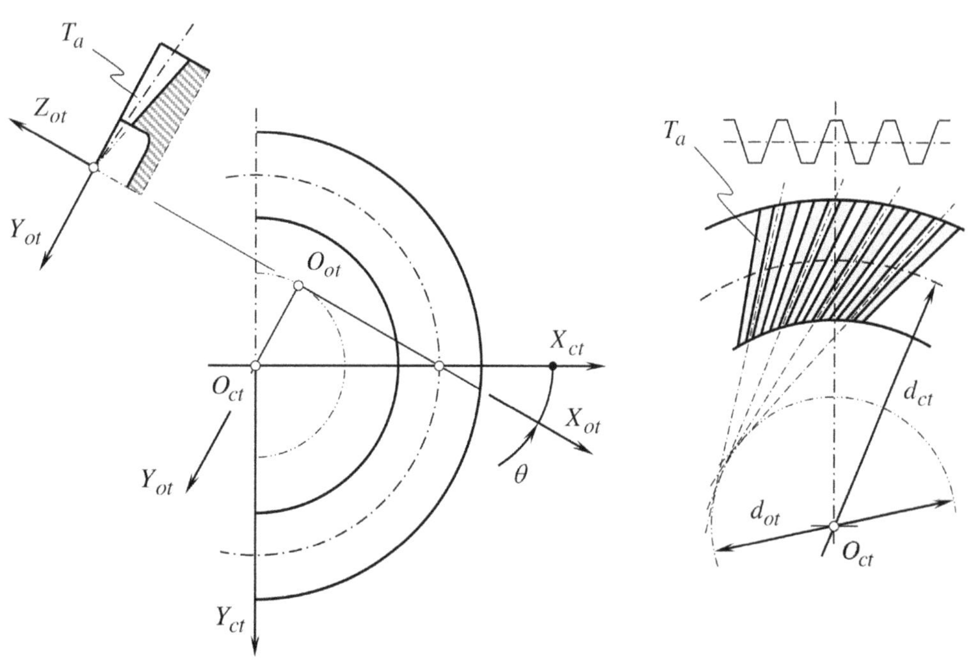


**FIGURE 12.11** Design parameters of the approximate generating surface $T_a$ for generating bevel gears with offset teeth.

### 12.3.1 Tooth Flank Geometry of the Gear with Offset Teeth

As it has been shown above, the expression [see Eq. (1.40)]

$$\mathbf{r}_g(U_g,\varphi_g) = \begin{bmatrix} (\sin\varphi_g - 2\varphi_g \tan\psi_{b.g} \sin\theta_g \sin\varphi_g)\cdot U_g \tan\theta_g \\ [\cos\varphi_g + 2\varphi_g \sin\varphi_g - \tan\psi_{b.g} \sin\theta_g \cos\varphi_g (\tan\theta_g - \varphi_g)]\cdot U_g \tan\theta_g \\ -U_g \\ 1 \end{bmatrix} \quad (12.17)$$

is valid for the position vector $\mathbf{r}_g$ of a point of the tooth flank $\mathscr{G}$ of an involute helical bevel gear.

Equation (12.17) allows for the calculation of all principal elements of the geometry of the involute straight bevel gear tooth flank $\mathscr{G}$ having offset teeth.

Without going into details of analysis here, it should be noted that helical bevel gears having a tooth flank geometry described by Eq. (12.17) are inconvenient for manufacturing purposes. As a result, the tooth flanks $\mathscr{G}$ with the desired geometry are approximated by tooth flanks $\mathscr{G}_a$, which are more convenient for the manufacturer. Straight bevel gears with offset teeth are a practical example of this type of approximation.

Straight bevel gears with offset teeth can be generated using the same process that is used to generate straight bevel gears. The principal kinematics of the generation of bevel helical gear tooth flanks is similar to that used to generate conventional straight bevel gear tooth flanks.

### 12.3.2 Generating Surface of the Gear-Cutting Tool

The generating surface must be determined before designing a gear-cutting tool for machining straight bevel gears with offset teeth. The generating surface of a gear-cutting tool is a type of virtual surface that is in correct mesh with the gear to be machined. Finishing cutting edges of the gear-cutting tool are within the generating surface.

Similar to generating ordinary straight bevel gears, the actual generating surface $T_a$ for generating straight bevel gears with offset teeth is not an envelope surface to consecutive positions of the desired tooth flank surface $\mathscr{G}$, which is given by Eq. (12.17). Instead, the approximate generating surface $T_a$ can be considered as a type of approximation to the desired generating surface $T$.

The approximate generating surface $T_a$ for generating straight bevel gears with offset teeth is depicted in Figure 12.11. In the case under consideration, the generating surface $T_a$ is shaped in the form of a crown rack. The diameter of the midsection of the face of the crown rack is denoted by $d_{ct}$, and the diameter of the concentric circle specifying the value of the offset is denoted by $d_{ot}$. Because of the offset, at points within the midsection, the teeth of the crown rack

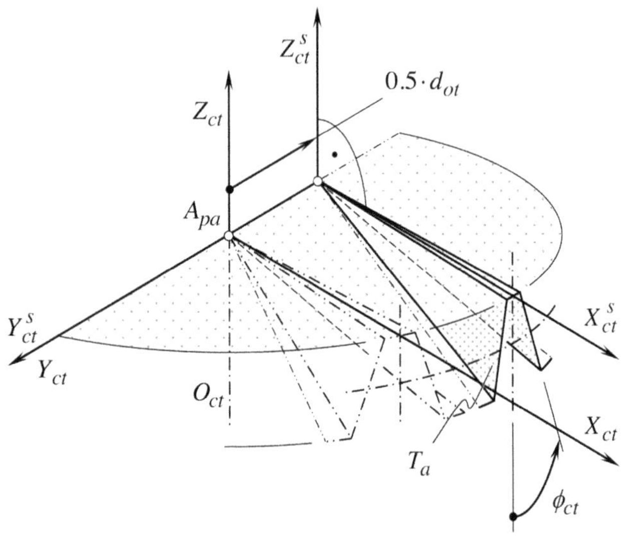


**FIGURE 12.12** Reference systems for analytical representation of the approximate generating surface $T_a$ for generating bevel gears having offset teeth.

$T_a$ are at a spiral angle $\theta$ with respect to the corresponding radial direction. Once the diameters $d_{ct}$ and $d_{ot}$ are specified, then the value of the spiral angle $\theta$ is calculated from the following formula:

$$\theta = \sin^{-1}\left(\frac{d_{ot}}{d_{ct}}\pi\right) \quad (12.18)$$

In the auxiliary reference system $X_{ot}Y_{ot}Z_{ot}$ (see Figure 12.11), the approximate generating surface $T_a$ is analytically described by an equation that is identical to Eq. (12.5). To represent the surface $T_a$ in the coordinate system $X_{ct}Y_{ct}Z_{ct}$ associated with the gear-cutting tool, a corresponding operator of the coordinate system transformation $\mathbf{Rs}(ot \mapsto c)$ must be composed. The coordinate system transformation can be much simplified if the reference systems are selected properly. For example (see Figure 12.12), in the coordinate system $X_{ct}^sY_{ct}^sZ_{ct}^s$, the approximate generating surface $T_a$ is analytically described by an equation that is identical to Eq. (12.5). Because the configuration of the reference surface $X_{ct}^sY_{ct}^sZ_{ct}^s$ relative to the reference surface $X_{ct}Y_{ct}Z_{ct}$ is convenient, the resultant coordinate system transformation can be described by the operator of elementary translation $\mathbf{Tr}[-0.5\cdot d_{ot}, Y_{ct}^s]$ of a simple structure.

Equation (12.5) and the operator of elementary translation $\mathbf{Tr}[-0.5\cdot d_{ot}, Y_{ct}^s]$ yield the expression

$$\mathbf{r}_{ta}^s = \mathbf{Tr}[-0.5\cdot d_{ot}, Y_{ct}^s]\cdot \mathbf{r}_{ta} \quad (12.19)$$

for the position vector $\mathbf{r}_{ta}^s$ of a point of the approximate generating surface $T_a$ for generating bevel gears with offset teeth.

Equation (12.19) is used for the derivation of the equation for the tooth flanks $\mathcal{G}_a$ of the generated bevel gears with offset teeth. It is also important for the calculation of the actual tooth flank surface $\mathcal{G}_a$ from the desired tooth flank $\mathcal{G}$ [see Eq. (12.17)] of the perfect gear.

## 12.4 GENERATION OF STRAIGHT BEVEL GEAR TEETH

The approximate generating surface of the gear-cutting tool must be reproduced by cutting edges of the cutting tool when machining straight bevel gears. We have shown above that the tooth flank of the approximate generating surface $T_a$ is shaped in the form of a plane surface in both cases of gear machining, that is, when machining straight bevel gears of conventional design [see Eq. (12.5)] and when machining bevel gears with offset teeth [see Eq. (12.19)].

The straight cutting edge of the gear-cutting tool is the simplest shape of the cutting edge to be used in the reproduction of the generating plane $T_a$. The straight motion of the cutting edge $CE$ and its rotation are the two motions to reproduce when cutting gears. The straight motion is performed in the direction that is parallel to the plane $T_a$. The rotation is performed about the axis of rotation, which is perpendicular to the plane $T_a$. Other possibilities to reproduce the plane $T_a$ also exist. However, not all of them are as simple as the two just mentioned.

### 12.4.1 Generation of the Plane $T_a$ by Straight Motion of the Cutting Edge

Reciprocation of the straight cutting edge $CE$ toward the axis of rotation $O_{ct}$ of the generation surface $T_a$, as shown in Figure 12.13, is the most reasonable means to reproduce the approximate generating surface $T_a$. Other directions of the reciprocation motion $\mathbf{V}_{cut}$ are possible, but they are not of practical importance to the gear-cutting tool designer. Under such a scenario, the plane $T_a$ is reproduced as the loci of consecutive positions of the straight cutting edge $CE$ when it is reciprocating toward axis $O_{ct}$ of the rotation $\boldsymbol{\omega}_{ct}$. The plane $T_a$ is defined by two directions through a point of the cutting edge $CE$. The direction of the cutting edge is one of two directions, and the direction of the reciprocation motion $\mathbf{V}_{cut}$ is the other one.

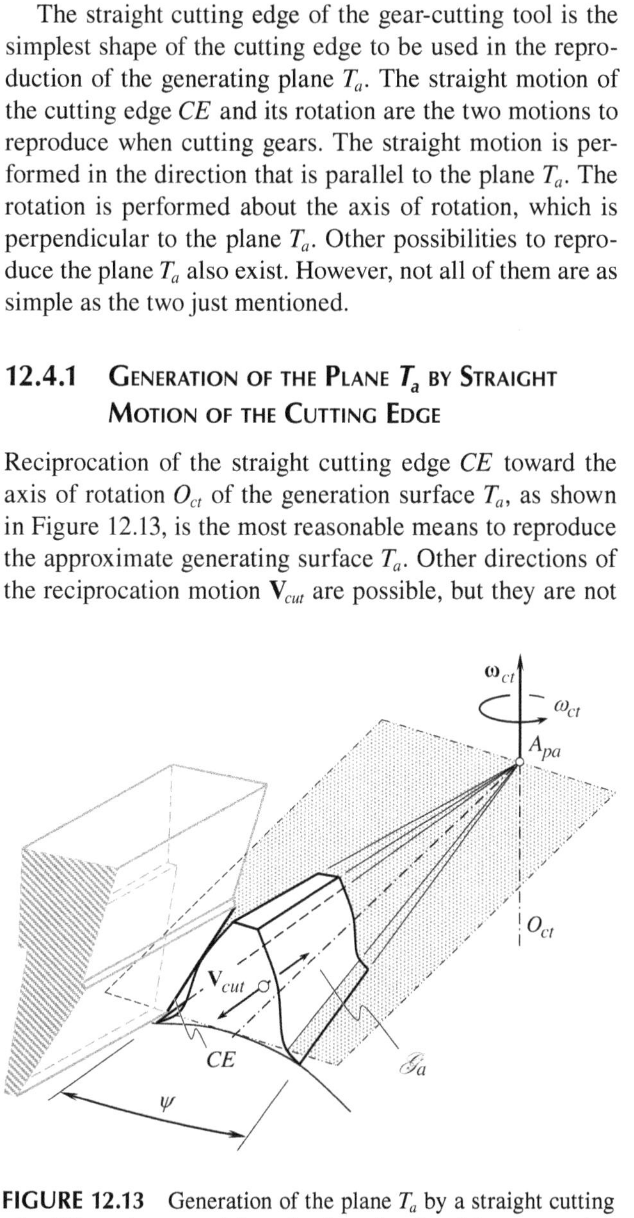


**FIGURE 12.13** Generation of the plane $T_a$ by a straight cutting edge $CE$ moving toward the axis of rotation $O_{ct}$ of the gear-cutting tool.

### 12.4.2 Machining of Straight Bevel Gears

The principles governing the generation of bevel gear teeth are analogous to the case of spur and helical gears; the difference lies in the fact that whereas spur and helical gears are generated by tools representing the teeth of the basic rack, the cutters used for bevel gear generation represent the teeth of the basic crown wheel. This is commonly called the *generating surface of the gear-cutting tool*. The motions resulting in generation are therefore those of rolling pitch cones instead of rolling pitch cylinders.

The cutters themselves must be given a form and motion which cause them to sweep out the surface of the basic crown wheel $T_a$, and the work gear is then given, relative to the cutters—the rolling motion that the finished gear would have when engaging with the crown wheel, which the cutters represent. Two distinct cases arise:

- That in which each of a pair of gears (both of which are to be generated) interacts with the same side of the surface of the imaginary crown wheel (which must therefore be symmetrical)
- That in which the mating gears interact with the opposite sides of the same basic crown wheel

The first case finds application in the cutting of bevel gears having straight (and "uncorrected") teeth, and the second in spiral bevel gears. Considering the generation of either of a pair of gears individually, however, both cases reduce to the same thing, the only difference being in the setting of the cutters.

Figure 12.14 illustrates the process diagrammatically for the case of straight bevel gears. Two cutters, which have straight-side cutting edges $CE$, are arranged to reciprocate $\mathbf{V}_{cut}$ along the radial lines, sweeping out the surfaces $T_a$ of the teeth of the imaginary crown wheel having its center at $O_{ct}$. The cutting edge $CE$ is viewed here as the line of intersection of the rake surface $R_s$ and of the clearance surface $C_s$ of the gear-cutting tool.

The work gear is arranged with its axis $O_g$ passing through $O_{ct}$ and with its pitch cone in contact with the pitch plane of the crown wheel $T_a$. It is then given a rotation $\boldsymbol{\omega}_g$ about its own axis, together with a rotation of the axis $O_{ct}$ of the crown wheel $T_a$, so related that the pitch cone of the work rolls over the pitch plane of the crown wheel. In passing through the zone in which the cutters operate, therefore, material is removed, and the result is a generated tooth flank $\mathcal{G}_a$ that interacts with that of the basic crown wheel.

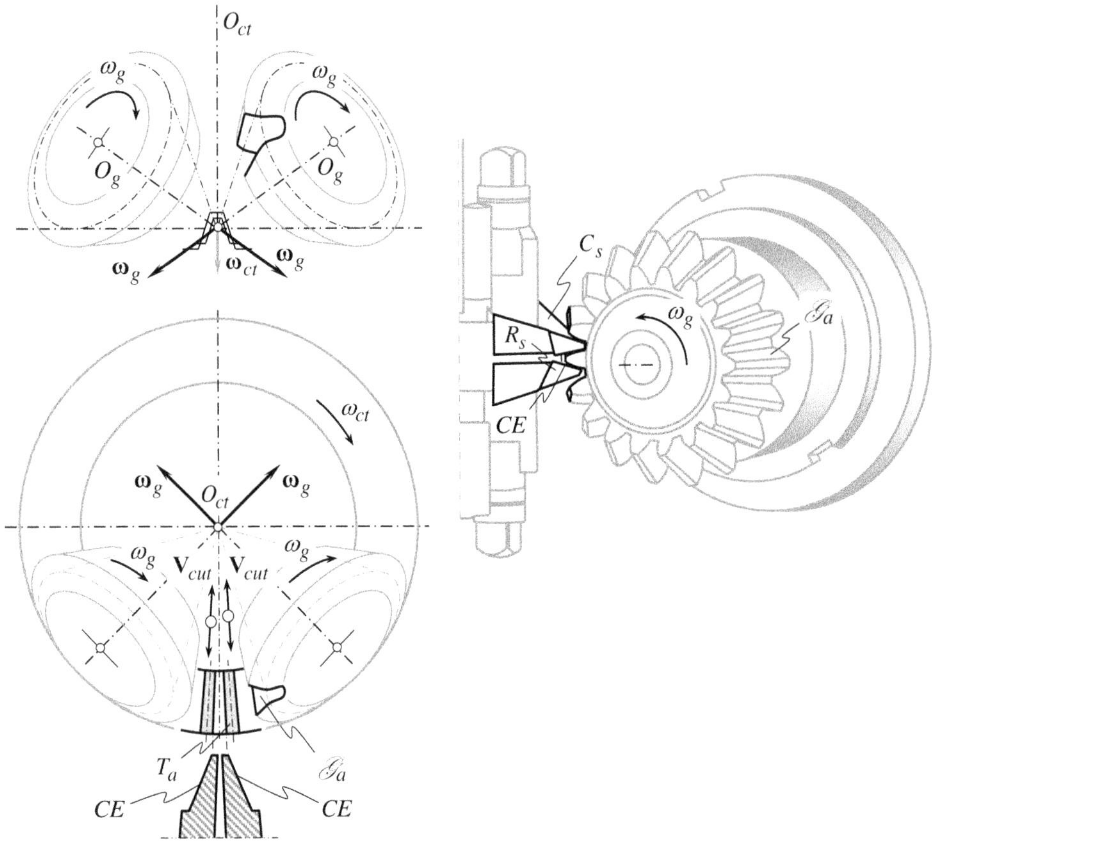


**FIGURE 12.14** Diagrammatic representation of straight bevel gear generation.

A close-up of the bevel gear planing operation is illustrated in Figure 12.15. A bevel gear generator unit that governs the reciprocation of the cutters is illustrated in Figure 12.16.

In practice, the component motions are rearranged as a matter of practical convenience, the work and the cutter each having only rotational motion about their respective axes.

### 12.4.3 Gear-Cutting Tools for Machining Straight Bevel Gears

A typical cutting tool for machining straight bevel gears is shown in Figure 12.17. Both roughing tools [see Figure 12.18(a)] and finishing tools [see Figure 12.18(b)] are specified by diametral pitch and profile angle. They are mounted in holders that are attached to the front of reciprocating slides on the face of the cradle (see Figure 12.16). Each tool can be adjusted relative to the holder so that the end of the tool is in the gear root plane and so that the tool edge travels along a line that intersects the cradle axis. This tool positioning is checked via gauging fixtures that are set on a proof block. The position of the toolholder on the face of the tool slide is adjusted to accommodate the cone distance of the work, and the length of stroke is adjusted so that the tool overtravels the face width by a small amount.

#### 12.4.3.1 Major Design Parameters of the Cutting Tool for Machining of Straight Bevel Gears

A close-up view of a cutting tool for machining straight bevel gears of both designs—gears of regular design, as well as gears with offset teeth—is shown in Figure 12.19. In addition to the major design parameters labeled in Figure 12.19, it is noteworthy that the required clearance angle $\alpha_o$ is due to the appropriate setup of the cutting tool.

Sharp corners of the tooth profile are usually rounded. The practical value of the radius $r$ of the roundness is about $r \approx 0.03m$. Here, module is denoted by $m$.

Tooth height $h$ (see Figure 12.19) must be sufficient for the purposes of the whole tooth height of the straight bevel gear. It is a common practice to assign $h = 2.5m$. Width $a$ of the top land must exceed half of the root land width at the heel side of the gear and must not exceed the root land width at the toe side of the gear. It is practical to assign $a \approx 0.4m$.

Cutting tools for machining straight bevel gears can be designed so as to have two working ends from the opposite ends of the cutting tool. After the cutting edges of the first working portion of the tool have been worn out, the tool setup is changed and the opposite working portion of the tool takes its turn to cut the stock. This makes it possible to use one rake face for planing bevel gears, and

**FIGURE 12.15** Planing of a straight bevel gear.

after it is worn out, to use the rake surface of the opposite end of the gear-cutting tool, thereby doubling the time interval between two consequent regrindings of the gear cutter.

Special duplet tools are used. These tools have two cutting edges in tandem—one edge roughs and the other finishes the bevel gear tooth flanks.

### 12.4.3.2 Cutting Tool Geometry

Cutting tools for machining straight bevel gears are designed so that the rake angle is in the range of $\gamma_o = 10° \ldots 25°$, depending on the material being worked. Standard gear-cutting tools have a rake angle $\gamma_o = 20°$. The lateral cutting edge is at the set angle $\beta = 12°$. Its orientation is also defined by the profile angle $\phi_{ct}$ of the cutting tool. Because the set angle $\beta = 12°$, an appropriate clearance angle is created. The clearance angle at the top cutting edge is equal to $\alpha_o \equiv \beta = 12°$.

The tilted orientation of the cutting tool makes it possible to create positive clearance angles at the top cutting edge ($\alpha_o > 0°$). However, clearance angles $\alpha_f$ at the lateral cutting edge *CE* are significantly smaller than the clearance angle $\alpha_o$. The actual value of the clearance angle $\alpha_f$ can be expressed in terms of the clearance angle $\alpha_o$ and the design parameters of the cutting tool.

**FIGURE 12.16** The bevel gear generator unit that governs the reciprocation of the cutters.

Because of the tilted orientation of the cutting tool, the inclination angle $\lambda$ at the lateral cutting edge does not have a zero value (that is, $\lambda \neq 0°$). Therefore, it is necessary to measure the clearance angle $\alpha_f$ in the section that is perpendicular to the cutting edge *CE* (see Figure 12.20).

The clearance angle $\alpha_f$ at the cutting edge *CE* can be calculated in the following manner. At a point of interest *m* within the cutting edge *CE* of the gear cutter, three vectors **A**, **b**, and **c** are constructed so as to comprise a set of coplanar vectors through *m*. The vectors **A**, **b**, and **c** need to be expressed in terms of the clearance angle $\alpha_f$ to be determined, and known parameters of the geometry of the cutting wedge of the cutting tool.

The vector **A** is pointed along the line of intersection of the clearance surface $C_s$ by normal section of the cutting wedge. In the Cartesian coordinate system $X_{ct}Y_{ct}Z_{ct}$, the following expression can be composed for the vector **A**

$$\mathbf{A} = \mathbf{i} \cdot \sin\alpha_f \cos\phi^*_{ct} + \mathbf{j} \cdot \sin\alpha_f \sin\phi^*_{ct} \\ -\mathbf{k} \cdot \cos\alpha_f \tag{12.20}$$

where $\phi^*_{ct}$ denotes the profile angle measured in the plane section that is perpendicular to the vector $\mathbf{V}_{cut}$ (see Figure 12.20)

**FIGURE 12.17** Typical cutting tool for machining straight bevel gears.

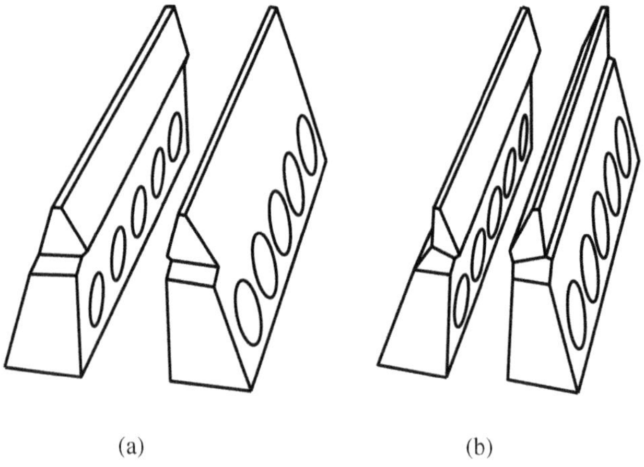


**FIGURE 12.18** Roughing cutters (a), and finishing cutters (b) for machining bevel gears with straight teeth.

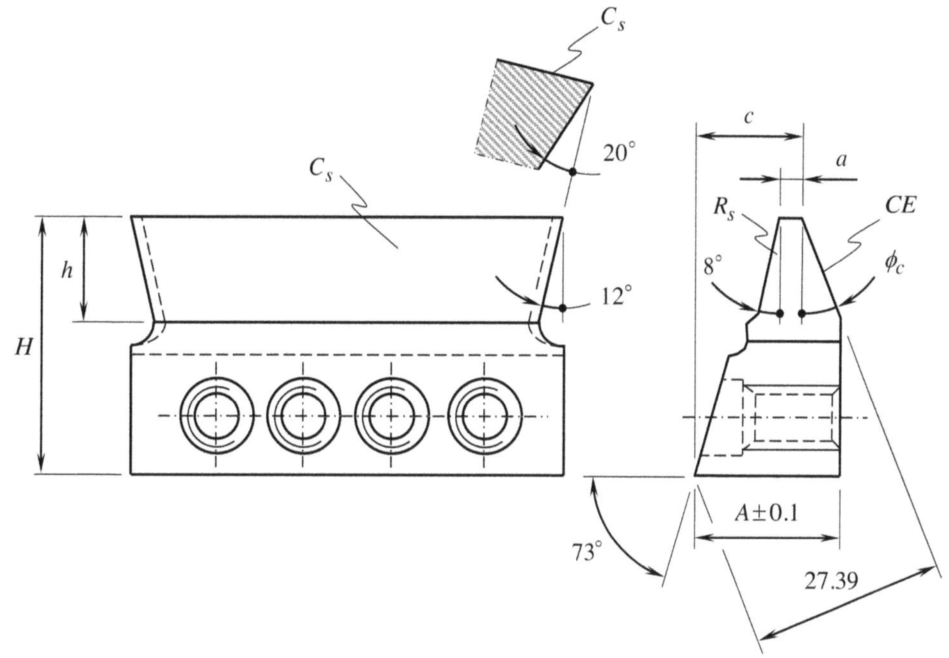


**FIGURE 12.19** Design parameters of the cutter for machining of bevel gears with straight teeth.

of the primary motion. It can be shown that for the calculation of the profile angle $\phi_{ct}^*$, the expression

$$\phi_{ct}^* = \tan^{-1}(\tan\phi_c \cdot \cos\alpha_o) \tag{12.21}$$

is valid.

For the calculation of the tooth profile angle in the plane section that is perpendicular to the generating straight line of the pitch cone, the formula

$$\phi_{ct}^{**} = \tan^{-1}(\tan\phi_{ct} \cdot \cos\alpha_o \cdot \cos\gamma_d) \tag{12.22}$$

can be derived (here $\gamma_d$ denotes the gear dedendum angle).

The unit vector **b** is pointed along the line of intersection of the clearance surface $C_s$ by a plane that is parallel to vector $\mathbf{V}_{cut}$. This vector is located entirely within the plane of Figure 12.20. The unit vector **b** can be expressed in terms of its projections onto the axes of the coordinate system $X_{ct}Y_{ct}Z_{ct}$:

$$\mathbf{b} = \mathbf{j}\cdot\sin\alpha_o - \mathbf{k}\cdot\cos\alpha_o \tag{12.23}$$

Ultimately, the unit vector **c** is pointed along the lateral cutting edge $CE$ of the gear cutter. This vector is analytically expressed as follows:

$$\mathbf{c} = -\mathbf{i}\cdot\sin\phi_c^* + \mathbf{j}\cdot\cos\phi_c^* \tag{12.24}$$

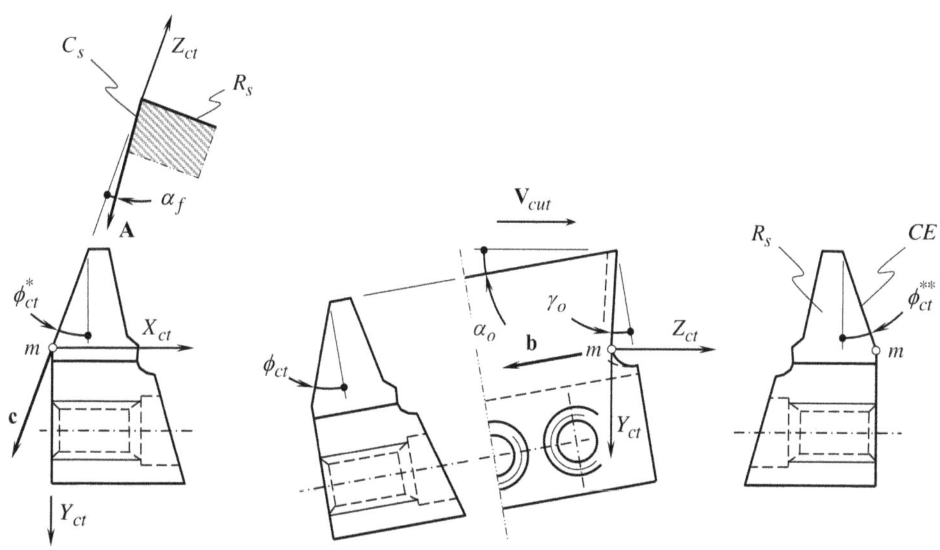


**FIGURE 12.20** Reference systems and vectors used for the calculation of the geometry of the cutting edge of the gear-cutting tool for machining straight bevel gears.

The vectors **A**, **b**, and **c** are constructed so that they comprise a set of coplanar vectors, all of which are located within the clearance plane $C_s$. Consequently, their triple scalar product is identical to zero ($\mathbf{A} \times \mathbf{b} \cdot \mathbf{c} \equiv 0$). Therefore, an expression

$$\mathbf{A} \times \mathbf{b} \cdot \mathbf{c}$$

$$\equiv \begin{vmatrix} \sin\alpha_f \cos\phi_{ct}^* & \sin\alpha_f \sin\phi_{ct}^* & -\cos\alpha_f \\ 0 & \sin\alpha_o & -\cos\alpha_o \\ -\sin\phi_{ct}^* & \cos\phi_{ct}^* & 0 \end{vmatrix} = 0 \quad (12.25)$$

is valid.

Equation (12.25) casts into the formula

$$\alpha_f = \tan^{-1}(\tan\alpha_o \cdot \sin\phi_{ct}^*) \quad (12.26)$$

for the calculation of the clearance angle $\alpha_f$.

The calculations reveal that for standard gear-cutting tools, the clearance angle $\alpha_f$ is approximately equal to $\alpha_f \cong 4°$. Because the angle $\alpha_f$ is obtained by tilting of the cutting tool, the rake angle is reduced by the corresponding value.

## 12.5 PECULIARITIES OF STRAIGHT BEVEL GEAR CUTTING

Production methods for straight bevel gears have received much attention, largely because of the substantial requirements of the automotive industry. In principle, the methods normally used for cutting spur and helical gears apply, but the generating process is modified to suit the rolling motions of two pitch cones instead of pitch cylinders. The important difference between conical and cylindrical gears is that teeth on the latter are generated by tools derived from the basic rack. In contrast, bevel gear teeth are produced by cutters representing the basic crown wheel (round rack). With this difference borne in mind, the many methods of cutting bevel gears have many similarities.

Straight bevel teeth are produced by a generating machine that reciprocates a cutting tool in a motion somewhat like that of a shaper cutter. The tools used do not resemble a pinion or a segment of a rack. The peculiar geometry of the bevel gear makes use of a special tool to cut each side of the bevel gear tooth. These tools each have a single inclined cutting edge that generates the bevel gear tooth on one side or the other. The machines achieve a generating motion by rolling the work gear and the cutter head at a slow rate, whereas the cutters reciprocate rapidly back and forth (see Figure 12.15).

In the production of straight bevel gears by planing, the cutters are straight sided, and they sweep out the surfaces of the basic crown wheel teeth, whereas the work gear is given the appropriate rolling motion.

A typical example of this method of cutting straight bevel gears features two straight-sided cutting tools that reciprocate across the face of a blank. The cutters are secured to the cradle of the machine, which rolls in correct relationship to the work gear. The blank rolls about the center of the crown wheel. Each cutter advances in turn as the combined rolling takes place and they remove material to form tooth space in a manner similar to that followed in planing a spur gear. Indexing is intermittent and takes place upon the completion of each tooth space.

This type of bevel gear generating machine is useful for cutting straddle-mounted pinions since the length of the cutting stroke can be limited so that generation over the full face can be completed without causing damage to the bearing seating. Another advantage is that during the cutting stroke, an additional motion is imparted to the work that produces a small amount of heel and toe end relief, thereby giving a barreled effect to the teeth. By providing a slightly excessive generating roll, tip and root easing are also obtained, and hard contact in the root and at the tip of gears meshing at their correct cone distance is prevented.

The latest models of straight bevel gear generators have a design feature that permits the tooth to be cut with a slight amount of crown. A cam can be set so that the cutting tool will remove a little extra metal at each end of the tooth. This makes it possible to secure a localized tooth bearing in the center of the face width.

In a few cases, cemented carbide cutters have been used to make gears instead of high-speed steel cutters. Much higher cutting speeds can be used with carbides, provided that the machine is rugged enough and powerful enough to drive the carbide cutter.

## 12.6 MILLING OF STRAIGHT BEVEL GEARS

Milling of straight bevel gears features a rotational primary motion (that is, a motion of cut). For the generation of the bevel gear tooth flank, the lateral tooth surface of the generating surface $T_a$ should be reproduced. In the bevel gear planing operation, the reproduction of the lateral tooth surface $T_a$ is achieved by reciprocation of the cutting tool in the radial direction of the generating surface. Reciprocation in the direction that is perpendicular to that just mentioned geometrically is feasible; however, it is extremely inconvenient for practical use. Rotation about an axis that is perpendicular to the plane $T_a$ is the ultimate possibility to reproduce the lateral tooth surface of the virtual crown gear. Mill cutters for machining bevel gears are designed based on this concept.

### 12.6.1 Peculiarities of Gear Machining Operation

A productive type of straight bevel generation is that which uses a pair of interlocking disk mill cutters having a multiplicity of blades that cut on two sides of a tooth space as shown in Figure 12.21 [191]. A set of two disk-type mill cutters (see Figure 12.22) is used for machining a straight bevel gear. As illustrated in Figure 12.23, an interlocking arrangement of the cutters allows a complete cutting

**FIGURE 12.21** Machining a straight bevel gear by two interlocking mill cutters.

process. The two interlocking cutters have to be adjusted independently during setup, which is complicated and time consuming.

The principal kinematics of straight bevel gear machining by interlocking mill cutters is illustrated in Figure 12.24. The profile angle $\phi_{ct}$ of the gears being cut is obtained via machine settings. The disk cutters rotate on axes $O_{ct}$, which are inclined to the face of the cradle. The cutting edges *CE* sweep out slightly internal conical surfaces.

The motions between the work gear $\omega_g$ and the cutting tools $V_{rol}$ are combined to give the minimum amount of rolling movement necessary to obtain fully generated teeth. No longitudinal traverse is provided, and the cutters feed into the work in a direction normal to the pitch cone. As a result, the tooth roots are slightly concave, but the tooth flanks could be slightly modified to give tip and root easing. Indexing takes place after each tooth space has been completed, and the machine is fully automatic in its motions. When a gear has been completed, the machine stops, the cutters withdraw, and the work gear can be changed with little delay.

In nearly all cases, gears cut using this method are completed from the solid blank in one operation that combines the roughing and finishing cuts. The method features high

FIGURE 12.22 A set of two disk-type mill cutters for machining a straight bevel gear.

productivity of straight bevel gear machining. The productivity is as much as four to five times greater than that for gear planing.

### 12.6.2 Design of Interlocking Mill Cutters

Straight bevel gears are cut with a circular mill cutter having a circumferential blade arrangement. As with other straight bevel gear generators, the tools form the blanks of a tooth simulating the basic crown wheel. Tooth flanks of the generating surface $T_a$ are identical to that used for planing of straight bevel gears [see Eq. (12.5) and Eq. (12.19)]. The cutter teeth are intermeshing, and the discs are inclined to one another at the required pressure angle, which is usually $\phi_{ct} = 20°$.

Mill cutters of two different designs are used for the machining of straight bevel gears.

***First***, lateral cutting edges *CE* of the mill cutter could be located within a plane that is perpendicular to the axis of rotation of the mill cutter. Mill cutters of this kind are used in the production of straight bevel gears of conventional design having no modification of the tooth flanks.

***Second***, lateral cutting edges *CE* of the mill cutter could make a certain angle $\delta$ in relation to the perpendicular to the axis of rotation of the mill cutter. Commonly, the angle $\delta$ is assigned in the range of $\delta = 1°...5°$. (Bevel gears having coarser teeth are machined with the mill cutters having a smaller angle $\delta$ and vice versa.)

Shown in Figure 12.25 is a disk-type milling cutter consisting of a body with slots. Cutting blades are bolted in the slots in the cutting tool body. The top of the cutting blade is wider than the width of the slots. This provides reliable fixturing of the blades in the cutting tool body. The cutting blade height $h$ exceeds the height of the gear tooth. It is practically equal to $h = 2.5 \cdot m$. Similar to bevel gear planing, the width of the top land of the cutting blades is assigned the value $a = 0.4 \cdot m$.

Rake angle $\gamma_f$ at the lateral cutting edge is commonly equal to $\gamma_f = 20°$. The clearance angle $\alpha_o$ at the top cutting edge is usually equal to $\alpha_o = 10°...12°$.

Standard mill cutters are available in two outer diameters $d_{o.ct}$. Milling cutters with $d_{o.c} = 150$ mm are used for machining of bevel gears of small module (up

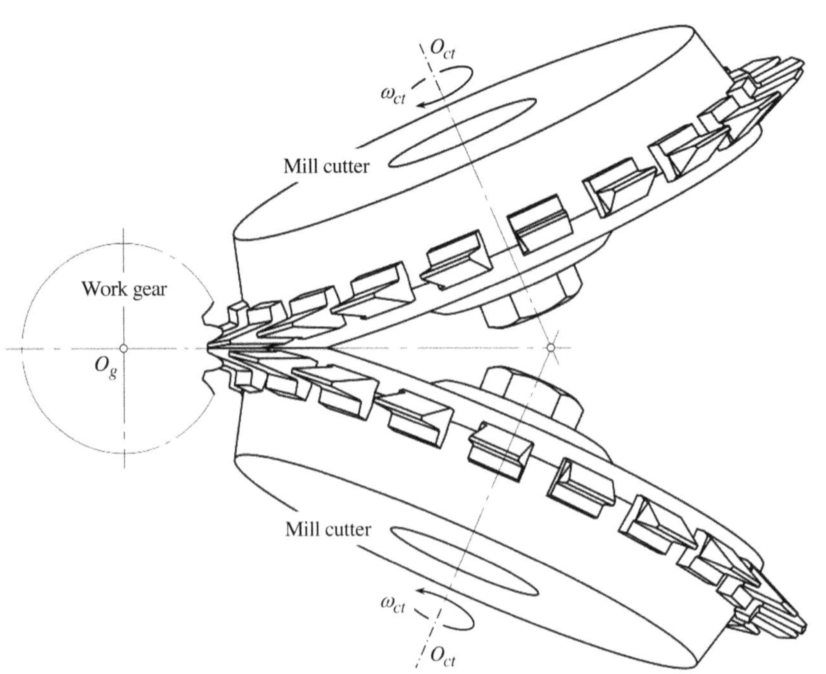


FIGURE 12.23 Schematic of straight bevel gear machining by interlocking mill cutters.

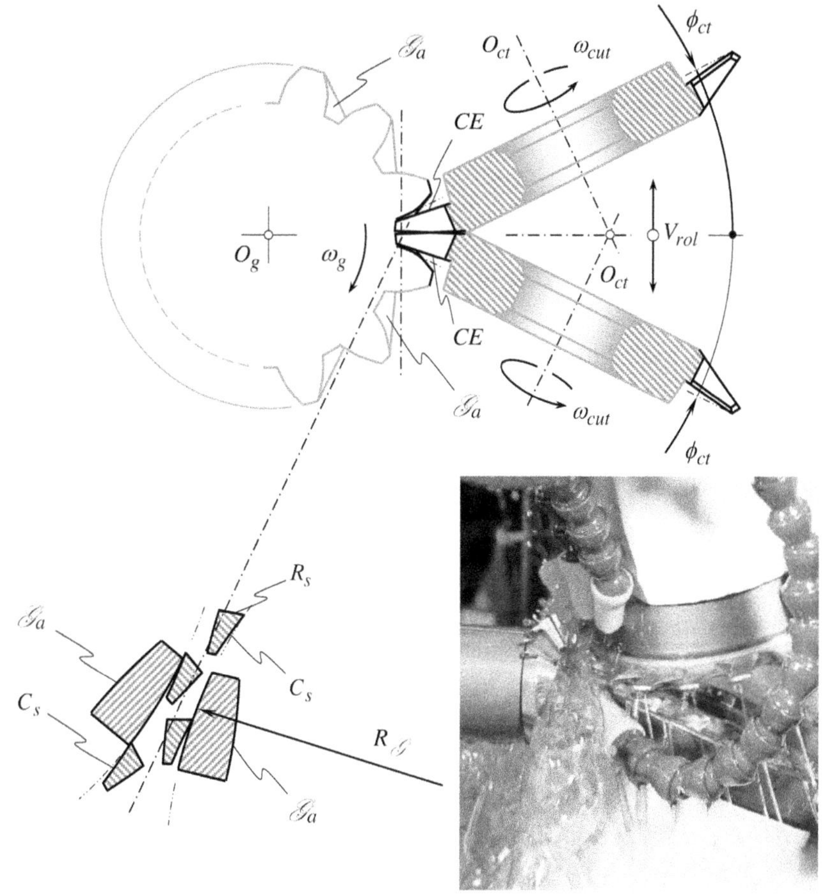


**FIGURE 12.24** Principal kinematics of straight bevel gear machining by interlocking mill cutters.

to $m = 2$ mm). For machining of bevel gears of coarse module (up to $m = 8$ mm), disk-type mill cutters having $d_{o.ct} = 275$ mm are commonly used. Custom-size mill cutters are also available.

The blades of the mill cutters are relief ground when manufactured and thus require sharpening only on the front face. Blade-to-blade spacing and cutter diameter, as well as angle and surface finish of the front face, must be closely controlled in sharpening. Special-purpose sharpeners, offered by the gear machine manufacturers, are recommended to assure consistency in sharpening, as well as the longest (cutter) tool life.

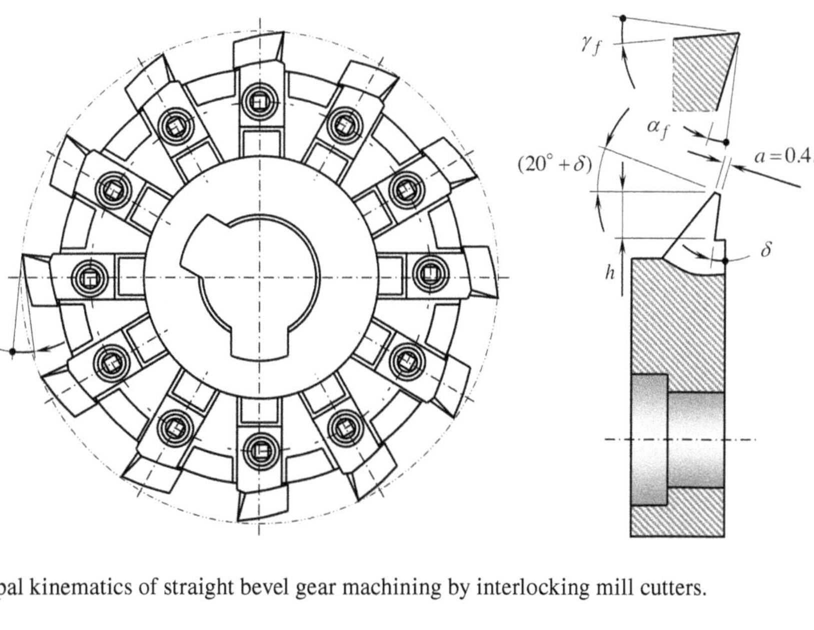


**FIGURE 12.25** Principal kinematics of straight bevel gear machining by interlocking mill cutters.

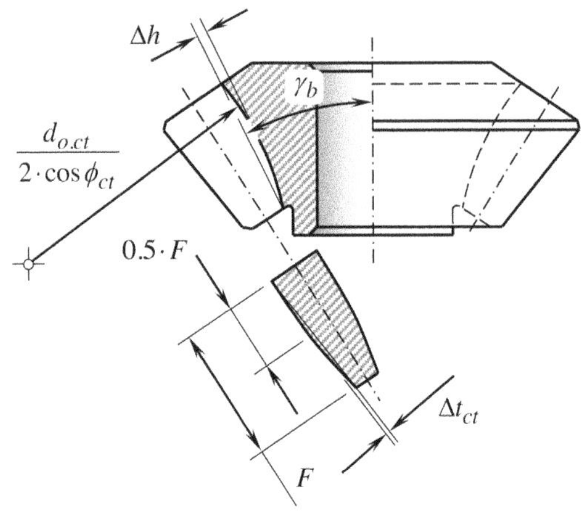


FIGURE 12.26 Specific features of the geometry of the finished bevel gear tooth flanks.

### 12.6.3 Specific Features of Shape of Finished Bevel Gear Tooth Flanks

Two cutters generate a combination of profile and length crowning in the flank surfaces of the bevel gear. A side effect of the cutter arrangement is a curved root line, depending on the cutter diameter. Slightly concave root land surfaces are produced because the cutters are not stroked in the direction of the teeth. The curve of the root line can be evaluated by the parameter $\Delta h$ (see Figure 12.26):

$$\Delta h = \frac{d_{o.c} \cdot \cos \phi_c - \sqrt{d_{o.c}^2 \cdot \cos^2 \phi_c - b^2}}{2} \quad (12.27)$$

When the mill cutter rotates, the cutting edges generate a conical surface. Mill cutters of this design are used in the production of bevel gears having longitudinal modification of their teeth. The value of the longitudinal modification $\Delta t_{ct}$ depends upon (1) the outer diameter $d_{o.ct}$ of the mill cutter, (2) the cutting edge angle $\delta$, and (3) the face width $F$ of the bevel gear (see Figure 12.26):

$$\Delta t_{ct} = \frac{F^2 \cdot \sin \delta}{4 \cdot d_{o.ct}} \quad (12.28)$$

The effect of the inner cone surface in connection with curved cutter blades is used to produce the crowning effect in straight bevel gears. As noted before, a pair of cutters is always required, one left-hand and one right-hand cutter, to realize the interlocking arrangement in the cutting machine in order to cut both flanks of a slot at the same time.

Again, milling of straight bevel gears with two mill cutters produces not the desired tooth flank $\mathcal{G}$ but a practical, suitable type of its approximation tooth flank $\mathcal{G}_a$.

## 12.7 MILLING OF BEVEL GEARS WITH CURVED TEETH

More opportunities in the design of gear-cutting tools appear due to the variety of permissible orientations of the rotation vector of the primary motion (the motion of cut). An example is shown in Figure 12.27: In this figure, a close-up of the machining process for a spiral bevel gear is illustrated. When milling a spiral bevel gear, the rotation vector $\omega_{cut}$ is pointed so that it is parallel to the axis of rotation $O_{ct}$ of the generating surface of the cutting tool, as shown in Figure 12.28. In this way, bevel gears with curved teeth having a certain spiral angle $\psi_g$ could be produced. Suitable spiral angles lie between about $\psi_g = 15°$ and 35°, but they are usually chosen between $\psi_g = 30°$ and 35° to provide adequate overlap, and it is normal practice to make spiral bevel gears with about 35% overlap.

When machining a bevel gear with curved teeth, the cutter is rotating about its axis with a certain angular velocity $\omega_{cut}$. The work gear and the generating surface $T_a$ roll over

FIGURE 12.27 Machining a spiral bevel gear.

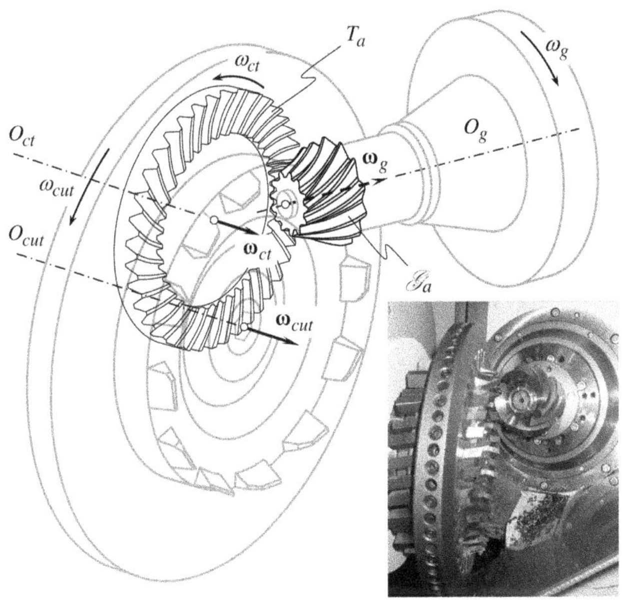


**FIGURE 12.28** Representation of the generating surface $T_a$ of the cutter head (of the face mill cutter).

one another. For this purpose, the rotations of the work gear $\omega_g$ and of the generating surface of the gear-cutting tool $\omega_{ct}$ are synchronized with one another in a timed manner. The generation of spiral bevel gears follows the same principle, and the only difference between the numerous types of spiral bevel gears lies in the motion of the cutter in relation to the work.

### 12.7.1 Peculiarities of the Gear Machining Operation

In the method of generating curved tooth bevel gears, the tooth spirals take the form of circular arcs. The straight-sided cutting tools represent the flanks of the basic crown wheel teeth, and the combined motions of the generating cutter and the work gear sweep out the surface of the imaginary crown wheel teeth (round rack teeth). Generation of the tooth flanks is obtained by giving the work gear a rolling motion relative to the cutter, similar to that which the finished gear would have when engaging with the crown wheel.

The toolholder is rotating to cause a cutting action while the work gear slowly rotates with the toolholder. The rotation of the work gear with respect to the toolholder causes a generating action to occur. After one tooth space is finished, the machine goes through an indexing motion to bring the cutter into the next tooth slot.

Pinions are cut as the reverse of wheels insofar as they are assumed to engage with the opposite side of the basic crown wheel surface. In practice, the axes of the generating cutter and the work gear are not inclined at the theoretical angle; the axes are arranged to accommodate the tapering depth of the tooth and also to provide deflection allowance in the tooth spirals.

In operation, the cutter is given a rotation speed and feed suitable for the material of the work gear and is fed into the full depth required while cutter and work gear roll together. A copious supply of cutting oil is fed to the cutting zone to act as lubricant and coolant. It is normal to expect a minimum of 100 gears to be cut between cutter sharpenings. As soon as a tooth space has been completed, the work gear and cutter roll out of engagement, the work gear is indexed to the next tooth space, and the cutting process continues.

Teeth are usually rough cut and finish cut as two separate operations, and if the quantities are sufficiently large, it is usual to carry out rough cutting on one machine and retain a second machine solely for finishing. Wheels being rough cut may be produced without tooth generation, and that refinement is deferred until the finish cutting operation. Pinions are invariably fully generated, and if they are intended to be run with form cut wheel, the full shape is applied to the pinion tooth flanks only.

Gears that require high quality are always provided with fully generated teeth on both the wheel and pinion.

To machine a bevel gear having a prescribed spiral angle $\psi_g$ at the central point $C$, the setup parameters of the cutting tool should meet the values that are calculated from the formulae (see Figure 12.29):

$$H = L - R_C \cdot \sin\psi_g \tag{12.29}$$

$$V = R_C \cdot \cos\psi_g \tag{12.30}$$

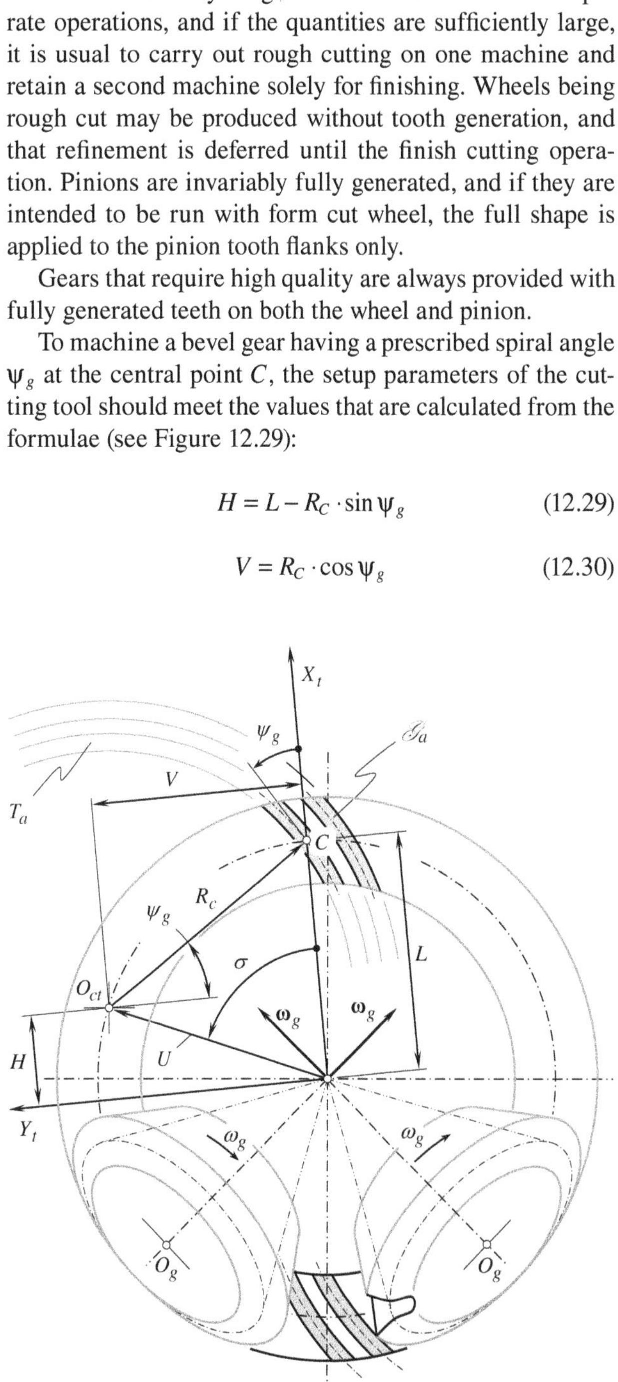


**FIGURE 12.29** Diagrammatic arrangement of spiral bevel gear generation.

**FIGURE 12.30** Cutter heads for machining spiral bevel gears.

Bevel gear generators are often designed so that they require setup parameters expressed in polar coordinates. The polar angle $\sigma$ and the offset distance $OP = U$ can be calculated from the expressions

$$U = \sqrt{H^2 + V^2} \tag{12.31}$$

$$\sigma = \tan^{-1}\left(\frac{V}{H}\right) \tag{12.32}$$

In rolling motion, the values of coordinates $H$ and $V$ change. However, the radial offset $U$ remains constant when the polar angle $\sigma$ varies.

### 12.7.2 Design of Gear Cutters

A variety of spiral bevel gear cutter heads used in industry is depicted in Figure 12.30. For machining bevel gears with circular arc teeth, the generating surface of the cutting tool is commonly chosen in the form of two cones of revolution having a common axis of rotation as shown in Figure 12.31. The equation of the approximate generating surface $T_a$ can be derived from an analysis of Figure 12.31.

The position vector of a point $\mathbf{r}_a^i$ for the inner portion $T_a^i$ of the approximate generating surface of the gear cutter head is expressed by an equation in the form:

$$\mathbf{r}_a^i(\varphi_{ct}, \theta_{ct}) = \frac{d_{ct} - t_{ct}}{2} \cdot \begin{bmatrix} \cos\varphi_{ct} \cdot \cos\theta_{ct} \\ \cos\varphi_{ct} \cdot \sin\theta_{ct} \\ \sin\varphi_{ct} \\ 1 \end{bmatrix} \cdot \frac{\cos\phi_{ct}}{\cos(\varphi_{ct} - \phi_{ct})} \tag{12.33}$$

Similarly, for the outer portion $T_a^o$ of the generating surface for the position vector of a point $\mathbf{r}_a^o$, the following formula

$$\mathbf{r}_a^o(\varphi_{ct}, \theta_{ct}) = \frac{(d_{ct} + t_{ct})}{2} \cdot \begin{bmatrix} \cos\varphi_{ct} \cdot \cos\theta_{ct} \\ \cos\varphi_{ct} \cdot \sin\theta_{ct} \\ -\sin\varphi_{ct} \\ 1 \end{bmatrix} \cdot \frac{\cos\phi_{ct}}{\cos(\varphi_{ct} - \phi_{ct})} \tag{12.34}$$

is derived.

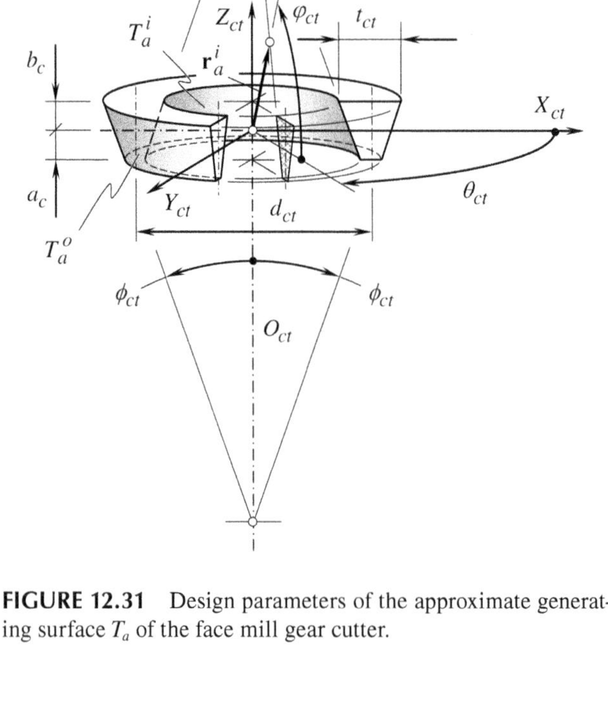

**FIGURE 12.31** Design parameters of the approximate generating surface $T_a$ of the face mill gear cutter.

**FIGURE 12.32** Solid spiral bevel gear cutter head of a small diameter (up to 50 mm).

Lateral cutting edges $CE$ of the face mill gear cutter are within the surfaces $T_a^i$ and $T_a^o$ [Eqs. (12.33) and (12.34)], correspondingly. In this way, the straight-sided cutting tools represent the flanks of the basic crown wheel teeth.

Spiral bevel gear cutter heads of small diameter (up to 50 mm) are made solid (see Figure 12.32). Cutting tools of larger diameter are often made assembled (see Figure 12.33). They consist of a tool body and a number of cutting blades bolted into the tool body (see Figure 12.34). The cutter blades themselves are bolted into slots in the body of the cutter and are arranged so that each flank of a tooth space is cut by alternate blades.

Cutting tools of small diameter have two cutting blades, whereas cutting tools of larger diameter usually have up to 32 cutting blades.

For roughing of coarse teeth, roughing face mill gear cutters are used. Finishing of bevel gears is performed by the finishing face mill gear cutters.

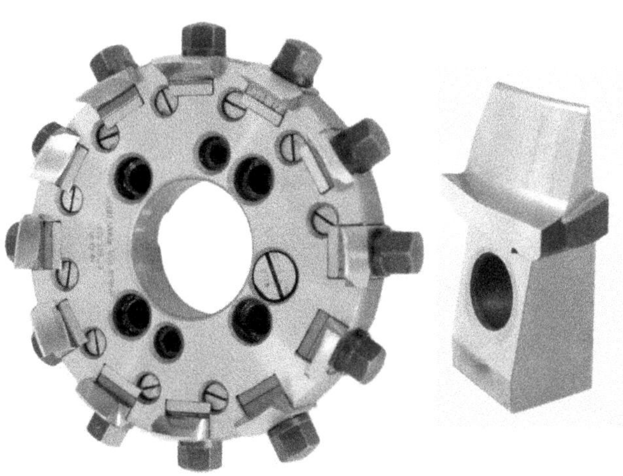

**FIGURE 12.33** Assembled spiral bevel gear cutter head of larger diameter.

It can be shown that the profile angle $\phi_{ct}^o$ of the outer cutting blades is smaller than its nominal value $\phi_{ct}$, whereas the profile angle $\phi_{ct}^i$ of the inner cutting blades exceeds its nominal value $\phi_{ct}$. Because of this, the cutting blades are designed with an asymmetrical tooth profile. The profile angles are as follows:

$$\phi_{ct}^o = \phi_{ct} - \Delta\phi_{ct} \tag{12.35}$$

$$\phi_{ct}^i = \phi_{ct} + \Delta\phi_{ct} \tag{12.36}$$

where

$$\Delta\phi_{ct} = \frac{\phi_{fr} + \phi_{fl}}{2} \cdot \sin\psi_g \tag{12.37}$$

where $\phi_{fr}$ and $\phi_{fl}$ denote the root profile angles, and $\psi_g$ is the spiral angle at the central point $C$.

Equations (12.35) through (12.37) are approximate. Calculations return the result of the calculation within 2%–3%.

The normal rake angle ranges from $\gamma_o = 10°\ldots15°$ to $\gamma_o = 22°\ldots27°$. For standard cutting tools, $\gamma_o = 20°$.

Clearance angle $\alpha_o = 11°...13°$ and $\alpha_f = 2°...5°$. Most designs of the cutting tools make it possible to increase the value of clearance angle $\alpha_f$.

## 12.8 GRINDING OF SPIRAL BEVEL GEAR CUTTER HEADS

Spiral bevel gear cutter teeth are ground on the clearance surfaces $C_s$ as well as on the rake surfaces $R_s$.

Relief grinding on the clearance surfaces $C_s$ is common in the production of spiral bevel gear cutter heads. For grinding purposes, the cutter head is installed on the spindle of the grinder as shown in Figure 12.35. When grinding, the cutter head performs a slow rotation about its axis. This motion of the cutter head is a feed motion in the cutter head grinding operation. The rotated grinding wheel performs a straight motion per each tooth of the cutter head. The straight motion of the grinding wheel is parallel to the axis of rotation of the cutter head.

If necessary, grinding on the clearance surfaces $C_s$ also can be used to resharpen worn spiral bevel gear cutter heads.

Grinding on the rake surfaces $R_s$ of the cutter head teeth is illustrated in Figure 12.36. This method of cutter head grinding is common when sharpening worn spiral bevel gear cutter heads. The accuracy of the ground cutter heads is checked by tracing the stylus connected to the indicator over the rake surface $R_s$, as shown in Figure 12.37. A deviation $\Delta\gamma_o$ of the rake angle $\gamma_o$ must be within a prespecified tolerance $[\gamma_o]$ for this deviation, that is, the inequality $\Delta\gamma_o \le [\gamma_o]$ must be valid after each regrinding of the gear cutter head teeth.

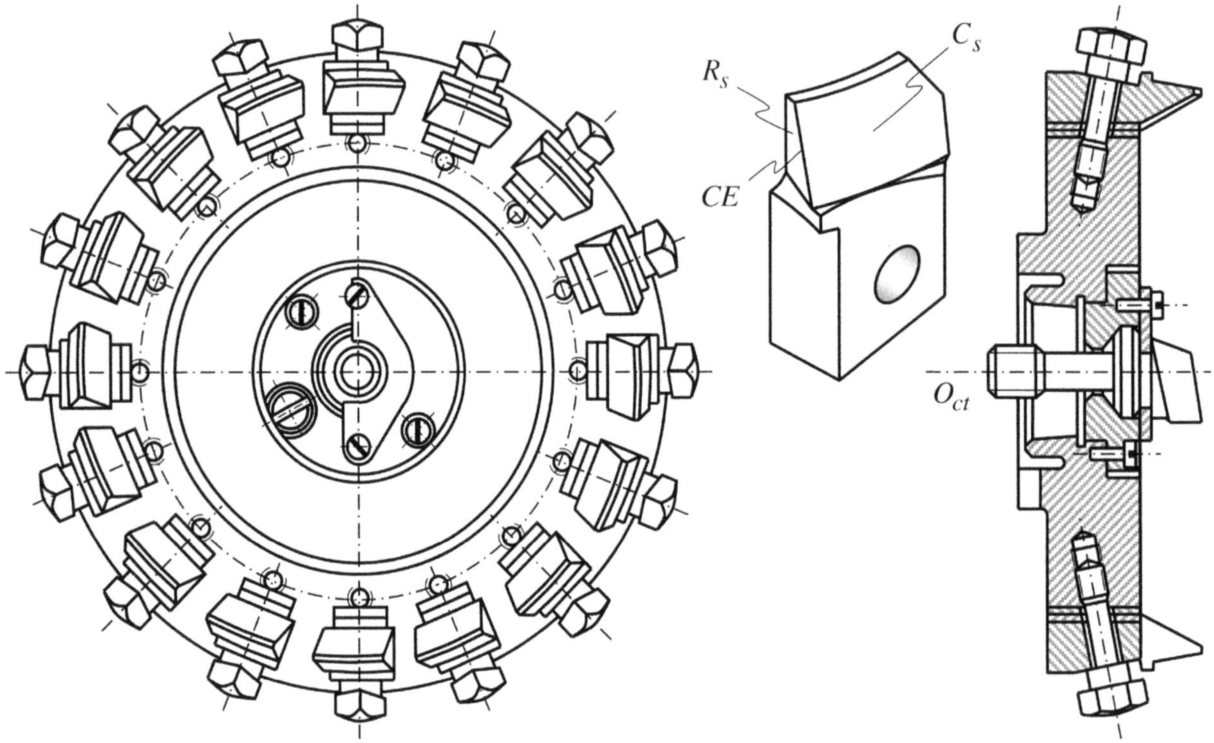


**FIGURE 12.34** Face mill gear cutter for machining bevel gears with circular-arc teeth.

**FIGURE 12.35** Grinding a spiral bevel gear cutter head on the clearance surface $C_s$ of the teeth.

## 12.9 CNC MILLING OF SPIRAL BEVEL GEAR TOOTH FLANK

Accomplishments in the field of sculptured part surface machining on multiaxis numerically controlled machines (NC machines) [151] make possible (and economical) the cutting of a large-scale spiral bevel gear in a low-volume production on NC machines. When this technology is

**FIGURE 12.36** Grinding a spiral bevel gear cutter head on the rake surface $R_s$ of the teeth.

**FIGURE 12.37** Inspection of the accuracy of grinding spiral bevel gear cutter head on the rake surface $R_s$ of the teeth.

**FIGURE 12.39** 5-axis CNC milling of a spiral bevel gear by the disk mill cutter.

used in gear production, both the gear tooth flank $\mathcal{G}$ and the pinion tooth flank $\mathcal{P}$ are viewed as a kind of sculptured part surface. The machining of the tooth flanks $\mathcal{G}$ and $\mathcal{P}$ is performed either by end mill cutters as illustrated in Figure 12.38 or by disk mill cutters as shown in Figure 12.39. In both cases, the generating surface of the gear-cutting tool and the gear $\mathcal{G}$ (and the pinion $\mathcal{P}$) tooth flank make line contact with one another. Ball-nose mill cutters are used for this purpose as well. In the latter case, the generating surface of the gear-cutting tool and the gear $\mathcal{G}$ (the pinion $\mathcal{P}$) tooth flank make point contact with one another.

The capabilities of multiaxis numerically controlled machines in cutting bevel gears are clearly illustrated in Figure 12.40. Practically, the tooth flanks $\mathcal{G}$ and $\mathcal{P}$ of any reasonable geometry can be reproduced using this technology.

**FIGURE 12.38** 5-axis CNC milling of a spiral bevel gear by the end mill cutter.

## 12.10 GRINDING OF SPIRAL BEVEL GEAR TOOTH FLANK

Rough-cut tooth flanks of spiral bevel gears can be finish-ground. Mostly heat-treated (hardened) spiral bevel gears are ground. Grinding of spiral bevel gear teeth is illustrated in Figure 12.41. The kinematics of the bevel gear grinding process features many common similarities with that of the spiral bevel gear-cutting process (see Section 12.1). For grinding the gear tooth flanks, the dressed grinding wheel is used. Dressing of the grinding wheel for grinding tooth flanks of a spiral bevel gear is depicted in Figure 12.42. Dressing of the grinding wheel is performed so as to shape it in compliance with generating surface $T$. The geometry of the generating surface $T$ that is used in the gear-grinding process is identical to that in the gear-milling process (see section 12.7.1).

## 12.11 LAPPING OF SPIRAL BEVEL GEAR TOOTH FLANK

All accurate spiral bevel gears, both in the green stage as well as heat treated, are lapped. Gear lapping is a finishing operation aiming at improving the surface finish and attaining an appropriate contact pattern in a gear pair. A close-up of a spiral bevel gear lapping process is shown in Figure 12.43.

For lapping, a spiral bevel gear and its mating pinion are engaged in a tight mesh and run together under a load. Lots of working fluid is supplied into the zone of interaction of the tooth flanks. The working fluid contains an abrasive material. Material removal occurs due to sliding of the tooth flanks of the gear $\mathcal{G}$ and of the pinion $\mathcal{P}$ in relation to one another. Two motions contribute to the tooth flank sliding. The profile sliding in the intersected-axes gear pair is one of the contributors, and the sliding

FIGURE 12.40 The capabilities of multiaxis numerically controlled machines for cutting bevel gears.

FIGURE 12.41 Grinding of tooth flanks of a spiral bevel gear.

FIGURE 12.42 Dressing of the grinding wheel for grinding the tooth flank spiral of a bevel gear.

**FIGURE 12.43** Lapping of tooth flanks of spiral bevel gear teeth.

in the lengthwise direction of the interacting tooth flanks $\mathscr{G}$ and $\mathscr{P}$ is the other contributor to the resultant tooth flank sliding.[3]

When lapping, the gear and the pinion are each considered a work gear, as there are no specially profiled gear-cutting tools used in this gear finishing operation. In a certain sense, the working fluid that contains abrasives can be considered the cutting tool in gear lapping.

Summarizing the discussion in this chapter, it must be clear to the reader that only bevel gears with a spherical involute tooth profile can be perfectly generated in gear machining processes based on intersected-axes vector diagrams of the gear machining mesh. This is attributed to the kinematics of the gear machining process and to violation of the condition of conjugacy of the tooth flanks $\mathscr{G}$ of a gear to be machined, and of the generating surface $T$ of the gear-cutting tool. In compliance with the condition of conjugacy of tooth flanks, at every point of contact of the surfaces $\mathscr{G}$ and $T$, the common perpendicular $\mathbf{n}$ must intersect the axis of instant rotation $P_{\text{ln}}$ of the gear and the gear-cutting tool in the gear machining mesh [171].

[3] Using the rotation vectors [171], of the gear, $\boldsymbol{\omega}_g$, of the pinion $\boldsymbol{\omega}_p$, of the plain of action, $\boldsymbol{\omega}_{pa}$, and the relative rotation, $\boldsymbol{\omega}_{pl}$, it can be readily shown that the sliding in the lengthwise direction of the interacting tooth flanks of the gear $\mathscr{G}$ and of the pinion $\mathscr{P}$ is always observed in any and all intersected-axes gear pairs.

# 13 Gear Shaper Cutters Having Tilted Axis of Rotation

## *Internal Gear Machining Mesh*

Gear machining operations that feature intersected axes of rotation of the work gear and the gear-cutting tool can be executed when an internal gear machining mesh is reproduced for machining the gear. Under such a scenario, either internal gears (ring gears) can be cut by gear-cutting tools having an external generating surface or external gears can be machined by enveloping gear-cutting tools.

### 13.1 PRINCIPAL KINEMATICS OF INTERNAL GEAR MACHINING MESH

The principal kinematics of the internal gear machining mesh can be illustrated by two rotation vectors.[1] One of the rotation vectors $\boldsymbol{\omega}_g$ represents rotation of the work gear, whereas another rotation vector $\boldsymbol{\omega}_{ct}$ represents the corresponding rotation of the cutting tool. The vectors $\boldsymbol{\omega}_g$ and $\boldsymbol{\omega}_{ct}$ of the rotations align with the intersecting axes of rotation $O_g$ and $O_{ct}$ of the work gear and the cutting tool. Moreover, the rotation vectors $\boldsymbol{\omega}_g$ and $\boldsymbol{\omega}_{ct}$ originate from the point of intersection of the axes $O_g$ and $O_{ct}$ as shown in Figure 13.1.

The internal gear machining mesh under consideration features the intersected-axes angle $\Sigma$ that is smaller than its critical value $\Sigma_{cr}$ (that is the inequality $\Sigma < \Sigma_{cr}$ is observed in the internal gear machining mesh).

Only the relative motions of the work gear and the gear-cutting tool are important for the consideration below.

The rotation of the cutter $\omega_{ct}$ can exceed the rotation of the work gear $\omega_g$. Such a configuration of the rotation vectors $\boldsymbol{\omega}_g$ and $\boldsymbol{\omega}_{ct}$ is schematically depicted in Figure 13.1,a, where the inequality $\omega_{ct} > \omega_g$ is observed. In this particular case the vector $\boldsymbol{\omega}_{pl}$ of the cutter's instant rotation can be constructed in compliance with the equality:

$$\omega_{pl} = \omega_{ct} - \omega_g \tag{13.1}$$

To construct the vector of instant rotation $\boldsymbol{\omega}_{pl}$, it is convenient to construct the rotation vector $-\boldsymbol{\omega}_g$ which is pointed opposite to the rotation vector $\boldsymbol{\omega}_g$ of the gear. After the rotation vector $-\boldsymbol{\omega}_g$ is applied to both to the work gear and the gear-cutting tool, then the work gear becomes motionless $(\omega_{ct} - \omega_g = 0)$. Instant rotation of the cutting tool relative to the work is represented by the rotation vector $\boldsymbol{\omega}_{pl}$.

[1] ***Reminder:*** Rotations by nature are not vectors. Therefore, special care is required when treating rotations as vectors.

For illustrative purposes, the axodes $\mathscr{A}_g$ and $\mathscr{A}_{ct}$ can be associated with the rotation vectors $\boldsymbol{\omega}_g$ and $\boldsymbol{\omega}_{ct}$ of the work gear and the cutting tool as shown in Figure 13.1,b. The relative motion of the gear-cutting tool with respect to the work gear allows for interpretation in the form of rolling with no sliding of the axode $\mathscr{A}_{ct}$ of the gear-cutting tool over the axode $\mathscr{A}_g$ of the work gear. The instant rotation $\boldsymbol{\omega}_{pl}$ is performed about the axis of instant rotation $P_{\text{ln}}$.

### 13.2 PECULIARITIES OF GEAR-CUTTING TOOL DESIGN

Derivation of the equation of the generating surface of a gear-cutting tool is one of the major purposes of analysis of the kinematics of the gear machining mesh. Then when the gear-cutting tool is designed, the gear machining mesh can be complemented with a primary motion of the cutting tool (that is, the motion of cut $\mathbf{V}_{cut}$). Ultimately, the kinematics of the gear machining process is composed in this manner.

#### 13.2.1 Shaping of Internal Gear

A shaper for machining an internal gear can be designed on the premise of the internal gear machining mesh. For design purposes, a Cartesian coordinate system $X_{ct}Y_{ct}Z_{ct}$ is associated with the gear-cutting tool to be designed as shown in Figure 13.2. The rotation vectors $\boldsymbol{\omega}_g$ and $\boldsymbol{\omega}_{ct}$ of the work gear and of the gear-cutting tool, respectively, form a certain intersected-axes angle $\Sigma$. The rotations $\omega_g$ and $\omega_{ct}$ are synchronized with each other in a timely manner.

The cutting tool reference system $X_{ct}Y_{ct}Z_{ct}$ can perform a certain motion in relation to the coordinate system $X_gY_gZ_g$ associated with the work gear. This relative motion can be employed as the primary motion (motion of cut). The motion of cut can be represented as the superposition of the elementary translations along the axes of the coordinate system $X_{ct}Y_{ct}Z_{ct}$ and the elementary rotations about axes of that same coordinate system. The number of elementary motions of the cutting tool is limited to a total of six. The elementary motions of the cutting tool can be considered separately as well as the combinations of the elementary motions by two, three, four, five, and six. In this way, all possible relative motions of the cutting tool with respect to the work gear are encompassed.

At this point, let us simplify the problem to be solved, and as an example let us consider not an arbitrary relative


DOI: 10.1201/9781003520528-17

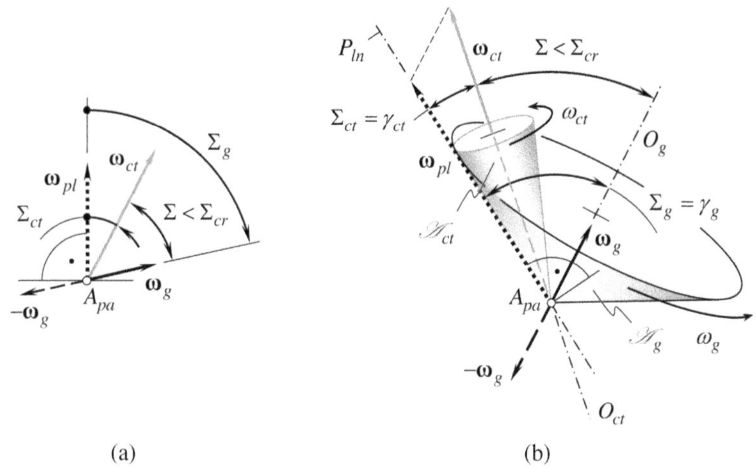


**FIGURE 13.1** Vector representation of the internal intersected-axes gear machining mesh for cutting internal gears: (a) the gear vector diagram, and (b) the corresponding axodes for the gear machining mesh.

motion of cut but a particular type of this motion instead. Let us assume that the motion of cut is a straight motion parallel to the work gear axis $O_g$. This means that the motion of cut $\mathbf{V}_{cut}$ can be represented in the form of vector summa

$$\mathbf{V}_{cut} = \mathbf{V}_x + \mathbf{V}_z \tag{13.2}$$

Here $\mathbf{V}_x$ and $\mathbf{V}_z$ designate projections of the vector $\mathbf{V}_{cut}$ onto the axes of the coordinate system $X_{ct}Y_{ct}Z_{ct}$.

The axodes $\mathscr{A}_g$ and $\mathscr{A}_{ct}$ can be associated with the work gear and the cutting tool, respectively. When machining a gear, the axodes $\mathscr{A}_g$ and $\mathscr{A}_{ct}$ roll over one another with no sliding between the rolling surfaces.

The actual shape of the operating pitch surfaces $\mathscr{P}_{sg}$ and $\mathscr{P}_{sc}$ of the work gear and the cutting tool, respectively, depends on the chosen type and direction of the primary motion. For the straight motion of cut $\mathbf{V}_{cut}$, the pitch surface of the work gear $\mathscr{P}_{sg}$ is shaped in the form of a cylinder of revolution. The pitch surface of the cutting tool $\mathscr{P}_{sc}$ is shaped in the form of the corresponding cone of revolution.

The kinematics of the gear-machining operation depicted in Figure 13.2 features the motion of cut $\mathbf{V}_{cut}$ parallel to the axis of rotation $O_g$ of the work gear. Such a kinematics can be used for the purpose of designing a shaper for machining an internal gear. Shaping an internal gear with the external shaper is illustrated in Figure 13.3.

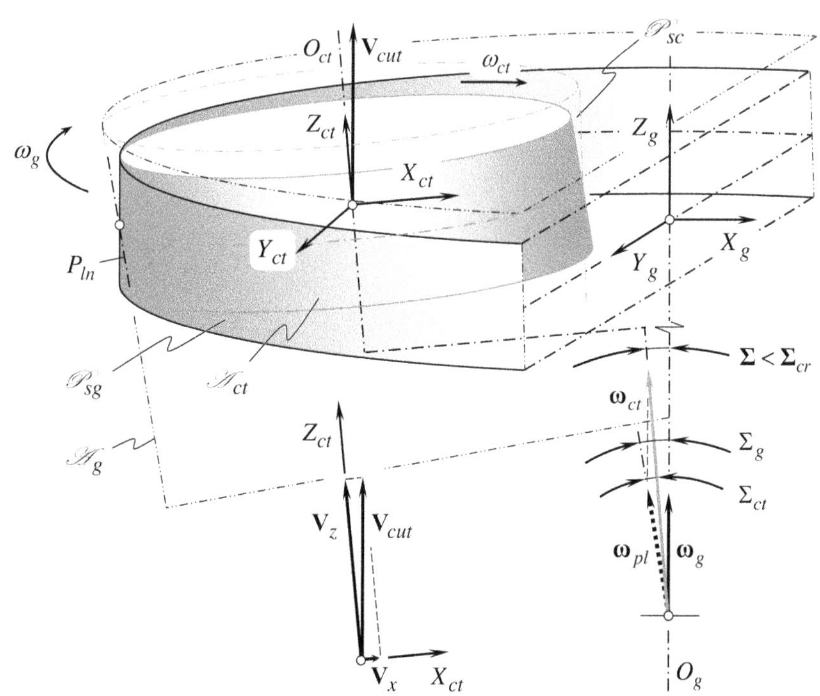


**FIGURE 13.2** Kinematics of the gear machining operation that features the primary motion (the motion of cut $\mathbf{V}_{cut}$) parallel to the axis of rotation $O_g$ of the work gear.

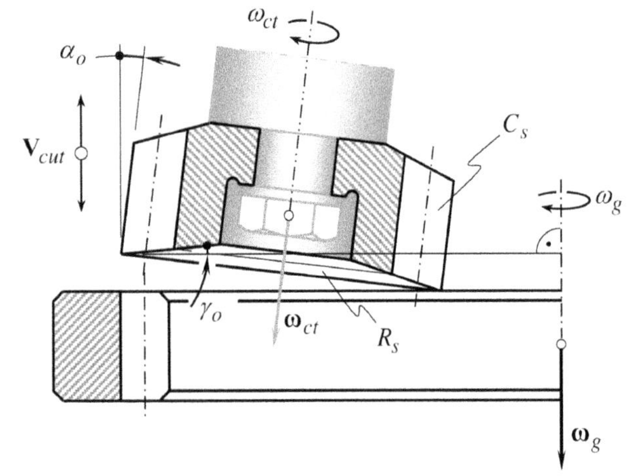


**FIGURE 13.3** Schematic of shaping of an internal gear by a shaper with a tilted axis of rotation.

The work gear rotates about its axis $O_g$. The shaper cutter rotates about its axis of rotation $O_{ct}$. The rotation of the work gear $\omega_g$ and the rotation of the shaper cutter $\omega_{ct}$ are synchronized with one another. The rotation vectors $\omega_g$ and $\omega_{ct}$ are aligned with the corresponding axes of rotation $O_g$ and $O_{ct}$. The rotation vectors $\omega_g$ and $\omega_{ct}$ form a certain angle, which in this case is equal to the clearance angle $\alpha_o$ at the top cutting edge of the shaper cutter. This makes it possible to shape the clearance surfaces $C_s$ in the form of nearly involute cylinders. Straight generating lines of the involute cylinders are parallel to the axis of rotation $O_{ct}$ of the shaper cutter. Finish grinding of the clearance surfaces $C_s$ in this case is much easier.

The rake surface $R_s$ of the shaper cutter teeth is shaped in the form of a cone of revolution. This cone is coaxial with the axis of rotation of the shaper cutter. As shown in Figure 13.3, the parameters of the conical rake surface $R_s$ can be expressed in terms of the rake angle $\gamma_o$ and the clearance angle $\alpha_o$ at the top cutting edge of the shaper cutter teeth.

Direction of the motion of cut $\mathbf{V}_{cut}$ is parallel to the axis of rotation $O_g$ of the work gear. It is easy to show that the motion of cut $\mathbf{V}_{cut}$ allows for representation in the form of vector summa, that is, as $\mathbf{V}_{cut} = \mathbf{V}_x + \mathbf{V}_z$.

### 13.2.2 Shaping of Spur Gear by Enveloping Shaper Cutter

Similar to machining an internal gear with the shaper cutter (see Figure 13.3), machining an external gear with the enveloping shaper cutter can be performed as well. Vector representation of the internal intersected-axes gear machining mesh in the process of cutting an external gear with an enveloping shaper cutter is schematically depicted in Figure 13.4. The vector diagram is constructed on the premise of the two rotation vectors $\boldsymbol{\omega}_g$ and $\boldsymbol{\omega}_{ct}$ of the work gear and of the gear-cutting tool. The vector $\boldsymbol{\omega}_{pl}$ of the instant rotation of the cutter is constructed in a manner similar to that illustrated in Figure 13.1. Again, solely for convenience, the rotation vector $-\boldsymbol{\omega}_g$ is applied to both the work gear and the gear-cutting tool.

A schematic of the machining process of a spur gear by the enveloping shaper cutter is shown in Figure 13.5. The work gear and the shaper cutter are rotating about their axes of rotation $O_g$ and $O_{ct}$, respectively. The rotations $\omega_g$ and $\omega_{ct}$ are synchronized with one another. The rotation vectors $\boldsymbol{\omega}_g$ and $\boldsymbol{\omega}_{ct}$ are aligned with the corresponding axes of rotation $O_g$ and $O_{ct}$. The rotation vectors $\boldsymbol{\omega}_g$ and $\boldsymbol{\omega}_{ct}$ form a certain angle. In this case, the angle is equal to the clearance angle $\alpha_o$ at the top cutting edge of the gear shaper

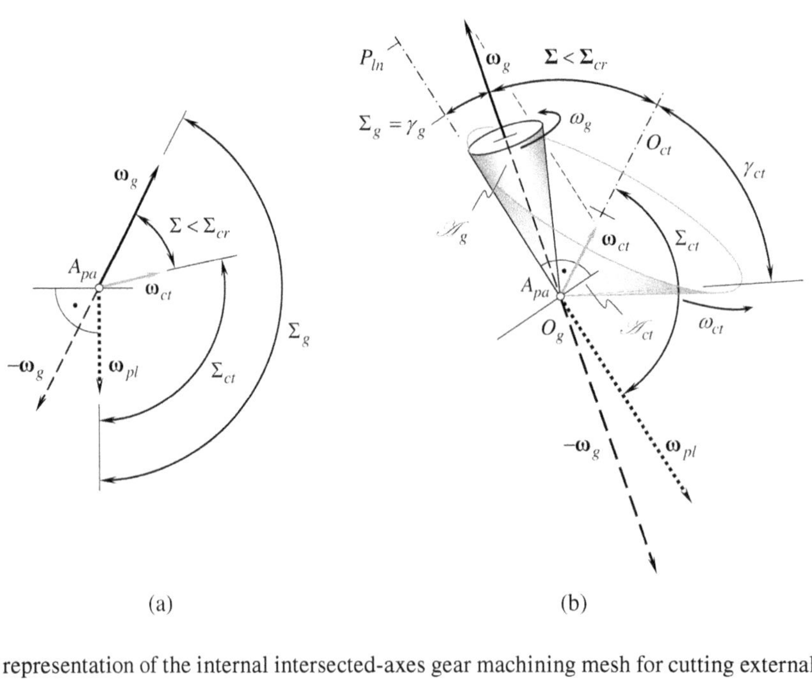


**FIGURE 13.4** Vector representation of the internal intersected-axes gear machining mesh for cutting external gear by the enveloping gear shaper cutter: (a) the gear vector diagram, and (b) the corresponding axodes for the gear machining mesh.

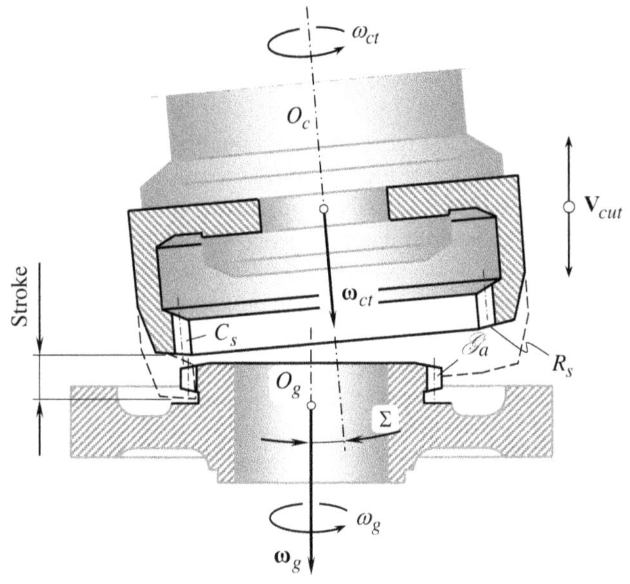


**FIGURE 13.5** Machining of a spur gear by the enveloping shaper cutter.

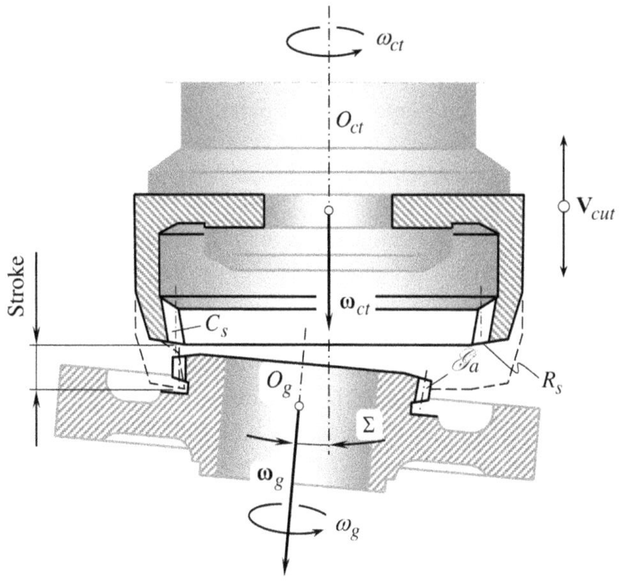


**FIGURE 13.6** Shaping of a recessed tooth form by an enveloping gear cutter. The tooth form is tapered by tilting the axis of the work gear relative to the axis of the gear shaper cutter.

cutter. This makes it possible to shape the clearance surfaces $C_s$ in the form of nearly involute cylinders. Straight generating lines of the involute cylinders are parallel to the axis of rotation $O_{ct}$ of the gear shaper cutter. Finish grinding of the clearance surfaces $C_s$ of the gear shaper cutter in this case is much easier.

The rake surface $R_s$ of the gear shaper cutter teeth is shaped in the form of a cone of revolution that is coaxial with the shaper cutter axis of rotation. The parameters of the conical rake surface $R_s$ can be expressed in terms of the rake angle $\gamma_o$ and of the clearance angle $\alpha_o$ at the top cutting edges of the gear shaper cutter teeth.

The direction of the primary motion (of the motion of cut $\mathbf{V}_{cut}$) is parallel to the axis of rotation $O_g$ of the work gear. It is easy to show that the motion of cut $\mathbf{V}_{cut}$ allows representation in the form of the vector summa $\mathbf{V}_{cut} = \mathbf{V}_x + \mathbf{V}_z$.

### 13.2.3 Shaping of External Recessed Tooth Forms by Enveloping Shaper Cutter

Implementation of the kinematic scheme of gear generating that is shown in Figure 13.4 can be extended to machining of external recessed tooth forms by an enveloping shaper cutter. Gears of this design are often referred to as *beveloids*. A perfect example of implementing gears of such a design can be found in [21].

The gear shown in Figure 13.6 has an external spline tooth that is recessed below the rim face of the gear.

The work gear and the gear shaper cutter rotate about their axes of rotation $O_g$ and $O_{ct}$. The rotations $\omega_g$ and $\omega_{ct}$ are synchronized with one another. The rotation vectors $\boldsymbol{\omega}_g$ and $\boldsymbol{\omega}_{ct}$ align with the corresponding axes of rotation $O_g$ and $O_{ct}$. The rotation vectors $\boldsymbol{\omega}_g$ and $\boldsymbol{\omega}_{ct}$ form a certain angle. In this case, the angle is equal to the angle of the recessed teeth.

The rake surface $R_s$ of the gear shaper cutter teeth is shaped in the form of a cone of revolution that is coaxial with the shaper cutter axis of rotation. The parameters of the conical rake surface $R_s$ can be expressed in terms of the rake angle $\gamma_o$ and of the clearance angle $\alpha_o$ at the top cutting edges of the gear shaper cutter teeth.

The direction of the motion of cut $\mathbf{V}_{cut}$ is parallel to the axis of rotation $O_g$ of the work gear. It can be shown that the motion of cut $\mathbf{V}_{cut}$ allows for representation in a form of the vector summa $\mathbf{V}_{cut} = \mathbf{V}_x + \mathbf{V}_z$.

With an external gear shaper cutter, the cutter diameter would be limited to a maximum 1.00 in. (25.4 mm), resulting in low generating feeds, multiple passes, long cycle times, and low cutter life. Tool life can be increased and cycle time reduced by using an enveloping shaper cutter with internal teeth [21]. The spline teeth can be cut straight or with back-taper as shown in Figure 13.6 on a tilting table mechanism. The accuracy of an enveloping cutter is inherently less because of the large manufacturing tolerances required for the internal teeth.

Along with machining of external beveloid teeth, shaping of internal gears having recessed tooth forms is feasible as well. For this purpose, gear shaper cutters of conventional design can be used.

In addition to the motion of cut that is parallel either to the axis of rotation of the work gear or to the axis of rotation of the gear shaper cutter, motions of cut of other types are also possible. To determine all possible motions of cut, it is necessary to investigate the possibility of implementation for this purpose of:

- All the elementary translational motions $\mathbf{V}_x$, $\mathbf{V}_y$, and $\mathbf{V}_z$
- All the elementary rotational motions $\omega_x$, $\omega_y$, and $\omega_z$
- All possible combinations of the elementary translations and elementary rotations by one, two, three, four, five, and six

The ratio of magnitudes of the elementary motions in every combination should be considered.

The potential capabilities of the kinematic scheme of gear generation (see Figure 13.1 and Figure 13.4) from the standpoint of gear-cutting tool design have not been thoroughly investigated yet. There is much room for research and development in this area of gear engineering. Novel methods of gear machining as well as novel designs of gear-cutting tools can be discovered by further analyzing the kinematic schemes shown in Figure 13.1 and Figure 13.4.

The means for cutting gears discussed in this section of the book allow for only approximate generation of the machined gear tooth flanks. This is attributed to the kinematics of the gear-cutting process and to violation of the condition of conjugacy of the tooth flanks $\mathscr{G}$ of a gear being machined and of the generating surface $T$ of the gear-cutting tool. In compliance with the condition of conjugacy of the tooth flanks at every point of contact of the surfaces $\mathscr{G}$ and $T$, the common perpendicular $\mathbf{n}$ must intersect the axis of instant rotation $P_{\text{ln}}$ of the gear and the gear-cutting tool in the gear machining mesh [171].

# *Subsection III-C*

# Crossed-Axes Gear Machining Mesh

Crossed-axes gear machining mesh (or spatial gear machining mesh, in other words) is extensively used when designing gear-cutting tools for continuous-indexing methods of generating gear tooth flanks. Hobs, shave cutters, and gear-cutting tools of other designs are perfect examples of gear-cutting tools designed on the premise of crossed-axes gear machining mesh.

Three main types of crossed-axes gear machining mesh are distinguished, depending on the actual values of the parameters of the mesh. They are

- External crossed-axes gear machining mesh.
- "Quasi-planar" crossed-axes gear machining mesh.
- Internal crossed-axes gear machining mesh.

Designing gear-cutting tools is considered separately for the gear machining mesh of each kind. This is solely because of the convenience of such an approach for readers. Gear-cutting tools of different designs are considered depending on the type of gear machining mesh that is simulated in a particular gear machining process. Therefore, discussion of the continuous-indexing gear-cutting tools design is split into three sections, each of which is devoted to

a. external crossed-axes gear machining mesh,
b. quasi-planar crossed-axes gear machining mesh, and
c. internal crossed-axes gear machining mesh.

Subsection III is the biggest of Subsections I to III. For the reader's convenience, this subsection is subdivided onto three parts. Part III-C 1 titled "External Gear Machining Mesh" is the first of three parts, III-C 1 to III-C 3, to be discussed in Subsection III.

# Part III-C.1

## Design of Gear-Cutting Tools: External Gear Machining Mesh

Gear-cutting tools of various types are designed on the premise of continuous-indexing external crossed-axes gear machining mesh. This group of gear-cutting tools is the most common in industry. The similarity of the kinematics of the gear machining process, together with the similarity of the geometry of generating surfaces of gear-cutting tools, make it reasonable to consider them in a common section.

This section of the book comprises four chapters, Chapter 14 through Chapter 17.

DOI: 10.1201/9781003520528-18

# 14 Generating Surface of Gear-Cutting Tool

The geometry of the tooth shape being machined, along with the kinematics of the gear machining operation, are the two components that are of critical importance for the generation of the generating surface of the gear-cutting tool.

Tooth flank geometry of the work gear is specified by the gear blueprint. It can also be described analytically, for example, by Eq. (1.2), Eq. (1.4), Eq. (1.32), Eq. (1.40), and others. Once the design parameters of a gear being machined are known from the blueprint, then the corresponding analytical description for tooth flanks can be derived from this data. Conventional technique is used for this purpose.

Gear vector diagrams are commonly used to investigate the kinematics of a gear machining mesh. An example of vector representation of *external crossed-axes gear machining mesh* is shown in Figure III-C 1.1.

The kinematics of spatial motion of the gear-cutting tool relative to the work gear is complex. It deserves to be discussed in greater detail.

## 14.1 KINEMATICS OF EXTERNAL CROSSED-AXES GEAR MACHINING MESH

External crossed-axes gear machining mesh is comprised of two rotations about crossing axes of rotation. One of the rotations is specified by the rotation vector $\boldsymbol{\omega}_g$. This is the rotation of the work gear.[1] Another one is specified by the rotation vector $\boldsymbol{\omega}_{ct}$. This is the rotation of the cutting tool being designed. The rotations $\boldsymbol{\omega}_g$ and $\boldsymbol{\omega}_{ct}$ are synchronized with one another in a timely manner so that when the cutting tool completes one full rotation about its axis of rotation $O_{ct}$, the work gear turns about the axis of rotation $O_g$ through the angle $2\pi(N_{ct}/N_g)$. Here, $N_g$ and $N_{ct}$ denote the tooth number of the gear being machined, and the tooth number (number of starts) of the gear-cutting tool, respectively.

An example of vector representation of external crossed-axes gear machining mesh is shown in Figure III-C 1.1. The vector diagram is specified in terms of

- Two rotation vectors $\boldsymbol{\omega}_g$ and $\boldsymbol{\omega}_{ct}$.
- The crossed-axes angle $\Sigma$.
- The center distance $C$.

The gear machining mesh shown in Figure III-C 1.1 is one of three possible types of crossed-axes gear machining meshes classified in Figure 3.8.

In Figure III-C 1.1(a), the rotation vectors $\boldsymbol{\omega}_g$ and $\boldsymbol{\omega}_{ct}$ are constructed so that they are parallel to the horizontal plane of projection $\pi_1$. Such a configuration of the rotation vectors $\boldsymbol{\omega}_g$ and $\boldsymbol{\omega}_{ct}$ is advantageous for two reasons. *First*, the actual value of the crossed-axes angle $\Sigma$ can be measured within the horizontal plane of projection $\pi_1$, and *second*, the actual value of the center distance $C$ can be measured within the vertical plane of projection $\pi_2$.

Let us suppose the axode of the work gear is motionless and the axode of the gear-cutting tool is free to roll over the axode of the work gear.[2] For this purpose, it is sufficient to assume that both the work gear and the gear-cutting tool are given an additional rotation $-\boldsymbol{\omega}_g$. Due to the additional rotation $-\boldsymbol{\omega}_g$, the work gear becomes stationary [that is, $\boldsymbol{\omega}_g + (-\boldsymbol{\omega}_g) = 0$], while the gear-cutting tool will be subjected to two rotations simultaneously, that is, it performs (a) the initial rotation $\boldsymbol{\omega}_{ct}$ and (b) the additional rotation $-\boldsymbol{\omega}_g$. Two rotations $\boldsymbol{\omega}_{ct}$ and $-\boldsymbol{\omega}_g$ together result in the instant rotation $\boldsymbol{\omega}_{pl} = \boldsymbol{\omega}_{ct} - \boldsymbol{\omega}_g$ of the gear-cutting tool about the axis of instant rotation $P_{\text{ln}}$. The vector $\boldsymbol{\omega}_{pl}$ of the instant rotation is applied at a certain point[3] $A_{pa}$ within the centerline ℄.

The rotation vector $\boldsymbol{\omega}_{pl}$ is parallel to the horizontal plane of projection $\pi_1$. Projections of the rotation vectors $\boldsymbol{\omega}_g$ and $\boldsymbol{\omega}_{ct}$ of the work gear and the cutting tool onto the vertical plane of projection $\pi_2$ are denoted by $\boldsymbol{\omega}_g^{rl}$ and $\boldsymbol{\omega}_{ct}^{rl}$, respectively. The components $\boldsymbol{\omega}_g^{rl}$ and $\boldsymbol{\omega}_{ct}^{rl}$ are parallel to the axis of instant rotation $P_{\text{ln}}$. The components $\boldsymbol{\omega}_g^{rl}$ and $\boldsymbol{\omega}_{ct}^{rl}$ result in pure rolling of the axodes $\mathscr{A}_g$ of the work gear and $\mathscr{A}_{ct}$ of the gear-cutting tool [see Figure III-C 1.1(b)]. For the magnitudes $\omega_g^{rl} = |\boldsymbol{\omega}_g^{rl}|$ and $\omega_{ct}^{rl} = |\boldsymbol{\omega}_{ct}^{rl}|$, the following ratio [see Eq. (14.3)] is valid

$$C_g \cdot \omega_{pl} = C \cdot \omega_{ct} \cdot \cos \Sigma_{ct} \tag{14.1}$$

$$C \cdot \omega_g \cdot \cos \Sigma_g = C_{ct} \cdot \omega_{pl} \tag{14.2}$$

$$C_g \cdot \omega_g \cdot \cos \Sigma_g = C_{ct} \cdot \omega_{ct} \cdot \cos \Sigma_{ct} \tag{14.3}$$

where $C_g$ denotes the distance from the apex point $A_{pa}$ to the axis of rotation $O_g$ of the work gear (this is the so-called gear

[1] ***Reminder:*** Rotations, by nature, are not vectors. Therefore, special care is required when treating rotations as vectors.

[2] The process of holding stationary different links of a kinematic chain is known as inversion.

[3] The designation $A_{pa}$ corresponds to the term *plane-of-action apex* $A_{pa}$. This term is widely used in the scientific theory of gearing (Radzevich, S.P., *Theory of Gearing: Kinematics, Geometry, and Synthesis*, 3rd edition, CRC Press, Boca Raton, Florida, 2022, 1154 pages) when intersected-axes gearing, as well as crossed-axes gearing, are discussed.


DOI: 10.1201/9781003520528-19

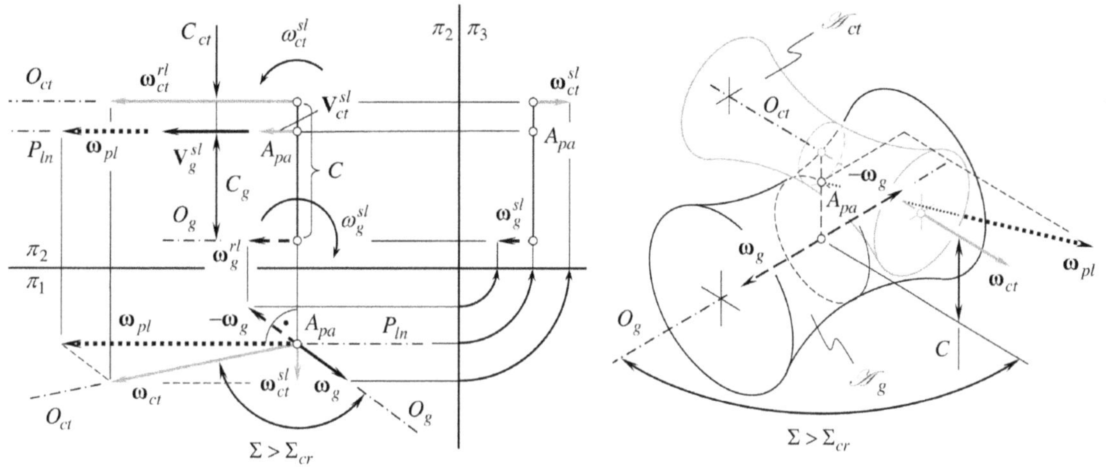


**FIGURE III-C1.1** Vector diagram of external crossed-axes gear machining mesh (a), and a set of corresponding axodes (b).

center distance $C_g$). Similarly, the distance from that same apex point $A_{pa}$ to the axis of rotation $O_{ct}$ of the gear-cutting tool is designated as $C_{ct}$ (this is the so-called gear-cutting tool center distance $C_{ct}$).

The cone angles $\Sigma_g$ and $\Sigma_{ct}$ in the gear and in the gear-cutting tool can be expressed in terms of the rotation vectors $\boldsymbol{\omega}_g$, $\boldsymbol{\omega}_{ct}$, and $\boldsymbol{\omega}_{pl}$. For this purpose, the formulas $\Sigma_g = \angle(\boldsymbol{\omega}_g, \boldsymbol{\omega}_{pl})$ and $\Sigma_{ct} = \angle(\boldsymbol{\omega}_{ct}, \boldsymbol{\omega}_{pl})$ apply.

Evidently, the equality

$$C_g + C_{ct} = C \tag{14.4}$$

is observed. This equality can be utilized for the purposes of determining the exact location of the apex point $A_{pa}$ within the centerline ℄. The condition of pure rolling is helpful for solving this particular problem.

The condition of pure rolling of the axodes can be employed for the purposes of determining the exact location of the apex point $A_{pa}$ within centerline ℄. The ratio

$$\frac{C_g}{C_{ct}} = \frac{\omega_{ct}^{rl}}{\omega_g^{rl}} \tag{14.5}$$

immediately follows from the condition of pure rolling of the axodes $\mathscr{A}_g$ and $\mathscr{A}_{ct}$. Here in Eq. (14.5) is designated $\omega_g^{rl} = |\boldsymbol{\omega}_g^{rl}|$ and $\omega_{ct}^{rl} = |\boldsymbol{\omega}_{ct}^{rl}|$.

Generally speaking, the magnitudes $\omega_g^{rl}$ and $\omega_{ct}^{rl}$ of the vectors of pure rolling $\boldsymbol{\omega}_g^{rl}$ and $\boldsymbol{\omega}_{ct}^{rl}$ are not equal to one another. The inequality $\omega_g^{rl} < \omega_{ct}^{rl}$ is commonly observed. The equality $\omega_g^{rl} = \omega_{ct}^{rl}$ is observed only in particular cases when the tooth number of the work gear $N_g$ and the tooth number of the cutting tool[4] $N_{ct}$ are equal to one another.

[4] In particular cases, the design parameter $N_{ct}$ could also mean the number of threads/starts of the worm-type gear-cutting tool.

The equality $C_g = C - C_{ct}$ immediately follows from Eq. (14.4). Substituting this value of the gear center distance $C_g$ into Eq. (14.5), the formula

$$C_{ct} = \frac{\omega_g^{rl}}{\omega_{ct}^{rl} + \omega_g^{rl}} \cdot C \tag{14.6}$$

for the calculation of the gear-cutting tool center distance $C_{ct}$ can be derived.

Then the gear center distance $C_g$ can be calculated from the above formula, $C_g = C - C_{ct}$. This expression for the gear center distance $C_g$ casts into

$$C_g = \frac{\omega_{ct}^{rl}}{\omega_{ct}^{rl} + \omega_g^{rl}} \cdot C \tag{14.7}$$

In the case under consideration, the apex point $A_{pa}$ is located within the centerline ℄ in between the gear cone apex $A_g$ and the gear-cutting tool cone apex $A_{ct}$.

The other two components $\boldsymbol{\omega}_g^{sl}$ and $\boldsymbol{\omega}_{ct}^{sl}$ of the rotation vectors $\boldsymbol{\omega}_g$ and $\boldsymbol{\omega}_{ct}$ are perpendicular to the axis of the instant rotation $P_{ln}$. With no distortion, these components are projected onto the frontal plane of projection $\pi_3$.

The rotation vectors $\omega_g^{sl}$ and $\omega_{ct}^{sl}$ are of the same magnitude and are pointed opposite to one another ($\boldsymbol{\omega}_g^{sl} = -\boldsymbol{\omega}_{ct}^{sl}$). The rotations $\omega_g^{sl}$ and $\omega_{ct}^{sl}$ result in pure sliding of the axodes of the work gear and the gear-cutting tool over one another.

One of two components $\mathbf{V}_g^{sl}$ of the vector of linear velocity of sliding of the axodes (that is, the component owing to the rotation of the work gear) is equal to

$$\mathbf{V}_g^{sl} = C_g \cdot \boldsymbol{\omega}_g^{sl} \tag{14.8}$$

The other component $\mathbf{V}_{ct}^{sl}$ of the vector of the linear velocity of the sliding of the axodes (that is, the component due to the rotation of the gear-cutting tool) can be calculated from the formula

$$\mathbf{V}_{ct}^{sl} = C_{ct} \cdot \boldsymbol{\omega}_{ct}^{sl} \tag{14.9}$$

Commonly, the inequality $C_g > C_{ct}$ is observed in a gear machining operation. Therefore, the component $\mathbf{V}_g^{sl}$ always exceeds the component $\mathbf{V}_{ct}^{sl}$ (and the inequality $|\mathbf{V}_g^{sl}| > |\mathbf{V}_{ct}^{sl}|$ is valid). This is because the magnitudes of the rotations $\boldsymbol{\omega}_g^{sl}$ and $\boldsymbol{\omega}_{ct}^{sl}$ are equal to each other $\left(|\boldsymbol{\omega}_g^{sl}| = |\boldsymbol{\omega}_{ct}^{sl}|\right)$.

The vectors $\mathbf{V}_g^{sl}$ and $\mathbf{V}_{ct}^{sl}$ of the linear velocities of sliding are pointed opposite from each other. Therefore, the vector $\mathbf{V}_{g-ct}^{sl}$ of the linear velocity of the resultant sliding of the axode of the work gear relative to the axode of the gear-cutting tool is calculated as the difference

$$\mathbf{V}_{g-ct}^{sl} = \mathbf{V}_g^{sl} - \mathbf{V}_{ct}^{sl} \tag{14.10}$$

The oppositely directed vector $\mathbf{V}_{ct-g}^{sl}$ of the resultant linear velocity of the sliding of the axode of the gear-cutting tool with respect to the axode of the work gear can be calculated from the equation

$$\mathbf{V}_{ct-g}^{sl} = -\mathbf{V}_{g-ct}^{sl} = \mathbf{V}_{ct}^{sl} - \mathbf{V}_g^{sl} \tag{14.11}$$

In the case under consideration, the magnitude $V_{\Sigma}^{sl}$ of the linear velocity of sliding of the axodes can be determined as summa

$$V_{\Sigma}^{sl} = V_g^{sl} + V_{ct}^{sl} \tag{14.12}$$

The component vectors $\boldsymbol{\omega}_g^{sl}$ and $\boldsymbol{\omega}_{ct}^{sl}$ comprise a pair of rotations. This is because the vectors $\boldsymbol{\omega}_g^{sl}$ and $\boldsymbol{\omega}_{ct}^{sl}$ are of the same magnitude and are pointed opposite to each other. A pair of rotations is equivalent to a translation with a linear velocity $\mathbf{V}_{sc}$. The vector of linear velocity $\mathbf{V}_{sc}$ of the translation is parallel to the vector of instant rotation $\boldsymbol{\omega}_{pl}$. The magnitude $V_{sc}$ of the vector $\mathbf{V}_{sc}$ is equal to

$$V_{sc} = |\mathbf{V}_{sc}| = C \cdot \omega_g \cdot \sin\Sigma_g = C \cdot \omega_{ct} \cdot \sin\Sigma_{ct} \tag{14.13}$$

It is worth pointing out here that the linear velocities $\mathbf{V}_{\Sigma}^{sl}$ and $\mathbf{V}_{sc}$ are of the same magnitude ($V_{\Sigma}^{sl} = V_{sc}$).

Two motions, namely, (a) the instant rotation $\boldsymbol{\omega}_{pl}$ and (b) the translation $\mathbf{V}_{sc}$ are performed simultaneously. For the calculation of parameter $p_{sc}$ of the resultant instant screw motion, the following formula applies:

$$p_{sc} = \frac{V_{sc}}{\omega_{pl}} = \frac{C \cdot \omega_g \cdot \sin\Sigma_g}{\omega_{pl}} = \frac{C \cdot \omega_{ct} \cdot \sin\Sigma_{ct}}{\omega_{pl}} \tag{14.14}$$

where $\omega_{pl} = |\boldsymbol{\omega}_{pl}|$.

Equations (14.1) through (14.3) yield an expression

$$\omega_{pl} = \frac{C \cdot \omega_g \cdot \cos\Sigma_g}{C_{ct}} = \frac{C \cdot \omega_{ct} \cdot \cos\Sigma_{ct}}{C_g} \tag{14.15}$$

for the calculation of magnitude $\omega_{pl}$ of the instant rotation $\boldsymbol{\omega}_{pl}$.

Therefore, the expression

$$p_{sc} = C_g \cdot \tan\Sigma_{ct} = C_{ct} \cdot \tan\Sigma_g \tag{14.16}$$

is valid. This immediately yields the ratio

$$\frac{C_g}{C_{ct}} = \frac{\sin\Sigma_{ct}}{\sin\Sigma_g} \tag{14.17}$$

Ultimately, in the case under consideration, two rotations $\boldsymbol{\omega}_g$ and $\boldsymbol{\omega}_{ct}$ about skew axes result in an instant screw motion. The resultant motion can be viewed as rolling with sliding of the hyperboloid $\mathscr{H}_{ct}$ of the gear-cutting tool over the motionless hyperboloid $\mathscr{H}_g$ of the work gear. The axode of the gear-cutting tool $\mathscr{H}_{ct}$ is represented as the locus of consecutive positions of the axis $P_{ln}$ in its rotation about the axis of rotation $O_{ct}$ of the gear-cutting tool. The stationary axode of the work gear $\mathscr{H}_g$ is represented as the locus of consecutive positions of the axis $P_{ln}$ in its rotation about the axis of rotation $O_g$ of the work gear. Rolling and sliding occur about and along a straight generating line when the lines align with each other at a certain instant of time.

Configuration of the rotation vectors $\boldsymbol{\omega}_g$ and $\boldsymbol{\omega}_{ct}$ relative to the axodes of the work gear $\mathscr{H}_g$ and the gear-cutting tool $\mathscr{H}_{ct}$ is illustrated in Figure III-C 1.1(b). Here we stress one more time that axodes $\mathscr{H}_g$ and $\mathscr{H}_{ct}$ themselves are not of prime importance and are not used in this text for the analysis of the kinematics of the gear machining operation. The vector diagrams are self-consistent and are preferred for the analysis of the kinematics of the gear machining processes.

## 14.2 AUXILIARY GENERATING SURFACE OF GEAR-CUTTING TOOL: OLIVIER PRINCIPLES

The auxiliary generating surface of the gear-cutting tool is utilized in cases when the first *Olivier principle*[5] [52] is used to determine the generating surface of the gear-cutting tool.

[5] In general cases of gear meshing, two principles proposed by Theodore Olivier (1842) are incorrect. Both the principles are valid only in some cases, when moving surfaces allow for sliding over themselves in the direction of the enveloping motion. In these cases the principles of Th. Olivier become useless. Therefore, there is no reason to apply Oliver principles for the generation of precise conjugate tooth flanks in a gear machining mesh. The principles by Th. Olivier are applicable only in cases when approximate gear-cutting tools are designed and used for cutting gears in continuous-indexing methods of gear machining (Radzevich, S.P., *Theory of Gearing: Kinematics, Geometry, and Synthesis*, *CRC* Press, Boca Raton, Florida, 2012, 743 pages).

The resultant motion of the gear-cutting tool in relation to the work gear is resolved into two elementary components. The auxiliary generating surface is an envelope to consecutive positions of the gear tooth flank when the gear performs one of the two component motions. Then, the generating surface is obtained as an envelope surface to consecutive positions of the auxiliary generating surface when the auxiliary surface performs the second of two elementary component motions.

For generating the auxiliary generating surface of the gear-cutting tool, consider a gear being machined that performs a certain motion relative to a Cartesian reference system $X_rY_rZ_r$. After having been generated, the auxiliary generating surface $\mathscr{R}_T$ will be associated with this Cartesian reference system $X_rY_rZ_r$.

In order to determine all possible types of auxiliary generating surfaces $\mathscr{R}_T$, it is necessary to consider all possible motions of the gear tooth flank $\mathscr{G}$ in relation to the coordinate system $X_rY_rZ_r$. An analysis is then performed determine whether or not an envelope surface to consecutive positions of the tooth flank $\mathscr{G}$ exists. If the envelope surface exists, then it can be used as an auxiliary generating surface of the gear-cutting tool. Actually, not only tooth flank $\mathscr{G}$ should be considered, but the surfaces forming the top and bottom lands of the gear tooth should be considered as well. Tooth flank $\mathscr{G}$ is taken into consideration here only as a representative example.

Consider a gear tooth flank $\mathscr{G}$ that is given by Eq. (1.4):

$$\mathbf{r}_g(U_g, V_g) = \begin{bmatrix} r_{b.g}\cos V_g + U_g\cos\lambda_{b.g}\sin V_g \\ r_{b.g}\sin V_g - U_g\sin\lambda_{b.g}\sin V_g \\ r_{b.g}\tan\lambda_{b.g} - U_g\sin\lambda_{b.g} \\ 1 \end{bmatrix} \quad \begin{matrix} V_g^{(l)} \le V_g \le V_g^{(a)} \\ 0 \le U_g \le [U_g] \end{matrix} \qquad (14.18)$$

The tooth flank $\mathscr{G}$ is used for the derivation of an equation of the auxiliary (virtual rack) $\mathscr{R}_T$ of the gear-cutting tool. The rack $\mathscr{R}_T$ must be conjugate to the gear tooth surface $\mathscr{G}$. Unfortunately, not all geometries of tooth flanks allow for a conjugate auxiliary rack $\mathscr{R}_T$. For example, spur and helical involute gears allow for a conjugate auxiliary rack $\mathscr{R}_T$, while gears with non-involute tooth profiles do not.

The auxiliary rack surface $\mathscr{R}_T$ can be generated as an envelope of consecutive positions of the gear tooth flank $\mathscr{G}$ in its motion relative to the coordinate system $X_rY_rZ_r$.

As shown in Figure 14.1(a), the gear being machined rotates about its axis of rotation $O_g$ with a certain angular velocity $\omega_g$. Simultaneously, the gear performs a straight motion with a certain linear velocity $V_g = |\mathbf{V}_g|$.

The relative motion of the work gear and the reference system $X_rY_rZ_r$ can be viewed as rolling of the gear pitch cylinder of a diameter $d_g = 2r_g$ over the pitch plane $W_r$ associated with the reference system $X_rY_rZ_r$. The auxiliary generating rack $\mathscr{R}_T$ [see Figure 14.1(b)] can be determined as the envelope of consecutive positions of the gear tooth flank $\mathscr{G}$ when the pitch plane $W_r$ rolls without sliding over the pitch cylinder of a diameter $d_g$. The linear velocity $V_g$ of the translational motion depends on the rotation $\omega_g$. It can be calculated from the equation $V_g = 0.5 \cdot \omega_g \cdot d_g$.

For the derivation of an equation of the auxiliary generating surfaces $\mathscr{R}_T$ of the gear-cutting tool, the following main coordinate systems are used (see Figure 14.2): (a) the coordinate system $X_gY_gZ_g$ associated with the gear being machined and (b) the coordinate system $X_rY_rZ_r$ which the auxiliary generating rack $\mathscr{R}_T$ will be associated with. A few more intermediate reference systems are used as well.

For the further analysis, it is necessary to represent the gear tooth flank $\mathscr{G}$ in the coordinate system $X_rY_rZ_r$. In this coordinate system, the position vector of a point $\mathbf{r}_g^r$ can be expressed in terms of the position vector of that same point $\mathbf{r}_g$ in the original reference system $X_gY_gZ_g$ and in terms of the operator $\mathbf{Rs}(\mathscr{G} \mapsto \mathscr{R}_T)$ of transition from the coordinate system $X_gY_gZ_g$ associated with the work gear to the coordinate system $X_rY_rZ_r$ [148, 158]. The expression

$$\mathbf{r}_g^r(U_g, V_g) = \mathbf{Rs}(\mathscr{G} \mapsto \mathscr{R}_T) \cdot \mathbf{r}_g(U_g, V_g) \qquad (14.19)$$

is valid for the position vector of a point $\mathbf{r}_g^r$.

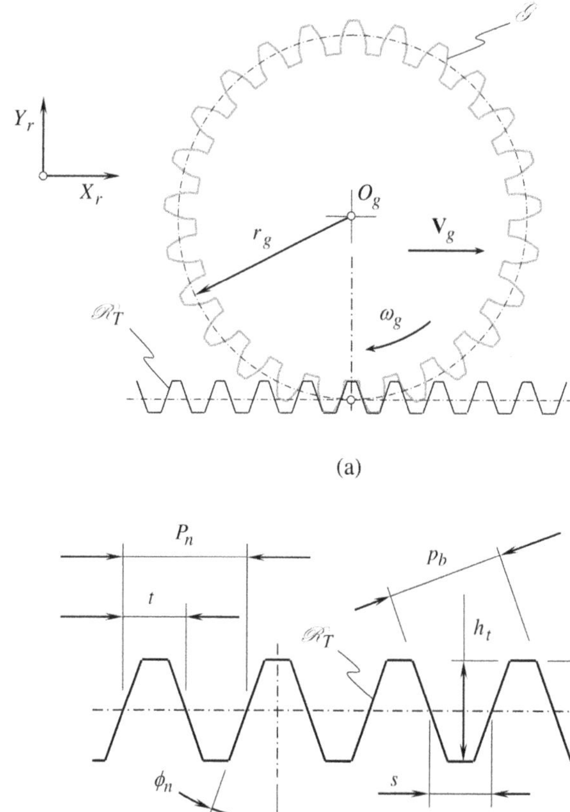


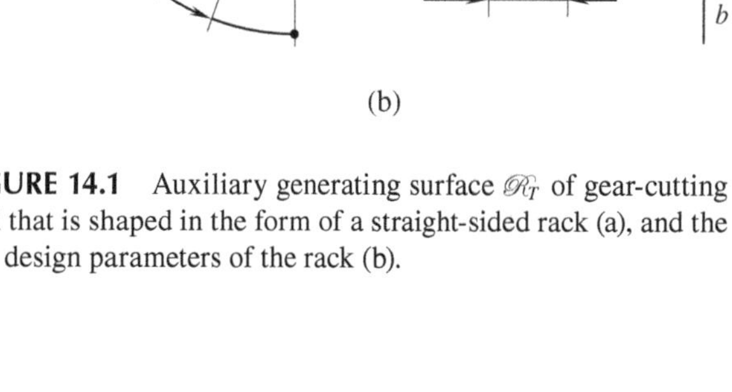


**FIGURE 14.1** Auxiliary generating surface $\mathscr{R}_T$ of gear-cutting tool that is shaped in the form of a straight-sided rack (a), and the key design parameters of the rack (b).

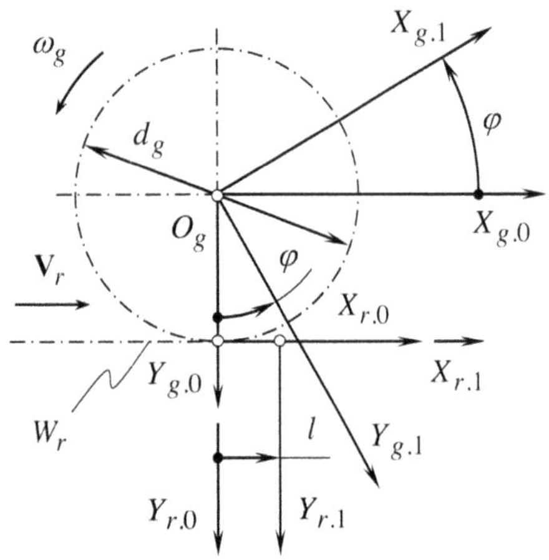


**FIGURE 14.2** The coordinate systems applied for the derivation of the equation of auxiliary generating surface $\mathscr{R}_T$ of a gear-cutting tool. (After: Radzevich, S.P., *ASME Journal of Mechanical Design*, 129, 334–345, 2007. With permission.)

The transformation operator $\mathbf{Rs}(\mathscr{G} \mapsto \mathscr{R}_T)$ analytically describes the resultant coordinate system transformation, that is, it describes the transitions from the Cartesian coordinate system $X_g Y_g Z_g$ to another Cartesian coordinate system $X_r Y_r Z_r$.

In the case under consideration, the resultant coordinate system transformation can be analytically represented as a superposition of several consequent elementary coordinate system transformations as illustrated in Figure 14.3. The required elementary linear transformations are described analytically by

- The operator $\mathbf{Rt}(\varphi, Z_g)$ of the rotation about the $Z_g$ – axis through an angle $\varphi$.
- The operator $\mathbf{Tr}(-0.5 \cdot d_g, Y_r)$ of the translation along the $Y_r$ – axis at a distance $-0.5 \cdot d_g$.
- The operator $\mathbf{Tr}(-l, X_r)$ of the translation along the $X_r$ – axis at a distance $-l$.

The above-mentioned operators of elementary coordinate system transformations yield an expression for the operator $\mathbf{Rs}(\mathscr{G} \mapsto \mathscr{R}_T)$ of the resultant coordinate system transformation:

$$\mathbf{Rs}(\mathscr{G} \mapsto \mathscr{R}_T) = \mathbf{Tr}(-l, X_r) \cdot \mathbf{Tr}(-0.5 \cdot d_g, Y_r) \cdot \mathbf{Rt}(\varphi, Z_g) \tag{14.20}$$

The value of the angle $\varphi$ is calculated from the formula $\varphi = \omega_g \cdot t$, and the distance $l$ is equal to $l = |\mathbf{V}_r| \cdot t$. Here time is denoted by $t$.

In the relative motion, the gear tooth flank surface $\mathbf{r}_g^r$ occupies various positions. Equation (14.18) is employed for the purpose of analytical representation of the screw involute surface $\mathbf{r}_{g.t}^r$ in its current location:

$$\mathbf{r}_{g.t}^r = \mathbf{r}_{g.t}^r(U_g, V_g, t) \tag{14.21}$$

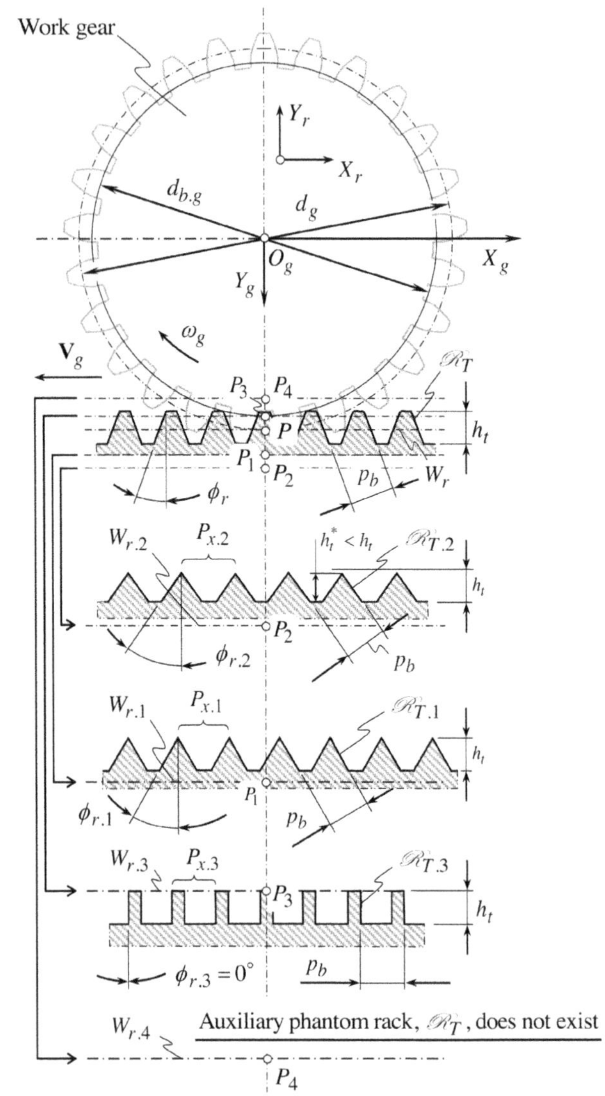


**FIGURE 14.3** Generation of auxiliary virtual racks $\mathscr{R}_T$ with different profile angles $\phi_r$. (After: Radzevich, S.P., *ASME Journal of Mechanical Design*, 129, 438–444, 2007. With permission.)

To eliminate time $t$, the equation $\mathbf{r}_{g.t}^r = \mathbf{r}_{g.t}^r(U_g, V_g, t)$ needs to be considered together with the Shishkov equation of contact, $\mathbf{n}_g \cdot \mathbf{V}_{g-r} = 0$ [204, 206]. The solution to the set of two equations

$$\begin{cases} \mathbf{r}_{g.t}^r = \mathbf{r}_{g.t}^r(U_g, V_g, t) \\ \mathbf{n}_g \cdot \mathbf{V}_{g-r} = 0 \end{cases} \tag{14.22}$$

returns[6] an equation of the tooth flank of the auxiliary rack $\mathscr{R}_T$. Here, the unit normal vector to the gear tooth

[6] It needs to be point out here that the solution to the set of Eq. (14.22) returns the equation of the tooth surface of the auxiliary rack $\mathscr{R}_T$ only in a case when the gear tooth flank $\mathscr{G}$ and the rack tooth flank $\mathscr{R}_T$ represent a pair of conjugate surfaces, as it is observed in the case of an involute tooth flank $\mathscr{G}$. Otherwise, when the gear tooth flank $\mathscr{G}$ and the rack tooth flank $\mathscr{R}_T$ are not conjugate, Eq. (14.22) returns an approximation, $\mathscr{R}_T^a$, to the desired auxiliary generating surface $\mathscr{R}_T$.

flank $\mathscr{G}_r$ (that is, in the coordinate system $X_rY_rZ_r$) is designated as $\mathbf{n}_g$, and $\mathbf{V}_{g-r}$ is the velocity vector of the motion of the gear tooth flank $\mathscr{G}$ relative to the coordinate system $X_rY_rZ_r$.

Expanding Eq. (14.22) and eliminating the enveloping parameter $t$, one can come up with an equation of the lateral tooth surface of the auxiliary generating surface $\mathscr{R}_T$ in matrix representation:

$$\mathbf{r}_r(U_r, V_r) = \begin{bmatrix} X_r(U_r, V_r) \\ Y_r(U_r, V_r) \\ Z_r(U_r, V_r) \\ 1 \end{bmatrix} \tag{14.23}$$

Eventually, the solution to the set of Eq. (14.22) reveals that

- The rack tooth flank surface $\mathscr{R}_T$ is a plane.
- This plane forms the base lead angle $\lambda_{b.g}$ with the gear axis of rotation $O_g$.
- The plane forms the profile angle $\phi_r$ with the pitch plane $W_r$ as depicted in Figure 14.1(b).

Just one tooth flank of the work gear can be generated by one tooth flank of the auxiliary rack $\mathscr{R}_T$. A corresponding number of planes $\mathscr{R}_T$ is required for simultaneous machining of several tooth flanks of the work gear. All of these planes are parallel to one another. Every two neighboring planes are at a distance that is equal to the normal base pitch $p_{nb}$ of the gear being machined. The distance $p_{nb}$ is measured along a perpendicular to the planes $\mathscr{R}_T$. The same is true with respect to the opposite side of the tooth profile of the gear, and of the auxiliary rack $\mathscr{R}_T$.

Hence, the auxiliary generating rack $\mathscr{R}_T$ of a gear-cutting tool for machining involute gears has a straight-sided tooth profile. The profile angle $\phi_r$ of the rack tooth $\mathscr{R}_T$ is identical to the gear profile angle $\phi_g$ that is measured in the transverse section on the gear pitch diameter ($\phi_r \equiv \phi_g$).

It is proved analytically [138] that for machining purposes the actual value of the gear pitch diameter $d_g$ can be chosen within a certain interval $d_{b.g} \le d_g \le d_g^*$. Here, $d_g^*$ designates the maximum permissible pitch diameter of the gear for which tooth pointing of the auxiliary rack $\mathscr{R}_{y.T}$ is avoided and, moreover, the width of the top land of teeth of the rack exceeds a certain minimum allowed value. The diameter $d_g^*$ can be determined on the premise of an analysis of the fulfillment of the set of necessary conditions of proper part surface generation [136, 148, 158] (see Appendix E). For different values $d_{y.g}$ of the gear pitch diameter, the conjugate rack surface $\mathscr{R}_{y.T}$ has the corresponding value of the profile angle $\phi_{y.r}$. The actual values of the pitch $P_{y.x}$ and the helix angle $\psi_{y.r}$ of the auxiliary rack $\mathscr{R}_{y.T}$ also depend on the pitch diameter $d_{y.g}$ of the gear. Therefore, as shown in Figure 14.3, variation in the gear pitch diameter $d_{y.g}$ results in corresponding alterations in

- Profile angle $\phi_{y.r}$.
- Pitch $P_{y.x}$.
- Helix angle $\psi_{y.r}$ of the auxiliary rack $\mathscr{R}_{y.T}$.

All the feasible auxiliary phantom racks $\mathscr{R}_{y.T}$ are always conjugate to the involute gear being machined.

A gear can be viewed as a certain number of the tooth-flank surfaces $\mathscr{G}$ that are evenly distributed around the gear pitch cylinder. Therefore, the auxiliary rack $\mathscr{R}_{y.T}$ can be considered as two sets of planes evenly distributed within the pitch plane $W_r$. The planes in each set are parallel to one another. The distance between every two neighboring planes of the auxiliary rack $\mathscr{R}_{y.T}$ (or the distance between every two neighboring straight-line profiles, which is the same) is measured along the pitch line. This distance equals the gear pitch

$$P_{y.x} = \frac{\pi \cdot d_g}{N_g} \tag{14.24}$$

which is measured on the pitch circle of the gear in a transverse section.

Thus, a variety of the auxiliary racks with different design parameters can be generated [140]. It is of critical importance to stress here that base pitch $p_b$ for all possible auxiliary racks $\mathscr{R}_{y.T}$ is of the same value and is equal to the base pitch of the gear being machined.

Variation of the pitch diameter $d_{y.g}$ within the interval $d_{b.g} \le d_g \le d_g^*$ results in variation of the gear pitch helix within the corresponding interval $\psi_g^* \le \psi_{y.g} \le \psi_{f.g}$.

The pitch diameter $d_g$ of the gear should not be too small; its actual value should not be less than the gear base diameter $d_{b.g}$ (see Figure 14.3), that is, the inequality $d_g \ge d_{b.g}$ has to be observed.

When the gear pitch diameter $d_g$ is smaller than the gear base diameter (that is, when the inequality $d_g < d_{b.g}$ is observed), then the corresponding auxiliary rack $\mathscr{R}_T$ does not exist; in such a scenario, no envelope can be generated for the consecutive positions of the gear tooth flank surface $\mathscr{G}$. This indicates that in the case under consideration, the first necessary condition of proper generation of the gear tooth flank is violated [136, 148, 158]. Therefore, in the event $d_g < d_{b.g}$, no auxiliary rack $\mathscr{R}_T$ can be generated.

The actual value of the profile angle $\phi_r$ of the auxiliary rack $\mathscr{R}_T$ depends on the actual timing of the rotation $\omega_g$ and the translation $V_g$ [see Figure 14.1(a)]. Two constraints are imposed on the actual value of the profile angle $\phi_r$.

*First*, the profile angle $\phi_r$ cannot be too small. Theoretically, the smallest permissible value of the profile angle equals $\phi_r = 0°$. Due to the necessity of creating an appropriate value of the normal clearance angle at the lateral cutting edges of the gear-cutting tool, the minimum value of the profile angle is commonly set to $\phi_r \cong 10°$ and not smaller.

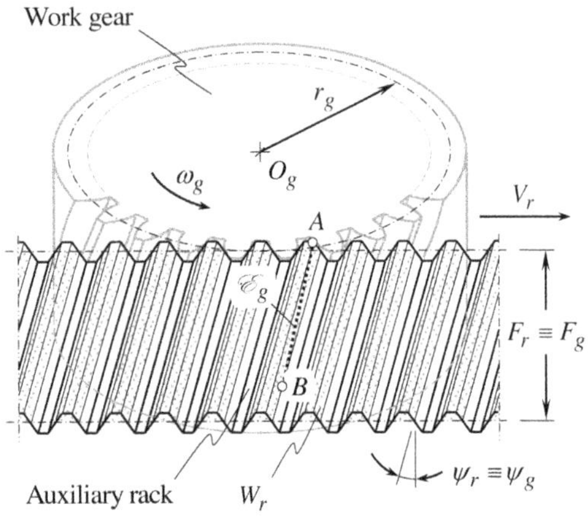


**FIGURE 14.4** The auxiliary generating rack $\mathscr{R}_T$ in mesh with the work gear. (After: Radzevich, S.P., *ASME Journal of Mechanical Design*, 124, 772–786, 2002. With permission.)

*Second*, the profile angle $\phi_r$ cannot be too big. A large profile angle $\phi_r$ results in a shorter top cutting edge of the gear-cutting tool. Because of violation of the fifth necessary condition of proper part surface generation [136, 148, 158], a large profile angle may even cause pointing of the top cutting edge as illustrated in Figure 14.3. The maximum allowed value for the profile angle $\phi_r$ is restricted by the inequality

$$\phi_r \le [\phi_r] \tag{14.25}$$

where $[\phi_r]$ designates the limit value of the tooth profile angle for which the width of the top land of the auxiliary rack tooth does not exceed a certain critical value.

The critical value of the length of the top land corresponds to the shortest allowed length of the top cutting edge of the gear-cutting tool.

The minimum required face width $F_r$ of the auxiliary generating rack $\mathscr{R}_T$ is equal to the face width $F_g$ of the work gear.

As shown in Figure 14.4, a tooth of the auxiliary generating rack $\mathscr{R}_T$ makes contact with the gear tooth flank $\mathscr{G}$ along a straight line. This straight line is commonly referred to as the *characteristic line* $\mathscr{E}_g$. The characteristic line $\mathscr{E}_g$ (that is, $\mathscr{E}_g \mapsto AB$) is in tangency with the base cylinder of the work gear and with the gear base helix (not shown in Figure 14.4). It is instructive to compare the gear tooth shown in Figure 14.4 with the screw involute surface depicted in Figure 1.13. The family of straight lines within the screw involute surface $\mathscr{G}$ in Figure 1.13 represents various configurations of the characteristic line $\mathscr{E}_g$ when the gear rolls relative to the reference system $X_rY_rZ_r$. At a certain instant of time, the characteristic $\mathscr{E}_g$ is aligned with the vector **C**. The ruled tooth flank surface $\mathscr{G}$ is generated as the family of consecutive positions of the characteristic $\mathscr{E}_g$.

In order to make sure the generation of the work gear tooth flanks is complete, the minimum face width $F_r$ of the auxiliary generating rack $\mathscr{R}_T$ has to be represented in full by the gear-cutting tool in the gear machining operation. For this purpose, the travel distance of the gear-cutting tool in the axial direction of the work gear must exceed the work gear face width ($F_r \ge F_g$).

It is important to stress here that the lateral plane of the auxiliary rack tooth is generated as an envelope to the consecutive positions of the gear tooth screw involute surface $\mathscr{G}$. This statement is evident and well-known. It will be helpful below for derivation of an equation of the generating surface of the gear-cutting tool.

## 14.3 EXAMPLES OF POSSIBLE TYPES OF AUXILIARY GENERATING SURFACES OF GEAR-CUTTING TOOL

The auxiliary generating surface $\mathscr{R}_T$ of the gear-cutting tool that is generated when the gear rotates $\omega_g$ about its axis of rotation, and simultaneously translates $V_g$ in the direction that is parallel to the $X_r$ – axis of the Cartesian reference system $X_rY_rZ_r$ [see Figure 14.1(a)], is the simplest example of the auxiliary generating surfaces. More examples of the auxiliary generating surfaces $\mathscr{R}_T$ can be derived on the basis of other types of kinematics of relative motion of the gear being machined and of the reference system $X_rY_rZ_r$.

The gear being machined can perform not just one translation in the direction that is parallel to the $X_r$ – axis, but it can perform two translations instead. The second translation can be performed along the $Y_r$ – axis of the coordinate system $X_rY_rZ_r$. In this case, the resultant translation $V_g$ is at an angle $\theta_r$ as depicted in Figure 14.5(a).

For machining of gears with a different profile angle at the opposite sides of the tooth profile, the auxiliary generating rack $\mathscr{R}_T$ also features different profile angles $\phi_{r.1}$ and $\phi_{r.2}$ at the opposite sides of its tooth profile [see Figure 14.5(b)]. The base pitch for the opposite sides of the tooth profiles of the rack $\mathscr{R}_T$ is not of the same value. This is because the teeth flank pitch that is measured within the pitch plane is the same for both sides of the tooth profile, whereas the profile angles $\phi_{r.1}$ and $\phi_{r.2}$ are not equal to each other.

Simultaneously with the rotation $\omega_g$, the gear being machined can perform an additional rotation $\omega_r$ with respect to the reference system $X_rY_rZ_r$. Depending on the parameters of the additional rotation $\omega_r$, the auxiliary generating rack $\mathscr{R}_T$ is shaped in the form either of an internal gear sector [see Figure 14.5(c)] or an external gear sector [see Figure 14.5(d)]. The radius of the pitch cylinder $R_r$ of the auxiliary rack $\mathscr{R}_T$ is negative ($R_r < 0$) in the first case [see Figure 14.5(c)], and it is positive ($R_r > 0$) in the second case [see Figure 14.5(d)]. Appropriate portions of the circular sectors can be employed as the auxiliary generating surfaces of the gear-cutting tools.

## 14.4 GENERATION OF THE GENERATING SURFACE OF GEAR-CUTTING TOOL

The generating surface of a gear-cutting tool is an envelope to consecutive positions of the auxiliary generating surface $\mathscr{R}_T$ in its motion relative to a reference system, which the

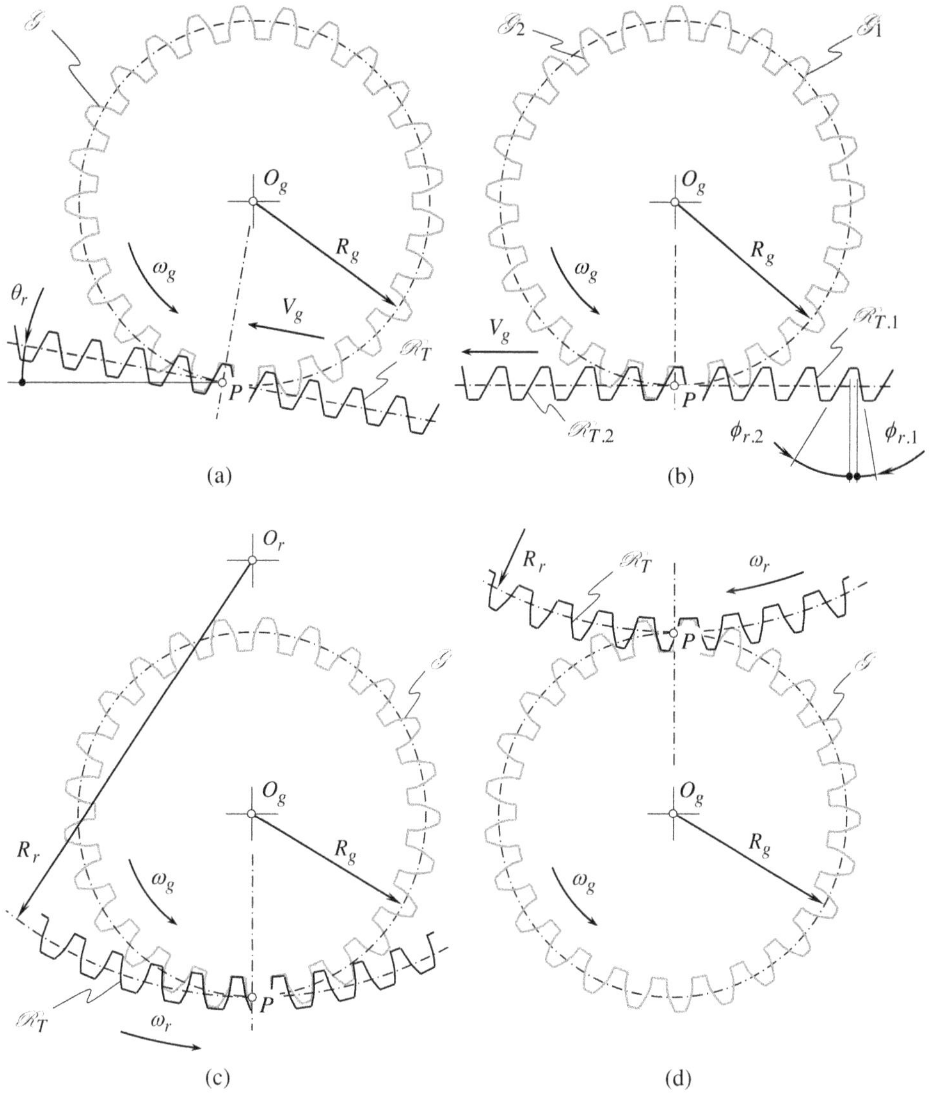


**FIGURE 14.5** Examples of possible types of auxiliary generating surfaces $\mathcal{R}_T$ of gear-cutting tools: (a) an inclined auxiliary rack; (b) auxiliary rack with asymmetric tooth profile; (c) enveloping auxiliary rack; and (d) convex auxiliary rack.

gear-cutting tool after being designed is associated with. Such an approach for determination of the generating surface of a gear-cutting tool is aligned with the first Olivier principle [52] of the generation of envelope surfaces. In addition to what is stated by the first Olivier principle, the generating surface of the gear-cutting tool has to be conjugate to the tooth flanks of the gear being machined, which is a must when accurate gears are machined.

For the specific case of a gear-cutting tool, the following definition for the generating surface $T$ can be derived from [136, 148, 158]:

**Definition 14.1** *Setting angle $\zeta_{ct}$ of the gear-cutting tool is the angle that complements (adds to 90°) the angle between (a) the projection of the axis of rotation of the gear-cutting tool onto the pitch plane $W_r$ of the auxiliary generating surface $\mathcal{R}_T$ and (b) the perpendicular to the auxiliary rack $\mathcal{R}_T$ tooth, which is constructed within the pitch plane $W_r$ of the auxiliary rack $\mathcal{R}_T$.*

Following the definition, corresponding formulae for calculation of the main design parameters of the generating surface of the gear-cutting tool can be derived.

### 14.4.1 Design Parameters of Generating Surface of Gear-Cutting Tool

For the purposes of creating the generating surface of the gear-cutting tool, an auxiliary generating rack $\mathcal{R}_T$ is used. The generating surface is an envelope to consecutive positions of the auxiliary rack $\mathcal{R}_T$ in its motion relative to the gear-cutting tool. At this point the generating surface is not constructed yet, so a Cartesian reference system $X_{ct}Y_{ct}Z_{ct}$ which the gear-cutting tool will be associated with is considered instead.

Permissible motion of the auxiliary generating surface $\mathcal{R}_T$ relative to the reference system $X_{ct}Y_{ct}Z_{ct}$ strongly depends on the geometry of the actual surface $\mathcal{R}_T$. For the auxiliary generating surfaces that are shaped in the form of straight-sided rack $\mathcal{R}_T$ [see Figure 14.1(a) and Figure 14.5(a,b)], a screw motion of the rack $\mathcal{R}_T$ about the axis of rotation $O_{ct}$ of the gear-cutting tool is permissible. For the auxiliary generating surfaces that are shaped in the form of a round rack $\mathscr{R}_T$ [see Figure 14.5(c,d)], a complex motion of the rack $\mathscr{R}_T$ about the axis of rotation $O_{ct}$ of the gear-cutting tool is anticipated to be permissible.[7] This motion can be interpreted as superposition of the rotation $\boldsymbol{\omega}_r$ of the auxiliary rack $\mathscr{R}_T$ about its axis of rotation $O_r$ and of the rotation $\boldsymbol{\omega}_{ct}$ of the rack $\mathscr{R}_T$ about the axis of rotation $O_{ct}$ of the gear-cutting tool. The resultant motion $\boldsymbol{\omega}_{scr} = \boldsymbol{\omega}_{ct} + \boldsymbol{\omega}_r$ is permissible when the rotations $\boldsymbol{\omega}_r$ and $\boldsymbol{\omega}_{ct}$ are synchronized with one another in a timely manner.

The scenario when a straight-sided generating rack $\mathscr{R}_T$ performs a screw motion about axis $O_c$ [see Figure 14.1(a)] is the simplest and the most common one. In the screw motion of the lateral planes of the rack $\mathscr{R}_T$, the teeth generate envelopes that are conjugate to the corresponding tooth flanks $\mathscr{G}$ of the work gear. For further analysis it is convenient to explore the geometry of an envelope to consecutive positions of a plane that performs a screw motion.

#### 14.4.1.1 Envelope to Successive Positions of a Plane with a Screw Motion

Consider a plane $\mathscr{R}_T$ that performs a screw motion. The plane $\mathscr{R}_T$ forms a certain angle $\tau_b$ with the $X_0-$ axis of a Cartesian reference system $X_0Y_0Z_0$. The reduced pitch $p$ of the screw motion is specified. The screw motion is performed about the axis $X_0$.

An auxiliary Cartesian coordinate system $X_1Y_1$ is rigidly associated with the plane $\mathscr{R}_T$ as illustrated in Figure 14.7. The plane $\mathscr{R}_T$ is perpendicular to the auxiliary coordinate system $X_1Y_1$.

The plane $\mathscr{R}_T$ is a plane through the straight line

$$Y_1 = X_1 \cdot \tan\tau_b \tag{14.26}$$

The auxiliary coordinate system $X_1Y_1Z_1$ performs the screw motion together with the plane $\mathscr{R}_T$. The screw motion is performing with respect to the motionless coordinate system $X_0Y_0Z_0$. In the coordinate system $X_1Y_1Z_1$, the unit normal vector $\mathbf{n}_T$ to the plane $\mathscr{R}_T$ is analytically expressed as

$$\mathbf{n}_T = \begin{bmatrix} 1 \\ -\tan\tau_b \\ 0 \\ 1 \end{bmatrix} \tag{14.27}$$

The position vector $\mathbf{r}_{\mathscr{R}}$ of arbitrary point $m$ of the plane $\mathscr{R}_T$ in the auxiliary coordinate system $X_1Y_1Z_1$ can be analytically described as

$$\mathbf{r}_{\mathscr{R}}(X_1, Z_1) = \begin{bmatrix} X_1 \\ X_1 \cdot \tan\tau_b \\ Z_1 \\ 1 \end{bmatrix} \tag{14.28}$$

[7] This particular kinematics of a gear machining process has not been thoroughly investigated yet.

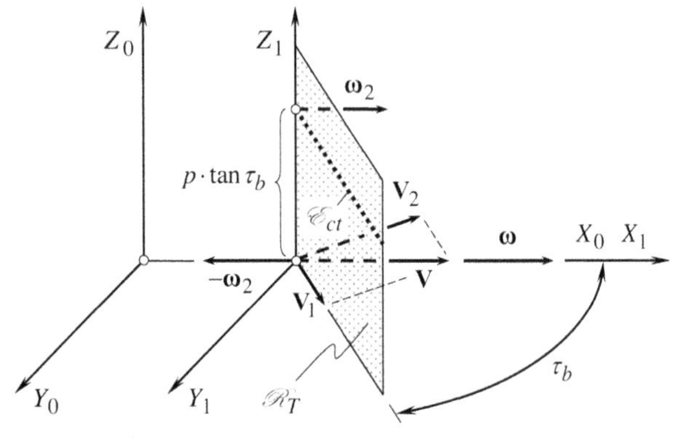


**FIGURE 14.6** Generation of screw involute surface as an envelope to consecutive positions of plane that performs screw motion.

The vector $\mathbf{v}_m$ of the speed of arbitrary point $m$ in the screw motion of the plane $\mathscr{R}_T$ is

$$\mathbf{v}_m = \mathbf{v} + [\boldsymbol{\omega} \times \mathbf{R}] \tag{14.29}$$

where:

- $\mathbf{v}$ is the vector of the speed of the translation.
- $\boldsymbol{\omega}$ is the vector of the speed of the rotation.
- $\mathbf{R}$ is the position vector of the point $m$ relative to the axis of the screw motion (magnitude of the vector $\mathbf{R}$ is equal to the distance of the point $m$ from the $X_0-$ axis, and the vector $\mathbf{R}$ is pointed from the axis $X_0$ toward $m$).

The direction of the vector $\mathbf{v}_m$ is of importance for determination of the characteristic line $\mathscr{E}_{ct}$, whereas the magnitude of the vector $\mathbf{v}_m$ is not of interest (it is important to stress here that, as shown in Figure 14.6, the characteristic

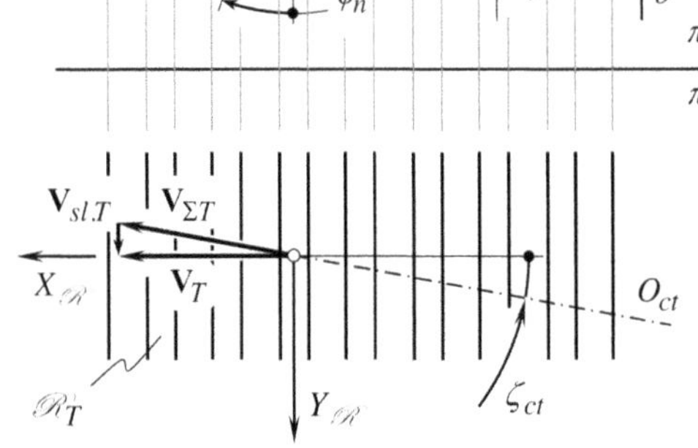


**FIGURE 14.7** A screw motion of the auxiliary generating rack $\mathscr{R}_T$ about the axis $O_{ct}$.

line $\mathscr{E}_{ct}$ and the characteristic line $\mathscr{E}_g$ in Figure 14.4 are two different characteristic lines). Because of this, it can be assumed that magnitude $\omega$ of the rotation vector $\boldsymbol{\omega}$ is equal $\omega = |\boldsymbol{\omega}| = 1$. Therefore:

$$\boldsymbol{\omega} = \mathbf{i} \tag{14.30}$$

$$\mathbf{v} = \mathbf{i} \cdot p \tag{14.31}$$

These yield

$$\mathbf{v}_m = \mathbf{i} \cdot p + \begin{vmatrix} \mathbf{i} & \mathbf{j} & \mathbf{k} \\ 1 & 0 & 0 \\ X_1 & Y_1 & Z_1 \end{vmatrix} \tag{14.32}$$

and

$$\mathbf{v}_m = \mathbf{i} \cdot p - \mathbf{j} \cdot Y_1 + \mathbf{k} \cdot Z_1 \tag{14.33}$$

The dot product of the unit normal vector $\mathbf{n}_T$ and of the vector of speed of the translation $\mathbf{v}_m$ is equal to

$$\mathbf{n}_T \cdot \mathbf{v}_m = p \cdot \tan \tau_b - Z_1 = 0 \tag{14.34}$$

Thus, the equation of contact can be represented in the form

$$Z_1 = p \cdot \tan \tau_b \tag{14.35}$$

The equation for the position vector of a point $\mathbf{r}_{\mathscr{E}}$ of the characteristic line $\mathscr{E}_{ct}$

$$\mathbf{r}_{\mathscr{E}}(t) = \begin{bmatrix} y \\ t \cdot \tan \tau_b \\ p \cdot \tan \tau_b \\ 1 \end{bmatrix} \tag{14.36}$$

is derived on the premise of simultaneous consideration of the Shishkov equation of contact $\mathbf{n}_T \cdot \mathbf{v}_m = 0$ together with the equation that describes the plane $\mathscr{R}_T$ in its current configuration with respect to the axis of the screw motion.

In Eq. (14.36), $\mathbf{r}_{\mathscr{E}}$ designates the position vector of a point of the characteristic line $\mathscr{E}_{ct}$, and the parameter of the characteristic line $\mathscr{E}_{ct}$ is denoted by $t$.

In the case under consideration, the characteristic line $\mathscr{E}_{ct}$ is a straight line of intersection of two planes. The plane $\mathscr{R}_T$ is the first of two planes. Another plane is parallel to the coordinate plane $X_1Z_1$ and is remote from this plane at the distanced $p \cdot \tan \tau_b$.

For a given screw motion, the location of the characteristic line $\mathscr{E}_{ct}$ within the plane $\mathscr{R}_T$ in the initial coordinate system $X_0Y_0Z_0$ remains the same.

The angle of rotation of the coordinate system $X_1Y_1Z_1$ about the $X_0$–axis is designated as $\varepsilon$. The translation of the coordinate system $X_1Y_1Z_1$ with respect to the stationary reference system $X_0Y_0Z_0$ that corresponds to the angle $\varepsilon$ is equal to $p \cdot \varepsilon$. This makes possible composing the operator $\mathbf{Rs}(1 \to 0)$ of the resultant coordinate system transformation:

$$\mathbf{Rs}(1 \to 0) = \begin{bmatrix} 1 & 0 & 0 & p \cdot \varepsilon \\ 0 & \cos\varepsilon & \sin\varepsilon & 0 \\ 0 & -\sin\varepsilon & \cos\varepsilon & 0 \\ 0 & 0 & 0 & 1 \end{bmatrix} \tag{14.37}$$

In order to represent analytically the envelope surface $T$, the equation $\mathbf{r}_{\mathscr{E}}(t)$ for the position vector of a point of the characteristic line $\mathscr{E}_{ct}$ has to be considered together with the operator $\mathbf{Rs}(1 \to 0)$ of coordinate system transformation:

$$\mathbf{r}_T(X_1, \varepsilon) = \mathbf{Rs}(1 \to 0) \cdot \mathbf{r}_E(t)$$
$$= \begin{bmatrix} X_1 + p \cdot \varepsilon \\ X_1 \cdot \tan \tau_b \cdot \cos\varepsilon + p \cdot \tan \tau_b \cdot \sin\varepsilon \\ -X_1 \cdot \tan \tau_b \cdot \sin\varepsilon + p \cdot \tan \tau_b \cdot \cos\varepsilon \\ 1 \end{bmatrix} \tag{14.38}$$

Consider the intersection of the envelope surface $T$ by the plane

$$X_0 = X_1 + p \cdot \varepsilon = 0 \tag{14.39}$$

The later equation yields representation in the form $X_1 = -p \cdot \varepsilon$. Therefore:

$$\mathbf{r}_{X_0}(\varepsilon) = \begin{bmatrix} 0 \\ p \cdot \tan \tau_b \cdot (\sin\varepsilon - p \cdot \varepsilon \cdot \cos\varepsilon) \\ p \cdot \tan \tau_b \cdot (\cos\varepsilon + p \cdot \varepsilon \cdot \sin\varepsilon) \\ 1 \end{bmatrix} \tag{14.40}$$

Equation (14.40) describes the involute of a circle.
The radius of the base circle of the involute curve

$$r_b = p \cdot \tan \tau_b \tag{14.41}$$

Therefore, a screw involute surface allows for interpretation in the form of the envelope to consecutive positions of a plane $\mathscr{R}_T$ that performs a screw motion. The reduced pitch of the involute screw surface is equal to $p$, and the radius of the base cylinder is equal to $r_b = p \cdot \tan \tau_b$. The screw involute surface shares common points with the base cylinder. These points are within a helix. The tangent to the helix forms an angle $\omega_b$ with the axis of the screw motion [1, 2]:

$$\tan \omega_b = \frac{r_b}{p} \tag{14.42}$$

From this, one may conclude that $\tan \omega_b = \tan \tau_b$, and therefore $\omega_b = \tau_b$.

The straight characteristic line $\mathscr{E}_{ct}$ is tangent to the base helix of the enveloping surface $T$. This means that

- If a plane $\mathscr{R}_T$ is tangent to the base cylinder of a diameter, $d_b = 2r_b$.
- A straight line $\mathscr{E}_{ct}$ within the plane $\mathscr{R}_T$ forms the angle $\tau_b$ with the axis of the screw motion.
- The plane $\mathscr{R}_T$ rolls over the base cylinder without sliding.

then the envelope surface $T$ can be represented as a locus of consecutive positions of the straight line $\mathscr{E}_{ct}$ that rolls with no sliding over the base cylinder together with the plane $\mathscr{R}_T$. The envelope surface is a screw involute surface.

The obtained screw involute surface (see Figure 14.6) is that shown in Fig. 1.14 and is analytically described by Eq. (14.16).

The preceding discussion can be summarized with the following theorem.

**Theorem 14.1** *If we screw, with pitch p, a plane about an axis fixed in space, which axis makes an angle* $\alpha$ *with the said plane, we describe in space a continuum of planes that successively intersect one another in a continuum of straight lines. These lines sweep out the involute helicoid* $(a, \alpha)$ *where* $a = p\tan\alpha$.

It might be correspondingly said that the screw of the plane sweeps in the same involute helicoid. The involute helicoid is, in any event, the so-called envelope of the continuum of planes.

This is a well-known theorem not found in most books about gearing. It may, however, be found in some books about kinematics [58]. See also page 335 in the book by Dr. J. Phillips [57].

Another solution to the problem of determining the envelope of a plane with a screw motion is given by P. Cormac [16].

#### 14.4.1.2 Principal Elements of the Geometry of Generating Surface of Gear-Cutting Tool

Figure 14.7 shows that the Cartesian reference system $X_{ct}Y_{ct}Z_{ct}$ is the reference system with which the gear-cutting tool is associated after being designed.

In the coordinate system $X_{ct}Y_{ct}Z_{ct}$, the relative motion of the auxiliary generating rack $\mathscr{R}_T$ can be represented as a superposition of two motions. The rotation $\omega_{ct}$ about the axis $O_{ct}$ of the gear-cutting tool is the first of two motions. The translation of the auxiliary rack $\mathscr{R}_T$ is the second motion. The vector of the speed of the translation is designated as $\mathbf{V}_{\Sigma T}$.

The vector of the speed $\mathbf{V}_{\Sigma T}$ of the translation can be resolved into two components (see Figure 14.7):

$$\mathbf{V}_{\Sigma T} = \mathbf{V}_T + \mathbf{V}_{sl.T} \tag{14.43}$$

The first component $\mathbf{V}_T$ is pointed along the axis $O_{ct}$ of the gear-cutting tool. The magnitude $V_T = |\mathbf{V}_T|$ of this component is equal to

$$V_T = \frac{V_{\Sigma T}}{\cos\zeta_{ct}} \tag{14.44}$$

where $\zeta_{ct}$ denotes the setting angle of the gear-cutting tool.[8]

The second component $\mathbf{V}_{sl.T}$ is along the straight generating line of the auxiliary rack $\mathscr{R}_T$. Sliding of the rack $\mathscr{R}_T$ over itself is a result of this motion. Due to the sliding, this motion does not affect the shape of the envelope surface, and thus it can be omitted from further consideration. Eventually, the resultant motion of the auxiliary generating rack $\mathscr{R}_T$ relative to the reference system $X_{ct}Y_{ct}Z_{ct}$ is a type of screw motion. The axis $O_{ct}$ is the axis of the screw motion. The parameter of the screw motion $p_{ct}$ is equal to

$$p_c = \frac{V_T}{\omega_{ct}} = \frac{V_{\Sigma T}}{\omega_{ct}\cos\zeta_{ct}} = \frac{\omega_g}{\omega_{ct}} \cdot \frac{r_g}{\cos\zeta_{ct}} \tag{14.45}$$

Owing to the ratio

$$\frac{\omega_{ct}}{\omega_g} = \frac{N_g}{N_{ct}} \tag{14.46}$$

the expression

$$p_{ct} = \frac{N_{ct} \cdot r_g}{N_g \cdot \cos\zeta_{ct}} \tag{14.47}$$

can be used for the calculation of the parameter of the screw motion $p_{ct}$.

A screw involute surface is the envelope to consecutive positions of the lateral plane of the auxiliary rack $\mathscr{R}_T$ in its screw motion about the axis $O_{ct}$. The radius of the base cylinder $r_{b.ct}$ of this screw involute surface is equal to

$$r_{b.ct} = p_{ct} \cdot \tan\varphi \tag{14.48}$$

where $\varphi$ denotes the angle that the axis $O_{ct}$ forms with the lateral plane $\mathscr{R}_T$ of the auxiliary rack tooth.

In the case under consideration, the characteristic line $\mathscr{E}_{ct}$ is a line within the lateral plane $\mathscr{R}_T$ of the rack tooth at which the perpendicular $\mathbf{n}_T$ to the plane $\mathscr{R}_T$ forms a certain angle $\varepsilon$ with the axis $O_{ct}$ of the screw motion. The value of the angle $\varepsilon$ can be calculated from *Ball's formula* [1, 2]:

$$\varepsilon = \tan^{-1}\left(\frac{\varphi}{r_i}\right) \tag{14.49}$$

where $r_i$ is the shortest distance of approach of the straight line along the normal vector $\mathbf{n}_T$ and the axis $O_{ct}$ of the screw motion.

[8] Commonly, the setting angle $\zeta_{ct}$ of a gear-cutting tool, e.g., of a hob, equals the hob pitch helix angle $\psi_c$, that is, the equality $\zeta_{ct} = \psi_{ct}$ is commonly observed. However, this is not a must; gear-cutting tools with the setting angle $\zeta_{ct} \neq \psi_{ct}$ are also used in industry [119].

When a plane performs a screw motion, the characteristic line is a straight line at a distance $r_i$ from the axis $O_{ct}$ of the screw motion. The characteristic line forms an angle $\varphi$ with the cutting tool axis $O_{ct}$. The value of the angle $\varphi$ can be determined analytically. For this purpose, the unit normal vector $\mathbf{n}_T$ to the lateral plane of the auxiliary rack tooth flank $\mathscr{R}_T$ is utilized.

Consider a Cartesian reference system $X_{\mathscr{R}}Y_{\mathscr{R}}Z_{\mathscr{R}}$ associated with the auxiliary generating rack $\mathscr{R}_T$, as shown in Figure 14.7. In the reference system $X_{\mathscr{R}}Y_{\mathscr{R}}Z_{\mathscr{R}}$, the unit normal vector $\mathbf{n}_T$ can be analytically expressed in the form

$$\mathbf{n}_T = \mathbf{i}\cos\phi_n + \mathbf{k}\sin\phi_n \tag{14.50}$$

(Within the characteristic line $\mathscr{E}_{ct}$, a perpendicular $\mathbf{n}_T$ is orthogonal to the linear velocity vector $\mathbf{V}_T$.)

The unit vector $\mathbf{a}$ is a vector along the axis $O_{ct}$ of the gear-cutting tool. (The unit vector $\mathbf{a}$ is aligned with the linear velocity vector $\mathbf{V}_{\Sigma T}$.) The vector $\mathbf{a}$ can be analytically represented in the form

$$\mathbf{a} = \mathbf{i}\cos\zeta_c - \mathbf{j}\sin\zeta_c \tag{14.51}$$

Because the vectors $\mathbf{n}_T$ and $\mathbf{a}$ are of unit length, then the equality

$$\cos(90° - \varphi) = \mathbf{n}_T \cdot \mathbf{a} \tag{14.52}$$

is valid.

This expression casts into the formula

$$\varphi = \sin^{-1}(\cos\phi_n \cdot \cos\zeta_{ct}) \tag{14.53}$$

for the calculation of the angle $\varphi$.

Equation (14.53) allows for an expression

$$d_{b.ct} = \frac{d_g \cdot N_{ct} \cdot \cos\phi_n}{N_g\sqrt{1-\cos^2\phi_n \cos^2\zeta_{ct}}} \tag{14.54}$$

for the calculation of the base diameter $d_{b.ct}$ of the generating surface $T$ of the gear-cutting tool.

For gear-cutting tools with a standard tooth profile (for which the tooth thickness $t$ and the space width $w$ on the pitch diameter $d_g$ are equal to each other), Eq. (14.54) reduces to

$$d_{b.ct} = \frac{m \cdot N_{ct} \cdot \cos\phi_n}{\sqrt{1-\cos^2\phi_n \cos^2\zeta_{ct}}} \tag{14.55}$$

The axial pitch of the screw involute surface can be calculated from the formula

$$P_{x.ct} = \frac{\pi d_g}{N_g \cos\zeta_{ct}} \tag{14.56}$$

For the calculation of the base pitch $p_{b.ct}$ of the gear-cutting tool, the formula

$$p_{b.ct} = P_{x.ct}\sin\varphi = \frac{\pi \cdot d_g \cdot \cos\phi_n \cdot \cos\zeta_{ct}}{N_g \cos\zeta_{ct}} = \frac{\pi \cdot d_{b.g}}{N_g} \equiv p_{b.g} \tag{14.57}$$

applies.

The preceding equations are used for the derivation of equations for the calculation of the design parameters of involute hobs, as well as gear-cutting tools of other types.

### 14.4.2 Equation of Generating Surface of Gear-Cutting Tool

Lateral cutting edges of a precise gear-cutting tool (that is, of a precise involute hob) has to be located within the generating surface $T$. Any displacements of the lateral cutting edges out of the generating surface $T$ cause tooth profile errors of the gear-cutting tool. The generating surface $T$ of the gear-cutting tool serves as the only reliable reference surface for the calculation of deviations of the actual machining surface $T_m$ of the gear-cutting tool from its desired shape $T$.

For the derivation of the equation of the generating surface $T$, consider an auxiliary generating surface $\mathscr{R}_T$ (see Figure 14.8). The auxiliary generating rack $\mathscr{R}_T$ is associated with a left-hand-oriented Cartesian reference system $X_rY_rZ_r$. The pitch plane of the rack is at a distance $0.5 \cdot d_{ct}$ from the $Z_r$ – axis, and it is parallel to the $X_rZ_r$ coordinate plane. Here, $d_{ct}$ denotes the pitch diameter of the gear-cutting tool. In the coordinate system $X_rY_rZ_r$, the equation of the lateral tooth surface of the auxiliary generating rack $\mathscr{R}_T$ yields representation in matrix form:

$$\mathbf{r}_r(U_r, V_r) = \begin{bmatrix} U_r \\ V_r \\ V_r \cdot \tan\phi_n \\ 1 \end{bmatrix} \tag{14.58}$$

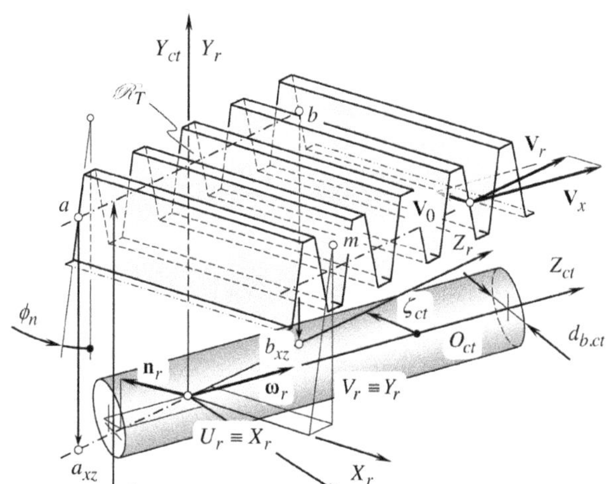


**FIGURE 14.8** Configuration of the auxiliary generating rack $\mathscr{R}_T$ relative to the axis of rotation $O_{ct}$ of the gear-cutting tool.

It is assumed in Eq. (14.58) that $U_r = X_r$ and $V_r = Y_r$. Therefore, the equality

$$V_r \cdot \tan\phi_n = Z_r \tag{14.59}$$

is valid.

In order to generate the generating surface $T$ of the gear-cutting tool, it is necessary to represent Eq. (14.58) in the left-hand-oriented Cartesian reference system $X_{ct}Y_{ct}Z_{ct}$ that is associated with the gear-cutting tool. The reference systems $X_rY_rZ_r$ and $X_{ct}Y_{ct}Z_{ct}$ share common axis $Y_r \equiv Y_{ct}$. They are turned about this axis through the setting angle $\zeta_{ct}$. Therefore, in the coordinate system $X_{ct}Y_{ct}Z_{ct}$, an expression for the position vector $\mathbf{r}_r^c$ of an arbitrary point $m$ of the auxiliary rack $\mathscr{R}_T$ can be represented in the form

$$\mathbf{r}_r^c(U_r, V_r) = \mathbf{Rt}(\zeta_{ct}, Y_r) \cdot \mathbf{r}_r(U_r, V_r) \tag{14.60}$$

Furthermore, consider a screw motion that the auxiliary rack $\mathscr{R}_T$ performs about the $Z_{ct}$ –axis. The reduced pitch (the screw parameter) of the screw motion is denoted by $p_{ct}$.

The screw motion of the auxiliary generating rack $\mathscr{R}_T$ can be resolved on two elementary motions. The translation along the $Z_{ct}$ – axis is the first of the two elementary motions. The translation of the rack $\mathscr{R}_T$ along the $Z_{ct}$ – axis through a certain distance $a_z$ is caused by this elementary motion. The coordinate system transformation due to the translation can be analytically described by the operator of translation $\mathbf{Tr}(a_z, Z_{ct})$. The rotation about the $Z_{ct}$ – axis is the second of the two elementary motions. This elementary motion causes a rotation of the auxiliary rack $\mathscr{R}_T$ about the $Z_{ct}$ – axis through a certain angle $\theta_z$. The coordinate system transformation due to the rotation is analytically described by the operator of rotation $\mathbf{Rt}(\theta_z, Z_{ct})$. Here, the actual values of the translation $a_z$ and of the rotation $\theta_z$ are not important. Only the ratio $a_z/\theta_z$ is of critical importance for the analysis that follows. The ratio $a_z/\theta_z$ is equal to the reduced pitch $p_{ct}$ of the screw motion of the auxiliary rack $\mathscr{R}_T$ in the case under consideration ($a_z/\theta_z = p_{ct}$).

The coordinate system transformation due to the screw motion of the auxiliary generating rack $\mathscr{R}_T$ can be analytically described by the operator of the resultant coordinate system transformation $\mathbf{Rs}(\mathscr{R} \mapsto ct)$. The operator $\mathbf{Rs}(\mathscr{R} \mapsto ct)$ yields representation in the form of the product

$$\mathbf{Rs}(\mathscr{R} \mapsto ct) = \mathbf{Rt}(\theta_z, Z_{ct}) \cdot \mathbf{Tr}(a_z, Z_{ct}) \tag{14.61}$$

The auxiliary generating rack $\mathscr{R}_T$ in its current location in the screw motion is described by the matrix equation

$$\overline{\mathbf{r}}_r^c(U_r, V_r, \theta_z) = \mathbf{Rs}(\mathscr{R} \mapsto ct) \cdot \mathbf{r}_r^c(U_r, V_r) \tag{14.62}$$

In Eq. (14.62), the position vector $\overline{\mathbf{r}}_r^c$ of a point depends on the rotation angle $\theta_z$. Due to the equality $a_z/\theta_z = p_{ct}$ being observed, the position vector $\overline{\mathbf{r}}_r^c$ can be expressed not in terms of the rotation angle $\theta_z$ but in terms of the translation $a_z$ instead.

Further, consider the family of the auxiliary rack tooth surfaces $\overline{\mathbf{r}}_r^c$ together with the equation of contact $\mathbf{n}_T \cdot \mathbf{V}_{r-c} = 0$. In the latter equation, the unit normal vector of the rack tooth surface $\overline{\mathbf{r}}_r^c$ is designated as $\mathbf{n}_T$, and the vector of instant relative motion of the auxiliary rack relative to the reference system $X_{ct}Y_{ct}Z_{ct}$ is denoted by $\mathbf{V}_{r-c}$.

The generating surface $T$ of the gear-cutting tool can be analytically described by Eq. (14.62) together with the Shishkov equation of contact $\mathbf{n} \cdot \mathbf{V}_\Sigma = 0$. Omitting bulky formulae transformation, the generating surface $T$ of the gear-cutting tool is represented in matrix form by the following equation:

$$\left.\begin{array}{r}\text{Generating surface, } T, \\ \text{of gear cutting tool}\end{array}\right| \Rightarrow \mathbf{r}_{ct}(U_{ct}, V_{ct})$$

$$= \begin{bmatrix} 0.5 \cdot d_{b.ct} \cdot \sin V_{ct} - U_{ct} \cdot \sin\psi_{b.ct} \cdot \cos V_{ct} \\ 0.5 \cdot d_{b.ct} \cdot \cos V_{ct} + U_{ct} \cdot \sin\psi_{b.ct} \cdot \sin V_{ct} \\ p_{b.ct} \cdot V_{ct} - U_{ct} \cdot \cos\psi_{b.ct} \\ 1 \end{bmatrix} \tag{14.63}$$

where:

- $U_{ct}$, $V_{ct}$ are the curvilinear coordinates (Gauss coordinates) of a point of the generating surface $T$.
- $d_{b.ct}$ is the base diameter of the screw involute surface $T$.
- $p_{b.ct}$ is the base pitch of the screw involute surface $T$.
- $\psi_{b.ct}$ is the base helix angle of the generating surface $T$.

In the case under consideration, the generating surface $T$ is a kind of screw involute surface that is conjugate to the gear tooth flanks $\mathscr{G}$. Figure 14.9 illustrates an example of the generating surface $T$ of the gear-cutting tool.

Equation (14.63) can be expressed in terms of the design parameters of the gear-cutting tool. For this purpose, the set of engineering gear equations is used (see Appendix F).

Equation (14.63) immediately produces an analytical representation of a perpendicular to the generating surface $T$ of the gear-cutting tool. The unit normal vector $\mathbf{n}_{ct}$ is calculated from Eq. (14.63) as the cross product $\mathbf{n}_{ct} = \mathbf{u}_{ct} \times \mathbf{v}_{ct}$ of the two unit tangent vectors $\mathbf{u}_{ct}$ and $\mathbf{v}_{ct}$. The unit tangent vector $\mathbf{u}_{ct}$ is aligned with the coordinate line $U_{ct}$ at the point of interest on the generating surface $T$. The unit vector $\mathbf{v}_{ct}$ is tangent to the coordinate curve $V_{ct}$ at the point of interest on the generating surface $T$. Unit tangent vectors $\mathbf{u}_{ct}$ and $\mathbf{v}_{ct}$ are equal to

$$\mathbf{u}_{ct} = \frac{\mathbf{U}_{ct}}{|\mathbf{U}_{ct}|} \tag{14.64}$$

$$\mathbf{v}_{ct} = \frac{\mathbf{V}_{ct}}{|\mathbf{V}_{ct}|} \tag{14.65}$$

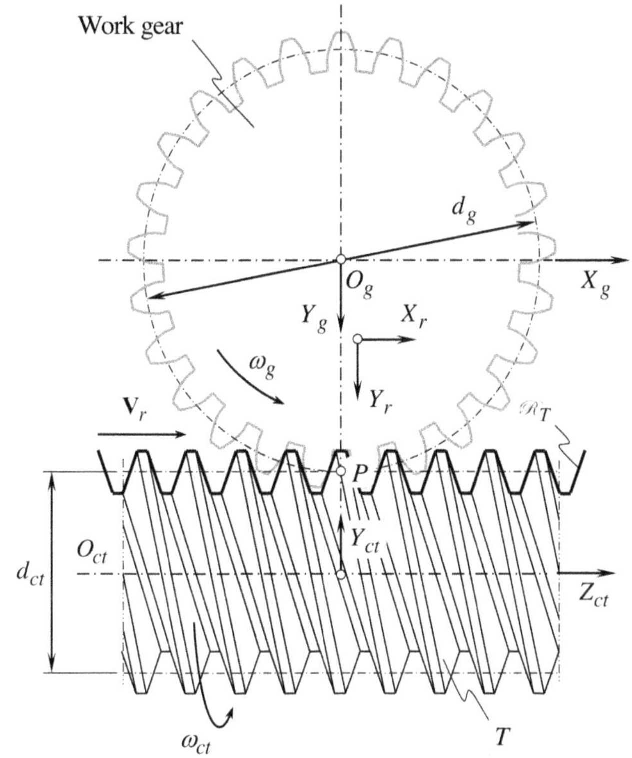


**FIGURE 14.9** Generating surface $T$ of cylindrical gear-cutting tool. (After: Radzevich, S.P., *ASME Journal of Mechanical Design*, 129, 334–345, 2007. With permission.)

respectively. Here, the equalities

$$\mathbf{U}_{ct} = \frac{\partial \mathbf{r}_c}{\partial U_{ct}} \tag{14.66}$$

$$\mathbf{V}_{ct} = \frac{\partial \mathbf{r}_{ct}}{\partial V_{ct}} \tag{14.67}$$

are valid.

For the case under consideration, the unit normal vector $\mathbf{n}_{ct}$ equals

$$\mathbf{n}_{ct} = \begin{bmatrix} \sin\psi_{b.ct} \cdot \sin V_{ct} \\ \sin\psi_{b.ct} \cdot \cos V_{ct} \\ \cos\psi_{b.ct} \\ 1 \end{bmatrix} \tag{14.68}$$

Equation (14.63) allows for calculation of the rest of the design parameters of the generating surface $T$ of the gear-cutting tool as well.

### 14.4.3 Setting Angle of the Gear-cutting Tool

The orientation of the auxiliary generating rack $\mathscr{R}_T$ with respect to the axis of rotation $O_{ct}$ of the gear-cutting tool is an important issue that needs to be considered in more detail. In the case considered above (see Figures 14.8 and 14.9), the axis of rotation $O_{ct}$ is parallel to the pitch plane $W_r$ of the auxiliary rack $\mathscr{R}_T$. In this way a cylindrical generating surface $T$ of the gear-cutting tool is obtained. However, the axis of rotation $O_{ct}$ can occupy various configurations within the pitch plane $W_r$. This variation of orientation of the auxiliary rack $\mathscr{R}_T$ with respect to the axis of rotation $O_{ct}$ is illustrated in Figure 14.10.

The auxiliary generating rack $\mathscr{R}_T$ forms a certain angle $\psi_r$ with a perpendicular to the axis of rotation $O_g$ of the gear. Commonly, this angle is referred to as the *helix angle* of the auxiliary generating rack $\mathscr{R}_T$. The axis of rotation $O_{ct}$ of the gear-cutting tool passes through the centerline ℄. In Figure 14.10, the center distance is depicted as point $C$ of the intersection of the centerline itself and the pitch plane $W_r$ of the auxiliary generating rack $\mathscr{R}_T$.

A straight line ${}^0O_{ct}$ through point $C$ is parallel to the pitch plane $W_r$. This straight line ${}^0O_{ct}$ is orthogonal to the line along the tooth of the auxiliary rack $\mathscr{R}_T$. For a spur gear, the straight line ${}^0O_{ct}$ is parallel to the gear face. The axis of rotation of the gear-cutting tool forms a certain

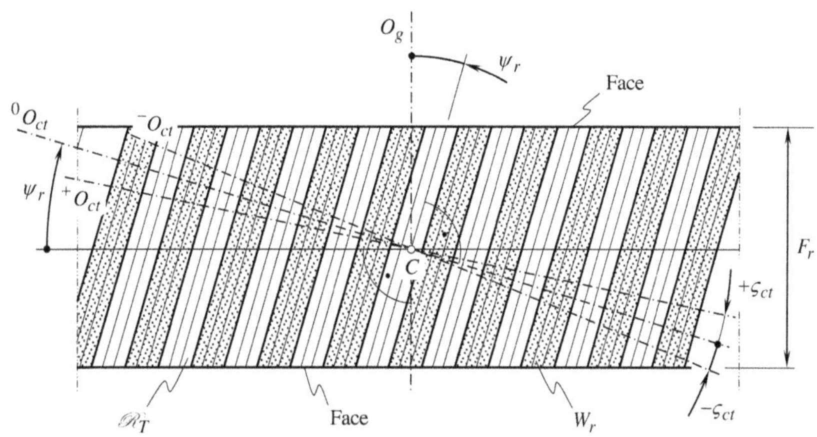


**FIGURE 14.10** Setting angle $\zeta_{ct}$ of gear-cutting tool. (Adapted from Radzevich, S.P., *ASME Journal of Mechanical Design*, 124, 772–786, 2002. With permission.)

angle with the straight line ${}^{0}O_{ct}$. The axis of rotation of the gear-cutting tool with a configuration that is denoted as ${}^{+}O_{ct}$ forms an angle $\zeta_{ct}$ of a positive value ($\zeta_{ct} > 0°$) with the straight line ${}^{0}O_{ct}$. The axis of rotation of the gear-cutting tool with a configuration that is designated as ${}^{-}O_{ct}$ forms an angle $\zeta_{ct}$ of a negative value ($\zeta_{ct} < 0°$) with the same straight line ${}^{0}O_{ct}$. It is convenient to use the designation $+\zeta_{ct}$ for positive ($\zeta_{ct} > 0°$) inclination of the axis of rotation of the gear-cutting tool and it is convenient to use the designation $-\zeta_{ct}$ for negative ($\zeta_{ct} < 0°$) inclination of the axis of rotation of the gear-cutting tool. Commonly, the angle $\zeta_{ct}$ is referred to as the *setting angle* of the gear-cutting tool [119, 161] and others. It is important to stress here that the setting angle $\zeta_{ct}$ is a design parameter of the gear-cutting tool and not a component of its setup parameters. An actual value of the setting angle $\zeta_{ct}$ is not mandatorily equal to the helix angle $\psi_{ct}$ of the gear-cutting tool.

The actual value of the setting angle $\zeta_{ct}$ is usually close or equal to the corresponding pitch helix angle $\psi_{ct}$ of the gear-cutting tool. This is the major reason why the angles $\zeta_{ct}$ and $\psi_{ct}$ are often not distinguished from one another. However, they are angles of completely different natures.

*Setting angle* $\zeta_{ct}$: The concept of the setting angle of the gear-cutting tool was introduced and discussed in detail earlier [111, 145], and others.

**Definition 14.2** *The generating surface of a gear-cutting tool is a screw surface that is conjugate to the tooth flank of the gear being machined.*

For a given gear-cutting tool, the cutting tool setting angle $\zeta_{ct}$ has a specified constant value ($\zeta_{ct} = const$). The actual value of the cutting tool setting angle $\zeta_{ct}$ does not depend on the current pitch diameter $d_{ct}$ of the gear-cutting tool, whereas the helix angle $\psi_{ct}$ of the gear-cutting tool is a function of the diameter $d_{ct}$ [that is, a relationship $\psi_{ct} = \psi_{ct}(d_c, \ldots)$ is observed]. Comparison of the equations $\zeta_{ct} = const$ and $\psi_{ct} = \psi_{ct}(d_{ct}, \ldots)$ makes evident the difference between the angles $\zeta_{ct}$ and $\psi_{ct}$.

The pitch diameter $d_{ct}$ changes after each resharpening of the gear-cutting tool. For example, its value reduces after each regrinding of a gear hob. The actual value of the setting angle $\zeta_{ct}$ also reduces after each redressing of the worm grinding worm. For gear-cutting tools with an enveloping generating surface $T$, an increase in the pitch diameter $d_{ct}$ is observed after each resharpening or redressing of the gear-cutting tool.

A clear understanding of the difference between the gear-cutting tool setting angle $\zeta_{ct}$ and the pitch helix angle $\psi_{ct}$ is of critical importance for the designer of gear-cutting tools. In order to emphasize the difference between the design parameters $\zeta_{ct}$ and $\psi_{ct}$, it is instructive to calculate the actual value of the helix angle $\psi_{ct}$ for a gear hob with a zero value of the setting angle ($\zeta_{ct} = 0°$). This calculation is performed below for a single-start *FETTE* gear hob (DIN8002A, Cat. No. 2022, Ident. No. 1202055) of module $m = 10$ mm, outer diameter $d_{o.ct} = 140$ mm, and normal pressure angle $\phi_n = 20°$ [22]. The calculations [140] reveal that the helix angle for the gear hob with a setting angle $\zeta_{ct} = 0°$ is equal to $\psi_{ct} = 2.490°$. It is evident that the pitch helix angle $\psi_{ct} = 2.490°$ is not identical to the setting angle $\zeta_{ct} = 0°$. The difference $\delta = |\zeta_{ct} - \psi_{ct}| = 2.490°$ of values between the angles $\zeta_{ct}$ and $\psi_{ct}$ is of significant value and cannot be neglected when calculating the design parameters of a precision gear hob. The difference $\delta$ is getting more significant for multi-start hobs. Therefore, multi-start hobs, and multi-start gear-cutting tools more generally, require a more careful treatment of the setting angle $\zeta_{ct}$.

Let us assume that that same involute hob is designed with setting angle not of a zero value, but with the setting angle of a value $\zeta_{ct} = 2.490°$. Calculations of the helix angle $\psi_{ct}$ for the involute hob with this set of design parameters return $\psi_{ct} = 2.492°$ [140]. The difference $\delta = |\zeta_h - \psi_{ct}| = 0.002°$ in this case is negligibly small and, therefore, can be neglected.

The consideration above makes it clear that the angles $\zeta_{ct}$ and $\psi_{ct}$ are two different design parameters of the gear-cutting tool. Moreover, it is shown below that the gear-cutting tool setting angle $\zeta_{ct}$ is the major design parameter, whereas the pitch helix angle $\psi_{ct}$ of the gear-cutting tool is a design parameter of secondary importance, the actual value of which can be expressed in terms of the angle $\zeta_{ct}$.

In the gear machining operation, the crossed-axes angle $\Sigma$ is equal to the difference $\Sigma = 90° - \zeta_{ct}$, regardless of the actual value of the helix angle $\psi_{ct}$ on the pitch cylinder of the gear-cutting tool.

### 14.4.4 Complementary Equations

The screw parameter (the reduced pitch) of the screw motion is

$$p_{ct} = \frac{d_g \cdot \omega_g}{2 \cdot \omega_{ct} \cdot \cos \zeta_{ct}} \tag{14.69}$$

For single-start involute hobs of conventional design, this equation reduces to

$$p_{ct} = \frac{d_g}{2 \cdot N_{ct} \cdot \cos \zeta_{ct}} \tag{14.70}$$

where the number of starts of the gear-cutting tool is denoted by $N_{ct}$.

When the axis of the screw motion $O_{ct}$ is parallel to the pitch plane $W_r$ of the auxiliary generating rack $\mathscr{R}_T$, the generating surface $T$ is shaped in the form of a cylindrical involute worm (see Figure 14.8). In the screw motion, the plane surfaces of the top land of the auxiliary rack $\mathscr{R}_T$ and the planes of the root land generate two coaxial surfaces of cylinders of revolution (see Figure 14.9). The first one is the inner cylinder of the gear-cutting tool, and the second one is the outer cylinder of the gear-cutting tool.

The envelope to consecutive positions of the lateral plane surfaces of the auxiliary rack $\mathscr{R}_T$ are shaped in the form of screw involute surfaces.

Equation (14.63) describes the generating surface $T$ of the involute gear-cutting tool. The position vector of a point $\mathbf{r}_{ct}$ is expressed here in terms of curvilinear $U_{ct}$ – and $V_{ct}$ – coordinates (Gauss' coordinates). It is easy to verify that a section of the generating surface $T$ by a plane $Z_{ct} = const$ is an involute curve having a base circle of radius $r_{b.ct}$.

Equation (14.63) yields the formula for the calculation of the base diameter $d_{b.ct}$ of the generating surface $T$ of the gear-cutting tool

$$d_{b.ct} = 2 \cdot r_{b.ct} = \frac{m \cdot N_{ct} \cdot \cos\phi_n}{\sqrt{1 - \cos^2\phi_n \cdot \cos^2\zeta_{ct}}} \tag{14.71}$$

where $m$ is the module of the gear-cutting tool.

Equation (14.71) indicates that the base diameter depends only on module $m$ of the gear-cutting tool, the number of its starts $Z_{ct}$, the normal profile angle $\phi_n$ of the auxiliary generating rack $\mathscr{R}_T$, and finally on the setting angle $\zeta_{ct}$ of the gear-cutting tool.

The base diameter of the generating surface of the gear-cutting tool can be expressed in terms of the pitch $P_{ct}$ of the generating surface $T$. Taking into account that the equality $m = 25.4/P_{cp}$ is observed, then Eq. (14.71) casts into the form

$$d_{b.cp} = \frac{25.4 \cdot N_{ct} \cdot \cos\phi_n}{P_{ct} \cdot \sqrt{1 - \cos^2\phi_n \cdot \cos^2\zeta_{ct}}} \tag{14.72}$$

Pitch helix angle $\psi_{ct}$ of the gear-cutting tool often (but not mandatorily) is approximately equal to the setting angle $\zeta_{ct}$ of the gear-cutting tool, that is, the approximate equality $\psi_{ct} \cong \zeta_{ct}$ is commonly observed. Under the assumption that the equality $\psi_{ct} \cong \zeta_{ct}$ is valid, Eq. (14.72) can be simplified to the form

$$d_{b.ct} \cong \frac{25.4 \cdot N_{ct} \cdot \cos\phi_n}{P_{ct} \cdot \sqrt{1 - \cos^2\phi_n \cdot \cos^2\psi_{ct}}} \tag{14.73}$$

Equation (14.63) also yields the formula for the calculation of base helix angle $\psi_{b.ct}$ of the generating surface $T$ of the gear-cutting tool:

$$\psi_{b.ct} = \tan^{-1}\left(\frac{\sqrt{\sin^2\phi_n + \tan^2\zeta_{ct}}}{\cos\phi_n}\right) \tag{14.74}$$

It is important to stress here that setting angle $\zeta_{ct}$ of the gear-cutting tool is incorporated into Eq. (14.74).

Equation (14.74) is equivalent to the equation

$$\psi_{b.ct} = \cos^{-1}(\cos\phi_n \cdot \cos\zeta_{ct}) \tag{14.75}$$

for the calculation of the base helix angle $\psi_{b.ct}$ that is proposed by Prof. S.P. Radzevich [138].

Equations (14.74) and (14.75) are used for the calculation of

1. The axial pitch $P_{x.ct}$ of the gear-cutting tool

$$P_{x.ct} = \pi \cdot d_{b.ct} \cdot \tan\psi_{b.ct} \tag{14.76}$$

2. The base pitch of the gear-cutting tool

$$p_{b.ct} = P_{x.ct} \cdot \cos\phi_n \cdot \cos\zeta_{ct} \tag{14.77}$$

as well as of many other design parameters of the gear-cutting tool.

The formula used for the calculation of the setting angle of the gear-cutting tool can also be derived from Eq. (14.63)

$$\zeta_{ct} = \tan^{-1}\left(\frac{m \cdot N_{ct}}{\sqrt{(d_{o.ct} - 2.5 \cdot m - \Delta d_{o.ct})^2 - m^2 \cdot Z_{ct}^2}}\right) \tag{14.78}$$

where:

- $d_{o.ct}$ is the outer diameter of the gear-cutting tool.
- $\Delta d_{o.ct}$ is the decrease of the outer diameter of the gear-cutting tool; it is equal to $\Delta d_{o.ct} = (d^n_{o.ct} - d^w_{o.ct})/2$.
- $d^n_{o.ct}$ is the outer diameter of the new gear-cutting tool.
- $d^w_{o.ct}$ is the outer diameter of the completely worn gear-cutting tool.

The following statements [119, 161, 175] immediately follow from the analysis of Eq. (14.63):

- The curvature of the involute profile of the generating surface $T$ at all points at the base cylinder (that is, at the start points of the screw involute surface) is equal to an infinity, and it is equal to zero at infinity.
- The principal curvatures of the involute generating surface of the gear-cutting tool at the points within the base helix are equal to $k_{1.ct} \to \infty$ and $k_{2.ct} = 0$ respectively.
- There is an infinite number of points at which the expressions $k_{1.ct} \to \infty$ and $k_{2.ct} = 0$ are valid.
- The first principal curvature of the generating surface of the gear-cutting tool is equal to zero ($k_{1.ct} = 0$) at points within the straight generating line of the surface $T$, whereas the second principal curvature is equal to infinity ($k_{2.ct} \to \infty$) at points within the base helix.
- The straight generating line (that is, the straight-line element of the involute generating surface of the gear-cutting tool) is tangent to the helix on the base cylinder. Normal vectors to the involute surface—those along the straight-line element of the generating surface of the involute gear-cutting tool—do not change their orientation; they are located within a common plane.

The above-listed statements are based on implementation of the formulae for the calculation of fundamental magnitudes of the first and the second order (see Appendix B) of generating surface $T$ of the gear-cutting tool.

## 14.5 USE OF *DG*-BASED METHODS FOR DETERMINING DESIGN PARAMETERS OF GENERATING SURFACES OF GEAR-CUTTING TOOL

In order to get an all-inclusive understanding of the geometry of the generating surface $T$ of the gear-cutting tool, in addition to the analytical investigation performed above, it is helpful to consider a *DG*-based (that is, a descriptive-geometry-based) approach for investigation of the geometry of the involute screw surface. Use of the *DG*-based approach provides a fruitful insight into the development and validation of the results of implementation of analytical methods of the analysis, including but not limited to implementation of the *DG/K*-based approach. Usually, results of implementation of the *DG*-based methods serve as a perfect filter for the eliminating of rough errors of the analysis. Both approaches, the *DG*-based as well as the *DG/K*-based, complement one another. Together, they provide the user with a profound understanding of the geometry of generating surfaces of the gear-cutting tool ([14, 98–100, 119, 138, 161] and others). As an example, the base helix angle and base diameter of the generating surface of the gear-cutting tool $T$ are determined below using the *DG*-based approach of analysis.

### 14.5.1 Base Helix Angle $\psi_{b.ct}$ of Generating Surface of Gear-Cutting Tool

To determine the base helix angle $\psi_{b.ct}$, the actual values of the normal pressure angle $\phi_n$ and of setting angle $\zeta_{ct}$ of the gear-cutting tool must be given.

The base helix angle $\psi_{b.ct}$ can be constructed in a system of planes of projection $\pi_1\pi_2\pi_3$. In case of necessity, additional auxiliary plane(s) of projection can be constructed as well.

At the beginning, it is necessary to construct the lateral tooth surface of the auxiliary generating rack $\mathscr{R}_T$. A possible way to construct the lateral plane of the auxiliary generating rack $\mathscr{R}_T$ is as follows.

Consider an arbitrary plane $A$ that is perpendicular to the axis of projection $\pi_1/\pi_2$ (see Figure 14.12). The plane $A$ is specified by the traces $A_1$ and $A_2$ within the horizontal $\pi_1$ and the vertical $\pi_2$ planes of projection. Next, turn the plane $A$ about the trace $A_2$ through the setting angle $\zeta_{ct}$ of the gear-cutting tool to the position of the plane $Q$. The plane $Q$ is specified by the traces $Q_1$ and $Q_2$, the second of which is aligned with the trace $A_2$ of the plane $A$. After that, the plane $Q$ is turned about the trace $Q_1$ through the normal

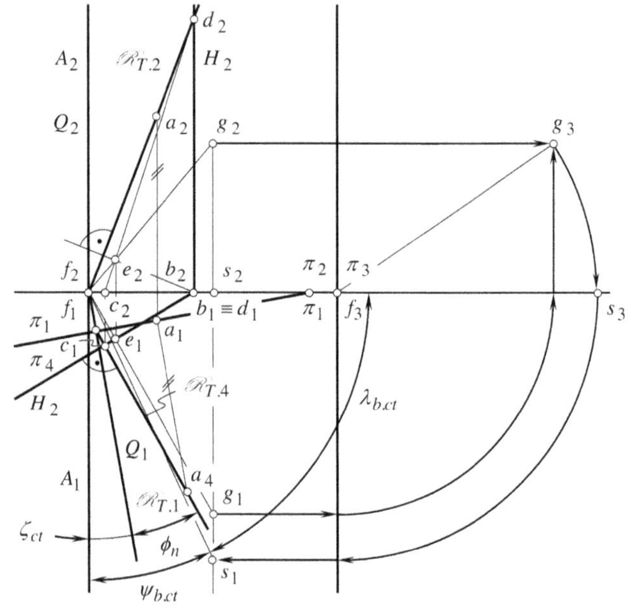


**FIGURE 14.11** Determination of base helix angle $\psi_{b.ct}$ of generating surface $T$ of involute gear-cutting tool. (Adapted from Radzevich, S.P., *ASME Journal of Mechanical Design*, 129, 334–345, 2007. With permission.)

profile angle $\phi_n$ of the gear-cutting tool. In this final location, the plane is designated as $\mathscr{R}_T$. The plane $\mathscr{R}_T$ is specified by the traces $\mathscr{R}_{T.1}$ and $\mathscr{R}_{T.2}$, respectively.

In order to construct the base helix angle $\psi_{b.ct}$ for this particular configuration (location and orientation) of the plane $\mathscr{R}_T$, an auxiliary plane of projection $\pi_4$ is constructed so that the axis of projection $\pi_1/\pi_4$ is perpendicular to the trace $\mathscr{R}_{T.1}$.

The base helix angle $\psi_{b.ct}$ of the gear-cutting tool is the angle that the lateral surface of the rack $\mathscr{R}_T$ forms with a plane, which is (a) orthogonal to the horizontal plane of projection $\pi_1$ and (b) orthogonal to the trace $\mathscr{R}_{T.4}$. Use of conventional rules—those developed in descriptive geometry—allows for construction of base helix angle $\psi_{b.ct}$ of the gear-cutting tool, as well as of base lead angle $\lambda_{b.ct}$ of the generating surface of the gear-cutting tool. The latter complements base helix angle $\psi_{b.ct}$ to the right angle (see Figure 14.11).

The derived *DG*-based solution to the problem of determining the base helix angle $\psi_{b.ct}$ gives an insight into how the expressions [see Eqs. (14.74) and (14.75)] can be derived analytically. It also serves as a perfect tool for verification of the results of the analytical solution to the problem.

### 14.5.2 Base Diameter $d_{b.ct}$ of Generating Surface of Gear-Cutting Tool

Let us again begin with the analysis of the property in accordance to which the generating surface $T$ of the gear-cutting

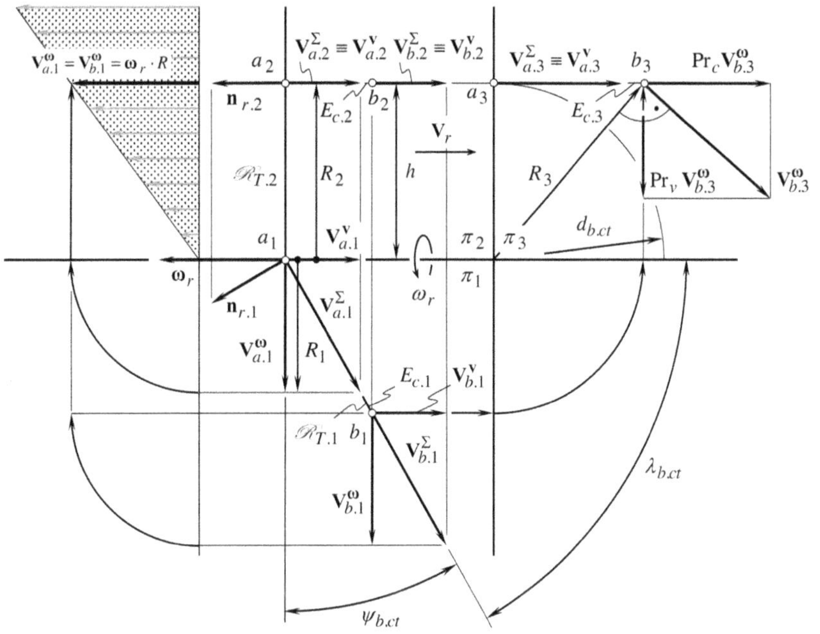

**FIGURE 14.12** Determination of the base diameter $d_{b.ct}$ of generating surface $T$ of involute gear-cutting tool. (Adapted from Radzevich, S.P., *ASME Journal of Mechanical Design*, 129, 334–345, 2007. With permission.)

tool is represented as an envelope surface to a family of consecutive positions of the lateral plane of the tooth of the auxiliary generating surface $\mathscr{R}_T$.

Consider a plane $\mathscr{R}_T$, configuration of which (location and orientation) is specified in a system of planes of projection $\pi_1\pi_2\pi_3$, as shown in Figure 14.12. The plane $\mathscr{R}_T$ performs a translation along the $\pi_1/\pi_2-$ axis of projection. The velocity of the translation is designated as $\mathbf{V}_r$. Simultaneously, the plane $\mathscr{R}_T$ rotates about that same axis $\pi_1/\pi_2$ with an angular velocity $\omega_r$.

The base diameter $d_{b.ct}$ of the gear-cutting tool is equal to the shortest distance of approach between the characteristic line $\mathscr{E}_{ct}$ and the axis $\pi_1/\pi_2$ of the screw motion of the rack. In order to determine the characteristic line $\mathscr{E}_{ct}$, it is necessary to select those points within the plane $\mathscr{R}_T$, the resultant velocity of which is perpendicular to the normal vector $\mathbf{n}_r$ to the plane $\mathscr{R}_T$ itself. To do that, the velocity vector $\mathbf{V}_r$ of the translation has to be considered together with the linear velocity of the rotation $\boldsymbol{\omega}_r$. Those points within the plane $\mathscr{R}_T$, the resultant instant velocity $\mathbf{V}^{\Sigma}$ of which is perpendicular to the normal vector $\mathbf{n}_r$, are the points that belong to the characteristic line $\mathscr{E}_{ct}$. In the case under consideration, the characteristic line $\mathscr{E}_{ct}$ is a straight line at a distance $d_{b.ct}/2$ from the axis of rotation $\pi_1/\pi_2$ of the gear-cutting tool. The characteristic line $\mathscr{E}_{ct}$ crosses the axis $\pi_1/\pi_2$ at the base helix, angle $\psi_{b.ct}$.

The *DG*-based solution to the problem of the determination of the base diameter $d_{b.ct}$ is depicted in Figure 14.12. The derived solution to the problem of determining the base diameter $d_{b.ct}$ of the generating surface $T$ of the gear-cutting tool is insightful for the derivation of Eq. (14.71) through Eq. (14.73).

Many other design parameters of the generating surface of the gear-cutting tool, as well as the design parameters of the gear-cutting tool itself, can be determined using the descriptive-geometry-based method.

## 14.6 POSSIBLE TYPES OF GENERATING SURFACES OF GEAR-CUTTING TOOLS

The generating surface of a gear-cutting tool is an envelope to consecutive positions of the auxiliary generating surface in its motion relative to a coordinate system with which the generating surface $T$ will be associated. Various types of the auxiliary generating surfaces $\mathscr{R}_T$ are depicted in Figures 14.1, 14.3, and 14.5. All of these, as well as auxiliary generating surfaces of those not shown above, can be used for the purpose of determining the generating surface of the gear-cutting tool. The motion of the auxiliary surface $\mathscr{R}_T$ relative to the reference system $X_{ct}Y_{ct}Z_{ct}$ should be appropriate in order for the surface $T$ to exist.

When the auxiliary generating rack $\mathscr{R}_T$ performs a screw motion about the $Z_{ct}-$ axis, and the axis is parallel to the pitch plane $W_r$ of the rack $\mathscr{R}_T$, the generating surface $T$ can be generated if and only if the velocity vector $\mathbf{V}_r$ of the translation, and the velocity vector $\boldsymbol{\omega}_g$ of the rotation, are timed with one another so that when the rack $\mathscr{R}_T$ turns through the angle of $2\cdot\pi$, it is simultaneously translated at a distance of $n\cdot P_x$. Here, $n$ designates an integer number (which actually is the number of starts of the generating surface $\mathscr{R}_T$), and the pitch of the rack's $\mathscr{R}_T$ teeth in the direction of the $Z_{ct}-$ axis is denoted by $P_x$. Ultimately, proper combination of the motions $\mathbf{V}_r$ and

$\omega_g$ results in the generation of the cylindrical surface $T$ shown in Figure 14.9.

### 14.6.1 Generating Surface of Gear-Cutting Tool Having Zero Profile Angle

In a particular case, a gear-cutting tool can be designed with zero normal profile angle $\phi_n$. An example of the auxiliary generating surface $\mathscr{R}_T$, having the profile angle of a zero value, is shown at the bottom of Figure 14.3 (the third case, labeled with a subscript 3). The generating surface of the gear-cutting tool is subject to significant changes when the equality $\phi_n = 0°$ is observed. Besides the reduced normal profile angle itself, an enormous increase of base diameter $d_{b.ct}$ of the gear-cutting tool occurs in this case. Following Eq. (14.71), the base diameter is equal to

$$d_{b.ct} = \frac{m \cdot N_{ct}}{\sqrt{1-\cos^2 \zeta_{ct}}} \tag{14.79}$$

when the equality $\phi_n = 0°$ is observed.

Taking into account that practical values of the setting angle $\zeta_{ct}$ of the gear-cutting tool are in the range of just a few angular degrees, the reduction of the normal profile angle to $\phi_n = 0°$ results in that the base diameter $d_{b.ct}$ in this case not only exceeds the outer diameter $d_{o.ct}$ of the gear-cutting tool, but it exceeds any reasonable values. Moreover, in the case of the setting angle of a zero value ($\zeta_{ct} = 0°$), the base diameter of the gear-cutting tool approaches infinity ($d_{b.ct} \to \infty$). No screw involute surface is feasible under this scenario, when the inequality $d_{o.ct} < d_{b.ct}$ is valid.

When the normal profile angle is zero, both the tooth addendum $a_c$, as well as the tooth dedendum $b_c$, of the auxiliary generating rack $\mathscr{R}_T$ shrink to zero ($a_c = 0$ and $b_c = 0$). The tooth profile of the gear-cutting tool is shrunk to a point—to the corner point of the auxiliary generating rack $\mathscr{R}_T$. Because of this feature, the work gear tooth profile is generated not by the corresponding conjugate profile of the gear-cutting tool (which does not exist for gear-cutting tools having the profile angle of zero value), but by the corner point of the generating rack $\mathscr{R}_T$ instead. This makes it possible to design a gear-cutting tool of coarse pitch having cutting edges made of superhard materials, such as manmade (artificial) diamond, borazon, and so forth (all of them are insufficiently small sizes).

Tooth height $h_{t.ct}$ of the generating rack $\mathscr{R}_T$ is not equal to the summa $h_{t.ct} \neq a_c + b_c$. Instead, it is approximately equal to tooth height $h_{t.g}$ of the gear being machined ($h_{t.ct} \approx h_{t.g}$).

However, it does not mean that a gear-cutting tool with a zero normal profile angle cannot be designed for machining an involute gear. The gear-cutting tool can be designed, using for this purpose a generating surface $T$ that is reduced in the case under consideration to two helices, which are located within the outer cylinder of the cutting tool (see Figure 14.13). For the calculation of axial pitch $P_{hx}$ of the helices, the formula

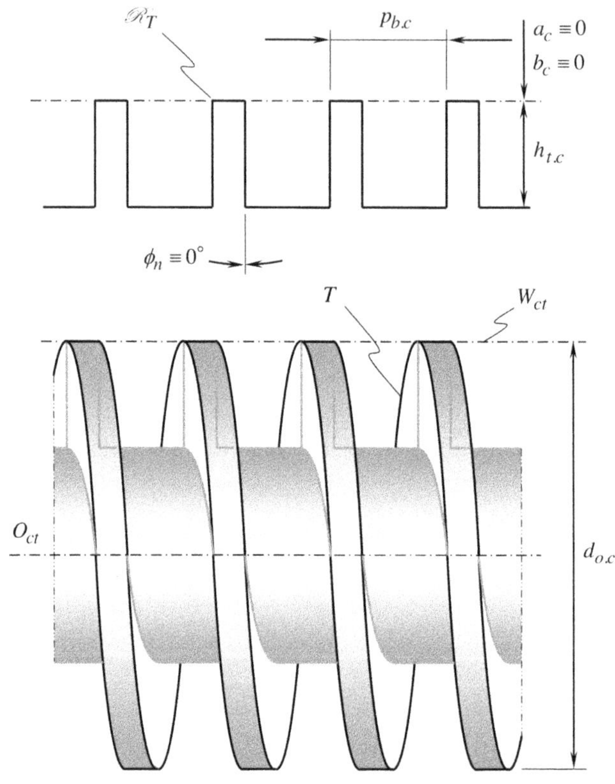


**FIGURE 14.13** Generating surface $T$ of a gear-cutting tool having normal profile angle $\phi_n$ of a zero value ($\phi_n = 0°$).

$$P_{hx} = p_{b.ct} \frac{N_{ct}}{\cos \zeta_{ct}} \tag{14.80}$$

is used.

When machining an involute gear, the gear-cutting tool has to be set up to allow the straight generating line $W_{ct}$ of the outer cylinder of the gear-cutting tool to be tangent to the base cylinder of the gear being machined.

For machining of involute gears with a low tooth count (for which the inequality $d_{l.g} \leq d_{b.g}$ is observed), a helical generating line $T$ is located within just one portion of the outer cylinder. For machining of involute gears with a greater tooth count (for which the inequality $d_{l.g} > d_{b.g}$ is observed), two portions of the helical generating line are required. The helices are located separately on two coaxial outer cylinders.

### 14.6.2 Conical Generating Surface of Gear-Cutting Tool

Permissible shapes of generating surfaces of gear-cutting tools are not limited only to cylindrical surfaces $T$. The auxiliary generating surface $T$ of other geometries are permissible as well.

The auxiliary generating rack $\mathscr{R}_T$ of a conical gear-cutting tool is tilted in relation to the axis of rotation $O_{ct}$

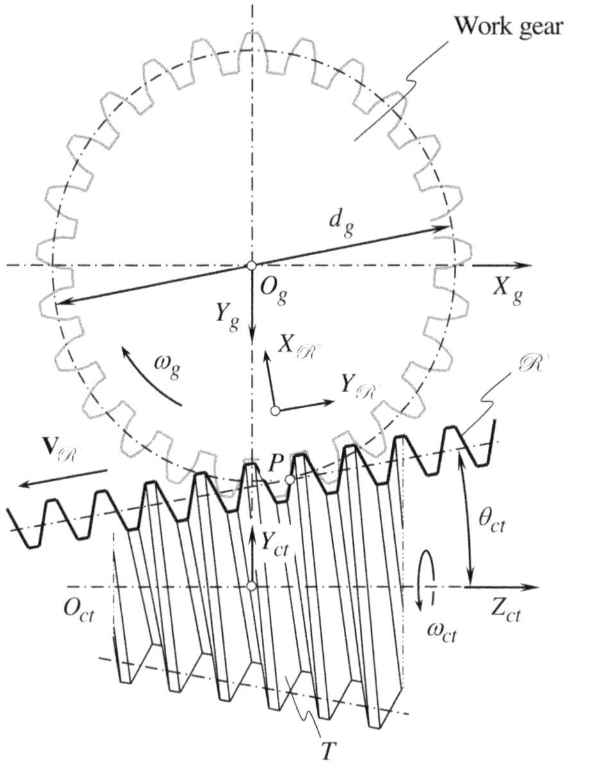


**FIGURE 14.14** Generation of conical generating surface $T$ of gear-cutting tool.

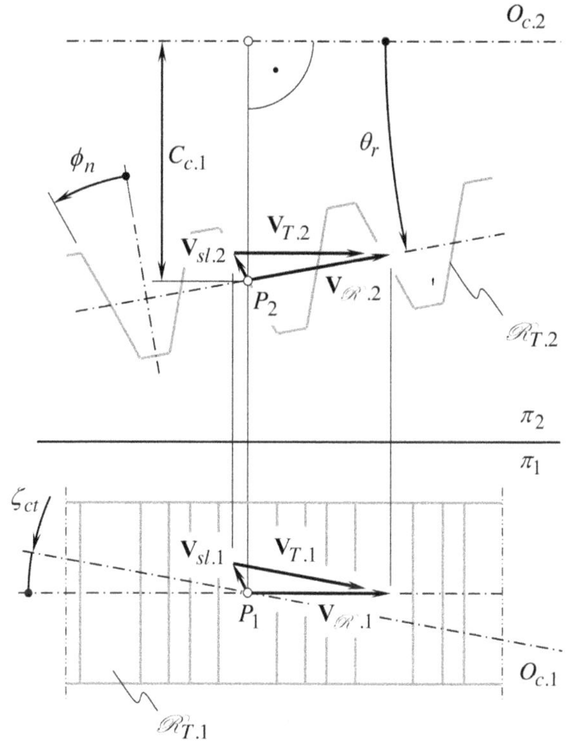


**FIGURE 14.15** Generation of generating surface $T$ of conical gear-cutting tool for machining cylindrical gears.

of the gear-cutting tool. The angle that the axis $O_{ct}$ forms with pitch plane $W_r$ of the auxiliary rack $\mathscr{R}_T$ is designated as $\theta_r$ [see Figure 14.5(a)]. The straight motion of the auxiliary generating surface $\mathscr{R}_T$ can be performed in a direction parallel to the pitch plane $W_r$ of the auxiliary generating rack $\mathscr{R}_T$. When the angles $\theta_r$ and $\phi_r$ correlate to one another properly (that is, when the inequality $-\phi_r < \theta_r < \phi_r$ is valid), then a conical generating surface $T$ of the gear-cutting tool is generated (see Figure 14.14).

That same auxiliary generating rack shown in Figure 14.1(b) can be used for the generation of a conical generating surface of the gear-cutting tool. When the rack $\mathscr{R}_T$ travels at an angle $\theta_r$ relative to the axis $O_{ct}$ of the gear-cutting tool, the generating surface $T$ is shaped in the form of a conical worm. The envelope surfaces to consecutive positions of the top lands and the bottom lands of the auxiliary rack $\mathscr{R}_T$ are shaped in the form of cones of revolution. The axes of rotation of the cones of revolution are aligned with axis $O_{ct}$ of rotation of the gear-cutting tool.

The straight motion of the auxiliary rack $\mathscr{R}_T$ should be investigated in more detail in order to get proper understanding of the geometry of the lateral surfaces of the conical generating surface $T$.

When generating the surface $T$ of the conical gear-cutting tool, the direction of the straight motion $\mathbf{V}_{\mathscr{R}}$ of the auxiliary rack $\mathscr{R}_T$ is perpendicular to the axis of rotation $O_g$ of the gear being machined. It is convenient to resolve the straight motion $\mathbf{V}_{\mathscr{R}}$ of the auxiliary rack $\mathscr{R}_T$ onto two components as shown in Figure 14.15. The first component $\mathbf{V}_{sl}$ is situated within the lateral plane of the auxiliary rack $\mathscr{R}_T$ tooth. The second component $\mathbf{V}_T$ is parallel to the axis of rotation $O_{ct}$ of the gear-cutting tool. Subscript 1 is assigned to the projections of the vectors $\mathbf{V}_{\mathscr{R}}$, $\mathbf{V}_T$, and $\mathbf{V}_{sl}$ onto the horizontal plane of projection $\pi_1$, and subscript 2 is assigned to the projections of those same vectors onto the vertical plane of projection $\pi_2$.

The equality

$$\mathbf{V}_{\mathscr{R}} = \mathbf{V}_T + \mathbf{V}_{sl} \tag{14.81}$$

is valid.

The translation $\mathbf{V}_{sl}$ results in the lateral plane of the auxiliary rack tooth *traveling on itself.* This motion does not affect the shape of the envelope to consecutive positions of the auxiliary rack $\mathscr{R}_T$ in its motion with respect to the cutting tool reference system $X_{ct}Y_{ct}Z_{ct}$. Therefore, the component $\mathbf{V}_{sl}$ can be neglected when determining conical generating surface $T$ of the gear-cutting tool.

The lateral plane travels along the axis $O_{ct}$ with a certain speed $V_T$. Simultaneously with the translation $V_T$, the rack $\mathscr{R}_T$ rotates about the $O_{ct}$ – axis with the rotation $\omega_r$. The ratio

$$\frac{V_T}{\omega_r} = \frac{P_{x.ct}}{2\pi} \tag{14.82}$$

is valid. Here, in Eq. (14.82), the axial pitch of the screw surface is denoted by $P_{x.ct}$.

Ultimately, the lateral surface of the conical generating surface $T$ can be viewed as the envelope to consecutive positions of the lateral plane of the auxiliary generating rack $\mathscr{R}_T$ that performs a screw motion. Two motions comprise the screw motion of the rack $\mathscr{R}_T$: (1) the rotation $\boldsymbol{\omega}_r$ of the rack $\mathscr{R}_T$ about the axis of rotation $O_{ct}$ of the gear-cutting tool and (2) the translation $\mathbf{V}_T$ along the $O_{ct}$ − axis.

When the linear velocity $V_{\mathscr{R}} = |\mathbf{V}_{\mathscr{R}}|$ is known, then the velocity $V_T = |\mathbf{V}_T|$ can be calculated from the equation

$$V_T = V_{\mathscr{R}} \cdot \frac{\cos\phi_n}{\cos\zeta_{ct} \cdot \cos(\phi_n \pm \theta_r)} \tag{14.83}$$

Owing to the fact that the generating surface is shaped in the form of a conical worm and not a cylindrical worm, two ways are possible to specify the configuration of the auxiliary generating rack $\mathscr{R}_T$ relative to the axis $O_{ct}$.

*First*, for the specification of a configuration of the auxiliary rack $\mathscr{R}_T$ relative to the axis $O_{ct}$, the setting angle $\zeta_{ct}$ of the gear-cutting tool can be utilized. The setting angle $\zeta_c$ complements to 90° the angle between the axis $O_{ct}$ and the straight generating lines of the rack $\mathscr{R}_T$. In Eq. (14.83), the linear velocity $V_T$ is expressed in terms of setting angle $\zeta_{ct}$ of the gear-cutting tool.

*Second*, an angle that complements to 90° the angle between the axes $O_g$ and $O_{ct}$ can be used for the specification of configuration of the auxiliary generating rack $\mathscr{R}_T$ relative to the axis $O_{ct}$. This angle is denoted as $\zeta^*_{ct}$. Evidently, the angle $\zeta^*_{ct}$ is not equal to the setting angle $\zeta_{ct}$ of the gear-cutting tool (that is, the inequality $\zeta^*_{ct} \neq \zeta_{ct}$ is valid). However, the angle $\zeta^*_{ct}$ can be expressed in terms of the setting angle $\zeta_{ct}$, and vice versa.

When the angle $\zeta^*_{ct}$ is used for the specification of a configuration of the auxiliary generating rack $\mathscr{R}_T$ relative to the axis $O_{ct}$, then the magnitude of the component $\mathbf{V}_T$ can be calculated from the equation

$$V_T = \frac{\sin\theta\tan(\theta+\phi_n)+\cos\theta}{\cos\zeta^*_{ct}} V_{\mathscr{R}} \tag{14.84}$$

This makes possible the conclusion that the lateral plane of the rack tooth performs a screw motion; the axis $O_{ct}$ is the axis of the screw motion, and the screw parameter of the screw motion is equal

$$p_{sc.1} = \frac{V_T}{\omega_r} = \frac{V_{\mathscr{R}}}{\omega_r} \cdot \underbrace{\frac{\sin\theta\tan(\theta+\phi_n)+\cos\theta}{\cos\zeta^*_{ct}}}_{a_1} = \frac{P_{x.ct}}{2\pi} \cdot a_1 \tag{14.85}$$

It is assumed there that $\omega_{\mathscr{R}} = 1$, and $V_T = P_{x.ct}$.

As already shown above (see Figure 14.6), the envelope to consecutive positions of the plane that performs a screw motion is a screw involute surface. By definition, the screw parameter $p_{ct}$ (the reduced pitch) of the screw involute surface is equal

$$p_{ct} = \frac{\mathbf{V}_T}{\omega_r} \tag{14.86}$$

where $\omega_r = |\boldsymbol{\omega}_r|$.

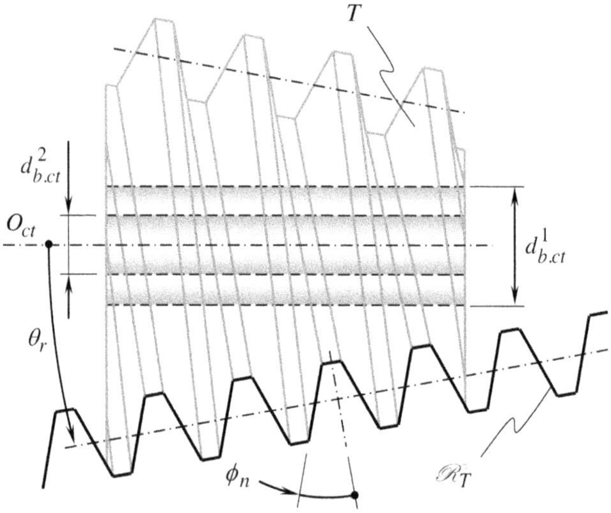


**FIGURE 14.16** Conical generating surface $T$ of gear-cutting tool for machining cylindrical gears.

Because of the lateral planes of the opposite sides of the auxiliary generating rack $\mathscr{R}_T$, tooth profiles are at different angles in relation to the axis of rotation $O_{ct}$, and the gear-cutting tool features not one but two base cylinders instead. One is the base cylinder of the screw involute surface of one side of the gear-cutting tool tooth profile, whereas the other is the base cylinder of the screw involute surface of the opposite side of the gear-cutting tool tooth profile. Schematically, the base cylinders of the opposite sides of the tooth profile of the gear-cutting tool are depicted in Figure 14.16.

For the purposes of determining the base diameters of conical generating surface $T$, it is convenient to consider a Cartesian reference system $X_{\mathscr{R}}Y_{\mathscr{R}}Z_{\mathscr{R}}$ associated with the auxiliary generating rack $\mathscr{R}_T$, as shown in Figure 14.15. The origin of the reference system $X_{\mathscr{R}}Y_{\mathscr{R}}Z_{\mathscr{R}}$ is at a distance $C_{ct}$ from the axis $O_{ct}$ of rotation of the gear-cutting tool. Another reference system $X_{ct}Y_{ct}Z_{ct}$ is associated with the gear-cutting tool.

The transition from the reference system $X_{\mathscr{R}}Y_{\mathscr{R}}Z_{\mathscr{R}}$ to the reference system $X_{ct}Y_{ct}Z_{ct}$ is analytically described by the operator $\mathbf{Rs}(\mathscr{R} \mapsto ct)$ of the result coordinate system transformation. In the case under consideration, the operator $\mathbf{Rs}(\mathscr{R} \mapsto ct)$ allows for matrix representation in the form

$$\mathbf{Rs}(\mathscr{R} \mapsto ct) = \begin{bmatrix} 0 & \sin\theta & \cos\theta & -C_{ct} \\ \cos\zeta^*_{ct} & \sin\zeta^*_{ct}\cos\theta & -\sin\zeta^*_{ct}\sin\theta & 0 \\ -\sin\zeta^*_{ct} & \cos\zeta^*_{ct}\cos\theta & -\cos\zeta^*_{ct}\sin\theta & 0 \\ 0 & 0 & 0 & 1 \end{bmatrix} \tag{14.87}$$

For the transition from the reference system $X_{ct}Y_{ct}Z_{ct}$ to the reference system $X_{\mathscr{R}}Y_{\mathscr{R}}Z_{\mathscr{R}}$, the operator $\mathbf{Rs}(ct \mapsto \mathscr{R})$ of the inverse transformation applies: $[\mathbf{Rs}(c \mapsto \mathscr{R}) = \mathbf{Rs}^{-1}(\mathscr{R} \mapsto c)]$.

In the coordinate system $X_{\mathscr{R}}Y_{\mathscr{R}}Z_{\mathscr{R}}$, the lateral plane of a tooth of the auxiliary rack $\mathscr{R}_T$ is analytically described by equation $Y_{\mathscr{R}} = -Z_{\mathscr{R}} \tan \phi_n$, or in matrix representation

$$Y_{\mathscr{R}}(X_{\mathscr{R}}, Z_{\mathscr{R}}) = \begin{bmatrix} X_{\mathscr{R}} \\ -Z_{\mathscr{R}} \tan \phi_n \\ -Z_{\mathscr{R}} \\ 1 \end{bmatrix} \tag{14.88}$$

Implementation of the operator $\mathbf{Rs}(\mathscr{R} \mapsto ct)$ of the linear transformation allows for an expression for the position vector of a point $\mathbf{r}_{\mathscr{R}}$ of that same plane represented in the coordinate system $X_{ct}Y_{ct}Z_{ct}$. Ultimately, for the calculation of the perpendicular $\mathbf{n}_{\mathscr{R}}$ to the lateral plane of the auxiliary rack $\mathbf{r}_{\mathscr{R}}$, the expression

$$\mathbf{n}_{\mathscr{R}} = \begin{bmatrix} \sin \zeta_c^* \\ \cos \zeta_c^* \\ \tan(\theta + \phi_n) \\ 1 \end{bmatrix} \tag{14.89}$$

is derived.

Let us determine the vector of the linear speed $\mathbf{V}_m$ of an arbitrary point $m$ within the lateral plane of the rack $\mathscr{R}_T$:

$$\mathbf{V}_m = \mathbf{i} \cdot Z_{ct} + \mathbf{j} \cdot p_{sc} - \mathbf{k} \cdot X_{ct} \tag{14.90}$$

Substituting the unit normal vector $\mathbf{n}_{\mathscr{R}}$ [see Eq. (14.89)] and the instant velocity vector $\mathbf{V}_m$ [see Eq. (14.90)] into the Shishkov equation of contact $\mathbf{n}_{\mathscr{R}} \cdot \mathbf{V}_m = 0$, and considering it together with the expression for the lateral plane in an arbitrary configuration of the auxiliary rack $\mathscr{R}_T$, allows for an equation

$$d_{b.c1} = 2 \frac{p_{sc.1} \cos \zeta_{ct}^*}{\sqrt{\sin^2 \zeta_{ct}^* + \tan^2(\theta + \phi_n)}} \tag{14.91}$$

for the base diameter of the screw involute generating surface $T$.

An equation

$$d_{b.c1} = 2 \frac{p_{sc.1} \cos \zeta_{ct}^*}{\sqrt{\sin^2 \zeta_{ct}^* + \tan^2(\theta - \phi_n)}} \tag{14.92}$$

similar to Eq. (14.91) is valid for the opposite side of the tooth profile of the rack $\mathscr{R}_T$.

Base diameters of a conical gear-cutting tool can be also expressed in terms of setting angle $\zeta_{ct}$:

$$d_{b.ct} = \frac{m \cdot N_{ct} \cdot \cos(\phi_n \pm \theta_r)}{\sqrt{1 - \cos^2(\phi_n \pm \theta_r) \cdot \cos^2 \zeta_{ct}}} \tag{14.93}$$

When the cone angle $\theta$ of the gear-cutting tool has a zero value (that is, when the equality $\theta = 0°$ is valid), then Eq. (14.91), Eq. (14.92) and Eq. (14.93) are applicable for cylindrical gear-cutting tools.

It is important to recall here that by definition the setting angle of the gear-cutting tool is measured within the pitch plane $W_{ct}$ (see Definition 14.1). Following from this statement, the setting angle $\zeta_{ct}$ of the conical gear-cutting tool does not complement to 90° the angle between the axes $O_g$ and $O_{ct}$. However, the angle $\overline{\zeta}_{ct}$ that complements to 90° the angle between the axis $O_g$ of the work gear and the axis $O_{ct}$ of the gear-cutting tool can be expressed in terms of the angles $\zeta_{ct}$ and $\theta_r$.

The angle that the axis of rotation $O_g$ of the work gear forms with the axis of rotation $O_{ct}$ of the gear-cutting tool is designated as $\Sigma$. The crossed-axes angle $\Sigma$ is equal to

$$\Sigma = 90° - \overline{\zeta}_{ct} \tag{14.94}$$

For the derivation of a formula for the calculation of the crossed-axes angle $\Sigma$, the operators of coordinate system transformations apply.

A Cartesian reference system $X_{ct}Y_{ct}Z_{ct}$ is associated with the generating surface of the conical gear-cutting tool as shown in Figure 14.17. Another Cartesian reference system

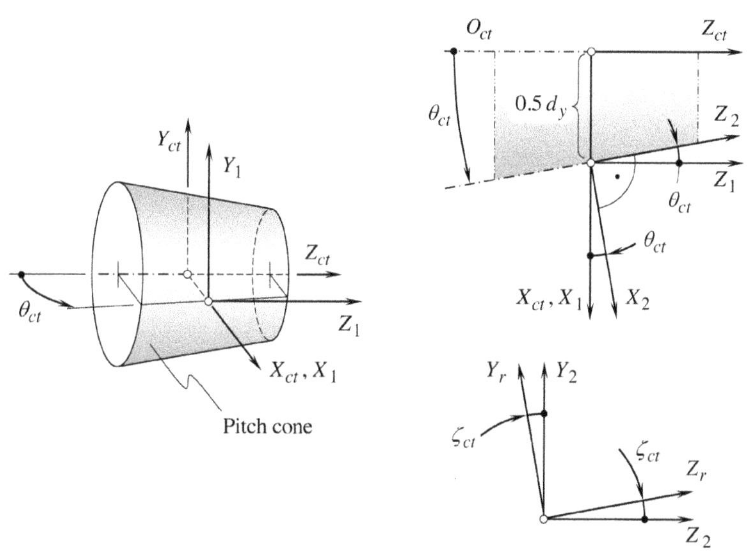


**FIGURE 14.17** One calculation of crossed-axes angle $\Sigma$, when machining a cylindrical gear with a conical gear-cutting tool.

$X_rY_rZ_r$ is associated with the auxiliary generating surface $\mathscr{R}_T$, so that (a) the $Y_rZ_r$ – coordinate plane is congruent with the pitch plane $W_r$ of the auxiliary rack $\mathscr{R}_T$ and (b) the $Y_r$ – axis is aligned with a straight generating line of the auxiliary rack tooth. The cone angle $\theta_{ct}$ is the angle that the axis $Z_{ct}$ of the gear-cutting tool reference system forms with the $Z_2$ – axis of the auxiliary coordinate system $X_2Y_2Z_2$. The setting angle $\zeta_{ct}$ of the gear-cutting tool is the angle that the axis $Y_2$ forms with the axis $Y_r$.

In such a scenario, the $Z_{ct}$ – axis and $Z_r$ – axis form the angle $\overline{\zeta}_{ct}$ to be determined, that is, $\overline{\zeta}_{ct} = \angle(Z_{ct}, Z_r)$.

In order to calculate the angle $\overline{\zeta}_{ct} = \angle(Z_{ct}, Z_r)$, both the axes $Z_{ct}$ and $Z_r$ have to be represented in a common reference system. The required coordinate system transformation can be analytically described by the operators of elementary coordinate system transformations.

For analytical representation of the transition of the reference system $X_{ct}Y_{ct}Z_{ct}$ along the $X_{ct}$ – axis through the distance $d_y/2$ to the position of the first auxiliary coordinate system $X_1Y_1Z_1$, the operator of translation $\mathbf{Tr}(d_y/2, X_{ct})$ is applied. Here, a current value of the pitch diameter of the conical pitch surface of the gear-cutting tool is designated as $d_y$.

For rotation of the auxiliary reference system $X_1Y_1Z_1$ about the $Y_1$ – axis through the cone angle $\theta_{ct}$ to the position of the second auxiliary coordinate system $X_2Y_2Z_2$, the operator of rotation $\mathbf{Rt}(\theta_{ct}, Y_1)$ is used.

Ultimately, the rotation of the reference system $X_2Y_2Z_2$ about the $X_2$ – axis through the setting angle $\zeta_{ct}$ of the gear-cutting tool to the position of the coordinate system $X_rY_rZ_r$ is analytically described by the operator of rotation $\mathbf{Rt}(\zeta_{ct}, X_2)$.

Use of the operators of elementary coordinate system transformations makes it possible to compose the operator $\mathbf{Rs}(ct \mapsto r)$ of the resultant coordinate system transformations. The expression

$$\mathbf{Rs}(ct \mapsto r) = \mathbf{Rt}(\zeta_{ct}, X_2) \cdot \mathbf{Rt}(\theta_{ct}, Y_1) \cdot \mathbf{Tr}(d_y/2, X_{ct}) \quad (14.95)$$

is used for calculating the operator $\mathbf{Rs}(ct \mapsto r)$ of this linear transformation.

The operator of translation $\mathbf{Tr}(d_y/2, X_{ct})$ does not affect the actual value of the crossed-axis angle $\Sigma$. Therefore, this operator can be omitted from further analysis.

The unit vector $\mathbf{k}_{ct}$ along the $Z_{ct}$ – axis of the reference system $X_{ct}Y_{ct}Z_{ct}$ can be represented in the coordinate system $X_rY_rZ_r$ of the auxiliary generating rack $\mathscr{R}_T$:

$$\mathbf{k}_{ct}^r = \mathbf{Rs}(ct \mapsto r) \cdot \mathbf{k}_{ct} \quad (14.96)$$

Equation (14.81) immediately yields an expression for the crossed-axes angle $\Sigma$

$$\Sigma = 90^\circ - \tan^{-1}\left(\frac{|\mathbf{k}_r \times \mathbf{k}_{ct}^r|}{\mathbf{k}_r \cdot \mathbf{k}_{ct}^r}\right) \quad (14.97)$$

where $\mathbf{k}_r$ denotes the unit vector along the $Z_r$ – axis of the reference system $X_rY_rZ_r$.

We stress again the principal difference between the helix angle $\psi_{ct}$ and the setting angle $\zeta_{ct}$ of the gear-cutting tool: Because the pitch diameter $d_{ct}$ varies within the face width of the conical gear-cutting tool, the pitch helix angle $\psi_{ct}$ is also varied within the face width of the gear-cutting tool. The value of the setting angle $\zeta_{ct}$ of the gear-cutting tool remains the same within the face width of the gear-cutting tool regardless of which portion of it is currently in mesh with the work gear.

### 14.6.3 Generating Surface of Gear-Cutting Tool with Asymmetric Tooth Profile

Gear-cutting tools with different profile angles at opposite sides of the tooth profile of the auxiliary generating rack are used for machining of spur and helical gears having different tooth profile angles. The auxiliary generating rack $\mathscr{R}_T$ with a corresponding conjugate profile is depicted in Figure 14.5(b). In the case under consideration, it is convenient to designate the opposite sides of the tooth of the auxiliary generating rack as $\mathscr{R}_{T.1}$ and $\mathscr{R}_{T.2}$.

When the auxiliary generating rack $\mathscr{R}_T$ performs a screw motion about the axis $O_{ct}$, then the generating surface of the gear-cutting tool that features an asymmetric tooth profile is generated (see Figure 14.18). Again, in the case under consideration it is convenient to designate the screw involute surfaces of the opposite sides of the thread profile as $T_1$ and $T_2$. The axial pitch of the opposite sides of the generating surface threads profile is of the same value. The setting angle $\zeta_{ct}$ of the gear-cutting tool is the same for the opposite sides of the threads profile.

Owing to a different profile angle, namely $\phi_{r.1}$ for one side and $\phi_{r.2}$ for the opposite side of the threads profile, base diameters $d_{b.ct}^1$ and $d_{b.ct}^2$ of the corresponding screw involute

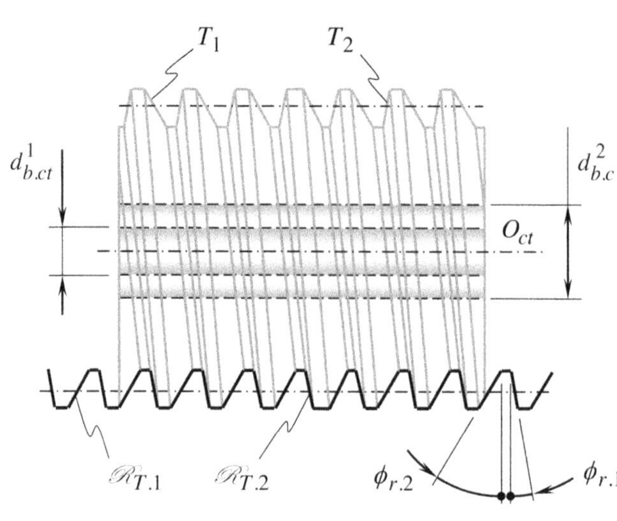


**FIGURE 14.18** Generating surfaces $T_1$ and $T_2$ of a gear-cutting tool for machining gears with asymmetric tooth profile.

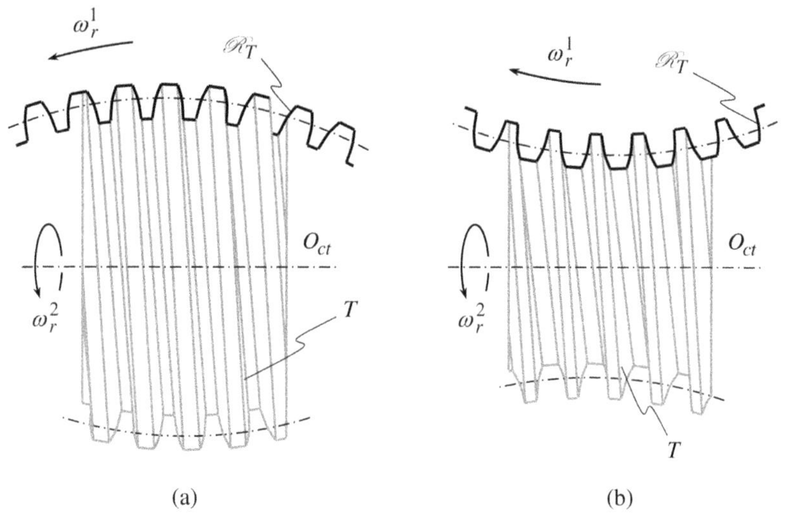


**FIGURE 14.19** Generating surface of gear-cutting tool with torus-shaped pitch surface, generated by a convex (a), and by a concave (b) auxiliary rack.

surfaces are also of a different value. For the calculation of the base diameters, the formulae

$$d_{b.ct}^1 = \frac{m \cdot N_{ct} \cdot \cos\phi_{r.1}}{\sqrt{1-\cos^2\phi_{r.1}\cdot\cos^2\zeta_{ct}}} \quad (14.98)$$

$$d_{b.ct}^2 = \frac{m \cdot N_{ct} \cdot \cos\phi_{r.2}}{\sqrt{1-\cos^2\phi_{r.2}\cdot\cos^2\zeta_{ct}}} \quad (14.99)$$

can be used.

When the inequality $\phi_{r.1} < \phi_{r.2}$ is observed, then the inequality $d_{b.ct}^1 > d_{b.ct}^2$ is valid as shown in Figure 14.18.

### 14.6.4 Generating Surfaces of Gear-Cutting Tools Featuring Torus-Shaped Pitch Surfaces

For the generation of the generating surfaces of the gear-cutting tool, the auxiliary generating surfaces in the form of either an internal gear sector [see Figure 14.5(c)] or an external gear sector [see Figure 14.5(d)] can be used. Generating surfaces of this type were investigated by Prof. S.P. Radzevich [144] as early as the 1970s.

When the auxiliary generating rack $\mathcal{R}_T$ rotates $\omega_r^1$ about its axis of rotation $O_r$ and when the reference system $X_rY_rZ_r$ simultaneously rotates $\omega_r^2$ about the axis of rotation of the gear-cutting tool $O_{ct}$, then the generating surface $T$ of the gear-cutting tool can be generated if the rotations $\omega_r^1$ and $\omega_r^2$ are properly synchronized with each other. Depending on the configuration of the auxiliary generating rack $\mathcal{R}_T$ and the axis of rotation $O_{ct}$ and on the timing of the rotations $\omega_r^1$ and $\omega_r^2$, either a convex [see Figure 14.19(a)] or a saddle-type [see Figure 14.19(b)] surface $T$ of the gear-cutting tool can be generated.

For the derivation of an equation for the position vector of a point of the generating surface $T$ of the gear-cutting tool, the kinematic method [2, 158, 206] is commonly used.

Consider two rotation vectors $\boldsymbol{\omega}_r$ and $\boldsymbol{\omega}_{ct}$, as shown in Figure 14.20. The rotation vectors $\boldsymbol{\omega}_r$ and $\boldsymbol{\omega}_{ct}$ are the vectors, one of which is pointed along the axis of rotation $O_r$ of the auxiliary generating rack $\mathcal{R}_T$, whereas the other is the vector along the axis of rotation $O_{ct}$ of the gear-cutting tool. The axes of rotation $O_r$ and $O_{ct}$ are positioned at a certain center distance $C_{r/ct}$. They cross each other at a certain crossed-axis angle $\Sigma = 90° - \zeta_{ct}$. The center distance $C_{r/ct}$ and the setting angle $\zeta_{ct}$ of the gear-cutting tool are considered as the given design parameters.

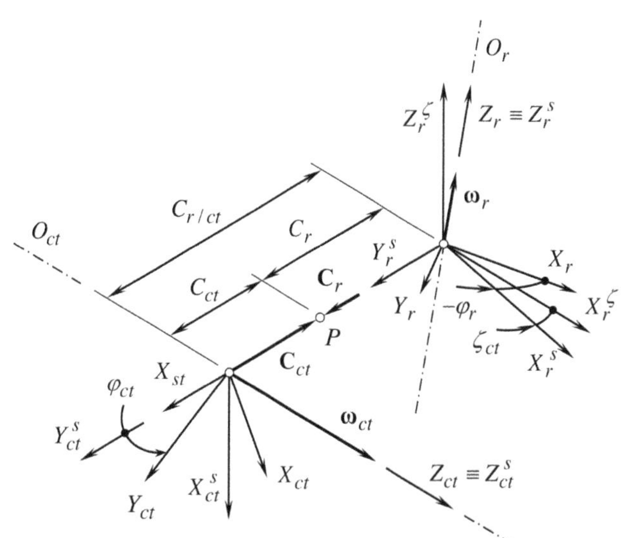


**FIGURE 14.20** Instant configuration of the reference systems $X_rY_rZ_r$ and $X_{ct}Y_{ct}Z_{ct}$ associated with the auxiliary generating rack $\mathcal{R}_T$ and with the gear-cutting tool that features a torus-shaped pitch surface.

Let us suppose that the auxiliary generating rack $\mathcal{R}_T$ of the gear-cutting tool is represented in the reference system $X_rY_rZ_r$ by the matrix equation

$$\mathbf{r}_r(U_r,V_r)=\begin{bmatrix} X_r(U_r,V_r)\\ Y_r(U_r,V_r)\\ Z_r(U_r,V_r)\\ 1\end{bmatrix} \quad (14.100)$$

For the derivation of an equation for the position vector of a point $\mathbf{r}_{ct}$ of generating surface $T$ of the gear-cutting tool, it is necessary to compose an equation of the auxiliary rack $\mathcal{R}_T$ in its current configuration when it is turned through an arbitrary angle of rotation $\varphi_r$. Then, it is necessary to represent the equation of the auxiliary rack $\mathcal{R}_T$ when the rack $\mathcal{R}_T$ occupies its current orientation in a reference system $X_{ct}Y_{ct}Z_{ct}$ associated with the gear-cutting tool. For this purpose, Eq. (14.74) has to be considered together with the operator $\mathbf{Rs}(r\mapsto ct)$ of the resultant coordinate system transformations. The operator $\mathbf{Rs}(r\mapsto ct)$ of the linear transformation can be expressed in terms of the operators of elementary coordinate system transformations:

$$\begin{aligned}\mathbf{Rs}(r\mapsto ct)&=\mathbf{Rt}(\varphi_{ct},Y_{ct}^s)\cdot\mathbf{Tr}(-C_{ct/r},Y_r^s)\\ &\cdot\mathbf{Rt}(-\zeta_{ct},Y_r^s)\cdot\mathbf{Rt}(-\varphi_r,Z_r)\end{aligned} \quad (14.101)$$

All the applied intermediate coordinate systems are depicted in Figure 14.20. Here, in Eq. (14.101), the rotation angle $\varphi_r$ of the auxiliary generating rack $\mathcal{R}_T$ and the rotation angle $\varphi_{ct}$ of the gear-cutting tool correlate to each other in compliance to the ratio $\varphi_r\cdot N_r=\varphi_{ct}\cdot N_{ct}$, where $N_r$ and $N_{ct}$ denote the number of teeth of the auxiliary rack $\mathcal{R}_T$ and the number of starts of the gear-cutting tool, respectively.

After the operator $\mathbf{Rs}(r\mapsto ct)$ of the resultant linear transformation is composed, then the auxiliary generating rack $\mathcal{R}_T$ in its current configuration with respect to the reference system $X_{ct}Y_{ct}Z_{ct}$ can be analytically described by the equation

$$\mathbf{r}_r^c(U_r,V_r,\varphi_r)=\mathbf{Rs}(r\mapsto ct)\cdot\mathbf{r}_r(U_r,V_r) \quad (14.102)$$

Using Eq. (14.102) together with the Shishkov equation of contact $\mathbf{n}_r^c\cdot\mathbf{V}_{r/ct}^c=0$, the following set of two equations is composed

$$\begin{cases}\mathbf{r}_r^c=\mathbf{r}_r^c(U_r,V_r,\varphi_r)\\ \mathbf{n}_r^c\cdot\mathbf{V}_{r/ct}^c=0\end{cases} \quad (14.103)$$

where $\mathbf{n}_r^c$ denotes the unit normal vector to the auxiliary rack surface $\mathcal{R}_T$ and $\mathbf{V}_{r/ct}^c$ denotes the vector of the resultant motion of the generating rack $\mathcal{R}_T$ with respect to the coordinate system $X_{ct}Y_{ct}Z_{ct}$. Both the vectors $\mathbf{n}_r^c$ and $\mathbf{V}_{r/ct}^c$ have to be represented in the coordinate system $X_{ct}Y_{ct}Z_{ct}$ with which the gear-cutting tool will be associated after it is designed.

A solution to the set of Eq. (14.103) returns an expression for the position vector of a point $\mathbf{r}_{ct}=\mathbf{r}_{ct}(U_{ct},V_{ct})$ of the generating surface $T$ of the gear-cutting tool.

It is important to stress here that the setting angle $\zeta_{ct}$ of the gear-cutting tool remains at a constant value regardless of which portion of threads of the surface $T$ is currently engaged in mesh with the work gear. This is one more piece of evidence for the difference between the setting angle $\zeta_{ct}$ and helix angle $\psi_{ct}$ of the gear-cutting tool.

For further analysis, it is convenient to introduce a stationary Cartesian reference system $X_{st}Y_{st}Z_{st}$. The origin of the reference system $X_{st}Y_{st}Z_{st}$ is located at the pitch point $P$ of the mesh of the auxiliary rack $\mathcal{R}_T$ and the generating surface $T$ of the gear-cutting tool. Due to lack of space, only the $X_{st}$ – axis of the newly introduced reference system $X_{st}Y_{st}Z_{st}$ is shown in Figure 14.20. The axes $Y_{st}$ and $Z_{st}$ are not shown here.

Similar to the vectors $\mathbf{C}_g$ and $\mathbf{C}_{ct}$ shown in Figure 3.6, the analogous vectors $\mathbf{C}_r$ and $\mathbf{C}_{ct}$ are depicted in Figure 14.20. The vectors $\mathbf{C}_r$ and $\mathbf{C}_{ct}$ can be expressed in terms of unit vector $\mathbf{c}$ along the $X_{st}$ – axis, and in terms of the radii $\mathcal{R}_r$ and $\mathcal{R}_{ct}$ of pitch surfaces of the auxiliary generating rack $\mathcal{R}_T$ and of the gear-cutting tool $\mathbf{C}_r=\mathbf{c}\cdot R_{w.r}$ and $\mathbf{C}_{ct}=-\mathbf{c}\cdot R_{w.ct}$. Depending on the actual values and configurations of (a) the rotation vectors $\boldsymbol{\omega}_r$ and $\boldsymbol{\omega}_{ct}$ and (b) the vectors $\mathbf{C}_r$ and $\mathbf{C}_{ct}$, either a convex (barrel-type) or saddle-type generating surface of the gear-cutting tool is generated (see Figure 14.19).

For gear-cutting tools with a pitch surface in the form of a torus, the specification of setting angle $\zeta_{ct}^*$ as the angle between the axes $O_g$ and $O_{ct}$ is the only possible specification of configuration of the gear-cutting tool relative to the work gear.

It is instructive to stress here that except for cylindrical and conical generating surfaces $T$ of the gear-cutting tools, fulfillment of the condition of conjugacy of the generating surfaces of all other geometries to the gear tooth flanks $\mathcal{G}$ have to be carefully checked. This is attributed to the kinematics of the gear machining process, and to violation (in many cases) of the condition of conjugacy of the tooth flanks $\mathcal{G}$ of a gear being machined, and of the generating surface $T$ of the gear-cutting tool. In compliance with the condition of conjugacy of the tooth flanks, at every point of contact of the surfaces $\mathcal{G}$ and $T$, the common perpendicular $\mathbf{n}$ must intersect the axis of instant rotation $P_{\text{ln}}$ of the gear and the gear-cutting tool in the gear machining mesh [171].

## 14.7 CONSTRAINTS ON DESIGN PARAMETERS OF GENERATING SURFACE OF GEAR-CUTTING TOOL

For machining of an involute gear, gear-cutting tools that feature the generating surface $T$ in the form of a screw involute surface are commonly used. The base cylinder is one of the key design features of the screw involute surface. The screw involute surface exists within the exterior of the base cylinder, not within the interior. The latter point imposes certain constraints onto values of the design parameters of generating surfaces of the gear-cutting tools.

Practically, the outer diameter of the gear-cutting tool is restricted by the design parameters of the gear generator at which the gear-cutting tool is applied. This means that the actual outer diameter of the gear-cutting tool $d_{o.ct}$ does not exceed its maximal permissible value $[d_{o.ct}]$ (that is, the inequality $d_{o.ct} \le [d_{o.ct}]$ is always valid). The working depth of the threads $h_k$ of the generating surface $T$ is equal to the summa of the addendum $a_c$ and the dedendum $b_c$ of the threads ($h_k = a_c + b_c$). Thus, the inner diameter of the gear-cutting tool $d_{f.ct}$ does not exceed the value of

$$d_{f.ct} \le [d_{o.ct}] - 2 \cdot h_k \tag{14.104}$$

From another side, the inner diameter of the gear-cutting tool $d_{f.ct}$ is equal to or exceeds the base diameter $d_{b.ct}$ (that is, the inequality $d_{f.ct} \ge d_{b.ct}$ must be fulfilled). In the most critical case, the latter inequality reduces to the equality $d_{f.ct} = d_{b.ct}$.

For a gear-cutting tool of the given module $m$ and the number of starts $N_{ct}$, the base diameter $d_{b.ct}$ depends only on the normal profile angle $\phi_n$ and on the setting angle $\zeta_{ct}$ of the gear-cutting tool. Equation (14.71) yields a graphical interpretation of the function $d_{b.ct}(\phi_n, \zeta_{ct})$. For a single-start gear-cutting tool ($N_{ct} = 1$) with a normalized module (that is, for $m = 1$), an example of the function $d_{b.ct}(\phi_n, \zeta_{ct})$ is plotted in Figure 14.21. As it immediately follows from Figure 14.21, the base diameter $d_{b.ct}$ approaches infinity ($d_{b.ct} \to \infty$) when the profile angle $\phi_n$ and the setting angle $\zeta_{ct}$ of the gear-cutting tool approach zero ($\phi_n \to 0°$, $\zeta_{ct} \to 0°$). The latter point makes possible a conclusion that when certain constraints are imposed on actual values of the diameters of the gear-cutting tool (outer diameter and inner root diameter), this immediately restricts the minimal value of the profile angle $\phi_n$ of the generating surface of the gear-cutting tool. The above-mentioned inequality $d_{f.ct} \ge d_{b.ct}$ along with Eq. (14.71)

$$d_{b.ct} = \frac{m \cdot N_{ct} \cdot \cos\phi_n}{\sqrt{1 - \cos^2\phi_n \cdot \cos^2\zeta_{ct}}} \tag{14.105}$$

for the calculation of the base diameter allow an expression

$$\phi_n \le [\phi_n] = \cos^{-1}\left(\frac{d_{f.ct}}{\sqrt{m^2 \cdot N_{ct}^2 + d_{f.ct}^2 \cdot \cos^2\zeta_{ct}}}\right) \tag{14.106}$$

for the calculation of the minimal permissible value of the normal pressure angle $\phi_n$ of the gear-cutting tool.

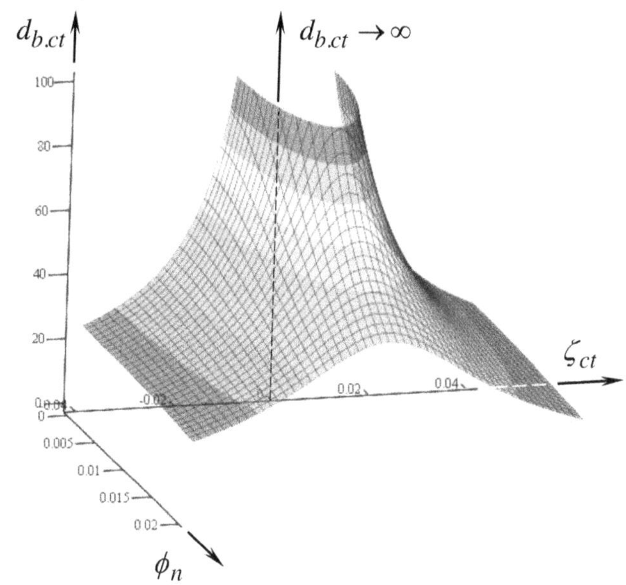


**FIGURE 14.21** Impact of normal profile angle $\phi_n$ and of setting angle $\zeta_{ct}$ on the value of the base diameter of the gear-cutting tool.

If normal profile angle $\phi_n$ exceeds its maximal allowed value $[\phi_n]$, then a corresponding portion of the generating surface $T$ of the gear-cutting tool cannot be reproduced by cutting edges. Ultimately, this results in violation of the conditions of proper part surface generation [136, 148, 158]. When conditions of proper part surface generation are not fulfilled, then the gear cannot be machined in compliance with the blueprint specification. Therefore, an accurate gear-cutting tool for machining of involute gears cannot be designed under a scenario when base helix angle $\psi_{b.ct}$ [see Eq. (14.75)] approaches zero ($\psi_{b.ct} \to 0°$), or its actual value is less than the minimal allowed value $[\psi_{b.ct}]$.

How is the latter statement correlated with the case depicted in Figure 14.3? In a scenario when the auxiliary generating rack $\mathscr{R}_{T.3}$ features a zero profile angle $\phi_n$ and setting angle $\zeta_{ct}$ of a small value, then the screw involute surface cannot be practically generated. No generating surface $T$ exists in this particular case. But auxiliary racks with zero profile exist.

However, the latter point does not mean that gear-cutting tools featuring zero profile angle $\phi_n = 0°$ are not possible. They are possible. However, in this particular case, the cutting edges (points) of the gear-cutting tools are located along the helix line on the outer cylinder of the gear-cutting tool.

# 15 Hobs for Machining Gears

Hobs are extensively used for machining gears, involute and straight-sided splines, sprockets, ratchets, and other form profiles. The automotive industry is the main consumer of hobs and hobbed gears. As an illustrative example, climb hobbing of a spur gear is illustrated in Figure 15.1.

Hobbing is usually and justifiably regarded as the accurate method of gear production, and more thought and mechanical skill have been expended on hobbing machines and their correct functioning than on any other type of gear-cutting machine. The process lends itself to the production of all types of gears, including those based on cones that constitute a special application, and it is unlikely to be superseded in the foreseeable future, particularly when large externally toothed cylindrical gears of high accuracy are required.

**FIGURE 15.1** Climb hobbing of spur gear.

## 15.1 DESIGN OF GEAR HOB ON PREMISE OF ITS GENERATING SURFACE

Consider a body that is made up of a cutting material, for example, of high-speed steel. The body is bounded by the generating surface $T$ of the gear-cutting tool. Such a body is referred to as the *generating body* of the gear-cutting tool.

In order to transform the generating body into the gear-cutting tool, it is necessary to make the body capable of removing the stock. Two ways to reach this goal are considered below.

In the first approach, the generating surface of the gear-cutting tool $T$ is entirely generated by cutting elements (that is, by the cutting edges). This is the easiest way to transform the generating body of the gear-cutting tool into a workable gear-cutting tool. Following this method, grinding worms for machining are designed.

As an example, consider the process of finishing a helical gear by the grinding worm as shown in Figure 15.2. The work gear rotates about its axis of rotation $O_g$, and the grinding worm rotates about its axis of rotation $O_{ct}$. The grinding worm is feeding in the axial direction of the work gear with a feed $F_{ct}$. Conversely, the work gear could be feeding in its axial direction. For surface generation, only the relative motions of the work gear and of the gear-cutting tool are of importance.

The rotations of the work gear $\omega_g$ and of the grinding worm $\omega_{ct}$ as well as the feed motion $F_{ct}$ are synchronized with one another in compliance with the design parameters of the gear to be machined (namely, with the tooth number $N_g$, pitch helix angle $\psi_g$, and the helix hand), as well as with the design parameters of the grinding wheel (namely, with the number of starts $N_{ct}$, pitch helix angle $\psi_{ct}$, and the helix hand).

The motions shown in Figure 15.2 are the same as those depicted in Figure 3.8. The feed motion $F_{ct}$ is the only difference. The need for the feed motion $F_{ct}$ is solely due to the fact that the generating surface $T$ of the gear-cutting tool makes point contact with the gear tooth flank surface $\mathscr{G}_g$. The characteristic line $\mathscr{E}_g$ (for the gear tooth flank $\mathscr{G}_g$ and the auxiliary generating surface $\mathscr{R}_T$) and the characteristic line $\mathscr{E}_{ct}$ (for the auxiliary generating surface $\mathscr{R}_T$ and the generating surface $T$ of the gear-cutting tool) intersect one another at the point of contact $K$ of the surfaces $\mathscr{G}_g$ and $T$. In rolling motion, the point of contact $K$ traces the involute tooth profile of the work gear. In order to spread the generated involute tooth profile along


DOI: 10.1201/9781003520528-20

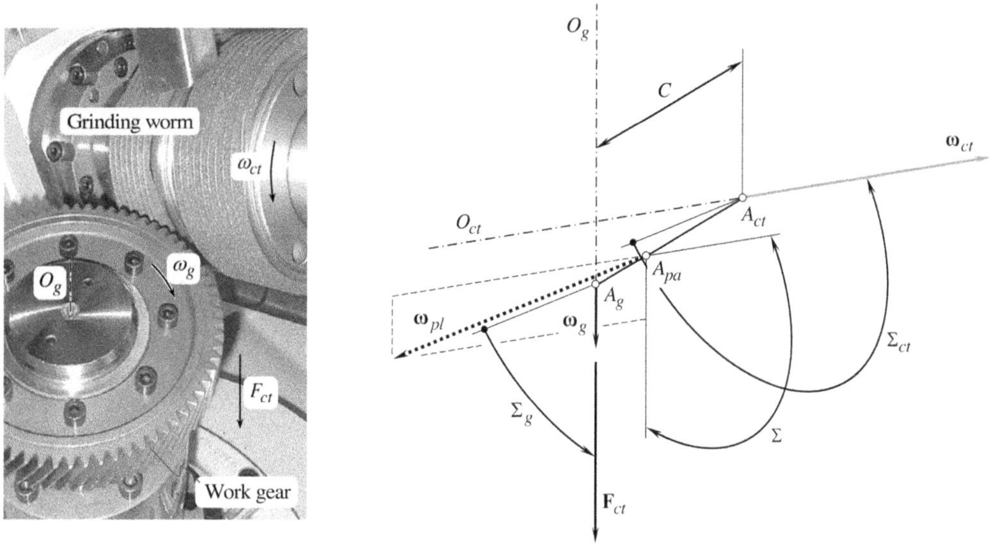


**FIGURE 15.2** Grinding a helical gear by the grinding worm.

the face width of the work gear,[1] the feed motion $\mathbf{F}_{ct}$ is required.

The kinematics of this type of part surface generation is used for semi-finishing and finishing of spur and helical gears. However, fine-pitch gears can be ground up from a solid blank—in such a scenario, gear grinding becomes not just a finishing operation, but also a roughing operation.

There are several possible ways to incorporate the feed motion $\mathbf{F}_{ct}$ into the kinematic scheme of the gear machining operation.

A helical gear can perform a screw motion $\mathbf{S}^g_{cr}$ under which the tooth flank $\mathcal{G}_g$ of the helical gear tooth moves on itself. The screw motion $\mathbf{S}^g_{cr}$ is the result of superposition of a rotation $\boldsymbol{\omega}^a_g$ and of a translation $\mathbf{V}^a_g$ of the work gear. These motions are timed with one another so as to keep the ratio $|\mathbf{V}^a_g| / |\boldsymbol{\omega}^a_g|$ equal to the reduced pitch $p_g$ of the tooth flank $\mathcal{G}_g$ (see Figure 15.3). The hand of the screw motion $\mathbf{S}^g_{cr}$ goes in the same direction as the direction of the hand of the gear tooth flank $\mathcal{G}_g$. For right-hand helical surface $\mathcal{G}_g$, the vectors of the rotation $\boldsymbol{\omega}^a_g$ and the translation $\mathbf{V}^a_g$ go in the same direction [see Figure 15.4(a,b)]. They can be pointed either in the same direction as the main rotation vector $\boldsymbol{\omega}_g$ of the work gear [see Figure 15.4(a)], or they can be pointed in the direction of the vector $\boldsymbol{\omega}_g$ [see Figure 15.4(b)]. Here, the vector $\boldsymbol{\omega}_g$ of the main rotation that is considered together with the vector $\boldsymbol{\omega}_{ct}$ of the main rotation of the gear-cutting tool result in rolling of the axodes of the work gear and of the gear-cutting tool.

For a left-hand-oriented helical surface $\mathcal{G}_g$, the vectors $\boldsymbol{\omega}^a_g$ and $\mathbf{V}^a_g$ are pointed in opposite directions (see Figure 15.4). While being pointed in opposite directions, the rotation vector $\boldsymbol{\omega}^a_g$ can be either pointing in the same direction as the main rotation vector $\boldsymbol{\omega}_g$ of the work gear [see Figure 15.4(c)] or it can be pointed in the opposite direction of the vector $\boldsymbol{\omega}_g$ [see Figure 15.4(d)].

Possible configurations of the vectors $\boldsymbol{\omega}^a_g$ and $\mathbf{V}^a_g$ with respect to one another, as well as with respect to the vector $\boldsymbol{\omega}_g$ of the main rotation of the work gear (see Figure 15.4), encompass both right-hand-oriented helices $\mathcal{G}_g$ and left-hand-oriented helices $\mathcal{G}_g$. They also encompass conventional methods of gear machining (that is, conventional gear hobbing operations) and climb methods of gear machining (that is, climb gear hobbing operations).

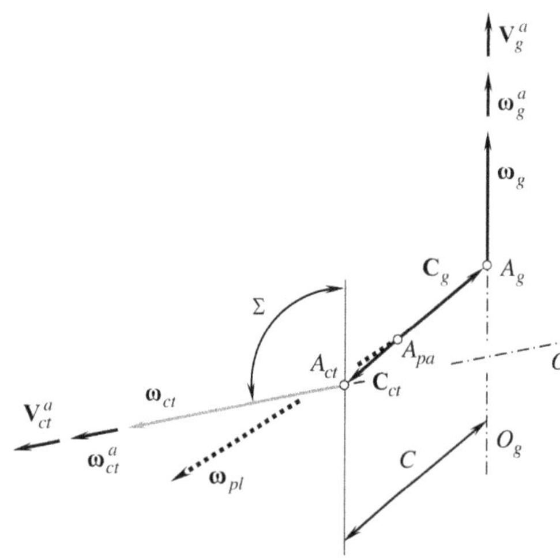


**FIGURE 15.3** Configuration of the motion vectors in the gear machining operation that features point contact of the gear tooth flank $\mathcal{G}_g$ and the generating surface $T$ of a gear-cutting tool.

[1] Only in a particular case when:

- The cutting edge of the gear-cutting tool is congruent to the characteristic line $\mathcal{E}_{ct}$.
- The characteristic line $\mathcal{E}_{ct}$ is congruent to the characteristic line $\mathcal{E}_g$.
- The characteristic line $\mathcal{E}_{ct} \equiv \mathcal{E}_g$ performs rolling motion about the axis $O_g$ of the work gear.

Then the feed motion $F_{ct}$ can be eliminated from the kinematics of the gear machining operation.

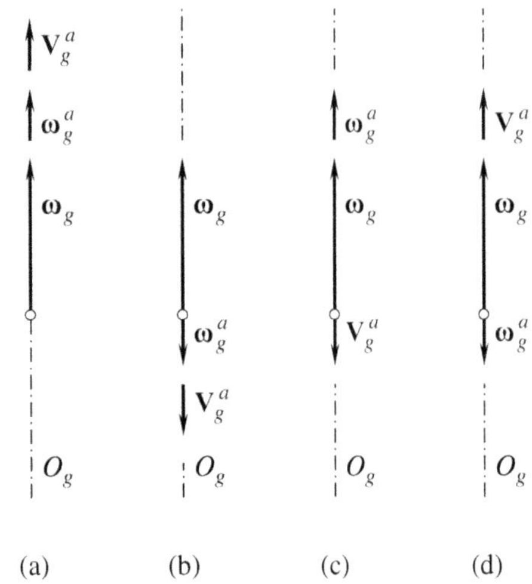


**FIGURE 15.4** Permissible configurations of the rotation vector $\boldsymbol{\omega}_g^a$ and the translation vector $\mathbf{V}_g^a$ in relation to the main rotation vector $\boldsymbol{\omega}_g$ of the work gear. The parts (a-d) are discussed in detail within the text body of the chapter.

In a specific case of gear machining, either rotation or translation degenerates to zero; that is, either the equality $\boldsymbol{\omega}_g^a = 0$ or the equality $\mathbf{V}_g^a = 0$ is observed. The first case corresponds to machining of spur gears, whereas the second case corresponds to machining of round racks.

A screw motion $\mathbf{S}_{cr}^c$ similar to the screw motion $\mathbf{S}_{cr}^g$ just considered can also be performed by the gear-cutting tool. When performing the screw motion $\mathbf{S}_{cr}^c$, the generating surface $T$ of the gear-cutting tool moves on itself. In the case under consideration, the screw motion $\mathbf{S}_{cr}^c$ is a kind of superposition of a rotation $\boldsymbol{\omega}_{ct}^a$ and a translation $\mathbf{V}_{ct}^a$ of the gear-cutting tool. The rotation $\boldsymbol{\omega}_c^a$ and the translation $\mathbf{V}_c^a$ are synchronized so that the ratio $|\mathbf{V}_c^a|/|\boldsymbol{\omega}_c^a|$ is equal to the reduced pitch $p_{ct}$ of the generating surface $T$ of the gear-cutting tool. The hand of the screw motion $\mathbf{S}_{cr}^c$ is going in the same direction as the hand of the generating surface $T$. For right-hand helical surface $T$, the vectors of the rotation $\boldsymbol{\omega}_{ct}^a$ and the translation $\mathbf{V}_{ct}^a$ are going in the same direction. They can be pointed either in the same direction as the main rotation vector $\boldsymbol{\omega}_{ct}$ of the gear-cutting tool, or they can be pointed in the opposite direction of the vector $\boldsymbol{\omega}_{ct}$.

Permissible configurations of the vectors $\boldsymbol{\omega}_{ct}^a$ and $\mathbf{V}_{ct}^a$ with respect to each other, as well as with respect to vector $\boldsymbol{\omega}_{ct}$ of the main rotation of the gear-cutting tool, encompass both right-hand-oriented helices $T$ and left-hand-oriented helices $T$. A screw motion $\mathbf{S}_{cr}^c$ of the gear-cutting tool is employed, for example:

- when hobbing worm gears,
- in the operation of diagonal hobbing of gears,
- in three main methods of gear shaving operation, and so forth.

Tools for grinding gears reproduce the generating surface $T$, which actually is a discontinuous surface. However, due to the large number of abrasive grains involved, it can be interpreted as a continuous one.

Reproduction of a continuous generating surface $T$ by cutting elements (that is, by abrasive grains) of the gear grinding tool is not the only possible way to transform the generating body into a workable gear-cutting tool. The surface $T$ can also be reproduced discretely. In this way, hobs for machining gears and gear-cutting tools of other designs can be designed. As illustrated in Figure 15.5, the generating surface $T$ of the gear-cutting tool can be reproduced by a certain number of discrete cutting edges $CE$. Geometrically, the cutting edge is viewed as the line of intersection of the generating surface $T$ by the rake surface $R_s$. Gashes are cut up in the generating body of the gear-cutting tool in order to form the rake surfaces $R_s$. After the cutting edges $CE$ and the rake surfaces $R_s$ are formed, then the clearance surfaces $C_s$ are created. The clearance surfaces $C_s$ pass through the cutting edges $CE$ and form a certain clearance angle $\alpha$ with the generating surface $T$. More details on the methods of transformation of the generating body into a workable gear-cutting tool can be found in [148, 158], as well as from other sources.

It is of critical importance to stress here that all the cutting edges, namely the cutting edges of both sides of the

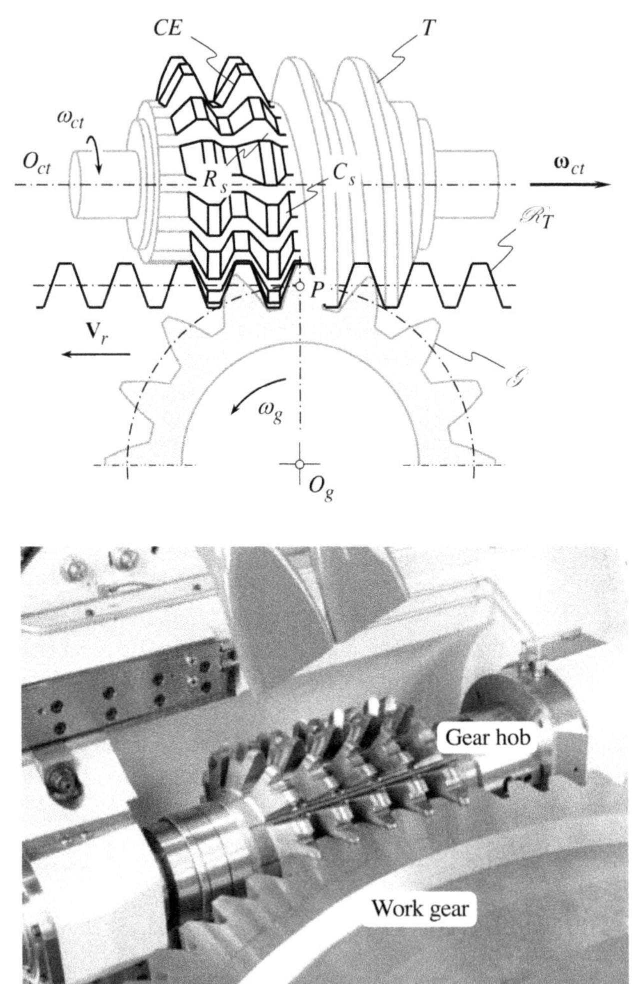


**FIGURE 15.5** Discrete reproduction of generating surface $T$ of a gear-cutting tool.

tooth profile of the gear-cutting tool, as well as the top cutting edges, are situated within the corresponding portions of the generating surface of the gear-cutting tool $T$. Any displacement of the cutting edge $CE$ from the generating surface $T$ causes deviations of the actually machined gear tooth surface from the desired gear tooth surface. The complexity of the problem is mostly due to all three surfaces:

- The generating surface $T$ of the gear-cutting tool.
- The rake surface $R_s$.
- The clearance surface $C_s$.

Must pass through the corresponding cutting edge $CE$.

In practical terms, it is often difficult to ensure that the three surfaces $T$, $R_s$, and $C_s$ pass through a cutting edge $CE$ (that is, it is a challenging problem to guarantee the three surfaces pass through a common line).

Rake surfaces $R_s$ and clearance surfaces $C_s$ of various geometries are used in the design of a gear-cutting tool. For the generation of the surfaces $R_s$ and $C_s$, numerous methods are used in the industry. Methods of roughing (that is, methods of machining by means of edge-cutting tools) of the surfaces $R_s$ and $C_s$ and finishing methods (that is, methods of machining by means of grinding wheels) are distinguished. The ultimate geometry of both the rake surfaces and the clearance surfaces is produced on finishing operations. This section of the book is focused on a detailed investigation of geometrical and kinematical aspects of finishing the surfaces $R_s$ and $C_s$.

## 15.2 GEOMETRY AND GENERATION OF RAKE SURFACE OF GEAR HOB

The cut chip flows out of the machined part surface over the rake surface of the hob tooth. The surface $R_s$ is one of the surfaces bounding the hob gashes. Often, the rake surface $R_s$ is orthogonal to the helix of the threads on the pitch cylinder of the generating surface $T$ of the hob.

### 15.2.1 Geometry of Rake Surface

Surfaces of various geometries can be implemented as the rake surface of a gear hob. A multiple-start involute hob with the rake surface $R_s$ in the form of a screw surface is depicted in Figure 15.6. A screw rake surface $R_s$ that features the helix angle $\psi_{rs}$ of an assigned value is considered below as a representative example of rake surfaces. An example of generating a rake surface $R_s$ of a precision involute hob is schematically depicted in Figure 15.7.

The rake angle $\gamma_o$ at the top cutting edge of the hob is often equal to zero, $\gamma_o = 0°$. However, for roughing of gears as well as for machining of low-accuracy gears, involute hobs having positive rake angle $\gamma_o > 0°$ are often used. For semi-finishing and/or for finishing of hardened gear, hobs with a negative rake angle $\gamma_o << 0°$ are used as well.

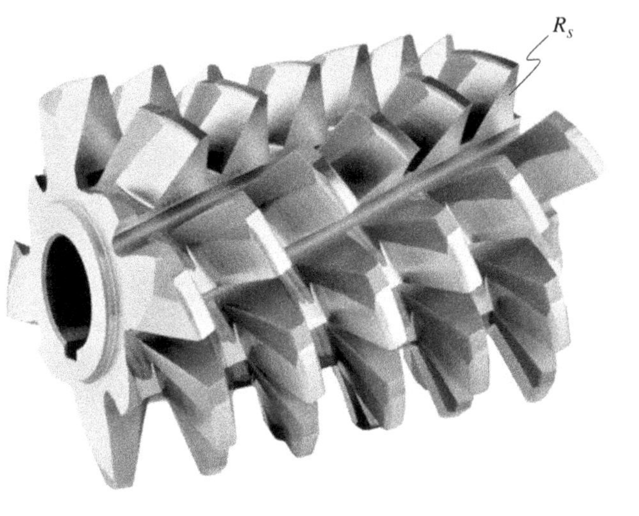


**FIGURE 15.6** A multiple-start involute hob with rake surface $R_s$ in the form of a screw surface.

The rake surface $R_s$ is represented here as a family of consecutive positions of a straight-line generator $G_{rs}$, which performs a screw motion about the hob axis $O_{ct}$. In the Cartesian coordinate system $X_{ct}Y_{ct}Z_{ct}$ associated with the gear-cutting tool, the straight-line generator $G_{rs}$ forms a rake angle $\gamma_o$ with the $Z_{ct}$ – axis. The position vector $\mathbf{r}_{rg}$ of a point of the straight-line generator $G_{rs}$ is located within the $Y_{ct}Z_{ct}$ – coordinate plane, as shown in Figure 15.6. The equation of the straight-line generator $G_{rs}$ yields vector representation in the form $\mathbf{r}_{rg} = \mathbf{r}_{rg}(U_{rs})$, where $U_{rs}$ denotes the parameter of the straight-line generator $G_{rs}$. For an involute hob with a rake angle of a certain value $\gamma_o$, the straight-line generator $G_{rs}$ is positioned at the distance $p_{rg}$ away from the axis $O_{ct} \equiv X_{ct}$. As it follows from the analysis of Figure 15.8, for a hob with an outer diameter $d_{o.ct}$, the parameter $p_{rg}$ can be calculated from the formula

$$p_{rg} = 0.5 \cdot d_{o.ct} \cdot \sin\gamma_o \tag{15.1}$$

The position vector of a point $\mathbf{r}_{rg}$ of the straight-line generator $G_{rs}$ forms a certain angle $U_{rs}$, with the shortest distance of approach of the generator $G_{rs}$ and the $X_{ct}$ – axis. Therefore, the position vector $\mathbf{r}_{rg}$ of a point of the straight-line generator $G_{rs}$ yields a matrix representation in the form [164]

$$\mathbf{r}_{rg}(U_{rs}) = \begin{bmatrix} 0 \\ p_{rg} \cdot \dfrac{\sin\gamma_o \cdot \cos(\gamma_o + U_{rs})}{\cos U_{rs}} \\ p_{rg} \cdot \dfrac{\sin\gamma_o \cdot \sin(\gamma_o + U_{rs})}{\cos U_{rs}} \\ 1 \end{bmatrix} \tag{15.2}$$

The generating line $G_{rs}$ performs a screw motion about the hob axis of rotation $O_{ct}$. The screw parameter of the screw motion is designated as $p_{rs}$.

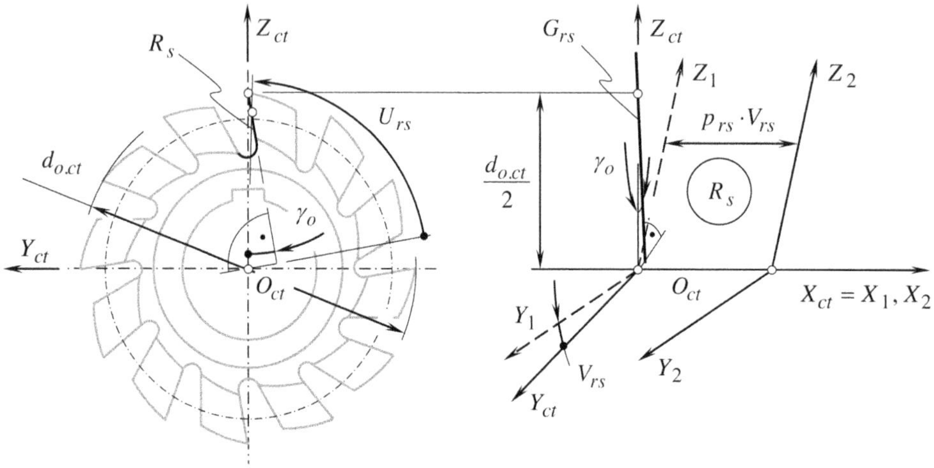


**FIGURE 15.7** Generation of rake surface $R_s$ of involute hob. (After: Radzevich, S.P., *SME Journal of Manufacturing Processes*, 9(2), 121–136, 2007. With permission.)

To make the analysis clear, two auxiliary reference systems, $X_1Y_1Z_1$ and $X_2Y_2Z_2$, are introduced as shown in Figure 15.7. The reference system $X_1Y_1Z_1$ rotates about the hob axis of rotation $O_{ct}$, whereas the reference system $X_2Y_2Z_2$ travels along the axis $O_{ct}$.

The screw motion of the generating line $G_{rs}$ can be decomposed onto the rotation together with the reference system $X_1Y_1Z_1$ about the $X_{ct}-$ axis and onto the translation together with the coordinate system $X_2Y_2Z_2$ along the $X_{ct}-$ axis. The rotation and the translation are synchronized with one another according to the parameter of the screw motion $p_{rs}$.

For the derivation of equation for the screw rake surface $R_s$ of the hob tooth, the operator $\mathbf{Rs}(2 \mapsto ct)$ of the resultant coordinate system transformation, say of the transition from the coordinate system $X_2Y_2Z_2$ to the gear hob coordinate system $X_{ct}Y_{ct}Z_{ct}$, is used. An analysis of Figure 15.7 reveals that the operator $\mathbf{Rs}(2 \mapsto ct)$ can be expressed in terms of the operator of rotation $\mathbf{Rt}(V_{rs}, Z_{ct})$ and the operator of translation $\mathbf{Tr}(p_{rs} \cdot V_{rs} \mapsto X_{ct})$, that is:

$$\mathbf{Rs}(2 \mapsto c) = \mathbf{Tr}(p_{rs} \cdot V_{rs} \mapsto X_{ct}) \cdot \mathbf{Rt}(V_{rs}, Z_{ct}) \tag{15.3}$$

The position vector of a point $\mathbf{r}_{rs}$ of the rake surface $R_s$, therefore, can be expressed by the equation

$$\mathbf{r}_{rs}(U_{rs}, V_{rs}) = \mathbf{Rs}(2 \mapsto ct) \cdot \mathbf{r}_{rg}(U_{rs}) \tag{15.4}$$

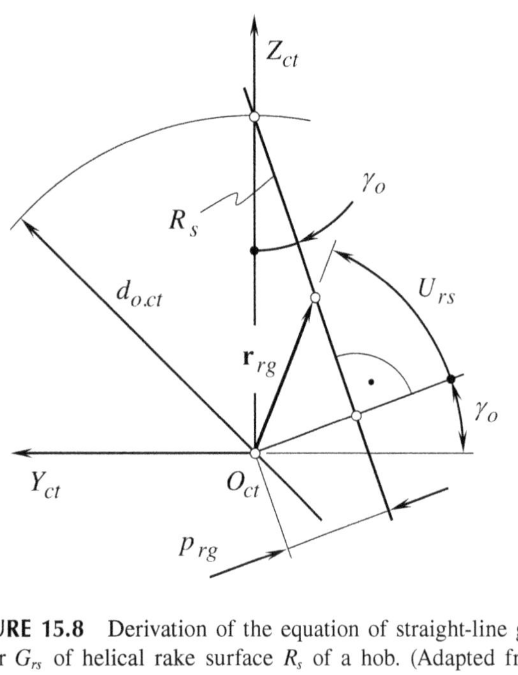


**FIGURE 15.8** Derivation of the equation of straight-line generator $G_{rs}$ of helical rake surface $R_s$ of a hob. (Adapted from: Radzevich, S.P., *SME Journal of Manufacturing Processes*, 9(2), 121–136, 2007. With permission.)

Equations (15.2) through (15.4) yield a matrix representation for the equation of the screw rake surface $R_s$ of an involute hob [164]:

$$\mathbf{r}_{rs}(U_{rs}, V_{rs}) = \begin{bmatrix} p_{rs} \cdot V_{rs} \\ \dfrac{\sqrt{2} \cdot p_{rg} \cdot \sin\gamma_o \cos(\gamma_o + U_{rs})}{\cos U_{rs}} \cdot \sin\left(\dfrac{\pi}{4} + V_{rs}\right) \\ \dfrac{\sqrt{2} \cdot p_{rg} \cdot \sin\gamma_o \sin(\gamma_o + U_{rs})}{\cos U_{rs}} \cdot \cos\left(\dfrac{\pi}{4} + V_{rs}\right) \\ 1 \end{bmatrix} \tag{15.5}$$

The parameters of the rake surface $R_s$ can be expressed in terms of the corresponding design parameters of the hob (see Appendix F).

Involute hobs with a small helix angle may have straight gashes. Under such a scenario, the rake surface $R_s$ is shaped in the form of a plane. An involute hob with the rake surface $R_s$ in the form of a plane is depicted in Figure 15.9.

The performed analysis of possible geometries of rake surfaces of gear hobs [164] reveals that all possible types of

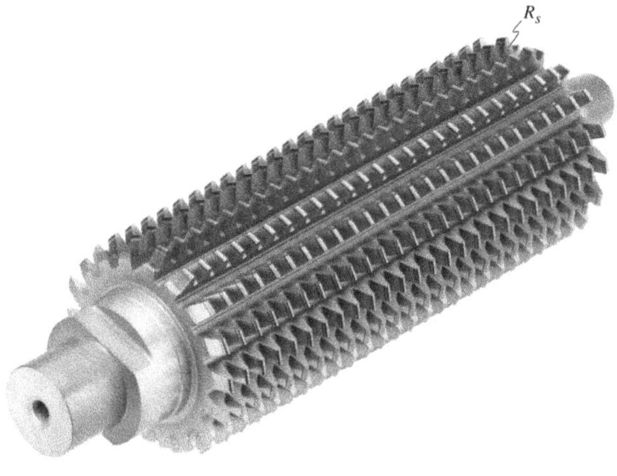


**FIGURE 15.9** Involute hob with rake surface $R_s$ in the form of a plane.

rake surfaces can be viewed as a specific case of the convolute surface (see Figure 15.10), and therefore an equation for each of them could be derived from Eq. (15.5).

Gear hobs of conventional design feature the rake surface $R_s$ in the form of an *Archimedean helicoid*, which is a specific case of a convolute surface with rake angle of a zero value ($\gamma_o = 0°$). For this particular case of gear hob design, an equation of the rake surface $R_s$ can be represented in matrix form:

$$\mathbf{r}_{rs}(Z_{ct}, V_{rs}) = \begin{bmatrix} p_{rs} \cdot V_{rs} \\ Z_{ct} \cdot \sin V_{rs} \\ Z_{ct} \cdot \cos V_{rs} \\ 1 \end{bmatrix} \tag{15.6}$$

The case of a gear hob having rake angle of a zero value $\gamma_o = 0°$ is a degenerate one. In the gear hob of such a design, the distance $p_{rg}$ reduces to zero ($p_{rg} = 0$).

For gear hobs with a negligibly small pitch lead angle $\lambda_{rs}$ of the rake surface $R_s$, the rake surface is often shaped in the form of a plane. In this particular case of gear hob design, the equation of the rake surface is described in matrix form:

$$\mathbf{r}_{rs}(U_{rs}, X_{ct}) = \begin{bmatrix} X_{ct} \\ \dfrac{\sin\gamma_o \cdot \cos(\gamma_o + U_{rs})}{\cos U_{rs}} \\ \dfrac{\sin\gamma_o \cdot \sin(\gamma_o + U_{rs})}{\cos U_{rs}} \\ 1 \end{bmatrix} \tag{15.7}$$

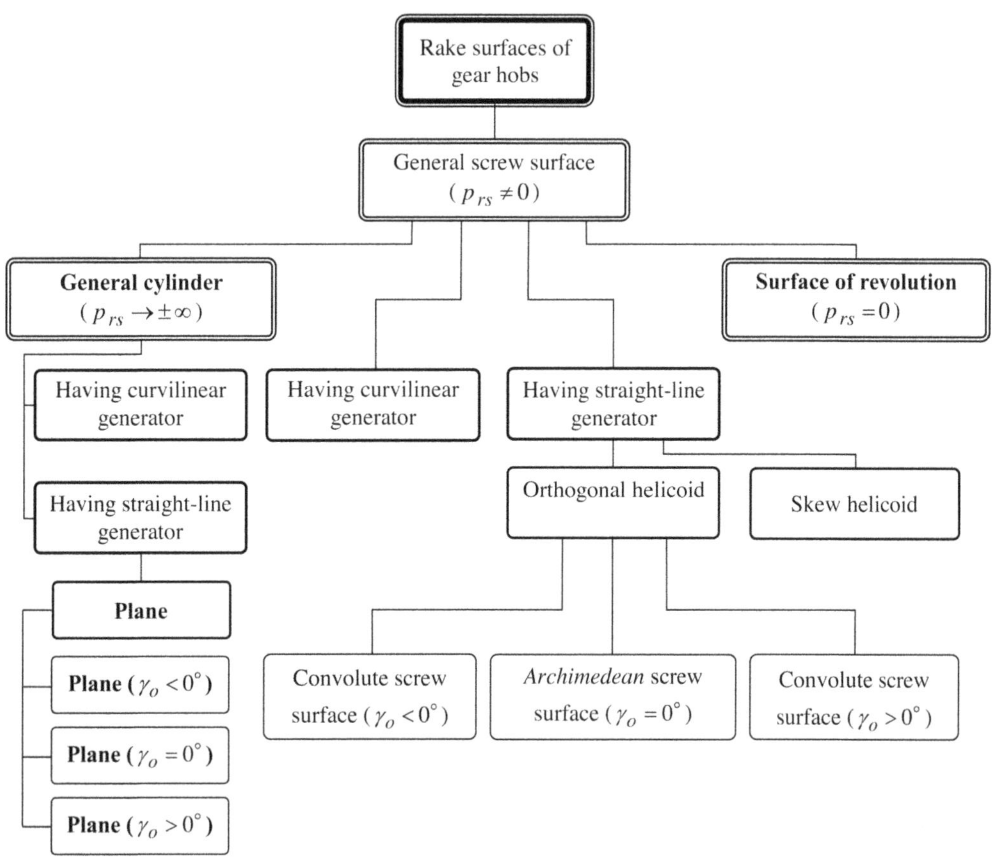


**FIGURE 15.10** Kinds of rake surfaces of a gear hob. (Adapted from: Radzevich, S.P., *SME Journal of Manufacturing Processes*, 9(2), 121–136, 2007. With permission.)

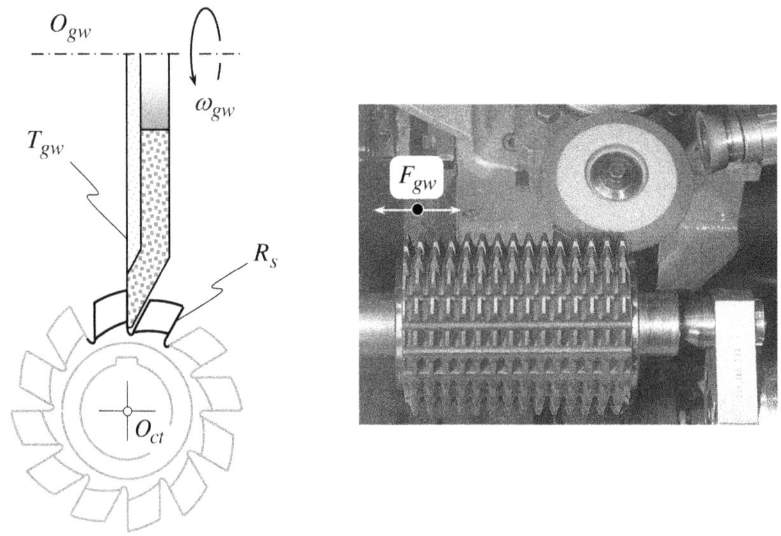


**FIGURE 15.11** Generation of the rake plane $R_s$ of a gear hob by a planar generating surface $T_{gw}$ of the grinding wheel.

The case considered above is also a degenerate one. This is because the screw parameter of the plane approaches infinity ($p_{rs} \to \infty$).

Ultimately, the rake angle $\gamma_o$ of the hob with a rake surface $R_s$ shaped in the form of a plane can also have a zero value ($\gamma_o = 0°$). This yields further simplification of the equation of the rake surface:

$$\mathbf{r}_{rs}(U_{rs}, X_{ct}) = \begin{bmatrix} X_{ct} \\ 0 \\ Z_{ct} \\ 1 \end{bmatrix} \tag{15.8}$$

It is important to note that the equation of the rake surface $R_s$ of any permissible geometry allows for representation in the form $\mathbf{r}_{rs} = \mathbf{r}_{rs}(U_{rs}, X_{ct})$. This form of representation of the rake surfaces is convenient for analyzing the geometry of gear hobs.

## 15.2.2 Generation of Rake Surface

Depending on the geometry of the actual rake surface of a hob, several methods for the generation of the surface $R_s$ are used. A comprehensive investigation of methods for generating screw surfaces is not a subject of this monograph. Therefore, only a few practical methods of generation of the rake surfaces are considered below as illustrative examples. The methods of rake surface generation are considered from the kinematic and geometric perspective.

### 15.2.2.1 Generation of Rake Surface in the Form of a Plane

The rake surface of the simplest geometry of the hob tooth is shaped in the form of a plane. The plane can be generated by means of another plane, which is congruent to the plane $R_s$ as depicted in Figure 15.11. Generation of the rake plane $R_s$ of the gear hob tooth is performed by means of the grinding wheel. The grinding wheel rotates $\omega_{gw}$ about its axis of rotation $O_{gw}$. The working plane $T_{gw}$ of the grinding wheel (in other words, the generating surface $T_{gw}$ of the grinding wheel) is perpendicular to the axis of rotation $O_{gw}$. The setup parameters of the grinding wheel are assigned so as to make the generating plane $T_{gw}$ congruent to the rake plane $R_s$. While rotating, the grinding wheel is reciprocating $F_{gw}$ in the axial direction of the hob being ground. In this way, the rake plane $R_s$ of the gear hob teeth is generated. The method is suitable for grinding gear hobs with a positive ($\gamma_o > 0°$) as well as with a negative ($\gamma_o < 0°$) rake angle $\gamma_o$. The method is also suitable for grinding gear hobs with a rake angle of zero ($\gamma_o = 0°$).

When grinding a hob on the rake surface $R_s$, the grinding wheel can be configured in relation to the hob so that its axis of rotation $O_{gw}$ is positioned vertically, as illustrated in Figure 15.12.

**FIGURE 15.12** Grinding of a hob with straight gashes on face surface $R_s$ by the grinding wheel with a vertical axis of rotation.

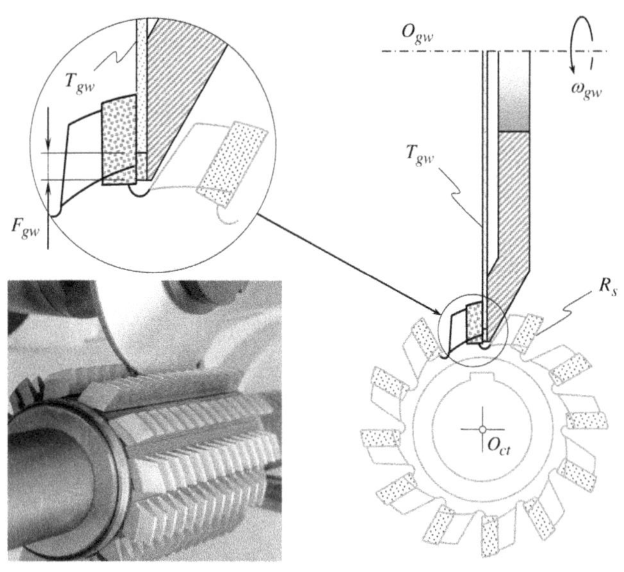


**FIGURE 15.13** Generation of rake plane $R_s$ by a narrow portion of plane of the generating surface $T_{gw}$ of the grinding wheel.

Geometrically, generation of the rake plane $R_s$ can be performed with a narrow strip of the generating surface $T_{gw}$ of the grinding wheel (see Figure 15.13). The width $F_{gw}$ of the working portion of the generating surface $T_{gw}$ is significantly smaller than the hob tooth height. This specific method of rake plane generation is used for grinding gear hobs featuring cutting elements made up of carbide alloys, ceramics, superhard alloys, and so forth.

### 15.2.2.2 Generation of Screw Rake Surface

For the purposes of generating a screw rake surface of a gear hob, a grinding wheel with a conical generating surface $T_{gw}$ is commonly used. The grinding wheel rotates $\omega_{gw}$ about its axis of rotation $O_{gw}$ (see Figure 15.14). The setup parameters of the grinding wheel are assigned so as to align the straight generating line of the conical generating surface $T_{gw}$ with the straight generating line of the screw rake surface $R_s$. While rotating, the grinding wheel reciprocates in the axial direction of the hob being ground (the reciprocation of the grinding wheel is not shown in Figure 15.14). The hob is spinning about its axis $O_{ct}$. The reciprocation of the grinding wheel is timed with the spinning of the hob. In this way, the screw rake surface $R_s$ of the hob teeth is generated.

In order to fulfil the set of necessary conditions of proper part surface generation [136, 148, 158], the setup parameters of the grinding wheel should have appropriate values. This problem can be solved using the descriptive geometry–based method (*DG – based method*). Then, on the premise of the derived *DG* – based solution to the problem, corresponding analytical expressions can be derived to determine the setup parameters of the grinding wheel.

Consider a screw rake surface $R_s$ of the hob having an outer diameter $d_{o.ct}$. In Figure 15.15, the screw rake surface $R_s$ is depicted in the system of planes of projection $\pi_1\pi_2\pi_3$. Assume that, in general, the surface $R_s$ is not an orthogonal helicoid but a skew helicoid, which features the straight generating line $L$ at an angle $\psi_{r.s}$ with respect to the axis $O_{ct}$. For the generation of the screw rake surface $R_s$, a conical grinding wheel of a diameter $D_{gw}$ applies. The generating surface $T_{gw}$ of the grinding wheel features a cone angle $\theta_{gw}$. It is necessary to determine the crossing angle $\delta_{gw}$ between the axis of rotation $O_{gw}$ of the grinding wheel and the axis of rotation $O_{ct}$ of the gear hob.

All straight generating lines of the rake surface $R_s$ are offset from the axis $O_{ct}$ at a distance $r_\gamma$. The distance $r_\gamma$ equals

$$r_\gamma = \frac{d_{o.c}}{2} \cdot \sin \gamma_o \tag{15.9}$$

This means that all straight generating lines of the rake surface $R_s$ are tangent to the round cylinder of a radius $r_\gamma$. The cylinder is shown in Figure 15.15.

A plane $P$ is tangent to the cylinder of radius $r_\gamma$. For convenience, the plane $P$ is chosen so that it is positioned parallel to the plane of projection $\pi_2$. An arbitrary point $a$

**FIGURE 15.14** Generation of a screw rake surface $R_s$ of a gear hob by conical generating surface $T_{gw}$ of the grinding wheel.

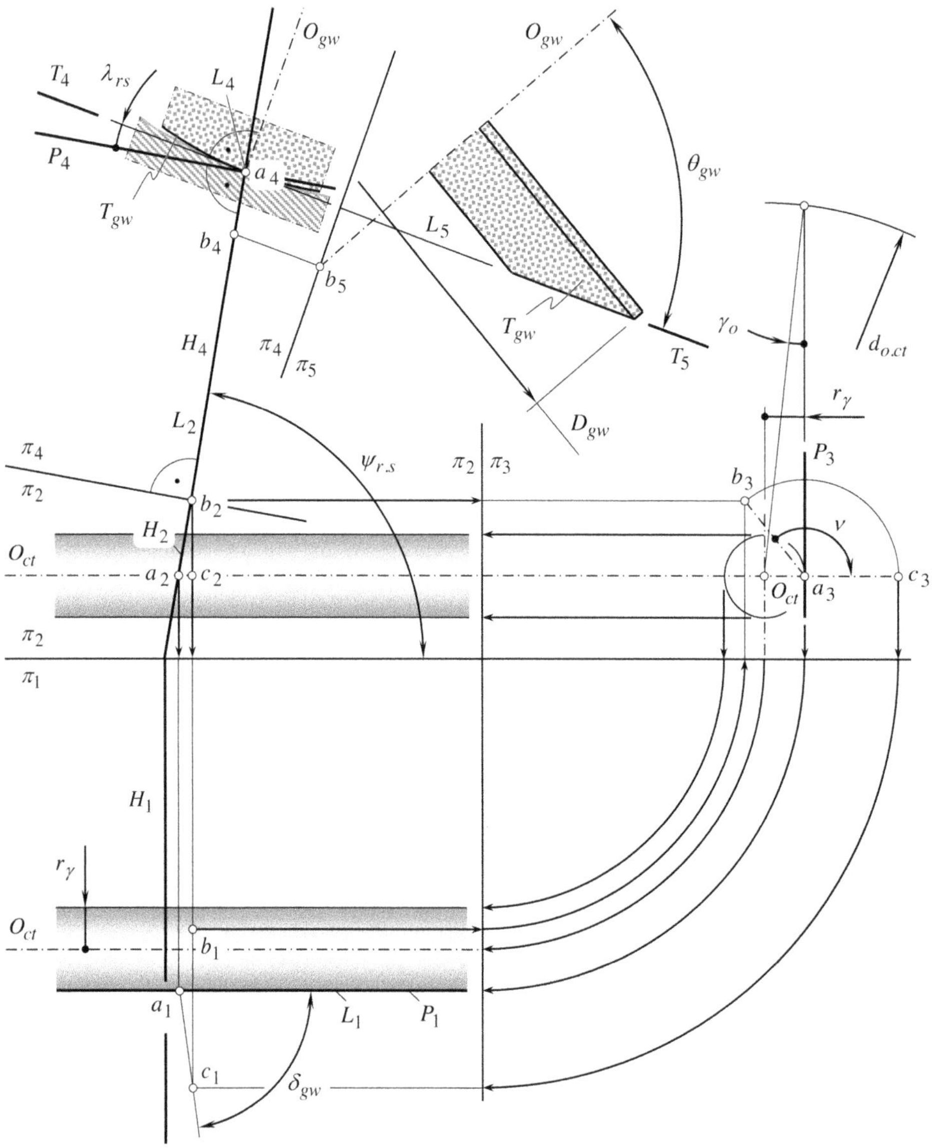


**FIGURE 15.15** Determination of setup parameters of conical grinding wheel for generation of screw rake surface $R_s$ of gear hob.

is located within the line of tangency of the cylinder and of the plane $P$. The straight generating line $L$ of the rake surface $R_s$ is the line through the point $a$ which is located within the plane $P$.

A plane $H$ passes through the straight line $L$ and is perpendicular to the plane of projection $\pi_2$. It forms the angle $\psi_{r.s}$ with the gear hob axis of rotation $O_{ct}$. The auxiliary axis of projection $\pi_2/\pi_4$ is perpendicular to the trace $H_2$ of the plane $H$. A plane $T$ is constructed so that it passes through the straight line $L$ and forms an angle $\lambda_{rs}$ with the plane $P$. On the plane of projection $\pi_4$, the plane $T$ is projected into the trace $T_4$. This allows for construction of the actual value of the angle $\lambda_{rs}$, and in this way constructs the trace $T_4$.

The auxiliary axis of projection $\pi_4/\pi_5$ is constructed so that it is perpendicular to the trace $T_4$. Onto the plane of projection $\pi_4$, the straight line $L$ is projected into the point $L_4$, whereas onto the plane of projection $\pi_5$, the straight line $L$ is projected with no distortion. In the plane of projection $\pi_5$, the projection $L_5$ aligns with the trace $T_5$.

Within the plane of projection $\pi_5$, the actual value of the cone angle $\theta_{gw}$ can be constructed. In this way, the configuration of the axis of rotation $O_{gw}$ of the grinding wheel is determined. The axis of rotation of the grinding wheel $O_{gw}$ intersects the axis of projection $\pi_4/\pi_5$ at point $b$. Point $b$ is constructed with the goal to specify the configuration of the axis $O_{gw}$ within the rest of the planes of projection.

Ultimately, spinning the projection $a_3b_3$ about the axis $O_{ct}$ through the angle $\nu$, the actual value of the crossing angle $\delta_{gw}$ can be constructed within the horizontal plane of projection $\pi_1$.

In most practical cases of generating the screw rake surface $R_s$ of the gear hob, the angle $\psi_{r.s}$ is equal to a right angle. The determination of the crossing angle $\delta_{gw}$ is simplified in cases when the equality $\psi_{r.s} = 90°$ is valid.

The method of generating the screw rake surface $R_s$ with a conical grinding wheel is suitable for grinding gear hobs with a positive ($\gamma_o > 0°$) and a negative ($\gamma_o < 0°$) rake angle $\gamma_o$, as well as for grinding gear hobs with a zero rake angle ($\gamma_o = 0°$).

#### 15.2.2.3 Peculiarities of Generation of Screw Rake Surface of Multiple-Start Gear Hob

Grinding wheels with a conical generating surface $T_{gw}$ can be used for generating screw rake surfaces of gear hobs either when the lead angle of the rake surface $R_s$ is reasonably small or when the accuracy of the gear hob is not a critical issue (that is, as it is observed for roughing gear hobs). This is due to the fact that the conical generating surface of the grinding wheel is capable of generating screw rake surface $R_s$ of the hob only approximately. Therefore, the screw rake surface that is actually generated features not a straight generating line, but a curved generating line (see Figure 15.16). The proof can be found in [138]. For this purpose, *DG* – based methods of analysis are used. After being generated by means of the conical generating surface $T_{gw}$, the profile of the transverse plane section of the screw rake surface $R_s$ becomes convex. The convexity of the transverse plane section of the rake surface $R_s$ is mostly due to the fact that the surfaces $R_s$ and $T_{gw}$ make line contact. The line of contact of the surfaces $R_s$ and $T_{gw}$ is not a planar curve, but a spatial 3*D* curve. Ultimately, the peculiarities of the geometry of the line of contact result in the screw rake surface $R_s$ of the hob deviating from what was desired.

Deviations of the actual parameters of the screw rake surface $R_s$ from its desired shape and geometry cause corresponding deviations in the tooth profile of the gear, of which the hob is used for machining.

When grinding a precision gear hob, especially when grinding a multiple-start gear hob, the deviations of the actually ground rake surface from the desired screw surface $R_s$ can exceed the tolerance for its shape and the parameters of the geometry. This particular source of inaccuracies of the rake surface $R_s$ can be eliminated if the screw rake surface of the gear hob teeth is generated by the grinding wheel having a form axial profile (see Figure 15.17). For a specific hob design, the generating surface of the form grinding wheel $T_{gw}$ can be found as an envelope for consecutive positions of the desired screw rake surface $R_s$ in its rotation about the axis of rotation of the grinding wheel. For the derivation of an equation of the generating surface $T_{gw}$ of the form grinding wheel, the kinematic method for determining enveloping surfaces can be used [204, 206].

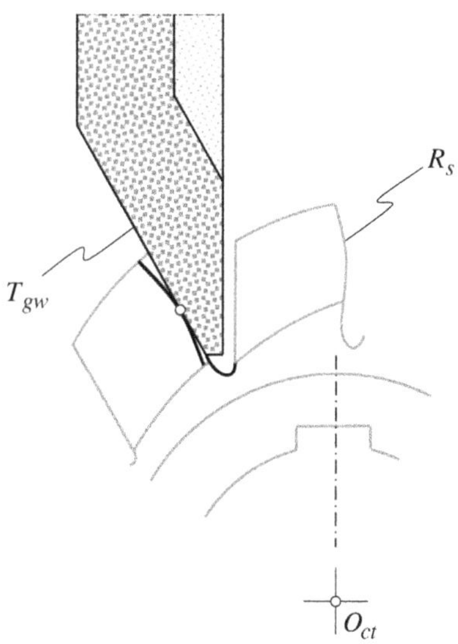


**FIGURE 15.16** Deviation of the transverse profile of a screw rake surface $R_s$ of a multiple-start gear hob.

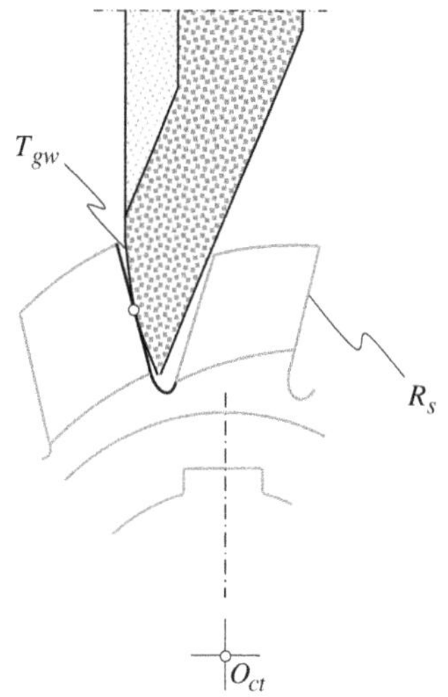


**FIGURE 15.17** Generation of a screw rake surface $R_s$ of a multiple-start hob by the form grinding wheel.

Then the desired axial profile of the form grinding wheel can be derived as the axial plane section of the surface $T_{gw}$. The axial profile can be used for dressing the form grinding wheel.

#### 15.2.2.4 Methods for Generating the "Intermittent Rake Surface" of a Special-Purpose Gear Hob

The methods considered for the generation of rake surfaces are applicable in cases when the rake surface of each tooth within a gash of the gear hob is a portion of a continuous (common) surface. The latter can be shaped either in the form of a common plane or in the form of a common screw surface. The rake surfaces of more complex geometries are used in the design of gear hobs as well.

Special-purpose gear hobs may have rake surfaces which do not belong to a common continuous surface. Instead, the rake surface of each tooth can be shifted in relation to one another along the pitch helix of the hob. Rake surfaces of this type are referred to as *intermittent rake surfaces* of the gear hob tooth. Special methods are developed for grinding the special-purpose gear hobs featuring intermittent rake surfaces. The methods make it possible to separately grind the rake plane of every single tooth of the gear hob.

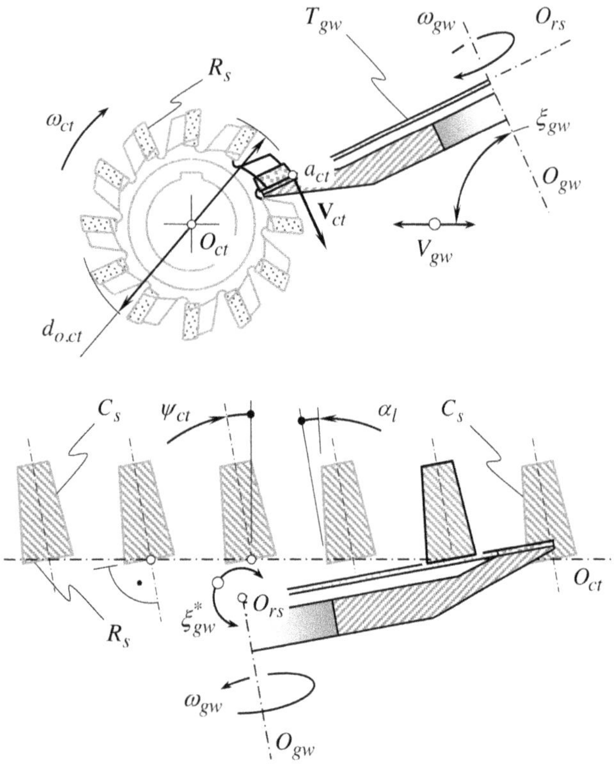


**FIGURE 15.18** Generation of an intermittent rake surface $R_s$ of a special-purpose gear hob.

As an example, the kinematics of a method[2] for generating an intermittent rake surface $R_s$ of a special-purpose gear hob is schematically illustrated in Figure 15.18 [86]. The hob being ground features the rake surfaces of each tooth that are shifted along the pitch helix with respect to one another at a certain distance. Owing to this, the rake surfaces that belong to a common gash of the hob are not situated within a common continuous surface, but represent the intermittent surface. When grinding, the hob rotates about its axis of rotation $O_{ct}$, as shown in Figure 15.18 [86]. The grinding wheel with the generating plane $T_{gw}$ is used for grinding the hob. The grinding wheel rotates $\omega_{gw}$ about its axis of rotation $O_{gw}$. The configuration of the axis of rotation $O_{gw}$ in relation to the gear hob is specified by two angles $\xi_{gw}$ and $\xi^*_{gw}$.

When grinding a rake plane $R_s$, the hob rotates with an angular velocity $\omega_{ct}$ of a constant value. The grinding wheel reciprocates back and forth with respect to the hob. The grinding wheel reciprocation features:

- Slow approaching to the gear hob.
- Instant interaction of the surfaces $R_s$ and $T_{gw}$.
- Fast departure of the grinding wheel from the outside of the gear hob.

In order to avoid the collision of the grinding wheel and the gear hob, the speed of the departure $V_{gw}$ and the angle

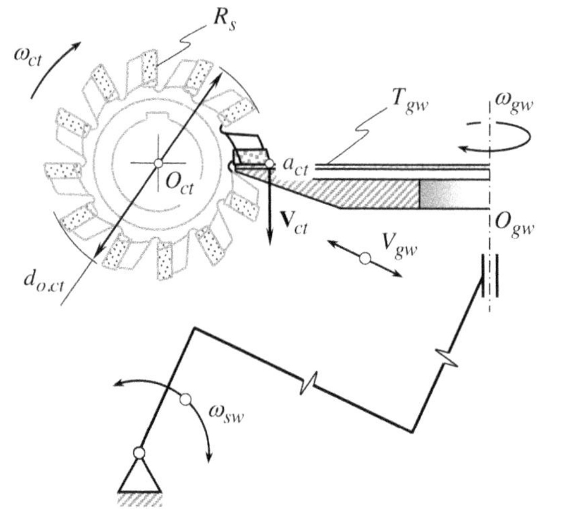


**FIGURE 15.19** Method of generating the intermittent rake surface $R_s$ of the special-purpose gear hob; the method features swivel motion $\omega_{sw}$ of the grinding wheel.

$\xi_{gw}$ have to be chosen so as to fulfill the following inequality (see Figure 15.18):

$$V_{gw} \geq |\mathbf{V}_{ct}| = 0.5 \cdot \omega_{ct} \cdot d_{o.ct} \cdot \cos \xi_{gw} \tag{15.10}$$

where:

- $\mathbf{V}_{ct}$ is the vector of linear velocity of rotation of the point, $a_{ct}$, about the hob axis of rotation.
- $d_{o.ct}$ is the outer diameter of the hob.
- $\xi_{gw}$ is an angle at which the axis of the grinding wheel is tilted in transverse cross-section of the hob.

In addition to the tilt angle $\xi_{gw}$, the axis of rotation of the grinding wheel $O_{gw}$ is turned about the axis $O_{rs}$ through an angle $\xi^*_{gw}$. The axis $O_{rs}$ is positioned perpendicular to the gear hob axis of rotation $O_{gw}$, and it is located within the rake plane $R_s$ of the gear hob tooth. In particular, the angle $\xi^*_{gw}$ could have the value $\xi^*_{gw} = \psi_{ct}$ under which the rake plane is perpendicular to the pitch helix of the hob. Here $\psi_{ct}$ denotes the pitch helix angle of the gear hob.

A method of grinding gear hobs similar to that discussed above is disclosed in [190]. The method[3] features a swivel motion $\omega_{sw}$ of the grinding wheel instead of a reciprocating motion. The method of rake surface generation is illustrated in Figure 15.19. The setup parameters of the grinding wheel and the motions $\omega_{ct}$ and $\omega_{sw}$ are determined so as to fulfil the inequality

$$|\mathbf{V}_{ct}| \leq pr_{\mathbf{V}ct} V_{gw} \tag{15.11}$$

where $pr_{\mathbf{V}ct} V_{gw}$ designates the projection of the vector $V_{gw}$ onto the direction of the vector $\mathbf{V}_{ct}$.

[2] SU Pat. No. 975339, *A Method of Grinding of Gear Hob*/S.P. Radzevich, Int. Cl. B24b 3/12, Filed: October 10, 1980.

[3] SU Pat. No. 1627375, *A Method of Grinding a Gear Hob*/S.P. Radzevich, I.V. Vinokurvov, Int. Cl. B24b 3/12, Filed: October 17, 1988.

Much room is available for further improvements in the area of grinding hobs that feature continuous as well as intermittent rake surfaces.

## 15.3 GEOMETRY AND GENERATION OF CLEARANCE SURFACES OF GEAR HOBS

Once the rake surface of the hob is generated, then the clearance surface of the hob tooth can be generated as well. As illustrated in Figure 15.20, the rake surface $R_s$ intersects the generating surface $T$ of the hob. Geometrically, the cutting edge is considered here as the line of intersection of the surface $T$ by the surface $R_s$. The desired clearance surface $C_s$ is a surface through the cutting edge $CE$, which forms clearance angle $\alpha$ of the required value with the surface of the cut.

Implementation of the surface of cut is inconvenient for the analysis below. The generating surface $T$ of the hob is a reasonable approximation to the surface of cut in the vicinity of the cutting edge $CE$. Therefore, approximately (however, with a sufficient degree of accuracy) the surface of the cut is substituted with the generating surface of the hob. Ultimately, the clearance surface $C_s$ is viewed as a surface through the cutting edge $CE$, which forms clearance angle $\alpha$ of the required value with the generating surface $T$ (and not with the surface of cut). In the ideal case, the cutting edge $CE$ is the line of intersection of three surfaces, that is, of the surfaces $T$, $R_s$, and $C_s$. It is of critical importance to notice here that the latter requirement is vital for new gear hobs as well as for reground hobs.

### 15.3.1 Equation of Desired Clearance Surface of Gear Hob Tooth

Three methods for the transformation of the generating body of the gear-cutting tool into a workable cutting tool are known [148, 158]. Below, the most common method for the transformation of the generating body of the gear-cutting tool into a workable cutting tool is followed. This approach features the clearance surface that passes through the cutting edge at the desired clearance angle. The cutting edge $CE$ is considered here as the line of intersection of generating surface $T$ of the hob by the rake surface $R_s$. Then for the derivation of an equation of the clearance surface $C_s$, a motion of the cutting edge $CE$ with respect to the generating surface $T$ is considered. The direction of this motion forms the desired clearance angle $\alpha$ with the surface $T$. Ultimately, the clearance surface of the hob tooth is represented as the set of consecutive positions of the cutting edge $CE$ when it performs the motion described above.

The geometry of the desired clearance surface $C_s$ of the involute gear hob tooth is illustrated in Figure 15.21. It is assumed below that the generating surface $T$ of the gear hob is determined, and the rake surface $R_s$ is specified. For simplicity, but without loss of generality of the consideration, the rake plane $R_s$ through the axis of rotation $O_{ct}$ of the hob is considered as an illustrative example.

The rake plane $R_s$ of the new gear hob intersects the generating surface $T$ of the hob. In Figure 15.21, the line of the intersection is denoted by $N$. The line $N$ starts at point $n$ within the base cylinder of a diameter $d_{b.ct}$ of the gear hob and goes outward toward the hob axis of rotation $O_{ct}$. Different cylinders of the new hob are intersected by the line $N$ at the following points:

- The cylinder of an outer diameter $d^n_{o.ct}$ at point $a$.
- The pitch cylinder of a diameter $d^n_{ct}$ at point $b$.
- The cylinder of limit diameter $d^n_{l.ct}$ at the point $c$.

The arc $abc$ of the intersection curve $N$ is chosen as the cutting edge $CE^n$ of the new gear hob.

Each time after the gear hob is reground, the diameters of the hob get smaller and smaller. This is due to the need for a clearance angle that produces a good cutting performance. Ultimately, after the last regrind, the diameters of the fully worn hob are designated in Figure 15.21 as $d^w_{o.ct}$, $d^w_{ct}$, and $d^w_{l.ct}$. These diameters correspond to the diameters $d^n_{o.ct}$, $d^n_{ct}$, and $d^n_{l.ct}$ of the new hob before it has been reground.

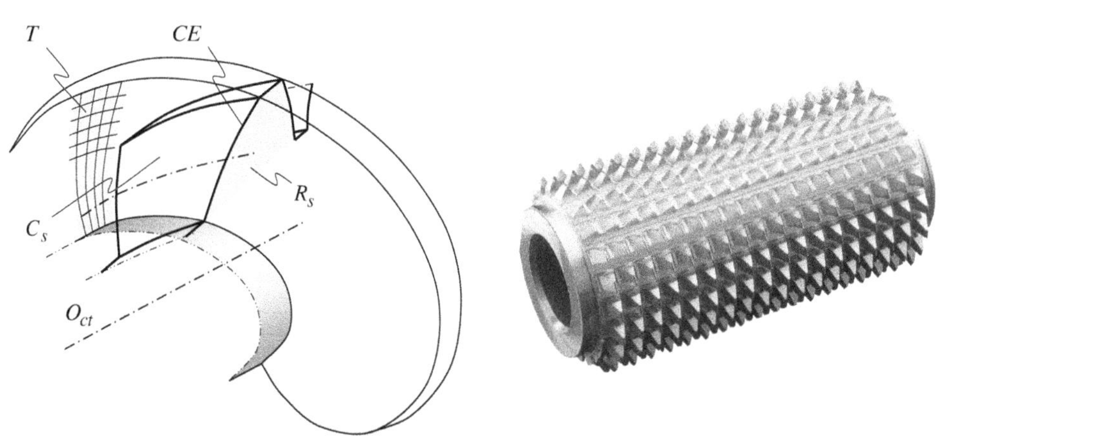


**FIGURE 15.20** Clearance surface $C_s$ of the hob tooth passing through the cutting edge $CE$ (the latter is interpreted as the line of intersection of the generating surface $T$ by the rake surface $R_s$ of the gear hob).

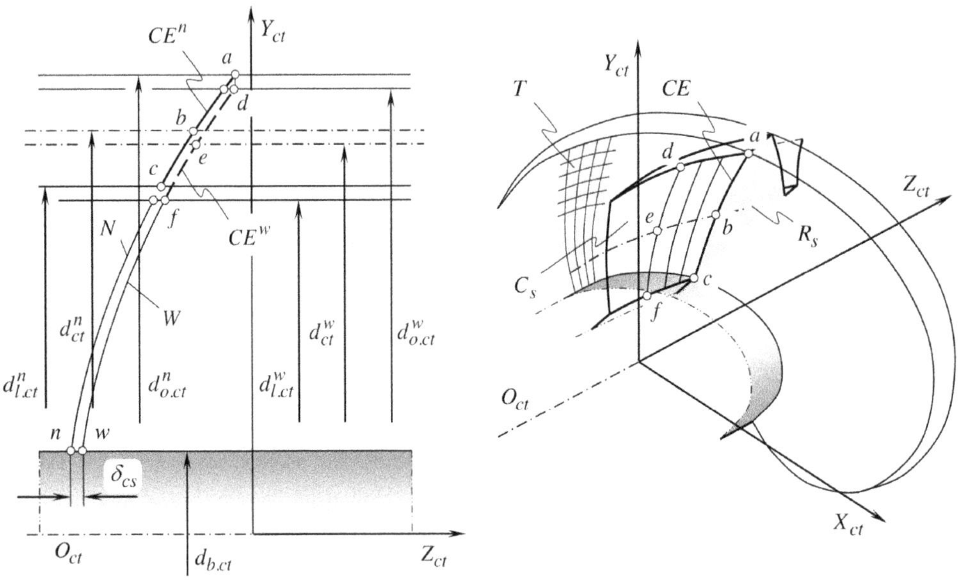


**FIGURE 15.21** Geometry of a desired clearance surface $C_s$ of a hob for machining involute gears.

The generating surface $T$ of the hob is intersected by the rake plane $R_s$ of the worn gear hob. In Figure 15.21, the line of the intersection is denoted as $W$. The line $W$ starts at point $w$ within the base cylinder of the diameter $d_{b.ct}$ of the hob and goes outward toward the hob axis of rotation $O_{ct}$. Different cylinders of the worn hob are intersected by the line $W$ at the following points:

- The cylinder of the outer diameter $d^w_{o.ct}$ at point $d$.
- The pitch cylinder of the diameter $d^w_{ct}$ at point $e$.
- The cylinder of the limit diameter $d^w_{l.ct}$ at the point $f$.

The arc $def$ of the intersection curve $W$ is chosen as the cutting edge $CE^w$ of the worn gear hob.

Similar to the approach above, any and all intermediate locations of the cutting edge $CE$ of the partially worn involute hob can be determined. The family of all of them is situated within the desired clearance surface $C_s$ of the involute gear hob.

The schematic discussed above (see Figure 15.21) can be employed for the purpose of deriving an equation of the clearance surface $C_s$ of the gear hob tooth.

Consider the equation [see Eq. (14.58)]

$$\left.\begin{array}{c}\text{Generating surface, } T,\\ \text{of involute gear hob}\end{array}\right| \Rightarrow \mathbf{r}_{ct}(U_{ct}, V_{ct})$$

$$= \begin{bmatrix} 0.5 \cdot d_{ct} \cdot \sin V_{ct} - U_{ct} \cdot \sin \psi_{b.ct} \cdot \cos V_{ct} \\ 0.5 \cdot d_{ct} \cdot \cos V_{ct} + U_{ct} \cdot \sin \psi_{b.ct} \cdot \sin V_{ct} \\ p_{b.ct} \cdot V_{ct} - U_{ct} \cdot \cos \psi_{b.ct} \\ 1 \end{bmatrix} \quad (15.12)$$

of the generating surface $T$ of a gear hob together with the equation

$$\left.\begin{array}{c}\text{Screw rake surface, } R_s,\\ \text{of involute gear hob}\end{array}\right| \Rightarrow \mathbf{r}_{rs}(U_{rs}, V_{rs})$$

$$= \begin{bmatrix} p_{rs} \cdot V_{rs} \\ \dfrac{\sqrt{2} \cdot p_{rg} \cdot \sin \gamma_o \cos(\gamma_o + U_{rs})}{\cos U_{rs}} \cdot \sin\left(\dfrac{\pi}{4} + V_{rs}\right) \\ \dfrac{\sqrt{2} \cdot p_{rg} \cdot \sin \gamma_o \sin(\gamma_o + U_{rs})}{\cos U_{rs}} \cdot \cos\left(\dfrac{\pi}{4} + V_{rs}\right) \\ 1 \end{bmatrix} \quad (15.13)$$

of a screw rake surface $R_s$ of the involute gear hob [see Eq. (15.5)].

The position vector of a point of the cutting edge $\mathbf{r}_{ce} = \mathbf{r}_{ce}(U_{ce})$ fulfils both Eq. (15.12) and Eq. (15.13) simultaneously. Two equations, Eq. (15.12) and Eq. (15.13), cast into an expression of the form

$$\mathbf{r}_{ce}(U_{ce}) = \begin{bmatrix} X_{ce}(U_{ce}) \\ Y_{ce}(U_{ce}) \\ Z_{ce}(U_{ce}) \\ 1 \end{bmatrix} \quad (15.14)$$

which can be derived for the calculation of the position vector $\mathbf{r}_{ce}$ of a point of the cutting edge $CE$ of an involute hob.

For the derivation of an equation of the desired clearance surface $C_s$ of the hob tooth, a screw motion of the cutting edge $\mathbf{r}_{ce}$ is considered. The reduced pitch of the screw motion of the cutting edge depends on (a) the reduced pitch $p_x$ of threads of the generating surface $T$ of the hob, and (b) the actual value of the clearance angle $\alpha$ at the lateral cutting edges of the hob tooth.

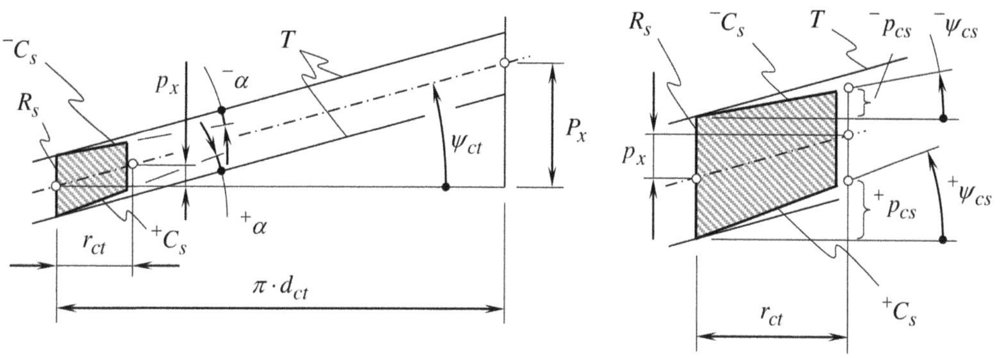


**FIGURE 15.22** Unfolded section of a thread of an involute gear hob by the pitch cylinder of a diameter $d_{ct}$.

The reduced pitch $p_x$ (the screw parameter, $p_x$) of threads of the generating surface $T$ is a signed value. It is positive for the right-handed involute hobs, and it is negative for the left-handed involute hobs. In the case under consideration, it is convenient to treat the clearance angle $\alpha$ as a signed value as well. The clearance angle $\alpha$ has a positive value for the clearance surface of the hob tooth when it increases the reduced pitch of the clearance surface (that is, it is added with the pitch helix angle $\psi_{ct}$). The clearance angle $\alpha$ has a negative value for the clearance surface of the hob tooth when it makes the reduced pitch of the clearance surface smaller (that is, it is subtracted from the pitch helix angle $\psi_{ct}$).

For further analysis, it is convenient to refer to Figure 15.22 where the unfolded section of a gear hob thread by the pitch cylinder of a diameter $d_{ct}$ is depicted. As shown in Figure 15.22, the section of the gear hob tooth is bounded by the rake surface $R_s$ and by two clearance surfaces $C_s$. The clearance surface that has a bigger reduced pitch, $^+p_{cs} > p_x$, is designated as $^+C_s$. The opposite clearance surface that features the smaller reduced pitch, $^-p_{cs} < p_x$, is designated as $^-C_s$.

The actual value of the reduced pitch $p_x$ for the involute gear hob with a given axial pitch $P_x$ of threads can be calculated from the following formula:

$$p_x = \frac{P_x}{2 \cdot \pi} \tag{15.15}$$

Pitch helix angle $\psi_{ct}$ can be expressed in terms of the axial pitch $P_x$ and of the pitch diameter $d_{ct}$ of the hob:

$$\psi_{ct} = \tan^{-1}\left(\frac{P_x}{\pi \cdot d_{ct}}\right) \tag{15.16}$$

For the clearance surface $^+C_s$, the corresponding helix angle $^+\psi_{cs}$ is equal:

$$^+\psi_{cs} = \psi_{ct} + \alpha \tag{15.17}$$

Once the helix angle $^+\psi_{cs}$ is calculated, then for the calculation of the reduced pitch $^+p_{cs}$, the following expression:

$$^+p_{cs} = 0.5 \cdot d_{ct} \cdot \tan {}^+\psi_{cs} \tag{15.18}$$

can be used.

Similarly, for the calculation of the reduced pitch $^-p_{cs}$ of the clearance surface $^-C_s$ of the opposite side of the hob tooth profile, the formula

$$^-p_{cs} = 0.5 \cdot d_{ct} \cdot \tan(^-\psi_{cs}) \tag{15.19}$$

is valid. Here, for the calculation of the helix angle $^-\psi_{cs}$, the formula

$$^-\psi_{cs} = \psi_{ct} - \alpha \tag{15.20}$$

can be used.

Clearance surface $^+C_s$ can be represented as a locus of consecutive positions of the cutting edge $CE$ [see Eq. (15.14)] when it performs the screw motion. The $Z_{ct}$ – axis of the coordinate system $X_{ct}Y_{ct}Z_{ct}$ associated with the gear hob is the axis of the screw motion, and $^+p_{cs}$ is the screw parameter (the reduced pitch) of the screw motion. Use of the operator of the linear transformation $\mathbf{Sc}_z(\varphi_{ce}, {}^+p_{cs})$ that analytically describes the screw motion [see Appendix D]:

$$\mathbf{Sc}_z(\varphi_{ce}, {}^+p_{cs}) = \begin{bmatrix} \cos\varphi_{ce} & \sin\varphi_{ce} & 0 & 0 \\ -\sin\varphi_{ce} & \cos\varphi_{ce} & 0 & 0 \\ 0 & 0 & 1 & {}^+p_{cs} \cdot \varphi_{ce} \\ 0 & 0 & 0 & 1 \end{bmatrix} \tag{15.21}$$

makes possible the following expression for the position vector $^+\mathbf{r}_{cs}$ of a point of the clearance surface $^+C_s$:

$$^+\mathbf{r}_{cs}(\varphi_{ce}, U_{ce}) = \mathbf{Sc}_z(\varphi_{ce}, {}^+p_{cs}) \cdot \mathbf{r}_{ce}(U_{ce}) \tag{15.22}$$

An equation for the position vector of a point $^-\mathbf{r}_{cs}$ of the clearance surface $^-C_s$ can be derived in a similar manner to how Eq. (15.22) is derived.

In practice, for many reasons, the actual clearance surface $C_s$ of the hob tooth deviates from the desired clearance surface. The desired clearance surface [see Eq. (15.22)] serves as the reference surface for the calculation of the deviations of the actual clearance surface from the desired clearance surface of the hob tooth.

### 15.3.2 Generating the Clearance Surface of the Gear Hob Tooth

Clearance surfaces of the hob teeth are commonly machined in a specific operation, which is often referred to as the *tooth relieving operation*. Relieving of the hob teeth is executed in the green (soft) stage of the gear hob and in the hard stage after heat treatment of the hob. Form tools are used in the first case, and grinding wheels are used in the second case.

#### 15.3.2.1 Cutting Clearance Surface of Gear Hob Teeth

The relieved clearance surface is a practical type of approximation of the desired clearance surface of the gear hob tooth. A properly relieved hob (a) features a clearance angle of a positive value, which is more or less close to its optimal value, and (b) has deviations of the cutting edges from their desired geometry within the tolerance for the accuracy of the gear hob.

Cutting of the clearance surface of gear hob teeth resembles the previously discussed cutting of the clearance surface of gear teeth in disk-type mill cutters (see Figure 6.13). Therefore, before we continue the discussion on cutting the clearance surface of gear hob teeth, it makes sense to refresh readers' memory on cutting the clearance surface of disk-type mill cutters for machining spur and helical gears (see Chapter 6 on details of cutting the clearance surface of disk-type mill cutters).

Cutting of relieved clearance surfaces of the gear hob teeth is used in two cases: (a) as a preliminary roughing operation of machining the clearance surfaces, and/or (b) as a finishing operation of machining the hob teeth. Basic methods of cutting the relieved clearance surface of the hob tooth are depicted in Figure 15.23.

When cutting the clearance surface $C_s$, the hob rotates $\omega_{ct}$ about its axis of rotation $O_{ct}$. The direction of the rotation is chosen so that the rake surface $R_s$ faces toward the rotation $\omega_{ct}$. The form cutter is reciprocated toward the axis of rotation $O_{ct}$. In Figure 15.23, the reciprocation motion of the cutter is denoted as $\mathbf{V}_{rc}$. In addition to the reciprocation motion $\mathbf{V}_{rc}$, the cutter travels $\mathbf{V}_{sc}$ in the axial direction of the hob. This motion $\mathbf{V}_{sc}$ of the cutter together with the rotation $\omega_{ct}$ provide the screw motion of the cutter with respect to the clearance surface $C_s$ to be cut.

The form cutter is designed so that it can cut either space between two adjacent hob teeth (case *I* in Figure 15.23), or cut the clearance surfaces of the opposite sides of the hob tooth (case *II* in Figure 15.23).

In the production of coarse pitch hobs, two more methods of cutting the clearance surfaces of the gear hob teeth are used as well. The first method features the reciprocation motion $\mathbf{V}_{rc}$ of the form cutter pointed at a certain angle with respect to the axis of rotation $O_{ct}$ of the hob. A schematic of the method is illustrated in Figure 15.23 (case *III*). The reciprocation motion $\mathbf{V}_{rc}$ in the second method of cutting the surfaces $C_s$ is parallel to the axis of rotation $O_{ct}$ of the hob (case *IV* in Figure 15.23).

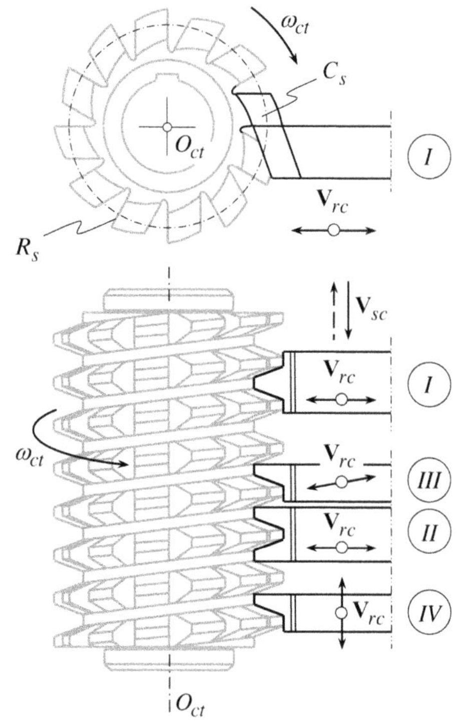


**FIGURE 15.23** Basic methods of cutting relieved clearance surface $C_s$ of gear hob tooth.

When the second method of cutting relieved surfaces of gear hob teeth is used, the gear hob manufacturer is often faced with the need to solve two problems.

The first problem is mostly due to the limited room between the adjacent teeth of the hob for the form cutter to be used.

The second problem is due to the following: The top land of the gear hob tooth is machined by the bottom-land cutting edge of the form cutter, and the bottom land of the hob tooth is machined by the top-land cutting edge of the form cutter. The distance between the top-land cutting edge and the bottom-land cutting edge of the form cutter must be chosen to be equal to half of the difference between the outer diameter $d^n_{o.ct}$ of the new hob and the smallest inner diameter $d^{\min}_{f.ct}$ of the hob. This means that the tooth height within the rake surface of the form cutter must be equal to

$$h_{f.ct} = \frac{d^n_{o.ct} - d^{\min}_{f.ct}}{2} \tag{15.23}$$

##### 15.3.2.1.1 *Elements of Geometry of Cutting Edge (the Clearance Surface of Relieved Hob Tooth)*

Normal clearance angle $\alpha_n$ and the clearance angle $\alpha_o$ that is measured in a section of the cutting edge *CE* by a plane that is perpendicular to the axis of rotation of the gear hob are the two major elements of the cutting edge geometry that depend on the geometry and the configuration of the clearance surface $C_s$ of the relieved hob tooth.

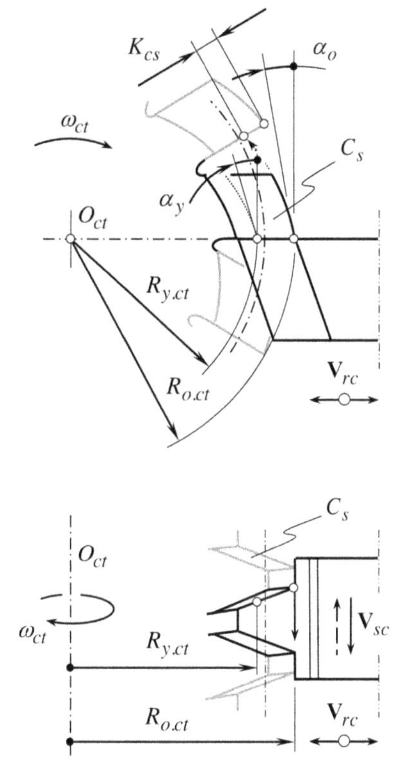


**FIGURE 15.24** Determination of major parameters of relieving operation in terms of the clearance angle $\alpha_o$ at the top cutting edge of gear hob tooth.

Without going into details of derivation, for gear hobs the clearance surface of which is shaped by means of an Archimedean spiral, Figure 15.24 yields an expression

$$\alpha_o = \tan^{-1}\left(\frac{K_{cs} \cdot Z_{gashes}}{2 \cdot \pi \cdot R_{o.ct}}\right) \tag{15.24}$$

for the calculation of the clearance angle $\alpha_o$.

Equation (15.24) reveals that the actual value of the clearance angle $\alpha_o$ depends on

- The outer diameter of the hob ($d_{o.ct} = 2 \cdot R_{o.ct}$).
- The number of gashes, $Z_{gashes}$. (For hobs featuring straight gashes, the parameter $Z_{gashes}$ is expressed in integer numbers; for hobs with helical gashes, the parameter $Z_{gashes}$ is expressed in numbers with fractions.)
- The parameter $K_{cs}$ of relieving (see Figure 15.24).

In practice, the calculations performed in the inverse order are important. Instead of the calculation of the clearance angle $\alpha_o$, calculation of the required value of the parameter $K_{cs}$ of relieving are performed:

$$K_{cs} = \frac{\pi \cdot d_{o.ct}}{Z_{gashes}} \cdot \tan\alpha_o \tag{15.25}$$

In this latter case, the value of the clearance angle $\alpha_o$ should be specified. For solid gear hobs (that is, for those made up of one piece of cutting material), the practically reasonable values of the clearance angle are within the interval $\alpha_o = 10...12°$.

Normal clearance angle $\alpha_n$ at the lateral cutting edge of the hob tooth depends on clearance angle $\alpha_o$ at the top cutting edge of the hob. This feature is inherited to relieved gear hobs of all designs. The actual value of the normal clearance angle $\alpha_n$ of gear involute hobs (featuring the rake angle $\gamma_o$ of a zero value) can be calculated from the following formula:

$$\alpha_n = \tan^{-1}\left(\tan\alpha_o \cdot \sin\phi_{rs}\right) \tag{15.26}$$

Here, in Eq. (15.26), $\phi_{rs}$ denotes a profile angle that is measured within the rake plane of the hob tooth (or, within a plane tangent to the rake surface of the hob tooth).

Gear hobs for machining involute gears feature the clearance angle $\alpha_o$, the actual value of which is almost of a constant value within the perimeter of the lateral cutting edge ($\alpha_o \cong const$). Only for coarse pitch hobs of small diameter, the variation of the clearance angle $\alpha_o$ within the lateral cutting edge could be of importance. For the calculation of the clearance angle $\alpha_y$ at an arbitrary point within the lateral cutting edge of the hob, the formula:

$$\alpha_y = \tan^{-1}\left(\frac{R_{o.ct}}{R_{y.ct}} \cdot \tan\alpha_o\right) \tag{15.27}$$

is commonly applied. Here in Eq. (15.27), the radius of a cylinder through an arbitrary point $m$ of the lateral cutting edge is denoted by $R_{y.ct}$ (see Figure 15.24).

Considered together, Eq. (15.26) and Eq. (15.27) allow for an expression

$$\alpha_{n.y} = \tan^{-1}\left(\frac{R_{o.ct}}{R_{y.ct}} \cdot \tan\alpha_o \cdot \sin\phi_{rs}\right) \tag{15.28}$$

for the calculation of normal clearance angle $\alpha_{n.y}$ at an arbitrary point $m$ within the perimeter of the lateral cutting edge $CE$ of the gear hob.

While the variation of the profile angle $\phi_{rs}$ along the lateral cutting edge of the gear hob for machining an involute gear is negligibly small and thus can be neglected, variation of the angle $\phi_{rs}$ is of critical importance for gear hobs for machining splines, and hobs for machining form profiles [for example, hobs for machining gears for conformal gear pairs (Novikov gearing), high-conformal gear pairs, sprockets, ratchets, etc.]. In this latter case, the angle $\phi_{rs}$ is measured within the rake plane of the gear hob tooth (it is assumed here that the equality $\gamma_o = 0°$ is valid) between the tangent through the point of interest $m$ within the cutting edge $CE$ and the perpendicular to the hob axis of rotation $O_{ct}$.

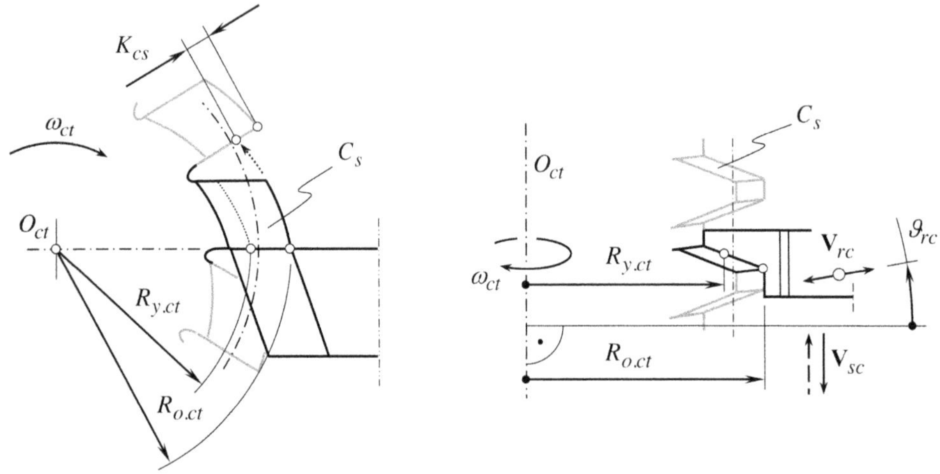


**FIGURE 15.25** More elements of geometry of cutting edge of relieved gear hob tooth.

Gear hobs with a clearance angle $\alpha_o = 10...12°$ feature very small normal clearance angles $\alpha_n$. Commonly, the actual value of the normal clearance angle at the lateral cutting edge of the gear hob tooth falls into the interval $\alpha_n = 2...4°$. Cutting performance of the gear hobs can be significantly improved by means of an increased actual value of the normal clearance angle $\alpha_n$.

The method for backing off a gear hob that features the reciprocation motion $\mathbf{V}_{rc}$ of the form cutter pointed at a certain angle relative to the axis of rotation $O_{ct}$ of the hob (see Figure 15.23, case *III*) is helpful for increasing the actual value of the normal clearance angle $\alpha_n$. When the reciprocation motion $\mathbf{V}_{rc}$ is at a certain angle $\vartheta_{rc}$ as shown in Figure 15.25, then the normal clearance angle $\alpha_{n.y}$ can be calculated from the formula

$$\alpha_{n.y} = \frac{K_{cs} \cdot Z_{gashes}}{2 \cdot \pi \cdot R_{y.ct}} \cdot \sin(\phi_{rs} + \vartheta_{rc}) \tag{15.29}$$

When hobbing splines and/or when machining parts with a form profile, it is of critical importance to calculate a reasonable value of the angle $\vartheta_{rc}$. A practical method for this is as follows.

Two points *a* and *b* can be chosen within the lateral cutting edge *CE*. The normal clearance angles $\alpha_{n.a}$ and $\alpha_{n.b}$ at the points *a* and *b* are equal to their critical values. They should be not less than $\alpha_{n.y} \geq 2...4°$. When the critical (the minimal) values $[\alpha_{n.a}]$ and $[\alpha_{n.b}]$ of the angles $\alpha_{n.a}$ and $\alpha_{n.b}$ are known, then a set of two equations:

$$\alpha_{n.a} = \frac{K_{cs} \cdot Z_{gashes}}{2 \cdot \pi \cdot R_{a.c}} \cdot \sin(\phi_{rs.a} + \vartheta_{rc}) \tag{15.30}$$

$$\alpha_{n.b} = \frac{K_{cs} \cdot Z_{gashes}}{2 \cdot \pi \cdot R_{b.c}} \cdot \sin(\phi_{rs.b} + \vartheta_{rc}) \tag{15.31}$$

can be composed.

The solution to the set of Eq. (15.30) and Eq. (15.31) returns a formula:

$$\vartheta_{rc} = \tan^{-1}\left( \frac{R_{a.ct} \cdot \tan\alpha_{n.a} \cdot \sin\phi_{rs.b} - R_{b.ct} \cdot \tan\alpha_{n.b} \cdot \sin\phi_{rs.a}}{R_{b.ct} \cdot \tan\alpha_{n.b} \cdot \cos\phi_{rs.a} - R_{a.ct} \cdot \tan\alpha_{n.a} \cdot \cos\phi_{rs.b}} \right) \tag{15.32}$$

for the calculation of the angle $\vartheta_{rc}$.

The corresponding value of the parameter $K_{cs}$ of relieving is calculated from the expression

$$K_{cs} = \frac{2 \cdot \pi \cdot R_{a.c} \cdot \tan\alpha_{n.a}}{Z_{gashes} \cdot \sin(\phi_{rs.a} + \vartheta_{rc})} = \frac{2 \cdot \pi \cdot R_{b.c} \cdot \tan\alpha_{n.b}}{Z_{gashes} \cdot \sin(\phi_{rs.b} + \vartheta_{rc})} \tag{15.33}$$

In particular, the angles $\alpha_{n.a}$ and $\alpha_{n.b}$ can be equal to one another, that is, the equality $\alpha_{n.a} = \alpha_{n.b}$ can be valid. In this, Eq. (15.30) and Eq. (15.31) can be significantly simplified.

The equations derived above are valid when the reciprocation motion $\mathbf{V}_{rc}$ is parallel to the axis of rotation of the hob to be relieved (see Figure 15.23, case *IV*). In this latter case, the angle $\vartheta_{rc}$ is equal to the right angle ($\vartheta_{rc} = 90°$).

When deriving Eq. (15.24) through Eq. (15.33), the impact of the translational motion $V_{sc}$ onto the elements of geometry of the hob cutting edge is neglected.

#### 15.3.2.1.2 Advanced Methods of Gear Hob Relieving Operation

In addition to the basic methods of the gear hob relieving operation (see Figure 15.23), a few more advanced methods of cutting the relieved clearance surface $C_s$ of the hob tooth are used.

Numerous methods of relieving coarse pitch hobs by shaper cutters have been developed.

A schematic of a method[4] of relieving a coarse pitch gear hob by a shaper cutter is depicted in Figure 15.26. In

[4] SU Pat. No. 155718, *A Method of Relieving of Gear Hob with the Shaper Cutter*/M.Ya. Geldfine, F.P. Denisenko, and A.L. Rabkin, Int. Cl. D24b 21/16, National Cl. B 23p, 49d, 7, Filed: September 25, 1961.

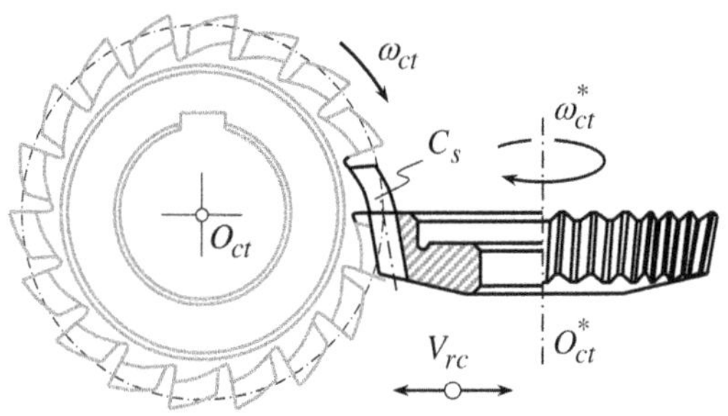


**FIGURE 15.26** Schematic of relieving process of coarse pitch hob by shaper cutter (SU Pat. No. 155718).

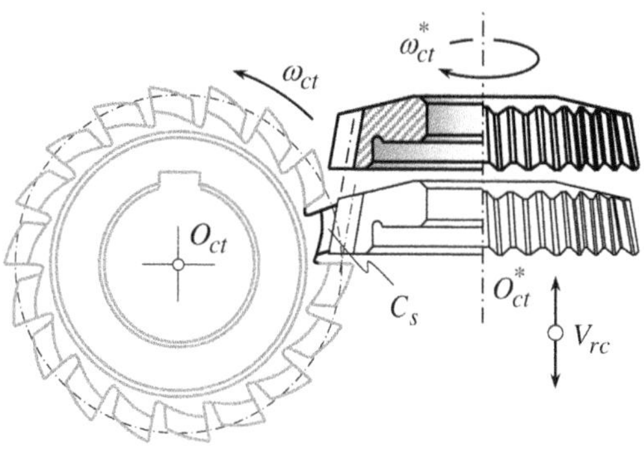


**FIGURE 15.27** Schematic of method of relieving of coarse pitch hob by shaper cutter (SU Pat. No. 751583).

this method, the hob to be relieved is rotated $\omega_{ct}$ about its axis of rotation $O_{ct}$ [55]. The shaper cutter rotates $\omega^*_{ct}$ about its axis of rotation $O^*_{ct}$. The shaper cutter is also performing the reciprocation motion $V_{rc}$ toward the axis of rotation of the hob $O_{ct}$. Simultaneously with the reciprocation $V_{rc}$, the shaper cutter travels $V_{sc}$ in the axial direction of the gear hob (this motion is not shown in Figure 15.26). The rotations $\omega_{ct}$ and $\omega^*_{ct}$ as well as the translation $V_{sc}$ are timed with each other, making it possible to relieve all the teeth of the coarse pitch gear hob in just one path of the shaper cutter along the axis of the hob. Multiple paths of the shaper cutter along the $O_{ct}$ – axis of the gear hob are no longer required when this method (see Figure 15.26) is used. In this way, the machining time is significantly reduced.

In another method for relieving a coarse pitch gear hob by the shaper cutter,[5] the reciprocation motion is tangential with respect to the clearance surface $C_s$ to be cut (see Figure 15.27). In this method, the gear hob to be relieved is rotated $\omega_{ct}$ about its axis of rotation $O_{ct}$ [82]. The shaper cutter rotates $\omega^*_{ct}$ about its axis of rotation $O^*_{ct}$. The shaper cutter is also performing the reciprocation motion $V_{rc}$. The direction of the reciprocation $V_{rc}$ is tangential to the clearance surface $C_s$ to be cut. The gear hob turns through one tooth for each reciprocation $V_{rc}$.

Simultaneously with the reciprocation $V_{rc}$, the shaper cutter travels $V_{sc}$ in the axial direction of the hob. (This motion is not shown in Figure 15.27.) The rotations $\omega_{ct}$ and $\omega^*_{ct}$, as well as the translation $V_{sc}$, are timed with one another, making it possible to relieve all the teeth of the coarse pitch hob in just one path of the shaper cutter along the axis of rotation of the gear hob. Multiple paths of the shaper cutter along the $O_{ct}$ – axis are not necessary when this method of hob relieving (see Figure 15.27) is applied. In this way, the machining time is significantly reduced. Better conditions of cutting are another advantage of the method of relieving schematically depicted in Figure 15.27. The better conditions of cutting result in a longer tool life for the shaper cutter.

The methods for relieving gear hobs shown in Figures 15.26 and 15.27 require implementation of shaper cutters of special design. This is because of the incompatibility of the cutting edge geometry of the top cutting edges of the

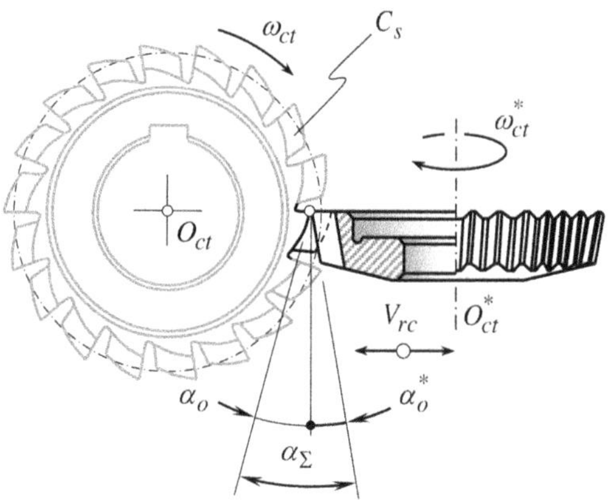


**FIGURE 15.28** Schematic of a method for relieving a coarse pitch gear hob with standard shaper cutter. (From Radzevich, S.P., USSR Pat. No. 1087309, Int. Cl. B23b 3/12, April 23, 1980.)

standard hob and of the standard shaper cutter. Clearance angle $\alpha_o$ at the top cutting edge of a standard hob is in the range of $\alpha_o = 10°...12°$. In special cases the clearance angle can be increased up to $\alpha_o = 18°$.

Clearance angle $\alpha^*_o$ at the top cutting edge of the standard shaper cutter is equal to $\alpha^*_o = 6°$.

Clearance angle $\alpha^*_o$ at the top cutting edge of a shaper cutter to be used on the relieving operation should exceed that angle $\alpha_o$ of the gear hob (that is, fulfillment of the inequality $\alpha^*_o > \alpha_o$ is a must when relieving hobs by shaper cutters). The latter requirement is because the actual clearance angle $\alpha_\Sigma$ in the relieving operation of the hob is equal to the difference $\alpha_\Sigma = \alpha^*_o - \alpha_o$, and it has to have a positive value ($\alpha_\Sigma > 0$). This means that implementation of special-purpose shaper cutters is required for machining clearance surfaces $C_s$ of the gear hob teeth. The shaper cutters should feature clearance angle $\alpha^*_o$ that meets the inequality $\alpha^*_o > \alpha_o$.

In order to make the relieving operation of standard gear hobs possible with standard shaper cutters, a method[6] for relieving hobs with the standard shaper cutter has been developed. A schematic of the method is illustrated in Figure 15.28 [83]. In this method, the hob to be relieved is rotated $\omega_{ct}$ about

[5] SU Pat. No. 751583, *A Method of Relieving of Gear Hob*/S.P. Radzevich, Int. Cl. B24b 3/12, B23b 5/42, Filed: April 4, 1977.

[6] SU Pat. No. 1087309, *A Method of Relieving of Gear Hob by the Standard Shaper Cutter*/S.P. Radzevich, Int. Cl. B24b 3/12, Filed: April 23, 1980.

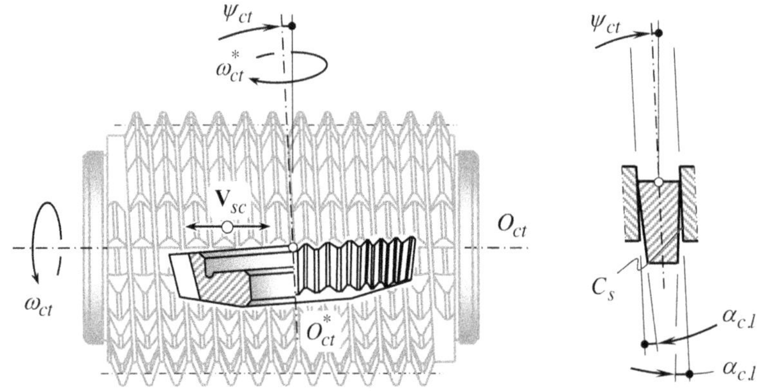


**FIGURE 15.29** Schematic of a method of hob relieving with a standard shaper cutter. (From Radzevich, S.P., USSR Pat. No. 1240548. Int. Cl. B24b 3/12, September 22, 1984.)

its axis of rotation $O_{ct}$. It is important to stress the readers' attention here on the direction of the rotation $\omega_{ct}$. The rotation $\omega_{ct}$ in the method, shown in Figure 15.28, is opposite to the direction of the rotation $\omega_{ct}$ in the method in [55]. The clearance surface of the hob tooth is faced toward the direction of the rotation.

The shaper cutter is rotated $\omega^*_{ct}$ about its axis of rotation $O^*_{ct}$. The shaper cutter is also reciprocating $V_{rc}$ toward the axis of rotation of the gear hob. Simultaneously with the reciprocation $V_{rc}$, the shaper cutter travels $V_{sc}$ in the axial direction of the hob. (This motion is not shown in Figure 15.28.) The rotations $\omega_{ct}$ and $\omega^*_{ct}$ as well as the translation $V_{sc}$ are synchronized with one another, making it possible to relieve all the teeth of the coarse pitch gear hob in a single path of the shaper cutter along the axis of rotation of the gear hob. The multiple paths of the shaper cutter along the $O_{ct}$ – axis are no longer necessary when using this method of hob relieving (see Figure 15.28). In this way, the machining time is significantly reduced.

Better conditions of cutting are another advantage of the method. The kinematics of the method of hob relieving results in the actual clearance angle $\alpha_\Sigma$ being equal to summa $\alpha_\Sigma = \alpha^*_o + \alpha_o$ and not to the difference ($\alpha_\Sigma = \alpha^*_o - \alpha_o$) of the clearance angles at the top cutting edges of the gear hob to be relieved, and of the shaper cutter to be used on the hob relieving operation. The use of this method has two important advantages: (a) shaper cutters of standard design can be used on the hob relieving operation, and (b) the increased clearance angle $\alpha_\Sigma$ allows for better quality of the machined surfaces, as well as longer tool life of the standard shaper cutter.

Further improvements to the methods of cutting of clearance surfaces of a gear hob are targeting the equalization of clearance angles $\alpha_{c.l}$ at the lateral cutting edges of the opposite sides of the tooth profile of the shaper cutter tooth.[7] A schematic of the method of relieving of a coarse pitch hob with the standard shaper cutter [84] is shown in Figure 15.29. The method of gear hob relieving is similar to the method just discussed (see Figure 15.28). The difference between the method shown in Figure 15.29 and the method schematically illustrated in Figure 15.28 is that the axis of rotation of the shaper cutter $O^*_{ct}$ is tilted at the pitch helix angle $\psi_{ct}$ of the hob. This makes it possible to have equal clearance angles $\alpha_{c.l}$ at the lateral cutting edges of the opposite sides of the tooth profile of the shaper cutter when relieving the hob.

The methods of hob relieving by shaper cutters discussed above (see Figures 15.26–15.28) feature that same translation motion $V_{sc}$ of the shaper cutter depicted in Figure 15.29.

Much room remains for further improvements to methods of cutting of clearance surfaces of gear hobs.

#### 15.3.2.2 Grinding of Relieved Clearance Surfaces of Gear Hob Teeth

Methods for relief grinding are used for finishing hardened clearance surfaces of gear hob.

##### *15.3.2.2.1 Relief Grinding of Solid Gear Hob*

Basic methods for relief grinding of the hob teeth are similar to those used for roughing of clearance surfaces (see Figure 15.23). When grinding (see Figure 15.30), the gear hob is rotated $\omega_{ct}$ about its axis of rotation $O_{ct}$. The direction of the rotation is chosen so that the rake surface $R_s$ of the hob tooth is facing toward the direction of the rotation. For relief grinding, a grinding wheel with a reasonably small outer diameter is used. The grinding wheel is rotated $\omega^*_{ct}$ about its axis of rotation $O^*_{ct}$. The axis $O^*_{ct}$ of the grinding wheel is tilted through a certain angle relative to the axis of rotation $O_{ct}$ of the gear hob. The axes $O_{ct}$ and $O^*_{ct}$ are crossed at the angle $(90° - \psi_{ct})$, where $\psi_{ct}$ denotes the pitch helix angle of the hob. The grinding wheel is reciprocating $\mathbf{V}_{rc}$ toward the axis of rotation $O_{ct}$ of the gear hob. In addition to the reciprocation $\mathbf{V}_{rc}$, the grinding wheel travels $\mathbf{V}_{sc}$ in the axial direction of the hob. This motion $\mathbf{V}_{sc}$ together with the rotation $\omega_{ct}$ provides the screw motion of the grinding wheel with respect to the clearance surface $C_s$ to be ground. The translational motion $\mathbf{V}_{sc}$ of the grinding wheel is not shown in Figure 15.30.

[7] SU Pat. No. 1240548, *A Method of Relieving of Gear Hob by the Standard Shaper Cutter*/S.P. Radzevich, Int. Cl. B24b 3/12, Filed: September 22, 1984.

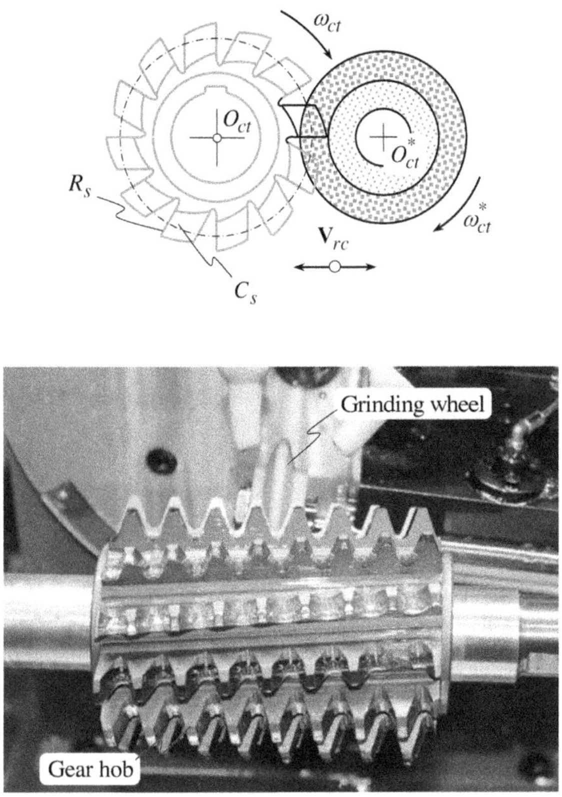


**FIGURE 15.30** Basic method for relief grinding of clearance surface $C_s$ of gear hob tooth.

The form grinding wheel is dressed so that it is capable of grinding either the space between two adjacent hob teeth (similar to case *I* in Figure 15.23), or it is grinding the clearance surfaces of the opposite sides of the hob tooth (similar to case *II* in Figure 15.23).

In the production of coarse pitch gear hobs, other methods for relief grinding feature reciprocation motion $\mathbf{V}_{rc}$ of the form grinding wheel pointed either (a) at a certain angle relative to the axis of rotation $O_{ct}$ of the hob (a schematic of the method is similar to that illustrated in Figure 15.23, case *III*), or (b) in the direction that is parallel to the axis of rotation $O_{ct}$ of the hob (a schematic of the method is similar to that illustrated in Figure 15.23, case *IV*). The application of the latter method is severely restricted due to the small amount of room for the grinding wheel, even in the case when the finger-type grinding wheel is used for this purpose.

An analysis of the contact geometry[8] of the clearance surface of the hob tooth $C_s$ and of the generating surface of the form grinding wheel $T_{gw}$ reveals that an opportunity for improvement to the basic methods of relief grinding is available. It is proven [89] that the accuracy of the relief-ground hob can be significantly increased if the path of reciprocation of the grinding wheel axis of rotation $O^*_{ct}$ is displaced

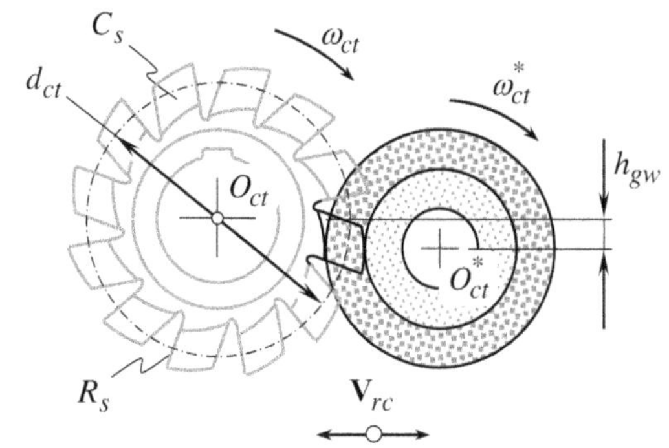


**FIGURE 15.31** Schematic of method for relief grinding of gear hob teeth. (From Radzevich, S.P., USSR Pat. No. 848290, Int. Cl. B24b 3/12, August 6, 1979).

down as shown in Figure 15.31. The recommended[9] value of the displacement $h_{gw}$ depends on the actual value of pitch diameter of the gear hob $d_{ct}$ and on the clearance angle $\alpha_w$ on the pitch diameter of the gear hob. The required displacement $h_{gw}$ can be calculated from the formula

$$h_{gw} = 0.5 \cdot d_{ct} \cdot \sin \alpha_w \tag{15.34}$$

Equation (15.27) yields an expression

$$\alpha_w = \tan^{-1}\left( \frac{d_{o.ct}}{d_{ct}} \cdot \tan \alpha_o \right) \tag{15.35}$$

for the calculation of the clearance angle $\alpha_w$.

In a relief grinding process for a gear hob, the clearance surface $C_s$ being ground and the generating surface of the form grinding wheel $T_{gw}$ make line contact. Every point of the line of contact traces a trajectory over the clearance surface $C_s$ of the hob tooth. Lines of this kind are commonly referred to as the *gear hob relief lines*.

When the grinding wheel is reciprocating, a point of the line of contact is moving from point $a$ to point $b$ along the corresponding *relief line*. The reduction of the pitch diameter from its value $d_{c.a}$ at point $a$ to the corresponding value $d_{c.b}$ at point $b$ is observed. Because the pitch diameter $d_{c.a}$ is larger compared to the pitch diameter $d_{c.b}$ (the inequality $d_{c.a} > d_{c.b}$ is observed), the pitch helix angle $\psi_{ct}$ also changes. Its actual value $\psi_{c.a}$ at point $a$ is smaller than that, $\psi_{c.b}$, at point $b$. In order to compensate for the change of pitch helix angle, an additional swivel motion $\omega^*_{rc}$ is performed by the grinding wheel in the method for relief grinding of gear hob teeth[10] as schematically illustrated in Figure 15.32 [90]. The additional swivel motion of the grinding wheel is performed for each tooth of the hob to be relief-ground.

This means that when the grinding wheel approaches the axis of rotation $O_{ct}$ of the gear hob (the reciprocation

[8] It is right to notice here that the *DG/K* – based method of surface generation [89], [148], [158], and [168], provides a perfect tool for the analysis of the contact geometry between the part surface to be machined and the generating surface of the cutting tool to be applied.

[9] SU Pat. No. 848290, *A Method of Relief Grinding of Gear Hob Teeth*/ S.P. Radzevich, Int. Cl. B24b 3/12, Filed: August 6, 1979.

[10] SU Pat. No. 1743810, *A Method of Relief Grinding of Gear Hob Teeth*/ S.P. Radzevich, Int. Cl. B24b 3/12, Filed: September 13, 1989.

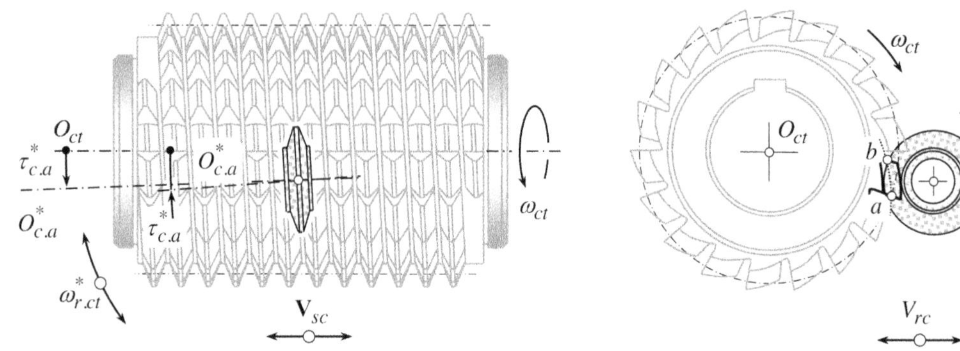


**FIGURE 15.32** Schematic of a method for relief grinding of gear hob teeth that features swivel motion of the grinding wheel. (From Radzevich, S.P., USSR Pat. No. 1743810, Int. Cl. B24b 3/12, Filed: September 13, 1989).

motion $V_{rc}$), it simultaneously performs the swivel motion $\omega^*_{rc}$ [92]. Owing to the swivel motion of the grinding wheel, the crossing angle between the axes $O_{ct}$ and $O^*_{ct}$ changes in time. The crossing angle $\tau^*_{c.a}$ at point $a$ is smaller than that $\tau^*_{c.b}$ at point $b$. This is obvious from the equalities $\tau^*_{c.a} = \psi_{c.a}$ and $\tau^*_{c.b} = \psi_{c.b}$ observed.

Because the method for relief grinding [92] features a swivel motion of the grinding wheel, this makes it possible to maintain proper conditions of contact between the clearance surface $C_s$ of the hob tooth and the generating surface $T_{gw}$ of the form grinding wheel. In this way, the accuracy of the ground gear hob is getting better.

Reciprocation $\mathbf{V}_{rc}$ of the grinding wheel can be replaced with an additional rotation of the hob being relieved. In the method[11] [91] of relief grinding, the axis of rotation $O_{ct}$ of the grinding wheel is motionless (that is, the grinding wheel is just rotating $\omega^*_{ct}$ about its axis; see Figure 15.33). The gear hob being relieved rotates $\omega_{ct}$ about its axis of rotation $O_{ct}$. In addition, the hob is rotated $\omega_e$ about an axis $O_e$, which is parallel to the hob axis of rotation $O_{ct}$ (that is, $O_e \parallel O_{ct}$). The axis $O_e$ is positioned at a distance $e_c$ from the axis $O_{ct}$ [90]. The distance $e_c$ is greater than the relieving parameter $K_{cs}$ (that is, the inequality $e_c > K_{cs}$ is observed).

The method for relief grinding can be executed when directions of the rotations $\omega_{ct}$ and $\omega_e$ are reversed. The inverse kinematics of the relief grinding operation is depicted in dashed lines.

Depending on the actual location of the axis $O_e$ of the additional rotation $\omega_e$, both the rotations $\omega_{ct}$ and $\omega_e$ can be either continuous or the rotation $\omega_e$ can be continuous while the rotation $\omega_{ct}$ is intermittent (that is, it serves only for indexing purposes).

Elimination of the reciprocation motion $\mathbf{V}_{rc}$ from the kinematics of the relief grinding operation improves the dynamics of the hob relieving process. Due to this, the accuracy of the relieved hobs improves.

Much room remains for further improvements to the methods of relief grinding of gear hobs.

[11] SU Pat. No. 1194612, *A Method of Relief Grinding of Gear Hob Teeth*/ S.P. Radzevich, Int. Cl. B23f 21/16, B24b 3/12, Filed: July 20, 1984.

#### *15.3.2.2.2 Relief Grinding of Assembled Gear Hobs*

Assembled gear hobs are relief-ground in one of two possible ways. First, conventional methods for relief grinding are also applicable to assembled hobs with inserted cutting racks. Second, because gear hobs are made assembled (and not solid), more opportunities for performing hob relieving operations appear in this regard.

At the beginning, it is necessary to stress that two different configurations of the cutting racks in the body of the assembled hob should be recognized. One of the configurations corresponds to that when the hob is capable of cutting the work gear. This configuration of the cutting racks is referred to as the *working position* of the cutting racks. Another configuration of the cutting racks corresponds to that at which the clearance surfaces of the hob teeth could be ground. The configuration of the cutting racks in the body of the assembled hob in the working position can differ from their configuration in the position when grinding clearance surfaces. The latter is referred to as the *position for machining* of the cutting racks.

For example, an assembled gear hob may feature bearing surfaces of the cutting racks shaped in the form of external cylinders of revolution as shown in Figure 15.34. These surfaces of the cutting racks interact with the corresponding internal surfaces in the form of cylinders of revolution in the body of the hob. This makes it possible to turn each

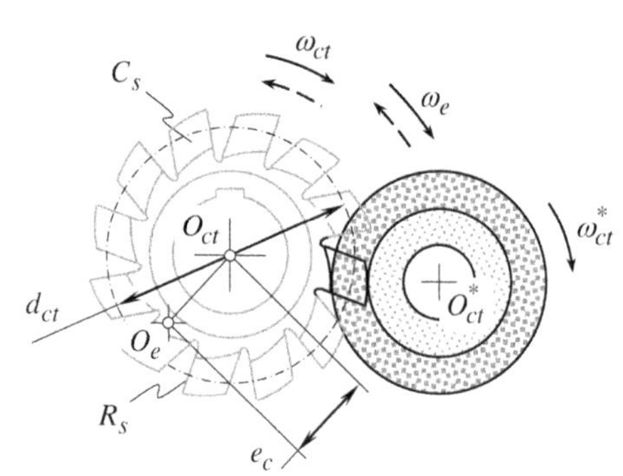


**FIGURE 15.33** Schematic of a method of relief grinding of gear hob teeth. (From Radzevich, S.P., USSR Pat. No. 1194612, Int. Cl. B24b 3/12, July 20, 1984).

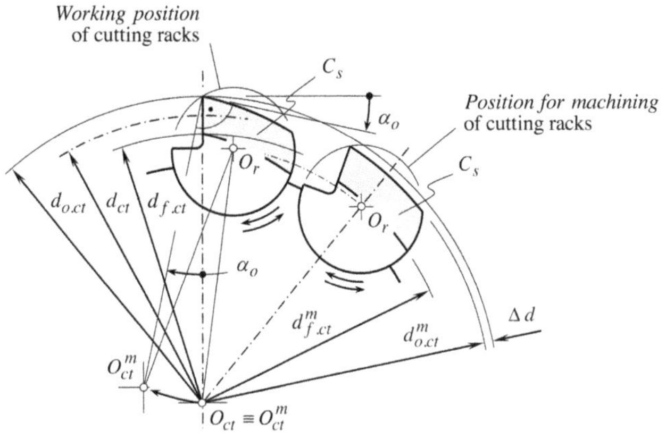


**FIGURE 15.34** Schematic of assembled gear hob for machining fine-pitch gears.

cutting rack about its axis $O_r$ through the clearance angle $\alpha_o$ from the working position to their position for machining, and vice versa.

When the cutting racks occupy their working positions, then the outer diameter of the hob, the pitch diameter, and the inner diameter are denoted as $d_{o.ct}$, $d_{ct}$, and $d_{f.ct}$, respectively. After the cutting racks are turned to their position for machining (or *technological position* of cutting racks, in other terminology), then the outer diameter and the inner diameter are designated as $d^m_{o.ct}$ and $d^m_{f.ct}$, respectively. It is important to note here that the outer diameter $d_{o.ct}$ is larger than the outer diameter $d^m_{o.ct}$, and the difference $\Delta d$ equals

$$\Delta d = \frac{d^m_{o.ct} - d_{o.ct}}{2} \tag{15.36}$$

In the position for machining of the cutting racks, the axis of rotation of the gear hob $O_{ct}$ and the technological axis $O^m_c$ align to each other. In working position of the cutting racks the axes, $O_{ct}$ and $O^m_{ct}$, do not align to each other; however, they remain parallel to one another. The location of the technological axis $O^m_{ct}$ in the working position of the cutting racks can be determined as illustrated in Figure 15.34.

When the cutting racks are turned about their axes $O_r$ from the working position to the position for machining, the clearance surfaces $C_s$ of all the teeth of all the cutting racks are located within a common continuous screw surface. This is owing to the fact that in the position for machining, the clearance angle $\alpha_o$ vanishes; it is getting equal to zero. It is evident that the continuous screw surface can be ground as threads of a worm using for this purpose conventional methods used for grinding worms. It is much more convenient to grind worm surfaces than to relief-grind clearance surfaces $C_s$ of the hob.

After being ground, the cutting racks turn back to their working position at which clearance angle $\alpha_o$ has a positive value ($\alpha_o > 0°$).

Implementation of the concept discussed above (see Figure 15.34) is practical when designing fine-pitch gear hobs. When cutting fine-pitch gears, the hob having round bearing surfaces is capable of withstanding cutting forces.

Another concept of assembled hobs can be implemented for machining coarse pitch gears.

Assembled gear hobs featuring rectangular-shaped slots for cutting racks can be relief-ground using a method similar to that depicted in Figure 15.34. As it follows from the analysis of Figure 15.35, the clearance surfaces $C_s$ of the gear hob teeth can be located within a common continuous screw surface when the cutting racks are displaced at a distance $h_{cs}$ from their working position to the position for machining. In the working positions of the cutting racks, all the clearance surfaces $C_s$ are ground as continuous thread surfaces of the worm. Again, it is much easier to grind worm surfaces rather than to relief-grind intermittent clearance surfaces $C_s$ of the gear hob. Moreover, the diameters of the cutting racks

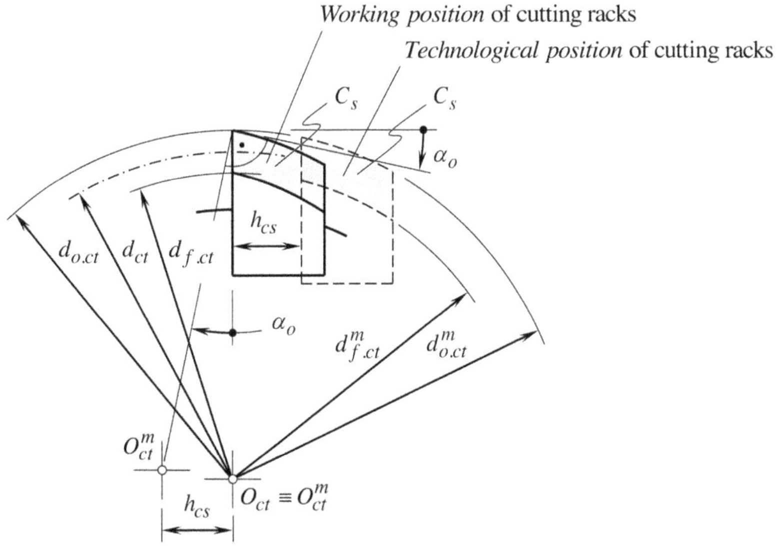


**FIGURE 15.35** Location and orientation of cutting racks in working position, and position for machining in special technological body of the hob.

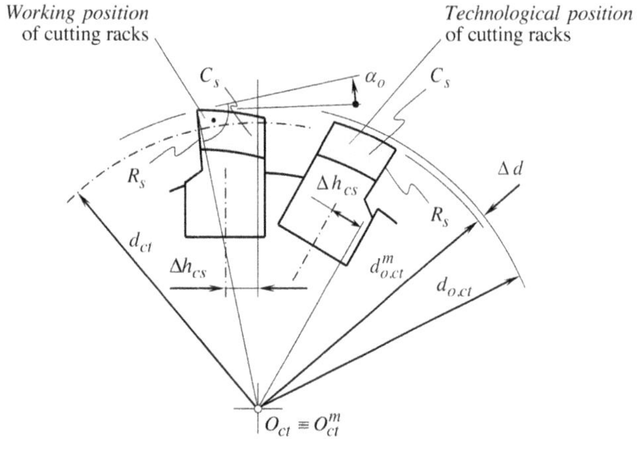


**FIGURE 15.36** Illustration of a possible implementation of a gear hob body for assembling the hob in working positions of the cutting racks, and in their positions for machining.

in their positions for machining can be kept equal to the corresponding diameters in the working positions of the cutting racks (that is, $d^m_{ct} = d_{ct}$, $d^m_{o.ct} = d_{o.ct}$ and $d^m_{f.ct} = d_{f.ct}$). This is because two different bodies are used: one of the bodies (the technological body of the gear hob) is used for the purposes of grinding the clearance surfaces $C_s$ of the hob teeth, and the other one is the working body of the hob.

Gear hobs can be relief-ground in the hob body, serving two purposes: grinding of the clearance surfaces of the hob teeth, and assembling the hob in the working positions of the cutting racks. Figure 15.36 reveals that under a certain value of the displacement $\Delta h_{cs}$ of the centerline of the cutting rack from the centerline of the gear hob, the hob body can be applied for two purposes: (a) for positioning the cutting racks in their working positions [in such cases, the orientation clearance angle $\alpha_o$ has a positive value ($\alpha_o > 0°$)], and (b) for the positioning of the cutting racks for the purposes of grinding of clearance surfaces [in such cases, the orientation the clearance angle $\alpha_o$ has a zero value ($\alpha_o = 0°$), and all clearance surfaces $C_s$ are located within a common continuous screw surface. This surface is commonly called the *technological worm*]. It is important to notice here that the outer diameter of the hob $d_{o.ct}$ is larger than that of the technological worm $d^m_{o.ct}$, that is, the inequality $d_{o.ct} > d^m_{o.ct}$ is observed ($d_{o.ct} - d^m_{o.ct} = 2 \cdot \Delta d$).

For designing assembled gear hobs with a hob body that is capable of serving both purposes simultaneously, a method of determining the location and the orientation of cutting rack slots in the hob body is developed.

Consider a simplified example of an assembled gear hob with a clearance angle of zero ($\alpha_o = 0°$). Assume also that the back face of the cutting rack is congruent to the bearing surface of the slot in the body of the gear hob as shown in Figure 15.37. Point $A$ is chosen at the top cutting edge of the hob tooth. The straight line $AB$ is located within the rake plane $R_s$. Point $B$ is at the outer cylinder of the hob body. Because the clearance angle is equal to zero ($\alpha_o = 0°$), the straight line $AB$ is the line through the axis of rotation $O_{ct}$ of the gear hob. It is assumed here that the transverse profile of the slot is aligned with the straight line $AB$.

Point $C$ within the outer cylinder of the gear hob body is chosen at a distance that is equal to the desired width of the slot. The opposite side of the transverse profile of the slot is through point $C$. The straight line $CD$ passes at a distance $r_{cs}$ from the axis $O_{ct}$ of the gear hob rotation, that is, the straight line $CD$ is tangent to a circle of radius $r_{cs}$. The distance $r_{cs}$ equals

$$r_{cs} = \frac{d_{o.ct}}{2} \cdot \sin \alpha_o \tag{15.37}$$

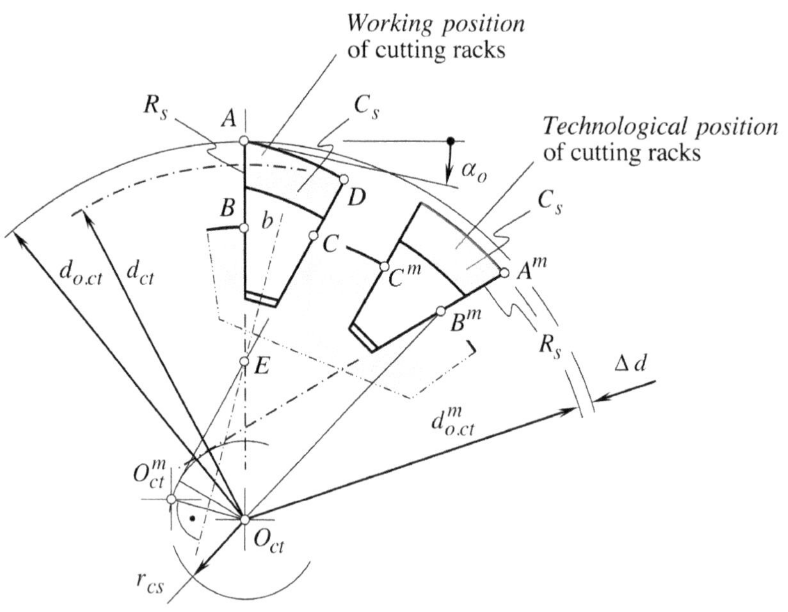


**FIGURE 15.37** Method of determination of design parameters of slots in the body of assembled gear hob.

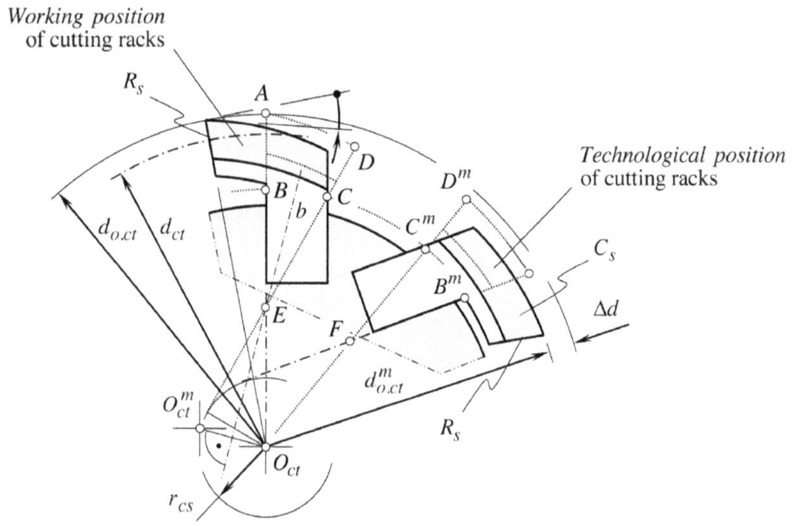


**FIGURE 15.38** An illustrative example of implementation of the method shown in Figure 15.36.

Equation (15.37) occurs because the cutting racks are turned through the angle of 180° from their working positions to their positions for machining of the clearance surfaces $C_s$.

In order to turn the cutting racks from working positions to their positions for machining, that is, to the technological position, the racks are turned through the angle of 180° about the bisector $b$ of the angle $\angle BEC$.

The axis $O^m_{ct}$ of the screw clearance surface $C_s$ is the mirror symmetry of the hob axis of rotation $O_{ct}$. The bisector $b$ serves as the axis of the symmetry.

In the technological position of the cutting racks, superscript $m$ is assigned to all the elements of the gear hob. In the technological position, the clearance surfaces $C_s$ of the cutting rack teeth are grinding as a continuous screw surface. No reciprocation of the grinding wheel is required in this case.

The importance of the disclosed solution to the problem of determining the location and orientation of the slots in the gear hob body (see Figure 15.37) is due to the fact that it can be used as a template when solving similar problems—for instance, when designing hobs with slots of any desired shape, and so forth. An illustrative example is shown in Figure 15.38.

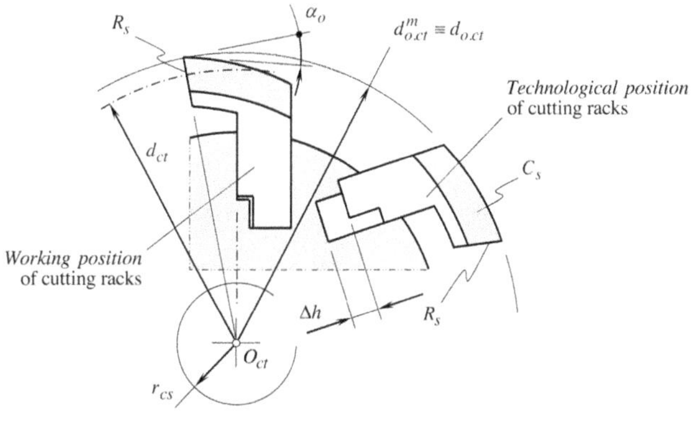


**FIGURE 15.39** Working position and position for machining of cutting rack in a gear hob body with slots of modified profile.

An assembled gear hob with slots in the hob body of a desired rectangular shape is depicted in Figure 15.38. Values of the rake angle $\gamma_o$ and of the clearance angle $\alpha_o$ are specified. The earlier constructed solution to the similar problem is shown in Figure 15.38 in dotted lines. It is used as the reference for the slots of the desired shape. Further, the cutting rack of the desired design is treated together with the cutting rack of the simplified geometry. In this way, the desired location and the orientation of the slots are determined.

It is important to notice here that the outer diameter $d_{o.ct}$ of the hob exceeds that of the technological worm $d^m_{o.ct}$, that is, the inequality $d_{o.ct} > d^m_{o.ct}$ is valid.

In a similar way to that described above, the slots of any desired shape in the hob body can be designed. The simplified solution to the problem (see Figure 15.37) works in all cases, and is applicable for gear hobs of any desired geometry of the cutting edges.

As it follows from the preceding discussion, the pitch diameter $d^m_{ct}$ of the technological worm (as well as other diameters of threads of the technological worm) is smaller than the pitch diameter $d_{ct}$ (and other corresponding diameters) of the hob itself. Just as the actual values of the diameters $d_{ct}$ and $d^m_{ct}$ are different, the pitch helix angles $\psi_{ct}$ and $\psi^m_{ct}$ are also of different values ($\psi^m_{ct} > \psi_{ct}$). The difference in the pitch helix angles is one of the root causes of the reduced accuracy of assembled hobs. In aiming for high accuracy of the gear hob, it is desirable to have identical pitch diameters for the hob and the technological worm. For this purpose, the slots in the hob body are shaped so that a ledge is formed at the bottom of every slot[12] [121]. When the cutting rack is in working position (see Figure 15.39), then the ledge

[12] SU Pat. No. 1505695, *An Assembled Gear Hob*/S.P. Radzevich, *et al*, Int. Cl. B23f 21/16, Filed: March 16, 1987.

surfaces do not interact with the corresponding surfaces of the cutting rack. After the cutting rack is turned to the technological position, then the lowest plane of the cutting rack contacts the top of the ledge. In this way the cutting rack is shifted outward about the hob axis of rotation $O_{ct}$. The ledge height $\Delta h$ is calculated so

$$\Delta h = \frac{d_{o.ct} - d^{*}_{o.ct}}{2 \cdot \cos(0.5 \cdot \alpha_o)} \tag{15.38}$$

to compensate for the change of diameter of the technological worm.

The slots with the ledge at their bottom can eliminate a source of deviations of the relief-ground gear hob from the desired geometry [121].

Deviation in the angular positions of the slots for cutting racks in the gear hob body is another source of inaccuracies in assembled gear hobs. The order of location of the cutting racks in the working position is the inverse to that in the position for grinding clearance surfaces. If one would decide to number the slots and the cutting racks in their working position, for example in a clockwise direction, then the order of the cutting racks in their technological position would be in the counterclockwise direction. It is clear that in the position for grinding, the cutting racks do not occupy their slots in their working positions. Therefore, deviation in the angular position of the slots for cutting racks in the hob body directly affects hob accuracy.

The desired clearance surface of the hob tooth [see Eq. (15.22)] can be approximated with a screw surface. In this case, it is possible to grind the clearance surfaces after they are displaced in the axial direction of the hob body to align with a screw surface, which in this configuration is continuous and common for all of the clearance surfaces $C_s$ [167].

In the working positions of the cutting racks of the assembled gear hob, the pitch helix angle $\psi_{ct}$ is equal [see Figure 15.40(a)]:

$$\psi_{ct} = \tan^{-1}\left(\frac{m \cdot N_{ct}}{d_{ct}}\right) \tag{15.39}$$

In Eq. (15.39), the number of starts of the hob is denoted by $N_{ct}$ (for a single-start hob the equality $N_{ct} = 1$ is valid).

The cutting racks can be displaced in the axial direction of the hob body to the technological positions as illustrated in Figure 15.40(b). The maximum displacement $\delta_{\max}$ of the cutting rack is equal to

$$\delta_{\max} = P_x \cdot m_{ct} \tag{15.40}$$

where the axial pitch of the hob is designated as $P_x$, and $m_{ct}$ is an integer number (usually $m_{ct} = \pm 1$).

The axial displacement $\delta_i$ of the $i-$th cutting rack depends on its number $i$

$$\delta_i = \frac{P_x \cdot m_{ct}}{n_{ct}} \cdot i \tag{15.41}$$

where the total number of the cutting racks of the hob is designated as $n_{ct}$.

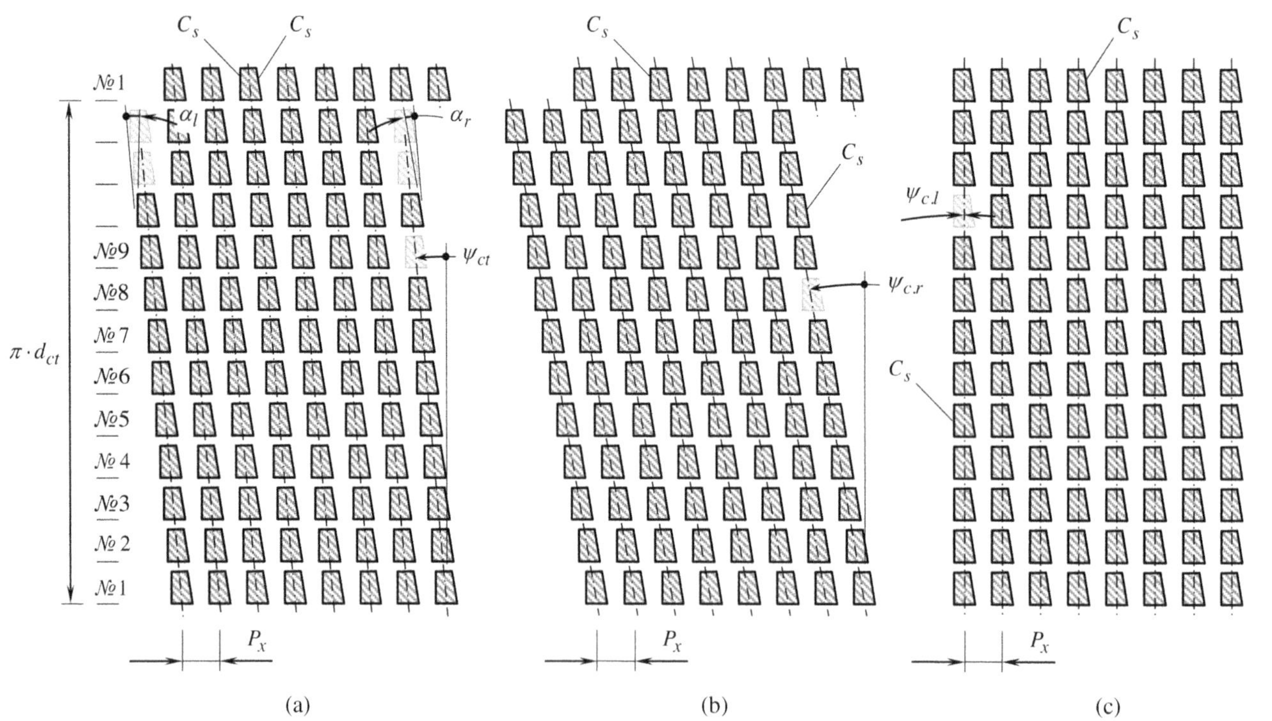


**FIGURE 15.40** Working positions and technological positions of the cutting racks in the gear hob body, (a) working positions of the cutting racks, (b) technological positions for grinding the right side of the cutting rack tooth profile, and (c) technological positions for grinding the left side of the cutting rack tooth profile. (The cutting racks are displaced in the axial direction of the hob.)

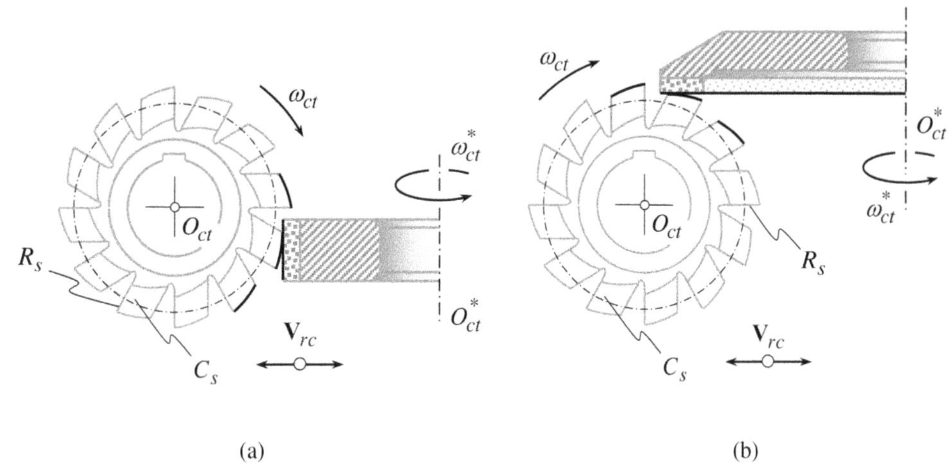


**FIGURE 15.41** Method of grinding clearance surfaces through the top cutting edge of the hob tooth by means of (a) the cylindrical surface of the grinding wheel, and (b) the plane surface of the grinding wheel. (From Radzevich, S.P. and Lempert V.I., USSR Pat. No. 1335425, Int. Cl. B24b 3/12, April 24, 1986).

In the technological position of the cutting racks, shown in Figure 15.40(b), the pitch helix angle $\psi_{c.r}$ of the right-hand-side clearance surfaces can be calculated from the formula

$$\psi_{c.r} = \tan^{-1}\left[\frac{m \cdot (N_{ct} + m_{ct})}{d_{ct}}\right] \quad (15.42)$$

In such a scenario, the right-hand-side clearance surfaces of the tooth profile of the gear hob are getting an inclination at the clearance angle $\alpha_r$:

$$\alpha_r = \psi_{c.r} - \psi_{ct} \quad (15.43)$$

Equations similar to those above are valid for the opposite side (that is, for the left-hand side) of the profile of the hob tooth [see Figure 15.40(c)]:

$$\psi_{c.l} = \tan^{-1}\left[\frac{m \cdot (N_{ct} - m_{ct})}{d_{ct}}\right] \quad (15.44)$$

$$\alpha_l = -\psi_{c.l} + \psi_{ct} \quad (15.45)$$

Implementation of the method of grinding of clearance surfaces of the assembled gear hobs (see Figure 15.40) makes possible precision generation of the clearance surfaces of the lateral sides of the tooth profile of the gear hob.

After clearance surfaces through the lateral cutting edges are ground, then simplified methods for relief grinding can be used[13] for grinding the clearance surfaces through the top cutting edges [176]. Two examples of kinematics of the simplified methods of relief grinding are shown in Figure 15.41.

For the relief grinding of clearance surfaces through the top cutting edges of the gear hob, standard grinding wheels, either cylindrical [see Figure 15.41(a)] or conical (dish-type) [see Figure 15.41(b)] are commonly used. The hob being relief-ground rotates about its axis of rotation $O_{ct}$ so that the rake surface $R_s$ is faced toward the direction of the hob rotation $\omega_{ct}$. The grinding wheel rotates about its axis of rotation $O^*_{ct}$. The rotation of the grinding wheel is designated as $\omega^*_{ct}$. The grinding wheel is reciprocating $V_{rc}$ toward the axis of the gear hob rotation $O_{ct}$ per each tooth of the hob being relief-ground. In addition, the grinding wheel travels $V_{sc}$ along the hob axis of rotation $O_{ct}$. There is no kinematical correlation between the translation $V_{sc}$ of the grinding wheel and the rotation $\omega^*_{ct}$ of the gear hob and its number of starts $N_{ct}$. Independence of the translation $V_{sc}$ from the rotation $\omega^*_{ct}$ and the starts number $N_{ct}$ in the method (see Figure 15.41) makes it possible to assign a favorable value to the speed $V_{sc}$ of the translational motion of the grinding wheel. The speed $V_{sc}$ can be set so that the relief grinding is performed in just a single path of the grinding wheel along the axis of rotation $O_{ct}$ of the hob.

For relief grinding of assembled hobs,[14] other methods are developed as well (see [93] and others).

## 15.4 ACCURACY OF HOBS FOR MACHINING OF INVOLUTE GEARS

The accuracy of gear hobs is an important consideration. Inaccuracies of the gear hobs directly affect the accuracy of the machined gears. Many efforts have been undertaken to investigate the accuracy of gear hobs, as well as investigating the accuracy of the hobbed gears. However, many of those issues related to the accuracy of the designed and manufactured gear hobs have not yet been properly investigated. The problem of the accuracy of gear hobs deserves to be comprehensively discussed separately in another book. A few problems are discussed next.

[13] SU Pat. No. 1335425, *A Method of Relief Grinding of Hob Teeth*/S.P. Radzevich, V.I. Lempert, Int. Cl. B23f 21/16, Filed: April 24, 1986.

[14] SU Pat. No. 961925, *A Method of Relief Grinding of an Assembled Hob Teeth*/S.P. Radzevich, Int. Cl. B23f 21/16, Filed: March 18, 1980.

### 15.4.1 Preliminary Remarks

The absence of the common line of intersection of three surfaces, that is, of

- the generating surface $T$ of the hob,
- the rake surface $R_s$ of the hob tooth, and
- the clearance surface $C_s$ of the hob tooth

is the major reason for the deviations of the cutting edges $CE$ of the hob tooth from their desired geometry and location. A few examples of the deviations of the rake surface $R_s$ from its desired position and geometry, along with the corresponding deviations in the profile of the hobbed gear tooth caused by the deviations, are illustrated in Figure 15.42.

No deviations of the tooth profile of the cut gear are observed when the rake surface of the hob is configured properly [see Figure 15.42(a)].

A slightly positive [see Figure 15.42(b)] or a slightly negative [see Figure 15.42(c)] rake angle at the top cutting edge of the hob tooth results in the corresponding reduction of tooth thickness either at the top of the tooth or at the bottom of the tooth of the hobbed gear.

The concavity of the rake surface of the hob [see Figure 15.42(d)] makes the machined gear tooth thicker, while convexity of the rake surface of the hob [see Figure 15.42(e)] makes the machined gear tooth thinner at the top and at the bottom.

Inaccuracies in the angular pitch of the hob teeth [see Figure 15.42(f)] result in a bigger height of the residual cusps on the machined surface of the gear tooth. No doubt more examples of this concern can be provided.

Taking into account the importance of proper shape and correct configuration of the rake surface, the parameters of shape and of the configuration of the rake surface of a hob tooth are carefully inspected for a new hob, as well as after every hob resharpening (see Figure 15.43).

Because the clearance surfaces of gear hob teeth are relief-ground, this makes corrective grinding of the hobs possible. The corrective grinding is aimed at reducing the inaccuracies of gear hobs. It is based on an analysis similar to that illustrated in Figure 15.42. More details on the corrective grinding of gear hobs can be found in the paper by Dr. Arkhangelskiy in [27].

The preliminary qualitative analysis of the deviations of the gear hob tooth, and their effect on the accuracy of the machined gear, make the necessity and importance of a more detailed analysis of gear hob accuracy clear.

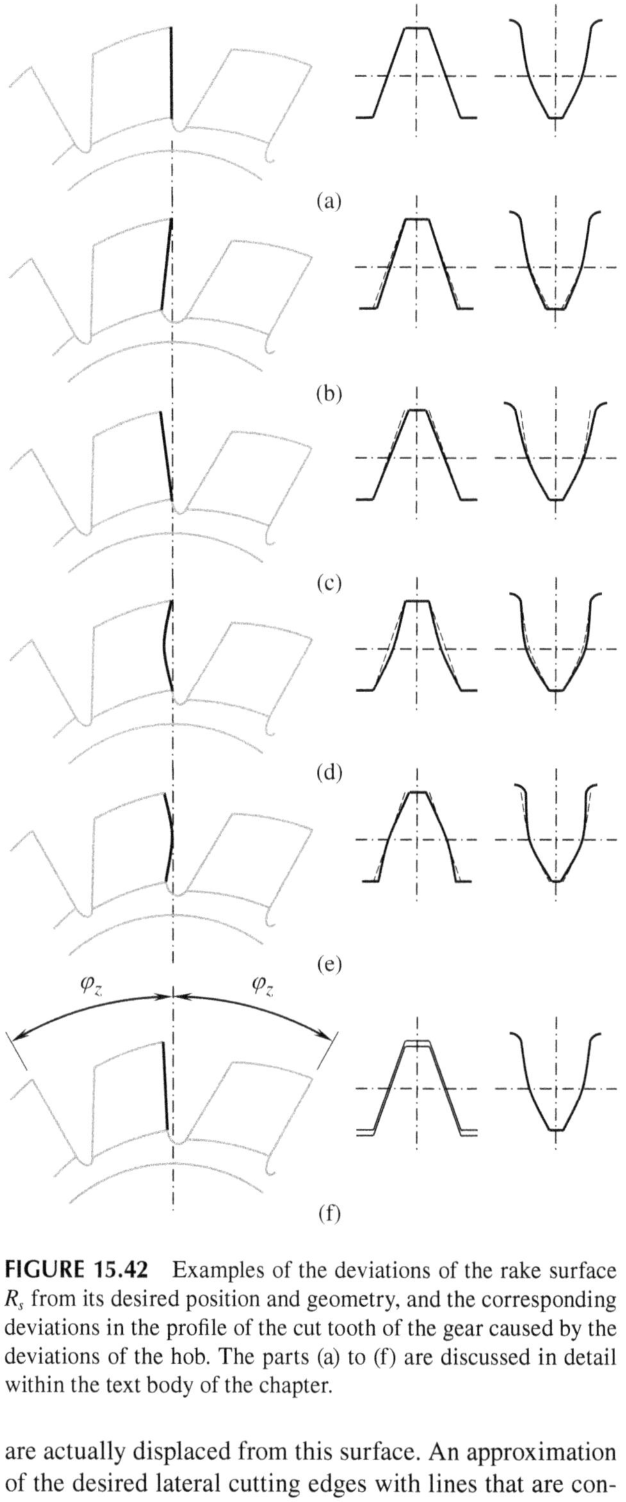


**FIGURE 15.42** Examples of the deviations of the rake surface $R_s$ from its desired position and geometry, and the corresponding deviations in the profile of the cut tooth of the gear caused by the deviations of the hob. The parts (a) to (f) are discussed in detail within the text body of the chapter.

### 15.4.2 Accuracy of an Involute Gear Hob as a Function of Its Design Parameters

The lateral cutting edges $CE$ of an involute hob are understood in this text as lines of intersection of the generating surface $T$ of the gear hob by the rake surface $R_s$. The third surface, namely the clearance surface $C_s$, must pass through the cutting edge $CE$ of the hob tooth. Though the lateral cutting edges $CE$ must be located within the surface $T$, they are actually displaced from this surface. An approximation of the desired lateral cutting edges with lines that are convenient in manufacturing of the hob is the chief reason for the displacement. Often the desired 3$D$ lateral cutting edges $CE$ of a hob tooth are approximated with a straight line. This results in inevitable deviation $\delta_c$ of the actual machining surface $T^m$ from the desired generating surface $T$ of the involute hob. The deviation $\delta_c$ is measured as a distance between the surfaces $T^m$ and $T$. The distance $\delta_c$ is measured along the perpendicular to the generating surface $T$ of the gear hob, and the perpendicular is through a point of interest within the lateral cutting edge $CE$ of the hob tooth.

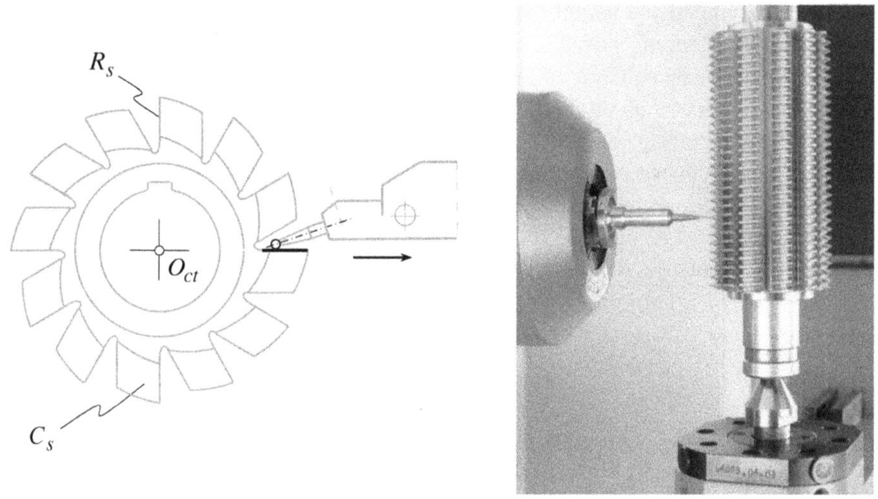


**FIGURE 15.43** Inspection of the geometry and configuration of the rake surface $R_s$ of a gear hob.

For the calculation of the deviation $\delta_c$, equations of the desired and of the actual lateral cutting edge of the gear hob tooth are required.

#### 15.4.2.1 Analytical Description of the Desired Lateral Cutting Edge

For the analytical description of the desired lateral cutting edges $CE$ of the involute gear hob, the equation $\mathbf{r}_{ct} = \mathbf{r}_{ct}(U_{ct}, V_{ct})$ of generating surface $T$ of the hob [see Eq. (15.12)] together with equation $\mathbf{r}_{rs} = \mathbf{r}_{rs}(U_{rs}, V_{rs})$ of the rake face $R_s$ [see Eqs. (15.4) through (15.8)] are convenient to use. Based on these equations, the position vector $\mathbf{r}_{ce}$ of the point of the desirable lateral cutting edge $CE$ can be analytically described by a matrix equation:

$$\mathbf{r}_{ce}(U_{ce}) = \begin{bmatrix} X_{ce}(U_{ce}) \\ Y_{ce}(U_{ce}) \\ Z_{ce}(U_{ce}) \\ 1 \end{bmatrix} \tag{15.46}$$

For the practical needs of designing an involute hob, the entire curve [Eq. (15.46)] is not of interest. The hob designer just needs an arc of the curve $\mathbf{r}_{ce} = \mathbf{r}_{ce}(U_{ce})$. This $3D$ arc through the points $a$, $b$, and $c$ is located within the outer hob diameter $d_{o.ct}$ and the limit hob diameter of the value $d_{l.ct}$. It is common practice to assume that

$$d_{l.ct} = d_{o.ct} - 2 \cdot 2.25 \cdot m \tag{15.47}$$

as it follows from Figure 15.44. Here $m$ designates the hob module.

The solution to Eq. (15.46) under $d_{o.ct}$ returns the position vector $\mathbf{r}_{a.c}$ of the point $a(X_a, Y_a, Z_a)$. Similarly, the solution to Eq. (15.46) under $d_{ct}$ returns the position vector $\mathbf{r}_{b.ct}$ of the point $b(X_b, Y_b, Z_b)$, while the solution to that same Eq. (15.46) under $d_{l.ct}$ returns the position vector $\mathbf{r}_{c.c}$ of the point $c(X_c, Y_c, Z_c)$.

#### 15.4.2.2 Analytical Description of the Actual Lateral Cutting Edge

Targeting a simplification of the manufacture of involute gear hobs, the desired lateral cutting edges $abc$ are often approximated with a straight-line segment through the points $a$ and $b$. The approximation causes a deviation of the actual lateral cutting edge ② with respect to the desired lateral cutting edge ① of the involute hob (see Figure 15.44).

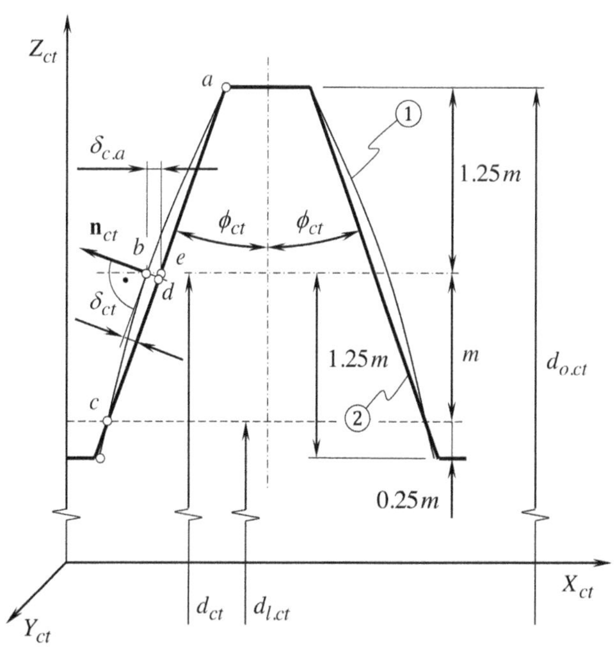


**FIGURE 15.44** Desirable ① and actual ② lateral cutting edges of an involute gear hob. (From Radzevich, S.P., *SME Journal of Manufacturing Processes*, 9(2), 121–136, 2007. With permission.)

For the calculation of deviations of the actual machining surface $T^m$ from the desired generating surface $T$ of the involute hob, it is necessary to derive an equation of the surface $T^m$. Such an equation can be drawn up on the premise of an equation of the actual lateral cutting edge of the hob.

Because the desired 3$D$ lateral cutting edge is approximated with a straight-line segment through the points $a$ and $b$, the equation for the position vector $\tilde{\mathbf{r}}_{ce}$ of the actual cutting edge of the hob can be represented in the form

$$(\tilde{\mathbf{r}}_{ce} - \mathbf{r}_{a.c}) \times (\tilde{\mathbf{r}}_{ce} - \mathbf{r}_{c.c}) = 0 \tag{15.48}$$

which immediately yields

$$\tilde{\mathbf{r}}_{ce}(\tilde{U}_{ce}) = \mathbf{r}_{a.c} + \tilde{U}_{ce} \cdot (\mathbf{r}_{c.c} - \mathbf{r}_{a.c}) \tag{15.49}$$

where $\tilde{U}_{ce}$ designates a parameter of the lateral cutting edge $\tilde{\mathbf{r}}_{ce} = \tilde{\mathbf{r}}_{ce}(\tilde{U}_{ce})$.

#### 15.4.2.3 Machining Surface of Involute Gear Hob

With the equation of the lateral cutting edge $\tilde{\mathbf{r}}_{ce}$ [see Eq. (15.49)], the machining surface $T^m$ of the gear hob can be represented as a set of consecutive positions of the cutting edge $\tilde{\mathbf{r}}_{ce}$ that performs a screw motion about the hob axis $O_{ct}$. The parameter of the screw motion of the cutting edge $\tilde{\mathbf{r}}_{ce}$ is equal to the reduced pitch $p_x$ of the involute hob [see Eq. (15.15)]. The schematic of the generation of the hob surface $T^m$ resembles that which is already discussed for the derivation of the rake face $R_s$ (see Figure 15.7). A Cartesian coordinate system $X_3Y_3Z_3$ is associated with the cutting edge $\tilde{\mathbf{r}}_{ce}$. The cutting edge $\tilde{\mathbf{r}}_{ce}$ performs a screw motion about the hob axis $O_{ct}$ with the parameter of the screw motion $p_x$. Recall that in Figure 15.7, the hob axis $O_{ct}$ is aligned with the $X_{ct}-$ axis of the coordinate system $X_{ct}Y_{ct}Z_{ct}$. For an analytical description of the screw motion, the operator $\mathbf{Rs}(3 \mapsto ct)$ of the resultant coordinate system transformation can be composed.

In the case under consideration, it is convenient to dissolve the screw motion of the cutting edge $\tilde{\mathbf{r}}_{ce}$ onto the rotation about the $X_{ct}-$ axis [this linear transformation is analytically described by the operator $\mathbf{Rt}(V_{rs}, X_{ct})$ of rotation], and onto the translation along the $X_{ct}-$ axis [this linear transformation is analytically described by the operator $\mathbf{Tr}(p_{rs} \cdot V_{rs} \mapsto X_{ct})$ of translation] [164]. The operators of the elementary coordinate system transformations yield the operator $\mathbf{Rs}(3 \mapsto ct)$ of the resultant coordinate system transformation:

$$\mathbf{Rs}(3 \mapsto ct) = \mathbf{Rt}(V_{rs}, X_{ct}) \cdot \mathbf{Tr}(p_{rs} \cdot V_{rs} \mapsto X_{ct}) \tag{15.50}$$

Skipping the routing formulae transformation, an equation of the actual machining surface $T^m$ of the involute gear hob can be represented in the following form:

$$\left.\begin{array}{r}\text{Machining surface, } T^m, \\ \text{of involute gear hob}\end{array}\right| \Rightarrow \tilde{\mathbf{r}}_{ct} = \mathbf{Rs}(3 \mapsto ct) \cdot \tilde{\mathbf{r}}_{ce} \tag{15.51}$$

Equation (15.51) can be cast into matrix form:

$$T^m : \Rightarrow \tilde{\mathbf{r}}_{ct}(\tilde{U}_{ct}, \tilde{V}_{ct}) = \begin{bmatrix} X_{ct}(\tilde{U}_{ct}, \tilde{V}_{ct}) \\ Y_{ct}(\tilde{U}_{ct}, \tilde{V}_{ct}) \\ Z_{ct}(\tilde{U}_{ct}, \tilde{V}_{ct}) \\ 1 \end{bmatrix} \tag{15.52}$$

Here, $\tilde{U}_{ct}$ and $\tilde{V}_{ct}$ designate curvilinear coordinates (*Gaussian coordinates*) of a point of the gear hob surface $T^m$.

#### 15.4.2.4 Deviation of the Actual Machining Surface from the Desired Generating Surface of an Involute Gear Hob

Calculation of the deviations of the actual machining surface $T^m$ from the desired generating surface $T$ of a precision involute gear hob is a challenging engineering problem. The discussion that follows is limited just to the calculation of the maximum deviation $\delta_{ct}$ of the gear hob surface $T^m$ from the desired surface $T$ of the involute hob.

##### *15.4.2.4.1 Definition of the Maximum Deviation $\delta_{ct}$ of the Hob Tooth Profile*

The maximum deviation $\delta_{ct}$ is observed on the gear hob pitch diameter (see Figure 15.44). The deviation $\delta_{ct}$ is measured in the direction of the outward unit normal vector $\mathbf{n}_{ct}$ to the generating surface $T$ of the hob. The deviation $\delta_{ct}$ is a signed value. It is positive when it is measured outward from the generating surface $T$ and it is negative if it is measured in the opposite direction.

An expression for the position vector $\mathbf{r}_\delta$ of a point of the straight line through point $b$ in the direction of the unit normal vector $\mathbf{n}_{ct}$ can be written in vector form:

$$(\mathbf{r}_\delta - \mathbf{r}_{b.c}) \times \mathbf{n}_{ct} = 0 \tag{15.53}$$

An equation for the position vector $\mathbf{r}_\delta(U_\delta)$ immediately follows from Eq. (15.53):

$$\mathbf{r}_\delta(U_\delta) = \mathbf{r}_{b.c} + U_\delta \cdot \mathbf{n}_c \tag{15.54}$$

Point $d$ is the point of interception of the straight line $\mathbf{r}_\delta(U_\delta)$ with the machining surface $T^m$ of the involute hob (see Figure 15.44). Coordinates $X_d$, $Y_d$, and $Z_d$ of point $d$ are calculated as a solution to the set of Eqs. (15.52) and (15.54). Ultimately, the expression

$$\delta_{ct} = -|\mathbf{r}_{b.c} - \mathbf{r}_{d.c}| \tag{15.55}$$

can be composed for the calculation of the deviation $\delta_{ct}$. Here, $\mathbf{r}_{d.c} = \mathbf{r}_{d.c}(X_d, Y_d, Z_d)$.

The deviation $\delta_{ct}$ of tooth profile has a negative value. This means that the deviation is directed toward the interior of the generating surface $T$. In other words, the deviation $\delta_{ct}$

is measured in a negative direction of the unit normal vector $\mathbf{n}_{ct}$ to the generating surface $T$.

It is important to point out here the perfect correlation between the deviations $\delta_{ct}$ and $\delta_{c.x}$. (The deviation $\delta_{c.x}$ is measured in the axial direction of the involute hob.) An approximate equality $\delta_{c.x} \cong \delta_{ct}/\cos\phi_n$ is observed. The accuracy of the approximation is commonly within the range of 5%.

#### 15.4.2.4.2 *Impact of Rake Angle $\gamma_o$ on Tooth Profile Deviation $\delta_{ct}$*

The deviation $\delta_{ct}$ of the gear hob tooth profile depends on the rake angle $\gamma_o$ of the involute hob. Numerical examples are convenient to illustrate the behavior of the function $\delta_{ct}(\gamma_o)$. The major design parameters of the involute hob for the illustrative calculations are summarized in Table 15.1.

An analytical expression for the calculation of the deviation $\delta_{ct}$ of the tooth profile of the gear hob in terms of the current value of the rake angle $\gamma_o$ can be derived from Eq. (15.55) for involute hobs with straight gashes that have various values of the module $m$, the normal pressure angle $\phi_n$, and the outer diameter $d_{o.ct}$.

A graphical interpretation of the function $\delta_{ct}(\gamma_o)$ for involute hobs having the design parameters (see Table 15.1) is shown in Figure 15.45 [164]. The functions $\delta_{ct}$ vs. $\gamma_o$ for the entering side of the hob tooth profile are plotted in Figure 15.45(a). Figure 15.45(b) illustrates the similar functions $\delta_{ct}$ vs. $\gamma_o$ for the exiting side of the hob tooth profile. The functions $\delta_{ct} = \delta_{ct}(\gamma_o)$ have extrema for both the entering side and the exiting side of the tooth profile. The extrema of the function $\delta_{ct} = \delta_{ct}(\gamma_o)$ corresponds to the tooth profile deviation that is equal to $\delta_{ct} = 0$. For the entering side of the hob tooth profile, the extremum $\delta_{ct} = 0$ is observed under a positive rake angle $\gamma_o > 0°$, while for the exiting side of the hob tooth profile the extremum $\delta_{ct} = 0$ is observed under a negative rake angle $\gamma_o < 0°$. The extremum value of the rake angle $\gamma_o$ is equal to $\gamma_o \cong +8.645°$ in the first case, and it is equal to $\gamma_o \cong -8.645°$ in the second case.

It is critical to stress here that the extreme values of the rake angle $\gamma_o$ have those values for which the lateral cutting edges of the hob tooth are tangent from the opposite sides to the hob base cylinder of a diameter $d_{b.ct}$ and to the base helices.

The impact of the gear hob module $m$ on the hob tooth profile deviation $\delta_{ct}$ is illustrated in Figure 15.46 [164] for the entering side of the hob tooth profile [see Figure 15.46(a)] and for the exiting side of the tooth profile [see Figure 15.46(b)]. Here the hob module is represented as a continuous parameter. This is valid because there are no physical restrictions on the module values with fractions. Following common practice, here module $m$ is understood as $m = 25.4/P$, where diametral pitch is designated as P.

It is natural to anticipate that involute gear hobs of a larger module $m$ have tooth profile deviation $\delta_{ct}$, which have different values for opposite sides of the tooth profile of the hob. Actually, they do (see Figure 15.46). Under similar initial conditions, the deviation $\delta_{ct}$ for the exiting side of the hob tooth profile is smaller compared to the deviation $\delta_{ct}$ for the entering side of the hob tooth profile. This is due to the additional impact of kinematics of the gear hobbing operation on the actual value of the deviation $\delta_{ct}$.

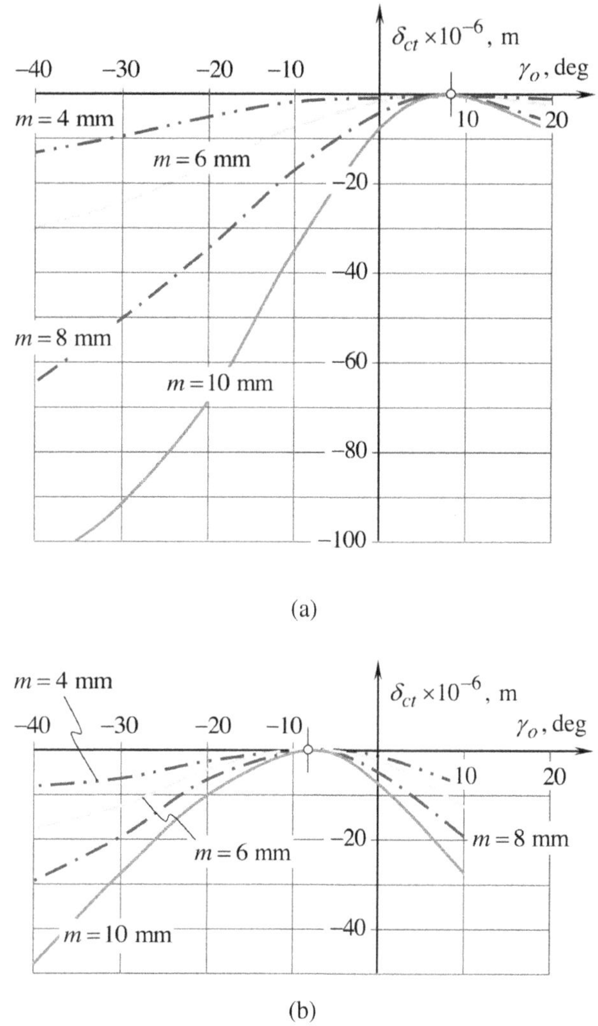


**FIGURE 15.45** Impact of the rake angle $\gamma_o$ on the tooth profile deviation $\delta_{ct}$ for involute hobs of various modulus with an outer diameter $d_{o.ct} = 160$ mm and the normal pressure angle of $\phi_n = 20°$: (a) entering side of the tooth profile and (b) exiting side of the tooth profile. (From Radzevich, S.P., *SME Journal of Manufacturing Processes*, 9(2), 121–136, 2007. With permission.)

**TABLE 15.1**
**Design Parameters of the Precision Involute Gear Hob**

| | |
|---|---|
| Module | $m = 10$ mm |
| Normal pressure angle | $\phi_n = 20°$ |
| Setting angle of the hob | $\zeta_{ct} = 4°$ |
| Outer diameter | $d_{o.ct} = 160$ mm |
| Number of starts | $N_{ct} = 1$ |

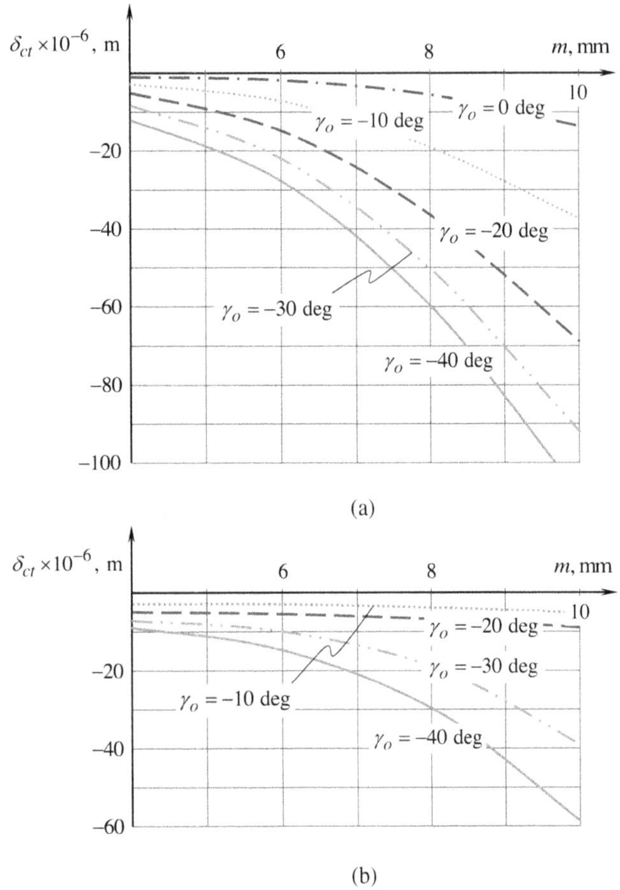


**FIGURE 15.46** Impact of the hob modulus $m$ on the tooth profile deviation $\delta_{ct}$ for involute hobs with various values of the rake angle $\gamma_o$, the outer diameter $d_{o.ct} = 160$ mm, and the normal pressure angle of $\phi_n = 20°$: (a) the entering side of the tooth profile and (b) the exiting side of the tooth profile. (From Radzevich, S.P., *SME Journal of Manufacturing Processes*, 9(2), 121–136, 2007. With permission.)

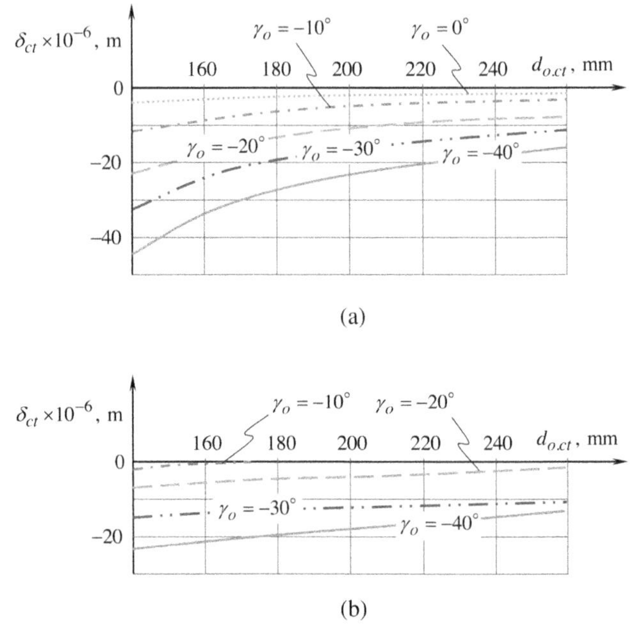


**FIGURE 15.47** Impact of the gear hob outer diameter $d_{o.ct}$ on the tooth profile deviation $\delta_{ct}$ for involute gear hobs with various values of the rake angle $\gamma_o$ and normal pressure angle of $\phi_n = 20°$: (a) the entering side of the tooth profile and (b) the exiting side of the tooth profile. (From Radzevich, S.P., *SME Journal of Manufacturing Processes*, 9(2), 121–136, 2007. With permission.)

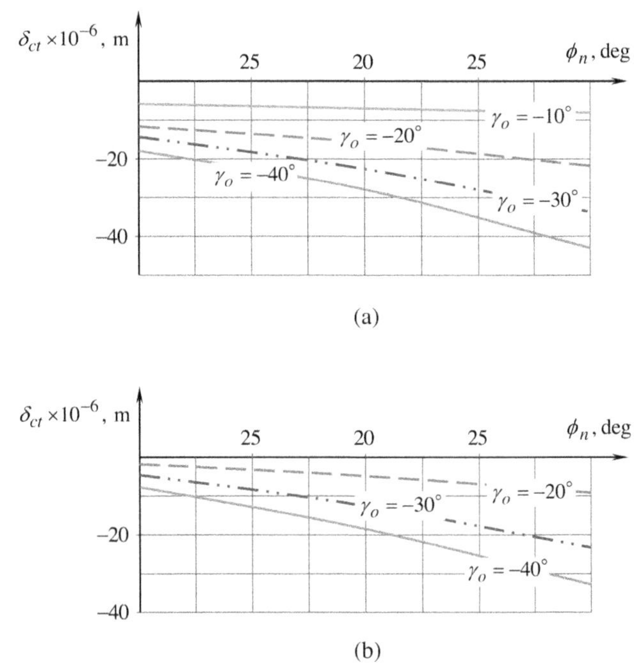


**FIGURE 15.48** Impact of the hob pressure angle $\phi_n$ on the tooth profile deviation $\delta_{ct}$ for involute gear hobs with various values of the rake angle $\gamma_o$ and the outer diameter of $d_{o.c} = 160$ mm: (a) the entering side of the tooth profile, and (b) the exiting side of the tooth profile. (From Radzevich, S.P., *SME Journal of Manufacturing Processes*, 9(2), 121–136, 2007. With permission.)

Two other issues are of critical importance for involute hob designers, as well as for manufacturers of precision gear hobs. They are the impact on the maximum value of the deviation $\delta_{ct}$ of (a) the gear hob outside diameter $d_{o.ct}$ and (b) the gear hob normal pressure angle $\phi_n$.

The larger the outer diameter $d_{ct}$ of the involute gear hob, the smaller the hob tooth profile deviations $\delta_{ct}$ for both, for the entering side of the hob tooth [see Figure 15.47(a)], as well as for the exiting side of the hob tooth profile [see Figure 15.47(b)]. Similarly, involute hobs with normal pressure angle $\phi_n$ also have larger tooth profile deviation $\delta_{ct}$ for the entering side of the tooth profile [see Figure 15.48(a)] as well as for the exiting side of the hob tooth profile [see Figure 15.48(b)]. Again, under similar initial conditions, the deviation $\delta_{ct}$ for the exiting side of the hob tooth profile is smaller than the deviations $\delta_{ct}$ for the entering side of the hob tooth profile. This is owing to the impact of the kinematics of the gear hobbing operation on the deviation $\delta_{ct}$.

### 15.4.2.4.3 *Impact of the Lead Angle $\lambda_{rs}$ of Screw Rake Surface on the Deviation $\delta_{ct}$ of Tooth Profile of an Involute Gear Hob*

The deviation $\delta_{ct}$ of the tooth profile of the involute gear hob depends on the actual value of the lead angle $\lambda_{rs}$ of the rake surface $R_s$ of the hob tooth. An equation

$$\delta_{ct} = \delta_{ct}(\lambda_{rs}, \zeta_{ct}, \phi_n, d_{b.ct}, d_{ct}, t_{ct}) \tag{15.56}$$

for the tooth profile deviation $\delta_{ct}$ can be derived. Then, the solution to the equation

$$\frac{\partial \delta_{ct}(\lambda_{rs}, \ldots)}{\partial \lambda_{rs}} = 0 \tag{15.57}$$

returns a formula for the calculation of an optimal value $\lambda_{rs}^{\text{opt}}$ of the lead angle of the rake surface $R_s$ under which the tooth profile deviation $\delta_{ct}$ vanishes.

Without going into the details of derivation, it should be mentioned here that the optimal value $\lambda_{rs}^{\text{opt}}$ of the lead angle $\lambda_{rs}$ of the rake face, under which the tooth profile deviation $\delta_{ct}$ vanishes, is calculated from the formula [164]

$$\lambda_{rs}^{\text{opt}} = -\tan^{-1}\left[\frac{d_{b.ct} \cdot \cos\zeta_{ct}}{\sqrt{(d_{ct} + t_{ct} \cdot \cot\phi_n)^2 - d_{b.ct}^2} \cdot \tan\phi_n + d_{b.ct} \cdot \sin\zeta_{ct}}\right] \tag{15.58}$$

The − sign in Eq. (15.58) indicates that the recommended direction of hand of the screw rake surface $R_s$ is opposite to the direction of hand for gear hobs of conventional design. Thus, the hand of the screw rake surface $R_s$ has the same direction as threads of the generating surface $T$ of the gear hob.

Equation (15.58) allows for the calculation of the optimal value $\lambda_{rs}^{\text{opt}}$ of the lead angle $\lambda_{rs}$ of the rake surface $R_s$ of the involute hob. For the single-start gear hob of module $m = 10$ mm, the outside diameter of $d_{o.ct} = 160$ mm, and the profile angle of $\phi_n = 20°$, and for Archimedean screw surface $R_s$, Eq. (15.58) returns $\lambda_{rs}^{\text{opt}} = -20.225°$ for the optimal value $\lambda_{rs}^{\text{opt}}$ of the lead angle $\lambda_{rs}$.

It is assumed in the calculation that the setting angle of the hob equals $\zeta_{ct} = 3.772°$. The chosen value of the hob-setting angle $\zeta_{ct}$ is close to the actual value of the pitch helix angle $\psi_{ct} = 3.699°$ of the gear hob. The latter is mostly for the purposes of avoiding pointing of the hob tooth.

Usually, the setting angle of the gear hob $\zeta_{ct}$ is either exactly or approximately equal to the hob helix angle $\zeta_{ct} \cong \psi_{ct}$. However, it is not mandatory to keep the equality $\zeta_{ct} \cong \psi_{ct}$. The actual value of the hob-setting angle $\zeta_{ct}$ affects the pitch helix angle $\psi_{ct}$ of the hob. This is due to the relationship (14.75):

$$\psi_{b.ct} = \cos^{-1}(\cos\phi_n \cdot \cos\zeta_{ct}) \tag{15.59}$$

for the base helix angle $\psi_{b.ct}$ is valid [138].

The thickness $t_{ct}$ of the hob tooth is calculated from the equation

$$t_{ct} = \frac{\pi \cdot m}{2} \tag{15.60}$$

and in the case under consideration, it is equal to $t_{ct} = 15.708$ mm.

For the calculation of the hob base diameter $d_{b.ct}$, Eq. (14.71)

$$d_{b.ct} = 2 \cdot r_{b.ct} = \frac{m \cdot N_{ct} \cdot \cos\phi_n}{\sqrt{1 - \cos^2\phi_n \cdot \cos^2\zeta_{ct}}} \tag{15.61}$$

is used [97, 119, 138, 143, 148]. In the illustrative numerical example, Eq. (15.61) returns $d_{b.ct} = 27.037$ mm.

The calculations reveal that deviations of the tooth profile of the involute gear hob with the above-listed design parameters (see Table 15.1) have a zero value ($\delta_{ct} = 0$) for both opposite sides of the tooth profile.

### 15.4.2.4.4 *Impact of Other Design Parameters of an Involute Gear Hob on the Tooth Profile Deviation $\delta_{ct}$*

As proven in our earlier work [160], minimum tooth profile deviation $\delta_{ct}$ is observed when the hob-setting angle $\zeta_{ct}$ is exactly equal to the pitch helix angle $\psi_{ct}$ of the involute hob. In other words, when the equality $\zeta_{ct} = \psi_{ct}$ is observed, then the condition $\delta_{ct} \mapsto \min$ is fulfilled. This statement is based on the premise of the solution to the equation $\partial\delta_{ct}(\zeta_{ct}, \ldots)/\partial\zeta_{ct} = 0$ [110, 140, 143, 159].

The tooth profile deviation $\delta_{ct}$ affects the base pitch of the involute hob.

Because of the difference between the actual value of the base pitch $\tilde{p}_{b.ct}$ and the desired value $p_{b.ct}$, the base pitch error

$$\delta p_b = p_{b.ct} - \tilde{p}_{b.ct} \tag{15.62}$$

becomes inevitable. The deviation $\delta p_b$ directly affects the conjugate action of the cut gear and the pinion tooth profiles, causing noise, whine, vibration, and other negative effects.

The results of the analytical investigation of the accuracy of involute gear hobs can be used for the purposes of increasing the accuracy of the hobbed gears.

Once the gear hobs are manufactured with straight lateral cutting edges, this feature of hob design allows for significant reduction or even for elimination of the deviation $\delta_{ct}$ of the tooth profile of the hob. Figure 15.45 reveals that the deviation $\delta_{ct}$ of tooth profile becomes the smallest possible under a specific value of the rake angle $\gamma_o$. This value of the rake angle $\gamma_o$ corresponds to the configuration of the lateral cutting edge $CE$ of the involute gear hob when it is aligned with the straight line that is tangent to the base helix angle of the generating surface $T$ of the hob (see Figure 15.49).

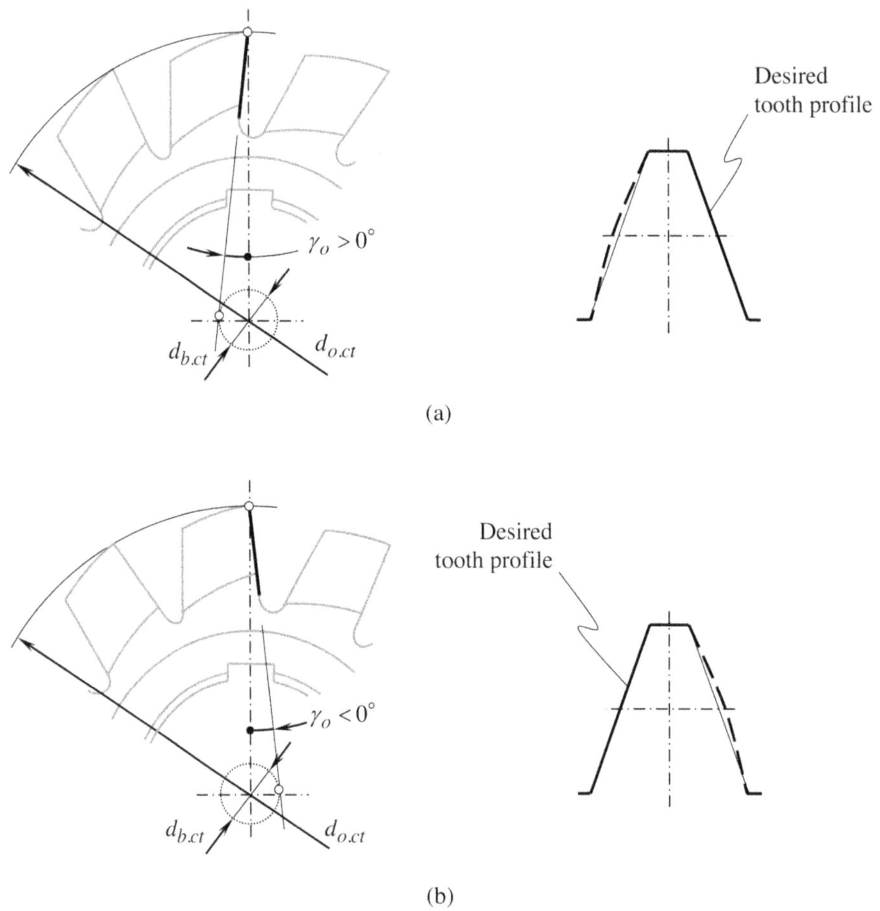


**FIGURE 15.49** Required configuration of the straight lateral cutting edge of a precision involute gear hob with (a) positive, and (b) negative rake angle $\gamma_o$. (From Radzevich, S.P., *SME Journal of Manufacturing Processes*, 9(2), 121–136, 2007. With permission.)

When the lateral cutting edge of one side of the tooth profile of a precision involute gear hob of conventional design is straight, the cutting edge of the opposite side of the tooth profile should be curved.

Precision single-start involute hobs featuring straight lateral cutting edges with tooth profile deviation of a zero value $\delta_{ct} = 0$ can be used for machining precision spur and helical gears of nonreversible gear trains, that is, for machining gears for helicopter transmissions, gear reducers for electric windmill stations, and turbine reducers. Precision involute hobs with a positive rake angle $\gamma_o > 0°$ [see Figure 15.49(a)] are recommended for gear hobs made of high-speed steel, while hobs having negative rake angle $\gamma_o < 0°$ [see Figure 15.49(b)] are recommended for carbide involute gear hobs. Positive rake angle $\gamma_o > 0°$ results in a corresponding positive value of the normal rake angle $\gamma_n > 0°$ at the hob tooth lateral cutting edges, which is suitable for high-speed steel hobs, while negative rake angle $\gamma_o < 0°$ causes a negative value of the normal rake angle $\gamma_n < 0°$ at the hob tooth lateral cutting edges, which is recommended for carbide hobs. In both cases, for single-start hobs of conventional design, the absolute value of the rake angle $\gamma_o$ is around $|\gamma_o| \approx 10°$.

Owing to the nonzero rake angle $\gamma_o$ on the top cutting edge of the gear hob tooth (see Figure 15.49), the lateral cutting edges of the hob tooth feature a corresponding value of the inclination angle $\lambda_{ce}$. The inclination angle $\lambda_{ce}$ has either a positive ($\lambda_{ce} > 0°$) or a negative ($\lambda_{ce} < 0°$) value depending on (a) the sign of the rake angle $\gamma_o$ and (b) the hand of the hob threads. Usually, the geometry of the lateral cutting edge is of critical importance for roughing and semi-roughing involute hobs. It is much less important for finishing and/or semi-finishing hobs [148, 156, 158].

The accuracy of precision skiving hobs with a rake angle $\gamma_o$ up to $\gamma_o \cong -60°$ [27] can also be improved using the method discussed above. The hob tooth profile deviation $\delta_{ct}$ can be eliminated not in the design of a single-start, but in the design of a multiple-start skiving hob [see Figure 15.50(a)] [141, 159, 168]. In this case, the required number of the hob starts $N_{ct}$ depends on the actual value of the rake angle $\gamma_o$ of the gear hob. To ensure the specified negative value of the rake angle $\gamma_o$, the precision involute skiving hob has to be of a right-hand helix in order to feature the entering lateral cutting edges of straight form [see Figure 15.50(b)], and it has to be of a left-hand helix to feature the entering lateral cutting edges of straight form [see Figure 15.50(c)]. Therefore, two different hobs are required for the machining of a precision hardened gear. Otherwise, the gear should be hobbed in two setups.

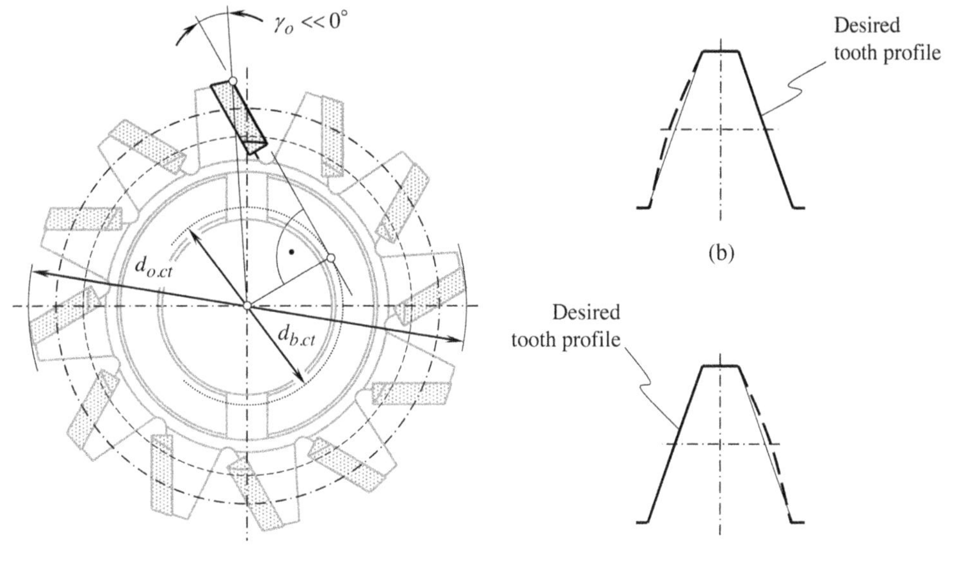


**FIGURE 15.50** Example of a precision skiving involute gear hob with the rake angle $\gamma_o$ of a prescribed value that features the desired straight lateral cutting edges, (a) side view of the skiving hob; (b) schematic of the skiving hob tooth with a straight right side of the tooth profile, and (c) schematic of the hob tooth with a straight right side of the tooth profile. (From Radzevich, S.P., *SME Journal of Manufacturing Processes*, 9(2), 121–136, 2007. With permission.)

### 15.4.3 Impact of Pitch Diameter on the Accuracy of Gear Hob

The kinematics of the relative motion of the work gear and of the involute hob depends on the current value of the pitch diameter of the gear hob. Changes to the kinematics of the relative motion can affect the resultant accuracy of the hobbed gear.

#### 15.4.3.1 Peculiarities of the Relative Motion of the Work Gear and the Gear Hob

The kinematics of meshing of a work gear and the involute gear hob includes rotations of the work gear and the hob about their axis of rotation. Simultaneously, the work gear and the gear hob are traveling relative to one another in the axial direction of the work gear with a feed motion.

The kinematics of meshing of the work gear and the gear hob resembles meshing that is observed in a worm gear set. A comprehensive analysis of the kinematics of worm gearing can be found in [14]. While the kinematics of meshing in both cases is similar, the kinematics of the relative motion of the work gear and the involute gear hob differs from relative motion of the worm and the worm gear in conventional worm gear set. The difference is mainly because of the alteration (of the reduction) in the pitch diameter after each regrinding of the gear hob. The reduction of the pitch diameter after the hob had been reground is illustrated in Figure 15.51.

After every regrinding, the clearance surface $C_s$ of the gear hob tooth is remained within a certain central angle $\sigma$ (see Figure 15.51). Practically, a portion of the clearance surface for regrinding of the hob tooth spans a central angle $\sim 0.6\sigma$. The outer diameter of the new hob is designated as $d_{o.ct}$, and the pitch diameter of the hob is denoted by $d_{ct}$. After being reground, the outer diameter of the hob reduces at a value of $\Delta d_{o.ct}$. The maximum reduction $\Delta d_{o.ct}$ of the gear hob outer diameter depends on the clearance angle $\alpha_o$ of the hob tooth (see Figure 15.51), and of the permissible value of the central angle $\sigma$. When the outer diameter $d_{o.ct}$ of the hob is getting smaller, a corresponding change to the pitch diameter $d_{ct}$ of the gear hob is observed [117, 138].

A change to the hob pitch diameter $d_{ct}$ causes a corresponding change to the pitch helix angle $\psi_{ct}$ of the hob. The pitch helix angle of a new hob can be calculated from the formula

$$\psi_{ct} = \sin^{-1}\left(\frac{N_{ct}\,P_{x.ct}}{d_{ct}}\right) \tag{15.63}$$

where $P_{x.ct}$ denotes the axial pitch of the involute hob.

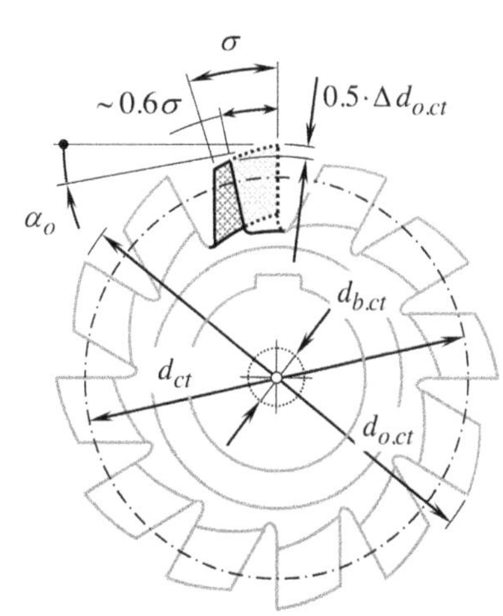


**FIGURE 15.51** Change of the pitch diameter of an involute gear hob after being reground. (From Radzevich, S.P., *ASME Journal of Mechanical Design*, 124, 772–786, 2002. With permission.)

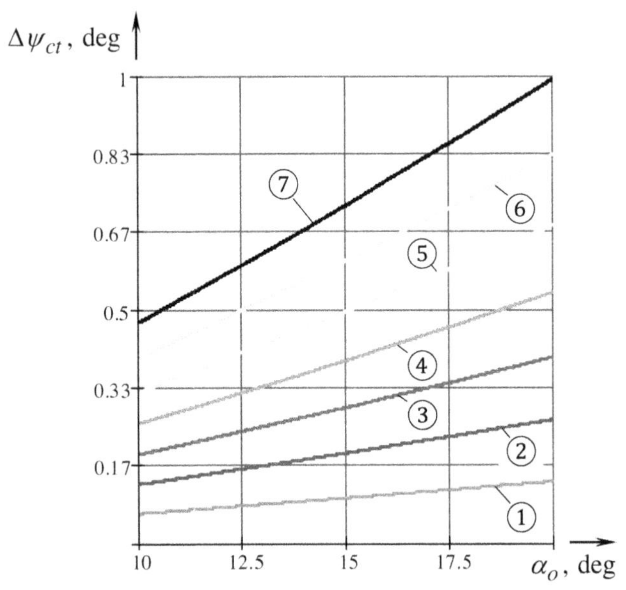


**FIGURE 15.52** Impact of the clearance angle $\alpha_o$ of the gear hob (module $m = 10$ mm, pitch diameter $d_{ct} = 160$ mm) on the pitch helix deviation $\Delta\psi_{ct}$ when the number of starts of the hob is ① – $N_{ct} = 1$, ② – $N_{ct} = 2$, ③ – $N_{ct} = 3$, ④ – $N_{ct} = 4$, ⑤ – $N_{ct} = 5$, ⑥ – $N_{ct} = 6$, and ⑦ – $N_{ct} = 7$. (From Radzevich, S.P., *ASME Journal of Mechanical Design*, 129, 1076–1085, 2007. With permission.)

After being reground, the pitch diameter $d_{ct}$ of the gear hob reduces to a certain value $d_{y.ct}$, while the hob axial pitch $P_{x.ct}$ and the number of starts $N_{ct}$ of the hob are remained at the same value and are identical to the values of the corresponding parameters of the new hob. Therefore, the pitch helix angle $\psi_{y.ct}$ for a partially worn gear hob increases to

$$\psi_{y.ct} = \sin^{-1}\left(\frac{N_{ct}\,P_{x.ct}}{d_{y.ct}}\right) > \psi_{ct} \tag{15.64}$$

The pitch helix angle $\psi_{y.ct}$ has its largest value when the pitch diameter $d_{y.ct}$ is the smallest permissible. This means that when the current value of the diameter equals $d_{y.ct} = d_{ct}^{\min}$, then the corresponding pitch helix angle equals $\psi_{y.ct} = \psi_{ct}^{\max}$. The difference in pitch helix angles $\Delta\psi_{ct} = -(\psi_{ct} - \psi_{y.ct})$ can be used to evaluate the range of change to the pitch helix angle $\psi_{ct}$ after each regrinding of the hob.

The difference $\Delta\psi_{ct}$ in the pitch helix angle of the gear hob strongly depends on the hob clearance angle $\alpha_o$. The larger the clearance angle $\alpha_o$, the larger the difference $\Delta\psi_{ct}$ in the pitch helix angle observed, and vice versa. Figure 15.52 illustrates the impact of the hob clearance angle $\alpha_o$ on the current value of the deviation $\Delta\psi_{ct}$ for involute gear hobs with various numbers of starts. It is assumed in Figure 15.52 that the maximum tooth depth at which the hob is reground is equal to the hob module, that is, the central angle ~ $0.6\sigma$ is equal to

$$\sim 0.6\sigma = \frac{2 \cdot m}{d_{o.ct}} \tag{15.65}$$

and thus, the equality $\sigma \cong 3.3 \cdot m/d_{o.ct}$ is valid. For the calculation of the deviation $\Delta\psi_{ct}$, the following formula [117, 138]

$$\Delta\psi_{ct} = \sin^{-1}\left(\frac{m \cdot N_{ct}}{d_{o.ct} - 2 \cdot m \cdot \tan\alpha_o}\right) - \sin^{-1}\left(\frac{m \cdot N_{ct}}{d_{o.ct}}\right) \tag{15.66}$$

applies.

Figure 15.52 reveals the strong impact of the clearance angle $\alpha_o$ on the deviation of the pitch helix angle $\Delta\psi_{ct}$ of a gear hob. The larger the number of starts $N_{ct}$ of the gear hob, the more severe the impact of the clearance angle $\alpha_o$ on the deviation $\Delta\psi_{ct}$.

The strong influence of the hob pitch diameter on the resultant accuracy of the hob is well recognized. The average value of pitch diameter $\bar{d}_{ct}$ is commonly used for calculation of the design parameters of an involute gear hob. Here,

$$\bar{d}_{ct} = \frac{d_{ct} - d_{ct}^{worn}}{2} \tag{15.67}$$

where $d_{ct}^{worn}$ designates the pitch diameter of the totally worn involute hob after its last permissible regrinding.

Figure 15.53 illustrates the desired change in value of the hob-setting angle $\zeta_{ct}$ vs. the current value of the gear hob pitch diameter $d_{y.ct}$. Involute hobs of smaller diameter and hobs with a greater number of starts require even more significant change to the hob-setting angle $\zeta_{ct}$. It is good to stress here that the maximum practical number of starts of an involute hob reaches $N_{ct} = 13$. This means that an appropriate change to the current value of the hob-setting angle $\zeta_{ct}$ is of critical importance for precision multiple-start involute gear hobs of small diameter [117, 138].

An inevitable change $\Delta\,p_{b.ct}$ to the base pitch $p_{b.ct}$ of the gear hob is observed if the setting angle of the hob $\zeta_{ct}$ changes in the proper manner, and no corresponding compensation

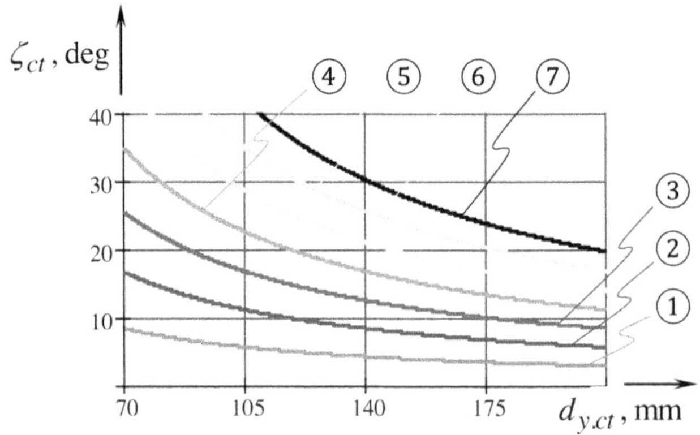


**FIGURE 15.53** Impact of the pitch diameter $d_{y.ct}$ of a hob of the module $m = 10$ mm on the desired value of the hob-setting angle $\zeta_{ct}$ when the number of starts of the hob is ① – $N_{ct} = 1$, ② – $N_{ct} = 2$, ③ – $N_{ct} = 3$, ④ – $N_{ct} = 4$, ⑤ – $N_{ct} = 5$, ⑥ – $N_{ct} = 6$, and ⑦ – $N_{ct} = 7$. (From Radzevich, S.P., *ASME Journal of Mechanical Design*, 129, 1076–1085, 2007. With permission.)

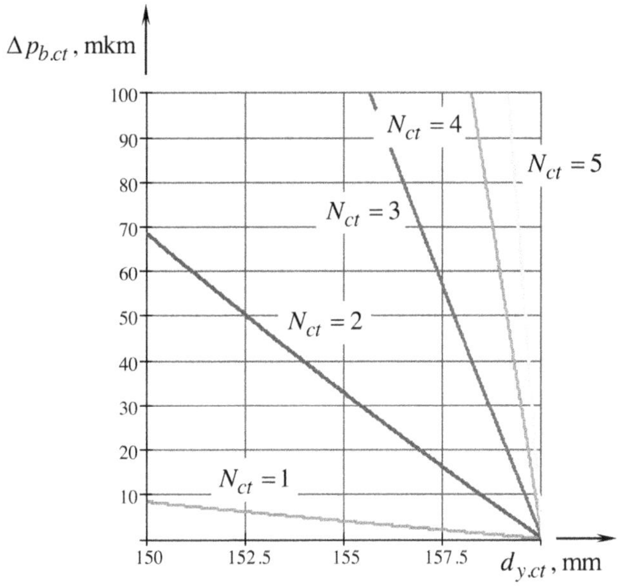


**FIGURE 15.54** Impact of the value of the reduction of the hob pitch diameter $d_{y.ct}$ on the deviations of the hob base pitch $\Delta\, p_{b.ct}$ for involute hobs with various numbers of starts $N_{ct}$. (From Radzevich, S.P., *ASME Journal of Mechanical Design*, 129, 1076–1085, 2007. With permission.)

is made to other design parameters of the hob. The impact of the hob pitch diameter $d_{ct}$ and the hob number of starts $N_{ct}$ on the deviation $\Delta\, p_{b.ct}$ in the base pitch of the hob are illustrated in Figure 15.54. The larger the alteration in the hob pitch diameter $d_{y.ct}$, the larger the deviation $\Delta\, p_{b.ct}$ of the actual base pitch of the hob from its nominal value $p_{b.ct}$. Involute gear hobs having larger numbers of starts $N_{ct}$ are more vulnerable to the changes to the hob pitch diameter (see Figure 15.54). The function

$$\Delta\, p_{b.ct} = \Delta\, p_{b.ct}(d_{y.ct}) \tag{15.68}$$

is nonlinear. However, the curves of the nonlinear function $\Delta\, p_{b.ct}(d_{y.ct})$ are close to straight lines. This is because the interval of alteration in the pitch diameter $d_{y.ct}$ is small. In the case under consideration, this interval is equal to just $d_{y.ct} = (150 - 140) = 10$ mm.

Any changes to the base pitch $p_{b.ct}$ of the gear hob are undesirable. The changes $\Delta\, p_{b.ct}$ are permissible only within a corresponding tolerance $[\Delta\, p_{b.ct}]$ for the alteration of the base pitch $p_{b.ct}$, that is, fulfillment of the inequality

$$\Delta\, p_{b.ct} \le [\Delta\, p_{b.ct}] \tag{15.69}$$

is a must. Otherwise, the accuracy of the hobbed involute gears gets unacceptably poor.

Evidently, an alteration $\Delta d_{ct}$ to hob diameter $d_{ct}$ results in a strongly undesirable change to the base pitch $p_{b.ct}$ of the gear hob.

#### 15.4.3.2 Principal Design Parameters of Involute Gear Hob

The following principal design parameters must be known before beginning to design an involute gear hob:

- The hob module $m$
- The normal profile angle $\phi_n$
- The number of starts $N_{ct}$ of the hob threads
- The desired value of the setting angle of the hob $\zeta_{ct}$

The rest of the important design parameters of an involute gear hob, namely

- The hob outside diameter $d_{o.ct}$
- The number of gashes
- The rake angle $\gamma_o$ of the hob tooth
- The clearance angle $\alpha_o$ of the hob tooth

are incorporated into the later stages of the design process.

The design parameters, that is, the gear hob module $m$ and the hob profile angle $\phi_n$, yield immediate calculation of the hob base pitch:

$$p_{b.ct} = \pi \cdot m \cdot \cos\phi_n \tag{15.70}$$

The axial pitch of the hob threads is calculated by the formula

$$P_{x.ct} = \frac{\pi \cdot m \cdot N_{ct}}{\cos\zeta_{ct}} \tag{15.71}$$

For the calculation of base diameter $d_{b.ct}$ of the hob, the following equation [110, 138, 159]

$$d_{b.ct} = \frac{m \cdot N_{ct} \cdot \cos\phi_n}{\sqrt{1 - \cos^2\phi_n \cdot \cos^2\zeta_{ct}}} \tag{15.72}$$

is applied.

Ultimately, the gear hob base helix angle $\psi_{b.ct}$ is calculated from the equation [138, 159]

$$\psi_{b.ct} = \cos^{-1}(\cos\phi_n \cdot \cos\zeta_{ct}) \tag{15.73}$$

Equation (15.73) also allows for representation in the form

$$\tan\psi_{b.ct} = \frac{\sqrt{\sin^2\phi_n + \tan^2\zeta_{ct}}}{\cos\phi_n} \tag{15.74}$$

which has proven to be convenient in certain applications.

The base lead angle $\lambda_{b.ct}$ is equal to $\lambda_{b.ct} = 90° - \psi_{b.ct}$.

Further, the base pitch of the hob $p_{b.ct}$ can be expressed in terms of the hob axial pitch $P_{x.ct}$ of the normal pressure angle $\phi_n$ and of the setting angle of the hob $\zeta_{ct}$:

$$p_{b.ct} = P_{x.ct} \cos\phi_n \cos\zeta_{ct} \tag{15.75}$$

This equation becomes useful when performing an analysis of the geometry of the involute gear hob and the kinematics of meshing in the gear hobbing operation.

If the axial pitch of the hob threads $P_{x.ct}$ is calculated from Eq. (15.71), then for the calculation of the hob pitch helix $\psi_{ct}$ angle, the equation

$$\psi_{ct} = \tan^{-1}\left(\frac{P_{x.ct}}{\pi \cdot d_{ct}}\right) \tag{15.76}$$

can be used.

Equation (15.71) yields a simple formula for the calculation of the pitch helix angle $\psi_{y.ct}$ of the gear hob:

$$\tan\psi_{y.ct} = \frac{P_{x.ct}}{\pi \cdot d_{y.ct}} = \frac{\pi \cdot m \cdot N_{ct}}{\pi \cdot d_{y.ct} \cdot \cos\zeta_{ct}} = \frac{m \cdot N_{ct}}{d_{y.ct} \cdot \cos\zeta_{ct}} \tag{15.77}$$

Equating the setting angle of the hob $\zeta_{ct}$ to the pitch helix angle $\psi_{y.ct}$ of the hob (that is, targeting the equality $\zeta_{y.ct} = \psi_{y.ct}$ to be valid), Eq. (15.77) yields the expression

$$\tan\zeta_{y.ct} = \frac{m \cdot N_{ct}}{d_{y.ct} \cdot \cos\zeta_{y.ct}} \tag{15.78}$$

Equation (15.78) reduces to

$$\sin\zeta_{y.ct} = \frac{m \cdot N_{ct}}{d_{y.ct}} \tag{15.79}$$

If the current value of the hob-setting angle $\zeta_{y.ct}$ is calculated from Eq. (15.79), then the equality $\zeta_{y.ct} = \psi_{y.ct}$ is observed for a new involute gear hob, and for the hob that has been reground as well.

The above equations are derived on the premise of the following.

Consider the auxiliary generating rack $\mathscr{R}_T$ of an involute gear hob. The tooth profile angle of the rack $\mathscr{R}_T$ is designated as $\phi_n$, and the base pitch of the rack is designated as $p_{b.\mathscr{R}}$. Other design parameters of the rack $\mathscr{R}_T$ (the tooth addendum $a_{\mathscr{R}}$, the tooth dedendum $b_{\mathscr{R}}$, the whole depth $h_{\mathscr{R}}$, the pitch $P_{n.\mathscr{R}}$, the tooth thickness $t_{\mathscr{R}}$, the space width $w_{\mathscr{R}}$) are also known. Commonly, most of the design parameters of the auxiliary generating rack $\mathscr{R}_T$ can be expressed in terms of module $m$.

Initially, the auxiliary rack $\mathscr{R}_T$ is described analytically in a left-hand-oriented Cartesian coordinate system $X_{ct}Y_{ct}Z_{ct}$ (see Figure 15.55). The pitch plane of the rack is positioned at a distance $d_{ct}/2$ from the $Z_{ct}$ – axis and is oriented parallel to the coordinate plane $X_{ct}Z_{ct}$. The axis of rotation $O_{ct}$ of the hob is aligned to the $Z_{ct}$ – axis. The auxiliary rack $\mathscr{R}_T$ is turned through the setting angle of the gear hob $\zeta_{ct}$ with respect to the hob axis of rotation $O_{ct}$.

In the coordinate system $X_{ct}Y_{ct}Z_{ct}$, the rack $\mathscr{R}_T$ performs a screw motion about the $Z_{ct}$ – axis. The screw parameter of the screw motion (the reduced pitch of the screw motion) of the rack $\mathscr{R}_T$ is designated as $p_{\mathscr{R}}$.

The screw motion of the auxiliary generating rack $\mathscr{R}_T$ can be dissolved onto two elementary motions, namely, onto (a) the rotation $\omega_{\mathscr{R}}$ of the rack $\mathscr{R}_T$ about the $Z_{ct}$ – axis, and (b) the translation $\mathbf{V}_{\mathscr{R}}$ of the rack $\mathscr{R}_T$ along the $Z_{ct}$ – axis.

Next, it is convenient to dissolve the translational motion $\mathbf{V}_{\mathscr{R}}$ onto two elementary motions:

$$\mathbf{V}_{\mathscr{R}} = \mathbf{V}_z + \mathbf{V}_0 \tag{15.80}$$

The translational motion with the speed $\mathbf{V}_z$ is pointed in the direction that is parallel to the $Z_{ct}$ – axis. The translational motion $\mathbf{V}_0$ is pointed in the direction of the generating line of the auxiliary rack $\mathscr{R}_T$. The velocity of the translation $\mathbf{V}_0$ results in sliding of the rack $\mathscr{R}_T$ on itself. This motion does not affect the shape of the enveloping surface $T$. Therefore, the translational motion $\mathbf{V}_0$ can be eliminated from further analysis.

The magnitude $V_z$ of the speed vector $\mathbf{V}_z$ equals

$$V_z = \frac{|\mathbf{V}_{\mathscr{R}}|}{\cos\zeta_{ct}} \tag{15.81}$$

Therefore, the screw parameter of the screw motion of the auxiliary rack $\mathscr{R}_T$ can be calculated from the formula

$$p_{\mathscr{R}} = \frac{V_z}{|\omega_g|} = \frac{d_{ct} \cdot \omega_{\mathscr{R}}}{2\omega_g \cdot \cos\zeta_{ct}} \tag{15.82}$$

For a single-start involute hob, Eq. (15.82) reduces to

$$p_{\mathscr{R}} = \frac{d_{ct}}{2N_g \cdot \cos\zeta_{ct}} \tag{15.83}$$

where $N_g$ denotes the number of teeth of the hobbed gear.

The design parameters of generating surface $T$ of the gear hob can be determined in the following way.

The characteristic line $\mathscr{E}_{\mathscr{R}}$ can be interpreted as a locus of points on the auxiliary generating surface $\mathscr{R}_T$, at which the unit perpendicular $\mathbf{n}_{\mathscr{R}}$ to the rack surface $\mathscr{R}_T$ forms a certain angle $\varepsilon$ with the axis of the screw motion $O_{ct}$. In other words, the characteristic line $\mathscr{E}_{\mathscr{R}}$ can be viewed as a locus of points at which the unit normal vector $\mathbf{n}_{\mathscr{R}}$ is perpendicular

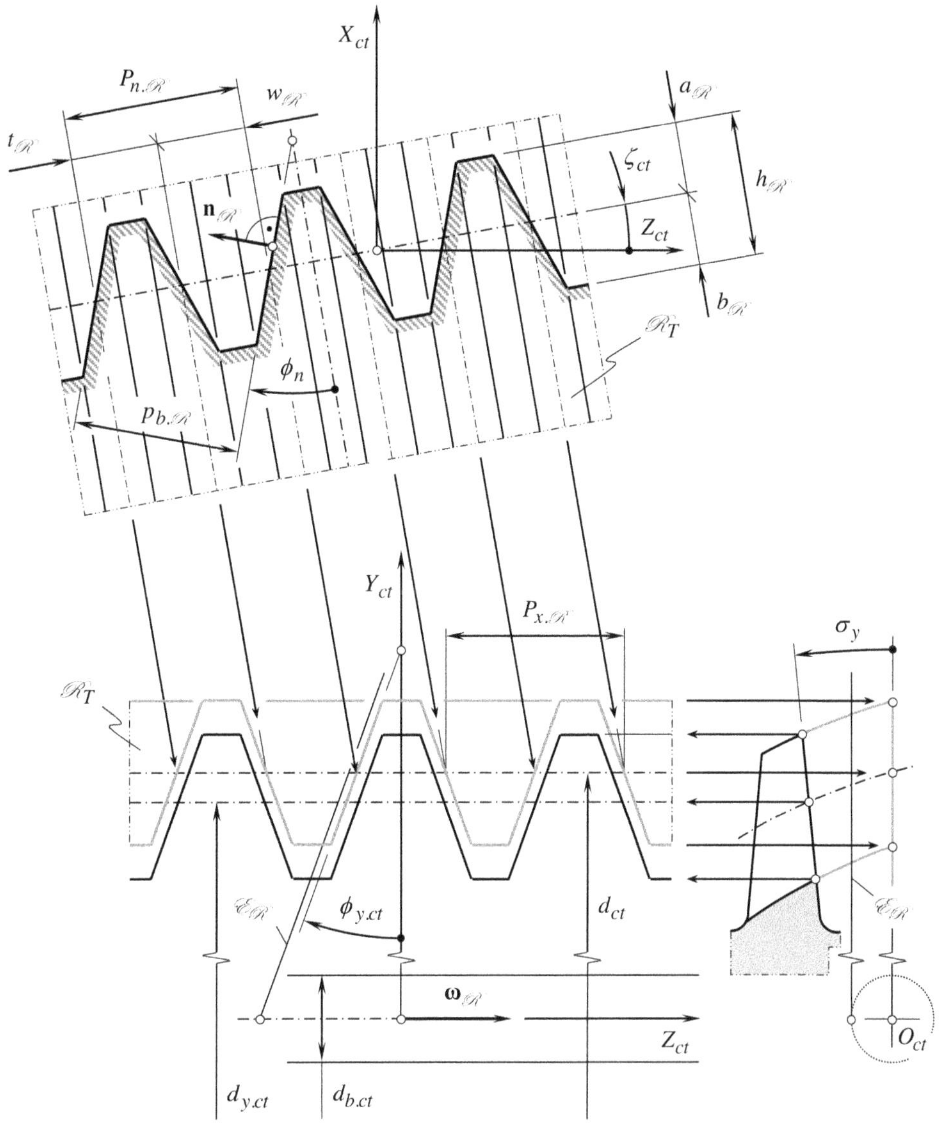


**FIGURE 15.55** Schematic for the calculation of the desired changes to the parameters of the kinematic geometry of an involute gear hob. (From Radzevich, S.P., *ASME Journal of Mechanical Design*, 129, 1076–1085, 2007. With permission.)

to the velocity vector $\mathbf{V}_{\mathscr{R}}$. For the calculation of the angle $\varepsilon$, the *Ball equation* [1, 77]

$$\tan\varepsilon = \frac{p_{\mathscr{R}}}{r_{y.ct}} \tag{15.84}$$

applies. Here $r_{y.ct}$ denotes the shortest distance of approach of the straight line along the unit normal vector $\mathbf{n}_{\mathscr{R}}$ and of the axis $Z_{ct}$ of the screw motion.

The angle that the tooth flank of the auxiliary rack $\mathscr{R}_T$ forms with the axis $Z_{ct}$ of the screw motion is designated as $\rho$. Thus, the unit normal vector $\mathbf{n}_{\mathscr{R}}$ forms the angle $(90° - \rho)$ with the axis $Z_{ct}$ of the screw motion. Owing to this, at points within the characteristic line $\mathscr{E}_{\mathscr{R}}$, the following equalities

$$\tan(90° - \rho) = \frac{p_{\mathscr{R}}}{r_{y.ct}} \tag{15.85}$$

$$r_{y.ct} = p_{\mathscr{R}} \cdot \tan\rho \tag{15.86}$$

are valid.

Equations (15.85) and (15.86) reveal that the characteristic line $\mathscr{E}_{\mathscr{R}}$ is a straight line that forms the angle $\rho$ with the $Z_{ct}$ – axis, and which is remote from the axis $Z_{ct}$ at a distance $r_{y.ct}$.

For further analysis, it is necessary to determine the actual value of the angle $\rho$. For this purpose, a unit vector $\mathbf{a}$ is constructed. The vector $\mathbf{a}$ is pointed along the axis $Z_{ct}$ of the screw motion. Therefore, in the Cartesian coordinate system $X_{ct}Y_{ct}Z_{ct}$, this vector can be represented in the form

$$\mathbf{a} = \mathbf{k} \tag{15.87}$$

In that same reference system $X_{ct}Y_{ct}Z_{ct}$, the unit normal vector $\mathbf{n}_{\mathscr{R}}$ can be represented as

$$\mathbf{n}_{\mathscr{R}} = \mathbf{i}\cdot\cos\phi_n\cdot\cos\zeta_{ct} + \mathbf{j}\cdot\sin\phi_n - \mathbf{k}\cdot\cos\phi_n\cdot\sin\zeta_{ct} \quad (15.88)$$

The vectors $\mathbf{a}$ and $\mathbf{n}_{\mathscr{R}}$ form the angle $\rho$. Therefore, the equality

$$\tan\rho = \frac{\mathbf{a}\cdot\mathbf{n}_{\mathscr{R}}}{|\mathbf{a}\times\mathbf{n}_{\mathscr{R}}|} \quad (15.89)$$

is valid.

After substituting Eq. (15.87) and Eq. (15.88), Eq. (15.89) casts into

$$\tan\rho = \frac{\cos\phi_n}{\sqrt{\sin^2\phi_n + \tan^2\zeta_{ct}}} \quad (15.90)$$

When the actual value of the angle $\rho$ is known, then the base diameter of the gear hob is calculated from

$$d_{b.ct} = 2\cdot p_{\mathscr{R}}\cdot\tan\rho = \frac{N_{ct}}{N_g}\cdot\frac{d_{ct}\cdot\cos\phi_n}{\sqrt{1-\cos^2\phi_n\cos^2\zeta_{ct}}} \quad (15.91)$$

In a specific case of the design of an involute gear hob for which tooth thickness $t_{\mathscr{R}}$ on the hob pitch diameter is equal to the hob space width $w_{\mathscr{R}}$, Eq. (15.91) reduces to

$$d_{b.ct} = \frac{m\cdot N_{ct}\cdot\cos\phi_n}{\sqrt{1-\cos^2\phi_n\cos^2\zeta_{ct}}} \quad (15.92)$$

In Eq. (15.92), the advantage of independence of the hob base diameter $d_{b.ct}$ from the hobbed gear tooth number $N_g$ is taken.

The equations derived above yield for the calculation of the hob axial pitch

$$P_{x.c} = 2\cdot\pi\cdot p_{\mathscr{R}} = \frac{\pi\cdot d_{ct}}{N_{ct}\cdot\cos\zeta_{ct}} \quad (15.93)$$

The normal pitch is measured on the unfolded base cylinder of the gear hob. It is equal to

$$p_{b.c} = p_{x.c}\cdot\sin\rho = \frac{\pi\cdot d_g}{N_g} = p_{b.g} \quad (15.94)$$

The fulfillment of the condition $p_{b.ct} = p_{b.g}$ is a must when designing involute gear hobs. This means that various hobs with different normal profile angles $\phi_n$ but with the same base pitch $p_{b.ct}$ can be used for hobbing of that same involute gear. However, the base pitch $p_{b.ct}$ of all of them must be the same, and it also must be equal to the base pitch of the involute gear being hobbed.

The set of equations derived above allow for the calculation of any and all principal design parameters of an involute gear hob.

#### 15.4.3.3 Elements of Kinematic Geometry of an Involute Gear Hob

The base pitch $p_{b.ct}$ of the hob must be exactly equal to the base pitch of the gear $p_{b.g}$ to be hobbed. The equality $p_{b.ct} = p_{b.g}$ reflects the principal condition for high accuracy of the hobbed gear. Because the equality $p_{b.ct} = p_{b.g}$ should always be observed, the involute hob base pitch $p_{b.ct}$ must remain the same value after regrinding of the hob. However, according to Eq. (15.75) and Eq. (15.78), change to the hob pitch diameter $d_{ct}$ causes a strongly undesired change to the hob base pitch $p_{b.ct}$. Ultimately, this results in the hob lateral cutting edges shifting from the generating surface $T$. In this way, the shift of the cutting edges causes errors of the hob tooth profile.

##### *15.4.3.3.1 Desired Change to Normal Profile Angle*

In order to maintain the high accuracy of the gear hob, we must keep the setting angle of the hob $\zeta_{ct}$ equal to the pitch helix angle $\psi_{ct}$ of the hob. This means that fulfillment of the equality $\zeta_{ct} = \psi_{ct}$ is one of the prerequisites for high accuracy of involute gear hobs [53, 128]. According to Eq. (15.75) and Eq. (15.78), the following equation for the calculation of the base pitch $p_{b.ct}$ is valid:

$$p_{b.ct} = \frac{d_{ct}\cdot P_{x.ct}}{\sqrt{d_{ct}^2 + m^2\cdot N_{ct}^2}}\cdot\cos\phi_n \quad (15.95)$$

where:

- $p_{b.ct}$ is the base pitch of the gear hob.
- $d_{ct}$ is the pitch diameter of the hob.
- $P_{x.ct}$ is the axial pitch of the hob.
- $m$ is the hob module.
- $N_{ct}$ is the number of starts of the hob (sometimes this design parameter is incorrectly referred to as *number of threads*).
- $\phi_n$ is the normal profile angle of the hob.

It is evident that the change to the pitch diameter after regrinding of the gear hob is inevitable. The hob diameters, namely (a) the pitch diameter $d_{ct}$, (b) the outer diameter $d_{o.ct}$, and (c) the limit diameter $d_{l.ct}$ get smaller after every regrinding of the hob.

Change to the pitch helix angle indicates that the parameters of relative motion of the worn hob with respect to the work gear differ from the corresponding parameters of relative motion of the new hob with respect to that same work gear. This means that the kinematics of meshing of the work gear with the new hob differ from the kinematics of meshing of the work gear with the worn hob. This also makes clear that the accuracy of an involute hob does not remain the same after the hob had been reground. The smaller the diameter of the gear hob, the larger the deviations of the lateral cutting edges from the generating surface $T$ of the hob. In order to accommodate for the deviations, a method for relieving involute hobs has been developed by Prof. S.P. Radzevich [92].

The proposed method for relief grinding of an involute gear hob[15] is a perfect example of implementation of the $DG/K$ – based method of surface generation [151] for solving a problem that pertains to hobbing of involute gears.

The reduction of the pitch diameter after regrinding of a gear hob with relieved teeth is inevitable. Therefore, the negative impact of this alteration to the hob base pitch must be compensated somehow. The required compensation can be performed by a corresponding change to another design parameter of the hob on the corresponding precalculated values.

Figure 15.51 illustrates the configurations of the auxiliary rack of the new gear hob and of the reground hob. The change of the configuration of the rake surface of the gear hob (as it is illustrated in Figure 15.55) is identical to that illustrated in Figure 15.51. The major parameters of kinematic geometry of the hob are indicated in Figure 15.55.

The straight-line generator (the characteristic line $\mathscr{E}_{\mathscr{R}}$) is tangent to the base cylinder of a diameter $d_{b.ct}$ of the gear hob. It also forms the base helix angle $\psi_{b.ct}$ with the axis $Z_{ct}$. Therefore, the characteristic line $\mathscr{E}_{\mathscr{R}}$ is tangent to the base helix of the hob. The profile angle of the auxiliary rack $\mathscr{R}_T$ in the axial section of the hob is designated as $\phi_{y.ct}$.

An analysis of Eq. (15.72) shows that the gear hob profile angle is the only design parameter that can be helpful for compensation of the inevitable alteration of the gear hob diameter. When the hob pitch diameter gets smaller, the hob pressure angle has to be properly changed in a manner that guarantees the hob base pitch $p_{b.ct}$ remains at a constant value.

An equation for the calculation of the required alteration to the hob profile angle $\phi_{y.ct}$ is derived below.

Configuration of the reground hob rake surface $R_s$ with respect to its configuration for the new hob can be expressed in terms of the current value of the central angle $\sigma_y$ (see Figure 15.55). Here, the angle $\sigma_y$ falls into the interval $0.6 \cdot \sigma \leq \sigma_y \leq 0°$. The design parameters $d_{y.ct}$ and $\phi_{y.ct}$ of the hob are functions of the angle $\sigma_y$. This means that the equations

$$d_{y.ct} = d_{y.ct}(\sigma_y) \tag{15.96}$$

$$\phi_{y.ct} = \phi_{y.ct}(\sigma_y) \tag{15.97}$$

are valid. The generalized form of these equations needs to be expressed in terms of the design parameters of the gear hob.

In order to maintain the gear hob base pitch at a constant value (and not changing after the hob regrinding), the equality $\partial p_{b.ct} / \partial \sigma_y = 0$ must be fulfilled. By differentiating Eq. (15.81) with respect to the angle $\sigma_y$, one can come up with the differential equation

$$\tan \phi_{y.ct}\, \partial \phi_{y.ct} = \left( \frac{1}{d_{y.ct}} - \frac{d_{y.ct}}{d_{y.ct}^2 + m \cdot N_{ct}} \right) \partial d_{y.ct} \tag{15.98}$$

Here, in Eq. (15.98), the variables [$\phi_{y.ct}(\sigma_y)$ and $d_{y.ct}(\sigma_y)$] are split. This is mostly due to both being dependent not on the angle $\sigma_y$, but on the pitch diameter variation $\Delta d_{y.ct}$. Therefore, separate integration of the left-hand side and of the right-hand side of Eq. (15.98) is valid:

$$\int_{\phi_n}^{\phi_{y.ct}} \tan \phi_{y.ct}\, \partial \phi_{y.ct} = \int_{d_{ct}}^{d_{y.ct}} \left( \frac{1}{d_{y.ct}} - \frac{d_{y.ct}}{d_{y.ct}^2 + m \cdot N_{ct}} \right) \partial d_{y.ct} \tag{15.99}$$

The intervals of integration (from $\phi_n$ to $\phi_{y.ct}$ and from $d_{ct}$ to $d_{y.ct}$) in Eq. (15.99) are semi-open.

Ultimately, the equation for the calculation of the desired current value of the profile angle $\phi_{y.ct}$ of the gear hob tooth can be represented in the form

$$\phi_{y.ct} = \arccos \left( \frac{d_{ct}}{d_{y.ct}} \cdot \sqrt{\frac{d_{y.ct}^2 + m^2 \cdot N_{ct}^2}{d_{ct}^2 + m^2 \cdot N_{ct}^2}} \cdot \cos \phi_n \right) \tag{15.100}$$

A 3D graph that illustrates the influence of the design parameters of the gear hob on the desirable value of the hob profile angle $\phi_{y.ct}$ is plotted in Figure 15.56. Because the deviations of the hob normal pressure angle are small, the 3D plot in Figure 15.56 represents the visualization only of the

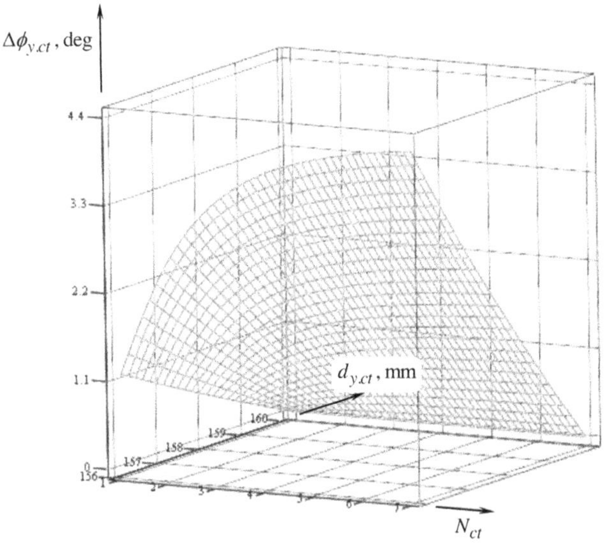


**FIGURE 15.56** Impact of the number of starts $N_{ct}$ of a gear hob, and of a hob current diameter $d_{y.ct}$ on the required precalculated value of the correction $\Delta\phi_{y.ct}$ to the hob profile angle $\phi_n$. (From Radzevich, S.P., *ASME Journal of Mechanical Design*, 129, 1076–1085, 2007. With permission.)

[15] SU Pat. No. 1743810, *A Method of Relief Grinding of Gear Hob Teeth*/ S.P. Radzevich, Int. Cl. B24b 3/12, Filed: September 13, 1989.

function "$\Delta\phi_{y.ct}$ vs. $d_{y.ct}$ and $N_{ct}$" (here $\Delta\phi_{y.ct} = \phi_n - \phi_{y.ct}$). The function "$\phi_{y.ct}$ vs. $d_{y.ct}$ and $N_{ct}$" is not shown in Figure 15.56.

The required alteration $\Delta\phi_n$ to the gear hob pressure angle $\phi_n$ can be expressed in terms of the current value of the pitch diameter $d_{y.ct}$ of the hob. The larger the alterations to the hob pitch diameter $d_{y.ct}$, the larger the precalculated desired alterations $\Delta\phi_n$ to the hob profile angle required. Again, for a multiple-start involute hob, the function $\Delta\phi_n = \Delta\phi_n(d_{y.ct})$ is much more vulnerable to the deviations of the actual value of pitch diameter compared to its nominal value (see Figure 15.56). As in the example considered above, the function $\Delta\phi_n = \Delta\phi_n(d_{y.ct})$ is a nonlinear one. Curves of the nonlinear function $\Delta\phi_n(d_{y.ct})$ are close to straight lines. This is because the interval of alteration of pitch diameter $d_{y.ct}$ is small, and it is only equal to $d_{y.ct} = 10$ mm. The increased interval of alteration of the diameter $d_{y.ct}$ makes the nonlinearity of the function $\Delta\phi_n(d_{y.ct})$ visible.

The following interpretation of the simultaneous change of the profile angle $\phi_n$ and of the hob-setting angle $\zeta_{ct}$ can be given.

Consider a lateral tooth surface of the auxiliary generating rack $\mathscr{R}_T$. The straight-line generator (that is, the characteristic line $\mathscr{E}_{\mathscr{R}}$) of the generating surface $T$ of the gear hob is located within this plane. Assume then that the lateral plane of the tooth turns through a certain angle about the straight-line generator $\mathscr{E}_{\mathscr{R}}$. This turn of the plane results in a corresponding change to the current value of the normal profile angle $\phi_n$ as well as to the setting angle of the hob $\zeta_{ct}$. The current values of the angles $\phi_n$ and $\zeta_{ct}$ are changing in the way that follows the rotation of the lateral plane of the rack $\mathscr{R}_T$ about the straight-line generator $\mathscr{E}_{\mathscr{R}}$. However, under such the rotation of the lateral plane of the auxiliary rack $\mathscr{R}_T$, the following design parameters of the hob, namely, (a) the base diameter $d_{b.ct}$, (b) the base helix angle $\psi_{b.ct}$, and (c) the base pitch $p_{b.ct}$, remain constant.

In order to accommodate for the deviations of base pitch $p_{b.ct}$ caused by the changes to the pitch diameter of the gear hob after every regrinding of the hob, a novel design of the gear hob is proposed by Prof. S.P. Radzevich [103]. The gear hob[16] features the ground relief clearance surfaces through the lateral cutting edges of the hob tooth, the geometry of which allows for fulfillment of the requirements imposed by Eq. (15.100).

The clearance surfaces of precision involute gear hobs of the proposed design can be ground. For this purpose, several methods for relief grinding of the hob were developed by Prof. S.P. Radzevich [87, 88, 197]. Use of the methods[17] makes it possible to keep the tooth profile angle $\phi_{y.ct} < \phi_n$ of the gear hob smaller after each regrinding operation in compliance with Eq. (15.100). The reduction of the gear hob profile angle on the precalculated value retains the constant value of the hob base pitch ($p_{b.ct} = const$). In this way, one of the major sources of the hob errors is eliminated. Ultimately, the gear hob accuracy is not reduced after each regrinding of the hob.

#### 15.4.3.3.2 Desired Value of Center Distance in Gear Hobbing Operation

Every regrinding of an involute gear hob causes a reduction of its diameter. Evidently, this reduction must be compensated by the corresponding reduction of center distance in the gear hobbing operation. At first glimpse, the required reduction $\Delta C_y$ of the center distance must be equal to half of the reduction of the hob diameter. However, because of the alteration into the gear hob normal profile angle, the desired value of the center distance depends on the current value of the hob profile angle $\phi_{y.ct}$.

The tooth ratio $u_{ct}$ in the gear hobbing operation is a constant parameter. This parameter is equal: $u_{ct} = N_g / N_{ct}$, and it does not depend on how significantly the gear hob diameter changes. From a diagram of meshing of the hob and of the work gear in the gear hobbing operation, the following formula for the calculation of the $\Delta C(\sigma_y)$ is derived:

$$\Delta C_y(\sigma_y) = \frac{1}{2}\left[ d_{ct} \cdot \sigma_y \cdot \tan\alpha_o + m \cdot N_g \cdot \left( 1 - \frac{\cos\phi_n}{\cos\phi_{y.ct}} \right) \right] \quad (15.101)$$

The performed analysis of Eq. (15.101) reveals that the required reduction $\Delta C_y(\sigma_y)$ of the center distance $C$ in the gear hobbing operation depends upon the tooth number of the work gear. Figure 15.57 illustrates an example of impact of the tooth number of the work gear $N_g$ and of

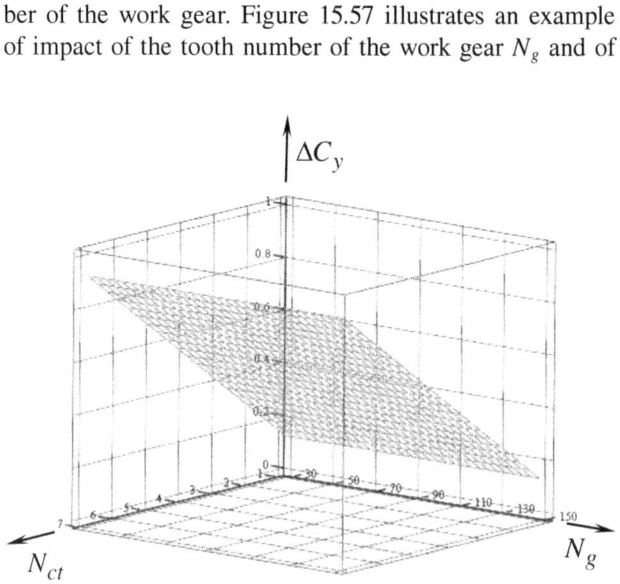


**FIGURE 15.57** Impact of the tooth number of the work gear $N_g$ and of the number of starts $N_{ct}$ of a gear hob on the required precalculated value of the correction $\Delta C_y$ to the center distance $C$. (From Radzevich, S.P., *ASME Journal of Mechanical Design*, 129, 1076–1085, 2007. With permission.)

[16] SU Pat. No. 1004029, *A Precision Gear Hob*/S.P. Radzevich, Int. Cl. B23f 21/16, Filed: November 6, 1979.

[17] SU Pat. No. 831546, *A Method of Relief Grinidg of Precision Gear Hob*/ Roding, P.R. and Radzevich, S.P., Int. Cl. B24b 3/00, Filed: July 2, 1979.

SU Pat. No. 965728, *A Method of Relief Grinidg of Precision Gear Hob*/S.P. Radzevich, Int. Cl. B24b 3/12, Filed: January 21, 1980.

SU Pat. 1365550, *A Method of Relief Grinding of Precision Gear Hob*/S.P. Radzevich, Int. Cl. B24b 3/12, Filed: April 29, 1986.

the hob number of starts $N_{ct}$ on the precalculated desired value of the correction $\Delta C_y(\sigma_y)$ to the center distance $C$. Because the alteration $\Delta C_c(\sigma_y)$ caused by the reduction to the gear hob diameter itself is small compared to the changes caused by reduction of the hob profile angle, the inequality $\Delta C_c(\sigma_y) > \Delta C_y(\sigma_y)$ is observed. Only one portion, $\Delta C_y(\sigma_y)$, of $\Delta C(\sigma_y)$ is plotted in Figure 15.57. For this purpose, Eq. (15.101) is represented in the form $\Delta C(\sigma_y) = \Delta C_c(\sigma_y) + \Delta C_y(\sigma_y)$:

$$\Delta C(\sigma_y) = \underbrace{\frac{1}{2} \cdot d_{ct} \cdot (\sigma_y) \cdot \tan\alpha_o}_{\Delta C_c(\sigma_y)} + \underbrace{\frac{m \cdot N_g}{2} \cdot \left(1 - \frac{\cos\phi_n}{\cos\phi_{y.ct}}\right)}_{\Delta C_y(\sigma_y)} \quad (15.102)$$

The portion $\Delta C_c(\sigma_y)$ depends on the reduction of the gear hob diameter. It does not depend on alteration of the profile angle after the hob had been reground.

Work gears with a larger tooth count require a slightly smaller reduction $\Delta C(\sigma_y)$ to the center distance. The impact of the hob number of starts onto the parameter $\Delta C(\sigma_y)$ is the strongest. The larger reduction $\Delta d_{y.ct}$ of the gear hob diameter causes the larger required reduction $\Delta C(\sigma_y)$ to the center distance.

The desired change to the parameters of the kinematic geometry of an involute gear hob can be achieved in one of two possible methods.

The first method is based on implementation of a precalculated modification of the clearance surface of the hob tooth. The modification of this type can be achieved on the hob relieving operation. Therefore, the first method for the improvement of the hob accuracy can be recommended for manufacturers of involute hobs.

The second method is based on the use of precalculated desired deviations of the rake surface of the gear hob tooth, and thus it can be used when regrinding the hob. This method for increasing the gear hob accuracy is briefly considered immediately below.

*15.4.3.3.3 Desired Correction to the Configuration of the Rake Surface of the Gear Hob*

As an example of the application of this approach, a method of corrective regrinding of an involute gear hob is discussed [117, 138].

Consider an involute gear hob with a plane rake surface that is parallel to the hob axis of rotation $O_{ct}$. The top cutting edge of the hob tooth has a positive rake angle $\gamma_o > 0°$. The clearance angle of that same cutting edge of the hob tooth equals $\alpha_o$. For a certain current value of the hob pitch diameter $d_{y.ct}$, a corresponding correction $\Delta\gamma_o$ to the rake angle can be calculated on the premise of the function $\phi_{y.ct} = \phi_{y.ct}(d_{y.ct})$ [see Eq. (15.100)]. Without going into details of derivation of equations, the implemented approach can be briefly summarized in the following way.

For the derivation of an equation for the calculation of the desired correction $\Delta\gamma_o$, consider three vectors **A**, **B**, and **c** through point of interest $m$ within the lateral cutting edge $CE$ of the hob after it has been reground (see Figure 15.58). The vectors **A**, **B**, and **c** are constructed so that they comprise a set of coplanar vectors. All the vectors are located within a certain plane that is tangent to the work gear tooth surface, and they can be expressed in terms of the known design parameters of the gear hob and the design parameter to be determined.

Three vectors **A**, **B**, and **c** can be expressed analytically as follows:

$$\mathbf{A} = -\mathbf{i} \cdot \tan\gamma_o - \mathbf{j} \cdot \cos\phi_n + \mathbf{k} \cdot \sin\phi_n \quad (15.103)$$

$$\mathbf{B} = -\mathbf{i} \cdot \tan(\gamma_o + \Delta\gamma_o) - \mathbf{j} \cdot \cos\phi_{y.ct} - \mathbf{k} \cdot \sin\phi_{y.ct} \quad (15.104)$$

$$\mathbf{c} = -\mathbf{i} \cdot \cos\alpha_o - \mathbf{j} \cdot \sin\alpha_o \quad (15.105)$$

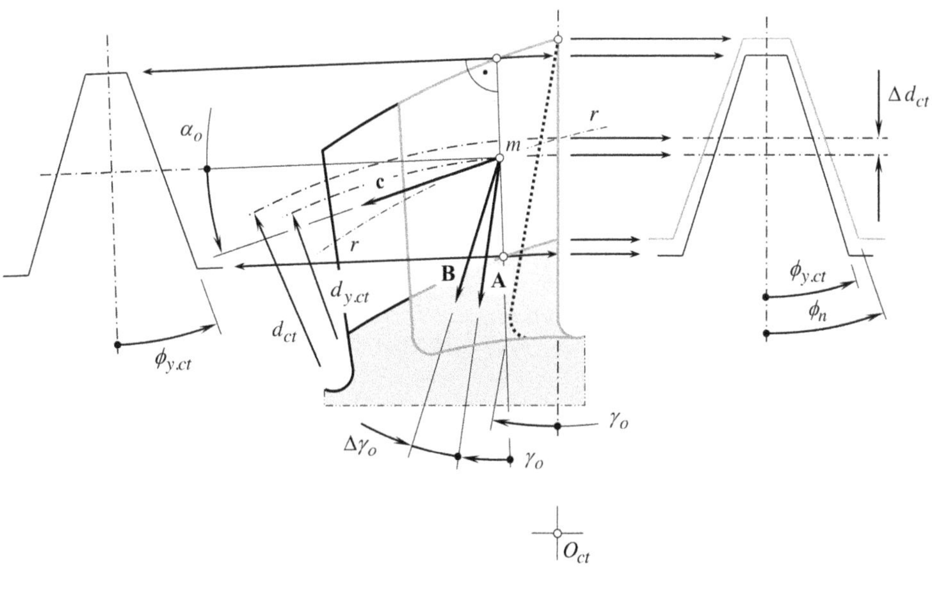


**FIGURE 15.58** Schematic to the calculation of the desired correction $\Delta\gamma_o$ to the hob rake angle $\gamma_o$. (From Radzevich, S.P., *ASME Journal of Mechanical Design*, 129, 1076–1085, 2007. With permission.)

By construction, the vectors **A**, **B**, and **c** are coplanar. Therefore, the triple scalar product of the vectors **A**, **B**, and **c** is identical to zero, that is, $\mathbf{A}\times\mathbf{B}\cdot\mathbf{c}\equiv 0$. This yields a determinant:

$$\mathbf{A}\times\mathbf{B}\cdot\mathbf{c}=\begin{vmatrix} -\tan\gamma_o & -\cos\phi_n & \sin\phi_n \\ -\tan(\gamma_o+\Delta\gamma_o) & -\cos\phi_{y.ct} & -\sin\phi_{y.ct} \\ -\cos\alpha_o & -\sin\alpha_o & 0 \end{vmatrix}=0 \tag{15.106}$$

After been expanded, the determinant casts into the formula for the calculation of the desired value of the correction $\Delta\gamma_o$:

$$\Delta\gamma_o=\arctan\left(\frac{\sin\phi_{y.ct}\tan\gamma_o+\sin(\phi_n-\phi_{y.ct})\cot\alpha_o}{\sin\alpha_o}\right)-\gamma_o \tag{15.107}$$

For involute gear hobs with the rake angle of zero ($\gamma_o = 0°$), Eq. (15.107) reduces to

$$\Delta\gamma_o=\arctan\left[\frac{\sin(\phi_n-\phi_{y.ct})\cdot\cot\alpha_o}{\sin\alpha_o}\right] \tag{15.108}$$

For an angle $\vartheta$ of relatively small value, the approximate equality $\tan\vartheta\cong\sin\vartheta\cong\vartheta$ is valid. (Here the angle $\vartheta$ is measured in radians.) Taking into consideration the latter equality, Eq. (15.108) can be simplified to

$$\Delta\gamma_o=\frac{(\phi_n-\phi_{n.y})}{\sin\phi_n\cdot\tan\alpha_o} \tag{15.109}$$

Figure 15.59 illustrates the influence of the current value of the pitch diameter $d_{y.ct}$ of the gear hob and of the number of starts of the hob $N_{ct}$ on the desired value of the correction $\Delta\gamma_o$ to the rake angle $\gamma_o$. The larger the change to the gear hob pitch diameter, and the larger the number of starts of the hob, the larger the precalculated desired value of the correction $\Delta\gamma_o$ to the rake angle $\gamma_o$.

Calculations reveal that for the difference in pressure angle of only $(\phi_n-\phi_{y.c})=10'$, the required value of the angular correction $\Delta\gamma_o$ can reach $\Delta\gamma_o=2.8°$. This numerical example clearly indicates the necessity and importance of implementing the precalculated corrections to the rake angle $\gamma_o$ when regrinding precision involute gear hobs. The larger the hob clearance angle $\alpha_o$, the smaller the precalculated value of the deviation $\Delta\gamma_o$ needed [117, 138].

*Numerical example.* Use of the disclosed approach allows for the calculation of the desirable change to the parameters of the kinematic geometry of the involute gear hob with the design parameters, summarized in Table 15.2.

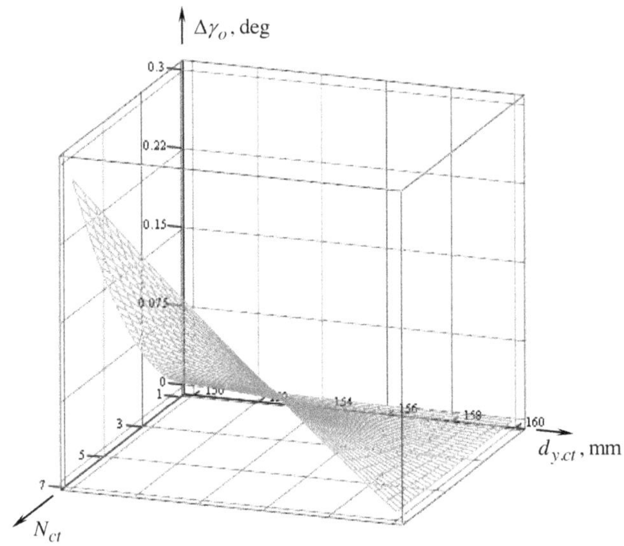


**FIGURE 15.59** Impact of the current diameter $d_{y.ct}$ of a gear hob and of the hob number of starts $N_{ct}$ on the desired precalculated value of the correction $\Delta\gamma_o$ to the hob rake angle $\gamma_o$. (From Radzevich, S.P., *ASME Journal of Mechanical Design*, 129, 1076–1085, 2007. With permission.)

The calculations reveal that in order to maintain the constant value of the base pitch of the involute gear hob, the corresponding design parameters of the hob have to be changed in accordance with the actual value of wear $VB_N$ of the hob clearance surface. (Here $VB_N$ designates the wear of the clearance surface of the hob tooth). Assume that the wear is equal to $VB_N = 0.6$ mm. This value of the wear $VB_N$ requires a reduction of the hob rake angle at $\Delta\gamma_o = 2.703°$ and an additional reduction of the center distance at $\Delta C_c = 0.621$ mm. These changes to the parameters of the kinematic geometry of the hob ensure that the base pitch of the hob remains the same value, $p_{b.ct} = 29.521$ mm, both for the new hob and for the hob that has been reground. The above-mentioned changes to the parameters of the kinematic geometry of the involute hob will also cause the increase of the pitch helix angle of the hob at $\Delta\psi_{ct} = 1.048'$. However, the change to the pitch helix angle $\psi_{ct}$ doesn't affect the hob accuracy.

**TABLE 15.2**
**Design Parameters of the Involute Gear Hob**

| | |
|---|---|
| Module | $m = 10$ mm |
| Normal pressure angle | $\phi_n = 20°$ |
| Pitch diameter | $d_{ct} = 160$ mm |
| Number of starts | $N_{ct} = 3$ |
| Work gear tooth number | $N_g = 100$ |
| Tooth clearance angle | $\alpha_o = 12°$ |

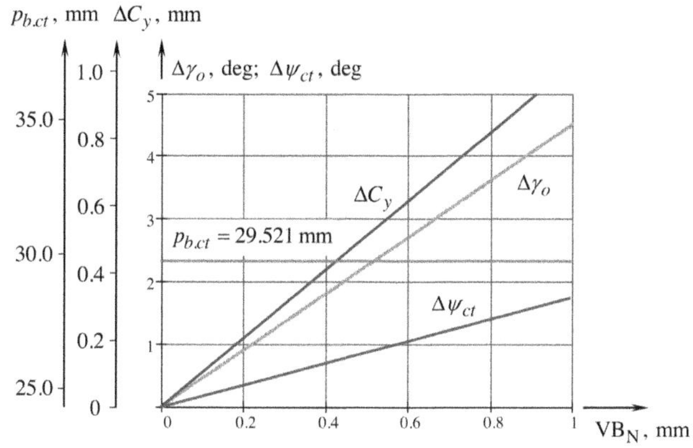


**FIGURE 15.60** A diagram for the corrective regrinding of an involute hob. (From Radzevich, S.P., *ASME Journal of Mechanical Design*, 129, 1076–1085, 2007. With permission.)

The obtained results of the calculation are plotted in Figure 15.60. In Figure 15.60, the following graphs are represented:

- The desired current value of the correction "$\Delta\gamma_o$ vs. $VB_N$."
- The desired increase to center distance $\Delta C_c = \Delta C_c(VB_N)$ in the mesh of the gear hob and of the work gear "$\Delta C_c$ vs. $VB_N$."
- The variation $\Delta\psi_{ct}$ of pitch helix angle $\psi_{ct}$ of the hob in terms of the hob wear $VB_N$.

The use of the method discussed above allows for the complete elimination of the deviation of the base pitch of the gear hob. As it follows from Figure 15.60, the base pitch $p_{b.ct}$ of the hob does not depend on the hob wear $VB_N$, and thus it remains the same value $p_{b.ct} = 29.521$ mm, regardless of how many times the hob has been reground. The same is valid with respect to the diameter of the base cylinder $d_{b.ct}$ of the gear hob. The base cylinder diameter does not depend on the hob wear, and it remains of constant value.

The derived Eq. (15.100) has been used for solving numerous gear-related engineering problems in the field regarding accuracy of hobbed involute gears. Implementation of the results obtained from the research made it possible to develop

- novel designs of precision involute hobs [131, 163],
- novel methods for relieving operations of an involute hob [8, 85, 88–90, 188, 197],
- novel designs of precision cutting tools for finishing of involute gears,
- novel methods for regrinding of precision cutting tools for finishing of involute gears [181],

and many others.

## 15.5 DESIGN OF GEAR HOBS

The design of gear hobs is mainly the problem of gear manufacturing organizations and gear-tool companies who handle the design of tools to machine gears. It is common for industry engineers to consider a gear hob as a cylindrical worm converted into a cutting tool. Cutting edges are formed by "gashing" the worm with a number of slots. The slots are usually either parallel with the hob axis of rotation, or perpendicular to the worm thread. The teeth of the hob are relieved back of the cutting edge to make an efficient cutting tool. The cutting face may be radial, or it may be given a slight hook to improve the cutting action.

### 15.5.1 Design Parameters of a Gear Hob

A close-up of typical shell-type hobs is shown in Figure 15.61. The lateral cutting edges of the hob tooth are usually made straight. Theoretically, the hob should have a slight curvature in its profile to cut a true involute gear. The curvature should correspond to that of an involute helicoid. Practically, though, the curvature required is so slight that it is disregarded. Only on multiple-start hobs of coarse pitch does it become necessary to grind an involute curve.

Hobs may be made with straight bores (see Figure 15.61), tapered bores, or integral with the hob arbor (see Figure 15.62). The shell-type hob with a straight bore is the most commonly used type. Hobs of this type feature a key slot. The key slot can be located either in the axial direction of the hob, as depicted

**FIGURE 15.61** Typical shell-type involute gear hobs.

FIGURE 15.62 Hob integral with the hob arbor.

FIGURE 15.64 Multiple-start gear hob.

in the right portion in Figure 15.61, or it can be located on one of the faces of the hob.

Taper bore hobs require more wall thickness that shell hobs. Hobs for machining helical gears require a taper when the gear helix angle exceeds $\psi_g > 30°$. Even below this angle, a taper is helpful if the gear tooth number is over $N_g > 150$. Some companies prefer the taper-bore hob because of the more rigid mounting that the taper provides.

Integral-shank hobs are expensive. They are used when the hob diameter has to be made so small that the hole of a big-enough diameter cannot be put through the hob. Very small hob diameters may be required when hob runout space or the gap between the helices is limited. Worm-gear hobs sometimes have to be very small to match the diameter of a small worm (see Figure 15.63).

In high-volume production of gears, the use of multiple-start gear hobs (see Figure 15.64) has proven to be effective.

In last decades of the twentieth century, there have been several important developments in gear hob design. The three most important developments to the gear industry are:

- Built-up hobs. Special hob blades are attached to a body made of less costly material than the blade material.
- Special roughing hobs. The special roughing hob is a very efficient tool to remove stock, but it does not produce gear teeth intended to run together.
- Skiving hobs [31]. Hobs of this design can finish-cut fully hardened gear teeth.

The built-up hobs may have blades brazed to an alloy steel body, or the blades may be mechanically attached. The blade material may be expensive high-speed steel, or it may even be a carbide material. The large sizes of built-up hobs may be re-bladed. The built-up hob tends to be less expensive (in large sizes) because the body material is not nearly as costly as the blade material.

The special roughing hobs cut very efficiently. They remove large chips, and the chips tend to cut and break away in an efficient manner. The K-Kut roughing hob is a perfect example of roughing hobs of this design. The design was developed by Barber-Colman (USA).

The skiving hob uses a special carbide blade and a negative rake angle of $\gamma_o = -30°$ (see Figure 15.65). Skiving hobs are used for finish-cutting a fully hardened gear that was finish-hobbed before hardening with a protuberance type of hob [31, 137, 141, 159]. It is a remarkable achievement, of course, to be able to finish a case-hardened gear by hobbing instead of grinding. In the case where very high accuracy and smooth finish are needed, the hard gear—finished by skiving—may be given a further honing operation or a very light final grinding. When a final grinding is used, the skiving serves to remove the bulk of the heat-treat

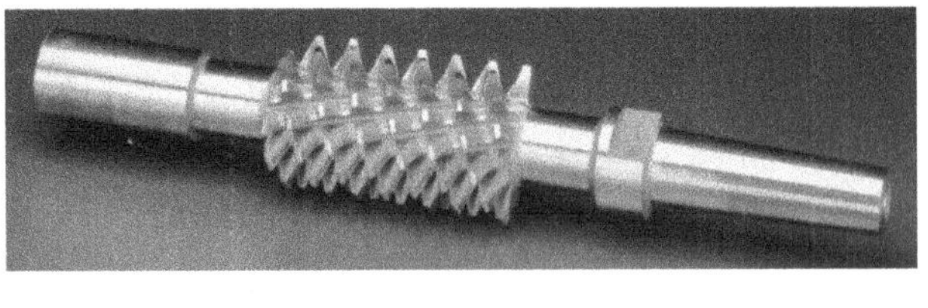

FIGURE 15.63 Worm-gear hob.

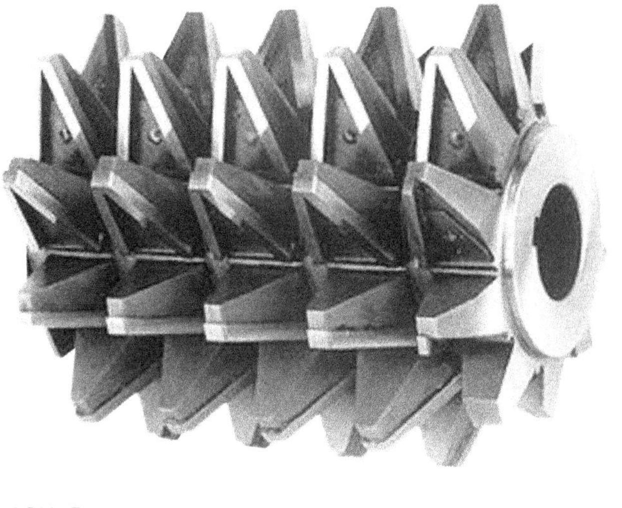

FIGURE 15.65 Skiving gear hob.

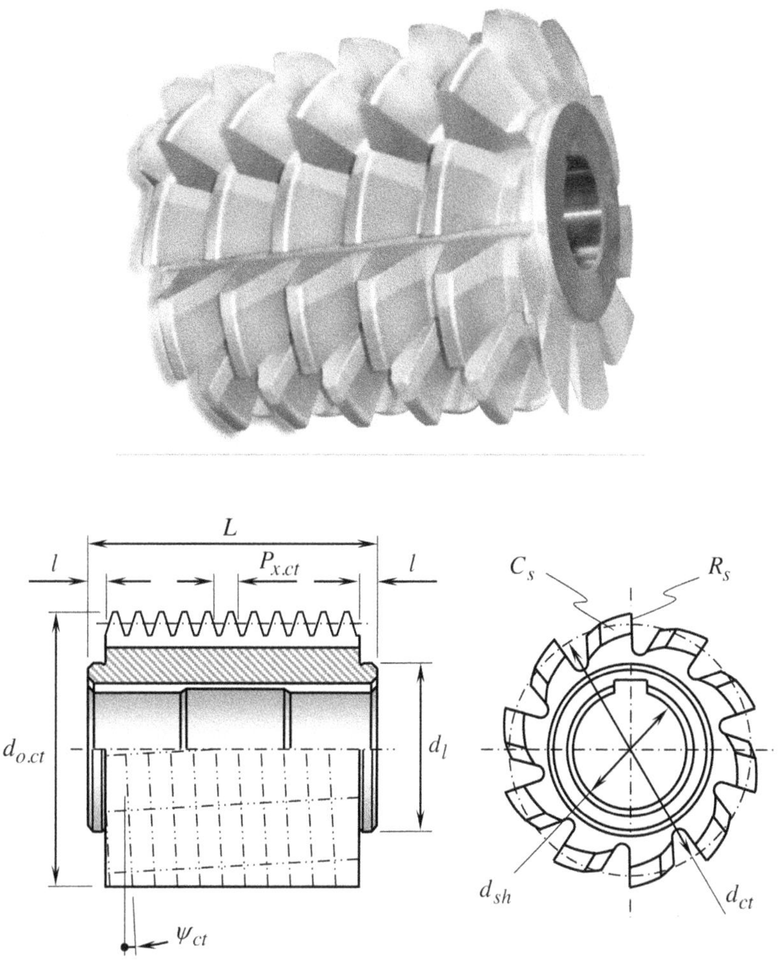


**FIGURE 15.66** Design parameters of a shell-type gear hob.

distortion, and to remove it in a quicker and more efficient manner than grinding.

The design parameters of a shell-type hob are schematically shown in Figure 15.66. The design parameters of a shell-type gear hob include but are not limited to

- Pitch diameter $d_{ct}$
- Outer diameter $d_{o.ct}$
- Whole length $L$

The diameter of the hob bore is designated as $d_{sh}$. The hob features two cylindrical surfaces of a diameter $d_l$ and length $l$ from both sides of the hob. They are used for checking the hob runout. The axial pitch of the hob threads and the pitch helix angle are designated in Figure 15.66 as $P_{x.ct}$ and $\psi_{ct}$, respectively.

The design of the hobs for machining gears for Novikov/conformal gearing is shown in Figure 15.67. Similar to the Novikov hob, the hob for machining sprockets is illustrated in Figure 15.68. It should be stressed here that neither gears for Novikov/conformal gearing, nor rotors with form rotor lobe profiles, as well as other components with non-involute profiles, can be precisely cut by hobs (i.e., non-involute profiles cannot be precisely cut in a continuous-indexing generation process). Only approximate generation of non-involute profiles is possible with hobs. The lower the tooth count of the work gear, the poorer the accuracy of the approximation, and vice versa.

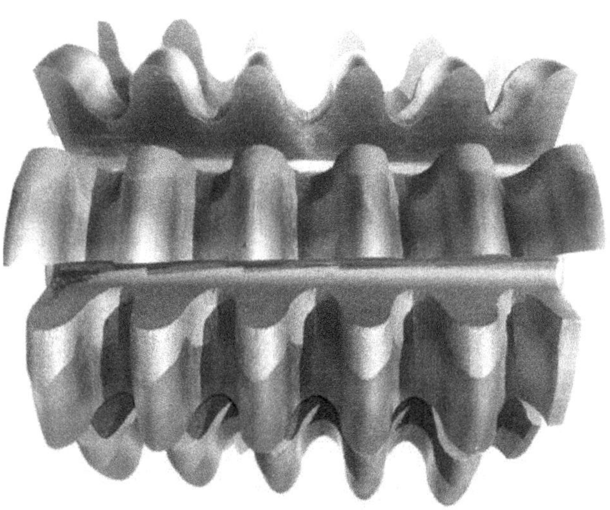

**FIGURE 15.67** Form gear hob: gear hob for machining gears for Novikov/conformal gearing.

**FIGURE 15.68** Form gear hob: gear hob for machining sprockets.

Shown in Figure 15.69, an involute hob with indexable inserts is used when the cost of the cutting tool material becomes critical. The hob body is made of a cheaper steel, while the indexable inserts are made of a cutting tool material that is more costly.

For hobbing of involute gears of medium module hobs with indexable inserts apply. The hob design is illustrated in Figure 15.70. The use of hobs of this design makes cost reduction on cutting tool material possible. The hob body can be used multiple times with several sets of the indexable inserts.

Medium module multiple-start involute hobs with indexable inserts are also used for cutting gears. A three-start gear hob of such a design is depicted in Figure 15.71. The rake surface of every hob tooth is oriented perpendicular to the hob thread. Owing to this, the geometry of the right-side cutting edge, and of the left-side cutting edge of the hob tooth profile, are identical to one another.

**FIGURE 15.69** Involute gear hob with indexable inserts.

**FIGURE 15.70** Involute gear hob of medium module having indexable inserts.

Indexable inserts of various designs are used in practice. A possible design of the indexable insert for a gear hob is shown in Figure 15.72. Cutting elements are bolted to the body of the indexable insert. Every cutting element of the indexable insert features a $V-$ shaped slot on its back surfaces. A corresponding protrusion is designed on the body of the indexable insert. The protrusion goes into the $V-$ shaped slot in the cutting element. These slots and protrusions are directed radially relative to the gear hob. Multiple $V-$ shaped slots and protrusions are also designed in the axial direction of the hob. This mean helps preventing the cutting element from displacements under unbalanced cutting forces exerting from the work gear, and from the chip when the gear hob operates.

Coarse module involute gear hobs with indexable inserts (see Figure 15.73) are designed so as to cut opposite gear

**FIGURE 15.71** Multiple-start (three-start) involute gear hob with indexable inserts for cutting gears of medium module.

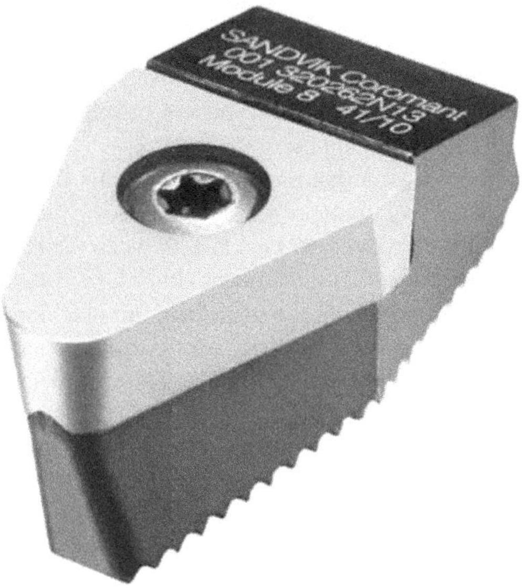


**FIGURE 15.72** Design of an indexable insert.

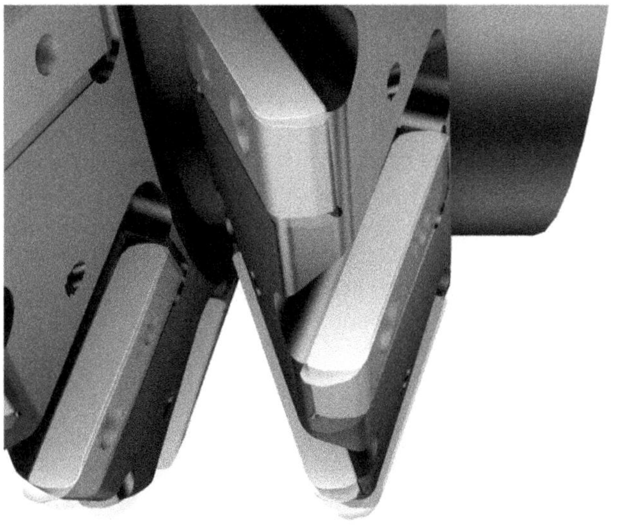

**FIGURE 15.74** Arrangement of the indexable inserts in the body of coarse pitch involute gear hob.

tooth profiles separately, that is, two separate indexable inserts cut the opposite sides of the work gear tooth profile. Figure 15.74 illustrates a possible arrangement of the indexable inserts in the body of a coarse pitch involute hob.

Coarse module involute hobs with indexable inserts are also designed as multiple-start gear hobs (see Figure 15.75). For a better understanding of principal features of the gear hob of this design, in Figure 15.75, one of two starts of the gear hob is marked up in blue.

A special-purpose coarse module involute hob with indexable inserts can be designed to feature an assembled gear hob housing as illustrated in Figure 15.76. This allows for the assembly of the hob of a required length (either longer, or shorter in length), depending on the tooth count of the work gear.

**FIGURE 15.75** Multiple-start (two-start) involute hob with indexable inserts for cutting gears of coarse module.

**FIGURE 15.73** Coarse module involute gear hob with indexable inserts.

**FIGURE 15.76** Special-purpose coarse module involute gear hob with indexable inserts.

### 15.5.2 Tooth Profile of Gear Hob

The tooth profile of the gear hob correlates, but is not identical to, the tooth profile of the gear being machined. As it follows from Figure 15.77, some of the design parameters of the gear hob could be identical to the corresponding design parameters of the gear, while others could not. The feasible difference between the design parameters of the hob and the gear is strongly restricted by the equality of base pitch of the hob $p_{b.ct}$ to base pitch of the gear $p_{b.g}$.

Fulfillment of the condition

$$p_{b.ct} = p_{b.g} = p_b \tag{15.110}$$

is a mandatory requirement for machining involute gears properly. In most practical cases of gear hob design, the profile angle $\phi_{n.ct}$ of the hob tooth and the axial pitch of the hob threads $P_{x.ct}$ are equal to the profile angle $\phi_n$ of the gear tooth and the axial pitch $P_x$ of the basic rack of the gear, respectively. When the equalities $\phi_{n.ct} = \phi_n$ and $P_{x.ct} = P_x$ are observed, then the hob tooth addendum $a_{ct}$ is equal to the gear tooth dedendum $b$; the hob tooth dedendum $b_{ct}$ is equal to the summa of the gear tooth addendum $a$, and the hob clearance $c_{ct}$, that is, the equality $b_{ct} = a + c_{ct}$, is valid.

The whole depth of the gear tooth $h_t$ is equal to $h_t = a + b$, and the working depth of the gear tooth can be calculated from the formula $h_k = h_t - c$, where $c$ designates clearance. Similar formulae are valid for the hob tooth. The whole depth $h_{t.ct}$ of the hob tooth is equal to $h_{t.ct} = a_{ct} + b_{ct}$, while the working depth of the hob tooth can be calculated from $h_{k.ct} = h_t$. For the calculation of the whole depth $h_{t.ct}$ of the hob tooth, an equality $h_{t.ct} = h_{t.ct} + c_{ct}$ is valid as well.

The normal tooth thickness $t$ and the space width $w$ are equal to each other ($t = w$) for gears with a standard tooth profile. The corresponding design parameters of the hob tooth correlate to the gear design parameters as follows: $t_{ct} = w$ and $w_{ct} = t$. The basic rack of the gear features a fillet of the radius $r_f$ at the bottom land. The basic rack of the hob features a similar roundness $r_{f.ct}$ at the top land of the tooth.

The cutting racks of an assembled gear hob (see Figure 15.78) have design parameters of the tooth profile similar to those of solid hobs.

One more reason that the hob tooth profile differs from the corresponding tooth profile of the gear to be machined is because of the necessity of backlash $B_n$ when the machined gears are assembled in the housing. In order to provide the gear pair with backlash, the gear teeth are machined thinner, and the gear space widths are machined wider compared to the corresponding parameters specified by the basic rack (see Figure 15.77).

Hobs with a modified tooth profile are used as well, with the aim of improving the cutting conditions. A few examples of the modified hob tooth profiles are schematically depicted in Figure 15.79. The modification of the hob tooth profile includes but is not limited to the following:

- sharp corner tooth profile [see Figure 15.79(a)],
- chamfered corner tooth profile [see Figure 15.79(b)],
- rounded corner tooth profile [see Figure 15.79(c)],
- full round (full capped) tooth profile [see Figure 15.79(d)],

and so forth.

Another reason for using hobs with a modified tooth profile is that they leave extra stock around the pitch line of the preshaved hobs. In this case, a special gear hob with a slight tooth hollow is implemented.

When designing a gear hob, much attention should be focused on increasing the cutting performance of the cutting tool. For this purpose, the wear of the hob tooth should be minimized by all means.

Figure 15.80(a) illustrates regular wear of the gear hob tooth. However, improper cutting conditions can result in intensive wear of the hob teeth. The interference of chip flows from adjacent cutting edges is the major reason for

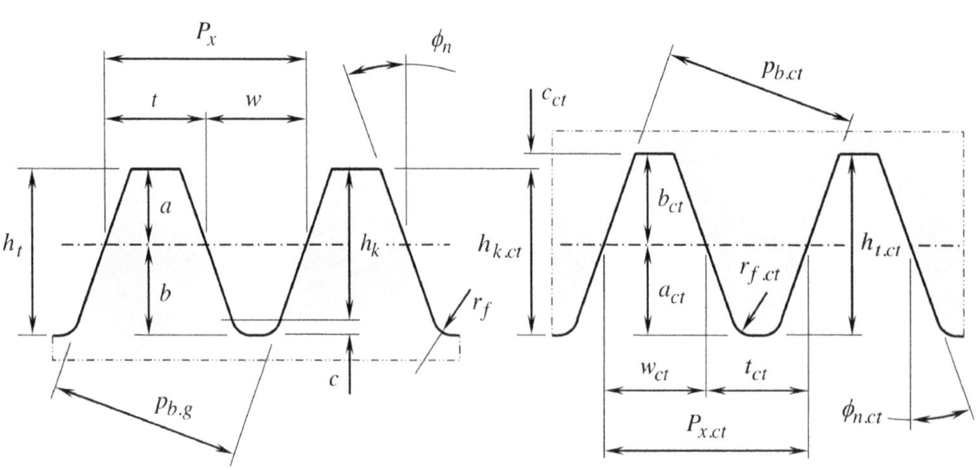


**FIGURE 15.77** Correlation among the design parameters of a work gear (at left) and the corresponding design parameters of the gear hob (at right).

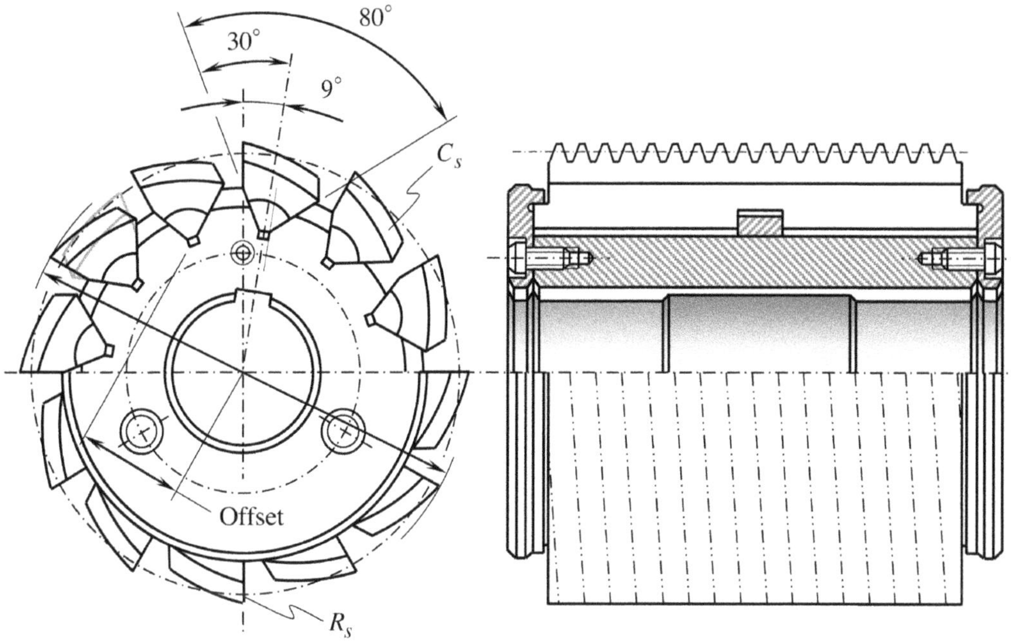


**FIGURE 15.78** Assembled gear hob with inserted cutting racks.

catastrophic wear of the gear hob of medium module, and, especially, of coarser module. Chip flows from the top cutting edge interact with chip flows from the lateral cutting edges [see Figure 15.80(b)]. The impact of interference of the chip flows on the cutting performance of the hob becomes greater for coarser pitch hobs. Possible types of wear of the hob tooth are summarized in Figure 15.80(c). In practice, not all of the possible types of tooth wear are observed simultaneously; only one or two commonly dominate.

In order to avoid catastrophic wear of the hob tooth [see Figure 15.80(d)], various cutting diagrams for improving cutting performance of the hob are implemented in the industry. The cutting diagrams eliminate the interference of chip flow from the adjacent cutting edges. They also result in a corresponding modification of the hob tooth profile. Some practical examples of tooth modification are shown in Figure 15.81.

A few illustrative examples are briefly presented below.

An involute gear hob can be modified to feature standard teeth and modified teeth. Every other tooth along the hob thread is modified, and thus it is the neighbor of two teeth of the standard profile. The modified teeth are shorter at $\Delta$ compared to the standard teeth as shown in Figure 15.81(a) (the modification does not exceed the radial clearance $\Delta \leq c(= 0.25m)$). Therefore, when machining gears, the modified teeth cut off chips only with their lateral cutting edges, while the standard teeth cut much thicker chips mainly with the top cutting edges. The chip cutoff by the lateral cutting edges is thin, and it does not affect the chip flow of the thick chips from the top cutting edge.

The hob tooth modification discussed here may be combined either with chamfers [as shown in Figure 15.81(b)] or with tooth roundness [see Figure 15.81(c)]. Grinding of the so-called mustache on the modified tooth profile of the gear hob [proposed by Dr. N.N. Smirnov, and illustrated in Figure 15.81(d)] is also efficient in this regard.

Many patents for inventions are granted on designs of gear hobs that feature improved cutting diagrams. The gear hobs in [171] and [179] are just a few examples.[18]

The modification of involute hobs is investigated in depth by Prof. S.N. Medveditskov [41]. More recent results of the research are available from Prof. S.N. Medveditskov et al. [42], Prof. S.P. Radzevich [159], Dr. N.N. Smirnov [208], M. Yu, R. Mertz, [222], and many others.

### 15.5.3 Precision Involute Hobs with Straight Lateral Cutting Edges

As proven in previous chapters, for machining of high-accuracy gears, the lateral cutting edges of the gear hob should be located within the generating surface, which is shaped in the form of an involute worm. The lateral cutting edges of the hob teeth align with the line of intersection of the generating surface $T$ of a gear hob by the rake surface $R_s$ of the hob tooth. These lines are a type of curved lines. For manufacturing purposes, the desirable curved lateral cutting edges of the hob teeth are usually approximated with the straight-line lateral cutting edges. Linear approximation of the lateral cutting edges of the hob tooth is one of the major sources of tooth deviations in a hobbed gear. Deviations of this kind are inevitable for gear hobs of conventional design.

[18] SU Pat. No. 1134318, *A Gear Hob*/S.P. Radzevich, S.N. Medveditskov, and N.N. Smirnov, Int. Cl. B23f 21/16, Filed: April 7, 1983.
SU Pat. No. 1187936, *A Gear Hob*/S.P. Radzevich, S.N. Medveditskov, and N.N. Smirnov, Int. Cl. B23f 21/16, Filed: June 28, 1984.

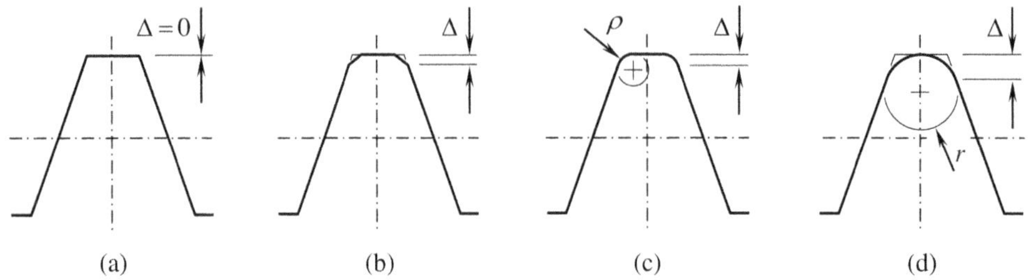


**FIGURE 15.79** Possible types of modification of the tooth profile of a gear hob, (a) non-modified hob tooth – reference, (b) the corner chamfers, (c) the corner roundness, and (d) the full top cutting edge roundness. (From Radzevich, S.P., *ASME Journal of Mechanical Design*, 128(4), 803–811, 2006. With permission.)

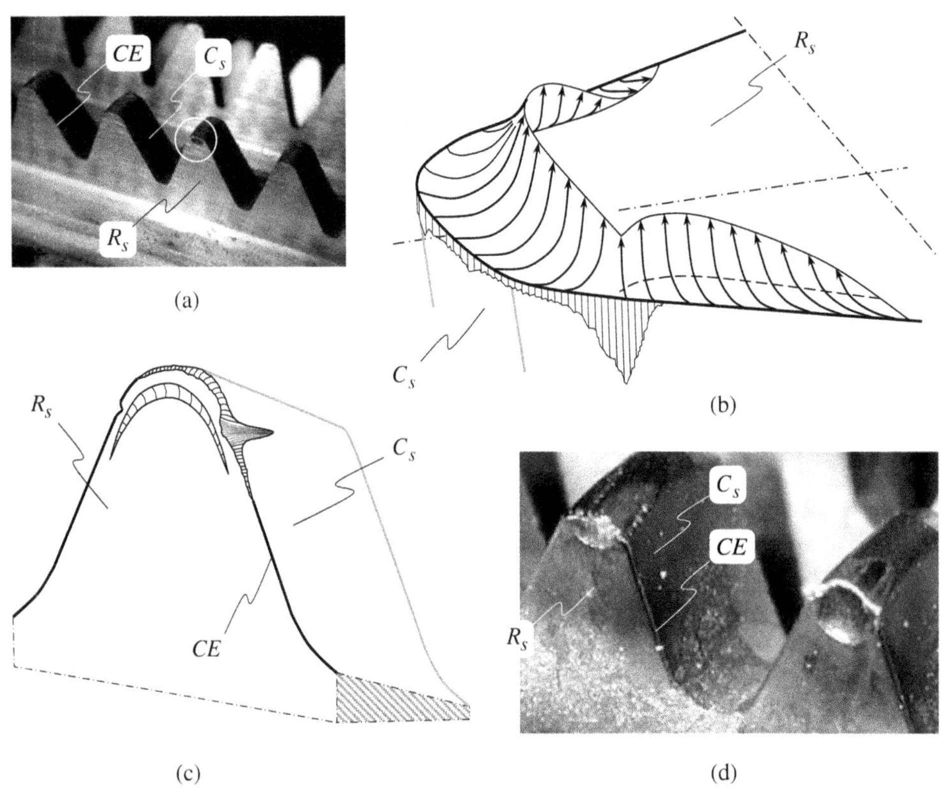


**FIGURE 15.80** Types of wear of a gear hob tooth, (a) close-up of a worn rake surface of a gear hob tooth, (b) traces of chip flow over the rake surface of the hog tooth, (c) location of the most worn portions of the hob tooth, and (d) close-up of critically worn gear hob tooth.

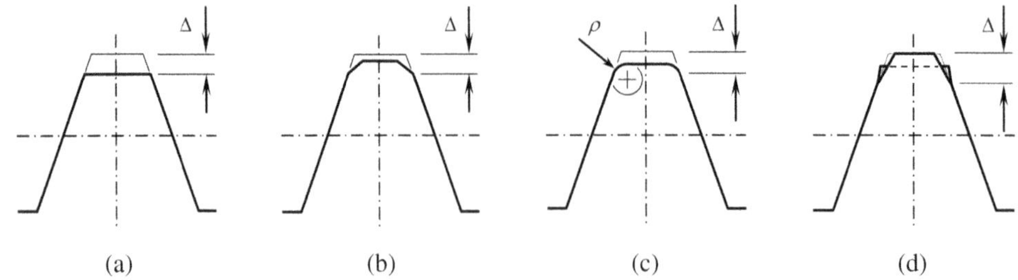


**FIGURE 15.81** Types of tooth profile modification of involute hobs, (a) a reduced height of the gear hob tooth, (b) a reduced height and corner chamfers of the gear hob tooth, (c) a reduced height and corner chamfers of the gear hob tooth, and (d) a reduced height and corner protrusions on the gear hob tooth. (From Radzevich, S.P., *ASME Journal of Mechanical Design*, 128(4), 803–811, 2006. With permission.)

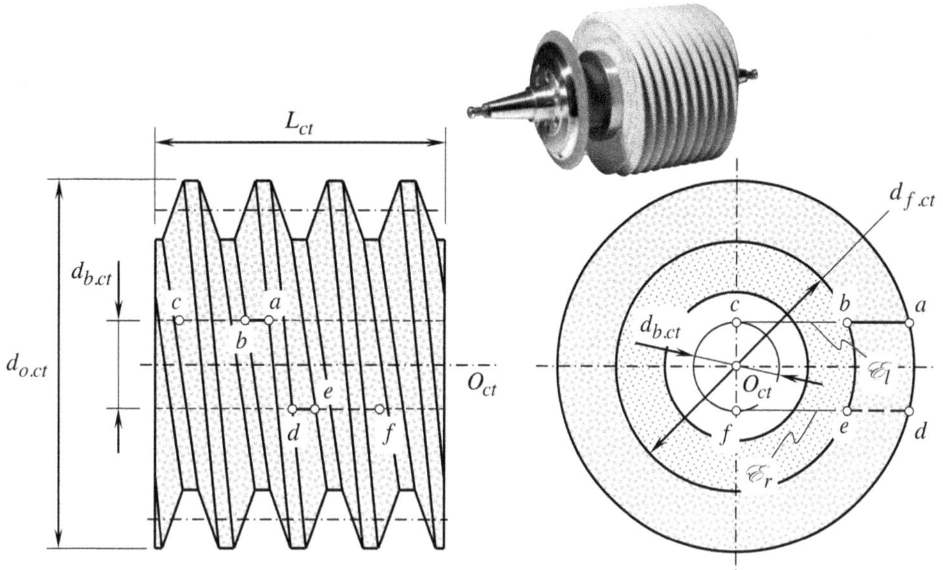


**FIGURE 15.82** Straight generating lines $\mathscr{E}_r$ and $\mathscr{E}_l$ within the screw involute surfaces of the opposite sides of the thread profile of the generating surface $T$ of a grinding worm.

The lateral portions of the generating surface of an involute gear hob are screw involute surfaces. A screw involute surface with any combination of the design parameters features a straight generating line which is entirely located within the generating surface $T$. This property of the screw involute surface is well known. It is often used, for example, for dressing of grinding worms for grinding of precision involute gears.

A grinding worm for finishing precision involute gears (see Figure 15.82) can be properly dressed if the dressing diamond reciprocates back and forth along the straight generating line. Two straight generating lines $\mathscr{E}_l$ and $\mathscr{E}_r$ for the opposite sides of the thread profile of the grinding worm are the straight lines through points $a$ and $b$ of one side of the thread surface, and through points $d$ and $e$ of the opposite side of the thread surface. The generating lines $\mathscr{E}_l$ and $\mathscr{E}_r$ are entirely located within the screw involute surfaces of the corresponding side of the thread profile. At points $c$ and $f$, the generating lines $\mathscr{E}_l$ and $\mathscr{E}_r$ are tangent to the base cylinder of a diameter $d_{b.c}$ of the grinding worm. Using the portions $ab$ and $de$ of the straight generating lines $\mathscr{E}_l$ and $\mathscr{E}_r$ for dressing the grinding worm makes the dressing operation much easier to perform.

The property of a screw involute surface that makes possible the precise generation of a surface with a straight line is utilized in many designs of gear finishing tools.[19] It can be enhanced and implemented for improving the design of a hob for machining involute gears.

The straight-line segments, similar to the straight-line segments $ab$ and $de$, can be used as the cutting edges of the lateral sides of the tooth profile of the precision gear hob. The possibility for increasing the accuracy of involute hobs is briefly discussed above (see Figure 15.49). However, for the manufacturer of precision involute hobs, as well as for the manufacturer of involute gears, it is much more convenient to have both lateral cutting edges of the hob tooth situated within a common rake plane. The possibility of designing precision involute hobs with straight lateral cutting edges is discussed immediately below.

It is assumed here that generating surface $T$ of the involute hob is described analytically by Eq. (14.63):

$$T:\Rightarrow \mathbf{r}_{ct}(U_{ct},V_{ct})=\begin{bmatrix} 0.5\cdot d_{b.ct}\cdot\sin V_{ct}-U_{ct}\cdot\sin\psi_{b.ct}\cdot\cos V_{ct} \\ 0.5\cdot d_{b.ct}\cdot\cos V_{ct}+U_{ct}\cdot\sin\psi_{b.ct}\cdot\sin V_{ct} \\ p_{b.ct}\cdot V_{ct}-U_{ct}\cdot\cos\psi_{b.ct} \\ 1 \end{bmatrix} \tag{15.111}$$

For the calculation of the hob base diameter $d_{b.ct}$ and the hob base helix angle $\psi_{b.ct}$, engineering formulae available in Appendix F can be used.

Equation (15.111) is of particular importance for the development of the design of a precision gear hob with straight lateral cutting edges. This is mainly owing to the important features of the geometry of the generating surface $T$. It is easy to verify that for a specific value of the $V_{ct}$ – parameter, Eq. (15.111) describes a straight generating line $\mathscr{E}_r$ (or a straight generating line $\mathscr{E}_l$) within the generating surface $T$. Further, the lateral cutting edge of the involute gear hob will be aligned with the straight-line characteristics $\mathscr{E}_l$ and $\mathscr{E}_r$. The straight characteristic lines $\mathscr{E}_l$ and $\mathscr{E}_r$ serve as a vital link between the geometry of the generating surface $T$ of the involute hob and the principal features of its design.

[19] SU Pat. No. 963653, *A Tool for Finishing Gears*/S.P. Radzevich, Int. Cl. B21h 5/00, Filed: October 2, 1980.

SU Pat. No. 1466861, *A Tool for Finishing Gears*/S.P. Radzevich, Int. Cl. B21h 5/00, Filed: February 2, 1987.

SU Pat. No. 1754302, *A Tool for Finishing Gears*/S.P. Radzevich, Int. Cl. B21h 5/02, Filed: September 27, 1989.

#### 15.5.3.1 Principal Design Parameters of Precision Involute Hob

The proposed concept of the design of the precision involute gear hob with straight lateral cutting edges[20] is based on the following considerations [97]:

1. The lateral cutting edges of one side of the tooth profile of an involute gear hob are located within the corresponding screw involute surface, say within the screw involute surface $T_r$.
2. The lateral cutting edges of the opposite side of the tooth profile of the involute gear hob are located within the screw involute surface of the opposite side of thread profile, say within the screw involute surface $T_l$.
3. The screw involute surfaces $T_r$ and $T_l$ of the opposite sides of the tooth profile of the involute hob intersect one another, and the line of the intersection is a helix.
4. Two characteristics, $\mathscr{C}_r$ and $\mathscr{C}_l$, pass through every point of the said helix.
5. Two characteristics, $\mathscr{C}_r$ and $\mathscr{C}_l$, through a common point of the helix intersect each other at that point, and thus a plane through the straight lines $\mathscr{C}_r$ and $\mathscr{C}_l$ is specified.
6. The plane through the straight lines $\mathscr{C}_r$ and $\mathscr{C}_l$ is used as the rake surface $R_s$ of cutting teeth of the precision involute gear hob.

Items 1 through 6 above allow for the determination of the desirable configuration of the rake surface $R_s$ of precision involute hob teeth.

##### 15.5.3.1.1 *DG – Based Approach to Determine Desirable Rake Plane Configuration*

The descriptive-geometry-based solution (or *DG – based solution*, for simplicity) to the problem of determining the desired configuration of the rake plane $R_s$ of the precision involute gear hob teeth is illustrated in Figure 15.83.

Before solving the problem of determining the desirable configuration of the rake plane of the hob tooth, the following design parameters of the hob must be known: modulus $m$ of the hob tooth, normal profile angle $\phi_n$, the setting angle of the hob $\zeta_{ct}$, the number of starts of the hob $N_{ct}$, the hob outer diameter $d_{o.ct}$, and the base diameter $d_{b.ct}$ of the hob.

The solution to the problem of determining the desirable configuration of the rake plane of the hob tooth is represented in the system of three planes of projection, namely, in the planes of projection $\pi_1$, $\pi_2$, and $\pi_3$, respectively. Two auxiliary planes of projection, that is, $\pi_4$ and $\pi_5$, are also used. The auxiliary axis of projection $\pi_1/\pi_4$ forms the angle $\zeta_{ct}$ with the axis of projection $\pi_1/\pi_2$. The axis $\pi_1/\pi_5$ is parallel to the axis $\pi_1/\pi_4$. The use of the plane of projection $\pi_5$ allows for significant reduction of the size of Figure 15.83.

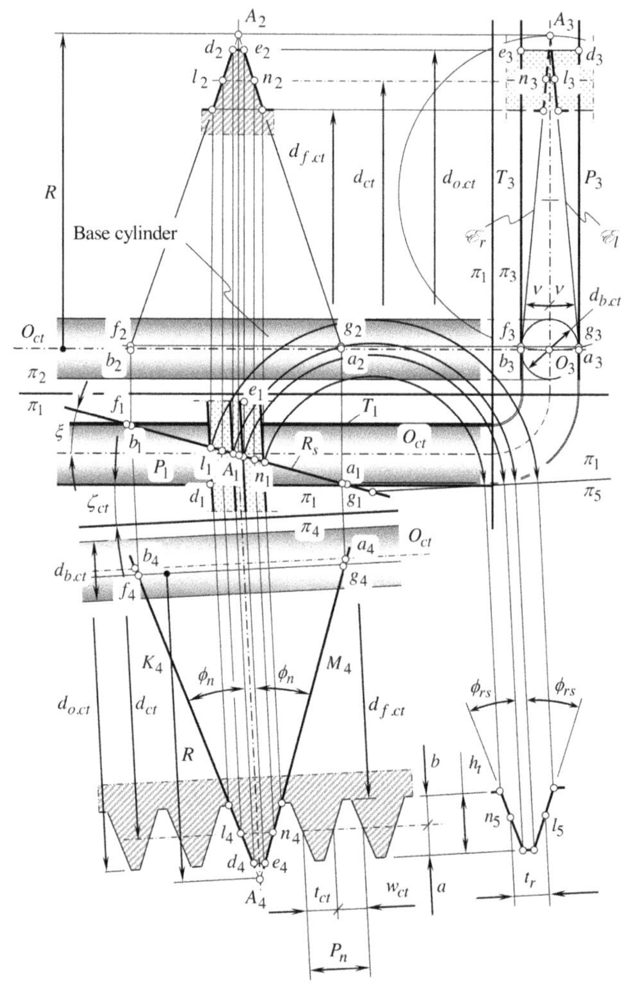


**FIGURE 15.83** An example of implementation of a *DG – based* approach to determine the desirable configuration of the rake plane $R_s$ of precision involute hob teeth. (From Radzevich, S.P., *ASME Journal of Mechanical Design*, 129, 334–345, 2007. With permission.)

In the system of planes of projection $\pi_1$, $\pi_2$, and $\pi_3$, the gear hob to be designed is oriented so that the hob axis of rotation $O_{ct}$ is positioned parallel to the axis $\pi_1/\pi_2$ (see Figure 15.83).

The normal plane section of the auxiliary generating rack $\mathscr{R}_T$ is shown in the plane of projection $\pi_4$.

Consideration of the images of the hob elements in the planes of projection $\pi_4$ and $\pi_3$ (that are connected to one another through the planes of projection $\pi_1$ and $\pi_2$) allows for the determination of the location of point $A$ within plane $\pi_3$. Two straight lines through point $A_3$ that are tangent from the opposite sides to the base cylinder of the diameter $d_{b.ct}$ of the hob represent straight lines with which the lateral cutting edges $CE$ of the hob will align. The rake plane of the hob tooth is the plane through these two straight lines. Conventional methods developed in descriptive geometry are used for the construction of the rest of the projections of these two straight lines that initially are constructed in the plane of projection $\pi_3$.

[20] SU Pat. No. 990445, *A Precision Involute Hob*/S.P. Radzevich, Int. Cl. B23f 21/16, Filed: October 8, 1981.

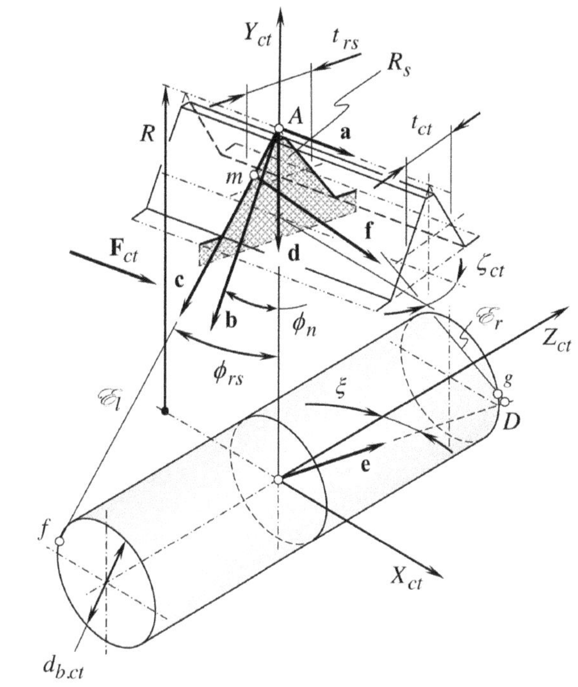


**FIGURE 15.84** Desirable orientation of the rake plane $R_s$ of the precision involute gear hob. (From Radzevich, S.P., *ASME Journal of Mechanical Design*, 129, 334–345, 2007. With permission.)

The angle $\xi$ that the rake plane $R_s$ forms with the axis of rotation $O_{ct}$ of the hob with no distortion is constructed in the plane of projection $\pi_1$. A profile of the hob tooth within the rake plane $R_s$ with no distortion is constructed in the plane of projection $\pi_5$.

##### 15.5.3.1.2 Analytical Determination of Desirable Rake Plane Configuration

In Figure 15.83, the solution to the problem of determining the main design parameters of the precision involute hob provides a clear understanding of the geometry of the precision involute gear hob with straight lateral cutting edges. The derived *DG* – based solution to the problem is helpful for deriving the corresponding formulae for the calculation of the design parameters of the gear hob.

Then consider the lateral surfaces of the auxiliary generating rack $\mathscr{R}_T$ whose teeth intersect one another (see Figure 15.84). The line of the intersection is the straight line through point $A$. This straight line is at a distance $R$ from the axis of rotation $O_{ct}$ of the hob. For the distance $R$, Figure 15.85 yields:

$$R = 0.5 \cdot (d_{ct} + t_{ct} \cdot \cot \phi_n) \tag{15.112}$$

The angle $\phi_{rs}$ between the lateral cutting edges is measured within the rake plane $R_s$ of the hob tooth. Prior to

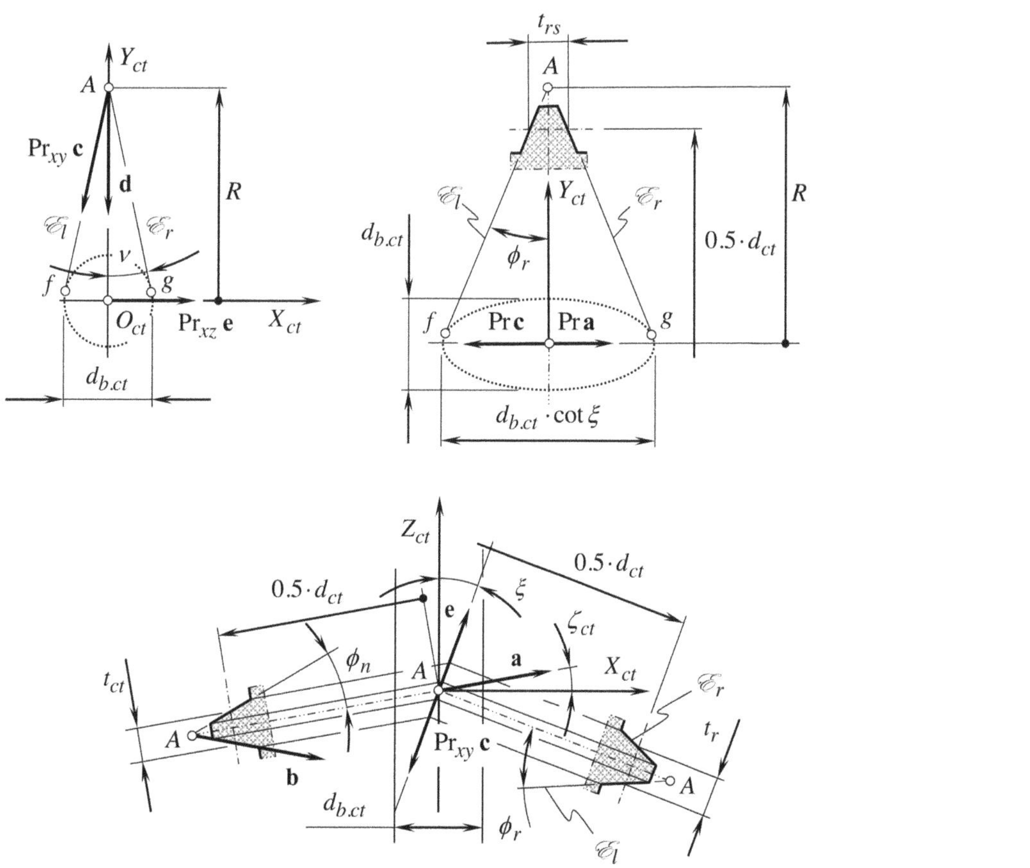


**FIGURE 15.85** Employed characteristic vectors. (From Radzevich, S.P., *ASME Journal of Mechanical Design*, 129, 334–345, 2007. With permission.)

deriving an equation for the calculation of the angle $\phi_{rs}$, it is convenient to derive an equation for the calculation of the projection $\nu$ of the angle $\phi_{rs}$ onto the coordinate plane $X_{ct}Y_{ct}$.

Projections of the lateral cutting edges of the involute hob onto the coordinate plane $X_{ct}Y_{ct}$ form a certain angle $\nu$. The actual value of the angle $\nu$ is calculated from the formula

$$\nu = \tan^{-1}\left(\frac{d_{b.ct}}{\sqrt{4\cdot R^2 - d_{b.ct}^2}}\right) \tag{15.113}$$

Then, consider three vectors of unit length **a**, **b**, and **c** (see Figures 15.84 and 15.85). The vector **a** is pointed along the line of intersection of the opposite tooth flanks of the auxiliary generating rack $\mathscr{R}_T$. The vector **b** is aligned with the lateral cutting edge *CE* of the involute hob. The vector **c** is pointed along the tooth profile of the auxiliary generating rack $\mathscr{R}_T$.

The vectors **a**, **b**, and **c** allow for analytical representation in vector form:

$$\mathbf{a} = \mathbf{i}\cdot\cos\zeta_{ct} + \mathbf{k}\cdot\sin\zeta_{ct} \tag{15.114}$$

$$\mathbf{b} = \mathbf{i}\cdot\sin\phi_n\sin\zeta_{ct} - \mathbf{j}\cdot\cos\phi_n - \mathbf{k}\cdot\sin\phi_n\cos\zeta_{ct} \tag{15.115}$$

$$\mathbf{c} = -\mathbf{i}\cdot\cos\phi_{rs}\tan\nu - \mathbf{j}\cdot\cos\phi_{rs} - \mathbf{k}\cdot\sin\phi_{rs}\cos\xi \tag{15.116}$$

where $\zeta_{ct}$ denotes the setting angle of the precision involute hob.

The actual value of the setting angle $\zeta_{ct}$ can be chosen by a designer of the gear hob. Usually, it is recommended to assign the actual value of the setting angle $\zeta_{ct}$ equal to the pitch helix angle $\psi_{ct}$ of the involute hob. In order to meet the equality $\zeta_{ct} = \psi_{ct}$ (which is the most favorable condition), the actual value of the setting angle of the hob must be calculated from the equation [110, 137]

$$\zeta_{ct} = \tan^{-1}\left(\frac{m\cdot N_{ct}}{\sqrt{(d_{o.ct} - 2\cdot 1.25\cdot m - \Delta d_{o.ct})^2 - m^2 N_{ct}^2}}\right) \tag{15.117}$$

where $\Delta d_{o.ct}$ designates the reduction of the outer diameter $d_{o.ct}$ of the gear hob caused by the resharpening of the worn gear hob (see Figure 15.51).

As the vectors **a**, **b**, and **c** are located by construction within the common lateral tooth flank of the auxiliary rack $\mathscr{R}_T$, therefore, the triple scalar product of the vectors is identical to zero, and thus the following identity $\mathbf{a}\times\mathbf{b}\cdot\mathbf{c}\equiv 0$ is observed. The latter expression yields a determinant:

$$\mathbf{a}\times\mathbf{b}\cdot\mathbf{c} = \begin{vmatrix} \cos\zeta_{ct} & 0 & \sin\zeta_{ct} \\ \sin\phi_n\sin\zeta_{ct} & -\cos\phi_n & -\sin\phi_n\cos\zeta_{ct} \\ -\cos\phi_{rs}\tan\nu & -\cos\phi_{rs} & -\sin\phi_{rs}\cos\xi \end{vmatrix} \equiv 0 \tag{15.118}$$

Equation (15.118) casts into the equation of two unknowns, namely, the angles $\phi_{rs}$ and $\xi$.

The rake plane $R_s$ of the involute hob is inclined relative to the hob axis $O_{ct}$ at a certain angle $\xi$. In order to determine the required value of the angle $\xi$, two more vectors **d** and **e**, also of unit length, are constructed (see Figure 15.85). The vector **d** is pointed along the $Y_{ct}$ – axis. The vector **e** is located within the rake plane $R_s$ and is perpendicular to the axis $Y_{ct}$.

For the vectors **d** and **e**, the following expressions are valid:

$$\mathbf{d} = -\mathbf{j} \tag{15.119}$$

$$\mathbf{e} = \mathbf{i}\cdot\sin\xi + \mathbf{k}\cdot\cos\xi \tag{15.120}$$

As the vectors **c**, **d**, and **e** are located within the rake plane $R_s$ of the hob tooth, therefore, the triple scalar product of the vectors is identical to zero, and thus, the identity $\mathbf{c}\times\mathbf{d}\cdot\mathbf{e}\equiv 0$ is observed. This yields the determinant:

$$\mathbf{c}\times\mathbf{d}\cdot\mathbf{e} = \begin{vmatrix} -\cos\phi_{rs}\tan\nu & -\cos\phi_{rs} & -\sin\phi_r\cos\xi \\ 0 & -1 & 0 \\ \sin\xi & 0 & \cos\xi \end{vmatrix} \equiv 0 \tag{15.121}$$

Equation (15.121) casts into the equation of two unknowns, namely, of the angles $\phi_{rs}$ and $\xi$.

Next, consider the set of two equations, say of Eq. (15.118) and Eq. (15.121) of the two unknowns, $\phi_{rs}$ and $\xi$. The solution to the set of the above equations can be represented in the form

$$\xi = \tan^{-1}\left(\frac{\cos\zeta_{ct}\cdot\tan\nu}{\tan\phi_n + \sin\zeta_{ct}\cdot\tan\nu}\right) \tag{15.122}$$

$$\phi_{rs} = \tan^{-1}\left(\frac{\tan\nu}{\sin\xi}\right) \tag{15.123}$$

The setting angle $\zeta_{ct}$ specifies an inclination of the axis of rotation $O_{ct}$ of the involute hob with respect to the auxiliary generating rack $\mathscr{R}_T$. It is necessary to point out here that the setting angle $\zeta_{ct}$ is a design parameter of the gear hob, and it is not a setup parameter in the gear hobbing operation.

Equations (15.122) and (15.123) return the required values of the angles $\phi_{rs}$ and $\xi$. These angles are indicated in the involute hob blueprint.

The precision involute gear hob of the discussed design [104] (a) features straight lateral cutting edges, and (b) is free of the major source of the tooth profile deviations. The concept of this design is enhanced to other designs of precision hobs[21] for the machining of involute gears; two more modifications of the gear hob design [105, 107] are developed as well. More details on the concept are available from [98–100, 102] and others.

[21] SU Pat. No. 1114505, *A Precision Involute Hob*/S.P. Radzevich, Int. Cl. B23f 21/16, Filed: October 18, 1982.
RU Pat. No. 2040376, *A Precision Involute Hob*/S.P. Radzevich *et al*, Int. Cl. B23f 21/16, Filed: January 03, 1992.

**TABLE 15.3**
**Calculating the Design Parameters of Involute Gear Hob with Straight Lateral Cutting Edges**

| Design Parameter of the Hob | Equation | Calculation | Result of the Calculation |
|---|---|---|---|
| Normal pitch | $P_n = \pi \cdot m$ | $P_n = \pi \cdot 10$ | $P_n = 31.4159$ mm |
| Setting angle of the hob | Eq. (15.117) | $\zeta_{ct} = \tan^{-1}\left(\frac{10 \cdot 1}{\sqrt{(180 - 2 \cdot 1.25 \cdot 10 - 3)^2 - 10^2 \cdot 1^2}}\right)$ | $\zeta_{ct} = 3.772$ deg |
| The hob base diameter | Eq. (15.70) | $d_{b.ct} = \frac{10 \cdot 1 \cdot \cos 20°}{\sqrt{1 - \cos^2 20° \cdot \cos^2 3.772°}}$ | $d_{b.ct} = 27.037$ mm |
| The hob base helix angle | Eq. (15.73) | $\psi_{b.ct} = \cos^{-1}(\cos 20° \cdot \cos 3.772°)$ | $\psi_{b.ct} = 20.338$ deg |
| Axial pitch | $P_x = \frac{P_n}{\cos \zeta_{ct}}$ | $P_x = \frac{31.4159}{\cos 3.772°}$ | $P_x = 31.484$ mm |
| Pitch diameter | $d_{ct} = d_{o.ct} - 2 \cdot 1.25 \cdot m$ | $d_{ct} = 180 - 2 \cdot 1.25 \cdot 10$ | $d_{ct} = 155$ mm |
| The hob pitch helix angle | $\tan \psi_{ct} = \frac{P_x \cdot N_{ct}}{\pi \cdot d_{ct}}$ | $\psi_{ct} = \tan^{-1}\left(\frac{31.484 \cdot 1}{\pi \cdot 155}\right)$ | $\psi_c = 3.699$ deg |
| Tooth thickness | $t_{ct} = \frac{P_n}{2}$ | $t_{ct} = \frac{31.4159}{2}$ | $t_c = 15.708$ mm |
| The auxiliary parameter $R$ | Eq. (15.112) | $R = 0.5 \cdot (155 + 15.708 \cdot \cot 20°)$ | $R = 99.079$ mm |
| Transverse projection of half of the angle that makes lateral cutting edges | Eq. (15.113) | $\nu = \tan^{-1}\left(\frac{27.037}{\sqrt{4 \cdot 99.709 - 27.037^2}}\right)$ | $\nu = 7.842$ deg |
| The angle that makes the rake face and the gear axis | Eq. (15.122) | $\xi = \tan^{-1}\left(\frac{\cos 3.772° \cdot \tan 7.842°}{\tan 20° + \sin 3.772° \cdot \tan 7.842°}\right)$ | $\xi = 20.225$ deg |
| Rake face pressure angle | Eq. (15.123) | $\phi_{rs} = \tan^{-1}\left(\frac{\tan 7.842°}{\sin 20.225°}\right)$ | $\phi_{rs} = 21.723$ deg |

*Example of the calculation*: A single-start involute hob with module $m = 10$ mm, outer hob diameter $d_{o.ct} = 180$ mm, and normal pressure angle $\phi_n = 20°$ is used as an example for the calculation of the design parameters of the precision involute hob with straight lateral cutting edges.

The results of the calculation are summarized in Table 15.3.

For accurate calculation of the design parameters of the precision involute hob, pitch diameter neither of the new hob nor of the completely worn hob can be entered into the equations. For accurate calculations, it is recommended to use the pitch diameter of the partly worn hob. This diameter correlates with the outer diameter of the cutting tool. The outer diameter of the new gear hob is equal to $d_{o.ct}$ (see Figure 15.51), while the outer diameter of the completely worn hob is calculated from the equation $(d_{o.ct} - \Delta d_{o.ct})$. For the calculation of the reduction $\Delta d_{o.ct}$ of the outer diameter $d_{o.ct}$, the following approximate equation:

$$\Delta d_{o.ct} \cong 2 \cdot \widehat{L} \cdot \tan \alpha_o = 5.585 \cdot \frac{\tan \alpha_o}{n_{ct}} \tag{15.124}$$

applies. Here it is designated:

- $\widehat{L}$ is the distance between two neighboring hob teeth that is measured along the helix on the outer cylinder of the hob (see Figure 15.51).
- $\alpha_o$ is the clearance angle at the top cutting edge of the hob tooth.
- $n_{ct}$ is the effective number of the hob teeth.

For involute hobs featuring straight slots, the effective number of the hob teeth $n_{ct}$ is always an integer number, and it is always equal to the actual hob tooth number $n_{ct}^a$, which usually falls into the range of $n_{ct}^a = 8 \sim 16$. This immediately yields a formula

$$\widehat{L} = \frac{\pi \cdot d_{o.ct}}{n_{ct}^a} \tag{15.125}$$

for the calculation of the distance $\widehat{L}$.

For gear hobs with helical slots, the effective hob teeth number $n_{ct}$ is always a number with fractions. Moreover, the actual value of this design parameter $n_{ct}$ depends on the hand of helix of the slots. This is owing to the fact that in the latter case the distance $\widehat{L}$ is calculated from equation [137, 138, 148] and others:

$$\widehat{L} = \frac{\pi \cdot d_{o.ct}}{n_{ct}^a} + P_{x.ct} \cdot N_{ct} \cdot \cos \lambda_{rs} \cdot \sin \lambda_{rs} \tag{15.126}$$

where:

- $d_{o.c}$ is the outside diameter of the gear hob.
- $P_{x.c}$ is the axial pitch of the hob.
- $N_c$ is the hob number of starts.
- $\lambda_{rs}$ is the lead angle of the hob rake surface $R_s$ (the lead angle $\lambda_{rs}$ is the signed value).

Equation (15.124) is a type of approximation that returns reasonably accurate results of the calculation.

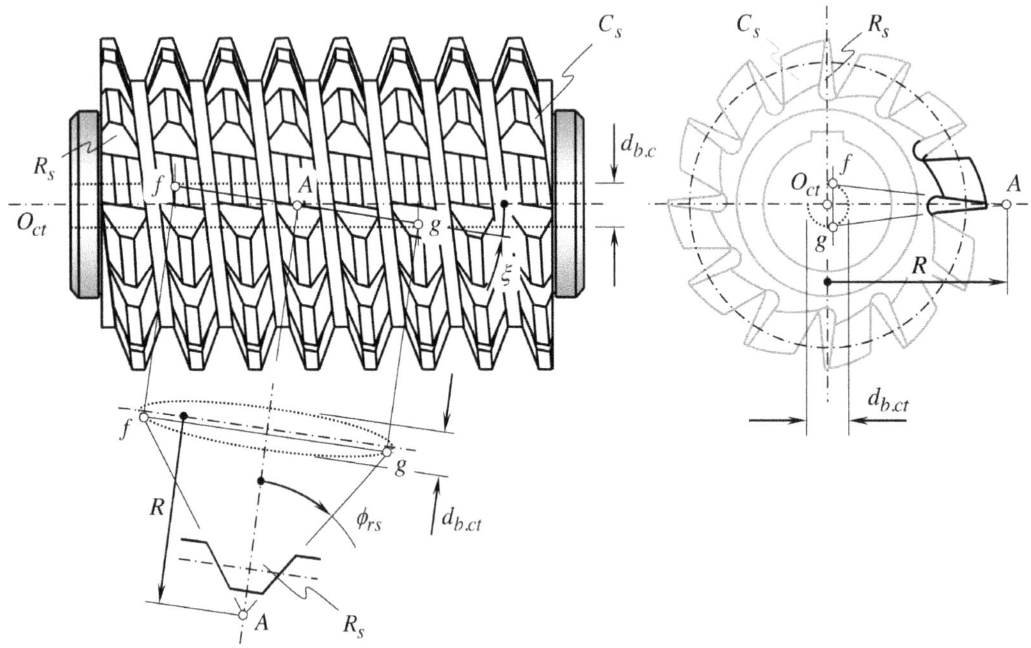


**FIGURE 15.86** Close-up of precision involute gear hob with straight lateral cutting edges. (From Radzevich, S.P., USSR Patent No. 990445, Int. Cl. B24b 3/12, October 8, 1981).

It is important to point out here the direction in which the angle $\xi$ of the rake plane orientation is measured. The angle $\xi$ is measured in the direction of positive pitch helix $\psi_{ct}$ of the precision involute hob. The direction to measure the angle $\xi$ is the same as the direction to measure the angle $\psi_{ct}$, thus the angle $\xi$ affects the cutting edge geometry of the lateral cutting edges of the precision involute hob.

Figure 15.86 illustrates a close-up of the precision involute gear hob with straight lateral cutting edges. For the readers' convenience, the designations of the main design parameters in Figure 15.86 are identical to those in Figures 15.84 and 15.85.

Analysis of the impact of the design parameters of the precision involute gear hob onto the value of the angle $\xi$ of the orientation of the rake surface $R_s$ reveals that a reduction of the angle $\xi$ is observed when

- The number of starts $N_{ct}$ of the hob is smaller, as it is illustrated in Figure 15.87.
- The profile angle $\phi_n$ of the hob tooth is greater, as shown in Figure 15.88.
- The setting angle of the hob $\zeta_{ct}$ is greater (see Figure 15.89).
- The hob pitch diameter $d_{ct}$ is greater (see Figure 15.87–15.89).

The rake planes of the involute hob teeth (see Figure 15.86) are not coplanar to one another. Figure 15.90(a) illustrates that a step is observed between the rake planes $R_s$ of each two neighboring teeth. The precision involute hob can be designed so that the rake planes $R_s$ of all the teeth in each row in the axial direction of the hob can be located in a common plane, as shown in Figure 15.90(b). Such an arrangement of the rake planes provides an insight into a practical type of approximation of the rake planes of the hob teeth within a common row.

An approximation of the rake plane of the involute gear hob by a surface that features geometry convenient for machining on machine tools is possible. For example, the set of the rake planes $R_s$ can be approximated by a screw surface of that same hand as the hand of the generating surface

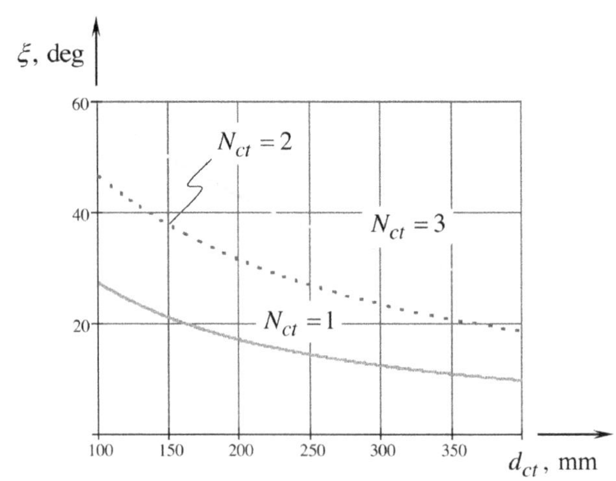


**FIGURE 15.87** Impact of the number of starts $N_{ct}$ of an involute gear hob on the actual orientation of the rake plane $R_s$ specified by the angle $\xi$ ($m = 10$ mm, $\phi_n = 20$ deg, $\zeta_{ct} = 3$ deg, $n_{ct} = 10$, $\alpha_o = 12$ deg). (From Radzevich, S.P., *ASME Journal of Mechanical Design*, 129, 334–345, 2007. With permission.)

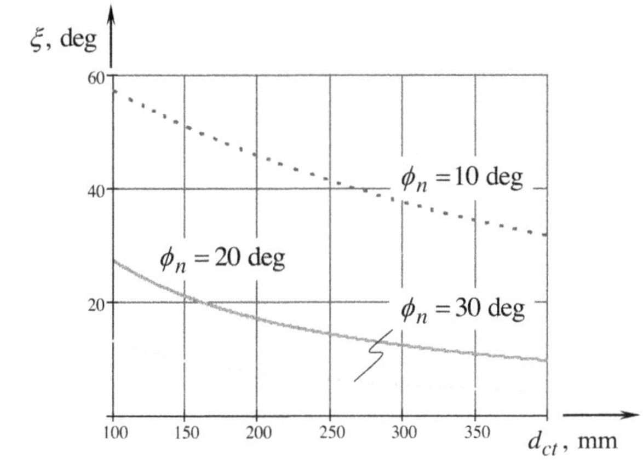


**FIGURE 15.88** Impact of the normal profile angle $\phi_n$ of an involute hob on the actual orientation of the rake plane $R_s$ specified by the angle $\xi$ ($m = 10$ mm, $N_{ct} = 1, \zeta_{ct} = 3$ deg, $n_{ct} = 10, \alpha_o = 12$ deg). (From Radzevich, S.P., *ASME Journal of Mechanical Design*, 129, 334–345, 2007. With permission.)

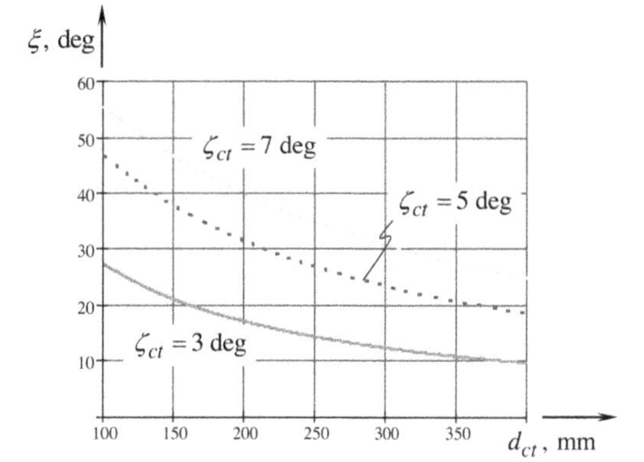


**FIGURE 15.89** Impact of the setting angle $\zeta_{ct}$ of a hob on the actual orientation of the rake plane $R_s$ specified by the angle $\xi$ ($m = 10$ mm, $\phi_n = 20$ deg, $N_{ct} = 1$, $n_{ct} = 10$, $\alpha_o = 12$ deg). (From Radzevich, S.P., *ASME Journal of Mechanical Design*, 129, 334–345, 2007. With permission.)

$T$ of the involute hob.[22] In such a scenario, the rake surface is shaped in the form of an Archimedean screw surface. The helix angle $\psi_{rs}$ of the screw rake surface $R_s$ is equal to the angle $\xi$, and thus the equality $\psi_{rs} = \xi$ is observed.

The involute gear hob with the approximated rake surface features zero rake angle at the top cutting edge ($\gamma_o = 0°$). The inclination angle $\lambda_o$ of the top cutting edge is approximately equal to the angle $\xi$.

### 15.5.3.2 Method for Resharpening Precision Involute Gear Hob

After an involute hob gets worn, either the rake surface $R_s$ or the clearance surface $C_s$ of the hob teeth can be reground in order to restore the cutting performance of the hob. For the purposes of resharpening of the worn precision involute hob, a method[23] of the hob regrinding can be employed [83]. A machine tool of conventional design for relief grinding of the cutting tools can be used for the purpose of grinding the clearance surfaces of the worn involute hob.

In the method schematically illustrated in Figure 15.91, the hob rotates about its axis of rotation $O_{ct}$ with steady rotation $\omega_{ct}$ [83]. The clearance surface $C_s$ of the hob tooth is faced toward the direction of the rotation $\omega_{ct}$. A dish-type grinding wheel is used in the hob resharpening operation. The grinding wheel rotates about its axis of rotation $O_{gw}$. The rotation of the grinding wheel is denoted by $\omega_{gw}$. Two angles $\theta_\phi$ and $\theta_\alpha$ together with two linear displacements $L_{gw}$ and $H_{gw}$ are the main setup parameters of the grinding wheel. The angles $\theta_\phi$ and $\theta_\alpha$ are used to accommodate mainly for the normal profile angle $\phi_n$ of the gear hob and for the clearance angle $\alpha_l$ at the lateral cutting edge of the hob tooth. The displacement $H_{gw}$ is specified by the design of the machine tool. Ultimately, the displacement $L_{gw}$ is a relatively free parameter, which makes possible optimization of the hob regrinding operation. The setup parameters $\theta_\phi$, $\theta_\alpha$, and $L_{gw}$ can be expressed in terms of the design parameters of the involute gear hob to be reground.

When resharpening an involute hob, the grinding wheel reciprocates toward the hob axis of rotation and in the opposite direction. The grinding wheel slowly approaches the hob. At a certain instant of time, the grinding plane $T$ of the grinding wheel touches the clearance surface $C_s$ of the hob tooth, and then it immediately moves outward on the hob. The speed of the latter motion is significantly higher compared to the speed of approach of the grinding wheel toward the hob axis of rotation. Simultaneously with the rotation $\omega_{ct}$ of the gear hob, the grinding wheel travels $V_{ax}$ in the axial direction of the hob being reground.

After the resharpening, the clearance surfaces $C_s$ of the hob teeth are shaped in the form of planes. Use of the method makes possible an increased clearance angle $\alpha_l$ at the lateral cutting edges of the hob. Normally, the method allows clearance angles up to $\alpha_l = 6...8°$ against $\alpha_l = 2...4°$ for the relief-ground hobs. The increased clearance angle $\alpha_l$ results in a significantly higher cutting performance of the involute hob.

Initially the method of resharpening of involute gear hobs [83] was developed [138] for the purposes of resharpening of semi-finishing and finishing built-up Azumi hobs (Japan), as well as of involute hobs of similar designs. However, the method also can be used for resharpening of the precision involute hobs of other designs.

### 15.5.3.3 Involute Hob for Machining Gear with a Modified Tooth Profile

The concept of the precision involute gear hob with straight lateral cutting edges can be evolved to a hob design for

[22] It is right to mention here that a similar approximation was proposed by Dr. Wildhaber [see: U.S. Pat. No. 1.924.476, *Method of Cutting Gears*, Filed: March 19, 1928, Patented: August 29, 1933]. Again, the solution proposed by Dr. Wildhaber to the problem under consideration is a kind of approximation, and not an accurate solution.

[23] SU Pat. No. 1087309, *A Method of Relieving of Gear Hob with the Standard Shaper Cutter*/S.P. Radzevich, Int. Cl. B24b 3/12, Filed: April 23, 1980.

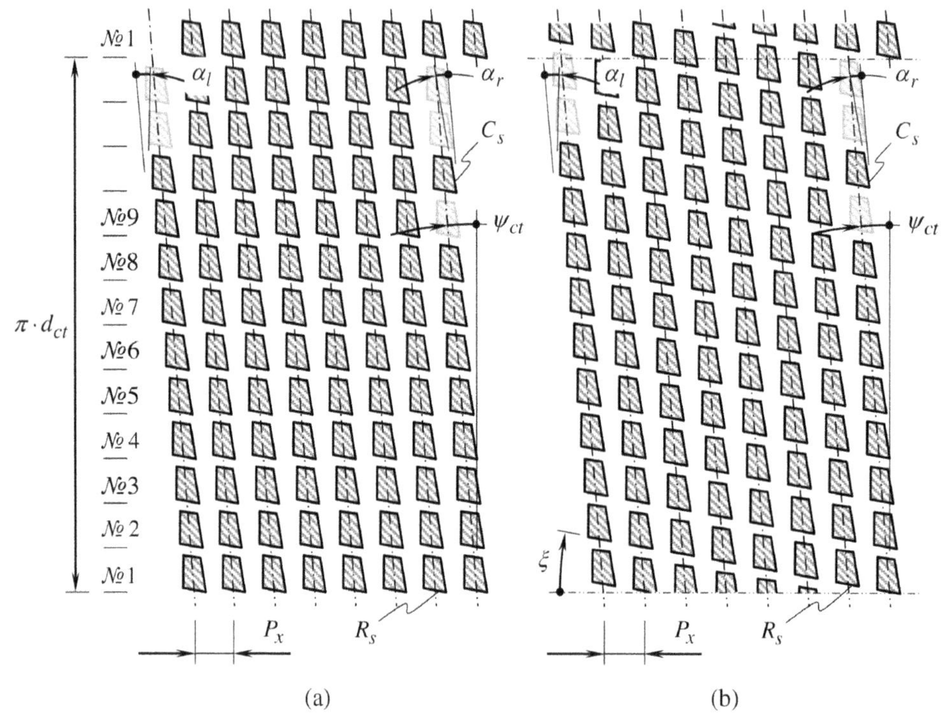


**FIGURE 15.90** Arrangement of the rake planes $R_s$ of a precision involute hob with straight lateral cutting edges so that they are located within a common plane, (a) involute hob cutting racks in their working positions, and (b) in technological positions for grinding over the rake surface. (From Radzevich, S.P., USSR Patent 990445, Int. Cl. B24b 3/12, October 8, 1981.)

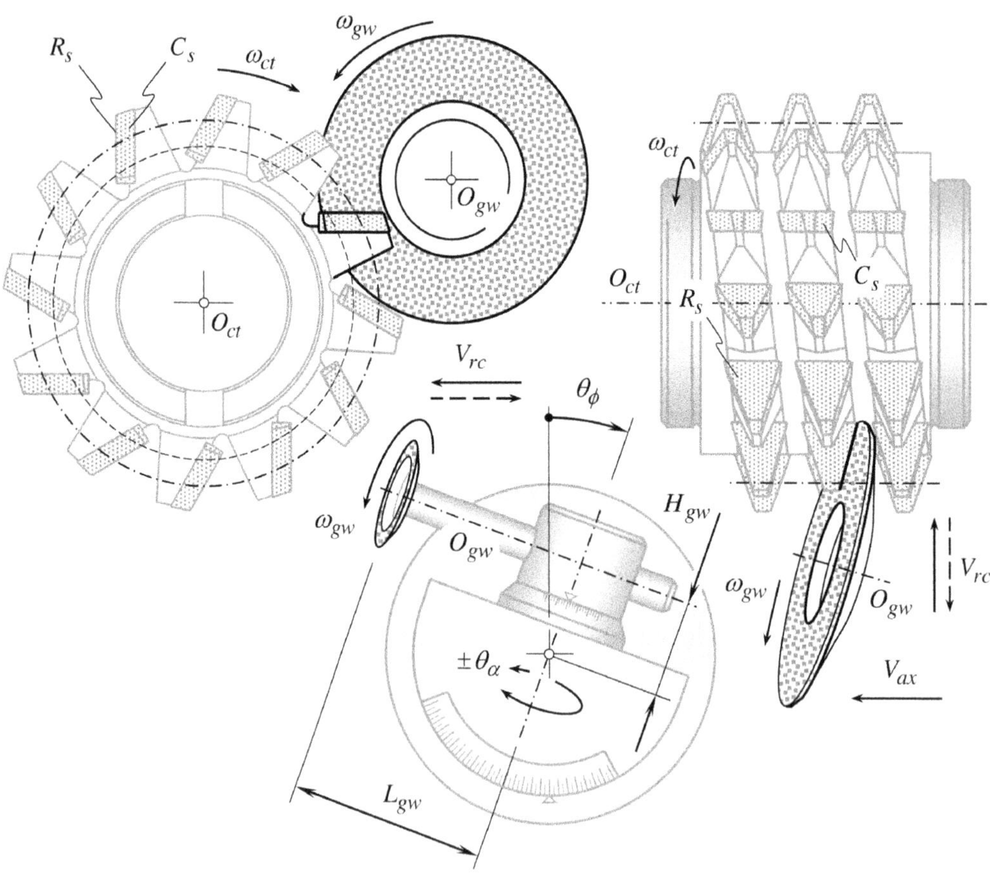


**FIGURE 15.91** Method for resharpening built-up involute gear hobs. (From Radzevich, S.P., USSR SU Patent 1087309, Int. Cl. B24b 3/12, April 23, 1980.)

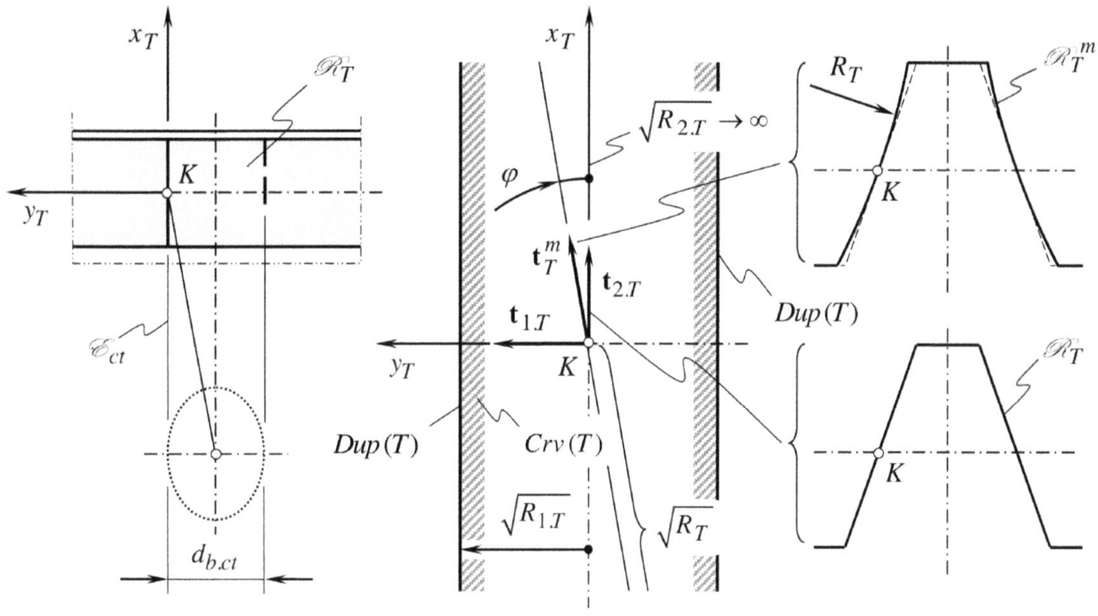


**FIGURE 15.92** Design concept of an involute hob for machining gears with a modified tooth profile (RU Pat. No. 2040376). (From Radzevich, S.P., *ASME Journal of Mechanical Design*, 129, 334–345, 2007. With permission.)

machining gears with a modified tooth profile. Such a possibility becomes evident from the analysis of Figure 15.92.

The straight lateral cutting edge *CE* of the precision involute hob is aligned with the straight line $\mathcal{E}_c$ which is entirely located within generating surface *T* of the hob. In searching for a possible approach to design a hob with a modified tooth profile, it sounds promising to turn our attention to the possibility of angular displacement of the characteristic line $\mathcal{E}_{ct}$ (and the lateral cutting edge *CE* of the gear hob as well) through a certain angle $\varphi$ about point *K* on the pitch line of the auxiliary generating rack $\mathcal{R}_T$ of the hob. Owing to the angular displacement of the straight line $\mathcal{E}_{ct}$, the lateral profile of the tooth of the auxiliary rack $\mathcal{R}_T^m$ becomes curved.[24] If the angular displacement $\varphi$ is determined properly, then a gear hob for machining gears with a modified tooth profile can be designed [107]. A design of the gear hob[25] is proposed for this purpose.

For the derivation of a formula for the calculation of the required value of the angular displacement $\varphi$, the use of the Euler formula

$$R_T(\varphi) = \left( \frac{\cos^2 \varphi}{R_{1.T}} + \frac{\sin^2 \varphi}{R_{2.T}} \right)^{-1} \qquad (15.127)$$

is helpful.

[24] In order to clarify the transition from the auxiliary rack $\mathcal{R}_T$ with a straight-line tooth profile to the auxiliary rack $\mathcal{R}_T^m$ with curved tooth profile, the following analogy can be considered. When a straight line rotates about an axis that is parallel to the line, a surface of a cylinder of revolution is generated. Then consider a straight line that crosses the axis of rotation. Consecutive positions of the straight line in the second case form a surface of a single-sheet hyperboloid of revolution.

[25] RU Pat. No. 2040376, *A Precision Involute Hob*/S.P. Radzevich, *et al*, Int. Cl. B23f 21/16, Filed: January 03, 1992.

Here, the principal radii of curvature $R_{1.T}$ and $R_{2.T}$ of the nonmodified generating surface *T* are equal to [148, 175]

$$R_{1.T} = \frac{\sqrt{d_{ct}^2 - d_{b.ct}^2} \cos \psi_{b.ct}}{2 \sin^2 \psi_{b.ct}} \qquad (15.128)$$

$$R_{2.T} \to \infty \qquad (15.129)$$

The approximate equation

$$\varphi \cong \cos^{-1} \sqrt{\frac{R_{1.T}}{R_T}} \qquad (15.130)$$

for the calculation of the desirable value of the angular displacement $\varphi$ immediately follows from the consideration of Eq. (15.127) together with Eq. (15.129).

Equation (15.130) is an approximate one because all points within the modified generating surface of the gear hob are of the hyperbolic type rather than the parabolic type, as observed for all points within the generating surface of the nonmodified hob. However, as the hob tooth profile modifications are small in nature, local approximation of saddle-type points (that is, hyperbolic-type points) on the modified rack $\mathcal{R}_T^m$ with parabolic points of the corresponding nonmodified rack $\mathcal{R}_T$ is permissible and practical.

To illustrate the feasibility of modification of the hob tooth profile, it is helpful to use the characteristic curves, namely, of the $\mathscr{I}n_R(T)$ – indicatrix of the first kind and the $\mathscr{I}n_k(T)$ – indicatrix of the second kind [101, 156].

The position vector $\mathbf{r}_{iR}$ of a point of the $\mathscr{I}n_k(T)-$ indicatrix of the first kind is given by the matrix equation

$$\mathscr{I}n_R(T) \Rightarrow \mathbf{r}_{iR}(\varphi) = \begin{bmatrix} R_T(\varphi)\cdot\cos\varphi \\ R_T(\varphi)\cdot\sin\varphi \\ 0 \\ 1 \end{bmatrix} \quad (15.131)$$

This characteristic curve indicates the alteration of normal radii of curvature $R_T(\varphi)$ of the generating surface $T$ of the gear hob in differential vicinity of the point $K$ (see Figure 15.93).

For the position vector $\mathbf{r}_{ik}$ of a point on the $\mathscr{I}n_k(T)-$ indicatrix of the second kind, the following matrix equation:

$$\mathscr{I}n_k(T) \Rightarrow \mathbf{r}_{ik}(\varphi) = \begin{bmatrix} k_T(\varphi)\cdot\cos\varphi \\ k_T(\varphi)\cdot\sin\varphi \\ 0 \\ 1 \end{bmatrix} \quad (15.132)$$

is known from [101, 148].

This characteristic curve indicates the alteration of normal curvature $k_T(\varphi)$ within the differential vicinity of the point $K$ of the generating surface $T$ of the involute hob (see Figure 15.93).

Figure 15.93 reveals that any desirable value of the hob tooth modification (that is, any desirable value of the radius of curvature $R_T$) is permissible [107].

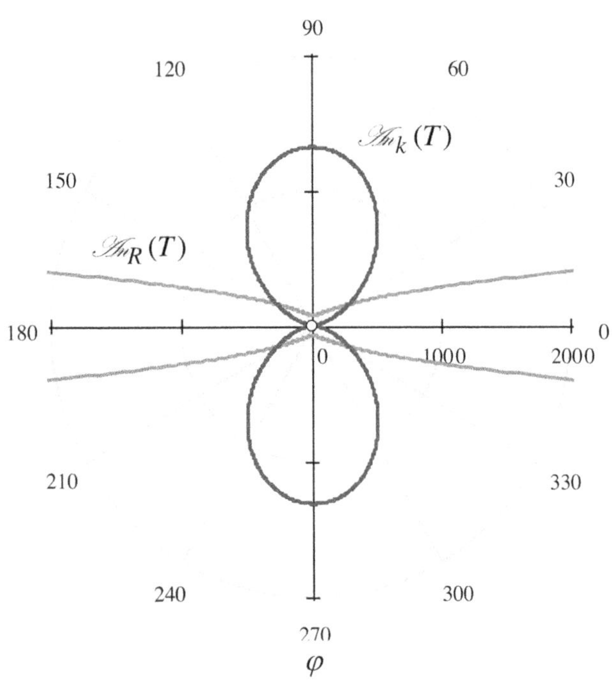


**FIGURE 15.93** Alteration of the normal curvature $k_T = k_T(\varphi)$ and the radii of curvature $R_T = R_T(\varphi)$ in the differential vicinity of a point on the generating surface $T$ of an involute hob. (From Radzevich, S.P., *ASME Journal of Mechanical Design*, 129, 334–345, 2007. With permission.)

Another way to illustrate the possibility of the modification of the involute gear tooth profile with the help of the angular displacement of the lateral cutting edges is based on the following consideration. The well-known equation $R_T = \Phi_{1.T}/\Phi_{2.T}$ for the calculation of the radii of normal curvature can be rewritten in extended form. Let us designate the ratio $dV_T/dU_T$ as $\upsilon = dV_T/dU_T$. Then, the expressions for $R_T(\varphi)$ and for $k_T(\varphi)$ can be replaced with the similar expressions $R_T(\upsilon)$ and $k_T(\upsilon)$ in terms of the $\upsilon-$ parameter:

$$R_T(\upsilon) = \frac{E_{ct} + 2\cdot F_{ct}\cdot\upsilon + G_{ct}\cdot\upsilon^2}{L_{ct} + 2\cdot M_{ct}\cdot\upsilon + N_{ct}\cdot\upsilon^2} \quad (15.133)$$

$$k_T(\upsilon) = R_T^{-1}(\upsilon) \quad (15.134)$$

The extreme values of the radii $R_{1.T}$ and $R_{2.T}$ of normal curvature, as well as of normal curvatures $k_{1.T}$ and $k_{2.T}$, occur at roots $\upsilon_1$ and $\upsilon_2$ of the equation

$$\det\begin{bmatrix} \upsilon^2 & -\upsilon & 1 \\ E_{ct} & F_{ct} & G_{ct} \\ L_{ct} & M_{ct} & N_{ct} \end{bmatrix} = 0 \quad (15.135)$$

where $E_{ct}$, $F_{ct}$, and $G_{ct}$ designate Gaussian coefficients of the first fundamental form $\Phi_{1.T}$ of the generating surface $T$ of the involute gear hob (see Chapter 1, Appendix B).

Gaussian coefficients of the second fundamental form $\Phi_{2.T}$ of the generating surface $T$ of the involute gear hob are designated as $L_{ct}$, $M_{ct}$, and $N_{ct}$ (see Chapter 1, Appendix B).

Both the characteristic curves, that is, the curves $R_T = R_T(\upsilon)$ and $k_T = k_T(\upsilon)$, are a type of graphical interpretation of the alteration of the radii of normal curvature, as well as the normal curvatures themselves in the differential vicinity of a point within the generating surface of the involute hob (see Figure 15.94). They perfectly correlate with the $\mathscr{I}n_R(T)-$ indicatrix of the first kind, and the $\mathscr{I}n_k(T)-$ indicatrix of the second kind (see Figure 15.93), of the generating surface $T$.

The calculation of the design parameters of the involute hob with a modified tooth profile is similar to that of the gear hob with a nonmodified tooth profile. The major difference occurs in the calculation of the parameter $R_m$ and the distance $d_m$. The parameter $R_m$ for the hob with a modified tooth profile differs from the corresponding parameter $R$ [see Eq. (15.112)], and the distance $d_m$ is not equal to the base diameter $d_{b.h}$ of the hob.

The desirable value of the distance $R_m$ must be expressed in terms of the modification $R_T$ of the hob tooth. For this purpose, it is convenient to solve an elementary geometrical problem, namely, to determine the coordinates of a certain point $S$ of the intersection of the circular arc of the radius $R_T$ (see Figure 15.92) with the centerline of the modified tooth profile. (Point $S$ is not shown in Figure 15.92). Then the parameter $R_m$ can be determined as a distance of point $S$ to

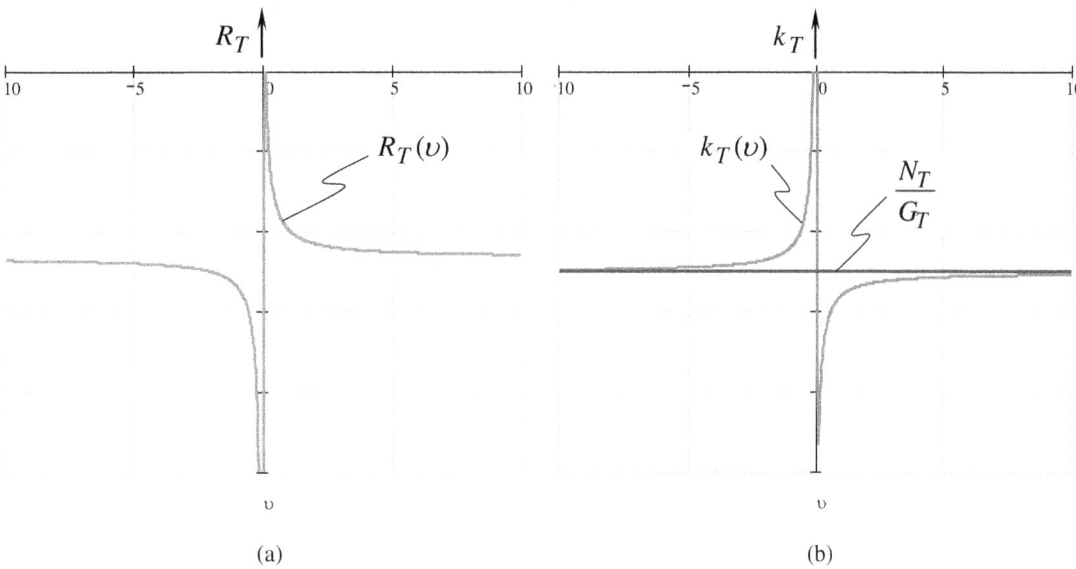


**FIGURE 15.94** (a) Alteration of the radii of normal radii $R_T(\upsilon)$ and (b) normal curvature $k_T(\upsilon)$ of the generating surface $T$ of an involute hob vs. $\upsilon = dV_T/dU_T$. (From Radzevich, S.P., *ASME Journal of Mechanical Design*, 129, 334–345, 2007. With permission.)

the axis of rotation $O_{ct}$ of the gear hob. Following the routine outlined above, the equation

$$R_m = \frac{R_T \sin\phi_n - t_r + 0.5d_{ct} - \sqrt{R_T^2 \sin^2\phi_n - R_T t_r \cos\phi_n - t_r^2}}{4R_T t_r \cos\phi_n + t_r^2} \tag{15.136}$$

for the calculation of the parameter $R_m$ is derived.

The corresponding equation for the distance $d_m$ can be represented in the form

$$d_m = d_{ct} \cdot \frac{\sin(\theta - \varphi)}{\cos\theta \cos\zeta_{ct}} \tag{15.137}$$

where for the calculation of the angle $\theta$, the earlier derived formula

$$\theta = \tan^{-1}\left(\frac{d_{b.ct}}{d_{ct}} \cdot \cos\psi_{b.ct}\right) \tag{15.138}$$

applies.

The generating surface $T_m$ of the gear hob with a modified tooth profile is different from the generating surface $T$ of the hob for machining nonmodified gears. However, it is of critical importance to stress here that the tooth profile angle $\phi_n$ and the setting angle of the hob $\zeta_{ct}$ do not change their values, and for the modified surface $T_m$ they are of that same value as for the generating surface $T$, which is evident.

Actually, the involute hob [107] provides a hyperbolic modification of the tooth profile of the hobbed gear. The hyperbolic modification enhances known types of modification of the gear tooth profile, namely, of a circular modification, parabolic modification, topological modification, and so forth.

*Illustrative example of the calculation*. Consider a single-start involute hob of module $m = 10$ mm, outer diameter $d_{o.ct} = 180$ mm, and normal pressure angle $\phi_n = 20°$. The desired modified tooth profile is a hyperbola through three points $a_3$, $a_k$, and $a_4$. The hyperbolic tooth profile is approximated with a circular arc through the same three points $a_3$, $a_k$, and $a_4$. It is assumed here that the approximation is reasonable, and it insignificantly affects the accuracy of the hobbed gear.

The initial tooth profile of the hob is not modified. It is a segment of straight line through points $a_1(x_1, y_1)$, $a_k(x_k, y_k)$, and $a_2(x_2, y_2)$ (see Figure 15.95). Actually, for the design of the involute gear hob under consideration, the coordinates of points $a_1$, $a_k$, and $a_2$ are as follows: $x_1 = 0$ mm, $y_1 = 0$ mm; $x_k = 3.64$ mm, $y_k = 10$ mm; and $x_2 = 8.189$ mm, $y_2 = 22.5$ mm.

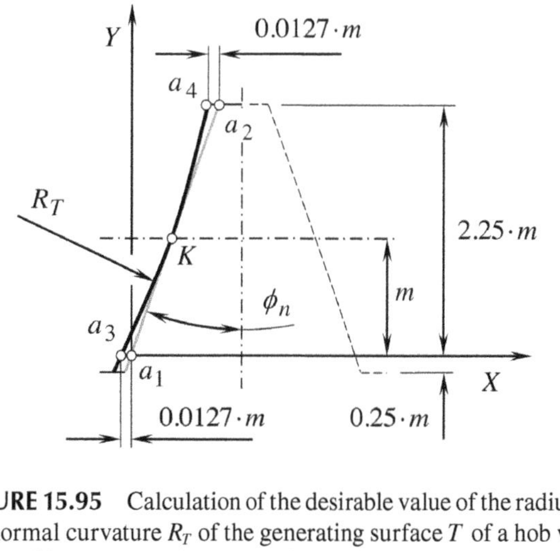


**FIGURE 15.95** Calculation of the desirable value of the radius of the normal curvature $R_T$ of the generating surface $T$ of a hob with the modified tooth profile. (From Radzevich, S.P., *ASME Journal of Mechanical Design*, 129, 334–345, 2007. With permission.)

The hob tooth profile has a modification of 0.127 mm. This allows for the calculation of the coordinates of points $a_3$ and $a_4$, which are as follows: $x_3 = -0.127$ mm, $y_3 = 0$ mm and $x_4 = 8.062$ mm, $y_4 = 22.5$ mm.

In order to provide the tooth of the involute hob with the modification in the range of 0.127 mm, the tooth profile radius of curvature $R_T$ has to be equal to

$$R_T = \sqrt{\frac{B^2 + C^2}{4 \cdot A^2} - \frac{D}{A}} = 593.625\,\text{mm} \qquad (15.139)$$

where all the parameters $A$, $B$, $C$, and $D$ are of constant value. They are calculated from the formulae immediately following:

$$A = \begin{vmatrix} x_3 & y_3 & 1 \\ x_k & y_k & 1 \\ x_4 & y_4 & 1 \end{vmatrix} = 2.858 \qquad (15.140)$$

$$B = -\begin{vmatrix} x_3^2 + y_3^2 & y_3 & 1 \\ x_k^2 + y_k^2 & y_k & 1 \\ x_4^2 + y_4^2 & y_4 & 1 \end{vmatrix} = 3.165 \times 10^3 \qquad (15.141)$$

$$C = \begin{vmatrix} x_3^2 + y_3^2 & x_3 & 1 \\ x_k^2 + y_k^2 & x_k & 1 \\ x_4^2 + y_4^2 & x_4 & 1 \end{vmatrix} = -1.224 \times 10^3 \qquad (15.142)$$

$$D = -\begin{vmatrix} x_3^2 + y_3^2 & x_3 & y_3 \\ x_k^2 + y_k^2 & x_k & y_k \\ x_4^2 + y_4^2 & x_4 & y_4 \end{vmatrix} = 401.864 \qquad (15.143)$$

The first principal radius of curvature $R_{1.T}$ of the generating surface $T$ of the hob [Eq. (15.115)] is

$$R_{1.T} = \frac{\sqrt{155^2 - 27.037^2} \cdot \cos 20.338^\circ}{2 \cdot \sin^2 20.338^\circ} = 592.357\,\text{mm} \qquad (15.144)$$

The required value of the angular displacement $\varphi$ [Eq. (15.130)] is

$$\varphi = \cos^{-1}\left(\sqrt{\frac{592.357\,\text{mm}}{593.625\,\text{mm}}}\right) = 3.746\,\text{deg} \qquad (15.145)$$

The projection of the lateral cutting edge onto the $X_{ct}Y_{ct}$ coordinate plane forms an angle $\nu_m$ with the centerline (see Figure 15.96). This angle has the value $\nu_m = 3.835$ deg.

The earlier calculated design parameters of the precision involute gear hob yield the calculation of the radius

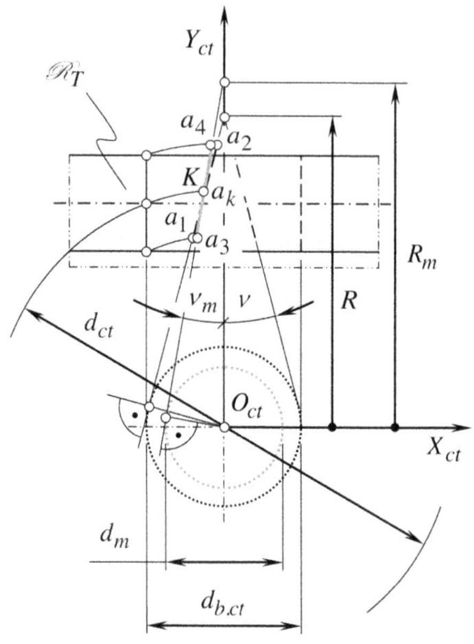


**FIGURE 15.96** On the calculation of the parameters $d^*_{b.h}$ and $R^*$ of the precision involute gear hob with a modified tooth profile. (From Radzevich, S.P., *ASME Journal of Mechanical Design*, 129, 334–345, 2007. With permission.)

$R_m = 123.175$ mm and the displacement $d_m = 16.478$ mm. These values are obtained from the solution of the triangles shown in Figure 15.96.

Finally, Eq. (15.122) yields for $\xi_m$:

$$\xi_m = \arctan \frac{\cos 3.772^\circ \cdot \tan 3.835^\circ}{\tan 20^\circ + \sin 3.772^\circ \cdot \tan 3.835^\circ} = 10.292^\circ \quad (15.146)$$

For the calculation of $\phi^m_{rs}$,

$$\phi^m_{rs} = \arctan \frac{\tan 3.835^\circ}{\sin 10.292^\circ} = 21.466^\circ \qquad (15.147)$$

Eq. (15.123) is applied.

The calculated design parameters allow for designing of the precision involute gear hob with a desired modification of the tooth profile.

### 15.5.4 Examples of Nonstandard Designs of Involute Gear Hobs

Although involute hobs of standard design are the most common cutting tools for machining gears, in many cases of gear production, gear hobs of improved designs are used as well.

#### 15.5.4.1 Cylindrical Gear Hobs of Nonstandard Design

Numerous designs of cylindrical hobs are developed that target the improvement of cutting performance, gaining the ability to cut hard stock, and so forth.

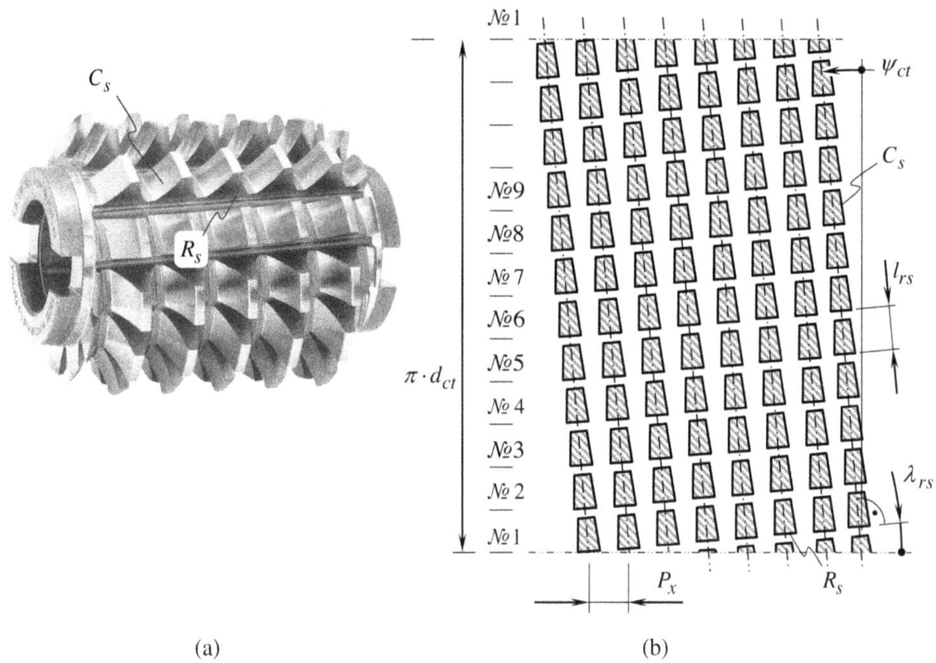


**FIGURE 15.97** Example of unfolded section of teeth of a standard involute gear hob by the pitch cylinder: (a) a hob, and (b) unfolded section by the pitch cylinder.

### *15.5.4.1.1 Gear Hob for Smooth Roughing of Coarse Pitch Gears*

An example of a standard involute gear hob for the machining of coarse pitch gears is depicted in Figure 15.97(a). Without going into the details of the description of the hob design, it is of interest to turn the reader's attention here to the configuration of the hob teeth within each row, that is, in the direction of the gashes.

The gashes in the body of the hob are machined so that the rake faces $R_s$ are orthogonal to the pitch helix through the centerline of each tooth [see Figure 15.97(b)]. This teeth arrangement equalizes the geometrical parameters of the lateral cutting edges of opposite sides of the hob tooth profile; the corresponding parameters of the geometrical parameters of the opposite lateral cutting edges are equal to each other. The gear hobs of standard design feature the pitch helix angle $\psi_{ct}$, identical to the lead angle $\lambda_{rs}$ of the screw rake surface $R_s$. For standard gear hobs, the identity $\psi_{ct} \equiv \lambda_{rs}$ is observed. In standard gear hobs, the hob setting angle $\zeta_{ct}$ is equal to the pitch helix angle $\psi_{ct}$, so the hob teeth of every row of teeth start the stock removal simultaneously (or nearly simultaneously). They also exit from cutting the stock simultaneously (or nearly simultaneously). Under such a scenario, intensive vibrations are generated. The vibrations negatively affect the gear machining process, which is especially critical when hobbing coarse pitch gears.

This particular problem of gear hob design is eliminated in the design of the gear hob[26] that is developed mostly for the purpose of roughing coarse pitch gears [76].

[26] SU Pat. No. 944828, *A Hob for Roughing Gears*/S.P. Radzevich, Int. Cl. B23f 21/16, Filed: September 4, 1980.

The frequency of the vibrations depends on the period of time between the instant of time when the cutting edges enter the stock and the instant of time when the cutting edges exit the stock. For gear hobs of a certain pitch diameter $d_{ct}$, this period of time $t$ can be expressed in the form

$$t = \frac{l_{rs}}{V_{ct}} = \frac{2 \cdot l_{rs}}{\omega_{ct} \cdot d_{ct}} \tag{15.148}$$

where:

- $l_{rs}$ is the distance between every two consecutive teeth along the thread of the hob.
- $V_c$ is the linear velocity of rotation of the point of the cutting edge at the pitch diameter of the hob.
- $\omega_c$ is the hob rotation.

For a given hob design, the distance $l_{rs}$ has a constant value, and it cannot be altered. The distance $l_{rs}$ can be expressed in terms of design parameters of the hob (see Figure 15.98):

$$l_{rs} = \frac{\sqrt{\pi^2 \cdot d_{ct}^2 + P_{x.ct}^2} - P_{x.ct} \cdot \sin \lambda_{rs}}{n_{ct}} \tag{15.149}$$

where:

- $P_{x.ct}$ is the axial pitch of the hob threads ($P_{x.ct} = \pi \cdot m \cdot N_{ct}$).
- $N_c$ is number of starts of the hob (sometimes this design parameter of the hob is referred to as the *number of hob threads*, which is not recommended).
- $\lambda_{rs}$ is the lead angle of the hob rake surface $R_s$.
- $n_{ct}$ is the number of teeth of the hob.

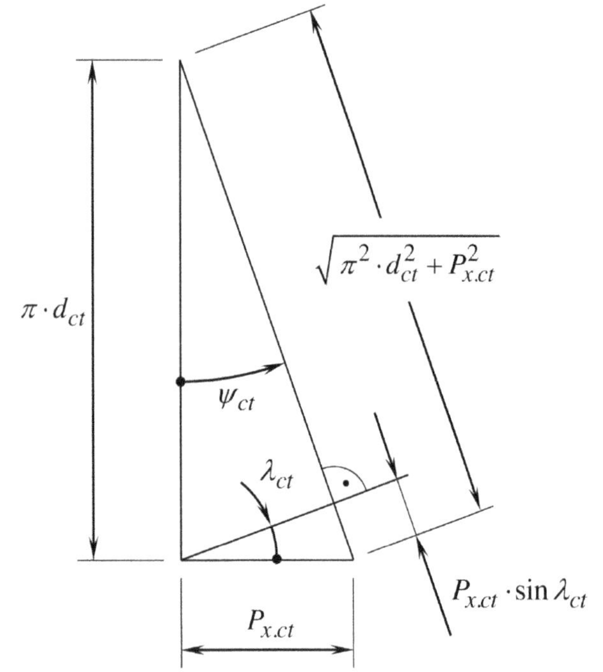


**FIGURE 15.98** Schematic for the calculation of the distance $l_{rs}$ between rake surfaces $R_s$ of a gear hob of a standard design.

When hobbing a gear, each cutting edge is loaded heavily. Heavy loading of the hob teeth is caused by the relatively large length of the distance $l_{rs}$. This results in every row of the hob teeth becoming loaded heavily as well. Alteration in the cutting load with the low frequency $t^{-1}$ causes undesirable vibrations.

In the hob design shown in Figure 15.99 [76], the rake surfaces $R_s$ are shifted at a certain distance $\delta_{rs}$ in the direction of the hob threads. After being shifted at the distance $\delta_{rs}$, the rake planes $R_s$ of the hob teeth align with a helix on the pitch cylinder of the hob. The helix angle that specifies the actual location of the rake planes is designated as $\xi_{rs}$.

The shift distance $\delta_{rs}$ can have either a positive value [in this case the angle $\xi_{rs}$ is greater than the lead angle $\lambda_{rs}$, as shown in Figure 15.99(a)] or a negative value (in this case $\xi_{rs} < \lambda_{rs}$), as depicted in Figure 15.99(b). In specific cases, the angle $\xi_{rs}$ can have a zero value ($\xi_{rs} = 0°$). The latter makes possible the design and manufacture of assembled gear hobs.

Each tooth of the hob (see Figure 15.99) withholds the cutting load of that same range as the standard gear hob does. A significant reduction of vibration is observed because the cutting teeth are entering into and exiting

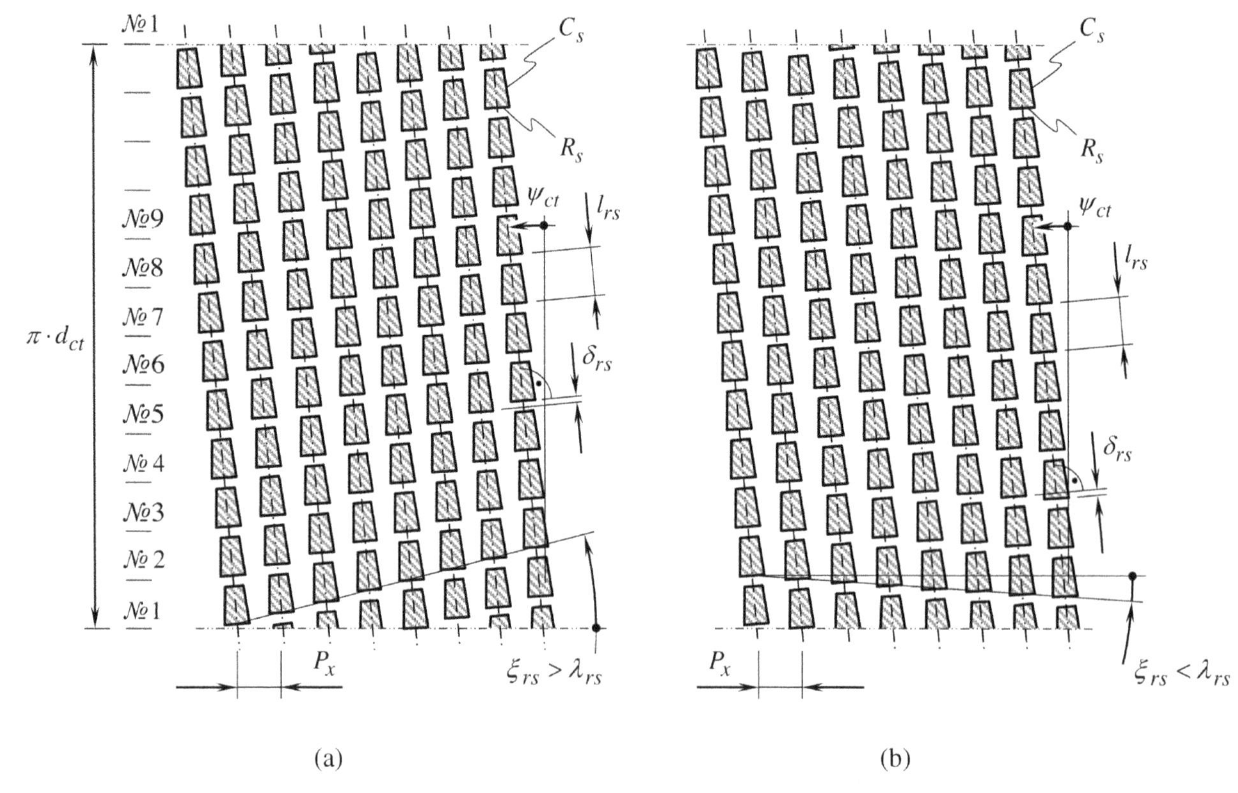


**FIGURE 15.99** Unfolded section (by the pitch cylinder) of a hob for roughing of coarse pitch gears: (a) left-handed, and (b) right-handed configurations of the rake surfaces. (From Radzevich, S.P., RU Pat. No. 944828, Int. Cl. B24b 3/12, Sept. 4, 1980.)

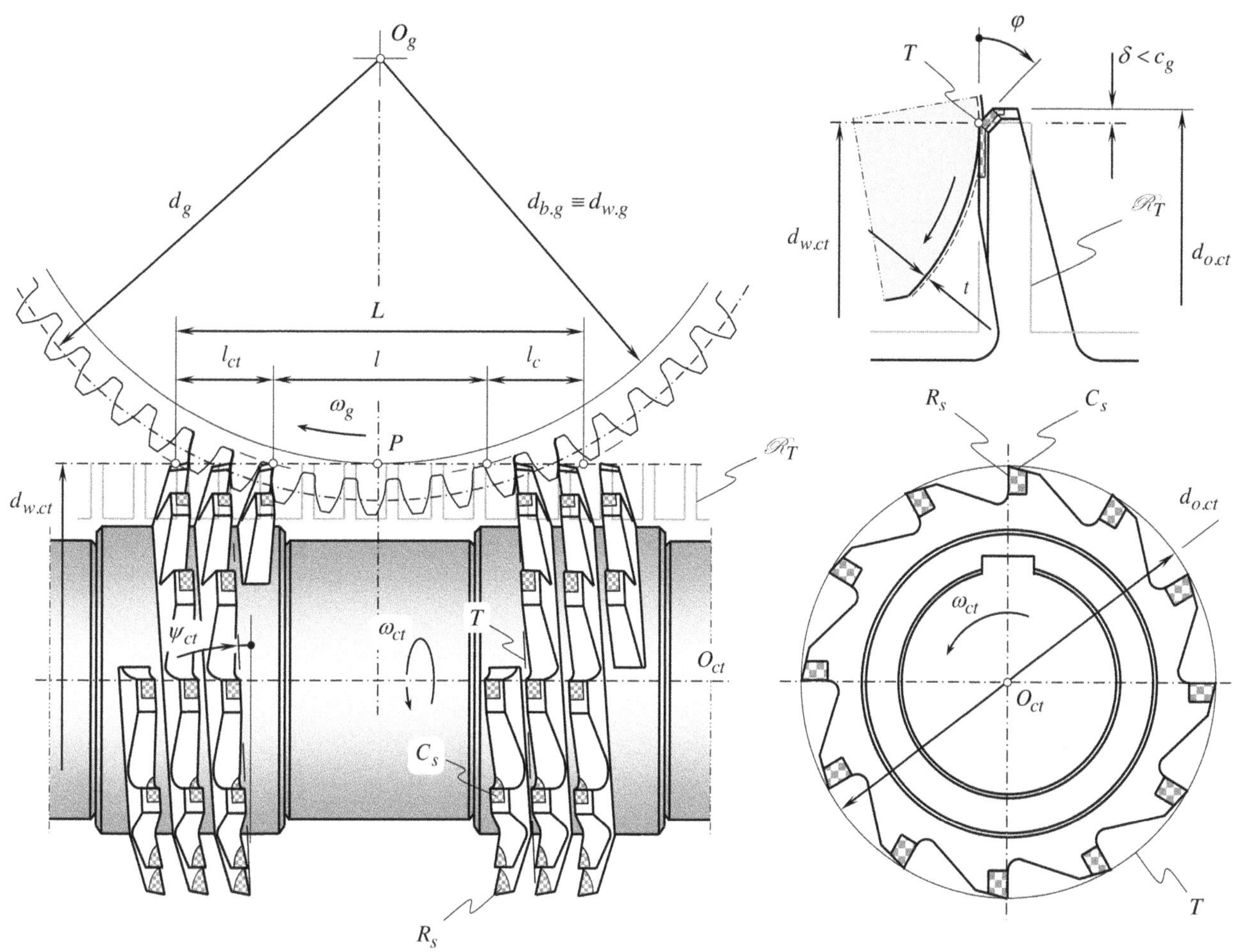


**FIGURE 15.100** Special-purpose hob for finishing and/or semi-finishing of hardened (Rockwell hardness up to $HR_C$ 62...64) gears.

from the chip removal process not simultaneously, but alternatively. The negative effect of the vibrations is eliminated mostly because of significant reduction of the variation in the cutting load and the increasing frequency of the vibrations.

For manufacturing of hobs for smooth roughing of coarse pitch gears (see Figure 15.99) [76], the methods[27] of hob grinding discussed above (see Figures 15.18, 15.19, 15.41, and 15.91) can be used [83, 86, 137, 138, 141, 159, 176, 190] and others.

### *15.5.4.1.2 Hob for Finishing Hardened (Rockwell Hardness up to $HR_C$ 62...64) Gears*

Hardened gears with surface hardness up to $HR_C$ 62...64 are used in the design of turbine reducers, helicopter transmissions, and so forth. For high production semi-finishing and/or finishing of hardened gears, special-purpose hobs are proposed [137, 141].

Chip removal from a hardened gear tooth surface is feasible if a built-up hob with cutting edges made of super-hard materials (borazon, human-made diamond, and so forth) is used for this purpose. Because of inevitable limitations imposed by the process of synthesizing super-hard materials, only small cutting elements can be synthesized. The available size of cutting elements is far from being implemented in the design of standard gear hobs of medium and/or coarse pitch. In order to make the use of super-hard materials in the design of gear-cutting tools possible, a special-purpose gear hob design is proposed[28] (see Figure 15.100).

The profile angle $\phi_n$ of the auxiliary generating rack $\mathscr{R}_T$ of the gear hob is equal to zero ($\phi_n = 0°$). The pitch line of the auxiliary rack $\mathscr{R}_T$ is tangent to the base cylinder $d_{b.g}$ of the gear being machined. When hobbing, the base cylinder $d_{b.g}$ serves as the pitch cylinder $d_{w.g}$ of the gear; therefore, the equality $d_{w.g} = d_{b.g}$ is observed in the gear hobbing process.

[27] SU Pat. No. 975339, *A Method of Grinding of Gear Hob*/S.P. Radzevich, Int. Cl. B24b 3/12, Filed: October 10, 1980.
SU Pat. No. 1087309, *A Method of Relieving of Gear Hob with the Standard Shaper Cutter*/S.P. Radzevich, Int. Cl. B24b 3/12, Filed: April 23, 1980.
SU Pat. No. 1335425, *A Method of Relief Grinding of Hob Teeth*/S.P. Radzevich, V.I. Lempert, Int. Cl. B23f 21/16, Filed: April 24, 1986.
SU Pat. No. 1627375, *A Method of Grinding of Gear Hob*/S.P. Radzevich, I.V. Vinokourvov, Int. Cl. B24b 3/12, Filed: October 17, 1988.

[28] In the 1970s, extensive research on design and implementation of special-purpose hobs for cutting hardened gears was undertaken by Dr. Oleg I. Moiseyenko (Kiev, Ukraine).

The hob is composed of two semi-hobs. The semi-hobs are separated from one another by a spacer (by a distance ring).

The total length $L$ of the active portion of the line of action is equal:

$$L = \sqrt{d_{o.g}^2 - d_{b.g}^2} \tag{15.150}$$

The similar expression

$$l = \sqrt{d_{l.g}^2 - d_{b.g}^2} \tag{15.151}$$

is valid for the inner portion $l$ of the line of action (see Figure 15.100).

Here, in Eq. (15.150) and Eq. (15.151), the outer diameter of the gear is designated as $d_{o.g}$, the base diameter of the gear is designated as $d_{b.g}$, and the limit diameter of the gear is designated as $d_{l.g}$.

Equations (15.150) and (15.151) allow for a formula

$$l_{ct} = \frac{L - l}{2} \tag{15.152}$$

for the calculation of the length $l_{ct}$ of the active part of the hob.

The pitch helix angle $\psi_{ct}$ of the hob can be specified in two ways.

*First*, the angle $\psi_{ct}$ is equal (see Figure 15.101):

$$\psi_{ct} = \tan^{-1}\left(\frac{P_{x.ct}}{\pi \cdot d_{w.ct}}\right) \tag{15.153}$$

*Second*, that same angle $\psi_{ct}$ can be expressed in terms of other design parameters of the hob and of the gear to be machined (see Figure 15.101):

$$\psi_{ct} = \cos^{-1}\left(\frac{p_{b.g}}{P_{x.ct}}\right) \tag{15.154}$$

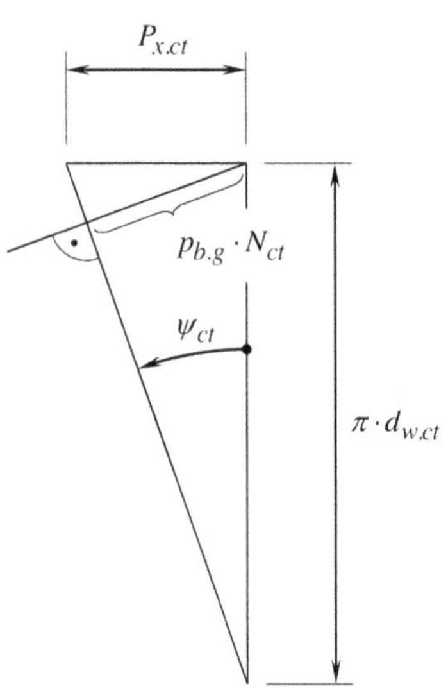


**FIGURE 15.101** Schematic for the calculation of the axial pitch $P_{x.ct}$ and pitch helix angle $\psi_{ct}$ of the hob shown in Figure 15.100.

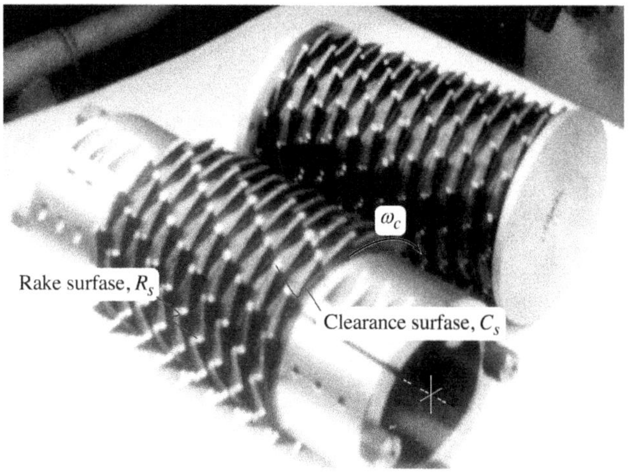


**FIGURE 15.102** Gear hob with a profile angle of a zero value for cutting involute gears (After Iziumov, S.N., Pankratov, Yu.M, and Sarafanov, I.N., of Leningrad Polytechnic Institute, USSR.)

Equations (15.153) and (15.154) specify that same pitch helix angle $\psi_{ct}$. The equations allow for a formula

$$P_{x.ct} = \frac{\pi \cdot p_{b.g} \cdot d_{w.ct}}{\sqrt{\pi^2 \cdot d_{w.ct}^2 - p_{b.g}^2}} \tag{15.155}$$

for the calculation of the axial pitch $P_{x.ct}$ of the hob helix.

Then, Eq. (15.153) considered together with Eq. (15.155) allow for an expression

$$\psi_{ct} = \tan^{-1}\left(\frac{p_{b.g}}{\sqrt{\pi^2 \cdot d_{w.ct}^2 - p_{b.g}^2}}\right) \tag{15.156}$$

for the calculation of the pitch helix angle $\psi_{ct}$.

As the hob profile angle is zero ($\phi_n = 0°$), in the case under consideration the setting angle of the hob $\zeta_c$ is equal to the pitch helix angle $\psi_{ct}$. This statement correlates with Eq. (14.75).

The design of the gear hob makes possible the use of cutting elements with short cutting edges for hobbing gears of any desired coarse pitch. Another advantage is that the geometry of the cutting edge is independent from the kinematics of meshing. Due to that, the gear hob can be designed with optimal geometry of the cutting edge. Taken as a whole, the use of the hob (see Figure 15.100) allows for high productive machining of hardened gears.

It should be mentioned here that in the past, there were many attempts to use a profile angle of zero in the design of gear hobs for cutting involute gears. The hob design[29] shown in Figure 15.102 is a good illustrative example in this regard.

[29] Proposed by Iziumov, S.N., Pankratov, Yu.M., and Sarafanov, I.N., of Leningrad Polytechnic Institute, USSR.

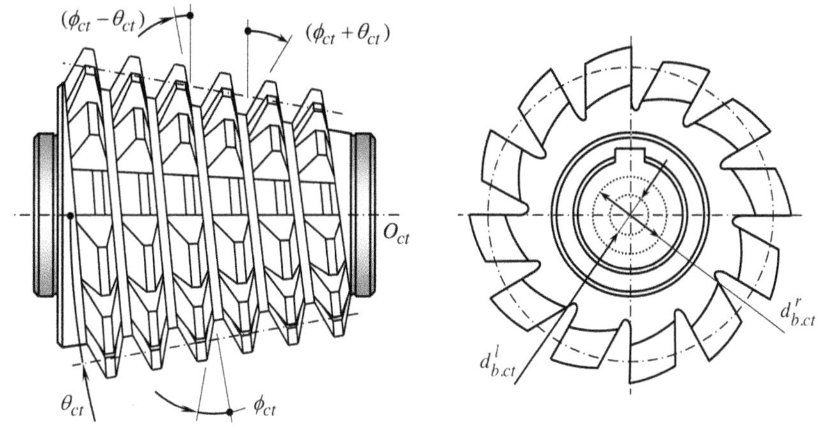


**FIGURE 15.103** Conical involute hob for machining spur and helical gears.

### 15.5.4.2 Conical Gear Hobs

Conical hobs can be used for machining of spur and helical gears. Conical gear hobs are designed on the premise of the generating surface $T$, which is shaped in the form of a conical worm conjugate to the gear being machined (see Figure 14.15). An example of a conical hob for machining spur and helical gears is schematically illustrated in Figure 15.103.

The profile angle of the auxiliary generating rack $\mathscr{R}_T$ tooth is designated in Figure 15.103 as $\phi_{ct}$. Owing to the cone angle $\theta_c$, the tooth flanks of the opposite sides of the tooth profile of the auxiliary rack $\mathscr{R}_T$ are at different angles relative to the axis of rotation $O_{ct}$ of the hob. The angle that the right-hand flank forms with the axis $O_{ct}$ is greater and is equal to $(\phi_{ct} + \theta_{ct})$. The angle that the left-hand flank forms with the axis $O_{ct}$ is smaller and is equal to $(\phi_{ct} - \theta_{ct})$.

Because the lateral tooth flanks of the auxiliary rack $\mathscr{R}_T$ have additional inclination through the cone angle $\theta_{ct}$ (see Figures 14.15 and 14.17), the conical hobs for machining spur and helical involute gears feature not one, but two base cylinders. The diameters of the base cylinders are designated as $d^r_{b.ct}$ and $d^l_{b.ct}$, respectively. The base cylinder of a bigger diameter $d^r_{b.ct}$ corresponds to the hob tooth side with a smaller angle $(\phi_{ct} - \theta_{ct})$, and the base cylinder of a smaller diameter $d^l_{b.ct}$ corresponds to the hob tooth side with a bigger angle $(\phi_{ct} + \theta_{ct})$.

Numerous designs of conical tools[30] for machining cylindrical (spur and helical) gears are known from [60, 105], and others. Numerous methods[31] for relief grinding, as well as for sharpening conical gear-cutting tools, have been developed [87, 88], and others. Comprehensive research in the field of design and implementation of conical hobs for machining cylindrical gears has been carried out in [15].

The researchers and users were particularly interested in conical hobs for machining of spur and helical gears, mainly for two reasons.

*First*, the use of conical hobs makes it possible to optimize the parameters of the *work gear-to-cutting tool* penetration curve (the *GT – penetration curve*, for simplicity).

The cutter's entering and exiting sides are distinguished when hobbing gears. Roughly speaking, the cutter's entering and exiting sides are separated from one another by the closest distance of approach of the work gear axis of rotation and the hob axis of rotation. The precutting zone of the hob is located mostly at the entering side of the hob, while the profile-generating zone is located from both sides of the hob.

The shape of the precutting zone depends on the external shape of the hob (see Figure 15.104). In the work gear/cutting tool penetration, the $GT$ – penetration curve can be understood as a spatial line of intersection of the cylindrical outer surface of the work gear by the outer surface of the hob. In Figure 15.104, the $GT$ – penetration curve is labeled as $G/C_{pc}$. The shape and position of the penetration curve $G/C_{pc}$ depends on

- The crossing angle of the two axes.
- The size of the hob.
- The type (geometry) of the outer surface of the hob.

It is clear from Figure 15.104 that when machining a work gear with the cylindrical hob [see Figure 15.104(a)], the shape and parameters of the $GT$ – penetration curve differ from those when machining a work gear with the conical hob [see Figure 15.104(b)]. The use of conical hobs makes it possible to optimize the parameters of the entering and exiting zones in a gear hobbing operation.

*Second*, a work gear can be machined by two conical hobs simultaneously. The hobs can be appropriately arranged on a common arbor as shown in Figure 15.105.

[30] SU Pat. No. 1028451, *A Conical Cutting Tool for Machining of Cylindrical Gears*/S.P. Radzevich, Int. Cl. B23f 21/16, Filed: April 26, 1982.

SU Pat. No. 1114505, *A Conical Hob for Machining of Cylindrical Gears*/S.P. Radzevich, Int. Cl. B23f 21/16, Filed: October 18, 1982.

[31] SU Pat. No. 965728, *A Method of Relief Grinding of Conical Hob for Machining of Cylindrical Gears*/S.P. Radzevich, Int. Cl. B24b 3/12, Filed: January 21, 1980.

SU Pat. No. 1365550, *A Method of Relief Grinding of Conical Hob for Machining of Cylindrical Gears*/S.P. Radzevich, Int. Cl. B24b 3/12, Filed: April 29, 1986.

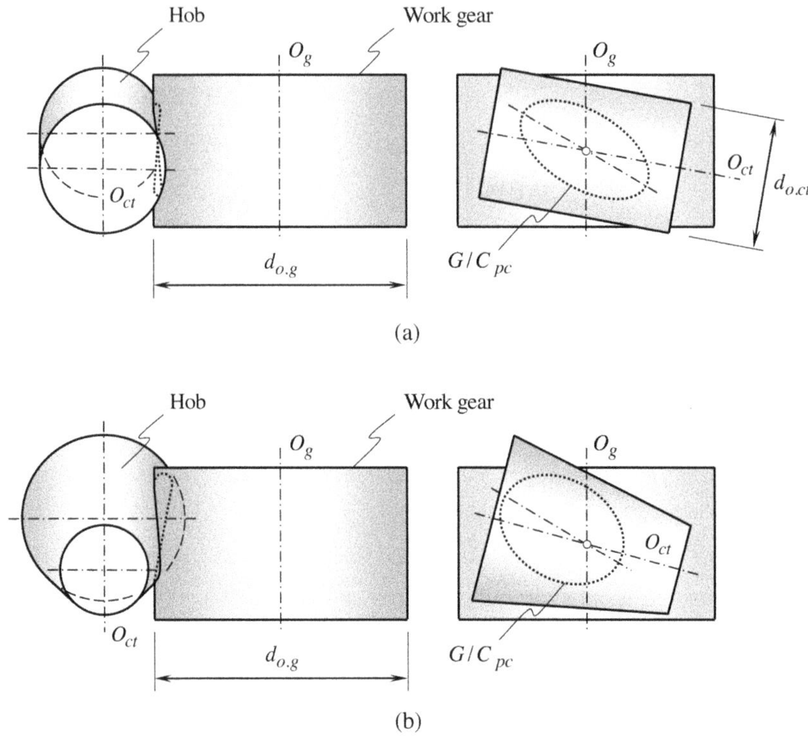


**FIGURE 15.104** Work gear-to-cutting tool penetration curve $G/C_{pc}$ for the cases of (a) machining a gear by the cylindrical hob and (b) machining that same gear by the conical hob.

Such an arrangement[32] of a pair of conical tools for machining cylindrical gears is known from [47, 61] and from other sources.

For machining involute gears with different values of the profile angle $\phi_g$ at the left-hand and right-hand sides of the tooth profile, cylindrical hobs of special design (the so-called involute hobs of asymmetric tooth profile) are used. An involute hob with asymmetric tooth profile is depicted in Figure 15.106(a). Details of the asymmetric tooth profile are illustrated in Figure 15.106(b).

The hob features different values of tooth profile angle $\phi_{ct}$ at the left-hand side (tooth profile angle $\phi_{ct.l}$) and the right-hand side (tooth profile angle $\phi_{ct.r}$) of the hob tooth (see Figure 15.107). Owing to this design feature, two base cylinders of different diameters $d^l_{b.ct}$ and $d^r_{b.ct}$ are used to construct the screw involute surfaces $T_l$ and $T_r$ of the opposite sides of the thread profile of the generating surface $T$ of the gear hob. The base cylinder with the bigger diameter $d^l_{b.ct}$ corresponds to the generating surface $T_l$ with a smaller tooth profile angle $\phi_{ct.l}$. Accordingly, the base cylinder with the smaller diameter $d^r_{b.ct}$ corresponds to the generating surface $T_r$ that features a bigger tooth profile angle $\phi_{ct.r}$.

The geometry of the generating surfaces $T_l$ and $T_r$ is identical to the geometry of corresponding surfaces of a conical hob for machining cylindrical gears (see Figure 15.103). This is why the cylindrical hob shown in Figure 15.107 is discussed together with conical hobs. The difference between the cylindrical hob (see Figure 15.107) and the conical hob (see Figure 15.103) is mainly because the first features the outer cylinder of a diameter $d_{o.ct}$ and the inner cylinder of a diameter $d_{f.ct}$, while the second features corresponding cone surface instead.

The pitch line of the cylindrical hob (see Figure 15.107) is parallel to the axis of rotation $O_{ct}$ of the hob.

Involute gears with asymmetric tooth profile are used in the design of gear transmissions when the conditions of transmission of a rotation in a direct and in a reverse direction are different.

In specific cases, the pitch cone angle $\theta_r$ could be equal to a right angle ($\theta_r = 90°$). Under such a scenario, the conical hob transforms to a disk-type gear-cutting tool that is often referred to as "face hob." "Butt-end hob" is another name for hobs of this design (see Figure 15.108) [102]. The conical

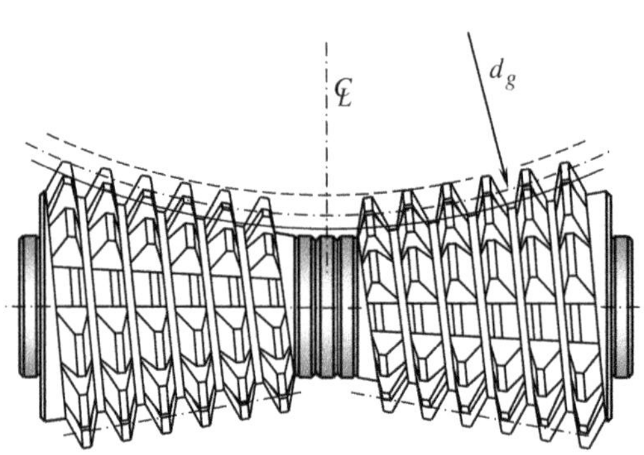


**FIGURE 15.105** Set of two conical hobs arranged on a common arbor for machining a cylindrical gear.

[32] SU Pat. No. 880590, *A Conical Tool for Finishing Cylindrical Gears*/ S.P. Radzevich, Int. Cl. B21h 5/02, Filed: November 6, 1979.

(a) (b)

**FIGURE 15.106** Involute gear hob with an asymmetrical tooth profile: (a) the gear hob and (b) the hob tooth profile. (Courtesy of Dr. A. Kapelevich, AKGears, LLC.)

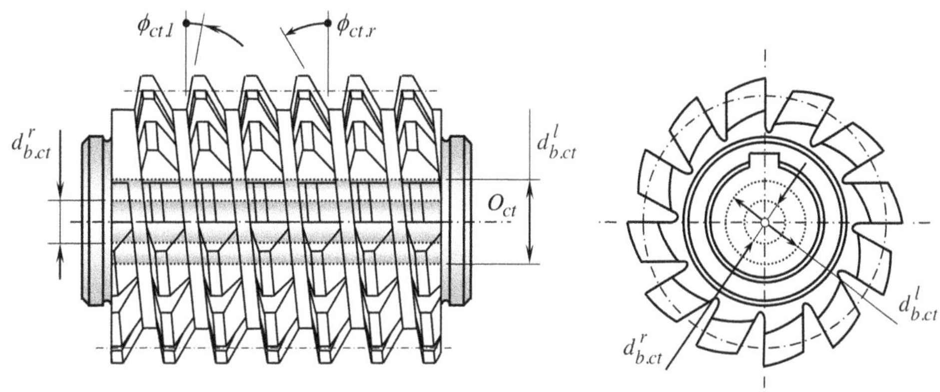


**FIGURE 15.107** Hob with different values of the profile angle at the left-hand side $\phi_{c.l}$ and at the right-hand side $\phi_{c.r}$ of the hob tooth.

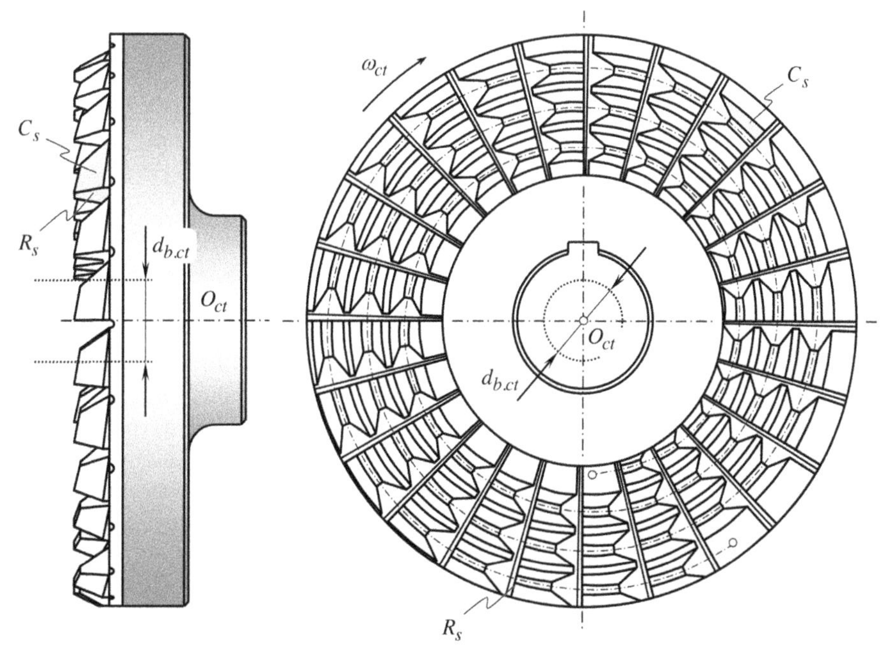


**FIGURE 15.108** Conical hob that features a cone angle of the value $\theta_{ct} = 90°$, the so-called face hob or but-end hob. (From Radzevich, S.P., USSR Patent 1038122, Int. Cl. B23f 21/04, May 4, 1981.)

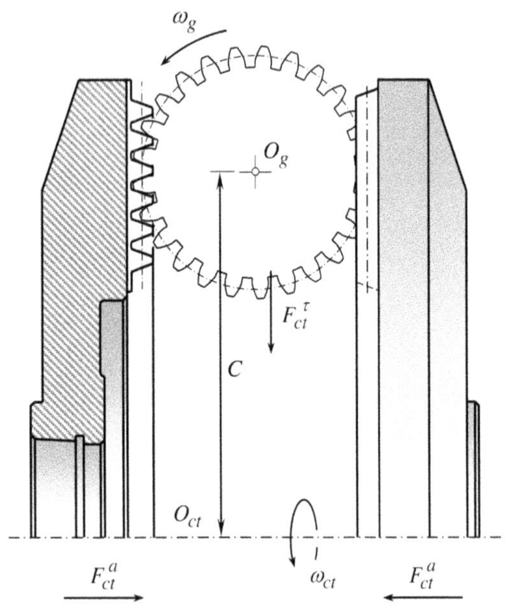


**FIGURE 15.109** Kinematics of machining of spur and helical gears by means of two face hobs. (From Radzevich, S.P. and Palaguta, V.A., USSR Patent 1328042, Int. Cl. B21h 5/00, Dec. 3, 1985.)

pitch surface of the butt-end hob[33] is converted to the pitch plane. The pitch plane is perpendicular to the axis $O_c$ of rotation of the hob. Under such a scenario, the pitch helix lines transform to the corresponding pitch spiral lines, which are planar curves and are located within the pitch plane.

The generating surfaces $T_l$ and $T_r$ of the face hob are constructed from two base cylinders of diameters $d_{b.ct}^{r}$ and $d_{b.ct}^{l}$. In the event when the tooth profile of the auxiliary generating rack $\mathscr{R}_T$ is symmetrical with respect to its centerline, the base cylinders become congruent to each other. Elementary analysis reveals that formally the diameters of the base cylinders can be interpreted as having opposite signs: the diameter of one of the base cylinders is of a positive value, while the diameter of another one is of a negative value.

Spur and helical gears can be machined either by one or by two[34] face hobs simultaneously, as shown in Figure 15.109 [182].

When machining a gear, the work gear rotates about its axis of rotation $O_g$. In Figure 15.109, the rotation of the work gear is denoted by $\omega_g$. The face hob is rotating about its axis of rotation $O_{ct}$. The rotation of the hob is denoted by $\omega_{ct}$. The axes $O_g$ and $O_{ct}$ of the rotations $\omega_g$ and $\omega_{ct}$ are positioned at a certain center distance $C$.

The rotations $\omega_g$ and $\omega_{ct}$ are timed so that the equality: $\omega_g \cdot N_g = \omega_{ct} \cdot N_{ct}$ is observed. Here, $N_g$ denotes the tooth count of the work gear, and the number of starts of the hob is denoted by $N_{ct}$. In order to fulfill the required ratio between the rotations $\omega_g$ and $\omega_{ct}$, and between the tooth numbers $N_g$ and $N_{ct}$, the hands of threads of both of the face hobs have to be the same.

The in-feed motions of two kinds are feasible.

*First*, after the work gear is set up at the desired center distance $C$, both face hobs travel in their axial direction toward the work gear with an axial feed $F_{ct}^{a}$. Before all the teeth of the work gear are hobbed, the work gear has to make more than one full rotation about its axis $O_g$ (when machining is performing only with one face hob). When two face hobs are used for machining a gear, then the rotation of the work gear through an angle over 180° is required.

*Second*, the face hobs are set at the desired distance between their pitch planes and do not travel in the axial direction. Instead, the work gear travels in its tangential direction with a tangential feed $F_{ct}^{\tau}$. In this second case, proper timing between the tangential feed $F_{ct}^{\tau}$ and the rotations of the face hobs is vital; for the in-feed period of time, the face hobs are rotating in the same direction, but with different angular velocities in order to make the feed $F_{ct}^{\tau}$ possible.

Using the face tools for machining spur and helical gears makes possible machining not only one work gear, but multiple work gears simultaneously.

### 15.5.4.3 Toroidal Gear Hobs

Toroidal hobs can be used for machining spur and helical gears. Gear hobs of this design feature the generating surface $T$ shown in Figure 15.20(b). The concept of toroidal tools[35] for machining spur and helical gears is known from [116]. Design features of the involute hob are shown in Figure 15.110. An example of a toroidal hob is illustrated in Figure 15.111.

When machining a spur or helical gear, the work gear rotates about its axis of rotation $O_g$. The hob rotates about its axis $O_{ct}$. The axes $O_g$ and $O_{ct}$ of the rotations $\omega_g$ and $\omega_{ct}$ cross with each other at a certain crossed-axes angle. The rotation of the work gear $\omega_g$ and the rotation of the toroidal hob $\omega_{ct}$ are synchronized so as to fulfill the equality $\omega_g \cdot N_g = \omega_{ct} \cdot N_{ct}$.

Toroidal hobs feature concave lateral cutting edges. This is because the pitch surface of the toroidal hob is a portion of a torus surface. Because the lateral cutting edges are concave, this reduces the height of the residual cusps on the tooth flanks of the machined gear. Ultimately, when radius $R_{w.\mathscr{R}}$ of the pitch surface of the auxiliary generating rack $\mathscr{R}_T$ becomes equal to the gear pitch radius $r_g = d_g/2$ (that is, when the equality $R_{w.\mathscr{R}} = r_g$ is observed), then the generating rack $\mathscr{R}_T$ becomes congruent to the gear tooth flank $\mathscr{G}$. In this particular case, the cusps vanish, and the true involute tooth profile can be machined. The waviness of the tooth flanks of the machined gear remains the only source of deviations of concern, if the consideration

[33] SU Pat. No. 1038122, *A Precision Gear Cutting Tool*/S.P. Radzevich, Int. Cl. B23f 21/04, Filed: May 4, 1981.

[34] SU Pat. No. 1328042, *A Tool for Finishing Cylindrical Gears*/S.P. Radzevich and V.A. Palaguta, Int. Cl. B21h 5/00, Filed: December 3, 1985.

[35] SU Pat. No. 965582, *A Toroidal Tool for Finishing Cylindrical Gears*/ S.P. Radzevich, Int. Cl. B21h 5/00, Filed: October 2, 1980.

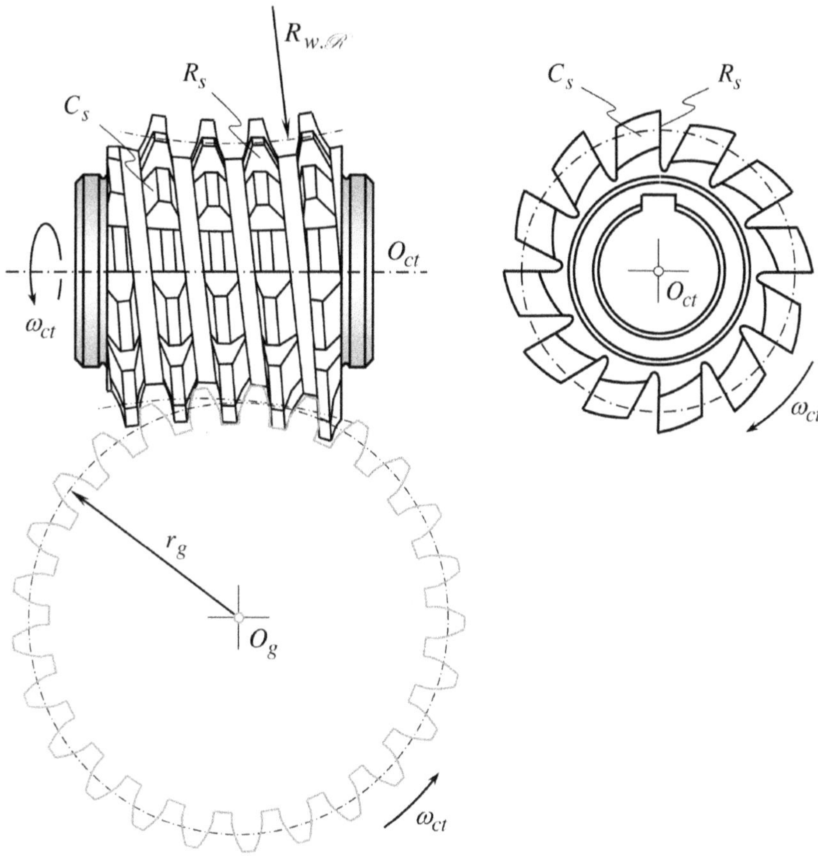


**FIGURE 15.110** Hob with toroidal pitch surface for machining spur and helical gears. (From Radzevich, S.P., USSR Patent 965582, Int. Cl. B21h 5/00, Oct. 2, 1980.)

**FIGURE 15.111** Hob shown in Figure 15.110 for machining spur and helical involute gears.

is limited only to the kinematics and geometry of the gear machining process.

For hobbing of internal gears, a special-purpose hob can be designed using for this purpose the toroidal generating surface $T$, schematically shown in Figure 15.20(a). Actually, a hob designed in this way is capable of cutting external gears as well. However, in the latter case of machining external gears, cusp height on tooth flanks of the cut gear is significantly greater.

## 15.6 CUTTING EDGE GEOMETRY OF GEAR HOB TOOTH

The geometry of the cutting edge of a gear hob tooth can be specified in terms of the rake angle, the clearance angle, the angle of inclination, and so forth. The specification of the geometry of the cutting edge of a gear hob tooth follows the requirements of the ISO Standard 3002.[36]

A certain correlation between the parameters of the geometry of the lateral and the top cutting edges of a gear hob tooth commonly is observed. This is mostly owing to the methods

[36] ISO Standard 3002. *Basic Quantities in Cutting and Grinding* – Part 1: *Geometry of the Active Part of Cutting Tools – General Terms, Reference Systems, Tool and Working Angles, Chip Breakers*, 1982, 52p.
The ISO 3002 standard has much room for improvement.

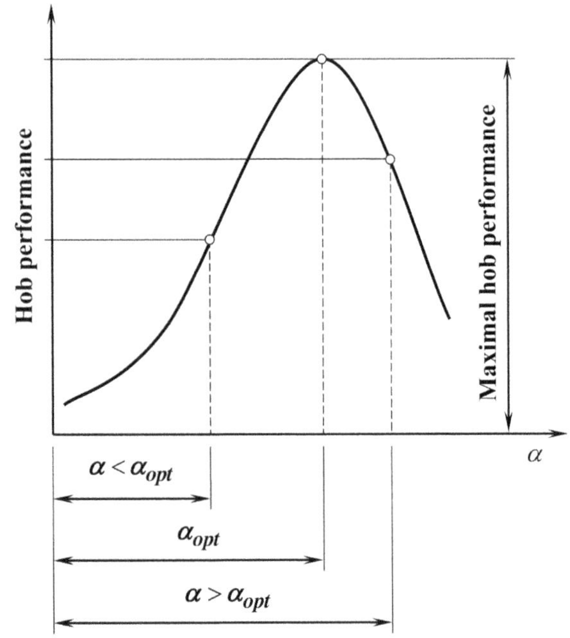


**FIGURE 15.112** Qualitative relation of a gear hob performance vs. the clearance angle α.

applied for the generation of the clearance surfaces of the hob teeth, as well as of the rake surfaces of the hob teeth.

It is proven [9, 10, 26, 28, 29, 39, 41, 43, 44, 53, 198, 200, 209, 210, 212, 213, 223, 224], and others, that there is an optimal combination of parameters of geometry of the cutting edge for cutting tools of any and all designs. This means that an optimal combination of the parameters of geometry of the cutting edge for a gear hob tooth exists, and thus it can be determined.

The nature of the optimal geometry of the cutting edge of a gear hob is not discussed here. At this point it is sufficient to realize that the optimal values of the geometry of the cutting edge can be determined somehow, and they are known. The optimal tool geometry for a given application is a matter of experimental finding. Figure 15.112 illustrates an example of the existence of an optimal value of the clearance angle α. The maximum the hob performance is observed when the clearance angle α is equal to its optimal value $\alpha_{opt}$. When the clearance angle α is either greater ($\alpha > \alpha_{opt}$) or smaller ($\alpha < \alpha_{opt}$) than the optimal value $\alpha_{opt}$, the hob performance becomes smaller. The same is true with respect to other parameters of the geometry of the cutting edge of the gear hob tooth.

It is of critical importance for the gear hob designer to express the desirable optimal geometry of the cutting edge in terms of the design parameters of the involute hob. The problem is not that easy to understand at first glance.

In practice, the value of the clearance angle at the top cutting edge of the hob tooth is close to the optimal value. The value of the clearance angle at the lateral cutting edge of the hob tooth depends on the value of the clearance angle at the top cutting edge. We often want to increase the clearance angle at the lateral cutting edge. However, the desirable increase entails a strongly unwanted increase of the clearance angle at the top cutting edge. There is a trade-off between the clearance angles at the top and at the lateral cutting edges of the hob tooth.

### 15.6.1 Penetration Curve and Machining Zone in Gear Hobbing Process

A zone within which the hob interacts with the work gear is commonly referred to as the *machining zone*. The stock removal and the gear tooth profile generation are observed only within the machining zone. The machining zone is bounded by the *penetration curve*. The penetration curve is an important consideration in hobbing of gears. It was briefly mentioned above and is depicted in Figure 15.104.

The relative motion of the work gear and the gear hob in the gear hobbing process is complex. The relative motion is composed of the elementary motions performed by the work gear and by the hob. The work gear is rotated about its axis $O_g$ with a certain angular velocity $\omega_g$ (see Figure 15.113). The hob is rotated about its axis $O_{ct}$ with the angular velocity $\omega_{ct}$. The axes of rotation $O_g$ and $O_{ct}$ are positioned at a certain center distance $C$. They are crossed at a crossed-axes angle Σ. The rotations $\omega_g$ and $\omega_{ct}$ are timed with each other so that the ratio $\omega_g/\omega_{ct} = N_{ct}/N_g$ is always fulfilled. The hob travels along the work gear axis of rotation $O_g$ with a feed $F_{ct}$. In a climb hobbing operation (see Figure 15.113), the gear hob travels upward. In a conventional hobbing

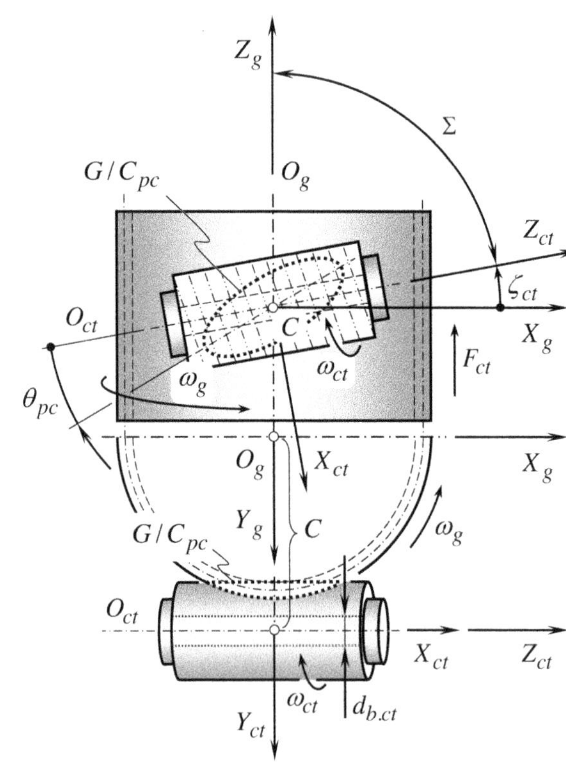


**FIGURE 15.113** Gear to hob penetration curve $G/H_{pc}$. (From Radzevich, S.P., *SME Journal of Manufacturing Science and Engineering*, 129(8), 750–759, 2007. With permission.)

operation, the direction of the gear hob travel is opposite to the direction of $F_{ct}$ shown in Figure 15.113.

Not all of the hob teeth are involved in the stock removal process simultaneously. Only a few of the hob teeth that pass through the machining zone remove chips and generate the gear tooth profile.

The machining zone is bounded by the *gear-to-hob penetration curve* $G/H_{pc}$. Therefore, in the *tool-in-use reference system*, the geometry of the cutting edge of the hob tooth can be determined only when a certain tooth of the hob is within the area bounded by the penetration curve $G/H_{pc}$. Otherwise, determining the geometry of the cutting edge loses any sense. Because of the above-mentioned issue, determination of the penetration curve $G/H_{pc}$ is needed prior to exploring the geometry of the cutting edge of the hob teeth.

#### 15.6.1.1 Parameters of the $G/H_{pc}$ Penetration Curve

Referring to Figure 15.113, the penetration curve $G/H_{pc}$ is understood as the line of intersection of the outer cylinder of the work gear $d_{o.g}$ and the outer cylinder of the hob $d_{o.ct}$. The penetration curve $G/H_{pc}$ is a type of spatial curve.

Targeting the derivation of an equation of the penetration curve, consider two cylinders of diameters $d_{o.g}$ and $d_{o.ct}$. The axes $O_g$ and $O_{ct}$ of the cylinders are positioned at a center distance $C$, and the axes form the crossed-axis angle $\Sigma$.

In a Cartesian coordinate system $X_g Y_g Z_g$ associated with the work gear, the position vector of a point $\mathbf{r}_{o.g}(\varphi_g, U_g)$ of the outer cylinder of the work gear can be expressed in the form of a column matrix:

$$\mathbf{r}_{o.g}(\varphi_g, U_g) = \begin{bmatrix} 0.5 \cdot d_{o.g} \cdot \cos\varphi_g \\ 0.5 \cdot d_{o.g} \cdot \sin\varphi_g \\ U_g \\ 1 \end{bmatrix} \tag{15.157}$$

where $\varphi_g$ and $U_g$ designate curvilinear coordinates (Gaussian coordinates) of a point within the surface $\mathbf{r}_{o.g}$.

In the Cartesian coordinate system $X_{ct} Y_{ct} Z_{ct}$ associated with the gear hob, an expression

$$\mathbf{r}_{o.ct}(\varphi_{ct}, U_{ct})\Big|_{ct} = \begin{bmatrix} 0.5 \cdot d_{o.ct} \cdot \cos\varphi_{ct} \\ 0.5 \cdot d_{o.ct} \cdot \sin\varphi_{ct} \\ U_{ct} \\ 1 \end{bmatrix}_{ct} \tag{15.158}$$

for the position vector $\mathbf{r}_{o.ct}(\varphi_{ct}, U_{ct})\Big|_{ct}$ of a point of the outer cylinder of the hob is similar to Eq. (15.157).

Here $\varphi_{ct}$ and $U_{ct}$ designate curvilinear coordinates (Gaussian coordinates) of a point within the surface of the cylinder $\mathbf{r}_{o.ct}$.

For the purposes of the derivation of an equation of the penetration curve $G/H_{pc}$, it is necessary to represent both the surfaces $\mathbf{r}_{o.g}$ and $\mathbf{r}_{o.ct}$ in a common reference system. In the specific case under consideration, it is convenient to employ the coordinate system $X_g Y_g Z_g$ and to represent the hob surface $\mathbf{r}_{o.ct}$ in the reference system associated with the work gear. For the coordinate system transformation, the operator $\mathbf{Rs}(ct \to g)$ of the resultant coordinate system transformation

$$\mathbf{Rs}(ct \to g) = \begin{bmatrix} \cos\Sigma & 0 & -\sin\Sigma & 0 \\ 0 & 1 & 0 & C \\ \sin\Sigma & 0 & \cos\Sigma & 0 \\ 0 & 0 & 0 & 1 \end{bmatrix} \tag{15.159}$$

is composed.

Use of the operator $\mathbf{Rs}(ct \to g)$ allows for the representation of Eq. (15.158) of the outer cylinder of the gear hob in the coordinate system $X_g Y_g Z_g$ (recall that initially the position vector $\mathbf{r}_{o.ct}$ is represented in the hob coordinate system $X_{ct} Y_{ct} Z_{ct}$):

$$\mathbf{r}_{o.ct}(\varphi_{ct}, U_{ct})\Big|_g = \mathbf{Rs}(ct \to g) \cdot \mathbf{r}_{o.ct}(\varphi_{ct}, U_{ct})\Big|_{ct} \tag{15.160}$$

Ultimately, the following expression:

$$GT_{pc} - \text{penetration curve} \Rightarrow \begin{cases} \mathbf{r}_{o.g}(\varphi_g, U_g) = \begin{bmatrix} 0.5 \cdot d_{o.g} \cdot \cos\varphi_g \\ 0.5 \cdot d_{o.g} \cdot \sin\varphi_g \\ U_g \\ 1 \end{bmatrix} \\ \mathbf{r}_{o.ct}(\varphi_{ct}, U_{ct})\Big|_g = \mathbf{Rs}(ct \to g) \cdot \mathbf{r}_{o.ct}(\varphi_{ct}, U_{ct})\Big|_{ct} \end{cases} \tag{15.161}$$

for the penetration curve can be obtained.

Equation (15.161) casts into the form of a column matrix:

$$G/H_{pc} \Rightarrow \mathbf{r}_{p.c.}(\varphi_g) = \begin{bmatrix} 0.5 \cdot d_{o.g} \cdot \cos\varphi_g \\ 0.5 \cdot d_{o.g} \cdot \sin\varphi_g \\ \dfrac{\sqrt{d_{o.ct}^2 - (d_{o.g} \cdot \sin\varphi_g - 2 \cdot C)^2} - d_{o.g} \cdot \cos\varphi_g \cdot \cos\Sigma}{2 \cdot \sin\Sigma} \\ 1 \end{bmatrix} \tag{15.162}$$

In the expanded form, Eq. (15.162) is bulky. It is better to compose a corresponding computer code for the calculations rather than use Eq. (15.162).

As seen in Figure 15.113, the direction of the view of the penetration curve is from the hob toward the work gear along the centerline ℄ of the axes $O_g$ and $O_{ct}$. The two cylinders $\mathbf{r}_{o.g}$ and $\mathbf{r}_{o.ct}$ penetrate into each other at a depth that is equivalent to the depth of the machining zone. The intersecting line between the two surfaces $\mathbf{r}_{o.g}$ and $\mathbf{r}_{o.ct}$ is a 3*D* curve, which follows both on the work gear cylinder and the gear hob cylinder.

For further analysis, projection of the penetration curve $G/H_{pc}$ onto the plane through the pitch point is important. This reference plane is perpendicular to the centerline ℄ (the shortest distance of approach $C$ of the axes $O_g$ and $O_{ct}$ is measured along the centerline ℄). When a reference is made below to the penetration curve, the projection of the intersecting line $G/H_{pc}$ onto the above-mentioned plane is understood. An equation of the projection of the penetration curve onto the plane through the pitch point (that is, the projection onto the plane that is parallel to the coordinate plane $X_gZ_g$) could be derived from Eq. (15.162):

$$\mathbf{Pr}_{xz}[\mathbf{r}_{p.c.}(\varphi_g)] = \begin{bmatrix} X_g(\varphi_g) \\ 0.5 \cdot d_g \\ Z_g(\varphi_g) \\ 1 \end{bmatrix} \tag{15.163}$$

where $d_g$ designates the pitch diameter of the work gear.

The shape and parameters of the penetration curve depend on

- The work gear outer diameter $d_{o.g}$.
- The gear hob outer diameter $d_{o.ct}$.
- The crossed-axis angle $\Sigma$.
- The center distance $C$ in the gear hobbing process.

All of the hob teeth not passing through the machining zone during the hob rotation (see Figure 15.113) do not make contact with the work gear. They are not, therefore, involved in the chip removal process. The geometry (in the tool-in-use reference system) of the cutting edge of the hob teeth located outside the area bounded by the penetration curve cannot be identified, which is evident.

A distinction must be drawn in hobbing between the precutting zone and the profile-generating zone. The greater portion of material is removed from the tooth space of the work gear in the precutting (roughing) zone. The precutting zone is at the end of the hob, which first enters the body of the work gear during gear hobbing. The standard hob must be positioned so as to completely cover the precutting zone.

#### 15.6.1.2 Partitioning of the Machining Zone

Not all of the hob teeth cut chips even when they are passing the machining zone. For a better understanding of the geometry of the cutting edge of the hob teeth, we must distinguish between the approach side and the recessing side of the hob teeth. The approaching side and the recessing side of the hob teeth are identical to the similar terms that are common for conventional gearing (and not only for gear hobbing).

Differentiation between the roughing sector and the gear tooth profile-generating sector must be understood as well.

For the differentiation, implementing the concept of the hob base cylinder is vital. The diameter of the base cylinder of the involute hob is calculated from Eq. (15.72):

$$d_{b.ct} = \frac{m \cdot N_{ct} \cdot \cos\phi_n}{\sqrt{1 - \cos^2\phi_n \cdot \cos^2\zeta_{ct}}} \tag{15.164}$$

The characteristic line $\mathscr{E}$ of the generating surface $T$ of the involute hob is a straight line that is tangent to the hob base cylinder of a diameter $d_{b.ct}$. The characteristic line $\mathscr{E}$ is at the base helix angle $\psi_{b.ct}$ to the hob axis of rotation $O_{ct}$. The base helix angle $\psi_{b.ct}$ can be calculated from Eq. (15.75):

$$\psi_{b.ct} = \cos^{-1}(\cos\phi_n \cos\zeta_{ct}) \tag{15.165}$$

Equations similar to that above are also valid for the involute tooth flank $\mathscr{G}$ of the gear being machined.

Three sides of the tooth profile of the involute hob are distinguished. They are

- The entering cutting edge.
- The recessing cutting edge.
- The top cutting edge.

Because of these, three corresponding areas of the machining zone are recognized. They are as follows (see Figure 15.114):

1. The area of cutting by the entering edges (roughing and profiling).
2. The area of cutting by the recessing edges (only profiling).
3. The area of cutting by the top edges of the involute hob.

Areas 1 to 3 are separated from one another by planes that are perpendicular to the plane of the drawing in Figure 15.114.

Area 1 of machining by the entering cutting edge is bounded by the plane though the straight-line segment *ab* at the bottom, and by the vertical plane through point *b* at the far left. The tilted bounding plane though the straight-line segment *ab* is tangent to the hob base cylinder at the top, and the vertical bounding plane through point *b* passes though the most remote left point of the entering line of action in the gear hobbing operation.

Area 2 of machining by the recessing cutting edge is bounded by the plane through the straight-line segment *cd* at the bottom, and by the vertical plane through point *d* at the far left. The tilted bounding plane through the straight-line segment *cd* is tangent to the hob base cylinder at the bottom, and the vertical bounding plane through point *d* passes through the most remote left point of the recessing line of action in the gear hobbing operation. In particular, the vertical bounding planes for the entering and recessing cutting edges can be coincidental with one another.

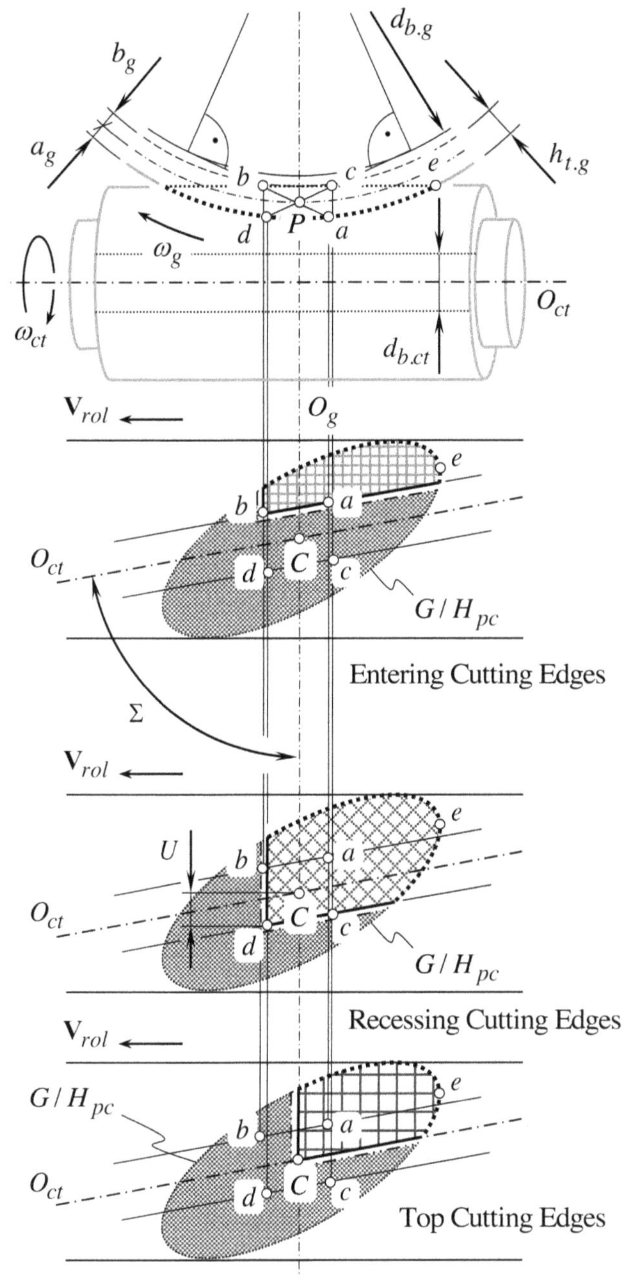


**FIGURE 15.114** Three areas within the machining zone in hobbing an involute gear. (From Radzevich, S.P., *SME Journal of Manufacturing Science and Engineering*, 129(8), 750–759, 2007. With permission.)

Area 3 of machining by the top cutting edge is bounded by the plane through the gear hob axis of rotation at the bottom, and by the vertical bounding plane though point $C$ at the far left. Point $C$ is the crossing point of the axes $O_g$ and $O_{ct}$.

Actually, all three tilted bounding planes are shifted down in the direction of the feed $\mathbf{F}_{ct}$ at a distance that is equal to half of the axial feed per revolution $0.5 \cdot F_{ct}$ of the work gear.

Point $e$ of the penetration curve is the far-right point of the machining zone, where entering cutting edges of the gear hob start cutting the stock.

The actual shape of the area of the entering cutting edge, as well as of the recessing cutting edge area, strongly depend on the location of the pitch point in the gear hobbing operation [156].

The investigation of the geometry of the cutting edge of an involute hob in the tool-in-use reference system makes sense only within the corresponding area of the machining zone (see Figure 15.114). This is because in the tool-in-use reference system, no geometry of the cutting edge can be determined outside of the machining zone. (The surface of cut $S_{ct}$, which is the principal reference surface, does not exist outside the machining zone).

### 15.6.2 Cutting Edge Geometry of Hob Tooth in the Tool-in-Use Reference System

Two major reference systems are commonly used for determining the geometry of cutting tools. The *tool-in-hand reference system* is one of them, and the *tool-in-use reference system* is the other one. The tool-in-use reference system is used for analysis of the geometry of the cutting edge of the active part of the hob tooth.

The interested reader is referred to Appendix G for the definitions and details on the basics of the geometry of cutting edges.

#### 15.6.2.1 Tool-in-Use Reference System in Gear Hobbing Operation

For the analysis of a gear hobbing process, it is convenient to place the origin of the tool-in-use reference system at a point of interest $m$ within the cutting edge $CE$ of the hob tooth. Three vectors are used for the purposes of constructing the tool-in-use reference system. These vectors are as follows [148]:

- The unit normal vector $\mathbf{n}_{ct}$ to the surface of cut.
- The velocity vector $\mathbf{V}_\Sigma$ of the resultant motion of the cutting edge $CE$ relative to the surface of cut $S_{ct}$.
- The unit vector $\mathbf{c}_e$ that is tangent to the cutting edge $CE$ at point $m$.

In addition to the vectors $\mathbf{n}_{ct}$, $\mathbf{V}_\Sigma$, and $\mathbf{c}_e$, a few complementary vectors are applied as well.

##### *15.6.2.1.1 Unit Normal Vector to the Surface of Cut*

The surface of cut $S_{ct}$ in a gear hobbing operation can be represented as a set of consecutive positions of the cutting edge $CE$ of the hob tooth that travels relative to the work gear. Three different cutting edges of the hob tooth—the entering cutting edge $CE_e$, the recessing cutting edge $CE_r$, and the top cutting edge $CE_t$—generate three different surfaces of cut: $S_{c.e}$, $S_{c.r}$, and $S_{c.t}$, respectively.

The resultant motion of the hob cutting edges is complex in nature. It can be represented as a superposition of three elementary motions. These elementary motions are created by (1) rotation of the work gear, (2) rotation of the

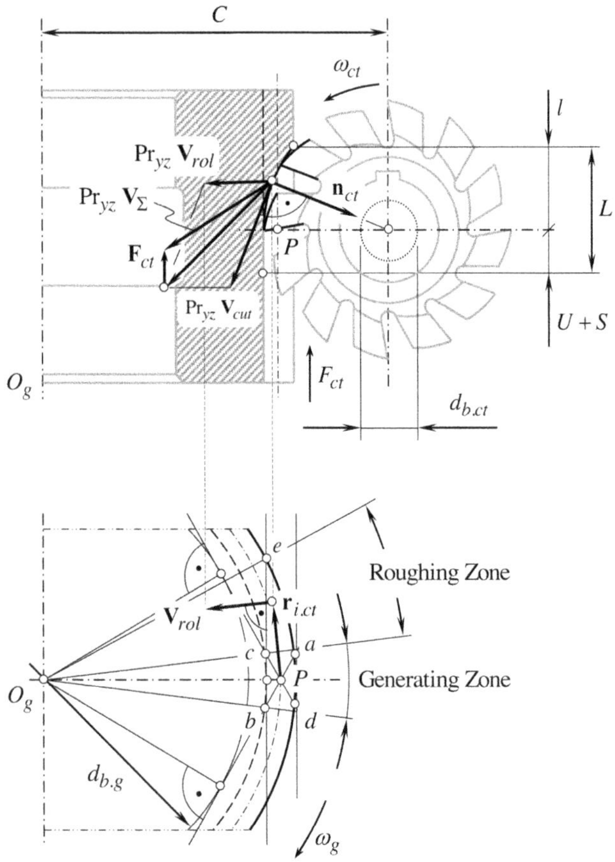


**FIGURE 15.115** Resultant speed in the gear hobbing process. (From Radzevich, S.P., *SME Journal of Manufacturing Science and Engineering*, 129(8), 750–759, 2007. With permission.)

gear hob, and (3) the feed motion of the hob. Ultimately, the complex resultant motion of the cutting edges results in complex and bulky equations for each of three surfaces of the cut. However, a reasonable simplification of the equations is possible. The simplification utilizes the property of generating surface $T$ of the involute gear hob according to which the surface $T$ is tangent to the surface of the cut $S_{ct}$.

This geometrical property of the generating surface $T$ allows a possible conclusion: The unit normal vector $\mathbf{n}_{ct}$ to the surface of the cut $S_{ct}$ is aligned with the unit normal vector $\mathbf{n}_T$ to the generating surface $T$ (that is, it is assumed here and below that the identity $\mathbf{n}_{ct} \equiv \mathbf{n}_T$ is valid).

The unit normal vector $\mathbf{n}_{ct}$ to the generating surface $T$ can be calculated from the equation

$$\mathbf{n}_{ct} = \mathbf{u}_T \times \mathbf{v}_T \tag{15.166}$$

where it is designated $\mathbf{u}_T = \mathbf{U}_T / |\mathbf{U}_T|$, $\mathbf{v}_T = \mathbf{V}_T / |\mathbf{V}_T|$ and $\mathbf{U}_T = \partial \mathbf{r}_T / \partial U_{ct}$, $\mathbf{V}_T = \partial \mathbf{r}_T / \partial V_{ct}$.

For the calculation of the derivatives, Eq. (15.63)

$$T: \Rightarrow \mathbf{r}_T(U_{ct}, V_{ct}) = \begin{bmatrix} 0.5 \cdot d_{b.ct} \cdot \sin V_{ct} - U_{ct} \cdot \sin \psi_{b.ct} \cdot \cos V_{ct} \\ 0.5 \cdot d_{b.ct} \cdot \cos V_{ct} + U_{ct} \cdot \sin \psi_{b.ct} \cdot \sin V_{ct} \\ p_{ct} \cdot V_{ct} - U_{ct} \cdot \cos s\psi_{b.ct} \\ 1 \end{bmatrix} \tag{15.167}$$

of the generating surface $T$ of the involute hob is used. In this equation, the negligible impact of the feed onto the parameters of the unit normal vector $\mathbf{n}_{ct}$ is ignored.

#### *15.6.2.1.2 Resultant Motion of Point of Interest Within the Cutting Edge*

The following three motions are performed simultaneously when hobbing an involute gear (see Figure 15.115):

- A rotation of the work gear $\omega_g$.
- A rotation of the hob $\omega_{ct}$.
- A feed motion $F_{ct}$.

Figure 15.116 illustrates the important details on the kinematics of the gear hobbing process. These three elementary motions ($\omega_g$, $\omega_{ct}$, and $F_{ct}$) cause the velocity vector $\mathbf{V}_\Sigma$ of the resultant motion of the cutting edge $CE$ with

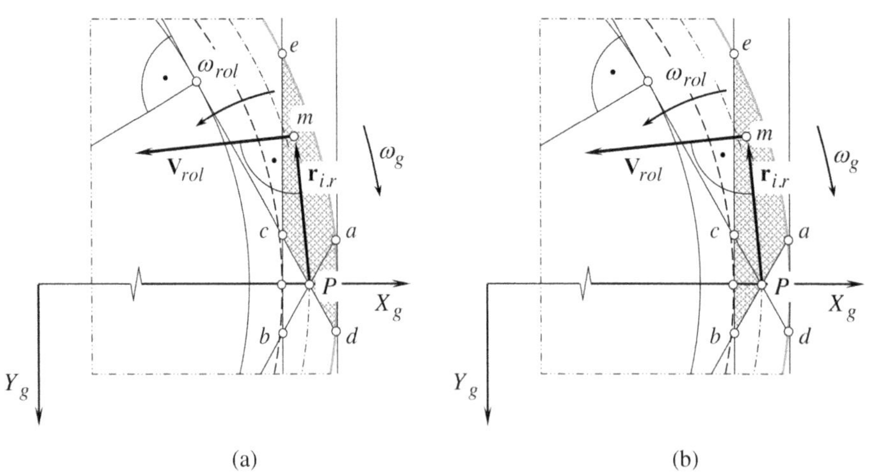


**FIGURE 15.116** Rolling motion of (a) the entering side of the hob tooth profile and (b) the recessing side of the hob tooth profile. (From Radzevich, S.P., *SME Journal of Manufacturing Science and Engineering*, 129(8), 750–759, 2007. With permission.)

respect to the surface of the cut $S_{ct}$. Evidently, the vector $\mathbf{V}_\Sigma$ can be expressed in terms of the elementary motions $\omega_g$, $\omega_{ct}$, and $F_{ct}$.

*First*, it is convenient to express the vector $\mathbf{V}_\Sigma$ as a summa of

$$\mathbf{V}_\Sigma = \mathbf{r}_{i.ct} \times \omega_{ct} + \mathbf{V}_{rol} + \mathbf{F}_{ct} \tag{15.168}$$

where $\mathbf{r}_{i.ct}$ designates a vector that is pointed from the axis of rotation $O_{ct}$ of the involute gear hob to the point of interest $m$ within the given cutting edge $CE$ at the specified angular position of the hob. (The vector $\mathbf{r}_{i.ct}$ is orthogonal to the hob axis of rotation $O_{ct}$, that is, $\mathbf{r}_{i.ct} \perp O_{ct}$). This vector can be also interpreted as the projection of a position vector of a point of interest $m$ onto the $X_{ct}Y_{ct}$ coordinate plane. The vector of the hob rotation is designated as $\omega_{ct}$. The rolling motion $\mathbf{V}_{rol}$ in the gear hobbing process can be expressed as the cross product:

$$\mathbf{V}_{rol} = \omega_{rol} \times \mathbf{r}_{i.rol} \tag{15.169}$$

Here, $\omega_{rol}$ is the vector of the rolling of the hob pitch plane over the work gear pitch cylinder, and $\mathbf{r}_{i.rol}$ designates the projection of the position vector of the cutting edge point of interest onto the $X_gY_g$ coordinate plane. The vector $\mathbf{F}_{ct}$ is a vector of the hob feed motion.

Figure 15.115 reveals that the following expressions for the vectors $\mathbf{r}_{i.ct}$, $\omega_{ct}$, $\mathbf{V}_{rol}$, and $\mathbf{F}_{ct}$ are valid:

$$\mathbf{r}_{i.ct} = \mathbf{i}\cdot|\mathbf{r}_{i.ct}|\cdot\cos\varepsilon + \mathbf{k}\cdot|\mathbf{r}_{i.ct}|\cdot\sin\varepsilon \tag{15.170}$$

$$\omega_{ct} = \mathbf{j}\cdot|\omega_{ct}| \tag{15.171}$$

$$\mathbf{V}_{rol} = \mathbf{i}\cdot|\mathbf{V}_{rol}|\cos\kappa + \mathbf{j}\cdot|\mathbf{V}_{rol}|\sin\kappa \tag{15.172}$$

$$\mathbf{F}_c = -\mathbf{i}\cdot|\mathbf{F}_{ct}|\cdot\cos\zeta_{ct} + \mathbf{k}\cdot|\mathbf{F}_{ct}|\cdot\sin\zeta_{ct} \tag{15.173}$$

Angle $\kappa$ is the angle that the vector of the rolling motion $\mathbf{V}_{rol}$ forms with the positive direction of the $X_g$ −axis. (The angle $\kappa$ is not shown in Figure 15.113).

Angle $\varepsilon$ in Eq. (15.170) defines the current angular position of the point of interest within the cutting edge of the rotating hob. The angle $\varepsilon$ is calculated from the equation $\varepsilon = |\omega_{ct}|\cdot t$, where $t$ designates time.

*Second*, all the vectors in Eq. (15.170) require being represented in a common reference system. For this purpose, a left-hand-oriented local coordinate system $xyz$ with the origin at the point of interest $m$ on the cutting edge $CE$ is used. The axis $z$ of the local coordinate system $xyz$ is aligned with the unit normal vector $\mathbf{n}_{ct}$ to the surface of the cut $S_{ct}$. This vector is pointed outward from the bodily side to the void side of the work gear tooth. The axis $y$ is directed along the velocity vector $\mathbf{V}_\Sigma$ of the resultant motion of the cutting edge. The axis $x$ complements the axes $y$ and $z$ to the left-hand-oriented Cartesian coordinate system $xyz$.

#### *15.6.2.1.3 Unit Tangent Vector to the Cutting Edge*

The lateral cutting edge of the precision involute gear hob is aligned with the line of intersection of the generating surface $T$ of the hob by the rake surface $R_s$ of the hob tooth. There is not much room to vary the parameters of the generating surface $T$. However, there is much room for variation of the shape and the parameters of the rake surface $R_s$ of the hob tooth. Because of this, the unit tangent vector $\mathbf{c}_e$ to the cutting edge $CE$ at the point of interest $m$ can be specified only for the hob of a given design. As an illustrative example, let us determine the unit tangent vector $\mathbf{c}_e$ for the precision involute hob, the design of which is proposed by the author [104].

The lateral cutting edge $CE$ is aligned with the straight-line characteristic line $\mathscr{E}$. The characteristic line $\mathscr{E}$ and the hob axis of rotation $O_{ct}$ are crossed at the hob base helix angle $\psi_{b.c}$ [see Eq. (15.165)] and are at a distance $0.5\cdot d_{b.ct}$ [see Eq. (15.164)] (see Figure 15.84). This yields an equation

$$\mathbf{c}_e = \frac{\mathbf{C}_e}{|\mathbf{C}_e|} \tag{15.174}$$

for the unit vector $\mathbf{c}_e$ where

$$\mathbf{C}_e = -\mathbf{i}\cdot\cos\phi_{rs}\tan\nu - \mathbf{j}\cdot\cos\phi_{rs} - \mathbf{k}\cdot\sin\phi_{rs}\cos\xi \tag{15.175}$$

and the angles $\nu$, $\xi$, and $\phi_{rs}$ are calculated from Eq. (15.113), Eq. (15.122), and Eq. (15.123) respectively.

The parameter $t_{ct}$ designates the tooth thickness within the pitch plane of the auxiliary rake $\mathscr{R}_T$ of the hob. In common practice, the equality $t_{ct} = 0.5\cdot\pi\cdot m$ is observed.

Here, $\zeta_{ct}$ designates the setting angle of the hob. The angle $\zeta_{ct}$ is measured within the pitch plane of the auxiliary rack $\mathscr{R}_T$. The required value of the hob-setting angle $\zeta_{ct}$ can be chosen by the designer of the hob. Usually (but not mandatorily), it is recommended to assign the actual value of the setting angle of the hob $\zeta_{ct}$ equal to the pitch helix angle $\psi_{ct}$ of the hob. The setting angle of the value $\zeta_{ct} = \psi_{ct}$ is preferred from many perspectives.

The derived equations for three vectors: $\mathbf{n}_{ct}$ (here $\mathbf{n}_{ct} = \mathbf{u}_T \times \mathbf{v}_T$), $\mathbf{V}_\Sigma$ [Eq. (15.168)], and $\mathbf{c}$ [Eq. (15.174)], comprise a trihedron of the tool-in-use reference system. In this reference system, the geometry of the cutting edge of the hob tooth can be determined.

### 15.6.2.2 Geometrical Parameters of Hob Cutting Edge in the Tool-in-Use Reference System

After all three principal vectors $\mathbf{n}_{ct}$, $\mathbf{V}_\Sigma$, and $\mathbf{c}_e$ are represented in a common reference system, the analysis of the cutting edge geometry of the hob tooth can be performed analytically.

The total number of geometrical parameters of cutting edges of the gear hob is limited by the specific features of the design of the involute hob. Ultimately, the following geometrical parameters of the cutting edge are considered:

- Rake angle $\gamma$
- Clearance angle $\alpha$
- Angle of inclination $\lambda$ of the cutting edge
- Cutting wedge angle $\beta$
- Angle of cutting $\delta$
- Cutting edge roundness $\rho$
- Cutting edge curvature $R_{ce}$

In special cases, the torsion $\tau_{ce}$ of the cutting edge could also be involved in the analysis of the geometry of the cutting edge. All of the above-listed geometrical parameters can be calculated for each of the three cutting edges of every tooth of the hob at any instant of the gear hobbing operation [148, 153, 155, 168]. For illustrative purposes, only the first three of the above-listed geometrical parameters of the cutting edge of the gear hob are discussed below in detail.

#### *15.6.2.2.1 Rake Angle of the Cutting Edge of the Hob Tooth*

The rake angle $\gamma$ is used to specify the inclination of the rake surface $R_s$ of the hob tooth with respect to the surface of the cut $S_{ct}$. For the analysis of the material removal process in a gear hobbing operation, it is convenient to specify the inclination of the rake surface $R_s$ by:

- Normal rake angle $\gamma_N$ that is measured in the normal plane-section of the cutting edge $CE$.
- The rake angle $\gamma_{\mathbf{V}}$ that is measured in the plane-section of the cutting edge $CE$ by main section plane.
- The rake angle $\gamma_{cf}$ that is measured in the plane-section of the cutting edge $CE$ by the plane of chip flow.

The normal rake angle $\gamma_N$ of the hob cutting edge is specified in the reference plane that is orthogonal to the cutting edge of the hob tooth at point of interest $m$. It is natural that this reference plane is perpendicular to the surface of the cut so it passes through the unit normal vector $\mathbf{n}_{ct}$. The normal rake angle $\gamma_N$ is specified as the angle that complements to 90° the angle between the unit vectors $\mathbf{n}_{ct}$ and $\mathbf{n}_{rs}$, that is, by definition, the normal rake angle equals

$$\gamma_N = 90° - \angle(\mathbf{n}_{ct}, \mathbf{n}_{rs}) \tag{15.176}$$

Here $\mathbf{n}_{rs}$ designates the unit normal vector to the rake surface $R_s$ of the gear hob tooth at point of interest $m$.

For the calculation of the normal rake angle $\gamma_N$, the equation

$$\gamma_N = 90° - \tan^{-1}\left(\frac{|\mathbf{n}_{ct} \times \mathbf{n}_{rs}|}{\mathbf{n}_{ct} \cdot \mathbf{n}_{rs}}\right) = \tan^{-1}\left(\frac{\mathbf{n}_{ct} \cdot \mathbf{n}_{rs}}{|\mathbf{n}_{ct} \times \mathbf{n}_{rs}|}\right) \tag{15.177}$$

applies.

The rake angle $\gamma_{\mathbf{V}}$ is measured in the direction of the vector $\mathbf{V}_\Sigma$ of velocity of the resultant motion of the cutting edge $CE$ relative to the surface of the cut $S_{ct}$. This reference plane is perpendicular to the surface of the cut as well. It also passes through the unit normal vector $\mathbf{n}_{ct}$. The rake angle $\gamma_{\mathbf{V}}$ is specified as the angle between the unit vectors $\mathbf{n}_{ct}$ and $\mathbf{b}$, that is, the rake angle $\gamma_{\mathbf{V}}$ is equal to:

$$\gamma_{\mathbf{V}} = \angle(\mathbf{n}_{ct}, \mathbf{b}) \tag{15.178}$$

Here $\mathbf{b}$ designates the unit vector that is tangent to the rake surface $R_s$ of the hob tooth. The vector $\mathbf{b}$ is located within the plane through the vectors $\mathbf{n}_{ct}$ and $\mathbf{V}_\Sigma$.

For the calculation of the rake angle $\gamma_{\mathbf{V}}$ the equation

$$\gamma_{\mathbf{V}} = \tan^{-1}\left(\frac{|\mathbf{n}_{ct} \times \mathbf{b}|}{\mathbf{n}_{ct} \cdot \mathbf{b}}\right) \tag{15.179}$$

can be used.

The rake angle $\gamma_{cf}$ is measured in the plane of chip flow over the rake face $R_s$ (*Stabler, G.V.*, 1951 [209]). The *chip-flow reference plane* is the plane through the vectors $\mathbf{V}_\Sigma$ and $\mathbf{V}_{cf}$. The velocity vector $\mathbf{V}_{cf}$ of the chip flow is located within the rake plane $R_s$ (or within the plane tangent to the rake surface $R_s$). The velocity vector $\mathbf{V}_{cf}$ forms an angle $\eta$ with the unit normal vector $\mathbf{n}_{ce}$ to the cutting edge $CE$. For involute hobs featuring the angle of inclination $\lambda \le 45°$, the approximate equality $\eta \cong \lambda$ is commonly valid. For involute hobs of special design, for example, for skiving gear hobs [137, 138] when the angle of inclination $\lambda > 45°$ (and it can even reach the value of $\lambda = 60°$), the angle $\eta$ can slightly deviate from $\lambda$. It is proven experimentally that for most cases of metal cutting, the deviation $\delta_\eta = \lambda - \eta$ does not exceed 5°...10° [148, 209].

The rake angle $\gamma_{cf}$ is defined as

$$\gamma_{cf} = \angle(\mathbf{V}_\Sigma, \mathbf{V}_{cf}) - 90° \tag{15.180}$$

The angle $\gamma_{cf}$ is calculated from the equation

$$\gamma_{cf} = \tan^{-1}\left(\frac{|\mathbf{V}_\Sigma \times \mathbf{V}_{cf}|}{\mathbf{V}_\Sigma \cdot \mathbf{V}_{cf}}\right) - 90° = \tan^{-1}\left(\frac{\mathbf{V}_\Sigma \cdot \mathbf{V}_{cf}}{|\mathbf{V}_\Sigma \times \mathbf{V}_{cf}|}\right) \tag{15.181}$$

The well-known formula for the calculation of the rake angle $\gamma_{cf}$ [Eq. (G.39) in Appendix G – Stabler, G.V. formula, 1951], can be derived from Eq. (15.181).

All the above-discussed rake angles, $\gamma_N$, $\gamma_{\mathbf{V}}$, and $\gamma_{cf}$, can have either a positive or a negative value. In specific cases they could be equal to zero as well.

#### *15.6.2.2.2 Clearance Angle of the Cutting Edge of the Hob Tooth*

Clearance angles are usually specified in two reference planes, say, in planes within which the rake angles $\gamma_N$ and $\gamma_{\mathbf{V}}$ are specified.

The normal clearance angle $\alpha_N$ is defined as an angle that complements to 180° the angle between the unit normal vectors $\mathbf{n}_{ct}$ and $\mathbf{n}_{cs}$. Here, $\mathbf{n}_{cs}$ designates the unit normal vector to the clearance surface $C_s$ at point of interest $m$ within the cutting edge $CE$. Therefore, the following equality is observed:

$$\alpha_N = 180° - \angle(\mathbf{n}_{ct}, \mathbf{n}_{cs}) \qquad (15.182)$$

This immediately yields an equation for calculation of the normal clearance angle:

$$\alpha_N = 180° - \tan^{-1}\left(\frac{|\mathbf{n}_{ct} \times \mathbf{n}_{cs}|}{\mathbf{n}_{ct} \cdot \mathbf{n}_{cs}}\right) = -\tan^{-1}\left(\frac{|\mathbf{n}_{ct} \times \mathbf{n}_{cs}|}{\mathbf{n}_{ct} \cdot \mathbf{n}_{cs}}\right) \qquad (15.183)$$

The clearance angle $\alpha_{\mathbf{V}}$ is defined as an angle that complements to 90° the angle that the unit normal vector $\mathbf{n}_{ct}$ forms with the vector $\mathbf{c}_s$. Here, $\mathbf{c}_s$ designates a unit tangent vector to the clearance surface $C_s$ of the cutting tool at point of interest $m$ within the cutting edge $CE$. The unit vector $\mathbf{c}_s$ is within the reference plane through the vectors $\mathbf{n}_{ct}$ and $\mathbf{V}_\Sigma$. Therefore, the equality

$$\alpha_{\mathbf{V}} = 90° - \angle(\mathbf{n}_{ct}, \mathbf{c}_s) \qquad (15.184)$$

is observed.

Ultimately, the equation

$$\alpha_{\mathbf{V}} = 90° - \tan^{-1}\left(\frac{|\mathbf{n}_{ct} \times \mathbf{c}_s|}{\mathbf{n}_{ct} \cdot \mathbf{c}_s}\right) = \tan^{-1}\left(\frac{\mathbf{n}_{ct} \cdot \mathbf{c}_s}{|\mathbf{n}_{ct} \times \mathbf{c}_s|}\right) \qquad (15.185)$$

is valid for the calculation of the clearance angle $\alpha_{\mathbf{V}}$.

Both the clearance angles $\alpha_N$ and $\alpha_{\mathbf{V}}$ have positive values. In specific cases, within a narrow land along the cutting edge, they could be equal to zero, or even have a negative value.

#### 15.6.2.2.3 *Angle of Inclination of the Cutting Edge of the Hob Tooth*

The inclination angle $\lambda$ is measured within the plane that is tangent to the surface of the cut at the point of interest $m$ of the cutting edge $CE$. The angle $\lambda$ complements to 90° the angle between the vectors $\mathbf{V}_\Sigma$ and $\mathbf{c}_e$, that is:

$$\lambda = \angle(\mathbf{V}_\Sigma, \mathbf{c}_e) - 90° \qquad (15.186)$$

Here $\mathbf{c}_e$ designates the unit vector that is tangent to the cutting edge $CE$. The unit vector $\mathbf{c}_e$ is situated within the plane that is tangent to the surface of the cut $S_{ct}$. The angle of inclination $\lambda$ has a positive value when the rotation from the unit vector $\mathbf{c}_e$ to the velocity vector $\mathbf{V}_\Sigma$ through a smaller angle is performed in a counterclockwise direction (looking from the far end of the unit normal vector $\mathbf{n}_{ct}$ toward the surface of the cut $S_{ct}$). Otherwise, the angle $\lambda$ has a negative value. The following equation

$$\lambda = \tan^{-1}\left(\frac{|\mathbf{c}_e \times \mathbf{V}_\Sigma|}{\mathbf{c}_e \cdot \mathbf{V}_\Sigma}\right) - 90° = \tan^{-1}\left(\frac{\mathbf{c}_e \cdot \mathbf{V}_\Sigma}{|\mathbf{c}_e \times \mathbf{V}_\Sigma|}\right) \qquad (15.187)$$

can be used for calculation of the inclination angle $\lambda$.

#### 15.6.2.2.4 *Cutting Wedge Angle* $\beta$

The cutting wedge angle $\beta$ is calculated from the known formula

$$\beta = 90° - (\alpha + \gamma) \qquad (15.188)$$

In this equation, all the angles must be taken in a common reference plane. The derived Eq. (15.177) and Eq. (15.183) reveal the impact of the design parameters of the gear hob on the current value of the cutting wedge angle $\beta$ at point of interest $m$ within the cutting edge $CE$.

#### 15.6.2.2.5 *Angle of Cutting* $\delta$

The angle of cutting $\delta$ is calculated from the known formula

$$\delta = 90° - \gamma \qquad (15.189)$$

Actually, this is the angle that is formed by the rake surface $R_s$ and by the machined surface of the work gear. The machined surface in the case under consideration is either the machined surface of the gear tooth flank, or it is a surface of stock, which is not cut yet.

In this equation, all the angles must be taken in a common reference plane. The derived Eq. (15.177) makes evident the influence of the design parameters of the hob onto the current value of the angle of cutting $\delta$ at point of interest $m$ within the cutting edge $CE$.

#### 15.6.2.2.6 *Effective Angle of Cutting* $\delta_{cf}$

The effective angle of cutting $\delta_{cf}$ is the angle that the inverse direction of the velocity vector $\mathbf{V}_\Sigma$ forms with the velocity vector $\mathbf{V}_{cf}$ of the chip flow over the rake surface $R_s$ of the involute hob tooth:

$$\left[\delta_{cf} = 180° - \angle(\mathbf{V}_\Sigma, \mathbf{V}_{cf})\right] \qquad (15.190)$$

For free oblique cutting, the effective angle of cutting $\delta_{cf}$ can be expressed in terms of the inclination angle $\lambda_{sc}$ and of the angle $\eta$ of the chip flow over the rake surface $R_s$ [148]

$$\delta_{cf} = \cos^{-1}(\cos\lambda \cdot \cos\eta \cdot \cos\delta_{\mathbf{V}} + \sin\lambda \cdot \sin\eta) \qquad (15.191)$$

where $\delta_{\mathbf{V}}$ designates the angle of cutting that is measured in the main reference plane.

This formula reveals that when the angle of inclination $\lambda$ and the angle of the chip flow $\eta$ decrease, then the effective angle of cutting $\delta_{cf}$ decreases as well. This relationship is in good correspondence with the results of experiments in [210].

The calculated effective angle of the cutting wedge can be used in a classic Merchant's formula

$$\tan\varphi = \frac{\xi \cdot \cos\gamma_{cf}}{1-\xi \cdot \sin\gamma_{cf}} \quad (15.192)$$

for the calculation of the effective shear angle $\varphi$. Here $\xi$ designates the chip thickness ratio $\xi = t_p/t_c$.

Further, the chip compression ratio $\zeta$ also depends on the effective rake angle $\gamma_{cf}$:

$$\zeta = \frac{L_p}{L_c} = \frac{\cos(\varphi - \gamma_{cf})}{\sin\varphi} \quad (15.193)$$

The effective rake angle $\gamma_{cf}$ is also used as an input parameter in other applications.

#### *15.6.2.2.7 Cutting Edge Roundness* $\rho$

The roundness $\rho$ of the cutting edge is of limited importance for a gear hobbing operation, as well as for wear of the working surfaces of the hob teeth. The hob cutting edge is usually subject to wear mostly of the clearance surface and the rake surface (crater wear of the rake face), as shown in Figure 15.80. The roundness $\rho$ of the cutting edge is not of prime importance from the standpoint of wear of the involute gear hobs.

#### *15.6.2.2.8 Radius of Curvature of the Cutting Edge* $R_{ce}$

The radius of curvature of the cutting edge $R_{ce}$ is almost out of control in a gear hobbing operation. The radius of curvature $R_{ce}$ (or the cutting edge curvature, $k_{ce} = R_{ce}^{-1}$) cannot be altered because of the following: Any change to the radius of curvature $R_{ce}$ immediately affects conditions of meshing of the gear hob and the work gear in the gear machining mesh, $p_b$ alterations of this sort are not base-pitch-preserving. Because of this, the accuracy of the hobbed gear would drop down drastically.

In a similar manner, any other required geometrical parameter of the cutting edge $CE$ of a gear hob tooth can be calculated.

#### *15.6.2.2.9 Calculating the Involute Gear Hob Cutting Edge Geometry*

The discussed approach for the calculation of the geometry of the cutting edge is applicable for a hob of any design. As an illustrative example, calculation of the geometrical parameters of the precision gear hob shown in Figure 15.86 is considered below.

*15.6.2.2.9.1 Calculation of the Inclination Angle* $\lambda$ *of the Lateral Cutting Edges* A hob of conventional design features the inclination angle of zero value at the lateral cutting edges. In specific cases, the inclination angle can be only a few angular degrees. In order to develop a design for a precision involute hob, alterations to the orientation of the rake surface $R_s$ with respect to the hob axis of rotation $O_{ct}$ have been introduced. The result in that the lateral cutting edges $CE$ of the precision gear hob (see Figure 15.86) [98, 104] do not pass through the hob axis of rotation $O_{ct}$, and therefore they are inclined relative to the velocity vector $\mathbf{V}_\Sigma$. For the case under consideration, Eq. (15.187) together with Eq. (15.164) for the calculation of the hob base diameter $d_{b.c}$ allows for the derivation of an expanded expression for the calculation of the inclination angle $\lambda_y$ at point of interest $m$ within the cutting edge $CE$ of the hob tooth:

$$\lambda_y = -\arcsin\frac{m \cdot N_{ct} \cdot \sin 2\phi_n}{2 \cdot d_{y.ct} \cdot \sqrt{1-\cos^2\phi_n \cdot \cos^2\zeta_{ct}}} \quad (15.194)$$

A diameter of the hob cylinder through the point of interest $m$ of the hob cutting edge is designated as $d_{y.ct}$. Actually, the diameter $d_{y.ct}$ is within an interval $d_{f.ct} \le d_{y.ct} \le d_{o.ct}$, where the diameters $d_{o.ct}$ and $d_{f.ct}$ are the outer diameter and the inner diameter of the gear hob, respectively.

Calculations reveal that the inclination angle $\lambda$ has opposite signs for the left-hand-side and for the right-hand-side profiles of the gear hob teeth [see Figure 15.117(a)]. The magnitude of the angle of inclination is in the range of $|\lambda_y| \cong 15° \div 20°$. The magnitude $|\Delta\lambda_y|$ of variation of the inclination angle $\lambda_y$ within the hob tooth height does not exceed $|\Delta\lambda_y| \le 5°$.

*15.6.2.2.9.2 Calculation of the Normal Rake Angle* $\gamma_N$ *of the Lateral Cutting Edges* Equation (15.177) allows for the derivation of an expanded expression for the calculation of the normal rake angle $\gamma_N$. It is instructive to mention here that in the specific case under consideration, an equation for the calculation of the normal rake angle $\gamma_N$ can be derived in a way that is easier than the straightforward approach, which is based on Eq. (15.177).

Consider three vectors **d**, **e**, and **f**, of unit length, as shown in Figure 15.84.

The unit vector **d** is aligned with the centerline of the hob tooth profile within the rake surface $R_s$. The equation for the vector **d** can be represented in the form

$$\mathbf{d} = -\mathbf{j} \quad (15.195)$$

The unit vector **e** is located within the plane that is congruent to the rake surface $R_s$. The vector **e** can be expressed by the equation

$$\mathbf{e} = \mathbf{i} \cdot \sin\xi + \mathbf{k} \cdot \cos\xi \quad (15.196)$$

Finally, the unit vector **f** is directed orthogonally to the cutting edge $CE$ at point of interest $m$ and is located within the rake plane $R_s$. The equation for the vector **f** can

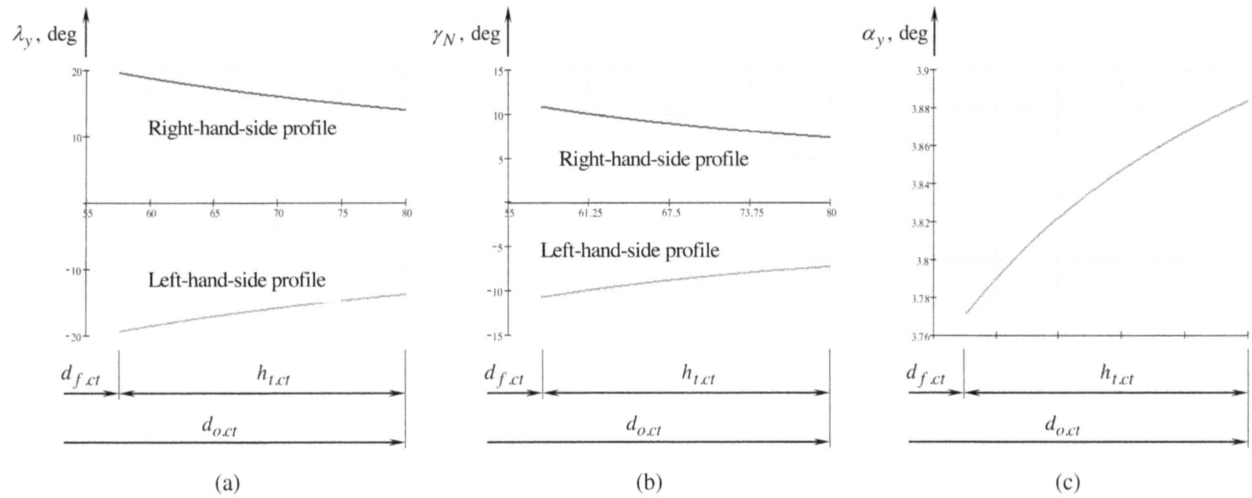


**FIGURE 15.117** Actual value of (a) the inclination angle $\lambda_y$, (b) the normal rake angle $\gamma_N$, and (c) the clearance angle $\alpha_y$ at the left-hand-side cutting edge *CE* at different points within the cutting edge of the hob shown in Figure 16.86. (From Radzevich, S.P., *SME Journal of Manufacturing Science and Engineering*, 129(8), 750–759, 2007. With permission.)

be represented in terms of the unknown yet normal rake angle $\gamma_N$:

$$\mathbf{f}(\gamma_N) = \mathbf{i}\cdot|\mathbf{f}(\gamma_N)|_x + \mathbf{j}\cdot|\mathbf{f}(\gamma_N)|_y + \mathbf{k}\cdot|\mathbf{f}(\gamma_N)|_z \tag{15.197}$$

By construction, the vectors **d**, **e**, and **f** are situated within the rake plane $R_s$ of the hob tooth. These vectors compose a set of coplanar vectors. Owing to that, a triple scalar product of the vectors **d**, **e**, and **f** is identical to zero, that is, the identity $\mathbf{d}\times\mathbf{e}\cdot\mathbf{f}\equiv 0$ is valid. This immediately yields an expression:

$$\mathbf{d}\times\mathbf{e}\cdot\mathbf{f} = \begin{vmatrix} 0 & -1 & 0 \\ \sin\xi & 0 & \cos\xi \\ |\mathbf{f}(\gamma_N)|_x & |\mathbf{f}(\gamma_N)|_y & |\mathbf{f}(\gamma_N)|_z \end{vmatrix} \equiv 0 \tag{15.198}$$

After the determinant [see Eq. (15.198)] has been expanded, and after the necessary formulae transformations are accomplished, an equation for the calculation of the normal rake angle $\gamma_N$ at the hob cutting edge point of interest[37] can be derived.

The function "$\gamma_N$ vs. $h_{t.c}$" is plotted in Figure 15.117(b). The calculations reveal that actual values of the normal rake angle $\gamma_N$ have opposite signs for the left-hand-side and the right-hand-side profiles of the gear hob teeth [see Figure 15.117(b)]. The magnitude of the normal rake angle $\gamma_N$ is in the range of $|\gamma_N| \cong 7 \div 11°$. The magnitude $|\Delta\gamma_N|$ of variation of the normal rake angle $\gamma_N$ within the hob tooth height does not exceed $|\Delta\gamma_N| \leq 4°$.

*15.6.2.2.9.3 Calculation of the Clearance Angle $\alpha_y$ at the Lateral Cutting Edges* Following the method discussed above, an equation for the calculation of the clearance angle $\alpha_y$ in the reference plane through the vectors $\mathbf{n}_c$ and $\mathbf{V}_\Sigma$ can be derived. Omitting the routing formulae transformation Figure 15.117(c) shows the results of the calculations. The clearance angle $\alpha_y$ has a positive value for both sides of the gear hob teeth. The magnitude of the clearance angle is in the range of $|\alpha_y| \cong 4°$. The magnitude $|\Delta\alpha_y|$ of variation of the clearance angle $\alpha_y$ within the hob tooth height does not exceed $|\Delta\alpha_y| \leq 1.5°$. The variation $\Delta\alpha_y$ of the clearance angle for the opposite cutting edges of the hob tooth is not shown in Figure 15.117(c) as this curve is almost congruent to the one that is depicted in Figure 15.117(c).

Slightly modified equations can also be applied for the calculation of the geometry of the cutting edge of the precision gear hobs [105, 107], as well as of the hobs of similar design.

The examples discussed above illustrate the possibility of calculating the geometry of the cutting edge of an involute hob. These results aim for further optimization of the hob tooth design. The derived equations make possible an optimization of all of the geometrical parameters of the hob cutting edges.

#### 15.6.2.3 Possibility of Improving Hob Design on the Premise of the Results of Investigation of the Geometry of Cutting Edge

Optimization of the geometry of the cutting edge of the gear hob tooth is a reliable way to improve the cutting performance of hobs for machining gears. It is strongly desired to develop a hob design that features optimal values of all cutting angles at every point of the cutting edge. Such a possibility is illustrated in the following with the improvement of the design of a hob for skiving of hardened gears.

The geometry of the active part of skiving hobs for finishing and/or semi-finishing of hardened gears significantly differs from that of the hobs of conventional

[37] This equation is not represented here as it is bulky.

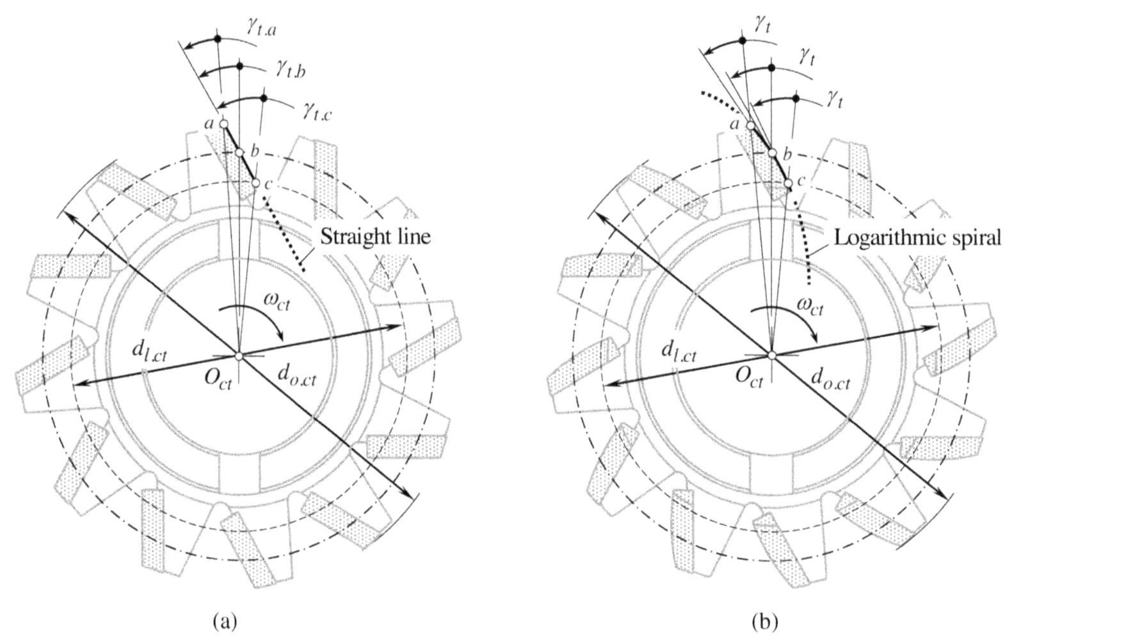


**FIGURE 15.118** Elements of the geometry of the active part of a skiving hob with (a) the rake angle at the top edge $\gamma_{o.a} < \gamma_{o.b} < \gamma_{o.c}$. (From *Kimura, K.*, and *Ainoura, M.*, US Patent 3.786.719) and (b) the skiving hob with $\gamma_o = Const$ (From Radzevich, S.P. USSR Patent 1.114.543, Int. Cl. B23f 21/16, Sept. 7, 1982.)

design. Only lateral cutting edges of the hob teeth remove the stock—the top edges do not cut the work material [74, 137, 138, 141, 158, 159]. This is an important feature of skiving hobs. The rake angle $\gamma_o$ of a negative value at the top edge of the hob tooth is commonly within the interval $\gamma_o = (-30° \div -60°)$.

Owing to the great negative rake angle $\gamma_o << 0°$ at the top edge of the skiving hob, the inclination angle $\lambda$ at the lateral cutting edges varies greatly. The estimation of the range of the variation of the inclination angle $\lambda$ can be derived from Figure 15.118(a). It is evident that the inequality $\gamma_{t.a} < \gamma_{t.b} < \gamma_{t.c}$ is observed. The difference between the values of the inclination angle $\lambda$ that is measured at the outer diameter of the hob $d_{o.ct}$ and at the limit diameter of the hob $d_{l.ct}$ is up to $\Delta\lambda \cong 30°$. It is known from many advanced sources that the deviation of the inclination angle of the cutting tool just on $\Delta\lambda \cong 5°$ from its optimal value can cause a double reduction of the cutting tool life. This means that the design of a skiving hob can be significantly improved if the inclination angle of the hob cutting edges would have a constant value along the cutting edge *CE* and the angle of inclination $\lambda$ would have optimal value at every point of the cutting edge of the gear hob.

For this purpose it is recommended[38] [74] to shape the rake surface $R_s$ with the help of a convex segment of a logarithmic spiral curve having a pole at the hob axis of rotation $O_{ct}$ as shown in Figure 15.118(b). (For hobs with a positive rake angle $\gamma_o$, a concave segment of the logarithmic spiral curve is used.)

[38] SU Pat. No. 1114543, *A Hob*/S.P. Radzevich, Int. Cl. B23f 21/16, Filed: September 7, 1982.

An equation of the generating curve of the hob rake surface $R_s$ can be derived on the premise of the well-known differential equation for isogonal trajectories:

$$\left(\frac{\partial\varphi}{\partial x}\cos\theta - \frac{\partial\varphi}{\partial y}\sin\theta\right)dx + \left(\frac{\partial\varphi}{\partial x}\sin\theta + \frac{\partial\varphi}{\partial y}\cos\theta\right)dy = 0 \tag{15.199}$$

In the case under consideration, the solution to Eq. (15.199) returns an equation of the *logarithmic spiral*:

$$\rho = \rho_0 \cdot e^{\varphi \tan \lambda_{opt}} \tag{15.200}$$

Actually, Eq. (15.200) describes a planar curve that is a reasonably good approximation to the transverse profile of the hob rake surface $R_s$ for which the requirement $\lambda_{opt} = Const$ is fulfilled. This is because the actual value of the inclination angle $\lambda$ is measured within the surface of the cut, and not in the transverse section of the hob. However, it is proven that the derived solution [see Eq. (15.200)] is practical.

The logarithmic spiral [see Eq. (15.200)] has many applications in the field of gear-cutting tool design.[39] Application of this curve includes but is not limited to the use (a) for the purposes of relief grinding of gear hobs, (b) for the purposes of dressing of form grinding wheels, (c) for inspection of form cutting tools, and so forth; see [65, 66, 122–129, 173, 174, 191, 192, 213] and others.

[39] SU Patents: No. 869978, No. 1009615, No. 1077706, No. 1117127, No. 1146134, No. 1199458, No. 1335426, No. 1371875, No. 1393527, No. 1433769, No. 1516738, No. 1629741, No. 1393527, No. 1754352, No. 1771937, No. 2009762 and others.

Modification of the rake surface of the hob teeth makes possible[40] the hobbing of the modified gear tooth profiles [68]. There is much room for improvement of hob design on the premise of in-depth analysis of the geometry of the cutting edge of the hob tooth.

## 15.7 CONSTRAINTS ON THE PARAMETERS OF MODIFICATION OF THE HOB TOOTH PROFILE

Involute hobs featuring a modified tooth profile are used for many purposes [97]. While modification aims at an increase of the hob cutting performance, modification of an involute hob tooth addendum often results in violation of the fifth necessary condition[41] of proper part surface generation (*conditions of proper PSG* for simplicity) (see Appendix E). The violation of the fifth condition of proper *PSG* can cause a significant reduction of the accuracy of the hobbed gears.

No modification is observed on the tooth profile of a standard gear hob [see Figure 15.79(a)]. Finishing and semi-finishing involute hobs (that is, the involute hobs for machining of hardened [up to *HRC* 60…62] gears) have shortened on Δ, the height of the tooth addendum [see Figure 15.81(a)] [31, 141, 159]. A similar tooth profile modification is used in the design of disk-type mill cutters [see Figure 6.22(c)]. The length of the lateral cutting edges of a hob with a chamfered tooth profile [see Figure 15.79(b)] is shortened compared to that for hobs with a standard tooth profile. Owing to shorter lateral cutting edges, generation of a gear tooth profile with a hob that features a chamfered hob tooth profile [see Figure 15.79(b)] is similar to that performed with the hob with a reduced addendum [see Figure 15.81(a)]. The addendum reduction [see Figure 15.81(a)] can be combined with the chamfered hob tooth profile [see Figure 15.79(b)]. This results in the modified hob tooth addendum [see Figure 15.81(b)].

The lateral cutting edges of the hob tooth also become shorter due to (a) the rounded ρ corners as shown in Figure 15.79(c), and (b) the full rounded *r* cap [see Figure 15.79(d)]. The roundness ρ as well as the full roundness *r* affect the hob tooth profile in a manner similar to how shortening the Δ height of the addendum [see Figure 15.81(a)] does. The rounded hob tooth profile [see Figure 15.79(c)] can be combined with the shortened tooth addendum [see Figure 15.67(a)]. This type of tooth modification results in the combined modification of the involute hob tooth profile depicted in Figure 15.81(c). A tooth profile modification of a complex nature, for example, as shown in Figure 15.81(d), also can entail violation of the fifth condition of proper part surface generation. Numerous other designs of the hob tooth modification are known [41, 42, 69, 70, 141, 159, 171, 208, 222], and so forth.

[40] SU Pat. No. 1114543, *A Gear Cutting Tool*/S.P. Radzevich, Int. Cl. B23f 21/16, Filed: April 26, 1982.

[41] In order to meet the fifth necessary condition of proper part surface generation, the intersection of adjacent portions of the generating surface *T* of a cutting tool must be eliminated.

The modification of the hob tooth profile can be the dominant factor that limits the accuracy of the hobbed gear. It is strongly desired to maintain a proper correspondence between the parameters of tooth profile modification of the hob and between the parameters of the work gear. The problem of the calculation of the min/max permissible value Δ of the tooth profile modification in terms of the tooth normal profile angle $\phi_{ct}$ for machining an involute gear with a prespecified tooth number is discussed below [97].

To calculate the constraints on the parameters of modification of the tooth profile of an involute hob, corresponding reference systems used for the analytical description of the kinematics and the geometry of the gear hobbing process must be established.

### 15.7.1 Applied Reference Systems

When hobbing a gear, the work gear is rotated $\omega_g$ about its axis $O_g$ (see Figure 15.119). The gear hob is rotated about its axis $O_{ct}$ with a certain angular velocity $\omega_{ct}$. The axes of rotation $O_g$ and $O_{ct}$ are positioned at a certain distance $C$ from one another. The gear and the hob rotate in a time-coordinated manner. The rotations $\omega_g$ and $\omega_{ct}$ are timed with each other in the way for which the ratio $\omega_g/\omega_{ct} = N_{ct}/N_g$ is observed. Here $N_{ct}$ designates the number of starts of the hob, and $N_g$ is the teeth count of the gear being hobbed. The hob climbs in the axial direction of the gear, that is, it travels up along the axis $O_g$ with a certain feed $\mathbf{F}_{ct}$. In a conventional gear hobbing operation, the gear hob travels down in the direction that is opposite to the direction of the gear hob climbing.

A Cartesian coordinate system $X_gY_gZ_g$ is associated with the work gear as shown in Figure 15.119. This reference system is rotated $\omega_g$ together with the work gear. The coordinate system $X_{ct}Y_{ct}Z_{ct}$ is associated with the hob. This reference system is rotated $\omega_{ct}$ and travels $\mathbf{F}_{ct}$ together with the hob. Numerous auxiliary reference systems are used in addition to the coordinate systems $X_gY_gZ_g$ and $X_{ct}Y_{ct}Z_{ct}$. They are not depicted in Figure 15.119.

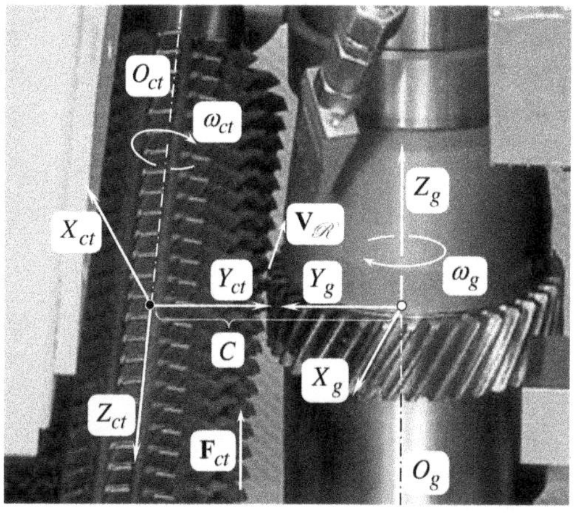


**FIGURE 15.119** Applied coordinate systems. (From Radzevich, S.P., *ASME Journal of Mechanical Design*, 128(4), 803–811, 2006. With permission.)

### 15.7.2 Kinematics of the Elementary Gear Drive

An elementary gear drive is composed of two gears and the housing, that is, it is composed of pinion 1 and of gear 2 (see Figure 15.120), which are assembled in the housing (not shown in Figure 15.120). The pinion 1 and the gear 2 rotate about the axes $O_{g.1}$ and $O_{g.2}$. The axes $O_{g.1}$ and $O_{g.2}$ of the rotations are positioned at a certain operating center distance.

The following two diameters are recognized for the pinion and the mating gear.

The *limit diameter* is considered the diameter of a gear at which the line of action intersects the maximum addendum circle of the mating gear. This is sometimes referred to as the start or end of contact of profile (*start of active profile SAP* or *end of active profile EAP*). The limit diameter $d_{l.g}$ is the lowest portion of a tooth that can actually come in contact with teeth of a mating gear. It is a calculated value and is not to be confused with form diameter. It is the boundary between the active profile and the fillet area of the tooth.

The *form diameter* is considered the diameter of a circle at which the transition curve produced by the tooling intersects, or joins, the involute or specified profile. This diameter cannot be less than the base circle diameter. Form diameter is a specified diameter on the gear, above which the transverse profile is to be in accordance with drawing specifications on profile. It is an inspection and should be placed at a somewhat smaller radius than the limit diameter to allow for shop tolerances.

In order to generate the desirable involute tooth profile between the outer diameter circle and the limit diameter circle, the actual modification of the involute hob tooth profile must not exceed the maximum permissible value [Δ]. It is necessary to determine the maximum allowed value of [Δ] in a way that provides the gear clearance $c$. The last is often equal to $0.25m$ (here $m$ designates the modulus of the involute hob).

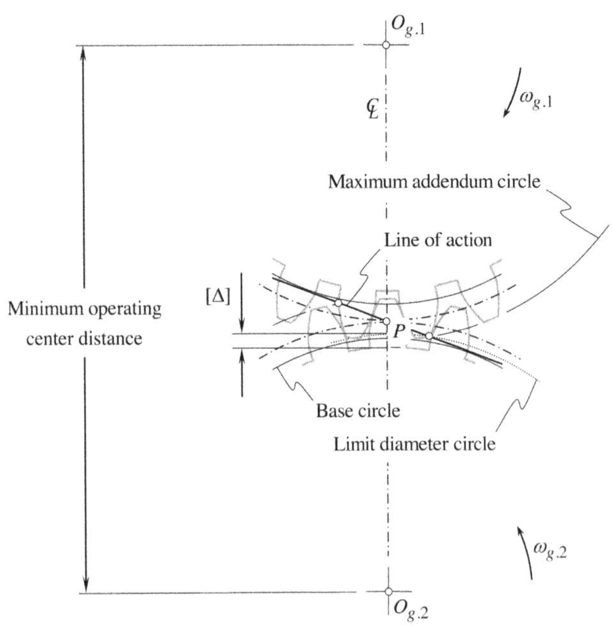


**FIGURE 15.120** Limit diameter and the maximum allowed value [Δ] of modification of a gear tooth profile. (From Radzevich, S.P., *ASME Journal of Mechanical Design*, 128(4), 803–811, 2006. With permission.)

Calculation of the maximum permissible value of the parameter [Δ] is a critical issue for designing a special-purpose involute hob.

### 15.7.3 Calculation of the Maximum Permissible Value of the Modification of Tooth Profile of Involute Hob

A convenient analytical approach to calculate the maximum permissible value of [Δ] is based on the analysis of fulfillment (or violation) of the fifth necessary condition of proper part surface generation. This condition requires avoiding intersection of the neighboring segments of the tooth profile of the generating surface $T$ of the involute gear hob.

A gear can be viewed as:

- A set of $N_g$ portions of a surface (portions of the cylinder of revolution) that form the top lands of the gear teeth.
- A set of $N_g$ portions of a surface (portions of the cylinder of revolution) that form the root lands of the gear teeth.
- A set of $N_g$ pairs of screw involute surfaces $\mathcal{G}$ through both sides of the gear tooth profile.

Owing to that, the generating surface $T$ of the involute hob is composed of $N_{ct}$ number of portions of a cylinder of revolution (root lands, as well as top lands), and that same number $N_{ct}$ of pairs of the screw involute surfaces through both sides of the hob tooth profile. When hobbing a gear, the surfaces composed the generating surface of the hob $T$ make contact with the corresponding portions of the gear tooth surface $\mathcal{G}$.

In order to make a conclusion whether the fifth necessary condition of proper part surface generation is fulfilled, or whether it is violated in a gear hobbing process, it is necessary to investigate the relative disposition of the neighboring portions of the generating surface $T$ of the gear hob.

In reality, three different kinds of relative disposition of the neighboring portions of generating surface $T$ of the hob are physically possible. The neighboring portions of the generating surface of a gear hob

- can share no common points, that is, they can be apart from one another,
- can be connected to each other at the end points (or the end point of one tooth profile and a point within another tooth profile), and
- can intersect each other.

In the first two cases, the fifth necessary condition of proper part surface generation is fulfilled. The corresponding dispositions of the neighboring portions of the generating

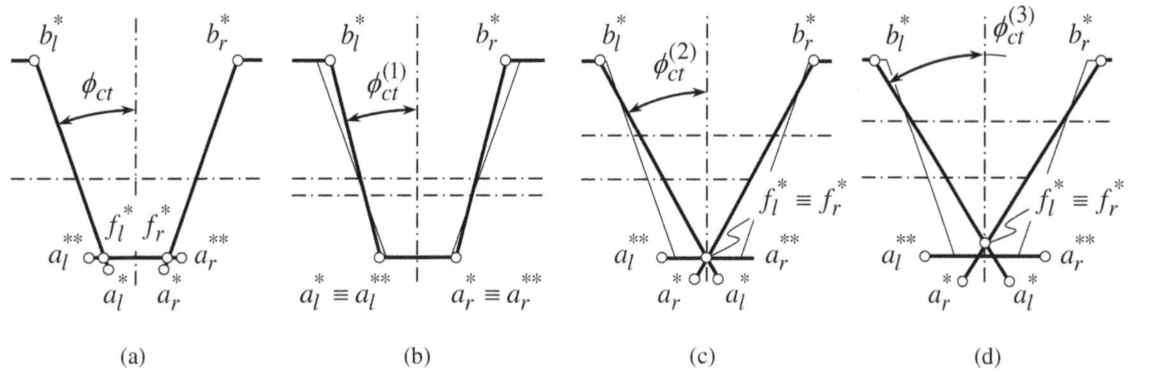


**FIGURE 15.121** Examples of violation of the fifth necessary condition of proper part surface generation. The parts (a) to (d) are discussed in detail within the text body of the chapter. (From Radzevich, S.P., *ASME Journal of Mechanical Design*, 128(4), 803–811, 2006. With permission.)

surface of the hob are not of interest from the perspective of satisfaction or violation of the fifth necessary condition of proper part surface generation.

When neighboring portions of the involute hob tooth profile intersect one another as observed in the third case, then violation of the fifth necessary condition of proper part surface generation occurs. As a result, the desired involute tooth profile is not generated in the bottom of the tooth space between the gear tooth profiles, but a transient curve is generated instead.

Consider an example of the generation of an involute profile of the work gear tooth space. The work gear tooth space is generated by a corresponding tooth of the auxiliary generating rack $\mathscr{R}_T$ of the hob. One side of the work gear tooth space is generated by the left-hand-side tooth profile $a_l^* b_l^*$ of the hob tooth [see Figure 15.121(a)]. The opposite side of the gear tooth space is generated by the right-hand-side tooth profile $a_r^* b_r^*$ of the hob tooth. The opposite sides $a_l^* b_l^*$ and $a_r^* b_r^*$ of the tooth profile of the auxiliary rack $\mathscr{R}_T$ form a certain profile angle $\phi_{ct}$ with the centerline of the auxiliary rack tooth. The bottom land of the gear tooth space is generated by the top land $a_l^{**} a_r^{**}$ of the hob tooth.

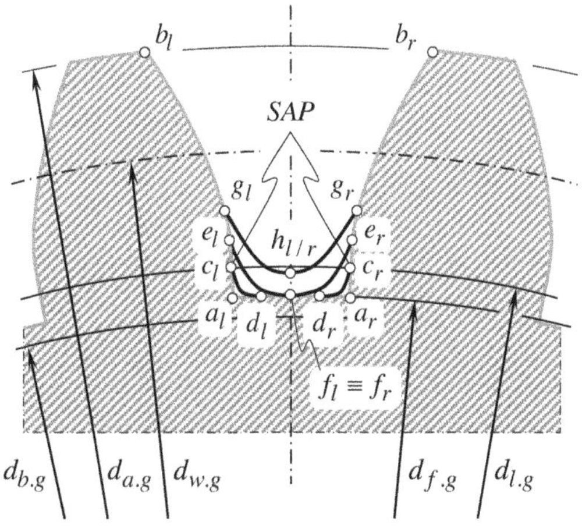


**FIGURE 15.122** Transient curves in the bottom of the tooth space of a hobbed gear. (From Radzevich, S.P., *ASME Journal of Mechanical Design*, 128(4), 803–811, 2006. With permission.)

The neighboring portions of tooth profile of the auxiliary rack $\mathscr{R}_T$ intersect each other. Point $f_l^*$ is the point of intersection of the straight-line segments $a_l^* b_l^*$ and $a_l^{**} a_r^{**}$. Similarly, point $f_r^*$ is the point of intersection of the straight-line segments $a_r^* b_r^*$ and $a_l^{**} a_r^{**}$. Because of the intersections, the portions $a_l^* f_l^*$ and $a_l^{**} f_l^*$ at one corner of the hob tooth, as well as the portions $a_r^* f_r^*$ and $a_r^{**} f_r^*$ at the opposite corner of the hob tooth, physically cannot exist. Instead of the absent portions $a_r^* f_r^*$ and $a_r^{**} f_r^*$ of the hob tooth profile, only the corners $f_l^*$ and $f_r^*$ physically exist in reality.

Because of the absence of portions of the hob tooth profile, corresponding portions of the work gear tooth space profile cannot be properly generated. Referring to Figure 15.122, the segments $a_l c_l$ and $a_r c_r$ of the involute profiles $a_l b_l$ and $a_r b_r$ of the gear tooth space, which should be generated by the absent portions $a_l^* f_l^*$ and $a_r^* f_r^*$ of the hob tooth profile, as well as the segments $a_l d_l$ and $a_r d_r$, which should be generated by the absent segments $a_l^{**} f_l^*$ and $a_r^{**} f_r^*$, actually are generated by the corners $f_l^*$ and $f_r^*$. The corners $f_l^*$ and $f_r^*$ generate not the desired portions $a_l c_l$ and $a_r c_r$ of the involute profiles, and not the portions $a_l d_l$ and $a_r d_r$ of the bottom land, but they generate the transient curves $c_l d_l$ and $c_r d_r$ instead. The transient curves are often referred to as fillets.[42]

The dimensions of the transient curves can be reduced to fit the tolerance by proper selection of the design parameters of the hob tooth. Moreover, if necessary, the transient curves can be eliminated [see Figure 15.121(b)].

Aiming for the elimination of the transient curves $c_l d_l$ and $c_r d_r$, the corresponding reduction of the normal profile angle of the hob tooth $\phi_{ct}$ is needed. When the hob normal profile angle $\phi_{ct}$ is reduced to the value of $\phi_{ct}^{(1)} = \phi_{l.g}$ (here $\phi_{l.g}$ designates the tooth profile angle on the limit diameter of the gear), then the transient curve vanishes. In such a scenario, the tooth profile of the hobbed gear is shaped exactly in the form of true involute curves $a_l b_l$ and $a_r b_r$, while the root land is shaped in the form of the arc $a_l a_r$ of a circle. The case of vanished transient curves is of interest mainly from the theoretical point of view, as a gear with sharp inner

[42] Reminder: The "*start of active profile*" (*SAP*) is the intersection of the limit diameter and the involute profile (see Figure 15.122). Points, $c_l$ and $c_r$, are the points of *SAP*.

corners at points $a_l$ and $a_r$ is impractical (an inner corner is a strong stress riser).

Two more examples of intersection of the adjacent segments of the tooth profile of the gear hob tooth are depicted in Figure 15.121(c,d). The corresponding transient curves are shown in Figure 15.122.

At a certain value of the normal profile angle $\phi_{ct}^{(2)} > \phi_{ct}$, the point of intersection $f_l^* \equiv f_r^*$ of the lateral sides $a_l^* b_l^*$ and $a_r^* b_r^*$ of the hob tooth profile is located within the top land profile $a_l^{**} a_r^{**}$ [see Figure 15.121(c)]. In this specific case, two transient curves $c_l d_l$ and $c_r d_r$ degenerate to a common continuous transient curve $e_l f_l (\equiv f_r) e_r$. The transient curve $e_l f_l (\equiv f_r) e_r$ (a) shares a common point $f_l \equiv f_r$ with the gear root land and (b) could spread above the limit circle of a diameter $d_{l.g}$ (see Figure 15.122).

When the point $f_l^* \equiv f_r^*$ of intersection of the lateral sides $a_l^* b_l^*$ and $a_r^* b_r^*$ of the gear hob tooth profile is shifted from the top land profile $a_l^{**} a_r^{**}$ toward the pitch line of the hob [see Figure 15.121(d)], then the transient curve in the tooth space of the hobbed gear has no common point with the bottom land, and the transient curve could spread far beyond the limit circle of a diameter $d_{l.g}$ (see Figure 15.122).

In the cases shown in Figure 15.121(c,d), the deviations of the desired profile of the gear tooth can exceed the tolerance for accuracy of the hobbed tooth profile.

Other types of the relative disposition of the neighboring portions of generating surface $T$ of a hob are possible in specific cases of gear hobbing. However, all of them can be interpreted as reduced cases of three major types of the relative dispositions listed above.

If we wish to prevent the impact of modification of the hob tooth profile on the accuracy of the hobbed gear, the transient curve must be located within the narrow space between the circle of a root diameter $d_{f.g}$ and the circle of a limit diameter $d_{l.g}$. This criterion can be analytically stated as:

$$[\Delta] \leq 0.5 \cdot (d_{l.g} - d_{f.g}) \tag{15.201}$$

The fifth necessary condition of proper part surface generation is fulfilled if and only if the inequality (15.201) is valid. In this case, a work gear can be hobbed in full compliance with the blueprint specification on the gear tooth profile. Otherwise, deviations of the actual gear tooth profile from the desired tooth profile are inevitable.

The inequality (15.201) can be expressed in terms of the design parameters of both the gear to be machined and the gear hob. For this purpose, consider the generation of a gear tooth profile with an involute hob (see Figure 15.123).

The left-hand-side of the gear tooth profile[43] is generated along the left-hand-side line of action $LA_l$, or to be more exact, within the portion $DE$ of the line of action $LA_l$. The involute hob tooth profile does not exceed the gear root circle of the diameter $d_{f.g}$. The hob tooth profile is bounded by the straight-line segment $AC$. Because of this,

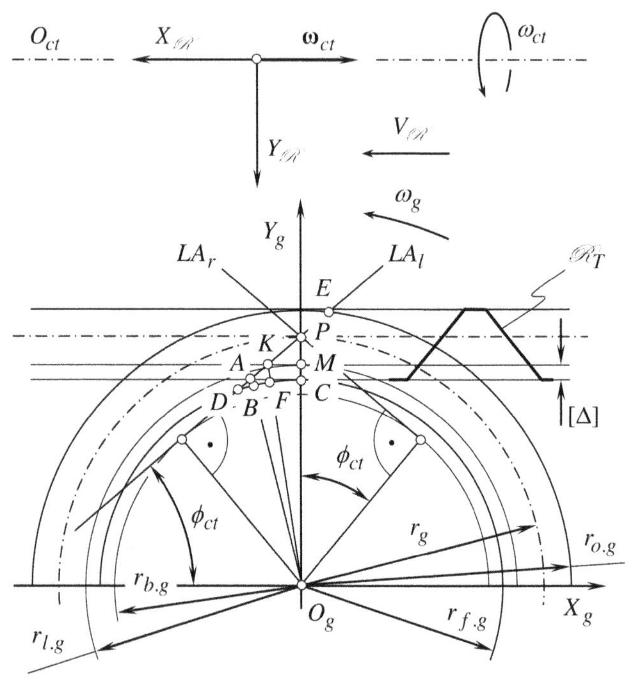


**FIGURE 15.123** Gear-to-hob diagram of meshing. (From Radzevich, S.P., *ASME Journal of Mechanical Design*, 128(4), 803–811, 2006. With permission.)

the portion $AD$ of the line of action $LA_l$ cannot be reproduced in the gear hobbing process. Therefore, within the portion $AB$ of the gear tooth height, that is, an overlap in the gear hobbing process by the portion $AD$ of the line of action $LA_l$, the transient curve of the height $AB = \Delta$ is observed. The parameter $\Delta$ can be analytically determined in the following way:

$$\Delta = AO_g - BO_g = AO_g - CO_g \tag{15.202}$$

The equation of the line of action $LA_l$ can be represented in the form

$$\text{Line of action, } LA_l \Rightarrow Y_g = X_g \cdot \tan\phi_{ct} + \frac{d_{b.g}}{2 \cdot \cos\phi_{ct}} \tag{15.203}$$

The $Y_A$ – coordinate of the point of intersection $A$ of the line of action $AL_l$ with the circle of a radius $r_{f.g}$ (see Figure 15.123) is equal $Y_A = 0.5 \cdot d_{f.g}$. This result makes possible the calculation of the $X_A$ – coordinate of point $A$. Equation (15.203) returns

$$X_A = 0.5 \cdot \left( d_{f.g} - \frac{d_{b.g}}{\cos\phi_{ct}} \right) \cdot \cot\phi_{ct} \tag{15.204}$$

Ultimately, Eq. (15.202) casts into

$$\Delta = 0.5 \cdot \left( \sqrt{\left( d_{f.g} - \frac{d_{b.g}}{\cos\phi_{ct}} \right)^2 \cot^2\phi_{ct} + d_{f.g}^2} - d_{f.g} \right) \tag{15.205}$$

[43] That same approach is applicable for the line of action, $LA_r$, of the right-hand side of the tooth profile.

The design parameters of standard gears satisfy the equalities

$$d_{b.g} = d_g \cdot \cos\phi_g = m \cdot N_g \cdot \cos\phi_g \qquad (15.206)$$

$$d_{f.g} = d_g - 2\cdot 1.25 \cdot m = m\cdot(N_g - 2.5) \qquad (15.207)$$

For the specific case of hobbing of a standard gear, Eq. (15.205) can be represented in the form

$$\Delta = \frac{m}{2}\left\{\sqrt{\left[(N_g - 2.5) - N_g \cdot \frac{\cos\phi_g}{\cos\phi_{ct}}\right]^2 \cot^2\phi_{ct} + (N_g - 2.5)^2} - (N_g - 2.5)\right\} \qquad (15.208)$$

The latter equation yields calculation of the actual value of the parameter $\Delta$ of the transient curve in the tooth space of an involute gear.

### 15.7.4 Normalized Deviation $\Delta_M$ of the Tooth Profile of Hobbed Gear

In specific cases of exploring the impact of gear hob design parameters on the parameters of the transient curve, it is convenient to use not the deviation $\Delta$ but instead the ratio of the deviation $\Delta$ to the gear modulus $m$. For this purpose, it is sufficient to divide both sides of the expression (15.208) by the modulus $m$. The ratio $\Delta/m$ is referred to as the *normalized deviation*, and it is designated as $\Delta_m$.

As it follows from Eq. (15.208), the normalized deviation $\Delta_m$ depends on the current value of the normal profile angle $\phi_{ct}$ of the hob tooth. The function $\Delta_m = \Delta_m(\phi_{ct})$ features an extremum [see Figure 15.124(a)]. This indicates that under certain conditions, say, for a certain combination of the design parameters of the gear hob, the deviation $\Delta_m$ can be minimized, and the accuracy of the hobbed gear can be improved.

The normalized deviation $\Delta_m$ also depends on the tooth count $N_g$ of the gear being machined. The function $\Delta_m = \Delta_m(N_g)$ also features an extremum [see Figure 15.124(b)]. The minimum normalized value of the deviation $\Delta_m$ for a hobbed gear with a given tooth count occurs when machining of a gear is executed by the gear hob with a certain value of normal pressure angle $\phi_{ct}$.

The results of the analysis make possible a convenient graphical interpretation of the area within which the fifth necessary condition of proper part surface generation in a gear hobbing operation is fulfilled, and it is violated outside of the area.

The maximum permissible value of the deviation $\Delta$ is designated as $[\Delta]$ (that is, the inequality: $\Delta \le [\Delta]$ has to be valid). The undesired increase of the deviation $\Delta$ beyond the tolerance $[\Delta]$ can be compensated by a corresponding change to the normal profile angle $\phi_{ct}$ of the gear hob. In order to keep the deviation $\Delta$ within the tolerance $[\Delta]$, the normal profile angle $\phi_{ct}$ of the hob tooth has to be within the interval (see Figure 15.123)

$$\cos^{-1}\left(\frac{d_{f.g}\cdot d_{b.g} + A_{ct}}{(d_{f.g} + 2\cdot[\Delta])^2}\right) \le \phi_{ct} \le \cos^{-1}\left(\frac{d_{f.g}\cdot d_{b.g} - A_{ct}}{(d_{f.g} + 2\cdot[\Delta])^2}\right) \qquad (15.209)$$

where $A_c = \sqrt{d_{f.g}^2 \cdot d_{b.g}^2 - (d_{f.g} + 2\cdot[\Delta])^2 \cdot (d_{b.g}^2 - 4\cdot d_{f.g}\cdot[\Delta] - 4\cdot[\Delta]^2)}$.

As an example, a zone within which the fifth necessary condition of proper part surface generation is met is

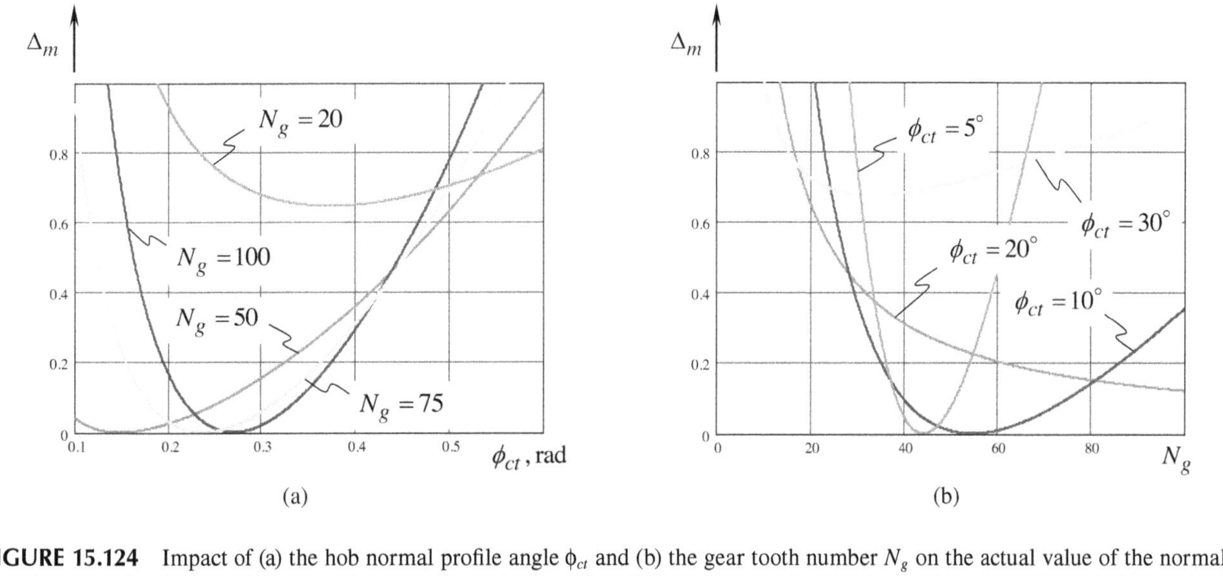


**FIGURE 15.124** Impact of (a) the hob normal profile angle $\phi_{ct}$ and (b) the gear tooth number $N_g$ on the actual value of the normalized deviation $\Delta_m$ of the work gear tooth profile. (From Radzevich, S.P., *ASME Journal of Mechanical Design*, 128(4), 803–811, 2006. With permission.)

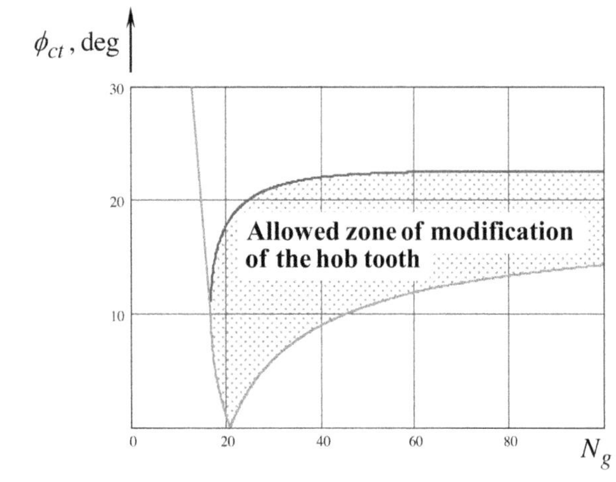


**FIGURE 15.125** Zone of fulfillment of the fifth necessary condition of proper part surface generation in the hobbing operation of involute gear. (From Radzevich, S.P., *ASME Journal of Mechanical Design*, 128(4), 803–811, 2006. With permission.)

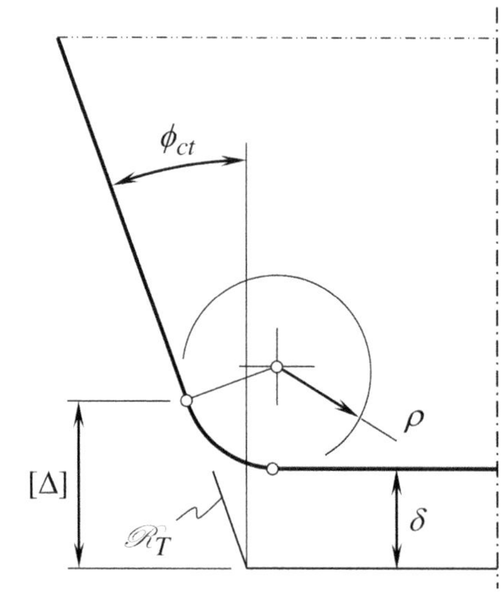


**FIGURE 15.126** Tooth profile of a finishing involute hob with a reduced δ addendum. (From Radzevich, S.P., *ASME Journal of Mechanical Design*, 128(4), 803–811, 2006. With permission.)

depicted in Figure 15.125. A spur involute gear with a profile angle $\phi_g = 20°$ can be hobbed by the gear hob with a corresponding normal profile angle $\phi_{ct}$. The deviation $\Delta$ falls into the tolerance band $0 \le \Delta \le [\Delta] = 0.25 \cdot m$. A similar boundary curve can be plotted for the spur and/or helical gear of any desirable design.

The application of gear hobs, design parameters of which correlate with the design parameters of the gear being machined [see inequality (15.209)], results in an increase in the accuracy of the hobbed gear.

### 15.7.5 Peculiarities of Involute Hobs with Reduced Addendum

To avoid an interaction of the top edge of the gear hob tooth with the completely finished bottom land of the precut gear, involute hobs with a shortened addendum $\delta$ are commonly applied (see Figure 15.126). Finishing and semi-finishing Azumi skiving hobs (see Figure 15.118), which feature the reduced addendum $\delta$, either combined with the rounded $\rho$ corners or not, are a good example in this regard [31, 141, 159]. Both the addendum reduction and the rounded corner cause the deviation $\Delta$.

In order to make possible an accurate hobbing of gears by hobs with either reduced tooth addendum and/or with rounded corners, the design parameters of the hob should fit the design parameters of the gear to be machined.

Referring to Figure 15.123, an expression

$$Y_g = X_g \cdot \tan\phi_{ct} + \frac{d_{b.g}}{2 \cdot \cos\phi_{ct}} \tag{15.210}$$

for the $Y_g$ – coordinate is derived.

For point $K$ at which the line of action intercepts the outer diameter of the hob, Eq. (15.210) yields expressions for the $X_K$ – and $Y_K$ – coordinates:

$$Y_K = 0.5 \cdot d_{f.g} + \delta \tag{15.211}$$

$$X_K = 0.5 \cdot \left( d_{f.g} - \frac{d_{b.g}}{\cos\phi_{ct}} + \delta \right) \cdot \cot\phi_{ct} \tag{15.212}$$

Therefore, the deviation $\Delta$ can be calculated from the formula

$$\Delta = 0.5 \cdot \left( \sqrt{\left( d_{f.g} - \frac{d_{b.g}}{\cos\phi_{ct}} + \delta \right)^2 \cot^2\phi_{ct} + (d_{f.g} + \delta)^2} - d_{f.g} \right) \tag{15.213}$$

On solving Eq. (15.213), the inequality

$$\delta \le [\Delta] = 0.5 \cdot \left( d_{b.g} \cdot \cos\phi_{ct} + \sqrt{(d_{f.g} + 2 \cdot [\Delta])^2 - d_{b.g}^2} \sin\phi_{ct} - d_{f.g} \right) \tag{15.214}$$

is obtained.

In order to meet the fifth necessary condition of proper part surface generation, the reduction of the involute hob tooth addendum must not exceed the maximum permissible value of $[\Delta]$. In the case under consideration, the analytical representation of the fifth necessary condition of proper part surface generation for the gear hobbing process casts

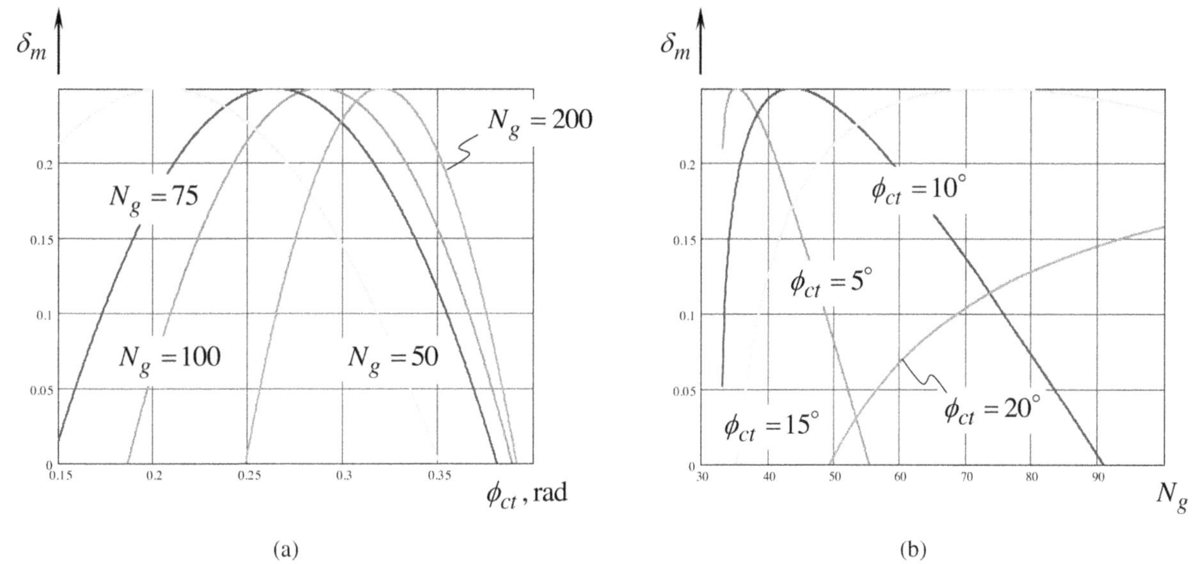


**FIGURE 15.127** Impact of (a) hob normal profile angle $\phi_{ct}$ and (b) work gear tooth count $N_g$ on the permissible value of the reduction of the tooth addendum of a hob. (From Radzevich, S.P., *ASME Journal of Mechanical Design*, 128(4), 803–811, 2006. With permission.)

into the following inequality:

$$d_{b.g} \cdot \cos\phi_{ct} + \sqrt{(d_{f.g} + 2 \cdot [\Delta])^2 - d_{b.g}^2} \cdot \sin\phi_{ct} - d_{f.g} - 2\delta \geq 0 \qquad (15.215)$$

In the particular case of hobbing of standard gears [for which the conditions of Eq. (15.206) and Eq. (15.207) are fulfilled], the inequality (15.214) casts into the form [97]

$$\delta \leq [\Delta] = \left[ N_g \cdot \cos\phi_{ct} \cdot \cos\phi_g + \sqrt{(N_g + 3)^2 - N_g^2 \cdot \cos^2\phi_g} \cdot \sin\phi_{ct} - (N_g + 2.5) \right] \qquad (15.216)$$

Examples of the functions $\delta = \delta(\phi_{ct})$ and $\delta = \delta(N_g)$ are plotted in Figure 15.127. All curves of the function $\delta = \delta(\phi_{ct})$ feature a maximum. This indicates that an increase of the allowed value of the parameter $\delta$ can be achieved by choosing a proper value of normal profile angle $\phi_{ct}$ [see Figure 15.127(a)]. A similar relationship is observed for the function $\delta = \delta(N_g)$ [see Figure 15.127(b)].

Equation (15.213) yields an analytical expression for the min/max boundary curves, which bound the area within which the fifth necessary condition of proper part surface generation is fulfilled:

$$\phi_{ct} = \cos^{-1}\left( \frac{(d_{f.g}d_{b.g} + 2\delta d_{b.g}) \pm \sqrt{(d_{f.g}d_{b.g} + 2\delta d_{b.g})^2 + (d_{f.g} + [\Delta])^2 \cdot \left\{(2d_{f.g} + \delta + 2[\Delta]) \cdot (2\Delta - \delta) - d_{b.g}^2\right\}}}{(d_{f.g} + [\Delta])^2} \right) \qquad (15.217)$$

For a specific case of hobbing of a standard involute gear [$\phi_g = 20°$, and when the conditions Eq. (15.206) and Eq. (15.207) are met], Eq. (15.217) casts into

$$\cos\phi_{ct} = \frac{(N_g - 2.4) \cdot N_g \cdot \cos\phi_g \pm \sqrt{(N_g - 2.4)^2 \cdot N_g^2 \cdot \cos^2\phi_g + (N_g - 2)^2 \cdot (0.9 \cdot N_g - N_g^2 \cos^2\phi_g - 1.9975)}}{(N_g - 2)^2} \qquad (15.218)$$

It is assumed here in Eq. (15.218) that the reduction $\delta$ of the tooth addendum of the hob is equal to $\delta = 0.05 \cdot m$ [that is, $\delta = 0.1 \cdot (0.5 \cdot m)$].

The area for possible combinations of the normal profile angle $\phi_{ct}$ of the hob tooth and the tooth count $N_g$ of the gear being machined is plotted in Figure 15.128 (the inner contour). The outer contour in Figure 15.128 corresponds to the case when no reduction $\delta$ of the tooth addendum of the hob is observed.

Referring to Figure 15.128, the reduction of the hob tooth addendum $\delta$ considerably narrows the zone within

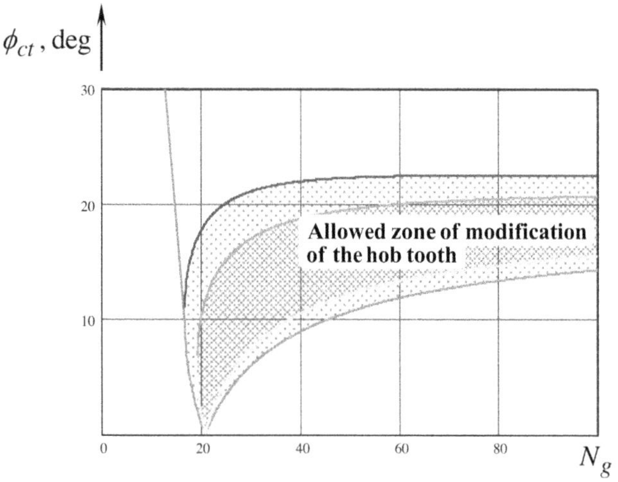


**FIGURE 15.128** Zone of fulfillment of the fifth necessary condition of proper part surface generation in the gear hobbing process by application of an involute hob with a reduced tooth addendum. (From Radzevich, S.P., *ASME Journal of Mechanical Design*, 128(4), 803–811, 2006. With permission.)

which the fifth necessary condition of proper part surface generation is fulfilled.

When the tolerance $[\Delta]$ is known, the reduction of the hob tooth addendum by the value $\delta < [\Delta]$ is permissible. Within the remaining portion $([\Delta] - \delta)$, the hob tooth can be either chamfered or rounded. For the calculation of the maximum permissible radius $\rho$ of rounding of the involute hob tooth, the formulae

$$\rho - \rho \cdot \sin\phi_{ct} = [\Delta] - \delta \tag{15.219}$$

$$\rho \leq \frac{[\Delta] - \delta}{1 - \sin\phi_{ct}} \tag{15.220}$$

are derived.

An analysis similar to that performed above can be done for hobbing helical gears as well.

### 15.7.6 Illustrative Examples of the Calculation

The following examples illustrate the capabilities of the approach discussed above for calculating the parameters of the modification of the tooth profile of the precision involute hob.

#### 15.7.6.1 Calculation of the Actual Value of the Deviation Δ

Consider the hobbing of a spur involute gear with a modulus $m = 10$ mm, normal profile angle $\phi_g = 20°$, tooth count $N_g = 75$, and clearance $c = 0.25m$. For the given involute gear, the rest of the design parameters can be calculated (see Appendix F):

- Pitch diameter: $d_g = m \cdot N_g = 750$ mm
- Outer diameter: $d_{a.g} = d_g + 2 \cdot m = 770$ mm
- Root diameter: $d_{f.g} = d_g - 2 \cdot 1.25 \cdot m = 725$ mm
- Base diameter: $d_{b.g} = d_g \cdot \cos\phi_g = 704.769$ mm

The gear is machined by a single-start $N_{ct} = 1$ involute hob with a normal profile angle $\phi_{ct} = 25°$. On solving Eq. (15.208), the actual value of the deviation $\Delta = 4.366$ mm is calculated.

The comparison of the actual deviation $\Delta = 4.366$ mm with the tolerance $[\Delta] \equiv c = 0.25 \cdot m = 2.5$ mm $[\Delta (= 4.366\,\text{mm}) > c (= 2.5\,\text{mm})]$ makes possible the conclusion that the given gear cannot be properly hobbed by the hob with a normal profile angle $\phi_{ct} = 25°$. In order to reduce the actual value of the deviation $\Delta$, an appropriate correction to the hob normal profile angle is required.

#### 15.7.6.2 Calculation of the Tolerance Band for the Hob Normal Profile Angle

For machining the spur involute gear of the same design as above, the expression (15.209) yields the following permissible interval for the involute hob normal profile angle: $8.398° \leq \phi_{ct} \leq 21.817°$. The calculated tolerance band for the normal profile angle $\phi_{ct}$ perfectly correlates with what is shown in Figure 15.125. Machining of the gear by the involute hob with a normal profile angle within the calculated interval $8.398° \leq \phi_{ct} \leq 21.817°$ results in the deviation $\Delta < c$ being within the tolerance band.

#### 15.7.6.3 Calculation of the Permissible Rounding of an Involute Hob Tooth Corner

The feasibility of the reduction of the hob tooth addendum was proved earlier. Based on these results as well as on the results discussed in the previous two examples of the calculations, for the gear hob with a normal profile angle $\phi_{ct} = 25°$, the calculated radius of the rounding of the hob tooth corner must not exceed $\rho \leq 2.598$ mm. Here $[\Delta] = c = 2.5$ mm, $\delta = 0.1 \cdot m = 1$ mm, and $\phi_{ct} = 25°$. Equations (15.219) and (15.220) are used in calculating the radius $\rho$ of the rounding.

## 15.8 APPLICATION OF HOBS FOR MACHINING GEARS

Spur, helical, and worm gears can be produced by hobbing. Double-helical and herringbone gears can be hobbed as well. Hobbing of gears of different types differs from each other.

### 15.8.1 Peculiarities of Gear Hobbing Operation

Spur and helical gears can be cut and finished by setting the hob to full depth and making only one cut. Medium and fine-pitch gears can be finishing cut in one step (see Figure 15.129). Hobs of various designs are used to cut gears in a single path of the hob past the work gear. As an illustrative example, Figure 15.130 shows climb hobbing of a spur involute gear by the hob with indexable inserts.

**FIGURE 15.129** Hobbing of a spur gear.

**FIGURE 15.130** Climb hobbing of a spur involute gear by the hob with indexable inserts.

Climb hobbing of a helical involute gear by the hob with indexable inserts is depicted in Figure 15.131.

For generating gears, gear generators with a horizontal spindle are often used. An example is illustrated in Figure 15.132. Horizontal configuration of the spindle is especially convenient when cutting gears with a wide face width, spline shafts (see Figure 15.133), and so forth. This also makes possible hobbing of multiple work gears in a single setup, as shown in Figure 15.134. Gears that feature a low tooth count, that is, *LTC – gears*, can also be hobbed on conventional gear generators (see Figure 15.135).

Where the highest accuracy is desired, it is customary to make a roughing and a finishing cut. Finish hobbing of a helical gear is illustrated in Figure 15.136. The finishing cut may remove from 0.25 to 1.0 mm (0.010 to 0.040 in.) of tooth thickness, depending on the size of the tooth.

Coarse pitch gears are customary roughed (see Figure 15.137) prior to the finishing cut. The roughing cut removes almost all the stock from the tooth space.

It is common practice to rough-cut gears on one gear generator, and then to finish-cut the gear on another gear generator. The productivity rate of the gear hobbing process can be significantly increased if two hobs are set up on a common spindle as illustrated in Figure 15.138. Initially, the work gear is roughly cut by the one hob (by the rough hob), then that same work gear is finish-cut by another hob (by the finish hob).

Coarse pitch gears are customary roughed prior to the finishing cut. The roughing cut removes almost all the stock from the tooth space.

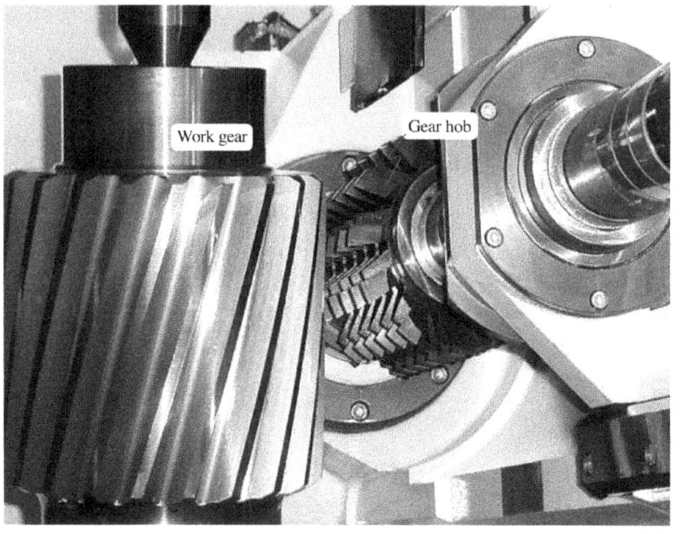


FIGURE 15.131 Climb hobbing of a helical involute gear by the hob with inserted cutting racks.

FIGURE 15.132 Hobbing a gear on a gear generator with a horizontal spindle.

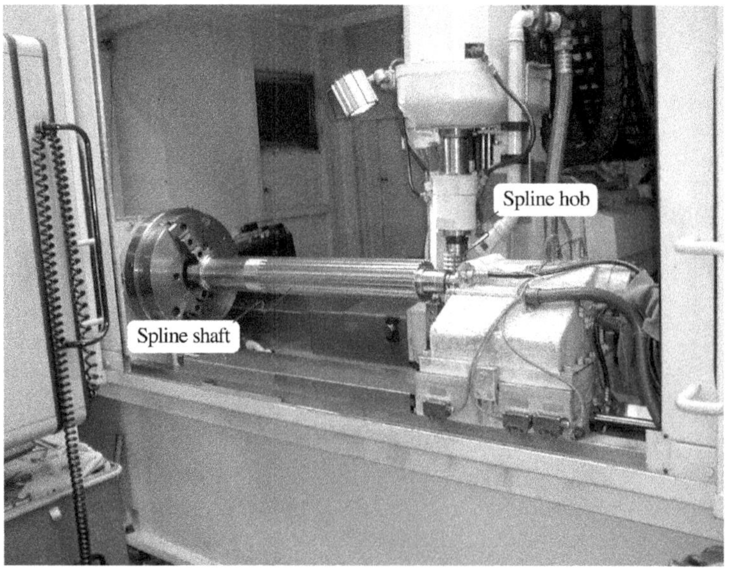


FIGURE 15.133 Horizontal hobbing of a spline shaft.

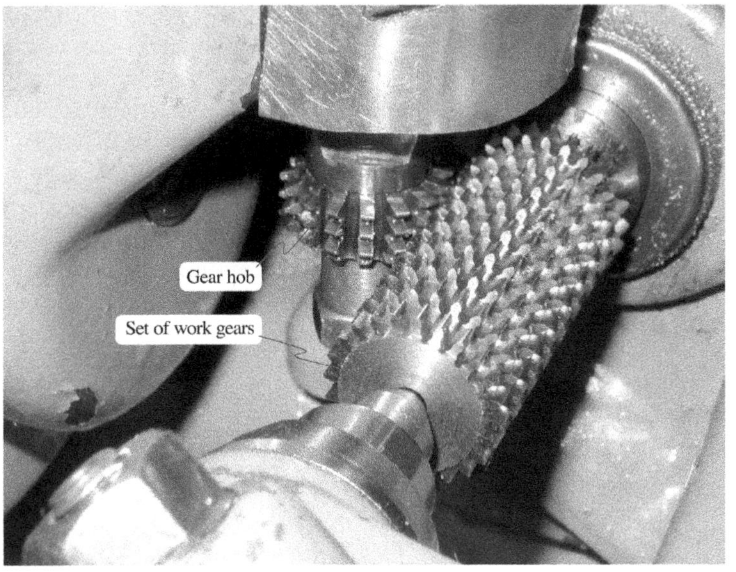


**FIGURE 15.134** Horizontal hobbing of multiple work gears in a single setup.

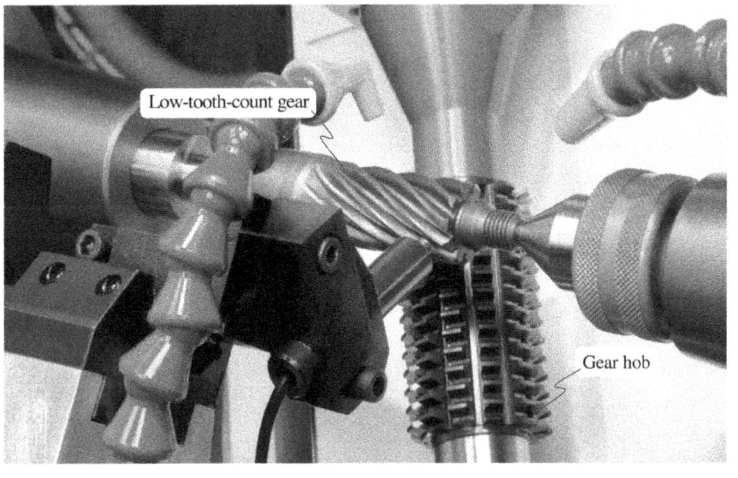


**FIGURE 15.135** Hobbing of a gear with a low tooth count (hobbing of *LTC – gear*).

**FIGURE 15.136** Finish hobbing of a helical involute gear.

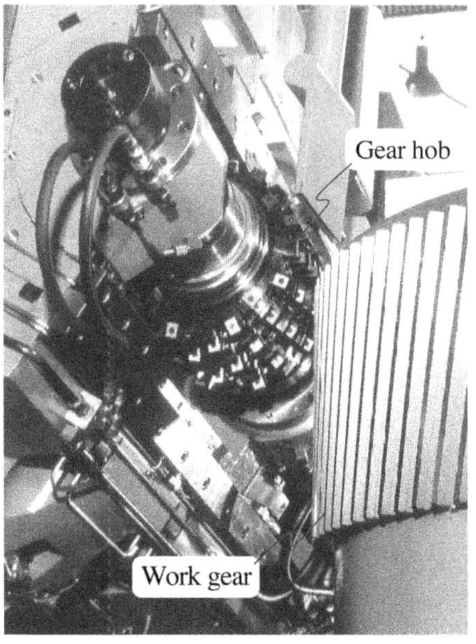


**FIGURE 15.137** Roughing of a coarse pitch helical gear.

When cutting coarse pitch gears, it is practical to use a gashing disk-type mill cutter for roughing tooth space in the work gear. Gashing of the work gear can be executed either by the single disk-type mill cutter as shown in Figure 15.139, or by two disk-type mill cutters as shown in Figure 15.140 (in the upward path, and the downward path of the cutting tool past the work gear).

Large-scale involute gears can be cut by the hob of a standard design. An illustrative example is depicted in Figure 15.141.

The hobbing process is quite advantageous in cutting gears with wide face widths or gears that have a toothed section which is integral with a long shaft. A high degree of tooth-spacing accuracy can be obtained with hobbing. High-speed marine and industrial gears with pitch-line speeds in the range of 15 to 100 m/s (3000 to 20,000 fpm) and diameters up to 5 m (200 in.) are often cut by hobbing. A few large mill gears up to 10 meters (400 in.) in diameter are hobbed. Hobbing of a giant gear by a hob with indexable inserts is illustrated in Figure 15.142.

When machining a gear, the generating surface *T* of the hob will make contact with the teeth on the work in a series of points as cutting proceeds. The points, in fact, become shallow elliptical slices in the metal being cut as each hob tooth makes contact with the work. Figure 15.143 shows the generation of a tooth space by the action of succeeding hob teeth. The envelope of the points or elliptical flecks approximates to an involute curve, and the extent to which the generated curve approaches a true involute theoretically depends on the number of teeth or gashes in the hob thread and the feed or distance the hob traverses across the face of the blank with each revolution of the work. In practice, it is not easy to generate a true involute tooth form by hobbing, but the continuous rolling of the gear being cut with cutting teeth of the hob ensures that every tooth on the work is progressively generated.

### 15.8.2 Peculiarities of Machining of Worm Gears and Worms

The hobbing of a worm gear is illustrated in Figure 15.144. The design parameters of the hob for cutting a worm gear are tightly connected to the corresponding design parameters of the mating worm. Hobs that are integral with the hob arbor (see Figure 15.63) are used for this purpose. Multiple-start hobs are used for machining a worm gear

**FIGURE 15.138** Rough hobbing and finish hobbing of a gear by two different hobs.

FIGURE 15.139 Gashing of a work gear by the disk-type mill cutter.

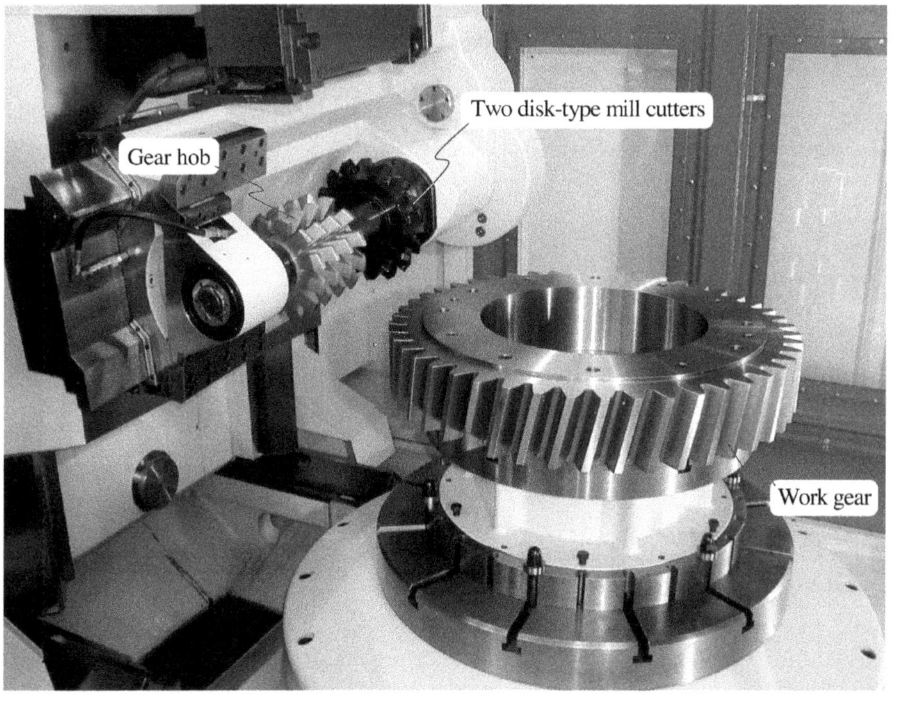


FIGURE 15.140 Gashing of the work gear by two disk-type mill cutters in the upward path, and the downward path of the cutting tool past the work gear.

**FIGURE 15.141** Hobbing of a large-scale helical involute gear by the hob of a standard design.

**FIGURE 15.142** Hobbing of a giant gear by a hob with indexable inserts.

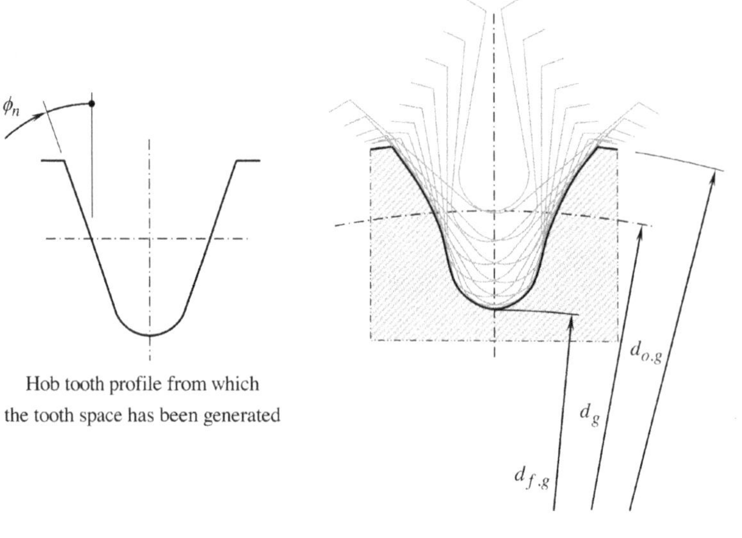


**FIGURE 15.143** Tooth space in 24-toothed spur gear generated by a hobbing process.

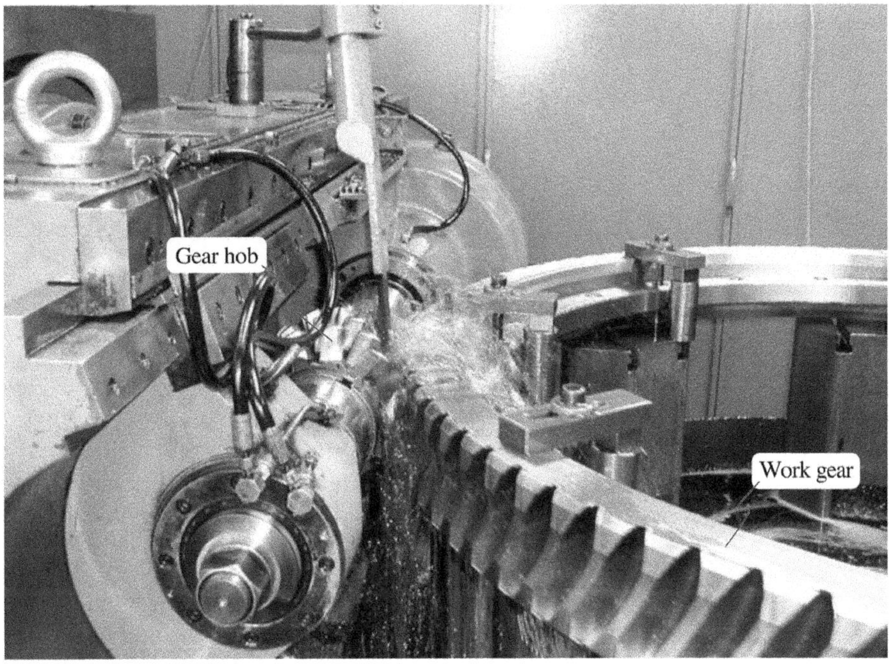


**FIGURE 15.144** Hobbing a worm gear by the hob of conventional design.

that is engaged in mesh with multiple-start worm as illustrated in Figure 15.145.

For hobbing worm gears with a large tooth count, hobs with a modified (conical) entering end are commonly used (see Figure 15.146). Fly-cutters can be used for cutting worm gears in a low-volume production of worm gearing (see Figure 15.147). When cutting worm gears with the fly-cutter, an additional motion of the gear generator spindle is required. This motion has to be synchronized with the rotation of the work gear, with the axial feed of the fly-cutter, and with the rotation of the gear generator spindle.

Coarse pitch worm gears can be cut on NC machines by end-mill cutters as illustrated in Figure 15.148. This process is especially efficient for roughing coarse pitch worm gears.

**FIGURE 15.145** Hobbing a worm gear by the multiple-start hob.

Various processes are used in the production of worms for worm gears. An illustrative example is depicted in Figure 15.149 of milling a worm with a disk-type milling cutter. For grinding of worms, disk-type grinding wheels are commonly used (see Figure 15.150). For this purpose, a face grinding wheel can also be used (see Figure 15.151).

Figure 15.152 illustrates a hob for cutting a worm gear that is engaged in mesh with a double enveloping (hourglass) worm. A multiple-start hob for cutting a worm gear

**FIGURE 15.146** Hobbing a worm gear by the hob with a modified (conical) entering end.

**FIGURE 15.147** Cutting of a worm gear by the fly-cutter.

that is engaged in mesh with a double enveloping (hourglass) worm is shown in Figure 15.153.

Other gear grinding processes are used for finishing worms and worm gears for worm gearings.

### 15.8.3 Feed Motion in Gear Hobbing Operation

All gears except worm gears are cut by feeding the hob across the face width of the gear. Referring to Figure 15.154, the feed motion is along the axis of rotation of the gear $O_g$. This axial feed motion is designated as $\mathbf{F}_{ag}$. Either the work gear travels along its axis of rotation or the hob travels along the axis of rotation $O_g$ of the work gear in the opposite direction. The latter depends on the design peculiarities of a particular gear hobber.

The positive direction of the feed motion $\mathbf{F}_{ag}$ corresponds to the positive direction of the rotation vector $\boldsymbol{\omega}_g$ of

**FIGURE 15.148** Milling of a worm gear on a numerically controlled machine.

**FIGURE 15.149** Milling a worm with a disk-type mill cutter.

**FIGURE 15.150** Grinding a worm with a disk-type grinding wheel.

the work gear. The method of hobbing gears that features the positive direction of the feed motion $\mathbf{F}_{ag}$ is customarily referred to as *climb hobbing.* Otherwise, when the direction of the feed motion $\mathbf{F}_{ag}$ is negative, the method of gear hobbing is often referred to as *conventional hobbing.*

The two possible methods of combining the direction of rotation $\omega_{ct}$ of the hob with the direction of the hob saddle feed $\mathbf{F}_{ag}$ are illustrated in Figure 15.155. In the first, the two motions $\boldsymbol{\omega}_{ct}$ and $\mathbf{F}_{ag}$ are in the same direction, and in the second, the direction of feed $\mathbf{F}_{ag}$ is opposite to the rotation of the hob $\boldsymbol{\omega}_{ct}$. The shapes of the uncut chip produced by the two methods are also schematically shown in Figure 15.155.

It has been found that the climb method of cutting has the beneficial effect of extending the life of a hob between

**FIGURE 15.151** Grinding a worm with a face grinding wheel.

**FIGURE 15.152** Hob for cutting a worm gear that is engaged in mesh with double enveloping (hourglass) worm.

**FIGURE 15.153** Multiple-start hob for cutting a worm gear that is engaged in mesh with double enveloping (hourglass) worm.

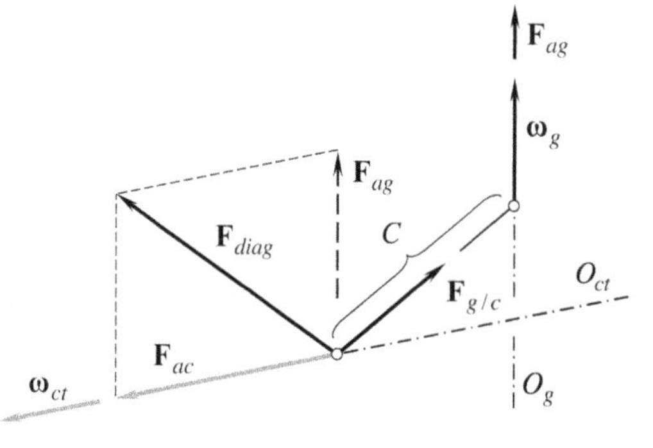


**FIGURE 15.154** Feed motions in a gear hobbing process.

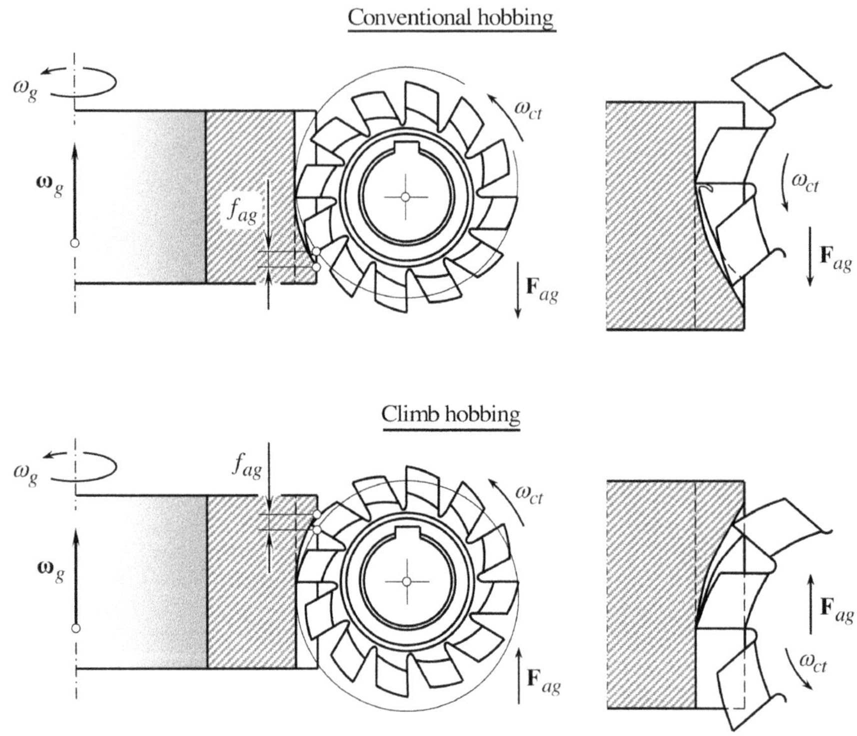


**FIGURE 15.155** Difference between climb gear hobbing ($\mathbf{F}_{ag} > 0$) and conventional gear hobbing ($\mathbf{F}_{ag} < 0$) processes.

grinds. This probably results from the hob teeth taking a maximum bite $f_{ag}$ at the commencement of cut, which gradually diminishes in thickness as the cut proceeds, and there is consequently less rubbing at the tip. In the conventional method of cutting, however, the cutter does not cut cleanly until the pressure between the cutter and the work is sufficient to bite into the metal. The cutter tends to rub before actually making chips.

Another advantage incidental to climb hobbing is that when cutting a double-helical gear and driving on the same flanks of the table worm wheel, it is possible to feed the hob downward on both helices. This results in a better surface finish on the teeth because the swarf can be washed away, and it does not fall back onto the work to produce scratches; in one direction climb cutting is necessary. If possible, it is an advantage to rough-hob precision gears on machines other than those used for finishing and to retain the highest-precision machines for finish hobbing only.

When helical gears are being hobbed, the relative motions of the hob and work are not quite the same as those applied to cutting spur gears because it is necessary to compensate for the advancement or regression which the hob would have moved owing to the helix angle, had it been meshing with a stationary gear.

When cutting a right-hand helix, it is necessary to use a right-hand hob and similarly, a left-hand hob should be used to cut a left-hand helix. This maintains a positive drive with the direction of loading on the master table always driving gear teeth in the same direction. When a left-hand helix is cut by a right-hand hob there is a tendency for the hob to control the work by pulling it away from the table drive. As a result, the tooth profile accuracy, pitch, and surface finish suffer in quality.

In the past it was customary to use single-started hobs for finishing and single- or double-started hobs for roughing. Single-started hobs give the best surface finish and tooth accuracy. In many cases, though, multiple-started hobs are used for both roughing and finishing. When gears are shaved or lapped after hobbing, the surface finish in hobbing is not quite so important. If the gear has enough teeth and they are not an even multiple of the number of starts of the hob, hobs with as many as five or seven starts may be quite profitably used. For the best results, there should be about 30 gear teeth for each hob start. This would mean that a five-start hob would not be used to cut fewer than 151 gear teeth. For commercial work of moderate accuracy, the number of gear teeth may be as low as 15 per hob start.

Worm gears are normally produced by hobbing (see Figure 15.145), and the process is basically the same as that used for cutting spur or helical gears. For that reason, worm gears may be cut in a standard gear hobbing machine. In the case of worm gears, the hob is fed either tangentially past the blank, or radially into the blank (see Figure 15.154).

In the first case, the hob travels along its axis of rotation $O_{ct}$. The axial feed motion of this kind is designated as $\mathbf{F}_{ac}$. The axial feed motion is positive when the vector $\mathbf{F}_{ac}$ is pointed in the same direction as the rotation vector $\boldsymbol{\omega}_{ct}$ of the hob. The feed motion is negative when the vector $\mathbf{F}_{ac}$ is pointed opposite to the direction of the rotation vector $\boldsymbol{\omega}_{ct}$.

In the second case, the hob travels toward the axis of rotation $O_g$ of the work gear. This feed motion is designated as $\mathbf{F}_{g/c}$. It is positive when the vector $\mathbf{F}_{g/c}$ is pointed toward the axis of rotation of the work gear, and negative when the vector $\mathbf{F}_{g/c}$ is pointed outward from the axis of rotation $O_g$ of the work gear.

With the introduction of tangential feed $\mathbf{F}_{ac}$, the process is slightly more complicated, but it has compensatory advantages. Simultaneously with the infeed, the hob is given additional feed in the direction of its axis, or, as an important cutting refinement to improve contact conditions, the axes of the hob and worm gear are at an angle that differs slightly from the axes of the worm and of the worm gear. This process is known as *swing hobbing*. The hob is made a bit larger in diameter than the worm but of the same normal pitch and pressure angle. The worm makes correct contact with the worm wheel, though the helix angles of the worm and hob differ by the amount of the swing.

When the tangential feed $\mathbf{F}_{ac}$ is applied, the hob should preferably move in a direction opposite to the rotation of the work so that backlash is taken up and there is no tendency for the hob to pull the wheel round. The rotary motions of the hob and the work gear are geared to the hob traversing motion to impart to the hob one more revolution than would be required by the ratio. This must operate at the same time that the hob is traversed one lead, to compensate for axial movement of the hob equal to the lead.

When the number of gashes in the hob is fewer than six or seven, the tangential feed $\mathbf{F}_{ac}$ should be applied in order to impart a satisfactory finish to the worm/wheel tooth profiles. This principle also applies when a fly hob is used to cut a worm wheel.

An important point to watch for when using radial infeed $\mathbf{F}_{g/c}$ for cutting worm gears by means of multiple-start hobs is to ensure that the number of starts and gashes in the hob do not contain a common factor greater than unit. When there is no common factor between the number of starts and gashes, all the teeth are effective in cutting the profile, which is then a much smoother curve.

A sharp hob is essential for cutting worm-wheel teeth to the required profile. Too often a poor tooth form is obtained because of the hob has been rubbing instead of cutting during the finishing pass. This defect can be rectified by using a really sharp hob for the final cut, and the practice has grown of employing separate hobs for roughing and finishing, particularly in quantity production. The fully relieved rounding hob is used to remove the bulk of material, leaving about 0.005 in. (0.12 mm) per flank for the finishing hob, often a serrated type made from case-hardened steel. When tungsten carbide is used for the hob material, a considerable increase in hob life between grindings is obtained, amounting to about 15-fold, and as a result, the quality of the work is improved.

In some cases, a gear hobber can perform two simultaneous feed motions, the axial feed $\mathbf{F}_{ag}$ along the axis of rotation of the work gear and the axial feed motion $\mathbf{F}_{ac}$ along the axis of rotation of the hob, as shown in Figure 15.156. Combination of the axial feed motions $\mathbf{F}_{ag}$ and $\mathbf{F}_{ac}$ results in the so-called diagonal hobbing of gears. Diagonal hobbing

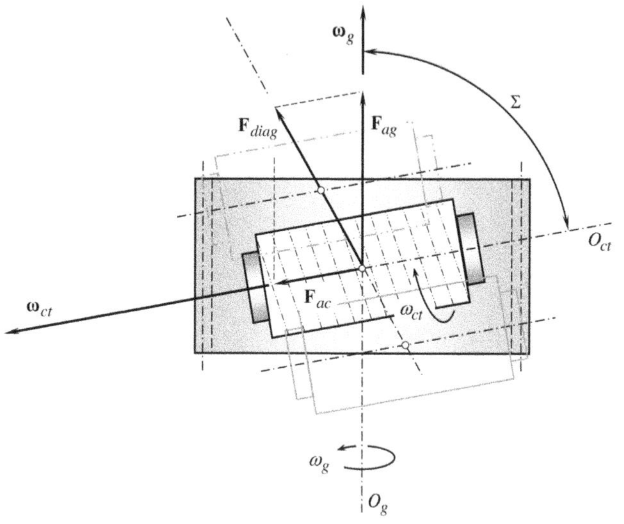


**FIGURE 15.156** Combining two feed motions $\mathbf{F}_{ag}$ and $\mathbf{F}_{ac}$ that are simultaneously performed results in the diagonal method of a gear hobbing process.

of gears can be specified by the resultant feed motion that is designated as $\mathbf{F}_{diag}$. The vector $\mathbf{F}_{diag}$ is equal

$$\mathbf{F}_{diag} = \mathbf{F}_{ag} + \mathbf{F}_{ac} \tag{15.221}$$

The tool life of hobs in a diagonal hobbing process is usually longer. In addition, the use of diagonal hobbing makes it possible to hob more accurate gears.

When hobbing gears, an extensive supply of coolant and/or lubricants is required (see Figure 15.157). Various designs of the nozzles are used to supply the coolant/lubricant in the gear hobbing process (see Figure 15.158).

### 15.8.4 Cycles of Gear Hobbing Operations

Many types of gear hobbing cycles have been developed for use in practice.

In the simplest case of hobbing a spur and/or a helical gear [see Figure 15.159(a)], during the cutting operation the hob is carried on a spindle, the axis of which is oriented at such an angle to the work gear axis that its thread meshes with the teeth of the gear being cut. The work gear and the gear hob are then caused to rotate with uniform motion and with relative rotational speeds in an inverse ratio to the number of teeth required in the work and the starts in the hob.

The total travel distance of the hob is composed of an approach distance $E$ of the face width of the gear $F_g$ and of the idle distance $U$. The clearance $E$ at the beginning of the cut is needed for run-in for the hob into the stock to be cut. The clearance $U$ is required to run out the hob at the end of the cut.

In the method schematically illustrated in Figure 15.159(a), the hob is set up at the nominal center distance $C$ from the work gear. Under such a scenario, the hob approach distance $E$ is greater than the idle distance $U$ (that is, the inequality

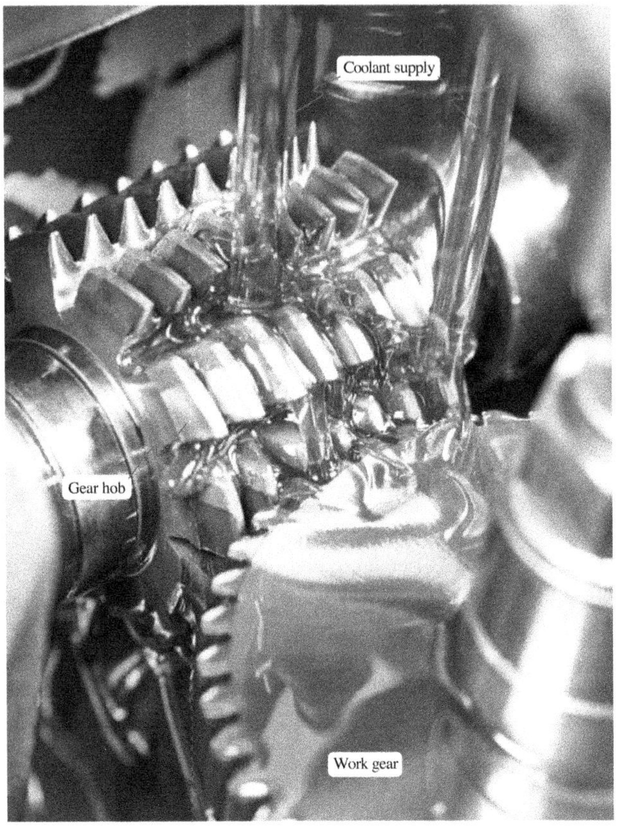


**FIGURE 15.157** Coolant supply when hobbing a gear.

$E > U$ is observed). The axial feed motion $\mathbf{F}_{ag}$ serves two purposes, (a) as the infeed motion, and (b) as the feed motion when the hob is generating tooth flanks of the gear.

To increase the productivity rate of a gear hobbing operation, it is possible to reduce the hob approach distance $E$ and to make it equal to the hob idle distance $U$ [see Figure 15.159(b)]. In order to make the equality $E = U$ possible, two feed motions are applied. The feed motion $\mathbf{F}_{g/c}$ is applied to infeed the hob into the stock to be cut, and the feed motion $\mathbf{F}_{ag}$ is applied for generating tooth flanks of the gear.

Precision gears and coarse pitch gears are often cut in two paths. When a gear is cut in two paths, the first and the second paths could be either a combination of the cycle shown in Figure 15.159(a,b), or it can be a doubled cycle that is depicted in Figure 15.159(b). Figure 15.159(c) illustrates the first case, while the second case is illustrated in Figure 15.159(d).

The axial feed $\mathbf{F}_{ag}$ can be combined with the continuous feed $\mathbf{F}_{g/c}$ in the direction of the center line ℄ of the axes of rotation $O_g$ and $O_{ct}$. The resultant feed motion $\mathbf{F}_\Sigma$ in this case is equal to vector summa

$$\mathbf{F}_\Sigma = \mathbf{F}_{ag} + \mathbf{F}_{g/c} \qquad (15.222)$$

The combined feed motion $\mathbf{F}_\Sigma$ makes possible the hobbing of the so-called beveloids that is depicted in Figure 15.159(e). For hobbing beveloid gears, hobbers of special design are required.

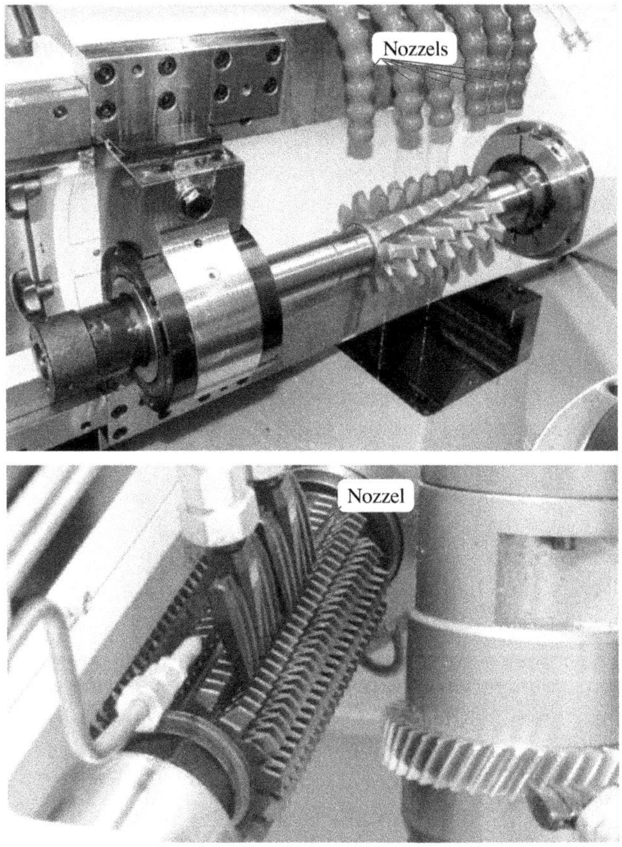


**FIGURE 15.158** Various designs of nozzles are used to supply the coolant/lubricant in the gear hobbing process.

There is one more advantage to hobbing gears in two separate paths. When roughing a gear, the productivity of the material removal process is a critical issue. The accuracy of the machining is an issue of secondary importance in the roughing process. In order to increase the productivity rate of the roughing of the work gear, it is recommended [40] to set the crossed-axis angle $\Sigma_r$ a bit bigger compared to its nominal value $\Sigma$. It is proven[44] that a significant increase in the productivity rate of roughing the gears can be achieved when the crossed-axes angle on the roughing path $\Sigma_r$ is set up in the range of

$$\Sigma_r = \Sigma + (0.5°...3.0°) \qquad (15.223)$$

On the finishing path, the crossed-axes angle is equal to its nominal value of $\Sigma$.

The axial feed $\mathbf{F}_{ag}$ as well as the feed motion $\mathbf{F}_{g/c}$ cannot be of constant value, but can be functions of time $t$, that is, $\mathbf{F}_{ag}(t)$ and/or $\mathbf{F}_{g/c}(t)$. In such a scenario, the resultant feed motion $\mathbf{F}_\Sigma$ also depends on time [$\mathbf{F}_\Sigma(t)$]:

$$\mathbf{F}_\Sigma(t) = \mathbf{F}_{ag}(t) + \mathbf{F}_{g/c}(t) \qquad (15.224)$$

[44] SU Pat. No. 1222442, *A Method of Multi-Paths Hobbing of Spur and Helical Gears*/S.N. Medveditskov *at al*, Int. Cl. B23f 5/22, Filed: July 26, 1984.

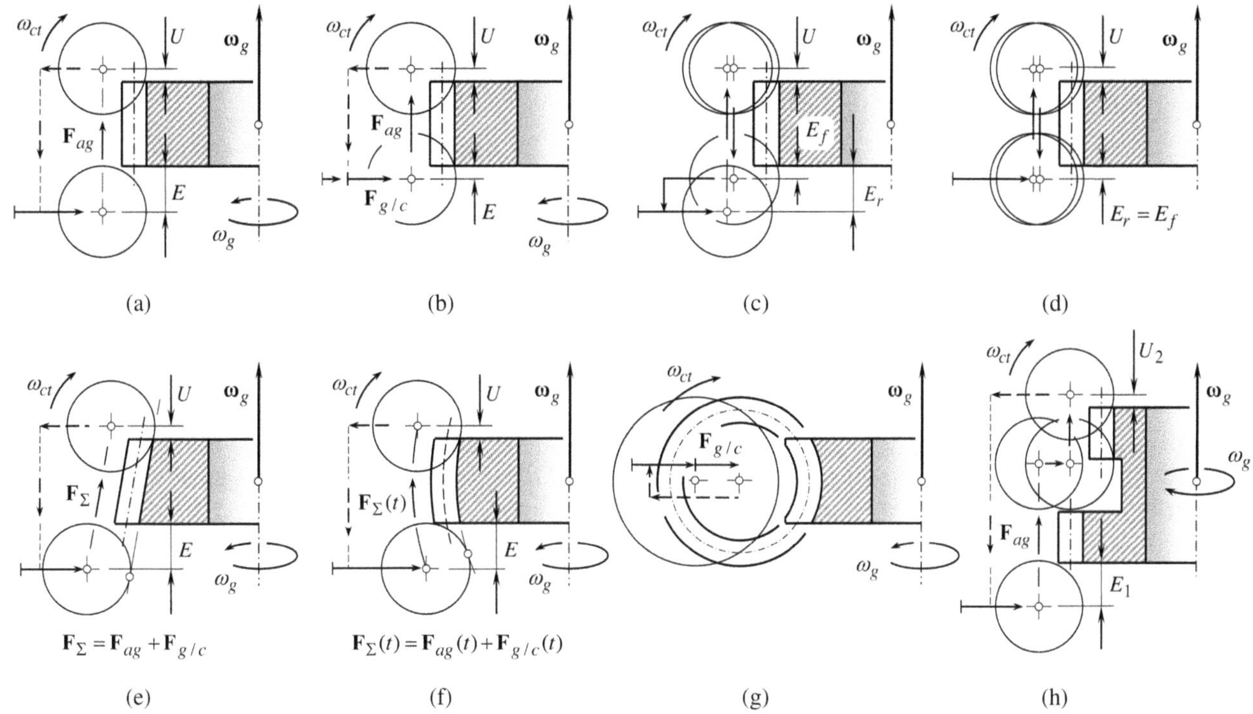


**FIGURE 15.159** Examples of cycles of the gear hobbing process. The parts (a) to (h) are discussed in detail within the text body of the chapter.

If the feed motions $\mathbf{F}_{ag}$ and $\mathbf{F}_{g/c}$ are properly timed with one another, this makes it possible to hob gears featuring a toroidal pitch surface. The possibility of hobbing a gear with a toroidal pitch surface is schematically illustrated in Figure 15.159(f).

A worm gear can be completely hobbed either when the feed motion $\mathbf{F}_{g/c}$ is applied [see Figure 15.159(g)], or axial feed $\mathbf{F}_{ac}$ is used instead (not shown in Figure 15.159).

In designing gears to be hobbed, especially in designing cluster gears, a number of things must be considered. Because a hob needs clearance to run in at the beginning of the cut and clearance to run out at the end of the cut, if the gear teeth come too close to a shoulder or other obstruction, it may be impossible to cut the part by hobbing [see Figure 15.159(h)]. If the gear is double-helical, a gap must be left between the helices for hob runout.

The optimal cycle of hobbing of a particular gear can be developed by varying values of the feed motions $\mathbf{F}_{ag}$, $\mathbf{F}_{ac}$, and $\mathbf{F}_{g/c}$. In addition, an appropriate variation of the crossed-axes angle can be implemented when the gear is hobbed in two or more paths.

### 15.8.5 Minimum Hob Travel Distance

The cutter's total travel distance is an important parameter in the gear hobbing process. The cutter's total travel distance $L = AD$ (see Figure 15.160) is equal to the sum of the hob approach distance $E$, of the face width of the gear to be machined $F_g$, and of the hob's idle distance (or hob overrun) $U$:

$$L = E + F_g + U \qquad (15.225)$$

The hob approach $E = AB$ is the distance that the hob should travel parallel to the gear axis, from the point of initial contact between the hob and the work to the point where the center distance reaches the first gear face (at the point $B$).

The gear face width is indicated on the part blueprint. When more than one blank per load are to be hobbed, the total face width of all the blanks must be considered.

The hob's idle distance $U = CD$ is the linear hob carriage travel beyond the second gear face $C$ that is required to complete generation of the teeth.

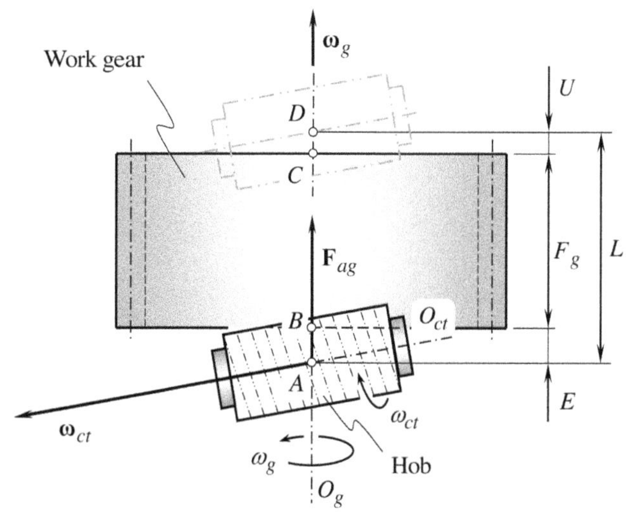


**FIGURE 15.160** Hob total travel distance in climb hobbing of a cylindrical gear.

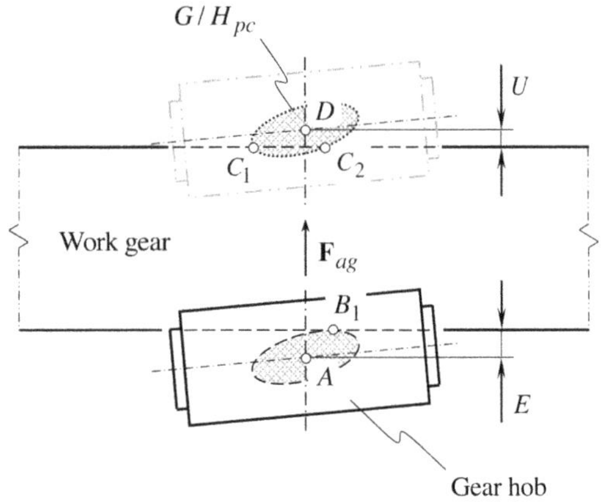


**FIGURE 15.161** Comparison of the hob approach distance $E$ and the hob idle distance $U$ in climb hobbing of a cylindrical gear. (From Radzevich, S.P., *ASME Journal of Mechanical Design*, 124, 772–786, 2002. With permission.)

There are two substantial reasons that make the hob total travel distance important to the development of an efficient gear hobbing process. They are (a) the hobbing time and (b) the axial size of a hobbed cluster gear. The smaller the approach distance and the hob overrun, the larger the productivity rate of the gear hobbing operation. It is necessary to calculate an exact value of the required total hob travel distance.

#### 15.8.5.1 Hobbing Time as a Function of the Hob Total Travel Distance

The actual calculation connected with finding the time that is required to hob a given part is relatively simple. The expression for hobbing time $T$ is [24, 211]

$$t = \frac{N_g \cdot L}{N_{ct} \cdot \omega_{ct} \cdot F_{ag}}, \text{min} \tag{15.226}$$

where:

- $N_g$ is the tooth number of the work gear.
- $L$ is the total hob travel distance, mm.
- $N_{ct}$ is the number of starts (threads) on the hob.
- $\omega_{ct}$ is the rotation of the hob, min$^{-1}$.
- $F_{ag}$ is the axial feed, mm per revolution of the work gear.

Equation (15.226) returns the time $t$ required for cutting the part. It can be easily shown that hobbing time $t$ can be represented as a function $t = t(E, F_g, U)$.

#### 15.8.5.2 Impact of the Hob Idle Distance on the Minimal Neck Width of the Hobbed Cluster Gear

Two distances, $E$ and $U$ [see Eq. (15.224)], make the total travel distance $L$ greater compared to the gear nominal face width $F_g$.

Referring to Figure 15.161, the length of the hob approach distance $E$ is specified as the height of the highest point $B_1$ on the penetration curve $G/H_{pc}$ above the horizontal plane through the center line—through the crossing point of the axis of rotation of the work gear $O_g$ and the axis of rotation of the hob $O_{ct}$.

When the hob is away from the upper face of the gear, the penetration curve $G/H_{pc}$ intersects the upper face at points $C_1$ and $C_2$. Because of the intersection, it is evident that in climb hobbing of cylindrical gears, the hob approach distance is greater than the hob idle distance ($E > U$). In specific cases of gear hobbing, the equality $E = U$ could be observed. However, under no circumstances does the hob idle distance exceed the hob approach distance.

In order to make the neck width of the cluster gear the shortest possible, hobbing of the cluster gear usually starts from the open end (where there is enough space for the hob approach distance) and finishes at the opposite end (where there is lack of space for the hob). This is because the minimum neck width of the cluster gear depends on the hob idle distance $U$. The greater the idle distance $U$ required, the wider the neck width $l_{min}$ of the cluster gear and vice versa (see Figure 15.162).

The hob idle distance must be the shortest possible to cut the time of gear hobbing and to minimize the neck width of the cluster gear. At that same time, the length of the hob's idle distance $U$ must be sufficient for the completion of the generation of the gear tooth flanks. The lack of length $\Delta U$ of the hob idle distance $U$ causes the deviation $f_{max}$ of the actual tooth flank $\mathscr{G}^{(a)}$ from the desirable shape $\mathscr{G}$ (see Figure 15.163).

It is important to point out here that the length of the distance $\Delta U$ depends only on the relative location of the gear

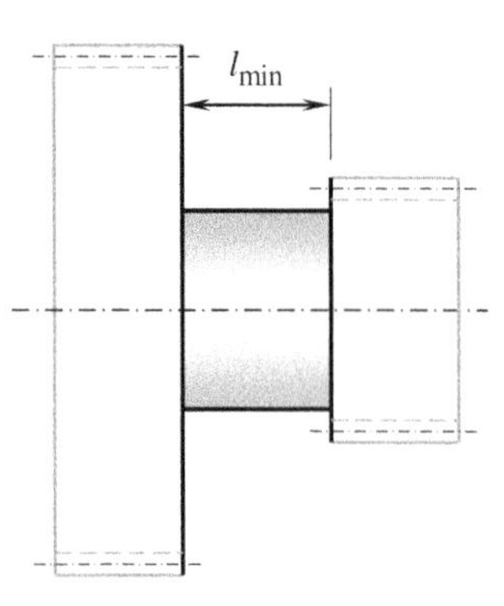


**FIGURE 15.162** Minimal neck width $l_{min}$ in a cluster gear.

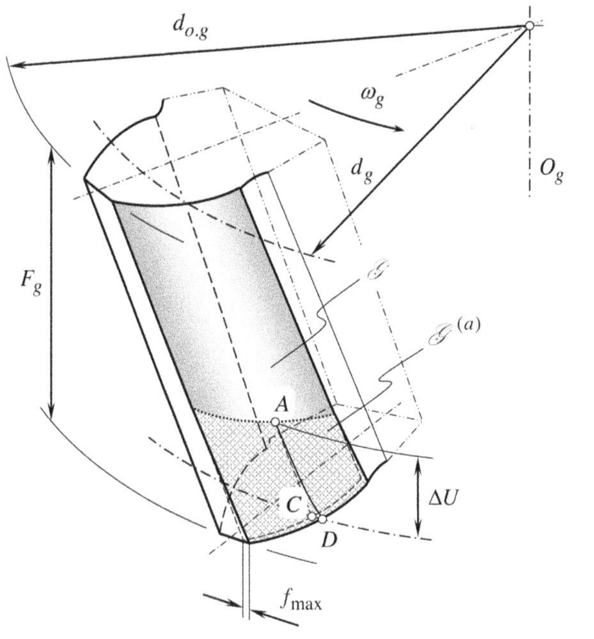


**FIGURE 15.163** Lack of the hob idle distance $\Delta U$ causes deviation $f_{max}$ of the actual gear tooth flank $\mathscr{G}^{(a)}$ from the desired gear tooth surface $\mathscr{G}$. (From Radzevich, S.P., *Gear Technology*, 20(4), 44–50, 2003. With permission.)

hob in the axial direction of the gear to be machined. It does not depend on the hob outer diameter, while the maximal deviation $f_{max}$ of the actual gear tooth flank $\mathscr{G}^{(a)}$ from the nominal gear tooth surface $\mathscr{G}$ does.

#### 15.8.5.3 Selection of a Proper Value of the Setting Angle of the Hob

The selection of the setting angle of the hob is of critical importance when hobbing a cluster gear. The importance of proper selection of the setting angle of the hob $\zeta_c$ is convenient to illustrate with the case of the hobbing of a spur gear (see Figure 15.164).

A hob of conventional design features a positive ($\zeta_{ct} > 0°$) setting angle as schematically shown in Figure 15.164(a). In this specific case, the axis of rotation of the hob is designated as ${}^{+}O_{ct}$. It is customary to assign a value of the setting angle $\zeta_{ct}$ of the hob equal to (or at least approximately equal to) the pitch angle $\psi_{ct}$ of the gear hob. In this case, the crossed-axes angle is obtuse ($\Sigma = 90° + \zeta_{ct}$ is greater than 90°, that is, $\Sigma > 90°$).

Gear hobs with a setting angle of zero, $\zeta_{ct} = 0°$, have been known for a long while. In Figure 15.164(b), the axis of rotation of the hob is designated as ${}^{0}O_{ct}$. Under such a scenario, the axis of rotation of the work gear $O_g$ and the cutter axis of rotation ${}^{0}O_{ct}$ cross at a right angle ($\Sigma = 90°$).

Ultimately, that same spur gear can be machined by a hob that features a negative setting angle ($\zeta_{ct} < 0°$) as schematically shown in Figure 15.164(e). In this particular case, the axis of rotation of the gear hob is designated in Figure 15.164(c) as ${}^{-}O_{ct}$. In this case, the crossed-axes angle is acute ($\Sigma = 90° - \zeta_{ct}$ is smaller than 90°, that is, $\Sigma < 90°$).

A different minimum idle distance is required when hobbing a gear by hobs featuring a different setting angle. For climb hobbing of a spur gear, an impact of the setting angle $\zeta_{ct}$ of the hob onto the required hob idle distance $U_{min}$ is illustrated in Figure 15.165.

The simplest possible configuration of the work gear and the hob is depicted in Figure 15.165(a). In this reduced case of gear hobbing, the hob setting angle is reduced to zero ($\zeta_{ct} = 0°$), the crossed-axes angle is equal to $\Sigma = 90°$, and the hob idle distance is equal to half the base diameter of the hob ($U = 0.5\,d_{b.ct}$).

Generating the gear tooth flanks with the hob begins at the instant of time when the straight generating line $E_1^r$ of the recessing flank of the hob thread profile in its horizontal position reaches the lower face of the work gear [see Figure 15.165(b)]. When the hob travel distance exceeds the length the of $l_1 = d_{b.c}$, the opposite flank of the hob tooth profile begins generating the opposite side of the gear tooth profile. From this moment, both flanks of the gear teeth are generating simultaneously.

Generating the gear tooth flanks is accomplished at the instant of time when the straight generating lines $E_1^a$ and $E_1^r$ of the opposite (the approaching) flank of the hob thread

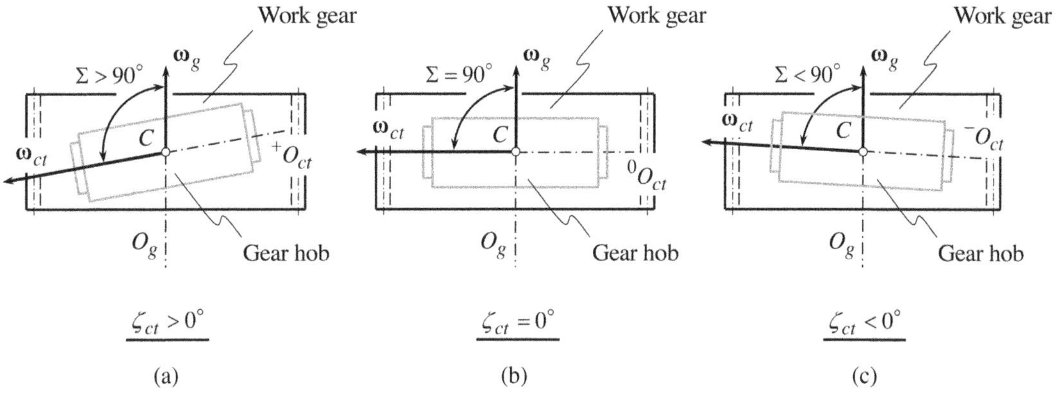


**FIGURE 15.164** Configuration of a work gear and the gear hob when hobbing a spur gear by the hob with (a) a positive $\zeta_{ct} > 0°$, (b) a zero $\zeta_{ct} = 0°$, and (c) a negative $\zeta_{ct} < 0°$ hob setting angle $\zeta_{ct}$. (From Radzevich, S.P., *ASME Journal of Mechanical Design*, 124, 772–786, 2002. With permission.)

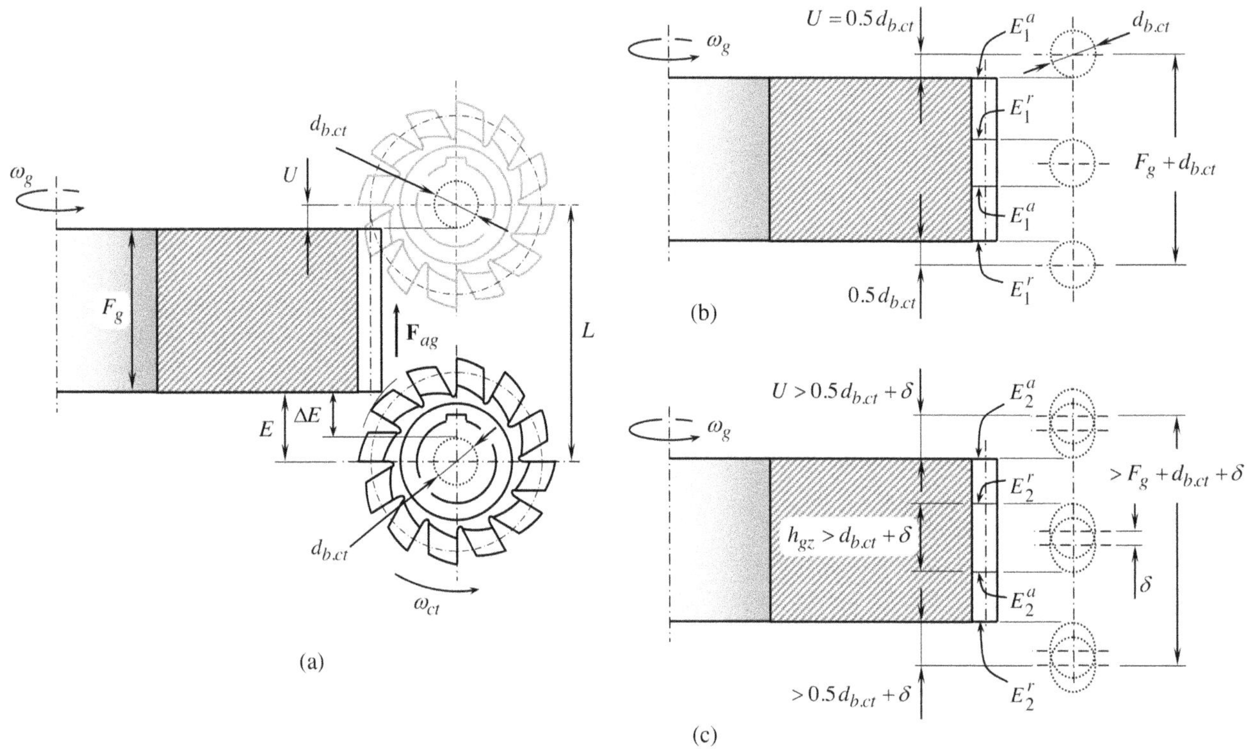


**FIGURE 15.165** Impact of the setting angle $\zeta_{ct}$ of a hob onto the hob idle distance $U$ in climb hobbing of a spur involute gear, (a) the total length of the gear hob travel distance, L, (b) determination of the hob idle distance, U, – case of a zero setting angle of the hob, and (c) determination of the hob idle distance, U, – case of non-zero setting angle of the hob. (From Radzevich, S.P., *ASME Journal of Mechanical Design*, 124, 772–786, 2002. With permission.)

profile in its horizontal position reaches the upper face of the work gear [see Figure 15.165(b)]. In the case under consideration, the hob idle distance $U$ is equal to half the base diameter of the hob ($U = 0.5d_{b.ct}$) and is the shortest possible. To accomplish generation of the gear tooth flanks, the cutter's total travel distance must be equal to or exceed the length of $L \geq F_g + d_{b.ct}$. It is of critical importance to stress here that the shortest possible hob idle distance cannot be less than $0.5d_{b.ct}$, that is, the inequality $U_{\min} \geq 0.5d_{b.ct}$ is always observed.

The above discussion can be extended to the case when the setting angle of the hob is not equal to zero.

Referring to Figure 15.165(c), consider the configuration when the setting angle of a hob is not zero, and, for example, this angle has a positive value. In this particular case of gear hobbing, the straight generating lines $E_2^a$ and $E_2^r$, which are similar to the straight generating lines $E_1^a$ and $E_1^r$ in Figure 15.165(b), are additionally shifted through a certain distance $\delta$ relative to one another along the axis of rotation $O_g$ of the work gear. The actual value of the shift $\delta$ depends on (a) the actual length of the generating zone of the gear tooth profile, and (b) the setting angle $\zeta_{ct}$ of the hob. This question is considered below in greater detail. The distance $\delta$ increases the length of the hob idle distance $U$ and the total hob travel distance, which is needed to generate the gear tooth profile.

### 15.8.5.4 Calculation of the Shortest Allowed Hob Idle Distance

The earlier discussion returns only qualitative understanding of the impact of the setting angle of the hob onto the shortest permissible hob idle distance [see Figure 15.165(c)]. Determination of the minimum permissible hob idle distance in quantities is the next step to be undertaken if we are to solve the problem under consideration. A conceptual solution to the problem under consideration that is accompanied by workable formulae for the calculation of $U_{\min}$ was derived by Prof. S.P. Radzevich[45] as early as 1987.

It is convenient to begin the consideration by determining the hob idle distance in the case of hobbing of a spur gear.

#### *15.8.5.4.1 Shortest Allowed Hob Idle Distance When Hobbing a Spur Gear*

For the development of an analytical solution to the problem under consideration, the use of methods developed in descriptive geometry (DG-based methods) is helpful.

Consider a gear to be machined that is specified by the following set of design parameters: (a) the outer diameter $d_{o.g}$, (b) the pitch diameter $d_g$, (c) the root diameter $d_f$, (d) the tooth addendum $a_g$, (e) the tooth dedendum $b_g$, and (f) the whole height $h_t$ of the tooth as shown in Figure 15.166.

The auxiliary generating rack $\mathscr{R}_T$ is conjugate to the gear being machined. The auxiliary rack $\mathscr{R}_T$ is specified by (a) the tooth addendum $a_{\mathscr{R}} = b_g$, (b) the tooth dedendum $b_{\mathscr{R}} = a_g$, (c) the whole tooth height $h_t$, and (d) the profile angle $\phi_{\mathscr{R}}$ that is equal to the profile angle of the gear tooth on the pitch cylinder (that is, $\phi_{\mathscr{R}} = \phi_g = \phi_n$).

[45] SU Pat. №4278973/08, *A Method of Hobbing Gear*/S.P. Radzevich, Int. Cl. B23f 5/22, Filed on: April 25, 1987.

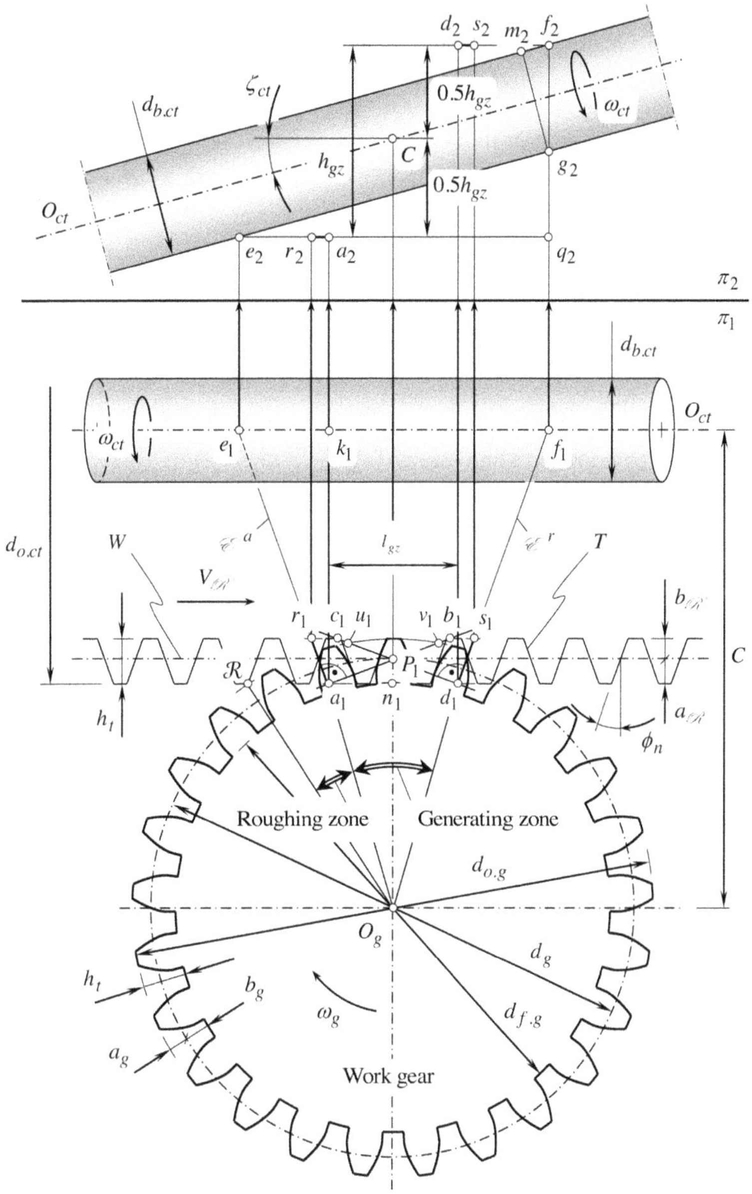


**FIGURE 15.166** Determination of the shortest permissible hob idle distance in hobbing of a spur involute gear. (From Radzevich, S.P., *ASME Journal of Mechanical Design*, 124, 772–786, 2002. With permission.)

The auxiliary generating rack $\mathscr{R}_T$ performs a straight steady motion with a linear velocity $V_{\mathscr{R}}$. The straight motion $V_{\mathscr{R}}$ of the generating rack $\mathscr{R}_T$ is synchronized with rotation $\omega_g$ of the gear.

The axis of rotation $O_{ct}$ of the hob is positioned at a center distance $C$ relative to the axis of rotation $O_g$ of the gear. The axis $O_{ct}$ forms a crossed-axes angle $\Sigma$ with the gear axis $O_g$. The crossed-axes angle $\Sigma$ is equal to $\Sigma = 90° - \zeta_c$. The rotation of the hob $\omega_c$ is synchronized with the rotation of the gear $\omega_g$.

It is customary to assign subscript 1 to projections of all elements (namely, of all points, of all lines, and so forth) onto the horizontal plane of projection $\pi_1$. Subscript 2 is assigned to the projections of the corresponding points, lines, and so forth onto the vertical plane of projection $\pi_2$.

The tooth profile of the gear being machined and the tooth profile of the auxiliary generating rack are conjugate to each other. In their relative motion, the line of action of the approaching side of the teeth profile is designated as $l_{LA}^a$. The line of action of the recessing side of the teeth profile is designated as $l_{LA}^r$. The lines of action $l_{LA}^a$ and $l_{LA}^r$ intersect one another at pitch point $P$. Projections of these elements onto the horizontal plane of projection $\pi_1$ are denoted as $P_1$, $l_{LA}^{(a)} = a_1 b_1$, and $l_{LA}^{(r)} = c_1 d_1$ (see Figure 15.166). Point $a$ is the point of intersection of the approaching portion of the line of action $l_{LA}^a$ with the hob outer diameter $d_{o.ct}$. It corresponds to the beginning of the generation zone. Point $d$ is the point of intersection of the recessing portion of the line of action $l_{LA}^r$ with the hob outer diameter $d_{o.c}$. It corresponds to the end of the generation zone. At that same time, point $a$ indicates the end of the roughing zone. The beginning of the roughing zone is denoted as $\mathcal{R}$. This point is located at the intersection of the outer diameter $d_{o.g}$ of the gear with the outer diameter $d_{o.c}$ of the hob.

At the beginning of the generation of the gear tooth profile, the straight generating line $ae$ of the generating surface of the gear hob (that is, the characteristic line $\mathcal{E}^a$) is in tangency with the hob base cylinder of a diameter $d_{b.c}$ at the point $e$. The characteristic line $\mathcal{E}^a$ remains orthogonal to the corresponding line of action $l_{LA}^a$ throughout the tooth profile generation. At the end of the generating zone, the straight generating line $df$ of the generating surface of the gear hob (that is, the characteristic line $\mathcal{E}^r$) is in tangency with the hob base cylinder at point $f$. The characteristic line $\mathcal{E}^r$ remains perpendicular to the corresponding line of action $l_{LA}^r$ throughout the tooth profile generation.

Figure 15.166 is helpful for a clear understanding of what the generating zone is in nature.

The height $h_{gz}$ of the generating zone is equal to

$$h_{gz} = f_2q_2 = f_2g_2 + g_2q_2 \tag{15.227}$$

Here $f_2g_2 = d_{b.ct} / \cos\zeta_{ct}$, and, therefore

$$h_{gz} = \frac{d_{b.ct}}{\cos\zeta_{ct}} + g_2q_2 \tag{15.228}$$

The length of the straight-line segment $g_2q_2$ can be determined from the $\Delta e_2g_2q_2$. In this triangle, $e_2q_2 = e_1f_1$, and therefore:

$$g_2q_2 = e_2q_2 \tan\zeta_{ct} = e_1f_1 \tan\zeta_{ct} \tag{15.229}$$

Ultimately, the length of the straight-line segment $e_1f_1$ can be represented as the summa $e_1f_1 = l_{gz} + 2e_1k_1$.

The length of the generating zone can be calculated from the simplified formula

$$l_{gz} = 2a_1n_1 = 2a_r / \tan\phi_n \tag{15.230}$$

that is evident from Figure 15.166. This simplified formula has been derived under the assumption that the gear tooth number is equal to infinity ($N_g = \infty$). The simplified formula is in wide use for performing approximate calculation. For more accurate calculation, the formula [119]

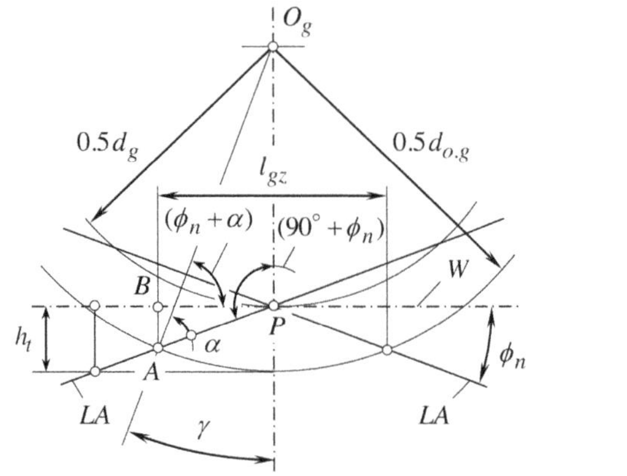


**FIGURE 15.167** Length of the generating zone in hobbing of an involute gear. (From Radzevich, S.P., *ASME Journal of Mechanical Design*, 124, 772–786, 2002. With permission.)

$$\begin{aligned} l_{gz} &= d_{o.g}\cos(\phi_n + \alpha) \\ &= \left(\sqrt{d_{o.g}^2 - d_g^2\cos^2\phi_n} - d_{o.g}\sin\phi_n\right)\cos\phi_n \end{aligned} \tag{15.231}$$

should be used instead. Figure 15.167 was used for the derivation of Eq. (15.231) [119].

The length of the straight-line segment $e_1k_1$ is equal to

$$e_1k_1 = 0.5d_{o.ct}\tan\phi_n \text{ (see } \Delta a_1e_1k_1) \tag{15.232}$$

Ultimately this yields the formula

$$\begin{aligned} U &= \frac{1}{2}h_{gz} = \frac{1}{2}\left(l_{gz} + d_{o.ct}\tan\phi_n\right) \\ &= \frac{1}{2}\left[\left(\sqrt{d_{o.g}^2 - d_g^2\cos^2\phi_n} - d_g\sin\phi_n\right)\cos\phi_n + d_{o.ct}\tan\phi_n\right] \end{aligned} \tag{15.233}$$

for the calculation of the hob idle distance when machining a spur gear.

In a specific case of hobbing of gears with a standard tooth profile (that is, when the equalities $a_g = m$ and $b_g = 1.25m$ are valid), Eq. (15.233) casts into:

$$U = \frac{1}{2}\left[\frac{m N_{ct}\cos\phi_n}{\sqrt{1 - \cos^2\phi_n\cos^2\zeta_{ct}}\,\cos\zeta_{ct}} + \left(d_{o.ct}\tan\phi_n + 2.25m\cot\phi_n\right)\tan\zeta_{ct}\right] \tag{15.234}$$

Other types of representation of Eq. (15.222) are known as well.

### 15.8.5.4.2 *Shortest Allowed Hob Idle Distance When Hobbing a Helical Gear*

The crossed-axes angle $\Sigma$ is equal to $\Sigma = 90° - \zeta_{ct} - \psi_g$ when hobbing a helical gear (see Figure 15.168). Because of that, the height of the generating zone $h_{gz}$ exceeds the length of the straight-line segment $f_4q_4$ that is similar to the straight-line segment $f_3q_3$ shown in Figure 15.166. The length of the straight-line segment $q_4u_4$ can be calculated from the formula (see Figure 15.168)

$$q_4u_4 = e_4u_4 \cdot \sin\psi_g = e_1f_1 \cdot \sin\psi_g \tag{15.235}$$

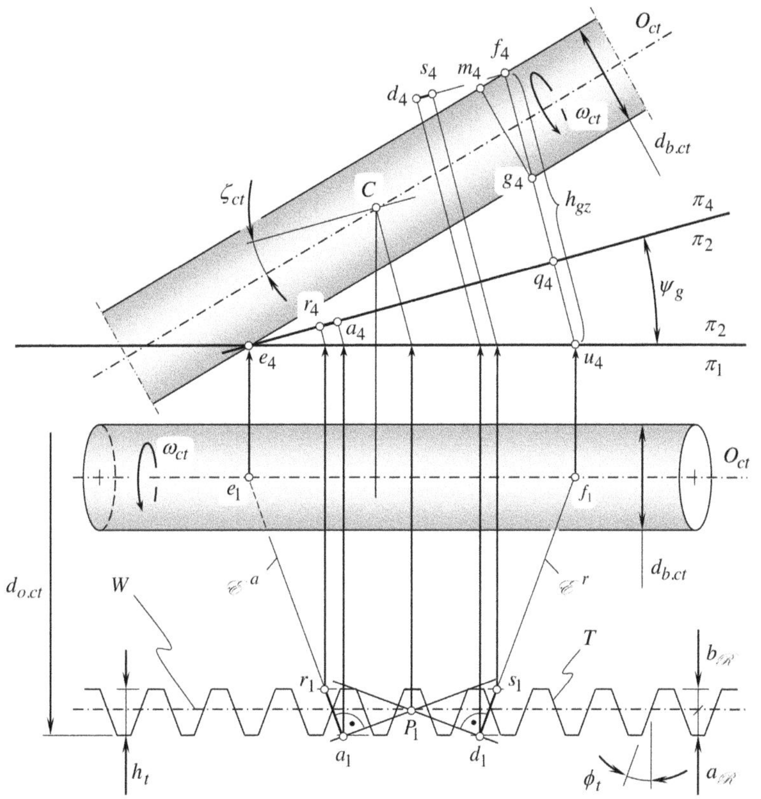


**FIGURE 15.168** Determination of the shortest permissible hob idle distance in hobbing of a helical involute gear. (From Radzevich, S.P., *ASME Journal of Mechanical Design*, 124, 772–786, 2002. With permission.)

Here, the length of the straight-line segment $e_1 f_1$ has been defined earlier.

A formula for the calculation of the hob idle distance when machining a helical gear cast into the form

$$
\begin{aligned}
U &= 0.5 h_{gz} = 0.5[l_{gz} + d_{o.ct}\tan\phi_n + (l_{gz} + d_{o.ct}\tan\phi_n)\sin\psi_g] \\
&= 0.5\left\{\left(\sqrt{d_{o.g}^2 - d_g^2\cos^2\phi_n} - d_g\sin\phi_n\right)\cos\phi_n + d_{o.ct}\tan\phi_n + \left[\left(\sqrt{d_{o.g}^2 - d_g^2\cos^2\phi_n} - d_g\sin\phi_n\right)\cos\phi_n + d_{o.ct}\tan\phi_n\right]\sin\psi_g\right\}
\end{aligned}
\tag{15.236}
$$

When the pitch helix angle $\psi_g$ of the work gear is zero ($\psi_g = 0°$), then Eq. (15.236) reduces to Eq. (15.233).

In the specific case of hobbing of gears with a standard tooth profile (that is, when the equalities $a_g = m$ and $b_g = 1.25m$ are valid), Eq. (15.236) casts into

$$
U = \frac{1}{2}\left[\frac{m N_{ct}\cos\phi_n}{\sqrt{1 - \cos^2\phi_n\cos^2\zeta_{ct}}\,\cos\zeta_{ct}} + (d_{o.ct}\tan\phi_n + 2.25m\cot\phi_n)\tan\zeta_{ct} \pm 1.25m\tan\psi_g\cot\phi_n\right] \tag{15.237}
$$

Other types of representation of the formula in Eq. (15.236) are known as well.

Examination of Eq. (15.233) and Eq. (15.236) reveals that the hob idle distance $U$ does not depend on the outer diameter $d_{o.ct}$ of the hob when the gear is machined by the hob with a zero setting angle $\zeta_{ct} = 0°$. In cases when the setting angle of the gear hob is not equal to zero ($\zeta_{ct} \neq 0°$), the hob idle distance $U$ depends on the hob's outer diameter.

#### 15.8.5.5 Impact of Tolerance on the Shortest Possible Hob Idle Distance

The generation of tooth flanks of both sides of the gear tooth is complete if and only if the hob passes beyond the upper face of the work gear at a distance that is equal to or exceeds the hob idle distance $U$. In cases when this requirement is not fulfilled, and the actual hob's idle distance $U^{act}$ is shorter at a certain distance $\Delta U$ than the nominal hob idle distance $U$ (that is, $U^{act} = U - \Delta U$), the actual profile $AD$ in

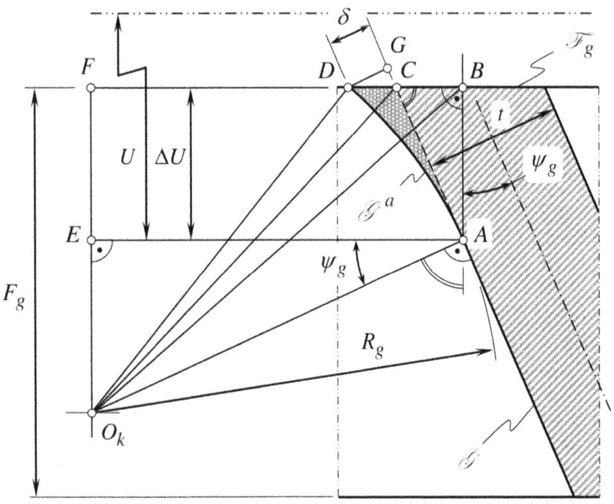


**FIGURE 15.169** Desirable $\mathscr{G}$ and actual $\mathscr{G}^a$ tooth flanks of a hobbed gear. (From Radzevich, S.P., *Gear Technology*, 20(4), 44–50, 2003. With permission.)

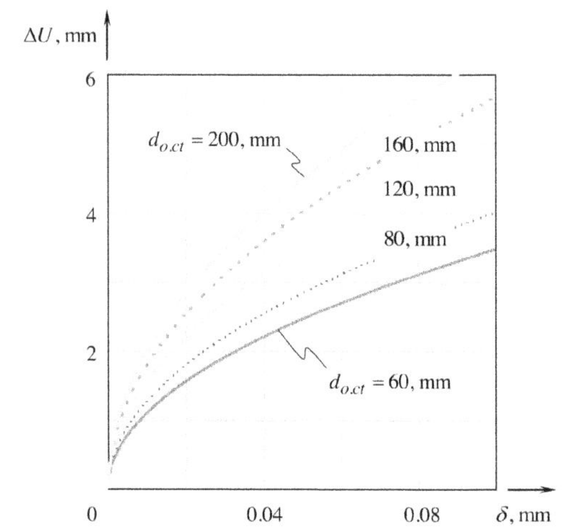


**FIGURE 15.170** Impact of the deviation δ onto the length $\Delta U$ (for cutting gears with hobs of different outer diameters $d_{o.ct}$). (From Radzevich, S.P., *ASME Journal of Mechanical Design*, 124, 772–786, 2002. With permission.)

the longitudinal direction of the gear tooth deviates from the desirable profile $AC$ (see Figure 15.169). The maximum deviation is observed at point $D$, at which the deviation is equal to δ (the maximum deviation is designated as $f_{\max}$ if measured in the transverse plane, and it is equal to $f_{\max} = CD$).

The maximum permissible deviation δ must not exceed the tolerance [δ] for the accuracy of the gear tooth flank.

Because of the tolerance [δ], the hob idle distance can be considerably reduced. Referring to Figure 15.169, the reduction $\Delta U$ of the minimum allowable hob idle distance $U$ can be expressed in terms of the tolerance [δ] [145]

$$\Delta U(\delta) = \frac{-[\delta]\cdot\cos\phi_n\cdot\tan\psi_g + \sqrt{[\delta]^2\cdot\cos^2\phi_n\cdot\tan^2\psi_g - \left[[\delta]^2\cdot\dfrac{\cos^2\phi_n}{\cos^2\psi_g} - 2\cdot R_g\cdot[\delta]\right]}}{\cos\psi_g} \tag{15.238}$$

For engineering applications, a simplified approximated formula

$$\Delta U(\delta) \cong \frac{\sqrt{2\cdot R_g\cdot[\delta]}}{\cos\psi_g} \tag{15.239}$$

for the calculation of the distance $\Delta U(\delta)$ is used.

It is assumed here in Eq. (15.239) that the radius $R_g$ is equal to $R_g \cong R_{1.ct}$, where $R_{1.ct}$ denotes the first principal radius of curvature of generating surface $T$ of the gear hob.

The first principal radius of curvature $R_{1.ct}$ is calculated from the expression [see Eq. (15.129) that is also available from Appendix F]

$$R_{1.ct} = \frac{\sqrt{d_{ct}^2 - d_{b.ct}^2}}{2\sin\psi_{ct}\sin\phi_n}\tan\psi_{b.ct} \tag{15.240}$$

The derivation of Eq. (15.240) can be found in [158].

For the calculation of the base helix angle $\psi_{b.ct}$, Eq. (14.74) and/or Eq. (15.74)

$$\psi_{b.ct} = \tan^{-1}\left(\frac{\sqrt{\sin^2\phi_n + \tan^2\zeta_{ct}}}{\cos\phi_n}\right) \tag{15.241}$$

can be used. The expression for the base helix angle $\psi_{b.ct}$ is also available from Appendix F.

An example of graphical interpretation of the function $\Delta U = \Delta U(\delta)$ is plotted in Figure 15.170. This graph indicates that incorporation of the deviation δ, even of a small value, makes possible a significant reduction $\Delta U$ of the hob idle distance $U$. The larger the outer diameter $d_{o.ct}$ of the hob, the larger the allowable distance $\Delta U$, and vice versa.

As shown by Prof. S.P. Radzevich [161], the neck width reduction up to 20% or so can be achieved by the appropriate calculation of the hob's idle distance $U$. A reduction of the neck width of a cluster gear entails the corresponding reduction in size and weight of a cluster gear as well as the weight reduction of the shafts and the gear housing.

The results of the calculation of the hob's idle distance can be entered into Eq. (15.226) for the calculation of machining time in a gear hobbing operation.

###### *15.8.5.5.1 Calculation of the Shortest Allowable Neck Width $l_{min}$ When the Outer Diameters $d_{o.g}^{(2)} \gg d_{o.g}^{(1)}$*

Consider the calculation of the shortest allowable neck width of a cluster gear that features a considerable difference between the outer diameter $d_{o.g}^1$ of the smaller gear and the outer diameter $d_{o.g}^2$ of the larger gear, as shown in Figure 15.171.

In the case under consideration, the collision of the hob of a certain outer diameter $d_{o.ct}$ and the upper face of the

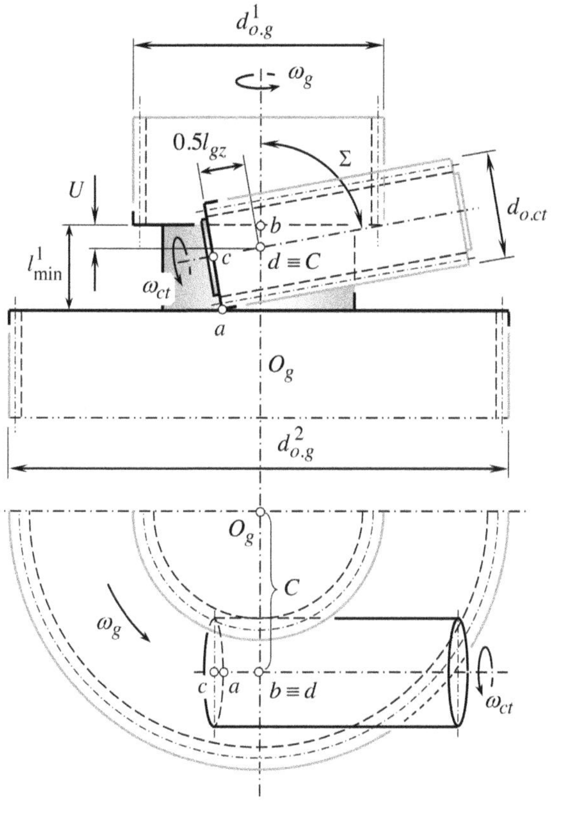


**FIGURE 15.171** Determination of the shortest allowable neck width $l^1_{min}$ of a cluster gear that features a considerable difference between the outer diameter $d^1_{o.g}$ of a smaller gear and the outer diameter $d^2_{o.g}$ of a larger gear. (From Radzevich, S.P., *ASME Journal of Mechanical Design*, 124, 772–786, 2002. With permission.)

larger gear at point $a$ can occur. The minimum allowable neck width $l^1_{min}$ of the cluster gear needs to be sufficient to avoid the collision while hobbing the smaller gear.

The minimum allowable neck width $l^1_{min}$ of the cluster gear can be represented as the summa of the straight-line segment $bd$ and of the projections of the straight-line segments $cd$ and $ac$ onto the axis $O_g$ of the cluster gear. The length of the straight-line segment $bd$ is equal to half of the hob's idle distance $U$. Projection of the straight-line segment $cd$ onto the cluster gear axis is equal to $0.5l_{gz}\cos\Sigma$. In addition, projection of the straight-line segment $ac$ onto the same axis is equal to $0.5d_{o.ct}\sin\Sigma$. The consideration immediately yields the formula

$$l^1_{min} = 0.5(l_{gz}\cos\Sigma + d_{o.c}\sin\Sigma + U) \qquad (15.242)$$

for calculation of the minimal permissible neck width $l^1_{min}$ of the cluster gear.

For a cluster gear of a given design, the shortest hob idle distance $U$ calculated from Eq. (15.233) allows for the shortest minimal allowable neck width $l^1_{min}$. As such, the cluster gear is shorter in its axial direction and its weight is lighter. It can be assembled with a shorter shaft that also is shorter in its axial direction and its weight is smaller. The gear train that is composed of such cluster gears needs a smaller gear housing that is lighter as well.

#### 15.8.5.5.2 Calculation of the Shortest Allowable Neck Width $l_{min}$ When the Outer Diameters $d^2_{o.g}$ and $d^1_{o.g}$ Are of Comparable Value

Consider the calculation of the permissible shortest neck width of a cluster gear that features no considerable difference between the outer diameter $d^1_{o.g}$ of the smaller gear and the outer diameter $d^2_{o.g}$ of the larger gear, as shown in Figure 15.172.

In the case under consideration, the collision of the gear hob with an outer diameter $d_{o.ct}$ and of the upper face of the larger gear at point $e$ can occur. The minimum allowable neck width $l^2_{min}$ of the cluster gear has to be sufficient to avoid the collision when hobbing the smaller gear.

To determine the minimum allowable neck width $l^2_{min}$ of a cluster gear, it is convenient to construct four vectors, namely, the vectors **A**, **B**, **C**, and **D**, shown in Figure 15.172. The vector **A** is located within the lower face of the smaller gear and is pointed out from gear axis $O_g$ parallel to the center line ℄. The vector **B** is pointed out from the endpoint $a$ of the vector **A**. It is parallel to the axis of rotation $O_g$ of the cluster gear. The length of the vector **B** is equal to the hob's idle distance $U$, to be determined. The vector **C** is pointed out from the endpoint $b$ of the vector **B**. It is directed along the hob axis of rotation $O_{ct}$ toward point $c$. The length of the

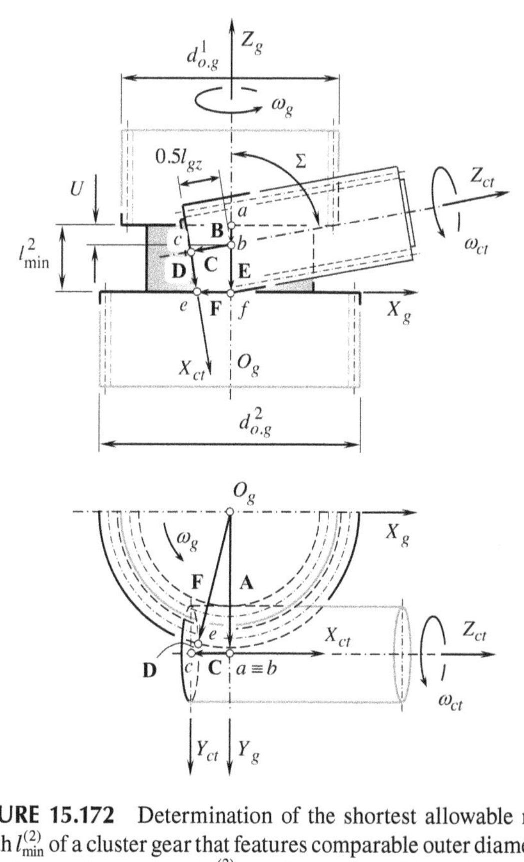


**FIGURE 15.172** Determination of the shortest allowable neck width $l^{(2)}_{min}$ of a cluster gear that features comparable outer diameters $d^{(1)}_{o.g}$ of a smaller gear and $d^{(2)}_{o.g}$ of a larger gear. (From Radzevich, S.P., *ASME Journal of Mechanical Design*, 124, 772–786, 2002. With permission.)

vector **C** is equal to the distance between the center line ℄ and the closest hob face *c*. Finally, the vector **D** is pointed out from the endpoint *c* of the vector **C**. It is directed along the hob face radius toward point *e* of contact of the upper face edge of a larger gear and the edge of closest face of the hob (see Figure 15.172).

Then, the minimal permissible neck width $l^2_{\min}$ of the cluster gear is represented as the projection of summa of the vectors $(\mathbf{A}+\mathbf{B}+\mathbf{C}+\mathbf{D})$ onto the axis of rotation $O_g$ of the cluster gear:

$$l^2_{\min} = \mathrm{Pr}_{Og}(\mathbf{A}+\mathbf{B}+\mathbf{C}+\mathbf{D}) \tag{15.243}$$

In this equation, the equality $|\mathbf{B}| = U$ takes place.

In order to complement the set of the vectors **A**, **B**, **C**, and **D** to a close loop of vectors, two complementary vectors **E** and **F** are constructed. The vectors **E** and **F** make possible the equality

$$(\mathbf{A}+\mathbf{B}+\mathbf{C}+\mathbf{D}) = (\mathbf{E}+\mathbf{F}) \tag{15.244}$$

The latter equality can be solved with respect to $U$ (remember that the equality $|\mathbf{B}| = U$ is valid). In this way the shortest permissible hob idle distance for machining of a given cluster gear with a given hob can be determined [111].

Another approach for the calculation of the minimum allowable neck width $l^2_{\min}$ of a cluster gear is based on the following consideration.

The round edge of the left-hand face of the hob collides with the round edge of the upper face of the larger gear at point *e* (see Figure 15.172). Once coordinates of the point *e* are expressed in terms of the minimum allowable neck width $l^2_{\min}$ of the cluster gear, then the neck width can be calculated as a solution to a set of two equations. These equations analytically describe (a) the round edge of the upper face of the larger gear and (b) the round edge of the left-hand face of the hob. It is important to point out here that both of the equations need to be represented in a common reference system.

Consider a Cartesian coordinate system $X_gY_gZ_g$ associated with the cluster gear as shown in Figure 15.172.

In the reference system $X_gY_gZ_g$, the equation for the round edge of the upper face of the larger gear allows for representation in matrix form

$$\mathbf{r}_{eg}(\varphi_g) = 0.5\,d_{o.g}\begin{bmatrix}\cos\varphi_g\\ \sin\varphi_g\\ 0\\ 1\end{bmatrix} \tag{15.245}$$

where:

- $\mathbf{r}_{eg}$ is the position vector of a point of the round edge of the upper face of the larger gear.
- $\varphi_g$ is the angular parameter of the round edge of the upper face of the larger gear.
- $d_{o.g}$ is the outer diameter of the larger gear.

Similarly, a Cartesian coordinate system $X_{ct}Y_{ct}Z_{ct}$ is associated with the hob as shown in Figure 15.172.

In the reference system $X_{ct}Y_{ct}Z_{ct}$, the round edge of the left-hand face of the hob allows for matrix representation as

$$\mathbf{r}_{ec}(\varphi_{ct}) = 0.5\,d_{o.ct}\begin{bmatrix}\cos\varphi_{ct}\\ \sin\varphi_{ct}\\ 0\\ 1\end{bmatrix} \tag{15.246}$$

where:

- $\mathbf{r}_{ec}$ is the position vector of a point of the round edge of the left-hand face of the hob.
- $\varphi_{ct}$ is the angular parameter of the round edge of the left-hand face of the hob.
- $d_{o.ct}$ is the outer diameter of the hob.

The set of two equations, that is, of Eq. (15.245) and Eq. (15.246), can be solved if both are represented in a common reference system. For example, the equations can be represented in the coordinate system $X_gY_gZ_g$. In such a case, Eq. (15.246) requires being rewritten in the coordinate system $X_gY_gZ_g$.

The transition from the coordinate system $X_{ct}Y_{ct}Z_{ct}$ to the coordinate system $X_gY_gZ_g$ is performed as a set of consequent translations along and rotations about the corresponding axis of coordinates.

*First*, the operator of translation $\mathbf{Tr}(0.5l_{gz}, Z_{ct})$ is used for the analytical description of the translation of the coordinate system $X_{ct}Y_{ct}Z_{ct}$, along the positive direction of the $Z_{ct}$ – axis at the distance $0.5l_{gz}$:

$$\mathbf{Tr}(0.5l_{gz}, Z_{ct}) = \begin{bmatrix}1 & 0 & 0 & 0\\ 0 & 1 & 0 & 0\\ 0 & 0 & 1 & -0.5l_{gz}\\ 0 & 0 & 0 & 1\end{bmatrix} \tag{15.247}$$

The new position of the coordinate system $X_{ct}Y_{ct}Z_{ct}$ is denoted as $X_1Y_1Z_1$.

*Second*, for the analytical description of the rotation of the intermediate reference system $X_1Y_1Z_1$ about the $Y_1$ – axis through the crossed-axes angle $\Sigma$ in the counterclockwise direction to the position, which is denoted as $X_2Y_2Z_2$, the operator of rotation $\mathbf{Rt}(\Sigma, Y_1)$ is used:

$$\mathbf{Rt}(\Sigma, Y_1) = \begin{bmatrix}\cos\Sigma & 0 & \sin\Sigma & 0\\ 0 & 1 & 0 & 0\\ \sin\Sigma & 0 & -\cos\Sigma & 0\\ 0 & 0 & 0 & 1\end{bmatrix} \tag{15.248}$$

*Third*, the operator of translation $\mathbf{Tr}\left\{-[l^2_{\min} - U], Z_2\right\}$ is implemented for the analytical description of the translation of the intermediate coordinate system $X_2Y_2Z_2$ at the distance $-[l_{\min}^2 - U]$ along the negative direction of the axis $Z_2$ to the position in which the axes of the reference system $X_2Y_2Z_2$ align with the corresponding axes of the reference system $X_gY_gZ_g$:

$$\mathbf{Tr}[-(l_{\min}^2 - U), Z_2] = \begin{bmatrix} 1 & 0 & 0 & 0 \\ 0 & 1 & 0 & 0 \\ 0 & 0 & 1 & -(l_{\min}^2 - U) \\ 0 & 0 & 0 & 1 \end{bmatrix} \quad (15.249)$$

The intermediate reference systems $X_1Y_1Z_1$ and $X_2Y_2Z_2$ are not shown in Figure 15.172.

The operator $\mathbf{Rs}(ct \rightarrow g)$ of the resultant coordinate system transformation, that is, of the transition from the coordinate system $X_{ct}Y_{ct}Z_{ct}$ to the coordinate system $X_gY_gZ_g$, is calculated as the product

$$\mathbf{Rs}(ct \rightarrow g) = \mathbf{Tr}[-(l_{\min}^2 - U), Z_2] \cdot \mathbf{Rt}(\Sigma, Y_1) \cdot \mathbf{Tr}(0.5l_{gz}, Z_{ct}) \quad (15.250)$$

The use of the derived operator $\mathbf{Rs}(ct \rightarrow g)$ of the resultant coordinate system transformation makes possible the analytical representation of the position vector of a point $\mathbf{r}_{ec}^g$ of the round edge of the left-hand face of the hob [see Eq. (15.246)] in the coordinate system $X_gY_gZ_g$:

$$\mathbf{r}_{ec}^g(\varphi_{ct}) = \mathbf{Rs}(ct \rightarrow g) \cdot \mathbf{r}_{ec}(\varphi_{ct}) \quad (15.251)$$

(Remember that initially the position vector of a point $\mathbf{r}_{ec}$ of the round edge of the left-hand face of the hob is specified in the reference system $X_{ct}Y_{ct}Z_{ct}$).

Ultimately, Eq. (15.251) for the position vector $\mathbf{r}_{ec}^g$ casts into matrix equation

$$\mathbf{r}_{ec}^g(\varphi_{ct}) = 0.5d_o \cdot \begin{bmatrix} \cos\Sigma\cos\varphi_{ct} \\ \sin\varphi_{ct} \\ \sin\Sigma\cos\varphi_{ct} - 0.5l_{gz} + l_{\min}^2 - U \\ 1 \end{bmatrix} \quad (15.252)$$

The round edge $\mathbf{r}_{eg}$ of the upper face of the larger gear [see Eq. (15.245)] and the round edge $\mathbf{r}_{ec}^g$ of the left-hand face of the hob [see Eq. (15.252)] meet each other at point $e$. At point $e$, the equality

$$\mathbf{r}_{eg}(\varphi_g)\Big|_e = \mathbf{r}_{ec}^g(\varphi_c)\Big|_e \quad (15.253)$$

is observed.

The set of three equations of three unknowns, that is, $\varphi_{ct}$, $\varphi_g$, and $l_{\min}^2$

$$0.5d_{o.ct}\cos\Sigma\cos\varphi_{ct} = 0.5d_{o.g}\cos\varphi_g \quad (15.254)$$

$$0.5d_{o.ct}\sin\varphi_{ct} = 0.5d_{o.g}\sin\varphi_g \quad (15.255)$$

$$0.5d_{o.ct}\sin\Sigma\cos\varphi_{ct} - 0.5l_{gz} + l_{\min}^2 - U = 0 \quad (15.256)$$

is derived from Eq. (15.253).

The solution to the set of Eq. (15.254) through Eq. (15.256) returns the shortest neck width $l_{\min}^2$ of a cluster gear that features only a minor difference between the outer diameter $d_{o.g}^1$ of the smaller gear and the outer diameter $d_{o.g}^2$ of the larger gear.

The calculated shortest neck width allows for the shortest possible axial size of the cluster gear. In this case, the cluster gear becomes smaller and lighter in weight. It can be assembled with a shorter shaft that is also lighter in weight. The gear train composed of such cluster gears requires a smaller gear housing and is lighter as well.

#### 15.8.5.6 Calculation of the Shortest Allowable Approach Distance of the Gear Hob

The solution to the problem of the calculation of the shortest allowable approach distance of the hob is given by E. Buckingham (see pages 167–169 in [14]). The problem to be solved is formulated by E. Buckingham as follows:

> Given the proportions of a helical involute gear and the setting of the hob of given proportions, to determine the minimum distance of the center of the hob from the face of the gear before the hob starts to cut.

Referring to Figure 15.173, when:

- $r_{o.g}$ is the outside radius of the gear.
- $r_{o.ct}$ is the outside radius of the hob.
- $C$ is the center distance between the centers of the hob and the gear.
- $\Sigma$ is the angle of the axis of the hob with the face of the gear.
- $\varepsilon_g$ is the angle on the gear to a point on the intersection line.
- $\varepsilon_{ct}$ is the angle on the hob to a point on the intersection line.

We first determine the projection of the form of the intersection of the outside cylinder of the hob with the outside cylinder of the work gear, on a plane parallel to the axes of the gear and of the hob. We use the intersection of these axes on the reference plane as the origin of the coordinate system, as shown in Figure 15.173.

When $x$ denotes the abscissa of projection of the intersection form, $y$ denotes the ordinate of projection of the intersection form, and we have the following from the conditions shown in Figure 15.173:

$$x = r_{o.g}\sin\varepsilon_g \quad (15.257)$$

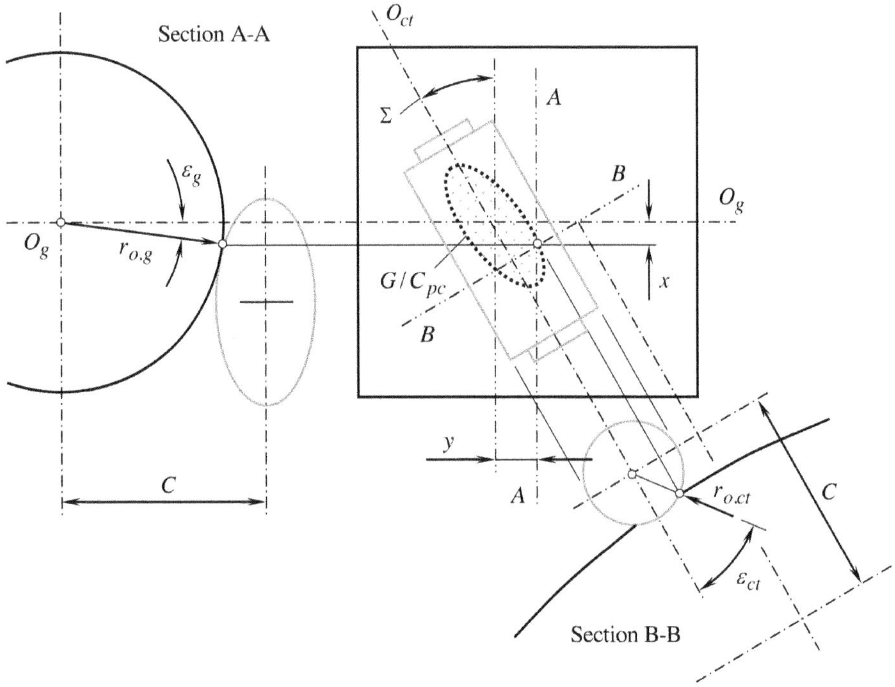


FIGURE 15.173 Calculation of the shortest permissible approach distance $E$ of the hob (after Buckingham, E., *Analytical Mechanics of Gears*, New York, NY: Dover Publications, Inc., 1988, 546 pages).

From

$$\sin\varepsilon_g = \frac{x}{r_{o.g}} \tag{15.258}$$

$$y = x\tan\Sigma + \frac{r_{o.ct}\sin\varepsilon_{ct}}{\cos\Sigma} \tag{15.259}$$

$$C = r_{o.g}\cos\varepsilon_g + r_{o.ct}\cos\varepsilon_{ct} \tag{15.260}$$

Rearranging and combining these equations to solve for values of $y$ in terms of $x$ and the other known values, we obtain

$$\cos\varepsilon_g = \sqrt{1-\sin^2\varepsilon_g} = \frac{\sqrt{r_{o.g}^2 - x^2}}{r_{o.g}} \tag{15.261}$$

$$C = r_{o.ct}\cos\varepsilon_{ct} + \sqrt{r_{o.g}^2 - x^2} \tag{15.262}$$

From

$$\cos\varepsilon_{ct} = \frac{C - \sqrt{r_{o.g}^2 - x^2}}{r_{o.ct}} \tag{15.263}$$

$$\sin\varepsilon_{ct} = \sqrt{1-\cos^2\varepsilon_{ct}} = \frac{\sqrt{r_{o.ct}^2 - \left(C - \sqrt{r_{o.g}^2 - x^2}\right)^2}}{r_{o.ct}} \tag{15.264}$$

Substituting this value of $\sin\varepsilon_{ct}$ into the equation for $y$, we obtain

$$y = \frac{x\tan\Sigma + \sqrt{r_{o.ct}^2 - \left(C - \sqrt{r_{o.g}^2 - x^2}\right)^2}}{\cos\Sigma} \tag{15.265}$$

This equation will give the coordinates of the projection of the intersection $G/C_{pc}$ of two cylinders on a plane parallel to their axes. The maximum value of $y$, that is, $y_m$, will be the minimum distance of the center of the hob from the face of the gear before the hob starts to cut, unless the hob is too short to cover the full intersection.

An examination of Eq. (15.265) shows that the maximum value of $x$, that is, $x_m$, will be reached when

$$C - \sqrt{r_{o.g}^2 - x_m^2} = r_{o.ct} \tag{15.266}$$

because the expression under the radical in Eq. (15.265) becomes minus and imaginary when the value of $x$ is greater than that given by this relationship. Solving this expression for $x_m$, we have

$$x_m = \sqrt{r_{o.g}^2 - (C - r_{o.ct})^2} \tag{15.267}$$

The value of $y$ when $x$ is equal to zero, that is, $y_0$, is given by the following:

$$y_0 = \frac{\sqrt{r_{o.ct}^2 - (C - r_{o.g})^2}}{\cos\Sigma} \tag{15.268}$$

The value of $x$ when the value of $y$ is a maximum, that is, $x_1$, is given very closely by the following equation:

$$x_1 = \frac{x_m^2 \tan\Sigma}{\sqrt{y_0^2 + x_m^2 \tan^2\Sigma}} \tag{15.269}$$

Then

$$y_m = x_1 \tan\Sigma + \frac{\sqrt{r_{o.ct}^2 - \left(C - \sqrt{r_{o.g}^2 - x_1^2}\right)^2}}{\cos\Sigma} \tag{15.270}$$

If the length of the hob is shorter than the distance to $x_1$, then the value of $x_1$ for use in Eq. (15.270) will be given very closely by the following:

When $x_{ct}$ is the extension of the end of the hob,

$$x_1 = x_{ct} \cos\Sigma \tag{15.271}$$

For the calculation of the shortest allowable approach distance of the hob, a corresponding computer code can be developed.

#### 15.8.5.7 Designing a Hob Featuring a Prescribed Value of the Setting Angle

The earlier discussed results of the investigation of the impact of design parameters of a hob onto the minimum allowable hob idle distance make it possible to design a special-purpose hob[46] with a prescribed value of the setting angle. Hobbing of a cluster gear with a minimum allowable neck width is the main purpose of implementing the hob of this design [140].

For hobbing a cluster gear with a minimum permissible neck width, use of the hobs with the setting angle $\zeta_{ct}$ equal to $\zeta_{ct} = -\psi_g$ is preferred. Hobs with the prescribed setting angle $\zeta_{ct}$ are referred to as $\zeta_{ct}$ – *hobs* [140]. In a specific case of hobbing of a spur gear, the use of $\zeta_{ct}$ – hobs featuring a zero setting angle ($\zeta_{ct} = 0°$) is recommended [see Figure 15.164(b)].

For the calculation of the design parameters of a $\zeta_{ct}$ – hob, a desired value of setting angle $\zeta_{ct}$ of the hob has to be known. Furthermore, the design parameters of the hob listed below, namely (a) module $m$, (b) normal profile angle $\phi_n = \phi_{\mathscr{R}}$, (c) pitch diameter $d_{ct}$ of the hob, and (d) number of the hob starts (threads) $N_{ct}$, should be specified as well.

Once the principal design parameters of the $\zeta_{ct}$ – hob are calculated, then the determination of the rest of the design parameters of the hob becomes trivial.

Two issues are of critical importance when designing a $\zeta_{ct}$ – hob. *First*, the crossed-axes angle $\Sigma$ has to be equal to $\Sigma = 90°$. *Second*, the base diameter of the hob has to be of the smallest possible value.

[46] SU Pat. №4278973/08, *A Method of Hobbing Gear*/S.P. Radzevich, Int. Cl. B23f 5/22, Filed on: April 25, 1987.

When hobbing a gear, the equality $\Sigma = \psi_g + \psi_{ct}$ is observed for the crossed-axes angle $\Sigma$. In order to keep the crossed-axes angle orthogonal ($\Sigma = 90°$), the pitch helix angle of the hob $\psi_{ct}$ (and, thus the setting angle $\zeta_{ct}$ of the hob) have to be equal to

$$\zeta_{ct} = \psi_{ct} = 90° - \psi_g \tag{15.272}$$

when designing a $\zeta_{ct}$ – hob.

Furthermore, the expression [see Eq. (14.75)]

$$\psi_{b.ct} = \cos^{-1}(\cos\phi_n \cdot \cos\zeta_{ct}) \tag{15.273}$$

allows for calculation of the base helix angle $\psi_{b.ct}$ of the $\zeta_{ct}$ – hob. Ultimately, Eq. (14.71) returns the desirable value of the base diameter $d_{b.ct}$ of the $\zeta_{ct}$ – hob.

The approximate value of the required number of starts $N_{ct}$ of the $\zeta_{ct}$ – hob can be calculated from Eq. (14.71), which in the case under consideration casts into

$$N_{ct} = \frac{d_{b.ct} \cdot N_g \cdot \sqrt{1 - \cos^2\phi_n \cdot \cos^2\zeta_{ct}}}{d_g \cdot \cos\phi_n} \tag{15.274}$$

Equation (15.274) returns a value $N_{ct}^c$ of the desired number of starts $N_{ct}$ of the $\zeta_{ct}$ – hob (it is common that the calculated value $N_{ct}^c$ is a number with fractions). The calculated value of $N_{ct}^c$ should be rounded to the nearest integer number $N_{ct}$.

Two options are available at this point. *First*, the calculated value $N_{ct}^c$ of number of starts $N_{ct}$ of the hob is rounded to $N_{ct}^{\max}$, which is the nearest integer that exceeds the calculated value of $N_{ct}^c$. In this case the inequality $N_{ct}^c < N_{ct}^{\max}$ is observed. *Second*, the calculated value $N_{ct}^c$ of the hob number of starts $N_{ct}$ is rounded to $N_{ct}^{\min}$, which is the nearest integer that is smaller than the calculated value of $N_{ct}^c$. In this case the inequality $N_{ct}^c > N_{ct}^{\min}$ is observed. Further calculations are split into two separate paths.

If the first option is chosen, the calculation of the minimum allowable neck width $l_{\min}$ has to be accomplished for $N_{ct}^{\min}$. When the second option is chosen, the calculation of the minimum allowable neck width $l_{\min}$ has to be performed for $N_{ct}^{\max}$. Practically, the solution for $l_{\min}$ under $N_{ct}^{\min}$ often makes the shorter minimum allowable neck width $l_{\min}$.

To minimize the critical neck width of a hobbed cluster gear, it is strongly recommended to use a hob with the smallest possible outer diameter $d_{o.ct}$ as well as the smallest possible base diameter $d_{b.ct}$. The minimum outer diameter of the hob $d_{o.ct}$ is limited to

$$d_{o.ct} = d_{b.ct} + 2 \cdot h_{ct} \tag{15.275}$$

where $h_{ct}$ designates the effective total height of the hob tooth ($h_{ct} = a_{\mathscr{R}} + b_{\mathscr{R}}$).

For hobbing a cluster gear, designing $\zeta_{ct}$ – hobs with the smallest possible outer diameter $d_{o.ct}$ is recommended.

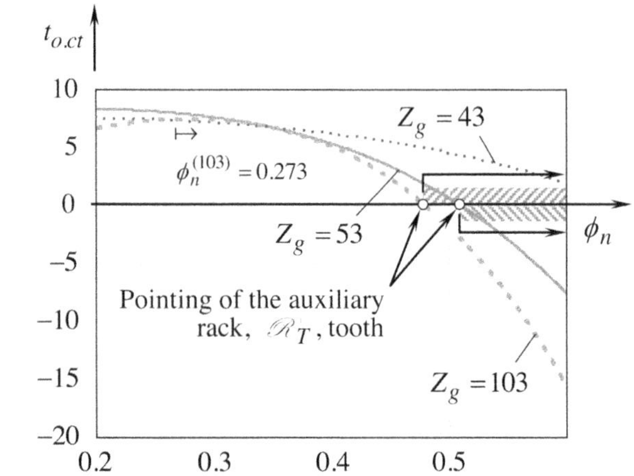


**FIGURE 15.174** Impact of the normal profile angle $\phi_n$ on the width of the top land $t_{o.ct}$ of the auxiliary generating rack $\mathscr{R}_T$ of a $\zeta_{ct}$ – hob. (From Radzevich, S.P., *ASME Journal of Mechanical Design*, 129, 438–444, 2007. With permission.)

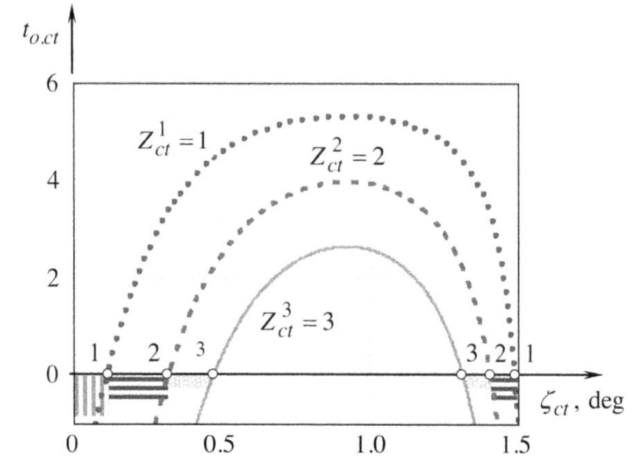


**FIGURE 15.175** Impact of the setting angle $\zeta_{ct}$ on the width of the top land $t_{o.ct}$ of the auxiliary generating rack $\mathscr{R}_T$ of a $\zeta_{ct}$ – hob. (From Radzevich, S.P., *ASME Journal of Mechanical Design*, 129, 438–444, 2007. With permission.)

This recommendation perfectly correlates with the earlier reported results of the research by the author [111].

The problem of minimization of the critical neck width $l_{\min}$ of a hobbed cluster gear is reduced at this point to the problem of minimization of the function $l = l(\phi_n, \zeta_{ct}, N_{ct}) \Rightarrow \min$. This problem is trivial. The calculated values of the parameters $\phi_n$, $\zeta_{ct}$, and $N_{ct}$ are used on further steps for the calculation of the design parameters of a particular design of the $\zeta_{ct}$ – hob.

When making a final decision on values of the design parameters of a $\zeta_{ct}$ – hob, it is important to bear in mind that both the normal profile angle $\phi_n$ and the setting angle $\zeta_{ct}$ of the hob affect the width of the top land $t_{o.ct}$ of the auxiliary generating rack $\mathscr{R}_T$ of the $\zeta_{ct}$ – hob. It is desirable to keep the length of the top cutting edge of the hob as long as possible. Therefore, the width of the top land $t_{o.ct}$ of the auxiliary rack $\mathscr{R}_T$ should also be the widest possible.

Examples of the function "$\phi_n$ vs. $t_{o.ct}$" are plotted in Figure 15.174. The maximum width $t_{o.ct}^{\max}$ of the top land $t_{o.ct}$ is observed when $\phi_n = \phi_{l.g}$ [140]. Here $\phi_{l.g}$ denotes the profile angle of the gear tooth at the limit diameter $d_{l.g}$.

A graphical interpretation of the impact of the setting angle $\zeta_{ct}$ of the $\zeta_{ct}$ – hob onto the width $t_{o.ct}$ of the top land is illustrated in Figure 15.175 [140]. The functions "$t_{o.ct}$ vs. $\zeta_{ct}$" have extremes. This means that there exists a combination of the principal design parameters of a particular design of the $\zeta_{ct}$ – hob under which the width $t_{o.ct}$ of the hob tooth top land is maximal $t_{o.ct}^{\max}$.

### 15.8.5.7.1 Example of the Calculation of the Design Parameters of a $\zeta_{ct}$ – Hob

It is assumed below that the design parameters of the gear to be machined are as follows: module $m = 10$ mm, normal profile angle $\phi_n = 20°$, pitch helix angle $\psi_g = 15°$, tooth count $N_g = 28$, pitch diameter $d_g = 280$ mm, outer diameter $d_{o.g} = 300$ mm, root diameter $d_{f.g} = 255$ mm, and face width $F_g = 24$ mm.

For a two-start $\zeta_{ct}$ – hob, Eq. (14.71) returns $d_{b.c} = 44.782$ mm for the base diameter of the hob. The use of the engineering formulae for the specification of the gear tooth flank (see Appendix F) allows for calculation of the hob pitch diameter $d_{ct} = 70$ mm, the hob outer diameter $d_{o.ct} = 95$ mm, and the pitch helix $\psi_{ct} = 15.993°$.

The difference $\Delta\psi = \psi_{ct} - \psi_g = 0.993°$ in helix angles is negligibly small for the case under consideration.

The implementation of the two-start $\zeta_{ct}$ – hob with the calculated design parameters makes it possible to cut a cluster gear with the neck width in the range of $l_{\min} \cong 70$ mm.

In a specific case of hobbing of a cluster gear, a problem, the nature of which is similar to that discussed earlier, could arise. The problem is illustrated in Figure 15.176. When special constraints are applied, it could be necessary to answer the question

- whether it is preferred to hob the cluster gear when the setting angle $\zeta_{ct}$ of the $\zeta_{ct}$ – hob is exactly equal to $\zeta_{ct} = -\psi_g$, but the number of the hob starts is greater, and hence, the base diameter of the hob is larger as well [see Figure 15.176(a)], or
- it is preferred to sacrifice the exact equality $\zeta_{ct} = -\psi_g$, and to use a hob with a slightly smaller setting angle and, instead, a smaller base diameter as shown in Figure 15.176(b).

This problem can be easily solved using the approach just disclosed.

The disclosed approach can be enhanced to design hobs for machining gears with a non-involute tooth profile, for example, for hobbing Novikov gears, as well as for hobbing splines, ratchets, sprockets, and so forth. When hobbing non-involute tooth profiles, it should be kept in mind that

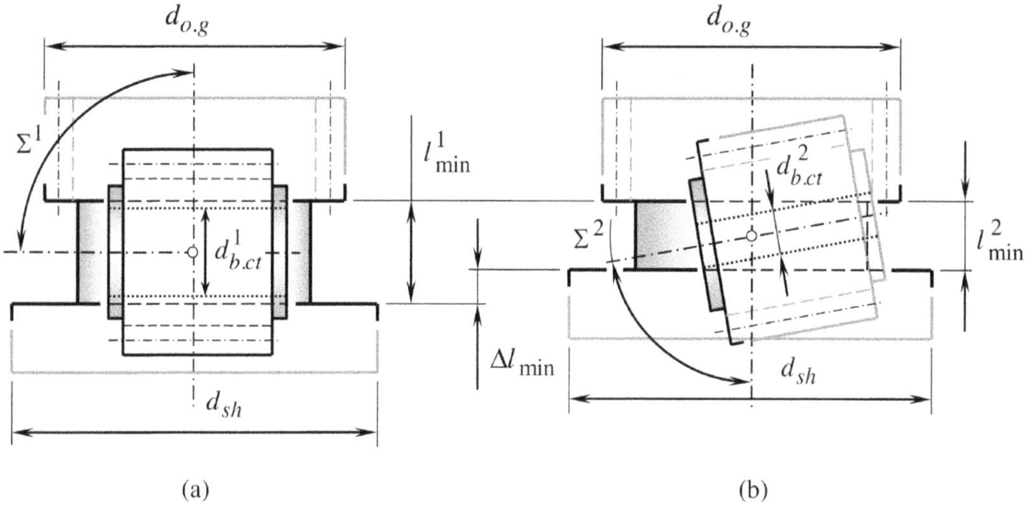


**FIGURE 15.176** One more problem that relates to designing of $\zeta_{ct}$ – hobs: (a) the case when the hob-setting angle is zero, and (b) the case when the hob-setting angle has a positive value.

only involute gears can be precisely generated in the gear hobbing process. No gears with other tooth flank geometry can be perfectly generated by the hob—only approximate non-involute gears can be cut this way. This is attributed to the kinematics of the gear hobbing process, and to violation of the condition of conjugacy of the tooth flanks $\mathscr{G}$ of a gear to be machined, and of the generating surface $T$ of the hob. In compliance with the condition of conjugacy of the tooth flanks, at every point of contact of the surfaces $\mathscr{G}$ and $T$, the common perpendicular **n** must intersect the axis of instant rotation $P_{ln}$ of the gear and the gear-cutting tool in the gear machining mesh [171].

### 15.8.6 Peculiarities of Generating Grinding of Gears

Gear grinding is a finishing process in gear production. Grinding allows for a finish grind of hardened gears and an improvement in gear accuracy. Many similarities can be found between the kinematics of the generating cutting of gears (that is, between gear hobbing) and the generating gear grinding process. Features of the kinematics of generating grinding of gears are briefly outlined below.

Generating grinding of a helical gear is depicted in Figure 15.177. Worm grinding wheels are used for this

**FIGURE 15.177** Worm grinding of a helical gear.

FIGURE 15.178 Worm-type grinding wheel.

process (see Figure 15.178). When grinding a gear, the work gear is rotated about its axis of rotation $O_{ct}$ with a steady angular velocity $\omega_{ct}$. The rotation of the grinding wheel is the primary motion in the gear grinding process. The work gear is rotated $\omega_g$ about its axis of rotation $O_g$. The rotations of the grinding wheel and the work gear are synchronized with one another so as to maintain the proportion $\omega_g/\omega_{ct} = N_{ct}/N_g$ (here $N_g$ and $N_{ct}$ are the tooth count [the number of starts] of the work gear and the grinding wheel, respectively). The closest distance of approach, that is, the distance between the work gear axis of rotation $O_g$ and the worm grinding wheel axis of rotation $O_{ct}$, is the center distance $C$. When rotated, the grinding wheel travels past the work gear along the axis $O_g$ with the feed motion.

The grinding wheel is redressed when it becomes worn. Various methods of dressing grinding wheels are used in industry. An illustrative example of the dressing process of a worm grinding wheel is shown in Figure 15.179.

FIGURE 15.179 Dressing of a worm grinding wheel.

FIGURE 15.180 Worm grinding wheel for grinding and polishing of gears: (*I*) grinding threads and (*II*) polishing threads of the worm-type grinding wheel.

**FIGURE 15.181** Coolant supply when grinding a helical gear with the worm grinding wheel.

Finish grinding of a work gear can be combined with work gear polishing (see Figure 15.180). For this purpose, worm grinding wheels of a special design are used.

A coolant supply can be used (see Figure 15.181) or not (dry grinding) when grinding gears with the worm grinding wheel.

# 16 Gear Shaving Cutters

Although many gears are satisfactorily produced by hobbing or gear shaping without subsequent gear finishing operations, finishing operations become a necessity when high load-carrying capacities, high speeds, long wear life, or quiet operation make surface finish and tooth accuracy of major design importance. For finishing of involute gears, the gear shaving process is used. In this process, finishing of gear tooth flanks is performed by shaving cutters. A close-up of the process of shaving of a spur gear is illustrated in Figure 16.1. The interaction between the shaving cutter tooth and the work gear teeth is shown in Figure 16.2.

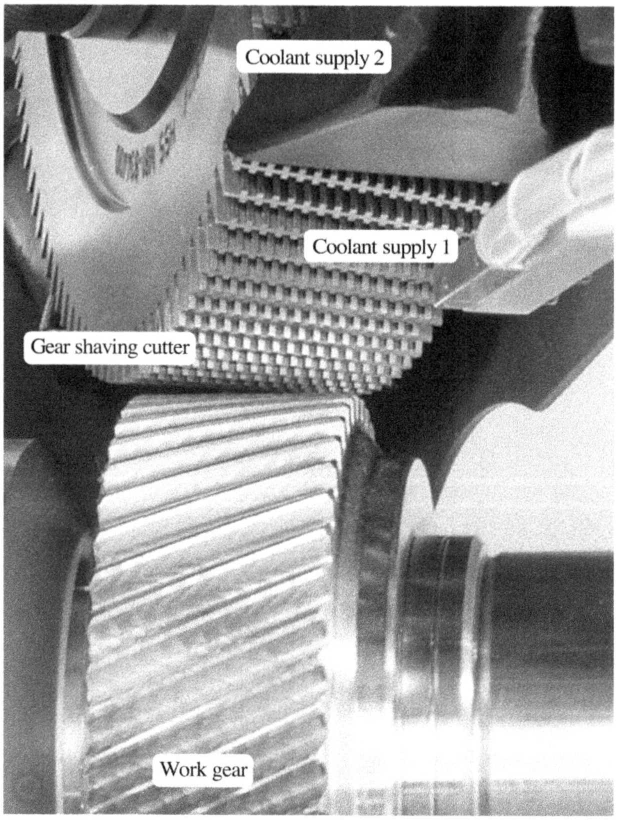


**FIGURE 16.1** Shaving of a spur gear.

## 16.1 DESIGN OF SHAVING CUTTER ON THE PREMISE OF ITS GENERATING SURFACE

The kinematics of the rotary shaving operation[1] is similar to that shown in Figure 15.3. Referring to Figure 16.3, the axis of rotation of the work gear $O_g$ and the axis of rotation of the shaving cutter $O_{ct}$ are shown at a certain center distance $C$. They cross one another at a certain crossed-axes angle $\Sigma$. The rotation of the work gear $\omega_g$ and the rotation of the shaving cutter $\omega_{ct}$ are synchronized with each other. In addition to the rotations, either the work gear or the shaving cutter reciprocates across the face of the shaving cutter in a transverse direction, and up-feeds an increment into the cutter with each stroke of the table. The direction of the reciprocation $\mathbf{F}_{ct}$ is perpendicular to the centerline ℄ between the work gear and the shaving cutter axes of rotations $O_g$ and $O_{ct}$. While remaining perpendicular to the centerline ℄, the actual direction of the reciprocation $\mathbf{F}_{ct}$ relative to the axes of rotation of the shaving cutter and of the work gear can vary. Ultimately, different kinds of gear shaving processes are recognized.

The up-feed motion $\mathbf{F}_{\text{up}}$ reduces the center distance $C$ and removes metal from the work gear tooth surfaces in the form of minute, hairlike chips. The up-feed motion $\mathbf{F}_{\text{up}}$ serves only for roughing purposes. If the motion $\mathbf{F}_{\text{up}}$ is assigned properly,

**FIGURE 16.2** Interaction between the shaving cutter tooth and the work gear teeth.

[1] The rotary gear shaving process was introduced to industry by *National Broach & Machine Division* (USA) in 1932. Originally the shaving process was applied (as early as in 1911) to worm wheels when serrated hobs were first used for finishing worm-wheel teeth, and the principle used then is applicable to shaving in general. After generation by a full hob in the normal manner, a serrated hob removes fine shavings of metal from the flanks of the teeth and imparts a smooth finish to the flanks of the worm-wheel teeth, which then conforms to the serrated hob shape.


DOI: 10.1201/9781003520528-21

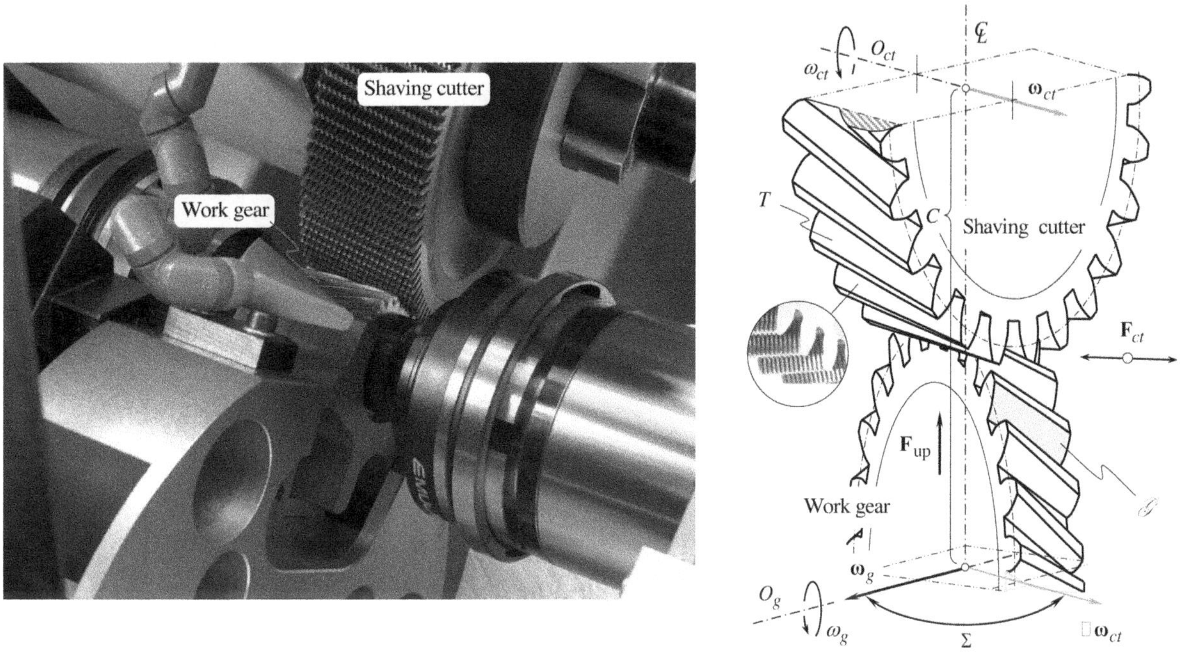


**FIGURE 16.3** Configuration of the work gear and the shaving cutter in the rotary shaving process.

it does not affect the shape and geometry of the machined tooth flank $\mathscr{G}$ of the shaved gear.[2]

### 16.1.1 Generating Surface of Shaving Cutter

Owing to similarity in the kinematics of surface generation, the equation of the generating surface of a shaving cutter for finishing of an involute gear resembles that of a hob for machining of that same involute gear. Omitting bulky derivation, an equation of the generating surface $T$ of a shaving cutter for finishing an involute gear can be represented in matrix form [see Eq. (14.63)]:

$$\left.\begin{array}{l}\text{Generating surface, } T, \\ \text{of a gear shaving cutter}\end{array}\right| \Rightarrow \mathbf{r}_{ct}(U_{ct}, V_{ct})$$

$$= \begin{bmatrix} 0.5 \cdot d_{b.ct} \cdot \sin V_{ct} - U_{ct} \cdot \sin \psi_{b.ct} \cdot \cos V_{ct} \\ 0.5 \cdot d_{b.ct} \cdot \cos V_{ct} + U_{ct} \cdot \sin \psi_{b.ct} \cdot \sin V_{ct} \\ p_{b.ct} \cdot V_{ct} - U_{ct} \cdot \cos \psi_{b.ct} \\ 1 \end{bmatrix} \quad (16.1)$$

The generating surfaces for both, for a hob and for a shaving cutter, are analytically described by similar equations, so the difference between them is mainly due to the different values of the base diameter $d_{b.ct}$ as well as of other design parameters of gear-cutting tool of this design.

[2] It is right to mention here that the kinematics in the rotary shaving process and in the honing of gears feature many similarities. This means that the term *abrasive shaving of gears* better fits the process that is nowadays referred to as *gear honing.* Both these processes feature the kinematics for which the main motion (the motion of cut) depends on the rotations of the work gear and of the gear hone. In the honing process, the main motion (the motion of cut) is independent of other motions of the machining process.

### 16.1.2 Rake Surface of the Cutting Teeth of a Shaving Cutter

Cutting edges are made up of all the shaving cutter teeth across every flank [see Figure 16.4(a)]. A slotting process is used for this purpose. After the slotting is complete, each flank of the shaving cutter tooth is grooved by serrations [5, 8]. In this way, every cutting tooth is provided with planar rake surfaces $R_s$. Wedges of intersection of the rake plane $R_s$ with the generating surface $T$ of the shaving cutter form the cutting edges $CE$ [see Figure 16.4(b)]. Under such a scenario, the cutting edges feature zero ($\gamma_c = 0°$) or nearly zero ($\gamma_c \cong 0°$) rake angle [see Figure 16.4(b)].

We mention here again that the chips to be removed in the gear shaving process are minute and hairlike. In practice, the uncut chip thickness $t$ is significantly below the roundness $\rho_{ct}$ of the shaving cutter cutting edge ($t \ll \rho_{ct}$). This fact is illustrated in Figure 16.5. Because the inequality $t \ll \rho_{ct}$ is observed, the actual value of the rake angle $\gamma_c$ does not affect the performance of the shaving process or the tool life of the shaving cutter. Regardless of the actual value of the rake angle (it can be either reasonably of a positive value $\gamma_c > 0°$ or reasonably of a negative value $\gamma_c < 0°$), the cutting performance of the gear shaving cutter practically remains in the same range. The cutting performance of a shaving cutter depends on the roundness $\rho_{ct}$ of the cutting edge of the shaving cutter rather than on the rake angle $\gamma_c$.

Because the design parameters of a shaving cutter are not restricted against the negative rake angle ($\gamma_c < 0°$), this

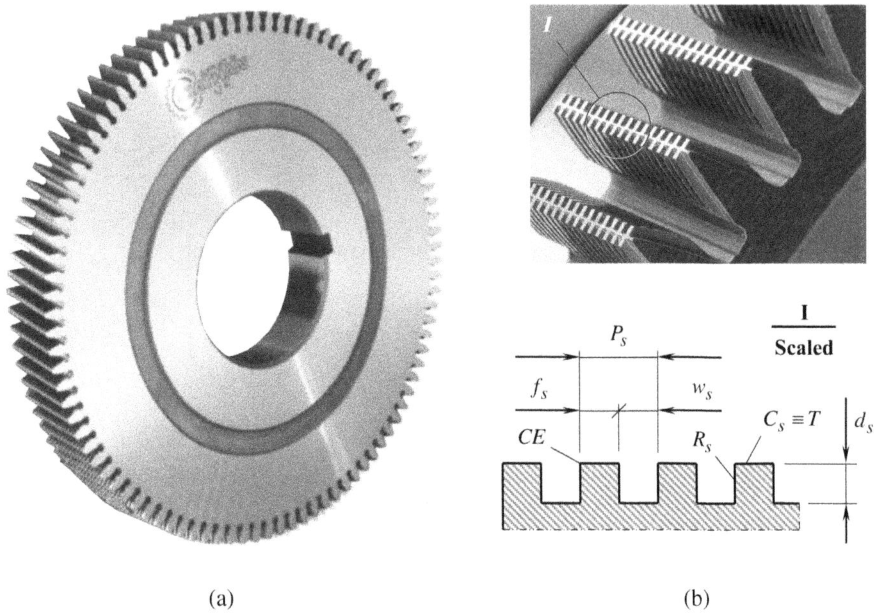


**FIGURE 16.4** Close-up of (a) a shaving cutter, and (b) serrations in a shaving cutter tooth. (From Radzevich, S.P., *ASME Journal of Mechanical Design*, 129(9), 969–980, 2007. With permission.)

makes it possible to design the serration with trapezoidal profile, as shown in Figure 16.5.

There is a significant advantage to the feasibility of the negative rake angle ($\gamma_c < 0°$): the serrations can be made of a trapezoidal profile. Slotting cutters of conventional design feature a clearance angle with a zero value ($\alpha_{sc} = 0°$). The cutting performance of slotting cutters with a zero clearance angle is low. It can be increased if a positive clearance angle can be created at the lateral cutting edges of the slotting cutter. If a shaving cutter is designed with a negative rake angle ($\gamma_c < 0°$), then the slotting cutters with a positive clearance angle ($\alpha_{sc} = 2 \div 4°$) at the lateral cutting edges can be used in the production of the shaving cutter. Slotting of the serrations with the slotting cutter with a positive clearance angle ($\alpha_{sc} > 0°$) is preferred; cutting conditions in this case are significantly smoother.

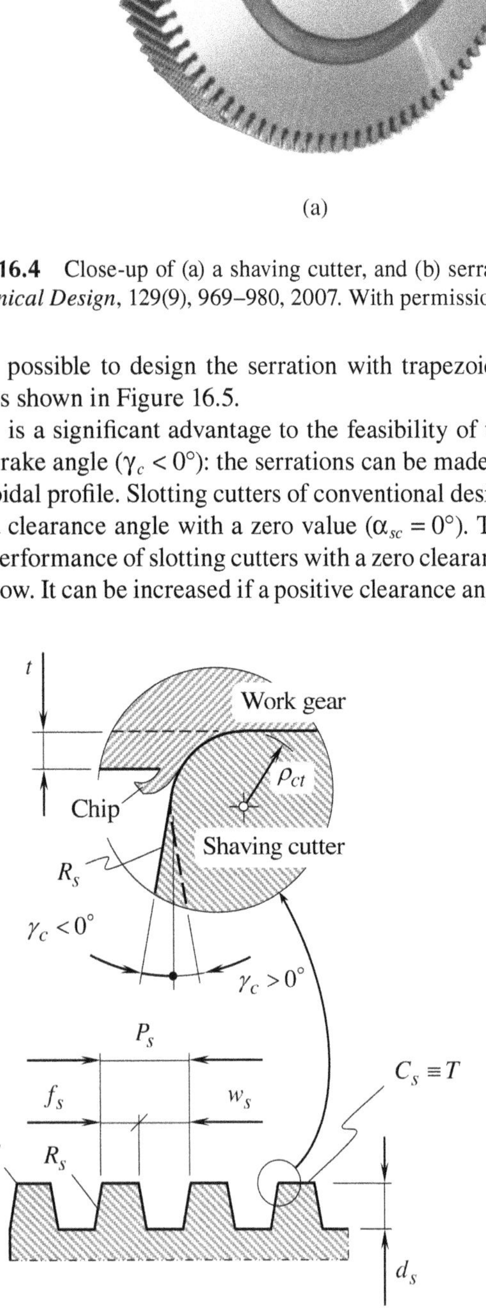


**FIGURE 16.5** Serrations in shaving cutter tooth flanks with a modified profile. (From Radzevich, S.P., *ASME Journal of Mechanical Design*, 129(9), 969–980, 2007. With permission.)

A trapezoidal serration profile also entails a variation of width of the serration land. This effect can be used to control contact pressure in the work gear-to-shaving cutter mesh, and in that way to compensate for the variation of the contact pressure caused by the variation of the number of teeth in contact in the mesh. This issue has not yet been comprehensively investigated.

### 16.1.3 Clearance Surface of the Cutting Teeth of a Shaving Cutter

Shaving cutters of most known designs feature a clearance surface $C_s$ that is congruent with the generating surface of the shaving tool [see Figure 16.4(b)]. The clearance surface $C_s$ is shaped in the form of a screw involute surface that is conjugate to the tooth flank $\mathscr{G}$ of the gear being machined. Because the surfaces $C_s$ and $T$ are congruent to one another, the clearance angle of zero value $\alpha_c = 0°$ at the cutting edges of the shaving cutter teeth is observed as a result.

### 16.1.4 Inclination Angle of the Cutting Edges of a Shaving Cutter

Since a shaving cutter is a finishing cutting tool, it is recommended to design it with a greater inclination angle $\lambda$ of the cutting edges. It is proven by many researchers and by

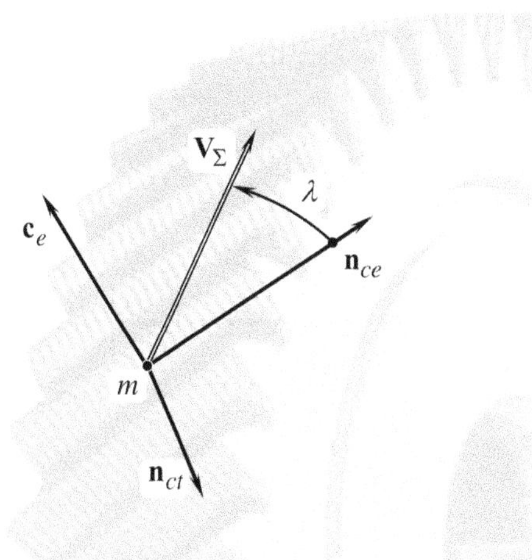


**FIGURE 16.6** Angle of inclination $\lambda$ of the cutting edge $CE$ of a gear shaving cutter.

practice that when thin chips are being cut off, the use of cutting tools with an increased inclination angle $\lambda$ is preferred.

At a point of interest $m$ within the cutting edge $CE$ of a shaving cutter, the angle of inclination $\lambda$ is measured within the plane of cut $P_{cv}$ (see Appendix G). The plane of cut $P_{cv}$ through the point of interest $m$ is tangent to the generating surface $T$ of the shaving cutter. The plane of cut $P_{cv}$ can also be specified as the plane that is perpendicular to the unit normal vector $\mathbf{n}_{ct}$ to the surface $T$ at point $m$.

The inclination angle $\lambda$ is the angle that the unit normal vector $\mathbf{n}_{ce}$ to the cutting edge $CE$ forms with the velocity vector of resultant motion $\mathbf{V}_\Sigma$ of the cutting edge relative to the surface of the cut [that is, by definition, the inclination angle equals $\lambda = \angle(\mathbf{n}_{ce}, \mathbf{V}_\Sigma)$], as shown in Figure 16.6.

Calculation is the only reasonable way to determine the actual value of the inclination angle $\lambda$ at a point of interest of the cutting edge of the shaving cutter tooth at a given instant of time. In the meantime, no special equipment for this particular purpose is available on the market.

Two vectors should be determined prior to calculation of the angle of inclination $\lambda$ of the cutting edge $CE$ of a shaving cutter tooth. They are (a) the unit normal vector $\mathbf{n}_{ce}$ and (b) the velocity vector of the resultant motion $\mathbf{V}_\Sigma$ of the cutting edge $CE$ relative to the surface of the cut $S_c$. Both the vectors must be determined at point of interest $m$ within the cutting edge. In addition, both the vectors have to be represented in a common reference system.

Referring to Figure 16.6, the unit normal vector $\mathbf{n}_{ce}$ to the cutting edge can be expressed in terms of the unit normal vector $\mathbf{n}_{ct}$ to the generating surface $T$ of the shaving cutter, and the unit vector $\mathbf{c}_e$ that is tangent to the cutting edge (that is, $\mathbf{n}_{ce} = \mathbf{c}_e \times \mathbf{n}_{ct}$). The earlier derived Eq. (15.68)

$$\mathbf{n}_{ct}(V_{ct}) = \begin{bmatrix} \sin\psi_{b.ct} \cdot \sin V_{ct} \\ \sin\psi_{b.ct} \cdot \cos V_{ct} \\ \cos\psi_{b.ct} \\ 1 \end{bmatrix} \tag{16.2}$$

is valid for the unit normal vector $\mathbf{n}_{ct}$.

The order of the multipliers in $\mathbf{n}_{ce} = \mathbf{c}_e \times \mathbf{n}_{ct}$ is of importance when determining the unit vector $\mathbf{n}_{ce}$.

An analytical expression for the unit vector $\mathbf{c}_e$ can be derived for a shaving cutter of a given design. Consider the rake plane $R_s$ of a shaving cutter that is perpendicular to the axis of rotation of the shaving cutter. The unit vector $\mathbf{c}_e$ is tangent to the involute profile of the shaving cutter tooth. This vector is located within the $X_{ct}Y_{ct}$ – coordinate plane and thus can be expressed just as a function of two $X_{ct}$ – and $Y_{ct}$ – parameters. Otherwise, if the rake plane $R_s$ is not perpendicular to the axis of rotation of the shaving cutter, the third $Z_{ct}$ – parameter is required to be incorporated into an expression for the unit tangent vector $\mathbf{c}_e$.

In a Cartesian coordinate system $X_{ct}Y_{ct}Z_{ct}$ associated with the shaving cutter, the equation for the unit vector $\mathbf{c}_e$ in the general case can be expressed in matrix form

$$\mathbf{c}_e(t_{ce}) = \begin{bmatrix} X_c(t_{ce}) \\ Y_c(t_{ce}) \\ Z_c(t_{ce}) \\ 1 \end{bmatrix} \tag{16.3}$$

where $t_{ce}$ designates the parameter of the cutting edge $CE$.

The cross product of the unit vector $\mathbf{n}_{ct}$ [see Eq. (16.2)] and the unit vector $\mathbf{c}_e$ [see Eq. (16.3)] returns the required expression for the unit normal vector: $\mathbf{n}_{ce} = \mathbf{c}_e \times \mathbf{n}_{ct}$.

The velocity vector $\mathbf{V}_\Sigma$ is also located within the plane of cut $P_{cv}$ (the plane of cut $P_{cv}$ is a plane that is tangent at the point of interest to the surface of cut $S_c$). This vector can be represented as superposition

$$\mathbf{V}_\Sigma = \mathbf{V}_{sl} + \mathbf{V}_{pr} + \mathbf{V}(\mathbf{F}_{ct}) \tag{16.4}$$

of three vectors, namely of (a) the velocity vector $\mathbf{V}_{sl}$ of sliding in the lengthwise direction of the work gear tooth, (b) the velocity vector $\mathbf{V}_{pr}$ of profile sliding, and (c) the vector $\mathbf{V}(\mathbf{F}_{ct})$ due to the reciprocation motion.

Determining the velocity vector $\mathbf{V}_\Sigma$ of the resultant motion of the shaving cutter tooth relative to the work gear is a bit more complicated than determining the unit vector $\mathbf{n}_{ce}$. For this purpose, refer to Figure 16.7, in which a schematic of a gear shaving process is depicted.

The work gear is rotated $\omega_g$ about its axis of rotation $O_g$. At the apex point[3] $A_{pa}$, the vector $\mathbf{V}_g$ of the linear velocity of the rotation can be expressed in terms of the rotation vector $\boldsymbol{\omega}_g$ and the radius $\mathbf{R}_g$ of the work gear pitch cylinder:

$$\mathbf{V}_g = \boldsymbol{\omega}_g \times \mathbf{R}_g \tag{16.5}$$

[3] The apex point, $A_{pa}$, is located within the centerline ℄. The point $A_{pa}$ is at a distance $R_g$ from the gear axis of rotation $O_g$ and at a distance $R_c$ from the shaving cutter axis of rotation $O_{ct}$. The apex point $A_{pa}$ divides the center distance $C$ reciprocally to the gear ratio $u$.

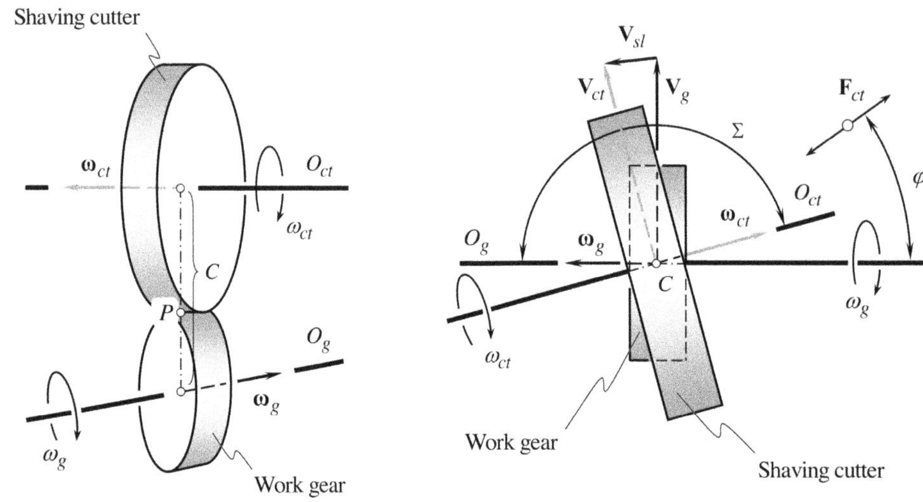


**FIGURE 16.7** Schematic of rotary gear shaving process.

Here the vector $\mathbf{R}_g$ is calculated from the formula $\mathbf{R}_g = \mathbf{k}_g(d_g/2)$. It is assumed here that the $Z_g$ – axis of the reference system $X_gY_gZ_g$ associated with the work gear is aligned with the centerline ℄ of the axes of rotation $O_g$ and $O_{ct}$. The $Z_g$ – axis is pointed from the work gear axis of rotation $O_g$ toward the shaving cutter axis of rotation $O_{ct}$.

The shaving cutter rotates $\omega_{ct}$ about its axis of rotation $O_{ct}$. At the apex point $A_{pa}$, the vector $\mathbf{V}_{ct}$ of the linear velocity of the rotation can be expressed in terms of the rotation vector $\boldsymbol{\omega}_{ct}$, and the radius $\mathbf{R}_{ct}$ of the shaving cutter pitch cylinder:

$$\mathbf{V}_{ct} = \boldsymbol{\omega}_{ct} \times \mathbf{R}_{ct} \tag{16.6}$$

Similar to the process above, for calculating the vector $\mathbf{R}_{ct}$, the formula $\mathbf{R}_{ct} = \mathbf{k}_{ct}(d_{ct}/2)$ can be used. It is assumed in the case under consideration that the $Z_{ct}$ – axis of the reference system $X_{ct}Y_{ct}Z_{ct}$ associated with the shaving cutter is aligned with the centerline ℄ of the axes $O_g$ and $O_{ct}$. The $Z_{ct}$ – axis is pointed from the shaving axis of rotation $O_{ct}$ toward the work gear axis of rotation $O_g$.

As the work gear axis of rotation $O_g$ and the shaving cutter axis of rotation $O_{ct}$ are crossed at a certain crossed-axes angle $\Sigma$, the velocity vectors $\mathbf{V}_g$ and $\mathbf{V}_{ct}$ do not align with each other, and they form a certain angle. The angle between the vectors $\mathbf{V}_g$ and $\mathbf{V}_{ct}$ is exactly equal to the crossed-axes angle $\Sigma$ (see Figure 16.6).

The sliding velocity vector $\mathbf{V}_{sl}$ is the difference between the vectors $\mathbf{V}_g$ and $\mathbf{V}_{ct}$:

$$\mathbf{V}_{sl} = \mathbf{V}_{ct} - \mathbf{V}_g \tag{16.7}$$

Both the vectors $\mathbf{V}_g$ and $\mathbf{V}_{ct}$ in Eq. (16.7) have to be expressed in a common reference system.

It can be shown that for the magnitude $V_{sl} = |\mathbf{V}_{sl}|$, Eq. (16.7) yields the formula

$$V_{sl} = 0.5 \cdot (\omega_g d_g \sin\psi_g + \omega_{ct} d_{ct} \sin\psi_{ct}) \tag{16.8}$$

where:

- $\omega_g$ is the rotation of the work gear.
- $\omega_{ct}$ is the rotation of the shaving cutter.
- $d_g$ is the pitch diameter of the work gear.
- $d_{ct}$ is the pitch diameter of the shaving cutter.
- $\psi_g$ is the pitch helix angle of the work gear.
- $\psi_{ct}$ is the pitch helix angle of the shaving cutter.

Note here in Eq. (16.8), pitch helix angles $\psi_g$ and $\psi_{ct}$ are signed values.

The velocity vector of the profile sliding $\mathbf{V}_{pr}$ of the shaving cutter tooth relative to the work gear tooth also contributes to the velocity vector $\mathbf{V}_\Sigma$ of the resultant motion. This contribution becomes more significant when the tooth number of the work gear and/or tooth number of the shaving cutter gets smaller. A tooth profile sliding of zero value is observed at the pitch point, and it becomes greater toward the top and bottom of the work gear teeth.

The feed motion $\mathbf{F}_{ct}$ of the shaving cutter is pointed at an angle $\varphi_{fc}$ with respect to the axis of rotation $O_g$ of the work gear (see Figure 16.7). The actual value of the angle $\varphi_{fc}$, as well as the magnitude of the feed motion $\mathbf{F}_{ct}$, is known. This makes possible the derivation of an expression for the vector $\mathbf{F}_{ct}$ in a reference system that is common for the work gear and the shaving cutter.

The components in Eq. (16.4) are not discussed in detail here. Analytical expressions for the components $\mathbf{V}_{sl}$, $\mathbf{V}_{pr}$, and $\mathbf{V}(\mathbf{F}_{ct})$ are derived below. The issue of critical importance at this point is that (a) the velocity vector $\mathbf{V}_\Sigma$ is located within the plane of cut $P_{cv}$ and (b) that the velocity vector $\mathbf{V}_\Sigma$ can be expressed in terms of (i) the design parameters of the work gear, and the shaving cutter, (ii) their configuration, and (iii) their relative motion in the gear shaving process.

The magnitude of the vector of up-feed motion $\mathbf{F}_{\text{up}}$ is negligibly small compared to the vectors $\mathbf{V}_{sl}$, $\mathbf{V}_{pr}$, and $\mathbf{V}(\mathbf{F}_{ct})$. Therefore, the vector $\mathbf{F}_{\text{up}}$ can be ignored when calculating the inclination angle $\lambda$.

After the velocity vector $\mathbf{V}_\Sigma$ is described analytically, it can be represented in matrix form as

$$\mathbf{V}_\Sigma = \begin{bmatrix} V_{\Sigma x} \\ V_{\Sigma y} \\ V_{\Sigma z} \\ 1 \end{bmatrix} \tag{16.9}$$

where $V_{\Sigma x}$, $V_{\Sigma x}$, and $V_{\Sigma x}$ are Cartesian components of the velocity vector $\mathbf{V}_\Sigma$ of resultant motion of the cutting edge *CE* relative to the plane of cut $P_{cv}$.

Having the unit normal vector $\mathbf{n}_{ce}$ along with the velocity vector $\mathbf{V}_\Sigma$ of the resultant motion of the cutting edge expressed in a common coordinate system, calculation of the inclination angle $\lambda$ becomes trivial:

$$\lambda = \tan^{-1}\left(\frac{|\mathbf{n}_{ce} \times \mathbf{V}_\Sigma|}{\mathbf{n}_{ce} \cdot \mathbf{V}_\Sigma}\right) \tag{16.10}$$

The inclination angle $\lambda$ is not of a constant value, but it varies along the cutting edge *CE* of a shaving cutter.

## 16.2 DESIGN OF GEAR SHAVING CUTTERS

The rotary shaving cutters (see Figure 16.8) are high-precision generating gear–cutting tools. They have ground tooth flanks and are held to Class A and AA tolerances in all principal elements. Precision shaving cutters are utilized for finishing gears by a rotary shaving process.

Shaving cutters are almost invariably made of high-speed steel. A hardness of 63 HRC minimum after tempering is considered satisfactory.

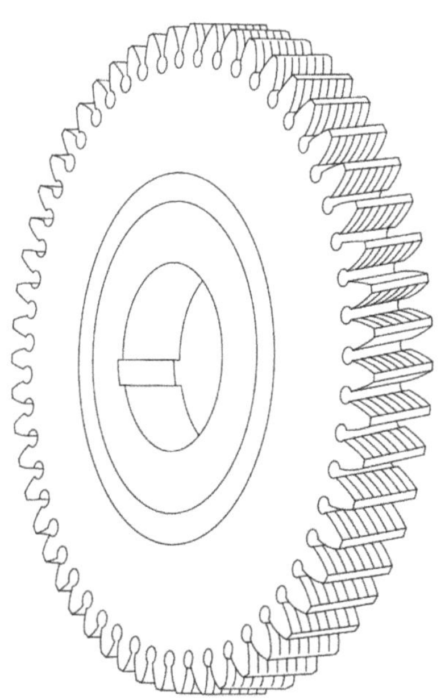

**FIGURE 16.8** Close-up of a shaving cutter for rotary shaving of gears.

### 16.2.1 Design Parameters of a Shaving Cutter

Just as the generating body of a gear shaving cutter is identical to the corresponding helical gear, many similarities are observed in the design of a gear shaving cutter and the helical gear.

The shaving cutter must have the same base pitch as the work gear it is intended to shave. To fulfill this critically important requirement, the normal diametral pitch and normal profile angle of the shaving cutter should be the same as those of the gears to be shaved [5, 8]. However, the mandatory equality of the base pitches of the shaving cutter and the work gear can be achieved under another combination of normal diametral pitch and normal pressure angle.

It is usual for a particular shaving cutter to shave a range of gears of the same base pitch. Unlike a hob, however, it is preferable to design a shaving cutter for a restricted range of gears, and if it is economically justified, a shaving cutter designed for shaving one particular gear can be expected to give the best results.

The average recommended shaving cutter pitch diameter is 8 in. for most external gears. A 9-in. pitch diameter shaving cutter is frequently used for low tooth numbers in the work gear.

The face widths of the shaving cutters vary from 1 to 4 in., depending on the requirements and type of application. Fine-pitch shaving cutters are furnished with face width of 3/8 to 5/8 in. and with a diameter of 3 to 4 in.

Helix angle is chosen so as to give a desirable crossed-axes angle between work gear and shaving cutter.

The shaving cutter tooth addendum is always calculated so that the shaving cutter will finish the gear profile slightly below the lowest point of contact with the mating gear. This addendum may be varied to suit the fillet (root radius, conditions of undercut, protuberance) and the size of the mating gear.

Better results are obtained from shaving cutters that have no common ratio between the numbers of teeth in the shaving cutter and the work gear. Therefore, a rotary shaving cutter is usually designed with a prime number of teeth. This *hunting-tooth condition* further refines spacing errors and increases the accuracy of the shaved gear. It is also beneficial in distributing small errors that may occur in the shaving cutter.

The dedendum area of the shaving cutter is supplemented with a drilled hole to form an undercut to allow the serrating tool to clear itself during manufacturing. The hole is the runout clearance for the serrating tool.

Holes are drilled between and at the base of the teeth of the shaving cutters to provide clearance for chips formed by the shaving operation. An unrestricted flow of coolant passes through the serrations and holes to keep the shaving cutter clean of these chips. The circle within which the holes are located is often referred to as a *hole circle*. Its diameter is chosen so as to allow for clearance between the tips of the gear teeth being shaved and the point of intersection between the shaving cutter profile and oil hole.

The holes are drilled at an angle ~ 1° smaller than the pitch helix angle of the shaving cutter. This is because the helix angle of a shaving cutter varies within the tooth height, and it gets smaller toward the axis of the shaving cutter.

### 16.2.2 Serrations on the Tooth Flanks of a Shaving Cutter

The tooth flanks of a rotary shaving cutter are serrated with multiple gashes [see Figure 16.9(a)]. The gashes (serrations) extend from top to bottom of the tooth, terminating in a clearance space at the bottom. These clearance spaces also provide unrestricted channels for a constant flow of coolant to promptly dispose of chips.

There is still considerable debate about the most suitable shape of the serrations. Commonly the serrations are shaped so that the cutting edges are located within a plane that is either perpendicular to the shaving cutter axis [see Figure 16.9(b)] or perpendicular to the line of tooth [see Figure 16.9(c)]. When the serrations are perpendicular to the line of tooth as shown in Figure 16.9(c), then the width of the end lands is not constant within the tooth height. In some cases the serrations could be designed with a trapezoidal profile [see Figure 16.9(d)]. Fine-pitch shaving cutters are usually made with the serrations cut through the teeth. These tend to be more fragile than when the teeth support the serrations [see Figure 16.9(e)]. In this latter case, the serrations with a trapezoidal profile are commonly used.

It is common practice to cut the serrations with a straight bottom that does not follow the tooth flank involute profile [see Figure 16.10(a)]. The serrations therefore do not have a constant depth $d_s$ along the whole tooth height. This peculiarity is not due to bad conditions of cutting when machining the serrations, but mostly to the complexity of the kinematics of machining of the serrations that feature equal the serration depth. Most slotting machines available on the market are capable of providing the slotting cutter with a rectilinear motion. In order to follow the involute profile, the machine kinematics would be greatly complicated without granting the shaving cutter a real advantage in performance.

However, when producing shaving cutters with a low tooth number, the convexity $d_s^*$ of the shaving cutter tooth

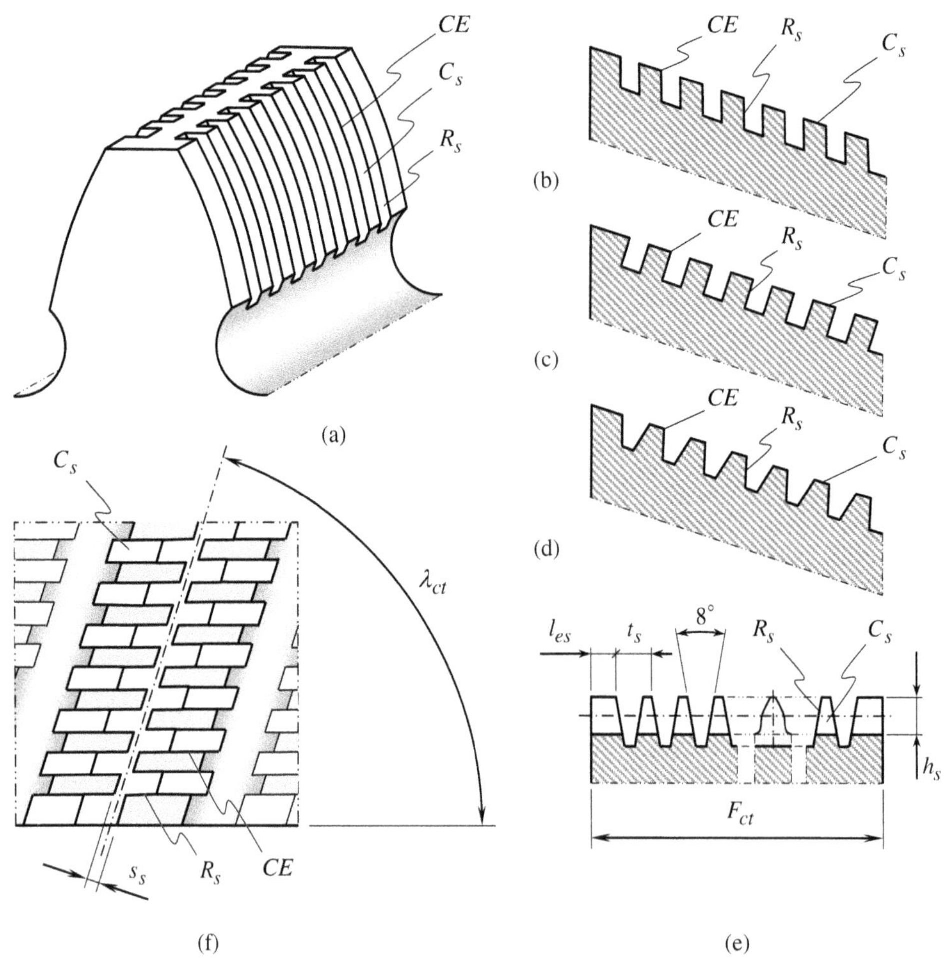


**FIGURE 16.9** Practical designs of serrations in a shaving cutter tooth flank: (a) gear shaving cutter tooth, (b) serrations perpendicular to the axis of rotation, (c) serrations perpendicular to the clearance surface, (d) trapezoidal serrations, (e) cut-through serrations, and (f) serrations on the opposite tooth flanks shifted in opposite directions.

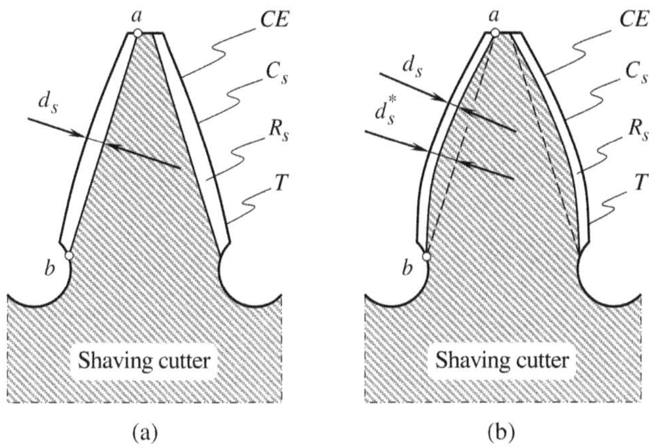


**FIGURE 16.10** Serrations of two types of shaving cutter tooth flanks: (a) serrations with straight bottom, and (b) serrations with convex bottom.

profile cannot be neglected. It is not allowed to approximate the bottom of the serrations with a straight-line segment *ab*. Therefore, the serrations have an involute form and a constant depth throughout [see Figure 16.10(b)].

The serrations in the shaving cutter tooth flank [see Figure 16.4(b)] can be specified by

- The serration pitch $P_s$.
- The serration land width $f_s$.
- The serration space width $w_s$.
- The serration depth $d_s$.

The serration pitch and the division between the land and space mainly depend on the module and also on the stock removal. The gashes are usually a little wider than the lands on a new shaving cutter. For standard shaving cutters, the serration pitch $P_s$ is in the range of $P_s = 1.80 \text{ mm} \div 2.20 \text{ mm}$. For these values of the serration pitch, the width of the land serration $f_s$ is in the range of $f_s = 1.05 \div 1.20$ mm, and the serration space $w_s$ is normally in the range of $w_s = 0.75 \div 1.00$ mm.

The serration depth value $d_s$ depends on the module and the profile angle of the shaving cutter. It is determined by the strength requirements, also taking into account an allowance for resharpening the shaving cutter. The serration depth is in the range of $d_s = 0.6...1.0$ mm.

The lands could be tapered on one side for strength and to facilitate washing the chips out.

The serrations in tooth flanks of opposite sides of the shaving cutter tooth profile must not interfere with one another.

A portion of the top land $s_s$ of the shaving cutter tooth within which no serrations occur must not be less than $s_s \geq 0.1$ mm [see Figure 16.9(f)]:

$$s_s = t_{oc} - 2 \cdot \frac{d_s}{\cos\phi_{t.o.c} \cos\psi_{o.ct}} \tag{16.11}$$

The width of the top land $t_{oc}$ of the shaving cutter can be calculated from the formula

$$t_{oc} = d_{o.ct}\left(\frac{t_{ct}}{d_{ct}} + \text{inv}\phi_{t.ct} - \text{inv}\phi_{t.o.ct}\right) \tag{16.12}$$

where the profile angle at the outer diameter of the shaving cutter is denoted by $\phi_{t.c.o}$.

The end lands are designed to be wider than the inner serrating lands as shown in Figure 16.11(a,b).

Shaving cutters are sometimes made with one flank unserrated so that only one flank of the work is shaved. This feature may be useful for nonreversible gears driving in one direction only.

Depending on the type of shaving, whether transverse, modified underpass, or full underpass, the serrations will be *annular pattern* or with a *staggered pattern*. The annular serration [see Figure 16.11(a)] is used for transverse or modified underpass methods of shaving. The staggered type of serration[4] [see Figure 16.11(b)] for full underpass shaving compensates for the reduced axial sliding and prevents the tracking of serration ends on the tooth flank of the shaved part. Therefore, the action of the cutting edges tends to be more nearly continuous.

The serrations are slotted (see Figure 16.12) by a serrating tool reciprocated vertically while the shaving cutter is rolled with its axis horizontal. In this way the shaving cutter tooth profile follows a desired involute curve.

The most convenient way to cut the serrations is to make them lie in planes perpendicular to the shaving cutter axis, and to follow approximately the involute profile. The serrations are made nearly parallel to the side faces of the shaving cutter with a small angular relief to facilitate cutting them. Unfortunately, they are not finished after hardening, apart from vapor blasting to remove deposits of hardening salt or light scale, and as a result, the edges of the lands lying between the serrations may not be ideally true.

### 16.2.3 Resharpening of Shaving Cutter

The shaving cutter, upon dulling, will react in a manner to warn that a tool change is needed. Keeping shaving cutters sharp is of critical importance for finishing high-quality gears. It is recommended to regrind the shaving cutter before using it for finishing gears.

Shaving cutters are sharpened on their flanks, which are screw involute surfaces. This makes the shaving cutter change quite appreciably in outside diameter during its life. Shaving cutters do not produce the same thickness on the work gear tooth throughout their lives. If the thickness of the work gear tooth is held constant, the depth that the shaving cutter is fed into the work gear will vary.

The shaving cutter throughout its life is designed to maintain the same *true form diameter* on the part. This is

[4] The staggered types of serrations are often referred to as *differential serrations*.

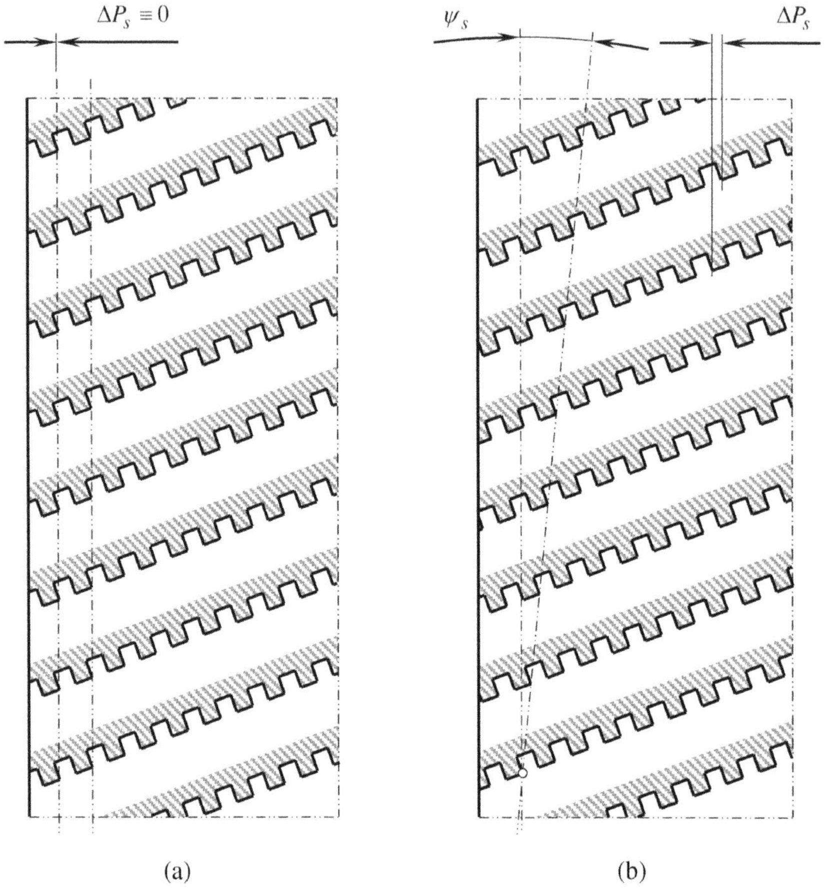


**FIGURE 16.11** Annular (a) and staggered (b) pattern of the serrations in tooth flanks of a shaving cutter. (From Radzevich, S.P., *ASME Journal of Mechanical Design*, 129(9), 969–980, 2007. With permission.)

accomplished by retaining the correct proportions of tooth addendum and tooth thickness. The tooth thickness of the shaving cutter is decreased with each resharpening of the shaving cutter. This causes a reduction in operating center distance and the consequent change in operating pressure angle. The addendum must be decreased to contact at the original *true involute form* (TIF) diameter [145–147]. This does not change in the same proportion as the change in center distance.

In practice, the base diameter is constant and remains so throughout the life of the shaving cutter. The cross-axes angle decreases slightly with each resharpening.

**FIGURE 16.12** Process of slotting the serrations in the tooth flank of a shaving cutter.

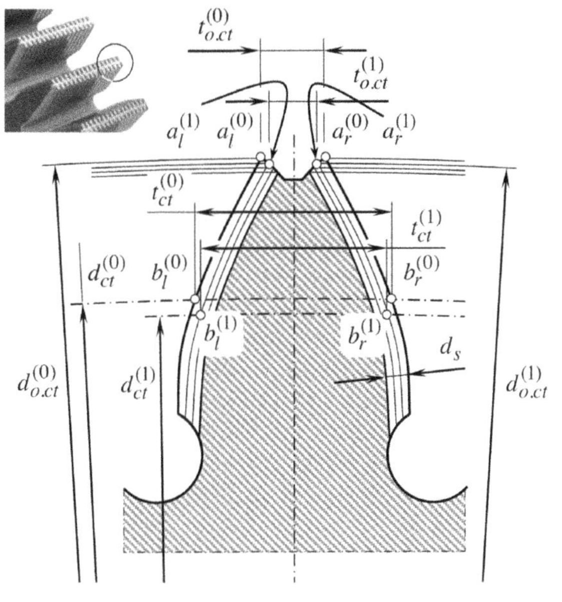

**FIGURE 16.13** Gear shaving cutter with a grooved top land. (From Radzevich, S.P., and Palaguta, V.A., USSR Patent 1632662, Int. Cl. B23f 21/28, March 30, 1989.)

The tendency of the shaving cutter to cut a deeper depth as the teeth are made thinner from sharpening may be controlled by reducing the outside diameter of the shaving cutter an appropriate amount each time it is sharpened [145–147]. Various techniques can be used for this purpose.

Grinding the outside diameter of the shaving cutter is inconvenient for the user of the shaving cutter. This undesirable process can be eliminated by using a gear shaving cutter of the design proposed by Prof. S.P. Radzevich and Dr. Palaguta [181].

The principal features of the gear shaving cutter of the proposed design[5] are illustrated in Figure 16.13.

Referring to Figure 16.13, a new shaving cutter can be specified by the set of the design parameters, namely by (a) a pitch diameter $d_{ct}^{(0)}$, (b) a tooth thickness $t_{ct}^{(0)}$, (c) an outer diameter $d_{o.ct}^{(0)}$, (d) a width of the top land $t_{o.ct}^{(0)}$, and others. The profile of the tooth addendum of the new shaving cutter is designated as $a_l^{(0)}b_l^{(0)}$ for the left side and $a_r^{(0)}b_r^{(0)}$ for the right side of the tooth profile.

After being reground for the first time, the pitch diameter of the shaving cutter becomes smaller and is designated now as $d_{ct}^{(1)}$. The tooth thickness of the resharpened shaving cutter $t_{ct}^{(1)}$ differs from the $t_{ct}^{(0)}$ of the new one. The reduction of the pitch diameter entails the need in the corresponding reduction of the outer diameter of the shaving cutter from the value of $d_{o.ct}^{(0)}$ to the value of $d_{o.ct}^{(1)}$. The corresponding width of the top land $t_{o.ct}^{(1)}$ is calculated for the given values of the design parameters of the resharpened shaving cutter. In this way the addendum profile $a_l^{(1)}b_l^{(1)}$ for the left side and $a_r^{(1)}b_r^{(1)}$ for the right side of the tooth of the resharpened shaving cutter can be determined.

These calculations are performed for a variety of different pitch diameters of the shaving cutter depending on the numbers of its resharpenings. Ultimately, in addition to points $a_l^{(0)}$, $a_l^{(1)}$, and $a_r^{(0)}$, $a_r^{(1)}$, more points $a_l^{(i)}$ and $a_r^{(i)}$ can be determined. The set of points $a_l^{(0)}$, $a_l^{(1)}$, ..., $a_l^{(i)}$ specifies the left-side profile, and the set of points $a_r^{(0)}$, $a_r^{(1)}$, and $a_r^{(i)}$ specifies the right-side profile of the groove along the top land of the shaving cutter tooth.

The shaving cutter that features grooves along the top lands of its teeth does not need regrinding over the outer cylinder after every resharpening [180], as is commonly practiced.

When resharpening a shaving cutter, the fifth necessary condition of proper part surface generation must be fulfilled. For the resharpened shaving cutter, as well as for the new cutter, zones of fulfillment of the fifth necessary condition of the proper part surface generation can be drawn similar to that which was done earlier for the gear hobs (see Figure 15.125, Figure 15.128).

The duration of a shaving cutter life depends primarily on the depth of the serration. This varies with the pitch.

The teeth are finished by final profile grinding and honing, and this tends to leave no feathering on the edges of the lands that may remain during the operative life of the shaving cutter.

Because the shaving cutter is designed for finishing gears, it is of critical importance to maintain a high accuracy throughout the lifetime of the shaving cutter. In other words, after the shaving cutter has been reground, it is highly desirable that it is capable of shaving gears with the same accuracy as before. For this purpose, a method of resharpening of shaving cutters is proposed by Prof. S.P. Radzevich and Dr. V.A. Palaguta [181]. The method of resharpening of shaving cutters[6] utilizes the concept of the constant base pitch of the new $[p_{b.ct}^{(0)}]$ and resharpened $[p_{b.ct}^{(i)}]$ shaving cutter. In other words, the use of this method makes possible maintaining the base pitch of the shaving cutter of a constant value ($p_{b.ct} = const$) regardless of how many times $i$ the shaving cutter is resharpened, and how much its pitch diameter $d_{ct}$ is reduced due to resharpening. The base pitch of a shaving cutter $p_{b.ct}$ must be (a) of a constant value throughout the lifetime of the shaving cutter and (b) equal to the base pitch $p_{b.g}$ of the gear intended to be shaved. Fulfillment of the equality $p_{b.ct} = p_{b.g}$ is a prerequisite to achieving high quality of the gear being shaved.

Grinding wheels with flat working surfaces are used for grinding shaving cutter tooth flanks (see Figure 16.14). A schematic of the prosed method of resharpening the gear shaving cutter is illustrated in Figure 16.15.

The generation of the shaving cutter tooth flank is executed by a plane. For this purpose, a dish-type grinding wheel commonly is used. The grinding wheel is rotated about its axis $O_{gw}$ with a certain rotation $\omega_{gw}$. The rotation

[5] SU Pat. No. 1632662, *A Gear Shaving Cutter/S.P. Radzevich* and *V.A. Palaguta*, Int. Cl. B23f 21/28, Filed: March 30, 1989.

[6] SU Pat. No. 1646725, *A Method of Re-Sharpening of Gear Shaving Cutter/S.P. Radzevich*, and *V.A. Palaguta*, Int. Cl. B23f 21/28, Filed: December 13, 1988.

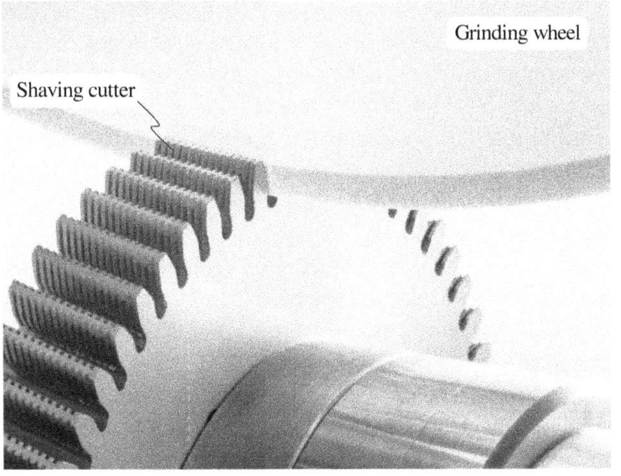


**FIGURE 16.14** Grinding the tooth flanks of a shaving cutter.

$\omega_{gw}$ serves as the primary motion of cutting in the shaving cutter grinding process.

In the relative motion of the shaving cutter and the grinding wheel, the swivel $\pm\omega_{ct}$ of the shaving cutter and the reciprocation $\pm V_{gw}$ of the grinding wheel are timed so that the shaving cutter pitch cylinder of a diameter $d_{ct}^{(0)}$ rolls with no sliding over the pitch plane $W_{gw}^{(0)}$ associated with the rotating grinding wheel, as shown in Figure 16.15. The pitch point in such a rolling motion is designated as $P^{(0)}$.

In an arbitrary configuration of the grinding wheel in relation to the shaving cutter, the perpendicular to the generating surface $T_{gw}^{(0)}$ of the grinding wheel through the point of contact $k^{(0)}$ of the plane $T_{gw}^{(0)}$, and the tooth flank is tangent to the base cylinder of the shaving cutter $d_{b.ct}^{(0)}$ at the corresponding point $a^{(0)}$. In such a scenario, the shaving cutter with a profile angle $\phi_{ct}^{(0)}$ is ground.

The shaving cutter needs to be resharpened when it gets dull. When the shaving cutter is resharpened for the $i$ – th time, its pitch diameter reduces to $d_{ct}^{(i)} < d_{ct}^{(0)}$. If no special action is taken, the reduction of the pitch diameter entails the corresponding changes in other design parameters of the shaving cutter as well as in the shaving cutter base pitch. The latter is not permitted as it violates the accuracy requirements.

The base pitch of the shaving cutter $p_{b.ct}$ can be expressed in terms of the axial pitch $P_{x.ct}$, pitch helix angle $\psi_{ct}$, and normal profile angle $\phi_{ct}$ of the shaving cutter [13]:

$$p_{b.ct} = P_{x.ct} \sin \psi_{ct} \cos \phi_{ct} \tag{16.13}$$

By using the engineering formulae for specification of the gear tooth flank (see Appendix F), Eq. (16.13) casts into an equivalent form:

$$p_{b.ct} = \frac{P_{x.ct}\, \pi d_{ct} \cos \phi_{ct}}{\sqrt{P_{x.ct}^2 + \pi^2 d_{ct}^2}} \tag{16.14}$$

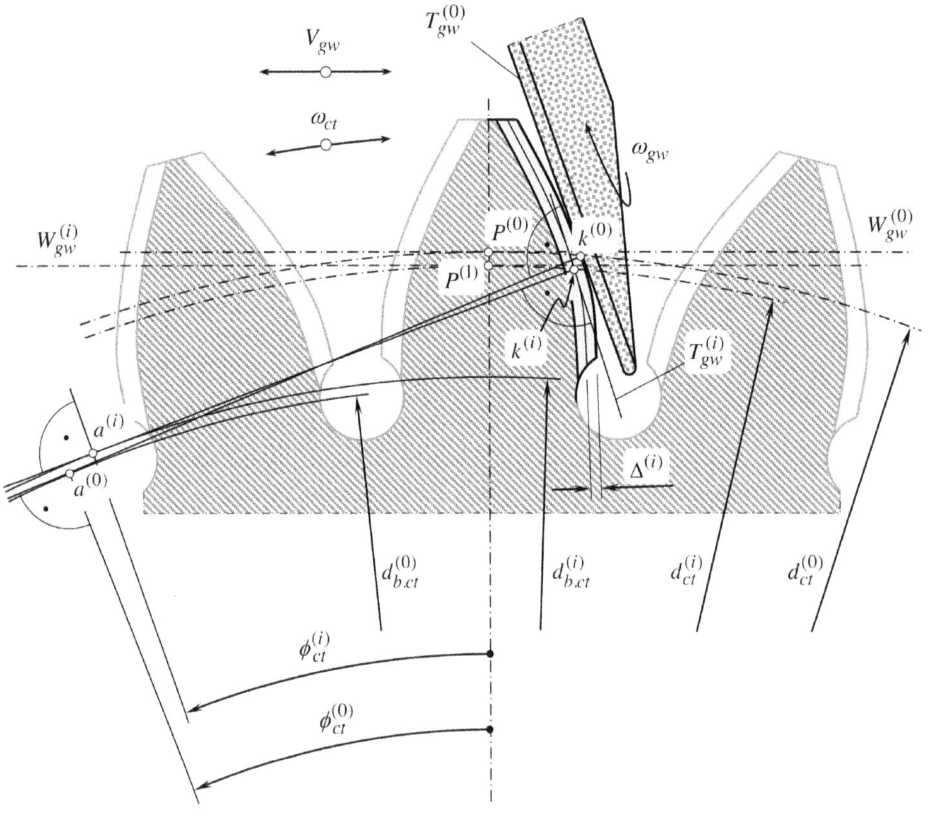


**FIGURE 16.15** Schematic of the method of resharpening a gear shaving cutter. (From Radzevich, S.P., and Palagura, V.A., USSR Patent 1646725, Int. Cl. B23f 21/28, Dec. 13, 1988; Radzevich, S.P., *SME Journal of Manufacturing Science and Engineering*, 127(4), 819–828, 2005. With permission.)

Equation (16.14) reveals that in the conventional methods of resharpening shaving cutters, the base pitch $p_{b.ct}$ is a function of pitch diameter $d_{ct}$ of the shaving cutter, that is, a dependence $p_{b.ct} = p_{b.ct}(d_{ct})$ is observed. The pitch diameter of the shaving cutter reduces from $d_{ct}^{(0)}$ to $d_{ct}^{(i)} < d_{ct}^{(0)}$ after each resharpening of the shaving cutter. Therefore, corresponding alterations to the base pitch $p_{b.ct}$ after resharpening of the shaving cutter are inevitable.

Any and all alterations to the base pitch of a shaving cutter are undesirable due to the accuracy requirements. As the alterations to diametral dimensions of the shaving cutter after resharpening are inevitable, their impact on the base pitch $p_{b.ct}$ can be compensated by corresponding changes to other design parameters of the shaving cutter. Ultimately, in this way it is possible to resharpen a shaving cutter so as to make its base pitch insensitive to the inevitable reduction of the pitch diameter.

The base pitch of a shaving cutter is not affected by any alterations to the pitch diameter if and only if the equality

$$\frac{\partial\, p_{b.ct}(d_{ct})}{\partial d_{ct}} = 0 \tag{16.15}$$

is valid. Equation (16.15) states that any and all alterations to the design parameters of the shaving cutter are allowable as long as the equality [see Eq. (16.15)] is fulfilled.

After substituting Eq. (16.14) in Eq. (16.15), the latter casts into a formula for the calculation of profile angle $\phi_{ct}^{(i)}$ of the shaving cutter after having been resharpened $i$ times:

$$\phi_{ct}^{(i)} = \cos^{-1}\left(\hat{C}\cdot\frac{P_{x.ct}^2 + \pi^2[d_{ct}^{(i)}]^2}{\pi^2[d_{ct}^{(i)}]^2}\right) \tag{16.16}$$

where $\hat{C}$ is the coefficient, which is calculated from the equation

$$\hat{C} = \frac{P_{x.ct}^2 + \pi^2\left[d_{ct}^{(0)}\right]^2}{\pi^2\left[d_{ct}^{(0)}\right]^2}\cdot\cos\phi_{ct}^{(0)} \tag{16.17}$$

Example. Design parameters of a new shaving cutter are collected in Table 16.1.

Equation (16.17) allows for calculation of the coefficient $\hat{C} = 0.06295$.

**TABLE 16.1**
**Design Parameters of a New Shaving Cutter**

| | |
|---|---|
| Module | $m = 5$ mm |
| Tooth number | $N_{ct} = 43$ |
| Pitch diameter | $d_{ct}^{(0)} = 222.584$ mm |
| Normal profile angle | $\phi_{ct}^{(0)} = 20°$ |
| Pitch helix angle | $\psi_{ct}^{(0)} = 15°$ |
| Base diameter | $d_{b.ct}^{(0)} = 208.288$ mm |
| Axial pitch | $P_{x.ct} = 2608.373$ mm |

In the case under consideration, the stock $\Delta^{(i)}$ for resharpening the shaving cutter is equal to $\Delta^{(i)} = 0.05$ mm.

The set of the design parameters of a new shaving cutter (see Table 16.1) along with the calculated value of the coefficient $\hat{C}$ and the specified value for the stock $\Delta^{(i)}$ to be reground, allows for calculation of the desirable profile angle $\phi_{ct}^{(i)}$ of the resharpened shaving cutter. Equation (16.16) returns $\phi_{ct}^{(i)} = 19°38'$ [181].

In order to obtain the desirable profile angle $\phi_{ct}^{(i)} = 19°38'$, the base diameter for the resharpened shaving cutter $d_{b.ct}^{(i)}$ must increase from $d_{b.ct}^{(0)} = 208.288$ mm to $d_{b.ct}^{(i)} = 209.390$ mm [181].

The recommended alterations to the design parameters of the resharpened shaving cutter compared to the corresponding design parameters of the new shaving cutter result in an increased accuracy of the finished gears.

The accuracy of the resharpened tooth profile of the shaving cutter is inspected (see Figure 16.16). For this purpose, either *CMM* or *GMM* is commonly used.

(a) (b)

**FIGURE 16.16** Inspection of a shaving cutter tooth profile after being resharpened: (a) close-up, and (b) the stylus probe tip in contact with the shaving cutter tooth.

## 16.3 AXIAL METHOD OF GEAR SHAVING PROCESS

The axial[7] method of gear shaving process is one of four basic methods of rotary shaving of external spur and helical gears. This method of shaving gears is extensively used in low- and medium-production operations. It is the most economical method for shaving wide-face-width gears.

### 16.3.1 Kinematics of Axial Method of Gear Shaving Process

When the axial method of gear shaving is used, the work gear and the shaving cutter are meshed in crossed-axes relationship (see Figure 16.17). The crossing axes of the work gear $O_g$ and the shaving cutter $O_{ct}$ are at a certain center distance $C$ apart from one another.

The shaving cutter is rotated in both directions during the work cycle. The work gear spindle and the shaving cutter spindle are driven in timed relation to each other by CNC-controlled motors. When a non-CNC machine is used, the rotation of the work gear spindle is not timed with that of the shaving cutter spindle. In Figure 16.17, the rotation of the work gear about its axis $O_g$ is designated as $\boldsymbol{\omega}_g$. The rotation of the shaving cutter about its axis $O_{ct}$ is designated as $\boldsymbol{\omega}_{ct}$. The rotations $\boldsymbol{\omega}_g$ and $\boldsymbol{\omega}_{ct}$ are properly timed with one another ($\omega_g \cdot N_g = \omega_{ct} \cdot N_{ct}$, here $N_g$ and $N_{ct}$ denote the tooth number of the work gear and the shaving cutter, respectively, and the equalities $\omega_g = |\boldsymbol{\omega}_g|$ and $\omega_{ct} = |\boldsymbol{\omega}_{ct}|$ are observed). The rotation vectors $\boldsymbol{\omega}_g$ and $\boldsymbol{\omega}_{ct}$ are at the angle $\Sigma = \angle(\boldsymbol{\omega}_g, \boldsymbol{\omega}_{ct})$. The angle $\bar{\Sigma}$ of crossing of the axes $O_g$ and $O_{ct}$ complements the angle $\Sigma$ to 180°, that is, $\bar{\Sigma} = \pi - \Sigma$.

In addition to the rotations $\boldsymbol{\omega}_g$ and $\boldsymbol{\omega}_{ct}$, the work gear is reciprocated $\mathbf{F}_{ct}$ in a path parallel to the work gear axis of rotation. It is customary to move the work gear past the shaving cutter by means of the feed slide. On some of the larger shaving machines, for obvious reasons, the shaving cutter is mounted on a feed slide. In the consideration below, the *relative motion* of the work gear and the shaving cutter is considered regardless of whether the worktable or the shaving cutter is actually reciprocated.

The length of traverse is determined by the face width of the work gear. In practice [211], for the best results, the length of traverse should be approximately 1.6 mm (1/16 in.) greater than the face width of the work gear, allowing for minimum over-travel at each end of the work gear face. In Figure 16.17, the left-end position and the right-end position of the reciprocating work gear is shown by dashed lines. The length of the reciprocating path depends on (a) the face width of the work gear, (b) the face width of the shaving cutter, and (c) the crossed-axes angle $\Sigma$.

During the work cycle, the center distance is being reduced in small controlled increments. In order to gradually reduce the center distance, the shaving cutter is fed into the work gear

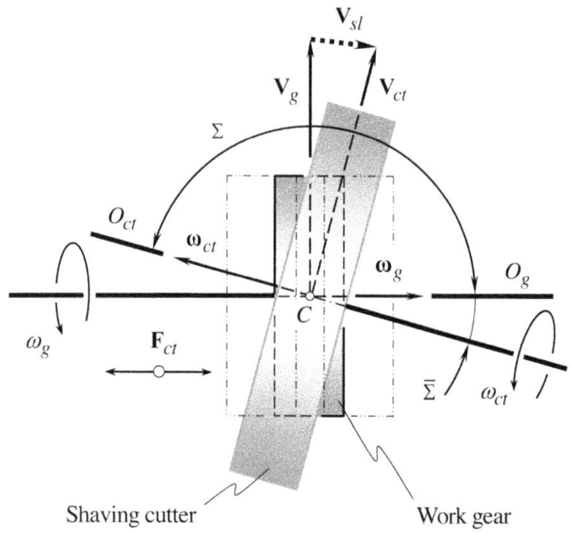


FIGURE 16.17 Schematic of the axial method of the gear shaving process.

to give the required radial tooth loading, and the machine is then run with the shaving cutter and the work gear in mesh for a predetermined number of passes. The automatic infeed of 0.001–0.002 in. (0.025–0.050 mm) is applied at each end of the shaving cutter traverse.

In the axial method of shaving, the traverse path is along the axis of the work gear. The number of strokes may vary due to the amount of stock to be removed. In this way the desirable tooth form, thickness, and finish are produced.

### 16.3.2 Cutting Speed in the Axial Method of Rotary Shaving of a Gear

The speed of cutting in the gear shaving process is created in a peculiar manner, which significantly differs from that, for example, in the turning or milling process. This issue was already briefly discussed in section 16.1.4.

The speed of the primary motion $\mathbf{V}_{cut}$ in rotary shaving of the gears is equal to the speed of the resultant sliding $\mathbf{V}_{\Sigma}$ of the shaving cutter tooth flank relative to the work gear tooth flank. As the teeth of the shaving cutter move across face width of the work gear, the fine chips are shaved up from flanks of the work gear teeth.

The nature of sliding of the shaving cutter tooth flanks relative to the work gear tooth flanks is the same as observed, for example, in the gear hobbing process. However, in rotary shaving the flank sliding is not as evident as it is when hobbing gears.

The speed of cut $\mathbf{V}_{cut}$ can be expressed in terms of (a) the sliding velocity $\mathbf{V}_{sl}$, in the lengthwise direction of the work gear tooth, (b) the velocity $\mathbf{V}_{pr}$ of the profile sliding, and (c) the velocity of sliding that is caused by the reciprocation motion $\mathbf{F}_{ct}$. Equation (16.4) reflects this functionality. Understanding the nature of the tooth flank sliding is the clue to determining the cutting speed in the rotary shaving of the gears. For this purpose, it is helpful to use the results

[7] The axial method of the gear shaving process is also referred to as either the *conventional* or *traverse* method of rotary shaving of the gears.

of investigation of the geometry and kinematics of crossed axes involute gearing [14].

In the rotary shaving process, geometrically, the work gear tooth flank $\mathscr{G}$ and the shaving cutter tooth flank $T$ make point contact at every instant of time.

#### 16.3.2.1 Impact of the Crossed-Axes Angle

When the axes of rotation $O_g$ and $O_{ct}$ of the work gear and the shaving cutter are not parallel to one another [see Figure 16.18(a)]—say they are at a certain crossing angle, $\bar{\Sigma}$—the pitch plane of the work gear $W_g$ travels $\mathbf{V}_g$ in the direction that is perpendicular to both, to the work gear axis $O_g$ and the centerline ℄. Similarly, the pitch plane of the shaving cutter $W_{ct}$ travels $\mathbf{V}_{ct}$ in the direction that is perpendicular to both, to the shaving cutter axis $O_{ct}$ and the centerline ℄.

The pitch cylinder of the work gear is driven by the pitch cylinder of the shaving cutter, while its own pitch plane is screwed along and engages the pitch cylinder of the shaving cutter, as indicated in Figure 16.18(a).

Thus, we have a closed circuit of action on the work gear and the shaving cutter. To summarize, we have the following: The helix of the shaving cutter screws its pitch plane $W_{ct}$ along its line of travel. This pitch plane is always tangent to the pitch cylinder of the work gear and causes it to rotate. The rotation of the work gear screws its pitch plane $W_g$ along its path, and this pitch plane is always tangent to the pitch cylinder of the shaving cutter. The rate of travel of this pitch plane must be the same as that of the circumference of the pitch cylinder of the shaving cutter. This completes the closed circuit.

The auxiliary generating surface $T$ is zero thickness film that is shaped in the form of the basic rack. The body of the rack that is conjugate to the gear intended to be machined is located on one side of the generating surface $T$, while the body of the rack that is conjugate to the shaving cutter is located on the opposite side of the generating surface $T$.

The forms of the conjugate racks of the work gear and of the shaving cutter match each other as indicated in Figure 16.18(b). When the rack of the shaving cutter is moved in the direction of motion of the pitch plane $W_g$ of the work gear, it acts as a cam on the rack of the work gear and forces it to move in the direction of motion of the pitch cylinder of the work gear, as indicated in Figure 16.18(b). Two intermediate conclusions follow from this consideration.

*First*, as the velocity vectors $\mathbf{V}_g$ and $\mathbf{V}_{ct}$ do not align with one another, this results in sliding velocity $\mathbf{V}_{sl} = \mathbf{V}_{ct} - \mathbf{V}_g$ [see Eq. (16.7)]. The sliding velocity $\mathbf{V}_{sl}$ is a component that most significantly benefits the speed of cut $\mathbf{V}_{cut}$, compared to the rest of the components in Eq. (16.4).

*Second*, because each of the velocity vectors $\mathbf{V}_g$ and $\mathbf{V}_{ct}$ is of a constant value, then the sliding velocity vector $\mathbf{V}_{sl}$ is also of a constant value, and it does not alter within tooth height either of the work gear or of the shaving cutter.

Increasing the angle between the shaving cutter and work gear axes increases the cutting action $\mathbf{V}_{cut}$, but as this reduces the width of the contact zone, the guiding action is sacrificed. Practically no cutting occurs at zero angle $\bar{\Sigma}$. Conversely, the guiding action can be increased by reducing the angle of crossed axes but at the expense of cutting action.

#### 16.3.2.2 Impact of the Traverse Motion

In the axial shaving of the gears, the direction of the traverse is parallel to the axis of rotation of the work gear. As the traverse motion is not tangent to the flank of the work gear tooth (except in the case of shaving the spur gear), the reciprocation speed affects the cutting speed in two ways.

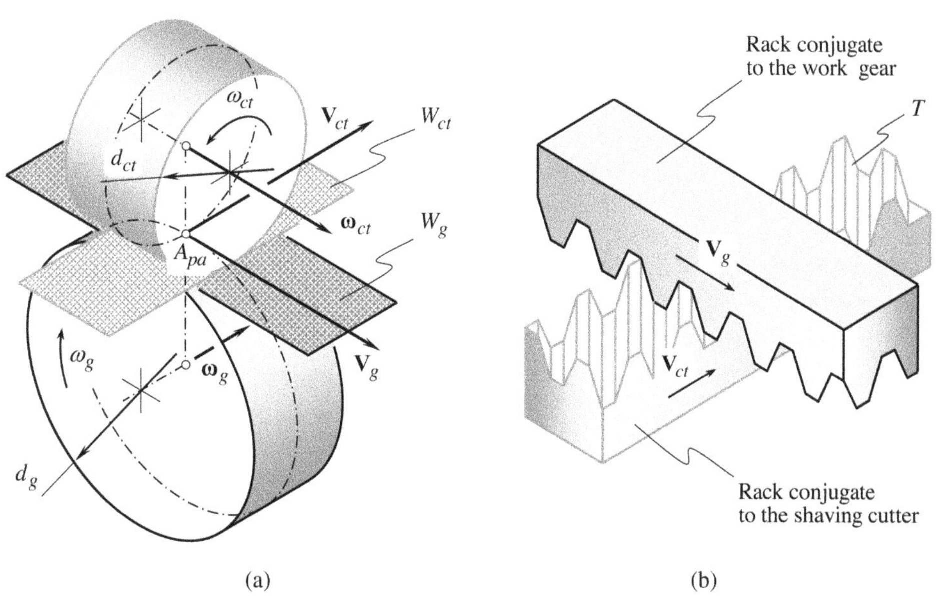


**FIGURE 16.18** Nature of the action between the work gear and the shaving cutter in the gear shaving process: (a) configuration of the pitch cylinders and (b) interaction of the auxiliary racks.

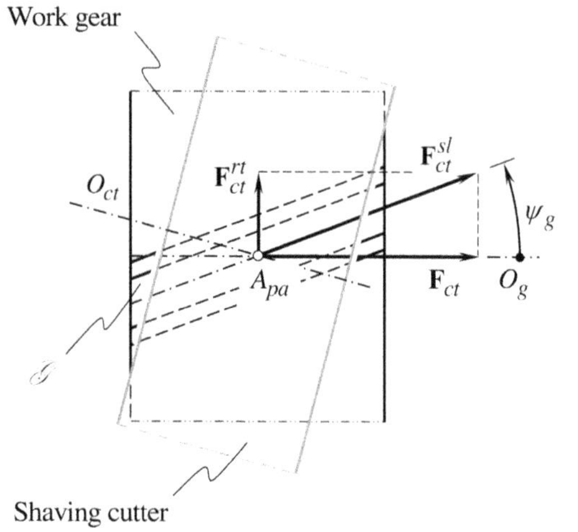


**FIGURE 16.19** Impact of the traverse motion on the cutting speed $\mathbf{V}_{cut}$ in the axial shaving of gears.

Let us assume that the shaving cutter is not rotating, and it is just reciprocating in the axial direction of the work gear. Under such a scenario, the straight reciprocation $\mathbf{F}_{ct}$ of the shaving cutter is at an angle to the work gear tooth flank, as indicated in Figure 16.19. Within the pitch plane this angle is equal to the work gear pitch helix angle $\psi_g$.

The traverse motion $\mathbf{F}_{ct}$ can be resolved in two directions.

1. One component, $\mathbf{F}_{ct}^{sl}$, is in the direction tangent to tooth flank $\mathcal{G}$ of the work gear. The magnitude of this component is equal:

$$F_{ct}^{sl} = \left|\mathbf{F}_{ct}^{sl}\right| = \frac{|\mathbf{F}_{ct}|}{\cos\psi_g} = \frac{F_{ct}}{\cos\psi_g} \tag{16.18}$$

   The component $\mathbf{F}_{ct}^{sl}$ represents the pure sliding of the tooth flanks of the work gear and the shaving cutter.

   It is important to point out here that the vector $\mathbf{F}_{ct}^{sl}$ of the sliding has a constant value, and it does not change with tooth height either of the work gear or of the shaving cutter.

2. Another component $\mathbf{F}_{ct}^{rt}$ of the reciprocation $\mathbf{F}_{ct}$ is located within the pitch plane and is perpendicular to the axis of rotation of the work gear $O_g$. For the calculation of the magnitude $F_{ct}^{rt}$ of the component $\mathbf{F}_{ct}^{rt}$, the following formula:

$$F_{ct}^{rt} = F_{ct}\tan\psi_g \tag{16.19}$$

   applies. The actual direction of the vector $\mathbf{F}_{ct}^{rt}$ depends on the hand of the pitch helix angle $\psi_g$.

The component $\mathbf{F}_{ct}^{rt}$ adds to the rotation of the work gear. This additional rotation $\omega_g^{ad}$ is equal:

$$\omega_g^{ad} = \frac{2F_{ct}\tan\psi_g}{d_g} \tag{16.20}$$

The rotation vector $\boldsymbol{\omega}_g^{ad}$ is aligned with the rotation vector $\boldsymbol{\omega}_g$. Depending on (a) the rotation of the shaving cutter $\boldsymbol{\omega}_{ct}$, (b) the direction of the hand and the value of the helix angle $\psi_g$, and (c) the actual direction of the reciprocation $\mathbf{F}_{ct}$, the rotation vector $\boldsymbol{\omega}_g^{ad}$ is pointed either in the same direction as the rotation vector $\boldsymbol{\omega}_g$ or opposite to $\boldsymbol{\omega}_g$. In the first case the rotation $\boldsymbol{\omega}_g^{ad}$ adds to the rotation $\boldsymbol{\omega}_g$. In the second case it subtracts from the rotation $\boldsymbol{\omega}_g$.

Because the additional rotation $\omega_g^{ad}$ of the work gear depends upon the pitch helix angle, the greater the pitch helix angle $\psi_g$ of the work gear, the greater the impact of the reciprocation motion $\mathbf{F}_{ct}$ of the shaving cutter on the cutting speed $\mathbf{V}_{cut}$.

### 16.3.2.3 Impact of Profile Sliding

Sliding of the shaving cutter tooth flank relative to the work gear tooth flank always occurs in the rotary shaving of gears. The speed of tooth profile sliding also contributes to the speed of cut in the gear shaving process.

The *work gear-to-shaving cutter mesh* is a kind of spatial mesh of cogged wheels. Implementation of 3*D* analysis is required to investigate the axial method of the gear shaving process in order to determine the sliding component of the speed of the cut.

Before undertaking a comprehensive investigation of the work gear-to-shaving cutter mesh in 3*D*, a preliminary simplified 2*D* consideration is helpful. In this case it is an advantage that frequently the crossed-axes angle in the rotary shaving process is reasonably small and does not exceed $\Sigma \leq 15 \div 20°$. This means that the results of the investigation of the work gear-to-shaving cutter mesh in 2*D* could be a good approximation to the accurate results of the investigation in 3*D*.

For the purposes of the preliminary analysis, consider the work gear-to-shaving cutter mesh that features the parallel axes $O_g$ and $O_{ct}$ of the rotation as shown in Figure 16.20 [165]. The base diameter of the work gear is designated as $d_{b.g}$, and the base diameter of the shaving cutter is designated as $d_{b.c}$.

The rolling action between the work gear and the shaving cutter is transmitted along the line of action *LA*. The sliding action between the work gear and the shaving cutter is observed in the direction perpendicular to the line of action.

The line of action in the work gear-to-shaving cutter mesh is tangent from the opposite sides to the base cylinders of the work gear $d_{o.g}$ and that of the shaving cutter $d_{o.ct}$. In Figure 16.20, the work gear and the shaving cutter are depicted in a position at which the line of action $N^g N^{ct}$ is horizontal. Points $P^g$ and $P^{ct}$ within the line of action *LA* correspond to the points of intersection of the line of action by the outer cylinder of the gear $d_{o.g}$ and by the outer cylinder of the shaving cutter $d_{o.ct}$, respectively. The segment $P^g P^{ct}$ of the line of action represents the active portion of the line of action.

The pitch cylinders of the work gear and the shaving cutter pass through the apex point $A_{pa}$ (through the pitch point $P$). Due to lack of space, the pitch cylinders are not shown in Figure 16.20.

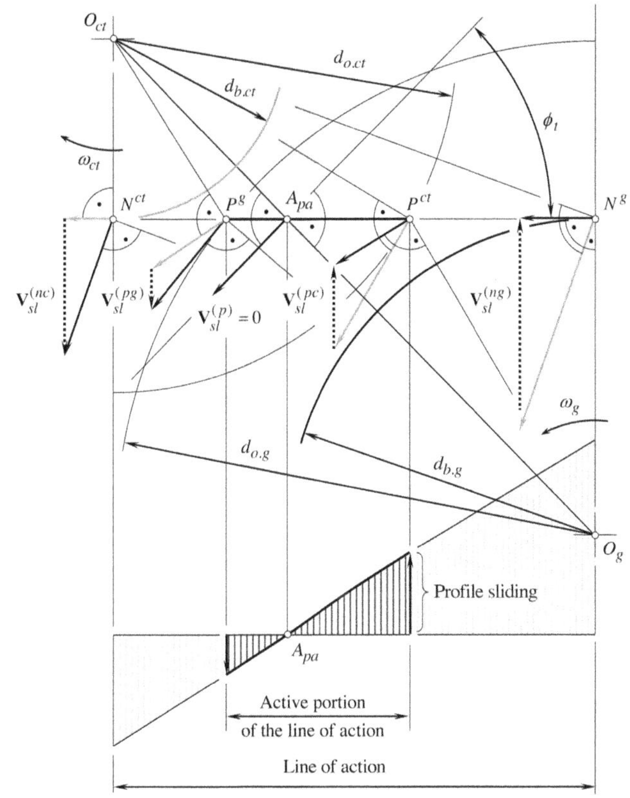


**FIGURE 16.20** Approximate 2*D* interpretation of the work gear-to-shaving cutter mesh in the rotary shaving of gears: Axes of the rotations of the work gear $O_g$ and the shaving cutter $O_{ct}$ are parallel to each other. (After Radzevich, S.P., *AGMA Technical Paper* 06FTM11, 2006, 22 pp.)

The points $N^g$, $P^{ct}$, $P$, $P^g$, and $N^{ct}$ within the line of action $N^g N^{ct}$ correspond to different instances of phases of meshing. For each of the points, the corresponding vector of linear velocity of the work gear $\mathbf{V}_g^{(i)}$ and the shaving cutter $\mathbf{V}_{ct}^{(i)}$ are depicted in Figure 16.20. Each of the vectors $\mathbf{V}_g^{(i)}$ is perpendicular to the radius to the corresponding point $N^g$, $P^{ct}$, $P$, $P^g$, and $N^{ct}$ from the gear axis of rotation $O_g$. Similarly, each of the vectors $\mathbf{V}_{ct}^{(i)}$ is perpendicular to the radius to the corresponding point $N^g$, $P^{ct}$, $P$, $P^g$ drawn from the shaving cutter axis of rotation $O_{ct}$. The vectors $\mathbf{V}_g^{(i)}$ and $\mathbf{V}_{ct}^{(i)}$ are not labeled in Figure 16.20 due to lack of space.

The magnitude of the vector $\mathbf{V}_g^{(i)}$ is equal to $V_g^{(i)} = \left|\mathbf{V}_g^{(i)}\right| = \omega_g R_g^{(i)}$. Here, the radius of the cylinder at which the $i-$ th point of contact is located is designated as $R_g^{(i)}$. For computing the magnitude of the vector $\mathbf{V}_{ct}^{(i)}$, the similar formula $V_{ct}^{(i)} = \left|\mathbf{V}_{ct}^{(i)}\right| = \omega_{ct} R_{ct}^{(i)}$ is valid. In this expression, $R_{ct}^{(i)}$ denotes the radius of the cylinder at which the $i-$ th point of contact is located.

Once the velocity vectors $\mathbf{V}_g^{(i)}$ and $\mathbf{V}_{ct}^{(i)}$ are expressed in a common reference system, the vector of the sliding velocity $\mathbf{V}_{sl}^{(i)}$ is equal to the difference:

$$\mathbf{V}_{sl}^{(i)} = \mathbf{V}_{ct}^{(i)} - \mathbf{V}_g^{(i)} \tag{16.21}$$

In this equation, the velocity vectors $\mathbf{V}_g^{(i)}$ and $\mathbf{V}_{ct}^{(i)}$ can be expressed in terms of the rotation vectors $\boldsymbol{\omega}_g$ and $\boldsymbol{\omega}_{ct}$ and of distances $\mathbf{R}_g^{(i)}$ and $\mathbf{R}_{ct}^{(i)}$ of the point of contact at a current instant of time:

$$\mathbf{V}_g^{(i)} = \boldsymbol{\omega}_g \times \mathbf{R}_g^{(i)} \tag{16.22}$$

$$\mathbf{V}_{ct}^{(i)} = \boldsymbol{\omega}_{ct} \times \mathbf{R}_{ct}^{(i)} \tag{16.23}$$

After being represented in a common coordinate system, the velocity vectors $\mathbf{V}_g^{(i)}$ and $\mathbf{V}_{ct}^{(i)}$ can be entered into Eq. (16.21) for calculating the sliding velocity magnitude $V_{sl}^{(i)}$

$$V_{sl}^{(i)} = 0.5\left(\omega_{ct}\sqrt{d_{y.ct}^2 - d_{b.ct}^2} - \omega_g\sqrt{d_{y.g}^2 - d_{b.g}^2}\right) \tag{16.24}$$

where $d_{y.g}$ and $d_{y.ct}$ are the diameters of the circles of the work gear and the shaving cutter passing through the $i-$ th point of contact.

The vector of the sliding velocity $\mathbf{V}_{sl}^{(i)}$ is always perpendicular to the line of action. Its magnitude, $V_{sl}^{(i)} = \left|\mathbf{V}_{sl}^{(i)}\right|$, varies within the tooth height of the work gear as well as within the tooth height of the shaving cutter. The larger the distance of the $i-$ th point of contact from the apex point $A_{pa}$ (through the pitch point $P$), the larger the sliding velocity $V_{sl}^{(i)}$, and vice versa. No sliding occurs at the apex point $A_{pa}$ (through the pitch point $P$), that is, the equality $\mathbf{V}_{sl}^{(p)} = 0$ is always valid at the apex point $A_{pa}$ (through the pitch point $P$).

Because the vectors $\mathbf{V}_g^{(i)}$ and $\mathbf{V}_{ct}^{(i)}$ of the linear velocity are not constant within the tooth height, the velocity vector $\mathbf{V}_{sl}^{(i)}$ of the profile sliding varies within tooth height of both the work gear and the shaving cutter teeth.

As the schematic depicted in Figure 16.20 features parallel axes of rotation $O_g$ and $O_{ct}$ of the work gear and the shaving cutter, the results of the analysis obtained on the premise of Figure 16.20 are a kind of approximation. However, the approximation is practically reasonable for three reasons. *First*, the crossing angle $\Sigma$ between the axes of rotation $O_g$ and $O_{ct}$ is relatively small. Commonly it does not exceed $\Sigma \le 20°$. *Second*, the speed of profile sliding $\mathbf{V}_{sl}^{(i)}$ contributes less significantly to the cutting speed. Only for work gears and/or shaving cutters with a low number of teeth does the component $\mathbf{V}_{sl}^{(i)}$ become more significant, while in the apex point $A_{pa}$ (through the pitch point $P$) it remains at zero value $[\mathbf{V}_{sl}^{(p)} = 0]$ regardless of the tooth numbers. *Third*, in reality, there is no need for absolutely accurate determination of the speed of the profile sliding. However, if necessary, an accurate calculation of the component $\mathbf{V}_{sl}^{(i)}$ can be performed as well.

For the purpose of accurately determining the speed of the profile sliding, the spatial, that is, three-dimensional and not planar (two-dimensional), mesh of the work gear-to-shaving cutter has to be considered instead. In this way it is possible to construct for the 3*D* mesh a diagram similar to that constructed for the 2*D* mesh (see Figure 16.20).

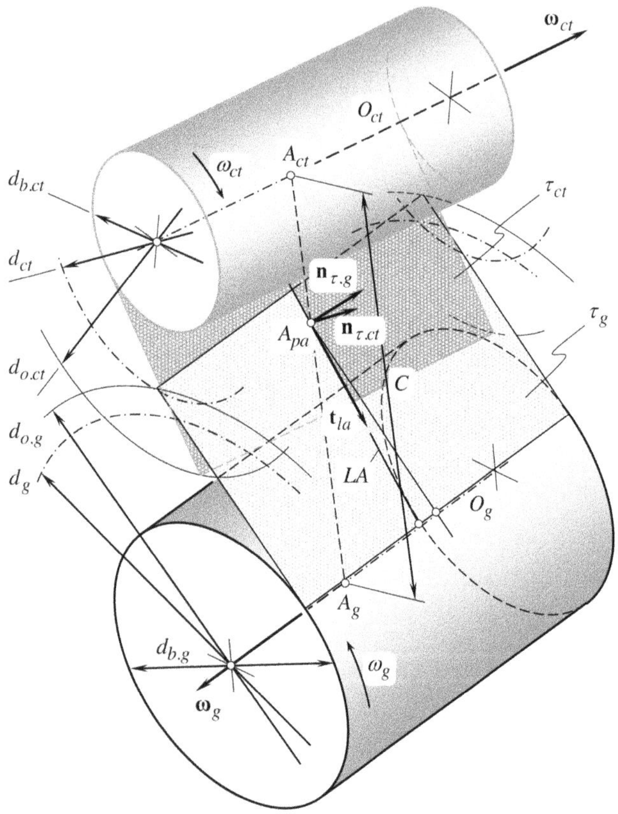


**FIGURE 16.21** Three-dimensional approximate schematic of the work gear-to-shaving cutter mesh.

Better results can be obtained if the approach immediately outlined in the following is used for the analysis.

Consider the disposition and relative orientation of the base cylinder of the work gear and the base cylinder of the shaving cutter shown in Figure 16.21.

The line of action *LA* is the line along which the shaving cutter interacts with the work gear. In the approach under consideration, the line of action *LA* is interpreted as the line of intersection of two planes $\tau_g$ and $\tau_{ct}$. The plane $\tau_g$ passes through the apex point $A_{pa}$. This plane is tangent to the base cylinder of the work gear. The plane $\tau_{ct}$ also passes through the apex point $A_{pa}$. This plane is tangent to the base cylinder of the shaving cutter. The line of action *LA* is the straight line of intersection of the planes $\tau_g$ and $\tau_{ct}$. It crosses each of the axes $O_g$ and $O_{ct}$ at the base helix angle. The line of action is tangent to both of the base cylinders. A segment of straight line between the points of tangency of the line of action *LA* and the base cylinders represents the entire line of action. The active portion of the line of action is terminated by the points of intersection of the straight line *LA* with the outer cylinder of the work gear and with the outer cylinder of the shaving cutter.

Further, as the axes of rotation $O_g$ and $O_{ct}$ cross each other at a certain angle $\bar{\Sigma}$, the equivalent base diameter $d^*_{b.g}$ of the work gear and the equivalent base diameter $d^*_{b.c}$ of the shaving cutter differ from that applied in the two-dimensional case.

Ultimately only three design parameters are required for the construction of the 3*D* mesh diagram. They are:

- The entire line of action *LA* measured between the points of tangency of the straight line of action with the base cylinders.
- The equivalent base diameter $d^*_{b.g}$ of the work gear.
- The equivalent base diameter $d^*_{b.ct}$ of the shaving cutter.

*Line of action*. Consider the tangent plane $\tau_g$ shown in Figure 16.22. In the Cartesian coordinate system $X_gY_gZ_g$ associated with the work gear, the position vector $\mathbf{r}_{\tau g}$ of a point of the plane $\tau_g$ can be expressed in matrix form as

$$\mathbf{r}_{\tau g}(X_g, Z_g) = \begin{bmatrix} X_g \\ 0.5d_{b.g} \\ Z_g \\ 1 \end{bmatrix} \qquad (16.25)$$

Similarly, in the Cartesian coordinate system $X_{ct}Y_{ct}Z_{ct}$ associated with the shaving cutter, the position vector $\mathbf{r}_{\tau c}$ of a point of the plane $\tau_c$ can be expressed by the equation

$$\mathbf{r}_{\tau c}(X_{ct}, Z_{ct}) = \begin{bmatrix} X_{ct} \\ -0.5d_{b.ct} \\ Z_{ct} \\ 1 \end{bmatrix} \qquad (16.26)$$

In order to be treated together, both of the position vectors $\mathbf{r}_{\tau g}$ and $\mathbf{r}_{\tau c}$ have to be represented in a common reference system. Neither the coordinate system $X_gY_gZ_g$ associated with the work gear nor the coordinate system $X_{ct}Y_{ct}Z_{ct}$ associated with the shaving cutter is convenient to serve as the common reference system. Therefore, in the case under consideration, an intermediate reference system is used as the common reference system.

The required coordinate system transformation can be performed in four consequent steps:

1. The rotation of the shaving cutter reference system $X_{ct}Y_{ct}Z_{ct}$ about the $Z_{ct}$ – axis through the angle $\vartheta$. This angle is calculated from the expression $\vartheta = \sin^{-1}(d_{b.ct}/d_{ct})$. This linear transformation is analytically described by the operator of rotation $\mathbf{Rt}(\vartheta, Z_{ct})$.
2. The translation along the $X_{ct}$ – axis at the center distance $C$. Analytically this linear transformation is described by the operator of translation $\mathbf{Tr}(C_{g/ct}, X_{ct})$.
3. The rotation about the $X_{ct}$ – axis through the crossed-axes angle $\bar{\Sigma}$. For this purpose, the operator of rotation $\mathbf{Rt}(\bar{\Sigma}, X)$ is applied.

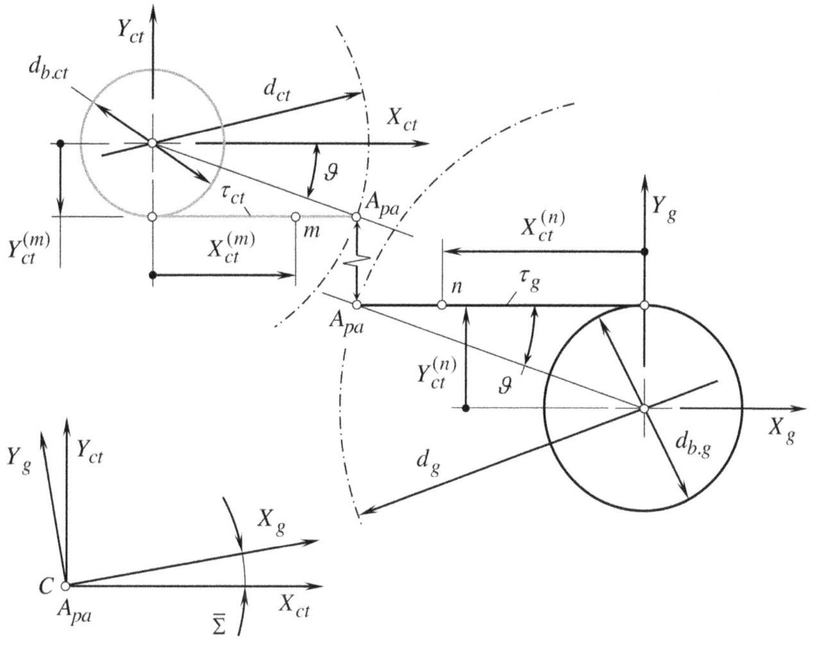


**FIGURE 16.22** Planes $\tau_g$ and $\tau_{ct}$ through the apex point $A_{pa}$ tangent to the base cylinders of a work gear and the shaving cutter.

4. The rotation of the work gear coordinate system $X_g Y_g Z_g$ about the $Z_g$ – axis through the angle $\vartheta = \sin^{-1}(d_{b.g}/d_g)$, which is equal to the angle $\vartheta$, used in the first step. This linear transformation is analytically described by the operator of rotation $\mathbf{Rt}(\vartheta, Z_g)$.

Steps 1 through 3 relate to transformations of the shaving cutter coordinate system $X_{ct}Y_{ct}Z_{ct}$ to the common coordinate system. In step 4, the work gear reference system $X_g Y_g Z_g$ is transformed into the common coordinate system.

In the common reference system, the position vector $\hat{\mathbf{r}}_{\tau.g}$ of a point of the plane $\tau_g$ can be expressed in terms of the position vector $\mathbf{r}_{\tau.g}$ and the operator of rotation $\mathbf{Rt}(\vartheta, Z_g)$:

$$\hat{\mathbf{r}}_{\tau.g} = \mathbf{Rt}(\vartheta, Z_g) \cdot \mathbf{r}_{\tau.g} \tag{16.27}$$

In that same reference system, the position vector $\hat{\mathbf{r}}_{\tau.ct}$ of a point of the plane $\tau_{ct}$ can be expressed by the equation

$$\hat{\mathbf{r}}_{\tau.ct} = \mathbf{Rt}(\Sigma, X) \cdot \mathbf{Tr}(C, X) \cdot \mathbf{Rt}(\vartheta, Z_{ct}) \cdot \mathbf{r}_{\tau.ct} \tag{16.28}$$

Equations (16.27) and (16.28) allow for the calculation of unit normal vectors $\mathbf{n}_{\tau.g}$ and $\mathbf{n}_{\tau.ct}$ to the planes $\tau_g$ and $\tau_c$, respectively. It is a routine procedure to calculate the unit normal vectors $\mathbf{n}_{\tau.g}$ and $\mathbf{n}_{\tau.ct}$. Ultimately, the equation of the line of action $LA$ can be represented in the form

$$LA \;\Rightarrow\; \mathbf{r}_{la}(\lambda) = \mathbf{r}_P + \lambda \cdot \mathbf{t}_{la} \tag{16.29}$$

where $\mathbf{t}_{la}$ designates a unit vector along the line of action $LA$, and $\lambda$ is the parameter of the line of action.

Once the equation of the line of action is derived, the following formula can be used for the calculation of length of the entire line of action

$$LA = \frac{d_{b.g}}{2} \frac{\tan \phi_{ot.g}}{\sin \psi_{b.g}} + \frac{d_{b.ct}}{2} \frac{\tan \phi_{ot.ct}}{\sin \psi_{b.ct}} \tag{16.30}$$

where:

- $d_{b.g}$ is the base diameter of the work gear.
- $d_{b.ct}$ is the base diameter of the shaving cutter.
- $\phi_{ot.g}$ is the transverse profile angle at the outer diameter of the work gear.
- $\phi_{ot.ct}$ is the transverse profile angle at the outer diameter of the shaving cutter.
- $\psi_{b.g}$ is the base helix angle of the work gear.
- $\psi_{b.ct}$ is the base helix angle of the shaving cutter.

Equation (16.30) returns a reasonably accurate value for the length of the line of action, which can be used for determining the accurate value of the speed of the profile sliding in the rotary shaving of gears.

For the specification of the location of the apex point $A_{pa}$ within the line of action, the tooth ratio $u$ is customarily used. By definition, the tooth ratio is equal to $u = \omega_{ct}/\omega_g$. The ratio of linear velocities $V_g$ and $V_{ct}$ can be expressed in terms of the pitch helix angles $\psi_g$ and $\psi_{ct}$ (see Figure 16.17):

$$\frac{V_g}{V_{ct}} = \frac{\cos \psi_{ct}}{\cos \psi_g} \tag{16.31}$$

With that stated, the expression

$$u = \frac{\omega_g}{\omega_{ct}} = \frac{d_{ct} \cos \psi_{ct}}{d_g \cos \psi_g} \quad (16.32)$$

is valid for the calculation of the tooth ratio $u$.

For specified values of the diameters $d_{y.g}$ and $d_{y.ct}$ of the circles of the work gear and the shaving cutter passing through the $i$ – th point of contact, the respective equivalent diameters $d^*_{y.g}$ and $d^*_{y.ct}$ are calculated by the formulae

$$d^*_{y.g} = \frac{d_{y.g}}{\cos \psi_{b.g}} \quad (16.33)$$

$$d^*_{y.ct} = \frac{d_{y.ct}}{\cos \psi_{b.ct}} \quad (16.34)$$

*Equivalent base diameters.* The line of action $LA$ crosses the work gear axis of rotation $O_g$ at the base helix angle $\psi_{b.g}$ of the work gear. Similarly, the crossing angle between the line of action $LA$ and the shaving cutter axis of rotation $O_{ct}$ is equal to the base helix angle $\psi_{b.ct}$ of the shaving cutter. At the point of tangency with the line of action, the radius of the normal curvature of the work gear base cylinder is equal to the equivalent base diameter $d^*_{b.g}$ of the work gear. Similarly, at the point of tangency with the line of action, the radius of the normal curvature of the shaving cutter base cylinder is equal to the equivalent base diameter $d^*_{b.ct}$ of the shaving cutter. It can be concluded from these statements that the equivalent diameters $d^*_{b.g}$ and $d^*_{b.ct}$ are doubled normal curvatures of the base helices of the work gear and the shaving cutter. Therefore, the *Euler equation* can be used for the calculation of the equivalent diameters:

$$d^*_{b.g} = \frac{d_{b.g}}{\cos^2 \psi_{b.g}} \quad (16.35)$$

$$d^*_{b.ct} = \frac{d_{b.ct}}{\cos^2 \psi_{b.ct}} \quad (16.36)$$

Having the parameters $d^*_{y.g}$, $d^*_{y.ct}$, $d^*_{b.g}$, and $d^*_{b.ct}$ calculated [Eqs. (16.33)–(16.36)] makes it possible to enter them into the equation

$$V^{(i)}_{sl} = 0.5\left[\omega_{ct}\sqrt{(d^*_{y.ct})^2 - (d^*_{b.ct})^2} - \omega_g\sqrt{(d^*_{y.g})^2 - (d^*_{b.g})^2}\right] \quad (16.37)$$

which is an analogy to Eq. (16.24).

#### 16.3.2.4 Resultant Formula for Cutting Speed in Axial Gear Shaving

It is good to summarize here the results of the investigation undertaken in this section, and to put together all the components of the speed of the cut.

*First*, while the rotation of the shaving cutter remains the same [see Eq. (16.6)], the rotation of the work gear is affected by the reciprocation motion. Owing to the reciprocation $\mathbf{F}_{ct}$, an additional component $\omega^{ad}_g$ of the rotation of the work gear [see Eq. (16.20)] should be incorporated into Eq. (16.5) for calculation of the velocity vector $\mathbf{V}_g$:

$$\mathbf{V}_g = (\omega_g + \omega^{ad}_g) \times \mathbf{R}_g \quad (16.38)$$

Therefore, the sliding velocity vector $\mathbf{V}_{sl}$ [see Eq. (16.7)] is equal:

$$\mathbf{V}_{sl} = \underbrace{\omega_{ct} \times \omega_{ct}}_{\mathbf{V}_{ct}} - \underbrace{(\omega_g + \omega^{ad}_g) \times \mathbf{R}_g}_{\mathbf{V}_g} \quad (16.39)$$

*Second*, the speed of the reciprocation $\mathbf{F}_{ct}$ adds a component $\mathbf{F}^{sl}_{ct}$ [see Eq. (16.18)] to the cutting speed $\mathbf{V}_{cut}$.

*Third*, due to the profile sliding, the component $\mathbf{V}^{(i)}_{sl}$ affects the cutting speed $\mathbf{V}_{cut}$ as well [see Eq. (16.21)].

Taking into account the results of the analysis disclosed in the previous three sections, Eq. (16.4) can be cast into an expression for the cutting speed $\mathbf{V}_{cut}$ in the axial gear shaving process:

$$\mathbf{V}_{cut} = \mathbf{V}_{sl} + \mathbf{F}^{sl}_{ct} + \mathbf{V}^{(i)}_{sl} \quad (16.40)$$

It is assumed in Eq. (16.40) that all the components are expressed in a common reference system.

Because vector $\mathbf{V}^{(i)}_{sl}$ of the profile sliding varies within the tooth height of the work gear as well as within the tooth height of the shaving cutter, then the resultant vector of the cutting speed varies within the tooth height of the shaving cutter as well.

The vector of the cutting speed $\mathbf{V}_{cut}$ is always within the tangent plane to the work gear tooth flank $\mathscr{G}$ at a point of its contact with the shaving cutter tooth flank.

An accurate solution to the problem of sliding in crossed-axes gearing (like the work gear-to-shaving cutter mesh) is discussed to the best possible extent in [172].

## 16.4 DIAGONAL METHOD OF THE GEAR SHAVING PROCESS

The direction of traverse of the shaving cutter may be obliquely inclined to the work gear axis. In this case the method of rotary shaving of the gears is commonly referred to as the *diagonal method* of gear shaving[8]. This is the most extensively used method of gear shaving.

The use of diagonal shaving makes it possible for finishing gears to have a wide range of face widths with the shaving cutter of a given design. It also permits shaving gears with face widths slightly in excess of the shaving cutter

[8] The diagonal method of the gear shaving process is also referred to as the *modified underpass* or *angular traverse* method of shaving the gears.

width. More even wear distribution across the face of the shaving cutter can be easily achieved with this method of rotary shaving.

Diagonal shaving is used primarily in medium- and high-production operations. Using this method, shaving times are reduced by as much as 50%. Its production time in shaving is high because of the short stroke of the feed slide. One shaving cycle is usually sufficient to complete the shaving.

Diagonal shaving not only increases production rates substantially but also has other advantages for high-production operations.

### 16.4.1 Kinematics of Diagonal Method of Gear Shaving Process

When the diagonal method of gear shaving is used, the work gear and the shaving cutter are meshed in a crossed-axes relationship (see Figure 16.23). The crossing axes of the work gear $O_g$ and of the shaving cutter $O_{ct}$ are positioned at a certain center distance $C$.

Referring to Figure 16.23, the rotation of the work gear about its axis of rotation $O_g$ is designated as $\boldsymbol{\omega}_g$. The rotation of the shaving cutter about its axis of rotation $O_{ct}$ is designated as $\boldsymbol{\omega}_{ct}$. The shaving cutter is rotated in both directions during the work cycle. The rotations $\boldsymbol{\omega}_g$ and $\boldsymbol{\omega}_{ct}$ are properly timed with one another ($\omega_g \cdot N_g = \omega_{ct} \cdot N_{ct}$, here $N_g$ and $N_{ct}$ denote tooth number of the work gear, and the shaving cutter, respectively, and the equalities $\omega_g = |\boldsymbol{\omega}_g|$ and $\omega_{ct} = |\boldsymbol{\omega}_{ct}|$ are observed). The rotation vectors $\boldsymbol{\omega}_g$ and $\boldsymbol{\omega}_{ct}$ are at the angle $\Sigma = \angle(\boldsymbol{\omega}_g, \boldsymbol{\omega}_{ct})$. The angle $\bar{\Sigma}$ of crossing of the axes $O_g$ and $O_{ct}$ complements the angle $\Sigma$ to 180°, that is, $\bar{\Sigma} = \pi - \Sigma$.

The work gear reciprocates $\mathbf{F}_{ct}$ obliquely in relation to its own axis while the work gear and the shaving cutter are in mesh (see Figure 16.23). The traverse angle $\varphi_{fc}$ is achieved either by positioning the workpiece table obliquely or by interpolating two machine axes. It is customary to move

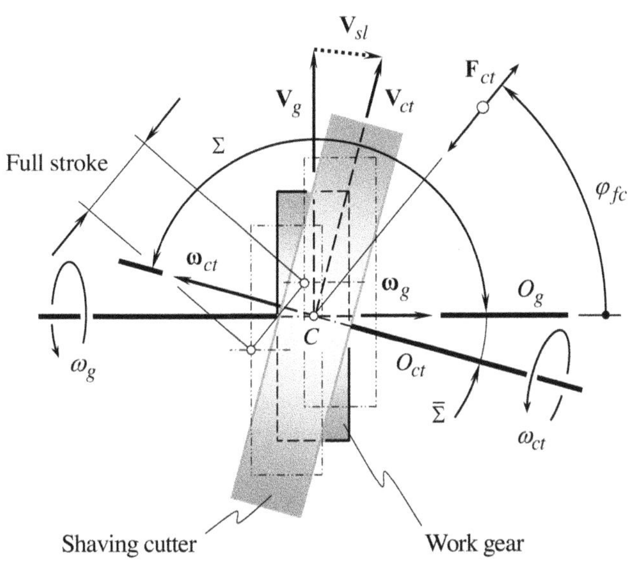


**FIGURE 16.23** Schematic of the diagonal method of the gear shaving process.

the work gear past the shaving cutter by means of the feed slide. On some of the larger shaving machines, for obvious reasons, the shaving cutter is mounted on a feed slide. In the consideration below, the relative motion of the work gear and the shaving cutter is considered regardless of whether the worktable or the shaving cutter actually reciprocates.

The productivity rate of rotary shaving of the gears directly depends on the length of the traverse path of the work gear. In Figure 16.23, the end positions of the reciprocating work gear are shown with dashed lines. The length of the traverse path in diagonal shaving depends on (a) the face width $F_g$ of the work gear, (b) the face width $F_{ct}$ of the shaving cutter, (c) the angle $\bar{\Sigma}$ between the axis of rotation $O_g$ of the work gear and the axis of rotation $O_{ct}$ of the shaving cutter, and (d) the traverse angle $\varphi_{fc}$. For calculation of the length of the traverse path in practice, an approximate formula

$$L \cong \frac{F_g \sin \bar{\Sigma}}{\sin \varphi_{fc}} + m \qquad (16.41)$$

is often used [166]. Here, in Eq. (16.41), the module is denoted by $m$.

With diagonal traverse shaving, the centerline ℄ of crossed axes is not restricted to a single position on the cutter as it is in conventional shaving, but is migrated across the shaving cutter face, evening out the wear. Consequently, shaving cutter life is extended.

*Multiple-stroke diagonal shaving*: An automatic up-feed mechanism on the shaving machine materially enlarges the scope of diagonal shaving by also making it available for multiple-stroke operations. This device feeds the work gear into the shaving cutter in a series of small increments, synchronized with table reciprocation.

Removal of stock from the work gear in a series of small increments instead of two large increments further increases shaving cutter life. It also makes the process feasible for gears requiring more stock removal than can be handled on a two-stroke cycle.

### 16.4.2 Traverse Angle in Diagonal Method of Rotary Shaving of Involute Gear

The traverse angle $\varphi_{fc}$ for diagonal shaving is the angle between the direction of traverse $\mathbf{F}_{ct}$ and the work gear axis of rotation $O_g$. To obtain conditions of cutting speed and work gear quality, in most cases the traverse angle will vary in the range of $\varphi_{fc} = 40 \ldots 70°$.

When shaving cutters with annular serrations are used in diagonal shaving, then the traverse angle $\varphi_{fc}$ is often limited to approximately $\varphi_{fc} \leq 55°$, unless differential-type serrations [see Figure 16.11(b)] are used. Otherwise, the serrations will track.

When gears are shaved by the shaving cutter traveling at an angle to the gear axis of rotation, the shaving cutter has to be wide enough to shave the whole gear face width in one stroke. The relative face width of the work gear and the shaving cutter has an important relationship with the traverse angle. A wide-faced work gear and a narrow-faced shaving cutter restrict the diagonal traverse to a small angle. Increasing the face width of the shaving cutter permits an increase in traverse angle.

For the calculation of the maximum permissible traverse angle $\varphi_{fc}$, the following formula is recommended [13]:

$$\varphi_{fc}^{\max} = \tan^{-1}\left(\frac{F_{ct} \cdot \sin\overline{\Sigma}}{F_g - F_{ct} \cdot \cos\overline{\Sigma}}\right) \tag{16.42}$$

where:

$F_g$ is the face width of the work gear.
$F_{ct}$ is the face width of the shaving cutter.
$\overline{\Sigma}$ is the angle between the axes of rotation $O_g$ and $O_{ct}$ of the work gear and the shaving cutter.

The traverse angle $\varphi_{fc}$ can be expressed in terms of the rotation vector $\omega_g$ of the work gear and the vector $\mathbf{F}_{ct}$ of the reciprocation motion:

$$\varphi_{fc} = -\tan^{-1}\left(\frac{|\omega_g \times \mathbf{F}_{ct}|}{\omega_g \cdot \mathbf{F}_{ct}}\right) \tag{16.43}$$

When the shaving cutter face width is increased to slightly greater than the work gear face width, traverse angles up to 90° are permissible. Diagonal shaving that features a large traverse angle is sometimes referred to as *diagonal-underpass shaving*. This shaving method is not only an application option, but is used particularly in the automotive industry. Since the traverse angle may be very large, this method is also suitable for shaving shoulder gears.

### 16.4.3 Cutting Speed in Diagonal Method of Rotary Shaving of Gear

Since the shaving cutter is mounted on a shaft that is not parallel to the gear axis of rotation, the teeth of the shaving cutter and the teeth of the work gear run together like a pair of *crossed-helical gears*. The choice of crossed-axes angle governs the cutting action of the shaving cutter. In general, the higher the crossed-axes angle, the faster the shaving cutter cuts. The best control over helix angle, though, is gained with a low crossed-axes angle.

A resultant formula for the calculation of the cutting speed $\mathbf{V}_{cut}$ in the diagonal gear shaving is similar to that [see Eq. (16.40)] earlier derived for the case of the axial method of gear shaving. As in diagonal shaving, the traverse motion $\mathbf{F}_{ct}$ is at the traverse angle $\varphi_{fc}$ with respect to the work gear axis of rotation, so the impact of the motion $\mathbf{F}_{ct}$ differs from that in the axial method of rotary shaving. Generally speaking, the traverse motion is not tangential to the flank of the work gear tooth. Because of this, the vector of the traverse motion affects the cutting speed in two different ways.

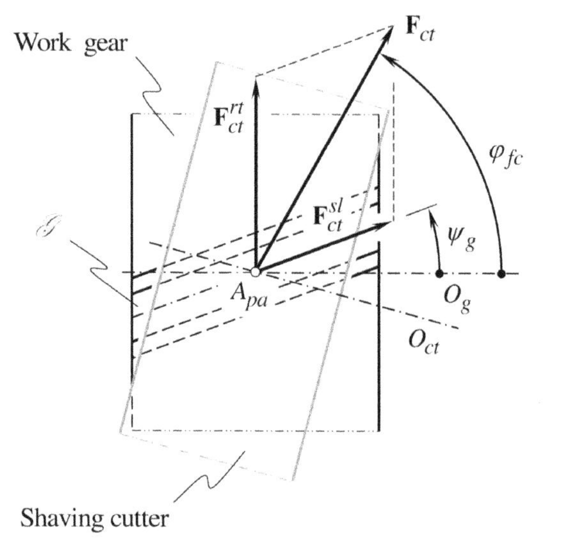


**FIGURE 16.24** Impact of the traverse motion on the cutting speed $\mathbf{V}_{cut}$ in the diagonal shaving of gears.

As was seen earlier (see Figure 16.19), let us assume again that the shaving cutter is not rotating and is only reciprocating at the traverse angle relative to the work gear axis $\varphi_{fc}$ as depicted in Figure 16.24. In such a scenario, the vector $\mathbf{F}_{ct}$ of the traverse motion of the shaving cutter is at an angle to the work gear tooth flank. Within the pitch plane, this angle is equal to $(\varphi_{fc} - \psi_g)$. The angle can be expressed in terms of the pitch helix angle $\psi_g$ of the work gear and the pitch helix angle $\psi_{ct}$ of the shaving cutter. In this case the angles $\psi_g$ and $\psi_{ct}$ should be considered as signed values depending on the direction of the hand of the helices.

The traverse motion $\mathbf{F}_{ct}$ can be decomposed in two directions. One of the components, $\mathbf{F}_{ct}^{sl}$, is pointed in the direction tangent to the tooth flank of the work gear. The magnitude of this component equals

$$F_{ct}^{sl} = |\mathbf{F}_{ct}^{sl}| = \frac{|\mathbf{F}_{ct}|\sin\varphi_{fc}}{\cos\psi_g} = \frac{F_{ct}\sin\varphi_{fc}}{\cos\psi_g} \tag{16.44}$$

The component $\mathbf{F}_{ct}^{sl}$ represents pure sliding of the tooth flanks of the work gear and the shaving cutter.

It is important to point out here that the sliding vector $\mathbf{F}_{ct}^{sl}$ is of a constant magnitude, and it does not change within the tooth height either of the work gear or within the tooth height of the shaving cutter.

Another component $\mathbf{F}_{ct}^{rt}$ of the traverse motion $\mathbf{F}_{ct}$ is within the pitch plane, and it is pointed perpendicular to the axis of rotation of the work gear $O_g$. For the calculation of the magnitude $F_{ct}^{rt}$ of the component $\mathbf{F}_{ct}^{rt}$, the formula

$$F_{ct}^{rt} = F_{ct}\frac{\sin(\varphi_{fc} - \psi_g)}{\cos\psi_g} \tag{16.45}$$

is applied. This formula is derived from Figure 16.24 on the premise of the sine law.

The component $\mathbf{F}_{ct}^{rt}$ contributes to the rotation of the work gear. This additional rotation, $\omega_g^{ad}$, is equal to

$$\omega_g^{ad} = F_{ct} \frac{2\sin(\varphi_{fc} - \psi_g)}{d_g \cos\psi_g} \tag{16.46}$$

The rotation vector $\boldsymbol{\omega}_g^{ad}$ is aligned with the rotation vector $\omega_g$. Depending on (a) the rotation of the shaving cutter $\boldsymbol{\omega}_{ct}$, (b) the direction of the hand and the value of the helix angle $\psi_g$, and (c) the actual direction of the reciprocation $\mathbf{F}_{ct}$ (the traverse angle $\varphi_{fc}$), the rotation vector $\boldsymbol{\omega}_g^{ad}$ is pointed either in that same direction as the rotation vector $\boldsymbol{\omega}_g$ or it is pointed opposite to $\omega_g$. In the first case, the rotation $\omega_g^{ad}$ adds to the rotation $\omega_g$. In the second case, it subtracts from the rotation $\boldsymbol{\omega}_g$.

Once Eqs. (16.44)–(16.46) are taken into account, then the resultant formula [see Eq. (16.40)] can be used for the calculation of the cutting speed vector $\mathbf{V}_{cut}$ in diagonal shaving of the gears.

As the additional rotation $\omega_g^{ad}$ of the work gear depends on the pitch helix angle, the greater the pitch helix angle $\psi_g$ of the work gear, the greater the impact of the traverse motion $\mathbf{F}_{ct}$ of the shaving cutter on the cutting speed $\mathbf{V}_{cut}$, and vice versa.

The tangential speed of the work gear is usually about 400 ft/min (2.0 m/s), so that the work speed is

$$\omega_g = \frac{400 \times 12}{d_g\,(\text{in.})} \text{ rev./min.} \tag{16.47}$$

With an angle between the shaving cutter spindle and the work gear spindle in the range of $\Sigma \cong 15°$, the cutting speed is about one-quarter of the circumferential speed of the work gear, in this case 110 ft/min. (0.55 m/s).

### 16.4.4 Optimization of the Kinematics in Diagonal Method of Rotary Shaving of Gears

The design parameters of a shaving cutter, as well as the parameters of the kinematics of the diagonal shaving process, commonly are chosen empirically. In this way, the choice is made only on the basis of accumulated practical experience.

An attempt to develop an analytical solution to the problem of determining the optimal design parameters of the shaving cutter along with the optimal parameters of the kinematics of the diagonal gear shaving process has been undertaken by Prof. S.P. Radzevich [54, 143, 184]. He proposed an analytical method for the synthesizing of the optimal gear shaving process. The method is based on an extensive implementation of the so-called *DG/K-based method* (**D**ifferential-**G**eometry/**K**inematics-*based method*) of part surface generation [64, 144, 148, 158]. Details on the proposed method of synthesis can also be found in [183]. The discussion below encompasses only preliminary results of the study, which was originally reported in [54, 143, 183, 184], and others.

The use of DG/K-based method is extremely helpful when optimizing the gear shaving process, as well as in many other applications [54, 143, 183, 184], and others.

#### 16.4.4.1 Concept of Synthesizing

The rate of metal removal and the surface finish strongly depend on the crossed-axes angle. Figure 16.25(a) shows diagrammatically the action between a shaving cutter and a helical gear. In Figure 16.25(a), the velocity vector of the total sliding motion is designated as $\mathbf{V}_{sl}^{\Sigma}$. The vector $\mathbf{V}_{sl}^{\Sigma}$ is equal to the summa of two components, namely (a) of the vector of sliding $\mathbf{V}_{sl}$ and (b) the vector $\mathbf{F}_{ct}^{sl}$ caused by the traverse motion $\mathbf{F}_{ct}$. Ultimately, the equality

$$\mathbf{V}_{sl}^{\Sigma} = \mathbf{V}_{sl} + \mathbf{F}_{ct}^{sl} \tag{16.48}$$

is observed.

The velocity vector $\mathbf{V}_{sl}^{\Sigma}$ is of a constant value at every point of contact between the tooth flank of the work gear and the tooth flank of the shaving cutter [see Figure 16.25(b)].

The smaller the crossed-axes angle $\Sigma$, the finer the shavings that are removed. In this way a very good finish can be produced. However, burnishing and work hardening of the

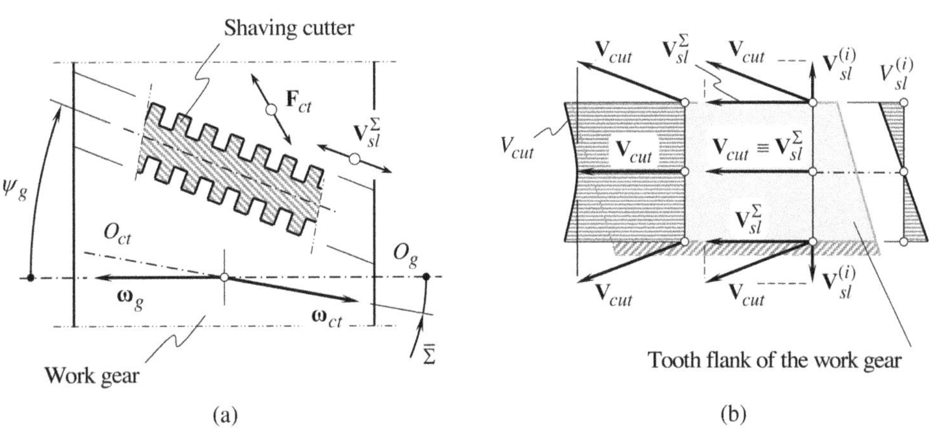


**FIGURE 16.25** Variation of the speed of cut vector $\mathbf{V}_{cut}$ within the tooth flank of a work gear: (a) configuration of the shaving cutter tooth and (b) variation of the linear velocity vector.

surface of the work gear teeth may occur unless the cutter is kept in a very sharp condition.

If the crossed-axes angle is larger, then the shaving action becomes much more positive, and it is easy to damage the work tooth surfaces unless a light cutter loading is applied. Consequently, the finished surface is of a rather less smooth texture than that produced by a smaller angle.

Since the choice of crossed-axes angle is largely a matter of judgment in weighing different variables, there are no fixed rules [211].

The tooth profile sliding $\mathbf{V}_{sl}^{(i)}$ varies within the tooth flank of the work gear as illustrated in Figure 16.25(b). The velocity vector $\mathbf{V}_{sl}^{(i)}$ affects both the value and the direction of speed of the cut vector $\mathbf{V}_{cut}$. The sliding velocity vector $\mathbf{V}_{sl}^{\Sigma}$ is predominant. However the impact of the profile sliding velocity $\mathbf{V}_{sl}^{(i)}$ on the speed of the cut vector $\mathbf{V}_{cut}$ could be significant when the tooth number of the work gear and/or of the shaving cutter is smaller.

Nominal contact between the work gear tooth flank and the shaving cutter tooth flank in diagonal shaving is a point. The point contact of the tooth flanks $\mathscr{G}$ and $T$ can be easily interpreted in the following way. Referring to Figure 16.26(a), the work gear tooth flank $\mathscr{G}$ and the auxiliary generating rack tooth flank $\mathscr{R}_T$ contact each other along the characteristic line $\mathscr{E}_g$. The auxiliary rack tooth flank $\mathscr{R}_T$ and the shaving cutter tooth flank $T$ contact one another along the characteristic line $E_{ct}$. Both straight lines $\mathscr{E}_g$ and $\mathscr{E}_{ct}$ are situated within the plane $\mathscr{R}_T$, and they intersect each other at contact

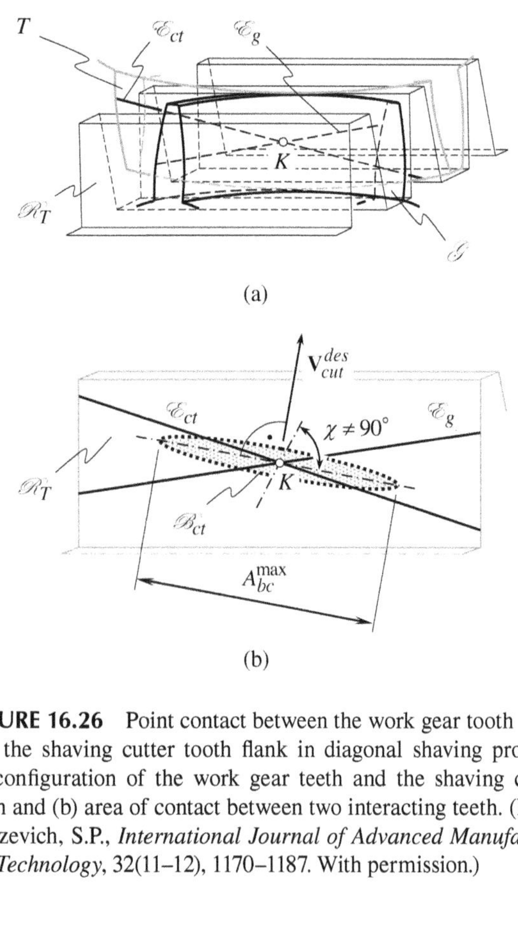


**FIGURE 16.26** Point contact between the work gear tooth flank and the shaving cutter tooth flank in diagonal shaving process: (a) configuration of the work gear teeth and the shaving cutter teeth and (b) area of contact between two interacting teeth. (From Radzevich, S.P., *International Journal of Advanced Manufacturing Technology*, 32(11–12), 1170–1187. With permission.)

point $K$. When the work gear and the shaving cutter rotate in tight mesh, the point of contact $K$ travels from the outer diameter $d_{o.g}$ of the work gear toward its limit diameter $d_{l.g}$ or in the opposite direction, depending on the direction of rotation of the work gear and the shaving cutter. In this way, a path of contact on the tooth flank $\mathscr{G}$ is traced. When the shaving cutter is traversing, then the path of contact travels across the face width of the work gear. Ultimately, any point of the work gear tooth flank can make contact with the shaving cutter tooth flank.

Under the load, the point of contact $K$ of the tooth flanks spreads over a certain area that is bounded by an ellipse-like boundary curve, as shown in Figure 16.26(b). The size and shape of the contact area depends upon the pitch helix angle $\psi_g$ of the work gear and the crossed-axes angle $\Sigma$ (or the angle $\bar{\Sigma}$).

Consider a surface $\mathscr{G}_{[h]}$ that is offset to the pinion tooth flank $\mathscr{G}$. When the offset distance is equal to the tolerance $[h]$ for the tooth flank $\mathscr{G}$ accuracy, then the offset surface is referred to as the *surface of tolerance* $\mathscr{G}_{[h]}$.

When tooth flank $\mathscr{G}$ of a work gear and tooth surface $T$ of the shaving cutter make contact at point $K$, the surface $T$ intersects the surface of tolerance $\mathscr{G}_{[h]}$. The line of the intersection of the surfaces is a certain closed 3-$D$ boundary curve $\mathscr{B}_{ct}$. This curve is often incorrectly referred to as a *contact ellipse*[9] (first of all, because an ellipse is a planar curve, while the boundary curve $\mathscr{B}_{ct}$ is a 3$D$ curve). The boundary curve $\mathscr{B}_{ct}$ bounds the spot of contact of the gear and the shaving cutter tooth. The contact area passes over the tooth flank of the work gear, and the asperities and protuberances are removed down to a basic smooth surface $\mathscr{G}$ as the shaving cutter rotates in contact with the work gear, and it is traversed $\mathbf{F}_{ct}$ across the face of the work gear at a feed rate.

The following issues are of critical importance for further consideration.

*First*, gear shaving is a kind of finishing processes. For accuracy reasons, forces that act on the work gear and the shaving cutter have to be at the lowest possible range. Low cutting forces cause small deflections of the shaving machine. A surface area that is bounded by the boundary curve $\mathscr{B}_{ct}$ should be as small as possible if we wish to achieve low cutting forces.

*Second*, the path of contact of the work gear tooth flank $\mathscr{G}$ and of the shaving cutter tooth flank $T$ should be the widest possible. For this purpose, the major axis $A_{bc}^{max}$ of the contact area has to be the longest possible. This makes possible a higher productivity rate of the rotary shaving process.

*Third*, as the productivity of the gear finishing process is important, the desired direction of the cutting speed $\mathbf{V}_{cut}^{des}$ has to be orthogonal to the major

[9] The boundary curve $\mathscr{B}_{ct}$ is a type of 3$D$ curve, while an ellipse is a type of planar curve.

axis $A_{bc}^{max}$ of the boundary curve $\mathcal{B}_{ct}$. It should be pointed out here that the minor axis $A_{bc}^{min}$ of the boundary curve $\mathcal{B}_{ct}$ [not shown in Figure 16.26(b)] is not perpendicular to the major axis $A_{bc}^{max}$. The axes $A_{bc}^{max}$ and $A_{bc}^{min}$ form a certain angle $\chi \neq 90°$.

Once the three items listed earlier are understood, then it becomes clear that varying the design parameters of the shaving cutter, and synchronizing the direction and the speed of traverse $\mathbf{F}_{ct}$ with the rotation of the work gear $\omega_g$ and of the shaving cutter $\omega_{ct}$, make control of the shape, size, and orientation of the contact area possible. The direction of the cutting speed $\mathbf{V}_{cut}$ can also be under control in this way. Ultimately, the accuracy of the finished gears can be increased and the shaving time reduced if the proper combination of design parameters of the work gear and shaving cutter is determined, and all the elementary motions of the rotary shaving process ($\omega_g$, $\omega_{ct}$, and $\mathbf{F}_{ct}$) are properly timed with one another. The crossed-axes angle $\Sigma$ is not listed here because the rotations of the work gear and the shaving cutter are represented in vector notation.

The local topology of the work gear tooth flank $\mathcal{G}$ and the shaving cutter tooth flank $T$ must be known in order to use the DG/K-based method of part surface generation to synthesize the most favorable process for the rotary shaving of gears.

### 16.4.4.2 Local Topology of the Contacting Tooth Flanks

Investigation of the local topology of the tooth flank of both the work gear $\mathcal{G}$ and the shaving cutter $T$ begins with the derivation of an equation of the tooth flank using the equation of the *Dupin indicatrix* at a surface point. Because the surfaces $\mathcal{G}$ and $T$ are similar, the required equations are considered in detail for the gear tooth flank $\mathcal{G}$ only. Corresponding equations for the generating surface $T$ of the shaving cutter can be written by an analogy with the equations derived for the tooth flank $\mathcal{G}$.

*16.4.4.2.1 Equation of Tooth Flank $\mathcal{G}$ of Involute Gear*

In a Cartesian reference system $X_gY_gZ_g$ associated with the gear, the equation of the tooth surface $\mathcal{G}$ of the work gear can be represented in the form of a column matrix [see Eq. (1.4)]:

$$\mathbf{r}_g(U_g, V_g) = \begin{bmatrix} r_{b.g}\cos V_g + U_g \cos\psi_{b.g} \sin V_g \\ r_{b.g}\sin V_g - U_g \sin\psi_{b.g} \sin V_g \\ r_{b.g}\tan\psi_{b.g} - U_g \sin\psi_{b.g} \\ 1 \end{bmatrix} \quad (16.49)$$

where:

- $r_{b.g}$ is the radius of the base cylinder of the work gear.
- $U_g, V_g$ are the curvilinear coordinates (Gaussian coordinates) of the tooth flank of the work gear.
- $\psi_{b.g}$ is the base helix angle of the work gear.

*16.4.4.2.2 First Fundamental Form of an Involute Gear Tooth Flank $\mathcal{G}$*

Fundamental magnitudes $E_g$, $F_g$, $G_g$ of the first fundamental form $\Phi_{1.g}$ of the work gear tooth flank $\mathcal{G}$ are calculated from Eq. (16.49) [150, 158]

$$E_g = 1 \quad (16.50)$$

$$F_g = -\frac{r_{b.g}}{\cos\psi_{b.g}} \quad (16.51)$$

$$G_g = \frac{U_g^2 \cos^4\psi_{b.g} + r_{b.g}^2}{\cos^2\psi_{b.g}} \quad (16.52)$$

Equations (16.50)–(16.52) yield an expression

$$\Phi_{1.g} \Rightarrow dU_g^2 - 2\frac{r_{b.g}}{\cos\psi_{b.g}} dU_g dV_g + \frac{U_g^2\cos^4\psi_{b.g} + r_{b.g}^2}{\cos^2\psi_{b.g}} dV_g^2 \quad (16.53)$$

for the first fundamental form $\Phi_{1.g}$ at a point of the work gear tooth flank $\mathcal{G}$.

A similar equation is also valid for the first fundamental form $\Phi_{1.T}$ of the generating surface $T$ of the shaving cutter.

*16.4.4.2.3 Second Fundamental Form of an Involute Gear Tooth Surface $\mathcal{G}$*

Equations (16.50)–(16.52) allow for the calculation of fundamental magnitudes $L_g$, $M_g$, $N_g$ of the second fundamental form $\Phi_{2.g}$ at a point of the work gear tooth flank $\mathcal{G}$ [150, 158]:

$$L_g = 0 \quad (16.54)$$

$$M_g = 0 \quad (16.55)$$

$$N_g = -U_g \sin\tau_{b.g} \cos\tau_{b.g} \quad (16.56)$$

Equations (16.54)–(16.56) yield an expression

$$\Phi_{2.g} \Rightarrow -U_g \sin\psi_{b.g}\cos\psi_{b.g} dV_g^2 \quad (16.57)$$

for the second fundamental form $\Phi_{2.g}$ of the work gear tooth flank $\mathcal{G}$.

A similar equation is also valid for the second fundamental form $\Phi_{2.T}$ of the generating surface $T$ of the shaving cutter.

*16.4.4.2.4 Curvature of an Involute Gear Tooth Flank $\mathcal{G}$*

Having calculated the first, $\Phi_{1.g}$, and the second, $\Phi_{2.g}$, fundamental forms [see Eq. (16.53) and Eq. (16.57)], normal curvature $k_g$ at a point of the work gear tooth flank $\mathscr{G}$ is calculated from the simple equation

$$k_g = \frac{\Phi_{2.g}}{\Phi_{1.g}} \tag{16.58}$$

For an involute gear tooth flank $\mathscr{G}$, this equation casts into

$$\begin{aligned} k_g = & -\frac{1}{U_g \sin\psi_{b.g} \cos\psi_{b.g}} \lambda^2 \\ & + 2\frac{r_{b.g}}{U_g \sin\psi_{b.g} \cos^2\psi_{b.g}} \lambda - \frac{U_g^2 \cos^4\psi_{b.g} + r_{b.g}^2}{U_g \sin\psi_{b.g} \cos^3\psi_{b.g}} \end{aligned} \tag{16.59}$$

where $\lambda$ denotes the ratio $\lambda = dU_g / dV_g$.

The first $k_{1.g}$ and second $k_{2.g}$ principal curvatures at a point of the work gear tooth flank $\mathscr{G}$ are equal

$$\begin{aligned} k_{1.g} &= \frac{\tan\psi_{b.g}}{U_g} = \frac{2\sin\psi_{b.g}}{\sqrt{d_{y.g}^2 - d_{b.g}^2}\cos\lambda_{b.g}} \\ &= \frac{2\sin\psi_g \sin\phi_n}{\sqrt{d_{y.g}^2 - d_{b.g}^2}\cos\lambda_{b.g}} \end{aligned} \tag{16.60}$$

$$k_{2.g} = 0 \tag{16.61}$$

correspondingly. It should be pointed out here that the inequality $k_{1.g} > k_{2.g}$ is always observed.[10]

##### 16.4.4.2.5 *Dupin Indicatrix of an Involute Gear Tooth Flank* $\mathscr{G}$

As tooth flank $\mathscr{G}$ of the work gear is an involute surface, its local patch in a differential vicinity of a point is a local patch of the parabolic type. In a local orthogonal coordinate system, the equation of the Dupin indicatrix $Dup(\mathscr{G})$ at a point of the gear tooth flank $\mathscr{G}$ can be represented in the form

$$Dup(\mathscr{G}) \Rightarrow \frac{U_g \sin\psi_{b.g} \cos^3\psi_{b.g}}{U_g^2 \cos^4\psi_{b.g} + r_{b.g}^2} y_g^2 = \mp 1 \tag{16.62}$$

##### 16.4.4.2.6 *Equations for the Generating Surface T of a Shaving Cutter*

A comparison of Eq. (16.1) for the generating surface $T$ of a shaving cutter and Eq. (16.49) for the tooth flank $\mathscr{G}$ of a work gear reveals that the equations are similar to one another in nature. Therefore, once the generating surface of a shaving cutter is represented in a Cartesian coordinate system $X_{ct}Y_{ct}Z_{ct}$ associated with the shaving cutter itself, then equations identical to Eq. (16.50) through Eq. (16.62) derived earlier are also valid for the generating surface $T$ of a shaving cutter. To be treated together, Eqs. (16.1) and (16.49) need to be represented in a common reference system.

#### 16.4.4.3 Applied Coordinate Systems

Several reference systems are used for analytical description of the surfaces $\mathscr{G}$ and $T$ of their configuration and relative motion. Three of them are of prime importance in the consideration below.

One of the coordinate systems is a left-hand-oriented Cartesian coordinate system $X_gY_gZ_g$ associated with the work gear. Another one is a left-hand-oriented Cartesian coordinate system $X_{ct}Y_{ct}Z_{ct}$ connected to the shaving cutter. The third one is a left-hand-oriented local Cartesian coordinate system $x_gy_gz_g$ with the origin at point $K$ of contact of the work gear tooth surface $\mathscr{G}$ and the shaving cutter tooth surface $T$ (see Figure 16.27). These three coordinate systems are referred to as *principal coordinate systems*. It is assumed below that the configuration of the principal reference system is known in a certain common coordinate system, for instance, in a coordinate system $X_{sm}Y_{sm}Z_{sm}$ associated with the shaving machine.

The axis $Z_g$ of the reference system $X_gY_gZ_g$ is aligned to the work gear axis of rotation $O_g$, and the $Z_{ct}$ – axis of the reference system $X_{ct}Y_{ct}Z_{ct}$ is aligned with the shaving cutter axis of rotation $O_{ct}$. The axes $X_g$ and $X_{ct}$ are parallel to the common tangent plane to the pitch cylinders of the work gear and the shaving cutter. Ultimately, the axes $Y_g$ and $Y_{ct}$ are perpendicular to the common tangent plane to the pitch cylinders of the work gear and the shaving cutter.

A local left-hand-oriented coordinate system $x_gy_gz_g$ is associated with the work gear tooth flank $\mathscr{G}$. The axis $x_g$ is aligned to the unit tangent vector $\mathbf{u}_g = \mathbf{U}_g / |\mathbf{U}_g|$, where $\mathbf{U}_g = \partial\mathbf{r}_g / \partial U_g$ (see Figure 16.27); the axis $z_p$ is aligned to the surface $\mathscr{G}$ unit normal $\mathbf{n}_g = \mathbf{u}_g \times \mathbf{v}_g$, where $\mathbf{v}_g = \mathbf{V}_g / |\mathbf{V}_g|$ and $\mathbf{V}_g = \partial\mathbf{r}_g / \partial V_g$; this axis is pointed outward from the bodily side of the work gear tooth surface $\mathscr{G}$; and the axis $y_g$ is aligned to the unit tangent vector $\mathbf{v}_g^* = \mathbf{n}_g \times \mathbf{u}_g$ (under such conditions, the axis $y_g$ is perpendicular to the $x_gz_g$ coordinate plane).

The similar local coordinate system $x_{ct}y_{ct}z_{ct}$, with the origin at point $K$, is associated with the involute tooth surface $T$ of the shaving cutter.

In addition to the coordinate systems just mentioned, several auxiliary reference systems are also employed.

#### 16.4.4.4 Geometry of Contact of the Tooth Flanks $\mathscr{G}$ and $T$

The geometry of contact of the tooth flanks $\mathscr{G}$ and $T$ of the work gear and the shaving cutter should be properly described analytically prior determining the optimal design parameters of a shaving cutter and the parameters of the kinematics of the rotary shaving operation. The indicatrix of conformity, $Cnf_R(\mathscr{G}/T)$, is used for the purpose of analytical description of the geometry of contact of two smooth

[10] Remember that algebraic values of the principal curvatures $k_{1.g}$ and $k_{2.g}$ relate to one another as $k_{1.g} > k_{2.g}$, and not as $k_{1.g} \geq k_{2.g}$. If one assumes that the equality $k_{1.g} = k_{2.g}$ is valid, then neither the first $k_{1.g}$ nor the second $k_{2.g}$ principal curvature can be defined. Therefore, the equality $k_{1.g} = k_{2.g}$ is never valid.

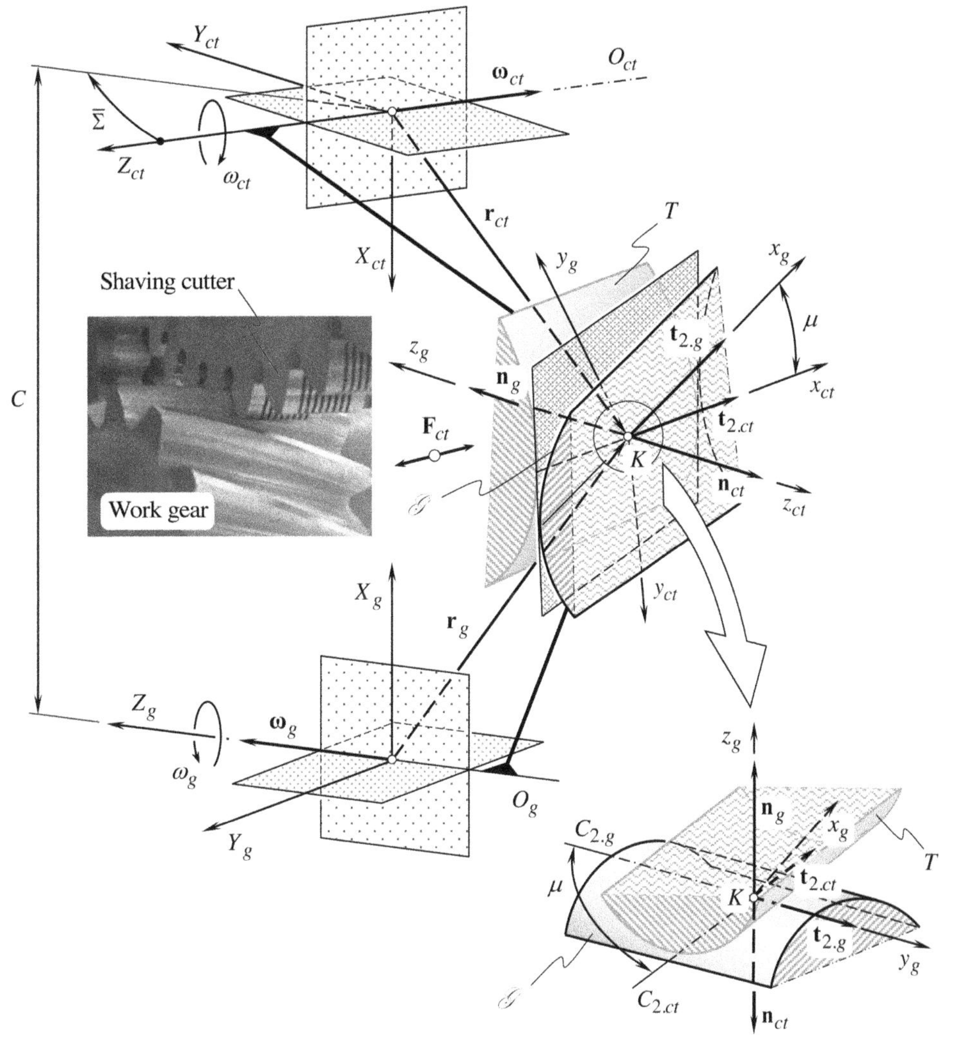


**FIGURE 16.27** Coordinate systems associated with the work gear $\mathcal{G}$ with the shaving cutter $T$ and with current point of contact $K$. (With kind permission from Springer Science+Business Media: *International Journal of Advanced Manufacturing Technology*, Diagonal shaving of an involute pinion: Optimization of parameters of the pinion finishing operation, 32, 2007, 1170–1187, Radzevich, S.P.)

regular surfaces $\mathcal{G}$ and $T$ in differential vicinity of point $K$ of their contact.

### 16.4.4.4.1 *Principal Directions on the Contacting Tooth Flanks $\mathcal{G}$ and $T$*

As the tooth flanks $\mathcal{G}$ and $T$ are smooth regular surfaces, two principal directions can be determined on each of the surfaces at point $K$ of their contact.

A direction on a work gear tooth flank $\mathcal{G}$ is uniquely specified by the ratio $\eta_g = dU_g/dV_g$. The directions $\mathbf{t}_{1.g}$ and $\mathbf{t}_{2.g}$ for which the equality

$$\begin{vmatrix} \eta_g E_g + F_g & \eta_g F_g + G_g \\ \eta_g L_g + M_g & \eta_g M_g + N_g \end{vmatrix} = 0 \tag{16.63}$$

is valid are referred to as the *principal directions* on the gear tooth flank $\mathcal{G}$.

Similarly, two principal directions $\mathbf{t}_{1.ct}$ and $\mathbf{t}_{2.ct}$ on the generating surface $T$ of a shaving cutter can be expressed in terms of the fundamental magnitudes $E_{ct}$, $F_{ct}$, $G_{ct}$ and $L_{ct}$, $M_{ct}$, $N_{ct}$, and of the ratio $\eta_{ct} = dU_{ct}/dV_{ct}$.

### 16.4.4.4.2 *Local Orientation of Tooth Flanks of a Work Gear and the Shaving Cutter*

The local orientation of the tooth flanks of a work gear $\mathcal{G}$ and the shaving cutter $T$ can be specified in terms of the angle $\mu$ of the surfaces $\mathcal{G}$ and $T$ in local relative orientation [64, 144, 148, 158]. The angle $\mu$ of the surfaces in local relative orientation is the angle that unit tangent vectors $\mathbf{t}_{1.g}$ and $\mathbf{t}_{1.ct}$ of the first principal directions (or unit tangent vectors $\mathbf{t}_{2.g}$ and $\mathbf{t}_{2.ct}$ of the second principal directions) of the surfaces $\mathcal{G}$ and $T$ form at contact point $K$. Thus, by definition, the angle $\mu$ of the surfaces' local relative orientation is equal:

$$\tan\mu = \frac{|\mathbf{t}_{1.g} \times \mathbf{t}_{1.ct}|}{\mathbf{t}_{1.g} \cdot \mathbf{t}_{1.ct}} \equiv \frac{|\mathbf{t}_{2.g} \times \mathbf{t}_{2.ct}|}{\mathbf{t}_{2.g} \cdot \mathbf{t}_{2.ct}} \tag{16.64}$$

In the case of rotary shaving of an involute gear, Eq. (16.64) is reduced to

$$\sin\mu = \frac{\sin\phi_n \sin\bar{\Sigma}}{\sqrt{(1-\cos^2\phi_n \sin^2\psi_g)\cdot(1-\cos^2\phi_n \sin^2\psi_{ct})}} \quad (16.65)$$

where:

- $\phi_n$ is the normal pressure angle in the work gear-to-shaving cutter mesh.
- $\psi_g$ is the pitch helix angle of the work gear.
- $\psi_{ct}$ is the pitch helix angle of the shaving cutter.
- $\bar{\Sigma}$ is the angle between the axes of rotation $O_g$ and $O_{ct}$ of the work gear and of the shaving cutter.

#### 16.4.4.4.3 Indicatrix of Conformity $Cnf_R(\mathcal{G}/T)$ at Point of Contact of a Work Gear Tooth Flank and Tooth Flank of the Generating Surface of the Shaving Cutter

The *indicatrix of conformity*[11] at point of contact of two smooth regular surfaces is a planar characteristic curve of the fourth order [64, 136, 144, 148, 158, 175]. This characteristic curve possesses the property of central symmetry, and, in particular cases, it also possesses the property of mirror symmetry. This curve quantitatively reflects the degree of conformity of the surfaces which make contact at a point or along a curve line. The characteristic curve $Cnf_R(\mathcal{G}/T)$ can be constructed in a common tangent plane (in the coordinate plane $x_g y_g$ of the local reference system $x_g y_g z_g$).

An equation of the indicatrix of conformity $Cnf_R(\mathcal{G}/T)$ at the point of contact of two smooth regular surfaces can be employed for synthesizing the most favorable process of rotary shaving of an involute gear. For this purpose, the expressions for the first $\Phi_{1.g}$, $\Phi_{1.ct}$ [see Eq. (16.53)] and second $\Phi_{2.g}$, $\Phi_{2.ct}$ [see Eq. (16.57)] fundamental forms have to be represented in a common reference system. The local coordinate system $x_g y_g z_g$ (see Figure 16.27) is used as the common reference system in the case under consideration.

For the purposes of linear transformation of surface fundamental forms, a formula

$$[\Phi_{1.g}]_K = \mathbf{Rs}^T(g \to K)\cdot[\Phi_{1.g}]_g \cdot \mathbf{Rs}(g \to K) \quad (16.66)$$

$$[\Phi_{2.g}]_K = \mathbf{Rs}^T(g \to K)\cdot[\Phi_{2.g}]_g \cdot \mathbf{Rs}(g \to K) \quad (16.67)$$

is proposed by Prof. S.P. Radzevich [154]. This formula is also known from other sources [64, 144, 148, 158].

[11] An equation of this characteristic curve has been published in:

a. SU Pat. No. 1185749, *A Method of Sculptured Surface Machining on Multi-Axis NC Machine*/S.P. Radzevich, Int. Cl. B23C 3/16, Filed: October 24, 1983, and in:
b. SU Pat. No. 1249787, *A Method of Sculptured Surface Machining on Multi-Axis NC Machine*/S.P. Radzevich, Int. Cl. B23C 3/16, Filed: December 27, 1984.

In Eq. (16.66) and Eq. (16.67), $[\Phi_{1.g}]_g$ and $[\Phi_{2.g}]_g$ denote the first and the second fundamental forms of the work gear tooth flank when the surface $\mathcal{G}$ is represented in the coordinate system $X_g Y_g Z_g$; $[\Phi_{1.g}]_K$ and $[\Phi_{2.g}]_K$ denote that same fundamental forms when the equation of the surface $\mathcal{G}$ is transformed to the local reference system $x_g y_g z_g$, and the operator of the resultant coordinate system transformation is designated as $\mathbf{Rs}(g \to K)$.

Equations similar to Eq. (16.66) and Eq. (16.67) can be composed for the tooth flank $T$ of the generating surface of the shaving cutter.

With the fundamental forms $[\Phi_{1.g}]_K$, $[\Phi_{2.g}]_K$, $[\Phi_{1.ct}]_K$, and $[\Phi_{2.ct}]_K$ calculated, this immediately yields an expression

Indicatrix of conformity, $Cnf_R(\mathcal{G}/T)$

$$\Rightarrow r_{cnf}(R_{1.g}, R_{1.ct}, \mu, \varphi) = \frac{\sqrt{R_{1.g}}}{\sin\varphi} + \frac{\sqrt{R_{1.ct}}}{\sin(\mu-\varphi)} \quad (16.68)$$

for the indicatrix of conformity $Cnf_R(\mathcal{G}/T)$ at point of contact of the tooth flanks $\mathcal{G}$ and $T$. Equation (16.61) is represented in the local coordinate system $x_g y_g z_g$.

Equation (16.68) is represented in polar coordinates with the origin at contact point $K$. The distance $r_{cnf}$ of a current point of the indicatrix of conformity $Cnf_R(\mathcal{G}/T)$ is expressed in terms of (a) the first principal radii of curvature $R_{1.g}$ and $R_{1.ct}$ of the contacting surfaces $\mathcal{G}$ and $T$ (here the equalities $R_{1.g} = k_{1.g}^{-1}$ and $R_{1.ct} = k_{1.ct}^{-1}$ are observed), (b) the angle $\mu$ of the tooth flanks $\mathcal{G}$ and $T$ in local relative orientation, and (c) the polar angle $\varphi$.

Equation (16.68) of the indicatrix of conformity $Cnf_R(\mathcal{G}/T)$ is often represented in vector notation

$$\mathbf{r}_{cnf}(R_{1.g}, R_{1.ct}, \mu, \varphi) = \mathbf{i}\cdot r_{cnf}\cos\varphi + \mathbf{j}\cdot r_{cnf}\sin\varphi \quad (16.69)$$

where $\mathbf{r}_{cnf}$ is the position vector of the point of the characteristic curve $Cnf_R(\mathcal{G}/T)$.

An illustrative example of the characteristic curve $Cnf_R(\mathcal{G}/T)$ is depicted in Figure 16.28. For the readers' convenience, the Dupin indicatrices $Dup(\mathcal{G})$ and $Dup(T)$ along with the curvature indicatrices $Crv(\mathcal{G})$ and $Crv(T)$ are shown in Figure 16.28 as well.

The degree of conformity at point of contact $K$ of the tooth flank $T$ of the shaving cutter to the tooth flank $\mathcal{G}$ of the work gear changes as the normal plane section of the surfaces $\mathcal{G}$ and $T$ turns about common perpendicular $\mathbf{n}_g$ through the point of contact [143]. In a certain normal plane section of the tooth flanks $\mathcal{G}$ and $T$, the degree of their conformity reaches the maximum value. This normal section $C_{cnf}^{max}$ of the surfaces $\mathcal{G}$ and $T$ is referred to as the plane section of the maximum degree of conformity of the surfaces. The maximum degree of conformity section of the surfaces $\mathcal{G}$ and $T$ is the plane section through the direction of minimum diameter $d_{cnf}^{min}$ of the indicatrix of conformity $Cnf_R(\mathcal{G}/T)$. Referring to Figure 16.29, this direction is specified by the unit tangent vector $\mathbf{t}_{cnf}^{max}$.

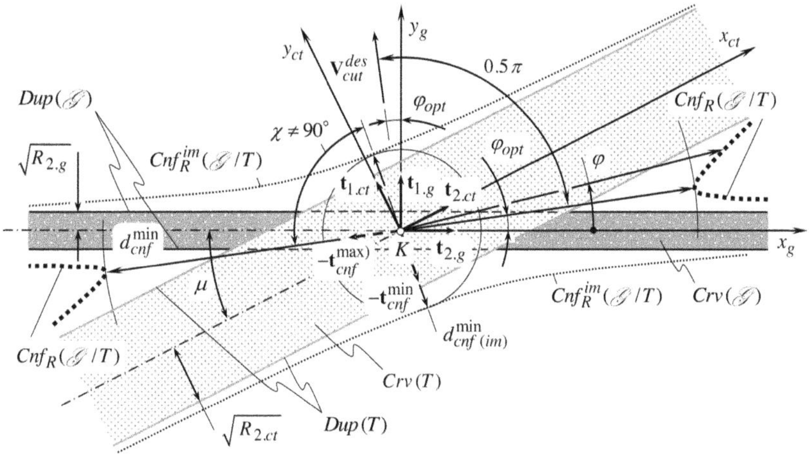


**FIGURE 16.28** Indicatrix of conformity $Cnf_R(\mathscr{G}/T)$ at a point of contact of the tooth flanks $\mathscr{G}$ of the work gear and $T$ of the shaving cutter, which is represented in the local reference system $x_g y_g z_g$, having a contact point as the origin. (With kind permission from Springer Science+Business Media: *International Journal of Advanced Manufacturing Technology*, Diagonal shaving of an involute pinion: Optimization of parameters of the pinion finishing operation, 32, 2007, 1170–1187, Radzevich, S.P.)

Equation (16.68) of the indicatrix of conformity $Cnf_R(\mathscr{G}/T)$ can be expressed in terms of the design parameters of the work gear and the shaving cutter and their relative orientation:

$$r_{cnf}(d_{y.g}, d_{y.ct}, \varphi, \mu) = \sqrt{\frac{2\sin\psi_g \sin\phi_n}{\sqrt{d_{y.g}^2 - d_{b.g}^2}\cos\lambda_{b.g}\sin^2\varphi}} + \sqrt{\frac{2\sin\psi_{ct}\sin\phi_n}{\sqrt{d_{y.ct}^2 - d_{b.ct}^2}\cos\lambda_{b.ct}\sin^2(\mu-\varphi)}} \tag{16.70}$$

All the information necessary for the calculation of the most favorable design parameters of the shaving cutter and the most favorable parameters of the kinematics of the rotary shaving process of a work gear can be extracted from Eq. (16.70).

The lateral tooth plane of the auxiliary generating rack $\mathscr{R}_T$ is tangent at $K$ to both to the tooth flank $\mathscr{G}$ of the work gear and the tooth flank $T$ of the shaving cutter. For the readers' convenience, in Figure 16.30 the characteristic

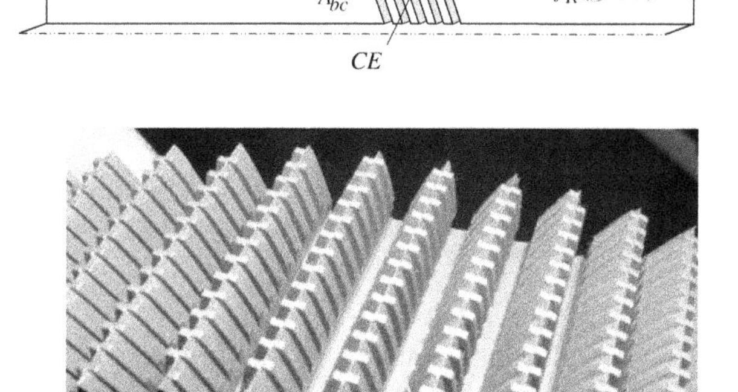


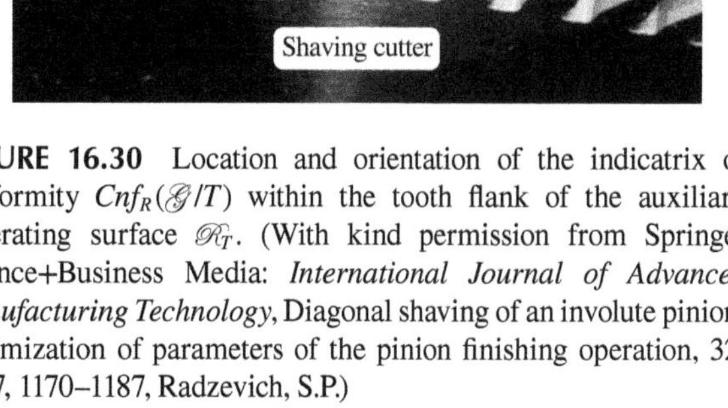


**FIGURE 16.30** Location and orientation of the indicatrix of conformity $Cnf_R(\mathscr{G}/T)$ within the tooth flank of the auxiliary generating surface $\mathscr{R}_T$. (With kind permission from Springer Science+Business Media: *International Journal of Advanced Manufacturing Technology*, Diagonal shaving of an involute pinion: Optimization of parameters of the pinion finishing operation, 32, 2007, 1170–1187, Radzevich, S.P.)

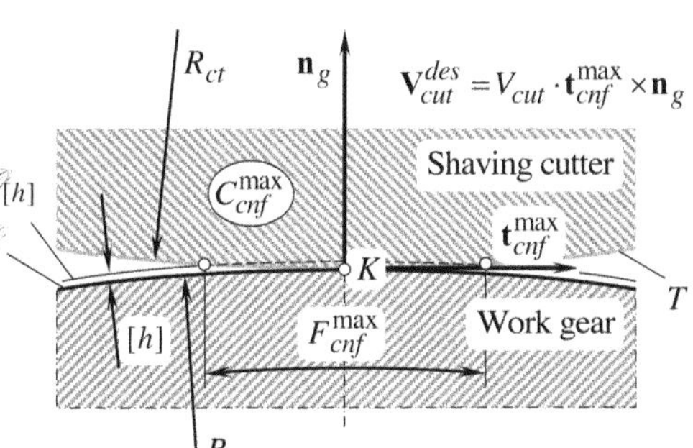


**FIGURE 16.29** Normal plane section $C_{cnf}^{max}$ at point of contact of the tooth flanks $\mathscr{G}$ and $T$ that corresponds to the direction $\mathbf{t}_{cnf}^{max}$ of the maximum degree of conformity of the surfaces $\mathscr{G}$ and $T$. (With kind permission from Springer Science+Business Media: *International Journal of Advanced Manufacturing Technology*, Diagonal shaving of an involute pinion: Optimization of parameters of the pinion finishing operation, 32, 2007, 1170–1187, Radzevich, S.P.)

curve $Cnf_R(\mathscr{G}/T)$ is depicted in the lateral plane $\mathscr{R}_T$ of the auxiliary generating rack.

The major axis $A_{bc}^{max}$ of the work gear-to-shaving cutter contact area is aligned with the direction of the minimum diameter $d_{cnf}^{min}$ of the indicatrix of conformity $Cnf_R(\mathscr{G}/T)$. The axis $A_{bc}^{max}$ is within the angle that the characteristic line $\mathscr{E}_g$ of the surface $\mathscr{G}$ and the characteristic line $\mathscr{E}_c$ of the surface $T$ form. It is important to stress here that the major axis $A_{bc}^{max}$ and the minor axis $A_{bc}^{min}$ of the boundary curve $\mathscr{B}_c$ are not orthogonal to each other, that is, in the general case the angle is not equal to a right angle $\chi \neq 90°$. The desirable direction of the speed of the cut $\mathbf{V}_{cut}^{des}$ at $K$ is perpendicular to the minimum diameter $d_{cnf}^{min}$ of the characteristic curve $Cnf_R(\mathscr{G}/T)$. The direction of the minimum diameter is also indicated with unit tangent vector $\mathbf{t}_{cnf}^{max}$ of the direction of the maximum degree of conformity at the point of contact of the tooth flanks $\mathscr{G}$ and $T$.

The angle of inclination of the cutting edge $\lambda$ is measured between the cutting speed vector $\mathbf{V}_{cut}^{des}$ and the normal unit vector $\mathbf{n}_{ce}$ to the cutting edge $CE$. The vector $\mathbf{n}_{ce}$ is located within the plane $\mathscr{R}_T$ (see Figure 16.30). Keeping this in mind, a shaving cutter that features an optimal inclination angle $\lambda_{opt}$ can be designed based on the features of the geometry of the indicatrix of conformity $Cnf_R(\mathscr{G}/T)$.

### 16.4.4.5 Favorable Design Parameters of a Shaving Cutter and of the Kinematics of the Rotary Shaving Operation

At this point, with the indicatrix of conformity $Cnf_R(\mathscr{G}/T)$ calculated, the problem of minimum shaving time can be expressed in terms of the design parameters of the shaving cutter along with the parameters of the kinematics of the rotary shaving process:

- The degree of conformity of the shaving cutter tooth flank $T$ to the work gear tooth flank $\mathscr{G}$ must be the maximum possible at every point of contact of the surfaces.
- An angle that the cutting speed vector $\mathbf{V}_{cut}$ forms with the vector $\mathbf{V}_{cut}^{des}$ of the desirable speed of the cut must be the smallest possible.

This is immediately discussed in more detail in the following.

#### 16.4.4.5.1 Necessary Conditions

Once the contact geometry of the tooth flanks $\mathscr{G}$ and $T$ is analytically described by means of the indicatrix of conformity $Cnf_R(\mathscr{G}/T)$, then the calculation of the favorable design parameters of a shaving cutter and the favorable parameters of the kinematics of a rotary shaving process turns to a routing problem.

A set of equations that specifies the necessary conditions for the minimum shaving time and maximum accuracy of the shaved tooth flank of a work gear immediately follows from Eq. (16.68) of the indicatrix of conformity $Cnf_R(\mathscr{G}/T)$:

The necessary conditions for the ***Minimum shaving time*** and the ***Maximum accuracy*** of the shaved gear

$$\Rightarrow \begin{cases} \dfrac{\partial r_{cnf}}{\partial R_{1.ct}} = \dfrac{1}{\sqrt{R_{1.ct}}\,\sin(\mu-\varphi)} = 0 \\ \dfrac{\partial r_{cnf}}{\partial \mu} = -\dfrac{\cos(\mu-\varphi)}{\sin^2(\mu-\varphi)}\sqrt{R_{1.ct}} = 0 \\ \dfrac{\partial r_{cnf}}{\partial \varphi} = -\dfrac{\cos\varphi}{\sin^2\varphi}\sqrt{R_{1.p}} - \dfrac{\cos(\mu-\varphi)}{\sin^2(\mu-\varphi)}\sqrt{R_{1.ct}} = 0 \end{cases} \tag{16.71}$$

The solution to the set of Eqs. (16.71) can be expressed in terms of the design parameters of the work gear and the shaving cutter.

#### 16.4.4.5.2 Sufficient Conditions

The sufficient conditions for the maximum of function $r_{cnf}(R_{1.ct}, \mu, \varphi)$ of three variables in this case are also fulfilled.

The solution to the set of Eq. (16.71) returns favorable values of three parameters, namely of (a) the first principal radius of curvature $R_{1.ct}^{opt}$ of the shaving cutter tooth flank $T$, (b) the angle $\mu_{opt}$ of the local relative orientation of the surfaces $\mathscr{G}$ and $T$, and (c) the angular parameter $\varphi_{opt}$. The calculated set of the parameters $R_{1.c}^{opt}$, $\mu_{opt}$, and $\varphi_{opt}$ is favorable for machining a given work gear. Two of the parameters (that is, the radius $R_{1.ct}^{opt}$ and the angle of local orientation $\mu_{opt}$) are sufficient for the calculation of the favorable values of the rest of the design parameters of a shaving cutter, while the third one, $\varphi_{opt}$, determines the desirable direction of the cutting speed vector $\mathbf{V}_{cut}^{des}$. Because of the constraints imposed by the design of shaving machines available on the current market, the actual values $R_{1.ct}^{act}$, $\mu_{act}$, and $\varphi_{act}$ of the parameters have be as close as possible to their calculated values $R_{1.ct}^{opt}$, $\mu_{opt}$, and $\varphi_{opt}$.

## 16.5 TANGENTIAL METHOD OF THE GEAR SHAVING PROCESS

In the extreme condition, a work gear is passed across the face of the shaving cutter at a right angle to the work gear axis of rotation. When the traverse is orthogonal to the axis of the work gear, the method of gear shaving is referred to as the *tangential method*. The tangential method is also often referred to as the *underpass method* of gear shaving. Because of the short feed stroke and the one cycle required, the tangential method of shaving is usually the fastest method of shaving. It is used primarily in high-production operations.

### 16.5.1 Kinematics of Tangential Method of Gear Shaving Process

Tangential shaving is basically the same as diagonal shaving, but the traverse path of the work gear is perpendicular to its axis (see Figure 16.31). With tangential shaving, there is no axial reciprocation of the worktable. Instead, the work gear reciprocates perpendicularly to its own axis. This method is usually dictated when shaving cluster gears, or gears with close adjacent shoulders. Usually, in order to avoid collision of the shaving cutter with the shoulder, the cross-axes angle falls into the range of 4° to 6°.

When the tangential method of gear shaving is used, the work gear and the shaving cutter are meshed in crossed-axes relationship (see Figure 16.31). The crossing axes of the work gear $O_g$ and the shaving cutter $O_{ct}$ are at a certain center distance $C$.

Referring to Figure 16.31, the rotation of the work gear about its axis of rotation $O_g$ is designated as $\omega_g$. The rotation of the shaving cutter about its axis of rotation $O_{ct}$ is designated as $\omega_{ct}$. The shaving cutter is rotated in both directions during the work cycle. The rotations $\omega_g$ and $\omega_{ct}$ are properly timed with one another ($\omega_g \cdot N_g = \omega_{ct} \cdot N_{ct}$; here $N_g$ and $N_{ct}$ denote the tooth number of the work gear and the shaving cutter, respectively, and the equalities $\omega_g = |\boldsymbol{\omega}_g|$ and $\omega_{ct} = |\boldsymbol{\omega}_{ct}|$ are observed). The rotation vectors $\boldsymbol{\omega}_g$ and $\boldsymbol{\omega}_{ct}$ are at an angle $\Sigma = \angle(\boldsymbol{\omega}_g, \boldsymbol{\omega}_{ct})$. The angle $\bar{\Sigma}$ of crossing of the axes $O_g$ and $O_{ct}$ complements the angle $\Sigma$ to 180°, that is, $\bar{\Sigma} = \pi - \Sigma$.

The work gear reciprocates $\mathbf{F}_{ct}$ tangentially in relation to its own axis of rotation while the work gear and the shaving cutter are in mesh (see Figure 16.31). It is customary to move the work gear past the shaving cutter by means of the feed slide. On some of the larger shaving machines, for obvious reasons, the shaving cutter is mounted on a feed slide. In the consideration below, the relative motion of the work gear and the shaving cutter is considered regardless of whether the worktable or the shaving cutter actually reciprocates.

The productivity rate of the rotary shaving of gears directly depends on the length of the traverse path of the work gear.

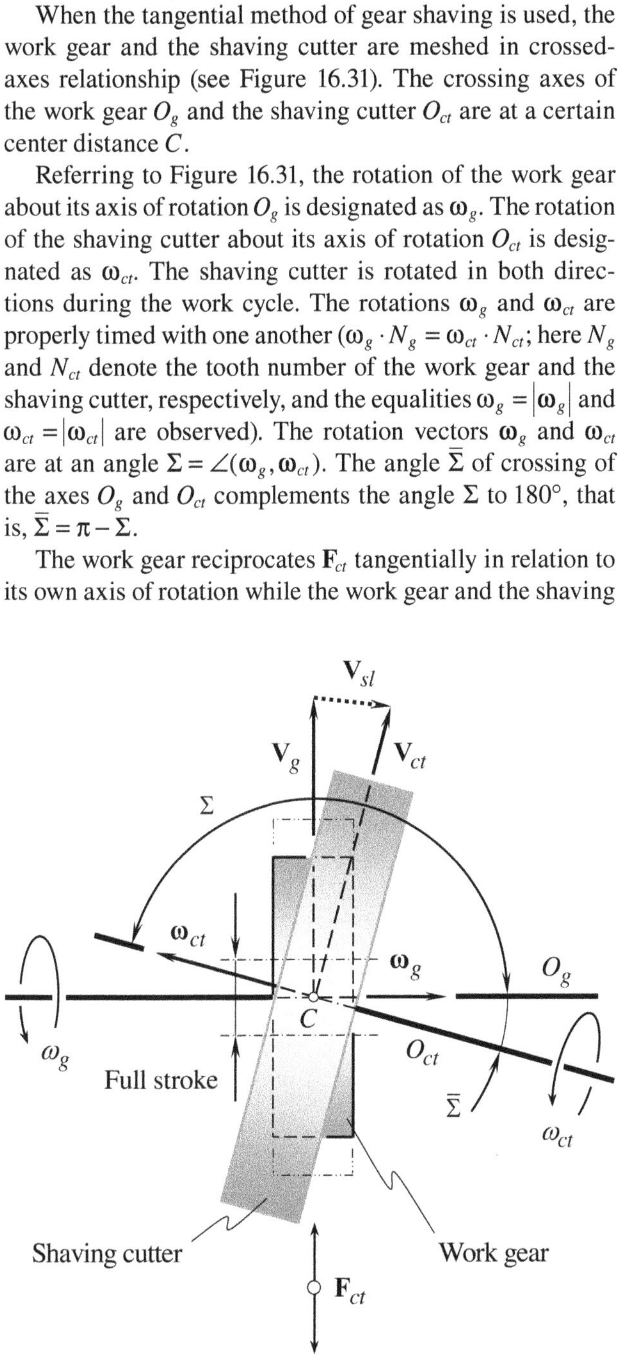


**FIGURE 16.31** Schematic diagram of the tangential method of the gear shaving process.

### 16.5.2 Cutting Speed in Tangential Method of Rotary Shaving of Gear

A formula for the calculation of the cutting speed $\mathbf{V}_{cut}$ in the diagonal shaving of gears is similar to that [see Eq. (16.40)] derived earlier for the case of the axial method of shaving of gears. As in tangential shaving, the traverse motion $\mathbf{F}_{ct}$ is perpendicular to the work gear axis of rotation, so the impact of the motion $\mathbf{F}_{ct}$ differs from that in the diagonal method of rotary shaving. The cutting speed is significantly affected by the transverse motion of the shaving cutter.

Assume again that the shaving cutter is not rotating and is just reciprocating with respect to the work gear shown in Figure 16.32. In such a scenario, the vector $\mathbf{F}_{ct}$ of the traverse motion of the shaving cutter is perpendicular to the work gear axis of rotation $O_g$.

The component $\mathbf{F}_{ct}^{sl}$ of the vector $\mathbf{F}_{ct}$ of the traverse motion in the direction tangent to the tooth flank of the work gear is equal to zero, $\mathbf{F}_{ct}^{sl} \equiv 0$. Another component $\mathbf{F}_{ct}^{rt}$ of the traverse motion $\mathbf{F}_{ct}$ is within the pitch plane and is perpendicular to the axis of the rotation of the work gear $\mathbf{F}_{ct}^{rt} \equiv \mathbf{F}_{ct}$. The component $\mathbf{F}_{ct}^{rt}$ contributes to the rotation of the work gear. This additional rotation $\omega_g^{ad}$ is equal to

$$\omega_g^{ad} = \frac{2F_{ct}}{d_g} \tag{16.72}$$

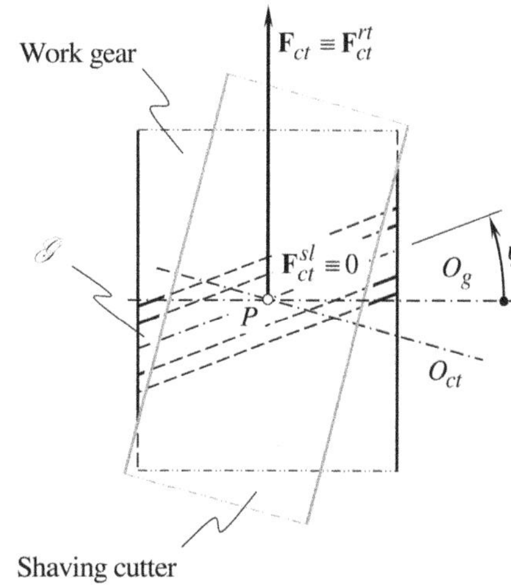


**FIGURE 16.32** Impact of the traverse motion on the cutting speed $\mathbf{V}_{cut}$ in the tangential shaving of gears.

The rotation vector $\boldsymbol{\omega}_g^{ad}$ is aligned with the rotation vector $\boldsymbol{\omega}_g$. Depending on (a) the rotation of the shaving cutter $\omega_c$ and (b) the direction of the hand and the value of the helix angle $\psi_g$, the rotation vector $\boldsymbol{\omega}_g^{ad}$ is pointed either in the same direction as the rotation vector $\boldsymbol{\omega}_g$ or it is pointed opposite to the vector $\boldsymbol{\omega}_g$. In the first case, the rotation $\boldsymbol{\omega}_g^{ad}$ adds to the rotation $\boldsymbol{\omega}_g$. In the second case, it subtracts from the rotation $\boldsymbol{\omega}_g$.

Once the identity $\mathbf{F}_{ct}^{sl} \equiv 0$ and Eq. (16.72) are taken into account, the resultant formula [see Eq. (16.40)] can be used for the calculation of the cutting speed $\mathbf{V}_{cut}$ in the tangential shaving of gears.

When the tangential method of gear shaving is applied, the shaving cutter has to be wider than the gear to be shaved, and its serrations must be placed on a helix in order to produce the relative tooth flank feed. Serrations of this type are shown in Figure 16.32(b). Serrations of this type are often referred to as *differential serration*, or *staggered pattern serrations*.

### 16.5.3 Tangential Shaving of Shoulder Gear: Descriptive-Geometry-Based Approach

Gears that are intended to be shaved should allow room for the shaving cutter to run out. The method of tangential shaving is ideally suited for shaving close-shoulder gears. We refer to a *shoulder* as any integral part of a gear that is close enough to the gear face, and diameter $d_{sh}$ of which exceeds outside diameter $d_{o.g}$ of the gear (see Figure 15.162). Sometimes a flange, a cam, or the tooling that holds the workpiece can be the interference. The neck width $l_{\min}$ or gap between the gear face and the face of the interfering shoulder should always be the minimum obtainable through the tolerance stackup of the component part.

It is necessary to prevent a shaving cutter at crossed axes from contacting the shoulder face. Therefore, when shaving a shoulder gear, the angle between the work gear axis and the shaving cutter axis is often reduced to $\Sigma = 3 \div 5°$. The chief use of a small crossed-axes angle is to operate in a restricted space near a shoulder (see Figure 16.33).

**FIGURE 16.33** Shaving of a cluster gear.

**FIGURE 16.34** Shaving a gear with a shoulder.

The following analysis is based on wide use of the powerful methods developed in descriptive geometry. This approach is referred to as the DG-based approach (descriptive-geometry-based approach).

#### 16.5.3.1 Maximum Allowed Outer Diameter of a Shaving Cutter

The following design parameters have to be known before determining a maximum permissible outer diameter of the shaving cutter[12]: (a) the work gear pitch helix angle $\psi_g$, (b) the face width of the work gear, (c) the width $l_{\min}$ of the shoulder gear neck, and (d) the shaving cutter pitch helix angle $\psi_{ct}$ [64].

Shaving of a gear with a shoulder is illustrated in Figure 16.34.

For solving the problem under consideration, a work gear with a shoulder is depicted in two planes of projection, namely in horizontal plane of projection $\pi_1$ and in vertical

[12] It is important to point out here that a few more geometrical problems arise in gear shaving of a shoulder gear. Examples are: (a) to determine a maximum permissible neck width $l_{\min}$ of the shoulder gear when the shaving cutter outer diameter $d_{o.ct}$, the work gear pitch helix angle $\psi_g$, and the shaving cutter pitch helix angle $\psi_{ct}$ are known, or (b) to determine an appropriate pitch helix angle $\psi_{ct}$ of the shaving cutter when its outer diameter $d_{o.ct}$, neck width $l_{\min}$ of the shoulder gear, and pitch helix angle $\psi_g$ of the work gear are given. Problems of this type are similar in nature to the problem of determining a maximum permissible outer diameter of the shaving cutter $d_{o.ct}$. Solutions to the problems (a), (b), and so forth can be drawn up from the solution to the problem of determining the shaving cutter outer diameter $d_{o.ct}$.

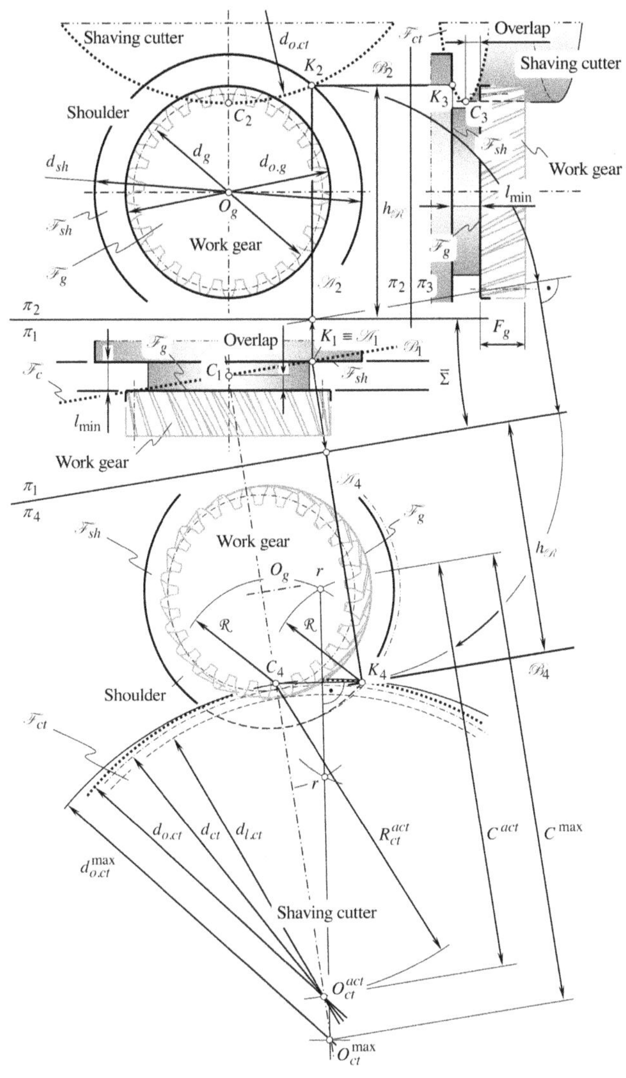


**FIGURE 16.35** Determination of the maximum permissible outer diameter $d_{o.ct}$ of a shaving cutter. (From Radzevich, S.P., *SME Journal of Manufacturing Science and Engineering*, 127(4), 893–900, 2005. With permission.)

plane of projection $\pi_2$ (see Figure 16.35) [64]. The work gear axis of rotation $O_g$ is at a right angle to the plane $\pi_2$.

The vertical plane $\mathscr{F}_{ct}$ coincides with the shaving cutter face in its limit closest to the shoulder. The vertical plane $\mathscr{F}_{sh}$ coincides with the shoulder face. Because both planes $\mathscr{F}_{ct}$ and $\mathscr{F}_{sh}$ are oriented vertically, in the plane of projection $\pi_1$ they are depicted as the straight-line traces $\mathscr{F}_{ct}$ and $\mathscr{F}_{sh}$ (see Figure 16.35).

The planes $\mathscr{F}_{ct}$ and $\mathscr{F}_{sh}$ intersect one another along a straight line $\mathscr{A}$. The line $\mathscr{A}$ is oriented vertically.

When the shaving cutter occupies the limit position at which it is closest to the shoulder, and no collision of the shaving cutter and the shoulder occurs, the outer cylinder of the shaving cutter and the shoulder cylinder contact each other at a point. This point is denoted as $K$. Point $K$ is located within the straight line $\mathscr{A}$. Because point $K$ belongs to a circle of a diameter $d_{sh}$, the vertical projection $K_2$ of point $K$ is located on the intersection of the vertical projection $\mathscr{A}_2$ of the straight line $\mathscr{A}$ and the vertical projection of the circle of a diameter $d_{sh}$.

An auxiliary plane of projection $\pi_4$ is constructed so that it is perpendicular to the plane of projection $\pi_1$, and the axis $\pi_1/\pi_4$ forms an angle $\overline{\Sigma}$ with the axis $\pi_1/\pi_2$. An image of the work gear in the plane of projection $\pi_4$ is located beyond the shoulder. Thus, the work gear is not visible in the plane of projection $\pi_4$. However, the work gear is shown in solid lines in the plane of projection $\pi_4$ for the readers' convenience.

The shaving cutter axis of rotation $O_{ct}$ is at a right angle with respect to the plane of projection $\pi_4$. Projection $K_4$ of point $K$ can be easily found with the help of the appropriate connection lines (see Figure 16.35), or by means of the horizontal straight line $\mathscr{B}$. Both projections $\mathscr{B}_2$ and $\mathscr{B}_4$ of the straight line $\mathscr{B}$ are at the same distance $h_{\mathscr{R}}$ from the horizontal plane of projection $\pi_1$.

The maximum permissible outer diameter $d_{o.ct}^{\max}$ of the shaving cutter can be determined in the auxiliary plane of projection $\pi_4$. The shaving cutter axis of rotation is located at the closest distance of approach $C$ of the axes $O_g$ and $O_{ct}$ (this straight line passes through the projection $C_4$). A circle of the maximum outer diameter $d_{o.ct}^{\max}$ passes through the projection $C_4$ and through the projection $K_4$. The straight-line segment $C_4K_4$ is subdivided into two equal portions by means of two arc segments of a certain radius $\mathcal{R}$. The most remote axis of rotation $O_{ct}^{\max}$ of the shaving cutter is located within the straight line $r-r$. Therefore, the center $O_{ct}^{\max}$ can be found as the intersection between the closest distance of approach of the axes $O_g$ and $O_{ct}$ and the straight line $r-r$. Point $O_{ct}^{\max}$ specifies the maximum permissible center distance $C^{\max}$. The actual shaving cutter outer diameter $d_{o.ct}$ should be the largest allowed in its design size class (3.5 in., 6.0 in., 8.0 in., 9.5 in., or 13.5 in.).

The actual shaving cutter axis of rotation $O_{ct}^{act}$ is located within the closest distance of approach of the axes of rotation $O_g$ and $O_{ct}$ at a distance $R_{ct}^{act} = 0.5 d_{o.ct}$. After being determined, the shaving cutter's actual outer diameter $d_{o.ct}$ allows for the determination of the actual center distance $C^{act}$ as well as the rest of the design parameters of the shaving cutter: the base diameter $d_{o.ct}$, the pitch diameter $d_{ct}$, the limit diameter $d_{l.ct}$, and so forth.

### 16.5.3.2 Minimum Required Overlap of the Work Gear and Shaving Cutter

When determining the maximum permissible outer diameter of the shaving cutter (see Figure 16.35), it has been assumed that the value of the overlap is known. However, usually the minimum required overlap has to be determined. Consideration of (a) the work gear auxiliary generating rack $\mathscr{R}_p$ of the minimum required face width $F_{\mathscr{R}.p}$ and (b) the shaving cutter auxiliary generating rack $\mathscr{R}_{ct}$ of the minimum required face width $F_{ct}$ is a key for determining the overlap.

This analysis follows immediately below.

### 16.5.3.2.1 Minimum Permissible Width of the Auxiliary Generating Rack $\mathscr{R}_g$ Conjugate to the Work Gear

A work gear with an involute tooth profile is depicted in three planes of projection $\pi_1$, $\pi_2$, and $\pi_3$, as shown in Figure 16.36 [58]. The axis of rotation $O_g$ of the work gear is at a right angle to the plane of projection $\pi_2$. The vertical auxiliary plane of projection $\pi_4$ is at a certain angle with respect to the plane of projection $\pi_2$. This angle is equal to the pitch helix angle $\psi_g$ of the work gear. Projection $\mathscr{R}_4$ of the infinitely wide auxiliary generating rack $\mathscr{R}$ is congruent to the section of the rack $\mathscr{R}$ by a normal plane. The generation of the screw involute surface of the work gear tooth flank occurs along two characteristic lines. The characteristic lines are labeled as $\mathscr{E}^1$ and $\mathscr{E}^2$. The characteristic lines $\mathscr{E}^1$ and $\mathscr{E}^2$ are located within two planes that are tangent from the opposite sides to the base cylinder of a diameter $d_{b.g}$ of the work gear. The active portions of the characteristic lines $\mathscr{E}^1$ and $\mathscr{E}^2$ are bounded with the endpoints $a$, $c$, and $d$, $f$, accordingly. The endpoints $c$ and $f$ are within the base cylinder. They are remote from the plane of projection $\pi_1$ at a distance $h_{ct}$. The straight-line segments $\mathscr{E}^1$ and $\mathscr{E}^2$ are tangent to the corresponding base helices of the work gear intended to be shaved. In order to get the work gear been completely shaved, not the entire length the characteristics $\mathscr{E}^1$ and $\mathscr{E}^2$ have to be generated in the rotary shaving process. It is sufficient to reproduce only their portions *ab* and *de*, that is, to reproduce their portions located within the work gear tooth height. Thus, for complete shaving of the work gear it is required to reproduce not the infinitely wide auxiliary generating rack $\mathscr{R}$ but only a portion $\mathscr{R}_g$ of it. This portion of the auxiliary generating rack is referred to as the *work gear auxiliary rack*. Within the plane of projection $\pi_1$, the portion $\mathscr{R}_g$ of the generating rack $\mathscr{R}$ is shaped in the form of a parallelogram $b_1g_1e_1h_1$. The straight-line segments $b_1g_1$ and $e_1h_1$ through points $b_1$ and $e_1$ are parallel to the work gear face $\mathscr{F}_g$. The straight-line segments $b_1h_1$ and $e_1g_1$ align

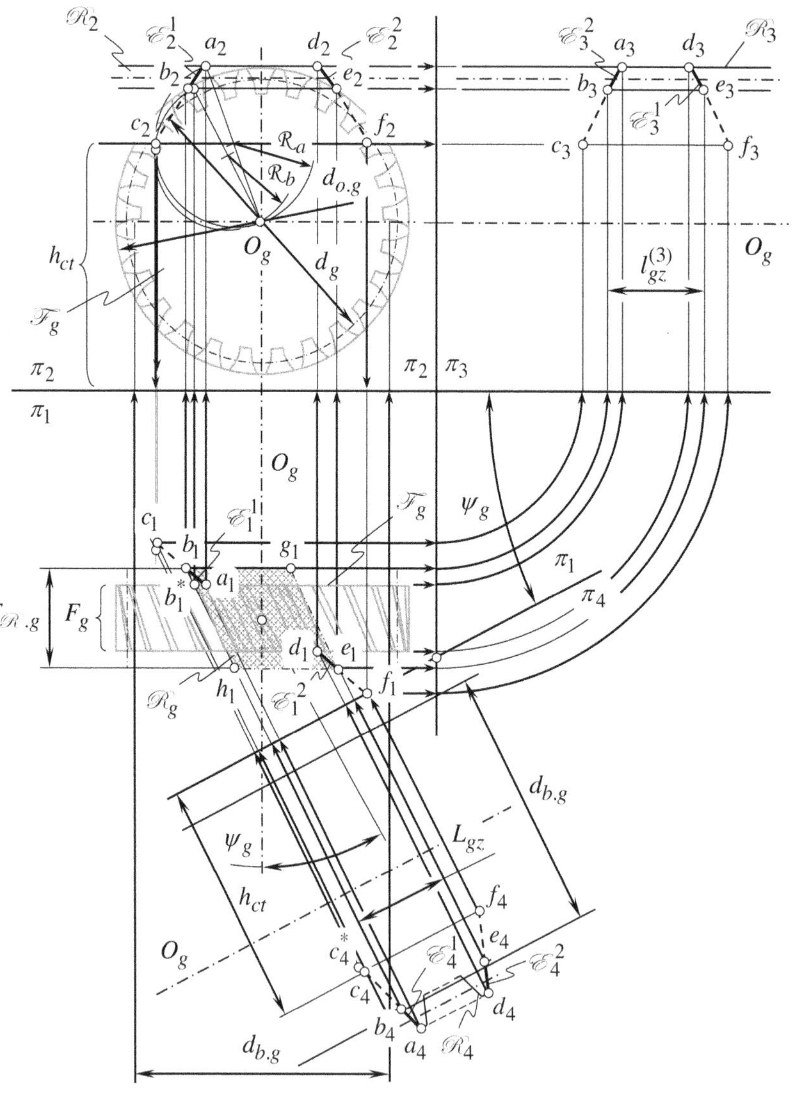


**FIGURE 16.36** Determining the minimum permissible width of the auxiliary generating rack $\mathscr{R}_g$ conjugate to a work gear. (From Radzevich, S.P., *SME Journal of Manufacturing Science and Engineering*, 127(4), 893–900, 2005. With permission.)

to the corresponding straight-line generatrix of the auxiliary rack $\mathscr{R}$. The minimum face width $F_{\mathscr{R}.g}$ of the auxiliary rack $\mathscr{R}_g$ of the work gear is equal to the length $L_{gz}$ of the generating zone as illustrated in Figure 16.36. It is important to point out here that, first, the straight-line segments $a_1b_1$ and $d_1e_1$ are within the tooth flanks of the rack $\mathscr{R}_g$ and, second, the rack $\mathscr{R}_g$ has to be of the smallest permissible size in the axial direction of the work gear.

Points $a$ and $b$ of the auxiliary rack $\mathscr{R}$ tooth are connected to the work gear axis of rotation $O_g$. Both the straight-line segments $aO_g$ and $bO_g$ were used as a diameter of a circle of the corresponding radii $\mathcal{R}_a$ and $\mathcal{R}_b$. The circles of the radii $\mathcal{R}_a$ and $\mathcal{R}_b$ meet the base cylinder $d_{b.g}$ at points $c$ and $c^*$. Because of the lack of space, only the projection $c_4^*$ of the point $c^*$ is depicted in Figure 16.36. That point of two points, $c$ and $c^*$, is chosen for whichever length $l_{gz}$ of the generating zone is longer [see the projection $l_{gz}^{(3)}$ of this straight-line segment].

The case is similar with respect to point $f$ of contact of the straight generating line $\mathscr{E}^2$ and the base cylinder of the work gear.

The auxiliary generating rack $\mathscr{R}_g$ has to be as short as possible and as narrow as possible. The length of the parallelogram $b_1g_1e_1h_1$ cannot be shorter than $L_{gz}(\cos\psi_g)^{-1}$, that is, the length of the parallelogram is specified by $L_{gz}$ and $\psi_g$. The required segments $ab$ and $de$ of the characteristics $\mathscr{E}^1$ and $\mathscr{E}^2$ have to be within the parallelogram. This restricts the shortest width of the parallelogram $b_1g_1e_1h_1$.

The length $l_{gz}$ of the generating zone is determined from the formula [64]

$$l_{gz} = d_{o.g}\cos(\phi_n+\alpha) = \left(\sqrt{d_{o.g}^2 - d_g^2\cos^2\phi_n} - d_g\sin\phi_n\right)\cos\phi_n \tag{16.73}$$

where:

$d_{o.g}$ is the outer diameter of the work gear.
$d_g$ is the pitch diameter of the work gear.
$\phi_n$ is the normal profile angle.

For the calculation of the angle $\alpha$, an expression

$$\alpha = \sin^{-1}\left(\frac{d_g}{d_{o.g}}\cdot\cos\phi_n\right) \tag{16.74}$$

has been derived by Prof. S.P. Radzevich [64].

The length $L_{gz}$ of the generating zone is equal to

$$L_{gz} = l_{gz} + 2\sqrt{(0.5d_{o.g})^2+(0.5d_g)^2+0.5d_{o.g}d_g\cos(\alpha+\phi_n)}\,\frac{\sin^2\phi_n}{\cos\phi_n} \tag{16.75}$$

The minimum required length $L_{gz}$ of the generating zone imposes certain constraints on the design parameter of the shaving cutter, as well as on the parameters of the kinematics of the rotary shaving process.

#### 16.5.3.2.2 *Minimum Permissible Width of the Auxiliary Generating Rack $\mathscr{R}_{ct}$ Conjugate to the Shaving Cutter*

The characteristics $E^1$ and $E^2$ for the work gear tooth flank $\mathscr{G}$ that is conjugate with the auxiliary generating rack $\mathscr{R}_g$ do not align to characteristics $\mathscr{E}^1$ and $\mathscr{E}^2$ for the auxiliary helix $\psi_p$ generating rack $\mathscr{R}_{ct}$ that is conjugate to the shaving cutter tooth surface $C_{ct}$. The characteristics $E^1$ and $\mathscr{E}^1$ are within the lateral tooth surface of the auxiliary generating rack $\mathscr{R}$ and also intersect each other at a certain point. The characteristics $E^2$ and $\mathscr{E}^2$ are within the opposite lateral tooth surface of the auxiliary rack $\mathscr{R}$ and also intersect each other at a certain point.

The tooth flank $C_{ct}$ of the shaving cutter can be generated as an envelope to a family of consecutive positions of the auxiliary generating rack $\mathscr{R}$ in its rolling about the shaving cutter axis of rotation $O_{ct}$. The minimum required portion of the auxiliary rack $\mathscr{R}$ of the shaving cutter is designated as $\mathscr{R}_{ct}$. The characteristics $\mathscr{E}^1$ and $\mathscr{E}^2$ can be found in a manner similar to how the characteristics $E^1$ and $E^2$ were reconstructed. This particular problem allows for interpretation in the form of the inverse to the problem considered earlier.

The characteristics $E^1$ and $\mathscr{E}^1$ form a certain angle $\mu$ with each other. The characteristics $E^2$ and $\mathscr{E}^2$ are at that same angle $\mu$ relative to one another. The angle $\mu$ is the angle of the tooth flanks' $\mathscr{G}$ and $C_{ct}$ local relative orientation. The straight-line characteristic $E^1$ is aligned with the first principal directions $\mathbf{t}_{1.g}$ of the work gear tooth flank $\mathscr{G}$. The first principal plane $C_{1.g}$ passes through the unit vector $\mathbf{t}_{1.g}$.

The same is observed for the opposite side of the work gear tooth flank and the tooth flank of the auxiliary generating rack $\mathscr{R}$.

In the rotary shaving of an involute gear, the rack surface $\mathscr{R}_g$ is an enveloping surface to a family of consecutive positions of the work gear tooth flank $\mathscr{G}$ in its motion relative to the rack $\mathscr{R}_g$. Referring to Figure 16.37, the straight-line segment $a_1b_1$ is a portion of the characteristic $E^1$, and the straight-line segment $d_1e_1$ is a portion of the characteristic $E^2$.

The tooth profile of the auxiliary generating rack $\mathscr{R}_c$ is not identical to the tooth profile of the auxiliary generating rack $\mathscr{R}_g$. The addendum $a_{\mathscr{R}c}$ exceeds the addendum $a_{\mathscr{R}g}$. The difference $(a_{\mathscr{R}c}-a_{\mathscr{R}g})$ is designated as $\Delta a$. For determining the required excess $\Delta a$, the fifth necessary condition of proper part surface generation (see Appendix E) can be used [136, 144, 148, 158].

The distance $h_{be}$ at which points $b$ and $e$ are remote from the plane of projection $\pi_5$ and the similar distance $h_{ad}$ at which points $a$ and $d$ are remote from the plane of projection $\pi_5$ (see Figure 16.37) are both equal to the distance at which the corresponding points $a$, $b$, $d$, and $e$ are remote from the plane of projection $\pi_4$.

The determination of the characteristics $\mathscr{E}^1$ and $\mathscr{E}^2$ of the auxiliary generating rack $\mathscr{R}$, which is conjugate to the shaving cutter tooth flank $C_{ct}$, is illustrated in Figure 16.37. Within the plane of projection $\pi_4$, the projections $E_4^1$ and $\mathscr{E}_4^1$,

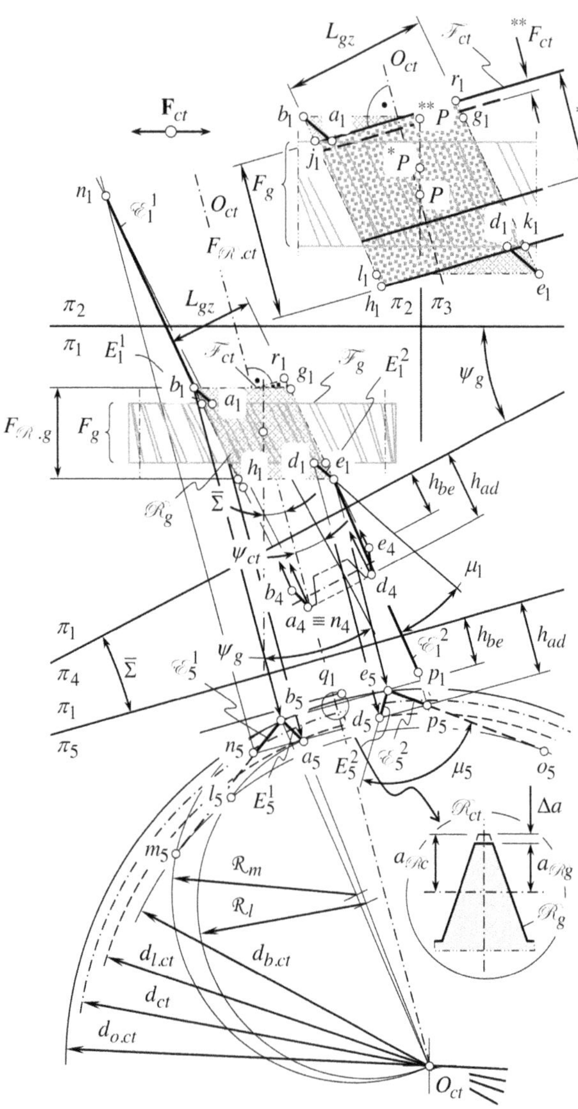

**FIGURE 16.37** Minimum required width of the auxiliary generating rack $\mathscr{R}_{ct}$ of a shaving cutter, and the minimum permissible overlap of the work gear and the shaving cutter. (From Radzevich, S.P., *SME Journal of Manufacturing Science and Engineering*, 127(4), 893–900, 2005. With permission.)

as well as the projections $E_4^2$ and $\mathscr{E}_4^2$, are congruent to one another. The straight-line segment $b_5m_5$ is tangent to the shaving cutter base cylinder of a diameter $d_{b.c}$. This allows for the construction of point $n$ and of projections of the characteristic $\mathscr{E}^1$ onto all others planes of projection. Similar to that just used, all projections of the characteristic $\mathscr{E}^2$ are constructed as well.

For the purpose of proper selection of one of two points $m$ and $l$, the semicircles of radii $\mathscr{R}_l$ and $\mathscr{R}_m$ (see the plane of projection $\pi_5$ in Figure 16.37) are used. The selection of a point is similar to the selection of a point of two points $c$ and $c^*$ in Figure 16.36.

The length $L_{gz}$ of the auxiliary generating rack $\mathscr{R}_{ct}$ of the shaving cutter is equal to (or exceeds) the length of the auxiliary generating rack $\mathscr{R}_g$ of the work gear. The width of the rack $\mathscr{R}_{ct}$ is specified by two straight lines through points $n_1$ and $p_1$. Both of these straight lines are parallel to the shaving cutter axis of rotation. The width of the auxiliary generating rack $\mathscr{R}_{ct}$ of the shaving cutter is equal to $F_{\mathscr{R}.ct}$.

Within the plane of projection $\pi_1$, the profile of the auxiliary rack $\mathscr{R}_{ct}$ is shaped in the form of a parallelogram $p_1q_1n_1r_1$. The straight-line segments $n_1r_1$ and $p_1q_1$ through points $n_1$ and $p_1$ are orthogonal to the shaving cutter axis of rotation $O_{ct}$, and thus they are parallel to the shaving cutter face $\mathscr{F}_{ct}$. The straight-line segments $n_1q_1$ and $r_1p_1$ are aligned with the corresponding straight-line generatrix of the auxiliary rack $\mathscr{R}$. The width of the shaving cutter auxiliary rack $\mathscr{R}_{ct}$ is equal to the length of the generating zone $L_{gz}$. Determination of the length $F_{\mathscr{R}.ct}$ is shown in Figure 16.37. It is important to point out here that the auxiliary generating rack $\mathscr{R}_{ct}$ of the shaving cutter has to be as short as possible, and it must be the narrowest permissible.

#### 16.5.3.3 Minimum Required Face Width of Shaving Cutter

The determination of the minimum required face width of a shaving cutter is based on the need to meet the two following requirements.

*First*, the entire work gear auxiliary rack $\mathscr{R}_g$ has to be generated in the rotary shaving process.
*Second*, the shaving cutter auxiliary rack $\mathscr{R}_{ct}$ must overlap the auxiliary rack $\mathscr{R}_g$.

In order to reproduce the characteristics $E^1$ and $E^2$ (see the straight-line segments $ab$ and $ed$), the corresponding characteristics $\mathscr{E}^1$ and $\mathscr{E}^2$ must be reproduced by the shaving cutter in the rotary shaving operation. Two straight lines through points $j_1$ and $k_1$, which are orthogonal to the shaving cutter axis of rotation $O_{ct}$, determine the minimum required width $F_{\mathscr{R}.ct}$ of auxiliary rack $\mathscr{R}_{ct}$ of the shaving cutter. For finishing of the entire tooth flanks of the work gear, the shaving cutter auxiliary rack of the width $F_{\mathscr{R}.ct}$ (and of length $L_{gz}$) has to be reproduced.

The minimum required width of the shaving cutter for the tangential shaving of a shoulder gear is equal to $F_{ct} = F_{\mathscr{R}.ct}$.

### 16.5.4 Tangential Shaving of Shoulder Gear: Analytical Approach

The descriptive-geometry-based analysis of tangential shaving of a shoulder gear gives an insight into how this problem can be solved analytically. On the premise of the derived graphical solution, a corresponding analytical solution to the problem is derived in the following.

#### 16.5.4.1 Optimal Design Parameters of Shaving Cutter

The design parameters of a shaving cutter, namely (a) the greatest outer diameter, and (b) the greatest pitch helix angle, under which no collision occurs between the adjacent shoulder and the shaving cutter can be determined analytically. An establishment of appropriate reference systems and the corresponding coordinate system transformations is the starting point for solving the problem under consideration.

#### 16.5.4.1.1 Applied Reference Systems

A left-hand-oriented Cartesian coordinate system $X_gY_gZ_g$ is associated with the work gear. The axis $X_g$ is aligned with the work gear axis of rotation $O_g$, as shown in Figure 16.38. In a gear shaving operation, the axis $Y_g$ is parallel in relation to the closest distance of approach $C$ between the axis of rotation $O_g$ of the work gear and the axis of rotation $O_{ct}$ of the shaving cutter.

A left-hand-oriented Cartesian coordinate system $X_{ct}Y_{ct}Z_{ct}$ is associated with the shaving cutter. The axis $X_{ct}$ is aligned with the shaving cutter axis of rotation $O_{ct}$. The axis $Y_{ct}$ is aligned with the closest distance of approach $C$ between the axis of rotation $O_g$ of the work gear and the axis of rotation $O_{ct}$ of the shaving cutter.

If necessary, auxiliary reference systems can be used in addition to the coordinate systems $X_gY_gZ_g$ and $X_{ct}Y_{ct}Z_{ct}$.

#### 16.5.4.1.2 Operators of the Linear Transformations

In order to be treated together, the work gear and the shaving cutter have to be represented in a common reference system. The coordinate system $X_gY_gZ_g$ can be used as the common reference system. If the coordinate system $X_gY_gZ_g$ is chosen, then an operator $\mathbf{Rs}(T \rightarrow \mathscr{G})$ of the resultant coordinate system transformation must be derived for the conversion of the tooth flank of the shaving cutter to the coordinate system $X_gY_gZ_g$ that is associated with the work gear.

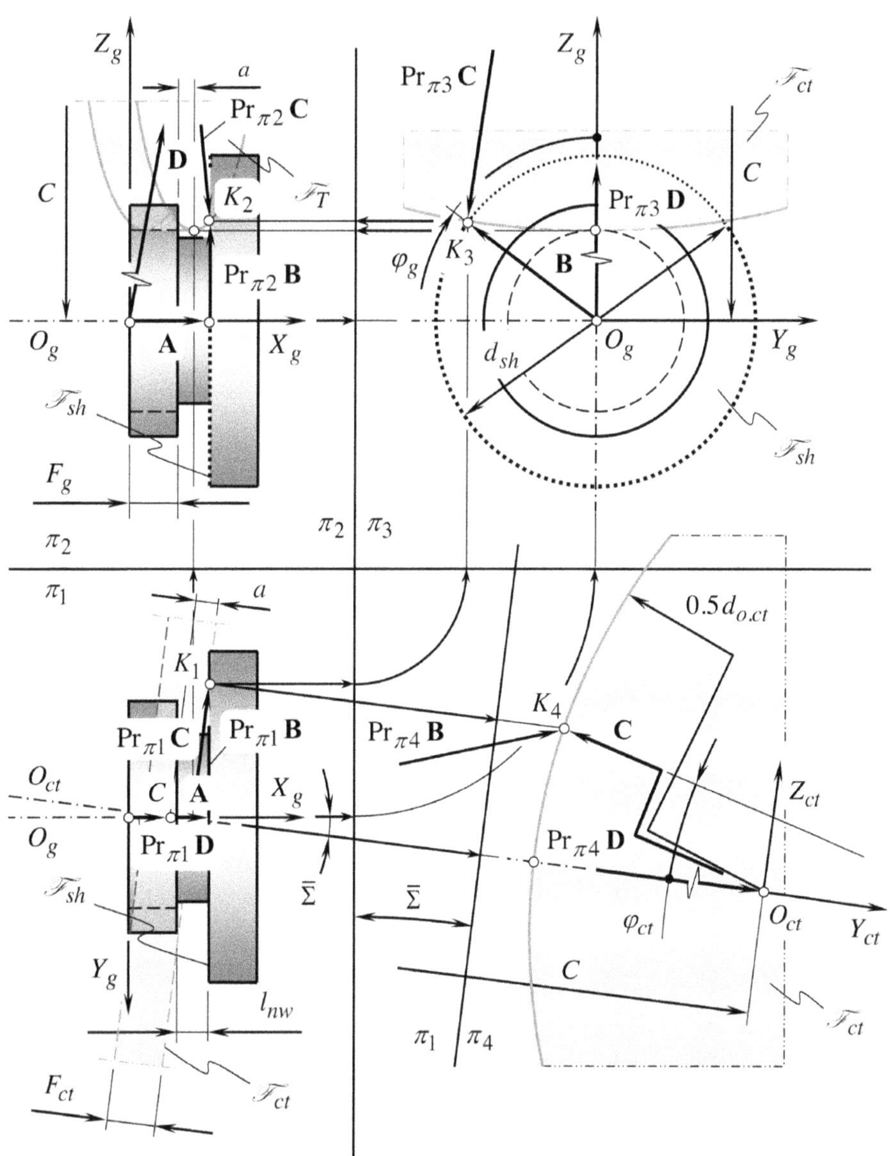

FIGURE 16.38 Applied coordinate systems and the closed vector polygon.

The operator $\mathbf{Rs}(T \to \mathscr{G})$ of the linear transformation is calculated as a product of the operators of the elementary coordinate system transformations, namely:

Operator $\mathbf{Tr}[(F_g + l_{nw}), X_{ct}]$ of translation along the $X_{ct}$ – axis through the distance $(F_g + l_{nw})$.
Operator $\mathbf{Tr}(C, Z_{ct})$ of translation along the $Z_{ct}$ – axis through the distance $C$.
Operator $\mathbf{Rt}(\bar{\Sigma}, Z_{ct})$ of rotation about the $Z_{ct}$ – axis through the angle $\bar{\Sigma}$.

With that said, the operator $\mathbf{Rs}(T \to \mathscr{G})$ of the resultant linear transformation can be calculated from the following formula:

$$\begin{aligned}\mathbf{Rs}(T \to \mathscr{G}) = \mathbf{Tr}[(F_g + l_{nw}),\, X_{ct}]\\ \cdot \mathbf{Tr}(C,\, Z_{ct}) \cdot \mathbf{Rt}(\bar{\Sigma},\, Z_{ct})\end{aligned} \qquad (16.76)$$

where:

$F_g$ is the face width of the gear.
$l_{nw}$ is the neck width of the work gear.
$C$ is the center distance.
$\bar{\Sigma}$ is the angle that is formed by the crossing axes of rotation $O_g$ and $O_{ct}$.

#### 16.5.4.1.3 Closed Vector Polygon

In the worst-case scenario, the shaving cutter is figured with respect to the shoulder gear so that the shaving cutter face $\mathscr{F}_T$ shares a common point $K$ with the shoulder face $\mathscr{F}_{sh}$ (see Figure 16.38).

In the Cartesian coordinate system $X_g Y_g Z_g$, the position vector $\mathbf{R}_K^{(g)}$ of point $K$ can be represented as the summa of two vectors

$$\mathbf{R}_K^{(g)} = \mathbf{A} + \mathbf{B} \qquad (16.77)$$

where vector $\mathbf{A}$ is equal to

$$\mathbf{A} = \begin{bmatrix} (F_g + l_{nw}) \\ 0 \\ 0 \\ 1 \end{bmatrix}_g \qquad (16.78)$$

and vector $\mathbf{B}$ yields representation in the form

$$\mathbf{B} = \begin{bmatrix} (F_g + l_{nw}) \\ -0.5 d_{sh} \sin\varphi_g \\ 0.5 d_{sh} \cos\varphi_g \\ 1 \end{bmatrix}_g \qquad (16.79)$$

The subscript $g$ at the brackets indicates that the vectors are represented in the coordinate system $X_g Y_g Z_g$ associated with the gear.

The position vector $\mathbf{D}$ that specifies the origin $O_{ct}$ of the coordinate system $X_g Y_g Z_g$ is equal to

$$\mathbf{D} = \begin{bmatrix} (F_g + l_{nw}) - (0.5 d_{o.ct} \cdot \sin\varphi_{ct} + a)\cos\bar{\Sigma} \\ 0 \\ C \\ 1 \end{bmatrix}_g \qquad (16.80)$$

Ultimately, for the position vector $\mathbf{R}_K^{(c)}$ of point $K$ in the coordinate system $X_{ct} Y_{ct} Z_{ct}$, the expression

$$\mathbf{R}_K^{(c)} \Rightarrow \mathbf{C} = \begin{bmatrix} a \\ -0.5 d_{o.ct} \cos\varphi_{ct} \\ 0.5 d_{o.ct} \sin\varphi_{ct} \\ 1 \end{bmatrix}_{ct} \qquad (16.81)$$

is valid. The subscript $ct$ at the brackets indicates that the vector $\mathbf{R}_K^{(c)}$ is represented in the coordinate system $X_{ct} Y_{ct} Z_{ct}$ associated with the shaving cutter.

In Eq. (16.73), the closest distance of approach of the shaving cutter to the shoulder is denoted as $a$ (see Figure 16.38).

Equations (16.76), (16.77), and (16.81) yield the expression

$$\begin{aligned}\mathbf{R}_K^{(g)} - \mathbf{R}_K^{(c)} \cdot \mathbf{Rs}(T \to \mathscr{G}) = (\mathbf{A} + \mathbf{B})\\ -[\mathbf{Rs}(T \to \mathscr{G}) \cdot \mathbf{C} + \mathbf{D}] \equiv 0\end{aligned} \qquad (16.82)$$

for the closed vector polygon [148].

After substituting Eq. (16.77)–Eq. (16.81) into Eq. (16.82), the following column matrix can be obtained:

$$\begin{bmatrix} F_g + l_{nw} - a \cdot \cos\bar{\Sigma} + 0.5 d_{o.ct} \sin\bar{\Sigma} \cos\varphi_{ct} - a + (0.5 d_{o.ct} \sin\varphi_{ct} + a) \cdot \cos\bar{\Sigma} \\ -0.5 d_{sh} \sin\varphi_g + a \cdot \sin\bar{\Sigma} + 0.5 d_{o.ct} \cos\bar{\Sigma} \cos\varphi_{ct} \\ 0.5 d_{sh} \cos\varphi_g - 0.5 d_{o.ct} \sin\varphi_{ct} - 2C_{g/ct} \\ 1 \end{bmatrix} = 0 \qquad (16.83)$$

It is necessary to eliminate two parameters, $\varphi_g$ and $\varphi_{ct}$, in order to solve Eq. (16.83). Upon elimination of the parameters $\varphi_g$ and $\varphi_{ct}$, Eq. (16.83) casts into an implicit equation of the form $F(l_{nw}, \bar{\Sigma}, d_{o.ct}) = 0$.

On solving the equation $F(l_{nw}, \overline{\Sigma}, d_{o.ct}) = 0$, a set of favorable design parameters of the shaving cutter for finishing a given work gear, as well as the closest distance of approach of the shaving cutter to the shoulder face $\mathscr{F}_{sh}$, can be obtained.

### 16.5.4.2 Influence of Overlap of the Shaving Cutter over the Work Gear onto the Accuracy of the Finished Tooth Flanks

A minimum required neck width $l_{nw}$ of a shoulder gear can be significantly reduced if the tolerance $[\delta]$ for the accuracy of the tooth flank $\mathscr{G}$ of the work gear exceeds the actual deviation $\delta$ of a finished gear. In such a scenario, the difference $\delta_\Delta$ between the tolerance $[\delta]$ and the actual deviation $\delta$ (that is, $\delta_\Delta = [\delta] - \delta$) allows for the corresponding reduction of the neck width $l_{nw}$. The reduction of the neck width $l_{nw}$ can be significant as the radii of the normal curvature $R_g$ and $R_{ct}$ of the work gear tooth flank $\mathscr{G}$ and generating surface $T$ of the shaving cutter are of great significant.

The problem of determining the minimum required overlap in rotary shaving of gears is similar in nature to the problem of determining the minimum required hob travel distance discussed earlier in Section 15.8.3.

In the event when the auxiliary rack $\mathscr{R}_{ct}$ of the shaving cutter overlaps the auxiliary rack $\mathscr{R}_g$ of the work gear insufficiently, then a portion of the gear tooth flank $\mathscr{G}$ remains unshaved. The lack of the overlap in the axial direction of the work gear is designated as $\Delta U$. In practice, the stock that is not removed from the unshaved portion of the tooth flank $\mathscr{G}$ causes the so-called "*tooth bias*" [64].

Referring to Figure 16.39, an expression for the function $\delta = \delta(\Delta U)$ can be represented in the form:

$$\delta(\Delta U) = \left(R_t \cos\psi_g - \sqrt{R_t^2 - (R_t \sin\psi_g + \Delta U)^2} - \Delta U \tan\psi_g\right)\cos\psi_g \tag{16.84}$$

where:

$R_t$ is the radius of normal curvature of the shaving cutter tooth surface $T$ in a specified direction.
$\psi_g$ is the pitch helix angle of the work gear.

It is important to stress here that the angle $\psi_g$ as well as angle $\psi_{ct}$ in the following both are a type of signed value.

The radius $R_t$ is measured in the direction of a unit tangent vector $\mathbf{u}_g$ (see Figure 16.40). The vector $\mathbf{u}_g$ is located within the tooth flank of the auxiliary generating rack $\mathscr{R}_T$ of the shaving cutter. It forms a certain angle $\varphi$ with the second principal direction $\mathbf{t}_{2.ct}$ on the shaving cutter tooth surface $T$. The angle $\varphi$ can be expressed in terms of (a) the angle $\mu$ of local relative orientation at point of contact of the surfaces $\mathscr{G}$ and $T$, (b) the base helix angle $\psi_{b.g}$ of the work gear, and (c) the base helix angle $\psi_{b.ct}$ of the shaving cutter:

$$\varphi = \mu - (\psi_{b.g} + \psi_{b.ct}) \tag{16.85}$$

For the calculation of the angle $\mu$, the formula [184]

$$\sin\mu = \frac{\sin\phi_n \sin\overline{\Sigma}}{\sqrt{(1-\cos^2\phi_n \sin^2\psi_g)\cdot(1-\cos^2\phi_n \sin^2\psi_c)}} \tag{16.86}$$

is applied.

Further, the Euler equation is utilized for the derivation of a formula for the calculation of the radius of normal curvature $R_t$:

$$R_t = \left[\underbrace{R_{1.ct}^{-1}}_{\equiv 0} \sin^2\varphi + R_{2.ct}^{-1}\cos^2\varphi\right]^{-1} \tag{16.87}$$

Ultimately, Eq. (16.87) casts to

$$R_t = \frac{R_{2.ct}}{\cos^2\varphi} \tag{16.88}$$

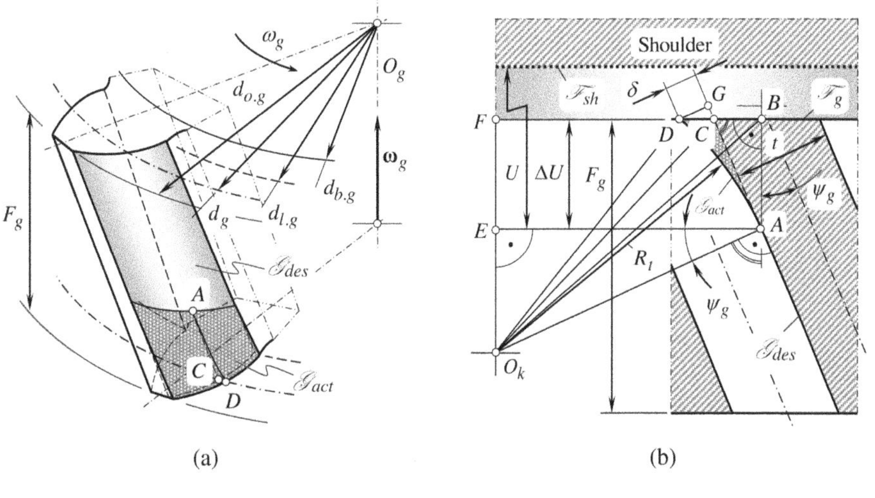


**FIGURE 16.39** Impact of the lack of the overlap $\Delta U$ on the deviation $\delta$ of the tooth flank $\mathscr{G}$ of a shaved gear: (a) a tooth of a finished gear and (b) gear with a shoulder. (From Radzevich, S.P., *Gear Technology*, 20(4), 44–50, 2003. With permission.)

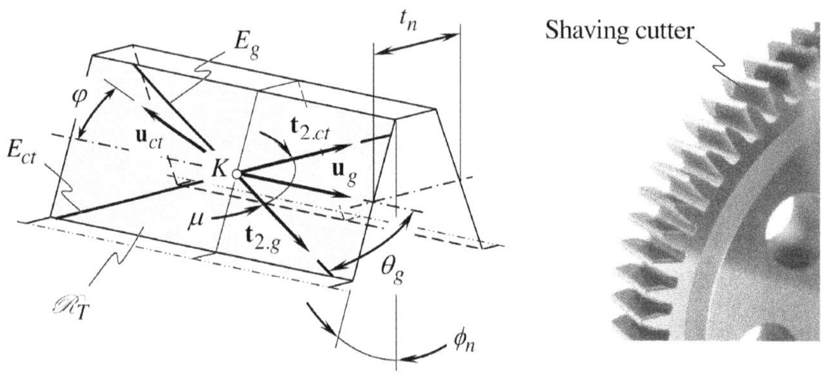


**FIGURE 16.40** Radius of normal curvature $R_t$ at a point of a shaving cutter tooth surface $T$ in a prespecified direction.

For the screw involute surface $T$ of the shaving cutter, the first principal radius of curvature $R_{1.ct}$ is equal to infinity ($R_{1.ct} \to \infty$). It is proven [175] that for the calculation of the second principal radius of curvature $R_{2.ct}$ of the shaving cutter surface $T$, the formula

$$R_{2.ct} = U_{ct} \tan \psi_{b.ct} = \frac{1}{2}\sqrt{\frac{d_{ct}^2 - d_{b.ct}^2}{1 - \sin^2 \psi_{ct} \cos^2 \phi_n}} \quad (16.89)$$

is valid.

The examination of the influence of the overlap of a shaving cutter over the face width of the work gear on the accuracy of the finished tooth flanks is illustrated with a graphical interpretation of the function $\delta = \delta(\Delta U)$, which is shown in Figure 16.41(a).

Figure 16.41(a) is important for the understanding of one of the major reasons for the tooth flank $\mathscr{G}$ bias, which appears when the gear is shaved improperly. The tooth flank $\mathscr{G}$ bias is inevitable when there is a lack of overlap, that is, when the work gear is shaved with a shaving cutter with insufficient face width. However, Figure 16.41(a) also reveals that if a gear can be shaved more accurately than specified by the tolerance $[\delta]$, the difference $\delta_\Delta = [\delta] - \delta$ makes possible a reduction of the minimum required neck width of the shoulder gear. The latter is important both for gear transmission designers and for manufacturers of gears.

The impact of the radius of normal curvature $R_t$ of the shaving cutter tooth flank on the deviation $\delta$ of the tooth flank $\mathscr{G}_{act}$ of the shaved gear with respect to the desirable true involute surface $\mathscr{G}_{des}$ is illustrated in Figure 16.41(b). The radius of normal curvature $R_t$ of the shaving cutter tooth flank $T$ is a function of the shaving cutter tooth number $N_{ct}$, that is, $R_t = R_t(N_{ct})$. The bigger the tooth number $N_{ct}$, the bigger the radius of the normal curvature $R_t$, and vice versa. The deviation $\delta$ becomes smaller when the radius of the normal curvature $R_t$ is larger. This immediately allows for a conclusion:

*Shaving a work gear with a shaving cutter that has a bigger tooth number $N_{ct}$ causes a reduction of the deviation $\delta$ of the actual tooth flank $\mathscr{G}_{act}$ with respect to the desirable true involute surface $\mathscr{G}_{des}$.*

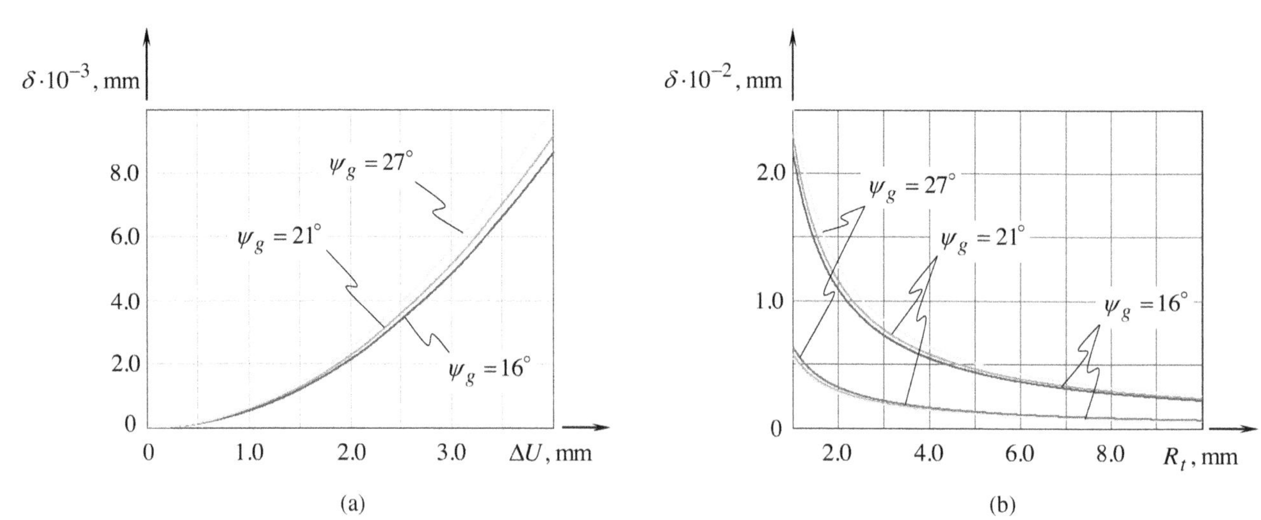


**FIGURE 16.41** Work gear tooth flank $\mathscr{G}$, deviation $\delta$, versus the parameter $\Delta U$ (a) and the deviation $\delta$ versus the radius of normal curvature $R_t$ (the upper curves for $\Delta U = 2$ mm, and lower curves for $\Delta U = 1$ mm) (b).

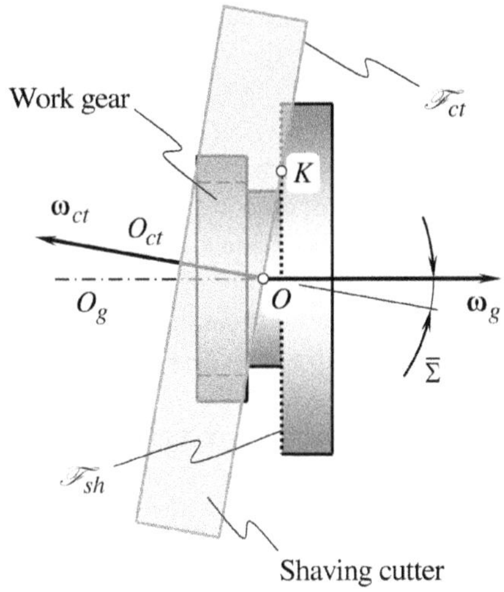


**FIGURE 16.42** Desirable posture of a shaving cutter in relation to the shoulder gear ($a \equiv 0$).

Once the difference $\delta_\Delta = [\delta] - \delta$ is known, then the permissible reduction $\Delta l_{nw}$ of the neck width can be expressed in terms of $\delta_\Delta$:

$$\Delta l_{nw} = \sqrt{R_t^2 - (R_t - [\delta])^2} - \sqrt{R_t^2 - (R_t - [\delta] + \delta_\Delta)^2} \qquad (16.90)$$

The distance between the face of a shoulder and the corresponding face of the shaving cutter is specified by the parameter $a$ (see Figure 16.38). It is desirable to keep the parameter $a$ as small as possible [64]. The smaller the distance $a$, the shorter the width of the neck of the shoulder gear that can be achieved. For the minimum deviations of the actual tooth flank of the shaved gear $\mathcal{G}_{act}$ from the desirable true screw involute surface $\mathcal{G}_{des}$, the parameter $a$ needs to fulfill the inequality

$$a \geq h_{t.\mathcal{R}} \cdot \cot \psi_{b.g} \cdot \cos \psi_{ct} \qquad (16.91)$$

as shown in Figure 16.42. Here $h_{t.\mathcal{R}}$ designates the whole depth of the auxiliary rack $\mathcal{R}$ tooth.

## 16.6 PLUNGE METHOD OF GEAR SHAVING PROCESS

With the plunge method of the gear shaving process, there is no worktable translation. Only a radial feed of the work gear against the shaving cutter is performed. Due to very short cycle time, plunge shaving is used in high-production operations. It is applied to gears, usually of narrow face widths. This method is particularly suited to shaving shoulder gears. The plunge method is not commonly used but is worth mentioning as a possible method.

### 16.6.1 Kinematics of Plunge Method of Gear Shaving Process

In plunge shaving, the work gear and the shaving cutter are meshed in crossed-axes relationship (see Figure 16.43). The crossing axes of the work gear $O_g$ and the shaving cutter $O_{ct}$ are at a certain center distance $C$.

Referring to Figure 16.43, the rotation of the work gear about its axis of rotation $O_g$ is designated as $\omega_g$. The rotation of the shaving cutter about its axis of rotation $O_{ct}$ is designated as $\omega_{ct}$. The rotations $\omega_g$ and $\omega_{ct}$ are properly timed with each other ($\omega_g \cdot N_g = \omega_{ct} \cdot N_{ct}$; here $N_g$ and $N_{ct}$ denote the tooth number of the work gear and the shaving cutter, respectively, and the equalities $\omega_g = |\boldsymbol{\omega}_g|$ and $\omega_{ct} = |\boldsymbol{\omega}_{ct}|$ are observed). The rotation vectors $\boldsymbol{\omega}_g$ and $\boldsymbol{\omega}_{ct}$ form the angle $\Sigma = \angle(\boldsymbol{\omega}_g, \boldsymbol{\omega}_{ct})$. The angle $\bar{\Sigma}$ of the crossing of the axes of rotation $O_g$ and $O_{ct}$ complements the angle $\Sigma$ to 180°, that is, $\bar{\Sigma} = \pi - \Sigma$.

Instead of feed-slide motion, as with the other types of shaving, the plunge method uses strictly an infeed to obtain the required tooth size of the finished gear. In the plunge method, the work gear travels along the centerline ℄. In this way the work gear is fed $\mathbf{F}_{ct}$ into the shaving cutter with no table reciprocation.

In all cases of plunge shaving, the face width of the shaving cutter has to be greater than that of the work gear. The shaving cutter must have the differential-type serrations, otherwise the cutting action will be impaired.

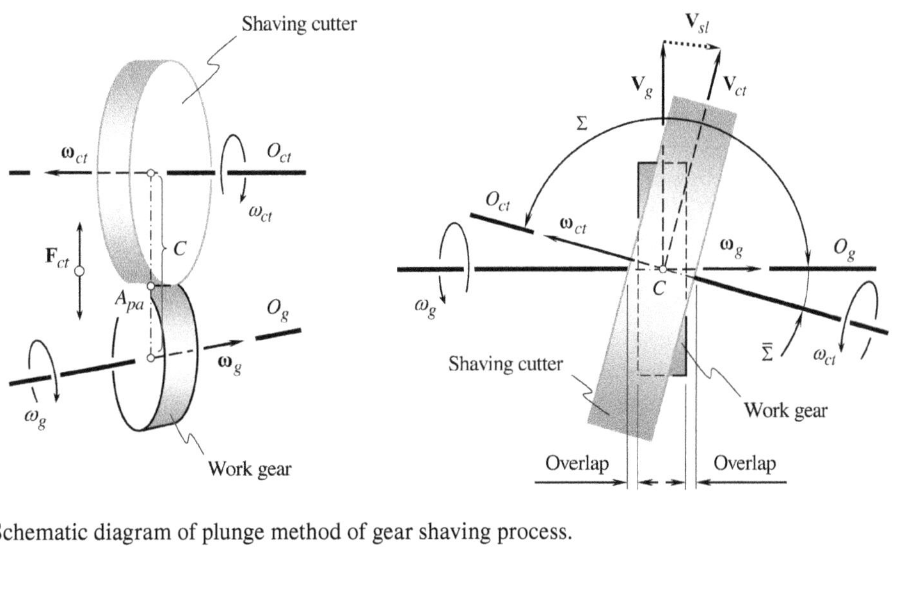


**FIGURE 16.43** Schematic diagram of plunge method of gear shaving process.

### 16.6.2 Cutting Speed in Plunge Method of Rotary Shaving of Gears

Since the shaving cutter is mounted on a shaft that is not parallel to the gear axis of rotation, the teeth of the shaving cutter and the teeth of the work gear run together like a pair of crossed-helical gears. The choice of crossed-axes angle governs the cutting action of the shaving cutter. In general, the higher the crossed-axes angle, the faster the shaving cutter cuts. The best control over helix angle, though, is gained with a low crossed-axes angle.

A formula for the calculation of the cutting speed vector $\mathbf{V}_{cut}$ in plunge shaving of gears can be derived from Eq. (16.4) under the assumption that $\mathbf{F}_{ct} = 0$:

$$\mathbf{V}_{cut} = \mathbf{V}_{sl} + \mathbf{V}_{pr} \tag{16.92}$$

The sliding velocity $\mathbf{V}_{sl} = \mathbf{V}_{ct} - \mathbf{V}_g$ is a component that most significantly contributes to the speed of the cutting speed vector $\mathbf{V}_{cut}$.

The magnitude $V_{sl} = |\mathbf{V}_{sl}|$ of the sliding velocity $\mathbf{V}_{sl} = \mathbf{V}_{ct} - \mathbf{V}_g$ is calculated from Eq. (16.8)

$$V_{sl} = 0.5 \cdot (\omega_g d_g \sin\psi_g + \omega_{ct} d_{ct} \sin\psi_{ct}) \tag{16.93}$$

The equality $V_{cut} = |\mathbf{V}_{cut}| \cong V_{sl}$ is valid for the calculation of the cutting speed in plunge shaving of gears.

### 16.6.3 Plunge Gear Shaving Process

The kinematics of the plunge gear shaving process is simple (see Figure 16.43). The geometry of the tooth flank $\mathscr{G}$ of the work gear is specified by Eq. (1.4):

$$\mathbf{r}_g(U_g, V_g) = \begin{bmatrix} r_{b.g}\cos V_g + U_g \cos\lambda_{b.g} \sin V_g \\ r_{b.g}\sin V_g - U_g \sin\lambda_{b.g} \sin V_g \\ r_{b.g}\tan\lambda_{b.g} - U_g \sin\lambda_{b.g} \\ 1 \end{bmatrix} \begin{matrix} V_g^{(l)} \le V_g \le V_g^{(a)} \\ 0 \le U_g \le [U_g] \end{matrix} \tag{16.94}$$

In the plunge shaving process, the tooth flank $\mathscr{G}$ is rotated about the axis of rotation $O_g$ of the work gear, while a reference system $X_gY_gZ_g$ is rotated about the axis of rotation $O_{ct}$ of the shaving cutter. The resultant motion of the work gear in relation to the reference system $X_{ct}Y_{ct}Z_{ct}$ associated with the shaving cutter is referred to as *epicyclic motion*.

The generating surface $T$ of a shaving cutter for the plunge gear shaving process can be determined as an envelope surface to a family of consecutive positions of the tooth flank surface $\mathscr{G}$ when the latter is performing the epicyclic motion relative to the reference system $X_{ct}Y_{ct}Z_{ct}$. The kinematic method for determining the generating surfaces of cutting tools, which is based on the Shishkov equation of contact $\mathbf{n}_g \cdot \mathbf{V}_\Sigma = 0$ [206], can be implemented to solve the problem of profiling of the shaving cutter. Ultimately, an equation of the generating surface $T$ of a plunge shaving cutter can be derived using the kinematic method.

The derived equation of the surface $T$ is bulky, and thus it is inconvenient for any practical needs. Therefore, the continuous analytical description of the surface $T$ is not used, but its discrete representation by a set of distinct points within the surface $T$ is used instead when developing a computer code for a numerically controlled (*NC*) machine.

The similar discrete representation of surfaces is used for determining the tooth flank of a shaving cutter for shaving of crowned gears, and, in the most general case, it is used for shaving gears with topologically modified tooth flanks. From the author's perspective it is reasonable to consider the problem of profiling of the shaving cutter for finishing topologically modified gears. Afterwards, the geometry of the tooth flank of a shaving cutter for plunge shaving of crowned gears can be obtained as a reduced case of the geometry of the shaving cutter for shaving of topologically modified gears. Ultimately, the geometry of the tooth flanks of the shaving cutter for plunge shaving of nonmodified gears is identical to that for the shaving cutter with topologically modified tooth flanks if it is assumed that the modification has a zero value.

### 16.6.4 Plunge Shaving of Topologically Modified Gears

It is customary to modify the gear tooth flank to accommodate for the axes displacement, the axes misalignment, and to minimize transmission error [170], which is the predominant cause of gear noise.

The problem of determining the geometry of the shaving cutter tooth flank may be informally referred to as: *Given a file of digitized gear tooth flank data, generate the corresponding file for the shaving cutter tooth flank.*

#### 16.6.4.1 Geometry of a Topologically Modified Gear Tooth Flank

The desirable topologically modified tooth flank deviates from the nominal tooth surface of the work gear. The screw involute surface $\mathscr{G}$ [see Eq. (16.94)] of a nonmodified tooth flank is used as the reference surface for the construction of the topologically modified tooth flank $\mathscr{G}_{\text{mod}}$. For the construction, a grid of points is specified within the tooth flank $\mathscr{G}$. For the construction of the grid, two sets of points are commonly used (see Figure 16.44).

The first set of $m$ points is chosen within the gear tooth profile. These $m$ points are evenly distributed either in the radial direction of the gear tooth or in the lengthwise direction of the tooth profile. The second set of $n$ points is chosen within the face width of the work gear. These $n$ points are evenly distributed within the face width of the gear. Ultimately, the gear tooth flank is covered by a grid of $m \times n$ points, each of which is located at a corresponding corner of elementary cells of the grid. For spur gears, the cells feature rectangular shape, and they are shaped in the form of a parallelogram for helical gears [139].

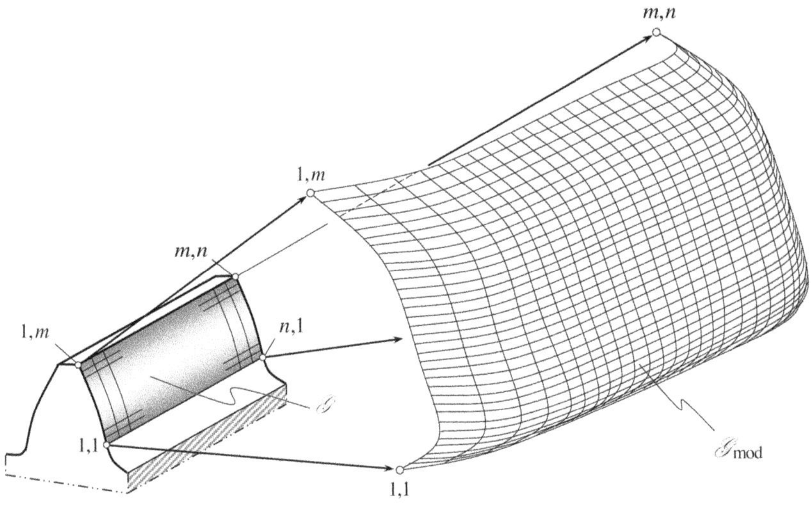


**FIGURE 16.44** Nominal screw involute surface $\mathscr{G}$ as the reference surface to the topologically modified tooth flank $\mathscr{G}_{\text{mod}}$ of a work gear.

Two numbers, $i = 1, 2, \ldots, m$ and $j = 1, 2, \ldots, n$, are assigned to each point of the grid of points. The deviation $\delta$ of the surface $\mathscr{G}_{\text{mod}}$ from the screw involute surface $\mathscr{G}$ at a point with the number $i, j$ is designated as $\delta_{i,j}$, correspondingly. The deviation $\delta_{i,j}$ is measured outwards of the work gear tooth body along the perpendicular $\mathbf{n}_g$ to the tooth flank $\mathscr{G}$.

For the particular case of the screw involute surface, an equation for the unit normal vector $\mathbf{n}_g$ can be derived not analytically as a cross product of two unit tangent vectors $\mathbf{u}_g$ and $\mathbf{v}_g$ [see Eq. (1.7) and Eq. (1.8)], but on the premise of two important features of the geometry of the screw involute surface $\mathscr{G}$.

*First*, the vector $\mathbf{n}_g$ is aligned to a line that is tangent to the base cylinder of the work gear.
*Second*, the line with which the vector $\mathbf{n}_g$ is aligned crosses the gear axis at the angle $(90° - \psi_{b.g})$.

Referring to Figure 1.14, this immediately yields an expression

$$\mathbf{n}_g = \begin{bmatrix} -\cos\psi_{b.g}\cos V_g \\ \cos\psi_{b.g}\sin V_g \\ \sin\psi_{b.g} \\ 1 \end{bmatrix} \tag{16.95}$$

for the unit normal vector $\mathbf{n}_g$.

Usually, the desirable deviations $\delta_{i,j}$ are given in matrix form [139]:

$$\left[\delta_g^{\text{mod}}\right] = \begin{bmatrix} \delta_{1,1} & \delta_{1,2} & \ldots & \delta_{1,j} & \ldots & \delta_{1,n} \\ \delta_{2,1} & \delta_{2,2} & \ldots & \delta_{2,j} & \ldots & \delta_{2,n} \\ \ldots & \ldots & \ddots & \ldots & \ldots & \ldots \\ \delta_{i,1} & \delta_{i,2} & \ldots & \delta_{i,j} & \ldots & \delta_{i,n} \\ \ldots & \ldots & \ldots & \ldots & \ddots & \ldots \\ \delta_{m,1} & \delta_{m,2} & \ldots & \delta_{m,j} & \ldots & \delta_{m,n} \end{bmatrix} \tag{16.96}$$

The matrix $\left[\delta_g^{\text{mod}}\right]$ is referred to as the *gear tooth flank modification matrix*.

For the determination of the desirable modified tooth flank $\mathscr{G}_{\text{mod}}$, the screw involute surface $\mathscr{G}$ [see Eq. (16.94)] is considered together with the gear modification matrix $\left[\delta_g^{\text{mod}}\right]$ [Eq. (16.96)]. For the further analysis, both the desirable tooth flank $\mathscr{G}_{\text{mod}}$ and the gear modification matrix $\left[\delta_g^{\text{mod}}\right]$ must be represented in a common reference system.

The origin of the local Cartesian reference system $x_g y_g z_g$ is at point $K_{i,j}$ of the grid of points. The axis $z_g$ is aligned with the unit normal vector $\mathbf{n}_g$. This makes it possible for representation of the position vector $\overline{\mathbf{r}}_{g.i,j}^{\text{mod}}$ of point $K_{i,j}$ in the form of column matrix

$$\overline{\mathbf{r}}_{g.i,j}^{\text{mod}} = \begin{bmatrix} 0 \\ 0 \\ \delta_{i,j} \\ 1 \end{bmatrix} \tag{16.97}$$

With the operator $\mathbf{Rs}_{i,j}(K_{i,j} \to \mathscr{G})$ of the resultant coordinate system transformation composed, the position vector $\mathbf{r}_{g.i,j}^{\text{mod}}$ of the $K_{i,j}$ – point can be represented in the coordinate system $X_g Y_g Z_g$ associated with the work gear [139]:

$$\mathbf{r}_{g.i,j}^{\text{mod}} = \mathbf{Rs}_{i,j}(K_{i,j} \to \mathscr{G}) \cdot \overline{\mathbf{r}}_{g.i,j}^{\text{mod}} \tag{16.98}$$

At a certain point $K_{i,j}$, the unit normal vector $\mathbf{n}_g^{\text{mod}}$ to the modified tooth flank $\mathscr{G}_{\text{mod}}$ does not align to the unit normal vector $\mathbf{n}_g$ to the screw involute surface $\mathscr{G}$. For the discretely specified tooth flank $\mathscr{G}_{\text{mod}}$, the unit normal vector $\mathbf{n}_g^{\text{mod}}$ can be determined under the assumptions that (a) the modified gear tooth flank $\mathscr{G}_{\text{mod}}$ is smooth, and (b) all the $K_{i,j}$ – points are distributed evenly (or at least, almost evenly) within the gear tooth flank. Both assumptions are reasonable.

Equation (16.97) allows for the position vectors $\mathbf{r}_{g.i,j}^{\text{mod}}$ and $\mathbf{r}_{g.i,(j+1)}^{\text{mod}}$ of points $K_{i,j}$ and $K_{i,(j+1)}$ (see Figure 16.45).

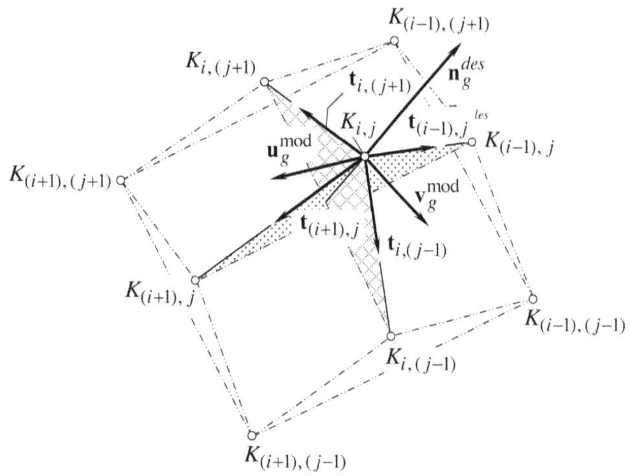


**FIGURE 16.45** Unit normal vector $\mathbf{n}_g^{\text{mod}}$ to the desirable topologically modified gear tooth flank $\mathcal{G}_{\text{mod}}$. (From Radzevich, S.P., *ASME Journal of Mechanical Design*, 125, 632–639, 2003. With permission.)

This immediately yields an expression for the vector $\mathbf{t}_{i,(j+1)}$ through point $K_{i,j}$ in the direction of the straight line through points $K_{i,j}$ and $K_{i,(j+1)}$:

$$\mathbf{t}_{i,(j+1)} = \mathbf{r}_{g.i,(j+1)}^{\text{mod}} - \mathbf{r}_{g.i,j}^{\text{mod}} \tag{16.99}$$

In that same way, an expression

$$\mathbf{t}_{i,(j-1)} = \mathbf{r}_{g.i,(j-1)}^{\text{mod}} - \mathbf{r}_{g.i,j}^{\text{mod}} \tag{16.100}$$

for the $\mathbf{t}_{i,(j-1)}$ through point $K_{i,j}$ in the direction of the straight line through points $K_{i,j}$ and $K_{i,(j-1)}$ is derived.

The cross product of the vectors $\mathbf{t}_{i,(j+1)}$ and $\mathbf{t}_{i,(j-1)}$ returns the tangent vector $\mathbf{u}_g^{\text{mod}}$ to the modified tooth flank $\mathcal{G}_{\text{mod}}$:

$$\mathbf{u}_g^{\text{mod}} = \mathbf{t}_{i,(j+1)} \times \mathbf{t}_{i,(j-1)} \tag{16.101}$$

Similarly, two vectors

$$\mathbf{t}_{(i+1),j} = \mathbf{r}_{g.(i+1),j}^{\text{mod}} - \mathbf{r}_{g.i,j}^{\text{mod}} \tag{16.102}$$

$$\mathbf{t}_{(i-1),j} = \mathbf{r}_{g.(i-1),j}^{\text{mod}} - \mathbf{r}_{g.i,j}^{\text{mod}} \tag{16.103}$$

are calculated through point $K_{i,j}$ (a) in the direction of the straight line through points $K_{i,j}$ and $K_{(i+1),j}$ and (b) in the direction of the straight line through points $K_{i,j}$ and $K_{(i-1),j}$.

Again, the cross product of the vectors $\mathbf{t}_{(i+1),j}$ and $\mathbf{t}_{(i-1),j}$ returns the tangent vector $\mathbf{v}_g^{\text{mod}}$ to the modified tooth flank $\mathcal{G}_{\text{mod}}$:

$$\mathbf{v}_g^{\text{mod}} = \mathbf{t}_{(i+1),j} \times \mathbf{t}_{(i-1),j} \tag{16.104}$$

Ultimately, the cross product of the two tangent vectors $\mathbf{u}_g^{\text{mod}}$ and $\mathbf{v}_g^{\text{mod}}$ allows for an expression [139]

$$\mathbf{n}_g^{\text{mod}} = \mathbf{u}_g^{\text{mod}} \times \mathbf{v}_g^{\text{mod}} \tag{16.105}$$

for the unit normal vector $\mathbf{n}_g^{\text{mod}}$ to the modified tooth flank $\mathcal{G}_{\text{mod}}$ at a specified point $K_{i,j}$.

### 16.6.4.2 Geometry of the Desirable Topologically Modified Tooth Flank of Shaving Cutter

Consider a Cartesian coordinate system $X_{ct}Y_{ct}Z_{ct}$ associated with the shaving cutter. In the reference system $X_{ct}Y_{ct}Z_{ct}$, the position vector $\overline{\mathbf{r}}_{g.i,j}^{\text{mod}}$ of point $K_{i,j}$ of the modified tooth flank $\mathcal{G}_{\text{mod}}$ is described by an expression

$$\overline{\mathbf{r}}_{g.i,j}^{\text{mod}} = \mathbf{Rs}(\mathcal{G} \rightarrow T) \cdot \mathbf{r}_{g.i,j}^{\text{mod}} \tag{16.106}$$

where the operator of the resultant coordinate system transformation is designated as $\mathbf{Rs}(\mathcal{G} \rightarrow T)$.

The operator $\mathbf{Rs}(\mathcal{G} \rightarrow T)$ of the linear transformation can be expressed in terms of the angle of rotation $\varphi_g$ of the work gear and the angle of rotation $\varphi_{ct}$ of the shaving cutter. However, the equality $\varphi_{ct} = \varphi_g \cdot (N_g / N_{ct})$ is valid. This is owing to the fact that in the rotary shaving process, the work gear and the shaving cutter are in tight mesh and are rotated about their axis in a properly timed manner. Here, the tooth number of the work gear is designated as $N_g$, and the tooth number of the shaving cutter is designated as $N_{ct}$. This means that in the coordinate system $X_{ct}Y_{ct}Z_{ct}$, the position vector $\overline{\mathbf{r}}_{g.i,j}^{\text{mod}}(\varphi_g)$ can be interpreted as a function of just one parameter, either of $\varphi_g$ or of $\varphi_{ct}$.

The timed rotation of the work gear and the shaving cutter makes it possible to determine a vector $\mathbf{V}_{\Sigma.i,i}$ of the resultant motion of point $K_{i,j}$ of the tooth flank $\mathcal{G}_{\text{mod}}$ relative to the coordinate system $X_{ct}Y_{ct}Z_{ct}$. In the coordinate system $X_{ct}Y_{ct}Z_{ct}$, this vector is represented as a function of the rotation angle, that is, $\mathbf{V}_{\Sigma.i,i} = \mathbf{V}_{\Sigma.i,i}(\varphi_g)$.

Ultimately, the position vector $\mathbf{r}_{c.i,j}^{\text{mod}}$ of a point of the modified tooth flank of the shaving cutter can be determined as the solution to the set of two equations [139]:

$$\mathbf{r}_{c.i,j}^{\text{mod}} \Leftarrow \begin{cases} \overline{\mathbf{r}}_{g.i,j}^{\text{mod}} = \overline{\mathbf{r}}_{g.i,j}^{\text{mod}}(\varphi_g) \\ \mathbf{n}_g^{\text{mod}} \cdot \mathbf{V}_{\Sigma.i,i}(\varphi_g) = 0 \end{cases} \tag{16.107}$$

On solving the set of Eqs. (16.107), that is, after the parameter $\varphi_g$ is excluded from Eqs. (16.107), an expression for the position vector $\mathbf{r}_{c.i,j}^{\text{mod}}$ of a point of the modified tooth flank of the shaving cutter can be obtained.

The discretely specified tooth flank of the topologically modified shaving cutter allows for representation in the form of the so-called shaving cutter tooth flank modification matrix [139]

$$\left[\delta_c^{\mathrm{mod}}\right]=\begin{bmatrix}\Delta_{1,1} & \Delta_{1,2} & \cdots & \Delta_{1,j} & \cdots & \Delta_{1,n}\\ \Delta_{2,1} & \Delta_{2,2} & \cdots & \Delta_{2,j} & \cdots & \Delta_{2,n}\\ \cdots & \cdots & \ddots & \cdots & \cdots & \cdots\\ \Delta_{i,1} & \Delta_{i,2} & \cdots & \Delta_{i,j} & \cdots & \Delta_{i,n}\\ \cdots & \cdots & \cdots & \cdots & \ddots & \cdots\\ \Delta_{m,1} & \Delta_{m,2} & \cdots & \Delta_{m,j} & \cdots & \Delta_{m,n}\end{bmatrix} \quad (16.108)$$

which is similar to that [see Eq. (16.96)] derived for the work gear tooth flank.

The points of the shaving cutter tooth flank modification matrix [see Eq. (16.108)] are in one-to-one correspondence with the points of the gear tooth flank modification matrix [see Eq. (16.96)].

#### 16.6.4.3 Grinding a Topologically Modified Tooth Flank of the Shaving Cutter

The tooth flank of the modified shaving cutter, specified by the shaving cutter modification matrix [see Eq. (16.108)], is a type of sculptured part surface. For grinding of a sculptured surface, a multiaxis NC machine is used. The tooth flank of the modified shaving cutter can be ground in a multiaxis NC grinder of an appropriate design (see Figure 16.46). Under such a scenario, the generating surface of the grinding wheel makes point contact with the modified tooth flank of the shaving cutter. The development of a computer code for a multiaxis NC grinder is a routine procedure that is beyond the scope of this book.

The grinding of the tooth flank of a modified shaving cutter can also be performed with a form grinding wheel that makes line contact with the modified tooth flank being ground. Such a machining process occurs when an NC grinder with a lack of articulation is used.

Precise generation of a topologically modified tooth flank of a shaving cutter is impossible when the grinding wheel and the shaving cutter tooth flank are in line contact. Approximation of the desirable generating surface of the grinding wheel with a surface of revolution is inevitable under such a scenario. The problem of profiling a form grinding wheel for grinding a tooth flank of the topologically modified shaving cutter using an NC grinder that features a lack of articulation is important for contemporary gear manufacturers.

##### 16.6.4.3.1 *Peculiarities of Kinematics of NC Grinders*

For grinding a topologically modified shaving cutter, an NC grinder featuring a lack of articulation can be used. A typical NC grinder has the following NC axes (see Figure 16.47):

$X-$ axis: Grinding wheel vertical motion
$Y-$ axis: Profile angle setting
$Z-$ axis: Grinding wheel motion
$A-$ axis: Helix angle setting
$B-$ axis: Shaving cutter rotation
$C-$ axis: Worktable feed

In addition, the $U-$ axis and $V-$ axis are used for the dresser (a) forward/backward and (b) vertical motions, and the $W-$ axis serves for the shaving cutter positioning.

The kinematics of grinding of a shaving cutter illustrated in Figure 16.47 is implemented, for example, in the design of the Mitsubishi ZA30CNC gear shaving cutter grinder.

The crossing axes of rotation of the shaving cutter $O_{ct}$ and the grinding wheel $O_{gw}$ are positioned at a certain center distance $C$. The shaving cutter swivels about its axis with a certain angular velocity. The swiveling of the

**FIGURE 16.46** Grinding the tooth flanks $\mathscr{G}_{\mathrm{mod}}$ of a shaving cutter.

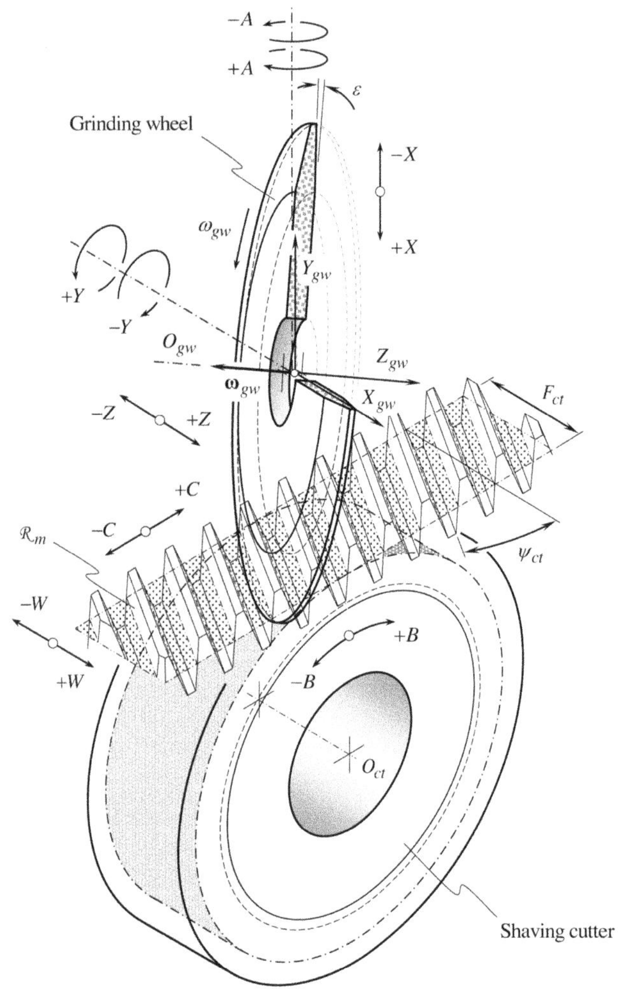


**FIGURE 16.47** Kinematics of grinding a shaving cutter with topologically modified tooth flanks. (From Radzevich, S.P., *SME Journal of Manufacturing Science and Engineering*, 127(4), 819–828, 2005. With permission.)

shaving cutter is properly timed with the reciprocation of the grinding wheel. In this way the shaving cutter generates the required auxiliary generating rack $\mathcal{R}_m$. The tooth flank of the approximate auxiliary rack $\mathcal{R}_m$ is a type of sculptured surface, and it is not a surface of revolution.

Following the way in which the tooth flank $\mathcal{G}_{\text{mod}}$ of the shaving cutter is determined [see Eq. (16.108)], a similar *auxiliary rack tooth flank modification matrix* [135, 142, 143, 147, 160]

$$[\varepsilon_{\mathcal{R}}] = \begin{bmatrix} \varepsilon_{1,1} & \varepsilon_{1,2} & \ldots & \varepsilon_{1,j} & \ldots & \varepsilon_{1,n} \\ \varepsilon_{2,1} & \varepsilon_{2,2} & \ldots & \varepsilon_{2,j} & \ldots & \varepsilon_{2,n} \\ \ldots & \ldots & \ddots & \ldots & \ldots & \ldots \\ \varepsilon_{i,1} & \varepsilon_{i,2} & \ldots & \varepsilon_{i,j} & \ldots & \varepsilon_{i,n} \\ \ldots & \ldots & \ldots & \ldots & \ddots & \ldots \\ \varepsilon_{m,1} & \varepsilon_{m,2} & \ldots & \varepsilon_{m,j} & \ldots & \varepsilon_{m,n} \end{bmatrix} \quad (16.109)$$

can be derived. A topologically modified tooth flank of the auxiliary generating rack $\mathcal{R}_m$ is specified by this matrix.

The generating surface of a grinding wheel $T_{gw}$ is a surface of revolution. When grinding a shaving cutter, the grinding wheel must generate the auxiliary rack $\mathcal{R}_m$. However, the sculptured surface $\mathcal{R}_m$ in no way can be congruent to the surface of revolution $T_{gw}$. A certain patch of the surface $\mathcal{R}_m$ can be *nearly* congruent to a patch of the surface $T_{gw}$. This means that the approximation of the tooth flank of the generating rack $\mathcal{R}_m$ by a portion of the surface of revolution is inevitable. The approximation entails corresponding errors in the tooth shape of the shaving cutter.

#### *16.6.4.3.2 Generating Surface of a Form Grinding Wheel*

The modification matrix [see Eq. (16.109)] is used for the purpose of transition from the auxiliary generating rack $\mathcal{R}_m$ to the generating surface $T_{gw}$ of the form grinding wheel.

Consider a Cartesian coordinate system $X_{gw}Y_{gw}Z_{gw}$ associated with the grinding wheel as shown in Figure 16.47. This coordinate system is rigidly connected to the grinding wheel and rotates together with the grinding wheel. Every $i, j-$ point of the tooth flank of the generating rack $\mathcal{R}_m$ traces in $X_{gw}Y_{gw}Z_{gw}$ a circle of radius $R_{i,j}$. The total number of the circles is equal to the total number of points $m \times n$ in the modification matrix [see Eq. (16.109)]. The circles intersect an axial plane of the grinding wheel, for example, they intersect the coordinate plane $X_{gw}Z_{gw}$ of the reference system $X_{gw}Y_{gw}Z_{gw}$. The points of the intersection form a cloud of points within the coordinate plane $X_{gw}Z_{gw}$ (see Figure 16.48). The axial profile of the grinding wheel can be derived from the cloud of points [135].

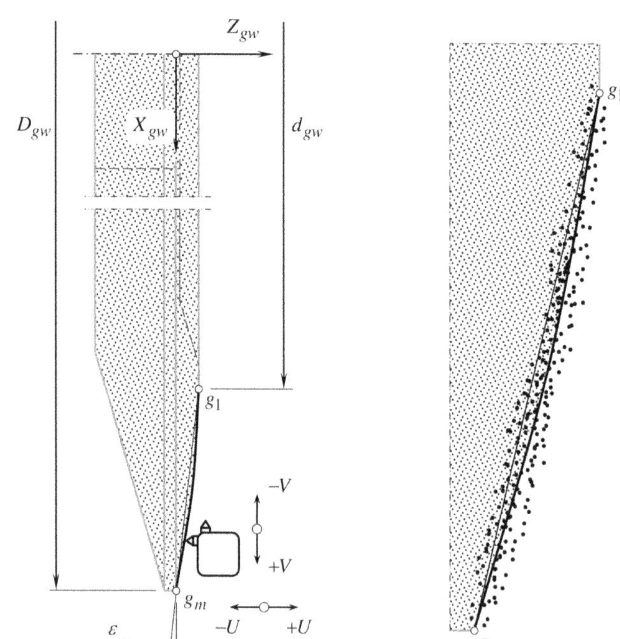


**FIGURE 16.48** Geometry of axial profile and kinematics of dressing of a form grinding wheel. (From Radzevich, S.P., *SME Journal of Manufacturing Science and Engineering*, 127(4), 819–828, 2005. With permission.)

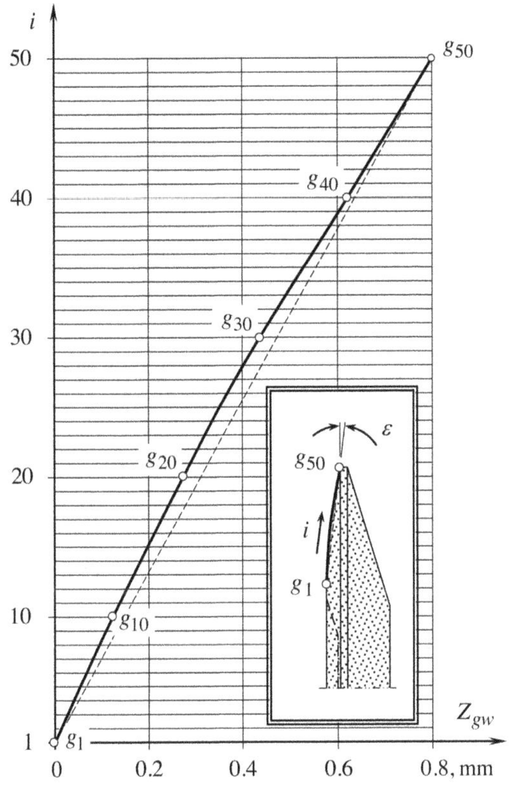


**FIGURE 16.49** Example of the axial profile of a form grinding wheel for grinding of a shaving cutter with a topologically modified tooth flank $\mathcal{G}_{mod}$. (From Radzevich, S.P., *SME Journal of Manufacturing Science and Engineering*, 127(4), 819–828, 2005. With permission.)

Referring to Figure 16.48, point $g_1$ of the axial profile of the grinding wheel is located within a circle of the diameter $d_{gw}$. In the grinding process, point $g_1$ generates the tooth profile at the outer diameter of the shaving cutter (see Figure 16.48). Accordingly, point $g_m$ generates the tooth profile at the inner diameter of the shaving cutter. The shaving cutter can be ground with the grinding wheel with an axial profile shown in Figure 16.49.

Once the axial profile is determined, the generating surface of the grinding wheel is shaped in the form of the surface of revolution with a predetermined axial profile. Further, this makes it possible for calculation of deviations because of the approximation of the topologically modified tooth flank $\mathcal{R}_m$ with a portion of a surface of revolution [135, 142, 143, 147, 160]. Ultimately, the predicted deviations of the tooth flank of the shaved gear can be calculated as well.

### 16.6.5 Fulfillment of Necessary and Sufficient Conditions of Proper Part Surface Generation When Designing a Shaving Cutter for Plunge Shaving of Gears

When designing a shaving cutter, the fulfillment of the set of necessary and sufficient conditions of proper part surface generation (conditions of proper part surface generating) is a must [136, 144, 148, 158, 168]. Fulfillment of the first and the fifth conditions of proper part surface generating is of critical importance when designing a shaving cutter for plunge shaving of either an involute helical gear or a gear with topologically modified tooth flanks (see Appendix E). An investigation into this problem was undertaken by Prof. S.P. Radzevich [162].

#### 16.6.5.1 Circular Mapping of Tooth Flanks of a Work Gear and the Shaving Cutter

The *circular mapping* (or simply $c-$ *mapping*) of the tooth flank $\mathcal{G}$ of the work gear and the tooth flank $T$ of the shaving cutter onto the mapping plane $\mathcal{M}$ [162] is employed in the analysis. The concept of $c-$ mapping is illustrated in Figure 16.50.

The plane of mapping $\mathcal{M}$ is a plane through the centerline ℄ of the work gear axis of rotation $O_g$ and the shaving cutter axis of rotation $O_{ct}$. At the apex point $A_{pa}$, the mapping plane $\mathcal{M}$ is tangent to both the tooth flank $\mathcal{G}$ and the tooth flank $T$. The unit normal vector $\mathbf{n}_{\mathcal{M}}$ to the mapping plane $\mathcal{M}$ is parallel to the common perpendicular $\mathbf{n}_g$ to the tooth flanks $\mathcal{G}$ and $T$ at the apex point $A_{pa}$.

Consider a plane through an arbitrary point $\mathcal{G}_{ij}$ within the tooth flank $\mathcal{G}$. The plane is orthogonal to the work gear axis of rotation $O_g$. The circle of a radius $R_{ij}$ with the center at $O_g$ intersects the mapping plane $\mathcal{M}$ at $\mathcal{M}_{ij}$. Point $\mathcal{M}_{ij}$ represents a $c-$ map of the point $\mathcal{G}_{ij}$. Similarly, all points of the tooth flank $\mathcal{G}$ can be mapped onto the mapping plane $\mathcal{M}$. In this way, the $c-$ map of the work gear tooth flank $\mathcal{G}$ is constructed. The $c-$ map of the shaving

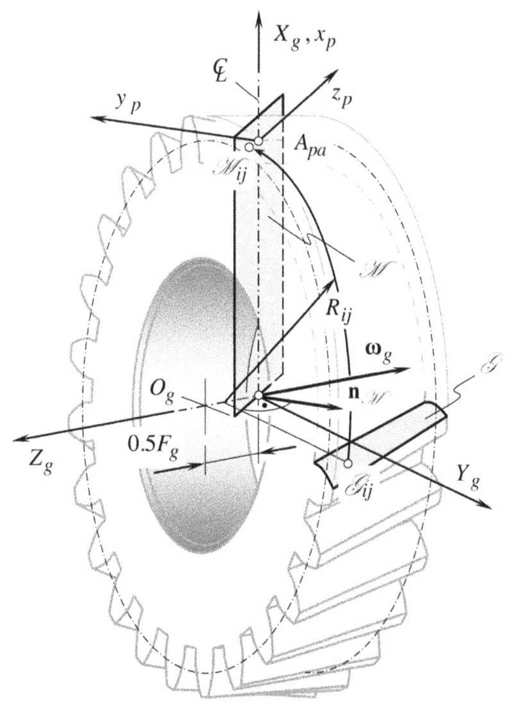


**FIGURE 16.50** Example of the $c-$ mapping of the work gear tooth flank $\mathcal{G}$. (From Radzevich, S.P., *ASME Journal of Mechanical Design*, 129(9), 969–980, 2007. With permission.)

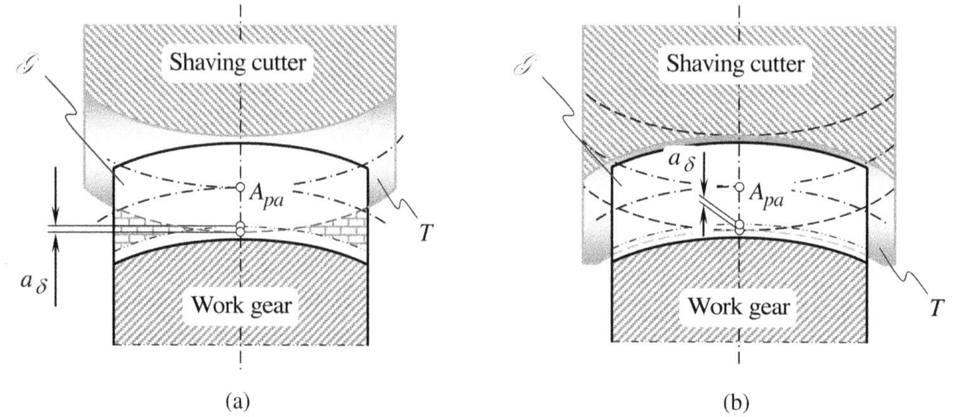


**FIGURE 16.51** Features of the $c-$ maps of the work gear $\mathcal{G}$ and shaving cutter $T$ tooth flanks, (a) in gear shaving operation by the shaver cutter of conventional design, and (b) in gear shaving operation by the enveloping shaver cutter. (From Radzevich, S.P., *ASME Journal of Mechanical Design*, 129(9), 969–980, 2007. With permission.)

cutter tooth flank can be constructed in the same way as discussed above.

The $c$-mapping of the tooth flanks $\mathcal{G}$ and $T$ yields an analytical representation. For this purpose, it is convenient to introduce a local Cartesian coordinate system $x_p y_p z_p$, with the origin at the apex point $A_{pa}$, as shown in Figure 16.50. The axis $x_p$ is aligned with the centerline ℄ of the axes of rotation $O_g$ of the work gear and $O_{ct}$ of the shaving cutter. The axis $y_p$ is parallel to the unit normal vector $\mathbf{n}_{\mathcal{M}}$ to the mapping plane $\mathcal{M}$. The axis $z_p$ complements the axes $x_p$ and $y_p$ to the left-hand-oriented local Cartesian coordinate system $x_p y_p z_p$.

For the calculation of the position vector of a point $\mathbf{r}_g^{(\mathcal{M})}$ of the $c-$ map of the work gear tooth flank $\mathcal{G}$ in the local reference system $x_p y_p z_p$, the following expression

$$\mathbf{r}_g^{(\mathcal{M})} = \begin{bmatrix} \left(\sqrt{X_g^2 + Y_g^2} - 0.5 \cdot d_g\right) \\ 0 \\ -\dfrac{Z_g}{\cos\psi_g} \\ 1 \end{bmatrix} \tag{16.110}$$

is valid. For the derivation of Eq. (16.110) for the position vector $\mathbf{r}_g^{(\mathcal{M})}$, Eq. (16.94) of the work gear tooth flank $\mathcal{G}$ is implemented together with Figure 16.50.

For the purposes of the analysis, the position vector $\mathbf{r}_g^{(\mathcal{M})}$ may be normalized by the module $m$. In this case the equality $\overline{\mathbf{r}}_g^{(\mathcal{M})} = \mathbf{r}_g^{(\mathcal{M})}/m$ specifies the correlation between the position vector of point $\overline{\mathbf{r}}_g^{(\mathcal{M})}$ of the normalized $\overline{c}-$ map with the corresponding position vector $\mathbf{r}_g^{(\mathcal{M})}$ of the original $c-$ mapping.[13]

Shown in Figure 16.51(a), the $c-$ maps of the tooth flank $\mathcal{G}$ of the work gear and the tooth flank $T$ of the shaving cutter reveal that the entire tooth flank $\mathcal{G}$ of the work gear cannot be finished by a shaving cutter of conventional design. Shadowed portions of the tooth flank $\mathcal{G}$ cannot be reached by the finishing tool, and thus they remain unshaved [185]. This means that a shaving cutter of a special design [see Figure 16.51(b)] is required for the plunge shaving process.

### 16.6.5.2 Shaving Cutter of a Special Design for Plunge Shaving of Precision Gears

The tooth design of a special-purpose plunge shaving cutter is schematically shown in Figure 16.52. Tooth flanks 1 feature a hollowed geometry. The top land 2 is also hollowed. The top land is represented with a patch of a surface of revolution. The effective face width of the shaving cutter is designated as $F_{ct}$. The total face width of the shaving cutter is designated as $F_{ct}^*$. The shaving cutter features two side lands 3 (from both faces). They are necessary to protect the shaving cutter teeth from damage. The face width of the side land is designated as $F_s$. The bottom land 4 has special shape and parameters.

Actually, the design of a plunge shaving cutter is specified in terms of conventional design parameters. Two groups

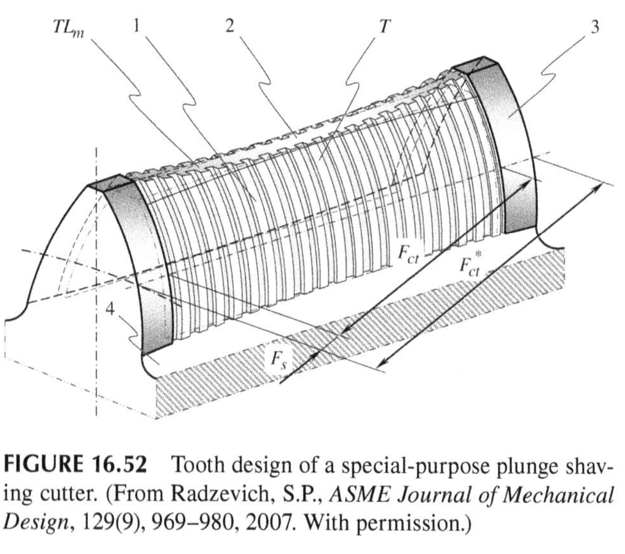


**FIGURE 16.52** Tooth design of a special-purpose plunge shaving cutter. (From Radzevich, S.P., *ASME Journal of Mechanical Design*, 129(9), 969–980, 2007. With permission.)

[13] To the best of the author's knowledge, a simplified type of mapping of the tooth flank of a gear is proposed by Gleason Works, Rochester, NY. The mapping plane through the axis of rotation of the helical gear is used in the simplified method of mapping of the tooth flank. This feature significantly limits the capabilities of the simplified type of the surface mapping.

of the design parameters can be distinguished in this regard. The following design parameters

1. Pitch diameter $d_{ct}$
2. Outer diameter $d_{o.ct}$
3. Root diameter $d_{f.ct}$
4. Addendum $a_{ct}$
5. Dedendum $b_{ct}$
6. Whole tooth height $h_{ct}$
7. Normal profile angle $\phi_n$
8. Pitch helix angle $\psi_g$
9. Functional face width $F_{ct}$

comprise the first group of the design parameters of the shaving cutter.

The second group of the design parameters comprises

1. Extended outer diameter $d^*_{o.ct}$
2. Addendum extension $a^*_{ct}$
3. Extended whole tooth height $h^*_{ct}$
4. Extended face width $F^*_{ct}$.

All points of the work gear tooth flank $\mathscr{G}$ can be reached by the shaving cutter tooth due to the addendum extension $a^*_c$. In this way the accuracy of the shaved work gear teeth is improved.

Fulfillment of two requirements is a must when determining the functional face width $F_{ct}$ of the shaving cutter.

*First*, the $c-$ map of the shaving cutter tooth flank $T$ must overlap the corresponding $c-$ map of the work gear tooth flank $\mathscr{G}$.

*Second*, the functional face width of the plunge shaving cutter must be the narrowest permissible.

Following these two requirements, a corresponding expression for the critical value of $F_{ct}$ is derived:

$$F_{ct} = F_g + 2 \cdot \frac{a}{\cos \Sigma} \tag{16.111}$$

where $a$ designates an excess of the plunge shaving cutter face width in relation to the gear face width $F_g$ [64].

In practice, the effective face width of the plunge shaving cutter can exceed its calculated value at a reasonable value.

For the calculation of the total face width of the plunge shaving cutter $F^*_{ct}$, the formula

$$F^*_{ct} = F_{ct} + 2 \cdot F_s \tag{16.112}$$

is valid. Here in Eq. (16.112), the width of the protective side land of the shaving cutter tooth is designated as $F_s$.

The $c-$ mapping is useful, and it can be implemented for determining other design parameters of the shaving cutter tooth, namely of (a) the desired axial profile of the modified top land $TL_m$ and (b) the desired axial profile of the modified bottom land as proposed in [162].

## 16.7 ADVANCES IN THE DESIGN OF THE SHAVING CUTTER

The productivity rate of the gear shaving process, as well as accuracy of the shaved gears, strongly depend on the design of the shaving cutter being applied for the finishing of a given work gear. Consideration in this section is limited only to those opportunities for the improvement of the design of a shaving cutter that can be drawn up from geometrical and kinematical analysis of a gear shaving process. Implementation of the methods developed in powder metallurgy technology, as well as in the production of cutting tools made of carbide alloys, provides the cutting tool designer with more freedom when designing shaving cutters. Ultimately, these technologies make it possible for the production of shaving cutters with serrations of complex geometry.

### 16.7.1 Elements of the Geometry of the Cutting Edges

A shaving cutter is a type of cutting tool for finishing gears. The cutting edges of a shaving cutter cut thin chips. When chips are thin, the use of cutting tools with a greater inclination angle is highly desired.

#### 16.7.1.1 Shaving Cutter with a Nearly Optimal Angle of Inclination of the Cutting Edges

The angle of inclination of cutting edges of a shaving cutter with annular serrations is insufficiently small. This becomes clear from the following consideration.

The derivation of a formula for the calculation of the angle of inclination $\lambda$ at a point of interest $m$ within the cutting edge $CE$ is briefly outlined here.

The tooth of a shaving cutter is depicted in Figure 16.53(a). At a point of interest $m$ within the cutting edge $CE$, three vectors are constructed: the unit normal vector $\mathbf{n}_{ct}$ to the generating surface $T$ of the shaving cutter, the cutting speed vector $\mathbf{V}_{cut}$, and the unit normal vector $\mathbf{n}_{ce}$ to the cutting edge $CE$. The vector $\mathbf{n}_{ce}$ is located within the plane of the cut. For convenience, the vectors $\mathbf{V}_{cut}$ and $\mathbf{n}_{ce}$ are also depicted in Figure 16.53(b) within the lateral plane of the auxiliary generating plane $\mathscr{R}_T$. By definition (see Appendix G), the inclination angle $\lambda$ is measured between the vectors $\mathbf{V}_{cut}$ and $\mathbf{n}_{ce}$. The angle of inclination $\lambda$ is positive when the vector of the cross product $\mathbf{V}_{cut} \times \mathbf{n}_{ce}$ is pointed outward from the body of the shaving cutter tooth. Figure 16.53(b) reveals that shaving cutters of conventional design feature an inclination angle of a negligibly small value.

Consider the vectors $\mathbf{V}_{cut}$ and $\mathbf{n}_{ce}$, referring to a certain reference system. The earlier introduced Cartesian coordinate system $X_{ct}Y_{ct}Z_{ct}$ (see Figure 16.27) can be employed in this case. The $X_{ct}-$ axis is pointed along the centerline $\mathbb{C}$ of the axes of rotation $O_g$ and $O_{ct}$ from the work gear axis of rotation $O_g$ toward the shaving cutter axis of rotation $O_{ct}$.

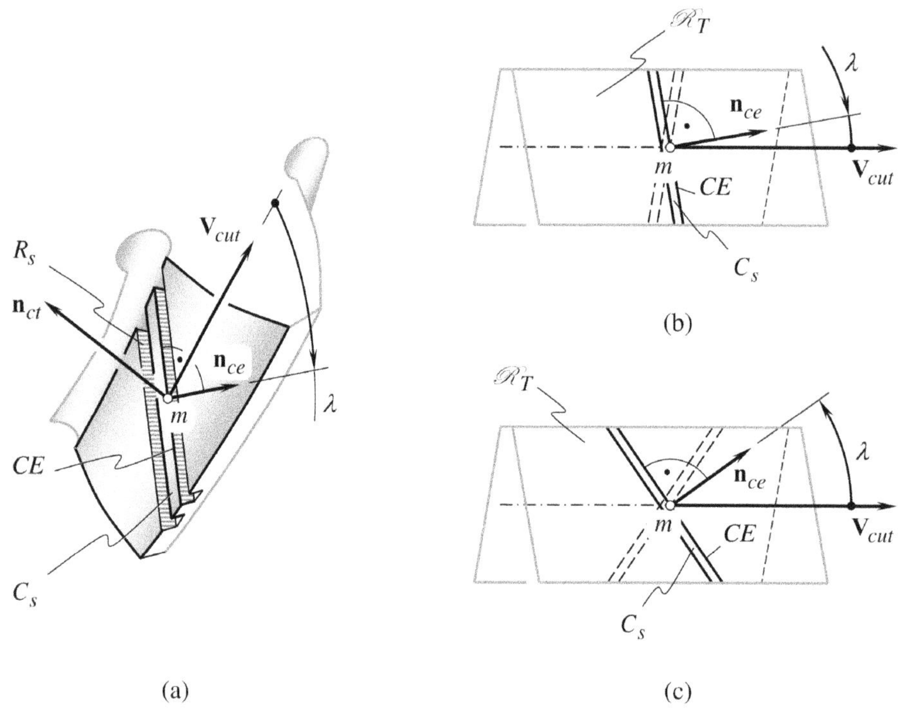


**FIGURE 16.53** Tooth design of a shaving cutter with an increased (with the near optimal) angle of inclination λ of the cutting edges: (a) gear shaving cutter tooth, (b) a case of annular serrations, and (c) a case of inclined serrations.

When the vectors $\mathbf{V}_{cut}$ and $\mathbf{n}_{ce}$ are through the apex point $A_{pa}$, then the following expressions are valid:

$$\mathbf{V}_{cut} = \mathbf{j}\cdot|\mathbf{V}_{cut}|\sin\psi_{ct} + \mathbf{k}\cdot|\mathbf{V}_{cut}|\cos\psi_{ct} \quad (16.113)$$

$$\mathbf{n}_{ce} = \mathbf{i}\cdot\cos\phi_t - \mathbf{j}\cdot\sin\phi_t \quad (16.114)$$

where:

$\psi_{ct}$ is the pitch helix angle of the shaving cutter.
$\phi_t$ is the transverse profile angle in the work gear-to-shaving cutter mesh.

By definition (see Appendix G), for the calculation of the angle of inclination, the formula

$$\lambda = \tan^{-1}\left(\frac{|\mathbf{V}_{cut}\times\mathbf{n}_{ce}|}{\mathbf{V}_{cut}\cdot\mathbf{n}_{ce}}\right) \quad (16.115)$$

is applied.

Substituting Eq. (16.113) and Eq. (16.114) into Eq. (16.115), a formula

$$\lambda = \tan^{-1}\left(\sqrt{\frac{1+\tan^2\phi_t\cos^2\psi_{ct}}{(1+\tan^2\phi_t)\sin^2\phi_t\sin^2\psi_{ct}}}\right) \quad (16.116)$$

for the calculation of the inclination angle λ is derived.

In order to express the inclination angle in terms of normal profile angle $\phi_n$, and not in terms of transverse profile angle $\phi_t$, an expression (see Appendix F)

$$\phi_t = \tan^{-1}\left(\frac{\tan\phi_n}{\cos\psi_{ct}}\right) \quad (16.117)$$

can be used.

The inclination angle of the cutting edges is significantly increased in the design of a shaving cutter featuring not annular but inclined serrations [see Figure 16.53(c)]. In this way the inclination angle λ can be of optimal or of nearly optimal value. The serrations at the opposite side of the shaving cutter tooth profile are inclined oppositely.

#### 16.7.1.2 Shaving Cutter with Inclination Angle of a Constant Value along the Cutting Edges

Once an optimal value of the inclination angle λ is determined, it is desirable to shape the cutting edges so as to have the optimal value $\lambda_{opt}$ of the angle of inclination at every point along the cutting edges. This problem is not of critical importance in cases when the tooth number of both the work gear and the shaving cutter is large enough. However, when the tooth number either of the work gear or the shaving cutter (or of both of them) is small, then variation of the direction of the cutting speed $\mathbf{V}_{cut}$ within the cutting edge [see Figure 16.25(b)] cannot be ignored. Variation of the direction of the cutting speed vector $\mathbf{V}_{cut}$ entails a corresponding alteration to the inclination angle λ. Once

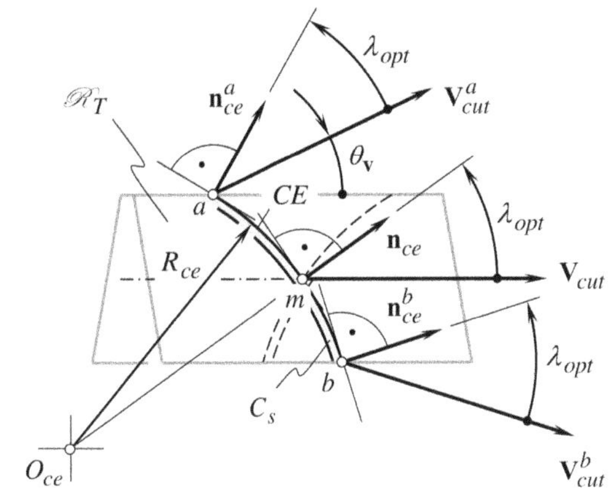


**FIGURE 16.54** Tooth design of a shaving cutter with an optimal inclination angle $\lambda_{opt}$ at every point of the cutting edge $CE$.

the inclination angle $\lambda$ varies along the cutting edge, it cannot have an optimal value at every point of the cutting edge of the shaving cutter. Ultimately this decreases the cutting performance of the shaving cutter.

The cutting performance of a shaving cutter can be increased if the inclination angle is kept at an optimal value at every point along the cutting edge $CE$. For this purpose, the serrations are made curvilinear in their lengthwise direction. A practical approximation to the desirable longitudinal shape of the serrations by a circular arc resolves the issue.

Consider a lateral tooth flank of the auxiliary generating rack $\mathscr{R}_T$ of the shaving cutter as shown in Figure 16.54. Within the plane $\mathscr{R}_T$, the cutting edge is shaped in the form a circular arc. Center $O_{ce}$ of the circular arc is located within a straight line that is aligned with the unit normal vector $\mathbf{n}_{ce}$ to the cutting edge $CE$ at the apex point $A_{pa}$. The radius $R_{ce}$ of the circular arc is calculated from the expression

$$R_{ce} = \frac{a_{ct}}{\sin(\theta_{\mathbf{V}} + \lambda_{opt}) - \sin \lambda_{opt}} \tag{16.118}$$

where:

$a_{ct}$ is the tooth addendum of the shaving cutter.

$\theta_{\mathbf{V}}$ is the angle within the lateral plane $\mathscr{R}_T$ that the cutting speed vector $\mathbf{V}_{cut}$ at the apex point $A_{pa}$ forms with the cutting speed vector $\mathbf{V}^a_{cut}$ at the top of the shaving cutter tooth $[\theta_{\mathbf{V}} = \angle(\mathbf{V}_{cut}, \mathbf{V}^a_{cut})]$.

$\lambda_{opt}$ is the desirable value of the inclination angle of the cutting edges of the shaving cutter.

After the cutting edge is constructed within the plane $\mathscr{R}_T$, then the plane $\mathscr{R}_T$ is folded over the screw involute surface $T$ of the shaving cutter tooth. Implementation of $c$-mapping of the circular arc cutting edge is helpful in this case.

In such a scenario, the inclination angle $\lambda$ has a constant value along the cutting edge, and it can be equal to an optimal value $\lambda_{opt}$ at every point of the cutting edge. In this way the cutting performance of the shaving cutter can be improved.

The concept of constant inclination angle is also used in the design of the tools for burnishing of gears with surface plastic deformation.[14]

### 16.7.2 Utilization of Features of the Generating Surface of Involute Shaving Cutter

The generating surface of a shaving cutter for machining of involute gears is shaped in the form of a screw involute surface [see Eq. (16.1)]. The screw involute surface $\mathscr{G}$ is conjugate to the work gear tooth flank $T$. The screw involute surface is a type of developable surface. It can be represented in the form of consecutive positions of a family of straight lines that are tangent to the base helix of the surface. This immediately allows for a conclusion: A straight line passes through every point of the generating surface $T$ of a shaving cutter. This straight line is often referred to as a *straight generating line*.

#### 16.7.2.1 Precision Involute Shaving Cutter with Straight Cutting Edges

A segment of the straight generating line can be employed as a cutting edge $CE$ of the shaving cutter.[15] The cutting edges of the precision shaving cutter align to the straight lines, which are tangent to the base helix of the generating surface $T$ of the shaving cutter (see Figure 16.55) [187].

The cutting speed vector $\mathbf{V}_{cut}$ is located within a plane tangent to the screw involute surface $T$. At the apex point $A_{pa}$, the cutting speed vector $\mathbf{V}_{cut}$ is tangent to the pitch cylinder, and it crosses the axis of rotation $O_{ct}$ of the shaving cutter at base helix angle $\psi_{b.ct}$.

The unit vector $\mathbf{n}_{ce}$ is perpendicular to the cutting edge $CE$ and is pointed opposite to the unit vector $\mathbf{v}^*_{ct}$. The latter vector is one of three direction vectors of the local Cartesian coordinate system. The origin of this reference system is at point of interest $m$ within the cutting edge $CE$.

Once the cutting speed vector $\mathbf{V}_{cut}$ and either the unit normal vector $\mathbf{n}_{ce}$ or the direction vector $\mathbf{v}^*_{ct}$ are calculated, then Eq. (16.115) is applied to calculate the inclination angle $\lambda$ at point of interest $m$.

Many advantages are inherent to a precision shaving cutter with straight cutting edges [187]. The possibility of the serrations of a constant depth is one of the advantages. The serrations can then be designed so that their depth gradually increases either toward the outer diameter or toward the base diameter of the shaving cutter. This feature in combination with the trapezoidal shape of the serrations allows for easier control over the width of the serration land.

---

[14] SU Pat. No. 1177016, *A Tool for Finishing Gears*/S.P. Radzevich, A.P. Radzevich, Int. Cl. B21h 5/02, Filed: July 8, 1983.
SU Pat. No. 1276408, *A Tool for Finishing Gears*/S.P. Radzevich, Int. Cl. B21h 5/02, Filed: December 25, 1984.

[15] SU Pat. No. 1537426, *A Shaving Cutter*/Radzevich, S.P., Palaguta, V.A., Radzevich, A.P., Tovmasian, M.A., Int. Cl. B23f 21/28, Filed: November 17, 1986.

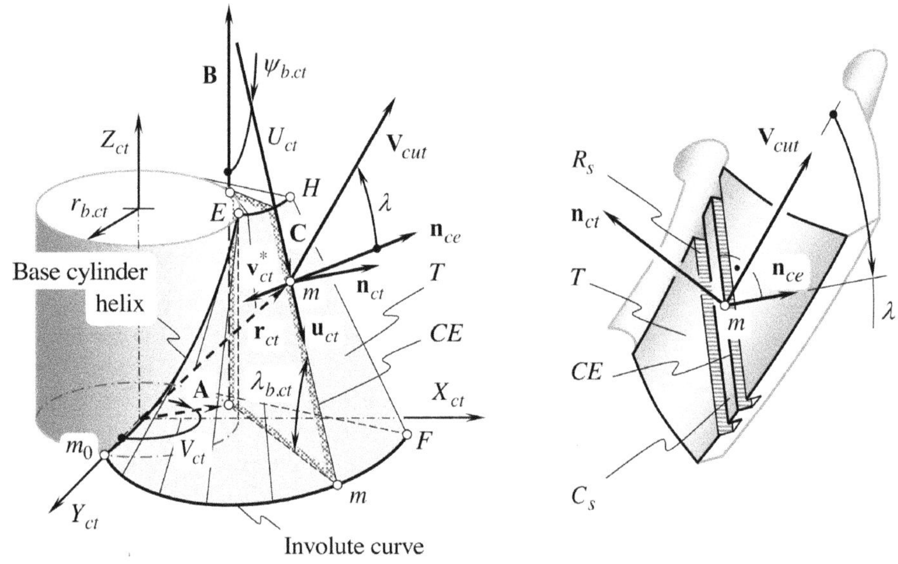


**FIGURE 16.55** Tooth design of a precision shaving cutter with straight cutting edges. (From Radzevich, S.P., *et al.*, USSR Patent 1537426, Int. Cl. D23f 21/28, Nov. 17, 1986.)

A precision worm-type shaving cutter can be designed with straight cutting edges as well.

#### 16.7.2.2 Shaving Cutter for Finishing Modified Tooth Flank of a Gear

Shaving cutters with straight cutting edges can be designed for finishing modified involute gears. As an example, consider a worm-type shaving cutter for finishing a modified work gear tooth flank. The worm-type shaving cutter features straight cutting edges.[16] That same concept for modification of the work gear tooth flanks can be implemented in the design of disk-type shaving cutters as well.

If the cutting edges of the shaving cutter are tangent to the base helix, then the tooth profile of the work gear is shaped in the form of an involute. In this particular case, the cutting edge of the shaving cutter is positioned at a distance of $0.5d_{b.ct}$ from the axis of rotation $O_{ct}$ of the shaving cutter. When cutting edge *CE* crosses the shaving cutter axis of rotation $O_{ct}$ at a distance of $0.5d^*_{b.ct}$ that exceeds the distance $0.5d_{b.ct}$, then the straight cutting edge is deviated from the generating surface *T* as shown in Figure 16.56. In this way a modified tooth profile of the work gear is generated in the gear finishing process. For the calculation of the desirable radius of the curvature $R_{ct}$ of the modified tooth profile, the Dupin indicatrix is applied.

For a screw involute surface, the Dupin indicatrix $Dup(T)$ of the auxiliary generating surface *T* is reduced to a pair of straight lines. These lines are parallel to the direction that is specified by the unit tangent vector $\mathbf{t}_{1.ct}$ of the first principal direction of the surface *T*. For a shaving cutter with given design parameters, the second principal radius of curvature $R_{2.ct} > 0$ can be calculated [the first principal radius of curvature $R_{1.ct}$ in the case under consideration is equal to infinity ($R_{1.ct} < 0$)]. Ultimately, the required angle $\theta_{ce}$ can be calculated from the equation

$$\theta_{ce} \approx \sin^{-1}\sqrt{\frac{R_{2.ct}}{R_{ct}}} \tag{16.119}$$

The direction of the straight cutting edge *CE* is specified by unit tangent vector $\mathbf{t}_{ct}$. The angle $\theta_{ce}$ is the angle that the cutting edge *CE* forms with the first principal direction $\mathbf{t}_{1.ct}$, or in other words, $\theta_{ce} = \angle(\mathbf{t}_{1.ct}, \mathbf{t}_{ct})$.

The approach illustrated in Figure 16.56 for the determination of the angle $\theta_{ce}$ is a type of approximation. The smaller the angle $\theta_{ce}$, the better the approximation is, and vice versa. Actually, the points of the generating surface *T* of the shaving cutter for finishing modified gears feature negative Gaussian curvature ($\mathcal{G}_{ct} < 0$). From this perspective, an exact expression for the calculation of angle $\theta_{ce}$ can be derived.

Once the angle $\theta_{ce}$ is calculated, then the calculation of the rest of the design parameters of the shaving cutter becomes a trivial procedure.

Investigation into the following:

- Cutting geometry of the cutting edges
- Shape and parameters of the generating surface of the shaving cutter
- Kinematics of the gear shaving process

may reveal new opportunities for the improvement of the design of shaving cutters[17] [186].

[16] SU Pat. No. 1683913, *A Method of Finishing of a Modified Gear Teeth*/S.P. Radzevich, S.P., Int. Cl. B23f 21/28, Filed: January 01, 1989.

[17] SU Pat. No. 1514516, *A Shaving Cutter*/S.P. Radzevich, V.A. Palaguta, A.P. Radzevich, V.A. Bulgakov, Int. Cl. B23f 21/28, Filed: November 27, 1986.

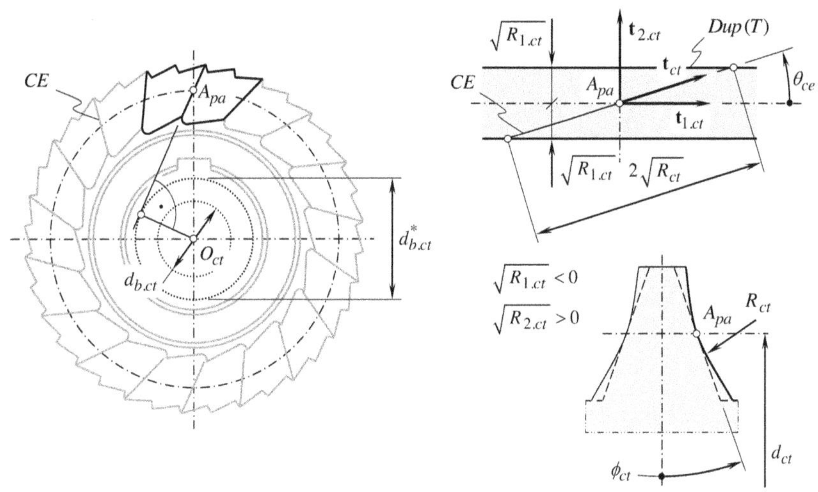


**FIGURE 16.56** Precision shaving cutter with straight cutting edges for finishing a modified tooth flank of the work gear. (From Radzevich, S.P., USSR Patent No. 1683913, Int. Cl. B23f 21/28, Jan. 1, 1989.)

## 16.8 PECULIARITIES OF THE GEAR SHAVING PROCESS

Rotary shaving is a low-pressure, free-cutting finishing operation that can be applied to finish preformed, hobbed, or shaped gear teeth before heat treatment. As the shaving cutter rotates with the work gear in tight mesh, the notches scrape off little shavings of material; hence the name *shaving.*

Gear shaving is the logical remedy for the inaccuracies inherent in gear cutting. It is not, therefore, a cure-all but a means of making a good gear better.

### 16.8.1 Shaving Cutter Selection

The equality of the base pitch of the shaving cutter to the base pitch of the work gear is a requirement of prime importance when selecting a design of a shaving cutter for finishing a given work gear. The base pitch of the shaving cutter must be equal to the base pitch of the work gear (and both of the pitches must be equal to the operating base pitch in the work gear-to-shaving cutter mesh—this is a must [172]).

Theoretically, one shaving cutter will shave all gears of a given pitch and profile angle regardless of the number of teeth. A slight involute variation due to the difference in sliding action may be noticeable on the gears in the coarser pitches with small numbers of teeth. Therefore, the range of gears that can be shaved with one shaving cutter will depend on the allowable involute tolerances.

Right- and left-hand shaving cutters are usually required for a range of helical gears. However, gears ranging from 5° to 18° helix angle, right- and left-hand, may be shaved with one straight spur shaving cutter. The range of helical angles that one shaving cutter can handle is determined by the crossed-axes angle between the shaving cutter and the work gear axes of rotation. The most desirable range of crossed-axes angles is from 5° to 15°. Thus, one shaving cutter having a 15° helix angle will shave spur gears and opposite-hand helical gears from 0° to 10° helix angle, and from 20° to 30°, approximately.

No particular requirements for the serration design need to be fulfilled when the axial method of shaving is applied. A shaving cutter with either annular serrations or serrations of differential type can be used in this case. That same is true with respect to a diagonal shaving process with reasonably small traverse angle. However, it is recommended to apply differential-type serrations in the design of shaving cutters for diagonal (with great traverse angle), tangential, and plunge shaving methods.

In general, it is desired to have a *hunting ratio* between the number of shaving cutter teeth and the number of work gear teeth. This tends to make it desirable to design shaving cutters with prime numbers of teeth. A prime number will hunt with all numbers of teeth except itself. Thus, numbers of teeth like 37, 41, 43, 47, 53, 59, 61, 67, 73, and so forth, are popular for shaving cutters.

If the gear and the shaving cutter are properly designed, it is possible to improve profile accuracy considerably.

### 16.8.2 Requirements for Pre-Shaved Work Gear

So much depends on the hobbed tooth form and the desirable final shaved tooth form that for the best results, both tools should be regarded as complementary, and the shaving cutter should be designed specifically to modify the hobbed tooth.

Of primary concern to the shaving cutter manufacturer is the fillet produced by the roughing operation. The tips of the shaving cutter teeth must not contact the gear root fillet during the shaving operation. If such contact does occur, excessive wear of the shaving cutter results, and the accuracy of the involute profile is affected.

Teeth intended to be shaved are often and preferably cut by a protuberance hob or shaper cutter. Gear-cutting tools with protuberance produce a slight undercut or relief near

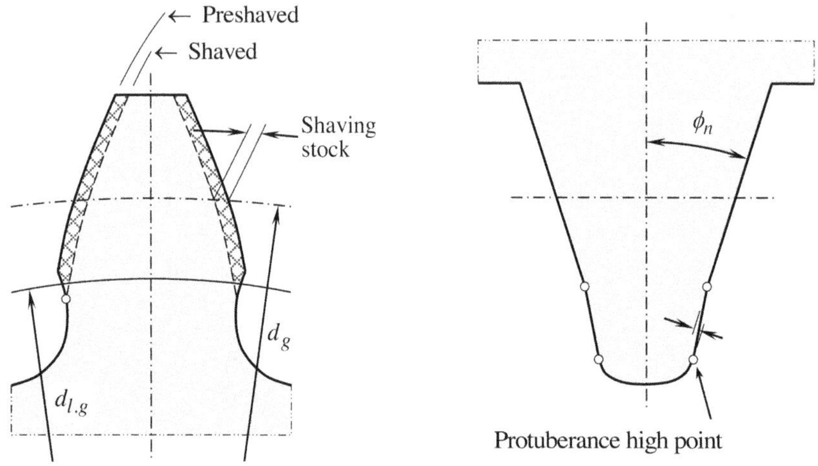


**FIGURE 16.57** Tooth of a pre-shaved work gear, and tooth of a protuberance-type hob.

the base of the gear tooth (see Figure 16.57). This method assumes a smooth blending of the shaved tooth profile and the unshaved tooth fillet, and reduces shaving cutter tooth tip wear. The amount of undercut produced by the protuberance-type gear-cutting tool should be made for the thin end of the tooth. The position of the undercut should be such that its upper margin meets the involute profile at a point below its limit diameter.

## 16.8.3 Manufacturing Aspects of Gear Shaving Operation

In rotary shaving of gears, the work gear and the shaving cutter rotate in tight mesh. There is no index gearing in the shaving process. Small parts are driven by the cutter that is shaving them, while large parts drive the shaving cutter instead.

When shaving, both, the work gear and the shaving cutter are set up on arbors. Examples of the arbors used in the gear shaving process are shown in Figure 16.58. Occasionally, especially when shaving low-tooth-count gears (the so-called *LTC −gears*), the work gear top-land surfaces are used instead of arbors. An example is depicted in Figure 16.59. When shaving an *LTC* −gear, the work gear is supported by means of top-land surfaces.

The shaving cutter is rotated at high speeds up to 122 m (400 and more surface ft) per minute. Feed is fine, and the shaving cutter tool contact zone is restricted. Shaving cutter life depends on several factors: operating speed, feed, material and hardness of the work gear, its required tolerances, type of coolant, and the size ratio of shaving cutter to work gear.

The desirable range of the crossed-axes angle is from 5° to 15° for spur gears and from 5° to 12° for helical gears. Experience has established these as the most desirable ranges, although in some cases it is permissible to extend them to from 3° to 18°. At the lower end of this extended range, cutting action is reduced, and at its upper end, guiding action is somewhat sacrificed.

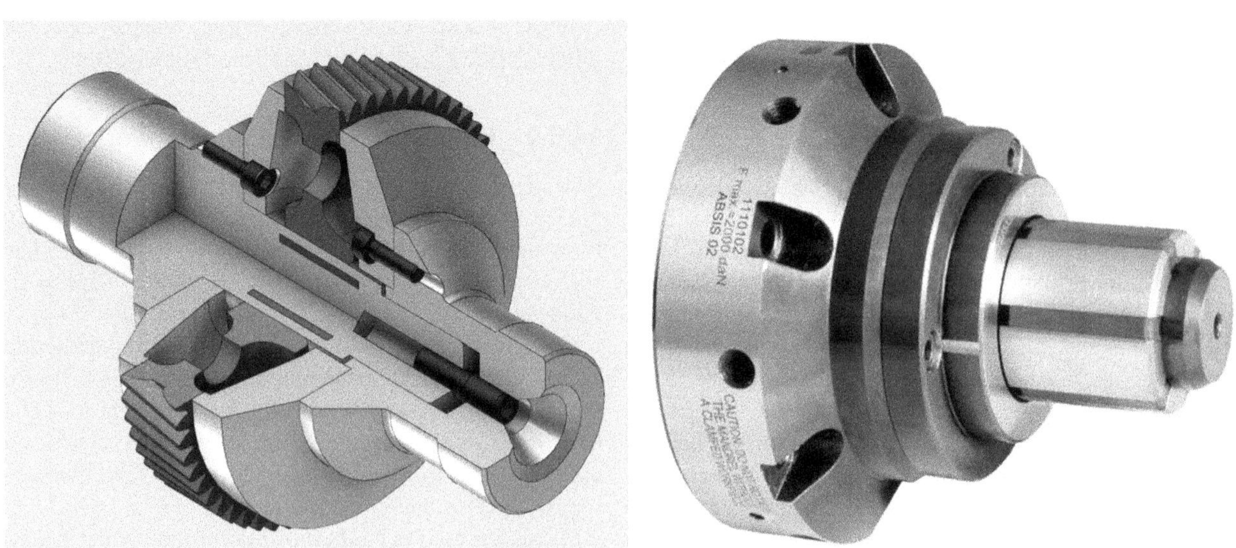

**FIGURE 16.58** Arbors that are used in the gear shaving process.

**FIGURE 16.59** Use of the top-land surfaces as the datum surface in the gear shaving process.

Shaving is a corrective process. The purpose of gear shaving is to

- Correct eccentricity and errors in index, helix angle, and tooth profile.
- Improve tooth surface finish.
- Maintain tooth size.
- Eliminate tooth end bearing conditions by producing a crowned tooth form.

The reason the shaving cutter can produce very high accuracy is that there is an averaging action as the shaving cutter rolls with the work. This tends to mask the effect of any slight indexing errors that may have been ground into the cutter.

Tooth-to-tooth spacing is improved by shaving, but accumulated spacing error is not changed by a large amount unless the accumulated error comes mostly from eccentricity effects. On narrow-face-width gears, helix errors can be controlled by shaving, but a narrow shaving cutter has little or no control on wide-face gears.

Gears with smaller numbers of teeth are more critical from a production and sound standpoint than those with greater numbers of teeth. Likewise, spur gears are generally more critical on both scores than helical gears.

Spur pinions with a small number of teeth are somewhat more difficult to shave. There are problems with obtaining good involute when shaving pinions with 10 teeth or fewer. It is also difficult to design a suitable shaving cutter for internal gears with fewer than 40 teeth. When internal gears have 25 teeth, it is just about impossible to get a shaving cutter inside that will shave on the crossed-axes principle.

The number of cuts required will depend on the amount of stock left for shaving and the amount taken per cut. Sometimes extra cuts may be needed to secure helix-angle correction. The amount of stock that may be removed in shaving is fairly limited. It is practical to put in only a certain amount of undercut in the pre-shave cutting. If too much stock is removed, the undercut allowance will be exceeded, and there will be trouble from cutter *bottoming.*

In shaving either spur or helical gears by the modified underpass shaving method, there is a choice of direction of feed and rotation. The direction of feed slide should be set to feed in the general direction of the cutter helix. The rotation of the cutter should be in a direction opposing the feed.

A shaving cutter becomes worn after a certain number of work gears are shaved. An example of the worn clearance surface $C_s$ of a shaving cutter is illustrated[18] in Figure 16.60. As shown in this figure, the shaving cutter tooth wear is spread over the clearance surface. Tooth wear of this kind is advantageous as a very thin stock must be ground off when resharpening the shaving cutter. After being reground, the shaving cutter is ready to finish the next set of work gears.

A conical spur or helical gears (tapered tooth gears that mesh on nonparallel axes) can be shaved by the high-speed diagonal shaving method.

### 16.8.4 Modification of Tooth Form and Shape

Modification of the tooth flank from its theoretical form has been found to have a marked effect on improving gear life and reducing operating sound levels by making allowances for deflection conditions under load. By far the most economical, and often the only way to achieve these important tooth modifications, is to make them by a finishing process.

Shaving cutters are capable of removing 0.0005 in. (0.012 mm) from the tip and root of tooth profiles to provide relief, and simultaneously they can remove undulations down a tooth profile.

In axial shaving as well as in diagonal shaving that features a small traverse angle, the rocking machine table is used in order to introduce the lead crown of the work gear teeth. For this purpose, the built-in crowning mechanism is used. The table supporting the part being shaved is rocking about a pivot point in timed relation with the feed. In this way, end relief, barreling, or crowning to the work gear teeth is imparted.

[18] Data received by Mitsubishi Heavy Industries, Ltd.

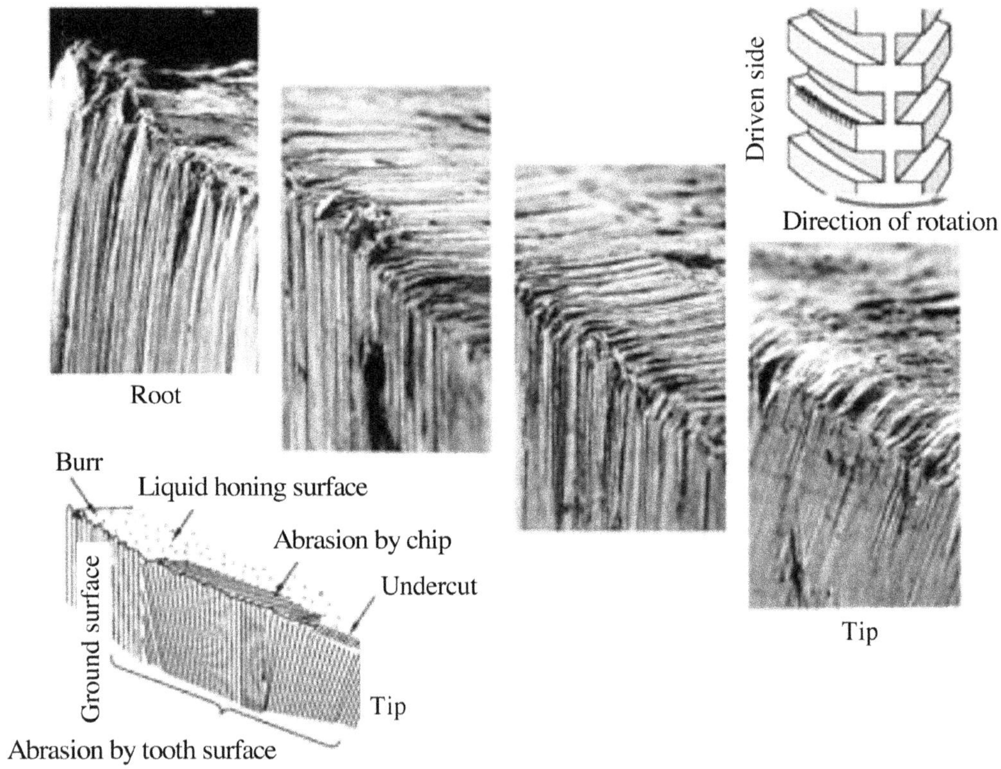


**FIGURE 16.60** Expending of the serration edge of shaving cutter (Adapted from: Mitsubishi Heavy Industries, Ltd.).

When traverse angles in diagonal shaving are from 60° to 90°, a shaving cutter with reverse-crowned teeth is used to provide the crowned tooth form. The same is observed when the tangential or plunge method of shaving is used. All tooth modifications in this case must be made to the shaving cutter as it will not be possible to realize them through axial movements on the machine. This method uses a shaving cutter with differential serrations and is applied to finish shoulder gears.

### 16.8.5 Shaving of Worm Gear

The object of worm-wheel shaving is to produce tooth surfaces more closely resembling those resulting from running in when small asperities and departures from the shape conjugate with the worm have been removed by wear or by plastic surface deformation.

In operation, the process is quite simple and logical. The lands on a serrated hob are not relieved, and when they engage with the surface of wheel teeth that have numerous small projections standing above the general surface level, these are removed in thin shavings. It is not feasible to take heavy cuts, and the closely spaced lands quickly remove the projections and produce a smoother finish on the wheel tooth flanks. If the pressure between the work gear and the hob is then slightly increased either by feeding the hob into the work gear radially or by driving the work gear by the hob against some resistance, the surface is locally elastically deformed. This permits the hob to shave off the surface of the work gear protruding between the lands of the hob and to remove metal surrounding hollows. About 0.005 to 0.010 in. (0.12–0.25 mm) of metal per flank is shaved off when finishing worm wheels.

Only involute gears can be precisely generated in the gear shaving process. No gears with other tooth flank geometry can be perfectly generated by the rack cutter—only approximate noninvolute gears can be cut this way [171]. This includes tooth flank modifications of any and all kinds. This is attributed to the kinematics of the gear shaving process, and to violation of the condition of conjugacy of the tooth flanks $\mathcal{G}$ of a gear intended to be machined, and of the generating surface $T$ of the shaving cutter. In compliance with the condition of conjugacy of the tooth flanks, at every point of contact of the surfaces $\mathcal{G}$ and $T$, the common perpendicular $\mathbf{n}$ must intersect the axis of instant rotation $P_{ln}$ of the gear and the shaving cutter in the gear machining mesh.

### 16.8.6 Abrasive Gear Shaving (Gear Honing)

Abrasive gear shaving (i.e., gear honing) is very similar to gear shaving by shaving cutters of a conventional design. Gear finishing is widely applied with abrasive shaving cutters (gear honing). This process is illustrated in Figure 16.61.

**FIGURE 16.61** Honing an involute gear by the enveloping abrasive shaver (by the enveloping hone).

**FIGURE 16.62** Ring gear and the abrasive shaver (gear hone) for finishing the teeth flanks.

When honing gears, it is customary to limit the crossed-axes angle by $\Sigma \leq 10°$. This is mostly because of accuracy issues.

Ring gears also can be finished by the abrasive shaving cutter (by the gear hone) as depicted in Figure 16.62. Tooth flanks of a special geometry (barrel type) are required for finishing ring gears.

**FIGURE 16.63** Honing a cluster gear.

**FIGURE 16.64** Dressing an abrasive shaving cutter (of a gear hone) by means of the diamond dressing tool.

Cluster gears can be finished by abrasive shaving as well. External gear hones are used for this purpose (see Figure 16.63).

Worn abrasive shaving cutters (worn gear hones) are redressed. Diamond dressing tools are used for this purpose (see Figure 16.64). When dressing/redressing, the abrasive shaving cutter and the dressing tool are rotated together in tight mesh under load. The process of dressing/redressing of worn gear hones is very fast.

# 17 Examples of Implementation of the Classification of the Gear Machining Meshes

Gear-cutting tools of all designs originate on vector diagrams of the corresponding gear machining mesh. It is not only the designs for hobs for machining gears and shaving cutters for finishing gears, discussed in previous chapters, that originate from the vector diagram of the external spatial gear machining mesh shown in Figure III-C 1.1(a). Other novel types of gear-cutting tools, and corresponding methods for machining cylindrical and conical gears, can be designed based on that same vector diagram of the external spatial gear machining mesh (see Figure 3.8). The following examples of novel designs of gear-cutting tools, as well as the methods for machining gears, demonstrate the potential capabilities of the classification of vector diagrams of gear machining meshes (see Figure 3.8).

## 17.1 HOB FOR TANGENTIAL HOBBING OF GEARS

Shown in Figure III-C 1.1(a), an external spatial gear machining mesh can be complemented with the feed motion $\mathbf{F}_{ct}$ that is pointed in the tangential direction relative to the work gear. Consequently, the principal kinematics of the tangential method for gear hobbing can be derived (see Figure 17.1). In this kinematics, the feed motion $\mathbf{F}_{ct}$ is pointed perpendicular to the work gear axis of rotation $O_g$.

Both spur and helical gears can be machined with the tangential method of gear hobbing. A hob of extended face width is used in this method as schematically shown in Figure 17.2. The length of the hob and the crossed-axis

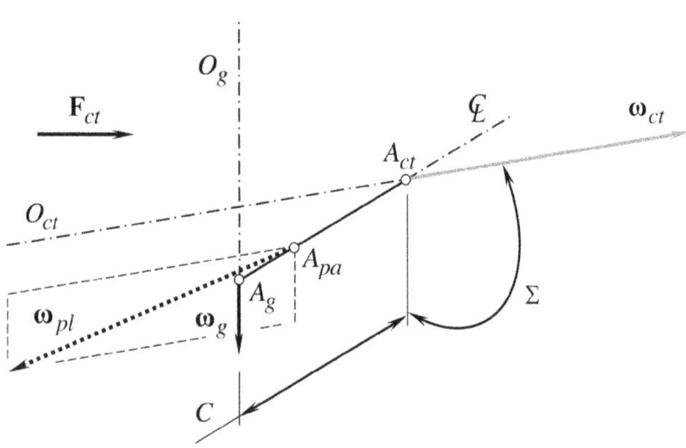


FIGURE 17.1 Schematic diagram of the kinematics of tangential method of gear hobbing.

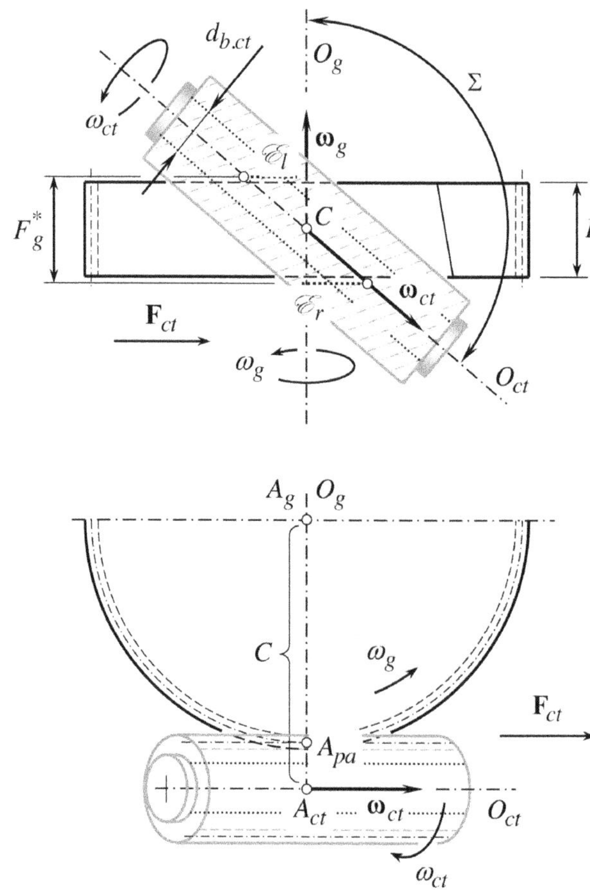


FIGURE 17.2 Configuration of the hob in tangential method of gear hobbing.

angle Σ need to be set to allow the required overlap of the work gear face by the hob. The approach that is illustrated in Figure 15.166 and Figure 15.168 can be employed for the determination the required combination of the crossed-axis angle Σ and the hob length. Two straight generating lines $\mathscr{E}_r$ and $\mathscr{E}_l$ of the generating surface $T$ of the hob are tangent to the base cylinder of the hob $d_{b.ct}$ and to the base helices as well. The straight lines $\mathscr{E}_r$ and $\mathscr{E}_l$ are configured so as to be parallel to the faces of the work gear. The crossed-axis angle Σ and the hob length should fulfil the condition under which the distance $F_g^*$ between the crossing straight generating lines $\mathscr{E}_r$ and $\mathscr{E}_l$ is equal to or exceeds the face width $F_g$ of the work gear, that is,



DOI: 10.1201/9781003520528-22

the equality $F_g^* \geq F_g$ must be valid. The work gear face width must not exceed the distance $F_g^*$. In such a scenario, the gear hob tangentially travels past the work gear. The whole work gear is machined just in one short pass of the hob. The travel distance of the hob does not depend upon the face width $F_g$ of the work gear.

Hobs for the tangential hobbing of gears are of the same design as conventional hobs and differ only in length.

## 17.2 HOB FOR PLUNGE GEAR HOBBING

Another opportunity arises in the event when the feed motion $\mathbf{F}_{ct}$ is pointed along the centerline ℄ of the axes of rotation $O_g$ and $O_{ct}$ of the work gear and of the gear-cutting tool, respectively. The principal kinematics of the gear hobbing process is illustrated in Figure 17.3. Because the feed motion is pointed toward the work gear axis of rotation $O_g$, this method of gear hobbing is referred to as the plunge method of gear hobbing.

Hourglass-type hobs are used in plunge gear hobbing (see Figure 17.4). The auxiliary generating surface $\mathscr{R}_T$ of the hob is congruent to the gear intended to be machined. The length of the hob should be sufficient to overlap the entire face width of the work gear. Again, the work cycle is very short—the whole work gear can be machined in one short path of the hob in the radial direction of the work gear.

Both spur and helical gears can be machined with the tangential method for gear hobbing.

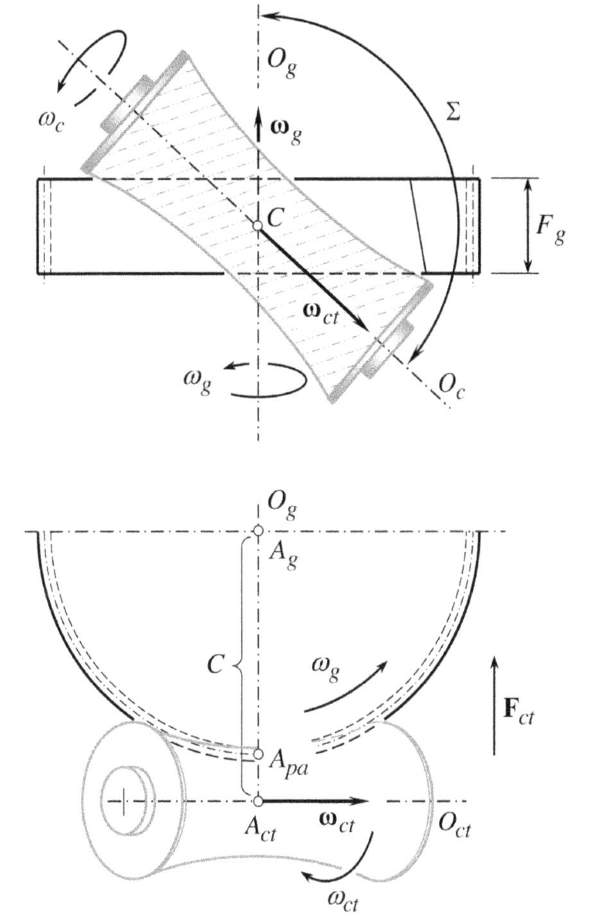


**FIGURE 17.4** Relative motions of the hob and the work gear in plunge method of hobbing gears.

## 17.3 HOBBING OF FACE GEAR

One of the possible directions of the feed motion $\mathbf{F}_{ct}$ of the gear-cutting tool corresponds to the principal kinematics of cutting a face gear with the standard hob. This possibility is illustrated in Figure 17.5. The work gear and the hob are rotated about their axes of rotation $O_g$ and $O_{ct}$, with axes crossing at a right angle. The rotation of the work gear $\boldsymbol{\omega}_g$ and the rotation of the hob $\boldsymbol{\omega}_{ct}$ are synchronized. The feed motion $\mathbf{F}_{ct}$ is pointed toward the work gear. The travel distance of the hob slightly exceeds the depth of the face gear tooth space. Consequently, the cycle time is very short.

This method of face-gear hobbing allows for its interpretation as an inversion of the method of gear cutting illustrated in Figure 15.109.

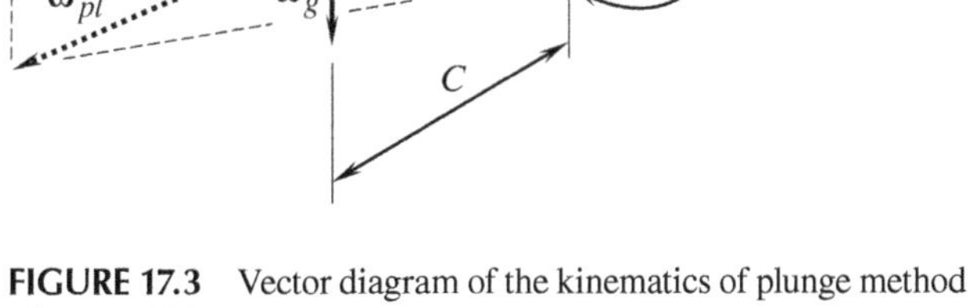


**FIGURE 17.3** Vector diagram of the kinematics of plunge method of gear hobbing process.

## 17.4 WORM-TYPE GEAR-CUTTING TOOL WITH CONTINUOUS HELIX-SPIRAL CUTTING EDGE

For most gear-cutting tools of conventional design, the rotation of the cutting tool contributes most to the cutting speed. From this standpoint, the rest of the elementary motions of the gear-cutting tool in relation to the work gear are of minor importance. The parameters of the external spatial gear machining mesh [see Figure III-C 1.1(a)] can be set up so as to make the rolling motion of the cutting tool the primary motion. In this particular case of special-purpose gear-cutting tool design, a minor motion can be the most beneficial. For example, consider a worm-type cutting tool having the continuous *helix-spiral cutting edge* (*HS – cutting edge* for simplicity). This design of gear-cutting tool is intended for the finishing of involute gears.

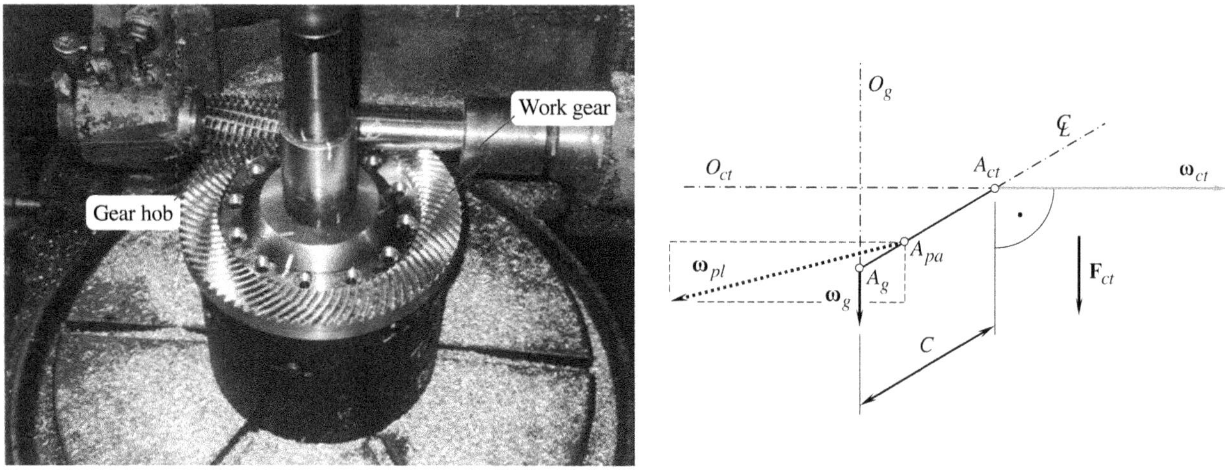


**FIGURE 17.5** Close-up and vector diagram of hobbing of a face gear by the standard hob.

Figure 17.6 depicts an example of a worm-type cutting tool for finishing an involute gear[1] [60, 67], and many others. In that case, the gear-cutting tool features a zero normal profile angle $\phi_n = 0°$. When the normal profile angle has a zero value, then the generating surface $T$ reduces to a helix located within the cylinder of outer diameter $d_{o.ct}$. The helix serves as the cutting edge $CE$ of the gear finishing tool.

The rake surface $R_s$ is shaped in the form of a screw surface through the helix $CE$. It forms the rake angle $\gamma$ with the perpendicular to the work gear tooth flank. No restrictions are imposed on the actual value of the rake angle $\gamma$. Therefore, it can be easily set equal to its optimal value.

The clearance surface $C_s$ is also a screw surface through the helix $CE$. It forms the clearance angle $\alpha$ with the surface of the cut. Again, no restrictions are imposed on the actual value of the clearance angle $\alpha$; thus, it can be easily set equal to its optimal value.

The inclination angle $\lambda$ is large. In this case, the actual value of the inclination angle $\lambda$ strongly depends upon the number of starts of the gear finishing tool. The greater the number of starts, the more reasonable the value of the inclination angle $\lambda$, and vice versa.

When machining a work gear, the gear finishing tool is set up so that the line of action is aligned with the top land of the auxiliary generating rack $\mathscr{R}_T$. Consequently, the pitch diameter of the gear finishing tool is equal to its outer diameter $d_{o.c}$.

Generation of the work gear tooth flank begins at a start point $a_{ce}$. The start point $a_{ce}$ is at a certain distance $l$ from the pitch point $P$. In Figure 17.6, the length of the active portion of the line of action is designated as $l_{ct}$. After the design parameters of the work gear are known, then the length of both straight-line segments $l$ and $l_{ct}$ can be calculated from Figure 17.6.

When finishing, the work gear is rotated $\omega_g$ about its axis of rotation $O_g$. The gear finishing tool is rotated $\omega_{ct}$ about its axis of rotation $O_{ct}$. Both the rotations are synchronized with one another. Either the work gear or the gear finishing tool is fed in the axial direction of the work gear (this motion is not shown in Figure 17.6).

The rolling motion of the gear finishing tool in relation to the work gear is used as the primary motion, that is, the rolling motion is used as the cutting speed motion $V_{cut}$. The use of a multiple-start gear finishing tool is preferable from this standpoint.

The reasonable value of stock $t$ can be cut from the work gear tooth flank in one tool path. The cut chip flows away from the work gear over the rake surface $R_s$ of the gear finishing tool.

High accuracy of the finished gears can be easily achieved due to peculiarities of design of the gear finishing tool. No profile errors are inherent because of profiling of the cutting tool. Sharpening the gear finishing tool does not affect its accuracy; the accuracy of the gear finishing tool remains constant regardless of how many times the tool has been sharpened. The total tool life of the gear finishing tool can be significantly extended because of the theoretically unlimited number of sharpenings allowed.

Additionally, the gear finishing tool is easy to be manufactured.

Applications of the gear finishing tool (see Figure 17.6) are limited to involute gears with large tooth counts. Only work gears having base diameter $d_{b.g}$ that is smaller than the form diameter $d_{l.g}$ can be finished with the gear finishing tool; that is, the inequality $d_{b.g} < d_{l.g}$ needs to be fulfilled. The greater the difference between the diameters $d_{b.g}$

---

[1] Radzevich, S.P., *A Tool for Machining Gears.* Pat. No. 933316, USSR, Int. Cl. B23F 21/00, Filed: October 18, 1979.
Radzevich, S.P., *A Gear Finishing Tool.* Pat. No. 1004030, USSR, Int. Cl. B23f 21/16, Filed: April 16, 1981.
Radzevich, S.P., *A Gear Finishing Tool.* Pat. No. 1106609, USSR, Int. Cl. B23f 21/16, Filed: September 27, 1982.
Radzevich, S.P., *A Gear cutting Tool.* Pat. No. 1028451, USSR, Int. Cl. B23F 21/16, Filed: April 26, 1982.
Radzevich, S.P., *A Gear Finishing Tool.* Pat. No. 1151392, USSR, Int. Cl. B23f 21/16, Filed: February 15, 1982.
Radzevich, S.P., *A Gear Finishing Tool.* Pat. No. 795677, USSR, Int. Cl. B21h 5/00, Filed: January 15, 1979.
Radzevich, S.P., *A Gear Finishing Tool.* Pat. No. 1230019, USSR, Int. Cl. B21h 5/00, Filed: December 25, 1984.

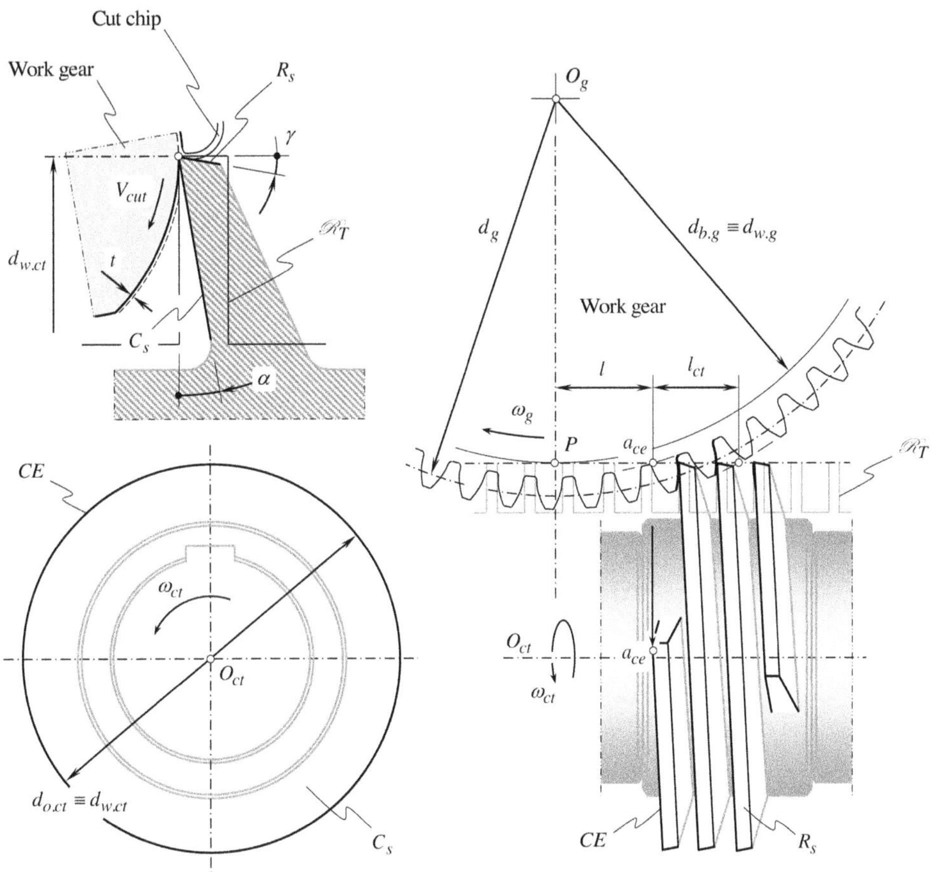


**FIGURE 17.6** Worm-type finishing gear-cutting tool with the continuous *HS* – cutting edge.

and $d_{l.g}$, the better the conditions for use of the gear finishing tool. This means that work gears with large tooth counts are preferable for finishing with the gear finishing tool.

The continuous *HS* – cutting edge of the gear finishing tool is in permanent contact with the work gear tooth flank at a distinct point. In the rolling motion, this point generates the tooth profile of the work gear. This schematic makes possible the finishing of both external and internal gears as well with the same gear finishing tool with no changes (or with limited changes) to the design parameters of the gear finishing tool.

In a transverse section of the work gear by a plane, the tooth profile is generated by a point, specifically by the point of intersection of the cutting edge with the transverse section.

The gear finishing tool is capable of finishing only one side of the work gear tooth profile. For finishing the opposite side of the tooth profile, the work gear needs to be inverted.

The concept of the worm-type gear finishing tool can be traced back to 1978, when a design of a worm-type finishing tool for burnishing of involute gears was proposed [118]. Gear finishing tools of this design[2] feature a working land and a calibrating land as shown in Figure 17.7. The gear finishing tool works in much the same way as other worm-type gear finishing tools (see Figure 17.6). The rolling motion of the gear burnishing tool contributes primarily to the machining speed $V_{mch}$. The working land deforms the stock while the calibrating land finishes the work gear tooth flank to the required size. This concept is widely used in numerous designs of gear finishing tools for burnishing tooth flanks of involute gears[3] [61, 85, 111, 118, 182], and others.

In summary, gear finishing tools (see Figure 17.7) can be used for burnishing of tooth flanks of work gears with a large tooth number (the requirement $d_{b.g} < d_{l.g}$ must be fulfilled). The use of multiple-start gear finishing tools is preferred.

These tools are easy to be manufactured, and a high degree of accuracy in the work gear tooth flank can be achieved with them.

It is instructive to note here that according to the author's recent findings, years ago Dr. Wildhaber was close to developing a gear-cutting tool for finishing involute gears. As early as 1930 he filed an invention[4] titled *Rotary Gear Cutting Tool*. In this invention, Dr. *Wildhaber* intended to use the rolling motion as the primary motion in the gear finishing process. The core idea of the invention is illustrated in Figure 17.8. In the invention, pitch point *P*

[2] SU Pat. No. 829280, *A Worm-Type Finishing Tool for Burnishing Involute Gear.*/ S.P. Radzevich, Int. Cl. B21h 5/00, Filed: May 5, 1978.

[3] SU Pat. No.: 795677, 795678, 829280, 880589, 880590, 923688, 956114, 965582, 969395, 980905, 996016, 1016004, 1031610, 1055578, 1174139, 1230019, 1328042 and others.

[4] U.S. Patent No. 1.996.987, *Rotary Gear Cutting Tool*, Ernst Wildhaber, Filed: June 23, 1930, Patented: April 9, 1935.

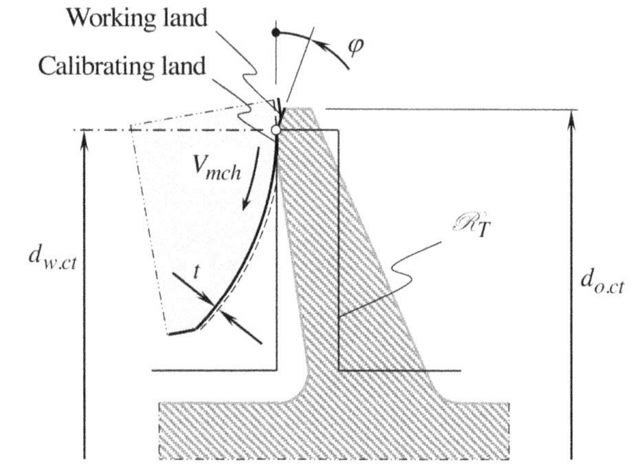


**FIGURE 17.7** Schematic of interaction of tooth flank of the work gear with the threads of the worm-type finishing tool for burnishing an involute gear. (From Radzevich, S.P., USSR Patent 829280, Int. Cl. B21h 5/00, May 5, 1978.)

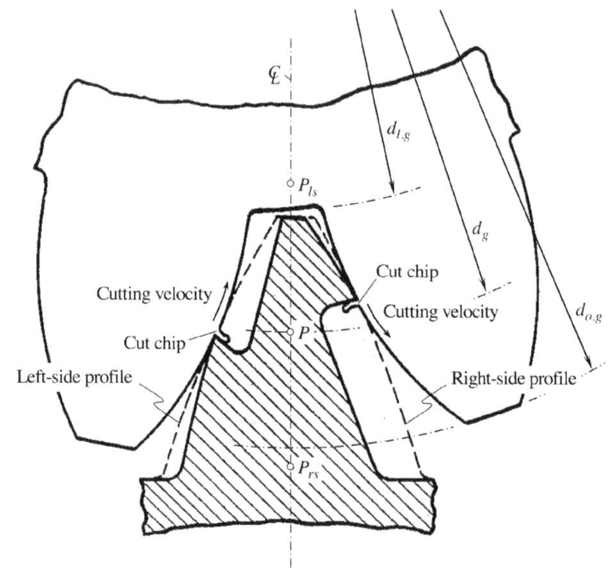


**FIGURE 17.8** Chip removal process when finishing a gear by the rotary gear-cutting tool. (U.S. Patent No. 1.996.987, *Rotary Gear Cutting Tool*, Ernst Wildhaber, Filed: June 23, 1930, Patented: April 9, 1935) [217].

in the gear finishing process is located on the intersection of the centerline ℄ by the pitch circle of a diameter $d_g$; that is, the pitch point $P$ is located in between the circles of the outer diameter $d_{o.g}$ and the end of the active profile diameter $d_{g.l}$ of the gear.

The pitch point $P_{ls}$ has to be located inside the circle of the diameter $d_{g.l}$ in order to cut chips with the left-side tooth profile of the rotary gear-cutting tool. Similarly, the pitch point $P_{rs}$ has to be located outside the circle of the outside diameter $d_{o.g}$ in order to cut chips with the left-side tooth profile of the rotary gear-cutting tool. Under no circumstances is the pitch allowed to be located between the circles of the diameters $d_{g.l}$ and $d_{o.g}$. This constraint is because at the pitch point, the cutting velocity changes to the opposite direction. No gear machining mesh with two pitch points $P_{ls}$ and $P_{rs}$ simultaneously is feasible. Therefore, the *Rotary Gear Cutting Tool* proposed by Dr. Wildhaber [217] is not workable. However, the concept of the invention is insightful.

## 17.5 CUTTING TOOLS FOR SCUDDING GEARS

The gear scudding process is a continuous cutting operation.[5] The kinematics of this method of gear machining is based on the vector diagram shown in [Figure III-C 1.1(a)]. Two rotation vectors—the rotation vector $\boldsymbol{\omega}_g$ of the work gear, and the rotation vector $\boldsymbol{\omega}_{ct}$ of the gear scudding tool—are at a certain center distance $C$ from one another. The rotation vectors cross at a crossed-axis angle $\Sigma = \angle(\boldsymbol{\omega}_g, \boldsymbol{\omega}_{ct})$. These two rotation vectors $\boldsymbol{\omega}_g$ and $\boldsymbol{\omega}_{ct}$ are the principal motion vectors in the gear scudding process. Thus, the scudding process has many similarities with processes such as gear hobbing, gear shaving, and so forth.

[5] The technology itself is a current slant on an older hard-gear finishing process developed in Germany called *walzschaelan* (translated as "hob-peeling"). The elements of hob-peeling are now being used in a different manner, providing the current green process called *scudding*.

### 17.5.1 Essentials of the Gear Scudding Process

In the scudding of external gears (see Figure 17.9), the cutter feeds directly along the workpiece axis as the cutter and workpiece spin in a synchronized fashion. The feed $\mathbf{F}_{ag}$ is

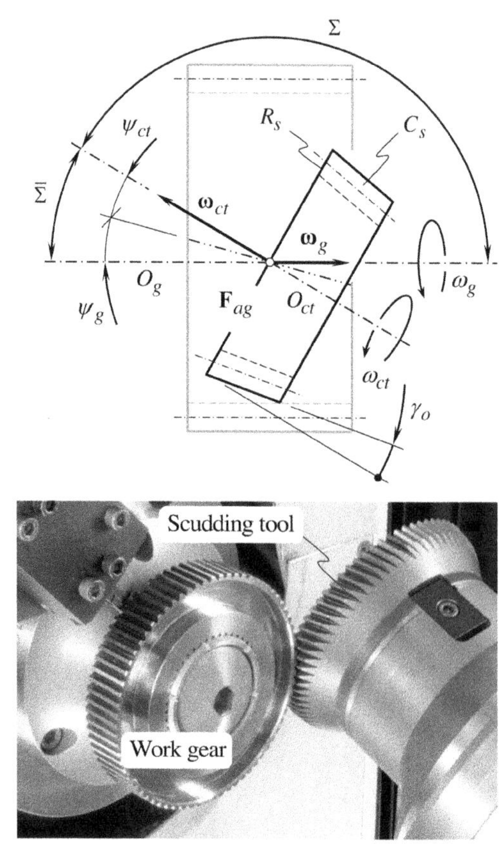


**FIGURE 17.9** Schematic of the gear scudding process. (Image courtesy of American Wera, Inc.)

**FIGURE 17.10** Close-up of scudding process of a ring gear.

directed parallel to the axis of rotation $O_g$ of the work gear. Cutting speed $V_{cut}$ is created as the axes of rotation $O_g$ and $O_{ct}$ of the work gear and the cutter cross at a certain angle $\bar{\Sigma}$ (the corresponding rotation vectors $\omega_g$ and $\omega_{ct}$ cross at angle $\Sigma = 180° - \bar{\Sigma}$).

Internal gears also can be successfully machined by the gear scudding process, as illustrated in Figure 17.10.

A gear-cutting tool of a design similar to a helical shaper cutter is used in the scudding of gears. A gear scudding tool can be viewed as a multiple-start hob with the same number of starts as the cutter has teeth. In other words, the gear scudding tool has only one cutting tooth at every start. Therefore, cutting tools for scudding can be designed in the way multiple-start gear hobs are designed. However, this approach to determining the design parameters of the gear scudding tool is applicable only when gears of very low accuracy need to be machined.

Profiling of gear scudding tools was investigated by Dr. Tsvis [213].

### 17.5.2 Design Concept of a Precision Cutting Tool for Gear Scudding Process

The generating surface of a precision gear scudding tool has to be determined using the auxiliary generating rack $\mathscr{R}_T$ of the enveloping type, as shown in Figure 14.20(b). Moreover, the auxiliary generating rack $\mathscr{R}_T$ must be congruent to the work gear $R_{w.g} \equiv R_{w.\mathscr{R}}$, similar to that of the gear hob in Figure 15.110. This methodology allows for the determination of the design parameters of the gear scudding tool.

As the auxiliary generating rack $\mathscr{R}_T$ of the gear scudding tool has to be congruent to the work gear, the equation of the tooth flank of the involute gear [see Eq. (1.4)] is valid for the analytical description of the tooth flank of the auxiliary surface $\mathscr{R}_T$. Therefore, the equation $\mathbf{r}_{\mathscr{R}} = \mathbf{r}_{\mathscr{R}}(U_{\mathscr{R}}, V_{\mathscr{R}})$ of the auxiliary rack tooth flank is known. This equation is represented in a Cartesian reference system $X_{\mathscr{R}}Y_{\mathscr{R}}Z_{\mathscr{R}}$ associated with the generating rack $\mathscr{R}_T$.

The position vector of a point $\mathbf{r}_{\mathscr{R}}$ of the auxiliary generating surface $\mathscr{R}_T$ must be represented in a Cartesian reference system $X_{ct}Y_{ct}Z_{ct}$ associated with the gear scudding tool. For this purpose, an operator $\mathbf{Rs}(\mathscr{R} \mapsto c)$ of the resultant coordinate system transformation is applied. Ultimately, the expression

$$\mathbf{r}_{\mathscr{R}}^{c} = \mathbf{Rs}(\mathscr{R} \mapsto ct) \cdot \mathbf{r}_{\mathscr{R}}(U_{\mathscr{R}}, V_{\mathscr{R}}) \tag{17.1}$$

can be derived for the position vector of a point $\mathbf{r}_{\mathscr{R}}^{c}$ of the auxiliary generating surface $\mathscr{R}_T$ in the coordinate system $X_{ct}Y_{ct}Z_{ct}$. The expression for the position vector of a point $\mathbf{r}_{\mathscr{R}}^{c}$ of the auxiliary rack $\mathscr{R}_T$ can be finally represented in the form

$$\mathbf{r}_{\mathscr{R}}^{c} = \mathbf{r}_{\mathscr{R}}^{c}(U_{\mathscr{R}}, V_{\mathscr{R}}, \varphi_{sc}) \tag{17.2}$$

In this expression, $\varphi_{sc}$ denotes the enveloping parameter because the operator $\mathbf{Rs}(\mathscr{R} \mapsto ct)$ of the resultant coordinate system transformation is a function of the angle $\varphi_{sc}$.

Equation (17.1) is considered together with the Shishkov equation of contact $\mathbf{n}_{\mathscr{R}} \cdot \mathbf{V}_{\Sigma} = 0$

$$\begin{cases} \mathbf{r}_{\mathscr{R}}^{c} = \mathbf{r}_{\mathscr{R}}^{c}(U_{\mathscr{R}}, V_{\mathscr{R}}, \varphi_{sc}) \\ \mathbf{n}_{\mathscr{R}} \cdot \mathbf{V}_{\Sigma} = 0 \end{cases} \tag{17.3}$$

where:

- $\mathbf{n}_{\mathscr{R}}$ is the unit normal vector to the auxiliary generating surface $\mathscr{R}_T$.
- $\mathbf{V}_{\Sigma}$ is the vector of the resultant motion of the generating rack $\mathscr{R}_T$ relative to the coordinate system $X_{ct}Y_{ct}Z_{ct}$.

After the enveloping parameter $\varphi_{sc}$ is eliminated, the set of Eqs. (17.3) casts into an expression

$$\mathbf{r}_T = \mathbf{r}_T(U_{\mathscr{R}}, V_{\mathscr{R}}) \tag{17.4}$$

for the position vector of a point $\mathbf{r}_T$ of the generating surface $T$ of the gear scudding tool.

A precision gear scudding tool is a special-purpose gear-cutting tool. It is designed for machining a particular work gear. Any alteration to the design parameters of the work gear tooth inevitably entails corresponding alterations to the design parameters of the gear scudding tool.

### 17.5.3 Application of the Gear Scudding Process

The scudding process can be used for a wide range of gear applications. Depending on the tolerances on the machined tooth profile, the scudding process can be used to cut involute gears like sprocket or ring gears, or on non-involute or nonsymmetrical gears, like belt pulleys or straight synchronic gears. It can also hard-finish internal gears with a carbide cutter.

Spur gears ($\psi_g = 0°$) as well as helical gears ($\psi_g \neq 0°$) can be machined by the gear scudding tool. For machining spur gears, cutters with helical teeth $\psi_{ct} \neq 0°$ are used. Helical work gears can also be machined with gear scudding cutters with straight teeth $\psi_{ct} = 0°$.

Gear scudding requires the tooth number of the work gear be in the range of $N_g = 80...100$ or even higher. Scudding of gears with lower tooth numbers cannot be performed currently on machine tools available on the market.

All cutting teeth of the gear scudding tool are loaded evenly. Consequently, the teeth wear evenly. Additionally, the design of the gear scudding tool allows the use of economical high-speed steel or other cutting materials.

The gear scudding process can cut a gear in nearly the same time as the hobbing process.

## 17.6 Shaper Cutter with Tilted Axis of Rotation for Shaping Cylindrical Gears

The rotation of the cutting tool is the prime motion by which cutting speed is created when using most gear-cutting tools. It is not, however, the only type of motion by which the cutting motion can be created; a straight motion of the cutting tool can also be employed.

### 17.6.1 Kinematics of Shaping a Helical Gear with the Straight-Tooth Shaper Cutter

Consider the kinematics of a gear-cutting process which is based on the vector diagram shown in Figure III-C 1.1(a) and featuring a straight cutting motion.

Assume that the cutting speed vector $\mathbf{V}_{cut}$ is at a certain angle $\vartheta_{ct}$ relative to the rotation vector $\boldsymbol{\omega}_g$ of the work gear. Machining of the work gear is possible if this angle is equal, for example, to the pitch helix angle $\psi_g$ of the work gear, that is, $\vartheta_{ct} \cong \psi_g$. Reasonable deviations of the angle $\vartheta_{ct}$ from the pitch helix angle $\psi_g$ are permitted in both directions $\vartheta_{ct} = \psi_g \pm \Delta\vartheta_{ct}$, where the reasonable deviation of the angle $\vartheta_{ct}$ from the pitch helix angle $\psi_g$ is designated as $\Delta\vartheta_{ct}$.

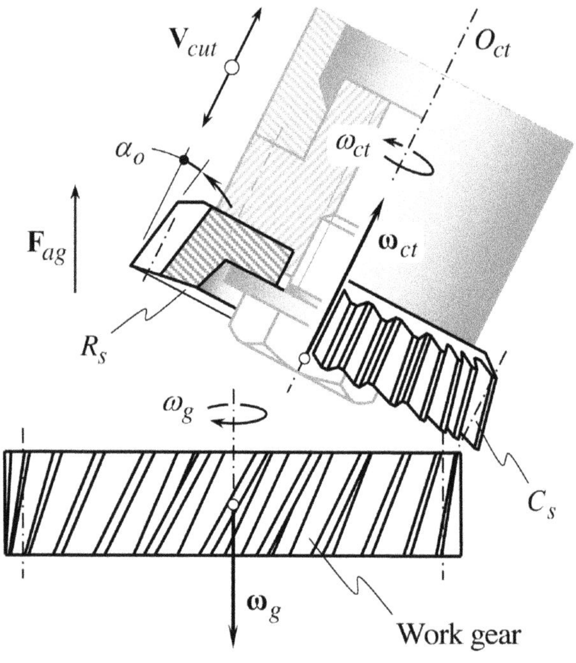


FIGURE 17.11 Schematic of a method of shaping a cylindrical gear. (From Radzevich, S.P., USSR Patent 1504903, Int. Cl. B23f 5/12, Dec. 2, 1987.)

In the scenario shown in Figure III-C 1.1(a), a vector diagram of the gear machining operation should be complemented with the cutting speed vector $\mathbf{V}_{cut}$. The vector $\mathbf{V}_{cut}$ forms the angle $\vartheta_{ct}$ with the axis of rotation $O_g$ of the work gear. Ultimately, the kinematics of the gear machining process appears as shown in Figure 17.11.

In addition, the work gear is fed $\mathbf{F}_{ag}$ in its axial direction. When the pitch helix angle $\psi_g$ of the work gear and the face width of the work gear are reasonably small, the feed motion $\mathbf{F}_{ag}$ can be unimportant.

### 17.6.2 Principal Elements of Design of the Gear-Cutting Tool

The relative motion of the work gear and the reference system in which the gear-cutting tool will be represented can be broken down into their elementary components in various ways. One method is to break down the relative motion so that the auxiliary generating surface is shaped in the form of the rack $\mathscr{R}_T$ conjugate to the work gear. The pitch plane of the auxiliary rack $\mathscr{R}_T$ is parallel to the axis of rotation of the gear-cutting tool [158].

Then the auxiliary generating rack $\mathscr{R}_T$ can be rolled with no sliding over the pitch cylinder of the gear-cutting tool. In such a scenario, the generating surface of the gear-cutting tool $T$ is shaped in a form identical to the generating surface of the standard gear shaper cutter. This means that a standard gear shaper cutter with straight teeth can be used for machining helical gears [158]. Such a possibility is schematically illustrated in Figure 17.9.

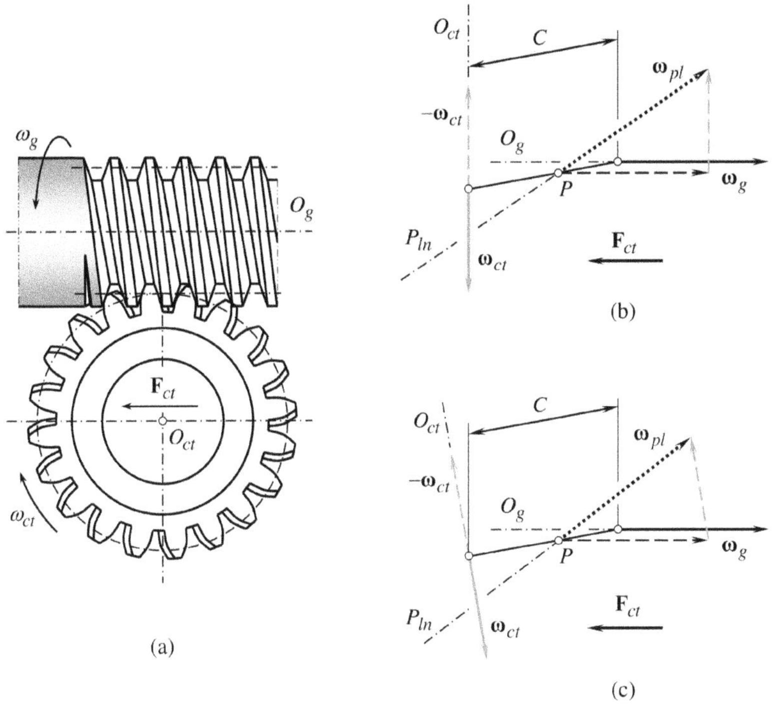


**FIGURE 17.12** Schematic and the principal kinematics of a method of generating cutting of a worm.

### 17.6.3 Possibility of Application of the Gear Shaper Cutter with a Tilted Axis of Rotation

The concept of machining helical gears using a straight-tooth shaper cutter was developed[6] with the goal of equalizing the clearance angles at lateral cutting edges on opposite sides of the tooth profile of the shaper cutter [94]. In this method[7] of gear cutting, the work gear and the shaper cutter rotate about their axes of rotation $O_g$ and $O_{ct}$ (see Figure 17.11) with certain angular velocities $\omega_g$ and $\omega_{ct}$ [94]. The rotations $\omega_g$ and $\omega_{ct}$ are synchronized in a timely manner. The shaper cutter axis of rotation $O_{ct}$ is inclined and forms the clearance angle $\alpha_o$ with the axis of rotation of the work gear $O_g$. The inclination of the axis $O_{ct}$ on $\xi_c = 1\ldots3°$ equalizes the clearance angle on the opposite flanks of the shaper cutter tooth. Such an inclination increases the tool life of the shaper cutter.

Eventually this concept can be enhanced for cutting helical gears with a shaper cutter that has straight teeth and features a tilted axis of rotation.

The method of shaping helical gears with a straight-tooth shaper cutter (see Figure 17.11) requires a machine tool of a special design, but using the method eliminates the need for two elements that are critical when cutting helical gears. Neither a helical guide nor a specially designed shaper cutter are necessary for shaping of helical gears. Additionally, no constraints are imposed on the value of the pitch helix angle. The actual value of the pitch helix angle of the work gear can be assigned any desired (optimal) value [158].

## 17.7 GEAR-CUTTING TOOL FOR MACHINING A WORM IN THE CONTINUOUSLY INDEXING METHOD

The vector diagram of the gear machining mesh (see Figure III-C 1.1) can be utilized for the purposes of cutting worms.

Worms can be cut by generating form cutters as shown in Figure 17.12(a). The use of generating form tools for cutting worms is particularly advantageous when a long worm (or a multiple-start worm) needs to be cut in only one path of the cutting tool in relation to the work.

In machining a worm, the work is rotated about its axis of rotation $O_g$ with a rotation $\omega_g$ as shown in Figure 17.12(a). The generating form cutter is rotating about its axis of rotation $O_{ct}$ with the rotation $\omega_{ct}$. Additionally, the cutter is fed $\mathbf{F}_{ct}$ in the axial direction of the work.

The axes $O_g$ and $O_{ct}$ of the rotations $\omega_g$ and $\omega_{ct}$ are either crossing at a right angle, as shown in Figure 17.12(b), or they are crossing at a certain angle [see Figure 17.12(c)] that is equal, or at least approximately equal, to the helix angle of the worm to be machined.

All the motions—specifically, the rotation of the work $\omega_g$, the rotation of the cutter $\omega_{ct}$, and the feed motion $\mathbf{F}_{ct}$—are synchronized with one another. When the number of starts of the work is denoted as $N_g$, and the number of teeth of the cutter is designated as $N_{ct}$, then the rotation of the cutter

[6] SU Pat. No. 1504903, *A Method of Shaping of Cylindrical Gear*/S.P. Radzevich, Int. Cl. B23f 5/12, Filed: December 2, 1987.

[7] SU Pat. No. 1504903, *A Method of Shaping of Cylindrical Gear*/S.P. Radzevich, Int. Cl. B23f 5/12, Filed: December 2, 1987.

$\omega_{ct}^{g}$ can be expressed in terms of tooth numbers $N_g$ and $N_{ct}$, as well as in terms of the rotation of the work $\omega_g$ by the expression

$$\omega_{ct}^{g} = \omega_g \frac{N_g}{N_{ct}} \quad (17.5)$$

When the cutter travels in the axial direction of the worm with the feed $F_{ct} = |\mathbf{F}_{ct}|$, then the cutter should rotate with the rotation $\omega_{ct}^{f}$ that is equal to

$$\omega_{ct}^{f} = \frac{2F_{ct}}{d_{w.ct}} \quad (17.6)$$

where $d_{w.ct}$ designates the pitch diameter of the cutter.

The resultant rotation of the cutter $\omega_{ct}$ can be calculated from

$$\omega_{ct} = \omega_g \frac{N_g}{N_{ct}} \pm \frac{2F_{ct}}{d_{w.ct}} \quad (17.7)$$

The choice of sign in Eq. (17.7) depends on two parameters: the direction of the hand of the worm to be cut, and the direction of the feed motion $\mathbf{F}_{ct}$.

Both single-start and multiple-start worms, as well as round racks, can be cut in just one path of the cutter regardless of the pitch of the work. A round rack can be viewed here as a worm having a zero number of starts.

The cutting tool and the method of machining are especially efficient when machining long, coarse pitch worms.

## 17.8 RACK SHAVING CUTTERS

The shaving of gears is commonly performed with a disk-type shaving cutter. The productivity of a gear shaving operation depends on the degree of conformity of the generating surface of the shaving cutter to the tooth flank of the work gear. The higher the degree of conformity of the generating surface $T$ to the tooth flank $\mathscr{G}$, the higher the productivity of the gear shaving process, and the higher the accuracy of the finished gears.

Examination of Eq. (16.70) reveals that the degree of conformity of the generating surface of a gear-cutting tool to the tooth surface of the work gear increases as the pitch diameter of the shaving cutter increases. In a gear shaving process, the degree of conformity at point of contact of the generating surface $T$ to the tooth flank $\mathscr{G}$ strongly depends on the tooth number of the shaving cutter.

Assume that the tooth number in a disk-type shaving cutter approaches infinity. Then the disk-type shaving cutter would transform into a rack-type shaving cutter. If an increase in the productivity of a gear shaving process is desired, then a rack-type shaving cutter is of particular interest; it has the greatest possible (infinite) number of teeth.

External spur and helical gears can be shaved by the rack-type shaving cutter.

### 17.8.1 Rack-Type Shaving Cutter

A rack-type shaving cutter can be viewed as a rack that is conjugate to the work gear. The lateral tooth surfaces of the rack are serrated. The rack-type shaving cutter is assembled from a certain number of teeth that are mounted in the shaving cutter body. The serration width is equal to 0.8…1.0 mm, the serration depth is ~1.0 mm, and the serration pitch is 1.6…2.0 mm.

The rack has to have a length and stroke longer than the work gear circumference. The length $L_{ct}$ of the rack-type shaving cutter is calculated from

$$L_{ct} = \frac{\pi m}{\cos\psi_{ct}}(N_g + 2) \quad (17.8)$$

where:

- $m$ is the module.
- $\psi_{ct}$ is the pitch helix angle of the shaving cutter.
- $N_g$ is the tooth number of the work gear.

The length of the shaving cutter calculated from Eq. (17.8) is sufficient for finishing all the teeth of the work gear.

The rack must be wider than the gear face width. In practice, the face width of the rack-type shaving cutter exceeds the face width of the work gear by three to four times. Therefore, different areas of the shaving cutter teeth can be involved in the gear finishing process. This extends tool life between regrindings of the shaving cutter.

It is worth pointing out here that rack-type shaving cutters were proposed before disk-type shaving cutters were invented.

### 17.8.2 Kinematics of the Rack Shaving Process

In the rack shaving process, the work gear is mounted on the arbor and is free to rotate about its axis $O_g$ (see Figure 17.13). The shaving cutter is mounted on the worktable of the shaving machine. It is engaged in mesh with the work gear. The worktable and the shaving cutter are reciprocated with a certain velocity $\pm V_{ct}$. The work gear is fed up toward the work gear after reciprocation of the worktable is complete. The feed of the shaving cutter is in the range of 0.025…0.080 mm per reciprocation. Spur gears are shaved with a shaving cutter with helical (inclined) teeth. The straight-tooth shaving cutter can be used for finishing helical gears.

The vector $\mathbf{V}_{ct}$ of the translation motion of the shaving cutter and the rotation vector $\omega_g$ are at a certain angle $\overline{\Sigma}$ to one another. This angle differs from the right angle $\overline{\Sigma} \neq 90°$. When a spur gear is shaved [see Figure 17.13(a)], the angle $\overline{\Sigma}$ is in the range $\overline{\Sigma} = 20…25°$. For the shaving of helical gears [see Figure 17.13(b)], the work gear pitch helix angle has to be taken into account.

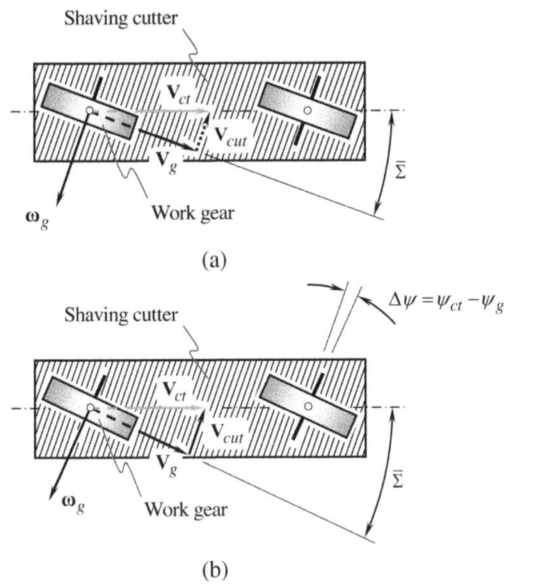


**FIGURE 17.13** Schematic of the rack shaving process of a gear: (a) shaving a spur gear and (b) shaving a helical gear.

Owing to rotation of the work gear, the linear velocity of the rotation is observed. This velocity is designated as

$$V_g = |\mathbf{V}_g| = 0.5 d_g \omega_g \tag{17.9}$$

The speed of the cut is equal to the difference

$$\mathbf{V}_{cut} = \mathbf{V}_{ct} - \mathbf{V}_g \tag{17.10}$$

The *speed-of-the-cut vector* $\mathbf{V}_{cut}$ is always within the tooth flank of the rack-type shaving cutter.

The vector diagram of the rack shaving process is illustrated in Figure 17.14.

As the work rolls with the reciprocating rack, it is also moved across a portion of the rack to equalize the wear of the rack. After every stroke of the rack, the gear is fed in a slight amount.

This method of shaving does not lend itself to shaving large parts. The racks used are quite expensive. The rack shaving process is very rapid, and a large number of parts are obtained per sharpening of the rack. The use of rack shaving is quite economical in high-production jobs. Rack-type shaving cutters have limited application in current industry mostly because of difficulties in their manufacturing and application.

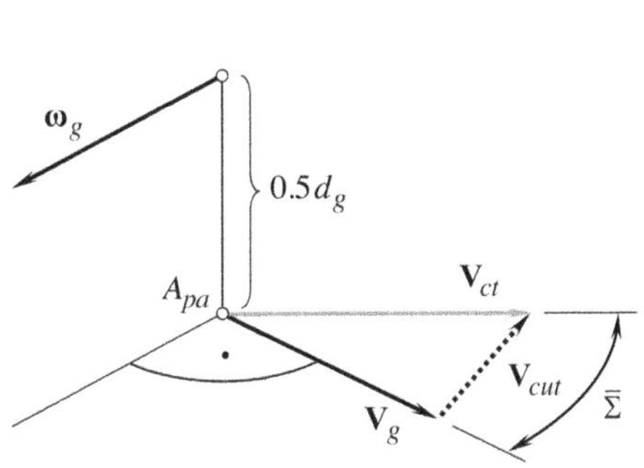


**FIGURE 17.14** Vector diagram of the rack shaving process.

It can be shown that there are no physical restrictions on the application of the vector diagram (see Figure 17.14) in the inverse manner, such as when the work is reciprocated while the cutting tool is rotated. Such a schematic corresponds, for example, to the shaving of a rack by the disk-type shaving cutter. With regard to the shaving process, the inverse application of the vector diagram (see Figure 17.14) is only of theoretical importance.

## 17.9 TOOL FOR BURNISHING GEAR TEETH

The sliding motion that is inherent to two rotations about two crossing axes is used to design a tool for burnishing gears. Figure 17.15 shows an example of a finishing tool[8] designed for burnishing of fillets either of bevel gears or of conical gears with skew teeth [113]. When a tool of this design is used, the work gear and the gear finishing tool are rotated in tight mesh about their axis of rotation. Free-rotating rollers are assembled along the teeth of the generating surface $T$ of the tool. Owing to the tooth sliding, the rollers roll over the fillet surfaces and deform thin surface layers of the work gear. Every set of rollers is shifted at a distance $\Delta R_r$ in the lengthwise direction of the teeth of the generating surface $T$ of the tool to overlap traces of the rollers. After having been shifted, the centers of the adjacent rollers are located within an Archimedean spiral curve.

No feed motion of the burnishing tool is required when using the tool for burnishing conical gears.

Because a lack of space for the rollers is an issue, hunting tooth number for the gear tool (see Figure 17.15) is of critical importance.

A few other designs of the gear finishing tool[9] employ this concept as well [23, 28, 114, 115, 198, 199], and others.

The sliding motion of the tooth flanks of the work gear and the cutting tool can be used for cutting bevel and conical gears, preferably with skew teeth. This concept of the gear-cutting tool is schematically illustrated in Figure 17.16.

Gear-cutting tools of this design have not yet been comprehensively investigated. Depending on the sign and magnitude of the helix of the work gear, and the sign and magnitude of the helix of the gear-cutting tool, the rake surface $R_s$ of the gear-cutting tool is faced either toward the axis of rotation $O_{ct}$ of the cutter (as shown in Figure 17.16), or it is faced in the opposite direction. The direction and magnitude of the offset of the work gear teeth, and the direction and magnitude of the offset of the cutting tool teeth, can be used instead of the signs and magnitudes of the helices.

[8] SU Pat. No. 846024, *A Tool for Burnishing of Conical Gears*/S.P. Radzevich, Int. Cl. B21h 5/04, Filed: November 15, 1978.

[9] SU Pat. No. 1094659, *A Tool for Burnishing of Conical Gears*/S.P. Radzevich, Int. Cl. B21h 5/04, Filed: March 11, 1983.
SU Pat. No. 1107943, *A Tool for Burnishing of Conical Gears*/S.P. Radzevich, Int. Cl. B21h 5/04, Filed: February 24, 1983.

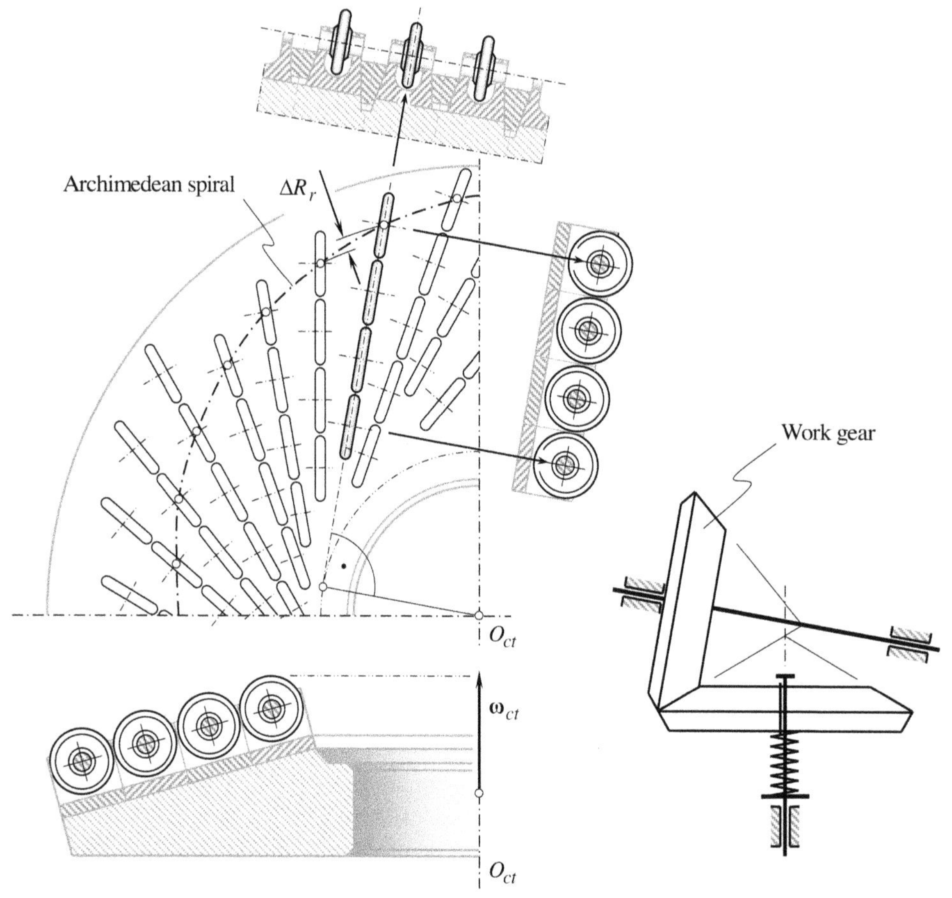


**FIGURE 17.15** Tool for burnishing of conical gears. (From Radzevich, S.P., USSR Patent 846024, Int. Cl. B21h 5/04, Nov. 15, 1978; Radzevich, S.P., USSR Patent 1094659, Int. Cl. B21h 5/04, March 11, 1983; Radzevich, S.P., USSR Patent 1107943, Int. Cl. B21h 5/04, Feb. 24, 1983).

## 17.10 CONICAL HOB FOR THE PALLOID METHOD OF GEAR CUTTING

A conical hob is used for machining conical gears with spiral teeth. The tooth profile of the machined conical gear is shaped in the form of an involute curve. The lengthwise direction of the gear tooth is also shaped in the form of an involute curve. The angle of the spiral of the work gear teeth is determined through the selection of the base circle of this second involute curve.

A conical hob for the *Palloid method of gear cutting* is an approximate type of gear-cutting tool.

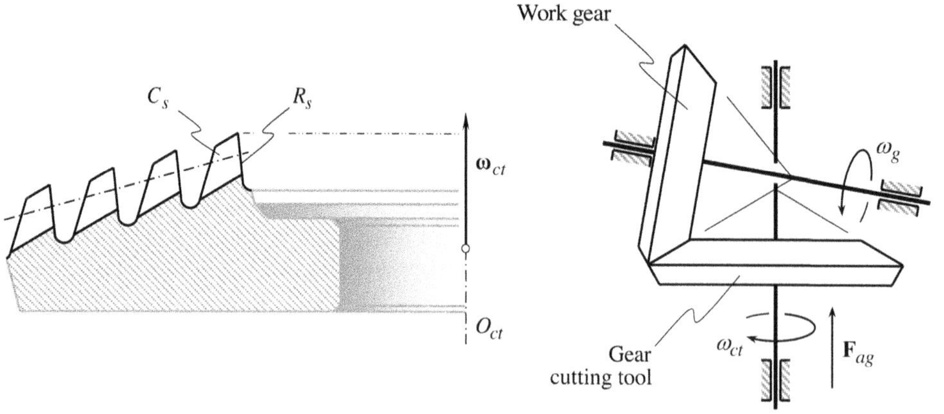


**FIGURE 17.16** Tool for generating conical gears.

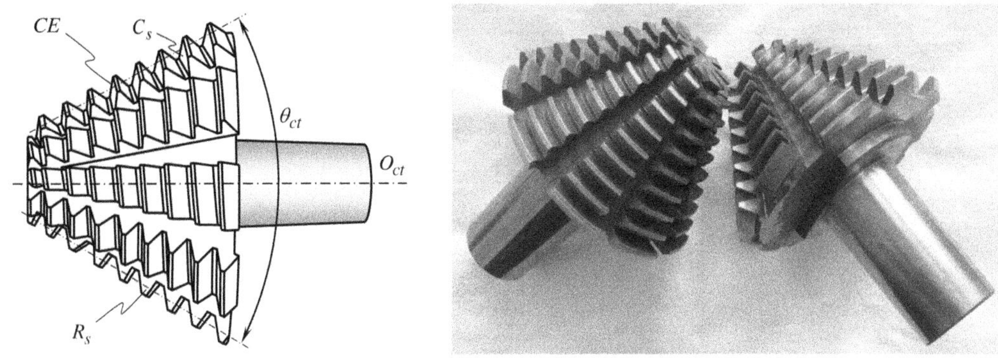


**FIGURE 17.17** Klingelnberg spiral bevel gear hob for cutting gears for Palloid gearing.

### 17.10.1 Preamble

In most cases, the solution to the problem of gear-cutting tool design begins with the specification of the tooth flank of the work gear. The generating surface of the gear-cutting tool to be designed is determined by the tooth flank of the work gear and by the parameters of the kinematics of the gear machining process. This allows for interpretation of the generating surface of the gear-cutting tool as a function of the tooth flank of the work gear and the kinematics of the gear machining operation.

In the Palloid method of gear cutting,[10] a conical hob is given instead of the work gear. The hob is designed independently from the gear to be machined. Afterwards, a gear is cut by the hob, and this cut gear is tested and a decision made about whether or not the machined gear fulfills the performance requirements. If it does, the hob is used for machining gears. If it does not, certain changes are introduced to the design parameters of the hob and a new gear is cut with a redesigned hob. Ultimately, after several runs, an appropriate set of design parameters of the conical hob can be derived. Therefore, in the Palloid method of gear cutting, the generating surface $T$ of the conical hob is not determined as a function of the desired work gear tooth flank, but a reasonable tooth flank of the work gear is determined as a function of the generating surface $T$ of the conical hob with practical values of its design parameters.

Approximate conical hobs are used primarily for two reasons. *First*, the desirable geometry of the tooth flank of the work gear is complex, and moreover it is not known yet.[11] The generating surface of the conical hob cannot be determined until the tooth flank geometry is determined. *Second*, the approximation makes possible the manufacture of the required design.

### 17.10.2 Design of the Conical Hob

A conical hob is illustrated in Figure 17.17 with a 60° included cone angle $\theta_{ct}$ for the basic pitch surface. A handed pair of hobs is needed to cut a pair of gears. The number of gashes is in the range of 8 to 10. The clearance angle at the top cutting edge is $\alpha_o = 4-6°$ at the hob end that has a bigger outer diameter and is $\alpha_o = 15-17°$ at the hob end that has a smaller outer diameter. The pitch helix angle $\psi_{ct}$ varies within the face width of the hob. The maximum value of the pitch helix angle $\psi_{ct} = \psi_{ct}^{\max}$ is observed at the hob end that has a smaller pitch diameter. The minimum value of the pitch helix angle $\psi_{ct} = \psi_{ct}^{\min}$ is observed at the opposite end of the hob. The variation of the pitch helix angle is owing to the difference between pitch diameters at both ends of the hob.

The core design parameters of the conical hob (see Figure 17.17) are similar to those of the conical hob for machining spur and helical gears (see Figure 15.103).

The hob design and manufacture are difficult and costly, which detracts from the usefulness of the system. Owing to the continuous change in the lead angle from one end of the hob to the other with the changing of diameter, a conical end-type grinding wheel is used to grind the hob tooth flanks. The use of an end-type grinding wheel is always more difficult than using a disk grinding wheel.

### 17.10.3 Kinematics of the Palloid Gear Hobbing Process

When machining a conical gear with spiral teeth (see Figure 17.18), the work gear is rotated about its axis of rotation $O_g$. Rotation of the work gear is designated as $\omega_g$. The conical hob with pitch cone angle $\theta_{ct}$ is rotated about its axis of rotation $O_{ct}$. The axes of rotation $O_g$ and $O_{ct}$ are at a certain center distance $C$ apart from each other (see Figure 17.19), and they cross at the crossed-axis angle $\bar{\Sigma}$. The crossing angle $\bar{\Sigma}$ is $\bar{\Sigma} = 180 - \Sigma$, where $\Sigma$ denotes the crossing angle between the rotation vectors $\boldsymbol{\omega}_g$ and $\boldsymbol{\omega}_{ct}$.

The pitch plane of the auxiliary generating surface $\mathscr{R}_T$ is perpendicular to the straight line that is aligned to the centerline $\mathbf{C\!\!\!\!L}$ of the axes of rotation $O_g$ and $O_{ct}$. The generating

[10] The Palloid method of gear hobbing was patented by the engineers Schlicht and Preis in the 1920s. This invention was based on the application of a conical hob. Use of the Palloid method makes it possible to machine a work gear having nearly uniform tooth height along the face width.

[11] So far, the monograph of 2012 [172] is the only source of information on a method to determine the geometry of a tooth flank of a gear like the Palloid gear.

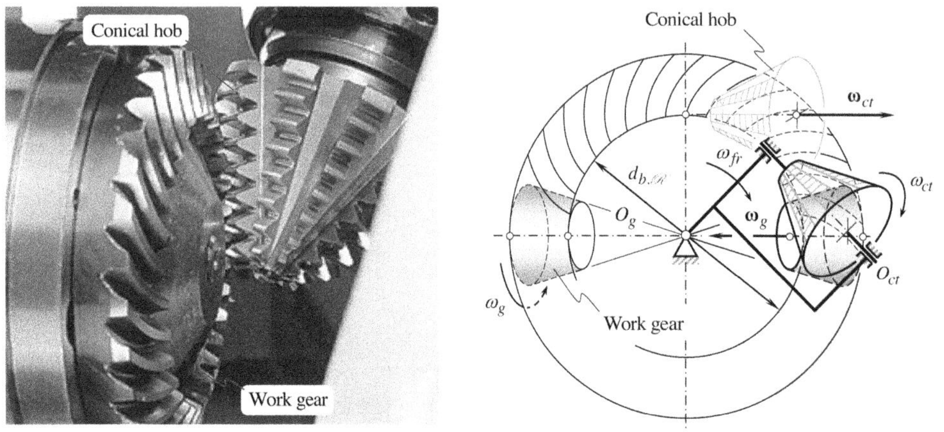


**FIGURE 17.18** Schematic of the Palloid method of hobbing of a conical gear with spiral teeth.

rack $\mathscr{R}_T$ is rotated about the axis of rotation $O_{\mathscr{R}}$ through the apex $A_{pa}$ of the pitch cone of the work gear, and it is perpendicular to the pitch plane of the surface $\mathscr{R}_T$.

In addition to the rotations $\omega_g$ and $\omega_{ct}$, the conical hob is rotated about the axis of rotation $O_{\mathscr{R}}$. This rotation $\omega_{fr}$ is the hob feed motion of the hob. The hob feed motion is a supplementary rotary movement of the imaginary crown wheel. Ultimately, the resultant motion of the hob relative to the work gear is a type of planetary (epicyclical) motion that is composed of three rotations: $\omega_g$, $\omega_{ct}$, and $\omega_{fr}$.

Cutting starts at the large end of the conical hob and finishes at the small end; as generation proceeds, the rotational speed of the hob is progressively increased to provide a more nearly equal peripheral speed for the hob teeth making the finishing cuts. This is achieved due to the rotation vectors $\boldsymbol{\omega}_g$, $\boldsymbol{\omega}_{ct}$, and $\boldsymbol{\omega}_{fr}$ (see Figure 17.19) being synchronized with one another. Because of the tiny slots available for the hob to pass on the small diameter, only very small chip removal magnitudes are achieved.

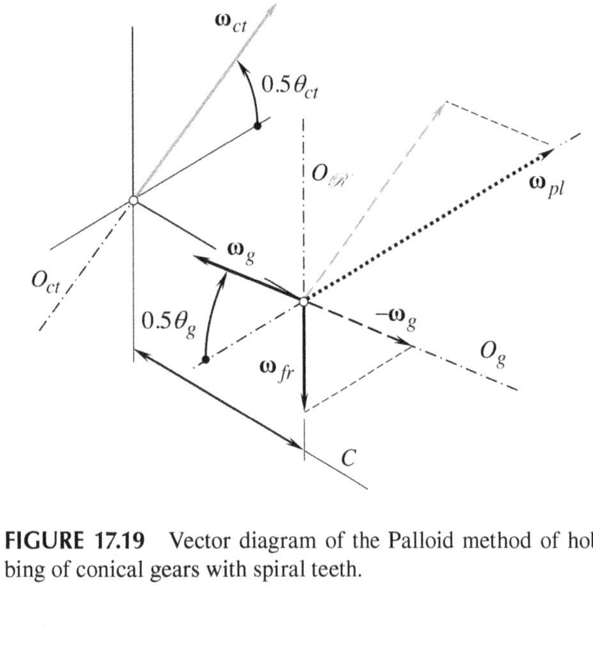


**FIGURE 17.19** Vector diagram of the Palloid method of hobbing of conical gears with spiral teeth.

The tooth profile of the work gear is of involute form in the transverse plane, and this is achieved by planetary feeding of the hob around the $O_{\mathscr{R}}$ – axis. Indexing is continuous as the work gear and hob rotate, and cutting is gradual for all teeth until the full depth of the tooth space has been attained. The depth of the tooth space is constant over the whole face width. The tooth profile from top to root of a portion of the work gear face width is generated by the hob. This is somewhat similar to the hobbing of spur and helical gears by means of tangential feeding.

The tooth spirals are of involute form as shown diagrammatically in Figure 17.18. Consequently, the normal pitch of the teeth is constant[12] for all positions across the face width of a work gear. All gears of a given normal pitch can be cut with a pair of standard hobs of the same pitch. When separate roughing and finishing hobs are utilized, the latter should preferably remove about 0.005 in. (0.12 mm) from each flank in a final cut to produce the best surface finish, taken in the same direction of rotation as that of the blank.

Advantages that even today occasionally lead to the application of the Palloid method are to be found in the involute tooth trace (insensibility to displacements along the centerline), flaky enveloping lines (smooth rolling and oil-pocket effect), and in the fully rounded root region (no grooving effect).

It can be shown based on [172] that no perfect Palloid gearing is possible at all. Perfect Palloid gears cannot be precisely cut by a conical hob. In nature, Palloid gearing is a kind of approximate gearing. Gears for $I_a$ – gearings, as well as for $C_a$ – gearings, can be generated only by face hobbing.

### 17.10.4 Peculiarities of Design of a Conical Hob for Machining Work Gears with Crowned Teeth

Originally, all the teeth of the conical hob have the same profile. This is because the generating surface $T$ of the conical hob is generated by the auxiliary generating rack

[12] This claim is not evident, and is not proven yet (see [172] for details).

$\mathcal{R}_T$ with a standard tooth profile (see also Figure 15.103). However, for the purpose of cutting a conical gear with a crowned tooth flank, the tooth profile of the conical hob is modified. Usually, the modification of the conical hob tooth profile results in the following.

The tooth thickness of the hob is not a constant value. Instead, it alternates within the face width of the hob. Starting from a certain tooth, the tooth thickness grows larger toward both ends of the hob. The thickness of the thinnest tooth is equal to half of the tooth pitch ($0.5\pi m$). This tooth is at a distance of one-third of the face width of the hob end that has a smaller diameter, and it is at a distance of two-thirds of the face width of the larger-diameter hob end. As a result, the thickness of the thickest tooth at the small diameter of the hob is equal to $(0.5\pi + 0.01) \cdot m$. Correspondingly, the thickness of the thickest tooth at the opposite end of the conical hob is equal to $(0.5\pi + 0.02) \cdot m$. Here, the conical hob module is designated as $m$.

The examples discussed clearly illustrate the capabilities of the gear machining mesh shown in Figure III-C 1.1. If explored properly, this vector diagram is capable of returning more novel designs of gear-cutting tools for machining gears.

Only involute gears can be precisely generated in a gear generating process. No gears with other tooth flank geometry can be perfectly generated by the rack cutter—only approximate non-involute gears can be generated. This is attributed to the kinematics of the gear generating process, and to violation of the condition of conjugacy of the tooth flanks $\mathcal{G}$ of a gear to be machined, and of the generating surface $T$ of the gear-cutting tool. In compliance with the condition of conjugacy of the tooth flanks, at every point of contact of the surfaces $\mathcal{G}$ and $T$ the common perpendicular $\mathbf{n}$ must intersect the axis of instant rotation $P_{ln}$ of the gear and the gear-cutting tool in the gear machining mesh [172].

# Part III-C 2

# Quasi-Planar Gear Machining Mesh

Being spatial, the gear machining meshes can feature a particular combination of the rotation vectors of the work gear and the gear-cutting tool for which the vector of instant rotation is perpendicular either to the rotation vector of the work gear or to the rotation vector of the gear-cutting tool. The component (either the work gear or the gear-cutting tool)—the rotation vector of which is perpendicular to the vector of instant rotation—resembles a planar gear machining mesh. For this reason, gear machining meshes that feature the vector of instant rotation perpendicular to one of the rotation vectors (either that of the work gear or of the gear-cutting tool) are referred to as *quasi-planar gear machining meshes*. The perpendicularity of two just mentioned angular velocity vectors[1] (rotation vectors) is a key feature of the gear vector diagrams under consideration.

In a quasi-planar gear machining mesh, the pitch angle of the generating surface of the gear-cutting tool reaches 90°. In such a scenario, the pitch cone transforms to a plane that is rotated about an axis perpendicular to the plane. The greater the pitch cone angle, the higher the degree of conformity of the generating surface of the gear-cutting tool to the tooth flank of the work gear. This statement directly follows from Eq. (16.70). This is one of many reasons why a generating surface in the form of a bevel gear having a 90° pitch cone angle is attractive for the designers of gear-cutting tools.

## KINEMATICS OF QUASI-PLANAR GEAR MACHINING MESH

A quasi-planar gear machining mesh is a particular case of spatial gear machining meshes. Perpendicularity either (1) of the rotation vector $\boldsymbol{\omega}_g$ of the work gear or (2) of the rotation vector $\boldsymbol{\omega}_{ct}$ of the generating surface of the gear-cutting tool to the vector $\boldsymbol{\omega}_{pl}$ of instant rotation is the major feature of quasi-planar gear machining meshes. In the first case, the equality

$$\angle(\boldsymbol{\omega}_g, \boldsymbol{\omega}_{pl}) = 90^\circ \qquad \text{(III-C 2.1)}$$

is observed. Similarly, the equality

$$\angle(\boldsymbol{\omega}_c, \boldsymbol{\omega}_{pl}) = 90^\circ \qquad \text{(III-C 2.2)}$$

is valid in the second case.

Figure III-C 2.1 is a sketch of a vector diagram for a quasi-planar gear machining mesh. In the case shown in Figure III-C 2.1, the cutting tool rotation vector $\boldsymbol{\omega}_{ct}$ is perpendicular to the vector $\boldsymbol{\omega}_{pl}$ of instant rotation ($\boldsymbol{\omega}_{ct} \perp \boldsymbol{\omega}_{pl}$). The pitch cone angle of the generating surface of the gear-cutting tool is equal to 90°, and thus the pitch cone is transformed to the pitch plane that is rotated about the $O_{ct}$ − axis. The axis of rotation $O_{ct}$ is aligned to the rotation vector $\boldsymbol{\omega}_{ct}$. In quasi-planar gear machining meshes, the pitch cone of the work gear rolls over the pitch plane of the generating surface of the gear-cutting tool.

In a way similar to that illustrated in Figure III-C 2.1, a vector diagram for a quasi-planar gear machining mesh can be constructed for the case when the work gear rotation vector $\boldsymbol{\omega}_g$ is perpendicular to the vector $\boldsymbol{\omega}_{pl}$ of instant rotation ($\boldsymbol{\omega}_g \perp \boldsymbol{\omega}_{pl}$).

The vector of instant rotation $\boldsymbol{\omega}_{pl}$ of the cutting tool with respect to the work gear is equal:

$$\boldsymbol{\omega}_{pl} = \boldsymbol{\omega}_{ct} - \boldsymbol{\omega}_g \qquad \text{(III-C 2.3)}$$

If the instant rotation of the work gear relative to the gear-cutting tool is considered instead of the relative motion of the cutting tool, then the vector of instant rotation is equal:

$$\boldsymbol{\omega}_{pl} = \boldsymbol{\omega}_g - \boldsymbol{\omega}_c \qquad \text{(III-C 2.4)}$$

This equality, along with the known property of the dot product of two vectors, allows for the expressions

$$\boldsymbol{\omega}_g \cdot (\boldsymbol{\omega}_g - \boldsymbol{\omega}_{ct}) = 0 \qquad \text{(III-C 2.5)}$$

$$\boldsymbol{\omega}_{ct} \cdot (\boldsymbol{\omega}_{ct} - \boldsymbol{\omega}_g) = 0 \qquad \text{(III-C 2.6)}$$

that are valid for a quasi-planar gear machining mesh.

A gear machining mesh that fulfills either Eq. (III-C 2.5) or Eq. (III-C 2.6) can be identified as a quasi-planar spatial mesh and vice versa; all quasi-planar gear machining meshes fulfill either Eq. (III-C 2.5) or Eq. (III-C 2.6).

A critical value of the crossed-axes angle $\Sigma$ corresponds to quasi-planar gear machining mesh $\Sigma = \Sigma_{cr}$.

[1] ***Reminder:*** Rotations, by nature, are not vectors. Therefore, special care is required when treating rotations as vectors.

DOI: 10.1201/9781003520528-23

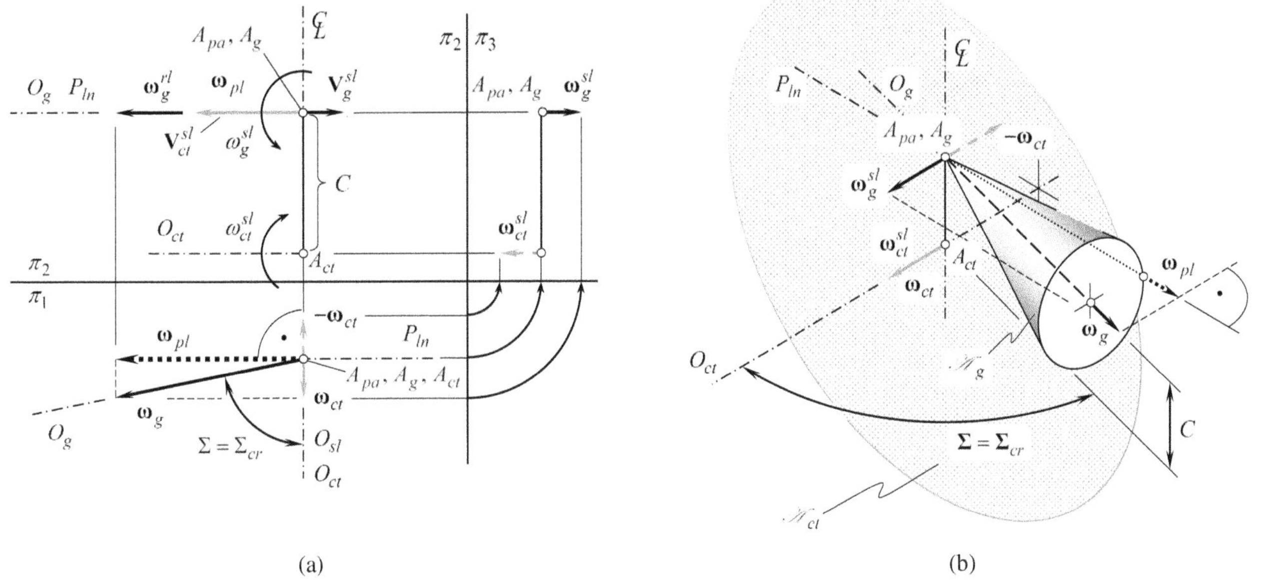


**FIGURE III-C 2.1** Vector diagram of a quasi-planar gear machining mesh, (a) shown in projections onto the planes of projections, and (b) close-up.

Within a plane through the centerline ℄ of the axes of rotation $O_g$ and $O_{ct}$, the vector of linear velocity of sliding $\mathbf{V}_g^{sl}$ of the axodes depends exclusively on the component $\boldsymbol{\omega}_g^{sl}$ of the rotation vector $\boldsymbol{\omega}_g$. Irrespective of whether the component $\boldsymbol{\omega}_{ct}^{sl}$ of the rotation vector $\boldsymbol{\omega}_{ct}$ is not zero ($\boldsymbol{\omega}_{ct}^{sl} \neq 0$), the vector of linear velocity of sliding of the axodes $\mathbf{V}_{ct}^{sl}$ is equal to zero $\mathbf{V}_{ct}^{sl} = 0$. This is because in the case under consideration, the equality

$$r_{w.g} = C \tag{III-C 2.7}$$

is observed.

Therefore, the pitch radius of the generating surface of the cutting tool is zero, $r_{w.ct} = 0$. Ultimately, the resulting sliding is given by $V_\Sigma = V_g^{sl}$.

A clear understanding of the vector diagram of a quasi-planar gear machining mesh (see Figure III-C 2.1) is the key to the successful design of gear-cutting tools based on this type of gear machining meshes, as well as to the design of gear-cutting tools for machining bevel gears in particular.

This section of the book comprises just one chapter, Chapter 18. There is much room for investigation into this area of gear-cutting tool design.

# 18 Gear-Cutting Tools for Machining Bevel Gears

The sliding of axodes is one of the major features of a quasi-planar gear machining mesh. Due to sliding of the axodes over one another, the sliding of the tooth flanks of the work gear and of the generating surface of the gear-cutting tool in gear machining meshes becomes inevitable. The sliding of the tooth flanks is similar to that observed in the gear-shaving process. The tooth flank sliding can be employed as the primary (cutting) motion for a gear-cutting tool; this is not physically prohibited.

## 18.1 DESIGN OF A GEAR-CUTTING TOOL FOR PLUNGE METHOD OF MACHINING OF BEVEL GEARS

Before designing a gear-cutting tool, the reader is advised to further study the kinematics of work gear tooth flank generation.

### 18.1.1 Kinematics

Figure 18.1 is a sketch of the simplest kinematics of generation of a tooth flank of a work gear. The simplest kinematics of work gear tooth flank generation is composed of just two rotations: a rotation of the work gear, and a rotation of the gear-cutting tool. The rotation vector $\boldsymbol{\omega}_g$ of the work gear and the rotation vector $\boldsymbol{\omega}_{ct}$ of generating surface of the gear-cutting tool are separated from each other at a center distance $C$ and are crossed at a certain crossed-axes angle $\Sigma$. The crossing angle $\Sigma_{ct}$ of the rotation vector $\boldsymbol{\omega}_{ct}$ and the vector of the instant rotation $\boldsymbol{\omega}_{pl}$ is a right angle.

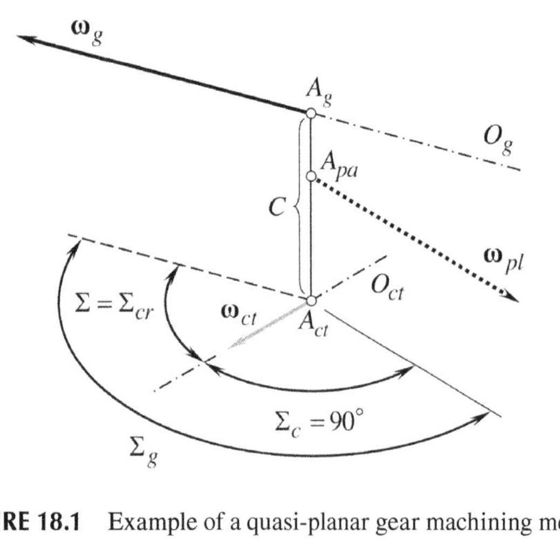


**FIGURE 18.1** Example of a quasi-planar gear machining mesh.

This is because, in a quasi-planar gear machining mesh, the crossed-axis angle $\Sigma$ is equal to its critical value $\Sigma = \Sigma_{cr}$.

Rotation of the work gear and of the gear-cutting tool about crossing axes $O_g$ and $O_{ct}$ makes it possible to utilize sliding velocity as the velocity of the primary (cutting) motion. No additional motion is required for the generation of a tooth flank of the work gear in this particular case of gear cutting. For a proper understanding of how the primary motion $\mathbf{V}_{cut}$ is created in the gear machining mesh, the rotation vectors $\boldsymbol{\omega}_g$, $\boldsymbol{\omega}_{ct}$, and $\boldsymbol{\omega}_{pl}$ are projected onto a plane of projection as shown in Figure 18.2. The projection plane is a plane through the apex point $A_{pa}$. It is perpendicular to the centerline ℄ of the axes of rotation $O_g$ and $O_{ct}$.

At the apex point $A_{pa}$, the linear velocity of a point created by the rotation $\boldsymbol{\omega}_g$ is designated as $\mathbf{V}_g$. The vectors $\boldsymbol{\omega}_g$ and $\mathbf{V}_g$ are perpendicular to one another ($\boldsymbol{\omega}_g \perp \mathbf{V}_g$). Similarly, the linear velocity of a point created by the rotation $\boldsymbol{\omega}_{ct}$ is designated as $\mathbf{V}_{ct}$. Again, the vectors $\boldsymbol{\omega}_{ct}$ and $\mathbf{V}_{ct}$ are perpendicular to each other ($\boldsymbol{\omega}_{ct} \perp \mathbf{V}_{ct}$). Because the rotation axes $O_g$ and $O_{ct}$ are crossed with one another at a certain crossed-axis angle $\Sigma$, the linear velocity vectors $\mathbf{V}_g$ and $\mathbf{V}_{ct}$ form a certain angle. The difference ($\mathbf{V}_{ct} - \mathbf{V}_g$) of the linear velocity vectors is equal to the vector $\mathbf{V}_{ax}^{sl}$ of axial sliding. When the pitch helix angle of the work gear is equal $\psi_g$, then projection of the vector $\mathbf{V}_{ax}^{sl}$ onto the pitch helix through the apex point $A_{pa}$ is equal to the velocity of sliding of the tooth flank of the work gear relative to the tooth flank of the cutting tool. The vector of linear velocity of this component $\mathbf{V}_{cut}$ can be utilized as the vector of primary motion in the gear machining process. For a given work gear, the design parameters (the pitch diameter $d_g$ and pitch helix angle $\psi_g$ at the apex point $A_{pa}$) have certain values. They are not allowed to be altered. The crossed-axis angle $\Sigma = \Sigma_{cr}$ also is not allowed to be changed. Therefore, an increase of the sliding velocity requires an alteration of the design parameters of the gear-cutting tool and the gear machining process that will result in an increase of the sliding vector $\boldsymbol{\omega}_g^{sl}$ (see Figure III-C 2.1), and an increase the distance of the interacting tooth flanks from the apex point $A_{pa}$. This is the only way to keep control over the speed of cut $\mathbf{V}_{cut}$ in the case under consideration. The greater the sliding vector $\boldsymbol{\omega}_g^{sl}$, and the greater the distance of the interacting tooth flanks from the apex point $A_{pa}$, the greater the velocity of cut $\mathbf{V}_{cut}$.

The vector diagram in Figure 18.2 also reveals that the direction of the linear velocity vector $\mathbf{V}_{cut}$ depends on the directions of the rotations $\boldsymbol{\omega}_g$ and $\boldsymbol{\omega}_{ct}$. Reversing the rotations $\boldsymbol{\omega}_g$ and $\boldsymbol{\omega}_{ct}$ results in the vector $\mathbf{V}_{cut}$ of the primary motion also being reversed.


DOI: 10.1201/9781003520528-24

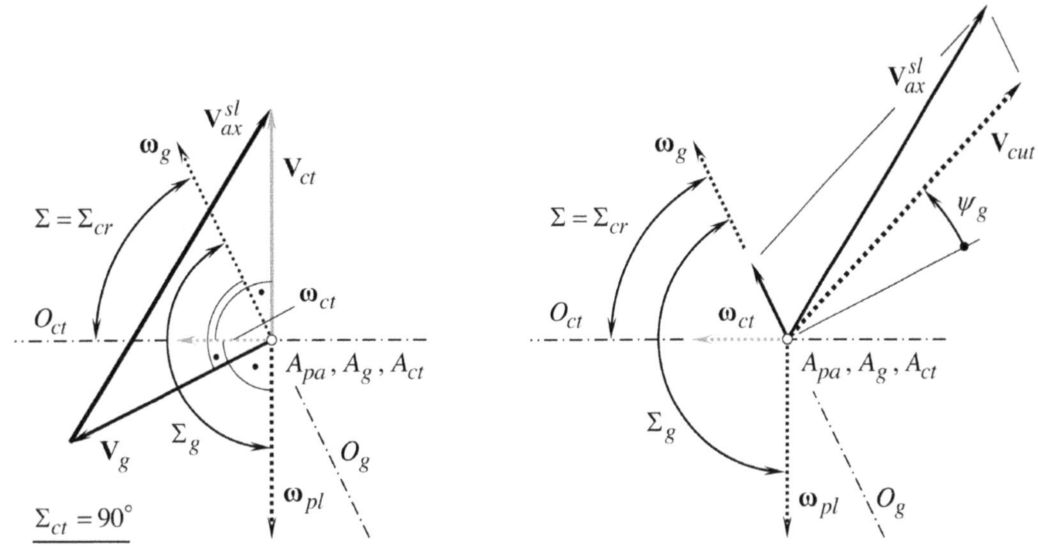


**FIGURE 18.2** Primary (cutting) motion $\mathbf{V}_{cut}$ in the plunge method of machining a bevel gear.

### 18.1.2 Possible Designs of Tools for Machining Bevel Gears

Many designs of gear-cutting/gear finishing tools can be developed based on the simplest kinematics of the work gear tooth flank generation shown in Figure 18.1.

The simple kinematics has been employed in the design of the gear-cutting tool depicted in Figure 18.3. The pitch angle of the gear-cutting tool is given as $\Gamma_{ct} = 90°$. As a result, the pitch cone is transformed to a pitch plane that is perpendicular to the axis of rotation $O_{ct}$.

When machining a bevel gear, the work gear is rotating $\omega_g$ about its axis of rotation $O_g$. The cutting tool is rotating $\omega_{ct}$ about its axis of rotation $O_{ct}$. The rotations $\omega_g$ and $\omega_{ct}$ are synchronized with one another in a timed manner. The gear-cutting tool is fed (plunged) $\mathbf{F}_{ac}$ in its axial direction toward the work gear.

Depending on the directions of the rotations $\boldsymbol{\omega}_g$ and $\boldsymbol{\omega}_{ct}$, the rake surface $R_s$ of the gear-cutting tool teeth should face either toward the axis $O_{ct}$ (as shown in Figure 18.3), or face oppositely, that is, outward from the axis of rotation $O_{ct}$ (not shown in Figure 18.3).

The sliding vector $\boldsymbol{\omega}_g^{sl}$ and the distance from the apex point $A_{pa}$ to the interacting tooth flanks of the work gear and the gear-cutting tool must be sufficient to create a reasonable speed of cut $\mathbf{V}_{cut}$. The gear-cutting tool can also be designed for finishing the work gear flanks. In this latter case, the cutting teeth can be shaped in a form similar to that of the shaving cutter feature. Abrasive finishing of the work gear tooth flanks is also possible.

The possibility of machining work gear tooth flanks so that they do not allow for sliding over themselves is a strong advantage of gear-cutting tools the design of which is based on the concept under consideration. This means that use of the simplest kinematics of work gear tooth flank generation (see Figure 18.1) makes possible the machining not only of approximate gears, but also of gears with correct tooth flank geometry—geometrically accurate gears.

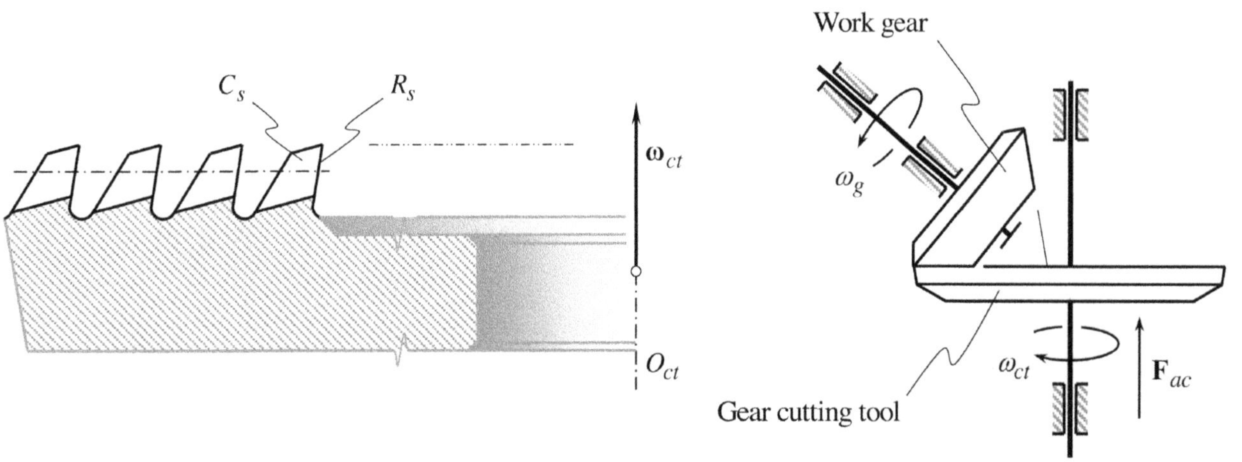


**FIGURE 18.3** Schematic of the plunge method of gear machining.

Tools for burnishing bevel gear fillets[2] similar to those previously developed by the author [113–115] can be designed based on the kinematics illustrated in Figure 18.1.

The simplest kinematics of the work gear tooth flank generation (see Figure 18.1) can be employed in inverse order. Conical gear-cutting tools for machining face gears can be designed on the premise of this kinematics. This opportunity is not discussed in detail here.

## 18.2 FACE HOB FOR CUTTING BEVEL GEAR

Approximate bevel gears can be cut by face hobs. A face hob can be viewed as a bevel hob (see Figure 17.17) featuring a pitch angle $\Gamma_{ct} = 90°$. An example of a face hob for cutting bevel gears is schematically illustrated in Figure 18.4. All teeth of the hob within the face width have the same profile. For this reason, only approximate bevel gears can be cut with a face hob of the design under consideration. A face hob can be designed either as a single-start or as a multiple-start gear-cutting tool. Many practical applications are based on the concept of the face hob shown in Figure 18.4. Some of the applications are briefly discussed below.

Face milling cutters for cutting bevel gears by the *Gleason* method are designed so that the cutter blades cut opposite flanks of the tooth space [219]. The *Oerlikon Spiromatic* spiral-bevel-gear generation process[3] bears considerable similarity to the Gleason method. There are, however, significant differences. The cutter blades are arranged in groups of two or three blades. One group, usually of two blades, cuts one tooth space while the next group follows in the next tooth space and so on until the complete gear is cut [219].

Similar to other methods of spiral-bevel-gear generation, the motions of the face hob, work gear, and machine cradle are such that the face hob sweeps out the imaginary crown wheel tooth flanks with which the work gear is in mesh. The crown wheel is concentric with the axis of rotation of the cradle. Spiral angles range between 30° and 45° but are usually made to lie between 35° and 40° and to give adequate overlap. When quiet running is important, it is customary to design to the larger angles.

The Oerlikon method employs continuous indexing. Generation of the tooth profile is obtained by giving the cutter cradle a slow rolling motion while the work gear is given a corresponding roll so that full involute profiles result. The rolling motion of the work gear is superposed on the indexing rotation through a differential. All the teeth are progressively cut in equal increments as generation proceeds. The Oerlikon gear is cut with a constant space width. All the teeth are finish-cut during one rolling motion of the cradle carrying the face hob.

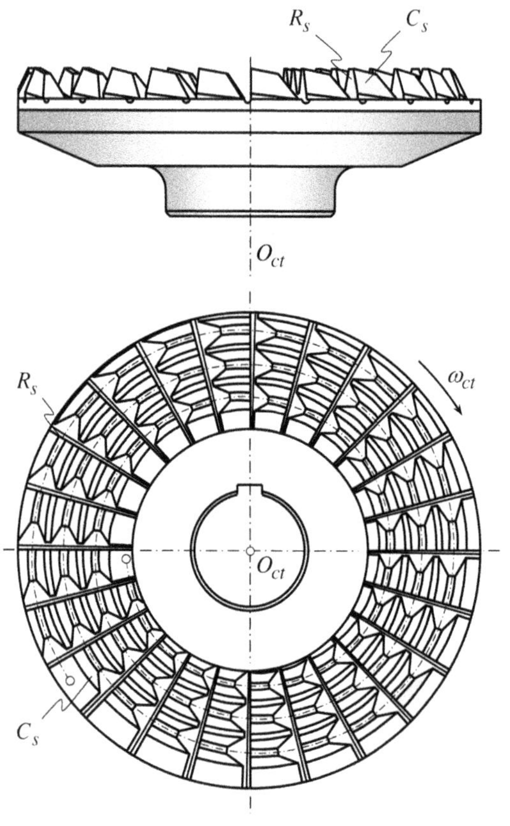


**FIGURE 18.4** Conical hob having pitch angle $\Gamma_{ct} = 90°$ (face hob).

It is often claimed that Oerlikon spiral-bevel and hypoid-type gears are of high quality and give satisfactory performance in service.

The *FIAT* spiral-bevel-gear generation process is another example of the application of a face hob that consists of one complete spiral of cutting blades [219]. Continuous indexing enables the cutter spiral to cut each consecutive tooth space. The tooth profiles are generated by the slow relative rolling between the face hob and the work gear as in the Oerlikon method of generation. The tooth depth is again constant.

There is much room for improvement in the design of cutting tools to generate bevel gears based on a quasi-planar gear machining mesh.

## 18.3 MORE POSSIBILITIES FOR DESIGNING GEAR-CUTTING TOOLS BASED ON A QUASI-PLANAR GEAR MACHINING MESH

The generating surface of a cutting tool for machining hypoid gears does not allow for sliding over itself. If approximate hypoid gears are machined, then it is practical to reproduce the generating surface by means of a surface of revolution. Rotation of the cutting tool in this case is superposed with the corresponding rolling motion. Ultimately,

[2] SU Pat. No. 846024, *A Tool for Burnishing of Bevel Gears*/S.P. Radzevich, Int. Cl. B21h 5/04, Filed: November 15, 1978.
SU Pat. No. 1094659, *A Tool for Burnishing of Bevel Gears*/S.P. Radzevich, Int. Cl. B21h 5/04, Filed: March 11, 1983.
SU Pat. No. 1107943, *A Tool for Burnishing of Bevel Gears*/S.P. Radzevich, Int. Cl. B21h 5/04, Filed: February 24, 1983.

[3] This is based on a method of producing spiral bevel gears developed by Dr. B. Mammano in Italy in 1936 [219].

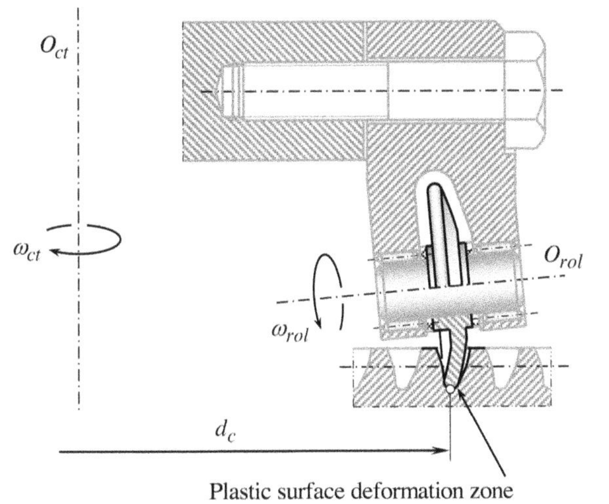


**FIGURE 18.5** Finishing tool for burnishing fillets of a hypoid gear.

either one tooth or one tooth space is machined in a single rolling stroke. The schematic of machining is similar to that of machining bevel gears.

Tools for machining bevel gears can also be designed for burnishing of the gear fillets. Numerous gear finishing tools have been developed [112] for this purpose.

The gear finishing tool[4] consists of a tool body with circumferentially located rollers (see Figure 18.5). The roller units are installed in the tool body in wedge-type slots. When the gear finishing tool is rotating $\omega_{ct}$ about its axis of rotation $O_{ct}$, the rollers are free to rotate $\omega_{rol}$ about their axes $O_{rol}$. The axis of rotation $O_{rol}$ of each roller is at a certain distance $\delta_{rol}$ from the axis of rotation $O_{ct}$ of the gear finishing tool. For the gear finishing tool of a diameter $d_{ct}$, the offset $\delta_{rol}$ is given by $\delta_{rol} = 0.5 d_{ct} \sin \psi_{ct}$, where $\psi_{ct}$ denotes the spiral angle of the work gear teeth.

When finishing hypoid gears, the rollers are pressed into the fillet and burnish thin surface layers of material of the gear.

In addition to gear-cutting tools and tools for burnishing hypoid gears, cutting tools for scudding bevel and hypoid gears can be developed. No limitations are observed for applying the concept of gear finishing tools that feature a continuous *HS*-cutting edge similar to that shown in Figure 17.6.

It is important to note that when the tooth number of the face gear approaches infinity, the schematic of the gear machining process transforms to a cylindrical gear-to-rack mesh. This kinematics of gear machining was discussed earlier (see Figures 17.13 and 17.14).

Only involute gears can be precisely generated in a quasi-planar gear machining mesh. No gears with other tooth flank geometry can be perfectly generated in a quasi-planar gear machining mesh—only approximate non-involute gears can be cut this way. This is attributed to the kinematics of the quasi-planar gear machining mesh and to violation of the condition of conjugacy of the tooth flanks $\mathscr{G}$ of a gear to be machined and of the generating surface $T$ of the gear-cutting tool. In compliance with the condition of conjugacy of the tooth flanks, at every point of contact of the surfaces $\mathscr{G}$ and $T$ the common perpendicular $\mathbf{n}$ must intersect the axis of instant rotation $P_{\text{ln}}$ of the gear and the gear-cutting tool in the gear machining mesh [172]. Otherwise, no geometrically accurate gears can be generated—only approximate gears can be machined instead.

[4] SU Pat. No. 721214, *Tool for Burnishing Bevel Gears*/S.P. Radzevich, Int. Cl. B21h 5/04, Filed: March 1, 1977.

# Part III-C 3

# Internal Gear Machining Mesh

An internal gear machining mesh is employed either when an internal work gear is cut by the external gear-cutting tool or when an external work gear is machined with an internal gear-cutting tool.

The internal gear machining mesh features an acute crossed-axis angle $\Sigma$, the actual value of which is less than the critical value $\Sigma_{cr}$ for this angle ($\Sigma < \Sigma_{cr}$). This means that the inequality $0° < \Sigma < \Sigma_{cr}$ is always observed in an internal gear machining mesh. The smaller the crossed-axis angle $\Sigma$, the higher the degree of conformity of the generating surface of the gear-cutting tool to the tooth flank of the work gear. This statement can be drawn from Eq. (16.70). The use of gear-cutting tools featuring a higher degree of conformity of their generating surface to the tooth flank of the work gear allows for a higher productivity rate. This is one of many reasons why an internal gear machining mesh is useful when designing gear-cutting tools. Another advantage of an internal gear machining mesh is that no waviness is observed on tooth flanks of the machined work gear.

## KINEMATICS OF INTERNAL GEAR MACHINING MESH

An internal gear machining mesh is a particular case of a spatial gear machining mesh. Both the rotation vector of the work gear $\boldsymbol{\omega}_g$ and the rotation vector of the gear-cutting tool $\boldsymbol{\omega}_{ct}$ form an acute angle with the vector of instant rotation $\boldsymbol{\omega}_{pl}$. The inequalities

$$\angle(\boldsymbol{\omega}_g, \boldsymbol{\omega}_{pl}) < 90° \qquad \text{(III-C 3.1)}$$

$$\angle(\boldsymbol{\omega}_{ct}, \boldsymbol{\omega}_{pl}) < 90° \qquad \text{(III-C 3.2)}$$

$$0° < \Sigma < \Sigma_{cr} \qquad \text{(III-C 3.3)}$$

are observed in any and all internal gear machining meshes.

Figure III-C 3.1 is a sketch of a vector diagram for an internal gear machining mesh. The case shown in Figure III-C 3.1 relates to the machining of an external gear by the cutting tool having an internal-type generating surface $T$. The work gear rotation vector $\boldsymbol{\omega}_g$ is at a certain acute angle $\Sigma_g$ in relation to the vector $\boldsymbol{\omega}_{pl}$ of instant rotation ($0° < \Sigma_g < 90°$). The cutting tool rotation vector $\boldsymbol{\omega}_{ct}$ forms an acute angle $\Sigma_c$ with the vector $\boldsymbol{\omega}_{pl}$ of instant rotation ($0° < \Sigma_c < 90°$). In the internal gear machining mesh, the external hyperboloid of revolution (of one sheet) associated with the work gear rolls over the internal hyperboloid of revolution (of one sheet) associated with the generating surface of the gear-cutting tool.[1]

Similar to the illustration in Figure III-C 3.1, a vector diagram for an internal gear machining mesh can be constructed for the case of machining an internal work gear with a cutting tool that has an external generating surface $T$.

The vector of instant rotation $\boldsymbol{\omega}_{pl}$ of the cutting tool relative to the work gear is equal to

$$\boldsymbol{\omega}_{pl} = \boldsymbol{\omega}_{ct} - \boldsymbol{\omega}_g \qquad \text{(III-C 3.4)}$$

If the instant rotation of the work gear relative to the cutting tool is considered instead of the cutting tool relative rotation, then the vector of instant rotation is equal to

$$\boldsymbol{\omega}_{pl} = \boldsymbol{\omega}_g - \boldsymbol{\omega}_{ct} \qquad \text{(III-C 3.5)}$$

These equalities along with the known property of the dot product of two vectors allow for the expressions

$$\boldsymbol{\omega}_g \cdot (\boldsymbol{\omega}_g - \boldsymbol{\omega}_{ct}) > 0 \qquad \text{(III-C 3.6)}$$

$$\boldsymbol{\omega}_{ct} \cdot (\boldsymbol{\omega}_{ct} - \boldsymbol{\omega}_g) > 0 \qquad \text{(III-C 3.7)}$$

which are valid for an internal gear machining mesh.

A gear machining mesh that meets either Eq. (III-C 3.6) or Eq. (III-C 3.7) is an internal spatial mesh, and vice versa. All internal gear machining meshes fulfil either Eq. (III-C 3.6) or Eq. (III-C 3.7).

Within a plane through the centerline ℄ of the axes of rotation $O_g$ and $O_{ct}$, a vector of the linear velocity of sliding $\mathbf{V}^{sl}$ of the axodes is created by two components $\mathbf{V}_g^{sl}$ and $\mathbf{V}_{ct}^{sl}$. The component $\mathbf{V}_g^{sl}$ is created by the component $\boldsymbol{\omega}_g^{sl}$ of the rotation vector $\boldsymbol{\omega}_g$, while the component $\mathbf{V}_{ct}^{sl}$ is created by the component $\boldsymbol{\omega}_{ct}^{sl}$ of the rotation vector $\boldsymbol{\omega}_{ct}$. Regardless of

[1] ***Reminder:*** *Axodes* (hyperboloids of revolution of one sheet) represent examples of obsolete categories, which are not recommended to operate with. Instead, it is recommended to operate with rotation vectors where appropriate.

DOI: 10.1201/9781003520528-25

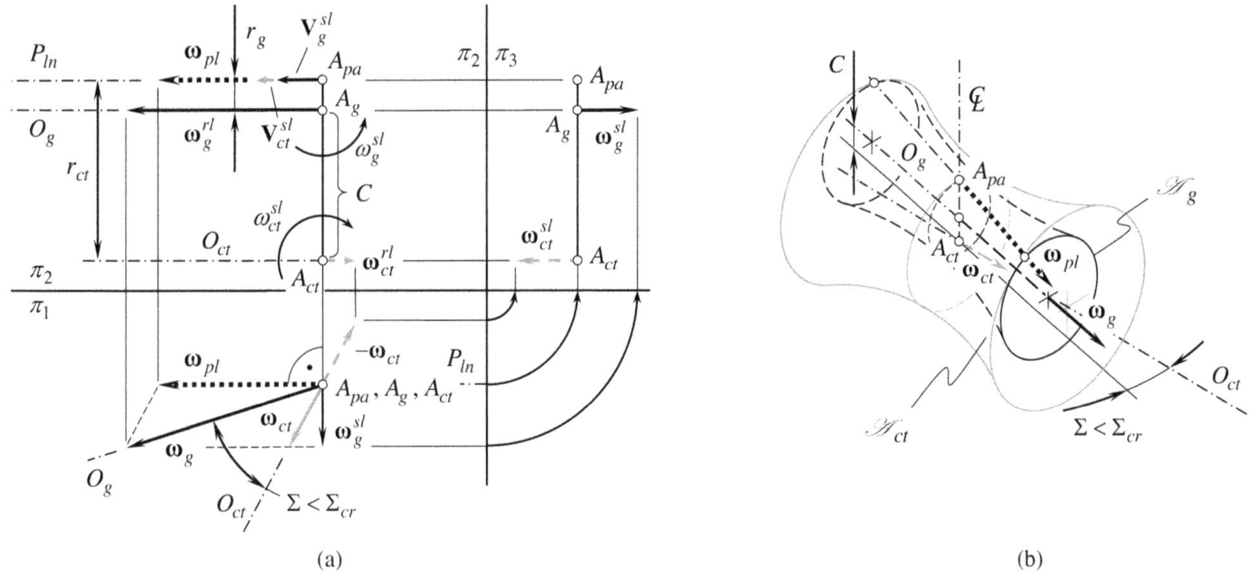


**FIGURE III-C 3.1** Vector diagram of an internal gear machining mesh.

whether the components $\omega_g^{sl}$ and $\omega_{ct}^{sl}$ of the rotation vectors are equal to each other, the components $\mathbf{V}_g^{sl}$ and $\mathbf{V}_{ct}^{sl}$ of the linear velocity of the sliding motion are not equal to each other. The difference $V_\Sigma$ between the components $V_g^{sl}$ and $\mathbf{V}_{ct}^{sl}$ is due to the difference between the distances of the axes of rotation $O_g$ and $O_{ct}$ from the apex point $A_{pa}$.

A clear understanding of the vector diagram of an internal gear machining mesh (see Figure III-C 3.1) is of critical importance when designing gear-cutting tools based on this type of gear meshing.

This section of the book comprises two chapters, Chapter 19 and Chapter 20.

# 19 Gear-Cutting Tools with an Enveloping Generating Surface

The sliding of axodes is one of the major features of internal work gear-to-cutting tool mesh. Because the axodes slide over one another, the sliding of tooth flanks of the work gear and the generating surface of the gear-cutting tool in a gear machining mesh becomes inevitable. The sliding of tooth flanks is similar to that observed in the gear shaving process. The tooth flank sliding can be utilized as the primary (cutting) motion for a gear-cutting tool to be designed as this is not physically prohibited.

## 19.1 GEAR-CUTTING TOOLS WITH A CYLINDRICAL GENERATING SURFACE

The determination of the design parameters of a work gear is the starting point when designing a gear-cutting tool. The design parameters of the work gear are the input, which are of prime importance when designing a high-performance gear-cutting tool.

### 19.1.1 Generating Surface of an Internal Cylindrical Gear-Cutting Tool

The kinematics of machining of a gear when the gear-cutting tool is in internal mesh with the work gear resembles that when the gear-cutting tool is in external mesh with the work gear. The geometry of the generating surface of the gear-cutting tool in both cases is of the same nature. The location of the body of the gear-cutting tool in relation to the generating surface $T$ is the only difference between the cases of the external and the internal meshing of the work gear and the gear-cutting tool.

For a gear-cutting tool that is externally meshed with the work gear (see Figure 14.9), the cutting tool body is located with respect to the generating surface $T$ so that the axis of rotation $O_{ct}$ passes through the bodily side of the cutting tool. In the case of a gear-cutting tool that is internally meshed with the work gear (see Figure 19.1), the cutting tool body is located with respect to the generating surface $T$ so that the axis of rotation $O_{ct}$ passes through the void side of the cutting tool. However, the location of the cutting tool body with respect to the generating surface $T$ does not affect the geometry of the generating surface itself. Therefore, without going into the details of derivation, Eq. (14.63) for the external generating surface $T$

$$\left.\begin{array}{r}\text{Generating surface, } T,\\ \text{of a gear cutting tool}\end{array}\right| \Rightarrow \mathbf{r}_{ct}(U_{ct}, V_{ct})$$

$$= \begin{bmatrix} 0.5 \cdot d_{b.ct} \cdot \sin V_{ct} - U_{ct} \cdot \sin \psi_{b.ct} \cdot \cos V_{ct} \\ 0.5 \cdot d_{b.ct} \cdot \cos V_{ct} + U_{ct} \cdot \sin \psi_{b.ct} \cdot \sin V_{ct} \\ p_{b.ct} \cdot V_{ct} - U_{ct} \cdot \cos \psi_{b.ct} \\ 1 \end{bmatrix} \quad (19.1)$$

is also valid for the case of the internal generating surface of the gear-cutting tool.

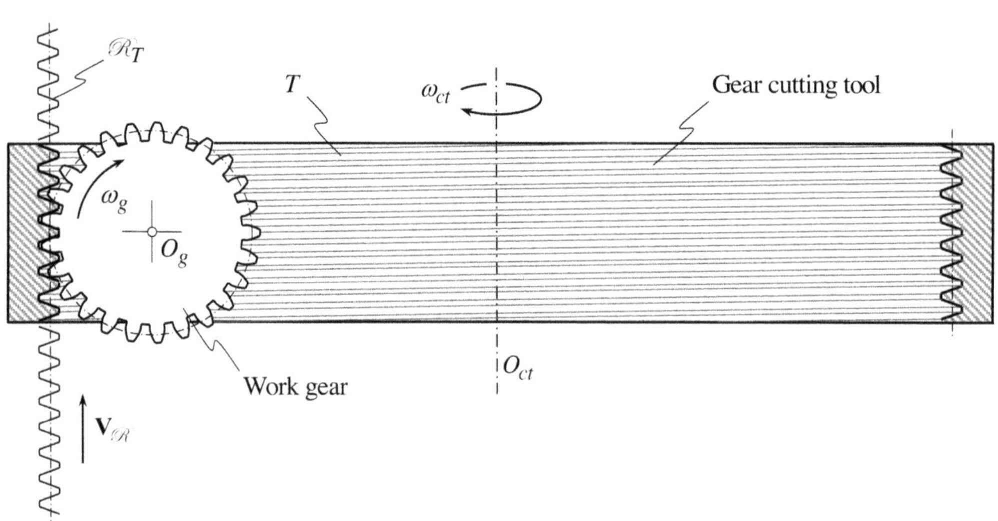


**FIGURE 19.1** Gear-cutting tool with a cylindrical enveloping generating surface $T$.


DOI: 10.1201/9781003520528-26

For an internal gear-cutting tool intended for machining of spur gears, the preferred design is with zero setting angle $\zeta_{ct} = 0°$ (however, this requirement $\zeta_{ct} = 0°$ is not a must). When helical gears are machined, then it is recommended that the equality $\zeta_{ct} = \psi_g$ be fulfilled (although this also is not a must). In such a scenario, the work gear axis of rotation $O_g$ and the cutting tool axis of rotation $O_{ct}$ are crossed with each other at right angles in both cases.

### 19.1.2 Solution to the Inverse Problem of Part Surface Generation

When a gear is machined by the gear-cutting tool having an internal generating surface $T$, no relative motion of the work gear and/or the gear-cutting tool in the axial direction of the work gear is permissible. Absence of the relative axial motion causes a violation of the conditions of proper part surface generation [133, 136, 145, 148, 158, 168] (see also Appendix E for details on the set of necessary and sufficient conditions of proper part surface generation). The violation of the conditions of proper part surface generation is inevitable in the case of the application of an internal gear-cutting tool for machining a work gear.

For the determination of whether or not a condition of proper part surface generation is violated, the inverse problem of part surface generation is commonly considered.

When seeking a solution to the inverse problem of part surface generation for machining a work gear with an internal gear-cutting tool, consider the generating surface $T$ that performs the motions relative to a reference system associated with the work gear. The motions are those that the cutting tool is performing relative to the work gear when machining the gear.

Consider a left-hand-oriented Cartesian coordinate system $X_gY_gZ_g$ associated with the work gear. This reference system is rotating with the work gear when the work gear is machined. Imagine that a left-hand oriented Cartesian coordinate system $X_{ct}Y_{ct}Z_{ct}$ is embedded in the gear-cutting tool. When the cutting tool is rotating, the reference system $X_{ct}Y_{ct}Z_{ct}$ is rotating together with the cutting tool. Once the reference systems $X_gY_gZ_g$ and $X_{ct}Y_{ct}Z_{ct}$ are chosen, this makes it possible to compose the operator $\mathbf{Rs}(ct \mapsto g)$ of the resultant coordinate transformation (see Appendix D). The operator $\mathbf{Rs}(ct \mapsto g)$ allows for representation of the generating surface $T$ of the gear-cutting tool in the reference system $X_gY_gZ_g$, associated with the work gear at every instant of time. Therefore, once the generating surface of the gear-cutting tool is represented in the form $\mathbf{r}_{ct} = \mathbf{r}_{ct}(U_{ct}, V_{ct})$ [see Eq. (19.1)], the position vector of a point $\mathbf{r}_{ct}^g$ of the generating surface $T$ in its current configuration relative to the work gear can be calculated from the equation

$$\mathbf{r}_{ct}^g(U_{ct}, V_{ct}, \varphi_{ct}) = \mathbf{Rs}(ct \mapsto g) \cdot \mathbf{r}_{ct}(U_{ct}, V_{ct}) \quad (19.2)$$

In addition to the curvilinear parameters $U_{ct}$ and $V_{ct}$, the position vector of a point $\mathbf{r}_{ct}^g$ of the generating surface $T$ in the gear reference system $X_gY_gZ_g$ also depends on the enveloping parameter. In the case of Eq. (19.2), the angle of rotation $\varphi_{ct}$ of the gear-cutting tool serves as the enveloping parameter solely for the purposes of convenience.

Equation (19.2), considered together with the Shishkov equation of contact $\mathbf{n} \cdot \mathbf{V} = 0$ (Prof. *V.A.* Shishkov, 1948), allows for an equation of the actually machined tooth flank $\mathcal{G}^a$ of the work gear. The Shishkov equation of contact $\mathbf{n} \cdot \mathbf{V} = 0$ is helpful for the purpose of eliminating the enveloping parameter $\varphi_{ct}$ from Eq. (19.2). Ultimately, the position vector of a point $\mathbf{r}_g^a$ of the actually machined tooth flank $\mathcal{G}^a$ of the work gear can be represented in the form

$$\mathbf{r}_g^a = \mathbf{r}_g^a(U_{ct}, V_{ct}) \quad (19.3)$$

At this point, the actually machined tooth flank of the work gear [see Eq. (19.3)] can be compared with the desirable tooth flank given by Eq. (1.4). To make a consistent comparison, the $U_{ct}$ – and $V_{ct}$ – parameters in Eq. (19.3) have to be compatible with the corresponding $U_g$ – and $V_g$ – parameters in Eq. (1.4). Depending on the kind of parameterization of the desired gear tooth flank [see Eq. (1.4)], and of the actually machined tooth flank [see Eq. (19.3)], it may be necessary to change the surface parameters (see Appendix C).

After completion, the comparison reveals that the actual tooth flank $\mathcal{G}^a$ [see Eq. (19.3)] differs from the desired tooth flank $\mathcal{G}$ [see Eq. (1.4)]. The difference is because of violation of the conditions of proper part surface generation caused by the absence of the axial relative motion of the work gear and of the gear-cutting tool in the axial direction of the work gear.

If the difference between the actual tooth flank $\mathcal{G}^a$ and the desirable tooth flank $\mathcal{G}$ of the work gear is within a reasonable range, then the gear-cutting tool can be used to machine the gears.

In practical terms, the inevitable violation of the conditions of proper part surface generation using enveloping gear-cutting tools has disadvantages as well as advantages. Approximate generation of the work gear tooth flank is the major disadvantage of using enveloping gear-cutting tools. However, if the pitch diameter of the enveloping gear-cutting tool is large enough, then the deviation of the actual tooth flank $\mathcal{G}^a$ from the desired tooth flank $\mathcal{G}$ can be negligibly small. In this case, the enveloping gear-cutting tool can be used for machining gears.

A large pitch diameter of the enveloping gear-cutting tool also has advantages and disadvantages. Although it is inconvenient to use gear-cutting tools of large diameters, most work gears can be machined simultaneously with them, as illustrated in Figure 19.2. In the case under consideration, every work gear is rotating $\omega_g$ about its axis of rotation. The gear-cutting tool is rotating $\omega_{ct}$ about the axis of rotation $O_{ct}$. Axes of rotation of the work gears are crossed with the gear-cutting tool axis $O_{ct}$ at a right angle. This is possible when the equality $\zeta_{ct} = \psi_g$ is observed. While rotating, every work gear travels in the axial direction of

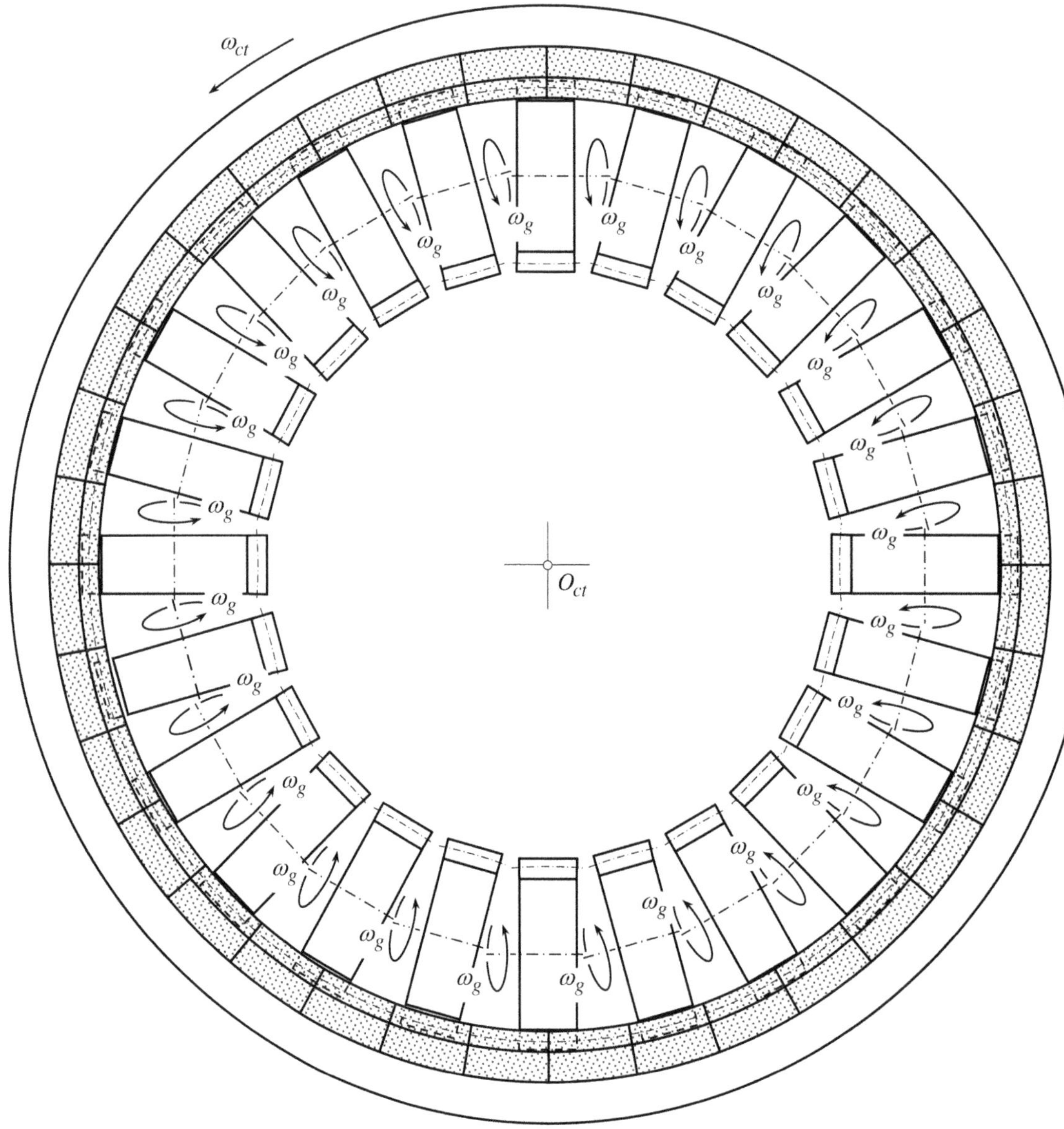


**FIGURE 19.2** Machining of multiple work gears simultaneously with an enveloping gear-cutting tool.

the gear-cutting tool $O_{ct}$. The rotations $\omega_g$ and $\omega_{ct}$ as well as the straight travel motion are synchronized with one another. It is convenient to load the work gears from one end of the gear-cutting tool and unload the machined gears from the opposite end. A gear-cutting tool designed on the principle illustrated in Figure 19.1 and Figure 19.2 could be practical in the high-volume production of gears.

### 19.1.3 Examples of Gear-Cutting Tools with an Enveloping Cylindrical Generating Surface

A vector representation of the kinematics of a gear machining process with a cylindrical enveloping gear-cutting tool (see Figure 19.1) is illustrated in Figure 19.3. The rotation vector of the work gear $\boldsymbol{\omega}_g$ and the rotation vector of the gear-cutting tool $\boldsymbol{\omega}_{ct}$ are pointed along the work gear axis of rotation $O_g$ and the cutting tool axis of rotation $O_{ct}$, respectively. The ratio $|\boldsymbol{\omega}_g|/|\boldsymbol{\omega}_{ct}|$ depends on tooth numbers $N_g$ and $N_{ct}$.

The axes of rotation $O_g$ and $O_{ct}$ are positioned at a distance $C$ from one another, and they are crossed at the crossed-axes angle $\Sigma$. The rotations $\omega_g$ and $\omega_{ct}$ are synchronized with each other.

The work gear and the gear-cutting tool travel $\mathbf{F}_{ct}$ relative to one another in the axial direction of the gear-cutting tool. The feed motion $\mathbf{F}_{ct}$ affects the timing of the rotations $\boldsymbol{\omega}_g$ and $\boldsymbol{\omega}_{ct}$. Ultimately, all three motions, $\boldsymbol{\omega}_g$, $\boldsymbol{\omega}_{ct}$, and $\mathbf{F}_{ct}$, have to be properly timed. No relative motion of the work gear and the gear-cutting tool is observed in the axial direction of the work gear.

Because no relative motion of the gear-cutting tool in the axial direction of the work gear is observed, then enveloping gear-cutting tools featuring cylindrical generating

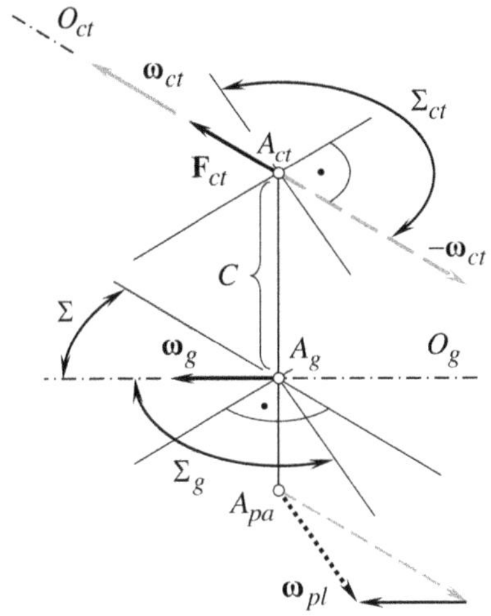


**FIGURE 19.3** Vector representation of the kinematics of a gear machining process with a cylindrical enveloping gear-cutting tool.

surfaces are perfectly suited for machining gears with crowned teeth. An example of a gear tooth with lengthwise flank modification is schematically depicted in Figure 19.4. The tooth flank modification results in the actual gear tooth flank deviating from the theoretical form. The modification is performed on a precalculated value $\delta$.

Numerous designs of gear-cutting tools can be developed based on the concept of an enveloping cylindrical generating surface $T$, similar to those considered in Chapter 14 for a conventional external generating surface. The possible types of tools for machining gears include, but are not limited to[2]

- enveloping hobs,
- enveloping shaving cutters,
- enveloping work gear grinding wheels, and
- gear broaching tools

[63, 77, 78, 81], and others.

A special method of hob relieving[3] has been developed for relieving enveloping hobs [188].

In addition to the machining of gear tooth flanks $\mathscr{G}$, enveloping gear-cutting tools can also be applied to machining chamfered gears. The design of a gear tooth with chamfers is schematically depicted in Figure 19.5.

For machining chamfered gears, multiple work gears are set up within the interior of the enveloping work gear as shown in Figure 19.6. The corresponding face of each work gear is displaced from the axis of rotation $O_{ct}$ of the enveloping gear-cutting tool at a certain distance $H$. The actual value of the displacement $H$ depends on the desirable parameters of the chamfers. When machining chamfered gears, the enveloping gear-cutting tool is rotating $\omega_{ct}$ about its axis of rotation $O_{ct}$. Every work gear is rotating $\omega_g$ about its axis $O_g$. Additionally, all work gears travel along the axis of rotation $O_{ct}$ of the gear-cutting tool.

The face width of the enveloping gear-cutting tool can be expressed in terms of the design parameters of the work gear along with the way the rotations $\omega_g$, $\omega_{ct}$, and the translation $F_{ct}$ are timed with one another. Under any circumstances, the face width of the gear-cutting tool has to be sufficient for machining all the gear teeth in a single path of the work gear through the gear-cutting tool.

It is easy to see that the productivity rate of gear chamfering by the enveloping gear-cutting tool (see Figure 19.6) will be extremely high.

It is important to note that enveloping gear-cutting tools featuring cylindrical generating surfaces can also be used for designing

- enveloping worm grinding wheels,
- gear-cutting tools for scudding gears,
- finishing gear-cutting tools featuring a continuous $HS$-cutting edge (similar to that shown in Figure 17.6),

as well as tools for burnishing tooth flanks of gears, and so forth.

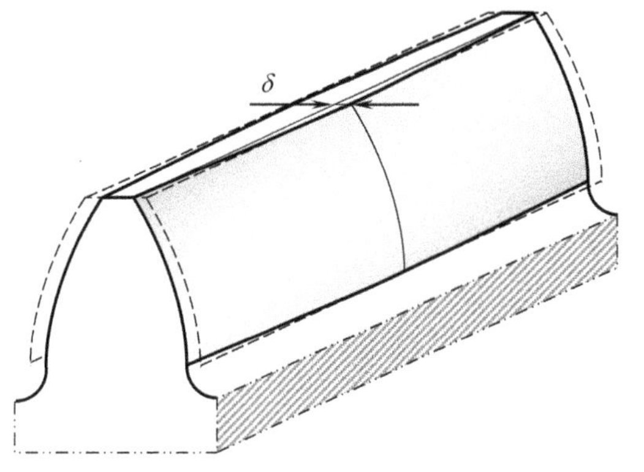


**FIGURE 19.4** Work gear tooth featuring lengthwise modification.

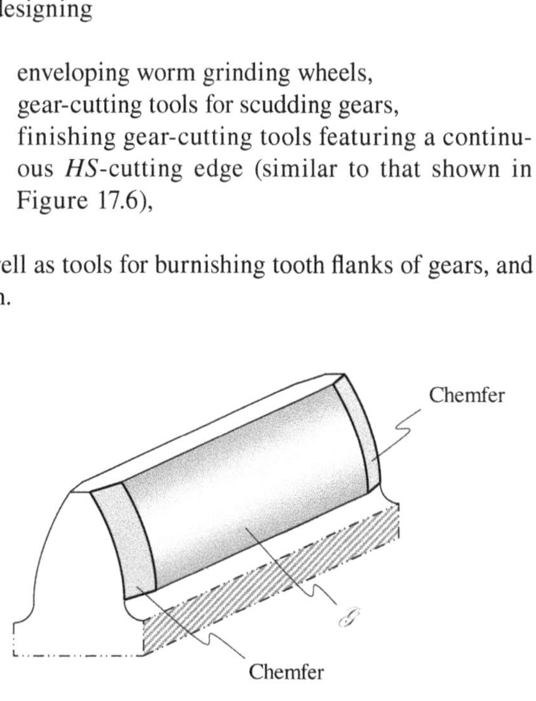


**FIGURE 19.5** Gear tooth featuring chamfers.

---

[2] SU Pat. No. 921727, *A Machine Tool for Finishing a Crowned Gear*/S.P. Radzevich, Int. Cl. B23F 19/00, Filed: September 26, 1980.
SU Pat. No. 984744, *A Machine Tool for Finishing a Crowned Gear*/ S.P. Radzevich, Int. Cl. B23F 19/00, Filed: April 16, 1981.
SU Pat. No. 1110566, *A Machine Tool for Finishing a Crowned Gear*/ S.P. Radzevich, Int. Cl. B23F 19/00, Filed: June 7, 1983.

[3] SU Pat. No. 1196232, *A Method for Relieving an Enveloping Hob*/S.P. Radzevich, Int. Cl. B24b 3/00, Filed: August 3, 1984.

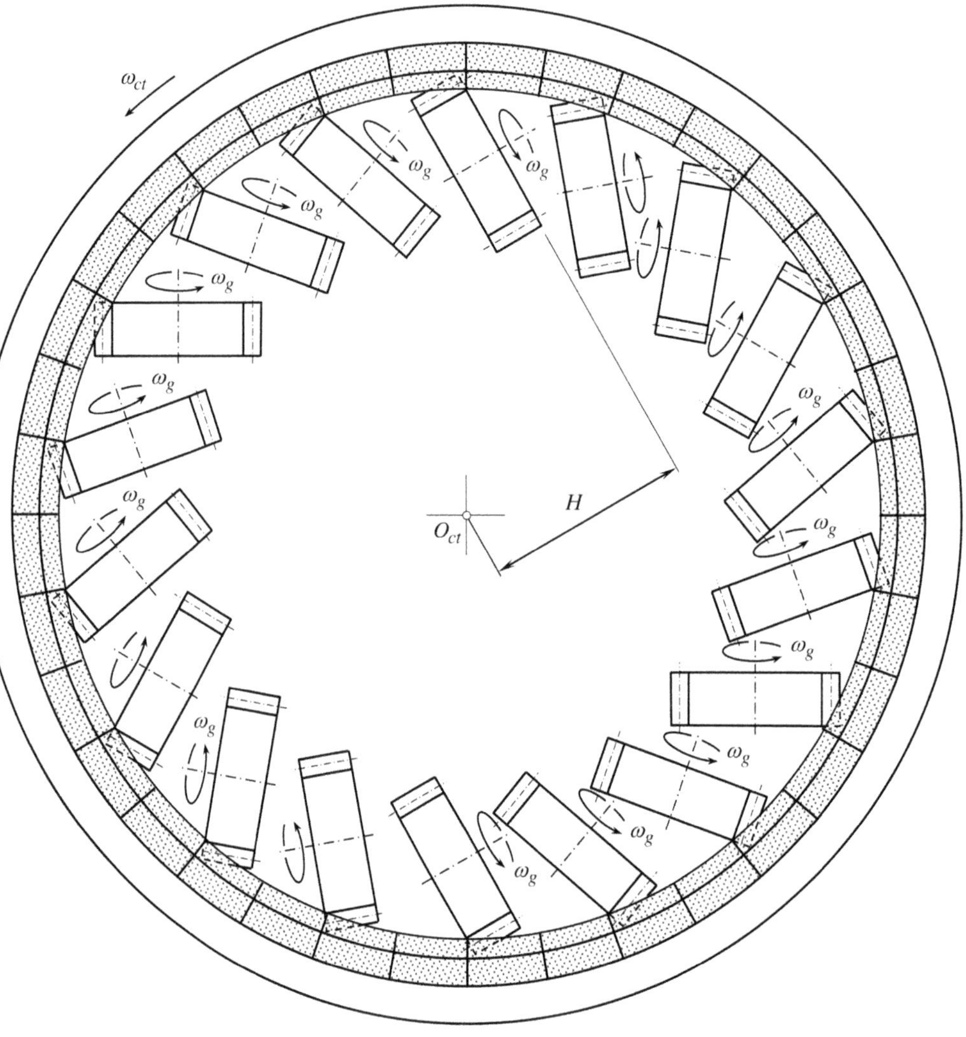


**FIGURE 19.6** Schematic of the chamfering process of spur and helical gears.

Pitch diameter of the enveloping gear-cutting tool increases as the cutting tool tooth number increases. Ultimately, when tooth number approaches infinity, the enveloping gear-cutting tool transfers into a rack similar to that discussed in Chapter 17 (see Figure 17.13).

## 19.2 GEAR-CUTTING TOOLS WITH A CONICAL GENERATING SURFACE

An enveloping conical generating surface of a gear-cutting tool can be created in a manner similar to the manner in which the external conical generating surface of a gear-cutting tool is created (see Figure 14.14) by means of a tilted auxiliary generating surface $\mathscr{R}_T$ [see Figure 14.5(a)].

### 19.2.1 Generating Surface of an Enveloping Conical Gear-Cutting Tool

The generating surface $T$ of an enveloping conical gear-cutting tool can be created as the enveloping surface to a family of consecutive positions of the auxiliary generating rack $\mathscr{R}_T$. For this purpose, a screw motion of the auxiliary rack $\mathscr{R}_T$ having the axis $O_{ct}$ of rotation of the gear-cutting tool as the axis of the screw motion is considered.

The rotation vector of the work gear $\boldsymbol{\omega}_g$ and the rotation vector of the gear-cutting tool $\boldsymbol{\omega}_{ct}$ are pointed along the work gear axis $O_g$ and the cutting tool axis of rotation $O_{ct}$, respectively. The axes of rotation $O_g$ and $O_{ct}$ are positioned at a distance $C$ from each other, and they are crossed at the crossed-axis angle $\Sigma$ (see Figure 19.7). The rotations $\omega_g$ and $\omega_{ct}$ are synchronized with each other.

In the case of a conical gear-cutting tool, the pitch plane of the auxiliary generating rack $\mathscr{R}_T$ is tilted through a certain angle $\gamma_{ct}$ relative to the axis of rotation $O_{ct}$. Translation $\mathbf{V}_{\mathscr{R}}$ of the auxiliary rack is performed at that same angle $\gamma_{ct}$ relative to the axis $O_{ct}$, as shown in Figure 19.7. When machining a gear with the conical enveloping gear-cutting tool, the feed motion $\mathbf{F}_{ct}$ is performed either in the direction of the translation $\mathbf{V}_{\mathscr{R}}$ or in the direction opposite to the linear velocity vector $\mathbf{V}_{\mathscr{R}}$.

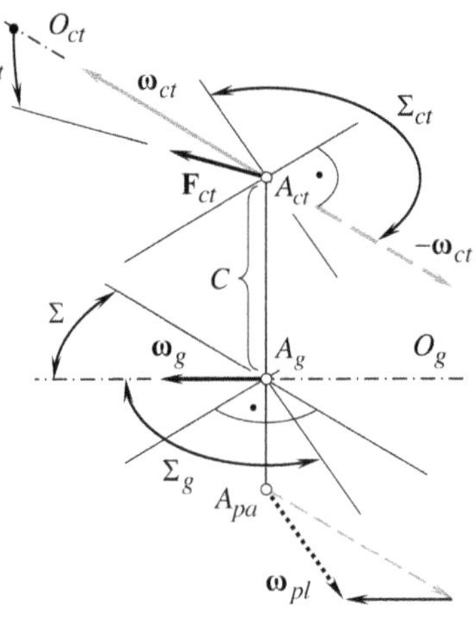


**FIGURE 19.7** Vector representation of the kinematics of a gear machining process with a conical enveloping gear-cutting tool.

The work gear and the gear-cutting tool travel $\mathbf{F}_{ct}$ in relation to each other in the direction that forms an angle $\gamma_{ct}$ relative to the axis of rotation $O_{ct}$ of the gear-cutting tool. The feed motion $\mathbf{F}_{ct}$ affects the timing of the rotations $\boldsymbol{\omega}_g$ and $\boldsymbol{\omega}_{ct}$. All three motions must be properly timed with one another.

The equation of the generating surface of an enveloping conical gear-cutting tool can be obtained immediately because of the similarity of generation of the enveloping conical generating surface $T$ to that of the external generating surface (see Chapter 14). The equation for the surface $T$ is identical to Eq. (19.1) except for one parameter. The base diameter $d_{b.ct}$ for the enveloping conical generating surface $T$ is calculated from Eq. (14.71):

$$d_{b.ct} = \frac{m \cdot N_{ct} \cdot \cos(\phi_n \pm \gamma_{ct})}{\sqrt{1 - \cos^2(\phi_n \pm \gamma_{ct}) \cdot \cos^2 \zeta_{ct}}} \tag{19.4}$$

The sign + here is valid for a screw involute surface of one side of the tooth profile of the generating surface $T$ (see Figure 19.8), while the sign – is valid for the opposite side of the tooth profile.

### 19.2.2 Examples of Gear-Cutting Tools with an Enveloping Conical Generating Surface

The conical enveloping generating surface $T$ (see Figure 19.8) is used for the design of enveloping hobs, grinding worms, shaving cutters, and so forth, similar to that for which a cylindrical enveloping generating surface is used (see Chapter 14). Examples of conical enveloping gear-cutting tools are known from many sources[4] [63, 79, 80, 158].

[4] SU Pat. No. 1000186, *A Machine Tool for Finishing a Crowned Gear*/ S.P. Radzevich, Int. Cl. B23F 19/00, Filed: November 6, 1981.
SU Pat. No. 1028450, *A Machine Tool for Finishing a Crowned Gear*/ S.P. Radzevich, Int. Cl. B23F 19/00, Filed: April 26, 1982.

Possible types of tools for machining of gears include, but are not limited to

- enveloping hobs,
- enveloping shaving cutters,
- enveloping worm grinding wheels,
- tools for scudding gears,
- gear broaching tools,
- finishing gear-cutting tools that feature a continuous *HS*-cutting edge (similar to that shown in Figure 17.6),

as well as the tools for burnishing of gear tooth flanks, and so forth.

As the tooth count of the gear-cutting tool increases, the enveloping conical gear-cutting tool degenerates into a corresponding rack-type gear-cutting tool (see Figure 17.13).

## 19.3 Gear-Cutting Tools with a Toroidal Generating Surface

For the purpose of design of enveloping gear-cutting tools, a generating surface with a round auxiliary generating rack $\mathcal{R}_T$ can be utilized. Examples of an auxiliary generating rack of this geometry is depicted in Figure 14.5(c) and (d). The use of a round auxiliary generating rack makes it possible to generate a toroidal enveloping generating surface $T$.

### 19.3.1 Generating Surface of an Enveloping Toroidal Gear-Cutting Tool

The generating surface $T$ of an enveloping toroidal gear-cutting tool can be created as the enveloping surface to a family of consecutive positions of the round auxiliary generating rack $\mathcal{R}_T$. For the purpose of generating of the surface $T$, the auxiliary generating rack $\mathcal{R}_T$ performs two motions simultaneously (see Figure 19.9). The motions are the rotation of the rack $\mathcal{R}_T$ about its axis of rotation $O_{\mathcal{R}}$ and the rotation of the rack $\mathcal{R}_T$ about the cutting tool axis of rotation $O_{ct}$. The rotations of the round auxiliary generating rack $\mathcal{R}_T$ are synchronized with each other. After the generation is completed, the generating surface of the enveloping toroidal gear-cutting tool is shaped in the form of a screw surface on the torus surfaces. The torus surface is specified by the radius $r_{\mathcal{R}}$ of the pitch circle of the auxiliary rack $\mathcal{R}_T$ and by the distance $C_{\mathcal{R}}$ of the rack axis of rotation $O_{\mathcal{R}}$ from the cutting tool axis of rotation $O_{ct}$. The equation of the generating surface $T$ of the enveloping toroidal gear-cutting tool can be derived in the form $\mathbf{r}_{ct} = \mathbf{r}_{ct}(U_{ct}, V_{ct})$, similar to that derived in Section 14.6.3 for the external toroidal gear-cutting tool (see Figure 14.19).

An example of the vector representation of the kinematics of a gear machining process with a toroidal enveloping gear-cutting tool is schematically illustrated in Figure 19.10.

The rotation vector of the work gear $\boldsymbol{\omega}_g$ and the rotation vector of the gear-cutting tool $\boldsymbol{\omega}_{ct}$ are along the work gear

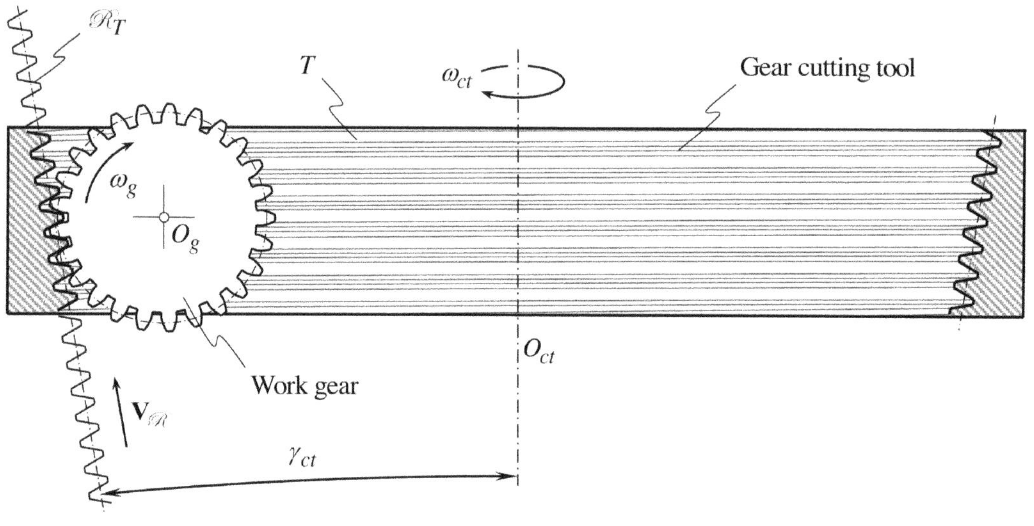


FIGURE 19.8 Gear-cutting tool with conical enveloping generating surface $T$.

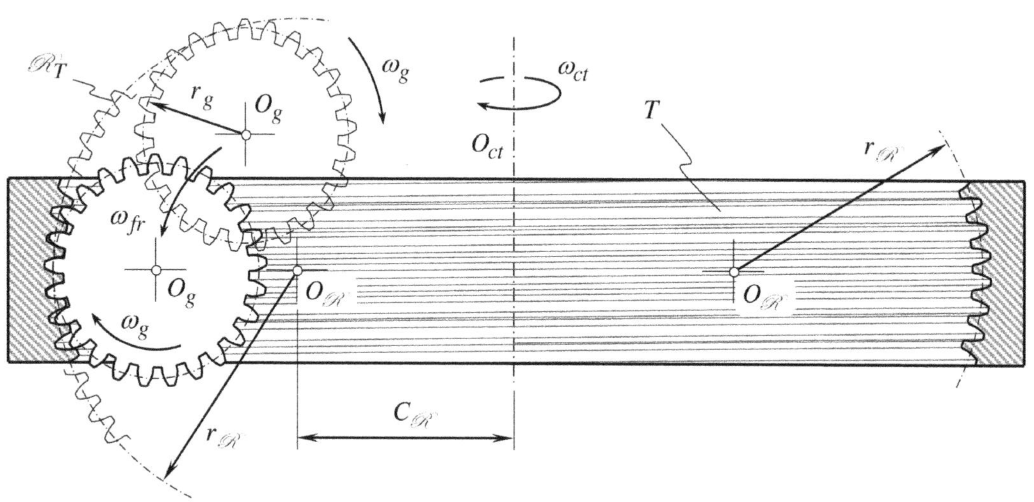


FIGURE 19.9 Gear-cutting tool with a toroidal enveloping generating surface $T$.

axis $O_g$ and the cutting tool axis of rotation $O_{ct}$, respectively. The axes of rotation $O_g$ and $O_{ct}$ are positioned at a distance $C$ from one another, and they form a certain crossed-axis angle $\Sigma$ (see Figure 19.10). The rotations $\omega_g$ and $\omega_{ct}$ are synchronized with each other.

While rotating, the work gear is fed into the gear-cutting tool. The rotation vector $\boldsymbol{\omega}_{fr}$ of the feed motion is pointed along the axis $O_{\mathscr{R}}$. The feed motion $\boldsymbol{\omega}_{fr}$ affects the timing of the rotations $\boldsymbol{\omega}_g$ and $\boldsymbol{\omega}_{ct}$. All three rotations—that is, $\boldsymbol{\omega}_g$, $\boldsymbol{\omega}_{ct}$, and $\boldsymbol{\omega}_{fr}$—have to be properly timed with one another.

A particular case of enveloping toroidal generating surfaces is of special interest to a gear-cutting tool designer.

No waviness on the tooth flanks of the machined work gear can be observed when an enveloping gear-cutting tool of any design (either cylindrical, conical, or toroidal) is used. The actual shape of the machined tooth profile features cusps, which are very undesirable. No cusps are observed in the lengthwise direction of the machined tooth

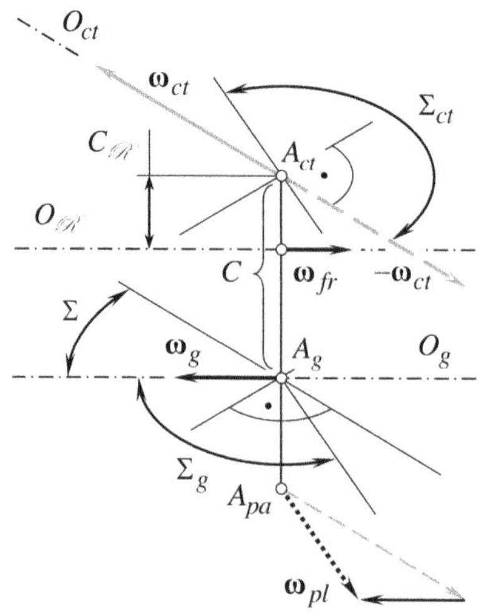


FIGURE 19.10 Vector representation of the kinematics of a gear machining process with a toroidal enveloping gear-cutting tool.

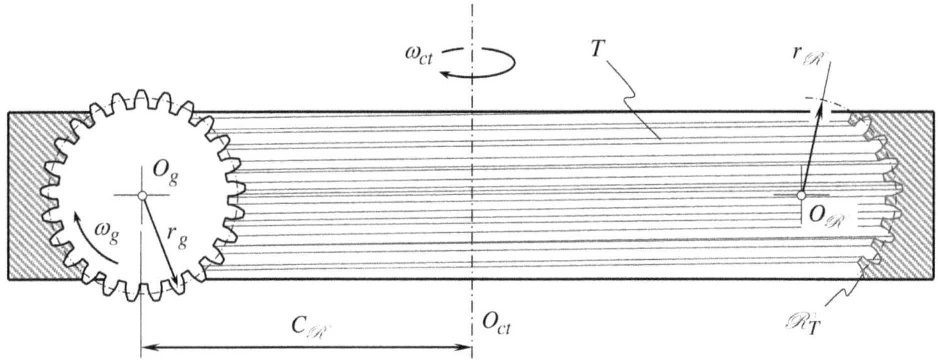


**FIGURE 19.11** Example of conformal toroidal enveloping generating surface $T$.

flank. At every point of the tooth profile of the work gear, the cusp height depends on the difference between the radius of curvature $R_g$ of the work gear tooth profile and the corresponding radius of curvature $R_{ct}$ of the generating surface $T$ of the gear-cutting tool. The smaller the difference[5] $(R_g + R_{ct})$, the smaller the cusp height on the machined tooth flank. If the difference $(R_g + R_{ct})$ approaches zero, then the cusp height approaches zero as well [133, 144, 148, 158, 168]. Cusps practically vanish when the difference $R_g + R_{ct} = 0$.

This concept of the enveloping toroidal generating surface of the gear-cutting tool (see Figure 19.9) can be enhanced to a point where it is possible to machine a work gear with no cusps on the tooth flanks. The generating surface $T$ of such a geometry is referred to as the *conformal* toroidal enveloping generating surface of the gear-cutting tool (see Figure 19.11). The equality $r_g = r_{\mathscr{R}}$ of the radii of pitch circles is the principal feature of the conformal toroidal generating surface $T$. Because the equality $r_g = r_{\mathscr{R}}$ is valid, the rotation vector $\boldsymbol{\omega}_{fr}$ of the feed motion is aligned with the rotation vector $\boldsymbol{\omega}_g$ of the work gear. The latter is schematically illustrated in Figure 19.12.

No cusps occur on the tooth flank of the work gear that is machined with a gear-cutting tool having the conformal toroidal enveloping generating surface $T$ (see Figure 19.11).

In contrast to the gear-cutting tools with an enveloping toroidal generating surface with a concave auxiliary rack $\mathscr{R}_T$ (see Figure 19.9), enveloping toroidal generating surfaces featuring a convex auxiliary generating rack $\mathscr{R}_T$ can be developed as well. An example is schematically illustrated in Figure 19.13. When a gear-cutting tool is designed on the premise of the generating surface $T$ shown in Figure 19.13, it is then possible to cut gears with no longitudinal modification (with no crowning). For this purpose, the setting angle $\zeta_{ct}$ has to be properly correlated with the crossed-axes angle $\Sigma$. This is a trivial engineering problem to assign a corresponding value of the setting angle $\zeta_{ct}$ of the gear-cutting tool.

[5] The involute tooth profile of an internal gear has concave geometry. Therefore, the corresponding radius of curvature of the tooth profile always has a negative value.

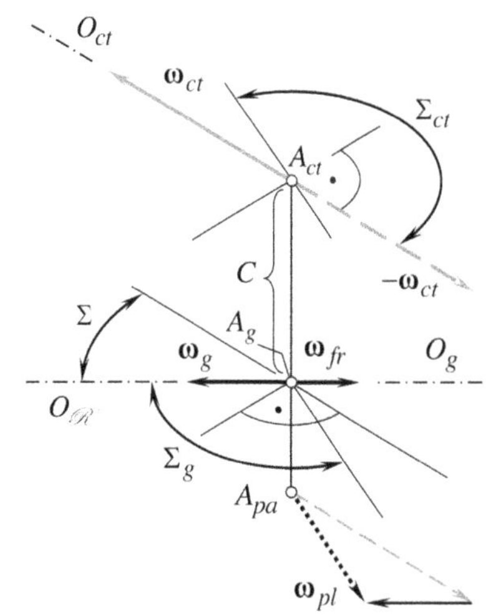


**FIGURE 19.12** Vector representation of the kinematics of a gear machining process with the conformal toroidal enveloping gear-cutting tool.

A vector representation of the kinematics of a gear machining process with a toroidal enveloping gear-cutting tool having a convex auxiliary generating rack $\mathscr{R}_T$ is shown in Figure 19.14. Possible types of the enveloping toroidal generating surfaces of gear-cutting tools were investigated by Prof. S.P. Radzevich [150].

### 19.3.2 Examples of Gear-Cutting Tools with Enveloping Toroidal Generating Surface

Gear-cutting tools having an enveloping toroidal generating surface (see Figure 19.11 and Figure 19.13) have no wide application in industry yet. An example is known from [67]. When machining a work gear, the enveloping toroidal gear-cutting tool[6] is rotated $\omega_{ct}$ about its axis of rotation $O_{ct}$. The work gear is rotated $\omega_g$ about its axis of rotation $O_g$. Simultaneous with the rotation $\omega_g$, the work gear performs another rotation $\omega_{fr}$. The work gear is fed into the gear-cutting tool by the feed rotation $\omega_{fr}$. All three rotations, $\omega_g$,

[6] SU Pat. No. 933316, *A Gear Cutting Tool*/S.P. Radzevich, Int. Cl. B23f 21/00, Filed: October 18, 1979.

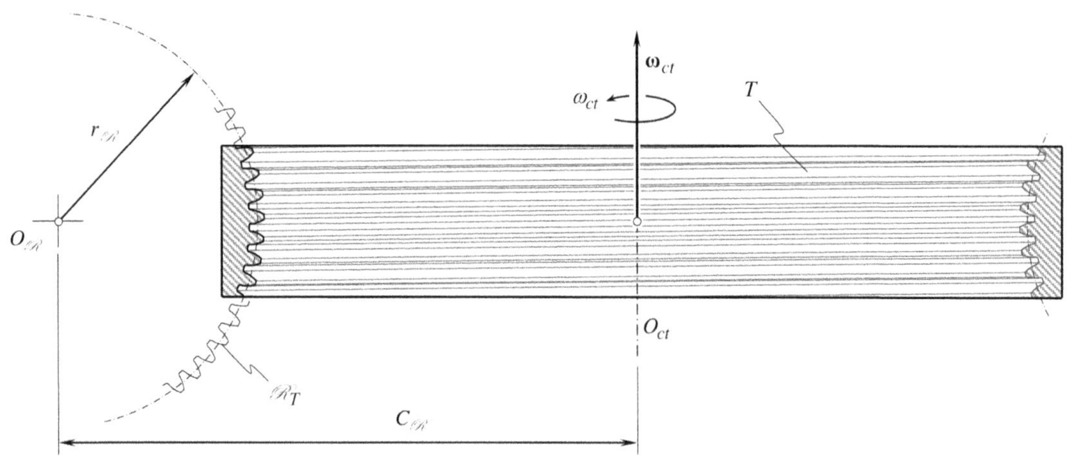


**FIGURE 19.13** Gear-cutting tool with toroidal enveloping generating surface $T$ having convex auxiliary generating rack $\mathscr{R}_T$.

$\omega_{ct}$, and $\omega_{fr}$, are synchronized with one another. The rotations are timed so that the work gear performs more than one full rotation while traveling within the face width of the gear-cutting tool for the complete generation of the tooth flanks of all the work gear teeth.

The gear honing process is a perfect example of the practicality of an application for the enveloping toroidal generating surface with a convex auxiliary generating rack $\mathscr{R}_T$ shown in Figure 19.13.

Gear honing is a hard-finishing method for eliminating the gear errors after the hardening and smoothing of the surface of the gear teeth. Gear honing is also referred to as *shave grinding*.[7] This is due to the fact that the machining kinematics and the tool geometry are similar to gear shaving, while the tool material and cutting speeds are similar to grinding.[8]

Gear honing is a continuous, chip-removing, precision machining process, and the kinematics is basically analogous to gear shaving. The driven honing tool and the generally nondriven work gear are in rolling contact with a specified radial pressure (see Figure 19.15).

The fact that the axes are crossed at a defined angle (crossed-axes angle) means that the superposition of the feed movements produces a resulting grinding movement in the material removal region, running diagonally from tip to root and producing a micro-cutting process with short cutting-point engagement. The large number of grains engaging in the tooth flank at the same time means there is a steady removal of chips, which in turn generates a surface texture on the tooth flank favorable for the gear operation.

The module range to be honed is between 0.5 and 8 mm, and the helix angle between 0° and 45°. Gear hardness is mostly below *HRC* 64.

Many designs of gear-cutting tools can be developed on the concept of enveloping toroidal generating surface $T$ similar to that for conventional external generating surfaces

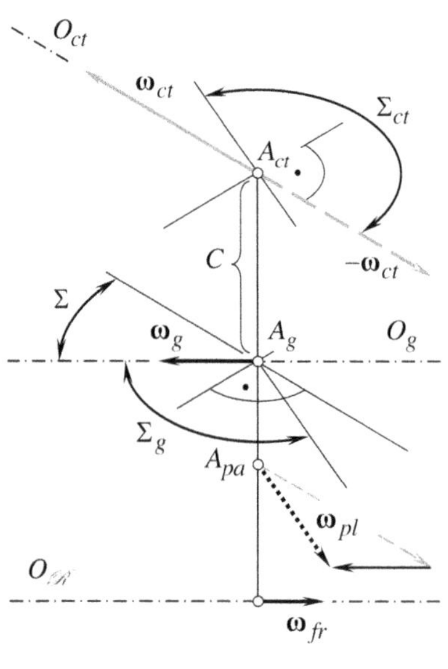


**FIGURE 19.14** Vector representation of the kinematics of a gear machining process with a toroidal enveloping gear-cutting tool having a convex auxiliary generating rack $\mathscr{R}_T$.

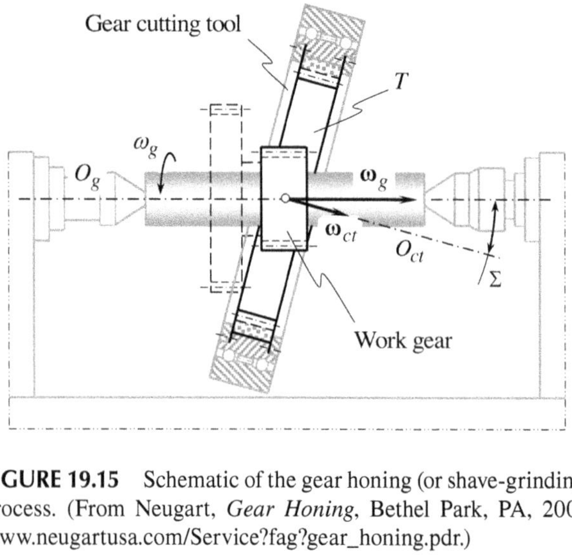


**FIGURE 19.15** Schematic of the gear honing (or shave-grinding) process. (From Neugart, *Gear Honing*, Bethel Park, PA, 2001. www.neugartusa.com/Service?fag?gear_honing.pdr.)

[7] *Abrasive shaving* is a more correct term for this gear finishing process rather than *shave grinding*.

[8] Neugart, *Gear Honing*, Bethel Park, PA, 2001, www.neugartusa.com/Service?fag?gear_honing.pdr

considered in Chapter 14. Possible types of tools for machining gears include, but are not limited to

- enveloping hobs,
- enveloping shaving cutters,
- enveloping work-grinding wheels,
- tools for scudding gears,
- gear broaching tools,
- finishing gear-cutting tools that feature the continuous *HS*-cutting edge (similar to that shown in Figure 17.6),

as well as tools for burnishing of gear tooth flanks, and so forth.

No geometrically accurate involute gears can be precisely generated in an internal gear machining mesh in the conventional gear honing machine (see Figure 19.15). Only approximate non-involute gears can be cut this way. This is attributed to the kinematics of the internal gear machining mesh and to violation of the condition of conjugacy of the tooth flanks $\mathscr{G}$ of a gear to be machined and of the generating surface $T$ of the gear-cutting tool. In compliance with the condition of conjugacy of the tooth flanks, at every point of contact of the surfaces $\mathscr{G}$ and $T$, the common perpendicular $\mathbf{n}$ must intersect the axis of instant rotation $P_{\text{ln}}$ of the gear and the gear-cutting tool in the gear machining mesh [171].

# 20 Gear-Cutting Tools for Machining Internal Gears

An internal work gear-to-cutting tool mesh can also be used to design a gear-cutting tool for machining an internal gear. Such a possibility can be directly drawn from the analysis of vector diagrams of the gear machining mesh and was investigated by Prof. S.P. Radzevich [150] as early as the early 1980s.

## 20.1 PRINCIPAL DESIGN PARAMETERS OF GEAR-CUTTING TOOL FOR MACHINING INTERNAL GEAR

The determination of a set of design parameters of a work gear is the starting point when designing a gear-cutting tool.

### 20.1.1 Geometry of Tooth Flank of Internal Gear

An example of an internal gear is illustrated in Figure 1.3. The location of the gear body with respect to the tooth surfaces is the major difference of the geometry of the internal gear shown in Figure 1.3 from the geometry of the external gear shown in Figure 1.5. Therefore, Eq. (1.4)

$$\mathbf{r}_g(U_g, V_g) = \begin{bmatrix} r_{b.g}\cos V_g + U_g \cos\lambda_{b.g}\sin V_g \\ r_{b.g}\sin V_g - U_g \sin\lambda_{b.g}\sin V_g \\ r_{b.g}\tan\lambda_{b.g} - U_g \sin\lambda_{b.g} \\ 1 \end{bmatrix} \begin{matrix} V_g^{(l)} \le V_g \le V_g^{(a)} \\ 0 \le U_g \le [U_g] \end{matrix} \qquad (20.1)$$

is valid for the internal gear as well.

The body of an external gear is located within the interior, which is bounded by the gear tooth flank $\mathcal{G}$. In contrast, the body of an internal gear is located outside the surface $\mathcal{G}$.

### 20.1.2 Kinematics of Machining an Internal Gear

The internal work gear and the gear-cutting tool are rotated about skew axes of rotation $O_g$ and $O_{ct}$ as depicted in Figure 20.1. The axes of rotation $O_g$ and $O_{ct}$ are positioned at a certain center distance $C$ from one another. The crossed-axis angle $\Sigma$ is the angle between the rotation vector $\boldsymbol{\omega}_g$ of the work gear and the rotation vector $\boldsymbol{\omega}_{ct}$ of the gear-cutting tool, that is, the crossed-axis angle is specified as $\Sigma = \angle(\boldsymbol{\omega}_g, \boldsymbol{\omega}_{ct})$.

The magnitudes $\omega_g$ and $\omega_{ct}$ of the rotation vectors $\boldsymbol{\omega}_g$ and $\boldsymbol{\omega}_{ct}$ relate to one another in the same inverse proportion that the corresponding tooth numbers $N_g$ and $N_{ct}$ relate to each other. When an internal work gear is machined, then the rotation $\omega_g$ is smaller compared to the cutting tool rotation $\omega_{ct}$. However, when the work gear tooth number $N_g$ is reduced, the work gear transforms into an internal worm. In such a scenario, the rotation $\omega_g$ exceeds the rotation $\omega_{ct}$.

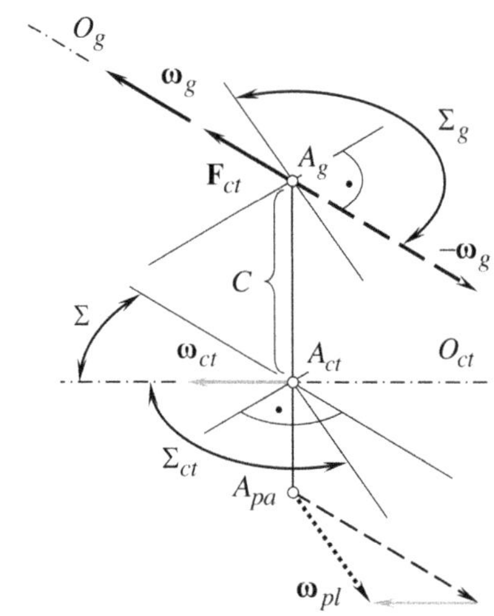


**FIGURE 20.1** Vector representation of the kinematics of a machining process of an internal cylindrical work gear by the external gear-cutting tool (the feed motion $\mathbf{F}_{ct}$ is pointed along the work gear axis $O_g$).

The feed motion $\mathbf{F}_{ct}$ of the gear-cutting tool is pointed along the work gear axis of rotation $O_g$. Either the work gear or the cutting tool can be fed along the axis $O_g$. This motion affects the resultant motion $\mathbf{V}_\Sigma$ of the work gear relative to the gear-cutting tool, and thus it should be taken into account when determining the generating surface of the gear-cutting tool.

Another schematic of machining of an internal gear employs feed motion $\mathbf{F}_{ct}$ pointed along the centerline ℄, as depicted in Figure 20.2. A feed motion of this type is referred to as *plunge feed motion* $\mathbf{F}_{ct}$. Plunge feed motion does not affect the resultant motion $\mathbf{V}_\Sigma$ of the work gear relative to the gear-cutting tool. Therefore, this motion can be neglected when determining the generating surface $T$ of the gear-cutting tool.

### 20.1.3 Determination of the Generating Surface of Gear-Cutting Tool for Machining Internal Gear

Two principal reference systems are used when determining the generating surface of the gear-cutting tool for


DOI: 10.1201/9781003520528-27

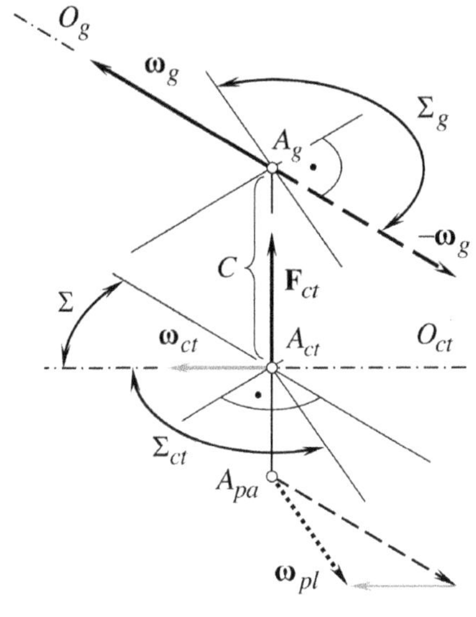


**FIGURE 20.2** Schematic of a machining process of an internal cylindrical work gear by the external gear-cutting tool that features plunge feed motion $\mathbf{F}_{ct}$.

**FIGURE 20.3** Hob for cutting an internal gear.

machining an internal gear. In many cases it is convenient to employ the Cartesian coordinate systems $X_gY_gZ_g$ and $X_{ct}Y_{ct}Z_{ct}$.

The first reference system $X_gY_gZ_g$ is associated with the work gear. This coordinate system is rotated with the work gear. The second reference system $X_{ct}Y_{ct}Z_{ct}$ is the coordinate system with respect to which the gear-cutting tool will be analytically described.

The generating surface of the gear-cutting tool is represented as the enveloping surface to a family of consecutive positions of the gear tooth flank $\mathscr{G}$ when considering the resultant motion of the coordinate system $X_gY_gZ_g$ relative to the coordinate system $X_{ct}Y_{ct}Z_{ct}$.

Use of the kinematics of a machining process of an internal cylindrical work gear by the external gear-cutting tool (shown in Figure 20.1) allows for the composition of the operator $\mathbf{Rs}(g \mapsto ct)$ of the resultant coordinate system transformation (see Appendix D for details). With the help of the operator of the linear transformation $\mathbf{Rs}(g \mapsto ct)$, the position vector of a point $\mathbf{r}_g$, the work gear tooth flank $\mathscr{G}$ can be represented in the coordinate system $X_{ct}Y_{ct}Z_{ct}$, which is associated with the gear-cutting tool

$$\mathbf{r}_g^c(U_g, V_g, \varphi_g) = \mathbf{Rs}(g \mapsto ct) \cdot \mathbf{r}_g(U_g, V_g) \tag{20.2}$$

The operator $\mathbf{Rs}(g \mapsto ct)$ of the resultant coordinate system transformation is function of the enveloping parameter $\varphi_g = \omega_g \cdot t$. Here $t$ denotes time.

To derive an equation of the generating surface $T$, the moving tooth flank surface $\mathbf{r}_g^c$ needs to be considered together with the Shishkov equation of contact $\mathbf{n}_g \cdot \mathbf{V}_\Sigma = 0$. Here in the equation of contact, the unit normal vector to the tooth flank $\mathscr{G}$ is designated as $\mathbf{n}_g$, and $\mathbf{V}_\Sigma$ is the vector of resultant motion of the tooth flank $\mathscr{G}$ with respect to the coordinate system $X_{ct}Y_{ct}Z_{ct}$. Considered together, this makes it possible to eliminate the enveloping parameter $\varphi_g$ from Eq. (20.2). Ultimately, the equation of the generating surface $T$ of the gear-cutting tool can be represented in the form of the vector equation $\mathbf{r}_{ct} = \mathbf{r}_{ct}(U_{ct}, V_{ct})$.

The generating surface of a gear-cutting tool for machining an internal gear is shaped in the form that resembles an external crowned gear.

## 20.2 EXAMPLES OF GEAR-CUTTING TOOLS FOR MACHINING INTERNAL GEAR

Hobs for machining internal gears are produced by the automotive industry, and others. An example of a hob for cutting internal gears is illustrated in Figure 20.3. The hob design is based on convex auxiliary generating rack $\mathscr{R}_T$. When cutting the work gear, the generating rack $\mathscr{R}_T$ is properly meshed with the work gear teeth. The rake surface $R_s$ of the hob tooth intersects the corresponding clearance surface $C_s$. The cutting edges $CE$ of the hob teeth align with the lines of intersection of the surfaces $R_s$ and $C_s$. Cutting edges of the hob teeth are entirely located within the generating surface $T$ of the hob.

A multiple-start hob for cutting an internal gear is illustrated in Figure 20.4.

A hob with indexable inserts for cutting an internal gear is depicted in Figure 20.5.

More examples of the design of hobs for cutting internal gears[1] are known from [75] and from other advanced sources.

[1] SU Pat. No. 946833, *A Hob for Machining Internal Gear*/S.P. Radzevich, Int. Cl. B23f 21/16, Filed: June 6, 1980.

FIGURE 20.4 Multiple-start hob for cutting an internal gear.

FIGURE 20.5 Hob with indexable inserts for cutting an internal gear.

The scudding process[2] can be used for a wide range of internal gear applications. In scudding, the cutter feeds directly through the work gear as the cutter and work gear spin in a synchronized fashion. Gear-cutting tools for scudding internal gears are a perfect example of cutting tools designed on the basis of internal work gear-to-cutting-tool mesh. Gear-cutting tools for scudding gears were investigated in [213].

The earlier considered concept of finishing gear-cutting tools having the continuous *HS*-cutting edge can also be extended to the area of gear-cutting tools for finishing internal gears. An example of a gear-finishing tool[3] of this design is schematically illustrated in Figure 20.6. The gear-finishing tool [111] is designed on the premise of a convex auxiliary generating rack $\mathscr{R}_T$, similar to that from which the hob (see Figure 20.3) is designed.

The line of action *LA* in the work gear-to-cutting-tool mesh is used to construct the continuous *HS* – cutting edge *CE* of the gear-finishing tool. When finishing the work gear, an imaginary (phantom) surface of revolution is generated by the line of action when the latter is rotated about the cutting tool axis of rotation $O_{ct}$. The line of intersection of the generating surface *T* by the surface of revolution is used as the continuous *HS* – cutting edge *CE* of the gear finishing tool (see Figure 20.6).

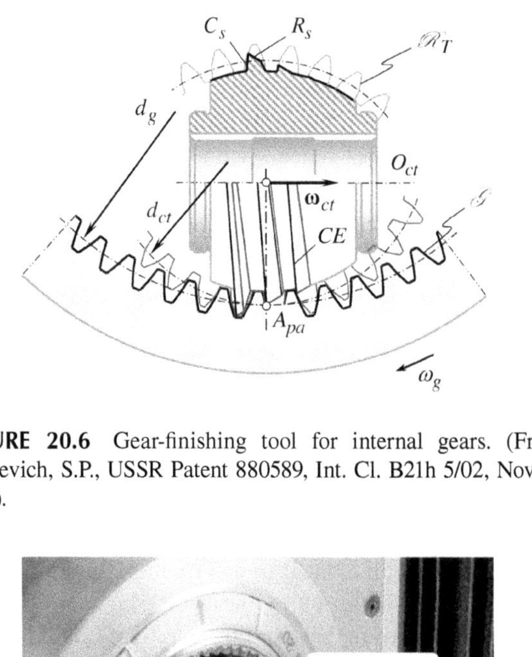


FIGURE 20.6 Gear-finishing tool for internal gears. (From Radzevich, S.P., USSR Patent 880589, Int. Cl. B21h 5/02, Nov. 6, 1979).

Owing to the peculiarities of the work gear generation of the tooth profile by the continuous *HS* – cutting edge, the gear-finishing tools (see Figure 17.6 and Figure 17.7) designed for finishing external gears can also be used for finishing internal gears. No changes to the design parameters of the gear-cutting tool are required in this regard.

Internal gears can be cut by the gear scudding tools (see Figure 20.7). They can also be finished with shaving cutters and/or they can be finished by a generating honing method.

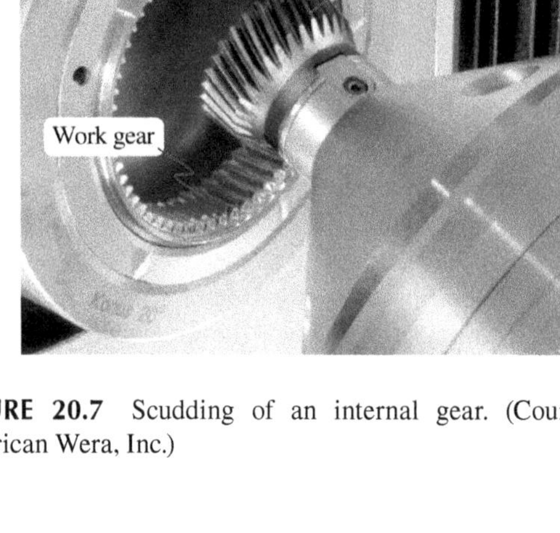


FIGURE 20.7 Scudding of an internal gear. (Courtesy of American Wera, Inc.)

[2] The technology itself is a current slant on an older hard-gear finishing process developed in Germany called *walzschaelan* (translated as "hob-peeling"). The elements of hob-peeling are now being used in a different manner, providing the current green process called *scudding*.

[3] SU Pat. No. 880589, *A Tool for Finishing Internal Gear*/S.P. Radzevich, Int. Cl. B21h 5/02, Filed: November 6, 1979.

Internal gears can be shaved on special machines in which the work drives the shaving cutter, or by an internal shaving cutter head attachment on external shaving machines.

Because of the crossed-axes relationship between the shaving cutter and the work gear in internal shaving, the shaving cutter requires a slight amount of crown in the teeth to avoid interference with the work gear teeth.

When internal gears are ¾-in. wide and less, or shoulder interferences limit the work gear reciprocation and crossed-axes angle, a plunge shaving method can be applied. Here, the shaving cutter is provided with differential serrations and plunge-fed upward into the work gear. If crowning is desired, a reverse-crowned shaving cutter is used with the plunge-feed shaving process.

When the work gear tooth count approaches infinity ($N_g \rightarrow \infty$), then the gear-cutting tool transforms into a cutting rack. Gear-cutting tools designed on the premise of the rack-to-gear-mesh were discussed earlier (see, for example, Figure 8.1).

Gear-cutting tools for machining bevel gears can also be designed using the work gear-to-cutting-tool mesh.

Only involute internal gears can be precisely generated in internal work gear-to-cutting tool mesh. No gears with other tooth flank geometry can be perfectly generated in internal work gear-to-cutting tool mesh—only approximate non-involute gears can be cut this way. This is attributed to the kinematics of the internal work gear-to-cutting tool mesh, and to violation of the condition of conjugacy of the tooth flanks $\mathscr{G}$ of a gear to be machined and of the generating surface $T$ of the gear-cutting tool. In compliance with the condition of conjugacy of the tooth flanks, at every point of contact of the surfaces $\mathscr{G}$ and $T$, the common perpendicular $\mathbf{n}$ must intersect the axis of instant rotation $P_{\text{ln}}$ of the gear and the gear-cutting tool in the gear machining mesh [171].

# 21 On the Lack of Understanding of the Scientific Theory of Gearing by the Majority of Gear Scientists and Engineers

The *scientific theory of gearing*, outlined in [172], is an advanced theory in the realm of gear science and gear art. The latest accomplishment in the theory and the most common mistakes that result in misunderstanding of the scientific theory of gearing are briefly discussed here.

The origin of the scientific theory of gearing can be traced back to the fundamental research carried out by Leonhard Euler and his predecessors in the eighteenth century, but the rapid growth of this theory began only in the early twenty-first century. Through the first two decades of the twenty-first century, most of the key problems in the proposed theory of gearing were solved. It is the right time now to study what has already been achieved in this area of mechanical engineering and to implement the results in everyday gear practice. Unfortunately, the latest achievements in the scientific theory of gearing are ignored by most gear scientists, gear engineers, and gear manufacturers. This is likely due to their lack of understanding of the scientific theory of gearing.

It takes time for many gear experts to become familiar with the theory, to study it, and to begin applying it in everyday practice. Obviously it cannot be done instantly.

To properly evaluate the advantages and benefits of using the modern scientific theory of gearing [172], one needs to acquire an in-depth knowledge of the theory along with reasonably extensive experience with putting it into practice. Otherwise, there is no chance to evaluate it properly and to realize its benefits. Moreover, a lack of knowledge of the theory in the gear community can lead to the formation of a toxic environment in response to its proposals.

Nowadays, some well-known gear experts who pose as gear theoreticians commit huge mistakes when trying to evaluate the theory due to their lack of qualifications to do such an evaluation.

As an illustrative example, an article published in recent years [210] in the *Gear Solutions* magazine can be considered. The author of this article has made an attempt to review some key accomplishments in the scientific theory of gearing and in this way has demonstrated a lack of understanding of the cornerstone ideas of gear science (see Figure 21.1).

The article [210] is instructive for gear theoreticians and other gear experts because it demonstrates how poor is the actual knowledge of the gear experts when they try to participate in a discussion on the theoretical aspects of the gear art in a general sense of this term.

Over a year after the article [210] was published, another article was published in *Gear Solutions* magazine [157]. The author's attention in the article [157] is focused mainly on the analysis of the principal mistakes committed in the article [210].

This monograph [172] makes an illustrative comparison of the outline of the scientific theory of gearing with the obsolete so-called established theory, on the development of which "… many scientists and gear engineers who worked on the theory and its improvement over approximately 100 years" is discussed. (See page 43 in [210].) It should be stressed here that the so-called established theory is *not a theory of gearing at all*, just a collection of recipes. Trial and error (and NOT a gear science!) is the tool used to derive such recipes.

This second article, which appeared in the August 2020 issue of *Gear Solutions* magazine, was available to the public for a very short time. The published article [157] was removed from the magazine almost immediately. This article is no longer available to the public, as this issue of the magazine no longer contains it. This is nonsense, but the already published article was withdrawn from that issue of the magazine![1] (Evidence for that can be found on page 6 of the August 2020 issue of *Gear Solutions*—see Figure 21.2, where the name of the author of this article was mistakenly not removed.) No scientific problems can be solved this way.

The article [157] is important, as many common mistakes committed by other gear experts are addressed there. However, despite the removal of this article from the August 2020 issue of *Gear Solutions*, it is still available to the public—a facsimile of this article can be found in Figure 21.3.

In this second article [157], the author's comments on all the mistakes and inconsistencies committed by the author of the article [210] pertaining to gear science are concisely outlined.

[1] The original version of the entire August 2020 Issue of *Gear Solutions* (which contains the article [157]) is available from the author upon request (and at no charge).


DOI: 10.1201/9781003520528-28

**FIGURE 21.1** (a–j) Facsimile of the article: Stadtfeld, H.J., "Why Are Today's Hypoids the Perfect Crossed-Axes Gear Pairs?" *Gear Solutions*, May 2019, pages 42–50 [210]. *(Continued)*

**FIGURE 21.1** *(Continued)*

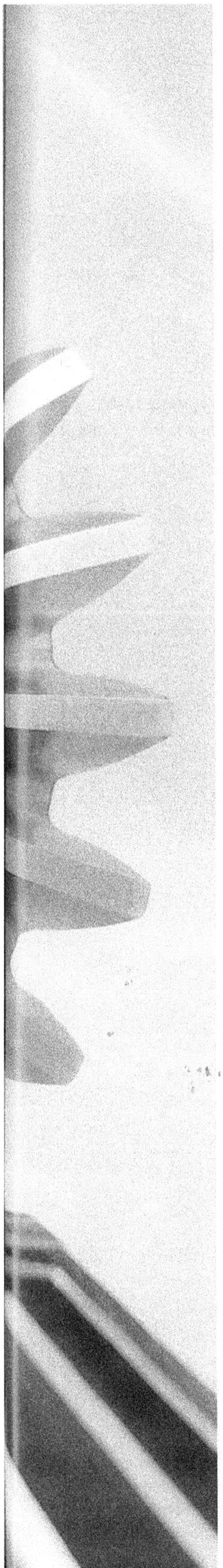

# Gear scientists, gear engineers, and gear manufacturers have worked successfully for many decades on finding the optimal flank forms and the optimal non-conjugate flank surface interaction.

By Dr. HERMANN J. STADTFELD

In 1924, Ernest Wildhaber, a well-known gear scientist, invented hypoid gearing. Compared to spiral bevel gears, hypoid gears provide an offset that allows lowering of the body of rear wheel drive vehicles by 50mm or more. This is possible because the propeller shaft between engine/transmission and the driving axle is not positioned at the center of the drive axle but is lowered by the offset amount (Figure 1). This allows the vehicle designer to lower the floor of the vehicle and subsequently the entire body by the same amount. Lowering the center of gravity of a passenger car by 50mm reduces the inertia responsible for sideways rolling by more than 10 percent, which provides better vehicle handling and more active safety. The lower body also reduces the CV coefficient for air resistance, providing higher gas mileage. Less than five years after the invention of hypoid gearing, all large automotive manufacturers around the world had converted their passenger cars and trucks to hypoid drive axles with a lower vehicle body.

Ernest Wildhaber emigrated from Switzerland to the U.S. in 1919 and was hired by The Gleason Works as a gear theoretician. Wildhaber received 279 patents, many of which changed the world of gearing. The cylindrical gear tooth profile that is today called *Wildhaber-Novikov* gearing was invented by Ernest Wildhaber in 1926. Mikhail Novikov, a Russian scientist with no access to western publications, invented the same tooth profile independent from Wildhaber in 1956. The contributions of both scientists are honored today by calling this system *Wildhaber-Novikov* gears.

Ernest Wildhaber is the father of modern gear theory. His pioneering contributions have been invaluable for the development of today's gear calculation and manufacturing processes.

## NEWS ABOUT HYPOID GEARS?

A recently published article by S.P. Radzevich [1] attempts to explain that hypoid gearsets designed and manufactured today lack good performance and are not perfect but approximate. The article seemingly questions the credibility of many scientists and gear engineers who worked on the theory and its improvement over approximately 100 years. However, reading the article carefully quickly reveals some key misperceptions and false assumptions. This write-up is meant to be a response on behalf of established gear theory and manufacturing.

In the following sections, certain statements and mathematical relationships of the published article are discussed and countering responses are presented and explained. Also, some assumptions about manufacturing and application procedures are discussed and compared to established practices today. The topics of this paper are structured accordingly in:

- The three fundamental laws of gearing.
- Perfect conjugacy.
- Real world applications.
- Transmission design.
- Heat treatment, lapping and grinding.

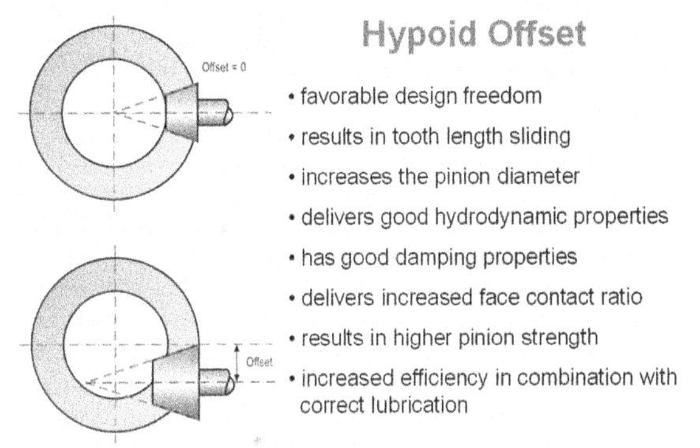


Figure 1: Features of pinion offset.

## THE THREE FUNDAMENTAL LAWS OF GEARING

The first fundamental law of gearing quoted in [1] $n_g \cdot V_\Sigma = 0$ also implies $|N_1 \times R_1| = i \cdot |N_2 \times R_2|$, where i is the constant transmission ratio. The three cases in Figure 2 visualize the problem of non-constant ratio and are noncompliant with the first gearing law because of a ratio change from case to case. This problem led Leonard Euler to discover the involute tooth profile. A simplistic mathematical approach teaches that the effective radius vector **R** remains unchanged while the contacting point between two mating flanks moves from $R_{b2}$ to $R_{b1}$ as shown to the



FIGURE 21.1 *(Continued)*

right in Figure 3 (movement along the line of action).

The line of action in parallel axes cylindrical gearing is a straight line, connecting the two base circles. If the surface normal vectors N1 and N2 are within the line of action, then the vector product $|\mathbf{N_i} \times \mathbf{R_i}|$ remains constant during a complete mesh cycle. The consequent application of this principle leads to the construction of an involute, as shown to the right in Figure 3.

The line that forms the tooth surface elements while traveling from position "a" to "f" along the line of engagement (line of action) is always perpendicular to the line of action. This principle implies that a tool, simply with straight cutting edges as shown in Figure 4, can be used to form the complex involute profile. Figure 4 also demonstrates the principle of profile shift while maintaining the first fundamental law of gearing.

The second fundamental law of gearing, which was proposed in 2017, $\mathbf{p_{ln}} \times \mathbf{V_m} + \mathbf{n_g} = 0$ [1], is a redundant relationship to the first gearing law and it is limited to cylindrical gears with parallel axes and straight bevel gears without hypoid offset. In this case, it adds nothing to the first gearing law; conjugacy is already given by the relationships required in the first gearing law.

The third fundamental gearing law is quoted in [1] in two different notations. The first notation covers only the special case of ratio = 1:

$$\varphi_{b.g} = \varphi_{b.p} = \varphi_{b.op}$$

However, the first notation is incorrect for all cases with different base pitch diameters because the circular pitch and not the angular pitch has to be equal between pinion, gear, and the operating base pitch. The second notation:

$$p_{b.g} = p_{b.p} = p_{b.op}$$

is consistent with the requirement of equal circular pitch, but the article makes no mention of the different notations.

## PERFECT CONJUGACY IN STRAIGHT BEVEL GEARSETS

Bevel gears with intersecting axes is the topic of a series of three papers published between October 2014 and January 2015 [2]. A straight bevel gearset with skew teeth was modeled, and a sample was manufactured. This publication addressed two points: the design of a gearset with low tooth count and the solution for perfect conjugacy. There is no mention in the article that the Coniflex® straight bevel gear system has been used since the 1940s for low tooth count gearsets such as differential gears.

Straight bevel gears such as Gleason Coniflex have tapered depth teeth, where the pitch cones roll on each other and the pitch apexes of pinion and gear match the crossing point of the axes. In the standard case, the face and root cone apexes match the crossing point. In such a standard Coniflex design, the base elements are also cones with cone apexes that match the crossing point of the axes. The involute development in Figure 3 can be applied to an infinite number of normal sections along the face width of a straight bevel gear which allows an involute development similar to that for cylindrical gears. The conical base elements of both members can be connected with a straight line (the line of action) in each section along the face width, whereas the plurality of all lines of action forms a plane (the plane of action).

This principle is shown in Figure 5. The two cones in Figure 5 are base cones of a straight bevel or a spiral bevel gearset. In the right two graphics, the view is directed such that the plane of action appears as a line that is tangential to the two cone-enveloping surfaces. The left side graphic in Figure 5 shows the plane of action three-dimensionally and how it connects the two base elements.

The plane of action cannot be extended beyond its tangential contacting line with the base element as shown incorrectly in Figure 1 in [2] and Figure 6 in [1]. The plane of action only exists where tooth engagement is possible and it is different than the generating gear plane (more specifically explained later).

There is, however, one difference from the true involute of cylin-

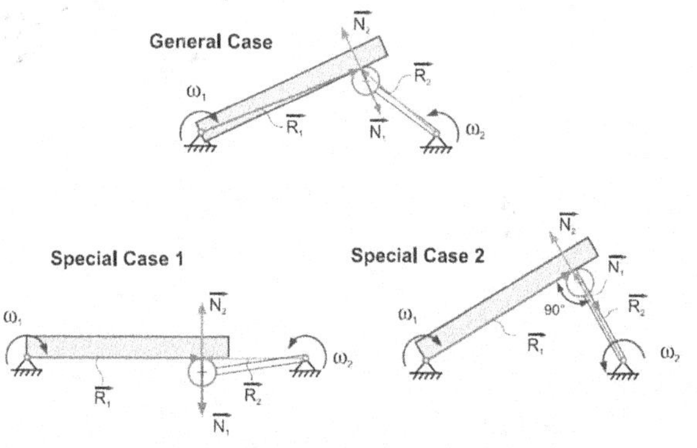


**Figure 2: Three general cases of motion transmission with a bar and a crank.**

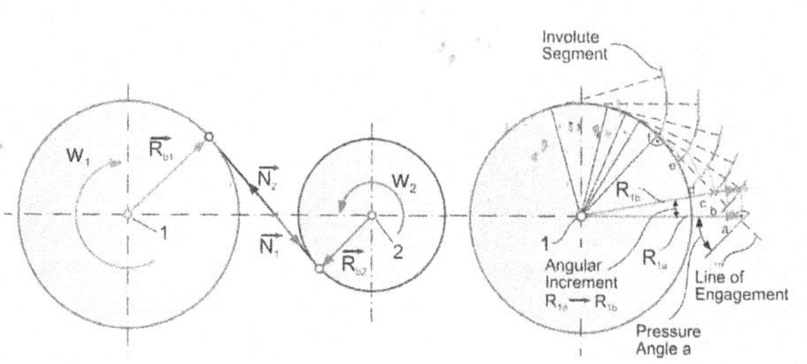


**Figure 3: Line of action between base circles (left) and subsequent involute development.**

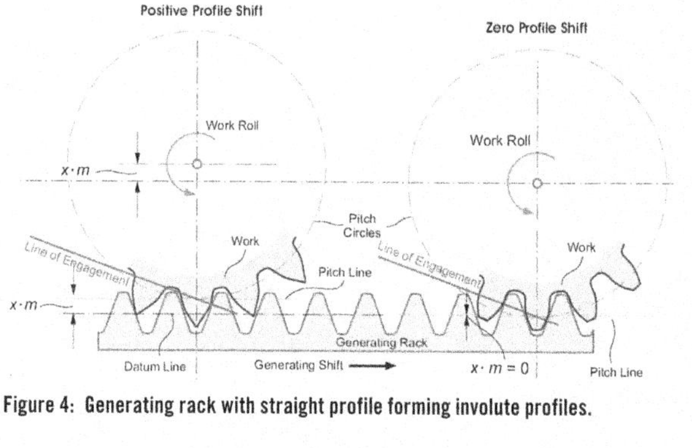


**Figure 4: Generating rack with straight profile forming involute profiles.**

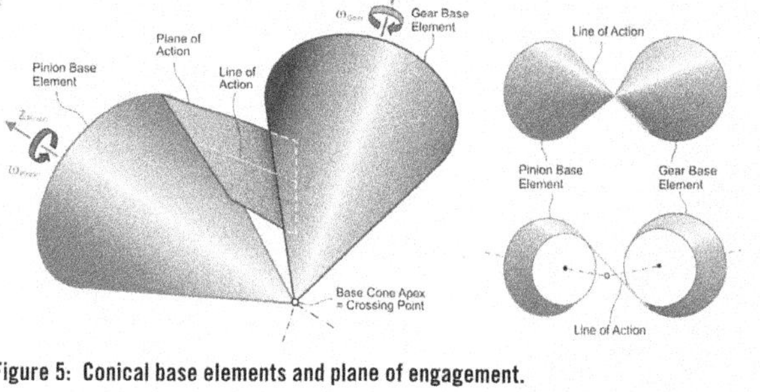


**Figure 5: Conical base elements and plane of engagement.**



**FIGURE 21.1** *(Continued)*

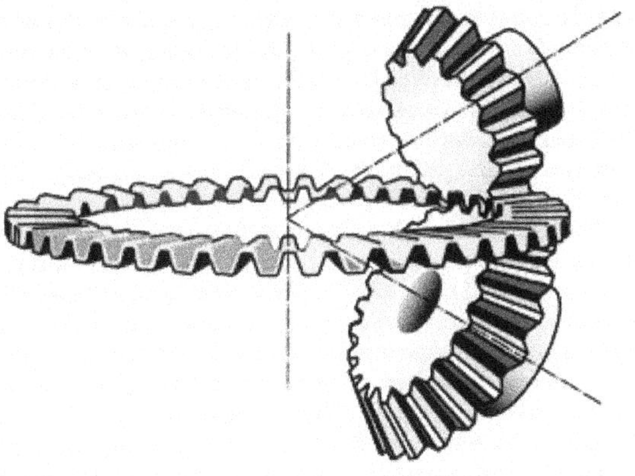

**Figure 6: Ring shaped generating rack with trapezoidal profile.**

**Figure 7: Solid model computer graphic of a perfectly conjugate straight bevel gearset.**

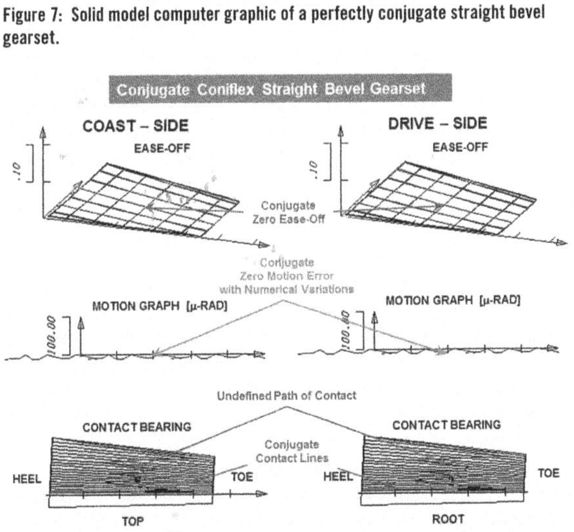


**Figure 8: Tooth contact analysis of the perfectly conjugate straight bevel gearset.**

drical gears. The rotation of pinion and gear does not occur in the normal plane but in the transverse plane. Because of this difference, the flank profile of straight bevel gears (and all other bevel gear types) is called Octoide. The Octoide is the analog function of an involute, and it provides to bevel gears the same advantages as an involute provides to cylindrical gears. Those advantages are constant ratio, center distance insensitivity, and ease of manufacturing.

Like cylindrical gears, bevel gears also have a trapezoidal generating profile. The straight rack of cylindrical gears becomes a ring, as shown in Figure 6. It is required to establish certain conditions in order to make the ring rack the generating gear for a pinion and a ring gear that will mesh perfectly together with zero transmission error and line contact identical to cylindrical gears. Those conditions are postulated in the kinematic coupling requirements:

- The flank surfaces of the generating gears of the two mating bevel gears are congruent (same shape but mirror images, as given in the example of Figure 6).
- The generating gears of the two mating bevel gears require identical axes of rotation (Top and bottom side of the generating gear in Figure 6 form the same generating gear, which rotates in both cases around the same axis and therefore satisfies condition 2).
- The surface of engagement of pinion and ring gear must be identical to the surface of engagement between pinion and generating gear and also to the one between ring gear and generating gear (without detailed knowledge of the surfaces of engagement, the global condition in Figure 6 seems to satisfy this requirement).

The generating gear principle has to be understood as the ultimate vehicle to form the teeth of two mating gears. The first fundamental law of gearing is fully executed by choosing trapezoidal profiles and by applying the kinematic coupling requirements. Gears are designed and manufactured in order to mesh with each other. What better way to manufacture them by way of a generating gear? The generating gear is represented by the manufacturing machine; it forms the teeth of the gear (at the bottom in Figure 6) while meshing with this gear perfectly. If the pinion is manufactured with the same generating gear but on the opposite side (at the top in Figure 6) and if the generating gear is imagined infinitely thin, then the result is a pinion that perfectly meshes with the gear having zero motion error. It is also given, in such a case, that line contact between the pinion and gear flank surfaces exists along the entire face width. The extraordinary effort shown in [2] to calculate straight bevel gear tooth surfaces that are skew with asymmetric teeth is based on the falsely assumed conjugate requirement. The sample gear pair shown in [2] was manufactured on a machining center with a ball nose endmill in a milling operation likely taking many hours to complete. It is rather simple and straightforward to manufacture a significantly better, perfectly conjugate straight bevel gearset in only a few minutes by using the generating gear method and a standard bevel gear manufacturing machine.

An example straight bevel gearset computer model is shown in Figure 7. The solid model in Figure 7 has been generated by using standard Gleason basic settings, based on the generating gear approach and by applying a standard Coniflex® Plus cutter head as used on Phoenix bevel gear manufacturing machines. The Coniflex straight bevel gear calculation for conjugate contact has to be conducted with a dish angle of 0° and no root angle correction (ΔGammaM = 0.). The dish angle is creating the length crowning and a ΔGammaM creates the profile crowning. With a dish angle of 0° and a ΔGammaM of 0° flank lead lines are straight lines and the profile is a true Octoide (involute equivalent).

A contact analysis of the gearset in Figure 7 is shown in Figure 8. The top of the figure shows the Ease-Offs of left and right flank (called coast and drive side in the graphic). The center of the figure shows the motion transmission errors of the pinion and gear flank pairs. The two bottom graphics are the representation of the tooth contact pattern. The contact pattern graphics are axial



**FIGURE 21.1** *(Continued)*

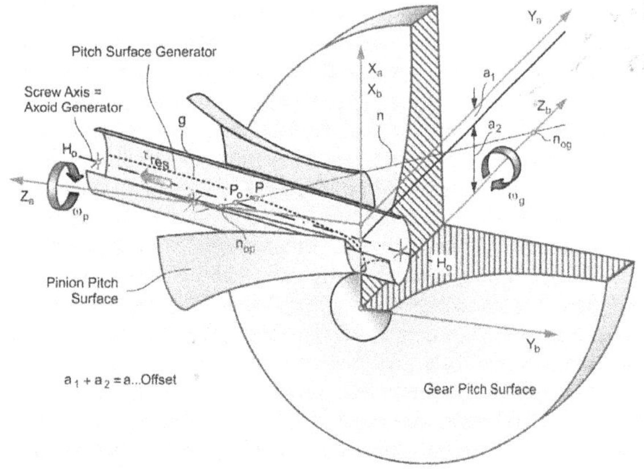


**Figure 9: The correct pinion and gear pitch surfaces.**

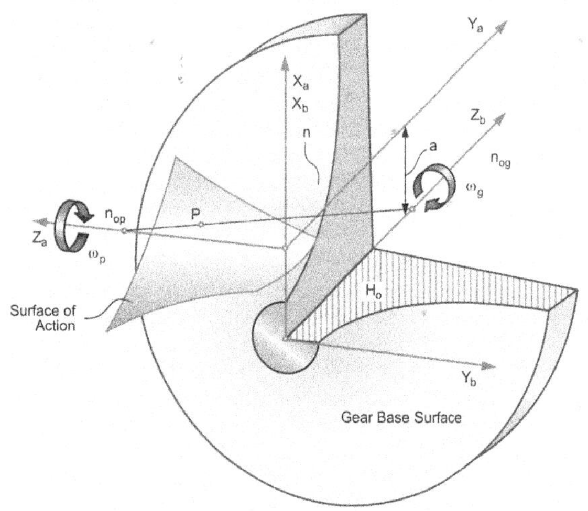


**Figure 10: Surface of action connecting the gear and pinion base surfaces.**

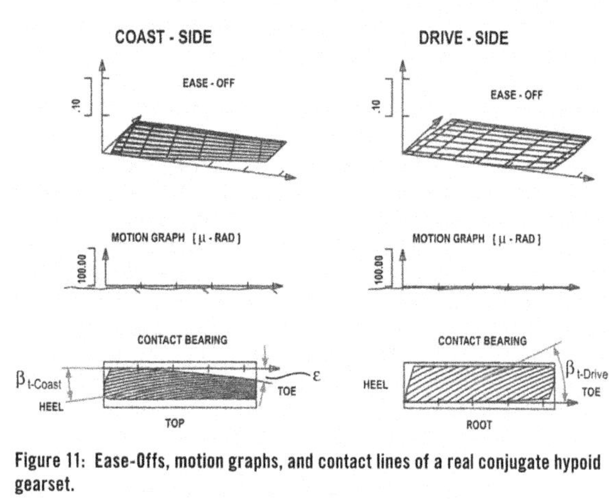


**Figure 11: Ease-Offs, motion graphs, and contact lines of a real conjugate hypoid gearset.**

projections of the flank surfaces and the contact lines in the same plane a two-dimensional part print would show the tooth area.

The contact analysis in Figure 8 confirms the full line contact in each roll position (lower graphics) as well as the zero motion error (only numerical static) in the center graphic which makes this example the perfectly conjugate straight bevel gearset. In the lower graphic the path of contact was calculated as a zig-zag line which indicates an undefined contact path. This means that due to the conjugacy, every point along each contact line is a path of contact point which makes the analysis program pick random points.

The Ease-Off base plane (top graphics in Figure 8) defines the conjugate state of a flank surface pair. Because the Ease-Off graph of the calculated flank pairs matches the presentation plane (base plane) precisely, that is the proof that a conjugate and precisely rolling gearset was the input of this contact analysis calculation.

The above experiment creating a conjugate straight bevel gearset is strictly academic. Conjugacy is the basis of all gearsets manufactured in high volume on dedicated manufacturing machines. A conjugate bevel gearset cannot be used for a power transmission because manufacturing tolerances and load affected deflections as well as material expansions and deformations under high operating temperatures will result in edge contact and high load concentrations. The load concentrations already start with moderate load and cause material damage and considerable noise emission. Although conjugacy is used as a reference for each design, predetermined amounts of length and profile crowning are applied. The right amount of crowning makes a gearset quiet and gives it the required load carrying capacity. The crowning is shown in the Ease-Off graphics with the conjugate reference always being present as the Ease-Off base plane. Several Ease-Off examples of a gearset with length and profile crowning are shown in the proceedings of this paper.

## PERFECT CONJUGACY IN HYPOID GEARSETS

It begins to become more problematic for hypoid gears. In [1] the pitch elements of crossed axes hypoid gears are drawn as cones. Even though the face cones of hypoid gears and pinions are machined conical, the pitch elements are hyperboloids.

Ernest Wildhaber and Arthur Stewart described their invention of hypoid gearing in 1926 [3]. Boris Shtipelman published in 1978 the relationships and derivations required to understand hypoid gears and their hyperbolic pitch elements [4]. Figure 9 offers a graphical interpretation of the hyperbolic pitch elements and their generator. The pitch surface generator is a line which winds on the surface of a cylinder beginning at the crossing point of the axes, equal to the first contact point of the pinion and gear pitch surfaces. The pitch surface generator is developed by the connecting line between the pinion and gear pitch surfaces ($\mathbf{n_{op}}$- $\mathbf{n_{og}}$) by shifting the connecting line along the pinion and gear axes. The connecting line ($\mathbf{n_{op}}$- $\mathbf{n_{og}}$) is normal to the pitch elements. Point **P** is one point of the pitch surface generator. If the division of the vector products of the distances between the axes and point **P** and their respective axis direction equals the ratio i of the hypoid gearset then one point of the pitch surface generator is found with:

$$\{(\mathbf{n_{op}} - \mathbf{P}) \times \mathbf{Z_a}\} \,/\, \{(\mathbf{P} - \mathbf{n_{og}}) \times \mathbf{Z_b}\} = \mathbf{i}$$

Although the pitch elements are hyperbolic and not conical, it is possible to use conical faces for the blanks of pinion and gear. If point **P** in Figure 9 was chosen at the center of the face width, then line ($\mathbf{n_{op}}$- $\mathbf{n_{og}}$) can be used as a normal vector to define the face angle a blank with straight face cones if the hypoid set was manufactured by face hobbing, which implies parallel depth teeth.

Straight face cones will merely influence the top root clearance of the gearset in the range of 30 to 60 microns. Using straight face cones will not change the form of the pitch surface nor will it influence the base surface. Those functional surfaces are given by the

46 gearsolutions.com

**FIGURE 21.1** *(Continued)*

kinematical relationships and have to be considered when thinking about the shape of the surface of action. No plane of action can exist between two hyperbolic base elements. The correct surface of action is curved and warped, as shown in Figure 10.

The conclusion is that the second fundamental law of gearing according to [1] does not exist and is not related in any way to the conjugacy of gears (as explained and demonstrated in the following section).

**CONJUGACY BETWEEN MESHING FLANKS**

The term conjugate is used in mathematics for two or more surfaces that contact each other along a line. Since the 1970s, the term conjugate has also been employed in gear technology literature to define the "exact" gear pair which presents a triple plurality of line contact between two gear flanks during the meshing process [5].

**DEFINITION OF THE CONJUGATE GEAR PAIR**

**1:** The flanks contact along a line (contact line), which is only limited by the boundaries of the teeth i.e. the working area.

**2:** The line contact between the flanks exists within the entire area of engagement in every mesh position.

**3:** Line contact is maintained in the entire area of engagement if pinion and ring gear are rotated by angular increments where: (angular pinion increment) / (angular ring gear increment) = transmission ratio.

The Ease-Off is a three-dimensional graphic of the flank deviations from a conjugate pair. It is calculated by transformation of a pinion flank "into" the gear coordinate system according to the first gearing law, resulting in a virtual gear flank that is conjugate to the actual pinion flank. This conjugate gear flank will then be compared to the present gear flank, where all differences in arc length are plotted point by point in ordinate direction into the Ease-Off graphic.

If both mating bevel gears have conjugate manufacturing data, then the Ease-Off graphic has no deviations in ordinate direction. Also, if the pinion flanks and the gear flanks have spiral angle and pressure angle errors of equal amounts, the Ease-Off graphic will not show any deviation. Although the individual gears are considered incorrect in this case, they will roll conjugate with each other, which subsequently leads to an Ease-Off without any ordinate values. Figure 11 shows the analysis results of a typical conjugate hypoid gearset. The Ease-Off graphics have zero crowning magnitudes in the ordinate direction. The motion graph has, next to some numerical entrance and exit variation, zero motion error. The contact bearings show line contact within the entire working area. The coast side contact ends at a toe root undercut (section ε).

Each spiral bevel and hypoid gearset with uniform tooth depth has a conjugate base design. This applies to all face hobbed and some face milled gearsets. A simple explanation of hypoid gearset conjugacy is possible with a non-generated gear that meshes with a generated pinion. For the calculation and manufacturing process, the hyperbolic pitch elements are calculated for the gear first. Then a suitable blade profile (gear cutter in Figure 12) is chosen and positioned in a face cutter head. The cutter head is positioned to create the desired spiral angle. With this procedure, a non-generated gear can be created by computer simulation and it can be manufactured with a bevel gear cutting machine [6].

A pinion cutter (see Figure 12) is positioned in a mathematical model or in a bevel gear cutting machine such that it represents one tooth of the non-generated gear by rotating around its axis. An additional simultaneous rotation around the pinion generating gear axis results in this pinion cutter becoming the generat-





**FIGURE 21.1** *(Continued)*

ing gear of a conjugate pinion. If the pinion is positioned with the same offset that was used to determine the pitch surfaces (Figure 9), then the cutter rotation around the pinion generating gear axis will form a pinion that is perfectly conjugate to the non-generated gear. The tooth contact analysis in Figure 11 has been obtained from such a non-generated hypoid gearset and therefore shows perfect conjugacy.

More complicated is the generation of a conjugate bevel or hypoid gearset with tapered depth teeth (see Figure 13). If the generating gear axes have identical axes of rotation that are perpendicular to the pitch line, the rotating cutter heads and their blades will not follow the root line of a tapered depth tooth. Tilting the cutter head in order to follow the root line would violate the first kinematic coupling requirement for teeth that are congruent to the slots of the mating member. The solution was developed in the 1940s [7].

If the cutting edges are adjusted in the cutting machine such that tooth reference profile and depth is matched at midface, and if an axial motion of the cradle is introduced that guides the blades along the tapered root line while the generating roll progresses along the face width, then the requirements of congruent teeth and slots are fulfilled with the result of perfect conjugacy.

The process configuration and kinematics in Figure 13 is called *Duplex Completing*. Today, all face-milled and ground spiral bevel and hypoid gears are manufactured with the Duplex Completing process. The axial cradle movement in this process is called Helical Motion and was first introduced with mechanical bevel gear machines in the late 1940s. The *Helical Motion* of the days of mechanical machines required an additional change gear box which actuated a cam that moved the sliding base during the generation process. Today's CNC controlled Phoenix® free form machines have the Helical Motion capability automatically by their interpolated axes movement.

## WHY IS CONJUGACY NOT DESIRABLE FOR REAL WORLD APPLICATIONS?

In 1926, Ernest Wildhaber [3, 8] was the first to propose applying surface crowning on hypoid gears. Wildhaber acknowledged that the slightest deviations in the gear housing and in the building position, as well as deflections affected by load and heat, will cause edge contact with peak stress levels of a multiple of the allowable values the gearset had been designed for. The conjugate gearset used for the TCA in Figure 11 was repeated with realistic displacement values of 50μm offset, 50μm pinion cone and 30' of shaft angle change. The results in Figure 14 show warped and tilted Ease-Offs and severe edge contact on heel and top. This edge contact will cause noisy operation followed by pittings and tooth fracture.

The theoretically conjugate hypoid set will not fulfill any of the fundamental gearing laws in case of the smallest gearbox inaccuracies or deflections. As mentioned above, already small deflections at moderate loads lead to load concentrations on the edges of conjugate flank pairs and can cause material damage and considerable noise emission. As such, the conjugate gear pair is not suitable for any task in a power transmission.

The introduction of 80μm length crowning and 15μm profile crowning to the conjugate hypoid design delivers the analysis results shown in Figure 15. The crowning makes the gearset insensitive to expected inaccuracies in the gear housing and load and heat affected deflections. Applying the same amounts of shaft displacements then used for the TCA of the conjugate hypoid set in Figure 14 moves the mean point slightly out of the initial position (see Figure 16), but a large contact area within the tooth boundaries is still maintained.

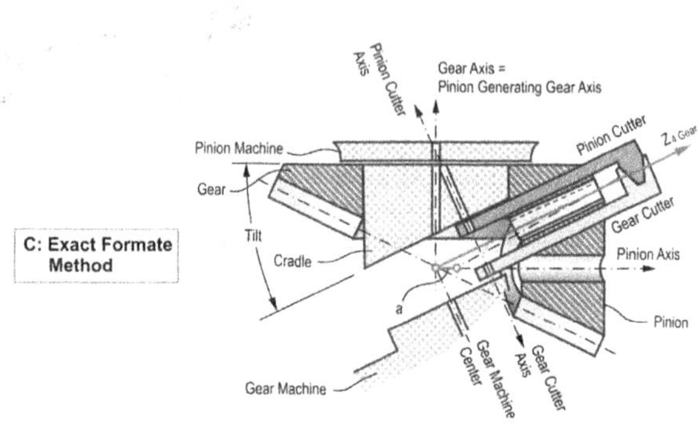


Figure 12: Generating of perfect conjugate hypoid pair.

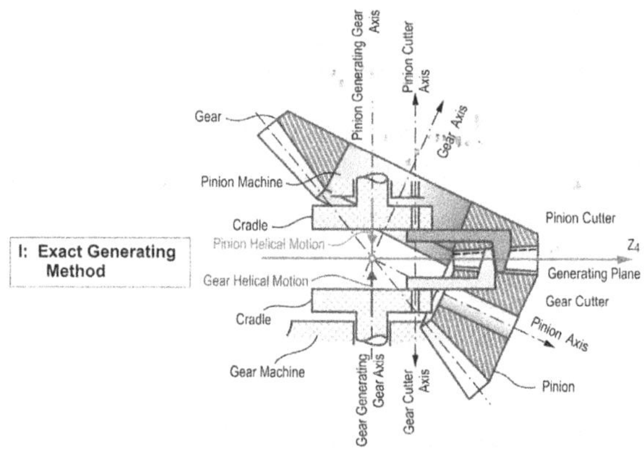


Figure 13: Generating of a conjugate tapered depth bevel gearset.

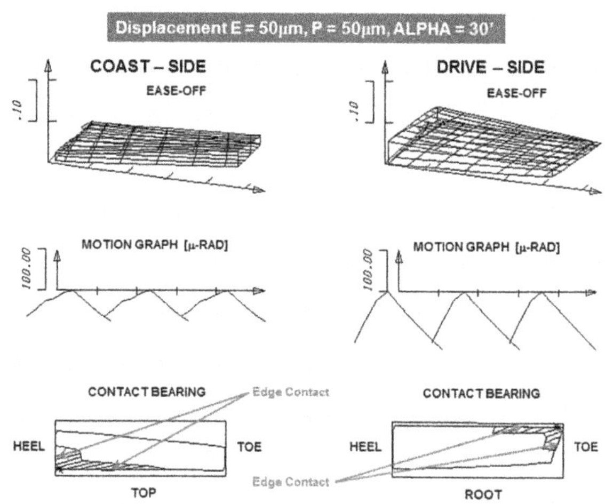


Figure 14: TCA of conjugate hypoid set from Figure 11 with shaft displacements.

It was demonstrated that the hypoid gearset with length and profile crowning in Figure 15 was developed based on a conjugate design. The first and third fundamental gearing law, mentioned in this paper, applies to the hypoid set in Figure 15 at the mean point roll position, if the load is zero. The first and third fundamental gearing law will apply in the area of contact as the load increases and the Hertzian contact spreads in contact line direction as well as in path of contact direction. This ideal condition can only be



**FIGURE 21.1** *(Continued)*

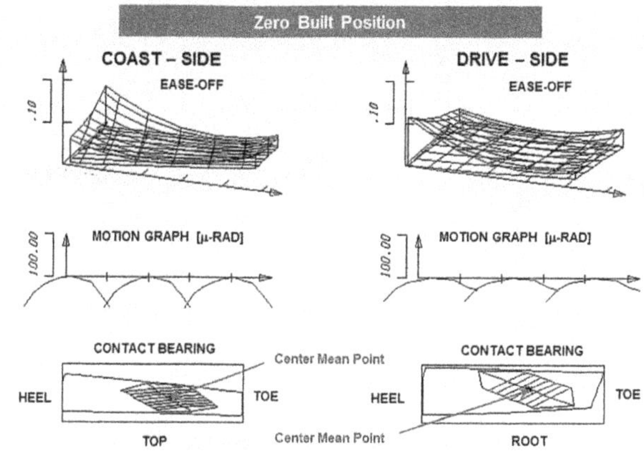


**Figure 15: TCA of hypoid set from Figure 11 without any displacements.**

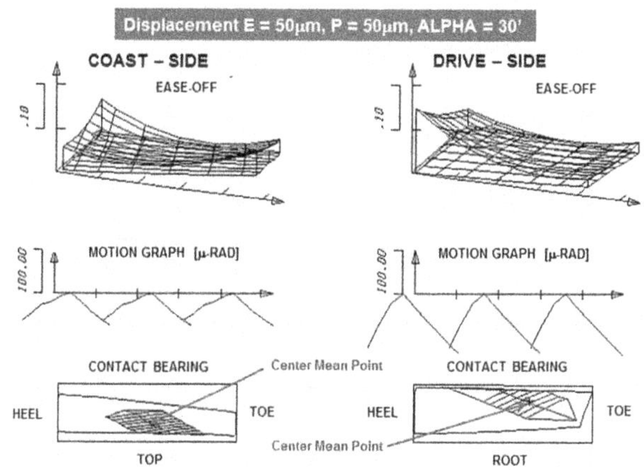


**Figure 16: TCA of hypoid set from Figure 11 with added crowning and shaft displacement.**

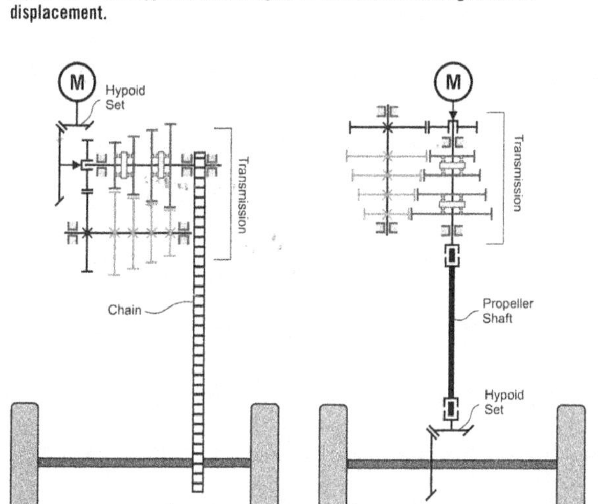


**Figure 17: Hypoid gearset locations in vehicles with rear wheel drive.**

achieved with correct amounts of crowning, adjusted to the operating displacements. It is interesting to mention that a hypoid gearset with crowning will fulfill the first and third fundamental gearing laws even in case of gearbox inaccuracies.

## TRANSMISSION ORIENTATIONS

The proposal in [1] to place the hypoid gearset between motor and transmission is not practical. Hypoid gears are not used as simple reducers, but their purpose is to redirect rotation and torque by a certain angle, commonly 90°. In case of hypoids, the second purpose of lowering the center of gravity of a vehicle body has become very important in the automotive and truck industry. The redirection of rotation and torque has to be done at the driving axle of a vehicle. The engine orientation of vehicles with a rear wheel drive is longitudinal. Because engines of cars and trucks are commonly in the front, a longitudinal oriented propeller shaft transmits the engine rotation to the rear drive axle (Figure 17, right side). The transmission, which is located between engine and propeller shaft, needs to have a gear shaft orientation identical to the direction of the engine crankshaft. The hypoid gearset redirects rotation and torque and provides the final reduction at the drive axle. The advantage of this concept is the lower torque in the complex shift or automatic transmission and the high torque only at the ring gear at the drive axle.

A concept as shown in the left side graphic in Figure 17 would require, for example, a chain connection from the transmission output to the drive axle. This solution has low efficiency, causes high noise, provides an unacceptable packaging, and is not very reliable.

## IS LAPPING AN ATTEMPT TO MAKE HYPOID GEARS CONJUGATE?

In [1, 2] it is mentioned that lapping of bevel and hypoid gearsets can be omitted if the gear geometry consists of a conjugate design. This reveals a misperception about the reason for lapping (and grinding). Lapping and grinding are hard-finishing operations. The soft manufactured bevel and hypoid gearsets have to be heat-treated, which in the most common case begins with a case carburizing of standard gear steels steel such as AISI 8620 or 16MnCr5. In order to give the low carbon steel a surface hardness in the 60 HRc range, a layer of carbon enrichment below the surface of 0.8 to 1,5mm depth is placed by a diffusion process. After the carburizing a quenching in oil and an additional tempering takes place. The result is a surface hardness that is commonly close to 60HRc and a core hardness in the 30HRc range. Case hardening provides an ideal transition between surface and core hardness that makes gears on the surface hard and wear resistant and in the core ductile. This makes shock loads and certain small plastic deformations tolerable without failure of the gearset. One major side effect of the heat-treatment process is the distortion of the gears that is caused by the carburizing, the re-crystallization of the steel, and the quenching. In order to make a gearset after heat-treatment suitable for power transmissions, for example in cars and trucks, a hard-finishing operation is required. The hard-finishing operating eliminates the heat-treatment distortions hereby providing the flank surfaces with the correct geometry from before the heat-treatment back. In addition, hard-finishing improves the surface finish to a low roughness and waviness, which enhances the hydrodynamic lubrication and reduces noise. Grinding and skiving are the preferred hard-finishing methods, creating a defined surface form that duplicates the original designed surfaces in the single micron range.

In case of face-hobbed bevel and hypoid gearsets, grinding is not possible, because of the epicyclical flank lead function. Skiving can generate epicyclical lead functions, but is not yet accepted for the high production volumes in the automotive and truck industry. This leaves only the lapping process for the hard-finishing of face-hobbed angular gearsets. However, the face-hobbed surface texture and the relative sliding between the flanks of hypoid gears make lapping an ideal alternative. Lapping can remove the surface scale



**FIGURE 21.1** *(Continued)*

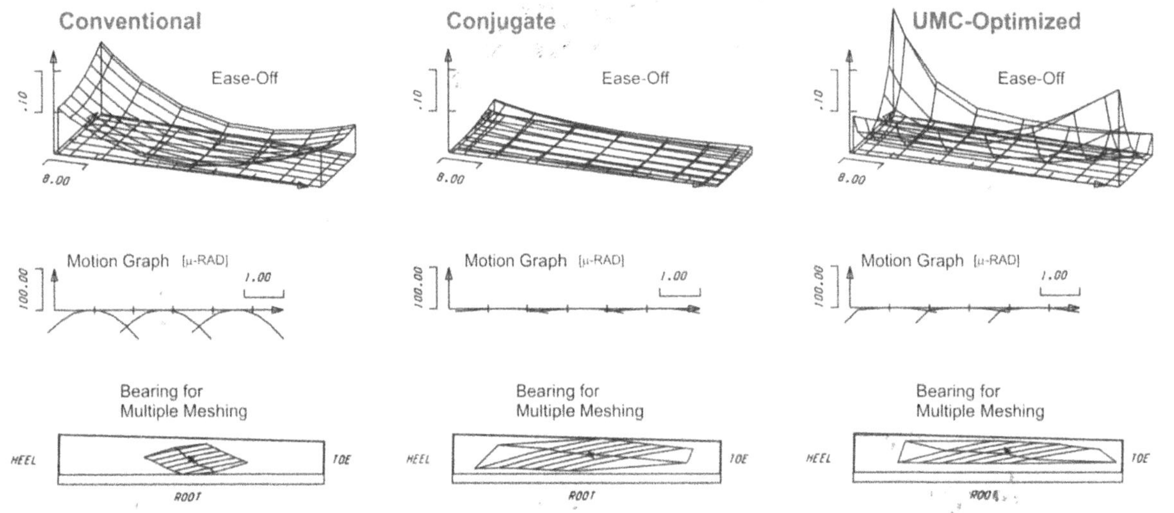


**Figure 18: From conventional crowning via conjugate to UMC-optimized.**

left from heat-treatment, and it re-matches two mating members by removing some runout and flank form distortions. Lapping can reduce the transmission error in many cases due to the fact that the major material removal is in the center region of the teeth where the tooth contact under light load is expected. In order for the lapping to work well, more crowning than required in the hard-finished gearset is used in the gearset design for the soft cutting. Lapping removes about 30 percent of this crowning, such that the length and profile crowning is just right after the lapping. Soft cutting of parts that are lapped after heat treatment considers a stock allowance of 0.03mm in the pinion and 0.01mm in the gear. If grinding is the hard finishing process (for face milled gearsets) then the design crowning is identical to the desired crowning after hard finishing. Between soft cutting and grinding, a uniform stock allowance of 0.10mm to 0.15mm is applied to the pinion and gear flanks.

## SUMMARY

Conjugacy between the members of straight bevel, spiral bevel, and hypoid gears was only the first step and goes back more than 100 years. Quickly, the early scientists and engineers found out that conjugacy only gives us an important basis, but not a solution for power transmissions. Angular gearsets under load experience deflections that move them away from their theoretical position by half a millimeter and more. Well-designed and manufactured bevel and hypoid gearsets today can live up to those requirements and still maintain a power density that is four times higher than it was 50 years ago.

Transmission errors of 50 to 150 micro-radiant that were normal in the 1970s are in today's high-power density gearsets only between 5 and 15 micro-radiant. All this was achieved by converting a global length and profile crowning (Figure 18, left) first back to conjugacy (Figure 18, center) and then into a UMC™-optimized selective crowning, which is limited to particular regions of the teeth as shown in the right graphic in Figure 18. It is notable that the flank center of the UMC-optimized Ease-Off is conjugate, and the transmission error is next to zero. In lapping, similar effects as in grinding are achieved by using low inertia spindles with rotational compliance and high-speed torque control (SmartLap™).

The dream of conjugate angular gearsets turned out to be a false objective. Gear scientists, gear engineers, and gear manufacturers worked successfully for many decades on finding the optimal flank forms and the optimal non-conjugate flank surface interaction. The conjugate tooth design is today considered simple compared to a sophisticated higher order surface modulation. There is still room for improvement, but this cannot be achieved by going back to antiquated conjugate designs.

## REFERENCES


[1] Radzevich, S.P.; Design Features of Perfect Gears for Crossed-Axes Gear Pairs; Gear Solutions Magazine, February 2019, Pages 36 to 43.

[2] Radzevich, S.P.; Preliminary Results of Testing of Low Tooth-Count Bevel Gears of a Novel Design, Part I, II and III, Gear Solutions Magazine, October 2014, December 2014 and January 2015.

[3] Wildhaber, E.; Stewart, A.J.; Design, production and application of the hypoid rear axle gear; The journal of the Society of Automotive Engineers Vol. 18, Issue 6, June 1926, Pages 575 to 589.

[4] Shtipelman, B.; Design and Manufacture of Hypoid Gears; John Wiley & Sons, Inc., New York, 1978.

[5] Stadtfeld, H.J.; Gleason Bevel Gear Technology; Expert Publishing, Renningen, Germany, May 2017.

[6] Schriefer, H.; Verzahnungsgeometrie und Laufverhalten bogenverzahnter; Kegelradgetriebe Dissertation RWTH Aachen, 1983.

[7] Wildhaber, E.; Relationships of Bevel Gears – Design for Duplex Cutting; American Machinist, October 1946.

[8] Wildhaber, E.; Relationships of Bevel Gears -Tooth Contact; American Machinist, February 1946.


ABOUT THE AUTHOR Dr. Hermann J. Stadtfeld is vice president of Bevel Gear Technology at The Gleason Works.



**FIGURE 21.1** *(Continued)*

# FROM THE EDITOR

## Industry keeps adapting to the new normal

The craziness that seems to be the driving force of 2020 has affected almost every part of our lives – from the tiniest things we may take for granted to the larger health issues that keep us up at night.

As we try to make sense and adapt to what has become a new normal, *Gear Solutions* has tried to adapt as well.

Our August issue and our September issue were originally scheduled to be a springboard for the International Manufacturing Technology Show. But like so many other 2020 tradeshows, the organizers were forced to shift gears due to the coronavirus pandemic.

I must applaud IMTS for working toward making the best lemonade out of the lemons COVID-19 has thrown at us. Even though the Chicago show is officially canceled, IMTS organizers are working to take much of the tradeshow virtual. A virtual experience means no face-to-face camaraderie as with past shows, but attendees will still get the benefit of IMTS' experts and presenters. And I suspect no one is going to really miss their annual soft pretzel food run. (OK, full confession: I do love a soft pretzel.)

It is somewhat of a relief to know that technology has been an awesome tool in keeping a lot of our business going in these weird times.

With physical tradeshows on hold, please take time to see how *Gear Solutions* can also be your ally in getting your message to your customers. We offer many ways in which to remind the industry that your products and services are available.

That's good news for your audience in search of the very services and products that you can provide every day. And with the world trying to cope with economic and medical hardships, the deep reach *Gear Solutions* can provide is more important than ever.

With that in mind, I hope you find the involute gearing and gear design articles in our August issue of interest.

In a fascinating article from Christian Weber, Thomas Tobie, and Karsten Stahl – all with the University of Munich's Gear Research Centre – the authors share their insights on the rapid and precise manufacturing of special involute gears for prototype testing.

Next up, frequent contributor Stephen Radzevich discusses the most critical accomplishments in the scientific theory of gearing.

And, as always, you'll find some expert and innovative advice from our columnists. I always appreciate the wisdom and time they volunteer for the magazine.

With all that in mind, a final reminder that *Gear Solutions* is here to serve you. If you have any suggestions or would like to contribute, please contact me. I'm always looking for exciting articles to share.

Stay safe and healthy out there, and, as always, thanks for reading!

**KENNETH CARTER, editor**
editor@gearsolutions.com
(800) 366-2185 x204

CALL FOR ARTICLES Have a technical paper or other work with an educational angle? Let *Gear Solutions* publish it. Contact the editor, Kenneth Carter, at editor@gearsolutions.com for how you can share your expertise with our readers.

6 gearsolutions.com

**GEARSolutions**

David C. Cooper
PUBLISHER

EDITORIAL
Kenneth Carter
EDITOR
Jennifer Jacobson
ASSOCIATE EDITOR
Joe Crowe
CONTRIBUTING EDITOR

SALES
Dave Gomez
NATIONAL SALES MANAGER
Tom McNulty
REGIONAL SALES MANAGER

CIRCULATION
Teresa Cooper
MANAGER
Jamie Willett
ASSISTANT

DESIGN
Rick Frennea
CREATIVE DIRECTOR
Michele Hall
GRAPHIC DESIGNER

CONTRIBUTING WRITERS
REBECCA BRINKLEY
E. BUDDY DAMM
BRIAN DENGEL
D. SCOTT MACKENZIE
ADRIAN NOWOISKY
STEPHEN RADZEVICH
KARSTEN STAHL
THOMAS TOBIE
CHRISTIAN WEBER

Media solutions

PUBLISHED BY MEDIA SOLUTIONS, INC.
P. O. Box 1987 • Pelham, AL 35124
(800) 366-2185 • (205) 380-1580 fax
David C. Cooper PRESIDENT
Teresa Cooper OPERATIONS

**FIGURE 21.2** Facsimile of page 6 in the August 2020 issue of *Gear Solutions* magazine, where the marks of the removed article [157] are still apparent.

**FIGURE 21.3** (a–h) Facsimile of the article: Radzevich, S.P., "Knowledge (of Gear Theory) is Power in the Design, Production, and Application of Gears," *Gear Solutions*, August 2020, pages 38–44 [157]. *(Continued)*

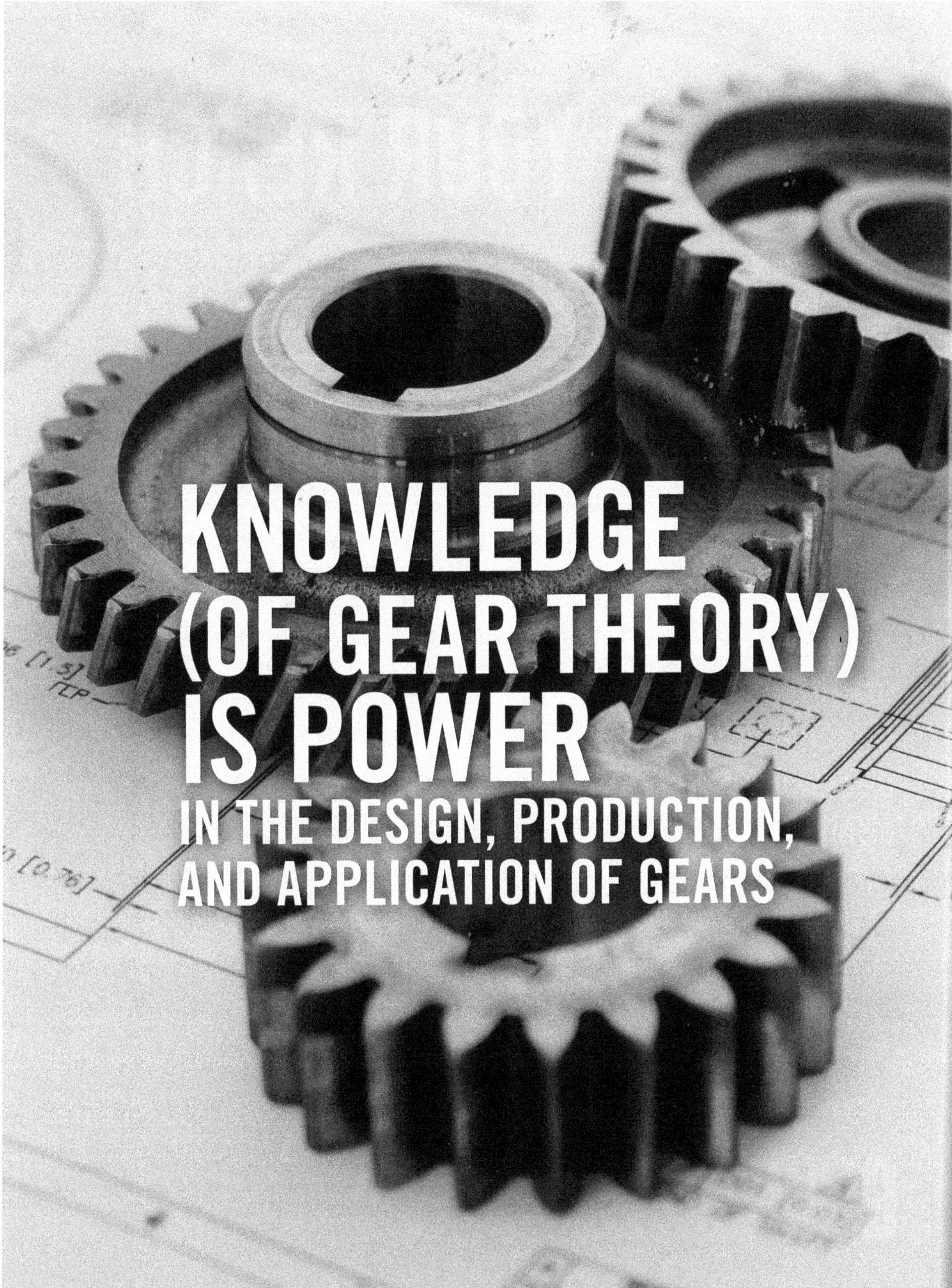


**FIGURE 21.3** *(Continued)*

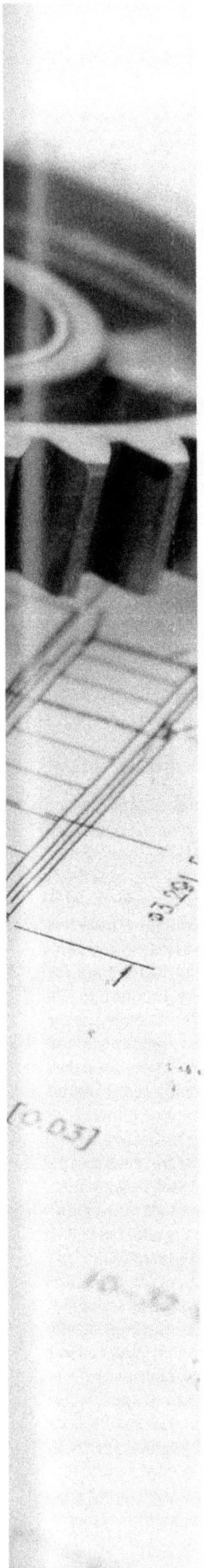

# In this article, a few of the most critical accomplishments in the scientific theory of gearing will be discussed.

By STEPHEN P. RADZEVICH

The now-famous equation, "knowledge is power" ("scientia potestas est"), was coined by Francis Bacon (January 22, 1561-April 9, 1626). Formulated as early as 1597, this famous equation was proven to be correct for centuries. Knowledge in any and all areas of human activity gives one a strong advantage over those less knowledgeable ones. Gearing is no exception. Knowledgeable and skilled gear engineers always have a greater chance for success doing business, rather than less knowledgeable and less skilled ones. This article is aimed at the promotion of knowledge (namely, the promotion of the "scientific theory of gearing") to gear practitioners, and, in this way, to equip them with a means to achieving success running a business in the gear industry.

## INTRODUCTION

The necessity of scientific foundations of gearing for design and manufacturing has long been realized by gear practitioners. Gears that operate on parallel axes of rotation of a gear and a mating pinion are the first to attract attention of mathematicians and gear theoreticians. It was Leonhard Euler who, in the mid-18th century, proved that an involute tooth profile fits the best gears that operate on parallel axes of rotation of a gear and a mating pinion.

Having started his own career in gearing in 1975, the author had many opportunities to get familiar with state-of-the-art innovations in both gear design and gear production. Shortly after graduating from college, the author was faced with: (a) numerous specific problems in gearing, and (b) the absence of an adequate means to solve these problems. The decades of research in the field of gearing theory by the author has resulted in a 2018 monograph titled "Theory of Gearing: Kinematics, Geometry, and Synthesis" [10]. It definitely makes sense to the gear community to get familiar with the capabilities that the proposed theory of gearing provides gear practitioners. In this article, a few of the most critical accomplishments in the scientific theory of gearing will be briefly discussed. The author concludes many in the gear community may be unfamiliar with the key accomplishments in the scientific theory of gearing. A recently published (2019) *Gear Solutions* article [15] (and not this article only) is taken as a representative example that proves this conclusion. Unfortunately, there are plenty of publications of this sort in current journals and magazines for gear engineers and scientists. Moreover, the article [15] (as well as other similar publications) may mislead gear practitioners, offering wrong approaches in the field of gear design and gear manufacturing. All the ambiguities in the practice of gear design and gear production should be eliminated. An old Chinese proverb, "The beginning of wisdom is to call things by their right names," is a good place to start.

## MOTIVATION

This article was necessitated by the recently published article: Stadtfeld, H.J., "Why are Today's Hypoids the Perfect Crossed-Axes Gear Pairs?" *Gear Solutions* magazine, May 2019, pp. 42-50 [15]. These comments pertain to the key accomplishments in the scientific theory of gearing. The comments could be confusing to inexperienced gear community members, especially those who attempt to study the gear theory on their own and apply the theory in their practical work.

Here and below in this text, the article [15], is used solely as a representative example of an improper understanding of the scientific theory of gearing by a majority of gear practitioners, including those who may call themselves "gear theoreticians." An additional reason to reference the article [15] is that it claims to be written on behalf of established gear theory and manufacturing (see page 43 in [15]): "This write-up is meant to be a response on behalf of established gear theory and manufacturing.") No reference is provided in [15] as to whether this "established gear theory" is available to the reader in the public domain.

Recommendation to the readers of the current article: Except for section 1, the rest of the sections in this article intentionally reflect the article [15]. This is to help the reader better understand what these two articles are about, their differences, and why the approach outlined in the current article is advantageous over the approach recommended in the article [15]. When reading the current article, it is recommended to have the article [15] for comparison.

## 1 THE INTRODUCTORY SECTION: NOVIKOV GEARING (CONFORMAL GEARING)

The evolution of gear art is heavily associated with Ernest Wildhaber. No doubt, Wildhaber is a gifted inventor and a famous gear engineer. Despite that, Wildhaber has never been recognized as a gear scientist, as is loosely claimed in [15] (page 43). Because of that, Wildhaber cannot be granted with the title of "the father of modern gear theory" (see page 43 in [15]). Both these statements are incorrect: Wildhaber was NOT a gear scientist at all, and, therefore, it is incorrect to refer to him as the father of modern gear theory. Wildhaber is credited with NO key accomplishments in the scientific theory of gearing (see [10] for details). A difference between the terms "a



FIGURE 21.3 (Continued)

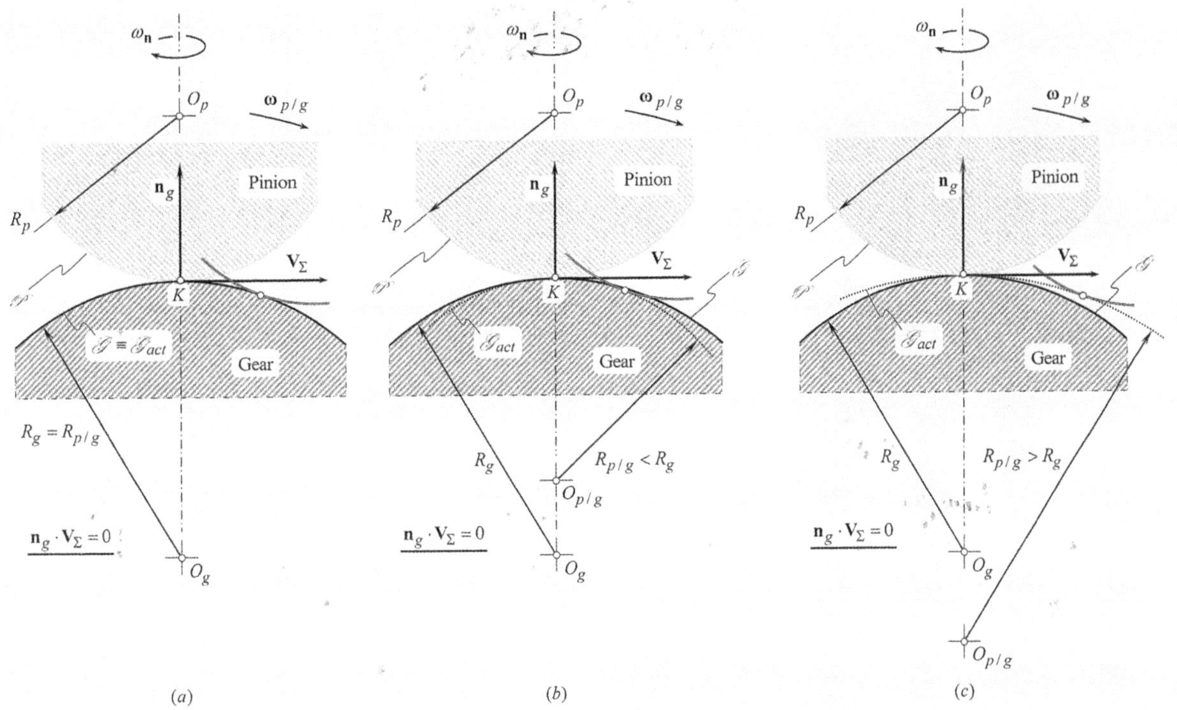


**Figure 1: Examples of (a) fulfillment, and violation: (b) interference, and (c) separation, of a gear, $\mathcal{G}$, and its mating pinion, $\mathcal{P}$, tooth flanks: the "Shishkov equation of contact, $\mathbf{n}_g \bullet \mathbf{V}_\Sigma = 0$" is fulfilled in all three cases (a) through (c), while the law of conjugacy is violated in the cases (b) and (v).**

gifted inventor and a famous gear engineer" and "a gear scientist" must be noted.

The earlier performed comparison of "Novikov gearing" [2], and Wildhaber "Helical gearing" [1] reveals these two gear systems are completely different, and, moreover, these gear systems are not compatible with one another [10, 7, 9]. On top of that, the term "Wildhaber-Novikov gearing" is incorrect by nature, as these two gear systems (namely, Wildhaber "Helical gearing" [1] and "Novikov gearing" [2]) cannot be combined into a common gear system (see [10, 7, 9], and others, for details). It is a huge mistake to adopt the term "Wildhaber-Novikov gearing."

"Novikov gearing" can be viewed as a reduced case of a parallel-axes involute gear pair, while Wildhaber "Helical gearing" cannot ([3, 6] and numerous other sources).

Today's gear professionals should know the difference between "Novikov gearing" [2] and Wildhaber "Helical gearing" [1].

## 2 NEWS ABOUT HYPOID GEARS?

The article [15] proves today's hypoid gearsets are approximate by nature, and, thus, they are not geometrically-accurate gearsets. No alterations can be even anticipated.

Further, no attempt is taken in the article [15] to "question the creditability of the scientists and gear engineers who worked on the theory and its improvement over approximately 100 years." No gear theory was developed during these 100 years, and no key scientific accomplishments are attained at this period of time. "Trial-and-error method" is the main – and perhaps only – tool used for "over approximately 100 years." But this is not a theory of gearing. Instead, this is a collection of more-or-less reasonable solutions to separate gear problems. Combined together, these solutions do not form a gear theory.

## 3 THE THREE FUNDAMENTAL LAWS OF GEARING

Leonhard Euler can be credited for originating geometrically-accurate gears (even though the term "geometrically-accurate gears" was introduced much later [10]). It is proven [10] that, in order to be referred to as "geometrically-accurate," the gears have to fulfill three fundamental laws of gearing. The fulfillment of three fundamental laws of gearing is necessary and sufficient to refer to a gearset as a "geometrically-accurate gearset." To the best possible extent, all three fundamental laws of gearing are discussed in [10]. These laws of gearing are briefly outlined here:

The first fundamental law of gearing: "At every point of contact of tooth flanks of a gear and a mating pinion, the vector of their instantaneous relative motion has to be orthogonal to the common perpendicular at every instant of time."

In this form, the first fundamental law of gearing is valid for gearing of all designs that operate on parallel axes, intersecting axes, and crossing axes of rotation of a gear and its mating pinion.

Numerous interpretations of the first fundamental law of gearing are known. The most practical and extensively adopted way of interpretation of the first fundamental law of gearing is to analytically describe this law of gearing in the form of the "Shishkov equation of contact, $\mathbf{n}_g \bullet \mathbf{V}_\Sigma = 0$." The "Shishkov equation of contact, $\mathbf{n}_g \bullet \mathbf{V}_\Sigma = 0$" is known to practically every gear engineer. Professor V.A. Shishkov was the first to express (~1948) the first fundamental law of gearing in the form of the dot product of a unit vector of a common perpendicular, $\mathbf{n}_g$, by the vector of the resultant linear velocity, $\mathbf{V}_\Sigma$, of the tooth flanks in relation to one another. Both the vectors, $\mathbf{n}_g$ and $\mathbf{V}_\Sigma$, are calculated at the contact point of a gear and its mating pinion tooth flanks.

The second fundamental law of gearing is often referred to as the "conjugate action law." When the "Shishkov equation of contact,



FIGURE 21.3 *(Continued)*

**Figure 2: Rotationally-positive external parallel-axes gear pair (Pat. No. 163857, USSR, A Helical Gearing./B.V. Shitikov, N.A. Bayazitov, Int. Cl. F06h, Filed: February 25, 1963, Published: July 22, 1964.).**

$n_g \bullet V_\Sigma = 0$" is fulfilled, the "conjugate action law" can either be fulfilled, or it can be violated. In Figure 1, an example of interaction of local patches of the tooth flanks, $\mathscr{G}$ and $\mathscr{P}$, is schematically illustrated. The tooth flanks, $\mathscr{G}$ and $\mathscr{P}$, make contact at point, $K$. The radii of curvature of the interacting tooth flanks at contact point, $K$, equal to $R_g$ and $R_p$, correspondingly (see Figure 1a). The centers of curvature of the tooth profiles, $\mathscr{G}$ and $\mathscr{P}$, are denoted by $0_g$ and $0_p$, correspondingly. In the instantaneous motion of the pinion, $\mathscr{P}$, in relation to the gear, $\mathscr{G}$, the pinion performs an instantaneous rotation, $\omega_{p/g}$, about the point $0_g$. The radius of curvature of the generated actual gear tooth flank, $\mathscr{G}_{act}$, equals to $R_{p/g} \equiv R_g$. In this scenario, the second fundamental law of gear is fulfilled, and the actual tooth flank, $\mathscr{G}_{act}$, is identical to the desirable gear tooth flank, $\mathscr{G}$, as shown in Figure 1a.

**Note:** In Figure 1a, the rotation, $\omega_n$, of the pinion, $\mathscr{P}$, in relation to the gear, $\mathscr{G}$, about the contact perpendicular, $n_g$, is not prohibited by the second fundamental law of gearing.

If the instantaneous rotation is performed either about the center $0_{p/g}$ (when $R_{p/g} < R_g$, see Figure 1b), or about the center $0_{p/g}$ (when $R_{p/g} > R_g$, see Figure 1c), the second fundamental law of gear is violated, and the actual tooth flank, $\mathscr{G}_{act}$, differs from the desirable gear tooth flank, $\mathscr{G}$.

**Reminder:** The first fundamental law of gearing (as well as the "Shishkov equation of contact, $n_g \bullet V_\Sigma = 0$") is fulfilled in all three cases shown in Figure 1, while the second fundamental law of gearing is fulfilled only in the first case illustrated in Figure 1a. The schematic in Figure 1 is helpful for understanding the difference between the first and second fundamental laws of gearing and prevents the making of unsubstantiated conclusions in this regard.

**Important:** When the second fundamental law of gearing is fulfilled, the first fundamental law of gearing is always fulfilled, and not vice versa.

With that said, it makes sense to go back to the article [15] and compare the results illustrated in Figure 1 with a loosely made statement (see page 44 in [15]): "The second fundamental law of gearing, ... , is a redundant relationship to the first gearing law, and it is limited to cylindrical gears with parallel axes and straight bevel gears without hypoid offset. In this case, it adds nothing to the first gearing law; conjugacy is already given by the relationships required in the first gearing law." It is correct to question whether a gear practitioner has a chance for success in the design and production of today's sophisticated gear systems if he or she strictly follows the afore mentioned quote (see page 44 in [15]).

In the most general case of gearing (namely, in the case of "crossed-axes gearing", or " $C_a$ – *gearing*", for simplicity) the second fundamental law of gearing is formulated as:

**The second fundamental law of gearing:** "In order to smoothly transmit a uniform rotary motion from a driving shaft to a driven shaft by means of gear teeth, perpendiculars to the tooth flanks of the interacting teeth at all points of their contact must intersect the axis of instantaneous rotation in the gear pair."

In a reduced case of gears that operate on parallel axes of rotation of a gear and its mating pinion (namely, in the case of "parallel-axes gearing", or "$P_a$ – *gearing*", for simplicity) this fundamental law of gearing is formulated as:

**The second fundamental law of (parallel-axes) gearing:** "In parallel-axes gearing, in order to smoothly transmit a uniform rotary motion from a driving shaft to a driven shaft by means of gear teeth, perpendiculars to the tooth flanks of the teeth at all points of their contact must pass through a stationary point located on the line of centers, namely, the pitch point $P$; the pitch point subdivides the center-distance reciprocal to the angular velocities of the gear and the pinion." (Robert Willis, 1841 [16])

These days, the second fundamental law of gearing (in a case of parallel-axes gearing) is commonly referred to as "Camus-Euler-Savari fundamental law of gearing" (or " fundamental law of gearing," for simplicity).

The second fundamental law of gearing is also often referred to as "conjugate action law" – and this is also correct.

The design of "Alpha Worm Gearing" (New Venture Gear, Syracuse, NY; US Pat. No. 6,148,683, Nov. 21, 2000) is a good example of the violation of the second fundamental law of gearing. Violation of the second fundamental law of gearing is the main reason for this design fail, and it has caused major financial damage to the gear industry.

The third fundamental law of gearing relates to the distribution of the gear teeth over the periphery of a gear and a mating pinion. In the most general case of gearing (in the case of crossed-axes gearing), the third fundamental law of gearing is formulated as:

**The third fundamental law of gearing:** "In order to smoothly transmit a uniform rotary motion from a driving shaft to a driven shaft by means of gear teeth, at every instant of time the angular base pitch of a gear, and that of a mating pinion, both have to be equal to the operating angular base pitch in the gear pair."

In a reduced case of "parallel-axes gearing," this fundamental law of gearing is formulated as:

**The third fundamental law of gearing (in parallel-axes gearing):** "In parallel-axes gearing, in order to smoothly transmit a uniform rotary motion from a driving shaft to a driven shaft by means of gear teeth, at every instant of time base pitch of a gear and that of a mating pinion, both must be equal to the operating base pitch in the gear pair."

August 2020 41

**FIGURE 21.3** *(Continued)*

All the geometrically-accurate gears (that is, all the precision gears – with no exception) meet all three fundamental laws of gearing – this is a must.

## 4 PERFECT CONJUGACY IN STRAIGHT BEVEL GEARSETS

The title of the section "Perfect Conjugacy in Straight Bevel Gearsets" in the article [15] may be confusing to readers, as "conjugacy" by nature cannot be "perfect" or "imperfect." Gearsets of all kinds are either conjugate, or they are not conjugate. The term "perfect conjugacy" could be considered analogous to saying "a bit pregnant."

It is loosely claimed (see page 44 in [15]) that: "The plane of action cannot be extended beyond its tangential contacting line with the base element as shown incorrectly .... The plane of action exists where tooth engagement is possible." Gearsets that feature the plane of action that extends far beyond its tangential contacting line with the base element have been known for decades. As an example, a parallel-axes gear set of this particular kind is depicted in Figure 2 [14]. Later on, this concept was evolved (~2008) by Prof. S.P. Radzevich to the cases of intersected-axes gearing, as well as to crossed-axes gearing [10].

In a set of articles [11], [12], and [13], the preliminary results of testing of low-tooth-count spiral bevel gears of a novel design are discussed. Without going into details of design and manufacture of spiral bevel gears of this design (proposed by Prof. S.P. Radzevich, ~2008, and designed using the earlier developed scientific theory of gearing [10]), it is right to stress the following:

**1.** As shown in Figure 3, the predicted contact patch is perfect (see Figure 7a on page 21 in [13]).

**2.** As shown in Figure 4, the predicted contact pattern is perfect (see Figure 9 on page 23 in [13]).

**3.** As shown in Figure 5, the contact pattern in the roll test is also perfect (see Figure 10 on page 23 in [13]), and it indicates excellent correlation with the predicted one.

**4.** No lapping was used to finish the gears.

**5.** No adjusting for the axial configuration of the gear and the mating pinion is required, and the gears are ready to run.

In today's production of precision gears, the use of the lapping process is inevitable. Those knowledgeable in the scientific theory of gearing can eliminate dirty and obsolete lapping processes from finishing tooth flanks of gears for precision gearsets.

Concerning the statements on page 45 in [15] that state: "The Octoid is the analog function of an involute, and it provides to bevel gears the same advantage as an involute provides to cylindrical gears. ... bevel gears have a trapezoidal generating profile. The straight rack ... becomes a ring ... as shown in Figure 6." It has been known for a while that the tooth profile of a crown rack for generating a geometrically-accurate bevel gear has to be shaped in the form of a curve illustrated in (see Figure 6) [10], and NOT in the form of a straight-sided crown rack. Only approximate bevel gears can be generated by the just-mentioned straight-sided crown rack "shown in Figure 6" [15]. No gearing with the so-called "octoidal" path of contact is possible at all: If one assumes the path of contact shaped in a form of an octoid curve, this inevitably entails violation of the second fundamental law of gearing – it is important to bear this in mind. The so-called "octoid gearing" is a mistake that often travels from one publication to another.

Concerning the statement on page 46 in [15] that states: "Conjugacy is the basis of all gearsets manufactured in high volume on dedicated manufacturing machines." All the bevel gears manufactured in high volume on dedicated manufacturing machines are generated by means of straight-sided crown rack (shown in Figure 6, page 45 in [15]), and, thus, all of them are approximate gears.

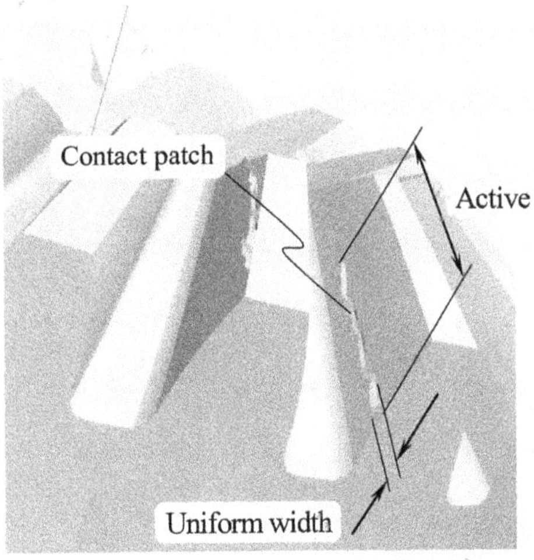


**Figure 3: Predicted contact patch in geometrically-accurate intersected-axes spiral bevel gears.**

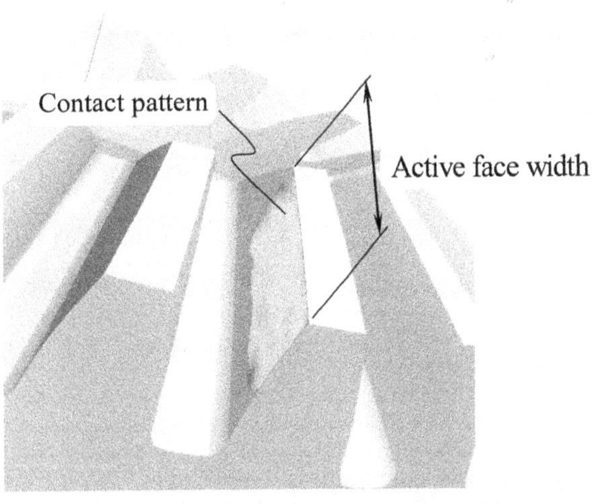


**Figure 4: The predicted contact pattern in geometrically-accurate intersected-axes gearing.**

Only approximate bevel gears can be generated by just mentioned straight-sided crown rack "shown in Figure 6" [15]. As the generated bevel gears are approximate, no conjugate bevel gear pairs can be composed of these gears.

Concerning the statement on page 46 in [15] that states: "Conjugate bevel gearsets cannot be used for a power transmission because manufacturing tolerances and ..." Today's gear industry is capable of manufacturing precision helical involute gears that are extensively used, for example, in the design of turbine reducers. The gear industry is NOT capable of manufacturing precision bevel gears, and this is the main reason for precision bevel gears NOT being extensively used in the industry. The bottom-line is: Precision bevel gears are NOT extensively used in the industry solely because the gear industry is not capable of producing (for a reasonable cost) these kinds of gears.

Concerning the statement on page 46 in [15] that states: "The right amount of crowning makes a gearset quiet and gives it the required load carrying capacity". How can the "right amount" be determined for a particular bevel gear application? The desirable amount of crowning can be determined by means of a "trial and error method" - this is the only tool developed by the "established



**FIGURE 21.3** *(Continued)*

**Figure 5: Contact pattern in geometrically-accurate spiral bevel gears in roll test.**

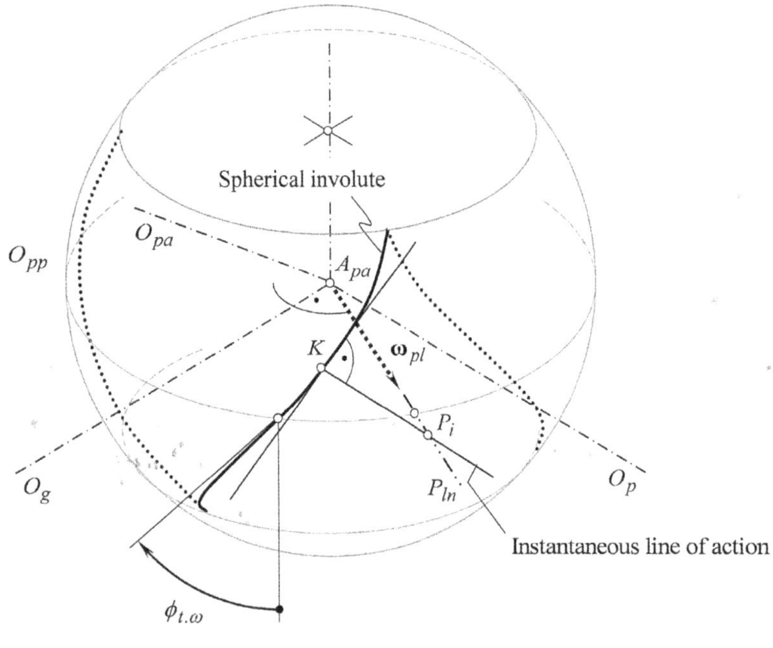


**Figure 6: Trace of contact point in geometrically-accurate intersected-axes gear pair.**

gear theory" in [15]. Today's scientific theory of gearing not only gives an answer to the question of how the "right amount" can be determined, but it also provides an in-detail specification of the entire modified tooth flank. Properly engineered gears (both design and production) do not need to use a "trial and error method."

## 5 PERFECT CONJUGACY IN HYPOID GEARSETS

The discussion on "perfect conjugacy in hypoid gearsets" in [15] is similar to "perfect conjugacy in straight bevel gearsets" discussed earlier (see Section 4). Again, the title of the section "Perfect Conjugacy in Straight Bevel Gearsets" in the article [15] may be confusing to readers as "conjugacy," by nature, can neither be "perfect"' nor "imperfect." Gearsets of all kinds are either conjugate, or they are not conjugate.

Concerning the statement on page 46 in [15] that states: "... the pitch elements of crossed axes hypoid gears are drawn as cones." It fails to describe one of the key features of the modern scientific theory of gearing [10]: In order to design gears for hypoid gearsets, no pitch elements are required at all. The term: "... the pitch elements are hyperboloids" is obsolete and is no longer used in the modern scientific theory of gearing [10].

Concerning the statement on page 47 in [15] that states: "No plane of action exists between two hyperbolic base elements." It is correct, as it proves there is no place for hyperboloids in the kinematics and the geometry of gears with crossing axes of rotation of a gear and its mating pinion.

Concerning the statement on page 47 in [15] that says: "The correct surface of action is curved and wrapped, as shown in Figure 10." Surface of action in hypoid gearing (as well as in gearing of all other kinds) is a plane through the axis of instantaneous rotation, or it is a plane through the common perpendiculars constructed at all points of the line of contact between the interacting tooth flanks of a gear and its mating pinion. The normal forces of interaction in gearing are commonly described by means of vectors. A vector is a quantity possessing both magnitude and direction. Vectors are described by means of straight-line segments with a specified direction. This leaves no room for "curved surface of action."

### 5.1 CONJUGACY BETWEEN MESHING FLANKS

Concerning the statement on page 47 in [15] that says: "The term conjugate is used in mathematics for two or more surfaces that contact each other along a line. Since the 1970s, the term conjugate has also been employed in gear technology literature to define..." Line contact between two surfaces (namely, between tooth flanks of two mating gears) is not sufficient (and is not always necessary) to refer to the pair of surfaces as to "conjugate surfaces." In addition, the second fundamental law of gearing has to be fulfilled for conjugate surfaces, which is a must. Moreover, it has been discovered over the last 10 years that two surfaces (two tooth flanks) can be conjugate to one another even when they are in point contact.

### 5.2 DEFINITION OF THE CONJUGATE GEAR PAIR

Concerning the statement on page 47 in [15] that says: "... hypoid gearset conjugacy is possible with a non-generated gear that meshes with a generated pinion." This reveals a poor understanding of the term "conjugacy," as no conjugate action is possible between non-generated tooth flank and generated tooth flank.

## 6 WHY IS CONJUGACY NOT DESIRABLE FOR REAL WORLD APPLICATIONS?

Most gear engineers don't doubt whether the conjugacy of inter-



**FIGURE 21.3** *(Continued)*

acting tooth flanks is desirable. Conjugacy is always desirable – especially in high-power-density transmissions and in high RPM-gearsets. Unfortunately, with no lapping process in finishing the tooth flanks, the gear industry is not capable of producing precision gears for real-world gearsets. For over a decade, a concept of crossed-axes gearsets that are insensitive to the mating gears axes displacements (both, the linear displacements, as well as the angular displacements) were introduced to the public [10]. Gears of this system are commonly referred to as "$S_{pr}$ – *gearing*." The "$S_{pr}$ – *gearing*" is a kind of point contact conjugate gearing capable of absorbing all the linear and angular displacements of a gear and its mating pinion axes of rotation, as long as the actual values of these displacements are within the prespecified tolerances for the accuracy of the gearset. At the same time, the "$S_{pr}$ – *gearing*" features the highest possible power density being transmitting by the gearset.

## 7 TRANSMISSION OPERATIONS

Only time will tell if it is practical "... to place the hypoid gearset between motor and transmission ...?" (see page 49 in [15]). Looking to the future, one can conclude that the vehicle powertrain is not the only potential application of gears with crossing axes of rotation of a gear and a mating pinion.

## 8 IS LAPPING AN ATTEMPT TO MAKE HYPOID GEARS CONJUGATE?

Eventually, the lapping process in the production of gears for bevel and hypoid gearsets will be eliminated [6]. The gear-grinding process is a No. 1 candidate for the future of gear-finishing operations. The sooner the key accomplishments in the scientific theory of gearing reaches gear practitioners, the sooner gear lapping will be replaced by gear grinding. This negates the statement on page 49 in [15] that says: "This reveals a misperception about the reason for lapping."

An in-depth familiarity with the key accomplishments in the scientific theory of gearing will make it possible to eliminate the gear-lapping process and replace it with the gear-grinding process. This is a reliable evolution of the gear design practice and the gear-finishing processes in particular.

## 9 SUMMARY

It is important to note that over the last 100 years, the gear industry has failed to:

- Eliminate lapping process.
- Avoid the necessity of adjustment of bevel/hypoid gears in pairs (that makes the gears not-self-replaceable).
- Eliminate the necessity of pairing of gears that operate on intersected, as well as on crossing axes of rotation of a gear and its mating pinion.
- Solve the problem of excessive gear noise excitation and vibration generation.

## CONCLUSION

In conclusion, it is likely the gear industry could end up wasting funds and time following the direction of evolution outlined in the article [15].

In the article [15] there is no evidence of:

- Whether the third fundamental law of gearing is understood by the author.
- The importance of the concept of the "operating base pitch" in a gear pair is realized.
- The concept of a gear, of a mating pinion base pitch in intersected-axes gearing, and in crossed-axes gearing, along with the concept of the "operating base pitch" in intersected-axes, and in crossed-axes gearsets as realized by the author.

Not all the inconsistences in [15] are addressed here – only some of the larger issues. However, a more detailed analysis (if necessary) can be undertaken.

## REFERENCES


[1] Pat. USA, No. 1,601,750, Helical Gearing, /E. Wildhaber, Patented: October 5, 1926, Filed: November 2, 1923.

[2] Pat. USSR, No. 109,113, Gear Pairs and Cam Mechanisms Having Point System of Meshing. /M.L. Novikov, National Classification 47h, 6; Filed: April 19, 1956, published in Bull. of Inventions No.10, 1957.

[3] Radzevich, S.P., "An Examination of High-Conformal Gearing", Gear Solutions, February, 2018, pages 31-39.

[4] Radzevich, S.P., "Conjugate Action Law in Intersected-Axes Gear Pairs and in Crossed-Axes Gear Pairs", Gear Solutions magazine, June 2020, pages 42-48.

[5] Radzevich, S.P., "Design Features of Perfect Gears for Crossed-Axes Gear Pairs", Gear Solutions, February, 2019, pp. 36-43.

[6] Radzevich, S.P., Gear Cutting Tools: Science and Engineering, 2nd Edition, CRC Press, Boca Raton, FL, 2017, 606 pages.

[7] Radzevich, S.P., Dudley's Handbook of Practical Gear Design and Manufacture, Third Edition, CRC Press, Boca Raton, FL, 2016, 629 pages.

[8] Radzevich, S.P., High-Conformal Gearing: Kinematics and Geometry, 2nd edition, Elsevier, Amsterdam, 2020, 506 pages.

[9] Radzevich, S.P., "On the Inconsistency of the Term "Wildhaber-Novikov Gearing": A New Look at the Concept of "Novikov Gearing"", Appendix A, pp. 487-501, in: Radzevich, S.P., (Editor), Advances in Gear Design and Manufacture, CRC Press, Boca Raton, Florida, 2019, 549 pages.

[10] Radzevich, S.P., Theory of Gearing: Kinematics, Geometry, and Synthesis, 2nd edition, revised and expanded, CRC Press, Boca Raton, FL, 2018, 934 pages.

[11] Radzevich, S.P., et al, "Preliminary Results of Testing of Low-Tooth-Count Bevel Gears of a Novel Design. Part 1", Gear Solutions, October 2014, pp. 25-26.

[12] Radzevich, S.P., et al, "Preliminary Results of Testing of Low-Tooth-Count Bevel Gears of a Novel Design. Part 2", Gear Solutions, December 2014, pp. 20-21.

[13] Radzevich, S.P., et al, "Preliminary Results of Testing of Low-Tooth-Count Bevel Gears of a Novel Design. Part 3", Gear Solutions, January 2015, pp. 20-23.

[14] Shitikov, B.V., Bayazitov, N.A., A Helical Gearing, Pat. No. 163857, USSR, Int. Cl. F06h, Filed: February 25, 1963, Published: July 22, 1964. [see also: Bayazitov, N.A., Helical Gears with a New Type of Gearing, Ph.D. Thesis, Kazan', Kazan' Technological & Chemical Institute, 1964.].

[15] Stadtfeld, H.J., "Why are Today's Hypoids the Perfect Crossed-Axes Gear Pairs?", Gear Solutions magazine, May 2019, pp. 42-50.

[16] Willis, R., Principles of Mechanisms, Designed for the Use of Students in the Universities and for Engineering Students Generally, London, John W. Parker, West Stand, Cambridge: J. & J.J. Deighton, 1841, 446 pages.


ABOUT THE AUTHOR Stephen P. Radzevich, Dr.(Eng.)Sci, can be reached at radzevich@usa.com or at (586) 292-7209.



**FIGURE 21.3** *(Continued)*

**Conclusion:** *The modern scientific theory of gearing is an efficient means for solving challenging problems in gear design, gear production, gear inspection, and gear application.*

Only proficient and knowledgeable gear experts can use the proposed theory of gearing in everyday practice. This theory is developed not for amateurs, and not for inexperienced gear experts.

# 22 Brief Notes on the History of Methods of Machining Gears and of Design of Gear-Cutting Tools[1]

Gears are the means by which power is transferred from source to application. The use of gears dates from time immemorial. The development of modern technology would be inconceivable without economically and accurately produced gears. A sizable section of industry and commerce in today's world depends on gearing for its economy, production, and livelihood. Today, as it has always been in the industry in the past, the design and manufacture of gears is an area of extremely specialized knowledge. Gear manufacturing practice is in a constant state of change. This results from the continuing search for novel and lower-cost methods of producing key components that are among the most difficult and expensive to produce.

Practical workers were able by various empirical means to produce gears adequate for their needs, at least until the early 19th century, when the mathematician's work was translated into practical language. Purely empirical solutions for the form of gear teeth can only be accounted for by the fact that gears operated at low speeds and under small loads.

The developments in the field of novel methods of gear cutting, novel designs of machines for cutting gears, and novel designs of gear-cutting tools, are tightly connected to one another, that is, it is common practice that a novel design of machine for cutting gears, as well as a novel design of the gear-cutting tool, are commonly initiated by the invention of a novel method of cutting gear teeth. The gear broaching process is a perfect illustration of this: The design of gear broaching machines, as well as the design of broaches for machining gears, are the consequences of the proposed method of gear broaching.

The main goal of this chapter is to briefly outline most of the fundamental accomplishments in the field of the methods of cutting gear teeth, and of the novel designs of special-purpose gear-cutting tools.

The following principles were adopted when writing this chapter:

- The principal contributions to the methods and means for cutting gear teeth should be identified, and a list of the principal accomplishments in the field should be composed.
- The names of the gear experts who are credited with certain accomplishments should be associated with their accomplishments.

For example,

a. The importance of the involute tooth profile to the gear-cutting tool designer is widely recognized. Academician Leonhard Euler[2] is credited with this accomplishment.
b. The condition of conjugacy of the work-gear tooth profile, and the gear-cutting tool tooth profile, is vital when designing cutting tools for cutting/finishing precision gears. This theorem is commonly referred to as the *fundamental theorem of parallel-axes gearing.* It is well known that the main theorem was proven by *Leonhard Euler* and (independently) by *Felix Savary.* It is also well known about the significant contributions made earlier by *Charles Camus* to the discovery of this theorem (but he failed to prove the theorem). Therefore, Charles Etienne Louis Camus,[3] Leonhard Euler, and Felix Savary should be credited with this accomplishment in the scientific theory of gearing, and it makes sense to refer to this theorem as the *Camus-Euler-Savary*[4] *fundamental theorem of parallel-axes gearing*, or just to the *CES – theorem*, for simplicity.
c. Analytical representation of the condition of contact of the tooth flank of the work-gear and the gear-cutting tool tooth flank in the form of dot product $\mathbf{n} \cdot \mathbf{V}_{\Sigma} = 0$ is of critical importance for the gear-cutting tool designer. Prof. V.A. Shishkov[5] is credited with this accomplishment. Commonly, the equation $\mathbf{n} \cdot \mathbf{V}_{\Sigma} = 0$ is referred to as the *Shishkov equation of contact,* $\mathbf{n} \cdot \mathbf{V}_{\Sigma} = 0$.

[1] Numerous stories on the design of gear-cutting tools and on methods of cutting gear teeth published earlier and available in the public domain were used when writing this chapter of the book. Most of the sources are listed either in the References or in the Bibliography.

[2] **Leonhard Euler** (April 15, 1707 – September 18, 1783) was a pioneering Swiss mathematician and physicist.

[3] **Charles Étienne Louis Camus** (August 25, 1699 – February 2, May 4, 1768) was a French mathematician and mechanician.

[4] **Felix Savary** (October 4, 1797 – July, 15, 1841) was a French mathematician and mechanician.

[5] **V.A. Shishkov** was a Soviet gear expert, doctor of engineering sciences, professor of mechanical engineering.

DOI: 10.1201/9781003520528-29

**FIGURE 22.1** Old-style gear transmission comprised of wooden gears.

In this manner, a corresponding name can be assigned to all principal accomplishments in the field of gear-cutting tool design.

Following this rule, not many gear experts are credited with the principal accomplishments in the field of gear machining—only those who were the first to discover a fundamental result of the research in the field. Unfortunately, in the meantime this approach is not applicable as not all the methods of gear cutting, as well as not all the designs of gear-cutting tools, can be associated with the corresponding names of the gear experts who were the first to propose a solution to a particular engineering problem that pertains to the methods and means of gear cutting. Therefore, this chapter of the book is limited just to brief notes on the subject. A full and complete history of the evolution of methods of cutting gear teeth, and of the corresponding designs of gear-cutting tools, is to be developed in the future.

## 22.1 EARLY ACCOMPLISHMENTS IN THE DESIGN OF TOOTHED WHEELS AND IN METHODS OF GEAR MANUFACTURE

The art and science of gearing have their roots before the Common Era. The original gear teeth were wooden pegs driven into the periphery of wooden wheels and driven by other wooden wheels of similar construction. An example of wooden gear transmission[6] is illustrated in Figure 22.1.

Neither special methods of cutting the gear teeth, nor gear-cutting tools of a special design were needed to manufacture wooden gears and gear transmissions. At that time, smart craftsmen were governed solely by common sense and by previously accumulated experience when designing and making wooden gears and gear trains. Primitive common-purpose cutting tools were used in gear production.

The oldest surviving relic containing gears is the *Antikythera mechanism*,[7] named for the Greek island near

[6] More examples of old-style wooden gears and gear transmissions can be found in: Dudley, D.W., *The Evolution of the Gear Art*, Published by AGMA, Washington DC, 1969, 93p. This rare Dudley's publication is reprinted as Appendix A, pages 483–510 in Radzevich, S.P., *Dudley's Handbook of Practical Gear Design and Manufacture*, 3rd ed., CRC Press, Boca Raton, Florida, 2016, 629 pages (also as Appendix A, pages 671–715 in the 2nd ed., CRC Press, Boca Raton, Florida, 2012, 878 pages). Many designs of old-style gears and gear transmissions are discussed in da Vinci, Leonardo, *The Madrid Codices*, Volume 1, 1493, Facsimile Edition of *Codex Madrid 1*, original Spanish title: *Tratado de Estatica y Mechanica en Italiano*, McGraw Hill Book Company, 1974.

[7] The artifact was recovered in 1900–1901 from the Antikythera shipwreck off the Greek island of Antikythera. Its significance and complexity were not understood until decades later. Believed to have been designed and constructed by Greek scientists, the instrument has been dated either between 150 and 100 BCE, or, according to a more recent view, at 205 BCE. The complexity grade of this precious example of *antique genius* was so high that artifacts of a similar complexity and workmanship did not reappear for a millennium and a half, when mechanical astronomical clocks were built in Europe.

(a) (b)

**FIGURE 22.2** Antikythera mechanism (100 BCE to 205 BCE), (a) the original artifact, and (b) that same mechanism overlapped by the nowadays replica.

which the mechanism was discovered in a sunken ship in 1900. The mechanism is not only the earliest relic of gearing, but it is also an extremely complex arrangement of epicyclic differential gearing. The mechanism is identified as a calendrical Sun and Moon computing mechanism and is dated to about 87 BCE.

The Antikythera mechanism (see Figure 22.2) is the oldest artifact consisting of gears known so far.[8] An image of the original Antikythera mechanism is shown in Figure 22.2(a). Figure 22.2(b) illustrates the Antikythera mechanism overlapped with the image of its replica. The device has more than 30 gears, although some scientists have suggested as many as 72 gears, with teeth formed through equilateral triangles. To be reproduced in metal, the simple geometry of the gear teeth used in the design of the Antikythera mechanism did not require any special methods for cutting gear teeth, or any gear-cutting tool of special design. This job was done by smart craftsmen governed solely by common sense and by previously accumulated experience. Primitive common-purpose cutting tools were used for cutting the gear teeth.

The history of gear-cutting mechanisms, as well as of gear-cutting tools, begins early in the 16th century, and may be even associated with some of *Leonardo da Vinci*'s drawings [20]. The first machines to cut gears were designed and used for machining gears at the end of the 17th century and at the beginning of the 18th century. *Robert Hooke*'s wheel cutting engine (circa 1672 CE), *Christopher Polhem*'s machine to cut clockwork gears[9] (circa 1729 CE), and a few others to be mentioned are perfect examples of early designs of machines to cut gear teeth.

Only common-purpose cutting tools of primitive design were used for cutting the gear teeth on these gear-cutting machines.

## 22.2 EARLY DESIGNS OF SPECIAL-PURPOSE CUTTING TOOLS TO PRODUCE GEAR TEETH

With progress in the use of mechanical devices came increased use of gears, and the form of the gear teeth changed to suit the application. The contacting sides or profiles of the teeth changed in shape until eventually they became parts of regular curves. The changes to the gear tooth profile entailed corresponding changes to the design of cutting tools for cutting the gear teeth.

The first known gear-cutting machine is described by Morales [18]. *Juanelo Turriano*[10] (see Figure 22.3) is credited with the invention of this early design of a gear-cutting machine. As shown by Woodbury [221], Turriano had a rotary file cutter not unlike those of 200 years later, but since it cuts iron, it must have been hardened. No details on the rotary file cutter tooth profile geometry are discussed by Woodbury [221].

The oldest known gear cutter (a rotary file) is shown in Figure 22.4. This gear-finishing tool was made in France,

[8] The earliest known reference to a gear was around 50 CE, *Hero of Alexandria*, though the *Book of Song* suggests that the south-pointing chariot may have employed differential gears as early as the reign of the Zhou Dynasty (1045–256 BCE) of China (Radzevich, S.P, *Dudley's Handbook of Practical Gear Design and Manufacture*, 3rd ed., Boca Raton, FL: CRC Press, 2016, 629 pages.). However, no artifact of the south-pointing chariot has been discovered so far. Only currently designed replicas are known. Therefore, in the meantime, the south-pointing chariot cannot be considered a relic of a mechanism with gears.

[9] See: (a) Robert Hooke's wheel-cutting engine, circa 1672 (Figure 35 on page 54 in [20]); (b) Christopher Polhem's machine to cut clockwork gears, circa 1729 (Figure 36 on page 55 in [20]), and others.

[10] **Juanelo Turriano** (Italian: Gianello Torriano; born Giovanni Torriani, c. 1500–1585) was an Italian and Spanish clockmaker, engineer, and mathematician.

**FIGURE 22.3** Juanelo Turriano (c. 1500–1585).

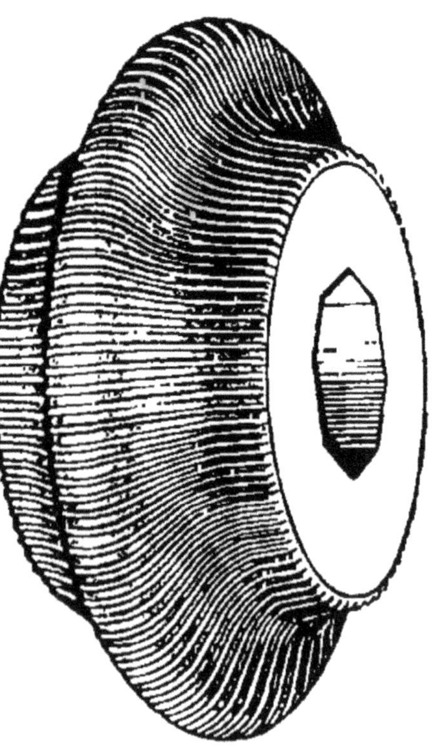

**FIGURE 22.4** Oldest known gear cutter (rotary file), circa 1782.

**FIGURE 22.5** Jacques de Vaucanson (1709–1782).

circa 1782. *Jacques de Vaucanson*[11] (see Figure 22.5), a famous French inventor, is credited with the invention of the form file for finishing wooden models of cylindrical gears. The gear models made of wood were used in finishing gears made of metal.

By the beginning of the 19th century, several other designs of gear-cutting tools (form gear mill cutters) were proposed. The concept of the design of form gear mill cutters can be traced back to Vaucanson's form rotary file (see Figure 22.4). Relieving the clearance surfaces of the teeth of form milling cutters became practical later on.

Originally, gear teeth were produced with form mill cutters on milling machines equipped with index heads.

Nowadays, designs of disk-type form gear mill cutters that originated from Vaucanson's form rotary file design are widely used in the low-volume production of gears.

## 22.3 GEAR-CUTTING TOOLS FOR THE FIRST PRODUCTION MACHINES

*Robert S. Woodbury* in his book on the history of gear-cutting machines [221] indicated that the first production machines to cut gears appeared in the first half of the 19th century (that is, between the year 1800 and the year 1855). At this approximate period of time, the first gear-cutting tools for production machines appeared on the market. These mechanisms and gear-cutting tools developed actively until 1930, when most of the primary types seem to have become stabilized. Thus, the entire sequence covers more than a century.

[11] **Jacques de Vaucanson** (February 24, 1709 – November 21, 1782) was a famous French inventor and artist who was responsible for the creation of impressive and innovative automated machines such as the first completely automated loom.

**FIGURE 22.6** John George Bodmer (1786–1864).

*Johann Bodmer* (see Figure 22.6) designed and manufactured spur gears with the ratio of 1:145 to 1:15, bevel gears, and helical gears. His achievements in the field of gears and in other areas of the machine industry were related to his remarkable ability to design tools.

In his patent[12] of 1839, Bodmer[13] described gear-cutting tools of two types. Those for cutting wooden patterns for cast gears were of the fly-cutter type. Those for cutting metallic gears were of the mill type, segmented for convenience in hardening. All were formed cutters. The problem of sharpening them gave great trouble and expense. Johann Bodmer was ahead of his time, and full recognition of his genius came later, and not in his lifetime.

As cast gears come more and more into use in the last half of the 18th century there were two trends: toward increased accuracy of gears, and toward increased size of gears.

It is likely the gear generating principle was utilized for the first time in *Saxton*'s gear generating machine (1842) [221]. The gear-cutting of a simple geometry was used to cut gears.

The molding-generating principle had already appeared in *Witworth*'s patent[14] of 1835 [221]. In this machine, the gear blank is geared by means of a train of gears with a worm-wheel hob in exact gear ratio without slippage. This was the first machine to generate involute teeth. In 1856, *Schile* took out a patent for a machine designed to cut spur gears with a hob,[15] also having gearing to rotate the blank and the hob at the correct relative speed.

[12] *Bodmer*'s British Patent No. 8070 of May 29, 1839.

[13] **John George Bodmer** (December 6, 1786 – May 29, 1864) was a prolific Swiss inventor, making contributions to areas ranging from weaponry to steam engines, textile manufacture (machinery for wool spinning), and railroad construction.

[14] *Witworth*'s British Patent No.6850 of June 11, 1835.

[15] *Schile*'s British Patent No. 2896 of December 6, 1856.

It seems clear that almost all types of gear-cutting tools for production-type gear-cutting machines that embody all the basic methods of forming the teeth had been developed by 1850.

## 22.4 EVOLUTION OF THE GEAR-CUTTING TOOLS FOR PRODUCTION MACHINES

Although in 1765 Leonhard Euler had already discovered in the involute curve a suitable tooth form for the kinematically correct transmission of rotation, there was still a long way to go before the design and production of gear-cutting machines that could generate an involute profile.

It had been recognized that the worm is a form of continuous rack and that all that was necessary to cut gears with it was to provide cutting edges on it—to make a hob (see Figure 22.7). According to [221], teeth had been cut by this method probably for the first time by *Ramsden*[16] in 1768. But this was only instrument work. The first application of the hobbing principle to a production machine was made in 1835 by *Joseph Whitehead*,[17] but his machine would cut only spiral gears. In 1839, *Pfaff* had used a hob to cut worm gears, as had Whitehead in 1853. In 1856, *Christian Schiele*[18] was granted a patent for a screw-shaped cutter to manufacture cylindrical gears, the forerunner of the modern hob. Schiele's patent specifications provided for gearing the blank to the hob in the proper ratio. This was the first hobbing machine to do this. Schiele used his hob 90° to cut spur and helical gears, and he also cut worm gears on his machine. Although Schiele's machine is clearly the beginning of the gear-hobbing type, it is not known if it ever appeared in practice (see Figure 22.8).

As early as 1854 the backed-off formed type of milling cutters (or *relieved milling cutters* in other terminology) had been invented by *Joseph R. Brown* (1843–1903). Backed-off formed milling cutters appeared on the market in 1867.

## 22.5 GOLDEN ERA OF PIONEERING DEVELOPMENTS IN GEAR MANUFACTURING

Modern gear-cutting techniques were established around 1900 by machine tools (generating machines) that produced the required tooth forms by means of mechanisms based on the geometry of gearing. Some of the outstanding engineering feats of those days have become closely linked with the names of eminent engineers, such as, in the field of generating machines for spur gears: *Edwin R. Fellows* (gear shaper, U.S. patent of 1896), *Sam Sunderland* (gear planer, British patent of 1908), *Max Maag* (gear planer, German patent of 1912; gear grinder, German patent of 1913), and also *Hermann Pfauter* (gear hobber, German patent of 1897).

[16] Ramsden J., *Description of an Engine for Dividing Mathematical Instruments*, London, 1777.

[17] *Whitehead*'s British Patent No. 6850 of June 11, 1835.

[18] *Schiele*'s British Patent No. 2896 of December 6, 1856.

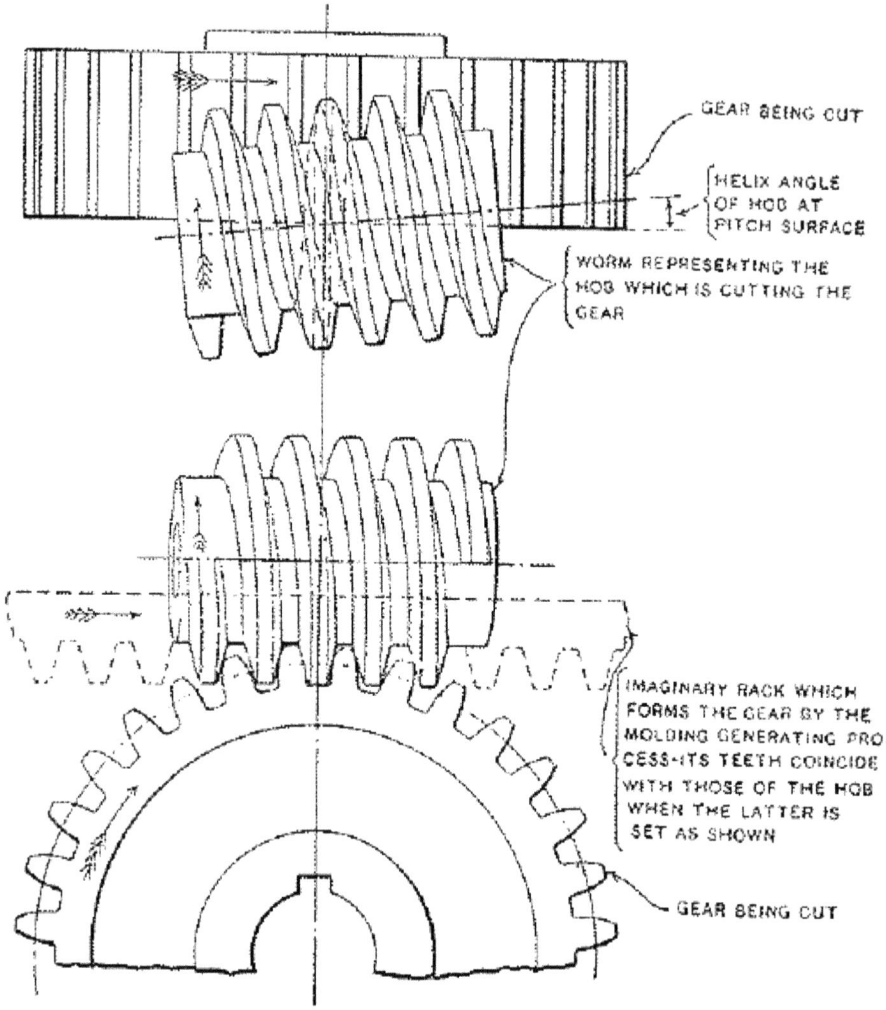


**FIGURE 22.7** Screw-shaped cutter to manufacture cylindrical gears by Christian Schiele.

**FIGURE 22.8** Sir David Brown.

Involute gearing alone offered the possibility of developing the first (basic) tool with straight flanks, thus ensuring accuracy, and of using it in the generating process. As a result, the turn of the 20th century saw the decision made in favor of the involute form for all spur and bevel gears produced on generating machines. From then on, the making of teeth by means of templates belonged to the past for the major portion of gear production.

Three inventors most significantly improved the gear-cutting process. They were pioneers in creating gear-producing automotive machines based on the continuous relation of motions between the generating tool and the gear being generated. Such machines provide continuous indexing and generate the gear tooth surface as the envelope to the family of tool surfaces.

These inventors are described next in chronological order. They are Robert Hermann Pfauter, Edwin R. Fellows, and Max Maag.

First, in 1897, *Robert Hermann Pfauter* invented the process (and machine) for generating spur and helical gears with hobs.

**FIGURE 22.9** Robert Hermann Pfauter (1854–1914).

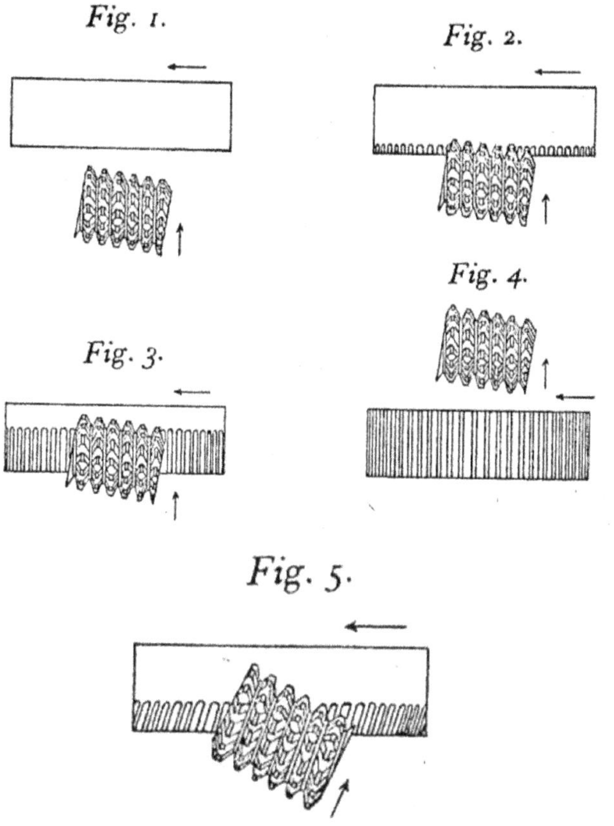


**FIGURE 22.10** Method and machine for hobbing helical gears by means of wormlike cutters, by Robert Hermann Pfauter (German Patent No. 112.082, Filed: September 2,1897. Patented: 1900).

For many years, Pfauter (see Figure 22.9) noted the key disadvantages of the existing manufacturing process for spur and helical gears:

- The necessity of indexing the gear for the generation of each tooth space.
- The generation of the tooth space only as a copy of the tool surface.

In 1897, he invented the process and machine for generating spur and helical gears with hobs. The 1897 patent for generating spur and helical gears was free of these disadvantages, and although it claimed to be limited to the generation of helical gears, it could actually be applied to the generation of spur and worm gears by hobs (see Figure 22.10). In 1900, Pfauter built a first experimental universal gear hobbing machine with a differential in accordance with his patent. Pfauter's machine was, of course, not the first that could produce gears of the correct form, but it added to those that used (or were to use) the geometry of gearing to achieve the same aim.

Pfauter was justifiably proud of his new method of generating gears, writing at the time of his patent application, "Taken as a whole the new method must be called perfect in itself, and rendering further developments unnecessary."

In a certain sense, Pfauter was the precursor of a group of inventors who later proposed other gear generation processes based on principles similar to those proposed in his patent.

Later on, the concept of Pfauter's hob design was evolved by *Nikola John Trbojevich* [also known as *Nicholas J. Terbo* (1886–1973)] to the design of globoidal hobs (see Figure 22.11).

The design of the hob shown in Figure 22.12 for forming double enveloping worm gears by *A. Macmann et al.* is another example of a hob of a similar design.

With the widespread use of the hobbing process, gear machining processes based on the application of templates is rapidly going out of use.

*Heinrich Schicht* took up the idea of hobbing cylindrical gears and transferred it to bevel gears, using a conical hob instead of a cylindrical hob to manufacture spiral bevel gears. Schicht and Preis applied for a patent with this idea in 1921.

The name of *Edwin R. Fellows* (see Figure 22.13) has to be mentioned as the second significant inventor. Fellows invented a revolutionary method to generate gears by shaping. This invention was based on the application of shapers. In the gear machining mesh, the interaction of two gears during their rotation about two parallel axes is simulated (see Figure 22.14). The manufacture of spur gears by shaping originated at the *Fellows Gear Shaper Company* in Springfield, Vermont, in 1896. Fellows' invention was a significant contribution because it enabled the automation of the internal and external gear manufacturing process. The automotive industry's broad application of the Fellows gear shaper process confirmed the importance of the invention.

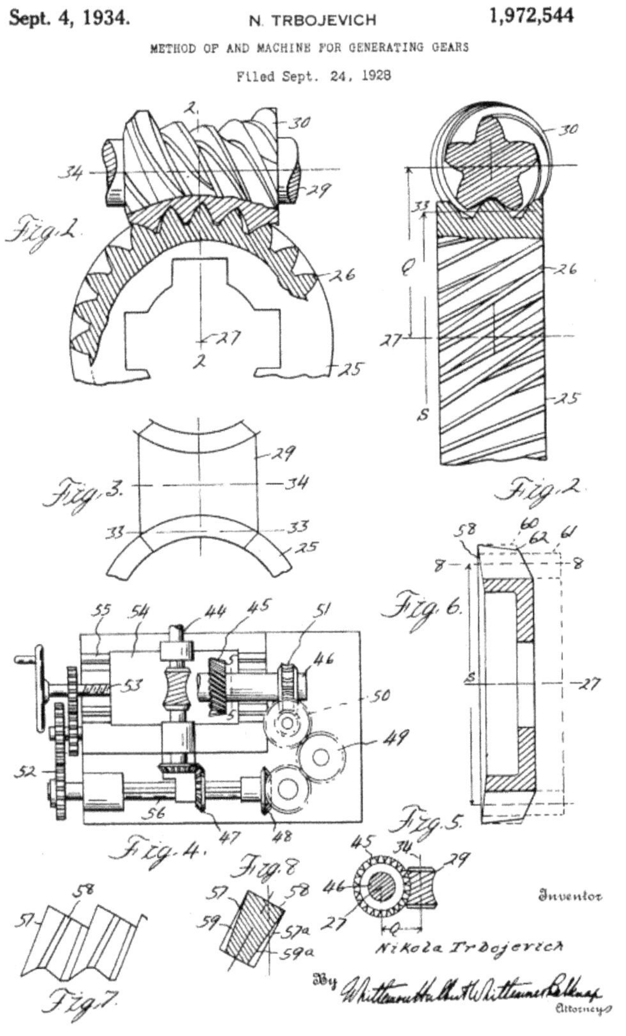


FIGURE 22.11 Method and machine for generating gears, by Nikola John Trbojevich (U.S. Pat. No. 1.972.544, Filed: September 24, 1928. Patented: September 4, 1934).

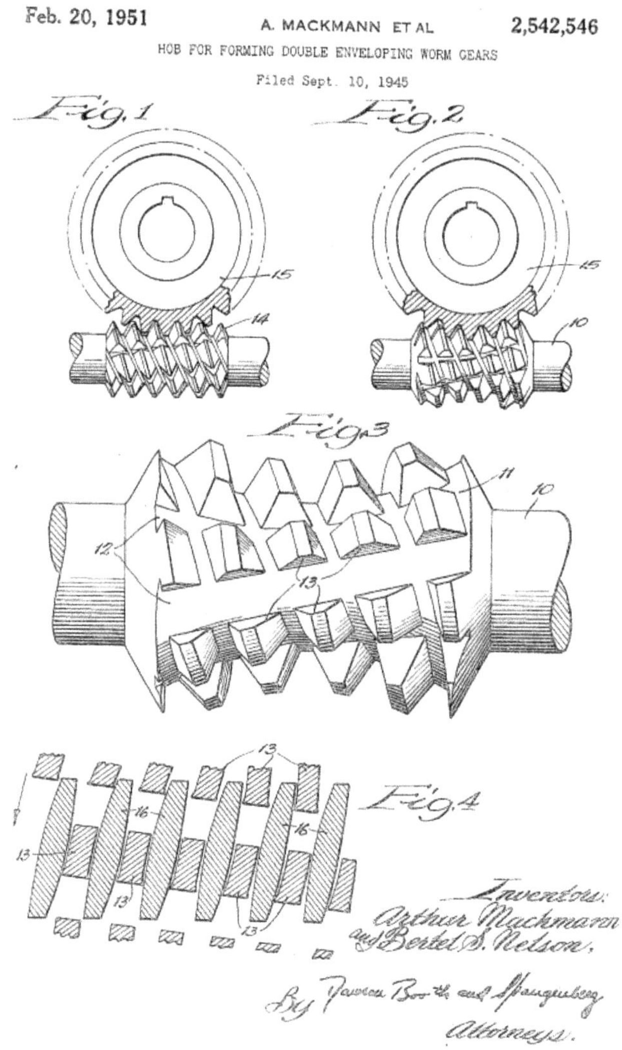


FIGURE 22.12 Hob for forming double enveloping worm gears, by A. Macmann et al. (U.S. Pat. No. 2.542.546, Filed: September 10, 1945. Patented: February 20, 1951).

Continued development produced machines capable of generating helical and face gears.

In 1897, the Fellows Company designed and built a machine for grinding involute profiles on gear shaper cutters. This process was later recognized for its application in grinding the various tooth profile cutters that were used to shape spur, helical, and special tooth profile gears.

In the early 1920s, *William Edwin Sykes* of Great Britain filed a patent for the design of a gear shaper cutter form generating cluster gears, gears with a closely located wall, and similar gears (see Figure 22.15).

It will be noted that in this patent the teeth 7 upon the periphery of the cutters are of helical formation and are of substantially the same shape in cross section as the teeth to be cut on the blank. The two are similar in shape, except that they are oppositely formed so as to provide right- and left-hand cutters, respectively, and are positioned opposite each other so that their cutting edges lie in opposed relation.

FIGURE 22.13 Edwin R. Fellows (1865–1945).

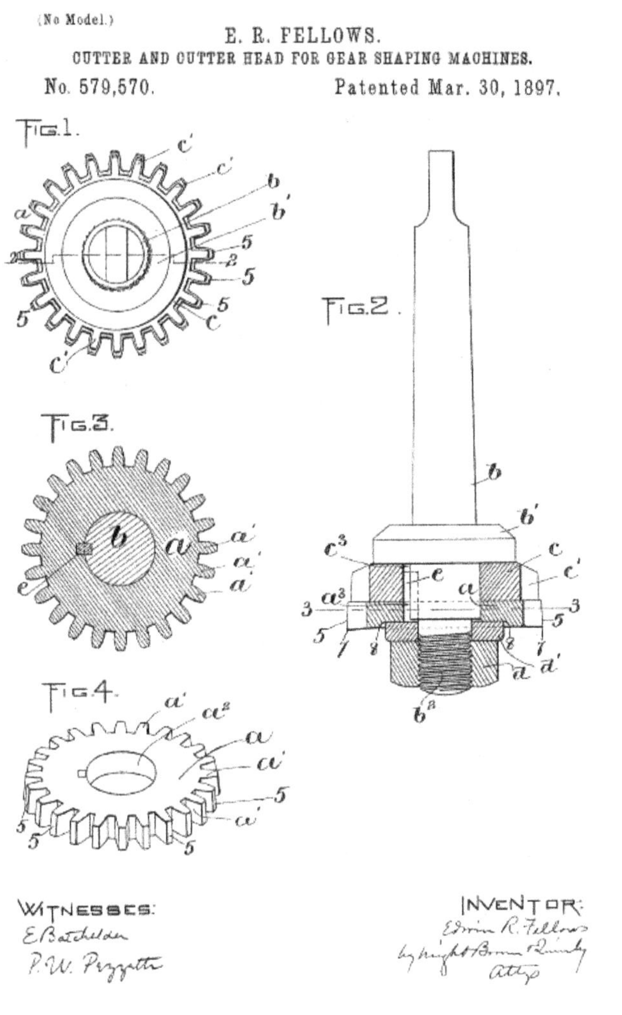


**FIGURE 22.14** Cutter and Cutter Head for Gear Shaping Machines, by Edwin R. Fellows (U.S. Patent No. 579.570, Filed: August 15, 1896. Patented: March 30, 1897).

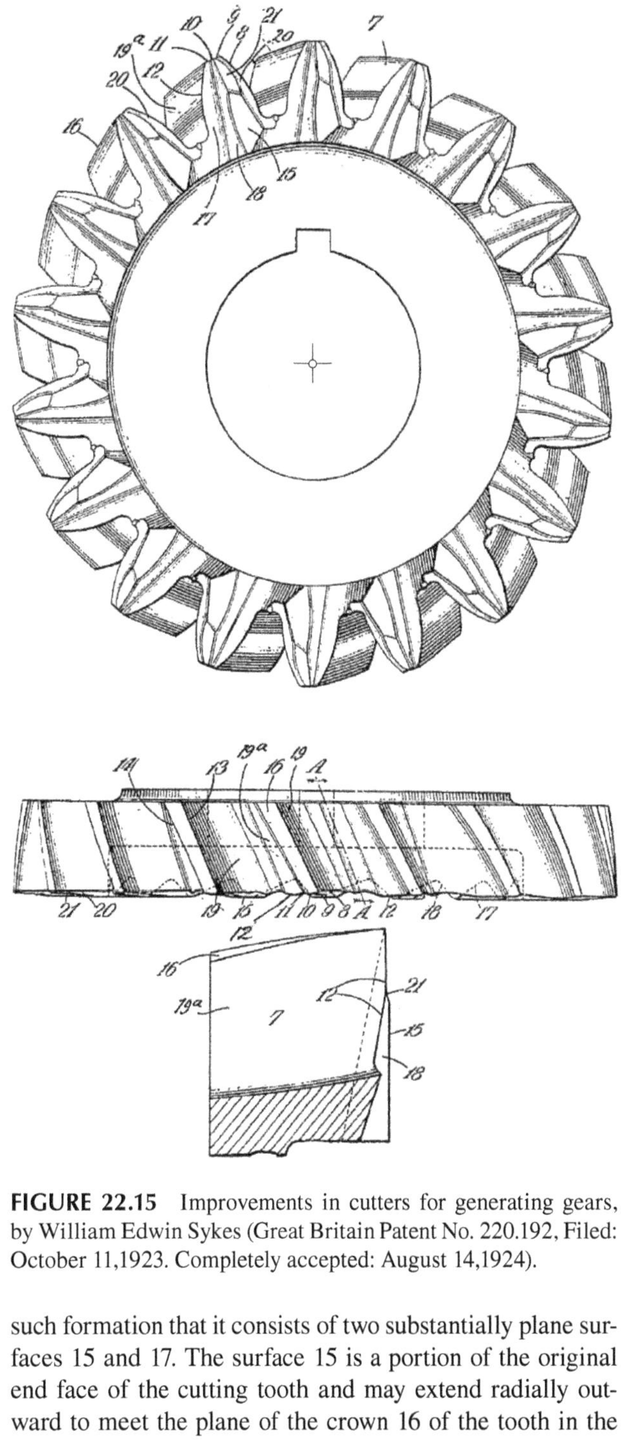

**FIGURE 22.15** Improvements in cutters for generating gears, by William Edwin Sykes (Great Britain Patent No. 220.192, Filed: October 11,1923. Completely accepted: August 14,1924).

The cutting edges of the teeth 7 which are represented in Figure 22.15 by the reference characters 8, 9, 10, 11, and 12, and constitute substantially the entire outline of the end face of the tooth, all lie substantially parallel to the side of the cutter pinion or at right angles to the shaft upon which the cutter is mounted, and make oblique angles with the longitudinal edges 13 and 14 of the cutter teeth. Such an arrangement will, as is obvious, obviate the disadvantages resulting from having one edge of the cutter in advance of another in its reciprocating movement relative to the face of the gear blank, as would be the case if the end face of the teeth were in a plane normal to the sides thereof.

It will be apparent, however, that if the entire end face of the tooth constituted a plane surface transverse to the axis of the cutter pinion, the edge 8 would be formed by two planes meeting at an angle considerably less than a right angle, the difference being determined by the helical angle of the cutter teeth, and the cutting edge 10 would be formed by two planes meeting at an angle at an angle considerably greater than a right angle. To obviate this disadvantage, the end face of each of the cutter teeth is, according to the invention, of such formation that it consists of two substantially plane surfaces 15 and 17. The surface 15 is a portion of the original end face of the cutting tooth and may extend radially outward to meet the plane of the crown 16 of the tooth in the edge 9. The surface 17 is cut away as explained below, the two surfaces 15, 17 being joined by a shoulder 18 extending more or less longitudinally of the tooth. The shoulder 18 is of considerable width adjacent to the base of the tooth and tapers toward the apex thereof until it finally runs out or nearly so at the edge 10. In order that the angle between the end face of the tooth and its side face 19 may be an angle only slightly less than a right angle, the outer marginal portion of the surface 15 is cut back from the face of the cutter pinion to some extent toward the outer edge 8, as shown at 20. The surface 17 is formed by cutting back the original surface 15 of the end face in a similar manner in order to form a similar

**FIGURE 22.16** Max Maag (1883–1960).

angle between the finished end face and the side face 19[a], the shoulder 18, which results from this latter cutting operation, being between the surfaces 17 and 15. This formation does not, however, interfere with the arrangement of having the cutting edges all in substantially the same plane transverse to the axis of the cutter. A relief cut is made at 21 in order to make the angle at the edge 9 somewhat acute.

The edge 8 is the forward cutting edge of the gear tooth, or that edge which first comes in contact with the blank during the generating rotary movement of the blank and cutters which take place during the cutting operation. It is this edge of the cutting teeth that moves into and cleans out the apices of the teeth on the blank. The advancing movement of the cutter ceases when this edge, which as stated is in a plane transverse to the blank axis, arrives at the centerline of the blank so as not to foul the other half of the gear tooth formed by the other cutter.

One more significant accomplishment in the field of cutting-tool design and manufacture is due to Dr. *Max Maag* (see Figure 22.16). In 1908, Dr. Maag developed the geometry of nonstandard involute spur and helical gears, which enabled designers to avoid undercutting and to increase the tooth thickness just by modifying the installation of the tool (the rack-cutter) with respect to the gear being generated. His research led to the development of the Maag system of generating nonstandard involute gears. The name of the Englishman Sam Sunderland can also be mentioned in this regard. In this system, he applied the method and machine invented by *Samuel Sunderland* of Kieghley, York, in Great Britain, and in 1909 received the first patent for cutting nonstandard involute gears based on the application of Sunderland's machine. The advantage in generating gears with a rack-cutter was that it was a simpler and more precise tool than a hob and a shaper.

A significant achievement of the Maag Company was the development of the method to grind spur and helical involute gears. The grinding tool was a saucer-shaped wheel with a 15 profile angle, a prototype of which was built in 1913. This invention was important because it provided the first automatic compensation for the wear of the grinding wheel, making it the first and most famous automatic control system in machine tool history.

In 1950, the method of grinding spur and helical gears by a plane was invented. The originality of the idea lay in the application of two grinding planes with a pressure angle of a zero value, each plane simulating the surface of a rack tooth with a pressure angle of zero.

## 22.6 DEVELOPMENT OF THE SKIVING INTERNAL GEARS PROCESS

On March 1, 1910, *Wilhelm von Pittler* registered a patent that aimed to significantly increase the productivity of the manufacture of internal gears. His patent for skiving, with the title *Procedure for the Cutting of Internal Gears by Means of an Internal Gear-Like Cutting Tool on which the Faces of the Teeth are Provided with Cutting Edges*, was initially reminiscent of the method for shaping internal gears that was already known at that time.

In order to produce rings with internal gearing, one is still forced to rely on broaching, form cutting, or shaping. It is precisely at this point that skiving opens up new opportunities. An internal gear with cutting on the abutting face is used as the tool. In contrast with shaping, however, the cutting movement is not generated by an oscillating stroke movement. It is far more the case that the intersecting axis alignment of the tool and workpiece creates an axial relative speed that makes the cutting movement possible. During the rotation of the tool, each cutting edge cuts through different tooth depths on the workpiece—a movement that is required for the cutting process. The intersection of the axes accounts for the fact that the helix angle of the tool and the helix angle of the internal gear to be manufactured differ by the amount of the axis intersection angle (see Figure 22.17 sketch). *Von Pittler* thus developed a continuously revolving gear teeth generating process, which drastically increases productivity for internal gearing. As a rule, skiving makes it possible to generate any kind of periodic structures on axially symmetrical lateral areas.

Figure 22.18 shows how Pittler's method of gear generation looks nowadays.

## 22.7 DEVELOPMENT OF THE ROTARY GEAR SHAVING PROCESS

The concept of rotary gear shaving can be traced back to the early years of the 20th century and can be associated with the gear shaving process by means of a rack-type gear shaver (see Figure 22.19). The work gear and the rack-type

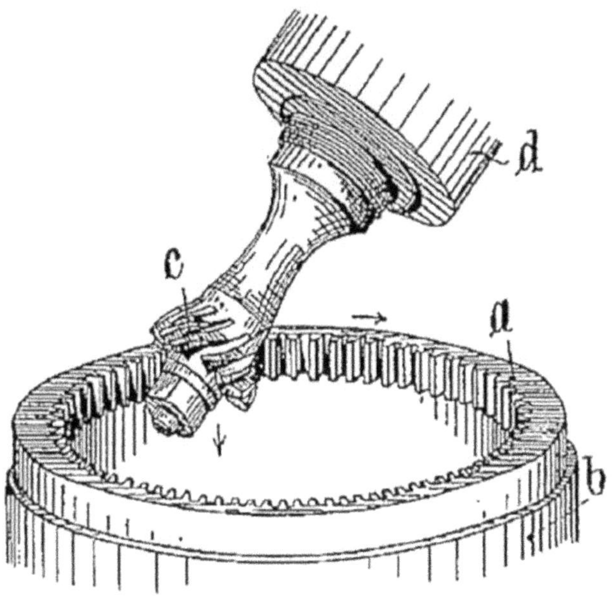


**FIGURE 22.17** Wilhelm von Pittler's 1910 sketch of the skiving process.

**FIGURE 22.18** Close-up of gear skiving process.

gear shave cutter are put in tight mesh with one another and reciprocate in the lengthwise direction of the cutter under light load. The work gear tooth flanks $\mathcal{G}$ and the rack-type gear shaver tooth flanks $\mathcal{P}$ are permanently in line contact. This is one of the key features of the gear shaving process due to which a high accuracy of finishing of involute gears is guaranteed.

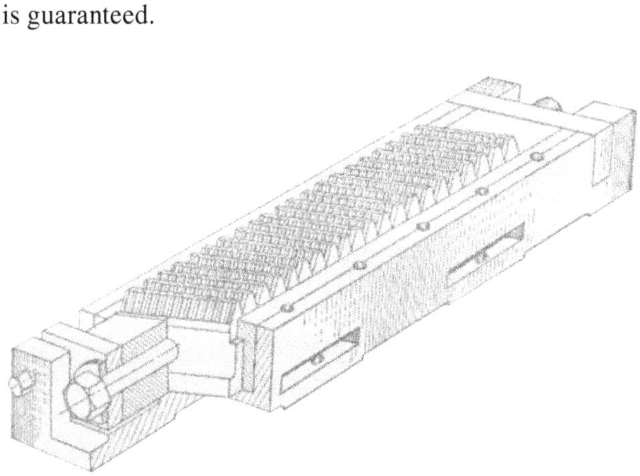

**FIGURE 22.19** Old-style rack-type gear shaver.

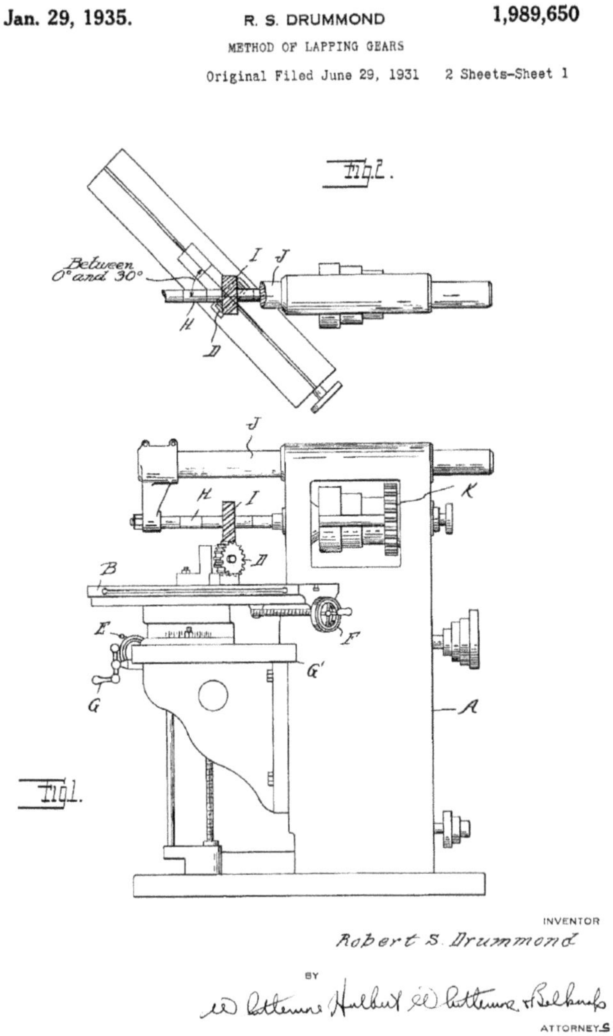


**FIGURE 22.20** A few of the illustrations to Drummond's U.S. Patent No. 1,989,650 on the method of lapping gears.

Gear shaving by a rack-type shaver can be considered the predecessor of the rotary gear shaving process. The development of the rotary gear shaving process can be traced back to the early 1930s when *Robert S. Drummond* was granted the U.S. patent[19] on a method of lapping gears (see Figure 22.20). The gear finishing tool is engaged in tight mesh with a gear to be finished at crossed axes of rotation of the gear finishing tool and the gear. When rotated, the gear finishing tool is reciprocating in the axial direction of the gear to be finished.

Evolution of the design of rotary gear shaving cutters is illustrated by an example shown in Figure 22.21.

Later on, about four dozen U.S. patents on inventions were granted to R. Drummond on the methods and means for rotary shaving of gears.

A method of shaving conical involute gears was invented by National Broach and Machine Co. (see Figure 22.22).

To provide a method of shaving conical involute gears which produces gears having teeth which are involutes of

[19] U.S. Pat. No. 1,989,650, *Method of Lapping Gears*, by *Robert S. Drummond*, Filed: June 29, 1931. Published: January 29, 1935.

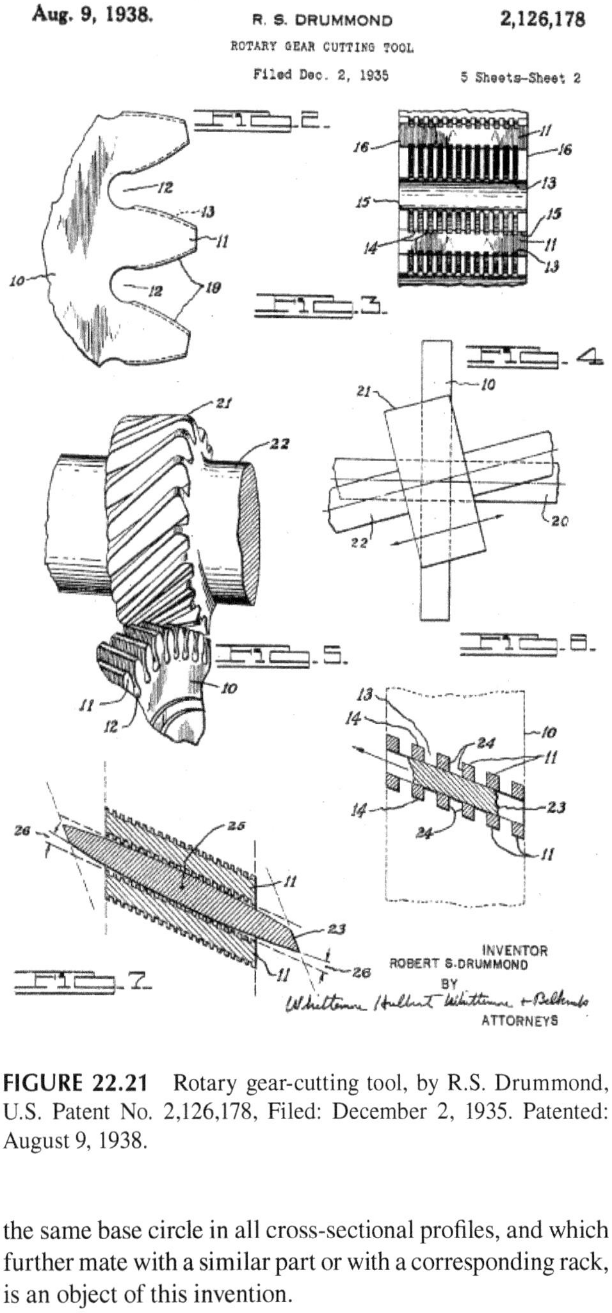


FIGURE 22.21 Rotary gear-cutting tool, by R.S. Drummond, U.S. Patent No. 2,126,178, Filed: December 2, 1935. Patented: August 9, 1938.

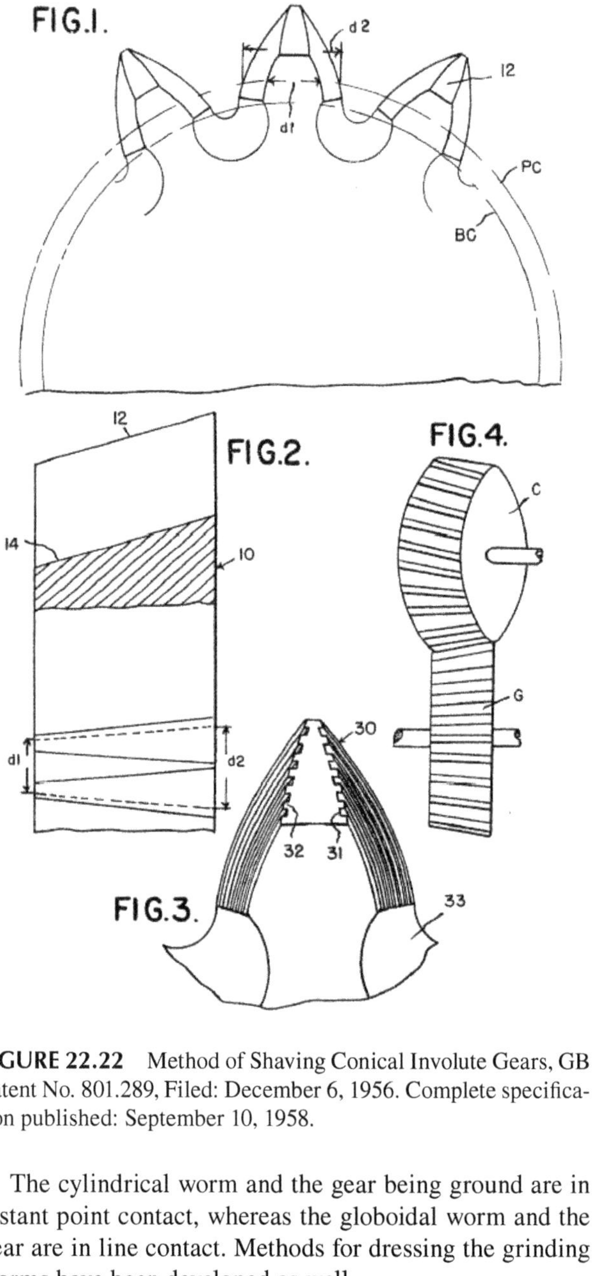


FIGURE 22.22 Method of Shaving Conical Involute Gears, GB Patent No. 801.289, Filed: December 6, 1956. Complete specification published: September 10, 1958.

the same base circle in all cross-sectional profiles, and which further mate with a similar part or with a corresponding rack, is an object of this invention.

## 22.8 GRINDING HARDENED GEARS

The first machine for grinding the hardened gears for automobiles was the work of *J.E. Reinecker* of Germany.

These methods of grinding high-precision involute spur and helical gears are based on the application of two types of grinding worms:

- A cylindrical worm whose design parameters depend mainly on the normal pressure angle and the normal module of the gear to be ground.
- A globoidal worm that is designed as the conjugate to the gear.

The cylindrical worm and the gear being ground are in instant point contact, whereas the globoidal worm and the gear are in line contact. Methods for dressing the grinding worms have been developed as well.

A device[20] by which the wear on the formed grinding wheel was corrected by two diamond points to maintain the correct form automatically, and the resulting wear on the grinding wheel was compensated for in the gear-grinding machine, also automatically, was proposed by *Ward* and *Taylor* of the *Gear Grinding Machine Company* of Detroit [221].

## 22.9 GEAR-CUTTING TOOLS FOR GENERATING BEVEL GEARS

The principle behind all the molding-generating machines grows out of *Sang*'s theoretical analysis, which showed how a rack-tooth-shaped cutter could be used in an intermeshing

[20] U.S. Pat. No. 888,675 of May 26, 1908.

**FIGURE 22.23** William Gleason (1836–1922).

method to produce interchangeable gears. The first machine of this type was the invention of *Hugo Bilgram* in 1884.

*William Gleason* (1836–1922), who in 1874 invented the straight bevel gear planer for the production of bevel gears with straight teeth, substantially advanced the progress of gear making (see Figure 22.23). Manufacturing early straight bevel gears was a difficult process because the gears were cast in metal and then were finished by manually filing each tooth.

The principle of the bevel gear generating method in the Gleason bevel gear planing machine is illustrated in Figure 22.24. Approximate (not perfect) straight bevel gears can be cut using the proposed method of generating tooth flanks of a bevel gear.

In 1898, *James E. Gleason* invented a bevel gear generating machine.[21] He also invented a two-tool bevel generator that enabled manufacturers to reduce the production time of straight bevel gears by half because two reciprocating blades could generate the opposite sides of the tooth space (1905), and he invented the Formate cut method by which spiral bevel gears and hypoid gears could be generated (1938).

As early as 1889, a gear generator using the describing generating principle was invented. The describing method of bevel gear generating was the first to be utilized—a single-point gear-cutting tool is used in this machine to cut straight bevel gears. It was *George Barnard Grant*[22] (1849–1917) (see Figure 22.25) who proposed (January 14, 1887) a correct method for the generation of tooth flanks in geometrically accurate intersected-axes gearing. The use of Grant's invention makes possible the generation of geometrically accurate tooth flanks of straight bevel gears. This is because in one of the possible applications of the invention, "... the rolling cone is increased in size until its center angle is ninety degrees, and it becomes a plane circle. Its element will form an epicycloidal surface as before, but it is now called an 'involute" surface' (see Figure 22.26). Therefore, the straight bevel gear tooth flanks are generated by the describing method adopted in the case of intersected-axes gearing, that is, bevel gearing in particular. This is a significant scientific achievement by G. Grant in the field of the scientific theory of gearing. There is no evidence that this achievement by G. Grant was properly valued by G. Grant himself.

The first spiral bevel gear generator, which operated using the single indexing method, was developed in the United States in 1913. The tool was a face cutter head for 5-cut method [5].

## 22.10 LATER ACCOMPLISHMENTS IN DESIGN OF GEAR-CUTTING TOOLS FOR GENERATING BEVEL GEARS

Until the automobile, there was little demand for new types of gears or gear trains. The appearance of the automobile produced a much greater demand for types of gears seldom used before 1900, and this was reflected in gear-cutting machines designed to cut helical gears, both spur and bevel, as well as skew-bevel gears. Machines utilizing the generating principle were produced to meet this demand, but the hobbing method was soon to dominate this field.

In the 1920s, automotive industry designers needed a gear drive to transform motions and power between crossed axes, and a lower location for the drive shaft.

The need for more accurate and quieter running gears became obvious with the advent of the automobile. Although the hypoid gear was within our manufacturing capabilities by 1916, it was not used practically until 1926, when it was used in the Packard automobile. The hypoid gear made it possible to lower the drive shaft and gain more

[21] U.S. Pat. No. 605,249. *Machine for Forming Gear Teeth*, Filed: February 23, 1897, Patented: June 7, 1899.

[22] **George Barnard Grant** (December 21, 1849 – August 16, 1917) is considered one of the founders of the gear-cutting industry in the United States. Grant established a gear-cutting machine shop in Charlestown, Massachusetts. When this business expanded, he moved the workshop to Boston, expanded it, and named it the *Grant Gear Works* (1877). From this extremely successful establishment evolved the *Philadelphia Gear Works* (1892) and the *Cleveland Gear Works* (*Boston Gear Works* and *Lexington Gear Works* are two more companies to mention here). George Grant even wrote several very successful books on the subject; for example, *A Treatise on Gear Wheels*; *A Handbook on the Teeth of Gears, Their Curves, Properties and Practical Construction*, and so forth].

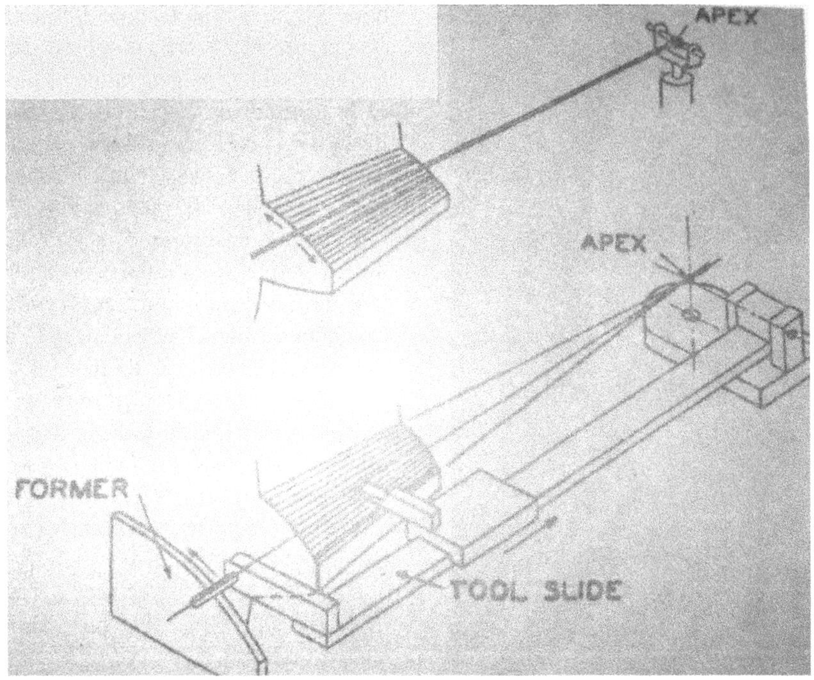


**FIGURE 22.24** Principle of Gleason bevel gear planing machine (1874) [20].

usable floor space. By 1937 almost all cars used hypoid-geared rear axles.

The success with the design, manufacture, and application of contemporary crossed-axes gearing is credited largely to two famous gear experts, *Nikola Trbojevich* (who is also known as *Nicholas Terbo*), and *Dr. Ernest Wildhaber.*

Nikola Trbojevich (see Figure 22.27), a world famous research engineer, mathematician, and inventor, was a

**FIGURE 22.25** George Barnard Grant (1849–1917).

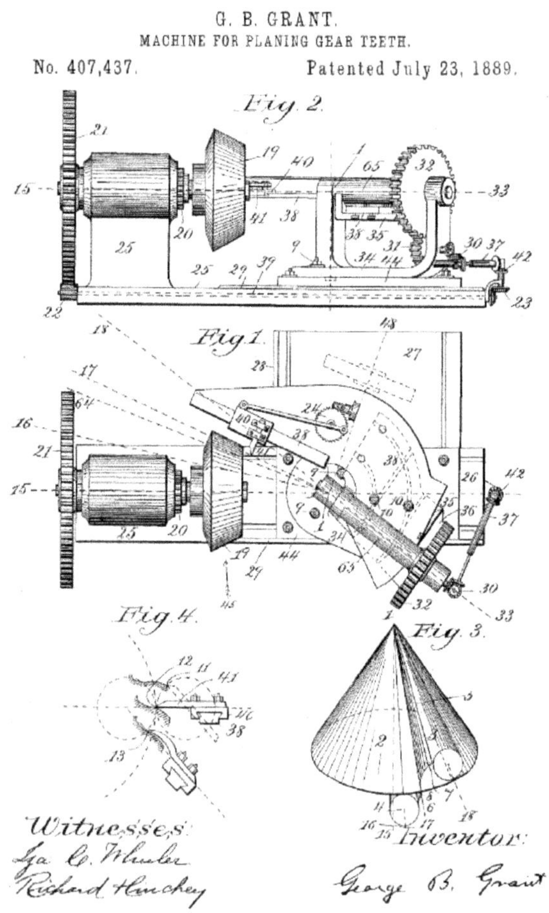


**FIGURE 22.26** *Machine for Planing Gear Teeth*, by G.B. Grant [Pat. No. 407.437, (U.S.). *Machine for Planing Gear Teeth.*/G.B. Grant, Filed: January 14, 1887 (serial No. 224,382). Patented: July 23, 1889].

**FIGURE 22.27** Nikola John Trbojevich, also known as Nicholas J. Terbo (1886–1973).

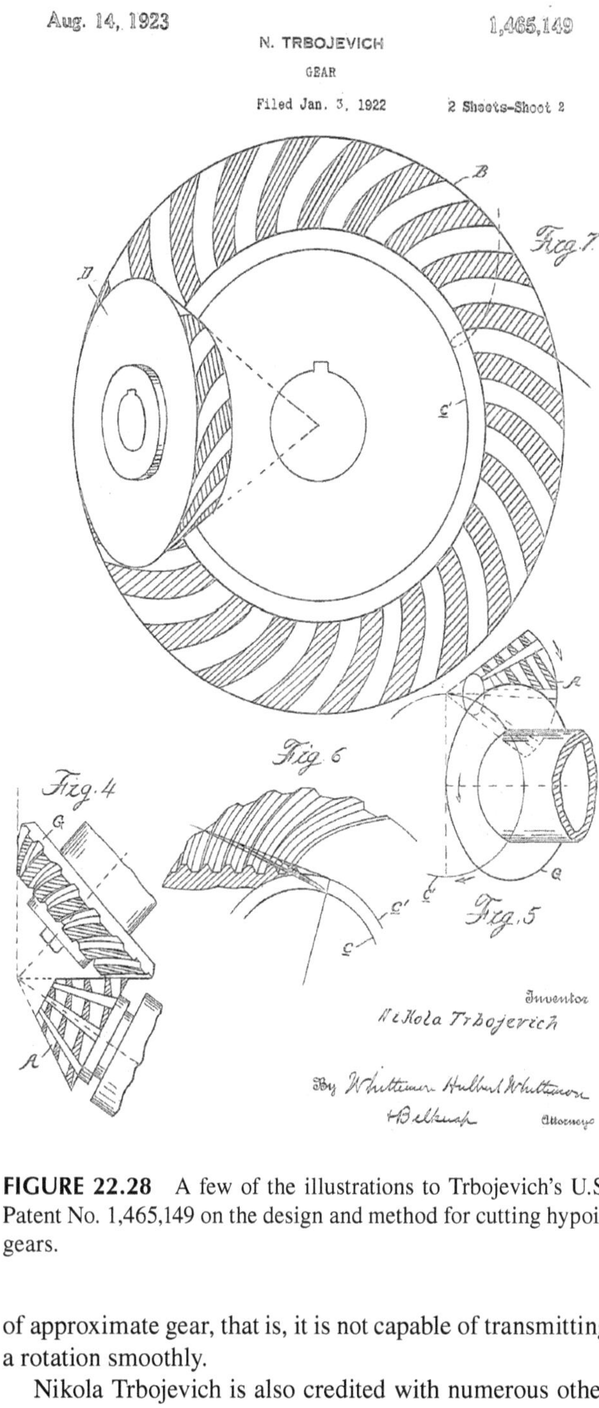


**FIGURE 22.28** A few of the illustrations to Trbojevich's U.S. Patent No. 1,465,149 on the design and method for cutting hypoid gears.

nephew and friend of *Nikola Tesla*. Mr. Trbojevich[23] was the first key developer of the methods and the means for machining hypoid gearing. He held nearly 200 U.S. and foreign patents, principally in the field of gear design.

Mr. Trbojevich's most notable work that brought him international recognition was the invention of the hypoid gear, along with a method and design of a cutting tool to cut gears of this novel design (see Figure 22.28). First published[24] in 1923, it was a new type of spiral bevel gear employing previously unexploited mathematical techniques. The hypoid gear is used on the great majority of all cars, trucks, and military vehicles today. Together with his invention of the tools and machines necessary for its manufacture, the hypoid gear became an integral part of the final drive mechanism of automobiles by 1931. Its effect was immediately apparent in that the overall height of rear-drive passenger automobiles was reduced by at least 4 inches. The accomplishments of N. Trbojevich in the design of hypoid gears and the methods and means for their production were recognized by his receipt of numerous U.S. and international patents on inventions.

It should be noted here that the hypoid gear produced by the methods and means proposed by N. Trbojevich is a kind of approximate gear, that is, it is not capable of transmitting a rotation smoothly.

Nikola Trbojevich is also credited with numerous other accomplishments in the field of methods for cutting gears, as well as of novel designs of gear-cutting tools for these purposes.

Dr. Ernest Wildhaber[25] (see Figure 22.29) is the second key developer of the methods and the means for machining hypoid gears. Wildhaber is one of the most famous inventors in the field of gear manufacture and design. He was granted 279 patents, some of which have a broad application in the gear industry. The hypoid gear drive is of the most famous inventions by Dr. Wildhaber. He proposed different

[23] **Nikola John Trbojevich** (May 21, 1886 – December 2, 1973), also known as Nicholas J. Terbo, was a world famous research engineer, mathematician, and inventor and held the basic patent for the hypoid gear.

[24] U.S. Pat. No. 1,465,149, *Gear*, by *Nikola Trbojevich*, Filed: January 3, 1922. Published: August 14, 1923.

[25] **Ernest Wildhaber** (1893–1979), doctor of engineering, h.c., inventor, and consultant for *The Gleason Works*.

FIGURE 22.29 Dr. Ernest Wildhaber (1893–1979).

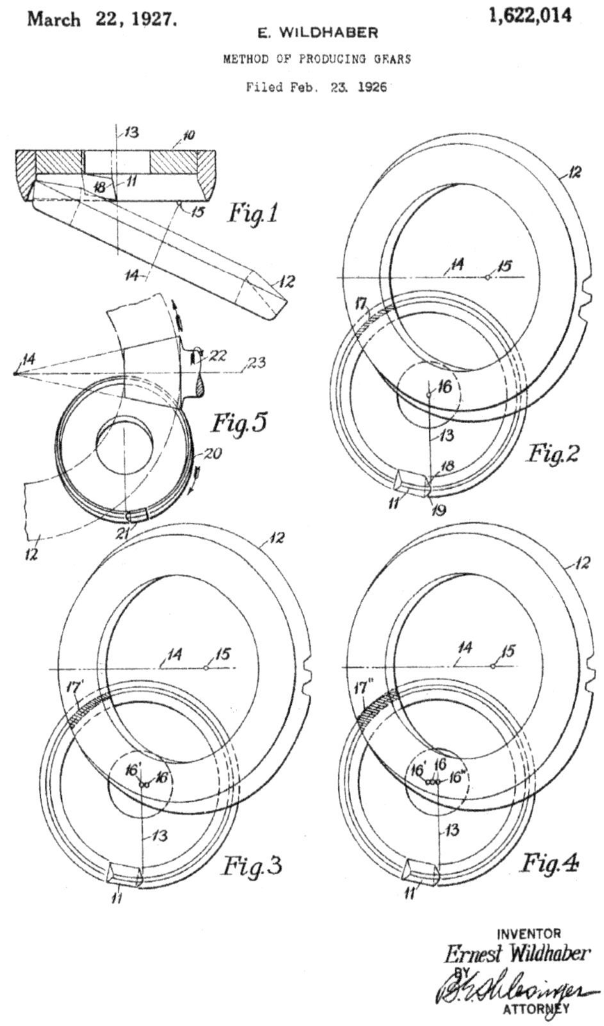


FIGURE 22.30 A few of the illustrations to Wildhaber's U.S. Patent No. 1,622,014 on the design of and method for cutting hypoid gears.

pressure angles for the driving and coast tooth sides of a hypoid gear, which allowed him to provide constancy of the tooth top land.

The proposed designs of crossed-axes gears in current industry are examples of approximate gearing because they are developed and manufactured based on the application of the imaginary crown gear with straight-sided profile (basic crown rack). Because of this, nowadays crossed-axes gears of all types, both face-milled and face-hobbed, do not meet all the requirements perfect gears need to meet.

The U.S. Patent No. 1,622,014 (1927) is the first patent granted to Dr. Wildhaber on the design of hypoid gears, as well as in the field of methods for cutting gears, and of novel designs of gear-cutting tools for these purposes.[26] Later on, dozens of U.S. patents were granted to Dr. Wildhaber in this particular area of gear design and production.

Teeth flanks of hypoid gears (see Figure 22.30) are generated by the cutter head. The gear generators of this type can produce high volumes of gears very efficiently. Due to economic factors, the majority of spiral and hypoid gear sets manufactured today are of the Wildhaber design. This is the main reason Wildhaber's methods and means for hypoid gear production are used widely today.

It should be stressed here that the hypoid gears produced by Wildhaber's methods are a type of approximate gear—that is, they are not capable of transmitting a rotation smoothly.

Dr. Wildhaber is also credited with one more milestone accomplishment in the production of bevel gears: that is, he invented the so-called *Revacycle* method of cutting straight bevel pinions, and a design of a rotary broach for this purpose.[27] The concept of the Revacycle method is illustrated in Figure 22.31. A few more U.S. patents on inventions [No. 2,271,753 (1942), No. 2,278,576 (1942), No. 2,288,058 (1942), No. 2,294,014 (1942), No. 2,315,147 (1943), No. 2,327,296 (1943), No. 2,357,153 (1944), No. 2,376,465 (1945), No. 2,392,278 (1946)] were granted for the Revacycle method of cutting straight bevel pinions.

One more name should be mentioned here in association with cutting gears for hypoid gearing.

The first continuous indexing generator for spiral bevel gears was presented in Germany in 1923 using a conical hob. *Heinrich Schicht*[28] (see Figure 22.32) was the key developer of the continuous indexing methods and the means for machining hypoid gears. Schicht took up the

[26] U.S. Pat. No. 1,622,014, *Method of Producing Gears*, by *Ernst Wildhaber*, Filed: February 23, 1926, Published: March 22, 1927.

[27] U.S. Pat. No. 2,267,182, *Cutter for Cutting Gears, Splined Shafts, and the Like*, by *Ernst Wildhaber*, Filed: December 31, 1937. Published: December 23, 1941.

[28] **Heinrich Schicht** (1886–1962), inventor of the Palloid system of gearing (Klingelnberg).

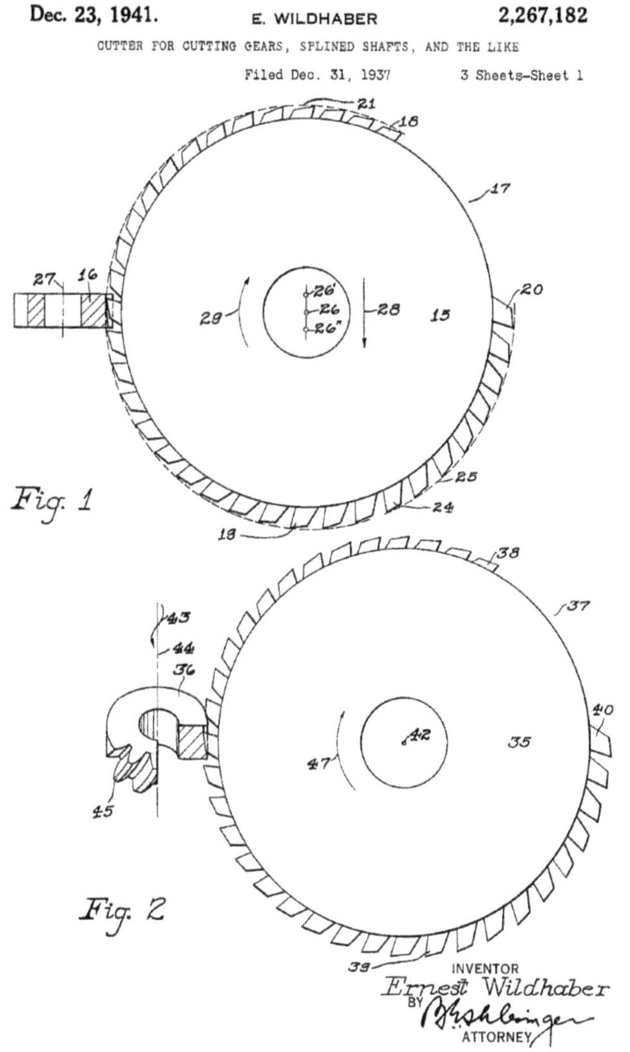


**FIGURE 22.31** A few of the illustrations to Wildhaber's U.S. Patent No. 2,267,182 on the design of a cutter for cutting gears, splined shafts, and the like.

**FIGURE 22.32** Heinrich Schicht (1886–1962).

idea of hobbing cylindrical gears and transferred it to bevel gears using a conical hob instead of a cylindrical hob to manufacture spiral bevel gears. Schicht and Preiss applied for a patent with this idea in 1921.

He is credited first of all with the invention of *Palloid gearing.*[29] In the Palloid® gear cutting method, two hobs with opposite threads are required to produce a pinion and wheel.

Tooth flanks in Palloid gears are generated using specially designed conical hobs as shown in Figure 22.33. Several Great Britain patents [No. 230885 (1925), and No. 234131 (1926)], as well as U.S. patents [No. 2,037,930 (1936), and No. 2,146,232 (1939)] were granted to H. Schicht for the Palloid gearing and the methods and means for its production.

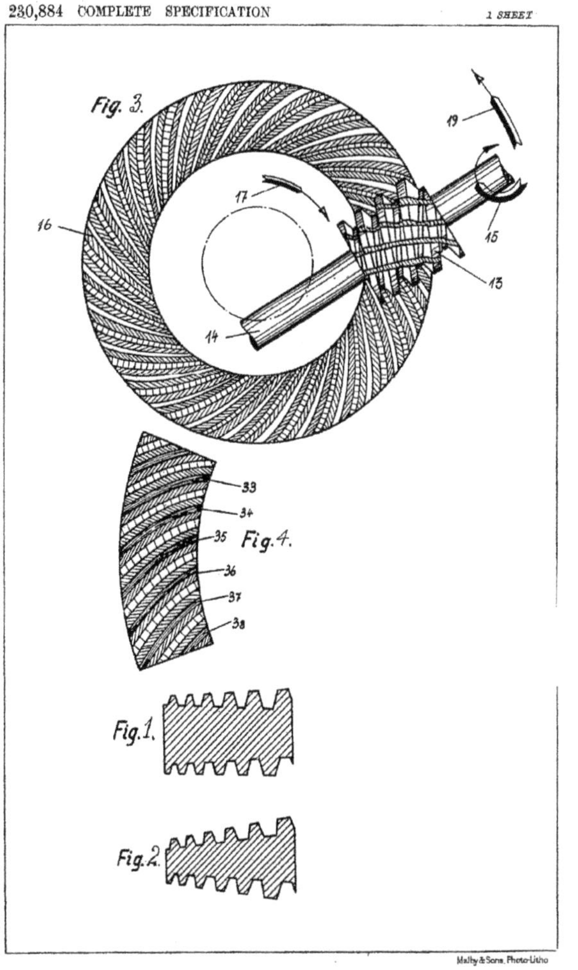


**FIGURE 22.33** A few of the illustrations to Schicht's GB Patent No. 230,884 on the improvements in the method of milling helicoidal bevel gears, and appliances therefor.

[29] GB Pat. No. 230,884, *Improvements in the Method of Milling Heliciodal Bevel Gears, and Appliances Therefore*, by *Heinrich Schicht*, Filed: November 17, 1923. Published: March 17, 1925.

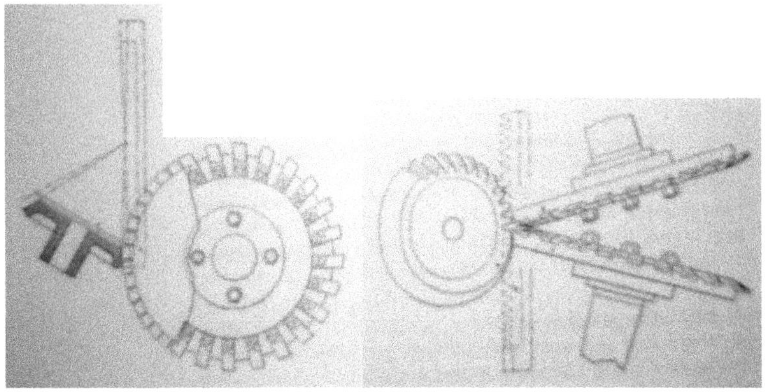

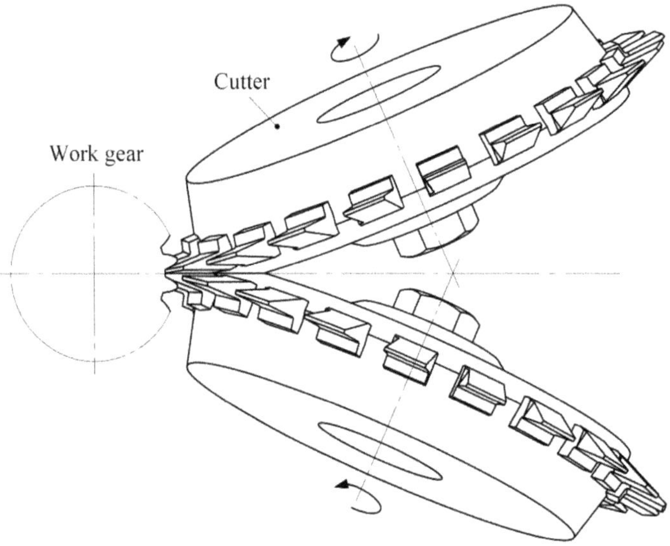


**FIGURE 22.34** Generating milling of straight bevel gears.

The first machine that used the Palloid method for generating spiral bevel gears and hypoid gears with a conical hob was built in 1923. This method was invented in the 1920s by engineers Schicht and Preis and was later replaced by the face-milled and face-hobbed methods.

It is often loosely claimed that the Palloid gears, which are cut by hobs, feature a tooth form that follows a true involute curve. This is incorrect. It should be stressed here that the Palloid gears produced by the methods proposed by H. Schicht also are a type of approximate gear, that is, they are not capable of transmitting a rotation smoothly.

In 1946, a face hobbing machine with a cutter head for the continuous indexing method was developed for the first time in Switzerland. Tooth depth was constant, and the tooth lengthwise shape was an elongated epicycloid.

## 22.11 GENERATING MILLING OF BEVEL GEARS

A different course was followed by *Oscar Beale*, who, around 1900, developed a generating method for the production of bevel gears using two disk-shaped cutters, which could machine both flanks at the same time. *Paul Böttcher* improved this concept and, in 1910, presented a face mill cutter system which produced spiral-shaped teeth on bevel gears.[30]

The molding-generating machine[31] which had the greatest commercial importance was the Brown & Sharpe machine, designed in 1900 by O. Beale[32] In this machine [221], both sides of a space are finished at the same time. This is done by using two toothed disks mounted on inclined axes with the teeth of one occupying the spaces of the other. The outer cutting faces of the disks act as the two sides of an involute worm-gear tooth. There is no motion of the axes of the cutters; all other motions—rotation and translation—are given to the blank. Provision is made for taking a finishing cut as well as to form bevel gears of various diameters, pitch, and pressure angle (see Figure 22.34).

Mostly common sense and accumulated practical experience (empirism), along with a limited use of the scientific approach, were the main features of designing gear-cutting tools and developing gear-cutting methods in the past.

[30] Krumme, W., *Klingelnberg Spiralkegelräder*, 2nd Edition, Springer, Heidelberg, 1950.

[31] Am. Mach., Jan. 29, 1903, pp. 145–150, and 1912, p. 1027.

[32] **Oscar James Beale** (1843–1911), an American designer of machines for cutting gears.

## 22.12 CONCISELY ABOUT THE RESEARCH ON GEAR-CUTTING TOOL DESIGN IN THE 20th CENTURY

In the 20th century, extensive research on gear-cutting tool design was carried out in the former USSR. The names of the key scientists are listed here in alphabetical order. They are: *Vasiliy S. L'ukshin*; *Peter R. Rodin* (1922–2008); *Georgiy N. Sakharov* (1910–1990); *Ivan I. Semenchenko* (1897–1964); *Galina I. Shevel'ova* (1929–2005); and a few more names could be mentioned here.

# Appendices

# Appendix A: Elements of Vector Calculus

The *vector*, the key to the entire theory of gearing, is a triple real number (in most computer languages, these are usually called *floating point numbers*) and is noted in a **bold** typeface, that is, **A** or **a**.

Care must be taken to differentiate between two types of vectors:

- **Position vector:** A position vector runs from the origin of coordinate $(0,0,0)$ to a point $(X,Y,Z)$ and its length gives the distance of the point from the origin. Its components are given by $(X,Y,Z)$. The essential concept to understand about a position vector is that it is anchored to specific coordinates (points in space). The set of points that are used to describe the shape of all part surfaces can be thought of as position vectors.
- **Direction vector:** A direction vector differs from a position vector in that it is *not* anchored to specific coordinates. Frequently, direction vectors are used in a form where they have unit length; in this case, they are said to be *normalized*. The most common application of a direction vector in the theory of part surface generation is to specify the orientation of a surface or ray direction. For this, we use a direction vector pointing away from the part surface at a right angle (normal). Such normal vectors are also the key in many calculations in the theory of gearing.

Vector calculus is a powerful tool for solving many geometrical and kinematical problems that pertain to the design and generation of gears and gearsets. In this book, vectors are understood as quantities that have magnitude and direction and obey the law of addition.

## A.1 FUNDAMENTAL PROPERTIES OF VECTORS

The distance-and-direction interpretation suggests a powerful way to visualize a vector as a directed line segment or arrow. The length of the arrow (at some predetermined scale) represents the magnitude of the vector, and the orientation of the segment and placement of the arrowhead (at one end of the segment or the other) represents its direction.

Vectors possess certain properties, the set of which is commonly interpreted as the set of fundamental properties of vectors.

### A.1.1 Addition

Given two vectors **a** and **b**, their sum $(\mathbf{a}+\mathbf{b})$ is graphically defined by joining the tail of **b** to the head of **a**. Then, the line from the tail of **a** to the head of **b** is the sum $\mathbf{c}=(\mathbf{a}+\mathbf{b})$.

### A.1.2 Equality

Two vectors are equal when they have the same magnitude and direction. The position of the vectors is unimportant for equality.

### A.1.3 Negation

The vector $-\mathbf{a}$ has the same magnitude as **a** but extends in the opposite direction.

### A.1.4 Subtraction

From the properties of addition and negation, the following $\mathbf{a}-\mathbf{b}=\mathbf{a}+(-\mathbf{b})$ can be defined.

### A.1.5 Scalar Multiplication

The vector $k\mathbf{a}$ has the same direction as **a**, with a magnitude $k$ times that of **a**. The magnitude $k$ is called a scalar, as it changes the scale of the vector **a**.

## A.2 MATHEMATICAL OPERATIONS OVER VECTORS

The following rules and mathematical operations can be determined from the fundamental properties of vectors listed in the previous section.

### A.2.1 Components of Vectors

Let's assume that a set of three vectors—**a**, **b**, and **c**—and two scalars—$k$ and $t$—are given. Then, vector addition and scalar multiplication have the following properties:

$$\mathbf{a}+\mathbf{b}=\mathbf{b}+\mathbf{a} \tag{A.1}$$

$$\mathbf{a}+(\mathbf{b}+\mathbf{c})=(\mathbf{a}+\mathbf{b})+\mathbf{c} \tag{A.2}$$

$$k(t\mathbf{a})=kt\mathbf{a} \tag{A.3}$$

$$(k+t)\mathbf{a}=k\mathbf{a}+t\mathbf{a} \tag{A.4}$$

$$k(\mathbf{a}+\mathbf{b})=k\mathbf{a}+k\mathbf{b} \tag{A.5}$$

Magnitude $a$ of a vector **a** is:

$$a=|\mathbf{a}|=\sqrt{a_x^2+a_y^2+a_z^2} \tag{A.6}$$

where $a_x$, $a_y$, and $a_z$ are the scalar components of **a**.

A unit vector $\bar{\mathbf{a}}$ in the direction of a vector $\mathbf{a}$ is:

$$\bar{\mathbf{a}} = \frac{\mathbf{a}}{|\mathbf{a}|} = \frac{\mathbf{a}}{a} \tag{A.7}$$

The components $\bar{a}_x$, $\bar{a}_y$, and $\bar{a}_z$ of a unit vector $\bar{\mathbf{a}}$ are also the direction cosines of the vector $\bar{\mathbf{a}}$:

$$\cos\alpha = \bar{a}_x \tag{A.8}$$

$$\cos\beta = \bar{a}_y \tag{A.9}$$

$$\cos\gamma = \bar{a}_z \tag{A.10}$$

It is common practice to denote the components $\bar{a}_x$, $\bar{a}_y$, and $\bar{a}_z$ by $l$, $m$, and $n$ accordingly.

### A.2.2 Scalar Product (or Dot Product) of Two Vectors

The formula:

$$\mathbf{a}\cdot\mathbf{b} = a_x b_x + a_y b_y + a_z b_z = |\mathbf{a}||\mathbf{b}|\cos\angle(\mathbf{a},\mathbf{b}) \tag{A.11}$$

is commonly used for calculation of scalar product of two vectors $\mathbf{a}$ and $\mathbf{b}$.

Equation (A.11) can also be represented in the form:

$$\mathbf{a}\cdot\mathbf{b} = [\mathbf{a}]^T\cdot[\mathbf{b}] = [\, a_x \;\; a_y \;\; a_z \,]\cdot\begin{bmatrix} b_x \\ b_y \\ b_z \end{bmatrix} \tag{A.12}$$

Angle $\angle(\mathbf{a},\mathbf{b})$ between two vectors $\mathbf{a}$ and $\mathbf{b}$ is calculated from:

$$\angle(\mathbf{a},\mathbf{b}) = \cos^{-1}\left(\frac{\mathbf{a}\cdot\mathbf{b}}{|\mathbf{a}||\mathbf{b}|}\right) \tag{A.13}$$

The scalar product of two vectors $\mathbf{a}$ and $\mathbf{b}$ features the following properties:

$$\mathbf{a}\cdot\mathbf{a} = |\mathbf{a}|^2 \tag{A.14}$$

$$\mathbf{a}\cdot\mathbf{b} = \mathbf{b}\cdot\mathbf{a} \tag{A.15}$$

$$\mathbf{a}\cdot(\mathbf{b}+\mathbf{c}) = \mathbf{b}\cdot\mathbf{a} + \mathbf{b}\cdot\mathbf{c} \tag{A.16}$$

$$(k\mathbf{a})\cdot\mathbf{b} = \mathbf{a}\cdot(k\mathbf{b}) = k(\mathbf{a}\cdot\mathbf{b}) \tag{A.17}$$

If $\mathbf{a}$ is perpendicular to $\mathbf{b}$, then:

$$\mathbf{a}\cdot\mathbf{b} = 0 \tag{A.18}$$

### A.2.3 Vector Product (or Cross Product) of Two Vectors

The vector product of two vectors can be calculated from the formula:

$$\begin{aligned}\mathbf{a}\times\mathbf{b} = (a_y b_z - a_z b_y)\,\mathbf{i} + (a_z b_x - a_x b_z)\,\mathbf{j}\\ + (a_x b_y - a_y b_x)\,\mathbf{k}\end{aligned} \tag{A.19}$$

Here, in Eq. (A.19), $\mathbf{i}$, $\mathbf{j}$, and $\mathbf{k}$ are unit vectors in the $X$, $Y$, and $Z$ directions of the reference system $XYZ$, in which the vectors $\mathbf{a}$ and $\mathbf{b}$ are specified.

The vector product possesses the following property: when $\mathbf{a}\times\mathbf{b} = \mathbf{c}$, vector $\mathbf{c}$ is perpendicular to a plane through vectors $\mathbf{a}$ and $\mathbf{b}$.

The vector product of two vectors $\mathbf{a}$ and $\mathbf{b}$ features the following properties:

$$\mathbf{a}\times\mathbf{b} = \begin{vmatrix} \mathbf{i} & \mathbf{j} & \mathbf{k} \\ a_x & a_y & a_z \\ b_x & b_y & b_z \end{vmatrix} \tag{A.20}$$

$$\mathbf{a}\times\mathbf{b} = |\mathbf{a}||\mathbf{b}|\,\mathbf{n}\sin\angle(\mathbf{a},\mathbf{b}) \tag{A.21}$$

where the unit normal vector to the plane through the vectors $\mathbf{a}$ and $\mathbf{b}$ is denoted by $\mathbf{n}$:

$$|\mathbf{a}\times\mathbf{b}| = |\mathbf{a}||\mathbf{b}|\sin\angle(\mathbf{a},\mathbf{b}) \tag{A.22}$$

Coordinates of the vector product $\mathbf{a}\times\mathbf{b}$ can also be expressed in the form:

$$|\mathbf{a}\times\mathbf{b}| = \begin{bmatrix} 0 & -a_z & a_y \\ a_z & 0 & -a_x \\ -a_y & a_x & 0 \end{bmatrix}\cdot\begin{bmatrix} b_x \\ b_y \\ b_z \end{bmatrix} = \begin{bmatrix} -a_z b_y + a_y b_z \\ -a_x b_z + a_z b_x \\ -a_y b_x + a_x b_y \end{bmatrix} \tag{A.23}$$

$$\mathbf{a}\times\mathbf{b} = -\mathbf{b}\times\mathbf{a} \tag{A.24}$$

$$\mathbf{a}\times(\mathbf{b}+\mathbf{c}) = \mathbf{a}\times\mathbf{b} + \mathbf{a}\times\mathbf{c} \tag{A.25}$$

$$(k\,\mathbf{a})\times\mathbf{b} = \mathbf{a}\times(k\,\mathbf{b}) = k(\mathbf{a}\times\mathbf{b}) \tag{A.26}$$

$$\mathbf{i}\times\mathbf{j} = \mathbf{k},\, \mathbf{j}\times\mathbf{k} = \mathbf{i},\, \mathbf{k}\times\mathbf{i} = \mathbf{j} \tag{A.27}$$

If $\mathbf{a}$ is parallel to $\mathbf{b}$, then:

$$\mathbf{a}\times\mathbf{b} = 0 \tag{A.28}$$

### A.2.4 Triple Scalar Product of Three Vectors

The product $(\mathbf{a}\times\mathbf{b})\cdot\mathbf{c}$ is commonly referred to as the *triple scalar product* of three vectors $\mathbf{a}$, $\mathbf{b}$, and $\mathbf{c}$.

The triple scalar product of three vectors **a**, **b**, and **c** features the following properties:

$$(\mathbf{a}\times\mathbf{b})\cdot\mathbf{c}=(\mathbf{b}\times\mathbf{c})\cdot\mathbf{a}=(\mathbf{c}\times\mathbf{a})\cdot\mathbf{b} \quad (A.29)$$

$$(\mathbf{b}\times\mathbf{c})\cdot\mathbf{a}=\mathbf{a}\cdot(\mathbf{b}\times\mathbf{c}) \quad (A.30)$$

$$(\mathbf{a}\times\mathbf{b})\cdot\mathbf{c}=\mathbf{a}\cdot(\mathbf{b}\times\mathbf{c}) \quad (A.31)$$

$$\mathbf{a}\cdot(\mathbf{b}\times\mathbf{c})=\begin{vmatrix} a_x & a_y & a_z \\ b_x & b_y & b_z \\ c_x & c_y & c_z \end{vmatrix} \quad (A.32)$$

### A.2.5 Triple Vector Product of Three Vectors

The product $(\mathbf{a}\times\mathbf{b})\times\mathbf{c}$ is commonly referred to as *triple vector product* of three vectors **a**, **b**, and **c**.

The product $(\mathbf{a}\times\mathbf{b})\times\mathbf{c}$ can be evaluated by two vector products. However, it also can be evaluated more simply using the identity:

$$(\mathbf{a}\times\mathbf{b})\times\mathbf{c}=(\mathbf{a}\cdot\mathbf{c})\mathbf{b}-(\mathbf{b}\cdot\mathbf{c})\mathbf{a} \quad (A.33)$$

It should be mentioned here that, in general, the triple vector products $(\mathbf{a}\times\mathbf{b})\times\mathbf{c}$ and $\mathbf{a}\times(\mathbf{b}\times\mathbf{c})$ are not equal:

$$(\mathbf{a}\times\mathbf{b})\times\mathbf{c}\neq\mathbf{a}\times(\mathbf{b}\times\mathbf{c}) \quad (A.34)$$

The analytical interpretation of many problems and results in the field of gear science have become much simpler through the use of vector calculus.

### A.2.6 Lagrange Equation for Vectors

The calculation of the mixed product of vectors **a** and **b** uses the equation:

$$(\mathbf{a}\times\mathbf{b})\cdot(\mathbf{a}\times\mathbf{b})=(\mathbf{a}\cdot\mathbf{a})(\mathbf{b}\cdot\mathbf{b})-(\mathbf{a}\cdot\mathbf{b})^2 \quad (A.35)$$

Equation (A.35) is attributed to *Lagrange*.[1]

## A.3 On the Similarity and Difference between Vectors and Matrices

A vector **a** is commonly represented in the form:

$$\mathbf{a}=\mathbf{i}a+\mathbf{j}b+\mathbf{k}c \quad (A.36)$$

That same vector **a** allows for matrix representation in one of the following forms:

$$\mathbf{a}=\begin{bmatrix} a \\ b \\ c \end{bmatrix} \quad (A.37)$$

$$\mathbf{a}=\begin{bmatrix} a \\ b \\ c \\ 1 \end{bmatrix} \quad (A.38)$$

$$\mathbf{a}=\begin{bmatrix} 1 & 0 & 0 & a \\ 0 & 1 & 0 & b \\ 0 & 0 & 1 & c \\ 0 & 0 & 0 & 1 \end{bmatrix} \quad (A.39)$$

Operations over vectors in a matrix representation [see the form Eq. (A.39)] are preferred, as multiple coordinate system transformations often need to be executed.

Vectors obey the commutative law—that is, the equalities $\mathbf{a}\times\mathbf{b}=\mathbf{b}\times\mathbf{a}$ and $\mathbf{a}\cdot\mathbf{b}=\mathbf{b}\cdot\mathbf{a}$ are valid for vectors. This is not always applicable to matrices.

Two kinds of products are valid for vectors, namely, a dot product $(\mathbf{a}\cdot\mathbf{b})$ and a cross product $(\mathbf{a}\times\mathbf{b})$ of vectors, but these are not valid with respect to matrices.

[1] Joseph-Louis Lagrange (January 25, 1736–April 10, 1813) was a famous French mathematician, astronomer, and mechanician.

# Appendix B: Elements of Differential Geometry of Surfaces

The discussion in this book is focused primarily on the elements of gearing theory.

The gear and the pinion tooth flank and their motion in space in relation to each other are analytically described in a reference system. An orthogonal Cartesian[1] reference system is a major kind of reference systems that is commonly used for this purpose. Mutually perpendicular coordinate axes of a Cartesian coordinate system are conventionally labeled as $X$, $Y$, and $Z$.

In a Cartesian reference system, the axes can have either a left- or right-handed orientation. A right-handed Cartesian reference system is preferred, and all algorithms and formulae used in this book assume right-handed orientation.

A coordinate system provides a numerical frame of reference for the three-dimensional space in which the theory is developed. Two coordinate systems are particularly useful: the ubiquitous Cartesian ($XYZ$) rectilinear system, and the spherical polar $(r, \theta, \varphi)$ or angular system. Cartesian coordinate systems are the most commonly used, but angular coordinates are often helpful as well.

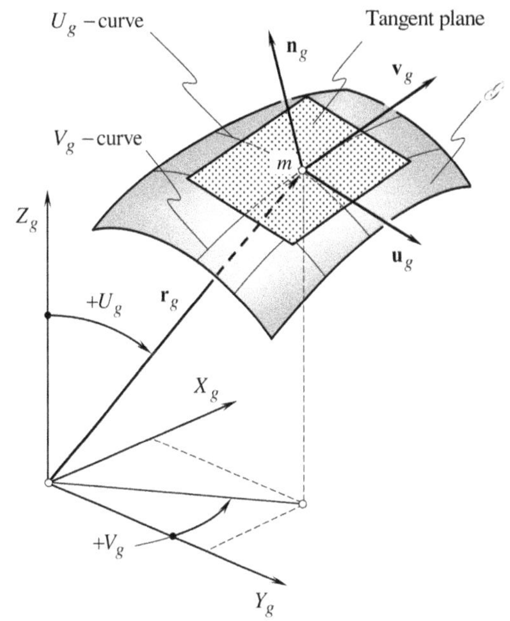


**FIGURE B.1** Principal parameters of local topology at point of gear tooth flank $\mathscr{G}$.

## B.1 SPECIFICATION OF A GEAR TOOTH FLANK

A gear tooth flank can be uniquely determined by two independent variables. Therefore, in most cases, we give a gear tooth flank $\mathscr{G}$ (see Figure B.1) by expressing its rectangular coordinated $X_g$, $Y_g$, and $Z_g$, as functions of two Gaussian[2] coordinates $U_g$ and $V_g$ in a certain closed interval[3]:

$$\mathscr{G} \Rightarrow \mathbf{r}_g = \mathbf{r}_g(U_g, V_g) = \begin{bmatrix} X_g(U_g, V_g) \\ Y_g(U_g, V_g) \\ Z_g(U_g, V_g) \\ 1 \end{bmatrix} \tag{B.1}$$

$$U_{1.g} \le U_g \le U_{2.g};\ V_{1.g} \le V_g \le V_{2.g}$$

where:

- $\mathbf{r}_g$ is the position vector of point of the gear tooth flank $\mathscr{G}$
- $U_g$ and $V_g$ are the curvilinear coordinates (Gaussian coordinates) of the gear tooth flank $\mathscr{G}$
- $X_g, Y_g, Z_g$ are the Cartesian coordinates of point of the gear tooth flank $\mathscr{G}$
- $U_{1.g}, U_{2.g}$ are the boundary values of the closed interval of the $U_g$ − parameter
- $V_{1.g}, V_{2.g}$ are the boundary values of the closed interval of the $V_g$ − parameter

The parameters $U_g$ and $V_g$ must enter into Eq. (B.1) independently, which means that the matrix:

$$\mathbf{M} = \begin{bmatrix} \dfrac{\partial X_g}{\partial U_g} & \dfrac{\partial Y_g}{\partial U_g} & \dfrac{\partial Z_g}{\partial U_g} \\ \dfrac{\partial X_g}{\partial V_g} & \dfrac{\partial Y_g}{\partial V_g} & \dfrac{\partial Z_g}{\partial V_g} \end{bmatrix} \tag{B.2}$$

has a rank of 2.

Positions where the rank is 1 or 0 are singular points; when the rank at all points is 1, then Eq. (B.1) represents a curve.

Other methods of a surface's specification are known as well. Among the most frequently used specifications of a gear tooth flank are:

- An equation in explicit form
- An equation in implicit form, and
- A set of parametric equations

[1] René Descartes (Latinized form: *Renatus Cartesius*; March 31, 1596–February 11, 1650) was a French mathematician, philosopher, and writer.
[2] Johan Carl Friedrich Gauss (April 30, 1777–February 23, 1855) was a famous German mathematician and physical scientist.
[3] Equations that are valid for the gear tooth flank $\mathscr{G}$ are also valid for the pinion tooth flank $\mathscr{P}$.

It is assumed here and in the following that any given kind of gear tooth flank $\mathscr{G}$ specification can be converted either into the vector form or into the matrix form of its specification, following from Eq. (B.1).

## B.2 TANGENT VECTORS, TANGENT PLANE, AND UNIT NORMAL VECTOR

The following notation is proven to be convenient in the subsequent consideration.

The first derivatives of $\mathbf{r}_g$ with respect to Gaussian coordinates $U_g$ and $V_g$ are designated as:

$$\frac{\partial \mathbf{r}_g}{\partial U_g} = \mathbf{U}_g \tag{B.3}$$

$$\frac{\partial \mathbf{r}_P}{\partial V_P} = \mathbf{V}_P \tag{B.4}$$

and for the unit tangent vectors:

$$\mathbf{u}_g = \frac{\mathbf{U}_g}{|\mathbf{U}_g|} \tag{B.5}$$

$$\mathbf{v}_g = \frac{\mathbf{V}_g}{|\mathbf{V}_g|} \tag{B.6}$$

respectively.[4]

The direction of the tangent line to the $U_g$ – coordinate line through a given point $m$ on the gear tooth flank $\mathscr{G}$ is specified by the unit tangent vector $\mathbf{u}_g$ (as well as by the tangent vector $\mathbf{U}_g$). Similarly, the direction of the tangent line to the $V_g$ – coordinate line through that same point $m$ on a gear tooth flank $\mathscr{G}$ is specified by the unit tangent vector $\mathbf{v}_g$ (as well as by the tangent vector $\mathbf{V}_g$).

The significance of the unit tangent vectors $\mathbf{u}_g$ and $\mathbf{v}_g$ becomes evident from the following considerations.

First, unit tangent vectors $\mathbf{u}_g$ and $\mathbf{v}_g$ yield an equation of the tangent plane to a gear tooth flank $\mathscr{G}$ at a specified point $m$:

$$\text{Tangent plane} \Rightarrow \begin{bmatrix} [\mathbf{r}_{t.p} - \mathbf{r}_g^m] \\ \mathbf{u}_g \\ \mathbf{v}_g \\ 1 \end{bmatrix} = 0 \tag{B.7}$$

where:

$\mathbf{r}_{t.p}$ is the position vector of point of the tangent plane to a gear tooth flank $\mathscr{G}$ at a specified point $m$

$\mathbf{r}_g^m$ is the position vector of point $m$ on a gear tooth flank $\mathscr{G}$

Second, tangent vectors yield an equation of the perpendicular $\mathbf{N}_g$ and of the unit normal vector $\mathbf{n}_g$ to a gear tooth flank $\mathscr{G}$ at a given point $m$:

$$\mathbf{N}_g = \mathbf{U}_g \times \mathbf{V}_g \tag{B.8}$$

and

$$\mathbf{n}_g = \frac{\mathbf{N}_g}{|\mathbf{N}_g|} = \frac{\mathbf{U}_g \times \mathbf{V}_g}{|\mathbf{U}_g \times \mathbf{V}_g|} = \mathbf{u}_g \times \mathbf{v}_g \tag{B.9}$$

When the order of the multipliers in Eqs. (B.8) and (B.9) is chosen properly, the unit normal vector $\mathbf{n}_g$ (as well as the normal vector $\mathbf{N}_g$) is pointed outward of the bodily side of the surface $\mathscr{G}$.

## B.3 LOCAL FRAME

Two unit tangent vectors $\mathbf{u}_g$ and $\mathbf{v}_g$, along with the unit normal vector $\mathbf{n}_g$, comprise a local frame $\mathbf{u}_g, \mathbf{v}_g, \mathbf{n}_g$, which has its origin at current point $m$ on a gear tooth flank $\mathscr{G}$. The unit tangent vector $\mathbf{u}_g$ is perpendicular to the unit normal vector $\mathbf{n}_g$ (that is, $\mathbf{u}_g \perp \mathbf{n}_g$), as is unit tangent vector $\mathbf{v}_g$ (that is, $\mathbf{v}_g \perp \mathbf{n}_g$). Generally, the unit tangent vectors $\mathbf{u}_g$ and $\mathbf{v}_g$ are not perpendicular to each other; rather, they form a certain angle $\omega_g$. In order to construct an orthogonal local frame, either the unit tangent vector $\mathbf{u}_g$ in the local frame $(\mathbf{u}_g, \mathbf{v}_g, \mathbf{n}_g)$ must be substituted with a unit tangent vector $\mathbf{u}_g^*$ or the unit tangent vector $\mathbf{v}_g$ in that same local frame $(\mathbf{u}_g, \mathbf{v}_g, \mathbf{n}_g)$ must be substituted with a unit tangent vector $\mathbf{v}_g^*$. For the calculation of the newly introduced unit tangent vectors $\mathbf{u}_g^*$ and $\mathbf{v}_g^*$, the following equations can be used:

$$\mathbf{u}_g^* = \mathbf{u}_g \times \mathbf{n}_g \tag{B.10}$$

$$\mathbf{v}_g^* = \mathbf{v}_g \times \mathbf{n}_g \tag{B.11}$$

It is convenient to choose that order of the multipliers in Eqs. (B.10) and (B.11), which preserves the orientation of the original local frame $(\mathbf{u}_g, \mathbf{v}_g, \mathbf{n}_g)$—that is, if the original local frame $(\mathbf{u}_g, \mathbf{v}_g, \mathbf{n}_g)$ is right-hand-oriented, then the newly constructed local frame [either the local frame $(\mathbf{u}_g^*, \mathbf{v}_g, \mathbf{n}_g)$ or the local frame $(\mathbf{u}_g^*, \mathbf{v}_g, \mathbf{n}_g)$] should also be a right-hand-oriented local frame, and vice versa.

It should be pointed out here that another possibility to construct an orthogonal local frame is also available. The local frame of this kind is commonly referred to as Darboux[5] frame and is briefly considered further in this section.

Unit tangent vectors $\mathbf{u}_g$ and $\mathbf{v}_g$ to a surface $\mathscr{G}$ at point $m$ are of critical importance when solving practical problems in the field of gearing. This statement is proven by numerous examples shown below.

---

4 It is necessary to underline here that the unit tangent vectors $\mathbf{u}_P$ and $\mathbf{v}_P$ are dimensionless values, following from Equations (B.5) and (B.6).

5 Jean Gaston Darboux (August 14, 1842–February 23, 1917) was a French mathematician.

## B.4 FUNDAMENTAL FORMS OF SURFACE

Consider two other important issues concerning the gear tooth flank geometry, both of which relate to intrinsic geometry in differential vicinity of surface point *m*.

### B.4.1 First Fundamental Form of a Surface

The first issue is the first fundamental form $\Phi_{1.g}$ of a gear tooth flank $\mathscr{G}$. The metric properties of a gear tooth flank $\mathscr{G}$ are described by the first fundamental form $\Phi_{1.g}$ of the surface. Usually, the first fundamental form $\Phi_{1.g}$ is represented as the quadratic form:

$$\Phi_{1.g} \Rightarrow ds_g^2 = E_g\, dU_g^2 + 2F_g\, dU_g\, dV_g + G_g\, dV_g^2 \tag{B.12}$$

In Eq. (B.12), the following designations apply:

$s_g$ is the linear element on a gear tooth flank $\mathscr{G}$ ($s_g$ is equal to the length of a segment of a certain curve on a gear tooth flank $\mathscr{G}$)
$E_g$, $F_g$, $G_g$ are the fundamental magnitudes of the first order at a surface point

Equation (B.12) for the first fundamental form $\Phi_{1.g}$ is known from many advanced sources. In the theory of gearing, another form of analytical representation of the first fundamental form $\Phi_{1.g}$ is proven to be useful:

$$\Phi_{1.g} \Rightarrow ds_g^2 = \begin{bmatrix} dU_g & dV_g & 0 & 0 \end{bmatrix} \cdot \begin{bmatrix} E_g & F_g & 0 & 0 \\ F_g & G_g & 0 & 0 \\ 0 & 0 & 1 & 0 \\ 0 & 0 & 0 & 1 \end{bmatrix} \cdot \begin{bmatrix} dU_g \\ dV_g \\ 0 \\ 0 \end{bmatrix} \tag{B.13}$$

This kind of analytical representation of the first fundamental form $\Phi_{1.P}$ is proposed by Prof. S.P. Radzevich (~2008).

The practical advantage of Eq. (B.13) is that it can easily be incorporated into computer programs when multiple coordinate system transformations are used. The last is vital for the theory of gearing.

Fundamental magnitudes of the first order $E_g$, $F_g$, and $G_g$ are calculated from the set of the following equations:

$$E_g = \mathbf{U}_g \cdot \mathbf{U}_g \tag{B.14}$$

$$F_g = \mathbf{U}_g \cdot \mathbf{V}_g \tag{B.15}$$

$$G_g = \mathbf{V}_g \cdot \mathbf{V}_g \tag{B.16}$$

Equations (B.14) through (B.16) can be represented in an expanded form:

$$E_g = \frac{\partial \mathbf{r}_g}{\partial U_g} \cdot \frac{\partial \mathbf{r}_g}{\partial U_g} = \frac{\partial X_g}{\partial U_g} \cdot \frac{\partial X_g}{\partial U_g} + \frac{\partial Y_g}{\partial U_g} \cdot \frac{\partial Y_g}{\partial U_g} + \frac{\partial Z_g}{\partial U_g} \cdot \frac{\partial Z_g}{\partial U_g} \tag{B.17}$$

$$F_g = \frac{\partial \mathbf{r}_g}{\partial U_g} \cdot \frac{\partial \mathbf{r}_g}{\partial V_g} = \frac{\partial X_g}{\partial U_g} \cdot \frac{\partial X_g}{\partial V_g} + \frac{\partial Y_g}{\partial U_g} \cdot \frac{\partial Y_g}{\partial V_g} + \frac{\partial Z_g}{\partial U_g} \cdot \frac{\partial Z_g}{\partial V_g} \tag{B.18}$$

$$G_g = \frac{\partial \mathbf{r}_g}{\partial V_g} \cdot \frac{\partial \mathbf{r}_g}{\partial V_g} = \frac{\partial X_g}{\partial V_g} \cdot \frac{\partial X_g}{\partial V_g} + \frac{\partial Y_g}{\partial V_g} \cdot \frac{\partial Y_g}{\partial V_g} + \frac{\partial Z_g}{\partial V_g} \cdot \frac{\partial Z_g}{\partial V_g} \tag{B.19}$$

Fundamental magnitudes of the first order $E_g$, $F_g$, and $G_g$ are functions of the $U_g$ – and $V_g$ – coordinates of point of a gear tooth flank $\mathscr{G}$. These relationships can be represented in the general form:

$$E_g = E_g(U_g, V_g) \tag{B.20}$$

$$F_g = F_g(U_g, V_g) \tag{B.21}$$

$$G_g = G_g(U_g, V_g) \tag{B.22}$$

It is important to point out here that fundamental magnitudes $E_g$ and $G_g$ are always positive (that is, $E_g > 0$, $G_g > 0$), and the fundamental magnitude $F_g$ can be equal zero ($F_g \geq 0$). This results in that the first fundamental form $\Phi_{1.g}$ at a point of a gear tooth flank $\mathscr{G}$ is always positively defined ($\Phi_{1.g} \geq 0$), and it cannot have a negative value.

By use of the first fundamental form $\Phi_{1.g}$, the following major parameters of geometry of a gear tooth flank $\mathscr{G}$ can be calculated:

a. The length of a curved line segment on a gear tooth flank $\mathscr{G}$
b. The area of a portion of gear tooth flank $\mathscr{G}$ that is bounded by a closed curve on the surface, and
c. The angle between any two directions on a gear tooth flank $\mathscr{G}$.

Length $s_g$ of a curved line segment:

$$U_g = U_g(t) \tag{B.23}$$

$$V_g = V_g(t) \tag{B.24}$$

on gear tooth flank $\mathscr{G}$ is given by the equation:

$$s_g = \int_{t_0}^{t} \sqrt{E_g \left(\frac{dU_g}{dt}\right)^2 + 2F_g \frac{dU_g}{dt}\frac{dV_g}{dt} + G_g \left(\frac{dV_g}{dt}\right)^2}\, dt \tag{B.25}$$

$$t_0 \leq t \leq t_1$$

For calculation of square $\mathcal{S}_g$ of gear tooth flank $\mathcal{G}$ patch $\Sigma$, which is bounded by a closed curve on the surface $\mathcal{G}$, the following equation can be used:

$$\mathcal{S}_g = \iint_{\Sigma} \sqrt{E_g G_g - F_g^2}\, dU_g\, dV_g \tag{B.26}$$

Ultimately, value of the angle $\omega_g$ between two given directions through a certain point $m$ on a gear tooth flank $\mathcal{G}$ can be calculated from one of the following equations:

$$\cos\omega_g = \frac{F_g}{\sqrt{E_g G_g}} \tag{B.27}$$

$$\sin\omega_g = \frac{H_g}{\sqrt{E_g G_g}} \tag{B.28}$$

$$\tan\omega_g = \frac{H_g}{F_g} \tag{B.29}$$

For the calculation of the discriminant $H_g$ of the first fundamental form $\Phi_{1.g}$, the following equation can be used:

$$H_g = \sqrt{E_g G_g - F_g^2} \tag{B.30}$$

It is assumed here that the discriminant $H_g$ is never negative—that is, $H_g = +\sqrt{E_g G_g - F_g^2}$.

The first fundamental form $\Phi_{1.g}$ represents the length of a curved line segment, and, thus, it always has a nonnegative value—that is, the inequality $\Phi_{1.g} \geq 0$ is always valid.

The first fundamental form $\Phi_{1.g}$ remains the same when the surface is banding. This is another important feature of the first fundamental form $\Phi_{1.g}$.

### B.4.2 Second Fundamental Form of a Surface

The second fundamental form $\Phi_{2.g}$ of a gear tooth flank $\mathcal{G}$ is another of the two aforementioned issues. Curvature at point of a smooth regular surface $\mathcal{G}$ is described by the second fundamental form $\Phi_{2.g}$.

Consider point $K$ on a smooth regular part surface $\mathcal{G}$ (see Figure B.2). The location of point $K$ is specified by two coordinates $U_g$ and $V_g$. A line through point $K$ is entirely located within the surface $\mathcal{G}$. A nearby point $m$ is located within the line through the point $K$. Location of the point $m$ is specified by the coordinates $U_g + dU_g$ and $V_g + dV_g$ as it is infinitesimally close to the point $K$. The closest distance of approach of the point $m$ to the tangent plane through the point $K$ is expressed by the second fundamental form $\Phi_{2.g}$. Torsion of the curve $Km$ is ignored. Therefore, the distance $a$ is assumed equal to zero ($a = 0$).

The curvature at point of a smooth, regular part surface $\mathcal{G}$ is described by the second fundamental form

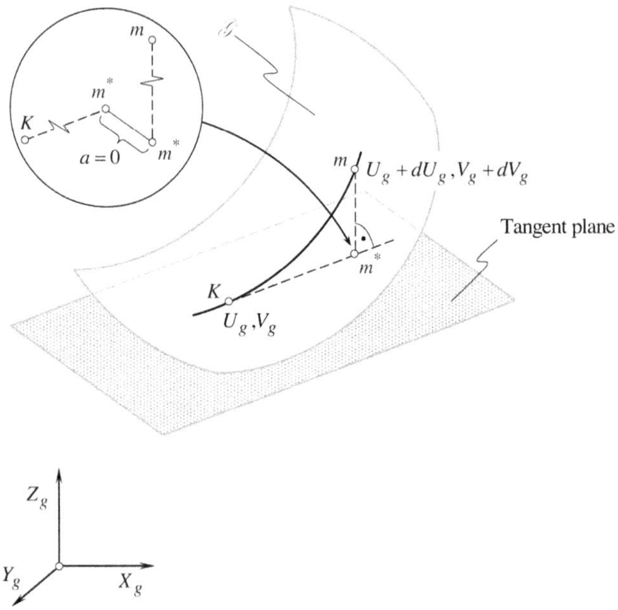


**FIGURE B.2** On the definition of the second fundamental form $\Phi_{2.g}$ at the point of smooth gear tooth flank $\mathcal{G}$.

$\Phi_{2.g}$. Usually, it is represented as the quadratic form (see Figure B.2):

$$\Phi_{2.g} \Rightarrow -d\mathbf{r}_g \cdot d\mathbf{n}_g = L_g\, dU_g^2 + 2M_g\, dU_g\, dV_g + N_g\, dV_g^2 \tag{B.31}$$

Equation (B.31) is known from many advanced sources.

In the theory of gearing, another analytical representation of the second fundamental form $\Phi_{2.g}$ is proven to be useful:

$$\Phi_{2.g} \Rightarrow \begin{bmatrix} dU_g & dV_g & 0 & 0 \end{bmatrix} \cdot \begin{bmatrix} L_g & M_g & 0 & 0 \\ M_g & N_g & 0 & 0 \\ 0 & 0 & 1 & 0 \\ 0 & 0 & 0 & 1 \end{bmatrix} \cdot \begin{bmatrix} dU_g \\ dV_g \\ 0 \\ 0 \end{bmatrix} \tag{B.32}$$

This analytical representation of the second fundamental form $\Phi_{2.P}$ was proposed by Prof. S.P. Radzevich (~2008).

Similar to Eq. (B.13), the practical advantage of Eq. (B.32) is that it can easily be incorporated into computer codes when multiple coordinate system transformations are used. The last is vital for the theory of gearing.

In Eq. (B.32), the parameters $L_g$, $M_g$, and $N_g$ designate fundamental magnitudes of the second order.

By definition, fundamental magnitudes of the second order are equal:

$$L_g = -\mathbf{U}_g \cdot \frac{\partial \mathbf{n}_g}{\partial U_g} = \mathbf{n}_g \cdot \frac{\partial \mathbf{U}_g}{\partial U_g} \tag{B.33}$$

$$M_g = -\frac{1}{2}\left(\mathbf{U}_g \cdot \frac{\partial \mathbf{n}_g}{\partial V_g} + \mathbf{V}_g \cdot \frac{\partial \mathbf{n}_g}{\partial U_g}\right) = \mathbf{n}_g \cdot \frac{\partial \mathbf{U}_g}{\partial V_g} = \mathbf{n}_g \cdot \frac{\partial \mathbf{V}_g}{\partial U_g} \quad \text{(B.34)}$$

$$N_g = -\mathbf{V}_g \cdot \frac{\partial \mathbf{n}_g}{\partial V_g} = \mathbf{n}_g \cdot \frac{\partial \mathbf{V}_g}{\partial V_g} \quad \text{(B.35)}$$

For the calculation of the fundamental magnitudes of the second order of a smooth regular gear tooth flank $\mathcal{G}$, the following equations are used:

$$L_g = \frac{\frac{\partial \mathbf{U}_g}{\partial U_g} \times \mathbf{U}_g \cdot \mathbf{V}_g}{\sqrt{E_g G_g - F_g^2}} \quad \text{(B.36)}$$

$$M_g = \frac{\frac{\partial \mathbf{U}_g}{\partial V_g} \times \mathbf{U}_g \cdot \mathbf{V}_g}{\sqrt{E_g G_g - F_g^2}} = \frac{\frac{\partial \mathbf{V}_g}{\partial U_g} \times \mathbf{U}_g \cdot \mathbf{V}_g}{\sqrt{E_g G_g - F_g^2}} \quad \text{(B.37)}$$

$$N_g = \frac{\frac{\partial \mathbf{V}_g}{\partial V_g} \times \mathbf{U}_g \cdot \mathbf{V}_g}{\sqrt{E_g G_g - F_g^2}} \quad \text{(B.38)}$$

Equations (B.36) through (B.38) can be represented in expanded form:

$$L_g = \frac{\begin{vmatrix} \frac{\partial^2 X_g}{\partial U_g^2} & \frac{\partial^2 Y_g}{\partial U_g^2} & \frac{\partial^2 Z_g}{\partial U_g^2} \\ \frac{\partial X_g}{\partial U_g} & \frac{\partial Y_g}{\partial U_g} & \frac{\partial Z_g}{\partial U_g} \\ \frac{\partial X_g}{\partial V_g} & \frac{\partial Y_g}{\partial V_g} & \frac{\partial Z_g}{\partial V_g} \end{vmatrix}}{\sqrt{E_g G_g - F_g^2}} \quad \text{(B.39)}$$

$$M_g = \frac{\begin{vmatrix} \frac{\partial^2 X_g}{\partial U_g \partial V_g} & \frac{\partial^2 Y_g}{\partial U_g \partial V_g} & \frac{\partial^2 Z_g}{\partial U_g \partial V_g} \\ \frac{\partial X_g}{\partial U_g} & \frac{\partial Y_g}{\partial U_g} & \frac{\partial Z_g}{\partial U_g} \\ \frac{\partial X_g}{\partial V_g} & \frac{\partial Y_g}{\partial V_g} & \frac{\partial Z_g}{\partial V_g} \end{vmatrix}}{\sqrt{E_g G_g - F_g^2}} \quad \text{(B.40)}$$

$$N_P = \frac{\begin{vmatrix} \frac{\partial^2 X_g}{\partial V_g^2} & \frac{\partial^2 Y_g}{\partial V_g^2} & \frac{\partial^2 Z_g}{\partial V_g^2} \\ \frac{\partial X_g}{\partial U_g} & \frac{\partial Y_g}{\partial U_g} & \frac{\partial Z_g}{\partial U_g} \\ \frac{\partial X_g}{\partial V_g} & \frac{\partial Y_g}{\partial V_g} & \frac{\partial Z_g}{\partial V_g} \end{vmatrix}}{\sqrt{E_g G_g - F_g^2}} \quad \text{(B.41)}$$

Fundamental magnitudes of the second order $L_g$, $M_g$, and $N_g$ are also functions of the $U_g$ –, and $V_g$ – coordinates of point of a gear tooth flank $\mathcal{G}$. In general form, these relationships can be represented in the form:

$$L_g = L_g(U_g, V_g) \quad \text{(B.42)}$$

$$M_g = M_g(U_g, V_g) \quad \text{(B.43)}$$

$$N_g = N_g(U_g, V_g) \quad \text{(B.44)}$$

The discriminant $T_g$ of the second fundamental form $\Phi_{2.g}$ can be calculated from the following equation:

$$T_g = \sqrt{L_g N_g - M_g^2} \quad \text{(B.45)}$$

We now come to the theorem, which is essential justification for considering the differential geometry of surfaces in connection with the six fundamental magnitudes. It was first proven by Bonnet (1867),[6] and may be enunciated as follows:

> **Theorem B.1:** *When six fundamental magnitudes $E_g, F_g, G_g$ and $L_g, M_g, N_g$ are given, and when they fulfill the Gaussian characteristic equation and the two Mainardi[7]–Codazzi[8] relations, the fundamental magnitudes uniquely determine a gear tooth flank $\mathcal{G}$ as to its position and orientation in space.*

This theorem is commonly referred to as the main theorem in the theory of surface, or the Bonnet theorem. According to the main theorem, two surfaces that have identical first and second fundamental forms must be either congruent or symmetrical to one another.

By use of six fundamental magnitudes, all the parameters of local geometry of a given part surface can be calculated.

---

[6] Pierre Ossian Bonnet (December 22, 1819–June 22, 1892) was a French mathematician.

[7] Gaspare Mainardi (June 27, 1800–March 9, 1879) was an Italian mathematician.

[8] Delfino Codazzi (March 7, 1824–July 21, 1873) was an Italian mathematician.

## B.5 PRINCIPAL DIRECTIONS AT POINT OF GEAR TOOTH FLANK

The direction of vectors $\mathbf{T}_{1.g}$ and $\mathbf{T}_{2.g}$ at a point on a gear tooth flank $\mathcal{G}$ can be specified in terms of the ratio $dU_g/dV_g$. For the vectors of the first $\mathbf{T}_{1.g}$ and the second $\mathbf{T}_{2.g}$ principal directions at point $m$ of a smooth, regular surface $\mathcal{G}$, the corresponding values of the ratio $dU_g/dV_g$ are calculated as roots of the quadratic equation:

$$\begin{vmatrix} E_g\,dU_g + F_g\,dV_g & F_g\,dU_g + G_g\,dV_g \\ L_g\,dU_g + M_g\,dV_g & M_g\,dU_g + N_g\,dV_g \end{vmatrix} = 0 \tag{B.46}$$

The first principal plane section $C_{1.g}$ is perpendicular to a gear tooth flank $\mathcal{G}$ at the current surface point $m$ and passes through the vector of the first principal direction $\mathbf{T}_{1.g}$. The second principal plane section $C_{2.g}$ is orthogonal to a gear tooth flank $\mathcal{G}$ at the current surface point $m$ and passes through the vector of the second principal direction $\mathbf{T}_{2.g}$.

The principal directions $\mathbf{T}_{1.g}$ and $\mathbf{T}_{2.g}$ can be identified at any and all points of a smooth regular gear tooth flank $\mathcal{G}$, except the umbilic points and flattened points of the surface. At both the umbilic points and the flattened points of a surface, principal directions cannot be identified.

In the theory of gearing, it is often preferable not to use the vectors $\mathbf{T}_{1.g}$ and $\mathbf{T}_{2.g}$ of the principal directions and instead use the unit vectors $\mathbf{t}_{1.g}$ and $\mathbf{t}_{2.g}$ of the principal directions. The unit tangent vectors $\mathbf{t}_{1.g}$ and $\mathbf{t}_{2.g}$ are calculated from the equations:

$$\mathbf{t}_{1.g} = \frac{\mathbf{T}_{1.g}}{|\mathbf{T}_{1.g}|} \tag{B.47}$$

$$\mathbf{t}_{2.g} = \frac{\mathbf{T}_{2.g}}{|\mathbf{T}_{2.g}|} \tag{B.48}$$

respectively.

Unit tangent vectors $\mathbf{t}_{1.g}$ and $\mathbf{t}_{2.g}$ of principal directions at point $m$ on a gear tooth flank $\mathcal{G}$ along with unit normal vector $\mathbf{n}_g$ at that same point $m$ comprise an orthogonal local frame $(\mathbf{t}_{1.g}, \mathbf{t}_{2.g}, \mathbf{n}_g)$. All three unit vectors—namely, vectors $\mathbf{t}_{1.g}$, $\mathbf{t}_{2.g}$, and $\mathbf{n}_g$—are mutually perpendicular to one another. The local frame $(\mathbf{t}_{1.g}, \mathbf{t}_{2.g}, \mathbf{n}_g)$ is commonly referred to as Darboux frame.

## B.6 CURVATURES AT A POINT OF THE GEAR TOOTH FLANK

The first $R_{1.g}$ and the second $R_{2.g}$ principal radii of curvature at a point of a gear tooth flank $\mathcal{G}$ are measured within the first and in the second principal plane sections $C_{1.g}$ and $C_{2.g}$, respectively. For the calculation of values of the principal radii of curvature, the following equation is commonly used:

$$R_g^2 - \frac{E_g N_g - 2F_g M_g + G_g L_g}{T_g} R_g + \frac{H_g}{T_g} = 0 \tag{B.49}$$

Remember that algebraic values of the radii of principal curvature $R_{1.g}$ and $R_{2.g}$ relate to each other as $R_{2.g} > R_{1.g}$. In certain cases, no principal curvatures, for example, at an umbilic point, gear tooth flank $\mathcal{G}$ can be identified, as all normal curvatures at an umbilic point of a surface are equal to one another.

Two other important parameters of local topology at a point of gear tooth flank $\mathcal{G}$ are:

- Mean curvature $\mathcal{M}_g$ and
- Intrinsic curvature $\mathcal{G}_g$ (Gaussian curvature or full curvature).

For the calculation of the curvatures $\mathcal{M}_g$ and $\mathcal{G}_g$, the following equations are commonly used:

$$\mathcal{M}_g = \frac{k_{1.g} + k_{2.g}}{2} = \frac{E_g N_g - 2F_g M_g + G_g L_g}{2 \cdot (E_g G_g - F_g^2)} \tag{B.50}$$

$$\mathcal{G}_g = k_{1.g} \cdot k_{2.g} = \frac{L_g N_g - M_g^2}{E_g G_g - F_g^2} \tag{B.51}$$

The expressions for the mean curvature $\mathcal{M}_g$ and for the Gaussian curvature $\mathcal{G}_g$ at a point of tooth flank $\mathcal{G}$:

$$\mathcal{M}_g = \frac{k_{1.g} + k_{2.g}}{2} \tag{B.52}$$

$$\mathcal{G}_g = k_{1.g} \cdot k_{2.g} \tag{B.53}$$

when considered together yield a quadratic equation with respect to principal curvatures $k_{1.g}$ and $k_{2.g}$:

$$k_g^2 - 2\mathcal{M}_g k_g + \mathcal{G}_g = 0 \tag{B.54}$$

The following formulae:

$$k_{1.g} = \mathcal{M}_g + \sqrt{\mathcal{M}_g^2 - \mathcal{G}_g} \tag{B.55}$$

$$k_{2.g} = \mathcal{M}_g - \sqrt{\mathcal{M}_g^2 - \mathcal{G}_g} \tag{B.56}$$

are the solutions to Eq. (B.54).

Here, in Eqs. (B.55) and (B.56), the first principal curvature of a gear tooth flank $\mathcal{G}$ at a current point $m$ is designated as $k_{1.g}$, and $k_{2.g}$ denotes the second principal curvature of gear tooth flank $\mathcal{G}$ at that same point $m$.

The principal curvatures $k_{1.g}$ and $k_{2.g}$ are the reciprocal to the corresponding principal radii of curvature $R_{1.g}$ and $R_{2.g}$:

$$k_{1.g} = \frac{1}{R_{1.g}} \tag{B.57}$$

$$k_{2.g} = \frac{1}{R_{2.g}} \tag{B.58}$$

The first principal curvature $k_{1.g}$ is always larger than the second principal curvature $k_{2.g}$ of a gear tooth flank $\mathcal{G}$ at current point $m$, meaning the inequality:

$$k_{1.g} > k_{2.g} \tag{B.59}$$

is always valid.

## B.7 BRIOSCHI FORMULA FOR GAUSSIAN CURVATURE

Gaussian curvature at surface $\mathcal{G}$, point $m$ can be expressed in terms of just three fundamental magnitudes $E_g$, $F_g$, $G_g$, of the first order, and their derivatives of the first, and of the second order, that is, with no application of three fundamental magnitudes $L_g$, $M_g$, $N_g$, of the second order. For this purpose, the Brioschi formula for the Gaussian curvature can be used:

$$\mathcal{G}_g = \frac{\begin{vmatrix} -\frac{1}{2}\frac{\partial^2 E_g}{\partial V_g^2} + \frac{\partial^2 F_g}{\partial U_g \partial V_g} - \frac{1}{2}\frac{\partial^2 G_g}{\partial U_g \partial V_g} & \frac{1}{2}\frac{\partial E_g}{\partial U_g} & \frac{\partial F_g}{\partial U_g} - \frac{1}{2}\frac{\partial E_g}{\partial V_g} \\ \frac{\partial F_g}{\partial V_g} - \frac{1}{2}\frac{\partial G_g}{\partial U_g} & E_g & F_g \\ \frac{1}{2}\frac{\partial G_g}{\partial V_g} & F_g & G_g \end{vmatrix} - \begin{vmatrix} 0 & \frac{1}{2}\frac{\partial E_g}{\partial V_g} & \frac{1}{2}\frac{\partial G_g}{\partial U_g} \\ \frac{1}{2}\frac{\partial E_g}{\partial V_g} & E_g & F_g \\ \frac{1}{2}\frac{\partial G_g}{\partial U_g} & F_g & G_g \end{vmatrix}}{(E_g G_g - F_g^2)} \tag{B.60}$$

For orthogonally parameterized surfaces (when the equality $F_g = 0$ is valid), the Gaussian curvature is:

$$\mathcal{G}_g = -\frac{1}{2\sqrt{E_g G_g}} \left( \frac{\partial}{\partial U_g} \frac{\frac{\partial G_g}{\partial U_g}}{\sqrt{E_g G_g}} + \frac{\partial}{\partial V_g} \frac{\frac{\partial E_g}{\partial V_g}}{\sqrt{E_g G_g}} \right) \tag{B.61}$$

The Brioschi[9] formula gives the Gaussian curvature solely in terms of the first fundamental form, as illustrated in Eqs. (B.60) and (B.61).

Other formats of representation of the Brioschi formula for the Gaussian curvature are also known.

## B.8 NATURAL FORM OF SURFACE REPRESENTATION

This brief consideration of major elements of part surface geometry necessitates the introduction of two definitions that are of critical importance for further discussion.

As mentioned earlier in this section of the book, it was proved by Bonnet that a unique surface is specified by the first and the second fundamental forms if the Gauss characteristic equation and the Mainardi-Codazzi relations of compatibility are satisfied, and those two surfaces have identical first and second fundamental forms that are congruent.[10] Six fundamental magnitudes uniquely determine a surface, except as to position and orientation in space.

The specification of a surface in terms of the first and second fundamental forms is usually called the natural kind of surfaces representation. In the general form, this kind of part surfaces representation can be expressed by a set of two equations:

$$\left.\begin{matrix}\text{Natural form} \\ \text{of surface } \mathcal{G} \text{ representation}\end{matrix}\right| \Rightarrow \mathcal{G} = \mathcal{G}(\Phi_{1.g}, \Phi_{2.g}) \begin{cases} \Phi_{1.g} = \Phi_{1.g}(E_g, F_g, G_g) \\ \Phi_{2.g} = \Phi_{2.g}(E_g, F_g, G_g, L_g, M_g, N_g) \end{cases} \tag{B.62}$$

Equation (B.62) can be derived from Eq. (B.1). A given gear tooth flank $\mathcal{G}$ can be expressed in both forms, either by Eq. (B.19) or by Eq. (B.1).

## B.9 A FEW MORE USEFUL EQUATIONS

Many calculations of the parameters of geometry can be significantly simplified by use of the first and the second fundamental forms of a smooth regular part surface $\mathcal{G}$.

[9] Francesco Brioschi (December 22, 1824–December 13, 1897) was an Italian mathematician.

[10] Two surfaces with the identical first and second fundamental forms might also be symmetrical. Refer to the literature—Koenderink, J.J., *Solid Shape*, The MIT Press, Cambridge, MA, 1990, 699 pages—on differential geometry of surfaces for details about this specific issue.

1. For the calculation of the actual value of radius $R_g$ of a normal curvature within a normal plane section through given point $m$ on a gear tooth flank $\mathcal{G}$ and at a given direction the following equation can be used:

$$R_g = \frac{\Phi_{1.g}}{\Phi_{2.g}} \tag{B.63}$$

2. The Euler formula for the calculation of normal curvature $k_{\theta.g}$ at point $m$ in a direction that is specified by the angle $\theta$ can be represented as follows:

$$k_{\theta.g} = k_{1.g}\cos^2\theta + k_{2.g}\sin^2\theta \tag{B.64}$$

Here, in Eq. (B.64), $\theta$ is the angle that the normal plane section $C_g$ makes with the first principal plane section $C_{1.g}$—in other words, $\theta = \angle(\mathbf{t}_g, \mathbf{t}_{1.g})$; here, $\mathbf{t}_g$ designates the unit tangent vector within the normal plane section $C_g$.

Equation (B.64) is also a perfect illustration of significant simplification of the calculations when fundamental magnitudes $E_g$, $F_g$, and $G_g$ of the first and $L_g$, $M_g$, and $N_g$ of the second order are used.

To develop a profound understanding of differential geometry of surfaces, the interested reader may wish to go to advanced monographs in the field. Systematic discussion of the topic is available from many sources. The author would like to turn the reader's attention, in particular, to the works of doCarmo, Eisenhart, and Stuik, among others.

## B.10 ILLUSTRATIVE EXAMPLE

Consider an example of how analytical representation of a surface in a Cartesian reference system can be converted into the natural representation of that same surface.

A Cartesian coordinate system $X_gY_gZ_g$ is associated with a gear tooth flank $\mathcal{G}$, as schematically shown in Figure B.3.

The position vector of point $\mathbf{r}_g$ of the gear tooth flank $\mathcal{G}$ can be represented as the summa of three vectors:

$$\mathbf{r}_g = \mathbf{A} + \mathbf{B} + \mathbf{C} \tag{B.65}$$

Each of the vectors **A**, **B**, and **C** can be expressed in terms of projections onto the axes of the reference system $X_gY_gZ_g$. Then, Eq. (B.65) casts into the equation:

$$\mathbf{r}_g(U_g, V_g) = \begin{bmatrix} r_{b.g}\cos V_g + U_g\cos\tau_{b.g}\sin V_g \\ r_{b.g}\sin V_g - U_g\sin\tau_{b.g}\sin V_g \\ r_{b.g}\tan\tau_{b.g} - U_g\sin\tau_{b.g} \\ 1 \end{bmatrix} \tag{B.66}$$

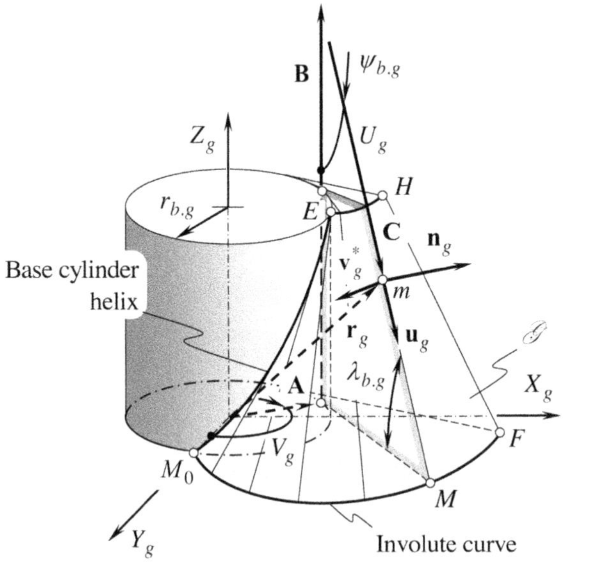


**FIGURE B.3** Derivation of the natural form of a representation of gear tooth flank $\mathcal{G}$.

This yields the calculation of two tangent vectors $\mathbf{U}_g(U_g, V_g)$ and $\mathbf{V}_g(U_g, V_g)$, which are respectively equal:

$$\mathbf{U}_g(U_g, V_g) = \begin{bmatrix} \cos\tau_{b.g}\sin V_g \\ -\cos\tau_{b.g}\cos V_g \\ -\sin\tau_{b.g} \\ 0 \end{bmatrix} \tag{B.67}$$

$$\mathbf{V}_g(U_g, V_g) = \begin{bmatrix} -r_{b.g}\sin V_g + U_g\cos\tau_{b.g}\cos V_g \\ r_{b.g}\cos V_g + U_g\cos\tau_{b.g}\sin V_g \\ r_{b.g}\tan\tau_{b.g} \\ 0 \end{bmatrix} \tag{B.68}$$

Substituting the derived vectors $\mathbf{U}_g$ and $\mathbf{V}_g$ into Eq. (B.14), one can create the formulae for the calculation of the fundamental magnitudes of the first order:

$$E_g = 1 \tag{B.69}$$

$$F_g = -\frac{r_{b.g}}{\cos\tau_{b.g}} \tag{B.70}$$

$$G_g = \frac{U_g^2\cos^4\tau_{b.g} + r_{b.g}^2}{\cos^2\tau_{b.g}} \tag{B.71}$$

These expressions can be substituted directly to Eq. (B.12) for the first fundamental form $\Phi_{1.g}$ of the gear tooth flank $\mathcal{G}$:

$$\Phi_{1.g} \Rightarrow dU_g^2 - 2\frac{r_{b.g}}{\cos\tau_{b.g}}dU_g dV_g + \frac{U_g^2\cos^4\tau_{b.g} + r_{b.g}^2}{\cos^2\tau_{b.g}}dV_g^2 \tag{B.72}$$

The derived expressions for the fundamental magnitudes $E_g$, $F_g$, and $G_g$ [see Eqs. (B.69) through (B.71)] can also be substituted to Eq. (B.13). In this way, a corresponding matrix representation of the first fundamental form $\Phi_{1.g}$ of the gear tooth flank $\mathcal{G}$ can be calculated. The interested reader may wish to complete these formulae on his or her own.

The discriminant $H_g$ of the first fundamental form of the gear tooth flank $\mathcal{G}$ can be calculated from the expression:

$$H_g = U_g \cos\tau_{b.g} \tag{B.73}$$

In order to derive an equation for the second fundamental form $\Phi_{2.g}$ at point of the gear tooth flank $\mathcal{G}$, parameters for the second derivatives of the position vector of a point $\mathbf{r}_g(U_g, V_g)$ with respect to $U_g$ – and $V_g$ – are necessary. The previous derived equations for the tangent vectors $\mathbf{U}_g$ and $\mathbf{V}_g$ [see Eqs. (B.67) and (B.68)] make the following expressions for the derivatives under consideration possible:

$$\frac{\partial \mathbf{U}_g}{\partial U_P} = \begin{bmatrix} 0 \\ 0 \\ 0 \\ 1 \end{bmatrix} \tag{B.74}$$

$$\frac{\partial \mathbf{U}_g}{\partial V_g} \equiv \frac{\partial \mathbf{V}_g}{\partial U_g} = \begin{bmatrix} \cos\tau_{b.g}\cos V_g \\ \cos\tau_{b.g}\sin V_g \\ 0 \\ 1 \end{bmatrix} \tag{B.75}$$

$$\frac{\partial \mathbf{V}_g}{\partial V_g} = \begin{bmatrix} -r_{b.g}\cos V_g - U_g\cos\tau_{b.g}\sin V_g \\ -r_{b.g}\sin V_g + U_g\cos\tau_{b.g}\cos V_g \\ 0 \\ 1 \end{bmatrix} \tag{B.76}$$

Further, substitute these expressions [see Eqs. (B.74) through (B.76)] into Eqs. (B.36) through (B.38). After the necessary formulae transformations are complete, Eqs. (B.36) through (B.38) cast into the set of formulae for the calculation of the fundamental magnitudes of the second order of the gear tooth flank $\mathcal{G}$. This set of formulae is as follows:

$$L_g = 0 \tag{B.77}$$

$$M_g = 0 \tag{B.78}$$

**TABLE B.1**
**Fundamental Magnitudes of the First and the Second Order at a Point of Involute Gear Tooth Flank $\mathcal{G}$.**

| | |
|---|---|
| $E_g = 1$ | $L_g = 0$ |
| $F_g = -\dfrac{r_{b.g}}{\cos\tau_{b.g}}$ | $M_g = 0$ |
| $G_g = \dfrac{U_g^2\cos^4\tau_{b.g} + r_{b.g}^2}{\cos^2\tau_{b.g}}$ | $N_g = -U_g\sin\tau_{b.g}\cos\tau_{b.g}$ |

$$N_g = -U_g \sin\tau_{b.g}\cos\tau_{b.g} \tag{B.79}$$

After substituting Eqs. (B.77) through (B.79) into Eq. (B.31), an equation for the calculation of the second fundamental form of the gear tooth flank $\mathcal{G}$ can be represented in the form:

$$\Phi_{2.g} \Rightarrow -d\mathbf{r}_g \cdot d\mathbf{N}_g = -U_g \sin\tau_{b.g}\cos\tau_{b.g} dV_g^2 \tag{B.80}$$

Similar to Eq. (B.72), the derived expressions for the fundamental magnitudes $L_g$, $M_g$ and $N_g$ of the second order can be substituted into Eq. (B.32) for the second fundamental form $\Phi_{2.g}$. By doing so, a corresponding matrix representation of the second fundamental form $\Phi_{2.g}$ of the surface $\mathcal{G}$ can be derived. The interested reader may wish to complete this formulae transformation independently.

For the calculation of the discriminant $T_g$ of the second fundamental form $\Phi_{2.g}$ of the gear tooth flank $\mathcal{G}$, the following expression is used:

$$T_g = U_g \sin\tau_{b.g}\cos\tau_{b.g} \tag{B.81}$$

The natural representation of the gear tooth flank $\mathcal{G}$ can be expressed in terms of the derived set of six equations for the calculation of the first fundamental magnitudes $E_g$, $F_g$, and $\mathcal{G}_g$ and second fundamental magnitudes $L_g$, $M_g$ and $N_g$ (see Table B.1).

All major elements of local geometry of the gear tooth flank $\mathcal{G}$ can be calculated based on the fundamental forms of magnitudes $E_g$, $F_g$, and $G_g$ of the first order $\Phi_{1.P}$ and $L_g$, $M_g$, and $N_g$ of the second order $\Phi_{2.g}$. The location and orientation of the gear tooth flank $\mathcal{G}$ are the two parameters that remain indefinite.

Once a part surface is represented in natural form—that is, it is expressed in terms of six fundamental magnitudes of the first and of the second order—further calculation of parameters of a gear tooth flank $\mathcal{G}$ becomes significantly easier.

# Appendix C: Change of Surface Parameters

When designing gears, it is often necessary to treat two or more surfaces simultaneously. For example, contact geometry at a common point of two tooth flanks $\mathcal{G}$ and $\mathcal{P}$ can be described analytically, if the tooth flanks are properly parameterized initially.

When it is necessary for two surfaces $\mathbf{r}_i$ and $\mathbf{r}_j$ to be treated simultaneously, not only must they be represented in a common reference system, but the $U_i-$ and $V_i-$ parameters of one of the surfaces $\mathbf{r}_i = \mathbf{r}_i(U_i, V_i)$ have to be synchronized with the corresponding $U_j-$ and $V_j-$ parameters of the other surface $\mathbf{r}_j = \mathbf{r}_j(U_j, V_j)$. The procedure of changing surface parameters is used for this purpose. Use of this procedure allows representation of one of the surfaces, for example of the surface $\mathbf{r}_j = \mathbf{r}_j(U_j, V_j)$, in the terms of the $U_i-$ and $V_i-$ parameters, say as $\mathbf{r}_j = \mathbf{r}_j(U_i, V_i)$.

If the parameterization of a surface is transformed by the equations $U^* = U^*(U,V)$ and $V^* = V^*(U,V)$, we obtain the new derivatives:

$$\frac{\partial \mathbf{r}}{\partial U^*} = \frac{\partial \mathbf{r}}{\partial U} \cdot \frac{\partial U}{\partial U^*} + \frac{\partial \mathbf{r}}{\partial V} \cdot \frac{\partial V}{\partial U^*} \tag{C.1}$$

$$\frac{\partial \mathbf{r}}{\partial V^*} = \frac{\partial \mathbf{r}}{\partial U} \cdot \frac{\partial U}{\partial V^*} + \frac{\partial \mathbf{r}}{\partial V} \cdot \frac{\partial V}{\partial V^*} \tag{C.2}$$

so that:

$$\mathbf{A}^* = \left[ \frac{\partial \mathbf{r}}{\partial U^*} \middle| \frac{\partial \mathbf{r}}{\partial V^*} \right] = \mathbf{A} \cdot \mathbf{J} \tag{C.3}$$

where:

$$\mathbf{J} = \begin{bmatrix} \dfrac{\partial U}{\partial U^*} & \dfrac{\partial U}{\partial V^*} \\ \dfrac{\partial V}{\partial U^*} & \dfrac{\partial V}{\partial V^*} \end{bmatrix} \tag{C.4}$$

which is called the Jacobian matrix of the transformation.

It can be shown that the new fundamental matrix $\mathbf{G}^*$ is given by:

$$\mathbf{G}^* = \mathbf{A}^{*T}\mathbf{A}^* = \mathbf{J}^T\mathbf{A}^T\mathbf{A}\mathbf{J} = \mathbf{J}^T\mathbf{G}\mathbf{J} \tag{C.5}$$

From this equation, we see by the properties of determinants that $|\mathbf{G}^*| = |\mathbf{J}|^2|\mathbf{G}|$. Using this result, as well as Eqs. (C.1) and (C.2), we can show that the unit surface normal $\mathbf{n}$ is invariant under the transformation, as we could expect.

The transformation of the second fundamental matrix can similarly be shown to be given by:

$$\mathbf{D}^* = \mathbf{J}^T\mathbf{D}\mathbf{J} \tag{C.6}$$

by differentiating Eqs. (C.1) and (C.2) and using the invariance of $\mathbf{n}$. From Eqs. (C.5) and (C.6), it can be shown that the principal curvatures and directions are invariant under the transformation.

We conclude that the unit normal vector $\mathbf{n}$ and the principal directions and curvatures are independent of the parameters used and are therefore geometric properties of the surface itself. They should be continuous if the surface is to be tangent and the curvature continuous.

# Appendix D: Applied Coordinate Systems and Linear Transformations

Consequent coordinate system transformations can easily be analytically described using matrices. The use of matrices for the coordinate system transformation[1] can be traced back to the mid-1940s,[2] when Dr. S.S. Mozhayev[3] began describing this process.

In this appendix, coordinate system transformation is briefly discussed from the standpoint of its implementation in the theory of gearing.

## D.1 COORDINATE SYSTEM TRANSFORMATION

Homogeneous coordinates utilize a mathematical trick to embed three-dimensional coordinates and transformations into a four-dimensional matrix format. As a result, inversions or combinations of linear transformations are simplified to inversions or multiplication of the corresponding matrices.

### D.1.1 Homogeneous Coordinate Vectors

Instead of representing each point $\mathbf{r}(x,y,z)$ in three-dimensional space with a single three-dimensional vector

$$\mathbf{r} = \begin{bmatrix} x \\ y \\ z \end{bmatrix} \tag{D.1}$$

homogeneous coordinates allow each point $\mathbf{r}(x,y,z)$ to be represented by any of an infinite number of four-dimensional vectors:

$$\mathbf{r} = \begin{bmatrix} T \cdot x \\ T \cdot y \\ T \cdot z \\ T \end{bmatrix} \tag{D.2}$$

The three-dimensional vector corresponding to any four-dimensional vector can be calculated by dividing the first three elements by the fourth, and a four-dimensional vector corresponding to any three-dimensional vector can be created by simply adding a fourth element and setting it equal to one.

### D.1.2 Homogeneous Coordinate Transformation Matrices of the Dimension 4 × 4

Homogeneous coordinate transformation matrices operate on four-dimensional homogeneous vector representations of traditional three-dimensional coordinate locations. Any three-dimensional linear transformation (translation, rotation, and so forth) can be represented by a $4\times4$ homogeneous coordinate transformation matrix. In fact, because of the redundant representation of three dimensions in a homogeneous coordinate system, an infinite number of $4\times4$ homogeneous coordinate transformation matrices are available to perform any given linear transformation. This redundancy can be eliminated to provide a unique representation by dividing all elements of a $4\times4$ homogeneous transformation matrix by the last element (which will become equal to one). By doing so, a $4\times4$ homogeneous transformation matrix can incorporate as many as 15 independent parameters. The generic format representation of a homogeneous transformation equation for mapping the three-dimensional coordinate $(x_1, y_1, z_1)$ to the three-dimensional coordinate $(x_2, y_2, z_2)$ is:

$$\begin{bmatrix} T^* \cdot x_2 \\ T^* \cdot y_2 \\ T^* \cdot z_2 \\ T^* \end{bmatrix} = \begin{bmatrix} T^* \cdot a & T^* \cdot b & T^* \cdot c & T^* \cdot d \\ T^* \cdot e & T^* \cdot f & T^* \cdot g & T^* \cdot h \\ T^* \cdot i & T^* \cdot j & T^* \cdot k & T^* \cdot m \\ T^* \cdot n & T^* \cdot p & T^* \cdot q & T^* \end{bmatrix} \cdot \begin{bmatrix} T \cdot x_2 \\ T \cdot y_2 \\ T \cdot z_2 \\ T \end{bmatrix} \tag{D.3}$$

If any two matrices or vectors of this equation are known, the third matrix (or vector) can be calculated, and the redundant $T$ element in the solution can then be eliminated by dividing all elements of the matrix by the last element.

Various transformation models can be used to constrain the form of the matrix to transformations with fewer degrees of freedom.

### D.1.3 Translations

Translation of a coordinate system is one of the major linear transformations used in the theory of gearing. Translations

1 Matrices were introduced into mathematics by A. Cayley in 1857. They provide a compact and flexible notation that is particularly useful in dealing with linear transformations, and they presented an organized method for the solution of systems of linear differential equations.

2 The application of matrices for the purposes of analytical representation of coordinate system transformation should be credited to Dr. S.S. Mozhayev [Mozhayev, S.S., *General Theory of Cutting Tools*, Doctoral Thesis, Leningrad, Leningrad Polytechnic Institute, 1951, 295 pages]. Since then, the matrix approach for coordinate system transformation has been used by Denavit & Hartenberg, as well as many other researchers.

3 S.S. Mozhayev is a soviet scientist known primarily for his accomplishments in the theory of cutting-tool design.

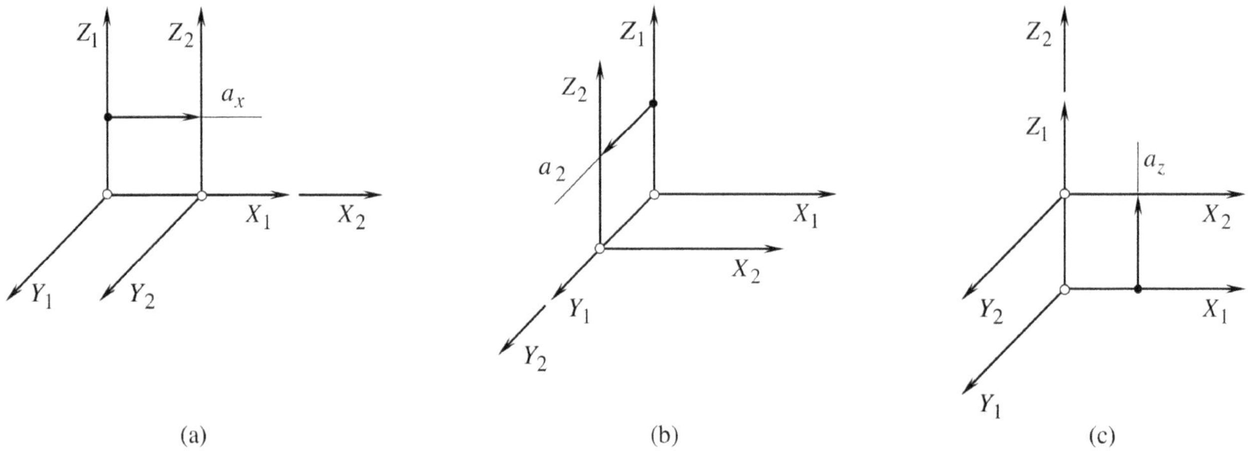


**FIGURE D.1** Analytical description of the operators of translations along (a) X-axis, $\mathbf{Tr}(a_x, X)$, along (b) Y-axis, $\mathbf{Tr}(a_y, Y)$, and along (c) Z-axis, $\mathbf{Tr}(a_z, Z)$ of a Cartesian reference system $XYZ$.

of the coordinate system $X_2Y_2Z_2$ along axes of the coordinate system $X_1Y_1Z_1$ are depicted in Figure D.1. Translations can be analytically described using a homogeneous $4 \times 4$ transformation matrix.

For an analytical description of translation along coordinate axes, the operators of translation $\mathbf{Tr}(a_x, X)$, $\mathbf{Tr}(a_y, Y)$, and $\mathbf{Tr}(a_z, Z)$ are used. These operators yield matrix representation in the form:

$$\mathbf{Tr}(a_x, X) = \begin{bmatrix} 1 & 0 & 0 & a_x \\ 0 & 1 & 0 & 0 \\ 0 & 0 & 1 & 0 \\ 0 & 0 & 0 & 1 \end{bmatrix} \quad \text{(D.4)}$$

$$\mathbf{Tr}(a_y, Y) = \begin{bmatrix} 1 & 0 & 0 & 0 \\ 0 & 1 & 0 & a_y \\ 0 & 0 & 1 & 0 \\ 0 & 0 & 0 & 1 \end{bmatrix} \quad \text{(D.5)}$$

$$\mathbf{Tr}(a_z, Z) = \begin{bmatrix} 1 & 0 & 0 & 0 \\ 0 & 1 & 0 & 0 \\ 0 & 0 & 1 & a_z \\ 0 & 0 & 0 & 1 \end{bmatrix} \quad \text{(D.6)}$$

In Eqs. (D.4) through (D.6), the parameters $a_x$, $a_y$ and $a_z$ are signed values that denote the distance of translation along the corresponding axis.

Consider two coordinate systems, $X_1Y_1Z_1$ and $X_2Y_2Z_2$, displaced along the $X_1$ –axis at a distance $a_x$, as schematically depicted in Figure D.1(a). A point $m$ in the reference system $X_2Y_2Z_2$ is given by the position vector $\mathbf{r}_2(m)$. In the coordinate system $X_1Y_1Z_1$, that same point $m$ can be specified by the position vector $\mathbf{r}_1(m)$. Then the position vector $\mathbf{r}_1(m)$ can be expressed in terms of the position vector $\mathbf{r}_2(m)$ using the equation:

$$\mathbf{r}_1(m) = \mathbf{Tr}(a_x, X) \cdot \mathbf{r}_2(m) \quad \text{(D.7)}$$

Equations similar to Eq. (D.7) are valid for the operators $\mathbf{Tr}(a_y, Y)$ and $\mathbf{Tr}(a_z, Z)$ of the coordinate system transformation. The latter is schematically illustrated in Figure D.1(b) and in Figure D.1(c).

Use of the operators of translation $\mathbf{Tr}(a_x, X)$, $\mathbf{Tr}(a_y, Y)$, and $\mathbf{Tr}(a_z, Z)$ makes the introduction of an operator $\mathbf{Tr}(a, \mathbf{A})$ of a combined transformation possible. Suppose that point $p$ on a rigid body goes through a translation describing a straight line from a point $p_1$ to a point $p_2$ with a change of coordinates of $(a_x, a_y, a_z)$. This motion of point $p$ is analytically described with a resultant translation operator $\mathbf{Tr}(a, \mathbf{A})$:

$$\mathbf{Tr}(a, \mathbf{A}) = \begin{bmatrix} 1 & 0 & 0 & a_x \\ 0 & 1 & 0 & a_y \\ 0 & 0 & 1 & a_z \\ 0 & 0 & 0 & 1 \end{bmatrix} \quad \text{(D.8)}$$

The operator $\mathbf{Tr}(a, \mathbf{A})$ of the resultant coordinate system transformation can be interpreted as the operator of translation along an arbitrary axis, with the vector $\mathbf{A}$ as the direction vector.

An analytical description of translation of the coordinate system $X_1Y_1Z_1$ in direction of an arbitrary vector $\mathbf{A}$ to the position of $X_2Y_2Z_2$ can be composed from Figure D.2. The operator of translation $\mathbf{Tr}(a, \mathbf{A})$ of that particular kind can be expressed in terms of the operators $\mathbf{Tr}(a_x, X)$, $\mathbf{Tr}(a_y, Y)$, and $\mathbf{Tr}(a_z, Z)$ of elementary translations:

$$\mathbf{Tr}(a, \mathbf{A}) = \mathbf{Tr}(a_z, Z) \cdot \mathbf{Tr}(a_y, Y) \cdot \mathbf{Tr}(a_x, X) \quad \text{(D.9)}$$

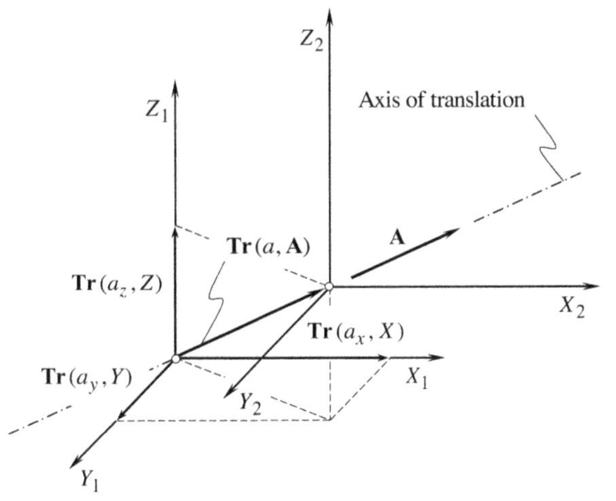


**FIGURE D.2** Analytical description of an operator **Tr** $(a, \mathbf{A})$ of translation along an arbitrary axis (vector **A** is the direction vector of the axis).

Evidently, the axis along the vector **A** is always the axis through the origins of both reference systems $X_1Y_1Z_1$ and $X_2Y_2Z_2$.

Any and all coordinate system transformations that do not change the orientation of a geometrical object are referred to as *orientation-preserving transformations*, or *direct transformations*. Transformation of translation is an example of a direct transformation.

## D.1.4 Rotation About a Coordinate Axis

Rotation of a coordinate system about a coordinate axis is another major linear transformation used in the theory of part surface generation. A rotation is specified by an axis of rotation and the angle of the rotation. It is a fairly simple trigonometric calculation to obtain a transformation matrix for a rotation about one of the coordinate axes.

Possible rotations of the coordinate system $X_2Y_2Z_2$ about the axis of the coordinate system $X_1Y_1Z_1$ are illustrated in Figure D.3.

For analytical description of rotation about a coordinate axis, the operators of rotation $\mathbf{Rt}(\varphi_x, X_1)$, $\mathbf{Rt}(\varphi_y, Y_1)$, and $\mathbf{Rt}(\varphi_z, Z_1)$ are used. These operators of linear transformations yield representation in the form of homogeneous matrices:

$$\mathbf{Rt}(\varphi_x, X_1) = \begin{bmatrix} 1 & 0 & 0 & 0 \\ 0 & \cos\varphi_x & \sin\varphi_x & 0 \\ 0 & -\sin\varphi_x & \cos\varphi_x & 0 \\ 0 & 0 & 0 & 1 \end{bmatrix} \quad \text{(D.10)}$$

$$\mathbf{Rt}(\varphi_y, Y_1) = \begin{bmatrix} \cos\varphi_y & 0 & \sin\varphi_y & 0 \\ 0 & 1 & 0 & 0 \\ -\sin\varphi_y & 0 & \cos\varphi_y & 0 \\ 0 & 0 & 0 & 1 \end{bmatrix} \quad \text{(D.11)}$$

$$\mathbf{Rt}(\varphi_z, Z_1) = \begin{bmatrix} \cos\varphi_z & \sin\varphi_z & 0 & 0 \\ -\sin\varphi_z & \cos\varphi_z & 0 & 0 \\ 0 & 0 & 1 & 0 \\ 0 & 0 & 0 & 1 \end{bmatrix} \quad \text{(D.12)}$$

Here, $\varphi_x$, $\varphi_y$, and $\varphi_z$ are signed values that denote the corresponding angles of rotations about a corresponding coordinate axis: $\varphi_x$ is the angle of rotation around the $X_1$ – axis (pitch) of the Cartesian coordinate system $X_1Y_1Z_1$; $\varphi_y$ is the angle of rotation around the $Y_1$ – axis (roll); and $\varphi_z$ is the angle of rotation around the $Z_1$ – axis (yaw) of that same Cartesian reference system $X_1Y_1Z_1$.

Rotation about a coordinate axis is illustrated in Figure D.3.

Consider two coordinate systems $X_1Y_1Z_1$ and $X_2Y_2Z_2$, which are turned about the $X_1$ – axis through an angle $\varphi_x$, as shown in Figure D.3(a). In the reference system $X_2Y_2Z_2$, point $m$ is given by a position vector $\mathbf{r}_2(m)$. In the coordinate system $X_1Y_1Z_1$, that same point $m$ can be specified by the position

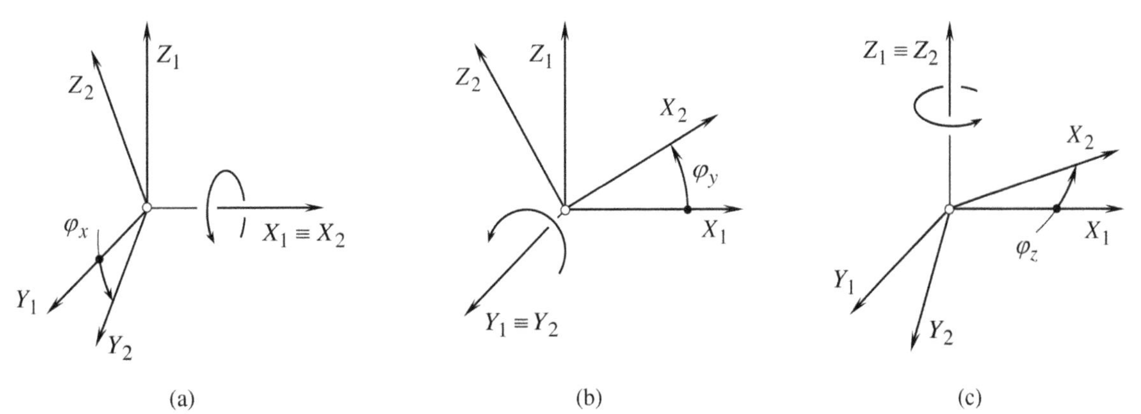


**FIGURE D.3** Analytical description of the operators of rotation (a) about X-axis, $\mathbf{Rt}(\varphi_x, X)$, (b) about Y-axis, $\mathbf{Rt}(\varphi_y, Y)$, and (c) about Z-axis, $\mathbf{Rt}(\varphi_z, Z)$ of a reference system $X_1Y_1Z_1$.

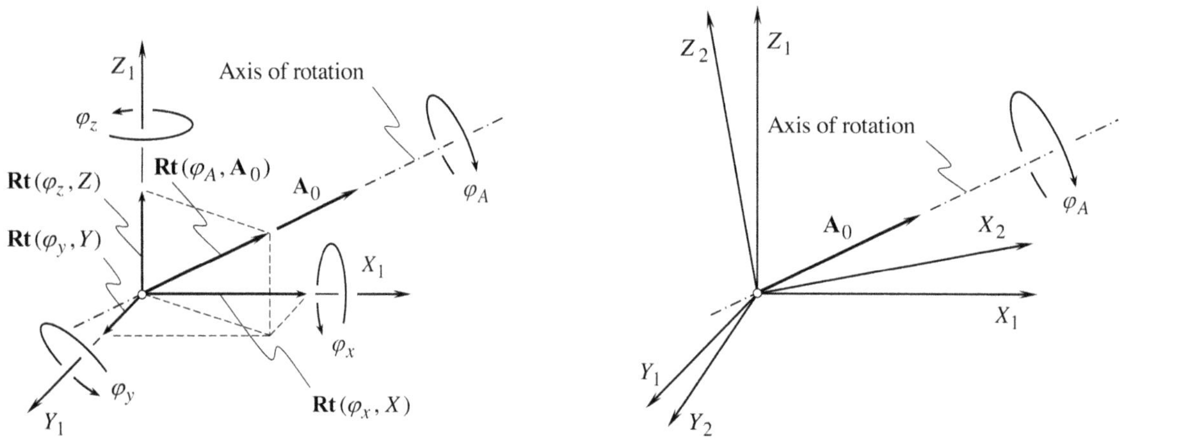


**FIGURE D.4** Analytical description of the operator $\mathbf{Rt}(\varphi_A, \mathbf{A})$ of rotation about an arbitrary axis through the origin of a Cartesian coordinate system $X_1Y_1Z_1$ (vector $\mathbf{A}$ is the direction vector of the axis of rotation).

vector $\mathbf{r}_1(m)$. Then, the position vector $\mathbf{r}_1(m)$ can be expressed in terms of the position vector $\mathbf{r}_2(m)$ using the equation:

$$\mathbf{r}_1(m) = \mathbf{Rt}(\varphi_x, X) \cdot \mathbf{r}_2(m) \qquad \text{(D.13)}$$

Equations similar to Eq. (D.13) are also valid for operators $\mathbf{Rt}(\varphi_y, Y)$ and $\mathbf{Rt}(\varphi_z, Z)$ of the coordinate system transformation. These elementary coordinate system transformations are schematically illustrated in Figure D.3(b) and (c), respectively.

## D.1.5 Rotation About an Arbitrary Axis Through the Origin

When a rotation is to be performed around an arbitrary vector based at the origin, the transformation matrix must be assembled from a combination of rotations about the Cartesian coordinate.

Two different approaches for analytical description of a rotation about an arbitrary axis through the origin are discussed in the following section.

### D.1.5.1 Conventional Approach

An analytical description of rotation of the coordinate system $X_1Y_1Z_1$ about an arbitrary axis through the origin to the position of a reference system $X_2Y_2Z_2$ is illustrated in Figure D.4. It is assumed here that the rotation is performed about the axis with vector $\mathbf{A}_0$ as the direction vector. The operator $\mathbf{Rt}(\varphi_A, \mathbf{A}_0)$ of rotation of that kind can be expressed in terms of the operators $\mathbf{Rt}(\varphi_x, X)$, $\mathbf{Rt}(\varphi_y, Y)$, and $\mathbf{Rt}(\varphi_z, Z)$ of elementary rotations:

$$\mathbf{Rt}(\varphi_A, \mathbf{A}_0) = \mathbf{Rt}(\varphi_z, Z) \cdot \mathbf{Rt}(\varphi_y, Y) \cdot \mathbf{Rt}(\varphi_x, X) \qquad \text{(D.14)}$$

Evidently, the axis of rotation (a straight line along vector $\mathbf{A}_0$) is always an axis through the origin.

The operators of translation and the operators of rotation also yield linear transformations of other kinds.

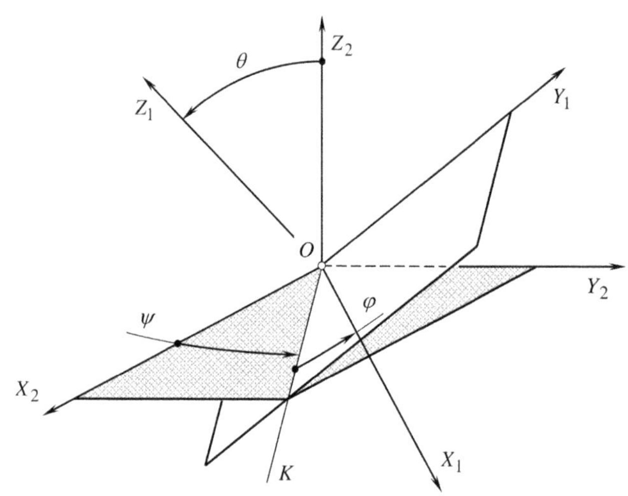


**FIGURE D.5** Euler angles.

### D.1.5.2 Euler Transformation

A Euler[4] transformation is a well-known kind of linear transformation used extensively in mechanical engineering. This kind of linear transformation is analytically described by the operator $\mathbf{Eu}(\psi, \theta, \varphi)$.

The operator $\mathbf{Eu}(\psi, \theta, \varphi)$ is expressed in terms of three *Euler angles* (or *Eulerian angles*) $\psi$, $\theta$, and $\varphi$. Configuration of an orthogonal Cartesian coordinate system $X_1Y_1Z_1$ in relation to another orthogonal Cartesian coordinate system $X_2Y_2Z_2$ is defined by the Euler angles $\psi$, $\theta$, and $\varphi$. These angles are shown in Figure D.5.

The line of intersection of the coordinate plane $X_1Y_1$ of the first reference system with the coordinate plane $X_2Y_2$ of the second reference system is commonly referred to as the *nodal line* (or *line of nodes*). In Figure D.5, the line $OK$ is the nodal line. It is assumed hereafter that the nodal line

[4] Leonhard Euler (April 15, 1707–September 18, 1783) was a famous Swiss mathematician and physicist who spent most of his life in Russia and Germany.

*OK* and the axes $Z_1$ and $Z_2$ form a frame of that same orientation as the reference systems $X_1Y_1Z_1$ and $X_2Y_2Z_2$.

The Euler angle $\varphi$ is referred to as the *angle of pure rotation*. This angle is measured between the $X_1-$ axis and the nodal line *OK*. The angle of pure rotation $\varphi$ is measured within the coordinate plane $X_1Y_1$ in the direction of shortest rotation from the axis $X_1$ to the axis $Y_1$.

The Euler angle $\theta$ is referred to as the *angle of nutation*. The angle of nutation $\theta$ is measured between the axes $Z_1$ and $Z_2$. The actual value of this angle never exceeds 180°.

The Euler angle $\psi$ is referred to as the *angle of precession*. The angle of precession $\psi$ is measured in the coordinate plane $X_2Y_2$. This is the angle between the line of nodes *OK* and the $X_2-$ axis. The direction of the shortest rotation from axis $X_2$ to axis $Y_2$ is the direction in which the angle of precession is measured.

In cases when the angle of nutation is either $\theta = 0°$ or $\theta = 180°$, the Euler angles are not defined.

The operator $\mathbf{Eu}(\psi,\theta,\varphi)$ of Eulerian transformation allows for the following matrix representation:

$$\mathbf{Eu}(\psi,\theta,\varphi) = \begin{bmatrix} -\sin\psi\cos\theta\sin\varphi + \cos\psi\cos\varphi & \cos\psi\cos\theta\sin\varphi + \sin\psi\cos\varphi & \sin\theta\sin\varphi & 0 \\ -\sin\psi\cos\theta\cos\varphi - \cos\psi\sin\varphi & \cos\psi\cos\theta\cos\varphi - \sin\psi\cos\varphi & \sin\theta\cos\varphi & 0 \\ \sin\theta\sin\varphi & -\cos\psi\cos\theta & \cos\theta & 0 \\ 0 & 0 & 0 & 1 \end{bmatrix} \tag{D.15}$$

It is important to stress here the difference between the operator $\mathbf{Eu}(\psi,\theta,\varphi)$ of the Eulerian transformation and the operator $\mathbf{Rt}(\psi_A,\mathbf{A}_0)$ of rotation about an arbitrary axis through the origin.

The operator $\mathbf{Rt}(\psi_A,\mathbf{A})$ of rotation about an arbitrary axis through the origin and the operator $\mathbf{Eu}(\psi,\theta,\varphi)$ of the Eulerian transformation can result in that same final orientation of the coordinate system $X_2Y_2Z_2$ in relation to the coordinate system $X_1Y_1Z_1$. However, the operators of linear transformations $\mathbf{Rt}(\psi_A,\mathbf{A}_0)$ and $\mathbf{Eu}(\psi,\theta,\varphi)$ are the operators of completely different nature. They can result in identical coordinate system transformation, but they are not equal to each other.

### D.1.6 Rotation About an Arbitrary Axis Not through the Origin

The transformation corresponding to the rotation of a coordinate system through an angle $\varphi$ around an arbitrary vector not through the origin cannot readily be written in a form similar to the rotation matrices about the coordinate axes.

The desired transformation matrix is obtained by combining a sequence of elementary translation and rotation matrices. (Once a single $4\times4$ matrix has been obtained representing the composite transformations, it can be used in the same way as any other transformation matrix).

The rotation of the coordinate system $X_1Y_1Z_1$ to a configuration, which the coordinate system $X_2Y_2Z_2$ possesses, can be performed about a corresponding axis that features an arbitrary configuration in space (see Figure D.6). The vector $\mathbf{A}$ is the direction vector of the axis of rotation. The axis of rotation is not a line through the origin.

The operator of this kind of linear transformation $\mathbf{Rt}(\psi_A,\mathbf{A})$ can be expressed in terms of the operator $\mathbf{Tr}(a,\mathbf{A})$ of translation along and of the operator $\mathbf{Rt}(\psi_A,\mathbf{A}_0)$ of rotation about an arbitrary axis through the origin:

$$\mathbf{Rt}(\varphi_A,\mathbf{A}) = \mathbf{Tr}(-b,\mathbf{B}^*)\cdot\mathbf{Rt}(\varphi_A,\mathbf{A}_0)\cdot\mathbf{Tr}(b,\mathbf{B}) \tag{D.16}$$

The following designations apply in Eq. (D.16):

$\mathbf{Tr}(b,\mathbf{B})$ is the operator of translation along the shortest distance of approach of the axis of rotation and origin of the coordinate system.

$\mathbf{Tr}(-b,\mathbf{B}^*)$ is the operator of translation in the direction opposite to the translation $\mathbf{Tr}(b,\mathbf{B})$ after the rotation $\mathbf{Rt}(\psi_A,\mathbf{A})$ is completed.

In order to determine the shortest distance of approach $B$ of the axis of rotation (that is, the axis along the direction vector $\mathbf{B}$) and origin of the coordinate system, consider the axis $\mathbf{B}$ through two given points $\mathbf{r}_{B.1}$ and $\mathbf{r}_{B.2}$.

The shortest distance between a certain point $\mathbf{r}_0$ and the straight line through the points $\mathbf{r}_{B.1}$ and $\mathbf{r}_{B.2}$ can be calculated from the following formula:

$$B = \frac{|(\mathbf{r}_2-\mathbf{r}_1)\times(\mathbf{r}_1-\mathbf{r}_0)|}{|\mathbf{r}_2-\mathbf{r}_1|} \tag{D.17}$$

For the origin of the coordinate system, the equality $\mathbf{r}_0 = 0$ is observed. Then,

$$B = |\mathbf{r}_1|\cdot\sin\angle[\mathbf{r}_1,(\mathbf{r}_2-\mathbf{r}_1)] \tag{D.18}$$

The matrix representation of the operators of translation $\mathbf{Tr}(a_x,X)$, $\mathbf{Tr}(a_y,Y)$, and $\mathbf{Tr}(a_z,Z)$ along the coordinate axes, together with the operators of rotation $\mathbf{Rt}(\varphi_x,X)$, $\mathbf{Rt}(\varphi_y,Y)$, and $\mathbf{Rt}(\varphi_z,Z)$ about the coordinate axes is convenient for implementation in gear science. Moreover, use of the operators is the simplest possible way to analytically describe the linear transformations.

### D.1.7 Resultant Coordinate System Transformation

The operators of translation $\mathbf{Tr}(a_x,X)$, $\mathbf{Tr}(a_y,Y)$, and $\mathbf{Tr}(a_z,Z)$ together with the operators of rotation $\mathbf{Rt}(\varphi_x,X)$,

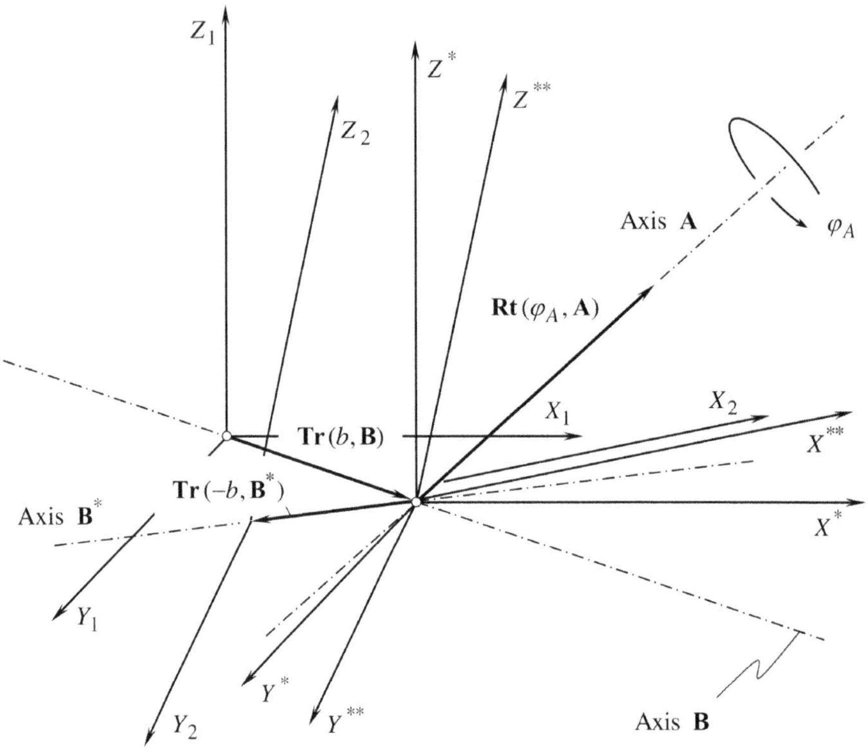


**FIGURE D.6** Analytical description of the operator $\mathbf{Rt}(\varphi_A, \mathbf{A})$ of rotation about an arbitrary axis not through the origin (vector **A** is the direction vector of the axis of rotation).

$\mathbf{Rt}(\varphi_y, Y)$, and $\mathbf{Rt}(\varphi_z, Z)$ are used for the purpose of composing the operator $\mathbf{Rs}(1 \mapsto 2)$ of the resultant coordinate system transformation. The transition from the initial Cartesian reference system $X_1Y_1Z_1$ to other Cartesian reference system $X_2Y_2Z_2$ is analytically described by the operator $\mathbf{Rs}(1 \mapsto 2)$ of the resultant coordinate system transformation.

For example, the expression:

$$\mathbf{Rs}(1 \mapsto 5) = \mathbf{Tr}(a_x, X) \cdot \mathbf{Rt}(\varphi_z, Z) \cdot \mathbf{Rt}(\varphi_x, X) \cdot \mathbf{Tr}(a_y, Y) \tag{D.19}$$

indicates, that the transition from the coordinate system $X_1Y_1Z_1$ to the coordinate system $X_5Y_5Z_5$ is executed in the following four steps (see Figure D.7):

- Translation $\mathbf{Tr}(a_y, Y)$ followed by
- Rotation $\mathbf{Rt}(\varphi_x, X)$, followed by
- Second rotation $\mathbf{Rt}(\varphi_z, Z)$, and followed by
- Translation $\mathbf{Tr}(a_x, X)$.

Ultimately, the equality:

$$\mathbf{r}_1(m) = \mathbf{Rs}(1 \mapsto 5) \cdot \mathbf{r}_5(m) \tag{D.20}$$

is valid.

When the operator $\mathbf{Rs}(1 \mapsto t)$ of the resultant coordinate system transformation is specified, the transition in the opposite direction can be performed by means of the operator $\mathbf{Rs}(t \mapsto 1)$ of the inverse coordinate system transformation. The following equality can be easily proven:

$$\mathbf{Rs}(t \mapsto 1) = \mathbf{Rs}^{-1}(1 \mapsto t) \tag{D.21}$$

In the example illustrated in Figure D.7, the operator $\mathbf{Rs}(5 \mapsto 1)$ of the inverse resultant coordinate system transformation can be expressed in terms of the operator $\mathbf{Rs}(1 \mapsto 5)$ of the direct resultant coordinate system transformation. Following Eq. (D.21), one can come up with the equation:

$$\mathbf{Rs}(5 \mapsto 1) = \mathbf{Rs}^{-1}(1 \mapsto 5) \tag{D.22}$$

It is easy to show that the operator $\mathbf{Rs}(1 \mapsto t)$ of the resultant coordinate system transformation allows for representation in the following form:

$$\mathbf{Rs}(1 \mapsto t) = \mathbf{Tr}(a, A) \cdot \mathbf{Eu}(\psi, \theta, \varphi) \tag{D.23}$$

The linear transformation $\mathbf{Rs}(1 \mapsto t)$ [see Eq. (D.23)] can also be expressed in terms of rotation about an axis $\mathbf{Rt}(\varphi_A, A)$, not through the origin [see Eq. (D.16)].

## D.2 COMPLEX COORDINATE SYSTEM TRANSFORMATION

In certain cases of complex coordinate system transformations that are repeatedly used in practice, special purpose

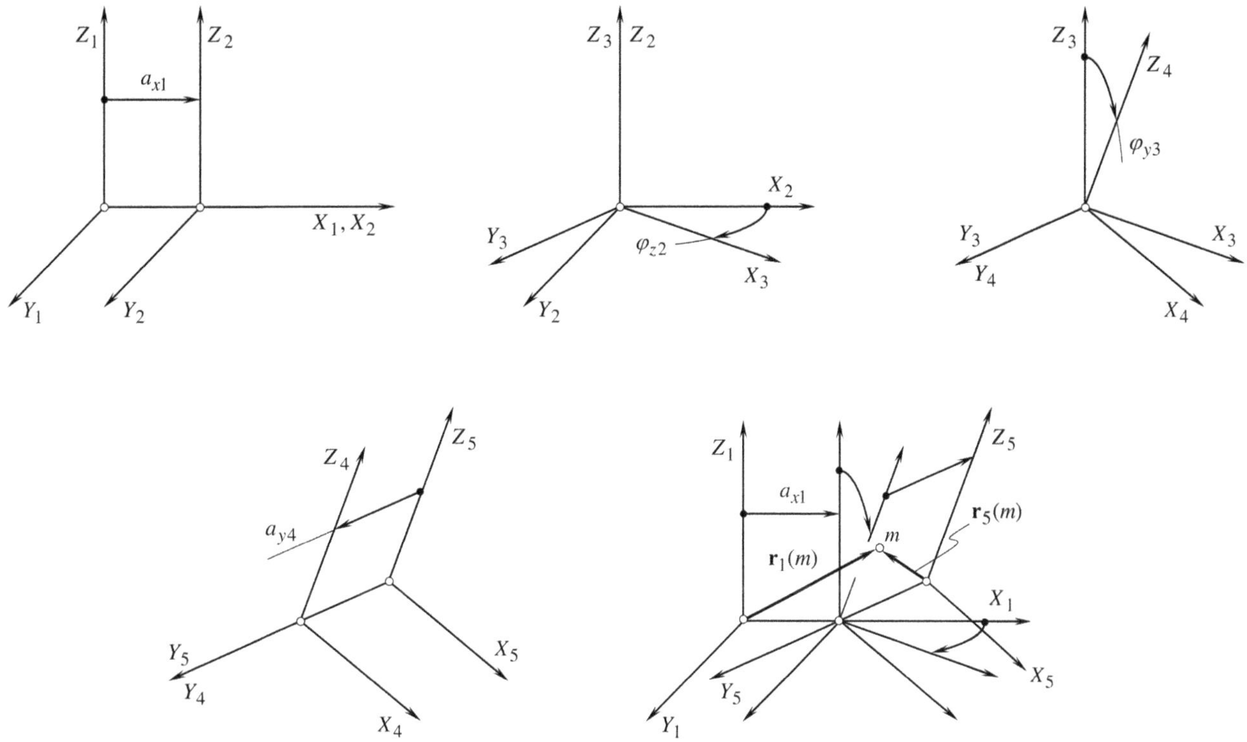


**FIGURE D.7** An example of the resultant coordinate system transformation, analytically expressed by the operator $\mathbf{Rs}(1 \mapsto 5)$.

operators of coordinate system transformation can be composed of elementary operators of translation, and operators of rotation.

### D.2.1 Linear Transformation Describing a Screw Motion About a Coordinate Axis

Operators for analytical description of screw motions about an axis of a Cartesian coordinate system are a particular case of the operators of the resultant coordinate system transformation.

By definition (see Figure D.8), the operator $\mathbf{Sc}_x(\varphi_x, p_x)$ of a screw motion about $X-$ axis of the Cartesian coordinate system $XYZ$ is equal to:

$$\mathbf{Sc}_x(\varphi_x, p_x) = \mathbf{Rt}(\varphi_x, X) \cdot \mathbf{Tr}(a_x, X) \tag{D.24}$$

After substituting the operator of translation $\mathbf{Tr}(a_x, X)$ and the operator of rotation $\mathbf{Rt}(\varphi_x, X)$ [see Eq. (D.10)], Eq. (D.24) casts into the expression:

$$\mathbf{Sc}_x(\varphi_x, p_x) = \begin{bmatrix} 1 & 0 & 0 & p_x \cdot \varphi_x \\ 0 & \cos\varphi_x & \sin\varphi_x & 0 \\ 0 & -\sin\varphi_x & \cos\varphi_x & 0 \\ 0 & 0 & 0 & 1 \end{bmatrix} \tag{D.25}$$

for the calculation of the operator of the screw motion $\mathbf{Sc}_x(\varphi_x, p_x)$ about $X-$ axis.

The operators of screw motions $\mathbf{Sc}_y(\varphi_y, p_y)$ and $\mathbf{Sc}_z(\varphi_z, p_z)$ about $Y-$ and $Z-$ axes, respectively, are defined similar to how the operator of the screw motion $\mathbf{Sc}_x(\varphi_x, p_x)$ is defined:

$$\mathbf{Sc}_y(\varphi_y, p_y) = \mathbf{Rt}(\varphi_y, Y) \cdot \mathbf{Tr}(a_y, Y) \tag{D.26}$$

$$\mathbf{Sc}_z(\varphi_z, p_z) = \mathbf{Rt}(\varphi_z, Z) \cdot \mathbf{Tr}(a_z, Z) \tag{D.27}$$

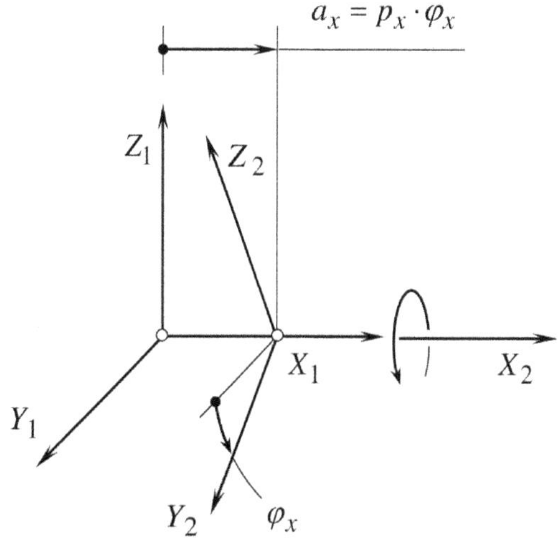


**FIGURE D.8** On analytical description of the operator of screw motion $\mathbf{Sc}_x(\varphi_x, p_x)$.

Using Eqs. (D.5) and (D.6) together with Eqs. (D.11) and (D.12), one can come up with the expressions:

$$\mathbf{Sc}_y(\varphi_y, p_y) = \begin{bmatrix} \cos\varphi_y & 0 & -\sin\varphi_y & 0 \\ 0 & 1 & 0 & p_y \cdot \varphi_y \\ \sin\varphi_y & 0 & \cos\varphi_y & 0 \\ 0 & 0 & 0 & 1 \end{bmatrix} \quad \text{(D.28)}$$

$$\mathbf{Sc}_z(\varphi_z, p_z) = \begin{bmatrix} \cos\varphi_z & \sin\varphi_z & 0 & 0 \\ -\sin\varphi_z & \cos\varphi_z & 0 & 0 \\ 0 & 0 & 1 & p_z \cdot \varphi_z \\ 0 & 0 & 0 & 1 \end{bmatrix} \quad \text{(D.29)}$$

for the calculation of the operators of the screw motion $\mathbf{Sc}_y(\varphi_y, p_y)$ and $\mathbf{Sc}_z(\varphi_z, p_z)$ about $Y-$ and $Z-$ axes.

Screw motions about a coordinate axis, as well as screw surfaces, are common in the theory of part surface generation. This makes it practical to use the operators of the screw motions $\mathbf{Sc}_x(\varphi_x, p_x)$, $\mathbf{Sc}_y(\varphi_y, p_y)$ and $\mathbf{Sc}_z(\varphi_z, p_z)$ in the theory of part surface generation.

When necessary, an operator of the screw motion about an arbitrary axis either through the origin of the coordinate system or not through the origin of the coordinate system can be derived following a method similar to that used for the derivation of the operators $\mathbf{Sc}_x(\varphi_x, p_x)$, $\mathbf{Sc}_y(\varphi_y, p_y)$ and $\mathbf{Sc}_z(\varphi_z, p_z)$.

### D.2.2 Linear Transformation Describing Rolling Motion of a Coordinate System

One more practical combination of a rotation and of a translation is often used in the theory of gearing.

Consider a Cartesian coordinate system $X_1Y_1Z_1$ (see Figure D.9). The coordinate system $X_1Y_1Z_1$ is traveling in the direction of the $X_1-$ axis. The velocity of the translation is denoted by $V$. The coordinate system $X_1Y_1Z_1$ rotates about its $Y_1-$ axis synchronous to the translation. The speed of rotation is denoted as $\omega$. Assume that the ratio $V/\omega$ is constant. Under such a scenario, the resultant motion of the reference system $X_1Y_1Z_1$ to its arbitrary position $X_2Y_2Z_2$ allows interpretation in the form of rolling with no sliding of a cylinder of a radius $R_w$ over the plane. The plane is parallel to the coordinate $X_1Y_1-$ plane, and it is remote from it at the distance $R_w$. For the calculation of the radius of the rolling cylinder, the expression $R_w = V/\omega$ can be used.

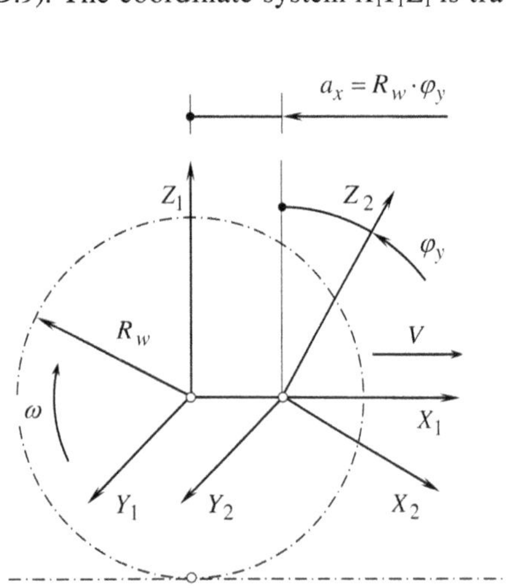


**FIGURE D.9** Illustration of the transformation of rolling $\mathbf{Rl}_x(\varphi_y, Y)$ of a coordinate system.

As the rolling of the cylinder of a radius $R_w$ over the plane is performed with no sliding, a certain correspondence between the translation and the rotation of the coordinate system is established. When the coordinate system turns through a certain angle $\varphi_y$, the translation of the origin of the coordinate system along $X_1-$ axis is equal to $a_x = \varphi_r \cdot R_w$.

Transition from the coordinate system $X_1Y_1Z_1$ to the coordinate system $X_2Y_2Z_2$ is analytically described by the operator of the resultant coordinate system transformation $\mathbf{Rs}(1 \mapsto 2)$. The operator $\mathbf{Rs}(1 \mapsto 2)$ is equal to:

$$\mathbf{Rs}(1 \mapsto 2) = \mathbf{Rt}(\varphi_y, Y_1) \cdot \mathbf{Tr}(a_x, X_1) \quad \text{(D.30)}$$

Here, $\mathbf{Tr}(a_x, X_1)$ designates the operator of the translation along the $X_1-$ axis, and $\mathbf{Rt}(\varphi_y, Y_1)$ is the operator of the rotation about the $Y_1-$ axis.

The operator of the resultant coordinate system transformation of this kind [see Eq. (D.30)] is referred to as the *operator of rolling motion over a plane*.

When the translation is performed along the $X_1-$ axis and the rotation is performed about the $Y_1-$ axis, the operator of rolling is denoted as $\mathbf{Rl}_x(\varphi_y, Y)$. In this particular case, the equality $\mathbf{Rl}_x(\varphi_y, Y) = \mathbf{Rs}(1 \mapsto 2)$ [see Eq. (D.30)] is valid. Based on this equality, the operator of rolling motion over a plane $\mathbf{Rl}_x(\varphi_y, Y)$ can be calculated from the equation:

$$\mathbf{Rl}_x(\varphi_y, Y) = \begin{bmatrix} \cos\varphi_y & 0 & -\sin\varphi_y & a_x \cdot \cos\varphi_y \\ 0 & 1 & 0 & 0 \\ \sin\varphi_y & 0 & \cos\varphi_y & a_x \cdot \sin\varphi_y \\ 0 & 0 & 0 & 1 \end{bmatrix} \quad \text{(D.31)}$$

While rotation remains about the $Y_1-$ axis, the translation can be performed not along the $X_1-$ axis, but along the $Z_1-$ axis instead. For rolling of this kind, the operator of rolling is equal to:

$$\mathbf{Rl}_z(\varphi_y, Y) = \begin{bmatrix} \cos\varphi_y & 0 & -\sin\varphi_y & -a_z \cdot \sin\varphi_y \\ 0 & 1 & 0 & 0 \\ \sin\varphi_y & 0 & \cos\varphi_y & a_z \cdot \cos\varphi_y \\ 0 & 0 & 0 & 1 \end{bmatrix} \quad \text{(D.32)}$$

For cases when the rotation is performed about the $X_1$ – axis, the corresponding operators of rolling are as follow:

$$\mathbf{Rl}_y(\varphi_x, X) = \begin{bmatrix} 1 & 0 & 0 & 0 \\ 0 & \cos\varphi_x & \sin\varphi_x & a_y \cdot \cos\varphi_x \\ 0 & -\sin\varphi_x & \cos\varphi_x & -a_y \cdot \sin\varphi_x \\ 0 & 0 & 0 & 1 \end{bmatrix} \quad \text{(D.33)}$$

for the case of rolling along the $Y_1$ – axis, and

$$\mathbf{Rl}_z(\varphi_x, X) = \begin{bmatrix} 1 & 0 & 0 & 0 \\ 0 & \cos\varphi_x & \sin\varphi_x & a_z \cdot \sin\varphi_x \\ 0 & -\sin\varphi_x & \cos\varphi_x & a_z \cdot \cos\varphi_x \\ 0 & 0 & 0 & 1 \end{bmatrix} \quad \text{(D.34)}$$

for the case of rolling along the $Z_1$ – axis.

Similar expressions can be derived for the case of rotation about the $Z_1$ – axis:

$$\mathbf{Rl}_x(\varphi_z, Z) = \begin{bmatrix} \cos\varphi_z & \sin\varphi_z & 0 & a_x \cdot \cos\varphi_z \\ -\sin\varphi_z & \cos\varphi_z & 0 & a_x \cdot \sin\varphi_z \\ 0 & 0 & 1 & 0 \\ 0 & 0 & 0 & 1 \end{bmatrix} \quad \text{(D.35)}$$

$$\mathbf{Rl}_y(\varphi_z, Z) = \begin{bmatrix} \cos\varphi_z & \sin\varphi_z & 0 & a_y \cdot \sin\varphi_z \\ -\sin\varphi_z & \cos\varphi_z & 0 & a_y \cdot \cos\varphi_z \\ 0 & 0 & 1 & 0 \\ 0 & 0 & 0 & 1 \end{bmatrix} \quad \text{(D.36)}$$

The use of the operators of rolling Eqs. (D.31) through (D.36) significantly simplifies the analytical description of the coordinate system transformations.

### D.2.3 Linear Transformation Describing the Rolling of Two Coordinate Systems

In the theory of gearing, combinations of two rotations about parallel axes are of particular interest.

As an example, consider the two Cartesian coordinate systems $X_1Y_1Z_1$ and $X_2Y_2Z_2$ shown in Figure D.10. The coordinate systems $X_1Y_1Z_1$ and $X_2Y_2Z_2$ are rotated about their axes, $Z_1$ and $Z_2$, respectively. The axes of the rotations are parallel to each other ($Z_1 \parallel Z_2$). The rotations $\omega_1$ and $\omega_2$ of the coordinate systems can be interpreted so that a circle of a certain radius $R_1$ that is associated with the coordinates system $X_1Y_1Z_1$ is rolling with no sliding over a circle of the corresponding radius $R_2$ that is associated with the coordinate system $X_2Y_2Z_2$. When the center-distance $C$ is known, then radii $R_1$ and $R_2$ of the circles (that is, of the *centrodes*) can be expressed in terms of the center-distance $C$ and of the given rotations $\omega_1$ and $\omega_2$. For these calculations, the following formulae:

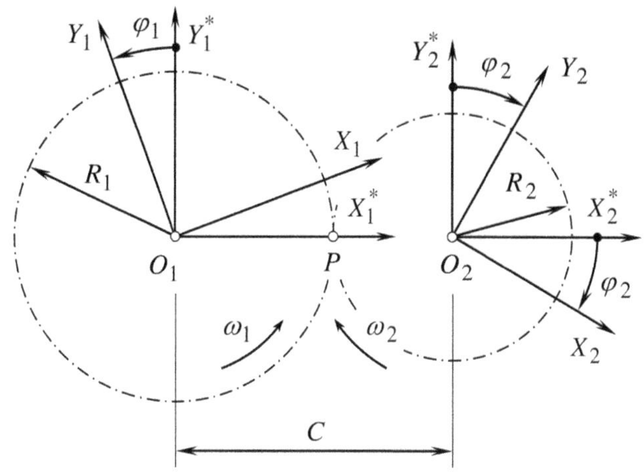


**FIGURE D.10** On derivation of the operator of rolling $\mathbf{Rr}_u(\varphi_1, Z_1)$ of two coordinate systems.

$$R_1 = C \cdot \frac{1}{1+u} \quad \text{(D.37)}$$

$$R_2 = C \cdot \frac{u}{1+u} \quad \text{(D.38)}$$

are used. Here, the ratio $\omega_1/\omega_2$ is denoted by $u$.

In the initial configuration, the $X_1$ –, and $X_2$ – axes align to each other. The $Y_1$ – and $Y_2$ – axes are parallel to each other. As shown in Figure D.10, the initial configuration of the coordinate systems $X_1Y_1Z_1$ and $X_2Y_2Z_2$ is labeled as $X_1^*Y_1^*Z_1^*$ and $X_2^*Y_2^*Z_2^*$, respectively.

When the coordinate system $X_1Y_1Z_1$ turns through a certain angle $\varphi_1$, the coordinate system $X_2Y_2Z_2$ turns through the corresponding angle $\varphi_2$. When the angle $\varphi_1$ is known, the corresponding angle $\varphi_2$ is equal to $\varphi_2 = \varphi_1/u$.

Transition from the coordinate system $X_2Y_2Z_2$ to the coordinate system $X_1Y_1Z_1$ can be analytically described by the operator of the resultant coordinate system transformation $\mathbf{Rs}(1 \mapsto 2)$. In the case under consideration, the operator $\mathbf{Rs}(1 \mapsto 2)$ can be expressed in terms of the operators of the elementary coordinate system transformations:

$$\mathbf{Rs}(1 \mapsto 2) = \mathbf{Rt}(\varphi_1, Z_1) \cdot \mathbf{Rt}(\varphi_1/u, Z_1) \cdot \mathbf{Tr}(-C, X_1) \quad \text{(D.39)}$$

Other equivalent combinations of the operators of elementary coordinate system transformations can result in the same operator $\mathbf{Rs}(1 \mapsto 2)$ of the resultant coordinate system transformation. The interested reader may wish to independently explore deriving the equivalent expressions for the operator $\mathbf{Rs}(1 \mapsto 2)$.

The operator of the resultant coordinate system transformations of this kind [see Eq. (D.39)] are referred to as the *operators of rolling motion over a cylinder.*

When rotations are performed around the $Z_1$ –, and the $Z_2$ –axes, the operator of rolling motion over a cylinder is designated $\mathbf{Rr}_u(\varphi_1, Z_1)$. In this particular case, the equality $\mathbf{Rr}_u(\varphi_1, Z_1) = \mathbf{Rs}(1 \mapsto 2)$ [see Eq. (D.39)] is valid. Based on this equality, the operator of rolling $\mathbf{Rr}_u(\varphi_1, Z_1)$ over a cylinder can be calculated from the equation:

$$\mathbf{Rr}_u(\varphi_1, Z_1) = \begin{bmatrix} \cos\left(\varphi_1 \cdot \dfrac{u+1}{u}\right) & \sin\left(\varphi_1 \cdot \dfrac{u+1}{u}\right) & 0 & -C \\ -\sin\left(\varphi_1 \cdot \dfrac{u+1}{u}\right) & \cos\left(\varphi_1 \cdot \dfrac{u+1}{u}\right) & 0 & 0 \\ 0 & 0 & 1 & 0 \\ 0 & 0 & 0 & 1 \end{bmatrix} \tag{D.40}$$

For the inverse transformation, the inverse operator of rolling of two coordinate systems $\mathbf{Rr}_u(\varphi_2, Z_2)$ can be used. This is equal to $\mathbf{Rr}_u(\varphi_2, Z_2) = \mathbf{Rr}_u^{-1}(\varphi_1, Z_1)$. In terms of the operators of the elementary coordinate system transformations, the operator $\mathbf{Rr}_u(\varphi_2, Z_2)$ can be expressed as:

$$\mathbf{Rr}_u(\varphi_2, Z_2) = \mathbf{Rt}(\varphi_1/u, Z_2) \cdot \mathbf{Rt}(\varphi_1, Z_2) \cdot \mathbf{Tr}(C, X_1) \tag{D.41}$$

Other equivalent combinations of the operators of elementary coordinate system transformations can result in that same operator $\mathbf{Rr}_u(\varphi_2, Z_2)$ of the resultant coordinate system transformation. The interested reader may wish to independently explore deriving the equivalent expressions for the operator $\mathbf{Rr}_u(\varphi_2, Z_2)$.

For the calculation of the operator of rolling of two coordinate systems $\mathbf{Rr}_u(\varphi_2, Z_2)$, the equation:

$$\mathbf{Rr}_u(\varphi_2, Z_2) = \begin{bmatrix} \cos\left(\varphi_1 \cdot \dfrac{u+1}{u}\right) & -\sin\left(\varphi_1 \cdot \dfrac{u+1}{u}\right) & 0 & C \\ \sin\left(\varphi_1 \cdot \dfrac{u+1}{u}\right) & \cos\left(\varphi_1 \cdot \dfrac{u+1}{u}\right) & 0 & 0 \\ 0 & 0 & 1 & 0 \\ 0 & 0 & 0 & 1 \end{bmatrix} \tag{D.42}$$

can be used.

Similar to the way the expression [see Eq. (D.40)] is derived for the calculation of the operator of rolling $\mathbf{Rr}_u(\varphi_1, Z_1)$ around the $Z_1$ – and $Z_2$ – axes, corresponding formulae can be derived for the calculation of the operators of rolling $\mathbf{Rr}_u(\varphi_1, X_1)$ and $\mathbf{Rr}_u(\varphi_1, Y_1)$ about parallel axes $X_1$ and $X_2$, as well as about parallel axes $Y_1$ and $Y_2$.

The use of the operators of rolling about two axes $\mathbf{Rr}_u(\varphi_1, X_1)$, $\mathbf{Rr}_u(\varphi_1, Y_1)$, and $\mathbf{Rr}_u(\varphi_1, Z_1)$ substantially simplifies analytical description of the coordinate system transformations.

### D.2.4 Coupled Linear Transformation

It worth noting that a translation $\mathbf{Tr}(a_x, X)$ along the $X$ – axis of a Cartesian reference system $XYZ$ and a rotation $\mathbf{Rt}(\varphi_x, X)$ about the axis $X$ of that same coordinate system $XYZ$ obey the commutative law; that is, these two coordinate system transformations can be performed in a different order equally. This means that the operators $\mathbf{Cp}_x(a_x, \varphi_x)$ of coupled linear transformations are invariant with respect to the order of the elementary transformations—namely, of the translation/rotation—and this is important. It makes no difference whether the translation $\mathbf{Tr}(a_x, X)$ is initially performed and followed by the rotation $\mathbf{Rt}(\varphi_x, X)$, or the rotation $\mathbf{Rt}(\varphi_x, X)$ is initially performed and followed by the translation $\mathbf{Tr}(a_x, X)$. This is because of the dot products $\mathbf{Rt}(\varphi_x, X) \cdot \mathbf{Tr}(a_x, X)$ and $\mathbf{Tr}(a_x, X) \cdot \mathbf{Rt}(\varphi_x, X)$ are identical:

$$\mathbf{Rt}(\varphi_x, X) \cdot \mathbf{Tr}(a_x, X) \equiv \mathbf{Tr}(a_x, X) \cdot \mathbf{Rt}(\varphi_x, X) \tag{D.43}$$

This means that the translation from the coordinate system $X_1Y_1Z_1$ to the intermediate coordinate system $X^*Y^*Z^*$ followed by the rotation from the coordinate system $X^*Y^*Z^*$ to the final coordinate system $X_2Y_2Z_2$ produces that same reference $X_2Y_2Z_2$ as when the rotation from the coordinate system $X_1Y_1Z_1$ to the intermediate coordinate system $X^*Y^*Z^*$ is followed by the translation from the coordinate system $X^*Y^*Z^*$ to the final coordinate system $X_2Y_2Z_2$.

The validity of Eq. (D.43) is illustrated in Figure D.11. The translation $\mathbf{Tr}(a_x, X)$ followed by the rotation $\mathbf{Rt}(\varphi_x, X)$, as shown in Figure D.11(a) is equivalent to the rotation $\mathbf{Rt}(\varphi_x, X)$ followed by the translation $\mathbf{Tr}(a_x, X)$, as shown in Figure D.11(b).

Therefore, the two linear transformations $\mathbf{Tr}(a_x, X)$ and $\mathbf{Rt}(\varphi_x, X)$ can become a coupled linear transformation:

$$\begin{aligned} \mathbf{Cp}_x(a_x, \varphi_x) &= \mathbf{Rt}(\varphi_x, X) \cdot \mathbf{Tr}(a_x, X) \\ &\equiv \mathbf{Tr}(a_x, X) \cdot \mathbf{Rt}(\varphi_x, X) \end{aligned} \tag{D.44}$$

The operator of linear transformation $\mathbf{Cp}_x(a_x, \varphi_x)$ can be expressed in matrix form [see Figure D.12(a)]:

$$\mathbf{Cp}_x(a_x, \varphi_x) = \begin{bmatrix} 1 & 0 & 0 & a_x \\ 0 & \cos\varphi_x & \sin\varphi_x & 0 \\ 0 & -\sin\varphi_x & \cos\varphi_x & 0 \\ 0 & 0 & 0 & 1 \end{bmatrix} \tag{D.45}$$

This expression is composed based on Eq. (D.4) for the linear transformation $\mathbf{Tr}(a_x, X)$ and on Eq. (D.10) for the linear transformation $\mathbf{Rt}(\varphi_x, X)$.

Two reduced cases of the operator of the linear transformation $\mathbf{Cp}_x(a_x, \varphi_x)$ are distinguished.

First, in a case where the component $a_x$ of the translation is zero, that is $a_x = 0$, the operator of linear transformation

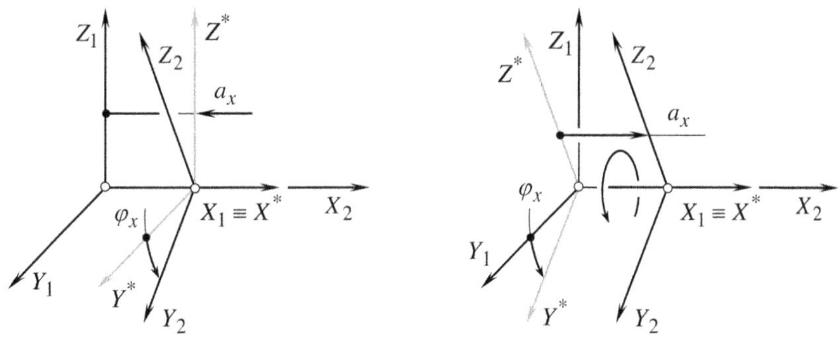


(a) (b)

**FIGURE D.11** On the equivalency of the linear transformations (a) $\mathbf{Rt}(\varphi_x, X) \cdot \mathbf{Tr}(a_x, X)$ and (b) $\mathbf{Tr}(a_x, X) \cdot \mathbf{Rt}(\varphi_x, X)$, which comprise the operator $\mathbf{Cp}_x(a_x, \varphi_x)$ of a coupled linear transformation of a Cartesian reference system *XYZ*.

$\mathbf{Cp}_x(a_x, \varphi_x)$ reduces to the operator of rotation $\mathbf{Rt}(\varphi_x, X)$, and the equality $\mathbf{Cp}_x(a_x, \varphi_x) = \mathbf{Rt}(\varphi_x, X)$ is observed.

Second, in a case where the component $\varphi_x$ of the rotation is zero, that is $\varphi_x = 0°$, the operator of linear transformation $\mathbf{Cp}_x(a_x, \varphi_x)$ reduces to the operator of translation $\mathbf{Tr}(a_x, X)$, and the equality $\mathbf{Cp}_x(a_x, \varphi_x) = \mathbf{Tr}(a_x, X)$ is observed.

These reduced cases are valid with respect to the translations and the rotations along and about the axes *Y* and *Z* of a Cartesian reference system *XYZ*. The corresponding coupled operators $\mathbf{Cp}_y(a_y, \varphi_y)$ and $\mathbf{Cp}_z(a_z, \varphi_z)$ for linear transformations of these kinds can also be composed [see Figure D.12(b) and (c)]:

$$\mathbf{Cp}_y(a_y, \varphi_y) = \begin{bmatrix} \cos\varphi_y & 0 & \sin\varphi_y & 0 \\ 0 & 1 & 0 & a_y \\ -\sin\varphi_y & 0 & \cos\varphi_y & 0 \\ 0 & 0 & 0 & 1 \end{bmatrix} \tag{D.46}$$

$$\mathbf{Cp}_z(a_z, \varphi_z) = \begin{bmatrix} \cos\varphi_z & \sin\varphi_z & 0 & 0 \\ -\sin\varphi_z & \cos\varphi_z & 0 & 0 \\ 0 & 0 & 1 & a_z \\ 0 & 0 & 0 & 1 \end{bmatrix} \tag{D.47}$$

In the operators of linear transformations $\mathbf{Cp}_x(a_x, \varphi_x)$, $\mathbf{Cp}_y(a_y, \varphi_y)$, and $\mathbf{Cp}_z(a_z, \varphi_z)$, the values of the translations $a_x$, $a_y$, and $a_z$, as well as values of the rotations $\varphi_x$, $\varphi_y$, and $\varphi_z$, are expressed in finite values rather than continuous values. The linear and angular displacements do not correlate to one another in time; thus, they do not comprise a screw. Rather, they are a kind of couples of a translation along and a rotation about a coordinate axis of a Cartesian reference system.

The introduction of the operators of linear transformation $\mathbf{Cp}_x(a_x, \varphi_x)$, $\mathbf{Cp}_y(a_y, \varphi_y)$, and $\mathbf{Cp}_z(a_z, \varphi_z)$ makes the linear transformations easier, as all the operators of the linear transformations become uniform.

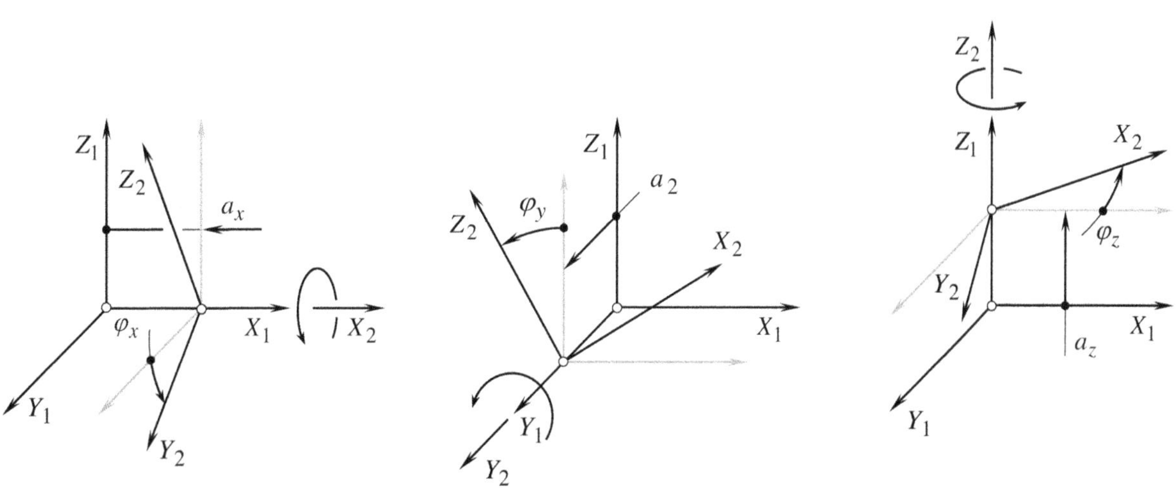


(a) (b) (c)

**FIGURE D.12** Representation of operators $\mathbf{Cp}_x(a_x, \varphi_x)$, $\mathbf{Cp}_y(a_y, \varphi_y)$, and $\mathbf{Cp}_z(a_z, \varphi_z)$ of coupled linear transformations of a Cartesian reference system *XYZ*, in terms of the operators of elementary linear transformations $\mathbf{Tr}(a_x, X)$ and $\mathbf{Rt}(\varphi_x, X)$, $\mathbf{Tr}(a_y, Y)$ and $\mathbf{Rt}(\varphi_y, Y)$, and, $\mathbf{Tr}(a_z, Z)$ and $\mathbf{Rt}(\varphi_z, Z)$, correspondingly.

The operators of linear transformation $\mathbf{Cp}_x(a_x, \varphi_x)$, $\mathbf{Cp}_y(a_y, \varphi_y)$, and $\mathbf{Cp}_z(a_z, \varphi_z)$ do not obey the commutative law. This is because rotation is not a vector by nature. Therefore, special care should be taken when treating rotations as vectors—particularly when implementing coupled operators of linear transformations.

The operators of coupled linear transformations $\mathbf{Cp}_x(a_x, \varphi_x)$, $\mathbf{Cp}_y(a_y, \varphi_y)$, and $\mathbf{Cp}_z(a_z, \varphi_z)$ [see Eqs. (D.45) through (D.47)] can be used for the analytical description of a resultant coordinate system transformation. In this scenario, the operator $\mathbf{Rs}(1 \mapsto t)$ of a resultant coordinate system transformation can be expressed in terms of all the operators $\mathbf{Cp}_x(a_x, \varphi_x)$, $\mathbf{Cp}_y(a_y, \varphi_y)$, and $\mathbf{Cp}_z(a_z, \varphi_z)$ using the following expression:

$$\mathbf{Rs}(1 \mapsto t) = \prod_{\substack{i=1 \\ j=x,y,z}}^{t-1} \mathbf{Cp}_j^i(a_j^i, \varphi_j^i) \tag{D.48}$$

In Eq. (D.48), only operators of coupled linear transformations are used.

### D.2.5 An Example of Non-Orthogonal Linear Transformation

Consider a non-orthogonal reference system $X_1Y_1Z_1$ having certain angle $\omega_1$ between the axes $X_1$ and $Y_1$. Axis $Z_1$ is perpendicular to the coordinate plane $X_1Y_1$. Another reference system $X_2Y_2Z_2$ is identical to the first coordinate system $X_1Y_1Z_1$ and is turned about the $Z_1$ – axis through a certain angle $\varphi$. The transition from the reference system $X_1Y_1Z_1$ to the reference system $X_2Y_2Z_2$ can be analytically described by the operator of linear transformation:

$$\mathbf{Rt}_\omega(1 \to 2) = \begin{bmatrix} \dfrac{\sin(\omega_1+\varphi)}{\sin\omega_1} & \dfrac{\sin\varphi}{\sin\omega_1} & 0 & 0 \\ -\dfrac{\sin\varphi}{\sin\omega_1} & \dfrac{\sin(\omega_1-\varphi)}{\sin\omega_1} & 0 & 0 \\ 0 & 0 & 1 & 0 \\ 0 & 0 & 0 & 1 \end{bmatrix} \tag{D.49}$$

In order to distinguish the operator of rotation in the orthogonal linear transformation $\mathbf{Rt}(1 \to 2)$ from the similar operator of rotation in a non-orthogonal linear transformation $\mathbf{Rt}_\omega(1 \to 2)$, the subscript "ω" is assigned to the latter.

When $\omega = 90°$, Eq. (D.49) casts into Eq. (D.12).

### D.2.6 Conversion of the Hand of a Coordinate System

The application of the matrix method of coordinate system transformation presumes that both of the reference systems "$i$" and "$(i \pm 1)$" are of the same hand. This means that it assumed from the very beginning that they are both either right-hand- or left-hand-oriented Cartesian coordinate systems. In the event the coordinate systems $i$ and $(i \pm 1)$ are oriented to opposite hands, one of the coordinate systems needs be converted.

For the conversion of a left-hand-oriented Cartesian coordinate system into a right-hand-oriented coordinate system, or vice versa, the operators of reflection are commonly used.

To reverse the direction of the $X_i$-axis of the initial coordinate system $i$ (in this case, in the new coordinate system $(i \pm 1)$, the equalities $X_{i\pm1} = -X_i$, $Y_{i\pm1} \equiv Y_i$, and $Z_{i\pm1} \equiv Z_i$ are observed), the operator of reflection $\mathbf{Rf}_x(Y_i\, Z_i)$ is applied. The operator of reflection yields a matrix representation:

$$\mathbf{Rf}_x(Y_i\, Z_i) = \begin{bmatrix} -1 & 0 & 0 & 0 \\ 0 & 1 & 0 & 0 \\ 0 & 0 & 1 & 0 \\ 0 & 0 & 0 & 1 \end{bmatrix} \tag{D.50}$$

Similarly, the use of the operators of reflections $\mathbf{Rf}_y(X_iZ_i)$ and $\mathbf{Rf}_z(X_iY_i)$ reverse the directions of the $Y_i$-axis and $Z_i$-axis. The operators of reflections $\mathbf{Rf}_y(X_iZ_i)$ and $\mathbf{Rf}_z(X_iY_i)$ can be expressed analytically in the form:

$$\mathbf{Rf}_y(X_iZ_i) = \begin{bmatrix} 1 & 0 & 0 & 0 \\ 0 & -1 & 0 & 0 \\ 0 & 0 & 1 & 0 \\ 0 & 0 & 0 & 1 \end{bmatrix} \tag{D.51}$$

$$\mathbf{Rf}_z(X_iY_i) = \begin{bmatrix} 1 & 0 & 0 & 0 \\ 0 & 1 & 0 & 0 \\ 0 & 0 & -1 & 0 \\ 0 & 0 & 0 & 1 \end{bmatrix} \tag{D.52}$$

A linear transformation that reverses the direction of the coordinate axis is an *opposite transformation*. A transformation of reflection is an example of *orientation-reversing transformation*.

## D.3 USEFUL EQUATIONS

The sequence of the successive rotations can vary depending on the intention of the researcher. Several special types of successive rotations are known, including Eulerian transformation, Cardanian transformation, two kinds of Euler-Krylov transformations, and so forth. The sequence of the successive rotations can be chosen from a total of 12 different combinations. Even though the Cardanian transformation is different from the Eulerian transformation in terms of the combination of the rotations, they both use a similar approach to calculate the orientation angles.

### D.3.1 *RPY*-Transformation

A series of rotations can be performed in the order *roll matrix* ($R$), by *pitch matrix* ($P$), by *yaw matrix* ($Y$). This kind of linear transformation is commonly referred to as an *RPY – transformation*. The resultant transformation can be represented by the homogenous coordinate transformation matrix:

$$\mathbf{RPY}(\varphi_x,\varphi_y,\varphi_z)=\begin{bmatrix} \cos\varphi_y\cos\varphi_z+\sin\varphi_x\sin\varphi_y\sin\varphi_z & \cos\varphi_y\sin\varphi_z-\sin\varphi_x\sin\varphi_y\cos\varphi_z & \cos\varphi_x\sin\varphi_y & 0 \\ -\cos\varphi_x\sin\varphi_z & \cos\varphi_x\cos\varphi_z & \sin\varphi_x & 0 \\ \sin\varphi_x\cos\varphi_y\sin\varphi_z-\sin\varphi_y\cos\varphi_z & -\sin\varphi_x\cos\varphi_y\cos\varphi_z-\cos\varphi_y\sin\varphi_z & \cos\varphi_x\cos\varphi_y & 0 \\ 0 & 0 & 0 & 1 \end{bmatrix} \tag{D.53}$$

The *RPY* – transformation can be used for solving problems in the field of gear science.

### D.3.2 Operator of Rotation About an Axis in Space

A spatial rotation operator for the rotational transformation of a point about a unit axis $\mathbf{a}_0(\cos\alpha,\cos\beta,\cos\gamma)$ passing through the origin of the coordinate system can be described as follows, with $\mathbf{a}_0=\mathbf{A}_0/|\mathbf{A}_0|$ designating the unit vector along the axis of rotation $\mathbf{A}_0$.

If the angle of rotation of the point about $\mathbf{a}_0$ is $\theta$, the *rotation operator* is expressed by:

$$\mathbf{Rt}(\theta,\mathbf{a}_0)=\begin{bmatrix} (1-\cos\theta)\cos^2\alpha+\cos\theta & (1-\cos\theta)\cos\alpha\cos\beta-\sin\theta\cos\gamma & (1-\cos\theta)\cos\alpha\cos\gamma+\sin\theta\cos\beta & 0 \\ (1-\cos\theta)\cos\alpha\cos\beta+\sin\theta\cos\gamma & (1-\cos\theta)\cos^2\beta+\cos\theta & (1-\cos\theta)\cos\beta\cos\gamma-\sin\theta\cos\alpha & 0 \\ (1-\cos\theta)\cos\alpha\cos\gamma-\sin\theta\cos\beta & (1-\cos\theta)\cos\beta\cos\gamma+\sin\theta\cos\alpha & (1-\cos\theta)\cos^2\gamma+\cos\theta & 0 \\ 0 & 0 & 0 & 1 \end{bmatrix} \tag{D.54}$$

Solutions to problems in the field of gearing can be significantly simplified through the implementation of the rotational operator $\mathbf{Rt}(\theta,\mathbf{a}_0)$ [see Eq. (D.54)].

### D.3.3 Combined Linear Transformation

Suppose point $p$ on a rigid body rotates with an angular displacement $\theta$ about an axis along a unit vector $\mathbf{a}_0$ passing through the origin of the coordinate system at first, followed by a translation at a distance $B$ in the direction of a unit vector $\mathbf{b}$. The linear transformation of this kind can be analytically described by the homogenous matrix:

$$\mathbf{Rt}(\theta_{\mathbf{a}_0},B_{\mathbf{b}})=\begin{bmatrix} (1-\cos\theta)\cos^2\alpha+\cos\theta & (1-\cos\theta)\cos\alpha\cos\beta-\sin\theta\cos\gamma & (1-\cos\theta)\cos\alpha\cos\gamma+\sin\theta\cos\beta & B\cos\alpha \\ (1-\cos\theta)\cos\alpha\cos\beta+\sin\theta\cos\gamma & (1-\cos\theta)\cos^2\beta+\cos\theta & (1-\cos\theta)\cos\beta\cos\gamma-\sin\theta\cos\alpha & B\cos\beta \\ (1-\cos\theta)\cos\alpha\cos\gamma-\sin\theta\cos\beta & (1-\cos\theta)\cos\beta\cos\gamma+\sin\theta\cos\alpha & (1-\cos\theta)\cos^2\gamma+\cos\theta & B\cos\gamma \\ 0 & 0 & 0 & 1 \end{bmatrix} \tag{D.55}$$

More operators of particular linear transformations can be found out in the literature.

## D.4 CHAINS OF CONSEQUENT LINEAR TRANSFORMATIONS AND CLOSED LOOP OF CONSEQUENT COORDINATE SYSTEMS TRANSFORMATIONS

Consequent coordinate system transformations form chains (circuits) of linear transformations. The elementary chain of coordinate system transformation is composed of two consequent transformations. Chains of linear transformations play an important role in the theory of gearing.

Two different kinds of chain of consequent coordinate system transformations are distinguished.

First is the transition from the coordinate system $X_gY_gZ_g$ associated with the gear tooth flank $\mathcal{G}$ to the local Cartesian coordinate system $x_gy_gz_g$ having the origin at contact point $K$ of the gear tooth flank $\mathcal{G}$ and the pinion tooth flank $\mathcal{P}$. This linear transformation is also made up of numerous operators of intermediate coordinate system transformations $(X_{in}Y_{in}Z_{in})$.

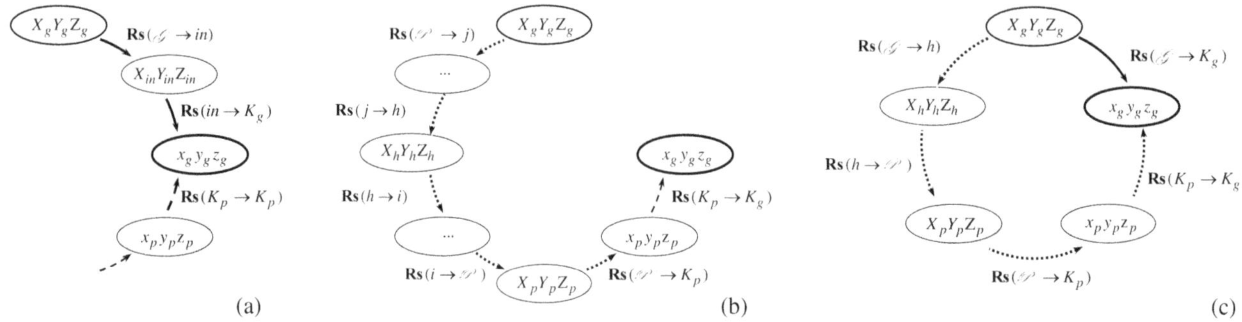


**FIGURE D.13** An example of direct chain (a), reverse chain (b), and a closed loop (c) of consequent coordinate system transformations.

It forms a chain of direct consequent coordinate system transformations, as illustrated in Figure D.13(a).

The local coordinate system $x_g y_g z_g$ is associated with the gear tooth flank $\mathscr{G}$. The operator $\mathbf{Rs}(\mathscr{G} \to K_g)$ of the resultant coordinate system transformation for a direct chain of linear transformations can be composed using a certain number of the operators of translation [see Eqs. (D.4) through (D.6)] and a corresponding number of operators of rotation [see Eqs. (D.10) through (D.12)].

Second is the transition from the coordinate system $X_g Y_g Z_g$ to the local Cartesian coordinate system $x_p y_p z_p$ with the origin at contact point $K$ of the tooth flanks $\mathscr{G}$ and $\mathscr{P}$. The local coordinate system $x_p y_p z_p$ is associated to pinion tooth flank $\mathscr{P}$. This linear transformation is also made up of numerous intermediate coordinate system transformations ($X_j Y_j Z_j$)—for example, transitions from the coordinate system $X_h Y_h Z_h$ associated with gear housing to intermediate coordinate system $X_i Y_i Z_i$. This kind of linear transformation forms a chain of inverse consequent coordinate system transformations, as shown in Figure D.13(b). The operator $\mathbf{Rs}\left(\mathscr{G} \to K_p\right)$ of the resultant coordinate system for the inverse chain of transformations can be composed using a certain number of operators of translation [see Eqs. (D.4), (D.5), and (D.6)] and a corresponding number of operators of rotation [see Eqs. (D.10) through (D.12)].

Chains of the direct and the reverse consequent coordinate system transformations, together with the operator of transition from the local coordinate system $x_p y_p z_p$ to the local coordinate system $x_g y_g z_g$ (or from the local coordinate system $x_g y_g z_g$ to the local coordinate system $x_p y_p z_p$), form a closed loop (a closed circuit) of the consequent coordinate system transformations, as depicted in Figure D.13(c).

If a closed loop of the consequent coordinate system transformations is complete, implementation of a certain number of the operators of translation [see Eqs. (D.4), (D.5), and (D.6)] and a corresponding number of the operators of rotation [see Eqs. (D.10) through (D.12)] returns a result that is identical to the input data. This means that the analytical description of a meshing process specified in the original coordinate system remains the same after implementation of the operator of the resultant coordinate system transformations. This condition is the necessary and sufficient condition for the existence of a closed loop of consequent coordinate system transformations.

The implementation of the chains, as well as of the closed loops of consequent coordinate system transformations, makes consideration of the meshing process of the tooth flanks $\mathscr{G}$ and $\mathscr{P}$ in any and all of reference systems that form up the loop possible. Therefore, for the consideration of a particular problem of part surface generation, the most convenient reference system can be chosen.

In order to complete the construction of a closed loop of consequent coordinate system transformations, an operator of transformation from the local coordinate system $x_p y_p z_p$ to the local coordinate system $x_g y_g z_g$ must be composed. Usually, the local reference systems $x_g y_g z_g$ and $x_p y_p z_p$ are semi-orthogonal coordinate systems. This means that the axis $z_P$ is always orthogonal to the coordinate plane $x_g y_g$, while the axes $x_g$ and $y_g$ can be either orthogonal or not orthogonal to each other. The same is valid with respect the local coordinate system $x_p y_p z_p$.

Two possible ways for performing the required transformation of the local reference systems $x_g y_g z_g$ and $x_p y_p z_p$ are to be considered.

Following the first way, the operator $\mathbf{Rt}_\omega(p \to g)$ of the linear transformation of semi-orthogonal coordinate systems (see Figure D.14) needs to be composed. The operator

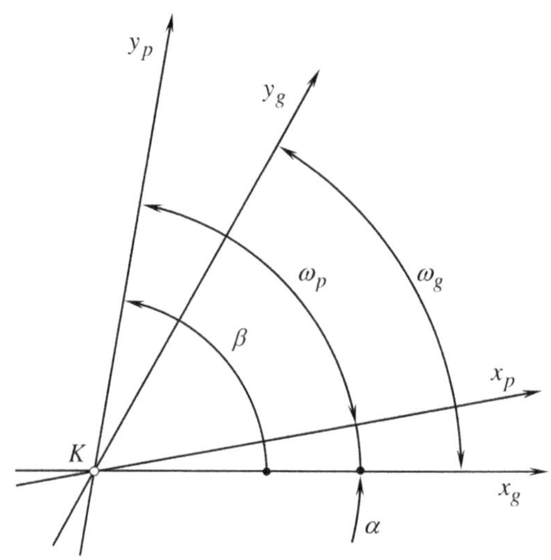


**FIGURE D.14** Local coordinate systems $x_g y_g z_g$ and $x_p y_p z_p$ that originate at contact point $K$.

$\mathbf{Rt}_\omega(p \rightarrow g)$ can be represented in the form of the homogenous matrix:

$$\mathbf{Rt}_\omega(p \rightarrow g) = \begin{bmatrix} \dfrac{\sin(\omega_p+\alpha)}{\sin\omega_p} & -\dfrac{\sin(\omega_g-\omega_p-\alpha)}{\sin\omega_p} & 0 & 0 \\ -\dfrac{\sin\alpha}{\sin\omega_p} & \dfrac{\sin(\omega_g-\alpha)}{\sin\omega_p} & 0 & 0 \\ 0 & 0 & -1 & 0 \\ 0 & 0 & 0 & 1 \end{bmatrix} \quad (D.56)$$

using the following designations:

- $\omega_g$ the angle that is formed by the coordinate lines $U_g$ and $V_g$ on the gear tooth flank $\mathcal{G}$
- $\omega_p$ the angle that is formed by the coordinate lines $U_p$ and $V_p$ on the pinion tooth flank $\mathcal{P}$
- $\alpha$ the angle that is formed by the axes $x_g$ and $x_p$ of the local coordinate systems $x_g y_g z_g$ and $x_p y_p z_p$

The auxiliary angle $\beta$ in Figure D.14 is equal to $\beta = \omega_T + \alpha$.

The inverse coordinate system transformation—that is, the transformation from the local coordinate system $x_g y_g z_g$ to the local coordinate system $x_p y_p z_p$ is analytically described by the operator $\mathbf{Rt}_\omega(g \rightarrow p)$ of the inverse coordinate system transformation. The operator $\mathbf{Rt}_\omega(g \rightarrow p)$ can be represented in the form of the homogenous matrix:

$$\mathbf{Rt}_\omega(g \rightarrow p) = \begin{bmatrix} \dfrac{\sin(\omega_g-\alpha)}{\sin\omega_g} & \dfrac{\sin(\omega_g-\omega_p-\alpha)}{\sin\omega_g} & 0 & 0 \\ \dfrac{\sin\alpha}{\sin\omega_g} & \dfrac{\sin(\omega_p+\alpha)}{\sin\omega_g} & 0 & 0 \\ 0 & 0 & -1 & 0 \\ 0 & 0 & 0 & 1 \end{bmatrix} \quad (D.57)$$

Following the second method for transformation of local coordinate systems, the auxiliary orthogonal local coordinate system must be constructed.

Let's consider an approach by which a closed loop (a closed circuit) of the consequent coordinate system transformations can be composed.

In order to construct an orthogonal normalized basis of the coordinate system $x_g y_g z_g$, an intermediate coordinate system $x_1 y_1 z_1$ is used. Axis $x_1$ of the coordinate system $x_1 y_1 z_1$ is pointed out along the unit vector $\mathbf{u}_g$ that is tangent to the $U_g$ – coordinate curve (see Figure D.15). Axis $y_1$ is pointed along vector $\mathbf{v}_g$ that is tangent to the $V_g$ – coordinate line on the gear tooth flank $\mathcal{G}$. The axis $z_1$ aligns with unit normal vector $\mathbf{n}_g$ and is pointed outward from the gear tooth body.

For an orthogonally parameterized gear tooth flank $\mathcal{G}$ (for which $F_g = 0$ and, therefore, $\omega_g = \pi/2$), analytical description of coordinate system transformations is significantly simpler. Further simplification of the coordinate system transformation is possible when the coordinate $U_g$ – and $V_g$ – lines are congruent to the lines of curvature on the part surface $\mathcal{G}$. In such a scenario, the local coordinate system is represented by the corresponding Darboux frame.

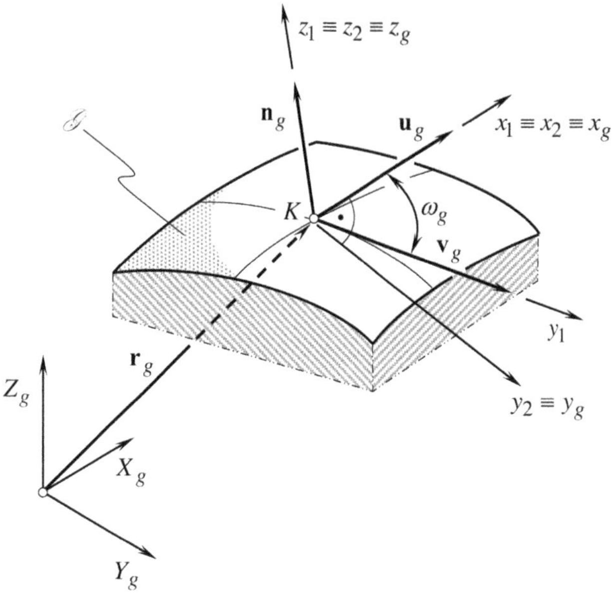


**FIGURE D.15** Local coordinate system $x_g y_g z_g$ associated with the gear tooth flank $\mathcal{G}$.

In order to construct a Darboux frame, the principal directions on the gear tooth flank $\mathcal{G}$ must be calculated. Determination of the unit tangent vectors $\mathbf{t}_{1.g}$ and $\mathbf{t}_{2.g}$ of the principal directions on the gear tooth flank $\mathcal{G}$ is considered in Appendix B.

In the common tangent plane, orientation of the unit vector $\mathbf{t}_{1.g}$ of the first principal direction on the gear tooth flank $\mathcal{G}$ can be uniquely specified by the included angle $\xi_{1.g}$ that the unit vector $\mathbf{t}_{1.g}$ forms with the $U_g$ – coordinate curve. This angle depends on both the gear tooth flank $\mathcal{G}$ geometry and the gear tooth flank $\mathcal{G}$ parameterization. Depicted in Figure D.16 is the relationship between the tangent vectors $\mathbf{U}_g$ and $\mathbf{V}_g$ and the included angle $\xi_{1.g}$. From the law of sine's,

$$\frac{\sqrt{G_g}}{\sin\xi_{1.g}} = \frac{\sqrt{E_g}}{\sin[\pi-\xi_{1.g}-(\pi-\omega_g)]} = \frac{\sqrt{F_g}}{\sin(\omega_g-\xi_{1.g})} \quad (D.58)$$

Here, $\omega_g = \cos^{-1}(F_g/\sqrt{E_g G_g})$.

Solving the expression for the included angle $\xi_{1.g}$ results in:

$$\xi_{1.g} = \tan^{-1}\frac{\sqrt{E_g G_g - F_g^2}}{E_g + F_g} \quad (D.59)$$

Another possible way of constructing of orthogonal local basis of the local Cartesian coordinate system $x_P y_P z_P$ is based on the following consideration.

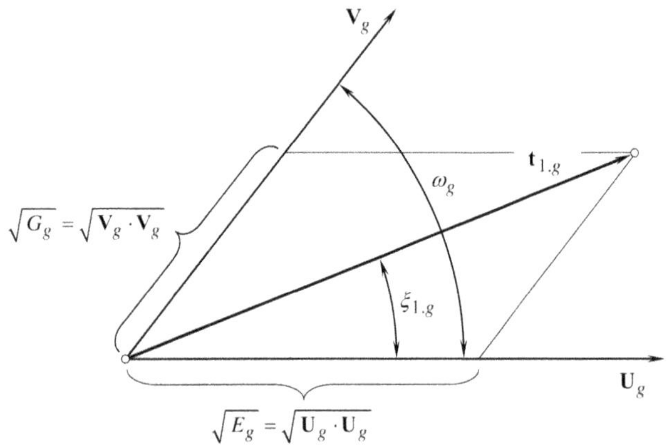


**FIGURE D.16** Differential relationships between the tangent vectors $\mathbf{U}_g$ and $\mathbf{V}_g$, the fundamental magnitudes of the first order, the included angle $\xi_{1.g}$, and the direction of the unit tangent vector $\mathbf{t}_{1.g}$ at the point of the gear tooth flank $\mathscr{G}$.

Consider an arbitrary non-orthogonal and not normalized basis $\mathbf{x}_1\mathbf{x}_2\mathbf{x}_3$ [see Figure D.17(a)]. Let's construct an orthogonal and normalized basis based on the initially given basis $\mathbf{x}_1\mathbf{x}_2\mathbf{x}_3$.

The cross product of any two of three vectors $\mathbf{x}_1$, $\mathbf{x}_2$, $\mathbf{x}_3$—for example, the cross product of the vectors $\mathbf{x}_1 \times \mathbf{x}_2$—determines a new vector $\mathbf{x}_4$ [see Figure D.17(b)]. Evidently, the vector $\mathbf{x}_4$ is orthogonal to the coordinate plane $\mathbf{x}_1\mathbf{x}_2$. Then, use the calculated vector $\mathbf{x}_4$ and one of two original vectors—for instance, vector $\mathbf{x}_2$—in the calculation of a new vector $\mathbf{x}_5 = \mathbf{x}_4 \times \mathbf{x}_2$ [see Figure D.17(c)]. The calculated basis $\mathbf{x}_1\mathbf{x}_4\mathbf{x}_5$ is orthogonal. In order to convert it into a normalized basis, vectors $\mathbf{x}_1$, $\mathbf{x}_4$, and $\mathbf{x}_5$ must each be divided by its magnitude:

$$\mathbf{e}_1 = \frac{\mathbf{x}_1}{|\mathbf{x}_1|} \tag{D.60}$$

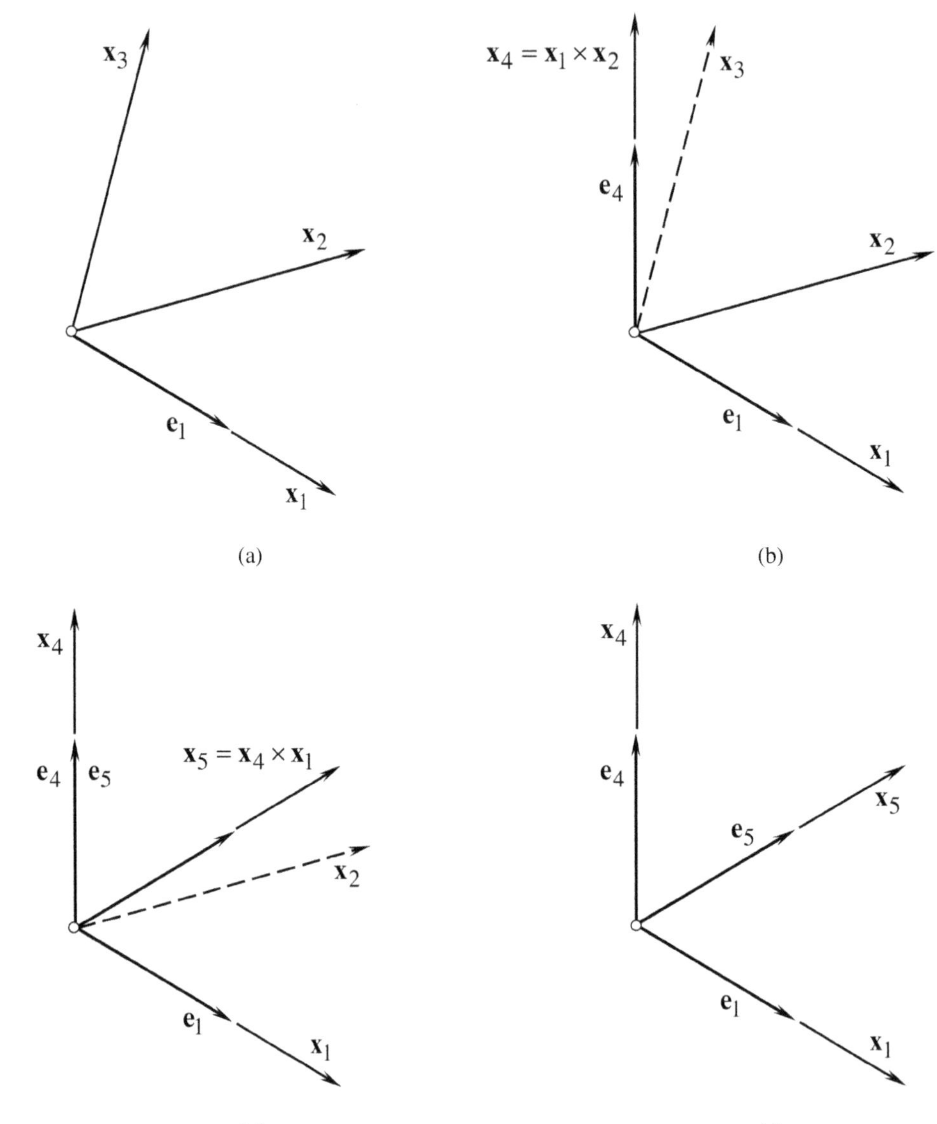


**FIGURE D.17** A normalized and orthogonally parameterized basis $\mathbf{e}_1\mathbf{e}_4\mathbf{e}_5$ that is constructed based on arbitrary basis $\mathbf{x}_1\mathbf{x}_2\mathbf{x}_3$. The parts (a) through (d) are discussed in detail in the text body of the chapter.

$$\mathbf{e}_4 = \frac{\mathbf{x}_4}{|\mathbf{x}_4|} \tag{D.61}$$

$$\mathbf{e}_5 = \frac{\mathbf{x}_5}{|\mathbf{x}_5|} \tag{D.62}$$

The resultant basis $\mathbf{e}_1\mathbf{e}_4\mathbf{e}_5$ [see Figure D.17(d)] is always both orthogonal and normalized.

In order to complete the analytical description of a closed loop of consequent coordinate system transformations, it is necessary to compose the operator $\mathbf{Rs}(K_p \to K_g)$ of transformation from the local reference system $x_p y_p z_p$ to the local reference system $x_g y_g z_g$ [see Figure D.13(c)].

In the case under consideration, the axes $z_g$ and $z_p$ align with the common unit normal vector, $\mathbf{n}_g$. The axis $z_g$ is pointed out from the bodily side to the void side of the gear tooth flank $\mathscr{G}$. The axis $z_g$ is pointed in the opposite direction. Due to this, the following equality is observed:

$$\mathbf{Rs}(K_p \to K_g) = \mathbf{Rt}(\varphi_z, z_p) \tag{D.63}$$

The inverse coordinate system transformation is analytically described by the operator:

$$\mathbf{Rs}(K_g \to K_p) = \mathbf{Rs}^{-1}(K_p \to K_g) = \mathbf{Rt}(-\varphi_z, z_p) \tag{D.64}$$

Implementation of the discussed results allows for:

a. The representation of the gear tooth flank $\mathscr{G}$ and the pinion tooth flank $\mathscr{P}$, as well as their relative motion, in a common coordinate system, and
b. Consideration of the meshing of the gear tooth flank $\mathscr{G}$ and the pinion tooth flank $\mathscr{P}$ in any desirable coordinate systems that is a component of the chain or the closed loop of consequent coordinate system transformations.

Transition from one coordinate system to another coordinate system can be performed in both of two feasible directions, say in direct as well as in inverse directions.

## D.5 THE IMPACT OF COORDINATE SYSTEM TRANSFORMATIONS ONTO FUNDAMENTAL FORMS OF SURFACES

Every coordinate system transformation entails a corresponding alteration to the equation of the gear tooth flank $\mathscr{G}$ or pinion tooth flank $\mathscr{P}$, or both. Because of this, it is often necessary to recalculate coefficients of the first $\Phi_{1.g}$ and of the second $\Phi_{2.g}$ fundamental forms of the gear tooth flank $\mathscr{G}$ (as many times as the coordinate system transformation is performed). This routing and time-consuming operation can be eliminated if the operators of coordinate system transformations are used directly on the fundamental forms $\Phi_{1.g}$ and $\Phi_{2.g}$.

After being calculated in an initial reference system, fundamental magnitudes $E_g$, $F_g$, and $G_g$ of the first $\Phi_{1.g}$ and fundamental magnitudes $L_g$, $M_g$, and $N_g$ of the second $\Phi_{2.g}$ order can be determined in any new coordinate system using operators of translation, operators of rotation, and operators of resultant coordinate system transformation. This kind of transformation of the fundamental forms $\Phi_{1.g}$ and $\Phi_{2.g}$ becomes possible due to implementation of the formula derived in the following.

Let's consider a gear tooth flank $\mathscr{G}$ given by equation $\mathbf{r}_g = \mathbf{r}_g(U_g, V_g)$, where $(U_g, V_g) \in \mathscr{G}$.

For the following analysis, it is convenient to use the equation of the first fundamental form $\Phi_{1.g}$ of the gear tooth flank $\mathscr{G}$ represented in matrix form:

$$[\Phi_{1.g}] = \begin{bmatrix} dU_g & dV_g & 0 & 0 \end{bmatrix} \cdot \begin{bmatrix} E_g & F_g & 0 & 0 \\ F_g & G_g & 0 & 0 \\ 0 & 0 & 1 & 0 \\ 0 & 0 & 0 & 1 \end{bmatrix} \cdot \begin{bmatrix} dU_g \\ dV_g \\ 0 \\ 0 \end{bmatrix} \tag{D.65}$$

Similarly, equation of the second fundamental form $\Phi_{2.g}$ of the gear tooth flank $\mathscr{G}$ can be given by:

$$[\Phi_{2.g}] = \begin{bmatrix} dU_g & dV_g & 0 & 0 \end{bmatrix} \cdot \begin{bmatrix} L_g & M_g & 0 & 0 \\ M_g & N_g & 0 & 0 \\ 0 & 0 & 1 & 0 \\ 0 & 0 & 0 & 1 \end{bmatrix} \cdot \begin{bmatrix} dU_g \\ dV_g \\ 0 \\ 0 \end{bmatrix} \tag{D.66}$$

The coordinate system transformation performed by the operator of linear transformation $\mathbf{Rs}(1 \to 2)$ transfers the equation $\mathbf{r}_g = \mathbf{r}_g(U_g, V_g)$ of the gear tooth flank $\mathscr{G}$ initially given in $X_1Y_1Z_1$ to the equation $\mathbf{r}_g^* = \mathbf{r}_g^*(U_g^*, V_g^*)$ of that same gear tooth flank $\mathscr{G}$ in a new coordinate system $X_2Y_2Z_2$. It is clear that $\mathbf{r}_g \neq \mathbf{r}_g^*$.

In the new coordinate system, the gear tooth flank $\mathscr{G}$ is analytically described by the following expression:

$$\mathbf{r}_g^*(U_g^*, V_g^*) = \mathbf{Rs}(1 \to 2) \cdot \mathbf{r}_g(U_g, V_g) \tag{D.67}$$

The column matrices of variables in Eqs. (D.65) and (D.66) casts the operator of the resultant coordinate system transformation $\mathbf{Rs}(1 \to 2)$ to the form:

$$\begin{bmatrix} dU_g^* & dV_g^* & 0 & 0 \end{bmatrix}^T = \mathbf{Rs}(1 \to 2) \cdot \begin{bmatrix} dU_g & dV_g & 0 & 0 \end{bmatrix}^T \tag{D.68}$$

Substitution of Eq. (D.68) into Eq. (D.65) and Eq. (D.66) makes the expressions for the fundamental forms $\Phi^*_{1.g}$ and $\Phi^*_{2.g}$ in the new coordinate system possible:

$$\left[\Phi^*_{1.g}\right] = \left[\mathbf{Rs}(1 \to 2) \cdot \left[\, dU_g \;\; dV_g \;\; 0 \;\; 0 \,\right]^T\right]^T \cdot \left[\Phi_{1.g}\right] \cdot \mathbf{Rs}(1 \to 2) \cdot \left[\, dU_g \;\; dV_g \;\; 0 \;\; 0 \,\right]^T \tag{D.69}$$

$$\left[\Phi^*_{2.g}\right] = \left[\mathbf{Rs}(1 \to 2) \cdot \left[\, dU_g \;\; dV_g \;\; 0 \;\; 0 \,\right]^T\right]^T \cdot \left[\Phi_{2.g}\right] \cdot \mathbf{Rs}(1 \to 2) \cdot \left[\, dU_g \;\; dV_g \;\; 0 \;\; 0 \,\right]^T \tag{D.70}$$

The following equation is valid for multiplication:

$$\left[\mathbf{Rs}(1 \to 2) \cdot \left[\, dU_g \;\; dV_g \;\; 0 \;\; 0 \,\right]^T\right]^T = \mathbf{Rs}^T(1 \to 2) \cdot \left[\, dU_g \;\; dV_g \;\; 0 \;\; 0 \,\right] \tag{D.71}$$

Therefore,

$$\left[\Phi^*_{1.g}\right] = \left[\, dU_g \;\; dV_g \;\; 0 \;\; 0 \,\right]^T \cdot \left\{\mathbf{Rs}^T(1 \to 2) \cdot \left[\Phi_{1.g}\right] \cdot \mathbf{Rs}(1 \to 2)\right\} \cdot \left[\, dU_g \;\; dV_g \;\; 0 \;\; 0 \,\right] \tag{D.72}$$

$$\left[\Phi^*_{2.g}\right] = \left[\, dU_g \;\; dV_g \;\; 0 \;\; 0 \,\right]^T \cdot \left\{\mathbf{Rs}^T(1 \to 2) \cdot \left[\Phi_{2.g}\right] \cdot \mathbf{Rs}(1 \to 2)\right\} \cdot \left[\, dU_g \;\; dV_g \;\; 0 \;\; 0 \,\right] \tag{D.73}$$

It can easily be shown that the matrices $\left[\Phi^*_{1.g}\right]$ and $\left[\Phi^*_{2.g}\right]$ in Eqs. (D.72) and (D.73) represent quadratic forms with respect to $dU_g$ and $dV_g$.

Transformation of the first $\Phi_{1.g}$ and the second $\Phi_{2.g}$ fundamental forms of the gear tooth flank $\mathscr{G}$ from the initial coordinate system $X_1Y_1Z_1$ to the new coordinate system $X_2Y_2Z_2$ is executed by the operator of transformation $\mathbf{Rs}(1 \to 2)$. As a result, in the new coordinate system, the corresponding fundamental forms are expressed as follows:

$$\left[\Phi^*_{1.g}\right] = \mathbf{Rs}^T(1 \to 2) \cdot \left[\Phi_{1.g}\right] \cdot \mathbf{Rs}(1 \to 2) \tag{D.74}$$

$$\left[\Phi^*_{2.g}\right] = \mathbf{Rs}^T(1 \to 2) \cdot \left[\Phi_{2.g}\right] \cdot \mathbf{Rs}(1 \to 2) \tag{D.75}$$

Equations (D.74) and (D.75) reveal that after the coordinate system transformation is completed, the first $\Phi^*_{1.g}$ and the second $\Phi^*_{2.g}$ fundamental forms of the gear tooth flank $\mathscr{G}$ in the coordinate system $X_2Y_2Z_2$ are expressed in terms of the first $\Phi_{1.g}$ and the second $\Phi_{2.g}$ fundamental forms, initially represented in the coordinate system $X_1Y_1Z_1$. In order to do this, the corresponding fundamental form (either $\Phi_{1.g}$ or $\Phi_{2.g}$) must be pre-multiplied by $\mathbf{Rs}(1 \to 2)$ and post-multiplied by $\mathbf{Rs}^T(1 \to 2)$.

Implementation of Eqs. (D.74) and (D.75) significantly simplifies formula transformations.

Equations similar to Eqs. (D.74) and (D.75) are valid with respect to pinion tooth flank $\mathscr{P}$, as well as to any and all quadratic forms.

In case of the use of the third $\Phi_{3.g}$ and the fourth $\Phi_{4.g}$ fundamental forms, their coefficients can be expressed in terms of the fundamental magnitudes of the first and the second order.

# Appendix E: Contact Geometry of Gear and Mating Pinion Tooth Flanks

In the theory of gearing, the kinematics of gearing is considered the prime element of the gear pair. Other important elements are considered the secondary elements of gearing, namely:

a. The shape and the geometry of the gear tooth flank $\mathcal{G}$ and
b. The shape and the geometry of the mating pinion tooth flank $\mathcal{P}$ (as well as numerous others)

This does not mean that the importance of the secondary elements is lower than that of the kinematics. It just means that the most favorable parameters of the secondary elements can be expressed in terms of the parameters of the prime element. Ultimately, the entire gear pair can be synthesized on the premise of just the prime elements—that is, it is possible to synthesize the rest of the design parameters of the gear pair based on the premise of the desirable kinematics of the gear pair to be designed. Only the kinematics of gearing is used for the purposes of synthesizing the best possible gear pair for transmitting the input rotation and torque.

In order to solve the problem of synthesizing the most favorable gear pair, an appropriate analytical description of contact geometry of the gear tooth flank $\mathcal{G}$ and the mating pinion tooth flank $\mathcal{P}$ is required. The problem of analytical description of contact geometry between two smooth regular surfaces in the first order of tangency is a sophisticated one.

Investigation into contact geometry of curves and surfaces can be traced back to the eighteenth century. In considerable detail, the study of the contact of curves and surfaces was undertaken by J.L. Lagrange[1] in his *Theorié des Fonctions Analytiques* (1797) and A.L. Cauchy[2] in his *Leçons sur les Applications du Calcul Infinitésimal á la Geometrie* (1826). In the twentieth century, an investigation in the realm of contact geometry of curves and surfaces was undertaken by J. Favard[3] in his *Course de Gèomètrie Diffèrentialle Locale* (1957). There will also be further discussion of other notable researchers in the field.

The results of the research on the contact geometry of two smooth regular surfaces are widely used in the theory of gearing. The problem of synthesizing the most favorable gear pair can be solved on the premise of the analysis of the topology of the contacting surfaces in the differential vicinity of their point of contact.

Various methods for analytical description of contact geometry between two smooth regular surfaces have been developed. The latest achievements in the field are discussed in a plurality of papers and monographs available in the public domain.

An in-depth analysis of the known methods of analytical description of the geometry of contact between two smooth regular surfaces uncovered the known methods' poor capability for solving problems in related to designing efficient gear pairs. Therefore, an accurate method for analytical description of contact geometry between two smooth regular surfaces $\mathcal{G}$ and $\mathcal{P}$ in the first order of tangency, which fits the needs of the theory of gearing, is necessary. Such a method is worked out in this appendix.

It is convenient to begin the discussion by starting with an analytical description of local relative orientation of the gear tooth flank $\mathcal{G}$ and the mating pinion tooth flank $\mathcal{P}$. The proposed analytical description is relevant to differential vicinity of the point of contact $K$ of the tooth flanks $\mathcal{G}$ and $\mathcal{P}$.

## E.1 LOCAL RELATIVE ORIENTATION AT THE POINT OF CONTACT OF GEAR AND MATING PINION TOOTH FLANKS

When gears rotate, a gear tooth flank $\mathcal{G}$ and the mating pinion tooth flank $\mathcal{P}$ are in permanent tangency with each other. The contacting surfaces $\mathcal{G}$ and $\mathcal{P}$ can be locally approximated by the corresponding quadrics, as schematically illustrated in Figure E.1. Certain restrictions on the relative configuration (location and orientation) of the tooth flanks $\mathcal{G}$ and $\mathcal{P}$ and on their instant relative motion are imposed by the requirement to be permanently in tangency to each other.

In the theory of gearing, a quantitative measure of relative orientation of the gear tooth flank $\mathcal{G}$ and of the mating pinion tooth flank $\mathcal{P}$ is established.

The relative orientation of the gear tooth flank $\mathcal{G}$ and of the mating pinion tooth flank $\mathcal{P}$ is specified by the angle $\mu$ of the local[4] relative orientation of the surfaces. By definition, angle $\mu$ is equal to the angle that the unit tangent vector $\mathbf{t}_{1.g}$ of the first principal direction of the gear tooth flank

[1] Joseph-Louis Lagrange (January 25, 1736–April 10, 1813) was a famous Italian-born [née Giuseppe Lodovico (Luigi) Lagrangia] French mathematician and mechanician.

[2] Augustin-Louis Cauchy (August 21, 1789–May 23, 1857) was a famous French mathematician.

[3] Jean Favard (August 28, 1902–January 21, 1965) was a French mathematician.

[4] The surfaces' orientation is *local* by nature, as it relates only to the differential vicinity of point $K$ to the tooth flanks $\mathcal{G}$ and $\mathcal{P}$.

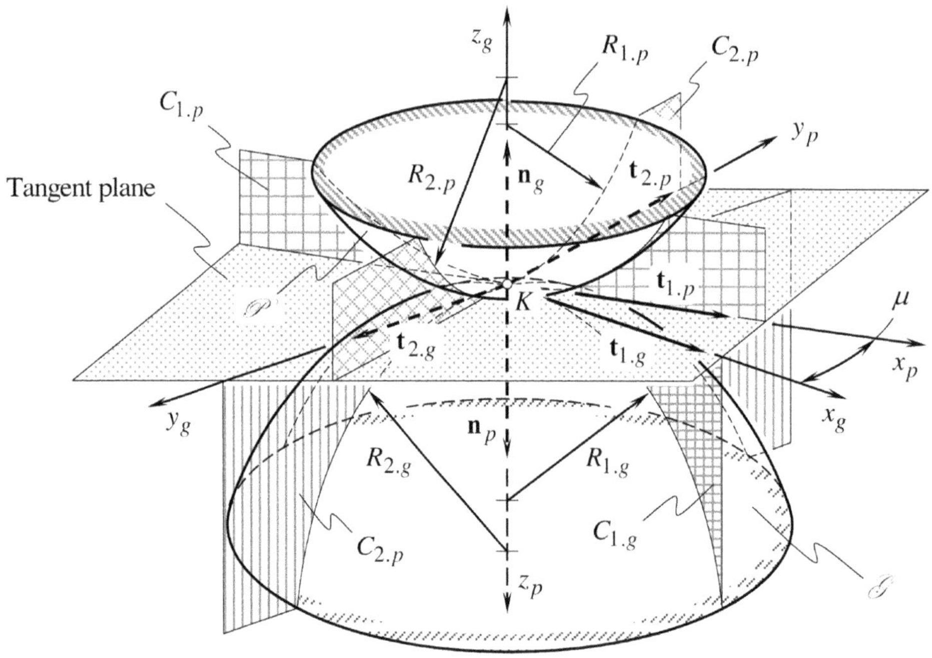


**FIGURE E.1** Local configuration of two quadrics that are tangent to a gear tooth flank $\mathscr{G}$ and to a mating pinion tooth flank $\mathscr{P}$ at contact point $K$. (After Prof. S.P. Radzevich, ~2008).

$\mathscr{G}$ forms with the unit tangent vector $\mathbf{t}_{1.p}$ of the first principal direction of the mating pinion tooth flank $\mathscr{P}$. That same angle μ can also be defined as the angle that makes the unit tangent vectors $\mathbf{t}_{2.g}$ and $\mathbf{t}_{2.p}$ of the second principal directions of the surfaces $\mathscr{G}$ and $\mathscr{P}$ at contact point $K$. This yields equations for the calculation of the angle μ:

$$\sin\mu = \left|\mathbf{t}_{1.g} \times \mathbf{t}_{1.p}\right| = \left|\mathbf{t}_{2.g} \times \mathbf{t}_{2.p}\right|, \tag{E.1}$$

$$\cos\mu = \mathbf{t}_{1.g} \cdot \mathbf{t}_{1.p} = \mathbf{t}_{2.g} \cdot \mathbf{t}_{2.p}, \tag{E.2}$$

$$\tan\mu = \frac{\left|\mathbf{t}_{1.g} \times \mathbf{t}_{1.p}\right|}{\mathbf{t}_{1.g} \cdot \mathbf{t}_{1.p}} \equiv \frac{\left|\mathbf{t}_{2.g} \times \mathbf{t}_{2.p}\right|}{\mathbf{t}_{2.g} \cdot \mathbf{t}_{2.p}} \tag{E.3}$$

with the following designations:

- $\mathbf{t}_{1.g}$, $\mathbf{t}_{2.g}$ are the unit vectors of principal direction on the gear tooth flank $\mathscr{G}$ measured at contact point $K$
- $\mathbf{t}_{1.p}$, $\mathbf{t}_{2.p}$ are the unit vectors of principal direction on the mating pinion tooth flank $\mathscr{P}$ at that same contact point $K$ of the tooth flanks $\mathscr{G}$ and $\mathscr{P}$

The directions of the unit tangent vectors of the principal direction $\mathbf{t}_{1.g}$ and $\mathbf{t}_{2.g}$ on the gear tooth flank $\mathscr{G}$ (as well as directions of the unit tangent vectors $\mathbf{t}_{1.p}$ and $\mathbf{t}_{2.p}$ of the principal directions on the pinion tooth flank $\mathscr{P}$) can be specified in terms of the ratio $dU_g/dV_g$ (or in terms of the ratio $dU_p/dV_p$ in case of the pinion tooth flank $\mathscr{P}$). The corresponding values of the ratio $dU_{g(p)}/dV_{g(p)}$ are calculated as roots of the quadratic equation:

$$\begin{vmatrix} E_{g(p)}\dfrac{dU_{g(p)}}{dV_{g(p)}} + F_{g(p)} & F_{g(p)}\dfrac{dU_{g(p)}}{dV_{g(p)}} + G_{g(p)} \\ L_{g(p)}\dfrac{dU_{g(p)}}{dV_{g(p)}} + M_{g(p)} & M_{g(p)}\dfrac{dU_{g(p)}}{dV_{g(p)}} + N_{g(p)} \end{vmatrix} = 0 \tag{E.4}$$

In the case of the contact point of the surfaces $\mathscr{G}$ and $\mathscr{P}$, the actual value of the angle μ is calculated at the contact point $K$. When the tooth flanks $\mathscr{G}$ and $\mathscr{P}$ are in line contact, the actual value of the angle μ can be calculated at every point within the line of contact.[5] The line of contact of the tooth flanks $\mathscr{G}$ and $\mathscr{P}$ is commonly referred to as the *characteristic line* $\mathscr{C}$, or just *characteristic* $\mathscr{C}$.

The determination of the angle μ of local relative orientation of the tooth flanks $\mathscr{G}$ and $\mathscr{P}$ of a gear and a mating pinion is illustrated in Figure E.1.

The unit vectors of the principal directions $\mathbf{t}_{1.g}$ and $\mathbf{t}_{1.p}$ are used to calculate the actual value of the angle μ of the local relative orientation of the tooth flanks $\mathscr{G}$ and $\mathscr{P}$.

Consider tooth flanks $\mathscr{G}$ and $\mathscr{P}$ in point contact, which are represented in a common reference system. The surfaces make contact at point $K$. For further analysis, an equation of the common tangent plane to the tooth flanks $\mathscr{G}$

[5] It is worth pointing out here that in a case of line contact, the relative orientation of two surfaces $\mathscr{G}$ and $\mathscr{P}$ is predetermined in a *global* sense. However, the actual value of the angle μ of the surface's *local* relative orientation at different points of the characteristic $\mathscr{C}$ is different.

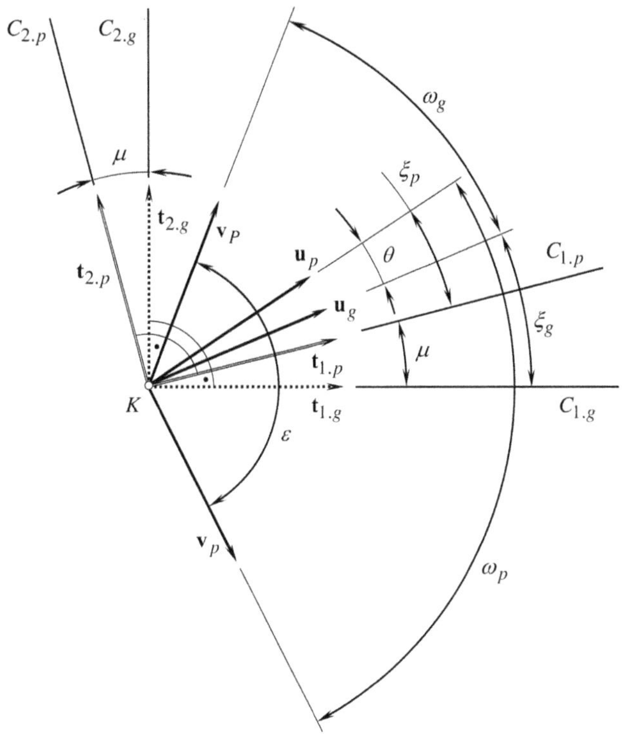


**FIGURE E.2** Local relative orientation at the contact point of tooth flanks of a gear $\mathcal{G}$ and a mating pinion $\mathcal{P}$ considered in a common tangent plane.

and $\mathcal{P}$ at the contact point $K$ is necessary (see Figure E.1 and Figure E.2).

$$(\mathbf{r}_{tp} - \mathbf{r}_K) \cdot \mathbf{u}_g \cdot \mathbf{v}_g = 0 \tag{E.5}$$

Here:

- $\mathbf{r}_{tp}$ is the position vector of point of the common tangent plane
- $\mathbf{r}_K$ is the position vector of the contact point $K$
- $\mathbf{u}_g$ and $\mathbf{v}_g$ are unit vectors that are tangent to $U_g$ – and $V_g$ – coordinate lines on the gear tooth flank $\mathcal{G}$ at the contact point $K$

The angle $\omega_g$ is the angle that is formed by the unit vectors $\mathbf{u}_g$ and $\mathbf{v}_g$. The actual value of the angle $\omega_g$ can be calculated from one of the following equations:

$$\sin \omega_g = \frac{\sqrt{E_g G_g - F_g^2}}{\sqrt{E_g G_g}} \tag{E.6}$$

$$\cos \omega_g = \frac{F_g}{\sqrt{E_g G_g}} \tag{E.7}$$

$$\tan \omega_g = \frac{\sqrt{E_g G_g - F_g^2}}{F_g} \tag{E.8}$$

Equations similar to Eqs. (E.6) through (E.8) are also valid for the calculation of the angle $\omega_p$ at the point on the pinion tooth flank $\mathcal{P}$.

Tangent directions $\mathbf{u}_g$ and $\mathbf{v}_g$ to the $U_g$ – and $V_g$ – coordinate lines at a point on the gear tooth flank $\mathcal{G}$, as well as tangent directions $\mathbf{u}_p$ and $\mathbf{v}_p$ to the $U_p$ – and $V_p$ – coordinate lines at a point on the pinion tooth flank $\mathcal{P}$, are specified in terms of the angles $\theta$ and $\varepsilon$. For the calculation of actual values of the angles $\theta$ and $\varepsilon$, the following equations are used:

$$\cos\theta = \mathbf{u}_g \cdot \mathbf{u}_p \tag{E.9}$$

$$\cos\varepsilon = \mathbf{v}_g \cdot \mathbf{v}_p \tag{E.10}$$

The angle $\xi_g$ is the angle that the first principal direction $\mathbf{t}_{1.g}$ on the gear tooth surface $\mathcal{G}$ forms with the unit tangent vector $\mathbf{u}_g$ (see Figure E.2). The equation for the calculation of an actual value of the angle $\xi_g$ was derived by Prof. *S.P. Radzevich*:

$$\sin \xi_g = \frac{\eta_g}{\sqrt{\eta_g^2 - 2\eta_g \cos\omega_g + 1}} \sin\omega_g \tag{E.11}$$

where $\eta_g$ designates the ratio $\eta_g = \dfrac{\partial U_g}{\partial V_g}$.

When $F_g = 0$, the equality $\tan \xi_g = \eta_g$ is observed. Here, the ratio $\eta_g$ is equal to the root of the quadratic equation:

$$\begin{aligned}(F_g L_g - E_g M_g)\eta_g^2 + (G_g L_g - E_g N_g)\eta_g \\ + (G_g M_g - F_g N_g) = 0\end{aligned} \tag{E.12}$$

which immediately follows from the equation:

$$\begin{vmatrix} E_g\, dU_g + F_g\, dV_g & F_g\, dU_g + G_g\, dV_g \\ L_g\, dU_g + M_g\, dV_g & M_g\, dU_g + N_g\, dV_g \end{vmatrix} = 0 \tag{E.13}$$

The equation for the calculation of the actual value of the angle $\xi_g$ allows for another representation. Following the chain rule, $d\,\mathbf{r}_g$ can be represented in the form:

$$d\,\mathbf{r}_g = \mathbf{U}_g\, dU_g + \mathbf{V}_g\, dV_g \tag{E.14}$$

By definition, $\tan \xi_g = \dfrac{\sin \xi_g}{\cos \xi_g}$. The functions $\sin \xi_g$ and $\cos \xi_g$ yield representation as:

$$\sin \xi_g = \frac{|\mathbf{U}_g \times d\,\mathbf{r}_g|}{|\mathbf{U}_g| \cdot |d\,\mathbf{r}_g|} \tag{E.15}$$

$$\cos \xi_{Pg} = \frac{\mathbf{U}_g \cdot d\,\mathbf{r}_g}{|\mathbf{U}_g| \cdot |d\,\mathbf{r}_g|} \tag{E.16}$$

The last expressions yield:

$$\tan\xi_g = \frac{\sin\xi_g}{\cos\xi_g} = \frac{|\mathbf{U}_g \times d\,\mathbf{r}_g|}{\mathbf{U}_g \cdot d\,\mathbf{r}_g} = \frac{|\mathbf{U}_g \times d\,\mathbf{r}_g|}{\mathbf{U}_g \cdot (\mathbf{U}_g dU_g + \mathbf{V}_g dV_g)} = \frac{|\mathbf{U}_g \times d\,\mathbf{r}_g| \cdot dV_g}{\mathbf{U}_g \cdot \mathbf{U}_g dU_g + \mathbf{U}_g \cdot \mathbf{V}_g dV_g} \tag{E.17}$$

By definition:

$$\mathbf{U}_g \cdot \mathbf{U}_g = E_g \tag{E.18}$$

$$\mathbf{U}_g \cdot \mathbf{V}_g = F_g \tag{E.19}$$

$$|\mathbf{U}_g \times \mathbf{V}_g| = \sqrt{E_g G_g - F_g^2} \tag{E.20}$$

Equation (E.14) through (E.20) yield the formula:

$$\tan\xi_g = \frac{\sqrt{E_g G_g - F_g^2}}{\eta_g \cdot E_g + F_g} \tag{E.21}$$

for the calculation of actual value of the angle $\xi_g$.

Equations similar to Eqs. (E.11) and (E.21) are also valid for the calculation of the actual value of the angle $\xi_p$ that the first principal direction $\mathbf{t}_{1.p}$ at a point on the pinion tooth flank $\mathscr{P}$ forms with the unit tangent vector $\mathbf{u}_p$.

The performed analysis makes it possible to derive the following equations for the calculation of the unit vectors of principal directions $\mathbf{t}_{1.g}$ and $\mathbf{t}_{2.g}$:

$$\mathbf{t}_{1.g} = \mathbf{Rt}(\xi_g, \mathbf{n}_g) \cdot \mathbf{u}_g \tag{E.22}$$

$$\mathbf{t}_{2.g} = \mathbf{Rt}\left[\left(\xi_g + \frac{\pi}{2}\right), \mathbf{n}_g\right] \cdot \mathbf{u}_g \tag{E.23}$$

for the gear tooth flank $\mathscr{G}$ and similar equations for the calculation of the unit vectors of principal directions $\mathbf{t}_{1.p}$ and $\mathbf{t}_{2.p}$:

$$\mathbf{t}_{1.p} = \mathbf{Rt}(\xi_p, \mathbf{n}_g) \cdot \mathbf{u}_p \tag{E.24}$$

$$\mathbf{t}_{2.p} = \mathbf{Rt}\left[\left(\xi_p + \frac{\pi}{2}\right), \mathbf{n}_g\right] \cdot \mathbf{u}_p \tag{E.25}$$

for the pinion tooth flank $\mathscr{P}$.

The operator of rotation $\mathbf{Rt}(\varphi_A, A_0)$ through an angle $\varphi_A$ about an axis $A_0$ is employed for the calculation of the operators of rotation in Eqs. (E.22) through (E.25).

## E.2 THE SECOND-ORDER ANALYSIS: PLANAR CHARACTERISTIC IMAGES

For a more accurate analytical description of contact geometry of the gear tooth flank $\mathscr{G}$ and the pinion tooth flank $\mathscr{P}$, consideration of the second-order parameters is prudent. The second-order analysis incorporates elements of the first-order analysis. When performing the second-order analysis, familiarity with the Dupin indicatrix is highly beneficial.[6] The Dupin indicatrix is a perfect starting point for consideration of the second-order analysis.

### E.2.1 Preliminary Remarks: Dupin Indicatrix

At any point of a smooth regular gear tooth flank $\mathscr{G}$ (as well as at any point of a smooth regular pinion tooth flank $\mathscr{P}$), a corresponding Dupin indicatrix can be constructed. The Dupin indicatrix $Dup(\mathscr{G})$ at a point of a gear tooth flank $\mathscr{G}$, as well as Dupin indicatrix $Dup(\mathscr{P})$ at a point of the pinion tooth flank $\mathscr{P}$, are planar characteristic curves of the second order. They are used for graphical interpretation of the distribution of normal radii of curvature of a surface in the differential vicinity of a surface point.

The Dupin indicatrix at a point of the tooth flank $\mathscr{G}$ (as well as at a point of the tooth flank, $\mathscr{P}$) is of critical importance in the theory of gearing. Generation of this planar characteristic curve is illustrated in the diagram shown in Figure E.3.

A plane $W$ through the unit normal vector $\mathbf{n}_g$ to the gear tooth flank $\mathscr{G}$ at a point $m$ is rotating about the unit normal vector $\mathbf{n}_g$. While rotating, the plane occupies consecutive positions, $W_1$, $W_2$, $W_3$, and so on. The radii of normal curvature of the line of intersection of the gear tooth flank $\mathscr{G}$ by normal planes $W_1$, $W_2$, and $W_3$ are equal to $R_{g,1}$, $R_{g,2}$, $R_{g,3}$, and so forth. The gear tooth flank $\mathscr{G}$ is intersected by a plane $Q$. The plane $Q$ is orthogonal to the unit normal vector $\mathbf{n}_g$. This plane is at a certain small distance $\delta$ from the point $m$. When the distance $\delta$ approaches a zero ($\delta \to 0$), and when the scale of the line of intersection of the gear tooth flank $\mathscr{G}$ by the plane $Q$ approaches infinity, then the line of intersection of the gear surface $\mathscr{G}$ by the plane $Q$ approaches the planar characteristic curve that is commonly referred to as the Dupin indicatrix $Dup(\mathscr{G})$.

In differential geometry, a surface is construed as a zero-thickness film. Because of this, the following five types Dupin indicatrices are distinguished in differential geometry of surfaces (see Figure E.4):

- Elliptic [see Figure E.4(a)]
- Umbilic [see Figure E.4(b)]
- Parabolic [see Figure E.4(c)]
- Hyperbolic [see Figure E.4(d)]
- Minimal [see Figure E.4(e)].

[6] Fransua Pier Charles Dupin (October 6, 1784–January 18, 1873) was a French mathematician.

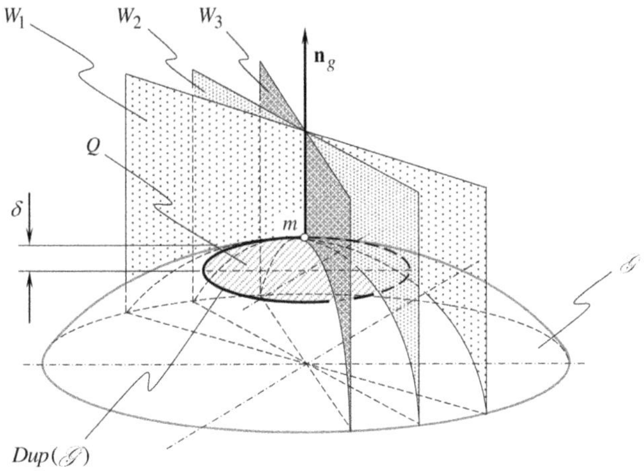


**FIGURE E.3** The Dupin indicatrix at a smooth regular point of gear tooth flank $\mathscr{G}$.

The Dupin indicatrix for a plane local surface patch doesn't exist. In the case of a plane, all points of the Dupin indicatrix are remote to infinity.

For local surface patches with negative full curvature ($\mathscr{G}_g < 0$), phantom branches—that is, the branches that are not intersected by a plane perpendicular to the unite normal vector $\mathbf{n}_g$ to the gear tooth flank $\mathscr{G}$ at point $m$—of the characteristic curve $Dup(\mathscr{G})$ are shown in dashed lines [see Figure E.4(d) and Figure E.4(e)].

An easy way to derive an equation of the characteristic curve $Dup(\mathscr{G})$ is as follows.

Euler's formula:

$$k_{1.g}\cos^2\varphi + k_{2.g}\sin^2\varphi = k_g \tag{E.26}$$

yields representation in the form:

$$\frac{k_{1.g}}{k_g}\cos^2\varphi + \frac{k_{2.g}}{k_g}\sin^2\varphi = 1 \tag{E.27}$$

Transition from polar coordinates to Cartesian coordinates is performed using well-known formulas:

$$x_g = \rho\cos\varphi \tag{E.28}$$

$$y_g = \rho\sin\varphi \tag{E.29}$$

These formulas make the following expressions for $\cos^2\varphi = x_g^2/\rho^2$ and $\sin^2\varphi = y_g^2/\rho^2$ possible. After substituting the formulas into Eq. (E.27), one can come up with the equation:

$$\frac{k_{1.g}}{k_g}\cdot\frac{x_g^2}{\rho^2} + \frac{k_{2.g}}{k_g}\cdot\frac{y_g^2}{\rho^2} = 1. \tag{E.30}$$

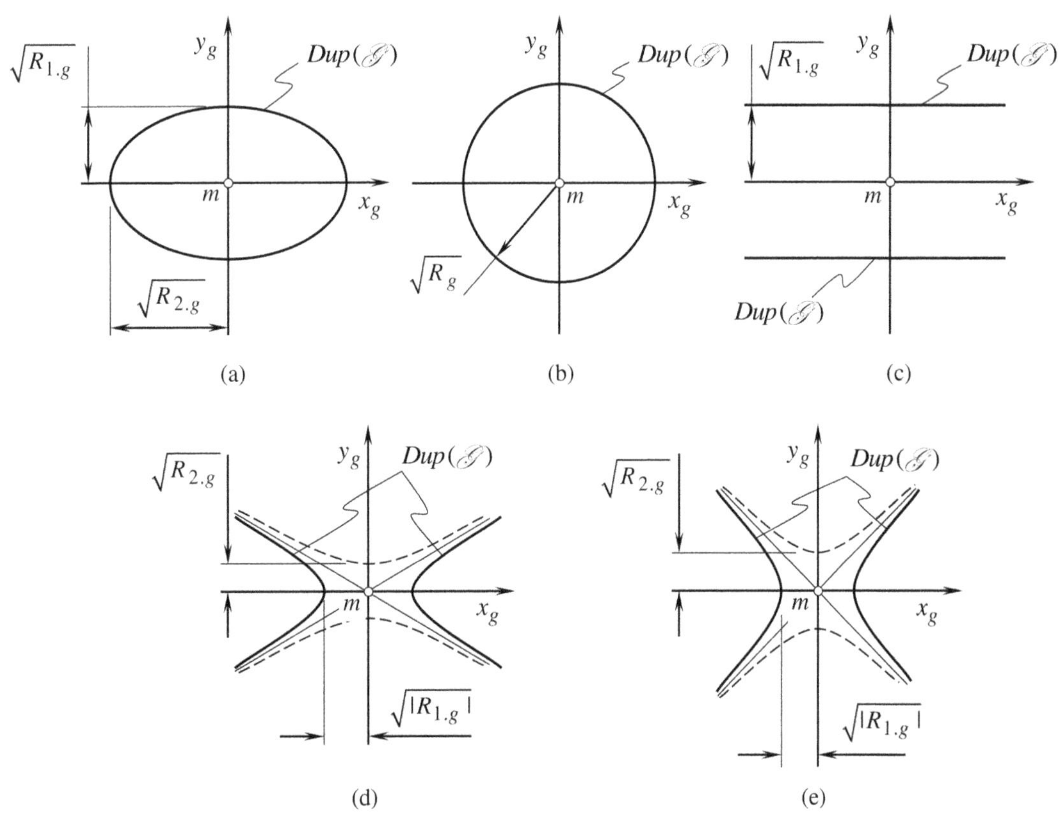


**FIGURE E.4** (a–e) Five kinds of Dupin indicatrices $Dup(\mathscr{G})$ at point $m$ of a smooth regular gear tooth flank $\mathscr{G}$.

It is convenient to designate $\rho = \sqrt{k_g^{-1}}$. Principal curvatures $k_{1.g}$ and $k_{2.g}$ are the roots of the quadratic equation:

$$\begin{vmatrix} L_g - E_g k_g & M_g - F_g k_g \\ M_g - F_g k_g & N_g - G_g k_g \end{vmatrix} = 0 \tag{E.31}$$

Substituting the calculated values of the principal curvatures $k_{1.g}$ and $k_{2.g}$ into Eq. (E.30), and after performing the necessary formulae transformations, an equation[7] for the Dupin indicatrix $Dup(\mathscr{G})$ can be represented in the form:

$$k_{1.g}\, x_g^2 + k_{2.g}\, y_g^2 = 1 \tag{E.32}$$

Equation (E.32) describes a particular case of the Dupin indicatrix, which is represented in a Darboux frame.[8]

The general form of the equation of the *Dupin indicatrix* at point $m$ of a gear tooth flank $\mathscr{G}$ is often represented as:

$$Dup(\mathscr{G}) \Rightarrow \frac{L_g}{E_g} x_g^2 + \frac{2M_g}{\sqrt{E_g G_g}} x_g\, y_g + \frac{N_g}{G_g} y_g^2 = 1 \tag{E.33}$$

In Eq. (E.33), the characteristic curve $Dup(\mathscr{G})$ is expressed in terms of the fundamental magnitudes $E_g$, $F_g$, and $G_g$ of the first order $\Phi_{1.g}$ and in terms of the fundamental magnitudes $L_g$, $M_g$, and $N_g$ of the second order $\Phi_{2.g}$ at a point of the gear tooth flank $\mathscr{G}$.

### E.2.2 Matrix Representation of Equation of *Dupin Indicatrix* at Point of Gear Tooth Flank

Like any other quadratic form, the Dupin indicatrix equation of the gear tooth flank $\mathscr{G}$ can be represented in matrix form:

$$Dup(\mathscr{G}) \Rightarrow \begin{bmatrix} x_g & y_g & 0 & 0 \end{bmatrix} \cdot \begin{bmatrix} \dfrac{L_g}{E_g} & \dfrac{2M_g}{\sqrt{E_g G_g}} & 0 & 0 \\ \dfrac{2M_g}{\sqrt{E_g G_g}} & \dfrac{N_g}{G_g} & 0 & 0 \\ 0 & 0 & \mp 1 & 0 \\ 0 & 0 & 0 & 1 \end{bmatrix} \cdot \begin{bmatrix} x_g \\ y_g \\ 0 \\ 0 \end{bmatrix} = \pm 1 \tag{E.34}$$

In a Darboux frame, this equation is reduced to:

$$Dup(\mathscr{G}) \Rightarrow \begin{bmatrix} x_g & y_g & 0 & 0 \end{bmatrix} \cdot \begin{bmatrix} L_g & M_g & 0 & 0 \\ M_g & N_g & 0 & 0 \\ 0 & 0 & \mp 1 & 0 \\ 0 & 0 & 0 & 1 \end{bmatrix} \cdot \begin{bmatrix} x_g \\ y_g \\ 0 \\ 0 \end{bmatrix} = \pm 1 \tag{E.35}$$

It is convenient to implement matrix representation of a Dupin indicatrix equation, for instance, when investigating spatial gearings—that is, crossed-axes gearing when multiple coordinate system transformations are required.

The Dupin indicatrix equation can be represented in the form:

$$r_{Dup}(\varphi) = \sqrt{|R_g(\varphi)|} \cdot \operatorname{sgn} \Phi_{2.g}^{-1} \tag{E.36}$$

The latter equation reveals that the position vector of a point of the Dupin indicatrix $Dup(\mathscr{G})$ in any direction is equal to the square root of the radius of curvature in that same direction.[9]

## E.3 The Degree of Conformity at the Point of Contact of Gear and Mating Pinion Tooth Flanks in the First Order of Tangency

For an accurate analytical description of the contact geometry of the gear and the mating pinion tooth flanks in the first order of tangency, a higher order analysis is necessary.

The method of a higher order analysis discussed in the following targets the development of an analytical description of degree of conformity of the pinion tooth flank $\mathscr{P}$ to the gear tooth flank $\mathscr{G}$ at a current contact point $K$. The higher the degree of conformity of the tooth flanks $\mathscr{G}$ and $\mathscr{P}$, the closer these surfaces are to each other in differential vicinity to point $K$. This qualitative (*intuitive*) definition of the degree of conformity of two smooth regular surfaces needs a corresponding quantitative measure.

### E.3.1 Preliminary Remarks

Implementation of the resultant deviation $l_{cnf}$ (see Figure E.5) of two smooth regular surfaces in contact for the analytical description of contact geometry is a type of straightforward

---

[7] The same equation for the Dupin indicatrix could be derived in another way. Coxeter considers a pair of conics obtained by expanding an equation in Monge's form $z = z(x, y)$ in a McLaurin series:

$$z = z(0,0) + z_1 x + z_2 y + \frac{1}{2}(z_{11}x_1^2 + 2z_{12}xy + z_{22}y^2) + \cdots = \frac{1}{2}(b_{11}x^2 + 2b_{12}xy + b_{22}y^2)$$

This gives the Dupin indicatrix equation as $(b_{11}x^2 + 2b_{12}xy + b_{22}y^2) = \pm 1$.

[8] Jean Gaston Darboux (August 13, 1842–February 23, 1917) was a French mathematician.

[9] Similar to the Dupin indicatrix $Dup(\mathscr{G})$, another type of planar characteristic curve can be introduced. This planar characteristic curve is referred to as a curvature indicatrix. The equation of this characteristic curve can be postulated in the form: $r_{Dup \cdot k}(\varphi) = \sqrt{|k_g(\varphi)|} \cdot \operatorname{sgn} \Phi_{2 \cdot g}^{-1}$. Application of a curvature indicatrix in the form $r_{Dup.k}(\varphi)$ makes it possible to avoid uncertainty in cases of plane surface. For plane surface, the characteristic curve $Dup(\mathscr{G})$ does not exist, while $r_{Dup.k}(\varphi)$ does exist. It shrinks to the point $m$ on the gear tooth flank $\mathscr{G}$.

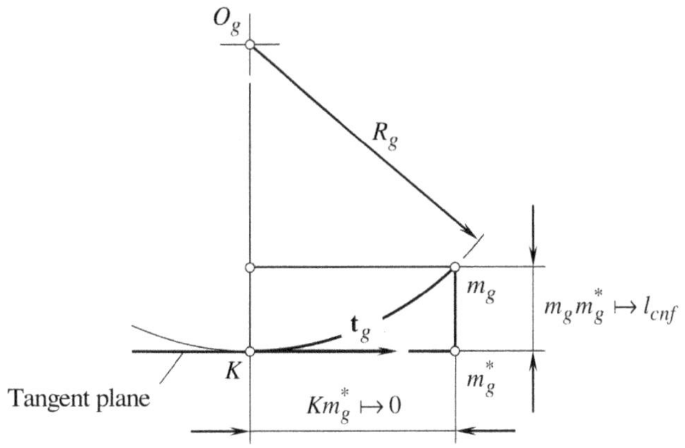


**FIGURE E.5** On transition from the resultant deviation $l_{cnf}$ to indicatrix of conformity $Cnf\,(\mathscr{G}/\mathscr{P})$ at contact point $K$ between two smooth regular tooth flanks $\mathscr{G}$ and $\mathscr{P}$.

solution to the problem under consideration. This approach has been proven to be computationally ineffective. However, the approach gives insight into how an effective method for solving the problem under consideration can be developed.

As seen in Figure E.5, three geometrical parameters are interrelated when a deviation of a surface from the tangent plane is considered in differential vicinity of surface point. These are:

a. The measure of the deviation $m_g m_g^*$ of a gear tooth flank $\mathscr{G}$ from the tangent plane $l_{cnf}$.
b. The distance $Km_g^*$ of current point $m_g$ from the contact point $K$.
c. The radius of normal curvature $R_g$ of the gear tooth flank $\mathscr{G}$ at the contact point $K$.

As a consequence of this relationship between the parameters $m_g m_g^*$, $Km_g^*$, and $R_g$, any one of them can be used for the purposes of quantitative evaluation of the degree of conformity of the contacting tooth flanks of the gear $\mathscr{G}$ and the mating pinion $\mathscr{P}$. Following Figure E.5:

$$m_g m_g^* = R_g - \sqrt{R_g^2 - \left(Km_g^*\right)^2}\,\Bigg|_{m_g \to K} \mapsto l_{cnf} \qquad \text{(E.37)}$$

Inversely, for the radius of normal curvature $R_g$ at a point of the gear tooth flank $\mathscr{G}$, the following expression is valid:

$$R_g = \frac{\left(m_g m_g^*\right)^2 + \left(Km_g^*\right)^2}{2 \cdot m_g m_g^*} \qquad \text{(E.38)}$$

Ultimately, one may conclude that any legitimate analytical function of normal radii of curvature $R_g$ and $R_p$ at a point of contact of the gear tooth flank $\mathscr{G}$ and the pinion tooth flank $\mathscr{P}$ can be used for this purpose.

Consider two smooth regular tooth flanks $\mathscr{G}$ and $\mathscr{P}$ in the first order of tangency that make contact at point $K$. Degree of conformity of the tooth flanks $\mathscr{G}$ and $\mathscr{P}$ can be construed as a function of the radii of normal curvature $R_g$ and $R_p$ of the contacting surfaces. The radii of normal curvature $R_g$ and $R_p$ of the tooth flanks $\mathscr{G}$ and $\mathscr{P}$ are taken in a common normal plane section through the point $K$. For a specified radius of normal curvature $R_g$ of the tooth flank $\mathscr{G}$, the degree of conformity of the tooth flanks depends on the corresponding value of the radius of normal curvature $R_p$ of the pinion tooth flank $\mathscr{P}$.

In gear meshing, the degree of conformity at the point of contact of the tooth flanks $\mathscr{G}$ and $\mathscr{P}$ is not of a constant value in most cases; it changes as the coordinates of the contact point change. The degree of the surfaces' conformity to one another depends on the orientation of the normal plane section through the contact point $K$ and alters as the normal plane section turns about the common perpendicular $\mathbf{n}_g$. This statement immediately follows from the previous conclusion, according to which the degree of conformity at the point of contact of the tooth flanks $\mathscr{G}$ and $\mathscr{P}$ yields an interpretation in terms of the radii of normal curvature $R_g$ and $R_p$.

The change in the degree of conformity of a gear tooth flank $\mathscr{G}$ and a mating pinion tooth flank $\mathscr{P}$ due to the turning of the normal plane section about the common perpendicular $\mathbf{n}_g$ is illustrated in Figure E.6 in two-dimensional examples, for which that same normal plane section of the gear tooth flank $\mathscr{G}$ makes contact with different plane sections $\mathscr{P}^i$ of the pinion tooth flank $\mathscr{P}$.

In the example shown in Figure E.6(a), the radius of normal curvature $R_p^1$ of the convex plane section $\mathscr{P}^1$ of the pinion tooth flank $\mathscr{P}$ is positive ($R_p^1 > 0$). The convex normal plane section of the pinion tooth flank $\mathscr{P}$ makes contact with the convex normal plane section ($R_g > 0$) of a gear tooth flank $\mathscr{G}$. The degree of conformity of the pinion tooth flank $\mathscr{P}$ to the gear tooth flank $\mathscr{G}$ in Figure E.6(a) is relatively low, as both the contacting surfaces are convex.

Another example is shown in Figure E.6(b). The radius of normal curvature $R_p^2$ of convex plane section $\mathscr{P}^2$ of the pinion tooth flank $\mathscr{P}$ is also of a positive value ($R_p^2 > 0$). However, its value exceeds the value of the radius of normal curvature $R_p^1$ in the first example ($R_p^2 > R_p^1$). This results in the degree of conformity of the pinion toot flank $\mathscr{P}$ to the gear tooth flank $\mathscr{G}$ [Figure E.6(b)] being greater compared to that shown in Figure E.6(a).

In the next example, depicted in Figure E.6(c), the normal plane section $\mathscr{P}^3$ of the pinion tooth flank $\mathscr{P}$ is represented with a locally flattened section. The radius of normal curvature $R_p^3$ of the flattened plane section $\mathscr{P}^3$ approaches infinity ($R_p^3 \to \infty$). Thus, the inequality $R_p^3 > R_p^2 > R_p^1$ is valid. Therefore, the degree of conformity of the pinion tooth flank $\mathscr{P}$ to the gear tooth flank $\mathscr{G}$ in Figure E.6(c) also increases.

Finally, for a concave normal plane section $\mathscr{P}^4$ of the pinion tooth flank $\mathscr{P}$, as illustrated in Figure E.6(d), the radius of normal curvature $R_p^4$ has a negative value ($R_p^4 < 0$). In this case, the degree of conformity of the pinion tooth flank $\mathscr{P}$ to the gear tooth flank $\mathscr{G}$ is the greatest of the four examples considered in Figure E.6.

The examples shown in Figure E.6 qualitatively illustrate what is already intuitively known, regarding the different

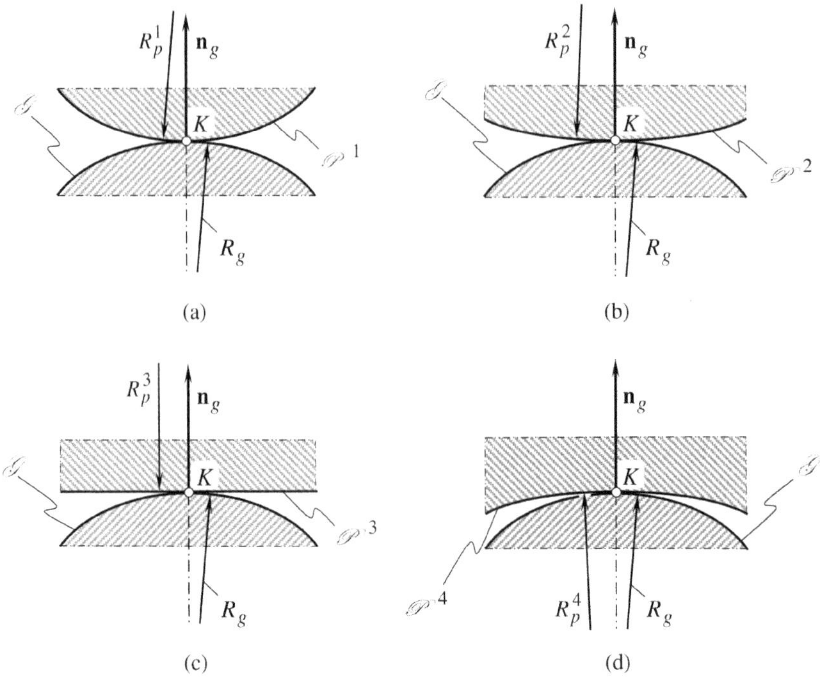


**FIGURE E.6** (a–d) Sections of two smooth regular tooth flanks $\mathscr{G}$ and $\mathscr{P}$ in contact by a plane through the common perpendicular $\mathbf{n}_g$.

degree of conformity of two smooth regular surfaces in the first order of tangency. Intuitively, one can realize that in the examples shown in Figure E.6(a) through Figure E.6(d), the degree of conformity at the points of contact of two tooth flanks $\mathscr{G}$ and $\mathscr{P}$ gradually increases.

A similar observation is made for a given pair of the tooth flanks $\mathscr{G}$ and $\mathscr{P}$ when different sections of the surfaces by a plane surface through the common perpendicular $\mathbf{n}_g$ are considered [see Figure E.7(a)]. When rotating the plane section about the common perpendicular $\mathbf{n}_g$, it can be observed that the degree of conformity of the gear and the pinion tooth flanks $\mathscr{G}$ and $\mathscr{P}$ is different in different configurations of the cross-sectional plane [see Figure E.7(b)].

These examples provide an intuitive understanding of what the degree of conformity at point of contact of two smooth regular tooth flanks $\mathscr{G}$ and $\mathscr{P}$ means. The examples cannot be employed directly for the purpose of quantitatively evaluating the degree of conformity at the point of contact of two smooth regular tooth flanks $\mathscr{G}$ and $\mathscr{P}$. The next step that is necessary is to introduce an appropriate quantitative evaluation of the degree of conformity of two smooth regular surfaces in the first order of tangency. In other words, how can a certain degree of conformity of two smooth regular surfaces be described analytically?

### E.3.2 Indicatrix of Conformity at the Point of Contact of Gear and Mating Pinion Tooth Flanks

This section aims to introduce a quantitative measure of the degree of conformity at the point of contact of two smooth regular surfaces. The degree of conformity at the point of contact of two tooth flanks $\mathscr{G}$ and $\mathscr{P}$ indicates how the pinion tooth flank $\mathscr{P}$ is close to the gear tooth flank $\mathscr{G}$ in the differential vicinity of contact point $K$, say how much

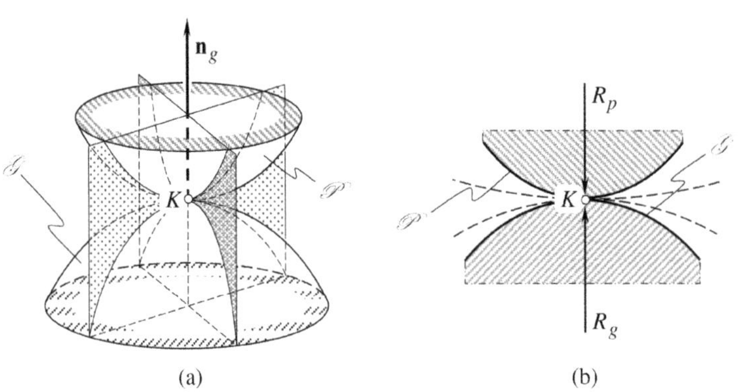


**FIGURE E.7** An analytical description of the contact geometry of two smooth regular tooth flanks of a gear $\mathscr{G}$ and a mating pinion $\mathscr{P}$: (a) a close-up, and (b) an example of section by normal plane through contact point, $K$.

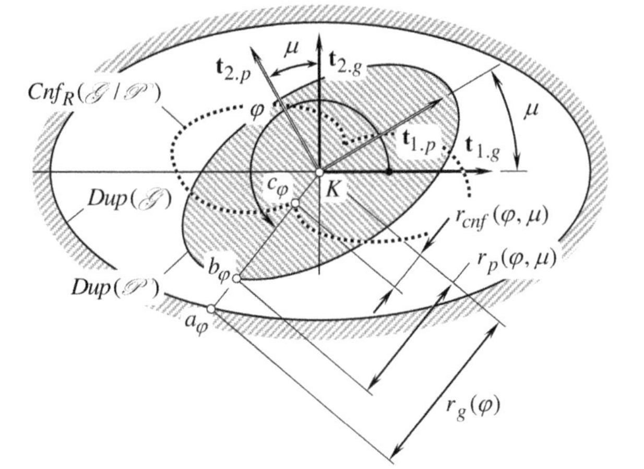


**FIGURE E.8** On the definition of the indicatrix of conformity $Cnf_R(\mathscr{P} \mapsto \mathscr{G})$ at the point of contact of tooth flanks $\mathscr{G}$ and $\mathscr{P}$. (See Radzevich, S.P., *Differential-Geometric Method of Surface Generation*, Dr.Sci. Thesis, Tula, Tula Polytechnic Institute, 1991, 300 pages).

the surface $\mathscr{P}$ is *congruent* to the surface $\mathscr{G}$ in differential vicinity of the contact point $K$. This type of congruency between the contacting surfaces $\mathscr{G}$ and $\mathscr{P}$ can also be construed as the *local congruency* of the contacting surfaces.

Quantitatively, the degree of conformity at the point of contact of a smooth regular surface $\mathscr{P}$ to another surface $\mathscr{G}$ can be expressed in terms of the difference between the corresponding radii of normal curvature of the contacting surfaces. In order to develop a quantitative measure of the degree of conformity of the tooth flanks $\mathscr{G}$ and $\mathscr{P}$, it is convenient to implement Dupin indicatrices $Dup(\mathscr{G})$ and $Dup(\mathscr{P})$ constructed at point of contact of the gear tooth flank $\mathscr{G}$ and the pinion tooth flank $\mathscr{P}$, respectively.

It is natural to assume that the smaller difference between the normal curvatures of the surfaces $\mathscr{G}$ and $\mathscr{P}$ in a common cross-section by a plane through the common normal vector $\mathbf{n}_g$ results in the greater degree of conformity at the point of contact of the tooth flanks $\mathscr{G}$ and $\mathscr{P}$.

The Dupin indicatrix $Dup(\mathscr{G})$ indicates the distribution of the radii of normal curvature at the point of the gear tooth flank $\mathscr{G}$ as it had been shown, for example, for a concave elliptic patch of the surface $\mathscr{G}$ (see Figure E.8). For a gear tooth flank $\mathscr{G}$, an equation of this characteristic curve in polar coordinates can be represented in the form:

$$Dup(\mathscr{G}) \Rightarrow r_g(\varphi_g) = \sqrt{|R_g(\varphi_g)|} \qquad \text{(E.39)}$$

where:

$r_g$ is the position vector of point of the Dupin indicatrix $Dup(\mathscr{G})$ at the point of the gear tooth flank $\mathscr{G}$

$\varphi_g$ is the polar angle of the indicatrix $Dup(\mathscr{G})$

This is similarly true with respect to the Dupin indicatrix $Dup(\mathscr{G})$ at the point of the pinion tooth flank $\mathscr{P}$, as was shown, for instance, for a convex elliptical patch of the pinion tooth flank $\mathscr{P}$ (see Figure E.8). An equation of this characteristic curve in polar coordinates can be represented in the form:

$$Dup(\mathscr{P}) \Rightarrow r_p(\varphi_p) = \sqrt{|R_p(\varphi_p)|} \qquad \text{(E.40)}$$

where:

$r_p$ is the position vector of point of the Dupin indicatrix $Dup(\mathscr{G})$ at the point of the pinion tooth flank $\mathscr{P}$

$\varphi_p$ is the polar angle of the indicatrix $Dup(\mathscr{P})$

In the coordinate plane $x_g y_g$ of the local reference system $x_g y_g z_g$, the equalities $\varphi_g = \varphi$ and $\varphi_p = \varphi + \mu$ are valid. Therefore, in the coordinate plane $x_g y_g$, Eqs. (E.39) and (E.40) cast into:

$$Dup(\mathscr{G}) \Rightarrow r_g(\varphi) = \sqrt{|R_g(\varphi)|} \qquad \text{(E.41)}$$

$$Dup(\mathscr{P}) \Rightarrow r_p(\varphi, \mu) = \sqrt{|R_p(\varphi, \mu)|} \qquad \text{(E.42)}$$

When the degree of conformity at the points of contact of the gear tooth flank $\mathscr{G}$ is greater, the difference between the functions $r_g(\varphi)$ and $r_p(\varphi, \mu)$ becomes smaller, and vice versa. The latter makes valid the following conclusion:

> **Conclusion E.1.** *The distance between the corresponding*[10] *points of the Dupin indicatrices $Dup(\mathscr{G})$ and $Dup(\mathscr{P})$ constructed at the point of contact of a gear tooth flank $\mathscr{G}$ and a mating pinion tooth flank $\mathscr{P}$ can be employed to indicate the degree of conformity at the point of contact of the gear tooth flank $\mathscr{G}$ and of the pinion tooth flank $\mathscr{P}$ at the contact point $K$.*

The equation of the indicatrix of conformity $Cnf_R(\mathscr{G}/\mathscr{P})$ at the point of contact of a gear tooth flank $\mathscr{G}$ and a mating pinion tooth flank $\mathscr{P}$ is defined by the following structure:

$$\begin{aligned} Cnf_R(\mathscr{G}/\mathscr{P}) &\Rightarrow r_{cnf}(\varphi, \mu) \\ &= \sqrt{|R_g(\varphi)|}\,\mathrm{sgn}\,R_g(\varphi) \\ &\quad + \sqrt{|R_p(\varphi, \mu)|}\,\mathrm{sgn}\,R_p(\varphi, \mu) \\ &= r_g(\varphi)\,\mathrm{sgn}\,R_g(\varphi) + r_p(\varphi, \mu)\,\mathrm{sgn}\,R_p(\varphi, \mu) \end{aligned} \qquad \text{(E.43)}$$

Because of the location of point $a_\varphi$ of the Dupin indicatrix $Dup(\mathscr{G})$ at the point of the gear tooth flank $\mathscr{G}$ is specified by the position vector $r_g(\varphi)$, and the location of point $b_\varphi$ of the Dupin indicatrix $Dup(\mathscr{P})$ at the point of the pinion tooth flank $\mathscr{P}$ is specified by the position vector

[10] Corresponding points of the Dupin indicatrices $Dup(\mathscr{G})$ and $Dup(\mathscr{P})$ share the same straight line through the contact point $K$ of the tooth flanks $\mathscr{G}$ and $\mathscr{P}$ and are located at the same side of point $K$.

$r_p(\varphi,\mu)$, the location of point $c_\varphi$ (see Figure E.8) of the indicatrix of conformity $Cnf_R(\mathcal{G}/\mathcal{P})$ at point of contact $K$ of the tooth flanks $\mathcal{G}$ and $\mathcal{P}$ is specified by the position vector $r_{cnf}(\varphi,\mu)$. Therefore, the equality $r_{cnf}(\varphi,\mu) = Kc_\varphi$ is observed, and the length of the straight-line segment $Kc_\varphi$ is equal to the distance $a_\varphi\, b_\varphi$.

In Eq. (E.43), the following are designated:

$r_g = \sqrt{|R_g|}$ is the position vector of point of Dupin indicatrix of the gear tooth flank $\mathcal{G}$ at point $K$ of contact with pinion tooth flank $\mathcal{P}$

$r_p = \sqrt{|R_p|}$ is the position vector of corresponding point of the Dupin indicatrix of the pinion tooth flank $\mathcal{P}$

In Eq. (E.43), the multipliers $\operatorname{sgn} R_g(\varphi)$ and $\operatorname{sgn} R_p(\varphi,\mu)$ are assigned to the functions $r_g(\varphi) = \sqrt{|R_g(\varphi)|}$ and $r_p(\varphi,\mu) = \sqrt{|R_p(\varphi,\mu)|}$, respectively, for the purpose of retaining the corresponding signs of the functions—that is, so that they have the same sign that the radii of normal curvature $R_g(\varphi)$ and $R_p(\varphi,\mu)$ have.

Ultimately, one can conclude that position vector $r_{cnf}$ of point of the indicatrix of conformity $Cnf_R(\mathcal{G}/\mathcal{P})$ can be expressed in terms of position vectors $r_g$ and $r_p$ of the Dupin indicatrices $Dup(\mathcal{G})$ and $Dup(\mathcal{P})$.

The calculation of a current value of the radius of normal curvature $R_g(\varphi)$ at the point of the gear tooth flank $\mathcal{G}$ uses the equality:

$$R_g(\varphi) = \frac{\Phi_{1.g}}{\Phi_{2.g}} \tag{E.44}$$

Similarly, the calculation of the current value of the radius of normal curvature $R_p(\varphi,\mu)$ at point of pinion tooth flank $\mathcal{P}$ can use the equality:

$$R_p(\varphi,\mu) = \frac{\Phi_{1.p}}{\Phi_{2.p}} \tag{E.45}$$

The use of the angle $\mu$ of local relative orientation of the tooth flanks $\mathcal{G}$ and $\mathcal{P}$ indicates that the radii of normal curvature $R_g(\varphi)$ and $R_p(\varphi,\mu)$ are taken in a common normal plane section through the contact point $K$.

Further, it is well-known that the inequalities $\Phi_{1.g} \geq 0$ and $\Phi_{1.p} \geq 0$ are always valid. Therefore, Eq. (E.43) can be rewritten in the following form:

$$r_{cnf} = r_g(\varphi)\operatorname{sgn}\Phi_{2.g}^{-1} + r_p(\varphi,\mu)\operatorname{sgn}\Phi_{2.p}^{-1} \tag{E.46}$$

For the derivation of an equation of the indicatrix of conformity $Cnf_R(\mathcal{G}/\mathcal{P})$, it is convenient to use Euler's equation for normal radius of curvature $R_g(\varphi)$ at a point of the gear tooth flank $\mathcal{G}$:

$$R_g(\varphi) = \frac{R_{1.g}\cdot R_{2.g}}{R_{1.g}\cdot\sin^2\varphi + R_{2.g}\cdot\cos^2\varphi} \tag{E.47}$$

Here, the radii of principal curvature $R_{1.g}$ and $R_{2.g}$ are the roots of the quadratic equation:

$$\begin{vmatrix} L_g\cdot R_g - E_g & M_g\cdot R_g - F_g \\ M_g\cdot R_g - F_g & N_g\cdot R_g - G_g \end{vmatrix} = 0 \tag{E.48}$$

Recall that the inequality $R_{1.g} < R_{2.g}$ is always observed.

Equations (E.47) and (E.48) allow for expression of the radius of normal curvature $R_g(\varphi)$ at the point of the gear tooth flank $\mathcal{G}$ in terms of the fundamental magnitudes of the first order $E_g$, $F_g$, and $G_g$ and of the fundamental magnitudes of the second order $L_g$, $M_g$, and $N_g$.

A similar consideration is applicable for the pinion tooth flank $\mathcal{P}$. Omitting routing analysis, one can conclude that the radius of normal curvature $R_p(\varphi,\mu)$ at point of the pinion tooth flank $\mathcal{P}$ can be expressed in terms of the fundamental magnitudes of the first order $E_p$, $F_p$, and $G_p$ and of the fundamental magnitudes of the second order $L_p$, $M_p$, and $N_p$.

Finally, on the premise of the performed analysis, the following equation for the indicatrix of conformity $Cnf_R(\mathcal{G}/\mathcal{P})$ at point of contact of the tooth flanks $\mathcal{G}$ and $\mathcal{P}$ can be derived:

$$\begin{aligned} r_{cnf}(\varphi,\mu) = &\sqrt{\left|\frac{E_gG_g}{L_gG_g\cos^2\varphi - M_g\sqrt{E_gG_g}\sin 2\varphi + N_gE_g\sin^2\varphi}\right|}\operatorname{sgn}\Phi_{2.g}^{-1} \\ &+ \sqrt{\left|\frac{E_pG_p}{L_pG_p\cos^2(\varphi+\mu) - M_p\sqrt{E_pG_p}\sin 2(\varphi+\mu) + N_pE_p\sin^2(\varphi+\mu)}\right|}\operatorname{sgn}\Phi_{2.p}^{-1} \end{aligned} \tag{E.49}$$

Equation (E.49) of the characteristic curve[11] $Cnf_R(P/T)$ was developed in the late 1970s and early 1980s.

[11] The equation of this characteristic curve is known from:

a. Pat. No. 1249787, USSR, *A Method of Sculptured Part Surface Machining on a Multi-Axis NC Machine*, S.P. Radzevich, B23C 3/16, Filed: December 27, 1984.

b. Pat. No. 1185749, USSR, *A Method of Sculptured Part Surface Machining on a Multi-Axis NC Machine*, S.P. Radzevich, B23C 3/16, Filed: October 24, 1983.

An analysis of Eq. (E.49) reveals that indicatrix of conformity $Cnf_R(\mathcal{G}/\mathcal{P})$ at point of contact of a gear tooth flank $\mathcal{G}$ and a mating pinion tooth flank $\mathcal{P}$ is represented by a planar centro-symmetrical curve of the fourth order. In certain cases, this characteristic curve also possesses a property of mirror symmetry. Mirror symmetry of indicatrix of conformity is observed, for example, when the angle $\mu$ of the local relative orientation of the tooth flanks $\mathcal{G}$ and $\mathcal{P}$ is equal to $\mu = \pm\pi\cdot n/2$, where $n$ designates an integer number.

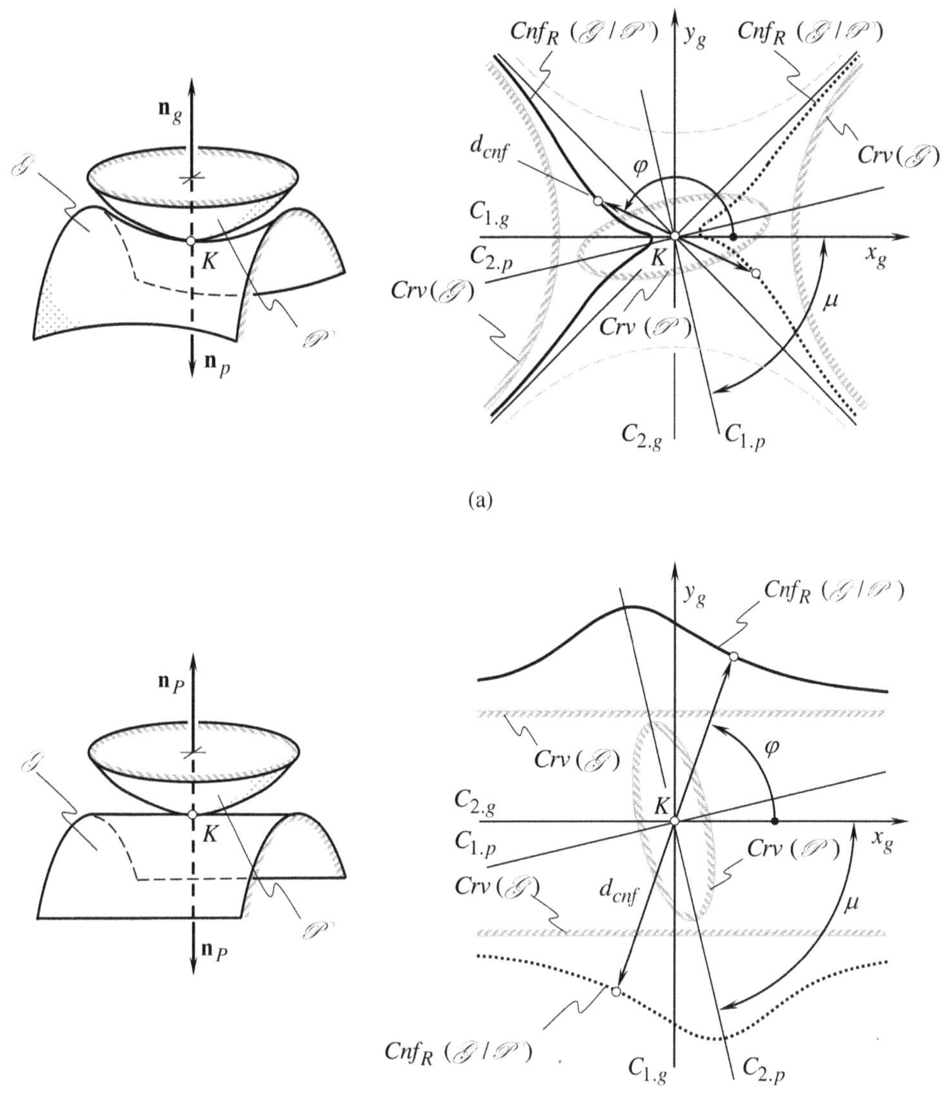


(a)



(b)

**FIGURE E.9** Examples of indicatrix of conformity $Cnf_R(\mathscr{G}/\mathscr{P})$ at point of contact $K$ of a smooth regular gear tooth flank $\mathscr{G}$ and a mating pinion tooth flank $\mathscr{P}$ in the cases of (a) elliptic-to-hyperbolic contact, and (b) elliptic-to-parabolic contact of the tooth flanks.

It is important to note here that even for the most general case of gearing, the position vector of point $r_{cnf}(\varphi,\mu)$ of indicatrix of conformity $Cnf_R(\mathscr{G}/\mathscr{P})$ is not dependent on the fundamental magnitudes $F_g$ and $F_p$. The independence of indicatrix of conformity $Cnf_R(\mathscr{G}/\mathscr{P})$ of the fundamental magnitudes $F_g$ and $F_p$ is due to the following.

Coordinate angle $\omega_g$ at point of the gear tooth flank $\mathscr{G}$ is calculated from the formula:

$$\omega_g = \arccos\frac{F_g}{\sqrt{E_g G_g}} \tag{E.50}$$

It is natural that the position vector $r_{cnf}(\varphi,\mu)$ of point of indicatrix of conformity $Cnf_R(\mathscr{G}/\mathscr{P})$ is not a function of the coordinate angle $\omega_g$. Although the position vector $r_{cnf}(\varphi,\mu)$ depends on the fundamental magnitudes $E_g$, $G_g$ and $E_p$, $G_p$, the performed analysis makes it evident why the position vector $r_{cnf}(\varphi,\mu)$ does not depend on the fundamental magnitudes $F_g$ and $F_p$.

Two illustrative examples of indicatrix of conformity $Cnf_R(\mathscr{G}/\mathscr{P})$ at point of contact of a gear tooth flank $\mathscr{G}$ and a mating pinion tooth flank $\mathscr{P}$ are shown in Figure E.9. The first example [see Figure E.9(a)] relates to contact of a hyperbolic paraboloid local patch of the tooth surface $\mathscr{G}$ and of a convex elliptical local patch of tooth surface $\mathscr{P}$. The second [see Figure E.9(b)] involves contact of a convex parabolic local patch of the tooth surface $\mathscr{G}$ and of a convex elliptical local patch of tooth surface $\mathscr{P}$. In both the cases, the corresponding curvature indicatrices $Crv(\mathscr{G})$ and $Crv(\mathscr{P})$ at the point of contact of the tooth flanks $\mathscr{G}$ and $\mathscr{P}$ are depicted in Figure E.9. The imaginary (phantom) branches of the Dupin indicatrix $Dup(\mathscr{G})$ [not labeled

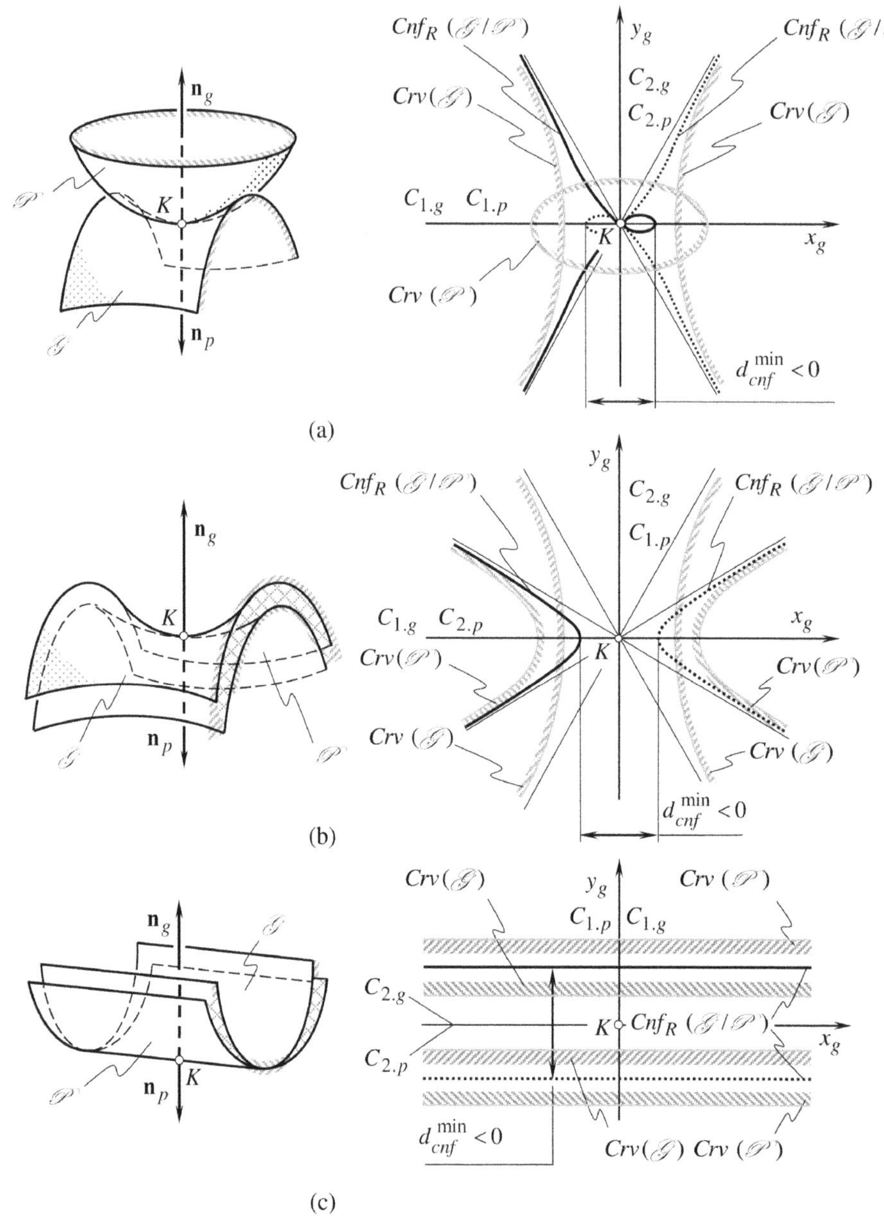


**FIGURE E.10** (a–c) Examples of violation of the physical condition of contact of a smooth regular gear tooth flank $\mathcal{G}$ and a mating pinion tooth flank $\mathcal{P}$.

in Figure E.9(a)] for the hyperbolic paraboloid local patch of the gear tooth flank $\mathcal{G}$ are shown in dashed lines [see Figure E.9(a)].

The gear tooth flank $\mathcal{G}$ and the pinion tooth flank $\mathcal{P}$ can make contact geometrically; however, physical conditions of their contact could be violated. Violation of the physical condition of contact results in the bodies bounded by the interacting surfaces $\mathcal{G}$ and $\mathcal{P}$ interfering into each other. Implementation of the indicatrix of conformity $Cnf_R(\mathcal{G}/\mathcal{P})$ uncovers the surfaces' interference, if any exists. Three illustrative examples of violation of the physical condition of contact are shown in Figure E.10. When correspondence between the radii of normal curvature of the contacting tooth flanks $\mathcal{G}$ and $\mathcal{P}$ is inappropriate, the indicatrix of conformity $Cnf_R(\mathcal{G}/\mathcal{P})$ either intersects itself [see Figure E.10(a)] or all of its diameters become negative [see Figures E.10(b) and E.10(c)].

Another interpretation of satisfaction and violation of the physical condition of contact of two smooth regular tooth flanks $\mathcal{G}$ and $\mathcal{P}$ is illustrated in Figure E.11. The condition of physical contact is fulfilled when all diameters of the indicatrix of conformity $Cnf_R(\mathcal{G}/\mathcal{P})$ have positive values. In this case, the gear tooth flank $\mathcal{G}$ and the mating pinion tooth flank $\mathcal{P}$ may contact each other as two rigid bodies do. An example of the indicatrix of conformity $Cnf_R(\mathcal{G}/\mathcal{P})$ for such a case is depicted in Figure E.11(a). In cases when this planar characteristic curve has negative diameters, as schematically shown in

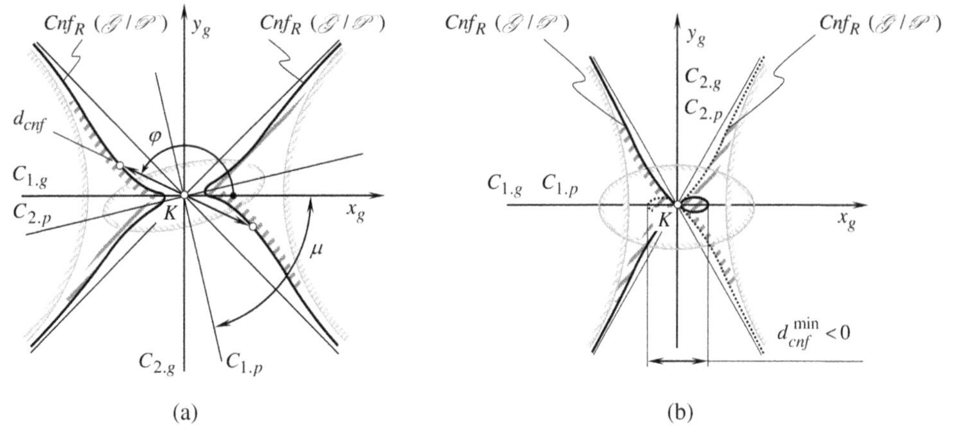


**FIGURE E.11** Another interpretation of satisfaction (a) and violation (b) of the condition of physical contact of a smooth regular gear tooth flank $\mathscr{G}$ and a mating pinion tooth flank $\mathscr{P}$.

Figure E.11(b), physical contact between the tooth flanks $\mathscr{G}$ and $\mathscr{P}$ is infeasible.

The actual value of current diameter[12] $d_{cnf}$ of the indicatrix of conformity $Cnf_R(\mathscr{G}/\mathscr{P})$ indicates the degree of conformity of the gear tooth flank $\mathscr{G}$ and the mating pinion tooth flank $\mathscr{P}$ to each other in a corresponding cross-section of the surfaces by a normal plane through the common perpendicular. The orientation of the normal plane section with respect to the tooth flanks $\mathscr{G}$ and $\mathscr{P}$ is specified by the corresponding central angle $\varphi$.

For the orthogonally parameterized gear tooth flank $\mathscr{G}$ and the mating pinion tooth flank $\mathscr{P}$ equation of the Dupin indicatrices $Dup(\mathscr{G})$ and $Dup(\mathscr{P})$ is simplified to:

$$L_g x_g^2 + 2M_g x_g y_g + N_g y_g^2 = \pm 1 \tag{E.51}$$

$$L_p x_p^2 + 2M_p x_p y_p + N_p y_p^2 = \pm 1 \tag{E.52}$$

After being represented in a common reference system, the use of Eqs. (E.51) and (E.52) makes it possible to create a simplified equation of the indicatrix of conformity $Cnf_R(\mathscr{G}/\mathscr{P})$ at a point of contact of the tooth flanks $\mathscr{G}$ and $\mathscr{P}$:

$$\begin{aligned} r_{cnf}(\varphi,\mu) = {} & (L_g\cos^2\varphi - M_g\sin 2\varphi + N_g\sin^2\varphi)^{-0.5}\,\mathrm{sgn}\,\Phi^{-1}_{2.g} \\ & + [L_p\cos^2(\varphi+\mu) - M_p\sin 2(\varphi+\mu) \\ & + N_p\sin^2(\varphi+\mu)]^{-0.5}\,\mathrm{sgn}\,\Phi^{-1}_{p.T} \end{aligned} \tag{E.53}$$

Equation (E.53) is valid for the orthogonally parameterized tooth flanks $\mathscr{G}$ and $\mathscr{P}$.

When designing gears, a reasonable degree of conformity $cnf_{\text{res}}$ at point of contact of the tooth flank of a gear $\mathscr{G}$ and the tooth flank of a mating pinion $\mathscr{P}$ is a trade-off between the desirable maximal degree of conformity $cnf_{\max}$ (max bearing capacity) and a minimal permissible degree of conformity $cnf_{\min}$ (min sensitivity to the axes misalignment)—that is, the inequality $cnf_{\min} < cnf_{\text{res}} < cnf_{\max}$ is observed.

[12] The diameter of a symmetrical planar curve that possesses a property of central symmetry can be defined as a distance between two points of the curve measured along the corresponding straight line through the center of symmetry of the curve.

## E.3.3 Vector and Matrix Representation of Equation of Indicatrix of Conformity

Equation (E.49) of indicatrix of conformity $Cnf_R(\mathscr{G}/\mathscr{P})$ at point of contact of tooth flank of a gear $\mathscr{G}$ and of a mating pinion $\mathscr{P}$ yields representation in both forms: in vector form and in matrix form.

Consider a unit tangent vector $\mathbf{t}_{cnf}(\varphi)$ through contact point $K$ in a common section of the contacting surfaces $\mathscr{G}$ and $\mathscr{P}$ by a normal plane. The unit tangent vector $\mathbf{t}_{cnf}(\varphi)$ is entirely situated in a common tangent plane through contact point $K$. In a common normal section, the radii of curvature of the tooth flanks $\mathscr{G}$ and $\mathscr{P}$ are designated as $R_g$ and $R_p$, respectively. For convenience, the following designation is used here: $r_g(\varphi) = \sqrt{R_g(\varphi)}$ and $r_p(\varphi+\mu) = \sqrt{R_p(\varphi+\mu)}$.

The position vector $\mathbf{r}_{cnf}$ of the indicatrix of conformity at the point of contact of the tooth flanks of a gear $\mathscr{G}$ and a mating pinion $\mathscr{P}$ can be expressed in terms of the parameters $r_g$ and $r_g$, as follows:

$$\mathbf{r}_{cnf}(\varphi+\mu) = [r_g(\varphi) - r_p(\varphi+\mu)]\cdot\mathbf{t}_{cnf}(\varphi) \tag{E.54}$$

In a conventional equation for $Cnf_R(\mathscr{G}\to\mathscr{P})$, the distance $r_{cnf}(\varphi)$ can be replaced with $\mathbf{r}_{cnf}$ after the corresponding expression for $Cnf_R(\mathscr{G}\to\mathscr{P})$ is multiplied by $\mathbf{t}_{cnf}(\varphi)$.

Equation (E.54) casts into:

$$\mathbf{r}_{cnf}(\varphi+\mu) = \mathbf{i}\cdot r_{cnf}(\varphi,\mu)\cos\varphi + \mathbf{j}\cdot r_{cnf}(\varphi,\mu)\sin\varphi \tag{E.55}$$

The unit vector $\mathbf{t}_{cnf}(\varphi)$ can be expressed in terms of the fundamental magnitudes of the first and the second order calculated at point of contact $K$ of tooth flanks of a gear $\mathcal{G}$ and a mating pinion $\mathcal{P}$.

Matrix representation of indicatrix of conformity $Cnf_R(\mathcal{G} \to \mathcal{P})$ follows immediately:

$$\mathbf{r}_{cnf}(\varphi, \mu) = \begin{bmatrix} r_{cnf}(\varphi, \mu)\cos\varphi \\ r_{cnf}(\varphi, \mu)\sin\varphi \\ 0 \\ 1 \end{bmatrix} \tag{E.56}$$

The equation of the indicatrix of conformity $Cnf_R(\mathcal{G} \to \mathcal{P})$ in matrix form [see Eq. (E.56)] is preferred when this characteristic curve ($Cnf_R(\mathcal{G} \to \mathcal{P})$) is involved in a multiple coordinate system transformation.

### E.3.4 Directions of Extremum Degree of Conformity at Point of Contact of Tooth Flanks of a Gear and a Mating Pinion

The directions along which the degree of conformity at the point of contact of a gear tooth flank $\mathcal{G}$ and a mating pinion tooth flank $\mathcal{P}$ is extremum—that is, the degree of conformity reaches either the maximal or the minimal of its value—are of prime importance for engineering applications. This issue is especially important for synthesizing a favorable gear pair.

The directions of the extremal degree of conformity of the interacting smooth regular tooth flanks $\mathcal{G}$ and $\mathcal{P}$—that is, the directions pointed along the extremal diameters $d_{cnf}^{\min}$ and $d_{cnf}^{\max}$ of indicatrix of conformity $Cnf_R(\mathcal{G}/\mathcal{P})$—can be found from the equation of the indicatrix of conformity $Cnf_R(\mathcal{G}/\mathcal{P})$. For the reader's convenience, the equation of this characteristic curve is represented in the form:

$$\begin{aligned} r_{cnf}(\varphi, \mu) &= \sqrt{|r_{1.g}\cos^2\varphi + r_{2.g}\sin^2\varphi|}\,\operatorname{sgn}\Phi_{2.g}^{-1} \\ &+ \sqrt{|r_{1.p}\cos^2(\varphi+\mu) + r_{2.p}\sin^2(\varphi+\mu)|}\,\operatorname{sgn}\Phi_{2.p}^{-1} \end{aligned} \tag{E.57}$$

Two directions within the common tangent plane are specified by the angles $\varphi_{\min}$ and $\varphi_{\max}$. These directions feature an extremum degree of conformity of the pinion tooth flank $\mathcal{P}$ to the gear tooth flank $\mathcal{G}$. Actually, the angles are the roots of the equation:

$$\frac{\partial}{\partial\varphi} r_{cnf}(\varphi, \mu) = 0. \tag{E.58}$$

It can easily be proven that, in a general case of contact of two smooth regular tooth flanks $\mathcal{G}$ and $\mathcal{P}$, the difference between the angles $\varphi_{\min}$ and $\varphi_{\max}$ is not equal to $0.5\pi$. This means that the equality

$$\varphi_{\min} - \varphi_{\max} = \pm 0.5\pi n \tag{E.59}$$

is not always observed, and in most cases the relationship:

$$\varphi_{\min} - \varphi_{\max} \neq \pm 0.5\pi n \tag{E.60}$$

is valid (here $n$ is an integer number). The condition [see Eq. (E.59)] $\varphi_{\min} = \varphi_{\max} \pm 0.5\pi n$ is fulfilled only in cases when the angle $\mu$ of the local relative orientation of the contacting surfaces $\mathcal{G}$ and $\mathcal{P}$ is equal to $\mu = \pm 0.5\pi n$, and thus the principal directions $\mathbf{t}_{1.g}$ and $\mathbf{t}_{2.g}$ of the gear tooth flank $\mathcal{G}$ and the principal directions $\mathbf{t}_{1.p}$ and $\mathbf{t}_{2.p}$ of the mating pinion tooth flank $\mathcal{P}$ are either aligned to each other or opposite.

This allows for the following statement:

> **Statement E.1.** *In a general case of contact of two smooth regular tooth flanks, the directions along which the degree of conformity of the tooth flanks $\mathcal{G}$ and $\mathcal{P}$ is extremal are not orthogonal to each other.*

This statement is important in engineering applications.

The solution to the equation $\partial r_{rel}(\varphi)/\partial\varphi = 0$ returns two extremal angles $\varphi_{\min}$ and $\varphi_{\max} = \varphi_{\min} + 90°$ [here $r_{rel}(\varphi)$ denotes the position vector of a point of the Dupin indicatrix at the point of the surface of relative curvature]. Eq. (E.58) allows for two solutions $\varphi_{\min}^{*}$ and $\varphi_{\max}^{*}$. Therefore, the extremal difference:

$$\Delta\varphi_{\min} = \varphi_{\min} - \varphi_{\min}^{*} \tag{E.61}$$

as well as the extremal difference:

$$\Delta\varphi_{\max} = \varphi_{\max} - \varphi_{\max}^{*} \tag{E.62}$$

can easily be calculated.

Generally speaking, neither the extremal difference $\Delta\varphi_{\min}$ nor the extremal difference $\Delta\varphi_{\max}$ is equal to zero. However, there are exceptions to this—say when the angle $\mu$ of local relative orientation of the surfaces $\mathcal{G}$ and $\mathcal{P}$ fulfills the relationship $\mu = \pm 0.5\pi n$.

Let us consider an example that illustrates Statement E.1.

Example. Let us analytically describe the contact geometry of two convex parabolic patches of the contacting tooth flanks $\mathcal{G}$ and $\mathcal{P}$ (see Figure E.12). The example pertains to finishing a helical involute gear using a disk-type shaving cutter. The design parameters of the gear and of the shaving cutter, along with the specified configuration of the gear and the shaving cutter, yield the following numerical data for the calculation. At the point $K$ of the tooth flanks' contact, the principal curvatures of the gear tooth flank $\mathcal{G}$ are equal to $k_{1.g} = 4mm^{-1}$ and $k_{2.g} = 0$. The principal curvatures of the mating pinion tooth flank $\mathcal{P}$ are equal to $k_{1.p} = 1mm^{-1}$ and $k_{2.p} = 0$. The angle $\mu$ of the local relative orientation of the tooth flanks $\mathcal{G}$ and $\mathcal{P}$ is equal to $\mu = 45°$.

Two approaches can be implemented for the analytical description of the contact geometry of the tooth flanks $\mathcal{G}$ and $\mathcal{P}$. The first one is based on implementation of the Dupin indicatrix of the surface of relative curvature. The

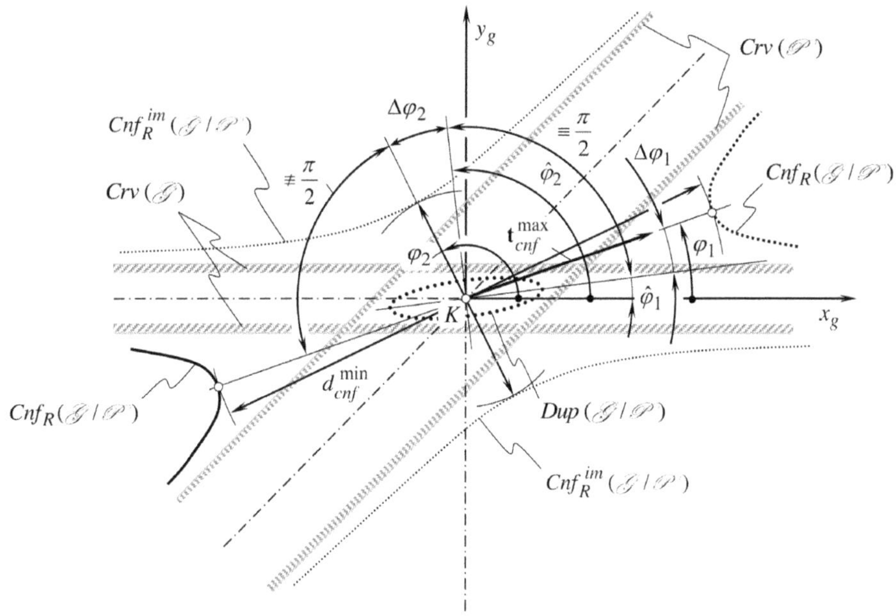


**FIGURE E.12** Example: Determination of the optimum instantaneous kinematics in a gear shaving operation.

second one is based on the application of the indicatrix of conformity $Cnf_R(\mathcal{G}/\mathcal{P})$ constructed at a contact point $K$ of the interacting tooth flanks $\mathcal{G}$ and $\mathcal{P}$.

The first approach. In the case under consideration, the normal curvature $k_{\mathcal{R}}$ of the surface of relative curvature $\mathcal{R}$ can be analytically expressed as:

$$k_{\mathcal{R}} = k_{1.g}\cos^2\varphi - k_{1.p}\cos^2(\varphi+\mu) \qquad \text{(E.63)}$$

Therefore, the following equality:

$$\frac{\partial k_{\mathcal{R}}}{\partial \varphi} = -2k_{1.g}\sin\varphi\cos\varphi + 2k_{1.p}\sin(\varphi+\mu)\cos(\varphi+\mu) = 0 \qquad \text{(E.64)}$$

is valid for the directions of the extremum degree of conformity of the tooth flaks $\mathcal{G}$ and $\mathcal{P}$ at every point of contact.

For the directions of the extremal degree of conformity at the point of contact of the gear tooth flank $\mathcal{G}$ and the mating pinion tooth flank $\mathcal{P}$, Eq. (E.64) yields a calculation of the extremal values $\varphi_{min} = 7°$ and $\varphi_{max} = \varphi_{min} + 90° = 97°$ for the angles $\varphi_{min}$ and $\varphi_{max}$.

The direction that is specified by the angle $\varphi_{min} = 7°$ indicates the direction of the minimal diameter of the Dupin indicatrix of the surface of relative curvature. That same direction corresponds to the maximal degree of conformity at the point of contact of the tooth flanks $\mathcal{G}$ and $\mathcal{P}$. Another direction specified by the angle $\varphi_{max} = 97°$ indicates the direction of the minimum degree of conformity of the contacting tooth flanks $\mathcal{G}$ and $\mathcal{P}$ at that same contact point.

The second approach. For the case under consideration, the use of Eq. (E.50) of the indicatrix of conformity $Cnf_R(\mathcal{G}/\mathcal{P})$ at the point of contact of the gear tooth flank $\mathcal{G}$ and the mating pinion tooth flank $\mathcal{P}$ makes the calculation of the extremal angles $\varphi^*_{min} = 19°$ and $\varphi^*_{max} = 118°$ possible.

Imaginary branches of the indicatrix of conformity $Cnf_R(\mathcal{G}/\mathcal{P})$ at the point of contact of the tooth flanks $\mathcal{G}$ and $\mathcal{P}$ in Figure E.12 are depicted in dashed lines.

It is important here to draw attention to two issues.

First, the extremal angles $\varphi_{min}$ and $\varphi_{max}$ calculated using the first approach are not equal to the corresponding extremal angles $\varphi^*_{min}$ and $\varphi^*_{max}$ calculated using the second approach. The relationships $\varphi_{min} \neq \varphi^*_{min}$ and $\varphi_{max} \neq \varphi^*_{max}$ are generally observed.

Second, the difference $\Delta\varphi^*$ between the extremal angles $\varphi^*_{min}$ and $\varphi^*_{max}$ is not equal to half of $\pi$. Therefore, the relationship $\varphi^*_{max} - \varphi^*_{min} \not\equiv 90°$ between the extremal angles $\varphi^*_{min}$ and $\varphi^*_{max}$ is observed. In a general case of contact of two sculptured surfaces, the directions of the extremal degree of conformity of the gear tooth flank $\mathcal{G}$ and the mating pinion tooth flank $\mathcal{P}$ are not orthogonal to each other.

The discussed example reveals that, in general cases of contact of two smooth regular tooth flanks, indicatrix of conformity $Cnf_R(\mathcal{G}/\mathcal{P})$ can be implemented for the purpose of accurate analytical description of the contact geometry of the surfaces. The Dupin indicatrix of the surface of relative curvature can be implemented for this purpose only in certain conditions of the relative orientation of surfaces $\mathcal{G}$ and $\mathcal{P}$. The application of the Dupin indicatrix of the surface of relative curvature enables only approximate analytical description of the geometry of the contact of the surfaces. The Dupin indicatrix of the surface of relative curvature could be equivalent to the indicatrix of conformity only in degenerate cases of surface contact. The advantages of the indicatrix of conformity over the Dupin indicatrix of the surface of relative curvature are due to this characteristic curve $Cnf_R(\mathcal{G}/\mathcal{P})$ being a curve of the fourth order.

### E.3.5 Important Properties of Indicatrix of Conformity $Cnf_R(\mathcal{G}/\mathcal{P})$ at the Point of Contact of the Tooth Flanks of a Gear and a Mating Pinion

The analysis of indicatrix of conformity $Cnf_R(\mathcal{G}/\mathcal{P})$ at the point of contact of a gear tooth flank and a mating pinion tooth flank performed in Eq. (E.50) reveals that this characteristic curve possesses the following important properties:

1. Indicatrix of conformity $Cnf_R(\mathcal{G}/\mathcal{P})$ at the point of contact of the tooth flanks $\mathcal{G}$ and $\mathcal{P}$ is a planar characteristic curve of the fourth order. It possesses the property of central symmetry and, in certain cases, the property of mirror symmetry.
2. Indicatrix of conformity $Cnf_R(\mathcal{G}/\mathcal{P})$ is closely related to the second fundamental forms $\Phi_{2.g}$ and $\Phi_{2.p}$ of surfaces $\mathcal{G}$ and $\mathcal{P}$, respectively. This characteristic curve is invariant with respect to the kind of parameterization of the tooth flanks $\mathcal{G}$ and $\mathcal{P}$, but its equation is not. A change in the parameterization of surfaces $\mathcal{G}$ and $\mathcal{P}$ leads to a change in the equation of indicatrix of conformity $Cnf_R(\mathcal{G}/\mathcal{P})$, while the shape and parameters of this characteristic curve remain unchanged.
3. The characteristic curve $Cnf_R(\mathcal{G}/\mathcal{P})$ is independent of the actual value of the coordinate angle $\omega_g$ that is formed by the coordinate lines $U_g$ and $V_g$ on the gear tooth flank $\mathcal{G}$. It is also independent of the actual value of the coordinate angle $\omega_p$ that is formed by the coordinate lines $U_p$ and $V_p$ on the mating pinion tooth flank $\mathcal{P}$. However, the parameters of indicatrix of conformity $Cnf_R(\mathcal{G}/\mathcal{P})$ depend on the angle $\mu$ of the local relative orientation of the tooth flanks $\mathcal{G}$ and $\mathcal{P}$. Therefore, for a given pair of the tooth flanks $\mathcal{G}$ and $\mathcal{P}$, the actual value of the degree of conformity of the surface varies correspondingly to variation of the angle $\mu$, while the pinion tooth flank $\mathcal{P}$ spins around the unit vector of the common perpendicular.

Additional properties of the indicatrix of conformity $Cnf_R(\mathcal{G}/\mathcal{P})$ at the point of contact of a gear tooth flank $\mathcal{G}$ and a mating pinion tooth flank $\mathcal{P}$ can be outlined.

### E.3.6 Converse Indicatrix of Conformity at Point of Contact of Tooth Flanks of a Gear and a Mating Pinion

For the Dupin indicatrix $Dup(\mathcal{G}/\mathcal{P})$ at the point of the surface of relative curvature $\mathcal{R}$, there exists a corresponding inverse Dupin indicatrix $Dup_k(\mathcal{G}/\mathcal{P})$. Similarly, for the indicatrix of conformity $Cnf_R(\mathcal{G}/\mathcal{P})$, at a point of contact of the tooth flanks $\mathcal{G}$ and $\mathcal{P}$, there exists a corresponding converse indicatrix of conformity $Cnf_k(\mathcal{G}/\mathcal{P})$. This characteristic curve can be expressed directly in terms of the normal curvatures $k_g$ and $k_p$ of surfaces $\mathcal{G}$ and $\mathcal{P}$:

$$Cnf_k(\mathcal{G}/\mathcal{P}) \Rightarrow r_{cnf}^{cnv}(\varphi,\mu) = \sqrt{|k_g(\varphi)|} \cdot \operatorname{sgn}\Phi_{2.g}^{-1} - \sqrt{|k_p(\varphi,\mu)|} \cdot \operatorname{sgn}\Phi_{2.p}^{-1} \quad \text{(E.65)}$$

For derivation of an equation of the converse indicatrix of conformity $Cnf_k(\mathcal{G}/\mathcal{P})$, Euler's formula for a surface's normal curvature is used in the following representation:

$$k_g(\varphi) = k_{1.g}\cos^2\varphi + k_{2.g}\sin^2\varphi \quad \text{(E.66)}$$

$$k_p(\varphi,\mu) = k_{1.p}\cos^2(\varphi+\mu) + k_{2.p}\sin^2(\varphi+\mu) \quad \text{(E.67)}$$

Here in Eqs. (E.66) and (E.67), the principal curvatures of the gear tooth flank $\mathcal{G}$ are designated as $k_{1.g}$ and $k_{2.g}$, while $k_{1.p}$ and $k_{2.p}$ designate the principal curvatures of the mating pinion tooth flank $\mathcal{P}$.

After substituting Eqs. (E.66) and (E.67) into Eq. (E.65), one can come up with the equation:

$$r_{cnf}^{cnv}(\varphi,\mu) = \sqrt{|k_{1.g}\cos^2\varphi + k_{2.g}\sin^2\varphi|}\operatorname{sgn}\Phi_{2.g}^{-1} - \sqrt{|k_{1.p}\cos^2(\varphi+\mu) + k_{2.p}\sin^2(\varphi+\mu)|}\operatorname{sgn}\Phi_{2.p}^{-1} \quad \text{(E.68)}$$

for the converse indicatrix of conformity $Cnf_k(\mathcal{G}/\mathcal{P})$ at the point of contact of the tooth flanks $\mathcal{G}$ and $\mathcal{P}$ in the first order of tangency.

In Eq. (E.68), principal curvatures $k_{1.g}$, $k_{2.g}$ and $k_{1.p}$, $k_{2.p}$ can be expressed in terms of the corresponding first-order fundamental magnitudes $E_g$, $F_g$, and $G_g$ and second-order fundamental magnitudes $L_g$, $M_g$, and $N_g$ of the gear tooth flank $\mathcal{G}$ and in terms of the corresponding first fundamental magnitudes $E_p$, $F_p$, and $G_p$ and second-order fundamental magnitudes $L_p$, $M_p$, and $N_p$ of the mating pinion tooth flank $\mathcal{P}$. Following this, Eq. (E.68) of the converse indicatrix of conformity $Cnf_k(\mathcal{G}/\mathcal{P})$ can be cast in a form that is similar to Eq. (E.50) of the ordinary indicatrix of conformity $Cnf_R(\mathcal{G}/\mathcal{P})$ at a point of contact of the tooth flanks $\mathcal{G}$ and $\mathcal{P}$.

It can be shown that, similar to the indicatrix of conformity $Cnf_R(\mathcal{G}/\mathcal{P})$, the characteristic curve $Cnf_k(\mathcal{G}/\mathcal{P})$ also possess the property of central symmetry and, in certain cases of the surfaces' contact, the property of mirror symmetry. The directions of the extremal degree of conformity of the gear tooth flank $\mathcal{G}$ and the mating pinion tooth flank $\mathcal{P}$ are orthogonal to each other only in degenerate cases of the surfaces contact.

Equation (E.68) of the converse indicatrix of conformity $Cnf_k(\mathcal{G}/\mathcal{P})$ is convenient for implementation when:

a. The gear tooth flank $\mathcal{G}$,
b. The mating pinion tooth flank $\mathcal{P}$, or
c. Both of them

feature point(s) or line(s) of inflection. In the point(s) or line(s) of inflection, the radii of normal curvature $R_{g(p)}$ of the surfaces $\mathcal{G}$ and $\mathcal{P}$ are equal to infinity. Points or lines of inflection cause indefiniteness when calculating the position vector $r_{cnf}(\varphi,\mu)$ of the point of the characteristic curve $Cnf_R(\mathcal{G}/\mathcal{P})$. Equation (E.68) of the converse indicatrix of conformity $Cnf_k(\mathcal{G}/\mathcal{P})$ is free of such disadvantages and it is therefore recommended for practical applications.

# Appendix F: The Closest Distance of Approach of Tooth Flanks of a Gear and a Mating Pinion

Generally, the problem of the calculation of the closest distance of approach between two smooth regular surfaces is a sophisticated and challenging one. As far as the author is aware, no general solution to the problem of calculation of the closest distance of approach between two smooth regular surfaces is available in the public domain. For the purpose of the calculation of the deviation $\delta_g$ of the actual gear tooth flank $\mathcal{G}_{ac}$ with respect to the desired (nominal) gear tooth flank $\mathcal{G}_{nom}$, the problem under consideration can be reduced to the problem of computation of the closest distance of approach between two torus surfaces $Tr_g$ and $Tr_p$.

Consider a gear tooth flank $\mathcal{G}$ and a mating pinion tooth flank $\mathcal{P}$ that initially are given in a common coordinate system $X_hY_hZ_h$ associated with the gear housing, as illustrated in Figure F.1. The tooth flanks $\mathcal{G}$ and $\mathcal{P}$ are locally approximated by portions of torus surfaces $Tr_g$ and $Tr_p$, respectively. Again, not all points of the torus surfaces $Tr_g$ and $Tr_p$ can be used for the local approximation of the gear tooth flanks $\mathcal{G}$ and the pinion tooth flanks $\mathcal{P}$. Only points that are located either within the greatest meridian or within the shortest meridian of the torus surface are employed for this purpose.

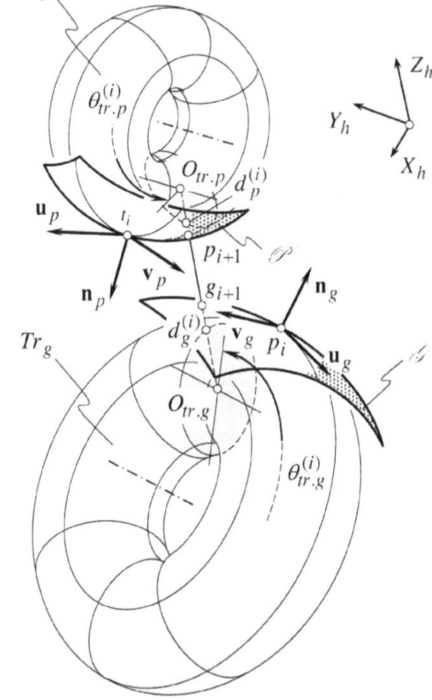

FIGURE F.1 Computation of the closest distance of approach between gear tooth flank $\mathcal{G}$ and mating pinion tooth flank $\mathcal{P}$.

The points $K_g$ and $K_p^*$ are chosen as the first guess points on the torus surfaces $Tr_g$ and $Tr_p$. For the following analysis, it is convenient to relabel the points $K_g$ and $K_p^*$ to $g_i$ and $p_i$, respectively.

For a given configuration of the torus surfaces $Tr_g$ and $Tr_p$, the closest distance of approach between these surfaces can be used as a first approximation to the closest distance of approach between the original gear tooth flanks $\mathcal{G}$ and the pinion tooth flanks $\mathcal{P}$.

The closest distance of approach between the torus surfaces $Tr_g$ and $Tr_p$ is measured along the common perpendicular to these surfaces. The following equations can be composed on the premise of this property of the closest distance of approach.

Unit normal vector $\mathbf{n}_{Tr.g}$ to the torus surface $Tr_g$ is located within a plane through the axis of rotation of the surface $Tr_g$. In the coordinate system $X_{tr.g}Y_{tr.g}Z_{tr.g}$ that is associated with the torus surface $Tr_P$, the equation of a plane through the axis of rotation of the torus surface $Tr_g$ can be expressed in the form:

$$[\mathbf{r}_{\tau g} - \mathbf{r}_{tr.g}^{(0)}] \times \mathbf{k}_{tr.g} \times \mathbf{R}_{tr.g} = 0 \tag{F.1}$$

with the following designations:

- $\mathbf{r}_{\tau g}$ is the position vector of the point of the plane through the axis of rotation of the torus $Tr_g$
- $\mathbf{r}_{tr.g}^{(0)}$ is the position vector of the point within the plane $\mathbf{r}_{\tau g}$ (it is assumed that this point coincides with the origin of the coordinate system $X_{tr.g}Y_{tr.g}Z_{tr.g}$)
- $\mathbf{k}_{tr.g}$ is the unit vector of the $Z_{tr.g}$ – axis

Equation F.1 is expressed in terms of the radius $\mathbf{R}_{tr.g}$. This indicates that the set of all planes through the fixed $Z_{tr.g}$ – axis forms a pencil of planes. The equation of the pencil of planes $\mathbf{r}_{\tau g}$ in the common coordinate system $X_hY_hZ_h$ can be represented in the form:

$$\mathbf{r}_{\tau g}(Z_{tr.g}, V_{\tau r.g}, \theta_{tr.g}) = \mathbf{Rs}(Tr_g \mapsto h) \cdot \begin{bmatrix} V_{tr.g} \cdot \cos\theta_{tr.g} \\ V_{tr.g} \cdot \sin\theta_{tr.g} \\ Z_{tr.g} \\ 1 \end{bmatrix} \tag{F.2}$$

The unit normal vector $\mathbf{n}_{Tr.p}$ to the torus surface $Tr_p$ is located within a plane through the axis of rotation of the surface $Tr_p$. In the coordinate system $X_{tr.p}Y_{tr.p}Z_{tr.p}$ that is associated with the surface $Tr_p$, the equation of a plane through the axis of rotation of the torus surface $Tr_p$ can be represented in the form:

$$[\mathbf{r}_{\tau p} - \mathbf{r}_{tr.p}^{(0)}] \times \mathbf{k}_{tr.p} \times \mathbf{R}_{tr.p} = 0 \qquad \text{(F.3)}$$

with the following designations:

- $\mathbf{r}_{\tau p}$ is the position vector of point of the plane through the torus $Tr_p$ axis of rotation
- $\mathbf{r}_{tr.p}^{(0)}$ is the position vector of point within the plane $\mathbf{r}_{\tau p}$ (it is assumed that this point coincides with the origin of the coordinate system $X_{tr.p}Y_{tr.p}Z_{tr.p}$)
- $\mathbf{k}_{tr.p}$ is the unit vector of the $Z_{tr.p}$ – axis

Equation F.3 is expressed in terms of the radius $\mathbf{R}_{tr.p}$. This indicates that the set of all planes through the fixed $Z_{tr.p}$ – axis forms a pencil of planes. The equation of this pencil of planes $\mathbf{r}_{\tau p}$ in the common coordinate system $X_hY_hZ_h$ can be represented in the form:

$$\mathbf{r}_{\tau p}(Z_{tr.p}, V_{tr.p}, \theta_{tr.p}) = \mathbf{Rs}(Tr_p \mapsto h) \cdot \begin{bmatrix} V_{tr.p} \cdot \cos\theta_{tr.p} \\ V_{tr.p} \cdot \sin\theta_{tr.p} \\ Z_{tr.p} \\ 1 \end{bmatrix} \qquad \text{(F.4)}$$

A straight line through the points $d_g^{(i)}$ and $d_p^{(i)}$, along which the shortest distance of approach $d_{gp}^{\min}$ of the torus surfaces $Tr_g$ and $Tr_p$ is measured, is the line of intersection of the planes $\mathbf{r}_{\tau g}$ and $\mathbf{r}_{\tau p}$. Therefore, this line $d_{gp}^{\min}$ must be aligned with both unit normal vectors $\mathbf{n}_{tr.g}$ and $\mathbf{n}_{tr.p}$.

In the coordinate system $X_hY_hZ_h$, the equation for the unit normal vector $\mathbf{n}_{tr.g}$ to the torus surface $Tr_g$ yields representation in matrix form:

$$\mathbf{n}_{tr.g} = \mathbf{Rs}(Tr_g \mapsto h) \cdot \begin{bmatrix} (C_{tr.g} + \cos\varphi_{tr.g}) \cdot \cos\varphi_{tr.g} \cdot \cos\theta_{tr.g} \\ (C_{tr.g} + \cos\varphi_{tr.g}) \cdot \cos\varphi_{tr.g} \cdot \sin\theta_{tr.g} \\ (C_{tr.g} + \cos\varphi_{tr.g}) \cdot \sin\varphi_{tr.g} \\ 1 \end{bmatrix} \qquad \text{(F.5)}$$

where $C_{tr.g}$ designates the parameter $C_{tr.g} = 1 - \dfrac{R_{2.g}}{R_{1.g}}$.

Similarly, in the coordinate system $X_hY_hZ_h$, the equation for the unit normal vector $\mathbf{n}_{tr.p}$ to the torus surface $Tr_p$ yields matrix representation in the form:

$$\mathbf{n}_{tr.p} = \mathbf{Rs}(Tr_p \mapsto h) \cdot \begin{bmatrix} (C_{tr.p} + \cos\varphi_{tr.p}) \cdot \cos\varphi_{tr.p} \cdot \cos\theta_{tr.p} \\ (C_{tr.p} + \cos\varphi_{tr.p}) \cdot \cos\varphi_{tr.p} \cdot \sin\theta_{tr.p} \\ (C_{tr.p} + \cos\varphi_{tr.p}) \cdot \sin\varphi_{tr.p} \\ 1 \end{bmatrix} \qquad \text{(F.6)}$$

where $C_{tr.p}$ designates the parameter $C_{tr.p} = 1 - \dfrac{R_{2.p}}{R_{1.p}}$.

Evidently, the points $O_{tr.g}$, $O_{tr.p}$, $d_g^{(i)}$, and $d_p^{(i)}$ (see Figure F.1), are located within the straight line through the centers, $O_{tr.g}$ and $O_{tr.p}$. The position vector $\mathbf{r}_{cd}$ of this straight line can be calculated from the equation:

$$(\mathbf{r}_{cd} - \mathbf{r}_{cg}) \times (\mathbf{r}_{cp} - \mathbf{r}_{cg}) = 0 \qquad \text{(F.7)}$$

where is designated:

- $\mathbf{r}_{cg}$ is the position vector of a point on the circle of a radius $R_{tr.g}$
- $\mathbf{r}_{cp}$ is the position vector of a point on the circle of a radius $R_{tr.p}$

It is necessary that the straight line $\mathbf{r}_{cd}$ be along the unit normal vectors $\mathbf{n}_{tg}$ and $\mathbf{n}_{tp}$ to the torus surfaces $Tr_g$ and $Tr_p$.

Considered together, Eqs. (F.2), (F.4), and (F.7) make the calculation of the closest distance of approach between the torus surfaces $Tr_g$ and $Tr_p$ possible. Then, the straight line $d_{gp}^{\min}$ intersects the gear tooth flank $\mathcal{G}$ and the pinion tooth flank $\mathcal{P}$ at the points $g_{i+1}$ and $p_{i+1}$, respectively. The points $g_{i+1}$ and $p_{i+1}$ serve as the second guess to the closest distance of approach between the surfaces $\mathcal{G}$ and $\mathcal{P}$.

The cycle of the recursive calculations is repeated as many times as necessary for making the deviation of the calculation of the closest distance of approach between the surfaces $\mathcal{G}$ and $\mathcal{P}$ smaller than the maximal permissible value.

There is an alternative approach for the calculation of the closest distance of approach between two torus surfaces. The direction of the unit normal vector to an offset surface to $Tr_g$ is identical to the direction of the unit normal vector $\mathbf{n}_{tr.g}$ to the torus surface $Tr_g$. This statement is also valid for the unit normal vector $\mathbf{n}_{tr.T}$ to the torus surface $Tr_p$. This property of the unit normal vectors $\mathbf{n}_{tr.g}$ and $\mathbf{n}_{tr.p}$ can be used for the modification of the method of calculation of the closest distance of approach between two torus surfaces.

The equation of the circle of radius $R_{tr.g}$ yields the matrix representation:

$$\mathbf{r}_{cg}(\theta_{tr.g}) = \mathbf{Rs}(\mathcal{G} \mapsto h) \cdot \begin{bmatrix} R_{tr.g} \cdot \cos\theta_{tr.g} \\ R_{tr.g} \cdot \sin\theta_{tr.g} \\ 0 \\ 1 \end{bmatrix} \qquad \text{(F.8)}$$

The equation of the circle of radius $R_{tr.p}$ can be analytically described in the similar way:

$$\mathbf{r}_{cp}(\theta_{tr.p}) = \mathbf{Rs}(\mathcal{P} \mapsto h) \cdot \begin{bmatrix} R_{tr.p} \cdot \cos\theta_{tr.p} \\ R_{tr.p} \cdot \sin\theta_{tr.p} \\ 0 \\ 1 \end{bmatrix} \qquad \text{(F.9)}$$

The distance $d_{gp}$ between two arbitrary points on the circles $\mathbf{r}_{cg}(\theta_{tr.g})$ and $\mathbf{r}_{cp}(\theta_{tr.p})$ equals:

$$d_{gp}(\theta_{tr.g}, \theta_{tr.p}) = |\mathbf{r}_{cg}(\theta_{tr.g}) - \mathbf{r}_{cp}(\theta_{tr.p})| \qquad \text{(F.10)}$$

The distance $d_{gp}$ is minimal for a specific (optimal) combination of the parameters $\theta_{tr.g}$ and $\theta_{tr.p}$. The favorable values of the parameters $\theta_{tr.g}$ and $\theta_{tr.p}$ can be calculated on solution of a set of two equations:

$$\frac{\partial}{\partial \theta_{tr.g}} \mathbf{r}_{cg}(\theta_{tr.g}) = 0 \quad \text{(F.11)}$$

$$\frac{\partial}{\partial \theta_{tr.p}} \mathbf{r}_{cp}(\theta_{tr.p}) = 0 \quad \text{(F.12)}$$

Following the solution of Eqs. (F.11) and (F.12), the optimal values $\theta_{tr.g}^{(opt)}$ and $\theta_{tr.p}^{(opt)}$ can be determined. These angels specify the direction of the closest distance of approach of the torus surfaces $Tr_g$ and $Tr_p$.

Following this method, the three-dimensional problem of calculation of the closest distance of approach of two torus surfaces is reduced to the problem of calculation of the closest distance of approach between two circles. In certain scenarios, the last approach could possess an advantage over the previous approach.

The convergence of the disclosed algorithms for the computation of the closest distance of approach between two smooth regular surfaces is illustrated in Figure F.2. The computation procedure is convergent regardless of the actual location of the first guess points on the surfaces $\mathscr{G}$ and $\mathscr{P}$.

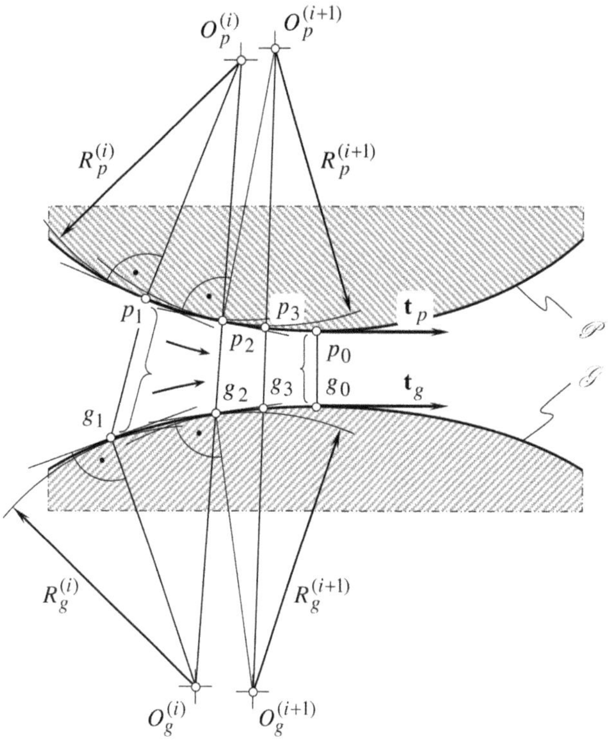


**FIGURE F.2** Convergence of the methods of calculation of the closest distance of approach of a gear tooth flank $\mathscr{G}$ and a mating pinion tooth flank $\mathscr{P}$.

It is instructive to draw attention here to the similarities between the disclosed iterative method for the computation of the closest distance of approach between two smooth regular surfaces and the Newton-Raphson method, the iterative method of chords, and so forth. Many similarities can be found out on this comparison.

# Appendix G: Engineering Formulae for Specification of Gear Tooth Flank

The engineering representation of a gear tooth flank can be converted into its scientific representation, and vice versa. For the conversion, it is convenient to use the so-called fundamental gear equations listed in Table G.1. More useful equations can be found out from many advanced sources.

**TABLE G.1**
**Fundamental Gear Equations**

| Name of the Parameter | Formulae |
|---|---|
| Transverse diametral pitch $P_t$ | $P_t = P_n \cos\psi$ |
| Pitch diameter $D$ | $D = \frac{N}{P_t}$ |
| Standard addendum $a$ | $a = \frac{1}{P_n}$ |
| Outside diameter $D_o$ | $D_o = D + 2 \cdot a$ |
| Transverse pressure angle $\phi_t$ | $\tan\phi_t = \frac{\tan\phi_n}{\cos\psi}$ |
| Base diameter $d_b$ | $d_b = D \cdot \cos\phi_t$ |
| Lead $L$ | $L = \pi \cdot D \cdot \cot\psi = \frac{\pi \cdot D}{\tan\psi}$<br>$L = \pi \cdot d_b \cdot \cos\psi_b$ |
| Normal circular pitch $p_n$ | $p_n = \frac{\pi}{P_n}$, $p_n = \frac{\pi \cdot D \cdot \cos\psi}{N}$ |
| Standard normal circular thickness $t_n$ | $t_n = \frac{p_n}{2}$, $t_n = t \cdot \cos\psi$ |
| Axial pitch $p_x$ | $p_x = \frac{\pi}{P_n \cdot \sin\psi} = \frac{p_n}{\sin\psi} = \frac{L}{N}$ |
| Transverse circular pitch $p_t$ | $p_t = \frac{\pi}{P_t} = \frac{p_n}{\cos\psi}$ |
| Helix angle, when the given center distance is standard $\psi$ | $\cos\psi = \frac{N_p + N_g}{2 \cdot p_n \cdot C}$ |
| Operating pitch diameter (pinion) $D_{rp}$; with non-standard center-distance $C$ | $D_{rp} = \frac{2 \cdot C \cdot N_p}{N_p + N_g}$ |
| Operating pressure angle, $\phi_{rt}$; with non-standard center-distance $C$, | $\sin\phi_{rt} = \frac{d_{bp} + d_{bg}}{2 \cdot C}$ |
| Normal diametral pitch $P_n$ | $P_n = P_t \cdot \sec\psi$ |
| Helix angle $\psi$ | $\cos\psi = \frac{N}{D \cdot P_n}$, $\sin\psi = \frac{\pi \cdot N}{P_n \cdot L}$ |
| Helix angle $\psi_y$ at any diameter $d_y$ | $\tan\psi_y = \frac{d_y}{d_1} \cdot \tan\psi_1$ |
| Transverse circular pitch $p_{t2}$ at any diameter $d_y$ | $p_{t2} = \frac{\pi \cdot d_y}{N}$ |
| Involute function of pressure angle | $\text{inv}\phi = \tan\phi - \breve{\phi}$ |
| Normal pressure angle $\phi_n$ | $\tan\phi_n = \tan\phi_t \cdot \cos\psi$<br>$\sin\phi_n = \sin\phi \cdot \cos\psi_b$<br>$\cos\phi_n = \sin\psi_b \cdot \csc\psi$ |

*(Continued)*

**TABLE G.1** ***(Continued)***

| Name of the Parameter | Formulae |
|---|---|
| Transverse pressure angle $\phi_{ty}$, at any diameter $d_y$ | $\cos\phi_{ty} = \frac{d_b}{d_y}$ |
| Base helix angle $\psi_b$ | $\cos\psi_b = \frac{\cos\psi \cdot \cos\phi_n}{\cos\phi_t} = \frac{\sin\phi_n}{\sin\phi_t}$ <br> $\sin\psi_b = \sin\psi \cdot \cos\phi_n$ <br> $\tan\psi_b = \tan\psi \cdot \cos\phi_t$ |
| Base pitch $p_b$ (here $\rho$ designates radius of profile curvature) | $p_b = \frac{\pi \cdot d_b}{N} = \frac{\pi \cdot D \cdot \cos\phi_t}{N} = \rho \cdot \cos\phi_t$ |

Formulae used for the conversion from the imperial (pitch) system to the metric system are summarized in Table G.2.

The formula $P = \frac{25.4}{m}$ is used to express diametral pitch $P$ in terms of module $m$. The expression:

$$m = \frac{25.4}{P} \tag{G.1}$$

is used for the inverse conversion. For the correspondence between millimeters and inches, the ratio:

$$\text{Millimeters } (mm) = \frac{\text{Inches}}{.03937} = 25.4 \text{ Inches} \tag{G.2}$$

$$\text{Inches} = .03937\,mm = \frac{mm}{25.4} \tag{G.3}$$

is valid.

A brief analysis of the types of gears used in industry, together with the analysis of local topology of gear tooth flanks to be machined, is necessary for the design of gear-cutting tools, especially gear-cutting tools for machining or finishing precision gears.

**TABLE G.2**
**Formulae for the Conversion from the Pitch System to the Metric System**

| Name of the Parameter | Imperial (inches) | Metric (millimeters) |
|---|---|---|
| Pitch diameter $D$ | $D = \frac{N}{P}$ | $D = m \cdot N$ |
| Addendum $a$ | $a = \frac{1}{P}$ | $a = m$ |
| Standard outside diameter $D_o$ | $D_o = D + 2 \cdot a$ | $D_o = D + 2 \cdot m$ |
| Base diameter $d_b$ | $d_b = D \cdot \cos\phi$ | |
| Circular pitch $p$ | $p = \frac{\pi}{P}$ | $p = \pi \cdot m$ |
| Standard circular tooth thickness $t$ | $t = \frac{p}{2}$ | |
| Average backlash per pair $B$ | $B = \frac{.040}{P}$ | $B = .040 \cdot m$ |

# Appendix H: On the Inadequacy of the Terms *Wildhaber-Novikov Gearing*, and *W-N Gearing*

*Wildhaber Helical Gearing* [100] and *Novikov Gearing* [122] are briefly discussed in this section, aiming to illustrate inadequacy of the terms *Wildhaber-Novikov Gearing* and *W-N Gearing*. As follows from the discussion, helical gearing proposed by Dr. E. Wildhaber must be referred to as Wildhaber Helical Gearing. Helical gearing proposed by Dr. M. Novikov must be referred to as Novikov Gearing. The extensively used terms *Wildhaber-Novikov Gearing* and *W-N Gearing* are meaningless and commonly are used by insufficiently knowledgeable people.

## H.1 *WILDHABER'S* HELICAL GEARING

Helical gearing with circular arc tooth profile [100] targets an improved power capacity of the gear pair. The invention relates to the tooth shape of gears, which run on parallel axes, and may be applied to helical gears, such as single helical gears and double helical gears or herringbone gears.

The purpose of the invention is threefold:

1. To provide helical gearing with an improved tooth contact, so as to lessen surface stresses and wear.
2. To provide helical gearing, which is capable of rapid and accurate production, and which may be ground without difficulty, if so desired.
3. To provide accurate gearing of circular tooth profile.

The invention is illustratively exemplified in the accompanying drawings, in which Figure H.1(a) is a side elevational view of the proposed gear showing parts thereof in section; Figure H.1(d) is a normal sectional view if Figure H.1(a) taken on the lines 2-2 of the latter figure; Figure H.1(g) is a side elevational view of a pair of gears constructed in accordance with the invention; Figure H.1(h) is a sectional view taken through a pair of gears; Figure H.1(b) and Figure H.1(c) are sectional views of milling cutters used in the manufacture of gear of the proposed design; Figure H.1(e) and Figure H.1(f) are elevational views of corresponding tools of rack shape, to be used in reciprocating machines for cutting helical gears in accordance with the invention; Figure H.1(k) and Figure H.1(l) are side elevational views of the improved gear showing a pair of grinding wheels in different operating positions, the wheels being set to grind opposite tooth surfaces; Figure H.1(m) is a view of a gear taken in normal section and showing the grinding wheels in operation position; Figure H.1(n) is a view of a mate pinion sowing the grinding wheels in operating position; Figure H.1(o) is a view of modified form of gear made in accordance with the invention; Figure H.1(j) is a sectional view taken through an internal gear and its pinion; Figure H.1 is a normal section through helical teeth of composite outline, constructed from the invention; Figure H.1(p) is a view of reciprocating tool of rack shape in operating position; and Figure H.1(s) is a view of a modified type of reciprocating tools, in position to start a cut on a herringbone gear.

Referring to the drawings, and particularly to Figure H.1(a) and Figure H.1(d), 1 denotes a helical gear having teeth 2 in contact with the teeth 3 of a mating pinion 4. In order to clearly illustrate the degree of contact between the teeth of the gear and pinion the tooth (4) is shown in section in Figure H.1(a).

It is customary to analyze helical gearing with reference to a normal section—that is, line 2-2 of Figure H.1(a), line 2-2 being normal to the helix of the pitch circle. Figure H.1(d) illustrates the said normal section 2-2 for both pinion 4 and gear 1.

It has been assumed as an example that the tooth profiles 6 of gear 1 are circular arcs of radii 7 and centers 8, in the shown normal section. Centers 8 are situated close to the pitch circle 9 of the gear.[1] The corresponding teeth of pinion 4 are so shaped as to allow rolling of the pitch circles 9 and 10 on each other, as well known to those skilled in the art.

When the gear tooth 2 is in the position shown, in Figure H.1(a) and Figure H.1(d), and its center is at 8, it contacts tooth 3 at point 11, which may be determined by a perpendicular to tooth 2 through point 12, point 12 being the contact point between the two pitch circles 9 and 10. The said perpendicular is in the present case the connecting line between point 12 and center 8 of the tooth profile.

Another position 2' of the gear tooth, and 3' of the corresponding pinion tooth are shown in dotted lines in Figure H.1(d). The tooth profiles contact here at a point 11', which can be determined like point 11. It will be noted that the contact point has traveled from 11 to 11' during a small angular motion of the gears.[2] The contact point has passed practically over the whole active profile during a turning

[1] Centers 8 need to be situated along the line of action *LA* that is a must. Otherwise, the condition of contact ($\mathbf{n} \cdot \mathbf{V}_\Sigma = 0$) is violated.

[2] Once the contact point has traveled from 11 to 11' during a small angular motion of the gears, this immediately reveals that the transverse contact ratio $m_p$ in the *Wildhaber's Helical Gearing* is greater than 0, that is $m_p > 0$ which is not permissible.

angle 13 of the gear, which angle corresponds to a fraction only of the normal pitch14, 14'. The said normal pitch equals the circle pitch of the shown normal section.

In gearing now in use, however, the tooth outline and the tooth proportions are so selected, that the contact of corresponding normal profiles lasts for an angle, which, as a rule, corresponds to more than one full pitch.

In gearing according to the invention, the contact point between two normal profiles passes over the whole active profile during a turning angle, which corresponds to less than one half the normal pitch, and usually to much less than that.

Gearing designed according to the invention, allows the teeth to come into better contact with each other, inasmuch as the tooth surfaces remain much closer to each in a direction perpendicular to the contact line between two mating gears. This is illustrated by a section taken in direction of lines 15, 15' of Figure H.1(d). In Figure H.1(a) the lateral profile 16 of tooth 3 and profile 17 of tooth 2 of said section are shown to contact at point 11, and to remain close to each other on their whole length. The same holds true for other sections, taken parallel to section 15, 15'.

Close contact between teeth is well known to reduce wear and to improve the efficiency of the gears.

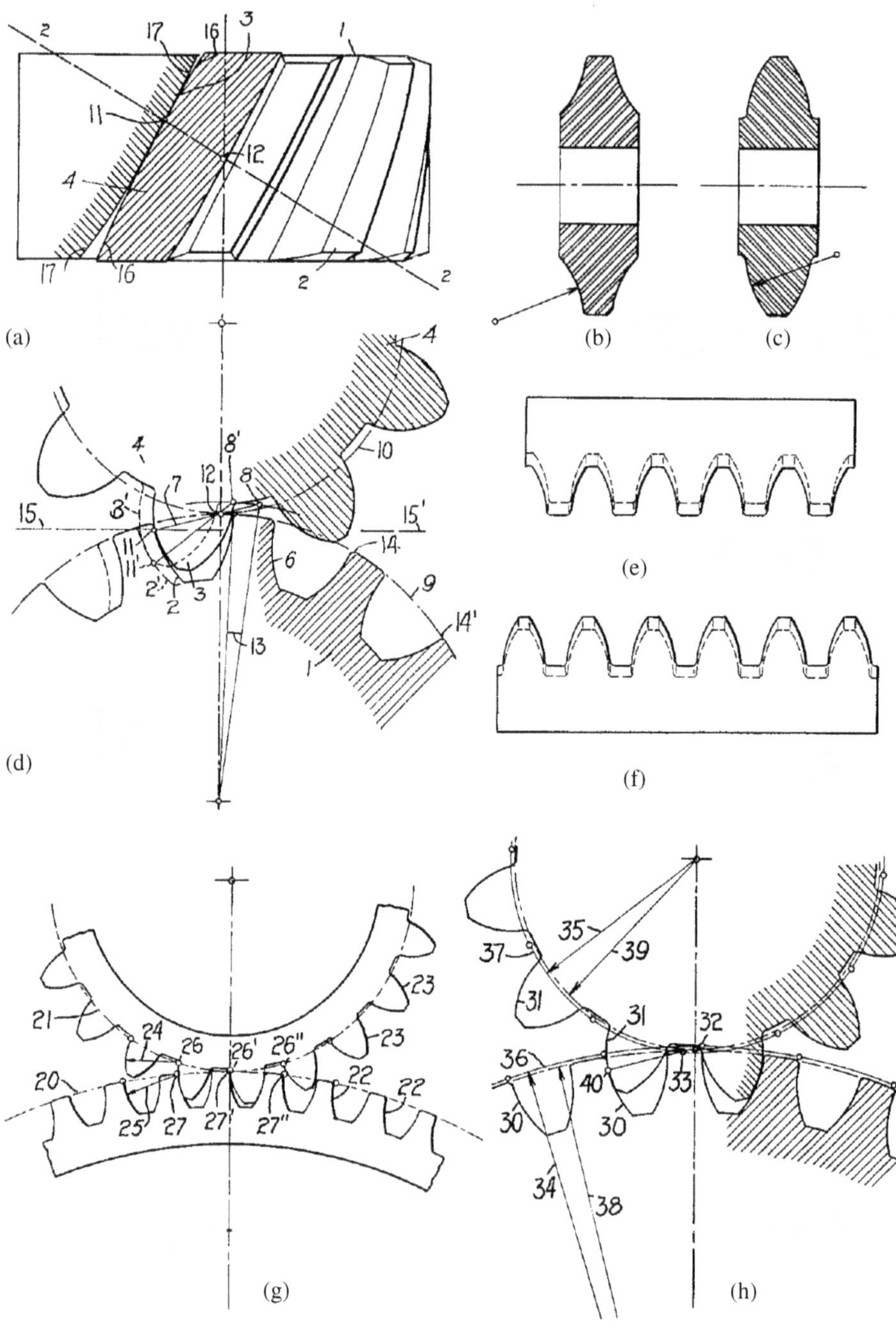


**FIGURE H.1** (a–s) *Helical Gearing.* (After E. Wildhaber., US Patent 1,601,750, Patented: October 5, 1926, Filed: November 2, 1923). *(Continued)*

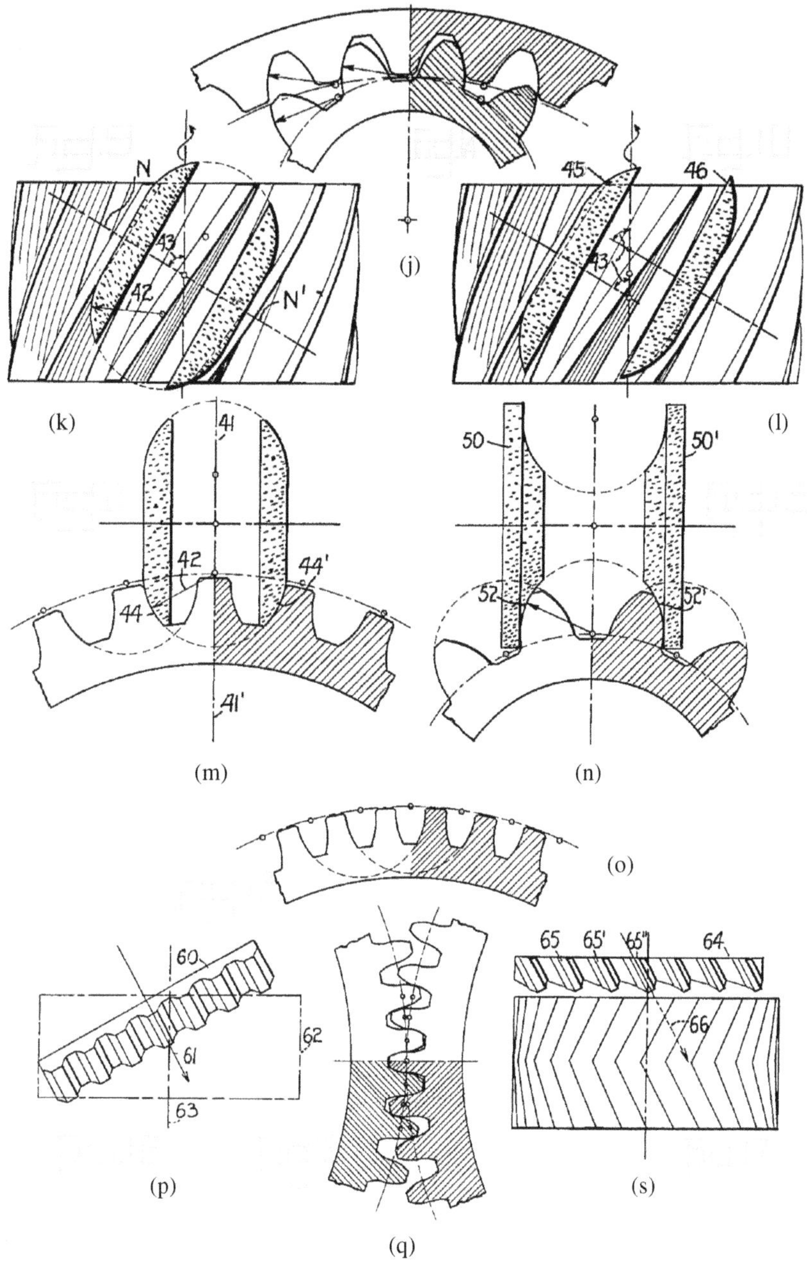


**FIGURE H.1** *(Continued)*

Although a circular arc is shown as the normal tooth profile of gear 1 in Figure H.1(d), it will be understood that this is not the only shape to affect the stated purpose of increasing the speed at which the contact point travels over the tooth profile of a normal section. As a rule, however, the shape can be approximated by a circle, whose center is close to the pitch center.

The gearing according to the present invention is strictly a gearing for helical teeth. It would not be advisable on straight teeth, on account of the explained short duration of contact between tooth profiles. This would cause intermittent action, whereas on helical gears similar parts of the teeth are always in contact, on account of the twisted nature of the tooth surfaces.

Figure H.1(g) may be considered as a view taken in the direction of the axes of a pair of gears. The tooth profiles are the circles in a section, which is perpendicular to the axes. The gear is provided with helical teeth, with working faces below the pitch circle 20, while the pinion teeth have working faces above the pitch circle 21 only. The working profiles 22 of the gear are concave and circular, and their centers are substantially situated on the pitch circle 20. The convex working profiles 23 of the pinion are also of circular shape. Their radii 24 are substantially the same as the radii 25 of the mate profiles. The centers 26, 26’, 26” are similarly situated on pitch circle 21. Profile centers 27, 27’, 27” of pitch circle 20, and profile centers 26, 26’, 26” of pitch circle 21 correspond to each other. They coincide during the mesh, which takes place on the whole tooth profile at once.

Figure H.1(g) may also be considered as a section perpendicular to the helical teeth and shows then the normal tooth profiles.

Figure H.1(h) shows a refinement of the preferred embodiments of the invention. It is a normal section through the helical teeth but can also be considered as a section perpendicular to the axes. Corresponding profiles 30 and 31 are circular, as in Figure H.1(g), but in this case the radius of the concave circular profile 30 is made a trifle larger than the radius of the convex circular profile 31. Consequently, the profile centers 32 and 33 do not exactly coincide during the mesh. The radii 34 and 35 of the circles 36 and 37, constituted by the profile centers 32 and 33, respectively, are not accurately identical with the pitch radii 38 and 39 of the two gears. The sum of the radii 34 and 35 is a trifle larger than the sum of the pitch radii. The radii 34 and 35 are so selected that the main tooth pressure runs about in a direction 33, 40.

The slight difference of the radii of profiles 30 and 31 facilitates the tooth contact, and allows for small errors in making and assembling.

Figure H.1(b) and (c) show a pair of milling cutters for milling gear teeth. The cutter may be applied in the usual manner, their axis being inclined in correspondence with the tooth inclination—that is, with the helix angle of teeth. It will be found that the cutters to be inclined for an angle, which is a trifle smaller than the helix angle in the pitch circle, for producing most accurate results.

In Figure H.1(e) and Figure H.1(f) a pair of rack shaped cutters is shown.[3] These cutters are for use in a reciprocating machine. The teeth of these tools are relieved inwardly, in the usual manner, as evident by the dotted lines.

The convex grinding wheels shown in Figure H.1(k) are illustrated in their operating positions, in a view which is taken perpendicular to the axis of the gear blank as well as to the axis of the grinding wheels, that is, in a view along the gear radii 41, 41' of Figure H.1(m). The wheels that are to produce concave circular teeth profiles in a normal section are of convex circular profile, with its radius 42 being the same as the radius of the concave circular profile. The grinding wheels are inclined for an angle 43, which equals the helix angle of the teeth, in the pitch circle. The wheels grind along their profiles indicated in dotted lines 44 and 44', which are located in a normal section. As shown in Figure H.1(k), the two grinding wheels are coaxially arranged with respect to each other.

The device shown in Figure H.1(l) corresponds to that shown in Figure H.1(k), with the exception that the grinding wheels 45 and 46 are not coaxially arranged. Although the arrangement shown in Figure H.1(k) imposes certain restrictions on the tooth design, it is frequently preferred. The arrangement of Figure H.1(l) is advantageous, when grinding wheels are not free to run out, for instance when they must clear against a shoulder, or in the case of herringbone teeth.

Referring particularly to Figure H.1(m), a normal section is illustrated and taken along lines N, N' of Figure H.1(k). In this view, the axis of the coaxially arranged grinding wheels is situated in the said normal section. The wheels grind along the profiles 44 and 44' of the shown normal section, while the blank performs a translator motion in the direction of its axis, and, in timed relation thereto, a turning motion about its axis. In other words, the blank is screwed past the grinding wheels.

Figure H.1(n) discloses a normal section through the teeth of the mating gear or pinion. Grinding wheels 50 and 50' are provided with concave circular profiles 52 and 52' with which they grind the convex gear teeth.

It will be understood that milling cutters might be used instead of the grinding wheels shown in Figure H.1(k) through Figure H.1(n); and also that grinding wheels of a shape shown in Figure H.1(b) and Figure H.1(c) might be used, if so desired.

The teeth ground according to Figure H.1(k), Figure H.1(m), and Figure H.1(n) are preferably so designed, that the centers of opposite tooth arcs 44 and 44', 52 and 52', respectively, in Figure H.1(n) coincide. In Figure H.1(m) and Figure H.1(n), the tooth arcs of every third tooth side have a common center.

The tooth arcs of every fifth tooth side have a common center in the normal section shown in Figure H.1(o).

In Figure H.1(m), the common center of opposite tooth arcs of alternate teeth is situated on the centerline of the intermediate tooth. The corresponding pinion shows convex circular profiles, of which opposite tooth sides of adjacent teeth have common centers in the middle of the intermediate tooth space.

The normal section shown in Figure H.1(j) shows an internal gear and its mate pinion, constructed in accord ace with the concave tooth profiles. Similarly, preference is given in external gears to providing the larger gear with concave tooth profiles.

The normal section through a pair of helical gears shown in Figure H.1(q) discloses opposite tooth profiles, the addendum being convex and the dedendum concave.

A rack shaped planing tool is illustrated in operating position in Figure H.1(p). Tools of this kind have been shown in another view in the Figure H.1(e) and Figure H.1(f). The reciprocal tool 60 moves in the direction 61, at an inclination, which equals the helix angle of the teeth. Gear 62, with its axis 63, is shown in dotted and dash lines. In order to cut the proper tooth shape, gear blank 62 after every cut is slightly fed in a rolling generating motion with respect to a rack which is embodied by tool 60.

Another reciprocal tool 64 is shown in Figure H.1(s), the tool in this case being provided with stepped teeth 65, 65', 65", which allow it to clear shoulders and herringbone teeth. The tool moves in the direction 66 of the helical teeth, which cuts.

[3] Conformal gears cannot be cut by rack cutters, and, more generally, they cannot be machined in any gear generating process.

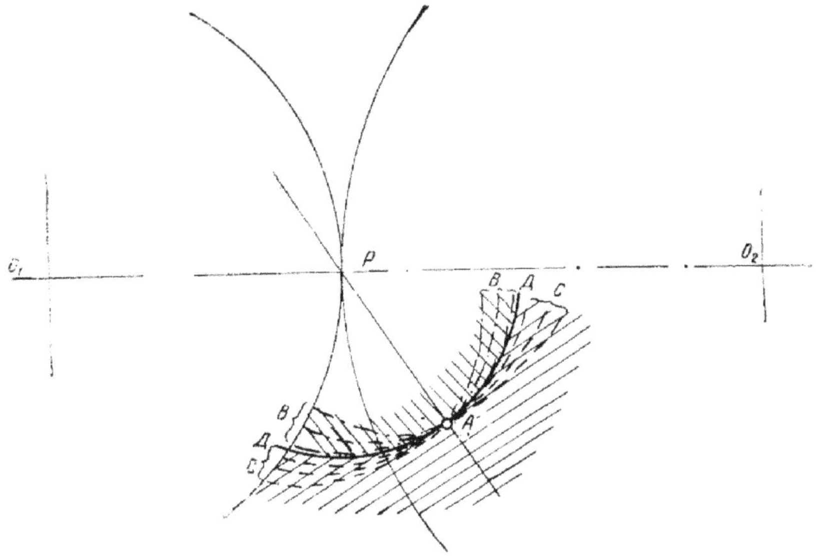


**FIGURE H.2** *Helical gearing* (After Dr. *M.L. Novikov*, Pat. No. 109,113, (USSR). *Gear Pairs and Cam Mechanisms Having Point System of Meshing*/National Classification 47h, 6; Filed: April 19, 1956, published in Bull. of Inventions No. 10, 1957).

Other ways of producing gearing according to the invention—that is, hobbing, planing with a pinion cutter, rolling, and casting—may be contemplated, but it is not deemed necessary at this time to give a detailed explanation of the mechanism used in connection therewith.

Briefly stated, the invention consists in providing helical gearing of such profile that the tooth contact passes rapidly over the normal profile of the teeth. This has been found to result in close contact between helical mate teeth. In a direction at right angles to the contact line, the mate teeth recede from each other only slightly, and thus provide a tooth contact, which is not very far from surface contact.

It is claimed,[4] the Wildhaber's gearing features a nonzero transverse contact ratio ($m_p > 0$) and a nonzero face contact ratio $m_F > 0$. The total contact ratio $m_t$ is greater than one ($m_t \equiv m_p + m_F > 1$).

The Wildhaber's gearing (see Figure H.1) does not meet the condition of contact. The condition of conjugacy of the tooth flanks, and the condition of equality of the base pitches of the gear and the pinion to the operating base pitch are not fulfilled in this design of gearing. This immediately yields a conclusion that Wildhaber's gearing is not workable at all.

The invention by Wildhaber is explained to the best possible extent in [155, 156].

## H.2 NOVIKOV GEARING

Novikov Gearing [122] is an example of conformal helical gearing. For a long while, Novikov's patent (S.U. Patent No. 109,113 of 1956) was not available to the most of gear experts.

Known designs of gearing, those featuring point system of meshing, feature low contact strength and are not widely used in practice.

The contact strength of known designs of gearing with a line system of meshing, including the widely used involute gearing, is limited.

The proposed design of gearing[5] [122] features higher contact strength due to favorable curvatures of the interacting tooth flanks. Under equivalent contact stress, similar dimensions, and comparable remaining design parameters, greater circular forces are permissible by the proposed gearing. Lower sensitivity to manufacturing errors and to deflections under the load is the other advantage of the proposed gearing.

The proposed design of gearing can be designed either with parallel, intersecting, or crossing axes of rotations of the gears. External gearing as well as internal gearing of the proposed system of meshing is possible. The tooth ratio of the proposed gearing can be either of constant value or it can be variable, and time dependent. The proposed concept of gearing can be utilized in the design of cam mechanisms.

In Figure H.2, possible tooth profiles in the cross-section of tooth flanks by a plane that is perpendicular to the instant axis of relative rotation through the current point of contact are illustrated.

Here, the point of intersection of the planar cross-section by the axis of instant relative rotation is denoted by $P$.

$O_1$ and $O_2$ are the points of intersection of the planar cross-section by the axes of the gear and the pinion.

$A$ is the point of meshing (in its current location).

$PA$ denotes the line of action.

[4] Wildhaber, E., *Helical Gearing*, US Patent 1,601,750, Patented: October 5, 1926, Filed: November 2, 1923.

[5] Pat. No. 109,113, (USSR). *Gear Pairs and Cam Mechanisms Having Point System of Meshing*. /M.L. Novikov, National Classification 47h, 6; Filed: April 19, 1956, published in Bull. of Inventions No. 10, 1957.

*ДАД* is the circle centered at the point *P* which corresponds to the limit case of the tooth profiles (in the case the profiles are aligned to each other).

Several curves *BAB* represent examples of the tooth profiles of one of the mating gears. The curves *BAB* are arbitrary smooth curves, which are located inside of the circular arc *ДАД*—that is, the arcs are located within the bodily side of the limit tooth flank of one of the gears. The curves *BAB* are located close to the circular arc *ДАД*, and they feature high degree of conformity to the circular arc.

Several curves *CAC* represent examples of the tooth profiles of the second of the mating gears. The curves *CAC* are arbitrary smooth curves, which are located outside of the circular arc *ДАД*—that is, the arcs are located within the bodily side of the limit tooth flank of another of two gears. The curves *CAC* are also located close to the circular arc *ДАД* and they feature high degree of conformity to the circular arc.

The entity of the invention is disclosed below in detail.

The location and orientation either of a straight path of contact or a smooth curved path of contact is specified in space in which the location and orientation of the axes of rotations of the gear and the pinion are given. The path of contact is located reasonably close to the axis of instant relative rotation of the gears. Either constant or time-dependent (smoothly varying in time) speed of motion of the contact point along the path of contact is assigned. A coordinate system is associated with the gear, and a corresponding coordinate system is associated with the pinion. In the coordinate systems, the moving contact point traces the so-called contact lines.[6] One of the paths of contact is associated with the gear and another one is associated with the pinion. Certain smooth regular surfaces through the pseudo-paths of contact can be used as the tooth flanks of the gear and the pinion. The following requirements should be fulfilled in order to the surfaces can be used as the tooth flanks:

- At every location of the contact point, the tooth flanks should have a common perpendicular, and thus, the requirements of the main theorem of meshing should be fulfilled.
- The curvatures of the tooth profiles should correspond to each other.
- No tooth flanks interference occurs within the working portions of the surfaces.

The tooth flanks of the proposed geometry fulfill these listed requirements and allow for high contact strength of the gear teeth.

Consider a plane through the current contact point that is perpendicular to the instant axis of relative rotation. Then, two circular arcs are constructed. The circular arcs are centered at points within the straight line through the pitch point and the contact point. The arcs centers are located close to the pitch point. The constructed circular arcs can be considered an example of the tooth profiles of the gear and the pinion. The tooth flanks are generated as loci of the tooth profiles constructed for all possible locations of the contact point. The working portion of one of two tooth flanks is convex, while the working portion of another tooth flank is concave (in the direction towards the axis of instant relative rotation). In a particular case, the radii of tooth profiles could be of the same magnitude and equal to the distance from the contact point to the axis of instant relative rotation. The centers of both profiles in this particular case are located at the axis of instant relative rotation. Under such a scenario, point meshing reduces to a special line meshing. This would require an extremely high accuracy of the center distance and independence of it from operation conditions, which is impractical. Point meshing is preferred when designing tooth profiles. A small difference between the radii of curvature of the tooth profiles is desirable. It should be kept in mind that during a run-in period of time, point meshing of the gear teeth will transform to the aforementioned locally line meshing of the tooth profiles. However, the theoretical point contact of the tooth flanks will be retained.

Tooth profiles can differ from the circular arcs. However, the tooth profiles of other geometries (those always passing through the contact point) should be located (for one gear) within the interior of the above-mentioned circular arc profile that centers at the point within the axis of instant relative rotation, as shown in the Figure H.2. For another gear, the tooth profile should be located outside the circular arc.

The law of motion of the contact point—that is, speed of the point and its trajectory—should be chosen so as to minimize the friction and wear losses. The friction and wear loss are proportional to the relative sliding velocity in the gear mesh. Therefore, it is desirable to reduce the sliding velocity as much as possible. For this purpose, the path of contact should be located not far from the axis of instant relative rotation. On the other hand, a location of the path of contact that is too close to the axis of instant relative rotation is also not desirable as that reduces the contact strength of the gear tooth flanks. In addition, it is recommended to ensure favorable angles between the common perpendicular (along which the tooth flanks of one of the gears acts against the tooth flank of another gear) and between the axes of rotations of the gears.

Opposite sides of tooth profiles are designed in a similar manner to that just discussed. Tooth thicknesses and angular pitch are assigned to ensure the required bending tooth strength.

The face width of the gear or the length of the gear teeth should correlate to their pitch to ensure the required value of the face contact ratio. Gear pairs can feature either one point of contact (when working portions of the tooth flank contact each other just in one point, excluding the phases of the teeth re-engagement), or they can feature multiple contact points when the tooth flanks contact each other at several points simultaneously.

---

[6] The *contact line* is a term used by Dr. *M.L. Novikov*. Actually, the *contact line* is the *pseudo-path of contact*.

For parallel-axes gear pairs, it is preferred to use a straight line as the path of contact, which is parallel to the axes of rotations of the gear and the pinion. The speed of the contact point along the straight path of contact can be of a constant value. In this particular case, the radii of curvature of the tooth profiles in all cross-sections by planes are equal to each other. Tooth flanks in this case are regular screw surfaces. Gears that feature tooth flanks of such geometry are easy for manufacture, and they can be cut on machine tools available on the market.

An example of parallel-axes gearing with limit geometry of the tooth profiles is illustrated in the Figure H.2. Point contact of the tooth flanks in this particular case is transformed to almost line contact of the tooth flanks. The curved contact line is located across the tooth profile. When axial thrust in the gear pair is strongly undesirable, herringbone gears can be used instead.

Novikov gearing (see Figure H.2) meets the condition of contact, as well as the condition of conjugacy of the tooth surfaces and the condition of equality of the base pitches of the gear and the pinion to the operating base pitch. Novikov gearing is a kind of perfect gearing.[7] Novikov gearing is the only feasible type of gearing with point contact of the tooth flanks that is capable of transmitting a rotation smoothly. As it is evident now, Novikov gearing is a reduced type of involute gearing.

[7] It needs to be noticed here that Dr. E. Wildhaber did not recognized the difference between Novikov gearing and the helical gearing proposed by him [100]. This conclusion immediately follows from Wildhaber's statement: "I may say also that it gives me satisfaction to see the original concept vindicated through the Russian reinvention and effort and through subsequent efforts and articles." Wildhaber's statement can be found on page 949 in the paper by T. Allan [Allan, T., "Some Aspects of the Design and Performance of Wildhaber-Novikov Gearing", *Proc. Inst. Mech. Engrs, Part I*, v. 179, n. 30, 1964/1965, pp. 931–954.] (see the *Communications* section). Other evidences in this concern are also known.

# Notation

$Cnf_R(\mathscr{G}/T)$ is the *indicatrix of conformity* at point of contact $K$ of the work gear tooth surface $\mathscr{G}$ and the generating surface $T$ of the gear cutting tool.

$C_{1.g}$, $C_{2.g}$ are the first and the second principal plane sections of the gear tooth flank $\mathscr{G}$.

$C_{1.T}$, $C_{2.T}$ are the first and the second principal plane sections of generation surface $T$ of the gear cutting tool.

$\mathscr{E}$ is the *characteristic line.*

$E_g$, $F_g$, $G_g$ are the *fundamental magnitudes of the first order* of the gear tooth surface $\mathscr{G}$.

$E_T$, $F_T$, $G_T$ are the *fundamental magnitudes of the first order* of generating surface $T$ of the gear cutting tool.

$\mathbf{Eu}(\psi,\theta,\varphi)$ is the operator of the *Eulerian transformation.*

$\mathscr{G}$ is the gear tooth flank.

$K$ is the point of contact of the tooth flanks $\mathscr{G}$ and $T$ (or a point within a line of contact of the tooth flanks $\mathscr{G}$ and $T$).

$L_g$, $M_g$, $N_g$ are the *fundamental magnitudes of the second order* of the gear tooth flank $\mathscr{G}$.

$L_T$, $M_T$, $N_T$ are the *fundamental magnitudes of the second order* of generating surface $T$ of the gear cutting tool.

$P_{ln}$ is the *pitch line* in the gear machining mesh.

$\mathbf{Rl}_x(\varphi_y, Y)$ is the *operator of rolling* over a plane ($Y$-axis is the axis of rotation, $X$-axis is the axis of translation).

$\mathbf{Rl}_z(\varphi_y, Y)$ is the *operator of rolling* over a plane ($Y$-axis is the axis of rotation, $Z$-axis is the axis of translation).

$\mathbf{Rl}_y(\varphi_x, X)$ is the *operator of rolling* over a plane ($X$-axis is the axis of rotation, $Y$-axis is the axis of translation).

$\mathbf{Rl}_z(\varphi_x, X)$ is the *operator of rolling* over a plane ($X$-axis is the axis of rotation, $Z$-axis is the axis of translation).

$\mathbf{Rl}_x(\varphi_z, Z)$ is the *operator of rolling* over a plane ($Z$-axis is the axis of rotation, $X$-axis is the axis of translation).

$\mathbf{Rl}_y(\varphi_z, Z)$ is the *operator of rolling* over a plane ($Z$-axis is the axis of rotation, $Y$-axis is the axis of translation).

$\mathbf{Rr}_u(\varphi, Z)$ is the *operator of rolling* of two coordinate systems.

$\mathbf{Rs}(A \mapsto B)$ is the *operator of the resultant coordinate system transformation*, say from the coordinate system $A$ to the coordinate system $B$.

$\mathbf{Rt}(\varphi_x, X)$ is the *operator of rotation* through an angle $\varphi_x$ about the $X$-axis.

$\mathbf{Rt}(\varphi_y, Y)$ is the *operator of rotation* through an angle $\varphi_y$ about the $Y$-axis.

$\mathbf{Rt}(\varphi_z, Z)$ is the *operator of rotation* through an angle $\varphi_z$ about $Z$-axis.

$R_{1.g}$, $R_{2.g}$ are the *first and second principal radii of curvature* of the work gear tooth flank $\mathscr{G}$.

$R_{1.T}$, $R_{2.T}$ are the *first and second principal radii of curvature* of the generating surface $T$ of the gear cutting tool.

$\mathbf{Sc}_x(\varphi_x, p_x)$ is the operator of screw motion about the $X$-axis.

$\mathbf{Sc}_y(\varphi_y, p_y)$ is the operator of screw motion about the $Y$-axis.

$\mathbf{Sc}_z(\varphi_z, p_z)$ is the *operator of screw motion* about the $Z$-axis.

$\mathbf{Tr}(a_x, X)$ is the *operator of translation* at a distance $a_x$ along the $X$-axis.

$\mathbf{Tr}(a_y, Y)$ is the *operator of translation* at a distance $a_y$ along the $Y$-axis.

$\mathbf{Tr}(a_z, Z)$ is the *operator of translation* at a distance $a_z$ along the $Z$-axis.

$U_g$, $V_g$ are the curvilinear (*Gaussian*) coordinates of a point on the work gear tooth flank $\mathscr{G}$.

$U_T$, $V_T$ are the curvilinear (*Gaussian*) coordinates of a point of generating surface $T$ of the gear cutting tool.

$\mathbf{U}_P$, $\mathbf{V}_P$ are the tangent vectors to curvilinear coordinate lines on the work gear tooth flank $\mathscr{G}$.

$\mathbf{U}_T$, $\mathbf{V}_T$ are the tangent vectors to curvilinear coordinate lines on generating surface $T$ of the gear cutting tool.

$\mathbf{V}_\Sigma$ is the vector of the resultant motion of generating surface $T$ of the gear cutting tool in relation to the tooth flank $\mathscr{G}$ of the work gear.

$k_{1.g}$, $k_{2.g}$ are the first and second principal curvatures of the work gear tooth flank $\mathscr{G}$.

$k_{1.T}$, $k_{2.T}$ are the first and second principal curvatures of generating surface $T$ of the gear cutting tool.

$\mathbf{n}_g$ is the unit normal vector to the work gear tooth flank $\mathscr{G}$.

$\mathbf{n}_T$ is the unit normal vector to generating surface $T$ of the gear cutting tool.

$r_{cnf}$ is the position vector of a point of the indicatrix of conformity $Cnf_R(G/T)$.

$\mathbf{r}_g$ is the position vector of a point of the work gear tooth flank $\mathscr{G}$.

$\mathbf{t}_{1.g}$, $\mathbf{t}_{2.g}$ are the *unit tangent vectors of principal directions* at point on the work gear tooth flank $\mathscr{G}$.

$\mathbf{t}_{1.T}$, $\mathbf{t}_{2.T}$ are the *unit tangent vectors of principal directions* at point on generating surface $T$ of the gear cutting tool.

$\mathbf{u}_g$, $\mathbf{v}_g$ are the *unit tangent vectors* to curvilinear coordinate lines on the work gear tooth flank $\mathscr{G}$.

$\mathbf{u}_T$, $\mathbf{v}_T$ are the *unit tangent vectors* to curvilinear coordinate lines on generating surface $T$ of the gear cutting tool.

$x_P y_P z_P$ is the *local Cartesian coordinate system* having origin at the point of contact of the surfaces $\mathscr{G}$ and $T$.

## GREEK SYMBOLS

$\Phi_{1.g}$, $\Phi_{2.g}$ are the first and second fundamental forms of the gear tooth flank $\mathscr{G}$.

$\Phi_{1.T}$, $\Phi_{2.T}$ are the first and second fundamental forms of generating surface $T$ of the gear cutting tool.

$\Sigma$ is the crossed-axis angle.

$\phi_n$ is the normal pressure angle of a gear cutting tool.

$\mu$ is the angle of the surfaces $\mathscr{G}$ and $T$ local relative orientation.

$\zeta_{ct}$ is the setting angle of a gear finishing tool.

$\boldsymbol{\omega}_c$ is the rotation vector of the gear cutting tool.

$\boldsymbol{\omega}_g$ is the rotation vector of the work gear.

$\boldsymbol{\omega}_{pl}$ is the vector of instant rotation of the gear cutting tool in relation to the work gear.

## SUBSCRIPTS

*cnf* is *conformity.*

max is *maximum.*

min is *minimum.*

opt is *optimal.*

*g* is *gear.*

*ct* is *cutting tool.*

# References


1. Ball, R.S., *A Treatise on the Theory of Screws*, Cambridge University Press, Cambridge, New York, 1998, 544 pages.
2. Ball, R.S., *The Theory of Screws*, Hodges & Foster, Dublin, 1876.
3. Basstein, G., "Calculation, Manufacturing and Applications of *Cylkro* Angular Face Gear Transmissions", In Proceedings of SME seminar: *Basic Gear Design and Manufacturing*, January 26–27, 2000, Troy (Detroit), 8 pages.
4. Bazhin, A.A., *Gear Cutting Tools*, ONTI NKTP, Moscow, 1935, 112 pages.
5. Klingelnberg, J., Editor, *Bevel Gear: Fundamentals and Applications*, Springer-Verlag GmbH Berlin Heidelberg, 2016, 328 pages.
6. Bianco, G., *Gear Hobbing*, Novaprint, Bologna, 2004, 201 pages.
7. Bianco, G., *Gear Shaving*, Jiuge, Beijing, 2000, 200 pages. Novaprint, Bologna, 2005, 174 pages.
8. Bianco, G.-F., Radzevich, S.P., *Precision Gear Shaving*, Nova Science Publishers, Inc, New York, 2010, 235 pages.
9. Bobrov, V.F., *Fundamentals of Metal Cutting*, Mashinostroyeniye, Moscow, 1975, 344 pages.
10. Bobrov, V.F., *Impact of the Angle of Inclination of the Main Cutting Edge of a Cutting Tool onto the Process of Machining of Metal Parts*, Mashgiz, Moscow, 1962, 152 pages.
11. Bonnet, P.O., *Journal de l'École polytechnique*, Vol. xiii, 1867, p. 31.
12. Börner, J., Humm, K., Joachim, F.J., "Developments of Conical Involute Gears (Beveloids) for Vehicle Transmissions", *Gear Technology*, November/December 2005, pp. 28–35.
13. Bregi, B.F., Erxleben, R.F., Tersch, R.W., *et al*, *Modern Methods of Gear Manufacture*, 4th edition, National Broach & Machine Division/Lear Siegler, Inc., 5600 St. Jean Ave., Detroit, MI 48213, 1972, 160 pages.
14. Buckingham, E., *Analytical Mechanics of Gears*, Dover Publications, Inc, New York, 1988, 546 pages.
15. Chkalova, O.N., *Conical Hobs for Machining Cylindrical Gears*, Ph.D. Thesis, Kiev, Kiev Polytechnic Institute, 1969, 26 pages.
16. Cormac, P., "A *Treatise on Screws and Worm Gear*", In *Their Mills and Hobs*, Chapman & Hall, Ltd, London, 1936, 138 page.
17. da Vinci, L., *The Madrid Codices*, Volume 1, 1493, Facsimile Edition of "*Codex Madrid* 1", original Spanish title: *Tratado de Estatica y Mechanica en Italiano*, McGraw Hill Book Company, 1974.
18. de Morales, A., *las Antigüedades de las ciudades de Espana*, Alcalá, 1575, 91–93 pages.
19. Denavit, J., Hartenberg, R.S., "A Kinematics Notation for Lower-Pair Mechanisms Based on Matrices", *ASME Journal of Applied Mechanics*, Vol. 77, 1955, pp. 215–221.
20. Dudley, D.W., *The Evolution of the Gear Art*, AGMA, Washington DC, 1969, 93 pages.
21. Endoy, R., *Gear Hobbing, Shaping and Shaving: A Guide to Cycle Time Estimating and Process Planning*, SME, Publication Development Department, Dearborn MI, 1990, 181 pages.
22. *FETTE Gear Cutting Tools: Hobbing, Gear Milling*, Leitz Metalworking Technology Group, Grand Rapids, USA, 2007.
23. Frayfeld, I.A., *Continuously Indexing Cutting Tools*, Mashgiz, Moscow, 1948, 252 pages.
24. *Gear Design: Manufacturing and Inspection Manual*, Society of Automotive Engineers, Inc, 1990, 643 pages.
25. Granovsky, G.I., *Kinematics of Machining*, Mashgiz, Moscow, 1948, 200 pages.
26. Granovski, G.I., Granovski, V.G., *Metal Cutting*, Visshaya Shcola, Moscow, 1985, 304 pages.
27. *Improvement of Gear Cutting Tools*, Collected papers, NIIMASh, Moscow, 1989, 571 pages.
28. Inozemtsev, G.G., *Design of Metal Cutting Tools*, Mashinostroyeniye, Moscow, 1984, 272 pages.
29. Inozemtsev, G.G., *Hobs With Rational Parameters of Design and of the Cutting Edge Geometry*, Saratov State University, Saratov, 1961, 224 pages.
30. Kau, B.-S., *Gear Shaper Cutters Having the Combined Rake Surface*, Ph.D. Thesis, Kiev Polytechnic Institute, Kiev, 1975, 215 pages.
31. Kimura, K., Ainoura, M., hobbing, C., U.S. Patent No. 3.786.719, AZUMI, 1974.
32. Kolchin, N.I., *Analytical Computation of Planar and Spatial Gearing: With Application to Profiling of the Gear Cutting Tools and to Computation of the Deviations in Gearing*, Mashgiz, Moscow, 1949, 210 pages.
33. Lashn'ev, S.I., *Fundamentals of the Theory of Surface Generation with Disk-Type, Rack-Type and Worm-Type Cutting Tools*, Doctoral Thesis, Tula Polytechnic Institute, Tula, 1968, 268 pages.
34. Lashn'ev, S.I., *Generation of Cogged Parts With Rack-Type and Worm-Type Cutting Tools*, Mashinostroyeniye, Moscow, 1971, 216 pages.
35. Lashn'ev, S.I., Yul'ikov, M.I., *Computation and Design of Metal Cutting Tools*, Mashinostroyeniye, Moscow, 1975, 392 pages.
36. Lashn'ev, S.I., Yul'ikov, M.I., *Computerized Design of Cutting Tools*, Mashinostroyeniye, Moscow, 1980, 208 pages.
37. *MAAG Gear Book: Calculation and Practice of Gears, Gear Drives, Toothed Couplings and Synchronous Clutch Couplings*, MAAG Gear Company Ltd, CH-8023 Zurich, Switzerland, 1990, 440 pages.
38. Margulis, D.K., *Broaches With Alterative Cutting Diagrams*, Mashgiz, Moscow, 1962, 270 pages.
39. Mat'ushin, V.M., *Fundamentals of Gear Cutting Tool Design (for Spur and Helical Gears)*, Doctoral Thesis, *STANKIN*, Moscow, 1957, 48 pages.
40. Medveditskov, S.N., *A Method of Multi-Paths Hobbing of Spur and Helical Gears*, Pat. No. 1222442, USSR, Int. Cl. B23f 5/22, Filed: July 26, 1984.
41. Medveditskov, S.N., *High-Productive Gear Hobbing*, Mashinostroyeniye, Moscow, 1981, 104 pages.
42. Medveditskov, S.N., Radzevich, S.P., Smirnov, N.N., "Gear Hobs With New Cutting Diagrams", *Mashinostroitel'*, 1985, No. 1, pp. 22–25.
43. Merchant, M.E., "Basic Mechanics of Metal-Cutting Process", *J. Appl. Mech*, Vol. 15, September 1944.
44. Merchant, M.E., "Mechanics of the Metal Cutting Process II, Plasticity Conditions in Orthogonal Cutting", *Journal of Applied Mechanics*, Vol. 16, June 1945.

45. Merritt, H.E., *Gear Engineering*, Pitman Publishing, London, 1971, 489 pages.
46. Michalec, G.W., *Precision Gearing: Theory and Practice*, John Wiley & Sons, Inc, New York, 1966, 620 pages.
47. Misevich, V.S., "A Method for Hobbing Cylindrical Gears With Radially Fed Hob", *Stanki i Instrument*, No. 1, 1964, pp. 16–18.
48. Mozhayev, S.S., *A Generalized Theory of Cutting Tools*, Doctoral Thesis, Leningrad Polytechnic Institute, Leningrad, 1953, 295 pages.
49. Mozhayev, S.S., *Analytical Theory of Twist Drills*, Mashgiz, Moscow, Leningrad, 1948, 136 pages.
50. Ngok, L.-M., *Investigation of the Geometry of the Rake Surface of Shaper Cutters*, Ph.D. Thesis, Kiev Polytechnic Institute, Kiev, 1972, 166 pages.
51. Oliphirenko, M.I., *Development and Investigation of the Methods to Improve the Accuracy of Gear Shaper Cutters*, Ph.D. Thesis, Kiev Polytechnic Institute, Kiev, 1980, 358 pages.
52. Olivier, T., *Theorie Geometrique Des Engrenages*, Bachelier, Paris, 1842.
53. Oxley, P.L.B., *Mechanics of Machining. An Analytical Approach to Assessing Machinability*, Ellis Horwood Ltd, Chichester, 1989, 242 pages.
54. Palaguta, V.A., *Investigation and Improvement of Productivity of Gear Shaving Process*, Ph.D. Thesis, DSTU, Dneprodzerzhinsk, 1995, 220 pages.
55. Pat. No. 155718 (USSR), *A Method of Relieving of Gear Hob with the Shaper Cutter*, Int. Cl. D24b 21/16, Filed: 1961.
56. Petrukhin, S.S., *Fundamentals of Cutting Tool Design: Kinematical Theory*, Mashgiz, Moscow, 1960, 163 pages.
57. Phillips, J., *General Spatial Involute Gearing*, Springer, Berlin Heidelberg, 2003, 498 pages.
58. Prudhomme, R., Lemasson, G., *Cinématique*, École Nationale Supérienre d'Arts et Métiers, École d'Engenieurs, Donod, Paris, 1906, 1955 pages.
59. Radzevich, S.P., "A Brief Overview on the Evolution of the Scientific Theory of Gearing: A Preliminary Discussion". In *Proceedings of International Conference on Gears 2015*, October 5–7, 2015, Technische Universität München (TUM), Garching (near Munich), Germany, 2015, pp. 1035–1946.
60. Radzevich, S.P., *A Conical Cutting Tool for Machining Cylindrical Gears*, Pat. No. 1028451, USSR, Int. Cl. B23f 21/16, Filed: April 26, 1982.
61. Radzevich, S.P., *A Conical Tool for Finishing Cylindrical Gears*, Pat. No. 880590, USSR, Int. Cl. B21h 5/02, Filed: November 6, 1979.
62. Radzevich, S.P., *A Continuous Indexing Method of Sharpening of Shaper Cutters*, Pat. No. 1604567, USSR, Int. Cl. B24B 3/34, Filed: January 12, 1988.
63. Radzevich, S.P., "A Crowning Achievements for Automotive Applications", *Gear Solutions*, December 2004, pp. 16–25.
64. Radzevich, S.P., "A Descriptive Geometry-Based Solution to a Geometrical Problem in Rotary Shaving of a Shoulder Pinion", *ASME Journal of Manufacturing Science and Engineering*, Vol. 127, No. 4, November 2005, pp. 893–900.
65. Radzevich, S.P., *A Device for Inspection of a Form Milling Cutter*, Pat. No. 1629741, USSR, Int. Cl. G01b 5/20, Filed: March 27, 1989.
66. Radzevich, S.P., *A Device for Inspection of a Form Milling Cutter*, Pat. No. 1754352, USSR, Int. Cl. B23c 9/00, Filed: June 29, 1990.
67. Radzevich, S.P., *A Gear Cutting Tool*, Pat. No. 933316, USSR, Int. Cl. B23f 21/00, Filed: October 18, 1979.
68. Radzevich, S.P., *A Gear Cutting Tool*, Pat. No. 1017444, USSR, Int. Cl. B23f 21/16, Filed: April 26, 1982.
69. Radzevich, S.P., *A Gear Hob*, Pat. No. 1009661, USSR, Int. Cl. B23f 21/16, Filed: October 21, 1981.
70. Radzevich, S.P., *A Gear Hob*, Pat. No. 1537425, USSR, Int. Cl. B23f 21/16, Filed: June 02, 1987.
71. Radzevich, S.P., "A Generalized Analytical Form of the Conditions of Proper Part Surface Generation. Part 1", In *Improvement of Efficiency of Metal Cutting*, VolgPI, Volgograd, 1987, pp. 70–79.
72. Radzevich, S.P., "A Generalized Analytical Form of the Conditions of Proper Part Surface Generation. Part 2", In *Improvement of Efficiency of Metal Cutting*, VolgPI, Volgograd, 1988, pp. 56–73.
73. Radzevich, S.P., *A Helical Shaper Cutter*, Pat. No. 956187, USSR, Int. Cl. B23F 21/10, Filed: September 26, 1980.
74. Radzevich, S.P., *A Hob*, Pat. No. 1114543, USSR, Int. Cl. B23f 21/16, Filed: September 7, 1982.
75. Radzevich, S.P., *A Hob for Machining Internal Gear*, Pat. No. 946833, USSR, Int. Cl. B23f 21/16, Filed: June 6, 1980.
76. Radzevich, S.P., *A Hob for Roughing Gears*, Pat. No. 944828, USSR, Int. Cl. B23f 21/16, Filed: September 4, 1980.
77. Radzevich, S.P., *A Machine Tool for Finishing a Crowned Gear*, Pat. No. 921727, USSR, Int.Cl. B23f 19/00, Filed: September 26, 1980.
78. Radzevich, S.P., *A Machine Tool for Finishing a Crowned Gear*, Pat. No. 984744, USSR, Int.Cl. B23f 19/00, Filed: April 16, 1981.
79. Radzevich, S.P., *A Machine Tool for Finishing a Crowned Gear*, Pat. No. 1000186, USSR, Int.Cl. B23f 19/00, Filed: November 6, 1981.
80. Radzevich, S.P., *A Machine Tool for Finishing a Crowned Gear*, Pat. No. 1028450, USSR, Int.Cl. B23f 19/00, Filed: April 26, 1982.
81. Radzevich, S.P., *A Machine Tool for Finishing a Crowned Gear*, Pat. No. 1110566, USSR, Int. Cl. B23f 19/00, Filed: June 7, 1983.
82. Radzevich, S.P., *A Method for Relieving of Gear Hob*, Pat. No. 751583, USSR, Int. Cl. B24b 3/12, B23b 5/42, Filed: April 4, 1977.
83. Radzevich, S.P., *A Method for Relieving of Gear Hob with the Standard Shaper Cutter*, Pat. No. 1087309, USSR, Int. Cl. B24b 3/12, Filed: April 23, 1980.
84. Radzevich, S.P., *A Method for Relieving of Gear Hob with the Standard Shaper Cutter*, Pat. No. 1240548, USSR, Int. Cl. B24b 3/12, Filed: September 22, 1984.
85. Radzevich, S.P., *A Method of Grinding and Re-Grinding of a Gear Finishing Tool*, Pat. No. 956114, USSR, Int. Cl. B21h 5/00, Filed: March 18, 1980.
86. Radzevich, S.P., *A Method of Grinding of Gear Milling Cutters*, Pat. No. 975339, USSR, Int. Cl. B 24 b 3/12, Filed: October 10, 1980.
87. Radzevich, S.P., *A Method of Hob Relieving*, Pat. No. 965728, USSR, Int. Cl. B24b 3/12, Filed: January 21, 1980.
88. Radzevich, S.P., *A Method of Relief Grinding of Conical Hob for Machining Cylindrical Gears*, Pat. No. 1365550, USSR, Int. Cl. B24b 3/12, Filed: April 29, 1986.
89. Radzevich, S.P., *A Method of Relief Grinding of Gear Hob*, Pat. No. 848290, USSR, Int. Cl. B 24 b 3/12, Filed: August 06, 1979.
90. Radzevich, S.P., *A Method of Relief Grinding of Gear Hob Teeth*, Pat. No. 1194612, USSR, Int. Cl. B23f 21/16, B24b 3/12, Filed: July 20, 1984.

91. Radzevich, S.P., *A Method of Relief Grinding of Gear Hob Teeth*, Pat. No. 1743810, USSR, Int. Cl. B24b 3/12, Filed: September 13, 1989.
92. Radzevich, S.P., *A Method of Relief Grinding of Precision Gear Hob*, Pat. No. 965728, USSR, Int. Cl. B24b 3/12, Filed: January 21, 1980.
93. Radzevich, S.P., *A Method of Relief Grinding of Teeth of an Assembled Hob*, Pat. No. 961925, USSR, Int. Cl. B24b 3/12, Filed: March 18, 1980.
94. Radzevich, S.P., *A Method of Shaping of Cylindrical Gear*, Pat. No. 1504903, USSR, USSR, Int. Cl. B23f 5/12, Filed: December 2, 1987.
95. Radzevich, S.P., *A Method to Control the Accuracy of Helical Shaper Cutter*, Pat. No. 1710231, USSR, Int. Cl. B23F 21/10, G01B 5/20, Filed: September 13, 1989.
96. Radzevich, S.P., "A New Angle on Cutting Bevel Gears", *Gear Solutions*, December, 2010, pp. 42–47.
97. Radzevich, S.P., "A Novel Approach for Computation of Constraints on Parameters of Modification of the Tooth Addendum of Precision Involute Hob", *ASME Journal of Mechanical Design*, Vol. 128, No. 4, July 2006, pp. 803–811.
98. Radzevich, S.P., "A Novel Design of Cylindrical Hob for Machining of Precision Involute Gears", *ASME Journal of Mechanical Design*, Vol. 129, 2007, pp. 334–345.
99. Radzevich, S.P., "A Novel Hob Design for Precision Involute Gears: Part I", *Gear Solutions*, 2007, pp. 34–43.
100. Radzevich, S.P., "A Novel Hob Design for Precision Involute Gears: Part II", *Gear Solutions*, 2007, pp. 38–43.
101. Radzevich, S.P., "A Possibility of Application of Plücker's Conoid for Mathematical Modeling of Contact of Two Smooth Regular Surfaces in the First Order of Tangency", *Mathematical and Computer Modeling*, Vol. 42, No. 9-10, November, 2005, pp. 999–1022.
102. Radzevich, S.P., *A Precision Gear Cutting Tool*, Pat. No. 1038122, USSR, Int. Cl. B23f 21/04, Filed: May 4, 1981.
103. Radzevich, S.P., *A Precision Gear Hob*, Pat. No. 1004029, USSR, Int. Cl. B23f 21/16, Filed: November 6, 1979.
104. Radzevich, S.P., *A Precision Involute Hob*, Pat. No. 990445, USSR, Int. Cl. B23f 21/16, Filed: October 08, 1981.
105. Radzevich, S.P., *A Precision Involute Hob*, Pat. No. 1114505, USSR, Int. Cl. B23f 21/16, Filed: October 18, 1982.
106. Radzevich, S.P., *A Precision Involute Hob*, Pat. No. 2040376, RU, Int. Cl. B23f 21/16, Filed: January 03, 1992.
107. Radzevich, S.P., *A Shaper Cutter*, Pat. No. 1039657, USSR, Int. Cl. B23F 21/10, Filed: March 16, 1982.
108. Radzevich, S.P., *A Shaper Cutter*, Pat. No. 1174187, USSR, Int. Cl. B23F 21/10, Filed: April 4, 1981.
109. Radzevich, S.P., *A Shaper Cutter*, Pat. No. 1172660, USSR, Int. Cl. B23F 21/10, Filed: February 13, 1984.
110. Radzevich, A.P., *A Study on Efficient Hobbing of Cylindrical Involute Gears*, Ph.D. Thesis, Kuybyshev Polytechnic Institute, Kuybyshev, 1988, 300 pages.
111. Radzevich, S.P., *A Tool for Finishing Internal Gear*, Pat. No. 880589, USSR, Int. Cl. B21h 5/02, Filed: November 6, 1979.
112. Radzevich, S.P., *A Tool for Burnishing of Bevel Gears*, Pat. No. 721214, USSR, Int. Cl. B21h 5/04, Filed: March 1, 1977.
113. Radzevich, S.P., *A Tool for Burnishing of Bevel Gears*, Pat. No. 846024, USSR, Int. Cl. B21h 5/04, Filed: November 15, 1978.
114. Radzevich, S.P., *A Tool for Burnishing of Bevel Gears*, Pat. No. 1094659, USSR, Int. Cl. B21h 5/04, Filed: March 11, 1983.
115. Radzevich, S.P., *A Tool for Burnishing of Bevel Gears*, Pat. No. 1107943, USSR, Int. Cl. B21h 5/04, Filed: February 24, 1983.
116. Radzevich, S.P., *A Toroidal Tool for Finishing Cylindrical Gears*, Pat. No. 965582, USSR, Int. Cl. B21h 5/00, Filed: October 2, 1980.
117. Radzevich, S.P., "A Way to Improve the Accuracy of Hobbed Involute Gears", *ASME Journal of Mechanical Design*, Vol. 129, October 2007, pp. 1076–1085.
118. Radzevich, S.P., *A Worm-Type Finishing Tool for Reinforcement of Involute Gear by Surface Plastic Deformation*, Pat. No. 829280, USSR, Int. Cl. B21h 5/00, Filed: May 15, 1978.
119. Radzevich, S.P., "About Hob Idle Distance in Gear Hobbing Operation", *ASME Journal of Mechanical Design*, Vol. 124, December 2002, pp. 772–786.
120. Radzevich, S.P., *An Advanced Technology of Machining of Hardened Gears*, VNIITEMR, Moscow, 1992, 72 pages.
121. Radzevich, S.P., *An Assembled Gear Hob*, Pat. No. 1505695, USSR, Int. Cl. B23f 21/16, Filed: March 16, 1987.
122. Radzevich, S.P., *An Attachment to a Grinding Machine for Dressing of a Form Grinding Wheel*, Pat. No. 1771937, USSR, Int. Cl. B24b 53/085, Filed: June 21, 1990.
123. Radzevich, S.P., *An Attachment to a Grinding Machining for Relief Grinding of Hobs*, Pat. No. 869978, USSR, Int. Cl. B23b 5/42, Filed: December 12, 1979.
124. Radzevich, S.P., *An Attachment to a Grinding Machining for Relief Grinding of Hobs*, Pat. No. 1009615, USSR, Int. Cl. B23b 5/42, Filed: September 26, 1980.
125. Radzevich, S.P., *An Attachment to a Grinding Machining for Relief Grinding of Hobs*, Pat. No. 1077706, USSR, Int. Cl. B23b 5/42, B24b 3/12, Filed: July 5, 1982.
126. Radzevich, S.P., *An Attachment to a Grinding Machining for Relief Grinding of Hobs*, Pat. No. 1117127, USSR, Int. Cl. B23b 5/42, B24b 3/12, Filed: June 15, 1983.
127. Radzevich, S.P., *An Attachment to a Grinding Machining for Relief Grinding of Hobs*, Pat. No. 1146134, USSR, Int. Cl. B23b 5/42, Filed: June 29, 1983.
128. Radzevich, S.P., *An Attachment to a Grinding Machining for Relief Grinding of Hobs*, Pat. No. 1199458, USSR, Int. Cl. B23b 5/42, Filed: May 29, 1984.
129. Radzevich, S.P., *An Attachment to a Grinding Machining for Relief Grinding of Hobs*, Pat. No. 1433769, USSR, Int. Cl. B24b 3/12, Filed: April 23, 1987.
130. Radzevich, S.P., *An Involute Hob*, Pat. No. 1004029, USSR, Int. Cl. B23f 21/16, Filed: November 6, 1979.
131. Radzevich, S.P., "Basic Conditions of Proper Part Surface Generating While Machining on Conventional Machine Tool", *The Scientific Journal: "Facta Universitatis,"* Series: *Mechanical Engineering*, University of Niš, Vol. 1, No. 6, 1998, pp. 637–651.
132. Radzevich, S.P., "Briefly on the Kinematic Method and on the History of the Equation of Contact in the Form of $\mathbf{nV} = 0$", *Theory of Mechanisms and Machines*, Vol. 15, No. 1, 2010, pp. 42–51. [in Russian].
133. Radzevich, S.P., *CAD/CAM of Sculptured Surfaces on Multi-Axis NC Machine: The DG/K-Approach*, M&C Publishers, San Rafael, California, 2008, 113 pages.
134. Radzevich, S.P., *Classification of Surfaces*, Monograph, Kiev, UkrNIINTI, No. 1440-Ук88, 1988, 185 pages. [in Russian].
135. Radzevich, S.P., "Computation of Parameters of a Form Grinding Wheel for Grinding of Shaving Cutter for Plunge Shaving of Topologically Modified Involute Pinion", *ASME Journal of Manufacturing Science and Engineering*, Vol. 127, No. 4, November 2005, pp. 819–828.
136. Radzevich, S.P., "Conditions of Proper Sculptured Surface Machining", *Computer-Aided Design*, Vol. 34, No. 10, September 2002, pp. 727–740.

137. Radzevich, S.P., *Cutting Tools for Machining of Hardened Gears*, VNIITEMR, Moscow, 1992, 60 pages.
138. Radzevich, S.P., *Design and Investigation of Skiving Hobs for Finishing of Hardened Gears*, Ph.D. thesis, Kiev Polytechnic Institute, Kiev, 1982, 298 pages.
139. Radzevich, S.P., "Design of Shaving Cutter for Plunge Shaving a Topologically Modified Involute Pinion", *ASME Journal of Mechanical Design*, Vol. 125, September 2003, pp. 632–639.
140. Radzevich, S.P., "Design of Special-Purpose Hob With Prescribed Value of the Hob-Setting Angle for Machining of Cluster Gear", *ASME Journal of Mechanical Design*, Vol. 129, 2007, pp. 438–444.
141. Radzevich, S.P., "Determination of Tooth Flank Geometry of Shaving Cutter for Plunge Shaving of Precision Gears", Part 1, *STIN*, 2008, No. 10, 9–13 pages.
142. Radzevich, S.P., "Determination of Tooth Flank Geometry of Shaving Cutter for Plunge Shaving of Precision Gears", Part 2, *STIN*, 2008, No. 11, 18–22 pages.
143. Radzevich, S.P., "Diagonal Shaving of an Involute Pinion: Optimization of Parameters of the Pinion Finishing Operation", *International Journal of Advanced Manufacturing Technology*, Vol. 32, No. 11–12, 2007, pp. 1170–1187.
144. Radzevich, S.P., *Differential-Geometrical Method of Surface Generation*, Doctoral Thesis, Tula, Tula Polytechnic Institute, 1991, 300 pages. [in Russian].
145. Radzevich, S.P., *Dudley's Handbook of Practical Gear Design and Manufacture*, 2nd Edition, CRC Press, Boca Raton, Florida, 2012, 878 pages.
146. Radzevich, S.P., *Dudley's Handbook of Practical Gear Design and Manufacture*, 3rd Edition, CRC Press, Boca Raton, Florida, 2016, 629 pages. [Radzevich, S.P., (Editor), *Dudley's Handbook of Practical Gear Design and Manufacture*, 4th edition, *CRC* Press, Boca Raton, FL, 2021, 1170 pages, 718 B/W Illustrations. ISBN-13: ISBN-10: 0367649020, 978-0367649029].
147. Radzevich, S.P., "Finishing Precise Gears for Low Noise/Noiseless Car Transmission", in: *Proceedings of the 1st Annual Manufacturing Technology Summit*, August 10–11, 2004, University of Michigan/Dearborn Fairlane Center, *SME*, 2004, 20 pages.
148. Radzevich, S.P., *Fundamentals of Surface Generation, Monograph*, Rastan, Kiev, 2001, 592 pages. [in Russian].
149. Radzevich, S.P., *Gear Cutting Tools: Fundamentals of Design and Computation*, CRC Press, Boca Raton, Florida, 2010, 786 pages.
150. Radzevich, S.P., "Generating Surfaces of Worm-Type Gear Cutting Tools", In: *Advance Processes in Production Technology*, VolgPI, Volgograd, 1985, pp. 64–78. [in Russian].
151. Radzevich, S.P., *Generation of Surfaces: Kinematic Geometry of Surface Machining*, CRC Press, Boca Raton, Florida, 2014, 747 pages.
152. Radzevich, S.P., *Geometry of Surfaces: A Practical Guide for Mechanical Engineers*, 2nd edition, Springer International Publishing, 2019, 466 pages. [1st edition: Radzevich, S.P., *Geometry of Surfaces: A Practical Guide for Mechanical Engineers*, Wiley, 2013, 264 pages].
153. Radzevich, S.P., "Geometry of the Active Part of Cutting Tools (in the Tool-in-Use System)", The *SME* MIDWEST 2005 Manufacturing Excellence Through Collaboration Conference, September 13–14, 2005, Rock Financial Showplace, Novi, Michigan, *SME* Paper No. TP06PUB36, 12 pages.
154. Radzevich, S.P., "Geometry of the Active Part of Cutting Tools (in the Tool-in-Hand System)", The *SME* Summit Conference: *Where Manufacturers, Technologies and Innovations Connect*, August 3-4, 2005, Olympia Resort and Conference Center, Oconomowoc (Milwaukee), WI. *SME* Paper No. TP06PUB37, 12 pages.
155. Radzevich, S.P., *High-Conformal Gearing: Kinematics and Geometry*, CRC Press, Boca Raton, Florida, 2015, 368 pages.
156. Radzevich, S.P., "Investigation of the Tooth Geometry of a Hob for Machining of Involute Gears (in the Tool-in-Use Reference System)", *ASME Journal of Manufacturing. Science and Engineering*, Vol. 129, No. 8, 2007, pp. 750–759.
157. Radzevich, S.P., "Knowledge (of Gear Theory) is Power in the Design, Production, and Application of Gears", *Gear Solutions* magazine, August 2020, pages 38–44.
158. Radzevich, S.P., *Kinematic Geometry of Surface Machining*, CRC Press, Boca Raton FL, 2008, 508 pages.
159. Radzevich, S.P., *Machining of Hardened Gears*, VNIITEMR, Moscow, 1985, 56 pages. [in Russian].
160. Radzevich, S.P., "Modified Shaving Cutters for Low Noise", p. 26, In Stott, W.R., "Software Bits", *Gear Technology*, January/February 2004, pp. 20–27.
161. Radzevich, S.P., "On a Possible Way of Size and Weight Reduction of a Car Transmission", *Gear Technology*, July/August 2003, pp. 44–50.
162. Radzevich, S.P., "On Satisfaction of the Fifth Necessary Condition of Proper Part Surface Generation in Design of Plunge Shaving Cutter for Finishing of Precision Involute Gears", *ASME J. of Mechanical Design*, Vol. 129, No. 9, 2007, pp. 969–980.
163. Radzevich, S.P., "On the Accuracy of Finishing Involute Hobs", in: *Cutting and Cutting Tools, Collected Papers*, Vol. 24, Vishcha Shkola, Kharkov, 1980, pp. 105–109. [in Russian].
164. Radzevich, S.P., "On the Accuracy of Precision Involute Hobs: An Analytical Approach", *SME Journal of Manufacturing Processes*, Vol. 9, No. 2, 2007, pp. 121–136.
165. Radzevich, S.P., "On Tooth Failure Analysis in Small-Teeth-Number Gearing: An Analytical Approach", *AGMA* Technical Paper 06FTM11, 2006, 22 pages.
166. Radzevich, S.P., "Part Surfaces Those Allowing Sliding Over Them", In *Research in the Field of Surface Generation*, S.P. Radzevich (ed.), UkrNIINTI, Kiev, No. 65–Uk89, 1989, pp. 29–53. [in Russian].
167. Radzevich, S.P., "Relief Grinding of Assembled Gear Hobs With Diamond Grinding Wheels", *Diamonds and Super-Hard Materials*, NIIMASh, Moscow, 1979, Vol. 4, pp. 13–14. [in Russian].
168. Radzevich, S.P., *Sculptured Surface Machining on Multi-Axis NC Machine*, Monograph, Kiev, Vishcha Schola, 1991, 192 pages. [in Russian].
169. Radzevich, S.P., "Technological Implementation of Shape Peculiarities of Involute Treads and Worms", *Vestnik Mashinostroyeniya*, 1984, No. 11, pp. 48–51. [in Russian].
170. Radzevich, S.P., "Technological Methods for Noise/Vibration Reduction in Driveline/Transmission of Trucks and All-Wheel-Drive Vehicle", *International Journal of Vehicle Noise and Vibration*, Vol. 2, No. 4, 2006, pp. 283–291.
171. Radzevich, S.P., *Theory of Gearing: Kinematics, Geometry, and Synthesis*, CRC Press, Boca Raton, Florida, 2012, 760 pages.

172. Radzevich, S.P., *Theory of Gearing: Kinematics, Geometry, and Synthesis, 3rd Edition, Revised and Expanded*, CRC Press, Boca Raton, FL, 2022, 1154 pages.
173. Radzevich, S.P., Chernishov, M.V., *An Attachment to a Grinding Machining for Relief Grinding of Hobs*, Pat. No. 1371875, USSR, Int. Cl. B24b 3/12, Filed: April 28, 1988.
174. Radzevich, S.P., Dashevskii, A.N., *An Attachment for Inspection of a Form Milling Cutter*, Pat. No. 1516738, USSR, Int. Cl. G01b 5/20, Filed: February 17, 1989.
175. Radzevich, S.P., Goodman, E.D., Palaguta, V.A., "Tooth Surface Fundamental Forms in Gear Technology", *University of Niš, the Scientific Journal Facta Universitatis, Series: Mechanical Engineering*, Vol. 1, No. 5, 1998, pp. 515–525.
176. Radzevich, S.P., Lempert, V.I., *A Method of Relieving of Gear Hobs*, Pat. No. 1335425, USSR, Int. Cl. B 24 b 3/12, Filed: April 24, 1986.
177. Radzevich, S.P., Medveditskov, S.N., *A Shaper Cutter*, Pat. No. 1641535, USSR, Int. Cl. B23F 21/10, Filed: June 26, 1988.
178. Radzevich, Medveditskov, S.P., Smirnov, A.P. N.N., *A Gear Hob*, Pat. No. 1134318, USSR, Int. Cl. B23f 21/16, Filed: April 7, 1983.
179. Radzevich, Medveditskov, S.P., Smirnov, A.P. N.N., *A Gear Hob*, Pat. No. 1187936, USSR, Int. Cl. B23f 21/16, Filed: June 28, 1984.
180. Radzevich, S.P., Palaguta, V.A., *A Gear Shaving Cutter*, Pat. No. 1632662, USSR, Int. Cl. B23f 21/28, Filed: March 30, 1989.
181. Radzevich, S.P., Palaguta, V.A., *A Method of Sharpening of Gear Shaving Cutter*, Pat. No. 1646725, USSR, Int. Cl. B23f 21/28, Filed: December 13, 1988.
182. Radzevich, S.P., Palaguta, V.A., *A Tool for Finishing Cylindrical Gears*, Pat. No. 1328042, USSR, Int. Cl. B21h 5/00, Filed: December 3, 1985.
183. Radzevich, S.P., Palaguta, V.A., *Advanced Methods for Gear Finishing*, VNIITEMR, Moscow, 1988, 52 pages.
184. Radzevich, S.P., Palaguta, V.A., "Synthesis of Optimal Gear Shaving Process", *Vestnik Mashinostroyeniya*, 1997, No. 8, pp. 36–41.
185. Radzevich, S.P., Palaguta, V.A., Goodman, E.D., *Geometry of Helical Gear and Tool Surfaces for Plunge Shaving of Involute Gears*, Final Data Report, Michigan State University/Chrysler Corp, 1998, 86 pages.
186. Radzevich, S.P., Palaguta, V.A., Radzevich, A.P., Bulgakov, V.A., *A Shaving Cutter*, Pat. No. 1514516, USSR, Int. Cl. B23f 21/28, Filed: November 27, 1986.
187. Radzevich, S.P., Palaguta, V.A., Radzevich, A.P., Tovmasian, M.A., *A Shaving Cutter*, Pat. No. 1537426, USSR, Int. Cl. B23f 21/28, Filed: November 17, 1986.
188. Radzevich, S.P., Pasechnik, M.F., *A Method of Hob Relieving*, Pat. No. 1196232, USSR, Int. Cl. B24b 3/00, Filed: August 3, 1984.
189. Radzevich, S.P., Petrenko, T.Y., "Part and Tool Surfaces those Allowing for Sliding Over Themselves", *Mechanika ta Mashinobuduvann'a*, Kharkiv, Ukraine, 1999, No. 1, pp. 231–240. [in Russian].
190. Radzevich, S.P., Vinokourov, I.V., *A Method of Grinding of Gear Milling Cutters having Straight Slots*, Pat. No. 1627375, USSR, Int. Cl. B 24 b 3/12, Filed: October 17, 1988.
191. Radzevich, S.P., Vinokourov, I.V., *An Attachment to a Grinding Machine for Dressing of a Form Grinding Wheel*, Pat. No. 1393527, USSR, Int. Cl. B23b 5/42, Filed: November 18, 1986.
192. Radzevich, S.P., Vinokourov, I.V., *An Attachment to a Grinding Machine for Relief Grinding of Hobs*, Pat. No. 1335426, USSR, Int. Cl. B24b 3/12, Filed: April 14, 1986.
193. Radzevich, S.P., Vinokourov, I.V., Radzevich, A.P., Androshchuk, G.A., Smol'nikov, N.Y., *A Method of Roughing of Coarse Pitch Gears*, Pat. No. 1815024, USSR, Int. Cl. B 23 f 5/20, 17/00, Filed: November 11, 1990.
194. Reuleaux, F., The Kinematics of Machinery: Outlines of a Theory of Machines, Translated and Edited by Alexander B. W. Kennedy, and With a New Introduction by Eugene S., Ferguson, Dover Publishers, Inc., New York, 1963, 622 pages. [Translated from: Reuleaux, F., *Theoretische Kinematik: Grundziige einer Theori des Maschinenwesens*, 1875].
195. Rodin, P.R., *Fundamentals of Cutting Tool Design*, Mashgiz, Moscow-Kiev, 1960, 160 pages. [in Russian].
196. Rodin, P.R., *Issues on Theory of Cutting Tool Design*, Doctoral Thesis, Odessa Polytechnic Institute, Odessa, 1961, 373 pages. [in Russian].
197. Rodin, P.R., Radzevich, S.P., *A Method of Relief Grinding of Precision Gear Hob*, Pat. No. 831546, USSR, Int. Cl. B24b 3/00, Filed: February 7, 1979.
198. Sakharov, G.N., *Continuously Indexing Cutting Tools*, Mashinostroyeniye, Moscow, 1983, 232 pages. [in Russian].
199. Sakharov, G.N., *Issues on Theory of Continuously Indexing Cutting Tools*, Doctoral Thesis, Stankin, Moscow, 1974, 320 pages. [in Russian].
200. Semenchenko, I.I., Mat'ushin, V.M., Sahharov, G.N., *Design of Metal Cutting Tools*, Mashgiz, Moscow, 1962, 952 pages. [in Russian].
201. Shaw, M.C., *Metal Cutting Principles*, Clarendon Press, Oxford, 1984.
202. Shevel'ova, G.I., *Theory of Surface Generation and of Contact of Moving Bodies*, Moscow University STANKIN, Moscow, 1999, 494 pages. [in Russian].
203. Shevchenko, N.A., *The Geometry of Cutting Edge and the Parameters of Uncut Chip*, Mashgiz, Moscow, 1957, 140 pages. [in Russian].
204. Shishkov, V.A., "Elements of Kinematics of Generating and Conjugating in Gearing", in: *Theory and Computation of Gears*, Vol. 6, LONITOMASH, Leningrad, 1948. [in Russian].
205. Shishkov, V.A., "Elements of Kinematics of Generating and Conjugating in Gearing", in: *Theory and Calculation of Gears*, Vol. 6, LONITOMASH, Leningrad, 1948, 123 pages.
206. Shishkov, V.A., *Generation of Surfaces Using Continuous Indexing Methods of Machining*, Mashgiz, Moscow, 1951, 152 pages. [in Russian].
207. Shishkov, V.A., *Generation of Surfaces in Continuous-Indexing Methods of Surface Machining*, Mashgiz, Moscow, 1951, 152 pages.
208. Smirnov, N.N., "Multi-Thread Hob With New Cutting Diagrams", *Gear Technology*, August 1991, pp. 27–31.
209. Stabler, G.V., "The Fundamental Geometry of Cutting Tools", *Proceedings of Institution of Mechanical Engineers*, Vol. 165, 1951, pp. 14–21. https://doi.org/10.1243/PIME_PROC_1951_165_008_02
210. Stadtfeld, H.J., "Why are Today's Hypoids the Perfect Crossed-Axes Gear Pairs?", *Gear Solutions* magazine, May 2019, pages 42–50.
211. Talantov, N.V., *Physical Foundations of Metal Cutting, Wear and Failure of Cutting Tools*, Mashinostroyeniye, Moscow, 1992, 240 pages. [in Russian].
212. Townsend, D.P., *Dudley's Gear Handbook. The Design, Manufacture, and Application of Gears.* 2nd Edition, McGraw Hill, Inc., NY, 1992.

213. Tsepkov, A.V., *Profiling of Gear Cutting Tools That Feature Relieved Clearance Surfaces*, Mashinostroyeniye, Moscow, 1979, 150 pages. [in Russian].
214. Tsvis, Y.V., *Profiling of Cutting Tools for Scudding of Gears*, Mashgiz, Moscow, 1961, 156 pages. [in Russian].
215. U.S. Pat. No. 407.437. *Machine for Planing Gear Teeth./* G.B. Grant, Filed: January 14, 1887 (serial No. 224,382), Patent issued: July 23, 1889.
216. US Pat. No. 1.913.865, *Cutting Tool*, Ernst Wildhaber, Filed: July 14, 1933, Published: June 13, 1930.
217. U.S. Pat. No. 1.924.476, *Method of Cutting Gears*, Filed: March 19, 1928, Patented: August 29, 1933.
218. U.S. Patent No. 1.996.987, *Rotary Gear Cutting Tool*, Ernst Wildhaber, Filed: June 23, 1930, Patented: April 9, 1935.
219. Vinokourov, I.V., Radzevich, S.P., *An Attachment to a Grinding Machining for Relief Grinding of Hobs*, Pat. No. 2009762, RU, Int. Cl. B23b 5/42, Filed: December 18, 1991.
220. Watson, H.J., *Modern Gear Production*, Pergamon Press, New York, 1970, 359 pages.
221. Willis, R., "Designed for the Use of Students in the Universities and for Engineering Students Generally", In *Principles of Mechanisms*, J. & J.J. Deighton, London, John W. Parker, West Stand, Cambridge, 1841, p. 446.
222. Woodbury, R.S., *History of the Gear-Cutting. A Historical Study in Geometry and Machine*, The M.I.T. Press, Massachusetts, USA, 1958, 135 pages.
223. Yu, M., Mertz, R., *An Evaluation of Smirnov Hob Cutter*, Technical Data Report, June 1, 1993, GMGC #0112, General Motors Gear Center, Romulus, MI 48174-1376, 1993, 73 pages.
224. Zorev, N.N., *Metal Cutting Mechanics*, Pergamon Press, New York, 1956. [in Russian].
225. Zorev, N.N., *Problems of Metal Cutting Mechanics*, ZNIITMash, Moscow, 1956, 367 pages. [in Russian].

# Bibliography


Anikin, Yu.V., *Sinusoidal Gearing*, Voronezh State University, Voronezh, 1975, 56 pages.

Babak, V.F., "Orientation of Instant Zone of Cutting in Shaving Process", Izvestiya VUZov, *Mashinostroyeniye*, 1969, 141–145 pages.

Barabol'a, S.Ya., *Hobbing Cylindrical Gears*, Technika, Kiev, 1992, 48 pages. [in Russian].

Baxter, M.L. Jr., "Basic Geometry and Tooth Contact of Hypoid Gears", *Industrial Mathematic*, Vol. 11, No. 2, 1961, pp. 19–42.

Baxter, M.L. Jr., *"Basic Theory of Gear-Tooth Action and Generation". In: Gear Handbook*, 1st edition, Darle Dudley, (Ed.), McGraw Hill, New York, 1962.

Bazhin, A.A., *Gear Cutting Tools*, ONTI NKTP, Moscow-Leningrad, 1935, 112 pages. [in Russian].

Bianco, G., *Gear Shaving*, Jiuge, Beijing, 2000, 200 pages.

Bianco, G., *Gear Hobbing*, Novaprint, Bologna, 2004, 201 pages.

Bianco, G., *Gear Shaving*, Novaprint, Bologna, 2005, 174 pages.

Bianco, G.-F., Radzevich, S.P., *Precision Gear Shaving*, Nova Science Publishers, Inc, New York, 2010, 235 pages.

Boguslaye, V.O., Ravs'ka, N.S., Kachan, O.Ya., Balushok, K.B., Mozgovii, V.F., and Okhrimenko, O.A., *Geretion of Hobs*, Zaporizhzh'a, VAT "Motor Sich", 2007, 176 pages. [in Russian].

Buckingham, E., Ryffel, H.H., *Design of Worm and Spiral Gears: A Manual for the Design and Manufacture of All-Redess-Action Worm and Spiral Gear Drives*, The Industrial Press, New York, 1960, 450 pages.

Chironis, N.P. (Ed.), *Gear Design and Application*, McGraw-Hill, New York, 1967, 374 pages.

Colbourne, J.R., *The Geometry of Involute Gears*, New York, Springer-Verlag, 1987, 532 pages.

Cormac, P., *A Treatise on Screws and Worm Gear, Their Mills and Hobs*, Chapman & Hall, Ltd, London, 1936, 1936, 138 pages.

Crosher, W.P., *Design and Application of the Worm Gear*, ASME Press, New York, 2002, 256 pages.

da Vinci, Leonardo, *The Madrid Codices*, McGraw Hill Book Company, Facsimile Edition of "*Codex Madrid,* 1974, 1493 pages.

Davis, J.R. (Ed.), *Gear Materials, Properties, and Manufacture*, ASM Press, 2005, 339 pages.

Derzhavets, Yu.A., Abramov, A.K., Pizh, O.L., *Hobs for Machining Ring Gears*, LDNTP, Leningrad, 1968, 24 pages. [in Russian].

Drago, R.J., *Fundamentals of Gear Design*, Butterworths, Boston, 1988, 560 pages. [ISBN 0-409-90127-X].

Dudley, D.W., *Practical Gear Design*, McGraw-Hill, New York, 1954.

Dudley, D.W., *The Evolution of the Gear Art*, AGMA, Washington DC, 1969, 93 pages.

Dudley, D.W., *Zahnräder-Berechnung, Entwurf und Herstellung nach Amerikanischen Erfahrungen*, Springer-Verlag, Berlin, 1961.

Dudley, D.W., *Gear Handbook: The Design, Manufacture, and Application of Gears*, McGraw-Hill, New York, 1962.

Dudley, D.W., *Handbook of Practical Gear Design*, McGraw-Hill, New York, 1984.

Dudley, D.W., *Handbook of Practical Gear Design*, CRC Press, Boca Raton, Florida, 1994, 688 pages.

Flanders, R.E., "Interchangeable Involute Gear Tooth Systems", *Journal A.S.M.E.*, Dec. 1908, 1501–1520 pages.

Flanders, R.E., *Am. Mach.*, Dec. 1910, p. 1064, and *Machinery*, Jan. 1911, p. 369, Mar. 1911, p. 569, June 1911, p. 798; H.H. Asbridge, *Am. Mach.*, Dec. 1910, p. 1211; E.J. Lewis, *Machinery*, Apr. 1911, p. 659; G.B. Grant, *Machinery*, June 1911, p. 813.

Flanders, R.E., *Bevel Gearing*, 4th edition, Machinery's Reference Series, Number 37, The Industrial Press, 1912, 48 pages.

Gibbs, J.W., *On the Form of the Teeth of Wheels in Spur Gearing*, Doctoral Dissertation, Yale University, New Haven, Connecticut, 1863.

Golembiyevskiy, A.I., *System Analysis of Gear Hobbing Process*, Navuka I Technika, Minsk, 1993, 152 pages. [in Russian].

Grant, G.B., *A Treatise on Gear Wheels*, 11th edition, Philadelphia Gear Works, Inc., Philadelphia, 1906, 105 pages.

Hertz, H., "On Contact of Solid Elastic Bodies and on Hardness", *Journal of Mathematics*, Vol. 92, 1881, pp. 156–171.

Hertz, H., "Über die Berührung Fester Elastischer Körper (The Contact of Solid Elastic Bodies)", *Journal für die Reine und Angewandte Mathematik (Journal for Pure and Applied Mathematic)*, 1981, pp. 156–171.; *Über die Berührung Fester Elastischer Körper und Über die Härte (The Contact of Solid Elastic Bodies and Their HarnessesThe Contact of Solid Elastic Bodies and Their HarnessesGesammelte Werke (Collected WorksCollected WorksMiscellaneous Papers*, translated by D.E. Jones and G.A. Schott), pp. 146–162, 163–183, London: McMillan and Co., Ltd., 1896.

Howes, M.A., *Source Book on Gear Design, Technology and Performance, A Comprehensive Collection of Outstanding Articles from the Periodical and Reference Literature*, Director, Metals Research, IIT Research Institute, American Society for Metals, Metals Park, Ohio, 1980, 417 pages.

Inozemtsev, G.G., *Hobs With Rational Geometrical and Design Parameters*, Saratov State University, Saratov, 1961, 224 pages. [in Russian].

Kapelevich, A.L., *Direct Gear Design*, CRC Press, Boca Raton, 2013, 299 pages.

Kedrinskiy, V.N., Pismanik, K.M., *Gear Generators for Cutting Bevel Gears*, Mashinostroyeniye, Moscow, 1967, 584 pages. (1st edition: 1958) [in Russian].

Klepikov, V.V., *Production of Gears*, Moscow State Industrial University, Moscow, 1997, 213 pages. [in Russian].

Klingelnberg, J., *Kegelräder: Grundlagen, Anwendungen*, Hrsg., Springer, 2008 edition (July 22, 2008), 381 pages.

Klingelnberg, J., (Ed.), *Bevel Gear: Fundamentals and Applications*, 1st edition, Springer Vieweg, 2016 edition (December 24, 2015), 325 pages.

Klocke, F., "Gear Cutting". In: The International Academy for Production Engineering, Laperriѐr, L., Reinhart, G., (eds.) *CIRP Encyclopedia of Production Engineering.* Springer, Berlin, Heidelberg. https://doi.org/10.1007/978-3-642-20617-7_6405

Kutzbach, K., "Bemerkungen zur Entwicklung der Verzahnung", ("Remarks on the Development of Gearing"), Appendix to: Matchoss, C., "Geschichte des Zahnrades" ("History of the Gear Whell"), Berlin, 1940.

Lopato, G.A., Kabatov, N.F., Segal', M.G., *Spiral Bevel and Hypoid Gearings*, 2nd edition, Moscow, Mashinostroyeniye, 1977, 423 pages. [in Russian].

L'ukshin, V.S., *Theory of Screw Surfaces With Application to Design of Cutting Tools*, Mashinostroyeniye, Moscow, 1968, 372 pages. [in Russian].

*MAAG Gear Book: Calculation and Manufacture of Gears and Gear Drives for Designers and Works Engineers*, MAAG Gear-Wheel Company Ltd, Zurich/Switzerland, 1963, 576 pages.

*MAAG Gear Book: Calculation and Practice of Gears, Gear Drives, Toothed Couplings and Synchronous Clutch Couplings*, MAAG Gear Company, Ltd, CH-8023 Zurich/ Switzerland, 1990, 440 pages.

*MAAG-Teshchenbuch*, 2nd edition, MAAG Gear-Wheel Company Ltd., Zurich/Switzerland, 1985, 727 pages.

Mark, W.D., *Performance-Based Gear Metrology: Kinematic-Transmission-Error Computation and Diagnosis*, Wiley, 2013, 269 pages.

Medveditskov, S.N., *High-Production Hobbing*, Mashinostroyeniye, Moscow, 1981, 104 pages. [in Russian].

Merritt, H.E., *Gear Engineering*, Putman Publishing, London, New York, Toronto, 1971, 489 pages.

Michalec, G.W., *Precision Gearing: Theory and Practice*, John Wiley & Sons, Inc, New York, 1966, 620 pages.

Minyang, Y., Heeduck, P. "The Prediction of Cutting Force in Ball-End Milling", *International Journal of Machine Tool & Manufacture: Design, Research and Application (Formerly: Machine Tool Design & Research)*, Vol. 31, No. 1, 1991, pp. 45–54.

Moiseyenko, O.I., *Gear Cutting Tools With Advanced Cutting Materials*, Vishcha Shkola, Kiev, 1986, 183 pages. [in Russian].

Moiseyenko, O.I., Tchkalova, O.N., *Hobs for High-Productive Cutting Hardened Gears*, Znaniye, Kiev, 1982, 20 pages. [in Russian].

Oberg, E., *Spur and Bevel Gearing*, The Industrial Press, New York, 1917, 274 pages.

Petrov, A.B., *et al.*, *Advanced Gear Shaving Processes*, Moscow, Moscow State Industrial University, 1997, 142 pages. [in Russian].

Phillips, J., *General Spatial Involute Gearing*, Springer, Berlin Heidelberg, 2003, 498 pages.

Romanov, V.F., *Calculations of Gear Cutting Tools*, Mashinostroyeniye, Moscow, 1969, 251 pages. [in Russian].

Saari, O.E., *Analytical Theory of Gear Tooth Surfaces*, Illinois Institute of Technology, 1961, 114 pages.

Shtipelman, B.A., *Design and Manufacture of Hypoid Gears*, John Wiley & Sons, New York, 1978, 394 pages.

Smirnov, N.N., "Multi-Thread Hob With New Cutting Diagrams", *Gear Technology*, July/August 1991, pp. 27–31.

Stadtfeld, H.J., *Handbook of Bevel and Hypoid Gears*, R.I.T, Rochester, NY, 1993, 251 pages.

Stadtfeld, H.J., *Gleason Bevel Gear Technology*, The Gleason Works, Division of Gleason Corporation, Rochester, NY, 1995, 202 pages.

Stadtfeld, H.J., *Advanced Bevel Gear Technology*, The Gleason Works, Division of Gleason Corporation, Rochester, NY, 2000, 343 pages.

Stadtfeld, H.J., *Gear Encyclopedia*, 2008 Edition, The Gleason Works, Division of Gleason Corporation, Rochester, NY, 2008, 462 pages.

Stadtfeld, H.J., *Gleason Kegelradtechnologie: Ingenieurwissenschaftliche Grundlagen und modernste Herstellunsverfahren für Winkelgetribe*, Expert-Verlag, Renningen, 2013, 491 pages.

Stadtfeld, H.J., *Gleason Bevel Gear Technology: Basics of Gear Engineering and Modern Manufacturing Methods for Angular Transmissions*, The Gleason Works, Rochester, NY, 2014, 503 pages.

Stadtfeld, H.J., *Gleason Bevel Gear Technology: The Science of Gear Engineering and Modern Manufacturing Methods for Angular Transmissions*, The Gleason Works, Rochester, New York, 2014, 491 pages.

Taits, B.A., *et al.*, *Production of Gears*, Mashinostroyeniye, Moscow, 1990, 463 pages. [in Russian].

*TCA: Tooth Contact Analysis Formulas and Calculation Procedure*, The Gleason Works, Rochester, New York, 1964.

Townsend, D.P., (1992), *Dudley's Gear Handbook. The Design, Manufacture, and Application of Gears.* 2nd Edition, McGraw Hill, New York, 1040 pages.

Tsepkov, A.V., *Profiling of Relieved Cutting Tools*, Mashinostroyeniye, Moscow, 1979, 150 pages. [in Russian].

*Understanding Tooth Contact Analysis*, The Gleason Works, Rochester, New York, 1981.

Watson, H.J., *Modern Gear Production*, Pergamon Press, New York, 1970, 359 pages.

Wildhaber, E., "Gear Tooth Curvature Treated Simply", *American Machinist*, Vol. 89, No. 18, August 30, 1945.

Wildhaber, E., "Basic Relations of Bevel Gears", *American Machinist*, Vol. 89, No. 20, September 7, 1945.

Wildhaber, E., "Relative Curvature Controls Gear Tooth Surface Strength", *American Machinist*, Vol. 89, No. 21, October 11, 1945.

Wildhaber, E., "Special Analysis of Gear Mesh Clarifies Curvature Conditions", *American Machinist*, Vol. 89, No. 22, October 25, 1945.

Wildhaber, E., "Basic Relationship of Hypoid Gears", *American Machinist*, Vol. 90, No. 4, February 14, 1946.

Wildhaber, E., "Basic Relationship of Hypoid Gears – II", *American Machinist*, Vol. 90, No. 5, February 28, 1946.

Wildhaber, E., "Basic Relationship of Hypoid Gears – III", *American Machinist*, Vol. 90, No. 6, March 14, 1946.

Wildhaber, E., "Tooth Contact", *American Machinist*, Vol. 90, No. 12, 1946.

Wildhaber, E., "Conjugate Tooth Surfaces", *American Machinist*, Vol. 90, No. 13, June 20, 1946.

Wildhaber, E., "Gear Tooth Sliding", *American Machinist*, Vol. 90, No. 15, July 18, 1946.

Wildhaber, E., "Skew Hypoid Gears", *American Machinist*, Vol. 90, No. 16, 1946.

Wildhaber, E., "Design for Duplex Cutting", *American Machinist*, Vol. 90, No. 17, August 15, 1946.

Wildhaber, E., *Fundamentals of Bevel and Hypoid Gearing*, Mashgiz, Moscow, 1948, 176 pages. [in Russian].

Wildhaber, E., *Foundations of Meshing of Bevel and Hypoid Gearings*, A collection of journal papers translated from English and comments by A.V. Slepak, Mashgiz, Moscow, 1948, 176 pages. [in Russian].

Wildhaber, E., "Surface Curvature", *Product Engineering*, May 1956.

Willis, R., "On the Teeth of Wheels", *Trans. Civ. Eng.*, Vol. II, 1838.

Willis, R., *Principles of Mechanisms, Designed for the Use of Students in the Universities and for Engineering Students*

*Generally*, J. & J.J. Deighton, London, John W. Parker, West Stand, Cambridge, 1841, 446 pages.

Winter, H., Stölzle, K., Placzek, T., *Topological Tooth Modifications and Contact Patterns of Spur and Helical Gears*, AGMA Paper 89FTM6, *AGMA*, 1989, 10 pages.

Woodbury, R.S., *History of the Gear-Cutting. A Historical Study in Geometry and Machine*, The M.I.T. Press, 1958, 1958, 135 pages.

*Worm Gearing*, Machinery's Reference Series, Number 1, 4th edition, The Industrial Press, New York, 1910, 48 pages.

Yeliseyev, Yu.S., *et al.*, *Production of Gears for Jet Engines*, Visshaya Shkola, Moscow, 2001, 593 pages. [in Russian].

Zhuravlev, V.L., *Production of Hypoid Gearings*, Mashinostroyeniye, Moscow, 1965, 152 pages. [in Russian].

# Index

**Note**: **Bold** page numbers refer to **tables** and *italic* page numbers refer to *figures*.

## H



## I

## Z

For Product Safety Concerns and Information please contact our EU
representative GPSR@taylorandfrancis.com
Taylor & Francis Verlag GmbH, Kaufingerstraße 24, 80331 München, Germany

www.ingramcontent.com/pod-product-compliance
Ingram Content Group UK Ltd.
Pitfield, Milton Keynes, MK11 3LW, UK
UKHW050059200926
475702UK00019B/2917
*9781032859262*

2ND EDITION

# Australian Mammals

## Biology and Captive Management

Editor: Stephen M. Jackson

CRC Press
Taylor & Francis Group
Boca Raton London New York

CRC Press is an imprint of the
Taylor & Francis Group, an **informa** business

Designed cover images: front cover: (top to bottom, left to right) Short-beaked Echidna puggle (photo by Stephen Jackson), Southern Hairy-nosed Wombat (photo by Stephen Jackson), Australian Sea Lion and pup (photo by Brad McKenzie), Dugong (photo by David Blyde), Koala (photo by Stephen Jackson), Quokka (photo by Stephen Jackson), Spotted-tailed Quoll (photo by Stephen Jackson), Dingo (photo by Stephen Jackson), Eastern tube-nosed Bat (photo by Stephen Jackson), Long-nosed Potoroo (photo by Stephen Jackson), pod of Common Bottlenose Dolphins (photo by David Blyde), Common Brush-tailed Possum (photo by Stephen Jackson)

This edition first published 2025
by CRC Press
2385 NW Executive Center Drive, Suite 320, Boca Raton FL 33431

and by CRC Press
4 Park Square, Milton Park, Abingdon, Oxon, OX14 4RN

*CRC Press is an imprint of the Taylor & Francis Group, an informa business*

First published by CSIRO Publishing. This edition published by arrangement with the Commonwealth Scientific and Industrial Research Organisation (CSIRO), Australia



ISBN: 9781041022916 (hbk)

Cover design by Cath Pirret

Typeset in Leo
by Envisage Information Technology

# Foreword

As past President of the Zoo and Aquarium Association and now running Taronga's two zoos, I have had the good fortune to work with Stephen Jackson on numerous topics and projects. Whether it be animal welfare, conservation or behavioural science, Stephen has always sought out the most contemporary research and practical experience available. Just as importantly, Stephen is always generous in sharing his knowledge for the improved welfare of wildlife.

In the second edition of this important book, Stephen has sought expertise from specific zoo keepers, veterinarians and field ecologists to provide us with deep insights into the care and management of our unique and fascinating Australian mammals.

Australia has been blessed with so many of the world's most distinctive and specialised animals, with no better an example than the platypus, featured in Chapter 2. Given that 87% of Australia's wildlife is not found anywhere else in the world, it is no wonder that Stephen's first edition of *Australian Mammals: Biology and Captive Management* quickly became the 'go to' for anyone caring for Australian wildlife. It has become an invaluable resource for those in zoos, in national parks, in research and conservation, as well as home wildlife carers.

As this new edition is published, we find ourselves in a situation where over 20% of Australia's native mammals are threatened, with numerous species on the brink of extinction. From the Regent Honeyeater to the iconic Koala, sadly I could rattle off too many examples of Australian species with a bleak outlook.

For many species, the only hope for their long-term return to robust populations in the wild will be intensive management and breeding programs. This second edition will be a critical support for professionals and volunteers alike.

On a positive note, the body of knowledge and understanding of the biology, behavioural science, husbandry and nutrition for Australian mammals has grown immensely since the first edition. Stephen has been meticulous in his attention to contacting those who are leading their specialisation to ensure this edition is comprehensive and reflects the latest understanding. This second edition will be a sought-after reference across the Australian wildlife community.

Cameron Kerr AO
Chief Executive, Taronga Conservation Society Australia

# Contents

# Introduction

Australian mammals have been held in captivity in Australia and throughout the world for over 230 years. Although they were originally kept as sources of curiosity, entertainment and study, today they are increasingly held to educate the public about their biology and threatened status, as part of captive breeding programs for conservation, for hand-rearing following the death of their mothers, for rehabilitation after injury or illness, for research and as pets.

When mammals are kept in captivity, they need to be held in facilities that optimise conditions for their welfare by providing high-quality nutrition, an appropriate physical environment, appropriate physical care to maintain good health, appropriate access to natural behavioural interactions, and these must result in a positive affective experience by the animals.

The husbandry of Australia's mammals in captivity is an expanding field, with earlier literature based largely on first-hand experiences of enclosure sizes, captive diet, behaviour and breeding. Although this information is highly valuable, there has been a need to bring together aspects of the wild biology (including wild diet, social behaviour, reproduction and nesting requirements) to maximise the husbandry provided in captivity. Publications such as the *Monotremes and Marsupials* (Collins 1973), *The Management of Australian Mammals in Captivity* (Evans 1982), *Care and Handling of Australian Native Animals* (Hand 1990), *Australian Mammals: Biology and Captive Management* (Jackson 2003a), *Haematology of Australian Mammals* (Clark 2004), *Medicine of Australian Mammals* (Vogelnest and Woods 2008), *Radiology of Australian Mammals* (Vogelnest and Allan 2015), *Current Therapy in Medicine of Australian Mammals* (Vogelnest and Portas 2025) and many published papers have made great advances in our knowledge of the husbandry of Australian mammals.

The aims of this second edition are to: (1) provide updated information on the current knowledge of the biology and husbandry of all Australia's native terrestrial and marine mammals and set out current accepted practice for captive management, and (2) utilise current knowledge of the wild biology and behaviour of the species or group to better understand how their animal husbandry can be improved and for more successful reproductive management in order to maximise the welfare of these animals in captivity. In order to achieve this each chapter includes as co-authors at least one zookeeper, veterinarian, and field ecologist in order to cover all aspects of the species or groups captive management requirements.

It is hoped that zookeepers, students, researchers, veterinarians, wildlife carers and the ever-expanding group of private individuals who keep Australian mammals as pets will find useful the information presented on the natural history, housing requirements, general husbandry, feeding requirements, handling and transport, health requirements, behaviour, breeding and artificial rearing. Although primarily focused on the management of Australian mammals in captivity, various aspects are of use to field biologists, including capture and restraint techniques, ageing techniques and behaviour and breeding information. It is also hoped that this edition will stimulate further improvement in the standard of husbandry of Australian mammals.

Despite attempts to incorporate as much published and unpublished information as possible, there are clearly numerous gaps in our knowledge that need to be filled. Areas of future development include fine-tuning diets, enclosure designs, area requirements, capture and restraint techniques, behavioural enrichment methods and population management techniques. The availability of animals within captive facilities also provides an opportunity to undertake significant research, including studies on taxonomy, ageing techniques, digestive physiology, social and reproductive behaviour, reproductive physiology such as oestrous cycles and gestation periods, artificial reproductive techniques, milk composition and growth and development. With this in mind, this book can be considered a consolidation of information for the start of a journey rather than an end and so readers are encouraged to further explore and publish their knowledge of the captive management of Australian mammals.

## ACKNOWLEDGEMENTS

Even more than for the first edition of this book, the development of this volume has relied on the assistance of numerous people from various institutions throughout Australia. In particular I would like to thank the 60 co-authors who have helped to significantly update this volume; I am very grateful to you all. In undertaking this project I have endeavoured to include the extraordinary knowledge that exists within the animal caretakers in the zoo industry and in numerous field biologists and veterinarians. This information has proved invaluable in making this work of greater quality and giving a broader perspective than a particular institution and therefore is more widely useful.

Many thanks also to Mark Hamilton and the team from CSIRO Publishing for all their hard work and patience in putting this work together. I also thank my family, including Kerstin, Olivia and James, for their support during the production of this book.

## OUTLINE

Each chapter covers a separate taxonomic group of Australian mammals and an effort has been made to make the scope of information covered as uniform as possible by using the husbandry manual outline described in Jackson (2003b). The common names and taxonomy used in this book follow Jackson and Groves (2015) and Baker and Gynther (2023) except where stated. The references throughout the text are located at the end of the relevant chapter. Given the large number of contributors, terminology may be inconsistent across all chapters, either because terminology may have changed over time or chapter authors may differ in which terminology they prefer. Nonetheless, efforts have been made to keep the terminology as consistent as possible between chapters.

## REFERENCES

Baker AM, Gynther IC (Eds) (2023) *Strahan's Mammals of Australia.* 4th edn. Reed New Holland Publishers, Sydney.

Clark P (2004) *Haematology of Australian Mammals.* CSIRO Publishing, Melbourne.

Collins LR (1973) *Monotremes and Marsupials: A Reference for Zoological Institutions.* Smithsonian Institution Press, Washington DC.

Evans DD (1982) *The Management of Australian Mammals in Captivity.* Zoological Board of Victoria, Melbourne.

Hand SJ (1990) *Care and Handling of Australian Native Animals.* Surrey Beatty & Sons in association with Royal Zoological Society of New South Wales, Sydney.

Jackson SM (2003a) *Australian Mammals: Biology and Captive Management.* CSIRO Publishing, Melbourne.

Jackson SM (2003b) Standardising husbandry manuals: guidelines for terrestrial vertebrates. *International Zoo Yearbook* **38**, 229–243. doi:10.1111/j.1748-1090.2003.tb02083.x

Jackson SM, Groves CP (2015) *Taxonomy of Australian Mammals.* CSIRO Publishing, Melbourne.

Vogelnest L, Allan G (2015) *Radiology of Australian Mammals.* CSIRO Publishing, Melbourne.

Vogelnest L, Portas T (Eds) (2025) *Current Therapy in Medicine of Australian Mammals.* Revised Edition. CSIRO Publishing, Melbourne.

Vogelnest L, Woods R (2008) *Medicine of Australian Mammals.* CSIRO Publishing, Melbourne.

# Preface to the second edition

Since the publication of the first edition of *Australian Mammals: Biology and Captive Management* in 2003 there has been an enormous increase in the knowledge of the biology and husbandry requirements of many species of Australian mammals - both terrestrial and marine. There have also been many advances in the understanding of animal welfare and the recognition that animals kept in captivity should not only be maintained in good health but also have a positive mental state. This lays down a fundamental challenge to animal managers to ensure that they maximise the physical and mental welfare of the animals they maintain in captivity. To reflect these changes, this edition contains a new chapter that specifically discusses the welfare of Australian mammals in human care. It also includes new chapters on the different groups of marine mammals, including dugong, seals and dolphins.

Advances in animal husbandry over the past 20 years include the more consistent breeding of platypus and short-beaked echidnas, the use of assisted breeding techniques such as cross-fostering in macropods, and artificial insemination in koalas. Despite these advances in our understanding of the biology and captive management of many species of Australian mammals, some species such as the marsupial mole remain an enigma due to the limited number of specimens discovered in the wild and the very few kept in captivity. As a result most chapters have changed considerably since the publication of the first edition, although some chapters have changed to a lesser degree. This typically reflects less effort being spent studying those animals in captivity and their requirements, which helps to highlight that more research is required to improve their management and welfare. Though many advances have been made in the nutritional requirements of Australian mammals held in captivity, there is still some disagreement and alternative approaches, so the diets provided here should be seen as a guide rather than definitive.

Genetic techniques have also dramatically increased the ability to diagnose disease, identify mating systems, identify the relatedness of individuals and understand the taxonomy of species. For example, since 2000 some 30 new species of Australian mammals have been recognised. The importance of correct taxonomy is highlighted by the feather-tailed glider, which was once recognised as one species but is now regarded as two species that partly overlap in their distribution. This may help explain why some facilities were able to readily breed this species while others were not as they may have unknowingly held both species together!

Sadly, since 2000 the decline and extinction of Australian mammals has continued at a rapid pace, despite the best efforts of many. This includes the extinction of the Bramble Cay melomys that has not been found since 2009–14, which appears to be the first species to vanish due to climate change, and the Christmas Island pipistrelle, which was also last seen in 2009. Many other once widely dispersed species have also collapsed in numbers and distribution in at least part of their range, including the koala, which appears to be suffering greatly from climate change. Various other species being severely affected by climate change include the mountain pygmy possum, lemuroid ring-tailed possum and several species of flying-fox. This highlights the potential importance of captive populations in the conservation of threatened species, though that should only be seen as a short-term solution due to the difficulties in maintaining captive populations over the long-term, including adaptation to captivity by individuals, inbreeding, and other genetic changes.

# List of contributors

**Guy Bedford**
Marine Mammal Specialist, Wildlife Consulting Services
Former Head Trainer of Marine Mammals, SeaWorld, Cold Coast, Qld, Australia.

**David J. Blyde** BVSc, MANZCVS (Zoo Medicine), MBA, MRCVS, GAICD, DipWildMed, GradCertCaptVertMgt
Veterinarian, Ex Sea World, Gold Coast, Qld, Australia.

**Rosemary J. Booth** BVSc
Wildlife Veterinarian, Manager Threatened Species Hidden Vale Wildlife Centre, Grandchester, Qld, Australia.
Associate Professor Wildlife Management, University of Southern Queensland, Toowoomba, Qld, Australia.

**Sara F.K. Brice** BASc
Registrar – Animal Populations, Taronga Zoo Conservation Society Australia, Mosman, NSW, Australia.

**Benn R. Bryant** BVSc, MVS
Senior Veterinarian, Taronga Western Plains Zoo, Taronga Conservation Society Australia, Dubbo, NSW, Australia.

**Kylie M. Cairns** BSc (Hons), PhD
Research Fellow, Centre for Ecosystem Science; School of Biological, Earth and Environmental Science, University of New South Wales, Sydney, NSW, Australia.

**Michelle Campbell-Ward** BSc, BVSc (Hons I), PGCertIAWEL, MANZCVS (Zoo Medicine), DZooMed (Mammalian), MRCVS
Associate Professor in Wildlife and Avian Medicine, Sydney School of Veterinary Science, University of Sydney, Camden, NSW, Australia.

**Lauren Clark** BAppSc (Animal Studies)
Senior Mammal Keeper, Currumbin Wildlife Sanctuary, Currumbin, Qld, Australia.

**Amy L. Coetsee** BSc (Hons), PhD
Threatened Species Biologist, Wildlife Conservation & Science, Zoos Victoria, Melbourne, Vic., Australia.

**Karen E Coomes** BSc (Hons), PhD
Director and Chair, Tree Roo Rescue and Conservation Centre Ltd., Malanda, Qld, Australia.

**Christine E. Cooper** BSc (Hons), PhD
Senior Lecturer, School of Molecular and Life Sciences, Curtin University, Perth, WA, Australia.

**Nick de Vos**
Former supervisor, Australian Mammal Precinct, Taronga Conservation Society Australia, Mosman, NSW, Australia.

**Chris R. Dickman** BSc (Hons), PhD
Professor in Terrestrial Ecology, School of Life and Environmental Sciences, University of Sydney, Sydney, NSW, Australia.

**Kerrie Dixon** CertIIICapAnimMan
Curatorial Supervisor, Sea Life Sydney Aquarium, Sydney, NSW, Australia.

**Sarah Eccleston** CertIIICaptivAnimMan
Koala Specialist Mammals Supervisor, Currumbin Wildlife Sanctuary, Currumbin, Qld, Australia.

**Bronwyn A. Fancourt** GradCertRes, BSc (Hons I), BCom, PhD.
Principal Conservation Officer, Queensland Parks and Wildlife Service, Department of Environment, Science and Innovation, Toowoomba, QLD, Australia.

Adjunct Research Fellow, School of Environmental and Rural Science, University of New England, Armidale, NSW, Australia.

**Arthur Ferguson** Cert III Zookeeping
Supervisor, Zoology Australian Fauna, Perth Zoological Parks Authority, Perth, WA, Australia.

**Peter J.S. Fleming** BSc (Hons), MResSc, PhD
Senior Principal Research Scientist, Vertebrate Pest Research Unit, NSW Department of Primary Industries, Orange, NSW, Australia.

**Claire M. Ford** BSc (Zoology)
Manager Population – Development and Welfare, Welfare Conservation and Science, Taronga Conservation Society Australia, Sydney, NSW, Australia.

**Danielle Fox**
Marine Mammals, Taronga Zoo, Taronga Conservation Society Australia, Sydney, NSW, Australia.

**Amber Gillett** BVSC (Hons), PhD
Senior Veterinarian and Research Coordinator, Australia Zoo Wildlife Hospital, Beerwah, Qld, Australia.

**Wendy Gleen** BSc, MSc (Res)
Senior Keeper, Australian Mammal Division, Taronga Zoological Society, Sydney NSW Australia.
Currently, Landcare Support Officer, Lake Macquarie City Council Landcare, Booragul, NSW, Australia.

**Ross L. Goldingay** BSc (Hons), PhD
Associate Professor, Faculty of Science and Engineering, Southern Cross University, Lismore, NSW, Australia.

**Peter H. Holz** BVSc, DVSc, MACVSc (Zoo Animal Medicine), PhD, Dipl.ACZM
Associate Veterinarian, Healesville Sanctuary, Zoos Victoria, Healesville, Vic., Australia.

**Frances A. Hulst** BVSc (Hons), MVetStud (WildMedHusb)
Veterinarian, Taronga Zoo Conservation Society Australia, Mosman, NSW, Australia.

**Stephen M. Jackson** BSc, MSc, PhD, FLS
Australian Museum Research Institute, Australian Museum, Sydney, NSW, Australia.
Adjunct Associate Professor, School of Biological, Earth and Environmental Sciences, University of New South Wales, Sydney, NSW, Australia.

**Tina Janssen** CertIVCaptAnim, Cert IVAssessWorkplTraining
Founder and Chairperson, Safe Haven - AACE (Australian Animals Care and Education Inc.) Sanctuary, Mt Larcom, Qld, Australia.

**Stephen D. Johnston** BSc (Zool), PhD, FSRB
Associate Professor, Reader in Reproduction (Wildlife), School of Agriculture and Food Science, University of Queensland, Gatton, Qld, Australia.

**Danielle Jose** BBiolEnvirSci (Hons)
Former Senior Technical Officer, Perth Zoo, Perth, WA, Australia.

**Belinda Laming** BSc, Cert III Zookeeping
Supervisor Zoology, Perth Zoological Parks Authority, Perth, WA, Australia.

**Lindy F. Lumsden** BSc, PhD, PSM
Principal Research Scientist, Arthur Rylah Institute for Environmental Research; Department of Energy, Environment and Climate Action, Heidelberg, Vic., Australia.

**Daniel Lunney** BSc, MSc, DSc (*honoris causa*) from University of Sydney and Murdoch University, Perth, WA, Australia.
Honorary Scientific Fellow; Science, Economics and Insights Division, Department of Planning and Environment, Sydney, NSW, Australia.
Adjunct Professor, School of Life and Environmental Sciences, University of Sydney, NSW, Australia.
Research Associate, Australian Museum, Sydney, NSW, Australia.

**Michael J. Lynch** BSc, BVSc, PhD, MANZCVS (Epidemiology)
Head of Veterinary Services, Life Sciences, Melbourne Zoo, Zoos Victoria, Melbourne, Vic., Australia.

**Rachel Lyons** BRTP, GradDipEnvManagement
Trainer and Species Coordinator, Wildcare Australia, Noosa, Qld, Australia.

**Paola Unda Marron** DVM
Veterinary Consultant, MarineLife Animal Management Advisors, Singapore.

**Helene Marsh** BSc, PhD
Emeritus Professor in Environmental Science and a Professorial Fellow, James Cook University, Townsville, Qld, Australia.

**Jennefer Mclean**
Director, Tolga Bat Rescue and Research, Atherton, Qld, Australia.

**David McLelland** BSc (Vet), BVSc, DVSc, MANZCVS(Zoo Medicine), Dipl.ACZM, DECZM (Zoo Health Management)
Veterinarian, Adelaide Zoo, Adelaide, SA, Australia.

**Justin Meager** BSc (Hons), PhD
Senior Marine Scientist (Marine Mammals), Threatened Species Operations, Department of Environment and Science, Brisbane, Qld, Australia.

**Stewart C. Nicol** BSc (Hons), PhD
Associate Professor, University of Tasmania, Hobart, Tas., Australia.

**Karen Nilsson** BSc (Zoology); GCCaptVertebrateManagement
Head of Koalas, Lone Pine Koala Sanctuary, Brisbane, Qld, Australia.

**Julie Old** BSc (Hons), PhD, MHEd, SFHEA
Associate Professor, School of Science, Western Sydney University, Penrith, NSW, Australia.

**Marissa L. Parrott** BSc (Hons), CertIISustEnvir, CertIIICaptAnim, PhD
Reproductive Biologist, Wildlife Conservation & Science, Zoos Victoria, Melbourne, Vic., Australia.

**Yvette Pauligk** BSc, BEd (Prim), CertIIICaptAnim
Natives Life Sciences Manager, Werribee Open Range Zoo, Zoos Victoria, Werribee, Vic., Australia.

**Benjamin J. Pitcher** BSc (Hons), PhD
Behavioural Biologist, Taronga Conservation Society Australia, Sydney, NSW, Australia.
School of Natural Sciences, Faculty of Science and Engineering, Macquarie University, North Ryde, NSW, Australia.

**Vicki L. Power** CertVetNurs, CertAnimSciTech, CertZookeeping, CertIVTrainAssess
Senior Technical Officer Zoology, Perth Zoological Parks Authority, Perth, WA, Australia.

**Peggy Rismiller** MSc, PhD, OAM
Environmental Physiologist, Echidna Researcher, Pelican Lagoon Research & Wildlife Centre, Kangaroo Island, SA, Australia.
University of Adelaide, School of Biological Sciences, Adelaide, SA, Australia.

**Alice Roser** CertIIICaptAnim, CertIVZookeeping
Native Mammal Supervisor, Currumbin Wildlife Sanctuary, Currumbin, Qld, Australia.

**Melody Serena** OAM, PhD
Conservation Biologist, Australian Platypus Conservancy, Campbells Creek, Vic., Australia.

**Bradley P. Smith** BPsych (Hons), PhD
Senior Lecturer in Psychology, School of Health, Medical and Applied Sciences, Central Queensland University, Adelaide, SA, Australia.

**Hayley Stannard** BAnimSc, BSc (Hons), GCLTHE, PhD
Senior Lecturer in Animal Anatomy and Physiology, School of Agricultural, Environmental and Veterinary Sciences, Charles Sturt University, Wagga Wagga, NSW, Australia.

**Louise Stockburger** DipLeadManag, DipRenewabResManag, CertIIICaptAnim, CertIVTrainAssess
Course Coordinator Cert III Wildlife and Exhibited Animal Care, Taronga Training Institute, Taronga Conservation Society Australia, Sydney, NSW, Australia.

**David A. Taggart** BSc, PhD
Adjunct Associate Professor of Wildlife Biology and Conservation Research, School of Animal and Veterinary Science, University of Adelaide (Waite Campus), Urrbrae, SA, Australia.
Principal Scientist, FAUNA Research Alliance, Kahibah, NSW, Australia.

**Peter D. Temple-Smith** PhD
Associate Professor, Department of Obstetrics and Gynaecology, School of Clinical Sciences at Monash Health, Monash University, Melbourne, Vic., Australia.

**Jessica L. Thomas** BSc, PhD, GDRS, MRS
Platypus Specialist, Healesville Sanctuary, Healesville, Vic., Australia.

**Joy S. Tripovich** GradDipBiolSci, BSc, PhD
Taronga Institute of Science and Learning, Taronga Conservation Society Australia, Sydney, NSW, Australia.
Centre for Ecosystem Science, School of Biological, Earth and Environmental Sciences, University of New South Wales, Sydney, NSW, Australia.

**Ludovica Valenza** BVSc (Hon), BSc (Hons), BAVBSc
Wildlife Hospital Manager & Veterinarian, Australia Zoo, Beerwah, Qld, Australia.

**Rebecca Vaughan-Higgins** BSc, BVMS, CertAVP (Zoo Medicine), PhD, DipECZM (Wildlife Population Health)
Veterinary Services Manager, Perth Zoo, Perth, WA, Australia.
Senior Lecturer, Wildlife and Conservation Medicine, Murdoch University, Perth, WA, Australia.

**Karl Vernes** BSc, MSc, PhD
Professor Mammal Ecology and Conservation, School of Environmental and Rural Science, University of New England, Armidale, NSW, Australia.

**Larry Vogelnest** BVSc (Hons), MVetStud (WildMedHusb), MANZCVS (Zoo Medicine), PSM
Senior Veterinarian, Taronga Zoo and NSW Sentinel Koala Population Monitoring Program, Taronga Conservation Society Australia, Sydney, NSW, Australia.

**John W. Winter** PhD
Wildlife Ecologist, Atherton, Qld, Australia.

# 1
# Welfare of Australian mammals in human care

*Claire Ford and Michelle Campbell-Ward*

## 1. INTRODUCTION

Australian native mammals may be cared for under the stewardship of humans in a variety of contexts, including (1) in zoos and fauna parks, (2) within large fenced semi-wild sanctuaries, (3) in wildlife rehabilitation facilities, (4) in research programs, (5) in conservation programs, and (6) as pets.

When animals are under direct or indirect human care, society expects them to be protected from harm and provided with conditions that support good animal welfare. This expectation stems from the moral perspective that all animals have intrinsic value. The maintenance of high standards of welfare is essential to meet professional and legal obligations and to retain the social licence to operate in an animal-related industry. Good animal welfare has been demonstrated scientifically to influence capacity to maintain self-sustaining zoo populations (Mason 2010) and in complex and multifaceted ways, there is increasing recognition that the welfare of animals in human care is inextricably connected with the wellbeing and culture of people (Wünderlich *et al.* 2021).

Animal welfare science relating to non-domesticated species is rapidly expanding. It explores how human actions and the environments created for animals directly influence welfare (Morgan and Tromborg 2007). The pursuit of higher welfare standards has driven remarkable advances in the husbandry, veterinary science and care of wild animals (Kirkwood 2003). This chapter outlines important animal welfare considerations relevant to the care of Australian mammals. It aims to define what animal welfare is, how it can be measured and assessed and how our actions can affect it.

## 2. DEFINITION OF ANIMAL WELFARE

Animal welfare is a complex concept without a universally agreed simple definition. One useful approach is to think of it as how animals experience the world in which they live encompassing both an animal's physical and mental state. An animal's welfare state is ultimately determined by the balance of positive and negative experiences. These experiences may be of varying intensity, duration and frequency and, rather than being fixed and unchanging, welfare exists on a continuum, shifting up and down over time (Mellor and Reid 1994; Melfi 2009).

Welfare is of most relevance at the individual animal level. It is subjective and self-determined, rather than determined by an observer or caregiver (Kagan and Veasey 2010) who can only infer an animal's welfare from physical, behavioural and/or situational indicators. To create conditions that support good welfare, caregivers should not just be minimising negative experiences associated with poor welfare, but actively promoting positive experiences (Melfi 2009; Mellor and Beausoleil 2015). Unlike free-ranging wildlife that are usually observed, studied and managed at the population level, animals cared for in facilities such as zoos, fauna parks and homes are typically managed as individuals. Consequently, there is capacity to examine animal welfare in these settings with far greater resolution.

## 3. SENTIENCE AND AUSTRALIAN MAMMALS

All vertebrates (including Australian mammals) and some invertebrates are considered sentient beings (Mellor 2019). Sentience is the capacity of an animal to perceive via the senses, to consciously feel emotions or to experience subjectively (Mellor 2019). These subjective experiences are termed 'affects' and encompass positive affects such as pleasure and comfort that are welfare enhancing and negative affects such as pain and fear that are welfare compromising (Yeates and Main 2008). These sensations, feelings and emotions play a role in modifying an animal's behaviour (Duncan and Petherick 1991; Fraser and Duncan 1998; Balcombe 2009; Denton *et al.* 2009; Mellor *et al.* 2009; Mellor and Beausoleil 2015; Mellor 2016). Formal and legislated recognition that some animals, including all mammals, are sentient beings is increasing internationally. From an ethical perspective, it is widely held that sentient beings are worthy of moral consideration.

Although specific research evaluating conscious perception is lacking for most Australian mammal species, the major features of brain processing of sensory inputs that are associated with subjective experiences appear to be common to all vertebrates (Baars 2001; Butler *et al.* 2005; Panksepp 2005; Boissy *et al.* 2007; Panksepp *et al.* 2011).

Given their short gestation and extreme immaturity at birth, marsupials have been included in studies evaluating the development of sentience. Research assessing differences in species that are profoundly immature at birth (wallabies), moderately immature at birth (rats) and mature at birth (lambs) reveals that the onset of conscious perception is not an 'on-off' phenomenon but rather more gradual (Diesch *et al.* 2008). The development of tammar wallaby (*Notamacropus eugenii*) joeys in this research revealed that:

- until around 120 days after birth they are not capable of conscious perception and hence cannot experience pain, based on a lack of behavioural or encephalographic (EEG) signs of conscious awareness
- behavioural signs of conscious perception are apparent by 160–180 days after birth, at which stage it would be expected that joeys are able to experience pain and, if it is noxious enough, to suffer
- the period between 120 and 160–180 days after birth is a transitional period during which major developmental changes in the nervous system occur (Diesch *et al.* 2010).

These observations suggest that for most marsupials the development of sentience occurs after approximately one-third to one-half of pouch life (Mellor *et al.* 2010).

The development of sentience in other groups of Australian mammals has not been specifically studied. Based on domestic animal research, however, it is thought that in the eutherian species with neurologically mature young at birth, there is fetal capacity for sentience in the last 20–25% of time spent in utero and neonatal consciousness within minutes or hours after birth (Mellor 2010).

## 4. MODELS AND ORIENTATIONS FOR ASSESSING AND UNDERSTANDING ANIMAL WELFARE

### 4.1 The five animal welfare domains

The Five Domains model is a well-recognised framework for assessing and managing animal welfare. It was originally developed in the 1990s (Mellor and Reid 1994) and has since undergone several refinements (Mellor and Beausoleil 2015; Mellor 2016; Mellor *et al.* 2020). Although there are various models for assessing and understanding welfare, the Five Domains model is widely utilised because it can be applied broadly across all taxa, it directs focus onto the animal's experience (rather than provisions), and it promotes continuous improvement through critical thinking (Beausoleil 2023).

The model recognises that animals can experience a range of feelings (positive and negative subjective experiences) and provides examples of some internal states and external circumstances from which these experiences can be inferred (Fig. 1.1). Each domain described in the model draws attention to specific areas relevant to welfare assessment and management (Mellor 2017).

The structure of the model includes four physical/functional domains:

- Domain 1: Nutrition – refers to the impact of nutrition and hydration.
- Domain 2: Physical environment – refers to the impact of the physical environment and atmospheric conditions (e.g. thermal conditions, humidity, space, noise, light, odours and substrates).
- Domain 3: Health – refers to the impact of health and functional state of the animal, including impacts of disease, injury and level of physical fitness.
- Domain 4: Behavioural interactions – refers to the impact of behavioural interactions on the state of the animal (i.e. interactions with the physical environment, with other animals and with humans).

The first three domains (nutrition, physical environment and health) refer to experiences that can influence the inner stability of the body and trigger behaviours associated with survival, whereas the fourth domain (behavioural interactions) directs attention to the external situation and the animals' interpretation of this. Combined, these four domains give rise to the fifth domain.

- Domain 5: Affective state – the emotions or feelings (affects) that ultimately determine an animal's welfare state, driven by what is important to the individual animal.

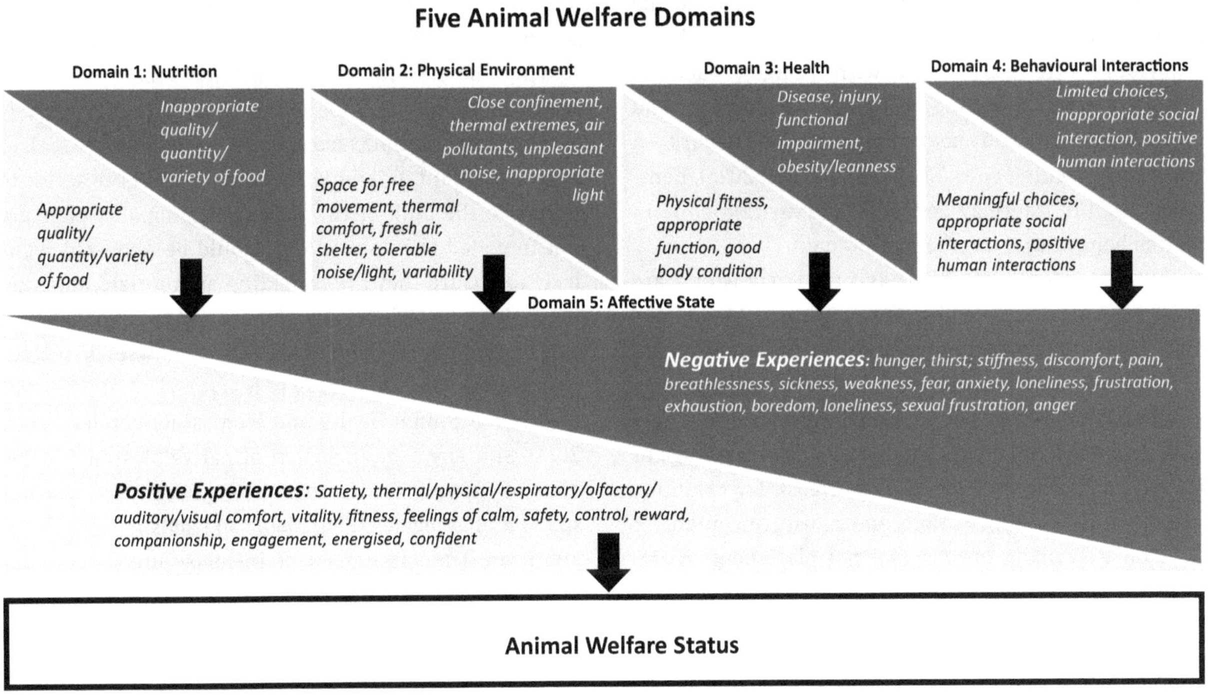


**Fig. 1.1.** The Five Domains Model of animal welfare. Adapted from Mellor and Beausoleil (2015) and Mellor *et al.* (2020).

The domains are explained in more detail in sections 4.1.1–4.1.5.

In general, the lives of free-ranging wild animals are significantly occupied by the first three domains: nutrition, physical environment and health. These are termed survival-critical factors (Mellor and Beausoleil 2015), with animals spending large proportions of their time dedicated to finding food, water and shelter and avoiding predation or injury. In a zoo or other animal care environment, domains 1–3 are generally addressed with food, water, shelter and health care provided (albeit with varying levels of satisfaction). Consequently, domain 4 (the behavioural interactions domain encompassing situation-related factors) is typically the focus for welfare enhancement in these settings (Mellor and Beausoleil 2015). This will be explored later in the chapter.

### 4.1.1 Domain 1 – Nutrition

Nutrition is the first of the survival-critical domains. Animals need access to food and water of sufficient quality and quantity to avoid prolonged hunger and thirst, and to experience satiation and quenching of thirst. An excessively negative hunger or thirst affect will impede all other states until this urgent need is satisfied (Shaw and Jarman 2025).

The nutritional composition and presentation of food and water need to be carefully considered for each animal or animal group to support good welfare. This does not necessarily mean feeding ad libitum. A well-balanced diet with a caloric density and nutritional composition similar to the natural diet should result in satiety when energy and nutrient needs are met, thus supporting overall health. This must be monitored by assessing and documenting feed intake, feeding behaviour, body condition/weight, faecal consistency and signs of gastrointestinal comfort or discomfort (Shaw and Jarman 2025). Some species are generalists that can eat a wide variety of foodstuffs, whereas others are specialists that eat a very narrow range of highly specific natural food items.

Obesity-related malnutrition can negatively impact welfare as much as underfeeding. Factors that may lead to overeating need to be understood, anticipated and prevented (Shaw and Jarman 2025). Some Australian mammals, including possums and gliders, are particularly predisposed to obesity and so excessive amounts of artificial nectars (typically high in simple sugars) must be avoided. This risk is compounded because the energy requirements of marsupials are typically lower than those of equivalent eutherians (Hume 2005). Diets that do not provide nutrients within the range of an animal's natural intake, such as a herbivore diet that is too high in carbohydrate and too low in fibre, may disrupt feedback mechanisms for satiation (Alcock *et al.* 2014). Boredom, dominance and displacement may be addressed with novel approaches to presenting feeds, offering more feeding stations and changing feeding schedules. Behaviours often associated with hunger, such as increased activity, pacing, begging, vocalisation and aggression may also be anticipatory behaviours in response to feeding cues (Shaw and Jarman 2025), demonstrating the interplay between the nutrition domain and the behavioural interactions domain.

Nutrient value, palatability and food acquisition are equal priorities for sound nutritional care. A diet may look, taste and feel natural yet lead to nutrient deficiencies, energy excess or behavioural issues if it is unbalanced and not prepared or presented appropriately (Shaw and Jarman 2025). The quality, harvesting and storage of browse for koalas (*Phascolarctos cinereus*), for example, influences the nitrogen, fibre and moisture content of the food eventually offered and pot placement in the enclosure affects the koalas' access to and selection of preferred leaf types (Freeman 2023). There may be a variety of textures, tastes and flavours in the natural diet, but even specialists such as those that feed exclusively on nectar or termites may have rewarding experiences generated through creative form and presentation of the diet (Hosey *et al.* 2013). The ability of an animal to exercise food preference is considered desirable, but does not mean allowing an animal to consume only favoured food items (Shaw and Jarman 2025). Identifying the higher value items in an animal's balanced diet and reserving them or a proportion of them for training, conditioning, enrichment and health programs is an important part of a successful nutrition management plan.

Where possible, allowing animals to access their food in a manner that is natural to them has welfare benefits. Grazing species, such as wombats and some macropods, should where possible, be provided with access to native grass pastures to feed on. The maintenance of these pastures is just as important as the sourcing of substitute foods such as hay and leafy greens.

The amount and variety of foods an individual consumes may be influenced by many factors, including age, reproductive status, time of year (thermal requirements), fitness, metabolism, health status and, if in a hierarchical group, the position of the individual. Therefore, in large groups, the presence of small amounts of a range of ingredients at each feeding is not necessarily sufficient to ensure all animals are consuming a varied and balanced diet. One strategy to address this is to vary items throughout the week, with a larger volume each time an item is offered, rather than giving a small amount of all food items every day. This prevents an individual from hoarding the same item every day, leaving the less favoured items for subordinates and resulting in malnutrition of the group. Another strategy is to include multiple feed stations so that one animal, or more dominant individuals, cannot restrict the access of others to the bulk of the ration. Additionally, diets are not intended to be static and should be reviewed regularly to ensure they are providing appropriate nutrition for each group or individual (Shaw and Jarman 2025).

The regular inclusion of native browse is recommended for some species (e.g. macropods and possums) (Fig. 1.2) (National Health and Medical Research Council 2014). Similarly, the inclusion of native blossoms in the diets of flying foxes (*Pteropus* sp.), gliders (e.g. *Petaurus* sp.) and phascogales (*Phascogale* sp.) adds seasonal variety. Even if a commercial or in-house mix is used for short-beaked echidnas (*Tachyglossus aculeatus*), access to termites, termite mounds and/or soil from around termite mounds is recommended to inoculate the gut and promote a natural microbiome (Shaw and Jarman 2025).

**Fig. 1.2.** The provision of native browse (lilly pilly, *Syzygium* sp.) to a swamp wallaby (*Wallabia bicolor*) is associated with a positive welfare state by satisfying nutritional and sensory needs. Photo by Louisa Todd/Taronga Conservation Society Australia.

When formulating diets, consideration should be given to the digestive physiology of the species, the importance of supporting a natural gastrointestinal microbiome and specific nutritional requirements. Commercial products may be suitable for some species, but careful attention to balance and supplementation requires professional advice, (e.g. from a veterinarian or animal nutritionist). Commercially produced insects require gut loading to supplement levels of calcium and various vitamins that would otherwise be deficient. Supplementation with thiamine and vitamin E is essential for species fed stored fish.

Access to diets designed for other species or food that is not part of the formal ration can introduce nutritional and health problems. For example, quokka (*Setonix brachyurus*) deaths due to mesenteric fat necrosis have been postulated to be due to ingestion of the meat-based diet provided for co-housed echidnas (Hulst *et al.* 2010). This prompted the redesign of echidna feeders to prevent non-target species being able to access the food.

There are some large gaps in our knowledge of the natural dietary composition of some species. For example, complete nutritional analyses of the seagrasses consumed by dugongs (*Dugong dugon*) and the range of prey items taken by platypuses (*Ornithorhynchus anatinus*) have not yet been conducted. For such species, variety and consideration of supplementation are necessary as is remaining abreast of emerging field research.

### 4.1.2 Domain 2 – Physical environment

The focus of domain 2 is the environmental factors that influence an animal's welfare. These include the space available for movement, the ambient temperature, the air and water quality, including the presence of noxious or pleasant odours, the light intensity and acoustics. The predictability or variability of any of these factors and the natural environments to which the species is adapted are also relevant. Monitoring how animals use the space in which they live is important, because environmental elements provided to support positive welfare are only meaningful if they are utilised. For instance, a spatially large enclosure area is of limited value if the animal confined within it is timid and only inhabits small densely vegetated areas to avoid feeling exposed in open areas. Natural ecology and the environment in which species evolved are recommended reference points for understanding the factors that are likely to be important for the physical environment of a given species in human care.

Some considerations for creating a physical environment that promotes positive animal welfare include:

- space – sufficient space supports positive affects associated with expression of natural behaviours (e.g. exploration) and meaningful choices with respect to social interactions and environmental conditions. It also enables the incorporation of a range of enclosure furnishings (Browning and Maple 2019) to promote complexity and environmental variability. For many species, particularly arboreal species, the three-dimensional space of the enclosure is highly relevant. Height should be considered together with floor area to maximise usable space. Space is likely to be especially important to animals that naturally have large home ranges and/or large flight distances.
- complexity of space – physical complexity within an enclosure includes components such as water features, substrates, climbing structures, tunnels, burrows, logs, pathways, refuges and climatic variables (e.g. light, temperature, humidity). These features support a diversity of significant behavioural, physical and cognitive opportunities and experiences for an animal or group of animals that promote positive affects such as physical comfort. The complexity and usability (quality) of space is often more relevant for animal welfare than the physical size of the space (quantity) (Browning and Maple 2019).
- microclimates – variable temperature, humidity and light gradient offers animals choice when seeking out the positive affects of thermal and solar comfort. In the context of global climate change, the provision of shade/shelter structures (natural or artificial), appropriate areas with access to the sun for basking and water sources for cooling down are increasingly relevant in outdoor enclosures. Koalas seek out cool tree trunks to enhance heat loss (Briscoe *et al.* 2014), highlighting the important role live plants and trees can play in supporting variable microclimates and consequently animal welfare.
- substrates – a variety of suitable substrates support physical comfort and rest. Key considerations include drainage, compaction and soiling (faecal build up). Substrates are likely to be particularly relevant for species that burrow or dig, such as bandicoots, bettongs and potoroos, and good quality substrates that support natural behaviours should be provided.

Substrates such as woodchips can also provide enrichment opportunities with buried invertebrates sought out through digging behaviours.

- refuge areas – opportunities for refuge are an important component of the physical environment. These include refuge from conspecifics or other animals, human caregivers, visitors, inclement weather and environmental elements that might be stressors (e.g. noisy human activities). The type of refuge will vary from species to species and examples include nest boxes, vegetation, perches, rock refuges, shelters and dens. Other animals of the same species can also be considered a refuge because some animals seek security within a social group structure. Whiptail wallabies (*Macropus parryi*) are considered one of the more social macropod species and derive benefits from social refuge, such as group vigilance (alert, monitoring of surrounds for predators) and warning (Kaufmann 1974).
- variability – judiciously changing the animals' physical environment can provide ongoing novelty and enrichment and supports good welfare. A dynamic or variable environment may include environmental enrichment devices (e.g. automatic feeders) or changeable environmental features (e.g. sprinklers). Enclosure and raceway design that facilitates rotation of animals between areas has been advocated as a mechanism to increase available space and provide stimulating behavioural opportunities (Coe 2004; Coe and Hoy 2020). These dynamic interventions encourage exploratory behaviours and their associated physical activation and mental stimulation benefits.
- safety –foundational for good animal welfare. Some considerations include prevention of physical injury (such as entanglement or other trauma through collision with barriers), exclusion of pests and potentially toxic plants and biosecurity measures to minimise disease risk.

The physical environments experienced by animals involved in mobile exhibitions or interactive displays that require them to leave their fixed housing should be carefully considered for their impact on welfare. This includes both transport containers and holding at temporary presentation sites. Potential environmental welfare risks include: frequent space confinement, uncontrolled environmental parameters (e.g. light, temperature, humidity, noise, vibration, olfaction) and reduced capacity to provide a species-appropriate, complex and safe space (Warwick *et al.* 2023). The duration, frequency and intensity of the experience, evaluation of species and individuals involved and the external or situational circumstances should all be factors in an assessment of the suitability of these types of environments.

Those responsible for designing, building and furnishing enclosures for animals should aim to create complex, dynamic environments (i.e. constructed habitats, hereafter referred to as enclosure habitats), with features relevant to the species' natural ecology and lifestage, and provide opportunities for animals to carry out appropriate behaviours, supported by environmental enrichment. The inference is that by utilising these features, animals experience positive affects and therefore wellbeing is enhanced.

### 4.1.3 Domain 3 – Health

The association between physical health and animal welfare is well recognised and the importance of biosecurity and minimising the risk of, and adequately treating, injury and disease is uncontested. This is the focus of domain 3, but it is supported by and overlaps with the other functional domains. Modern veterinary, nutritional and enrichment programs for Australian mammals aim to prevent health issues and promote physical wellness. It is crucial that daily routines for caregivers include monitoring the physical fitness and body condition of their animals, checking for any signs of injury or disease and seeking veterinary advice as appropriate to address acute health issues and manage chronic ones. Minimising disease and injury promotes comfort and functional capacity while reducing negative experiences such as pain and debility. Additionally, it is hypothesised that a positive affective state boosts immunity and improves physical health (Melfi 2009).

The prevention of many common diseases in Australian mammals relies heavily on the provision of good husbandry. Some examples are given below.

- Macropod progressive periodontal disease (MPPD), previously termed 'lumpy jaw', is a debilitating and painful dental condition that develops when dietary, environmental and social factors affect the physical integrity of and microbial populations in the oral cavity of kangaroos and wallabies (McLelland 2025). Healthy dentition is necessary to enable animals to eat normally. The resulting satisfaction of hunger, maintenance of good body condition and strength

and intake of essential nutrients give rise to the positive affects associated with the comfort of satiation, good health and functional capacity (Mellor *et al.* 2020). Conversely, an animal with advanced MPPD is more likely to experience the negative affects of hunger, pain and the frustration of impaired foraging ability.
- Diseases of the eye associated with exposure to ultraviolet radiation and subpar water quality are common and frequently progressive or recurrent in captive Australian pinnipeds (Bodley 2025). Vision is an important sense for pinnipeds and their ocular anatomy is adapted to function well both in the air and under water (Hanke *et al.* 2009). Ocular disease can compromise welfare by causing discomfort, pain, abnormal light sensitivity and vision loss (Gage 2012).
- Many species that are flighty (readily startled) in nature necessitate the use of appropriately trained, skilled and experienced staff, careful enclosure design and planning for handling events to avoid traumatic injuries, such as those that can result from colliding with enclosure barriers and obstacles.

Some physical features of Australian mammals serve a multitude of purposes and their integrity or impairment can impact their welfare in a variety of ways. For instance:

- The wing membranes of bat species are critical for normal mobility (flight) and maintenance of fitness, but they also play a role in thermoregulation, as a cradle for nursing pups and (for some species) camouflage or capturing prey (Olsson and Woods 2008). This means that damage to a wing membrane can adversely impact welfare by restricting flight as well as interfering with mothering, hunting and feelings of security.
- Platypuses, with their relatively poor vision, use a combination of mechanoreceptors and electroreceptors in the bill to locate their invertebrate prey (Pettigrew 1999). In addition to these specialised sensory functions, the bill houses the nostrils necessary for normal respiratory function and is used for prehending food. The integrity of this unique anatomical structure is therefore critical to platypus welfare.

### 4.1.4 Domain 4 – Behavioural interactions

The primary focus of domain 4 is evidence of hindered and/or enhanced behavioural interactions with the environment, other non-human animals and human beings. It captures the behavioural outputs that reflect an animal's perception of external circumstances or situational-related affects. Behavioural interactions are influenced by genetics, learning and engagement with the physical, biological and social environment beyond an animal's immediate needs for survival.

Positive experiences are generated by stimulating environments that offer animals opportunities to engage in voluntary and purpose-driven behaviours that they find rewarding. Typically, this may include motivated environmental exploration, foraging, bonding, sexual activity, maternal reward from rearing young and play (Mellor *et al.* 2015; Mellor 2017). Conversely, a barren, small environment that restricts options to interact in a species-appropriate fashion is likely to result in negative affects. Such negative affects may be experienced through:

- daily thwarting of normal long-duration grazing motivation in species such as wombats and kangaroos without access to native grasses
- predators not provided with a hunting substitute
- social species being kept isolated
- solitary species housed in close quarters with other individuals
- transient human company for species or individuals that are bonded to a particular person or people in general.

For example, a study examining the prevalence and cause of stereotypic behaviour (repetitive and unvaried behaviour with no apparent goal or function) in zoo-based bare-nosed wombats (*Vombatus ursinus*) revealed that this species is susceptible to the development of locomotor stereotypies. The study suggested that frustration of the foraging motivational system was the underlying reason for the development of the abnormal behaviour (Hogan and Tribe 2007). In addition to enrichment and careful enclosure design, conditioning programs and favourable carer/animal relationships can assist in the prevention or management of stress-induced abnormal repetitive behaviours (Bacon 2018).

Multispecies enclosures are one way of creating stimulating behavioural interactions and are increasingly popular in zoos and wildlife parks. Through promoting more space, increased social complexity (both within and between species) and higher activity levels, such enclosures can improve animal welfare. However, it is important these potential welfare benefits are balanced against the potential welfare risks. Careful planning and management is necessary to minimise resource guarding, competition for refuges, negative interactions such as aggression and disease risk (Veasey and Hammer 2010; Kaandorp 2012).

Humans may also exert influence over an animal's external circumstances and the interactions between animals and caregivers or unfamiliar persons have the potential to result in welfare-enhancing positive affects or welfare-compromising negative affects (Mellor *et al.* 2020). Aspects of interactions with animals such as caregiver attitude (e.g. degree of confidence and patience), voice (tone and volume), skill level and handling approach can all factor into the animal's welfare experience (Mellor *et al.* 2020). Furthermore, the response to caregivers or visitors can be influenced by characteristics such as the animal's previous experience, genetics, species biology and individual predisposition (Hosey 2008; Rault *et al.* 2020). There is increasing recognition that positive human-animal relationships can benefit welfare (Rault *et al.* 2020). Factors that can assist in developing these include frequent and familiar human interactions associated with a positive experience; predictable interactions that incorporate choice and control (when and how interaction occurs); and utilisation of positive reinforcement techniques. The life stage of the animal may also influence the success of these interactions, as younger animals tend to have experienced fewer negative interactions with humans and display greater learning ability, curiosity and exploration than adult animals (Rault *et al.* 2020). Human-animal relationships such as training and conditioning and animal-visitor interactions are discussed further in section 8.

### 4.1.5 Domain 5 – Affective experience

The first four domains can be observed and measured, but the fifth domain is not directly quantifiable and must be inferred from the evidence provided by the other four domains. Physiological, behavioural and pathological changes are interpreted by the nervous system and generate the animal's subjective mental experiences, which can be negative, neutral or positive (Beausoleil 2020), and it is the balance of these that determine an animal's welfare state.

Positive nutritional, environmental and health conditions are likely to give rise to the respective positive affects of satiety, comfort and good health. Varied, novel, engaging environmental and social challenges allow an animal to make choices, to avoid threats, to explore and to rest freely and these are expected to translate into the animal feeling safe/secure, calm, companionable, interested, rewarded, in control and engaged. On the other hand, negative nutritional, environmental, health and behavioural conditions are likely to give rise to negative affects such as hunger, breathlessness, pain and frustration (Mellor *et al.* 2020). Furthermore, if a highly motivated behaviour is prevented (domain 4: e.g. grazing), this may result in feelings of fear, anxiety, boredom, loneliness, frustration or anger (domain 5). An example of a condition that gives rise to a negative experience of hunger and thirst (domain 5) is the restriction of food and water intake (domain 1). Conversely, satiety and pleasure (domain 5) from drinking and experiencing good tastes and smells contribute to a positive animal welfare state as a consequence of enjoying sufficient quality food and water (domain 1). As demonstrated by these examples, the five domains are inherently linked, interact dynamically and are not discrete packages to be viewed in isolation or applied to a restricted species type. Rather, the model provides generalised and comprehensive welfare insights that are applicable for a wide range of taxa including Australian mammals.

## 4.2 Other models/orientations

In attempts to gain a holistic understanding of animal welfare, several authors have proposed alternative approaches or frameworks to the Five Domains model for assessing animal welfare. These are not mutually exclusive of the Five Domains Model and share many common themes.

Fraser's (2003) three orientations of animal welfare include (1) biological function (health and physical fitness), (2) affective state (integrated mental experience), and (3) natural living (the concept that the natural environment is an important reference point for establishing what is important to animals, such as highly motivated and natural species-specific behaviours) (Koene 2013; Yeates 2018).

Hosey *et al.* (2009) describe the following four overlapping approaches to the evaluation of wild animal welfare: (1) the subjective experience including emotions (the mind), (2) the ability to thrive and survive in the environment (the body), (3) the use of wild animal behaviour as a template (nature), and (4) the drawing of analogies from other animals, including humans. Similarly, Sherwen and Hemsworth (2019) suggest that a comprehensive welfare assessment should include behavioural, physiological and, if possible, emotional measures.

Fundamentally, what these models and frameworks have in common is that they infer that both physical and psychological wellbeing matter and that natural history, behavioural ecology, sensory capacity and adaptations of a species can act as valuable reference points for any assessment. The application of this approach to Australian mammals is complicated by the lack of comparative field-based observations for many species (Kagan and Veasey 2010), highlighting the need for further research.

## 5. HOW DO WE MEASURE ANIMAL WELFARE?

### 5.1 Principles

Animal welfare scientists are challenged to identify objective, valid and largely quantitative methods for measuring a state that is subjective, self-determined, variable over time and influenced by individual and species-specific factors. Assessment techniques were initially developed and later refined in the agricultural sector (Hill and Broom 2009) with subsequent spillover into research settings and other animal industries.

Animals are complex organisms and the factors that contribute to their welfare are numerous (Campbell-Ward 2023). Even in well-studied species such as domestic cattle, pigs and poultry, there is no single universal measure that accurately describes an individual's welfare. Given the lack of capacity for direct communication with animals about how they feel, assessment relies on selecting indicators that reveal empirical but indirect information about animal wellbeing (Hosey *et al.* 2009). Indicators may be systematically evaluated by direct observation of the animals and their environments and/or by reviewing available records. They may be measured in a binary fashion (e.g. presence/absence), via a scoring system, or simply quantified and recorded at a single point in time or tracked longitudinally (Campbell-Ward 2023).

Important challenges and limitations of using indicators from domestic animal welfare science to accurately evaluate the welfare of Australian mammals include the wide variety of species, their relative unfamiliarity to some observers and the limited research on the needs and husbandry of some taxa (Maple and Perdue 2013). Additionally, if handling is required to measure a particular indicator, this may pose risks to the animal and/or handler and adversely influence results (Campbell-Ward 2023).

There are two broad categories of welfare indicators:

- resource-based measures, which attempt to infer the welfare status of animals from the inputs and logistical practices that go into caring for them, such as the quality of substrate, stocking density, ventilation, feed trough size, food provided, lighting, heating/cooling, shelter, hygiene and staff training (see section 5.2)
- output-based measures, which aim to gather more direct evidence regarding the welfare state of an animal. As they relate to Australian mammals these may be divided into those that evaluate behaviour, physical health and physiology (see sections 5.3, 5.4 and 5.5).

To overcome the significant limitations and caveats of using any single indicator, most established animal welfare assessment protocols include a variety of resource-based (indirect) and/or output-based (direct) measures to infer welfare state.

### 5.2 Resource-based measures

Resource-based measures tend to focus on carer provisions for animals both in terms of resources (diet, housing) and management practices (health care, enrichment). Such parameters are often easy to define and measure. However, they lack an animal-centric perspective (Yeates and Main 2008). The provision of preferred and motivating resources and an appropriate level of environmental complexity are vital, but individual differences may exist and keeper ratings of preference do not always correlate with the animal's actual preference (Gaalema *et al.* 2011). Input-based measures are therefore best combined with output-based measures.

### 5.3 Output-based measures: behaviour

Behavioural observations are often the focus of output-based welfare assessments of wildlife in human care (Watters *et al.* 2009). The frequency, duration and intensity of key behaviours (e.g. vocalisation, foraging, aggression, exploration) may be monitored via species-specific ethograms (inventories detailing the behaviours exhibited by animals) or time (activity budgets) and space (enclosure usage) metrics can be developed (Manteca Vilanova 2020; Rose and Riley 2021). Normal behaviour implies not only natural behaviours, including those behaviours that an animal is highly motivated to perform (McPhee and Carlstead 2010), but also appropriate behavioural range and context for a given individual (Bacon 2018). For example, locomotor activity may be appropriate if hunting for food but inappropriate if in the form of repetitive pacing; mating may be appropriate if consistent with seasonal patterns but abnormal if excessive (Campbell-Ward 2023). There is evidence that behavioural variability correlates with improved welfare via increased control over the environment (Asher *et al.* 2009).

Activity patterns of a species in the wild can serve as an important baseline for interpreting and understanding behavioural needs, preferences and observations of the same species in a care environment. Differences in time budgets between wild animals and those in human care may indicate problems with animal management (Melfi and Feistner 2002), but not all wild behaviours are considered necessary for optimal welfare (e.g. fighting behaviour) (Mellor 2015). Ethograms and activity patterns have been studied for a diversity of Australian mammals in the wild, including:

- platypus (*Ornithorhynchus anatinus*; Bethge 2002)
- Tasmanian devil (*Sarcophilus harrisii*; Andersen *et al.* 2020)
- koala (*Phascolarctos cinereus*; Nagy and Martin 1985; Logan and Sanson 2003; Ryan *et al.* 2013)
- southern hairy-nosed wombat (*Lasiorhinus latifrons*; Finlayson *et al.* 2010)
- bare-nosed wombat (*Vombatus ursinus*; Evans 2008; Story *et al.* 2014)
- mahogany glider (*Petaurus gracilis*; Jackson and Johnson 2002)
- squirrel glider (*Petaurus norfolcensis*; Holland *et al.* 2007; Sharpe and Goldingay 2019)
- greater glider (*Petauroides volans*; Comport *et al.* 1996)
- brush-tailed rock-wallaby (*Petrogale penicillata*; Botma *et al.* 2020)
- agile wallaby (*Notamacropus agilis*; Stirrat 2004)
- black flying-fox (*Pteropus alecto*; Markus and Blackshaw 2002)
- dingo (*Canis familiaris*; Thomson 1992; Allen *et al.* 2013; McNeill *et al.* 2016; Déaux *et al.* 2017).

Abnormal behaviours, most obviously stereotypies, have been correlated with poor welfare and maladaptation (Campbell-Ward 2023). However, such behaviours may, in some cases, be satisfying or soothing to perform and can be considered a coping mechanism (Mason 1991). Other potential indicators of poor welfare include improper vocalisation, extreme timidity, hyperaggression, excessive sleep, inactivity, escape behaviours, self-mutilation, fur plucking and decreased performance of behaviours critical to survival and reproduction. These need to be interpreted in the broader situational context (Kagan and Veasey 2010).

Other behaviours may indicate positive affects, including play, affiliative behaviours (i.e. friendly, peaceful interactions with other individuals), approach, allogrooming (social grooming) and decreased vigilance (McPhee and Carlstead 2010; Manteca Vilanova 2020).

Experimentally, more complex behavioural assessments may help define what animals want or need but can be impractical for real-world assessments (Campbell-Ward 2023). An animal's choice between alternatives can measure relative preference (Fraser *et al.* 1993), while motivation can be tested via operant tasks (tasks accompanied by reinforcement, where an animal performs a behaviour for access to a desirable resource, e.g. a food reward) (Kirkden and Pajor 2006). Researchers have demonstrated that a change in a dolphin's willingness to participate in training sessions is an early indicator of an altered health status (Clegg *et al.* 2019) and have evaluated how much of an aversive stimulus an animal will endure for a given resource (Kirkden and Pajor 2006). The application of investigations into animal preferences can be complex, as a preference for something does not necessarily equate to poor welfare in its absence and animals do not consistently select resources that are in their individual best interests (Kagan and Veasey 2010). Additionally, the history and past experiences of animals (e.g. wild-origin versus captive-born) may influence behaviour and coping strategies (Campbell-Ward 2023).

Recent developments in behavioural outputs to inform welfare include:

- measuring cognitive bias to identify positive animal emotions. This is aligned with the 'glass half full or glass half empty' paradigm. Animals that experience positive affects or emotions tend to make more optimistic judgements when presented with ambiguous stimuli (Brydges *et al.* 2011; Clegg 2018).
- applications of qualitative behavioural assessments, whereby animal body language is examined as a whole rather than as discrete individual behaviours (Fleming *et al.* 2015). In particular, subjective caregiver ratings are potentially meaningful wellbeing indicators. They reflect the relationship and accumulated experience

animal care staff have with the animals in their care and their capacity to intuitively detect subtle changes in behaviour, posture, attitude, expression or movement to characterise an animal's mental state (e.g. tense, calm, frustrated, excited) (Whitham and Wielebnowski 2009).

- social network analyses to better understand social bonds and associated positive affects. By mapping social relationships, animal care and management can be tailored to ensure these are supported and not unknowingly disrupted, resulting in unintended consequences for animal welfare (Rose and Croft 2015; Rose and Riley 2021).
- quantifying behavioural diversity (the number and frequency of behaviours) leading to the inference that high behavioural diversity equates to animal needs being met, whereas low behavioural diversity signals possible welfare compromise, particularly if associated with abnormal repetitive behaviours and/ or lethargy (Miller *et al.* 2020).

## 5.4 Output-based measures: physical health

Physical health is a commonly used welfare indicator that aims to identify deviations from 'normal'. Measurable physical factors include body condition, cleanliness, the presence and severity of lesions and evidence of pain or injury. Physical proxies for behavioural issues can be observed (e.g. wounds or fur loss from aggressive interactions), but the age and cause of lesions may be unclear (Campbell-Ward 2023). In some circumstances, direct handling is necessary for measurement, although there is a trend to make use of conditioning programs that facilitate handling-free health evaluations (Hellmuth *et al.* 2012). Group findings may only be valid if a representative cohort is inspected.

Veterinary and necropsy reports can provide retrospective health data for individuals or groups, but are reliant on the quality of record-keeping (Campbell-Ward 2023). Longevity and mortality rates are sometimes used as indirect and crude physical health measures, but are insensitive and often controversial (Kagan and Veasey 2010), especially when used in isolation. High mortality rates may signal poor welfare but there are many confounding factors to consider, including population demography, life cycle and euthanasia/quality-of-life protocols. For example, even under optimal husbandry and welfare conditions, some deaths, including euthanasia, would be expected annually in a large population of older (over 3 years of age) spinifex hopping-mice (*Notomys alexis*).

## 5.5 Output-based measures: physiology

Physiological indicators have the potential to demonstrate an animal's response to a situation and are mostly aimed at identifying negative welfare states. Examples include body temperature, respiratory rate, the functioning of the sympathetic-adrenal-medullary axis or the hypothalamic-pituitary-adrenal axis, levels of other hormones (e.g. prolactin, oxytocin) and measures of immunity (e.g. neutrophil-lymphocyte ratio) (Campbell-Ward 2023). Variables such as cost, time, training of personnel and animals, invasiveness, biological fluctuations, social grouping of animals and handling for sampling or direct measurement determine their real-world utility (Wolfensohn *et al.* 2018). Research evaluating the correlation of physiological parameters with welfare has produced mixed results (Maple and Perdue 2013), with some measures, such as glucocorticoids, increasing not only with stress but also with general arousal (Whitham *et al.* 2020) such as via copulation or play.

Physiological indicators of chronic stress may be particularly significant in assessing the level of everyday welfare of animals in human care. They must be combined with multiple concurrent behavioural observations for context (Campbell-Ward 2023). The relatively recent development of non-invasive techniques (e.g. hormone levels in faeces, urine or hair) has permitted the incorporation of physiological parameters into welfare assessments of animals, but lack of validation of tests remains a significant limitation for many Australian mammal species. In their attempts to monitor adrenocortical activity in 13 species of marsupial (including the numbat (*Myrmecobius fasciatus*), Tasmanian devil, koala, greater bilby (*Macrotis lagotis*), hairy-nosed wombats, possums and macropods), Fanson *et al.* (2017) found that the assay that performed best varied between species, even those that are closely related.

By evaluating time budget behavioural differences and correlating them with salivary cortisol levels, researchers have been able to compare the welfare implications of open and closed pool facilities for bottlenose dolphins (*Tursiops truncatus*) (Ugaz *et al.* 2013). The dolphins that participated in this research were trained to emerge from the water and open their mouths voluntarily. This allowed non-invasive sample collection of saliva via swabs introduced into the mouth, up to the base of the tongue.

## 5.6 Aggregating welfare measures

Welfare scientists advocate utilising a range of welfare measures (such as those discussed in sections 5.1–5.5) to gain the best understanding of an animal's welfare state.

The precise composition of a welfare assessment protocol and how heavily each indicator contributes to the cumulative result is context-dependent. The time available, type of operation and purpose of the assessment can determine the format. Scientific robustness relies on species-relevant parameters with high levels of interobserver reliability, trained assessors and measures that have been validated against other known welfare-relevant indicators. A balance must be struck between cost-effective feasibility, impositions on the animals and caregivers and meaningful results (Campbell-Ward 2023). Early versions of aggregated welfare assessment tools were heavily reliant on resource-based measures with a focus on the presence or absence of welfare harms. There have been concerted efforts recently to increase the proportion of output-based measures and to include both negative and positive welfare indicators (Richmond *et al.* 2017).

Scientific validation of welfare indicators continues to be investigated. Consulting experts and research literature and using existing tools for analogous species and the species' natural ecology as a reference point is likely to yield best results (Manteca Vilanova 2020).

Several individual animal-specific and species-specific welfare assessment tools have been developed (Sherwen *et al.* 2018) and these can be valuable aids in animal welfare management. They may be used in situations such as monitoring animal introductions, movement of animals to a new enclosure, to assist in providing life stage-appropriate (e.g. geriatric) animal care or simply to better understand the welfare experience of animals for which humans have care responsibilities.

An example of a basic welfare checklist is provided in Addendum 1. This can be used as a starting point for caregivers new to welfare assessment and can be tailored to the species and situation in question. It does not require access to scientific assays or specialised equipment. Such checklists are not designed to definitively determine the welfare state of an animal, but rather help identify a range of criteria that may assist in delivering positive animal welfare outcomes.

## 5.7 Accreditation and online tools

Some industry associations (e.g. Australasian Zoo and Aquarium Association) provide welfare assessment tools as a key component of their accreditation schemes. This may involve staff completing self-assessments and providing evidence in combination with validation by external auditors.

Several online tools are now available for the practical assessment and visual tracking of animal welfare in zoological settings, including:

- Zoological Information Management System (ZIMS) Care and Welfare module (Species360): utilising the Five Domains framework, this program allows monitoring of user-determined input and output welfare indicators and the assignment of expected values or value ranges to different taxa.
- The Animal Welfare Assessment Grid (AWAG): software designed to assess and monitor the welfare and cumulative lifetime experience of animals. The tool encompasses welfare needs across four parameters, namely physical, behavioural/psychological, environmental and procedural.
- Zoo Monitor: software developed by Lincoln Park Zoo designed to record behaviour and space/enclosure use and can accommodate focal and group sampling.

# 6. CONTEMPORARY ANIMAL WELFARE CONCEPTS: KEY CONSIDERATIONS IN SUPPORTING POSITIVE ANIMAL WELFARE

To create conditions that support positive animal welfare, which can then be substantiated by a framework such as the Five Domains, there are some important, progressive and intrinsically linked concepts caregivers should become familiar with:

- competence,
- choice,
- agency and control
- 24 hours/day, 7 days/week (24/7) welfare.

In combination, these four concepts provide a creative toolbox from which to draw inspiration when designing enclosure habitats, daily routines and enrichment programs. Skilled and experienced personnel and ready access to appropriate husbandry and veterinary expertise are prerequisites.

## 6.1 Competence

Competence refers to the capacity of an animal to effectively behave, respond and function in line with species evolutionary adaptations. Creating the conditions that

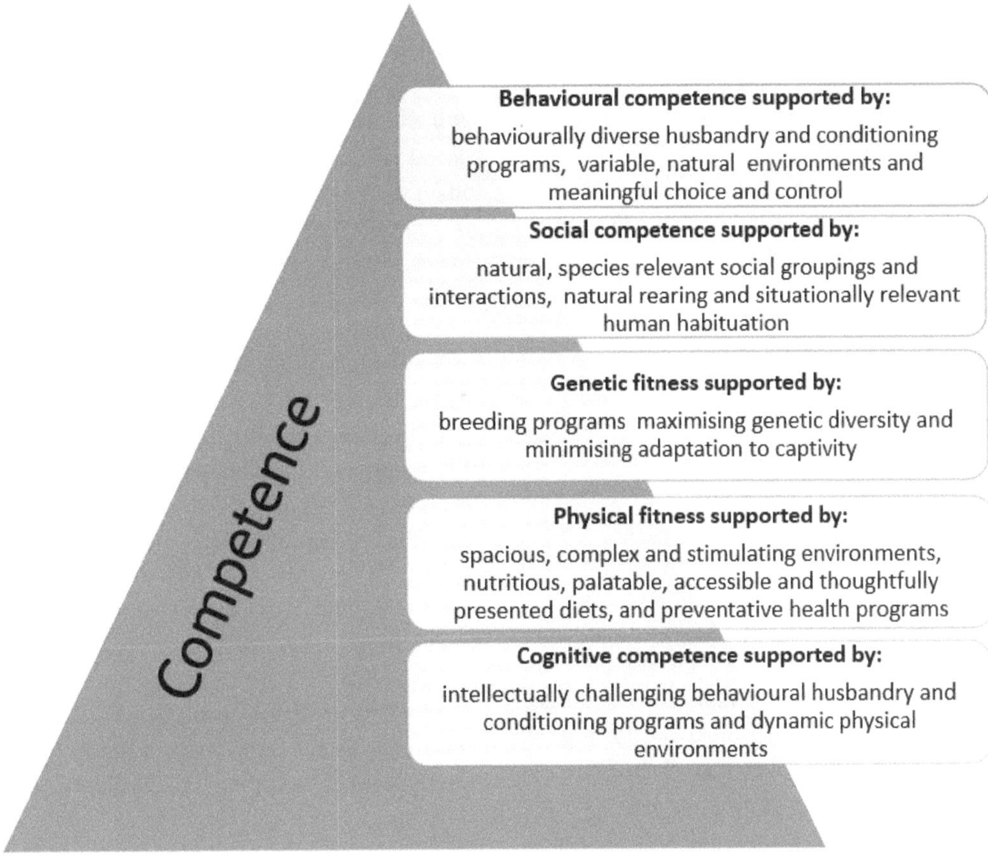


**Fig. 1.3.** Summary of the various forms of competence relevant to animal wellbeing and the husbandry elements that sustain each type.

support physical, behavioural, social and cognitive competence is therefore important for promoting the positive affects that align with an animal's intrinsic nature.

There are many forms of competence that are relevant to animal wellbeing (Fig. 1.3) and. these will be discussed in detail in sections 6.1.1.–6.1.6.

### 6.1.1. Behavioural competence

Behavioural competence refers to the capacity of an animal to display a diversity of natural behaviours, with appropriate frequency, intensity and duration, as well as the ability to manage environmental stressors to which they are ordinarily exposed. Species-relevant behavioural diversity is widely recognised as important for both conservation and animal wellbeing (McPhee and Carlstead 2010; Miller *et al.* 2020) and aligns with the concept of naturalness being an important welfare consideration. Behavioural diversity is a measure of both the richness of behaviour (number of behaviours) as well as the evenness (frequency of each behaviour) (Miller *et al.* 2020). The retention of a varied and complex behavioural repertoire is considered essential for competency in a novel or changing environment.

In addition to supporting positive animal welfare, natural behaviours are essential to deliver key advocacy objectives of zoos and wildlife parks, such as fostering an appreciation of wildlife and inspiring conservation-related actions. Visitors tend to correlate the display of natural behaviours with contented and well-cared-for animals (Godinez and Fernandez 2019). The display of non-natural, unwanted or abnormal repetitive behaviours detracts from the intended message or purpose of an exhibit and if the behaviours are perceived as negative or harmful, this can erode an organisation's reputation as a care facility and ultimately the social licence to operate (McPhee and Carlstead 2010).

Individual animals can adjust their behaviour to respond to a specific need, such as conforming to a feeding schedule or to a group structure, as an adaptive coping mechanism (McPhee and Carlstead 2010). For example, reduced time spent foraging due to a plentiful supply of food may elicit an activity budget that does not necessarily reflect that of wild counterparts. To avoid this adjustment leading to a loss of behavioural diversity, behavioural competence can be promoted through enrichment programs that support opportunities to carry out highly motivated behaviours (see section 7) and thoughtfully designed enclosure habitats that respect natural ecology, facilitate natural behaviours and promote choice and control (see sections 6.2 and 6.3).

Animal behaviours may be influenced by genetic and environmental factors, including past experiences. Wild-born Tasmanian devils brought into care can develop abnormal behaviours such as repetitive pacing more readily than those born in captivity. This is also a notable behaviour in males housed near, but unable to access, females in oestrus. Once established, such behaviours can be difficult to eradicate, even with increased behavioural enrichment (Izzard *et al.* 2019).

Behavioural competence is also a key goal for rehabilitation and conservation breeding for release programs (McPhee and Carlstead 2010). Animals may undergo targeted behavioural training and/or assessment before release to ensure they are behaviourally adept, can find shelter and have the appropriate foraging, hunting and predator awareness skills to maximise post-release survival. Published examples of behavioural training or assessment for rehabilitation and conservation purposes include:

- hand-reared brushtail possums (*Trichosurus vulpecula*) that demonstrated more exploratory behaviours and less human habituation had better post-release survival (Mella *et al.* 2022). This study also revealed that the process of hand-rearing itself and the use of a 'buddy' system can affect the development of personality and behavioural traits.
- exposing predator-naive greater bilbies to feral cats under controlled conditions before release resulted in improved survival in the first 40 days following translocation when compared with non-predator exposed bilbies. Additionally, predator-aware bilbies spent less time moving and more time in cover, suggesting predator training resulted in behavioural wariness (Ross *et al.* 2019).
- conditioning northern quolls (*Dasyurus hallucatus*) in a wildlife park to regard cane toads as aversive through exposure to toad sausages (containing a non-lethal quantity of cane toad and laced with a nausea-inducing chemical). Wild quolls similarly formed an aversion to the toad sausages, although this appeared to wane over time (Indigo *et al.* 2018).

### 6.1.2 Social competence

Social competence refers to the capacity of animals to adapt their behaviour according to their social setting. It is relevant for behaviours associated with selecting mates, rearing offspring, establishing dominance hierarchies, resolving conflicts or forming social bonds. Affiliative behaviours such as grooming, bonding and play are likely to be of significance to some Australian native mammals (Russell 1984). For instance, sniffing nose-to-nose, also sniffing pouch, cloaca, and scrotum are the more common positive bonding interactions observed in marsupials (Russell 1984). Allogrooming is described in many macropods, with most grooming occurring between dam and offspring (Higginbottom and Croft 1999). Play has been observed most frequently in species of macropods, larger dasyurids and wombats (Byers 1999).

When animals are in human care, the key risks to social competence are unnatural social grouping and humanisation, including imprinting. As demonstrated in sugar gliders (*Petaurus breviceps*), differences in kin social relationships have been observed between wild animals and those maintained in captivity, although the relevance of these differences to animal welfare remains unclear (Klettenheimer *et al.* 1997; Sadler and Ward, 1999).

Although many marsupials are not considered highly social (Russell 1984), social learning may occur in some species, particularly in relation to predator recognition and avoidance, use of habitat in species with a stable home range and recognition of conspecifics and dominant individuals (Higginbottom and Croft 1999). For longer-lived species and those that have an extended association between female young and mother, social learning is thought to be related to food selection and distribution, avoidance of obstacles, identification of refuge locations and development of mothering skills (Higginbottom and Croft 1999).

Larger dasyurids, such as quolls, that spend extended periods of time foraging with their mothers, may use social learning to confer appropriate predator responses. Northern quolls translocated to a predator-free area lost recognition of dingoes (*Canis familiaris*) as a natural predator after 13 generations (Jolly *et al.* 2020). It is possible this loss of predator awareness is a consequence of altered social learning conditions or, alternatively, may represent a population level shift in behaviour due to a change in selection pressures.

Some marsupials such as the brushtail possum appear to also use social learning to enhance their problem-solving ability. A study compared the capacity of naive possums to solve a food-based puzzle that they had not previously encountered, with the capacity of possums that had repeatedly observed other possums solving the puzzle before attempting it themselves. The possums that had observed the food puzzle being solved prior outperformed the naive possums (Godfrey *et al.* 2023).

The significance of social development, learning and competence mechanisms in Australian mammals appears to vary between species and life circumstances and has been the subject of only limited studies to date. The potential welfare and conservation impacts of social competence and their interaction with other competencies are promising areas for future research.

### 6.1.3 Genetic fitness

*Ex situ* conservation breeding programs typically focus on retaining genetic diversity and minimising inbreeding to promote individual fitness in the short term and to secure future population level adaptive potential in the long term (Frankham *et al.* 2010). Selection pressures that promote behaviours suited to the *ex situ* environment, also known as adaptation to captivity, is also an important consideration. Generally, biologists advocate suspending evolution while animals are in a zoo-based breeding program, so that if descendants are returned to the wild in the future, they genetically resemble their forebears. Inevitably though, as the number of generations in captivity increases, the greater the chance of behavioural (or even morphological or physiological) changes occurring as a result of traits conferring greater survivorship in a zoo-based environment. Therefore, after many generations, the behavioural repertoire of zoo animals may diverge from that of wild animals. A comparison of wild and zoo-bred feathertail gliders (*Acrobates pygmaeus*) demonstrated that the latter experienced less frequent, shorter and shallower torpor, higher metabolic rates during rest and torpor and slower rates of rewarming. Zoo-bred gliders also became hypothermic at low temperatures whereas wild gliders did not. The number of generations in zoo-based care (in this case, 4–5) may have contributed to a population level behavioural and physiological shift (Geiser and Ferguson 2001). Similarly, a study of Tasmanian devils demonstrated that as the number of generations in intensive breeding facilities increased, so too did vulnerability to vehicle strike after release. This behavioural outcome was attributed to long-term captive breeding (Grueber *et al.* 2017), although both studies were unable to determine if the patterns observed were the result of genetic changes in captivity or accumulated non-genetic changes in behaviour and biology. Nonetheless, both examples highlight the value of preserving genetically influenced competencies and furthering research into adaptation to captivity.

### 6.1.4 Physical competence or fitness

Being physically fit, hosting a healthy microbiome and maintaining an appropriate bodyweight and muscle tone contribute to normal biological functioning and positive affects. Physical competence is supported by:

- providing a physical and social environment that is sufficiently spacious, complex and stimulating to encourage species-appropriate exercise
- ensuring diets are nourishing, palatable, accessible, safe and thoughtfully presented
- implementing a preventative health management program in consultation with a veterinarian, including regular health assessment and monitoring of weight, body condition and mobility (Fig. 1.4)
- encouraging early reporting of suspected health issues or alterations in fitness.

These elements are important for animals at all life stages, from preweaning through to the geriatric period.

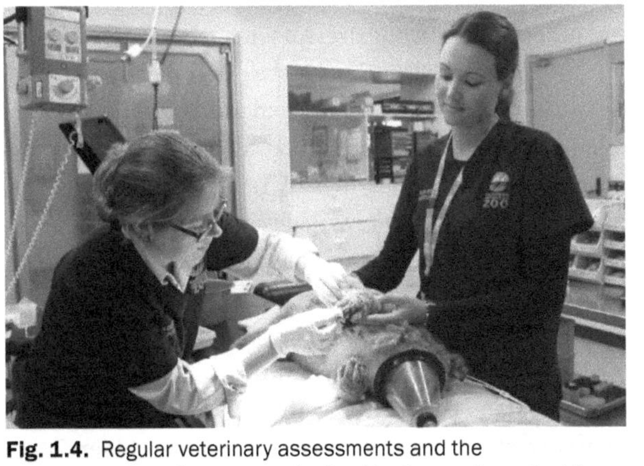

**Fig. 1.4.** Regular veterinary assessments and the implementation of a preventative health plan are important for monitoring and supporting physical competence. The early detection of health issues favours effective treatment and management and minimises longer term welfare implications. Photo by Taronga Conservation Society Australia.

### 6.1.5 Cognitive competence

Cognitive competence, or intellectual fitness, refers to behavioural adjustment through individual patterns of learning, which can influence response to future events. Appropriate cognitive challenges, such as problem solving, decision making and cognitive skill development, provide animals with the opportunity to express motivating behaviours and are associated with positive welfare (Meehan and Mench 2007).

In the wild, animals use their cognitive skills to obtain and respond to information and to draw on their problem-solving capacity to navigate safely through their environment. For instance, scent markings may indicate the presence of a predator, a competitor or a mate. An animal moving through a given environment must decide whether to continue or alter their route after detecting such scents. Those animals that respond appropriately to challenges are more likely to have a competitive advantage than those that do not. Other examples of common challenges in the wild include way-finding, food recognition and extraction, appropriate social responses, attracting a mate, finding suitable shelter and protecting territory or other resources.

In a zoo-based environment, the challenges faced by animals are typically different to those present in the wild. Nevertheless, it is important to consider the wild cognitive challenges a species would face, as they provide an indication of adaptative capacities. The use of such cognitive capacities may remain highly relevant and important to the animal even in a different setting. Of equal importance is the management of challenges they would not experience in the wild such that they do not become significant stressors. Examples include unusual sensory inputs such as noise generated by machinery or people, pungent odours and unnatural light levels. Such potential stressors must be considered in light of a species' sensory capabilities. A review of social behaviour and organisation revealed that the most significant avenues of communication for marsupials appear to be olfactory and auditory, with visual communication more significant to diurnal macropods (Russell 1984).

The welfare benefits of cognitive challenges through the provision of tasks can be diminished by either over- or under stimulation (Špinka and Wemelsfelder 2011). Tasks that are too challenging are likely to cause stress or frustration and tasks that are not challenging enough are likely to cause boredom or apathy. It is therefore important that investigations are carried out to ensure the right balance is achieved so that tasks are possible and positive affects realised.

### 6.1.6 Balancing competence, conservation and welfare

The importance of the relative preservation of different forms of competence must be viewed in the context of the species' and/or individual's role in a given care environment. For instance, for species or individuals in a zoo-based environment whose role is to generate awe, wonderment and connection with visitors, a docile demeanour unsuitable for survival in the wild (i.e. altered behavioural competence) may be associated with improved welfare outcomes. In the wild, character traits such as boldness, behavioural flexibility and being non-migratory may confer resilience to anthropogenic disturbance and these traits can predict likely welfare outcomes for such animals in human care (Mason 2010). Conversely, for threatened species cared for under a recovery action or insurance setting and destined for wild release, preservation of wild traits and prevention of adaptation to captivity are essential. Macropods, for example, can be naturally flighty and animals with calm natures may have better welfare outcomes in a walk-through type enclosure, whereas individuals in a breed-for-release program should retain a flight distance from people.

Typically, in a zoo-based or other care environment, a balance must be struck between acclimatisation to human presence and the display of natural behaviours. Fear responses should be tempered and yet animals must maintain the capacity to recognise and relate appropriately to conspecifics.

## 6.2 Choice

When contemplating ways to promote positive experiences for animals, the concept of choice is increasingly referenced. Choice allows an animal to meaningfully react to the environment or situation, through having options and an ability to select preferences. This may be as simple as finding physical comfort through environmental alternatives (e.g. different substrates), thermal comfort via access to a variety of shaded and unshaded areas or visual comfort due to a light gradient (Špinka 2019). Choice can also be promoted through the social environment, by empowering animals to choose with whom, how often and for how long to interact with another animal or human (Špinka 2019) (Fig. 1.5). Facilitating the ability of animals to make choices over aspects

**Fig. 1.5.** To promote choice, designated retreat areas in walk-through enclosures provide animals with the ability to choose when and for how long they wish to be in close proximity to visitors. Photo by Michelle Campbell-Ward.

of their daily lives and to partake in meaningful behaviours (e.g. foraging and exploration) has the potential to increase animal wellbeing (Kagan *et al.* 2015). It is important to bear in mind, however, that the benefit of choice can be diminished by dilemma. For instance, high-value food placed in an exposed area for a cryptic species creates a dilemma for the animal – whether to be anxious and get the food reward or to feel safe but miss the food opportunity.

Determining which choices are meaningful for animals can be guided by preference and cognitive bias testing. These tests infer what an animal 'wants' to be doing for pleasure, comfort or satisfaction and have demonstrated that many animals prefer challenges and engagement over passive rewards (Mench 1998).

Animals are individuals and have personal preferences. Thus, it is impossible to define an optimal environment for a species, but rather individually distinct, optimal environments that can be created by flexibly applying tailored solutions to standard species-specific designs. Choice is important from an animal welfare perspective and must be evaluated in the context of the individual's history, characteristics and internal drivers.

## 6.3 Agency and control

Together with choice, the concepts of agency and control are oft-cited focus areas for animal wellbeing. Agency (or the feeling of control over actions and their consequences) is supported when animals can engage in voluntary, self-directed, goal-directed behaviours (Wemelsfelder 1997) which, in turn, generate positive affects (e.g. feelings of being in control, energised or interested) (Mellor *et al.* 2020). Having the capacity to make engaging choices supports animals to exercise agency. From an adaptive perspective, animals need to behave proactively and gather knowledge and enhance skills for future use to be successful; being only reactive is an inferior approach from a competitive perspective

(Špinka 2019). Exercising agency, therefore, is likely to be important to animals, at least most mammals.

Agency includes the ability to engage in activities to achieve an immediate outcome, such as food acquisition, securing safety and feeding young. It is also important for promoting behavioural competence through building capacity and resources for coping with future challenges, such as via social learning and exploration. Suppression of agency in a human-care setting can be associated with boredom, depression or helplessness and may lead to underdeveloped competence in comparison with the wild, where animals are constantly making choices (Špinka 2019).

Control empowers an animal to influence the course of events (i.e. the likelihood of something occurring is influenced by the animal's actions). An animal empowered by control can predict an outcome. Hence, a related concept to control is predictability. There are different types of predictability: spatial (refers to the location of an event), temporal (refers to the timing of an event) and signalled predictability (refers to an event being signalled by a cue, typically visual or auditory) (Bassett and Buchanan-Smith 2007; Buchanan-Smith 2021). Although spatial and temporal predictability have varied effects on welfare and, in some cases, can encourage unwanted behaviours (e.g. anticipatory behaviours preceding a positive event such as feeding), signalled predictability has been shown to be beneficial for welfare in several cases. Signalled predictability lets animals prepare for an event or to perceive an element of control over what follows the signal. For example, caregivers alerting animals they are about to enter their enclosure by using a reliable predictable signal (e.g. knocking) can reduce anxiety-related behaviours caused by unexpected sounds of doors opening and closing (Rimpley and Buchanan-Smith 2013). Signalled predictability is thought to be most beneficial when it precedes a negative event. By being able to predict what is coming next, an animal can anticipate and avoid heightened vigilance. For positive or enjoyable events, predictability appears not to be as significant (Rimpley and Buchanan-Smith 2013).

Promoting control can facilitate animal care activities. This concept encourages animal independence (i.e. is there something caregivers are doing/providing for animals that they could be doing for themselves?) (Coe and Hoy 2020).

New technologies can assist in giving animals control or empowering them to improve their own welfare. Examples include:

- managing the ambient environment (e.g. via motion sensors that adjust light levels, temperature and humidity or activate underwater jets)
- choosing when and what to eat through animal-controlled feeders
- accessing spaces via microchip-controlled doors (e.g. nest boxes that exclude non-target species or individuals) to provide selective access to refuge or other resources
- managing social and refuge opportunities (e.g. through mate choice) (Coe and Hoy 2020).

Control and predictability can empower animals in a care environment to exhibit purpose and independence, whether it be preparing for an upcoming negative event (e.g. transport) or setting the course for more positive events. Interspecies and interindividual differences regarding the value or meaning of these should be expected; however, they are valuable concepts for promoting positive welfare.

### 6.4 24/7 welfare

The concept of 24/7 animal welfare is a philosophy that adapts the environment and care throughout the animal's life cycle, taking into account species natural history, as well as variations across day and night, weekdays, weekends and seasons (Brando and Buchanan-Smith 2018).

Although direct animal care and husbandry typically occur during daytime hours, these are not necessarily biologically relevant times for animals, particularly the nocturnal and crepuscular species of which there are numerous Australian mammal examples. The 24/7 animal welfare framework also considers variations over the animal's entire lifespan from birth through to end of life. This acknowledges that animals have differing habitat requirements and associated needs at different life stages. Additional factors, including lunar cycles, seasons, visitor presence/absence and climatic differences, can hold differential relevance for animals and their wellbeing. In a human-care environment, welfare can be enhanced by providing more opportunities for animals to exercise choice and control (e.g. to forage and obtain food at individually- and species-relevant times, rather than only when convenient for the caregiver).

## 7. ENVIRONMENTAL ENRICHMENT

Environmental enrichment aims to improve animals' mental and physical wellbeing through a goal-driven process by changing or enhancing their environment. Enrichment or behavioural husbandry programs are designed to promote choice and control, integrate complexity, and support behavioural, physical and social competence. They are most meaningful when they reflect the animals' behavioural biology and natural history (Shepherdson 2010). Environmental enrichment goals may include increasing frequency and/or duration of desirable behaviours, decreasing frequency and/or duration of undesirable behaviours, increasing behavioural diversity, extending enclosure utilisation and activating animals to improve physical health and agility (Shepherdson 2010). Additionally, enrichment may ameliorate negative affects by improving coping mechanisms and decreasing the aversive response to stressful situations (Carlstead and Shepherdson 2000). The diversity of environmental enrichment is vast and is commonly categorised as food-related, sensory, cognitive, social and physical environment-related (Fig. 1.6).

Ultimately, enrichment should provide stimulation but not frustration, create engaging choices for animals, provide animals with purpose and promote behaviours that are meaningful, motivating and relevant to the species' biology. Successful enrichment practices support animal welfare through promoting positive experiences such as exploration, foraging, increased activity, feelings of reward, satisfaction, novelty and decreased stress and anxiety.

Attempts to introduce enrichment practices may be subject to implementation impediments (Tuite *et al.* 2022). Changes to policies and procedures may be difficult to fulfil for a range of reasons and managers and caregivers need to be cognisant of barriers and enablers to proposed enrichment ideas. Attention should be given to the benefits of enrichment for both animal physical and mental health, alongside increased transparency around enrichment objectives.

Despite the wealth of resources relating to environmental enrichment for zoo-based animals (Shape of Enrichment Inc. 2022), information specifically related to Australian native mammals is currently limited and needs further research. A six-step framework affectionately known as 'S.P.I.D.E.R.' is widely accepted as critical to the development and sustainability of best practice enrichment programs (Alligood and Leighty 2015) and is recommended for those looking to fill knowledge gaps. The six steps are: (1) setting goals, (2) planning, (3) implementing, (4) documenting, (5) evaluating, and (6) readjusting (Mellen and MacPhee 2001).

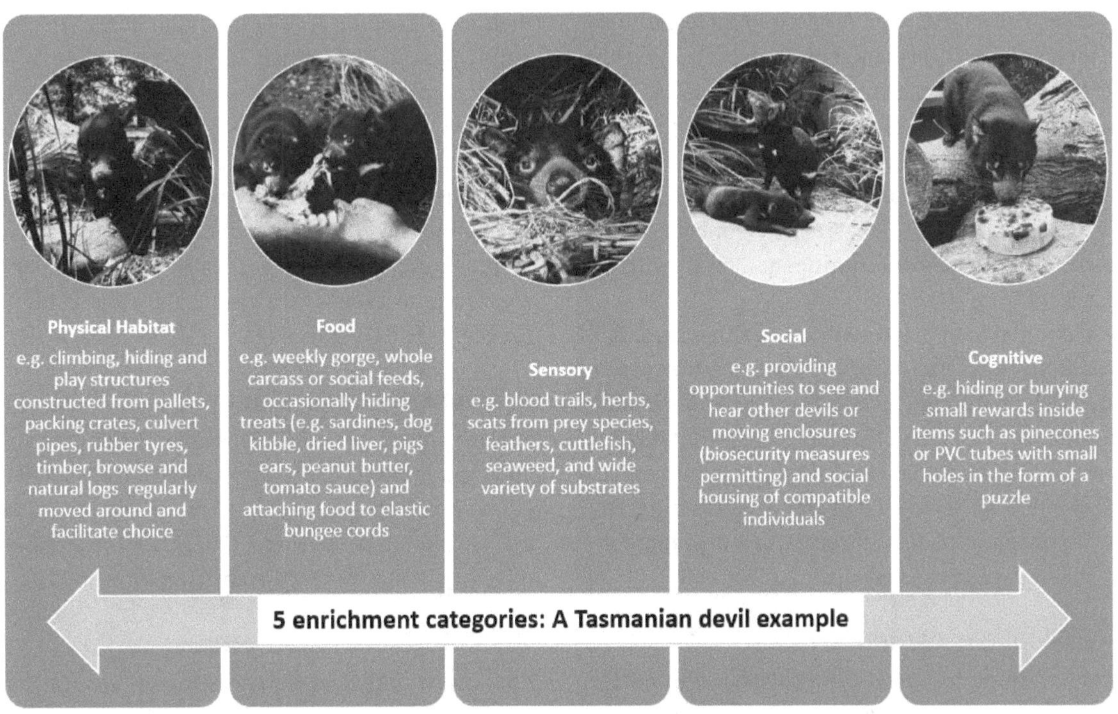


**Fig. 1.6.** In the wild, Tasmanian devils are very active and replicating this in an area as small as 70 $m^2$ is challenging. To increase enclosure habitat complexity, Tasmanian devil caregivers plan enrichment around five main principles that are not mutually exclusive. Adapted from Izzard *et al.* (2019). Photos by Bridget Kaitler, Louisa Todd and Emma Lloyd used with permission of Taronga Conservation Society Australia.

## 7.1 Food-related enrichment

Many species spend a significant portion of their natural activity budget foraging. Consequently, food-related enrichment generally aims to extend the time spent foraging to closer reflect wild activity and increase the diversity of natural foraging-related behaviours (Hosey *et al.* 2009). Providing food in a bowl, trough or confined space is a common practice in care environments, but typically does not support replicating natural foraging behaviours for most species.

Hiding food (e.g. in termite mounds) and using timed feeders are widely used strategies across many taxa for enrichment purposes. For carnivores, feeding or foraging devices such as feeding poles and rope pulls that encourage hunting behaviours combined with olfactory enrichment and scatter-feeds have been shown to diminish pacing. A study of Tasmanian devils concluded that pacing behaviour reduced following the introduction of bungee carcass feeding and koala scats (O'Neal 2011).

An evaluation of four enrichment items provided to southern hairy-nosed wombats was conducted by Hogan *et al.* (2010). It included three feed-related items (browse, grass and buried food) and one olfactory item (a log scented with aniseed, peppermint, vanilla, lemon, coconut or rose). The enrichment resulted in increased time spent foraging and exploring, but did not change time spent in abnormal pacing and being inactive. The study concluded that foraging and olfaction were unlikely to be causal factors relating to the locomotor repetitive behaviours, but that welfare was still improved by providing stimulus diversity, greater choice in behavioural options and additional opportunities to interact with the environment.

To avoid nutritional and physical health issues, it is important that food offered as part of an enrichment program, as well as any animal food given to visitors for animal feeding activities, is accounted for in the total ration and part of a balanced diet. Equally, to avoid unintended harms, visitors must be discouraged from feeding animals with unapproved items through the use of appropriate signage, education and adequate supervision.

## 7.2 Physical enrichment

An animal care environment is at risk of becoming static over time and consequently physical enrichments are typically included to create a dynamic and varied environment that keeps animals interested and activated and provides them with options.

**Fig. 1.7.** A quokka (*Setonix brachyurus*) seeking shade and a hiding spot under a live tussock grass. Photo by Louisa Todd with permission of Taronga Conservation Society Australia.

Substrates such as dirt, litter, mulch and dust hollows encourage foraging and exploration through food-finding (e.g. invertebrates), promote olfactory-stimulated digging and burrowing behaviours and provide environmental variability and complexity that supports information gathering (Carlstead and Shepherdson 2000). Live plants are another form of physical enrichment and have the added benefit of producing variable microclimates, influencing temperature and humidity and providing natural hides (Fig. 1.7).

Other physical components such as water features (e.g. sprinklers, ponds, waterfalls), climbing structures, perches and rock mounds can be incorporated into enclosure habitats. Wallabies will utilise browse for feed and resting, sugar gliders are encouraged to leap or glide by providing a series of vertical trunks with few horizontal connections and diggable substrate is important for echidnas (Hawkins 2005).

## 7.3 Sensory enrichment

Sensory inputs can generate interest, stimulate animals and even manage stress. They are best considered in the context of a species' natural history, dominant senses and perception of the world. Categories include auditory, olfactory, visual and tactile enrichment.

### 7.3.1 Auditory enrichment

Auditory stimuli may be enriching or aversive for animals. Playing sounds reminiscent of the natural environment may be pleasant for an animal and classical music has been shown to enhance affiliative behaviour in bottlenose dolphins (Guérineau *et al.* 2022). Conversely, care environments can be unnaturally loud (Wells 2009) and reduction of ambient noise may be necessary to promote good welfare. The impact of human-generated noise such as machinery, fireworks and concerts must be considered and steps taken to reduce negative consequences for animals. Sound control barriers may be appropriate for noisy environments and efforts should be made to remedy problematic enclosure aspects such as noisy gates. In addition to sound level, tempo and gradual versus non-gradual introduction of novel sounds are relevant auditory considerations.

### 7.3.2 Olfactory enrichment

As for auditory stimuli, there are mixed results relating to the use of odours for enrichment. Differences likely reflect the natural and historical sensory environment, species' characteristics and individual preferences. Olfaction is noted to be the most important of the sensory communication avenues for marsupials (Higginbottom and Croft 1999). Essential oils, scents and pheromones have been used to stimulate arousal and excitement in various taxa and some aromatic herbs (e.g. lavender) have been shown to reduce anxiety and aggression and promote other positive affects (Wells 2009). For Tasmanian devils, blood trails and scattered goat or camel fur can make engaging olfactory enrichment. Care must be taken to ensure that products or materials are safe for use in relation to potential toxicities or exposure to pathogens and that attempts to provide this type of enrichment or excessive cleaning of enclosures with chemicals does not disrupt natural scent marking communication.

### 7.3.3 Visual enrichment

Visual enrichment encompasses changes to the physical environment, even relatively simple ones, as well as the use of audiovisual media or computer screens, mirrors for self-recognition and colours to benefit mood (Reiss and Marino 2001; Wells 2009). It should be noted, however, that colour is unlikely to be meaningful for nocturnal species. Visual enrichment has been shown to be the least frequently provided type of enrichment to mammals in human care (Hoy *et al.* 2010).

Dingoes may benefit from visual enrichment if they can access a high area (e.g. a rockpile or platform that provides a vista to observe the activity of other animals and/or humans). Some zoos make use of so-called 'howdy doors' that provide visual access between animals who do not have physical access to one another. This involves aspects of social enrichment and potentially auditory and olfactory stimuli too, but the visual information may be important to some animals (Maple and Perdue 2013). For many species, having the choice to retreat from visual contact or viewing by visitors or other animals is similarly important.

### 7.3.4 Tactile enrichment

Different textures, such as varied substrates for moving on or interacting with, can provide engaging sensory experiences. Piles of bark, mulch, feathers, grass, browse, shredded paper, hay, pebbles and sand may promote positive affects associated with exploration, play, rubbing and digging. Such tactile enrichment can be combined with foraging (e.g. crickets or mealworms hidden in substrates for insectivores; shellfish detected by the vibrissae of pinnipeds in sand). For aquatic or semi-aquatic animals, sprinklers, jets or bubble walls can provide meaningful water-based tactile enrichment (Fig. 1.8).

## 7.4 Cognitive enrichment

Cognitive enrichment is broadly defined as enrichment that allows animals to solve problems or to control their environment (Hall *et al.* 2021). Examples include puzzle feeders and the learning associated with training and conditioning sessions (Melfi 2013). Most enrichment for Australian native mammals is directed towards opportunities for foraging and exploratory behaviours. A species' life history, including natural cognitive experiences, and the individual history are important considerations in the design of cognitive enrichment plans, as their success may be influenced by genetics, environmental complexity, past experience and social dynamics (Hall *et al.* 2021).

**Fig. 1.8.** An Australian sea lion (*Neophoca cinerea*) interacting with the directional flow of water jet enrichment. The animal can choose to move into or out of the stream of water in various ways so that different parts of the body can experience the sensation. Photo by Brad McKenzie with permission of Taronga Conservation Society Australia.

The cognitive abilities of Australian native mammals should not be underestimated. The short-beaked echidna is often considered primitive from an evolutionary perspective, but is capable of carrying out a range of cognitive tasks, such as black and white discrimination, circle/triangle discrimination and same/different categorisation (Russell 2014).

A possible trade-off between memory capacity and learning speed has been identified in a study involving quokkas and fat-tailed dunnarts (*Sminthopsis crassicaudata*) (Bonney and Wynne 2004). Quokkas were determined to have relatively good memories, capable of retaining long-term information such as location of resources. In contrast, dunnarts' memories were inferior but they excelled in behavioural flexibility and capacity to adapt quickly to changing circumstances. This may be an evolutionary consequence of quokkas being relatively long-lived, with stable food sources and fewer natural predators, whereas dunnarts are short-lived, must evade a wide variety of predators, experience fluctuating resources and must hunt to secure food. Studying and understanding these types of natural cognitive differences may offer insights into meaningful cognitive enrichment opportunities.

Although food and other resources are usually in abundance when animals are in care, cognitive challenge may be rewarding for some animals and, in such cases, important to wellbeing. To be enriching, cognitive challenges should be relevant, motivating, controllable and possible to master (Clark 2013). There are numerous documented cases where animals have been observed preferring to work for a food reward utilising motivated behaviours, even in the presence of identical freely available food items. This is often referred to as contrafreeloading and is consistent with the idea that for many species, the processing of cues, decision making and cognitive challenge, along with foraging and exploration are highly motivated behaviours. The potential of contrafreeloading in Australian native mammals is an area that would benefit from further research.

## 7.5 Social enrichment

Social enrichment refers to intentional efforts to facilitate species-appropriate social interactions. This can include housing naturally social species in demographically appropriate groups, carefully designing multispecies enclosure habitats and promoting meaningful human-animal interactions. The opportunity for social encounters that include feeding, exploration, play, bonding, sexual activity, rearing offspring, defending a territory, cooperative problem solving, affiliative behaviours and even safety through taking refuge within a group, are likely to be rewarding and are aided by favourable social groupings.

Some Australian native mammal groups include naturally gregarious species, such as bats, gliders, and many larger species of macropod. Social enrichment is likely to be highly relevant for them. Even species considered solitary are unlikely to spend their entire life without social interaction and this should be considered in the animals' full life experience.

There is a trend in zoos and wildlife parks to house multiple species together, such as in macropod walk-through enclosures. There are important considerations for this approach, including the social impact of different individuals and species on each other. As discussed in section 4.1.1, it is also important that the welfare of one species is not significantly compromised by a desire to improve the welfare of another. Some individuals or species may dominate the optimal refuge or shade areas, or best food and water resources, necessitating ongoing close monitoring and management.

Visitors may also affect the behaviour and enclosure space use in walk-through areas. One study demonstrated that fewer quokkas were visible when the enclosure was open to visitors (Learmonth *et al.* 2018). Conversely, a different study found that, unlike other macropods, quokkas were not affected by zoo visitors and postulated that appropriate housing and husbandry may have adequately compensated for the visitor effect or that evolutionary differences result in the species being less sensitive to visitor pressure than others (Meade *et al.* 2021). Observations of Kangaroo Island (KI) kangaroos (*Macropus fuliginosus fuliginosus*) and red kangaroos (*Osphranter rufus*) revealed that when visitor numbers increased these macropods increased their time spent engaged in visitor-directed vigilance and KI kangaroos also increased their time spent locomoting and rested less (Sherwen *et al.* 2015). In the same study, there was no effect of visitor number on the distance kangaroos positioned themselves from the visitor pathway or on faecal glucocorticoid concentration in either species. These results suggest that it is difficult to draw general conclusions regarding animal welfare in walk-through enclosures and assessments should be made on a case-by-case basis.

Training and conditioning using positive reinforcement may be considered a form of social enrichment due to the positive affects generated through behavioural interactions experienced by animals with their human caregiver(s). Voluntary cooperation for husbandry and veterinary procedures can result in welfare benefits through improved stress management, socialisation of species and the provision of novel situations and cognitive challenges. This is outlined further in section 8.1.

## 8. HUMAN-ANIMAL RELATIONSHIPS

Depending on context, wild mammals in care usually have daily contact with a variable number of people, both familiar and unfamiliar (Claxton 2011). Responsible caregivers and managers must monitor the effect of visitor and staff presence on the welfare of animals in their care (Davey 2007). Human-animal interactions, which may lead to human-animal relationships (HARs), may be positive (enriching), neutral (no effect) or negative (stressful) in nature. The response of animals to caregivers has been described in three dimensions: affinity to carer, fear of people and sociable/curious nature (Carlstead 2009). Fear of people is significantly and positively correlated with poor welfare and there can be significant species differences in fear and affinity responses. The quality of the HAR can be influenced by whether caregivers enter the enclosure with the animals, the frequency and time of feeding, carer visibility to the animal and the method of rearing. Certain husbandry styles, attitudes and habits (such as the making of unexpected noises) can be aversive to animals, although this has not been studied in Australian mammals to date. Good HARs established between animals and their carers can have positive effects on animal welfare. Benefits include stress resilience, socialisation/bonding and physical and mental stimulation (Hosey and Melfi 2015)

### 8.1 Training and conditioning

Training and conditioning have the potential to deliver animal welfare benefits, particularly with the shift away from coercive methods and greater focus on empowering animals through positive reinforcement. Positive reinforcement training promotes trust, desirable behaviours and a stimulating environment (Brando 2012). It can also improve welfare by providing an opportunity for animals to voluntarily participate in their own care, reducing unnecessary fear, anxiety, hypervigilance, aggressive behaviours and abnormal repetitive behaviours that may be associated with some husbandry and veterinary procedures (Brando 2012). These procedures include being handled or weighed, the collecting of blood samples and being moved to a different enclosure, including holding areas or into pet packs or crates. Health assessments without sedation can be facilitated by training and conditioning, such as eye, mouth and joint examinations in seals (Bodley 2025) and ultrasonography and endoscopy in dolphins (Dold 2015; Blyde 2025).

The concept of training as a form of enrichment has been gaining traction over the past decade. It achieves this by:

- giving animals more control over their environment (Westlund 2014)
- expanding behavioural choices (Westlund 2014)
- promoting species-appropriate behavioural repertoires (Westlund 2014)
- empowering the animal to deal adequately with challenges (Westlund 2014)
- offering learning opportunities (Melfi 2013)
- facilitating goal-driven enrichment usage and more complex behaviours (Fernandez 2022; Melfi 2013)
- encouraging the development of favourable human-animal interactions (Melfi 2013)
- promoting more favourable interactions with conspecifics (Fernandez 2022).

Engagement with training and conditioning programs varies depending on species, individual history, current environment and the skill of the trainer. To ensure positive welfare benefits are achieved, ongoing assessment and monitoring of behaviour at the level of the individual animal is essential (Fernandez and Martin 2021).

### 8.2 Interactive experiences with Australian native mammals

Personalised and spatially close experiences with ambassador wildlife are popular with visitors to zoos, aquariums, mobile exhibitions and wildlife organisations (Spooner *et al.* 2021). A variety of interactive experiences are offered to visitors, ranging from indirect contact (e.g. in a walk-through enclosure) through to more direct contact such as feeding and directly touching, petting or holding animals. A global review of more than 1000 members of zoo and aquarium associations revealed that 75% offered at least one form of animal visitor interaction (AVI), with petting of wild animals being the most prevalent (D'Cruze *et al.* 2019). Common examples in Australia include photographic opportunities with koalas, macropod walk-through enclosures, walking with dingoes and personalised meet-and-greet interactions with native mammals including echidnas, wombats and possums.

Organisations that offer animal encounters have an important role to play in fostering respect and appreciation for wildlife and this needs to occur without compromising animal welfare and broader conservation efforts for the species in the wild. Balancing welfare and intended conservation advocacy goals is complex and it is important that those offering animal encounters duly consider the ethics of the practice and the welfare of the individual animals involved.

Some studies demonstrate that facilitated close interactions can help people build connections with nature and these interactions have the potential to inspire human behaviour change that benefits species in the wild (Dunstan *et al.* 2016). Conversely, there are several studies that show that wildlife interactions can inadvertently be harmful not only to the individual animals involved, but to species in the wild. Broadly shared social media posts of some of these experiences without appropriate context can promote illegal wildlife trade and encourage welfare-compromising tourism encounters (Lenzi *et al.* 2020). They can also increase the desirability of wildlife as pets and give the impression that the species is not imperilled in the wild (Leighty *et al.* 2015). Particular or exacerbated welfare risks associated with mobile exhibitions include the higher frequency of handling, cross-handling, sensory and circadian disturbance, repeated transport stress, suboptimal housing, restricted access to food and water and disease risk (Warwick *et al.* 2023). Careful attention to enclosure and encounter design, species selection and visitor education can reduce the potential welfare risks of AVIs while still facilitating engagement goals (Fernandez *et al.* 2009).

There is a relative paucity of published literature relating to the animal welfare impact of interactive experiences involving Australian native mammals, but research in this area is increasingly being undertaken. Several recent studies include:

- red kangaroos in walk-through enclosures – potential visitor effects on activity levels, social proximity, space use patterns and foraging behaviour have been observed in relation to crowd presence or size (Sherwen *et al.* 2015; Jones *et al.* 2021; Meade *et al.* 2021). Context and individual differences in response to visitor presence are noted, highlighting the importance of assessing facility design and species and individual behavioural suitability for AVIs (Jones *et al.* 2021).
- swamp wallabies in walk-through enclosures – a UK study comparing zoo closure (visitor absence) and open periods (visitor presence) demonstrated no difference in zone usage or the proportion of time spent near visitor pathways between the two periods (Williams *et al.* 2021). Nevertheless, when visitor numbers increased, another study in an Australian zoo revealed this species demonstrated greater visitor-directed vigilance, prompting the authors to suggest operators consider restricting the number of visitors entering an enclosure at one time to mitigate any potentially related welfare issues (Meade *et al.* 2021).
- koalas – visitor proximity and noise have been shown to be correlated with increased koala vigilance (Larsen *et al.* 2014) and handling often interrupts resting or feeding times (Benesch 2007). Previous unpublished industry data raising concerns validated by these studies was used by the New South Wales government to ban the holding of koalas by zoo visitors (Jackson and Crane 2008).

A welfare-focused AVI checklist is provided for reference in Addendum 2 identifying features for consideration in developing interactive experiences with Australian native mammals.

## 9. END-OF-LIFE-CARE, AGED ANIMAL ASSESSMENTS AND EUTHANASIA

With advances in welfare provisions, veterinary care and husbandry, and the reduced risk of natural threats such as predation, trauma and disease, Australian mammals in human care may live well beyond the average age of their wild counterparts. Physical and psychological impairments due to advanced age affect multiple body systems, including degeneration of the skeleton, dentition, muscle strength and mental capacity and changes in body composition and reproductive capability. This decline can present welfare concerns for elderly animals, as degenerative processes and changes in social structure may correlate with reduced quality of life. The care of wildlife presents some unique challenges in managing the geriatric period in light of welfare needs. Recognition of age-related conditions can be difficult due to the cryptic nature of some species and their ability to mask the signs of illness. In consultation with veterinarians, caregivers may implement a range of management strategies for older animals, including dosing with prescribed medications, increasing opportunities to seek comfort, modifying diets to meet individual needs, altering enclosure habitats to accommodate changing functional capacity and providing age-appropriate enrichment. In deteriorating animals of advanced age, euthanasia is often warranted to alleviate unmanageable pain or distress and to avoid a prolonged death for an animal in an unacceptable welfare state. For companion animals, veterinary hospice options may be available before euthanasia or death, but for wildlife in human care such palliative options may be more restricted, difficult or stressful and despite good intentions may actually negatively impact welfare. If there are pressing welfare concerns, prolonged debate and delayed decision making over end-of-life planning and euthanasia must be avoided.

An important step in geriatric animal management is the education and engagement of all care staff in the process, preparing them for an animal's death and improving the shared understanding of decision-making factors. To facilitate these discussions, the development of criteria and objective protocols for regular welfare assessment of geriatric animals can be helpful (Campbell-Ward 2023). Such protocols, often consolidated into an Aged Animal Assessment tool or Quality-of-Life template, enable veterinarians, keepers, managers and curators to assess and longitudinally track an animal's health and quality of life to facilitate welfare-focused end-of-life management and guide the timing of euthanasia decisions (Vogelnest and Talbot 2019). The implementation of these processes helps justify and explain decisions for euthanasia by promoting compassionate animal care and welfare as key principles.

It is helpful if record-keeping practices facilitate the identification of animals that are approaching or have reached expected longevity for their species. At a predetermined time (e.g. starting when an animal reaches 80% of expected longevity), such tools can be used to objectively score criteria in comparison with an animal of the same species, or even better the same individual, in its prime. This can be repeated at regular intervals to detect a decline. Possible criteria include:

- physical assessment factors (e.g. body condition score, mobility, activity levels, changes in urine and faecal output, medications)
- psychological assessment factors (e.g. self-directed behaviour, self-injurious behaviour, response to stimuli, spatial awareness)
- social assessment factors (e.g. changes in hierarchy, change in human–animal relationship, role of the individual within the social structure)
- other considerations (e.g. staff or visitor concern over the animal's physical appearance, genetic value to the regional population).

Each criterion can be scored, with some elements weighted more heavily to reflect the greater significance on the animal's quality of life (Campbell-Ward 2023). For example, joint health for a dingo and tooth wear for a koala are considered highly welfare relevant for these species and should contribute meaningfully to an assessment. Total scores are calculated by adding together all components of the assessment and can provide an indication of quality of life. This helps direct decision-making when a particular threshold is reached. A sample Aged Animal Assessment template is provided in Addendum 3.

Euthanasia and death are sensitive topics that can also affect the wellbeing of people. It is inevitable that animal caregivers form emotional attachments with animals in their care (Vogelnest and Talbot 2019). Empathy and compassion for staff are important considerations during this end-of-life period. Appropriate support and counselling should be provided when animals decline significantly, die unexpectedly or euthanasia is carried out.

## 10. STANDARDS AND LEGISLATIVE FRAMEWORKS

Animal welfare legislation, including expected standards of care, is a means to ensure the community's expectations are met and that those responsible for animals comply with good management practices. Although concepts of animal welfare assessment have become increasingly sophisticated and abstract, there is recognition that the use of clearly articulated and practical guidance documents is necessary and worthwhile (Green and Mellor 2011).

Animal welfare may be regulated at the national or the state/territory/provincial level, depending on the jurisdiction. A complex milieu of legislation, codes of practice and guidelines relate to the keeping of Australian mammals in permanent human care and to the rescue and rehabilitation of injured, orphaned and diseased wildlife. Despite some consistency in purpose, these legislative instruments and their implementation can vary, in some cases quite considerably, between jurisdictions. Mandatory standards related to animals cared for or accessible for visitor viewing exist in some localities and these may include not only animal welfare considerations, but also human safety, education of zoo visitors and facilities. Some Australian mammal species may be kept legally as pets in some locations and elsewhere such practices are prohibited or restricted to licence or permit holders. Additionally, national approaches have been developed for animals used for scientific purposes (with compulsory adherence required in all Australian states and territories).

Generally, standards provide the basis for the legal requirements for animal welfare and include the word 'must'. Guidelines often include the word 'should' and complement the standards by providing advice and/or recommendations to achieve desirable animal welfare and security outcomes but non-compliance with them may not constitute an offence under law.

In addition to locally relevant legislation, regional associations or individual institutions may develop their own set of accreditation or self-regulating standards to demonstrate their commitment to animal welfare and safety (Watters *et al.* 2021).

## 11. NATIVE MAMMALS AS PETS

Traditionally, pet animals kept in homes and at private residences have been largely limited to domesticated species; however, there is a global trend for the increasing popularity of non-traditional pets, including Australian wildlife (Chomel *et al.* 2007; Grant *et al.* 2017). Owners typically cite companionship and enjoyment as reasons for the acquisition of a pet, but for unusual animals the motivations can be expanded to include enthusiasm and interest in the species, recreation, novelty and amateur study. Rather than having an intrinsic respect for animals or their wellbeing, some research also indicates that unusual pet acquisition can be significantly influenced by status factors, social recognition and materialistic indulgence (Warwick *et al.* 2018). Regardless of the individual drivers that lead to pet ownership, there are a myriad of factors influencing an animal's welfare in domestic settings that typically lack the professional management and facilities of other care contexts.

A wide variety of Australian birds, reptiles and frogs have been kept recreationally for many decades (Johnson 2008). Around the turn of the 21st century, the push for legalising more wildlife species as pets in Australia, including mammals, was argued in the scientific literature (Oakwood and Hopwood 1999; Archer 2002; Hopwood 2002; Cooney *et al.* 2010). The concept continues to be debated and variably implemented globally, with attention drawn not only to animal welfare considerations but also conservation, genetics, biosecurity and public safety (Schuppli *et al.* 2014; Warwick *et al.* 2018). These issues relate to where and how animals are procured, the experience and knowledge of prospective owners, the suitability of the home-based environment, selective breeding, disease risks and the potential for escape (Schuppli *et al.* 2014). Concerns are countered by favourable views that in Australia native animals could replace domestic pets that are ecologically harmful (Lilith *et al.* 2006; Hughes and Macdonald 2013) and that this may raise the public profile of native species (Cooney *et al.* 2010), thereby contributing to conservation efforts. It has also been suggested that the pet industry could encourage *ex-situ* propagation of native animals (Hess 2011); however, species that are selectively bred for particular traits such as docility or novel colouration are non-natural and do not serve a conservation purpose. Opinions vary regarding which, if any, of Australia's mammals are considered appropriate as pets (Viggers and Lindenmayer 2002). Detailed analysis of risk factors has determined that some species of native mammals may be suitable for private keeping (Cooney *et al.* 2010) depending on, among other things, adaptability to the human environment and availability of adequate biological and husbandry information. Generally, the keeping of native mammals as pets in Australia is tightly regulated and more so than for other groups of local fauna. Licensing requirements, such as permits, and lists of allowable species vary significantly between jurisdictions (Cooney *et al.* 2010). Keeping, breeding and trade for the domestic market is banned in some locations, while in others there are variable restricted species licensing arrangements, including some permit exemptions. Outside of Australia, markets for selected Australian mammal pets have developed considerably. The most notable example is the legal trade in pet sugar gliders in North America, and to a lesser extent Asia and the United Kingdom, that has burgeoned since the 1990s (Campbell *et al.* 2019). However, despite assumptions to the contrary, recent genetic provenance studies suggest the US sugar glider population originated from Indonesia, where wild trapping via a quota system is reported (Lyons and Natusch 2011; Lyons and Natusch 2012), rather than Australia (Campbell *et al.* 2019). There is also a sizeable niche macropod pet industry servicing North America, Europe and Japan (Cooney *et al.* 2010).

To safeguard the welfare of Australian mammals, the natural histories and biological needs of species must be respected, regardless of their ownership arrangements. As with animals in other care settings, pets should be able to function well biologically, be free from negative psychological states, be able to experience normal pleasures and lead reasonably natural lives (Schuppli *et al.* 2014). Keeping animals in settings to which they are poorly adapted is a threat to their welfare. Despite the increasing availability and accessibility of quality resources on established care protocols and disease prevention in the form of hard copy and electronic textbooks, open access journal articles, online professional forums and veterinary-endorsed advice sheets, there is considerable evidence that husbandry deficiencies are a common cause of clinical issues in pet Australian mammals (Gamble 2004).

To function well, animals must be well nourished and free from disease and injury. The nutritional requirements of a species must be adequately known and understood by prospective owners and appropriate foodstuffs be readily available to avoid issues. For example, improper nutrition is one of the most common causes of illness in pet sugar gliders, predisposing to metabolic bone disease, neurological disorders, obesity, dental disease, anaemia and cataracts (Ness and Johnson-Delaney 2012; Johnson 2020). This has led to the development of several commercial diets. Disease and injury prevention requires that owners are able to house and handle their pets safely and seek out appropriate veterinary expertise for general advice and when problems arise. The health management, medicine and surgery of unusual pets is a rapidly expanding discipline, with several veterinary professional associations encouraging research, disseminating educational materials to members and accrediting or certifying those who undertake postgraduate study in this area.

To feel well, animals must not be unduly stressed by home environments. Australian mammals are not domesticated and unless they are hand-raised, some species are flighty, difficult to handle and prone to stress in situations from which they cannot escape proximity to humans and other pets. Stress-inducing physical and social deficiencies readily result in pet sugar gliders developing behavioural disorders such as self-mutilation, overgrooming, aggressiveness and cannibalism (Johnson-Delaney 2010). Disturbance during biologically determined rest periods (e.g. during the day for nocturnal species such as possums and quolls) may be stressful. Furthermore, these species have periods of highest activity at times that may not suit an owner's preference or lifestyle.

To permit a life that respects naturalness, pet owners must have a good knowledge of a species' natural behaviour. Some species require high levels of exercise, social or psychological stimulation or key stimuli in the environment (Schuppli and Fraser 2000). For example, although some native rodents can be kept relatively easily in cages or aquaria that can be readily accommodated within a home, care must be taken to ensure there is appropriate substrate, social groupings and hide areas to allow the animals to feel safe and to satisfy their innate motivation to dig and interact with conspecifics.

Finally, animal welfare may be jeopardised if an owner loses interest in, or commitment to, the animal. This can happen if the animal grows too large, aggressive or destructive and becomes difficult to house or costly to keep (Schuppli and Fraser 2000). For example:

- sexually mature hand-reared male kangaroos can become highly aggressive towards people
- adult wombats are incredibly powerful and can easily damage home and garden structures. They may become aggressive and can inflict serious injury
- many of the personality traits of dingoes differ from those typically considered ideal characteristics for a successful and rewarding companion dog–owner relationship (Smith 2014).

Pet dissatisfaction can arise due to a misalignment of expectations and may lead to repeated rehoming (Grant *et al.* 2017). Owners may find it unsatisfactory if an animal is inactive, nocturnal or prefers solitude to human company. Similarly, if species have qualities that reduce, or fail to enhance, the welfare of the owner, the animal may become neglected. In this respect, the animal's suitability as a pet depends greatly on the owner.

Mitigating the welfare (and other) risks of keeping native animals as pets requires careful assessment of a species' suitability and a range of educational, regulatory and industry structure measures (Cooney *et al.* 2010).

# 12. WELFARE OF WILD AUSTRALIAN MAMMALS

## 12.1 Rescue and rehabilitation

Research indicates that the numbers of injured, orphaned and diseased wild animals in need of rescue in Australia is increasing and that the causes are overwhelmingly anthropogenic (Taylor-Brown *et al.* 2019). Common reasons for mammalian rescues include motor vehicle collision, entanglement or entrapment, orphaning or abandonment, introduced predator attack and being found in an unsuitable environment (Kwok *et al.* 2021). It is estimated that 4 million wild mammals are killed by motor vehicle trauma in Australia annually (Englefield *et al.* 2018a) and that 143 million mammals in forests and woodlands were impacted by the 2019–20 Australian bushfires (WWF 2020). In intense and widespread natural disasters, the animals that are unable to flee or find refuge typically perish, but of those that do survive, large mammals such as koalas are the most likely to be rescued and presented to triage units, wildlife carers and veterinarians (Parrott *et al.* 2021; Sherwen *et al.* 2023).

Wildlife rescue and rehabilitation in the Australian context is carried out almost exclusively by volunteers, numbering in the order of 20 000 – as individuals and groups of assorted sizes and organisational structures (Englefield *et al.* 2019). The work undertaken is largely self-funded and often conducted in a home or private property environment. Each state and territory has its own legislative framework and codes of conduct and policy (Englefield *et al.* 2018b). The variation in these regulatory structures means that there is no common welfare framework that all carers in the country work under, which creates inconsistent approaches and gaps.

Wildlife rehabilitators are supported by a network of veterinarians (some working exclusively with wildlife in purpose-built hospitals and others through general practices). A significant amount of this work is provided pro bono and there can be challenges in resourcing this demand (Orr and Tribe 2018).

One of the primary considerations when contemplating the welfare implications of intervening when a wild animal casualty is discovered is that proximity to humans and handling is inherently stressful. Furthermore, signs of pain and distress are often readily masked or difficult for observers to detect, particularly in species that are unfamiliar. There is likely to be significant health compromise if a wild animal can be approached without attempting to flee.

For wild animals the process of capture, transport, handling, examination, veterinary intervention, hospitalisation and rehabilitation can be physically and/or psychologically harmful. Treatments that are well tolerated by domestic pets or farm animals may be impossible or ethically questionable in a wild species. Unless they are juveniles, wild animals typically do not habituate to the presence of people and if they do, this renders them unsuitable for release according to many codes of practice. Although some animals appear to cope better than others, there has been little in the way of robust objective assessment of wild Australian mammal welfare in rehabilitation settings (Campbell-Ward 2020). Due to their sedentary and arboreal nature, housing a wild koala short-term can be achieved much easier than housing an adult macropod or a fur seal, but that does not mean their needs are any less complex or that they are assured good welfare in care. Managing a rescued koala's nutritional intake to maintain or establish an appropriate gastrointestinal microbiota to support feeding behavioural competence post-release has been identified as just one potentially critical but poorly understood factor in rehabilitation success (Goldenberg *et al.* 2022).

Triage for wildlife casualties in any scenario must consider the nature of the injury or illness, as well as a range of other factors, including the availability of resources and facilities. The ultimate goal is ensuring optimal welfare outcomes for animals alongside broader conservation objectives. For animals removed from the wild for assessment, options that may be variably available include emergency medical stabilisation, transfer to temporary care with a rehabilitator, immediate release back to the wild or euthanasia. Decision makers must consider the species status, legislation, human health and safety and a range of prognostic factors (e.g. body condition, age/stage of development, disease status, threat to wider population health if released, expected hospitalisation and recovery times and availability of suitable habitat for release).

Assessing the prognosis for wildlife with respect to their short- and long-term welfare cannot be directly compared to that for domestic animals with similar injuries or degrees of illness. If a dog or horse sustains a severe injury and requires intensive care and/or surgery, the assumption is that even if recovery is slow or incomplete, the animal will remain in long-term human care and can be reassessed regularly, receiving medication or other therapies as needed. With wild animals, however, veterinarians, wildlife carers and managers can only deem an animal fit for release if chronic effects are not anticipated and no further intervention or assessment is expected post-recovery. Even if the initial prognosis is deemed fair to good, long-term welfare should be continuously re-evaluated throughout the various stages of the rehabilitation process (from emergency care, hospitalisation and intermediate stages, through to the often-overlooked pre-release stage) (Sherwen *et al.* 2023). If it becomes clear that an animal will not be physically and behaviourally capable of independent life in the wild, whether that is due to severity of injury or illness, progression of an earlier problem, an inability to cope with the expected duration of rehabilitation required or an issue that has developed in care, euthanasia is appropriate (with the rare exception of licensing for retention in long-term care). Multiple data reviews indicate that the majority of rescued mammals do not survive the rehabilitation process (Burton and Tribe 2016; Kwok *et al.* 2021).

The release of hand-reared orphans is contentious and policies and practices vary throughout Australia (Englefield *et al.* 2018b). The nature of typical marsupial hand-rearing techniques and the length of time animals are in care make it very challenging to avoid habituation to humans. Humanisation is recognised generally as a poor indicator of future welfare for free-ranging animals but robust post-release studies are lacking for many species that undergo rehabilitation.

The welfare of rescued animals can be improved by promoting swift triage processes that ensure medical assessment of injuries and disease is not delayed, that suffering is promptly relieved, that rehabilitators and treating veterinarians undergo appropriate training and are mentored by those with species experience, that housing facilities are fit for purpose and that research evaluating welfare indices of these animals is supported and funded.

## 12.2 Conservation translocations

Conservation translocations refer to purposefully moving threatened animals (or plants) from one site to another to achieve conservation outcomes; they are recognised as a valuable tool that supports population restoration through reintroduction or reinforcement, assisted colonisation or ecological replacement (IUCN/SSC 2013). Animal welfare is an integral consideration for translocation feasibility, alongside other key factors such as environmental suitability, resource availability, quality of source populations (genetic, demographic and behavioural suitability), social impact and effective program governance (IUCN/SSC 2013). Breed-for-release programs are a common source of conservation translocation animals and have added complexity for welfare considerations compared with wild-to-wild translocations. This is especially true with respect to determining whether the animals bred can develop an ecologically appropriate behavioural repertoire (McPhee 2004).

A global review of 293 translocations by Berger-Tal *et al.* (2020) demonstrated that animal behaviour-related issues account for many frequently encountered challenges. They found that problems associated with dispersal and movement are common (45% of all behaviour-related issues reported), with animals typically dispersing long distances from the release site. Rapid dispersal may be attributed to a lack of conspecifics, the presence of aggressive conspecifics or competitors, a lack of desired resources, stress associated with a novel environment and animals dispersing in search of habitat that has the same features as their natal habitat. To mitigate potential dispersal issues, release site fidelity can be encouraged by:

- providing release site cues while still in care
- providing supplementary food near the release site
- using playback calls of conspecifics near the release site and/or
- choosing sites with desirable resources.

Other problems encountered in translocations are those associated with learning, when animals have trouble adapting to a novel environment, and when inefficient feeding habits negatively impact fitness. Suggested prerelease mitigations include training, the use of natural diets, promoting natural methods of foraging and minimising human interactions to avoid habituation. Competitive behavioural and social issues are reported when translocated animals are unable to adequately compete with existing populations or do not demonstrate appropriate social behaviours. For Australian native mammals, one of the major factors impacting translocation success is the effectiveness of cat and fox control (Short 2009). The apparent lack of predator awareness and appropriate anti-predator behaviour explains, at least in part, why introduced cats and foxes remain a key threat for many species. Many Australian native mammal translocations occur in either introduced predator-free sanctuaries or in areas with introduced predator control programs to manage this risk. Predator avoidance training has also been trialled with mixed success for numbats, bilbies, eastern barred bandicoots (*Perameles gunnii*), quokkas, northern quolls and tammar wallabies (Robertson 2007; Edwards *et al.* 2021; Taylor *et al.* 2021).

Environmental enrichment can be a valuable tool to support reintroduction by ensuring relevant competencies develop (see section 6.1). Enrichment for release candidates has a different objective to enrichment for animals in permanent human care. It occasionally requires animals to be subject to aversive or stressful stimuli, which is not usually synonymous with positive welfare (Freeman 2023). However, exposure to manageable stressors if delivered in short bursts can promote desirable behavioural skills and the development of appropriate stress responses. It must be remembered that welfare is the sum of an animal's life experiences and good welfare in an *ex-situ* environment should not be at the expense of welfare in the post-release phase. Enrichment has the potential to benefit post-release outcomes through improved behavioural skill development such as predator avoidance, foraging, social skills, courtship, mating, habitat selection, physical condition and learning movement and migration routes (Reading *et al.* 2013).

Despite attempts to manage behavioural competencies before release, translocation represents a stressful event via capture, handling, transport and release to a novel area (Dickens *et al.* 2010). Stress-mitigating strategies should be included in translocation planning and can include avoiding unnecessary handling, minimising transit time and allowing for acclimatisation and provision of supportive resources in transit and at release (Tarszisz *et al.* 2014). Additionally, there is a growing interest in improved monitoring pre- and post-release, which should support a better understanding of the impact of stressors on release success. Success is often defined as survival or population persistence, but that is not a good measure of the welfare of individually translocated animals. More robust monitoring of key physiological and behavioural welfare indicators and further research to understand sensory ecology and how both negative and positive affects operate in wild settings should help refine release protocols in the future (Swaisgood 2010; Tarszisz *et al.* 2014; Browning and Veit 2023).

## 13. ANIMAL WELFARE AND ETHICS

The terms 'animal welfare' and 'animal ethics' often get confused and used interchangeably in common language. However, they are distinct concepts, so understanding each of them and how they relate to each other is important.

'Ethics' is a philosophical discipline that attempts to discern how people ought to act, make decisions and place differential value on things. So, animal ethics encompasses how people ought to act around and with animals, how people make decisions that affect animals and what differential value is placed on individual animals or species in a wide range of situations.

A plethora of ethical theories are applicable in describing how people feel animals should be treated and on what grounds decisions affecting them should be made. With respect to Australian mammals in human care, 'animal rights' is perhaps the best known and controversial ethical theory but there are others. A description of selected theories is presented in Table 1.1. People or organisations may align strongly with one theory, or more commonly, may hold opinions that represent a complicated mixture of more than one, and these may differ with context.

In contrast to ethics and ethical theories, as discussed in section 2, 'animal welfare' encompasses both an animal's physical and mental state. It is a multidimensional concept (Miele *et al.* 2011; Ohl and van der Staay 2011) describing an intrinsic, internal state that is peculiar to an individual animal and may be thought of broadly as 'the quality of life as perceived by the animal itself' (Bracke *et al.* 1999). So simplistically, animal welfare is about animals and animal ethics is actually about people. As such, ethics can influence animal welfare but is not synonymous with it.

A person's commitment to achieving the highest possible standards of animal welfare is ethically driven (Mellor *et al.* 2015). Conflicts and moral distress can arise in communities, management structures and among animal care staff when the ethical tendencies of individuals do not align with those of their organisation, colleagues or those that influence regulation.

Although contemporary assessment of animal welfare is approached scientifically, values and ethical judgements are involved in deciding how animals should be treated. Every definition of animal welfare, and there are many, is influenced by the moral or ethical standards of the society in which it was proposed. Scientists, whether they are aware of it or not, are affected by their own ethical viewpoint. This places animal welfare at an interesting convergence of science and philosophy. In practice what this means is that animal welfare issues cannot simply be investigated and addressed by means of objective biological measurements of an animal's welfare state under certain circumstances (Ohl and van der Staay 2011). Interpretation of welfare status and active management of perceived welfare issues are influenced by context and cultural and societal values.

**Table 1.1.** Selected ethical theories and their application to animals.

| Ethical theory | Description |
|---|---|
| **Animal rights** | The conventional animal rights position is that beings described as '"subjects-of-a-life" have rights and should never be treated merely as an ends to others' means' (Gjerris *et al.* 2013). Animal rights incorporates the welfare-related right to freedom from pain and suffering, but also includes ideas about the right to liberty, right to life and the right to freedom from ownership, irrespective of the impact on welfare. Anything that disrespects an individual animal's rights is irrefutably considered to be morally wrong in this traditional rights view. Although many animal rights advocates oppose the keeping of animals in captivity altogether, others such as Gunasekera (2018) argue that a more moderate rights view points to the obligation to compensate animals for their loss of freedom in a captive setting by maximally promoting the remainder of their interests, including the provision of reliable food, shelter, safety, companionship, mental stimulation and a relatively long and painless life. |
| **Utilitarianism** | Provides an approach that attempts to balance the competing interests of all (animals and people) involved in or affected by a decision. The welfare of all interested parties is morally important to the utilitarian. There are no limits on how any individual animal may be treated; an action will be considered right by a utilitarian only if it maximises welfare. Effectively it holds that it is morally justifiable to trade off the fundamental interests of one or more animals in order to secure an aggregative increase in welfare for other animals and/or humans influenced by the action. |
| **Respect for nature** | Respect for nature proponents hold overlapping views about protecting the value of 'naturalness' and the preservation of species, biodiversity and/or ecosystems. The respect for nature standpoint suggests that animals should live 'natural' lives, according to species norms, and that welfare is concerned with the functioning or fitness of animals (Appleby and Sandøe 2002). This view appeals to many ordinary citizens who consider the expression of natural behaviours and living a life appropriate for the species as being conducive to good animal welfare (Lassen *et al.* 2006; Miele *et al.* 2011). For some there may be a focus on populations rather than individuals. |
| **Contractarian** | Ethical obligations originate in mutual agreements or contracts between people. Indirect ethical obligations exist towards animals only if they matter to other people. |
| **Relational** | The emphasis is on the relationships between animals and humans and between and among humans. In this view people's duties to animals depend on whether the individual or group has a close relationship with people and/or whether the way a person treats an animal might affect their treatment of humans. |
| **Deontology** | Universal laws, defining good and bad, govern behaviour and decisions. Intentions must be good regardless of the outcome. |
| **Consequentialism** | An action is judged by its consequences: the better the consequences of an action, the more moral it is. |

The way in which management practices are viewed by society is affected markedly by both public understanding and community attitudes. Solutions to identified problems may not be straightforward and may need to be tailored to the societal context. This chapter has provided a contemporary understanding of animal welfare expectations for native mammals in Australia at the time of publication. Some of the concepts discussed are relatively recent developments and their application is anticipated to evolve further.

From a practical perspective, ethics can influence the answers to the following questions (Veissier *et al.* 2011).

- What is the baseline standard for how animals should be treated?
- Where do the thresholds exist between unacceptable, acceptable and good welfare?
- What kind of compromises are acceptable in a less-than-perfect world?
- Can a deficiency in one contributor to welfare be compensated for by the meeting of others?

Research projects involving Australian mammals (in the wild and in human care) must be assessed and approved by an Animal Ethics Committee (AEC) in accordance with a recognised code of practice. In Australia, this is the Australian Code of Practice for the Care and Use of Animals for Scientific Purposes (National Health and Medical Research Council 2013). The aim of this document is to ensure that studies meet humane, ethical and responsible standards and balance the proposed scientific insights to be gained against the potential animal welfare harms. One of the governing principles used by researchers and AECs in designing and assessing projects is the so-called '3Rs' (reduction, replacement and refinement). Projects can only be justified when the minimum number of animals is used to obtain meaningful results, non-animal alternatives are considered and used where appropriate and techniques are refined to minimise the welfare impact of the interventions. Although there is no formal mechanism to apply this concept to Australian mammals in non-research settings, the 3Rs can provide a valuable framework for the review of care and management protocols. For example, the use of training and conditioning is a refinement to physical capture for routine veterinary examination or sampling to reduce stress and the potential for injury.

## 14. ACKNOWLEDGEMENTS

The authors thank Matthew Crane for his thoughtful and constructive review of the draft manuscript.

## ADDENDUM 1

An example of a welfare assessment checklist.

This simple collection of criteria was designed for macropods, but can be adapted for other species. Rather than being a comprehensive evaluation, it is designed to flag areas of concern and to act as a guide for triggering more thorough welfare investigations such as tailored intensive welfare monitoring and assessments. Derived from: Mellor *et al.* 2015; Sherwen *et al.* 2018; Shaw and Jarman 2025 and ZAA 2021 Accreditation Welfare Assessment Tool (Zoo and Aquarium Association Australasia pers. comm.).

| **Nutrition – Goal: sufficient food and water leading to feelings of satiety, hydration** | |
|---|---|
| Weight of animal ________<br>Normal species weight range (at same life stage) _____________ | Is weight within normal range?<br>Y/N<br>Action: |
| Review of water and food quality and safety | Is water or food quality compromised (e.g. freezing, overheated, stagnant, stale, contaminated, fouling)?<br>Are there any known species-specific toxicity risks?<br>Y/N<br>Action: |
| Review animal diet | Is diet similar to wild diet in nutrient/energy/moisture content and variety?<br>Is there sufficient food for the number of animals?<br>Y/N/Partially/Unsure<br>Action: |
| Review water and food availability | Are there sufficient watering and feeding stations to avoid competition/guarding and are they accessible to all animals?<br>Y/N/Partially/Unsure<br>Action: |
| Review food presentation, opportunity for natural foraging behaviours | Food presented in species-appropriate manner (e.g. browse) and provided at species relevant time of day?<br>Y/N/Partially/Unsure<br>Action: |
| **Environment – Goal: comfortable environment that is sufficiently spacious and complex to support natural behaviours and minimise sensory disturbance (e.g. unpleasant noise/scents) leading to feelings of comfort, relaxation, calmness** | |
| Assess enclosure's spatial dimensions and enclosure furniture, referring to species' natural ecology | Is there sufficient space and complexity of space to promote a variety of natural behaviours likely to be meaningful for the animal?<br>Y/N/Partially/Unsure<br>Action: |
| Examine environmental factors that are likely to affect physical comfort and sensory experience:<br>• Tactile<br>• Auditory<br>• Scent<br>• Taste<br>• Visual/light<br>• Temperature<br>• Dryness/dampness | Does the environment support comfort for the animal (e.g. species-relevant variety of substrates such as sand or mulch; thermal cline where animal can utilise shade, sun, refuges, heating/cooling, misters and other related features; auditory; scents.<br>Y/N/Partially/Unsure<br>Action: |

| | |
|---|---|
| Examine safety aspects of the enclosure, containment, potential for pest incursion, boundary and obstacle collision, entanglement | Is the environment considered safe?<br>Y/N/Unsure<br>Action: |
| Examine opportunities for animals to take refuge (e.g. move away from potential stressors (visitors, activities, other animals)) | Does the animal utilise environmental refuges?<br>Y/N/Partially/Unsure<br>Action: |
| Examine variability and novelty in environment and how the environment is made interesting and engaging | Does the animal utilise environmental enrichment and does it achieve the intended goals?<br>Y/N/Partially/Unsure<br>Action: |
| Review capacity for meaningful choice and control | Does the animal have meaningful choice and control over its daily activities?<br>Y/N//Partially/Unsure<br>Action: |
| Examine how the animal uses the space provided and identify unused space | Carry out enclosure usage review |
| **Health – Goal: good health leading to feelings of vitality and being free from pain** | |
| Veterinary and keeper visual assessment for signs of good health and absence of disease, injury, parasites | Does the animal look healthy, with good fur condition and without signs of disease, parasites, abnormalities?<br>Y/N/Partially/Unsure<br>Action: |
| Examine animal for evidence of functional impairment preventing natural behaviour (e.g. lameness, coordination issues) | Does the animal look functionally fit?<br>Y/N/Partially/ Unsure<br>Action: |
| Review animal for species-specific health concerns in consultation with veterinarian (e.g. any facial swellings?) | Are there any species-specific health concerns not captured above?<br>Y/N<br>Action: |
| Body Condition Score (BCS).<br>Value: 1 (emaciated) -<br>5 (obese)_________ | Is BCS within an optimal range (2.5–3.5/5)?<br>Normal range: Y/N<br>Action: |
| **Behavioural interactions – Goal: positive interactions with the environment, other animals and humans that generate feelings of being engaged, energised, affectionate, social, rewarded, secure, playful, calm, confident** | |
| Examine opportunities to carry out species relevant natural behaviours – an ethogram can help determine highly motivated and meaningful behaviours | Is exploration observed in line with natural ecology?<br>Y/N/Partially/Unsure<br>Action:<br>Is foraging observed in line with natural ecology?<br>Y/N/Partially/Unsure<br>Action:<br>Are highly motivated behaviours observed?<br>Y/N/Partially/Unsure<br>Action:<br>Is play observed in line with natural ecology?<br>Y/N/Partially/Unsure<br>Action:<br>Are abnormal behaviours (e.g. repetitive behaviours) observed?<br>Y/N/Partially/Unsure<br>Action:<br>Has a change in aggression level been observed recently?<br>Y/N/Partially/Unsure<br>Action: |

**Addendum 1.** continued

| | |
|---|---|
| Examine species relevant social opportunities, including assessment of social make up (e.g. numbers in group, sexes, other species) | Has species-relevant social behaviour been observed (e.g. sexual activity, grooming, rearing, and affiliative behaviours)?<br>Y/N/Partially/Unsure<br>Action: |
| Positive human–animal interactions | Have positive human–animal interactions been observed with caregiver/s?<br>Y/N/Partially/Unsure<br>Action: |
| Subjective animal carer assessment | What is the animal carer's subjective assessment of the animal's welfare state?<br>Poor/Intermediate/Good<br>Action: |
| Duration and frequency of natural behaviours; behavioural diversity | Carry out an activity budget assessment or behavioural diversity assessment and determine the degree to which it aligns with wild/semi-wild scenarios, if relevant.<br>Action: |

## ADDENDUM 2

An example of a checklist outlining considerations for Animal-Visitor Interactions (AVIs)

Derived from: ZAA (2018); WAZA (2020); TCSA (2022).

**Considerations for developing higher welfare and respectful AVIs**

| | **Investigative point** | **Achieved Yes/No/Partially** | **Recommend modification** |
|---|---|---|---|
| 1 | Individual animal(s) (personality/temperament, age, sex, previous life experiences, method of rearing) and species (natural history traits) assessed as behaviourally appropriate for the experience | | |
| 2 | Location of the experience reviewed and options explored for carrying out experience in animal's enclosure or within an area the animal finds comfortable, safe and secure | | |
| 3 | The maximum number, density and proximity of visitors defined; minimising risk of negative visitor effect | | |
| 4 | The animal's experience or interpretation of the encounter assessed (humans are usually quite tactile, but other species may be more influenced by other senses such as scent or taste); how animals perceive unfamiliar humans should be considered, as it may be aversive (e.g. a human may be perceived as an enemy, prey, symbiote, conspecific, no consequence or friend (Hosey 2013)) | | |
| 5 | Potential stressors identified and mitigated | | |
| 6 | Animal presented in a natural setting, with minimal disturbance or interruption to natural behavioural patterns | | |
| 7 | Animal has genuine choice to participate or move away and seek refuge if desired (contained but not restrained) | | |
| 8 | Animal has a range of food opportunities offered outside the AVI and the impact of any food offered during the AVI has been considered from a nutritional, social and choice perspective | | |
| 9 | Training and conditioning program in place that supports good animal welfare and natural behaviours that matter to the animal | | |
| 10 | The experience frequency, duration and time of day has been assessed for animal welfare implications (e.g. is it within the animal's natural activity period?) | | |
| 11 | The impact of the experience on an animal's ability to carry out natural activities such as resting, grooming, bonding or playing with minimal disturbance has been assessed | | |

**Considerations for developing higher welfare and respectful AVIs**

| | Investigative point | Achieved Yes/No/Partially | Recommend modification |
|---|---|---|---|
| 12 | Regular monitoring and/or welfare assessments carried out to ensure good welfare is maintained | | |
| 13 | The AVI adds value to the animal's life experience (e.g. as part of an enrichment program) | | |
| 14 | When not participating in the encounter the animal is afforded the same level of husbandry care, complexity of space, social interactions and opportunities as other animals (e.g. those on display) | | |
| 15 | Staff/volunteers are trained to recognise signs of stress and discomfort and empowered to act accordingly (adapt the experience or remove the animal from the situation) | | |
| 16 | The AVI benefits the species by delivering a conservation, animal welfare or sustainability message | | |
| 17 | The experience does not inadvertently encourage damaging animal encounters elsewhere or promote wildlife trade or unsuitable pet ownership | | |
| 18 | Safety risk assessment carried out that considers visitors, animals and staff, actual and perceived risks, and zoonoses | | |

## ADDENDUM 3

An example of an Aged Animal Assessment template

This tool can assist with end-of-life planning and decision making for older animals and can be adapted to suit different species or taxonomic groups. Age-related conditions may impact animal welfare and consequently quality of life. Aged animal assessments assist with objectively assessing animal welfare and other relevant factors to inform the development of plans to ensure older animals are adequately supported later in life and to emphasise that quality of life is given priority over length of life. Repeated assessments monitor changes over time and prompt appropriate intervention strategies to respond to observed declines or quality-of-life concerns. Adapted with permission from the Taronga Conservation Society 2022 Aged Animal Assessment Template for Monotremes and Marsupials.

**Animal information**

| | |
|---|---|
| **Species** | **Expected longevity*** |
| **Date of birth** | **Maximum age in the wild** |
| **Sex** | **Maximum age in captivity** |
| **Animal ID#** | |
| **Assessment completed by** | **Assessment Date** |

**Age score**

| *Note: Expected longevity is calculated as the age at which 90% of a population has died. Typically determined through analysis of studbook information or the Zoological Information Management System (ZIMS) database. | | **Score** |
|---|---|---|
| < 80% expected longevity | Assign score −2 | |
| 80–100% expected longevity | Assign score −1 | |
| 101–120% expected longevity | Assign score 2 | |
| > 120% expected longevity | Assign score 4 | |

**Physical criteria**

Note: criteria are scored based on a comparison with an animal of the same species in its prime. Selected criteria are weighted: for them, the raw score is multiplied by the number in the multiply score column to generate their row subtotal

| Criteria | Multiply score | Score | Subtotal |
|---|---|---|---|
| Body condition | × 2 | 1 = excellent, 5 = poor (either thin or obese) | |
| Lameness in 1 limb | | 1 = mild, 5 = severe | |

**Addendum 3.** continued

| Criteria | Multiply score | Score | Subtotal |
|---|---|---|---|
| Lameness in > 1 limb | × 2 | 1 = mild, 5 = severe | |
| Stiffness in 1 limb | | 1 = mild, 5 = severe | |
| Stiffness in > 1 limb | × 2 | 1 = mild, 5 = severe | |
| Overgrown nails | | If present = 3 | |
| Deformity/swelling of joints | × 2 | 1 = mild, 5 = severe | |
| Lethargy | × 2 | Listless, drowsy, sleepy<br>1 = mild, 5 = severe | |
| Mobility | × 2 | 1 = reduced, 5 = severe immobility | |
| Activity | × 2 | Time spent being active (e.g. exploring, foraging)<br>1 = reduced, 5 = very inactive | |
| Weakness | | Lack of muscle strength<br>1 = mild, 5 = severe | |
| Vomiting | | 1 = rarely, 5 = often | |
| Diarrhoea | | 1 = rarely, 5 = often | |
| Constipation | | 1 = rarely, 5 = often | |
| Reduced appetite | | 1 = rarely, 5 = often | |
| Coughing | × 2 | 1 = mild, 5 = severe | |
| Abnormal breathing | × 2 | 1 = mild, 5 = severe | |
| **Discharges** | | | |
| Eyes | | 1 = mild, 5 = severe | |
| Ears | | 1 = mild, 5 = severe | |
| Mouth | | 1 = mild, 5 = severe | |
| Nasal | | 1 = mild, 5 = severe | |
| Urogenital sinus/penis | | 1 = mild, 5 = severe | |
| Cloaca | | 1 = mild, 5 = severe | |
| Pouch | | 1 = mild, 5 = severe | |
| Glands | | 1 = mild, 5 = severe | |
| Dribbles saliva | | 1 = mild, 5 = severe | |
| Red, swollen gums | | 1 = mild, 5 = severe | |
| Broken, worn or lost teeth | | 1 = few, 5 = many | |
| Bill/beak abnormalities | | 1 = mild, 5 = severe | |
| Difficulty eating | | 1 = rarely, 5 = often | |
| Food presentation | | 1 = food presentation normal; 5 = significant modifications to manage impairments (e.g. food chopped or softened) | |
| Loss of vision | | 1 = mild, 5 = blind | |
| Other ocular abnormalities | | 1 = mild, 5 = severe | |
| Excessive urination | | 1 = mild, 5 = severe | |
| Decreased urination | | 1 = mild, 5 = severe | |
| Excessive thirst | | 1 = mild, 5 = severe | |
| Poor coat/skin condition | | Dry coat/skin, fur loss<br>1 = mild, 5 = severe | |
| Skin ulcers, swellings, wounds, infections | | 1 = mild, 5 = severe | |
| Loss of coordination | | 1 = mild, 5 = severe | |
| Urinary incontinence | | 1 = mild, 5 = severe | |

| Criteria | Multiply score | Score | Subtotal |
|---|---|---|---|
| Faecal incontinence | | 1 = mild, 5 = severe | |
| Vagueness | | 1 = mild, 5 = severe | |
| Reaction time | | 1 = slow, 5 = very slow | |
| **Radiographic changes present at last examination** | | | |
| Degenerative joint disease (DJD) in 1 joint | | 1 = mild, 5 = severe | |
| DJD > 1 joint | × 2 | 1 = mild, 5 = severe | |
| Ankylosed joint | | Single joint = 5 | |
| Ankylosed joint | × 2 | > 1 joint = 5 | |
| Other (specify) | | | |
| **Clinical abnormalities at last veterinary examination** | | | |
| Anaemia | | 1 = mild, 5 = severe | |
| Renal impairment | × 2 | 1 = mild, 5 = severe | |
| Liver impairment | | 1 = mild, 5 = severe | |
| Neoplasia | | 1 = mild, 5 = severe | |
| Reproductive pathology | | 1 = mild, 5 = severe | |
| Other | | | |
| **Medication** | | | |
| Current treatments | | 1 = nutraceutical alone<br>2 = anti-inflammatory alone<br>3 = nutraceutical + anti-inflammatory<br>4 = opioid alone<br>5 = all of them together | |
| Other (specify) | | 1 per additional medication | |
| Duration of treatment | | 1 = 3 months<br>2 = 3–6 months<br>3 = 6–12 months<br>4 = > 12 months<br>If currently on ≥ 1 drug, score for the most longstanding one | |
| **Behaviour** | | | |
| Self-directed behaviour | | Directing behaviour towards self (licking, scratching, biting) at one area for a prolonged period.<br>Time spent directing behaviour to be rated as:<br>1 = low, 5 = high | |
| Self-injurious behaviour | × 2 | When self-directed behaviour leads to damage:<br>1 = minor damage (e.g. hair loss),<br>5 = major damage (e.g. loss of digits) | |
| Response to stimuli | | Stimuli include weather, visitors, enclosure structures:<br>1 = mild impairments, 5 = not responsive | |
| Spatial awareness | | Ability to navigate around enclosure and use all enclosure furnishings:<br>1 = mild impairment, 5 = limited enclosure use | |
| Sensory awareness | | Ability to respond appropriately to thermal conditions, sounds, lights, smells, tactile stimuli:<br>1 = mild impairments, 5 = not responsive | |
| Activity pattern | | Change in behavioural time budget (e.g. foraging, exploring):<br>1 = mild reduction in activity, 5 = severe reduction in activity | |

**Addendum 3.** continued

| | |
|---|---|
| Behavioural diversity | Degree to which individual and species appropriate behavioural repertoire reflects diversity expected:<br>1 = mildly reduced repertoire, 5 = severely reduced repertoire |
| Engagement with environment | Degree to which the animal interacts with the environment and can navigate around enclosure habitat and utilise furnishings:<br>1 = minimal change, 5 = significant change |
| Engagement with conspecifics | Degree to which the animal responds to conspecifics (if applicable):<br>1 = minimal change, 5 = significant change |
| Engagement with keepers | Degree to which the animal responds to the presence or cues provided by keepers:<br>1 = minimal change, 5 = significant change |
| Apathy | Presence of apathetic behaviours (e.g. depressed, disinterested):<br>1 = mild, 5 = severe |
| Self-care behaviours | Change in self-care behaviours such as bathing, grooming:<br>1 = mild change, 5 = severe reduction in self-care |
| Subjective keeper assessment | 1 = mostly in a positive affective state<br>5 = mostly in a negative affective state |
| Other (specify) (e.g. any stressors identified) | Score 1 per additional issue identified |

**Social (only applies to social species)**

| | |
|---|---|
| Changes in hierarchy | Monitored by change in access to food, shelter, grooming opportunities, contact with others (e.g. huddling):<br>1 = minimal change, 5 = significant changes |
| Target for aggression/ostracised | Monitored by change in access to food, shelter, grooming opportunities, contact with others (e.g. huddling):<br>1 = minimal change, 5 = significant changes |
| Beneficial role within social structure | 1 = significant, 5 = not very beneficial |

**Other species-specific considerations**

**Results**

| | |
|---|---|
| **Total score** | **Increase or decrease since last assessment** |
| **Previous score** | **Date of previous assessment** |

**Geriatric care plan**

| | |
|---|---|
| Medications | Action: |
| Enclosure modifications | Action: |
| Enrichment | Action: |
| Dietary changes | Action: |
| Animal watch requirements | Action: |
| Other: | Action: |
| Next health check date | |
| Next aged assessment date | |

| Medications | Action: |
|---|---|

**Recommended outcome:**

[ ] Monitor
[ ] Treatment
[ ] End-of-Life Planning
[ ] Euthanasia Recommended

## REFERENCES


Alcock J, Maley CC, Aktipis CA (2014) Is eating behavior manipulated by the gastrointestinal microbiota? Evolutionary pressures and potential mechanisms. *BioEssays* **36**, 940–949. doi:10.1002/bies.201400071

Allen BL, Goullet M, Allen LR, Lisle A, Leung LKP (2013) Dingoes at the doorstep: preliminary data on the ecology of dingoes in urban areas. *Landscape and Urban Planning* **119**, 131–135. doi:10.1016/j.landurbplan.2013.07.008

Alligood C, Leighty K (2015) Putting the "E" in SPIDER: evolving trends in the evaluation of environmental enrichment efficacy in zoological settings. *Animal Behavior and Cognition* **2**, 200–217. doi:10.12966/abc.08.01.2015

Andersen GE, McGregor HW, Johnson CN, Jones ME (2020) Activity and social interactions in a wide-ranging specialist scavenger, the Tasmanian devil (*Sarcophilus harrisii*), revealed by animal-borne video collars. *PLoS One* **15**, e0230216. doi:10.1371/journal.pone.0230216

Appleby MC, Sandøe P (2002) Philosophical debate on the nature of well-being: implications for animal welfare. *Animal Welfare (South Mimms, England)* **11**, 283–294. doi:10.1017/S0962728600024866

Archer M (2002) Confronting crises in conservation: a talk on the wild side. In *A Zoological Revolution – Using Native Fauna to Assist in its Own Survival*. (Eds D Lunney and C Dickman) pp. 12–52. Royal Zoological Society of New South Wales and Australian Museum, Sydney.

Asher L, Collins LM, Ortiz-Pelaez A, Drewe JA, Nicol CJ, Pfeiffer DU (2009) Recent advances in the analysis of behavioural organization and interpretation as indicators of animal welfare. *Journal of the Royal Society, Interface* **6**, 1103–1119. doi:10.1098/rsif.2009.0221

Baars AJ (2001) There are no known differences in brain mechanisms of consciousness between humans and other mammals. *Animal Welfare (South Mimms, England)* **10**, S31–S40. doi:10.1017/S0962728600023496

Bacon H (2018) Behaviour-based husbandry: a holistic approach to the management of abnormal repetitive behaviours. *Animals (Basel)* **8**, 103. doi:10.3390/ani8070103

Balcombe J (2009) Animal pleasure and its moral significance. *Applied Animal Behaviour Science* **118**, 208–216. doi:10.1016/j.applanim.2009.02.012

Bassett L, Buchanan-Smith H (2007) Effects of predictability on the welfare of captive animals. *Applied Animal Behaviour Science* **102**, 223–245. doi:10.1016/j.applanim.2006.05.029

Beausoleil N (2020) *EAZA Animal Welfare Webinars: The Five Domains model: history value and updates*. <https://www.youtube.com/watch?v=t6v8A1l6zT4>

Beausoleil N (2023) *2023 ZAA Conference: The Five Domains model: why and how should we use it in our zoo?* <https://www.youtube.com/watch?v=nAI4quNxvQA&list=PLshDDGUYOTCFylL55DLmEAJMlTFkE0d7X&index=32>

Benesch AR (2007) Chronoethological assessment of well-being and husbandry in captive koalas *Phascolarctos cinereus*, Goldfuss 1817. PhD Thesis. Johann Wolfgang Goethe-Universität, Germany.

Berger-Tal O, Blumstein DT, Swaisgood RR (2020) Conservation translocations: a review of common difficulties and promising directions. *Animal Conservation* **23**, 121–131. doi:10.1111/acv.12534

Bethge P (2002) Energetics and foraging behaviour of the platypus *Ornithorhynchus anatinus*. PhD thesis. University of Tasmania, Hobart.

Blyde D (2025) Cetaceans. In *Current Therapy in Medicine of Australian Mammals*. Revised Edition. (Eds L Vogelnest and T Portas) pp. 635–647. CSIRO Publishing, Melbourne.

Bodley K (2025) Pinnipeds. In *Current Therapy in Medicine of Australian Mammals*. Revised Edition. (Eds L Vogelnest and T Portas) pp. 621–634. CSIRO Publishing, Melbourne.

Boissy A, Manteuffel G, Jensen MB, Moe RO, Spruijt B, Keeling LJ, Bakken M (2007) Assessment of positive emotions in animals to improve their welfare. *Physiology & Behavior* **92**, 375–397. doi:10.1016/j.physbeh.2007.02.003

Bonney KR, Wynne CDL (2004) Studies of learning and problem solving in two species of Australian marsupials. *Neuroscience and Biobehavioral Reviews* **28**, 583–594. doi:10.1016/j.neubiorev.2004.08.005

Botma KL, Shapland F, Hoy JM, Pritchard T, Murray PJ (2020) Habitat influence on activity patterns of brush-tailed rock-wallabies in Queensland. *Australian Journal of Zoology* **68**, 136–144. doi:10.1071/ZO20090

Bracke MBM, Spruijt BM, Metz JHM (1999) Overall animal welfare reviewed – part 3: welfare assessment based on needs and supported by expert opinion. *Netherlands Journal of Agricultural Science* **47**, 307–322. doi:10.18174/njas.v47i3.468

Brando S (2012) Animal learning and training: implications for animal welfare. *The Veterinary Clinics of North America. Exotic Animal Practice* **15**, 387–398. doi:10.1016/j.cvex.2012.06.008

Brando S, Buchanan-Smith HM (2018) The 24/7 approach to promoting optimal welfare for captive wild animals. *Behavioural Processes* **156**, 83–95. doi:10.1016/j.beproc.2017.09.010

Briscoe N, Handasyde K, Griffiths S, Porter W, Krockenberger A, Kearney M (2014) Tree-hugging koalas demonstrate a novel thermoregulatory mechanism for arboreal mammals. *Biology Letters* **10**, 20140235. doi:10.1098/rsbl.2014.0235

Browning H, Maple TL (2019) Developing a metric of usable space for zoo exhibits. *Frontiers in Psychology* **10**, 791. doi:10.3389/fpsyg.2019.00791

Browning H, Veit W (2023) Positive wild animal welfare. *Biology & Philosophy* **38**, 14. doi:10.1007/s10539-023-09901-5

Brydges NM, Leach M, Nicol K, Wright R, Bateson M (2011) Environmental enrichment induces optimistic cognitive bias in rats. *Animal Behaviour* **81**, 169–175. doi:10.1016/j.anbehav.2010.09.030

Buchanan-Smith H (2021) Evolutionary principles, learning from the wild, and making a case for positive change for animals. *WeBuzz by Animal Concepts Podcast* Ep **27** <https://podcasts.apple.com/au/podcast/ep27-hannah-buchanan-smith-on-evolutionary-principles/id1514809055?i=1000530815794>

Burton E, Tribe A (2016) The rescue and rehabilitation of koalas (*Phascolarctos cinereus*) in Southeast Queensland. *Animals (Basel)* **6**, 56. doi:10.3390/ani6090056

Butler AB, Manger PR, Lindahl BIB, Arhem P (2005) Evolution of the neural basis of consciousness: a bird-mammal comparison. *BioEssays* **27**, 923–936. doi:10.1002/bies.20280

Byers J (1999) The distribution of play behaviour among Australian marsupials. *Journal of Zoology* **247**, 349–356. doi:10.1111/j.1469-7998.1999.tb00997.x

Campbell CD, Pecon-Slattery J, Pollak R, Joseph L, Holleley CE (2019) The origin of exotic pet sugar gliders (*Petaurus breviceps*) kept in the United States of America. *PeerJ* **7**, e6180. doi:10.7717/peerj.6180

Campbell-Ward M (2020) Ethics and wild animal welfare in natural disasters: a case study of the 2019/2020 summer bushfires. In *33rd AWMS Annual Conference Book of Abstracts*. 8–10 December 2020, Online. p.6. Australasian Wildlife Management Society.

Campbell-Ward M (2023) Quality-of-life assessments in zoo animals: not just for the aged and charismatic. *Animals (Basel)* **13**, 3394. doi:10.3390/ani13213394

Carlstead K (2009) A comparative approach to the study of keeper-animal relationships in the zoo. *Zoo Biology* **28**, 589–608. doi:10.1002/zoo.20289

Carlstead K, Shepherdson D (2000) Alleviating stress in zoo animals with environmental enrichment. In *The Biology of Animal Stress: Basic Principles and the Implications for Animal Welfare* (Eds GP Moberg and JA Mench) pp. 337–354. CABI Publishing, New York.

Chomel BB, Belotto A, Meslin F-X (2007) Wildlife, exotic pets and emerging zoonoses. *Emerging Infectious Diseases* **13**, 6–11. doi:10.3201/eid1301.060480

Clark FE (2013) Marine mammal cognition and captive care: a proposal for cognitive enrichment in zoos and aquariums. *The Journal of Zoo and Aquarium Research* **1**, 1–6.

Claxton AM (2011) The potential of the human–animal relationship as an environmental enrichment for the welfare of zoo-housed animals. *Applied Animal Behaviour Science* **133**, 1–10. doi:10.1016/j.applanim.2011.03.002

Clegg ILK (2018) Cognitive bias in zoo animals: an optimistic outlook for welfare assessment. *Animals (Basel)* **8**, 104. doi:10.3390/ani8070104

Clegg ILK, Rödel HG, Mercera B, van der Heul S, Schrijvers T, de Laender P, Gojceta R, Zimmitti M, Verhoeven E, Burger J, Bunskoek PE, Delfour F (2019) Dolphins' willingness to participate (WtP) in positive reinforcement training as a potential welfare indicator, where WtP predicts early changes in health status. *Frontiers in Psychology* **10**, 2112. doi:10.3389/fpsyg.2019.02112

Coe J (2004) Mixed species rotation exhibits. In *Proceedings of the Australasian Regional Association of Zoological Parks and Aquaria*. Presented at Christchurch, New Zealand. <https://joncoe.net/wp-content/uploads/2021/12/MxdSpRo.pdf>

Coe J, Hoy J (2020) Choice, control and computers: empowering wildlife in human care. *Multimodal Technologies and Interaction* **4**, 92. doi:10.3390/mti4040092

Comport SS, Ward SJ, Foley WJ (1996) Home ranges, time budgets and food tree use in a high density tropical population of greater gliders, *Petauroides volans minor* (Pseudocheiridae: Marsupialia). *Wildlife Research* **23**, 401–419. doi:10.1071/WR9960401

Cooney R, Chapple RS, Doornbos S, Jackson S (2010) A feasibility study into the use of Australian native mammals as pets. Rural Industries Research and Development Corporation Publication No. 10/072, Canberra.

D'Cruze N, Khan S, Carder G, Megson D, Coulthard E, Norrey J, Groves GA (2019) Global review of animal–visitor interactions in modern zoos and aquariums and their implications for wild animal welfare. *Animals (Basel)* **9**, 332. doi:10.3390/ani9060332

Davey G (2007) Visitors' effects on the welfare of animals in the zoo: a review. *Journal of Applied Animal Welfare Science* **10**, 169–183. doi:10.1080/10888700701313595

Déaux E, Crowe T, Charrier I (2017) Recreational fishing alters dingo foraging behavior on Fraser Island. *The Journal of Wildlife Management* **82**, 85–92. doi:10.1002/jwmg.21340

Denton DA, McKinley MJ, Farrell M, Egan GF (2009) The role of primordial emotions in the evolutionary origin of consciousness. *Consciousness and Cognition* **18**, 500–514. doi:10.1016/j.concog.2008.06.009

Dickens MJ, Delehanty DJ, Romero LM (2010) Stress: an inevitable component of animal translocation. *Biological Conservation* **143**, 1329–1341. doi:10.1016/j.biocon.2010.02.032

Diesch TJ, Mellor D, Johnson C, Lentle R (2008) Responsiveness to painful stimuli in anaesthetised newborn and young animals of varying neurological maturity (wallaby joeys, rat pups and lambs). *Alternatives to Animal Testing and Experimentation : AATEX* **14**, 549–552.

Diesch TJ, Mellor DJ, Johnson CB, Lentle RG (2010) Developmental changes in the electroencephalogram and responses to a noxious stimulus in anaesthetized tammar wallaby joeys (*Macropus eugenii eugenii*). *Laboratory Animals* **44**, 79–87. doi:10.1258/la.2009.009045

Dold C (2015) Cetacea (whales, dolphins, porpoises). In *Fowler's Zoo and Wild Animal Medicine*. Vol. 8. (Eds RE Miller and ME Fowler) pp. 422–435. Elsevier Saunders, St Louis.

Duncan IJH, Petherick JC (1991) The implications of cognitive processes for animal welfare. *Journal of Animal Science* **69**, 5017–5022. doi:10.2527/1991.69125017x

Dunstan E, Fairbrother B, Van Sluys M (2016) Empowering citizens to effect change: a case study of zoo-based community conservation. *Pacific Conservation Biology* **22**, 90–97. doi:10.1071/PC15035

Edwards M, Ford C, Hoy J, FitzGibbon S, Murray P (2021) How to train your wildlife: a review of predator avoidance training. *Applied Animal Behaviour Science* **234**, 105170. doi:10.1016/j.applanim.2020.105170

Englefield B, Starling M, McGreevy P (2018a) A review of roadkill rescue: who cares for the mental, physical and financial welfare of Australian wildlife carers? *Wildlife Research* **45**, 103–118. doi:10.1071/WR17099

Englefield B, Blackman SA, Starling M, McGreevy PD (2018b) A review of Australian animal welfare legislation, regulation, codes of practice, and policy, and their influence on stakeholders caring for wildlife and the animals for whom they care. *Animals (Basel)* **9**, 335. doi:10.3390/ani9060335

Englefield B, Candy S, Starling M, McGreevy P (2019) The demography and practice of Australians caring for native wildlife and the psychological, physical and financial effects of rescue, rehabilitation and release of wildlife on the welfare of carers. *Animals (Basel)* **9**, 1127. doi:10.3390/ani9121127

Evans MC (2008) Home range, burrow-use and activity patterns in common wombats (*Vombatus ursinus*). *Wildlife Research* **35**, 455–462. doi:10.1071/WR07067

Fanson KV, Best EC, Bunce A, Fanson BG, Hogan LA, Keeley T, Narayan EJ, Palme R, Parrott ML, Sharp TM, Skogvold K, Tuthill L, Webster KN, Bashaw M (2017) One size does not fit all: monitoring faecal glucocorticoid metabolites in marsupials. *General and Comparative Endocrinology* **244**, 146–156. doi:10.1016/j.ygcen.2015.10.011

Fernandez E (2022) Training as enrichment: a critical review. *Animal Welfare (South Mimms, England)* **31**, 1–12. doi:10.7120/09627286.31.1.001

Fernandez EJ, Martin AL (2021) Animal training, environmental enrichment, and animal welfare: a history of behaviour analysis in zoos. *Journal of Zoological and Botanical Gardens* **2**, 531–543. doi:10.3390/jzbg2040038

Fernandez EJ, Tamborski MA, Pickens SR, Timberlake W (2009) Animal–visitor interactions in the modern zoo: conflicts and interventions. *Applied Animal Behaviour Science* **120**, 1–8. doi:10.1016/j.applanim.2009.06.002

Finlayson GR, White CR, Dibben RM, Shimmin GA, Taggart DA (2010) Activity patterns of the southern hairy-nosed wombat (*Lasiorhinus latifrons*) (Marsupialia:Vombatidae) in the South Australian Murraylands. *Australian Mammalogy* **32**, 39–46. doi:10.1071/AM09024

Fleming PA, Clarke T, Wickham SL, Stockman CA, Barnes AL, Collins T, Miller DW (2015) The contribution of qualitative behavioural assessment to appraisal of livestock welfare. *Animal Production Science* **56**, 1569–1578. doi:10.1071/AN15101

Frankham R, Ballou J, Briscoe D (2010) *Introduction to Conservation Genetics*. 2nd edn. Cambridge University Press, Cambridge.

Fraser D (2003) Assessing animal welfare at the farm and group level: the interplay of science and values. *Animal Welfare (South Mimms, England)* **12**, 433–443. doi:10.1017/S0962728600026038

Fraser D, Duncan IJH (1998) "Pleasures", "pains" and animal welfare: towards a natural history of affect. *Animal Welfare (South Mimms, England)* **7**, 383–396. doi:10.1017/S0962728600020935

Fraser D, Phillips PA, Thompson BK (1993) Environmental preference testing to access the well-being of animals: an evolving paradigm. *Journal of Agricultural & Environmental Ethics* **6**, 104–114.

Freeman M (2023) The behavioural biology of marsupials and monotremes. In *The Behavioural Biology of Zoo Animals* (Ed P Rose) pp. 119–132. CRC Press, Boca Raton.

Gaalema DE, Perdue BM, Kelling AS (2011) Food preference, keeper ratings, and reinforcer effectiveness in exotic animals: the value of systematic testing. *Journal of Applied Animal Welfare Science* **14**, 33–41. doi:10.1080/10888705.2011.527602

Gage LJ (2012) Ocular disease and suspected causes in captive pinnipeds. In *Fowler's Zoo and Wild Animal Medicine Current Therapy*. Vol 7. (Eds RE Miller and ME Fowler) pp. 490–494. Elsevier Saunders, St Louis.

Gamble KC (2004) Marsupial care and husbandry. *The Veterinary Clinics of North America. Exotic Animal Practice* **7**, 283–298. doi:10.1016/j.cvex.2004.01.002

Geiser F, Ferguson C (2001) Intraspecific differences in behaviour and physiology: effects of captive breeding on patterns of torpor in feathertail gliders. *Journal of Comparative Physiology. B, Biochemical, Systemic, and Environmental Physiology* **171**, 569–576. doi:10.1007/s003600100207

Gjerris M, Nielsen MEJ, Sandøe P (2013) The good, the right and the fair: an introduction to ethics. *Texts in Philosophy*, nr. 22, College Publications.

Godfrey EJ, Cameron EZ, Hickling GJ (2023) Social learning in a nocturnal marsupial: is it a possum-ability? *Biology Letters* **19**, 20220460. doi:10.1098/rsbl.2022.0460

Godinez AM, Fernandez EJ (2019) What is the zoo experience? How zoos impact a visitor's behaviors, perceptions, and conservation efforts. *Frontiers in Psychology* **10**, 1746. doi:10.3389/fpsyg.2019.01746

Goldenberg SZ, Parker JM, Chege SM, Greggor AL, Hunt M, Lamberski N, Leigh KA, Nollens HH, Ruppert KA, Thouless C, Wittemyer G, Owen MA (2022) Revisiting the 4 R's: improving post-release outcomes for rescued mammalian wildlife by fostering behavioral competence during rehabilitation. *Frontiers in Conservation Science* **3**, 910358. doi:10.3389/fcosc.2022.910358

Grant RA, Montrose VT, Wills AP (2017) ExNOTic: Should we be keeping exotic pets? *Animals (Basel)* **7**, 47. doi:10.3390/ani7060047

Green TC, Mellor DJ (2011) Extending ideas about animal welfare assessment to include 'quality of life' and related concepts. *New Zealand Veterinary Journal* **59**, 263–271. doi:10.1080/00480169.2011.610283

Grueber CE, Reid-Wainscoat EE, Fox S, Belov K, Shier DM, Hogg CJ, Pemberton D (2017) Increasing generations in captivity is associated with increased vulnerability of Tasmanian devils to vehicle strike following release to the wild. *Scientific Reports* **7**, 2161. doi:10.1038/s41598-017-02273-3

Guérineau C, Lõoke M, Ganassin G, Bertotto D, Bortoletti M, Cavicchioli L, Furlati S, Mongillo P, Marinelli L (2022) Enrichment with classical music enhances affiliative behaviours in bottlenose dolphin. *Applied Animal Behaviour Science* **254**, 105696. doi:10.1016/j.applanim.2022.105696

Gunasekera CA (2018) The ethics of killing "surplus" zoo animals. *Journal of Animal Ethics* **8**, 93–102. doi:10.5406/janimalethics.8.1.0093

Hall BA, McGill DM, Sherwen SL, Doyle RE (2021) Cognitive enrichment in practice: a survey of factors affecting its implementation in zoos globally. *Animals (Basel)* **11**, 1721. doi:10.3390/ani11061721

Hanke FD, Hanke W, Scholtyssek C, Dehnhardt G (2009) Basic mechanisms in pinniped vision. *Experimental Brain Research* **199**, 299. doi:10.1007/s00221-009-1793-6

Hawkins M (2005) Enrichment for Australian mammal species: a review. In *Proceedings of the 7th International Conference on Environmental Enrichment*. 31 July–5 August, New York. (Eds N

Clum, S Silver and P Thomas) pp. 79–87. Wildlife Conservation Society, New York, USA.

Hellmuth H, Augustine L, Watkins B, Hope K (2012) Using operant conditioning and desensitization to facilitate veterinary care with captive reptiles. *The Veterinary Clinics of North America. Exotic Animal Practice* **15**, 425–443. doi:10.1016/j.cvex.2012.06.003

Hess L (2011) Exotic animals: appropriately owned pets or inappropriately kept problems? *Journal of Avian Medicine and Surgery* **25**, 50–56. doi:10.1647/1082-6742-25.1.50

Higginbottom K, Croft DB (1999) Social learning in marsupials. In *Mammalian Social Learning: Comparative and Ecological Perspectives* (Eds HO Box and KR Gibson) pp. 80–101. Cambridge University Press, Cambridge.

Hill SP, Broom DM (2009) Measuring zoo animal welfare: theory and practice. *Zoo Biology* **28**, 531–544. doi:10.1002/zoo.20276

Hogan LA, Tribe A (2007) Prevalence and cause of stereotypic behaviour in common wombats (*Vombatus ursinus*) residing in Australian zoos. *Applied Animal Behaviour Science* **105**, 180–191. doi:10.1016/j.applanim.2006.06.006

Hogan LA, Johnston SD, Lisle A, Horsup AB, Janssen T, Phillips CJC (2010) Stereotypies and environmental enrichment in captive southern hairy-nosed wombats, *Lasiorhinus latifrons*. *Applied Animal Behaviour Science* **126**, 85–95. doi:10.1016/j.applanim.2010.05.009

Holland GJ, Bennett AF, van der Ree R (2007) Time-budget and feeding behaviour of the squirrel glider (*Petaurus norfolcensis*) in remnant linear habitat. *Wildlife Research* **34**, 288–295. doi:10.1071/WR06104

Hopwood P (2002) Native Australian mammals as pets: an overview. In *A Zoological Revolution. Using Native Fauna to Assist in Its Own Survival.* (Eds D Lunney and C Dickman) pp. 77–83. Royal Zoological Society of New South Wales and Australian Museum, Sydney.

Hosey G (2008) A preliminary model of human–animal relationships in the zoo. *Applied Animal Behaviour Science* **109**, 105–127. doi:10.1016/j.applanim.2007.04.013

Hosey G (2013) Hediger revisited: how do zoo animals see us? *Journal of Applied Animal Welfare Science* **16**, 338–359. doi:10.1080/10888705.2013.827916

Hosey G, Melfi V (2015) Are we ignoring neutral and negative human–animal relationships in zoos? *Zoo Biology* **34**, 1–8. doi:10.1002/zoo.21182

Hosey G, Melfi V, Pankhurst S (2009) *Zoo Animals: Behaviour, Management and Welfare.* Oxford University Press, Oxford.

Hosey G, Melfi V, Pankhurst S (2013) *Zoo Animals: Behaviour, Management and Welfare.* 2nd edn. Oxford University Press, Oxford.

Hoy JM, Murray PJ, Tribe A (2010) Thirty years later: enrichment practices for captive mammals. *Zoo Biology* **29**, 303–316. doi:10.1002/zoo.20254

Hughes J, Macdonald DW (2013) A review of the interactions between free-roaming domestic dogs and wildlife. *Biological Conservation* **157**, 341–351. doi:10.1016/j.biocon.2012.07.005

Hulst F, Allan G, Vogelnest L (2010) Mesenteric fat necrosis in the quokka (*Setonix brachyurus*). In *Proceedings of the Wildlife Disease Association, Australasian Section.* December 13–17, Dover, Tasmania.

Hume ID (2005) Nutrition of marsupials in captivity. *International Zoo Yearbook* **39**, 117–132. doi:10.1111/j.1748-1090.2005.tb00011.x

Indigo N, Smith J, Webb J, Phillips B (2018) Not such silly sausages: evidence suggests northern quolls exhibit aversion to toads after training with toad sausages. *Austral Ecology* **43**, 592–601. doi:10.1111/aec.12595

IUCN/SSC (2013) *Guidelines for Reintroductions and Other Conservation Translocations*. Version 1.0. IUCN Species Survival Commission, Gland.

Izzard S, Barnard O, Schaap D (2019) Managing and maintaining wild temperament and behaviours in captivity. In *Saving the Tasmanian Devil: Recovery Through Science-Based Management* (Eds CJ Hogg, S Fox, D Pemberton and K Belov) pp. 207–221. CSIRO Publishing, Melbourne.

Jackson SM, Crane MD (2008) Getting close to animals: changing attitudes and the regulation of the zoo industry in New South Wales. In *Too Close for Comfort: Contentious Issues in Human-Wildlife Encounters* (Eds D Lunney, A Munn and W Meikle) pp. 149–157. Royal Zoological Society of New South Wales, Sydney.

Jackson SM, Johnson CN (2002) Time allocation to foraging in the mahogany glider *Petaurus gracilis* (Marsupialia, Petauridae) and a comparison of activity times in exudivorous and folivorous possums and gliders. *Journal of Zoology* **256**, 271–277. doi:10.1017/S0952836902000304

Johnson DH (2020) Sugar glider medicine and common diseases. Proceedings of the 92nd Western Veterinary Conference. 16–19 February, Las Vegas. Viticus Group, Las Vegas, Nevada.

Johnson R (2008) Thirty years of rehab: experiences of a wildlife veterinarian. In *Too Close for Comfort: Contentious Issues in Human-Wildlife Encounters.* (Eds D Lunney, A Munn and W Meikle) pp. 15–21. Royal Zoological Society of New South Wales, Sydney.

Johnson-Delaney C (2010) Marsupials. In *BSAVA Manual of Exotic Pets.* 5th edn. (Eds A Meredith and C Johnson-Delaney) pp. 103–126. British Small Animal Veterinary Association, Gloucester.

Jolly CJ, Webb JK, Gillespie GR, Phillips BL (2020) Training fails to elicit behavioral change in a marsupial suffering evolutionary loss of antipredator behaviors. *Journal of Mammalogy* **101**, 1108–1116. doi:10.1093/jmammal/gyaa060

Jones M, Gartland KN, Fuller G (2021) Effects of visitor presence and crowd size on zoo-housed red kangaroos (*Macropus rufus*) during and after a COVID-19 closure. *Animal Behavior and Cognition* **8**, 521–537. doi:10.26451/abc.08.04.06.2021

Kaandorp J (2012) Veterinary challenges of mixed species exhibits. In *Fowler's Zoo and Wild Animal Medicine Current Therapy*. Vol 7. (Eds RE Miller and ME Fowler) pp. 24–31. Elsevier Saunders, St Louis.

Kagan R, Veasey J (2010) Challenges of zoo animal welfare. In *Wild Mammals in Captivity: Principles and Techniques for Zoo Management*. 2nd edn. (Eds DG Kleiman, KV Thompson and CK Baer). pp. 11–21. The University of Chicago Press, Chicago.

Kagan R, Carter S, Allard S (2015) A universal animal welfare framework for zoos. *Journal of Applied Animal Welfare Science* **18**, S1–S10. doi:10.1080/10888705.2015.1075830

Kaufmann JH (1974) Social ethology of the whiptail wallaby, *Macropus parryi* in northeastern New South Wales. *Animal Behaviour* **22**, 281–369. doi:10.1016/S0003-3472(74)80032-1

Kirkden RD, Pajor EA (2006) Using preference, motivation and aversion tests to ask scientific questions about animals' feelings. *Applied Animal Behaviour Science* **100**, 29–47. doi:10.1016/j.applanim.2006.04.009

Kirkwood JK (2003) Welfare, husbandry and veterinary care of wild animals in captivity: changes in attitudes, progress in knowledge and techniques. *International Zoo Yearbook* **38**, 124–130. doi:10.1111/j.1748-1090.2003.tb02072.x

Klettenheimer BS, Temple-Smith PD, Sofronidis G (1997) Father and son sugar gliders: more than a genetic coalition? *Journal of Zoology* **242**, 741–750. doi:10.1111/j.1469-7998.1997.tb05823.x

Koene P (2013) Behavioral ecology of captive species: using behavioral adaptations to assess and enhance welfare of nonhuman zoo animals. *Journal of Applied Animal Welfare Science* **16**, 360–380. doi:10.1080/10888705.2013.827917

Kwok ABC, Haering R, Travers SK, Stathis P (2021) Trends in wildlife rehabilitation rescues and animal fate across a six-year period in New South Wales, Australia. *PLoS One* **16**, e0257209. doi:10.1371/journal.pone.0257209

Larsen MJ, Sherwen SL, Rault J-L (2014) Number of nearby visitors and noise level affect vigilance in captive koalas. *Applied Animal Behaviour Science* **154**, 76–82. doi:10.1016/j.applanim.2014.02.005

Lassen J, Sandøe P, Forkman B (2006) Happy pigs are dirty! Conflicting perspectives on animal welfare. *Livestock Science* **103**, 221–230. doi:10.1016/j.livsci.2006.05.008

Learmonth MJ, Sherwen S, Hemsworth PH (2018) The effects of zoo visitors on Quokka (*Setonix brachyurus*) avoidance behavior in a walk-through exhibit. *Zoo Biology* **37**, 223–228. doi:10.1002/zoo.21433

Leighty KA, Valuska AJ, Grand AP, Bettinger TL, Mellen JD, Ross SR, Boyle P, Ogden JJ (2015) Impact of visual context on public perceptions of non-human primate performers. *PLoS One* **10**, e0118487. doi:10.1371/journal.pone.0118487

Lenzi C, Sperian S, Grasso C (2020) "Let me take a selfie": implications of social media for public perceptions of wild animals. *Society & Animals* **31**, 64–83. doi:10.1163/15685306-BJA10023

Lilith M, Calver M, Styles I, Garkaklis M (2006) Protecting wildlife from predation by owned domestic cats: application of a precautionary approach to the acceptability of proposed cat regulations. *Austral Ecology* **31**, 176–189. doi:10.1111/j.1442-9993.2006.01582.x

Logan M, Sanson GD (2003) The effects of lactation on the feeding behaviour and activity patterns of free-ranging female koalas (*Phascolarctos cinereus* Goldfuss). *Australian Journal of Zoology* **51**, 415–428. doi:10.1071/ZO03017

Lyons JA, Natusch DJD (2011) Wildlife laundering through breeding farms: illegal harvest, population declines and a means of regulating the trade of green pythons (*Morelia viridis*) from Indonesia. *Biological Conservation* **144**, 3073–3081. doi:10.1016/j.biocon.2011.10.002

Lyons JA, Natusch DJD (2012) Over-stepping the quota? The trade in sugar gliders in West Papua, Indonesia. *Traffic Bulletin* **24**, 5–6.

Manteca Vilanova X (2020) *EAZA Animal Welfare Webinars: The Fundamentals of Animal Welfare Assessments*. <https://www.youtube.com/watch?v=8TT6xBLcggU>

Maple TL, Perdue BM (2013) *Zoo Animal Welfare*. Springer, Heidelberg.

Markus N, Blackshaw JK (2002) Behaviour of the black flying fox *Pteropus alecto*: 1. An ethogram of behaviour, and preliminary characterization of mother–infant interactions. *Acta Chiropterologica* **4**, 137–152. doi:10.3161/001.004.0203

Mason GJ (1991) Stereotypies: a critical review. *Animal Behaviour* **41**, 1015–1037. doi:10.1016/S0003-3472(05)80640-2

Mason GJ (2010) Species differences in welfare responses to captivity: stress, welfare and the comparative method. *Trends in Ecology & Evolution* **25**, 713–721. doi:10.1016/j.tree.2010.08.011

McLelland D (2025) Macropod progressive periodontal disease ('lumpy jaw'). In *Current Therapy in Medicine of Australian Mammals*. Revised Edition. (Eds L Vogelnest and T Portas) pp. 483–494. CSIRO Publishing, Melbourne.

McNeill AT, Leung LKP, Goullet MS, Gentle MN, Allen BL (2016) Dingoes at the doorstep: home range sizes and activity patterns of dingoes and other wild dogs around urban areas of north-eastern Australia. *Animals (Basel)* **6**, 48. doi:10.3390/ani6080048

McPhee ME (2004) Generations in captivity increases behavioral variance: considerations for captive breeding and reintroduction programs. *Biological Conservation* **115**, 71–77. doi:10.1016/S0006-3207(03)00095-8

McPhee ME, Carlstead K (2010) The importance of maintaining natural behaviors in captive mammals. In *Wild Mammals in Captivity: Principles and Techniques for Zoo Management*. 2nd edn. (Eds DG Kleiman, KV Thompson and CK Baer) pp. 303–313. The University of Chicago Press, Chicago.

Meade J, Melfi VA, Keith M, Burns AL (2021) The effect of visitor number on the behavior of zoo-housed macropods. *Anthrozoos* **34**, 477–489. doi:10.1080/08927936.2021.1914432

Meehan CL, Mench JA (2007) The challenge of challenge: can problem solving opportunities enhance animal welfare. *Applied Animal Behaviour Science* **102**, 246–261. doi:10.1016/j.applanim.2006.05.031

Melfi V (2009) There are big gaps in our knowledge, and thus approach, to zoo animal welfare: a case for evidence-based zoo animal management. *Zoo Biology* **28**, 574–588. doi:10.1002/zoo.20288

Melfi V (2013) Is training zoo animals enriching? *Applied Animal Behaviour Science* **147**, 299–305. doi:10.1016/j.applanim.2013.04.011

Melfi VA, Feistner ATC (2002) A comparison of the activity budgets of wild and captive Sulawesi crested black macaques (*Macaca nigra*). *Animal Welfare (South Mimms, England)* **11**, 213–222. doi:10.1017/S0962728600028165

Mella VSA, Gillies C, McArthur C, Webb E, Herbert CA (2022) Factors influencing the success of hand-reared juvenile brushtail possums (*Trichosurus vulpecula*) released into the wild. *Wildlife Research* **50**, 96–107. doi:10.1071/WR22056

Mellen J, MacPhee MS (2001) Philosophy of environmental enrichment: past, present, and future. *Zoo Biology* **20**, 211–226. doi:10.1002/zoo.1021

Mellor DJ (2010) Galloping colts, fetal feelings, and reassuring regulations: putting animal-welfare science into practice. *Journal of Veterinary Medical Education* **37**, 94–100. doi:10.3138/jvme.37.1.94

Mellor DJ (2015) Positive animal welfare states and reference standards for welfare assessment. *New Zealand Veterinary Journal* **63**, 17–23. doi:10.1080/00480169.2014.926802

Mellor DJ (2016) Updating animal welfare thinking: moving beyond the 'Five Freedoms' towards 'A Life Worth Living'. *Animals (Basel)* **6**, 21. doi:10.3390/ani6030021

Mellor DJ (2017) Operational details of the Five Domains Model and its key applications to the assessment and management of animal welfare. *Animals (Basel)* **7**, 60. doi:10.3390/ani7080060

Mellor DJ (2019) Welfare-aligned sentience: enhanced capacities to experience, interact, anticipate, choose and survive. *Animals (Basel)* **9**, 440. doi:10.3390/ani9070440

Mellor DJ, Beausoleil NJ (2015) Extending the 'Five Domains' model for animal welfare assessment to incorporate positive welfare states. *Animal Welfare (South Mimms, England)* **24**, 241–253. doi:10.7120/09627286.24.3.241

Mellor DJ, Reid CSW (1994) Concepts of animal well-being and predicting the impact of procedures on experimental animals. In *Improving the Well-being of Animals in the Research Environment* (Eds RM Baker, G Jenkin and DJ Mellor) pp. 3–18. Australian and New Zealand Council for the Care of Animals in Research and Teaching, Glen Osmond.

Mellor DJ, Patterson-Kane E, Stafford KJ (2009) *The Sciences of Animal Welfare.* Wiley-Blackwell Publishing, Oxford.

Mellor DJ, Diesch TJ, Johnson CB (2010) Legal and animal welfare implications of when consciousness first appears in developing young and of the potential for delayed onset of increased pain sensitivity. In *Australian Animal Welfare Strategy Conference* <https://citeseerx.ist.psu.edu/document?repid=rep1&type=pdf&doi=1a60338b93e77aaa7c06350c144fe42247474b0f>

Mellor DJ, Hunt S, Gusset M (2015) *Caring for Wildlife: The World Zoo and Aquarium Animal Welfare Strategy.* WAZA Executive Office, Gland, Switzerland.

Mellor DJ, Beausoleil NJ, Littlewood KE, McLean AN, McGreevy PD, Jones B, Wilkins C (2020) The 2020 Five Domains model: including human–animal interactions in assessments of animal welfare. *Animals (Basel)* **10**, 1870. doi:10.3390/ani10101870

Mench J (1998) Why it is important to understand animal behaviour. *ILAR Journal* **39**, 20–26. doi:10.1093/ilar.39.1.20

Miele M, Veissier I, Evans A, Botreau R (2011) Animal welfare: establishing a dialogue between science and society. *Animal Welfare (South Mimms, England)* **20**, 103–117. doi:10.1017/S0962728600002475

Miller LJ, Vicino GA, Sheftel J, Lauderdale LK (2020) Behavioral diversity as a potential indicator of positive animal welfare. *Animals (Basel)* **10**, 1211. doi:10.3390/ani10071211

Morgan KN, Tromborg CT (2007) Sources of stress in captivity. *Applied Animal Behaviour Science* **102**, 262–302. doi:10.1016/j.applanim.2006.05.032

Nagy KA, Martin RW (1985) Field metabolic rate, water flux, food consumption and time budget of koalas, *Phascolarctos cinereus* (Marsupialia: Phascolarctidae) in Victoria. *Australian Journal of Zoology* **33**, 655–665. doi:10.1071/ZO9850655

National Health and Medical Research Council (2013) *Australian Code for the Care and Use of Animals for Scientific Purposes.* 8th edn. National Health and Medical Research Council, Canberra.

National Health and Medical Research Council (2014) *A Guide to the Care and Use of Australian Native Mammals in Research and Teaching.* National Health and Medical Research Council, Canberra.

Ness RD, Johnson-Delaney CA (2012) Sugar gliders. In *Ferrets, Rabbits and Rodents: Clinical Medicine and Surgery.* 3rd edn. (Eds KE Quesenberry and JW Carpenter) pp. 393–410. Elsevier Saunders, St Louis.

O'Neal T (2011) Animal enrichment strategies for promoting natural behaviors in captive populations of Tasmanian devils (*Sarcophilus harrisii*). Independent Study Project (ISP) Collection. 1129. <https://digitalcollections.sit.edu/isp_collection/1129>

Oakwood M, Hopwood P (1999) A survey of the attributes and requirements of quolls that may affect their suitability as household pets. *Australian Zoologist* **31**, 365–375. doi:10.7882/AZ.1999.038

Ohl F, van der Staay FJ (2011) Animal welfare: at the interface between science and society. *Veterinary Journal (London, England)* **192**, 13–19. doi:10.1016/j.tvjl.2011.05.019

Olsson A, Woods R (2008) Bats. In *Medicine of Australian Mammals* (Eds L Vogelnest and R Woods) pp. 465–402. CSIRO Publishing, Melbourne.

Orr B, Tribe A (2018) Animal welfare implications of treating wildlife in Australian veterinary practices. *Australian Veterinary Journal* **96**, 475–480. doi:10.1111/avj.12765

Panksepp J (2005) Affective consciousness: core emotional feelings in animals and humans. *Consciousness and Cognition* **14**, 30–80. doi:10.1016/j.concog.2004.10.004

Panksepp J, Fuchs T, Iacobussi P (2011) The basic neuroscience of emotional experience in mammals: the case of subcortical FEAR circuitry and implications for clinical anxiety. *Applied Animal Behaviour Science* **129**, 1–17. doi:10.1016/j.applanim.2010.09.014

Parrott ML, Wicker LV, Lamont A, Banks C, Lang M, Lynch M, McMeekin B, Miller KA, Ryan F, Selwood KE, Sherwen SL, Whiteford C (2021) Emergency response to Australia's Black Summer 2019–2020: the role of a zoo-based conservation organisation in wildlife triage, rescue, and resilience for the future. *Animals (Basel)* **11**, 1515. doi:10.3390/ani11061515

Pettigrew JD (1999) Electroreception in monotremes. *The Journal of Experimental Biology* **202**, 1447–1454. doi:10.1242/jeb.202.10.1447

Rault JL, Waiblinger S, Boivin X, Hemsworth P (2020) The power of a positive human–animal relationship for animal welfare. *Frontiers in Veterinary Science* **7**, 590867. doi:10.3389/fvets.2020.590867

Reading RP, Miller B, Shepherdson D (2013) The value of enrichment to reintroduction success. *Zoo Biology* **32**, 332–341. doi:10.1002/zoo.21054

Reiss D, Marino L (2001) Mirror self-recognition in the bottlenose dolphin: a case of cognitive convergence. *Proceedings of the National Academy of Sciences of the United States of America* **98**, 5937–5942. doi:10.1073/pnas.101086398

Richmond SE, Wemelsfelder F, de Heredia IB, Ruiz R, Canali E, Dwyer CM (2017) Evaluation of animal-based indicators to be used in a welfare assessment protocol for sheep. *Frontiers in Veterinary Science* **4**, 210. doi:10.3389/fvets.2017.00210

Rimpley K, Buchanan-Smith H (2013) Reliably signalling a startling husbandry event improves welfare of zoo-housed capuchins *(Sapajus apella). Applied Animal Behaviour Science* **147**, 205–213. doi:10.1016/j.applanim.2013.04.017

Robertson H (2007) Wildlife conservation and Perth Zoo. In *Proceedings of the National Wildlife Rehabilitation Conference.* 5–9 August, Fremantle. Australian Wildlife Rehabilitation Conference. <https://www.awrc.org.au/uploads/5/8/6/6/5866843/robertson_helen_zoo.pdf>

Rose P, Croft DP (2015) The potential of Social Network Analysis as a tool for the management of zoo animals. *Animal Welfare (South Mimms, England)* **24**, 123–138. doi:10.7120/09627286.24.2.123

Rose PE, Riley LM (2021) Conducting behavioural research in the zoo: a guide to ten important methods, concepts and theories. *Journal of Zoological and Botanical Gardens* **2**, 421–444. doi:10.3390/jzbg2030031

Ross A, Letnic M, Blumstein D, Moseby K (2019) Reversing the effects of evolutionary prey naiveté through controlled predator exposure. *Journal of Applied Ecology* **56**, 1761–1769. doi:10.1111/1365-2664.13406

Russell E (1984) Social behaviour and social organisation of marsupials. *Mammal Review* **14**, 101–154. doi:10.1111/j.1365-2907.1984.tb00343.x

Russell FA (2014) Conditional same/different discrimination learning in the short-beaked echidna *(Tachyglossus aculeatus)*. PhD thesis. University of Wollongong, Wollongong, Australia.

Ryan MA, Whisson DA, Holland GJ, Arnould JPY (2013) Activity patterns of free-ranging koalas (*Phascolarctos cinereus*) revealed by accelerometry. *PLoS One* **8**, e80366. doi:10.1371/journal.pone.0080366

Sadler L, Ward S (1999) Coalitions in male sugar gliders: are they natural? *Journal of Zoology* **248**, 91–96. doi:10.1111/j.1469-7998.1999.tb01026.x

Schuppli CA, Fraser D (2000) A framework for assessing the suitability of different species as companion animals. *Animal Welfare (South Mimms, England)* **9**, 359–372. doi:10.1017/S0962728600022958

Schuppli CA, Fraser D, Bacon HJ (2014) Welfare of non-traditional pets. *Revue Scientifique et Technique (International Office of Epizootics)* **33**, 221–231. doi:10.20506/rst.33.1.2287

Shape of Enrichment Inc (2022) *Resources to Download*. <https://theshapeofenrichmentinc.wildapricot.org/Complimentary-Resources-to-Download-(PDF)>

Sharpe D, Goldingay RL (2019) Time budget of the squirrel glider (*Petaurus norfolcensis*) in subtropical Australia. *Australian Journal of Zoology* **66**, 251–260. doi:10.1071/ZO18049

Shaw M, Jarman A (2025) Nutrition. In *Current Therapy in Medicine of Australian Mammals*. Revised Edition. (Eds L Vogelnest and T Portas) pp. 229–255. CSIRO Publishing, Melbourne.

Shepherdson DJ (2010) Principles of and research on environmental enrichment for mammals. In *Wild Mammals in Captivity: Principles and Techniques for Zoo Management*. 2nd edn. (Eds DG Kleiman, KV Thompson and CK Baer). pp. 62–67. The University of Chicago Press, Chicago.

Sherwen SL, Hemsworth PH (2019) The visitor effect on zoo animals: implications and opportunities for zoo animal welfare. *Animals (Basel)* **9**, 366. doi:10.3390/ani9060366

Sherwen SL, Hemsworth PH, Butler KL, Fanson KV, Magrath MJ (2015) Impacts of visitor number on kangaroos housed in free-range exhibits. *Zoo Biology* **34**, 287–295. doi:10.1002/zoo.21226

Sherwen SL, Hemsworth LM, Beausoleil NJ, Embury A, Mellor DJ (2018) An animal welfare risk assessment process for zoos. *Animals (Basel)* **8**, 130. doi:10.3390/ani8080130

Sherwen SL, Campbell-Ward M, Woods R (2023) Wildlife welfare and the 2019–2020 wildfires. In *Australia's 2019–2020 Megafires: Biodiversity Impacts and Lessons for the Future* (Eds L Rumpff, S Legge, S van Leeuwen, B Wintle and J Woinarski) pp. 343–356. CSIRO Publishing, Melbourne.

Short J (2009) 'The characteristics and success of vertebrate translocations within Australia'. A progress report to Department of Agriculture, Fisheries and Forestry. <https://researchportal.murdoch.edu.au/esploro/outputs/book/The-characteristics-and-success-of-vertebrate/991005542576007891>

Smith BP (2014) Living with wild dogs: personality dimensions in captive dingoes (*Canis dingo*) and implications for ownership. *Anthrozoös* **27**, 423–433. doi:10.2752/175303714X14023922797869

Species360 (2023) Zoological Information Management System (ZIMS). Minneapolis, USA. <www.species360.org>

Špinka M (2019) Animal agency, animal awareness and animal welfare. *Animal Welfare (South Mimms, England)* **28**, 11–20. doi:10.7120/09627286.28.1.011

Špinka M, Wemelsfelder F (2011) Environmental challenge and animal agency. In *Animal Welfare*. pp. 27–44. CAB International, Wallingford.

Spooner SL, Farnworth MJ, Ward SJ, Whitehouse-Tedd KM (2021) Conservation education: are zoo animals effective ambassadors and is there any cost to their welfare? *Journal of Zoological and Botanical Gardens* **2**, 41–65. doi:10.3390/jzbg2010004

Stirrat SC (2004) Activity budgets of the agile wallaby, *Macropus agilis*. *Australian Journal of Zoology* **52**, 49–64. doi:10.1071/ZO04005

Story G, Driscoll D, Banks S (2014) What can camera traps tell us about the diurnal activity of the nocturnal bare-nosed wombat (*Vombatus ursinus*)? In *Camera Trapping: Wildlife Management and Research* (Eds P Meek, P Fleming, G Ballard, P Banks, A Claridge, J Sanderson and D Swann) pp. 35–43. CSIRO Publishing, Melbourne.

Swaisgood R (2010) The conservation-welfare nexus in reintroduction programmes: a role for sensory ecology. *Animal Welfare (South Mimms, England)* **19**, 125–137. doi:10.1017/S096272860000138X

Tarszisz E, Dickman CR, Munn AJ (2014) Physiology in conservation translocations. *Conservation Physiology* **2**, cou054. doi:10.1093/conphys/cou054

Taylor R, Coetsee AL, Doyle RE, Sutherland DR, Parrot ML (2021) Sniffing out danger: rapid antipredator training of an endangered marsupial. *Australian Mammalogy* **44**, 109–116. doi:10.1071/AM20048

Taylor-Brown A, Booth R, Gillett A, Mealy E, Ogbourne SM, Polkinghorne A, Conroy GC (2019) The impact of human activities on Australian wildlife. *PLoS One* **14**, e0206958. doi:10.1371/journal.pone.0206958

TCSA (2022) Dignity and respect for wildlife. Taronga Conservation Society Australia, Sydney, <https://taronga.org.au/sites/default/files/2022-03/Taronga_Animal_Welfare_Dignity_Respect_Charter_2022.pdf>

Thomson PC (1992) The behavioural ecology of dingoes in north-western Australia. II. Activity patterns, breeding season and pup rearing. *Wildlife Research* **19**, 519–530. doi:10.1071/WR9920519

Tuite EK, Moss SA, Phillips CJ, Ward SJ (2022) Why are enrichment practices in zoos difficult to implement effectively? *Animals (Basel)* **12**, 554. doi:10.3390/ani12050554

Ugaz C, Valdez RA, Romano MC, Galindo F (2013) Behavior and salivary cortisol of captive dolphins (*Tursiops truncatus*) kept in open and closed facilities. *Journal of Veterinary Behavior* **8**, 285–290. doi:10.1016/j.jveb.2012.10.006

Veasey J, Hammer G (2010) Managing captive mammals in mixed-species communities. In *Wild Mammals in Captivity: Principles and Techniques for Zoo Management*. 2nd edn. (Eds DG Kleiman, KV Thompson and CK Baer) pp. 151–161. The University of Chicago Press, Chicago.

Veissier I, Jensen KK, Batreau R, Sandøe P (2011) Highlighting ethical decisions underlying the scoring of animal welfare in the Welfare Quality scheme. *Animal Welfare (South Mimms, England)* **20**, 89–101. doi:10.1017/S0962728600002463

Viggers KL, Lindenmayer DB (2002) Problems with keeping native Australian mammals as companion animals. In *A Zoological Revolution. Using Native Fauna to Assist in Its Own Survival*. (Eds D Lunney and C Dickman) pp. 130–151. Royal Zoological Society of New South Wales and Australian Museum, Sydney.

Vogelnest L, Talbot JJ (2019) Quality-of-life assessment and end-of-life planning for geriatric zoo animals. In *Fowler's Zoo and Wild Animal Medicine Current Therapy*. Vol. 9. (Eds RE Miller, N Lamberski and P Calle) pp. 83–91. Elsevier, St Louis.

Warwick C, Steedman C, Jessop M, Arena P, Pilny A, Nicholas E (2018) Exotic pet suitability: understanding some problems and utilizing a labelling system to aid animal welfare, environment and consumer protection. *Journal of Veterinary Behavior* **26**, 17–26. doi:10.1016/j.jveb.2018.03.015

Warwick C, Pilny A, Steedman C, Howell T, Martinez-Silvestre A, Cadenas V, Grant R (2023) Mobile zoos and other itinerant handling events: current status and recommendations for future policies. *Animals (Basel)* **13**, 214. doi:10.3390/ani13020214

Watters JV, Margulis SW, Atsalis S (2009) Behavioral monitoring in zoos and aquariums: a tool for guiding husbandry and directing research. *Zoo Biology* **28**, 35–48. doi:10.1002/zoo.20207

Watters J, Krebs B, Eschmann C (2021) Assessing animal welfare with behavior: onward with caution. *Journal of Zoological and Botanical Gardens* **2**, 75–87. doi:10.3390/jzbg2010006

WAZA (2020) WAZA guidelines for animal-visitor interactions. World Association of Zoos and Aquariums, Barcelona <https://www.waza.org/wp-content/uploads/2020/05/ENG_WAZA-Guidelines-for-AVI_FINAL_-April-2020.pdf>

Wells DL (2009) Sensory stimulation as environmental enrichment for captive animals: a review. *Applied Animal Behaviour Science* **118**, 1–11. doi:10.1016/j.applanim.2009.01.002

Wemelsfelder F (1997) The scientific validity of subjective concepts in models of animal welfare. *Applied Animal Behaviour Science* **53**, 75–88. doi:10.1016/S0168-1591(96)01152-5

Westlund K (2014) Training is enrichment: and beyond. *Applied Animal Behaviour Science* **152**, 1–6. doi:10.1016/j.applanim.2013.12.009

Whitham JC, Wielebnowski N (2009) Animal-based welfare monitoring: using keeper ratings as an assessment tool. *Zoo Biology* **28**, 545–560. doi:10.1002/zoo.20281

Whitham JC, Bryant JL, Miller LJ (2020) Beyond glucocorticoids: integrating dehydroepiandrosterone (DHEA) into animal welfare research. *Animals (Basel)* **10**, 1381. doi:10.3390/ani10081381

Williams E, Carter A, Rendle J, Ward SJ (2021) Impacts of COVID-19 on animals in zoos: a longitudinal multi-species analysis. *Journal of Zoological and Botanical Gardens* **2**, 130–145. doi:10.3390/jzbg2020010

Wolfensohn S, Shotton J, Bowley H, Davies S, Thompson S, Justice WSM (2018) Assessment of welfare in zoos: towards optimum quality of life. *Animals (Basel)* **8**, 110. doi:10.3390/ani8070110

Wünderlich NV, Mosteller J, Beverland MB, Downey H, Kraus K, Lin M-H, Syrjälä H (2021) Animals in our lives: an interactive well-being perspective. *Journal of Macromarketing* **41**, 646–662. doi:10.1177/0276146720984815

WWF (2020) 'Australia's 2019–2020 bushfires: the wildlife toll'. World Wide Fund for Nature, <https://assets.wwf.org.au/image/upload/Animals_Impacted_Interim_Report_24072020_final?_a=ATO2Ba20>

Yeates J (2018) Naturalness and animal welfare. *Animals (Basel)* **8**, 53. doi:10.3390/ani8040053

Yeates JW, Main DCJ (2008) Assessment of positive welfare: a review. *Veterinary Journal (London, England)* **175**, 293–300. doi:10.1016/j.tvjl.2007.05.009

ZAA (2018) Position statement: animal visitor interactions. Zoo and Aquarium Association Australasia, Sydney.

# 2
# Platypus

*Stephen M. Jackson, Jessica L. Thomas, Melody Serena and Peter Temple-Smith*

## 1. INTRODUCTION

The platypus (*Ornithorhynchus anatinus*) and short-beaked echidna (*Tachyglossus aculeatus*) comprise the Australian monotremes. Due to their egg-laying ability, distinctive appearance and biology, they are of enormous public and scientific interest. These unusual features, along with its secretive and semi-aquatic behaviour, have made the platypus a longstanding focus of worldwide attention.

Platypuses appear to have been first held in captivity by Lauderdale Maule (1832) who captured a female and two young in 1831 that lived for 2 weeks on a diet of worms, bread and milk. In 1832–33 George Bennett (1834a, 1834b) kept several animals, including two nestlings, that survived for 5 weeks on moistened bread, chopped egg and finely minced meat. Subsequently, platypuses were reportedly held by Jules Verreaux (1848) on a diet of rice and egg yolk and by Harry Burrell (1927), who was the first person to construct a specialised facility for their public display in 1910. Burrell's animals survived for 3 months at the Sydney Zoological Gardens when it was located at Moore Park, before being moved to Mosman where it was renamed Taronga Zoological Park.

Few platypuses have been successfully exported. Budapest Zoo was the first overseas zoo to receive two animals in 1913 (Collins 1973). In 1922 a platypus was transferred to the New York Zoological Park where it was on display for 1 hour per day and survived for 49 days (Joseph 1922; Burrell 1927). In addition, one male and two females were sent overseas in both 1947 and 1958 to the New York Zoological Society (Fleay 1980). More recently, in 2019 two platypuses were flown from Taronga Zoo in Sydney to San Diego Zoo in California (San Diego Zoo 2022).

Early attempts to keep platypuses in captivity resulted in them dying after only a few weeks or months, until 1932 when Robert Eadie maintained three captive platypuses for between 6 months and more than 2 years at the Sir Colin Mackenzie Sanctuary (now Healesville Sanctuary) in Victoria, Australia (Eadie 1935). Little effort was made to breed platypuses until David Fleay (1944) bred them successfully in the summer of 1943–44 at Healesville Sanctuary. After Fleay, it would be another 47 years until platypuses were bred at the semi-wild facility at Warrawong Sanctuary, South Australia, in the summer of 1990–91 (Gemmell *et al.* 1992, 1995; T. Grant pers. comm.) and 55 years until they bred again at the fully artificial captive facility at Healesville Sanctuary in 1998–99 (Holland and Jackson 2002). Platypuses have subsequently been bred on multiple occasions at Healesville Sanctuary and Taronga Zoo, facilitating further research into the husbandry and breeding requirements of this species (Hawkins and Battaglia 2009; Thomas *et al.* 2017).

Captive management of platypuses has an essential role to play in biological research as well as conservation-based educational displays. Accordingly, there is a need to ensure that impeccable standards for the captive management of platypuses are developed. Although many current standards may be universally applicable, no definitive 'recipe' for exhibiting, maintaining and breeding platypuses in captivity has been developed.

## 2. TAXONOMY

### 2.1 Nomenclature

The platypus was originally described as *Platypus anatinus* by Shaw (1799). However, as the word

*Platypus* was already being used for a genus of beetles, the term *Ornithorhynchus* was instead adopted. This name was first used by Blumenbach (1800) to describe the platypus when he called it *Ornithorhynchus paradoxus*.

Class: Mammalia
Subclass: Prototheria
Order: Monotremata
Family: Ornithorhynchidae
Genus and species: *Ornithorhynchus anatinus*

## 2.2 Subspecies

None.

## 2.3 Recent synonyms

Synonyms of the platypus can be found in Jackson and Groves (2015).

## 2.4 Other common names

In the past it has been called the duckbill and water mole.

# 3. NATURAL HISTORY

## 3.1 Bodyweight and morphometrics

The platypus is one of Australia's most easily recognisable animals and is distinguished from other mammals by its soft-skinned bill, webbed feet and semi-aquatic lifestyle. It is covered by dense, waterproof fur (except on the bill, feet and tail) and the bill is soft, pliable around the edges and covered in sensory receptors used to navigate underwater and locate prey (Proske and Gregory 2003).

Adult male platypuses weigh 1379–2393 g on average and females weigh 888–1365 g on average, with bodyweight varying greatly by geographic location (Table 2.1). The front feet terminate in a broad band of webbing; adult males possess a venomous spur on the inner hindleg.

There is a general north to south cline in body size, with the smallest animals found in Queensland and the largest in Tasmania (Carrick 1995; Connolly and Obendorf 1998; Furlan *et al.* 2011). Bodyweight can also vary significantly with season (Temple-Smith 1973; Grant and Temple-Smith 1983; Bino *et al.* 2015), implying that length may be a more reliable indicator of size (P. Temple-Smith pers. obs.). There is a distinct sexual dimorphism, with males being longer (≈25%; Grant, 2015), larger and heavier than females.

## 3.2 Distribution and habitat

Platypuses live in freshwater habitats along the east coast of mainland Australia from mid-north Queensland to the Victorian border with South Australia (including some river systems running west of the Great Dividing Range) as well as in Tasmania, on King Island and on Kangaroo Island, where they were introduced. They occupy streams, creeks, lakes and ponds from sea level to near the top of the Australian Alps, including some human-made reservoirs and waterways that can be brackish during periods of low flows. When resting or sleeping out of the water, platypuses mainly occupy burrows that are dug into the banks of a water body. Burrows are usually relatively short and simple in construction, with the entrance located either above or below water level, often hidden under tree roots or overhanging vegetation (Serena 1994; Gardner and Serena 1995; Serena *et al.*1998; Thomas *et al.* 2019). The average temperature inside burrows has

**Table 2.1.** Mean (± standard deviation) measurements of platypus body length (bill tip to tail tip) and bodyweight in different parts of Australia.

Values for Queensland and New South Wales from Carrick (1995); values for Tasmania from Connolly and Obendorf (1998).

| Location | Bodyweight (g) | | Total length (cm) | |
|---|---|---|---|---|
| | Males | Females | Males | Females |
| North Queensland | 1018 ± 208 | 704 ± 49 | 44.1 ± 3.1 | 41.0 ± 1.8 |
| South-east Queensland | 1556 ± 194 | 1222 ± 94 | 49.3 ± 2.7 | 43.8 ± 1.6 |
| New South Wales – East of Great Dividing Range | 1434 ± 218 | 857 ± 107 | 50.5 ± 2.4 | 41.5 ± 2.0 |
| New South Wales – On Great Dividing Range | 1379 ± 132 | 888 ± 92 | 47.4 ± 3.5 | 40.3 ± 2.0 |
| New South Wales – West of Great Dividing Range | 2215 ± 364 | 2000 | 54.9 ± 29 | 47.0 |
| Tasmania | 2393 ± 358 | 1365 ± 202 | 57.8 ± 2.8 | 48.2 ± 2.0 |

been found to be 18°C in summer and 14°C in winter (Grant and Dawson 1978a; Bethge *et al.* 2004).

## 3.3 Conservation status

The platypus remains widespread and relatively common in many places, but is officially classified as Near Threatened in Australia; however, it is currently listed as Endangered in South Australia and Vulnerable in Victoria. Threats include altered water flow regimens associated with the use of large dams to support irrigation, inappropriate fishing practices, litter entanglement, habitat degradation, climate change, water pollution and the activities of introduced species such as foxes and carp (Grant and Temple-Smith 2003; Woinarski *et al.* 2014; Bino *et al.* 2019).

## 3.4 Diet in the wild

Platypuses feed on a wide variety of freshwater adult and larval invertebrates such as worms, mayflies, dipterans and caddisflies (Table 2.2), generally taking whatever small prey is available to them, including small freshwater crayfish and yabbies (Munks *et al.* 2000; McLachlan-Troup *et al.* 2010; Marchant and Grant 2015; Klamt *et al.* 2016; Hawke *et al.* 2022). Platypuses are believed to locate their food mainly by detecting the weak electrical impulses created when invertebrate muscles contract (Proske *et al.* 1998). The platypus has a complex bill structure that is also used to expel inedible material and water from the mouth when feeding (Booth and Connolly 2008). Prey items are stored in cheek pouches while the platypus forages underwater, and thoroughly masticated as the animal floats on the water surface. Fragments of exoskeleton are largely discarded and the soft parts of prey ingested, making it impossible to conduct standard faecal dietary analysis.

**Table 2.2.** Example of food consumed by platypuses in the wild, from a study at the upper Shoalhaven River, New South Wales.

Derived from Faragher *et al.* (1979) and Grant (1995).

| Food | % in winter | % in summer |
|---|---|---|
| Horsehair worms | 17 | - |
| Freshwater shrimps | 12 | - |
| Caddisfly larvae | 41 | 64 |
| Midge and fly larvae | 12 | 18 |
| Mayfly larvae | 18 | - |
| Stonefly larvae | - | 9 |
| Dragonfly larvae | - | 9 |

## 3.5 Longevity

### 3.5.1 Wild

Mark-recapture studies in the upper Shoalhaven River in New South Wales have revealed that platypuses may occasionally survive to 21 years of age in the wild, but most survive from 7 to 14 years, with male longevity being less than that of females (Grant 2004; Serena *et al.* 2014). During a more recent study in Melbourne's south-eastern suburbs (Serena *et al.* 2024), a male was captured when nearly 24 years of age.

### 3.5.2 Captivity

Platypuses routinely live more than 10 years in captivity, and often over 20 years, with the oldest platypuses on record being a 30-year-old female held at Healesville Sanctuary and a 27-year-old female at Taronga Zoo (J. Thomas pers. obs.; Carrick *et al.* 1982; Serena *et al.* 2024). Other long-lived captive platypuses include an animal at Sydney Aquarium that survived to over 23 years of age (W. Meikle pers. comm.), an animal that lived to 21 years of age at Lone Pine Koala Sanctuary,Queensland, and an animal that lived to 18 years of age at the Australian Reptile Park, New South Wales.

### 3.5.3 Methods used to determine the age of adults

Platypuses are difficult to age once they have achieved adult bodyweight. The size and appearance of male spurs can be used to identify first-year juveniles and second-year subadults with reasonable reliability (Fig. 2.1). Most juvenile females retain a tiny vestigial spur sheath (≤ 4 mm long) on each ankle until the age of ≈9 months; in older females, a small pit marks the site of the vestigial sheath (Temple-Smith 1973; Grant 2007; Williams *et al.* 2013; Fig. 2.1). A small percentage of young subadults (4% of males and 2% females) may be misclassified as juveniles due to the sheaths being retained for longer than normal. Spur length is greatest when the males are 19–24 months of age and decreases linearly with age thereafter (Williams *et al.* 2013).

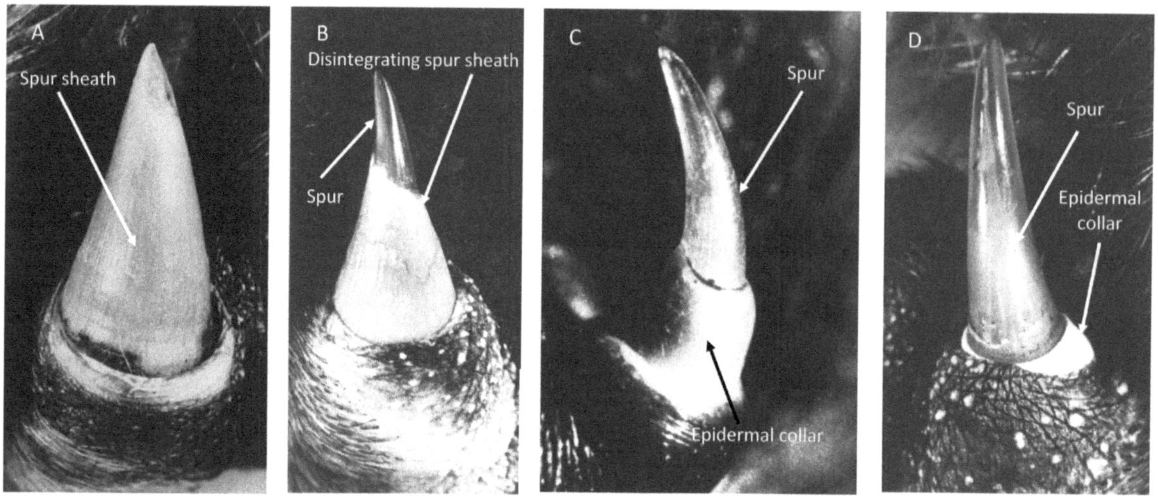


**Fig. 2.1.** Spur morphology and age classes of the male platypus. (a) Recently emerged juvenile with spur sheath intact. (b) Later juvenile with spur sheath beginning to disintegrate. Both of these animals are in their first year of life. (c) Subadult spur with no spur sheath present and the epidermal collar extending up the spur. This animal is in its second year of life. (d) Adult spur with epidermal collar no longer prominent. Photos by Peter Temple-Smith (A,B and D) and Tom Grant (C). See Williams *et al.* (2013) and Grant *et al.* 2024 for further details of age-related change in spur morphology.

## 4. HOUSING REQUIREMENTS

### 4.1 Enclosure design

In 1848, the French biologist Jules Verreaux reported that the species was 'no trouble' to maintain, but provided little documentation about his methods. However, since the early 1900s many dedicated and determined platypus enthusiasts have progressively addressed the challenges associated with platypus husbandry, culminating in the first platypus being born and raised successfully in captivity in 1943.

**Table 2.3.** Captive facilities holding platypus past and present.

| Facility | Years held on display |
|---|---|
| Brisbane Forest Park, Qld, Aust. | 1992–2009 |
| Bronx Zoo, New York, USA. | 1947–58 |
| Budapest Zoo, Hungary. | 1913–17? |
| Fleay's Fauna Park, Qld, Aust. | 1952–present |
| Australian Reptile Park, NSW, Aust. | 1968–present |
| Healesville Sanctuary, Vic., Aust. | 1933–present |
| Lone Pine Koala Sanctuary, Qld, Aust. | 1972–88<br>2010–present |
| Melbourne Zoo, Vic., Aust. | 1937–present |
| SEA LIFE, Sydney Aquarium, NSW, Aust. | 1997–present |
| Taronga Zoo, NSW, Aust. | 1934–present |
| San Diego Safari Park, CA, USA | 2019–present |
| Warrawong Sanctuary, SA, Aust. | 1987-present |
| Walkabout Creek, Qld, Aust | 2007–present |

Currently, 10 zoological institutions hold platypuses (Table 2.3). Most of these facilities are fundamentally quite similar, with the animals being displayed within buildings. However, the platypuses held at Warrawong Sanctuary are semi-wild and held within large, open, fully constructed enclosures (stream, ponds and a small wetland) in a previously permanently dry creek line within a degraded and abandoned dairy farm, surrounded by a predator-proof fence (T. Grant pers. comm.; Gemmell *et al.* 1995).

Platypuses are semi-aquatic, spending similar time active in the water and sleeping in burrows. Therefore, natural soil banks in an enclosure provide opportunities for important behaviours such as digging a burrow and also platforms for grooming and travel to other areas. The following components should be considered for inclusion when constructing an artificial river bank within a platypus enclosure:

- the soil should be clay-rich or compacted to reduce risk of burrow collapse
- the bank should run along the full length of the water source, preferably on both sides in some sections
- provision of plants and tree roots to stabilise the soil
- minimum depth of 0.5 m, and minimum width of 0.5 m

- have a base or underwire to prevent animals digging out
- provide sections where the bank overhangs the water
- any digging by the animal should not send soil into the water.

The captive environment for platypuses should include:

- natural foraging opportunities of live food items
- a selection of protected feeding and grooming sites in secure locations, designed so animals can retreat to their own space
- water movement
- water and nest box temperatures maintained within the range normally experienced by platypuses in the wild
- three burrows per platypus
- tunnel systems (used to connect feeding tanks, display areas and nest boxes) modelled on tunnels in the wild with respect to length and internal dimensions
- nest boxes that contain dry and clean nesting material. Timber boxes have good longevity. Recycled plastic is durable but slippery inside and retains condensation. PVC pipe although circular in shape requires non-slip matting glued inside for traction but it becomes unstuck quickly. Hatches in the roof of the pipe also warp their shape easily, providing an escape risk and leaking daylight.
- a variety of natural objects such as logs, rocks, soil and live plants
- natural earthen banks or earth-filled tanks to allow breeding females to construct nesting burrows
- provision for breeding pairs to forage and sleep separately.

The aquatic environment should be maintained to a high standard of clarity and cleanliness. Tank water should be changed frequently if recirculating filters are not incorporated into the system. Filter inlets should be shielded to prevent platypuses becoming trapped.

Facilities should be insulated from electrical currents, excessive noise and vibration (e.g. associated with pumps and filtering equipment).

Tunnels used by platypuses should conform to the following:

- measure at least 6–8 cm (H) and 9–13 cm (W) internally, which corresponds to the minimum documented dimensions of tunnels in the wild (Burrell 1927; Grant 1983a; Serena 1994). In practice, it is recommended that tunnels be at least 15 cm (H) × 15 cm (W) internally to reduce the potential for fur loss due to rubbing
- be constructed of materials that minimise abrasion to feet, bills and fur while providing traction underfoot
- microchip weighbridge can be fitted into tunnels so each animal can be identified and weighed as it travels along the designated tunnel
- slope at an angle of 25° or less
- ideally extend 1–4 m in length from the water's edge to burrow chambers, but should not exceed 4 m. Ideally it should be ~1.5 m (Serena *et al.* 1998)
- enable water that enters tunnels with platypuses to escape through drainage and/or evaporation
- have interiors that are accessible via secure lids/hatches for inspection or cleaning purposes
- be rainproof (if located outdoors)
- be sufficiently well shaded and/or insulated so the interior does not exceed ambient air temperatures on sunny days (and ideally does not exceed 25°C)
- be fitted with sliding internal doors (or comparable devices) at both the nest box and tank ends. More than one entry/exit point should be available to avoid dead ends if housing more than one animal.

Breeding females require access to a mound of earth so they can dig their own burrows and nesting chambers and plug tunnels with backfilled soil, known as a 'pug' (Thomas *et al.* 2018a). Adequate humidity during incubation is maintained by both the plugged entrances and the wet vegetation used by the female to construct the nest (Temple-Smith and Grant 2001; Thomas *et al.* 2019). The soil should be clay-rich to ensure that it holds its structure and does not collapse. Soil structure can be enhanced by making it open to the weather and allowing vegetation to grow on top so roots help to bind the soil together.

Breeding females also require a substantial volume of floating nesting material in tanks adjoining potential nesting areas. Appropriate nesting materials includes grasses, rushes and mat rushes (e.g. *Lomandra longifolia* and *Poa ensiformis*) and the bark and leaves of eucalypts (e.g. *E. viminalis* and *E. ovata*) (Burrell 1927; Fleay 1944; Gilfedder *et al.* 1992; Holland and Jackson 2002; Thomas *et al.* 2018a). *Lomandra* has been found to be the most highly utilised platypus nesting material at Healesville Sanctuary (Thomas *et al.* 2018a).

A dry ledge or section of bank (at least 0.6 m long × 0.25 m wide) should be provided directly below the external entrance of nesting tunnels for resting and grooming purposes. Additional dry resting sites (ledges, emergent rocks or logs, sections of bank) should be available in every tank. Enough dry area should be provided that all platypuses using the tank can rest out of the water simultaneously without having to rest next to each other.

Each tank should have at least one section of overhanging bank or comparable cover (at least 3 m (L) × 0.25 m (W)) under which platypuses can float comfortably while consuming food at the water's surface. At least one additional protected resting site per tank should be provided underwater.

## 4.2 Holding area design

The layout and husbandry of a holding facility should minimise disturbance to animals (e.g. staff not stepping on or over the tunnel system; water pipes and drainage pipes should be insulated to minimise noise or vibration transfer). An example of a platypus holding facility and associated artificial burrow system, illustrating all fundamental requirements, is shown in Figs 2.2 and 2.3.

The facility should include at least one feeding tank with associated tunnels and nest boxes and adequate storage, sinks, taps and hoses for cleaning. Taps and hoses should be positioned for ease of use, as should drains and sumps. Feeding tanks should be as large as is practicable and the tunnels should meet minimum requirements for internal dimensions and lengths as described previously.

A choice of nest box sizes and construction is desirable to give the platypuses variety. Nest boxes can be made of plywood, terracotta or recycled plastic and may contain soil as well as fibrous nesting material.

## 4.3 Spatial requirements

Relevant factors to consider when providing tanks for platypuses include how many animals share the enclosure, how many tanks they share and tank size and shape. At least one tank measuring at least 10 $m^2$ in surface area and at least 1 m deep should be provided. Tanks should

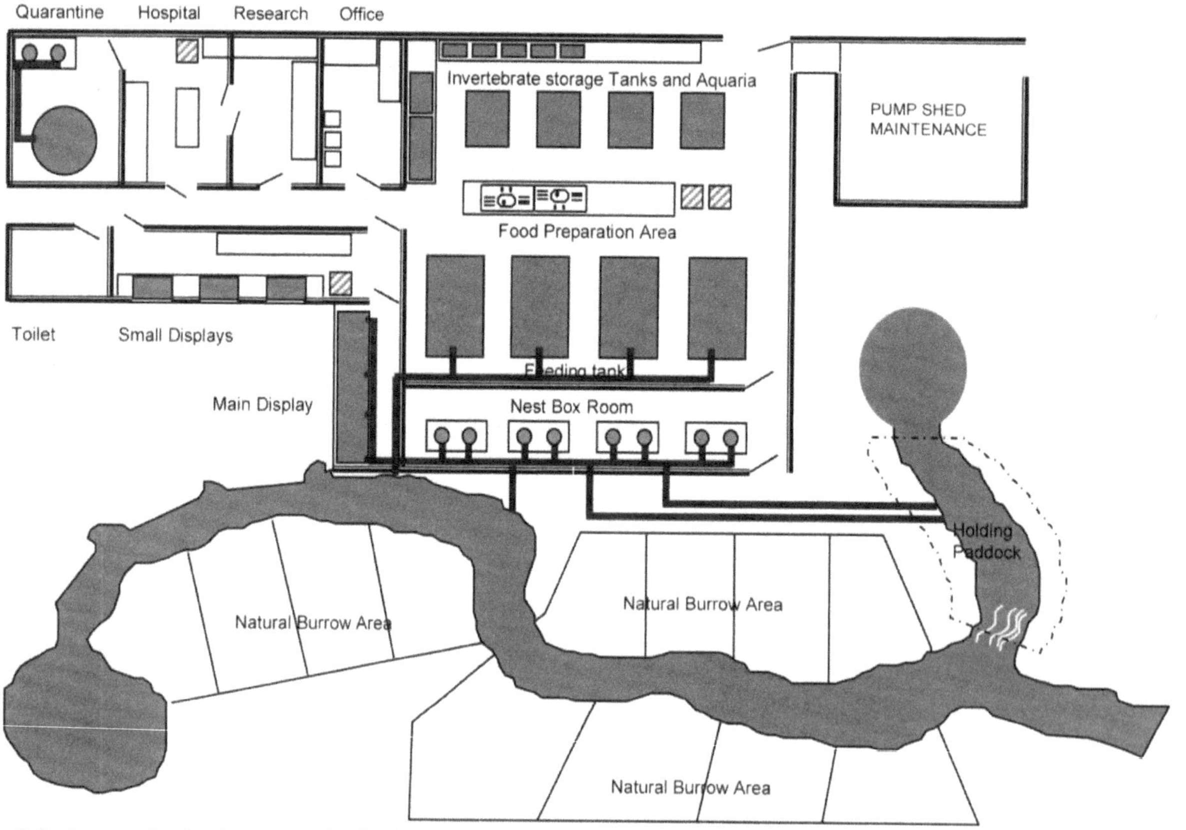


**Fig. 2.2.** An example of a platypus captive facility. Taken from McKee *et al.* (1995), via Booth and Connolly (2008). Used with permission of Jeff McKee.

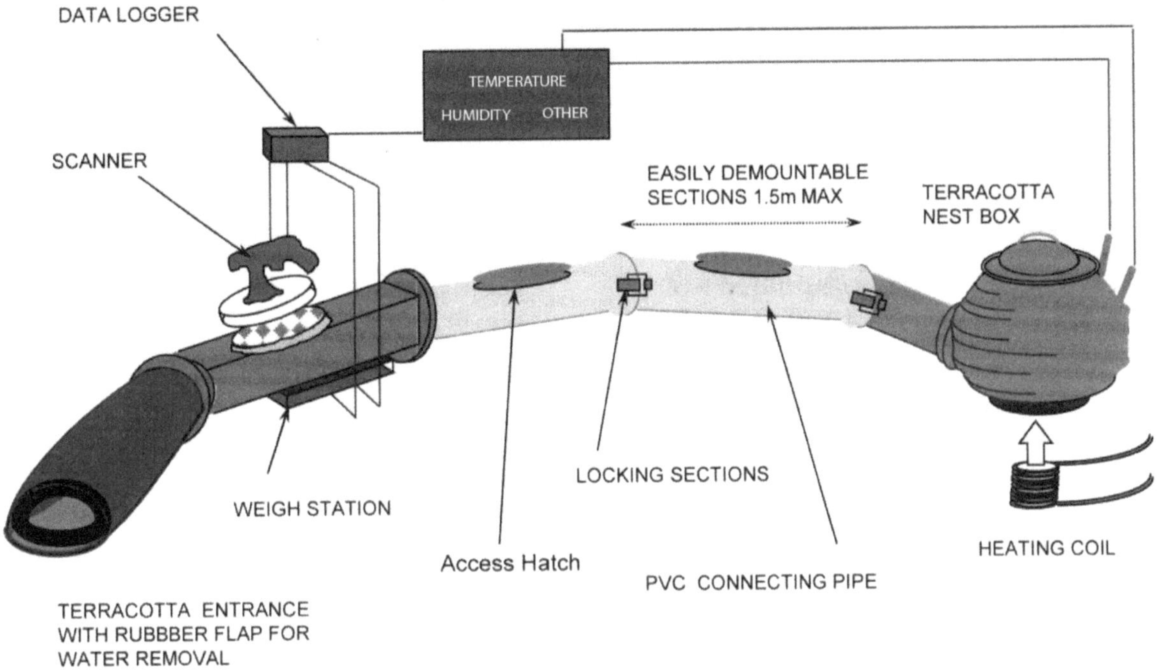


**Fig. 2.3.** Example of an artificial burrow system. Taken from McKee *et al.* (1995), via Booth and Connolly (2008). Used with permission of Jeff McKee.

also include shallow riffles, refuge areas where animals can hide both at the water's edge and underwater, and digging areas. If two animals are held together, at least one adjoining tank with a surface area of at least 4 $m^2$ should be provided. Holding two males or more than two female platypuses together is not recommended.

## 4.4 Position of enclosures

If feeding tanks are located outdoors, they should be in an area with substantial overhead cover (ideally provided by shrubs or trees). Providing plenty of shade will also help to moderate summer temperatures in the enclosure.

## 4.5 Weather protection

Where possible, some tunnels and nest boxes should be either under a veranda or inside a building to prevent rain entering the tunnel and nest box system. Any rainwater entering the tunnels or nest boxes must be drained away quickly. It is also important to protect the animals from high temperatures (> 25°C) to prevent heat stress. Because of the highly insulative qualities of platypus fur, which allows animals to remain warm while foraging for hours in near freezing water (Grigg *et al.* 1992), low ambient temperatures should not normally be a problem as long as dry nest boxes and nesting material are provided. Though platypuses routinely remain active throughout the year, even in very cold climates (Grigg *et al.* 1992), animals have occasionally been documented to retire to burrows in winter and become inactive for periods up to 156 h in captivity (Eadie 1934; Fleay 1980) and 154 h in the wild (Serena 1994). However, this behaviour is not typically seen in modern captive facilities where a high-quality diet is consistently available and temperatures are controlled.

## 4.6 Temperature requirements

Water and nest box temperatures should be maintained at ~16°C. Because the thermoneutral zone for resting platypuses on land is 20–25°C (Grant and Dawson 1978a, 1978b), ambient temperatures should not exceed 25°C.

Platypuses do not pant or have sweat glands under their dense fur on the body (Augee 1976), and sweat glands are largely confined to the bill region (Martin 1902; Brice 2009). Animals may turn onto their back and 'faint' when exposed to air temperatures of 35°C for as little as 17 min (Martin 1902). Similarly, a platypus maintained at ambient temperatures > 30°C did not tolerate it well and was observed to roll over and lose consciousness (Martin 1902; Robinson 1954; Smyth 1973).

Studies of wild platypuses in Tasmania found their mean burrow temperature was 14.2°C in winter and 17.5°C in summer, with a range of 12.2–22.8°C (Bethge *et al.* 2004). Studies of captive platypuses by Krueger *et al.* (1992) found they spent more time resting than active in water at low temperatures (< 16°C) and the maximum ratio of active to resting behaviours in the water occurred when the water temperature was 16–18°C. It was also found that the proportion of time that platypuses spent resting on land rose steadily as temperatures increased from 10–12°C to 22–26°C. Some modern facilities have water cooling systems to maintain the pool water temperature at ~16°C (L. Vogelnest pers. obs.).

## 4.7 Life support systems

Water quality is the basis for the health of platypuses. Therefore, the life support system must not only maintain good water quality through removal of biological waste, but facilitate temperature controls and sterilise the water from bacteria. This is often one of the more expensive aspects of building a new facility, but it is also the most important. It is recommended that an aquarist or water technician be consulted to ensure the system includes some of the following features:

- mechanical filtration such as a sand/bead filter
- temperature control that heats and cools water
- sterilisation: ultraviolet (UV) is common for freshwater and easy to use; ozone is used at some facilities but needs specialist support
- biological filtration or medium-pressure UV (removes ammonia and sterilises)
- programmable logic controller (PLC)
  - variable speed pumps
  - fault logs
  - program for weather events: rain, flood, rapids
  - water level sensors (e.g. turn off pump if too low (fail safe))
  - automate solenoid top up after backwash (can select between recycled or potable based on availability)
- backwash recycling or water storage tank for cleaning
- vacuum system
  - recycles water back through system
  - can remove old food
  - does not cause too much vibration due to the sensitivity of the platypus bill (Proske *et al.* 1998).

  A vacuum can be installed as part of the existing filtration system, which will not cause any additional vibrations, or can be purchased as a portable unit with a standard pool pump that sits outside the enclosure during use. Vacuuming should typically occur when the animal is out of the water.

## 4.8 Substrate

Substrate in the tanks should be non-porous, easily cleaned and diverse enough to provide behavioural enrichment for foraging platypuses and shelter for live prey. The type of substrate will be partly determined by whether or not water filters are used and the type of filter being used. Sand, fish tank gravel, large river rocks, tree bark or even a fibreglass or concrete base with no other substrate have been used with varying success. However, abrasive surfaces must be avoided as these can damage the animal's feet and nails. Leftover food is more difficult to remove from sand and fish tank gravel; a concrete or fibreglass floor is easy to clean but looks unnatural.

## 4.9 Nest boxes

Several nest boxes that are large enough to accommodate two adult platypuses and a substantial volume of nesting material should be provided per non-breeding adult. Breeding nest boxes should be large enough to accommodate one adult platypus and as many as three well-grown juveniles.

Nest boxes should have a non-abrasive internal surface and be fitted with a securely latching hinged lid to facilitate access when inspecting animals or replacing nesting material. About 75% of the nest box volume should be filled with nesting material. Seagrass was historically used as nesting material, but is expensive and often not easily obtained. Sphagnum moss holds moisture better and does not get soggy like sea grass when damp. Dry leaf litter and polar fleece blankets have also been successfully used as bedding material. Polar fleece blankets do not become saturated when wet and allow the animal to tuck up tightly (as within a snug burrow). They are also easy to clean and quick to dry.

## 4.10 Enclosure furnishings

All furnishings should be non-toxic, non-abrasive and should not present any risk of entanglement or injury. Enclosure furnishings can include plant material such as logs, fern fronds, tree branches, rocks or river pebbles. All of these items should be cleaned as needed and checked to ensure they have not come in contact with any toxins (e.g. weedkiller). Branches should be weathered or previously soaked to remove tannins before they are put into the tank. Fresh branches should be changed weekly.

# 5. GENERAL HUSBANDRY

## 5.1 Hygiene and cleaning

It is critical to support animal health by applying high standards of hygiene and effective quarantine procedures. The senior platypus animal caretaker should have a minimum of 5 years' experience working with captive wildlife. Other animal caretakers should have a minimum of 2 years' experience working with captive wildlife (or equivalent) and undergo a training period under the direct supervision of the senior platypus animal caretaker.

All tanks used by platypuses should be routinely cleaned and filled with mains water, rainwater, pumped spring water or creek water to ensure the water quality is maintained. Off-exhibit feed tanks without filtration are typically cleaned daily. Water in display tanks should be filtered at the rate of one complete water change for at least every 6 h that animals are on display. Display tanks should be drained and thoroughly cleaned (tank and furniture scrubbed and hosed down) at least once per week, depending on the filtering system and water quality. Water sterilisation such as UV is recommended to control coliform bacteria.

Additional protocols that should be applied are:

- chemical agents should not be used to clean tanks or tunnels used by platypuses
- base substrate in display tanks should be replaced when soiled
- off-display tanks should be drained, cleaned and refilled daily if not equipped with a system for filtering recirculated water. Cleaning other off-display facilities such as natural ponds may not be necessary
- nesting material in non-breeding nest boxes should be checked regularly. The presence of damp nesting material is not of major concern, as long as it is not soaking wet and each platypus has access to dry material
- tunnels and nest boxes should not be scrubbed or hosed unless fouling occurs. Every effort should be made to allow excess water to drain or evaporate from the tunnel and nest environment. Drainage holes need to be checked regularly for blockages
- whether originating from the wild or another institution, newly arrived platypuses should be thoroughly examined upon arrival by an experienced veterinarian to determine quarantine requirements (see section 8).

Animals very quickly habituate to a daily routine, thereby facilitating husbandry, moving animals on and off display and even hand feeding. However, if a platypus is prone to abnormally repetitive behaviour, varying the routine may be beneficial.

## 5.2 Record keeping

It is important to establish a system whereby the health, condition and reproductive status of captive platypuses are routinely monitored and recorded:

- identification details (platypuses should be individually identifiable. e.g. microchip; see section 5.3)
- results of veterinary examinations and treatments
- behavioural changes or problems
- reproductive behaviour or condition
- bodyweight, tail fat index (see section 8.2.2) and linear measurements
- changes in diet and food consumption
- movements of individuals between enclosures or institutions
- husbandry and water quality changes
- births with dam and sire if known
- deaths with necropsy results.

Collecting information on the physical, physiological and behavioural patterns of individual platypus can contribute greatly to the husbandry of this species. It also allows individual histories to be shared with other institutions and facilitates a cooperative approach to data collection among institutions. ZIMS, the Zoo Information Management System (Species360 2023), can be used for recording general information on age, sex, parentage, births, identification, bodyweights, transfers and deaths, with separate ZIMS modules for veterinary information and studbooks. ZIMS interfaces with programs such as PMx, which provides demographic and genetic analysis tools to assist with managing breeding programs (Lacy *et al.* 2011).

### 5.2.1 Daily records

The following information should be recorded daily where possible:

- observations of behaviour
- relevant environmental and water temperatures including maximum and minimum
- location of the animal in the nest box/tunnel system (early in the morning) and identity of tank(s) used for display/feeding if more than one is available.

### 5.2.2 Weekly/monthly records

Female platypuses can be conditioned to allow physical examination and tail fat scoring to occur without manual restraint, reinforced with tactile stimulus (i.e. rubbing under the bill or belly). Physical examination of males needs to be done with extreme care, especially during the breeding season, to avoid risk of being spurred.

The following data should be recorded on a monthly basis or more frequently if needed:

- dry bodyweight of animal to the nearest 10 g (or more accurately if feasible)
- Tail Fat Index (TFI) (see section 8.2.2)
- water pH and conductivity in display and off-display tanks
- animal's general demeanour
- temperatures of water, air and nest boxes
- activity budgets and behaviour should be monitored (e.g. by means of automated, tunnel-mounted activity recording system, time-lapse video monitoring or records from trained volunteers). This is important to provide information relevant to assessing the health or reproductive status and behaviour of targeted individuals.

All information should be recorded in a standardised format to facilitate retrieval, analysis and the preparation of internal reports and technical papers.

## 5.3 Methods of individual identification

Passive integrated transponder (PIT) tags (microchips) are an excellent method of identification because they are permanent. For platypuses, the tag is implanted subcutaneously between the scapulae (Grant and Whittington 1991; Booth and Connolly 2008). Sealing the entry wound with tissue glue (Vetbond®) will ensure that tags stay in place immediately after implantation. For identification purposes, individuals can be scanned with a hand-held PIT tag reader or a reader can be fitted into the artificial tunnel system (Booth and Connolly 2008; Macgregor *et al.* 2014).

The following identification methods used in the past are no longer recommended: tattoos on the inside of the upper or lower frontal shield; visual identification based on the skin pigment pattern on the underside of the bill; freeze branding; plastic cattle tags or metal leg bands (Grant and Carrick 1974; Grant and Whittington 1991).

# 6. FEEDING REQUIREMENTS

## 6.1 Captive diet

Captive platypuses need a diet that caters for individual preferences in quantity and type of food, so it should be varied, nutritious and encourage natural foraging behaviours (Shaw and Jarman 2025). The diet should include as high a proportion of live food as possible (Carrick *et al.* 1982). The next best alternative is food that is freshly frozen for storage purposes and thawed just before being offered to platypuses.

Yabbies (freshwater crayfish) should be fed live. Mean daily food consumption by platypuses in captivity has been found to vary between 14.9% and 27.8% of bodyweight throughout the year (Krueger *et al.* 1992). These data are similar to those estimated for platypuses in the wild in Tasmania where the mean food intake was estimated to be 19% of bodyweight (Munks *et al.* 2000). A lactating female was found to consume up to 36.4% of her bodyweight daily in the final month of lactation (Thomas *et al.* 2020). The daily captive diet for adults and weaned juveniles should normally include an adequate selection from the following items (contingent upon availability and palatability):

- earthworms (*Oligochaeta* sp.)
- yabbies (live) (*Cherax destructor* and *C. albidus*)
- mealworms (*Tenebrio molitor*)
- fly pupae (*Musca domestica*)
- freshwater shrimp (*Paratya australiensis*)
- tubifex worms (*Tubifex* sp.)
- crickets (*Acheta domesticus*)
- cockroaches (e.g. *Blatella* sp.; *Periplaneta* sp.)
- blackworms (*Lumbriculus variegatus*)
- aquatic insect larvae.

The amount of food offered should be:

- Earthworms:
  - fed daily (50–150 g per animal)

- earthworms are generally supplied in a plastic container with some soil; worms are removed from the container with as little soil as possible and weighed on an electronic balance
- without further washing, worms are placed into the feeding tank
- earthworms are collected and weighed the following morning as for mealworms
- caution needs to be taken in comparing weights of earthworms before and after feeding as they gain weight in water.

- Yabbies:
  - fed daily (10–70 per animal depending on size of yabbies, allowing ad libitum access to 4 kg)
  - yabbies are counted and washed; water is drained away through a colander for ~5 min; yabbies are then weighed on an electronic balance and placed into the feeding tank
  - feeding tanks are drained and hosed out the following morning
  - remaining food items are collected in a fine mesh basket placed under the outlet drainpipe; the number of yabbies eaten is calculated by counting the number of thoraxes present
  - all leftover yabby material is collected, washed, drained (for 5 min) and weighed to determine the weight of food consumed.
- Mealworms:
  - fed daily (50–120 g per animal)
  - mealworms are weighed on an electronic balance and placed into the feeding tank
  - remaining mealworms are collected (as for yabbies) and quantity estimated; occasionally, remaining mealworms are weighed to determine accuracy of estimate.
- Fly pupae:
  - fed daily (20–40 g per animal)
  - pupae are weighed on an electronic balance and placed in the tank
  - at the end of the feeding period the remaining fly pupae are estimated
  - occasionally remaining pupae are netted out and weighed to determine the accuracy of the estimate.

Food should be provided at times that best suit individual activity patterns, with the amounts altered based on seasonal levels of consumption. In general, this will involve providing most food in late afternoon for overnight consumption ad libitum. Ad libitum feeding does not result in overweight animals and are the most successful feeding method for maintaining healthy captive platypuses.

Consideration should be given to systematically varying the proportion of different foods offered to platypuses on different days, thereby providing a temporal dimension of variability to their diet. Studies of the captive diet at Healesville Sanctuary revealed there was no significant seasonal change in preference for different dietary items, with yabbies comprising the greatest proportion of the diet (Thomas *et al.* 2017). The highest amount of kilojoules was consumed in the post-breeding season (November–March) and the least amount during the breeding season (August–October) (Table 2.4; Thomas *et al.* 2017).

## 6.2 Supplements

No supplements are required if a variety of food items are offered.

**Table 2.4.** Macronutrients in dietary items (Thomas *et al.* 2017).

| | Fly pupae | Mealworms | Earthworms | Freshwater crayfish viscera | Freshwater crayfish tail |
|---|---|---|---|---|---|
| Energy (kJ/100 g) | 560 | 960 | 220 | 650 | 280 |
| Fats (g/100 g) | 4.7 | 13.2 | 1.1 | 13.3 | 0.7 |
| Protein (g/100 g) | 18.9 | 20.6 | 7.4 | 9.2 | 14.9 |
| Carbohydrate (g/100 g) | 4 | 7 | 3 | < 1 | < 1 |
| Saturated fatty acids (%) | 37.8 | 24.0 | 40.2 | 26.9 | 41.5 |
| Mono-unsaturated fatty acids (%) | 39.2 | 47.5 | 17.9 | 48.9 | 35.0 |
| Polyunsaturated fatty acids (%) | 22.7 | 28.2 | 35.7 | 22.6 | 22.6 |
| Moisture (g/100 g) | 70.2 | 56.6 | 87.7 | 74.9 | 79.2 |

### 6.3 Presentation of food

Food is scattered in the water in either the display tank or separate feeding tanks (if water clarity or filtration is a problem). As with any animal, food should be fresh and varied and leftovers need to be removed every 1–2 days. The length of time before removal depends on the filtration system and the water temperature. Dead yabbies and mealworms will decompose and putrefy if left in water.

## 7. HANDLING AND TRANSPORT

### 7.1 Timing of capture and handling

Platypuses are generally best caught while in their nest box or tunnel system. To minimise stress, they should generally not be caught while they are swimming in a tank.

### 7.2 Capture equipment

Captive platypuses are best contained in a soft cotton bag, ~500 mm (W) × 800 mm (L), such as a pillowcase. Use of rougher materials such as hessian/burlap sacking is not recommended, as they can be quite abrasive to the feet and bill. Removable nest boxes can also be used to reduce handling when transporting and weighing, as used successfully at Healesville Sanctuary (J. Thomas pers. obs.).

### 7.3 Capture and physical restraint

Platypuses are generally caught by the tail from within the nest box or tunnel system. When handling an adult male platypus (or animal of unknown sex) make sure to avoid the ventral part of the body between the hindlimbs and along the abdomen (Vogelnest 2025a). Because of the male's venomous spur, they should be only held by the tail, with the palm against the tail's dorsal surface (Fig. 2.4 left), or by grasping the tail with the thumb on the underside of the tail and other fingers on the top. This puts all digits and wrist out of potential contact. Female platypuses are also best held by the tail, as they may squirm free if held any other way. Tail restraint should only be maintained for a short period due to tension on the spine. Females can be supported with the other hand under the front feet to support their bodyweight.

Platypus envenomation causes immediate extreme pain, swelling and debilitation of the limb that has been spurred (Jamieson 1818; Spicer 1877; Tidswell 1906; Calaby 1968; Sutherland 1983; Fenner *et al.* 1992; Tonkin and Negrine 1994). Though no human fatalities have been recorded, a male platypus can inflict deep punctures and the spurs may be difficult to remove. To date

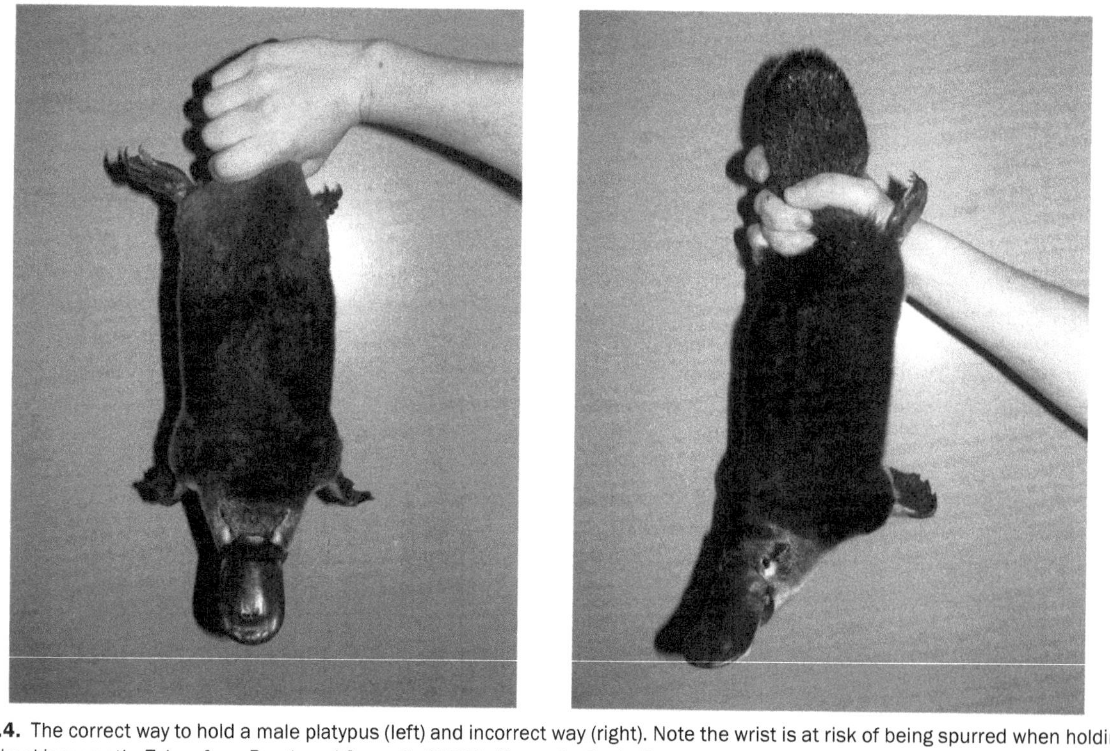

**Fig. 2.4.** The correct way to hold a male platypus (left) and incorrect way (right). Note the wrist is at risk of being spurred when holding the animal incorrectly. Taken from Booth and Connolly (2008). Photos by Rosie Booth.

there is no antivenom available (Vogelnest and Whinfield 2025). The wound should be washed with soap and water to remove free venom and contamination, then lightly bandaged and medical attention sought as soon as possible. The application of ice is not recommended as it will greatly increase pain (Fenner *et al.* 1992). Symptoms are treated with parenteral opiates or regional nerve blockade, with analgesic support required for several days (Fenner *et al.* 1992). In one well-documented case, the patient spent 6 days in hospital and the envenomated area remained painful, swollen and with little movement for 3 weeks; significant functional impairment of the hand persisted for 3 months (Fenner *et al.* 1992; Tonkin and Negrine 1994).

Routine handling of platypuses should be restricted to that required for the management and veterinary needs of individual animals and should only be undertaken by trained staff. Veterinary protocols should make provision for sedation and/or anaesthesia when this is deemed necessary to minimise stress to the animals.

## 7.4 Weighing and examination

When animals have been captured they should be weighed and a brief examination undertaken to check for cuts or fur loss and assess body condition.

## 7.5 Release

To release a platypus into an enclosure for the first time, place it in the burrow system, so it can more easily find the burrow entrance when leaving the water and feel secure in returning to the burrow. Place the platypus in its new home immediately before its normal activity period (i.e. just before dark).

## 7.6 Transport requirements

### 7.6.1 Transport box design

Transporting a platypus over a short distance (e.g. within the zoo) is easiest by securing it in a cloth bag, tied at the top, that is placed inside a sturdy box (e.g. plywood). In warm weather the bag should be moistened with water and placed in a box with good ventilation to prevent overheating.

The transport of platypuses over longer distances (e.g. interstate and particularly overseas) requires significant preparation and discussion with veterinarians and others with experience working with platypus as it is a highly complex operation. The platypus must be conditioned to a specific nest box, which is placed within a well-ventilated transport box with foam padding to prevent the inner box shifting around. During transport, the box should be held in an air-conditioned environment to ensure it is maintained at a temperature of ~18–20°C. A thermometer recording the internal temperature of the transport box must be available outside the box.

An animal caretaker must accompany the platypus and ideally have access to the animal during transport. Transportation time and external noise and vibration from the time of crating to release at the destination must be kept to a minimum. Further specific details of an appropriate box design can be found in the *Conditions for the Overseas Transfer of Platypus* (Department of Environment and Energy 2019; Addendum 1).

### 7.6.2 Furnishings

Box should contain nesting material. See Addendum 1.

### 7.6.3 Water and food

No water or food should be required during transport as long as the animal has a minimum Tail Fat Index = 2 (see section 8.2.2) and it arrives at the destination within 36 h for overseas flights or within 6–8 h within Australia.

### 7.6.4 Number of animals per transport box

Only one animal should be held in a given box.

### 7.6.5 Timing of transportation

It may be better to transport platypuses during the day for longer distances when the animals would normally be resting/not feeding. In the case of overseas transport, consideration should be given to the platypus being in the box for 36 h. Transported animals must be kept in the dark in an air-conditioned environment that maintains the temperature at ~8–20°C. It is important, particularly in warmer weather, to check the ambient temperature of the transport box to make sure it does not overheat.

### 7.6.6 Release from the transport box

Upon arrival at the new location, the animal inside the nest box should be allowed to explore its new environment. Ideally, the new enclosure will be similar to the old one to help the animal become familiar with the facility. Moving to a new facility is likely to be very stressful to the platypus so anything that can be done to assist it settling in is likely to be beneficial. Significant follow-up monitoring will be required to ensure the platypus is not losing weight, is feeding normally and is not showing signs of stress such as abnormal repetitive behaviour.

# 8. HEALTH REQUIREMENTS

*Larry Vogelnest*

## 8.1 Daily health observations

Each platypus should be observed daily when active and undisturbed for signs of injury or illness, especially after introductions to new animals or enclosures. An appropriate time to do this is when platypuses are swimming and foraging soon after being fed. Observing the undisturbed animal both while swimming and while dry in the nest boxes is important. Each animal should be checked for the following:

- coat/skin condition: check for alopecia, crusting, thickening of skin or wounds. Pay particular attention to the skin and webbing of the feet and the nails. Any fur loss, lesions, deformed or brittle nails may be signs of infection (e.g. fungal) or mechanical trauma and should be reported to a veterinarian. Seasonal mechanical alopecia of the tail from conspecific biting, plugging the burrow entrance with a pug or overgrooming is more common than infectious causes (Booth and Connolly 2008). Note that platypuses moult seasonally and may lose large areas of guard hairs from the dorsal, especially, and ventral surfaces during late spring/early summer after the breeding season (presumably in response to warmer weather or possibly a change in circulating hormones) (Grant and Carrick 1978; P. Temple-Smith pers. obs.).
- discharges from the eyes, ears, nares, mouth or cloaca
- appetite
- faeces (if possible)
- changes in demeanour, activity level and behaviour
- injuries.

All animals should be weighed and have body condition assessed on a regular basis (see section 8.2). Each individual should be accounted for each day. Any signs of injury or illness or unusual changes in behaviour should be documented and reported to a veterinarian. If an animal is found dead, make note of the circumstances and anything unusual or abnormal on the body or surrounding area. Carefully collect the body (using PPE where appropriate) and submit it for necropsy.

## 8.2 Bodyweight and body condition score

### 8.2.1 Bodyweight

Trends in bodyweight are a good indicator of general health. Some seasonal variation in bodyweight and percentage body fat has been documented in wild platypuses (Hulbert and Grant 1983; Grant and Temple-Smith 1983; Handasyde *et al.* 2003; Bino *et al.* 2015) and may also be seen in captive animals. However, a persistent decline in bodyweight should be reported to a veterinarian.

### 8.2.2 Body condition score

Tail Fat Index (TFI; or Tail Volume Index, TVI) – see Fig. 2.5) has been widely used to assess the body condition of platypuses (since Grant and Carrick 1978). More recently, three condition indices have been developed by Macgregor *et al.* (2016): (1) Relative Tail Fat Volume (RTFV) based on cross-sectional area measurements taken from detailed ultrasound images compared with total body length (RTFVTBL); (2) Body Condition Index (BCI) based on bodyweight ($m_b$) and bill width (BW) (BCIBW); and (3) BCI based upon ultrasonographic measurement of tail fat depth and BW (Relative Fat Depth: RFDBW). They suggest that their techniques may provide a better assessment of body condition than TVI, but they also point to significant correlations between TVI assessment and their RTFVTBL, and that their relative fat depth RFDBW only appears to outperform TVI. The BCIBW is possibly the most appropriate condition index for platypuses, but it needs further testing against at least TVI and preferably also one or both of the other tail fat protocols. Regardless, the TVI remains a useful and practical index of the body condition of platypuses (Vogelnest and Whinfield 2025). The current scoring system of 1–5, with 1 being excellent and 5 as worst condition, is contrary to all other animal body condition scoring systems where the lowest number is very poor condition and the highest numbers indicates excellent condition or obesity. We have followed the suggestion of Macgregor *et al.* (2016) in reversing the numbering of the TVI, so that animals with lower body condition are given a lower number and those with greater body condition are given a higher number (Fig. 2.5).

| Tail volume index | Description and interpretation | Images |
|---|---|---|
| 1 | Tail flaccid and strap-like and outlines of vertebrae are apparent when viewed ventrally. Emaciated condition. | 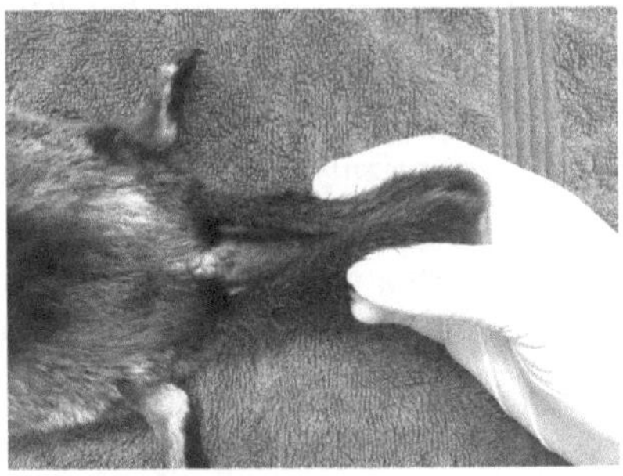 |
| 2 | Tail fairly flaccid. At the upper end of this fat index class, lateral edges can be bent inwards until they touch. At the lower end, the two lateral tail halves can be pressed entirely against each other along the ventral midline. Poor/below average condition. | 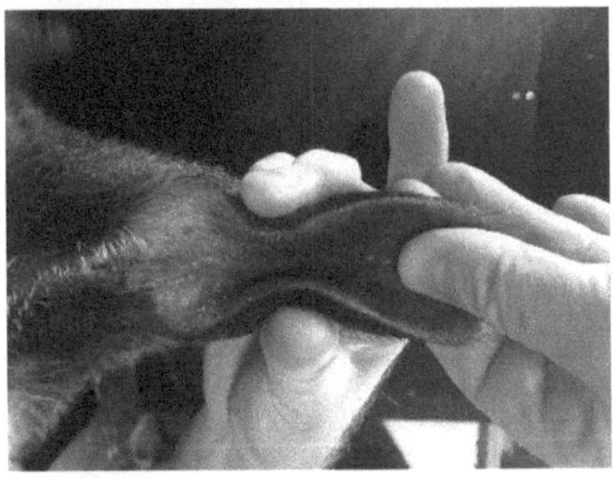 |
| 3 | Tail moderately turgid; about half to two-thirds of tail width can be bent inwards with creases evident. Ventral tail is slightly concave. Fair/average condition. | 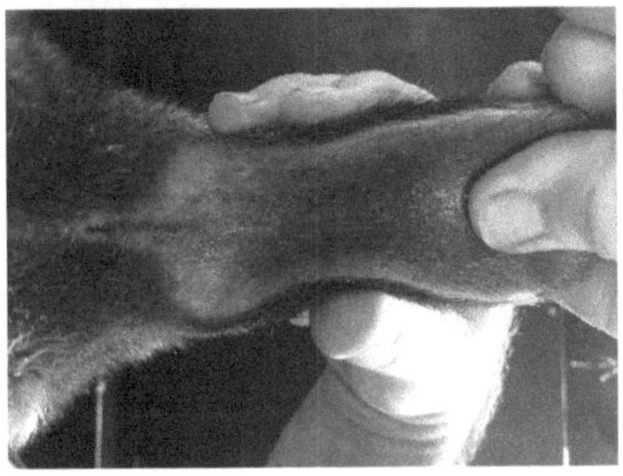 |
| 4 | Tail moderately turgid; lateral edges at the middle of the tail can be bent slightly inwards (though no 'creases' evident). Ventral tail is flat. Good/above average condition. | 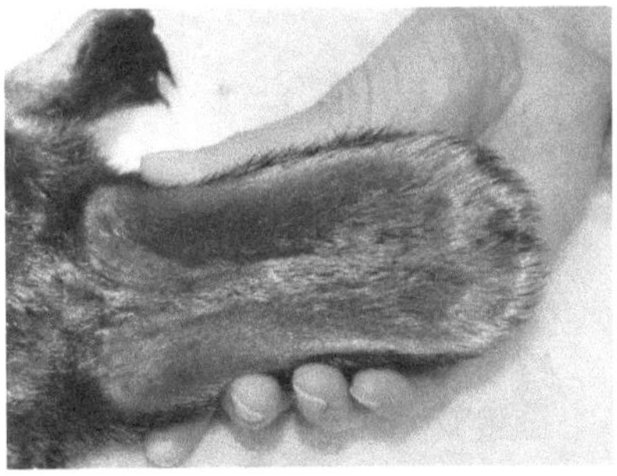 |
| 5 | Tail turgid; lateral edges resists curling inwards when gently pressed together. Ventral tail is convex. Excellent condition. | 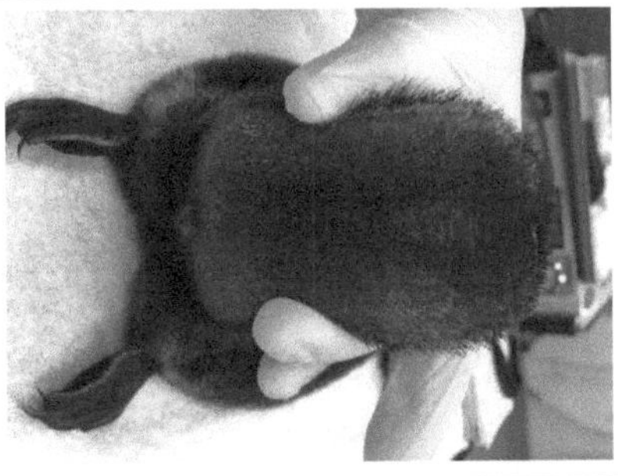 |

**Fig. 2.5.** Tail volume index (or tail fat index) as a measure of body condition in platypuses. Note the change from the traditional 1–5 score where 1 is excellent and 5 is emaciated. Photos by Larry Vogelnest.

## 8.3 Physical examination

Most platypuses can be readily handled (see Booth and Connolly 2008, Vogelnest 2025a and section 7.3) and a basic physical examination can be conducted by keepers (or a veterinarian) without the need for chemical restraint. Male platypuses are potentially dangerous, especially during the breeding season, and should only be handled by experienced animal caretakers (see section 7.3). Examinations should be conducted on a weekly or monthly basis, with geriatric platypuses or those with health concerns or injuries examined more frequently. In addition to the list in section 8.1, the following can be assessed:

- body temperature – normal 32.5°C (range < 30–34.3°C); taken using a digital thermometer inserted deep into the rectum via the cloaca (usually only possible with the animal under anaesthesia)
- abnormal swellings, masses or deformity
- cloaca – should be clean and free of faeces
- spurs of males – swelling, ulceration or discharges at the base of the spur
- penis – if indicated the penis should be extruded and examined (Fig. 2.6; Temple-Smith and Grant 2001). Penises are more easily extruded during the breeding season, without anaesthesia, and will often hang out while the animal is being handled. At other times this is usually only possible with the animal under anaesthesia
- eyes – can be difficult to examine, except under anaesthesia, as the palpebral fissure and globe are small. The cornea should be clear, the pupil should respond to light and there should be no swelling or discharge.

If required, a detailed physical examination should be conducted by a veterinarian, with the animal under anaesthesia. Veterinarians should refer to Vogelnest and Woods (2008) and Vogelnest and Portas (2025) for detailed information on examination and medicine of platypuses. All body areas and systems should be evaluated: cardiovascular; respiratory; oral cavity (secateuring ridges and keratinous grinding pads, oral mucosa and pharynx), abdomen; musculoskeletal; genitourinary; ocular; and skin and coat. Important parameters and vital signs should be measured and recorded: bodyweight and condition (Fig. 2.5), heart rate, respiratory rate and body temperature (Booth and Connolly 2008; McLelland 2025).

## 8.4 Chemical restraint

Chemical restraint is required for a detailed physical examination, sample collection other clinical or surgical procedures and measurements and reduces the risk of envenomation when examining male platypuses.

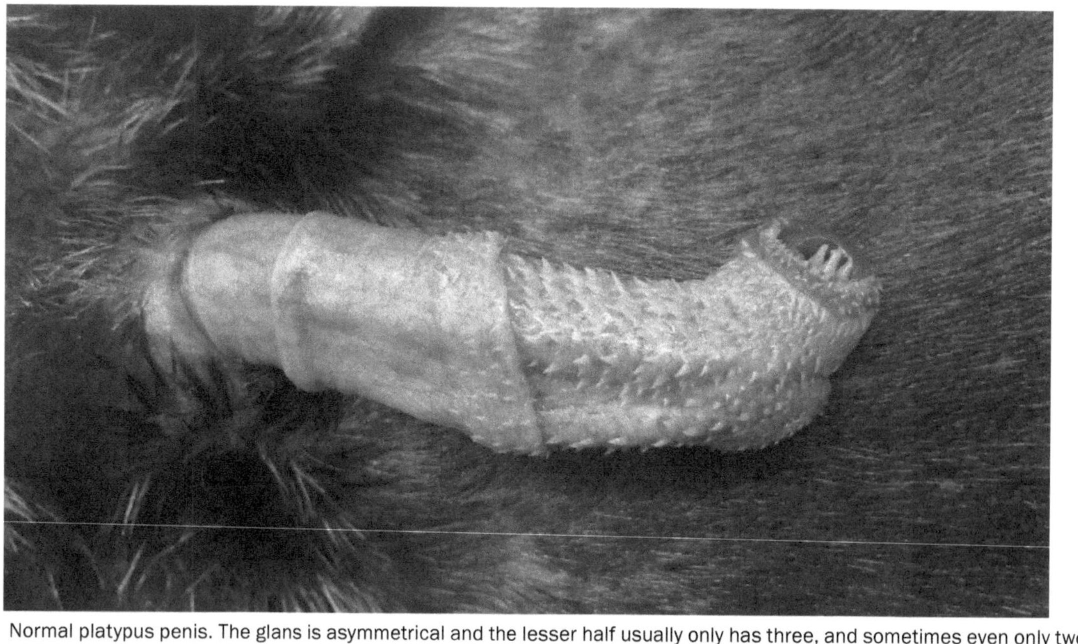

**Fig. 2.6.** Normal platypus penis. The glans is asymmetrical and the lesser half usually only has three, and sometimes even only two, evertible foliate papillae. Photo by Larry Vogelnest.

Ideally, chemical restraint is performed by a veterinarian and relevant chapters in Vogelnest and Woods (2008) and Vogelnest and Portas (2025) should be referred to. Sedation (using diazepam) is rarely required, but has been used for minor procedures and transportation and at low doses has been useful in controlling anxiety and encouraging feeding in platypuses undergoing rehabilitation (Vogelnest 2025a, 2025b).

Prior to general anaesthesia, pre-anaesthetic fasting is not usually required as platypuses are not prone to regurgitation under anaesthesia. If the animal has been foraging just before capture the caudal oral cavity and buccal pouches may be full of ground prey that could potentially obstruct the airway. It is generally preferable that the platypus has no access to food for at least 1 h before anaesthesia so the oral cavity is free of food (although food may remain in the buccal pouches beyond this timeframe). This can be achieved by locking the animal in its nest box or placing the animal in a bag, which also allows time for the fur to dry.

Volatile (isoflurane or sevoflurane) inhalational agents delivered in oxygen are the anaesthetic of choice for platypuses. Induction may be via a face mask or induction chamber. Induction chambers are useful for adult males to avoid the need for physical restraint and thereby reduce the risk of spurring. The animal is usually left in a cloth bag in the chamber. Once anaesthetised the animal is removed from the chamber and bag and anaesthesia is maintained via a mask. Induction using a facemask requires a brief period of physical restraint, which is best accomplished with the animal in a cloth bag that has a hole cut out of one corner just wide enough for the bill or by manipulating the position of the animal within an intact bag so only the bill is exposed. Gently restrain the animal within the bag on a foam mat or towel while placing the mask over the bill. Anaesthesia is always maintained via a mask as it is very difficult to intubate platypuses. Captive platypuses can be returned to their tunnel system and allowed access to water after 1 h. In the field, platypuses are best held in a bag for 30–60 min after full recovery before returning them to the water (Vogelnest 2025a).

## 8.5 Common health problems

Several surveys of free-ranging platypus populations have found evidence of exposure to infectious disease, but clinical disease associated with these infections is rare. However it can be challenging to assess the health of free-ranging platypuses and there is a risk of underestimating disease processes to which platypuses may be susceptible. Infectious diseases in captive platypuses are uncommon. A few infectious agents and diseases have been documented in debilitated, rescued platypuses. The majority of reported non-infectious causes of morbidity and mortality in platypuses are directly or indirectly anthropogenic (Serena and Williams 2010, 2022). There have been few reports of neoplasia and very few other non-infectious diseases reported. Table 2.5 lists infectious diseases and Table 2.6 lists non-infectious diseases of platypuses.

## 8.6 Preventative medicine

The cornerstones of preventative medicine are good biosecurity; optimal husbandry, enclosure design, life support systems and environment; good nutrition; disease screening; vaccination; routine preventative treatments; and conducting necropsies on all animals that die. All promote good health and welfare. Captive platypuses suffer from few infectious diseases. All new platypuses entering a captive population should undergo a period of quarantine and relevant health examinations (e.g. routine blood work, screening for parasites and treatment for parasites if indicated). There are no other specific preventative medicine recommendations for platypuses (e.g. vaccination, routine screening and treatments). A necropsy must be conducted on any platypus that dies to facilitate a diagnosis and, if relevant, inform preventative action to protect other platypuses in a captive population.

## 8.7 Zoonoses

There are few specific zoonotic diseases of concern that can be contracted from platypuses. Fungal skin infections (ringworm) are seen in platypuses and some species of these fungi may be transmissible to humans. Appropriate PPE and hand hygiene are indicated if handling platypuses with suspected or confirmed ringworm. Male platypuses are considered dangerous during handling, with most accounts of envenomation describing spurring to the hand or wrist. Envenomation results in immediate onset of intense pain and swelling. No antivenom is available. Treatment is symptomatic, requiring powerful analgesics for prolonged periods (Booth and Connolly 2008; Vogelnest and Whinfield 2025).

**Table 2.5.** Infectious diseases or agents of captive and rescued platypuses of clinical significance (adapted from Vogelnest and Whinfield 2025) (see Booth and Connolly 2008 and Vogelnest and Whinfield 2025 for detailed information).

PCR = polymerase chain reaction; SC = subcutaneous.

| Disease/ infectious agent | Cause | Clinical signs | Diagnosis | Treatment | Prevention/Comments | Reference |
|---|---|---|---|---|---|---|
| **Ectoparasites** | | | | | | |
| Platypus tick | *Ixodes ornithorhynchi* | None to mild chronic active dermatitis | Observation. Generally found on the less densely haired parts of the limbs. Small ticks may be found in large numbers, particularly on the hindlimbs of wild platypuses; large ticks are mostly found in the axillae and deep in the fur over the dorsal pelvic region | Generally not required Manual removal. Ivermectin SC | Rarely seen in captive platypuses. Ticks are intermediate host for blood parasites | 1, 2 |
| Fleas and mites | *Pygiopsylla hopli* and *P. zethi*, two species of trombiculid mite | Skin irritation n including the external ear canal | Observation | Ivermectin SC | Rarely seen in captive platypuses | 1–3 |
| External nematodes | *Cercopithifilaria johnstoni* | None. Mild thickening of the superficial skin layers, mild inflammation. Ticks are known intermediate host for *C. johnstoni* | Skin scrapings, histopathology | Not warranted | Not seen in captive platypuses | 5 |
| **Endoparasites** | | | | | | |
| Intestinal worms | Trematodes: *Mehlisia ornithorhynchi, Maritrema ornithorhynchi, Moreauia mirabilis* | Mild enteritis | Observation at necropsy (incidental finding) | Not warranted | Not seen in captive platypuses | 1, 4 |
| | Nematode: *Angiostrongylus cantonensis* | Neurological signs, including circling; death Verminous meningoencephalitis | Necropsy, histopathology, *A. cantonensis*-specific qPCR and then confirmation by sequencing | None reported | | 6, 7 |
| Sparganosis | *Spirometra erinacei* plerocercoids | Dependant on location of spargana | Radiographs, ultrasound examination, surgical exploration, necropsy | Surgical removal but generally very challenging | Not seen in captive platypuses | 8, 9 |

| Disease/ infectious agent | Cause | Clinical signs | Diagnosis | Treatment | Prevention/Comments | Reference |
|---|---|---|---|---|---|---|
| **Protozoal** | | | | | | |
| Toxoplasmosis | *Toxoplasma gondii* | Cysts observed in the heart muscle of a platypus that had been held in captivity for 4 years | Necropsy | None | Ensure cats do not have access to any bedding provided to the platypus | 5, 10 |
| Coccidiosis | Coccidia *Eimeria* sp. | Generally non-pathogenic. Can cause disease in debilitated rescued dispersing juveniles | Faecal examination, histopathology | Only warranted if considered clinically significant | Not seen in captive platypuses | 4, 5 |
| Blood parasites | *Theileria ornithorhynchi* | Considered non-pathogenic. Associated with anaemia in compromised, wild juveniles. Ixodid ticks as intermediate host | Blood smear, PCR | No specific treatment. Usually resolves with supportive care and treatment of comorbidities | Parasites occasionally seen as incidental finding in blood of captive platypuses | 11, 12 |
| | *Trypanosoma binneyi* | Non-pathogenic. Thought to be transmitted by the tick *Ixodes ornithorhynchi* and leeches | Blood smear, PCR | None | Not seen in captive platypuses | 3, 5, 13 |
| | *Ehrlichia ornithorhynchi* | Non-pathogenic. Associated with the tick *Ixodes ornithorhynchi* | Blood smear, PCR | None | Not seen in captive platypuses | 14 |
| **Viral** | | | | | | |
| Cutaneous papilliform lesions | Unknown (trauma, papilloma or herpesvirus) | Papilloma-like lesions associated with the nails of the forefeet | Histopathology | Surgical debridement | Seen in captive platypuses and may be associated with chronic trauma from enclosure and tunnel surfaces | 4 |
| **Bacterial** | | | | | | |
| Cutaneous ulcers | *Corynebacterium ulcerans* | Skin ulcers similar to mucormycosis | Culture, histopathology | Not attempted | Not seen in captive platypuses | 22 |
| Leptospirosis | *Leptospira interrogans* var Hardjo | Subclinical, chronic interstitial nephritis, seropositivity | Serology, histopathology, PCR | Not warranted | | 3, 16, 17 |
| Dermatophilosis | *Dermatophilus congolensis* | Multifocal, mild to severe scab formation not protruding above the fur | Gram stain of scabs, histopathology, culture | Not attempted | Not seen in captive platypuses | 18 |
| Pneumonia | *Aeromonas hydrophila, Escherichia coli* | Unknown | Necropsy | Not attempted | Not seen in captive platypuses | 19 |

**Table 2.5.** continued

| Disease/ infectious agent | Cause | Clinical signs | Diagnosis | Treatment | Prevention/Comments | Reference |
|---|---|---|---|---|---|---|
| **Fungal** | | | | | | |
| Mucormycosis | *Mucor amphibiorum* | Granulomatous and often ulcerative dermatitis. May disseminate to internal organs, particularly the lungs, leading to death | Histopathology, molecular diagnostics | None reported | Only seen in wild platypuses in Tasmania. Not seen in captive platypuses | 3, 19–21 |
| *Mucor circinelloides* | *Mucor circinelloides* | Ulcerative dermatitis; white, irregular, crumbling keratin of the mandibular and maxillary secateuring ridges; encephalitis | Culture, histopathology | None reported | Ulcerative mycosis seen in a wild platypus in Tasmania. Oral keratinopathy and encephalitis seen in a captive platypus | 4, 22 |
| Dermatophytosis (ringworm) | *Trichophyton mentagrophytes*<br>*Microsporum gypseum* | Tail alopecia, multifocal alopecia, crusting and thickened skin | Culture, histopathology | Long-term topical and systemic antifungal medication | Only seen in rescued, debilitated juveniles. Usually immunosuppressed individuals with comorbidities such as theileriosis | 4 |
| Paronychia (nail infections) | *Fusarium oxysporum*<br>*Fusarium solani* complex<br>*Microsporum gypseum* | Elongated, brittle, split, flaking nails, inflamed and swollen nail beds of the forefeet | Culture | Long-term topical antifungal medication | Seen in captive platypuses and may be secondary to chronic nail and nail base trauma from enclosure and tunnel surfaces | 4 |

References: (1) Whittington and Spratt (1989); (2) Whittington (1992); (3) Munday *et al.* (1998); (4) Vogelnest and Whinfield (2025); (5) McColl (1983); (6) Whinfield (2024); (7) Australia Zoo Wildlife Hospital records; (8) Whittington *et al.* (1992); (9) Booth (1999); (10) Whittington (1988); (11) Collins *et al.* (1986); (12) Kessell *et al.* (2014); (13) Paparini *et al.* (2014); (14) Gofton *et al.* (2018), (15) Macgregor *et al.* (2010); (16) McColl and Whittington (1985); (17) Loewenstein *et al.* (2008); (18) Lunn *et al.* (2016); (19) Munday and Peel (1983); (20) Connolly *et al.* (1999); (21) Connolly (2015); (22) Stewart *et al.* (1999).

**Table 2.6.** Common non-infectious diseases and causes of mortality of captive and rescued platypuses.

| Disease/ agent | Cause | Clinical signs | Diagnosis | Treatment | Prevention/ Comments | Reference |
|---|---|---|---|---|---|---|
| Starvation | Maladaptation, weaning-related challenges, failure to thrive of dispersing juveniles, drought | Emaciation, death | Signalment, clinical signs | Euthanasia, supportive care, treatment of specific infections | Not relevant to captive platypuses | 1 |
| Fur loss (primarily dorsal tail) | Unknown. Moulting has been correlated with higher water temperatures. On the tail, is seen with intraspecific biting particularly males biting females during breeding; excessive grooming, mechanical wear on enclosure surfaces | Patchy fur loss, primarily on the dorsal tail, mostly towards the tip and base of tail. An estimation of the moult on the dorsal and ventral surfaces of the body has been developed:<br>None: guard hairs of pelage intact.<br>Very light: small groups of guard hairs missing. Noticeable only on dry animal.<br>Light: patches of guard hairs missing. Often noticeable only on dry animal.<br>Moderate: large patches of guard hairs missing. Usually noticeable on wet animal.<br>Heavy: pelage almost devoid of guard hairs. Always noticeable in wet animal | Observation. Negative findings on skin scrapings and cultures | None | There may be a seasonal correlation with fur loss/moulting and periods of warmer water temperatures | 2 |
| Predation | Dogs, foxes, feral cats, Tasmanian devil, spot-tailed quoll, raptors, eels, goannas, pythons, Murray cod, crocodiles | Wounds, death | Observation/ history, physical examination, necropsy | Medical and surgical treatment if not fatal | Not relevant to captive platypuses | 1 |
| Extreme climatic events | Flood and drought | Emaciation, death | Observation, necropsy | None | Not relevant to captive platypuses | 1 |
| Anthropogenic threats | Forestry and logging; mining; land clearing and agriculture; urban development; invasive species; fisheries; dams and weirs; salinity; vehicle strike; pollution; human malice; climate change | Emaciation, trauma, dehydration, entanglement, entrapment, drowning, death | Observation/ history, physical examination, necropsy | Medical and surgical treatment if not fatal | Increased conservation action, climate change action, policy changes. Not relevant to captive platypuses | 1, 3–9 |

References: (1) Vogelnest and Whinfield (2025); (2) Grant and Carrick (1978); (3) Booth and Connolly (2008); (4) Ladds (2009); (5) Serena and Williams (2010); (6) Gust and Griffiths (2010); (7) Macgregor (2015); (8) Scheelings (2016); (9) Bino *et al.* (2019).

# 9. BEHAVIOUR

## 9.1 Activity

Wild platypuses are mostly nocturnal, but can be active during the day. Several studies suggest they typically forage for 7–18.8 h per day, although they can feed continuously for up to nearly 30 h (Grant 1983b; Grant *et al.* 1992; Serena 1994; Gust and Handasyde 1995; Otley *et al.* 2000; Bethge *et al.* 2009). Females also tend to be more diurnally active in spring and summer as a result of the higher energy demands associated with lactation or as a strategy to avoid resident males until they are ready to mate (Grant 1995; Otley *et al.* 2000; Thomas *et al.* 2018b). The incidence of diurnality on the Australian mainland has been found to range from 6.5% to 25%, whereas in Tasmania, 50% of the platypuses tracked during winter and 20% of those tracked during summer were diurnal (Grigg *et al.* 1992; Serena 1994; Gardner and Serena 1995; Gust and Handasyde 1995; Otley *et al.* 2000).

In captivity, a solitary platypus or two females held together generally follow(s) the nocturnal light cycle, with activity periods ranging from 3.5 to 11 h (average = 8 h) (Hawkins 1998). When mixed sex pairs are held together, a dominance relationship may be established (with adult males dominant to adult females and adult females dominant to juvenile males) in which the dominant animal's activity pattern is similar to that when kept alone whereas the subordinant animal's activity becomes shortened, more fragmented and more diurnal, thereby helping to segregate the two animals temporally (Hawkins 1998).

## 9.2 Social behaviour

In the wild, home ranges overlap between adult and juvenile individuals, but these individuals appear to seldom interact with each other (Serena 1994). Sexually mature males appear to be territorial and may use temporal avoidance as their home ranges can have considerable overlap in larger systems such as the Shoalhaven River in New South Wales (Gust and Handasyde 1995). Overlap in movements of both males and females in the latter part of the breeding season (September–November) was described in the Snowy River downstream of Jindabyne Dam (Hawke *et al.* 2021). Individuals generally have multiple burrows where they usually rest alone (Serena 1994; Gardner and Serena 1995; Serena *et al.* 1998), although burrow sharing has been observed between a subadult and adult male, two adult females, a grown female and an independent first-year female, and two independent first-year females (Serena 1994).

## 9.3 Reproductive behaviour

During the breeding season the female changes her behaviour from typically avoiding to initiating contact with the male (Hawkins and Battaglia 2009). Females will use three strategies to avoid males before being receptive: avoidance, flight and resistance (Thomas *et al.* 2018b). Mating behaviour appears to occur only in the water, where the male repeatedly follows the female and bites her tail. Mating also occurs in the water where the male curls up behind the female to mount (Figs 2.7, 2.8). Detailed descriptions of behaviours associated with courtship and mating can be found in Strahan and Thomas (1975), Hawkins and Fanning (1992), De-La-Warr and Serena (1999) and Thomas *et al.* (2018b). The time between the onset of courtship behaviour and mating behaviour is ~40 days, with females showing behavioural receptivity for up to ~30 days (Thomas *et al.* 2018b).

At 1–2 weeks after successfully mating, a female commences nesting burrow preparation by digging or modifying a dirt burrow, followed by intensively collecting wet nesting material for 2–5-days before retiring to the nesting chamber, now filled with damp vegetation, to lay eggs (Thomas *et al.* 2018b). Eggs are laid in the first 60 h of retirement (Thomas *et al.* 2018b). The first period the females spend in their burrow to lay and incubate eggs lasts ~80–94 h (Hawkins and Battaglia 2009). Subsequent retirement intervals range from ~32 to 50 h until hatching, after which the in-burrow stay intervals gradually shorten (Hawkins and Battaglia 2009). Females with nestlings do not appear to resume full feeding activity (> 8 h) until the nestlings are 23–36 days old (Thomas *et al.* 2018a). In captivity, females have been documented to engage in breeding activity between the ages of 2 and 12 years (Table 2.7).

A summary of the time lines for events related to the first two breeding successes at Healesville Sanctuary is given in Table 2.8.

## 9.4 Grooming

Although water is required for feeding, platypuses should be given opportunities to groom by providing rocks or logs near the water edge or rising above the water where they can rest and preen their fur.

## 9.5 Behavioural problems

Like any animal, platypuses can display strong abnormal repetitive behaviour, such as swimming to and fro along

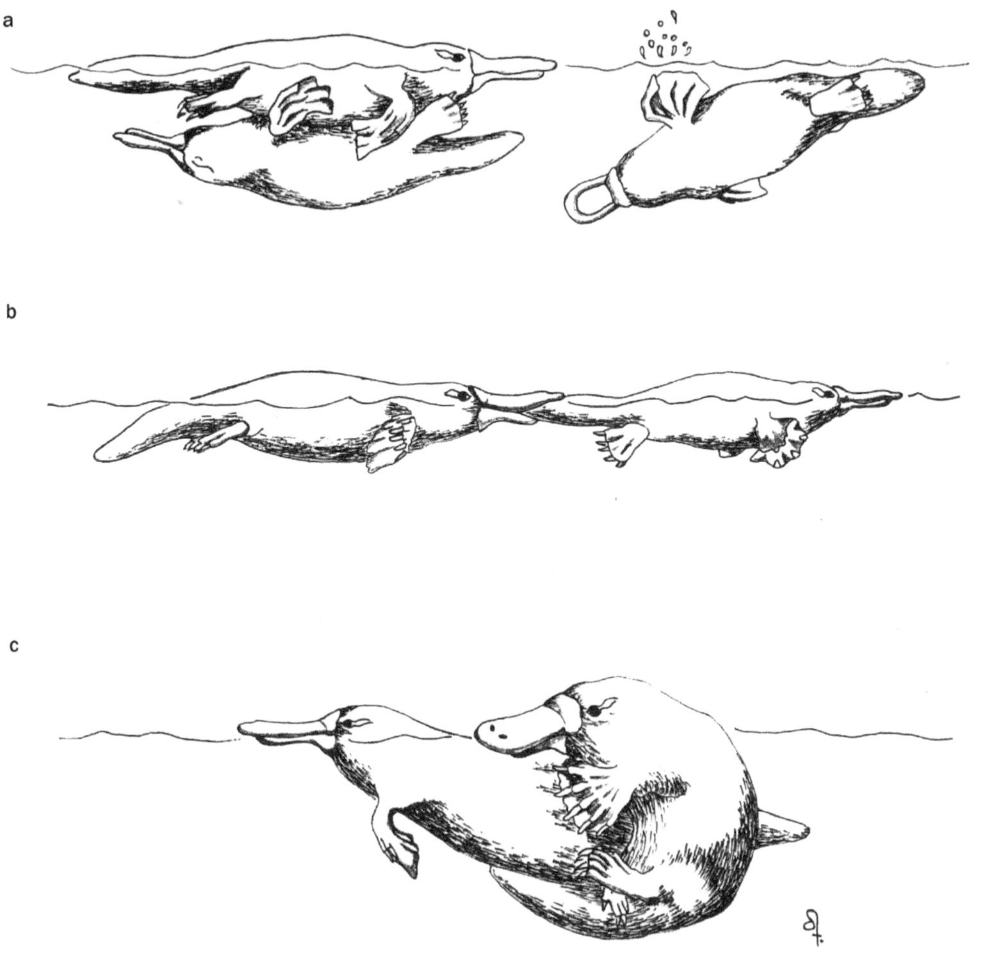


**Fig. 2.7.** Mating behaviour of the platypus. (a) male and female passing and touching each other; (b) male grasping the female's tail; and (c) apparent copulation position. Taken from Grant (2007) with permission of the illustrator Dominic Fanning.

**Fig. 2.8.** Circling behaviour observed in wild platypuses. Photo used with the permission of Zoos Victoria.

**Table 2.7.** Reproductive events recorded for captive platypuses from 1943 to 2020.

Derived from Fleay (1980), Holland and Jackson (2002), A. Battaglia and M. Hawkins (pers. comm.), J. Thomas (unpubl. data). Note that the year range reflects the summer when mating occurred (first year) and young were observed (second year).

| Year | Event | Dam name | Age[1] | Relationship time[2] | Institution |
|---|---|---|---|---|---|
| 1943–44 | 1 Birth | Jill | 6 | 4 years | Healesville Sanctuary |
| 1944–45 | Eggs | Jill | 7 | 5 years | Healesville Sanctuary |
| 1953–54 | Nesting | Penelope | 6 | 5 years | Bronx Zoo, New York |
| 1971–72 | Eggs | Unknown | 2 | 2 years? | Taronga Zoo |
| 1972–73 | Birth[†] | Penny | 8 | 2 years | Fleay's Fauna Park |
| 1974–75 | Eggs | Penny | 10 | 4 years | Fleay's Fauna Park |
| 1975–76 | Eggs | Penny | 11 | 5 years | Fleay's Fauna Park |
| 1976–77 | Eggs | Penny | 12 | 6 years | Fleay's Fauna Park |
| 1985–86 | Mating | Darkbill | 2 | 1.5 years | Taronga Zoo |
| 1986–87 | Nesting | Lightbill | 2 | 2 years | Taronga Zoo |
| 1987–88 | Eggs | Lightbill | 3 | 3 years | Taronga Zoo |
| 1988–89 | Mating? | Darkbill | 5 | 5 years | Taronga Zoo |
| 1990–91 | Mating | Darkbill | 7 | 7 years | Taronga Zoo |
| 1990–91 | 2 Births[#] | Undetermined | 3 | 2 years | Warrawong Sanctuary |
| 1991–12 | Nesting | Darkbill | 8 | 8 years | Taronga Zoo |
| 1992–93 | 1 Birth[@] | Undetermined | 5 | Up to 4 years | Warrawong Sanctuary |
| 1994–95 | Mating behaviour | Koorina | 4 | 3 months | Healesville Sanctuary |
| 1995–96 | 2 Births* | Undetermined | 9 | Up to 8 years | Warrawong Sanctuary |
| 1995–96 | Mating behaviour | Koorina | 5 | 3 months | Healesville Sanctuary |
| 1996–97 | Mating behaviour | Koorina | 6 | 1 year | Healesville Sanctuary |
| 1998–99 | 2 Births | Koorina | 8 | 3 years | Healesville Sanctuary |
| 2000–01 | 1 Birth | Koorina | 10 | 5 years | Healesville Sanctuary |
| 2002–03 | 2 Births | Maryanne | 8 | 3 years | Taronga Zoo |
| 2003–04 | 2 Births | Maryanne | 9 | 4 years | Taronga Zoo |
| 2004–05 | 1 Birth | Maryanne | 10 | 5 years | Taronga Zoo |
| 2007–08 | 2 Births | Binarri | 5 | 1 years | Healesville Sanctuary |
| 2008–09 | 2 Births | Binarri | 6 | 2 years | Healesville Sanctuary |
| 2009–10 | 1 Birth | Binarri | 7 | 3 years | Healesville Sanctuary |
| 2011–12 | 2 Births | Binarri | 9 | 1 years | Healesville Sanctuary |
| 2012–13 | 2 Births | Binarri | 10 | 2 years | Healesville Sanctuary |
| 2014–15 | 1 Birth | Binarri | 12 | 4 years | Healesville Sanctuary |
| 2015–16 | 2 Births | Binarri | 14 | 6 years | Healesville Sanctuary |
| 2019–20 | 1 Birth | Waddirrang | 12 | 5 years | Healesville Sanctuary |

Notes:
(1) Age was estimated by the development of spurs and the bodyweight of animals that suggested they were born during a particular breeding season.
(2) Relationship time is the length of time the pair has been known to each other either continuously or if given access to each other only during the breeding season.
[†]Young was found dead.
[#]The three juvenile females were introduced to the sanctuary in early 1987 from Kangaroo Island, followed by two juvenile males in early 1988. During the 1990–91 breeding season, two juvenile males were captured at 3 and 8 months after birth. Genetic testing determined the dam for both young (Gemmell *et al.* 1992; 1995; T. Grant pers. comm.).
[@]During the 1992–93 breeding season one juvenile male was captured 4 months after birth. Genetic testing determined the dam for the young (T. Grant pers. comm.).
*During the 1995–96 breeding season one juvenile female and one juvenile male were captured (both 8 months after birth). Genetic testing determined the dam for both young (T. Grant pers. comm.).

**Table 2.8.** Comparison of breeding activity between the two occasions in which young have successfully been raised in captivity at Healesville Sanctuary.

From Fleay (1944) and Holland and Jackson (2002).

| Activity | Jill (dates) Fleay (1944) | Day no. | Koorina (dates) Holland and Jackson (2002) | Day no. |
|---|---|---|---|---|
| Inactivity | 28/5/43 | -151 | | |
| | 31/5/43 to 1/6/43 | -148 to -147 | | |
| | 4/6/43 | -144 | | |
| | 7/6/43 | -141 | | |
| | 12/6/43 to 19/6/43 | -136 to -129 | | |
| | 24/6/43 to 26/6/43 | -124 to -122 | | |
| | 28/6/43 to 3/7/43 | -120 to -115 | | |
| | 6/7/43 to 12/7/43 | -112 to -106 | | |
| | 17/7/43 to 21/7/43 | -101 to -97 | | |
| | 28/7/43 to 2/8/43 | -90 to -85 | 25/8/98 to 30/8/98 | -90 to -85 |
| | 8/8/43 to 13/8/43 | -79 to -74 | 10/9/98 to 13/9/98 | -74 to -71 |
| | 25/8/98 | -62 | | |
| | 29/8/98 | -58 | | |
| | 1/9/43 to 3/9/43 | -55 to -53 | | |
| | | | 21/10/98 | -33 |
| Mating | 11/10/43 | -15 | 6/11/98 to 8/11/98 | -17 to -15 |
| Nest building | 23/10/43 | -3 | 14/11/98 to 18/11/98 | -9 to -5 |
| Male separated | 18/10/43 | -8 | 20/11/98 | -3 |
| In nest all night | 26/10/43 to 30/10/43 | 0 to 4 | 23/11/98 to 27/11/98 | 0 to 4 |
| | 2/11/43 | 7 | 29/11/98 | 6 |
| | 4/11/43 to 5/11/43 | 9 to 10 | 1/12/98 to 3/12/98 | 8 to 10 |
| | 8/11/43 | 13 | 6/12/98 | 13 |
| | | | 12/12/98 | 19 |
| Young emerge | 26/2/44 | 123 | 3/4/99 | 131 |
| | | | 8/4/99 | 136 |

the glass of the tank, circling or escape behaviour (trying to climb up the walls). It is always good practice to allow the animals to exhibit as much natural behaviour as possible. Abnormal repetitive behaviour can be exhibited because of stress, which can be in varying forms, or illness. Animals should undergo a health check as a first step by a veterinarian if they start showing an abnormal repetitive behaviour to rule out a medical condition. Changes in their routine or environment, new animal caretakers, noise or vibration or even introducing two animals together can cause stress. Sometimes it can be difficult to find or resolve the problem, which could be as simple as that a particular animal is just not coping with being in a captive environment. Temporary strategies for reducing the behaviour include moving to a novel tank for a day and dramatically increasing the water flow.

## 9.6 Signs of stress

Typical signs of stress include inappetence with hyperactivity or lethargy (Munday *et al.* 1998). Hyperactive animals attempt to dig out of their enclosure or transport box, often resulting in abrasions on the bill and feet (Munday *et al.* 1998). Abnormal behaviour can also include moving the nesting material frequently or frenetic feeding behaviour. Stressed platypuses can be detected because they lose bodyweight, do not dry themselves properly after leaving the water so the nest becomes wet and often lay quietly and are unresponsive to stimulation (Munday *et al.* 1998). When being handled, males will often try to spur the handler, especially during the breeding season.

It is critically important to understand the signs of stress, particularly when bringing an animal in from the wild or when introducing platypuses to each other (Booth and Connolly 2008; see section 11.12). Indicators that suggest a failure of the platypus to adapt to captivity include:

- abnormal repetitive behaviour
- reversal of neutrophil to lymphocyte ratio
- escalating or persistent escape behaviour
- failure to eat the expected daily food intake
- > 10% bodyweight loss within 2 weeks of capture.

If an individual does not adapt appropriately to captivity, it should be released immediately (see section 11.11).

### 9.6.1 Short-term stress

Short-term anxiety or annoyance may be communicated via a growl-like vocalisation that animals sometimes make when disturbed, especially when being caught or held by someone new. Animals that struggle incessantly when contained in a bag or box will in most cases settle quickly when placed in a container in a dark and quiet place. However, these animals are not necessarily unstressed and cases of sudden death within a few hours of capture have been reported.

### 9.6.2 Longer term stress

Several distress behaviours have been reported, alone or in combination, that indicate a recently captured animal should be released back into the wild:

- quiet retirement to sleeping chamber and not feeding
- excessive movement in and out of the water and not feeding, even while spending considerable periods in the water
- failure to groom water from fur or to keep sleeping chamber dry
- entering water whenever disturbed
- abnormal repetitive swimming
- prolonged periods of submergence under an object followed by return to the surface only to breathe
- defaecation inside the nesting chamber.

These behaviours may also occur in combination with weight loss and abrasions to the feet, bill and tail.

A response to stress is signalled by an increase in cortisol, reflected in increased numbers of circulating neutrophils and decreased circulating lymphocytes. Cortisol causes neutrophils to move from the tissues into the circulatory system, resulting in lymphocytolysis (Schalm *et al.* 1975). Measurement of neutrophils is considered a better indicator of stress than measuring cortisol because cortisol rises rapidly when an animal is restrained (Booth and Whittington 1993; Booth and Connolly 2008).

## 9.7 Behavioural enrichment

Different methods of providing behavioural enrichment include:

- providing live food
- creating water movement with artificial streams and waterfalls (platypuses in captivity seem to enjoy playing in waterfalls)
- housing more than one animal in an enclosure (e.g. a female with other females, or a male with a female). Moving a piece of bedding from one system to another can simulate the social presence of another animal without the risk of its actual presence
- providing some land to move around on or foraging ponds of different depths
- changing the length of burrows or the geometry of intersections and providing soil in which to construct burrows
- floating objects that can be manipulated such as paperbark, tree fern trunks, dry logs, kickboards
- foraging challenges such as plant pots for pin point feeding or floating baskets with food
- varying the captive environment by broadcasting nocturnal or rain sounds, using sprinklers or misting the surface of the tanks to simulate rain, raising or dropping the water level to simulate drought or flood

## 9.8 Training and conditioning

Platypuses can be trained with positive reinforcement to accept an unrestrained physical examination, hand feeding (Fig. 2.9), stationing and crate training. The most important prerequisites for a successful training program are a positive relationship to be established between the platypus and animal caretaker (someone they know) and for the program to set the platypus up for success by making it feel safe and comfortable.

Favoured food items to use for positive reinforcement include blackworms, glass shrimp and dragonfly nymphs, fed in the water where a platypus feels most secure. Other positive reinforcement may include providing a favoured enrichment item, such as a tree fern trunk.

No food items should be excluded from the diet nor should food be withheld. Platypuses should have the choice to participate or not at all, with times and sessions done in their own tank. Animal caretakers getting into the water and sitting at water level with some shelter nearby will give the platypuses a sense of security while they investigate the animal caretaker.

## 9.9 Introductions and removals

A platypus can usually be readily introduced back into its home enclosure (including one shared with another individual) without any problems, although the time spent out of the enclosure should be minimised. Addendum 1 contains an outline of the appropriate protocols to follow.

## 9.10 Intraspecific compatibility

The home ranges of platypuses belonging to different sex and age classes characteristically overlap, though individuals are mostly solitary in their habits apart from family groups or pairs associating for the purpose of mating (Serena 1994; Gardner and Serena 1995; Gust and Handasyde 1995). Sexually mature males appear to be much more strongly territorial than females: males occupying overlapping areas typically avoid each other through staggered temporal use of shared areas (Gardner and Serena 1995), with greater spatial separation evident during the breeding season than at other times of year (Gust and Handasyde 1995). Therefore, individuals are best kept either solitarily, as two females, a male–female pair or potentially a male with two females. Males cannot be housed together, particularly during the breeding season, as they interact aggressively and there are records of males killing other males in captivity by spurring (Burrell 1927; Fleay 1980). Take care

**Fig. 2.9.** Hand feeding a captive platypus. Photo used with the permission of Zoos Victoria.

when introducing animals for the first time, as there can be some aggression and some individuals appear to be inherently incompatible. Even if there is no aggression, at times one animal will not cope with sharing space with another and can fret or go off its food.

The size and design of the captive environment can affect how well animals get along when housed together. Animals ideally will share several tanks of water so that each animal can feed, swim and groom alone. Enclosure size is to some extent arbitrary as variables such as tank shape and furnishings can make even a small enclosure more interesting. Addendum 2 contains more information on the protocol for introducing platypus to each other.

### 9.11 Interspecific compatibility

Platypuses have been successfully housed with several other species including:

- water dragons ( e.g. *Intellagama lesueurii*)
- turtles (e.g. *Chelodina longicollis*)
- ducks (e.g. wood ducks (*Chenonetta jubata*) and plumed whistling ducks (*Dendrocygna eytoni*)).

### 9.12 Suitability for captivity

A review of captive platypus deaths between 1934 and 1988 (228 platypuses) suggested they did not generally live longer than 1 year (Whittington 1991, 1993). These results caused controversy at the time, with more current data from 1987 to 1991 suggesting that most animals lived for > 5 years (Williams 1992; Serena and Williams 1993a, 1993b). Great advances in platypus husbandry and breeding in the past 30 years have clearly highlighted that they can be held for many years in captivity and breed regularly if managed appropriately.

## 10. BREEDING

### 10.1 Mating system

Platypuses in captivity can be both polygynous and polyandrous, with both males and females potentially having several partners.

### 10.2 Ease of breeding

Platypus have historically been very difficult to breed and remain so for most institutions. The first captive birth of a platypus occurred at Healesville Sanctuary, in the summer of 1943–44 when a single young was born and raised (Fleay 1944). After this time there were numerous records of breeding activity at several institutions; however, these did not lead to the successful rearing of young (Table 2.7). It was not until the summer of 1998–99 when twins were born (though one later died) at Healesville Sanctuary that the key components to breeding this species in captivity were unlocked (Holland and Jackson 2002).

Since the summer of 1998–99, platypuses have been bred regularly by both Healesville Sanctuary and Taronga Zoo (Hawkins and Battaglia 2009; Thomas *et al.* 2017), but not any other institution. These successful breeding events have included the production of 11 young that survived to maturity, at Healesville Sanctuary between 2007 and 2019 (Thomas *et al.* 2017, 2020) and 5 young at Taronga Zoo between 2003 and 2005.

### 10.3 Oestrous cycle, gestation and incubation periods

The length of the oestrous cycle of the platypus is unknown, although gestation is thought to last ~16 days (range 13–23 days) based on the timing of observed matings and the subsequent retirement of a female to a burrow in captive studies (Holland and Jackson 2002; Hawkins and Battaglia 2009; Thomas *et al.* 2020). After laying eggs, the female incubates them by holding them against her belly with her tail while she lies curled up in the nest chamber of the burrow. Hatching takes place ~10–12 days after the eggs are laid (Carrick 1995; Hawkins and Battaglia 2009; Thomas *et al.* 2020). It is possible that female platypuses may be induced ovulators as only mated females appear to produce eggs (J. Thomas pers. obs.).

### 10.4 Reproductive status

#### 10.4.1 Females

The best method to determine a female's reproductive status is via observation of behaviour, including:

- courtship behaviour, including willingness to be around the male
- mating behaviour
- preparing a nesting burrow in an earthen bank
- disappearing for several days after the estimated gestation period based on observed mating
- greatly increased food consumption indicating lactation (which can be confirmed by injecting oxytocin).

Tail fat stores also decrease towards the end of the mating period (around October: Temple-Smith 1973), indicative of high-energy demands during the mating period (Handasyde *et al.* 2003). Lower bodyweight has also been reported in lactating female platypuses in the upper Shoalhaven River in New South Wales at the end of the breeding season (February–March) (Bino *et al.* 2015).

If young are found in or after emerging from the nest, several developmental stages and measurements can be recorded and compared with existing growth curves (see section 10.16) or to establish curves for future reference.

- Developmental attributes recorded on a weekly basis:
  - sex: observe spur morphology
  - eyes open
  - fur visible: slightly tinged, medium length or well developed
  - eating solids
  - self-feeding
  - independent.
- Measurements recorded on a weekly basis:
  - bodyweight (g)
  - dorsal bill length (mm) exclusive of the dorsal shield
  - maximum bill width (mm)
  - interocular width (mm)
  - interaural width (mm)
  - head length: from occiput to snout tip (mm)
  - head width: maximum width across the zygomatic arches (mm)
  - crown-rump length (mm): primarily for very small neonates
  - body length (mm): from the snout tip to cloaca
  - tail length (mm): from the cloaca to the end of the last vertebra of the tail tip
  - total length (mm): from snout tip to tail tip
  - maximum tail thickness (mm) at the level of the cloaca
  - spur length (mm) and development in both males and females
  - length of pes (mm), not including the toenail.

### 10.4.2 Males

The size of the venom-producing crural glands and testes show a clear increase in size during the breeding season (Temple-Smith 1973; Grant and Temple-Smith 1998a). Males also are more aggressive during the breeding season (Temple-Smith 1973), are more likely to vocalise and spur if handled and often become more restless. They can also show increased ranging behaviour at this time, moving around all areas of the enclosure.

## 10.5 Reproductive management

Physical separation of animals is the most effective technique to control breeding in platypuses. There are no records of captive facilities using contraceptives or other methods to control reproduction and the difficulties of successfully breeding platypuses means that there normally is little or no desire to prevent this activity.

## 10.6 Occurrence of hybrids

None.

## 10.7 Timing of breeding

Platypuses have a distinct breeding season that varies to some extent with latitude. Platypus mating behaviour in both captivity and the wild has been recorded from August to early November in Victoria (De-La-Warr and Serena 1999; Holland and Jackson 2002; Easton *et al.* 2008) and New South Wales (Burrell 1927; Hawkins and Battaglia 2009), and is believed to begin and end somewhat earlier in Queensland and later in Tasmania. Young emerge from the nesting burrow for the first time between mid-January and early March (Carrick 1995; Grant 1995; Williams *et al.* 2013). Captive studies suggest female platypuses have a receptive period of 4–6 days (Hawkins and Battaglia 2009).

## 10.8 Age at first and last breeding

### 10.8.1 Wild

Observations of wild platypuses suggest that females can first breed when they are ~2 years of age (although some do not breed until their third or fourth year) and can continuing breeding to the age of at least 21 years (Grant *et al.* 1983; Grant and Griffiths 1992; Grant 1995; Grant 2004; T. Grant pers. comm.).

### 10.8.2 Captivity

The age at which platypuses show reproductive behaviour in captivity appears to be much older than in the wild (Table 2.7). The occasions where reproductive behaviour has been observed have had several common factors:

- the individuals were brought into captivity at a very early age, generally as soon as possible after weaning and leaving the burrow for the first time in February–March
- they were housed together with their mate all their lives

- they were kept in the same enclosure all of their lives
- females appear to show reproductive behaviour after 6 years of age
- there was a continuity of handlers. Platypuses recognise individual people and behave differently when someone else takes over while one of the animal caretakers is absent. For this reason, most institutions have a policy of having the same animal caretaker working with their platypuses.

## 10.9 Ability to breed every year

Evidence from the wild suggests that female platypuses do not normally breed each year (Grant *et al.* 1983). Data collected over many years from wild populations suggests that the birth interval of individual females ranges from 1 to 6 years (Temple-Smith and Grant 2001; Grant 2004; Grant *et al.* 2004).

Records of captive births also suggest that breeding typically occurs every second or more years. However, observations at Taronga Zoo and Healesville Sanctuary suggest that females can breed in consecutive years on occasion, though there appears to be an increased likelihood of a single young being raised (Hawkins and Battaglia 2009). This may reflect the enormous energy investment by the female during lactation during the previous year.

## 10.10 Ability to breed more than once per year

Platypuses are only able to raise young once per year.

## 10.11 Nesting requirements

The length of the nesting burrow in captivity has been shown to vary from 3.2 m to 10.4 m (Thomas *et al.* 2018a). After mating occurs, gravid females should be provided with plenty of nesting material in the water to allow her to build a nest within the nesting chamber. Timing and placement of presentation of the nesting material appears to be critical. Nesting material should be provided as close as possible to the dirt mound and should include small terminal branches of eucalypts (with leaves ~20 cm in length) for ease of transport by the female, along with leaves of *Lomandra* and *Poa*, tree fern fronds and strips of tree bark. The female transfers these materials to the nesting chamber using her curled-up tail (Fig. 2.10), which she then uses to construct a spherical nest (Fig. 2.11). The vegetation is thought to maintain the humidity required to prevent desiccation of the eggs during incubation and neonates during the early post-hatching period (Burrell 1927; Grant 1995; Temple-Smith and Grant 2001). On average, a female makes 24 ± 4 trips to carry nesting material into the nest (Thomas *et al.* 2018a), with ~80 L of nesting material transferred in one well-documented instance (Holland and Jackson 2002). However, females differ in their skill level: some completely this task easily while others struggle to transport material. In some cases, females only manage to bring 31–52% of collected material into the nest, with the remainder strewn along the burrow (Thomas *et al.* 2018a). Other platypuses should be separated from the nesting female as they can impede her success collecting and transporting material.

Nesting burrows are highly variable in design and include sections of backfilled soil known as pugs that may provide security from rising flood waters, predators and disturbance by other platypuses (Burrell 1927; Thomas *et al.* 2017). In captivity, female platypuses should be provided with an earth mound in which to dig a nesting tunnel and chamber of at least 2–3 m in length. The soil should have a significant clay component to enhance stability and incorporate roots and branches to further prevent tunnel collapse. The mound should also be open to the weather because rain also helps to keep the soil stabilised, keeps the female cool during hot weather, maintains humidity and minimises the likelihood of the eggs drying out. Ideally, vegetation should also be allowed to grow on the mound to help stabilise the soil over the burrow and nesting chamber.

The ideal system for breeding appears to include at least two separate systems, comprising at least two feed tanks, two earth banks and two sets of nest boxes and associated tunnels. During the 1998–99 breeding season at Healesville Sanctuary, after mating was observed and no further interest was shown between Koorina and her mate, he was separated from Koorina to ensure that he did not interfere with her while she built her nest, laid her eggs and fed and raised the young (Holland and Jackson 2002). Males with access to a female and her young may try to enter the nesting burrow and will consume large portions of the female's food.

## 10.12 Breeding diet

In captivity the diet remains the same during the breeding season; however, the female becomes voracious in her appetite, particularly during late lactation, when energy demands are at their highest. When Koorina bred at Healesville Sanctuary in the 1998–99 breeding season

**Fig. 2.10.** Female collecting nesting material in her tail. Photo used with permission of Zoos Victoria.

**Fig. 2.11.** An excavated nesting chamber (left) and a nest removed from the nesting chamber (right) with material comprised of mostly of *Lomandra longifolia* and *Poa ensiformis*. Reproduced from Thomas *et al.* (2018a) with permission from CSIRO Publishing. Photographs used with permission of Zoos Victoria.

her food consumption increased to up to 800 g of food per day late in lactation (Fig. 2.12) as compared to ~200–300 g of food per day before and after lactation (Holland and Jackson 2002).

Another study found that the energy intake of a lactating female platypus was more than twice that of non-lactating females in the last month of lactation (Fig. 2.13). More specifically, she consumed comparable kilojoules to non-lactating platypuses in the first month of lactation, increasing by a factor of 1.54-fold in the third month and 2.10-fold in the fourth and final month. Initially, the female spends the majority of her time in the nest; this then decreases until she is only feeding the young every second day until weaning, which allows her more time to feed in order to gain the additional nutrients she requires for milk production (Thomas *et al.* 2019).

## 10.13 Number of young born

Of 70 active nests examined by Burrell (1927), 16% contained one egg or juvenile, 77% contained two eggs or twins and 7% contained three eggs or triplets.

## 10.14 Age at weaning

The young suckle milk secreted from the mother's abdominal surface. The duration of lactation in the wild and captivity has been recorded to be between 114 and 127 days (Grant and Griffiths 1992; Carrick 1995; Holland and Jackson 2002; Hawkins and Battaglia 2009) or, in a more recent captive study, 128 ± 1 days (Thomas

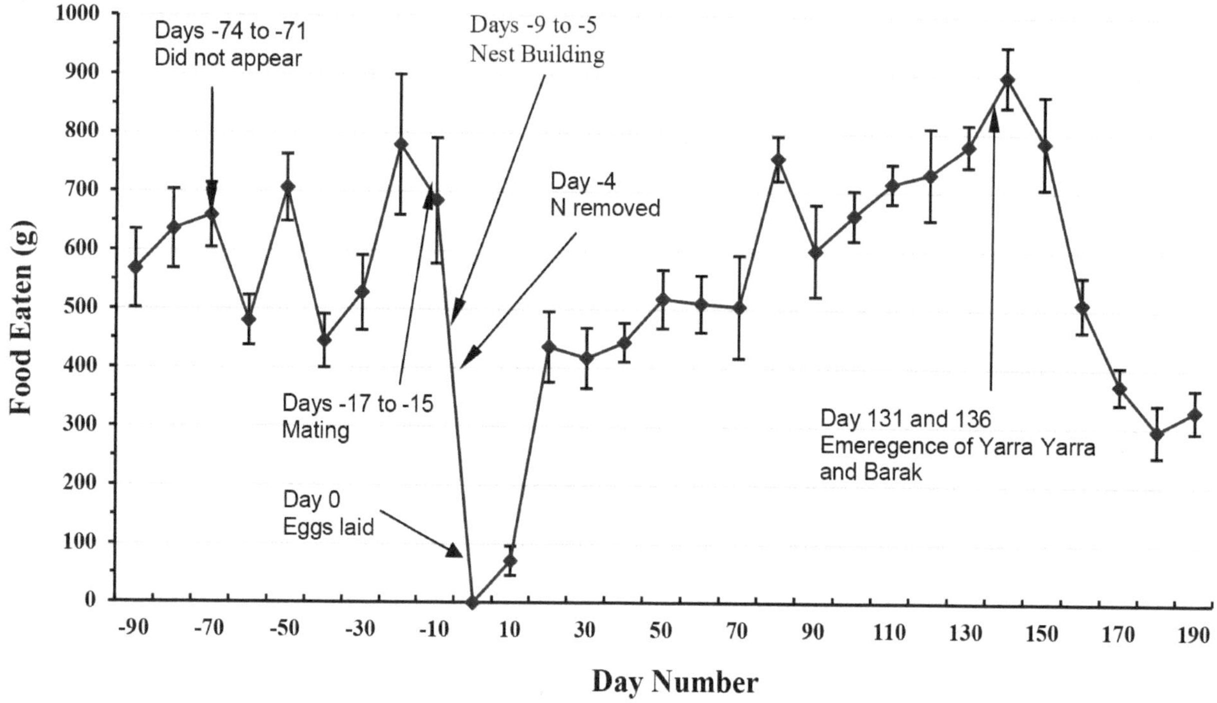


**Fig. 2.12.** Average food consumption of a female platypus during the 1998–99 breeding season. Standard error bars are shown. Prior to day 4 the food consumption was for both N and Koorina, with some food consumed by the young following their emergence. Taken from Holland and Jackson (2002).

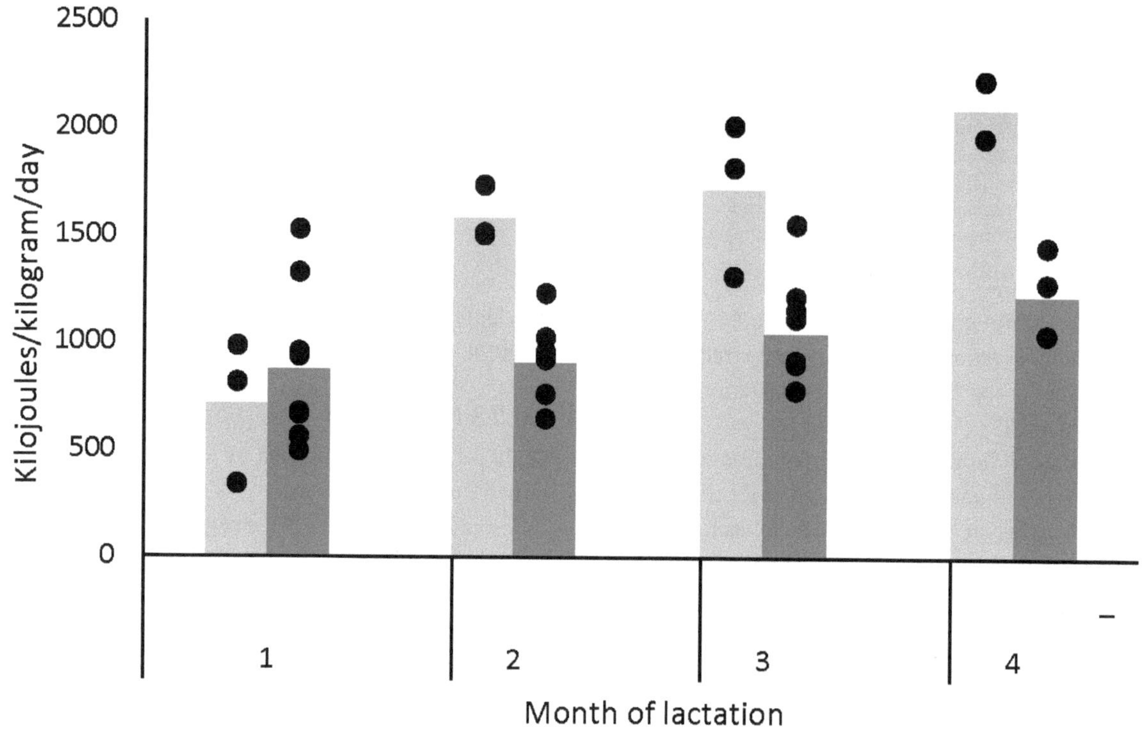


**Fig. 2.13.** Mean kilojoules consumed per day from November to February by a lactating female (light grey) compared with five non-lactating females (dark grey). Reproduced from Thomas *et al.* (2020) with permission from CSIRO Publishing.

*et al.* 2019). Juveniles have been recorded to emerge from the nesting burrow between 128 and 133 days after the female retired to lay eggs (Hawkins and Battaglia 2009; Thomas *et al.* 2020). Thomas *et al.* (2019) reported that 60% of captive nestlings emerged after weaning, suggesting that the transition from a completely milk-based diet to a weaned diet occurs rapidly within days.

## 10.15 Age at removal from the mother

Young have been separated from their mother at ~155 days of age or 1 month after they have emerged from the burrow (Holland and Jackson 2002) or from a few days after emergence up to 6 months after emergence (J. Thomas pers. comm.). Observations obtained in the wild suggest that some young disperse from the maternal home range in their first autumn (Serena 1994) but others may remain longer, with 12% of juvenile males and 25% of juvenile females eventually recruited into their birth populations near Melbourne, Victoria (Serena *et al.* 2014). It is unknown if any maternal care ever occurs following weaning. However, there is a record of a captured juvenile platypus in the upper Shoalhaven River in New South Wales, caught in late January, that regurgitated milk during sampling of its cheek pouches that also contained macroinvertebrate material (Grant 2007).

## 10.16 Growth and development

The potential for age assessment based on bodyweight is complicated by the considerable size differences that exist among populations of adult platypuses (Grant and Temple-Smith 1983). Some information on the growth of nestlings has been obtained from museum specimens; however, the age of these animals was estimated (Grant and Temple-Smith 1998b). In nestlings < 3 months old, bill width was greater than length, but this was reversed in older nestlings and juveniles of both sexes.

Presently, the best method of assigning age classes in males and females is by examining the size and appearance of spurs and/or spur sheaths found on the hindlimbs, which can be used to distinguish reliably between two female age classes (juveniles up to 9 months of age; all other females) and three male age classes (juveniles up to 12 months old; subadults 13–24 months old; all other males) (see section 3.5.3; Temple-Smith 1973; Grant 1995; Williams *et al.* 2013).

An extensive review of nestling platypuses of different ages from museum collections has been used to examine changes in growth and development with age (Manger *et al.* 1998). That review has detailed drawings of the changes in the external anatomy of developing platypuses and includes various measurements of some

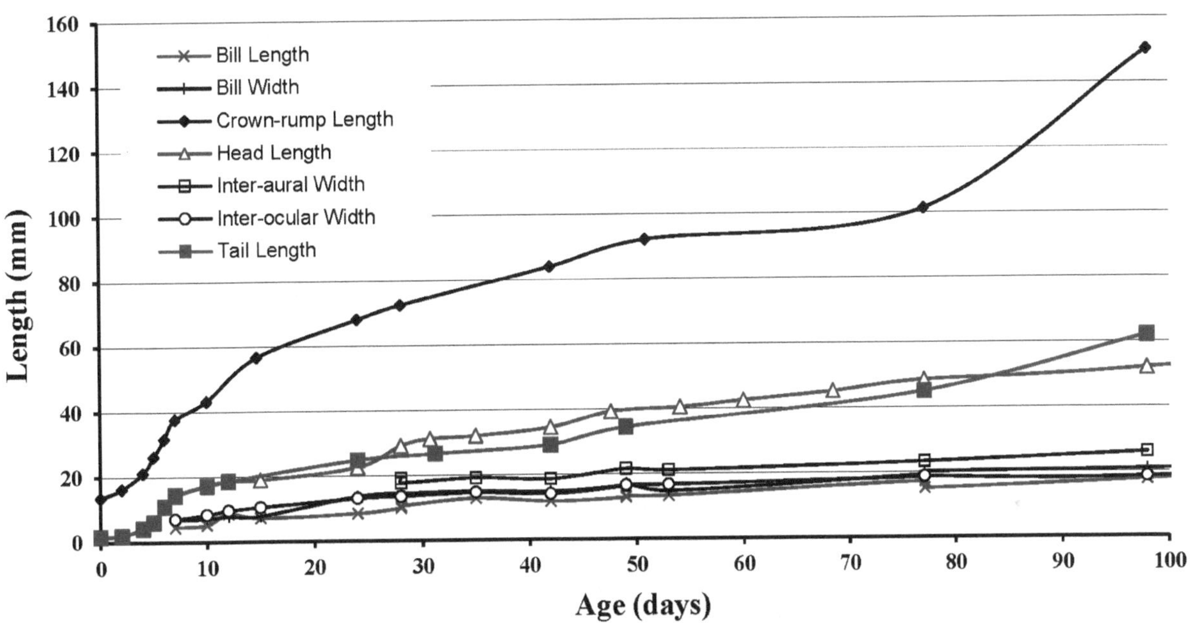


**Fig. 2.14.** Growth of different morphological features of platypuses using museum specimens. Derived from Manger *et al.* (1998).

physical features (Fig. 2.14). Caution needs to be taken with these estimates as the derivation of the ages is based on the 'Burrell' collection housed at the National Museum of Australia. It is assumed that Harry Burrell himself aged these animals and although they are the best approximation available, they may not be accurate (Manger *et al.* 1998).

More recently, a growth chart has been developed from captive platypuses at Healesville Sanctuary after egg laying (Fig. 2.15; Thomas *et al.* 2020). The age of developmental milestones, including the presence of fur on the tail, ventral and dorsal surfaces, opening of the eyes, commencement of grooming and moving, can be found in Fig. 2.16.

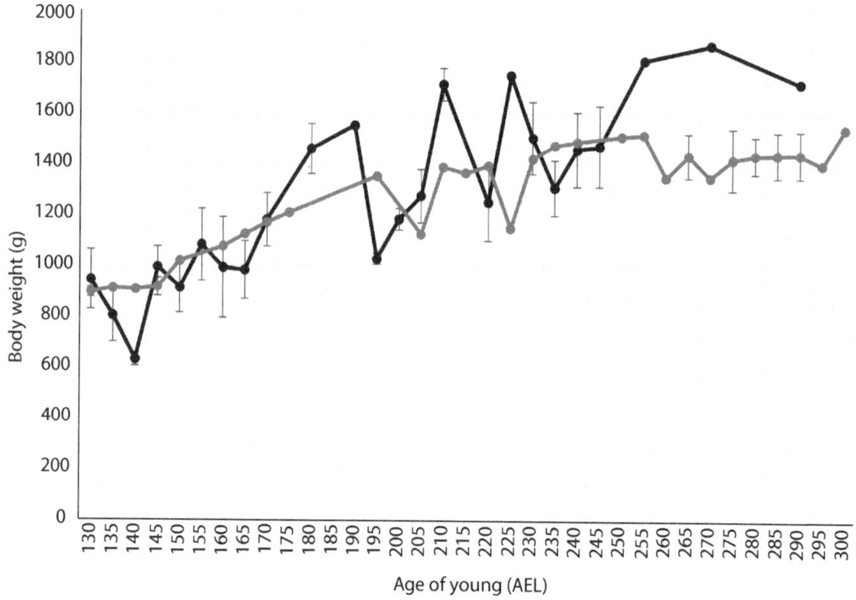


**Fig. 2.15.** Bodyweight of juvenile platypuses. The black line shows the pooled bodyweight data of males and the grey line for females. Note that 'AEL' refers to after egg laying. Reproduced from Thomas *et al.* (2020) with permission from CSIRO Publishing.

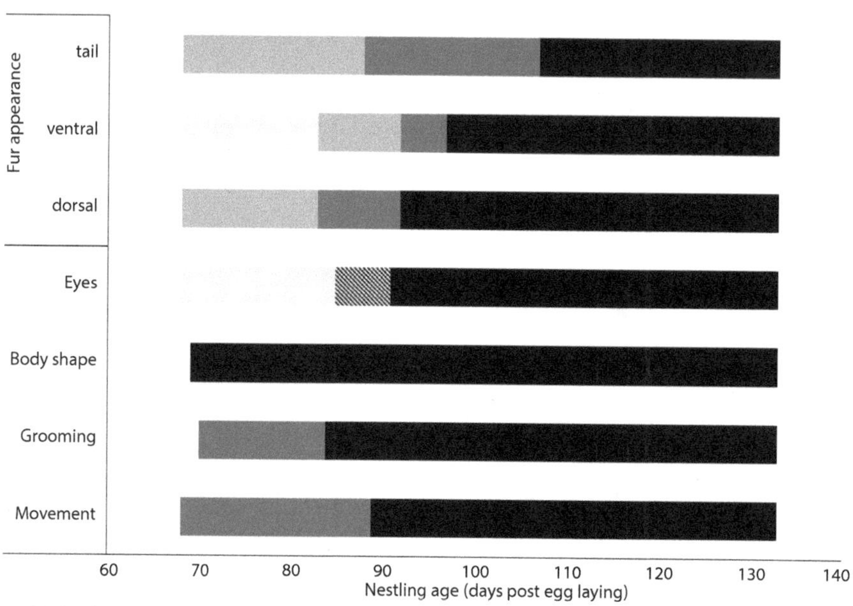


**Fig. 2.16.** Physical and behavioural development of nestlings as recorded from first visibility at day 68–70 of age until emergence at day 133 of age. Fur appearance: white = no fur present, light grey = velvet fur, dark grey = short fur, black = long or glossy fur; eyes: white = eyes closed, patterned = partially open, black = eyes open; body shape: black = straight; grooming: dark grey = slow foot motion, black = fast and full foot motion; movement: dark grey = slow crawl, black = full walk. Reproduced from Thomas *et al.* (2020) with permission from CSIRO Publishing.

# 11. ARTIFICIAL REARING

## 11.1 Housing

As with all native mammals that have been taken into care, minimising stress is a major consideration. Choosing suitable housing can help to create a stress-free environment. To achieve this, several factors should be considered:

- securing the area from children and other animals
- maintaining high standard of hygiene
- escape-proofing the area
- clearing the area of obstacles and hazards
- offering shelter from extreme weather and noise.

Facilities in which young are being hand-reared should consist of the following:

- an indoor feed tank ~1–2 m × 1–2 m
- a nest box divided into two chambers of equal size by a partition bisecting the box's long axis and reaching up to the lid. Provision should be made for platypuses to move freely from one chamber to the other through an aperture 0.2 m in diameter, located 5 cm above the nest box floor
- a tunnel connecting the nest box and tank above the water.

Other criteria governing the facility and its operation are:

- the nest box and feed tank environs should be quiet
- lighting should normally be subdued
- ambient temperatures should be maintained in the range 15–25°C
- soft, dry, clean substrate should be provided in the nest box
- water depth of the feed tank should initially be maintained in the range 5–10 cm, and gradually increased as animals become stronger and more adept at swimming/feeding
- water should be changed at least daily (more frequently if required).
- Two 11-week-old platypuses raised by Robertson (1989) were housed in a solid timber box measuring 600 (L) × 500 (W) × 500 (H) cm and half filled with dried kangaroo grass (*Themeda australis*). It was heated to 22–25°C with a 250-Watt dull emitter heat globe suspended over the box.
- An 8–10-week-old animal hand-raised by Taronga Zoo (Beavan 1997) was held in a wooden box (62 × 62 × 129 cm) throughout the rearing process. The box was lined with sphagnum moss and divided into four compartments. The platypus had access to all four compartments and was kept in the box at all times other than feeding sessions. The temperature within the box was ~22°C. It appears weaning is rapid so there is no need for a slow transition from milk to solids.

## 11.2 Temperature requirements

Robertson (1989) and Beavan (1997), when hand-rearing platypuses, held the ambient temperature at 22–25°C, using a minimum/maximum temperature gauge with a plastic-coated probe that was placed next to the juvenile to ensure the temperature could be monitored. The water temperature for the individual raised by Taronga Zoo was kept above 32°C initially, as it was found that the animal would not swim or feed in water at a lower temperature (Beavan 1997). However, observations by Grant (pers. comm.) suggest that adults can become very lethargic in water temperatures above 30°C. The platypus releases heat from exercise to the water and if the water is the same temperature, this heat transfer does not occur readily, as conduction and convection need a temperature gradient for evaporative cooling to occur, except at the surface of the water. Therefore, the water should probably be at slightly lower temperatures. Platypuses must also be fully furred and waterproof before being given access to water; otherwise animals should only be fed milk and not permitted to swim. Temperature measurements taken over 1 week at Healesville Sanctuary, in a natural burrow where the nestling was parent-reared, at ~8 weeks of age, the temperature was 26°C when measured with the nestling alone in the chamber (J. Thomas pers. obs.). This may have been due to the respiration of the nestling and the burrow chamber being sealed by the mother at her departure. Wild and captive-bred juveniles leaving the nest, normally at the end of lactation, are usually between 750 and 1000 g (or more) bodyweight. Once weaned, animals can be housed in water at ~19–21°C.

## 11.3 Specific requirements

When first brought in for hand-rearing, animals may be dehydrated. If so, they should be provided with plain boiled water, with 5 g (1 teaspoon) of glucose to 100 mL of water or 1 g of electrolyte replacer if available (Austin 1997). Vytrate® can also be given at a ratio of 20 mL Vytrate® to 250 mL water. It is important to warm the juvenile platypus before feeding, to lessen the risk of inhalation pneumonia. If this takes some time, give fluids subcutaneously. If the nestling is really cold, place it in a warm water bath and dry it off rather than putting it in a hot box.

Stress is a major problem in the successly rearing native mammals and can be fatal. Therefore, it is important to keep noise to a minimum, not overhandle them and maintain high standards of hygiene.

## 11.4 Diet and feeding routine

### 11.4.1 Natural milk

The best way to determine if a platypus is still milk-dependent is if it has a square rather than rectangular bill and a pronounced sucking action if presented with milk in a cupped hand (Booth and Connolly 2008). There may still be milk in the check pouches on arrival. The time of year the animal comes into care will also suggest if the platypus needs milk (i.e. in Victoria, juveniles emerge mid-January–March, so an animal in care before that will likely require milk). Platypus milk typically contains 391 g/L total solids, 222 g/L crude lipid, 82 g/L crude protein, 33 g/L total hexose, 0.43% sialic acid and 0.5% minerals (Griffiths *et al.* 1984; Gibson *et al.* 1988) (Table 2.9). The major fatty acids of their milk are palmitic, stearic, palmitoleic, oleic, linoleic, linolenic and arachidonic, with the other fats comprising long-chain polyunsaturated fatty acids (Griffiths *et al.* 1984). When considering carbohydrates, the milk has a mean hexose content of 33%, of which nearly half is L-fucose. Of the total monosaccharides present in acid hydrolysates of the water-soluble carbohydrates, L-fucose constitutes 33%, D-galactose 29%, glucosamine 20%, D-glucose 11% and sialic acid 7% (Messer and Kerry 1973; Messer *et al.* 1983). Free lactose has been only found in trace amounts (Messer *et al.* 1983).

### 11.4.2 Milk formulas

The main commercial formulas for hand-rearing platypuses include:

- Wombaroo® Echidna Milk – recommended milk formula for a platypus nestling is to make 100 mL of Wombaroo echidna late lactation formula as per packet instructions (36 g of powder plus 70 mL of preboiled warm water), then add a further 10 mL of Wombaroo® Good Oil for Animals and 8 mL of water (total volume = 118 mL). This formula best resembles the macronutrient content of platypus milk. Although the protein is slightly higher, the nestlings will simply use this for muscle and tissue growth and it is unlikely to be an issue
- Biolac® – this formula can be used with the addition of lactose-free cream
- Di-Vetelact® (Lillelund Pty Ltd) – a widely used, low-lactose milk formula. Due to its low energy concentration when prepared as directed, some groups advise the addition of mono- and polyunsaturated fats such as canola oil.

Robertson (1989) made a mixture of 15 g Di-Vetelact® and 100 g water, which was mixed to a creamy consistency. Feeding occurred twice per day, morning and night, with a maximum of 20 mL given per day. From day 45, after care commenced, the platypus was noticeably reluctant to take the milk mixture; however, it would take a blended egg and crayfish mixture wetted with a few millilitres of Di-Vetelact®.

Beavan (1997) made up a formula containing Di-Vetelact® (1 scoop in 50 mL water), vegetable oil (0.8 mL), and thickened cream (1.8 mL) mixed together. Olive oil is an excellent source of palmitic acid and oleic acid and can be added to the milk (Booth and Connolly 2008).

Palatability of the milk can be improved by adding live blackworms to the mix (J. Thomas pers. comm.) and also increases the iron content.

### 11.4.3 Feeding apparatus

Initially, small quantities of milk can be poured into the palm of the hand and the platypus encouraged to suck up the mixture (Robertson 1989). If this method is used, ensure the nostrils are not submerged (Booth and Connolly 2008). The cupped hand method can be messy and not very efficient, as well as not hygienic. An alternative method of feeding utilises a 'feed brush' constructed from a camera cleaning brush adapted to fit on the end of a 20-mL disposable syringe, which has proved very successful. Milk flow can be controlled better and the

**Table 2.9.** Concentrations of the major constituents of platypus milk.

Derived from Griffiths *et al.* (1984) and Gibson *et al.* (1988).

| Total solids (%) | Carbohydrates (%) | Lipids (%) | Protein (%) | Calcium (mg/L) | Iron (mg/L) |
|---|---|---|---|---|---|
| 39.1 | 3.3 | 22.2 | 8.2 | 1910 | 21.1 |

bristles could potentially stimulate the sensitive nerve endings in the bill. An individual at Taronga Zoo was successfully tube fed using a 2.5-mL endotracheal tube due to the difficulty in passing the tube beyond the pharyngeal area into the oesophagus. After the platypus gained strength, live food was given (Beavan 1997). A small rubber bowl has also been used successfully at Healesville Sanctuary. If the feeding response is poor, a paste can be prepared from 1 scoop of Di-Vetelact® and 40 mL of water and placed directly into the mouth using a syringe (Booth and Connolly (2008).

### 11.4.4 Feeding routine

Parenteral fluid therapy should be commenced during periods of inanition, under the supervision of the veterinarian.

The volume of milk that should be offered per feed should not exceed 10 mL, in keeping with this species very small stomach. Taronga Zoo provided six feeding sessions each day, beginning at 0600 h and ending shortly after 1800 h (Beavan 1997). The room in which the animal was kept had reverse cycle lighting. This location was chosen because of the isolation of the room and not because of the light regimen. Each session lasted between 15 min and 120 min. Other authors suggest offering 10–15 mL of milk per feed and fed every 6–8 h with the amount offered varying based on the animal's enthusiasm for feeding and weight gain. The total daily intake per day should be ~10–20% of the bodyweight using 3–6 feeds (Booth and Connolly 2008).

When Taronga Zoo hand-raised a juvenile female platypus, it was taken from its box at the beginning of the feed session and allowed to wake up thoroughly in the animal caretaker's lap (Beavan 1997). It was then placed on a rock in a tray of water so that it could make its own way to the feed tray where it was provided solid food. This usually occurred within 10–15 s. The animal was removed from the water once it started to scratch at the sides of the tray and try to pull herself out. The young animal was lifted from the tray and towel dried, as it did not have adequate waterproofing, and placed back in the box until the next feeding session. All food was either counted or measured in and out of the water.

In preparation for weaning the milk can be mixed with a small amount of Small Carnivore Food (Wombaroo® Food Products) or Insecta-Pro (Vetafarm) in slowly increasing proportions. An alternative is to make a slurry of milk, mealworms (with heads removed) and insectivore mix that is offered in a small dish held in the hand (Booth and Connolly 2008).

Taronga Zoo offered a juvenile platypus brine shrimp, bloodworms, mosquito larvae, mealworms, fly pupae, maggots, crickets, grasshoppers, earthworms, cockroaches and small yabbies (Beavan 1997). The young platypus indicated it was ready for solid foods when it started grabbing at them. Solid foods should be offered by hand and in the water. Young platypuses can consume ~10 mealworms at a time from approximately week 14. These and other insects may be chewed for hours and expulsion of the exoskeletons from the mouth may be facilitated by their removal with fine forceps. The young platypus will steadily increase its intake of insects but will still require milk until 21 weeks of age (Booth and Connolly 2008).

The young platypus should urinate and defecate each day and can either be stimulate with moistened cottonwool or preferably by placing the cloaca and hindlimbs in a shallow container of water (Booth and Connolly 2008).

The insects provided by Taronga Zoo were placed in the water tray, which was 40 × 25 × 15 cm, with a water level of 10 cm. Water temperature was kept above 30°C initially. A resting area (smooth rock, 1/8 of the tray area) was placed in the tray to allow the platypus to climb in and out of the water. Larger food items such as crickets and yabbies were initially offered by hand. Preferences for any particular food sources changed from day to day. Initially, the platypus showed little interest in the yabbies. To overcome this, their legs and exoskeletons were removed, which proved successful (Beavan 1997).

The 8–10-week-old platypus raised at Taronga Zoo initially would not swim or feed in water temperature lower than 31.5–32°C (Beavan 1997). Instead, it would scratch and climb out of the tub, which suggested a lack of insulation.

## 11.5 Data recording

When an animal is first brought in for hand-rearing, its sex and approximate age, using growth charts, should be recorded. During the hand-rearing process several important pieces of information should be recorded on a daily basis:

- date when the information is recorded
- time when the information is recorded
- dry bodyweight to the nearest 1 g if possible
- Tail Fat Index (see section 8 for method). Care should be taken using this method for juveniles as the majority of wild juveniles have a TFI of 3, with very

few having a TFI higher than this, but they survive and are recaptured (T. Grant pers. comm.). Therefore, decisions on transportation, release and rehabilitation of juveniles should not rely too heavily on this crude index (T. Grant pers. comm.)
- general activity and demeanour
- characteristics and frequency of defecation and urination
- amount (g) of different food types offered
- food consumption at each feed
- veterinary examinations and results.

This information serves several purposes, including providing important background information such as food consumption, which will assist a veterinarian to reach a diagnosis if the animal becomes sick or fails to grow or gain weight. It also allows a comparison with growth curves to assess progress (section 10.16) and to establish new growth curves that do not exist for other measurements. Developmental attributes and measurements should be recorded on a weekly basis (see section 10.4.1).

Photographic records should be used to supplement morphometric data in providing information on growth and development. Areas of particular interest are the bill, feet, spurs and pelage. Both still and video technology may be used as deemed appropriate by the veterinarian and a scale included to assist in demonstrating changing dimensions and characteristics.

## 11.6 Identification methods

See section 5.

## 11.7 Hygiene

Maintaining a high standard of hygiene is critical to the survival of the juvenile platypus. Emphasis needs to be placed on the following:

- maintaining a clean nest box; using a new liner for juvenile platypus after each feed
- mainataining personal hygiene – wash and disinfect hands before and after handling juveniles. Use antibacterial solution for washing hands with furless juveniles, as their immune system is not well developed
- using boiled water when making up formulas
- cleaning spilt milk formula, faeces and urine from the animal's skin and fur as soon as possible, and then drying the animal
- after feeding, cleaning the nasal area with a soft tissue and regularly washing it with fresh water to prevent milk entering the nasal passage and lungs
- washing all feeding equipment in warm soapy water and sterilising it in a suitable agent such as Milton® Anti-bacterial Solution, or boiling it for 10 min. After sterilising, rinsing the equipment in cold water
- only heating the milk once and discarding leftovers
- avoiding contact with other animals unless certain that they pose no health risk
- stimulating the animal to toilet before or after feeding.▪

Taronga Zoo's juvenile was found to suffer from the fungus *Trichosporum beigelii* (Beavan 1997). She scratched continuously at her stomach, undertail and underlimbs causing hair on these areas to fall out. An iodine (betadine) solution was then applied to the affected skin daily with cotton wool. An antifungal drug, itraconozol, was also used to treat the fungal problem. Sporanox® (itraconozol) beads were removed from a capsule and injected into a cricket, which was the first to be fed during the 0600 h feeding session and the animal was monitored to ensure that the cricket was totally ingested.

The Taronga Zoo platypus was commonly flatulent and often defecated and urinated at each feed (Beavan 1997). The urine was bright yellow and the faeces had a toothpaste consistency. Hiccupping was also observed.

## 11.8 Behavioural considerations

The rearing of the Taronga Zoo platypus has resulted in a platypus that was not imprinted (Beavan 1997). Previous attempts at artificially rearing juvenile platypus have resulted in significant imprinting.

## 11.9 Use of foster animals

Using foster species is not an option.

## 11.10 Weaning

The estimated 75-day-old platypus hand-reared by Robertson (1989) weaned itself 80 days after it arrived at Fleay's Fauna Park. Complete weaning from milk took place over a 2-week period after ~155 days. Robertson (1989) suggested that it is critical to provide large quantities of food at this time due to the animal's ravenous appetite and being very wasteful feeders. This animal remained quite 'affectionate' and showed no disturbance at any human activity. The Taronga Zoo platypus began weaning after 96 days in captivity (Beavan 1997). At this time, all hand feeding stopped, forcing the platypus to catch the food herself. This decision was made due to the animal's good weight gain.

Wild and captive-bred juveniles normally leaving the nest at the end of lactation are usually between 750 and 1000 g bodyweight.

## 11.11 Rehabilitation considerations

The decision to transfer young juveniles to larger outdoor facilities should be based on a combination of the following criteria:

- loss of interest in milk
- success in foraging on solid food items in water
- ability to swim and dive
- pelage dries efficiently after emergence from water
- restless behaviour in juvenile facility
- consistent positive weight gain
- ability to thermoregulate
- able to find its own shelter.

It is important to ensure that platypus releases back into the wild are handled with due consideration for the welfare of the released animal and minimum impact on wild populations. When releasing platypus back into the wild, several issues should be considered.

- Platypuses qualifying for release should include mis-rescued individuals, animals that have fallen accidentally into human hands or animals not adapting to captivity.
- Only animals judged by an experienced wildlife veterinarian to be healthy at the time of release should be considered for release, if deemed in the animal's best interests.
- Animals should normally be released shortly before dusk at a suitable water body as close as possible to where they were captured or rescued.
- If circumstances prevent the release occurring at the exact point of capture/rescue, then the nearest suitable site should be chosen. This should be determined by facility staff in consultation with the rescuer and local wildlife officers.

## 12. ACKNOWLEDGEMENTS

Sincere thanks also to Sara Brice and Nick De Vos for providing details of the platypus transport and to Tom Grant for editing the manuscript and providing details on spur development.

## ADDENDUM 1: PLATYPUS CRATE DESIGN FOR OVERSEAS TRANSPORT BY NICK DE VOS AND SARA BRICE

The platypus crate developed by Taronga Zoo was designed for lengthy air transportation (e.g. 24 h) to comply with the *Conditions for the Overseas Transfer of Platypus* (Department of Environment and Energy (2019) and the IATA (2022) standards. The design considered the safety and welfare of platypus and aimed to minimise stress from heat, manage potential impacts from high noise levels and provide adequate ventilation.

This description should not be taken as an absolute template for all future transfers as we consider this crate design to be a prototype that can be adapted after more platypus transfers are completed.

*Conditions for the Overseas Transfer of Platypus* require:

- the nest box is either placed inside a transport box or is robust enough to be the travel crate
- the temperature in the box must be maintained at ~18–20°C for the duration of the journey
- a thermometer monitoring the internal crate temperature must be readable from the outside of the crate
- an internal camera (with external monitoring) and sustainable illumination to facilitate viewing of the platypus while inside the crate.

The crates met those conditions as follows.

### Robust nest box

The crate was constructed from three layers of materials with external dimensions 85–90 cm (L) × 61 cm (W) × 55 cm (H) and weighed ~42 kg (Fig. 2.17). The design included a removable panel at one end of the crate that was the same size of the platypus' enclosure tunnel system so that the crate could act as a nesting box pre and post transfer to assist with acclimation to the crate/crate training and conditioning. Ventilation panels (60.0 × 6.0 cm long sides and 30.0 × 6.0 cm on the short side) were incorporated into three sides of the crate, which together amounted to 10% of the crate area. Handles were placed on each side of the crate and acted as spacers when placed in the aircraft hold so nothing could be placed hard up against the crate and impede ventilation.

### Temperature stability

The preferred environmental land temperature for platypus ranges from 10°C to 26°C. The *Conditions for the Overseas Transfer of Platypus* stipulate that the

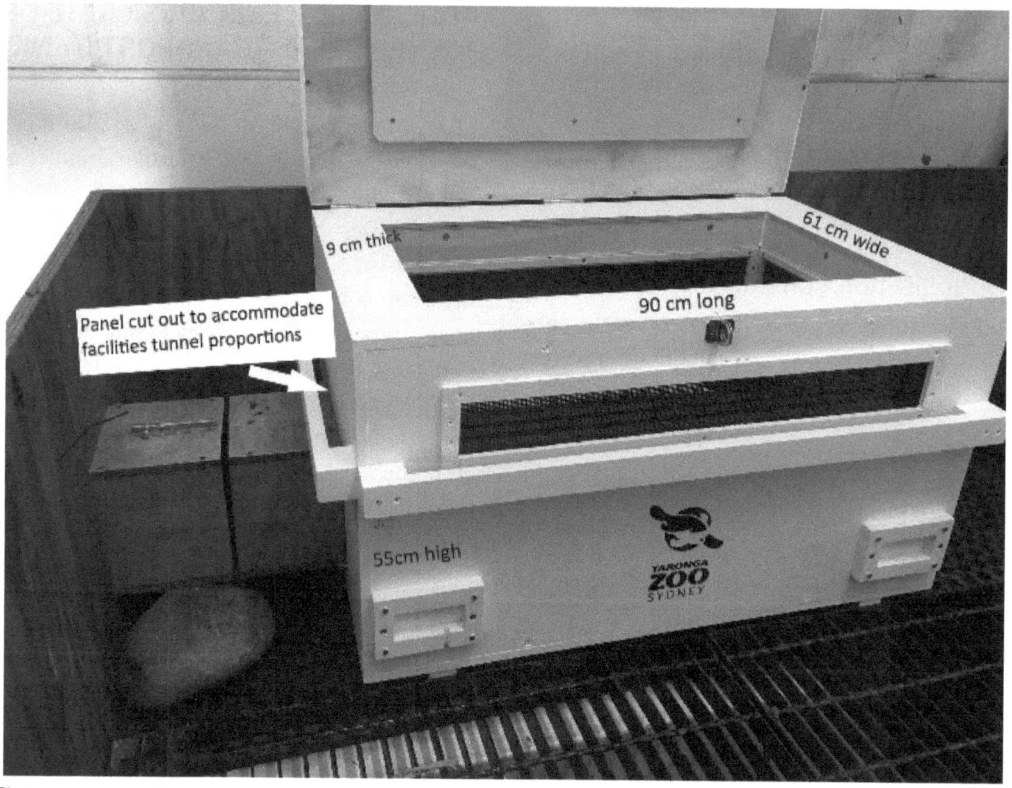


**Fig. 2.17.** Platypus crate designed to act as a nest box as well as a transport crate. Photo by Nick de Vos.

**Fig. 2.18.** Crate during testing stage and showing Infinite-R™ Phase Change Material in the lid. Photo by Sara Brice.

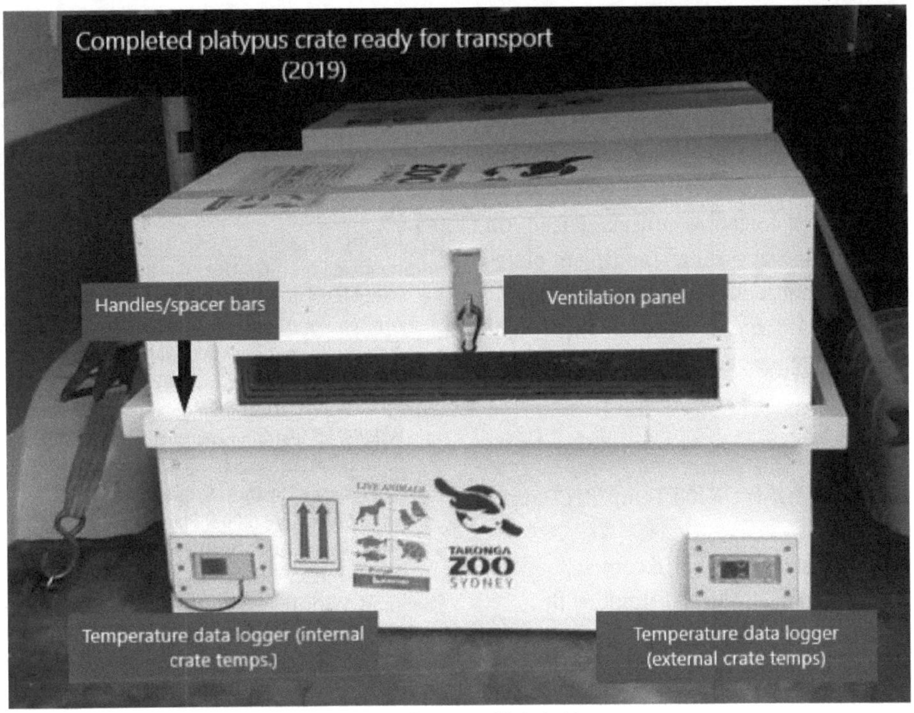


**Fig. 2.19.** Completed crate showing temperature data logger. Photo by Nick de Vos.

temperature within the crate must be maintained between 18°C and 20°C, largely as a precautionary approach due to the data deficiency for platypus transportation. To meet this criterion the design of the crate incorporated materials that enhanced temperature stability as well as providing sound insulation.

The walls of the crate were thick (9 cm wide) and made from three layers with airspace between each layer to allow airflow. The inner and outer walls were constructed from plywood, but the internal layer sandwiched between the plywood was made from FOILBOARD® (polystyrene with foil on one side) to regulate the temperature.

Infinite-R™ Phase Change Material was placed into the lid and bottom of the crate (Fig. 2.18). The material is made of a combination of clay, salt and water and effectively works as a temperature stabiliser by absorbing and releasing thermal energy to maintain a regulated temperature that can be set by the supplier, in this case 19°C.

Button data loggers (HOBO UX100 Temp/RH Data Logger) were installed to monitor the temperature inside and outside the crate (Fig. 2.19).

Inspection cameras (Signet 9mm Inspection Camera with 2.4-inch LCD Monitor) were installed to detect and can monitor animals in tight spaces because the camera has a 1.0-m flexible cable that can be inserted into the crate and a 2.4-inch display console that allowed users to view the images from outside the crate.

## Crate testing

A series of tests were undertaken to measure carbon dioxide levels produced by a platypus while contained in the crate and to determine the effectiveness of the crate's ventilation. This was considered necessary because the crate's side walls were designed to be well insulated from temperature changes and were significantly thicker than those of crates used for most other animal transports. Accumulating carbon dioxide levels would indicate lack of air flow.

As the result of the ventilation testing of the prototype crate, the final version had increased ventilation panels installed in three sides of the crate (through all three layers of the crate) and eight vent holes (2.5 cm diameter) in the base of the crate.

Results from temperature data loggers indicated that both internal and external temperatures dropped lower than 18°C during the flight. Both platypuses arrived at San Diego Zoo Safari Park in excellent condition despite being exposed to lower temperatures than specified by the *Overseas Conditions for Transport of Platypus*. This was not considered an issue because it is generally considered that warmer temperatures are more of a problem for the welfare of platypuses (i.e. temperatures > 25°C) than those at the lower end of the range. The narrow temperature range set by *Conditions for the Overseas Transfer of Platypus* was prescribed as a precautionary measure where concern is directed more towards higher than lower temperatures. Based on these results and contemporary understanding of platypus biology/husbandry, subsequent updates to the *Conditions for the Overseas Transfer of Platypus* might consider expanding the temperature range for travel to below 18°C.

## Suppliers of materials

Button data loggers - HOBO UX100 Temp/RH Data Logger
Web address: www.instrumentchoice.com.au/
Infinite-R™ Phase Change Material supplied by Environmental Technology Solutions.
Web address: https://infiniterpcm.com/
FOILBOARD® supplied by Foilboard Insulation Systems.
Web address: https://www.foilboard.com.au
Signet 9 mm Inspection Camera with 2.4 inch LCD Monitor
Web address: www.jaycar.com.au/

## REFERENCES


Augee M (1976) Heat tolerance in monotremes. *Journal of Thermal Biology* **1**, 181–184. doi:10.1016/0306-4565(76)90011-5

Austin MA (1997) *A Practical Guide to the Successful Handrearing of Tasmanian Marsupials*. Regal Publications, Melbourne.

Beavan M (1997) Hand rearing of a juvenile platypus. *Proceedings of the ASZK/ARAZPA Conference*. 16–20 March. Auckland Zoo, New Zealand.

Beeh P (1995) Piecing together the platypus killer puzzle. *Geo Australasia* **17**(3), 88–98.

Bennett GF (1834a) Notes on the natural history and habits of the *Ornithorhynchus paradoxus*, Blum. *Transactions of the Zoological Society of London* **1**, 229–258. doi:10.1111/j.1096-3642.1835.tb00622.x

Bennett G (1834b) Notes on the natural history and habits of the *Ornithorhynchus paradoxus*. *Proceedings of the Zoological Society of London* **2**, 141–146.

Bethge P, Munks S, Otley H, Nicol S (2004) Platypus burrow temperatures at a subalpine Tasmanian lake. *Proceedings of the Linnean Society of New South Wales* **125**, 272–276.

Bethge P, Munks S, Otley H, Nicol S (2009) Activity patterns and sharing of time and space of platypuses, *Ornithorhynchus anatinus*, in a subalpine Tasmanian lake. *Journal of Mammalogy* **90**, 1350–1356. doi:10.1644/08-MAMM-A-355R.1

Bino G, Grant T, Kingsford R (2015) Life history and dynamics of a platypus (*Ornithorhynchus anatinus*) population: four decades of mark-recapture surveys. *Scientific Reports* **5**, 16073. doi:10.1038/srep16073

Bino G, Kingsford RT, Archer M, Connolly JH, Day J, Dias K, Goldney D, Gondora J, Grant T, Griffiths J, Hawke T, Klamt M, Lunney D, Mijangos L, Munks S, Sherwin W, Serena M, Temple-Smith P, Thomas J, Williams G, Whittington C (2019) The platypus: evolutionary history, biology and an uncertain future. *Journal of Mammalogy* **100**, 308–327. doi:10.1093/jmammal/gyz058

Blumenbach JF (1800) Über das Schnabelthier (Ornithorhynchus paradoxus) ein neuentdecktes Geschlecht von Säugthieren des fünften Welttheils. *Magazin für den Neuesten Zustand Naturkunde mit Rücksicht auf die dazu Gehörigen Hülfswissenschaften* **2**, 205–214.

Booth R (1999) Care and medical management of monotremes. In *Wildlife in Australia: Healthcare & Management*. Post Graduate Foundation in Veterinary Science. (Ed. DI Bryden) pp. 41–50. University of Sydney, Sydney.

Booth R, Connolly J (2008) Platypuses. In *Medicine of Australian Mammals*. (Eds L Vogelnest and R Woods) pp. 103–132. CSIRO Publishing, Melbourne.

Booth RJ, Whittington RJ (1993) Indirect assessment of the health of platypus using haematology. Symposium on comparative biochemistry and physiology of marsupial and monotreme blood. Department of Biochemistry, University of Sydney, Sydney.

Brice PH (2009) Thermoregulation in monotremes: riddles in a mosaic. *Australian Journal of Zoology* **57**, 255–263. doi:10.1071/ZO09039

Burrell H (1927) *The Platypus*. Rigby Ltd, Adelaide.

Calaby JH (1968) The platypus (*Ornithorhynchus anatinus*) and its venomous characteristics. *Venomous Animals and their Venoms. Vol. 1. Vertebrates*. (Eds W Bucherl, EE Buckley and V Deulofeu) pp. 15–29. Academic Press, New York

Carrick FN (1995) Platypus. In *The Mammals of Australia*. (Ed. R Strahan) pp. 36–38. Reed Books, Sydney.

Carrick FN, Grant TR, Williams R (1982) Platypus *Ornithorhynchus anatinus*: its captive maintenance. In *The Management of Australian Mammals in Captivity*. (Ed. DD Evans) pp. 4–12. Zoological Board of Victoria, Melbourne.

Collins GH, Whittington RJ, Canfield PJ (1986) *Theileria ornithorhynchi* Mackerras, 1959 in the platypus, *Ornithorhynchus anatinus* (Shaw). *Journal of Wildlife Diseases* **22**, 19–24. doi:10.7589/0090-3558-22.1.19

Collins LR (1973) *Monotremes and Marsupials: A Reference for Zoological Institutions*. Smithsonian Institution Press, Washington DC.

Connolly JH (2009) A review of mycormycosis in the Platypus (*Ornithorhynchus anatinus*). *Australian Journal of Zoology* **57**, 235–244. doi:10.1071/ZO09043

Connolly JH (2015) Mucormycosis in the platypus and amphibians caused by Mucor amphibiorum. *Microbiology Australia* **36**, 83–87. doi:10.1071/MA15027

Connolly JH, Obendorf DL (1998) Distribution, capture and physical characteristics of the platypus (*Ornithorhynchus anatinus*) in Tasmania. *Australian Mammalogy* **20**, 231–237. doi:10.1071/AM98231

Connolly JH, Obendorf RJ, Muir DB (1998) Causes of morbidity and mortality in platypus (*Ornithorhynchus anatinus*) from Tasmania, with particular reference to *Mucor amphibiorum* infection. *Australian Mammalogy* **20**, 177–187. doi:10.1071/AM98177

Connolly JH, Obendorf DL, Whittington RJ (1999) Haematological, serum biochemical and serological features of platypuses with and without mycotic granulomatous dermatitis. *Australian Veterinary Journal* **77**, 809–813. doi:10.1111/j.1751-0813.1999.tb12952.x

Connolly JH, Obendorf DL, Whittington RJ (2001) *Mucor amphibiorum* infections in the platypus. In A. Martin and L. Vogelnest (2001) Veterinary conservation biology wildlife health and management in Australasia. *Proceedings of International Joint Conference*. Taronga Zoo, Sydney, Australia. 1–6 July 2001, pp. 253–259.

De-La-Warr M, Serena M (1999) Observations of platypus *Ornithorhynchus anatinus* mating behaviour. *Victorian Naturalist* **116**, 172–174.

Department of Environment and Energy (2019) *Conditions for the Overseas Transfer of Platypus*. Australian Government, Canberra.

Eadie R (1934) The platypus in captivity. *Victorian Naturalist* **1**, 3–11.

Eadie R (1935) *The Life and Habits of the Platypus*. Stillwell & Stephens, Melbourne.

Easton L, Williams G, Serena M (2008) Monthly variation in observed activity of the platypus Ornithorhynchus anatinus. *Victorian Naturalist* **125**, 104–109.

Faragher RA, Grant TR, Carrick FN (1979) Food of the platypus, *Ornithorhynchus anatinus*, with notes on the food of the brown trout, *Salmo trutta*, in the Shoalhaven River, New South Wales. *Australian Journal of Ecology* **4**, 171–179. doi:10.1111/j.1442-9993.1979.tb01207.x

Fenner PJ, Williamson JA, Meyers D (1992) Platypus envenomation: a painful experience. *The Medical Journal of Australia* **157**, 829–832. doi:10.5694/j.1326-5377.1992.tb141302.x

Fleay D (1944) *We Breed the Platypus*. Robertson and Mullens, Melbourne.

Fleay D (1980) *Paradoxical Platypus: Hobnobbing with Duckbills*. Jacaranda Press.

Furlan E, Griffiths J, Gust N, Armistead R, Mitrovski P, Handasyde KA, Serena M, Hoffman AA, Weeks AR (2011) Is body size variation in the platypus (*Ornithorhynchus anatinus*) associated with environmental variables? *Australian Journal of Zoology* **59**, 201–215. doi:10.1071/ZO11056

Gardner JL, Serena M (1995) Spatial organisation and movement patterns of adult male platypus, *Ornithorhynchus anatinus* (Monotremata: Ornithorhynchidae). *Australian Journal of Zoology* **43**, 91–103. doi:10.1071/ZO9950091

Gemmell NJ, Grant TR, Western PS, Watson JM, Murray ND, Graves JAM (1992) Preliminary molecular studies of platypus family and population structure. In *Platypus and Echidnas*. (Ed. ML Augee). pp. 277–284. Royal Zoological Society of NSW, Sydney.

Gemmell NJ, Grant TR, Western PS, Walmsley J, Watson JM, Murray ND, Graves JAM (1995) Determining platypus relationships. *Australian Journal of Zoology* **43**, 283–291. doi:10.1071/ZO9950283

Gibson RA, Neumann M, Grant TR, Griffiths M (1988) Fatty acids of the milk and food of the platypus (*Ornithorhynchus anatinus*). *Lipids* **23**, 377–379. doi:10.1007/BF02537355

Gilfedder L, Whinam J, Harris S (1992) An observation of apparent platypus nesting behaviour. *Tasmanian Naturalist* **109**, 4.

Gofton AW, Loh S, Barbosa AD, Paparini A, Gillett A, Macgregor J, Oskam CL, Ryan UM, Irwin PJ (2018) A novel *Ehrlichia* species in blood and *Ixodes ornithorhynchi* ticks from platypuses (*Ornithorhynchus anatinus*) in Queensland and Tasmania, Australia. *Ticks and Tick-Borne Diseases* **9**, 435–442. doi:10.1016/j.ttbdis.2017.12.011

Grant TR (1983a) The behavioural ecology of monotremes. In *Advances in the Study of Mammalian Behaviour*. (Eds JF Eisenberg and DG Kleiman) pp. 360–394. American Society of Mammalogists Special Publication No. 7.

Grant TR (1983b) Body temperatures of free ranging platypuses, *Ornithorhynchus anatinus* (Monotremata) with observations on their use of burrows. *Australian Journal of Zoology* **31**, 117–122. doi:10.1071/ZO9830117

Grant TR (1995) *The Platypus*. New South Wales University Press, Sydney.

Grant TR (2004) Captures, capture mortality, age and sex ratios of platypuses, *Ornithorhynchus anatinus*, during studies over 30 years in the upper Shoalhaven River in New South Wales. *Proceedings of the Linnean Society of New South Wales* **125**, 217–226.

Grant T (2007) *Platypus*. 4th edition. CSIRO Publishing, Melbourne.

Grant TR (2015) Family Ornithorhynchidae (Platypus). In *Handbook of the Mammals of the World. Vol. 5 Monotremes and Marsupials* (Eds DE Wilson and RA Mittermeier). pp. 58–67. Lynx Edicions, Barcelona.

Grant TR, Carrick FN (1974) Capture and marking of the platypus, *Ornithorhynchus anatinus*, in the wild. *Australian Zoologist* **18**, 133–135.

Grant TR, Carrick FN (1978) Some aspects of the ecology of the platypus, *Ornithorhynchus anatinus*. in the upper Shoalhaven River, New South Wales. *Australian Zoologist* **20**, 181–199.

Grant TR, Dawson TJ (1978a) Temperature regulation in the platypus, *Ornithorhynchus anatinus*: maintenance of body temperature in air and water. *Physiological Zoology* **51**, 1–6. doi:10.1086/physzool.51.1.30158659

Grant TR, Dawson TJ (1978b) Temperature regulation in the platypus, *Ornithorhynchus anatinus*: production and loss of metabolic heat in air and water. *Physiological Zoology* **51**, 315–332. doi:10.1086/physzool.51.4.30160956

Grant TR, Griffiths M (1992) Aspects of lactation and determination of sex ratios and longevity in a free-ranging population of platypuses, *Ornithorhynchus anatinus*, in the Shoalhaven River, NSW. In *Platypus and Echidnas*. (Ed. ML Augee) pp. 80–89. Royal Zoological Society of New South Wales, Sydney.

Grant TR, Temple-Smith PD (1983) Size, seasonal weight change and growth in platypuses, *Ornithorhynchus anatinus* (Monotremata: Ornithorhynchidae), from rivers and lakes of New South Wales. *Australian Mammalogy* **6**, 51–60. doi:10.1071/AM83008

Grant TR, Temple-Smith PD (1998a) Field biology of the platypus (*Ornithorhynchus anatinus*): historical and current perspectives. *Philosophical Transactions of the Royal Society of London* **353**, 1081–1091. doi:10.1098/rstb.1998.0267

Grant TR, Temple-Smith PD (1998b) Growth of nestling and juvenile platypuses (*Ornithorhynchus anatinus*). *Australian Mammalogy* **20**, 221–230. doi:10.1071/AM98221

Grant T, Temple-Smith P (2003) Conservation of the platypus, *Ornithorhynchus anatinus*: threats and challenges. *Aquatic Ecosystem Health & Management* **6**, 5–18. doi:10.1080/14634980301481

Grant TR, Whittington RJ (1991) The use of freeze drying-banding and implanted transponder tags as a permanent marking method for Platypus, *Ornithorhynchus anatinus* (Monotremata: Ornithorhynchidae). *Australian Mammalogy* **14**, 147–150. doi:10.1071/AM91020

Grant TR, Griffiths M, Leckie RMC (1983) Aspects of lactation in the platypus, *Ornithorhynchus anatinus* (Monotremata), in waters of eastern New South Wales. *Australian Journal of Zoology* **31**, 881–889. doi:10.1071/ZO9830881

Grant TR, Grigg GC, Beard LA, Augee ML (1992) Movement and burrow use by platypuses, *Ornithorhynchus anatinus*, in the Thredbo River, New South Wales. In *Platypus and Echidnas*. (Ed ML Augee) pp. 263–267. Royal Zoological Society of New South Wales, Sydney.

Grant TR, Griffiths M, Temple-Smith PD (2004) Breeding in a free-ranging population of platypuses, *Ornithorhynchus anatinus*, in the upper Shoalhaven River in NSW: a 27 year study. *Proceedings of the Linnean Society of New South Wales* **125**, 217–236.

Grant T, Serena M, Williams G, Temple-Smith P (2024) Age determination in the platypus (*Ornithorhynchus anatinus*) using spur sheath and spur developmental stages: a review. *Australian Mammalogy* **46**, AM24020. doi:10.1071/AM24020

Griffiths M, Green B, Leckie RMC, Messer M, Newgrain KW (1984) Constituents of platypus and echidna milk, with particular reference to the fatty acid complement of the triglycerides. *Australian Journal of Biological Sciences* **37**, 323–329. doi:10.1071/BI9840323

Grigg G, Beard L, Grant T, Augee M (1992) Body temperature and diurnal activity patterns in the platypus (*Ornithorhynchus anatinus*) during Winter. *Australian Journal of Zoology* **40**, 135–142. doi:10.1071/ZO9920135

Gust N, Griffiths J (2010) *Tasmanian Platypus Management Plan*. Department of Primary Industries, Parks, Water and Environment, Hobart.

Gust N, Handasyde K (1995) Seasonal variation in the ranging behaviour of the platypus (*Ornithorhynchus anatinus*) on the Goulburn River, Victoria. *Australian Journal of Zoology* **43**, 193–208. doi:10.1071/ZO9950193

Handasyde KA, McDonald IR, Evans BK (2003) Plasma glucocorticoid concentrations in free-ranging platypuses (*Ornithorhynchus anatinus*): response to capture and patterns in relation to reproduction. *Comparative Biochemistry and Physiology* **136A**, 895–902. doi:10.1016/S1095-6433(03)00280-0

Hawke T, Bino G, Kingsford RT, Iervasi D, Iervasi K, Taylor MD (2021) Fine-scale movements and interactions of platypuses, and the impact of an environmental flushing flow. *Freshwater Biology* **66**, 177–188. doi:10.1111/fwb.13628

Hawke T, Bino G, Shackleton ME, Ross AK, Kingsford RT (2022) Using DNA metabarcoding as a novel approach for analysis of platypus diet. *Scientific Reports* **12**, 2247. doi:10.1038/s41598-022-06023-y

Hawkins MR (1998) Time and space sharing between platypuses (*Ornithorhynchus anatinus*) in captivity. *Australian Mammalogy* **20**, 195–205. doi:10.1071/AM98195

Hawkins M, Battaglia A (2009) Breeding behaviour of the platypus (*Ornithorhynchus anatinus*) in captivity. *Australian Journal of Zoology* **57**, 283–293. doi:10.1071/ZO09090

Hawkins MR, Fanning D (1992) Courtship and mating behaviour of captive platypuses at Taronga Zoo. In *Platypus and Echidnas*. (Ed. ML Augee) pp. 106–114. Royal Zoological Society of New South Wales, Sydney.

Holland N, Jackson SM (2002) Reproductive behaviour and food consumption associated with the captive breeding of platypus (*Ornithorhynchus anatinus*). *Journal of Zoology* **256**, 279–288[London]. doi:10.1017/S0952836902000328

Hulbert AJ, Grant TR (1983) A seasonal study of body condition and water turnover in a free-living population of platypuses, *Ornithorhynchus anatinus* (Monotremata). *Australian Journal of Zoology* **31**, 109–116. doi:10.1071/ZO9830109

International Air Transport Association (IATA) (2022) *Live Animals Regulations*. 48th edition. International Air Transport Association, Geneva Switzerland.

Jackson SM, Groves CP (2015) *Taxonomy of Australian Mammals*. CSIRO Publishing, Melbourne.

Jamieson J (1818) Poisonous effects of blow from spur of *Ornithorhynchus*. *Transactions of the Linnean Society of London* **12**, 584–585.

Joseph ES (1922) My experience with the platypus in captivity. *Bulletin of the New York Zoological Society* **25**(5), 105–111.

Kessell AE, Boulton JG, Dutton GJ, Woodgate R, Shamsi S, Peters A, Connolly JH (2014) Haemolytic anaemia associated with Theileria sp. in an orphaned platypus. *Australian Veterinary Journal* **92**, 443–449. doi:10.1111/avj.12225

Klamt M, Davis JA, Thompson RM, Marchant R, Grant TR (2016) Trophic relationships of the platypus: insights from stable isotope and cheek pouch dietary analyses. *Marine and Freshwater Research* **67**, 1196–1204. doi:10.1071/MF15004

Krueger B, Hunter S, Serena M (1992) Husbandry, diet and behaviour of platypus *Ornithorhynchus anatinus* at Healesville Sanctuary. *International Zoo Yearbook* **31**, 64–71. doi:10.1111/j.1748-1090.1991.tb02364.x

Lacy RC, Balou JD, Pollak JP (2011) PMx: software package for demographic and genetic analysis and management of pedigreed populations. *Methods in Ecology and Evolution* **3**, 433–437. doi:10.1111/j.2041-210X.2011.00148.x

Ladds P (2009) *Pathology of Australian Native Wildlife*. CSIRO Publishing, Melbourne.

Loewenstein L, McLachlan-Troup T, Hartley M, English A (2008) Serological survey for evidence of *Leptospira interrogans* in free living platypuses (*Ornithorhynchus anatinus*). *Australian Veterinary Journal* **86**, 242–245. doi:10.1111/j.1751-0813.2008.00305.x

Lunn T, Macgregor J, Munks S, Carver S (2016) Dermatophilus congolensis infection in platypus (*Ornithorhynchus anatinus*), Tasmania, Australia. *Journal of Wildlife Diseases* **52**, 965–967. doi:10.7589/2016-02-030R

Macgregor JW (2015) Conservation of the platypus (*Ornithorhynchus anatinus*): Development of a framework to assess the health of wild platypus populations. PhD thesis. Murdoch University, Perth.

Macgregor JW, Holyoake CS, Munks SA, Robertson ID, Warren KS (2010) Preliminary investigation into the prevalence of mucormycosis in the platypus (*Ornithorhynchus anatinus*) in three catchments in north-west Tasmania. *Australian Veterinary Journal* **88**, 190–196. doi:10.1111/j.1751-0813.2010.00568.x

Macgregor JW, Holyoake C, Munks S, Connolly J, Robertson I, Fleming P, Warren K (2014) Novel use of in-stream microchip readers to monitor wild platypuses. *Pacific Conservation Biology* **20**, 376–384. doi:10.1071/PC140376

Macgregor JW, Holyoake C, Munks S, Conolly JH, Robertson ID, Fleming PA, Lonsdale RA, Warren K (2016) Assessing body condition in the platypus (*Ornithorhynchus anatinus*): a comparison of new and old methods. *Australian Journal of Zoology* **64**, 421–429. doi:10.1071/ZO16071

Manger PR, Hall LS, Pettigrew JD (1998) The development of the external features of the platypus (*Ornithorhynchus anatinus*). *Philosophical Transactions of the Royal Society of London. Series B, Biological Sciences* **353**, 1115–1125. doi:10.1098/rstb.1998.0270

Marchant R, Grant TR (2015) The productivity of the macroinvertebrate prey of the platypus in the upper Shoalhaven River, New South Wales. *Marine and Freshwater Research* **66**, 1128–1137. doi:10.1071/MF14301

Martin CJ (1902) Thermal regulation and respiration exchange in monotremes and marsupials: a study in the development of homeothermism. *Philosophical Transactions of the Royal Society of London. Series B, Biological Sciences* **195**, 1–37.

Maule L (1832) On the habits and economy of the *Ornithorhynchus*. *Proceedings of the Committee of Science and Correspondence of the Zoological Society of London* **2**, 145–146.

McColl KA (1983) Pathology in captive platypus (*Ornithorhynchus anatinus*) in Victoria, Australia. *Journal of Wildlife Diseases* **19**, 118–122. doi:10.7589/0090-3558-19.2.118

McColl KA, Whittington RJ (1985) Leptospiral titres in wild platypuses (*Ornithorhynchus anatinus*) in New South Wales. *Australian Veterinary Journal* **62**, 66–67. doi:10.1111/j.1751-0813.1985.tb14242.x

McKee J, Booth RJ, Spittal D, Romer E (1995) Design of a semi-free range platypus breeding and research facility. Unpublished report.

McLachlan-Troup TA, Dickman CR, Grant TR (2010) Diet and dietary selectivity of the platypus in relation to season, sex and macroinvertebrate assemblages. *Journal of Zoology* **280**, 237–246. doi:10.1111/j.1469-7998.2009.00645.x

McLelland D (2025) Appendix 1. Clinical pathology and physiological values. In *Current Therapy in Medicine of Australian Mammals*. Revised Edition. (Eds L Vogelnest and T Portas) pp. 656–719. CSIRO Publishing, Melbourne.

Messer M, Kerry KR (1973) Milk carbohydrates of the echidna and the platypus. *Science* **180**(4082), 201–203. doi:10.1126/science.180.4082.201

Messer M, Gadiel PA, Ralston GB, Griffiths M (1983) Carbohydrates of the milk of the platypus. *Australian Journal of Biological Sciences* **36**, 129–137. doi:10.1071/BI9830129

Munday BL, Peel BF (1983) Severe ulcerative dermatitis in platypus (*Ornithorhynchus anatinus*). *Journal of Wildlife Diseases* **19**, 363–365. doi:10.7589/0090-3558-19.4.363

Munday BL, Whittington RJ, Stewart NJ (1998) Disease conditions and subclinical infections of the platypus (*Ornithorhynchus anatinus*). *Philosophical Transactions of the Royal Society of London. Series B, Biological Sciences* **353**, 1093–1099. doi:10.1098/rstb.1998.0268

Munks SA, Otley HM, Bethge P, Jackson J (2000) Reproduction, diet and daily energy expenditure of the platypus in a sub-alpine Tasmanian Lake. *Australian Mammalogy* **21**, 260–261.

Obendorf DL, Peel BF, Minday BL (1993) Mucor amphibiorum infection in platypus (*Ornithorhynchus anatinus*) from Tasmania. *Journal of Wildlife Diseases* **29**, 485–487. doi:10.7589/0090-3558-29.3.485

Otley HM, Munks SA, Hindell MA (2000) Activity patterns, movements and burrows of platypuses (*Ornithorhynchus anatinus*) in a sub-alpine Tasmanian lake. *Australian Journal of Zoology* **48**, 701–713. doi:10.1071/ZO99010

Paparini A, Macgregor J, Irwin PJ, Warren K, Ryan UM (2014) Novel genotypes of *Trypanosoma binneyi* from wild platypuses (*Ornithorhynchus anatinus*) and identification of a leech as a potential vector. *Experimental Parasitology* **145**, 42–50. doi:10.1016/j.exppara.2014.07.004

Proske U, Gregory E (2003) Electrolocation in the platypus: some speculations. *Comparative Biochemistry and Physiology. Part A, Molecular & Integrative Physiology* **136**, 821–825. doi:10.1016/S1095-6433(03)00160-0

Proske U, Gregory JE, Iggo A (1998) Sensory receptors in monotremes. *Philosophical Transactions of the Royal Society of London. Series B, Biological Sciences* **353**, 1187–1198. doi:10.1098/rstb.1998.0275

Robertson G (1989) Aspects of the hand rearing of a captive platypus. *Proceedings of the ASZK Conference*. 5–8 May. Melbourne Zoo, Melbourne.

Robinson KW (1954) Heat tolerance of Australian monotremes and marsupials. *Australian Journal of Biological Sciences* **7**, 348–360. doi:10.1071/BI9540348

San Diego Zoo (2022) Platypus *Ornithorhynchus anatinus*. <https://animals.sandiegozoo.org/animals/platypus#:~:text=The%20two%20platypuses%20were%20flown,by%20pollution%20and%20climate%20change> (accessed 3 June 2022).

Schalm OW, Jain NC, Carroll ED (1975) *Veterinary Haematology*. Lea & Febiger, Philadelphia.

Scheelings T (2016) Morbidity and mortality of monotremes admitted to the Australian Wildlife Health Centre, Healesville Sanctuary, Australia. *Australian Veterinary Journal* **94**, 121–124. doi:10.1111/avj.12417

Serena M (1994) Use of time and space by the platypus (*Ornithorhynchus anatinus*) along a Victorian stream. *Journal of Zoology* **232**, 117–131[London]. doi:10.1111/j.1469-7998.1994.tb01563.x

Serena M, Williams GA (1993a) The survival of platypuses in captivity: a reappraisal with recommendations for veterinary management and future research. *Australian Veterinary Journal* **70**, 63–65. doi:10.1111/j.1751-0813.1993.tb15143.x

Serena M, Williams GA (1993b) Survival of platypuses in captivity. *Australian Veterinary Journal* **70**, 234–235. doi:10.1111/j.1751-0813.1993.tb03317.x

Serena M, Williams G (2010) Factors contributing to platypus mortality in Victoria. *Victorian Naturalist* **127**, 178–183.

Serena M, Williams GA (2022) Factors affecting the frequency and outcome of platypus entanglement by human rubbish. *Australian Mammalogy* **44**, 81–86. doi:10.1071/AM21004

Serena M, Thomas JL, Williams GA, Officer RCE (1998) Use of stream and river habitats by the platypus, *Ornithorhynchus anatinus*, in an urban fringe environment. *Australian Journal of Zoology* **46**, 267–282. doi:10.1071/ZO98034

Serena M, Williams GA, Weeks AR, Griffiths J (2014) Variation in platypus (*Ornithorhynchus anatinus*) life-history attributes and population trajectories in urban streams. *Australian Journal of Zoology* **62**, 223–234. doi:10.1071/ZO13079

Serena M, Snowball G, Thomas JL, Williams GA, Danger A (2024) Platypus longevity: a new record in the wild and information on captive life span. *Australian Mammalogy* **46**, AM23048.

Shaw G (1799) *The Naturalist's Miscellany: or coloured figures of natural objects, drawn and described immediately from nature*. Volume 10. F.P. Nodder & Co., London.

Shaw M, Jarman A (2025) Nutrition. In *Current Therapy in Medicine of Australian Mammals*. Revised Edition. (Eds L Vogelnest and T Portas) pp. 229–255. CSIRO Publishing, Melbourne.

Smyth DM (1973) Temperature regulation in platypus, *Ornithorhynchus anatinus* (Shaw). *Comparative Biochemistry and Physiology* **45**, 705–715. doi:10.1016/0300-9629(73)90074-1
Species360 (2023) Zoological Information Management System (ZIMS). Minneapolis, USA. <www.species360.org>
Spicer WW (1877) On the effects of wounds on the human subject inflicted by the spurs of the platypus. *Papers and Proceedings of the Royal Society of Tasmania* **1876**, 162–167.
Stewart NJ, Munday BL, Hawkesford T (1999) Isolation of *Mucor circinelloides* from a case of ulcerative mycosis of platypus (*Ornithorhynchus anatinus*), and a comparison of the response of *Mucor circinelloides* and *Mucor amphibiorum* to different culture temperatures: isolation of *Mucor circinelloides* from a case of ulcerative mycosis of platypus. *Medical Mycology* **37**, 201–206. doi:10.1111/j.1365-280X.1999.00221.x
Strahan R, Thomas DE (1975) Courtship of the platypus, *Ornithorhynchus anatinus*. *Australian Zoologist* **18**, 165–178.
Sutherland SK (1983) *Australian Animal Toxins: The Creatures, Their Toxins and Care of the Poisoned Patient*. Oxford University Press, Melbourne.
Temple-Smith PD (1973) Seasonal breeding biology of the platypus, *Ornithorhynchus anatinus* (Shaw 1799), with special reference to the male. PhD Thesis. Australian National University, Canberra.
Temple-Smith P, Grant T (2001) Uncertain breeding: a short history of reproduction in monotremes. *Reproduction, Fertility and Development* **13**, 487–497. doi:10.1071/RD01110
Thomas JL, Handasyde KA, Temple-Smith P, Parrott ML (2017) Seasonal changes in food selection and nutrition of captive platypuses (*Ornithorhynchus anatinus*). *Australian Journal of Zoology* **65**, 319–327. doi:10.1071/ZO18004
Thomas JL, Handasyde K, Parrott ML, Temple-Smith P (2018a) The platypus nest: burrow structure and nesting behaviour in captivity. *Australian Journal of Zoology* **65**, 347–356. doi:10.1071/ZO18007
Thomas JL, Parrott ML, Handasyde KA, Temple-Smith P (2018b) Female control of reproductive behaviour in the platypus (*Ornithorhynchus anatinus*), with notes on female competition for mating. *Behaviour* **155**, 27–53. doi:10.1163/1568539X-00003476
Thomas JL, Parrott ML, Handasyde KA, Temple-Smith P (2019) Burrow use by juvenile platypuses (*Ornithorhynchus anatinus*) in their natal home range. *Journal of Mammalogy* **100**, 1182–1190. doi:10.1093/jmammal/gyz078
Thomas JL, Parrott ML, Handasyde KA, Temple-Smith P (2020) Maternal care of platypus nestlings (*Ornithorhynchus anatinus*). *Australian Mammalogy* **42**, 283–292. doi:10.1071/AM19019
Tidswell F (1906) *Researches on Australian Venoms – Snake-bite, Snake-Venom and Antivenine, The Poison of the Platypus, and The Poison of the Red-Spotted Spider*. Department of Public Health, Sydney.
Tonkin MA, Negrine J (1994) Wild platypus attack in the antipodes: a case report. *The Journal of Hand Surgery* **19**, 162–164. doi:10.1016/0266-7681(94)90155-4
Verreaux J (1848) Observations sur l'Ornithorhynque. *Revue Zoologique* **11**, 127–134.
Vogelnest L (2025a) Capture, physical and chemical restraint. In *Current Therapy in Medicine of Australian Mammals*. Revised Edition. (Eds L Vogelnest and T Portas) pp. 122–151. CSIRO Publishing, Melbourne.
Vogelnest L (2025b) Appendix 3. Suggested chemical restraint agents, regimens and doses. In *Current Therapy in Medicine of Australian Mammals*. Revised Edition. (Eds L Vogelnest and T Portas) pp. 725–731. CSIRO Publishing, Melbourne.
Vogelnest L, Portas T (Eds) (2025) *Current Therapy in Medicine of Australian Mammals*. Revised Edition. CSIRO Publishing, Melbourne.
Vogelnest L, Whinfield J (2025) Platypus. In *Current Therapy in Medicine of Australian Mammals*. Revised Edition. (Eds L Vogelnest and T Portas) pp. 423–434. CSIRO Publishing, Melbourne.
Vogelnest L, Woods R (Eds) (2008) *Medicine of Australian Mammals*. CSIRO Publishing, Melbourne.
Whittington RJ (1988) The monotremes in health and disease. In *Australian Wildlife*. (Ed. DI Bryden) pp. 727–787. Proceedings 104. Post Graduate Committee in Veterinary Science, University of Sydney, Sydney.
Whittington RJ (1991) The survival of platypuses in captivity. *Australian Veterinary Journal* **68**, 32–35. doi:10.1111/j.1751-0813.1991.tb09842.x
Whittington RJ (1992) The role of infectious diseases in the population biology of monotremes. In *Platypus and Echidnas* (Ed. ML Augee) pp. 285–292. The Royal Zoological Society of New south Wales, Sydney.
Whittington RJ (1993) Survival of platypuses in captivity. *Australian Veterinary Journal* **70**, 234. doi:10.1111/j.1751-0813.1993.tb03316.x
Whittington RJ, McColl KA (1983) Aspiration pneumonia in a wild platypus Ornithorhynchus anatinus. *Australian Veterinary Journal* **60**, 277. doi:10.1111/j.1751-0813.1983.tb07107.x
Whittington RJ, Spratt DM (1989) Lesions associated with metazoan parasites of wild platypuses (*Ornithorhynchus anatinus*). *Journal of Wildlife Diseases* **25**, 521–526. doi:10.7589/0090-3558-25.4.521
Whittington R, Middleton D, Spratt DM, Muntz F, Carmel B, McCracken HE, Strackosch MR, Stephenson-Shaw J, Harper PAW, Hartley WJ (1992) Sparganosis in the monotremes *Tachyglossus aculeatus* and *Ornithorhynchus anatinus* in Australia. *Journal of Wildlife Diseases* **28**, 636–640. doi:10.7589/0090-3558-28.4.636
Whittington RJ, Connolly J, Obendorf D, Emmins J, Grant TR, Handasyde K (2001) Serological responses against the pathogenic dimorphic fungus Mucor amphibiorum in populations of platypus (*Ornithorhynchus anatinus*) with and without ulcerative dermatitis. In *Veterinary Conservation Biology Wildlife Health and Management in Australasia. Proceedings of International joint Conference* (Eds A Martin and L Vogelnest). pp. 261. Taronga Zoo, Sydney Australia. 1–6 July 2001.
Williams G (1992) Management of platypus in Australian zoos, 1987–1991. *Thylacinus* **17**(3), 6–8.
Williams GA, Serena M, Grant TR (2013) Age-related change in spurs and spur sheaths of the platypus (*Ornithorhynchus anatinus*). *Australian Mammalogy* **35**, 107–114. doi:10.1071/AM12011
Woinarski JCZ, Burbidge AA, Harrison P (Eds) (2014) *The Action Plan for Australian Mammals 2012*. CSIRO Publishing, Melbourne.

# 3
# Short-beaked echidna

*Stephen M. Jackson, Stewart Nicol, Alice Roser, Lauren Clark, Stephen Johnston, Arthur Ferguson, Belinda Laming and Peggy Rismiller*

## 1. INTRODUCTION

The short-beaked echidna (*Tachyglossus aculeatus*) is a highly unusual species because of its long spines that are reminiscent of hedgehog or porcupine quills, its reproduction (which involves laying eggs similar to platypuses) and its diet, which consists of invertebrates, primarily termites, ants and the larvae of beetles. Its closest relatives are the long-beaked echidnas (*Zaglossus* sp.) of New Guinea (Flannery and Groves 1998; Nicol 2015) that eat earthworms and a variety of arthropods in the soil and rotting logs (Nicol 2021). Short-beaked echidnas attract great interest in the zoological world and the general public because of their unique combination of mammalian and reptilian characteristics.

Short-beaked echidnas have had a long history in captivity. London Zoo held an echidna as early as 1845 (Owen 1845). It was offered bread, milk and some mealworms, but only consumed the milk using its long tongue, unfortunately, dying unexpectedly only 4 days after its arrival in London (Flower 1929). Subsequently, animals were kept by Philadelphia Zoo in 1903, with one living for an amazing 49 years and 5 months (Crandall 1964). A male short-beaked echidna transferred from Australia to San Diego Zoo in 1956 survived for a further 56 years (Rismiller and Grutzner 2019; SDZWA 2021). Amsterdam Zoo kept animals in 1903 (Flower 1931), New York Zoo in 1913 and St Louis Zoo received an animal in 1938 (Crandall 1964; Fieseler and Junge 1997), with most of these surviving to 12–20 years of age. Short-beaked echidnas are relatively common in zoological institutions across Australia and have been displayed in several zoos throughout the world (Rismiller 1999).

## 2. TAXONOMY

### 2.1 Nomenclature

It was long believed that there were only two species of echidna: the short-beaked echidna and the long-beaked echidna from New Guinea (*Zaglossus bruijnii*) (Table 3.1). The most recent revision of the long-beaked echidnas suggested there are two additional species: the eastern long-beaked echidna (*Z. bartoni* with four subspecies including the nominate subspecies) and Attenborough's long-beaked echidna (*Z. attenboroughi*) (Flannery and Groves 1998). The discovery of a museum specimen of *Z. bruijnii* labelled as having been collected in the West Kimberley in Western Australia suggested that this species may have also occurred in Australia until very recently (Helgen *et al.* 2012), although Burbidge (2018) has disputed this claim. As no living animals have been recorded in Australia, this species is not considered within this chapter.

Class: Mammalia
Order: Monotremata
Family: Tachyglossidae
Genus and species: *Tachyglossus aculeatus*

**Etymology**

*Tachyglossus* – swift tongue.
*Aculeatus* – furnished with spines.

### 2.2 Subspecies

The short-beaked echidna has five subspecies (Jackson and Groves 2015):

*T. a. aculeatus* – coastal regions and Great Dividing Range of Queensland, New South Wales, Victoria and South Australia

*T. a. acanthion* – Western Australia, Northern Territory, most of arid and semi-arid Australia
*T. a. lawesii* – North Queensland and south-eastern New Guinea
*T. a. multiaculeatus* – Kangaroo Island
*T. a. setosus* – Tasmania and Bass Strait Islands.

Based on functional cranial morphology, particularly the relative length of the beak, the Australian short-beaked echidnas can be divided into two main groups: those from south-east Eastern Australia and Tasmania (*T. a. aculeatus* and *T. a. setosus*) and those from arid and semi-arid Australia and Kangaroo Island (*T. a. acanthion* and *T. a. multiaculeatus*) (Nicol 2021). Differences in beak length are associated with differences in diet and the associated feeding behaviour: the main dietary items of *T. a. acanthion/multiaculeatus* are ants and termites, while for *T. a. aculeatus/setosus*, the main dietary items are ants and scarab larvae (Nicol 2021). There are also differences between the two groups of Australian short-beaked echidnas in thermoregulation and metabolism (Nicol 2017), reproductive behaviour and maternal care (Morrow and Nicol 2012; Rismiller and Grutzner 2019), and so, when appropriate, details are given separately for the two clades in this chapter.

### 2.3 Recent synonyms

Synonyms of the short-beaked echidna can be found in Jackson and Groves (2015).

### 2.4 Other common names

The short-beaked echidna has been called the spiny ant-eater, and incorrectly, porcupine (Augee 1995).

## 3. NATURAL HISTORY

### 3.1 Bodyweight and morphometrics

Short-beaked echidnas are easily recognisable by their spines and pointed snouts. They typically weigh 2–7 kg and have head and body length of 30–45 cm (Augee 1995). Sexually mature male and female short-beaked echidnas show seasonal cycles in bodyweight related to reproduction and hibernation (Rismiller and McKelvey 2000; Nicol *et al.* 2019a).

### 3.2 Distribution and habitat

Short-beaked echidnas are found throughout mainland Australia, Tasmania and major offshore islands, apart from Melville Island, and also in New Guinea. Within this extensive range they occur in all habitat types from deserts to grasslands, forests and alpine areas (Augee 1995; Flannery 1995; Rismiller and Grutzner 2019; Nicol 2021).

### 3.3 Conservation status

Throughout its distribution the short-beaked echidna is relatively common and considered Least Concern of extinction. However, the subspecies on Kangaroo Island is considered Endangered due to predation by cats and pigs, fires, habitat fragmentation and road deaths (Woinarski *et al.* 2014).

### 3.4 Diet in the wild

Short-beaked echidnas eat a variety of small invertebrates, including ants, termites, small beetles, crickets, and many types of soil-living moth and beetle larvae (Griffiths 1968; Abensperg-Traun 1988; Rismiller 1999; Spencer and Richards 2009). Australia-wide, their overall diet includes ants, but in arid and semi-arid areas termites may be an equally or more important dietary item (Griffiths 1978, 1989). In temperate areas where termites are much less common or virtually absent, scarab larvae form a major part of the diet (Nicol 2021).

### 3.5 Longevity

#### 3.5.1 Wild

There are very few records of the longevity of short-beaked echidnas in the wild; however, one wild animal, with only three legs, lived for over 45 years and there is another record of 48 years (Rismiller 1999; Augee *et al.* 2006; Hulbert *et al.* 2008). In Tasmania, male survival was found to be longer than that of females: at 10 years, survival was 73% for females and 90% for males; at 17 years, survival was 42% and 64% respectively (Nicol *et al.* 2019a).

#### 3.5.2 Captivity

Short-beaked echidnas are relatively long-lived species that typically live 14–30 years in captivity; however, a specimen at the Philadelphia Zoo lived for 49 years and 6 months (Crandall 1964; Collins 1973). Another animal that arrived at the San Diego Zoo in 1956 lived for more than 56 years (Rismiller and Grutzner 2019).

#### 3.5.3 Methods used to determine the age of adults

It is very difficult to determine the age of adult short-beaked echidnas because growth peaks at ~2 years of age. However, growth curves in bodyweight data and regression analysis from short-beaked echidnas may provide keepers with an approximate estimate of age until they are adult (see section 10.16). Juvenile echidnas of both sexes have spurs (Griffiths 1989) covered by a spur sheath; however, Rismiller and McKelvey (2000) observed that the spur sheaths were completely lost in animals between 14 and up to 48 months of age, providing a tool for ageing young animals. In the Tasmanian subspecies all females had lost their spurs between 1 and 2.6 years of age, while males lost their juvenile spurs between 2.1 and 3.2 years of age (Nicol *et al.* 2019a).

## 4. HOUSING REQUIREMENTS

### 4.1 Enclosure design

Enclosures for short-beaked echidnas must be well built, because they can use their very strong forelimbs to pull apart wire and other materials to escape from their enclosure. For example, an individual at Taronga Zoo pulled a wooden sliding door off its hinges and another pulled apart wire at the front of its enclosure to escape (S. Jackson pers. obs.). This species has also been observed to force out wire netting that has been well attached and to move large stones weighing 14 kg (Bennett 1881). Short-beaked echidnas are also remarkably good climbers; however, a solid, smooth wall that is at least 1.2 m high will readily contain them. If chain link fencing is used then a smooth metal barrier at least 60 cm high is needed to prevent climbing, as they have been observed to climb up chain link fencing over 2 m high and if they fall while climbing, they are prone to injury (Augee and Gooden 1997).

The enclosure should be provided with a good vegetation cover (such as shrubs and grasses), logs and branches on the ground and lots of areas for the animals to dig downwards to submerge themselves. Concrete or wire floors should only be used if they are covered with a substrate such as leaf litter, sand, soil and/or mulch to a depth of at least 15–50 cm because concrete or wire can cause injuries to their feet (Figs 3.1, 3.2).

An alternative to a concrete or wire substrate is to extend the walls at least 1 m below the surface to prevent the short-beaked echidnas from digging underneath and escaping. Enclosures that are intended to house a female for breeding are thought more likely to encourage better use of provided nursery burrow boxes for the deposit of puggles if the substrate depth is no deeper than 15 cm (A. Ferguson pers. obs.). Extra care must be taken when moving through enclosures in which substrate depth enables the animal to be concealed, to prevent animal injury and/or burrow collapse under foot.

Short-beaked echidna enclosures should have a variety of different substrates and furniture that replicate the species' wild habitat. Enclosure provisions should provide opportunities for the animals to engage in behaviours that complement their preferences, including sufficient space to maintain fitness through walking, foraging and digging and access to hide/shelter options they can wedge into or under to gain a sense of security. A water body such as a stream or pond can also be incorporated into the enclosure, but not used as barrier, as the short-beaked echidna can easily move through water (e.g. Rismiller 1999).

The area provided for the short-beaked echidnas to dig needs to be well drained so that when they dig down into the soil during rainy weather it is not waterlogged. A dry substrate is also important because persistent moisture favours survival and build-up of coccidia oocysts. A minimum of at least one suitably designed and installed

**Fig. 3.1.** Example of a short-beaked echidna exhibit at Perth Zoo. Photos by Arthur Ferguson.

**Fig. 3.2.** Example of a short-beaked echidna enclosure. Note the vegetation, hollow logs, heat lamp, nest box (with the lid lifted) and sand and leaf mulch for the echidnas to dig within. Photo by Alice Roser.

dry burrow option should be available for short-beaked echidnas to access as a hide/shelter. The design of the burrow should also consider ease of animal caretaker access to the animal and for routine cleaning.

Short-beaked echidnas are readily held outdoors, even in cold climates, as long as they have adequate shelter and substrate in which they can dig. They can also be held within nocturnal houses with reverse lighting. If this option is chosen, appropriate diurnal and nocturnal lighting should be provided and customised to replicate seasonal photoperiod. Opportunities for basking under infrared heat lamps should be provided.

As short-beaked echidnas are often sensitive to noise, it is generally advisable to provide soundproofing to minimise visitor noise (Collins 1973). This is most relevant for animals that have recently come into captivity from the wild. Initially, these animals are likely to demonstrate a stress response to novel audio stimuli, but over time will become accustomed to the general soundscape in their area. Captive-bred animals are generally more comfortable with the soundscape they have grown up with as opposed to a wild animal coming into a new/foreign soundscape (A. Ferguson pers. obs.). Where short-beaked echidnas are exhibited in a nocturnal house setting, measures to limit noise, including from visitors, should be incorporated.

An example of a dedicated short-beaked echidna research facility has been constructed at Currumbin Wildlife Sanctuary, Queensland. This off-exhibit facility consists of 10 enclosures, 5 on each side of an aisle (Wallage *et al.* 2015). All enclosures are uniform in size and shape (3.8 (L) × 4.8 m (W)), with a 0.9-m high brick wall surrounding the enclosures and nylon netting extending above the wall to prevent entry by scavenging birds. A partial metal roof runs down the centre of the facility covering the aisle and one-quarter of each enclosure; one half of the remaining roof consists of clear alsynite sheeting leaving the final one-quarter composed of 2-cm wire mesh that allows exposure to the weather. All enclosures are filled with a sand-based substrate to aid in drainage and this is maintained at a depth of 20 cm.

## 4.2 Holding area design

Holding area designs can be relatively simple, but should include the various attributes mentioned above. At Currumbin Wildlife Sanctuary, a minimum 450 L plastic storage tub is typically used to hold short-beaked echidnas requiring medical attention or for overnight holding. Hospital tubs need to be round. Other substrate options for hospitalised echidnas include clean leaf litter, shredded newspaper in large strips or blankets and towels.

## 4.3 Spatial requirements

The minimum floor area for up to two animals is 16 $m^2$, with an additional floor area of at least 4.0 $m^2$ for each extra animal. Abnormal repetitive pacing is regularly reported for animals in smaller enclosures, especially those with basic furnishing. Short-beaked echidnas are intelligent and inquisitive (Russell and Burke 2016), with a strong innate drive for exploratory and locomotory behaviours. The opportunity for individuals to engage in natural behaviours can be better achieved with spatial provision greater than the prescribed minimum standard, together with the provision of environmental complexity as described throughout this chapter.

If breeding is intended, multiple enclosures are required: one enclosure dedicated to the female that facilitates egg incubation and rearing young and a second enclosure for males when not housed with the female for mating. A third enclosure is also preferred to support the separation (dispersal) and independence of weaned young (A. Ferguson pers. obs.).

## 4.4 Position of enclosures

The enclosure should be situated to allow short-beaked echidnas to seek shade during the warmer months of the year (in summer) and also to bask in the sun (especially during the cooler months of the year). On sunny days, especially in cooler weather, they may sun bask by lying on their stomach with their legs spread out. Therefore, the opportunity to access sun for basking in cooler weather is likely beneficial for their wellbeing and should be considered in the positioning of the enclosures.

## 4.5 Weather protection

Field studies suggest that hollow logs and depressions under the roots of fallen trees are commonly used by wild echidnas as daily shelter sites (Wilkinson *et al.* 1998). Hollow logs appear to be less effective at providing an appropriate microhabitat compared with burrows because the logs are often open at both ends. As a result, rabbit burrows have been observed to be commonly used as hibernacula because they offer much better thermal buffering than hollow logs (Wilkinson *et al.* 1998).

Short-beaked echidnas may choose security over protection from rainy weather, such as burrowing into substrate in an exposed position or hiding in or under an exposed log. It is therefore important to provide short-beaked echidnas with options so they can maintain their sense of security while also providing good protection from rain. This can be achieved by positioning preferred hides/boxes/logs under a solid roof section or providing burrow boxes that remain dry during rainy weather. Additional cover options in the form of grasses, shrubs and logs should also be incorporated.

The artificial burrows provided for captive short-beaked echidnas at Perth Zoo have been used by females as year-round refuges (Fig. 3.3). These were placed under shelter and filled with a 10–15 cm layer of substrate such as dry sandy loam (Ferguson and Turner 2013). Similar artificial burrows fitted with video camera and temperature sensors have been used successfully at Currumbin Wildlife Sanctuary, along with the provision of multiple half (split longitudinally) hollow logs (Wallage *et al.* 2015). On extremely hot days in south-east Queensland, the lid of the artificial burrow may be covered with wet towels or ice bricks to reduce the internal burrow temperature for the nesting puggle when it exceeds 25°C. If the enclosure becomes hot for adults, then water misting with adequate ventilation, or fans if needed, can be used to cool the area.

As short-beaked echidnas do not appear to tolerate high temperatures well (see discussion in section 4.6) it is important to provide adequate shading such as caves, burrows and logs, and deep substrate so that they can behaviourally thermoregulate and seek refuge during the heat of the day when required. A water body that is large enough to allow each animal to submerge and/or sprinklers should also be provided during hot weather (Brice *et al.* 2002a, 2002b; Pierce *et al.* 2006).

## 4.6 Temperature requirements

Short-beaked echidnas do not typically require heating if they are displayed outdoors, as they prefer colder rather than hotter temperatures. However, when displayed indoors (and even outdoors in cool weather) they should be provided with a heat lamp (e.g. 230 V Red or 240 V Red) placed ~550–700 mm off the substrate to allow them to bask (Ferguson and Turner 2013; Wallage *et al.* 2015; S. Jackson pers. obs.). For example, Wallage *et al.* (2015) provided short-beaked echidnas with the use of artificial burrow boxes and infrared heat lamps in their outdoor enclosures. The lamps were turned on at 15:00 h

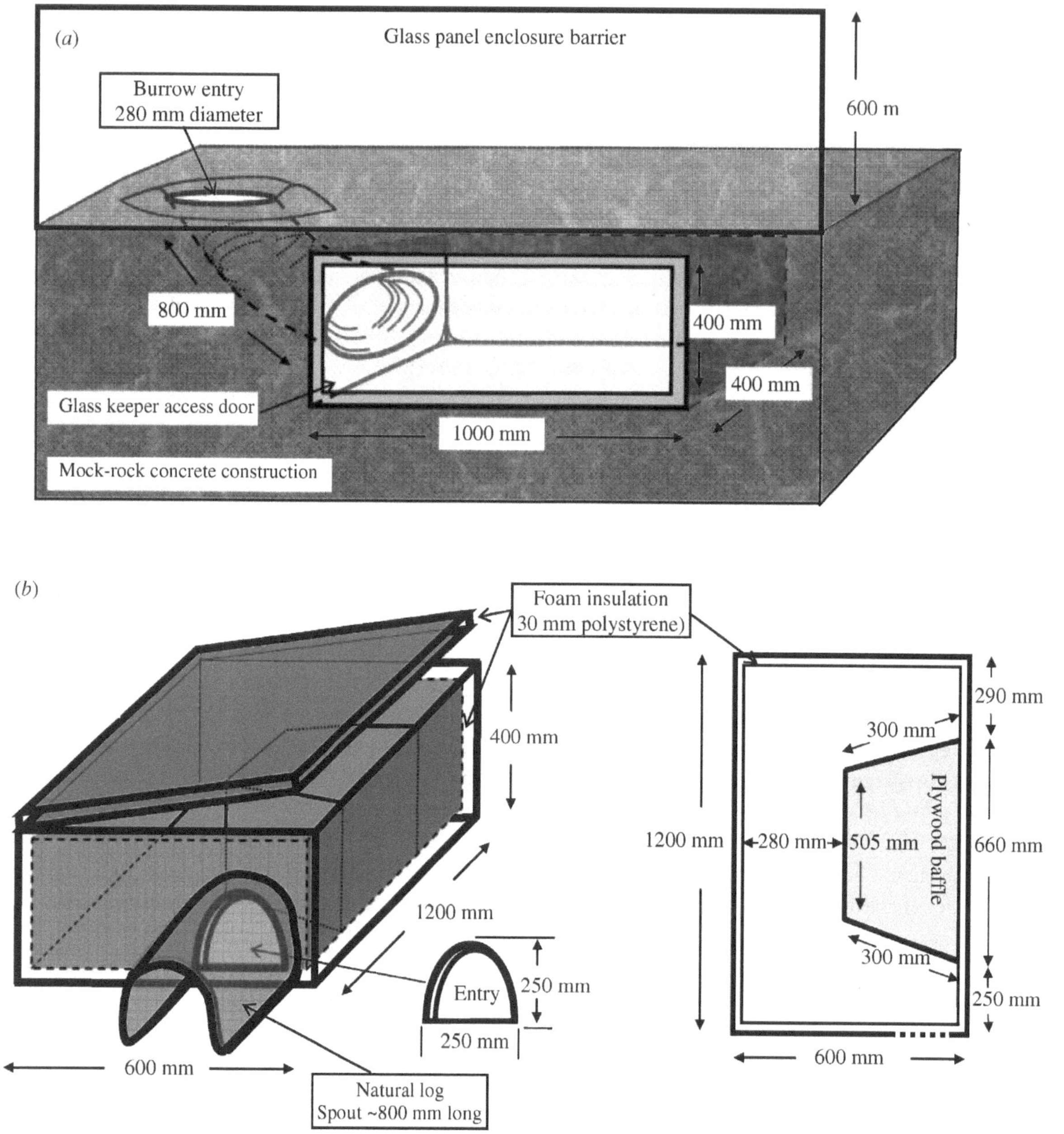


**Fig. 3.3.** Artificial burrows that have been provided to short-beaked echidnas to hide/shelter and egg incubation. Reproduced from Ferguson and Turner (2013) with permission from CSIRO Publishing.

and off the following morning at 07:00 h to provide additional heat during the night if needed during cooler weather. These additions coincided with 13 young being born to four females, although no direct causal effect was attributed to these changes (Wallage *et al.* 2015).

Short-beaked echidnas can be active over a range of temperatures, but appear to have a thermal preference of 20–30°C (Holz 2015). Although they have some capacity for increasing evaporative water loss during warm weather (Barker *et al.* 2016), they do not tolerate well ambient temperatures above 35°C for long periods and are known to die when exposed to temperatures approaching 40°C (Martin 1903; Robinson 1954; Schmidt-Nielsen *et al.* 1966; Augee 1976; Whittington 1988; Brice *et al.* 2002a). However, other observations of wild short-beaked echidnas have shown individuals to tolerate temperatures of 35–40°C in a shelter during bushfires for up to 10 h (Brice *et al.* 2002a) and routinely tolerate ambient temperatures exceeding 35–42°C for several hours in some of their daytime retreats (Brice 2007).

Short-beaked echidna enclosures should be monitored closely during summer heatwaves to ensure the animals remain comfortable and do not overheat. At Perth Zoo, nursery burrow temperature parameters were obtained during nursery burrow life of five young and showed they had an average temperature of 20–25°C between October and February (Ferguson and Laming 2022).

## 4.7 Substrate

Various substrates have been used, including earth, peat, gravel and leaves, sand, weldmesh over earth, wood and concrete (Collins 1973; S. Jackson pers. obs.). Soil or leaf litter, which is most often used, should be at least 20 cm deep to allow individuals to bury themselves and feel secure. The use of sand substrates of different particle sizes in the enclosures of short-beaked echidnas has been examined, noting the importance of adequate drainage to reduce the incidence of disease (e.g. coccidiosis).

## 4.8 Nest boxes

Artificial burrows or burrow boxes have been used by the facilities with successful breeding of short-beaked echidnas (Ferguson and Turner 2013; Wallage *et al.* 2015). However, if burrow boxes are not designed or installed correctly, the animals sometimes dig under them to rest rather than go inside the box, which may pose a risk of injury to the animal. When artificial burrows are provided, they should be positioned directly against the lower substrate barrier mesh or concrete so there is no room for a short-beaked echidna to burrow under. If they are designed and set up correctly, these boxes may be well used as they appear to mimic refuge opportunities in the wild.

Short-beaked echidnas in the wild are known to rest in hollow logs, at the base of hollow trees and under rocky outcrops, dig under dense piles of dead branches, leaf litter, grass tussocks and depressions under the roots of fallen trees and burrow into the side of termite mounds (Augee and Ealey 1968; Augee *et al.* 1975; Abensperg-Traun 1988, 1991; Wilkinson *et al.* 1998).

Captive short-beaked echidnas have used dedicated nursery burrow boxes to deposit their young, with females back-filling the entrance of the burrow chamber with soil, to maintain security for the young while she is away feeding and replenishing her milk supply (Wallage *et al.* 2015). These burrow boxes can also be fitted with thermometers to monitor the temperature within and CCTV cameras can be focused on the entrance or positioned inside the box to monitor the behaviour of the female and young. The lid of the box can be lifted by animal caretakers for easy routine access to monitor the burrow environment, as well as the growth and development of the young without the need to disturb the burrow entrance substrate back-fill created by the female.

## 4.9 Enclosure furnishings

To promote and encourage typical behaviours, the enclosure should contain obstacles that allow the animals to utilise different senses. Freshly and regularly replaced branches for smelling, leaf litter or soil to dig in and hollow logs to take refuge should be provided. Termite mounds can also be included to provide enrichment for the animals as they dig into them and pull them apart. It is important to ensure that nest boxes and enclosure furnishings are at a suitable distance from the perimeter fence to prevent escape and that placement of furnishings considers the risk of crush injury should an animal excavate the surrounding substrate.

# 5. GENERAL HUSBANDRY

## 5.1 Hygiene and cleaning

Uneaten food should be removed daily and the substrate under areas where the short-beaked echidnas typically defecate ('latrine') should be a focus for daily cleaning. Water bowls should also be cleaned and refilled daily. Full substrate changes of enclosures with concrete bases are ideally completed every 6–12 months depending on the size of the enclosure. Ponds or water bodies should be cleaned weekly, or as required, to maintain clean water. Enclosures with a concrete base should be scrubbed out and fresh topsoil provided before short-beaked echidnas are introduced.

## 5.2 Record keeping

It is important to establish a system whereby the health, condition and reproductive status of the captive short-beaked echidna population are routinely monitored and recorded:

- identification numbers (all individuals should be identifiable)
- results of veterinary examinations and treatments

- behavioural changes or problems
- reproductive behaviour or condition
- bodyweight and measurements
- changes in diet and food consumption
- movements of individuals between enclosures or institutions
- births, with dam and sire if known
- deaths with necropsy results.

Collecting information on the physical, physiological and behavioural patterns of individual short-beaked echidnas can contribute greatly to the husbandry of this species. It also allows individual histories to be shared with other institutions and facilitates a cooperative approach to data collection among institutions. ZIMS, the Zoo Information Management System (Species360 2023) can be used for recording general information on age, sex, parentage, births, identification, bodyweights, transfers and deaths, with separate ZIMS modules for veterinary information and studbooks. These systems have been developed by Species360, which is part of the Conservation Planning Specialist Group (CPSG) software. As these are standardised there is a high degree of efficiency in transferring information between institutions. ZIMS also interfaces with programs such as PMx, which provides demographic and genetic analysis tools to assist with the management of breeding programs (Lacy *et al.* 2011).

## 5.3 Methods of individual identification

### 5.3.1 Passive Integrated Transponder (PIT) tags

PIT tags are an excellent method of identification because they are permanent. The standard recommended site of insertion of a PIT tag in captive and rehabilitation short-beaked echidnas is subcutaneously on the dorsal midline between the scapulae, ideally under general anaesthetic but can be done conscious if necessary (Middleton 2008). Sealing the entry wound with tissue glue (Vetbond®) will ensure that tag stays in place immediately after implantation. The ventral side at the interface between spines and ventral hair is also used by field researchers (e.g. Rismiller and McKelvey 2000), but less commonly for zoo short-beaked echidnas.

### 5.3.2 Plastic tubing on spines

Short-beaked echidnas can be easily marked by gluing different colour combinations of polyvinylchloride (PVC) or insulation tubing over several spines on the back (Augee *et al.* 1975; Rismiller 1992; S. Jackson pers. obs.). Several spines should be covered (as the spines are periodically shed) on different parts of the body, including the tail, rump, back, shoulder or head. In these locations they can also be placed on the right, centre or left portion of the body (Rismiller 1992). Colour codes can be developed that utilise three or four letters to describe first the colour, second the body part and third the location on the body part (Rismiller 1992). For example, RTL indicates Red, Tail Left and YGBR refers to Yellow Green Back Right. Colour coding in the wild has been known to work for up to 10 years (Rismiller 1992). If CCTV monitoring is used for animals involved with breeding introductions, a small flag made of reflective safety tape (e.g. 3M Scotchlite) can be fitted onto the colour identification tags of males to distinguish them from females when reviewing recorded video footage (Ferguson and Turner 2013).

# 6. FEEDING REQUIREMENTS

## 6.1 Captive diet

- Ad lib
  - water
- Daily diet (per animal) – provided by Perth Zoo (B. Laming pers. obs.). Blend together:
  - 56.25 g beef mince (lean; raw)
  - 0.5 g calcium carbonate powder
  - 16.8 g egg (hard-boiled and peeled)
  - 7 g cellulose powder
  - 2.8 g mealworms (chopped)
  - 0.6 g banana (peeled)
  - 0.14 mL Pentavite® (multivitamin + iron liquid)
  - 43.75 mL water (may be adjusted slightly to attain required consistency)

Perth Zoo offers 120-g portion per animal per day August–February. The recipe is increased by 20–25% to yield 150-g portion per animal per day March–July. Portions may be adjusted in response to monthly animal bodyweights, to maintain optimal seasonal bodyweight. Lactating females remain on larger portions until young are weaned. Live termites are offered in a separate tray on an ad hoc basis.

Alternative diets

- Daily diet (per animal) – provided by Currumbin Sanctuary (Wallage *et al.* 2015). Blend together:
  - 750 g beef mince (lean)
  - 3 eggs
  - 160 g dextrose ('Cattlekare' brand)
  - 50 g fly pupae ('Pisces' brand; increased 2 × leading into and during breeding season)

- 20 g vitamin E ('White E' brand)
- 5 g calcium powder (Vetasearch international balanced cal' brand)
- 100 mL olive oil (increased 2 × leading into and during breeding season)
- 5 mL Soluvet (Vetafarm)
- 60 g wheat bran

Currumbin Wildlife Sanctuary gives 100 g as maintenance diet, 150 g as increased portion for breeding females 1 month before the breeding season commencing and lactating females are offered 200 g as soon as their activity increases (once the puggle has hatched at day 10 of incubation) and continues until the puggle has weaned. Termites can be also separated from their dirt and sprinkled over the food if encouragement to eat is required, particularly in adults that are unwell or juveniles when starting to wean (A. Roser pers. obs.).

Both 'Echidna Food' by Wombaroo® Food Products and the 'Complete Echidna Diet' (Vetafarm) products are being used widely in zoos in Australia and overseas and by those who rehabilitate animals (Stannard *et al.* 2017; Shaw and Jarman 2025; Perry *et al.* 2022; Shaw 2022).

## 6.2 Supplements

See diets described in section 6.1.

## 6.3 Presentation of food

Food is presented in a dish, at a time appropriate to the activity of the individual animals and in a manner that prevents spoilage and contamination. To reduce competition for food in enclosures with multiple short-beaked echidnas, or from wild rats and birds, it may be necessary for the food to be presented in plastic containers with holes drilled in the sides. Not only does this exclude other species from eating the food, it gives an increased degree of foraging enrichment and promotes an equally shared distribution of food among the short-beaked echidnas. The feed dishes need to be thoroughly cleaned daily.

A vermin exclusion feeder has been developed by Perth Zoo to keep the food secure for the short-beaked echidnas (Figs 3.4, 3.5). The feeder has a lid that is

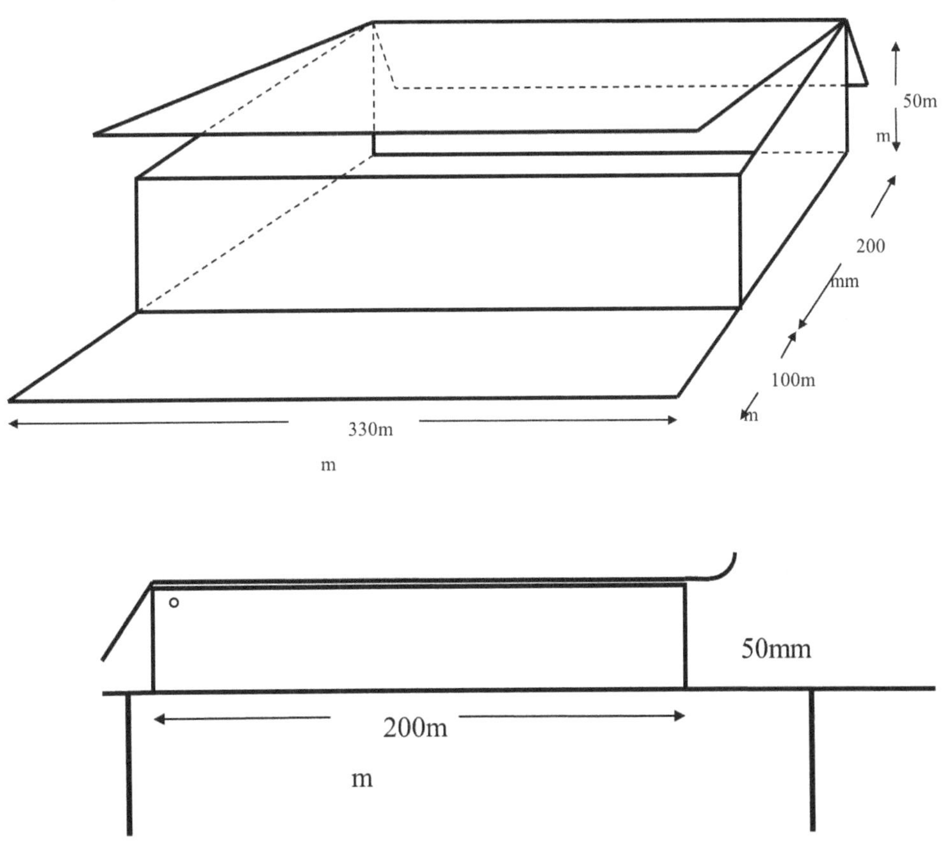


**Fig. 3.4.** Dimensions of a short-beaked echidna feeder developed by Perth Zoo.

Fig. 3.5. Vermin exclusion feeder being used at Perth Zoo to prevent wild birds from eating short-beaked echidna food. Photos by Perth Zoo.

330 mm (L) × 220 mm (W) plus 40 mm angled at the back to prevent the lid staying open when a short-beaked echidna has finished feeding. There is a 20-mm overhang at the front of the feeder that is curved up for the short-beaked echidnas to lift the lid to gain access to the food. A 5-mm overhang on both sides of the lid is included to keep rain out of feeder. The food tray compartment is 200 mm (W) × 320 mm (L) × 50 mm (D). The base plate is 350 × 330 mm with four 100-mm long metal spikes fitted to the bottom of feeder to fix it into the ground.

An alternative vermin exclusion feeder consists of a steel frame with sheet metal roof, clear acrylic sides and swinging clear acrylic flap doors for short-beaked echidnas to push their way through. This feeder is 60 × 60 cm (W) and 30 cm, with an A-frame roof extending a further 13 cm high (Fig. 3.6; A. Ferguson pers. obs.).

When short-beaked echidnas are displayed within an enclosure the food can be placed inside an artificial termite mound that can be protected from rain and predators to simulate feeding on a termite colony (Fig. 3.7).

Potential enrichment opportunities include:

- live termite nests and live ants and ant larvae
- slow feeder bowl and/or mat (e.g. LickiMat) that can be used both for enrichment or to slow a fast eater (A. Roser pers. obs.)
- PVC pipe with screw-on caps either end and holes drilled in the sides and mounted on a wooden stand.

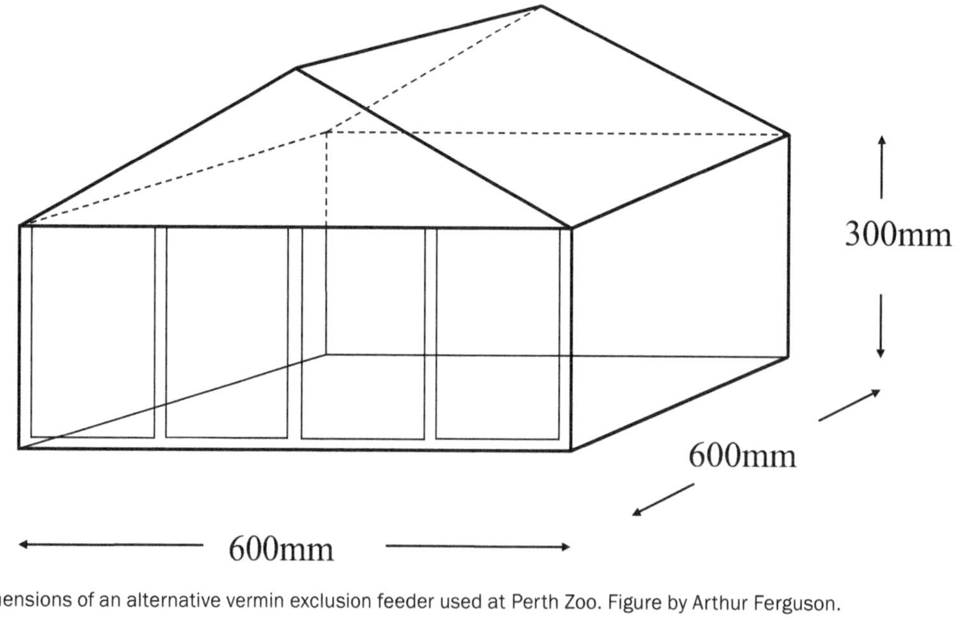


Fig. 3.6. Dimensions of an alternative vermin exclusion feeder used at Perth Zoo. Figure by Arthur Ferguson.

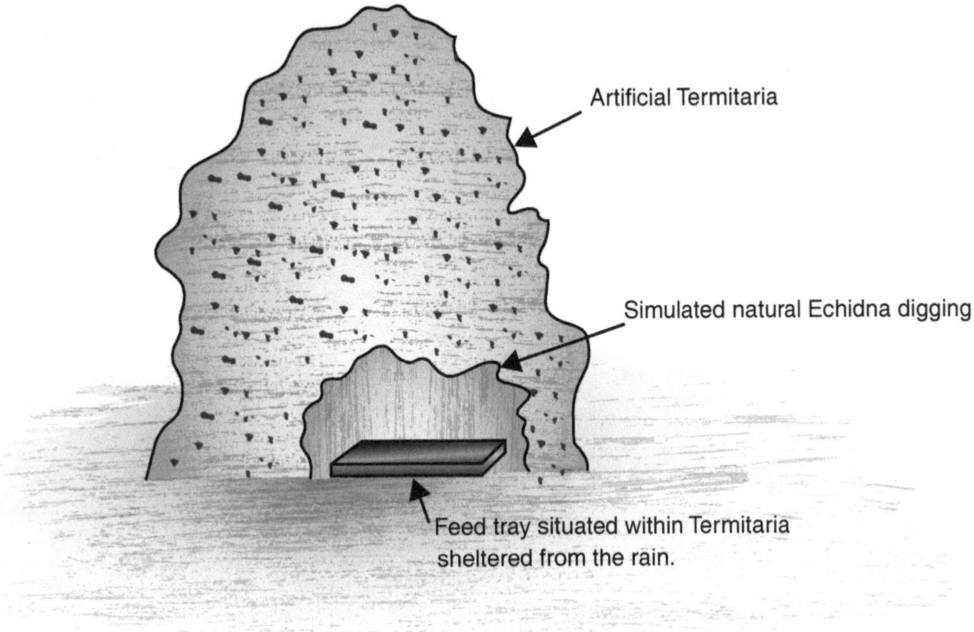


**Fig. 3.7.** Artificial termite mound used to offer food for short-beaked echidnas. Redrawn based on a figure by Arthur Ferguson.

Make sure the holes drilled in plastic containers are smoothed over with sandpaper to prevent abrasions of the beak and the pipe feeder should not be used for short-beaked echidnas with beak injuries (A. Roser pers. obs.).

## 7. HANDLING AND TRANSPORT

### 7.1 Timing of capture and handling

Short-beaked echidnas can readily be caught at any time, but may be at higher risk of heat stress when handled during high ambient temperatures. It is not recommended to handle females that have an egg or neonate in pouch (Ferguson and Turner 2013).

### 7.2 Capture equipment

Catching equipment can include soft leather gloves or towels to protect the hands. Once captured the short-beaked echidnas can be held in a tightly woven jute bag or inside a plastic garbage bin, which is ideal for short-term holding or transport. Note: most catching bags are unsuitable for short-beaked echidnas because their spines make it difficult to get them out of the bag once they are inside.

### 7.3 Capture and physical restraint

Several techniques have been used to catch short-beaked echidnas. Despite the short-beaked echidna's sharp spines, many people prefer not to use gloves because they decrease the handler's sensitivity to the animal's movements. Light-weight safety gloves are a useful compromise because they prevent most scratches and puncture wounds, but allow gentle handling. The best option for catching short-beaked echidnas is to slide a hand under each side of the animal between the front and back legs until the soft underbelly is felt and then lift (Fig. 3.8). With a firm grip behind the front legs, the animal usually relaxes, enabling it to be lifted, taking care not to cause injury to the sensitive beak area. When the animal is lifted its natural reaction is to roll forward and curl itself around the hands, which allows the handler to gain a firm grip while protecting the animal from spiking itself (Rismiller and McKelvey 2000; Scott *et al.* 2018).

Experienced animal caretakers may place the curled up echidna on its back, rolling it onto its side (which will encourage the animal to uncurl, providing access to the hindfeet), then quickly and carefully grasping the upper hindfoot first as the animal starts to uncurl, followed by the other hindfoot and then lifting it up slightly while allowing the front feet to rest on the ground. This enables the short-beaked echidna to support its bodyweight and will also encourage it to stretch out and expose its ventral surface for visual inspection. This method is used at Perth Zoo to monitor the area anterior to the cloaca for the level of bulging that becomes prominent in

**Fig. 3.8.** Method used to pick up a short-beaked echidna. Photos by Lauren Clark, used with permission of Currumbin Wildlife Sanctuary.

males during the mating period (July–August) and for pouch changes of females (A. Ferguson pers. obs.).

Tools of any type, such as shovels, must not be used to extract a buried echidna as they can cause injuries to the animal's snout and feet. Instead be patient, dig along one side with your hand until your fingers are under the belly, then can work on getting out the rest of the body. If handling of puggles is necessary, they should be cupped in both hands to hold them safely. Ensure hands are clean or wear disposable gloves when handling puggles. It is not recommended to remove puggles from their mothers' pouch unless there are health reasons to do so.

Short-beaked echidnas are surprisingly strong and have often been known to dig or squirm into the substrate during capture attempts. They can also curl into a tight ball and grip onto even the smoothest of surfaces, making it virtually impossible to get hands underneath. In these cases, thick gloves or towels are useful. Great care needs to be taken when short-beaked echidnas are in this situation and it may be best to attempt capture at a later time. Animals that have not been stressed by forceful capture are generally quieter and easier to handle while those that have had a bad experience usually respond by digging and becoming harder to extract (P. Rismiller pers. comm.).

## 7.4 Weighing and examination

Short-beaked echidnas are easily weighed by either placing them in a plastic bucket and using hanging scales or directly on a balance scale (Fig. 3.9). An alternative method is to use a large square made out of heavy fabric material ~65 × 65 cm, with loops or buttonholes at each corner for reinforcement. The corners are picked up and the animal is weighed using hanging scales (Rismiller and McKelvey 2003). The advantages of this method over the bucket method include (1) the weighing square being easily folded for storing and transport and easy to keep clean and (2) the short-beaked echidna tends to go quiet in a dark and confined space. When weighing is finished, simply put down the square so its sides fall open and the animal can walk away with no

**Fig. 3.9.** Methods used to weigh short-beaked echidnas. Photos by Alice Roser.

tipping or shaking required. Some institutions weigh their animals monthly as there are significant seasonal changes in wild echidna bodyweight (Rismiller and McKelvey 2000), which is sometime observed in captive animals (A. Roser pers. obs.).

Weighing of nursery burrow young was undertaken daily for eight young bred at Perth Zoo (performed whenever the female was not in the nursery burrow with her young to avoid disturbing the nursing process). Daily weighing of young after their nursery burrow emergence through weaning and post separation from the dam was also undertaken to track weight gain and loss closely to achieve successful weaning and rearing to independence (Ferguson and Laming 2022).

## 7.5 Release

Short-beaked echidnas are easily released back into their enclosures. When using a plastic rubbish bin for transport, gently place the bin on its side and the animal can exit when ready. Always allow for extra time and patience to catch or release a short-beaked echidna as this will result in a higher chance of a successful event and calmer circumstances for both the animal and animal caretaker.

## 7.6 Transport requirements

### 7.6.1 Transport box design

Transport boxes for short-beaked echidnas need to be of robust construction and have no gaps or joins to prevent the echidna pulling it apart. Further specific details of box design can be found in IATA (2022). Plastic stacker boxes (640 × 410 × 390) with a clip lid have been used successfully to transport short-beaked echidnas (P. Rismiller pers. comm.). Leaf litter is put in the box for them to burrow into and a bolt with wing nut is used to secure the lid. A row of air holes is also drilled at each end of the box (P. Rismiller pers. comm.). For within facility transport (e.g. from enclosure to veterinary surgery) or for local transport (journeys of 1–2 h), short-beaked echidnas can be transported in an air-conditioned vehicle in a soft plastic rubbish bin with ventilation holes in the lid; a towel or shredded paper is typically placed in the bin in case of urine contamination and to provide comfort and security. It is recommended that the bin lid is secured with a strap during transport in the vehicle to prevent escape.

### 7.6.2 Furnishings

No furnishings are required except for shredded paper, which may be added to the box for hiding.

### 7.6.3 Water and food

Water and food are generally not required during transport.

### 7.6.4 Animals per transport box

Only one short-beaked echidna should be placed in each box.

### 7.6.5 Timing of transportation

It is important that the short-beaked echidna does not overheat; the temperature should not be higher than 25°C and optimally should be ~22°C (Whittington 1988).

### 7.6.6 Release from the transport box

Short-beaked echidnas are easily released from the box. For boxes with a door positioned on the side, open the box and allow the short-beaked echidna to move out of the box in its own time. Some animals can be shy and will want to stay hidden in the security of the box so it may take some time for them to emerge.

## 8. HEALTH REQUIREMENTS

*Larry Vogelnest*

### 8.1 Daily health observations

Each short-beaked echidna should be observed daily for signs of injury or illness if possible. This is particularly important after events such as introductions with other animals (conspecific or other species) or new enclosures and breeding. An appropriate time to do this is when short-beaked echidnas are being fed and in the period after feeding when they tend to be more active. Distant observation of the undisturbed animal is also important. During colder months short-beaked echidnas may bury themselves deep in the substrate when going into torpor or hibernation and may be difficult to find. Digging them up or disturbing them is not recommended. When appropriate, each animal should be checked for the following:

- appetite
- faeces – faecal consistency grading is described in Table 3.1. Faecal appearance and consistency are variable and depend on factors such as diet, access to soil and different soil types. Free-ranging and captive short-beaked echidnas on ideal diets with access to soil typically pass Grade 2–3 faeces. Captive short-beaked echidnas on gruel-type diets without access to dirt typically pass Grade 3–4 faeces (Tobias 2025a). Persistent Grade 4, and particularly Grade 5 faeces, is an indication for investigation and should be discussed with a veterinarian. A sample may need to be submitted for diagnostic assessment
- demeanour, activity level and behaviour
- injuries
- gait, particularly lameness
- fur and spines – check for alopecia, broken spines, crusting or flaking of the skin, inflammation or wounds
- discharges – from the eyes, ears, nose or mouth. Also check for the presence of blood within the enclosure or in faeces.

All animals should be weighed and have body condition assessed on a regular basis (see section 8.2). Any signs of injury or illness or unusual changes in behaviour should be documented and reported to a veterinarian. If an animal is found dead, make note of the circumstances and anything unusual or abnormal on the body or surrounding area. Carefully collect the body (using PPE where appropriate) and submit it for necropsy.

**Table 3.1.** Faecal consistency grading for the short-beaked echidna (adapted from Stannard *et al.* 2017).

| Grade | Consistency | Description |
|---|---|---|
| 1 | Extremely dry | ▪ Often passed in several clumps of individual stools instead of one complete unit<br>▪ No indentations are made when stool is picked up<br>▪ Maintains original shape after collection but may break into smaller pieces (or dry powder) |
| 2 | Firm (ideal) | ▪ Firm, cylindrical shape, can range from straight to horseshoe shaped; ~2 cm in diameter<br>▪ Slightly moist surface<br>▪ May leave residue on surfaces and gloves<br>▪ Often defecated as one unit and maintains shape after collection |
| 3 | Soft with shape | ▪ Soft, with a definite cylindrical shape<br>▪ Will break apart easily and will definitely leave residue on surfaces and gloves<br>▪ Often loses original shape after collection |
| 4 | Soft without shape | ▪ Soft, has texture but no cylindrical shape (often described as 'cow pat')<br>▪ Leaves residue on surfaces and gloves<br>▪ Piles may spread over an area of several centimetres |
| 5 | Liquid (diarrhoea) | ▪ Always resembles liquid and may or may not be particulate matter present<br>▪ Mucus is often present<br>▪ Difficult to collect and leaves residue on surfaces and gloves<br>▪ Often defecates in many pools instead of one complete unit and may spread over wide area |

## 8.2 Bodyweight and body condition score

### 8.2.1 Bodyweight

Short-beaked echidna bodyweight can vary markedly throughout the year. Males may lose up to 30% of their bodyweight during the breeding season in the wild (Rismiller and McKelvey 2000) and are at their lowest bodyweight by the end of the breeding season (August). Captive short-beaked echidnas may have normal weight variations of up to 2 kg. Adult weights are higher in autumn–early winter and lower in spring–early summer (Tobias 2025a), which may be related to preparation for torpor/hibernation followed by an increased metabolic rate during the warmer months. Seasonal bodyweight variation in free-ranging short-beaked echidnas differs significantly depending on geographical region (Nicol *et al.* 2019a). Short-beaked echidnas in zoos should ideally be weighed at least monthly to indicate trends. Trends in bodyweight are a good indicator of general health. A persistent decline in bodyweight outside of the normal seasonal variation is an indication for more detailed examination and should be reported to a veterinarian.

### 8.2.2 Body condition score

A specific body condition scoring guide has been developed at Taronga Conservation Society Australia for the short-beaked echidna and is yet to be published (M. Shaw pers. comm).

Similar to bodyweight, there is seasonal variation in body condition. However, in rescued or sick short-beaked echidnas, body condition may be an important clinical consideration. Body condition can be subjectively assessed as follows (Tobias 2025a; M. Shaw pers. comm.):

- circumference shape (viewed from above): ranges from distinct hourglass shape indicating under-condition to oval indicating over-condition
- dorsal profile (viewed in the dorsal plane either from head to tail or vice versa): ranges from a pointed to narrow dome indicating under-condition, to a wide dome indicating over-condition
- degree of abdominal concavity and epipubic and pelvic bone protrusion when in dorsal recumbency: concave and prominent bone protrusion indicating under-condition and flat or convex with minimal bone protrusion indicating over-condition
- lateral skin fold thickness assessed under anaesthesia: a wide, non-compressible fold indicates over-condition, a narrow fold indicates under-condition. This measurement is particularly useful because it is a subjective measure of body fat, which in short-beaked echidnas is predominately stored in the dorsolateral subcutaneous space.

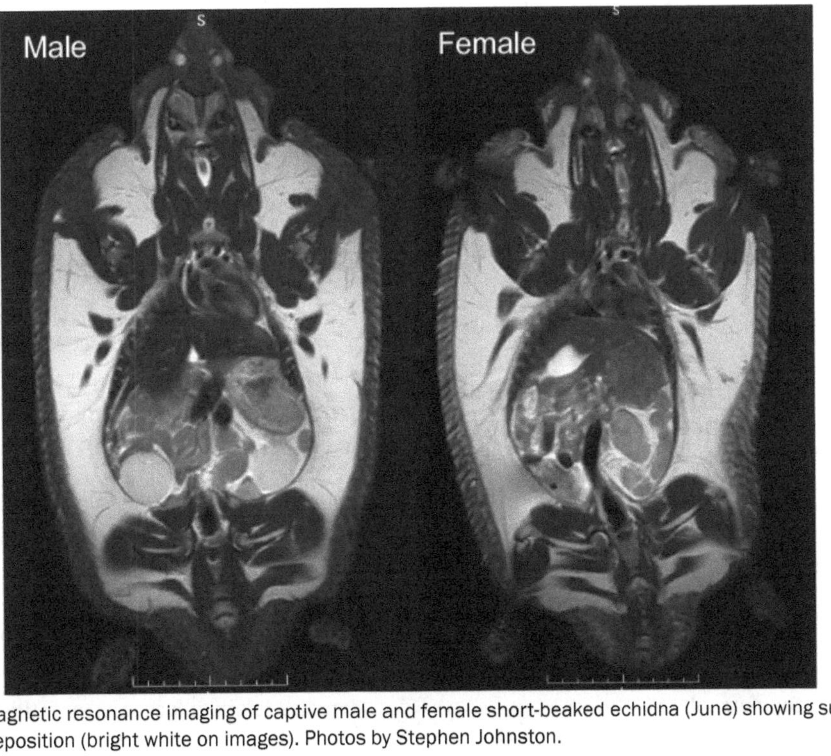


**Fig. 3.10.** Coronal magnetic resonance imaging of captive male and female short-beaked echidna (June) showing subcutaneous and intra-abdominal fat deposition (bright white on images). Photos by Stephen Johnston.

Fat deposition in the short-beaked echidna can be substantial, with a subcutaneous fat layer up to ~3 cm leading into the breeding season (Fig. 3.10; S. Johnston unpubl. data); there are also significant deposits of intra-abdominal (mesenteric and gonad) adipose tissue. The significance of fat depots in captive short-beaked echidnas is still uncertain, but may be related to breeding success (S. Johnston unpubl. data).

## 8.3 Physical examination

Some captive short-beaked echidnas can be readily handled (see Middleton 2008; Vogelnest 2025a) and a basic physical examination can be conducted by animal caretakers (or a veterinarian) without the need for chemical restraint. The extent and ease of the examination will, however, be dependent on the temperament of the individual animal. Physical examinations should be conducted from weekly to monthly depending on the individual, health status and age. Geriatric short-beaked echidnas or those with health concerns or injuries should be examined, weighed and body condition scored more frequently. In addition to the list above (section 8.1) the following can be assessed:

- body temperature – basal body temperature range is 30–33°C (Middleton 2008; Morrow and Nicol 2009), however, it can be lower depending on ambient temperature. Torpor or hibernation typically occurs at ambient temperatures less than 12°C and body temperatures less than 27°C. During these periods, body temperature can decrease to 4°C. Short-beaked echidnas are intolerant of high ambient temperatures and above 28°C heat loss is minimal and hyperthermia begins to develop (Barker *et al.* 2016). Echidnas behaviourally thermoregulate (rather than sweat or pant as a means of heat loss) and are adept at finding shelter, restricting activity to cooler times of the day or at night or making use of water bodies (swimming and/or drinking) for cooling (Middleton 2008). Body temperature can be taken using a digital thermometer inserted deep into the rectum via the cloaca, but is usually only possible with the animal under anaesthesia (Tobias 2025a).
- abnormal swellings, masses or deformities
- cloaca – should be clean and free of discharge or faeces; occasionally males housed together or male-female pairs may show evidence of cloacal trauma associated with excessive probing/investigation with the beak (S. Johnston pers. obs.)
- nails – check for excessively overgrown or broken nails
- penis – if indicated the penis should be extruded and examined (Johnston *et al.* 2006), but is only possible with the animal under anaesthesia.
- ventral skin and pouch area – should be clean, dry and without any areas of inflammation, crusting or alopecia. Both males and females have a 'pouch' area. If the pouch is prominent in a female and moisture or milk can be expressed from the milk patches, this may indicate the presence of a puggle in a burrow.
- eyes – can be difficult to examine because the palpebral fissure and globe are small and is usually only possible under anaesthesia. The cornea should be clear, the pupil should respond to light and there should be no swelling or discharge. The eyelids should be clean, dry to slightly moist and free of inflammation or swelling.

If required, a detailed physical examination should be conducted by a veterinarian, with the animal under anaesthesia. Veterinarians should refer to Vogelnest and Woods (2008) and Vogelnest and Portas (2025) for detailed information on examination and medicine for short-beaked echidnas. All body areas and systems should be evaluated: cardiovascular; respiratory; oral cavity (tongue, oral mucosa), abdomen; musculoskeletal; genitourinary; ocular; and skin, fur and spines. Important parameters and vital signs should be measured and recorded: bodyweight and condition; heart rate; respiratory rate; and body temperature (Middleton 2008; McLelland 2025).

Sex determination in short-beaked echidnas can be challenging because they lack external genitalia, males are testicond (the testes are located within the abdominal cavity) and both sexes have a pouch-like area that can become more pronounced when the longitudinal muscles on the abdomen contract (Rismiller 1993). Various methods of determining sex have been described (Johnston *et al.* 2006) including the following:

- spur morphology – the only method that can be used in conscious short-beaked echidnas. The length of the spurs in adult males typically ranges from 4.5 to 15 mm while spurs of females are much shorter (juvenile) or absent (Fig. 3.11; Johnston *et al.* 2006; Nicol *et al.* 2019a). However, on Kangaroo Island, 25% of adult females retain at least one spur (Rismiller 1993).

- palpation/extrusion of the penis - requires general anaesthesia. In subadult and adult males, a longitudinal ridge of firm tissue can be palpated cranial to the cloaca, but in juveniles this may not be discernible. In adults the penis can be extruded externally (Fig. 3.12). It is not advisable to attempt this in juveniles or subadults because the preputial opening can be very small. During the breeding season (July-August) when mature males are most reproductively active, a noticeable bulge of the penis in the cloaca is evident cranial to the cloaca when compared with the same area in females (A. Ferguson pers. obs.). This requires the animal to be handled to visualise (refer to sections 7.3-7.4)
- testicular palpation - requires general anaesthesia. May give inconclusive results in young males and in males with inactive or immature testes. Requires experience and only possible in adult males during the breeding season
- ultrasound examination of the testes - requires general anaesthesia. May give inconclusive results in young males and in males with inactive or immature testes. Requires some sonographic experience (Fig. 3.13)
- faecal steroid analysis - two methods involve the measurement of the oestradiol:androgen ratio (E:A ratio) and the androgen concentration. One study, with a small sample size found that the mean E:A ratio for two females was $18 \pm 6$ and $23 \pm 8$, respectively, compared with only $1.3 \pm 0.6$ for a male. Similarly, the concentration of androgen in dried faeces was $1.7 \pm 0.5$ and $2.2 \pm 0.6$ ng/g, respectively, for the two

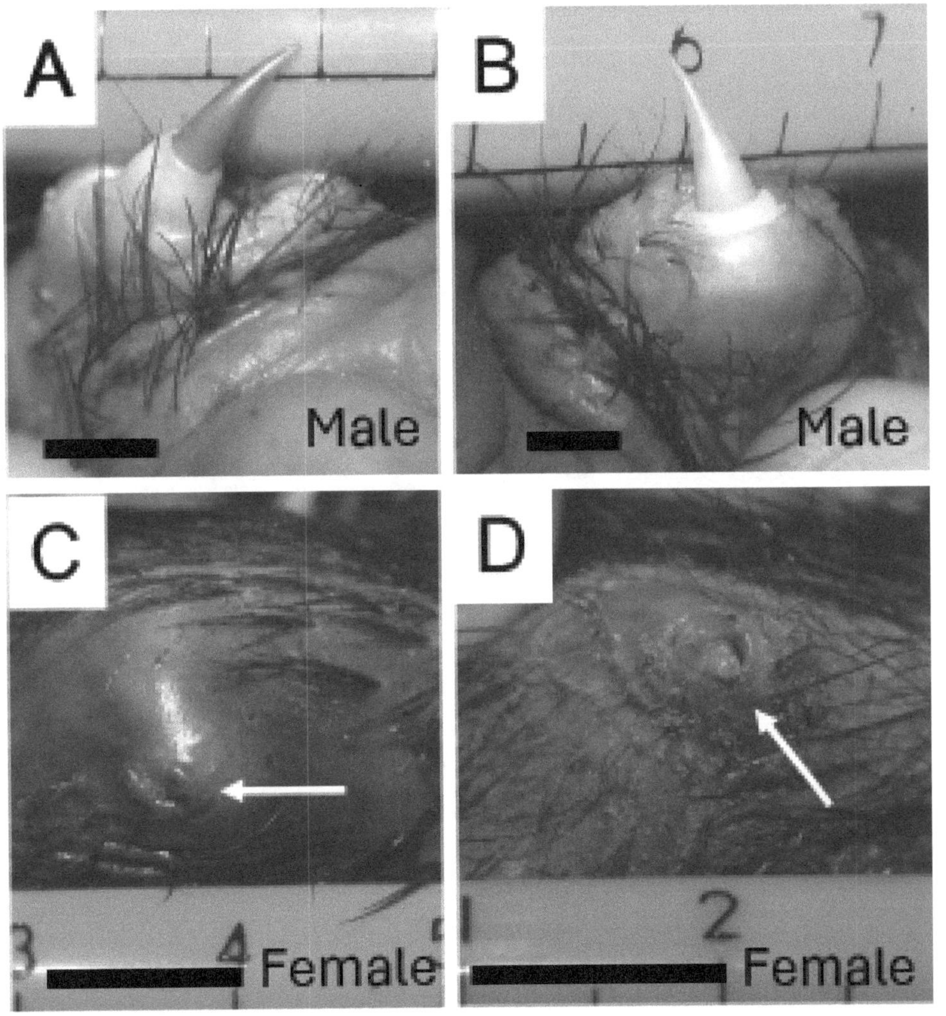


**Fig. 3.11.** Hindlimb spurs of the short-beaked echidna male (A, B) and female (C, D). Taken from Johnston *et al.* (2006), with approval from the publisher. ©2006 John Wiley & Sons Inc.

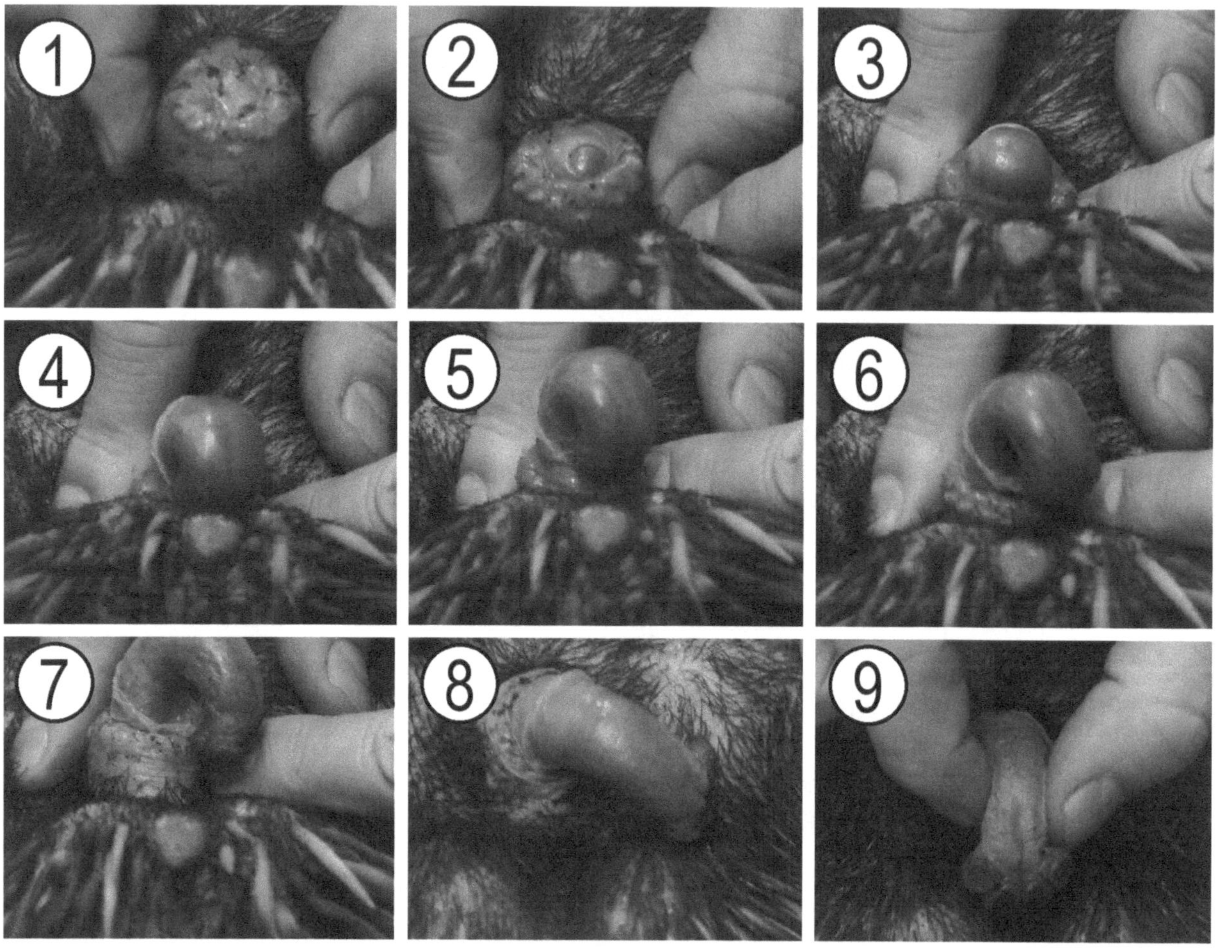


**Fig. 3.12.** Extrusion of short-beaked echidna penis. Taken from Johnston *et al.* (2006) with approval from the publisher. ©2006 John Wiley & Sons Inc.

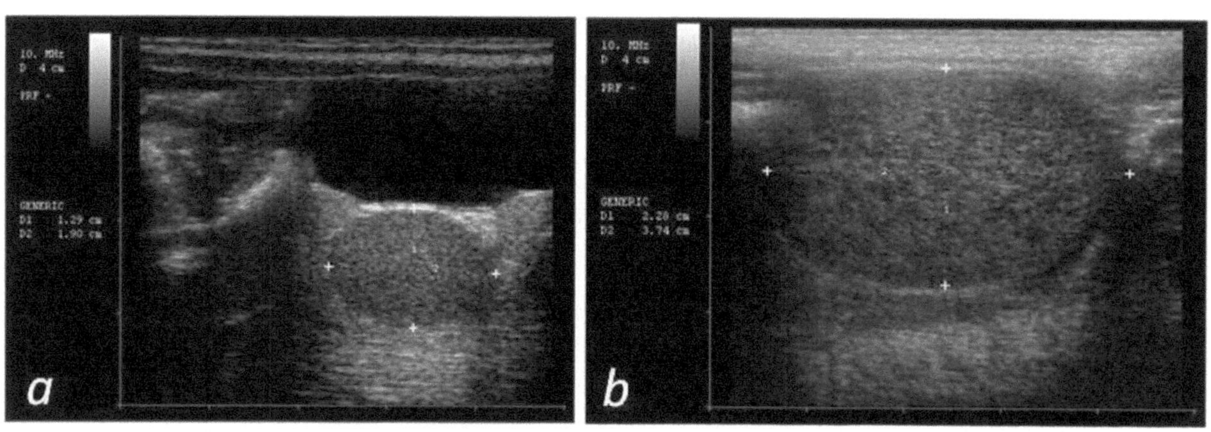


**Fig. 3.13.** Ultrasound images of left testis in December before testes recrudescence being initiated (a), left testis in June at the beginning of the breeding season (b). Images were taken at the same scale and the white crosses indicate linear measurements used to calculate volume. Taken from Morrow *et al.* (2016) by permission of Oxford University Press.

females and 10.1 ± 2.3 ng/g for the male (Oates *et al.* 2002)

- non-invasive genetic sexing of juvenile short-beaked echidnas - technique based on polymerase chain reaction detection of two Y chromosome (male-specific) genes (CRSPY and AMHY) and X chromosome gene (AMHX) using genomic DNA extracted from hair follicles (Perry *et al.* 2019; Keating *et al.* 2022).

## 8.4 Chemical restraint

Chemical restraint is generally required for a detailed physical examination, sample collection other clinical or surgical procedures and measurements (Fig. 3.14). Chemical restraint should be performed by a veterinarian and relevant chapters in Vogelnest and Woods (2008) and Vogelnest and Portas (2025) should be consulted. Prior to anaesthesia, fasting the animal for 4–6 h is recommended. Short-beaked echidnas can vomit, although they are not prone to regurgitation under anaesthesia. Additionally, anaesthesia after feeding may result in gut stasis and predispose to gastric yeast overgrowth and/or bloating. Injectable agents can be used for sedation, anaesthetic induction before maintenance with gaseous agents, brief examination, sample collection including venepuncture (Portas 2025), minor treatments or when an echidna is difficult to extract from a surface or space. Intramuscular injections can be given in the hindlimbs or epaxial muscles (useful in unrestrained short-beaked echidnas that cannot be readily extracted from a space for physical restraint) (Vogelnest 2025a). Volatile inhalational anaesthetic agents such as isoflurane or sevoflurane delivered in oxygen are frequently used as induction agents or for maintenance after injectable induction. Induction is typically via a face mask, although an induction chamber can be used to avoid the need to physically restrain the animal and thus may be less stressful. The chamber must be ruggedly constructed to withstand the echidna's physical strength (Middleton 2008; Holz 2015). Once anaesthetised, the animal is removed from the chamber and maintained via a face mask.

## 8.5 Common health problems

Captive short-beaked echidnas are susceptible to a range of infectious diseases. The same is not true for free-ranging short-beaked echidnas, suggesting that captive environmental, husbandry, nutritional or other management factors are likely contributing factors. Detailed information on infectious and non-infectious diseases

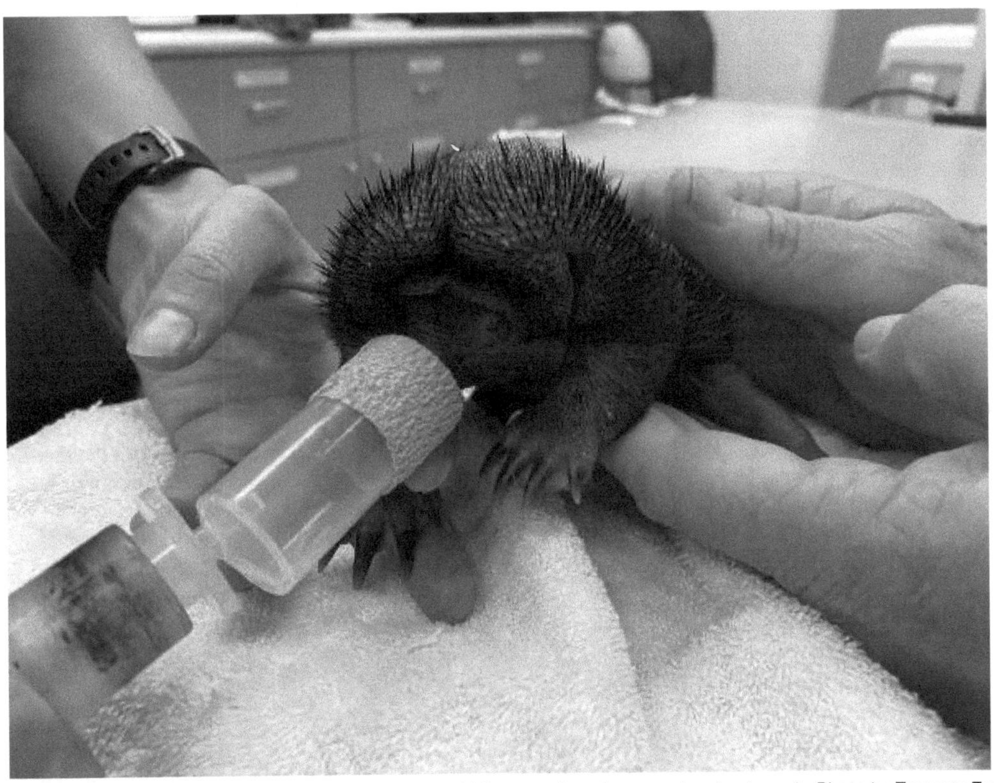

**Fig. 3.14.** Gaseous anaesthetic induction of a short-beaked echidna puggle using a customised mask. Photo by Taronga Zoo.

of short-beaked echidnas is given in Middleton (2008) and Tobias (2025a). The most significant disease of captive short-beaked echidnas is coccidiosis and this is discussed in detail in Tobias (2025b). Important diseases are summarised in Tables 3.2–3.5.

## 8.6 Preventative medicine

There are no specific preventative medicine recommendations for short-beaked echidnas (e.g. vaccination, routine screening and treatments). However, appropriate and optimal enclosure design, husbandry, quarantine and optimal nutrition all promote good health and welfare. In captive populations where particular pathogens or parasites are prevalent, routine screening is warranted (e.g. faecal wet preparations and floatation). A necropsy must be conducted on any short-beaked echidna that dies to facilitate a diagnosis and, if relevant, inform preventative action to protect other short-beaked echidnas in a captive population.

## 8.7 Zoonoses

There are no known zoonotic diseases of concern that can be contracted from short-beaked echidnas.

**Table 3.2.** Selected infectious diseases or agents of captive and rescued short-beaked echidnas.

| Disease/infectious agent | Cause | Clinical signs | Diagnosis | Treatment | Prevention/ comments | Reference |
|---|---|---|---|---|---|---|
| **Ectoparasites** | | | | | | |
| Mites | *Odontacarus echidnus* | May be associated with dermatitis | Skin scrape, sticky tape cytology | Rarely required but various acaricides including ivermectin, fipronil and selamectin are likely to be effective | Generally not required | 1–5 |
| Ticks | *Bothriocroton concolor, Ixodes* spp., *Haemaphysalis* spp. | None or dermatitis, anaemia with heavy burdens | Direct observation | Various acaricides | Generally not required | 1, 6, 7 |
| Fleas | *Echidnophaga ambulans, Bradiopsylla echidnae* | None or mild pruritus | Direct observation | Rarely required but flea products available for domestic animals are likely to be effective | Generally not required | 1, 8 |
| **Endoparasites** | | | | | | |
| Intestinal worms | Cestoda: *Echidnotaenia tachyglossi* *Linstowia echidnae* Nematoda: *Parastrongyloides* sp., *Nicollina* sp. *Tachynema* sp. *Tasmanema* sp. | None or diarrhoea | Faecal examination and the presence of eggs or proglottids (segments of tape worms) | Anthelmintics | Generally not required; however, maintaining good sanitation and hygiene of enclosures are important | 1, 2, 5, 9–14 |

| Disease/infectious agent | Cause | Clinical signs | Diagnosis | Treatment | Prevention/ comments | Reference |
|---|---|---|---|---|---|---|
| Sparganosis | Plerocercoids of *Spirometra ernaceieuropaei* | Subcutaneous, intrathoracic, abdominal masses to fatal disseminated disease | Clinical signs, radiography, ultrasound examination, surgical exploration, necropsy | Surgical removal but generally very challenging | Rarely seen in captive short-beaked echidnas. Infection via consumption of copepods infected with larvae or other hosts containing a plerocercoid | 1, 14 |
| **Protozoa** | | | | | | |
| Coccidiosis | *Eimeria echidnae* and *E. tachyglossi*. Combination of pathogen–host–environment factors influence the development of clinical coccidiosis (see Table 3.3) | Infection without disease is common (asymptomatic) Weight loss; inappetence; failure to thrive; diarrhoea; lethargy; vomiting; reduced activity; hunched posture; weakness; collapse; neurological signs | History of recent exposure factors. Presence or numbers of faecal oocysts is not necessarily indicative of presence of disease. Presence of *Eimeria* in blood cells, abnormal haematology, necropsy | Early treatment is important. Anticoccidial drugs, supportive care | Common and significant disease of captive short-beaked echidnas. Disease is seen in recently rescued free-ranging short-beaked echidnas, and is more common in subadults, but also seen in long-term captive adults. Prevention: anticoccidial drugs to reduce shedding, manage risk factors | 1, 15–19 |
| Blood parasites | *Theileria tachyglossi* | Considered non-pathogenic | Blood smear | Not required | Parasites occasionally seen as incidental finding in blood of captive short-beaked echidnas | 20 |
| Toxoplasmosis | *Toxoplasma gondii* | Neurological abnormalities, death in both free-ranging and captive short-beaked echidnas | Necropsy, histopathology and specific test for toxoplasma | None reported | Toxoplasma is carried by cats. Infection can be prevented by excluding cats from enclosures or substrate | 1, 5 |
| **Bacterial** | | | | | | |
| Salmonellosis | Various *Salmonella* spp. | None, grade 5 diarrhoea ± blood, lethargy, weakness, death | Faecal culture, necropsy | Fluids, antibiotics | Good hygiene and husbandry | 4, 5, 11, 21, 22 |

**Table 3.2.** continued

| Disease/infectious agent | Cause | Clinical signs | Diagnosis | Treatment | Prevention/ comments | Reference |
|---|---|---|---|---|---|---|
| Botryomycosis (staphylococcal pododermatitis) | *Staphylococcus* spp., *Streptococcus* spp. Secondary to injury or nail bed infection | Soft tissue swelling, ulceration of foot/digits (usually a single foot affected) | Biopsy, culture | Antibiotics, surgery, analgesics and anti-inflammatory drugs | | 4, 5, 11, 20, 22 |
| **Fungal** | | | | | | |
| Mycotic stomatitis | *Candida albicans* (occasionally *C. glabrata* or *C. parapsilois*) | Excessive salivation; excessive time to consume food; oral mucosa bleeds easily; tongue discoloured (red or pale), swollen, ulcerated | Examination under anaesthesia, cytology of oral swabs | Systemic and topical antifungal therapy, supportive care | May be a primary condition or with concurrent disease or general debilitation. May be associated with poor environmental conditions, stress and immune suppression | 4, 11 |
| Ringworm (dermatophytosis) | *Microsporum gypseum*, occasionally other fungi | Fur loss, crusting, skin redness, brittle spines | Clinical examination, skin sampling (scrapings, biopsy) | Systemic and topical antifungal therapy | | 4, 11 |

References: (1) Tobias (2025a); (2) Middleton (2008); (3) Domrow (1991); (4) Whittington (1988); (5) McOrist and Smales (1986); (6) Roberts (1970); (7) Loh *et al.* (2016); (8) Dunnet and Mardon (1974); (9) Whittington and Obendorf (2012); (10) Spratt *et al.* (1991); (11) Whittington (1992); (12) Durette-Desset and Cassone (1983); (13) Spratt and Beveridge (2016); (14) Whittington *et al.* (1992); (15) Barker *et al.* (1985); (16) Backhouse and Bolliger (1959); (17) Mackerras (1959); (18) Tong (2019); (19) Debenham *et al.* (2012); (20) Priestley (1915); (21) Gilruth *et al.* (1911); (22) Finnie (1988).

**Table 3.3.** Risk factors that may influence development of clinical coccidiosis in short-beaked echidnas (adapted from Tobias 2025b).

| Pathogen factors | Host factors | Environmental factors |
|---|---|---|
| ▪ Introduction of a novel coccidian species or strain shed into an established captive group<br>▪ Increased environmental parasite load<br>▪ Infective dose | ▪ Immune compromise (e.g. due to concurrent disease, inadequate immunity from dam, poor nutrition)<br>▪ Lack of immunity due to lack of exposure (e.g. naïve adult)<br>▪ Stress (see environmental factors) | ▪ Poor hygiene<br>▪ Conditions favouring development of the parasites in the environment (e.g. mild/cool ambient temperature, moisture/humidity)<br>▪ Sudden diet change<br>▪ Confinement in small areas<br>▪ Overcrowding<br>▪ Noise, disturbance<br>▪ Transportation<br>▪ Extreme weather<br>▪ Change in routine |

**Table 3.4.** Non-infectious diseases of captive and rescued short-beaked echidnas.

| Disease/agent | Cause | Clinical signs | Diagnosis | Treatment | Prevention/comments | Reference |
|---|---|---|---|---|---|---|
| Ocular and eyelid disease | Trauma, unknown, allergens | Conjunctivitis, corneal opacity and ulceration, discharge, crusting, cataracts, swollen eyelids | Ophthalmic examination | Topical treatment, systemic anti-inflammatory drugs | Treatment for cataracts not required; blind captive short-beaked echidnas cope well. Application of topical medications can be difficult | 1, 2 |

| Disease/agent | Cause | Clinical signs | Diagnosis | Treatment | Prevention/comments | Reference |
|---|---|---|---|---|---|---|
| Dorsal dermatopathy (hyperkeratotic dermatitis) | Seasonal, nutritional, unknown | Varying degrees of patchy fur and sometimes spine loss, spine fragmentation, brittle spines, crusts, scale, secondary infections. Dorsal trunk usually affected, extending to the ventrum in some animals | Clinical examination, skin sampling (scrapings, biopsy) | Often self-limiting, may respond to topical or systemic omega fatty acid and zinc supplement | Dietary insufficiencies, low or high humidity are likely causes. The role of infectious agents and multifactorial causes is possible | 3 |
| Seborrhoea | Unknown, environmental factors | Dandruff, redness of skin, ulceration, poor spine quality, secondary infections | Clinical examination, skin sampling (scrapings, biopsy) | Keratolytic/ antiseptic shampoo | Mostly hand-reared puggles | 4 |
| Burns | Bush fires | Burnt spines, foot pads, skin, inappetence, lethargy, malodour | Clinical examination | Topical burn treatment, analgesia, fluids | Severe cases require euthanasia | 2 |
| Gastric impaction | Feeding of sinewy meat; ingestion of fur or indigestible plant fibres | Poor or no appetite, abdominal distension, lethargy, vomiting, death | Clinical examination, radiography, necropsy | Surgery | Feed commercial diet; or use lean good quality mince in diet mixes. Prognosis is often poor | 2 |
| Hyperthermia | Low heat tolerance, high ambient temperatures, lack of opportunity for behavioural thermoregulation | Incoordination, dull mentation, muscle fasciculation, increased heart and respiratory rates, body temperature > 37°C | History, clinical signs | Cooling with water, subcutaneous fluids | Avoid exposure to high ambient temperatures, provide appropriate shelter and refuges | 2 |
| Trauma | Motor vehicles, dog attacks | Beak wounds and fractures (common), broken spines, skin wounds, lameness, paralysis | Clinical examination, full-body radiography | Analgesia, antibiotics as indicated, surgery (e.g. repair of abdominal hernias) | | 1 |
| Neoplasia | Fibroma, lymphosarcoma, chondroma, leukaemia, adenoma, leiomyosarcoma, soft tissue sarcoma | Dependent on type and location of tumour | Clinical examination, imaging, haematology | Excision, euthanasia | Neoplasia is rare in short-beaked echidnas | 1, 5 |

References: (1) Middleton (2008); (2) Tobias (2025a); (3) Vogelnest (2025b); (4) Whittington (1993); (5) Tong Marwood (2025).

**Table 3.5.** Diseases of mixed cause in captive and rescued short-beaked echidnas (Tobias 2025a; Tong 2019).

| Disease | Cause | Clinical signs | Diagnosis | Treatment | Prevention/ comments |
|---|---|---|---|---|---|
| Bronchopneumonia | Bacteria, fungi, secondary to other diseases | Nasal discharge, death | Culture of discharges, radiography, computed tomography | Antifungal drugs, antibiotics, supportive care | |
| Ulcerative gastritis | Mixed bacteria and fungi that are not normal gastric flora in the short-beaked echidna. May be associated with poor or inappropriate diet (particularly hand-reared juveniles), other diseases (e.g. coccidiosis) | Weight loss, failure thrive or gain weight (juvenile short-beaked echidnas), capricious appetite, lethargy | Examination of gastric samples | Antifungal and antibacterial drugs, supportive care, treat concurrent disease | Appropriate diet, improve hygiene and husbandry |
| Diarrhoea | Inappropriate diet, change in diet, poor husbandry, stress, bacterial or other enteric pathogens (coccidia, yeasts) | Diarrhoea, lethargy, weight loss | Faecal examination (wet preparation, floatation, culture), clinical examination under anaesthesia including radiography | Fluids, anti-diarrhoeal preparations, transfaunation, specific antimicrobial drugs as indicated by culture results | Addition of soil to diet, feed commercial diet, reduce stress, improved husbandry and hygiene |
| Failure to thrive in hand-reared short-beaked echidnas | Inappropriate milk formula (particularly in relation to stage of development/weaning), inadequate transfer of immunity from dam, enteric infections, stress, inexperienced carer, poor husbandry | Poor weight gain, weight loss, poor body condition | Signalment, history | Appropriate diet for stage of development/ weaning, specific treatment of any concurrent problems | For wild puggles: correct determination if weaned or unweaned. Appropriate hand-rearing formula and methodology (see section 11) |

# 9. BEHAVIOUR

## 9.1 Activity

Short-beaked echidnas are semi-fossorial, typically solitary (when not engaged in breeding) and cryptic animals. Although short-beaked echidnas are one of the few Australian mammals that may be commonly encountered during daytime in the wild, they can also be active at night. Because of this behavioural flexibility, providing food only during the day may encourage short-beaked echidnas to increase daytime activity. In captivity, some have suggested that activity can be negatively influenced by noise levels from visitors (Collins 1973). A 2-year study of wild short-beaked echidnas during the summer and spring seasons in Western Australia found, on average, that they spent over 60% of their active hours engaged in foraging and digging behaviours, with the remainder of active time spent walking and climbing (Clemente *et al.* 2016).

During cool weather, short-beaked echidnas will often sun bask with their abdomen on the ground and their fore- and hindlegs stretched out (S. Jackson pers. obs.; Coleman 1934, 1935). Animals rewarming from hibernation may speed up the process by basking (Nicol *et al.* 2008). Observations of captive animals have shown that this behaviour peaks when the proximal body temperature spiked above 30°C (Ferguson and Turner 2013). Some individuals may prefer basking under heat lamps either during the day or night (Ferguson and Turner 2013; see section 4.6).

As noted earlier, short-beaked echidnas in the wild show reduced activity for several months preceding mating. In *T. aculeatus/setosus* this is manifests as deep hibernation, which may begin as early as late summer, whereas *T. acanthion/multiaculeatus* shows reduced activity and short periods of torpor. Hibernation is not due to absence of food or simple avoidance of the cold: in Tasmania the males in best condition enter hibernation earliest and come out of hibernation first in order to be the first to mate, with some males emerging as early as May (Nicol *et al.* 2019b). Male emergence is well before the winter solstice (i.e. before the coldest weather). Females enter hibernation about 1 month after males and emerge in June–September, with the majority of female arousals being due to disturbance by males. If females mate early in the season they may return to hibernation even if pregnant (Nicol *et al.* 2019b). Adult males and females with low bodyweight (i.e. low-fat reserves) may hibernate through until September–October and do not mate. Unlike most hibernators, short-beaked echidnas do not dig or construct a hibernaculum. Instead, they may use existing rabbit or wombat burrows, hollow logs or dig under tree stumps or into grass tussocks, piles of bark or leaves (Wilkinson *et al.* 1998) or simply burrow into the substrate and thus are often more exposed to ambient conditions (Nicol 2017).

Although wild echidnas in south-east Queensland are known to hibernate (Beard and Grigg 2000), echidnas kept over an extended period in captivity in south-east Queensland have never been observed to hibernate and rarely torpor (S. Johnston pers. obs.).

## 9.2 Social behaviour

Wild short-beaked echidnas are solitary, except during the breeding season. In captivity, observations of groups up to 33 have shown them to live readily together and to generally ignore each other except for the occasional sniff of another's body and when feeding on provided food, which they readily share without noticeable aggression (although they push their way towards the food) (Brattstrom 1973; Augee *et al.* 1978, S. Jackson pers. obs.). Despite their normally solitary behaviour in the wild, Brattstrom (1973) suggested there is a loose dominance hierarchy in captive short-beaked echidnas that is based on body size. In contrast, Augee *et al.* (1978) observed a dominance hierarchy of animals based on the same sex using observations of encounters at their feeding dish (though no aggressive behaviour was observed); however, if anything, there was an inverse relationship between dominance and body size, with smaller individuals generally being successful in encounters.

Enclosure companions during the breeding season may be selected based on their eating and sleeping patterns to ensure that individuals do not consume other animals' food while they are in torpor or hibernation. The degree of inactivity will greatly affect the body condition of the animals if some are eating and others are not. Issues with food sharing appear to be relatively uncommon, as multiple animals can feed from the same dish and often share dishes; however, close monitoring is required to prevent under- or over-feeding. Male homosexual behaviours have also been observed, which has caused injuries to the cloacal region of their house mate (S. Johnston pers. obs.).

One observation of aggressive behaviour that was observed in captivity involved a female, subsequently known to have been carrying a pouch young, placing her snout under another female and pushing her across the enclosure on several occasions, resulting in the pushed animal adopting a defensive position with its head under its body (Augee *et al.* 1978). The animal being pushed was only touching the ground with its left legs. The female with pouch young then walked away and was followed by the second animal sniffing towards her, resulting in the pouch young female sniffing under the second animal for several minutes. The roles were then reversed with the second animal pushing its snout under the female with the pouch young, which attempted to turn away to present her spines to the second animal. The young of the female with a pouch was found dead 5 days later, so it is possible the young may have been dislodged or injured during one of these encounters, though there was no direct evidence.

## 9.3 Reproductive behaviour

Field studies of reproductive behaviour have been carried out on short-beaked echidna populations on Kangaroo Island (subspecies *multiaculeatus*), Tasmania (subspecies *setosus*), Mount Kosciuszko (*aculeatus*) and south-east Queensland (*aculeatus*). Short-beaked echidnas are seasonal breeders and males have very large testes, reaching 1% of bodyweight in some males during the breeding season (Morrow *et al.* 2016), strongly suggestive of sperm competition (Jones *et al.* 2004; Parker 2016). There are significant differences in aspects of wild short-beaked echidna reproductive behaviour, probably reflecting differences between the *acanthion/multiaculeatus* clade and the *aculeatus/setosus* clade as well as climatic effects. During the breeding season there is intense male-male competition and males may form 'trains' with a female being followed by several males. Short-beaked echidna mating trains with 2–5 males have been reported from several areas of mainland Australia (Augee *et al.* 1975; Griffiths 1978), but females were never found with more than one male in Kosciuszko National Park (Beard *et al.* 1992) or more than two males in south-east Queensland (Beard and Grigg 2000). There is photographic evidence of mating trains in wild short-beaked echidnas during the breeding season in Chinchilla, on the western downs of Queensland (S. Johnston pers. obs.). On Kangaroo Island trains of up to 10 males have been reported, although the most common numbers are 1–3 males (Rismiller 1992; Rismiller and McKelvey 2000). In Tasmania, trains are much less commonly seen, particularly early in the breeding season, and mating usually occurs in hollow logs, basal tree hollows and wombat burrows. Most females were found with only one male at a time and the maximum recorded number of males in a group was four, although over the whole mating season Tasmanian short-beaked echidnas were found with similar numbers of males to those on Kangaroo Island (Nicol *et al.* 2019b).

On Kangaroo Island the short-beaked echidna trains have been recorded to continue for as many as 36 days, with all males dispersing within 48 h of one copulating with the female. Rismiller and McKelvey (2000) described four stages of courtship and mating: (1) males stay close to the female and often forage or travel in a loose group within 3–5 m of each other; (2) males walk beside and behind the female with bodies touching and those males that are not flanking the females try to dislodge those that are; (3) males remain close to the female at all times and their beaks nudge and prod her or they stroke her with their front paws, which results in the female digging in and erecting her spines causing the males to retreat; (4) mating occurs. The female lies flat on the ground when prodded by the male with the spines in a relaxed position. If one male is present, he strokes her spines with a forefoot and lifts her tail with his hindfoot. When several males are present, they all dig beside her and try to push the others out of the way, resulting in a trench around her. When one animal remains, usually the largest, he continues to dig until his tail is placed under the female's so that his cloaca is on hers. Mating then occurs for 30–180 min.

In a captive facility in south-east Queensland, male 'train' behaviour is not necessary for successful breeding in captivity. Generally, one male will be introduced to ensure genetic management. Observations show that the male will follow closely behind the female when he is sexually interested and he will persist until the female shows receptivity. The male will dig behind and slightly underneath her at the tail, then curls on his side into a 'C' shape and places his tail under hers. The female will lift her tail slightly so his cloaca is near hers for mating (Wallage *et al.* 2015; Dutton-Regester *et al.* 2021, 2022a). These captive studies also showed that in breeding pair behaviours, including 'urogenital sniffing' (male sniffs the female urogenital region), 'rolling' (male rolls on his side at the caudal region of the female) and 'copulation attempt' (male attempts to mate with female and holds position), showed a significant positive increase over time before copulation and a negative rate of change after copulation. The frequency of these behaviours by males towards females can be used to indicate that copulation has likely occurred. The time females spent 'pacing' (along lengths of enclosure) significantly increased and both the 'time eating' and the 'quantity of food eaten' significantly decreased in pregnant females on the day of oviposition. These studies were not able to identify oestrus from specific behaviours; however, they suggested that female echidnas are receptive for less than 24 h (Dutton-Regester *et al.* 2023).

In Tasmania the breeding season follows a period of deep hibernation, with males in the best condition emerging first in early May and the earliest matings occurring 1 month later. Males use olfactory cues to find females, even though they are still hibernating, and males appear to be able to then use these cues to distinguish between females that are in good reproductive condition and those in poorer condition without actually disturbing them (Harris *et al.* 2014, 2016). Females that mate in a given year are significantly heavier than those that do not mate (Nicol *et al.* 2019b). Males stay with the female until she has at least partially aroused from hibernation, (a period of 8–12 h) and then mate. Early in the mating season females will re-enter torpor after mating, even when pregnant. Although most females become pregnant in their first mating group, males continue to seek them out and mate with them. This is costly for females as it is energetically expensive, disrupts their feeding and sometimes leads to abrasions around the cloaca that may become infected (Nicol *et al.* 2019b). Males may even dig out and mate with females in a nursery burrow, leading to loss of the egg or death of the young (Harris and Nicol 2014). Because of this, in captive breeding programs the male should be removed several days after the first observed copulation. Currumbin Wildlife Sanctuary observed that male Queensland short-beaked echidnas can be left in the enclosure for up to 14–15 days after mating as interest subsides after mating and the two animals do not cross paths. The male can stay with the female until she is nearing her egg lay (S. Johnston pers. obs.). In contrast this does not appear to be the case in Tasmania (S. Nicol pers. obs.) or on Kangaroo Island, and probably not so in Victoria, so the general advice would be to remove males after mating, particularly if there is more than one male.

Rismiller and McKelvey (2000) observed Kangaroo Island females closely during the courtship period and no female was seen to copulate more than once, which contrasts to those in Tasmania, where fresh sperm on cloacal swabs taken from females unambiguously demonstrated that multiple copulations had occurred (Morrow *et al.* 2017; Nicol *et al.* 2019b).

Throughout Australia, short-beaked echidna mating occurs during the period June–September, with little difference between climatic regions (Beard *et al.* 1992; Rismiller and McKelvey 1996; Beard and Grigg 2000; Rismiller and McKelvey 2000; Morrow *et al.* 2009; Nicol *et al.* 2019b). A small number of matings observed in October in south-east Queensland and Tasmania have involved females that lost their first young and mated a second time (Beard and Grigg 2000; Harris and Nicol 2014; Nicol *et al.* 2019b).

In some regions, mating follows a period of reduced activity. Short-beaked echidnas in Mount Kosciuszko National Park have shown periods of deep hibernation between April and August, with body temperatures falling as low as 4°C (Grigg *et al.* 1989, 1992). At Stanthorpe in south-east Queensland, short-beaked echidnas showed periods of hibernation between April and September, with a minimum body temperature of 12°C, although not all animals hibernated in a given year (Kuchel 2003). In contrast, captive animals in coastal south-east Queensland rarely display torpor, and there is no evidence of hibernation, before the breeding season (S. Johnston pers. obs.).

On Kangaroo Island males and female showed reduced activity and periods of several days of torpor between April and August, with body temperatures as low as 12°C (Rismiller and McKelvey 1996). In Tasmania, males may enter hibernation as early as late January and hibernate until the end of August, while females may enter hibernation in mid-February and hibernate until mid-October. Short-beaked echidnas that mated in a given year emerged earlier than non-mating individuals: males from early May to mid-July, females from early June to early September (Nicol *et al.* 2019b). The differences in use of torpor and hibernation may be due to physiological differences between the *aculeatus/setosus* and *acanthion/multiaculeatus* clades as well as climatic differences. Generally, short-beaked echidnas from the warmer parts of Australia seem to have more variable body temperature ($T_b$) when active than animals from cooler climates (Nicol 2017). During entry into hibernation, $T_b$ falls towards the temperature of the substrate that the short-beaked echidna is buried in and Tasmanian animals appear to be unable to maintain a prolonged hibernation when soil temperature and thus body temperature is above 17°C (Nicol 2017); they also appear to seek out cooler areas to hibernate (Nicol and Andersen 2007). Early in the hibernation season, Tasmanian short-beaked echidnas may show short periods of torpor and then rewarm when the ambient conditions warm slightly, similar to the pattern seen in winter in warmer parts of Australia.

Although the reproductive physiology of wild and captive short-beaked echidnas is likely to be similar, other than geographical differences with respect to the occurrence and use of hibernation, there are nevertheless also differences in behaviour related to their captive reproductive management, especially associated with the appropriating timing of joining (e.g. detection of oestrus and the introduction of the male). The specifics of captive short-beaked echidna reproductive management are outlined in section 10, an important technical consideration in this process is the need for high resolution video monitoring of the enclosure. At Currumbin Wildlife Sanctuary's dedicated off-exhibit breeding facility, video cameras provide 24-h surveillance of breeding activity , which has greatly contributed to the monitoring of reproductive status and breeding success (Wallage *et al.* 2015; Dutton-Regester *et al.* 2021). Infrared cameras are set up to monitor the entire enclosure, including cameras within the burrow and nest boxes. As most captive short-beaked echidna breeding activity is crepuscular and/or nocturnal, afterhours monitoring of reproductive activity via playback is essential. It also means time and effort must be dedicated to monitoring the video recording during the next day to carefully map key reproductive events such as courtship behaviour, mating, egg lay and incubation. These behaviours can then be mapped with reproductive hormone profiles using blood (Dutton-Regester *et al.* 2021) or faecal samples, which allows non-invasive monitoring (Dutton-Regester *et al.* 2022b, 2022c); monitoring of faecal progesterone metabolite concentrations, although suitable for the assessment of the luteal phase of the oestrous cycle/pregnancy, are not as informative as plasma samples, especially in the early stages of the luteal phase.

The reproductive behaviour of captive short-beaked echidnas has been described by Wallage *et al.* (2015), together with supplementary videos of the key behaviours, which include courtship behaviour with the male closely following the female for a period of up to 6 h and the male probing the female's urogenital sinus with his snout. The infrared video cameras occasionally also capture the act of copulation, which revealed the female positioned in the corner of the burrow box with only her tail visible and with the male at her caudal extremity using his forelimbs to lift her to gain access to her urogenital region. At the same time, the male then curled himself around the female, forming a 'C' shape, and positioned his urogenital region underneath her, allowing his erect penis (Johnston *et al.* 2007a) access the female's urogenital opening. Although it was not possible to observe actual intromission, once the male had appeared to penetrate the female there was a subsequent period of thrusting (in this case) for 60 min, although other observations of copulation have lasted from 30 min up to 90 min, including multiple copulations on the same night or over multiple days (3–7 days). The majority of copulations (53/59) observed by Wallage *et al.* (2015) occurred between 18:00 and 06:00 h, with 23 of these documented between 22:00 and 02:00 h. Post-copulation or disengagement was defined by a sudden lack of courtship or interest by the female when approached by the male, suggesting that only receptive females are likely to breed. At Currumbin Wildlife Sanctuary there is generally a single strong mating in the evening and if successful, the male will slowly lose interest over the following 4–5 days. A more recent study of reproductive behaviour of the captive short beaked echidna over a 3-year period (Dutton-Regester *et al.* 2022a) took a quantitative approach to exploring oestrus, pregnancy and oviposition. This study found for breeding pairs that there was a significant rate of change over time before and after copulation for the behaviours of 'urogenital sniffing', 'rolling' and 'copulation attempts'. For pregnant females, time engaged in 'pacing' significantly increased while 'time eating' and the 'quantity of food eaten' significantly decreased on the day of oviposition. While this study was not able to identify oestrus from specific behaviours, it suggested that in contrast to what was seen in Tasmanian echidnas in the wild (Morrow *et al.* 2017), in these northern females the period of sexual receptivity may be less than 24 h.

The reproductive anatomy of the short-beaked echidna penis has been described by Johnston *et al.* (2007b) and more recently by Fenelon *et al.* (2021). Johnston *et al.* (2007b) also reports on the unique erection and ejaculation behaviour of the short-beaked echidna in which the flaccid penis with four definable glans penis rosettes transforms into a fully erect penis with only two rosettes in which the semen pools (Fig. 3.15).

Although there appear to be no specific gestational behaviours per se, there have been recent observations of egg-laying (oviposition) behaviour recorded at the Currumbin Wildlife Sanctuary short-beaked echidna research colony (Dutton-Regester *et al.* 2022a). The following is a description of one such observation from the animal caretaker's notes.

'Females seem to have no interest in food the evening of egg lay; they are active as normal until a few hours prior and they then become what appears to be uncomfortable or agitated, walking back and forth over a repetitive path of very short distance. It is shortly after this time that the female will generally lay in lateral recumbency in a 'C' shape to position her cloaca near her pouch. She will lay the egg into the pouch area and assist the egg with her beak. Although the tongue is active during egg lay behaviour, it is not used to physically manipulate egg deposition' (A. Roser and H. Meer pers. obs.).

Perhaps the most obvious reproductive behaviour in the captive short beaked is that when the female is incubating an egg in her temporary pouch (Wallage *et al.* 2015; Dutton-Regester *et al.* 2021). During incubation, which typically lasts for ~10 days, the female remains in her burrow box or is secluded under a log or other furniture item within the enclosure and does not move, defecate or eat. Once hatched from the egg, and the puggle is secured within the pouch, the mother resumes her normal active behaviour (feeding and movement). The timing of these behaviours (gestation, oviposition and incubation) has recently been confirmed by direct observations of females that were repeatedly anaesthetised, bled and pouch inspected for a study that documented the luteal phase of the short-beaked echidna (Dutton-Regester *et al.* 2021). The duration and lack of activity during this period was

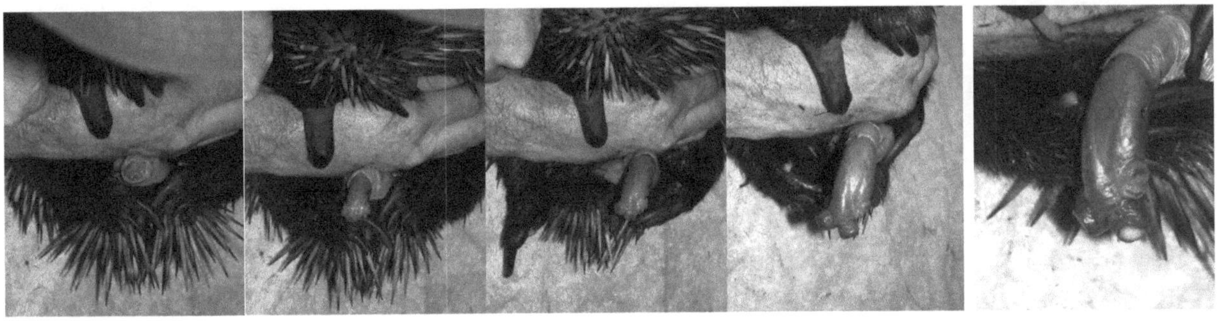

**Fig. 3.15.** Erection of the penis of the short-beaked echidna. Modified from Johnston *et al.* (2007b).

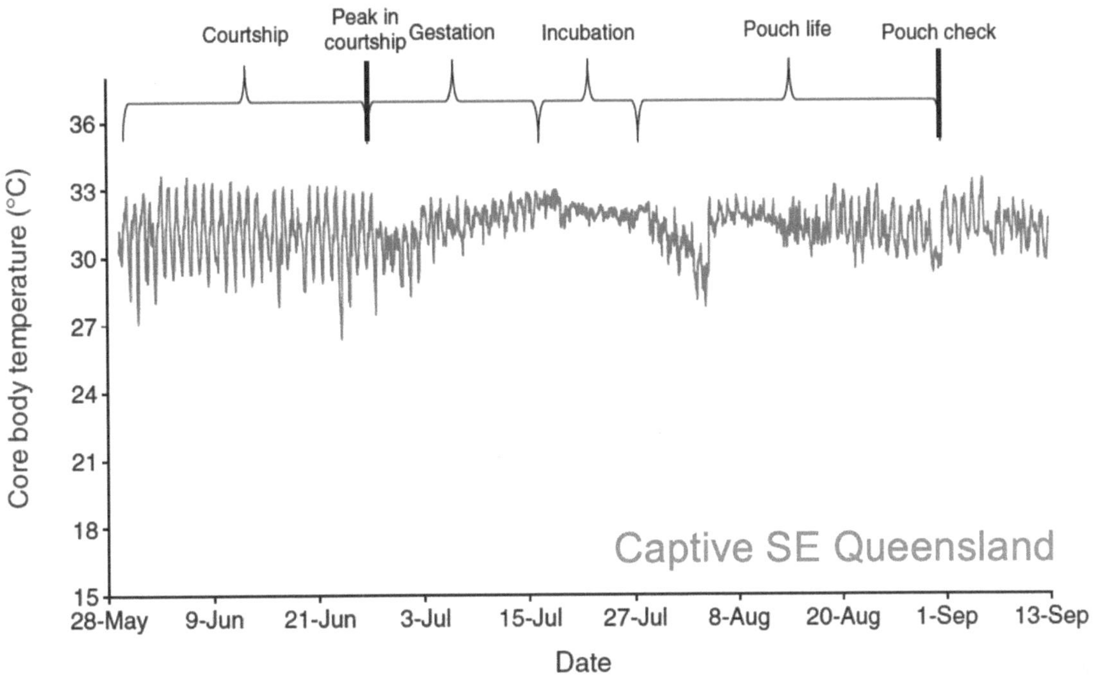


**Fig. 3.16.** Changes in core body temperature of a south-east Queensland captive female short-beaked echidna during key reproductive events. Reproduced from Wallage *et al.* (2015) with permission from CSIRO Publishing.

also mapped by a consistent period of homothermic regulation of the mother's body temperature; outside of this period the short-beaked echidna shows a significant degree of heterothermy (Beard *et al.* 1992; Nicol and Andersen 2006; Wallage *et al.* 2015; Fig. 3.16). Although this change in body temperature is diagnostic of incubation behaviour, the implantation of thermal transmitters into the body cavity is not practical for routine monitoring, or even necessary, given the distinctive change in the activity pattern of the incubating mother.

## 9.4 Bathing

Short-beaked echidnas will utilise water: they will readily walk through it and are good swimmers, being able to swim across creeks, rivers and dams (Rismiller 1999). At Currumbin Wildlife Sanctuary bathing tubs/pools ~60 × 30 cm are provided during the warmer months, but not to those who have puggles to prevent possible drowning. The short-beaked echidnas spend a significant amount of time bathing each day and through the evening.

## 9.5 Behavioural problems

Abnormal repetitive pacing behaviour has been long known in captive short-beaked echidnas, with records of it as early as 1961 (Heniger and Kummer 1961, in Brattstrom 1973). Animals housed in enclosures of small size and low complexity; and bachelor males housed within sensory range of, but excluded from, reproductive females during the breeding season, appear to show the highest incidence of pacing.

## 9.6 Signs of stress

Signs of acute stress in captive short-beaked echidnas include increased rate and depth of respiration that leads to more pronounced 'snuffing' (serous nasal discharge) (Spielman 1994). Other signs include defensive mechanisms such as curling up and withdrawing limbs. Signs associated with chronic stress include decreased appetite, weight loss and increased escape behaviour, leading to stupefaction/inactivity if the stress continues long enough (Spielman 1994). Some animals may shake, urinate or defecate when being handled. Russell *et al.* (2022) have recently reported the validation of a faecal glucocorticoid assay for the captive short-beaked echidna, which should prove useful for the non-invasive assessment of stress in this species.

## 9.7 Behavioural enrichment

Behavioural enrichment should aim to complement the evolutionary features and behavioural traits of short-beaked echidnas in the wild. It should include: 1) opportunities to pull apart termite mounds or rotten decaying logs, substrate for digging and changing/adding new substrates periodically to provide renewed interest and exploration; 2) areas of leaf litter mulch, areas of softer sand/loam or harder gravel will increase interest and exploration; 3) moving the location of feeders periodically to encourage exploration and foraging behaviour and varying the type of feeding devices used to offer food also increase interest; and 4) half hollow logs and tight spaces for short-beaked echidnas to wedge into so they have places that give an increased sense of security. Heat lamps or natural sun basking opportunities should be provided and a living enclosure, including live plants such as grasses, shrubs and trees, will promote invertebrate activity in the enclosure substrate layer for exploration and interest.

There are several behavioural enrichment activities that can be pursued to stimulate short-beaked echidnas in captivity:

- making the enclosure surface as varied as possible with the soil profile and the addition of furniture such as hollow logs and rock piles that allow animals to investigate. Take care to ensure that the rocks are of adequate size and properly placed so they are not dislodged by the short-beaked echidnas' digging, as this may cause injury
- adding piles of soil, sticks and leaf litter to investigate
- providing freshly cut browse (old eucalyptus fodder from the koala enclosure) for olfactory stimulus
- providing a water body such as a pond or small stream for bathing
- periodic 'free walks' under supervision throughout the zoo
- food enrichment in the form of live termite mounds and ant colonies or the use of food containers in which short-beaked echidnas need to use cognition to gain access to a food reward (Russell 2014)
- providing scent trails and inanimate objects to investigate (e.g. cardboard boxes, balls, ramps, soft toys), changing enclosure furniture or moving it within the enclosure
- addition of new enclosure mates
- induction of foraging behaviour via the addition of bark piles or the use of slow feeders.

## 9.8 Training and conditioning

Training has typically not been used in captive short-beaked echidnas, though a male animal at Currumbin Wildlife Sanctuary has been trained to ejaculate on demand (see Johnston *et al.* 2007b). A hand-reared short-beaked echidna was conditioned to enter a crate, move along a specified pathway on a staged area and exit the stage. She was also conditioned to step onto scales for weighing (B. Laming pers. obs.). At another facility short-beaked echidnas were trained to weigh themselves on a scale within their enclosure so that they did not need to be caught (Cable 2021). This was achieved by placing a large electronic scale within the enclosure that was low enough for the animals to climb on. The scale had an aluminium canopy over it so that the animal had to position fully on the scale before it was weighed. Logs placed at each side of the scale made it easier for each animal to get onto the scales. The animals were encouraged onto the scales by placing a small amount of food within a plastic feed dish on the scales and theanimals were slowly desensitised to the food bowls and scales over a period of time. Some animals were observed to quickly adapt to feeding on the scales while others took several months (Cable 2021).

Captive short-beaked echidnas have also been conditioned/habituated to wake for their food either to administer oral treatments in the food or for voluntary participation in encounters with visitors or research activities. Assessed on an individual basis, some will choose not to participate; some may choose to eat then return to sleep until dark. Conditioning with a cue (such as a tap or click noise) can be rewarded with food as a positive reinforcer (A. Roser pers. obs.).

### 9.9 Introductions and removals

Both introductions and removals are easily undertaken, with no noticeable adverse effects on any of the animals involved. However, introductions should be closely observed with particular attention given to relative food consumptions of each individual to avoid one of the enclosure mates from dominating the food.

### 9.10 Intraspecific compatibility

Although large numbers of short-beaked echidnas of either sex can be held together with few problems, it is still important to have enough shelter per animal and to ensure that all animals are able to receive sufficient food (Brattstrom 1973). Although males and females can be housed together out of the breeding season, Currumbin Wildlife Sanctuary generally houses males and females separately all year round, except for breeding. Juveniles can be housed together and also, once successfully weaned, can be housed with adults.

Caution should be taken when housing juveniles with adults as a general rule. There is much variation in the development of young and some of the smaller animals may benefit from more cautious housing options until they are of sufficient weight and vigour to compete with larger animals for food and shelter.

Dam aggression towards puggles has been observed. On two occasions at Currumbin Wildlife Sanctuary puggles were either found dead or had sustained fatal injuries with no conclusive reasons for the occurrence (i.e. no video evidence). On another occasion, a video recorded a female 'attacking' (digging/gouging) her puggle with clear intent to harm. The puggle was subsequently removed and hand raised (A. Roser pers. obs.). There are several factors that may have caused this behaviour, including the dam being a first-time mother, her inability to confine her puggle in the burrow or poor health or development of the puggle that was not detected by the veterinary or keeper staff.

### 9.11 Interspecific compatibility

Short-beaked echidnas have been held with a wide number of species, including koalas (*Phascolarctos cinereus*), long-beaked echidnas, bettongs (*Bettongia* and *Aepyprymnus rufescens*) sugar gliders (*Petaurus breviceps*), tree shrews (*Tupai* sp.), lesser pandas (*Ailurus fulgens*)and various species of birds (S. Jackson pers. obs.; Fieseler and Junge 1997). Short-beaked echidnas can potentially be placed with numerous different species; however, due to their generally 'shy disposition' they should not be housed with species that make frequent abrupt noises when moving or calling, or with species that may interrupt their rest periods. It is also not recommended to co-house with other species during the breeding season in order to avoid any disturbance during breeding/incubation/puggle in burrow. Recently, short-beaked echidnas have been recorded predating the eggs of two species of freshwater turtle eggs (Robinson *et al.* 2024), so holding echidnas with these species is not advisable if intending to breed the turtles.

## 10. BREEDING

### 10.1 Mating system

Although there appear to be differences in some aspects of reproductive behaviour between the short-beaked echidna subspecies (see above), wild animals all appear to have a roaming male mating system (Clutton-Brock 1989) characterised by intense competition (including the suggestion of sperm competition) between males to mate with females (Jones *et al.* 2004; Nicol *et al.* 2019b; Rismiller and McKelvey 2000).

### 10.2 Ease of breeding

Historically, short-beaked echidnas bred very poorly in captivity, with records of them being bred on only a few occasions. Berlin Zoo bred this species in 1908 (Heck 1908), Basel Zoo bred them in 1955 and 1967 (Lang 1958, in Crandall 1964; Augee *et al.* 1978) and Taronga Zoo bred them in 1977 (Augee *et al.* 1978); however, on each occasion the offspring did not survive to maturity (Collins 1973). Philadelphia Zoo bred them in 1983 and 1985 (Anon 1986, 1988), Taronga Zoo in 1988 (George 1990), Adelaide Zoo in 1988 and 1989 (Campbell 1989; Muirhead 1989), St Louis Zoo successfully bred and reared a short-beaked echidna in 1996 (Fieseler and Junge 1997) and Oklahoma City Zoo bred them in 1987 (Boisvert and Grisham 1988). More recently, there has been a significant improvement in captive breeding success to the point that they are now routinely bred by several institutions, including Currumbin Wildlife Sanctuary, which bred 13 young to 4 females between 2011 and 2014

(Clark 2013; Wallage *et al.* 2015; Dutton-Regester *et al.* 2021). At Perth Zoo 10 young were bred and reared to independence between 2007 and 2016, with 4 of these offspring being F2 generation from Perth Zoo-bred females (Ferguson and Turner 2013; Ferguson and Laming 2022; A. Ferguson and B. Laming unpubl. data).

It has been speculated that the failure of many breeding attempts may well be due to the female having nowhere appropriate to place the young when it is time for them to leave the pouch, as in the wild the young spend all or most (depending on the subspecies, see above) of their life before weaning in a nursery burrow. High rates of offspring mortality during the early stages of the incubation occurred in the Tasmania, with only 20% of young surviving to weaning (Morrow and Nicol 2012). The same study also estimated that 60% of young died either before hatching or in the first 2 weeks of lactation, which may be due to male disturbance of the nursery burrow (Morrow and Nicol 2012). On Kangaroo Island the eggs of all females monitored hatched (Rismiller and McKelvey 2000, 2009). Therefore, it appears that in captivity the successful rearing of young is greatly facilitated by the provision of artificial burrows (section 4.5; Ferguson and Turner 2013; Wallage *et al.* 2015). Breeding success may have also been facilitated in part by removing males after the cessation of receptivity by females, preventing disturbance by the males (Harris and Nicol 2014).

## 10.3 Oestrous cycle, gestation and incubation periods

Although there is circumstantial evidence that ovulation in the short-beaked echidna may be induced (coitus and or seminal) there is currently no direct experimental evidence to support this. Given that the female will typically come back into oestrus if she loses an egg or puggle, it is possible to broadly estimate the length of an oestrous cycle by calculating the period from when she loses the egg until she mates and lays an egg in her pouch; however, this estimate will also depend on how soon the male is introduced following the previous loss. Dutton-Regester *et al.* (2021, 2022c) have recently reported that of four of five females that they observed entering a second gestation period or luteal phase after the loss of an egg or pouch young, the shortest interval was 15 days (range 16–18 days); this then is currently the best estimate of the length of the follicular phase of the short-beaked echidna oestrous cycle. If a 17-day gestation is assumed, this suggests an oestrous cycle length of ~32 days, which is remarkably similar to that suggested by Higgins *et al.* (2004) and Harris and Nicol (2014). Measurement of oestrogen secretion has proven challenging in the short-beaked echidna, both in plasma (Dutton-Regester *et al.* 2021) and in faeces (Dutton-Regester *et al.* 2022b, 2022c), so the full endocrine characterisation of the cycle remains to be documented. Figure 3.17 from Dutton-Regester *et al.* (2021) summarises, by way of two examples, what is currently known about the endocrine profile of the captive short-beaked echidna's oestrous cycle, gestation and incubation.

Except in Tasmania, where females may re-enter hibernation in the first few days after mating (Morrow *et al.* 2017), the gestation period has been calculated to be ~17 days (range 15– 24 days (Morrow *et al.* 2009; Rismiller and Grutzner 2019; Dutton-Regester *et al.* 2021). On Kangaroo Island, *T. a. multiaculeatus* females ($n$ = 10) were routinely checked after copulation was sighted and eggs were found 22–24 days later (Rismiller and McKelvey 2000). The exact timing of gestation is difficult because it depends on precise characterisation of ovulation, which is not yet possible. In *T. a. setosus* females that re-enter hibernation after fertilisation, the embryo may remain at the blastocyst stage until the mother rewarms, giving a gestation period of up to 80 days (Morrow *et al.* 2017); this interesting phenomenon has not been reported for any captive short-beaked echidna in south-east Queensland and appears to be unique to Tasmanian short-beaked echidnas. For Tasmanian echidnas that do not re-enter hibernation the gestation period is 21–23 euthermic days (Nicol and Morrow 2012; Morrow *et al.* 2017). Therefore, the best estimate of gestation period is 20–24 days.

## 10.4 Reproductive status

### 10.4.1 Females

There are clear stages that the female goes through during reproduction (Griffiths *et al.* 1969; Dutton-Regester *et al.* 2021; Figs 3.18, 3.19).

(1) Non breeding – the pouch is poorly developed and flattened with no lateral margins or with slightly raised and turgid skin (Fig. 3.18A–C).
(2) Pregnancy – pronounced swelling of the lateral margins of the pouch creates an increase in pouch depth. The skin is deep pink in colour 3–5 days before the egg is laid into the pouch (Fig. 3.18D–F). Pregnancy can be positively confirmed by detection

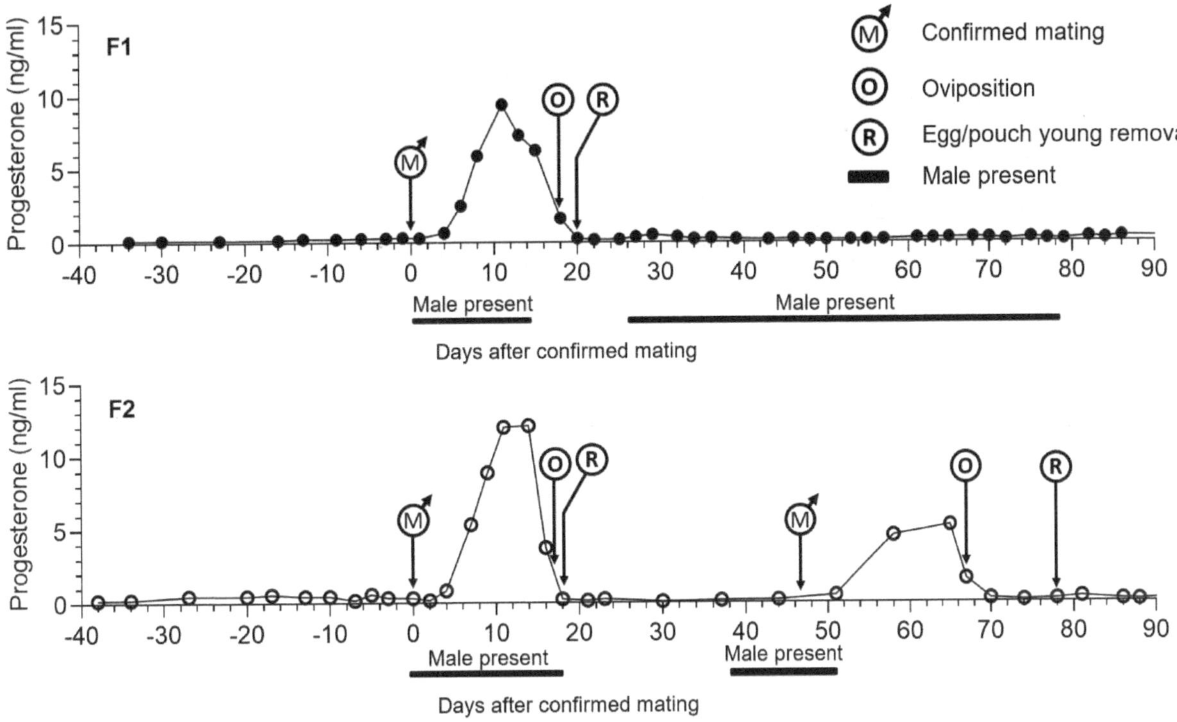


**Fig. 3.17.** Progesterone secretion of two short-beaked echidna following a successful mating, removal of egg/pouch young and the subsequent oestrus (F2 only). Taken from Dutton-Regester *et al.* (2021).

of an egg, via ultrasonography, and is confirmed by an elevated secretion of plasma progesterone (Fig. 3.18; Dutton-Regester *et al.* 2021).

(3) Egg presence - well-developed pouch and mammary glands with an egg inside. Lateral pouch margins draw together tightly to close the pouch (Fig. 3.18G–I).

(4) Pouch young - egg has hatched and puggle is present.

(5) Lactation - pouch empty, large mammary glands and young in the nest.

(6) Regression - lateral margins of the pouch remain but margins appear loose (having lost their turgidity) and the pouch is soft pink in colour. This occurs 12–30 days after the egg or pouch young has been removed (Fig. 3.18J–L).

If pouch young are present, several developmental stages and measurements can be recorded and compared with existing growth curves (see section 10.16) or curves established for future reference.

- Developmental stages
  - sex - not distinguishable due to the testes in males being inguinal
  - ear slit visible
  - eyes open - usually at 60 days of age
  - fur visible - slight tinge, medium or well developed
  - spines visible
  - eating solids
  - self-feeding
  - independent
- Measurements (see Appendix 3)
  - bodyweight (g)
  - head length - from occiput to snout tip (mm)
  - head width - maximum width across the zygomatic arches (mm)
  - crown-rump length (mm) - primarily for very small neonates
  - body length (mm) - from the snout tip to cloaca
  - tail length (mm) - from the cloaca to the end of the last vertebra of the tail tip
  - total length (mm) - from snout tip to tail tip
  - spur length (mm) - and development in both males and females
  - pes length (mm) - not including the toenail; measure the same pes each time.

### 10.4.2 Males

Although the penis is obvious when erect, it is normally retracted into the preputial sheath, which lies closely opposed to the ventral body wall, just above the opening

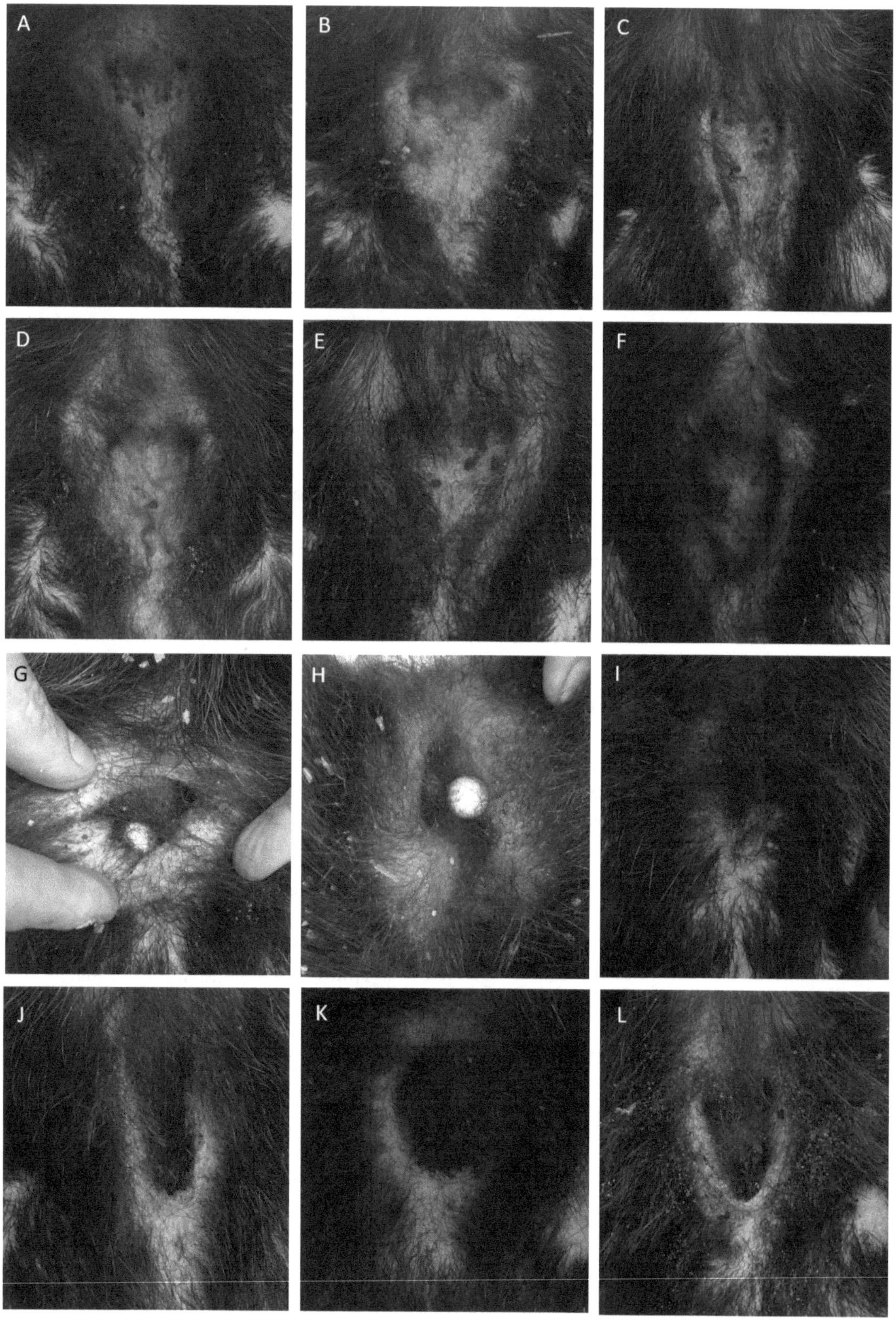


**Fig. 3.18.** Stages of pouch development in a female short-beaked echidna. Taken from Dutton-Regester *et al.* (2021).

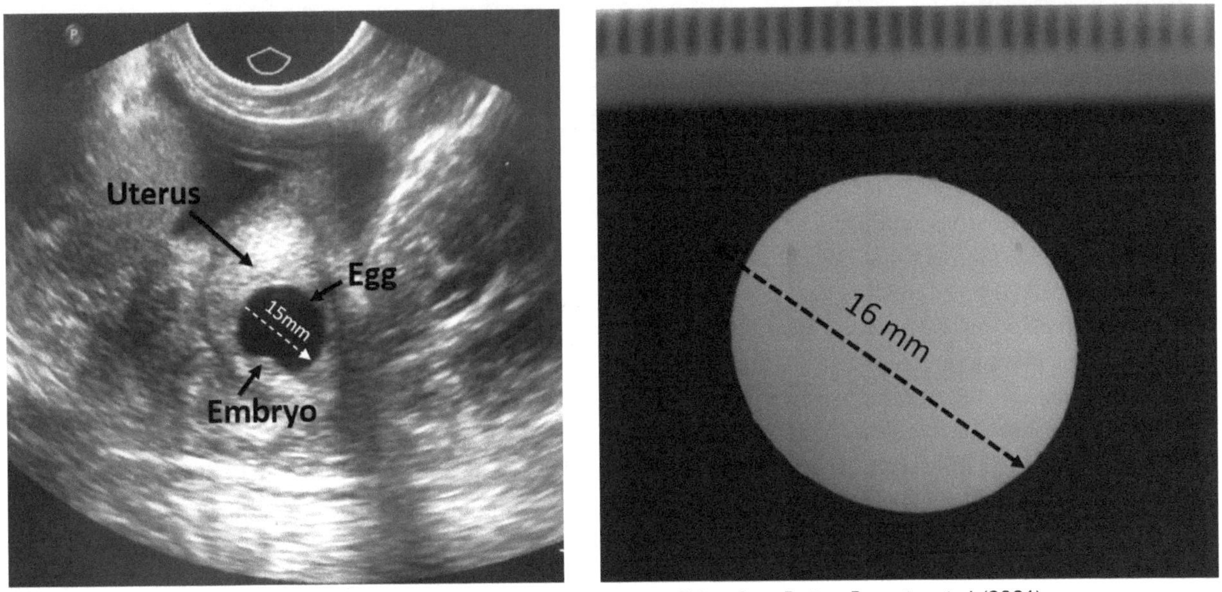


**Fig. 3.19.** Egg detected in utero by ultrasonography and a freshly laid egg. Taken from Dutton-Regester *et al.* (2021).

of the cloaca (Fenelon *et al.* 2021). In reproductively active males, the retracted penis produces a visible bulge around the cloacal region and can be easily palpated and everted (Johnston *et al.* 2006); the penis is particularly noticeable during the breeding season. Johnston *et al.* (2007a) has described several techniques for the assessment of reproductive status in the male short-beaked echidna, which should be conducted at the start of the breeding season. The procedures are performed while the animal is under general anaesthesia and includes the recovery of spermatozoa flushed from the urethra via application of gentle pressure on the bladder while the animal is in dorsal recumbency. Spermatozoa recovered in this way (known as spermatorrhoea) can be used to simply confirm the production of sperm by the testis. The spermatozoa recovered are individual, typically non-motile and not in their normal bundle arrangement as occurs when semen is ejaculated (Johnston *et al.* 2007b). Changes in testicular and Cowper's gland volume can be also readily assessed using ultrasonography. Johnston *et al.* (2007a) have also described the use of a GnRH or hCG (human chorionic gonadotropin) challenge as a reliable measure of the androgenic capacity (testosterone) of the testis. The male short-beaked echidna shows a strong seasonal reproductive cycle (Johnston *et al.* 2007a; Morrow *et al.* 2016).

## 10.5 Reproductive management

Up until quite recently institutions have had limited success in breeding short-beaked echidnas, but improved knowledge of their physiology and behaviour has led to increasing success. St Louis Zoo found that over a 15-year period when three males and a female were housed together, no mating was observed. However, after the death of two of the males and a change in the diet offered (see section 6.1) the single male and female successfully produced and weaned a young (Fieseler and Junge 1997). Fieseler and Junge (1997) suggested that the presence of the three males confined in close proximity to the female may have resulted in a lack of normal oestrous cycles and conception, infant care or a combination of these. Additionally, the enclosure was made more secure to decrease disturbance from the public. The depth of the substrate may also be important in allowing the male to get below the female to gain access to her cloaca for mating. Contrary to observations by Fieseler and Junge (1997), a short-beaked echidna was bred at Oklahoma City Zoo where the sex ratio was 2:2 (Boisvert and Grisham 1988).

At Currumbin Wildlife Sanctuary several measures have resulted in high breeding success. Leading into the breeding season (April) both male and female short-beaked echidnas are fed a diet rich in additional olive oil and fly pupae (see section 6.1). In early June, all short-beaked echidnas to be used in the breeding program are scheduled for a full veterinary health check and the males assessed as per section 10.3.2. In mid-June females are established in their own enclosure fitted with an infrared light heat source, burrow and nest box. The enclosure is also fitted with video cameras that allow the recording of behaviour over 24 h and which is monitored (in fast forward mode) the following morning by staff, volunteers or university students.

As distinctive oestrus behaviour is not normally present without the presence of the male, a suitable genetically compatible, sexually mature, proven male is then placed in the enclosure and closely monitored. It is important to note here that multiple males are not used to elicit 'train' behaviour and that this behaviour is not needed for successful breeding. It is not uncommon for mating to take place the day/evening after the male is introduced and mating and repeated mating can still occur within days of the male's first entry. Courtship and copulation behaviour are described in section 9.3. Once the female ceases to be receptive (and/or) the male loses sexual interest, the male is removed from the enclosure. As estimates of gestation in captivity are from 16.5 to 20 days (Wallage *et al.* 2015; Dutton-Regester *et al.* 2021), video is carefully monitored from day 14 post first mating for signs of egg lay or a sudden change in the female's behaviour that would indicate incubation (see section 10.3.2).

From a research perspective, it is possible to remove the female for pouch inspection for the presence of the egg, but for routine breeding purposes this is not recommended. Although incubation behaviour is typically strong evidence of an egg in the pouch, from time to time the female may produce a degenerative egg that fails to develop; in these circumstances, it is difficult to ascertain whether this phenomenon is associated with a lack of oocyte fertilisation or embryonic loss. It is still unknown whether an egg is ovulated spontaneously or induced by coitus (Dutton-Regester *et al.* 2021), but circumstantial evidence would appear to favour the latter, based on the observation that females during the breeding season without access to the male have not been observed to produce eggs. In wild short-beaked echidnas, in the majority of cases, fertilisation occurs at or soon after the initial mating. The end of the incubation period coincides with the hatching of an egg and demarcates a significant change and increase in the female's activity pattern, including recommencement of feeding (Rismiller and McKelvey 2009). Although the pouch can be checked post hatching while the short-beaked echidna is anaesthetised, there is always a risk that capture of the female or the pouch check procedure itself may result in loss of the pouch young.

## 10.6 Occurrence of hybrids

None.

## 10.7 Timing of breeding

Short-beaked echidnas are strongly seasonal breeders and throughout Australia mating occurs during the period June to mid-September (see above). Captive bred short-beaked echidnas at Perth Zoo have been strongly seasonal breeders, with reproductive activity being observed between June and September. At Currumbin Wildlife Sanctuary, captive animals have bred and produced young from June through to early November. Successful captive breeding of animals at Berlin Zoo and Basel Zoo in the northern hemisphere occurred in March and May (Collins 1973).

## 10.8 Age at first and last breeding

The minimum age at first breeding in the wild is 5 years for both sexes (Nicol *et al.* 2019a; Rismiller and McKelvey 2003). For captive short-beaked echidnas in Western Australia, zoo-born females as young as 4 years old have bred and a zoo-born male successfully sired offspring at 3 years of age (Ferguson and Laming 2022). At Currumbin Wildlife Sanctuary, a zoo-born female known to be 3 years old successfully produced young. To date, F2 males at Currumbin Sanctuary have not been reproductive until 4.5 years of age. Because short-beaked echidnas of both sexes appear to live until over 30 years of age it is likely that they can breed throughout this time. An individual male was estimated to be at least 36 years of age at Currumbin Wildlife Sanctuary and sired at 35 years of age. At Brookfield Zoo in the USA a pair of *T. a. multiaculeatus* successfully bred at 47 years of age (D. Vinci pers. comm.).

## 10.9 Ability to breed every year

The ability of short-beaked echidnas to breed depends on their condition; wild animals in good condition are capable of breeding on successive years (Nicol *et al.* 2019b). In some geographical regions wild females that lose their egg or pouch young may breed in the following year, although it is less likely that females that have taken a young through to weaning will breed in the following season because they usually do not have time to rebuild their energy stores. When mating takes place, it is almost always successful in producing a viable egg (Rismiller and McKelvey 2000; Nicol *et al.* 2019b). Approximately 20% of pregnancies of wild Tasmanian short-beaked echidnas resulted in a young surviving to weaning (Morrow and Nicol 2012), compared with 50% of Kangaroo Island young that survived to weaning (Rismiller and McKelvey 2000).

Captive short-beaked echidnas in the breeding program at Currumbin Wildlife Sanctuary have been shown to produce eggs and/or puggles on successive years, although there has been variability noted in which some years being more successful than others, despite consistency in husbandry. At Perth Zoo one female successfully raised young to independence over 3 consecutive years while four other females that successfully reared young to independence did not breed in consecutive years (Ferguson and Turner 2013; Ferguson and Laming 2022; A. Ferguson and B. Laming unpubl. data). In wild Tasmanian short-beaked echidnas, adult females that mated in a given season were significantly heavier at the end of the preceding summer than females that did not mate (Nicol *et al.* 2019b). Short-beaked echidnas have large deposits of subcutaneous and intra-abdominal adipose tissue and the large seasonal variation in the bodyweights of wild animals are associated with changes in these fat stores (Nicol *et al.* 2019b). Research into the building up of these stores at the appropriate time of the year is likely to further improve the success of captive breeding. The relationship between nutrition and reproduction is poorly understood in most Australian wildlife.

## 10.10 Ability to breed more than once per year

Wild short-beaked echidnas typically breed only once per year; however, as noted above a small number of matings observed in October in south-east Queensland and Tasmania have involved females that lost their first young and mated a second time (Beard and Grigg 2000; Harris and Nicol 2014; Nicol *et al.* 2019b), although very few of these resulted in young that weaned.

Captive female short-beaked echidnas at Currumbin Wildlife Sanctuary have been observed to regularly cycle again and produce an egg after losing an egg or puggle. One female has even been able to produce three eggs in the same breeding season (Johnston, Roser and Meers, unpubl. data); these observations clearly confirm that short-beaked echidna are polyoestrus and that lactation results in anoestrus. The ability to successfully breed in the same breeding season following the loss of a puggle is related to the energy stores available to the female.

## 10.11 Nesting requirements

Provision of an area for female short-beaked echidnas to dig a burrow appears to be very important for the successful rearing of puggles, as a significant part of their preweaning growth normally occurs in a nursery burrow.

Soon after a captive female has successfully mated, she will enter an artificial burrow (Fig. 3.3) for egg incubation. Unlike their wild counterparts the burrow entrances of captive short-beaked echidnas remained open during the incubation process. Three females were observed to remain within a burrow for 11–13 days without feeding or drinking, which resulted in the hatching of five young (Ferguson and Turner 2013). At Currumbin Wildlife Sanctuary the incubating females have shown no particular preference as to where they incubate. They have incubated under hollow logs, under nest boxes, in nest boxes (these have minimal sand ~10 cm), in leaf litter; one female incubated out in the open next to a rock in her enclosure. Incubation was 10–11 days and if the female remained in her incubation position, video footage showed that she was physically awake with movement or readjustment of the body position but not changing her location. During this time the females do not eat, drink or defecate and this can continue for a few days after incubation; it is also not unusual for the females to go 14–15 days without food but generally they will emerge to eat once the egg has hatched at day 10. Raking the area surrounding her incubation location from day 10 will show the animal caretaker if the female has moved overnight and provide a clue that she is out of the incubation stage via the presence of tracks. After 10 days, most females will come out, defecate, eat and then return to the incubation location. This can continue for days following the hatching, and even up to day 20 of puggle development, until her behaviour returns to her new normal (carrying the puggle). Although each female shows variable behaviour once out of incubation, there is always a striking increase in her activity post incubation.

If possible (depending on the burrow) puggles/dams need to be monitored carefully (most easily done with cameras), to ensure they are feeding regularly and to observe any aggression from the dam. Assistance may be required with some dams (particularly first-time breeders) that may not back-fill their burrow sufficiently thus allowing burrow temperatures to vary outside appropriate limits and allowing rodents to enter the burrow. Poorly sealed burrows may also allow the puggle to roam when it is not old enough to do so. Dams will feed puggles in the burrow anywhere from 1- to 10-day intervals, with 4–5 days being the average. After 10 days of not feeding, intervention will most likely be required for hand-rearing or supplemental feeding.

Once the puggle is left in the burrow/box and has been confined by the mother, cover the box with foil insulation sheeting and browse, monitor the temperature closely and add ice bricks under the insulation and on top of the box lid to keep the temperature below 25°C on hot summer days. Double-washed river sand is the preferred substrate at Currumbin Wildlife Sanctuary and this is moistened with a water spray bottle multiple times during the day to keep temperatures down and to allow the female to mould the sand when she is constructing and deconstructing the burrow to feed the puggle. The female can take 1 h to deconstruct the burrow before feeding and then up to 4 h to reconstruct it and confine her puggle securely until the following feed, which may be days later.

Prior to the female creating a burrow in which to leave her puggle (this occurs usually from 2 to 2.5 months depending on the age, the experience of the mother and her pouch size), the animal caretaker should watch for signs of the female digging around the enclosure. If it is too early, do not add additional sand or a second box because this can inadvertently encourage her to 'dump' her puggle early. Video monitoring and thorough checking of the enclosure first thing the following morning to look for a puggle around this time is highly recommended. If the female is digging and the puggle has reached 60 days of age, an additional nest box is added to the enclosure and both boxes are two-thirds filled with sand/loam, giving the dam a choice to burrow and leave her puggle. She may still choose to burrow elsewhere, but this provides a structured option to support close monitoring.

## 10.12 Breeding diet

Taronga Zoo and Currumbin Wildlife Sanctuary use a diet that contains a higher level of fat (see section 6.1) during the breeding season for short-beaked echidnas, though most institutions do not use a specific diet for the breeding season and have still had success in breeding. Portion sizes may need to be increased for lactating females, to account for their higher energy requirements.

## 10.13 Number of young born

Typically, a single egg is laid, though there are rare records of twins being raised (Griffiths 1978; Rismiller and McKelvey 2000; Pierce *et al.* 2006; Wallage *et al.* 2015; Dutton-Regester *et al.* 2021). Johnston *et al.* (unpub. data) have also reported the occurrence of triplets, but none of them survived within the pouch. Wallage *et al.* (2015) reported the successful rearing of two young by a captive female that placed one young in an artificial burrow that was provided and the other in a self-made underground burrow. The young were fed on different nights. One of the puggles was markedly smaller than the other and ultimately the smaller offspring required some intervention whereby the larger puggle was weaned slightly earlier to leave the smaller one with the mother to continue feeding. The smaller puggle was in an underground burrow where it was not possible to monitor its health until it started leaving the burrow. Clearly, captive females are prone to produce multiple eggs and young, but survival of these puggles is likely to be low; Johnston, Roser and Clark (unpubl. obs.) have also observed that the tendency for captive short-beaked echidnas to produce multiple offspring appears to be a characteristic of individual females.

## 10.14 Age at weaning

Weaning time varies greatly between the subspecies: in Tasmania and south-eastern Australia weaning occurs at 130–160 days and on Kangaroo Island and Western Australia at 204–210 days (Rismiller and McKelvey 2000, 2003; Morrow *et al.* 2009). In Tasmania puggles begin to leave the burrow for several hours at ~12 days before weaning (Morrow and Nicol 2012). During this time the digestive system is transitioning from an all-milk diet to a much less digestible invertebrate diet. On Kangaroo Island weaning seems more abrupt, as puggles are not observed to leave the burrow until the mother ceases returning (Rismiller and McKelvey 2000). It appears that bodyweight is not related to age or sexual maturity (Rismiller and McKelvey 2003).

Puggles produced at Currumbin Wildlife Sanctuary have solid food introduced from 7 months of age and in the majority of cases have weaned by 9 months of age. Once the consumption of the captive diet is consistent by the puggle, it is separated from the female to allow her to regain condition before the following breeding season. Although further confirmatory evidence is needed in support of the impact of limiting nutrition (energy) on reproductive success, given the length of the weaning period it is possible that the dam may not be able to regain sufficient body condition for the subsequent breeding season, particularly if the egg is laid and hatched late in the breeding season.

At Perth Zoo, suckling from the dam by some individuals was observed post nursery burrow emergence (Ferguson and Laming 2022). Nine parent-raised young were provided with access to solid food from the time they began to emerge from the nursery burrow (Ferguson and Laming 2022; A. Ferguson and B. Laming unpubl. data). Weights of young were monitored daily (when possible, without disturbing the female) to track weight gain/loss during the weaning stage transition from milk onto solid food (A. Ferguson and B. Laming unpubl. data). After burrow emergence, the animal caretakers were mindful not to disturb the female (especially when the young were with her) as the young may be suckling and disturbance could stress the female.

## 10.15 Age at removal from mother

Captive puggles have been observed to explore outside the burrow well before weaning. To assist the weaning of juvenile short-beaked echidna, artificial food can be placed on a flat dish and live termites (separated from their dirt) are sprinkled over the top. Puggles appear to take to eating live food more readily than to the captive diet, but as they eat the termites, they will ingest small amounts of the diet and gradually become accustomed to it. Some puggles will feed from their mothers for an extended period and not consume much solid food, so may need to be removed to help the process. Other puggles can readily live with their mothers after weaning without problems. Some females have been observed 'offering' themselves to the puggle for feeding before weaning.

At weaning the weight of the puggle will usually drop when first removed from its mother. It is also important to monitor the young at this time for signs of coccidiosis, which appears to be prevalent at this stage of development in captivity.

At Perth Zoo the young short-beaked echidnas were separated from their mothers at 233–251 days of age weighing 1912–3185 g. All young lost weight following separation from their mothers, with the time taken to regain separation weight ranging from 1.5 to 5 weeks (Ferguson and Laming 2022). It is possible the degree of weight loss at separation could be reduced by leaving the young with the female for longer and then separating the female out of the enclosure initially to allow the young to gain full independence in the familiarity of its enclosure until it is ~10–12 months of age. The juvenile could then be moved to a new enclosure that may better replicate more natural dispersal behaviour reported to occur at ~12 months of age by Abensperg-Traun (1991).

## 10.16 Growth and development

Observations of wild short-beaked echidnas in Tasmania (*T. a. setosus*) and Kosciuszko National Park (*T. a. aculeatus*) have revealed that females dig a nursery burrow 2–3 days before egg laying and then back-fill the entrance with soil (Beard and Grigg 2000; Morrow and Nicol 2012). Burrows are ~1–1.5 m long, quite shallow with nursery chambers never more than 40 cm below the surface, and are always newly dug and never re-used in subsequent years (Griffiths 1965,1978; Abensperg-Traun 1989; Beard *et al.* 1992; Beard and Grigg 2000; Rismiller and McKelvey 2000). On Kangaroo Island nursery burrows are not dug until the female is ready to deposit her young at age ~40–45 days. Lengths of tunnels to the nursery chamber are 1.25–2 m (Rismiller and McKelvey 2009). Nursery burrow temperatures between 15°C and 20°C have been recorded in Tasmania (Morrow and Nicol 2012), with similar temperatures documented on Kangaroo Island (Rismiller 2014).

Unlike platypus burrows, the nursery chambers are not lined with grass, probably because short-beaked echidnas have a pouch and platypuses do not. The entrances are always back-filled with a mound of earth (Beard *et al.* 1992; Rismiller and McKelvey 2000).

The egg is laid directly into a pouch that forms on the female's abdomen. The timing of egg laying has been estimated in Tasmanian short-beaked echidnas using implanted data loggers. There is a significant reduction in body temperature variability when the female enters the nursery burrow and a small spike in temperature ~2 days later marks egg laying (Nicol and Andersen 2006). Timing of egg laying in Kangaroo Island short-beaked echidnas was determined by examining the pouch of females from day 15 after observed or suspected mating (Rismiller and McKelvey 2000).

In Tasmania and south-east Australia females are only very rarely found outside the burrow with an egg or young in the pouch and when this does occur the female is in the act of moving to a new burrow due to disturbance or other problems with the original burrow; in south-east Queensland and Kangaroo Island, females are often found foraging with young in the pouch (Beard and Grigg 2000) and on Kangaroo Island females also leave the burrow with the egg in the pouch (Rismiller and McKelvey 1996, 2009).

The egg is incubated in the pouch and hatches 10–11 days after being laid (Griffiths 1968; Morrow *et al.* 2009). In south-east Queensland and on Kangaroo Island the female echidnas forage with the young in the pouch for 45–50 days before depositing the ~200 g young in a burrow (Beard and Grigg 2000; Rismiller and McKelvey 2009). In the cooler south-east of Australia the female remains in the burrow with the young in the pouch (~3 weeks in Kosciuszko and 5–7 weeks in Tasmania) before beginning to leave the burrow (Morrow *et al.* 2009; Beard and Grigg 2000). The female always plugs the burrow while away and the time between return visits increases from ~2 days initially to 6–7 days in late lactation (Beard and Grigg 2000; Rismiller and McKelvey 2000; Nicol and Andersen 2007; Ferguson and Laming 2022). Weaning occurs after ~21 weeks of lactation in Tasmanian echidnas, slightly longer in south-east Queensland and 30 weeks on Kangaroo Island and in Western Australia (Table 3.6; Abensperg-Traun 1989; Beard and Grigg 2000; Nicol and Andersen 2007; Rismiller and McKelvey 2009; Morrow and Nicol 2012; Ferguson and Laming 2022).

Growth curves have been developed for several subspecies of short-beaked echidnas including *T. a. acanthion* (captive) (Fig. 3.20), *T. aculeatus aculeatus* (captive) (Fig. 3.21), *T. a. multiaculeatus* (wild) (Fig. 3.22) and *T. a. setosus* (wild) (Fig. 3.23).

Morrow and Nicol (2012) divided the lactation period for *T. a. setosus* into four stages corresponding to the time spend by the mother in the nursery burrow per suckling event and the stage of short-beaked echidna development. During stage 1 (Table 3.7) young are in the nursery burrow with their mother, presumably in her pouch, and experience a constant ambient temperature because their mother maintains a stable body temperature and actively keeps burrow conditions warm and stable. During stages 2 and 3 the young may become hypothermic when left alone in the nursery burrow. Stage 2 of lactation begins when the young is first left alone in the burrow and during this phase the female spends long periods of time with the young, which is only very lightly furred. In stage 3 the hair is denser and the mother is only spending ~5 h each visit with the young and may be away from the burrow for 4 or more days. By stage 4 of lactation the young has developed a thick covering of fur and remain euthermic while the mother is away. As lactation progresses the time spent by the mother on foraging trips increases and time spent in the burrow with the young decreases (< 4.5 h per suckling event) (see Table 3.7). The longest foraging trips by the mother generally occurs just before the young is weaned. The longest recorded foraging trips, and hence longest intervals between young being fed, were 319 and 281 h (~13 and 12 days) during stage 4 of lactation, but in both of these cases the young had begun to emerge from the burrow (Morrow and Nicol 2012).

**Table 3.6.** Developmental milestones and events for captive Western Australian short-beaked echidnas (*T. a. acanthion*).

Taken from Ferguson and Laming (2022).

| Developmental milestone | Average ± SD | Range | *n* |
|---|---|---|---|
| Age (days) of young when deposited in nursery burrow | 61 ± 2 | 58–63 | 8 |
| First measured weight (g) of young after deposited in nursery burrow | 295 ± 64 | 215–412 | 8 |
| Age (days) at which young first venture out of nursery burrow | 169 ± 21 | 146–201 | 6 |
| Bodyweight (g) of young at first venture out of nursery burrow | 1674 ± 511 | 1154–2460 | 6 |
| Age (days) of young when separated from mother | 245 ± 7 | 233–251 | 6 |
| Bodyweight (g) of young when separated from mother | 2518 ± 531 | 1912–3185 | 6 |
| Time (weeks) taken for young to regain separation weight | 3.1 ± 1.5 | 1.5–5 | 6 |
| Age (days) when weight of young reaches 3 kg | 348 ± 97 | 212–445 | 8 |

In contrast to the observation seen in Tasmanian short-beaked echidnas, female echidnas from Kangaroo Island that have produced young spend much less time in the nursery burrow (Table 3.8). In stage 1 the mother spends no time in the nursery burrow because the young remain in the mother's pouch while the mother is free roaming. In stage 2, the young is deposited in the nursery burrow at 45–60 days after hatching and the mother spends only short periods there of ~2.68 h. During the main lactation period of stage 3, from 61–180 days after hatching, the mothers spend only 1.99 h in the burrow. In late lactation of stage 4, from 181–210 days after hatching, the female spends only 1.44 h in the nursery burrow (P. Rismiller unpubl. data).

In Tasmania, young short-beaked echidnas emerge from their nursery burrow for several hours to forage at ~135 days of age and this exploratory foraging period lasts for ~12 days during which they return to the nursery where they are also fed by the mother. Weaning occurs at ~130–163 days when the young weigh 1.49 ± 0.16 kg (SD) and there appears to be no further interaction with the mother (Table 3.9; Morrow and Nicol 2012).

In south-east Queensland *T. a. aculeatus* has not been observed to dig new nursery burrows, but use rabbit burrows, plugging the entrance and remaining in the burrow for 2–3 weeks while incubating the egg and then the hatchling for the first week or so of its life (Beard and Grigg 2000). Nursery burrows have also been located under flat rocks, rotten vegetation, under a termite mound and even using garden refuse (Griffiths *et al.* 1988; Rismiller 1999). After this time south-east Queensland short-beaked echidna mothers were observed to leave the nursery burrow with the young in the pouch, until it is ~45–50 days old and weighing ~200 g. Although the young do not have erupted spines by this stage, they are too large to be accommodated in the pouch and the mother then leaves the young in a nursery burrow, returning to feed it over the next 4 months, with weaning occurring at 150–165 days (Beard and Grigg 2000).

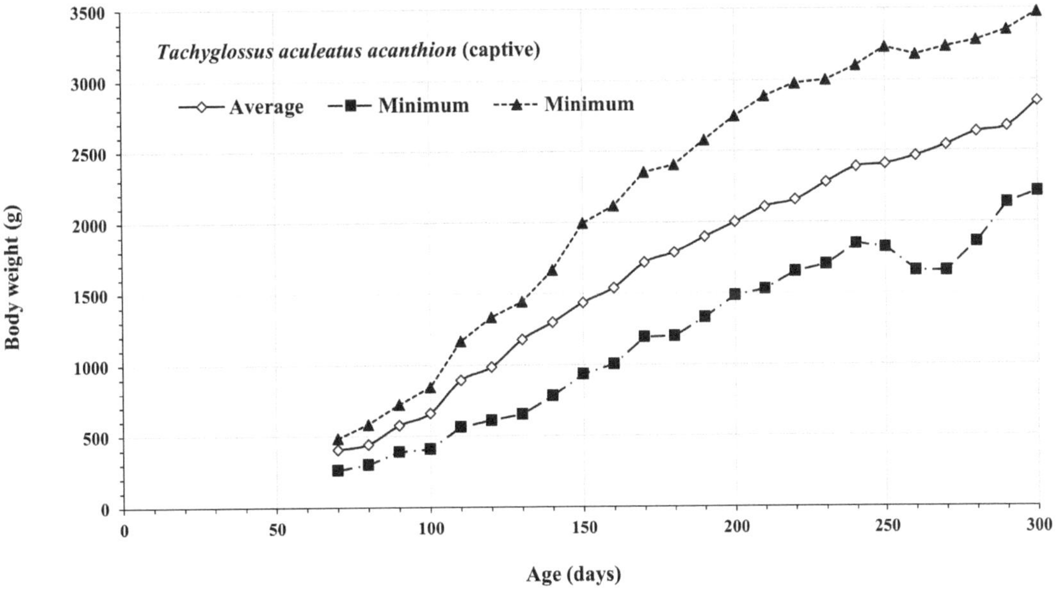


**Fig. 3.20.** Growth curve of a captive short-beaked echidna *T. a. acanthion*. Reproduced from Ferguson and Laming (2022) with permission from CSIRO Publishing.

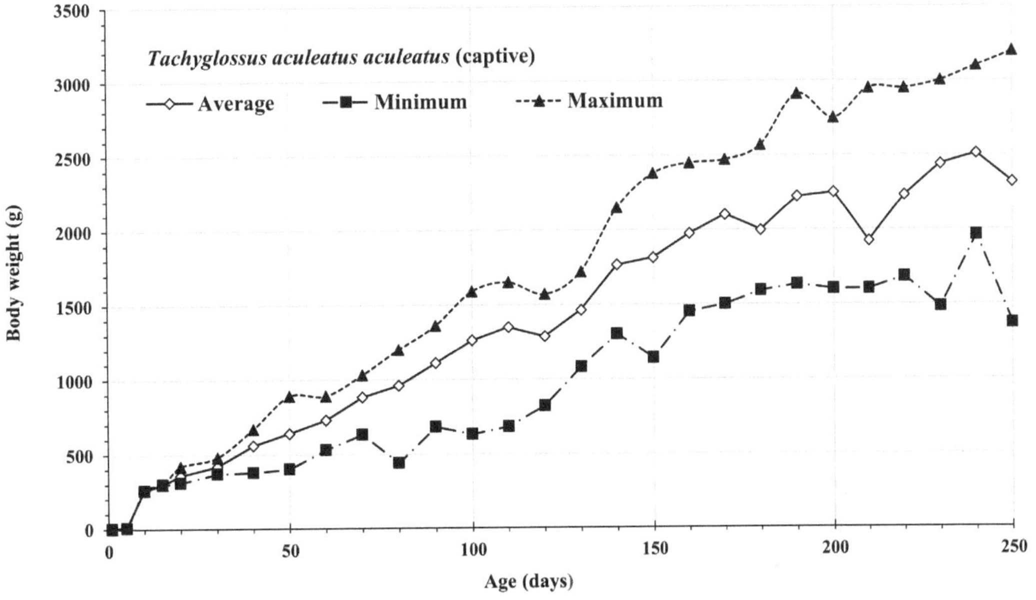


**Fig. 3.21.** Growth curve of a captive short-beaked echidna *T. a. aculeatus*. Data taken from Alice Roser and Stephen Johnston unpubl. data.

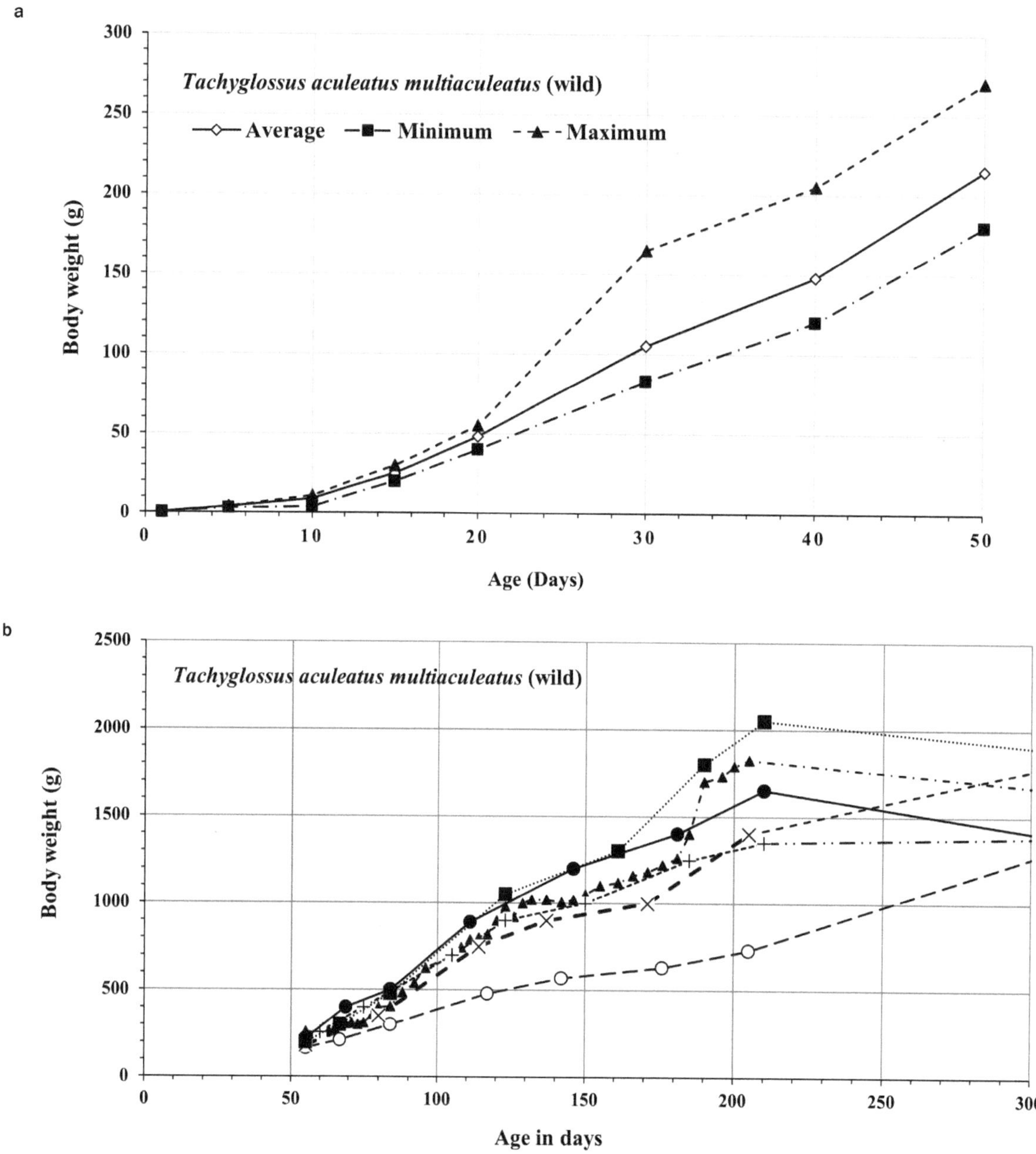


**Fig. 3.22.** (a) Growth curve of a wild short-beaked echidna *T. a. multiaculeatus*. Data taken from Rismiller and McKelvey (2003). (b) Growth curve of six wild short-beaked echidnas *T. a. multiaculeatus*. Data taken from P. Rismiller unpubl. data.

On Kangaroo Island (*T. a. multiaculeatus*) some females remain in the incubation burrow until the egg hatches, but Rismiller and McKelvey (2009) found 80% of females foraging with the egg in the pouch. After the egg hatches females forage with the young in the pouch until the puggle is ~55 days of age, when the young weighs ~215 g (Rismiller and McKelvey 2003). At this time the female digs a nursery burrow and deposits the young there, returning at 5-day intervals to feed it. Once the young is in the nursery burrow the mother forages over a larger area. The young are weaned at ~204–210 days, with a weight of ~1.44 kg (i.e. the lactation period for *T. a. multiaculeatus* is 2 months longer than that of *T. a. setosus*, although the weaning weights are the same (Morrow *et al.* 2009)).

Observations of one individual of the subspecies *T. a. acanthion* from the Western Australian wheatbelt gave an estimated weaning bodyweight of 1.7 kg at

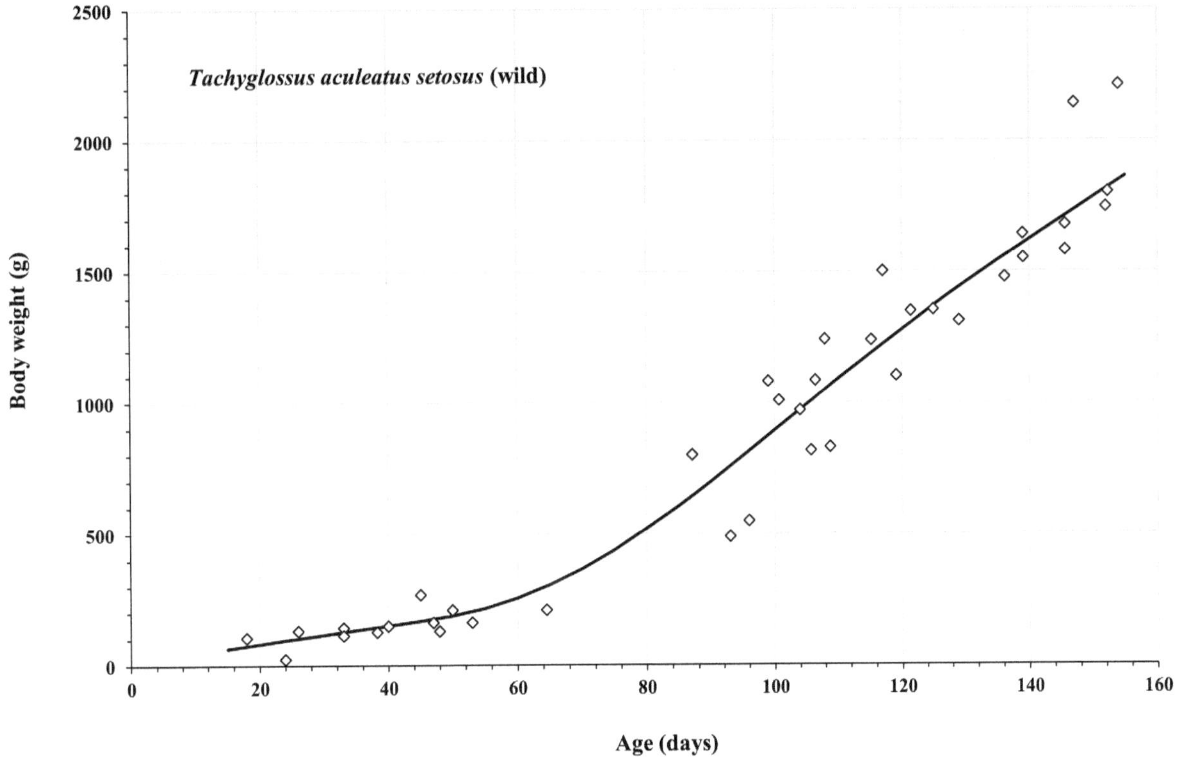


**Fig. 3.23.** Growth curve of a wild short-beaked echidna *T. a. setosus*. Data taken from Morrow and Nicol (2012); S. Nicol unpubl. data. Line of best fit to field data from 17 puggles. Weaning occurs at about 150 days.

**Table 3.7.** Time that Tasmanian short-beaked echidna (*T. a. setosus*) mothers spent inside nursery burrows per suckling event and time between suckling events (mean ± SD).

Derived from Morrow and Nicol (2012).

| Lactation stage | Days after hatching | Mean time spent in burrow (h) | Mean time between burrow visits (h) |
|---|---|---|---|
| 1 | 0-37 | 100% | |
| 2 | 38-66 | 29.9 ± 24.2 | 52.8 ± 34.6 |
| 3 | 67-100 | 5.2 ± 4.1 | 109.5 ± 44.1 |
| 4 | 101-150 | 2.6 ± 1.9 | 153.3 ± 71.0 |

**Table 3.8.** Time that Kangaroo Island short-beaked echidna (**T. a. multiaculeatus**) mothers spent inside nursery burrows per suckling event and time between suckling events (mean ± SD).

Derived from Rismiller and McKelvey (2009).

| Lactation stage | Days after hatching | Mean time spent in burrow (h) | Mean time between burrow visits (h) |
|---|---|---|---|
| 1 | 0-45 | 0 | |
| 2 | 45-60 | 2.68 ± 0.28 | 88 ± 4 |
| 3 | 61-180 | 1.99 ± 0.02 | 120 |
| 4 | 181-210 | 1.44 ± 0.24 | 144 |

**Table 3.9.** Growth and development of young Tasmanian short-beaked echidnas (*T. a. setosus*).

Derived from Morrow and Nicol (2012).

| Age (days) | Mean weight | Hindlegs | Hair | Spines | Eyes |
|---|---|---|---|---|---|
| < 2 | Not weighed (*n* = 4) | Not developed | No hair | No spines | Eyespot |
| 39–66 | 201 g, *n* = 3 | Developed | Fine hair 3 mm in length | Spines beginning to erupt the skin | Fused shut |
| 103–118 | 995 g, *n* = 5 | Developed | Thick dense hair 2 cm in length | Present, shorter than hair | Open |
| 119–164 | 1490 g, *N* = 10, *n* = 5 | Developed | Thick dense hair 3 cm in length | Present, longer than hair | Open |

**Table 3.10.** Nursery burrow life parameters for captive Western Australian short-beaked echidnas (*T. a. acanthion*).

Taken from Ferguson and Laming (2022).

| Short-beaked echidna nursery life parameters | Average ± SD | Range | *n* |
|---|---|---|---|
| Total time (days) that young were in the nursery burrow | 109 ± 21 | 84–138 | 6 |
| % gain in bodyweight during time in nursery burrow | 432 ± 102 | 311–601 | 6 |
| Total no. of nursery burrow nursing visits by mother | 31.3 ± 5.5 | 22–37 | 6 |
| Total combined time (h) that mother was in nursery burrow during burrow life | 135.1 ± 66.3 | 72.7–246.2 | 5 |
| Maximum recorded weight loss by young in one night during burrow life (%) | 5.57 ± 0.8 | 4.6–6.7 | 7 |
| Maximum weight loss by young between feeds (%) | 10.7 ± 2.1 | 7.3–13.5 | 6 |
| Maximum weight loss (g) by young between feeds | 71.3 ± 33.1 | 40–116 | 6 |
| Maximum daily weight gain (%) by young | 24 ± 5.8 | 16–31 | 7 |
| Age (days) at maximum daily weight gain by young | 99.4 ± 18.4 | 82–135 | 7 |
| Biggest gain in daily weight (g) by young following a feed | 174 ± 152.0 | 118–255 | 5 |
| Age (days) at biggest weight gain by young following a feed | 129.2 ± 130.0 | 109–150 | 5 |

195–200 days (Abensperg-Traun 1989). More recent observations by Ferguson and Laming (2022), have provided significant details of the growth in short-beaked echidnas from Western Australia (Table 3.10; Figs 3.24–3.26).

These observations highlight that there are significant differences between the different subspecies of short-beaked echidnas for the age at weaning, where *setosus* and south-east *aculeatus* are similar and contrast with the more arid zone/western subspecies (Nicol 2021). There is no specific information on the lactation period for *T. a. lawesii* or any other aspect of its reproduction.

While the mother is in the burrow, the young consume up to 40% of the maternal bodyweight in milk (Griffiths 1978; Griffiths *et al.* 1988; Rismiller and McKelvey 2009; Rismiller 1999). Females have been known to move their young to a new nursery burrow as a result of disturbance by males at the old burrow (Harris and Nicol 2014). Nursery disturbance by wild male short-beaked echidnas in Tasmania has been suggested as a form of infanticide in order to force the female to breed again because males do not seem to disturb the burrows of females they have mated with (Harris and Nicol 2014).

Female Tasmanian short-beaked echidnas lose a lot of bodyweight during the nursery period of lactation, greater than the weight loss during hibernation (Nicol and Morrow 2012). Despite this weight loss they continue to feed their young and gain body condition in preparation for the following hibernation season (Nicol and Morrow 2012). Female Kangaroo Island short-beaked echidnas lose most bodyweight while carrying pouch young and weight stabilises after placing the young in a nursery burrow (Rismiller and McKelvey 2000).

In the case of a successfully weaned short-beaked echidna at St Louis Zoo, the female dug a burrow when the puggle was ~60 days of age (Fieseler and Junge 1997). The burrow consisted of a split hollow log, under which she placed the infant and then buried the entire log to entomb the young. She returned every 4 or 5 days to unearth the young and nurse it. Burrow activity and nursing occurred only at night and the time spent with the young ranged

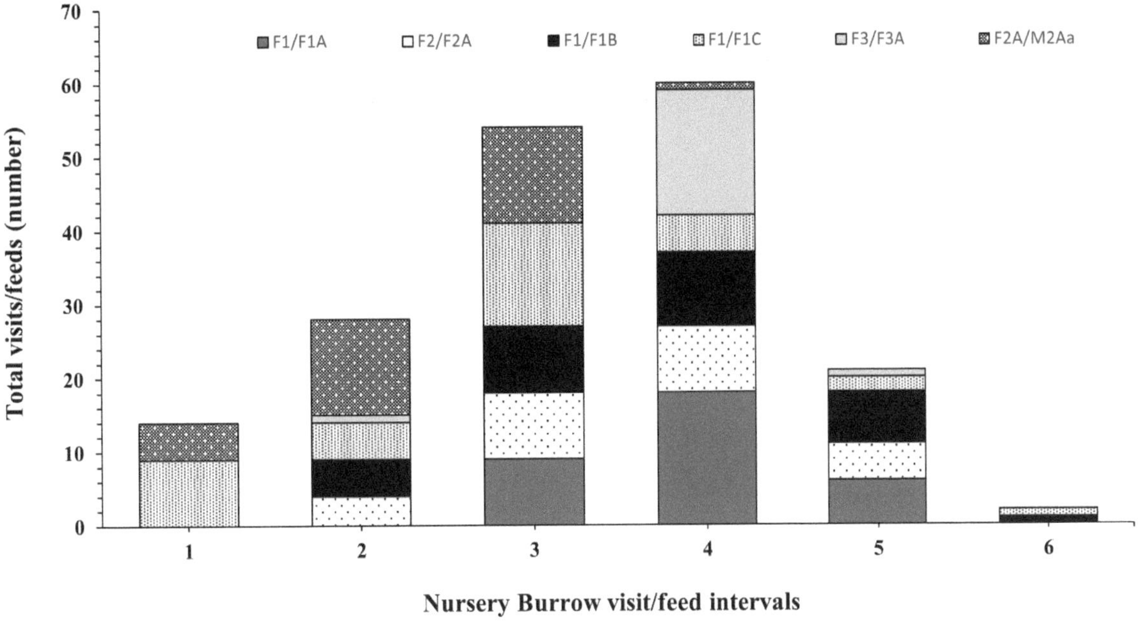


**Fig. 3.24.** Short-beaked echidna females F1, F2, F3 and F2A nursery burrow visits/feeding young intervals days, to feed six young F1A (dark grey), F1B (black), F2A (white dots), F1C (lots of dots), F3A (light grey) and M2Aa (grey with white dots). Reproduced from Ferguson and Laming (2022) with permission from CSIRO Publishing.

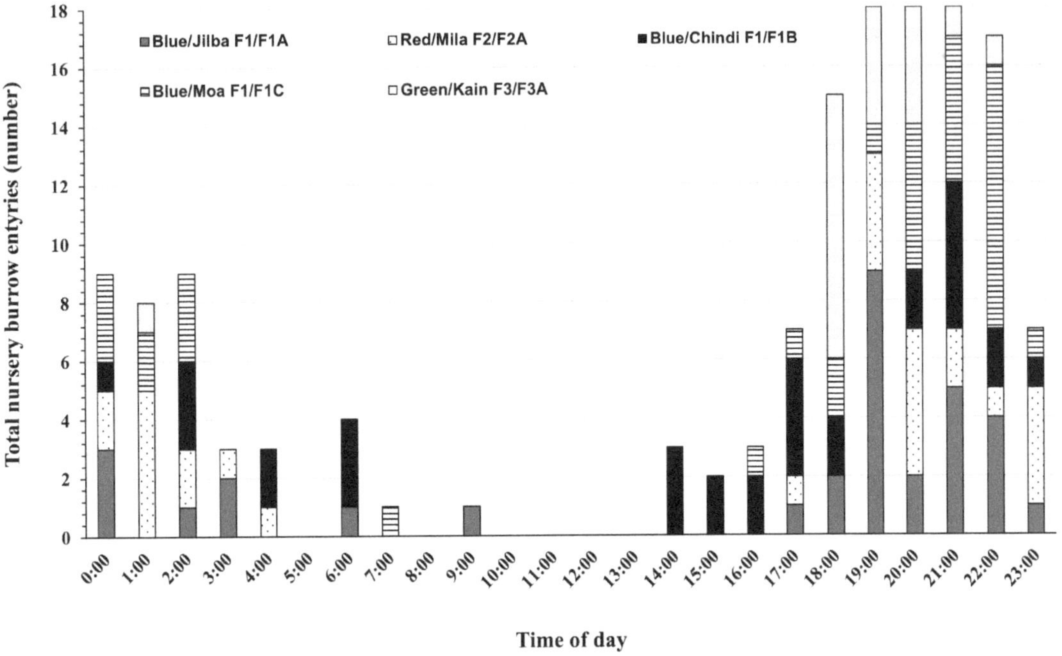


**Fig. 3.25.** Short-beaked echidna nursery burrow entry times by females F1, F2 and F3 to feed five young F1A (dark grey), F1B (black), F2A (white dots), F1C (lots of dots) and F3A (light grey), during nursery burrow life. Reproduced from Ferguson and Laming (2022) with permission from CSIRO Publishing.

from 20 min to 2 h. By 4 months of age, the puggle was more active and was frequently seen out of the burrow during the day. By 6 months of age the puggle weighed 1058 g and the female unearthed it and destroyed the burrow. Nursing bouts became sporadic and the female was reluctant to allow nursing, at which time the young short-beaked echidna was removed from the mother. Until this point it had never been observed to eat solid food, so

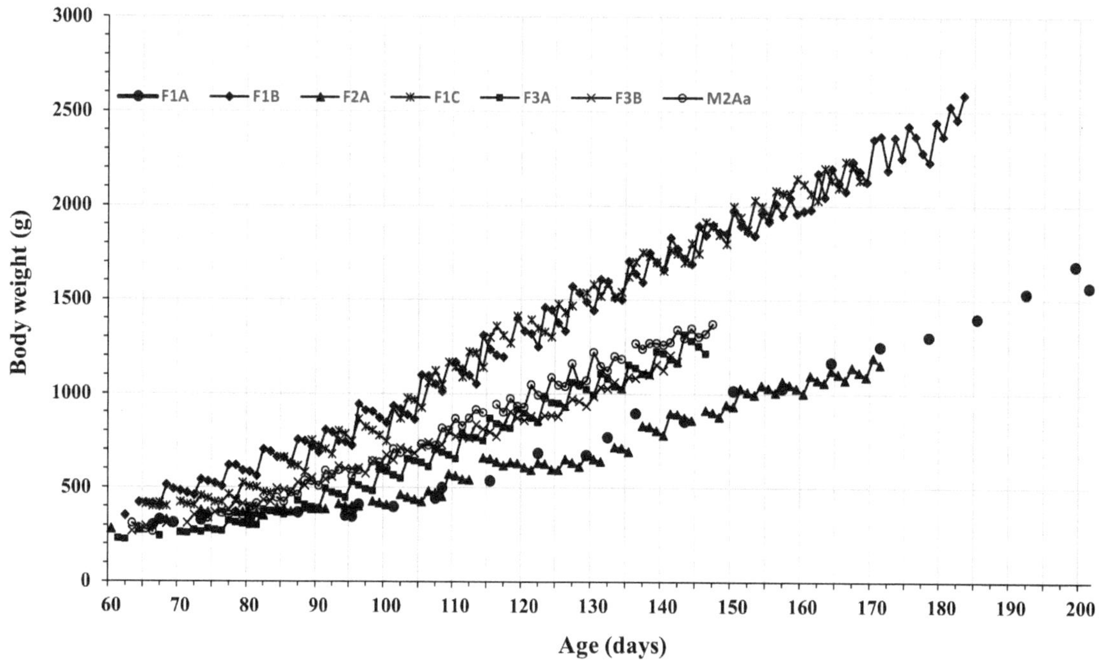


**Fig. 3.26.** Growth in nursery burrow of seven parent-reared young at Perth Zoo. Reproduced from Ferguson and Laming (2022) with permission from CSIRO Publishing.

logs with ants in them were provided and a grub paste (ground insects and water) was spread on logs. The short-beaked echidna lost ~16% of its bodyweight (1233–1030 g), but by 28 days after removal from its mother it had gained 21 g and steadily gained weight subsequently.

Observations on Kangaroo Island suggest that weaning is an abrupt process; however, this is not the case in other areas. Weaning bodyweight and age is highly variable among locations. In Tasmania short-beaked echidnas can be weaned from 139 to 154 days of age and weighing 1.5–2.2 kg, while on Kangaroo Island they can be weaned at 204–210 days of age and weighing 0.8–1.3 kg (Abensperg-Traun 1989; Rismiller 1999; Pierce *et al.* 2006; Morrow and Nicol 2012; Morrow *et al.* 2009).

There is some variability in whether the young are fully weaned by the time they leave the burrow. In Tasmania some young were observed to make short exploratory trips of up to several hours out of the burrow over a period of 12 days or so while the mother was absent, but were fed when she returned to the burrow, before they finally abandoned it. The mother returned to feed the young but abandoned the burrow herself when the young did not return. Later in the breeding season the young did not leave the burrow before the mother abandoned it. (Morrow and Nicol 2012). On Kangaroo Island young were considered weaned when the female no longer backfilled the entrance to the nursery and never returned. In 5 years, only 2 of 15 females returned to suckle their young once the puggles had emerged from the nursery (Rismiller and McKelvey 2009). Pierce *et al.* (2006) reported a rare case of twin young that first emerged from the nursery burrow 8 days apart, but then stayed outside the burrow for several days before returning to the burrow. While they were out of the burrow they were suckled by the mother. This behaviour lasted for 60 days before they abandoned the burrow. After weaning the young appear to have no further interactions with their mother, though the young may remain within their mother's home range until ~12 months of age (Morrow and Nicol 2012).

## 11. ARTIFICIAL REARING

### 11.1 Housing

As with all orphan native mammals that have been taken into care, minimising stress is a major consideration. Choosing suitable housing can help to create a stress-free environment. To achieve this, several factors should be considered:

- providing the appropriate temperature
- providing a space that is free from noise or vibration

- securing the area from children and other animals
- maintaining the area in a hygienic manner
- escape-proofing the area
- clearing the area of obstacles and hazards
- offering shelter from light (pouch and burrow life stages), weather and noise.

Pouch-stage/furless puggles (i.e. 30–50 days) can be held in a soft or hard type of cooler with ventilation holes in the top and shredded paper to nestle in (Bellamy 1994), or placed in a sterile pouch within the cooler, for an added feeling of security.

Larger furred/spined puggles will become increasingly active within the burrow environment and in need of opportunities for strength-building and sensory enrichment. Burrow young, more than 50 days of age can be housed in large (60 (L) × 40 (W)× 40 cm (H)) smooth-walled containers with small air holes near the top and on opposite walls of the container. As the young are usually underground in a nursery burrow with no daylight, an artificial hollow can be formed using half a terracotta pot, wooden box or similar dark hide. The floor of this burrow can be covered in deep substrate and dry leaf litter to mimic the nursery burrow. Puggles at this age have also been held successfully in artificial burrows in a 205-L drum and kept in a roofed outside area. The puggle is also provided with a mixture of clean soil and leaf litter to nestle into, which works well because it better simulates what it would have in a burrow (P. Rismiller pers. obs.).

The puggle can be housed outdoors from 5 months of age and should be in a small, fully enclosed pen with solid or mesh flooring (sufficiently strong to prevent digging out), covered in deep soil substrate with natural burrow opportunities (e.g. hollow logs and leaf litter).

## 11.2 Temperature requirements

Short-beaked echidnas generally experience an ambient temperature between 15°C and 21°C in nursery burrows (Morrow and Nicol 2012; Rismiller 2014; R. Rismiller pers. obs.). Hand-reared puggles should also be housed within this range (Whittington 1988).

Use a minimum/maximum temperature gauge with a plastic-coated probe that can be placed next to the puggle to monitor the temperature. Suckling short-beaked echidnas will become torpid if the temperature falls to ~12°C and they are not suckled (Whittington 1988). It is important to note that puggles will often appear cold, sluggish and withered just before the next feed is due and then warm, wriggly and engorged after their feed – this is natural and should only be of concern if paired with other symptoms (e.g. prolonged constipation or severe dehydration).

Like adults, puggles suffer from heat stress at high temperatures and the ideal temperature range is ~15–22°C (Rismiller 2008). They should not be given artificial heat, even unfurred young, as long as they are housed as described above (Bellamy 1994).

## 11.3 Specific requirements

It is normal for the skin of a young hairless short-beaked echidna more than 15 days old to be dry. When the burrow young is housed in cool soil/leaf litter it does not dry out. Spraying the burrow lightly with water every few days helps maintain a healthy atmosphere. The temperature should never go above 21°C (Rismiller 2008).

When first brought in for hand-rearing, the short-beaked echidna may be dehydrated. Dehydration can be determined by checking the skin for looseness or dryness. If the animal is dehydrated it can be given plain boiled water, with 5 g (1 teaspoon) of glucose to 100 mL of water or 1 g of electrolyte replacer if available (Austin 1997). Lectade or any electrolyte replacer (e.g. Vytrate®) can also be used at a ratio of 20 mL to 250 mL water and offered in a cupped hand but should not be forced (P. Rismiller pers. comm.; Staker 2007). It is important to gently warm the puggle before rehydrating to avoid the risk of aspiration pneumonia. If this is taking time, give fluids subcutaneously. If the puggle is really cold (though they are often cool to touch anyway – so be careful), cup the puggles in warmed hand (P. Rismiller pers. obs.). Placing it in a shallow warm water bath (taking care not to submerge the beak) and drying it rather than putting it in a hot box has also been suggested.

Common problems associated with hand-reared short-beaked echidnas include failure to thrive (Tobias 2025a, 2025b) and constipation and diarrhoea, which is generally due to *Candida* or *Salmonella* (Woods 1999). Although they do not defecate as frequently as marsupials, the frequency of defecation should be monitored and constipation treated with an enema (e.g. Microlax®).

It is important to remember that there is some variation in adult bodyweight of the different subspecies of short-beaked echidna (Nicol *et al.* 2019a) and differences in growth rates between individual puggles of the same subspecies as well as between subspecies (see Figs 3.20–3.23). Any concern over the bodyweight of the puggle should be interpreted in conjunction with other physical and behavioural indicators of health.

## 11.4 Diet and feeding routine

### 11.4.1 Natural milk

Short-beaked echidna milk is more concentrated than platypus milk; it contains 48·9% solids, 31.0% lipid, 12.4% protein and has high concentrations of iron (33.33 mg/L) (Table 3.11; Griffiths *et al.* 1984, 1988).

This high energy content is due to the high lipid content, with oleic acid being the dominant fatty acid. The milk fat contains only small amounts of free fatty acids, cholesterol esters and phospholipids. Approximately 88% of the milk fats are triglycerides and it is very low in polyunsaturated fatty acids (Griffiths *et al.* 1984, 1988).

As in other mammals, the main proteins in short-beaked echidna milk are casein and α-globulin proteins, but these occur at higher concentrations than in most other mammals (Skibiel *et al.* 2013) and are highest in early lactation (Teahan *et al.* 1991). Short-beaked echidna milk contains three specific caseins (Lefèvre *et al.* 2009) and the whey proteins include albumin, antimicrobial protein (EchAMP) (Kumar *et al.* 2019) and the iron-binding protein transferrin (not lactoferrin as in most other mammals (Teahan *et al.* 1991).

An important point for hand-rearing is that short-beaked echidna milk contains only trace amounts of lactose and only in early lactation. Instead, short-beaked echidna milk contains a multitude of oligosaccharides, with the major neutral oligosaccharide being the short-chain fucosyllactose and sialyllactose as the major acidic oligosaccharide, with the number of oligosaccharides decreasing during lactation (Oftedal *et al.* 2014). The sialyllactose is in a form resistant to microbial degradation and this and EchAMP are likely to be important in protecting both mother and young from pathogenic bacteria growing in milk leakage when the puggle is drinking from the nipple-less mammary gland (Oftedal *et al.* 2014).

The calcium levels in short-beaked echidna milk are similar to those in eutherian milks, but lower than found in marsupials (Table 3.11; Griffiths *et al.* 1984), while the iron concentrations are very high. Because they are so underdeveloped at birth, monotremes (and marsupials) rely on the iron in milk for haemoglobin, cytochromes and myoglobin (Stannard *et al.* 2020) and the iron bound to transferrin gives short-beaked echidna milk its pink colour.

### 11.4.2 Milk formulas

The main commercial formulas for hand-rearing short-beaked echidnas include:

- Biolac® – M100 for furless puggles; M150 is a transitional milk for feeding when dense fur has developed. When the animal produces solid, dark pellet droppings the M200 formula is used because it contains increased lipid in the form of canola oil. When the juvenile is nearing weaning, add 2–5 mL of canola oil per 100 mL of formula. Mixing the formulas helps the transition between the two formulas.
- Di-Vetelact® (Lillelund Pty Ltd) – a widely used, low-lactose milk formula. Due to its low energy concentration when prepared as directed, some groups advise the addition of mono- and polyunsaturated fats such as canola oil. Add 2–5 mL canola oil to 100 mL milk during the last phase of milk feeding (Messer and Walker 1992).
- Wombaroo® Echidna Milk Replacer – two milk formulas are available, one for animals < 30 days old and another for animals older than 30 days. Once the puggle is nearing weaning (depending on subspecies – see section 10.16), solid food can be introduced by adding small amounts of Wombaroo® Small Carnivore Food into the milk and slowly increasing the amount. Over ~4 weeks the amount of solids is increased and the amount of milk decreased so the feed changes from a thin porridge to a thick paste. Note: cases of metastatic mineralisation associated with excess vitamin D in the diet have been reported in hand-raised short-beaked echidnas given Wombaroo® Echidna Milk Replacer (Scheelings *et al.* 2017). As a result of this investigation the vitamin D level of Wombaroo® Echidna Milk Replacer was reduced to 41 IU/MJ (Scheelings *et al.* 2017).

**Table 3.11.** Concentrations of the major constituents of mid-lactation milk of short-beaked echidnas.

From Griffiths *et al.* (1984, 1988).

| Total solids (%) | Carbohydrates (%) | Lipids (%) | Protein (%) | Calcium (mg/L) | Iron (mg/L) | Energy (kJ/g) |
|---|---|---|---|---|---|---|
| 30–55 | 2.3 | 31.0 | 12.4 | 1170 | 33.3 | 15.2 |

### 11.4.3 Feeding apparatus

As female short-beaked echidnas have no teat or nipple, teaching the puggle to suckle milk from the palm of a clean cupped hand is similar to pushing against the milk patch and the most natural feeding option (Rismiller 2008). A small rubber bowl similar to that used for platypuses has also been used successfully.

### 11.4.4 Feeding routine

Short-beaked echidnas weighing only 50 g have been successfully reared (Whittington 1988). In preparation for feeding the young should be warmed in the palm of the hands, 5–10 min beforehand until it becomes quite active when the milk is offered. Pouch young are fed once per day with 15–20% of the bodyweight in milk being offered, which can be consumed in less than 30 min.

Once the female has left the young puggle in its burrow at ~45–55 days of age, she normally returns every 3–5 days to feed it (Rismiller and McKelvey 2009), when it will take in 10–35% of its bodyweight (Whittington 1988; Rismiller 2008, 2010). After coming into care, the young should be offered milk every day until it starts to suckle. After that, a natural feeding regime of once every 3–5 days should be established. (P. Rismiller pers. obs.). Hand-warm milk (i.e. milk temperature feels only slightly warm when tested on the skin of your wrist) can be dribbled onto the hand using a syringe (Fig. 3.27) or pipette. Place the young on the lap with it body curled around the hand and it can be suckled in this position.

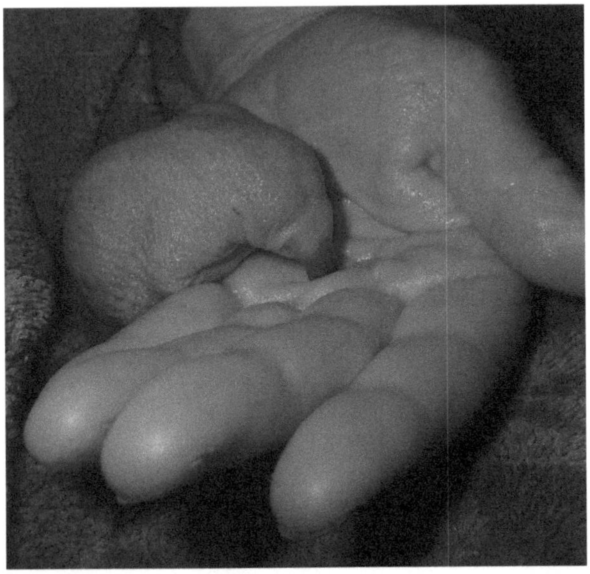

**Fig. 3.27.** Hand-feeding a short-beaked echidna. Photo by Peggy Rismiller.

It has been suggested that feeding at longer intervals is better than shorter intervals due to the potential of the milk not being fully digested if the feeding interval is too short, which can result in the death of the puggle (Staker 2007). As the animal grows the interval between feeds can be changed to 3 and then 4 days (Staker 2007). Indicators of hunger in puggles include activity (especially if it lies on its side or back) or vocalising such as squeaking, though not all puggles do this.

An alternative feeding method is to use a small plastic medicine cup. Hold the short-beaked echidna with the snout in a small amount of milk at the base of the cup, without submerging the nostrils. Allow suckling of 3–5 mL at a time (Bellamy 1994).

Force-feeding with milk formula using an infant gastric feeding tube may be indicated if other feeding methods fail. The procedure has been used on furless puggles by waiting for the animal's tongue to poke out, inserting the catheter into the animal's mouth and letting it draw the catheter back into its mouth/throat. The catheter was then moved into the stomach, the animal was fed and then left alone until the milk could no longer be seen inside its transparent stomach (usually 2–3 days). This procedure was repeated until the animal was self-feeding. Despite some successes in using force-feeding techniques, some authorities do not recommend it because any tube introduced into the mouth of a short-beaked echidna can damage the tiny mouth opening as well as the lower jaw and beak (P. Rismiller pers. comm.). Tube feeding is a complex procedure and veterinary support and the advice of an experienced carer is strongly recommended before undertaking this method of feeding.

Pouch and burrow young generally do not have to be stimulated to urinate and defecate, but should be closely monitored for signs of constipation or diarrhoea, as these could indicate underlying illness. For pouch-stage puggles, urination is typically observed every 3–4 days, often when the puggle is handled. Defecation at this stage is less frequent, expected every 1–2 weeks.

For burrow-stage puggles, urination and defecation are slightly more frequent, but unpredictable. It is common for a puggle to not urinate or defecate following a feed. Faeces are a yellow/mustard colour of toothpaste consistency, changing to a firmer and darker consistency once solid food is introduced.

At 5–6 months of age, once the puggle is standing and moving strongly, milk and subsequent solid foods may be presented in a shallow bowl to allow self-feeding. The bowl should deep enough to discourage the puggle from stepping in, but shallow enough that the food can be reached without standing. It should be positioned to prevent tipping over or spillage. It is recommended that the carer hold the bowl for the first few feeds while the method is introduced and then progress to monitoring the feed only. Some puggles may still require cleaning after self-feeding (see Hygiene section below).

Ants and termites can be added after several months, closer to weaning, though this is not required. Preboiled water, cooled to milk temperature, can be offered following a milk feed from 3 to 5 months onwards. From 5 months onward, fresh water should be supplied ad libitum.

## 11.5 Data recording

When an animal is first brought in for hand-rearing, its sex and approximate age, using growth charts, should be recorded. During the hand-rearing process several important pieces of information should be recorded on a daily basis:

- date when the information is recorded
- time when the information is recorded
- bodyweight to the nearest 1 g if possible
- eyes open or date when eyes open
- ear slits open or date when open
- date when spines erupt through skin
- general activity and demeanour
- characteristics and frequency of defecation and urination
- amount (g) of different food types offered
- food consumption at each feed
- veterinary examinations and results.

This information serves several purposes, including providing important background information such as food consumption, which will assist a veterinarian to reach a diagnosis if the animal becomes sick or fails to grow or gain weight. It also allows a comparison with growth curves to assess progress (section 10.16) and to establish new growth curves that do not exist for other measurements. The developmental stages and measurements outlined in section 10.3.1 should also be recorded on a weekly basis if possible.

## 11.6 Identification methods

These are generally not required.

## 11.7 Hygiene

Maintaining a high standard of hygiene is critical to the survival of the puggle. Emphasis needs to be placed on the following:

- maintaining a clean nest area at all times; using a new liner for the pouch young after each feed
- personal hygiene – washing and disinfecting hands before and after handling the puggle. Use antibacterial solution for washing hands because their immune systems are not well developed
- using boiled water when making up formulas for very young puggles
- cleaning spilt milk formula, faeces and urine from the puggle's skin and fur as soon as possible, and then drying the animal
- washing all feeding equipment in warm soapy water and sterilising it in a suitable agent such as Milton® Anti-bacterial Solution or boiling it for 10 min. After sterilising, rinsing all equipment in cold water
- heating up milk only once and discarding leftovers
- avoiding contact with other animals unless certain they pose no health risk.

## 11.8 Behavioural considerations

Short-beaked echidna are typically a solitary species and socialisation with conspecifics should be avoided during hand-rearing, although housing with others of similar health status at weaning age is unlikely to be detrimental. Exposing the puggle to a range of stimuli that it may encounter in its future environment will support its transition to a new environment post-weaning.

The most important factor to consider for short-beaked echidnas being released is an intact healthy natural habitat that contains an abundant variety of invertebrates. In most cases, hand-reared individuals are kept in captivity as part of an educational display, which means they become used to noises and people. They generally make very good display animals as they are usually more active around visitors.

## 11.9 Use of foster animals

No foster species are known to be used.

## 11.10 Weaning

The lactation period varies among the different subspecies (see details above). In the wild the mother simply leaves the young and does not return, which means the young may not feed for a week or more before it starts to go out and explore for food. It is therefore the age of the animal and not the bodyweight that is most important for the timing of release. In the wild on Kangaroo Island, the young never leave the nursery and feed only on mother's milk until weaning (Rismiller and McKelvey 2000). In Tasmania, young make some short excursions out of the burrow during the last 2 weeks of burrow life (Morrow and Nicol 2012). Once released at 7 months of age the young do not need additional milk or food, because they start exploring and feeding on their own. It is normal for young short-beaked echidnas to lose bodyweight over the following year (Rismiller 2008).

Puggles destined to remain in the captive environment may be gradually offered a small, less-concentrated portion of the adult diet, in addition to the milk feed, from 6 months of age, where health and behaviour indicates readiness. The amount of milk can be gradually decreased and the adult diet portion increased, until the puggle is reliably consuming the adult diet (typically between 7 and 10 months of age).

## 11.11 Release procedures

It is important to ensure that releases back into the wild are handled with due consideration for the welfare of the released animal and that they entail minimum impact upon wild populations. When releasing short-beaked echidnas back into the wild there are several issues that should be considered.

- Short-beaked echidnas qualifying for release include those that have been hand-reared or animals not adapting to captivity.
- Only animals judged by an experienced wildlife veterinarian to be behaviourally competent and healthy at the time of release should be considered for release, if deemed in the animal's best interests.
- Animals should normally be released shortly before dusk as close as possible to where they were captured or rescued.
- If circumstances prevent the release occurring at the exact point of capture/rescue, then the nearest suitable site should be chosen as determined by institution staff in consultation with the rescuer and local wildlife officers.
- When releasing, a useful technique involves leaving the short-beaked echidna in an artificial burrow tipped on its side so that it can start to come and go naturally (P. Rismiller pers. comm.). It will eventually wander out, often staying in the area for several months before beginning its travels in search of a home range.
- Relocating short-beaked echidnas should be carried out with great care, because if they have been brought in from the wild, they will attempt to return to their original home range, at great risk (Rismiller and McKelvey 1995) if they attempt to cross busy roads.

## REFERENCES


Abensperg-Traun MA (1988) Food preference of the echidna, *Tachyglossus aculeatus* (Monotremata: Tachyglossidae), in the wheatbelt of Western Australia. *Australian Mammalogy* **11**, 117–123. doi:10.1071/AM88015

Abensperg-Traun MA (1989) Some observations on the duration of lactation and movements of a *Tachyglossus aculeatus acanthion* (Monotremata: Tachyglossidae) from Western Australia. *Australian Mammalogy* **12**, 33–34. doi:10.1071/AM89005

Abensperg-Traun MA (1991) A study of home-range, movements and shelter use in adult and juvenile echidnas, *Tachyglossus aculeatus* (Monotremata: Tachyglossidae), in Western Australian wheatbelt reserves. *Australian Mammalogy* **14**, 13–21. doi:10.1071/AM91002

Anon (1986) Species of mammals bred in captivity in 1982 and 1983 and multiple generation births. *International Zoo Yearbook* **24/25**, 506–564.

Anon (1988) Species of mammals bred in captivity in 1985 and multiple generation births. *International Zoo Yearbook* **27**, 395–435.

Augee ML (1976) Heat tolerance of monotremes. *Journal of Thermal Biology* **1**, 181–184. doi:10.1016/0306-4565(76)90011-5

Augee ML (1995) Short-beaked echidna *Tachyglossus aculeatus*. In *The Mammals of Australia*. (Ed. R Strahan) pp. 40–43. Reed Books, Sydney.

Augee ML, Ealey EHM (1968) Torpor in the echidna, *Tachyglossus aculeatus*. *Journal of Mammalogy* **49**, 446–454. doi:10.2307/1378202

Augee ML, Gooden B (1997) *Echidnas of Australia and New Guinea*. University of New South Wales Press, Sydney.

Augee ML, Ealey EHM, Price IP (1975) Movements of echidnas, *Tachyglossus aculeatus*, determined by marking-recapture and radio-tracking. *Australian Wildlife Research* **2**, 93–101. doi:10.1071/WR9750093

Augee ML, Bergin TJ, Morris C (1978) Observations on behaviour of echidnas at Taronga Zoo. *Australian Zoologist* **20**, 121–129.

Augee ML, Gooden B, Musser A (2006) *Echidna: Extraordinary Egg-Laying Mammal*. CSIRO Publishing, Melbourne.

Austin MA (1997) *A Practical Guide to the Successful Hand Rearing of Tasmanian Marsupials*. Regal Publications, Melbourne.

Backhouse TC, Bolliger A (1959) *Babesia tachyglossi* n. sp. from the echidna *Tachyglossus aculeatus*. *The Journal of Protozoology* **6**, 320–322. doi:10.1111/j.1550-7408.1959.tb04377.x

Barker IK, Beveridge I, Munday BL (1985) Coccidia (*Eimeria tachyglossi* n. sp., *E. echidnae* n. sp. and *Octosporella hystrix* n.

sp.) in the echidna, *Tachyglossus aculeatus* (Monotremata: Tachyglossidae). *The Journal of Protozoology* **32**, 523–525. doi:10.1111/j.1550-7408.1985.tb04055.x

Barker MM, Cooper CE, Withers PC, Nicol SC (2016) Reexamining echidna physiology: the big picture for *Tachyglossus aculeatus acanthion*. *Physiological and Biochemical Zoology* **89**, 169–181. doi:10.1086/686716

Beard LA, Grigg GC (2000) Reproduction in the short-beaked echidna, *Tachyglossus aculeatus*: field observations at an elevated site in south-east Queensland. *Proceedings of the Linnean Society of New South Wales* **122**, 89–99.

Beard LA, Grigg GC, Augee ML (1992) Reproduction by echidnas in a cold climate. In *Platypus and Echidnas*. (Ed. ML Augee) pp. 93–100. Royal Zoological Society of New South Wales, Sydney.

Bellamy T (1994) Hand rearing echidnas. In *Wildlife*. Proceedings 233. Post Graduate Committee in Veterinary Science. (Ed. DI Bryden) pp. 23–27. University of Sydney, Sydney.

Bennett GJ (1881) Observations on the habits of the *Echidna hystrix* of Australia. *Proceedings of the Zoological Society of London* **49**, 737–739. doi:10.1111/j.1096-3642.1881.tb01330.x

Boisvert M, Grisham J (1988) Reproduction of the short-nosed echidna *Tachyglossus aculeatus* at the Oklahoma City Zoo. *International Zoo Yearbook* **27**, 103–108. doi:10.1111/j.1748-1090.1988.tb03202.x

Brattstrom BH (1973) Social and maintenance behaviour of the echidna, *Tachyglossus aculeatus*. *Journal of Mammalogy* **54**, 50–70. doi:10.2307/1378872

Brice P (2007) Heat tolerance and cold indifference in the short-beaked echidna, *Tachyglossus aculeatus*: body temperature management in a mammalian constitutional eurytherm. PhD thesis, School of Integrative Biology, The University of Queensland.

Brice PH, Grigg GC, Beard LA, Donovan JA (2002a) Heat tolerance of short-beaked echidnas (*Tachyglossus aculeatus*) in the field. *Journal of Thermal Biology* **27**, 449–457. doi:10.1016/S0306-4565(02)00015-3

Brice PH, Grigg GC, Beard LA, Donovan JA (2002b) Patterns of activity and inactivity in echidnas (*Tachyglossus aculeatus*) free-ranging in a hot dry climate: correlates with ambient temperature, time of day and season. *Australian Journal of Zoology* **50**, 461–475. doi:10.1071/ZO01080

Burbidge AA (2018) Did *Zaglossus bruijnii* occur in the Kimberley region of Western Australia? *Australian Mammalogy* **40**, 315–318. doi:10.1071/AM17053

Cable B (2021) Voluntary scale training in the short-beaked echidna (*Tachyglossus aculeatus*). *Thylacinus* **45**(1), 3–7.

Campbell B (1989) Echidnas breed at Adelaide Zoo. *Thylacinus* **14**(4), 3.

Clark L (2013) Currumbin Wildlife Sanctuary's short-beaked echidna reproductive program. *Thylacinus* **37**(4), 2–3.

Clemente C, Cooper C, Withers P, Freakley C, Singh S, Terrill P (2016) The private life of echidnas: using accelerometry and GPS to examine field biomechanics and assess the ecological impact of a widespread, semi-fossorial monotreme. *The Journal of Experimental Biology* **219**, 3271–3283. doi:10.1242/jeb.143867

Clutton-Brock TH (1989) Mammalian mating systems. *Proceedings of the Royal Society of London. Series B, Biological Sciences (1934–1990)* **236**(1285), 339–372.

Coleman E (1934) The echidna under domestication. *Victorian Naturalist* **51**, 12–21.

Coleman E (1935) The echidna under domestication. *Victorian Naturalist* **52**, 151–154.

Collins LR (1973) *Monotremes and Marsupials: A Reference for Zoological Institutions*. Smithsonian Institution Press, Washington DC.

Crandall LS (1964) *The Management of Wild Mammals in Captivity*. Chicago University Press, Chicago.

Debenham JJ, Johnson R, Vogelnest L, Phelan DN, Whittington R, Šlapeta J (2012) Year-long presence of *Eimeria echidnae* and absence of *Eimeria tachyglossi* in captive short-beaked echidnas (*Tachyglossus aculeatus*). *The Journal of Parasitology* **98**, 543–549. doi:10.1645/GE-2982.1

Domrow R (1991) *Acari prostigmata* (excluding Trombiculidae) parasitic on Australian vertebrates: an annotated checklist, keys and bibliography. *Invertebrate Taxonomy* **4**, 1283–1376. doi:10.1071/IT9901283

Dunnet GM, Mardon DK (1974) A monograph of Australian fleas (Siphonaptera). *Australian Journal of Zoology Supplementary Series* **22** (30), 1–273.

Durette-Desset MC, Cassone J (1983) A taxonomic revision of the trichostrongyloid nematode parasites of the echidna, *Tachyglossus aculeatus* (Monotremata). *Australian Journal of Zoology* **31**, 257–284. doi:10.1071/ZO9830257

Dutton-Regester KJ, Keeley T, Fenelon JC, Roser A, Meer H, Hill A, Pyne M, Renfree MB, Johnston S (2021) Plasma progesterone secretion during gestation of the captive short-beaked echidna. *Reproduction (Cambridge, England)* **162**, 267–275. doi:10.1530/REP-21-0110

Dutton-Regester KJ, Roser A, Meer H, Renfree MB, Phillips C, Johnston SD (2022a) Reproductive behaviour before and after oestrus and oviposition in the captive short-beaked echidna (*Tachyglossus aculeatus*). *Reproduction, Fertility and Development* **34**, 920–932. doi:10.1071/RD22092

Dutton-Regester KJ, Roser A, Meer H, Renfree MB, Johnston SD, Keeley T (2022b) Optimizing captive short-beaked echidna (*Tachyglossus aculeatus*) fecal sample identification and hormonal analysis. *Zoo Biology* **42**, 322–327. doi:10.1002/zoo.21732

Dutton-Regester KJ, Roser A, Meer H, Russell FA, Pyne M, Renfree MB, Johnston SD, Keeley T (2022c) Investigating the utility of using fecal hormone metabolites as a reproductive management tool for captive short-beaked echidnas (*Tachyglossus aculeatus*). *General and Comparative Endocrinology* **330**, 114142. doi:10.1016/j.ygcen.2022.114142

Dutton-Regester KJ, Roser A, Meer H, Renfree MB, Phillips C, Johnston SD (2023) Reproductive behaviour before and after oestrus and oviposition in the captive short-beaked echidna (*Tachyglossus aculeatus*). *Reproduction, Fertility and Development* **34**, 920–932. doi:10.1071/RD22092

Fenelon JC, McElrea C, Shaw G, Evans AR, Pyne M, Johnston SD, Renfree MB (2021) The unique penile morphology of the short-beaked echidna, *Tachyglossus aculeatus*. *Sexual Development: Genetics, Molecular Biology, Evolution, Endocrinology, Embryology, and Pathology of Sex Determination and Differentiation* **15**, 262–271. doi:10.1159/000515145

Ferguson A, Laming B (2022) Maternal care and juvenile development of captive-bred short-beaked echidnas (*Tachyglossus aculeatus acanthion*) at Perth Zoo, Western Australia. *Australian Journal of Zoology* **70**, 43–55. doi:10.1071/ZO22013

Ferguson A, Turner B (2013) Reproductive parameters and behaviour of captive short-beaked echidna (*Tachyglossus aculeatus acanthion*) at Perth Zoo. *Australian Mammalogy* **35**, 84–92. doi:10.1071/AM12022

Fieseler CA, Junge RE (1997) The management and husbandry of the short-beaked echidna (*Tachyglossus aculeatus*) at the St. Louis Zoo. St Louis Zoo, St Louis, Missouri. Unpublished manuscript.

Finnie EP (1988) Diseases and injuries of other Australian mammals. In *Australian Wildlife*. Proceedings 104. Post Graduate Committee in Veterinary Science. (Ed. DI Bryden) pp. 91–93. University of Sydney, Sydney.

Flannery TF (1995) *Mammals of New Guinea*. 2nd edn. Australian Museum/Reed Books, Sydney.

Flannery TF, Groves CP (1998) A revision of the genus *Zaglossus* (Monotremata, Tachyglossidae), with description of new species and subspecies. *Mammalia* **62**, 367–396. doi:10.1515/mamm.1998.62.3.367

Flower SS (1929) List of vertebrated animals exhibited in the gardens of the Zoological Society of London 1828–1927. Zoological Society of London, London.

Flower SS (1931) Contributions to our knowledge of the duration of life in vertebrate animals. *Proceedings of the Zoological Society of London* **101**, 145–234. doi:10.1111/j.1469-7998.1931.tb06192.x

George G (1990) Monotreme and marsupial breeding programs in Australian Zoos. In *Mammals from Pouches and Eggs*. (Eds JA Marshall Graves, RM Hope and DW Cooper) pp. 39–63. CSIRO, Melbourne.

Gilruth JA, Sweet G, Dodd S (1911) Observations on the occurrence in the blood of various animals (chiefly monotremes and marsupials) of bodies apparently identical with *Anaplasma marginale* Theiler, 1910. *Parasitology* **4**, 1–6. doi:10.1017/S0031182000002389

Griffiths M (1965) Rate of growth and intake of milk in a suckling echidna. *Comparative Biochemistry and Physiology* **16**, 383–392. doi:10.1016/0010-406X(65)90304-X

Griffiths M (1968) *Echidnas*. Pergamon Press, London.

Griffiths M (1978) *The Biology of the Monotremes*. Academic Press, New York.

Griffiths M (1989) Tachyglossidae. In *Fauna of Australia Volume 1B Mammalia*. (Eds DW Walton and BJ Richardson) pp. 407–435. Australian Government Publishing Service, Canberra.

Griffiths M, McIntosh DL, Coles REA (1969) The mammary gland of the echidna, *Tachyglossus aculeatus'* with observations on the incubation of the egg and on the newly-hatched young. *Journal of Zoology* **158**, 371–386[London]. doi:10.1111/j.1469-7998.1969.tb02155.x

Griffiths M, Green B, Leekie RMC, Messer M, Newgrain K (1984) The constituents of platypus and echidna milk with particular reference to the fatty acid complement of the triglycerides. *Australian Journal of Biological Sciences* **37**, 323–330. doi:10.1071/BI9840323

Griffiths M, Kristo F, Green B, Fogerty AC, Newgrain K (1988) Observations on free-living lactating echidnas *Tachyglossus aculeatus* (Monotremata: Tachyglossidae) and sucklings. *Australian Mammalogy* **11**, 135–143. doi:10.1071/AM88017

Grigg GC, Beard LA, Augee ML (1989) Hibernation in a monotreme, the echidna (*Tachyglossus aculeatus*). *Comparative Biochemistry and Physiology* **92A**, 609–612. doi:10.1016/0300-9629(89)90375-7

Grigg GC, Augee ML, Beard LA (1992) Thermal relations of free-living echidnas during activity and in hibernation in a cold climate. In *Platypus and Echidnas*. (Ed. ML Augee) pp. 160–173. Royal Zoological Society of New South Wales, Sydney.

Harris RL, Nicol SC (2014) Observations of breeding behaviour and possible infanticide in a wild population of Tasmanian echidnas (*Tachyglossus aculeatus setosus*). *Australian Mammalogy* **36**, 108–112. doi:10.1071/AM13011

Harris RL, Holland BR, Cameron EZ, Davies NW, Nicol SC (2014) Chemical signals in the echidna: differences between seasons, sexes, individuals and gland types. *Journal of Zoology* **293**, 171–180. doi:10.1111/jzo.12133

Harris RL, Cameron EZ, Davies NW, Nicol SC (2016) Chemical cues, hibernation and reproduction in female short-beaked echidnas (*Tachyglossus aculeatus setosus*): Implications for sexual conflict. In *Chemical Signals in Vertebrates 13*. (Eds BA Schulte, TE Goodwin and MH Ferkin) pp. 145–166. Springer International Publishing, Cham.

Heck L (1908) Echidna-Züchtung im Berliner Zoologischen Garten. *Sitzungsbericht der Gesellschaft Naturforschender Freunde* **1908**, 187–189.

Helgen KM, Miguez RP, Kohen JL, Helgen LE (2012) Twentieth century occurrence of the long-beaked echidna *Zaglossus bruijnii* in the Kimberley region of Australia. *ZooKeys* **255**, 103–132. doi:10.3897/zookeys.255.3774

Heniger H, Kummer H (1961) Das Verhalten der Schnabeligel (Tachyglossidae). *Handbuch der Zoologie* **8**, 1–8.

Higgins DP, Tobias G, Stone GM (2004) Excretion profiles of some reproductive steroids in the faeces of captive female short-beaked echidna (*Tachyglossus aculeatus*) and long-beaked echidna (*Zaglossus* sp.). *Proceedings of the Linnean Society of New South Wales* **125**, 279–286.

Holz P (2015) Monotremes. In *Zoo and Wild Animal Medicine*. 7th edn. (Eds RE Miller and ME Fowler) pp. 517–521. Elsevier Saunders, St Louis.

Hulbert AJ, Beard LA, Grigg GC (2008) The exceptional longevity of an egg-laying mammal, the short-beaked echidna (*Tachyglossus aculeatus*) is associated with peroxidation-resistant membrane composition. *Experimental Gerontology* **43**, 729–733. doi:10.1016/j.exger.2008.05.015

International Air Transport Association (IATA) (2022) *Live Animals Regulations*. 48th edn. International Air Transport Association, Geneva Switzerland.

Jackson SM, Groves C (2015) *Taxonomy of Australian Mammals*. CSIRO Publishing, Melbourne.

Johnston SD, Madden C, Nicolson V, Pyne M (2006) Identifying the sex of short-beaked echidnas. *Australian Veterinary Journal* **84**, 63–65. doi:10.1111/j.1751-0813.2006.tb13131.x

Johnston SD, Nicholson V, Madden C, Logie S, Pyne M, Roser A, Lisle AT, D'Occhio M (2007a) Assessment of reproductive status of male echidnas. *Animal Reproduction Science* **97**, 114–127. doi:10.1016/j.anireprosci.2005.12.016

Johnston SD, Smith B, Pyne M, Stenzel D, Holt WV (2007b) One-sided ejaculation of echidna sperm bundles. *American Naturalist* **170**, E162–E164. doi:10.1086/522847

Jones RC, Djakiew D, Dackeux JL (2004) Adaptations of the short-beaked echidna *Tachyglossus aculeatus* for sperm production, particularly on an arid environment. *Australian Mammalogy* **26**, 199–204. doi:10.1071/AM04199

Keating SE, Fenelon JC, Pyne M, Pinto BJ, Guzmán-Méndez IA, Johnston SD, Renfree MB, Gamble T (2022) Genetic sex test for the short-beaked echidna (*Tachyglossus aculeatus*). *Conservation Genetics Resources* **14**, 271–278. doi:10.1007/s12686-022-01258-3

Kuchel LJ (2003) The energetics and patterns of torpor in free-ranging *Tachyglossus aculeatus* from a warm-temperate climate. PhD thesis, The University of Queensland.

Kumar A, Parveen S, Sharma I, Pathak H, Deshmukh MV, Sharp JA, Kumar S (2019) Structural and mechanistic insights into EchAMP: a antimicrobial protein from the Echidna milk. *Biochimica et Biophysica Acta (BBA)-Biomembranes* **1861**, 1260–1274. doi:10.1016/j.bbamem.2019.03.020

Lacy RC, Balou JD, Pollak JP (2011) PMx: software package for demographic and genetic analysis and management of pedigreed populations. *Methods in Ecology and Evolution* **3**, 433–437. doi:10.1111/j.2041-210X.2011.00148.x

Lang EM (1958) Neves vom Schnabeligel (*Echidna aculeata*). *Bulletin des Zoologischen Gartens Basel* **19**, 5–8.

Lefèvre CM, Sharp JA, Nicholas KR (2009) Characterisation of monotreme caseins reveals lineage-specific expansion of an ancestral casein locus in mammals. *Reproduction, Fertility and Development* **21**, 1015–1027. doi:10.1071/RD09083

Loh SM, Gofton AW, Lo N, Gillett A, Ryan UM, Irwin PJ, Oskham CL (2016) Novel *Borrelia* species detected in echidna ticks, *Bothriocroton concolor*, in Australia. *Parasites & Vectors* **9**, 339. doi:10.1186/s13071-016-1627-x

Mackerras MJ (1959) The haematozoa of Australian mammals. *Australian Journal of Zoology* **7**, 105–135. doi:10.1071/ZO9590105

Martin CJ (1903) Thermal adjustment and respiratory exchange in monotremes and marsupials: a study in the development of homeothermism. *Philosophical Transactions of the Royal Society of London. Series B, Biological Sciences* **195**, 1–37.

McLelland D (2025) Appendix 1. Clinical pathology and physiological values. In *Current Therapy in Medicine of Australian Mammals*. Revised Edition. (Eds L Vogelnest and T Portas) pp. 656–719. CSIRO Publishing, Melbourne.

McOrist S, Smales L (1986) Morbidity and mortality of free-living and captive echidnas, *Tachyglossus aculeatus* (Shaw) in Australia. *Journal of Wildlife Diseases* **22**, 375–380. doi:10.7589/0090-3558-22.3.375

Messer M, Walker DM (1992) Milk substitutes for marsupial: In theory how good (or bad) are they? In *Urban Wildlife*. Proceedings 204. Post Graduate Committee in Veterinary Science. (Eds L Vogelnest, D Spielman, T Bellamy, M Messer and DM Walker) pp. 99–124. University of Sydney, Sydney.

Middleton D (2008) Echidnas. In *Medicine of Australian Mammals*. (Eds L Vogelnest and R Woods) pp. 77–101. CSIRO Publishing, Melbourne.

Morrow G, Nicol SC (2009) Cool sex? Hibernation and reproduction overlap in the echidna. *PLoS One* **4**, e6070. doi:10.1371/journal.pone.0006070

Morrow GE, Nicol SC (2012) Maternal care in the Tasmanian echidna (*Tachyglossus aculeatus setosus*). *Australian Journal of Zoology* **60**, 289–298. doi:10.1071/ZO12066

Morrow G, Andersen NA, Nicol S (2009) Reproductive strategies of the short-beaked echidna: a review with new data from a long-term study on the Tasmanian subspecies (*Tachyglossus aculeatus setosus*). *Australian Journal of Zoology* **57**, 275–282. doi:10.1071/ZO09037

Morrow GE, Jones SM, Nicol SC (2016) Interaction of hibernation and male reproductive function in wild Tasmanian echidnas *Tachyglossus aculeatus setosus*. *Journal of Mammalogy* **97**, 852–860. doi:10.1093/jmammal/gyw013

Morrow GE, Jones SM, Nicol SC (2017) Frozen embryos? Torpor during pregnancy in the Tasmanian short-beaked echidna *Tachyglossus aculeatus setosus*. *General and Comparative Endocrinology* **244**, 139–145. doi:10.1016/j.ygcen.2015.11.006

Muirhead L (1989) Zoo News: Adelaide Zoo. *Thylacinus* **14**(3), 28.

Nicol SC (2015) Family Tachyglossidae (Echidnas). In *Handbook of the Mammals of the World. Volume 5. Monotremes and Marsupials*. (Eds DE Wilson and RA Mittermeier) pp. 34–56. Lynx Edicons, Barcelona.

Nicol SC (2017) Energy homeostasis in monotremes. *Frontiers in Neuroscience* **11**(195), 1–17.

Nicol SC (2021) Diet, feeding behaviour and echidna beaks: a review of functional relationships within the tachyglossids. *Australian Mammalogy* **44**, 39–50. doi:10.1071/AM20053

Nicol SC, Andersen NA (2006) Body temperature as an indicator of egg-laying in the echidna, *Tachyglossus aculeatus*. *Journal of Thermal Biology* **31**, 483–490. doi:10.1016/j.jtherbio.2006.05.001

Nicol SC, Andersen NA (2007) Cooling rates and body temperature regulation of hibernating echidnas (*Tachyglossus aculeatus*). *The Journal of Experimental Biology* **210**, 586–592. doi:10.1242/jeb.02701

Nicol SC, Morrow GE (2012) Sex and seasonality: reproduction in the echidna (*Tachyglossus aculeatus*). In *Living in a Seasonal World*. (Ed. T Ruf) pp. 143–153. Springer-Verlag, Berlin.

Nicol SC, Morrow G, Andersen NA (2008) Hibernation in monotremes: a review. In *Hypometabolism in Animals: Torpor, Hibernation and Cryobiology*. (Eds BG Lovegrove and AE McKechnie) pp. 251–262. University of KwaZulu-Natal, Pietermaritzburg.

Nicol SC, Anderson NA, Morrow GA, Harris RL (2019a) Spurs, sexual dimorphism and reproductive maturity in Tasmanian echidnas (*Tachyglossus aculeatus setosus*). *Australian Mammalogy* **41**, 161–169. doi:10.1071/AM18005

Nicol SC, Morrow GE, Harris RL (2019b) Energetics meets sexual conflict: the phenology of hibernation in Tasmanian echidnas. *Functional Ecology* **33**, 2150–2160. doi:10.1111/1365-2435.13447

Oates JE, Bradshaw FJ, Bradshaw SD, Lonsdale RA (2002) Sex identification and evidence of gonadal activity in the short-beaked echidna (*Tachyglossus aculeatus*) (Monotremata: Tachyglossidae): non-invasive analysis of faecal sex steroids. *Australian Journal of Zoology* **50**, 395–406. doi:10.1071/ZO02017

Oftedal OT, Nicol SC, Davies NW, Sekii N, Taufik E, Fukuda K, Saito T, Urashima T (2014) Can an ancestral condition for milk oligosaccharides be determined? Evidence from the Tasmanian echidna (*Tachyglossus aculeatus setosus*). *Glycobiology* **24**, 826–839. doi:10.1093/glycob/cwu041

Owen R (1845) Observations on the living *Echidna* exhibited at the Menagerie of the Society in May 1845. *Proceedings of the Zoological Society of London* **13**, 80–82.

Parker GA (2016) The evolution of expenditure on testes. *Journal of Zoology* **298**, 3–19. doi:10.1111/jzo.12297

Perry T, Toledo-Flores D, Kang WX, Ferguson A, Laming B, Tsend-Ayush E, Lim SL, Grutzner F (2019) Non-invasive genetic sexing technique for analysis of short-beaked echidna (*Tachyglossus aculeatus*) populations. *Reproduction, Fertility and Development* **31**, 1289–1295. doi:10.1071/RD18142

Perry T, West E, Eisenhofer R, Stenhouse A, Wilson I, Laming B, Rismiller P, Shaw M, Grützner F (2022) Characterising the gut microbiomes in wild and captive short-beaked echidnas reveals diet-associated changes. *Frontiers in Microbiology* **13**, 687115.

Pierce F, Mattiske J, Menkhorst P (2006) A probable case of twins in the short-beaked echidna *Tachyglossus aculeatus* (Tachyglossidae: Monotremata), with observations on suckling young after their emergence from the nursery burrow. *Victorian Naturalist* **123**, 332–340.

Portas T (2025) Appendix 2. Blood collection sites. In *Current Therapy in Medicine of Australian Mammals*. Revised Edition. (Eds L Vogelnest and T Portas) pp. 720–724. CSIRO Publishing, Melbourne.

Priestley H (1915) *Theileria tachyglossi* n. sp. A blood parasite of *Tachyglossus aculeatus*. *Annals of Tropical Medicine and Parasites* **9**, 233–238. doi:10.1080/00034983.1915.11687681

Rismiller PD (1992) Field observations on Kangaroo Island echidnas (*Tachyglossus aculeatus multiaculeatus*) during the breeding season. In *Platypus and Echidnas*. (Ed. ML Augee) pp. 101–105. Royal Zoological Society of New South Wales, Sydney.

Rismiller PD (1993) Overcoming a prickly problem. *Australian Natural History* **24**(6), 22–29.

Rismiller PD (1999) *The Echidna: Australia's Enigma*. Hugh Lauter Leven Associates.

Rismiller PD (2008) *Biology, Rescue and Rehab of Short-beaked Echidnas*. The author, Penneshaw, Kangaroo Island. South Australia. Revised edition 2019.

Rismiller PD (2010) Echidnas, the different mammal. National Wildlife Rehabilitation Conference. Adelaide. www.awrc.org.au/uploads/5/8/6/6/5866843/rismiller_echidnas.pdf, accessed 13 May 2023.

Rismiller PD (2014) *Tachyglossus*: Survival strategies for successful rescue and hand rearing. National Wildlife Rehabilitation Conference, Hobart.

Rismiller PD, Grutzner F (2019) Tachyglossus aculeatus. *Mammalian Species* **51**(980), 75–91. doi:10.1093/mspecies/sez012

Rismiller PD, McKelvey MW (1995) Orientation and location in short-beaked echidnas *Tachyglossus aculeatus*. In *Reintroduction Biology of Australian and New Zealand Fauna*. (Ed. M Serena) pp. 227–234. Surrey Beatty & Sons, Sydney.

Rismiller PD, McKelvey MW (1996) Sex, torpor and activity in temperate climate echidnas. In *Adaptations to the Cold: Tenth International Hibernation Symposium*. (Eds F Geiser, AJ Hulbert and SC Nicol) pp. 23–30. University of New England, Armidale.

Rismiller PG, McKelvey MW (2000) Frequency of breeding and recruitment in the short-beaked echidna, *Tachyglossus aculeatus*. *Journal of Mammalogy* **81**, 1–17. doi:10.1644/1545-1542(2000)081<0001:FOBARI>2.0.CO;2

Rismiller PD, McKelvey MW (2003) Body mass, age and sexual maturity in short-beaked echidnas, *Tachyglossus aculeatus*. *Comparative Biochemistry and Physiology. Part A, Molecular & Integrative Physiology* **136**, 851–865. doi:10.1016/S1095-6433(03)00225-3

Rismiller PD, McKelvey MW (2009) Activity and behaviour of lactating echidnas (*Tachyglossus aculeatus multiaculeatus*) from hatching of egg to weaning of young. *Australian Journal of Zoology* **57**, 265–273. doi:10.1071/ZO09031

Roberts FHS (1970) *Australian Ticks*. CSIRO, Melbourne.

Robinson KW (1954) Heat tolerance of Australian monotremes and marsupials. *Australian Journal of Biological Sciences* **7**, 348–360. doi:10.1071/BI9540348

Robinson KJ, Limpus DJ, Crosbie B, Limpus CJ, Fabbro L (2024) Depredation of eggs of threatened freshwater turtles by the short-beaked echidna (*Tachyglossus aculeatus* (Shaw, 1792)). *Australian Journal of Zoology* **71**, ZO23029.

Russell FA (2014) Conditional same/different discrimination learning in the short-beaked echidna (*Tachyglossus aculeatus*). PhD thesis. The University of Wollongong, Wollongong.

Russell F, Burke D (2016) Conditional same/different concept learning in the short-beaked echidna (*Tachyglossus aculeatus*). *Journal of the Experimental Analysis of Behavior* **105**, 133–154. doi:10.1002/jeab.185

Russell FA, Johnston SD, Hill A, Roser A, Meer H, Fenelon JC, Renfree MB, Keeley T (2022) Validation of a non-invasive assessment technique for quantifying faecal glucocorticoid metabolite concentrations in the short-beaked echidna (*Tachyglossus aculeatus*). *General and Comparative Endocrinology* **327**(114092), 1–8. doi:10.1016/j.ygcen.2022.114092

San Diego Zoo Wildlife Alliance (SDZWA) (2021) Short-beaked echidna (*Tachyglossus aculeatus*) fact sheet: Managed care. https://ielc.libguides.com/sdzg/factsheets/short-beaked-echidna/managedcare, accessed 22 April 2023.

Scheelings TF, Slocombe FF, Bayley C (2017) Metastatic mineralisation caused by vitamin D intoxication in hand-raised short-beaked echidnas (*Tachyglossus aculeatus*). *Australian Veterinary Journal* **95**, 427–430. doi:10.1111/avj.12640

Schmidt-Nielsen K, Dawson TJ, Crawford EC, Jr (1966) Temperature regulation in the echidna (*Tachyglossus aculeatus*). *Journal of Cellular Physiology* **67**, 63–71. doi:10.1002/jcp.1040670108

Scott K, Toomey V, Griffiths S, Stenner R (2018) Principles of Echidna Rehabilitation: For Veterinarians, Veterinary Technicians and Volunteer Wildlife Rehabilitators. Wildcare Australia Inc., Nerang, Queensland.

Shaw ME (2022) Echidna nutrition. In *Fowler's Zoo and Wild Animal Medicine Current Therapy*. Volume 10 (Eds. E Miller, N Lamberski and P Calle) pp. 697–708. Elsevier, Amsterdam.

Shaw M, Jarman A (2025) Nutrition. In *Current Therapy in Medicine of Australian Mammals*. Revised Edition. (Eds L Vogelnest and T Portas) pp. 229–255. CSIRO Publishing, Melbourne.

Skibiel AL, Downing LM, Orr TJ, Hood WR (2013) The evolution of the nutrient composition of mammalian milks. *Journal of Animal Ecology* **82**, 1254–1264. doi:10.1111/1365-2656.12095

Species360 (2023) Zoological Information Management System (ZIMS). Minneapolis, USA. www.species360.org.

Spencer C, Richards K (2009) Observations on the diet and feeding habits of the short-beaked echidna (*Tachyglossus aculeatus*) in Tasmania. *Tasmanian Naturalist* **131**, 36–41.

Spielman D (1994) First aid emergency care for Australian native mammals. In *Wildlife*. Proceedings 233. Post Graduate Committee in Veterinary Science. (Ed. DI Bryden) pp. 339–366. University of Sydney, Sydney.

Spratt D, Beveridge I (2016) Helminth parasites of Australian monotremes and marsupials. *Zootaxa* **4123**, 1–198. doi:10.11646/zootaxa.4123.1.1

Spratt DM, Beveridge I, Walter EL (1991) A catalogue of Australasian monotremes and marsupials and their recorded helminth parasites. *Records of the South Australian Museum. Monograph Series* **1**, 1–105.

Staker L (2007) Hand rearing echidnas. Unpublished manuscript. http://www.lyndastaker.com/freestuff/echidnas.pdf, accessed 22 August 2020.

Stannard HJ, Bekkers JM, Old JM, McAllan BM, Shaw ME (2017) Digestibility of a new diet for captive short-beaked echidnas

(*Tachyglossus aculeatus*). *Zoo Biology* **36**, 56–61. doi:10.1002/zoo.21347

Stannard HJ, Miller RD, Old JM (2020) Marsupial and monotreme milk: a review of its nutrient and immune properties. *PeerJ* **8**, e9335. doi:10.7717/peerj.9335

Teahan CG, McKenzie HA, Griffiths M (1991) Some monotreme milk "whey" and blood proteins. *Comparative Biochemistry and Physiology Part B* **99**, 99–118. doi:10.1016/0305-0491(91)90014-5

Tobias G (2025a) Short-beaked echidna. In *Current Therapy in Medicine of Australian Mammals*. Revised Edition. (Eds L Vogelnest and T Portas) pp. 435–456. CSIRO Publishing, Melbourne.

Tobias G (2025b) Coccidiosis in short-beaked echidnas. In *Current Therapy in Medicine of Australian Mammals*. Revised Edition. (Eds L Vogelnest and T Portas) pp. 457–464. CSIRO Publishing, Melbourne.

Tong LJ (2019) Pathology of echidna. In *Proceedings: Wildlife Health and Pathology Short Course*. (Ed. J Hall) pp.174–197. Taronga Zoo, February 2019. Sydney, Taronga Conservation Society Australia.

Tong Marwood L (2025) Neoplasia. In *Current Therapy in Medicine of Australian Mammals*. Revised Edition. (Eds L Vogelnest and T Portas) pp. 299–320. CSIRO Publishing, Melbourne.

Vogelnest L (2025a) Capture, physical and chemical restraint. In *Current Therapy in Medicine of Australian Mammals*. Revised Edition. (Eds L Vogelnest and T Portas) pp. 122–151. CSIRO Publishing, Melbourne.

Vogelnest LJ (2025b) Dermatology. In *Current Therapy in Medicine of Australian Mammals*. Revised Edition. (Eds L Vogelnest and T Portas) pp. 186–211. CSIRO Publishing, Melbourne.

Vogelnest L, Portas T (Eds) (2025) *Current Therapy in Medicine of Australian Mammals*. Revised Edition. CSIRO Publishing, Melbourne.

Vogelnest L, Woods R (Eds) (2008) *Medicine of Australian Mammals*. CSIRO Publishing, Melbourne.

Wallage A, Clarke L, Thomas L, Pyne M, Beard L, Ferguson A, Lisle A, Johnston S (2015) Advances in the captive breeding and reproductive biology of the short-beaked echidna (*Tachyglossus aculeatus*). *Australian Journal of Zoology* **63**, 181–191. doi:10.1071/ZO14069

Whittington RJ (1988) The monotremes in health and disease. In *Australian Wildlife*. Proceedings 104. Post Graduate Committee in Veterinary Science. (Ed. DI Bryden) pp. 727–787. University of Sydney, Sydney.

Whittington RJ (1992) The role of infectious diseases in the population biology of monotremes. In *Platypus and Echidnas*. (Ed. ML Augee) pp. 285–292. Royal Zoological Society of New South Wales, Sydney.

Whittington RJ (1993) Diseases of monotremes. In *Zoo and Wild Animal Medicine. Current Therapy 3*. (Ed. ME Fowler) pp. 269–275. WB Saunders, Philadelphia.

Whittington RJ, Obendorf D (2012) Monotremes. In *Diagnostic Pathology of Aquatic, Aerial and Terrestrial Wildlife. Wildlife Pathology Short Course*. 21–24 August Taronga Zoo, Sydney. p. 214. Australian Registry of Wildlife Health, Sydney.

Whittington RJ, Middleton D, Spratt DM, Muntz F, Carmel B, Mc Cracken HE, Strakosch MR, Stephanson-Shaw J, Harper PAW, Hartley WJ (1992) Sparganosis in the monotremes *Tachyglossus aculeatus* and *Ornithorhynchus anatinus* in Australia. *Journal of Wildlife Diseases* **28**, 636–640. doi:10.7589/0090-3558-28.4.636

Wilkinson DA, Grigg GC, Beard LA (1998) Shelter selection and home range of echidnas, *Tachyglossus aculeatus*, in the highlands of south-east Queensland. *Wildlife Research* **25**, 219–232. doi:10.1071/WR97072

Woinarski JCZ, Burbidge AA, Harrison P (Eds) (2014) *The Action Plan for Australian Mammals 2012*. CSIRO Publishing, Melbourne.

Woods R (1999) Prevention of disease in hand reared native wildlife. In *Wildlife in Australia: Healthcare & Management*. (Ed. DI Bryden) pp. 455–470. Proceedings 327. Post Graduate Foundation in Veterinary Science. University of Sydney, Sydney.

# 4
# Carnivorous marsupials

*Stephen M. Jackson, Marissa L. Parrott, Chris R. Dickman, Bronwyn A. Fancourt, Julie Old, Peter Holz and Hayley Stannard*

## 1. INTRODUCTION

The carnivorous marsupials are a highly diverse group that includes 88 species in the order Dasyuromorphia with three families: the Dasyuridae (antechinuses, quolls, Tasmanian devil (*Sarcophilus harrisii*) and their relatives), the Thylacinidae (extinct thylacine) and the Myrmecobiidae (numbat (*Myrmecobius fasciatus*), see Chapter 5) (Baker *et al.* 2023). The two species of marsupial moles of the Order Notoryctemorphia are also carnivorous. The members of this group range in body size from only 3 g for the Pilbara ningaui (*Ningaui timealeyi*) up to 35 kg for the thylacine (*Thylacinus cynocephalus*). Species occur in desert to rainforest habitats and the group includes fossorial species such as the marsupial mole and relatively arboreal species such as phascogales and spotted-tailed quolls (*Dasyurus maculatus*). The carnivorous marsupials are distributed throughout Australia and New Guinea, with several cryptic species remaining to be formally classified (Baker *et al.* 2023).

The carnivorous marsupials have a long history of being held in captivity. The Tasmanian devil was kept in captivity in Hobart from before 1807 (Harris 1807), at which time it was considered 'untameably savage; biting severely, and uttering at the same time a low yelling growl'.

Tasmanian devils were held in captivity at London Zoo as early as 1833 and the New York Zoo in 1909 (Crandall 1964; Collins 1973; Oglesby 1978) and have since been held in zoos throughout Australia. They were first bred at the Beaumaris Zoo in 1915, with the first mainland breeding occurring at Healesville Sanctuary, Victoria, in 1952 (Oglesby 1978).

Other species of carnivorous marsupial such as brush-tailed phascogales (*Phascogale tapoatafa*) were held at London Zoo in 1905, first bred in captivity in 1932 (Fleay 1950) and have been held by several zoos including Healesville Sanctuary, Taronga Zoo and Perth Zoo, as well as several private breeders. Yellow-footed antechinus (*Antechinus flavipes*) were held at London Zoo in 1908 and have since been held in zoos in Australia (Collins 1973). Eastern quolls (*Dasyurus viverrinus*) were held at London Zoo in 1911, planigales at Taronga Zoo in 1972 and kowaris (*Dasyuroides byrnei*), dunnarts, antechinuses, red-tailed phascogales (*Phascogale calura*) (Foster *et al.* 2006; Stannard *et al.* 2013; Old *et al.* 2021), kultarrs (*Antechinomys laniger*) (Stannard and Old 2010, 2014) and dunnarts have been held at various zoos, universities and museums (Mitchell 1911; Crandall 1964; Woolley 1971a; Collins 1973; Au *et al.* 2010; Parrott *et al.* 2019a). Over the years, many species have been held in captivity in zoos and research institutions throughout Australia and the world. Due to the lack of facilities such as nocturnal houses, the smaller species have often been poorly held in zoos and mostly held in private facilities as subjects for research. Some dasyurids such as the phascogales, quolls, dunnarts, kowari and the Tasmanian devil have bred relatively well in captivity, but other species such as the kultarr and antechinuses have been a challenge to breed over multiple generations.

The thylacine was held in Australian and overseas zoos as early as at least 1850 when London Zoo exhibited it on several occasions (Gunn 1863; Crandall 1964), with other zoos holding it including Adelaide Zoo (1886–1903), Antwerp Zoo (1914), Beaumaris Zoo in Hobart (1910–36), Berlin Zoo (1902–08), Cologne Zoo (1909), Melbourne Zoo (1875–1925), New York Zoo (1902–19), National Zoo in Washington (1902–09) and Taronga Zoo in Sydney (1918) (Crandall 1964; Guiler 1985). Sadly, the thylacine has been considered extinct since the last animal died in Beaumaris Zoo in Hobart in 1936, so it is not included in this chapter. The marsupial moles have never been held on exhibit and only very rarely have been held in captivity (and even then, they have fared very poorly), so the information on them is very limited and any information that will assist in keeping them alive for extended periods will be of significant use. The endangered status of various carnivorous marsupials is a reminder of the important role that zoos and other captive facilities have in maintaining viable populations via their captive breeding programs and using their facilities for breeding and research (e.g. Parrott *et al.* 2019b) of threatened species.

## 2. TAXONOMY

### 2.1 Nomenclature

The extant carnivorous marsupials comprise three separate families and two orders (Table 4.1). The Dasyuridae is the largest family, containing 85 species in total, of which 65 occur in Australia, 22 in New Guinea and 2 in both regions (Flannery 1995a, 1995b; AMTC 2022; Baker and Gynther 2023). The Thylacinidae contains only the one extinct species; the two species of marsupial mole, which are very different from all other marsupials, are included within their own Order.

**Australian carnivorous marsupials**

Class: Mammalia
Supercohort: Marsupialia
Cohort: Australidelphia
Order: Dasyuromorphia
Family: Dasyuridae
Genus and species: 65 species in 13 genera
Family: Myrmecobiidae
Genus and species: 1 species (see Chapter 5)
Family: Thylacinidae
Genus and species: 1 species
Order: Notoryctemorphia
Family: Notoryctidae
Genus and species: 2 species in 1 genus

**Etymology**

See Strahan and Conder (2007).

### 2.2 Subspecies

See Jackson and Groves (2015).

### 2.3 Recent synonyms

Synonyms of the Australian carnivorous marsupials can be found in Jackson and Groves (2015).

### 2.4 Other common names

See Baker and Gynther (2023) and AMTC (2024).

## 3. NATURAL HISTORY

### 3.1 Bodyweight and morphometrics

The carnivorous marsupials range in body size from only 3 g to over 21 kg (Table 4.1). The morphometrics of each species can be found in Baker and Gynther (2023).

### 3.2 Distribution and habitat

There is great variation in the distribution and habitats of the Australian species of carnivorous marsupials. They range from the arid deserts (marsupial moles, kowari, mulgaras, ningauis, kultarr and various species of dunnarts) to woodlands (quolls, phascogales, Tasmanian devil, thylacine (extinct) and other species of dunnarts) to rainforest (species of dunnarts and populations of quolls) and even the Australian alpine zone for some species such as *Antechinus*. More specific details of the distribution and habitats occupied by the Australian carnivorous marsupials can be found in Baker and Dickman (2018) and Baker and Gynther (2023).

### 3.3 Conservation status

Many of the Australian carnivorous marsupials have declined in number and range over the past 230 years, including the thylacine (extinct in 1936); six species are Endangered, and six species are classified as Vulnerable to extinction on the IUCN Red List (Woinarski *et al.* 2014; Table 4.1).

### 3.4 Diet in the wild

The smallest species, such as planigales, ningauis and dunnarts, are largely insectivorous. Larger species, such as the marsupial moles, antechinuses, mulgaras, kowari, phascogales and smaller quolls, eat insects and small vertebrates such as lizards, while the larger quolls and Tasmanian devils are primarily carnivorous. Tasmanian devils and quolls will also scavenge on carrion (Table 4.2). A strong positive correlation has been observed between body size of dasyurids and invertebrate and vertebrate prey, with prey handling time

**Table 4.1.** Species of extant carnivorous marsupials within Australia, their bodyweights and conservation status.

*Also occurs in New Guinea and/or surrounding islands. EN = endangered, EX = extinct, LC = least concern, NT = near threatened, UNK = unknown, VU = vulnerable.
Derived from Woinarski *et al.* (2014), Jackson and Groves (2015), Aplin (2015), Baker (2015), Mason *et al.* (2017), Rovinsky *et al.* (2020), Baker and Gynther (2023), AMTC (2024), Umbrello *et al.* (2023) and J. Newman-Martin (pers. comm.).

| Common name | Scientific name | Weight (g) | IUCN status |
|---|---|---|---|
| **Order Dasyuromorphia** | | | |
| **Family Dasyuridae** | | | |
| Southern mulgara | *Dasycercus archeri* | Unknown | EX |
| Brush-tailed mulgara | *D. blythi* | 60–110 | LC |
| Crest-tailed mulgara | *D. cristicauda* | c. 75 | EX |
| Ampurta | *D. hillieri* | 65–185 | NT? |
| Little mulgara | *D. marlowi* | c. 40 | EX |
| Northern mulgara | *D. woolleyae* | c. 75 | EX |
| Kaluta | *Dasykaluta rosamondae* | 20–40 | LC |
| Kowari | *Dasyuroides byrnei* | 70–175 | VU |
| Western quoll (*) | *Dasyurus geoffroii* | 595–2185 | NT |
| Northern quoll | *D. hallucatus* | 240–1120 | EN |
| Spotted-tailed quoll | *D. maculatus* | 900–5000 | NT |
| Eastern quoll | *D. viverrinus* | 700–2225 | EN |
| Dibbler | *Parantechinus apicalis* | 40–125 | EN |
| Sandstone pseudantechinus | *Pseudantechinus bilarni* | 15–40 | LC |
| Fat-tailed pseudantechinus | *P. macdonnellensis* | 20–45 | LC |
| Carpentarian pseudantechinus | *P. mimulus* | 14–25 | NT |
| Ningbing pseudantechinus | *P. ningbing* | 15–25 | LC |
| Woolley's pseudantechinus | *P. woolleyae* | 30–50 | LC |
| Tasmanian devil | *Sarcophilus harrisii* | 5000–14 000 | EN |
| Rusty antechinus | *Antechinus adustus* | 18–45 | LC |
| Agile antechinus | *A. agilis* | 16–40 | LC |
| Silver-headed antechinus | *A. argentus* | 16–52 | EN? |
| Black-tailed antechinus | *A. arktos* | 38–120 | EN? |
| Fawn antechinus | *A. bellus* | 25–83 | VU |
| Yellow-footed antechinus | *A. flavipes* | 28–67 | LC |
| Atherton antechinus | *A. godmani* | 42–123 | NT |
| Cinnamon antechinus | *A. leo* | 32–124 | LC |
| Mainland dusky antechinus | *A. mimetes* | 30–178 | LC |
| Swamp antechinus | *A. minimus* | 36–129 | LC |
| Buff-footed antechinus | *A. mysticus* | 15–46 | LC |
| Brown antechinus | *A. stuartii* | 17–71 | LC |
| Subtropical antechinus | *A. subtropicus* | 16–61 | LC |
| Tasmanian dusky antechinus | *A. swainsonii* | 30–115 | LC |
| Tasman Peninsula antechinus | *A. vandycki* | 46–93 | VU? |
| Red-tailed phascogale | *Phascogale calura* | 38–68 | NT |
| Northern phascogale | *P. pirata* | 108–234 | VU |
| Brush-tailed phascogale | *P. tapoatafa* | 106–311 | NT |

**Table 4.1.** continued

| Common name | Scientific name | Weight (g) | IUCN status |
|---|---|---|---|
| Giles' planigale | *Planigale gilesi* | 5–16 | LC |
| Long-tailed planigale | *P. ingrami* | 2.6–6.6 | LC |
| Orange-headed Pilbara planigale | *P. kendricki* | 4.1–12.5 | UNK |
| Common planigale | *P. maculata* | 5.5–22 | LC |
| Cracking-clay Pilbara planigale | *P. tealei* | 2.9–6.1 | UNK |
| Narrow-nosed planigale | *P. tenuirostris* | 4–9 | LC |
| Kultarr | *Antechinomys laniger* | 10–30 | LC |
| Long-tailed dunnart | *A. longicaudata* | 15–25 | LC |
| Wongai ningaui | *Ningaui ridei* | 6.5–11.5 | LC |
| Pilbara ningaui | *N. timealeyi* | 3.5–9.5 | LC |
| Southern ningaui | *N. yvonneae* | 6–14 | LC |
| Chestnut dunnart (*) | *Sminthopsis archeri* | 12–16 | NT |
| Kakadu dunnart | *S. bindi* | 12–20 | NT |
| Butler's dunnart | *S. butleri* | 15–30 | VU |
| Fat-tailed dunnart | *S. crassicaudata* | 10–21 | LC |
| Little or Lesser dunnart | *S. dolichura* | 10–21 | LC |
| Julia Creek dunnart | *S. douglasi* | 40–70 | NT |
| Grey-bellied dunnart | *S. fuliginosa* | 10–25 | LC |
| Gilbert's dunnart | *S. gilberti* | 14–25 | LC |
| White-tailed dunnart | *S. granulipes* | 18–37 | LC |
| Hairy-footed dunnart | *S. hirtipes* | 13–19.5 | LC |
| White-footed dunnart | *S. leucopus* | 16–32 | LC |
| Stripe-faced dunnart | *S. macroura* | 13–25 | LC |
| Common dunnart | *S. murina* | 10–28 | LC |
| Ooldea dunnart | *S. ooldea* | 8–17 | LC |
| Sandhill dunnart | *S. psammophila* | 25–55 | VU |
| Red-cheeked dunnart (also *) | *S. virginiae* | 18–58 | LC |
| Lesser hairy-footed dunnart | *S. youngsoni* | 8.5–12 | LC |
| **Family Thylacinidae** | | | |
| Thylacine | *Thylacinus cynocephalus* | 15 000–35 000 | EX |
| **Order Notoryctemorphia** | | | |
| **Family Notoryctidae** | | | |
| Northern marsupial mole | *Notoryctes caurinus* | 30–50 | LC |
| Southern marsupial mole | *N. typhlops* | 40–70 | LC |

increasing exponentially with invertebrate prey size (Dickman 1988a; Fisher and Dickman 1993; Woolnough and Carthew 1996; Chen *et al.* 1998; Jones and Stoddart 1998). Larger dasyurids, such as the spotted-tailed quoll and the Tasmanian devil, can take prey several times their own body size (Jones 2003). Fungal material has also been reported in the diets of three species of Antechinus, the common dunnart and brush-tailed phascogale (Elliott *et al.* 2022) and may form a small but hitherto poorly recognised part of the diet of other dasyurids.

Marsupial moles have been poorly studied due to both their highly cryptic habits and the very few specimens that have been found over the years. Stomach contents of individuals have revealed ants, ant eggs, termites, grass seed material (which appears to be the result of eating

**Table 4.2.** Wild diet of the different genera of carnivorous marsupials.

Derived from Baker and Gynther (2023) and Terry and Golden (2019).

| Genus | Diet |
|---|---|
| **Dasyuromorphia** | |
| **Dasyuridae** | |
| *Dasycercus* | Insects, other arthropods and small vertebrates (e.g. lizards and small birds) |
| *Dasykaluta* | Insects, other arthropods and small vertebrates (e.g. lizards) |
| *Dasyuroides* | Insects, other arthropods and small vertebrates (e.g. lizards) |
| *Dasyurus* | Insects, other arthropods, birds, frogs, lizards, small to medium-sized mammals up to own body size, or larger for spotted-tailed quolls, including carrion. Some vegetation matter such as berries, grasses and fruit |
| *Parantechinus* | Insects, other arthropods and small vertebrates (e.g. lizards) |
| *Pseudantechinus* | Insects, other arthropods and small vertebrates (e.g. lizards) |
| *Sarcophilus* | Insects, frogs, crustaceans, birds, fish and mammals, including carrion. |
| *Antechinus* | Insects, other arthropods and small vertebrates (e.g. lizards, birds, eggs) |
| *Phascogale* | Insects, other arthropods and small vertebrates (e.g. lizards, birds, eggs) |
| *Planigale* | Insects, other arthropods and small vertebrates (e.g. lizards) |
| *Antechinomys* | Insects, other arthropods and small vertebrates (e.g. lizards) |
| *Ningaui* | Insects, other arthropods and small vertebrates |
| *Sminthopsis* | Insects, other arthropods and small vertebrates (e.g. lizards) |
| **Notoryctemorphia** | |
| **Notoryctidae** | |
| *Notoryctes* | Ants, ant pupae, chrysomelid beetles, sawfly larvae, beetle larvae, small lizards |

grass-eating ants), weevil larvae, beetles and spiders (Stirling 1891; Winkel and Humphrey-Smith 1988; Johnson and Walton 1989; Pavey *et al.* 2012). Other observations in captivity suggest they also eat small lizards such as skinks and geckoes, and centipedes.

## 3.5 Longevity

### 3.5.1 Wild

In the wild, most dasyurids live for a relatively short time. The smaller species, such as the dunnarts and ningauis, live for only 1–2 years at most. The males of antechinuses, kaluta and phascogales in the wild die after each breeding season, so live for up to 11 months only, with females generally living into a second year, although larger species such as the mainland dusky antechinus can live to 3 years (Braithwaite and Lee 1979; Bradley 1987; Parrott *et al.* 2007a). Dibblers can live up to 2–3 years in the wild on the mainland (though island populations experience male die-off in some years), with the occasional animal (both males and females) living until 3.5 years of age. In some years there has been male die-off of dibblers after breeding, but it does not appear to be an annual event in mainland populations (Mills and Bencini 2000). The larger quolls and Tasmanian devils live for up to 4–7 years, with longevity generally increasing with the size of the species (e.g. Serena and Soderquist 1988, 1989a; Hughes 1999a; Belcher 2003; Baker and Gynther 2023). However, due to the contagious cancer, devil facial tumour disease, many devils in diseased populations are not surviving past 2 years of age, significantly changing the age structure in wild populations (Lazenby *et al.* 2018). Unique among the quolls, the northern quoll appears to have a similar life history to antechinuses and phascogales in which the males die after mating at the end of their first year (Dickman and Braithwaite 1992; Braithwaite and Griffiths 1994). Subsequent observations on the testes histology of northern quolls suggest that they do not become reproductively senile after their first breeding season and therefore the phenomenon is different to that seen in antechinuses, phascogales and the kaluta (P. Woolley pers. comm.). Details of the marsupial moles' longevity in the wild are unknown.

### 3.5.2 Captivity

In captivity, all species of dasyurids live longer than they do in the wild (Table 4.3). The smaller species such as dunnarts and ningauis generally live for 2–3 years,

**Table 4.3.** Longevity (months) of different genera of carnivorous marsupials in the wild and in captivity.

| Species | Wild | Captivity | Reference |
|---|---|---|---|
| **Dasyuridae** | | | |
| *Dasycercus* | 24–42 | 67–72 | 1–3 |
| *Dasyurus* | 11–60? | 12–60 (82) | 4-6 |
| *Parantechinus* | 24–36 | 36–54 | 7, 8 |
| *Sarcophilus* | 60–72 | 60–96+ | 1, 5, 9–11 |
| *Antechinus* | 11 (M), 36(F) | 22–72+ | 2, 4, 12, 13 |
| *Phascogale* | 11 (M), 24–36(F) | 24–60 | 10, 14, 15 |
| *Planigale* | – | 36–60 | 1, 2 |
| *Antechinomys* | – | 47–67 | 2, 16 |
| *Ningaui* | 12 | 19–23 | 1, 2, 17 |
| *Sminthopsis* | 24–36 | 29–51 | 1, 2, 17 |
| **Notoryctidae** | | | |
| *Notoryctes* | – | 1–2 | 18, 19 |

References: (1) Baker and Gynther (2023); (2) Aslin (1982); (3) Carnio (1993); (4) Mitchell (1911); (5) Flower (1931); (6) Gaikhorst (1999); (7) Mills and Bencini (2000); (8) Lambert (2000); (9) Guiler (1978); (10) Slater (1993); (11) Loh *et al.* pers. obs.; (12) Braithwaite and Lee (1979); (13) M. Parrott pers. obs.; (14) Bradley (1997); (15) Stannard *et al.* (2013); (16) Stannard and Old (2010); (17) Dickman *et al.* (2001); (18) Howe (1975); (19) Withers *et al.* (2000).

kultarrs > 5 years (Stannard and Old 2010), with the antechinuses and phascogales living for 3–6 years for both sexes, though the males are sterile after their first year and often appear to have visibly rough fur (Stannard *et al.* 2013; Letendre *et al.* 2018). One agile antechinus is known to have lived to 6 years of age at Healesville Sanctuary (M. Parrott pers. obs.). If male antechinuses are taken into captivity approximately 3 or more months before field matings they usually survive well; however, if brought in shortly before mating they usually do not survive longer than their wild counterparts (Dickman 1985).

The larger species of carnivorous marsupials, such as the quolls, live for 3–5 years and the Tasmanian devil generally has a longevity of between 6 and 8 years (Slater 1993; M. Parrott pers. obs.), though animals have survived to 9 or even 10 years old on rare occasions (Loh *et al.* pers. obs.). Male northern quolls usually die after the first breeding season. However, they can live an extra year in captivity, but do not breed (Hellingham 1999), which is not necessarily due to sterility (P. Woolley pers. comm.). The marsupial moles have not done well in captivity; the few records show that animals survived only a few months (Wood Jones 1923; Stirling 1891; Howe 1975; Johnson and Walton 1989). One marsupial mole died after 4 weeks when it was left out in the cold (Howe 1975), and another stopped feeding after 5 weeks and died (Withers *et al.* 2000).

### 3.5.3 Methods used to determine the age of adults

There are presently few methods of accurately determining the age of adult carnivorous marsupials; however, given their generally short life spans it is most important to know that they are either an adult or juvenile (which can be derived from growth curves where they exist) and their reproductive strategy. The use of molar eruption and wear has been used with success in Tasmanian devils to divide them into year classes. This species has continuous growth of the canine teeth with age, with an average additional 1.02 mm growth visible from the original dentine line per year of age (Pollock *et al.* 2021a; Loh *et al.* pers. obs.). Devils without fully erupted fourth molars can confidently be identified as < 12 months of age. The use of tooth wear involves examining the wear of cusps on the molars on the lower jaw (in which the first molar is bicuspid and the others are tricuspid), which are pointed in early life, but wear down to gums in old age (Table 4.4). Although tooth wear can assist in the determination of age, the overlap of wear classes between ages means that this method is best used in conjunction with growth measurements (Pemberton 1990). Additionally, there is no difference in the type of tooth wear in captive and wild devils, but the rate of tooth wear is slower in captive devils, occurring at older ages (which is likely due to higher quality diet items; Pollock *et al.* 2021a). In species such as those of the genus *Antechinus*, females in their first year (until about 1 month before breeding have uniform coloured fur on the belly and nipples are scarcely visible. Second-year females

**Table 4.4.** Wear characteristics of the lower molars (molars 1 and 2) used to identify age classes in wild Tasmanian devils.

Derived from Pemberton (1990).

| Tooth wear score | Wear characteristics for molar | Approx. age (months) |
|---|---|---|
| 0 | No wear | 0–12 |
| 1 | Tip worn | 0–12 |
| 2 | Dentine showing | 12–24 |
| 3 | All cusps worn but still distinguishable | 12–36 |
| 4 | No distinguishable cusps | 24–36 |
| 5 | Flat and worn to gums | > 36 |

have stained fur in the area around the open 'pouch' and prominent nipples, especially after they have reared young. Older individuals may show small patches of grey hair.

Another technique for adult dasyurids utilises the relative incisor wear on western quolls (Serena and Soderquist 1989a). When viewed anteriorly, incisor cusps typically appear sharply angular until 11–14 months of age, becoming progressively rounded thereafter; by 23–33 months the cutting surfaces of the incisors are worn flat. With this technique, individuals can be reliably assigned to an annual cohort up to the age of ~18 months (Serena and Soderquist 1989a). With females, minimum age can be gauged by looking at the condition of the teats; if minute, they are nulliparous whereas if swollen, red or encrusted they are parous (C. Dickman pers. obs.; B. Fancourt pers. obs.).

## 4. HOUSING REQUIREMENTS

### 4.1 Enclosure design

Small dasyurids are very good climbers, and some are arboreal, so they should be contained in fully enclosed enclosures or holding facilities with solid partitions or wire, with a maximum aperture of 5–10 mm between, depending on the species. The enclosure should be made to look natural by adding soil or leaf litter with grass tussocks, hollow logs, large branches for climbing, small branches with leaves and/or flowers (e.g. eucalypts), rocks, and bark over the top of the substrate to provide stimulation, shelter and reduce stress. The walls of the enclosure need to be strongly constructed, as brush-tailed phascogales have been known to chew a 5-cm hole through plasterboard and escape from the enclosure (W. Gleen pers. comm.).

If metal is used, it needs to be thick due to urine 'marking' that corrodes enclosure metal frames, wire, latches and window locks (red-tailed phascogale) (J. Old and H. Stannard pers. obs.). Red-tailed phascogales are also known to have chewed through galvanised wire mesh used to cover air vents (H. Stannard pers. obs.).

Quolls and juvenile Tasmanian devils are difficult to contain in open enclosures due to their ability to climb and dig (Godsell 1982a), so a minimum 1.5-m high smooth wall around the outside is needed to stop animals escaping. Tin strips, 0.3–0.5-m wide, added to the top of tin sheets at an angle of 45° have also been used with success to contain eastern quolls (Godsell 1982a). If wire is used, the mesh size is important as juvenile and adult female eastern quolls, for example, can squeeze through a mesh size larger than 5 × 5 $cm^2$ (Godsell 1982a). Double layers of mesh have not been recommended due to animals entangling their limbs and hanging themselves (Godsell 1982a). A double gate system is highly recommended, particularly in outdoor enclosures, due to the potential for animals to escape while people are entering the enclosure (Williams 1990).

The floor should ideally have wire mesh buried ~30–60 cm below the soil surface for Tasmanian devils and quolls, although this may not be needed if the footings to the outside of the enclosure descend at least 30 cm. Care also needs to be taken to ensure no vegetation grows over the edge, as the animals will use it to escape. Further, trees in the enclosure should be secured with tree guards, or tin strips, added to trunks to stop animals climbing trees and jumping out. Ideally, the enclosure should be fully enclosed to minimize any chance of escape. A pond, shrubs, tree ferns and hollow logs should be provided. The tree ferns may not survive animals climbing on them, so protect them with tree guards. Large enclosure furnishings should be secured to the floor, so animals cannot dig them up. Higher points within the centre of Tasmanian devil enclosures (such as on climbing furniture or on top of dens) are often used by animals to look out and provide a greater perception of space. Devils have been successfully maintained in individual enclosures (which can allow for greater genetic management for breeding animals) and in larger free-ranging group enclosures of 0.3–400 ha, which allow for greater expression of natural behaviours.

In the wild and in captivity, all species of carnivorous marsupials appear to bask in the sun, so access to natural light, heat lamps and/or heat rocks is highly recommended. Ideally, flat rocks should also be provided to bask upon, especially for smaller species (e.g. Slater 1993; Geiser 2004; Pavey and Geiser 2008; Warnecke *et al.* 2008; Warnecke and Geiser 2010; Stannard *et al.* 2015; S. Jackson pers. obs.). Exposure to natural light or lighting cycles that closely mimic natural cycles also appears to be important in the breeding success of many species of dasyurids (see section 10.4).

The marsupial moles have never been exhibited; however, a specimen has been held in a container 120 × 90 cm, which had one side of plate glass so that it could be observed under the soil (Howe 1975). Another specimen was held in an aquarium that contained ~15 cm of loose soil, taken from the Great Sandy Desert of Western

Australia where the animal was collected (Withers *et al.* 2000). Marsupial moles could potentially be exhibited by providing heating on the front glass of the enclosure (as they are known to rest against the glass here), by mimicking rain occasionally as they are known to come to the surface after rain (Stirling 1891) and by placing food on the surface of the sand, where they often feed.

## 4.2 Holding area design

Holding enclosures for small dasyurids can be of relatively simple design. Enclosures 45 cm (L) × 30 cm (W) × 25 cm (H) with a glass front and a solid wood (not less than ~10 mm thick) or stainless-steel back, sides and a mesh top have been used successfully for holding pairs of a range of small species (Woolley 1982). However, size standards have increased over time with increased understanding of animal behaviour and welfare requirements. Nest areas can be added by placing coconuts with a hole cut in the end, toilet rolls with one end covered over (S. Jackson pers. obs.) or small wooden boxes (~10–12 cm (L) × 10–12 cm (W) × 10–12 cm (H)) that can be either located within the enclosure or attached to one or both sides (Woolley 1982) (Fig. 4.1a; Table 4.5). A mesh partition can be used to separate the male from females so that access can be allowed after both visual and smell access (Woolley 1982; Carnio 1993). Alternatively, a single unit can be used in which the nest box is incorporated into the enclosure (Fig. 4.1b).

a

b

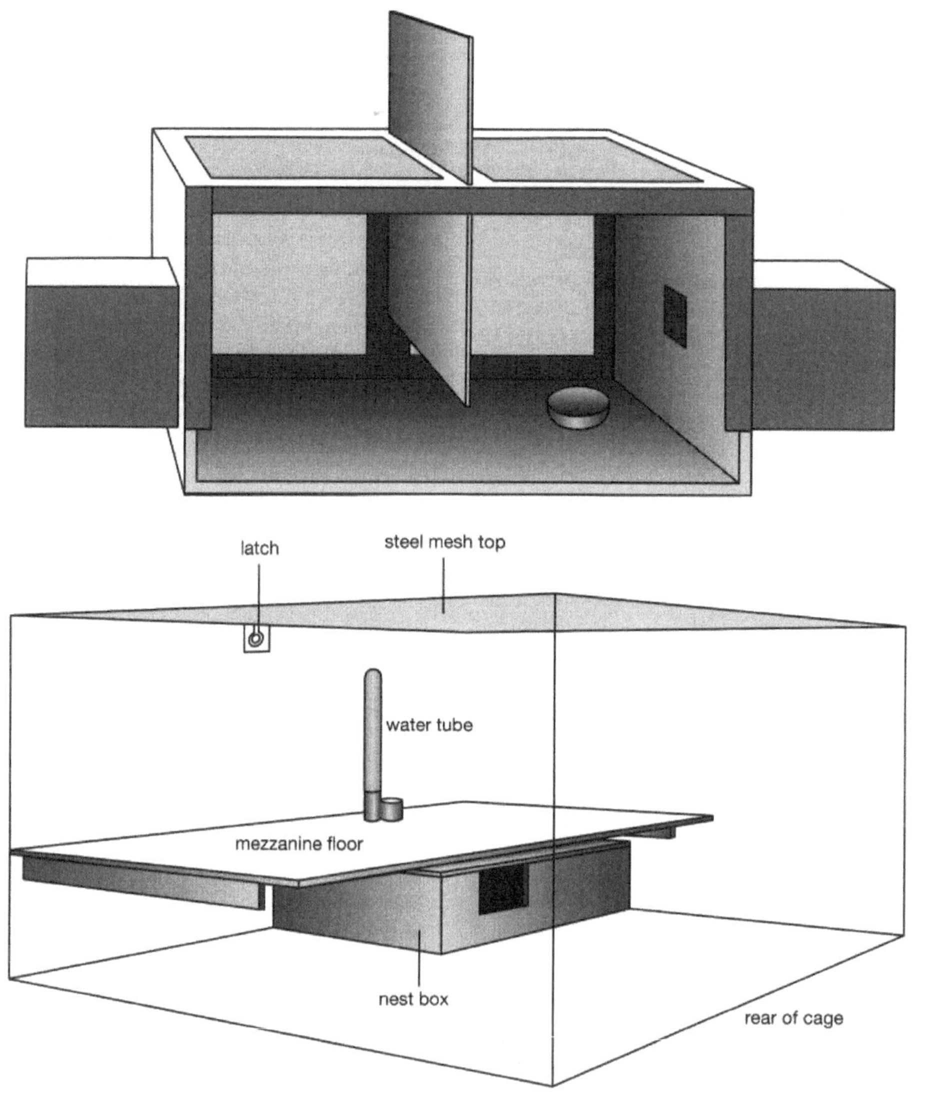


**Fig. 4.1.** Examples of enclosure to hold small dasyurids. (a) With subdivision in middle and nest boxes on each side) and (b) with a mezzanine level and single nest area. Derived from Woolley (1982) and Woolley (1993).

**Table 4.5.** Minimum areas of enclosures recommended for pairs of animals of different genera of Australian carnivorous marsupials.

Derived from Howe (1975), Aslin (1980), Collins *et al.* (1993), Slater (1993), Hogg and Hockley (2019) and S. Jackson (pers. obs.).

| Genus | Area ($m^2$) × (H m) | Additional floor area for each extra animal ($m^2$) |
|---|---|---|
| **Dasyuridae** | | |
| *Dasycercus* | 2.0 × 0.5 | 0.25 |
| *Dasykaluta* | 2.0 × 0.5 | 0.25 |
| *Dasyurus hallucatus* | 10.0 × 2.4 | 4.00 |
| *Dasyurus viverrinus & D. geoffroii* | 15.0 × 2.4 | 6.00 |
| *Dasyurus maculatus* | 30.0 × 3.5 | 10.00 |
| *Parantechinus* | 2.0 × 0.5 | 0.25 |
| *Pseudantechinus* | 2.0 × 0.5 | 0.25 |
| *Sarcophilus* | 100.0 × 1.2 | 20.00 |
| *Antechinus* | 2.0 × 0.5 | 0.25 |
| *Phascogale* ( e.g. *P. tapoatafa, pirata*) | 10.0 × 2.0 | 1.00 |
| *Phascogale* (e.g. *P. calura*) | 10.0 × 2.0 | 0.50 |
| *Planigale* | 1.0 × 0.5 | 0.25 |
| *Antechinomys* | 1.0 × 0.5 | 0.25 |
| *Ningaui* | 1.0 × 0.5 | 0.25 |
| *Sminthopsis* | 1.0 × 0.4 | 0.25 |
| **Notoryctidae** | | |
| *Notoryctes* | 1.0 × 1.0 | 0.5 |

Permanent indoor research enclosures for red-tailed phascogales and kultarrs have been made of wood and lined with glass (1 m (L) × 1 m (W) × 0.58 m (H)) (Stannard *et al.* 2015). The door to each enclosure was made of glass with heavy-duty metal vents. The glass-lined enclosures made for very easy cleaning and prevented phascogales chewing through the enclosure and escaping. The indoor rooms were maintained at 21°C ± 5°C and a ventilation system supplied heated/cooled air from outside to each indoor enclosure. The substrate was a recycled paper-based product (Breeders Choice kitty litter (Fibre-Cycle Pty Ltd, Queensland)). Each enclosure contained at least two wooden nest boxes containing shredded paper and enclosure furniture (e.g. eucalypt, casuarina and melaleuca branches).

Woolley and Watson (1984) used large outdoor enclosures to hold a colony of fat-tailed dunnarts. These enclosures were 5 m (L) × 10 m (W) with the walls being made from 120-cm wide metal sheeting sunk 30 cm into the ground. Above this sheeting, 5 cm cyclone mesh was used. Large 26 L plastic garbage bins filled with polystyrene foam and foam plastic sheeting were used as nesting areas. Although laboratory bred and wild caught animals bred in these enclosures (at different times of the year, July and September/October respectively) none of the young were raised successfully and the population declined considerably, possibly because of human disturbance and escapes, so the enclosures were abandoned.

Red-tailed phascogales have also been housed in outdoor enclosures (1.5 m (L) × 1.6 m (W) × 1.9 m (H)) (Stannard *et al.* 2013). Outdoor enclosures had concrete bases (for ease of cleaning and noting that phascogales rarely go on the ground and prefer to hide on the branches in the leaves), wire sides and a clear PVC roof. Enclosures contained large amounts of furniture (browse and large branches for climbing). Entry into the enclosures included an antechamber to prevent exit of phascogales when staff were entering the outdoor enclosures.

## 4.3 Spatial requirements

Area requirements for the different genera of carnivorous marsupials should be large enough to allow breeding, social behaviour and long-term survival, and typically range from 45 × 30 × 25 cm for small species up to 600 × 300 × 150 cm for larger species (Table 4.5).

## 4.4 Position of enclosures

Access to natural light to allow natural circadian rhythms is highly recommended, particularly for breeding. Wherever possible, access should also be given to direct sun as most species of carnivorous marsupials like to sun bask (although adequate areas to seek shade should also be supplied – see below).

## 4.5 Weather protection

Ideally, the enclosures should be outdoors for large species such as quolls and Tasmanian devils or, if indoors, should have large windows and/or skylights to allow adequate natural light and preferably the ability to sun bask, but also not overheat the enclosure and animals. Excessive exposure to sunlight has resulted in sunburn, heat stress, premature loss of vision and solar dermatitis resulting in hair loss on the lower back in Tasmanian devils (Kelly 1993).

Marsupial moles appear to be highly sensitive to cold (Lydekker 1894; Howe 1975), so they should never be held outdoors, never exposed to temperature below ~16°C and should be held at ~22°C (Howe 1975).

## 4.6 Temperature requirements

In the wild, most species of dasyurids sun bask (including the thylacine historically), although not much is known of the marsupial moles (Baker and Gynther 2023). Enclosed room temperatures for species such as dunnarts and antechinuses should be ~22 ± 1°C. Heating and natural light should be provided wherever possible with the use of 100–150 W infrared lamps for at least several hours per day (Woolley 1982). Each lamp should be 20–50 cm above the floor of the enclosure with the temperature at floor level being as high as 35–48°C so that a thermal gradient is provided from cooler temperatures to high temperatures (Woolley 1982).

Electric basking rocks can also be provided to small dasyurids, including red-tailed phascogales, which have been observed to use them on a strictly nocturnal basis (1900–0700 h), and kultarrs have been observed for up to 2.5 h on such rocks at all hours of the day (Stannard *et al.* 2015). The rocks should be placed directly onto flat ground, so animals cannot dig underneath, and the cord should be wrapped in heavy duty plastic (slicing clear vinyl plastic tubing longitudinally and placing over the cord should suffice). Although kultarrs and red-tailed phascogales have not been observed to chew on the cords of the electric heat rocks, all cords should be covered as a precaution (H. Stannard pers. obs.).

All small wild dasyurids use daily torpor to conserve energy. Even though they may not be short of food in captivity, it may be beneficial for them to enter torpor in captivity by allowing enclosure temperatures to drop below 22°C at night by switching off basking rocks for a few hours.

As early as 1891, records of First Nations people suggested that marsupial moles were not seen during cold weather. Two moles held in captivity died after a frosty night, even though their box was well protected and they had access to sand in which to burrow (Stirling 1891). Subsequent observations showed them to be highly sensitive to cold as shivering spasms were observed at 15.6°C and one died when no heating was supplied and the temperature dropped to 1.7°C (the animal was found in the middle of the soil) (Howe 1975). Heating has been achieved in marsupial mole enclosures by placing the container near a window with a northerly aspect to receive maximum solar heating and by providing supplementary heating with an electric radiator (Howe 1975) or with the use of a heating pad under one end to achieve a thermal gradient from ~22°C to 32°C (Withers *et al.* 2000).

## 4.7 Substrate

Various substrates can be used for the enclosures of small species, including sand, soil, paper, cardboard and wood shavings (which should be spot cleaned daily). Nesting material can include dry eucalyptus leaves, melaleuca bark, wood shavings, shredded paper, soft hay or dry sea grass (Woolley 1982; C. Dickman pers. obs.). Paper cut in the shape of eucalyptus leaves has been successfully used to weave ball-shaped nests by agile antechinus (Parrott 2006). Larger species such as quolls and Tasmanian devils should have a mulch, soil or leaf litter substrate. A mixture of two parts Solite® (Solite Corporation), two parts pine bark (untreated and not artificially coloured) and one part peat moss to a depth of 45 cm has been used to allow spotted-tailed quolls to dig and show good reproductive success (Collins *et al.* 1993). Concrete floors for larger dasyurids have resulted in sore feet so are not recommended (Williams 1990).

Most species of dasyurids appear to undergo an annual moult, particularly after the breeding season (Woolley 1982). It appears that some species such as the mulgaras may require moist sand (e.g. bricklayer's sand that is unwashed and has a high clay content) to burrow in and keep their fur in good condition (Woolley 1982). Several species, such as the kowari and kultarr, appear to keep their fur clean and scent mark by sand bathing, so the provision of fine sand is recommended (Aslin 1974; Woolley 1982). Similar behaviour has been observed with the fat-tailed dunnart, with sand rolling linked to toilet behaviour (Ewer 1968a).

Marsupial moles have been kept successfully, albeit briefly, in a substrate of red sand (Howe 1975; Withers *et al.* 2000). When river loam was used, the moles' claws, nose and cloacal area became heavily encrusted with clay (Howe 1975). Withers *et al.* (2000) found that the mole was unable to burrow into air-dried sand, so the sand was kept slightly moistened. The enclosure should have sand to a depth of at least 40 cm to allow the moles to dig into the soil.

## 4.8 Nest boxes

Nest boxes should be provided for all species of carnivorous marsupials except the marsupial moles (Table 4.6). They provide shelter, warmth and security and can assist in capture, so they should have a hinged or removable roof. They are usually made of thin plywood for small species and stronger timber, or even concrete, for spotted-tailed quolls and Tasmanian devils. Fat-tailed dunnarts and stripe-faced dunnarts held off-exhibit

**Table 4.6.** Minimum nest box sizes recommended for different species of carnivorous marsupials.

Derived from Woolley (1971a), Halley (1992), Carnio (1993), Slater (1993), Woolley (1993) and Hogg and Hockley (2019).

| Genus | L × W × H (cm) | Entrance diameter (cm) |
|---|---|---|
| **Dasyuridae** | | |
| *Dasycercus* | 20 × 20 × 15 | 8 |
| *Dasykaluta* | 10 × 10 × 10 | 5 |
| *Dasyurus* – smaller species | 40 × 50 × 40 | 25 |
| *Dasyurus maculatus* | 70 × 60 × 70 | 30 |
| *Parantechinus* | 23 × 15 × 15 | 6 |
| *Pseudantechinus* | 10 × 10 × 10 | 5 |
| *Sarcophilus* | 120 × 90 × 70 | 40 |
| *Antechinus* | 10 × 10 × 10 | 5 |
| *Phascogale* ( *e.g. P. tapoatafa)* | 40 × 25 × 25 | 6 |
| *Phascogale* ( *e.g. P. calura)* | 30 × 20 × 20 | 5 |
| *Planigale* | 10 × 10 × 10 | 5 |
| *Antechinomys* | 10 × 10 × 10 | 5 |
| *Ningaui* | 10 × 10 × 10 | 5 |
| *Sminthopsis* | 10 × 10 × 10 | 5 |

have been provided with small cardboard boxes (e.g. small tea boxes, cereal boxes, tissue boxes, and card boxes) as nest boxes (H. Stannard pers. obs.).

Uncontaminated natural nesting material, such as grasses, dry eucalyptus leaves, washed and dried sea grass, soft hay and stringybark, should also be provided in the enclosure for the animals' use (C. Dickman pers. obs.). Kultarrs and red-tailed phascogales held off-exhibit have been provided with shredded paper (J. Old pers. obs.). The Tasmanian devil nest box area can double as a feed and capture area, where the animals can be briefly locked up (if they are aggressive) while the enclosure is cleaned or repaired (Hogg and Hockley 2019). Generally this is not necessary, as they usually keep out of the way. The entrance of the nest box can be shut with a slide and the inside of the box can be viewed through a wire mesh roof (Slater 1993; S. Jackson pers. obs.). Nesting material for Tasmanian devil nest boxes can include straw, fresh grass, ferns, bracken or similar vegetation and be replaced weekly (Hogg and Hockley 2019).

Well-constructed wooden nest boxes can be used in captivity and then for transporting and releasing

**Fig. 4.2.** Nest box used for wild agile antechinus. Photo by Marissa Parrott.

animals to the wild, providing a familiar area for nesting and acclimatisation post-release (Fig. 4.2). Agile antechinus, for example, were bred in captivity, with females and pouch young successfully released by attaching their familiar nest boxes 2–3 m high on suitable tree trunks (using metal strips pre-attached to the nest boxes to minimise noise and disturbance while nests were in use). Most females continued to use the nest boxes after release, allowing researchers to easily monitor them and their young (Parrott 2006).

## 4.9 Enclosure furnishings

To minimise abnormal repetitive behaviour, it is advisable to provide several different furnishings within the enclosure. These can range from small branches, running wheels (see Woolley 1993 for details; off-exhibit only), rocks, hollow logs, climbing branches, pieces of bark, PVC pipes (off-exhibit only) and climbing branches for small species, to planted trees and shrubs, large logs and branches for larger species. Running wheels should be provided sparingly to ensure that abnormal repetitive behaviour does not occur. Any heavy items such as rocks, logs and nest boxes must be placed directly on the base of the enclosure rather than on the leaf litter, because small dasyurids may dig under them and be crushed (J. Old and H. Stannard pers. obs.).

Numerous tussocks of grass (e.g. Mitchell grass *Astrebla* spp.) generally need to be added because phascogales, and particularly quolls and Tasmanian devils, often dig them up and use them for nesting material (W. Gleen pers. comm.). Paperbark branches will be stripped of bark for use as nesting material by red-tailed phascogales (J. Old and H. Stannard pers. obs.). The addition of enclosure furnishings such as branches helps to encourage activity, scenting behaviour and nest building.

# 5. GENERAL HUSBANDRY

## 5.1 Hygiene and cleaning

All enclosures should be cleaned daily to remove faecal matter and uneaten food. Small enclosures can be spot cleaned daily and given a full substrate clean weekly, or more often if required. Larger dasyurids use latrines so most of the faeces are in the one place and are therefore generally easy to find and remove. Shallow drinking water dishes should be cleaned and refilled daily. As phascogales have been known to drown easily, water bottles are preferable and easier to clean. When all individuals permanently leave an enclosure, it should be thoroughly cleaned (e.g. F10® SC Veterinary Disinfectant), rested and scrubbed out if possible before new animals arrive.

## 5.2 Record keeping

It is important to establish a system whereby the health, condition and reproductive status of captive carnivorous marsupials are routinely monitored and recorded:

- identification numbers (all individuals should be identifiable)
- results of veterinary examinations and treatments
- behavioural changes or problems
- reproductive behaviour or condition
- bodyweights and measurements
- changes in diet and food consumption
- movements of individuals between enclosures or institutions
- births with dam and sire if known
- deaths with necropsy results.

Collecting information on the physical, physiological and behavioural patterns of individual animals can contribute greatly to the husbandry of these species. It also allows individual histories to be shared with other institutions and facilitates a cooperative approach to data collection among institutions. ZIMS, the Zoo Information Management System (Species360 2023) can be used for recording general information on age, sex, parentage, births, identification, bodyweights, transfers and deaths, with separate ZIMS modules for veterinary information and studbooks. These systems have been developed by Species360, which is part of the Conservation Planning Specialist Group (CPSG) software. ZIMS also interfaces with programs such as PMx, which provides demographic and genetic analysis tools to assist with the management of breeding programs (Lacy *et al.* 2011).

## 5.3 Methods of individual identification

### 5.3.1 Passive Integrated Transponder (PIT) tags

PIT tags (microchips) are an excellent method of identification because they are permanent. They are implanted under the skin between the scapulae of individuals > 10 g in bodyweight and can be used on most species of carnivorous marsupials (Pavey *et al.* 2022; Waudby *et al.* 2022). Sealing the entry wound with tissue glue (Vetbond®) will ensure that tag stays in place immediately after implantation. Animals generally need to be caught to confirm identification with a PIT tag reader, though some can be checked by passing the reader over the animal in a nest box or with a remote chip reader in the enclosure, which records the animal when it passes by. Transponders such as those of Trovan produce sizes including the Microchip (11.5 × 2.12 mm) and Midichip (8 × 1.4 mm). It should be noted that the smaller the chip size, the shorter the read range.

### 5.3.2 Tattoos

Numbers can be tattooed on the pinnae (ears) of species or individuals in which the pinna is sufficiently large and relatively hairless (Woolley 1982). Tattoos have been used successfully on rodents ~20 g in bodyweight (Lindner and Fuelling 2002). Pinna tattooing has also been used on pouch-young western quolls (Soderquist and Dickman 1988). However, this technique is rarely used now and has largely been replaced by PIT tags.

### 5.3.3 Visual identification

Most individuals within the species of carnivorous marsupials are difficult to tell apart; however, Tasmanian devils can generally be identified by the markings of their fur, especially the white markings near the chest and rump. Similarly, quolls can be identified by their unique spot patterns (Hohnen *et al.* 2013; Rowland *et al.* 2020).

The use of non-toxic permanent markers for short-term identification has also been successfully used in the field to identify animals that have been recaptured during short (< 5 day) trap surveys, with a different colour mark on the ear used each day to determine recapture success (Pavey *et al.* 2022; Waudby *et al.* 2022). In captivity, the use of thick black bands and thin black stripes on tails has allowed identification of individual agile antechinus held in groups during mate choice trials (Parrott *et al.* 2015). The tail markings fade and disappear after ~5 days.

### 5.3.4 Ear tags

Small metal ear tags have previously been used on several species, including the brown antechinus and the kowari; however, they are unsuitable because the ear often tears and the wound from the tag may become infected (Woolley 1982; Salamon and Klettenheimer 1994).

### 5.3.5 Ear notching

Ear notching is one of the few ways to individually identify small species of dasyurids, including dunnarts, antechinuses, ningauis and mulgaras (Fig. 4.3; C. Dickman pers. obs.) and can have the additional benefit of collection of tissue for genetic analysis (Waudby *et al.* 2022). It has also been used on larger species, such as dibblers, where a small notch is taken out of one ear for females and the opposite ear for males to help identify behaviours during mating. Notches should be placed in the ear margin to allow a partial, rather than fully enclosed, punch hole. Fully enclosed holes can catch on obstacles, including branches and needle-shaped spinifex leaves, which could tear the ear easily.

When notching the ears, ear notching equipment should be sterilised (dipped in ≥ 70% ethanol and flamed or allowed to dry naturally) and used to excise a small piece of tissue (~1–2 $mm^2$ per clip) in the appropriate places around the ear margin (Fig. 4.3; Waudby *et al.* 2022; C. Dickman pers. obs). Make sure to always avoid cartilage and any obvious blood vessels. The clipping process takes ~30 s (quicker for a single clip) and rarely results in bleeding or obvious distress in the animal being processed (C. Dickman pers. obs.).

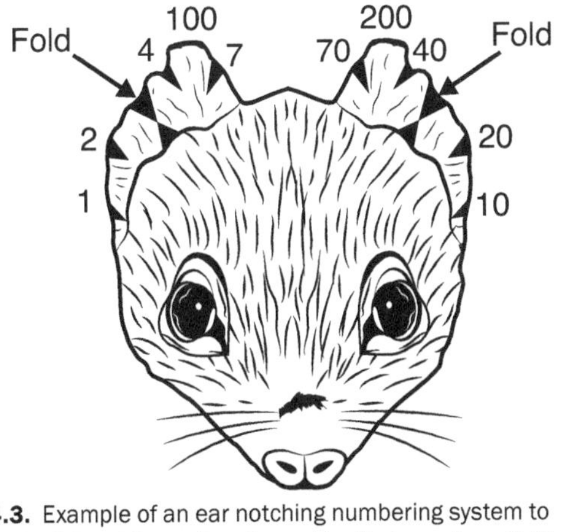


**Fig. 4.3.** Example of an ear notching numbering system to identify individual Australian dasyurids. Redrawn from C. Dickman (pers. obs.).

### 5.3.6 Toe clipping

Toe clipping has been used to individually identify small animals (e.g. Aslin 1982; Woolley 1982; Wood and Slade 1990; Braude and Ciszek 1998), but is no longer used for adults of any species. However, it has been used successfully for ecological and genetic research on brown and agile antechinuses, with no detectable negative effects on the growth or survival of the pouch young that were toe-bud clipped (Parrott *et al.* 2007a; Fisher and Blomberg 2009). This technique involves cutting the very tip of the developing toe-bud at an extremely immature stage, so the area that is cut is very small and the memory of pain is unlikely to be great (Fisher and Blomberg 2009). For genetic research (i.e. to collect samples from the mother and her full litter while still in the pouch), the very tip of the developing tail (< 1 mm skin) has also been taken with no issues noted (Parrott *et al.* 2007a, 2007b). Clipping of the toe-bud or tail tip typically must be done while the young are still attached to the mother's nipples, requiring one operator to immobilise the mother and a highly dexterous second operator to complete the clipping (C. Dickman pers. obs.). Toe-bud clipping has been used on several taxa of wild dasyurids, including antechinuses, dibbler, dunnarts, ningauis, and mulgaras. Despite this, the use of this technique on captive populations is not recommended, as young can have samples taken at a later age.

### 5.3.7 Hair bleaching

Hair bleaching has been used to identify brown antechinus in captivity (Rigby 1972), but care must be taken to avoid contact between the bleach and skin.

## 6. FEEDING REQUIREMENTS

### 6.1 Captive diet

Diets for captive dasyurids should consider the age, activity level, sex and reproductive state of the individual, as well as climatic and seasonal conditions, and be adjusted accordingly. For example, older animals, or animals in warmer climates, may require less food than younger animals housed in cooler climates. If housed outdoors in cooler climates, an increased food allocation may be required in winter. All animals should be carefully monitored for signs of obesity (Woolley 1982). There is a need to better identify which diets are most appropriate for breeding rather than just maintenance. Protein has been noted as especially important for breeding, as diets too low in protein can cause reduced breeding success and potential cannibalism of young.

Female dasyurids with pouch young, particularly in late lactation, should be fed 2–3 times their normal maintenance amount to ensure adequate nutrition for the mother and the young and reduce the potential for cannibalism (Aslin 1980; Woolley 1982). Energy requirements of reproducing females have only been studied in two dasyurid species, thus far. Daily energy requirements of the eastern quoll and red-tailed phascogale increase during late lactation to 205–222% above non-lactating requirements (Green *et al.* 1997; Stannard and Old 2015). Litter size does not appear to influence nutritional needs and female red-tailed phascogales with smaller litter sizes ($n = 4$) produce young that are heavier than those in larger litters ($n = 8$) (Stannard and Old 2015). After weaning, when the young are removed, the female's diet will need to be reduced to a maintenance level.

As with observations in macropods, some carnivorous marsupials are known to undertake regurgitation or merycism of their food. This process involves convulsive regurgitation of the stomach contents into the mouth before re-swallowing with limited or no further chewing. This behaviour has been observed in the crest-tailed mulgara (Fleay 1961), western quoll (Archer 1974a), red-tailed phascogale (Archer 1974a) and agile antechinus (M. Parrott pers. obs.).

Some species of dasyurids, particularly the mulgaras, do not appear to tolerate a high fur or feather content in their diet, as feeding 1–4-day-old chicks has led to feather balls, resulting in fatal intestinal obstruction (Booth 1994; Woolley 1971a), which is unusual considering the natural diet of the mulgaras includes small vertebrates (Chen *et al.* 1998; Masters 1998). Nonetheless, they should be provided with unfurred mice and insects (Booth 1994). However, the inclusion of food items such as bone, fur, feathers can be important in the diet of larger species, such as Tasmanian devils, to maintain gut and dental health and ensure high calcium intake. This is especially important for lactating mothers to ensure that rickets does not occur in joeys. Alternatively, the food can be dusted with calcium powder.

All meat should be frozen for at least 4 weeks at −20°C before feeding to minimize the chance of introducing *Toxoplasma* or other parasites to the colony (Attwood *et al.* 1975; Attwood and Woolley 1982) (see section 8.4). Some food items, such as live insects, eggs, smaller vertebrates (e.g. day-old chicks for Tasmanian devils or quolls), bones and blood smears can be provided as behavioural enrichment to encourage natural foraging behaviours and increased activity.

### 6.1.1 *Antechinus, Pseudantechinus* and *Sminthopsis*

- Ad lib
  - water
- Daily diet (per animal)
  - 2 g Small Carnivore Food (Wombaroo®)
  - 9 g mixed invertebrates (e.g. mealworms, crickets, cockroaches, fly pupae, wild-caught insects)
  - 1 pinkie mouse fed weekly
- Supplement
  - 2 crickets - 3–4 times per week as available
  - 2 earthworms - 1–2 times per week as available.
  - 2 moths - 3–4 times per week as available
  - blossoms as available
- Alternative diet provided by Healesville Sanctuary
  - 6 g daily: meat mix finely crumbled
  - g daily: Wombaroo® Small Carnivore Food (weight is powder before water is added)
  - g daily: mealworms, crickets, cockroaches, fly pupae, wild-caught insects
  - 1 weekly: pinkie mouse whole

The proportion of food offered depends on the species and weight of the animals. Small dasyurids (including antechinuses, dunnarts, kultarrs, red-tailed phascogales) consume between 17% and 60% of their bodyweight in food per day depending on the type of food offered (e.g. they tend to have lower intakes when offered mealworms and Small Carnivore Food) (Cowan *et al.* 1974; Nagy *et al.* 1978; Stannard and Old 2011; Stannard *et al.* 2014; Stannard and Old 2015). Fat-tailed dunnarts will consume up to 81% of their bodyweight in food per day when fed tinned cat food (Stannard *et al.* 2014). Therefore, they should be fed live insects such as crickets, mealworms, cockroaches, moths and spiders for behavioural enrichment and to increase the natural calcium in the diet (Booth 1994). Note that earthworms and cockroaches are known to ingest and transmit *T. gondii* oocysts (Wallace 1972; Bettiol *et al.* 2000a, 2000b), so should be raised in sterile soil or on an artificial substrate. Some antechinuses and dunnarts selectively hunt wolf spiders in the wild (Potter *et al.* 2018a, 2018b), suggesting that these common and widespread invertebrates could be included as supplements in the diets of captive animals.

Small Carnivore Food is not recommended by some establishments. Some dasyurids such as red-tailed phascogales will not eat it, whereas kultarrs and dunnarts will (Stannard and Old 2011; Stannard *et al.* 2014).

### 6.1.2 *Antechinomys*

- Ad lib
  - water
- Daily diet (per animal)
  - Monday: 8 g crickets*
  - Tuesday: 1 adult mouse cut into 1-cm pieces
  - Wednesday: 8 g mealworms*
  - Thursday: 8 g cockroaches*
  - Friday: 8 g mealworms*
  - Saturday: starve day
  - Sunday: 8 g pinkie mice*

*Dusted with calcium and vitamin E powder.

### 6.1.3 *Phascogale, Dasycercus, Dasyuroides*

- Ad lib
  - water
- Daily diet (per animal)
  - ½ day-old chick or pinkie mouse (mouse only for mulgara)
  - 20 g Pet health food (10-mm cubes)
  - 1 g fly pupae
  - 5 g Small Carnivore Food (Wombaroo®)
  - 5 mealworms
- Supplement
  - 0.2 g pollen grains weekly
  - 5 g fruit (apple, orange, pear or fruit in season) - 10 mm cube twice per week
  - 4-5 mealworms, crickets or woody cockroaches dusted with calcium and vitamin E several times per week
  - 6 g egg and cheese - 3-4 times per week
  - 3 moths - 3-4 times per week as available
  - fresh flowers of eucalypts, melaleucas and other plants

### 6.1.4 *Parantechinus*

(Non-breeding diet)

- Ad lib
  - water
- Daily diet (per animal)

(Morning feed: insects)

  - mealworms (*Tenebrio molitor*)
  - crickets (*Acheta domestica*)
  - wood cockroaches (*Panesthia laevicollis*)
  - fruit-fly pupae (*Ceratitis capitata*)
  - termites (*Coptotermes* spp. and *Nasutitermes* spp.)
  - moths

(Afternoon feed: diced meats)

  - chicken - cooked flesh only
  - kangaroo (*Macropus*, *Notamacropus* or *Osphranter*)
  - rats (*Rattus norvegicus* unfurred)
  - SF52 (vitamin mineral supplement) 1 teaspoon per 100 g meat
- Supplement (a couple of times per week for each)
  - fruit, such as dried figs and sultanas
  - boiled egg
  - Small Carnivore Food (Wombaroo®)
  - premium-quality cat kibble (e.g. Hill's Science Diet®)
  - branches of fresh flowers of Proteaceae and Myrtaceae, including *Grevillea* and *Callistemon*, to provide nectar

A 60 g dibbler should receive 4 g of invertebrates and 8 g of meat daily. Typically, all the live food is eaten and most or all of the meat is consumed each day (Lambert and Mills 2006).

Note: crickets, lightly dusted with calcium powder, cockroaches and moths are scattered around the tank to provide enrichment and promote natural foraging behaviour (Lambert and Mills 2006).

(Breeding diet)

- Invertebrates are increased to ~50% of the total diet for both males and females. Following mating the males are returned to the non-breeding diet. Females are then given 10 g live food and ad lib meat each day until the young are eating solid food. Once the young are eating solids, the female's insect ration is decreased to 4 g per day, the young dibblers are given 2 g and the whole group is fed meat ad lib (Lambert 2000).

### 6.1.5 *Planigale* and *Ningaui*

- Ad lib
  - water
- Daily diet (per animal)
  - 5-8 g Pet health food (5 mm grated)
  - 5 g egg and cheese
  - 2 mealworms

  - < 1 g Advance™ Puppy kibble (¼–½ of a cube)
  - 1/8 day-old chick or mouse
  - 1 g fly pupae
- Supplement
  - 1 cricket dusted with calcium and vitamin E powder given 3–4 times per week as available
  - 1 earthworm: 1–2 times per week as available

## 6.1.6 Western quoll

- Ad lib
  - water
- Daily diet (per animal)
  - rats
  - mice
  - mealworms
  - chicks
  - fish
  - Advance™ Puppy kibble
  - doves
  - insects including cockroaches, crickets, mealworms

(Total food = 105 g for males and 70 g for females)

The amount of food provided is regulated according to the animal (particularly as they are prone to obesity) (Gaikhorst 1999). During the breeding season the male's food is increased to 140 g per day and the female's diet is increased to 105 g per day once young are produced. When the young are 30 days old, the food amount is increased to 140 g and at 110 days an extra 35 g is given daily to each young (as they start eating). At 170 days of age the young are each given 70 g of food. It is important to monitor the health of the mother and young, as underfeeding can result in cannibalism or self-mutilation (Gaikhorst 1999). Once the young are separated from their mother, they can be fed ad lib until they reach their optimal bodyweight of 1500 g for males and 900 g for females (Gaikhorst 1999).

## 6.1.7 Northern quoll

- Ad lib
  - water
- Diet provided by Healesville Sanctuary
  - Monday: 25 g pilchards (whole or cut)
  - Tuesday: 30 g quail (cut piece)
  - Wednesday: 40 g hard-boiled egg with shell cut in half
  - Thursday: 20 g weaner rat or mouse, whole or cut
  - Friday: 10 g Advance™ Puppy kibble
  - Saturday: 25 g day-old chick
  - Sunday: 15 g, mixed invertebrates – cockroaches or moths (50%), crickets or mealworms (50%)
- Enrichment
  - mealworms: 15 daily
  - cooked vegetables (e.g. pumpkin, sweet potato, beetroot, carrot), smear, 3 × weekly
  - peanut butter, smear only, 1 × weekly
  - blood, smear only, 1 × weekly
  - tinned cat food, smear only, 1 × monthly
  - novel invertebrates (e.g. worms, maggots, fly pupae, grass hopers, woodroaches, moths). May be replaced with other invertebrates. Feed on a weight for weight basis as available

## 6.1.8 Spotted-tailed quoll

- Ad lib
  - water
- Daily diet (per animal) – provided by Healesville Sanctuary
  - Monday: 120 g pilchards
  - Tuesday: 120 g rat
  - Wednesday: 120 g rabbit, possum or wallaby, 34 g mealworms, 30 g mealworms activity feed
  - Thursday: 120 g rabbit, 34 g Advance™ dog kibble
  - Friday: 60 g meat mix
  - Saturday: 120 g day-old chick, 25 g crickets
  - Sunday: 60 g mice adult, 33 g crickets, 60 g day-old chick
- Enrichment
  - mealworms: 15 daily
  - cooked vegetables (e.g. pumpkin, sweet potato, beetroot, carrot), smear, 3 × weekly
  - peanut butter, smear only, 1 × weekly
  - blood, smear only, 1 × weekly
  - tinned cat food, smear only, 1 × monthly
  - novel invertebrates (e.g. worms, maggots, fly pupae, grass hopers, woodroaches, moths). May be replaced with other invertebrates. Feed on a weight-for-weight basis as available
  - pig's ear: monthly

## 6.1.9 Eastern quoll

- Ad lib
  - water
- Daily diet (per animal) – provided by Healesville Sanctuary
  - Monday: 100 g pilchards
  - Tuesday: 80 g day-old chick

- Wednesday: 65 g hardboiled egg with shell, cut in half
  - Thursday: 80 g weaner rat or mouse
  - Friday: 100 g rabbit
  - Saturday: 100 g day-old chick
  - Sunday: 20 g mixed invertebrates (cockroaches/moths 50%, crickets/mealworms 50%)
- Enrichment
  - mealworms: 15 daily
  - cooked vegetables (e.g. pumpkin, sweet potato, beetroot, carrot), smear, 3 × weekly
  - peanut butter, smear only, 1 × weekly
  - blood, smear only, 1 × weekly
  - tinned cat food, smear only, 1 × monthly
  - novel invertebrates (e.g. worms, maggots, fly pupae, grass hopers, woodroaches, moths. May be replaced with other invertebrates). Feed on a weight-for-weight basis, as available

### 6.1.10 Tasmanian devil

In the wild, devils often gorge themselves and can eat up to 40% of their bodyweight in food in a single sitting (Pemberton and Renouf 1993). They would then not eat again for several days. To mimic this natural behaviour, most facilities feed a variety of large carcasses, smaller amounts and have starve days to mimic gorge feeding in the wild (excepting lactating females or those being paired to breed to allow for identification of oestrus when females eat a reduced amount). Devils should have at least one starve day per week, which can assist with their digestion.

Adult devils should be provided with approximately 5–6% of their bodyweight in food per day (adjusted for age, sex, activity level, reproduction and season as required). This can be calculated across a fortnightly diet to allow for larger items on some days (gorge days) and smaller or no food on other days (starve days). It should be noted that weaning and juvenile devils often require 8–10% of their bodyweight of food per day while growing (Healesville Sanctuary pers. comm.).

- Ad lib
  - water
- Daily diet (per non-breeding animal) – provided by Healesville Sanctuary

(Fortnight 1)

  - Monday: rabbit (triple amount)
  - Tuesday: starve day
  - Wednesday: starve day
  - Thursday: fish (e.g. pilchards)
  - Friday: kangaroo/beef (double amount)
  - Saturday: starve day
  - Sunday: quail (double amount)
  - Monday: starve day
  - Tuesday: wallaby (triple amount)
  - Wednesday: starve day
  - Thursday: starve day
  - Friday: possum (double amount)
  - Saturday: starve day
  - Sunday: brisket bone plus 1 egg

(Fortnight 2)

  - Monday: wallaby (triple amount)
  - Tuesday: starve day
  - Wednesday: starve day
  - Thursday: fish (e.g. mackerel)
  - Friday: rabbit (triple amount)
  - Saturday: starve day
  - Sunday: starve day
  - Monday: quail (double amount)
  - Tuesday: starve day
  - Wednesday: possum (triple amount)
  - Thursday: starve day
  - Friday: starve day
  - Saturday: wallaby legs and tails (double amount)
  - Sunday: starve day
- Enrichment
  - feeds and scatters such as day-old chicks, blood smears and live insects (see below for other food-based enrichment items).

Various food items can be offered to Tasmanian devils, including large carcasses that can last 3–4 days in the enclosure such as possum, rabbit, wallaby and large chickens, and smaller carcasses and pieces that can last 1-2 days in the enclosure such as kangaroo, quail, small chickens, beef, horse, and wallaby and possum legs and tails. Kangaroo and horse should only be fed once per fortnight. Day and enrichment feeds can consist of smaller items such as chicken wings and necks, day-old chicks, rabbit feet, wallaby and possum legs and tails, fish (pilchards and mackerel), live insect scatters, raw and hard-boiled eggs, brisket bones, meat and blood smears, bones and offal (liver, hearts, kidneys). Due to its richness, offal should only be provided once per month (Healesville Sanctuary pers. comm.).

During oestrus females either eat very little or stop eating (Kelly 1993; Stannard *et al.* 2017). This can be used to determine if a female is entering oestrus and thus the daily ration should be provided and recorded for females that will be paired for breeding. When lactating, females should be provided with a daily diet as the young grow larger. Once joeys are 90 days old, the fat content in the females' diet should be increased (using items such as egg or mince, M. Zabinskas pers. comm.). However, care must be taken to avoid obesity in females. Joeys start eating solid food, in addition to drinking milk, from 150 days old, when additional food should be provided. As the young grow this food portion can increase to that of 2–3 adult devils to ensure the female and her joeys are receiving adequate food. The amount of food required will vary based on the number of joeys and the activity level of the joeys and mother (Healesville Sanctuary pers. comm.).

Tasmanian devils are prone to obesity, which can be seen in the tail, lower abdomen, neck and back; rolls of fat can also develop around the neck and under the chin (Kelly 1993). High-fat items in the diet such as chicken (42% fat) and quail (36% fat) should be alternated with high-protein items such as kangaroo (80% protein) and/or rabbit (76% protein) (Stannard *et al.* 2017) and the quantities of high fat items should be reduced for animals prone to obesity.

### 6.1.11 Marsupial moles

Very few marsupial moles have been held in captivity, but several different food items have been found in the stomach contents or are accepted while in captivity (Stirling 1891; Spencer 1896; Wood Jones 1923; Howe 1975; Winkel and Humphrey-Smith 1988; Johnson and Walton 1989; Johnson 1995; Withers *et al.* 2000).

- Recommended foods to be offered in captivity
  - ant eggs, larvae and pupae, including *Rhytidoponera* spp.
  - longicorn beetle larvae
  - lepidoptera larvae
  - sawfly larvae
  - scarabaeid beetle larvae
  - gekkonid lizards
  - spiders
  - fly pupae
  - mealworms
  - dog kibble is crushed finely and mixed with the ant eggs, which have been mashed, and mincemeat. The mix can be frozen in serving sizes in plastic bags and should not be refrozen after thawing.

### 6.1.12 Mealworms and crickets as food

Mealworms (*Tenebrio molitor*), which are not worms but larvae of the darkling beetle, are frequently fed to many species of carnivorous marsupials. Mealworms and crickets are useful food items because they are convenient and, if produced properly, nutritional. The typical nutrient composition of several species of invertebrates, including mealworms and crickets, that are fed to insectivores has been determined (Finke 2003a; Stannard and Old 2011; Stannard *et al.* 2014). Insects that are fed to carnivorous marsupials should be gut-loaded to enhance their nutrient content (Finke 2003b).

The disadvantage of using mealworms and crickets is that they are high in fat and typically have low calcium content and low calcium to phosphorus (Ca:P) ratios (Jones *et al.* 1972; Allen and Oftedal 1989). These ratios can range from 1:2 up to 1:14, when they should be ~1.2:1 (Zwart and Rulkens 1979). Therefore, feeding a diet primarily of mealworms or crickets can lead to obesity and a deficiency in calcium (Martin *et al.* 1976; Anderson 2000). This problem can be minimised by maintaining the mealworms on a commercial high-calcium diet for as little as 24 h before animals are fed (Anderson 2000). A study of field crickets found the calcium content and Ca:P ratio increased during the initial period of feeding but remained stable after 48 h (Allen and Oftedal 1989). Further details on mealworms can be found in Chapter 13, section 6.1.6. The practice of dusting or dipping insects in calcium supplements may adversely affect their palatability, so has inconsistent results and therefore provide inadequate levels of calcium. Additionally, if the insects are not consumed immediately, self-grooming or other activity may significantly reduce or eliminate the supplement (Bernard *et al.* 1997).

Mealworm media of different composition have been developed to maintain the insects, including a wheat bran/pollard medium that can include vegetables such as cabbage, carrots, lettuce, potato and pumpkin, as well as wholegrain bread. Gut-load the mealworms 1–2 days before feeding the carnivorous marsupials or coat them immediately before feeding is needed. Examples of gut-loading or coating media are given below.

(1) 50:50 mix of Insect Booster (Wombaroo®/Passwell®) and wheat bran in addition to vegetables such as leafy greens, sweet potato and carrots (Lyons 2020).

(2) Blended chick-starter (60%), wheat bran (38.5%) and calcium carbonate powder (1.5%; Balanced Calcium Powder® Mavlab™ Animal Health) (Lyons 2020)
(3) 50:50 mix of Wombaroo® Small Carnivore Food and bran, or generously coat the mealworms with Wombaroo® Small Carnivore Food immediately before feeding to the carnivorous marsupials. Mealworms are supplemented with banana peel when in the bran mix (L. Lumsden pers. obs.).
(4) The Missing Link® (Skin and Coat Supplement Powder for Dogs) can be used as a coating of the mealworms immediately before feeding them to the carnivorous marsupials. Sprinkle 1 heaped teaspoon over ~100 g mealworms or pro-rata (R. Lyon pers. obs.). Alternatively, in the preparation for each feed, more worms than are needed can be placed in a container and The Missing Link® powder added at the rate of ~1 teaspoon per 100 worms. Between the morning and afternoon feeds the mealworms are left in the powder for ~7 h and those placed in it overnight for ~17 h will consume most of the powder (J. Mclean pers. obs.).

Mealworms can be placed in a shallow container with a fine mesh floor. The waste falls through the mesh, keeping the media clean. A large-sized bran media can be used so that it does not readily go through the floor sieve (e.g. McKenzie's Natural Process Bran) and then a sieve can be used to collect the mealworms to be fed to the carnivorous marsupials.

House crickets (*Acheta domesticus*) can be held in various sized containers depending on whether the intention is to breed them or maintain them after purchase until they are fed to carnivorous marsupials (e.g. see von Hackewitz 2020). For example, plastic containers that have cardboard egg cartons added can be used, with the size of the container depending on the number of crickets being held. If adult crickets are purchased, they can be fed lettuce and carrot. A high-protein pellet food such as chick starter can also be provided. Open water dishes may result in deaths because crickets are not good swimmers if they jump in the water; the lettuce and carrot may provide the necessary moisture. If a water dish is added, include a sponge in the bowl to reduce the risk of drowning. The food should be replaced, and any dead crickets removed, every few days. The container should be kept at ~25–26°C, but not too humid, so ensure there is adequate ventilation. Similar to mealworms, crickets should be gut-loaded or dusted before being fed to carnivorous marsupials.

### 6.1.13 Meat mix

| Ingredient | Percentage by weight (%) |
|---|---|
| Kangaroo mince | 41 |
| Advance™ Puppy kibble (crushed) | 14 |
| Whole boiled egg including shell | 24 |
| Fly pupae | 10 |
| Wombaroo® Insectivore Rearing Mix | 7 |
| Bran | 2 |
| Calcium carbonate powder | 2 |

- Dry ingredients (crushed kibble, insectivore powder, calcium carbonate and bran) may be prepared separately and stored for up to 2 weeks in air-tight containers.
- Kangaroo mince, egg fly pupae should be added fresh before feeding out the following day.

### 6.2 Supplements

A variety of different food items can be provided to increase activity and for behavioural enrichment (see individual diets above). Some species eat soft nuts and dried fruit (Woolley 1982).

### 6.3 Presentation of food

The food for the smaller species can be supplied in a metal or plastic dish, but larger species such as quolls and Tasmanian devils are fed on the ground or the food can be hung from bungee cords. Carcasses or larger feeds can be staked to the ground or attached to safe bungee systems to facilitate pulling behaviours and activity of larger carnivores such as devils. Small feeds to increase activity for public exhibit (activity feeds) are generally provided by randomly throwing in food, ideally at different times of the day, to any part of the floor, or up branches for arboreal species, to encourage foraging and hunting behaviour. Blood or a food mash can be smeared around the enclosure to encourage activity and foraging behaviours. Live insects in the mulch or leaf litter work well as a way of encouraging natural feeding behaviours. Take care that the use of such feeds does not lead to obesity.

## 7. HANDLING AND TRANSPORT

### 7.1 Timing of capture and handling

Carnivorous marsupials are generally best caught during the day while they are asleep in their nest box. If held in a nocturnal house they can often be caught first thing in the morning before the lights go out. Alternatively, they can be netted or trapped in the enclosure.

## 7.2 Capture equipment

Smaller species (ningauis, dunnarts and red-tailed phascogales) can be easily held in calico bags or click-seal plastic bags (with small air holes) for short periods during weighing. Larger species such as quolls and Tasmanian devils should be placed in larger, thick canvas, cotton, calico or hessian bags, taking care that the animal does not bite the handler through the bag. Take care with hessian bags because dust can fall into the animal's eyes and nose, and coarsely woven hessian bags allow the animal to see movement through the bag, which can distress and overexcite it making it difficult to contain. Specially made canvas sacks with air holes in the corners and airhole eyelets at regular intervals around the base are another option. Bags made from flexible thin neoprene/wetsuit type material or microfleece have also been used successfully for quolls because the dense fabric maintains a dark interior that keeps the animal calm. If using fabric bags with seams, ensure they are used hem-side/seam-side out because toes and claws can be caught and limbs can get wound up in cotton/string and cause injuries.

## 7.3 Capture and physical restraint

### 7.3.1 Small dasyurids

Small species can move very quickly and the best way to catch them is by tipping them from the nest box into a cloth bag. If they are out in the open within their enclosure, position them into a corner using a hand inside a cloth bag and then grab them through the bag and turn up the edges. An alternative to hand catching is an Elliott small live mammal trap, which is particularly good for medium-sized species such as antechinuses, mulgaras, kowari and phascogales. Once in the trap, the animals can be slid directly into a catching bag. Once in a bag, the animal can be handled in the bag while keeping the eyes covered to promote calm behaviour or by holding its head and upper body down with one hand and using the other hand to grab it over the shoulders so that the head is held securely between the index and middle fingers (Fig. 4.4a).

Small dasyurids, such as dunnarts, may be captured by experienced handlers by cupping a hand over them and holding them in the palm of the hand (Fig. 4.4b). The handler's thumb is placed firmly (but not tightly) over the animal's chest to under the chin and the fingers support the animal's back and neck (Scicluna *et al.* 2025). Scruffing, holding the animal by the back of the neck with the thumb

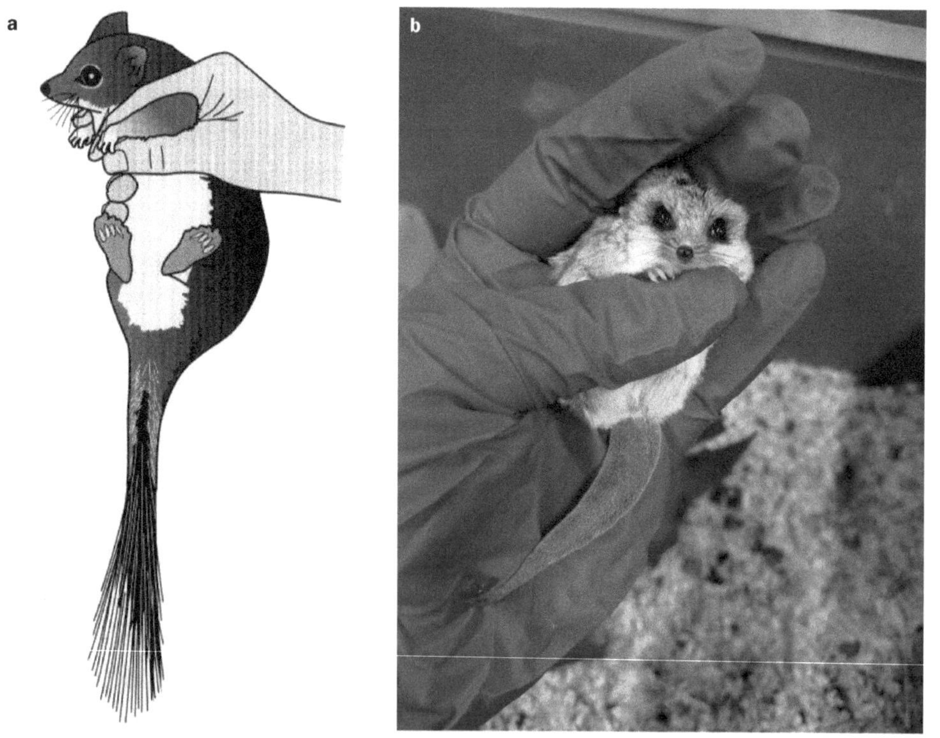


**Fig. 4.4.** Two techniques for holding small dasyurids: (a) between the index and middle fingers and (b) cupped around the body with the thumb under the chin (from Scicluna *et al.* 2025 with permission).

and index fingers, has been used for small dunnarts, ningauis and planigales. However, it is not recommended for larger species such as antechinuses, mulgaras and kowaris as they are too strong and brush-tailed phascogales readily lose their fur (C. Dickman pers. obs.). Care must be taken to not injure smaller animals or cause suffocation when holding animals using this method, so this technique should not be used by inexperienced handlers.

Another way to capture antechinuses, dunnarts and planigales is to put a hand inside a calico bag like a glove and then place that hand over the whole animal. However, some species such as the Atherton antechinus are known to give deep and painful bites (C. Dickman pers. obs.). Greater caution should be taken with medium-sized dasyurids such as phascogales, mulgaras and kowari as their bites can hurt considerably. Do not catch them by the tail because the outside sheath of skin may come off. If the animal is to be held for some time, the animal in the bag should be placed in a secure box because potentially it can chew out of a bag, given time.

### 7.3.2 Large dasyurids

Quolls and the Tasmanian devil are very strong animals that can bite very hard. Quolls and young devils can also be very fast and agile climbers. The best method of catching them is to trap them in their nest box by first blocking the entrance so they cannot escape and then carefully opening a slide built into the side of the box so the animal drops directly into a bag. Blowing a sharp puff of air on their rump can sometimes be used with success and is faster and less stressful than shaking them out.

If the box is large and too cumbersome to move, the roof should be hinged so that the animal can be caught by holding a large net over the entrance to the nest box and chasing it out of the box. An alternative method of catching Tasmanian devils is to place a catching bag or any heavy piece of material over the devil's head and front quarter of the body. By moving the hessian bag over the head, then removing it, the animal usually moves around to avoid the bag, giving the handler the opportunity to manoeuvre it to the best position. Simply touching a devil on the rump is sometimes enough to make it spin around and present the rump and tail. Lean forward and with one quick motion grab the animal firmly by the base of the tail, lifting it clear at arms' length from the body - or any other objects that it might grab or bite - and into a bag being held by a second person (Fig. 4.5). Tasmanian devils do not have the flexibility to climb up and bite the handler. However, inexperienced handlers should not use this technique as injuries and breaks to tails have been occasionally observed. Devils should not be held by their tails after a large feed or if senescent, and the time being held by the tail should be as brief as possible to avoid the possibility of tail fracture. Holding quolls by the tail is not recommended as their tail bones are known to fracture and they can twist up and bite the handler's hand or arm.

The devil can be transferred to a canvas or heavy cotton sack by placing the sack on the ground while still holding the top of the cuff as this will make the bag form a tunnel. With the other hand lower the animal towards the ground so its front feet can touch the ground. Run the animal into the bag by letting it run itself while you gently push it along. The animal may at this point grab the bag, stopping itself from moving into it. Simply pull them apart and try again. Usually, the dark inside of the bag is enough incentive for the animal to want to go into it. Once the animal is at least three-quarters of the way in, lift the bag off the ground and at the same time release your grip on the tail. Tie the neck of the bag at

**Fig. 4.5.** Techniques for briefly holding large dasyurids such as spotted-tailed quolls and Tasmanian devils.

least seven-eighths of the way up, and do not carry the bag in your arms or on your lap - keep it clear of your body. The stitching can be unpicked in one of the bottom corners to ~70 mm to allow the animal to poke its nose through the bag, making it much easier for the vet to place an anaesthetic mask over its mouth/nose.

An alternative technique involves conditioning the quoll or devil to take food from the back of a net (held open by the animal caretaker) or within a box for several days before trying to catch it. If in a larger enclosure, devils and quolls can also be caught in large PVC pipe traps set overnight and checked first thing in the morning.

If the quoll or Tasmanian devil is out in the enclosure, the best way to catch it is with a net on a long pole and a second person (with a net or plastic rake in case the animal turns on them and tries to bite their legs) who herds it towards the net. Once it is in, quickly lift the net off the ground and twist it so the animal cannot climb out.

Once captured, smaller quolls, such as eastern and northern quolls, can be restrained by holding their head firmly between index and middle fingers, in a similar way to that suggested for smaller dasyurids and controlling the body by holding the tail. Larger species such as the spotted-tailed quoll, especially males, and the Tasmanian devil have very thick necks so are best held by strongly holding the head between the thumb and index finger. Alternatively (and more strongly recommended) they should be kept in the catching bag and held firmly on the ground. The position of the head should be monitored so that the quoll does not bite through the bag. When required, the limb or part of the body required for examination should be brought out of the bag, keeping the eyes and head covered at all times unless they are to be examined. The inclusion of two-way, centre-opening zippers along the length of the handling bag can facilitate access to a limb, pouch or other required body part, without the risk of the animal escaping.

During the capture process, care needs to be taken to minimise stress, which may be manifested by loud vocalisations, defecation and urination (Spielman 1994). If these occur over a prolonged period then the capture should be halted and resumed at a later stage or using other techniques.

### 7.3.3 Marsupial mole

Marsupial moles have been located within their enclosure while under the sand with the use of a stethoscope to detect movement. They are attracted to the heat of a radiator when confined to a small surface area so this can facilitate capture (Howe 1975). The marsupial mole is readily held and does not appear to show any intention of biting. Some observers suggest that initially they resent handling and continually try to wriggle out of the hand, but after 1 week they will lie quietly in the hand and will even climb into the hand to go to sleep (Howe 1975). They always appear to resent being restrained (Howe 1975).

## 7.4 Weighing and examination

The smaller species (weighing less than ~10 g) can be weighed in plastic bags (that are not sealed or have several small holes at the top to allow air flow) using a fine-scale spring balance as the weights will be more accurate and easier to see. They can also be weighed using hanging or electric scales with 1 g (ideally 0.1 g with digital scales for species less than 10–15 g) increments.

The muscles of the pouch of larger dasyurids are strong enough to deter a person using only one hand trying to look inside, so both hands are usually needed, which means two or three handlers are recommended. One person restrains the animal with its dorsal surface closest to the ground and its head region is restrained (an additional person may restrain the hindfeet or rump area), while the other person uses two or three fingers to open the pouch and investigate inside. Take care to reduce the amount of cold air entering the pouch if the young are unfurred, especially when they are very small. This means opening the pouch as little as possible and if measurements are required, inserting the tip of the callipers into the pouch. The pouches of most species of small to medium-sized dasyurids are poorly developed, so the young are firmly attached but can be readily seen hanging from the nipples (Tyndale-Biscoe and Renfree 1987).

Spotted-tailed quolls have been trained to enter a clear Perspex® tube 60 cm long × 10 cm in diameter when it was placed directly in front of them and more readily if it was covered to darken the interior (Collins *et al.* 1993). Once the anterior half of the body is in the tube, the quoll can be restrained and manipulated easily because it cannot move and bite the handler. This technique allows the posterior half of the body to be examined and facilitates pouch checking (Collins *et al.* 1993).

## 7.5 Release

The different species are generally best released either directly into a nest box or onto the ground.

## 7.6 Transport requirements

The various species of carnivorous marsupials are relatively easily transported. For short distances (e.g. several hours' duration) they are readily transferred in a catching bag, although they should ideally be placed inside a nest box with the entrance sealed, which acts as a secondary barrier if they escape from the bag. The nest box also provides protection from other objects that may crush them, particularly if they are small. Care must be taken that they do not become tangled in the bag, or overheat, which is more likely in a bag compared with being loose in a box. Animals have also been successfully transferred from the wild to captivity in their familiar traps (e.g. Elliott traps) that are placed inside a larger second box in case of escape. Ensure adequate bedding is available in the trap to give the animal comfort and stability.

Whenever carnivorous marsupials are transported by aircraft, they should be placed in a wooden box recommended by the International Air Transport Association (IATA 2022). Inside the box the animal can be held directly inside a bag (for shorter distances, e.g. several hours) and provided with nesting material so that it does not roll around inside the box. For longer distances, the animal need only be placed inside the box, not in a bag, and provided with adequate nesting material.

### 7.6.1 Transport box design

For smaller species, such as those below the size of a phascogale or even the smaller species of quoll, the box can be divided into two or more compartments for convenience. Further specific details of the box design can be found in IATA (2022).

### 7.6.2 Furnishings

Wood shavings or shredded paper should be provided to minimise the animal being moved around during transport and to provide insulation against heat and cold.

### 7.6.3 Water and food

Water should be provided on all except very short trips (e.g. less than 1–2 h) in cool weather. A small water bottle can be used, but make sure the animal is familiar with using it before shipment. Although they often do not eat during transport, provide a small dish of food for the journey, particularly for small species that have higher metabolic rates and feeding requirements.

### 7.6.4 Animals per transport box

Only one individual should be placed per compartment within a box. Females with pouch young should not be transferred unless only recently born and still attached to the teat. Even then there is some risk that a stressed female will remove her young (C. Dickman pers. obs.). Female agile antechinus have been successfully transferred in a familiar (natal) nestbox with large, but permanently attached, young of 3–4 weeks old, and released to the wild without any loss of young (Parrott 2006). Similarly, devils with permanently attached pouch young have been translocated in the wild with no loss of young.

### 7.6.5 Timing of transportation

Wherever possible, transport should be done either in the early morning or overnight so that animals do not become overheated.

### 7.6.6 Release from the transport box

Smaller species are generally released by catching them in the box and placing them in the enclosure, while larger species can be released by opening the box and letting them self-release into the enclosure in their own time.

# 8. HEALTH REQUIREMENTS

## 8.1 Daily health observations

Each individual should be observed daily for any signs of injury or illness. The most appropriate time to do this is generally when the enclosure is being cleaned or when animals are being fed, as many of the larger species, especially in nocturnal houses, will approach to be fed. During these times, each animal within the enclosure should be checked for the following:

- coat condition, fur loss
- discharges: from the eyes, ears, nose, mouth or cloaca

- appetite
- faeces: amount, appearance and consistency
- eyes: for cloudiness, vision
- changes in demeanour
- injuries
- presence and development of pouch young by observation of the bulge in the pouch
- abnormal repetitive behaviour
- movement: unsteady gait, hindquarters swaying, stumbling

## 8.2 Bodyweight and body condition score

### 8.2.1 Bodyweight

Record and compare with previous weights. Trends in bodyweight of carnivorous marsupials give a good general indication of the animal's state of health, provided age and sex are considered. Animals in captivity should be weighed at least monthly for larger species (unless females have young and should not be disturbed) and, ideally, weekly in smaller species due to their fast metabolism, to gain an indication of trends. It should be noted that some female carnivorous marsupials can lose up to 15% of their bodyweight via the loss of oedematous fluid during birth/pseudo-birth, which should be considered in individuals with fluctuating bodyweights.

### 8.2.2 Body condition score

There is no agreed method for determining a body condition score in carnivorous marsupials. However, the generic nine-point scale for body condition scoring zoo animals defined by the Zoological Information Management System (ZIMS, Species360) could be used, where 1 = emaciated, 5 = ideal condition, and 9 = obese. A specific body condition index has been developed for the Tasmanian devil and can be found within Table 4.7 (Hogg and Hockley 2019).

## 8.3 Physical examination

Whenever an animal is caught and is being examined, look for wounds and the presence of lumps throughout

**Table 4.7.** Body condition score for Tasmanian devils.

Derived from Hogg and Hockley (2019).

| Condition score | Neck | Shoulder, forearms | Rib, back | Pelvis, hindlimbs |
|---|---|---|---|---|
| 1. Emaciated | Negligible fat and muscle present. Cervical vertebrae palpable. Jugular furrow visible on extension of neck | Scapula prominent. Muscle wastage from scapula > scapula spine visible or palpable as a sharp edge<br>Forearms show muscle wastage and are thin | Individual ribs easily visible. Skin furrows between ribs<br>Lumbar spinous processes easily visible | Ischial tuberosity (pin bone) and epipubic bone very prominent – jutting out<br>Advanced muscle wastage from thigh. Stifle prominent |
| 2. Moderately thin | Muscle and fat present. Discernible indentation from shoulder to neck | Muscle palpable on scapula > scapula spine just palpable | Ribs easily palpated, last three may be just visible if coat thinning<br>Hollowing of abdomen after final rib | Epipubic bone prominent but with fleshy covering<br>Adequate thigh musculature |
| 3. Average | Neck well covered, contour tapering from shoulders | Difficult to palpate scapula spine | Ribs covered with tissue, last three easily palpated. Minimal 'step down' from ribs to abdomen | Epipubic bone well covered and obvious |
| 4. Good | Neck thick and well muscled, blends into head and shoulder region | Forearms and shoulders well-muscled | Cranial ribs challenging to palpate. Caudal ribs palpable through thick layer of tissue. Barely discernible step down from ribs to abdomen | Epipubic bone visible as two peaks in the pubic area |
| 5. Obese | As above. Roll of flesh at base of neck sitting against shoulder | As above | Difficult to palpate all ribs. Loose abdominal tissue > gives 'pan-cake' appearance if laid on back | Roll of fat above base of tail |

the body, which may be tumours. The eyes should also be checked closely for cloudiness and general clarity. Bodyweight is a useful indicator of condition.

The physical examination may include the following:

- body condition score - best assessed by muscle palpation in the area over the scapula, spine and temporal fossa
- temperature - normally 35-36.5°C, can be measured via the cloaca
- pulse rate - varies greatly with the species; rate decreases with increasing body size. Taken over the femoral artery under anaesthesia as the rate will be increased immediately after the animal is caught
- respiratory rate - varies greatly with the species; rate decreases with increasing body size. It should be taken under anaesthesia or observed from a distance without disturbing the animal as it will increase after the animal is caught
- fur - check for alopecia, ectoparasites, fungal infections or trauma
- eyes
  - should be clear, bright and alert
  - normal bilateral pupillary light response
  - normal corneal reflex
  - should not have any discharges
- check for the presence of lumps over body and auscultation of the lungs
- teeth: can be checked for wear or tooth fracture, particularly in older animals or those fed on harder objects such as bone
- cloaca
  - should be clean
  - check for faeces around the cloaca
- females
  - condition of the pouch
  - check whether lactation is occurring by milking teats, although in some small species it is very difficult to express milk; redness (vascularisation) of the pouch is a good indicator, especially if the pouch is swollen (C. Dickman pers. obs.)
  - if pouch young are present, record sex, stage of development, bodyweight if detached from the teat and measure the head to determine age from growth curves, if available.
- males
  - check testes - size (length, width, depth) and consistency (firm - not squishy)
  - extrude penis and assess in larger species only
  - check the size and activity of the sternal gland

## 8.4 Chemical restraint

Although most dasyurids are not prone to regurgitation under anaesthetic, pre-anaesthetic fasting for 6-8 h is recommended (Vogelnest 1999). Minor manipulative procedures and transportation can be performed using prior sedation with intramuscular diazepam (Valium®) at 1-2 mg/kg (Vogelnest 1999).

Inhalation anaesthesia is the preferred method in all dasyurids, by restraining the animal and inducing with a mask over the face, even through the fabric of a bag or net (Vogelnest 1999). Isoflurane is preferred for inhalation anaesthesia and it can be maintained via a mask or the animal can be intubated if it is large enough (Vogelnest 1999). In larger dasyurids, such as Tasmanian devils and quolls, an inhalation induction is not always possible. These individuals can be induced with alfaxalone (Alfaxan®) 2-3 mg/kg given intramuscularly (Vogelnest 1999).

## 8.5 Common health problems

Carnivorous marsupials suffer few problems in captivity. Most of the parasites and diseases that have been recorded are presented below in Tables 4.8 and 4.9. Although some parasites and diseases are species-specific, others are less so and easily transferred between species, hence caution must be exercised when staff and animals move between enclosures, using the same catching bags or other apparatus.

## 8.6 Preventative medicine

All dasyurids should be quarantined for at least 30 days before entering a collection. A minimum of two faecal floats, taken at least 14 days apart, should be examined during this period. Preventative medicine focusses on daily monitoring and, ideally, an annual physical examination. No routine vaccinations are required. Necropsies should be performed on any dasyurids that die.

## 8.7 Zoonoses

Several zoonotic diseases have been associated with carnivorous marsupials (Table 4.10). For many zoonotic diseases, the dasyurid host will show no signs of disease. If zoonotic disease is suspected, seek medical advice.

**Table 4.8.** Common infectious and parasitic diseases of zoo-based carnivorous marsupials.

| Disease | Cause | Clinical signs | Diagnosis | Treatment | Prevention | Comments | Reference |
|---|---|---|---|---|---|---|---|
| **Ectoparasites** | | | | | | | |
| Ectoparasitism | Various species of fleas, ticks, mites and lice. Includes fleas of the genera *Echidnophaga*, *Uropsylla* and *Stephanocircus*; ticks of the genera *Ixodes Haemaphysalis*; and mites of the genera *Diabolicoptes*, *Satanicoptes*, *Demodex* and *Dasyurochirus*. Also includes lice of the genus *Boopia* | The flea, *Uropsylla tasmanica*, has been reported in the Tasmanian devil, eastern quoll, western quoll and spotted-tailed quoll where heavy infestations have been recorded in the distal limbs, ears, groin, scrotum and face. These can cause considerable irritation and result in severe scratching and hair loss. *Demodex* has been found in various dasyurids. It results in skin nodules filled with mites, granulomatous dermatitis and folliculitis | Generally by visual signs and skin scrapings for mites with microscopic examination to identify the parasites | Controlled using carbaryl, fipronil or imidacloprid topically or injectable ivermectin or moxidectin. Nodules can be surgically removed | Reduce ectoparasite burden by changing the nest material and washing the nest boxes frequently. Mite infestations have been known to occur after their transfer from laboratory rodents used as food. Placing leaf litter in the freezer for 4 days has been effective in eliminating the incidence of mites introduced from the bush | | 1–13 |
| **Endoparasites** | | | | | | | |
| Endoparasitism | Various species of endoparasites have been found in dasyurids including nematodes (e.g. Ascarids, *Cylicospirura*, *Strongyle* spp., *Mackerra strongylus*, *Trichinella spiralis*) and cestodes (e.g. *Taenia ovis* and *Anoplotaenia*). *Coelomotrema antechinomes*, a trematode, has been identified in a kultarr. In wild populations, antechinuses have been found to carry numerous species of endoparasites, the number of which rise sharply during the breeding season in males, compared with females, and are considered directly involved with the seasonal mortality of males (see section 10.1.1) | Often subclinical but may cause diarrhoea | Faecal flotation and the presence of eggs or proglottids (segments that make up the cestodes) | Nematodes can be treated with ivermectin, moxidectin or similar parasiticides. Tapeworms can be treated with praziquantel | Remove faeces daily. Regular monitoring via faecal floats | | 5, 8, 14–16 |

| Disease | Cause | Clinical signs | Diagnosis | Treatment | Prevention | Comments | Reference |
|---|---|---|---|---|---|---|---|
| **Protozoal** | | | | | | | |
| Toxoplasmosis | Caused by *Toxoplasma gondii*. Toxoplasmosis has been observed in kowari, antechinuses, dunnarts, mulgaras and kultarr and has caused significant mortality in captive populations. *T. gondii* infection in eastern quolls does not affect their reproduction or survival, and infection does not advance to clinical toxoplasmosis | Abnormalities include altered behaviour, blindness, incoordination, paralysis or death without prior signs. Clouding of the cornea or lens of one or both eyes and destructive retinochoroiditis have also been observed in several kowaris and a white-footed dunnart | Antemortem diagnosis of toxoplasmosis is confirmed by serological testing to detect rising IgG *T. gondii* titres. Necropsy findings include tissue cysts, inflammation and necrosis in multiple organs but especially in the heart, brain, lungs, adrenals, skeletal muscle and liver | Treatment is usually unsuccessful but can be attempted using clindamycin, trimethoprim/ sulfonamide or atovaquone | Cats are the definitive host. Infection is avoided by preventing access to cat faeces or food or water that has been contaminated with cat faeces. Infectious cysts may be found in the muscles of any mammalian or avian food animals such as sheep and rodents. Prevent transmission by cooking meat to at least 66°C or freezing it for 4 weeks | | 17–30 |
| Mycobacteriosis | Usually caused by *Mycobacterium avium* but other mycobacteria have also been implicated | Weight loss, lameness, areas of thickened skin with secondary ulceration, abscesses and neurological signs | Based on clinical signs, especially chronic weight loss, as there are no accurate antemortem tests | Treatment is usually unsuccessful | Opportunistic disease, often affecting animals that may be chronically stressed or have other comorbidities. Prevention is difficult as the organisms are normal soil and water inhabitants | | 10 |
| Salmonellosis | A range of *Salmonella* species has been isolated from dasyurids | Individuals are usually asymptomatic | Faecal culture will identify the presence of the bacteria but shedding is intermittent | Treatment is not warranted as infections are generally asymptomatic | | Zoonosis. Hand washing and appropriate personal hygiene are important | 3, 10 |

**Table 4.8.** continued

| Disease | Cause | Clinical signs | Diagnosis | Treatment | Prevention | Comments | Reference |
|---|---|---|---|---|---|---|---|
| Tasmanian devil facial tumour disease | Transferable cancer. The presence of the cancer in wild populations of Tasmanian devils was first recorded in 1996 and since then has resulted in a reduction of the wild population of > 80% across their wild range. Infected animals rarely live beyond 2 years rather than their usual 5 years of age. Transmission of the neuroendocrine tumours occurs via direct inoculation of tumour cells from one Tasmanian devil to another by biting. In 2014, a second transmissible cancer (DFTD2) was detected in wild devil populations | Large solid masses that ulcerate and appear predominantly on the head and neck. Death usually occurs within a few months of masses appearing | Presence of tumours and confirmation via biopsy | No treatment is available and the disease is consistently fatal within months. Research on vaccines and immunotherapies is progressing | Ensure that animals with DFTD are kept separate from those that do not have it. The captive population of Tasmanian devils is free from DFTD with any new wild animals undergoing lengthy quarantine periods before inclusion in the program. Disease-free insurance populations have been established on Maria Island, the Forestier Peninsula and in a multi-institutional captive insurance program across Tasmania and mainland Australia | | 31–39 |

References: (1) Nutting and Woolley (1965); (2) Woolley (1982); (3) Obendorf (1993); (4) Booth (1994); (5) Cunningham (1994); (6) Holz (1998); (7) Lambert (2000); (8) Oakwood and Spratt (2000); (9) Desch and Holz (2006); (10) Holz (2008); (11) Vilcins *et al.* (2008); (12) Pearse (1981); (13) Fancourt (2010); (14) Angel (1970); (15) Beveridge and Barker (1976); (16) Arundel *et al.* (1977); (17) Attwood and Woolley (1970); (18) Attwood and Woolley (1973); (19) Dubey (1974); (20) Attwood *et al.* (1975); (21) Attwood and Woolley (1980); (22) Attwood and Woolley (1982); (23) Canfield *et al.* (1990); (24) Obendorf (1993); (25) Innes (1997); (26) Booth (1999); (27) Bettiol *et al.* (2000a); (28) Bettiol *et al.* (2000b); (29) Miller *et al.* (2000); (30) Fancourt *et al.* (2014); (31) Dennis (2006); (32) Hawkins *et al.* (2006); (33) Jones *et al.* (2008); (34) Hamede *et al.* (2013); (35) Kreiss *et al.* (2015); (36) Tovar *et al.* (2017); (37) Hogg *et al.* (2019a); (38) Pyecroft (2025); (39) Pye *et al.* (2016).

**Table 4.9.** Common non-infectious diseases of zoo-based carnivorous marsupials.

| Disease/agent | Cause | Clinical signs | Diagnosis | Treatment | Prevention | Reference |
|---|---|---|---|---|---|---|
| Nutritional osteodystrophy or hind limb paralysis | Calcium deficiency is sometimes called metabolic bone disease. It is caused by inappropriate diets deficient in calcium and/or an incorrect ratio of calcium to phosphorus, which should be 1.2.1 | Abnormalities including growth defects, lameness in gait or posture and the dragging of limbs in extreme cases | An incorrect Ca:P ratio can cause osteodystrophy or softening of the bones. Radiography of the vertebral, pelvic, and long bones can show low bone density | Cases identified early may respond to a high-calcium, high-vitamin D3 diet and strict enclosure rest | Animals fed only meat will develop calcium deficiency. Feeding a nutritionally complete diet will prevent this disease occurring. Calcium in the form of calcium carbonate must be added to any meat preparations | 1–3 |
| Neoplasia | Non-inflammatory growth of tissue that is outside the normal control mechanisms of the body. Tumours are generally age-related and usually occur after reproductive senescence at an age when individuals would normally have died in the wild. There are two broad types: (a) malignant tumours that tend to grow and spread rapidly, destroying neighbouring tissues and infiltrating the healthy structures nearby, and (b) benign tumours that grow slowly in one site, press neighbouring parts aside, but neither invade not destroy them | Clinical signs are variable depending on the body system affected | Clinical signs and biopsies of tumours | Depends on the tumour type but they are generally untreated. Affected animals are given supportive care or euthanased | Difficult to prevent | 4–12 |

**Table 4.9.** continued

| Disease/agent | Cause | Clinical signs | Diagnosis | Treatment | Prevention | Reference |
|---|---|---|---|---|---|---|
| Male die-off in antechinuses and phascogales | Antechinuses, phascogales, kaluta, and some populations of northern quolls and dibblers are unusual in that after mating in the wild all the males die. This condition is termed semelparity | In the antechinuses, phascogales and kaluta, the deaths are generally the result of an uncontrolled increase in plasma corticosteroids and associated adrenal weight increases. This results in a decrease in bodyweight, negative nitrogen balance and immunosuppression that results in gastrointestinal haemorrhage, marked neutrophilia and lymphopenia, and anaemia associated with heavy parasitemia by *Babesia* sp. | In captivity the males can live for more than 1 year, and up to 4 years, although they all show testicular failure and are sterile after their first breeding season. Quolls appear to have a different mechanism causing their die-off. Dibbler males are unusual in that they can sire litters in three consecutive seasons, although during the third season their litters are of smaller size | None. Natural event | None. Natural event | 5,13–21 |
| Central nervous system degeneration | Unknown cause. Occurs in older quolls and devils | Blindness, ataxia, swaying gait, hindquarter collapse | Based on age and clinical signs. Definitive diagnosis requires necropsy | None | None | 10 |
| Intervertebral disc disease | Degenerative intervertebral discs resulting in rupture and/or prolapse. Mostly seen in older quolls and devils | Ataxia, partial or complete hindlimb paralysis | Clinical signs and radiography | None | None. Maintain optimum bodyweight | 22 |
| Heart disease | Unknown cause. Seen in older Tasmanian devils | Weakness, lethargy, coughing | Clinical signs, ECG, radiography | Supportive care | None | 23 |

References: (1) Zwart and Rulkens (1979); (2) Woolley (1982); (3) Obendorf (1993); (4) Attwood and Woolley (1973); (5) Arundel *et al.* (1977); (6) Griner (1979); (7) Twin and Pearse (1986); (8) Hine (1988); (9) Reece and Hartley (1994); (10) Holz and Little (1995); (11) Boden (1998); (12) Stannard and Old (2014); (13) Braithwaite and Lee (1979); (14) Barnett (1973); (15) Lee *et al.* (1977); (16) Barker *et al.* (1978); (17) Braithwaite and Lee (1979); (18) Lee and Cockburn (1985); (19) Halley (1992); (20) Oakwood *et al.* (2001); (21) Letendre *et al.* (2018); (22) Holz (2008); (23) Eden (2025).

**Table 4.10.** Zoonotic diseases associated with carnivorous marsupials.

| Disease | Cause | Clinical signs in humans | Transmission | Prevention |
|---|---|---|---|---|
| Mycobacteriosis | Environmental bacteria. Numerous different species | Pneumonia, skin lesions. Disease usually only occurs in immunocompromised people | Infected lesions | Hand washing and appropriate hygiene |
| Assorted bacteria including *Salmonella, Campylobacter, Pasteurella, Leptospira* | Numerous different species have been isolated from dasyurids | Diarrhoea, abdominal pain, flu-like symptoms, wound infections | Faeces, urine, bites | Hand washing and appropriate hygiene |
| Giardiasis | *Giardia* spp. | Diarrhoea, abdominal pain | Faeces | Hand washing and appropriate hygiene |

# 9. BEHAVIOUR

## 9.1 Activity

Cold temperatures can induce torpor in all species of carnivorous marsupials weighing less than ~1000 g (Geiser 1994, 2003). The period during which animals can be seen on exhibit can be maximised by reducing the amount of time the lights are off (though this may cause issues with circadian rhythms and reproduction) and providing activity feeds. For example, a study of the fat-tailed dunnart found that activity lasted proportionally longer with 8 h darkness than with 20 h darkness (Holloway and Geiser 1996). In the wild, food shortage is also an important determinant of activity as it promotes torpor (Holloway and Geiser 1996). However, as food is readily available in captivity, this factor should not be of significance. Studies of the stripe-faced dunnart showed a peak of activity shortly after the lights were turned off, which steadily declined until they retired before the lights were turned on again (O'Reilly *et al.* 1984). Although most species of dasyurids are considered nocturnal, many species sun bask and some, such as the dusky antechinuses, kaluta and spotted-tailed quoll, are active in both the day and night (Stannard *et al.* 2015; Baker and Gynther 2023; C. Dickman pers. obs.).

To survive cold temperatures and a lack of food in the wild, many dasyurids less than ~1000 g (Geiser 1994) use several techniques to reduce heat loss, including growing thicker fur (Wallis 1982), using nests (Frey 1991), nest sharing (Morton 1978a; Frey 1991), reducing activity (Frey 1991), sunbasking (Frey 1991), storing fat in the tail and undergoing torpor (Morton 1978b, 1978c, 1980). Torpor generally occurs at low ambient temperatures (e.g. < 6°C in fat-tailed dunnarts, < 13°C for stripe-faced dunnarts, < 15°C in common dunnarts, < 11°C for the kultarr, and < 18.3°C for red-tailed phascogales) (Wallis 1982; Geiser *et al.* 1984; Geiser and Baudinette 1985; Geiser 1986; Frey 1991). Among the dasyurids that undergo torpor there is a remarkable uniformity in response to cold temperatures in body temperature (> 14°C), and maximum duration (< 10 h) (Wallis 1982). However, as noted by Geiser and Cooper (2023), torpor can be an almost routine occurrence in small dasyurids, and in the numbat, and may occur in response to a wide range of stimuli other than just temperature.

## 9.2 Social behaviour

In many species of dasyurids, groups can be successfully kept together in captivity in single species colonies. However, they generally show aggression towards each other, with fighting intensity usually increasing with body size. Fighting can result in significant injury or death in larger species such as quolls and Tasmanian devils if adequate space is not provided. Groups of breeding and non-breeding devils have been successfully maintained in large free-ranging type enclosures and groups of newly weaned devils are often housed together to promote activity and learning of social behaviours.

It is often difficult to detect interactions between individuals as most species are very small and nocturnal, except for quolls and Tasmanian devils. Infrared cameras can be used to review behaviours of all sizes of carnivorous marsupial, but cameras must have an infrared wavelength > 920 nm to avoid the light negatively influencing circadian rhythms or behaviour. One technique that has been used successfully in free-living dibblers and grey-bellied dunnarts is the use of fluorescent pigments and a portable ultraviolet light source (Dickman 1988b). Different coloured pigments are brushed onto the ventral surface, flanks and dorsal and ventral surfaces of the tail. Through the use of this technique various behaviours have been identified including:

- mounting: males transferred pigment to the posterior dorsal fur and base of the tail of the female and females transferred pigments to the posterior ventral surface, scrotal sac and underside of the tail of males
- huddling: transfer of pigments from a donor animal to the sides, all over the back and underbelly of other animals
- fighting: transfer of pigment to the muzzle of the attacking animal and occasionally to the fore- and hindclaws of the attacker
- social grooming: characterised by licking or rubbing the face, muzzle and body of another with pigment being transferred to the groomer's forearms and forepaws.

### 9.2.1 *Dasycercus*

Little is known about the social behaviour of the mulgara and kowari in the wild, although excavated burrows of brush-tailed mulgara were found to include only single animals and offspring. The kowari appears to be solitary with overlapping home ranges of several square kilometres in which they come together briefly to mate (Woolley 1990a; Aslin and Lim 1995). Observations of mulgaras in the wild by First Australians suggest that the male is never found in the same burrow with the female when they have young, but returns later (Spencer 1896). Despite being largely solitary in the wild, both species in captivity show little aggression towards each other and that which does occur generally constitutes threat postures rather than physical contact (Ewer 1968b; Sorenson 1970). In contrast, other observations have shown kowaris in captivity to be very aggressive towards each other when the female is in oestrus. Observations on captive kowari have found 90% of the females' active periods occurred during the night, but only 52% of the males' active time occurred then (Aslin 1974). The females were most active around midnight, but could be found active throughout the night, while the males were sporadically active during the night and day. They show little contact-promoting behaviour, which suggests they do not form groups in the wild. Despite this, they do not inflict serious injuries on each other in captivity (Aslin 1974). They display various postures and can produce a range of vocalisations, including hissing, chattering, snorting and grating, that are associated with various agonistic and other social behaviours (Aslin 1974).

### 9.2.2 *Dasyurus*

Quolls are generally solitary and actively defend core home ranges, although they have home ranges that overlap considerably (Dempster 1995; Godsell 1995). The home ranges of eastern quolls overlap and they occasionally share dens. Adults usually avoid one another except during the short breeding season when fights between males become more frequent (Bryant 1988; Godsell 1995).

Both sexes of the western quoll are solitary, with female home ranges showing little or no overlap, suggesting intrasexual territoriality. In contrast to females, males' home ranges are much larger and overlap with other males and females. In captivity, females deposit scent by cloacal dragging in response to other individuals (Serena and Soderquist 1989a).

Female northern quolls are typically visited by up to four males per night, with males adopting a roving strategy and regularly visiting several widely spaced females in rapid succession, presumably to monitor the onset of oestrus. There is also an increase in scat deposition in prominent positions and an increase in sternal gland activity in males, which suggest the importance of olfactory communication (Oakwood 2002).

Field studies of spotted-tailed quolls have found that adult females show intrasexual territoriality throughout the year but seem to tolerate the presence of female offspring. In contrast, male spotted-tailed quolls were not territorial and their home ranges overlapped both with other males and females throughout the year (Belcher and Darrant 2004). Female spotted-tailed quolls with offspring have been observed in maternity dens located within burnt-out tree stumps, under large boulders and in old rabbit and wombat burrows. Nesting material such as dry grass is carried in their mouth to the dens. They used these dens for up to 39 days before moving to a new den, which was typically within a few hundred metres of the old one (Körtner *et al.* 2019).

Unlike their wild relatives, a highly unusual behaviour has been observed in which a spotted-tailed quoll was observed to bring food to a female with young with whom he was housed (Settle 1978). However, other observations suggest that males and females are more likely to kill each other if housed together with young.

### 9.2.3 *Parantechinus* and *Pseudantechinus*

Little is known of the social behaviour of these species. Dibblers are largely crepuscular, but often carry out some foraging by day and it has been suggested that they are solitary and only come together to breed (Lambert 2000). In captivity, two similar-sized adult females can be housed together permanently for exhibit purposes (Lambert 2000). During the breeding season one male can be held with two females. Aggression can occur between pairs and the aggressor is usually the female if she is larger than the male, with problems occurring immediately after introduction (Lambert 2000). Therefore, the male should be larger than the female to reduce any chance of incompatibility (Lambert 2000). At Perth Zoo they have been held and bred successfully as monogamous pairs and polygynous groupings (one male and two females) (Lambert and Mills 2006).

### 9.2.4 *Sarcophilus*

The Tasmanian devil is normally solitary but promiscuous during the breeding season, with home ranges that overlap extensively with others (Guiler 1970; Pemberton and Renouf 1993; Jones 1995). Despite their generally solitary nature, devils often congregate when feeding on large prey such as kangaroos, with as many as 22 being observed feeding on a dead cow (Pemberton and Renouf 1993; Jones 1995). They often show aggressive behaviour when first introduced and over food. This behaviour is highly ritualised and has many elements including posture (20 types), vocal signals (11 forms) and cloacal dragging. Both sexes carry out cloacal dragging, which involves bending the hindlegs under the body and dragging the rump along the ground by pulling with the forelimbs. Both prey size and group size influence devil behaviours, with different techniques used to collaboratively pull apart whole carcasses compared with a more individual focus on smaller pieces. Larger group sizes increase the frequency of social pulling behaviours (Pollock *et al.* 2021b).

Physical clashes, particularly between male Tasmanian devils, can lead to significant injuries (especially the muzzle and rump) (Eisenberg *et al.* 1975; Buchmann and Guiler 1977; Pemberton and Renouf 1993). This is also a key transmission time for DFTD (see section 8.5). Aggressive interactions generally result in the establishment of a stable dominance and then a progressive decline in the frequency of aggressive interactions due to the lower-ranking animal's reluctance to challenge the superior enclosure mate (Buchmann and Guiler 1977). If the dominant animal is the female, this usually results in unsuccessful breeding due to the male's inability to successfully mate. If this occurs, the female needs to be partnered with a more dominant animal or be removed for several weeks just before the breeding season, to allow the male to establish himself, before she is reintroduced. Devils can be housed in single sex groups and fed a single large meal in captivity. They will show some aggressive behaviours and generally this does not affect an individual's time spent feeding (Skelton and Stannard 2018). Most group-housed devils have a dynamic shifting social hierarchy; however, feeding single large meals may result in subordinate males not meeting their nutrient requirements and being stressed (Skelton and Stannard 2018).

Although they are generally considered solitary, and it is recommended that the male be removed after mating; there are records of males washing and cleaning the young (Turner 1970). In the wild, weaned and dispersing juveniles are believed to socially interact, which may be important to build social behaviours and skills around other devils. In captivity, weaned devils from different litters can be housed together, with larger groups eliciting greater activity levels and longer active times. There are no increases in aggression, despite larger group sizes. Young devils appear to form relationships with others and frequently den together, especially if introduced at a young age or from the same litter (Rosewarne 2011). Scent is important in communication, with devils using communal latrine sites, which are believed to facilitate scent sharing. Devils can recognise individuals based on scent and show differing levels of interest in different individuals and sexes (Nightingale 2013). Male devils in captivity show increased interest in the scent of females and can display searching, digging and pacing behaviours if a female in a nearby enclosure is in oestrus.

### 9.2.5 *Antechinus*

If held as a colony outside the breeding season, antechinuses show relatively little aggression while still forming a linear hierarchy. However, during the breeding season the males become much more aggressive towards each other as the need to find and defend mates increases greatly (Wood 1970; Rigby 1972; Braithwaite 1974). During the mating season both sexes are promiscuous; males in some species spend no more than 8 h in a communal nest and travel extensively between nests, probably to increase reproductive success (Lazenby-Cohen and Cockburn 1988; Lazenby-Cohen 1991). Females choose to mate with multiple males (Parrott *et al.* 2015) and almost all wild litters (92–96%) have mixed paternity with up to five fathers per litter (Holleley *et al.* 2006; Parrott 2006). Lactating females nest solitarily and increase the number of foraging trips, but do not reduce the time in the nest (Lazenby-Cohen 1991). Prior to nest entry, females utter soft 'chit-chit' sounds that appear to reassure the young. The agile antechinus is mostly nocturnal, with males nesting communally and always with females, with a maximum aggregation size of 18 being observed during the breeding season. This number falls after the breeding season has ended (Cockburn and Lazenby-Cohen 1992). Other observations on wild subtropical antechinuses found juveniles to be communal when plenty of food was available and solitary when food availability diminished (Braithwaite 1979). Aggregations are not stable and during one winter a single nest attracted 28 females and 24 males. Not all antechinuses are nocturnal or communal; dusky antechinuses are often diurnal (as are yellow-footed antechinus) and nest solitarily (Hall 1980; Green and Crowley 1989; Baker and Dickman 2018). Scent is important for communication in antechinuses. Males will scent mark pointed or hard objects (e.g. branches, food bowls) in their enclosures using urine and cutaneous sternal and paracloacal glands (Parrott 2006).

## 9.2.6 *Phascogale*

Adult male brush-tailed phascogales very rarely nest with other males or females, except during the breeding season, suggesting that they are largely solitary in the wild (Soderquist and Ealey 1994). One study over multiple years in south-western Australia, however, found that during a single year 56% of animals shared nest boxes with 2–4 individuals of any age/sex cohabiting. The nest sharing during this event coincided with a prolonged drought and was thought to indicate an energetic crisis before the population crashed the following year (Rhind 2003).

Both sexes are promiscuous, with females occupying home ranges that are intrasexually exclusive, whereas the males' home ranges overlap extensively with females and other males and expand during the breeding season (Soderquist 1995; Millis *et al.* 1999). As a result, both brush-tailed and red-tailed phascogales can demonstrate multiple paternity within litters (Foster 2008). In captivity, however, brush-tailed and red-tailed phascogales will often nest together (Fleay 1934; Cuttle 1982a; J. Old and H. Stannard pers. obs.). Phascogales (common brush-tailed, northern brush-tailed and red-tailed) are highly arboreal, nesting above the ground, and nocturnal; their nightly activity interspersed between periods of inactivity (Cuttle 1982a). The activity bouts also appear to have some degree of social content, as observations of animals in captivity show them to often share the same nest box and synchronise periods of feeding and activity (Cuttle 1982a). Interactions between wild animals rarely include contact and generally only involve chases (Soderquist and Ealey 1994). Wild animals drum their forefeet on branches and other substrates if alarmed, producing a sound that is audible to human observers at distances up to 50 m (C. Dickman pers. obs.). This has been presumed to act as a warning to other phascogales that a predator has been detected, but it could be (more likely) a signal to the predator that it has been spotted; any intraspecific function of foot-tapping thus remains to be confirmed (C. Dickman pers. obs.).

Red-tailed phascogales housed in captivity can be kept in groups and will readily share nest boxes. The group composition can include females only, one male with a group of females or a combination of males from the same litter that are raised in the same enclosure from a young age and females. However, after breeding groups have been established and females are pregnant, they do not tolerate males sharing the nest box, hence during the breeding season a minimum of two next boxes should be provided in each enclosure. When males are placed in with females, they will initially hiss and straighten their tail. Males will actively mark their territories with urine and sternal gland secretions (J. Old pers. obs.). On introduction of a male to a female's enclosure, the male will chase the female and may bite and hold her by the tail. If the female stands still the male will sniff the female's cloacal area (H. Stannard pers. obs.).

## 9.2.7 *Planigale*

Little is known about the social behaviour of planigales in the wild due to their highly cryptic nature; however, they are known to sometimes occupy communal nests and olfactory marking has been observed (Morrison 1975; Van Dyck 1979; Andrew and Settle 1982; Read 1984a, 1984b, 1987). In captivity they demonstrate a dominance hierarchy, although they generally do not show much aggression towards each other except over refuges when females are in oestrus (females were found to be dominant to males), when an animal is added to the enclosure or after food thieving (Morrison 1975; Van Dyck 1979; Andrew and Settle 1982). Despite aggressive behaviour, threat and appeasement calls usually inhibit serious fighting (Van Dyck 1979; Andrew and Settle 1982). Planigales have periods of intense activity interspersed with periods of torpor (Van Dyck 1979).

### 9.2.8 *Ningaui*

Little is known about the social behaviour of the ningauis; however, radio-tracking observations suggest that individuals of this genus do not have fixed home ranges and have an extended breeding season (Coventry and Dixon 1984; Kitchener *et. al.* 1986). Captive adults appear to be intolerant of one another outside the breeding season (Fanning 1982). Males have a sternal gland and its activity appears to be closely associated with the reproductive season. Males actively mark bark and other surfaces as they pass along them (Fanning 1982; Dickman *et al.* 2001).

### 9.2.9 *Sminthopsis*

In the wild, fat-tailed dunnarts have been found to nest solitarily in the breeding season (though they do not appear to defend a territory), but up to 70% of individuals share nests in groups from two to eight in the non-breeding season (Morton 1978a; Dickman *et al.* 2001). The nest sharing groups appear to be non-permanent and made up of a random aggregation of individuals during the non-breeding season; however, in the breeding season if nest sharing occurs it involves a male with an oestrus female (Morton 1978a).

Male common dunnarts have been observed to be aggressive towards each other and may be wounded in combat (Fox 1995). Both male and female fat-tailed dunnarts have large overlapping home ranges during the breeding and non-breeding seasons (Morton 1978a). This contrasts with trapping results for the white-footed dunnart that suggest female home ranges do not overlap (Lunney 1995). In captivity, male fat-tailed dunnarts have been observed to live together without trouble, unless a female in oestrus is present (Ewer 1968a). When the female has pouch young, she becomes increasingly aggressive towards the male, who defends himself by fleeing. Hostility between females does not normally occur unless one of the females is reproducing (Ewer 1968a). Both hairy-footed and lesser hairy-footed dunnarts show similar patterns of home range and behaviour to fat-tailed dunnarts (C. Dickman pers. obs.).

In the stripe-faced dunnart, aggression between males and females can occur when introduced to pair (from minor fur loss or bites to the tail, to death of one or both animals). Captive mate choice trials (see section 9.3.1) indicate that females do not choose males based on familiarity. However, non-familiar pairs show higher aggression levels and rates of minor injuries (e.g. fur loss and bites to the tail) compared with familiar pairs, which receive no injuries (Parrott *et al.* 2019a). This suggests that familiarity between mates could significantly decrease fighting, injuries and deaths in captive breeding programs. Aggression can be diminished by housing the animals next to each other with restricted contact before introduction (Parrott *et al.* 2019a).

Julia Creek dunnarts held in pairs, with a running wheel, over a 6-week period showed that one of eight pairs had to be separated due to aggressive interactions. The other paired animals displayed more abnormal repetitive pacing and self-grooming than animals kept by themselves (Phillips *et al.* 2012).

Dunnarts typically emerge at dusk and return to their refuge shortly after dawn. They are not active the entire night, having several feeding bouts between periods of rest (Ewer 1968a). Although dunnarts are generally nocturnal, the photoperiod has been found to influence both the duration and intensity of their activity (Holloway and Geiser 1996). Shorter photoperiods (L:D 8:16) produce a longer duration of activity that is confined to darkness. When the light cycle was L:D 16:8, dunnarts usually did not become active until 20:00 h, but were regularly active up to 4 h after the lights turned on. The level of food available also had an effect, with activity increasing from 150–200 movements/30 min with food to a maximum of 255 movements/30 min without food. When food was returned the number of movements decreased greatly to 50–60/30 min (Holloway and Geiser 1996). Unlike some rodents, the photoperiod had no effect on torpor. Torpor was apparent at all times of the year, although more so in winter, and appears to be more associated with food restriction and low ambient temperatures (Holloway and Geiser 1996).

### 9.2.10 *Antechinomys*

Kultarrs are generally solitary, although they can be housed in single-sex pairs. Males housed together are not aggressive towards each other outside the breeding season, but will become aggressive if a female in oestrus is present or has been recently removed (Valente 1984). Males will perform cloacal dragging if another animal (male or female) is introduced into their enclosure (Valente 1984). Males make a high pitched 'chir-ir-up' sound during the breeding season when investigating new substrate or enclosure mate (male or female) (Valente 1984). On the introduction of a male to a female's enclosure, the female may chase and bite the male, usually the male will run to escape or turn to confront the female and bite her around the neck while she also bites him. This aggressive behaviour only lasts a short period of time and within 24 h kultarrs will usually eat together and share a nest box. Kultarrs make a 'hiss-like' noise when they feel threatened and a particular individual kultarr was heard to make a 'grunt-like' noise when being restrained for health checks (Stannard and Old 2010; H. Stannard pers. obs.). Kultarrs are most vocal during the breeding season, with males and females vocalising to each other even when housed in separate enclosures (Happold 1972). Young will also vocalise to alert the female when they become dislodged from the pouch (Happold 1972).

### 9.2.11 *Notoryctes*

The marsupial moles are very cryptic species and little is known of their biology or behaviour. They have no eyes and live in the sand plain and dune regions of central Australia where they swim through the sand rather than forming permanent burrows (Johnson 1991). Although they spend most of the time below ground, they will come to the surface and move short distances (Johnson 1991). Little is known of their social behaviour: they appear to be solitary, as permanent burrows where communication would be relatively easy are not formed (Johnson and Walton 1989).

## 9.3 Reproductive behaviour

There is an extraordinary uniformity in the behaviour of dasyurids (Croft 1982; Tyndale-Biscoe and Renfree 1987; Wolfe *et al.* 2000). The female may show agonistic responses to the approach of the male, who chases her for a long time in the period leading up to oestrus. There can be significant injuries to the female in captivity if she cannot escape. During oestrus the female will show behavioural signs that she is receptive to males, including a loss of appetite and allowing the male to approach and investigate her mouth and genital region and groom her flank. The male then grasps the female by the scruff of the neck with his mouth and clasps her abdomen, which results in intromission (which can last 1-12 h or more) if not resisted. Females can actively pursue males of interest, approaching them and displaying mating postures (e.g. in the agile antechinus). If in a group, a dominant female may actively chase another female away from a male (Parrott *et al.* 2015).

Because behavioural oestrus of some species can be observed, this can potentially be used instead of collecting urine samples (which provides information before the onset of behavioural oestrus; P. Woolley pers. comm.) to determine oestrus. The kowari, for example, shows behavioural signs of oestrus, as the females normally does not tolerate males (although they will tolerate other females) and begins to tolerate them only when in oestrus (Ganslosser and Meissner 1984). The male sniffs the female's head, fur and anal region and the female will even approach the male and lift her tail to allow anogenital sniffing, which may involve the male lifting the female's hindquarters off the ground. During this sniffing, the female performs a head bobbing behaviour (Ganslosser and Meissner 1984).

Male antechinuses increase their activity greatly during the breeding season by being active for longer periods and travelling much further searching for females (Lee *et al.* 1977). Although generally nocturnal, they become increasingly active during the day in search of females. Males have been observed to mate for long periods in captivity; bouts usually last for 5–6 h and can continue for 12 h (Marlow 1961; Woolley 1966). In captivity the females often lose hair on the back of their neck, and on their hips, due to repeated matings, whereas in the wild this normally does not occur to the same extent except in high-density populations, which indicates a lower frequency of copulations in the wild (Lee *et al.* 1977).

At the onset of breeding activity, and particularly during oestrus, both males and females of many small species of dasyurids make characteristic calls. This is the best time to introduce the male in with the female (Croft 1982; Collins *et al.* 1993). Species in which the females call to attract mates include:

- spotted-tailed quolls and western quolls ('clucking' sound) (Settle 1978; Collins *et al.* 1993; Gaikhorst 1999)
- Tasmanian devils: when allowed contact through a fence (wire or even semi-solid palings) will solicit to

males using a growling whine, sometimes quite high pitched, and short, soft barks. The male is generally silent or responds with snorts

- ningauis: mate-seeking calls (from one-to-many syllables) by both males and females (Fanning 1982)
- common dunnarts, red-cheeked dunnarts and lesser hairy-footed dunnarts (using a repetitive 'chit-chit-chit-chit-chit') (Fox and Whitford 1982; Whitford *et al.* 1982).
- Giles' planigale (soft clucking that seems to develop into a high-pitched clicking like that of insectivorous bats) (Whitford *et al.* 1982; Read 1984a)
- narrow-nosed planigale (makes a 'tsst tsst' only when deprived of males) (Read 1984a)
- common planigale (wheezy loud 'tsz tsz') (Van Dyck 1979)
- red-tailed phascogales: make a chit-chit-chit-chit-chit' noise when introduced at the start of the breeding season (H. Stannard pers. obs.)
- brush-tailed phascogales: appear to call males using a chirp call (Soderquist and Ealey 1994).

Oestrus in Tasmanian devils coincides with decreased appetite, development of a loose (fluid-filled) roll on the back of the neck, an increase in nesting behaviour, lethargy, changes in pouch condition and being more submissive to the male (whereas for the rest of the year she may be dominant to the male) (Kelly 1993; Slater 1993; Smith 1993; Hesterman *et al.* 2008). Males can show increased activity and sniffing and searching behaviours if in olfactory contact with a female nearby. They may dig at the wall of an adjacent female's enclosure to try to reach the female. During mating the male may grip the loose skin of the neck roll with his teeth and drag or escort her around the enclosure until he finds an appropriate den in which to mate (Kelly 1993; Slater 1993). Once mating has commenced the male guards the female and may continue to copulate with her for up to 10 days (Kelly 1993).

### 9.3.1 Mate choice

Allowing a female her choice of mate can significantly increase breeding success in captive marsupial breeding colonies. Studies of female mate choice in the agile antechinus showed it is an important factor in the mating system. In captivity the females choose males that are genetically dissimilar to themselves via olfactory cues (Parrott *et al.* 2006; 2007b; 2015) and male siring success is significantly influenced by genetic relatedness between mates in wild populations (M. Parrott unpubl. data). Male body size did not affect female mate choice, but in the wild, larger males had greater siring success of litters, suggesting a level of reproductive coercion (Kraaijeveld-Smit *et al.* 2002; Holleley *et al.* 2006; Parrott *et al.* 2006; M. Parrott unpubl. data). Females will actively choose males and approach them to breed, but avoid males that are very vocal or aggressive. Males can also show mate choice, choosing to mate with some females multiple times, though, on rare occasions, rejecting other females (Parrott *et al.* 2015).

In wild spotted-tailed quolls, the sires of litters are less related to dams than expected by chance, suggesting that female mate choice based on genetic dissimilarity may occur. Further, litters with multiple paternity show lower levels of relatedness than litters sired by a single male (Glen *et al.* 2009). Successful males had a higher bodyweight than males that were not assigned paternity in known litters of young.

Mate choice trials with Tasmanian devils suggest that females show a distinct preference for one male over another. This suggests that in every species of dasyurid a female may not like an individual male that has been put in her enclosure. Therefore, providing her with a choice may be desirable, perhaps allowing her to contact males through a fence or partition first. This technique maximises the likelihood of successful breeding.

The application of female mate choice using male scent in a long-term breeding colony of the stripe-faced dunnart significantly increased breeding success, with more than double the number of matings (75% *v.* < 35% of females with sperm detected in their urine following mating) and pregnancies (72% *v.* 32%) compared with pairing based on pedigree alone (Parrott *et al.* 2019a). Females paired with their chosen male also became pregnant significantly more quickly than females paired based on pedigree. However, pairing with a chosen male did not affect litter size, which was high in both mate choice and pedigree pairings. The age of the male (using 2-year-old males compared with their 1-year-old sons) did not influence female mate choice. The same techniques elicited similar behaviours and were successful in breeding sandhill dunnart, red-tailed phascogale and kultarr (M. Parrott unpubl. data). Allowing mate choice in the fat-tailed dunnart did not increase the number of pregnancies or litter size, but female mate choice influenced the litter sex ratio, with females paired with their preferred mate having more daughters. Additionally, females paired with a male in better body condition produced more sons (Gill 2019). Although genetic dissimilarity was a predictor of mate choice in the smaller carnivorous marsupials (e.g. antechinuses), studies of Tasmanian devil mate choice and genes of the major histocompatibility complex (MHC) showed that MHC-based heterozygosity does not influence reproductive outcomes in captive females or males, or in wild female Tasmanian devils (Day *et al.* 2019).

Studies of captive fat-tailed dunnarts have revealed that female mate choice influences the offspring sex ratio, but not fecundity (Gill *et al.* 2023). The study by Gill *et al.* (2023) revealed that although females spent some 67% of their time with a preferred male's scent, the reproductive success was not influenced by assigned mate choice. The results revealed that the sex ratio of captive populations of fat-tailed dunnarts may be manipulated to produce more female offspring by pairing them with their preferred males, whereas more male offspring can be produced by pairing females with males in better body condition (Gill *et al.* 2023).

Mate choice behaviours, including sniffing, licking and pushing scent-marked objects (see below) and grooming of the face and genitals are comparable between a variety of species, including the agile antechinus, stripe-faced dunnart, sandhill dunnart, red-tailed phascogale and kultarr (as well as non-carnivorous marsupials, such as the mountain pygmy-possum; Parrott *et al.* 2015, 2019a; M. Parrott unpubl. data). Techniques for mate choice can be easily modified to suit different species (see Parrott *et al.* 2019a and Hartnett *et al.* 2018 for techniques). However, care must be taken to maintain appropriate genetic diversity in captive populations as free female mate choice (where females may choose the same restricted number of males from the suite of available males) may result in lower retention of genetic diversity of the population (e.g. in group-housed devils – see Gooley *et al.* 2018). To avoid this, females could be offered a small number of potential mates to choose from (e.g. based on mean kinship/pedigree or genetic analyses) from the pool of possible males to ensure appropriate pairings and sire contributions are maintained (see Parrott *et al.* 2019a; Hartnett *et al.* 2018).

### 9.3.2 Scent marking

Many species of carnivorous marsupial scent mark their enclosures using urine, faeces and facial, sternal and paracloacal glands (Toftegaard and Bradley 2003; Parrott *et al.* 2007b, 2019a; J. Old pers. obs.). Tasmanian devils and spotted-tailed quolls use latrine sites to communicate both in the wild and in captivity. Faeces of males and females can be moved between enclosures before pairings to build familiarity and interest between pairs. During collections of scent for stripe-faced dunnart mate choice trials, males chose hard-edged and vertical objects over rounded and horizontal objects for scent marking, with new items all marked on the first night (Parrott *et al.* 2019a). Females accepted and reacted strongly to objects during mate choice trials (see section 9.3.1). In wild brush-tailed phascogales, scent experiments suggested that sternal gland marking by males provided intersexual communication and may serve to familiarise females with future mates (Soderquist and Ealey 1994).

Experiments freezing scent-marked objects for 7, 10, 25 and 40 days showed that freezing the male scent does not affect a female's ability to choose mates and that scents can be stored for up to 40 days without significantly affecting the scent. This may be a powerful tool when determining mate preference before animals travel between captive breeding institutes (Parrott *et al.* 2019a).

## 9.4 Bathing

Tasmanian devils will often bathe their feet or even swim in water during warm weather, so a large body of water should be provided in the enclosure to allow them to at least submerge their legs up to the knees.

Smaller species, including fat-tailed dunnarts (Ewer 1968a), ningauis (Fanning 1982), mulgaras and kowari (Aslin 1974), often bathe in fine sand to maintain the condition of their fur, so either the whole substrate or a sand bath of fine sand should be provided (Booth 1994).

## 9.5 Behavioural problems

Most species of dasyurids can suffer from abnormal repetitive behaviour that can be very difficult to stop once established. Dibblers can pace and continuously jump from their nest box to the lid of their enclosure and back down (Lambert 2000). Pacing has also been observed in many species, including fat-tailed dunnarts that continually run up and down alongside the enclosure wall (Ewer 1968a), and can range from wanting to feed and waiting for the animal caretaker to walk by to feed them (e.g. Tasmanian devils) or pacing, apparently due to looking for a mate (e.g. male dunnarts). Male red-tailed phascogales have shown abnormal repetitive behaviour by perpetual jumping from branch to branch in a uniform pattern when housed singularly. Ensuring that branches are replaced and moved in different configurations regularly, and enclosure mates are present, all but removes this activity (J. Old pers. obs.).

## 9.6 Signs of stress

Signs of short-term and chronic stress in dasyurids include increased irritability (e.g. vocalising, fighting or attacking more), listlessness and signs of depression; rough fur and alopecia and abnormal repetitive behaviour (Spielman 1994). It is important to remember that semelparous marsupials show many signs associated with stress (e.g. alopecia) as part of their natural ageing process.

## 9.7 Behavioural enrichment

Several steps can be taken to stimulate individual animals and minimise behavioural problems such as repetitive behaviour becoming established. Providing different food and non-food-based enrichments elicits different behaviours, including foraging, climbing, digging, pulling, sniffing and increased activity levels (O'Neil 2011). The frequency with which items are provided also affects the levels of interest and behaviours; thus a calendar of enrichment can be followed to vary the types of enrichment items being provided. Enrichment items include the following.

- Making the enclosure surface as variable as possible by changing the soil profile and adding hollow logs and branches to climb through and up; this may include obstructing the chosen abnormal behaviour path with obstacles.
- Scattering or hiding small pieces of food so that it takes them time to hunt for it. Take care to ensure animals are not overfed by adding in too many 'treat' items or additional food items. All food items provided need to be calculated into the daily diet formulation to prevent animals from overeating and obesity.
- Feeding at different times of the day so the animals do not time their activity with feeding time.
- Raw bones for quolls and Tasmanian devils to chew on can provide significant enrichment as the animals gain access to bone marrow and also keep their teeth clean.
- Providing live insects and other food types (e.g. fish), and hiding them within the enclosure, helps to promote hunting behaviour and reduce abnormal repetitive behaviours.
- Using passive insect feeders and suitable logs as temporary refuge for the insects (Hawkins 1998).
- Varying the way food is fed, such as feeding a lot of food for large species (e.g. Tasmanian devils) to let them gorge and then not feeding them for several days, rather than feeding the same types of food each day.
- Use of a 'bungee feeder' for Tasmanian devils (and quolls), which is a heavy-duty, elasticised cord with a brass clip on either end. One end is attached to a suitable fixture point within the enclosure, while the carcass is attached to the other end. This provides resistance and exercise when pulled (Schaap 2002). Large carcasses can also be pegged to the ground to increase enrichment and elicit natural pulling behaviours.
- Freezing blocks of blood or food items such as rats, rabbits or chickens inside blocks of ice to stimulate olfactory and taste senses (Markus 1996; Schaap 2002).
- Providing running wheels for some small species (e.g. Martin 1965; Nelson and Smith 1971; Breckon and Hulse 1972; Woolley 1993), which helps them to remain fit and decreases the likelihood of obesity. However, these need to be regularly taken out of the enclosure as they can lead to abnormal repetitive behaviours if left in all the time (J. Old and H. Stannard pers. obs.).
- Using various scents, including prey faeces such as wombat, possum or kangaroo; dragging food items around the enclosure and hiding them; or other novel smells such as spices or aromatherapy oils.
- Varying the diet throughout the week (see section 6.1).
- Planting various species of plants such as grass tussocks and bushes.

## 9.8 Training and conditioning

Captive-bred northern quolls have been fed 'toad sausages' before release to train them to avoid eating wild cane toads (*Rhinella marina*). The animals that consumed a single toad sausage showed reduced interest in cane toads and showed reduced attack behaviour (Indigo *et al.* 2018).

A parent-raised spotted-tailed quoll that was aggressive towards animal caretakers was trained to station on a log within its enclosure after three knocks on the door. This animal also allowed a harness to be attached to it, leave the enclosure and to climb onto the shoulder of an animal caretaker for talks to the public (Costelloe 2014).

## 9.9 Introductions and removals

Males should ideally be introduced into the female's enclosure (Woolley 1990b) as it results in lower aggression and increased pregnancies. Smaller dasyurids, particularly those in holding enclosures with a partition in the middle, can be introduced after first having had visual, auditory and olfactory communication with the potential mate and can be watched through the glass front or mesh sides of the enclosure (see section 4.2). Allowing familiarity between mates can significantly decrease aggression and instances of minor injury (such as fur loss and bites to the tail) from 44% in unfamiliar stripe-faced dunnart pairings to 0% when dunnarts are familiarised through enclosure mesh walls (Parrott *et al.* 2019a).

Male spotted-tailed quolls and northern quolls have been known to kill and partially consume the intended mate during breeding encounters, so it may be advisable to increase the food by up to 20% just before introductions and monitor them closely (Collins *et al.* 1993; Hellingham 1999). Male quolls should be introduced (under supervision) just before the breeding season, particularly once the females start calling. If the female appears uninterested, the male can be moved to the next prospective mate's enclosure and the process repeated until a receptive female is found (Collins *et al.* 1993). This procedure can be made easier by placing the male quoll in a portable cat cage and placing it in the female's enclosure. If the female starts to show interest by making clucking vocalisations then the male is released; otherwise the male is moved to another female's enclosure (Collins *et al.* 1993). As females enter oestrus for only 3–4 days (which may occur only once or be repeated several times, depending on the species) during the breeding season, it is very important if using this technique to rotate the males every 1–2 days to maximise the opportunity of successful breeding (Collins *et al.* 1993; Edgar and Belcher 1995).

Tasmanian devils should ideally be introduced to each other only after they have had a period of olfactory and visual contact, and once introduced should be carefully monitored for at least several days afterward (Slater 1993). Great care must be taken during introductions as incompatible animals (which may be because the female is not in oestrus) can become highly aggressive towards each other and fighting can result in serious injury or death (Slater 1993; S. Jackson pers. obs.). Mate choice trials suggest the animals must be familiar with each other by spending at least 1 or 2 days together before the breeding season starts. This suggests that females do not severely injure males that they know, even if they are not in oestrus, because they have a pre-established dominance hierarchy. Familiarity between devils can be increased by moving faeces between prospective mates' enclosures before pairing.

Red-tailed phascogale females should be moved to separate enclosures when young appear in the pouch, as young can easily be mis-mothered/swapped if multiple females are sharing nest boxes with young that have recently detached from the teat (J. Old pers. obs.).

## 9.10 Intraspecific compatibility

Although not much is known of the social behaviour of many of the carnivorous marsupials, they are generally not very social species and usually only come together during mating, after which the males seek further mates. The exception is species such as the agile

**Table 4.11.** Nesting behaviour of carnivorous marsupials and the suggested sex ratio of different genera when held in captivity except when breeding.

| Genus | Nesting behaviour | Suggested sex ratio |
|---|---|---|
| **Dasyuridae** | | |
| *Dasycercus* | Solitary | 1:1 |
| *Dasyurus* | Solitary | 1:1 |
| *Parantechinus/ Pseudantechinus* | Solitary | 1:1 or female pairs |
| *Sarcophilus* | Solitary | 1:1 |
| *Antechinus* | Groups* | 1:1–3:3 |
| *Phascogale – P. tapoatafa* | Solitary | 1:1 or solitary |
| *Phascogale – P. calura* | Solitary–Groups | 1:6 |
| *Planigale* | Solitary | 1:1 |
| *Ningaui* | Solitary? | Solitary – 1:1? |
| *Sminthopsis* | Solitary? Groups – non-breeding | 1:1–10:10 |
| *Antechinomys* | Solitary | 1:1 |
| **Notoryctidae** | | |
| *Notoryctes* | Solitary | 1:1 – solitary? |

* After the breeding season all males die and females then nest solitarily.

antechinus, which nest communally before the breeding season, and some species of dunnart, which can huddle to increase heat and conserve energy. Therefore, in most cases they should be held in pairs or, ideally, the males should be introduced to the females shortly before they are in oestrus. An outline of the number of individuals that are known to nest together and suggested sex ratios in captivity can be found in Table 4.11.

Western Sydney University successfully bred red-tailed phascogales by keeping one male with up to six females during the breeding season. When the females became aggressive towards the male he was removed from the enclosure. When the females had pouch young, they were also moved to individual enclosures to reduce the chances of mis-mothering. During the non-breeding season female red-tailed phascogales were held together without any problems. Juvenile males and adult males could also be held together if they were siblings and grew up in the one enclosure together. Male adult red-tailed phascogales were never put together if they were not siblings or siblings that been separated at any time from one another. One solitary male showed abnormal repetitive behaviours; hence suggest housing as mixed groups or sibling pairings. Some males were housed singularly, but were able to smell, see and hear other males and females in adjacent enclosures (J. Old pers. obs.).

## 9.11 Interspecific compatibility

Dasyurids are generally incompatible with most species due to their carnivorous nature, which results in them eating most things their own size or smaller, as well as their ability to escape and the difficulty of properly managing them in large enclosures. Species such as quolls and Tasmanian devils are not recommended to be held with any other species. Species of dasyurids that have been held successfully with other animals are the red-tailed phascogale and brush-tailed phascogale, which have been held with larger ground-dwelling mammals including long-nosed potoroos (*Potorous tridactylus*), long-footed potoroos (*Potorous longipes*) and rufous bettongs (*Aepyprymnus rufescens*) (Halley 1992; M. Parrott pers. obs.).

# 10. BREEDING

## 10.1 Mating system

All dasyurids are polygamous, although they may use different reproductive strategies including polygyny, polyandry or promiscuity. Many species can also store sperm, with sperm competition known to occur in several species, so several males can father individuals from the same litter (Taggart and Temple-Smith 1991; Millis *et al.* 1999; Shimmin *et al.* 1999, 2000; Taggart *et al.* 2003; Foster 2008; Parrott *et al.* 2015).

Some dasyurids, including antechinuses, are promiscuous, with females actively choosing to mate with multiple males (Parrott *et al.* 2015). This may have benefits including: (1) allowing females to 'trade up' to obtain better genes for offspring; (2) promoting sperm competition/sperm choice to obtain better genes; (3) enabling females to avoid genetically incompatible gene combinations such as with inbreeding; (4) enabling females to avoid infertility, and (5) increasing survival of the young (Fisher *et al.* 2006a, 2006b).

### 10.1.1 Reproductive strategies

Within the dasyurids, there are at least six different reproductive strategies used (Table 4.12) (Lee *et al.* 1982; Lee and Cockburn 1985).

Strategy 1: The female is mono-oestrous (one oestrus period per year) and there is an abrupt death of males at

**Table 4.12.** Key factors in the classification of dasyurid life histories.

Taken from Lee *et al.* (1982), Lee and Cockburn (1985) and Krajewski *et al.* (2000).

| Strategy | Oestrus pattern | Seasons per male | Duration of mating period | Seasonality of breeding | Age of sexual maturity (months) |
|---|---|---|---|---|---|
| 1 | Mono-oestrus | Annual | Restricted | Seasonal | 11 |
| 2 | Mono-oestrus | Perennial | Restricted | Seasonal | 11 |
| 3 | Mono-oestrus | Perennial | Restricted | Seasonal | 11 |
| 4 | Polyoestrus | Perennial | Extended | Seasonal | 6 |
| 5 | Polyoestrus | Perennial | Extended | Seasonal | 8–11 |
| 6 | Polyoestrus | Perennial | Extended | Aseasonal | 8–11 |

the conclusion of the mating period, which is highly synchronised and shorter than the gestation period (semelparity). The mortality or die-off of antechinuses is because of the effect of high free testosterone and cortisol levels on many organ systems (Naylor *et al.* 2008). Another example is the males of the mainland dusky antechinus that begin to die at the time of ovulation and complete die-off within 5–10 days. This occurrence in dusky antechinus is very consistent within populations and between years (e.g. 20–30 July in the Otway Ranges in Victoria) (McDonald *et al.* 1981). Births occur 2–3 weeks after the male die-off. This semelparous strategy is used by the antechinuses, phascogales, kaluta and some populations of dibblers and northern quolls (Table 4.13). However, weather and resource availability can affect timing in several species. In the brush-tailed phascogales, births can occur earlier in a year with good rainfall, but it has been suggested that drought causes a delay in births potentially due to an increased period of sperm storage (Rhind 2002). In the agile antechinus, during an extreme drought year, young were left in nests at least 2 weeks later than in the previous non-drought year, suggesting that oestrus, ovulation or birth dates may have been 2 weeks later, or lactation 2 weeks longer, in the drought year (Parrott *et al.* 2007a).

Strategies 2 and 3: Can be mono-oestrous or if poly-oestrous reproduce synchronously and once yearly and so behave as if mono-oestrous (facultative mono-oestrous). In these species, some individuals of both sexes survive to reproduce in at least two years and include both large and small dasyurids (12–7000 g). Strategy 2, known as iteroparity, is similar to Strategy 1 except that males may survive beyond the end of the mating season and species include dibblers, pseudantechinuses, western quoll and

**Table 4.13.** Life history traits of dasyurid marsupials.

Derived from Lee *et al.* (1982) and Krajewski *et al.* (2000). *Some populations have different strategies.

| Strategy | Species | Oestrus | Seasons per male | Duration of breeding | Seasonality | Maturity (months) | Reference |
|---|---|---|---|---|---|---|---|
| 1 | *Dasykaluta rosamondae* | Mono | 1 | Restricted | Seasonal | 11 | 1, 2 |
| | *Dasyurus hallucatus** | Mono | 1 | Restricted | Seasonal | 11 | 3 |
| | *Parantechinus apicalis** | Mono | 1 | Restricted | Seasonal | 11 | 3 |
| | *Antechinus agilis* | Mono | 1 | Restricted | Seasonal | 11 | 4–6 |
| | *A. bellus* | Mono | 1 | Restricted | Seasonal | 11 | 7, 8 |
| | *A. flavipes* | Mono | 1 | Restricted | Seasonal | 11 | 4, 9 |
| | *A. godmani* | Mono | 1 | Restricted | Seasonal | 11 | 5, 10 |
| | *A. leo* | Mono | 1 | Restricted | Seasonal | 11 | 11, 12 |
| | *A. minimus* | Mono | 1 | Restricted | Seasonal | 11 | 5, 13 |
| | *A. stuartii* | Mono | 1 | Restricted | Seasonal | 11 | 5 |
| | *A. mimetes* | Mono | 1 | Restricted | Seasonal | 11 | 5, 14 |
| | *A. subtropicus* | Mono | 1 | Restricted | Seasonal | 11 | 15, 16 |
| | *Phascogale calura* | Mono | 1 | Restricted | Seasonal | 11 | 17 |
| | *P. tapoatafa* | Mono | 1 | Restricted | Seasonal | 11 | 18–20 |
| 2 | *Dasyurus hallucatus** | Mono | > 1 | Restricted | Seasonal | 11 | 21, 22 |
| | *Parantechinus apicalis** | Mono | > 1 | Restricted | Seasonal | 11 | 23 |
| | *Pseudantechinus bilarni* | Mono | > 1 | Restricted | Seasonal | 11 | 7, 24 |
| | *P. macdonnellensis* | Mono | > 1 | Restricted | Seasonal | 11 | 14, 25 |
| | *P. ningbing* | Mono | > 1 | Restricted | Seasonal | 11 | 26 |
| | *P. woolleyae* | Mono | > 1 | Restricted | Seasonal | 11 | 27 |
| | *Sarcophilus harrisii* | Fac. poly | > 1 | Restricted | Seasonal | 12 | 28–33 |
| | *Sminthopsis fuliginosa* | Mono | > 1 | Restricted | Seasonal | 11 | 34 |
| 3 | *Dasycercus blythi/hillieri* | Fac. poly | > 1 | Restricted | Seasonal | 11 | 35–39 |

| Strategy | Species | Oestrus | Seasons per male | Duration of breeding | Seasonality | Maturity (months) | Reference |
|---|---|---|---|---|---|---|---|
| | *Dasyurus geoffroii* | Fac. poly | > 1 | Restricted | Seasonal | 11 | 40–42 |
| | *D. maculatus* | Fac. poly | > 1 | Restricted | Seasonal | 11 | 43, 44 |
| | *D. viverrinus* | Fac. poly | > 1 | Restricted | Seasonal | 11 | 29, 45–48 |
| | *Sminthopsis leucopus* | Fac. poly | > 1 | Restricted | Seasonal | 11 | 49–51 |
| 4 | *S. bindi* | ? | ? | Extended | ? | ? | 6 |
| | *S. crassicaudata* | Poly | > 1 | Extended | Seasonal | 6 | 14, 52, 53 |
| | *S. dolichura* | Poly | > 1 | Extended | Seasonal | 6 | 54 |
| | *S. douglasi* | Poly | > 1 | Extended | Seasonal | 6 | 6, 27 |
| | *S. gilberti* | Poly | > 1 | Extended | Seasonal | ? | 6, 27 |
| | *S. macroura* | Poly | > 1 | Extended | Seasonal | 6 | 55 |
| | *S. murina* | Poly | > 1 | Extended | Seasonal | 6 | 56 |
| | *S. v. nitela* | Poly | > 1 | Extended | Seasonal | 6 | 6, 57 |
| 5 | *Dasyuroides byrnei* | Poly | > 1 | Extended | Seasonal | 8-11 | 14, 58–60 |
| | *Planigale gilesi* | Poly | > 1 | Extended | Seasonal | 8-11 | 61–63 |
| | *P. ingrami* | Poly | > 1 | Extended | Seasonal | 8-11 | 27, 64–67 |
| | *P. m. maculata* | Poly | > 1 | Extended | Seasonal | 8-11 | 67–70 |
| | *P. tenuirostris* | Poly | > 1 | Extended | Seasonal | 8-11 | 61, 63 |
| | *Ningaui ridei* | Poly | > 1 | Extended | Seasonal | 8-11 | 38, 71, 72 |
| | *N. timealeyi* | Poly | > 1 | Extended | Seasonal | 8-11 | 6, 71 |
| | *N. yvonneae* | Poly | > 1 | Extended | Seasonal | 8-11 | 6, 27 |
| | *Antechinomys laniger* | Poly | > 1 | Extended | Seasonal | 8-11 | 5, 14, 73, 74 |
| | *S. ooldea* | Poly | > 1 | Extended | Seasonal | 8-11 | 74 |
| | *S. longicaudata* | Poly | > 1 | Extended | Seasonal | 8-11 | 75 |
| | *S. psammophila* | ? | ? | Extended | ? | ? | 6 |
| | *S. youngsoni* | ? | ? | Extended | Seasonal | ? | 6, 39 |
| | *Planigale m. sinualis* | Poly | > 1 | Extended | Aseasonal | ? | 76, 77 |
| | *Sminthopsis v. virginiae* | Poly | ? | Extended | Aseasonal | 6 | 6, 78 |

References: (1) Woolley (1982); (2) Woolley (1991a); (3) Dickman and Braithwaite (1992); (4) Woolley (1966); (5) Lee *et al.* (1982); (6) Baker and Gynther (2023); (7) Calaby and Taylor (1981); (8) Woolley (1981); (9) Fleay (1949); (10) Van Dyck (1982); (11) Van Dyck (1980); (12) Leung (1999); (13) Wainer (1976); (14) Woolley (1973); (15) Wood (1970); (16) C. Dickman pers. obs.; (17) Bradley (1997); (18) Fleay (1934); (19) Cuttle (1982b); (20) Soderquist (1993a); (21) Fleay (1962); (22) Begg (1981a); (23) Woolley (1971b); (24) Begg (1981b); (25) Woolley (1991b); (26) Woolley (1988); (27) Krajewski *et al.* (2000); 28) Fleay (1935a); (29) Green (1967); (30) Guiler (1970); (31) Buchmann and Guiler (1977); (32) Hughes (1982); (33) Jones *et al.* (2008); (34) Crowther *et al.* (1999); (35) Wood Jones (1923); (36) Fleay (1961); (37) Michener (1969); (38) Woolley (1971a); (39) Dickman *et al.* (2001); (40) Arnold and Shield (1970); (41) Archer (1974b); (42) Arnold (1976); (43) Fleay (1940); (44) Settle (1978); (45) Hill and O'Donoghue (1913); (46) Hill and Hill (1955); (47) Fletcher (1977); (48) Godsell (1982b); (49) Read *et al.* (1983); (50) Woolley and Ahern (1983); (51) Woolley and Gilfillan (1990); 52) Godfrey and Crowcroft (1971); (53) Morton (1978b); (54) Friend *et al.* (1997); (55) Godfrey (1969a); (56) Fox and Whitford (1982); (57) Morton *et al.* (1987); 58) Mack (1961); (59) Aslin (1974); (60) Hutson (1976); (61) Denny (1982); (62) Whitford *et al.* (1982); (63) Read (1984a); (64) Fleay (1965); (65) Heinsohn (1970); (66) Woolley (1974); (67) Archer (1976); (68) Morrison (1975); (69) Van Dyck (1979); (70) Denny *et al.* (1979); (71) Fanning (1982); (72) Kitchener *et al.* (1986); (73) Woolley (1984); 74) Aslin (1983); (75) Woolley and Valente (1986); (76) Davies (1960); (77) Taylor *et al.* (1982); (78) Taplin (1980).

some populations of the northern quoll. Recent research on dibblers suggests that they may warrant inclusion in a new category of life history for species experiencing facultative male die-off (Mills and Bencini 2000). Strategy 3 is similar to Strategy 2 except that females are facultatively polyoestrus and can undergo a second oestrous if unmated or if they have prematurely lost the first litter. Species include eastern quolls, spotted-tailed quolls, mulgaras, and the white-footed dunnart.

Strategies 4, 5 and 6: These species are polyoestrous and usually produce at least two litters in a breeding season. In Strategy 4 species, the females are polyoestrus, the males are perennial, the breeding seasons are extended and maturity is attained in 6 months or less. Strategy 5 is similar to Strategy 4 except that maturity occurs at 8–11 months. Examples of Strategy 4 species include the fat-tailed dunnart, striped-faced dunnart and common dunnart, with Strategy 5 species including ningauis, common dunnarts, planigales, kultarr and the kowari.

Strategy 6 is similar Strategy 5 except that breeding is aseasonal with year-round reproduction. Species that use this strategy include northern populations of the common planigale and the New Guinean habbema dasyure (*Murexia habbema*) and long-nosed dasyure (*Murexia naso*).

The breeding of some species is known to shift slightly in captivity. For example, observations indicate that the breeding season of the red-tailed phascogale has shifted to earlier in the season (May/June rather than July). Therefore, there is a need to monitor signs of oestrus earlier in captivity as the season may shift, even under natural photocycles (Foster *et al.* 2006; Old *et al.* 2021).

Beyond the dasyurids, other reproductive strategies may be expected in the marsupial moles and perhaps also the thylacine. The marsupial moles have two nipples and have been reported to carry 1–2 pouch young, a litter size that is smaller than that of any of the dasyurids. The life history of the thylacine is poorly known, but its larger size than any extant dasyurid and placement in its own family make it difficult to commit with any confidence to any of the six dasyurid reproductive strategies.

### 10.1.2 Sperm competition and multiple paternity

Sperm competition and multiple paternity occur in some genera of dasyurids. The potential of species for sperm competition and multiple paternity in litters can be estimated by examining the relative size of the testis to the bodyweight, as those with large testis compared with their body size are likely to have higher sperm competition (Dickman 1993; Taggart *et al.* 1998). The genera most likely to use sperm competition include *Antechinus*, *Phascogale*, some *Sminthopsis*, *Parantechinus*, *Pseudantechinus* and *Sarcophilus* (Millis 1995; Shimmin *et al.* 2000; Kraaijeveld-Smit *et al.* 2002; Foster *et al.* 2008; Sale *et al.* 2013). This has implications for captive-breeding programs, as it can increase the genetic diversity of the population, although it also means that the exact paternity is not known unless paternity tests are undertaken. One method used with good success with brush-tailed phascogales is rotation of males through the females' enclosures twice per week during the breeding season (Halley 1992). Male stripe-faced dunnarts have also been rotated through females' enclosures, with three males provided sequentially over the three nights of the female's receptive period, with high breeding success (Parrott *et al.* 2019a).

## 10.2 Ease of breeding

A detailed knowledge of the life-cycle of most or all species is essential if successful breeding groups are to be maintained (Williams 1990). Most species are kept as pairs until young are born and pairing of animals is often required until a compatible pair is found (Williams 1990). Males may also be cycled through the females' enclosures and removed following ovulation, if cycles are well monitored (e.g. in the stripe-faced dunnart). Good breeding success has been achieved when the male is introduced when the female is in oestrus, which is determined by the presence of cornified epithelial cells in urine (see section 10.5).

Many of the smaller species of dasyurid are relatively easy to breed if they are housed with access to natural light cycles (see section 4.3). To date, several species have bred and reared young successfully in captivity, including the brush-tailed mulgaras, ampurtas, kaluta, kowari, yellow-footed antechinus, agile antechinus, brown antechinus, mainland dusky antechinus, brush-tailed phascogale, red-tailed phascogale, western quoll, eastern quoll, spotted-tailed quoll, dibbler, fat-tailed pseudantechinus, Tasmanian devil, fat-tailed dunnart, Julia Creek dunnart, stripe-faced dunnart and kultarr (Woolley 1982; Dickman 1985; Bennett *et al.* 1990; Gaikhorst 1999; Lambert 2000; Foster *et al.* 2008; Stannard and Old 2010; Parrott *et al.* 2015; Old *et al.* 2021).

The species that have bred with the greatest success and are considered the most suitable for establishing long-term colonies include kowari, kaluta, yellow-footed antechinus, fat-tailed dunnart, stripe-faced dunnart, brush-tailed phascogale, red-tailed phascogale, Tasmanian devil, eastern quoll and western quoll (Aslin 1980; Woolley 1982; Halley 1992; Gaikhorst 1999; Selwood and Cui 2006; Old *et al.* 2021). The routine introduction of wild-caught animals is typically required every 4–5 years to maintain genetic diversity and reproductive output (Selwood and Cui 2006).

Although many species of dasyurid have bred successfully in captivity, the long-term viability of some species has been problematic due to irregular breeding. Observations of some carnivorous marsupials suggest that wild founders tend to breed well in captivity, but the subsequent captive progeny sometimes do not breed as well over the medium to long-term, resulting in the long-term decline of the population (Woolley 1971a; Aslin 1982). Examples of species that have proven difficult to breed consistently include dibbler, ningbing pseudantechinus, kultarr, agile antechinus, and white-footed dunnart (Aslin 1982; Woolley 1982).

Common planigales can breed rapidly in captivity initially; however, success appears to decline over successive generations. Animals more than two generations from the wild generally do not breed (Aslin 1982). Similar observations have been made in early colonies of the stripe-faced dunnart, where a captive population produced 109 young in the first two generations and in the third breeding season the females had irregular oestrous cycles and in the few instances where copulation was recorded there was nearly 100% prenatal mortality. This population subsequently became extinct without the cause being definitely established (Godfrey 1969a). However, in later colonies of the stripe-faced dunnart, and following increased research, long-term colonies of over 20 years were successfully maintained with consistent breeding success (Au *et al.* 2010; Parrott *et al.* 2019a).

Some species, such as dibblers, breed well for 1–2 years, then the rate of reproduction decreases in the third year (Lambert and Mills 2006). The incidence of cannibalism in captive-bred dibblers can be high – they have been known to eat their entire litter (Lambert 2000). Therefore, it is important that captive-raised dibblers are exposed to minimum stress by not handling, creating visual barriers between adjacent enclosures and feeding ad lib pinkie rats and more invertebrates (Lambert 2000).

Even those species that have been bred successfully do not have a 100% success rate (i.e. every adult female in the population successfully breeds each year). Although based on small sample sizes, the success rate has been observed by Aslin (1980, 1982) to range from 14% and 30% for fat-tailed dunnarts/Ooldea dunnarts; 12.5–33% for fawn antechinus/yellow-footed antechinus; 50–83% for kowaris; and 50–78% for common planigales. In the stripe faced dunnart, long-term success rates of 35% breeding success per oestrous cycle were more than doubled by allowing females to choose their mates (see section 9.3.1). Therefore, to maximise the likelihood of long-term captive population viability it is important to hold several animals, rotate them in order to maximise the opportunities for breeding, provide mate choice where possible and ideally supplement the population with wild stock.

As early as 1915, Tasmanian devils were considered difficult to breed in captivity (Roberts 1915; Fleay 1935a, 1952). Since that time, they have been bred routinely in captivity; however, recent observations of this species suggest that 90% of breeding females were wild-caught, with most producing young from the first oestrous cycle of the season. In contrast, 74% of unsuccessfully bred females were predominantly captive born. Records of the overall breeding success rate of the mainland devil insurance population was 53% in 2009 and 40.5% in 2010, with the population steady across the insurance program (Keeley *et al.* 2012). However, the success of devil breeding often increases with the experience of the holding institution, with the breeding success of devils increasing from an estimated 19.8% in the first year of holding devils to 75.2% by the eighth year across all insurance population sites (Hogg *et al.* 2019b).

## 10.3 Oestrous cycle and gestation period

Oestrous cycles range widely and can vary from only 10 days for the dibbler to 60 days for the kowari (Table 4.14). The gestation period for the various species of carnivorous marsupials is short, like other marsupials and ranges from only 11–12 days in the common dunnart, stripe-faced dunnart and kultarr to more than 60 days in the kaluta (Table 4.14).

## 10.4 Reproductive status

### 10.4.1 Females

Carnivorous marsupials are generally placed in several categories depending on their reproductive status. The examination of reproductive status in small to medium-sized species can be facilitated by putting them inside a transparent plastic tube and examining the pouch with an otoscope (Roberts and Kohn 1991). Small species can also be pouch checked while in a bag and exposing the pouch or in the hand (Fig. 4.6). For females the categories are:

- non-parous (females that have never bred) – pouch small with no skin folds, clean and dry; teats very small, surrounded by fur
- parous (females that have bred previously but not presently) – pouch is small but distinct, dry and sometimes dirty; the teats are slightly elongated and obvious even if surrounded by fur
- pregnant – hair in pouch area becoming sparse, pouch pink in colour and glandular in appearance; skin folds and long white hairs may be observed on the lateral margins of the pouch
- pouch young present – attached to the teats

**Table 4.14.** Duration of oestrous cycle and gestation (days) of carnivorous marsupials for which it is known.

M, mono-oestrous.

| Species | Oestrous cycle | Duration of oestrus | Gestation | Diapause | Ovulation number | Reference |
|---|---|---|---|---|---|---|
| *Dasycercus hillieri* | – | – | 35–44 | N | – | 1 |
| *Dasyuroides byrnei* | 60 | – | 30–35 | N | 11–18 | 1-6 |
| *Dasykaluta rosamondae* | – | – | 38–62 | – | – | 7, 8 |
| *Dasyurus geoffroii* | – | – | 16–18 | N | 50 | 9, 10 |
| *D. maculatus* | 36–58 (49.5) | – | 20–21 | N | – | 11, 12 |
| *D. viverrinus* | 34–37 | 5 | 19–24 | N | 7-35 | 13–15 |
| *Parantechinus apicalis* | 10–20 | 1–3 | 41–48 | – | – | 16–18 |
| *Pseudantechinus bilarni* | – | – | c. 38 | – | 10 | 6, 19 |
| *P. ningbing* | – | – | 45–52 | – | 12–15 | 20 |
| *P. macdonnellensis* | – | – | 45–55 | – | 12 | 6, 7 |
| *Sarcophilus harrisii* | – | – | 19–23(21 ave) | N | 40–56 | 6, 12, 21, 22 |
| *Antechinus agilis* | M | 11 | 27 | – | ≥ 19 | 6, 12, 23–27 |
| *A. bellus* | M | – | – | – | 16 | 12 |
| *A. leo* | M | – | 31 | – | – | 28 |
| *A. mimetes* | M | – | 29–35 | – | – | 29 |
| *A. minimus* | M | – | 29–31 | – | – | 29 |
| *A. stuartii* | M | 4 | 26–31 | Y | 11–19 | 30,* 31 |
| *Phascogale tapoatafa* | 40 M | – | 24–33 | Y? | – | 32 |
| *P. calura* | – | – | 28–30 | ? | 10–13 | 33 |
| *P. gilesi* | 15–22 | 3–5 | 14.5–16.5 | N | – | 34, 35 |
| *P. maculata* | – | – | 19–20 | – | 17 | 6, 12 |
| *P. tenuirostris* | 31–34 | 1–2 | 18.5–19.5 | N | – | 35 |
| *Ningaui ridei* | – | – | 13.5–21.25 | N | – | 36 |
| *Antechinomys laniger* | – | – | 12 | – | – | 37, 38 |
| *Sminthopsis crassicaudata* | 31 | 1–3 | 13–16 | N | 14 | 6, 39–41 |
| *S. archeri* | 31–36 | – | 13–16 | – | – | 42, 43 |
| *S. douglasi* | 28 | – | – | – | – | 44 |
| *S. longicaudata* | 34.4 | – | 17–19 | N | – | 45 |
| *S macroura* | 23.25–26 | 1–2 | 9.5–11 | N | 20–40 | 6, 46–49 |
| *S. murina* | 24 | – | 10.5–13.5 | N | – | 50 |
| *S. virginiae* | 30 | – | 15 | – | – | 12 |

References: (1) Woolley (1971a); (2) Mack (1961); (3) Aslin (1980); (4) Close (1983); (5) Fletcher (1983); (6) Taggart *et al.* (2003); (7) Woolley (1988); 8) Woolley (1991b); 9) Morton *et al.* (1989); 10) Gaikhorst (1999); (11) Collins *et al.* (1993); (12) Baker and Gynther (2023); (13) Hill and O'Donoghue (1913); (14) Fletcher (1985); (15) Tyndale-Biscoe and Renfree (1987); (16) Woolley (1971b); (17) Woolley (1973); (18) Lambert (2000); (19) Woolley (1995); (20) Woolley (1991a); (21) Guiler (1970); (22) Hughes (1999b); (23) Taggart *et al.* (1999); (24) Selwood (1982a); (25) Selwood (1982b); (26) Selwood (1983); (27) Selwood (1985); (28) Leung (1999); (29) Williams and Williams (1982); (30) Wilson (1986); (31) Marlow (1961); 32) Millis *et al.* (1999); 33) Bradley (1997); (34) Whitford *et al.* (1982); (35) Read (1984a); (36) Fanning (1982); (37) Woolley (1984); (38) Stannard and Old (2010); (39) Smith and Godfrey (1970); (40) Godfrey and Crowcroft (1971); (41) Morton (1978a); (42) Woolley (2007); (43) Woolley (2015); (44) Pollock *et al.* (2010); (45) Woolley and Valente (1986); (46) Godfrey (1969a); (47) Woolley (1990b); (48) Taggart *et al.* (1997); (49) Selwood and Woolley (1991); (50) Fox and Whitford (1982).

*Animals referred to in this paper were named *A. flavipes* but were *A. stuartii* (Woolley 1966), those of Marlow were *A. stuartii*, and those of Selwood were *A. agilis*.

- lactating (young absent from the pouch but still suckling) – pouch area large, skin folds flaccid, hair sparse and stained, skin smooth and dark pink; teats elongated
- post lactation – teats expressing only clear liquid and/or regressing.

For larger species such as Tasmanian devils and spotted-tailed quolls, a description of the pouch condition can be found in Table 4.15.

If pouch young are present, several developmental stages and measurements can be recorded and compared to existing growth curves (see section 10.16) or

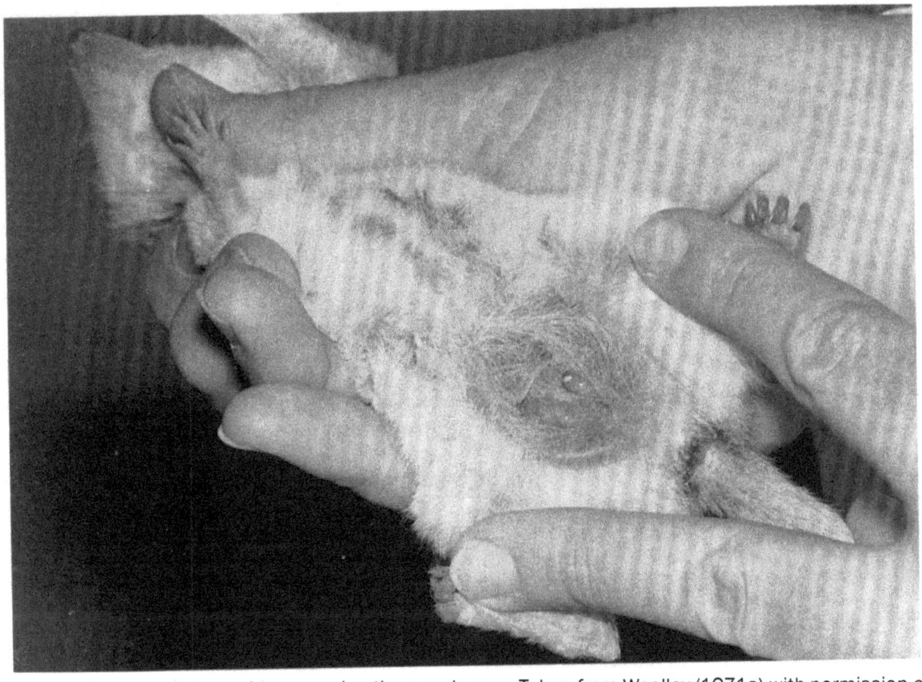

**Fig. 4.6.** Method for holding a small dasyurid to examine the pouch area. Taken from Woolley (1971a) with permission of the publisher. Reprinted with permission from John Wiley & Sons Inc.

**Table 4.15.** Pouch development in female Tasmanian devils and spotted-tailed quolls during the breeding season.

Taken from Hesterman *et al.* (2008).

| Pouch score/condition | Pouch description |
|---|---|
| 0 (immature) | Round shape. Pale, small and shallow |
| 1 (adult) | Oval shape. Pale and clean, teats may still be elongated if young only recently weaned; occasional yellow 'wax' |
| 2 | Skin flushes pink and interior secretes a sticky pink/red, grease-like substance |
| 3 | Skin very flushed, marked/copious volumes of thin, red oily secretions produced; area deepens slightly |
| 4 | Enlarged: interior very deep and pouch edge thickened; secretion becomes clear and watery leaving interior damp<br>Pouch hairs thicken and lengthen<br>White 'studs' develop after 1–2 weeks, and persist several weeks in females without pouch young |
| ? Post-reproductive (> 6 years old) | Secretes low-moderate levels of brown, thin/watery liquid |

new curves can be established for future reference. These include:

- developmental stages
  - sex distinguishable
  - tips of ears free
  - papillae of facial vibrissae evident
  - eyelashes visible
  - eyes open
  - fur visible – slight tinge, medium or well developed
  - tips of first incisors through the gums
  - on back of mother or in nest
  - eating solids
  - self-feeding
  - independent
- measurements (see Appendix 3)
  - bodyweight (g) – if not on teat
  - head length (mm) – from the occiput to snout tip
  - head width (mm) – maximum width across the zygomatic arches
  - crown-rump length (mm) – primarily for very small neonates still in pouch
  - body length (mm) – from snout tip to cloaca

- tail length (mm) – from the cloaca to the end of the last vertebra of the tail tip
- total length (mm) – from snout tip to tail tip
- tibia length (mm) – from the hip to the bottom of the pes
- pes length (mm) – from the heel to the base of the longest toe, not including the claw.

### 10.4.2 Males

The males of some species, including antechinuses and phascogales, have a sternal gland that develops with age and reaches maximal development during the breeding season. The sternal gland is usually bare of hair with surrounding fur stained yellow to brown (Baker and Gynther 2023). The activity of the gland can be measured from the following scale (Woolley 1966; Millis 1995; Millis and Bradley 2001):

(1) little or no activity – little or no staining of the surrounding hair; little or no hair loss over the gland area; no obvious glandular product.
(2) medium level activity – some staining of the surrounding hair; some loss of hair over the gland area; waxy glandular products visible.
(3) high activity – much staining of the surrounding hair; total hair loss over gland area; waxy glandular product prominent.

In the males of most species, the size of testes can be measured as they increase during the breeding season due to the onset of spermatogenesis. In antechinuses, testes are maximal up to 6 weeks before mating, at which point spermatogenesis stops (C. Dickman pers. obs.). Measure the length, width and depth of the testes in millimetres. Testis volume can be calculated by using the equation $V = \pi/6 \times (\text{length}) \times (\text{width})^2$ (Spencer 1996). The width of the scrotum has been used as a simple way of measuring increases in testis size successfully in captive studies and in the field (Dickman *et al.* 2001).

## 10.5 Reproductive management

The timing of breeding can be determined by examining the urine of females for cornified epithelial cells, sperm and, in some species, a marked increase in bodyweight is observed (Godfrey 1969a; Woolley 1971a, 1971b; Close 1983; Gaikhorst 1999). The urine can easily be collected with a disposable plastic pipette or by pressing a clean slide against the cloaca immediately after capture (as subsequent urine will be much more dilute in sperm and epithelial cells) (Aslin 1980; Close 1983) or by placing the animal in a plastic tub (with holes in the lid) for a short period of time (e.g. < 5 min) and collecting urine from the tub if present. Cells in the sample can then be examined immediately under ×100 magnification (even without a coverslip) (Godfrey 1969a). Although some authors have used air-drying and staining with methylene blue (which stains the nuclei of the parabasal epithelial cells so they can be distinguished from anucleated epithelial cells) to determine oestrus (Close 1983), it is usually not required. Once cells are detected, urine samples should be taken every 1–2 days and the number of cells per field of view can be scored as one of three scores: + cells present (Fig. 4.7a), ++ many cells present (Fig. 4.7b), and +++ cells are abundant (Fig. 4.7c) (Selwood 1982a). The males can also be examined to see if they are reproductively active by looking for sperm in their urine (Selwood 1982a). When the number of epithelial cells reaches ++ or +++ (which usually corresponds with a decline in bodyweight) the female should be mated with a male. Successful mating can be determined by direct observation or sperm present in the urine of the female (Fig. 4.7d) (Selwood 1982a; Taggart and Temple-Smith 1991). Close (1983) found oestrus in kowaris to be correlated with 20–50 anucleate epithelial cells and leucocytes per ×100 field, 1–2 days before the appearance of sperm in the urine.

Ovulation and mating have been induced in fat-tailed dunnarts by injecting gonadotrophins, but no fertilisation occurred (Smith and Godfrey 1970). This failure was attributed to an unexpected period between ovulation and mating when ova were retained in the oviduct. Ovulation of non-breeding and non-cycling female stripe-faced dunnarts was successfully induced using equine serum gonadotrophin. This resulted in successful pregnancies and the production of fertile young (Au *et al.* 2010).

Despite the polygynous mating system of the dasyurids (Taggart *et al.* 2003), and most likely the marsupial mole in the wild, the sexes should only be put together as pairs during breeding and sometimes feeding, as they are generally solitary at other times. If held together throughout the year, they often will not mate as they appear to become too accustomed to their enclosure partner. In order to increase the potential of successful matings and encourage mating behaviour, males should ideally be rotated through the female enclosures every few days for species such as phascogales and antechinuses that have very short breeding seasons of only several weeks. In species with longer breeding seasons,

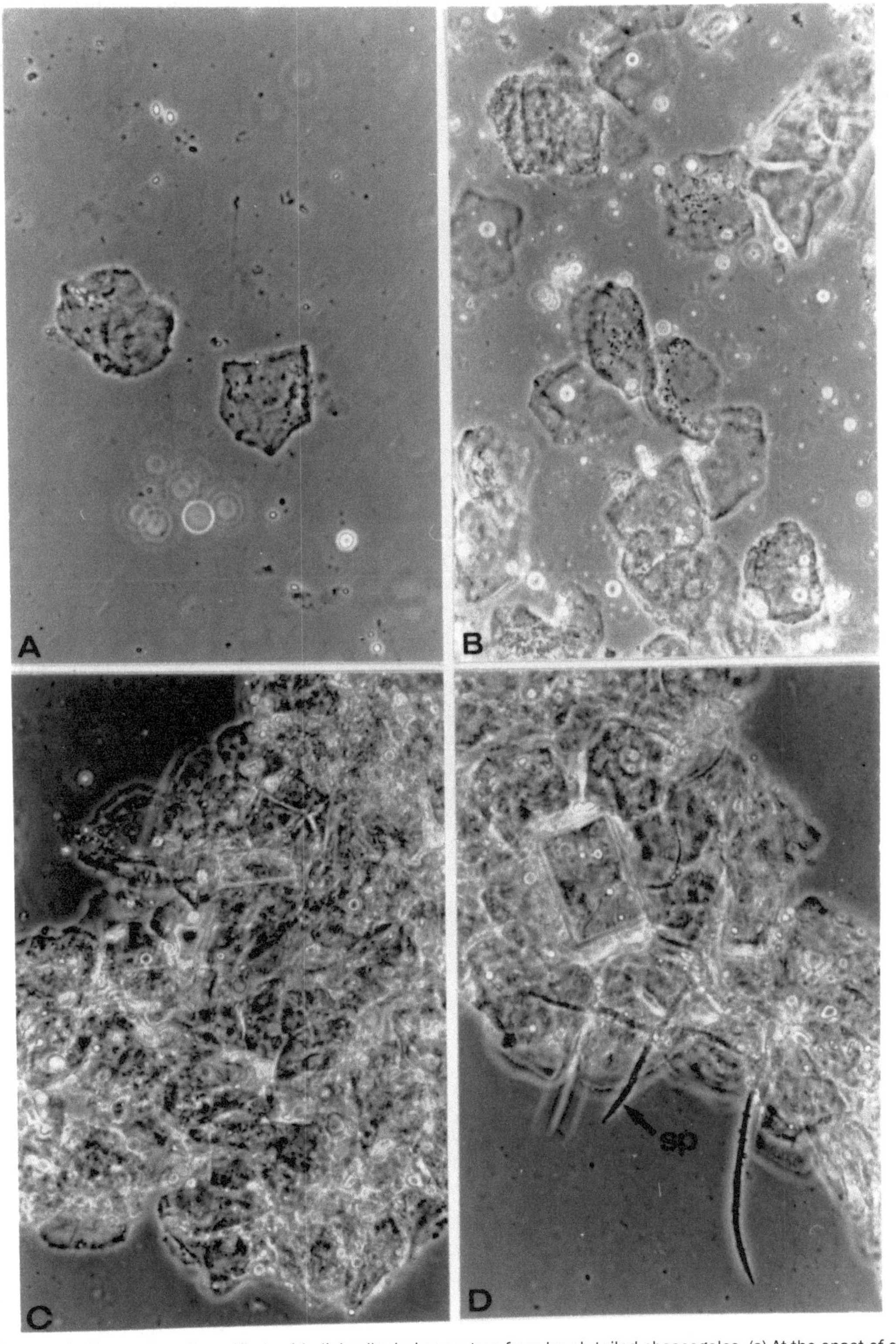


**Fig. 4.7.** Changes in the number of cornified epithelial cells during oestrus from brush-tailed phascogales. (a) At the onset of oestrus the number of cornified epithelial cells (CEC) is low and they are generally separate. (b) As oestrus progresses, the number of CEC increases and cells begin to occur in groups. (c) Large numbers of CEC that occur in clumps or sheets. (d) Spermatozoa may often appear after mating within the clumps of epithelial cells. Taken from Millis (1995), reprinted with permission from Monash University.

such as quolls and the Tasmanian devil, they should, if possible, be rotated every few weeks or, ideally, the technique described in sections 9.3 and 9.9 should be used. Once the females have been observed to have successfully mated with one or more males or when pouch young are observed, the male should generally be removed (Aslin 1982). However, contrary to other research, kowaris have been held in family groups throughout the year, with young of several litters being raised together and in the presence of the male without trouble (Meissner and Ganslosser 1985). Deaths that were observed (21% mortality) were caused by negligence or mistreatment by the mother and not overt aggression by the male (Meissner and Ganslosser 1985). This compares with 39% mortality observed in a colony in which the mothers were isolated from other adults and kept in a relatively small enclosure (Aslin 1980).

The fat-tailed dunnart is a strictly seasonal breeder with two peaks of birth, one in August and another in December, which coincide with a seasonal abundance of terrestrial invertebrates (Tyndale-Biscoe and Renfree 1987). The earliest births occur in the third week of July and the last in February (Godfrey and Crowcroft 1971; Morton 1978b). Therefore, the onset of breeding, both in the wild and in captivity, begins shortly after the winter solstice and ends sometime after the summer solstice (Tyndale-Biscoe and Renfree 1987). Female fat-tailed dunnarts respond to increasing photoperiod from 12L:12 D to 15L:9 D by returning to breeding condition (Godfrey 1969b). The signal to begin breeding appears to be a response to a change from short to long day, as animals maintained in 16L:8D for most of the time and then exposed to 8L:16D for 3 weeks ceased oestrous cycles during this period and resumed 20–30 days after the return to 16L:8D (Smith *et al.* 1978). Bennett *et al.* (1982) used artificial lighting with success by maintaining fat-tailed dunnarts in 16L:8D for 6 months, then a period of three weeks of 8L:16D followed by a return to 16L:8D. In these situations, the females can produce as many as five successive litters whereas in the wild females are not known to produce more than two litters nor produce in a second season (Morton 1978b; Smith *et al.* 1978). Breckon and Hulse (1972) bred fat-tailed dunnarts in 12D:12L conditions although this was only over 1 year, and weaning success was very low. Long-tailed dunnarts have been bred successfully with artificial light in which the daylength was adjusted every 10 days (Woolley and Valente 1986).

The fat-tailed dunnart has been bred with great success by holding males with 2–4 females in artificial lighting, checking the females' pouches 1–2 times per week and moving the female to a separate enclosure if she has pouch young (Bennett *et al.* 1990). As soon as the litter is weaned (~65 days) the female is re-paired so that she can mate during the post-weaning oestrus (Bennett *et al.* 1990). If pairs have not produced young within 5 weeks they are re-paired with another animal. The male is not reused if he has not sired more than one litter in 3 months of his initial pairing or within 6 weeks if he has previously mated successfully (Bennett *et al.* 1990). Importantly, these animals were held in artificial lighting in which it was found they had significantly more litters if 'daylength' was varied by having long days (16L: 8 D) and short days (8L:16D) for approximately 3 weeks every 6 months (Bennett *et al.* 1990).

Observations on agile antechinus under various day/night lengths have also failed to result in animals breeding. Scott (1986) placed animals in short daylength enclosures (10L:14D) to mimic breeding season but it had no effect on reproductive condition. If they were placed in short day enclosures, the reproductive condition appeared to be completely suppressed in both sexes (Scott 1986). In antechinuses and phascogales the females are mono-oestrous with ovulation occurring synchronously each year. The timing in antechinus appears to be defined by an endogenous circannual clock with ovulation being determined by the absolute photoperiod and a precise threshold rate of change in photoperiod which acts as the major zeitgeber (Dickman 1985; Selwood 1985; McAllan and Dickman 1986; Bradley 1987; McAllan *et al.* 1991, 2006; Halley 1992; McAllan and Geiser 2006). Therefore, exposure to natural light cycles (i.e. sunlight) or the artificial light cycles that mimic it, which is unlikely to be achieved, is highly recommended to facilitate breeding of dasyurids in captivity. It is important to use light globes with a high colour rendering index which mimics the wavelengths found in sunlight to ensure the correct cues are provided to animals under artificial lighting.

Brush-tailed phascogales are generally housed separately outside the breeding season, at which time several techniques can be used to mate animals. First, a single male can be introduced into the female's enclosure and kept there until pouch young are noticed in the female (Halley 1992; Slater 1993). Second, a round robin system can be used in which males are rotated through the females every few days while the females are in oestrus (with oestrus being determined by urine samples and generally occurring in the first two weeks of May in Victoria) (Halley 1992; Millis *et al.* 1999).

An analysis of a colony of fat-tailed dunnart breeding records showed that a female that does not produce a litter after being paired with a male for two or more oestrus cycles is more likely to reproduce if paired with a different male than if she stays with the first male (Smith *et al.* 1978). Similar observations have been made with kultarrs (J. Old and H. Stannard pers. obs.).

It appears that many species of dasyurids, particularly quolls and Tasmanian devils, do not mate well with individuals that they are regularly housed with, even during the receptive period (Settle 1978; Gaikhorst 1999). Female quolls also do not appear to tolerate being re-mated by the same male and can become extremely aggressive, unless they have since been mated by a second male (Settle 1978). This may reflect that in the wild they would meet and socialise with several different males before the breeding season, but den alone.

## 10.6 Occurrence of hybrids

None known at this stage.

## 10.7 Timing of breeding

Oestrus is highly synchronised in most species of dasyurids and can occur at any time of the year, depending on the species (Table 4.16). Behavioural oestrus in agile antechinus lasts up to 11 days, 7 days in yellow-footed and mainland dusky antechinus, 2-3 days in the dunnarts,

**Table 4.16.** Reproduction and development of the carnivorous marsupials.

| Species | Litter size (mean) | First detach (days) | Permanent pouch exit (days) | Weaning (days) | Sexual maturity F (months) | Sexual maturity M (months) | Mating period | Birth season | Reference |
|---|---|---|---|---|---|---|---|---|---|
| **Dasyuridae** | | | | | | | | | |
| *Dasycercus blythi/hillieri* | 2-6 (5) | 55-60 | 88 | 100-120 | 10-11 | 10-11 | Aug-Sep | Sept-Oct | 1-6 |
| *Dasyuroides byrnei* | 4-6 (5) | 55 | 70-78 | 100-120 | 8-9 | 8-9 | May-Oct | Jun-Dec | 4, 7-11 |
| *Dasykaluta rosamondae* | 6-8 | - | - | 100-120 | 10 | 10 | Sept | Nov | 12 |
| *Dasyurus geoffroii* | 1-6 | - | 61 | 110-54 | 12 | 12 | Apr-Jul | May-Sept | 13-15 |
| *D. hallucatus* | 6-8 (7) | 60-70 | 56-70 | 125-50 | 10-11 | - | May-Aug | Jun-Sept | 16- 21 |
| *D. maculatus* | 4-6 (5) | 35-49 | 96 | 125-150 | 11 | 11 | Apr-Aug | Jun-Aug | 22-25 |
| *D. viverrinus* | 1-6 (6) | 49-65 | 91 | 135-140 | 11 | 11 | May-Jun | May-Aug | 17, 26-30 |
| *Parantechinus apicalis* | 8 | 60 | 64 | 90- 120 | 10-11 | 10-11 | Mar-Apr | Apr-May | 18, 31-33 |
| *Pseudantechinus bilarni* | 4-6 | - | c. 30-45 | c. 90 | 12 | 12 | May-Jul | Aug-Sept | 14, 34 |
| *P. ningbing* | - | - | - | 112 | 10-11 | 10-11 | Jun | Jul-Aug | 35 |
| *P. macdonnellensis* | 5-6 | - | - | 98 | 12 | 12 | Jun-Jul | Jul-Oct | 36, 37 |
| *P. woolleyae* | 4-6 | - | - | - | 10 | 10 | | | 14 |
| *Sarcophilus harrisii* | 1-4 (3) | 90-105 | 105 | 150-280 | 12 | 12 | Mar-Apr | Apr-May | 28, 38, 39 |
| *Antechinus agilis* | 6-10 | 35 | 35 | c. 90 | 10-11 | 10-11 | Aug | Sept | 14, 50, 51 |
| *A. bellus* | 1-10 | 28-35 | | c. 100 | 10-11 | 10-11 | Aug | Sept-Oct | 40 |
| *A. flavipes* | 1-12 (7) | 42 | | 90-120 | 10-11 | 10-11 | Jun-Sept | Jul-Oct | 18, 41, 42 |
| *A. godmani* | 1-6 | - | - | - | 10-11 | 10-11 | Jun-Aug | Jul-Sept | 43 |
| *A. leo* | 8-10 (9) | - | - | - | 10-11 | 10-11 | Sept-Oct | Oct-Nov | 44 |

**Table 4.16.** continued

| Species | Litter size (mean) | First detach (days) | Permanent pouch exit (days) | Weaning (days) | Sexual maturity F (months) | Sexual maturity M (months) | Mating period | Birth season | Reference |
|---|---|---|---|---|---|---|---|---|---|
| *A. mimetes* | 4–8 | 33–43 | 56 | 90–95 | 10–11 | 10–11 | Jun–Jul | Jul–Aug | 45–48 |
| *A. minimus* | – | – | – | – | 10–11 | 10–11 | May–Jul | Jun–Aug | 49, 52 |
| *A. subtropicus* | 6–8 | < 35 | 35 | c. 90 | 10–11 | 10–11 | Sept | Oct | 53 |
| *A. stuartii* | 6–8 | 35–45 | 35 | 90–110 | 9–10 | – | Jul–Aug | Aug–Sept | 54, 55 |
| *Phascogale calura* | 6–8 (7) | 44 | | 90–110 | 12 | 11.5 | Jul | Jun–Aug | 56–58 |
| *P. tapoatafa* | 3–8 (6) | 49–54 | 49–54 | 120–140 | 7.5 | 11 | May–Jul | Jun–Sept | 59–62 |
| *P. gilesi* | 6–12 (7) | 37 | | 65–70 | – | – | Jul–Dec | Aug–Jan | 63–65 |
| *P. ingrami* | 4–10 (7) | 35–40 | | 90 | – | – | Aug–Feb | Dec–Mar | 66–71 |
| *P. maculata* | 4–12 (8) | 28 | 45 | 70 | 10 | – | Jan–Dec | Jan–Dec | 72, 73 |
| *P. tenuirostris* | 4–12 (6) | 40 | 40 | c. 95 | – | – | Jul–Jan | Aug–Feb | 63, 65 |
| *Ningaui ridei* | 5–7 | 42–44 | 48–49 | 70–81 | 10 | 6–11 | Sept–Dec | Oct–Jan | 6, 74, 75 |
| *N. timealeyi* | 4–6 (5) | – | – | – | – | < 12 | Aug–Jan | Sept–Mar | 75, 76 |
| *N. yvonneae* | – | – | – | – | – | 8 | Aug–Sept | – | 14 |
| *Antechinomys laniger* | 6–8 | 30–48 | 30–48 | 80–90 | 11.5 | 11.5 | Jul–Oct | Jun–Feb | 18, 77–79 |
| *Sminthopsis archeri* | – | – | – | – | – | – | Oct–Feb | Oct–Feb– | 80 |
| *S. crassicaudata* | 3–10 | 43 | 59–63 | 65–68 | 4 | 5 | Jun–Jan | Jul–Feb | 18, 81–83 |
| *S. dolichura* | 4–8 (7) | – | – | – | – | – | Aug–Feb | Sept–Mar | 84 |
| *S. douglasi* | 1–8 | – | – | 70 | 4–7 | 7–8 | Aug–Mar | Aug–Mar | 14, 85 |
| *S. fuliginosa* | 1–8 | 35 | 28–35 | 70 | 12 | 12 | Jul | Aug | 14, 77, 86 |
| *S. gilberti* | 1–8 | – | – | – | – | – | Oct–Nov | Oct–Dec | 14 |
| *S. leucopus* | 1–10 | 56 | 56 | 86 | – | – | Jul–Aug | Aug–Sept | 78, 79, 87, 88 |
| *S. macroura* | 6–8 | 40 | 40 | 70 | 3–9 | 5–11 | Jun–Feb | Jul–Mar | 80–82, 89–91 |
| *S. murina* | 4–10 | 34 | – | 65 | 5 | 5 | Aug–Jan | Aug–Mar | 83, 92 |
| *S. ooldea* | 7–8 | 30 | 45 | 70 | – | – | – | – | 14 |
| *S. virginiae* | 6–8 | – | – | 65–70 | 6 | 6 | All year? | All year? | 14 |
| *S. youngsoni* | 5–6 | – | 30+ | – | 8 | 8 | Aug–Sept | Sept–Oct | 6, 14 |
| | – | – | – | – | – | – | – | – | |
| **Thylacinidae** | – | – | – | – | – | – | – | – | |
| *Thylacinus cynocephalus* | 1–4 | – | – | – | – | – | – | May Sept | 93 |

References: (1) Fleay (1961); (2) Michener (1969); (3) Sorenson (1970); (4) Woolley (1971a); (5) Gibson and Cole (1992); (6) Dickman *et al.* (2001); (7) Mack (1961); (8) Aslin (1974); (9) Aslin (1980); (10) Fletcher (1983); (11) Meissner and Ganslosser (1985); (12) Woolley (1991a); (13) Soderquist and Serena (1990); (14) Baker and Gynther (2023); (15) Gaikhorst (1999); (16) Fleay (1962); (17) Nelson and Smith (1971); (18) Woolley (1973); (19) Begg (1981a); (20) Braithwaite and Griffiths (1994); (21) Nelson and Gemmell (2003); (22) Fleay (1940); (23) Collins (1973); (24) Settle (1978); (25) Green and Scarborough (1990); (26) Hill and O'Donaghue (1913); (27) Hill and Hill (1955); (28) Green (1967); (29) Fleay (1935b); (30) Bryant (1988); (31) Woolley (1971b); (32) Lambert (2000); (33) Lambert and Mills (2006); (34) Begg (1981b); (35) Woolley (1988); (36) Woolley (1991b); (37) Gilfillan (2001); (38) Fleay (1935a); (39) Guiler (1970); (40) Friend (1985); (41) Fleay (1949); (42) Smith (1984); (43) Watt (1997); 44) Leung (1999); (45) Fleay (1932); (46) Wakefield and Warneke (1963); (47) Dickman (1982); (48) Williams and Williams (1982); (49) Wilson and Bourne (1984); 50) Woolley (1966); (51) Selwood (1982a); (52) Wilson (1986); (53) Wood (1970); (54) Horner and Taylor (1959); (55) Marlow (1961); (56) Kitchener (1981); (57) Bradley (1997); (58) Foster *et al.* (2008); (59) Fleay (1934); 60) Cuttle (1982b); (61) Millis *et al.* (1999); 62) Rhind (2002); 63) Denny (1982); 64) Whitford *et al.* (1982); 65) Read (1984a);( (66) Davies (1960); (67) Fleay (1965); (68) Heinsohn (1970); (69) Woolley (1974); (70) Archer (1976); (71) Woolley and Elliott (2014); (72) Aslin (1975); (73) Van Dyck (1979); (74) Fanning (1982); (75) Kitchener *et al.* (1986); (76) Dunlop and Sawle (1982); (77) Woolley (1984); (78) Lee and Cockburn (1985); (79) Stannard and Old (2010); (80) Woolley (2007); (81) Fleay (1929); (82) Godfrey and Crowcroft (1971); (83) Morton (1978b); (84) Friend *et al.* (1997); (85) Woolley (2015); (86) Crowther *et al.* (1999); (87) Woolley and Ahern (1983); (88) Lunney and Ashby (1987); (89) Godfrey (1969a); (90) Woolley (1990b); (91) Taggart *et al.* (1997); (92) Fox and Whitford (1982); (93) Guiler (1961).

1-3 days in the kowari, 3 days in quolls, 5 days in phascogales and 1-3 days in ningauis and planigales, so several bouts of copulation can occur (Marlow 1961; Woolley 1966; Tyndale-Biscoe and Renfree 1987; Taggart *et al.* 2003).

Almost all Australian dasyurids are seasonal breeders, with the period of late spring and early summer generally coinciding with late lactation and pouch emergence in temperate Australia and with the post-monsoonal period in northern Australia (Tyndale-Biscoe and Renfree 1987; Fisher *et al.* 2013). The timing of breeding seasons for various carnivorous marsupials is shown in Table 4.16.

## 10.8 Age at first and last breeding

Although male antechinuses and phascogales die or are reproductively sterile after mating, the females can survive to reproduce to a second, or rarely a third, year (Lee *et al.* 1982; Lee and Cockburn 1985; Stannard *et al.* 2013) (Table 4.17). Most dasyurids reach reproductive senescence before they die in captivity, so careful management is required of the captive population. If space is limited and breeding is stopped, it will result in an ageing population and ultimately the demise of the colony. Therefore, if space is limited, euthanasia of reproductively finished stock may be considered rather than stopping breeding. Breeding should be allowed to begin as soon as the individuals are sexually mature and not stopped, as this invariably results in no breeding when attempts are made to commence again. This may negatively affect the longevity of the colony.

## 10.9 Ability to breed every year

All species can breed every year; however, males of the species with Strategy 1 reproduction, such as the antechinuses and phascogales, are limited to the males

**Table 4.17.** Reproductive life (months) of carnivorous marsupials for which it is known in captivity.

| Species | Reproductive life (months) | | Reference |
|---|---|---|---|
| | Males | Females | |
| *Dasycercus hillieri* | 12-72 | 12-72 | 1, 2 |
| *Dasyuroides byrnei* | 7-48 | 8-49 | 3-5 |
| *D. geoffroii* | 12-36 | 12-36 | 6, 7 |
| *D. hallucatus* | 11-12 | 11-24 | 8 |
| *D. maculatus* | 12-60 | 11-36 | 9 |
| *Parantechinus apicalis* | 12-36 | 12-36 | 10 |
| *Pseudantechinus bilarni* | 12-24 | 12-36 | 2 |
| *P. macdonnellensis* | 12-36 | 12-48 | 11 |
| *P. ningbing* | 12-24 | 12-24 | 2 |
| *P. woolleyae* | 12-48 | 12-48 | 2 |
| *Sarcophilus harrisii* | 12-60 | 12-48 | 12, 13 |
| *Antechinus* spp. | 11-11.5 | 11-24 | 14 |
| *Phascogale* spp. | 11-11.5 | 11-48 | 15, 16 |
| *Planigale maculata* | 8-38 | - | 3 |
| *Antechinomys laniger* | 12-31 | 12-31 | 17, 18 |
| *Sminthopsis archeri* | 24 | 24 | 19 |
| *S. crassicaudata* | 4-30 | - | 20, 21 |
| *S. dolichura* | 4-24 | 8-24 | 2 |
| *S. fuliginosa* | 12-24 | 12-24 | 22 |
| *S. longicaudata* | < 12-24 | < 12-24 | 23 |
| *S. youngsoni* | - | > 20 | 24 |

References: (1) Woolley (1971a); (2) Baker and Gynther (2023); (3) Aslin (1980); (4) Aslin (1982); (5) Carnio (1993); (6) Serena and Soderquist (1988); (7) Serena and Soderquist (1989a); (8) Oakwood (1997); (9) Collins *et al.* (1993); (10) Lambert (2000); (11) Woolley (1991b); (12) Hughes (1982); (13) C. Srb pers. comm.; (14) Slater (1993); (15) Bradley (1997); 16) Stannard *et al.* (2013); (17) Woolley (1984); (18) Stannard and Old (2010); (19) Woolley (2015); (20) Smith *et al.* (1978); (21) Bennett *et al.* (1982); (22) Crowther *et al.* (1999); (23) Woolley and Valente (1986); (24) Dickman *et al.* (2001).

having only one breeding season (Lee *et al.* 1977; Lee and Cockburn 1985). In captivity, the males of these species can live longer than 1 year, however, normally they are sterile (Woolley 1966; Dickman 1993; Slater 1993). In brush-tailed phascogales and red-tailed phascogales the sterile scrotum appears blue because of hair loss from the scrotal skin and the black pigment in the tunica vaginalis, which is visible through the skin (S. Jackson pers. obs.). Apart from these species, all the others should be encouraged to breed every year, as they will generally not breed again if stopped for ≥ 1 year.

Many dasyurids often do not breed at all if they do not breed in their first year (Carnio 1993), so attempts should be made to breed them in their first year to maximise their reproductive output. However, this is not always the case; hence if there are genetically important animals in the colony it is worth the effort to see if they will breed in a subsequent year, such as female red-tailed phascogales (Stannard *et al.* 2013). Tasmanian devils that do not breed in their first or second year can breed in subsequent years, though success may be lower.

## 10.10 Ability to breed more than once per year

Most species of dasyurid are polyoestrous, except antechinuses and phascogales; that is, they can undergo oestrus if the first young are lost or they fail to become pregnant in earlier cycles (Tyndale-Biscoe and Renfree 1987). The only true mono-oestrous dasyurids are the antechinuses, phascogales and the kaluta where the males die after mating (Kitchener 1981; Cuttle 1982b; Lee *et al.* 1982; Slater 1993).

Fat-tailed dunnarts can breed twice per year as the female can go into oestrus 1–2 days after the young finish weaning and the second litter can be born 82–90 days after the first litter was born (Bennett *et al.* 1982). Similarly, kowaris and planigales can produce two litters per year (Woolley 1971b; Woolley 1973; Fletcher 1983; Baker and Gynther 2023). The lesser hairy-footed dunnart breeds irregularly in the wild, but could potentially breed more often in captivity due to the greater availability of food and water (Dickman *et al.* 2001). Stripe-faced dunnarts, for example, are known to produce three litters per year in captivity (P. Woolley pers. comm.). In contrast, observations on the wild brush-tailed mulgara and wongai ningaui showed them not to breed more than once per year (Dickman *et al.* 2001).

## 10.11 Nesting requirements

Nest boxes and/or hollow logs should be provided for all species of dasyurids, similar to their behaviour in the wild.

- Antechinuses build their nests in tree hollows or fallen logs from eucalypt leaves and grass (Dickman 1991).
- Brush-tailed phascogales make their nests in tree hollows with small entrance holes and composed of material including stringybark strips, bark, feathers and fur (Soderquist 1993b). Similarly, red-tailed phascogales use tree hollows with a small entrance for shelter and nesting. They construct large nests of wool, bark, feathers and grass (Short *et al.* 2011). Red-tailed phascogales readily use nest boxes in captivity and will build nests from shredded paper and melaleuca bark in a ball-like shape (J. Old pers. obs.).
- Dunnarts make nests under rocks or within holes of bird-eating spiders, scorpions, rodents or lizards, into which they drag some leaves or grass (Morton 1978c; Dickman 1996). In areas with hard substrates animals may simply use cracks in the soil (Waudby and Petit 2017). Ningauis make use of similar pre-existing burrows, cracks in the soil or under the protection of spinifex hummocks (Baker and Dickman 2018).
- Kowari and mulgaras dig their own burrows, with those of mulgaras ranging from shallow (15 cm) and temporary single burrows to complex systems that descend to over 0.5 m with many entrances and exit holes (Baker and Dickman 2018).
- Western quolls have been observed to den within root channels exposed after fire and old rabbit burrows or under the stumps of fallen trees or boulders. Nursery nests may contain eucalypt leaves and some feathers (Serena and Soderquist 1989b).
- Spotted-tailed quolls den in wombat and rabbit burrows, rock crevices, tree hollows and artificial structures (Glen and Dickman 2006; Körtner *et al.* 2019).
- Tasmanian devils den in burrows, which may be underground, under rocks or rocky overhangs, and in caves, fallen logs or dense vegetation. Devils may create burrows themselves or use dens that have been dug and vacated by wombats. They are, on occasion, found in other unusual areas such as underneath houses. They make nests from soft vegetation and grasses and in captivity can be seen taking nesting material into their dens.
- Kultarrs nest in hollow logs and stumps, spinifex hummocks and burrows of other animals (Lidicker

and Marlow 1970). In captivity kultarrs will build a shallow burrow (10 cm) if provided with compacted sloped sand and cover the entrance with native grasses (Valente 1984). If provided with a nest box lined with shredded paper or straw kultarrs will make a shallow depression in one corner and may line it with leaves provided in the enclosure (H. Stannard pers. obs.).

- Marsupial moles do not appear to build any form of nest as they make deep burrows, although nothing definite is known about their breeding biology (Troughton 1973).

## 10.12 Breeding diet

There are no data that specifically show that the dietary requirements for any dasyurid change during breeding; however, additional food should be provided to males before breeding (especially for quolls), as they have been known to kill and partially eat their proposed mates, and for lactating females, as a shortage will often result in cannibalism of the young (see section 6.1).

## 10.13 Number of young born

With the exception of the species discussed above (see section 10.10), only one litter per year is generally produced by dasyurids, but most species can generally produce a second litter (except antechinuses, phascogales and the kaluta) if the first litter is lost (Tyndale-Biscoe and Renfree 1987).

Most dasyurids are superovulators; they produce supernumerary young, which means there are more neonates than there are teats. Overproduction of neonates maximises the chance of all the teats being occupied (Tyndale-Biscoe and Renfree 1987). The excess young that do not attach to a teat either drop off and die or are consumed by their mothers. The litter size often changes with female age and bodyweight. Litter size in the Tasmanian devil is inversely related to the weight of the mother, with the lighter and presumably younger females tending to have larger litters than heavier older females (Guiler 1970). The sex ratio of litters can vary among individuals; for example, female antechinuses and dunnarts in good condition produce a higher proportion of sons, while females in poor condition produce more daughters (Dickman 1988b, 1988c). Female fat-tailed dunnarts that were allowed to choose their mates had a higher proportion of daughters (Gill 2019; see section 9.3.1). There are also shifts with age in some species (Cockburn *et al.* 1985), which could be of value if more males or females are needed at any time in a breeding colony (C. Dickman pers. obs.). Red-tailed phascogales show a bias in females surviving to weaning (Foster *et al.* 2008).

## 10.14 Age at weaning

Despite the large variation in body size there is remarkably little variation in the age of weaning, with dunnarts being weaned at 65–85 days, the antechinuses and brush-tailed phascogales at 90–140 days, the quolls at 110–155 days and the Tasmanian devil at 150–280 days (Table 4.16).

## 10.15 Age at removal from the mother

In captivity, Tasmanian devils are usually removed at 8–10 months of age and transferred to another enclosure so that a male can be introduced to the female in January for the start of the breeding season in late February (Smith 1993). Separation of mothers and weaned young is generally based on when the mother has ceased lactating or is in the final stages of lactation. If young are removed closer to 8 months old, 1–2 young may be left with the mother while lactation ceases to avoid mammary gland issues when suckling ceases. In other species there is a brief period (1–4 weeks) when the young accompany the mother and learn about different types of prey and hunting techniques before the young become fully independent and begin to forage alone (Baker and Dickman 2018). Separation of mothers and offspring can be safely carried out at this point.

## 10.16 Growth and development

Figure 4.8 shows the growth and development of various species of quoll and the Tasmanian devil, with further figures shown in Bach (1998). Additional references for growth and development are given in Table 4.18.

# 11. ARTIFICIAL REARING

## 11.1 Housing

As with all native mammals that have been taken into care, minimising stress is a major consideration. Choosing suitable housing can help to create a stress-free environment. To achieve this, several factors should be considered:

- securing the area from children and animals (particularly domestic dogs and cats)
- maintaining high standard of hygience
- escape-proofing the area
- clearing the area of obstacles and hazards
- offering shelter from extreme weather and noise.

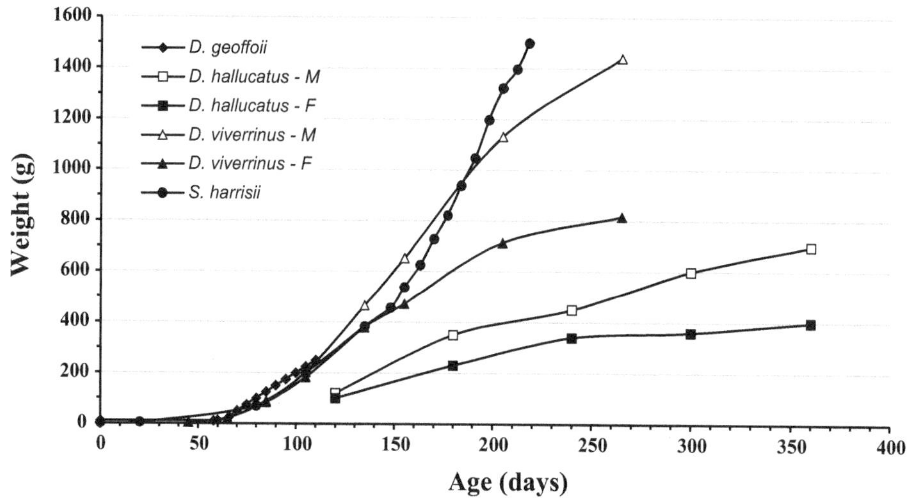


**Fig. 4.8.** Growth in bodyweight of quolls and the Tasmanian devil. Derived from Begg (1981a), Merchant *et al.* (1984), Serena and Soderquist (1988), Austin (1997), Gaikhorst (1999), Phillips and Jackson (2003).

**Table 4.18.** Growth curve measurements that have been developed for different species of carnivorous marsupials.

AR = forearm length, CR = crown to rump length, EA = ear length, HB = head-body length, HE = head length, HW = head width, LE = hindleg length, PE = pes length, TA = tail length, TO = total length, WT = bodyweight.

| Species | Measurements | Reference |
|---|---|---|
| *Dasycercus hillieri* | WT | 1, 2 |
| *Dasyuroides byrnei* | WT | 2, 3 |
| *Dasykaluta rosamondae* | WT, CR, HE, PE | 4 |
| *Dasyurus geoffroii* | WT, CR, HW | 5, 6 |
| *D. hallucatus* | WT, TO | 2, 7, 8 |
| *D. maculatus* | CR, HB, HE | 2, 9, 10 |
| *D. viverrinus* | WT, AR, CR, HB, HE, LE, PE, TO | 2, 11-14 |
| *Sarcophilus harrisii* | WT, CR, HE, HW, LE, PE | 2, 15-21 |
| *Parantechinus apicalis* | WT, HE, HW, PE, TA | 22 |
| *Antechinus mimetes* | WT, CR, HE, HB, PE, TA | 23 |
| *A. stuartii* | HB, PE, TO | 24 |
| *Phascogale calura* | WT, AR, CR, HE, HW, PE | 25 |
| *P. tapoatafa* | WT, CR, HW | 2, 26-28 |
| *P. gilesi* | WT, HE, HW, CR, PE, HB, TA | 29 |
| *P. tenuirostris* | WT, CR, HE, HW, HB, TA | 30, 31 |
| *Ningaui ridei* | WT, CR, HB, HE, HW, PE, TA | 32 |
| *Sminthopsis crassicaudata* | WT, CR | 2, 33, 34 |
| *S. macroura* | WT, CR, HB, HE, TA | 35 |
| *S. murina* | WT, CR, HB, HE, PE, TA | 36 |
| *S. ooldea* | WT, EA, HB, PE, TA | 37 |
| *S. psammophila* | WT, HE | 38 |
| *S. virginiae* | CR, HE | 39 |

References: (1) Michener (1969); (2) Collins (1973); (3) Aslin (1974); 4) Woolley (1991a); (5) Serena and Soderquist (1988); (6) Gaikhorst (1999); 7) Begg (1981a); (8) Nelson (1992); (9) Green and Scarborough (1990); (10) Collins *et al.* (1993); (11) Fleay (1935b); (12) Hill and Hill (1955); (13) Merchant *et al.* (1984); (14) Bryant (1988); (15) Fleay (1935a); (16) Guiler (1970); (17) Guiler (1978); (18) Smith (1993); (19) Austin (1997); (20) Phillips and Jackson (2003); (21) Pemberton (1990); (22) Mills *et al.* (2000); (23) Williams and Williams (1982); (24) Marlow (1961); (25) Foster *et al.* (2006); (26) Fleay (1950); (27) Cuttle (1982b); (28) Soderquist (1993a); (29) Whitford *et al.* (1982); (30) Read (1985); 31) Read (1987); (32) Fanning (1982); 33) Godfrey and Crowcroft (1971); (34) Morton (1978a); (35) Frigo and Woolley (1997); (36) Fox and Whitford (1982); (37) Aslin (1983); 38) Lambert *et al.* (2011); (39) Taplin (1980).

Unfurred or just furred joeys are best kept inside artificial pouches made of non-synthetic fibres such as cotton as they retain the warmth better. Use a cotton pouch liner and put it inside the woollen pouch for warmth (brushed cotton is softer on the skin) and keep it in a warm environment. Small furless dasyurids are unlikely to be successfully reared as they are normally attached to the mother's teat. Pouches need to be washed and disinfected every day as bacteria and yeasts thrive in these warm conditions (Bellamy 1992). Pouches should have the corners sown out so that joeys do not get their heads caught. Ensure there are no loose fibres that can wrap around their toes.

## 11.2 Temperature requirements

The temperature of the bag should be 34–36°C if the joey is furless. As the joey grows fur the temperature can be reduced to 30°C (Bellamy 1992). Use a minimum/maximum temperature gauge with a plastic-coated probe that can be placed next to the joey, as this will ensure that the temperature can be monitored.

## 11.3 Specific requirements

The skin of unfurred and slightly furred young should be kept moist with the use of Sorbolene cream (not with added glycerine) so that the skin does not become dry and cracked (George *et al.* 1995). Baby oil does not appear to be properly absorbed and tends to stay on the skin surface where it rubs off and is absorbed by the liner bag fabric (George *et al.* 1995).

When first brought in for hand-rearing, the joey may be dehydrated. Vytrate® can be given at a ratio of 20 mL Vytrate® to 250 mL water. It is important to warm the joey before feeding, otherwise there is a greater risk of inhalation pneumonia. It needs to be a sterile isotonic solution such as Hartmann's administered at up to 10% bodyweight subcutaneously.

Stress is a major problem in successfully rearing native mammals and can be fatal. Therefore, it is important to keep noise to a minimum, not to overhandle the animals and maintain high standards of hygiene.

## 11.4 Diet and feeding routine

### 11.4.1 Natural milk

The composition of carnivorous marsupials' milk has so far only been determined for the eastern quoll and Tasmanian devil, but nutrient composition probably undergoes significant changes throughout lactation in all marsupials. The solids, lipids and protein appear to peak in mid- to late lactation and the carbohydrates reach a peak in mid-lactation. The levels of calcium reach a peak in late lactation whereas iron peaks early and then decreases towards late lactation (Table 4.19). Milk consumption by young eastern quolls increases linearly from 7 to 14 weeks of age, which corresponds with the increasing energetic needs of lactating females (Green *et al.* 1997).

### 11.4.2 Milk formulas

The main commercial formulas for hand-rearing carnivorous marsupials are as follows.

- Biolac® – there are three formulas: M100 for furless joeys; M150, which is a transitional milk to use when dense fur has developed and M200, which is used when the animal produces solid dark pellet droppings, as it contains elevated lipid in the form of canola oil. When the joey is nearing weaning, 2–5 mL of canola oil is added per 100 mL of formula. Mixing the formulas eases the transition from one formula to another. Animals should be fed 10–15% of their bodyweight per day.
- Di-Vetelact® (Lillelund Pty Ltd) –a widely used, low-lactose milk formula. Due to its low energy concentration when prepared as directed, some groups advise the addition of mono- and polyunsaturated fats such as canola oil, as with Wombaroo® diets. The addition of saturated fats in the form of cream has been suggested, but cream is too highly saturated and can lead to the malabsorption of calcium. Di Vetelact should be fed at ~20% bodyweight, except for very small joeys

**Table 4.19.** Concentrations of major constituents of milk of two species of carnivorous marsupial.

| Species | Total solids (%) | Carbohydrates (%) | Lipids (%) | Protein (%) | Calcium (mg/L) | Iron (mg/L) | Reference |
|---|---|---|---|---|---|---|---|
| *Dasyurus viverrinus* | 15.0–34.0 | 2.0–7.4 | 4.0–16.0 | 3.0–10.0 | 2100–3200 | 12–22 | 1, 2 |
| *Sarcophilus harrisii* | 45.0 | 5.0 | 30.0 | 6.0 | 3200 | 12 | 1 |

References: (1) Green (1984); (2) Green *et al.* (1987).

(< 100 g). Some people also add a tablespoon per litre of high-protein baby cereal for furred joeys.
- Wombaroo® – charts are provided to assist in determining the type and volume to be fed. Though Wombaroo® does not have a dasyurid-specific milk replacer, they suggest using Kangaroo Milk Replacer >0.7 with the addition of 1 mL of The Good Oil for Animals or canola oil per 10 mL milk.

### 11.4.3 Feeding apparatus

Very small joeys can be fed using a syringe fitted with a bicycle tyre rubber valve, plastic intravenous catheter or 1-inch length of infant gastric feeding tube (Bellamy 1992). The larger carnivorous marsupials can be fed with plastic feeder bottles, 50 or 100 mL, with a special Type C teat for Tasmanian devils and Type D for quolls (Austin 1997). The teat should be punctured with a hot needle.

### 11.4.4 Feeding routine

Milk should be fed at ~36°C. When feeding, it is important not to feed the milk formula too quickly: the rate at which the milk is squeezed into the mouth should not be faster than the rate at which it is swallowed. Ensuring the hole in the teat is not too large will help (it should only be the size of a pinhole). Too much milk results in an accumulation in the pharynx, which is suddenly sneezed or coughed out the nostrils. To avoid this, be very careful of the rate at which milk is released to the joey and use a smaller hole on the teat if required.

The number of daily feeds changes as the joey develops (Bellamy 1992). Very young unfurred joeys should be fed every 2–3 h around the clock. When furred, the number of feeds is decreased to 5 per day and the volume increased per feed. At full emergence the number of feeds is reduced to 2–3 feeds per day.

## 11.5 Data recording

When an animal is first brought in for hand-rearing, its sex and approximate age, using growth charts, should be recorded. During the hand-rearing process several important pieces of information should be recorded on a daily basis:

- date when the information is recorded
- time when the information is recorded
- bodyweight to the nearest 1 g if possible (very important for small species)
- general activity and demeanour
- characteristics and frequency of defecation and urination
- amount (g) of different food types offered
- food consumption at each feed
- veterinary examinations and results.

This information serves several purposes, including providing important background information such as food consumption, which will assist a veterinarian to reach a diagnosis if the animal becomes sick or fails to grow or gain weight. It also allows a comparison with growth curves to assess progress (section 10.16) and to establish new growth curves that do not exist for other measurements. The developmental stages and measurements outlined in section 10.16 should also be recorded on a weekly basis if possible.

## 11.6 Individual identification methods

Generally not required, but when furred different coloured nail polish may be applied to the claws. Larger species can be given a PIT tag under the skin (see section 5.3.1). Species such as Tasmanian devils and quolls can be individually identified by their markings, which are visible as soon as fur fuzz starts to grow.

## 11.7 Hygiene

Maintaining a high standard of hygiene is critical to the survival of the joey. Emphasis needs to be placed on the following:

- maintaining a clean pouch lining at all times. Older joeys can be trained to urinate on newspaper by keeping a piece of newspaper with the smell of urine on it
- maintaining personal hygiene by washing and disinfecting hands before and after handling the joey. Use antibacterial solution for washing hands with furless joeys, as their immune system is not well developed
- washing hands between feeding different joeys
- using boiled water when making up formulas for joeys of all ages
- cleaning spilt milk formula or paste, faeces and urine from the joey's skin and fur as soon as possible and the fur dried to prevent bacterial or fungal skin infections
- washing all feeding equipment in warm soapy water and sterilise it in a suitable agent such as a microwavable baby steam steriliser or F10 ® disinfectant. After sterilising, rinse the equipment in cold water

- storing the teats and bottles in the fridge in an air-tight container after they have been disinfected
- only heating up milk once and discarding leftovers.
- avoiding contact with other animals unless certain they pose no health risk and are behaviourally suitable to be in contact
- stimulate the joey to toilet before or after feeding. As with other marsupials, toileting can be done by the application of warm water to the cloaca using cotton wool to stimulate urination and defecation, which allows the animal to keep drier and warmer in its artificial pouch
- if furless, cover the joey's body with Sorbolene cream after each feed, only until the fur appears.

## 11.8 Behavioural considerations

Larger species such as quolls and Tasmanian devils can become bonded with their rearer, so if they are to be released, contact should be minimised between feeding times. Ideally, they should also be raised in a group, so they are familiar with other members of their species.

## 11.9 Use of foster animals

Fostering has not been widely used for any of the carnivorous marsupials; however, there is a record of a dominant female kowari adopting a subordinate female's five young so that she had a total of 10 young in the nest (Carnio 1993). Three of the young (two males and a female) were then placed with another female (who already had four young) held elsewhere and all were accepted, and all were raised successfully (Carnio 1993). Meissner and Ganslosser (1985) have also made similar observations on kowaris. Julia Creek dunnarts have been observed to rear the young of other females as well as their own (Woolley *et al.* 1998). Females with young 50 days old, reared young that were 12 days younger and 2 days older than their own young (Woolley *et al.* 1998). Fat-tailed pseudantechinus will also accept young of another female (P. Woolley pers. comm.). Brush-tailed phascogales are known to have raised the young of other individuals on several occasions. One female that had three young ~12 weeks of age was given a further seven young of approximately 9 weeks of age (after the scent of the original young was rubbed on them), which were raised successfully (Hunter 1991). A further example of cross-fostering in brush-tailed phascogales involved females with young of 95 and 98 days of age being given young 68 days old, which resulted in them still heavily lactating at 174 days after giving birth, when they normally would have ceased lactating (Hunter 1991; Soderquist 1993a).

Although not recommended due to the very different milk composition and the potential of predation, there is a record of three abandoned juvenile Tasmanian devils being fostered to a domestic cat that reared them successfully (Turner 1970).

## 11.10 Weaning

Once the teeth begin to emerge in larger species such as Tasmanian devils and quolls, small amounts of lean beef with 1 teaspoon of calcium powder per 500 g can be offered (Austin 1997). A mixture of beef or chicken mince in with the milk allows the animal to lap it and get a taste for solids.

The type of food offered should simulate the wild diet; that is, the smaller dasyurids should be offered a variety of insects such as mealworms, crickets, moths, beetles and other insects as available. Care needs to be taken with feeding mealworms, as they are generally deficient in calcium (unless maintained in a high-calcium medium). The natural diet may be supplemented with Wombaroo® Small Carnivore Food. Once the teeth are well developed in quolls, they can be given large insects, day-old chickens, mice and rabbit or wallaby meat. Tasmanian devils can be given these and rats, whole rabbits and large bones to chew on (Austin 1997). At weaning, fresh water should be supplied.

## 11.11 Release procedures

In preparation for release, human contact should be minimised, especially for the larger species such as quolls and Tasmanian devils that are most likely to become imprinted. The provision of natural prey items is important so that hunting skills can be developed. The animal should also be fit and healthy and should not be overweight or have any injuries.

Carnivorous marsupial releases back into the wild must be handled with due consideration for the welfare of the released animal to ensure minimum impact upon wild populations. When releasing carnivorous marsupials back into the wild, the following issues should be considered.

- Individuals qualifying for release include mis-rescued individuals, animals that have fallen accidentally into human hands or animals not adapting to captivity.
- Only animals judged by an experienced wildlife veterinarian to be healthy at the time of release

should be considered for release, if deemed in the animal's best interests.

- Animals should normally be released shortly before dusk and as close as possible to where they were captured or rescued.
- Ensure that the released animal has appropriate predator avoidance skills.
- Larger species, such as quolls and Tasmanian devils, may require the provision of additional food until their hunting skills are developed.
- Soft release, where animals are maintained in large pens well away from human smells and sounds, can be used.
- If circumstances prevent the release occurring at the exact point of capture/rescue then the nearest suitable site should be chosen. The exact site should be determined by institution staff in consultation with the rescuer and local wildlife officers.

A good example of how animals can be returned successfully to the wild after a period in captivity is provided by the study of Cooper *et al.* (2009) on yellow-footed antechinus. Studies on the release of Tasmanian devils as part of the conservation breeding program suggest that various factors are important when releasing this species (Sinn *et al.* 2014) including:

- stress – including an understanding of the welfare requirements and how to manage them (Teixeira *et al.* 2007)
- behaviour – including unnatural rearing or selection towards captivity (Bremner-Harrison *et al.* 2004)
- early experiences before release – including habitat selection behaviour (Stamps and Swaisgood 2007)
- fitness of the released animals (Mathews *et al.* 2005)
- habitat quality at the release site (Sinn *et al.* 2014)
- release method (i.e. hard or soft release) (Sinn *et al.* 2014)
- consideration of individual variation in behaviour.

A study of Tasmanian devils released from captivity to the wild on Maria Island found very high survival of released devils, with 96% surviving more than 1 year post-release (Rogers *et al.* 2016). There was no difference in survival or bodyweight of devils raised in intensive captive facilities versus larger free-ranging enclosures. Given that the survival rate and bodyweights were the same for both groups of Tasmanian devils, it suggests that those raised in intensive captive facilities retained the innate behaviours required to scavenge and hunt prey and maintained their bodyweight post-release (Rogers *et al.* 2016).

## ACKNOWLEDGEMENTS

Sincere thanks to Jocelyn Hockley for providing comments on artificial rearing rearing and Jake Newman-Martin for providing several body weights of new *Dasycercus*.

## REFERENCES

Allen ME, Oftedal OT (1989) Dietary manipulation of the calcium content of feed crickets. *Journal of Zoo and Wildlife Medicine* **20**, 26–33.

AMTC (2024) The AMTC Australian Mammal Species List. Version 4.0. https://australianmammals.org.au/publications/amtc-species-list (accessed 21 July 2024).

Anderson SJ (2000) Increasing calcium levels in cultured insects. *Zoo Biology* **19**, 1–9. doi:10.1002/(SICI)1098-2361(2000)19:1<1::AID-ZOO1>3.0.CO;2-F

Andrew DL, Settle GA (1982) Observations on the behaviour of species of *Planigale* (Dasyuridae, Marsupialia) with particular reference to the narrow-nosed Planigale. In *Carnivorous Marsupials*. (Ed. M. Archer) pp. 311–324. Royal Zoological Society of New South Wales, Sydney.

Angel M (1970) *Coelomotrema antechinomes*, gen. et. sp. nov., a prosthogonimid trematode from *Antechinomys* and *Antechinus* (Marsupialia: Dasyuridae) from Australia. *Australian Journal of Zoology* **18**, 119–124. doi:10.1071/ZO9700119

Aplin K (2015) Family Notoryctidae (Marsupial moles). In *Handbook of the Mammals of the World. Monotremes and Marsupials*. Vol. 5. (Eds DE Wilson and RA Mittermeier). pp. 210–219. Lynx Edicions, Barcelona.

Archer M (1974*a*) Regurgitation or merycism in the western native cat, *Dasyurus geoffroii*, and the red-tailed wambenger, *Phascogale calura* (Marsupialia, Dasyuridae). *Journal of Mammalogy* **55**, 448–452. doi:10.2307/1379014

Archer M (1974*b*) Some aspects of reproductive behaviour and the male erectile organs of *Dasyurus geoffroii* and *D. hallucatus* (Dasyuridae: Marsupialia). *Memoirs of the Queensland Museum* **17**, 63–67.

Archer M (1976) Revision of the marsupial genus *Planigale* Troughton (Dasyuridae). *Memoirs of the Queensland Museum* **17**, 341–365.

Arnold J (1976) Growth and development of the chuditch, *Dasyurus geoffroii*. PhD thesis. University of Western Australia, Perth.

Arnold J, Shield J (1970) Growth and development of the chuditch (*Dasyurus geoffroii*). *Bulletin of the Australian Mammal Society* **2**, 198.

Arundel JH, Barker IK, Beveridge I (1977) Diseases of marsupials. In *The Biology of Marsupials*. (Eds B Stonehouse and D. Gilmore) pp. 141–154. Macmillan, London.

Aslin HJ (1974) The behaviour of *Dasyuroides byrnei* (Marsupialia) in captivity. *Zeitschrift für Tierpsychologie* **35**, 187–208. doi:10.1111/j.1439-0310.1974.tb00443.x

Aslin HJ (1975) Reproduction in *Antechinus maculatus* Gould (Dasyuridae). *Australian Wildlife Research* **2**, 77–80. doi:10.1071/WR9750077

Aslin HJ (1980) Biology of a laboratory colony of *Dasyuroides byrnei* (Marsupialia: Dasyuridae). *Australian Zoologist* **20**, 457–471.

Aslin HJ (1982) Small dasyurid marsupials: their maintenance and breeding in captivity. In *The Management of Australian*

*Mammals in Captivity*. (Ed. DD Evans) pp. 22–26. Zoological Board of Victoria, Melbourne.

Aslin HJ (1983) Reproduction in *Sminthopsis ooldea* (Marsupialia: Dasyuridae). *Australian Mammalogy* **6**, 93–95. doi:10.1071/AM83014

Aslin HJ, Lim L (1995) Kowari *Dasyuroides byrnei*. In *The Mammals of Australia*. (Ed. R Strahan) pp. 59–61. Reed Books, Sydney.

Attwood HD, Woolley PA (1970) Toxoplasmosis in dasyurid marsupials. *Pathology* **2**, 77–78. doi:10.1016/S0031-3025(16)39577-0

Attwood HD, Woolley PA (1973) Spontaneous malignant neoplasms in dasyurid marsupials. *Journal of Comparative Pathology* **83**, 569–581. doi:10.1016/0021-9975(73)90014-5

Attwood HD, Woolley PA (1980) Pancreatic pathology in dasyurid marsupials. *Journal of Wildlife Diseases* **16**, 245–250. doi:10.7589/0090-3558-16.2.245

Attwood HD, Woolley PA (1982) Histopathology of captive dasyurid marsupials. In *The Management of Australian Mammals in Captivity*. (Ed. DD Evans) pp. 27–30. Zoological Board of Victoria, Melbourne.

Attwood HD, Woolley PA, Rickard MD (1975) Toxoplasmosis in dasyurid marsupials. *Journal of Wildlife Diseases* **11**, 543–551. doi:10.7589/0090-3558-11.4.543

Au PCK, Nation A, Parrott M, Selwood L (2010) Induced ovulation mimics the timetable of natural development in the stripe-faced dunnart, *Sminthopsis macroura*, and results in the birth of fertile young. *Reproduction (Cambridge, England)* **139**, 419–425. doi:10.1530/REP-09-0203

Austin MA (1997) *A Practical Guide to the Successful Handrearing of Tasmanian Marsupials*. Regal Publications, Melbourne.

Australasian Mammal Taxonomy Consortium (AMTC) (2022) The AMTC Australian Mammal Species List. Version 2.0. https://australianmammals.org.au/publications/amtc-species-list (accessed 14 January 2023).

Bach C (1998) *Birth Date Determination in Australasian Marsupials*. Australasian Regional Association of Zoological Parks and Aquaria, Sydney.

Baker AM (2015) Family Dasyuridae (Carnivorous marsupials). In *Handbook of the Mammals of the World. Monotremes and Marsupials*. Vol. 5. (Eds DE Wilson and RA Mittermeier). pp. 232–348. Lynx Edicions, Barcelona.

Baker AM, Dickman C (2018) *Secret Lives of Carnivorous Marsupials*. CSIRO Publishing, Melbourne.

Baker AM, Gynther IC (eds) (2023) *Strahan's Mammals of Australia*. 4th edn. Reed New Holland Publishers, Sydney.

Baker AM, Eldridge MDB, Fisher DO, Frankham G, Helgen K, Jackson SM, Potter S, Travouillon KJ, Umbrello L (2023) Taxonomy and diversity of living Australasian marsupials. In *American and Australasian Marsupials* (Eds NC Cáceres and CR Dickman). pp. 163–247. Springer, Switzerland.

Barker IK, Beveridge I, Bradley AJ, Lee AK (1978) Observations of spontaneous stress related mortality among males of the dasyurid marsupial *Antechinus stuartii* Macleay. *Australian Journal of Zoology* **26**, 435–447. doi:10.1071/ZO9780435

Barnett JL (1973) A stress response in *Antechinus stuartii* (Macleay). *Australian Journal of Zoology* **21**, 501–513. doi:10.1071/ZO9730501

Begg RJ (1981a) The small mammals of Little Nourlangie Rock, N.T. III: ecology of *Dasyurus hallucatus*, the northern quoll (Marsupialia: Dasyuridae). *Australian Wildlife Research* **8**, 73–85. doi:10.1071/WR9810073

Begg RJ (1981b) The small mammals of Little Nourlangie Rock, N.T. II: ecology of *Antechinus bilarni*, the sandstone antechinus (Marsupialia: Dasyuridae). *Australian Wildlife Research* **8**, 57–72. doi:10.1071/WR9810057

Belcher CA (2003) Demographics of the tiger quoll (*Dasyurus maculatus maculatus*) populations in south-eastern Australia. *Australian Journal of Zoology* **51**, 611–626. doi:10.1071/ZO02051

Belcher CA, Darrant JP (2004) Home range and spatial organisation of the marsupial carnivore, *Dasyurus maculatus maculatus* (Marsupialia: Dasyuridae) in south-eastern Australia. *Journal of Zoology* **262**, 271–280. doi:10.1017/S0952836903004631

Bellamy T (1992) Marsupial handrearing. In *Urban Wildlife*. Proceedings 204. Post Graduate Committee in Veterinary Science. (Eds L Vogelnest, D Spielman, T Bellamy, M Messer and DM Walker) pp. 49–67. University of Sydney, Sydney.

Bennett JH, Smith MJ, Hope RM, Chesson CM (1982) Fat-tailed dunnart *Sminthopsis crassicaudata*: establishment and maintenance of a laboratory colony. In *Management of Australian Mammals in Captivity*. (Ed. DD Evans) pp. 38–44. The Zoological Board of Victoria, Melbourne.

Bennett JH, Breed WG, Hayman DL, Hope RM (1990) Reproductive and genetical studies with a laboratory colony of the dasyurid marsupial *Sminthopsis crassicaudata*. *Australian Journal of Zoology* **37**, 207–222. doi:10.1071/ZO9890207

Bernard JB, Allen ME, Ullrey DE (1997) *Feeding Captive Insectivorous Animals: Nutritional Aspects of Insects as Food*. AZA Nutrition Advisory Group. Fact Sheet 003. American Zoo Association.

Bettiol SS, Obendorf DL, Nowarkowski M, Milstein T, Goldsmid JM (2000a) Earthworms as paratenic hosts of toxoplasmosis in eastern barred bandicoots in Tasmania. *Journal of Wildlife Diseases* **36**, 145–148. doi:10.7589/0090-3558-36.1.145

Bettiol SS, Obendorf DL, Nowarkowski M, Goldsmid JM (2000b) Pathology of experimental toxoplasmosis in eastern barred bandicoots in Tasmania. *Journal of Wildlife Diseases* **36**, 141–144. doi:10.7589/0090-3558-36.1.141

Beveridge I, Barker IK (1976) The parasites of *Antechinus stuartii* Macleay from Powelltown, Victoria, with observations on seasonal and sex-related variations in numbers of helminths. *Australian Journal of Zoology* **24**, 265–272. doi:10.1071/ZO9760265

Boden E (1998) *Black's Veterinary Dictionary*. A & C Black, London.

Booth R (1994) Medicine and husbandry: Dasyurids, possums and bats. In *Wildlife*. Proceedings 233. Post Graduate Committee in Veterinary Science. (Ed. DI Bryden) pp. 423–41. University of Sydney, Sydney.

Booth R (1999) Macropods: Hand raising, husbandry, diseases and rehabilitation. *Wildlife Veterinary Notes*. Unpublished.

Bradley AJ (1987) Stress and mortality in the red-tailed phascogale, *Phascogale calura* (Marsupialia: Dasyuridae). *General and Comparative Endocrinology* **67**, 85–100. doi:10.1016/0016-6480(87)90208-5

Bradley AJ (1997) Reproduction and life history in the red-tailed phascogale, *Phascogale calura* (Marsupialia: Dasyuridae): the adaptive-stress senescence hypothesis. *Journal of Zoology* **241**, 739–755. doi:10.1111/j.1469-7998.1997.tb05745.x

Braithwaite RW (1974) Behavioural changes associated with the population cycle of *Antechinus stuartii* (Marsupialia). *Australian Journal of Zoology* **22**, 45–62. doi:10.1071/ZO9740045

Braithwaite RW (1979) Social dominance and habitat utilisation in *Antechinus stuartii* (Marsupialia). *Australian Journal of Zoology* **27**, 517–528. doi:10.1071/ZO9790517

Braithwaite RW, Griffiths AD (1994) Demographic variation and range contraction in the northern quoll *Dasyurus hallucatus* (Marsupialia: Dasyuridae). *Wildlife Research* **21**, 203–217. doi:10.1071/WR9940203

Braithwaite RW, Lee AK (1979) A mammalian example of semelparity. *American Naturalist* **113**, 151–156. doi:10.1086/283372

Braude S, Ciszek D (1998) Survival of naked mole-rats marked by implantable transponders and toe-clipping. *Journal of Mammalogy* **79**, 360–363. doi:10.2307/1382873

Breckon G, Hulse EV (1972) Difficulties in the management of *Sminthopsis crassicaudata* due to iodine deficiency and thyroid disease. *Laboratory Animals* **6**, 109–118. doi:10.1258/002367772781006329

Bremner-Harrison S, Prodohl PA, Elwood RW (2004) Behavioural trait assessment as a release criterion: boldness predicts early death in a reintroduction programme of captive-bred swift fox (*Vulpes velox*). *Animal Conservation* **7**, 313–320. doi:10.1017/S1367943004001490

Bryant S (1988) Maintenance and breeding of the eastern quoll *Dasyurus viverrinus*. *International Zoo Yearbook* **27**, 119–124. doi:10.1111/j.1748-1090.1988.tb03204.x

Buchmann OLK, Guiler ER (1977) Behaviour and ecology of the Tasmanian devil, *Sarcophilus harrisii*. In *The Biology of Marsupials*. (Eds B Stonehouse and D Gilmore) pp. 155–168. Macmillan, London.

Calaby JH, Taylor JM (1981) Reproduction in two marsupial mice, *Antechinus bellus* and *A. bilarni* (Dasyuridae), of tropical Australia. *Journal of Mammalogy* **62**, 329–341. doi:10.2307/1380709

Canfield PJ, Hartley WJ, Dubey JP (1990) Lesions of toxoplasmosis in Australian Marsupials. *Journal of Comparative Pathology* **103**, 159–167. doi:10.1016/S0021-9975(08)80172-7

Carnio J (1993) Captive management and breeding of the kowari (*Dasyuroides byrnei*) at the Metropolitan Toronto Zoo. In *The Biology and Management of Australasian Carnivorous Marsupials*. (Eds M Roberts, J Carnio, G Crawshaw and M Hutchins) pp. 61–72. Metropolitan Toronto Zoo, Ontario and AAZPA Monotreme and Marsupial Advisory Group, Washington DC.

Chen X, Dickman CR, Thompson MB (1998) Diet of the mulgara, *Dasycercus cristicauda* (Marsupialia: Dasyuridae), in the Simpson Desert, central Australia. *Wildlife Research* **25**, 233–242. doi:10.1071/WR97087

Close RL (1983) Detection of oestrous in the kowari *Dasyuroides byrnei* (Marsupialia: Dasyuridae). *Australian Mammalogy* **6**, 41–43. doi:10.1071/AM83006

Cockburn A, Lazenby-Cohen KA (1992) Use of nest trees by *Antechinus stuartii*, a semelparous lekking marsupial. *Journal of Zoology* **226**, 657–680. doi:10.1111/j.1469-7998.1992.tb07508.x

Cockburn A, Scott MP, Dickman CR (1985) Sex ratio and intrasexual kin competition in mammals. *Oecologia* **66**, 427–429. doi:10.1007/BF00378310

Collins LR (1973) *Monotremes and Marsupials: A Reference for Zoological Institutions*. Smithsonian Institution Press, Washington DC.

Collins L, Flicker L, Heath A, Peterson G, Pojeta K, Rodden M, Rodden R, Watson-Jones J, Baird MW (1993) Captive management and husbandry procedures for tiger quolls (*Dasyurus maculatus*). In *The Biology and Management of Australasian Carnivorous Marsupials*. (Eds M Roberts, J Carnio, G Crawshaw and M Hutchins) pp. 47–54. Metropolitan Toronto Zoo, Ontario and AAZPA Monotreme and Marsupial Advisory Group, Washington DC.

Cooper CE, Vernes K, Cooper T (2009) Fate of wild-caught *Antechinus flavipes* released after physiological experiments. *Australian Mammalogy* **31**, 143–145. doi:10.1071/AM09013

Costelloe E (2014) 'The curious art of quoll quieting' [video]. Wild Life Sydney Zoo. https://www.youtube.com/watch?v=cxklRrrLO6w (accessed 18 December 2022).

Coventry AJ, Dixon JM (1984) Small native mammals from the Chinaman Well area of north-western Victoria. *Australian Mammalogy* **7**, 111–115. doi:10.1071/AM84012

Cowan IMcT, O'Riordan AM, Cowan JSMcT (1974) Energy requirements of the dasyurid marsupial mouse *Antechinus swainsonii* (Waterhouse). *Canadian Journal of Zoology* **52**, 269–275. doi:10.1139/z74-033

Crandall LS (1964) *The Management of Wild Mammals in Captivity*. University of Chicago Press, Chicago.

Croft DB (1982) Communication in the Dasyuridae (Marsupialia): a review. In *Carnivorous Marsupials*. (Ed. M Archer) pp. 291–309. Royal Zoological Society of New South Wales, Sydney.

Crowther MS, Dickman CR, Lyman AJ (1999) *Sminthopsis griseoventer boullangerensis* (Marsupialia: Dasyuridae), a new subspecies in the *S. murina* complex from Boullanger Island, Western Australia. *Australian Journal of Zoology* **47**, 215–243. doi:10.1071/ZO98062

Cunningham M (1994) Parasites of native animals observed at Taronga Zoo from 1979–1994. In *Wildlife*. Proceedings 233. Post Graduate Committee in Veterinary Science. (Ed. DI Bryden) pp. 291–295. University of Sydney, Sydney.

Cuttle P (1982a) A preliminary report on aspects of the behaviour of the dasyurid marsupial *Phascogale tapoatafa*. In *Carnivorous Marsupials*. (Ed. M Archer) pp. 325–332. Royal Zoological Society of New South Wales, Sydney.

Cuttle P (1982b) Life history strategy of the dasyurid marsupial *Phascogale tapoatafa*. In *Carnivorous Marsupials*. (Ed. M Archer) pp. 13–22. Royal Zoological Society of New South Wales, Sydney.

Davies SJJF (1960) A note on two small mammals of the Darwin area. *Journal of the Proceedings of the Royal Society of Western Australia* **43**, 63–66.

Day J, Gooley RM, Hogg CJ, Belov K, Whittington CM, Grueber CE (2019) MHC-associated mate choice under competitive conditions in captive versus wild Tasmanian devils. *Behavioral Ecology* **30**, 1196–1204. doi:10.1093/beheco/arz092

Dempster ER (1995) The social behaviour of captive northern quolls, *Dasyurus hallucatus*. *Australian Mammalogy* **18**, 27–34. doi:10.1071/AM95027

Dennis C (2006) Time to raise the devil. *Nature* **439**, 530. doi:10.1038/439530a

Denny MJS (1982) Review of planigale (Dasyuridae, Marsupialia) ecology. In *Carnivorous Marsupials*. (Ed. M Archer) pp. 131–138. Royal Zoological Society of New South Wales, Sydney.

Denny MJS, Gibson D, Read D (1979) Some results of field observations on planigales. *Bulletin of the Australian Mammal Society* **5**, 27.

Desch CE, Holz PH (2006) *Demodex tortellinioides* n. sp. (Acari: Demodecidae) from the agile antechinus, *Antechinus agilis* Dickman, Parnaby, Crowther and King, 1998 (Marsupialia: Dasyuridae) in Australia. *International Journal of Acarology* **32**, 81–84. doi:10.1080/01647950608684445

Dickman CR (1982) Some ecological aspects of seasonal breeding in *Antechinus* (Dasyuridae: Marsupialia). In *Carnivorous Marsupials*. (Ed. M Archer) pp. 139–150. Royal Zoological Society of New South Wales, Sydney.

Dickman CR (1985) Effects of photoperiod and endogenous control on timing of reproduction in the marsupial genus *Antechinus*. *Journal of Zoology* **206**, 509–524[London]. doi:10.1111/j.1469-7998.1985.tb03555.x

Dickman CR (1988a) Body size, prey size, and community structure in insectivorous mammals. *Ecology* **69**, 569–580.

Dickman CR (1988b) Detection of physical contact interactions among free-living mammals. *Journal of Mammalogy* **69**, 865–868. doi:10.2307/1381651

Dickman CR (1988c) Sex ratio variation in response to interspecific competition. *American Naturalist* **132**, 289–297. doi:10.1086/284851

Dickman CR (1991) Use of trees by ground-dwelling mammals: implications for management. In *Conservation of Australia's Forests*. (Ed. D Lunney) pp. 125–136. Zoological Society of NSW, Sydney.

Dickman CR (1993) Evolution of semelparity in male dasyurid marsupials: a critique and an hypothesis of sperm competition. In *The Biology and Management of Australian Carnivorous Marsupials*. (Eds M Roberts, J Carnio, G Crawshaw and M Hutchins) pp. 25–38. Metropolitan Toronto Zoo, Ontario and AAZPA Monotreme and Marsupial Advisory Group, Washington DC.

Dickman CR (1996) Vagrants in the desert. *Nature Australia* **25**, 54–62.

Dickman CR, Braithwaite RW (1992) Postmating mortality of males in the dasyurid marsupials, *Dasyurus* and *Parantechinus*. *Journal of Mammalogy* **73**, 143–147. doi:10.2307/1381875

Dickman CR, Haythornthwaite AS, McNaught GH, Mahon PS, Tamayo B (2001) Population dynamics of three species of dasyurid marsupials in arid central Australia: a 10-year study. *Wildlife Research* **28**, 493–506. doi:10.1071/WR00023

Dubey JP (1974) Effect of freezing on the infectivity of Toxoplasma cysts to cats. *Journal of the American Veterinary Medical Association* **165**, 534–536.

Dunlop JN, Sawle M (1982) The habitat and life history of the Pilbara ningaui *Ningaui timealeyi*. *Records of the Western Australian Museum* **10**, 47–52.

Eden P (2025) Dasyurids and the numbat. In *Current Therapy in Medicine of Australian Mammals*. Revised Edition. (Eds L Vogelnest and T Portas) pp. 570–577. CSIRO Publishing, Melbourne.

Edgar R, Belcher C (1995) Spotted-tailed quoll *Dasyurus maculatus*. In *The Mammals of Australia*. (Ed. R Strahan) pp. 67–69. Reed Books, Sydney.

Eisenberg JF, Collins LR, Wemmer C (1975) Communication in the Tasmanian devil (*Sarcophilus harrisii*) and a survey of auditory communication in the Marsupialia. *Zeitschrift für Tierpsychologie* **37**, 379–399. doi:10.1111/j.1439-0310.1975.tb00887.x

Elliott TF, Truong C, Jackson SM, Zuniga CL, Trappe JM, Vernes K (2022) Mammalian mycophagy: a global review of ecosystem interactions between mammals and fungi. *Fungal Systematics and Evolution* **9**, 99–159.

Ewer RF (1968a) A preliminary survey of the behaviour in captivity of the dasyurid marsupial, *Sminthopsis crassicaudata* (Gould). *Zeitschrift für Tierpsychologie* **25**, 319–365. doi:10.1111/j.1439-0310.1968.tb00019.x

Ewer RF (1968b) *Ethology of Mammals*. Elek Science, London.

Fancourt BA (2010) Spatial and temporal variation in declining eastern quoll (*Dasyurus viverrinus*) populations in Tasmania. BSc Honours thesis, University of Tasmania, Hobart.

Fancourt BA, Nicol SC, Hawkins CE, Jones ME, Johnson CN (2014) Beyond the disease: is *Toxoplasma gondii* infection causing population declines in the eastern quoll (*Dasyurus viverrinus*)? *International Journal for Parasitology. Parasites and Wildlife* **3**, 102–112. doi:10.1016/j.ijppaw.2014.05.001

Fanning FD (1982) Reproduction, growth and development in *Ningaui* sp. (Dasyuridae, Marsupialia) from the Northern Territory. In *Carnivorous Marsupials*. (Ed. M Archer) pp. 23–37. Royal Zoological Society of New South Wales, Sydney.

Finke MD (2003*a*) Complete nutrient composition of commercially raised invertebrates used as food for insectivores. *Zoo Biology* **21**, 269–285. doi:10.1002/zoo.10031

Finke MD (2003*b*) Gut loading to enhance the nutrient content of insects as food for reptiles: a mathematical approach. *Zoo Biology* **22**, 147–162. doi:10.1002/zoo.10082

Fisher DO, Blomberg SP (2009) Toe-bud clipping of juvenile small marsupials for ecological field research: No detectable negative effects on growth or survival. *Austral Ecology* **34**, 858–865. doi:10.1111/j.1442-9993.2009.01991.x

Fisher DO, Dickman CR (1993) Body size-prey relationships in insectivorous marsupials: tests of three hypotheses. *Ecology* **74**, 1871–1883. doi:10.2307/1939944

Fisher DO, Double MC, Moore BD (2006a) Number of mates and timing of mating affect offspring growth in the small marsupial *Antechinus agilis*. *Animal Behaviour* **71**, 289–297. doi:10.1016/j.anbehav.2005.03.041

Fisher DO, Double MC, Blomberg SP, Jennions MD, Cockburn A (2006b) Post-mating sexual selection increases lifetime fitness of polyandrous females in the wild. *Nature* **444**, 89–92. doi:10.1038/nature05206

Fisher DO, Dickman CR, Jones ME, Blomberg SP (2013) Sperm competition drives the evolution of suicidal reproduction in mammals. *Proceedings of the National Academy of Sciences of the United States of America* **110**, 17910–17914. doi:10.1073/pnas.1310691110

Flannery TF (1995a) *Mammals of New Guinea*. Reed Books, Sydney.

Flannery TF (1995b) *Mammals of the South-west Pacific & Moluccan Islands*. Reed Books, Sydney.

Fleay D (1929) The fat-tailed pouched mouse. *Victorian Naturalist* **45**, 278–280.

Fleay D (1932) Swainson's phascogale (The Bush Mouse). *Victorian Naturalist* **49**, 132–134.

Fleay D (1934) The brush-tailed phascogale: first record of breeding habitat. *Victorian Naturalist* **51**, 89–100.

Fleay D (1935*a*) Notes on the breeding of Tasmanian devils. *Victorian Naturalist* **52**, 100–105.

Fleay D (1935*b*) Breeding of *Dasyurus viverrinus* and general observations on the species. *Journal of Mammalogy* **16**, 10–16. doi:10.2307/1374524

Fleay D (1940) Breeding of the tiger cat. *Victorian Naturalist* **56**, 159–163.

Fleay D (1949) The yellow-footed marsupial mouse. *Victorian Naturalist* **65**, 273–277.

Fleay D (1950) Experiences with Australia's brush-tailed tuan. *Animal Kingdom* **53**(5), 152–157.

Fleay D (1952) The Tasmanian or marsupial devil: its habits and family life. *Australian Museum Magazine* **10**, 275–280.

Fleay D (1961) Breeding the mulgara. *Victorian Naturalist* **78**, 160–167.

Fleay D (1962) The northern quoll, *Satanellus hallucatus*. *Victorian Naturalist* **78**, 288–293.

Fleay D (1965) Australia's 'needle-in-a-haystack' marsupial: vicissitudes in the pursuit and study of Ingram's planigale, the smallest pouch bearer. *Victorian Naturalist* **82**, 195–204.

Fletcher TP (1977) Reproduction in the native cat, *Dasyurus viverrinus* (Shaw). BSc (Hons) thesis. University of Tasmania, Hobart.

Fletcher TP (1983) Endocrinology of reproduction in the dasyurid marsupial *Dasyuroides byrnei* (Spencer). PhD thesis. La Trobe University, Melbourne.

Fletcher TP (1985) Aspects of reproduction in the male eastern quoll, *Dasyurus viverrinus* (Shaw) (Marsupialia: Dasyuridae), with notes on polyoestry in the female. *Australian Journal of Zoology* **33**, 101–110. doi:10.1071/ZO9850101

Flower SS (1931) Contributions to our knowledge of the duration of life in vertebrate animals. *Proceedings of the Zoological Society of London* **101**, 145–234. doi:10.1111/j.1469-7998.1931.tb06192.x

Foster WK (2008) Reproductive strategies of the red-tailed phascogale (*Phascogale calura*). PhD thesis. University of Adelaide, Adelaide.

Foster WK, Bradley AJ, Caton W, Taggart DA (2006) Comparison of growth and development of the red-tailed phascogale (*Phascogale calura*) in three captive colonies. *Australian Journal of Zoology* **54**, 343–352. doi:10.1071/ZO06033

Foster WK, Caton W, Thomas J, Cox S, Taggart DA (2008) Timing of births and reproductive success in captive red-tailed phascogales, *Phascogale calura*. *Journal of Mammalogy* **89**, 1136–1144. doi:10.1644/08-MAMM-A-014.1

Fox BJ (1995) Common dunnart *Sminthopsis murina*. In *The Mammals of Australia*. (Ed. R Strahan) pp. 150–151. Reed Books, Sydney.

Fox BJ, Whitford D (1982) Polyoestry in a predictable coastal environment: reproduction, growth and development in *Sminthopsis murina* (Dasyuridae, Marsupialia). In *Carnivorous Marsupials*. (Ed. M Archer) pp. 39–48. Royal Zoological Society of New South Wales, Sydney.

Frey H (1991) Energetic significance of torpor and other energy-conserving mechanisms in free-living *Sminthopsis crassicaudata* (Marsupialia: Dasyuridae). *Australian Journal of Zoology* **39**, 689–708. doi:10.1071/ZO9910689

Friend GR (1985) Ecological studies of a population of *Antechinus bellus* (Marsupialia: Dasyuridae) in tropical northern Australia. *Australian Wildlife Research* **12**, 151–162. doi:10.1071/WR9850151

Friend GR, Johnson BW, Mitchell DS, Smith GT (1997) Breeding, population dynamics and habitat relationships of *Sminthopsis dolichura* (Marsupialia: Dasyuridae) in semi-arid shrublands of Western Australia. *Wildlife Research* **24**, 245–262. doi:10.1071/WR96070

Frigo L, Woolley PA (1997) Growth and development of pouch young of the stripe-faced dunnart, *Sminthopsis macroura* (Marsupialia: Dasyuridae), in captivity. *Australian Journal of Zoology* **45**, 157–170. doi:10.1071/ZO97002

Gaikhorst G (1999) *Animal Management Guidelines for the Western Quoll Chudditch Dasyurus geoffroii*. Perth Zoo, Western Australia.

Ganslosser U, Meissner K (1984) Behavioural signs of oestrous in *Dasyuroides byrnei* (Marsupialia: Dasyuridae). *Australian Mammalogy* **7**, 223–224. doi:10.1071/AM84025

Geiser F (1986) Thermoregulation and torpor in the kultarr, *Antechinomys laniger* (Marsupialia: Dasyuridae). *Journal of Comparative Physiology Series B* **156**, 751–757. doi:10.1007/BF00692755

Geiser F (1994) Hibernation and daily torpor in marsupials: a review. *Australian Journal of Zoology* **42**, 1–16. doi:10.1071/ZO9940001

Geiser F (2003) Thermal biology and energetics of carnivorous marsupials. In *Predators with Pouches*. (Eds M Jones, C Dickman and M. Archer) pp. 238–253. CSIRO Publishing, Melbourne.

Geiser F (2004) The role of torpor in the life of Australian arid zone mammals. *Australian Mammalogy* **26**, 125–134. doi:10.1071/AM04125

Geiser F, Baudinette RV (1985) The influence of temperature and photophase on daily torpor in *Sminthopsis macroura* (Dasyuridae: Marsupialia). *Journal of Comparative Physiology. A, Neuroethology, Sensory, Neural, and Behavioral Physiology* **156**, 129–134.

Geiser F, Cooper CE (2023) Daily torpor, hibernation, and heterothermy in marsupials. In: *American and Australasian Marsupials: An Evolutionary, Biogeographical, and Ecological Approach*. (Eds NC Cáceres and CR Dickman) pp. 1221–1248. Springer Nature, Cham .

Geiser F, Augee ML, McCarron HCK, Raison JK (1984) Correlates of torpor in the insectivorous dasyurid marsupial *Sminthopsis murina*. *Australian Mammalogy* **7**, 185–191. doi:10.1071/AM84020

George H, Parker G, Coote P (1995) Common wombats: rescue rehabilitation release. Unpublished manuscript.

Gibson DF, Cole JR (1992) Aspects of the ecology of the mulgara, *Dasycercus cristicauda* (Marsupialia: Dasyuridae) in the Northern Territory. *Australian Mammalogy* **15**, 105–112. doi:10.1071/AM92013

Gilfillan SL (2001) An ecological study of a population of *Pseudantechinus macdonnellensis* (Marsupialia: Dasyuridae) in central Australia. I: invertebrate food supply, diet and reproductive strategy. *Wildlife Research* **28**, 469–480. doi:10.1071/WR99062

Gill B (2019) Can captive breeding programs be improved by incorporating mate choice? Investigating female mate choice in fat-tailed dunnarts, *Sminthopsis crassicaudata*. Honours thesis. La Trobe University.

Gill BP, Edwards AM, Parrott ML, Scicluna EL, Robert KA (2023) Female mate choice in the fat-tailed dunnart (*Sminthopsis crassicaudata*) influences offspring sex ratio, but not fecundity. *Australian Journal of Zoology* doi:10.1071/ZO22028.

Glen AS, Dickman CR (2006) Home range, denning behaviour and microhabitat use of the carnivorous marsupial *Dasyurus maculatus* in eastern Australia. *Journal of Zoology* **268**, 347–354[London]. doi:10.1111/j.1469-7998.2006.00064.x

Glen AS, Cardoso MJ, Dickman CR, Firestone KB (2009) Who's your daddy? Paternity testing reveals promiscuity and multiple paternity in the carnivorous marsupial *Dasyurus maculatus* (Marsupialia: Dasyuridae). *Biological Journal of the Linnean Society. Linnean Society of London* **96**, 1–7. doi:10.1111/j.1095-8312.2008.01094.x

Godfrey GK (1969a) Reproduction in a laboratory colony of the marsupial mouse *Sminthopsis larapinta* (Marsupialia: Dasyuridae). *Australian Journal of Zoology* **17**, 637–654. doi:10.1071/ZO9690637

Godfrey GK (1969b) The influence of increased photoperiod on reproduction in the dasyurid marsupial, *Sminthopsis crassicaudata*. *Journal of Mammalogy* **50**, 132–133. doi:10.2307/1378643

Godfrey GK, Crowcroft P (1971) Breeding the fat-tailed marsupial mouse *Sminthopsis crassicaudata* in captivity. *International Zoo Yearbook* **11**, 33–38. doi:10.1111/j.1748-1090.1971.tb01839.x

Godsell J (1982a) Notes on housing the eastern quoll *Dasyurus viverrinus* in outdoor enclosures. In *The Management of Australian Mammals in Captivity*. (Ed. DD Evans) pp. 45. Zoological Board of Victoria, Melbourne.

Godsell J (1982b) The population ecology of the eastern quoll (Dasyuridae: Marsupialia), in southern Tasmania. In *Carnivorous Marsupials*. (Ed. M Archer) pp. 199–207. Royal Zoological Society of New South Wales, Sydney.

Godsell J (1995) Eastern quoll *Dasyurus viverrinus*. In *The Mammals of Australia*. (Ed. R Strahan) pp. 70–71. Reed Books, Sydney.

Gooley RM, Hogg CJ, Belov K, Grueber CE (2018) The effects of group versus intensive housing on the retention of genetic diversity in insurance populations. *BMC Zoology* **3**, 2. doi:10.1186/s40850-017-0026-x

Green B (1984) Composition of milk and energetics of growth in marsupials. *Symposia of the Zoological Society of London* **51**, 369–387.

Green B, Merchant J, Newgrain K (1987) Milk composition in the eastern quoll, *Dasyurus viverrinus* (Marsupialia: Dasyuridae). *Australian Journal of Biological Sciences* **40**, 379–387. doi:10.1071/BI9870379

Green B, Merchant J, Newgrain K (1997) Lactational energetics of a marsupial carnivore, the eastern quoll (*Dasyurus viverrinus*). *Australian Journal of Zoology* **45**, 295–306. doi:10.1071/ZO97003

Green K, Crowley H (1989) Energetics and behaviour of active subnivean insectivores *Antechinus swainsonii* and *A. stuartii* (Marsupialia: Dasyuridae) in the Snowy Mountains. *Australian Wildlife Research* **16**, 509–516. doi:10.1071/WR9890509

Green RH (1967) Notes on the devil (*Sarcophilus harrisii*) and the quoll (*Dasyurus viverrinus*) in north-eastern Tasmania. *Records of the Queen Victoria Museum* **27**, 1–13.

Green RH, Scarborough TJ (1990) The spotted-tailed quoll *Dasyurus maculatus* (Dasyuridae, Marsupialia) in Tasmania. *Tasmanian Naturalist* **100**, 1–15.

Griner LA (1979) Neoplasms in Tasmanian devils (*Sarcophilus harrisii*). *Journal of the National Cancer Institute* **62**, 589–595. doi:10.1093/jnci/62.3.589

Guiler ER (1961) The breeding season of the thylacine. *Journal of Mammalogy* **42**, 396–397. doi:10.2307/1377040

Guiler ER (1970) Observations on the Tasmanian devil, *Sarcophilus harrisii* (Marsupialia: Dasyuridae). II: reproduction, breeding, and growth of pouch young. *Australian Journal of Zoology* **18**, 63–70. doi:10.1071/ZO9700063

Guiler ER (1978) Observations on the Tasmanian devil, *Sarcophilus harrisii* (Marsupialia, Dasyuridae), at Granville Harbour, 1966–75. *Papers of the Royal Society of Tasmania* **112**, 161–188. doi:10.26749/rstpp.112.161

Guiler E (1985) *Thylacine: The Tragedy of the Tasmanian Tiger*. Oxford University Press, Melbourne.

Gunn R (1863) Letter announcing the shipment of living thylacines, with remarks on their habits. *Proceedings of the Zoological Society of London* **31**, 103–104.

Hall S (1980) Diel activity of three small mammals coexisting in forest in southern Victoria. *Australian Mammalogy* **3**, 67–79. doi:10.1071/AM80008

Halley M (1992) Maintenance and captive breeding of the brush-tailed phascogale *Phascogale tapoatafa*. *International Zoo Yearbook* **31**, 71–78. doi:10.1111/j.1748-1090.1991.tb02365.x

Hamede RK, McCallum H, Jones M (2013) Biting injuries and transmission of Tasmanian devil facial tumour disease. *Journal of Animal Ecology* **82**, 182–190. doi:10.1111/j.1365-2656.2012.02025.x

Happold M (1972) Maternal and juvenile behaviour in the marsupial jerboa *Antechinomys spenceri* (Dasyuridae). *Australian Mammalogy* **1**, 27–37. doi:10.1071/AM72005

Harris GP (1807) Description of two new species of *Didelphis* from Van Diemen's Land. *Transactions of the Linnean Society of London* **9**(1), 174–178. doi:10.1111/j.1096-3642.1818.tb00336.x

Hartnett CM, Parrott ML, Mulder RA, Coulson G, Magrath MJL (2018) Opportunity for female mate choice improves reproductive outcomes in the conservation breeding program of the eastern barred bandicoot (*Perameles gunnii*). *Applied Animal Behaviour Science* **199**, 67–74. doi:10.1016/j.applanim.2017.10.008

Hawkins CE, Baars C, Hesterman H, Hocking GJ, Jones ME, Lazenby B, Mann D, Mooney N, Pemberton D, Pyecroft S, Restani M, Wiersma J (2006) Emerging disease and population decline of an island endemic, the Tasmanian devil *Sarcophilus harrisii*. *Biological Conservation* **131**, 307–324. doi:10.1016/j.biocon.2006.04.010

Hawkins MR (1998) Effects of olfactory enrichment on Australian marsupial species. In *Proceedings of the Third International Conference on Behavioural Enrichment*. Sea World, Orlando, Florida. 12–17 October 1997. pp. 135–149. Shape of Enrichment Inc., San Diego, California.

Heinsohn GE (1970) World's smallest marsupial: the flat-headed marsupial mouse. *Animals (Basel)* **13**, 220–222.

Hellingham H (1999) Detecting oestrous cycles in the northern territory quoll (*Dasyurus hallucatus*) at Melbourne Zoo. ASZK/ARAZPA Conference, Alice Springs Desert Park.

Hesterman H, Jones SM, Schwarzenberger F (2008) Pouch appearance is a reliable indicator of the reproductive status in the Tasmanian devil and the spotted-tailed quoll. *Journal of Zoology* **275**, 130–138. doi:10.1111/j.1469-7998.2008.00419.x

Hill JP, Hill WCO (1955) The growth stages of the pouch young of the native cat (*Dasyurus viverrinus*) together with observations on the anatomy of the new-born young. *Transactions of the Zoological Society of London* **28**, 349–352. doi:10.1111/j.1096-3642.1955.tb00003.x

Hill JP, O'Donoghue CH (1913) The reproductive cycle in the marsupial *Dasyurus viverrinus*. *The Quarterly Journal of Microscopical Science* **59**, 133–174.

Hine RS (1988) *Concise Veterinary Dictionary*. Oxford University Press, Oxford.

Hogg CJ, Hockley J (Eds) (2019) *DPIPWE/ZAA Husbandry Guidelines for the Tasmanian Devil,* Sarcophilus harrisii. 4th edn. Zoo and Aquarium Association, Australia.

Hogg CJ, Fox S, Pemberton D, Belov K (Eds) (2019a) *Saving the Tasmanian Devil: Recovery Through Science-based Management*. CSIRO Publishing, Melbourne.

Hogg CJ, Wright B, Morris KM, Lee AV, Ivy JA, Grueber CE, Belov K (2019b) Founder relationships and conservation management: empirical kinships reveal the effect on breeding programmes when founders are assumed to be unrelated. *Animal Conservation* **22**, 348–361. doi:10.1111/acv.12463

Hohnen R, Ashby J, Tuft K, McGregor H (2013) Individual identification of northern quolls (*Dasyurus hallucatus*) using remote cameras. *Australian Mammalogy* **35**, 131–135. doi:10.1071/AM12015

Holleley CE, Dickman CR, Crowther MS, Oldroyd BP (2006) Size breeds success: multiple paternity, multivariate selection and male semelparity in a small marsupial, *Antechinus stuartii*. *Molecular Ecology* **15**, 3439–3448. doi:10.1111/j.1365-294X.2006.03001.x

Holloway JC, Geiser F (1996) Effect of photoperiod on torpor and activity of *Sminthopsis crassicaudata* (Marsupialia). In *Adaptations to Cold: Tenth International Hibernation Symposium*. (Eds F Geiser, AJ Hulbert and SC Nicol) pp. 95–102. University of New England Press, Armidale.

Holz PH (1998) Nodules caused by demodex-like mites in the skin of tiger quolls (*Dasyurus maculatus*). *Australian Veterinary Journal* **76**, 59–60. doi:10.1111/j.1751-0813.1998.tb15689.x

Holz PH (2008) Dasyurids. In *Medicine of Australian Mammals*. (Eds L Vogelnest and R Woods) pp. 359–382. CSIRO Publishing, Melbourne.

Holz PH, Little PB (1995) Degenerative leukoencephalopathy and myelopathy in dasyurids. *Journal of Wildlife Diseases* **31**, 509–513. doi:10.7589/0090-3558-31.4.509

Horner BE, Taylor JM (1959) Results of the Archbold Expeditions. No. 80. Observations on the biology of the marsupial mouse, *Antechinus flavipes flavipes*. *American Museum Novitates* **1972**, 1–24.

Howe D (1975) Observations on a captive marsupial mole *Notoryctes typhlops*. *Australian Mammalogy* **1**, 361–365. doi:10.1071/AM75007

Hughes RL (1982) Reproduction in the Tasmanian devil *Sarcophilus harrisii* (Dasyuridae, Marsupialia). In *Carnivorous Marsupials*. (Ed. M Archer) pp. 49–63. Royal Zoological Society of New South Wales, Sydney.

Hughes RL (1999a) Reproduction in wild caught female Tasmanian devils *Sarcophilus harrisii*. Abstract. Australian Mammal Society Newsletter.

Hughes RL (1999b) Gestational development and placentation in the Tasmanian devil *Sarcophilus harrisii*. Abstract. *Australian Mammal Society Newsletter.*

Hunter S (1991) Fostering tuans. *Thylacinus* **16**, 10.

Hutson GD (1976) Grooming behaviour and birth in the dasyurid marsupial *Dasyuroides byrnei*. *Australian Journal of Zoology* **24**, 277–282. doi:10.1071/ZO9760277

Indigo N, Smith J, Webb JK, Phillips B (2018) Not such silly sausages: evidence suggests northern quolls exhibit aversion to toads after training with toad sausages. *Austral Ecology* **43**, 592–601. doi:10.1111/aec.12595

Innes EA (1997) Toxoplasmosis: comparative species susceptibility and host immune response. *Comparative Immunology, Microbiology and Infectious Diseases* **20**, 131–138. doi:10.1016/S0147-9571(96)00038-0

International Air Transport Association (IATA) (2022) *Live Animals Regulations*. 48th edn. International Air Transport Association, Geneva Switzerland.

Jackson SM, Groves C (2015) *Taxonomy of Australian Mammals*. CSIRO Publishing, Melbourne.

Johnson K (1991) The mole who comes in from the sun. *Wildlife Australia* (Spring), 8–9.

Johnson KA (1995) Marsupial mole. *Notoryctes typhlops*. In *The Mammals of Australia*. (Ed. R Strahan). pp. 409–411. Reed Books, Sydney.

Johnson KA, Walton DW (1989) Notoryctidae. In *Fauna of Australia. Mammalia*. Vol. 1B. (Eds DW Walton and BJ Richardson) pp. 591–602. Australian Government Publishing Service, Canberra.

Jones LD, Cooper RW, Harding RS (1972) Composition of mealworm *Tenebrio molitor* larvae. *Journal of Zoo Animal Medicine* **3**(4), 34–41. doi:10.2307/20094161

Jones M (1995) Tasmanian devil *Sarcophilus harrisii*. In *The Mammals of Australia*. (Ed. R Strahan) pp. 82–84. Reed Books, Sydney.

Jones ME, Stoddart DM (1998) Reconstruction of the predatory behaviour of the extinct marsupial thylacine (*Thylacinus cynocephalus*). *Journal of Zoology* **246**, 239–246[London]. doi:10.1111/j.1469-7998.1998.tb00152.x

Jones ME (2003) Convergence in ecomorphology and guild structure among marsupial and placental carnivores. In *Predators with Pouches: The Biology of Carnivorous Marsupials*. (Eds ME Jones, CR Dickman and M Archer) pp. 285–296. CSIRO Publishing, Melbourne.

Jones ME, Cockburn A, Hamede R, Hawkins C, Westerman H, Lachish S, Mann D, McCallum H, Pemberton D (2008) Life-history change in disease-ravaged Tasmanian devil populations. *Proceedings of the National Academy of Sciences of the United States of America* **105**, 10023–10027. doi:10.1073/pnas.0711236105

Keeley T, O'Brien JK, Fanson BG, Masters K, McGreevy PD (2012) The reproductive cycle of the Tasmanian devil (*Sarcophilus harrisii*) and factors associated with reproductive success in captivity. *General and Comparative Endocrinology* **176**, 182–191. doi:10.1016/j.ygcen.2012.01.011

Kelly A (1993) Management of Tasmanian devils (*Sarcophilus harrisii*) at the Tasmanian Wildlife Park. In *The Biology and Management of Australasian Carnivorous Marsupials*. (Eds M Roberts, J Carnio, G Crawshaw and M Hutchins) pp. 79–84. Metropolitan Toronto Zoo, Ontario and AAZPA Monotreme and Marsupial Advisory Group, Washington DC.

Kitchener DJ (1981) Breeding, diet and habitat preferences of *Phascogale calura* (Gould, 1844) (Marsupialia: Dasyuridae) in the southern wheatbelt, Western Australia. *Records of the Western Australian Museum* **9**, 173–186.

Kitchener DJ, Cooper N, Bradley A (1986) Reproduction in the male *Ningaui* (Marsupialia: Dasyuridae). *Australian Wildlife Research* **13**, 13–25. doi:10.1071/WR9860013

Körtner G, Claridge A, Ballard G (2019) Denning behaviour of female spotted-tailed quolls during the breeding season. *Australian Journal of Zoology* **67**, 145–152. doi:10.1071/ZO20012

Kraaijeveld-Smit FJL, Ward SJ, Temple-Smith PD (2002) Multiple paternity in a field population of a small carnivorous marsupial, the agile antechinus, *Antechinus agilis*. *Behavioral Ecology and Sociobiology* **52**, 84–91. doi:10.1007/s00265-002-0485-z

Krajewski C, Woolley PA, Westerman M (2000) The evolution of reproductive strategies in dasyurid marsupials: implications of molecular phylogeny. *Biological Journal of the Linnean Society. Linnean Society of London* **71**, 417–435. doi:10.1111/j.1095-8312.2000.tb01267.x

Kreiss A, Brown GK, Tovar C, Lyons AB, Woods GM (2015) Evidence for induction of humoral and cytotoxic immune responses against devil facial tumor disease cells in Tasmanian devils (*Sarcophilus harrisii*) immunized with killed cell preparations. *Vaccine* **33**(26), 3016–3025. doi:10.1016/j.vaccine.2015.01.039

Lacy RC, Balou JD, Pollak JP (2011) PMx: software package for demographic and genetic analysis and management of pedigreed populations. *Methods in Ecology and Evolution* **3**, 433–437. doi:10.1111/j.2041-210X.2011.00148.x

Lambert C (2000) *Animal Management Guidelines for Dibbler* (Parantechinus apicalis). Perth Zoo, Western Australia.

Lambert C, Mills H (2006) Husbandry and breeding of the dibbler *Parantechinus apicalis* at Perth Zoo. *International Zoo Yearbook* **40**, 290–301. doi:10.1111/j.1748-1090.2006.00290.x

Lambert C, Gaikhorst G, Matson P (2011) Captive breeding of the sandhill dunnart, *Sminthopsis psammophila* (Marsupialia: Dasyuridae): reproduction, husbandry and growth and development. *Australian Mammalogy* **33**, 21–27. doi:10.1071/AM10004

Lazenby BT, Tobler MW, Brown WE, Hawkins CE, Hocking GJ, Hume F, Huxtable S, Iles P, Jones ME, Lawrence C, Thalmann S, Wise P, Williams H, Fox S, Pemberton D (2018) Density trends and demographic signals uncover the long-term impact of transmissible cancer in Tasmanian devils. *Journal of Applied Ecology* **55**, 1368–1379. doi:10.1111/1365-2664.13088

Lazenby-Cohen KA (1991) Communal nesting in *Antechinus stuartii* (Marsupialia: Dasyuridae). *Australian Journal of Zoology* **39**, 273–283. doi:10.1071/ZO9910273

Lazenby-Cohen KA, Cockburn A (1988) Lek promiscuity in a semelparous mammal, *Antechinus stuartii* (Marsupialia: Dasyuridae)? *Behavioral Ecology and Sociobiology* **22**, 195–202. doi:10.1007/BF00300569

Lee AK, Cockburn A (1985) *The Evolutionary Ecology of Marsupials*. Cambridge University Press, Cambridge.

Lee AK, Bradley AJ, Braithwaite RW (1977) Corticosteroid levels and male mortality in *Antechinus stuartii*. In *The Biology of Marsupials*. (Eds B Stonehouse and D Gilmore) pp. 209–220. Macmillan, London.

Lee AK, Woolley P, Braithwaite RW (1982) Life history strategies of dasyurid marsupials. In *Carnivorous Marsupials*. (Ed. M Archer) pp. 1–11. Royal Zoological Society of New South Wales, Sydney.

Letendre C, Sawyer E, Young LJ, Old JM (2018) Immunosenescence in a captive semelparous marsupial, the red-tailed phascogale (*Phascogale calura*). *BMC Zoology* **3**, 10. doi:10.1186/s40850-018-0036-3

Leung LKP (1999) Ecology of Australian tropical rainforest mammals. I: the Cape York antechinus, *Antechinus leo* (Dasyuridae: Marsupialia). *Wildlife Research* **26**, 287–306. doi:10.1071/WR96042

Lidicker W, Marlow B (1970) A review of the dasyurid marsupial genus *Antechinomys* Krefft. *Mammalia* **34**, 212–227. doi:10.1515/mamm.1970.34.2.212

Lindner E, Fuelling O (2002) Marking methods in small mammals: ear-tattoo as an alternative to toe-clipping. *Journal of Zoology* **256**, 159–163. doi:10.1017/S0952836902000195

Lunney D (1995) White-footed dunnart *Sminthopsis leucopus*. In *The Mammals of Australia*. (Ed. R Strahan) pp. 143–145. Reed Books, Sydney.

Lunney D, Ashby E (1987) Population changes in *Sminthopsis leucopus* (Gray) (Marsupialia): Dasyuridae), and other small mammal species in forest regenerating from logging and fire near Bega, New south Wales. *Australian Wildlife Research* **14**, 275–284. doi:10.1071/WR9870275

Lydekker R (1894) *A Handbook of the Marsupialia and Monotremata*. W.H. Allen & Co., London.

Lyons R (2020) *Introduction to the Care and Rehabilitation of Microbats (Focussing on species of south east Queensland)*. Version 6.0. Wildcare Australia, Nerang, Queensland.

Mack G (1961) Mammals of south-western Queensland. *Memoirs of the Queensland Museum* **13**, 213–229.

Markus N (1996) Methodology of enrichment and applications on Tasmanian devils. *Thylacinus* **21**(3), 3–8.

Marlow BJ (1961) Reproductive behaviour in the marsupial mouse, *Antechinus flavipes* (Waterhouse)(Marsupialia) and the development of the pouch young. *Australian Journal of Zoology* **9**, 203–218. doi:10.1071/ZO9610203

Martin PG (1965) The potentialities of the fat-tailed marsupial mouse, *Sminthopsis crassicaudata* (Gould) as a laboratory animal. *Australian Journal of Zoology* **13**, 559–562. doi:10.1071/ZO9650559

Martin RD, Rivers JPW, Cowgill UM (1976) Culturing mealworms as food for animals in captivity. *International Zoo Yearbook* **16**, 63–70. doi:10.1111/j.1748-1090.1976.tb00130.x

Mason ED, Firn J, Hines HB, Baker AM (2017) Breeding biology and growth in a new, threatened carnivorous marsupial. *Mammal Research* **62**, 179–187. doi:10.1007/s13364-016-0303-z

Masters P (1998) The Mulgara *Dasycercus cristicauda* (Marsupialia: Dasyuridae) at Uluru. *Australian Mammalogy* **20**, 403–407. doi:10.1071/AM98403

Mathews F, Orros M, McLaren G, Gelling M, Foster R (2005) Keeping fit on the ark: assessing the suitability of captive-bred animals for release. *Biological Conservation* **121**, 569–577. doi:10.1016/j.biocon.2004.06.007

McAllan BM, Dickman CR (1986) The role of photoperiod in the timing of reproduction in the dasyurid marsupial *Antechinus stuartii*. *Oecologia* **68**, 259–264. doi:10.1007/BF00384797

McAllan BM, Geiser F (2006) Photoperiod and the timing of reproduction in *Antechinus flavipes* (Dasyuridae: Marsupialia). *Mammalian Biology* **71**, 129–138. doi:10.1016/j.mambio.2006.01.005

McAllan BM, Joss JM, Firth BT (1991) Phase delay of the natural photoperiod alters reproductive timing in the marsupial *Antechinus stuartii*. *Journal of Zoology* **225**, 633–646[London]. doi:10.1111/j.1469-7998.1991.tb04330.x

McAllan BM, Crowther MS, Dickman CR (2006) Photoperiod as a reproductive cue in the marsupial genus *Antechinus*: ecological and evolutionary consequences. *Biological Journal of the Linnean Society. Linnean Society of London* **87**, 365–379. doi:10.1111/j.1095-8312.2006.00571.x

McDonald IR, Lee AK, Bradley AJ, Than KA (1981) Endocrine changes in dasyurid marsupials with differing mortality patterns. *General and Comparative Endocrinology* **44**, 292–301. doi:10.1016/0016-6480(81)90004-6

Meissner K, Ganslosser U (1985) Development of young in the kowari, *Dasyuroides byrnei* Spencer, 1896. *Zoo Biology* **4**, 351–359. doi:10.1002/zoo.1430040405

Merchant JC, Newgrain K, Green B (1984) Growth of the eastern quoll *Dasyurus viverrinus* (Shaw), (Marsupialia) in captivity. *Australian Wildlife Research* **11**, 21–29. doi:10.1071/WR9840021

Michener GR (1969) Notes on the breeding and young of the crest-tailed marsupial mouse, *Dasycercus cristicauda*. *Journal of Mammalogy* **50**, 633–635. doi:10.2307/1378801

Miller DS, Mitchell GF, Biggs B, McCracken H, Myroniuk P, Hewish M (2000) Detection of agglutinating antibodies to *Toxoplasma gondii* in sera from captive mainland Australian eastern barred bandicoot (*Perameles gunnii*). *Journal of Wildlife Diseases* **36**, 213–218. doi:10.7589/0090-3558-36.2.213

Millis AL (1995) Reproductive and genetic studies of the brush-tailed phascogale, *Phascogale tapoatafa* (Marsupialia: Dasyuridae). BSc Honours thesis. Monash University, Melbourne.

Millis AL, Bradley AJ (2001) Reproduction in the squirrel glider, *Petaurus norfolcensis* (Petauridae) in south-east Queensland. *Australian Journal of Zoology* **49**, 139–154. doi:10.1071/ZO01003

Millis AL, Taggart DA, Bradley AJ, Phelan J, Temple-Smith PD (1999) Reproductive biology of the brush-tailed phascogale, *Phascogale tapoatafa* (Marsupialia: Dasyuridae). *Journal of Zoology* **248**, 325–335. doi:10.1111/j.1469-7998.1999.tb01032.x

Mills H, Bencini R (2000) New evidence for facultative male die-off in island populations of dibblers, *Parantechinus apicalis*. *Australian Journal of Zoology* **48**, 501–510. doi:10.1071/ZO00025

Mills H, German RZ, Lambert C, Bradley MP (2000) Growth and sexual dimorphism in the dibbler, *Parantechinus apicalis* (Marsupialia, Dasyuridae). *Australian Mammalogy* **21**, 239–243. doi:10.1071/AM00239

Mitchell PC (1911) On longevity and relative viability in mammals and birds; with a note on the theory of longevity. *Proceedings of the Zoological Society of London* **81**, 425–548. doi:10.1111/j.1096-3642.1911.tb01942.x

Morrison RGB (1975) Emergence of the pygmy Antechinus. *Australian Natural History* **18**, 164–167.

Morton SR (1978*a*) An ecological study of *Sminthopsis crassicaudata* (Marsupialia: Dasyuridae). III: reproduction and life history. *Australian Wildlife Research* **5**, 183–211. doi:10.1071/WR9780183

Morton SR (1978*b*) An ecological study of *Sminthopsis crassicaudata* (Marsupialia: Dasyuridae). II: behaviour and social organisation. *Australian Wildlife Research* **5**, 163–182. doi:10.1071/WR9780163

Morton SR (1978*c*) Torpor and nest-sharing in free-living *Sminthopsis crassicaudata* (Marsupialia) and *Mus musculus* (Rodentia). *Journal of Mammalogy* **59**, 569–575. doi:10.2307/1380234

Morton SR (1980) Ecological correlates of caudal fat storage in small mammals. *Australian Mammalogy* **3**, 81–86. doi:10.1071/AM80009

Morton SR, Armstrong MD, Braithwaite RW (1987) The breeding season of *Sminthopsis virginiae* (Marsupialia: Dasyuridae) in the Northern Territory. *Australian Mammalogy* **10**, 41–42. doi:10.1071/AM87009

Morton SR, Dickman CR, Fletcher TP (1989) Dasyuridae. In *Fauna of Australia. Mammalia*. Vol. 1B. (Ed. DW Walton) pp. 560–582. Australian Government Publishing Service, Canberra.

Nagy KA, Seymour RS, Lee AK, Braithwaite R (1978) Energy and water budgets in free-living *Antechinus stuartii* (Marsupialia: Dasyuridae). *Journal of Mammalogy* **59**, 60–68. doi:10.2307/1379875

Naylor R, Richardson SJ, McAllan BM (2008) Boom and bust: a review of the physiology of the marsupial genus *Antechinus*. *Journal of Comparative Physiology. B, Biochemical, Systemic, and Environmental Physiology* **178**, 545–562. doi:10.1007/s00360-007-0250-8

Nelson JE (1992) Developmental staging in a marsupial, *Dasyurus hallucatus*. *Anatomy and Embryology* **185**, 335–354. doi:10.1007/BF00188546

Nelson JE, Gemmell RT (2003) Birth in the northern quoll *Dasyurus hallucatus* (Marsupialia: Dasyuridae). *Australian Journal of Zoology* **51**, 187–198. doi:10.1071/ZO02016

Nelson JE, Smith G (1971) Notes on the growth rates in native cats of the family Dasyuridae. *International Zoo Yearbook* **11**, 38–41. doi:10.1111/j.1748-1090.1971.tb01840.x

Nightingale R (2013). Understanding social behaviour of Tasmanian devils (*Sarcophilus harrisii*) in captivity: Implications for conservation and management. Honours thesis. Deakin University.

Nutting W, Woolley P (1965) Pathology in *Antechinus stuartii* (Marsupialia) due to *Demodex* sp. *Parasitology* **55**, 383–389. doi:10.1017/S0031182000068852

O'Neil T (2011) *Animal Enrichment Strategies for Promoting Natural Behaviors in Captive Populations of Tasmanian Devils (*Sarcophilus harrisii*)*. Healesville Sanctuary, Victoria.

O'Reilly HM, Armstrong SM, Coleman GJ (1984) Response to variations in lighting schedules on the circadian activity rhythms of *Sminthopsis macroura froggatti* (Marsupialia: Dasyuridae). *Australian Mammalogy* **7**, 89–99. doi:10.1071/AM84009

Oakwood M (1997) The ecology of the northern quoll *Dasyurus hallucatus*. PhD thesis. Australian National University, Canberra.

Oakwood M (2002) Spatial and social organisation of the carnivorous marsupial *Dasyurus hallucatus* (Marsupialia: Dasyuridae). *Journal of Zoology* **257**, 237–248. doi:10.1017/S0952836902000833

Oakwood M, Spratt DM (2000) Parasites of the northern quoll, *Dasyurus hallucatus* (Marsupialia: Dasyuridae) in tropical savanna, Northern Territory. *Australian Journal of Zoology* **48**, 79–90. doi:10.1071/ZO99056

Oakwood M, Bradley AJ, Cockburn A (2001) Semelparity in a large marsupial. *Proceedings. Biological Sciences* **268**(1465), 407–411. doi:10.1098/rspb.2000.1369

Obendorf DL (1993) Diseases of dasyurid marsupials. In *The Biology and Management of Australasian Carnivorous Marsupials*. (Eds M Roberts, J Carnio, G Crawshaw and M Hutchins) pp. 39–46. Metropolitan Toronto Zoo, Ontario and AAZPA Monotreme and Marsupial Advisory Group, Washington DC.

Oglesby R (1978) Notes on breeding the Tasmanian devil in captivity. *Thylacinus* **3**(1–4), 17–20.

Old JM, Ong OTW, Stannard HJ (2021) Red-tailed phascogales: a review of their biology and importance as model marsupial species. *The Journal of Experimental Zoology* **335**, 217–227. doi:10.1002/jez.2438

Parrott ML (2006) Female choice and breeding success in a small carnivorous marsupial, the agile antechinus (*Antechinus agilis*). PhD thesis, University of Melbourne.

Parrott ML, Ward SJ, Temple-Smith PD (2006) Genetic similarity, not male size, influences female mate choice in the agile antechinus (*Antechinus agilis*). *Australian Journal of Zoology* **54**, 319–323. doi:10.1071/ZO06030

Parrott ML, Ward SJ, Temple-Smith PD, Selwood L (2007*a*) Effects of drought on weight, survival and breeding success of agile antechinus (*Antechinus agilis*), dusky antechinus (*A. swainsonii*) and bush rats (*Rattus fuscipes*). *Wildlife Research* **34**, 437–442. doi:10.1071/WR07071

Parrott ML, Ward SJ, Temple-Smith PD (2007*b*) Olfactory cues, genetic relatedness and female mate choice in the agile antechinus (*Antechinus agilis*). *Behavioral Ecology and Sociobiology* **61**, 1075–1079. doi:10.1007/s00265-006-0340-8

Parrott ML, Ward SJ, Temple-Smith PD, Selwood L (2015) Females choose mates based on genetic relatedness in a small dasyurid marsupial, the agile antechinus (*Antechinus agilis*). *PLoS One* **10**(4), e0122381. doi:10.1371/journal.pone.0122381

Parrott ML, Nation A, Selwood L (2019a) Female mate choice significantly increases captive breeding success, and scents can be frozen to determine choice, in the stripe-faced dunnart. *Applied Animal Behaviour Science* **214**, 95–101. doi:10.1016/j.applanim.2019.03.006

Parrott ML, Dowling E, Faulkner T, Hughes C, Keeley T, Kelly A, Miller KA, Pohl B, O'Brien KA, Hogg CJ (2019b) Captive research: working together for the common good. In *Saving the Tasmanian Devil: Recovery Through Science-Based Management*. (Eds CJ Hogg, S Fox, D Pemberton and K Belov) pp. 251–265. CSIRO Publishing, Melbourne.

Pavey CR, Geiser F (2008) Basking and diurnal foraging in the dasyurid marsupial (*Pseudantechinus macdonnellensis*). *Australian Journal of Zoology* **56**, 129–135. doi:10.1071/ZO08032

Pavey CR, Burwell CJ, Benshemesh J (2012) Prey selection and diet of marsupial moles. *Journal of Zoology* **287**, 115–123. doi:10.1111/j.1469-7998.2011.00889.x

Pavey CR, Molyneux J, Körtner G, Friend JA, Vaughan-Higgins R, Fancourt BA, Heiniger J, Waudby HP (2022) Marsupial carnivores. In *Wildlife Research in Australia: Practical and Applied Methods* (Eds BP Smith, HP Waudby, C Alberthsen and JO Hampton) pp. 389–400. CSIRO Publishing: Melbourne.

Pearse AM (1981) Aspects of the biology of *Uropsylla tasmanica* Rothschild (Siphonaptera). MSc thesis. University of Tasmania, Hobart.

Pemberton D (1990) Social organisation and behaviour of the Tasmanian devil, *Sarcophilus harrisii*. PhD thesis. Zoology Department, University of Tasmania, Hobart.

Pemberton D, Renouf D (1993) A field study of communication and social behaviour of the Tasmanian devil at feeding sites. *Australian Journal of Zoology* **41**, 507–526. doi:10.1071/ZO9930507

Phillips BT, Jackson SM (2003) Growth and development of the Tasmanian devil *Sarcophilus harrisii* at Healesville Sanctuary. *Zoo Biology* **22**, 497–505. doi:10.1002/zoo.10092

Phillips CJC, Davies E, Lisle A (2012) Housing systems to meet the behavioural needs of a solitary mammal with an extensive home range: the Julia Creek dunnart (*Sminthopsis douglasi*). *Applied Animal Behaviour Science* **141**, 36–42. doi:10.1016/j.applanim.2012.07.007

Pollock K, Booth R, Wilson R, Keeley T, Grogan K, Kennerley P, Johnston SD (2010) Oestrus in the Julia Creek dunnart (*Sminthopsis douglasi*) is associated with wheel running behaviour but not necessarily changes in body weight, food consumption or pouch morphology. *Animal Reproduction Science* **117**, 135–146. doi:10.1016/j.anireprosci.2009.03.005

Pollock TI, Parrott ML, Evans AR, Hocking DP (2021a) Wearing the devil down: rate of tooth wear varies between wild and captive Tasmanian devils. *Zoo Biology* **40**, 444–457. doi:10.1002/zoo.21632

Pollock TI, Evans AR, Hunter D, Parrott ML, Zabinskas M, Hocking DP (2021b) Torn limb from limb: the ethology of prey-processing in Tasmanian devils (*Sarcophilus harrisii*). *Australian Mammalogy* **44**, 126–138. doi:10.1071/AM21006

Potter TI, Greenville AC, Dickman CR (2018a) Assessing the potential for intraguild predation among taxonomically disparate micro-carnivores: marsupials and arthropods. *Royal Society Open Science* **5**, 171872. doi:10.1098/rsos.171872

Potter TI, Stannard HJ, Greenville AC, Dickman CR (2018b) Understanding selective predation: are energy and nutrients important? *PLoS One* **13**, e0201300. doi:10.1371/journal.pone.0201300

Pye RJ, Pemberton D, Tovar C, Tubio JM, Dun KA, Fox S, Darby J, Hayes D, Knowles GW, Kreiss A, Siddle HV, Swift K, Lyons AB, Murchison EP, Woods GM (2016) A second transmissible cancer in Tasmanian devils. *Proceedings of the National Academy of Sciences of the United States of America* **113**, 374–379. doi:10.1073/pnas.1519691113

Pyecroft SB (2025) Devil facial tumour disease. In *Current Therapy in Medicine of Australian Mammals*. Revised Edition. (Eds L Vogelnest and T Portas) pp. 578–587. CSIRO Publishing, Melbourne.

Read DG (1984a) Reproduction and breeding season of *Planigale gilesi* and *P. tenuirostris* (Marsupialia: Dasyuridae). *Australian Mammalogy* **7**, 161–173. doi:10.1071/AM84017

Read DG (1984b) Movements and home ranges of three sympatric dasyurids, *Sminthopsis crassicaudata, Planigale gilesi* and *P. tenuirostris* (Marsupialia), in semi-arid western New South Wales. *Australian Wildlife Research* **11**, 223–234. doi:10.1071/WR9840223

Read DG (1985) Development and growth of *Planigale tenuirostris* (Marsupialia: Dasyuridae) in the laboratory. *Australian Mammalogy* **8**, 69–78. doi:10.1071/AM85005

Read DG (1987) A comparison of growth rates in dependent juveniles of *Planigale gilesi* and *P. tenuirostris* (Marsupialia: Dasyuridae). *Australian Journal of Zoology* **35**, 161–171. doi:10.1071/ZO9870161

Read DG, Fox BJ, Whitford D (1983) Notes on breeding in *Sminthopsis* (Marsupialia: Dasyuridae). *Australian Mammalogy* **6**, 89–92. doi:10.1071/AM83013

Reece R, Hartley B (1994) The pathology registry and some interesting cases. In *Wildlife*. Proceedings 233. Post Graduate Committee in Veterinary Science. (Ed. DI Bryden) pp. 217–233. University of Sydney, Sydney.

Rhind SG (2002) Reproductive demographics among brush-tailed phascogales (*Phascogale tapoatafa*) in south-western Australia. *Wildlife Research* **29**, 247–257. doi:10.1071/WR01013

Rhind SG (2003) Communal nesting in the usually solitary marsupial, *Phascogale tapoatafa*. *Journal of Zoology* **261**, 345–351. doi:10.1017/S0952836903004308

Rigby RG (1972) A study of the behaviour of caged *Antechinus stuartii*. *Zeitschrift für Tierpsychologie* **31**, 15–25.

Roberts MG (1915) The keeping and breeding of Tasmanian devils (*Sarcophilus harrisii*). *Proceedings of the Zoological Society of London* **85**, 575–581. doi:10.1111/j.1469-7998.1915.00575.x

Roberts M, Kohn F (1991) A technique for obtaining early life history data in pouched marsupials. *Zoo Biology* **10**, 81–86. doi:10.1002/zoo.1430100111

Rogers T, Fox S, Pemberton D, Wise P (2016) Sympathy for the devil: captive-management style did not influence survival, body-mass change or diet of Tasmanian devils 1 year after wild release. *Wildlife Research* **43**, 544–552. doi:10.1071/WR15221

Rosewarne T (2011). Understanding animal behaviours in captive insurance populations: the Tasmanian Devil as a case study. Honours thesis. Deakin University.

Rovinsky DS, Evans AR, Martin DG, Adams JW (2020) Did the thylacine violate the costs of carnivory? Body mass and sexual dimorphism of an iconic Australian marsupial. *Proceedings. Biological Sciences* **287**, 20201537. doi:10.1098/rspb.2020.1537

Rowland J, Hoskin CJ, Burnett S (2020) Camera traps are an effective method for identifying individuals and determining the sex of spotted-tailed quolls (*Dasyurus maculatus gracilis*). *Australian Mammalogy* **42**, 349–356. doi:10.1071/AM19017

Salamon M, Klettenheimer B (1994) A new technique for marking and later recognising small mammals in the field. *Journal of Zoology* **233**, 314–317. doi:10.1111/j.1469-7998.1994.tb08592.x

Sale MG, Kraaijeveld-Smit FJL, Arnould JPY (2013) Multiple paternity in the swamp antechinus (*Antechinus minimus*). *Australian Mammalogy* **35**, 227–230. doi:10.1071/AM12039

Schaap D (2002) Enriching the devil: the Tasmanian devil. *The Shape of Enrichment* **11**(1), 1–4.

Scicluna EL, Newton AH, Hutchison JC, Dimovski AM, Fanson KV, D'Souza G, Whitehead S, Pask AJ (2025) Breeding fat-tailed dunnarts (*Sminthopsis crassicaudata*) in captivity: revised practices to minimize stress whilst maintaining considerations of wild biology. *Developmental Biology*. doi:10.1002/dvdy.755

Scott MP (1986) The timing and synchrony of seasonal breeding in the marsupial, *Antechinus stuartii*: interaction of environmental and social cues. *Journal of Mammalogy* **67**, 551–560. doi:10.2307/1381287

Selwood L (1982a) Brown antechinus *Antechinus stuartii*: management of breeding colonies to obtain embryonic material and pouch young. In *The Management of Australian Mammals in Captivity*. (Ed. DD Evans) pp. 31–37. Zoological Board of Victoria, Melbourne.

Selwood L (1982b) A review of maturation and fertilisation in marsupials with special reference to the dasyurid: *Antechinus stuartii*. In *Carnivorous Marsupials*. (Ed. M Archer) pp. 65–76. Royal Zoological Society of New South Wales, Sydney.

Selwood L (1983) Factors influencing pre-natal fertility in the brown marsupial mouse, *Antechinus stuartii*. *Journal of Reproduction and Fertility* **68**, 317–324. doi:10.1530/jrf.0.0680317

Selwood L (1985) Synchronisation of oestrous, ovulation and birth in female *Antechinus stuartii* (Marsupialia: Dasyuridae). *Australian Mammalogy* **8**, 91–96. doi:10.1071/AM85007

Selwood L, Cui S (2006) Establishing long-term colonies of marsupials to provide models for studying developing mechanisms and their application to fertility control. *Australian Journal of Zoology* **54**, 197–209. doi:10.1071/ZO05052

Selwood L, Woolley PA (1991) A timetable of embryonic development, and ovarian and uterine changes during pregnancy, in the stripe-faced dunnart, *Sminthopsis macroura* (Marsupialia: Dasyuridae). *Journal of Reproduction and Fertility* **91**, 213–227. doi:10.1530/jrf.0.0910213

Serena M, Soderquist TR (1988) Growth and development of captive and wild pouch young of *Dasyurus geoffroii* (Dasyuridae: Marsupialia). *Australian Journal of Zoology* **36**, 533–543. doi:10.1071/ZO9880533

Serena M, Soderquist TR (1989a) Spatial organisation of a riparian population of the carnivorous marsupial *Dasyurus geoffroii*. *Journal of Zoology* **219**, 373–383[London]. doi:10.1111/j.1469-7998.1989.tb02586.x

Serena M, Soderquist TR (1989b) Nursery dens of *Dasyurus geoffroii* (Marsupialia: Dasyuridae), with some notes on nest building behaviour. *Australian Mammalogy* **12**, 35–36. doi:10.1071/AM89006

Settle GA (1978) The quiddity of tiger quolls. *Australian Natural History* **19**, 164–169.

Shimmin GA, Jones M, Taggart DA, Temple-Smith PD (1999) Sperm transport and storage in the agile antechinus (*Antechinus agilis*). *Biology of Reproduction* **60**, 1353–1359. doi:10.1095/biolreprod60.6.1353

Shimmin GA, Taggart DA, Temple-Smith PD (2000) Sperm competition and genetic diversity in the agile antechinus (Dasyuridae: *Antechinus agilis*). *Journal of Zoology* **252**, 343–350. doi:10.1111/j.1469-7998.2000.tb00629.x

Short J, Hide A, Stone M (2011) Habitat requirements of the endangered red-tailed phascogale, *Phascogale calura*. *Wildlife Research* **38**, 359–369. doi:10.1071/WR10220

Sinn DL, Cawthen L, Jones SM, Pukk C, Jones ME (2014) Boldness towards novelty and translocation success in captive-raised, orphaned Tasmanian devils. *Zoo Biology* **33**, 36–48. doi:10.1002/zoo.21108

Skelton CJA, Stannard HJ (2018) Preliminary investigation of social interactions and feeding behavior in captive group-housed Tasmanian devils (*Sarcophilus harrisii*). *Journal of Applied Animal Welfare Science* **21**, 295–303. doi:10.1080/10888705.2017.1418667

Slater G (1993) Husbandry strategies for the Tasmanian devil and brush-tailed phascogale at the Healesville Sanctuary. In *The Biology and Management of Australasian Carnivorous Marsupials*. (Eds M Roberts, J Carnio, G Crawshaw and M Hutchins) pp. 73–78. Metropolitan Toronto Zoo, Ontario and AAZPA Monotreme and Marsupial Advisory Group, Washington DC.

Smith GC (1984) The biology of the yellow-footed antechinus, *Antechinus flavipes* (Marsupialia: Dasyuridae), in a swamp forest on Kinaba Island, Cooloola, Queensland. *Australian Wildlife Research* **11**, 465–480. doi:10.1071/WR9840465

Smith H (1993) Breeding of Tasmanian devils (*Sarcophilus harrisii*) at Adelaide Zoo. In *The Biology and Management of Australasian Carnivorous Marsupials*. (Eds M Roberts, J Carnio, G Crawshaw and M Hutchins) pp. 85–90. Metropolitan Toronto Zoo, Ontario and AAZPA Monotreme and Marsupial Advisory Group, Washington DC.

Smith MJ, Godfrey GK (1970) Ovulation induced by gonadotrophins in the marsupial, *Sminthopsis crassicaudata* (Gould). *Journal of Reproduction and Fertility* **22**, 41–47. doi:10.1530/jrf.0.0220041

Smith MJ, Bennett JH, Chesson CM (1978) Photoperiod and some other factors affecting reproduction in female *Sminthopsis crassicaudata* (Gould) (Marsupialia: Dasyuridae) in captivity. *Australian Journal of Zoology* **26**, 449–463. doi:10.1071/ZO9780449

Soderquist TR (1993a) Maternal strategies of *Phascogale tapoatafa* (Marsupialia: Dasyuridae). I: breeding seasonality and maternal investment. *Australian Journal of Zoology* **41**, 549–566. doi:10.1071/ZO9930549

Soderquist TR (1993b) Maternal strategies of *Phascogale tapoatafa* (Marsupialia: Dasyuridae). II: juvenile thermoregulation and maternal attendance. *Australian Journal of Zoology* **41**, 567–576. doi:10.1071/ZO9930567

Soderquist TR (1995) Spatial organisation of the arboreal carnivorous marsupial *Phascogale tapoatafa. Journal of Zoology* **237**, 385–398. doi:10.1111/j.1469-7998.1995.tb02769.x

Soderquist TR, Dickman CR (1988) A technique for marking marsupial pouch young with fluorescent pigment tattoos. *Australian Wildlife Research* **15**, 561–563. doi:10.1071/WR9880561

Soderquist TR, Ealey L (1994) Social interactions and mating strategies of a solitary carnivorous marsupial, *Phascogale tapoatafa*, in the wild. *Wildlife Research* **21**, 527–542. doi:10.1071/WR9940527

Soderquist TR, Serena M (1990) Occurrence and outcome of polyoestry in wild western quolls *Dasyurus geoffroii* (Marsupialia: Dasyuridae). *Australian Mammalogy* **13**, 205–208. doi:10.1071/AM90021

Sorenson MW (1970) Observations on the behaviour of *Dasycercus cristicauda* and *Dasyuroides byrnei* in captivity. *Journal of Mammalogy* **51**, 123–130. doi:10.2307/1378538

Species360 (2023) Zoological Information Management System (ZIMS). Minneapolis, USA. www.species360.org.

Spencer PBS (1996) Coping with a naturally fragmented environment: a genetic and ecological study of the allied rock wallaby, *Petrogale assimilis*. PhD thesis. James Cook University, Townsville.

Spencer WB (1896) Mammalia. In *Report on the Horn Expedition to Central Australia. Part 2. Zoology*. (Ed. WB Spencer) pp. 1–52. Dulau, London.

Spielman D (1994) First aid emergency care for Australian native mammals. In *Wildlife*. Proceedings 233. Post Graduate Committee in Veterinary Science. (Ed. DI Bryden) pp. 339–366. University of Sydney, Sydney.

Stamps JA, Swaisgood RR (2007) Someplace like home: experience, habitat selection and conservation biology. *Applied Animal Behaviour Science* **102**, 392–409. doi:10.1016/j.applanim.2006.05.038

Stannard HJ, Old JM (2010) Observation of reproductive strategies of captive kultarrs (*Antechinomys laniger*). *Australian Mammalogy* **32**, 179–182. doi:10.1071/AM10011

Stannard HJ, Old JM (2011) Digestibility of feeding regimes of the red-tailed phascogale (*Phascogale calura*) and the kultarr (*Antechinomys laniger*) in captivity. *Australian Journal of Zoology* **59**, 257–263. doi:10.1071/ZO11069

Stannard HJ, Old JM (2014) Biology, life history and captive management of the kultarr (*Antechinomys laniger*). *Zoo Biology* **33**, 157–165. doi:10.1002/zoo.21128

Stannard HJ, Old JM (2015) Changes to food intake and nutrition of female red-tailed phascogales (*Phascogale calura*) during late lactation. *Physiology & Behavior* **151**, 398–403. doi:10.1016/j.physbeh.2015.08.012

Stannard H, Borthwick C, Ong O, Old J (2013) Longevity and breeding in the red-tailed phascogale (*Phascogale calura*). *Australian Mammalogy* **35**, 217–219. doi:10.1071/AM12042

Stannard HJ, McAllan BM, Old JM (2014) Dietary composition and nutritional outcomes in two marsupials, *Sminthopsis macroura* and *S. crassicaudata. Journal of Mammalogy* **95**, 503–515. doi:10.1644/13-MAMM-A-071

Stannard H, Fabian M, Old J (2015) To bask or not to bask: behavioural thermoregulation in two species of dasyurid, *Phascogale calura* and *Antechinomys laniger. Journal of Thermal Biology* **53**, 66–71. doi:10.1016/j.jtherbio.2015.08.012

Stannard HJ, Tong L, Shaw M, Van Sluys M, McAllan B, Raubenheimer D (2017) Nutritional status and functional digestive histology of the carnivorous Tasmanian devil (*Sarcophilus harrisii*). *Comparative Biochemistry and Physiology. Part A, Molecular & Integrative Physiology* **205**, 1–7. doi:10.1016/j.cbpa.2016.12.008

Stirling EC (1891) Further notes on the habits and anatomy of *Notoryctes typhlops. Transactions of the Royal Society of South Australia* **14**, 283–291.

Strahan R, Conder P (2007) *Dictionary of Australian and New Guinean Mammals*. CSIRO Publishing, Melbourne.

Taggart DA, Temple-Smith PD (1991) Transport and storage of spermatozoa in the female reproductive tract of the brown marsupial mouse, *Antechinus stuartii* (Dasyuridae). *Journal of Reproduction and Fertility* **93**, 97–110. doi:10.1530/jrf.0.0930097

Taggart DA, Selwood L, Temple-Smith PD (1997) Sperm production, storage, and the synchronization of male and female reproductive cycles in the iteroparous, strip-faced dunnart (*Sminthopsis macroura*; Marsupialia): relationships to reproductive strategies within the Dasyuridae. *Journal of Zoology* **243**, 725–736. doi:10.1111/j.1469-7998.1997.tb01972.x

Taggart DA, Breed WG, Temple-Smith PD, Purvis A, Shimmin G (1998) Reproduction, mating strategies and sperm competition in marsupials and monotremes. In *Sperm Competition and Sexual Selection*. (Eds TR Birkhead and AP Moller) pp. 623–665. Academic Press, London.

Taggart DA, Shimmin GA, McCloud P, Temple-Smith PD (1999) Timing of mating, sperm dynamics, and ovulation in a wild population of agile antechinus (Marsupialia: Dasyuridae). *Biology of Reproduction* **60**, 283–289. doi:10.1095/biolreprod60.2.283

Taggart DA, Shimmin GA, Dickman CR, Breed WG (2003) Reproductive biology of carnivorous marsupials: Clues to the likelihood of sperm competition. In *Predators with Pouches: The Biology of Carnivorous Marsupials*. (Eds ME Jones, CR Dickman and M Archer) pp. 358–375. CSIRO Publishing, Melbourne.

Taplin LE (1980) Some observations on the reproductive biology of *Sminthopsis virginiae* (Tarragon) (Marsupialia: Dasyuridae). *Australian Zoologist* **20**, 407–418.

Taylor JM, Calaby JH, Redhead TD (1982) Breeding in wild populations of the marsupial-mouse *Planigale maculata sinualis* (Dasyuridae, Marsupialia). In *Carnivorous Marsupials*. (Ed. M Archer) pp. 83–87. Royal Zoological Society of New South Wales, Sydney.

Teixeira CP, Schetini de Azevedo C, Mendl M, Cipreste CF, Young RJ (2007) Revisiting translocation and reintroduction programmes: the importance of considering stress. *Animal Behaviour* **73**, 1–13. doi:10.1016/j.anbehav.2006.06.002

Terry W, Golden A (2019) An observation of nest robbing and bird predation by wild brush-tailed phascogales in central Victoria. *Australian Zoologist* **40**, 237–240. doi:10.7882/AZ.2019.003

Toftegaard CL, Bradley AJ (2003) Chemical communication in dasyurid marsupials. In *Predators with Pouches*. (Eds M Jones,

C Dickman and M Archer) pp 347–357. CSIRO Publishing, Melbourne.
Tovar C, Pye RJ, Kreiss A, Cheng Y, Brown GK, Darby J, *et al.* (2017) Regression of devil facial tumour disease following immunotherapy in immunised Tasmanian devils. *Scientific Reports* 7, 43827. doi:10.1038/srep43827
Troughton ELeG (1973) *Furred Animals of Australia*. 9th edn. Angus and Robertson, Sydney.
Turner K (1970) Breeding Tasmanian devils *Sarcophilus harrisii* at Westbury Zoo. *International Zoo Yearbook* **10**, 65. doi:10.1111/j.1748-1090.1970.tb01279.x
Twin JE, Pearse AM (1986) A malignant mixed salivary tumour and a mammary carcinoma in a young wild eastern spotted native cat *Dasyurus viverrinus* (Marsupialia). *Journal of Comparative Pathology* **96**, 301–306. doi:10.1016/0021-9975(86)90050-2
Tyndale-Biscoe H, Renfree M (1987) *Reproductive Physiology of Marsupials*. Cambridge University Press, Cambridge.
Umbrello LS, Cooper NK, Adams M, Travouillon KJ, Baker AM, Westerman M, Aplin KP (2023) Hiding in plain sight: two new species of diminutive marsupial (Dasyuridae: Planigale) from the Pilbara, Australia. *Zootaxa* **5330**, 1–46 doi:10.11646/zootaxa.5330.1.1.
Valente A (1984) Reproductive biology of *Antechinomys laniger* (*spenceri* form) (Marsupialia: Dasyuridae). PhD thesis. La Trobe University, Melbourne.
Van Dyck S (1979) Behaviours in captive individuals of the dasyurid marsupial *Planigale maculata* (Gould 1851). *Memoirs of the Queensland Museum* **19**, 413–431.
Van Dyck S (1980) The cinnamon antechinus, *Antechinus leo* (Marsupialia: Dasyuridae), a new species from the vine forests of Cape York Peninsula. *Australian Mammalogy* **3**, 5–17. doi:10.1071/AM80001
Van Dyck S (1982) The status and relationships of the Atherton antechinus, *Antechinus godmani* (Marsupialia: Dasyuridae). *Australian Mammalogy* **5**, 195–210. doi:10.1071/AM82019
Vilcins IM, Old JM, Körtner G, Deane EM (2008) Ectoparasites and skin lesions in wild-caught spotted-tailed quoll (*Dasyurus maculatus*) (Marsupialia: Dasyuridae). *Comparative Parasitology* **75**, 271–277. doi:10.1654/4339.1
Vogelnest L (1999) Chemical restraint of Australian fauna. In *Wildlife in Australia: Healthcare & Management*. (Ed. DI Bryden) pp. 149–186. Graduate Foundation in Veterinary Science. University of Sydney, Sydney.
von Hackewitz L (2020) Sustainable and cost-efficient feed ingredients for optimum breeding of house crickets (*Acheta domesticus*) for human consumption in Thailand. Master of Food Science thesis. Swedish University of Agricultural Sciences, Uppsala.
Wainer JW (1976) Studies of an island population of *Antechinus minimus* (Marsupialia: Dasyuridae). *Australian Zoologist* **19**, 1–7.
Wakefield NA, Warneke RM (1963) Some revision in *Antechinus* (Marsupialia). *Victorian Naturalist* **80**, 194–219.
Wallace GD (1972) Experimental transmission of *Toxoplasma gondii* by cockroaches. *The Journal of Infectious Diseases* **126**, 545–547. doi:10.1093/infdis/126.5.545
Wallis RL (1982) Adaptations to low environmental temperatures in the carnivorous marsupials. In *Carnivorous Marsupials*. (Ed. M Archer) pp. 285–290. Royal Zoological Society of New South Wales, Sydney.
Warnecke L, Geiser F (2010) The energetics of basking behaviour and torpor in a small marsupial exposed to simulated natural conditions. *Journal of Comparative Physiology. B, Biochemical, Systemic, and Environmental Physiology* **180**, 437–445. doi:10.1007/s00360-009-0417-6
Warnecke L, Turner J, Geiser F (2008) Torpor and basking in a small arid zone marsupial. *Naturwissenschaften* **95**, 73–78. doi:10.1007/s00114-007-0293-4
Watt A (1997) Population ecology and reproductive seasonality in three species of *Antechinus* (Marsupialia: Dasyuridae) in the wet tropics of Queensland. *Wildlife Research* **24**, 531–547. doi:10.1071/WR96033
Waudby HP, Petit S (2017) Thermoregulatory value of cracking-clay soil shelters for small vertebrates during extreme desert conditions. *Integrative Zoology* **12**, 237–249. doi:10.1111/1749-4877.12225
Waudby HP, Burns PA, Jensen MA, Hampton JO, Hunter D, McKnight DT, Pestell AJ, Bengsen AJ, Bower D, Coulson G, Heard G, de Laive A, Jolly CJ, Hale S, Fancourt BA, Kelleher SR, Petit S, Behrendorff L, Lumsden LF, McGregor M (2022) Wildlife marking methods. In *Wildlife Research in Australia: Practical and Applied Methods* (Eds BP Smith, HP Waudby, C Alberthsen and JO Hampton). pp. 150–179. CSIRO Publishing: Melbourne.
Whitford D, Fanning FD, White AW (1982) Some information on reproduction growth and development in *Planigale gilesi* (Dasyuridae, Marsupialia). In *Carnivorous Marsupials*. (Ed. M Archer) pp. 77–81. Royal Zoological Society of New South Wales, Sydney.
Williams R (1990) Carnivorous marsupials. In *Care and Handling of Australian Native Animals*. (Ed. SJ Hand) pp. 67–73. Surrey Beatty & Sons and the Royal Zoological Society of New South Wales. Sydney.
Williams R, Williams A (1982) The life cycle of *Antechinus swainsonii* (Dasyuridae, Marsupialia). In *Carnivorous Marsupials*. (Ed. M Archer) pp. 89–95. Zoological Society of New South Wales, Sydney.
Wilson BA (1986) Reproduction in the female dasyurid *Antechinus minimus maritimus* (Marsupialia: Dasyuridae). *Australian Journal of Zoology* **34**, 189–197. doi:10.1071/ZO9860189
Wilson BA, Bourne AR (1984) Reproduction in the male dasyurid *Antechinus minimus maritimus* (Marsupialia: Dasyuridae). *Australian Journal of Zoology* **32**, 311–318. doi:10.1071/ZO9840311
Winkel K, Humphrey-Smith I (1988) Diet of the marsupial mole, *Notoryctes typhlops* (Stirling 1889) (Marsupialia: Notoryctidae). *Australian Mammalogy* **11**, 159–161. doi:10.1071/AM88020
Withers PC, Thompson GG, Seymour RS (2000) Metabolic physiology of the north-western marsupial mole, *Notoryctes caurinus* (Marsupialia: Notoryctidae). *Australian Journal of Zoology* **48**, 241–258. doi:10.1071/ZO99073
Woinarski JCZ, Burbidge AA, Harrison P (Eds) (2014) *The Action Plan for Australian Mammals 2012*. CSIRO Publishing, Melbourne.
Wolfe KM, Robertson H, Bencini R (2000) The mating behaviour of the dibbler, *Parantechinus apicalis*, in captivity. *Australian Journal of Zoology* **48**, 541–550. doi:10.1071/ZO00030

Wood DH (1970) An ecological study of *Antechinus stuartii* (Marsupialia) in south-east Queensland rain forest. *Australian Journal of Zoology* **18**, 185–207. doi:10.1071/ZO9700185

Wood MD, Slade NA (1990) Comparison of ear-tagging and toe-clipping in prairie voles, *Microtus ochrogaster. Journal of Mammalogy* **71**, 252–255. doi:10.2307/1382178

Wood Jones F (1923) *The Mammals of South Australia*. Government Printer, Adelaide.

Woolley P (1966) Reproduction in *Antechinus* sp. and other dasyurid marsupials. In *Comparative Biology of Reproduction in Mammals*. (Ed. IW Rowlands) pp. 281–294. Academic Press, New York.

Woolley PA (1971a) Maintenance and breeding of laboratory colonies: *Dasyuroides byrnei* and *Dasycercus cristicauda. International Zoo Yearbook* **11**, 351–354. doi:10.1111/j.1748-1090.1971.tb01932.x

Woolley PA (1971b) Observations on the reproductive biology of the dibbler, *Antechinus apicalis* (Marsupialia: Dasyuridae). *Journal of the Royal Society of Western Australia* **54**, 99–102.

Woolley PA (1973) Breeding patterns, and the breeding and laboratory maintenance of dasyurid marsupials. *Jikken Dobutsu [Experimental Animals]* **22**, Supplement, 161–172.

Woolley PA (1974) The pouch of *Planigale subtilissima* and other dasyurid marsupials. *Journal of the Royal Society of Western Australia* **57**, 11–15.

Woolley PA (1981) *Antechinus bellus*, another dasyurid marsupial with post-mating mortality of males. *Journal of Mammalogy* **62**, 381–382. doi:10.2307/1380717

Woolley PA (1982) Laboratory maintenance of dasyurid marsupials. In *The Management of Australian Mammals in Captivity*. (Ed. DD Evans) pp. 13–21. Zoological Board of Victoria, Melbourne.

Woolley PA (1984) Reproduction in *Antechinomys laniger* ('spenceri' form) (Marsupialia: Dasyuridae): field and laboratory investigations. *Australian Wildlife Research* **11**, 481–489. doi:10.1071/WR9840481

Woolley PA (1988) Reproduction in the ningbing antechinus (Marsupialia: Dasyuridae): field and laboratory observations. *Australian Wildlife Research* **15**, 149–156. doi:10.1071/WR9880149

Woolley PA (1990*a*) Mulgaras, *Dasycercus cristicauda* (Marsupialia: Dasyuridae): their burrows and attempts to collect live animals between 1966 and 1979. *Australian Mammalogy* **13**, 61–64.

Woolley PA (1990*b*) Reproduction in *Sminthopsis macroura* (Marsupialia: Dasyuridae). I. The female. *Australian Journal of Zoology* **38**, 187–205. doi:10.1071/ZO9900187

Woolley PA (1991*a*) Reproduction in *Dasykaluta rosamondae* (Marsupialia: Dasyuridae): field and laboratory observations. *Australian Journal of Zoology* **39**, 549–568. doi:10.1071/ZO9910549

Woolley PA (1991*b*) Reproduction in *Pseudantechinus macdonnellensis* (Marsupialia: Dasyuridae): field and laboratory observations. *Wildlife Research* **18**, 13–25. doi:10.1071/WR9910013

Woolley PA (1993) Collection and laboratory maintenance of New Guinean Dasyurid Marsupials. In *The Biology and Management of Australasian Carnivorous Marsupials*. (Eds M Roberts, J Carnio, G Crawshaw and M Hutchins) pp. 91–97. Metropolitan Toronto Zoo, Ontario and AAZPA Monotreme and Marsupial Advisory Group, Washington DC.

Woolley PA (1995) Observations on reproduction in captivity of *Parantechinus bilarni* (Marsupialia: Dasyuridae). *Australian Mammalogy* **18**, 83–85. doi:10.1071/AM95083

Woolley PA (2007) Observation on the oestrous cycle and breeding season of the chestnut dunnart *Sminthopsis archeri*, a little known dasyurids marsupial. *Australian Mammalogy* **29**, 89–95. doi:10.1071/AM07011

Woolley PA (2015) The Julia Creek dunnart, *Sminthopsis douglasi* (Marsupialia: Dasyuridae): breeding of a threatened species in captivity and in wild populations. *Australian Journal of Zoology* **63**, 411–423. doi:10.1071/ZO15056

Woolley PA, Ahern LD (1983) Observations on the ecology and reproduction of *Sminthopsis leucopus* (Marsupialia: Dasyuridae). *Proceedings of the Royal Society of Victoria* **95**, 169–180.

Woolley PA, Elliott C (2014) Breeding in wild populations of a small dasyurid marsupial, *Planigale ingrami*, in north-western Queensland using a novel method for collection of specimens. *Australian Mammalogy* **36**, 81–89. doi:10.1071/AM13027

Woolley PA, Gilfillan SL (1990) Confirmation of polyoestry in captive white-footed dunnarts, *Sminthopsis leucopus* (Marsupialia: Dasyuridae). *Australian Mammalogy* **14**, 137–138. doi:10.1071/AM91016

Woolley PA, Valente A (1986) Reproduction in *Sminthopsis longicaudata* (Marsupialia: Dasyuridae): laboratory observations. *Australian Wildlife Research* **13**, 7–12. doi:10.1071/WR9860007

Woolley PA, Watson MR (1984) Observations on a captive outdoor breeding colony of a small dasyurid marsupial, *Sminthopsis crassicaudata. Australian Wildlife Research* **11**, 249–254. doi:10.1071/WR9840249

Woolley PA, Beckman JL, Coleman ME (1998) Rearing of foster young by the Julia Creek dunnart. Abstract. Australian Mammal Society Annual Conference.

Woolnough A, Carthew S (1996) Selection of prey by size in *Ningaui yvonneae. Australian Journal of Zoology* **44**, 319–326. doi:10.1071/ZO9960319

Zwart P, Rulkens RJ (1979) Improving the calcium content of mealworms. *International Zoo Yearbook* **19**, 254–255. doi:10.1111/j.1748-1090.1979.tb00574.x

# 5
# Numbat

*Vicki Louise Power, Danielle Jose, Christine E. Cooper and Rebecca Vaughan-Higgins*

## 1. INTRODUCTION

The numbat (*Myrmecobius fasciatus*) is an unusual marsupial with distinct dorsal striping. It is also unusual in that it is one of few diurnal Australian mammals and feeds exclusively on termites. Numbats have been maintained in captivity by only a few institutions due to the difficulty of supplying adequate numbers of termites. Healesville Sanctuary in Victoria was one of the first zoos to keep numbats when it received a female in 1941, but the animal lived only 2 months (Fleay 1942). Taronga Zoo acquired a pair of numbats in 1968, which lived until at least 1978, and although they bred, no young were raised (Purse 1972; Strahan 1978; Anon 1979; Barlow 1998). Perth Zoo first received numbats in 1968, but did not receive additional animals until December 1986 when four individuals were obtained from a captive colony at the Western Australian Wildlife Research Centre at Woodvale; at both institutions numbats bred successfully (Anon 1969; Hume 1987). In recent years, Alice Springs Desert Park, Healesville Sanctuary and Wildlife Sydney Zoo have obtained post-breeding numbats for display and maintained them to end of life. Numbats are active, captivating display animals that provide visitors with an interesting experience. Importantly, they educate the public about threatened species and the threatening processes they face, including the impacts of introduced predators on Australian wildlife.

## 2. TAXONOMY

### 2.1 Nomenclature

The numbat was first described by Waterhouse (1836) as *Myrmecobius fasciatus*.

Class: Mammalia
Order: Dasyuromorphia
Family: Myrmecobiidae
Genus and species: *Myrmecobius fasciatus*

### 2.2 Subspecies

The type specimen of *Myrmecobius fasciatus* originated from near Brookton in the south-west of Western Australia (Waterhouse 1836). On the basis of its reddish pelage, Wood Jones (1923) described numbats from South Australia as a separate species *Myrmecobius rufus*, but later authors (Finlayson 1933; Tate 1951; Strahan 1995, Groves 2005; Friend 2008) regarded this form as a subspecies, which is now extinct (Woinarski *et al.* 2014). The taxonomic history of the numbat is reviewed by Cooper (2011) and Jackson and Groves (2015).

### 2.3 Recent synonyms

None.

### 2.4 Other common names

The current common name numbat is derived from the Australian Indigenous terms noombat, nombat, nyoombot and nambart (Wood Jones 1923; Troughton 1967; Strahan and Conder 2007). Other terms used by Indigenous people include wai-hoo, wai-hao, weeoo, weeou, wee-u, weeu, wi-u, wiu, walpurti, mutjurarranypa and parrtjilaranypa (Abbott 2001). European names initially included the marsupial anteater, banded anteater and white-banded bandicoot; *M. f. rufus* was distinguished from the nominate form as the south-eastern or rusty numbat, reflecting its redder pelage (Troughton 1967; Friend *et al.* 1982).

## 3. NATURAL HISTORY

### 3.1 Bodyweight and morphometrics

Numbats are unmistakable; their bottle-brush tail, long snout with black and white face markings and bold stripes on the rear of the body are distinctive (Cooper 2011). Adult numbats have a total length of 470 mm, with a head and body length of ~270 mm, tail length of ~200 mm and an adult body mass of 500–700 g. There is some sexual dimorphism, with males usually having greater body mass than females (Friend 1995). The skull is elongated (~60 mm) and the dentition is variable and highly modified; unusually among land mammals, numbats have at least 7 post-canine teeth for a typical total of 50 teeth (Friend 1993).

### 3.2 Distribution and habitat

Numbats once occurred across southern Australia from south-west Western Australian into north-west Victoria and western New South Wales. Following the arrival of Europeans this distribution contracted, with only two naturally occurring populations persisting in the south-west of Western Australia, one at Dryandra Woodland, 170 km east of Perth and the other at Perup Nature Reserve, 280 km south-east of Perth (Friend 1995). These areas consist of eucalypt forest and woodland dominated by wandoo (*Eucalyptus wandoo*) or jarrah (*E. marginata*), with an understory of poison pea (*Gastrolobium* spp.). Habitats inhabited by numbats throughout their historical distribution included eucalypt woodland, mulga (*Acacia aneura*) woodland, spinifex (*Triodia* spp.) grassland and sand dune country (Finlayson 1933; Friend *et al.* 1982; Bester and Rusten 2009).

Numbats have been reintroduced to several sites within their historical distribution over the past 30 years. These include Boyagin, Karroun Hill, Dragon Rocks and Tutanning nature reserves, Batalling State Forest, Cocanarup Timber Reserve, Stirling Range National Park and Mount Dale Conservation Park and the fenced Karakamia and Mount Gibson Sanctuaries in Western Australia (Friend and Thomas 2003; Morris *et al.* 2015; Harley *et al.* 2018; Perth Zoo unpubl. data), with varying success in establishing self-sustaining populations. In addition, numbats have been released at Yookamurra Sanctuary and Secret Rocks in South Australia and Scotia Sanctuary and Mallee Cliffs National Park in New South Wales.

Habitat requirements include an abundance of termites in the soil and protection from predators, especially introduced predators such as the European red fox (*Vulpes vulpes*) and feral cat (*Felis catus*). This explains the persistence of numbats in areas with abundant fallen timber and poison pea (*Gastrolobium* spp.), which can result in secondary poisoning of feral predators (Cooper 2011). Translocations have been most successful in discrete remnant patches, and especially fenced reserves, with considerable understory where post-release dispersal is limited and predation is minimised (Friend and Thomas 2003).

### 3.3 Conservation status

The numbat is listed as Endangered by the IUCN Red List, Western Australian state legislation and Australian federal legislation.

### 3.4 Diet in the wild

Numbats feed exclusively on termites (Isoptera). Numbats eat ~15 000 per day in captivity and more than 21 500 termites per day in the wild (Calaby 1960; Friend 1998; Cooper and Withers 2004a). Genera of termites known to be consumed include *Amitermes*, *Coptotermes*, *Heterotermes*, *Microcerotermes*, *Nasutitermes*, *Paracapritermes*, *Termes*, *Tumulitermes* and *Occasitermes*, with the proportion in the diet depending on their relative abundance (Calaby 1960). The species most commonly consumed within the current natural distribution of the numbat are *Coptotermes acinaciforms* and *Amitermes obeuntis* (Calaby 1960; Christensen *et al.* 1984). Small predatory ants and a considerable quantity of soil debris, which sticks to the tongue during feeding, are also consumed (Cooper 2011).

### 3.5 Longevity

#### 3.5.1 Wild

In the wild, both male and female numbats typically live 4–5 years (Friend 2008).

#### 3.5.2 Captivity

In captivity, the oldest female numbat recorded lived to 11 years of age and male numbats can live for up to 10 years (Perth Zoo unpubl. data).

#### 3.5.3 Methods used to determine the age of adults

There are currently no methods of accurately determining the age of adult numbats due to the morphological simplicity of their dentition. They are generally classed as either adult (> 1 year old) or juvenile.

## 4. HOUSING REQUIREMENTS

### 4.1 Enclosure design

Numbats have been successfully housed in display enclosures 11 m (L) × 7 m (W) × 2 m (H) covered with 25 × 12 mm welded mesh to prevent escape (Hume 1987). Perth Zoo's current display enclosure was designed in 1987, consisting of a large semi-enclosed natural area

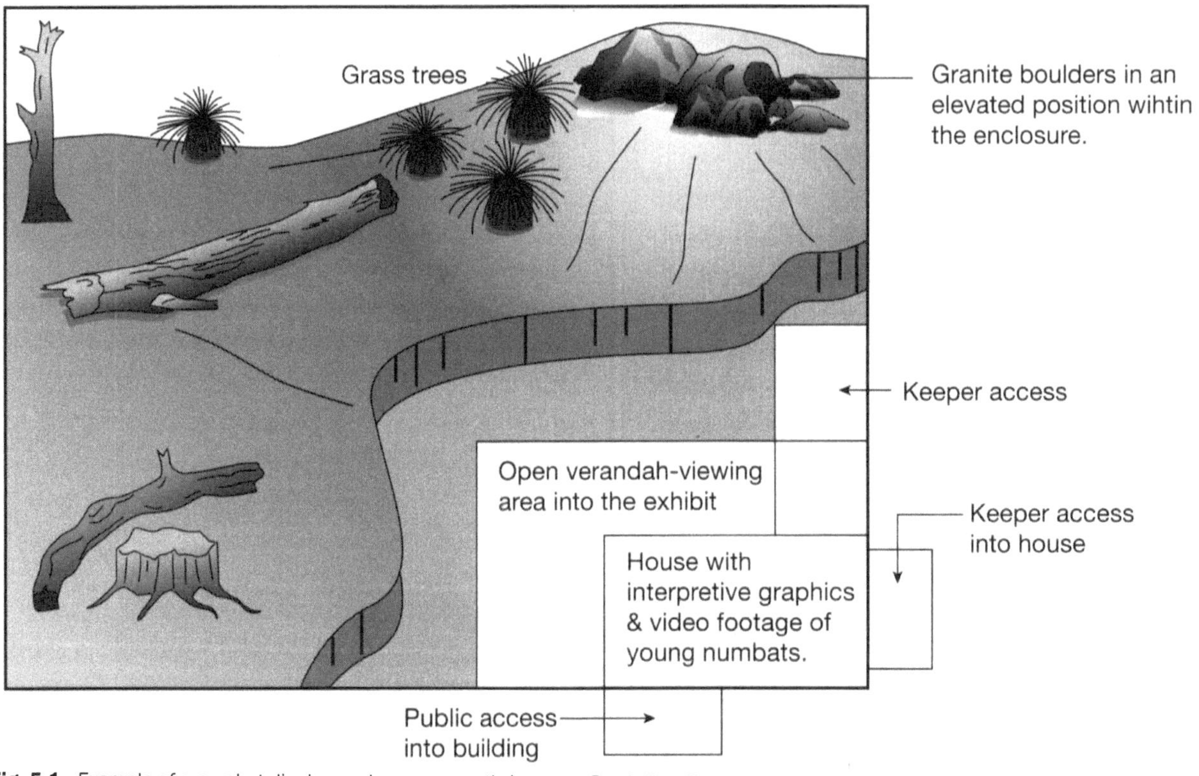


**Fig. 5.1.** Example of a numbat display enclosure currently in use at Perth Zoo. The external area is 15 × 13 × 6 m and the internal house within the enclosure is 4.5 × 3 m. Drawing by Perth Zoo.

with an emphasis on public viewing and education. The enclosure is situated in an elevated north-facing location with shade trees for protection from the summer heat (Fig. 5.1). The 'farmhouse' attached to the enclosure offers interpretive graphics and a screen that plays information about numbats for visitors to view. The display is covered with an aviary mesh 25 × 12.5 mm and the wall mesh is buried to a depth of 1 m to avoid potential predator incursions and numbat escape.

Numbat display enclosures ideally mimic the numbats' natural environment, including environmental enrichment and aesthetic elements such as native shrubs, horizontal and vertical hollow logs (with hollows of 70–120 mm diameter) and elevated boulders (see Table 5.1). Numbats will dig burrows, which provide shelter from extreme weather. Nests are constructed from finely shredded bark, grass and eucalypt leaves in both hollow logs and burrows and play a crucial role in thermoregulation (Christensen *et al.* 1984; Cooper and Withers 2005), so providing material to construct nests is important. Reticulation to support enclosure plantings should be used sparingly to minimise substrate moisture; planting of drought-tolerant plants is advisable. Misting the substrate may aid thermoregulation during hot weather, with some numbats lying under a shrub in damp areas during hot periods.

## 4.2 Holding area design

Perth Zoo is the only institution currently breeding numbats. The breeding complex at Perth Zoo consists of a series of 24 interconnecting enclosures, each 5 × 3 × 2 m with 25 × 12.5 mm mesh used for the walls, floor and roof (Fig. 5.2). A 1.5-m wide safety race surrounds the entire complex. The mesh in the floor is buried to a depth of 1 m, allowing the numbats to excavate burrows without risk of escape. Each enclosure has small gates at ground level (600 mm (H) × 1450 mm (W)) that can be opened to allow access to adjoining enclosures. For most of the year, each numbat has access to two enclosures. As with the exhibit enclosures, holding/breeding enclosures are ideally naturalistic with a variety of plants, hollow logs and climbing logs to provide environmental enrichment; a simulated natural environment is particularly desirable for individuals that are intended for release to the wild.

**Table 5.1.** Recommended plant species for numbat enclosures.

| Scientific name | Common name |
|---|---|
| **Family Asteraceae** | |
| *Rhodanthe manglesii* | Pink sunray |
| **Family Cyperaceae** | |
| *Lepidosperma pubisquameum* | Sedge |
| **Family Fabaceae** | |
| *Acacia extensa* | Wiry wattle |
| *Acacia nervosa* | Rib wattle |
| *Acacia pulchella* | Prickly moses |
| *Acacia squamata* | Wattle |
| *Gastrolobium bilobum* | Heart leaf poison |
| *Gompholobium tomentosum* | Hairy yellow pea |
| *Hovea trisperma* | Common hovea |
| **Family Iridaceae** | |
| *Patersonia occidentalis* | Purple flag |
| **Family Myrtaceae** | |
| *Kunzea micromera* | Kunzea |
| *Hypocalymma angustifolium* | White myrtle |
| **Family Poaceae** | |
| *Neurachne alopecuroidea* | Fox-tailed mulga grass |
| **Family Proteaceae** | |
| *Banksia lindleyana* | Porcupine banksia |
| *Banksia nivea* | Honeypot dryandra |
| *Banksia nobilis* | Golden dryandra |
| *Banksia polycephala* | Many-headed dryandra |
| *Banksia sessilis* | Parrot bush |
| *Banksia squarosa* | Pingle |
| *Hakea amplexicaulis* | Prickly hakea |
| *Hakea lissocarpha* | Honey bush |
| *Hakea stenocarpa* | Narrow-fruited hakea |
| *Hakea undulata* | Wavy-leaved hakea |
| **Family Xanthorrhoeaceae** | |
| *Xanthorrhoea preissii* | Grass tree |

## 4.3 Spatial requirements

An appropriate enclosure size for a pair of numbats is 135 $m^2$, with each additional individual allocated a further 2.25 $m^2$ of floor area.

## 4.4 Position of enclosures

The position of the enclosure is important to facilitate behavioural thermoregulation. An aspect that allows numbats exposure to sunlight throughout the day is

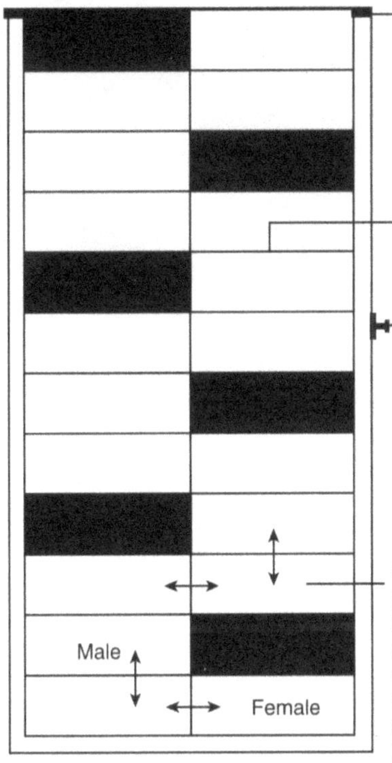


**Fig. 5.2.** Design of Perth Zoo's numbat breeding enclosures. Filled in spaces are empty enclosures. Taken from Power (2002).

important to reduce energy expenditure, especially during winter (Friend and Whitford 1988; Cooper *et al.* 2003a). Shaded areas allow numbats respite from incident radiation at high environmental temperatures. A gentle slope or sump wells in the substrate allow for good drainage. Ponds or flowing water streams are not recommended as high moisture levels may cause skin issues.

## 4.5 Weather protection

Protection from excessive heat and rain can be provided with (UV blocked) polycarbonate sheeting over part of the roof and 90% density shade cloth can be used to cover portions of the roof between November and May to provide extra protection during hot summer months while still allowing access to sunlight and escape of hot air via the uncovered portion of the roof. Numbats may also dig burrows to escape the heat or during cold, rainy winter periods. Half a plastic pipe (e.g. diameter 30 cm) or log can also be used to provide weather protection (Fig. 5.3).

## 4.6 Temperature requirements

Heating or cooling is not required in Mediterranean or similar climates, particularly if numbats have excavated

**Fig. 5.3.** Plastic weather protection for numbats that may also be placed over an above-ground nest box. Photo by Perth Zoo.

a burrow and are provided with a dry nest box with adequate nest material (e.g. dried grass or eucalypt bark). Females with young are more easily managed in a nest box than in a burrow, but females may be persistent in digging their own burrow. Sick, injured or debilitated numbats can be provided with an external heat source at low temperatures to allow them to maintain body temperature at reduced metabolic cost.

## 4.7 Substrate

Numbat enclosures should have a well-draining substrate; for example, a heavy sand with some laterite/soil blend and very little clay to limit soil moisture content. The substrate should be soft enough for the numbat to excavate a burrow but firm enough for a burrow to hold its shape and not collapse. Plants grown in the enclosure will reinforce burrow stability, as numbats generally dig along the side of a log, fence or an established root system. River sand was historically used as a buffer layer in numbat enclosures, but excessive ingestion while digging and foraging can lead to sand impaction in the gut. Finer grade whitewash sand has proved more suitable.

## 4.8 Nest boxes

Artificial nest boxes of various designs provide shelter for numbats in enclosures. The entrance to above-ground boxes is best shaped as an archway so attached young are not bumped on the bottom upon entry or exit of the nest box. A hinged lid provides access by keeping staff (Fig. 5.4a). Boxes with an outer chamber allow for provision of a heat source (Fig. 5.4b) and may be useful for compromised individuals. Ideally, each numbat is provided with access to multiple nest boxes placed throughout the enclosure to provide a variety of environmental conditions. Hollow logs do not allow for easy access to the animals by staff. Shelter from the elements can be provided by a Perspex cover on a semicircle aluminium frame placed over the box (Fig. 5.3).

Underground nest chambers can simulate a burrow (Fig. 5.4c). Constructing the chamber portion (400 mm high × 185 mm square) from 12 × 12 mm mesh reduces condensation and a plastic fly-screen lining will prevent surrounding soil from entering. A foam block on the top of the chamber provides insulation and can be covered with a solid aluminium lid to protect it from the animals. A hinged lid provides staff access. Animal access is achieved by 800 mm plastic agricultural drainage pipe (Fig. 5.4c). It is important to ensure the drainage pipe does not rise up, which can potentially cause harm to mothers with pouch young on rapid entry, as the pouch young could be knocked off. The chamber should be buried at least 500 mm below ground.

## 4.9 Enclosure furnishings

Various enclosure furnishings provide for enrichment, including hollow logs, rocks and tussocks. Landscaping ideally includes native vegetation to provide shade and cover (Table 5.1). Tussock grasses in particular provide good cover. However, it can become difficult to sight animals for monitoring in an overgrown enclosure. Logs can be arranged in such a way to allow animals to climb and tree stumps cut off into flat steps and large stone pieces make excellent sun basking platforms. *Melaleuca* logs provide additional value as numbats will often tear strips of bark for nesting material. Fresh eucalypt branches propped up against walls or in the corners of enclosures so the leafy end rests on the ground can also provide shade and shelter. Shade cloth along the walls of enclosures can prevent animals becoming startled from outside activity and is particularly beneficial if the area is frequented by staff.

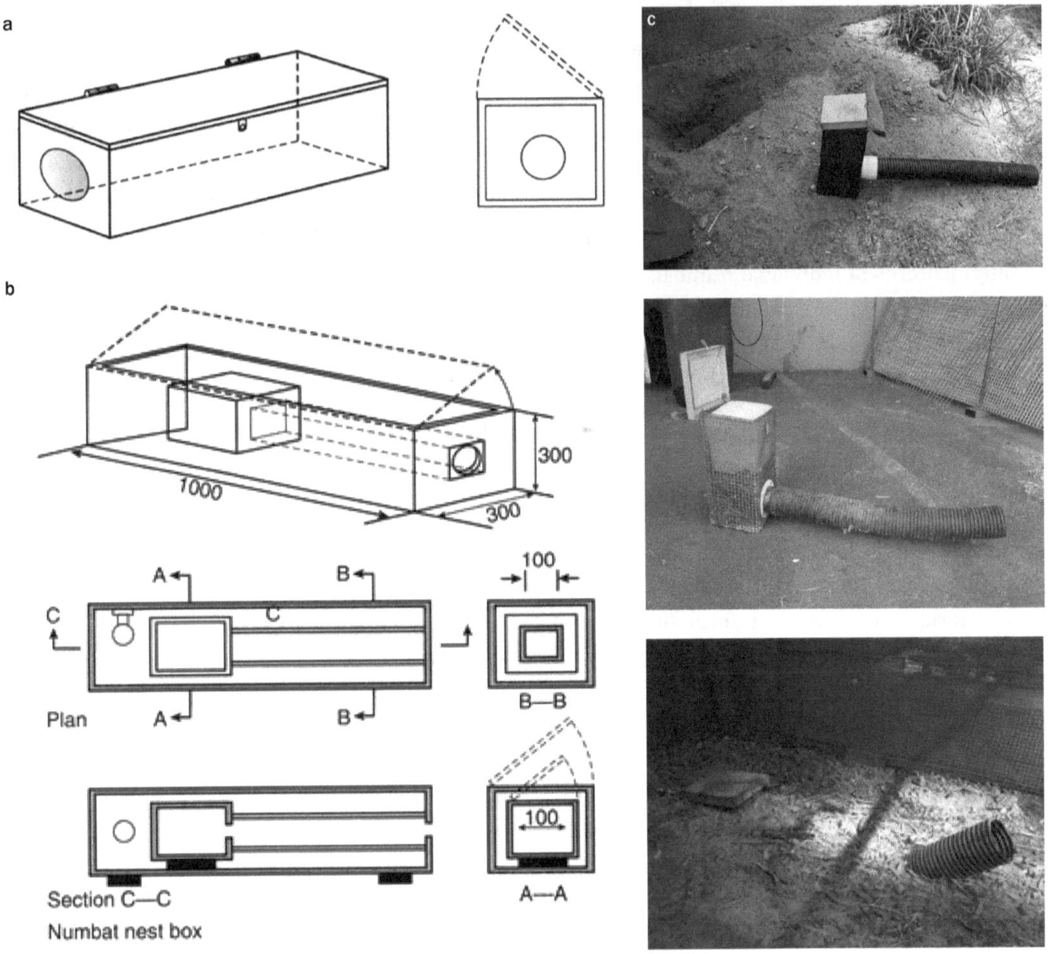


**Fig. 5.4.** Artificial nest boxes with (a) hinged lid, (b) internal heat source and (c) underground nest chamber and access tunnel before deployment (middle right) and placed in situ (bottom right). Photos by Vicki Power.

## 5. GENERAL HUSBANDRY

Sighting each individual numbat daily by the keeper responsible allows for monitoring for any health issues or injuries. CCTV cameras are useful for observing animals' activity and mating behaviour generally, but are not as useful for detecting unwell animals or actual matings. Motion sensor cameras can also be used to monitor activities of mothers with young

### 5.1 Hygiene and cleaning

Water bowls and food dishes should be cleaned daily and disinfected with F10® solution and sun dried. Enclosures are cleaned daily to remove a build up of faeces, with more intensive cleaning during weaning due to larger numbers of animals per enclosure. For breeding enclosures, scheduled maintenance and a deep clean is best carried out during early December each year, after the previous young are independent and before animals are paired for breeding. Enclosures should also be thoroughly cleaned before new animals are admitted. Enclosure furniture such as logs should be replaced when they are showing signs of damage or rotting. Eucalypt branches used for shelter should be replaced every 2 weeks before they degrade.

### 5.2 Record keeping

It is important to establish a system whereby the health, condition and reproductive status of captive numbats are routinely monitored and recorded:

- identification details (e.g. Species360 GAN and microchip numbers) – all individuals should be identifiable; this may be achieved by microchipping and/or their individually unique stripe pattern
- results of veterinary examinations and treatments
- behavioural changes or problems
- reproductive behaviour or condition
- bodyweight and measurements

- changes in diet and food consumption
- movements of individuals between enclosures or institutions
- births, with dam and sire if known
- deaths with necropsy results
- enrichment introductions/additions.

Collecting information on the physical, physiological and behavioural patterns of individual numbats can contribute greatly to the husbandry of this species. It also allows individual histories to be shared with other institutions and facilitates a cooperative approach to data collection among institutions. ZIMS, the Zoo Information Management System (Species360 2023), can be used for recording general information on age, sex, parentage, births, identification, bodyweights, transfers and deaths, with separate ZIMS modules for veterinary information and studbooks. These software systems have been developed by the International Species Information System (ISIS), which is part of the Conservation Breeding Specialist Group (CBSG) of the IUCN. As these programs are standardised, there is a high degree of efficiency in transferring information between affiliated institutions. ZIMS also interfaces with programs such as PMx, which provides demographic and genetic analysis tools to assist with the management of breeding programs (Lacy *et al.* 2011).

## 5.3 Methods of individual identification

### 5.3.1 Passive integrated transponder (PIT) tags

PIT tags (microchips) are typically implanted in the intrascapular area and are an excellent method of individual identification. They should not be implanted in the back of neck of females as males often latch on to this area during mating. Females should instead be implanted in the left inguinal region. PIT tags can be implanted when the animals are ~10 months of age (150–200 g). Sealing the entry wound with tissue glue (Vetbond®) or similar fast-setting adhesive (e.g. Superglue®) or by placing a couple of stitches across the site will prevent the tag tracking out along the insertion wound. Routine microchip scanning will detect if the PIT tag has been lost.

### 5.3.2 Visual identification

Each numbat has a unique stripe pattern that can be used to identify individuals. A photographic library can maintain a record of each individual's stripe pattern. A small drop of different coloured nail polish or food colouring on the head of young numbats can serve for temporary individual identification.

# 6. FEEDING REQUIREMENTS

## 6.1 Captive diet

The development of an artificial diet for numbats (Hume 1987; Friend and Whitford 1988; Barlow 1998), coupled with the development of a reliable termite trapping technique (see Addendum 1) to supplement the artificial diet, has facilitated captive breeding and display of numbats. The artificial diet, consisting of lactose-free milk powder and eggs (Table 5.2), is used for general maintenance of males and non-breeding females, supplemented with fresh or frozen termites (Friend and Whitford 1988). The proportion of termites in the diet is increased for breeding individuals (see section 10.12).

## 6.2 Termites as a food source

*Coptotermes* and *Nasutitermes* termite species are most readily harvested, at least in the Perth region of Western Australia. Termite harvesting is described in detail in

**Table 5.2.** Artificial numbat diet (recipe sufficient for 2–3 individual numbats).

| Ingredients | Amount | Instructions |
|---|---|---|
| **Cooked custard** | | |
| Warm water | 145 g | The water, eggs and Di-Vetelact® (Lillelund Pty Ltd) are mixed together using an electric beater and then |
| Eggs | 75 g | cooked to the consistency of scrambled eggs. This can be stored in the fridge overnight to be used the |
| Di-Vetelact® | 30 g | following day. Note: ensure there are no small eggshell fragments |
| **Custard mix** | | The cold, cooked custard is placed in a bowl with the rest of the ingredients and mixed thoroughly using an |
| Cooked custard | 186 g | electric beater |
| Termite mound | 20 g | SF52® (Speciality Feeds, WA) |
| Cellulose powder | 5 g | |
| Fish oil | 1 g | |
| Vitamin B12 | 0.5 g | |
| $CaCO_3$ (Calcium powder) | 0.9 g | |
| SF52® | 0.2 g | |

Addendum 1. *Coptotermes* are preferred by some numbats over *Nasutitermes*; *Coptotermes* have a higher digestibility (81% *v.* 64%) and better facilitate maintenance of body mass (Cooper and Withers 2004a). Blending the two termite species into the artificial diet in varying amounts has been successful. To ensure a reliable termite supply year-round, termites can be bulk frozen for up to 6 months, but they must be used soon as possible after defrosting as they deteriorate within hours. Captive numbats typically consume 70 g of artificial diet with 10 g of termites daily, or 58–63 g of pure termite. Water is typically provided ad lib; although numbats do not require drinking water in the wild (Cooper and Withers 2004a), they do drink when water is available in captivity.

### 6.3 Presentation of food

Small plastic bowls 200 mm in diameter and 40 mm in height are suitable for feeding numbats. Feeding bowls that have a water moat protect the food from ants are useful as numbats will not eat if their food is contaminated with ants. The food should be thoroughly homogenised and added to a dish then the termites added; live termites must be quickly stirred into the mixture to prevent them escaping and the mixture presented to the numbats immediately after the addition of the termites. The daily food allowance is ideally split into a morning and afternoon feed to ensure freshness. Feeding times should correspond with the seasonal activity of termites in the wild, with later morning and earlier afternoon feeds in winter compared with summer (Cooper and Withers 2004b).

## 7. HANDLING AND TRANSPORT

### 7.1 Timing of capture and handling

Captive numbats are most easily caught first thing in the morning before they become active and leave the nest box or in the evening after they retire for the night. During the heat of the day numbats often return to their nest to rest and this provides an additional opportunity to catch them. If they have a burrow, however, they will be inaccessible until they emerge and can be encouraged to retreat into a nest box.

### 7.2 Capture equipment

Pillowcases or cloth bags are useful for blocking the entrance to a nest box containing a numbat during capture and for transporting numbats. Mesh nets are not recommended as numbats can become stressed if chased with a net and they may also become tangled in the net with a risk of damaging their claws.

### 7.3 Capture and physical restraint

Numbats can be easily caught in a nest box with a hinged lid; the entrance to the box is blocked with a bag, the lid lifted slightly and a hand slipped over the animal's shoulders to restrain it, while restraining the hindlegs with the other hand. The animal can then be removed and placed into a cloth bag. Numbats, like most other small mammals, are best handled in a cloth bag whenever possible to keep them calm. To remove a numbat from a bag and restrain it by hand, the animal is pinned down by the shoulders with one hand while the other hand is slid into the bag to restrain the shoulder area. Then the hindfeet should be held together firmly with the other hand to prevent kicking. The numbat can then be lifted out of the bag (Fig. 5.5). Numbats have small, reduced teeth and rarely bite, at least not sufficiently to cause any serious damage, and although they can scratch with their hind claws, they generally make little attempt to escape when being handled (Fleay 1942; Cooper 2011).

### 7.4 Weighing and examination

Numbats can be weighed easily while in a cloth bag by placing them on an electronic or spring balance. Examination can be readily undertaken while manually restraining the animal. Monthly weighing and examination are recommended and claws should be trimmed as required.

### 7.5 Release

Numbats are best released directly into a nest box or log inside their enclosure. If they are released straight onto open ground, they may to injure themselves by running into objects in the enclosure or wire mesh in an attempt to flee from the handler.

### 7.6 Transport requirements

#### 7.6.1 Transport box design

Numbats can be safely transported by air in a wooden box (350 (W) × 400 (L) × 200 (H) mm), made from 18-mm marine ply or similar (Fig. 5.6), as suggested by the International Air Transport Association (2022), or in a small commercial container (e.g. PetPak) (550 mm (L) × 350 mm (W) × 350 mm (W)).

#### 7.6.2 Furnishings

No specialised furnishings are required when transporting numbats. Numbats are always transported inside a cloth bag or pillowslip tied at the top and placed within their transport container.

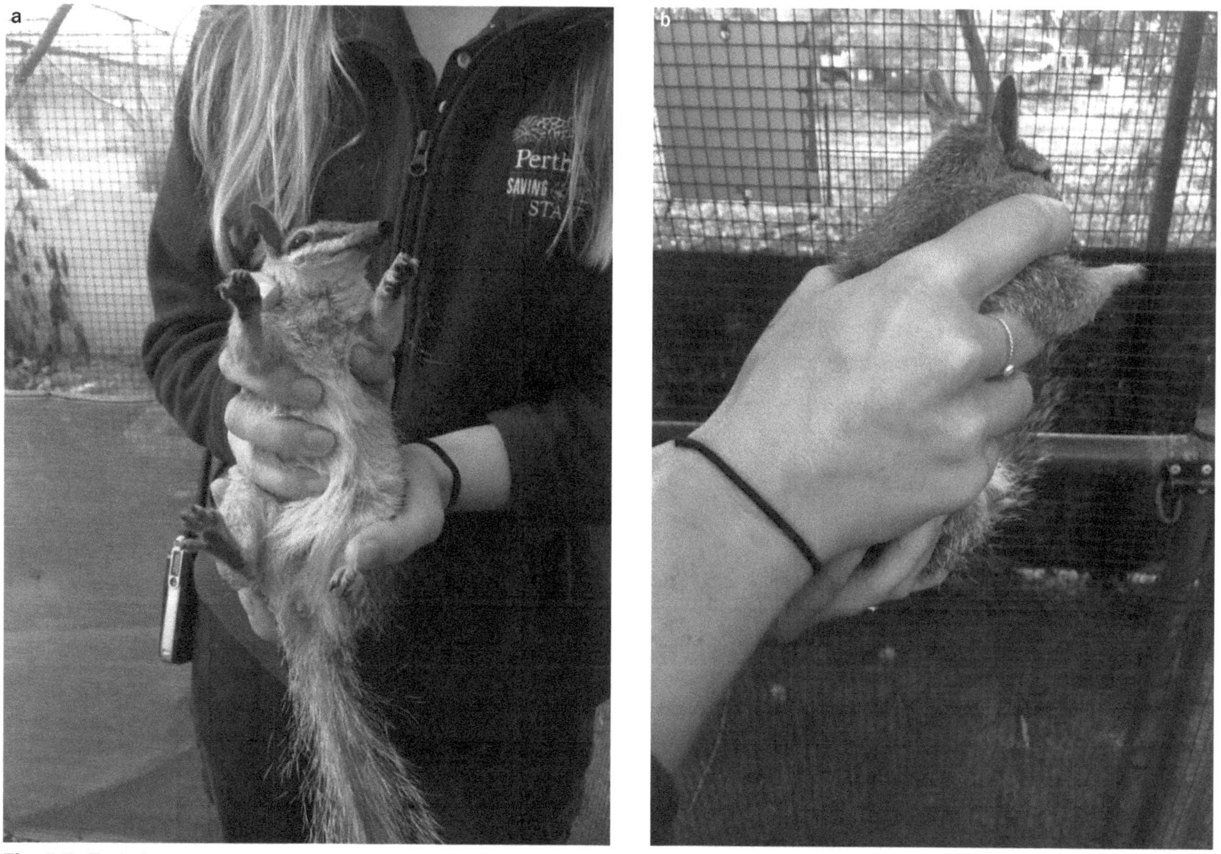

**Fig. 5.5.** Technique used to hold numbats: (a) ventral and (b) dorsal views. Photos by Perth Zoo.

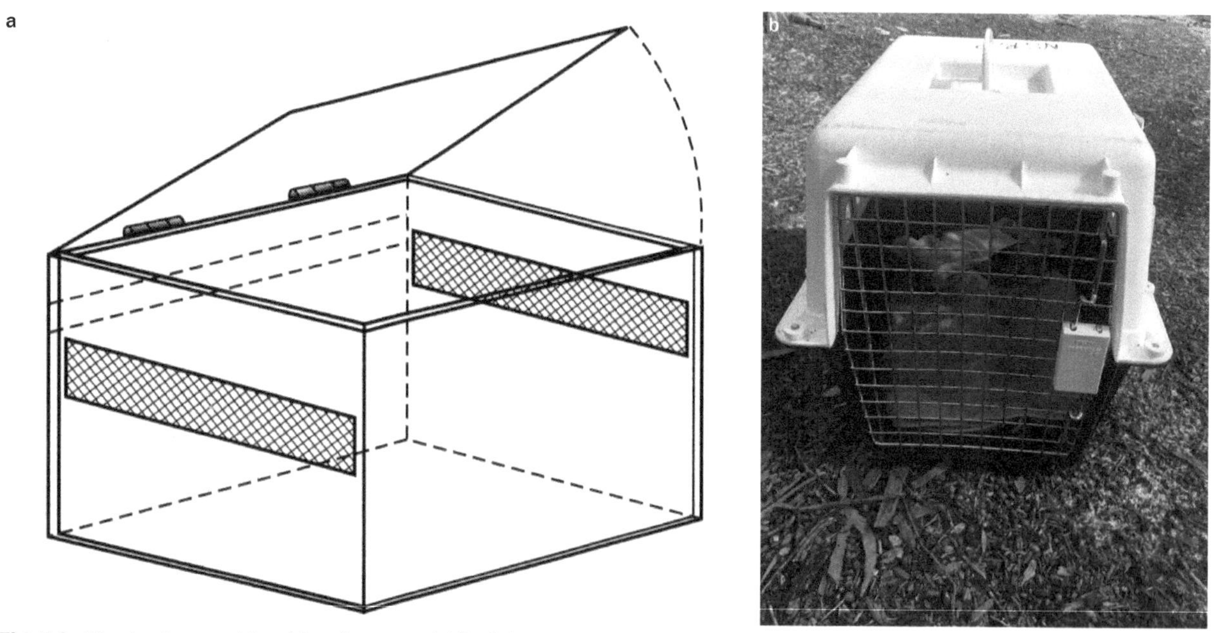

**Fig. 5.6.** Wooden transport box (a) and commercial PetPak (b) used for transporting numbats. Drawing from Power (2002) and photo by Perth Zoo.

Shredded paper or a towel should be placed on the bottom of the box for cushioning and two rolled up towels on either side prevent the numbat from moving around too much while in the box.

### 7.6.3 Water and food

Food and water are typically not required for short-term (< 24 h) transport.

### 7.6.4 Animals per transport box

Multiple animals can be transported within the same transport box, provided they are secured in separate bags and divided by a barrier or rolled up towel. It is not advisable to transport multiple adult males in the same transport box as they are territorial and not usually in close proximity to one another.

### 7.6.5 Timing of transportation

Wherever possible, transport should occur early in the morning or overnight, avoiding periods of high temperature and minimising disruption to feeding times. The maximum transportation period should be less than 24 h as significant weight losses have been noted for animals transported without food and water over a 36-h period. (S. Haigh pers. comm.). Internal transport within the institution can occur at any time of the day, although preferably not during the heat of the day in summer.

### 7.6.6 Release from the transport box

Numbats do not require sedation for transport. They can be released directly into a log or nest box, so that they can emerge on their own and slowly adjust to their new enclosure/release site.

## 8. HEALTH REQUIREMENTS

### 8.1 Daily health observations

Each animal should be observed daily when active for signs of injury or illness. This is particularly important in quarantine and after events such as introduction to other animals or to new enclosures. The most appropriate time to do this is generally when the enclosure is being cleaned or when animals are being fed; however, remote (e.g. CCTV) observation of the undisturbed animal is also beneficial. Each animal should be checked for the following:

- coat/skin condition - check for alopecia, crusting, thickening of skin or wounds
- body condition score - typically graded from 0 (emaciated) to 5 (obese) (see section 8.2.2)
- discharges - from the eyes, ears, nose, mouth or cloaca
- appetite
- faeces - number and consistency
- eyes - cloudiness, redness, open or closed
- changes in demeanour, activity level and behaviour
- presence and development of young by observation of the pouch region
- injuries.

### 8.2 Bodyweight and body condition score

#### 8.2.1 Bodyweight

Ideally, bodyweight and assessment of body condition should be recorded regularly. The frequency of weight measurements will depend on the logistics required to weigh particular individuals and any underlying health concerns.

#### 8.2.2 Body condition score

There is no agreed method for determining a body condition score in numbats. The generic nine-point scale for body condition scoring of zoo animals defined by ZIMS (Species360 2023) could be used, where 1 = emaciated, 5 = ideal condition, and 9 = obese.

### 8.3 Physical examination

Numbats can be manually restrained for physical examination and/or restrained in a cloth bag, exposing the extremities as required by keepers or a veterinarian. Monthly weighing is also facilitated by restraining the animal in a bag. If required, detailed physical examination should be conducted by a veterinarian with wildlife expertise. Veterinarians should refer to Vogelnest (2025) and Vogelnest and Portas (2008) for detailed information on examination and medicine for Australian mammals. All body areas and systems should be evaluated: cardiovascular, respiratory, oral cavity (dentition, oral mucosa and pharynx), abdomen, musculoskeletal, genitourinary, ocular, skin and coat, ears, teats and mammary glands, and lymph nodes. Important parameters and vital signs should be measured and recorded such as body mass and condition; heart/pulse rate; respiratory rate and effort; and body temperature.

### 8.4 Chemical restraint

Pre-anaesthetic fasting is not essential, but an 8-h fasting period is recommended for planned procedures on numbats maintained in captivity. Anaesthesia is most commonly induced and maintained with isoflurane gas (Vogelnest 2025). Induction is achieved with 5% isoflurane in oxygen administered via an induction chamber or tube attached to an isoflurane anaesthetic circuit. Numbats often breath hold during induction, which can slow the process. Once induced they can be easily maintained on a mask for short procedures. For longer procedures numbats can be intubated using an uncuffed 2-mm endotracheal tube. A unidirectional anaesthetic circuit (e.g. Bain or T-piece) should be used and the numbat maintained on 3% isoflurane and 2 L/min (Bain) or 3 L/min (T-piece) oxygen. Recovery is often rapid following cessation of isoflurane anaesthesia. Sedation of numbats is not commonly performed, although presumably the principles would be similar to those for small dasyurids (Vitali and Monaghan 2008). Zoletil® has been used for wild numbats at a dose rate of 5–9 mg/kg intramuscularly (Vogelnest 1999) and 'topping up' is not recommended (Haigh and Friend 1999). Isoflurane gaseous anaesthesia is preferable.

General anaesthesia may be better for blood sample collection and microchip insertion owing to the thick and tough skin of numbats. Blood collection is difficult due to the thick skin, lack of visible vessels and poor blood flow if the individual is hypothermic. Active warming of extremities with warming devices may be required to minimise peripheral vasoconstriction. The best blood collection sites are the femoral, jugular and brachiocephalic veins (Vitali and Monaghan 2008). The choice of therapeutic agents and dosage, route, frequency and duration of administration are typically based on those administered to domestic cats and dogs, but the low metabolic rate of numbats (Cooper and Withers 2002; Cooper *et al.* 2003b) should be considered when calculating drug doses. There have been no pharmacokinetic studies to date for numbats (Bodley 2025).

## 8.5 Common health problems

No species-specific health issues have been reported for numbats. Numbat records from wild caught and captive bred individuals from the Native Species Breeding Program at Perth Zoo (Tables 5.3–5.6) were sourced through a ZIMS (Species360 2023) medical search and cross-referenced with Vitali and Monaghan (2008). For wild-caught individuals, trauma is the most commonly reported issue.

## 8.6 Preventative medicine

A 30-day quarantine period is recommended for all wild numbats entering a facility, with three faecal examinations for parasite detection. During this time, numbats will need to adapt to an artificial diet. Treatment for acanthocephalan infection is necessary if detected (see Table 5.2) and a standard physical examination, which includes examination of the eyes, ears, mouth, dentition, feet, nails, integument, abdomen, thorax and limbs, pouch region and sternal gland, is advised. Blood is typically collected for routine complete blood count and biochemical analysis and screening radiographs are taken for baseline health monitoring.

## 8.7 Zoonoses

Many numbat cultures are positive for *Salmonella* spp. and other enteric pathogens such as *Escherichia coli* and *Campylobacter* spp., which have caused gastrointestinal disease in people. Excellent hand hygiene and disinfection is recommended when handling numbats, their food and when cleaning enclosures.

# 9. BEHAVIOUR

## 9.1 Activity

The numbat is the only strictly diurnal marsupial (Dickman and Calver 2023). Wild numbats spend their active period foraging for termites and retreat to hollow logs when they are alarmed (Christensen *et al.* 1984). The activity patterns of numbats reflect the accessibility of termites in subsurface soil galleries and local weather conditions. They are most active in summer, which is also the breeding season, when they forage in the morning and late afternoon and rest during the heat of the day in hollow logs or burrows. During winter, activity declines and is restricted to the period from mid-morning to mid-afternoon (Christensen *et al.* 1984; Friend and Burrows 1983). Patterns of activity for captive numbats are broadly similar to those of wild numbats. Captive numbats are exclusively diurnal and are typically active for ~20% of a 24-h period. They are more active in summer than winter, avoid activity during rain and are less active during periods of low light when they are unable to exploit solar radiation to aid thermoregulation (Cooper and Withers 2004b). Numbats may become torpid overnight during periods of cold weather (Cooper and Withers 2004c).

## 9.2 Social behaviour

Numbats are largely solitary, with burrows or logs typically containing only a single individual or female with young, although breeding pairs may share a nest (Calaby 1960; Christensen *et al.* 1984). In the wild, adult numbats

**Table 5.3.** Common infectious diseases of captive numbats.

| Disease | Cause | Clinical signs | Diagnosis | Treatment | Prevention | Reference |
|---|---|---|---|---|---|---|
| **Ectoparasites** | | | | | | |
| | Fleas *Echidnophaga myrmecobii* and *E. perilis* | Sometimes alopecia, pruritus, erythematous skin (but not always), often visible with a quick scan of the haircoat | Careful examination of the pelage | Small quantities of Fipronil® or Frontline® spray that can be applied with a gauze swab | Quarantine all new arrivals, monitoring, treatment of infected individuals and regular substrate changes | 1, 2 |
| | Mites *Mesolaelaps australiensis* | Sometimes alopecia, pruritus, erythematous skin (but not always), often visible with a quick scan of the haircoat | Superficial and deep skin scrapings | Small quantities of Fipronil® or Frontline® spray that can be applied with a gauze swab | Quarantine all new arrivals, monitoring, treatment of infected individuals and regular substrate changes | 2 |
| | Ticks *Ixodes vestites, I. holocyclus* and *Amblyomma triguttatum* | Sometimes alopecia, pruritus, erythematous skin (but not always), often visible with a quick scan of the haircoat | Careful examination of the pelage | Physical removal and small quantities of Fipronil® or Frontline® spray that can be applied with a gauze swab | Quarantine all new arrivals, monitoring, treatment of infected individuals and regular substrate changes | 2 |
| **Endoparasites** | | | | | | |
| | Nematodes - phylum Acanthocephala *Multisentis myrmecobius* (thorny-headed worm)<br>The lifecycle of acanthocephalans is indirect and involves ingestion of an intermediate arthropod host (termites) by the definitive host. It is assumed that the termites ingest eggs that develop to the infective stage within the termite over 1–3 months. The numbat is infected when it ingests the infected termite | Diarrhoea is a possible presenting sign for all endoparasitic infections.<br>Obvious antemortem signs have not been recorded. Has been associated with gastrointestinal pathology, including inguinal hernias and gastrointestinal torsion in wild and captive individuals | Difficult to detect on routine faecal float<br>The presence of whole worms in the small intestine may be observed with associated pathology on histopathology<br>Whole worms can be expelled from the cloaca by living individuals | Ivermectin® at 200 ug/kg PO or SC. It is recommended to treat animals 2–3 times if infection detected and feed 'clean' termites during this time to break the lifecycle<br>'Clean' termites are sourced from the wild where numbat populations do not exist | Provision of 'clean' termites. Faecal parasite screening conducted annually and pre-export to minimise infection | 2–5 |
| | Two species of trichostrongyloid nematodes (*Beveredgilla calabyi* and *B. inglisi*) have been reported | Colonisation of the gastrointestinal tract is generally not associated with illness; asymptomatic carriers | Zinc sulfate faecal flotation and direct faecal examination | Thought to be commensal, no treatment indicated | Not associated with disease; monitor load | 6 |

**Table 5.3.** continued

| Disease | Cause | Clinical signs | Diagnosis | Treatment | Prevention | Reference |
|---|---|---|---|---|---|---|
| **Viral** | | | | | | |
| | No reports of viral disease were found. | | | | | |
| **Bacterial** | | | | | | |
| *Campylobacteriosis* | *Campylobacter coli* detected on faecal culture from several clinically normal numbats | Colonisation of the gastrointestinal tract by *Campylobacter* spp. is generally not associated with illness; may be asymptomatic carriers Typically only show signs of gastrointestinal disease with concurrent infections or if immunocompromised | Faecal sample submitted for microbial culture and sensitivity | No treatment or control has been initiated unless clinical signs of gastroenteritis. Most *Campylobacter* sp. infections are self-limiting. Fluid support is suggested if diarrhoea is present. Antimicrobial therapy may be used to treat the severely ill or if at risk of severe disease | Good hygiene and disinfection practices, and low host density. In other species, outbreaks are typically associated with contaminated water, and food. Humans can become infected from contact with infected faeces | 2 |
| Clostridial necrotic enteritis (NE) | *C. perfringens* type A Clostridial NE associated with three sudden deaths in captivity surrounding weaning<br>Type A is a known cause of NE in domestic poultry, cattle, sheep, pigs, dogs and possibly horses. It is part of the normal intestinal flora of all endothermic animals and under certain conditions can overgrow and produce potent toxins that cause circulatory shock and rapid death. Necrotic enteritis in livestock is usually associated with type A, B and C toxins | Three weanling numbats presented with bloated abdomens, haemorrhagic diarrhoea, collapse and sudden death | Histopathology and microbial culture and sensitivity of tissues. The small intestines are typically dark red, haemorrhagic and filled with haemorrhagic liquid. A rapid presumptive diagnosis can be made by demonstrating large rod-shaped bacteria in gram-stained mucosal impression smears | Treatment is often futile as lesions are usually irreversible at the onset of diarrhoea<br>Resuscitation, anti-bacterial agents such as amoxicillin, clavulanic acid initially then based on microbial culture and sensitivity results, fluid support | In domestic animals, risk factors for NE include factors that may upset the normal gastrointestinal flora and allow the organism to proliferate. This includes sudden diet change, feeding non-starch polysaccharides in cereals, high-protein diets, antibiotic use or gastroenteritis due to other causes. Cases most commonly occur in numbats at weaning; therefore attention to minimising stress around weaning and a slow diet transition seem warranted<br>*Clostridium perfringens* Type A Toxoid Cattle Vaccine is not available in Australia and has not been trialled for numbats. However, ongoing research on vaccine development for NE using the NetB gene may prove promising in the future | 7–9 |

| Disease | Cause | Clinical signs | Diagnosis | Treatment | Prevention | Reference |
|---|---|---|---|---|---|---|
| Collibacillosis | *Escherichia coli* | One case report associated with septicaemia with ascending infection of kidneys with non-haemolytic *E. coli.* causing a suppurative pyelonephritis and marked, acute pyometra | Histopathology and microbial culture and sensitivity, faecal culture and sensitivity | Antibacterial therapy based on results of culture and sensitivity, hydration and supportive care | Good hygiene and disinfection practices, and low host density | 10 |
| Mycobacteriosis | *Mycobacterium* spp. | *Mycobacterium intracellulare* associated with foot lesions and *M. chelonae* abscesses associated with pustular skin lesions on the ventrum in captive numbats<br>Thought to be associated with primary skin trauma | Biopsy of lesions for histopathology and microbial culture and sensitivity | Treatment is often difficult, prolonged and usually unsuccessful Amikacin 7.5 mg/kg IM b.i.d. for 6 days with a repeat 10-day course 6 weeks later has been reported | Control is problematic, as asymptomatic animals are hard to identify until lesions develop, and the organisms are found normally in soil and water | 2, 11 |
| Salmonellosis | **Salmonella* of varying serotypes including Muenchen, *S. cholerae-suis* (var. Australia and var. decatur), *S. wandsbek* and *S.i indiana* diagnosed on routine faecal culture in wild numbats | Cultured in animals with mucoid or malodorous faeces and shedding thought to be associated with periods of stress | Faecal collection for microbial culture and sensitivity, which can take 12 weeks | Treatment with oral antibacterial agents, such as trimethoprim-sulfamethoxazole and amoxycillin clavulanic acid, has been attempted on occasions when infected with known pathogenic strains. However, due to intermittent shedding and lack of clinical evidence of disease the success of treatment is often difficult to determine | Reducing stress may reduce *Salmonella* sp. shedding | 2, 12 |
| **Mixed aetiology** | | | | | | |
| | Skin disease | In captivity, obesity and excessive moisture in enclosure are risk factors for development. Many cases resolve spontaneously, but unresponsive cases require investigation. Bacterial dermatitis may include exudative lesions, hair loss, scurf and scale production | Skin scrapes for mites, biopsy and culture | Cases usually respond to oral or topical antibiotic preparations based on culture and sensitivity | Good nutrition, weight management and drainage of enclosures to prevent excessive moisture build up | 2 |

**Table 5.3.** continued

| Disease | Cause | Clinical signs | Diagnosis | Treatment | Prevention | Reference |
|---|---|---|---|---|---|---|
| | Cardiopulmonary disease | Often isolated incidents; for example, Pleurisy with pyothorax (possibly secondary to oesophageal perforation) | Thoracic radiography, echocardiography, thoracocentesis | Antibiotics based on results on microbial culture and sensitivity, often futile | Often isolated incidents difficult to prevent | 2 |
| | | Epicarditis and possible carcinoma leading to chylothorax and atherosclerosis of the aorta and carotid vessels | Thoracic radiography, echocardiography, thoracocentesis Necropsy, microbial culture and sensitivity and histopathology | Antibiotics based on results of microbial culture and sensitivity, often futile | Often isolated incidents difficult to prevent | 2 |
| | | Noisy respiration and mucoid nasal discharge and bubbling | Clinical signs. Thoracic auscultation | Typically resolves without treatment | Thought to be a stress or handling response; therefore try to minimise such stressors | 2 |

References: (1) Calaby (1960), (2) Vitali and Monaghan (2008), (3) Smales (1997), (4) Haigh and Friend (1999), (5) Haigh *et al.* (1994), (6) Beveridge and Spratt (1996), (7) Zaragoza *et al.* (2019), (8) Van Immerseel *et al.* (2004), (9) Rood *et al.* (2016), (10) Perth Zoo unpubl. data; (11) Gaynor *et al.* (1990), (12) Power and Monaghan (2003).

**Table 5.4.** Common non-infectious diseases of captive numbats.

NSAIDs = non-steroidal anti-inflammatory drugs

| Disease/agent | Cause | Clinical signs | Diagnosis | Treatment | Prevention | Reference |
|---|---|---|---|---|---|---|
| Degenerative joint disease | Osteoarthritis of the hips and hock in captive aged individuals | Weakness, slow in movements, shuffling gait, lameness, poor mobility | Physical examination of musculoskeletal system noting range of motion, and presence of crepitus, radiographs of musculoskeletal system, joint taps may rule out infectious causes | NSAIDs such as meloxicam, tramadol | Consider modification of enclosure for aged individuals to minimise potential for injury and ease of movement (e.g. use of ramps, low branches, minimise angles) | 1 |
| Gastrointestinal disease | Diarrhoea | Lethargy, inappetence, dehydration | Repeated faecal culture and sensitivity, parasite examination | Most cases are transitory and resolve with supportive care. *Salmonella* culture is recommended for all recurrent or persistent cases | Minimise dietary change and stressors | 1 |
| | Sand impaction typically secondary to another disease altering gut motility | Salivation, lethargy, inappetence, distended abdomen, cloacal prolapse or swelling, vomiting or tenesmus. Voiding large amounts of sand in faeces Common predisposing disease includes lower motor neurone injury and subsequent defecation abnormalities, intermittent *Salmonella* infection and chronic proctitis | Sand can be palpated in loops of intestine in the abdomen. Radiographs plain ± contrast to define the presence of sand | Dietary supplementation with psyllium, administration of oral laxative gel such as Cat Laxative (Aristopet Animal Health, Australia), warm water enemas administered under general anaesthesia and supportive fluid therapy | Daily monitoring of faeces | 1 |
| Neonatal disease | Usually secondary to feeding difficulties or complications of hand-rearing such as aspiration pneumonia and hypothermia | Typically poor weight gain, lethargy, poor feeding response | Thorough review of history and husbandry<br>Faecal culture to rule out infectious causes<br>Blood profile to establish if infectious/inflammatory and organ involvement | Affected pouch young die very quickly and intervention in these cases has generally been unsuccessful<br>If pouch young appear dehydrated or debilitated on initial presentation it may be necessary to rehydrate with oral electrolytes and subcutaneous fluids for 12–24 h before feeding milk replacers | Poor lactation appears to be associated with environmental stress so it is critical that lactating females be maintained in a quiet warm environment and provided with excellent nutrition | 1 |

**Table 5.4.** continued

| Disease/agent | Cause | Clinical signs | Diagnosis | Treatment | Prevention | Reference |
|---|---|---|---|---|---|---|
| Trauma | Mating related trauma | Hair loss, crusting and ulceration in the dorsal cervical region | Characteristic site of trauma | Topical antibacterial and or wound healing products may be used but often resolve spontaneously | If one individual causing repeated trauma, consider resting that animal | 1 |
| | Sudden-onset neurological and/or musculoskeletal conditions, often the result of climbing injuries, including toenail avulsion, in captivity | If startled while climbing on the wire they may injure themselves, breaking nails or injuring their tail or a limb. A numbat presenting with loss of tail tone and hindlimb ataxia was diagnosed radiographically as having ventral luxations of the first and sixth coccygeal vertebrae. This injury was thought to result from jumping behaviour | History<br>Radiography to rule out bone pathology | Local wound dressing products for toenail avulsions ± removal of nail and bandaging. Rest and supportive care. NSAIDs and antibiotics may be indicated depending on severity | Minimise access to high wire enclosures, provide 'safe' climbing structures, stress reduction | 1 |
| | Corneal ulcers due to trauma have been observed in several individuals | Characteristic 'blueness of cornea' often with discharge, redness, photophobia and blepharospasm | Fluorescein staining of eye. The uptake of fluorescein stain can be highlighted by the use of a cobalt blue light. The fluorescein stain fluoresces under this light.<br>Schirmer tear test – using canine reference result typically ≥15 mm in 1 min not consistent with dry eye<br>Swab for microbial culture and sensitivity | Correct any predisposing cause. Atropine is indicated if the pupil is small or if uveitis is present<br>NSAIDs indicated when uveitis is associated with the corneal ulceration.<br>Antibacterial choice dependent on results of swab and consideration of numbats temperament (e.g. topical and or systemic). Deep or non-responsive ulcers may require referral to an ophthalmologist | Minimise access to low, spiky shrubs | 1 |

Reference: (1) - Vitali and Monaghan (2008).

**Table 5.5.** Other infrequent infectious reported diseases reported for captive numbats.

| Disease | Cause | Clinical signs | Diagnosis | Treatment | Prevention | Reference |
|---|---|---|---|---|---|---|
| **Ectoparasites** | | | | | | |
| | *Prostigmata* ear mite | Bilateral grey crusting of the ears in one captive numbat | Superficial and deep skin scrapings | Small quantities of Fipronil® or Frontline® spray that can be applied with a gauze swab | Quarantine all new arrivals, monitoring, treatment of infected individuals and regular substrate changes | 2 |
| **Protozoal** | | | | | | |
| *Toxoplasma gondii* yet to be diagnosed in captivity over a 27-year period (Perth Zoo unpubl. data), but presumed susceptibility | *Toxoplasma gondii* is a small intracellular protozoan parasite that can affect any endothermic animal. Numbats are considered moderately susceptible to toxoplasmosis | Presumably ataxia, in-coordination, poor hair coat and diarrhoea | Demonstrating either an elevated IgM titre in conjunction with a low or negative IgG titre (acute disease) or rising/elevated IgG and IgM titres (chronic disease) combined with supporting clinical signs and history. A definitive diagnosis can be made postmortem using a combination of histopathology and immunohistochemistry with or without molecular diagnostic techniques | There are few published reports of treatment of Australian marsupials with toxoplasmosis. Treatment is therefore often extrapolated from therapeutic regimens in domestic animals | No access to cats, which are the definitive host of the parasite. Reduce access of cats to enclosures, food and nesting material | 1-3 |
| **Bacterial** | | | | | | |
| Erysipelas | *Erysipelothrix rhusiopathiae* isolated from a wild numbat on necropsy | Typically results in septicaemia and acute death.<br>In an adult female the external pouch was palpably fluctant and the nipples oozed a purulent material | Necropsy, microbial culture and sensitivity and histopathology. Histopathology reported widespread fibrin thrombi containing bacterial microcolonies within interstitial vessels of the mammary gland with surrounding necrotic tissue. Bacterial microcolonies were present throughout the kidney, intestine, lung and mammary tissue and culture produced a moderate growth of *E. rhusiopathiae* | If detected antemortem, suggested treatment regimens have included therapy with penicillins, potentiated sulfonamides, and tetracyclines although no cases of successful treatment have been reported | Regular soil substrate changes, minimising exposure to reservoir hosts and daily health monitoring are likely to reduce disease risk. Vaccination (Eryvac, Pfizer Animal Health) is common in production animals and has been practiced in captive cetaceans; however, vaccines are generally not administered in captive marsupials owing to the sporadic nature of the disease | 4,5 |

**Table 5.5.** continued

| Disease | Cause | Clinical signs | Diagnosis | Treatment | Prevention | Reference |
|---|---|---|---|---|---|---|
| Listeriosis | *Listeria monocytogenes* | Clinically progressive sudden onset head tilt, tremors and ataxia in an 8-month-old numbat. Elected to euthanase on welfare grounds | Necropsy, microbial culture and sensitivity and histopathology Histopathology of the brain reported severe, multifocal, subacute heterophilic and lymphoplasmacytic encephalitis and ventriculitis. Listeria immune-histochemistry was positive and culture of the brain was positive for *L. monocytogenes* | Antibacterial therapy based on culture and sensitivity | Attention to food hygiene and disinfection. *Listeria* spp. are ubiquitous in the environment and frequently contaminate the outside of fruits and vegetables. In addition, *Listeria* spp. are psychrophilic and can proliferate in a refrigerated environment (e.g. pre-made artificial diet which is refrigerated and contains milk powder). Avoid decaying vegetation in numbat enclosures as it may harbour *Listeria* spp. | Unpubl. data |
| **Fungal** | | | | | | |
| | Cryptococcal pneumonia diagnosed at necropsy in a captive individual with co-infection of *Mycobacterium intracellulare* | Lethargy | Radiography<br>Latex Cryptococcal antigen test<br>Nasal cytology, culture and genotypin. | Not attempted. Consider fluconazole and amphotericin B protocol as used for a long-nosed potoroo | Need to determine species causing infection and then likely environmental sources (e.g. numbats are ground dwellers and climbers and will likely come into contact with both *Eucalyptus* sp. associated with *Cryptococcus gattii* and bird guano associated with *Cryptococcus neoformans*) | 6, 7 |

References: (1) Vitali and Monaghan (2008); (2) Vogelnest and Portas (2025); (3) Wildlife Health Australia (2019); (4) Vaughan-Higgins *et al.* (2013); (5) Gaynor *et al.* (1990); (6) Krockenberger *et al.* (2005); (7) Vaughan *et al.* (2007).

**Table 5.6.** Other infrequent non-infectious reported diseases reported for captive numbats.

NSAIDs = non-steroidal anti-inflammatory drugs.

| Disease/agent | Cause | Clinical signs | Diagnosis | Treatment | Prevention | Reference |
|---|---|---|---|---|---|---|
| Genito-urinary disease | Urine changes, including poor concentrating ability and proteinuria reported in an individual that died of renal failure | Lethargy, inappetence, weight loss, wetness around cloaca,. | Urinalysis: bladder is frequently difficult to palpate and many individuals urinate during handling for capture and induction of anaesthesia Manual expression of the bladder and cystocentesis are possible under anaesthesia Dipstick urinalysis parameters are typical of most mammals Isosthenuria and proteinuria have been noted in several individuals with renal disease Urine culture and sediment examination. Necropsy, microbial culture and sensitivity and histopathology | Supportive care but ultimately futile | Difficult, often age related. | 1 |
| | Penis prolapse – post mating | Swollen, erythematous everted penis | Clinical examination | Responded to lubrication, topical concentrated sugar solution to reduce oedema and manual reduction under anaesthesia | Difficult – isolated event | 1 |
| | Preputial tear | Cloacal bleeding post mating | Clinical examination | Responded to NSAIDs and clavulanic acid–amoxycillin and topical antibiotic ointment | Difficult – isolated event | 1 |
| | Testicular capsular fibroma | Increased size and firmness of the affected testis | Biopsy and histopathology | Hemicastration was performed and the animal recovered uneventfully | Difficult – isolated event | 1 |
| Gastrointestinal disease | Oesophageal perforation from food items and associated ingesta in mediastinum | Inappetence and weight loss; sudden death | Necropsy and histopathology | Usually futile | Ensure captive diet does not contain sharp items | |
| Tumours | Adenocarcinoma, squamous cell carcinomas (mandibular), histiocytic sarcoma | More commonly reported in aged numbats | Biopsy for histopathology and microbial culture and sensitivity | Superficial skin tumours can be resected under general anaesthetic. Large or invasive tumours are assessed on a case-by-case basis and supportive care or euthanasia may be considered | Not applicable | 1 |

Reference: (1) Vitali and Monaghan (2008).

maintain home ranges of ~50 ha; home ranges of males and females may overlap, but other individuals of the same sex are excluded (Cooper 2011).

Calaby (1960) reported that numbats, when handled or disturbed, often produce a low throaty growl, with the mouth closed, and this type of vocalisation may also be produced by a female rejecting the advances of a male during the breeding season. Male and female numbats vocalise with a series of soft clicks during the breeding season and a soft chirping sound may facilitate communication between mother and young, and occasionally is heard from females in oestrus. Like other marsupials, numbats also produce hissing noises when disturbed (Fleay 1942).

## 9.3 Reproductive behaviour

Breeding numbats respond to one another in a variety of ways when first introduced. Extroverted individuals exit their nest boxes promptly to explore the new enclosure. Generally, males spend the first day scent marking the females' enclosure; during the breeding season the male's sternal gland becomes active and its exudate is smeared on sticks, rocks and logs (Friend 1998). When the pair first encounter each other there can be a loud altercation with growling vocalisations by one or both animals, usually the female. Males will often attempt to mount a female during these first encounters, which often leads to them tumbling together on the ground. Some pursuit or chasing by the male is acceptable, but if chasing is excessive and for extended periods, this may indicate an incompatible pairing and it is recommended that the male is removed and reintroduced again the following day. If the same chasing behaviour is noted the male should be replaced with another male. A good indication that a compatible pairing has been established is the animals feeding from the same food bowl, sleeping together and generally spending a lot of time in close proximity to each other.

Pairings should be determined by unrelatedness of individuals using PMx® (a demographic and genetic analysis tool; Lacy *et al.* 2011; Ballou *et al.* 2020) and appropriate body mass. Suitable males generally need to be larger than or evenly matched with female body mass. Mating has been observed over a 48-hour period and copulation times range from a few minutes to an hour. A white, gelatinous seminal plug containing sperm and cornified epithelial cells may be passed from the female's cloaca up to 48 h after mating and typically indicates a successful mating. Post-mating, females may have a wet neck from biting by the male and ruffled fur around the cloacal area near the tail base.

The first observation of a female mounting a male during breeding was caught on film, with her mounting the male several times and biting his neck. This has not been observed in captivity before and is considered not normal. This female later went on to conceive pouch young with the same male from a later oestrus cycle (K. Cavanough and V. Power pers. obs.).

## 9.4 Bathing

Numbats do not appear to bathe even if sufficient free water is available.

## 9.5 Behavioural problems

Numbats can engage in abnormal repetitive behaviour, including pacing. This can be ameliorated by the provision of fresh leafy branches to break up the path being paced to encourage the animal to climb and explore their environment, although pacing is difficult to stop in front of doors where objects cannot be placed. The provision of any new natural items will stimulate the animals to investigate their environment.

Numbats are naturally very inquisitive and have been observed on occasions to climb the wire mesh, scanning adjoining enclosures. This has been observed mostly during the warmer months, which corresponds with increased activity before and during the breeding season. In captivity, this can be difficult to prevent. Great care should be taken not to frighten an animal if it is on the mesh as they can suffer nail injuries or broken bones if they are startled and jump off. Regular trimming of nails helps reduce the severity of nail injuries, which are more commonly seen in wild-caught animals during the first 6 months in captivity.

## 9.6 Signs of stress

Inappetence and associated weight loss, along with inactivity and hiding are the primary signs of stress.

## 9.7 Behavioural enrichment

Behavioural enrichment suitable for numbats includes modifying the enclosure and/or diet. Making the enclosure as variable as possible by modifying the soil profile and adding rocks, logs and climbing branches wired to enclosure walls encourages numbats to explore and climb. Frequently changing or moving objects in the enclosure stimulates the highly inquisitive numbat. Nest material from another numbat's nest box is a great enrichment item. Providing fresh termite crumb (leftover debris after extracting harvested termites, see Addendum 1) so numbats can forage for remaining termites encourages natural foraging behaviour and will help develop or maintain feeding skills of individuals scheduled for release into the wild. Likewise, an artificial termite log (Fig. 5.7) filled with live termites and crumb provides considerable enrichment and further encourages natural foraging behaviour. The addition of *Amitermes* mounds before release given to weaner young provides diversity of termite species and stimulates foraging behaviour.

### 9.8 Training and conditioning

Numbats are not trained but are encouraged to undertake natural behaviour that would be necessary for survival in the wild, such as foraging for termites.

### 9.9 Introductions and removals

Introductions of males to females should be done after the morning feed, initially under supervision when the female is in proestrus. The male is moved to an adjoining enclosure ahead of time so the dividing gates between the male and female enclosures can be opened to pair them. The male generally spends the first day scent marking within the female's enclosure, with brief bouts of chasing of the female. Although aggression is not normally observed, some males will chase females rather than engage in reproductive behaviour. If this behaviour lasts longer than a day, the male should be replaced with another male.

### 9.10 Intraspecific compatibility

Numbats are solitary animals and are housed separately most of the year. Males and females are usually compatible with each other during the breeding season, although this needs to be closely monitored. Two females have been successfully housed together for display purposes when introduced simultaneously to the new enclosure to minimise development of any dominance behaviour. Some aggression between females has been observed, particularly over food, so feeding in separate areas of the enclosure is advisable. Male numbats can become aggressive towards each other (through the fence) during the breeding season. To avoid this, they should be housed in enclosures that are separated by either a female's enclosure or an empty enclosure and shade-cloth screens erected on the enclosure walls.

### 9.11 Interspecific compatibility

Numbats are normally housed alone, but can be held as part of a mixed species enclosure with small insectivorous birds including rainbow bee-eaters (*Merops ornatus*), rufous whistlers (*Pachycephala rufiventris*), welcome swallows (*Hirundo neoxena*) and splendid fairy wrens (*Malurus splendens*).

## 10. BREEDING

### 10.1 Mating system

Numbats are polygynous, with males having the ability to mate with more than one female.

**Fig. 5.7.** Artificial termite log being used by a numbat. Photo by Vicki Power.

## 10.2 Ease of breeding

Numbats are seasonal breeders and readily breed in captivity.

## 10.3 Oestrous cycle and gestation period

Females are receptive to males for approximately 2 days in each oestrous cycle. The gestation period is 14 days, so the pouch region can be checked for young after a further 1–2 days (Friend and Whitford 1988). At Perth Zoo, checks of young occur 2 days after birth and 8 weeks post birth, with no further checks until weaning.

## 10.4 Reproductive status

### 10.4.1 Females

Females can be categorised into several reproductive stages including:

- non-breeding
  - non-parous (never bred) – pouch region small with barely any indentation, teats very small, clean and dry
  - parous (bred previously) – there is an indentation in the pouch area with all four nipples clearly visible and the pouch region may appear dirty with a dry exudate
- breeding
  - pro-oestrous – nipples becoming pink with slight swelling on abdomen around mammary area
  - oestrous – a thick and creamy discharge from the cloaca may be observed (Friend and Whitford 1988)
- pregnant – pouch pink in colour, with a glandular and moist appearance. Swelling on the lateral margins of the pouch region (the abdomen just in front of the mammary area and caudal to the inner surfaces of the thighs), so that the teats are enclosed within a depression. The nipples also appear to increase in size slightly and deepen in colour to red; slight swelling around the nipples forms a ridge up to the nipples
- lactation
  - early lactation (young attached to the teat) – long guard hairs hang down over the pouch area, offering some warmth and protection. As the young develop, the skin folds become flaccid, hair sparse and stained, skin smooth and dark pink
  - late lactation (young absent from the pouch but still suckling) – teats dark pink and elongated (~15 mm in length). Heavy, engorged mammary glands
  - post lactation – teats expressing clear liquid and/or regressing. Mammary area can have a lumpy appearance while milk supply is drying up. Takes 4–6 weeks for nipples to fully regress to non-breeding size.

### 10.4.2 Males

Male numbats go through clear reproductive changes from October to February (Calaby 1960; Friend and Whitford 1988; Friend 1998). These include:

- sternal scent gland becoming active and exudes an oily, sticky liquid that stains the fur on the chest and stomach a red-brown colour. The secretions begin in late September reaching a maximum extent in late December and declining during January–February
- testes increasing in size during early October with a maximum volume attained in late December–January. This is maintained until February, when they regress rapidly
- cloacal region swells noticeably with associated glandular enlargement.

## 10.5 Reproductive management

The current breeding protocol at Perth Zoo is to pair females mid-December before oestrus. Direct behavioural observation has been successful in recent years to determine breeding activity so swabbing of females to determine oestrus is no longer necessary. Oestrous cycles for individuals may occur within a few days to 1 week of their previous years' oestrus dates. This allows a certain predictability of each oestrus event with all females except new wild-caught animals. Opportunistic collection of any mucus passed by females in enclosures can be examined under a microscope to confirm either oestrus or post-mating sperm. Use staining procedures listed below.

The following laboratory procedure can be used to detect the presence of sperm in the urine of males.

- Place the numbat in a wire cage trap with a fly wire floor over a tray to collect urine and faeces.
- Label and refrigerate urine sample immediately.
- Centrifuge urine samples at 6400 rpm for 2 min.
- Resuspend the pellet in PBS (pH-buffered saline) solution at 0.5–1.0 mL and agitate.
- Pipette up a small amount (two drops) not removing any sediment and drop onto slide and air dry. Stain according to DIFF QUIK® protocol:
- stain slide as soon as air-dried
- fixative 5 × 1 s dips, drain on paper between each of the three parts of the process
  - Sol. 1 5 × 1 s dips, drain

- Sol. 2 5 × 1 s dips, drain
- gently rinse with deionised water
- air dry
- mount with DePGX® mounting medium if required for long-term storage.

On occasion, urine samples may not show sperm. This does not mean that males cannot sire young; sometimes the sperm does not show in the urine sample, but if a male with no sperm in his urine fails to produce young, further investigation is required. Individuals can remain paired until pouch young are found or if no mating has occurred males may be replaced before the next oestrous cycle.

## 10.6 Occurrence of hybrids

None.

## 10.7 Timing of breeding

Patterns of faecal steroid secretion of captive numbats reveal a highly synchronised seasonal pattern of reproduction (Hogan *et al.* 2012). The mating period extends from late December to early February, although there have been instances of mating in late February. Births occur from early January to the start of March, though most occur in the second half of January (Fig. 5.8). In the Perth Zoo breeding programme, numbats are paired during mid- to late December. Males generally have sperm visible in the urine from mid-December with some year-to-year variability. Sperm are usually absent by the end of February. New wild-caught males have their urine checked for sperm production just before pairing in December. Female numbats can undergo 2–3 oestrous cycles in a breeding season, ~28 days apart. Most mating occurs in January; however, if a litter is lost within the first few weeks after birth or a mating produces no young and it is early enough in the season, females may be able to produce a litter from their next oestrus cycle.

## 10.8 Age at first and last breeding

Males reach sexual maturity at 2 years of age and females at 1 year (Hume 1987; Friend and Whitford 1988). Females are able to breed past 5 years of age, but the number of young produced declines with age. Currently, a breeding management model is in place at Perth Zoo where the youngest female breeding is 1 year old and male is 2 years old. Animals are represented in each age class, breeding for a total of 4 years and are then retired out of the breeding program as 4-year-old females and 5-year-old males. These older animals are rehomed to other institutions, released into the wild or used for display, allowing for regular importation of new young animals to optimise potential breeding output.

## 10.9 Ability to breed every year

Females and males can breed every year.

## 10.10 Ability to breed more than once per year

Numbats are poly-oestrous, so if a litter is lost early in the breeding season or the female fails to mate, it is possible to produce a litter at the next oestrous cycle.

## 10.11 Nesting requirements

Several types of nest boxes are provided (see section 4.8).

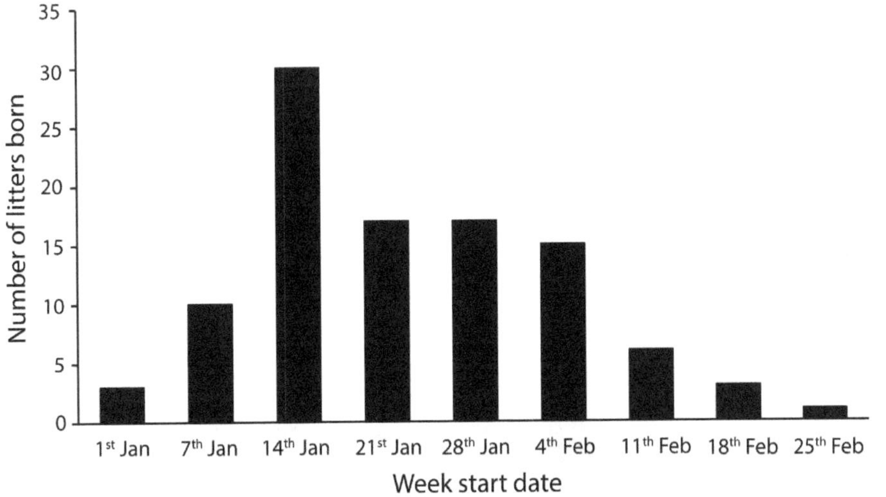


**Fig. 5.8.** Timing of births for captive numbats at Perth Zoo. $N = 102$ litters. Data from Perth Zoo.

## 10.12 Breeding diet

In early December before the onset of the breeding season the proportion of termites in the diet of captive numbats is increased to a minimum of 50% for females and 12% for males, with additional termite crumb fed ad lib during the day. Females with pouch young are provided with a higher proportion of termites in the diet until they wean their young in October or November. After this time the termite percentage in their diet can be gradually reduced to 12% (over a period of 1–2 weeks). Having a reliable supply of termites during the breeding season is critical to success.

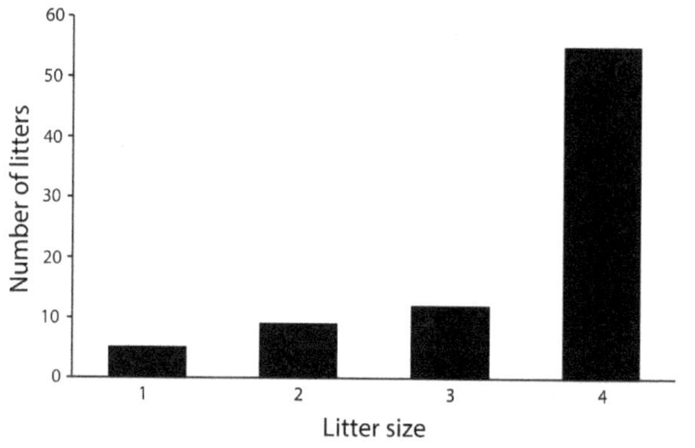


**Fig. 5.10.** Litter size of captive numbats bred at Perth Zoo between 1997 and 2018. *n* = 81 litters. Data from Perth Zoo.

## 10.13 Number of young born

A maximum of four young can attach to the four teats, though potentially more than four young are born (Friend 1998). In the wild, females almost always carry three or four pouch young (AJ Friend. pers. comm.). The average litter size for numbats in captivity is between three and four (average ± SD 3.5 ± 0.85, $n = 81$; Fig. 5.9). It is rare for females to give birth to only one or two young. On average females successfully wean 2.2 ± 1.8 ($n = 81$) pouch young with a survival rate of 64.9% (Fig. 5.10).

## 10.14 Age at weaning

In the wild, numbat young are independent at ~11–12 months. The natal dispersal time is quite short, rarely taking more than 1 week from departure to establishment of a territory in the area where the numbat will spend the rest of its life (Friend and Burrows 1983; Friend 1998). The weaning process for captive born young is commenced in late October/early November when the young are 10 months of age. At that stage the young weigh 200–300 g and they are gradually introduced to an artificial diet (see section 6.1) one or two young at a time to allow the mother's milk supply to diminish and the nipples regress in readiness for the next breeding season. Wombaroo® Impact Colostrum Supplement (1 g per 35 g of artificial diet once a day) can be added to the diet of young and mothers before weaning in September and it is discontinued by late December.

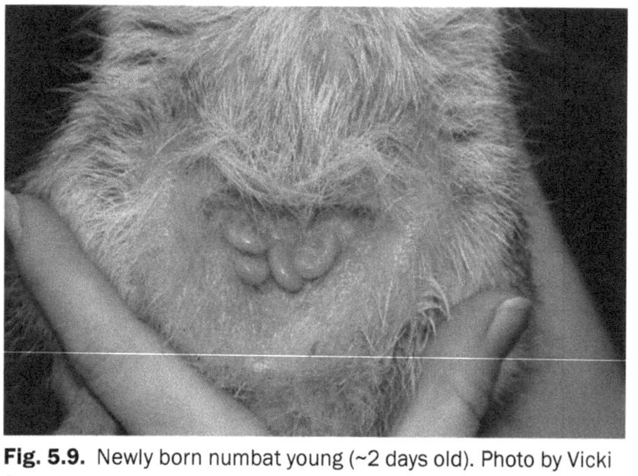

**Fig. 5.9.** Newly born numbat young (~2 days old). Photo by Vicki Power.

## 10.15 Age at removal from the mother

Young can be removed once they are weaned.

## 10.16 Growth and development

Numbat young have a relatively slow growth rate (Fig. 5.11; Table 5.7) compared with other similar-sized marsupials, which has been attributed to their dependence on a

**Table 5.7.** Developmental stages in the growth of the numbat. Data from five litters born at Perth Zoo.

| Developmental stage | Mean age in days (range) | Date (approximate) |
|---|---|---|
| Sex can be identified visually | n/a (90–100) | 30/04 - 30/05 |
| Soft fur growing | 122.6 (116–137) | 30/05 - 10/06 |
| First deposited | 190.6 (186–193) | 05/08 - 25/08 |
| Eyes still closed | 190.0 (186–193) | 05/08 - 11/08 |
| Long guard hairs growing | 194 (194) | 27/08 |
| First vocalisations heard | 200 (194–206) | 06/08 - 26/08 |
| Eyes open first time | 201.8 (192–215) | 06/08 - 11/09 |
| Young attached to nipple again | 212.7 (199–230) | 23/08 - 11/09 |
| First emerges from box | 216.4 (203–226) | 15/08 - 25/09 |
| First eats artificial diet | 252.6 (239–262) | 02/10 - 03/11 |
| Weaned from mother | 273 (269–279) | 30/10 - 22/11 |

n/a – data not available.

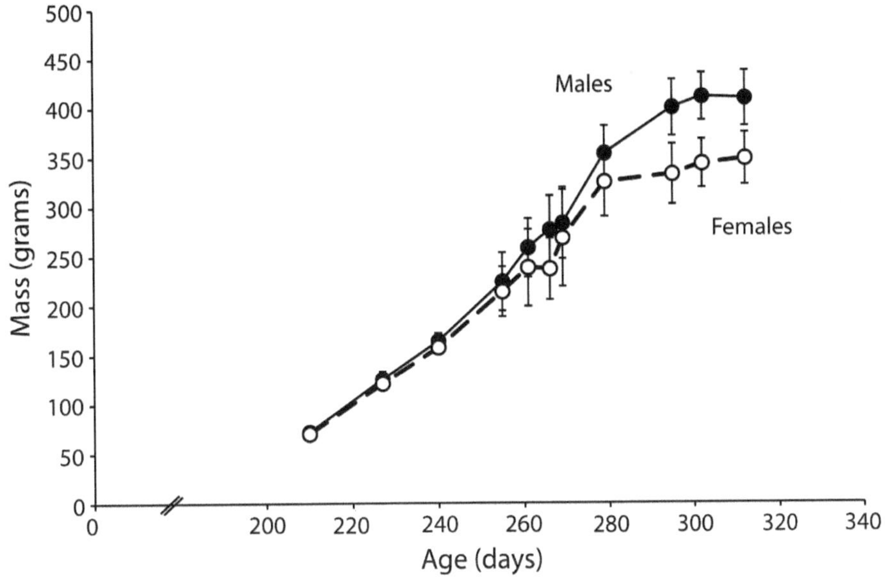


**Fig. 5.11.** Body mass (mean ± standard deviation; *N* = 22) of developing numbat young born at Perth Zoo. From 0 to 200 days pouch young are attached to the mother's nipples and cannot be weighed individually. Data from Perth Zoo.

termite diet (Friend 1998). Young are born in the second half of January and carried until late July or early August when they are ~45 mm long and weighing 18–27 g. They are then deposited in a nest, which in the wild is usually in a burrow (Friend 1998). In captivity, this is the first time the young can be weighed and sexed, which is usually done within 2 weeks of when the dam deposits them in the nest. Within 1 week of deposit, they should increase their mass by 30–50%. In early September the young begin to emerge from the burrow each morning and remain outside after the mother has departed for the day. By mid-October the young supplement their mother's milk with their own foraging efforts (Friend 1998) and in the wild the young disperse from the maternal (natal) range from November (Friend 1998).

## 11. ARTIFICIAL REARING

Several numbat young required assisted rearing at Perth Zoo between 1999 and 2020. A reliable feeding and husbandry routine has been established, based on experience gained from these animals. Requirements for hand-rearing young are variable and will depend on the age and circumstances. For example, a neonate with a suddenly dead mother will be more developed and in better body condition than a neonate of the same age that is removed because of mismothering or lack of milk supply.

### 11.1 Housing

As with all native mammals that are taken into care, minimising stress is a major consideration.

Housing requirements depend on the age and development of the young. In general, very small young require more heat as they have less developed thermoregulatory capacity. The following is a guide based on various litters that have been reared from different ages:

- poorly furred/very young neonate (eyes and ears closed):
  - kept together in a flannel pouch (within a woollen pouch) - tied shut and kept at a temperature of 34°C
  - the woollen pouch is to be kept in a thermostat-controlled hot box with water tray underneath the shelf, so the air is not too dry
  - care must be taken that the numbats do not get too cold when removed from their hot box for feeding.
- For uncompromised young of older ages (i.e. numbats with eyes open; Fig. 5.12a):
  - kept locked in a nest box lined with grass (ideally from mother's nest box initially) at 30°C
  - nest box kept within thermostat controlled hot box with water tray underneath.
- Emerging age (numbats trying to climb out of nest box - more active at feeds - eyes and ears fully open and well furred (Fig. 5.12):

- in a wooden nest box lined with grass (door open) – this is set up within a plastic tub with a heat mat placed underneath (temperature probe to be kept in nest box) – temperature range of 20–28°C)
- substrate cotton (pillow slips or towels are good) – newspaper is a bit slippery at this stage of development as emerging numbats are quite unsteady on their feet
- the ambient room temperature should be maintained at ~22°C
- a heat lamp on top of the tub lid, or a microwaveable heat disc inside a material cover are options for providing additional warm spots.

If the young are old enough and are able to lap milk, it is advisable to try and keep them in their existing heat box with familiar scents. This may minimise stress and the young are more likely to come out and explore their environment and use the basking lamps.

## 11.2 Temperature requirements

During assisted rearing the temperature in the incubator can be held at 22–34°C. A heated nest box (Fig. 5.12) can be used if an incubator is not available.

## 11.3 Specific requirements

The skin of unfurred, or slightly furred, young can be kept moist with Sorbolene cream so it does not become dry and cracked. When first brought in for hand-rearing the joey may be dehydrated. Rehydration can be achieved with boiled water with 5 g (1 teaspoon) of glucose to 100 mL of water or 1 g of electrolyte replacer if available (Austin 1997). Vytrate® can also be used at a ratio of 20 mL Vytrate® to 250 mL water (J. Cowey pers. comm.). It is important to warm the joey before feeding to minimise the risk of inhalation pneumonia. Fluids can be given subcutaneously if the individual is very dehydrated and hydration via bottle-feeding will take too long. No unfurred pouch young have ever been successfully hand-reared. Only pouch young at least 6 months of age, which are fully furred and eyes open, have been successfully reared to independence.

## 11.4 Diet and feeding routine

### 11.4.1 Natural milk

Development of the young in a pouch during a relatively long lactation is a characteristic of marsupial reproduction. Marsupial lactation can be divided into three phases (Tyndale-Biscoe and Janssens 1988). Phase one occurs during pregnancy when the mammary gland is prepared for suckling. Phase two starts with birth and is completed when the joey is physiologically mature. In the early stage of phase two, the joey is continuously attached to the teat and in the later part the young can detach from the teat but is still wholly dependent upon milk for nourishment. At this stage the joey is deposited in the nest but may re-attach for a few days. Phase three spans the period of weaning to the adult diet. As the joey grows and develops, the composition of the milk changes to accommodate the joey's energetic requirements (see Green and Merchant 1988 for review).

**Table 5.8.** Concentrations of major constituents of numbat milk during late-stage lactation.

Data from Griffiths *et al.* (1988).

| Total Solids (%) | Carbohydrates (%) | Lipids (%) | Protein (%) | Calcium (mg/L) | Iron (mg/L) |
|---|---|---|---|---|---|
| 28.0 | 2.0 | 12.0 | 14.0 | – | – |

The composition of milk from early lactation in numbats has not been determined due to the long-term attachment of the young. The composition of late lactation milk of wild numbats (Table 5.8) has been determined (Griffiths *et al.* 1988) and is similar to that described for other marsupials at a similar stage of lactation. Carbohydrate content is low and is mostly present as free hexose. Nearly all the lipids are triglycerides with a very high level of oleic acid, probably a consequence of the numbats' exclusively termitivorous diet (Griffiths *et al.* 1988). Numbats fed an artificial diet have lower levels of oleic acid in the milk fat.

### 11.4.2 Milk formulas

Numbat young have been successfully reared on a Di-Vetelact® (Lillelund Pty Ltd) formula with egg yolk added for extra protein and fat. All neonates should be started on Vytrate®, then moved to a 50:50 Vytrate®/Di-Vetelact® mix the following day. A day or two later when they are well hydrated and more stable, young should be moved to 100% Di-Vetelact®.

When stable on this dilution (1 scoop to 40 mL water) slowly start adding egg yolk to the formula in increments of 0.5 mL. Aim to increase by 0.5 mL every 3–4 days to a maximum of 2 mL egg yolk.

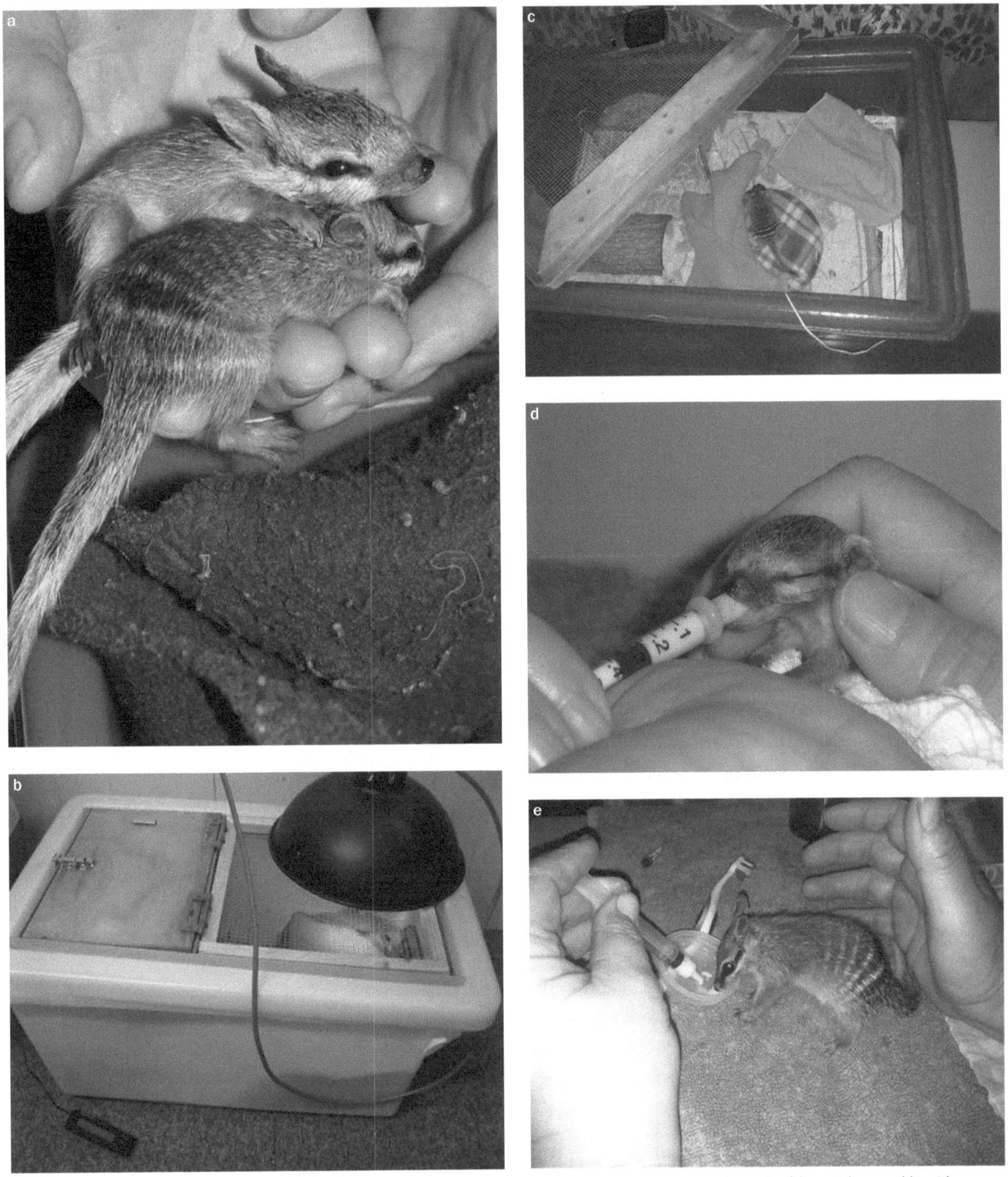

**Fig. 5.12.** Hand-rearing numbat young in captivity. (a) Furred young, eyes just open. (b) Hand-rearing tub with nest box and heat lamp. (c) Sleeping tub with heat pad. (d) Syringe-feeding a young numbat. (e) A young numbat lapping milk. Photos from Perth Zoo.

## Formula intake goals

Some individuals will willingly take over 20% of their body mass in a single day, but this can result in bloating and very rapid weight gain. Appropriate weight gain can be achieved when food intake is capped at 18% of body mass (e.g. an average intake of 17–18% body mass produces average daily gains of 4% of body mass and an intake of 15% body mass produced a daily gain of 3.5% body mass).

### Feeding frequency

Feeding frequency depends on the age of the neonates and their health (degree of debilitation, dehydration etc). As a general rule of thumb, the younger/less developed numbats (under 40 g) will require six feeds within a 24-h period with a maximum 5-h break between feeds, whereas those heavier than 40 g can be maintained on five feeds with an 8–10 h break overnight.

Feed frequency should only be reduced when:

- there has been a steady and consistent weight gain in the days leading up to the change
- the numbat can hold the required volume for each feed without excessive stretching/bloating of the stomach.

## 11.4.3 Feeding apparatus

Neonates have been successfully raised using a 1–3-mL syringe and teats (Bushbaby™ teats, which are long thin latex teats designed to fit a 1-mL syringe are recommended; Barlow 1998; Fig. 5.12). Due to the length of these teats, a plastic catheter insert placed inside the teat will give it more stability and allow for easier placement inside the numbat joey's mouth.

## 11.4.4 Feeding routine

Once the numbats have built up sufficient strength, they can switch from syringe feeding to lapping (Fig. 5.12), which involves less handling and encourages more independence. The recommended method is as follows.

- Add milk in small increments (no greater than 0.5 mL at a time) to a small plastic bowl.
- Keep milk warm and pre-warm the bowl before adding milk.
- Hold the bowl tilted under the numbat's mouth until there is brief contact with the milk. There may be some spluttering at first until the numbat learns how to lap.

### Feeding method

Syringe/teat feeding is necessary when the neonate's eyes and ears are still closed and they are unable to stand or sit for a few minutes without tiring.

- It may be difficult to insert a teat into the numbat's mouth. An often successful approach is to go in under the top lip (possibly even on the side of the mouth), then work the teat tip around to the front, where the mouth should open slightly.
- The teat should only partially be inserted to avoid the numbat gagging on it or its long tongue.
- Milk should be fed very slowly – one drop at a time, making sure the numbat swallows it. If they do not swallow the milk, there is the danger of milk inhalation into the lungs. Gentle stimulation of the throat area may encourage swallowing. If milk is fed too fast, it may also seep out the sides of the mouth.

### Handling during feeding

Whether teat feeding or lapping, human contact should be minimised.

- For teat feeding wrap the numbat in a bag with the eyes covered.
- When lapping, keep the numbat under a cloth as much as possible. Some of the young prefer no contact when lapping. In this case the cloth should be draped between the numbat and keeper feeding them.
- Numbats should not be restrained with bare hands for feeding – use a cloth.
- If numbats are out while waiting (while milk warms etc) they should be placed in a bag to keep them from getting too cold and from being exposed to those feeding them.

If the neonates are still being syringe-fed after half an hour or are compromised in any way, then heat should be supplied in the form of a microwaveable heat disc. Stress is a major problem in the rearing of numbats. Therefore, it is important that noise is kept to a minimum and they are not over-handled.

## 11.5 Data recording

When an animal is first brought in for hand-rearing, its sex and approximate age, or developmental stage, should be recorded. During the hand-rearing process several important pieces of information should be recorded on a daily basis.

- date when the information is recorded
- time when the information is recorded
- body mass
- general activity and demeanour
- characteristics and frequency of defecation and urination
- amount (g) and different food types offered.
- food consumption at each feed
- veterinary examinations and results.

This information serves several purposes, including providing important background information such as food consumption, which will assist a veterinarian to reach a diagnosis if the animal becomes sick or fails to grow or gain weight. It also allows a comparison with growth curves to assess progress (section 10.16) and to establish new growth curves that do not exist for other measurements. The developmental stages and measurements outlined in Table 5.7 should also be recorded on a weekly basis if possible.

## 11.6 Identification methods

A small drop of different coloured nail polish or food colouring on the head of young numbats can serve for temporary individual identification.

## 11.7 Hygiene

As with any animal, maintaining a high standard of hygiene is critical to the survival of the young. Feeding utensils should be cleaned and sterilised after each use. Formula should be freshly prepared each day and kept refrigerated between feeds.

## 11.8 Behavioural considerations

Young numbats may learn some behaviours from their mother such as nest building and foraging, and possibly also predator avoidance. It may be possible to place the young numbat(s) with a non-lactating female or a surrogate female that has recently weaned her young. Young should have access to natural sunlight and outside stimulation once they are able to regulate their body temperature. Periods of exercise are essential to starting the weaning process. Trays of termite crumb and small water dishes can be provided. Habituation to humans during the hand-rearing process can be difficult to reverse. This means that very few hand-raised individuals end up as suitable candidates for release to the wild. However, they have been proven to be good captive breeders (particularly females). They can also make excellent display animals due to their relaxed nature.

## 11.9 Use of foster animals

Numbats have been fostered successfully onto other lactating mothers whose own litter has been recently weaned.

## 11.10 Weaning

Introduction of the artificial diet should coincide with the natural weaning age. Numbats need to first be accustomed to lapping, then a pea-sized piece of the plain egg custard can be mixed with the milk. The proportion of custard to milk must be increased very slowly (usually every few days), monitoring closely for any lack of acceptance, gut pain and ease of defecation. It is important that steady weight gain is maintained. Once young are eating full custard meals, a sprinkle of termites can be added and mixed in. The termites can be increased to a couple of grams per meal once the weanlings are willingly eating this mixture. A weaned numbat should not receive more than 40–50 g of custard/termite mixture by the time they are 10–11 months old (by November–December). At weaning, fresh water should be supplied.

## 11.11 Release procedures

There is a recovery plan for the numbat in Western Australia (Department of Parks and Wildlife 2015), managed by the WA Department of Biodiversity, Conservation and Attractions. Translocation of numbats to areas within the species' former range where introduced predators are controlled is one of the actions of the plan. Young numbats thus require preparation before their release. Termite mound is provided daily for enrichment and to stimulate natural foraging activity; termite-infested logs are ideal if available. Body mass should range between 350 g and 400 g, to facilitate the appropriate fitting of radio-collars for monitoring the animals once they are released. All young should undergo a prerelease physical (see section 8). Numbats have many predators, including foxes, cats, pythons, birds of prey and quolls. Captive-bred numbats have comparable rates of survival on release to wild-born numbats of the same age (Palmer *et al.* 2020).

# ADDENDUM 1. SUSTAINABLE TERMITE HARVESTING TECHNIQUES

Numbats are obligate termitivores, so a reliable termite supply is essential. *Nasutitermes exitiosus* is Australia's best-known species of termite (Watson and Gay 1991). It occurs across southern mainland Australia, from the southern half of Western Australia east to the Pacific coast, but is absent from Tasmania (Watson and Abbey 1985). *Coptotermes acinaciformis* is a common and widely distributed termite species in Western Australia and is certainly the most abundant and destructive species in the southern part of the state (Calaby and Gay 1956). Both these species are found in the Perth metropolitan area, but neither is trapped in this area due to the potential risk from past use of pesticides and the risk of transfer of residual chemicals to the numbats. In summer the trapping efforts are directed towards *Coptotermes* as larger quantities are consistently trapped. During the winter months, trapping *Nasutitermes* can often be more productive than trapping *Coptotermes* if a large enough nest can be located. However, *Coptotermes* have a considerably higher digestibility (81%) than *Nasutitermes* (64%; Cooper and Withers 2004a), so fewer *Coptotermes* are required to meet the numbat's average daily metabolic rate, and they also appear to be more palatable.

## 1 Termite harvesting

A method of collecting termites has been developed to supply termites for the numbat breeding program at Perth Zoo. The basic principle is to lure termites into an open-topped drum packed with karri (*Eucalyptus diversicolor*) wood (one of the termites' preferred foods). Once the wood has been infested with termites, the drum is retrieved and brought to a place where the termites can be separated from the wood. Nests are not destroyed and when yields start to diminish, the termite traps are removed, allowing the nests to recover. Perth Zoo has established collection sites that are mainly within water catchment areas with limited public access or on private land. It is essential to have multiple sites established in the event of a natural disaster (bush fire) or destruction of the drums by public interference.

## 2 Finding termites

The mounds of *N. exitiosus* (Fig. 5.13) found on the Swan coastal plain in sandy low rainfall areas are low, 200–400 mm and 'thin-skinned', but smooth and dome-shaped with most of the nest underground (Eutick 1983). *Coptotermes* have underground nests, often at the base of trees or in dead or dying trees. These are usually detected by locating their feeding galleries, which radiate out from the nest (Hadlington 1987). The galleries can be found by turning fallen deadwood over and looking for termite activity, keeping disruption to a minimum, or by placing wet pieces of karri wood on the ground and returning several weeks later to check for activity. It is important to bury the bait wood into the ground well so they make good contact with the soil and avoid black ant infestation. Blocks of wood can be stacked on top of each other to create a platform for the drum to sit on once termite infiltration has occurred. Searching for evidence of termite activity on large tree bases can lead to the primary nest and the source of galleries to which drums are set on.

## 3 Preparing the termite traps

Open-topped metal drums (15–20 L) are used with 8 × 12 mm diameter holes drilled in the bases. Untreated karri wood battens of size 40 × 25 mm, free of debris and fungus and cut to 400 mm long are packed into the drums. A densely packed drum creates a more attractive environment for termites.

Care must be taken when setting drums to avoid leaving the nests or galleries exposed to infiltration by predatory black ants. The soldier termites can repel a few ants, but whole nests can be destroyed if ant numbers are high. The period of exposure of termites in their mound or in the soil should be minimised.

## 4 Trapping methods

A sharp square-mouthed spade can be used cut the top off the mound and flatten the surface for *Nasutitermes*, while for *Coptotermes* the drum can be placed directly on the exposed galleries. The drums are put in place and all then gaps around the base of the drum sealed with dirt. Approximately 0.5–1.0 L of water should be poured over the wood in the drum and a heavy-duty black plastic garbage bag used to firmly seal the top (Fig. 5.13). The moisture and plastic bag combine to create a humid environment, which is attractive to termites.

## 5 Replacing the drums when full

Termites are very active in warmer weather and a good site may yield a full drum within 2–3 weeks. In winter, a drum on the same nest may take several months to fill. When the original drums are collected, a replacement drum should be immediately placed back onto the nest lining up the holes of the new drum with the holes in the nest where the termites are emerging from.

## 6 Storing full drums

Drums of termites may be stored for up to 1 week before separating the termites from the wood. However, they must be kept out of the sun and wind and stored on antproof tables. To prevent drums from drying out they should be stored inside a tightly-sealed plastic bag on top of damp karri slats inside a plastic tray. Pesticides should never be used for ant control as the risk of contaminating the termites is too high.

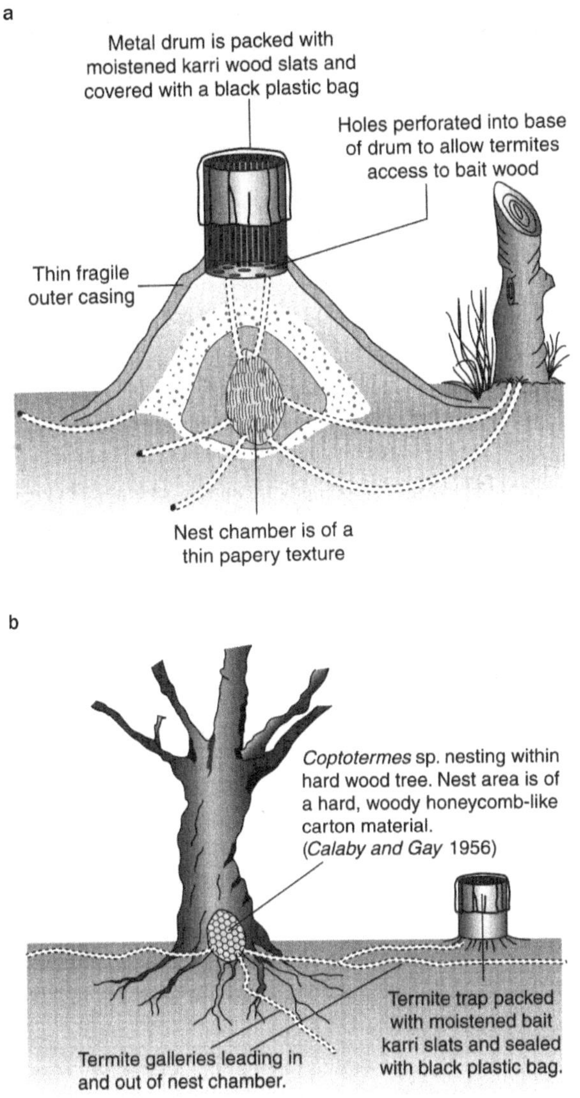


**Fig. 5.13.** Trapping techniques for (a) *Nasutitermes* and (b) *Coptotermes* sp. Drawing by Vicki Power.

## 7 Separating the termites

Termites are very delicate and will crush easily if roughly handled and will dehyrdrate quickly when separated from their mound. Dust masks should be worn during termite separation. Termites can be removed from the drums intact by removing one karri slat at a time and hitting with a tool such as a crowbar, so the termites fall into a large plastic tray. A combination of termites and dirt (mound material) will collect in the plastic tray. Termites can be then separated from the dirt by placing the mixture onto a platform sitting in another large tray (Gay *et al.* 1955). The termites are photophobic and will move to the edge of the 'platform' and fall into the plastic tray beneath. Any remaining dirt/termite mix can be provided to numbats as supplemental food and enrichment.

## 8 Freezing the termites

Termites can be frozen for future use without losing their nutritional value or palatability to numbats. The sooner they are frozen after separating the better and if necessary, they can be refrigerated for a day before freezing in zip lock bags. All air should be expelled from the bag before sealing, and they should be labelled with the species identification, date of processing and weight.

## ACKNOWLEDGEMENTS

Many thanks to Dr Tony Friend for all his valuable advice and support since the establishment of the numbat breeding program at Perth Zoo. A sincere thank you to all the staff at the Perth Zoo and Department of Biodiversity, Conservation and Attraction's Native Species Breeding Program who have contributed to the program over the years. Dr Peter Mawson and Dr Harriet Mills facilitated the production of this manuscript and provided input and editorial assistance.

## REFERENCES

Abbott I (2001) Aboriginal names of mammal species in south-west Western Australia. *CALMscience* **3**, 433–486.

Anon (1969) Census of rare animals living in zoos and other institutions in 1968. *International Zoo Yearbook* **9**, 274–306.

Anon (1979) Census of rare animals living in zoos and other institutions in 1978. *International Zoo Yearbook* **19**, 384–423.

Austin MA (1997) *A Practical Guide to the Successful Hand Rearing of Tasmanian Marsupials*. Regal Publications, Melbourne.

Ballou JD, Lacy RC, Pollak J (2020) PMx: Software for demographic and genetic analysis and management of pedigreed populations (Version 1.6.2.20200110). Chicago Zoological Society, Brookfield, Illinois, USA. http://www.scti.tools (accessed 17/01/2022)

Barlow S (1998) Notes on the assisted rearing of three juvenile *Myrmecobius fasciatus* at Perth Zoo. *Thylacinus* **22**, 47–52.

Bester AJ, Rusten K (2009) Trial translocation of the numbat (*Myrmecobius fasciatus*) into arid Australia. *Australian Mammalogy* **31**, 9–16. doi:10.1071/AM08104

Beveridge I, Spratt D (1996) The helminth flora of Australian marsupials, origins and evolutionary biology. *Advances in Parasitology* **37**, 135–254. doi:10.1016/S0065-308X(08)60221-3

Bodley K (2025) Appendix 4. Drug formulary. In *Current Therapy in Medicine of Australian Mammals*. Revised Edition. (Eds L Vogelnest and T Portas) pp. 732–761. CSIRO Publishing, Melbourne.

Calaby JH (1960) Observations on the banded anteater *Myrmecobius fasciatus* Waterhouse (Marsupialia), with particular reference to its food habits. *Proceedings of the*

*Zoological Society of London* **135**, 183–207. doi:10.1111/j.1469-7998.1960.tb05841.x
Calaby JH, Gay FJ (1956) The distribution and biology of the genus *Coptotermes* (Isoptera) in Western Australia. *Australian Journal of Zoology* **4**, 19–39. doi:10.1071/ZO9560019
Christensen P, Maisey K, Perry DH (1984) Radiotracking the numbat, *Myrmecobius fasciatus*, in the Perup Forest of Western Australia. *Australian Wildlife Research* **11**, 275–288. doi:10.1071/WR9840275
Cooper CE (2011) *Myrmecobius fasciatus* (Dasyuromorphia: Myrmecobiidae). *Mammalian Species* **43**, 129–140. doi:10.1644/881.1
Cooper CE, Withers PC (2002) Metabolic physiology of the numbat (*Myrmecobius fasciatus*). *Journal of Comparative Physiology* **172**, 669–675. doi:10.1007/s00360-002-0294-8
Cooper CE, Withers PC (*2004*a) Termite digestibility and water and energy contents determine the water economy index of numbats (*Myrmecobius fasciatus*) and other Myrmecophages. *Physiological and Biochemical Zoology* **77**, 641–650. doi:10.1086/421750
Cooper CE, Withers PC (2004b) Influence of season and weather on activity patterns of the numbat (*Myrmecobius fasciatus*) in captivity. *Australian Journal of Zoology* **52**, 475–485. doi:10.1071/ZO04038
Cooper CE, Withers PC (2004c) Patterns of body temperature variation and torpor in the numbat, *Myrmecobius fasciatus* (Marsupialia: Myrmecobiidae). *Journal of Thermal Biology* **29**, 277–284. doi:10.1016/j.jtherbio.2004.05.003
Cooper CE, Withers PC (2005) Physiological significance of the microclimate in night refuges of the numbat *Myrmecobius fasciatus*. *Australian Mammalogy* **27**, 169–174. doi:10.1071/AM05169
Cooper CE, Walsberg GE, Withers PC (2003a) Biophysical properties of the pelt of a diurnal marsupial, the numbat (*Myrmecobius fasciatus*), and its role in thermoregulation. *The Journal of Experimental Biology* **206**, 2771–2777. doi:10.1242/jeb.00484
Cooper CE, Withers PC, Bradshaw SD (2003b) Field metabolic rate and water turnover of the numbat (*Myrmecobius fasciatus*). *Journal of Comparative Physiology. B, Biochemical, Systemic, and Environmental Physiology* **173**, 687–693. doi:10.1007/s00360-003-0380-6
Department of Parks and Wildlife (DPW) (2015) Numbat (*Myrmecobius fasciatus*) Recovery Plan. Wildlife Management Program No. 60. Prepared by J.A. Friend and M.J. Page, Department of Parks and Wildlife, Perth.
Dickman CR, Calver MC (2023) Food habits and activity patterns of Australasian marsupials. In *American and Australasian Marsupials* (Eds NC Cáceres, CR Dickman) pg 1–37. Springer, Cham.
Eutick ML (1983) Collecting and maintaining termites of the Sydney region. In *Apicultural Review*. (Ed. KC Parson) pp. 1–12. Avicultural Society of N.S.W.
Finlayson HH (1933) On the Eremian representative of *Myrmecobius fasciatus* (Waterhouse). *Transactions of the Royal Society of South Australia* **57**, 203–205.
Fleay D (1942) The numbat in Victoria. *Victorian Naturalist* **59**, 3–7.
Friend JA (1993) Myrmecobiidae. In *Fauna of Australia*. Vol. 1B. *Mammalia* (Eds GJ Glasby, GJB Ross and PL Beesley) pp. 583–590. Australian Government Publishing Service, Canberra.
Friend JA (1995) Numbat *Myrmecobius fasciatus*. In *The Mammals of Australia*. (Ed. R Strahan) pp. 160–162. Reed Books, Sydney.
Friend JA (1998) Numbats on a junk food diet. *Nature Australia* (Winter), 40–49.
Friend JA, Burrows RG (1983) Bringing up young numbats. *Swans* **13**, 3–9.
Friend JA, Thomas ND (2003) Conservation of the numbat. In *Predators with Pouches: The Biology of Carnivorous Marsupials*. (Eds M Jones, C Dickman and M Archer) pp. 452–463. CSIRO Publishing, Sydney.
Friend JA, Whitford D (1988) Captive breeding of the numbat *Myrmecobius fasciatus*. Unpublished report to the WWF Australia and the Department of Conservation and Land Management, Western Australia.
Friend JA, Fuller PJ, Davis JA (1982) The numbat in central Australia. *Swans* **12**, 20–26.
Friend T (2008) Numbat, *Myrmecobius fasciatus*. In *The Mammals of Australia*. 3rd edn. (Eds S Van Dyck and R Strahan) pp 163–1165. Reed New Holland, Sydney.
Gay FJ, Greaves T, Holdaway FG, Wetherley AH (1955) Standard laboratory colonies of termites for evaluating the resistance of timber, timber preservatives, and other materials to termite attack. *CSIRO Bulletin* **277**, 1–60.
Gaynor WT, Cousins DV, Friend JA (1990) Mycobacterial infection in numbats (*Myrmecobius fasciatus*). *Journal of Zoo and Wildlife Medicine* **21**, 476–479.
Green B, Merchant JC (1988) The composition of marsupial milk. In *The Developing Marsupial: A Model for Biomedical Research*. (Eds CH Tyndale-Biscoe and PA Janssens) pp. 41–54. Springer-Verlag, London.
Griffiths M, Friend JA, Whitford D, Fogerty AC (1988) Composition of the milk of the numbat, *Myrmecobius fasciatus* (Marsupialia: Myrmecobiidae), with particular reference to the fatty acids of the lips. *Australian Mammalogy* **11**, 59–62. doi:10.1071/AM88006
Groves CP (2005) Order Dasyuromorphia. In *Mammal Species of the World: A Taxonomic and Geographic Reference*. 3rd edn. (Eds DE Wilson and DM Reeder) pp. 23–37. Johns Hopkins University Press, Baltimore, Maryland.
Hadlington P (1987) *Termites and Other Common Timber Pests*. University of New South Wales Press, Sydney.
Haigh SA, Friend JA (1999) Veterinary aspects of several threatened Western Australian mammal species. *AAVCB Newsletter* **16**.
Haigh SA, Friend JA, Morris KD (1994) An investigation of the health status and the occurrence of disease agents in several threatened Western Australian species. Internal DBCA document. https://library.dbca.wa.gov.au/static/FullTextFiles/015851.pdf
Harley D, Mawson PR, Olds L, MacFadden M, Hogg C (2018) The contribution of zoo-based captive breeding programs towards the conservation of Australian native fauna. In *Hope and Beyond: Recovering Australia's Threatened Species*. (Eds S Garnett, P Latch, D Lindenmayer and JZ Woinarski) pp 281–294. CSIRO Publishing, Melbourne.
Hogan LA, Johnston SD, Lisle AT, Robertson H, Valentine L (2012) Non-invasive monitoring of male and female numbat (*Myrmecobius fasciatus*: Myrmecobiidae) reproductive activity. *Animal Reproduction Science* **133**, 237–245. doi:10.1016/j.anireprosci.2012.07.005

Hume D (1987) The numbat as a zoological exhibit. *Thylacinus* **12**, 4–6.

International Air Transport Association (IATA) (2022) *Live Animals Regulations*. 48th edn. International Air Transport Association, Geneva, Switzerland.

Jackson SM, Groves C (2015) *Taxonomy of Australian Mammals*. CSIRO Publishing, Melbourne.

Krockenberger M, Stalder K, Malik R, Canfield P (2005) Cryptococcosis in Australian Wildlife. *Microbiology Australia* **26**, 69–73. doi:10.1071/MA05069

Lacy RC, Balou JD, Pollak JP (2011) PMx: software package for demographic and genetic analysis and management of pedigreed populations. *Methods in Ecology and Evolution* **3**, 433–437. doi:10.1111/j.2041-210X.2011.00148.x

Morris K, Page M, Kay R, Renwick J, Desmond A, Comer S, Burbidge A, Kuchling G, Sims C (2015) Forty years of fauna translocations in Western Australia: lessons learned. In *Advances in Reintroduction Biology of Australian and New Zealand Fauna*. (Eds DP Armstrong, MW Hayward, D Moro and PJ Seddon) pp 217–234. CSIRO Publishing, Melbourne.

Palmer N, Smith MJ, Ruykys L, Jackson C, Volck G, Riessen N, Thomas A, Moir C, Palmer B (2020) Wild-born versus captive-bred: a comparison of survival and refuge selection by translocated numbats (*Myrmecobius fasciatus*). *Wildlife Research* **47**, 217–223. doi:10.1071/WR19105

Power VL 2002 *Numbat Animal Management Guidelines*. Perth Zoo, Perth.

Power VL, Monaghan C (2003) Numbats. In *Australian Mammals: Biology and Captive Management*. (Ed. SM Jackson). pp. 99–125. CSIRO Publishing, Melbourne.

Purse B (1972) The numbat: Taronga's special trust. *Koolewong* (June), 11–12.

Rood JI, Keyburn AL, Moore RJ (2016) NetB and necrotic enteritis: the hole movable story. *Avian Pathology* **45**, 295–301. doi:10.1080/03079457.2016.1158781

Smales LR (1997) *Multisentis myrmecobius*, Gen. et. Sp. Nov (Acanthocephala: Oligocanthorhynchidae), from the numbat, *Myrmecobius fasciatus*, and a key to genera of the Oligacanthorhynchidae. *Invertebrate Systematics* **11**, 301–307. doi:10.1071/IT95023

Species360 (2023) Zoological Information Management System (ZIMS). Minneapolis, USA. zims.Species360.org.

Strahan R (1978) Status and husbandry of Australian monotremes and marsupials. In *Breeding Endangered Species in Captivity*. (Ed. RD Martin) pp. 171–182. Academic Press, London.

Strahan R (1995) *The Mammals of Australia*. Reed Books, Sydney.

Strahan R, Conder P (2007) *Dictionary of Australian and New Guinean Mammals*. CSIRO Publishing, Melbourne.

Tate GHH (1951) The Banded Anteater, *Myrmecobius fasciatus* Waterhouse (Marsupialia). *American Museum Novitates* **1521**, 1–8.

Troughton E (1967) *Furred animals of Australia*. Angus and Robertson, Sydney.

Tyndale-Biscoe CH, Janssens PA (1988) Introduction. In *The Developing Marsupial*. (Eds CH Tyndale-Biscoe and PA Janssens) pp. 1–7. Springer-Verlag, London.

Van Immerseel F, De Buck J, Pasmans F, Huyghebaert G, Haesebrouck F, Ducatelle R (2004) *Clostridium perfringens* in poultry: an emerging threat for animal and public health. *Avian Pathology* **33**, 537–549. doi:10.1080/03079450400013162

Vaughan RJ, Vitali SD, Eden PA, Payne KL, Warren KS, Forshaw D, Friend JA, Horwitz AM, Main C, Krockenberger MB, Malik R (2007) Cryptococcosis in Gilbert's and long-nosed potoroo. *Journal of Zoo and Wildlife Medicine* **38**, 567–573. doi:10.1638/2007-0005R.1

Vaughan-Higgins RJ, Bradfield K, Friend JA, Riley TV, Vitali SD (2013) Erysipelas in a numbat (*Myrmecobius fasciatus*). *Journal of Zoo and Wildlife Medicine* **44**, 208–211. doi:10.1638/1042-7260-44.1.208

Vitali S, Monaghan C (2008) Numbats. In *Medicine of Australian Mammals*. (Eds L Vogelnest and R Woods) pp. 383–394. CSIRO Publishing, Melbourne.

Vogelnest L (1999) Chemical restraint of Australian native fauna. In *Wildlife in Australia: Healthcare and Management*. (Ed. DI Bryden) pp. 149–181. Proceedings 327. Post Graduate Foundation in Veterinary Science. University of Sydney, Sydney.

Vogelnest L (2025) Capture, physical and chemical restraint. In *Current Therapy in Medicine of Australian Mammals*. Revised Edition. (Eds L Vogelnest and T Portas) pp. 122–151. CSIRO Publishing, Melbourne.

Vogelnest L, Portas T (2008) Macropods. In *Medicine of Australian Mammals*. (Eds L Vogelnest and R Woods) pp. 133–225. CSIRO Publishing, Melbourne.

Vogelnest L, Portas T (Eds) (2025) *Current Therapy in Medicine of Australian Mammals*. Revised Edition. CSIRO Publishing, Melbourne.

Waterhouse GR (1836) Description of a new genus (*Myrmecobius*) of mammiferous animals from New Holland, probably belonging to the marsupial type. *Proceedings of the Zoological Society of London* **4**, 69–70.

Watson JAL, Abbey HM (1985) Seasonal cycles of the *Nasutitermes exitiosus* (Hill) (Isoptera: Termitidae). *Caste Developmental Sociobiology* **10**, 73–92.

Watson JAL, Gay FJ (1991) Isoptera. In: *Insects of Australia* (Ed. Division of Entomology, CSIRO). Pp 330–347. Melbourne University Press, Melbourne.

Wildlife Health Australia (2019) Toxoplasmosis in Australian mammals. https://wildlifehealthaustralia.com.au/Portals/0/Documents/FactSheets/Mammals/Toxoplasmosis_of_Australian_Mammals.pdf

Woinarski JCZ, Burbidge AA, Harrison PL (2014) *Action Plan for Australian Mammals 2012*. CSIRO Publishing, Melbourne.

Wood Jones F (1923) *The Mammals of South Australia*. Parts 1–3. Government Printer, Adelaide.

Zaragoza NE, Orellana CA, Moonen GA, Moutafis G, Marcellin E (2019) Vaccine production to protect animals against pathogenic clostridia. *Toxins* **11**, 525. doi:10.3390/toxins11090525

# 6 Bandicoots

*Stephen M. Jackson, Marissa L. Parrott, Michael J. Lynch, Yvette Pauligk and Amy L. Coetsee*

## 1. INTRODUCTION

The bandicoots are a diverse group of marsupials that includes 31 species in the Order Peramelemorphia within four families: the Chaeropodidae (pig-footed bandicoots), the Peroryctidae (echymiperas), the Peramelidae (bandicoots) and the Thylacomyidae (bilbies) (Baker *et al.* 2023). The bandicoots' common name is derived from the bandicoot rats (*Bandicota* species) that occur in Asia, but they are marsupials with a well-developed pouch.

Despite the appeal of some species, especially the very popular greater bilby, bandicoots were historically poorly represented in zoos, with most captive animals generally held in research institutions up until the 1970s for reproductive studies (e.g. Lyne 1971, 1974). Western barred bandicoots (*Perameles bougainville*) were held and bred as early as 1924 (Wood Jones 1924), long-nosed bandicoots (*Perameles nasuta*) have been held in the McMaster Laboratory in Sydney since 1954 (Collins 1973) and southern brown bandicoots (*Isoodon obesulus*) and eastern barred bandicoots (*Perameles gunnii*) have been held by the University of California (Heinsohn 1966). Rufous spiny bandicoots (*Echymipera rufescens*) have rarely been held in captive facilities, although there are records of them being kept at the National Zoo in Washington in 1972 (Collins 1973). Greater bilbies (*Macrotis lagotis*) were exhibited as early as 1848 at London Zoo and subsequently in the late 1800s to early 1900s at Frankfurt Zoo and Philadelphia Zoo in 1904 (Collins 1973). Since the early 1980s, however, multiple Australian zoos have been heavily involved in conservation efforts for greater bilbies and some bandicoot species, many of which are endangered. This has led to significant developments in the captive management of these species, including their breeding. Greater bilbies have been held at Monarto Zoo, Taronga Zoo, Western Plains Zoo in Dubbo, Adelaide Zoo, Currumbin Sanctuary, Healesville Sanctuary and the Alice Springs Desert Park in the Northern Territory, with releases to the wild and fenced locations. Other species, such as golden bandicoots (*Isoodon auratus*), have been held at Perth Zoo and Territory Wildlife Park. Mainland eastern barred bandicoots have been held at Healesville Sanctuary, Melbourne Zoo, Werribee Open Range Zoo, Kyabram Fauna Park, Serendip Sanctuary, Adelaide Zoo, Western Plains Zoo in Dubbo and Taronga Zoo as part of a successfully completed 33-year recovery program for insurance and release of this endangered species into fox-free sites (Parrott *et al.* 2017). Tasmanian eastern barred bandicoots have been held and bred in a new program at Trowunna Wildlife Sanctuary, in Tasmania, since 2018. The western barred bandicoot has been bred at the Kanyana Wildlife Rehabilitation Centre (Western Australia) since 1994 and the Peron Captive Breeding Centre (Western Australia) since 1998 as part of a captive breeding program. Monarto Zoo has also held this species (Bennett 2008).

## 2. TAXONOMY

### 2.1. Nomenclature

The bandicoots comprise two separate families within the Order Peramelemorphia and consist of 31 species. Of these, 19 are found in Australia, 14 are found in New Guinea and surrounding islands and 2 are found in both regions (Seebeck *et al.* 1990; Flannery 1995a, 1995b; Baker and Gynther 2023; Baker *et al.* 2023; Table 6.1).

**Australian Bandicoots**

Class: Mammalia
Supercohort: Marsupialia
Cohort: Australidelphia
Order: Peramelemorphia
Family: Chaeropodidae
Genus and species: 1 genus, 2 species
Family: Peroryctidae
Genus and species: 1 genus, 2 species
Family: Peramelidae

**Table 6.1.** Species of bandicoots within Australia, their bodyweight and conservation status.

*Also occurs in New Guinea and/or surrounding islands. EN = Endangered, EX = Extinct, LC = Least Concern, VU = Vulnerable.
Derived from Seebeck *et al.* (1990), Flannery (1995a, 1995b), Woinarski *et al.* (2014), Baker and Gynther (2023) and Baker *et al.* (2023).

| Common name | Scientific name | Weight (g) | IUCN status |
|---|---|---|---|
| **Family Chaeropodidae** | | | |
| Southern pig-footed bandicoot | *Chaeropus ecaudatus* | c. 280–600 | EX |
| Northern pig-footed bandicoot | *C. yirratji* | c. 360–600 | EX |
| **Family Peroryctidae** | | | |
| Rufous spiny bandicoot* | *Echymipera rufescens* | 525–2225 | LC |
| **Family Peramelidae** | | | |
| Golden bandicoot | *Isoodon auratus* | 100–670 | VU |
| Dusky-bellied bandicoot or quenda | *I. fusciventer* | 556–2000 | |
| Northern brown bandicoot* | *I. macrourus* | 500–3100 | LC |
| Southern brown bandicoot | *I. obesulus* | 400–1850 | LC |
| Cape York brown bandicoot | *I. peninsulae* | c. 555.690 | LC |
| Western barred bandicoot | *Perameles bougainville* | 165–379 | VU |
| Desert bandicoot | *P. eremiana* | ~250 | EX |
| Liverpool Plains striped bandicoot | *P. fasciata* | c. 450–585 | EX |
| Eastern barred bandicoot | *P. gunnii* | 500–1450 | VU |
| Marl | *P. myosuros* | c. 540–885 | EX |
| Long-nosed bandicoot | *P. nasuta* | 520–1450 | LC |
| South-eastern striped bandicoot | *P. notina* | c. 570–850 | EX |
| Northern long-nosed bandicoot | *P. pallescens* | 500–1050 | LC |
| Nullarbor barred bandicoot | *P. papillon* | 367–551 | EX |
| **Family Thylacomyidae** | | | |
| Greater bilby | *Macrotis lagotis* | 600–2500 | VU |
| Lesser bilby | *M. leucura* | 310–435 | EX |

Genus and species: 2 genera, 14 species
Family: Thylacomyidae
Genus and species: 1 genus, 2 species

## 2.2 Subspecies

See Jackson and Groves (2015).

## 2.3 Recent synonyms

Synonyms for Australian bandicoots can be found in Jackson and Groves (2015).

## 2.4 Other common names

See Baker and Gynther (2023).

# 3. NATURAL HISTORY

## 3.1 Bodyweight and morphometrics

Australian bandicoots range in size from ~150 g to 2500 g (Table 6.1) (Flannery 1995a, 1995b; Baker and Gynther 2023). The morphometrics of Australian species can be found in Baker and Gynther (2023).

## 3.2 Distribution and habitat

Bandicoots occur throughout mainland Australia and New Guinea and surrounding islands in a wide range of habitat types. Pig-footed bandicoots once occurred in the southern grasslands, golden bandicoots occur in the Top End and Kimberley tropics of Australia, desert bandicoots were strictly desert inhabitants, bilbies are also desert dwellers, although they once occurred in both arid and semi-arid regions of central Australia. The eastern barred bandicoot prefers grasslands and brown bandicoots live in more closed forest (Baker and Gynther 2023). Members of the family Peroryctidae all occupy more tropical environments of north-eastern Australian and New Guinea. More specific details of the distribution and habitats occupied by the Australian bandicoots can be found in Seebeck *et al.* (1990) and Baker and Gynther (2023).

## 3.3 Conservation status

Overall, the bandicoots and bilbies have not done well since European settlement in Australia, with eight species having gone extinct and four species considered vulnerable out of only 19 species. On the mainland, the eastern barred bandicoot was listed as Extinct in the Wild until 2021 when it was reclassified in Victoria to Endangered. The decline and extinction of these marsupials has primarily been a result of predation from introduced feral cats (*Felis catus*) and European red foxes (*Vulpes vulpes*) on the mainland of Australia, habitat destruction and fragmentation and habitat degradation from introduced livestock and rabbits (Woinarski *et al.* 2014; Table 6.1). However, the mainland eastern barred bandicoot population is behaviourally flexible in its habitat requirements and can survive in the presence of feral cats. The eastern barred bandicoot has been the focus of a greater than 33-year recovery program including captive breeding and insurance, research, reintroduction to predator-free fenced reserves and introductions to three fox-free islands (Parrott *et al.* 2017; Rendall *et al.* 2018). Populations are continuing to expand and the captive breeding and insurance program was phased out in 2021 as it was no longer required (Parrott and Edwards 2023; A. Coetsee pers. obs.).

## 3.4 Diet in the wild

Bandicoots are typically omnivorous, though specific diets are known for few species. They typically eat a wide range of invertebrates, bulbs and grasses, with fungi also contributing to the diet of at least some species (Claridge and May 1994; Baker and Gynther 2023). The long-nosed bandicoot has been observed to predate upon the nest of a spotted pardalote (*Pardalotus punctatus*) and eat three of the young birds (Guppy and Guppy 2018). Southern-brown bandicoots and eastern barred bandicoots have been observed eating from a wallaby carcass in the wild, though it is unknown whether the bandicoots were eating the wallaby or parasites such as maggots that may have been found in its stomach or on the carcass (M. Parrott and Y. Pauligk pers. obs.). Further details of the diet of different genera can be found in Table 6.2 and Baker and Gynther (2023).

## 3.5 Longevity

### 3.5.1 Wild

Bandicoots are typical of species that breed rapidly and die at a relatively young age. Longevity for *Perameles* is 2–3 years in the wild (Mallick *et al.* 2000). *Isoodon* species may live up to 4 years of age (Table 6.3).

### 3.5.2 Captivity

As with other species, bandicoots live longer in captivity, with most species typically living at least 4–5 years. Although there are records of bilbies living up to 10 years in captivity, they typically live 5 years (Table 6.3). Eastern barred bandicoots in captivity can continue to breed up to 4.8 years old for females and > 5.8 years old for males (Hartnett *et al.* 2018; M. Parrott unpubl. data), though fecundity reduces in bandicoots over 3 years of age (Pauligk 2021). They often live to 4–5 years old and have been recorded living up to 6 years old in captivity (Parrott *et al.* 2017).

**Table 6.3.** Longevity (months) of different genera of bandicoots in the wild and in captive facilities.

Number in brackets is the average longevity.

| Genus | Wild | Captivity | Reference |
|---|---|---|---|
| *Isoodon* | 42–48 | 23+ | 1, 2 |
| *Perameles* | 24–36 | 42–72 | 3–6 |
| *Macrotis* | 12+ | 48–120 (60) | 7, 8 |

References: (1) Gemmell (1990); (2) Lobert and Lee (1990); (3) Heinsohn (1966); (4) Dufty (1991); (5) Mallick *et al.* (2000); (6) Pauligk (2021); (7) Flower (1931); (8) Southgate *et al.* (2000).

**Table 6.2.** Diet of different genera of bandicoots.

| Genus | Food types | Reference |
|---|---|---|
| *Chaeropus* | Grasses | 1, 2 |
| *Echymipera* | Invertebrates, fungi, fruits, seeds, plant matter | 3–5 |
| *Isoodon* | Various invertebrates: worms, snails, ants, larvae, insects; frogs, grasses, seeds, mosses, fungi | 6–8 |
| *Perameles* | Various invertebrates: worms, snails, insects, spiders; fungi, seeds, slugs, frogs, skinks, birds | 6, 8–13, 15 |
| *Macrotis* | Various invertebrates: insect larvae, other insects, termites, ants; bulbs, seeds, plant fibre, rodents? | 2, 8, 14, 16 |

References: (1) Wright *et al.* (1982); (2) Dixon (1988); (3) Flannery (1995a); (4) Baker and Gynther (2023); (5) Shevill and Johnson (2008); (6) Heinsohn (1966); (7) Quin (1985); (8) Claridge and May (1994); (9) Reimer and Hindell (1996); (10) Claridge (1993); (11) Guppy and Guppy (2018); (12) O'Hara *et al.* (2012); (13) Thums *et al.* (2005); (14) Southgate (1990); (15) Loeffler (2018); (16) Gibson (2001).

### 3.5.3 Methods used to determine the age of adults

There appears to be no useful technique for determining the age of adult bandicoots. An examination of various parameters in long-nosed bandicoots included various skull measurements, time of eruption of teeth, wear of the teeth, rate of deposition of dentine and cementum in the teeth, size of the lenses and time of epiphyseal fusion (Kingsmill 1962). None of these measures were found to provide satisfactory results (Kingsmill 1962). The only age classes that could be distinguished were 0–4 months and > 4 months, using the degree of fusion of the epiphyses of the limb bones (Kingsmill 1962).

## 4. HOUSING REQUIREMENTS

### 4.1 Enclosure design

Due to their nocturnal behaviour, bandicoots are often exhibited in nocturnal houses. Most species are readily held in enclosures with a deep layer of soil, leaf litter and/or mulch, at least 10 cm deep so that they can dig a nest depression. Various tussocks, hollow logs and branches should also be added. Care needs to be taken when providing logs to ensure the entry/exit points are not too small that they get stuck (or cannot be retrieved) or so the animals do not dig underneath and get crushed. All furniture provided should be light and placed so that it cannot be dug under. Some species, such as members of the genera *Isoodon* and *Perameles*, generally prefer enclosures with good vegetation cover; efforts should be made to provide cover, but it should not be so thick that they cannot be seen. For individuals not held within a nocturnal setting, sand pads can be added to provide key indirect information on activity.

### 4.2 Holding area design

Holding enclosures should have sheet metal lining to a minimum of 1.5 m because some bandicoots, such as the eastern barred bandicoot, can jump 1.2 m and climb mesh. Without the sheet metal, bandicoots can catch their paws in the mesh and break claws and digits or may fall and injure themselves, including breaking their tails. Vegetation and furniture should be positioned to discourage the individual using it to gain additional height up the walls (Fig. 6.1).

The enclosure should be mouse-proof through the use of a concrete base or fine underwire that is submerged 15 cm (most species) to 25 cm (bilbies) below the surface to avoid damage to claws and snouts (Kingston 1998; Pauligk 2021). The mesh should be bird netting or fabric mesh with a hole of 1.2 cm (Williams 1990) or galvanised weldmesh with a gap of 10 × 10 mm to exclude rodents (Pauligk 2021). At least one end of the enclosure should be covered to provide shelter from rain and with multiple dry nesting opportunities (Kingston 1998; Pauligk 2021). A 100–200 cm area of sand at the front of the enclosure is useful for monitoring the movements of the elusive and nocturnal animals. If the sand area is raked daily, animal caretakers can monitor the footprints and gaits of animals and determine their levels of activity.

### 4.3 Spatial requirements

Bandicoots have been held in relatively small enclosures, though minimum size standards have increased over time to improve animal welfare (Table 6.4). If possible, enclosures should be made larger to further reduce the chance

**Fig. 6.1.** Example of an eastern barred bandicoot holding enclosure showing nesting areas including structural plywood panels and half rubbish bins. Also note the smooth tin on the wall up to a height of ~1.5 m. Photo used with permission of Zoos Victoria.

**Table 6.4.** Minimum areas of enclosures recommended for pairs of different genera of Australian bandicoots.

From Mackerras and Smith (1960), McCracken (1986), Williams (1990), Kingston (1998), Pauligk (2021) and Animal Welfare Victoria (2021).

| Genus | Area (L × B × H) (m) | Additional floor area for each extra animal ($m^2$) |
|---|---|---|
| *Isoodon* | 4 × 5 × 1.5 | 10 |
| *Perameles* | 4 × 5 × 1.5 | 10 |
| *Macrotis* | 5 × 5 × 1.5 | 10 |
| *Echymipera* | 4 × 5 × 1.5 | 10 |

of aggression and stress in females with young. The Victorian Code of Practice for the public exhibition of animals (Animal Welfare Victoria 2021) suggests a minimum floor space of 20 m, with an increased floor space of 10 m per additional animal, and a fence height of 1.5 m.

## 4.4 Position of enclosures

The enclosure should be away from significant traffic and other noises and positioned so that it is protected from wind and rain (Pauligk 2021). Exposure to heat may be a factor in hot regions of Australia.

## 4.5 Weather protection

Protection from extremes in weather, especially exposure to very hot weather or heavy rain and strong wind, is necessary. It is recommended that at least one-third of the enclosure (particularly the end facing the direction of inclement weather) should be covered. Shade cloth can also be used to increase the level of shade in very exposed enclosures (Kingston 1998; Pauligk 2021).

Bandicoots should ideally be provided with at least three shelter sites that have nesting material included within them (Pauligk 2021). These can include:

- structural plywood panels (~50–60 cm (L) × 40 cm (H)) that are leant against the walls of the enclosure at an angle of ~45–60° (to make a tent-like structure)
- plastic bins that have been cut in half and have both ends open
- large tussocks for nest sites. Grasses of the genera *Poa* and *Lomandra* grasses have been successfully used, but serrated edge tussocks should be avoided because they may cause eye damage. Nests can be well-developed for some species, so it is important to provide adequate grasses to allow these to be constructed (Fig. 6.2)
- hollow logs that are relatively light and (40–80 cm long) can be used as functional nest sites.

**Fig. 6.2.** Eastern barred bandicoot nest. Photo used with permission of Zoos Victoria.

## 4.6 Temperature requirements

Heating is generally not needed for species that live in cooler, more temperate climates, provided adequate dry bedding and cover are available (Kingston 1998). Species from more arid regions, such as bilbies, should be provided with a heat lamp if held in temperate regions and allowed to retreat from the heat in hot weather. In warmer areas, misters can be used to cool enclosures in the summer or when temperatures exceed 28°C. However, for species of bandicoot from non-tropical regions, misters should only be used briefly or in short bursts to cool the area to avoid increased humidity, which can promote bacterial growth.

## 4.7 Substrate

Bandicoots and bilbies should be maintained on well-drained sand or soil, though other substrates, including leaf litter and mulch, have been used. If the enclosure does not have a concrete base, galvanised weldmesh (10 × 10 mm) should be laid down and crushed granite of ~10 mm diameter should be compacted to a depth of 80–100 mm over it, with mulch, sand and/or soil then added as a top layer (Pauligk 2021). Mulch and soil, of at least 15 cm, with turf and tussock grasses have been very successful for housing eastern barred bandicoots. The depth of the substrate is important to reduce the incidence of damage to the bandicoots' feet and snouts. Bandicoots may also develop foot issues, calluses or cracking of feet if held on inappropriate or damp substrate. The provision of soil and/or sand allows bandicoots to dig nesting areas and to forage for invertebrates. Bilbies in particular should be provided with sand, and/or soil, in which to dig. Bilbies in the wild dig to 25 cm while foraging (Pavey 2006); thus at least 25 cm depth of substrate is recommended for housing captive bilbies.

### 4.8 Nest boxes

If natural nesting locations are not provided, sturdy nest boxes should be provided. Boxes with dimensions 35 cm (L) × 24 cm (W) × 24 cm (H) have been used successfully for *Isoodon* and *Perameles* (Lyne 1982). Nest boxes of 40 cm (L) × 40 cm (W) × 40 cm (H) have been used successfully with bilbies (Hulbert 1982). They should be filled with a nesting material such as fine rye grass, meadow hay or straw, and nesting materials can be provided in the enclosure, which the animals will collect to create their own nest. Structural plywood panels (~50–60 cm (L) × 40 cm (H)) leant against walls to create a tent-like structure with bedding underneath have also been used successfully for bandicoots (Pauligk 2021).

### 4.9 Enclosure furnishings

Various enclosure furnishings should be supplied, including grass tussocks and sedges (e.g. *Lomandra longifolia*), hollow logs, branches of eucalypts laid on the ground (especially in corners), large pieces of bark and nest boxes. Fresh grass or hay should also be supplied in each of the nesting sites to assist in nest building (Krake and Halley 1993).

## 5. GENERAL HUSBANDRY

### 5.1 Hygiene and cleaning

Each enclosure should be cleaned every 1–2 days to remove faecal matter and uneaten food. Small enclosures can be spot-cleaned daily and given a full substrate clean weekly or more often if required. Drinking water dishes should be cleaned daily and water bottles should be checked daily to make sure the nozzle is working properly and that the bottle is at least two-thirds full. When an enclosure is emptied it should be scrubbed out before new animals are introduced.

### 5.2 Record keeping

It is important to establish a system whereby the health, condition and reproductive status of the captive bandicoot population are routinely monitored and recorded:

- identification numbers (all individuals should be identifiable)
- results of veterinary examinations and treatments administered
- behavioural changes or problems
- reproductive behaviour or condition
- bodyweight and measurements
- changes in diet and food consumption
- movements of individuals between enclosures or institutions
- births, with dam and sire if known
- deaths with necropsy results.

Collecting information on the physical, physiological and behavioural patterns of individual bandicoots can contribute greatly to the husbandry of this species. It also allows individual histories to be shared with other institutions and facilitates a cooperative approach to data collection among institutions. ZIMS, the Zoo Information Management System (Species360 2023), can be used for recording general information on age, sex, parentage, births, identification, bodyweights, transfers and deaths, with separate ZIMS modules for veterinary information and studbooks. As these are standardised, there is a high degree of efficiency in transferring information between institutions. The information can be directly transferred into management software, such as PMx, to assist with planning, reports and recommendations (Lacy *et al.* 2011). The genetic management and breeding recommendations of both bandicoots (e.g. Parrott *et al.* 2017) and bilbies (e.g. Miller *et al.* 2015) can be aided via the use of microsatellite genetic markers and other molecular techniques.

### 5.3 Methods of individual identification

#### 5.3.1 Passive Integrated Transponder (PIT) tags

PIT tags (microchips) are implanted between the scapulae of individuals and can be used on all bandicoots. PIT tags are an excellent and permanent method of identification, but can track out along the injection site. This can be avoided by sealing the entry wound with tissue glue (Vetbond®) or similar fast setting adhesive, but is not essential. In some species with very tough skin a small hole is cut in the shoulder region to allow the insertion of the tag needle, after which the hole is sutured with 2–3 stiches. A disadvantage of the use of PIT tags is that they generally require the animal to be caught to confirm identification with a PIT tag reader, though remote tag readers baited with peanut butter are an effective way to monitor tagged bandicoots where required.

### 5.3.2 Ear tattoos

Tattooing was used for the identification of bandicoots in the past, due to their large, thinly furred ears (*Isoodon* has small dark ears). However, the use of tattoos has been replaced by PIT tags.

### 5.3.3 Ear tags

Metal ear tags have been used, but were generally unsatisfactory as they are often lost within several months (Lyne 1982). Fingerling ear tags have been successfully used on bilbies in captivity, but are no longer recommended because they pull out causing ear damage and the sites can easily become infected.

### 5.3.4 Ear notching

Although not generally used on adults, ear notching of pouch young has been used for identification (Lyne 1982). It is not recommended because it damages the ears and can be difficult to distinguish from other wear and tear of the ears over time for animals in the wild, particularly in males that frequently have torn ears due to fighting. However, ear notching is useful for the collection of tissue for genetic analyses. Tissue should only ever be collected from the ear margin, via removal of a thin sliver (< 1 mm) of skin or leaving an open small notch at the edge. Complete holes (fully encircled by skin) should not be left because they can snag on branches or similar, leading to tearing of the ear.

## 6. FEEDING REQUIREMENTS

### 6.1 Captive diet

The diet of captive bandicoot species should approximate the composition of their diet in the wild and be appropriately balanced for vitamins and minerals. For most species this can be achieved by offering a diet that roughly consists of:

- 45% dried dog kibble
- 30% mixed vegetables (chosen from carrot, pumpkin, mushroom, broccoli, sweet potato and sweet corn)
- 15% invertebrates (crickets, fly pupae, mealworms, moths, earthworms, grasshoppers or cockroaches)
- 10% seeds (dry and sprouted). Some species such as eastern barred bandicoots are not offered seeds, but instead more vegetables and invertebrates.

Young growing animals need to be supplemented with calcium to achieve a calcium content of the diet of ~0.7% on a dry matter basis.

The amount of food fed is dependent on the size of the animal. Setting a target bodyweight range and monitoring the weight of the animal is the most reliable method of not over- or underfeeding.

### 6.2 Supplements

Di-Vetelact® (125 g powder and 900 mL water) has been used to increase the bodyweight of aged bilbies. They were given 50 mL in a bowl to gain weight after it was found their teeth were worn almost to the gum line (Gamble and Blyde 1992).

### 6.3 Presentation of food

Most of the food is supplied in a dish; however, live food can be scattered around the enclosure to promote foraging behaviour. Food dishes positioned on sand pads can assist with indirect monitoring of movement and faecal output. Multiple food dishes should be provided when multiple individuals are within the same enclosure (Pauligk 2021).

## 7. HANDLING AND TRANSPORT

### 7.1 Timing of capture and handling

Capture and handling are best undertaken during the day or just before the lights go on in the case of nocturnal houses, when the bandicoots are in their nest or nest box.

### 7.2 Capture equipment

Large calico bags and strong good-quality pillowcases (otherwise they will be easily ripped by the animals' feet) are useful in holding bandicoots and depending on the species should be 40–60 cm deep or 30–40 cm deep. A strong cotton net ~60 cm deep and 45–50 cm can be used as required.

### 7.3 Capture and physical restraint

Larger species such as bilbies are generally readily caught from their nest box, often simply by placing a bag over them and scooping them up. If the animal to be caught is out in the enclosure, a cotton net can be used to scoop them up easily.

When capturing smaller species, if the animal is in a nest depression (which is usually thin and long with an entry and exit point), its location can generally be identified by a slight bulge at the base of a tussock or other areas where they are known to nest. Bedding material, such as hay, can often be seen dragged to nesting locations and emerging from under the substrate where their nest is. Different techniques can be used, usually involving two people, but with practice one person can capture them. If two people are present, both hold their nets over each end of the nest, with the bottom of the net touching the ground in case the bandicoot shoots out.

Gently use a hand to disturb one end of the nest, to encourage the bandicoot to exit in the other direction and enter the awaiting net. With practice, one person can carry out this technique, but two people should be present, if possible, to ensure a quick capture process. If the bandicoot escapes into the enclosure, then it should be netted as soon as possible. The catching net should have a diameter of ~30–40 cm. The net ring should be padded, with a short handle attached and can be lined with a simple material such as a flannel pillowcase or a more solid material (Pauligk 2021).

Bandicoots tend to run very rapidly in a zig-zag pattern and can jump to over 1.2 m, which can result in injury from impact when they fall to the ground. Body position and furniture can be used to encourage the bandicoot down a wall and then intercepted with a net (Pauligk 2021). Care also needs to be taken to ensure that pouch young have not been thrown when catching females. Ideally, females with large pouch young should not be caught. For eastern barred bandicoots, for example, once pouch young are confirmed, the female and young are not caught again for 60 days (unless required for veterinary treatment).

Once caught, bandicoots are best transferred to calico bags for examination. When their eyes are covered, they should stop kicking, resulting in less fur loss, more control and less stress. However, if required, they can be held firmly between the index and middle fingers of one hand, in a similar method to that used for possums and rodents. The other hand holds the rump, with the tail held between the index and middle finger or middle and fourth finger (Fig. 6.3). A slight modification of this technique involves placing the index finger of the second hand between the hindlegs and holding them in a tighter semicircle shape to reduce the likelihood of the bandicoot kicking out and potentially damaging its spine (Pauligk 2021).

Never hold a bandicoot by the tail, because it may cause severe trauma to the tail base vertebrae. However, unlike *Perameles* and *Isoodon*, you can readily hold bilbies by the tail.

## 7.4 Weighing and examination

Weighing is easily undertaken in a handling bag with spring or balance scales by one person, making sure to cover the eyes, while areas of interest on the animal are

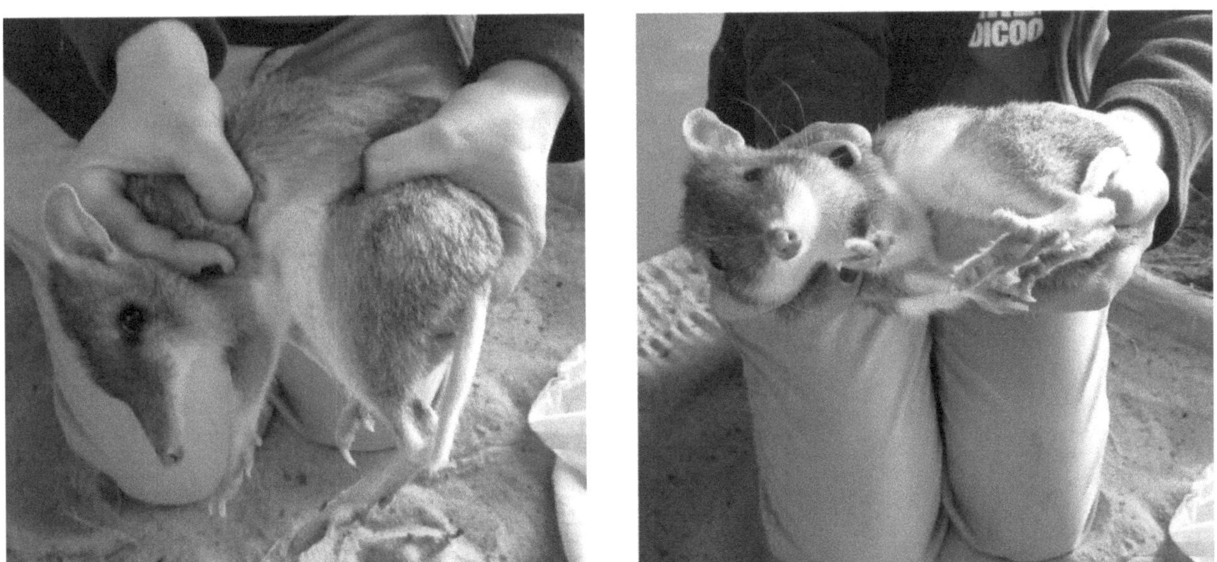

**Fig. 6.3.** Handling technique used for bandicoots. Photo used with permission of Zoos Victoria.

checked. Generally, only the area to be examined is exposed from the bag while the rest of the animal stays securely inside the bag. Individuals not readily observed should be weighed and examined under manual restraint once a month to ensure they are in good health.

Pouch checking and other examinations should be undertaken in an enclosed space and with the animal in a catching bag to keep the eyes covered (they tend to try and escape if they can see). One or more people can carry out the examination in an enclosed space while sitting down. When examining a bandicoot that is manually restrained, sit and place the bag in your lap with the animal on its back and its head facing away from the handler. Some handlers may also place the head towards them. The bandicoot is restrained between the thighs by putting pressure on the bandicoot's bottom half and less on the head. This allows good access to the pouch, feet, tail base and head (Fig. 6.4; Pauligk 2021).

Examination can commence at the hindquarters of the body, which includes pouch checks. The pouch can be examined by pulling the hindlegs apart with the outer part of your hand, keeping your fingers free to open the pouch. Examination can continue up the body to include the feet and head. With the bandicoot sitting the right way up in your lap, body condition, parasites and eyes can easily be checked.

## 7.5 Release

As most species of bandicoots are very flighty and will often run around the enclosure and jump up the walls and/or climb the walls after release, it is often easier to place the bandicoot in a transport box lined with bedding material pulled over the entrance, within the enclosure (Pauligk 2021). After 15 min the bandicoot will have settled and the entrance to the box can be opened, allowing the animal to emerge in its own time. Bilbies generally handle being released much better and can be released into a nest box or open enclosure. In either case, the enclosure should be as free as possible of obstacles to minimise the opportunity for the animal to run into any of them.

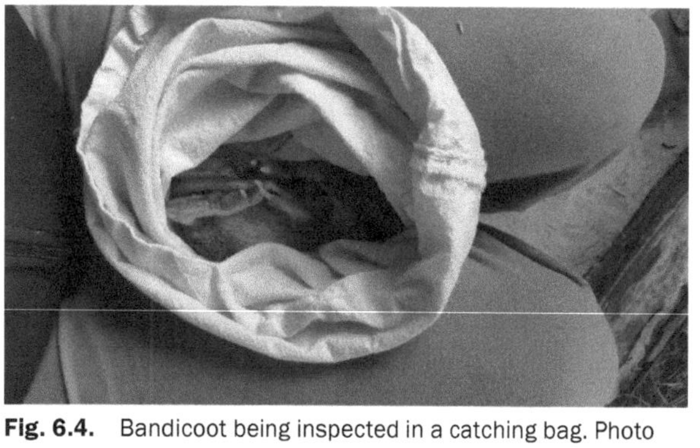

**Fig. 6.4.** Bandicoot being inspected in a catching bag. Photo used with permission of Zoos Victoria.

## 7.6 Transport requirements

### 7.6.1 Transport box design

Strongly built boxes with sliding doors or plastic PetPaks are recommended. It is important to have adequate ventilation holes and to ensure they do not get blocked during transport. Other considerations for wooden box design include carrying handles and spacer bars on the outside to allow ventilation. Wooden boxes are best made from 7 mm plywood and a 15 × 15 mm pine framework (Kingston 1998). Further specific details of box design can be found in IATA (2022).

### 7.6.2 Furnishings

Nesting material such as shredded paper should be provided. Hay or grass should not be used in boxes transporting animals being released to the wild to avoid the spread of seeds or weeds at release sites.

### 7.6.3 Water and food

Water and food are normally not required during travel, as the animals usually do not eat it as they are transported during the day while they are sleeping. On longer journeys or when animals have been caught at night, food items such as mealworms and kibble can be offered (Kingston 1998; Pauligk 2021).

### 7.6.4 Animals per transport box

Only one animal per box. Females with pouch young should not be transferred, unless unavoidable, as there is always a risk that pouch young can detach from the teat. If a transfer is required, the pouch young should be less than 25 days of age (i.e. when they are still permanently attached to the teat and before they start developing whiskers).

### 7.6.5 Timing of transportation

Ideally, animals should be transported during the day in an air-conditioned vehicle.

### 7.6.6 Release from the transport box

Most species of bandicoots are very flighty, so care must be taken when releasing them. Generally, the best way to release a bandicoot is to gently place the box on the ground, remove or fully open the box door and leave the enclosure to allow the bandicoot to emerge from the box in its own time. Bilbies generally cope with being handled quite well and often respond to being released, even by hand, by digging and ambling around the enclosure.

# 8. HEALTH REQUIREMENTS

## 8.1 Daily health observations

Each bandicoot should be observed daily for any signs of injury or illness, especially after they have been introduced, as they often show aggression towards each other. The most appropriate time to do this is generally when the enclosure is being cleaned or when they are being fed, as many of the larger species, especially in nocturnal houses, will approach to be fed. Each animal should be checked for the following:

- coat - condition: in particular, fur missing around the rump
- discharges - from the eyes, ears, nose, mouth or cloaca
- appetite
- faeces - number and consistency
- eyes - cloudiness
- changes in demeanour, activity level and behaviour
- presence and development of pouch young by observation of the bulge in the pouch
- injuries.

If a front sand patch is present in the enclosure, the area can be examined for footprints showing signs of limping or drag marks, blood, fur or other concerns.

For individuals held under a diurnal light cycle, direct observation can be difficult. Handling should occur after 1 week within a new environment to monitor bodyweight and health. If concerns arise prior, remote infrared cameras can be used to observe the animals during periods of natural activity. Cameras with high infrared wavelengths, > 920 nm, should be used to ensure the animal is not disrupted by the infrared light. Following the transfer and collection of animals from the wild, some individuals can take 1-3 weeks to adapt to the captive environment (Pauligk 2021).

## 8.2 Bodyweight and body condition score

### 8.2.1 Bodyweight

Bodyweight is a very useful indicator of health status and weight loss often reflects reduced intake or illness. Bandicoots in the captive environment tend to have quite stable bodyweights, so monitoring individuals for this parameter, food intake and activity via sand pads or night cameras is recommended. Animals in captivity should be weighed at least monthly, unless they are females with pouch young in which case they should be left undisturbed.

### 8.2.2 Body condition score

Determining a body condition score in bandicoots is similar to the method used in many other mammals where the musculature of the back and hindquarters is palpated to assess the prominence of the bones of the vertebral column and pelvis. Ideally, the 9-point scale for body condition scoring defined by the Zoological Information Management System (ZIMS, Species360 2023) is used, where 1 = emaciated, 5 = ideal condition, and 9 = obese.

## 8.3 Physical examination

If bandicoots are being caught up and examined, look for wounds. The presence of open wounds or lumps throughout the body, especially around the face and rump, suggests aggression problems. Also check the eyes closely for cloudiness and general clarity.

The physical examination may include the following:

- temperature - normally 34.5–35.5°C; can be measured through the anus via the cloaca
- bodyweight - record and compare with previous weights. Trends in bodyweight of bandicoots give a good general indication of the animal's state of health, provided age, sex and geographical location are taken into account. Animals in captivity should be weighed monthly to indicate trends
- fur - check for alopecia, ectoparasites, fungal infections or wounds
- eyes
  - should be clear, bright and alert
  - normal bilateral pupillary light response
  - normal corneal reflex
  - should not have any discharges
- also check for the presence of lumps over body and auscultate the lungs
- cloaca
  - should be clean
  - check for faeces around the cloaca
- pouch
  - condition of the pouch
  - check whether lactation is occurring by milking teats

    - if pouch young are present, record sex, stage of development, bodyweight if detached from the teat and measure to determine age from growth curves if available. Pouch young should not be removed from the pouch as this may result in the loss of young. Once a female has young it is advisable to leave her undisturbed for ~60 days
- males
    - check testes: size (length, width, depth) and consistency (firm – not squishy)
    - extrude penis and assess
    - check the size and activity of the sternal gland.

### 8.4 Chemical restraint

Pre-anaesthetic fasting is not necessary for adult animals as they are not prone to regurgitation (Vogelnest 1999). Hand-reared animals, however, should not be fed for at least 1 h before anaesthesia, as they may regurgitate the milk formula (Vogelnest 1999). Bilbies can be sedated to facilitate subsequent anaesthesia induction using intramuscular midazolam at 0.1–0.2 mg/kg) (Vogelnest 1999).

Anaesthesia is best undertaken by inhalation, as bandicoots are readily handled and induction with a conical face mask is straightforward. Induction and recovery are rapid and muscle relation is excellent (Vogelnest 1999). Injectable agents are rarely used (Booth 1994). Volatile agents such as sevoflurane and isoflurane in oxygen have been extensively used in bilbies and a range of bandicoot species and are reliable and safe when anaesthesia is adequately monitored and managed. Intubation is straightforward using a bladed laryngoscope and a 2 mm uncuffed endotracheal tube.

### 8.5 Common health problems

The majority of the parasites and diseases reported from bandicoots and bilbies are presented in Tables 6.5 and 6.6.

### 8.6 Preventative medicine

For captive bandicoots it is important to regularly monitor bodyweight and appetite as ongoing indicators of health. Faecal samples should be collected twice yearly for parasite examination. Positive results do not always mean treatments should be administered. The exceptions to this are the presence of eggs/larvae showing the animal is infected with round worm or strongyloides species. Captive animals live much longer lives than their wild counterparts so it is important that as animals age there is proactive investigation of the development of debilitating arthritis. This should be by a combination of scheduled screening radiography and the use of night-cameras to observe gait.

### 8.7 Zoonoses

Bandicoot species are not distinguished as the carriers of specific zoonoses. However, it is good practice when handling animals and their excretory products to wear disposable gloves and protective clothing plus practice good hygiene (e.g. hand washing) to reduce the risk of disease transfer to or from the animals.

## 9. BEHAVIOUR

### 9.1 Activity

All species of bandicoots are primarily nocturnal (Heinsohn 1966). Observations in captivity have shown bilbies to have several activity phases, usually of several hours duration (Aslin 1982). Bilbies range widely overnight; females have been recorded travelling up to 2.2 km, and males 4.6 km, per night while foraging (McRae 2004). Males have also been recorded occupying burrows more than 2 km away from each other on consecutive days (Southgate and Possingham 1995).

Species such as eastern barred bandicoots are frequently seen in paddocks after emergence at dusk until several hours before sunrise, whereas southern brown bandicoots prefer more covered habitat (Heinsohn 1966). In captivity, observations of southern brown bandicoots in nocturnal conditions showed them to spend 36% of their time in the nest, 28% stationary, 16% moving, 9% digging, 4% grooming, 3% standing upright, 3% feeding and 1% wall running (Garling 1982). Eastern barred bandicoots spend a large amount of time digging for food. On Churchill Island during winter, it was estimated that one eastern barred bandicoot undertakes ~487 foraging digs per night, which displaces ~13.15 kg of soil per night, equating to ~400 kg of soil per winter month per bandicoot (Halstead *et al.* 2020).

### 9.2 Social behaviour

#### 9.2.1 *Isoodon*

Southern brown bandicoots are solitary and appear to be strongly territorial (Heinsohn 1966). In captivity, animals of this species are usually very aggressive towards each other. One animal will often dominate and attack another, which usually does not defend itself but

**Table 6.5.** Common infectious diseases of zoo-based bandicoots.

| Disease | Cause | Clinical signs | Diagnosis | Treatment | Prevention | Reference |
|---|---|---|---|---|---|---|
| **Ectoparasites** | | | | | | |
| | Fleas: *Pygiopsylla* spp. | Can occur in seasonal infestations, especially during the warmer months of spring and summer | Visual examination | Selamectin or imidacloprid percutaneously or fipronil sprayed lightly over the entire coat; all have been successful in treating flea burdens | Continual monitoring, especially if the animals are in a natural habitat enclosure. Change the bedding regularly | 1–4 |
| | Mites: *Odontacarus* spp., *Ornithonyssus bacoti* and *Haemolaelaps* spp. have been seen in bandicoots including bilbies<br>Trombiculid mites have been observed in wild eastern barred bandicoot populations. *Sarcoptes*-like mites have also been observed in southern brown bandicoots. Mites of the genus *Demodex* spp. have been observed in eastern-barred bandicoots, northern brown bandicoots, and other species | May be associated with alopecia | Visual examination. Trombiculid mites in females around the pouch. In males around the stalk of the scrotum | Selamectin percutaneously can be used for heavy mite burdens | Continual monitoring, especially if the animals are in a natural habitat enclosure. Change the bedding regularly | 4, 5 |
| | Ticks: *Ixodes* spp. and *Haemaphysalis* spp. | Heavy tick burdens have been observed in eastern barred bandicoots. Generally thought that bandicoots are immune to the toxins of the *I. holocyclus*. However, heavy loads are associated with deaths | Visual inspection | Manual removal. Ivermectin can be used subcutaneously for heavy tick burdens | Continual monitoring, especially if the animals are in a natural habitat enclosure. Change the bedding regularly | 1, 4, 6–10 |

Table 6.5. continued

| Disease | Cause | Clinical signs | Diagnosis | Treatment | Prevention | Reference |
|---|---|---|---|---|---|---|
| **Endoparasites** | | | | | | |
| | Roundworm of the phylum Nematoda include the genera *Nicollina* spp., *Asymmetracantha* spp. *Austrostrongylus* spp., *Labiobulura* spp., *Mackerrastongylus* spp., *Moniliformis* spp. *Parastrongyloides* spp., *Physaloptera* spp., *Pilarinema* spp., *Peramelistrongylus* spp., *Strongyloides* spp., *Trichurus* spp. | If burdens of 20–50 are present, they can cause severe ulcerative granulomatous gastritis sufficient to cause debility | Faecal flotation and the presence of eggs or proglottids (segments that make up the worms) | Ivermectin at 200–400 ug/kg by mouth repeated in 14 days or 25 mg/mL fenbendazole at 10–30 mg/kg by mouth daily for 3 consecutive days and repeated in 14 days | Generally not required but could be by routine treatment with anthelminthics. It is also important to remove faeces from the enclosure and to maintain good hygiene | 1, 11–13 |
| **Fungal** | | | | | | |
| | *Trichophyton* spp. has been observed in captive bilbies | Hair loss | Dermatophyte test media. Fungal culture | Terbinofine 20 mg/kg once daily orally | Clean environment very important once diagnosed | 14 |
| **Viral** | | | | | | |
| Papilloma or polyoma virus | Papilloma-like virus observed in captive and free-ranging western barred bandicoots | Multicentric proliferative lesions and hair loss involving the skin and mucosal surfaces. Formation of nodular growths that can be anywhere, but often around the face, feet and pouch | Detection of large intranuclear inclusion bodies ,in some cases. Indirect immunohistochemistry using antibovine papillomavirus antibody has been positive in some animals | Face and feet nodules can be very debilitating and result in the animal needing to be euthanased | | 15, 16 |
| **Bacterial** | | | | | | |
| Mycobacteriosis | *Mycobacteria* spp. | Poor condition and death. Has been observed in northern brown bandicoots | Affects the lungs and gastrointestinal tract | Cytology, culture and PCR of lung wash samples. Histopathological diagnosis on tissues collected at postmortem | Multimodal and prolonged antimicrobial treatment is possible, but likely to be not feasible | 4 |
| Septicaemia | Mixture of the bacteria *Escherichia coli, Proteus* spp., *Vibrio metschinkovii* and *Erysipelothrix rhusiopathiae.* Due to infection of wounds. Has been observed in eastern-barred bandicoots, bilbies, northern brown bandicoots and southern brown bandicoots | Clinical signs are reduced appetite and activity progressing to severe debilitation and death | Combination of clinical signs and altered blood parameters (especially white blood cell parameters) | Appropriate antimicrobial therapy based on wound and/or blood culture | Appropriate wound treatments to prevent establishment and spread of bacteria from the wound site | 4, 17 |

| Disease | Cause | Clinical signs | Diagnosis | Treatment | Prevention | Reference |
|---|---|---|---|---|---|---|
| *Chlamydia* | *Chlamydia* in western barred bandicoots | Clinical signs of chlamydia infection include purulent ocular discharge, corneal opacity, conjunctivitis, ruptured eyeball, and blepharitis | Swabbing of discharge for microbiological culture | Successful treatment of Chlamydiales has included isolation of individuals and administration of systemic oxytetracyline (20 mg/kg IM once weekly for 6 weeks). The eyes were administered with topical oxytetracycline and neomycin daily for 4 months | | 17 |
| Protozoans | *Toxoplasma gondii*. Bandicoots become infected as a result of ingesting sporulated oocytes contaminating food, soil and invertebrate paratenic hosts. They can come into contact directly as a result of soil or plant matter put in their enclosure or secondarily via earthworms that have ingested some of the oocysts in the soil | Toxoplasmosis can cause cataracts and retinal disease, uncharacteristic docility, incoordination, apparent blindness, erratic staggering and circling movements, unnatural daytime activity and death in both wild and captive bandicoots. Eastern barred bandicoots have died within 11–14 days from toxoplasmosis after eating infected worms | Antemortem diagnosis of toxoplasmosis is confirmed by serological testing to detect rising IgG *T. gondii* titres. Direct agglutination test or modified agglutination test using the commercial kit Antigene Toxo-AD and microtiter plate reagents (bioMerieux SA, Marcy l'Etoile, France) are useful. Agglutination tests have been used to monitor the development of specific *T. gondii* antibodies | Once infected, individuals usually die. Could potentially medicate with antiprotozoal drugs such as sulfonamides including amprolium; toltrazuril can be used to treat coccidiosis | Keep all bedding material and food away from cats | 4, 12, 18–26 |
| | *Cryptosporidium muris* Isolated from a colony of the greater bilby | | | *Cryptosporidium* within the captive colony of bilbies reduced when additional enclosures became available and the stocking rate was reduced. Some bilbies were actively treated with dimetridazole soluble powder at a dose rate of 25 g/20 L (0.6 g/500 mL) administered over 5 days in the drinking water | Isolated from a colony of the greater bilby, apparently due to stress associated with overstocking | 26 |

References: (1) Lenghaus *et al.* (1990); (2) Thomas (1990); (3) Booth (1994); (4) Lynch (2008); (5) Buchecker and Kleinig (2002); (6) Roberts (1970); (7) Presidente (1982); (8) Gemmell *et al.* (1991); (9) Norman (1991); (10) Stone (1992); (11) Mawson (1960); (12) Kingston (1998); (13) Bodley (2025); (14) Wilson (1984); (15) Friend (2002); (16) O'Hara *et al.* (2004); (17) Warren *et al.* (2005); (18) Pope *et al.* (1957a); (19) Pope *et al.* (1957b); (20) Desmonts and Remington (1980); 21) Obendorf and Munday (1990); (22) Booth (1999); (23) Bettiol *et al.* (2000a); (24) Bettiol *et al.* (2000b); (25) Miller *et al.* (2000); (26) Warren *et al.* (2003).

**Table 6.6.** Common non-infectious diseases of zoo-based bandicoots.

| Disease/agent | Cause | Clinical signs | Diagnosis | Treatment | Prevention | Reference |
|---|---|---|---|---|---|---|
| Skin and hair conditions | Poor coat quality, including alopecia and hyperkeratosis in bilbies | Occurs on the head of male bilbies and tail of female bilbies. In bilbies lesions are often more extensive in hotter weather | Visual inspection | Environmental alteration | Providing a cool and humid burrow | 1 |
| | Aggression | Fur loss as a result of fighting, especially among male bilbies | Visual inspection | Early intervention and removal of aggressive individuals | Careful paring and monitoring of animals | 1 |
| Ulceration and abscesses of the feet | May be caused by excessive moisture and/or a build-up of faecal matter in the enclosure substrate | Ulcers and abscesses observed on the plantar pads of the tarsus in captive bilbies. Often seen in aged bilbies | Clinical observation. *Staphylococcus aureus* has been obtained from two cases | Removal of damaged tissue, bandaging the foot, antibiotics and anti-inflammatory treatment | More frequent substrate changes may help. Daily removal of faecal matter | 1–3 |
| Periodontal disease | | Appears to be diet-related and observed in captive bilbies and eastern barred bandicoots. Due to poor diets not mimicking the abrasive qualities of the wild diet (e.g. a banana-based diet in one institution) | Calculus formation on the lateral surfaces of the teeth, gingivitis and tooth loss. Periodontal disease has been associated with severe bronchopneumonia and pleurisy in free-ranging eastern barred bandicoots | | Reduction in soft food items and introducing commercial dog kibble. Including browse such as *Acacia* spp. gum inside bamboo sticks | 1, 3 |
| Trauma | Trauma is the most common reason for treatment in captive eastern barred bandicoots. It can result from aggression by other individuals, from climbing up the walls and falling or running into obstacles, especially during capture or transport | Fur loss, especially around the rump or damage to the ears and face, is usually caused by aggression from conspecifics. Injuries can occur to the claws and toes (frantic digging on hard surfaces) and the base of tail (from repeated jumping and ground impact) in stressed animals trying to escape cages or enclosures | Clinical signs and radiography | Depends on the injury sustained | Great care needs to be taken when introducing bandicoots into enclosures. Mixing animals of similar size may result in less aggression, as will introducing animals into larger enclosures. Enclosure walls should have a metal skirting ~1.5 m high to decrease the opportunity of climbing up them and falling, or breaking toes in the wire. A compatibility check should be conducted 1 week after introduction if the individual cannot be readily observed | 4 |

References: (1) Lynch (2008); (2) Blyde (1993); (3) Buchecker and Kleinig (2002); (4) Kingston (1998).

crouches and exposes its rump to the attacker (Heinsohn 1966). In captivity, male southern brown bandicoots have been observed to attack and kill females (Heinsohn 1966). Females will sometimes attack males, which may be due to their relatively larger size or because they are introduced into a foreign territory (Heinsohn 1966). Therefore, only animals of similar size should be introduced . Northern brown bandicoots have been kept successfully with one male and several females. Although these species are generally not considered to dig burrows (Heinsohn 1966), a captive southern brown bandicoot has been observed to dig a burrow (Kirsch 1968). Observations of the northern brown bandicoot in captivity have also shown them to be solitary and when males and females were housed together, aggressive interactions were not uncommon, with the female the aggressor on most occasions (Meritt 1970).

### 9.2.2 *Perameles*

Wild eastern barred bandicoots are promiscuous (Dufty 1994a). They demonstrate very limited social behaviour and typically use mutual avoidance and feed separately. However, they will fight, chase each other and make snorting sounds, which appears to reflect a dominance hierarchy (Heinsohn 1966; Dufty 1994a). Male eastern barred bandicoots can be very aggressive towards one another; injuries frequently seen on wild male bandicoots include torn ears, fur loss, scratches across the rump and back and tail amputations. Eastern barred bandicoots build nests by digging shallow depressions in the ground, but are also known to use rabbit warrens. They are quite resourceful, using whatever they can find to line their nest, but usually favour grass. Eastern barred bandicoots have several nests throughout their home range, with different bandicoots known to occupy the same nest at different times (Heinsohn 1966; Winnard *et al.* 2013). Observations of captive animals show pairs generally nest apart, although they will nest together, and will abandon the nest if disturbed and build another one in a different location (Heinsohn 1966). The nests can be either a shallow depression in the soil ~10 cm deep under thick shrubbery or in grass that covers a lined shallow depression (Heinsohn 1966). Other captive observations suggest that there is a hierarchy in both male and female eastern barred bandicoots. They can show significant aggression towards each other, especially in a confined space and, as a result, are generally held separately. If paired for breeding, animals should be checked carefully for signs of aggression as both males and females can attack and potentially injure their mate. Observations of captive male eastern barred bandicoots show them to be highly aggressive, with a male chasing a second male continuously after it was introduced for 4 nights until it was removed (Heinsohn 1966; Seebeck 1979). Similar observations have been made when long-nosed bandicoots were introduced to each other, resulting in one of them eventually dying from injuries inflicted by the other male (Stodart 1966). Sex ratios that have been used successfully with eastern barred bandicoots (and should be adequate for other species) include 1:1, 0:2 and 1:2 in an enclosure that was 10–12 m (L) × 34 m (W) × 2.6 m (H) (Krake and Halley 1993). Retired groups of up to four female eastern barred bandicoots have been held together with at least 1 ha per female.

### 9.2.3 *Macrotis*

Groups of bilbies can be maintained in captivity without serious fighting, with aggressive behaviour seldom being observed (Aslin 1982; Johnson and Johnson 1983). Bilbies have proven to be relatively passive in captivity in comparison with other bandicoots and a rigid dominance hierarchy amongst males is usually maintained without serious fighting (Johnson and Johnson 1983). When aggression occurs it usually consists of one animal directing loud threat hisses at another, which usually retreats immediately (Aslin 1982; Johnson and Johnson 1983). If the second animal does not retreat it generally results in the two circling one another briefly nose to tail and hissing loudly and on rare occasions this is followed by one animal leaping on the other attempting to bite its rump or flank, which results in fur loss (Aslin 1982). Unlike other bandicoots that generally excavate only shallow depressions in which to nest, the bilby digs deep spiral burrows up to 3 m deep (Aslin 1982; Johnson and Johnson 1983). Dominant males chase subordinate males out of and away from burrows and the alpha male maintains priority of access to all the well-used burrows in the enclosure, which is assisted by scent marking them (Johnson and Johnson 1983). Males share burrows freely with females (at which time their ears normally fall down due to decreased blood flow) and copulation appears to take place in the burrow (Aslin 1982; Johnson and Johnson 1983). Females also appear to have a less intense hierarchy and will share their burrows with each other (Johnson and Johnson 1983). Up to 30% of burrows have been recorded being reused by both females and males (Moseby and O'Donnell 2003).

Although the majority of burrows have a single entrance (84%; McRae 2004) and are isolated, complex systems of interconnecting burrows have been recorded (Watts 1969). In one 30-day radiotelemetry study, males used an average of 7.2 burrows, and females an average of 3.6 burrows (McRae 2004).

## 9.3 Reproductive behaviour

Reproductive behaviour appears to be fairly consistent between the different species of bandicoots. It involves the male following the female for up to several hours, sniffing her rump and making numerous attempts to mount her (Stodart 1977; Coulson 1990). Mating attempts can be as short as 3–7 s, but can be up to 30 s or longer (Heinsohn 1966). During mating attempts, the male rests his head on the female's back and grips her with his forelegs. Copulations are rapid and may be repeated intermittently for up to 45 min (Dufty 1994a). Male eastern barred bandicoots have been observed scent marking their enclosures and female enclosures following introductions, using cloacal scent glands and scent glands behind their ears.

Mate choice occurs in a wide variety of taxa and can significantly affect reproductive success both in the wild and in captivity. Allowing a female her choice of mate can significantly increase captive breeding success, while also decreasing potential aggression and the time taken to produce young (Parrott *et al.* 2019). In a range of Australian mammals, mate choice can be used to improve captive breeding outcomes and potentially the quality of young produced (see further information in Chapter 4 Carnivorous Marsupials and Chapter 9 Possums and Gliders). In the eastern barred bandicoot, research into mate choice showed that females were interested in both the scent of, and interactions with, male bandicoots and showed different preferences between potential mates. Based on female preference displayed during interactions with two potential and genetically suitable males (based on microsatellite analysis; see Hartnett *et al.* 2018), females paired with their preferred males were significantly more likely to produce young (82%), than those paired with non-preferred males (33%). These preferred pairings also resulted in significantly earlier conception of young than non-preferred pairings (13 *v.* 43 days respectively). Thus, allowing mate preference, while ensuring genetic diversity targets are met, could improve the productivity of bandicoots in captive-breeding programs (Hartnett *et al.* 2018).

## 9.4 Bathing

Bandicoots are not known to bathe, although they have been observed splashing through shallow puddles.

## 9.5 Behavioural problems

The major behavioural problems of captive bandicoots are aggression and cannibalism by females of their young. Aggression has been recorded in multiple hand-reared male eastern barred bandicoots (Pauligk 2021). Aggressive behaviour may be triggered by several factors including: overcrowding, incompatible pairing of animals, inadequate number of nest sites, inadequate vegetation cover and inadequate number of feed stations (Pauligk 2021).

Both aggression and cannibalism appear to be greatly lessened by increasing the size of the enclosure and providing adequate nest sites. Abnormal repetitive behaviour of southern brown bandicoots has been observed; however, it did not occur frequently and may have been due to the animal caretakers walking past (Garling 1982). Repetitive behaviours, such as pacing, have also been observed in eastern barred bandicoots, with an increase during the optimum breeding season, likely associated with trying to find or reach a mate. These can be mitigated through increased enrichment or movement to a larger enclosure or release location (Pauligk 2021). Major aggression problems are rare in bilbies.

## 9.6 Signs of stress

Signs of acute stress include escape attempts and aggression towards other bandicoots. Bilbies tend to hiss, but other species tend not to vocalise (Spielman 1994). Acute stress generally results in reduced food intake, reduced bodyweight, poor coat condition and alopecia (Spielman 1994). Bilbies, especially females after raising young, can experience hair loss, which does not appear to debilitate them, but does not look good when on display. The measurement of faecal glucocorticoid metabolites has been used successfully in free-ranging bandicoots to determine the level of stress (Dowle *et al.* 2013), with assays validated in captive bilbies and 12 other marsupial species (Fanson *et al.* 2017).

## 9.7 Behavioural enrichment

Various behavioural enrichment activities can be provided for bandicoots in captivity (Pauligk 2021). These include:

- addition of fresh tussocks and branches (removal of old branches)
- moving enclosure furniture such as nesting areas

- live insects scattered throughout the enclosure to promote foraging
- sensory and cogitative activities such as rattle balls, stuffed pinecones, smears, mealworm barbells
- cardboard boxes and tubes stuffed with bedding material and food
- rotten logs, untreated timber, with lots of insects
- pine cones with a small amount of jam or other treat lodged in them.

## 9.8 Training and conditioning

Bandicoots and bilbies have been successfully trained with recall to a sound cue, weigh scale training, crate training and target training. Bilbies have been trained to enter Perspex boxes adjoining their enclosure for treats, to be weighed and to interact with people (M Parrott pers. obs.). Conditioning to be handled has also been successful, especially with hand-reared females. As with some other marsupial species, males can become aggressive once they reach sexual maturity. Bilbies (Moseby *et al.* 2012) and eastern barred bandicoots (Taylor *et al.* 2021) have been trained for predator avoidance of cats before release to the wild. Wild-caught northern brown bandicoots have been trained in captivity to use microchip-automated doors to access safe refuges, before release back to the wild (Edwards *et al.* 2020).

## 9.9 Introductions and removals

Pairs of eastern barred bandicoots can be placed together and are usually compatible. However, when they are first introduced, they will generally take some time to settle down. Temperament of both individuals should be considered, as aggression can occur from either sex. Ideally, the dominant individual (often the male) should be introduced into the submissive individual's territory (Pauligk 2021). Where possible, pairing can occur via slide access when territories are already established. Introduced individuals should be checked daily when on a nocturnal cycle or examined under manual restraint once a week (Pauligk 2021). Injuries have been caused by both sexes, typically involving fur loss off the rump and sides leaving superficial wounds. Males have been known to cause significant trauma and open wounds to some females, followed by significant weight loss. Males will make advances on the female following introduction. Initially, the female may show aggression; however, after a short period of time will often accept the male's advances (Y. Pauligk pers. obs.). Other observations of eastern barred bandicoots have included a male scaling a partition and biting and scratching a female on the other side, partly mutilating her ear, scratching her back and rump and tearing out large patches of fur. After being separated they were later reintroduced and were compatible (Heinsohn 1966).

Similar aggression has been observed in southern brown bandicoots; both males and females are aggressive to members of the same or opposite sex, which can result in death from injuries sustained (Heinsohn 1966). Given the dominance hierarchy that exists, it is best to place the larger animal into the small animal's established enclosure rather than the other way around. It has also been successful to place the male into the female's enclosure or to house animals next to each other and then open slides between enclosures to provide additional space. With this in mind, successful introductions appear to be those in which the body size of the individuals is similar (Stoddart and Braithwaite 1979). To assist in the introduction there should be at least 2–3 nest areas per individual once introduced and daily checks should be made, if possible, to check for signs of aggression such as fur, blood or an animal out in the open. Raked sand pads can also be helpful to facilitate these checks. Other considerations include making sure there is adequate cover, feed stations (one each and widely separated, ideally with a visual barrier), and make sure the enclosure is not overcrowded (Kingston 1998).

Southern brown bandicoots require a large area and it is important to choose animals of a similar size; otherwise, the smaller one may be maimed or killed (Mackerras and Smith 1960). A method used to successfully introduce a captive bandicoot to a group, and reduce fighting, is to remove all bandicoots and place them in a new enclosure with the newcomer.

## 9.10 Intraspecific compatibility

Most species of bandicoots are generally unsocial and frequently highly aggressive if placed with their own sex. Species within the genera *Perameles* and *Isoodon* are usually highly aggressive to one another and so should be kept either by themselves or in pairs. The bilby is comparatively tolerant of other individuals and can readily be kept in pairs or small groups if the enclosure is large enough. However, bilbies can be aggressive towards each other and their own offspring if the density of animals is too high (Warren *et al.* 2003; Table 6.7). It is very important that animals of similar size are introduced, the enclosure is as large as possible and they are monitored, at least initially, for signs of aggression.

**Table 6.7.** Social behaviour of bandicoots and the suggested sex ratio of different genera when held in captivity.

| Genus | Social behaviour | Suggested sex ratio |
|---|---|---|
| **Peramelidae** | | |
| *Isoodon* | Solitary | Solitary or 1:1 |
| *Perameles* | Solitary | Solitary, 1:1, 1:2 or female pairs |
| *Macrotis* | Solitary/Can nest together? | 1:1–1:2 |

## 9.11 Interspecific compatibility

Bandicoots have been held with various species of other animals, especially arboreal species such as the yellow-bellied glider (*Petaurus australis*), sugar glider (*Petaurus breviceps*), squirrel glider (*Petaurus norfolcensis*), Leadbeater's possum (*Gymnobelideus leadbeateri*), eastern ring-tailed possum (*Pseudocheirus peregrinus*), common brush-tailed possum (*Trichosurus vulpecula*), brush-tailed phascogale (*Phascogale tapoatafa*) and tawny frogmouth (*Podargus strigoides*) (Garling 1982; Krake and Halley 1993; Pauligk 2021). Bilbies have been held with other terrestrial mammals, including mulgara (*Dasycercus cristicauda*) (Lee 1990).

# 10. BREEDING

## 10.1 Mating system

All species of bandicoots appear to exhibit a promiscuous mating system.

## 10.2 Ease of breeding

Bandicoots generally breed well in captivity and several species, if not all, will re-enter oestrus if the young are lost. Long-nosed bandicoots, for example, can re-enter oestrus as early as 5–10 days (Close 1977).

Mate choice studies undertaken on eastern barred bandicoots revealed that females placed in enclosures adjacent to unrelated males (kept on either side) and allowed to evaluate them simultaneously via interaction through perforated clear acrylic barriers and choose their preferred mate were significantly more likely to produce young and conceive earlier (Hartnett *et al.* 2018; see section 9.3 Reproductive behaviour). In the eastern barred bandicoot, if a pair has not conceived within 50 days of pairing, they are unlikely to conceive and produce young. In this instance, it is suggested to remove the male, wait 2 weeks to ensure young are not produced (encompassing the 12.5-day gestation), and repair the female with a different male (Hartnett *et al.* 2018; Pauligk 2021).

Studies of wild greater bilbies found that they have a promiscuous mating system and that reproductive success was fairly randomly distributed among males and that a larger body size did not provide males with a reproductive advantage (Miller *et al.* 2010). They also found that within any particular year nearly 60% of males failed to produce any offspring and that of those that did sire young ~70% fathered one offspring and 30% 2–3 offspring (Miller *et al.* 2010).

## 10.3 Oestrous cycle and gestation period

Bandicoots have relatively short oestrous cycles that typically last 12–37 days. Their gestation periods are close to the shortest known of any mammal group and typically range from 12.5 to 14 days (Table 6.8). When born, the neonates are very unusual among marsupials in that the placenta can be seen stretched between the cloaca and the pouch. Bandicoots (and wombats and the koala) have a chorioallantoic placenta more similar to eutherian or placental mammals, which allows high nutrient transfer, whereas other marsupials have a less invasive placenta called a choriovitelline (or yolk sac) placenta (Tyndale-Biscoe and Renfree 1987). Interestingly, bandicoots have an 'alternate nipple strategy', generally using fewer than half their nipples per litter

**Table 6.8.** Duration of oestrous cycle and gestation (days) for bandicoots.

Numbers in brackets are mean values.

| Species | Oestrous cycle | Gestation | Diapause | Reference |
|---|---|---|---|---|
| *Isoodon macrourus* | 14–30 (22) | 12.5 | N | 1–3 |
| *Perameles nasuta* | 17–34 (26) | 12.5 | N | 3–8 |
| *Perameles gunnii* | ~26 | 12.5 | N | 5 |
| *Macrotis lagotis* | 12–37 (21) | 13–17 (14) | N | 8–10 |

References: (1) Lyne (1974); (2) Lyne (1976); (3) Gemmell (1988a); (4) Hughes (1962); (5) Lyne (1964); (6) Stodart (1966); (7) Close (1977); (8) McCracken (1986); (9) McCracken (1990); (10) Ballantyne *et al.* (2009).

(Heinsohn 1966). This allows females to produce litters in quick succession without waiting for the used teats to regress to a small enough size for the next litter of young. Female eastern barred bandicoots, for example, have an oestrous cycle and can mate again during their late lactation phase, ~1 week before her litter is weaned. Females will give birth just 12.5 days later with the new litter of young attached to the smaller teats, while the larger, recently used teats have time to regress. It is important for animal caretakers to note that litters can be produced in quick succession and remove males if multiple litters from the same parents are not required.

## 10.4 Reproductive status

### 10.4.1 Females

Bandicoots are generally placed in several categories depending on their reproductive status. Females can be categorised into several reproductive stages including:

- non-parous (females that have never bred) - pouch small with no skin folds, clean and dry, teats very small
- parous (females that have bred previously but not presently) - pouch is small but distinct, dry and dirty, the teats are slightly elongated
- oestrus - the female's urogenital opening changes with swelling of the lips, which corresponds with the presence of cornified epithelial cells (Lyne 1976)
- pregnant - pouch pink in colour and glandular in appearance, skin folds may be observed on the lateral margins of the pouch
- pouch young present - attached to the teat
- lactating (young absent from the pouch but still suckling) - pouch area large, skin folds flaccid, hair sparse and stained, skin smooth and dark pink, teats elongated
- post lactation - with teats expressing only clear liquid and/or regressing.

If pouch young are present, several developmental stages and measurements can be recorded and compared with existing growth curves (see section 10.16) or new curves established for future reference. These include:

- developmental stage
  - sex distinguishable
  - tips of ears free
  - papillae of facial vibrissae evident
  - eyelashes visible
  - eyes open
  - fur visible - slight tinge, medium or well developed
  - tips of first incisors through the gums
  - at foot
  - eating solids
  - self-feeding
  - independent
- measurements (see Appendix 3)
  - bodyweight (g) - if not on teat
  - head length (mm) - from the occiput to snout tip
  - head width (mm) - maximum width across the zygomatic arches
  - crown-rump length (mm) - primarily for neonates
  - body length (mm) - from snout tip to cloaca
  - tail length (mm) - from the cloaca to the end of the last vertebra of the tail tip
  - total length (mm) - from snout tip to tail tip
  - tibia length (mm) - from the hip to the bottom of the pes
  - pes length (mm) - from the heel to the base of the longest toe, not including the claw.

### 10.4.2 Males

Sexual maturity of male northern brown bandicoots (where sperm first appears in the reproductive tract) corresponds with pigmentation of the scrotum, though these signs precede sexual maturation by at least 150 days and are not good indicators of reproductive ability in male bandicoots (Gemmell 1987). Male eastern barred bandicoots are able to reproduce from 4 months of age, though are unlikely to be competitive against larger older males, if present.

## 10.5 Reproductive management

Although the detection of cornified epithelial cells has been used as a technique to determine oestrus in bandicoots (Lyne 1976), it is not generally necessary due to the ease with which most species breed. Further information on this technique is given in Chapter 4. The male can generally be left in with the female after mating without adverse effects on the female or young, if enough space is available to avoid fighting. However, as above, females can breed and become pregnant again before weaning their first litter of young; thus males should be removed if further litters from the same parents are not desired.

Although many species of bandicoots can breed almost continuously throughout the year, some species of bandicoots show evidence of annual cycles in their reproductive activity, with peaks in spring and early summer, while others have peaks when food availability may increase due to increased rainfall making soil easier to dig and soil invertebrates more accessible. Environmental variables of day length, rainfall and temperature were assessed for northern brown bandicoots, southern brown bandicoots and eastern barred bandicoots and it was discovered that reproduction was most related to the rate of change in minimum temperature, although there were additional associations with rainfall and day length (Barnes and Gemmell 1984). Other observations of captive northern brown bandicoots have suggested that day length may influence breeding activity because captive, well-fed, northern brown bandicoots were not continuous breeders (Gemmell 1982). Therefore, the manipulation of these factors could potentially increase reproductive output in captivity.

To maximise breeding success in eastern barred bandicoots, breeding pairs have been kept together continuously with juveniles removed after 10–12 weeks (Krake and Halley 1993). However, to manage genetic diversity of the captive population and ensure appropriate genetics for release (i.e. with minimal relatedness in founder groups), this is no longer recommended and does not hinder breeding success.

## 10.6 Occurrence of hybrids

There are no known occurrences of hybrids between bandicoot and bilby species. However, there is a successful genetic rescue program underway that is breeding mainland eastern barred bandicoots with the closely related Tasmanian subspecies of eastern barred bandicoot. This will provide an increase in genetic diversity and future adaptability of the mainland eastern barred bandicoot, with no issues observed to date in out-bred offspring or later generations (DELWP 2021).

## 10.7 Timing of breeding

Most species of bandicoots can breed throughout the year, although there may be a peak during spring and summer in some species, or winter in others, when food availability is greatest. The northern brown bandicoot, for example, can produce young at all times of the year, but appears to have a period of non-breeding each year in late autumn to early winter (FitzGibbon 2015). The western-barred bandicoot is known to have a distinct breeding season between March and April (Table 6.9).

The mainland eastern barred bandicoot has the highest breeding success from autumn to spring, with a peak in winter in Victoria, and generally reduced or no breeding over summer depending on weather conditions, population density and food availability. Bilbies can produce young year-round in captive populations and in the arid zone, but reintroduced populations to temperate areas did not produce young in autumn (between February and mid-May). Male bilbies in these areas had a significantly smaller testis size in late summer, when breeding activity ceased (Berris *et al.* 2019).

## 10.8 Age at first breeding and last breeding

Bandicoots reach sexual maturity early, with some species, such as eastern barred bandicoots and southern brown bandicoots, being able to breed 3–4 months after

**Table 6.9.** Reproduction and development for different species of bandicoots.

Numbers in brackets are mean values.

| Species | Litter size (mean) | First detach (days) | Permanent pouch exit (days) | Weaning (days) | Sexual maturity M (months) | Sexual maturity F (months) | Mating period | Reference |
|---|---|---|---|---|---|---|---|---|
| *Echymipera rufescens* | 1–4 (3) | – | 65 | – | – | – | Nov–May | 1 |
| *Isoodon macrourus* | 1–6 (3) | 42 | 55 | 60 | 350 | 250 | All year | 2–10 |
| *Isoodon obesulus* | 1–6 (2.5) | – | – | 60–70 | 120 | 90–120 | All year | 11–15 |
| *Perameles bougainville* | 1–3 (2.5) | – | – | – | – | 90–150 | Mar–Nov | 16 |
| *Perameles gunnii* | 1–5 (2.4) | ~35 | 55 | 60–65 | 150 | 90–120 | All year | 12, 17,–21 |
| *Perameles nasuta* | 1–5 (2) | – | 50–54 | 62–63 | 150 | 120 | All year | 11, 22 |
| *Macrotis lagotis* | 1–3 (1.5) | – | 67–82 | 90 | 270–420 | 175–220 | All year | 23–25 |

References: (1) Shevill and Johnson (2008); (2) Mackerras and Smith (1960); (3) Gemmell (1982); (4) Gemmell *et al.* (1984a); (5) Gemmell (1986); (6) Gemmell (1987); (7) Gemmell (1988b); (8) Gemmell (1989a); (9) Friend (1990); (10) Vernes and Pope (2009); (11) Lyne (1964); (12) Heinsohn (1966); (13) Stoddart and Braithwaite (1979); (14) Lobert and Lee (1990); (15) Mallick *et al.* (1998); (16) Short *et al.* (1998); (17) Dufty (1991); (18) Dufty (1994b); (19) Kingston (1998); (20) Hartnett *et al.* (2018); (21) Pauligk (2021); (22) Stodart (1966); (23) McCracken (1986); (24) McCracken (1990); (25) Southgate *et al.* (2000).

birth and other species (for which it is known) being able to breed within 270–420 days (Table 6.9). Climatic conditions appear to be important in allowing breeding, as female eastern barred bandicoots have been found to delay sexual maturity by up to 6 months under severe drought conditions. In captivity, bandicoots can generally breed until shortly before they die and bilbies have been known to breed until more than 4 years of age (Southgate *et al.* 2000). In captive eastern barred bandicoots, individuals can live to over 5 years of age and can breed up to 4 years old for females and > 5 years old for males (Hartnett *et al.* 2018; M. Parrott unpubl. data), though they generally do not reproduce after 3.5 years of age.

## 10.9 Ability to breed every year

Bandicoots can breed every year. Northern brown bandicoots can produce up to five litters per year, with an average interval between litters of 90 days (range 51–108) (Friend 1990; Kemper *et al.* 1990; FitzGibbon 2015). Similar observations have been made of eastern barred bandicoots, which can produce an average of 3.8 litters per season, (Heinsohn 1966) though can have up to 5 litters per year when food availability is high and climatic conditions are favourable. Southern brown bandicoots can have up to four litters per year (Copley *et al.* 1990; Lobert and Lee 1990), as can bilbies (Southgate *et al.* 2000; Berris *et al.* 2019). One of the techniques that bandicoots appear to use to achieve this is by using an alternative nipple strategy, where young attach to up to half the teats for one litter and the other teats for the second litter (Heinsohn 1966).

## 10.10 Ability to breed more than once per year

All species of bandicoots appear to be able to breed more than once per year.

## 10.11 Nesting requirements

Nest boxes, plywood panels leant against the walls of the enclosure and grass tussocks should be provided, as outlined in sections 4.5 and 4.8.

## 10.12 Breeding diet

The amount of food supplied can be slightly increased towards late lactation, and increased further if all the food is being eaten, or can be provided ad libitum. In eastern barred bandicoots, crushed kibble can be provided from 45 days so that emerging pouch young can begin to forage (Pauligk 2021).

## 10.13 Number of young born

The litter size of bandicoots often decreases with the age of the young (e.g. northern brown bandicoot litters can be as high as seven at birth and decrease to the time of weaning) (Gemmell *et al.* 1984b; Gemmell 1989b). In suburbia, the average northern brown bandicoot litter size fell from 3.7 young per litter in early lactation to 1.7 young per litter in late lactation (FitzGibbon 2015). The loss of pouch young occurred throughout pouch life and there was no obvious period of highest risk (Gemmell 1989b). In eastern barred bandicoots, litter size appears to vary with season and age; females may begin with a small litter and have larger ones as they mature. Over 33 years of captive breeding of the eastern barred bandicoot, there has been no decrease in fecundity or litter size and average litter size is comparable across time (Parrott *et al.* 2017).

It is important that once the young are born the female is not put in stressful situations (e.g. excess noise or overcrowding), as most bandicoots (rarely, if at all, in bilbies) are known to be cannibalistic in captivity. Northern brown bandicoots, long-nosed bandicoots and eastern barred bandicoots, for example, are known to kill and eat their young when stressed, which appears to be more common in small enclosures or if handled too frequently (Mackerras and Smith 1960; Lyne 1971; Lyne 1982; Gemmell 1989b; Pauligk 2021; S. Jackson pers. obs.).

## 10.14 Age at weaning

Weaning in bandicoots is very quick as the young grow rapidly and the females breed almost continuously, with often only several months between litters (Stodart 1966). In both captive and wild animals, the greatest losses are immediately after weaning (Gordon 1974; Hall 1983; Gemmell 1989b).

## 10.15 Age at removal from the mother

Young should be removed soon after weaning as there is a high chance of inbreeding or aggression if they are left in the enclosure. Northern brown bandicoots can be removed after 60 days as they do not suckle and have no need of their mother. Eastern barred bandicoots, for example, are removed at 75–80 days. This is particularly important in the case of young females, which reach sexual maturity in 3–4 months (Krake and Halley 1993; Table 6.9).

## 10.16 Growth and development

The growth and development curves are known for several of the Australian species of bandicoots (Fig. 6.5). Graphs of most of these can be seen in Bach (1998).

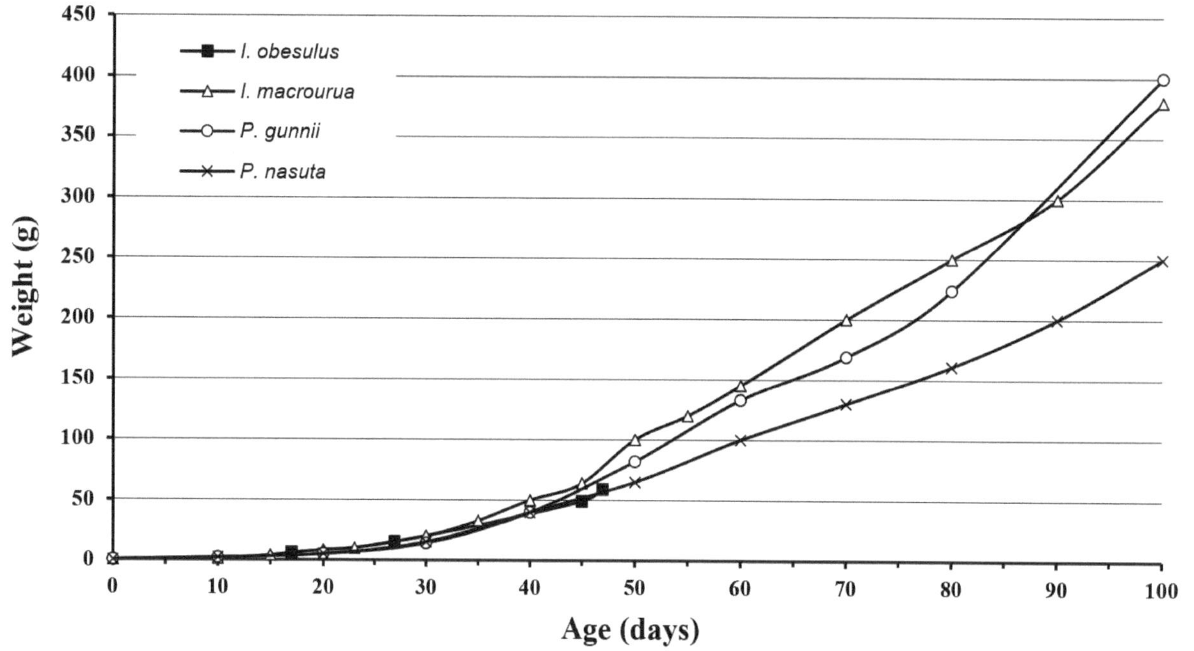


**Fig. 6.5.** Growth in bodyweight of several species of bandicoots. From Mackerras and Smith (1960), Lyne (1964), Gemmell *et al.* (1984a); Austin (1997).

Further references on the growth and development of bandicoots are given in Table 6.10.

## 11. ARTIFICIAL REARING

### 11.1 Housing

As with all native mammals that have been taken into care, minimising stress is a major consideration. Choosing suitable housing can help to create a stress-free environment. To achieve this, several factors should be considered including:

- securing the area from children and other animals (particularly domestic dogs and cats)
- maintaining high standard of hygiene
- escape-proofing the area
- clearing the area of obstacles and hazards
- ensuring the area offers shelter from extreme weather and noise.

If possible, use an ICU humidicrib or a secure box. First, line the box with a towel, then supply warm pouches for the juvenile to nest in. Unfurred or just furred animals should be placed in multiples layers of artificial pouches that simulates, as closely as possible, the security and warmth of the mother's pouch.

**Table 6.10.** Growth curve measurements that have been developed for different species of bandicoots.

CR = crown-rump length; EA = ear length; HB = head-body length; HE = head length; LE = leg length; MA = manus length; PE = pes length; TA= tail length; WT = bodyweight.

| Common name | Measurements | Reference |
|---|---|---|
| *Isoodon macrourus* | WT, CR, HE, HB, PE, TA | 1-6 |
| *I. obesulus* | WT, EA, HE, LE, PE, TA | 7-10 |
| *Perameles gunnii* | WT, CR, EA, HE, PE, TA | 2, 7, 8, 11, 12 |
| *P. nasuta* | WT, EA, HE, MA, PE, TA | 13 |
| *Macrotis lagotis* | WT, HE | 14-16 |

References: (1) Mackerras and Smith (1960); (2) Collins (1973); (3) Gemmell *et al.* (1984a); (4) Hall (1990); (5) Gemmell and Hendrikz (1993); (6) Attard and McKillup (1998); (7) Heinsohn (1966); (8) Austin (1997); (9) Hale (2000); (10) Duffy and Rose (2007); (11) Dufty (1995); 12) Pauligk (2021); (13) Lyne (1964); (14) Hulbert (1972); (15) McCracken (1990); (16) Southgate *et al.* (2000).

Natural-fibre products such as cotton or wool are required as they retain warmth better and prevent the young from overheating.

### 11.2 Temperature requirements

The temperature is generally kept at 28–30°C for furred animals and 32–34°C for unfurred animals (Table 6.11; Austin 1997; Kingston 1998; Pauligk 2021). An ICU humidicrib is highly recommended for unfurred and

**Table 6.11.** Heating and suggested housing of bandicoots at different ages for hand-rearing bandicoots.

ICU = intensive care unit.

| Age (in days) | Suggested housing | Heating |
|---|---|---|
| < 50 (furless - fine fur appearing) | ICU - in closed pouch | 30–32°C |
| 50–55 (fur short and velvety) | ICU - in opened pouch | 28–30°C |
| 55–60 (fine furred) | ICU - in opened pouch in colder climates. Otherwise, a 100 L plastic storage tub with a well-ventilated lid in a thermostatically controlled room-. | 24–28°C |
| 60–75 (fully furred) | 250 L plastic tub with ventilated lid set up in a secure indoor room. Set up tub to encourage foraging and nest building with access to open pouch in a box in natural substrate (e.g. leaf litter, mulch, clean soil) | 20–24°C |
| 75 + | Outdoor enclosure - set up to encourage foraging and nest building with access to multiple nest sites (e.g. hollow logs) | Ambient temperature with heat lamp (if required) |

just furred joeys as it greatly increases their chance of survival. Furred joeys can be maintained at an appropriate temperature with the use of an electric heating pad or SnuggleSafe® Microwave Heat Pad, though the latter is not temperature controlled. Use a minimum/maximum temperature gauge with a plastic-coated probe that can be placed in the pouch layers next to the joey, as this will ensure that the temperature can be monitored. Do not put the probe in the pouch with the joey in case they chew on it. Unfurred joeys will remain in a closed pouch and do best in a humidicrib at the required temperature. Once fully furred, external heating is generally not necessary, although in colder climates placing a SnuggleSafe® Microwave Heat Pad in the enclosure at night will provide gentle warmth if required.

## 11.3 Specific requirements

The success rate of rearing unfurred joeys is typically low unless the nutritional and environmental requirements of the immature animal are precisely met. The skin of unfurred and slightly furred young should be kept moist with the use of Sorbolene cream (not with added glycerine) so that the skin does not become dry and cracked (George *et al.* 1995). However, if using an ICU humidicrib with adequate humidity, there should be no need to moisturise the skin if the crib is maintained at the correct temperature.

When first brought in for hand-rearing, the animal may be dehydrated. If so, give it boiled water that has cooled to a tepid temperature, with 5 g (1 teaspoon) of glucose to 250 mL of water or 1 g of electrolyte replacer if available (Austin 1997). It is important to warm the joey before feeding to avoid the risk of inhalation pneumonia.

Stress is a major problem in rearing native mammals successfully and can be fatal. It is important to minimise noise, not to overhandle and to maintain high standards of hygiene. The success of hand-raising furred joeys of species such as eastern barred bandicoots has been high (> 80%; M. Parrott unpubl. data), with causes of death in a small number of animals including bone disease, and dehydration on arrival for veterinary assessment. It should be noted that no unfurred eastern barred bandicoots have survived past 10 days of hand-raising (M. Parrott unpubl. data) and the raising of unfurred young is not recommended.

## 11.4 Diet and feeding routine

### 11.4.1 Natural milk

The milk of northern brown bandicoots during lactation has been found to be similar to other marsupials, with milk solids increasing from 8% to > 40% over 55 days. Carbohydrate concentration increased from ~2% initially to 7% in mid-lactation, declining to 1% in late lactation. Concentration of lipids reached 25% by 55 days and protein increased from 2% to 14% in mid-lactation and then to 2% in late lactation (Merchant and Libke 1988; Merchant 1990). Milk intake by pouch young increases from 2 mL/day after 20 days to 18 mL by day 55 (Merchant 1990). The concentrations of the major constituents are shown in Table 6.12.

Eastern barred bandicoot milk steadily becomes more concentrated with solids, fat and protein during lactation. For example, lipids rise from 3.1% in early lactation, to 8.4% in mid- lactation and increases to 17.4% by late stages (Ikonomopoulou *et al.* 2005). Energy similarly increases throughout lactation from 2.25 kJ/mL at

**Table 6.12.** Concentrations of major constituents of the milk of several species of bandicoot.

From Merchant and Libke (1988), Duffy and Rose (2007), Pauligk (2021) and Ikonomopoulou *et al.* (2005).
*Measured in grams per 100 mL. **From weeks 4–8 of lactation in the eastern barred bandicoot.

| Species | Total solids (g/100 mL) | Carbohydrates (g/100 ml) | Lipids (g/100 mL) | Protein (g/100 mL) | Calcium (mg/L) | Iron (mg/L) |
|---|---|---|---|---|---|---|
| *Isoodon macrourus* | 80–400 | 10–70 | 20–250 | 20–140 | – | – |
| *I. obesulus* * | 27–41 | 0.5–2.0 | 10–26 | 9–14 | – | – |
| *Perameles gunnii*** | 28–54 | 2.0–4.5 | 3.1–17.4 | 4.5–12.1 | – | – |

4 weeks of lactation to 5.48 kJ/mL at 6 weeks and 8.86 kJ/mL at 8 weeks lactation (Ikonomopoulou *et al.* 2005), allowing rapid growth (Pauligk 2021).

## 11.4.2 Milk formulas

The main commercial formulas for hand-rearing bandicoots are listed.

- Biolac® – the three formulas are M100 for furless joeys, M150, which is a transitional milk to feed when dense fur has developed, and M200. When the joey is nearing weaning, 2–5 mL of canola oil is added per 100 mL of formula. Mixing the formulas assists in the transition from one formula to another. The manufacturer recommends that animals should be fed 10–15% of their bodyweight per day.
- Di Vetelact® (Lillelund Pty Ltd) – widely used low-lactose milk formula with a low energy concentration when prepared as directed. Should be fed at ~20% bodyweight, except for very small joeys, which may require up to 30% bodyweight.
- Wombaroo® > 0.7 Kangaroo Milk – different formulas are used for the different stages of development to mimic the changes that occur in the female's milk during lactation. Charts are provided to assist in determining the volume to be fed. Wombaroo® 0.7 Kangaroo Milk Replacer has been used with good success (Kingston 1998). The formula is made up according to the instructions, with canola oil added incrementally for the first three days of care to increase the fat content (Pauligk 2021) and subsequently the amount added remains stable until weaning.

## 11.4.3 Feeding apparatus

Very small joeys can be fed using a silicone or latex syringe teat plastic intravenous catheter (cannula) or 25 mm length of infant gastric feeding tube (Bellamy 1992; Pauligk 2021). Larger bandicoot joeys will readily lap from the end of a cannula and syringe before they progress to lapping from a shallow bowl. Syringe teats can be punctured with a hot needle or cut a 1–2 mm slit in the tip with a small sharp pair of scissors. From 40 days old, eastern barred bandicoots can lap from a shallow dish with minimal assistance. A soft tissue should be used to clean any spilled milk, particularly from the mouth of the bandicoot (Pauligk 2021).

## 11.4.4 Feeding routine

When deciding on a feeding regime for the bandicoot joey several things need to be considered.

(1) Read and understand the manufacturers' guidelines for making the milk formula.
(2) Decide how much the animal can comfortably consume in one feed.
(3) The age of the joey.
(4) Is the joey eating natural food (e.g. invertebrates) yet?
(5) Is the joey dehydrated and needs more frequent fluid intake?

Milk should be warmed to ~36°C before feeding. Older animals (if furred) will generally lap readily from a shallow saucer or the end of a syringe. Milk formula should be made up fresh every 24 h, but can be prepared in advance in small batches and frozen for ease..

The bandicoot joey is given 10–30% of its bodyweight in millilitres per day. Bandicoots do not generally have a tendency to overfeed; when they are full they will stop feeding. The number of daily feeds changes as the joey develops (Bellamy 1992). Very young, unfurred joeys should be fed every 3 h (i.e. eight feeds per day). When furred, the number of feeds is reduced to five and the volume increased per feed. At full emergence, the number of feeds is reduced to 2–3 per day and by this stage they should be lapping well from a shallow bowl. Once fine fur begins to emerge, infant cereal can be added to the formula and invertebrates such as earthworms, mealworms, grubs, moths, crickets, snails and grasshoppers can be offered, mixed in with Wombaroo® Small Carnivore or Insectivore Mix (Austin 1997). Milk can be offered from a small dish from 40 days of age, as bandicoots quickly learn to feed themselves.

A slightly different method used with the eastern barred bandicoot is to offer 6-8 feeds spaced out over 18 h for pouch young under 40 days old. Milk is presented in a shallow dish and lapped up by the pouch young. If this does not occur, the use of an intravenous catheter (cannula cut to a ~10-15 mm length) may be required. At 40-50 days old the schedule reduces to 5-6 feeds over 8-12 h. From 50 days old, there is the addition of live food items (such as crickets and mealworms) and 4-5 milk feeds over an 8 h period. Once insects are being consumed, crushed kibble, vegetables and water can be added ad lib. After 60 days old, the milk feeds slowly reduce to one per day, with live food and continued access to crushed kibble, vegetables and water (Pauligk 2021).

If the young develops diarrhoea, the milk formula should be replaced with Vytrate®, an electrolyte solution, until the faeces are firm (Kingston 1998). Vytrate® can be used at a ratio of 20 mL to 250 mL water. Once firm faeces have been established, the milk formula can be reintroduced by providing 75% water and 25% milk for 24 h, increasing to 100% milk over the following 3 days. Diarrhoea in any hand-reared animal should be investigated by a veterinarian to rule out bacterial infection or candidiasis.

It is important not to feed the milk formula too quickly. The rate at which the milk is squeezed into the mouth should not be faster than the rate at which it is swallowed. Ensuring the hole in the teat is an appropriate size for the joey will help regulate the flow. Too much milk too rapidly results in it accumulating in the pharynx and then suddenly sneezed or coughed out the nostrils. It is essential to prevent aspiration into the lungs. For small joeys, encouraging them to lap from a syringe, syringe/cannula or a shallow bowl allows them to feed at their own preferred rate and will prevent aspiration.

## 11.5 Data recording

When an animal is first brought in for hand-rearing, its sex and approximate age, using growth charts, should be recorded. During the hand-rearing process the following important information should be recorded on a daily basis:

- date when the information is recorded
- time when the information is recorded
- bodyweight to the nearest 1 g
- general activity and demeanour
- characteristics and frequency of defecation and urination
- amount (g) of different food types offered
- food consumption at each feed
- veterinary examinations and results.

This information serves several purposes, including providing important background information such as food consumption that will assist a veterinarian to reach a diagnosis if the animal becomes sick or fails to grow or gain weight. It also allows a comparison with growth curves to assess progress (section 10.16) and to allow new growth curves to be established for measurements in which they presently do not occur. The developmental stages and measurements outlined in section 10.3.1 should also be recorded on a weekly basis if possible.

## 11.6 Identification methods

Young can be temporarily marked via nail polish on a claw or a small dot of non-toxic marker if they are not easily distinguishable. Once furred, other tagging methods can be used (see section 5.3.1).

## 11.7 Hygiene

Maintaining a high standard of hygiene is critical to the survival of the bandicoot joey. Emphasis needs to be placed on:

- maintaining a clean pouch lining at all times; older joeys can be trained to urinate on newspaper by keeping a piece of newspaper with the smell of urine on it
- maintaining personal hygiene by washing and disinfecting hands before and after handling the joey. Use antibacterial solution for washing hands when caring for furless joeys, as their immune system is not well developed
- washing hands between feeding joeys from different litters
- using boiled water when making up formulas during all stages of development
- cleaning spilt milk, faeces and urine from the joeys' skin and fur as soon as possible, drying the animals and replacing their pouch liners
- washing all feeding equipment in warm water and dishwashing detergent and sterilise it in a microwavable baby steam steriliser or F10® disinfectant
- storing the teats and bottles in the fridge in an airtight container after they have been disinfected

- heating up milk once only and discarding leftovers
- making up fresh milk every 24 h
- avoid contact with other animals unless certain that they pose no health risk
- stimulating the joey to toilet before or after feeding. As with other marsupials, toileting can be done by the application of warm water to the cloaca using a soft tissue to stimulate urination and defecation, which prevents the animal from becoming wet and cold in its pouch
- if furless, covering the joey's body with Sorbolene cream after each feed until fur appears (if not using an ICU humidicrib).

## 11.8 Behavioural considerations

There are no specific considerations as these species do not generally appear to show any bonding, although if they are to be released, they should not be exposed to other species of animals, with which they may become accustomed. Reducing handling as the bandicoot gets closer to release age is important to prevent humanisation. Joeys seem to do much better if paired with another bandicoot of approximately the same age. Many carers find that single unfurred or just furred joeys have a much lower chance of survival if raised on their own. Hand-reared animals have been successfully rehabilitated to the wild and included in captive-breeding programs where they have produced and raised young. For the eastern barred bandicoot, there is no difference in the longevity or breeding success of hand-reared young compared with parent-reared young.

## 11.9 Use of foster animals

The use of cross-fostering does not appear to have been actively pursued for bandicoots. Nonetheless, pouch young are known to move between females. Three young northern brown bandicoots were found to move through a wire fence from their mother's enclosure into an adjacent enclosure and into the pouch and onto the teat of a female that had two young (Gemmell 1988b). The young appeared to move between day 44 and 50 after birth and were successfully weaned (Gemmell 1988b). Despite these observations, cross-fostering should not be relied upon because bandicoots can lose or eat their young when disturbed.

## 11.10 Weaning

Solid food can first be offered to eastern barred bandicoots at ~50 days of age. Commence live feeds by offering 1–2 crickets or mealworms after each milk feed and then gradually increase the variety (earthworms, fly pupae, moths) and number of invertebrates offered (Pauligk 2021). Additional crushed kibble and diced vegetables can be offered overnight from 50 days. Once the bandicoot is lapping water readily, the amount of solid food is slowly increased over the next 2–3 weeks. Finely diced lean meat mixed with Wombaroo® insectivore powder can also be given. Most species of bandicoots should be weaned by 60–70 days of age (Austin 1997). At weaning, fresh water should be supplied.

## 11.11 Release procedures

Prior to release, they should be placed in an outside enclosure that is as large as possible and given as much live food as possible, including earthworms, fly pupae, beetles and moths, to encourage foraging behaviour. Releasing animals older than 4 months and bodyweight > 400 g to wild-type sites has been successful for the eastern barred bandicoot. Single animals, reared correctly, do not imprint on their carer. Males can be particularly aggressive when defending their 'captive' territory. This is important for them in the wild to claim territories. Wildlife groups generally release northern brown and long-nosed bandicoots at ~250 g, or slightly less if they came into care as a juvenile.

## ACKNOWLEDGEMENTS

Sincere thanks to Karen Scott and Alice Roser for making numerous edits to the hand-rearing section.

## REFERENCES

Animal Welfare Victoria (2021) Code of Practice for the Public Display of Exhibition of Animals. Department of Jobs, Precincts and Regions. https://agriculture.vic.gov.au/livestock-and-animals/animal-welfare-victoria/pocta-act-1986/victorian-codes-of-practice-for-animal-welfare/code-of-practice-for-the-public-display-of-exhibition-of-animals (accessed 22 October 2021).

Aslin HJ (1982) Notes on captive rabbit bandicoots (*Macrotis lagotis*). *Western Australian Naturalist (Perth)* **15**, 67–71.

Attard SM, McKillup SC (1998) Reproduction and growth of the bandicoot *Isoodon macrourus* at four sites in Rockhampton, Queensland. *Australian Mammalogy* **20**, 411–414. doi:10.1071/AM98411

Austin MA (1997) *A Practical Guide to the Successful Hand Rearing of Tasmanian Marsupials.* Regal Publications, Melbourne.

Bach C (1998) *Birth Date Determination in Australasian Marsupials.* Australasian Regional Association of Zoological Parks and Aquaria, Sydney.

Baker AM, Gynther IC (eds) (2023) *Strahan's Mammals of Australia*. 4th edn. Reed New Holland Publishers, Sydney.

Baker AM, Eldridge MDB, Fisher DO, Frankham G, Helgen K, Jackson SM, Potter S, Travouillon KJ, Umbrello L (2023) Taxonomy and diversity of living Australasian marsupials. In *American and Australasian Marsupials*. (Eds NC Cáceres and CR Dickman). pp. 1–85. Springer, Switzerland.

Ballantyne K, Matson P, Noakes N, Nicolson V, Johnson SD (2009) Reproductive physiology of the female greater bilby (*Macrotis lagotis* Thylacomyidae): evidence for a male-induced luteal phase. *Reproduction, Fertility and Development* **21**, 274–282. doi:10.1071/RD08134

Barnes A, Gemmell RT (1984) Correlation between breeding activity in the marsupial bandicoots and some environmental variables. *Australian Journal of Zoology* **32**, 219–226. doi:10.1071/ZO9840219

Bellamy T (1992) Marsupial handrearing. In *Urban Wildlife*. Proceedings 204. Post Graduate Committee in Veterinary Science. (Eds L Vogelnest, D Spielman, T Bellamy, M Messer and DM Walker) pp. 49–67. University of Sydney, Sydney.

Bennett MD (2008) Western barred bandicoots in health and disease. PhD thesis. Murdoch University, Perth.

Berris KK, Cooper SJB, Breed WG, Carthew SM (2019) Timing of breeding and female fecundity of the greater bilby (*Macrotis lagotis*) in the temperate zone of South Australia: implications for translocations of a previously widespread species. *Wildlife Research* **46**, 444–453. doi:10.1071/WR18159

Bettiol SS, Obendorf DL, Nowarkowski M, Goldsmid JM (2000*a*) Pathology of experimental toxoplasmosis in eastern barred bandicoots in Tasmania. *Journal of Wildlife Diseases* **36**, 141–144. doi:10.7589/0090-3558-36.1.141

Bettiol SS, Obendorf DL, Nowarkowski M, Milstein T, Goldsmid JM (2000*b*) Earthworms as paratenic hots of toxoplasmosis in eastern barred bandicoots in Tasmania. *Journal of Wildlife Diseases* **36**, 145–148. doi:10.7589/0090-3558-36.1.145

Blyde D (1993) Staphylococcal infections in the greater bilby (*Macrotis lagotis*). In Proceedings of the Wildlife Diseases Association (Australasian Section) pp. 25–27. North Stradbroke Island, Queensland.

Bodley K (2025) Appendix 4. Drug formulary. In *Current Therapy in Medicine of Australian Mammals*. Revised Edition. (Eds L Vogelnest and T Portas) pp. 732–761. CSIRO Publishing, Melbourne.

Booth R (1994) Medicine and husbandry: monotremes, wombats and bandicoots. In *Wildlife*. Proceedings 233. Post Graduate Committee of Veterinary Science. (Ed. DI Bryden) pp. 395–421. University of Sydney, Sydney.

Booth R (1999) Macropods: Hand raising, husbandry, diseases and rehabilitation. *Wildlife Veterinary Notes*.

Buchecker J, Kleinig S (2002) Husbandry manual for the greater bilby *Macrotis lagotis lagotis*. Monarto Zoological Park, South Australia. Unpublished manuscript.

Claridge AW (1993) Fungal diet of the long-nosed bandicoot (*Perameles nasuta*) in south-eastern Australia. *Victorian Naturalist* **110**, 86–91.

Claridge AW, May TW (1994) Mycophagy among Australian mammals. *Australian Journal of Ecology* **19**, 251–275. doi:10.1111/j.1442-9993.1994.tb00489.x

Close RL (1977) Recurrence of breeding after cessation of suckling in the marsupial *Perameles nasuta*. *Australian Journal of Zoology* **25**, 641–645. doi:10.1071/ZO9770641

Collins LR (1973) *Monotremes and Marsupials: A Reference for Zoological Institutions*. Smithsonian Institution Press, Washington, DC.

Copley PB, Read VT, Robinson AC, Watts CHS (1990) Preliminary studies of the Nuyts Archipelago bandicoot *Isoodon obesulus nauticus* on the Franklin Islands, South Australia. In *Bandicoots and Bilbies*. (Eds JH Seebeck, RR Brown, RL Wallis and CM Kemper) pp. 345–56. Surrey Beatty & Sons, Sydney.

Coulson G (1990) Applied behaviour: its role in conservation biology of the eastern barred bandicoot. In *Management and Conservation of Small Populations*. (Eds TW Clark and JH Seebeck) pp. 77–88. Chicago Zoological Society, Brookfield, IL.

DELWP (Department of the Environment Land, Water and Planning) (2021) *Draft National Recovery Plan for the Mainland Eastern Barred Bandicoot* Perameles gunnii *(named subspecies)*. Victoria State Government, Melbourne, and Australian Government, Canberra.

Desmonts G, Remington JS (1980) Direct agglutination test for diagnosis of Toxoplasmosis infection: method for increasing sensitivity and specificity. *Journal of Clinical Microbiology* **11**, 562–568. doi:10.1128/jcm.11.6.562-568.1980

Dixon JM (1988) Notes on the diet of three mammals presumed to be extinct: the pig footed bandicoot, the lesser bilby and the desert rat kangaroo. *Victorian Naturalist* **105**, 208–210.

Dowle M, Webster KN, Deane E (2013) Faecal glucocorticoid metabolite concentrations in the free-ranging bandicoots (*Perameles nasuta* and *Isoodon obesulus*) of northern Sydney. *Australian Mammalogy* **35**, 1–7. doi:10.1071/AM11033

Duffy DJ, Rose R (2007) Milk composition and growth in the southern brown bandicoot, *Isoodon obesulus* (Marsupialia: Peramelidae). *Australian Journal of Zoology* **55**, 323–329. doi:10.1071/ZO07036

Dufty AC (1991) Some population characteristics of *Perameles gunnii* in Victoria. *Wildlife Research* **18**, 355–366. doi:10.1071/WR9910355

Dufty AC (1994a) Field observations of the behaviour of free-ranging eastern barred bandicoot *Perameles gunnii*, at Hamilton, Victoria. *Victorian Naturalist* **111**, 54–59.

Dufty AC (1994b) Population demography of the eastern barred bandicoot (*Perameles gunnii*) at Hamilton, Victoria. *Wildlife Research* **21**, 445–457. doi:10.1071/WR9940445

Dufty AC (1995) The growth and development of the eastern barred bandicoot *Perameles gunnii* in Victoria. *Victorian Naturalist* **112**, 79–85.

Edwards MC, Hoy JM, FitzGibbon SI, Murray PJ (2020) Bandicoot bunkers: training wild-caught northern brown bandicoots (*Isoodon macrourus*) to use microchip-automated safe refuge. *Wildlife Research* **47**, 239–243. doi:10.1071/WR19151

Fanson KV, Best EC, Bunce A, Fanson BG, Hogan LA, Keeley T, Narayan EJ, Palme R, Parrott ML, Sharp TM, Skogvold K, Tuthill L, Webster KN, Bashaw M (2017) One size does not fit all: monitoring faecal glucocorticoid metabolites in marsupials. *General and Comparative Endocrinology* **244**, 146–156. doi:10.1016/j.ygcen.2015.10.011

FitzGibbon SI (2015) Reproductive ecology of the northern brown bandicoot (*Isoodon macrourus*) in habitat fragments of urban Brisbane. *Australian Mammalogy* **37**, 253–259. doi:10.1071/AM14032

Flannery TF (1995a) *Mammals of New Guinea*. Reed Books, Sydney.

Flannery TF (1995b) *Mammals of the South-West Pacific & Moluccan Islands*. Reed Books, Sydney.
Flower SS (1931) Contributions to our knowledge of the duration of life in vertebrate animals. *Proceedings of the Zoological Society of London* **101**, 145–234. doi:10.1111/j.1469-7998.1931.tb06192.x
Friend GR (1990) Breeding and population dynamics of *Isoodon macrourus* (Marsupialia: Peramelidae): studies from the wet-dry tropic of northern Australia. In *Bandicoots and Bilbies*. (Eds JH Seebeck, RR Brown, RL Wallis and CM Kemper) pp 357–365. Surrey Beatty & Sons, Sydney.
Friend T (2002) Workshop on disease in western barred bandicoot populations, 16 July 2002. Department of Conservation and Land Management, Perth.
Gamble J, Blyde D (1992) Artificial milk diets as a supplement for aged and infirmed marsupials. *International Zoo Yearbook* **31**, 90–92. doi:10.1111/j.1748-1090.1991.tb02368.x
Garling D (1982) Activity and relationships in the short-nosed bandicoots at Taronga Zoo. *Thylacinus* (December), 8–18.
Gemmell RT (1982) Breeding bandicoots in Brisbane (*Isoodon macrourus*; Marsupialia, Peramelidae). *Australian Mammalogy* **5**, 187–193. doi:10.1071/AM82018
Gemmell RT (1986) Sexual maturity in the female bandicoot, *Isoodon macrourus* (Gould) in captivity. *Australian Journal of Zoology* **34**, 199–204. doi:10.1071/ZO9860199
Gemmell RT (1987) Sexual maturity in the male bandicoot *Isoodon macrourus*. *Australian Journal of Zoology* **35**, 433–441. doi:10.1071/ZO9870433
Gemmell RT (1988a) The oestrous cycle length of the bandicoot *Isoodon macrourus*. *Australian Wildlife Research* **15**, 633–635. doi:10.1071/WR9880633
Gemmell RT (1988b) A composite litter of young in the pouch of the bandicoot, *Isoodon macrourus* (Marsupialia: Peramelidae). *Australian Mammalogy* **11**, 157–158. doi:10.1071/AM88019
Gemmell RT (1989a) Breeding season and litter size of the bandicoots *Isoodon macrourus* (Marsupialia: Peramelidae) in captivity. *Australian Mammalogy* **12**, 77–79. doi:10.1071/AM89013
Gemmell RT (1989b) Survival of pouch young and juvenile bandicoots *Isoodon macrourus* (Marsupialia: Peramelidae). *Australian Mammalogy* **12**, 73–76. doi:10.1071/AM89012
Gemmell RT (1990) Longevity and reproductive capability of *Isoodon macrourus* in captivity. In *Bandicoots and Bilbies*. (Ed JH Seebeck, PR Brown, RL Wallis and CM Kemper) pp. 213–217. Surrey Beatty & Sons, Sydney.
Gemmell RT, Hendrikz JK (1993) Growth rates of the bandicoot *Isoodon macrourus* and the brushtail possum *Trichosurus vulpecula*. *Australian Journal of Zoology* **41**, 141–149. doi:10.1071/ZO9930141
Gemmell RT, Johnston G, Barnes A (1984a) The uniformity of growth within the litter of the marsupial *Isoodon macrourus*. *Growth* **48**, 221–233.
Gemmell RT, Walker MT, Johnston G, Cepon G (*1984b*) The number of young present in sequential litters of the marsupial bandicoot, *Isoodon macrourus*, in captivity. *Australian Journal of Zoology* **32**, 623–629. doi:10.1071/ZO9840623
Gemmell RT, Cepon G, Green PE, Stewart NP (1991) Some effects of tick infestations on northern brown bandicoots (*Isoodon macrourus*) juveniles. *Journal of Wildlife Diseases* **27**, 269–275. doi:10.7589/0090-3558-27.2.269
George H, Parker G, Coote P (1995) Common wombats: Rescue rehabilitation release. Unpublished manuscript.
Gibson LA (2001) Seasonal changes in the diet, food availability and food preferences of the greater bilby (*Macrotis lagotis*) in south-western Queensland. *Wildlife Research* **28**, 121–134. doi:10.1071/WR00003
Gordon G (1974) Movement and activity of the short-nosed bandicoot *Isoodon macrourus* Gould (Marsupialia). *Mammalia* **38**, 405–431. doi:10.1515/mamm.1974.38.3.405
Guppy M, Guppy S (2018) The long-nosed bandicoot (*Perameles nasuta*) as a nest-predator. *Australian Mammalogy* **40**, 106–108. doi:10.1071/AM16048
Hale ML (2000) Inheritance of geographic variation in body size and countergradient variation in growth rates in the southern brown bandicoot *Isoodon obesulus*. *Australian Mammalogy* **22**, 9–16. doi:10.1071/AM00009
Hall LS (1983) Observations in body weights and breeding of the northern brown bandicoot *Isoodon macrourus* trapped in south-east Queensland. *Australian Wildlife Research* **10**, 467–476. doi:10.1071/WR9830467
Hall LS (1990) Growth and a description of the development of external features of pouch young of captive *Isoodon macrourus*. In *Bandicoots and Bilbies*. (Eds JH Seebeck, PR Brown, RL Wallis and CM Kemper) pp. 123–133. Surrey Beatty & Sons, Sydney.
Halstead LM, Sutherland DR, Valentine LE, Rendall AR, Coetsee AL, Ritchie EG (2020) Digging up the dirt: quantifying the effects on soil of a translocated ecosystem engineer. *Austral Ecology* **45**, 97–108. doi:10.1111/aec.12833
Hartnett CM, Parrott ML, Mulder RA, Coulson G, Magrath MJL (2018) Opportunity for female mate choice improves reproductive outcomes in the conservation breeding program of the eastern barred bandicoot (*Perameles gunnii*). *Applied Animal Behaviour Science* **199**, 67–74. doi:10.1016/j.applanim.2017.10.008
Heinsohn GE (1966) Ecology and reproduction of the Tasmanian bandicoots (*Perameles gunnii* and *Isoodon obesulus*). *University of California Publications in Zoology* **80**, 1–96.
Hughes RL (1962) Role of the corpus luteum in marsupial reproduction. *Nature* **194**, 890–891. doi:10.1038/194890a0
Hulbert AJ (1972) Growth and development of pouch young in the rabbit-eared bandicoot, *Macrotis lagotis* (Peramelidae). *Australian Mammalogy* **1**, 38–39. doi:10.1071/AM72006
Hulbert AJ (1982) Notes on the management of a captive breeding colony of the greater bilby *Macrotis lagotis*. In *The Management of Australian Mammals in Captivity*. (Ed. DD Evans) pp. 53. Zoological Board of Victoria, Melbourne.
Ikonomopoulou MP, Smolenski AP, Rose RW (2005) Changes in milk composition during lactation in the eastern barred bandicoot (*Perameles gunnii*) (Marsupialia: Peramelidae). *Australian Journal of Zoology* **53**, 59–65. doi:10.1071/ZO04044
International Air Transport Association (IATA) (2022) *Live Animals Regulations*. 48th edn. International Air Transport Association, Geneva Switzerland.
Jackson SM, Groves C (2015) *Taxonomy of Australian Mammals*. CSIRO Publishing, Melbourne.
Johnson CN, Johnson KA (1983) Behaviour of the bilby *Macrotis lagotis* (Reid), (Marsupialia: Thylacomyidae) in captivity. *Australian Wildlife Research* **10**, 77–87. doi:10.1071/WR9830077
Kemper C, Kitchener DJ, Humphreys WF, How RA, Schmitt LH, Bradley A (1990) The biology of the northern brown bandicoot, *Isoodon macrourus* (Marsupialia: Peramelidae) at Mitchell Plateau, Western Australia. *Australian Journal of Zoology* **37**, 627–644. doi:10.1071/ZO9890627

Kingsmill E (1962) An investigation of criteria for estimating age in the marsupials *Trichosurus vulpecular* Kerr and *Perameles nasuta* Geoffroy. *Australian Journal of Zoology* **10**, 597–616. doi:10.1071/ZO9620597

Kingston J (1998) *Handbook for the Captive Management of the Eastern Barred Bandicoot* Perameles gunnii. Zoological Parks and Gardens Board, Melbourne.

Kirsch JAW (1968) Burrowing in the quenda (*Isoodon obesulus*). *Western Australian Naturalist (Perth)* **10**, 178–180.

Krake D, Halley M (1993) Maintenance and captive breeding of the eastern barred bandicoot *Perameles gunnii* at Healesville Sanctuary. *International Zoo Yearbook* **32**, 216–221. doi:10.1111/j.1748-1090.1993.tb03538.x

Lacy RC, Balou JD, Pollak JP (2011) PMx: software package for demographic and genetic analysis and management of pedigreed populations. *Methods in Ecology and Evolution* **3**, 433–437. doi:10.1111/j.2041-210X.2011.00148.x

Lee L (1990) Behaviour of the bilby in captivity at the Territory Wildlife Park, N.T. *Thylacinus* **15**(1), 9–11.

Lenghaus C, Obendorf DL, Wright FH (1990) Veterinary aspects of *Perameles gunnii* biology with special reference to species conservation. In *Management and Conservation of Small Populations.* (Eds TW Clark and JH Seebeck) pp. 89–108. Chicago Zoological Society, Brookfield, IL.

Lobert B, Lee AK (1990) Reproduction and life history of *Isoodon obesulus* in Victorian heathland. In *Bandicoots and Bilbies.* (Eds JH Seebeck, PR Brown, RL Wallis and CM Kemper) pp. 311–318. Surrey Beatty & Sons, Sydney.

Loeffler E (2018) The foraging ecology of Eastern barred bandicoots in newly established island populations. Honours thesis. Deakin University, Melbourne.

Lynch M (2008) Bandicoots and bilbies. In *Medicine of Australian Mammals.* (Eds L Vogelnest and R Woods) pp. 439–464. CSIRO Publishing, Melbourne.

Lyne AG (1964) Observations on the breeding and growth of the marsupial *Perameles nasuta* Geoffroy, with notes on other bandicoots. *Australian Journal of Zoology* **12**, 322–339. doi:10.1071/ZO9640322

Lyne AG (1971) Bandicoots in captivity. *International Zoo Yearbook* **11**, 41–43. doi:10.1111/j.1748-1090.1971.tb01841.x

Lyne AG (1974) Gestation period and birth in the marsupial *Isoodon macrourus. Australian Journal of Zoology* **22**, 303–309. doi:10.1071/ZO9740303

Lyne AG (1976) Observations on oestrus and the oestrous cycle in the marsupials, *Isoodon macrourus* and *Perameles nasuta. Australian Journal of Zoology* **24**, 513–521. doi:10.1071/ZO9760513

Lyne AG (1982) The bandicoots *Isoodon macrourus* and *Perameles nasuta*: their maintenance and breeding in captivity. In *The Management of Australian Mammals in Captivity.* (Ed. DD Evans) pp. 47–52. Zoological Board of Victoria, Melbourne.

Mackerras MJ, Smith RG (1960) Breeding the short-nosed marsupial bandicoot *Isoodon macrourus* (Gould) in captivity. *Australian Journal of Zoology* **8**, 371–382. doi:10.1071/ZO9600371

Mallick SA, Driessen MM, Hocking GJ (1998) Biology of the southern brown bandicoot (*Isoodon obesulus*) in south-eastern Tasmania. II: demography. *Australian Mammalogy* **20**, 339–347. doi:10.1071/AM98339

Mallick SA, Driessen MM, Hocking GJ (2000) Demography and home range of the eastern barred bandicoot (*Perameles gunnii*) in south-eastern Tasmania. *Wildlife Research* **27**, 103–115. doi:10.1071/WR97006

Mawson PM (1960) Nematodes belong to the Trichostrongylidae, Subuluridae, Rhabdiasidae, and Trichuridae from bandicoots. *Australian Journal of Zoology* **8**, 261–284. doi:10.1071/ZO9600261

McCracken HE (1986) Observations on the oestrous cycle and gestation period of the greater bilby, *Macrotis lagotis* (Reid) (Marsupialia: Thylacomyidae). *Australian Mammalogy* **9**, 5–16. doi:10.1071/AM86001

McCracken HE (1990) Reproduction in the greater bilby, *Macrotis lagotis* (Reid): a comparison with other perameloids. In *Bandicoots and Bilbies.* (Eds JH Seebeck, PR Brown, RL Wallis and CM Kemper) pp. 199–204. Surrey Beatty & Sons, Sydney.

McRae PD (2004) Aspects of the ecology of the greater bilby, *Macrotis lagotis*, in Queensland. MSc thesis. University of Sydney, Sydney.

Merchant JC (1990) Aspects of lactation in the northern brown bandicoot *Isoodon macrourus.* In *Bandicoots and Bilbies.* (Eds JH Seebeck, PR Brown, RL Wallis and CM Kemper) pp. 219–228. Surrey Beatty & Sons, Sydney.

Merchant JC, Libke JA (1988) Milk composition in the northern brown bandicoot, *Isoodon macrourus* (Peramelidae: Marsupialia). *Australian Journal of Biological Sciences* **41**, 495–505. doi:10.1071/BI9880495

Meritt DA, Jr (1970) Observations on the short-nosed bandicoot at Lincoln Park Zoo, Chicago. *International Zoo Yearbook* **10**, 130–131. doi:10.1111/j.1748-1090.1970.tb01308.x

Miller DS, Mitchell GF, Biggs B, McCracken H, Myroniuk P, Hewish M (2000) Detection of agglutinating antibodies to *Toxoplasma gondii* in sera from captive mainland Australian eastern barred bandicoot (*Perameles gunnii*). *Journal of Wildlife Diseases* **36**, 213–218. doi:10.7589/0090-3558-36.2.213

Miller EJ, Eldridge MDB, Thomas N, Marlow N, Herbert CA (2010) The genetic mating system, male reproductive success and lack of selection on male traits in the greater bilby. *Australian Journal of Zoology* **58**, 113–120. doi:10.1071/ZO09092

Miller EJ, Eldridge MDB, Morris K, Thomas N, Herbert CA (2015) Captive management and the maintenance of genetic diversity in a vulnerable marsupial, the greater bilby. *Australian Mammalogy* **37**, 170–181. doi:10.1071/AM14009

Moseby KE, O'Donnell E (2003) Reintroduction of the greater bilby, *Macrotis lagotis* (Reid) (Marsupialia: Thylacomyidae), to northern South Australia: survival, ecology and notes on reintroduction protocols. *Wildlife Research* **30**, 15–27. doi:10.1071/WR02012

Moseby KE, Cameron A, Crisp HA (2012) Can predator avoidance training improve reintroduction outcomes for the greater bilby in arid Australia? *Animal Behaviour* **83**, 1011–1021. doi:10.1016/j.anbehav.2012.01.023

Norman RJ (1991) *Physaloptera peramelis* infection in the eastern barred bandicoot (*Perameles gunnii*): taxonomy and pathogenicity. Diploma of Wildlife Medicine and Husbandry thesis. University of Sydney, Sydney.

O'Hara AJ, Warren KS, Swan RA, Sims C, Friend JA (2004) Cutaneous papillomatosis and carcinmatosis in the highly endangered western barred bandicoot (*Perameles bougainville*). *Veterinary Dermatology* **15**(Suppl. 1), 25. doi:10.1111/j.1365-3164.2004.411_17.x

O'Hara PJ, Murray PJ, Klieve AV (2012) A review of the nutrition of Australian peramelid marsupials. *Australian Mammalogy* **34**, 133–144. doi:10.1071/AM11008

Obendorf DL, Munday BL (1990) Toxoplasmosis in wild eastern barred bandicoots, *Perameles gunnii*. In *Bandicoots and Bilbies*. (Eds JH Seebeck, PR Brown, RL Wallis and CM Kemper) pp. 193–197. Surrey Beatty & Sons, Sydney.

Parrott ML, Edwards AM (2023) Reproductive strategies and biology of the Australasian marsupials. In *American and Australasian Marsupials: An Evolutionary, Biogeographical, and Ecological Approach*. (Eds NC Caceras and CR Dickman). pp 931–979. Springer, Cham.

Parrott ML, Coetsee AL, Hartnett CM, Magrath MJL (2017) New hope for the Eastern barred bandicoot *Perameles gunnii* after 27 years of recovery effort. *International Zoo Yearbook* **51**, 154–164. doi:10.1111/izy.12157

Parrott ML, Nation A, Selwood L (2019) Female mate choice significantly increases captive breeding success, and scents can be frozen to determine choice, in the stripe-faced dunnart. *Applied Animal Behaviour Science* **214**, 95–101. doi:10.1016/j.applanim.2019.03.006

Pauligk Y (2021) *Eastern Barred Bandicoot (*Perameles gunnii*) Captive Husbandry Manual*. Zoos Victoria, Melbourne.

Pavey C (2006) *National Recovery Plan for the Greater Bilby* Macrotis lagotis. Northern Territory Department of Natural Resources, Environment and the Arts, Alice Springs.

Pope JH, Bicks VA, Cook I (1957a) Toxoplasma in Queensland. I: observations on a strain of *Toxoplasma gondii* from a bandicoot, *Thylacis obesulus*. *Australian Journal of Experimental Biology* **35**, 467–480. doi:10.1038/icb.1957.49

Pope JH, Bicks VA, Cook I (1957b) Toxoplasma in Queensland. II: natural infections in bandicoots and rats. *Australian Journal of Experimental Biology* **35**, 481–490. doi:10.1038/icb.1957.50

Presidente PJA (1982) Common ringtail possum *Pseudocheirus peregrinus*: maintenance in captivity, blood values and diseases. In *The Management of Australian Mammals in Captivity*. (Ed. DD Evans) pp. 75–81. Zoological Board of Victoria, Melbourne.

Quin DG (1985) Observations on the diet of the southern brown bandicoot, *Isoodon obesulus* (Marsupialia: Peramelidae), in southern Tasmania. *Australian Mammalogy* **11**, 15–25. doi:10.1071/AM88002

Reimer AB, Hindell MA (1996) Variation in body condition and diet of the eastern barred bandicoot (*Perameles gunnii*) during the breeding season. *Australian Mammalogy* **19**, 47–52. doi:10.1071/AM96047

Rendall AR, Coetsee AL, Sutherland DR (2018) Predicting suitable release sites for assisted colonisations: a case study of eastern barred bandicoots. *Endangered Species Research* **36**, 137–148. doi:10.3354/esr00893

Roberts FHS (1970) *Australian Ticks*. CSIRO Publishing, Melbourne.

Seebeck JH (1979) Status of the barred bandicoot *Perameles gunnii* in Victoria: with a note on husbandry of a captive colony. *Australian Wildlife Research* **6**, 255–264. doi:10.1071/WR9790255

Seebeck JH, Brown PR, Wallis RL, Kemper CM (1990) *Bandicoots and Bilbies*. Surrey Beatty & Sons, Sydney.

Shevill DI, Johnson CN (2008) Diet and breeding of the rufous spin bandicoot *Echymipera rufescens australis*, Iron Range, Cape York Peninsula. *Australian Mammalogy* **29**, 169–175. doi:10.1071/AM07021

Short J, Richards JD, Turner B (1998) Ecology of the western barred bandicoot (*Perameles bougainville*) (Marsupialia: Peramelidae) on Dorre and Bernier Islands, Western Australia. *Wildlife Research* **25**, 567–586. doi:10.1071/WR97131

Southgate RI (1990) Habitats and diet of the greater bilby *Macrotis lagotis* Reid (Marsupialia: Peramelidae). In *Bandicoots and Bilbies*. (Eds JH Seebeck, PR Brown, RL Wallis and CM Kemper) pp. 303–309. Surrey Beatty & Sons, Sydney.

Southgate RI, Possingham H (1995) Modelling the reintroduction of the greater bilby *Macrotis lagotis* using the metapopulation model analysis of the likelihood of extinction (ALEX). *Biological Conservation* **73**, 151–160. doi:10.1016/0006-3207(95)90041-1

Southgate RI, Christie P, Bellchamber K (2000) Breeding biology of captive, reintroduced and wild greater bilbies, *Macrotis lagotis* (Marsupialia: Peramelidae). *Wildlife Research* **27**, 621–628. doi:10.1071/WR99104

Species360 (2023) Zoological Information Management System (ZIMS). Minneapolis, USA. www.species360.org.

Spielman D (1994) First aid emergency care for Australian native mammals. In *Wildlife*. Proceedings 233. Post Graduate Committee of Veterinary Science. (Ed DI Bryden) pp. 339–366. University of Sydney, Sydney.

Stodart E (1966) Management and behaviour of breeding groups of the marsupial *Perameles nasuta* Geoffroy in captivity. *Australian Journal of Zoology* **14**, 611–623. doi:10.1071/ZO9660611

Stodart E (1977) Breeding and behaviour of Australian bandicoots. In *The Biology of Marsupials*. (Eds B Stonehouse and D Gilmore) pp. 179–191. Macmillan, London.

Stoddart DM, Braithwaite RW (1979) A strategy of utilisation of regenerating heathland habitat by the brown bandicoot (*Isoodon obesulus*; Marsupialia, Peramelidae). *Journal of Animal Ecology* **48**, 165–179. doi:10.2307/4107

Stone BF (1992) Ticks and wildlife: interactions, particularly of the Australian paralysis tick *Ixodes holocyclus* with koalas, bandicoots and fruit bats. In Proceedings of Wildlife Disease Association (Australasian Section). p. 17. Warrumbungle National Park, NSW.

Taylor R, Coetsee AL, Doyle RE, Sutherland DR, Parrott ML (2021) Sniffing out danger: rapid antipredator training of an endangered marsupial. *Australian Mammalogy* **44**, 109–116. doi:10.1071/AM20048

Thomas LN (1990) Stress and population regulation in *Isoodon obesulus* (Shaw and Nodder). In *Bandicoots and Bilbies*. (Eds JH Seebeck, PR Brown, RL Wallis and CM Kemper) pp. 335–343. Surrey Beatty & Sons, Sydney.

Thums M, Klaassen M, Hume ID (2005) Seasonal changes in the diet of the long-nosed bandicoot (*Perameles nasuta*) assessed by analysis of faecal scats and of stable isotopes in blood. *Australian Journal of Zoology* **53**, 87–93. doi:10.1071/ZO04030

Tyndale-Biscoe H, Renfree M (1987) *Reproductive Physiology of Marsupials*. Cambridge University Press, Cambridge.

Vernes K, Pope CC (2009) Reproduction in the northern brown bandicoot (*Isoodon macrourus*) in the Australian Wet Tropics. *Australian Journal of Zoology* **57**, 105–109. doi:10.1071/ZO09019

Vogelnest L (1999) Chemical restraint of Australian native fauna. In *Wildlife in Australia: Healthcare & Management*. Proceedings 327. Post Graduate Foundation in Veterinary Science. (Ed. DI Bryden) pp. 149–186. University of Sydney, Sydney.

Warren KS, Swan RA, Morgan-Ryan UM, Friend JA, Elliot A (2003) *Cryptosporidium muris* infection in bilbies (*Macrotis lagotis*). *Australian Veterinary Journal* **81**, 739–741. doi:10.1111/j.1751-0813.2003.tb14602.x

Warren KS, Swan R, Bodetti T, Friend T, Hill S, Timms P (2005) Ocular Chlamydiales infections of western barred bandicoots (*Perameles bougainville*) in Western Australia. *Journal of Zoo and Wildlife Medicine* **36**, 100–102. doi:10.1638/02-067

Watts CHS (1969) Distribution and habits of the rabbit bandicoot. *Transactions of the Royal Society of South Australia* **93**, 135–141.

Williams R (1990) Bandicoots. In *Care and Handling of Australian Native Animals*. (Ed. SJ Hand) pp. 75–81. Surrey Beatty & Sons & the Royal Zoological Society of New South Wales. Sydney.

Wilson GS (1984) The diseases and parasite status of the bilby (*Macrotis lagotis*) and rufous (western) hare wallaby (*Lagorchestes hirsutus*) in captivity. Diploma of Wildlife Health and Medicine thesis. University of Sydney, Sydney.

Winnard AL, di Stefano J, Coulson G (2013) Habitat selection of a critically endangered species in a predator-free but degraded reserve. *Wildlife Biology* **19**, 429–438. doi:10.2981/12-116

Woinarski JCZ, Burbidge AA, Harrison P (eds.) (2014) *The Action Plan for Australian Mammals 2012*. CSIRO Publishing, Melbourne.

Wood Jones F (1924) *The Mammals of South Australia. Part II: the bandicoots and herbivorous marsupials*. Government Printer, Adelaide.

Wright W, Sanson GD, McArthur C (1982) The diet of the extinct bandicoot *Chaeropus ecaudatus*. In *The Fossil Vertebrate Record of Australasia*. (Eds PV Rich and EM Thompson) pp. 229–245. Monash University, Melbourne.

# 7
# Koala

*Stephen M. Jackson, Karen Nilsson, Sarah Eccleston, Amber Gillett, Stephen Johnston, Daniel Lunney and Rosie Booth*

## 1. INTRODUCTION

The koala, the large kangaroos and the platypus are among Australia's most popular and iconic mammals. The empathy created by the koala appears to some observers to be because it is one of the few mammals that has a face rather than a muzzle, a trait it shares with humans (Lee and Martin 1988; Jackson 2007). The koala is neither a danger to people nor a pest to farmers, so it is not only tolerated across the landscape, but welcomed, In the past 30 years it has become the subject of serious conservation efforts (McAlpine *et al.* 2015), which include knowing how to care for a captive koala, whether on exhibit or in a rehabilitation facility (Lunney *et al.* 2022a, 2022b; 2023). The role of koala rescue and rehabilitation (i.e. captive husbandry) was highlighted by the extensive Black Summer fires of 2019–20 in eastern Australia (Lunney *et al.* 2022c).

Koalas are crepuscular and nocturnal; they are often seen moving at dusk, or more alarmingly, on roads in the evening or at night where they are at serious risk of injury or death (Lunney *et al.* 2022d, 2022e). The koala is one of the largest arboreal mammals (4.1–14.9 kg), resting in trees without building nests (Martin & Handasyde 1995). The koala's feet and hands are well developed and possess long, sharp claws, which are essential in climbing branches and tree trunks. After koalas were rescued following the major fires in coastal New South Wales in 1994, many remained in care for up to 9 months until their claws regrew (Lunney *et al.* 2004, 2007). Without the claws, the koalas cannot climb, find food or escape from predatory dogs. The fur is thick and dense and has some of the best insulating properties found in marsupials, verging on those of some arctic mammals. Dawson *et al.* (2014) have shown that koalas and polar bears have high and comparable levels of fur insulation. The colour and length of koala fur differs among locations, ranging from light grey and relatively short fur in Queensland, darker grey in New South Wales and chocolate brown and relatively long furred in Victoria and South Australia, and combinations at locations in between, with brown also found in the western part of its range in New South Wales. At times, this has caused confusion, with brown-coloured koalas appearing to be sick to people who are familiar with grey-coloured koalas.

Koalas initially did poorly in captivity, with most dying after only a very short period. An early record of them being held in captivity was by a soldier in who kept a female with 'two' young, which lived for over 1 month on gum leaves and bread soaked in milk or water (Stanbury & Phipps 1980). London Zoo maintained its first koala received in 1880 for 14 months by feeding it dried eucalypt leaves brought from Australia and later fresh leaves (Flower 1880). Several other koalas subsequently purchased by London Zoo did not make the journey or survived less than 1 year after being offered food items that included bread, milk and honey and even *Eucalyptus* throat pastilles (Crandall 1964).

In Australia, the first records of koalas in captivity are from Taronga Zoo in Sydney, which began keeping koalas in 1914, 2 years before the zoo officially opened. Koalas were sent to New York Zoo in 1920, but died only 5 days after refusing both dried and refrigerated leaves (Crandall 1964). San Diego received two koalas in 1925, of which one died after 5 months, while the other survived nearly 2 years on the local introduced eucalypts that occur there (Crandall 1964; Jackson 2007). Due to widespread hunting of koalas throughout their distribution in the latter part of the 19th century and early 20th century, Lone Pine Koala Sanctuary in Brisbane and Koala Park in Sydney were established in 1927. These were soon followed by Healesville Sanctuary (initially known as the Sir Colin Mackenzie Sanctuary) in Victoria, which opened in 1934 (Jackson 2001, 2007). The first breeding in captivity was at Melbourne Zoo in 1925, followed by the Koala Farm at Adelaide in 1937 and in 1938 at Taronga Zoo (Fleay 1937; Minchin 1937; Crandall 1964). The records on this subject are likely to be incomplete, given the testimony that Koala Park had the 'first and only attempt to propagate native bears in captive yet natural conditions' (Morgan *et al.* 1933).

San Diego received four more koalas in 1952 and they lived 5–6 years being fed the leaves of the abundant eucalypt trees in the area (Crandall 1964). Further koalas received by San Diego Zoo and San Francisco Zoo produced young in 1960 (Pournelle 1961). Koalas were sent from Taronga Zoo to Tama Zoo and Nagoya Higashimyama Zoo, Japan, in 1984. They were also sent from Lone Pine Koala Sanctuary to San Diego Zoo in 1976 and 1981, to Kagashima Zoo in Japan in 1984, San Francisco Zoo in 1986 and Yokohama Zoo in 1987 (Jackson 2001; K. Nilsson unpubl. data).

Taronga Zoo lent several individuals to the Singapore Zoological Gardens for 6 months in 1991 and food had to be packed and flown to Singapore daily to ensure that the specialised dietary needs of the animals would be met (Jackson 2001). More recently, koalas have been held in numerous zoos throughout Australia, the USA, Africa, Asia and Europe (Jackson 2001, 2007), as well as in rescue and rehabilitation centres within the koala's range.

Due to their popularity, koalas are an excellent educational tool and flagship species for increasing public awareness of species conservation for both children and adults (Finnie 1990). Koalas in zoos are ambassadors for conservation, particularly as the major factors affecting the long-term survival of wild koalas, and many other species, are the availability of suitable habitat, disease, their interactions with humans and climate change, especially drought, heatwaves and fire (Lunney *et al.* 2012, 2020; Reckless *et al.* 2017; Beale *et al.* 2022a, 2022b; DAWE 2022a, 2022b).

## 2. TAXONOMY

### 2.1 Nomenclature

The koala was first described by Goldfuss (1817) as *Lipurus cinereus*. The present genus name *Phascolarctos* was described by de Blainville (1816). The koala is the only extant member of the once-diverse family Phascolarctidae (Black *et al.* 2014; Jackson and Groves 2015).

Class: Mammalia
Supercohort: Marsupialia
Cohort: Australidelphia
Order: Diprotodontia
Suborder: Vombatiformes
Family: Phascolarctidae
Genus and species: *Phascolarctos cinereus*

### 2.2 Subspecies

Previously there were thought to be three subspecies of koalas. The nominate subspecies *P. c. cinereus* was considered to occur along eastern New South Wales. The northern subspecies *P. c. adustus* (Thomas 1923) was to be found in eastern Queensland, with the type specimen coming from O'Bil Bil near Mundubbera. The third subspecies *P. c. victor* (Troughton 1935) was found in much of Victoria with the type locality being Booral (Jackson and Groves 2015).

Genetic research by Houlden *et al.* (1999) found limited genetic distinction between geographically distant populations along the koala's distribution, suggesting a single evolutionary significant unit. Consequently, its range has been recognised as a cline or continuum that follows Bergmann's rule, whereby animals from high latitudes (i.e. koalas from Victoria and South Australia are larger, have thicker fur and are darker (to conserve heat) than those closer to the equator). In their study of the genomic comparisons in koalas, Kjeldsen *et al.* (2019) found biogeographic and anthropogenic impacts, but no support for a taxonomic classification of three koala subspecies, with only a single evolutionary significant unit supported. In a groundbreaking study of adaptation and conservation insights from the koala genome, Johnson *et al.* (2018) produced a complete and contiguous genome of the koala, thereby revealing, for example, that the koala's ability to detoxify eucalypt foliage may be due to a particular gene family and its ability to smell, taste and moderately ingest plant secondary metabolites may be due to expansions in the vomeronasal and taste receptors.

In a subsequent study linking genomic and environmental data for managing koalas, Lott *et al.* (2022) used next-generation sequencing to investigate the mechanisms underlying population structure and adaptive genetic variation. They concluded that the conservation of genome-wide genetic variation must be aligned with the protection of core koala habitat to increase the resilience of vulnerable populations. Specifically, they called for the five major genetic clusters identified in their study to be taken into account in future koala actions, such as translocations, because current management divisions in Queensland and New South Wales do not reflect historic or contemporary population structure.

As a result of the genetic studies, all populations should be referred to under the scientific name of *Phascolarctos cinereus* with no recognised subspecies (Jackson 2015; Jackson and Groves 2015).

## 2.3 Recent synonyms

Synonyms of koalas can be found in Jackson and Groves (2015).

## 2.4 Other common names

The word 'koala' comes from the Dharuk language (Sydney region) (Dixon 2019). Koala bear or native bear, and occasionally monkey bear, have historically been used, but despite these names, they are not related to bears. When searching the historical record for koalas, native bear is one of the best search terms, although in their account of the initial discovery by Europeans of the koala in the Sydney region at the end of the 18th, and beginning of the 19th centuries, Lunney *et al.* (2010) report three similar words used at the time for the common name of the koala: Coola, Cola and Cullawine.

# 3. NATURAL HISTORY

## 3.1 Bodyweight and morphometrics

Adult koalas can weigh between 4.1 and 14.9 kg and reach 648–820 mm in body length depending on sex and latitude (Table 7.1) (Martin *et al.* 2023).

Sexual size difference is evident, with males being up to ~50% larger than females (Lee and Martin 1988). It is suggested that this sexual dimorphism is associated with a polygynous mating system and a male dominance hierarchy (Tyndale-Biscoe and Renfree 1987). The koala has small eyes in comparison with the size of the head, with the slits of the pupils being vertical rather than horizontal as in other marsupials. The fur of the koala is thick and woolly and thicker and longer on the back than on the belly. Both the inside and outside of the ears are heavily furred (Lee and Martin 1988). Koalas have a shorter coat the further north they progress within their distribution; presumably due to the increase in average temperatures. The Victorian koala also has a heavier, shaggier coat with more fur in the ears and around the face while the Queensland koala is much smaller in size with a sleeker coat.

The koala has several morphological features, some of which it shares with wombats, which distinguish it from the other diprotodont marsupials. In contrast to the wombats, which only have one pair of upper and lower incisors, the koala has three incisors on either side of the upper jaw and the teeth have roots. The manus (hand) is forcipate with digits I and II opposed to the remaining three digits, with each digit terminating in a strongly curved claw (Lee and Carrick 1989).

The females have a pouch that contains two teats and opens centrally and downwards when not occupied and backwards when there is a large pouch young present. Males have a prominent sternal gland, which normally stains the fur around this area orange to dark-brown (Allen *et al.* 2010). Male koalas can be distinguished from females by the shape of the head (Fig. 7.1). The head of

**Table 7.1.** Average bodyweight and size for koalas in Queensland and Victoria.

Martin *et al.* (2023).

| Location | Weight (range)(kg) | Head and body length (range)(mm) |
|---|---|---|
| **Queensland** | | |
| Males | 6.5 (4.2–9.1) | 705 (674–736) |
| Females | 5.1 (4.1–7.3) | 687 (648–723) |
| **Victoria** | | |
| Males | 12.0 (9.5–14.9) | 782 (750–820) |
| Females | 8.5 (7.0–11.0) | 716 (680–730) |

a

b

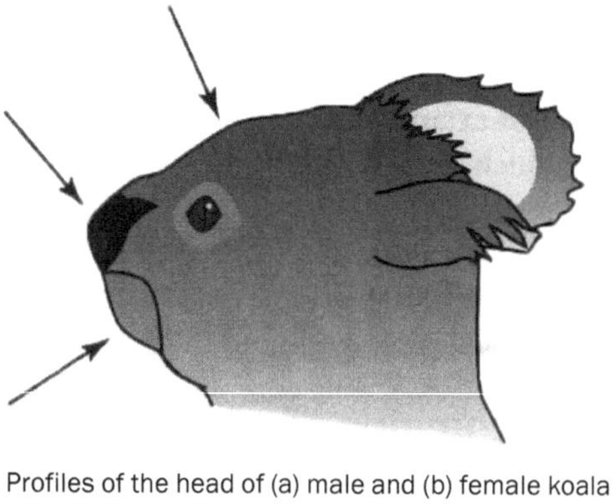

**Fig. 7.1.** Profiles of the head of (a) male and (b) female koala illustrating the distinctive 'roman' nose of the male and the straighter nose of the female.

adult males is relatively larger than that of females and appears broader and squared off in profile. Males also have a broad rather than pointed chin, relatively small ears and a pendulous scrotum.

## 3.2 Distribution and habitat

The koala is endemic to Australia. It has a wide but patchy distribution in eucalypt forests and woodlands of eastern and south-eastern Australia, including Queensland, New South Wales, the Australian Capital Territory, Victoria and South Australia, as well as many islands where it has often been introduced (Jackson 2007). Its distribution includes the coastal and inland areas of Queensland, north to Herberton, and extends westwards into the semi-arid climates of central Queensland and New South Wales. The koala is widespread in Victoria where its distribution extends to the south-east corner of South Australia. Reintroductions in the 1920s expanded the range into South Australia, notably to Kangaroo Island and the Adelaide Hills. The natural range of the koala is determined by its specialist diet, habitat and environmental requirements. Typically, this includes forests and woodlands dominated by *Eucalyptus* species (Melzer *et al.* 2000). A detailed description of the distribution of the koala in each state can be found in the Commonwealth conservation advice for the koala (DAWE 2022a) with a useful map in the National Recovery Plan for the koala (DAWE 2022b).

## 3.3 Conservation status

The conservation status of the koala varies among the states. In 2012 the koala was listed as Vulnerable to Extinction under the Commonwealth *Environment Protection and Biodiversity Conservation Act 1999* (EPBC Act) in New South Wales, Queensland and the Australian Capital Territory (DAWE 2022a). It was not listed for Victoria or South Australia. The koala population was subdivided in 2012 because of substantial differences in management and conservation status across the species' range. Under Section 517 of the EPBC Act the combined koala populations of Queensland, New South Wales and the Australian Capital Territory were declared to be a species for the purposes of the Act. This entity was listed as Vulnerable. In 2022, the EPBC Act listing for New South Wales, Queensland and the ACT was increased to Endangered. In Victoria and South Australia, the koala is not listed as a threatened species under Commonwealth legislation.

In considering the status of the koala in the *Action Plan for Australian Mammals 2012*, Woinarski *et al.* (2014) and Woinarski and Burbidge (2020) identified that the cause of its decline includes habitat loss, fragmentation and degradation; predation by dogs; deaths from vehicle collision; disease; climate change; and inbreeding. The *National Recovery plan for the Koala* (DAWE 2022b) identifies the overarching threats to the listed koala (i.e. the threats in Queensland, New South Wales and the Australian Capital Territory) are land use change and climate change. The recovery plan then lists other direct threats as disease, dogs and vehicles. The plan further notes that the threats interact and further notes that, because of the listed koala's large distribution, their relative importance varies at local scales.

## 3.4 Diet in the wild

Study of the diet of koalas in the wild is relevant to the management of captive koalas, as it helps explains why koalas choose the leaves of some species of eucalypts over others and further, the leaves from some individual trees of some species, but not the leaves of other individual trees. The chemical ecology of leaf choice is complicated, but it is intriguing to grasp at last what lies behind the well-known observations of so many managers and zookeepers that koalas are 'fussy eaters'.

Koalas feed predominantly on the foliage of eucalypts, with some non-eucalypts occasionally contributing to the diet, but more importantly, acting as shelter trees. Over the past 2 decades there has been an increase in the understanding of the trees and leaves selected as browse by koalas. One good example is a novel description of koala diets and habitat quality by Moore *et al.* (2023) who used faecal molecular analysis. They analysed the diets of koalas throughout Queensland, using undigested plant DNA in koala dung, also known as scats or faecal pellets. They found that koalas fed overwhelmingly from trees of the *Eucalyptus* subgenus *symphyomyrtus*. The genus *Corymbia* made a limited contribution to koala diets, but *Angophora* was scarcely used. Non-eucalypts comprised only small part of the diet, with *Lophostemon confertus*, *L. suaveolens* and *Melaleuca* species present, but the genera *Allocasuarina* or *Casuarina* were not present. Blyton *et al.* (2023a) further describe this new genetic method for diet determination from faeces that provides species-level resolution.

Although there are many species of trees that koalas typically select, they frequently feed on the leaves of some trees and not others of the same species. There is some evidence that soil type influences palatability. Braithwaite (2004) found a positive association between leaf nutrients and soil quality, while Moore *et al.* (2004) took the next step and reviewed the role of nutrition in the conservation of the marsupial folivores of eucalypt forests. They noted that it seems to be accepted that good koala habitat occurs on fertile soils with favoured food species.

In a study of habitat selection by koalas in the far north coast of New South Wales, McAlpine *et al.* (2023) identified that the dominant factor influencing habitat use and koala occurrence was the distribution of *E. propinqua*, *E. robusta*, *E. tereticornis*, *E. microcorys* and *E. grandis*. They found that koalas were more likely to use medium-sized trees of these species where they occurred on soils with high levels of Colwell phosphorous, which is a measure of the phosphorus available for plant uptake. The study also confirmed that non-preferred species of eucalypts and non-eucalypts are extensively used by koalas and form important components of koala habitat, thereby lending support to restoring a mosaic of koala-preferred tree species and other species recognised for their value as shelter.

Moore *et al.* (2004) identified the 'anti-nutritional properties of foliage', particularly plant secondary metabolites (PSMs). This field of study continues to expand, and it is also important in understanding feed for captive koalas. Moore *et al.* pointed out that the association between dense koala populations and eucalypts growing on fertile soils may be driven as much by water availability as it is by nutrients.

More recently there have been studies of the impact of drought on koala populations, especially towards their western edge (i.e. the more arid edge of their range) (Davies *et al.* 2013, 2014; Seabrook *et al.* 2014; Smith *et al.* 2013a, 2013b; Lunney *et al.* 2012, 2020). These have revealed the importance of water, with photos of wild koalas drinking water that was offered to them during drought. Other koalas suffering dehydration were revived with an injection of saline solution (Lunney *et al.* 2012). Mella *et al.* (2019, 2020) demonstrate the use of water provided to wild koalas. Moore *et al.* (2004) were among the first researchers to formally recognise the potential impact of climate change on the distribution of eucalypts and their chemistry and thus the nutritional resources of marsupial folivores.

Intraspecific variation in palatability of eucalypts for koalas is controlled by the variable concentrations of PSMs, including formylated phloroglucinol compounds (FPCs) (Moore *et al.* 2004) and tannins, also possibly unsubstituted B-ring flavanones, which are found in many subgenus *Eucalyptus* species and which also strongly deter common brush-tailed possums (*Trichosurus vulpecula*). They reduce the availability of nitrogen and DeGabriel *et al.* (2008) developed a method of measuring nitrogen in leaf samples to evaluate how tannins influence the nutritional value of forage. Among the important findings were that digestible nitrogen concentrations of leaves varied significantly more than the total nitrogen concentrations. Thus, predictions of the nutritional suitability of foliage for herbivores, based on total nitrogen rather than available nitrogen, can be misleading. Consequently, the variability in the nutritional quality of eucalypt forests may explain the distribution and numbers of mammalian herbivores, including the koala. DeGabriel *et al.* (2009) expanded on this research by stating that this new knowledge was just beginning to emerge. As DeGabriel *et al.* (2008) stated, their central finding was that the availability of in vitro digestible nitrogen, resulting from the interaction between tannin and nitrogen concentrations, correlated with reproductive success in a population of common brush-tailed possums.

Marsh *et al.* (2018) tackled another perennial problem, that of leaf age and nutritional composition. They noted that average differences may not identify the nutritional trade-offs, such that the amount of PSMs that can be ingested may depend on the nutritional context. In some cases, higher concentrations of nitrogen or other nutrients may allow herbivores to tolerate higher PSM intakes. Thus, the same concentration of PSMs may be less of a deterrent in young than in mature leaves. Crucially, they concluded that PSMs and nutrient concentrations can differ widely among different trees of the same species, meaning that some individual plants may not produce enough PSMs to deter herbivores, even if young leaves contain higher concentrations of PSMs than mature leaves from the same tree. The opposite is that animals may prefer higher nutrient concentrations in young leaves only when PSM concentrations are low. Thus, young leaves will more likely be selected on individual plants with low concentrations of PSMs.

A study of wild koalas, taken into captivity for feeding trials, revealed that fundamental dietary specialisation explained the different use of eucalypt resources within a koala population (Marsh *et al.* 2021). The researchers asked whether differential use of two eucalypt species

(*E. obliqua* and *E. viminalis*) by individual koalas was due to individual specialisation or to extrinsic factors. They commented that many subgenus *Eucalyptus* species tend to be classified as secondary rather than primary food tree species for koalas, but Marsh *et al.* (2021) suggest that the relative importance of some secondary food species may depend on scale, such that plant species may be eaten relatively little by the population as a whole, but may be a critical resource for some individuals within the population. One implication being that these individuals may buffer populations from extinction, such as during periods of dramatic habitat alteration. Another implication is individual specialisation in a captive setting, meaning that not all individuals may benefit from being fed the same diet.

Koalas have been observed to consume large quantities of the bark of specific *E. mannifera* in subalpine regions. It was hypothesised that the consumption of bark in this setting was utilised due to a deficiency of sodium in leaves at high elevations (Au *et al.* 2017). In captivity koalas are also known to consume bark while undergoing rehabilitation, though the tree species is not consistent. The consumption of bark appears to be associated with gastrointestinal disease and it is thought that a more fibrous component in the diet is required either via a shift in the microbiome or for increased bulk in the gastrointestinal tract (A. Gillett pers. obs.).

When eucalypt leaves have a high-water content (~60–80%), koalas normally do not need to drink, but obtain sufficient water from their food. Water requirements are driven as much by water demands (varying with thermoregulatory demands). In cool conditions, 50% water is adequate (J. DeGabriel pers. comm.). However, recent studies have found that, during hot weather and drought, koalas do need additional water (Mella *et al.* 2019, 2020). In the millennium drought, koalas died from dehydration and a population of koalas in Warrumbungle National Park declined from abundant to rare (Lunney *et al.* 2012, 2020). Eucalypt leaves have a high fibre and low protein content. They contain strong-smelling oils, phenolic compounds that make them unpalatable or even poisonous to most mammals. To cope with this diet, the koala has numerous adaptations including their molar teeth that finely comminute the leaves; a large caecum; a capacity to tolerate the toxic compounds in their food; and a low metabolism/slow digestion rate. Oils and many phenolic compounds are detoxified in the liver. There is no evidence for the widespread belief that eucalyptus oils intoxicate koalas, rendering them lethargic.

The hindgut of the koala is well developed in the caecum and proximal colon. The caecum is used as a fermentation chamber and with the aid of bacteria partially breaks down the cellulose. Apart from the enormous size of the hindgut the stomach of the koala also contains a cardiogastric gland, similar to that in the wombat, although in the koala it is branched and more complex (Hume 1999). At the commencement of weaning, the joey eats semiliquid faeces from the cloaca of the mother (Minchin 1937; Osawa *et al.* 1993; Martin and Handasyde 1999). This substance is commonly called 'pap' and has a higher microbial density than normal faeces.

There has been much recent interest in documenting the koala microbiome (see Alfano *et al.* 2015; Blyton *et al.* 2019, 2023b; Brice *et al.* 2019; Chong *et al.* 2020). A better understanding of the organisms in the koala gut will provide new insights into both dietary preference and energy utilisation. There is an ever-sharper focus on different aspects of koala diet, such as the study by Blyton *et al.* (2023b) in which they found that the koala gut microbiome was largely unaffected by dietary change and thus supported previous findings that the koala gut microbiome influences the tree species chosen for feeding. Their study was based on the recognition that koalas inherit their gut microbiomes maternally by feeding on pap. The study also concluded that animal microbiomes may not be affected by translocations.

## 3.5 Longevity

### 3.5.1 Wild

There are few records of longevity in the wild. A female tagged on French Island, Victoria, was still breeding at 10 years of age and a male at Walkerville, Victoria, was estimated to have died at 16 years of age (Lee and Martin 1988). There are also records of wild females living 17–18 years (Martin and Handasyde 1995). A study in north-east New South Wales of post-release survival of koalas rescued and rehabilitated to the wild recorded that, of the top 10 post-release survivors with the longest time in the wild, the shortest total time was 7 years and the longest was a remarkable 13.6 years for a koala that was already 5 years old at first admission to rehabilitation (Lunney *et al.* 2022a, 2022b).

Determining the major causes of death in past koala populations is difficult because of a paucity of records, so constructing an ecological history of change from the time of European settlement is challenging. Nevertheless, besides the findings of Cahir *et al.* (2020, 2021) a series of studies in New South Wales has shown that it is possible to construct an ecological interpretation of changes in koala populations after European settlement. Among the features to emerge are the considerable fluctuations in population sizes at most locations, marked regional differences, the importance of land clearing, logging, stock and the impact of predation by dogs, drought, heatwaves and, for a few decades, the fur trade from the latter part of the 19th century (Lunney and Leary 1988; Knott *et al.* 1998; Lunney *et al.* 2010, 2016, 2020).

Fires kill koalas, but the records are scant as to the effect on numbers. Koalas have no means of escaping high-severity fires that sweep through the crowns of eucalypts and the few that survive have little hope of avoiding starvation or dehydration before the trees produce epicormic growth. However, detailed satellite studies of the severity of the Black Summer bushfires in 2019–20 found that there was a range of severity levels within the mapped fire grounds (i.e. the perimeter of the fire). It is now appreciated that there is a higher level of survival at the lower levels of fire severity, such as when the canopy remains intact and the fire is confined to the ground level (Gibson *et al.* 2020). In the previous major fire in 1994 in coastal New South Wales, a radio-tracking study in Port Stephens found that koalas were back in the burnt bush within 3 months (Mathews *et al.* 2007).

In the early 1980s, a severe drought in central Queensland caused browning and a loss of leaves in eucalypt forests, resulting in substantial numbers of deaths of koalas (Gordon *et al.* 1988). In the agriculturally rich lands of the Liverpool Plains of north-west New South Wales, an estimated one-quarter of the koala population died during heatwaves at the end of 2009, just before the breaking of the millennium drought (Lunney *et al.* 2012). Again, this drought reduced an abundant koala population in Warrumbungle National Park to rarity by 2009 and following a fire in 2013, no koalas were seen (Lunney *et al.* 2020).

Other factors that influence the longevity of wild populations include disease (particularly *Chlamydia*, related pathology, leukaemia and neoplasia) that lead to death and the rate of wear of the teeth, which ultimately results in an inability to masticate sufficient food to meet the animal's nutritional needs (Gonzalez-Astudillo *et al.* 2017; 2019a).

### 3.5.2 Captivity

The average longevity in captivity is ~14–16 years for females and 12–14 for males; however; a male and female from Tennoji Zoo in Osaka, Japan, lived for 22 and 21 years of age respectively. Lone Pine Sanctuary held the previous record for the longest surviving koala, which survived 23 years, before it died in 2001; however, a koala in Japan held at the Awaji Farm Park England Hill on Awaji Island has lived for over 24 years (K. Nilsson pers. obs.; Anon 2021). In captivity the major cause of death is generally old age. A survey of koalas housed in Australia by Gillett (2014) reported that the primary causes of death in captive koalas were neoplasia (lymphoma and leukaemia), cryptococcosis and dermatitis in older animals. Several captive facilities in south-east Queensland have also noted an increase in pouch-young deaths, the aetiology of which is still under investigation (S. Johnston pers. obs.).

Koala retrovirus (KoRV)-related illnesses are still poorly defined are the most common cause of captive koala deaths in south-east Queensland and are most often observed in animals ~7 years of age (Gillett 2014, 2023; Gillett and Hanger 2025). KoRV-related illness include neoplasia, bone marrow dysplasias and conditions associated with immune dysfunction (Gillett 2023). Long-term captivity records are not available from rehabilitation groups because the objective is to release koalas back into the wild upon recovery.

### 3.5.3 Methods used to determine the age of adults

Techniques have been developed to determine the approximate age of koalas, although there can be wide variation in the rate of tooth wear between individuals, so wear classes may not align with age between locations (Gordon 1991; Melzer *et al.* 2011). Therefore, tooth wear age class classification should ideally be verified with local data whenever possible (Melzer *et al.* 2011). One method used to estimate koala age uses only the fourth upper premolar (P4), which has been correlated with age (Fig. 7.2; Martin 1981; Martin and Handasyde 1999). A second method examines the tooth wear of the upper premolar (P4) and molars (especially M1) to determine the Tooth Wear Class, which in turn is used to estimate the approximate age (Fig. 7.2; Gordon 1991).

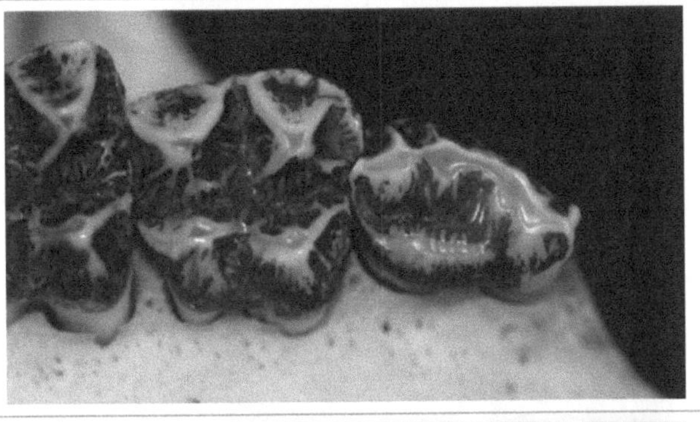

Tooth Wear Class - ~3 years of age. Lateral rim of premolar and four pyramidal cusps of M1 just in wear with the first hint of brown dentine just appearing. Black pigment on unworn enamel is prominent

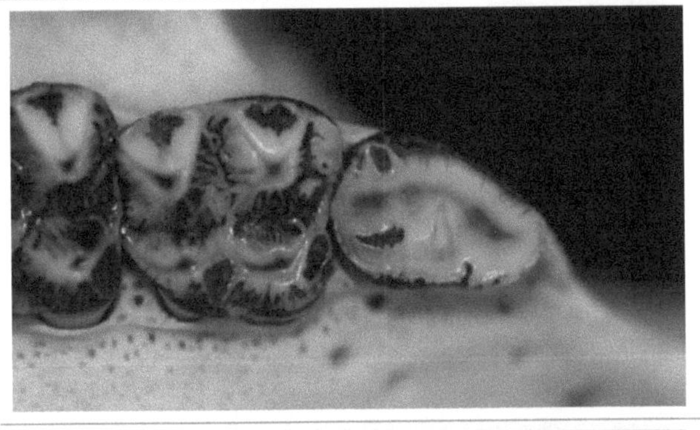

Tooth Wear Class - ~4 years of age. The entire lateral and medial ridge of the premolar is worn to a flat plateau. There is still significant black stain on the enamel. All four pyramids of M1 are in wear with dark squares of dentine exposed at the peaks

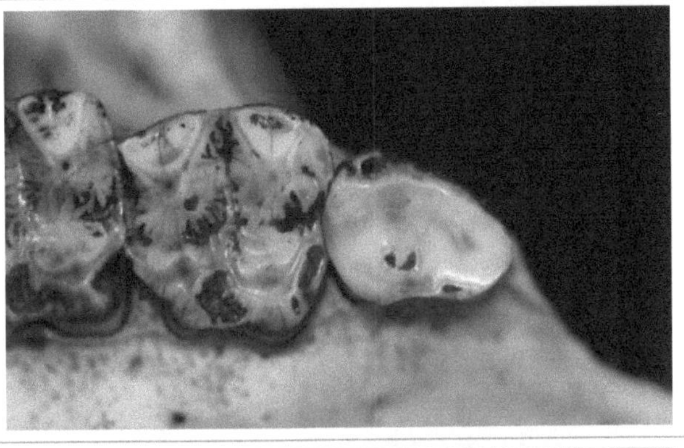

Tooth Wear Class - ~5 years of age. There is a remnant of black enamel on the smooth surface of the premolar. There is a small brown circle of dentine in a crater on the peak of each lingual pyramid on M1

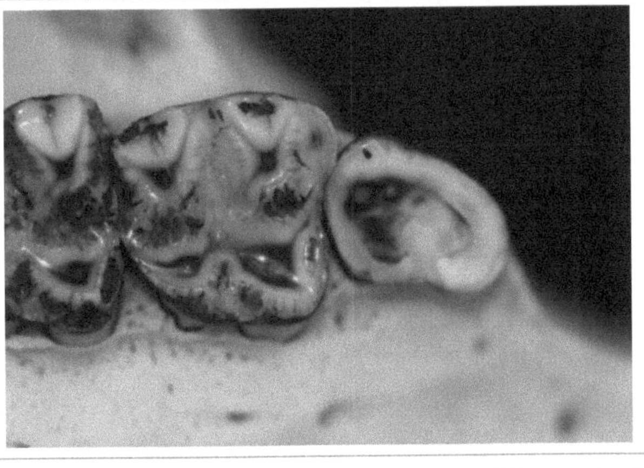

Tooth Wear Class - ~6 years of age. The premolar is worn smooth. Brown dentine remnants are visible on the lateral edge.
M1 still has four distinguishable pyramids but the lingual pair are worn nearly flat. Shrinking areas of black-stained enamel remain at the base of the pyramidal cusps

**Fig. 7.2.** Examples of tooth wear on the right upper left premolar (P4) and molars 1–2 of known-age captive koalas from a facility in south-east Queensland. Note that this figure does not represent average wear classes of koalas across captive facilities or the wild population, which may vary considerably. As such these images should only be used as a guide to estimate the age of koalas. Photos and captions by Rosie Booth unpubl. data.

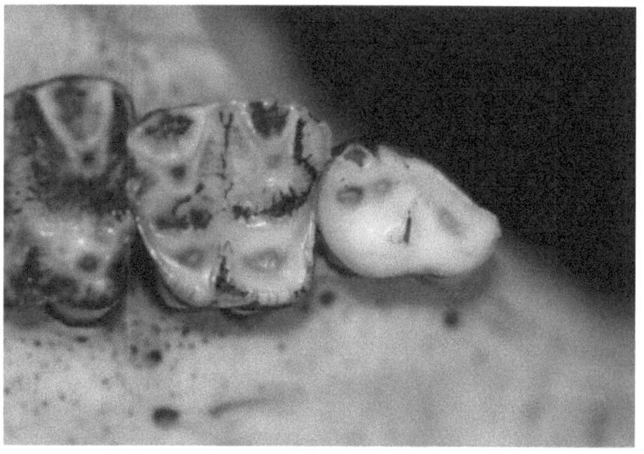

Tooth Wear Class – ~7 years of age. The premolar is worn flat with a lateral brown chain of dots and variable minute remnants of black enamel. The two lateral pyramids of M1 remain, but the lingual pyramids are worn further to reveal two dark circles of dentine with central white cores being the root bases

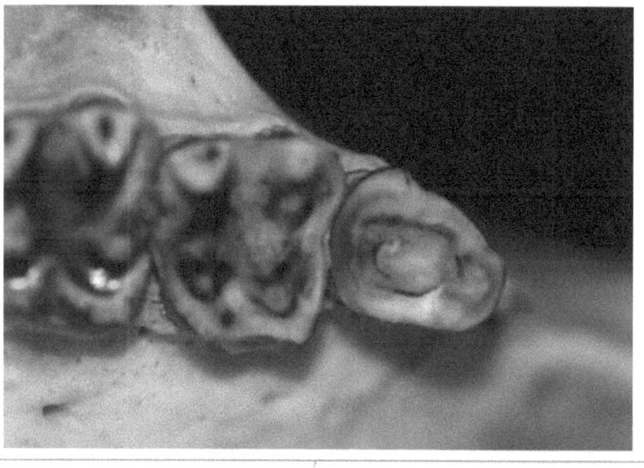

Tooth Wear Class – ~8 years of age. The premolar is worn smooth with some wave and a circular line of dentine. The pyramids of M1 are worn to a smooth wave and there are four dark circles of dentine with central pale circles being the root bases

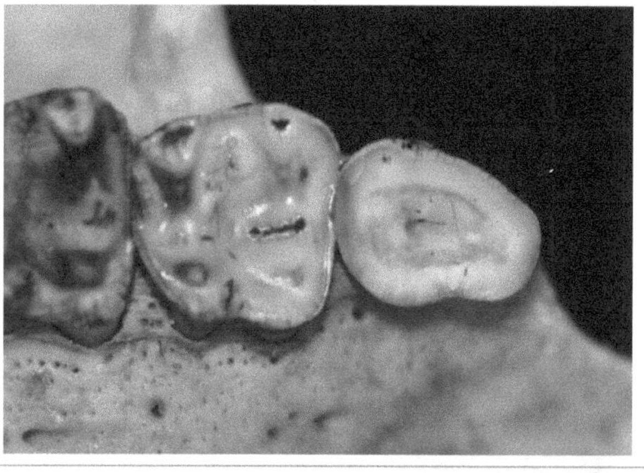

Tooth Wear Class – ~9 years of age. The premolar is worn smooth and flat with a distinct even enamel border around an ellipse of brown dentine. The premolar is becoming more crescent-shaped. The pyramids of M1 and now indistinct, with each having a central brown circle indicating the roots and very small remnants of black-stained enamel

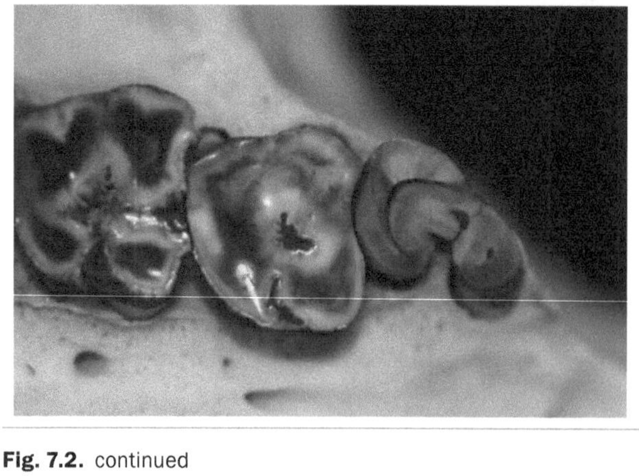

Tooth Wear Class – ~10 years of age. The premolar is worn down to a crescent with root canal exposure and roots starting to separate. M1 has lost most of the wave structure and retains just a central remnant of black-stained enamel. In life the gingiva may separate the premolar into anterior and posterior segments from this age onwards

**Fig. 7.2.** continued

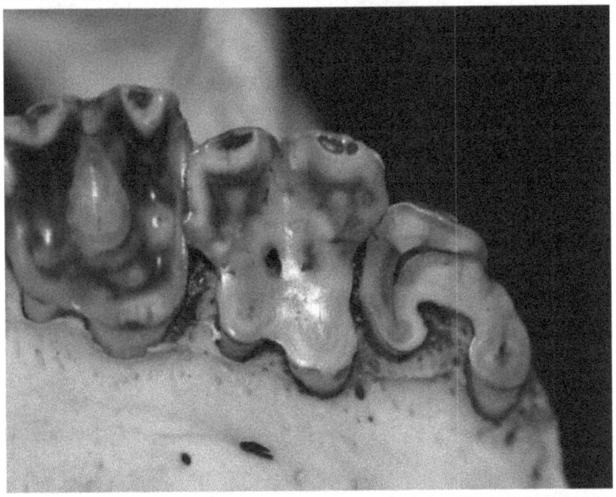

Tooth Wear Class - ~11 years of age. The premolar is worn closer to the gum line and there is a thinner bridge of enamel connecting the roots.
M1 is worn past the low troughs in the original four pyramidal cusps and gaps appear as the four roots separate.
The outline of the molar and premolar now have a more 4 lobed irregular shape as the wear continues into the tooth roots.

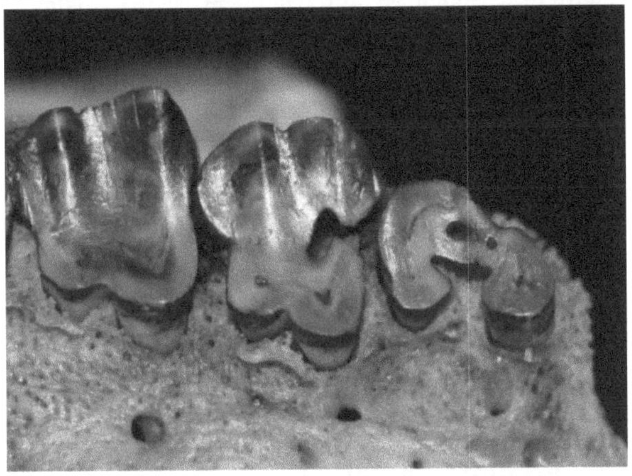

Tooth Wear Class - ~16 years of age. The premolar is worn to two smooth roots visible at the level of the gum line.
M1 is dividing into separate roots.
The four lobes of M1 are now more distinct. There is still a wave of wear on the surface of the molars.

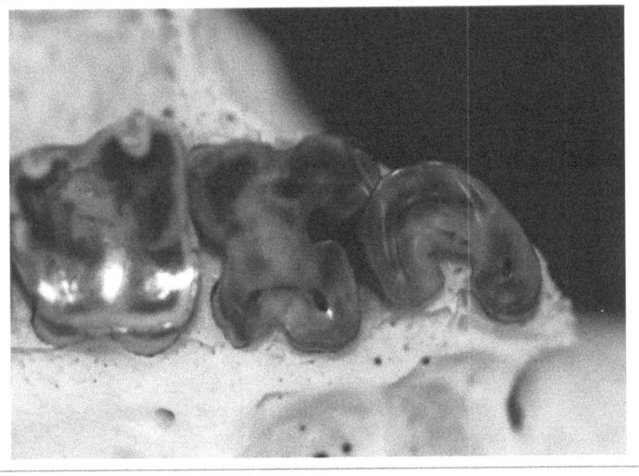

Tooth Wear Class - ~18 years of age. Similar to above. The premolar is still a crescent but there is more pulp cavity exposure in a crescent caudally and in a circle anteriorly. The outline of M1 is more divided into lobes (four-leaf clover shape) and the pulp cavity of the two medial roots is exposed

## 4. HOUSING REQUIREMENTS

### 4.1 Enclosure design

Koala enclosures should follow several general principles in order to satisfy minimum conditions for the keeping of koalas in captivity and those outlined here can be adapted for research facilities. Formalised standards have been developed for exhibiting koalas in New South Wales (Anon 2006) require the enclosures to:

- provide shelter from wind, rain and extremes in temperature and allow sufficient access to shade during the hot periods of the day
- provide freedom of movement both vertically and horizontally

- be well drained and have either a readily cleanable substrate or be of a material that can be replaced to avoid the accumulation of faeces and urine.

Nationally recognised standards have been drafted for koalas (Anon 2019). These standards require:

- enclosure barriers to be unclimbable and have a minimum height of 1.2 m
- enclosures have security features such as a secure roof, sheet metal panel and other deterrents to prevent escape of the enclosed koalas
- each koala be provided a minimum of two resting forks, with one at least 1.5 m above the ground and the other at least 1.8 m
- each branch to have sufficient traction for koalas to climb easily and safely
- that there are no climbable features inside or outside the enclosure that could facilitate koala escape or entry by wild koalas.

Various enclosure designs that have been used by institutions to exhibit koalas are open, semi-enclosed or totally enclosed in design. They include a comparatively simple design with a circular or oval wall at least 1.2 m high, with a floor that is grassed or made of sand or concrete. If enclosure fence heights of less than 2 m are used, then a perimeter fence of the facility needs to be predator proof and be at least 2 m high due to the potential for predation by cats and foxes (e.g. Mella *et al.* 2018).

Within each enclosure, rough-barked trees (e.g. *Eucalyptus obliqua*) should be provided that are ideally greater than 3 m in height and have at least two natural forks provided per koala that are at least 1.5 m above the ground for the animals to sit in. Trees should be placed ~2–3 m apart to encourage the koalas to jump from tree to tree. The forks may also be joined by lateral branches to facilitate movement from tree to tree without descending to the ground.

Larger, naturally vegetated enclosures have been used at some facilities where the ground is covered with natural grasses and various native shrubs and trees are provided. The koalas are viewed from elevated pathways several metres off the ground to bring the viewer closer to the animals. A sheet of metal can be placed around the fence of the enclosure and branches of trees near the fence line are trimmed from fence to prevent the animals from escaping. The enclosure can be furnished with large branching trees (such as stringy bark) that are least 6 m high for the animals to rest in. The trees can be inserted into the ground and replaced approximately every 12 months as they lose their bark and become slippery over time. Metal stirrups concreted into the ground can be used to prevent the base of poles rotting (K. Nilsson pers. obs.). The enclosure can be watered by a sprinkler system and water may also be sprayed from a removable bayonet system elevated off the ground on 2 m poles to help keep the foliage fresh on hot windy days.

## 4.2 Holding area design

Holding areas for koalas need only be of a simple design. They should be totally roofed and can be constructed of chain or welded mesh of a size to ensure that koalas are not able to get any part of their body stuck. Some facilities are also known to use insulated roofing panels in their off-exhibit areas.

A cement floor, well drained to a good sized sump with a grate is the easiest way to maintain off-exhibit koalas on a long-term basis, as there is not the need to replace dolerite substrates regularly. The floor is swept, hosed and scrubbed daily to remove any algal growth and ensure a safe and hygienic surface. Metal bases can be used as pole stands to allow the pole configuration to be easily changed when necessary.

## 4.3 Spatial requirements

Areas of enclosures typically range from 30 $m^2$ to 100 $m^2$ for 2–4 animals (see sections 4.1 and 4.2 for more details). An additional area of 2.5 × 2.5 m should be provided for each extra animal.

The floor area required by the *Standards for Exhibiting Australian Mammals in New South Wales* (Anon 2006) for up to two animals is at least 12 $m^2$ with a height of 2.4 m (assuming the enclosure has walls up to 2.4 m and a roof). If the facility has a predator-proof external fence and the koala enclosure is not roofed, the enclosure fence may be 1.2 m high. An additional area of 4 $m^2$ is required for each additional animal.

The proposed national standards for exhibiting koalas require an area of 9 $m^2$ for a single koala, with an additional 3 $m^2$ for each additional adult koala of the same sex. These standards also require a floor area of at least 25 $m^2$ for mixed sex enclosures per male koala that is 30 months of age or older. An additional area of 3 $m^2$ is required for each additional adult female koala. Koalas must be able to perch in the highest fork without being restricted by the roof of the enclosure.

## 4.4 Position of enclosures

The aspect of an enclosure is important because the koalas need to be provided with sufficient shelter from wind, rain and extreme heat, but have an opportunity to warm themselves during cold weather. Shelter can be provided in the form of partial or total overhead coverage by placing the trees within the enclosure in a compact arrangement to provide shelter.

## 4.5 Weather protection

The trees in the enclosure can be placed in a compact arrangement to help insulate the animals against the cold (Drake 1982). Alternatively, enclosure walls can be placed on the side of the prevailing weather. In open enclosures that do not include artificial shade, taller trees, including non-food species, with shadier/denser foliage need to be provided for shelter, so the koalas can adapt their behaviour and seek shade during hot weather (Ellis *et al.* 2010; Crowther *et al.* 2014). Koalas in the wild also rest at different heights in the trees depending on season and temperature, with koalas positioning themselves high in the trees to catch the sun in winter and sheltering well down in the tree, at shrub level, in hot weather (Matthews *et al.* 2007). Shade sails have been used by some exhibitors where koalas are exposed to long periods of midday sun and these also provide protection from hail during storms.

## 4.6 Temperature requirements

No heating is generally required for koalas unless they are held at temperatures that are consistently low (0°C or less), in which case they may need to be held indoors with a source of heating. Koalas are able to tolerate very low temperatures (as low -10°C) for many weeks. However, there is a growing body of research demonstrating evidence of heat stress in koalas (e.g. Mella *et al.* 2019, 2020). Therefore, the provision of water bowls and the use of water mister fans and sprinklers, not directly on or at animals, should be applied during extreme temperatures. The sprinklers can be left on day and night if required. It is important that size of the water droplets of the sprinkler are sufficient to penetrate through the fur and wet the skin in order for evaporative cooling to be effective (Adam *et al.* 2021). A comfortable ambient temperature range for koalas is ~8–28°C, though southern koalas can cope better with colder temperatures due to their longer coat and northern koalas can cope better with heat above 30°C (R. Booth pers. obs.). It should also be noted, however, that high humidity can reduce the koala's capacity to deal with high ambient temperatures (A. Gillett pers. obs.).

By feeding mostly at night, free-ranging koalas are mostly active when daily ambient temperature is lowest (Ryan *et al.* 2013), potentially offsetting the impact of heat production from activity and metabolic heat production from digestion (although digestion is continuous in koalas) on their overall body temperature profile (Adam *et al.* 2021). Adam *et al.* (2021) found that koalas exhibit a diurnal rhythm in body temperature regardless of the time of year. They also revealed that captive koalas fed in the morning for the purposes of public exhibit had a higher body temperature (0.2–0.7°C higher) than those fed in the evening, speculating that this phenomenon may be of significance particularly for koalas held in the north of their normal range or during period of extreme heat. These authors have also suggested the use of larger droplets of water from sprinklers designed to wet the skin in combination with morning 'wetting' could be of benefit in heat wave conditions to lower the body temperature of koalas in the morning so that they are able to cope with the rise during the day.

## 4.7 Substrate

The base of the enclosure can be made of any number of materials, from concrete to various types of soil, sand, leaf litter or dolerite. Dolerite is especially good under perches, as it is easily raked and drains well, leaving a dry, compacted and attractive surface. For exhibit enclosures it is better to use soil as it is more aesthetically pleasing while off-exhibit enclosures are easily maintained with smooth-finished concrete. The base of the enclosure must drain readily to ensure that, in the event of rain, the koalas are able to move between trees without having to wade through water.

## 4.8 Enclosure furnishings

Each enclosure needs to include the eucalypt resting forks, as described above, no closer than 0.9 m to each other. The tree trunk with forks should be very sturdy and ideally have a base diameter of a least 10–15 cm and have rough bark (e.g. ironbark species) to assist in the climbing mobility of the koalas. Though additional shade should also be provided, field studies suggest trunk thickness is important because during hot weather they seek out shade against the bark of live, smooth-barked tree trunks, as they are cooler than the surrounding ambient air temperature (Briscoe *et al.* 2014).

Cross branches can also be supplied to link each of the tree trunks to assist the females to escape from male aggressive behaviour. These may not be required if the trees are placed reasonably closely together with limbs from adjacent trees that come within 1–2 m of each other, as the koalas will be able to jump from one tree to another. If the koalas are unable to jump from one tree to another, they will readily descend to move from tree to tree. Ideally a garden, and other trees, should also surround the enclosure to provide additional shade from the sun, shelter from the wind and rain. Shade trees should be watered regularly and may need the protection of metal guards to prevent the koalas climbing them. Depending on the size and position of the shade trees, it may be adequate to let the koalas climb in them, particularly as this may voluntarily bring them closer to the public. Care should be taken that the trees surrounding the enclosure do not allow the animals to escape because they will make use of any overhanging limbs, and even fern fronds if given the opportunity. As a general rule, maintaining a 1.8-m gap should stop any koalas from escaping.

Generally, koala branches are long lasting, but will need to be replaced when the surfaces of the trunk begin to wear smooth, which is usually every 12 months. The ideal species for tree trunks are ironbarks as they typically keep their outer bark longer, whereas other species such as stringy bark wear more quickly and become smooth. Ideally, enclosure tree trunks should be at least 6 m high and at least 40–60 cm in diameter.

# 5. GENERAL HUSBANDRY

## 5.1 Hygiene and cleaning

All enclosures should be cleaned daily to remove faecal matter and uneaten food. Soil substrates should be raked daily and concrete substrates should be hosed daily to remove all faecal matter. If the koalas are in a grassed enclosure that is large enough to allow natural decomposition of the faeces, then manual removal of them may not be required. All faecal material should be removed from the tree trunks on a scheduled basis. All leaf pots should be emptied and refilled daily to keep the water fresh. Drinking water dishes should be cleaned and refilled daily. When a koala permanently leaves an enclosure with a concrete floor, the floor should be thoroughly disinfected and scrubbed in preparation for the next arrival.

## 5.2 Record keeping

It is important to establish a system whereby the health, condition and reproductive status of each captive koala are routinely monitored and recorded:

- identification (all individuals should be identifiable)
- results of veterinary examinations and treatments
- changes in defecation and urination
- behavioural changes or problems
- reproductive (oestrus) behaviour and condition (cycling, lactation, sexually immature)
- bodyweights and measurements
- body condition score
- changes in diet and food consumption
- movements of individuals between enclosures or institutions
- births, with dam and sire if known
- deaths with necropsy results.

The collection of information on physical, physiological, and behavioural patterns of individual koalas can contribute greatly to the husbandry of this species. It also allows individual histories to be transferred to other institutions if required and facilitates a cooperative approach to data collection amongst institutions. ZIMS, the Zoo Information Management System (Species360 2023) can be used for recording general information on age, sex, parentage, births, identification, changes in bodyweights, transfers and deaths, with separate ZIMS modules for veterinary information and studbooks. These systems have been developed by Species360, which is part of the Conservation Planning Specialist Group (CPSG) software. ZIMS interfaces with programs, such as PMx, which provides demographic and genetic analysis tools to assist with the management of breeding programs (Lacy *et al.* 2011). Genetic management of captive koala populations is an increasing issue, so an understanding of its genetic diversity will assist its captive management and conservation (Johnson *et al.* 2018). Therefore captive institutions are encouraged to actively manage their captive koala populations, not just based on a studbook, but on molecular genetics.

When new facilities seek to exhibit koalas, they may be required to maintain additional records on diet to provide an index of appetite and feed preference. This precaution may be necessary because of local and seasonal differences in digestibility and palatability of leaves. New exhibitors (especially those outside Australia) may be required to demonstrate access to adequate fresh supplies of leaves from at least three species of koala food trees in their local region that are considered most favoured.

## 5.3 Methods of individual identification

Each animal should be individually identified, with a separate record card for each animal.

### 5.3.1 Passive Integrated Transponder (PIT) tags

PIT tags are implanted between the scapulae of the individual koalas and are an excellent, permanent method of identification. To prevent the tags tracking out along the injection site, the entry point can be sealed with a small drop of tissue glue (e.g. Vetbond®) or similar fast-setting adhesive, but is generally not necessary, and if used in excess can leave a messy plug of glue on the fur. When the koala scratches the glue-matted fur, it can re-open the insertion point (A. Gillett pers. obs). PIT tags generally require the animal to be caught to confirm identification with a PIT tag reader. Further details can be found in Vogelnest (1998).

### 5.3.2 Ear tattoos

Tattoos work well in koalas and are best placed below the centre of the ear because the edge of the ear becomes pigmented with age.

### 5.3.3 Visual identification

Generally, koalas have quite distinctive faces, which can be readily distinguished with practice. Potentially, a photo could be taken of their faces and kept as a record. Research is currently being conducted on artificial intelligence for facial recognition. Nonetheless, a form of permanent marking is still recommended. Unique rump markings can also be used for identification of animals. Nose markings are generally unique and widely used. Koalas also have unique fingerprints.

### 5.3.4 Ear tags

Ear tags are no longer routinely used with captive koalas as they have been replaced by PIT tags. However, they are used on wild animals for ease of identification post release for both research and from rehabilitation (Lunney *et al.* 2022b). More recently, ear tags have been fitted with radio-transmitters for tracking and observation post release (A. Gillett pers. obs.). Large, coloured ear tags such as sheep tags (e.g. Leader or MS Schippers Tags) are effective in determining individuals without catching them; however, these can be hidden by the fur on the ears, particularly in the Victorian koalas (this can be overcome by trimming the fur of the ears).

Ear tag numbers make it easy to report recapture, research observations and the death or illness of rehabilitated koalas. When applying the tags, care needs to be taken to avoid veins when punching through the ear. Very small (~8 mm/2 mm in size) metal tags are also used by some research groups as a backup to the larger coloured plastic tags (A. Gillett pers. obs.). These tags are uniquely numbered and are inserted into the opposite ear to the larger plastic tags. Note that large metal tags can be clamped down onto ears during fights, causing subsequent pressure necrosis and tag loss.

# 6. FEEDING REQUIREMENTS

## 6.1 Captive diet

- Ad lib: branches of eucalypt leaves

Usually, two large or three smaller branches of at least three different species are provided per koala per day. Fresh branches of browse should always be available. At least one leaf pot should be provided for each individual koala as this will allow plenty of room for them to move and reduce the incidence of aggressive encounters, particularly during the breeding season. Note that during peak lactation, female koalas increase their food intake by ~35% to meet the nutritional demands of reproduction (Krockenberger and Hume 2007).

- Ad lib: water

It has become increasingly evident that koalas will routinely use free water in the wild if provided, even during winter, with the time spent drinking increasing with rising temperature and the time since the last rain (Mella *et al.* 2019; 2020).

### 6.1.1 Species of eucalypts preferred

To ensure that the koalas held by an institution are kept in good physical condition it is important to ensure that the diet is varied and of high quality. A koala typically eats between 400 and 800 g of fresh leaves per day or ~10% of its bodyweight (Hawkes 1978; Nagy and Martin 1985; Zoidis and Markowitz 1992; Jackson 2007). The amount of food also appears to vary seasonally, with more food being eaten in winter than summer (Moore and Foley 2000).During periods of drought, the water content can reduce greatly and has resulted in deaths of koalas in wild populations (e.g. Clifton 2010).

Koalas show a definite preference towards some eucalypt species, with some being highly preferred, others eaten occasionally and some are rarely, if ever, eaten. Of ~900 species of eucalypts (which now include three genera *Eucalyptus*, *Corymbia* and *Angophora*), koalas feed primarily upon the leaves of the subgenus *Symphyomyrtus* that includes some 400 species. Approximately 70 species have been recorded as having been eaten by koalas at some time (Table 7.2).

Note that as the number of studies of tree preferences increases, the list of species they utilise increases, especially as new locations are examined in detail. For example, none of the tree species for north-east New South Wales (McAlpine *et al.* 2023) are even present in north-west New South Wales (Crowther *et al.* 2014), and the list for Queensland has common features with New South Wales, but has its own set of characteristics (Moore *et al.* 2023). Not only does each koala's choice vary according to locality, it also varies with season (Crowther *et al.* 2014) and Crowther *et al.* (2022) found that, in the fragmented landscape of north-west New South Wales, food quality and shelter drive koala tree re-visits. Scattered, isolated trees with nutrient-rich leaves are valuable resource patches for koalas despite the movement costs to the koala to reach them.

In captivity, the staple browse species provided to koalas depends on where the captive facility is located within Australia and the availability of nearby trees and/or the species of trees that are being grown in a plantation. The most common species of eucalypts and other trees eaten by koalas throughout their distribution are listed in Table 7.2.

### 6.1.2 Choice of eucalypt branches to be cut

Part of the skill of successfully maintaining koalas involves the knowledge of which species of eucalypt is the most appropriate to feed at different times of the year, as the preference within and between species varies considerably throughout the year.

Depending on the species and time of year, branches should be chosen that possess new growth tips. Some species of eucalypts are almost wholly eaten, adult leaves included, while others have only the new growth eaten. It is important to highlight that koalas do not always prefer young tips, as Pratt (1937) found koalas will reject the young leaves of juvenile trees (especially *E. viminalis*) while another study found young foliage accounted for 5–35% of the diet and mature foliage 50–90% of the diet of four rehabilitated koalas (Nyo Tun 1993 in Moore and Foley 2000). Young koalas or unwell animals are commonly observed consuming sticks/twigs and fibrous mature leaf (S. Eccleston pers. obs.; A. Gillett per. obs.).

Branches cut for koalas are normally at least 1 m long and contain as much fresh new growth as possible. The diameter of the branches cut is generally around 2–5 cm and is best cut at an angle of 45° to facilitate maximum coppice regrowth.

The frequency of leaf collection varies from 1–2 times per week to every day. A minimum of 3–4 species of eucalypt is collected each trip and cuttings should be kept out of direct sun and wind (particularly during summer) to prevent dehydration of the tips.

### 6.1.3 Storage of leaves

The leaves should not be stored beyond a maximum of 1 week, though ideally no more than 2–3 days or less if the condition of the leaves has deteriorated. They should be stored in an enclosed shaded area with an overhead sprinkler system to keep the leaves moist, particularly in hot dry weather, as the tips can brown very quickly, making them unpalatable.

Although browse can be kept in a refrigerated unit at 4–5°C, observation of leaves held in refrigerators has shown that the tips can dry out, so the use of refrigeration is not always recommended (it is likely that some refrigerators are better than others).

The leaves should be stored in large bins filled with water (changed at least once per week) that are ~60–80 cm high with a diameter of ~60 cm. Plastic bins will not rust and are easy to clean. It is important to clean them weekly to remove the build-up of any algae and other contaminants in the water. Other containers such as troughs, with partitioning to keep the branches upright, or large plastic upright bins with wheels (wheelie bins) that are ~1000 mm high and a have an opening of 500 mm × 560 mm and a hinged lid (that is left open or removed) can also be used. Some facilities cover the stems and leaves of koala branches in the bins with thick black plastic bags to stop the tips from drying out.

### 6.1.4 Eucalypt plantations

The maintenance of captive koalas increasingly requires the establishment of plantation-grown trees as the native local eucalypt stands are often inadequate or, in the case of overseas zoos, not available, particularly as koalas require several species. In order to establish a plantation, the size of the population to be fed and the daily requirements will need to be determined. For each koala there needs to be ~500–1000 trees planted (made up of at least

**Table 7.2.** Species of eucalypts and other tree species eaten by koalas throughout their distribution in Australia.

Derived from Taronga Zoo, Lone Pine Sanctuary, Currumbin Wildlife Sanctuary, Melbourne Zoo, Healesville Sanctuary, Tidbinbilla Nature Reserve, Martin and Handasyde (1999) and Phillips and Callaghan (2000), Crowther *et al.* (2014, 2022), Moore *et al.* (2023).
***Most preferred species. Initials for the different tree genera include *C* = *Corymbia*, *E* = *Eucalyptus*, and *L* = *Lophostemon*.

| Scientific name | Common name | Queensland | New South Wales | Victoria |
|---|---|---|---|---|
| *C. citriodora* | Lemon-scented gum | *** | * | – |
| *C. eximia* | Yellow bloodwood | – | * | – |
| *C. gummifera* | Red bloodwood | – | * | – |
| *C. henryi* | Large-leaved spotted gum | * | – | – |
| *C. maculata* | Spotted gum | * | * | – |
| *C. tessellaris* | Moreton Bay ash | * | – | – |
| *E. albens* | White box | | ** | |
| *E. acaciiformis* | Wattle-leaved peppermint | – | * | – |
| *E. acmenoides* | White mahogany | – | * | – |
| *E. agglomerata* | Blue-leaved stringybark | – | * | – |
| *E. amplifolia* | Cabbage gum | * | * | – |
| *E. botryoides* | Southern mahogany | – | * | *** |
| *E. bridgesiana* | Apple box | – | * | * |
| *E. camaldulensis* | River red gum | *** | *** | *** |
| *E. camphora* | Mountain swamp gum | – | * | * |
| *E. canaliculata* | Large-fruited grey gum | – | * | – |
| *E. cephalocarpa* | Silver-leaved stringybark | – | – | * |
| *E. coolabah* | Coolabah tree | * | – | – |
| *E. creba* | Narrow-leaved red ironbark | * | * | – |
| *E. cypellocarpa* | Mountain grey gum | – | * | * |
| *E. dealbata* | Tumbledown red gum | | *** | |
| *E. drepanophylla* | Grey ironbark | * | – | – |
| *E. dunnii* | Dunn's white gum | * | – | – |
| *E. exserta* | Queensland peppermint | * | – | – |
| *E. eugenoides* | Thin-leaved stringybark | – | * | – |
| *E. fastigata* | Brown barrel | – | * | – |
| *E. globulus* | Tasmanian blue gum | – | *** | *** |
| *E. globoidea* | White stringybark | – | * | |
| *E. goniocalyx* | Long-leaved box | – | *** | *** |
| *E. grandis* | Flooded gum | * | * | * |
| *E. haemostoma* | Scribbly gum | – | * | – |
| *E. leucoxylon* | Pink flowering gum | – | – | *** |
| *E. macrorhyncha* | Red stringybark | – | – | * |
| *E. maidenii* | Maiden's gum | – | * | – |
| *E. major* | Brittle or red spotted gum | * | – | – |
| *E. mannifera* | Mottled gum | – | * | – |
| *E. melliodora* | Yellow box | * | ** | * |
| *E. microcorys* | Tallowwood | *** | *** | – |
| *E. moluccana* | Grey box | * | * | – |
| *E. nicholii* | Narrow-leaved peppermint | * | * | * |
| *E. obliqua* | Messmate | – | *** | *** |

**Table 7.2.** continued

| Scientific name | Common name | Queensland | New South Wales | Victoria |
|---|---|---|---|---|
| *E. oblonga* | Narrow-leaved stringybark | – | * | – |
| *E. ochrophloia* | Yapunyah | * | – | – |
| *E. orgadophila* | Mountain coolibah | * | – | – |
| *E. ovata* | Swamp gum | – | *** | *** |
| *E. paniculata* | Grey ironbark | – | * | – |
| *E. parramattensis* | Parramatta or drooping red gum | – | * | – |
| *E. pellita* | Large-fruited red mahogany | * | – | – |
| *E. pilularis* | Blackbutt | * | *** | – |
| *E. piperita* | Sydney peppermint | – | * | – |
| *E. polyanthemos* | Silver dollar gum or red box | – | – | * |
| *E. populnea* | Poplar box or bimble box | ** | ***– | – |
| *E. propinqua* | Small-fruited grey gum | *** | *** | – |
| *E. punctata* | Large-fruited grey gum | *** | *** | *** |
| *E. racemosa* | Southern scribbly gum | – | * | – |
| *E. radiata* | Narrow-leaved peppermint | – | *** | *** |
| *E. regnans* | Mountain ash | – | – | * |
| *E. resinifera* | Red mahogany | * | * | – |
| *E. robusta* | Swamp mahogany | *** | *** | * |
| *E. rossii* | Scribbly gum | – | * | – |
| *E. rubida* | Candle bark, ribbon or white gum | – | – | * |
| *E. saligna* | Sydney blue gum | * | * | * |
| *E. scoparia* | Wallangarra white gum | – | * | – |
| *E. seeana* | Narrow-leaved grey gum | * | – | – |
| *E. sideroxylon* | Red ironbark | * | * | – |
| *E. signata* | Scribbly gum | * | – | – |
| *E. tereticornis* | Forest red gum | *** | *** | *** |
| *E. umbra* | Broad-leaved white mahogany | * | – | – |
| *E. viminalis* | Manna gum | - | *** | *** |
| *L. confertus* | Brush box | * | – | – |
| *L. suaveolens* | Swamp box | *** | – | – |

5-6 preferred species) ~4–6 years ahead of acquiring koalas. In Queensland, ~500 trees per koala are required with a planting and harvesting turn over period of 2-3 years. A brief outline is covered below for the establishment of a plantation. A more detailed description can be found in Knight (2004) and Addendum 1. Beale *et al.* (2022a) have produced a valuable set of guidelines for revegetating koala habitat that incorporates current knowledge of their requirements and nutritional ecology from peer-reviewed literature. While the authors state that it is not intended to provide instruction for growing or planting vegetation per se, it is relevant and does supply further reading and resources.

When first considering a plantation, various factors need to be considered and include location, tenure, accessibility, construction of capital improvements, harvesting regimen and other factors regarding the future management of the plantation. Once planted the trees should typically be left alone until they reach at least 4–5 m high and have a closed canopy. This occurs at different years of age depending on the species and the site where they have been planted. At this stage, coppicing a proportion of the stems, ~20% of the stand, allows continuing growth on the crowns of the more vigorous browse trees and provide for manipulation of the stocking of browse trees. When collecting branches, it is important for the long-term management of the plantation not to over crop or coppice individual trees. If overcutting is occurring, more trees should be planted. Eucalypt plantations are usually established for forestry purposes with 1 tree per 4 m square, giving an initial stocking of 625 trees per hectare (up to a maximum of 1000 trees per hectare). After treatment the stocking rate can be reduced to 500 plants per hectare. Ingrowth from the coppiced stumps will replace those stems removed with the effect of introducing a second age class of leaf growth into the plantation (Hawkes 1978). If all the coppices are removed from a stump, foliage can generally be harvested from *E. viminalis* and *E. goniocalyx* at intervals of 12–14 months (Drake *et al.* 1991). Some species such as *E. ovata* can be harvested every 6–8 months because the koalas do not reject juvenile leaves and regrowth is rapid (Drake *et al.* 1991).

### 6.1.5 Artificial diets

The use of an artificial diet was tested with koalas many years ago in the form of a thin, flexible biscuit and a thick paste (Pahl and Hume 1990). The moisture, nitrogen and fibre contents of the biscuits are similar to those observed in leaves preferred by koalas. Using dry bodyweights, the biscuit cell wall content was 24%, cellulose content 16%, lignin content 3%, ash content 6%, nitrogen content of 1.9% and a moisture content of 62%. The biscuit form of the artificial diet was 2 (D) × 15 (W) × 60 (L) mm in size (Pahl and Hume 1990). The thick paste consisted of 'Presbo' powder, a constant amount of ground eucalyptus foliage and water. The biscuits were dipped into the paste by hand before presentation to the koalas, but the paste was also administered orally with a syringe (Pahl and Hume 1990). Although the use of artificial diets in conjunction with fresh foliage is able to sustain the bodyweight of koalas for a limited period of time, it is very time consuming and koalas appear to become less willing to eat artificial diets over time. Therefore, it is not recommended for long-term maintenance of koalas and is not currently used by any institution.

### 6.1.6 Supplementary milk diets for aged and sick koalas

As koalas age, their teeth become increasingly worn so the cutting edge decreases and becomes less efficient in grinding food. This results in larger particles to digest and less efficient digestion, thus significant increases occur in the amount of time spent feeding and the number of leaves chewed, but ultimately ill-health occurs due to malnutrition. The larger particles can also lead to gut impaction due to poor absorption (Lanyon and Sanson 1986; Gamble and Blyde 1992; Logan and Sanson 2002). Older koalas with advanced tooth wear may show fibrous strands in their faeces that link the pellets together. These coarse pieces of fibre can cause ulceration of the small intestine, ileum or colon, which can ultimately lead to sepsis and death (R. Booth pers. obs.).

Supplementary feeding with high-energy milk powders can greatly help to maintain sick or old koalas that have unweaned young and are losing weight, as it provides additional nutrition to compensate for the increased energetic demands during lactation or the poor digestion of food due to teeth wear (Osawa and Carrick 1990).

Several milk powders have been used successfully to supplement the nutrition of koalas (Blanshard and Bodley 2008). However, there is a trend away from soy-based supplements and more recently Crittacare Koala (Vetafarm) has been the supplement of choice by several institutions (K. Nilsson pers. obs.).

Other supplements that have been used include (Blanshard and Bodley 2008):

- Di-Vetelact® (Lillelund Pty Ltd) - low-lactose milk supplement.
- Karicare+ Soy Milk Formula (Nutricia)
- Portagen® (Mead Johnson Nutrition)
- ProSobee® (Enfamil)
- EmerAid Intensive Care Herbivore®.

Koalas that weigh 5–8 kg and are eating some leaf can be provided with ~60 mL of paste per day that is divided into 2–3 feeds throughout the day to maintain body condition. Finely chopped eucalyptus leaves can be added to the paste mixture of animals not eating leaf to help maintain the gut bacteria. Koalas that are eating some leaves can be given 20–40 mL per day to maintain health or recover from illness (Blanshard and Bodley 2008). Feeding paste on its own is not recommended for koalas that are not eating any leaf; blended leaf alone is preferred (Gillett and Hanger 2025). Nutritional supplementation for koalas under intensive care can be found in Table 7.3.

**Table 7.3.** Nutritional supplementation for koalas under intensive care.

Derived from Gillett and Hanger (2025).
*Yes if still eating some leaf; #not required if feeding paste combined with blended leaf.

| Condition | Paste formula | Blended leaf (no paste) | Paste combined with blended leaf | Leaf provisions and feeding requirements | Transfaunation |
|---|---|---|---|---|---|
| Inappetence | No | Yes | No | Wide variety of mature and soft-tip eucalypt species (mixture of varieties once or twice daily). Non-eucalypt species. Hand-feed twice daily | No |
| Ileus or caecocolic dysbiosis/typhlocolitis syndrome (CDTS) | No | Yes | No* | Wide variety of mature and soft-tip eucalypt species (mixture of varieties offered daily). Non-eucalypt species. Hand-feed twice daily | Yes – if dysbiosis involved, but avoid if complete gastrointestinal stasis |
| Eating small amounts of leaf | No | Yes# | Yes | Identify preferred leaf. Non-eucalypt species. Hand-feed twice daily | Yes – if dysbiosis involved |
| Eating approximately 50% of daily leaf requirement | Yes | Not required | Yes | Identify preferred leaf, slowly introduce other species. Non-eucalypt species | Yes – if dysbiosis involved |
| Eating normal leaf quantity, but underweight or with healing fractures | Yes | Not required | Not required | Wide variety of eucalypt species. Occasionally offer non-eucalypt species | Not required |

Some koalas appear to learn to like the taste of the milk formula and may actively seek the top of the catheter-tipped syringe into their mouths, but other koalas do not accept the syringe so readily but with persistence will willingly swallow most of the paste. After the paste has been given, it is common to provide ~10 mL of water in the syringe to cleanse any sticky residue in the koala's mouth. Any paste that goes on the face or fur needs to be removed and cleaned up after feeding (Blanshard and Bodley 2008).

## 6.2 Supplements

There are several reports of koalas eating soil, gravel and termite mound. These observations suggest they may assist in digestion by providing trace minerals or digestive enzymes in the case of termite mounds, associated with secretions by the termites (Troughton 1941; Smith 1979a; Melzer *et al.* 2014; Wong *et al.* 2014). These observations are supported by necropsies of various koalas in one study that found varying amounts of pebbles and gravel, totalling up to 7 g (Bolliger 1962), in the caecum of all animals studied (Bolliger 1962). Therefore, soil or termite mound should be provided to all koalas (Gillett and Hanger 2025).

Providing leaves of non-eucalypt species such as *Lophostemon*, *Melaleuca* and *Casuarina* sp. is important for koalas that have received antibiotic therapy or are suffering from any gastrointestinal tract disturbances (Gillett and Hanger 2025).

## 6.3 Presentation of food

A surplus of food should always be provided. Generally, at least three branches (though sometimes two larger branches of favourite species can be used) of at least two species should be provided to each koala per day, in long thin pots (~10–15 cm in diameter and 60–70 cm long) that are filled with water; however, sometimes only one species can be fed depending on availability and preference. The provision of different species of eucalypt leaves allows the koalas to always have at least two species that they will eat. All branches should be free of foreign material such as dirt, insects and bird droppings (hose off the branches before offering). Captive koalas cannot leave their setting, unlike their wild counterparts, so cannot move away from a contaminated branch. The browse should be fed as fresh as possible, with no obvious signs of wilting. The branches should be positioned so that all tips are within easy access for the koalas.

The leaf pots should be connected to the tree trunk just below the tree forks so the koala can sit in the tree fork and feel secure while feeding. The leaf pots should be rinsed and refilled daily with water before the new leaves being added. The leaf pots should be placed in the shade to minimise desiccation of the leaves. There should be at least one leaf pot (preferably 2–3) per individual and they are spread out among the various branch forks in the enclosure to reduce competition over food.

Usually, the leaf pots are be made of plastic or stainless steel. Plastic pots are lighter in weight, but do not last as long, as the bottoms often fall out if they are dropped. Metal containers are more durable, but they are heavier to handle. Powder-coated aluminium pots and brackets are suitable because they are lighter in weight, have less heat absorption properties, are easier to clean and aesthetically more appealing for exhibit enclosures. The pots should be attached to the trunk by drilling a hole at the top of the leaf pot and attaching it to a bolt that has been screwed into the tree trunk. A second method used to attach the pots to the trunk consists of an arm with an elbow, which is connected to the tree with the use of a sleeve (Fig. 7.3). The height of the leaf pot is usually where the top of the pot is ~1200 mm off the ground.

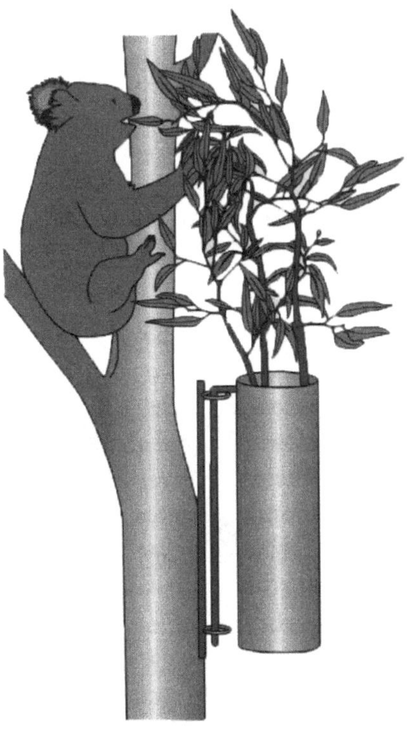

**Fig. 7.3.** A koala leaf pot and its attachment to the tree trunk. Not drawn to scale.

The uneaten leaves in each enclosure should be changed daily, preferably in the afternoon to stop the leaves from drying out during the day. This is particularly important in hot weather, because the koalas generally will not eat the leaves until it is dark. Some institutions change the leaves in the morning and in the afternoon. In this case, the morning feed is minimal due to the koalas' general inactivity during the day.

Consumption of the leaves provided to the koalas should be monitored to check if they are also being eaten by wild common brush-tailed possums, which may compete with the koalas particularly during winter.

## 7. HANDLING AND TRANSPORT

### 7.1 Timing of capture and handling

The best time to capture koalas for examination is in the morning when the temperature is cooler, especially important during warmer weather. In enclosures with tall trees, it may be more convenient to capture the animals when they are fed in the afternoon as they will often come down from the higher branches to feed. However, koalas are unlikely to come down to feed in the heat.

### 7.2 Capture equipment

Catching bags should be made of thick cotton or other strong material. Although hessian bags have often been used, there have been reports of the koala's eyes being irritated when these are used for capture. The bag opening should be wide, with a diameter of ~45–60 cm and the depth should be ~60–90 cm.

Specific equipment used to capture koalas includes long poles and traps. The configuration of these is discussed further in section 7.3.

### 7.3 Capture and physical restraint

#### 7.3.1 Capture

Koalas can be captured using several techniques depending on how easily they can be accessed.

- 'Flagging' technique – a plastic bag, fabric-covered halo or loop, rag or hessian sack is attached to the end of an extendable pole (Fig. 7.4). When the flag, halo, sack, rag or bag is waved just above the head of the koala it will typically begin to descend the tree. The bag is kept slightly above the head of the koala as it descends. This can be performed by a ground team if koalas can be reached from the ground or when koalas are high up, trained tree-climbing personnel can 'flag' the koala from within the tree. When within reach, unconditioned animals can generally be coaxed down further by placing a hand firmly on its head and pushing it gradually down the branch and into a catching bag. Koalas can also be removed from a tree by putting the catching bag over the animal's head (which helps to calm the koala) and then push the edge of the bag over the back towards the rump and finally unhook the feet (Fig. 7.5). A broom can also be used if koalas are sitting low enough to reach and specialised catching equipment is not available.

- 'Halo' technique – a fabric-covered halo or loop attached to the end of an extendable pole (Fig. 7.4) can be held above the koala's head to encourage it to descend the tree in a passive manner or can be placed on top of the koala's head and downward pressure applied to forcibly/actively encourage the koala to descend. The Halo technique can be performed by a ground team or from within the tree by a trained tree-climber as with the flagging technique.
- Noosing technique – using a long pole with a loop with a stop knot to prevent overtightening has been widely used to capture koalas. The loop is tightened, then stops at the stop knot to make it a non-tightening loop and the koala is gently restrained from climbing up the tree and encouraged to descend the tree by waving a flag above the koala (i.e. it is flagged down the tree) (Figs 7.6, 7.7). However, this technique carries inherent risk of injury and death to the koala and should only ever be used by experienced teams under strict safety guidelines to prevent koala injury (Hanger and Gillett 2025).
- Using a koala trap -placing a circular pool fence or corflute ~2 m away from the tree containing the koala that is to be caught (Fig. 7.8). A smooth collar of tin may be placed around the tree ~1.5 m above the ground (not always required), so that when the koala climbs down it cannot climb back up. A collapsible treadle trap component should be placed on the outer edge of the fence that is ~45 cm (L) × 45 cm (W) × 120 cm (H) (Fox Trap, Wiretainers) (Hasegawa and Carrick 1995; Phillips 2011; Hanger and Gillett 2025; Ashman and Whisson 2020). This method may only be needed rarely for animals that have escaped from their enclosure.

**Fig. 7.4.** Halo technique used to catch koalas. Photo by Amber Gillett.

**Fig. 7.5.** Restraint of a koala from a tree using a catching bag.

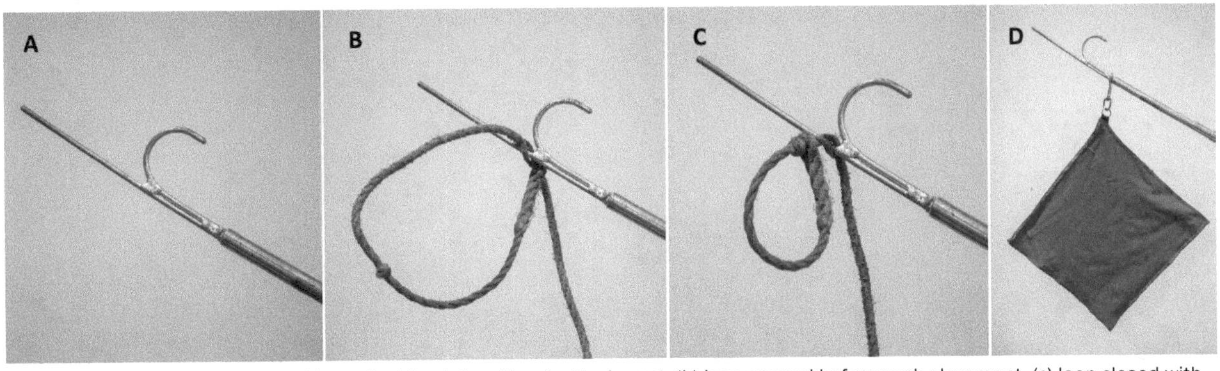


**Fig. 7.6.** Components of the catching pole: (a) point and hook attachment, (b) loop opened before neck placement, (c) loop closed with stopper knot in place to prevent tightening and asphyxiation, (d) flag attachment. Photographs by George Madani. Taken from Madani *et al.* 2020 with permission of the photographer.

**Fig. 7.7.** Example of a koala catch with the loop and flag method: (a) placing the noose over the koala's head, (b) positioning the noose around the koala's neck, (c) flagging the koala down the tree, (d) placing the koala into a handling bag. Reproduced from Madani *et al.* 2020 with permission from CSIRO Publishing. Photographs by George Madani.

**Fig. 7.8.** Koala trap showing the external fence and a trap placed on the outer edge. Taken from Phillips (2011) with permission from Stephen Phillips.

## 7.3.2 Handling

Although koalas appear to be docile and cute, they can be both agile and aggressive if disturbed. Their teeth and claws are very strong and sharp and when handled, they will tend to grab at anything within their reach. Care needs to be taken to avoid being scratched or bitten. One way of carrying a koala is to hold both arms and one leg of the koala (Fig. 7.9a). Another way to restrain a koala is to use a blanket from behind the animal, grab one arm and the opposite rear leg, supporting the rear end (Fig. 7.9b). The animal is then placed within a large 'catch bags' made of a soft but strong material. In this way it may be safely carried facing away, at arm's length, or effectively restrained, reducing the possibility of injury by the koala.

Non-compliant (i.e. semi-wild/wild) koalas may be restrained for examination by placing them in a closely woven sack or bag (such as made of cotton or nylon). The koala's head can be exteriorised for examination or procedures such as ear tagging, fitting tracking collars and morphometric head measurements. Examining the rest of the body usually requires exteriorising part or all of the koala. The head can be covered and only the parts required for examination brought out at any one time; for example, ear for ear tag identification, leg or arm for examination. It may be necessary to use more than one person for restraint. When done this way the claws are kept prevented from scratching. Be careful while the animal is in the sack because if given the chance, they will bite hard through the sack.

Pouch young can be examined by holding the mother on the ground or on a bench while in a sack, exposing the

**Fig. 7.9.** Restraint techniques for a koala. Larger individuals can be held by (a) holding both arms and one leg or (b) restraint using a towel/blanket. Note that all four legs should normally be held as a free leg can injure the handler. Photos courtesy of Australia Zoo.

lower half of the mother for examination (i.e. the head and upper arms are held firmly in the sack). While one person is keeping the forearms under control, a second is holding the legs down, allowing a third person to open the pouch to check for pouch young. If required, measurements can be taken of the pouch young to estimate its age and to chart its growth. Back young should be removed from the mother, measured and then returned.

If an animal has been hand-reared and is accustomed to handling they can generally be carried on the body with the arms gripping clothing and the rump supported so that the koala does not need to hold on as strongly. Animals that are habituated to humans can be easily held for procedures including venepuncture and artificial insemination (Fig. 7.10). It is an often advisable to wear a heavy-duty clothing with long sleeves when carrying a koala in this way as the claws will easily go through a thin shirt, particularly if it becomes frightened. Young animals can be easily held by giving them a large stuffed toy or 'teddy' to hold onto.

## 7.4 Weighing and examination

Weighing should typically occur at the same time each day to account for gut fill. During an examination, the animal should be weighed by placing them in a sack or bag and weighing them with either a set of hanging spring scales (usually 5 kg, 10 kg or 20 kg; Fig. 7.11) or preferably on electronic scales as they are more accurate. If using spring scales use the same set every time to reduce the differences between scales. Koalas can be weighed using various methods including placing them in a bag and hanging them from scales; placing them in a pet carrier (particularly for quiet animals); holding them and stepping on the scales together, then deducting person's bodyweight; and placing a koala on a branch positioned on larger scales and placing the koala on the branch. Some institutions have also had success with the koala sitting on a flatbed scale long enough for the bodyweight to be recorded. If housing multiple koalas, it is advised to weigh all on the same day to account for differences in leaf species each day, weather events and time of the day, which might affect gut fill.

**Fig. 7.11.** Method used to weigh koala using hanging scales and a postal bag. Photo by Dan Lunney.

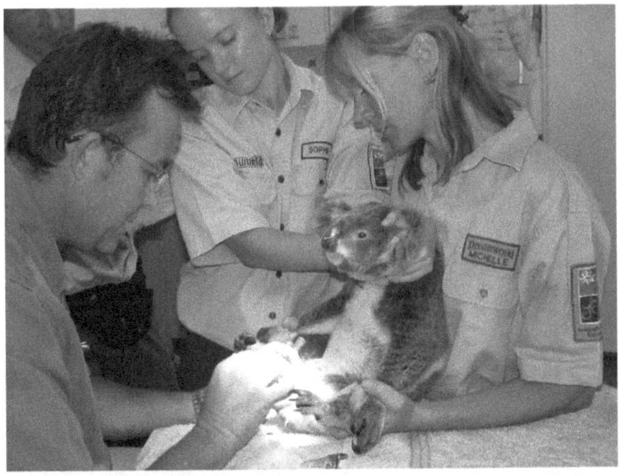

**Fig. 7.10.** Holding position (with all four limbs secured) for female koala being prepared for artificial insemination. Photo by Stephen Johnston.

## 7.5 Release

Release of the koala from the cotton or nylon bag or sack is ideally done with the opening facing the base of a tree trunk that is 1–3 m away. When the koala leaves the sack it should run straight to the tree, although it may choose another trunk. Animals being held are best released by standing next to the tree for them to go to, ideally at a fork, and moving the koala's arm closest to the tree on to the tree. Move the other arm across to the tree while at the same time lifting the koala's bottom over to the tree.

Often the koala will begin moving itself onto the branch by leaning over to the branch and reaching out either before or after the first arm has been moved across. If this occurs, keep supporting the animal's rump and carefully lift the koala over to the branch or fork.

## 7.6 Transport requirements

The conditions for the transport of koalas have been formulated to maximise the welfare of koalas involved in overseas transactions. The conditions set by Environment Australia provide the framework for the transport of koalas. A full list of the conditions for the overseas transfer of koalas can be obtained from the Australia Government's *Conditions for the Overseas Transfer of Koalas* (Anon 2009).

### 7.6.1 Transport box design

Each koala should be transported internationally in a solid framed box with inside measurements of 1000 mm (L) × 820 mm (W) × 1040 mm (H). The box should have a removable, leakproof metal drop tray fitted to the base. The side walls and top must be stout wire mesh and fitted with light hessian or shade cloth covers. Further specific details of the box design can be found in IATA (2022). More recently, for short travels, there are purpose-built transport containers, which have a plastic base and are wire-topped, allowing the top to be opened, which helps overcome a lot of the problems associated with getting a koala in and out of a side-opening transport container (Fig. 7.11).

### 7.6.2 Furnishings

At least 1-2 fork branches are required so the koala is able to sit during its transportation. These branches need to be securely fixed to the box to prevent them becoming dislodged during handling and shipment. Local transport, within 1-3 h, does not require furnishings.

### 7.6.3 Water and food

Although koalas generally do not drink, a stable dish of water should be placed in the box. Three or four shortened branches of tips should also be placed in a modified shortened leaf pot that is filled with water. Depending on the length of the flight, the leaves may need to be changed at least once during the flight. Water dishes are not usually required for short journeys (1-3 h).

### 7.6.4 Animals per transport box

One koala should be transported per box. Females with pouch young should generally not be transported.

### 7.6.5 Timing of transportation

Overnight is preferable as it is generally cooler.

### 7.6.6 Release from the transport box

When releasing the koala from the transport box, the box should be placed in the enclosure next to the base of a tree (Fig. 7.12). The door to the box should be completely opened or removed and the animal allowed to leave the box in its own time, thereby giving the koala the opportunity to explore its surroundings and climb a tree at its leisure. When the animal is up a tree the box can be removed from the enclosure. On arrival after transport, there should be minimal noise and disturbance that might cause distress to the koala.

# 8. HEALTH REQUIREMENTS

## 8.1 Daily health observations

Each koala should be observed daily for any signs of injury or illness (Gillett and Hanger 2025). The most appropriate time to do this is generally when the enclosure is being cleaned or when the branches are replaced,

**Fig. 7.12.** Purpose-built koala transport container for short-term transportation. Photo by Rosie Booth.

which is often when the koalas are more active. During these times, each animal within the enclosure should be checked for the following:

- coat condition
- body condition score - see section 8.2
- discharges from the eyes, ears, nose, mouth or cloaca (diarrhoea)
- appetite
- faeces -number of pellets, size, colour and consistency
- reduced skin elasticity, suggestive of dehydration
- dirt around the mouth, suggestive of dirt eating
- changes in demeanour ( e.g. evidence of agitation when koalas are mostly inactive)
- climbing ability using all four limbs
- wetness of the cloaca and rump - evidence of cystitis (dripping bladder), stained fur around cloaca
- physical injuries
- presence and development of pouch young by observation of the bulge in the pouch.

During the summer months, koalas should be checked for signs of heat stress, which include panting, laying along the branch, sweating from the pads on the hands and feet, lethargy (A. Gillett pers. obs.) and the presence of loose, very dry skin on the nose (Drake 1982). It may also include sitting on the floor at the base of the tree or even lying face down on the floor (K. Nilsson per. obs.). Heat stress tends to occur at temperatures of ~35°C or higher, and at high humidity (> 75%), so they need to be checked regularly during high temperature events for signs of stress. The monitoring of heat stress (temperature and humidity) is particularly important for koalas held in the north of their range or in subtropical or tropical environments outside the tolerance of their thermal niche; in these circumstances the koala is likely to require specialised housing and a period of acclimation when first introduced. Heat stress can be reduced in hot weather by placing a sprinkler where it will spray the trees of approximately one-third of the enclosure (Drake 1982) and be positioned so that koalas can move in or out of the spray zone. Care needs to be taken with sprinklers in humid environments that they do not cause the humidity to increase or lead to fungal lesions on the skin (K. Nilsson pers. obs.).

The faeces can be checked for the numbers of pellets dropped (normally between 75 and 150), particularly in the case of sick animals and animals that are solitary. The consistency of the faeces should also be noted for diarrhoea or soft faeces. Although the presence of runny faeces can indicate potential problems, very high-quality leaves and leaves of particular species can also cause it, particularly after a period when the leaf quality has not been optimal. Small dry faecal pellets are commonly indicative of dehydration and often associated with candida infection in the gastrointestinal tract disease (A. Gillett pers. obs.).

It is important that the same animal caretakers regularly inspect and weigh the koalas as they are more able to determine the subtle changes in the health of individuals, including behavioural changes, which may indicate a health-related problem.

The abdomen can be palpated in koalas to determine stomach fill and caecal tone. The stomach is hard when it is full with leaf and is palpable in the upper left quadrant of the abdomen, under or below the rib cage. The remainder of the abdomen can be assessed for tone and should be soft but spring back when gently compressed from each side of the koala. If the abdominal tone is hard it may indicate a build-up of gas. To assess if a koala is dehydrated, the skin around the scruff of the neck (over the neck and scapular area) can be pinched and gently pulled up. In dehydrated koalas the skin will remain pinched up for > 3 s instead of returning to its normal position (A. Gillett pers. obs.). The pouch should be examined to check for any pouch young and the body checked over for any wounds. Although the nails of koalas are generally quite long, they do not require trimming as they must be sharp to climb trees efficiently. However, older animals can develop arthritic changes in joints that can result in a change in how they grip the branches and this can lead to overgrowth of claws, which should be monitored to prevent the claws curling into the koala's pads. The claws will need to be clipped/reshaped in such cases (K. Nilsson pers. obs.).

## 8.2 Bodyweight and body condition score

### 8.2.1 Bodyweight

Record and compare with previous measurements. Trends in the bodyweight of koalas give a good general indication of the animal's state of health, provided age, sex and geographical location are taken into account. Animals in captivity should ideally be weighed fortnightly (at the most monthly), due to the speed with which they can deteriorate and to indicate trends in bodyweight over time. This may range from monthly for healthy animals to several times a week for sick or injured animals. Fluctuations of up to 200–400 g may result from variable gut fill. However, if a consistent decline in bodyweight occurs, a veterinarian should be consulted and supplementary feeding may be required (e.g. daily doses of plant (Prosobee®) or milk protein-based formulas (Portagen®)) (Handasyde *et al.* 1988).

### 8.2.2 Body condition score

The body condition score is best assessed by palpation of the muscle on either side of the scapula spine. A standardised condition score of 1–5 (or sometimes 1–10) based on muscle quality in this region is used (Table 7.4; Carrick *et al.* 1990).

## 8.3 Physical examination

The physical examination may include the following:

- temperature – normally 34.2–36.9°C (Connolly 1999; Adam *et al.* 2018, 2020); can be measured in the cloaca using a rectal thermometer
- pulse rate – normally 70–100 beats/min in adults at rest (Connolly 1999; A. Gillett pers. obs.) measured over the femoral artery.
- respiratory rate – normally 10–15 breaths/min at rest (Connolly 1999). Panting (rapid, shallow breathing with the mouth closed) is normal in stressed or excited animals
- fur – check for alopecia, ectoparasites, fungal infections or trauma
- eyes
  - should be clear, bright and alert
  - normal bilateral pupillary light response (koalas that are stressed do not exhibit this light response)
  - normal corneal reflex
  - should not have any discharges
- also check for the presence of lumps over body and auscultate the lungs
- cloaca
  - should be clean.
  - check for urine staining and faecal contamination around the cloaca
- pouch
  - condition of the pouch.
  - check whether lactation is occurring by milking teats.
  - if pouch young are present, record the sex, stage of development, bodyweight if detached from the teat and measure to determine age from growth curves if available
- males
  - check testes: size (length, width, depth) and consistency (firm – not squishy), evidence of cryptorchidism
  - check fullness of the cauda epididymis
  - extrude penis and assess for evidence of trauma
  - check the size and activity of the sternal gland (see Allen *et al.* 2010).

## 8.4 Chemical restraint

Pre-anaesthetic fasting for at least 30 min is desirable to reduce the risk of ground leaf being in the mouth or pharynx. Regurgitation is rare in koalas. Alfaxalone (Alfaxan®) is the injectable induction agent of choice (3–5 mg/kg intramuscular) followed by isoflurane/oxygen via mask and subsequently an endotracheal tube. Recovery times with Alfaxan® are faster than with Zoletil® 100 (tiletamine hydrochloride and zolazepam hydrochloride). Other agents can also be used and have been described by Vogelnest (1999, 2019). Very quiet koalas or small koalas can be masked down with isoflurane and oxygen without prior sedation.

**Table 7.4.** Condition index used for koalas.

From A. Reiss. pers. comm.

| Score | Rating | Definition |
|---|---|---|
| 5 | Excellent | Strong muscle tone. Obviously convex muscle weights on either side of the scapula. Scapula spine detectable on careful palpation |
| 4 | Good | Good muscle tone. Slightly convex muscles on either side of the scapula. Scapula spine easily palpable |
| 3 | Fair | Flat to slightly convex muscles on either side of the scapula. Scapula spine prominent on palpation |
| 2 | Poor | Slight dishing or concave muscles on either side of scapula. Scapula spine very obvious on palpation. Edges of scapula palpable |
| 1 | Emaciated | Noticeable dishing of muscles on either side of scapula. In very emaciated animals there may be almost no muscles palpable on either side of the scapula spine. In these cases, the entire scapula will be palpable through the skin |

The cephalic vein is usually used for blood collection, catheter placement or intravenous injections and the thigh muscles are used for intramuscular injections (Vogelnest 1999). It is also important that koalas should not be given access to trees or other climbing apparatus until fully recovered from anaesthesia (Vogelnest 1999). When a deeper plane of anaesthesia is required (e.g. surgery or procedures such as electroejaculation for semen collection), it is advisable to intubate the koala. Given the complicated soft palate of the koala, intubation can aided by means of a laryngoscope or fine rigid endoscope (Blanshard and Bodley 2008).

## 8.5 Common health problems

Several diseases are known to occur in koalas with the most common listed in Tables 7.5 and 7.6. More detailed information can be found in excellent reviews compiled by Blanshard and Bodley (2008), Gillett and Hanger (2025), Higgins and Maher (2019), and Jelocnik *et al.* (2019).

Other diseases found in koalas include tubulointerstitial nephrosis, neoplasia, oxalate nephrosis, dermatomycosis, gastritis, enteritis, typhlocollitis, dermoid cysts, rhinitis and necrobacillosis of the jaw (Brown and Woolcock 1988; Finnie 1988; Booth and Blanshard 1999; Gillett and Hanger 2025). There is also a record of an outbreak of sarcoptic mange in a colony of koalas that resulted in the death of several koalas because the mange was difficult to see under the thick fur. Two treatments of amitraz (as a 0.025% aqueous suspension) applied topically 10 days apart effectively controlled the outbreak (Brown *et al.* 1981).

## 8.6 Preventative medicine

Captive koalas can suffer from several significant life-threatening infectious diseases that require rigorous quarantine of new specimens and routine vaccinations. All new koalas, particularly if from the wild, entering a captive population should undergo a period of quarantine and relevant health examinations (e.g. routine haematology, chlamydial PCR, screening and treatment for parasites if indicated). A necropsy must be conducted on any koala that dies to facilitate diagnosis and, if relevant, inform preventative action to protect other koalas in a zoo population.

## 8.7 Zoonoses

There are no confirmed cases of zoonotic disease transmission from koalas to humans (Blanshard and Bodley 2008).

## 8.8 Stress

It is increasingly evident that stress increases the negative impacts of diseases such as chlamydia and koala retrovirus (Speight 2022). Stressors associated with captive koalas may include handling, disruption of feeding times, disruption of sleeping, overcrowding, separation of the sexes, noise, controlled mating and weaning. Stress in wild koalas can also arise from severe drought and heatwaves (Lunney *et al.* 2012).

Sick or injured animals can be challenging to treat (Finnie 1988) and there may be underlying contributory conditions (Gillett 2023). In any treatment or care setting, stress should be kept to a minimum and it may be necessary to hand-feed or use multivitamin therapy to stimulate the appetite. Although numerous studies have examined the levels of glucocorticoids in koalas (reviewed in Fanson 2019), the precise relationship and usefulness of these measures for koala captive management have been the subject of current research. A recent study found that diseased and injured koalas had significantly higher faecal cortisol metabolites (FCM) than clinically healthy animals, with females showing a more elevated response to stress manifested by injury and disease than males (Santamaria *et al.* 2023).

## 8.9 Quarantine requirements

The quarantine protocol for koalas is primarily aimed at preventing the introduction of chlamydia. However, following the protocol is also likely to prevent the introduction of other diseases and will establish a comprehensive set of baseline parameters for each animal. Quarantine is also important to help prevent introduction of respiratory issues such as *Bordetella* and *Cryptococcus* (K. Nilsson pers. obs.).

### 8.9.1 Procedures

A standard 30-day quarantine is recommended. Before entering animal facilities, all imported koalas should have the following checks (in priority order).

(1) Thorough clinical examination, including a full clinical history if available. The following should be recorded: bodyweight, body condition, identification, age, pouch check, physical abnormalities, teeth condition.
(2) Chlamydial testing of conjunctival and urogenital swabs (see Hulse *et al.* 2018, 2019, 2020).

**Table 7.5.** Common diseases and parasites of koalas of clinical significance.

Ig = immunoglobulin, KoRV = koala retrovirus, PCR = polymerase chain reaction.

| Disease/ infectious agent | Cause | Clinical signs | Diagnosis | Treatment | Prevention/comments | Reference |
|---|---|---|---|---|---|---|
| **Ectoparasites** | | | | | | |
| Ticks | *Ixodes holocyclus*<br>*I. cornuatus*<br>*I. hirsti*<br>*I. tasmani*<br>*Haemophysalis bancrofti*<br>*H. longicornis* | Local swelling and pruritis | Observation. Fully engorged ticks are mostly found around the jaw, in the axillae and deep in the fur over the thorax | Generally not required.<br>Manual removal.<br>Ivermectin, cydectin oral, injectable or pour-on | Ticks can be present in moderate numbers without causing harm.<br>Heavy burdens may cause anaemia.<br>Ticks can transmit a range of blood parasites | 1–4 |
| Mites | *Koalachirus perkinsi* | Can cause irritation, with significant scratching along the back and flanks | Observation within the fur or under microscopy on a fur pluck.<br>Magnification required | Subcutaneous injection of ivermectin<br>Frontline spray | Small host-specific fur-dwelling mite.<br>No records of associated pathology | 3–6 |
| | *Sarcoptes scabiei* | Mange on hands and face | Skin scraping | Cydectin repeated at weekly intervals for 4–6 weeks or until negative scraping | Quarantine; close contact is usually required for transmission.<br>Severe mange probably indicates underlying immunocompromise and poor prognosis for recovery | 3, 4 |
| | *Demodex* sp. | Hair loss and dermatitis | Skin scraping | Cydectin repeated at weekly intervals for 4–6 weeks or until negative scraping | Quarantine; close contact is usually required for transmission<br>Severe mange is likely to indicate underlying immunocompromise and poor prognosis for recovery | 3, 4 |
| **Endoparasites** | | | | | | |
| Cestodes | *Bertiella obesa* | Commensal. Mostly non-pathogenic unless the host is compromised | Observation; white tapeworm segments in faecal pellets are extremely common<br>Adult worms are often an incidental finding in the small intestine on ultrasound | Generally not required.<br>Oral praziquantel at a dose of 10 mg/kg<br>Grind up in paste | Very common host-specific tapeworm that has been sometimes associated with enteritis and debility when present in large numbers | 3, 4 |
| Nematodes | *Ophidascaris robertsi* | Likely subclinical | Incidental finding at necropsy | Not applicable | Transmission is from the faeces of carpet pythons, contamination of soil or leaves possible | 7 |
| | *Durikainema phascolarcti* | Dyspnoea, coughing, lethargy and general pulmonary disease | Bronchoalveolar lavage and cytological examination, radiography of lungs.<br>Histopathology postmortem | Can be self-limiting but treatment with corticosteroids (prednisolone) has been successful | | 4, 6, 8 |

| Disease/ infectious agent | Cause | Clinical signs | Diagnosis | Treatment | Prevention/comments | Reference |
|---|---|---|---|---|---|---|
| **Protozoal** | | | | | | |
| Cryptosporidiosis | *Cryptosporidia* sp. | Severe refractory diarrhoea, sudden death | Faecal gram stain or acid-fast staining techniques | Metronidazole can be used with caution | Good hygiene to prevent build-up in the environment or contamination of leaves with the faeces of other species | 3, 4 |
| Toxoplasmosis | *Toxoplasma gondii* | Respiratory or neurological signs | Serology IgG and IgM titres Histopathology postmortem | Unlikely to be successful but antiprotozoal regimens used in macropods could be attempted | Avoid allowing koala browse to lie on the ground where contamination with cat faeces may occur | 3, 4, 9 |
| Trypanosomiasis | *Trypanosoma irwini, T. gilletti, T. copemani, T. vegrandis, T. noyesi* | Largely non-pathogenic though some have been associated with clinical disease. Thought to be transmitted by ticks | Blood smear, PCR | In the presence of anaemia, blood transfusions and antimalarial treatment may be required | Not reported in captive koalas | 3, 4, 6, 10–12 |
| **Viral** | | | | | | |
| KoRV | | Clinical signs are not specific to KoRV, but are putatively associated with infection. Pathophysiology is poorly understood Putative KoRV-associated diseases are leukaemia, lymphoma and other neoplasia, immunocompromise or immune dysfunction predisposing to other pathogens | There are several subtypes of KoRV (A–M) and virus has both endogenous (n the genome) and exogenous (infectious) forms | Animals invariably die despite any attempts at symptomatic treatment. Research into vaccination is underway, but is currently not available. Antivirals have been trialled but are not effective | | 6 |

**Table 7.5.** continued

| Disease/ infectious agent | Cause | Clinical signs | Diagnosis | Treatment | Prevention/comments | Reference |
|---|---|---|---|---|---|---|
| Herpesvirus | Herpesvirus, possibly two koala-specific viruses implicated so far:PhaHV-1 and PhaHV-2 | Dyspnoea, nasal discharge, sneezing, sudden death | Specialist laboratory PCR of respiratory tract swabs | Experimental treatment in marsupials with vancyclovir orally | Quarantine. Pathogenicity is poorly understood in koalas. Likely spread by droplet aerosol. Concurrent infection with *Chlamydia pecorum* | 2, 13–15 |
| **Bacterial** | | | | | | |
| Chlamydiosis | *Chlamydia pecorum*<br>*C. pneumonae* | Conjunctivitis<br>Cystitis<br>Reproductive tract disease<br>Pneumonia | Clinical signs are highly suggestive of diagnosis<br>Quantitative PCR of conjunctival and urogenital swabs is the gold standard for diagnosis before and after treatment | Doxycycline or chloramphenicol are effective antichlamydial drugs in koalas. Treatment duration depends on the severity of clinical signs but is usually 2–4 weeks | *Chlamydia pecorum* is primarily sexually transmitted<br>Chlamydial pneumonia with either species can be transmitted by droplet aerosol.<br>Maintain chlamydia-free captive colonies where possible and use two-way koala proof fencing to prevent infected wild visitors | 1, 3–5, 16, 17 |
| Sepsis | *Salmonella typhimurium*<br>*Pseudomonas aeruginosa*<br>*Aeromonas hydrophila*<br>*Pseudomonas pseudomallei*<br>*Chromobacterium violaceum* | Phlebitis, oedema, petechial haemorrhages, sudden death | Blood culture or tissue | Rarely successful. Intravenous antibiotics based on culture and sensitivity, koala plasma transfusion | Risk with overcrowding, poor hygiene and environmental exposure, particularly during wet weather | 1, 4, 5, 18 |
| Rhinitis/ pneumonia | *Bordatella bronshiseptica* | Nasal discharge<br>Increased respiratory rate<br>Weight loss | Culture and sensitivity of nasal discharge and/or bronchoalveolar lavage<br>Radiography<br>Necropsy | Antibiotics based on culture and sensitivity<br>Nebulisation with antibiotics may improve treatment success | Vaccination with killed canine Bordatella vaccine every 6–12 months depending on institutional incidence | 1, 4, 5, 18 |

| Disease/ infectious agent | Cause | Clinical signs | Diagnosis | Treatment | Prevention/comments | Reference |
|---|---|---|---|---|---|---|
| Pneumonia | Usual suspects include a variety of bacterial pathogens or via aspiration of milk or leaf matter | Nasal discharge<br>Increased respiratory rate<br>Weight loss | Culture and sensitivity of nasal discharge and/or bronchoalveolar lavage<br>PCR test of lavage or nasal secretions<br>Radiography<br>Necropsy | Antibiotics based on culture and sensitivity<br>Nebulisation with antibiotics may improve treatment success | Quarantine if bacterial origin and likely spread by droplet aerosol | 1, 4, 18 |
| Cutaneous ulcers | *Mycobacterium ulcerans* | Large under-run skin ulcers or ulcers on the hands and feet | History, culture, histopathology.<br>PCR test | Poorly responsive to antimicrobial therapy | Has been identified in koalas from Raymond Island and mainland Victoria | 1, 4, 18 |
| **Fungal** | | | | | | |
| Cryptococcosis | *Cryptococcus neoformans* and *C. gattii* | Nasal discharge, facial asymmetry, increased respiratory rate, audible breathing, neurological signs | Cytology of discharge<br>LCAT<br>Lateral flow<br>Radiography | Respiratory disease can be treated if detected before granulomas form, with prolonged itraconazole or fluconazole.<br>Neurological disease likely fatal so euthanasia recommended | Cryptococcal disease is likely to be precipitated by stress, including poor nutrition, starvation, illness due to other causes and transport<br>Heavy environmental load precipitates disease risk<br>Environmental contamination is the source of the yeast, strong association with *Eucalyptus camaldulensis*<br>Not transmitted from animal to animal or from animal to humans | 1, 4, 16 |
| Candidiasis | *Candida albicans* | Dry, small faecal pellets, yellow plaques in mouth, cloaca, inappetence, diarrhoea, weight loss | Budding yeasts can be detected on a wet-preparation or gram stain of a mouth swab or faecal sample | Can be treated with nystatin, fluconazole or amphotericin B. Drug selection and duration of treatment depend on severity of clinical signs | Usually occurs as an opportunistic infection in koalas being treated with broad-spectrum antibiotic therapy for chlamydiosis or other bacterial infections. Or as a common finding in hand-reared joeys | 4, 5, 17 |
| Dermatophytosis (ringworm) | *Trichophyton mentagrophytes*<br>*Microsporum gypseum*<br>*M. canis*<br>*Acremonium* sp.<br>*Penicillium* sp. | Sparse fur on fingers, crusting, circular alopecia | Fungal culture, fur pluck, sticky tape impression or skin swab cytology, histopathology | Long-term topical medication<br>Sometime systemic antifungal medication is required | Reduce shade and moisture in enclosure, usually occurs in immunosuppressed individuals with comorbidities such as chlamydiosis | 1, 3, 4 |
| Paronychia (nail infections) | Any of the above fungal species | Elongated, brittle, split, deformed nails, inflamed and swollen nail beds of hands and feet | Culture | Long-term topical antifungal medication | Associated with nail bed trauma, partial avulsion of nail, and damp dark enclosures. Improve hygiene, ventilation, sunlight | 14 |

References: (1) Stone (1992); (2) Stone and Carrick (1996); (3) Booth and Blanshard (1999); (4) Blanshard and Bodley (2008), (5) Gillett and Hanger (2025); (6) Gillett (2023); (7) Gonzalez-Astudillo *et al.* (2019b); (8) Spratt and Gill (1998); (9) Hartley *et al.* (1990); (10) McInnes *et al.* (2011a); (11) McInnes *et al.* (2011b); (12) Barbosa *et al.* (2016); (13) Vaz *et al.* (2012); (14) Vaz *et al.* (2011); (15) Stalder *et al.* (2015); (16) Booth and Nyari (2020); (17) Palmieri *et al.* (2018); (18) Blanshard (1994).

**Table 7.6.** Common non-infectious diseases and causes of mortality of captive and rescued koalas.

Derived from Gillett and Hanger (2025).

| Disease/agent | Cause | Clinical signs | Diagnosis | Treatment | Prevention/comments |
|---|---|---|---|---|---|
| Starvation | Drought, after fire, land clearing | Emaciation, death | History of environmental issue, muscle wasting and dehydration are suggestive | Supportive care with intravenous fluids and oral energy supplements may allow individuals to recover<br>Euthanasia of severely affected individuals | Not relevant to captive koalas |
| Climate change | High ambient temperatures, wildfire, drought | Incineration, smoke inhalation, burns, starvation, death | History of environmental issue, muscle wasting and dehydration are suggestive | Euthanasia of severely affected individuals | Not relevant to captive koalas |
| Predation | Domestic and wild dogs, raptors, owls, goannas, pythons | Puncture wounds, saliva on fur | Observation/history, physical examination, abdominocentesis useful to detect caecal trauma from dog mauling, necropsy | Medical and surgical treatment if not fatal | Back young in captivity are vulnerable to predation from owls, eagles and pythons unless precautions are taken with enclosure design |
| Anthropogenic threats | Forestry and logging, land clearing and agriculture, urban development, swimming pools, vehicle strike | Emaciation, trauma, dehydration, entanglement, entrapment, drowning, death | Observation/history, physical examination, necropsy | Medical and surgical treatment if not fatal | Increased conservation action, climate change action, policy changes. Captive koalas are ambassadors to educate the public about the anthropogenic threats that face wild koalas |

(3) Blood sample taken for blood cell count and biochemistry.
(4) Blood sample taken for cryptococcal antigen serology (serum gel tube 2 mL minimum) if coming from areas where *Cryptococcus* is endemic.
(5) Faecal flotation.

All animals should be held in quarantine until all test results are negative. Koalas should be monitored daily by animal caretakers. All animals should be checked and cleared by a veterinarian before introduction to resident populations. All imports should be considered infective until proven otherwise. All handling of animals and their uneaten feed and excreta is done by one person. Uneaten feed should be wrapped in plastic and removed from the premises. Gloves, coat and rubber boots should be worn while handling quarantined koalas and their feed. These items of protective clothing are to remain in the quarantine area. The hands and forearms of animal caretakers should be scrubbed with Hibitane® or iovone scrub after boots; gloves and coat have been removed.

## 8.9.2 Maintenance cleaning of quarantine facilities

Enclosure floors should be cleaned with veterinary disinfectant such as F10® (Meadow's Animal Healthcare) at least once weekly, but ideally daily, and should be spelled for as long as possible between animals.

## 8.9.3 Cleaning up after positive Chlamydia results

Koala forks used by koalas with only an antibody response to *Chlamydia* and with negative swabs can be disinfected by thoroughly drenching with Microcide or Virkon at antiviral concentration (20 m/L). Contaminated forks from positive swab enclosures are to be removed and disposed of. All enclosure support beams, furniture, walls and hard floors should be saturated with Microcide at antiviral concentrations. Once a movable enclosure has been disinfected, it should be moved to another site. The exposed soil should be sprayed with Avisafe and allowed to dry out. Sand around koala forks should be raked into bags and removed. After underlying material has dried out, spray with Microcide then apply new sand around all the forks. All staff involved in the clean-up procedure should wear PPE consisting of overalls, rubber boots and disposable gloves. A facemask and protective goggles should be worn during disinfectant spraying for personal protection. All PPE should be cleaned and disinfected or disposed after use.

## 9. BEHAVIOUR

### 9.1 Activity

Koalas generally rest and sleep in the forks of trees but are occasionally found stretched along a branch. On hot days their limbs are extended and they often lie free on either side of the trunk. The animal may recline along the limb and the head is held free of the chest, exposing the belly. On cold, wet and windy days they sit with their backs to the wind, with their arms folded against the chest and legs drawn against the belly. These changes in posture have an important role in temperature regulation (Lee and Martin 1988; Ryan *et al.* 2013). Koalas may regularly move to seek the sun during winter and shade in summer (Matthews *et al.* 2007). In fact, habitat selection in the wild varies with the season in inland New South Wales (Crowther *et al.* 2014) and koalas can hug the bark of trees to thermoregulate (Briscoe *et al.* 2014), thereby identifying the importance of temperature in tree and habitat choice.

It has been consistently shown that wild koalas spend ~18–20 h of each day resting or asleep (with the least activity typically between 7 a.m. and 3 p.m.), 1–3 h spent feeding and the remaining time spent moving between branches or trees, grooming or in social behaviour (Fig. 7.13; Benesch *et al.* 2010; Ryan *et al.* 2013). Feeding episodes normally last from 20 min to 2 h in duration with 4–6 of these bouts per day. Although feeding can occur at any time of the day or night, there is a tendency for wild koalas to feed immediately before or after dusk or dawn (Lee and Martin 1988; Ryan *et al.* 2013). In contrast, captive koalas tend to habituate to the feeding routine and feed when the food is supplied.

A study that examined the effect of visitor number and visitor noise as they moved along a boardwalk within a 5-m radius of koalas suggested there is an impact. The research revealed that an increase in the number of nearby visitors (but not the total daily number of visitors) resulted in an increase in the time koalas spent vigilant. It also showed that an increase in visitor noise resulted in an increase in the time koalas spent vigilant (Larsen *et al.* 2014). Therefore, the design of koala enclosures and any public interaction with them should take this into account.

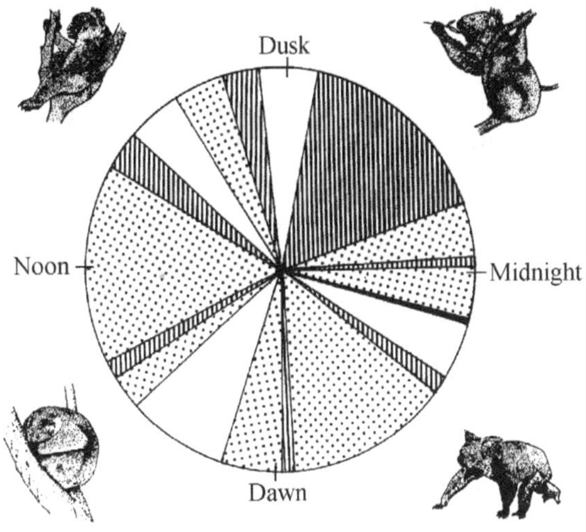


**Fig. 7.13.** The koala's daily cycle of activity. Vertical hatched areas signify periods of feeding; stippled areas, periods of sleeping; unshaded areas, periods of resting; and areas shaded black, periods moving between trees. Taken from Lee and Martin (1988).

It also appears that handling koalas by members of the public during photographic sessions can have an impact (Webster *et al.* 2017). A study that used a faecal cortisol metabolite (FCM) enzyme-immunoassay (EIA) to monitor the physiological stress revealed that males that were handled had a 200% higher FCM level than animals that were not handled. Female koalas were not affected by handling as long they were not undergoing lactation (Narayan *et al.* 2013).

### 9.2 Social behaviour

In the wild, koalas are generally a solitary species and their movement patterns are mostly confined to dusk and evening. In a study of home ranges at Port Stephens, New South Wales, the movements of wild koalas showed a wide variation in patterns, from highly localised movements to long-range dispersal over 20 km. The daily movement distances of males (mean = 278.3 m) were higher than for females (141.0 m). The monthly displacement ranged from less than 10 m to more than 5 km. Home ranges of males (95% kernel) were significantly larger than those of females (mean for males = 58.9 ± 10.5 ha; mean for females = 25.7 ± 8.6 ha) (Matthews *et al.* 2016). Another study on the Liverpool Plains, in north-west New South Wales, found that distance travelled was strongly associated with habitat fragmentation, sex and year. Further, the connectivity metric had the highest relative importance among all the variables. In this fragmented rural landscape, the distance travelled increased with decreasing patch connectivity and in the most fragmented habitat, koalas travelled over 30 km, which was 3-fold more than in the least fragmented habitat, and male koalas moved further than females (Rus *et al.* 2021).

The vocalisations of koalas are diverse. Calls include the bellow, which is primarily used by the males, although some females occasionally bellow, often when in oestrus. Field studies have revealed that the number of bellow vocalisations coincided with increased breeding activity, with nearly all males bellowing during the peak of the mating season (Ellis *et al.* 2011). The peak in koala bellowing and distance travelled appears to occur around midnight in wild populations and the study authors suggested that bellowing was used by males to attract females rather than to repel males (Ellis *et al.* 2011). It has also been suggested appears that koalas may be able to recognise other individuals by their distinctive bellow up to 50 m away (Charlton *et al.* 2012). Koala bellows are now being used to determine the presence of koalas in the wild, which represents a new survey technique (Law *et al.* 2021)

Other vocalisations include repeated squeaks of joeys, which may serve to attract the mother's attention. The wails, squawk, low grunt, snarls and screams of females probably serve as a defensive threat or as rejection to the male's advance during attempted copulation (Smith 1980a; Lee and Martin 1988). The bellowing calls of female koalas are significantly shorter in duration than those of males and had a higher frequency than male bellows. It may be that these calls identify the animal's age and sex (Charlton 2015). Males also emit wailing, squeaking vocalisations (similar to females) if approached by another male in an aggressive way (thought to show submissiveness) or if distressed by other factors (K. Nilsson pers. obs.). Overtly stressed males and females may display a high-pitched 'bleating' behaviour, particularly during capture or at release (A. Gillett pers. obs.).

Fighting between koalas is relatively infrequent, with most agonistic interactions being relatively minor and typically are intensified squabbles. Severe fighting can involve savage biting, wrestling and chasing and occurs mostly between males that are unfamiliar to each other or by being aroused by other aggressive interactions. Males may also attack females and back young, which may include pulling or knocking them from the tree. It is not advisable to have all male enclosures because of fighting that can result in physical injuries, especially to the ears and arms. Females can also become aggressive towards other koalas, but this typically occurs at the end of lactation (Smith 1980b). Introduction of the male to a female enclosure typically results in bellowing behaviour, urination and sternal gland scent marking. Strong bellowing behaviour also typically follows a successful mating.

Male koalas have been held in groups that have grown up together; however, they need to be monitored closely over their lifetime for any signs of disharmony. If aggression occurs, either the antagonist or the one being targeted needs to be removed and housed alone. New males are typically not introduced into a harmonious adult male group, so knowledge of the group's dynamics is essential so as to not create even more disharmony (K. Nilsson pers. obs.). Sometimes females from female-only groups will also need to be removed due to them displaying unusually aggressive tendencies towards the others. In this case it is often one female targeting another, so moving her to another group that does not include the target is usually sufficient (K. Nilsson pers. obs.). Sometimes females may also need to be removed from harem groups because of consistent aggression towards the male. This aggression can manifest whether the female has been in oestrus or not (K. Nilsson pers. obs.).

## 9.3 Reproductive behaviour

The beginning of the breeding season is heralded by an increase in the frequency of bellowing by males (Lee 1988). Male koalas scent mark trees by grasping the tree trunk and rubbing their chest (containing the sternal gland) up and down against the base of the tree trunk and branches as they climb. They also scent mark using urine, with both sexes occasionally urinating on the trunk or on the ground close to the tree. The use of scent marking may help to establish the dominance of the male and the reproductive status of the female (Smith 1980c, 1980d; Lee 1988; Thompson and Fadem 1989).

Encounters between males are sometimes aggressive and can result in injuries. During these fights the males endeavour to wrap their forelimbs around the other's shoulders while trying to bite the other's neck and grunting. Sometimes the males rear up during these bouts on their hindlimbs and wrestle. Bite and scratch wounds are also inflicted to the ears, nose, elbow or knee (Smith 1980b, 1980d).

Koalas are induced ovulators (Johnston *et al.* 2004) (i.e. females ovulate when induced by the act of coitus and possibly in combination with seminal stimulation). When female koalas are housed separately in non-pregnant or lactating (i.e. cycling) cohorts they may show overt signs of oestrus that can include an increase in activity (movement and pacing), jerking behaviour, homosexual pseudo-copulation, bellowing and body-weight loss (Johnston *et al.* 2000a; Feige *et al.* 2007; Takahashi *et al.* 2009). These oestrous behaviours can be facilitated and intensified in the presence of a 'teaser' male (Johnston and Holt 2019). A teaser male is used for routine reproductive management and can help the animal caretaker gain a precise understanding of the reproductive status of the females in the collection. Typically, the male teaser is held by the animal caretaker and introduced into the female enclosure and presented (without physical contact) to each individual female in the enclosure so that they are aware of his presence. The male teaser is then placed on the floor of the female's enclosure where he is allowed to scent mark objects (base of poles) with the sternal glands, bellow and urinate.

The presence and sound of the male teaser enhances the female's behavioural excitement, typified by extremely overt signs of behavioural oestrus that include most or all of the following behaviours: 'jerking' or convulsive behaviour that resembles violent hiccoughing, homosexual pseudo-copulation behaviour of females by other females, a dramatic change in the female's behaviour from passive lack of interest to acute attention in response to the male, urination and occasionally bellowing (Johnston *et al.* 2000a; Johnston and Holt 2019); non-oestrus females do not typically bellow. It is best to use a different teaser male each time for subsequent oestrus detections (particularly if conducted daily) to reduce the possibility of either the male and female becoming habituated to each other's presence.

Oestrus in the non-mated koala typically lasts for ~10 days and has been mapped to elevated concentrations of plasma oestradiol 17B (Johnston *et al.* 2000a). Although the length of oestrus is variable among individuals (0–19 days), once ovulation has been induced (pre-ovulatory follicle ruptures) the source of oestradiol is removed and the oestrous behaviour typically subsides 24 h post mating. As a result, male presence is not tolerated by the female once oestrous signs have stopped. Most females will reject the male and strike out aggressively as soon as mating has occurred and if another male is introduced to the enclosure the following day, she will again reject him (K. Nilsson per. obs.).

In contrast, non-oestrous females will typically show distinctive rejection behaviour to the advances of the male, often emitting violent squeaks or squawks and striking the male with the forearm as he approaches. The significance of such distinctive oestrous behaviour in captive koalas, while an advantage for their reproductive management, has not commonly been described or observed in wild koalas, such that the behaviour may be an artefact of captivity. It has been proposed that the sexual excitement associated with elevated levels of oestradiol 17B may be linked the physio-behavioural mechanism that causes wild female koalas to seek out males in their normal home range during the breeding season (Feige *et al.*, 2007; Ellis *et al.*, 2011).

Once mated, the female presumably ovulates and oestrus subsides. These activities generally decline as summer progresses and are not typically observed during the cold months (Lee 1988; Lee and Martin 1988; Martin and Handasyde 1991). Mating of females with back young should not be encouraged as the young can become dislodged and fall to the ground causing injury and she will no longer allow the joey access to milk (K. Nilsson pers. obs.).

## 9.4 Bathing

Although koalas should always be given access to water they never bathe in water.

## 9.5 Behavioural problems

Some animals that are hand-reared can become overly attached to human company, resulting in longer weaning times and the urge to climb on anyone who passes by (such as during cleaning).

## 9.6 Signs of stress

Visual signs of stress, especially if acute, include restlessness, stupefaction, frequent urination or defecation (S. Eccleston pers. obs.), low whining vocalisations or high-pitched bleating (A. Gillett pers. obs.) and typical alarm posture (wide eyes, ears forward and vertical spine). Other signs include loud vocalisations, aggressive defence such as threatening, biting or scratching (this may be before, during or after a stressful event), head shaking or ear flicking (Spielman 1994). Stress may also manifest as trembling when separated from the mother, diarrhoea (soft pellets or watery) within 24 h and usually lasting 24 h (Spielman 1994). If the number of pellets produced each day falls below 100, it may be indicative of dehydration, gastrointestinal dysbiosis or early ileus (A. Gillett pers. obs.) and should be monitored closely (Gillett and Hanger 2025). At a minimum, faecal assessment by a veterinarian who is familiar with wildlife or koala medicine is recommended (A. Gillett pers. obs.; Spielman 1994). Further information on stress in koalas can be found in sections 7.7.1–7.7.2.

A non-invasive assessment of stress in koala has been developed using a faecal cortisol metabolite (FCM) enzyme-immunoassay (EIA) to monitor physiological stress in both wild and captive koalas (Narayan *et al.* 2013; Santamaria *et al.* 2023). The FCM levels of captive koalas differed significantly by sex, handling and reproductive condition (lactating *v.* non-lactating). Handled male koalas had 200% higher FCM levels than their non-handled counterparts, while females were not affected.

## 9.7 Behavioural enrichment

Koalas generally do not display abnormal repetitive behaviour as they do not have the energy to spare while they rest 18–20 h per day (see section 9.6 for signs of stress). Some individuals will pace near the animal caretakers' entrance before the set feed times, but this may also indicate mate seeking, breeding or oestrous behaviour. Movement of individual koalas can be maximised and conflict minimised by ensuring adequate forks for feeding and resting and several cross branches.

## 9.8 Training and conditioning

Limited training has been undertaken with koalas; young animals that are hand-reared can be held by trained staff.

### 9.8.1 Koala handling and photographing by the public

Koala handling by the public is not permitted in some jurisdictions and exhibitors needs to be aware of the requirements in their state. Whether a koala is being held by a member of the public or the visitor stand next to the koala that is sitting in a tree fork for photos to be taken, it is imperative the animals used are chosen based on their temperament. These animals must be conditioned from weaning for handling or close human positioning. Even after this conditioning process many koalas are not suitable for consistent use for interactions with members of the public, so those facilities that permit public interactions must have enough koalas to allow for a proportion that is not suitable. The experience of the animal caretakers involved in conditioning is important to accurately interpret behavioural cues and not force a koala to interact with humans. Koalas must be monitored constantly and if their responsiveness to humans changes (can often occur at sexual maturity), they should no longer be included in public handling experiences (K. Nilsson pers. obs.).

It is also important that constant attention is paid to the koalas, particularly if being held by the visitor, to assess their behaviour. Conditioning for handling by staff should occur to reduce the stress of handling for weighing, moving and reproductive husbandry. Any koalas used for handling sessions must be routinely weighed and the time and dates that it is used should be recorded (Anon 2019). When using koalas for photographic opportunities, it is recommended that they are placed in tree forks that are approximately chest or shoulder height.

### 9.8.2 Signs of stress during handling

When koalas are used for handling and/or photography, the signs of stress when undertaking this activity should be monitored at all times (Booth 1989). These include:

- being restless (including not sitting on a branch, keeps coming to the ground and walking around)
- being listless (completely flaccid and tractable)
- being aggressive (including biting and/or scratching)
- urinating and defecating often
- continuously ear flicking
- showing signs of anxiety including a low whining vocalisation, and a typical alarm posture (wide eyes, ears forward, spine very vertical)
- excessive yawning and neck stretching.

### 9.8.3 Minimising stress during handling

Stress of koalas can be minimised during handling or photographic sessions by (Booth 1989; Anon 2019):

- only using koalas with suitable temperaments
- choosing captive-bred koalas that have been habituated to handling from weaning or koalas that were rescued but unable to be released
- not using koalas that are not fully weaned and independent or have pouch young
- using responsible and experienced animal caretakers who will respond to the needs of each koala
- using minimal restraint to get the most cooperation
- monitoring of bodyweight of koalas and giving the individuals a 'holiday' whenever a drop or insufficient gain in weight (if growing) is noted

- daily recording each koala's handling time for photography and comparing with bodyweights recorded as least fortnightly (ideally weekly)
- repositioning the tourist, not the koala, for a photo if needed
- having clear stress parameters that trigger the withdrawal of the koala from public interactions (see section 9.10.1)
- ensuring the total interactive handling time (including transport and exhibit) does not exceed 4 h per day
- ensuring koalas are not handled more than (i) 30 min per day, (ii) 180 min in a 7-day period or (iii) on 3 consecutive days.

### 9.9 Introductions and removals

Animal introductions to an enclosure with resident koalas are normally done early in the morning to minimise any public reaction to aggressive confrontations between koalas and allow the whole day for animals to be observed before being left together overnight. After the introduction of a new koala into an enclosure it should be watched to check for any agonistic behaviour, which should decrease as the new animal works out its position in the group. If the aggression continues after several hours, the new animal should be removed.

The introductions of koalas should ideally be customised according to the age and sex of the individuals involved and the reason for their move. Best practice is to avoid placing adult males together, as excessive aggression, likely injury, and stress leading to stress-related illness is common. Female to female introductions are generally uneventful, unless they are displaying oestrous behaviours, but even in these circumstances, injuries are unlikely (S. Eccleston pers. obs.).

### 9.10 Intraspecific compatibility

Female koalas can readily be held with each other and one or more males. If more than one male is held together, the dominance hierarchy means that the most dominant male is likely (but not always) to do most or all the breeding. However, in most institutions, paternity is important for breeding program, so only one male is introduced and thus given access to a female at any one time. This is particularly important during the breeding season, although if adequately separated and depending on the nature of the individuals, 2–3 male koalas can be housed together.

### 9.11 Interspecific compatibility

Koalas have been exhibited with other species including echidnas (*Tachyglossus aculeatus*), bettongs (*Bettongia* spp.), pademelons (*Thylogale* spp.), quokkas (*Setonix brachyurus*), parma wallabies (*Notamacropus parma*), various species of lizards and large birds such as magpie geese (*Anseranus semipalmata*). They can potentially be exhibited with other small macropods such as bettongs and potoroos (*Potorous* sp.). Wombats are generally not compatible with koalas (or most species) as they have been known to cause injuries, and potentially fatalities, by biting koalas. With the exception of the echidna, these species are generally nocturnal and are unlikely to be seen regularly by the public in enclosures with natural light cycles.

## 10. BREEDING

### 10.1 Mating system

In the wild, the koala is normally polygynous, with a male mating with more than one female during a single breeding season.

### 10.2 Ease of breeding

Koalas readily breed naturally in captivity. Long-term analysis of breeding success at San Diego Zoo suggested that male bodyweight did not affect reproductive success, but reproductive output was greatest when males were slightly older than females (Bercovitch *et al.* 2006).

### 10.3 Oestrous cycle and gestation period

Although there may be individual variation, female koalas start to show evidence of reproductive cycles and are capable of producing offspring around 2–3 years of age (though some are capable of breeding at between 18 months and 2 years of age (A. Gillett pers. obs.). The reproductive physiology of the koala has been described in detail elsewhere (Johnston *et al.* 2000a, 2000b, 2004; Johnston and Holt 2019), so only a brief overview will be given here. Given that the koala is an induced ovulator (stimulated by coitus and seminal biochemistry), there are three possible oestrous cycle scenarios.

(1) Female in oestrus is not mated, or not successfully mated, by the male so ovulation does not occur. In this situation the female does not experience a luteal phase (no ovulation, formation of a corpus luteum

(CL) or secretion of progesterone) and she will typically return to oestrus ~33 (25–48) days later.

(2) Female is successfully mated and ovulation occurs (typically a lutenising hormone surge occurs 28 h after natural mating (Johnston *et al.* 2004), but the pregnancy fails (lack of oocyte fertilisation or embryonic loss or the joey fails to attach to a teat); in this scenario, the female experiences a luteal phase, but will return to oestrus ~50 (47–53) days after the mating.

(3) Pregnancy occurs and the female will typically give birth in a narrow window of 34–36 days. Presumably, the sucking stimulus of the pouch young results in lactational anoestrus and the prevention of further cyclical activity. Koala behavioural oestrus can last on average for 10 days (Johnston *et al.* 2000a), but Allen *et al.* (2008) have shown that successful pouch young may result if females are mated on day 8 (50% success) and even day 10 (30% success) of oestrus.

## 10.4 Reproductive status

### 10.4.1 Females

Koalas are generally placed in several categories depending on their reproductive status. For females these include:

(1) non-parous (never bred) – pouch small with no skin folds, clean and dry; teats very small
(2) parous (bred previously but not presently) – pouch is small but distinct, dry and dirty; teats are slightly elongated
(3) pregnant – pouch roomy, pink in colour and glandular in appearance; skin folds may be observed on the lateral margins of the pouch in those near birthing
(4) pouch young present – attached to the teat
(5) lactating (young absent from the pouch but still suckling) – pouch area large, skin folds flaccid, hair sparse and stained, skin smooth and dark pink; teats elongated
(6) post lactation – teats expressing only clear liquid and/or regressing.

If pouch young are present there are several developmental stages and measurements that can be recorded and compared with existing growth curves (see section 10.16) or to establish curves for future reference. These include:

- developmental stages
  - sex distinguishable
  - tips of ears free
  - papillae of facial vibrissae evident
  - eyelashes visible
  - eyes open
  - fur visible – slightly tinged, medium or well developed
  - tips of first incisors through the gums
  - emergence and pap feeding
  - eating solids
  - riding on tummy and back
  - self-feeding
  - independent
- measurements (see Appendix 3)
  - bodyweight (g) – if not on teat
  - head length (mm) – from the occiput to snout tip
  - head width (mm) – maximum width across the zygomatic arches
  - crown-rump length (mm) – primarily for very small neonates
  - body length (mm) – from snout tip to cloaca
  - tibia length (mm) – from the hip to the bottom of the pes
  - pes length (mm) – from the heel to the base of the longest toe, not including the claw.

### 10.4.2 Males

The male's scent gland over the sternum starts to develop when quite young and may become bare. Most likely resulting from elevated levels of androgen, sexually mature males develop a collar of cuticular spines at the base of the glans penis (S. Johnston unpubl. obs.).

Studies of spermatogenesis, sperm and seminal biology are beyond the scope of this chapter, so the reader is referred to the most recent review of koala reproductive biology by Johnston and Holt (2019) in which male anatomy and physiology are described. A recent study has reported on koala seminal proteomics (Skerrett-Byrne *et al.* 2021).

## 10.5 Reproductive management

### 10.5.1 Selective breeding method

This method of breeding involves placing a single male and female koala together. This system appears to be most effective when the male and female koalas are kept separate and placed together when the female is in oestrus (O' Callaghan and Blanshard 1991). Prior to the male entering the female enclosure, the females are typically teased with a different male to the one that is being used for mating. This form of oestrus detection allows the animal caretaker to identify the females suitable for mating. It is prudent to conduct daily oestrus detection during the breeding season to carefully monitor the day of oestrus. A reproductive ethogram can be established to record the occurrence of oestrous behaviours.

Females can be mated with the chosen male on days 1-10 of their behavioural oestrus, although on days 2–4 is most typical. The sire can either be brought into the female's enclosure for mating if there is a suitable pole or both the male and female can be brought into a dedicated mating enclosure where the copulatory behaviour can be monitored. Best practice is to take the female to the male enclosure because if females are co-housed, the rejection behaviours from other females in the vicinity can deter the male from a successful copulation. The selective breeding method is encouraged to allow precise genetic management, but requires the use of a cohort of cycling females, which is not always possible in institutions with small populations.

A successful mating involves placing the female on a dedicated mating pole that is cut off at human head height so that she is unable to escape. The male is then placed underneath the female and allowed to establish a neck bite. If the female is in oestrus, she will typically be in standing oestrus, in which she will push out her rump and be receptive to the male's advance. Once the male has established a neck bite (nape of neck) and secures the female around her chest with his forearms, his erect penis is directed towards the female's urogenital region. The male will then attempt to find the urogenital opening with his glans penis and once detected he will commence thrusting behaviour; typically, the male penile thrusts an average of 40 times in 20 s. During this period, the female will remain upright, still and continue to display a lordosis or arching of her back. Following thrusting, the male then remains still as he ejaculates, but the female then commences to jerk and ear flick, still in a lordosis position.

Manipulative studies using an artificial vagina for semen collection have revealed that during thrusting the male produces a seminal plug fraction (presumably from the bulbourethral glands) and that ejaculation of the sperm-rich fraction (including prostatic secretions) occurs once the male has completed thrusting. The final thrusting behaviour is normally characterised by two prolonged final ejaculatory thrusts (Johnston *et al.* 2000b; Johnston and Holt 2019). If mating is not observed directly, the presence of a seminal copulatory (dried semen) on the fur of the female urogenital sinus is strong evidence of copulation.

Mating on average lasts for 60 s (~30 s of thrusting followed by ~30 s of ejaculation) and is terminated by the male releasing the neck, then biting the shoulder region of the female, which initiates disengagement of the pair. When the male releases the neck bite, the animal caretaker can intervene and remove the male before he bites the shoulder region again. This prevents unnecessary injury to the female. It is important where possible to observe the mating closely, to ensure the penis is inserted into the female's urogenital sinus and that the male applies sufficient copulatory stimulus to ensure induction of ovulation; the role of sufficient semen deposition to the ovulatory process has still to be determined.

From time to time, inexperienced males may fail to insert their penis and it possible for the animal caretaker to provide assistance to the male to locate the urogenital opening. It is also very typical for the male to bellow vigorously post mating and once this has ceased the male should be placed back in his enclosure. Successful ovulation can be confirmed with the cessation of oestrous behaviour 24 h later or by detecting elevated levels of progesterone in the plasma or faeces.

The use of levonorgestrel implants in female koalas has proved effective contraception and the removal of the implants resulted in successful reversal of the contraceptive effect in the next breeding season (Hynes *et al.* 2010). Levonorgestrel implants have also been used with variable success to synchronise oestrus in the koala, but further studies are required to refine this procedure for the purposes of reproductive management (Ballantyne *et al.* 2015a).

### 10.5.2 Harem breeding method

This system involves a single male being placed with a group of female koalas, which allows the paternity of any offspring to be established for several females. Care needs to be taken not to let any particular male overbreed, otherwise a loss of genetic variation can occur over the long-term together with habituation leading to a loss of sexual interest by the male.

### 10.5.3 Random breeding method

This method involves placing several male and female koalas in the one enclosure. There is little control over which animals mate and the paternity of most of the young will not be known, particularly as many matings will be unobserved (O' Callaghan and Blanshard 1991). Determining the pedigree is therefore almost impossible, except with the use of expensive techniques such as DNA fingerprinting (profiling). Female koalas housed in an enclosure with several males often get little rest during the breeding season as they are repeatedly harassed by different males. This can result in trauma and/or death of pouch young if present. Another disadvantage is that subadult females can be overpowered and mated, and give birth at a small size and bodyweight (O' Callaghan and Blanshard 1991).

### 10.5.4 Breeding group sex ratio

Generally, a single male is placed with one or more females, as discussed with the selective and harem breeding methods, to ensure the paternity of the young is known and to avoid fighting by males trying to achieve alpha status and mating rights (Smith 1980b). Lactating and non-lactating females are normally separated to avoid accidental adoption of young by non-lactating females. Some females that have had their young removed for hand-rearing are often found with the young of other females, seemingly in an attempt to replace the loss of their own.

Housing lactating females with similarly aged joeys has proved very successful (K. Nilsson pers. obs.). Females will readily feed and welcome joeys that are not their own into their pouch (even if another joey is already present). This method is also beneficial if a joey falls or becomes separated from its mother - other females will intervene and collect the joey. Animal caretakers need to be familiar with each joey/dam combination before the young emerge, so knowledge of genetics is maintained if joeys venture onto another dam. Older joeys should not be housed with dams that have permanent pouch young in case they dislodge the younger joey in their search for milk (K. Nilsson pers. obs.).

The ideal grouping arrangement is to have all cycling (non-pregnant and non-lactating) females in the same enclosure so that detection of oestrus can be made efficiently with a single teaser male and in order to enhance female–female interactions such as pseudo-copulation. However, it is not always possible to house koalas in this arrangement in institutions with smaller koala numbers, so in those circumstances the harem method is preferred.

### 10.5.5 Artificial breeding

Over the past 2 decades there has been significant research into the basic science and application of the use of assisted breeding technology in the koala (reviewed by Johnston *et al.* 1999; Johnston and Holt 2019), which has resulted in the production of 34 pouch young born by artificial insemination (Johnston *et al.* 2003; Allen *et al.* 2010). Johnston (2019) has identified that successful artificial insemination involves three fundamental components.

(1) Semen collection using either electro-ejaculation (e.g. Johnston *et al.* 1994) or an artificial vagina (Johnston *et al.* 1997) and semen storage (chilled [Johnston *et al.* 2000c; Johnston *et al.* 2012] or cryopreservation [e.g. Johnston *et al.* 2006; Zee *et al.* 2007, 2008, 2009])
(2) Determining the most appropriate time for insemination associated with the physiological and behavioural characterisation of oestrus, the oestrous cycle and pregnancy (Johnston *et al.* 2000a, 2000b, 2004).
(3) Determining the most appropriate site for semen deposition (Johnston 1999).

There have also been attempts to develop strategies that allow for controlled timing of oestrus (oestrus synchronisation; Ballantyne *et al.* 2015a, 2015b) and induction of ovulation using human chorionic gonadotrophin (Johnston *et al.* 2000b) and gonadotrophin-releasing hormone (GnRH) (Allen *et al.* 2008). Although the koala clearly represents the most successful implementation of an artificial insemination program in any marsupial species (Rodger *et al.* 2009), no offspring have yet been born following artificial insemination with frozen-thawed semen; this likely to be associated with the propensity of the koala sperm DNA to decondense following cryopreservation (Johnston 2019). Apart from the successful production of offspring, the primary achievement from developing artificial breeding technology in the koala has been the acquisition of basic reproductive information on anatomy, physiology and behaviour, all of which have not only contributed to captive breeding enhancement but also instructed our understanding of koala behavioural ecology and provided us with valuable tools for assessing fertility.

It is also becoming clear that small, fragmented wild koala populations will require a level of human intervention and management to maintain their genetic integrity and disease-free (*Chlamydia* and koala retrovirus) status. Assisted breeding technology in association with a thorough understanding of koala reproductive husbandry, in conjunction with koala movement patterns and the nature of the fragmentation of the landscape, will not only help secure the welfare of captive koala colonies, but is also likely to play an important role in the management and conservation of wild koala populations in regions where they are rare and endangered. The reader is referred to the concept of the Live Koala Genome Bank in Johnston and Holt (2019), which highlights the role of captive populations in what they describe as the practice of genetic connectivity, genetic capture, genetic rescue and genetic propagation - captive koalas are the central tenet of this paradigm.

### 10.5.6 Contraception

Although it is rare that contraception is applied to captive koala populations, fertility control with the use of slow-release hormone implants has been used in over-abundant populations of southern wild koalas (Menkhorst 2008). Studies suggest that implants containing either levonorgestrel or oestradiol have been successful in significantly reducing the fertility of koalas for at least four breeding seasons after they have been treated (Middleton *et al.* 2003; Hynes *et al.* 2010, 2019). The use of levonorgestrel implants in female koalas has also proven effective in preventing the production of young and the removal of the implants successfully resulted in reversal of the contraceptive effect in the next breeding season (Hynes *et al.* 2010, 2019). To date the use of GnRH agonists appears to be ineffective for male marsupial contraception (AZA Reproductive Management Center, Saint Louis Zoo), including koalas (S. Johnston pers. obs.).

## 10.6 Occurrence of hybrids

None.

## 10.7 Timing of breeding

It is generally accepted that the koala is a seasonal breeder, with a birth season in wild populations lasting from September-October to March-April (Tyndale-Biscoe and Renfree 1987; Allen *et al.* 2010). However, more recent studies by Ballantyne *et al.* (2015c) have suggested that although non-lactating captive koalas in south-east Queensland showed reproductive cycles during September-April, nevertheless there was a proportion of female koalas that continued to cycle through late autumn and winter, suggesting that with improved husbandry and nutrition, it might be possible to extend the breeding season in captive koalas. Given that koalas are weaned at ~11-12 months, the breeding season schedule apparent in most zoos could simply be a consequence of rebreeding 12 months after the previous birth.

There is some variation in the timing of the breeding seasons in zoo populations within Australia.

- Northern Australia (includes Queensland and northern New South Wales)
  Mating period: July-April (O'Callaghan 1996)
  Birth period: August-May (O'Callaghan 1996)
- Southern Australia (includes southern New South Wales, Victoria and South Australia)
  Mating period: September - February (Lee and Martin 1988)
  Birth period: October - April (Lee and Martin 1988)
- North America
  Mating season: March - May (Thompson 1987)
- Eastern China
  Breeding season: March-August (S. Eccleston pers. obs.).

Despite the male koala also being recognised as a seasonal breeder (semen quality, testosterone secretion, sternal gland size, bulbourethral glands) in the wild and captivity, Allen *et al.* (2010) have shown that semen quality (motility - at least 70%, rate of sperm movement of 3 and sperm concentration $10 \times 10^6$/mL) is still sufficiently suitable for artificial insemination and therefore presumably natural conception throughout the year. These authors and Hulse *et al.* (2021) have also noted that, in general, the semen quality of captive koalas tends to be better than that of wild koalas; the reason for this difference remains unresolved but may be a consequence of better nutrition and energy availability. Interestingly, Allen *et al.* (2010) have also shown that koala semen collected in winter for cryopreservation resulted in better post-thaw survival, albeit that koala semen has proven very difficult to freeze successfully (Johnston and Holt 2019).

## 10.8 Age at first and last breeding

### 10.8.1 Males

Males are capable of reproducing at 18 months of age, but most are prevented from gaining access to females by older and larger males in the wild (Martin and Handasyde 1991). In captivity, males as young as 16 months have been observed attempting to mate mount and have successfully produced young at 18 months of age (Thompson 1987). Interestingly, spermatozoa have been recovered postmortem from the epididymis of a deceased 11-month-old male (S. Johnston pers. obs.). Currumbin Wildlife Sanctuary had a male produce multiple offspring in one breeding season at 17 years of age (S. Eccleston pers. obs.).

### 10.8.2 Females

Females occasionally have their first young when ~18–24 months of age in the wild (though births have been observed in a female from 12 months per season in captivity (O'Callaghan 1996), when they approach adult size, but these young rarely survive pouch life. Most females breed first towards the end of their second year/beginning of their third year in the wild and may produce one young each year up until 10–15 years of age (Gall 1980; Thompson 1987; McLean and Handasyde 2006).

## 10.9 Ability to breed every year

Males are able to breed every year. The largest number of matings observed in a season is eight, although the libido of the males generally drops after 4–5 matings in captivity (O'Callaghan 1996). A breeding male at Currumbin Wildlife Sanctuary had 13 copulations in one season before rejecting females. Towards the end of the breeding season males that have had multiple matings/breeding access appear to reject females and are unlikely to attempt copulation (K. Nilsson pers. obs.).

Females are able to breed every year. If the female remains unmated, she will typically cycle every 33 days throughout the breeding season and sometimes beyond (Johnston *et al.* 2000a; Ballantyne *et al.* 2015c). A female that is mated and ovulation is successfully induced, but loses her pregnancy, will cycle again ~50 days after the first mating and can be mated again (Johnston *et al.* 2000a). As koalas have a relatively short anovulatory and ovulatory oestrous cycle, there will be numerous opportunities for breeding koalas throughout the breeding season if the initial attempts are unsuccessful.

## 10.10 Ability to breed more than once per year

Koalas can only raise one young per year. Twins are rare. When twins are born at least one usually dies or requires hand-rearing at ~4–6 months of age (A. Gillett pers. obs.).

## 10.11 Nesting requirements

None needed.

## 10.12 Breeding diet

There is no specific diet required during breeding, though even greater attention should be given to the quality of eucalypt leaves provided. Lactating females typically consume more leaf than non-lactating females (Moore and Foley 2000).

## 10.13 Number of young born

Approximately 65–73% of adult females breed per year in the wild, each producing a single young. Very rarely two young are produced (Martin and Handasyde 1991). In captivity 50–80% of females breed each year, depending on their age, with conception rate decreasing once > 7 years old where it reduces to 20% for females > 13 years of age (O'Callaghan 1996; S. Eccleston pers. obs.). It is presumed that the female is unable to rear two young at the same time successfully to independence as per the comments above (Lee 1988; Lee and Martin 1988; Martin and Handasyde 1991; K. *Nilsson pers. obs.).*

Koalas have approximately 70–90% juvenile survival rate in captivity; however, this does vary with age and is lowest for females < 9 years old and highest for animals 11–12 years old (O'Callaghan 1996). The time of highest pouch young mortality is in the first 3 months of life, before it is obvious from the physical appearance of the pouch that it contains a joey, so some joey losses may go undetected (O'Callaghan 1996). If the young is lost, the female has a 70% chance of losing any subsequent joeys (O'Callaghan 1996). More recent information on koalas in south-eastern Australia found survival rates for dependent young are 86–96% for the period from birth to permanent pouch emergence and 88–100% for the period from first permanent emergence to the completion of weaning (Martin and Handasyde 1991).

The loss of pouch young and back young has significantly increased in recent breeding seasons in south-east Queensland. The causes are currently unknown, but may include environmental condition changes, microclimate changes in enclosures, pathogens or pouch (Maidment *et al.* 2023). Necropsy details have revealed little information; some results have indicated malnutrition, exposure and mismothering. Losses have been observed from early pouch young to 11–13 months old (S. Eccleston pers. obs.).

Several practices can reduce the death rate of koalas (O'Callaghan 1996), including (1) separating males from females, as the males can increase the level of stress in the females and dislodge the joey while trying to mate; (2) grouping females with same age young together, as some joeys can die when the joey climbs onto a female that is not lactating or has a very small pouch young; (3) isolating females with pouch young so that neither the mother nor pouch young are interfered with by other koalas; (4) observing the pouch exterior for signs of discharge or excessive grubbiness, which can alert animal caretakers to pouch infections or illness of the joey that can be treated successfully if caught early enough; (5) observing the dam: if she is seen repeatedly licking or scratching at her pouch it may indicate infection; (6) observing the young, looking for ruffled fur, head tilt, cloudy eyes or discharge from the eyes; and (7) transferring pouch or back young, if the mother dies, to another female that is lactating.

Some facilities inspect the pouch of females by inserting an index finger into the pouch from between the animals' hindlegs while she is walking along a branch. This method is gentler and is less traumatic than if the female has to be caught (Drake 1982). In contrast to this approach some facilities avoid putting anything inside a female's pouch to prevent introducing bacteria or potentially dislodging the joey from the teat (especially in the first 5 months). In these facilities, observations by animal caretakers are used as the primary monitoring technique for any unusual discharges, odours from the pouch, and size of the pouch from outside are recorded (K. Nilsson pers. obs.).

If a female is mated, she should be pouch checked on days 34–36. Excessive pouch checking may interfere with a successful birth by stressing the female. Checking on day 35 is ideal. Behaviour changes in the following days and timing of the next oestrus will identify if the female was pregnant or if a pouch young was present or lost.

If a female has not given birth, then she can be mated again, but she is unlikely to return to oestrus until ~50 days after the initial mating. The koala can be pouch checked using two people and the koala is restrained in the all-fours position by one person while the other checks the pouch and progress of the pouch young and the teat on which the joey is attached. The bodyweight and body condition of the joey can also be checked when it is fully emerged from the pouch and independent from the mother (K. Nilsson pers. obs.).

## 10.14 Age at weaning

The age of weaning, for management purposes, is typically at 12 months and ~2.2 kg for northern koalas, whichever occurs later (K. Nilsson pers. obs.). The adult male should be removed when the female is first observed to have a pouch young. The young koala should be removed from the female when independent to avoid metabolic drain and prolonged lactation, and to allow the female time to recover condition before giving birth again. Young can be housed together, especially males, so they are more likely to be compatible as they get older.

Ballantyne *et al.* (2015d) has described the changes in prolactin secretion of lactating captive female koalas, noting that plasma prolactin concentrations were highest, coinciding with maturing young, between day 189 and day 231. Females do not exhibit any signs of oestrus while lactating until day 269, some 100 days before pouch young are weaned (Ballantyne *et al.* 2015d). Plasma luteinising hormone concentrations of the females were relatively high throughout lactation (Ballantyne *et al.* 2015d).

Young koalas in the wild usually stay in the general vicinity of the mother for another year. Some females settle in a home range nearby the mother. These females may be mated by their father in subsequent breeding seasons. Young males usually disperse from their mother's home range at ~2 years of age and may roam for the next 2–3 years before settling (Martin and Handasyde 1999).

## 10.15 Age at removal from the mother

The joey remains with its mother until ~12 months of age, at which time it weighs ~2 kg for northern koalas and ~2.5 kg for southern koalas (Smith 1979b; Lee and Martin 1988; O' Callaghan 1996). If another young is born at about this time, the bond between the yearling and mother abruptly breaks down. Yearlings attempting to suckle are treated aggressively by their mother and although often found in the same tree as the mother, they are no longer tolerated on her back. Yearlings are sometimes found with a surrogate mother, an adult female without back young and even occasionally with a male courting the mother.

## 10.16 Growth and development

A young neonate is 0.5 g and 19 mm long when born and bears a strong superficial appearance to those of other marsupials (Lee and Martin 1988; Lee 1988; Martin and Handasyde 1991). The forelimbs, shoulders and lips are well developed and the digits are equipped with deciduous claws. By contrast, the toes of the hindlimbs are buds.

The relationship of head length versus age in days was developed for Queensland koalas by Blanshard (1991) and for Victorian koalas by Martin and Handasyde (1991) (Fig. 7.14). Smith (1979b), Thompson (1987) and O'Callaghan (1996) examined the relationship between age and bodyweight for Queensland koalas and Martin and Handasyde (1991) developed a growth curve for weight with age for Victorian koalas (Fig. 7.15a). It should be noted that these figures contain data for both Queensland and Victorian animals and though the koalas

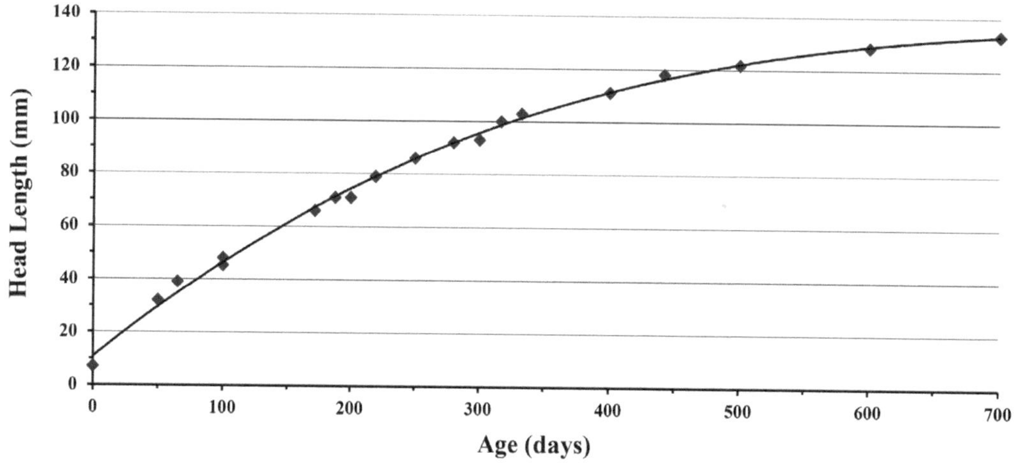


**Fig. 7.14.** Growth in head length for koalas. Derived from Blanshard (1991) and Martin and Handasyde (1991).

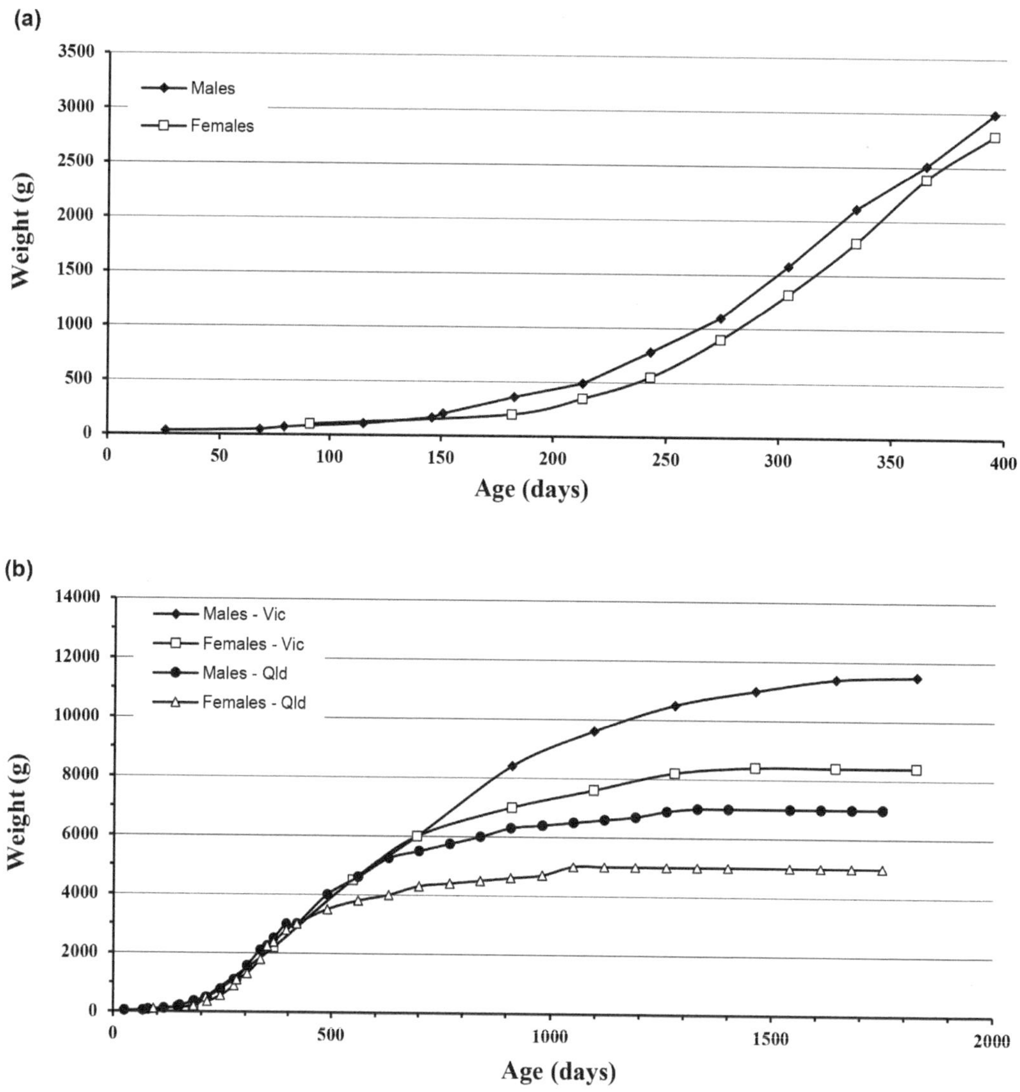


**Fig. 7.15.** Growth in bodyweight of the koala: (a) up to 400 days and (b) up to 1825 days. Derived from Thompson (1987), Blanshard (1991), Martin and Handasyde (1991) and O'Callaghan (1996).

remain relatively similar while juveniles, their bodyweights diverge greatly as adults (from ~400 days), with Victorian koalas being considerably larger (Fig. 7.15b). Woods (1999) uses head length and bodyweight, but the location (and hence adult size) is not known.

Development of pouch young is slow, with the young remaining in the pouch for 5-6 months and relying only on the mother's milk. When the joey is ~5-6months of age (170-210 days; Thompson 1987), the female produces a special type of faeces (known as pap), which the joey eats over several weeks. Koalas will feed on pap several times over a period of several weeks from the beginning of emergence and can continue up to the joey becoming a permanent back young at ~8 months of age. The process is always initiated by the joey nuzzling the mother's cloaca (the mother does not spontaneously start producing pap) and does not occur day after day (there is usually 2-3 days between feeds) (K. Nilsson pers. obs.). Pap introduces the appropriate gut flora and bacteria into the developing juvenile's stomach and caecum so that it can begin to digest the eucalyptus leaves and wean from the mother (Table 7.7). There is also emerging research that suggests the coprophilous fungi in the gut of koalas may represent a novel source of antimicrobials that warrant further exploration (Peterson *et al.* 2009; Hayhoe 2016). The joey commences eating eucalyptus leaves at 5-6 months of age and will progressively consume greater amounts of leaf until it is weaned at ~11-12 months after birth.

The composition of the gut microbiome is of critical importance, so herbivorous species need to ensure their offspring are inoculated with an appropriate microbial assemblage. The koala is a prime example of inoculating the infants with gut microbes, via pap, with the faecal microbiome differing among individual koalas feeding on different eucalypts. Pap provides joeys with a higher microbial density than regular faeces. Blyton *et al.* (2022) sampled faeces from captive joeys to track changes in the composition and function of their gut microbiomes and found that their microbiomes matured from pap feeding until independence, reaching an adult composition at ~1 year. They also found that the change in microbial composition increases the joey's microbiome capacity to degrade fibrous plant cell walls, which the researchers expect facilitates a successful shift to the nutritionally poor, high-fibre diet of eucalyptus leaves.

**Table 7.7.** Timetable of major developmental stages in juvenile koalas.

Derived from Thompson (1987) and Blanshard (1991).

| Stage of development | Age range (days) |
|---|---|
| Head out of pouch | 162-203 |
| First total emergence | 166-224 |
| Pap eaten | 171-213 |
| Eucalypt leaves eaten | 192-232 |
| First seen off mother | 214-275 |

Stages ages and bodyweight of koala joey development (approximate guidelines) (S. Eccleston pers. obs.) are given below.

- day 1: ~1.5-2.0 cm long in pouch, 35 days post mating
- 6-13 days: skin pink, no pigment on nose, hands or feet
- 28 days: light-brown pigment on dorsum of nose, eyes closed, ears open, pads slightly pigmented, extremely fine hair on nose
- 35 days: faint pigmentation on parts of face, hands and feet
- 64 days: pads dark, grey skin pigmentation
- 69-80 days: nose showing pigment, eyes still closed
- 53-126 days: eyes begin to open and bodyweight ~200-250 g
- 91-123 days: long whisker guard hairs develop on chin and fine hair begins to develop on upper body
- 105-140 days: claws are amber brown
- 163-182 days: darker hair on the head and shoulders, very fine pale hair on lower body, thicker hair beginning to develop on hindlegs
- 184-189 days: dense, smooth fur over lower half of body, longer fur over shoulders
- 160-203 days: head emerges from pouch
- 186-211 days: first total pouch emergence
- 180-215 days: pap consumed
- 300-450 g: may continue this process for much longer; all joeys seem to differ
- 190-232 days: eucalyptus leaves eaten, weighs 300-450 g
- 189-212 days: whorls of fur apparent over shoulders and fur growth still more prominent on upper half of the body; weighs 450-700 g at 6-7 months
- 214-300 days: first seen moving independently of mother.

## 11. ARTIFICIAL REARING

### 11.1 Housing

As with all native mammals that have been taken into care, minimising stress is a major consideration.

Choosing suitable housing can help to create a stress-free environment. To achieve this, several factors should be considered, including:

- securing the area from children and other animals (particularly domestic dogs and cats)
- maintaining a high standard of hygiene
- escape-proofing the area
- clearing the area of obstacles and hazards
- ensuring the area offers shelter from extreme weather and noise.

The type of accommodation that needs to be provided to the juvenile koala depends on the age of the animal. There are three main stages (Wildcare Australia 2018).

- Unfurred or just furred – should be placed in multiples layers of artificial pouches that simulate, as closely as possible, the security and warmth of the mother's pouch. Natural-fibre products such as cotton are required as they retain warmth better. In unfurred joeys, cotton pouches must be soft to reduce the risk of causing rubbing sores to the hip/pelvic region. The joey should be placed in 2–3 layers of cotton pouch liners (i.e. one inside the other), which are then placed within a thicker, warm outer pouch, such as a feather or woollen pouch. (Wildcare Australia 2018). The koala joey can be contained in a plastic rescue basket (for a small inactive pouch joey) or a deep cane basket with suitable heating, clean pouches and blankets, and the provision of a soft teddy bear for the joey to hold onto provides an ideal environment.
  Gentle warmth can be provided to the koala with the use of a SnuggleSafe® Microwave Heat Pad (Wildcare Australia 2018) or a neonatal ICU humidicrib if available as these are portable and regulate temperature much better (A. Gillett pers. obs.). A thermometer is also recommended to monitor the temperature in the 'pouch' via a probe and the reader on the outside measures the ambient temperature. An example of a koala pouch with several linings and a temperature probe is shown in Fig. 7.16. Note that the thermometer probe should be placed in one of the layers of the pouch beyond the reach of any the joey. If it is placed in same actual pouch, there is a risk of the joey chewing on it.
- Fully furred pouch young to front young – can be placed in a larger pouch similar to that described for stage 1 and housed in a deep cane basket with several soft baby blankets and teddy bears. (Wildcare Australia 2018). The joey can be given a smaller teddy bear, which is placed in the pouch with them. At this age, they will still require gentle warmth, particularly at night, and a SnuggleSafe® Microwave Heat Pad is a popular option because it is waterproof and easily cleaned. The basket can be placed either on the ground close to the perch or hung securely from a branch of the perch. At this age the koala will explore and actively climb the perches. The placement of the basket at this stage is to familiarise them visually with larger structures. The basket is also intended to be the middle ground for easy transition to small perches for exploring as they become more confident. Most koalas at this size just climb around on the edge of the basket or on its handle until they feel confident enough to grip a perch, but even then will not venture far from the basket setup until a bit older. During periods of cold weather, young joeys of this age may benefit from being housed in a humidicrib ICU at night. Once the joey has pap fed and is starting to nibble on browse, a small bunch of soft tips should be placed in a small water bottle within easy reach of the joey (Fig. 7.17). The use of a soft toy as a substitute for the mother koala has been found to be very successful. From the onset of hand-rearing the joey, it is placed onto the soft toy, which it clings to eagerly. This helps minimise stress on the joey and allows the rearer to handle the joey more easily for feeding and cleaning. The soft toy also makes the introduction of the koala into new environments less stressful as the soft toy

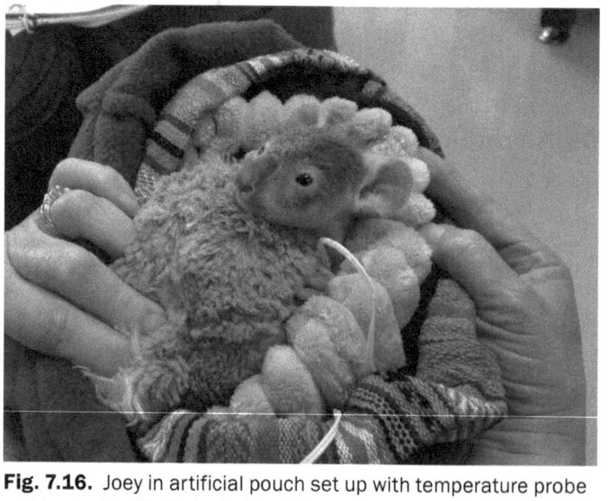

**Fig. 7.16.** Joey in artificial pouch set up with temperature probe and toy teddy. Photo by Amber Gillett.

**Fig. 7.17.** Joey in artificial pouch set up with toy teddy and leaves for feeding. Photo by Karen Scott.

can go as well. Even when hand-reared, koalas that are fully-grown often readily accept climbing onto a soft toy, which makes it easy to carry them.

- Fully furred back young – should be housed with easy access to climbing perches (both horizontal and vertical), which should include secure browse holders in which to place fresh leaf each day. Small-furred young should be offered a deep cane basket with soft baby blankets and several soft teddy bears to sleep in during the day and then utilise the perches when active at night (Wildcare Australia 2018).

A variety of pouch liner sizes will be needed to accommodate the growing koala joey (Wildcare Australia 2018). For example, unfurred joeys can be placed within pouches 15 × 20 cm; just furred animals can be placed within 20 × 25 cm; and furred animals placed within 25 × 30 cm pouches (Wildcare Australia 2018).

When the joey gets older, two thick-barked branches should be placed in an upright position, making sure the branches have a good fork in them at approximately chest height of the animal caretaker for the koala to sit in. A horizontal branch connecting the two uprights will create a good climbing structure. By 12 months of age the koala should be in an outdoor enclosure with various climbing structures (Fig. 7.18). When the joey is old enough to go into a larger enclosure with forked branches it can be placed with a soft toy in the fork until it is confident enough to leave the soft toy.

## 11.2 Temperature requirements

Furless joeys need special attention to ensure they maintain a constant temperature. A temperature of ~32–34°C is recommended for unfurred koala joeys, which is reduced to 28–30°C for furred animals. The use of a

**Fig. 7.18.** Example of climbing structure that can be provided to a young koala. Photo by Karen Scott.

humidicrib ICU is highly recommended for unfurred and just furred joeys as it greatly increases their chance of survival. Furred joeys can be maintained at an appropriate temperature with the use of a heating pad or SnuggleSafe® Microwave Heat Pad. Use a minimum/maximum thermometer that has a plastic-coated probe that can be placed next to the joey to monitor the temperature. Heat pads should be thermostatically controlled to avoid overheating. Furred joeys can be given a heat pad in an upright position in a corner of their basket, which will allow the joey to move to and from the heat source to adjust its own temperature. By 8–9 months of age, heating should only be needed at night and by 11 months of age the joey should be completely weaned from heating.

## 11.3 Specific requirements

When the koala first arrives:

- minimise handling
- have it examined by a veterinarian
- get a history of the animal
- record initial bodyweight and measurements
- estimate its age
- do not attempt to feed it until it is warm as it may aspirate milk
- organise bedding or a heat pad
- organise the appropriate diet.

The animal should initially be weighed daily to ensure bodyweight gain. Once fully furred and approaching weaning it can be weighed every 2–3 days to ensure continuing weight gain. If there is a failure to gain weight, or a change in the rate of gain, this should be investigated by a veterinarian and the diet investigated if necessary.

The skin of unfurred and slightly furred young should be kept moist with the use of Sorbolene cream (not with added glycerine) so that the skin does not become dry and cracked (George *et al.* 1995). Baby oil does not appear to be properly absorbed and tends to stay on the skin surface where it rubs off and is absorbed by the liner bag fabric (George *et al.* 1995). When using a humidicrib ICU at this stage of development, there is generally no need to use any artificial moisturises as the humidity produced ensures that the skin does not become dry.

When first brought in for hand-rearing the joey may be dehydrated; if so, it can be given plain boiled water, with 5 g (1 teaspoon) of glucose to 100 mL of water or 1 g of electrolyte replacer if available (Austin 1997). It is important to warm the joey before feeding otherwise there is a greater risk of inhalation pneumonia. If this is taking time give fluids subcutaneously and bottle-feed later.

Stress may be a major factor in the success of rearing native mammals. Therefore, it is important that noise is kept to a minimum, do not over-handle and hygiene should be maintained as stringent as possible.

## 11.4 Diet and feeding routine

### 11.4.1 Natural milk

Koala milk composition shows deviations from the general marsupial pattern, with the exception of the common brush-tailed possum and eastern ring-tailed possum (*Pseudocheirus peregrinus*), which are also folivorous (Krockenberger 1996). Milk solids decrease from a peak of 33.6% at 220 days to 28.3% at 300 days, after which they remain relatively constant, in contrast to the solids of most marsupials, which rise at this time. Lipids provide a major source of energy in early to mid-lactation, but do not rise at pouch exit, unlike lipid levels in other marsupials that rise sharply at pouch exit to high levels in late lactation (Krockenberger 1996). Total carbohydrates of koala milk range from 8.8% at day 140 and decrease steadily to 1.1% at 380 days (Table 7.8). Oligosaccharides comprised > 90% of the total carbohydrates during most of the lactation period, but decrease to < 5% towards the end of lactation when there is an increase in lactose (Marshall *et al.* 1990; Krockenberger 1996). Concentrations of sodium, potassium, calcium, and phosphorus are within those typically found in marsupials; however, copper was not detected, suggesting its concentration is < 2 mg/L (Marshall *et al.* 1990).

### 11.4.2 Milk formulas

The main commercial formulas for hand-rearing koalas are listed.

- Biolac® Formula Milk – the three formulas are M100 for furless joeys, M150, which is a transitional milk to feed when dense fur has developed, and M200when the animal produces solid dark pellet droppings as it contains elevated lipid in the form of canola oil. Mixing the formulas assists in the transition between the two formulas. Unfurred joeys may take up to 30% of their bodyweight per day, just furred up to 20% of their bodyweight, decreasing to 10–20% as they become more active and are eating more leaf, until finally decreasing to 5–10% of their bodyweight as their diet consists mostly of their natural food. It has been recommended that Wombaroo® Impact should be added to Biolac® at the rate of 1 g to 100 mL of made-up milk formula (Wildcare Australia 2018).
- Di-Vetelact ® (Lillelund Pty Ltd) – a widely used low-lactose milk formula. Due to its low energy concentration when prepared as directed, some groups advise the addition of mono- and polyunsaturated fats such as canola oil, as with the Wombaroo® diets. The addition of saturated fats in the form of cream has been suggested, but it is too highly saturated and can lead to the malabsorption of calcium. Di-Vetelact® should be fed at ~20% bodyweight, except for very small joeys (< 100 g). It has been recommended that Wombaroo® Impact should be added to Di-Vetelact® at the rate of 1 g to 100mL of made-up milk formula (Wildcare Australia 2018).
- Wombaroo® Koala Milk Replacer – the most widely used milk formula for koalas. Charts are provided to assist in determining the type and volume to be fed. There are two formulas: 'Early Lactation', which is given until the head first appears out of the pouch at

**Table 7.8.** Concentrations of the major constituents of koala milk.

Derived from Green (1984), Marshall *et al.* (1990) and Krockenberger (1996).

| Total Solids (%) | Carbohydrates (%) | Lipids (%) | Protein (%) | Calcium (mg/L) | Iron (mg/L) |
|---|---|---|---|---|---|
| 28.3–33.6 | 1.0–8.8 | 10.1–18.0 | 5.5–12.5 | 4000–5100 | 13.0 |

~160 days, and then after a transition period from 160 to 180 days, the 'Late Lactation' formula is given up to 250 days, when the young would normally be on the mother's back, and another transitional period until 270 days during which the late lactation formula is used. It has been recommended that Wombaroo® Impact should be added to Biolac® at the rate of 1 g to 100mL of made-up milk formula (Wildcare Australia 2018).

- Karicare+ Soy Milk Formula ® (Nutricia) - a human infant formula commonly used in koala rehabilitation as a 'paste', typically used when a fully furred joey is not receiving adequate nutrition from the mother due to her poor health, is extremely malnourished or require supplementary nutrition. It is also beneficial for feeding older orphaned koalas that will not tolerate handling for normal milk feeds. Two level scoops of powder are added to 10 mL preboiled warm water to forms a 'custard' consistency (Wildcare Australia 2018).

## 11.4.3 Feeding apparatus

Very small koala joeys that are fed < 10 mL per feed may be fed with a glass or plastic syringe and a teat to allow the milk flow to be controlled and reduce the chance of aspiration. A standard koala teat will fit on the end of a 10-mL glass or plastic syringe for older joeys (Wildcare Australia 2018).

The milk formula should be heated until it is ~36°C (not too hot that it burns the mouth of the joey). Each day ~10–20% of the bodyweight of the individual should be given over the 24-h period (or the amount specified on the Wombaroo® chart for a joey of that age). This amount is divided up into the number of feeds given per day. Care needs to be taken not to over feed Wombaroo® as diarrhoea can occur. Initially only one person should feed and handle the koala. Once settled, two people can be used if required.

## 11.4.4 Feeding routine

When deciding on a feeding regimen for the koala joey several things need to be considered.

(1) Read and understand the manufacturers' guidelines for making the milk formula.
(2) Decide how much the animal can comfortably consume in one feed.
(3) The age of the joey.
(4) Is the joey eating leaves yet?
(5) Is the joey dehydrated and need more frequent fluid intake?

When a young koala comes into care it is important to assess its age and hence its food requirements. An outline of the milk feeding schedule for koalas is given in Table 7.9. Feeding is initially required every 3–4 h, around the clock, regardless of the age of the joey, until the animal is settled and feeding well.

Once the joey's routine has been established and it is feeding well, the time between feeds can be increased if the joey is more than 180 days old. The number of feeds should also be varied depending on the health and keenness of the joey for feeding. For example, a poor feeder may require more frequent small feeds than an animal that feeds well of the same age.

A small koala joey that is fed < 10 mL per feed should be fed with a syringe and teat, but as it grows and is consuming more than 10 mL of milk per feed it should be transitioned to a glass bottle and teat (Wildcare Australia 2018). Joeys often feed better initially if their eyes are covered, as it helps keep them focussed on feeding. This technique tends to make the feeding process a lot easier and less stressful.

When koalas are ~400–600 g, their behaviour begins to change and most will start to refuse feeds between mid-morning to early afternoon (Wildcare Australia 2018) and they can be transitioned to demand feeding. Once

**Table 7.9.** The number of feeds per day for different aged koala joeys in Queensland.

Derived from Wildcare Australia (2018).

| Weight (g) | Development | Feeds per day |
|---|---|---|
| < 100 | Unfurred, skin transparent; eyes closed | 6–7 |
| 100–200 | Unfurred, pink skin; eyes closed | 6 |
| 200–300 | Unfurred, skin starting to turn grey Eyes open or close to opening | 6 |
| 300–450 | Short sleek fur; eyes open | 6 |
| 450–700 | Lightly furred | 6 |
| 700–1100 | Fluffy furred all over | 4 |
| 1100–1500 | - | 3 |
| 1500–2000 | - | 2 |
| 2000–2500 | - | 1 |
| > 2500 | - | Weaned |

this occurs, joeys will be eager to feed from mid- to late afternoon through to early morning. It is recommended not to feed a joey if it is unwilling to take it as it causes significant stress and can lead to aspiration pneumonia. If a joey is unwilling to take more than two consecutive feeds, they should be assessed by a wildlife veterinarian as soon as possible. Behaviour that indicates a young koala is ready for feeding includes (Wildcare Australia 2018):

- climbing towards the top of the pouch and looking around
- climbing out of the pouch and onto a teddy
- starting to explore its surrounds and may turn its attention to their leaf pot
- starting to vocalise
- eagerly taking the teat and commencing suckling when offered
- consuming a reasonable amount of milk before contentedly falling asleep.

Generally, if a young koala has started eating leaves, it is assumed that it has already eaten pap from its mother. A way of determining if a koala has consumed leaves (and therefore pap) is by checking for brown staining on the erupted cheek teeth or plant cell walls in faecal smears. Soft faeces or pap is a special type of faeces produced by the female when her joey is ~6 months old. If giving the joey crushed up faeces (pap), first screen the faeces for parasites.

If the joey has not consumed pap before hand-rearing, it is important to be aware of the signs a joey may display when it requires pap, generally at ~6 months of age, but can be any time from when the joey is starting to cut fur to being fully furred. Signs that an orphaned joey may require pap include refusing milk feeds, being unsettled (appearing hungry but won't accept milk) and nuzzling at their soft toy or the rearer. Collecting pap (caecal content) from any freshly deceased subadult or adult koala (either male or female) suffice. Two types of pap have been described: 'clean' pap that obtained from the caecum of disease-free koalas and 'dirty' pap from koalas that showing clinical signs of disease, especially chlamydiosis (Wildcare Australia 2018). The use of 'clean' pap is the preferred option; however, 'dirty' pap has been used when there is no other option available.

If pap is not available, then collect fresh faeces from an adult koala and mix it into a slurry. Offer the joey as much as it wants until the joey no longer willingly accepts it. It can also be included in the milk for bottle feeds. Pap substitute should be offered regularly to all hand-reared joeys from 5–6 months of age to stimulate the natural pap feeding process. Pap can also be fed regularly to adult koalas that have gastrointestinal disturbances, mainly when the flora of their gastrointestinal tract has been affected by antibiotics.

As the joey begins to start eating eucalypt leaves, the quantity of milk formula will remain the same, but the number of feeds will vary depending on how many leaves have been eaten. During more advanced stages of leaf eating (300–365 days), when the joey is eating a lot of leaves (or should be), it is important to make sure they get a good milk feed in the morning and at night. Hand-reared joeys are often fed on milk for slightly longer than a parent-raised animal to ensure they can cope with any extra stresses and have a good bodyweight before weaning.

## 11.5 Data recording

When an animal is first brought in for hand-rearing, its sex and approximate age, using growth charts, should be recorded. During the hand-rearing process the following important information should be recorded on a daily basis:

- date when the information is recorded
- time when the information is recorded
- bodyweight to the nearest 1 g if possible, weighing the koala while on its soft toy
- general activity and demeanour
- characteristics and frequency of defecation and urination
- amount (g) of milk offered
- milk consumption at each feed
- species of leaves offered and eaten
- veterinary examinations and results
- date when pap or faeces was offered.

This information serves several purposes, including providing important background information such as food consumption which will assist a veterinarian to reach a diagnosis if the animal becomes sick or fails to grow or gain bodyweight. It also allows a comparison with growth curves to assess progress (section 10.16) and to allow new growth curves to be established for measurements in which they presently do not occur. The developmental stages and measurements outlined in section 10.3.1 should also be recorded on a weekly basis.

## 11.6 Identification methods

Generally not needed; however, an implant chip can be used once furred. In some states, orphaned wild koalas are required to be microchipped as small back young.

## 11.7 Hygiene

Maintaining a high standard of hygiene is critical to the survival of the koala joey. Emphasis needs to be placed on the following:

- maintaining a clean pouch lining at all times. Older joeys may be able to be trained to urinate on newspaper by keeping a piece of newspaper with the smell of urine on it. The joey is placed on the paper while clinging to its toy. Slowly turn the toy until the koala's bottom is touching the paper. They usually place their rear legs on the paper, urinate and jump back on their toy. When joeys are ~7 months old, they tend to come to the ground on their own accord to urinate and continue to do so for several months
- personal hygiene - wash and disinfect hands before and after handling the joey. Use antibacterial solution for washing hands with furless joeys, as their immune system is not well developed
- washing hands between feeding different joeys
- using boiled water when making up formulas for joeys of all ages
- cleaning up spilt milk formula or paste, faeces and urine from the joey's skin and fur as soon as possible and the fur dried to prevent bacterial or fungal skin infections
- washing all feeding equipment in warm water and dishwashing detergent and sterilising it in a microwavable baby steam steriliser or F10® disinfectant
- storing the teats and bottles in the fridge in an air-tight container after they have been disinfected.
- only heating the milk once and discarding leftovers
- avoiding contact with other animals unless certain they pose no health risk
- stimulating the joey to toilet before or after feeding. As with other marsupials toileting can be done by the application of warm water to the cloaca using a soft tissue to stimulate urination and defecation, which allows the animal to keep drier and warmer in its pouch. Koala joeys will start to urinate and defecate on their own at an early age, generally by the time they become a small, back young
- keeping a furless joey's skin hydrated with Sorbolene cream applied after each feed until fur appears, unless using an ICU humidicrib
- changing the water in the leaf pots daily.

## 11.8 Behavioural considerations

Koalas become highly attached to the person who is hand-rearing them and can resist weaning, resulting in it taking much longer than it would if the mother was raising the young. Therefore, it is important not to allow the koala to become overly dependent on humans. The use of the teddy bear minimises the bonding of the koala to the rearer as the joey becomes more attached to its scent on the teddy and should willingly climb onto it for anyone. When koalas are being reared for release, carers should spend the least amount of time possible (feeding and cleaning purposes only) with a joey in the later stages of hand-rearing to reduce their reliance on the carers for comfort.

## 11.9 Use of foster animals

Not recommended.

## 11.10 Weaning

At ~6–6.5 months of age, after the consumption of pap, the joey should be feeding on eucalyptus leaves. A general rule is to decrease the formula by 5% per week as long as the joey continues to gain bodyweight at a minimum of 5–10% of bodyweight per day. The joey should be given the fresh, light-green tips of leaves from at least three species of eucalypt per day and these should be changed at least twice per day (and should be sprayed with water several times per day, especially in hot weather). The leaves should be placed in leaf pots filled with water to keep them fresh. Make sure the leaf pots always have browse in them, as juvenile koalas may fall into the pot and get stuck, and the mouth of the pots should be small enough so that a joey cannot fall into it and drown.

The joey should be completely weaned by ~12 months of age. Often the koala will start to refuse the formula at this time and start weaning itself. The basic rule is to decrease the milk content as the leaf intake increases. Water should be available once koalas are starting to climb on the perches.

## 11.11 Release procedures

When it is getting closer to the release stage, start collecting leaves from the future release site so that the koala is familiar with those species. The release site should be in an area with a healthy residential population of koalas and be away from roads, residential areas and dogs.

Some studies of the release of wild koalas after rehabilitation suggest that in the first 30 days after release females tended to stay in the release site while males tended to progressively move away (Ellis *et al.* 1990). However, this is not always the case, as each animal behaved differently. Some stay in a small area, others immediately wander large distances, often heading back to the site where they were rescued as adults (presumably returning to their known home range) (A. Gillett pers. obs.).

Post-release monitoring of rehabilitated koalas has been limited, so long-term success rates are unknown. Gauging the success of rehabilitation has largely been limited to using release as the measure of success, though some long-term post release survival studies in recent years have provided an additional measure of success (Gillett and Fitzgibbon 2024). Using ear tag records as unique identifiers of individual koalas, Lunney *et al.* (2022c) investigated a 31-year set of 5051 koala admission records (1989–2020) of a koala rehabilitation group, Friends of the Koala, in Lismore, north-east New South Wales, searching for records of koalas that had returned to rehabilitation for a second, third or fourth time. Of the 1771 koalas that were released, 270 were admitted to rehabilitation two or more times. Re-admission figures mostly fitted the pattern of the reasons for first admissions, meaning that the released koalas had become part of the local koala population. Of the 270 re-admitted koalas, 66% had remained for more than 6 months in the wild, and 33% had remained in the wild for more than 2 years. The conclusion drawn was that rehabilitated and released koalas can survive in the wild long-term, even though some koalas were re-admitted after a very brief period post-release. This is a more robust view of success for koala survival after rehabilitation than simply the proportion of released versus non-released koalas. What this study has shown is that koalas fit within an international pattern of release, rejoining the wild population, including long-term survival, that justifies the effort and care required to rehabilitate each individual koala.

Another study monitored 36 koalas post-release in the greater Sydney region in New South Wales and assessed if the koalas were fed appropriately while in care (Leigh *et al.* 2023). The koalas were allocated to one of three categories of medical intervention (none, minor, major) during rehabilitation. The 'minor' category was assigned if the koala received minor interventions such as water for dehydration or topical creams, whereas 'major' was assigned if the koala received major medical intervention including surgery, intravenous fluids for dehydration, oral antibiotics or chlamydia treatment. After release the koalas were tracked and their survival, movements post-release and variation between the species of browse fed in care versus browse selected by koalas in-situ were evaluated. Leigh *et al.* (2023) found that the post-release survival rate of koalas was 58.5%, with only koalas that received medical intervention not surviving. They discovered a critical threshold for death occurred at 2 weeks post-release, with death being related to low body condition and poor climbing ability at the time of release. They found there was a poor overlap between the tree species that koalas were fed in care and suggested that the captive versus wild dietary gap is an important consideration (Leigh *et al.* 2023). Their results revealed a high number of tree species that koalas used that were not fed in the rehabilitation facility, including important food trees. They recommended further research, given the potential impact of diet on koala health. This, along with so many other aspects of koala biology, both in and out of a captive environment, is an important area for further study.

A recent study by Gillett and FitzGibbon (2024) reported a survival rate of 92% post release in rehabilitated koalas monitored for more than 3 months. This study included animals that had been treated and rehabilitated at a wildlife hospital for serious trauma, illness and orphaning, with time in care ranging from 56 to 571 days. On average, orphaned koala spent 371 + 98 days in care and koalas admitted for injury or disease spent 187 + 47 days in care prior to release. Length of time in care did not appear to affect post-release survival and high rehabilitation success was largely attributed to high-quality care during rehabilitation and providing access to outdoor enclosures during rehabilitation to ensure sufficient fitness, strength and behavioural competence prior to release.

## ACKNOWLEDGEMENTS

Sincere thanks to Jane DeGabriel, Bill Ellis, Jo Loader, Ben Moore and Karen Scott for providing valuable comments on the chapter. Karen Scott also provided several useful images.

## ADDENDUM 1. THE MANAGEMENT OF EUCALYPTUS PLANTATIONS FOR KOALA FODDER

Derived from O'Callaghan (1999) and Knight (2004)

Qualified horticulturists, experienced with eucalypt plantation management, should be consulted at all stages of plantation development. The growth rates of plantation eucalypts vary depending on the climate, soil conditions, insect pests and harvesting rates.

## General principles

Successful plantations aim to produce high-quality, palatable leaf for koalas by removing factors that adversely affect the growth of the tree.

### 1. Why plantations?

- Although costing more initially in capital expenditure, plantations are generally cheaper than the costs associated with roadside collection or purchase from commercial sources.
- Plantations provide a reliable and controlled supply of feed for koalas, as well as other animals.
- Zoos and wildlife parks have an obligation to be sustainable in their use of natural resources.
- Plantations allow better control of the quality of food being given to the koalas.

### 2. Site selection

Factors to consider when purchasing or selecting land to be used for koala leaf plantations include:

- distance to the facility holding the koalas, as the travelling distance can affect leaf quality
- soil quality:
  - soil type, which should suit 80% of the species intended to grow
  - soil fertility - high soil fertility will save costs through less use of fertiliser and will generally produce highly palatable leaves
  - salinity level - low salt levels are generally preferred
  - pH - should be ~6.0–6.5.
  - drainage - most *Eucalyptus* species prefer well-drained soils
  - avoid soils with a dominant clay or sand content as they are often unsuitable, but a sand or clay-based loam with a depth of ~50–100 cm is generally suitable
- existing vegetation:
  - do not plant under a canopy of any species, as competition for light and water will hinder growth
  - existing vegetation will give an indication of soil conditions (e.g. ironbarks and bloodwoods = poor soil; redgums = fertile soils)
- access to the site in inclement weather (e.g. clays stay wet a long time and may make vehicle access not possible)
- surrounding land use that may cause damage to the plantation or leaf quality (e.g. fires, use of chemicals, smoke and ash)
- aspect of the land (avoid slopes that do not face the sun and slopes facing strong prevailing winds as these can greatly reduce the growth of the trees in the plantation)
- access to good, year-round, quality and quantity of water for irrigation as part of the plantation management. Maintaining the availability of moisture to the plants will enhance their productivity and the quality of the browse produced, so plantations should ideally include some form of irrigation.

### 3. Site preparation

- Considerable preparation is required before the establishment of the plantation. This includes an assessment of the existing vegetation cover, condition of the soil, level of infrastructure required. This stage is important in maximizing the productivity of the plantation.
- Site preparation aims to provide optimum conditions for tree growth by reducing competition for nutrients, light and water and any hindrance to root development.
- All existing vegetation should be removed except in areas that are not going to be planted.
- Retention of vegetation surrounding the plantation is encouraged as it attracts birds into the area and reduces eucalyptus-specific pest opportunities.
- Rows should be formed according to land contours to prevent erosion.
- Ripping of rows to a depth of 0.5–1.0 m helps to breakup compacted soils so the plant roots will be able to more easily move through the soil. Poorly developed roots can result in stunted growth and death of trees.
- All rows should be mounded to increase aeration and prevent water logging and erosion.
- Rotary hoeing may be required to provide friable soil for root/soil contact, particularly when the trees are small.

### 4. Species selection

- Consider the eucalypt species that the animals are being fed at present and which are the preferred species. Preferred species should make up at least 70–80% of the tree species planted.
- Other species and new species not fed before should be planted in small 'trial' numbers to gauge animal's response, novelty value and the effect of plantation palatability versus wild eucalyptus palatability.

- Plant at least five staple (i.e. preferred) species and three other species. The greater number of species planted the greater the flexibility in feeding the animals.
- New tree species and those from other climatic or geographical areas should be trialled first in small numbers as they may not grow well or be palatable.
- Select species according to soil type and conditions (e.g. if wet soil, plant water-loving tree species).
- Different eucalypt 'types' have different growth rates that generally relate to the amount of leaf on the tree (large amount of leaf infers faster growth rates). Iron-barks and boxes tend to be slower growers than the gums and stringy-barks.

## 5. Planting density

- Density of trees planted depends on the size and age of browse preferred (e.g. tips *v.* leaves to be collected) and the growth habit of each tree species.
- Plant spacing will depend on the type of browse the koala will be provided, the size of the plantation, and the number of trees needed. The distance between trees typically ranges from 1.5 to 3.0 m. Another option to calculate the area required is to include one tree per 4 $m^2$.
- A koala eats 600–800 g (10% of its bodyweight) per day. The quantity will depend on the age of the leaf, moisture content, season, activity level of koalas (e.g. breeding, lactating) and the age and weight of the koala. A general ratio of trees to plant is 1000 trees per koala.

## 6. Planting

- Trees should be planted as tube stock.
- The optimal time to plant is in late winter (after the last frost) or autumn when the climate is still relatively warm, there is moisture in the soil and there is the prospect of further rainfall. Low rainfall can be compensated by the use of irrigation.
- The layout of the plantation needs to be designed to enhance the productivity of the plant's growth and the efficiency the browse collection. To achieve this, plantations are typically designed in blocks or rows.
- Space between rows needs to be large enough for vehicle access to collect the branches and for maintenance without causing damage to the trees. A distance of 2–4 m between rows is typically required.
- Trees should be planted after being grown in tubes to promote straight root growth. This ensures tree establishes faster because the roots are facing downwards instead of having formed a ball.
- Trees must be removed from the tube pots as soon as possible because failure to do so has been shown to cause poor root growth and stem rot.
- Staking trees should generally be avoided as it may result in poorly developed root systems. Staking can artificially support the tree and weaken root growth. Rubbing on stakes can also cause stem rot.
- Consideration should be given to protection from vertebrate pests, including macropods, deer, rabbits and hares. Tree guards also help to provide a microclimate for the plants and protect them from dry winds and frosts.
- Tree fertiliser may be useful when planting to ensure adequate levels of nutrition, especially when planting in soils with lower fertility.

## 7. Soil fertility and the relationship to palatability

- Soil type and the amount of available nutrients in the soil determine leaf palatability, with koala leaf preference shown to be directly related to the percentage of nutrients to anti-nutrients (e.g. tannins) in the leaf.
- Trees produce fewer anti-nutrients in fertile soils. Anti-nutrients include tannins and cineoils that make the leaf taste bad and protect it from leaf insects.
- Important nutrients for good *Eucalyptus* growth are nitrogen, phosphorus, potassium, calcium, magnesium, copper, sulfur and sodium.
- It is important to have high nitrogen, high potassium and low phosphorus.
- Fertiliser should be applied twice per year.
- Quantity of fertiliser used is determined by harvesting rates.
- It is preferable to use organic fertilisers; however, chemical fertilisers can be used if they are checked to ensure they do not affect leaf palatability.
- Soil test at least once every 3 years to determine effects on soil fertility.
- Feeder roots occur in the top 10 cm of soil so avoid excess watering after fertiliser application.
- Do not clump fertiliser as this may cause burning and tree death.
- Avoid fertilising immediately before or after spraying with herbicides, as this may cause a chemical reaction and affect tree growth.

## 8. Irrigation

- The ability to provide water to the trees is essential to eliminate drought as a limiting factor to tree growth.
- Dripper lines, T-tape or overhead irrigation can be used. Overhead is preferable as there are fewer problems and the sprinklers wash the leaves regularly, thus removing any foreign substances.
- Irrigation frequency will depend on soil type and natural weather conditions.
- When trees are planted, they should be irrigated until well established (e.g. noticeable development or growth, usually ~2 weeks).
- After trees are established, they should be stressed to encourage root growth. Stress the trees over the first year to the point where new growth starts to droop. At this point trees should be well watered. This will only need to be done two or three times in the first year.
- Irrigation on mature trees can be used to promote new growth in blocks.

## 9. Weed control

- Competing vegetation growth is a major limiting factor to the growth and survival of plants through competition for light and nutrients.
- Grass is a major consumer of nutrients due to its fast growth rates.
- Weed control should ensure a 1 $m^2$ area around the base of the tree free of vegetation.
- The most effective preventative weed controllers are weed mat, carpet underfelt, mulch and mowing and slashing. Mulch can include straw, garden mulch, newspapers, lawn clippings, wood shavings and sawdust
- When spraying weeds, use a non-residual herbicide such as Bioactive®.
- Avoid sprays with surfactants as they are damaging to the environment.
- Care must be taken when spraying young trees as green stems absorbs poison at one-third the rate less than foliage.

## 10. Disease

Eucalypts are susceptible to various diseases:

- foliage diseases, including fungal diseases such as *Eucalyptus* shoot blight and powdery mildew
- root rot diseases such as cinnamon fungus (*Phytophthora cinnamomi*) and *Armillaria* sp.
- stem diseases such as fungal stem canker
- wood rot, which generally infects older plants
- damping off, which is common in young seedlings in warm humid conditions. Often caused by fungi of the genera *Fusarium*, *Rhizoctonia*, *Phytophthora* and *Pythium*.

## 11. Pest control

- Eucalypts are susceptible to a range of adult and larval insect pests, including Christmas beetles (*Anoplognathus* sp.), sawfly larvae (e.g. *Pterygophorus facielongus*) and various psyllid, lerp and scale insects.
- Avoid using sprays as it may affect the palatability of the leaf and kill useful bugs.
- Where practical, manual removal of most pests can quickly achieve results.
- Insect deterrents may be beneficial (e.g. permaculture).
- Planting mixed species in stands will attract less insect pests.

## 12. Harvesting techniques

- New trees are initially pruned after planting when at least 1.5–2 m in height and have had good seasonal growth. At this time the top of the plant is pruned so the height is reduced to 1.3–1.5 m. When the lateral growth results from the pruning, two of these are selected on opposite sides of the trunk that have strong bushy growth. Regular follow-up pruning will encourage these branches to be bushy.
- Pollarding trees (cutting up from the base) appears to be suitable, stimulating branching faster than coppicing trees (cutting close to the ground). It also appears that pollarding results in fewer dead trees.
- Trees react to the stimulus of having branches removed, which allows the harvester to shape the tree as preferred. Plantation managers often shape plants into a Y-shape, where 5–7 stems per plant can be produced. The base of the Y-shape can vary in height, but is often ~1.5 m high.
- Removal of a branch on a tree can often stimulate the production of young growth on all the remaining branches.
- Trees are competing between themselves for light and this affects the extent of growth of the lateral branches.
- Individual pieces on a tree are also competing with each other and thinning loads are required to reduce timber production and increase leaf production on the tree.

- Trees should be harvested at a height that is comfortable for the person harvesting and above the height that maintenance equipment will cause damage.
- All harvesting cuts should attempt to be on an angle to allow maximum exposure to sun and to encourage water runoff.
- Saws or similar equipment should always be used to harvest browse to prevent splitting of the trunk.

## 13. General plantation maintenance

- Lower branches and shoots should be removed to allow air movement at ground level throughout the plantation.
- Replanting all gaps and replacing all dead trees every 2 years will ensure continued leaf production.
- A major coppice of individual trees may be required if the bole at harvesting level becomes too large or if large amounts of dead timber are present.
- Because of the effect of constant, young growth, production and removal, the expected tree life will be greatly reduced (e.g. some tree species that usually live over 200 years may be old at 10 years in a plantation). Signs of this may be reduced leaf production, increased insect attack and death.

## 14. Climate controlled facilities

In regions where the climate may be unsuitable for the year-round production of eucalypt leaves, it may be appropriate to grow the plantation within a climate-controlled facility such as glasshouses and plastic-covered greenhouses. Within these facilities there is a greater ability to control the soil quality, plant nutrition, pests and diseases, climate and air movement (See Knight 2004).

## 15. Browse storage

Once the branches have been cut they can be stored in large tubs of clean water (e.g. 60–80 cm in height and a diameter of ~60 cm) for several days (max. 1 week). The tubs should ideally be placed in a cool area that is maintained at 5-10°C. Alternatively, they can be placed in a shaded area (such as a structure covered in shade cloth) that has a fine mist to keep the leaves damp and prevented from drying out and turning brown, which can occur very quickly. The stems can be recut near the base to enhance the uptake of water.

## REFERENCES


Adam D, Beard L, Johnston SD, Nicolson V, Lisle A, McKinnon A, Larkin R, Theilemann P, Gillett A, Brackin K, FitzGibbon S, Barth B, Ellis W (2018) Recording body temperature in koalas (*Phascolarctos cinereus*): a comparison of techniques. *Australian Veterinary Journal* **96**, 308–311. doi:10.1111/avj.12719

Adam D, Johnston SD, Beard L, Nicolson V, Gaughan JB, Lisle AT, FitzGibbon S, Barth BJ, Gillett A, Grigg G, Ellis W (2020) Body temperature of free-ranging koalas (*Phascolarctos cinereus*) in south-east Queensland. *International Journal of Biometeorology* **64**, 1305–1318. doi:10.1007/s00484-020-01907-y

Adam D, Johnston SD, Beard L, Nicolson V, Lisle AT, Gaughan JB, Larkin R, Theilemann P, Ellis W (2021) Temporal effect of feeding on the body temperature and behaviour of captive koalas (*Phascolarctos cinereus*). *Australian Mammalogy* **44**, 16–23. doi:10.1071/AM20024

Alfano N, Courtiol A, Vielgrader H, Timms P, Roca AL, Greenwood AD (2015) Variation in koala microbiomes within and between individuals: effect of body region and captivity status. *Scientific Reports* **5**, 10189. doi:10.1038/srep10189

Allen CD, Burridge M, Mulhall S, Chafer ML, Nicolson VN, Pyne M, Zee YP, Jago SC, Lundie-Jenkins G, Holt WV, Carrick FN, Curlewis JD, Lisle AT, Johnston SD (2008) Successful artificial insemination in the koala (*Phascolarctos cinereus*) using extended and extended-chilled semen collected by electroejaculation. *Biology of Reproduction* **78**, 661–666. doi:10.1095/biolreprod.107.064824

Allen CD, de Villiers DL, Manning BD, Dique DS, Burridge M, Chafer ML, Nicolson VN, Jago SC, McKinnon AJ, Booth RJ, McKee JJ, Pyne MJ, Zee YP, Lundie-Jenkins G, Theilemann P, Wilson RJ, Carrick FN, Johnston SD (2010) Seasonal reproduction in wild and captive male koala (*Phascolarctos cinereus*) populations in south-east Queensland. *Reproduction, Fertility and Development* **22**, 695–709. doi:10.1071/RD09113

Anon (2006) *Standards for Exhibiting Australian Mammals in New South Wales*. NSW Department of Primary Industries, Orange.

Anon (2009) Conditions for the overseas transfer of koalas. Department of the Environment, Water, Heritage and the Arts. Australian Government, Canberra. https://www.environment.gov.au/biodiversity/wildlife-trade/publications/conditions-overseas-transfer-australian-native-animals (accessed 14 October 2021).

Anon (2019) *Australian Animal Welfare Standards and Guidelines. Exhibited Animals: Koala*. Agriculture Ministers Forum, Canberra.

Anon (2021) Koala at Japanese zoo becomes world's oldest in captivity. *The Japan Times*. https://www.japantimes.co.jp/news/2021/03/02/national/oldest-koala (accessed 27 July 2021).

Ashman KR, Whisson DA (2020) Developing guidelines for the use of traps to capture koalas. *Australian Mammalogy* **42**, 380–384. doi:10.1071/AM20018

Au J, Youngentob KN, Clark RG, Phillips R, Foley WJ (2017) Bark chewing reveals a nutrient limitation of leaves for a specialist folivore. *Journal of Mammalogy* **98**, 1185–1192. doi:10.1093/jmammal/gyx045

Austin MA (1997) *A Practical Guide to the Successful Hand Rearing of Tasmanian Marsupials*. Regal Publications, Melbourne.

Ballantyne K, Anderson ST, Pyne N, Nicolson V, Mucci A, Lisle A, Johnston SD (2015a) The use of a synthetic progesterone, levonorgestrel (LNG), to control the oestrous cycle in the koala. *Reproduction, Fertility and Development* **28**, 1526–1532. doi:10.1071/RD14387

Ballantyne K, Anderson ST, Pyne N, Nicolson V, Mucci A, Lisle A, Johnston SD (2015b) Use of the gonadotrophin-releasing hormone antagonist azaline B to control the oestrous cycle in the koala (*Phascolarctos cinereus*). *Reproduction, Fertility and Development* **28**, 1686–1694. doi:10.1071/RD14349

Ballantyne K, Lisle A, Mucci A, Johnston SD (2015c) Seasonal oestrous cycle activity of captive female koalas in south-east Queensland. *Australian Mammalogy* **37**, 245–252. doi:10.1071/AM14018

Ballantyne K, Anderson ST, Mucci A, Nicolson V, Johnston SD (2015d) Plasma prolactin concentrations during lactation, pouch young development and the return to behavioural oestrus in captive koalas (*Phascolarctos cinereus*). *Reproduction, Fertility and Development* **28**, 1838–1843. doi:10.1071/RD14384

Barbosa A, Austen J, Gillett A, Warren K, Paparini A, Irwin P, Ryan U (2016) First report of *Trypanosoma vegrandis* in koalas (*Phascolarctos cinereus*). *Parasitology International* **65**, 316–318. doi:10.1016/j.parint.2016.03.004

Beale P, Marsh KJ, Youngentob KN (2022a) *Revegetating Koala Habitat*. Report prepared for the Department of Climate Change, Energy, the Environment and Water, Canberra, October.

Beale P, Youngentob K, Marsh K (2022b) *Effects of Fire on Koalas and their Habitat*, DCCEEW (Department of Climate Change, Energy, the Environment and Water), Canberra. www.dcceew.gov.au/environment/epbc/publications (accessed 1 December 2023)

Benesch AR, Munro U, Krop T, Fleissner G (2010) Seasonal changes in the behaviour and circadian rhythms in activity and behaviour of captive koalas *Phascolarctos cinereus*. *Biological Rhythm Research* **41**, 289–304. doi:10.1080/09291010903018164

Bercovitch FB, Tobey JR, Andrus CH, Doyle L (2006) Mating patterns and reproductive success in captive koalas (*Phascolarctos cinereus*). *Journal of Zoology* **270**, 512–516. doi:10.1111/j.1469-7998.2006.00169.x

Black KH, Price GJ, Archer M, Hand SJ (2014) Bearing up well? Understanding the past, present and future of Australia's koalas. *Gondwana Research* **25**, 1186–1201. doi:10.1016/j.gr.2013.12.008

Blanshard WH (1991) Growth and development of the koala from birth to weaning. In *Biology of the Koala*. (Eds AK Lee, KA Handasyde and GD Sanson) pp. 193–202. Surrey Beatty & Sons, Sydney.

Blanshard WH (1994) Medicine and husbandry of koalas. In *Wildlife*. Proceedings 233. Post Graduate Committee in Veterinary Science. (Ed. DI Bryden) pp. 547–626. University of Sydney, Sydney.

Blanshard W, Bodley K (2008) Koalas. In *Medicine of Australian Mammals*. (Eds L Vogelnest and R Woods) pp. 227–327. CSIRO Publishing, Melbourne.

Blyton MDJ, Soo RM, Whisson D, Marsh KJ, Pascoe J, Le Pla M, Foley W, Hugenholtz P, Moore BD (2019) Faecal inoculations alter the gastrointestinal microbiome and allow dietary expansion in a wild specialist herbivore, the koala. *Animal Microbiome* **1**, 6. doi:10.1186/s42523-019-0008-0

Blyton MDJ, Soo RM, Hugenholtz P, Moore BD (2022) Maternal inheritance of the koala gut microbiome and its compositional and functional maturation during juvenile development. *Environmental Microbiology* **24**, 475–493. doi:10.1111/1462-2920.15858

Blyton MDJ, Brice KL, Heller-Uszynska K, Pascoe J, Jaccoud D, Leigh KA, Moore BD (2023a) A new genetic method for diet determination from faeces that provides species level resolution in the koala. *bioRxiv* preprint doi:10.1101/2023.02.12.528172; this version posted February 13, 2023.

Blyton MDJ, Pascoe J, Hynes E, Soo RM, Hugenholtz P, Moore BD (2023b) The koala gut microbiome is largely unaffected by host translocation but rather influences host diet. *Frontiers in Microbiology* doi:10.3389/fmicb.2023.1085090.

Bolliger A (1962) Gravel in the caecum of the koala (*Phascolarctos cinereus*). *Australian Journal of Science* **24**, 416–417.

Booth R (1989) *The Effects of Handling on Koalas*. Lone Pine Koala Sanctuary, Queensland. ASZK Proceedings. Melbourne Zoo and Healesville Sanctuary.

Booth RJ, Blanshard WH (1999) Diseases in koalas. In *Zoo and Wild Animal Medicine: Current Therapy 4*. (Eds ME Fowler and RE Miller) pp. 321–333. W.B. Saunders, Philadelphia.

Booth RJ, Nyari S (2020) Clinical comparison of five anti-chlamydial antibiotics in koalas (*Phascolarctos cinereus*). *PLoS One* **15**(7), e0236758. doi:10.1371/journal.pone.0236758

Braithwaite W (2004) Do current forestry practices threaten forest fauna? A perspective. In *Conservation of Australia's Forest Fauna*. 2nd edn. (Ed D Lunney) pp. 513–536. Royal Zoological Society of New South Wales, Sydney.

Brice KL, Trivedi P, Jeffries TC, Blyton MD, Mitchell C, Singh BK, Moore BD (2019) The Koala (*Phascolarctos cinereus*) faecal microbiome differs with diet in a wild population. *PeerJ* **7**, e6534. doi:10.7717/peerj.6534

Briscoe NJ, Handasyde KA, Griffiths SR, Porter WP, Krockenberger A, Kearney MR (2014) Tree-hugging koalas demonstrate a novel thermoregulatory mechanism for arboreal mammals. *Biology Letters* **10**, 20140235. doi:10.1098/rsbl.2014.0235

Brown AS, Seawright AA, Wilkinson GT (1981) An outbreak of sarcoptic mange in a colony of koalas. In *Wildlife Diseases of the Pacific Basin and Other Countries*. (Ed. ME Fowler) pp. 111. Proceedings of the 4th International Conference of the Wildlife Disease Association, Sydney.

Brown S, Woolcock J (1988) Epidemiology and control of chlamydial disease in koalas. In *Australian Wildlife*. Proceedings 104. Post Graduate Committee in Veterinary Science. (Ed. DI Bryden) pp. 495–502. University of Sydney, Sydney.

Cahir F, Schlagloth R, Clarke ID (2020) The historic importance of the koala in Aboriginal society in New South Wales, Australia: an exploration of the archival record. *Journal of Indigenous Studies and First Nations and First Peoples' Cultures* **3**(2). Pennsylvania State University, University Park.

Cahir F, Schlagloth R, Clarke ID (2021) The importance of the koala in Aboriginal society in nineteenth-century Queensland (Australia): A reconsideration of the archival record. *Anthrozoos* **35**, 75–89. doi:10.1080/08927936.2021.1963544

Carrick FN, Wood AD, Fyfe L (1990) Standards for treatment of koalas. In *Koalas: Research for Management*. (Ed. G Gordon) pp. 148–153. World Koala Research Incorporated, Brisbane.

Charlton BD (2015) The acoustic structure and information content of female koala vocal signals. *PLoS One* **10**(10), e0138670. doi:10.1371/journal.pone.0138670

Charlton BD, Reby D, Ellis WAH, Brumm J, Tecumseh Firch W (2012) Estimating the active space of male koala bellows: propagation of cues to size and identity in a *Eucalyptus* forest. *PLoS One* **7**(9), e45420. doi:10.1371/journal.pone.0045420

Chong R, Cheng Y, Hogg CJ, Belov K (2020) Marsupial gut microbiome. *Frontiers in Microbiology* **11**, 1058. doi:10.3389/fmicb.2020.01058

Clifton ID (2010) High koala mortality associated with low browse moisture in tropical environments. *Australian Mammalogy* **32**, 157–159. doi:10.1071/AM10015

Connolly JH (1999) Emerging diseases of koalas and their medical management. In *Wildlife in Australia: Healthcare & Management.* Proceedings 327. Post Graduate Foundation in Veterinary Science. (Ed. DI Bryden) pp. 15–38. University of Sydney, Sydney.

Crandall LS (1964) *The Management of Wild Mammals in Captivity.* University of Chicago Press, Chicago.

Crowther MS, Lunney D, Lemon J, Stalenberg E, Wheeler R, Madani G, Ross KA, Ellis M (2014) Climate-mediated habitat selection in an arboreal folivore. *Ecography* **37**, 336–343. doi:10.1111/j.1600-0587.2013.00413.x

Crowther MS, Rus AI, Mella VSA, Krockenberger MB, Lindsay J, Moore BD, McArthur C (2022) Patch quality and habitat fragmentation shape the foraging patterns of a specialist folivore. *Behavioral Ecology* **33**, 1007–1017. doi:10.1093/beheco/arac068

Davies N, Gramotnev G, McAlpine C, Seabrook L, Baxter G, Lunney D, Rhodes J, Bradley A (2013) Physiological stress in koala populations near the arid edge of their distribution. *PLoS One* **8**(11), e79136. doi:10.1371/journal.pone.0079136

Davies N, Gramotnev G, McAlpine C, Seabrook L, Baxter G, Lunney D, Rhodes J, Bradley A (2014) Climate-driven changes in diet composition and physiological stress in an arboreal folivore at the semi-arid edge of its distribution. *Biological Conservation* **172**, 80–88. doi:10.1016/j.biocon.2014.02.004

Dawson TJ, Webster KN, Maloney SK (2014) The fur of mammals in exposed environments: do crypsis and thermal needs necessarily conflict? The polar bear and marsupial koala compared. *Journal of Comparative Physiology* **184**, 273–284. doi:10.1007/s00360-013-0794-8

de Blainville HMD (1816) Prodrome d'une nouvelle distribution systematique du regne animal. *Bulletin des Sciences (Societe de Philomathique de Paris)* **1816**, 113–124.

DeGabriel JL, Wallis IR, Moore BD, Foley WJ (2008) A simple, integrative assay to quantify nutritional quality of browses for herbivores. *Oecologia* **156**, 107–116. doi:10.1007/s00442-008-0960-y

DeGabriel JL, Moore BD, Foley WJ, Johnson CN (2009) The effects of plant defensive chemistry on nutrient availability predict reproductive success in a mammal. *Ecology* **90**, 711–719. doi:10.1890/08-0940.1

Department of Agriculture Water and the Environment (DAWE) (2022a) *Conservation Advice for* Phascolarctos cinereus *(Koala),* Canberra. Department of Agriculture, Water and the Environment, Canberra.

Department of Agriculture Water and the Environment (DAWE) (2022b) *National Recovery plan for the Koala:* Phascolarctos cinereus *(Combined Populations of Queensland, New South Wales and the Australian Capital Territory).* Department of Agriculture, Water and the Environment, Canberra.

Dixon RMW (2019) *Australia's Original Languages: An Introduction.* Allen & Unwin, Sydney.

Drake B (1982) Koala *Phascolarctos cinereus*: its husbandry at the Royal Melbourne Zoological Gardens. In *Proceedings of the Scientific Meeting of the Australian Mammal Society.* (Ed. DD Evans) pp. 129–131. Zoological Board of Victoria, Melbourne.

Drake B, Miller M, Morley NW (1991) Management of koalas in captivity. In *Biology of the Koala.* (Eds AK Lee, KA Handasyde and GD Sanson) pp. 323–329. Surrey Beatty & Sons and the World Koala Research Corporation, Sydney.

Ellis WAH, White NA, Kunst ND, Carrick FN (1990) Response of koalas (*Phascolarctos cinereus*) to re-introduction to the wild after rehabilitation. *Australian Wildlife Research* **17**, 421–426. doi:10.1071/WR9900421

Ellis WAH, Melzer A, Clifton ID, Carrick F (2010) Climate change and the koala *Phascolarctos cinereus*: water and energy. *Australian Zoologist* **35**, 369–377. doi:10.7882/AZ.2010.025

Ellis WAH, Bercovitch F, FitzGibbon S, Roe P, Wimmer J, Melzer A, Wilson R (2011) Koala bellows and their association with the spatial dynamics of free-ranging koalas. *Behavioral Ecology* **22**, 372–377. doi:10.1093/beheco/arq216

Fanson KV (2019) Stress physiology of Australian mammals. In *Current Therapy of Medicine of Australian Mammals.* (Eds L Vogelnest and T Portas) pp. 495–506. CSIRO Publishing, Melbourne.

Feige S, Nilsson K, Phillips CJC, Johnston SD (2007) Heterosexual and homosexual behavior and vocalizations in captive female koalas (*Phascolarctos cinereus*). *Applied Animal Behaviour Science* **103**, 131–145. doi:10.1016/j.applanim.2006.03.014

Finnie EP (1988) Diseases and injuries of other Australian mammals. In *Australian Wildlife.* Proceedings 104. Post Graduate Committee in Veterinary Science. (Ed. DI Bryden) pp. 91. University of Sydney, Sydney.

Finnie T (1990) *The Role of Zoos in the Conservation of Koalas.* In *Koala Summit: Managing Koalas in NSW.* (Eds D Lunney, CA Urquhart and P Reed) pp. 109. National Parks and Wildlife Service, Sydney.

Fleay D (1937) Observations on the koala in captivity: successful breeding in Melbourne Zoo. *Australian Zoologist* **9**, 68–80.

Flower WH (1880) Additions to the menagerie. *Proceedings of the Zoological Society of London* **48**, 355–358. doi:10.1111/j.1469-7998.1880.tb06565.x

Gall BC (1980) Aspects of the ecology of the koala, *Phascolarctos cinereus* (Goldfuss), in Tucki Tucki Nature Reserve, New South Wales. *Australian Wildlife Research* **7**, 167–176. doi:10.1071/WR9800167

Gamble J, Blyde D (1992) Artificial milk diets as a supplement for aged and infirmed marsupials. *International Zoo Yearbook* **31**, 90–92. doi:10.1111/j.1748-1090.1991.tb02368.x

George H, Parker G, Coote P (1995) *Common Wombats: Rescue Rehabilitation Release.* Unpublished manuscript.

Gibson R, Danaher T, Hehir W, Collins L (2020) A remote sensing approach to mapping fire severity in south-eastern Australia using sentinel 2 and random forest. *Remote Sensing of Environment* **240**, 111702. doi:10.1016/j.rse.2020.111702

Gillett AK (2014) An examination of disease in captive Australian koalas *(Phascolarctos cinereus)* and potential links to koala retrovirus (KoRV). *Technical Reports of the Australian Museum* **24**, 39–45. doi:10.3853/j.1835-4211.24.2014.1612

Gillett AK (2023) Defining putative koala retrovirus-associated disease in koalas. In *Proceedings of the Second Koala Retrovirus Workshop*. (Eds DE Alquezar-Planas, DP Higgins, CL Singleton and AD Greenwood). *Technical Reports of the Australian Museum Online* **38**, 23–29.

Gillett A, Fitzgibbon S (2024) *Monitoring the Survival of Rehabilitated Koalas Post-release: Assessing the conservation benefits of koala rehabilitation at Australia Zoo Wildlife Hospital*. Report prepared for the Queensland Department of Environment, Science and Innovation, Brisbane.

Gillett AK, Hanger J (2025) Koala. In *Current Therapy in Medicine of Australian Mammals*. Revised Edition. (Eds L Vogelnest and T Portas) pp. 495–520. CSIRO Publishing, Melbourne.

Goldfuss GA (1817) *Die Säugethiere, in Abbildungen nach Natur mit Beschreibungen von Johan Christian Daniel von Schreiber*. Part 65. No text. Plate 155, Aa, Ab.

Gonzalez-Astudillo V, Allavena R, McKinnon A, Larkin R, Henning J (2017) Decline causes of koalas in South East Queensland, Australia: a 17-year retrospective study of mortality and morbidity. *Scientific Reports* **7**(1), 42587. doi:10.1038/srep42587

Gonzalez-Astudillo V, Henning J, Valenza L, Knott L, McKinnon A, Larkin R, Allavena R (2019a) A necropsy study of disease and comorbidity trends in morbidity and mortality in the koala (*Phascolarctos cinereus*) in South-East Queensland, Australia. *Scientific Reports* **9**(1), 17494. doi:10.1038/s41598-019-53970-0

Gonzalez-Astudillo V, Knott L, Valenza L, Henning J, Allavena R (2019b) Parasitism by *Ophidascaris robertsi* with associated pathology findings in a wild koala (*Phascolarctos cinereus*). *Veterinary Record Case Reports* **7**, e000821. doi:10.1136/vetreccr-2019-000821

Gordon G (1991) Estimation of the age of the koala *Phascolarctos cinereus* (Marsupialia: Phascolarctidae) from tooth wear and growth. *Australian Mammalogy* **14**, 5–12. doi:10.1071/AM91001

Gordon G, Brown AS, Pulsford T (1988) A koala (*Phascolarctos cinereus* Goldfuss) population crash during drought and heatwave conditions in south-western Queensland. *Australian Journal of Ecology* **13**, 451–461. doi:10.1111/j.1442-9993.1988.tb00993.x

Green B (1984) Composition of milk and energetics of growth in marsupials. *Symposia of the Zoological Society of London* **51**, 369–387.

Handasyde KA, Martin RW, Lee AK (1988) Field investigations into chlamydial disease and infertility in koalas in Victoria. In *Australian Wildlife*. Proceedings 104. Post Graduate Committee in Veterinary Science. (Ed. DI Bryden) pp. 505–515. University of Sydney, Sydney.

Hanger J, Gillett A (2025) Veterinary involvement in the research and management of free-ranging koala populations. In *Current Therapy in Medicine of Australian Mammals*. Revised Edition. (Eds Vogelnest and T Portas) pp. 543–550. CSIRO Publishing, Melbourne.

Hartley WJ, Dubey JP, Spielman DS (1990) Fatal toxoplasmosis in koalas (*Phascolarctos cinereus*). *The Journal of Parasitology* **76**, 271–272. doi:10.2307/3283032

Hasegawa M, Carrick F (1995) First catch your koala! Use of a trap to capture koalas *Phascolarctos cinereus* for ecological studies. *Australian Zoologist* **30**, 68–70. doi:10.7882/AZ.1995.010

Hawkes NH (1978) Identification and Management of Koala Eucalypt Trees in New South Wales. In *The Koala: Proceedings of the Taronga Symposium on Koala Biology* (Ed. TJ Bergin) pp. 89–96. Zoological Parks Board of NSW, Sydney.

Hayhoe E (2016) Coprophilous fungi from koala faeces: a novel source of antimicrobial compounds. PhD thesis. Swinburne University of Technology, Melbourne.

Higgins D, Maher I (2019) Koala retrovirus. In *Current Therapy of Medicine of Australian Mammals* (Eds L Vogelnest and T Portas) pp 487–506. CSIRO Publishing, Melbourne.

Houlden BA, Costello BH, Sharkey D, Fowler EV, Melzer A, Ellis W, Carrick F, Baverstock PR, Elphinstone MS (1999) Phylogeographic differentiation in the mitochondrial control region in the koala, *Phascolarctos cinereus* (Goldfuss 1817). *Molecular Ecology* **8**, 999–1011. doi:10.1046/j.1365-294x.1999.00656.x

Hulse LS, Hickey D, Mitchell JM, Beagley KW, Ellis W, Johnston SD (2018) Development and application of two multiplex real-time PCR assays for detection and speciation of bacterial pathogens in the koala. *Journal of Veterinary Diagnostic Investigation* **30**, 523–529. doi:10.1177/1040638718770490

Hulse LS, McDonald S, Johnston SD, Beagley KW (2019) Rapid point-of-care diagnostics for the detection of *Chlamydia pecorum* in koalas (*Phascolarctos cinereus*) using loop-mediated isothermal amplification without nucleic acid purification. *MicrobiologyOpen* **8**(12), e916. doi:10.1002/mbo3.916

Hulse LS, Beagley K, Ellis W, Fitzgibbon S, Gillett A, Barth B, Robbins A, Pyne M, Larkin R, Johnston SD (2020) Epidemiology of chlamydia-induced reproductive disease in male koalas (*Phascolarctos cinereus*) from Southeast Queensland, Australia as assessed from penile urethral swabs and semen. *Journal of Wildlife Diseases* **56**, 82–92. doi:10.7589/2019-03-062

Hulse L, Beagley K, Larkin R, Nicolson V, Gosálvez J, Johnston S (2021) The effect of Chlamydia infection on koala (*Phascolarctos cinereus*) semen quality. *Theriogenology* **167**, 99–110. doi:10.1016/j.theriogenology.2021.03.016

Hume ID (1999) *Marsupial Nutrition*. Cambridge University Press, New York.

Hynes EF, Handasyde KA, Shaw G, Renfree MB (2010) Levonorgestrel, not etonogestrel, provides contraception in free-ranging koalas. *Reproduction, Fertility and Development* **22**, 913–919. doi:10.1071/RD09253

Hynes EF, Shaw G, Renfree MB, Handasyde KA (2019) Contraception of prepubertal young can increase cost effectiveness of management of overabundant koala populations. *Wildlife Research* **46**, 317–325. doi:10.1071/WR18109

International Air Transport Association (IATA) (2022) *Live Animals Regulations*. 48th edn. International Air Transport Association, Geneva, Switzerland.

Jackson SM (2001) Koalas. In *Encyclopedia of the World's Zoos*. (Ed. C Bell) pp. 687–690. Fitzroy Dearborn, Chicago.

Jackson SM (2007) *Koala: Origins of an Icon*. Allen & Unwin, Sydney.

Jackson SM (2015) Family Phascolarctidae (Koala). In *Handbook of the Mammals of the World*. Vol. 5. *Monotremes and Marsupials*. (Eds DE Wilson and RA Mittermeier) pp. 400–417. Lynx Editions, Barcelona.

Jackson SM, Groves C (2015) *Taxonomy of Australian Mammals*. CSIRO Publishing, Melbourne.

Jelocnik M, Gillett A, Hanger J, Polkinghorne A (2019) Chlamydiosis in koalas. In *Current Therapy of Medicine of Australian Mammals*. (Eds L Vogelnest and T Portas) pp. 495–506. CSIRO Publishing, Melbourne.

Johnson RN, O'Meilly D, Chen Z, Etherington GJ, Ho SYW, *et al.* (2018) Adaptation and conservation insights from the koala genome. *Nature Genetics* **50**, 1102–1111. doi:10.1038/s41588-018-0153-5

Johnston SD (1999) Studies towards the development of an artificial insemination protocol in the koala, *Phascolarctos cinereus*. PhD thesis. The University of Queensland, Brisbane.

Johnston SD (2019) Challenges associated with the development and transfer of assisted breeding technology in marsupials and monotremes: lessons from the koala, wombat and short-beaked echidna. *Reproduction, Fertility and Development* **31**, 1305–1314. doi:10.1071/RD18113

Johnston SD, Holt WV (2019) Using the koala (*Phascolarctos cinereus*) as a case study to illustrate the development of artificial breeding technology in marsupials: An update. In *Reproductive Sciences in Animal Conservation*. (Eds P Comizzoli, JL Brown and WV Holt) pp. 327–362. Springer Nature, Cham, Switzerland.

Johnston SD, McGowan MR, Carrick FN, Cameron RDA (1994) Seminal characteristics and spermatozoal morphology of captive Queensland koalas (*Phascolarctos cinereus adustus*). *Theriogenology* **42**, 501–511. doi:10.1016/0093-691X(94)90688-F

Johnston SD, O'Callaghan P, McGowan MR, Phillips NJ (1997) Characteristics of koala (*Phascolarctos cinereus adustus*) semen collected by artificial vagina. *Journal of Reproduction and Fertility* **109**, 319–323. doi:10.1530/jrf.0.1090319

Johnston SD, McGowan MR, O'Callaghan P (1999) Assisted breeding technology for the conservation and propagation of *Phascolarctos cinereus* or how to make a koala pouch young. In *Wildlife in Australia: Healthcare & Management*. Proceedings 327. Post Graduate Foundation in Veterinary Science. (Ed. DI Bryden) pp. 199–216. University of Sydney, Sydney.

Johnston SD, McGowan MR, O'Callaghan P, Cox R, Nicolson V (2000a) Studies of the oestrous cycle, oestrus and pregnancy in the koala (*Phascolarctos cinereus*). *Journal of Reproduction and Fertility* **120**, 49–57. doi:10.1530/reprod/120.1.49

Johnston SD, McGowan MR, O'Callaghan P, Cox R, Nicolson V (2000b) Natural and artificial methods for inducing the luteal phase in the koala (*Phascolarctos cinereus*). *Reproduction (Cambridge, England)* **120**, 59–64. doi:10.1530/reprod/120.1.59

Johnston SD, McGowan MR, Phillips N, O'Callaghan P (2000c) Optimal physiochemical conditions for the manipulation and short-term preservation of koala (*Phascolarctos cinereus*) spermatozoa. *Journal of Reproduction and Fertility* **118**, 273–281. doi:10.1530/reprod/118.2.273

Johnston SD, McGowen MR, O'Callaghan P, Cox R, Houlden B, Haig S, Taddeo G (2003) Birth of Koalas *Phascolarctos cinereus* at Lone Pine Koala Sanctuary following artificial insemination. *International Zoo Yearbook* **38**, 160–172. doi:10.1111/j.1748-1090.2003.tb02076.x

Johnston SD, O'Callaghan P, Nilsson K, Tzipori G, Curlewis JD (2004) Semen-induced luteal phase and identification of a LH surge in the koala (*Phascolarctos cinereus*). *Reproduction (Cambridge, England)* **128**, 629–634. doi:10.1530/rep.1.00300

Johnston SD, MacCallum C, Blyde D, McClean R, Lisle A, Holt WV (2006) An investigation into the similarities and differences governing the cryopreservation success of koala (*Phascolarctos cinereus*: Goldfuss) and common wombat (*Vombatus ursinus*: Shaw) spermatozoa. *Cryobiology* **53**, 218–228. doi:10.1016/j.cryobiol.2006.06.001

Johnston SD, Zee YP, Lopez-Fernandez C, Gosalvez J (2012) The effect of chilled storage and cryopreservation on the sperm DNA fragmentation dynamics of a captive population of koalas. *Journal of Andrology* **33**, 1007–1015. doi:10.2164/jandrol.111.015248

Kjeldsen SR, Raadsma HW, Leigh KA, Tobey JR, Phalen D, Krockenberger A, Ellis WA, Hynes E, Higgins DP, Zenger KR (2019) Genomic comparisons reveal biogeographic and anthropogenic impacts in the koala (*Phascolarctos cinereus*): a dietary-specialist species distributed across heterogeneous environments. *Heredity* **122**, 525–544. doi:10.1038/s41437-018-0144-4

Knight C (2004) Phascolarctos cinereus *Captive Husbandry Guidelines: Koala Browse Plantation Guide*. Australasian Regional Association of Zoological Parks and Aquaria, Sydney.

Knott T, Lunney D, Coburn D, Callaghan J (1998) An ecological history of Koala habitat in Port Stephens Shire and lower Hunter on the central coast of New South Wales, 1801–1998. *Pacific Conservation Biology* **4**, 354–368. doi:10.1071/PC980354

Krockenberger AK (1996) Composition of the milk of the koala, *Phascolarctos cinereus*, an arboreal folivore. *Physiological Zoology* **69**, 701–718. doi:10.1086/physzool.69.3.30164223

Krockenberger AL, Hume ID (2007) A flexible digestive strategy accommodates the nutritional demands of reproduction in a free-living folivore, the koala (*Phascolarctos cinereus*). *Functional Ecology* **21**, 748–756. doi:10.1111/j.1365-2435.2007.01279.x

Lacy RC, Balou JD, Pollak JP (2011) PMx: software package for demographic and genetic analysis and management of pedigreed populations. *Methods in Ecology and Evolution* **3**, 433–437. doi:10.1111/j.2041-210X.2011.00148.x

Lanyon JM, Sanson GD (1986) Koala (*Phascolarctos cinereus*) dentition and nutrition. II: implications of tooth wear in nutrition. *Journal of Zoology* **209**, 169–181. doi:10.1111/j.1469-7998.1986.tb03573.x

Larsen MJ, Sherwin SL, Rault JL (2014) Number of nearby visitors and noise level affect vigilance in captive koalas. *Applied Animal Behaviour Science* **154**, 76–82. doi:10.1016/j.applanim.2014.02.005

Law B, Gonsalves L, Burgar J, Brassil T, Kerr I, Wilmott L, Madden K, Smith M, Mella V, Crowther M, Krockenberger R, Rus A, Pietsch R, Truskinger A, Eichinski P, Roe P (2021) Estimating and validating koala *Phascolarctos cinereus* density estimates from acoustic arrays using spatial count modelling. *Wildlife Research* **49**, 438–448. doi:10.1071/WR21072

Lee AK (1988) Life histories of marsupials, with particular reference to the life history of the koala. *Australian Wildlife*. Proceedings 104. Post Graduate Committee in Veterinary Science. (Ed. DI Bryden) pp. 613–621. University of Sydney, Sydney.

Lee AK, Carrick FN (1989) Phascolarctidae. In *Fauna of Australia*. Vol. 1B. (Eds DW Walton and BJ Richardson) pp. 740–754. Australian Government Publishing Service, Canberra.

Lee AK, Martin R (1988) *The Koala: A Natural History*. New South Wales University Press, Sydney.

Leigh KA, Hofweber LN, Sloggett BK, Inman VL, Pettit LJ, Sriram A, Haering R (2023) Outcomes for an arboreal folivore after rehabilitation and implications for management. *Scientific Reports* **13**, 6542. doi:10.1038/s41598-023-33535-y

Logan M, Sanson GD (2002) The effect of tooth wear on the feeding behaviour of free-ranging koalas (*Phascolarctos cinereus*, Goldfuss). *Journal of Zoology* **256**, 63–69. doi:10.1017/S0952836902000080

Lott MJ, Wright BR, Neaves LE, Frankham GJ, Dennison S, Eldridge MDB, Potter S, Alquezar-Planas DE, Hogg CJ, Belov K, Johnson RN (2022) Future-proofing the koala: synergising genomic and environmental data for effective species management. *Molecular Ecology* **31**, 3035–3055. doi:10.1111/mec.16446

Lunney D, Leary T (1988) The impact on native mammals of land-use changes and exotic species in the Bega District (New South Wales) since settlement. *Australian Journal of Ecology* **13**, 67–92. doi:10.1111/j.1442-9993.1988.tb01417.x

Lunney D, Gresser SM, Mahon PS, Matthews A (2004) Post-fire survival and reproduction of rehabilitated and unburnt koalas. *Biological Conservation* **120**, 567–575. doi:10.1016/j.biocon.2004.03.029

Lunney D, Gresser S, O'Neill LE, Matthews A, Rhodes J (2007) The impact of fire and dogs on koalas at Port Stephens, New South Wales, using population viability analysis. *Pacific Conservation Biology* **13**, 189–201. doi:10.1071/PC070189

Lunney D, Close R, Crowther MS, Bryant J, Shannon I, Madden K, Ward S (2010) Campbelltown's koalas: their place in the natural history of Australia. In *The Natural History of Sydney*. (Eds D Lunney, P Hutchings and D Hochuli) pp. 319–325. Royal Zoological Society of New South Wales, Sydney.

Lunney D, Crowther MS, Wallis I, Foley WJ, Lemon J, Wheeler R, Madani G, Orscheg C, Griffith JE, Krockenberger M, Retamales M, Stalenberg E (2012) Koalas and climate change: a case study on the Liverpool Plains, north-west NSW. In *Wildlife and Climate Change: Towards Robust Conservation Strategies for Australian Fauna*. (Eds D Lunney and P Hutchings) pp. 150–168. Royal Zoological Society of NSW, Sydney.

Lunney D, Wells A, Miller I (2016) An ecological history of the Koala in Coffs Harbour and its environs, on the mid-north coast of New South Wales, c1861–2000. *Proceedings of the Linnean Society of New South Wales* **138**, 1–48.

Lunney D, Sonawane I, Wheeler R, Tasker E, Ellis M, Predavec M, Fleming M (2020) An ecological reading of the history of the koala population of Warrumbungle National Park. *Proceedings of the Linnean Society of New South Wales* **141** (Supplement), S131–S154.

Lunney D, Cope H, Sonawane I, Stalenberg E, Haering R (2022a) An analysis of the long-term trends in the records of Friends of the Koala in north-east New South Wales. I: cause and fate of koalas admitted for rehabilitation (1989–2020). *Pacific Conservation Biology* **29**, 177–196. doi:10.1071/PC22008

Lunney D, Cope H, Sonawane I, Stalenberg E, Haering R (2022b) An analysis of the long-term trends in the records of Friends of the Koala in north-east New South Wales. II: post-release survival. *Pacific Conservation Biology* **29**, 197–222. doi:10.1071/PC21077

Lunney D, Cope H, Sonawane I, Haering R (2022c) A state-wide picture of koala rescue and rehabilitation in New South Wales during the 2019–2020 bushfires. *Australian Zoologist* **42**, 243–255. doi:10.7882/AZ.2022.013

Lunney D, Moon C, Sonawane I, Predavec M, Rhodes JR (2022d) A 6-year study of mitigating koala roadkill during an upgrade of the Pacific Highway at Lindsay's cutting, Coffs Harbour New South Wales. *Australian Mammalogy* **44**, 305–318. doi:10.1071/AM21032

Lunney D, Moon C, Sonawane I, Predavec M, Rhodes JR (2022e) Factors that drive koala roadkill: an analysis across multiple scales in New South Wales, Australia. *Australian Mammalogy* **44**, 328–337. doi:10.1071/AM21040

Lunney D, Cope H, Griffith J, Orscheg C, Bryant J, Haering R (2023) Trends in the rescue and fate of koalas in NSW (1973–2020), with a focus on disease and trauma. *Australian Zoologist* (In press). doi:10.7882/AZ.2023.026

Madani GF, Ashman KR, Mella VSA, Whisson DA (2020) A review of the 'noose and flag' method to capture free-ranging koalas. *Australian Mammalogy* **42**, 341–348. doi:10.1071/AM19064

Maidment TI, Bryan ER, Pyne M (2023) Characterisation of the koala (*Phascolarctos cinereus*) pouch microbiota in a captive population reveals a dysbiotic compositional profile associated with neonatal mortality. *Microbiome* **11**, 75. doi:10.1186/s40168-023-01527-9

Marsh KJ, Ward J, Wallis IR, Foley WJ (2018) Intraspecific variation in nutritional composition affects the leaf age preferences of a mammalian herbivore. *Journal of Chemical Ecology* **44**, 62–71. doi:10.1007/s10886-017-0911-3

Marsh KJ, Blyton MDJ, Foley WJ, Moore BD (2021) Fundamental dietary specialisation explains differential use of resources within a koala population. *Oecologia* **196**, 795–803. doi:10.1007/s00442-021-04962-3

Marshall V, Carrick F, Doherty MD, Maclean DJ (1990) Aspects of the composition of koala milk. In *Biology of the Koala*. (Eds AK Lee, KA Handasyde and GD Sanson) pp. 229–241. Surrey Beatty & Sons, Sydney.

Martin RW (1981) Age specific fertility in three populations of the koala, *Phascolarctos cinereus* Goldfuss, in Victoria. *Australian Wildlife Research* **8**, 275–283. doi:10.1071/WR9810275

Martin RW, Handasyde K (1991) Population dynamics of the koala (*Phascolarctos cinereus*) in southern Australia. In *Biology of the Koala*. (Eds AK Lee, KA Handasyde and GD Sanson) pp. 75–84. Surrey Beatty & Sons and the World Koala Research Corporation, Sydney.

Martin RW, Handasyde K (1995) Koala *Phascolarctos cinerea*. In *Mammals of Australia*. (Ed. R Strahan) pp. 196–198. Reed Books, Sydney.

Martin RW, Handasyde K (1999) *The Koala: Natural History, Conservation and Management*. Australian Natural History Series, UNSW Press, Sydney.

Martin RW, Handasyde KA, Krockenberger A, Ellis W, Lunney D (2023) Koala *Phascolarctos cinereus*. In *Strahan's Mammals of Australia*. 4th edn. (Eds AM Baker and IC Gynther) pp. 201–203. Reed New Holland, Sydney.

Matthews A, Lunney D, Gresser S, Maitz W (2007) Tree use by koalas (*Phascolarctos cinereus*) after fire in remnant coastal forest. *Wildlife Research* **34**, 84–93. doi:10.1071/WR06075

Matthews A, Lunney D, Gresser S, Maitz W (2016) Movement patterns of koalas in remnant forest after fire. *Australian Mammalogy* **38**, 91–104. doi:10.1071/AM14010

McAlpine CA, Lunney D, Melzer A, Menkhorst P, Phillips S, Phalen D, Ellis W, Foley W, Baxter G, de Villiers D, Kavanagh R, Adams-Hosking C, Todd C, Whisson D, Molsher R, Walter M, Lawler I, Close R (2015) Conserving koalas: a review of the contrasting regional trends, outlooks and policy challenges. *Biological Conservation* **192**, 226–236. doi:10.1016/j.biocon.2015.09.020

McAlpine CA, Callaghan J, Lunney D, Rhodes JR, Goldingay R, Goulding W, Adams-Hosking C, Fielding K, Hetherington SB, Brace A, Hopkins M, Caddick L, Taylor E, Vass L, Swankie L (2023) The influences on koala habitat selection across four local government areas on the far north coast of NSW. *Austral Ecology* **48**, 928–951. doi:10.1111/aec.13313

McInnes LJ, Hanger J, Simmons G, Reid SA, Ryan UM (2011a) Novel trypanosome *Trypanosoma gilletti* sp. (Euglenozoa: Trypanosomatidae) and the extension of the host range of *Trypanosoma copemani* to include the koala (*Phascolarctos cinereus*). *Parasitology* **138**, 59–70. doi:10.1017/S0031182010000971

McInnes LJ, Gillett A, Hanger J, Reid SA, Ryan UM (2011b) The potential impact of native Australian trypanosome infections on the health of koalas (*Phascolarctos cinereus*). *Parasitology* **138**, 873–883. doi:10.1017/S0031182011000369

McLean N, Handasyde KA (2006) Sexual maturity, factors affecting the breeding season and breeding in consecutive seasons in populations of overabundant Victorian koalas (*Phascolarctos cinereus*). *Australian Journal of Zoology* **54**, 385–392. doi:10.1071/ZO06015

Mella VSA, McArthur C, Frend R, Crowther MS (2018) Foxes in trees: a threat for Australian arboreal fauna? *Australian Mammalogy* **40**, 103–105. doi:10.1071/AM16049

Mella VSA, McArthur C, Krockenberger MB, Frend R, Crowther MS (2019) Needing a drink: rainfall and temperature drive the use of free water by a threatened arboreal folivore. *PLoS One* **14**(5), e0216964. doi:10.1371/journal.pone.0216964

Mella VSA, Orr C, Hall L, Velasco S, Madani G (2020) An insight into natural koala drinking behaviour. *Ethology* **126**, 858–863. doi:10.1111/eth.13032

Melzer A, Carrick F, Menkhorst P, Lunney D, St John B (2000) Overview, critical assessment, and conservation implications of koala distribution and abundance. *Conservation Biology* **14**, 619–628. doi:10.1046/j.1523-1739.2000.99383.x

Melzer A, Ellis W, Gordon G, Tucker G, Kindness R, Carrick F (2011) Unusual patterns of tooth wear among koalas *Phascolarctos cinereus* from St Bees Island, Queensland, require re-evaluation of criteria for aging koalas by tooth-wear class. *Australian Zoologist* **35**, 550–554. doi:10.7882/AZ.2011.006

Melzer A, Cristescu R, Ellis W, FitzGibbon S (2014) The habitat and diet of koalas (*Phascolarctos cinereus*) in Queensland. *Australian Mammalogy* **36**, 189–199. doi:10.1071/AM13032

Menkhorst P (2008) Hunted, marooned, re-introduced, contracepted: a history of koala management in Victoria. In *Too Close for Comfort: Contentious Issues in Human-wildlife Encounters*. (Eds D Lunney, A Munn and W Meikle) pp. 73–92. Royal Zoological Society of New South Wales, Sydney.

Middleton DR, Walters B, Menkhorst P, Wright P (2003) Fertility control of the koala, *Phascolarctos cinereus*: the impact of slow-release implants containing levonorgestrel or oestradiol on the production of pouch young. *Wildlife Research* **30**, 207–212. doi:10.1071/WR02052

Minchin AK (1937) Notes on the weaning of a young koala (*Phascolarctos cinereus*). *Records of the South Australian Museum* **6**, 1–3.

Moore BD, Foley WJ (2000) A review of feeding and diet selection in koalas (*Phascolarctos cinereus*). *Australian Journal of Zoology* **48**, 317–333. doi:10.1071/ZO99034

Moore BD, Wallis IR, Marsh KJ, Foley WJ (2004) The role of nutrition in the conservation of the marsupial folivores of eucalypt forests. In *Conservation of Australia's Forest Fauna*. 2nd edn. (Ed. D Lunney) pp. 549–575. Royal Zoological Society of New South Wales, Sydney.

Moore B, Blyton M, Nicole D, Whale J (2023) *What Do Koalas Really Eat? Novel Characterisation of Koala Diets and Habitat Quality Using Faecal Molecular Analysis*. Hawkesbury Institute for the Environment, Western Sydney University, Hawkesbury Campus, Richmond, NSW.

Morgan S, Ward P, Burnet N (1933) *The Native Bear of Australia*. W A Pepperday and Co. Ltd, George St, Sydney.

Nagy KA, Martin RW (1985) Field metabolic rate, water flux, food consumption and time budgets of koalas, *Phascolarctos cinereus* (Marsupialia: Phascolarctidae) in Victoria. *Australian Journal of Zoology* **33**, 655–665. doi:10.1071/ZO9850655

Narayan EJ, Webster K, Nicolson V, Mucci A, Hero JM (2013) Non-invasive evaluation of physiological stress in an iconic Australian marsupial: the Koala (*Phascolarctos cinereus*). *General and Comparative Endocrinology* **187**, 39–47. doi:10.1016/j.ygcen.2013.03.021

Nyo Tun U (1993) Re-establishment of rehabilitated koalas in the wild and their use of habitat in Sheldon, Redland Shire, southeast Queensland with particular reference to dietary selection. MSc thesis. University of Queensland, Brisbane.

O' Callaghan P (1996) *Growth and Mortality of Koala Pouch and Back Young*. Australian Koala Foundation Conference Proceedings, Greenmount Coolangatta, Queensland.

O' Callaghan P (1999) The management of *Eucalyptus* plantations for koala fodder. *Thylacinus* **23**(2), 22–24.

O' Callaghan P, Blanshard W (1991) *Breeding Koalas in Captivity*. ASZK Conference Proceedings, Lone Pine Koala Sanctuary, Queensland.

Osawa R, Carrick FN (1990) Use of a dietary supplement in koalas during systematic antibiotic treatment in chlamydial infection. *Australian Veterinary Journal* **67**, 305–307. doi:10.1111/j.1751-0813.1990.tb07805.x

Osawa R, Blanshard WH, O'Callaghan PG (1993) Microbiological studies of the intestinal microflora of the koala, *Phascolarctos cinereus*. II: pap, a special maternal faeces consumed by juvenile koalas. *Australian Journal of Zoology* **41**, 611–620. doi:10.1071/ZO9930611

Pahl LI, Hume I (1990) Preferences for *Eucalyptus* species of the New England Tablelands and initial development of an artificial diet for koalas. In *Biology of the Koala*. (Eds AK Lee, KA Handasyde and GD Sanson) pp. 123–128. Surrey Beatty & Sons and the World Koala Research Corporation, Sydney.

Palmieri C, Hulse L, Pagliarani S, Larkin R, Higgins DP, Beagley K, Johnston S (2018) *Chlamydia pecorum* infection in the male reproductive system of koalas (*Phascolarctos cinereus*). *Veterinary Pathology* **56**, 300–306. doi:10.1177/0300985818806963

Peterson RA, Bradner JR, Roberts TH, Nevalainen KMH (2009) Fungi from koala (*Phascolarctos cinereus*) faeces exhibit a broad range of enzyme activities against recalcitrant substrates. *Letters in Applied Microbiology* **48**, 218–225. doi:10.1111/j.1472-765X.2008.02513.x

Phillips S (2011) Development of a lightweight, portable trap for capturing free-ranging koalas (*Phascolarctos cinereus*). *Australian Zoologist* **35**, 747–749. doi:10.7882/AZ.2011.025

Phillips S, Callaghan J (2000) Tree species preferences of koalas (Phascolarctos cinereus) in the Campbelltown Area south-west Sydney, New South Wales. *Wildlife Research* **27**, 509–516. doi:10.1071/WR98087

Pournelle GH (1961) Notes on reproduction of the koala. *Journal of Mammalogy* **42**, 396. doi:10.2307/1377039

Pratt A (1937) *The Call of the Koala*. Robertson & Mullins, Melbourne.

Reckless HJ, Murray M, Crowther MS (2017) A review of climatic change as a determinant of the viability of koala populations. *Wildlife Research* **44**, 458–470. doi:10.1071/WR16163

Rodger JC, Paris DBBP, Czarny NA, Harris MS, Molinia FC, Taggart DA, Allen CD, Johnston SD (2009) Artificial insemination in marsupials. *Theriogenology* **71**, 176–189. doi:10.1016/j.theriogenology.2008.09.006

Rus AI, McArthur C, Mella VSA, Crowther MS (2021) Habitat fragmentation affects movement and space use of a specialist folivore, the koala. *Animal Conservation* **24**, 26–37. doi:10.1111/acv.12596

Ryan MA, Whisson DA, Holland GJ, Arnould JPY (2013) Activity patterns of free-ranging koalas (*Phascolarctos cinereus*) revealed by accelerometry. *PLoS One* **8**(11), e80366. doi:10.1371/journal.pone.0080366

Santamaria F, Schlagloth R, Valenza L, Palme R, de Villiers D, Henning J (2023) The effect of disease and injury on faecal cortisol metabolites, as an indicator of stress in wild hospitalised koalas, endangered Australian marsupials. *Veterinary Sciences* **10**, 65. doi:10.3390/vetsci10010065

Seabrook L, McAlpine C, Baxter G, Rhodes J, Bradley A, Lunney D (2014) Determining range edges: habitat quality, climate or climate extremes? *Diversity & Distributions* **20**, 95–106. doi:10.1111/ddi.12152

Skerrett-Byrne DA, Anderson AL, Hulse L, Wass C, Dun MD, Bromfield EG, De Iuliis GN, Pyne M, Nicolson V, Johnston SD, Nixon B (2021) Proteomic analysis of koala (*Phascolarctos cinereus*) spermatozoa and prostatic bodies. *Proteomics* **21**(19), 2100067. doi:10.1002/pmic.202100067

Smith AG, McAlpine C, Rhodes J, Seabrook L, Lunney D, Baxter G (2013a) Are there habitat thresholds in koala occupancy in the semiarid landscapes of the Mulgalands Bioregion? *Wildlife Research* **40**, 413–426. doi:10.1071/WR13010

Smith AG, McAlpine C, Rhodes J, Seabrook L, Baxter G, Lunney D (2013b) At what spatial scales does resource selection vary? A case study of koalas in a semi-arid region. *Austral Ecology* **38**, 230–240. doi:10.1111/j.1442-9993.2012.02396.x

Smith M (1979a) Behaviour of the koala, Phascolarctos cinereus Goldfuss, in captivity. I: non-social behaviour. *Australian Wildlife Research* **6**, 117–129. doi:10.1071/WR9790117

Smith M (1979b) Notes on reproduction and growth in the koala, *Phascolarctos cinereus* (Goldfuss). *Australian Wildlife Research* **6**, 5–12. doi:10.1071/WR9790005

Smith MTA (1980a) Behaviour of the koala, *Phascolarctos cinereus* (Goldfuss), in captivity. III: vocalisations. *Australian Wildlife Research* **7**, 13–34. doi:10.1071/WR9800013

Smith MTA (1980b) Behaviour of the koala, *Phascolarctos cinereus* (Goldfuss), in captivity. VI: aggression. *Australian Wildlife Research* **7**, 177–190. doi:10.1071/WR9800177

Smith MTA (1980c) Behaviour of the koala, *Phascolarctos cinereus* (Goldfuss), in captivity. IV: scent-marking. *Australian Wildlife Research* **7**, 35–40. doi:10.1071/WR9800035

Smith MTA (1980d) Behaviour of the koala, *Phascolarctos cinereus* (Goldfuss), in captivity. V: sexual behaviour. *Australian Wildlife Research* **7**, 41–51. doi:10.1071/WR9800041

Species360 (2023) Zoological Information Management System (ZIMS). Minneapolis, USA. www.species360.org.

Speight N (2022) Health and diseases of koalas. *Animals (Basel)* **12**(1005), 1–2.

Spielman D (1994) First aid emergency care for Australian native mammals. In *Wildlife*. Proceedings 233. Post Graduate Committee in Veterinary Science. (Ed. DI Bryden) pp. 339–366. University of Sydney, Sydney.

Spratt DM, Gill PA (1998) *Durikainema phascolarcti* n. sp. (Nematoda: Muspiceoidea: Robertdollfusidae) from the pulmonary arteries of the koala Phascolarctos cinereus with associated pathological changes. *Systematic Parasitology* **39**, 101–106. doi:10.1023/A:1005957809179

Stalder K, Vaz PK, Gilkerson JR, Baker R, Whiteley P, Ficorilli N, Tatarczuch L, Portas T, Skogvold K, Anderson GA, Devlin JM (2015) Prevalence and clinical significance of herpesvirus infection in populations of Australian marsupials. *PLoS One* **10**(7), e0133807. doi:10.1371/journal.pone.0133807

Stanbury P, Phipps G (1980) *Australia's Animals Discovered*. Pergamon Press Sydney.

Stone BF (1992) Ticks and wildlife: interactions, particularly of the Australian paralysis tick Ixodes holocyclus with koalas, bandicoots and fruit bats. In *Proceedings of Wildlife Disease Association (Australasian Section) Annual Conference*. Warrumbungle National Park. p. 17.

Stone BF, Carrick FN (1996) Some effects of ticks on koalas, especially tick paralysis. In *Koalas: Research for Management. Proceedings of the Brisbane Koala Symposium*. 22–23 September 1990. (Ed. G Gordon) pp. 65–69. World Koala Research Inc., Brisbane.

Takahashi M, Tobey JR, Pisacane CB, Andrus CH (2009) Evaluating the utility of an accelerometer and urinary hormone analysis as indicators of estrus in a zoo-housed koala (*Phascolarctos cinereus*). *Zoo Biology* **28**, 59–68. doi:10.1002/zoo.20212

Thomas O (1923) On some Queensland Phalangeridae. *Annals & Magazine of Natural History* **11**(62), 246–250. doi:10.1080/00222932308632848

Thompson VD (1987) Parturition and development in the Queensland Koala *Phascolarctos cinereus cinereus* at San Diego Zoo. *International Zoo Yearbook* **26**, 217–222. doi:10.1111/j.1748-1090.1987.tb03162.x

Thompson VD, Fadem BH (1989) Scent marking in the koala (Phascolarctos cinereus): related behaviour and sex differences. *Zoological Garten* **59**, 157–165.

Troughton E le G (1935) The southern race of the koala. *Australian Naturalist* **9**, 137–140.

Troughton E le G (1941) *Furred Animals of Australia*. Angus & Robertson, Sydney.

Tyndale-Biscoe H, Renfree M (1987) *Reproductive Physiology of Marsupials*. Cambridge University Press, Cambridge.

Vaz P, Whiteley PL, Wilks CR, Duignan PJ, Ficorilli N, Gilkerson JR, Browning GF, Devlin JM (2011) Detection of a novel gammaherpesvirus in koalas (*Phascolarctos cinereus*). *Journal of Wildlife Diseases* **47**, 787–791. doi:10.7589/0090-3558-47.3.787

Vaz P, Whiteley PL, Wilks CR, Browning GF, Gilkerson JR, Ficorilli N, Devlin JM (2012) Detection of a second novel gammaherpesvirus in a free-ranging koala (*Phascolarctos cinereus*). *Journal of Wildlife Diseases* **48**, 226–229. doi:10.7589/0090-3558-48.1.226

Vogelnest L (1998) Use of transponder identification implants in koalas *Phascolarctos cinereus*. *Thylacinus* **22**(1), 77–78.

Vogelnest L (1999) Chemical restraint of Australian fauna. In *Wildlife in Australia: Healthcare & Management*. Proceedings 327. Post Graduate Foundation in Veterinary Science. (Ed. DI Bryden) pp. 149–186. University of Sydney, Sydney.

Vogelnest L (2019) Capture, physical and chemical restraint. In *Current Therapy of Medicine of Australian Mammals*. (Eds L Vogelnest and T Portas) pp. 121–150. CSIRO Publishing, Melbourne.

Webster K, Narayan E, de Vos N (2017) Fecal glucocorticoid metabolite response of captive koalas (*Phascolarctos cinereus*) to visitor encounters. *General and Comparative Endocrinology* **244**, 157–163. doi:10.1016/j.ygcen.2016.07.012

Wildcare Australia (2018) *Hand-raising Orphaned Koalas*. Version 1. Wildcare Australia Inc., Nerang, Queensland.

Woinarski JCZ, Burbidge AA (2020) *Phascolarctos cinereus* (amended version of 2016 assessment). *The IUCN Red List of Threatened Species* 2020: e.T16892A166496779. doi:10.2305/IUCN.UK.2020-1.RLTS.T16892A166496779.en (accessed 11 October 2021).

Woinarski JCZ, Burbidge AA, Harrison P (Eds) (2014) *The Action Plan for Australian Mammals 2012*. CSIRO Publishing, Melbourne.

Wong LJ, H'ng PS, Wong SY, Lee SH, Lum WC, Chai EW, Wong WZ, Chin KL (2014) Termite digestomes as a potential source of symbiotic microbiota for lignocelluloses degradation: a review. *Pakistan Journal of Biological Sciences* **17**, 956–963. doi:10.3923/pjbs.2014.956.963

Woods R (1999) Prevention of disease in hand reared native wildlife. In *Wildlife in Australia: Healthcare & Management*. Proceedings 327. Post Graduate Foundation in Veterinary Science. (Ed. DI Bryden) pp. 455–489. University of Sydney, Sydney.

Zee YP, Holt WV, Allen CD, Nicolson V, Burridge M, Lisle A, Carrick FN, Johnston SD (2007) Effects of cryopreservation on mitochondrial function and heterogeneity, lipid raft stability and phosphatidylserine translocation in koala (*Phascolarctos cinereus*) spermatozoa. *Reproduction, Fertility and Development* **19**, 850–860. doi:10.1071/RD07084

Zee YP, Holt WV, Gosalvez J, Allen CD, Nicolson V, Pyne M, Burridge M, Carrick FN, Johnston SD (2008) Dimethylacetamide can be used as an alternative to glycerol for the successful cryopreservation of koala (*Phascolarctos cinereus*) spermatozoa. *Reproduction, Fertility and Development* **20**, 724–733. doi:10.1071/RD08036

Zee YP, Holt WV, Nicolson V, Pyne M, Johnston SD (2009) Individual variability in post-thaw sperm survival in a captive koala population. *Cryobiology* **59**, 69–74. doi:10.1016/j.cryobiol.2009.04.010

Zoidis AM, Markowitz H (1992) Findings from a feeding study of the koala (*Phascolarctos cinereus adustus*) at the San Francisco Zoo. *Zoo Biology* **11**, 417–431. doi:10.1002/zoo.1430110608

# 8
# Wombats

*Stephen M. Jackson, Stephen Johnston, Benn Bryant, Tina Janssen and David Taggart*

## 1. INTRODUCTION

The family Vombatidae contains three species of wombats: the bare-nosed (or common) wombat (*Vombatus ursinus*), the southern hairy-nosed wombat (*Lasiorhinus latifrons*) and the northern hairy-nosed wombat (*Lasiorhinus krefftii*). Wombats are large stocky, powerful marsupials that weigh up to 40 kg and are adapted for digging extensive underground burrow systems. Bare-nosed wombats and southern hairy-nosed wombats are still relatively common, though many are killed by cars or shot each year under mitigation permits in some regions. The northern hairy-nosed wombat, although once widely distributed between central Queensland and Southern New South Wales (Horsup 2014; Swinbourne *et al.* 2016), remains one of the world's most endangered species. At the last census in 2017, the population consisted of ~315 individuals at Epping Forest in central Queensland (Taylor and Fewster 2017), with a further 14 individuals at Richard Underwood Nature Refuge near St George in southern Queensland (Horsup *et al.* 2021), the result of a recent introduction.

Despite being considered unattractive and sometimes truculent (Crandall 1964), bare-nosed wombats have been held in numerous zoos throughout Australia and the world. Collins (1973) reported that they had been held in some 41 zoos up until 1969. The first bare-nosed wombats to be held in a zoo in Europe were sent to Newcastle, England, by Captain John Hunter, Governor of New South Wales in 1798. In 1803 the zoological gardens attached to the Natural History Museum in Paris received five bare-nosed wombats that were taken back to France alive by the Nicholas Baudin expedition on the board the ship *Le Naturaliste* (Jackson *et al.* 2021). Everard Home (1808) held another animal in London, which lived for 2 years in captivity and was fed vegetables and hay. London Zoo had both bare-nosed wombats and southern hairy-nosed wombats as early as 1862 (Gray 1863; Flower 1929, 1931). Today bare-nosed wombats are held in many Australian zoos (Collins 1973; Lees and Johnson 2002). The northern hairy-nosed wombat is not currently held by any institution and the only known captive individual was a female animal known as 'Joan' that was held by a farmer on a property adjacent to Epping Forest National Park where the remaining animals live. A second animal, an adult male, 'Solstice' was brought into captivity in June 1996 at Western Plains Zoo in Dubbo but died after ~7months due to difficulty adjusting to captivity. Consequently, there is an urgent need to better understand the husbandry of the northern hairy-nosed wombat in order to establish a sustainable captive population.

## 2. TAXONOMY

### 2.1. Nomenclature

Wombats belong to the family Vombatidae. There are two genera and three extant species of wombats within this family (Table 8.1). The first species to be described was the common wombat (Shaw 1800), followed by the southern hairy-nosed wombat (Owen 1845). The northern hairy-nosed wombat was not described until 1872 by Sir Richard Owen.

Class: Mammalia
Supercohort: Marsupialia
Cohort: Australidelphia
Order: Diprotodontia
Suborder: Vombatiformes
Superfamily: Vombatoidea
Family: Vombatidae
Genus and species: *Lasiorhinus krefftii* (northern hairy-nosed wombat)
*Lasiorhinus latifrons* (southern hairy-nosed wombat)
*Vombatus ursinus* (bare-nosed wombat)

## 2.2 Subspecies

The hairy-nosed wombats do not have any recognised subspecies, but the bare-nosed wombat has three subspecies, which include *Vombatus ursinus ursinus* from Flinders Island, *V. u. hirsutus* from the mainland and *V. u. tasmaniensis* from Tasmania (Jackson and Groves 2015).

## 2.3 Recent synonyms

Synonyms can be found in Jackson and Groves (2015).

## 2.4 Other common names

- Northern hairy-nosed wombat: Queensland wombat, Queensland hairy-nosed wombat or Moonie River wombat (Baker and Gynther 2023).
- Southern hairy-nosed wombat: hairy-nosed wombat (Baker and Gynther 2023).
- Bare-nosed wombat: common wombat, naked-nosed wombat, coarse-haired wombat, island wombat, forest wombat (Baker and Gynther 2023).

# 3. NATURAL HISTORY

## 3.1 Bodyweight and morphometrics

Depending on the species, wild adult wombats range in bodyweight from ~19 to 40 (typically 20–35) kg and are ~900–1150 mm in body length (Van Dyck and Strahan 2008), although in captivity, weights of common wombats of up to 50 kg have been reported (Table 8.1). The wombats are distinctive due to their short powerful limbs, stocky body, blunt and broad head, small eyes and a backward ventrally facing pouch.

**Table 8.1.** Species of wombats, their adult bodyweights, body lengths (without tail) and conservation status.

CR = critically endangered, LR = lower risk, NR = near threatened. Derived from Van Dyck and Strahan (2008), Woinarski *et al.* (2014), Baker and Gynther (2023) and Taggart and Kleemann (unpubl. data).

| Species | Weight (g) | Head and body length (mm) | Status |
|---|---|---|---|
| *Lasiorhinus krefftii* | 25–40 | 965–1195 | CR |
| *Lasiorhinus latifrons* | 18–36 | 840–1100 | NR |
| *Vombatus ursinus* | 20–45 | 840–1150 | LR |

## 3.2 Distribution and habitat

Bare-nosed wombats are found predominantly in south-eastern Australia, mostly in temperate forests and grasslands, although they also occur above the snowline in mountainous areas during winter. The most northerly extent of their range extends up into border regions of Queensland near Stanthorpe in Girraween National Park (McIlroy 2008). Although the natural existing wild population of northern hairy-nosed wombat is only found in central Queensland at Epping Forest National Park, where it occupies semi-arid grasslands, there have been attempts to establish a secondary semi-free-range colony in suitable habitat near St George in southern Queensland. The southern hairy-nosed wombat occurs in southern South Australia west of the Murray River and in south-eastern Western Australia as disjunct populations in semi-arid chenopod grasslands.

## 3.3 Conservation status

The northern hairy-nosed wombat is one of the rarest mammals in the world and is Critically Endangered under the IUCN, with a population size that has been as low as 65 individuals but is now predicted by the Queensland Government, based on genetic data from hair censuses, to be ~315 individuals (Hoyle *et al.* 1995; Horsup 2014; Queensland Government 2021, 2022). The southern hairy-nosed wombat is currently listed as Near Threatened, although populations have rebounded somewhat after the millennium drought (Taggart *et al.* 2020). The bare-nosed wombat is considered at low risk of extinction apart from the South Australian populations that are regarded as vulnerable (Temby 1998; Woinarski *et al.* 2014; Baker and Gynther 2023).

## 3.4 Diet in the wild

Wombats are grazing herbivores that feed on a variety of grasses, herbs, sedges, forbs, roots, shrubs and bulbs and eat these largely proportional to their availability. Grasses most utilised by bare-nosed wombats for food include tussock grass (*Poa* sp.), kangaroo grass (*Themeda australis*), spear grass (*Stipa* sp.) and wallaby grass (*Danthonia penicillata*), with others such as oats (*Avena sativa*), Australian salt grass (*Distichlis distichophylla*), perennial rye grass (*Lolium perenne*), club rushes (*Scirpus* sp.), sedges (*Carex* sp.) and mat-rushes (*Lomandra* sp.) also eaten at times (Mallett and Cooke 1986; Rishworth *et al.* 1995; Triggs 1996; Woolnough 1998).

Common dietary items eaten by the southern hairy-nosed wombat include: the native spear grasses (*Atriplex* spp. and *Astrebla* spp.); corrugated sida (*Sida corrugate*); blue bush and salt bush (*Chenopodiaceae* spp. and *Camphorosmeae* spp.); and introduced weeds including thread iris (*Moraea setifolia*); wards weed (*Carrichtera annua*); onion weed (*Asphodelus fistulosus*); and burr medic (*Medicago minima*) (Camp *et al.* 2020; Taggart and

Kleemann unpubl. data). Grazing halos can often be seen surrounding a southern hairy-nosed wombat warren complex where the palatable vegetation has been eaten down compared with adjacent vegetation. Due to the highly abrasive nature of the silica in the grasses that wombats eat, their incisors and cheek teeth grow continuously, a feature that is unique among the marsupials. Common dietary items eaten by the northern hairy-nosed wombat include the native *Aristida* spp., *Cenchrus ciliaris*, *Enneapogan* spp., and other unidentifiable monocotyledons (Woolnough 1998; Woolnough and Johnson 2000).

## 3.5 Longevity

### 3.5.1 Wild

In the wild southern hairy-nosed wombats have been known to live for > 18 years (Walker *et al.* 2020; Taggart and Temple-Smith 2023; Taggart and Kleemann unpubl. data). Comparatively little is known of the longevity of bare-nosed wombats in the wild; however, it appears that they can live for more than 15 years (Table 8.2) (Triggs 1996).

### 3.5.2 Captivity

Bare-nosed wombats typically live 12–15 years in captivity (e.g. Schmidt 1880; Fleay 1957; Crandall 1964; Collins 1973); however, there are records of them living to 26 years of age in London Zoo (Flower 1931). A further record at the National Zoo and Aquarium was of a bare-nosed wombat over 31 years of age (ABC News 2017). Similar to bare-nosed wombats, southern hairy-nosed wombats typically live 10–15 years in captivity, but life spans of 25 and 27 years have been recorded (Flower 1931; Treby 2005). The only record of longevity of a captive northern hairy-nosed wombat is that of the female wombat 'Joan', which was caught as an adult and lived for some 27 years in captivity with the Dennis family who live next to Epping Forest National Park; as the individual was an adult when brought into captivity it must have been at least 30 years old at death (Woolnough 1998).

### 3.5.3 Methods used to determine the age of adults

Growth charts based on head length, and bodyweight exist for juvenile and subadult southern hairy-nosed wombats up to 2.2 years of age (Taggart *et al.* 2007), but once wombats reach adult size there is no reliable technique for aging them. Patterns of tooth wear cannot be used for aging wombats as their teeth grow continuously.

**Table 8.2.** Longevity (years) of different genera of wombats in the wild and in captivity.

The longest recorded longevity is in brackets.

| Genus | Wild | Captivity | Reference |
|---|---|---|---|
| *Lasiorhinus latifrons* | 14–15 (18) | 10–15 (27) | 1–5 |
| *Lasiorhinus krefftii* | – | 30 | 6 |
| *Vombatus* | 15 | 12–15 (31) | 1, 7–11 |

References: (1) Mitchell (1911); (2) Flower (1931); (3) Wells (1989); (4) Taggart and Temple-Smith (2023); (5) D. Taggart and S. Kleemann unpubl. data; (6) Woolnough (1998); (7) Schmidt (1880); (8) Fleay (1957); (9) Crandall (1964); (10) Collins (1973); (11) Triggs (1996).

## 4. HOUSING REQUIREMENTS

### 4.1 Enclosure design

Enclosure design needs to consider that wombats are powerful and destructive. They will dig and chew in attempts to escape, pulling up plants and digging under logs and other furnishings. The floor should be soil with a mesh underlay or concrete layer ~1–1.5 m below the surface. The walls should be smooth to preclude chewing and scrabbling at mesh fences, causing damage to the teeth, gums and feet (Booth 1999). Reflective metal fencing should be avoided as it may cause excessive elevated enclosure temperatures (Williams 1990).

All wombats live in burrows, which, for hairy-nosed wombats, are often found within large warren complexes comprising many burrows. Burrows help protect wombats from physical threats, dehydration and environmental extremes (Taggart *et al.* 2020; Taggart and Temple-Smith 2023). Wombats can live in extreme environments where the temperatures can reach 35–48°C and relative humidity reach as low as 2–5%. Despite this, they are relatively intolerant of heat, so shelter in burrows during the day where the corresponding temperature is ~10–27°C with 60–70% humidity (Wells 1971; Shimmin *et al.* 2001, 2002). Wombats do not sweat, which is useful for conserving water, but also contributes to their susceptibility to heat stress (Wells 1978a). Common wombats can develop hyperthermia when ambient temperatures exceed 24°C (Brown 1964) and for hairy-nosed wombats > 30°C, with deaths known to occur at temperatures > 38°C (Ride 1970; Wells 1971; Gaughwin 1982; Williams 1990). Southern hairy-nosed wombats will salivate profusely when they get hot, saturating the fur around their lower jaw and upper chest

(Taggart and Kleemann unpubl. data). In their burrows southern hairy-nosed wombats will sometimes create earth plugs to help reduce moisture loss through increased humidity (Steele and Temple-Smith 1998; Shimmin *et al.* 2002).

If captive animals are kept outdoors where they cannot construct extensive burrows, they must have the opportunity to behaviourally thermoregulate within an artificial burrow, be provided adequate soil to construct one or have access to air-conditioned dens. Burrow analogues can be constructed using mock rock caves, pipes or hollow tree trunks. Overstorey planting will provide shade and should be included. Sprinklers and adequate shading during warm weather should be provided where environmental temperatures exceed 24°C (Gaughwin 1982; S. Jackson pers. obs.). Artificial burrows should be regularly checked for temperature to ensure they meet welfare standards and are below 26°C, particularly during hot weather in the summer months.

Water can be provided via a water feature, a stainless-steel bowl (secured so it cannot be tipped over) or an automatic filling device.

Containment walls should be ~1.2–1.5 m high and continue below the surface to a depth of at least 1 m, or until it reaches the buried mesh, so there are no points that can allow the wombats to escape, above or below the ground.

Wombats have also been successfully held indoors or where there is an indoor and outdoor component to the enclosure, which enables better control of the ambient temperature. Wombats are generally content to lay on cool concrete flooring, but if soil and leaf litter is preferred in the den, then this needs to be renewed on a regular basis to prevent contamination and disease. The indoor enclosure allows the use of reverse lighting to exhibit them during their nocturnal phase when they are most active (Crowcroft and Soderlund 1977).

Several zoos such as Taronga Western Plains Zoo (bare-nosed wombats), Rockhampton Zoo and the non-government Australian Animals Care and Education (AACE) facility located at Mount Larcom in central coastal Queensland (southern-hairy-nosed wombats) have maintained research colonies of wombats and have adapted their enclosure designs for better animal caretaker access, sample collection and breeding management. These designs are less about the visitor experience per se and more about allowing greater facilitation of animal husbandry and sample collection. The enclosures at Taronga Western Plains Zoo were based on a design for four cycling female bare-nosed wombats that included a central but separate internal den system with animal caretaker access from an elevated ceiling and which allowed periodic access to an external open vegetated pen system; daily movement of wombats was controlled by opening or closing a series of tunnel doors, which also facilitated the exclusion of wombats for cleaning of the den system (Fig. 8.1; MacCallum 2004). Rockhampton Zoo developed a similar enclosure design for their research colony of southern hairy-nosed wombats (Hogan *et al.* 2009, 2010a; Descovich *et al.* 2012a; Swinbourne *et al.* 2015, 2018a, 2018b).

A captive research colony of up to 22 southern hairy-nosed wombats has been held at the AACE facility, initially at Marlborough and later at Mt Larcom on the central coast of Queensland. More recently Janssen and her research colleagues from AACE (2022) have established a very successful breeding colony of southern

**Fig. 8.1.** Bare-nosed wombat research facility housing based at Taronga Western Plains Zoo. (Top) Purpose-built facility showing central den system and surrounding open pens; (Bottom) ceiling door of the den system that allows animal caretaker access. Photos by Jodie Lardner-Smith.

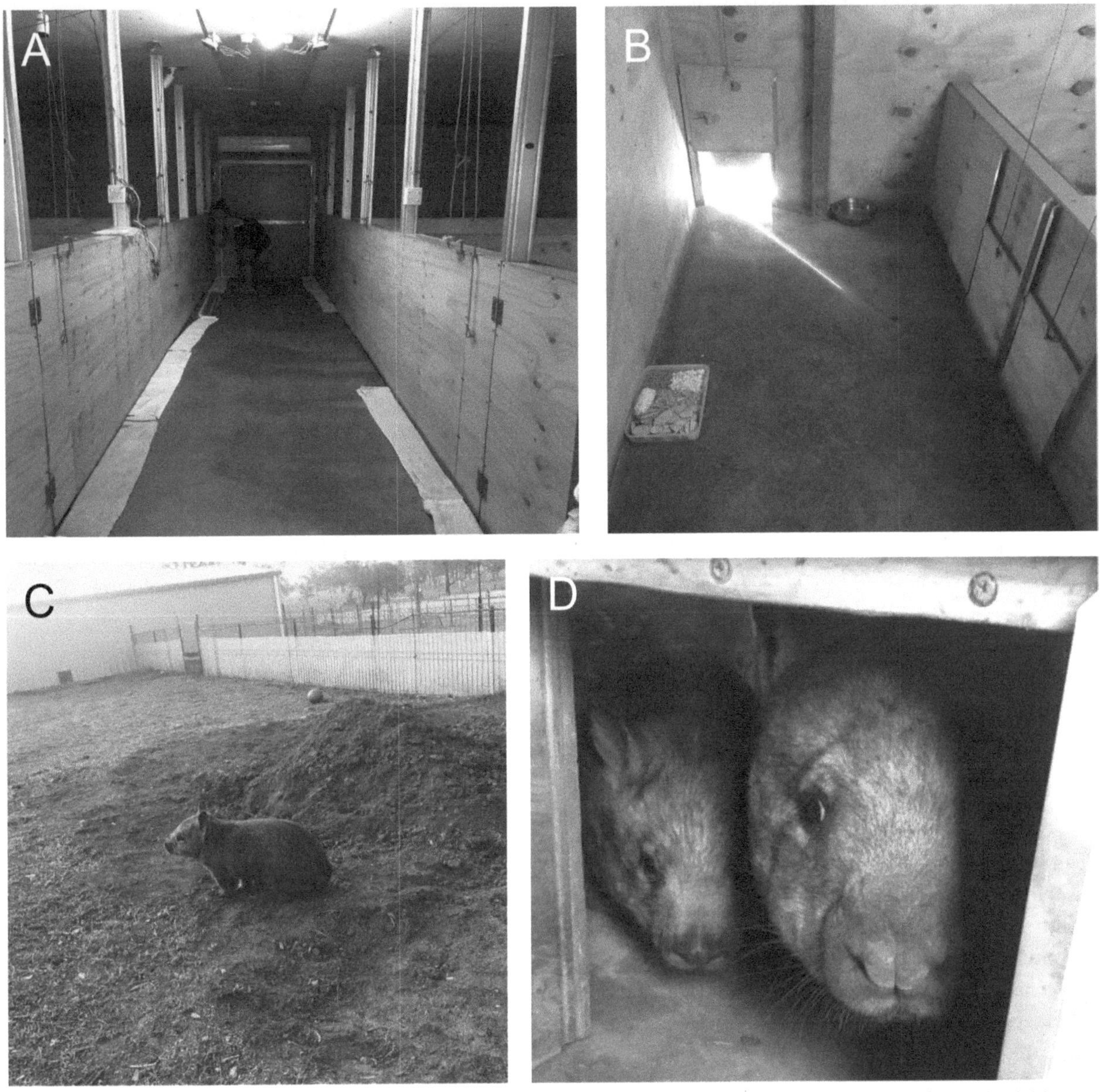


**Fig. 8.2.** Research colony facility based at Australian Animals Care and Education, Mount Larcom, central coastal Queensland. (a) Air-conditioned ground level internal den system with easy access for zoo animal caretakers. Note the concrete floor for ease of cleaning. (b) Individual animal internal den showing internal and external doors with pulley access to avoid contact with wombats. (c) Large external pen area with dirt mounds for excavation and for providing enrichment. Note also the bowling ball up against the fence line, which has been used for enrichment. (d) Mother and joey typically share the same den system and enclosure until weaning. Photos by Tina Janssen.

hairy-nosed wombats in which the enclosure design has been optimised for reproductive research (Fig. 8.2). These animals typically entered the research program as hand-reared pouch young from South Australia, the colony of which was established as a reproductive model to better understand the captive husbandry and breeding of the nearby northern hairy-nosed wombat.

Critical to the success of the program has been the design of enclosures (discussed below), which has incorporated the use of air-conditioners that typically maintain the temperature of the artificial den system around 24°C. It is recommended that air-conditioners should be an integral component of any wombat artificial den system, but especially in hotter and humid climates (T. Janssen and S. Johnston pers. obs.).

The AACE facility also utilises an internal den system with concrete floors for ease of cleaning and the recovery of faecal and urine samples for hormone analysis

(Swinbourne *et al.* 2017a, 2018a, 20186b; Du *et al.* 2017, 2018). Each den system opens to a large open grassed/dirt pen area with regularly maintained dirt mounds for excavation by the wombats. Typically, wombats are held singly in each enclosure, except for breeding purposes or mothers with young at foot. The whole internal den system and outdoor pen is monitored 24 h by video surveillance. Pulley doors to the outside pens and between sleeping areas in the internal pens are used to move the wombats for cleaning and reproductive management.

Male common and southern hairy-nosed wombats can be very aggressive both to their conspecifics and animal caretakers, so that males are typically housed separately.

## 4.2 Holding area design

The holding area design is based on a similar principle to the enclosure design and only needs to be basic if the wombats are held in small numbers. If held as pairs, and the animals are compatible, provision should be made to minimise the effect of aggression by providing separate resting areas.

## 4.3 Spatial requirements

A pair of wombats requires at least 45 $m^2$, with an additional 10 $m^2$ for each additional animal. Larger enclosures are preferable and enclosures up to 400 $m^2$ have been suggested to reduce the likelihood of pacing, climbing and other attempts to dig out of the enclosure (Booth 1999). An alternative enclosure design for a pair of wombats consists of an indoor temperature controlled (air-conditioned) area (5.8 $m^2$) with individual sleeping chambers (3.0 × 0.6 $m^2$ each) and an adjoining outdoor area (~100 $m^2$) consisting of soil substrate, partial grass vegetation and a dirt mound for enrichment (Swinbourne *et al.* 2017a).

Although common wombats are usually not kept with more than two individuals together, southern hairy-nosed wombats can be held in groups of 4–6 individuals. When holding multiple animals together, at least 10 $m^2$ should be provided for each additional animal above two. Studies of southern hairy-nosed wombats revealed that when a male and two females were housed in small (75.5 $m^2$), medium (151 $m^2$), and large (224 $m^2$) enclosures there were significant differences in observed social conflict behaviour (Descovich *et al.* 2012b). The wombats in the largest enclosures displayed less biting behaviour, visual scanning, retreated less from conspecifics and dug less along the fence line (which may represent escape behaviour). The wombats in the largest enclosures also showed less self-directed grooming than the animals housed in the smaller enclosures (Descovich *et al.* 2012b).

**Fig. 8.3.** Southern hairy-nosed wombat basking in the sunshine in the wild at Kooloola Station, Murraylands, South Australia. Photo by David Taggart.

## 4.4 Position of enclosures

Enclosures should be situated in an area that provides shade during hot weather and sunny areas during cooler weather, as wombats like to sunbathe when it is cold (Fig. 8.3).

## 4.5 Weather protection

The enclosures can be open, semi-enclosed or fully enclosed, but as wombats do not sweat and rely on the burrow to help modulate their body temperature and energy expenditure (Wells 1978a; Taggart *et al.* 2020), combined indoor (air conditioned) and outdoor enclosures work best. If the enclosure is open outdoors only, then the animals need to have adequate shade and sprinkler systems to cool the wombats in hot weather.

## 4.6 Temperature requirements

Although heating is not required, a clean, dry shelter should always be available. Wombats should have access to areas that are lower than 24°C at all times due to their susceptibility to heat stress, unless they have access to extensive burrow systems. Internal den systems of wombats held in hotter and humid climates should include air-conditioning.

## 4.7 Substrate

The substrate should be soil, leaf litter (that is regularly renewed) or sand that is well-drained so that flooding does not occur. Straw and hay in den systems can cause respiratory and fungal problems and should be avoided. Digging behaviour allows the wombats to wear down their claws, which may otherwise grow too long (Williams 1990). If wombats are permitted to excavate their own burrows, a sandy loam or clay works well because wombats use this for burrowing in the wild (Steele and Temple-Smith 1998; Marshall *et al.* 2018; Swinbourne *et al.* 2021). For example, a study of wild southern hairy-nosed wombats found the average clay content of the soil where they occur was 23% (range 9–40%) and the average sand content was 63% (range 41–88%) (Swinbourne *et al.* 2021). Sand substrates may also allow the wombats to dust bath. Internal den systems are preferably furnished with a concrete floor for ease of cleaning, but these areas should be complemented with the addition of outdoor soil-based substrates to allow digging.

## 4.8 Den sites

Wombats need areas to retreat from high temperatures. A suitable den should be positioned in the shade and ideally be ~1.2 m (L) × 1.2 m (W) × 1.0 m (H) with a hinged lid. It should be lined with dense rubber to allow for easy cleaning and to protect the box from urine (Fig. 8.4). The entrance to the den box should be ~40 cm (H) × 40 cm (W) to allow the wombat to rub its back against the doorframe (Gaughwin 1982; Williams 1990; Steele and Temple-Smith 1998). Ideally, a small den system with a lockdown drop door is preferable for the sleeping area and feeding, with a small doorway to allow the animal to scratch while entering or exiting the den (Fig. 8.4). If fed in the den area, the wombat is more likely to use this area and this can facilitate containment by dropping the door. Facilities housing larger research colonies of wombats have typically incorporated separate sleeping dens and tunnel systems into the overall design, with various door systems to facilitate daily husbandry and movement of animals (MacCallum 2004; Hogan *et al.* 2009, 2010b; Descovich *et al.* 2012b).

## 4.9 Enclosure furnishings

Furnishings need to be strong or wombats will destroy them. Large hollow logs, branches, large rocks and terracotta or concrete pipes that are buried in the soil and large enough to sleep inside are useful. Heavy furnishings need to be carefully and thoughtfully placed in order to avoid the risk of wombats excavating under them and being crushed. A suitably positioned scratching/rubbing post is also recommended. Branches and rocks can also be used for this purpose. Wombats will

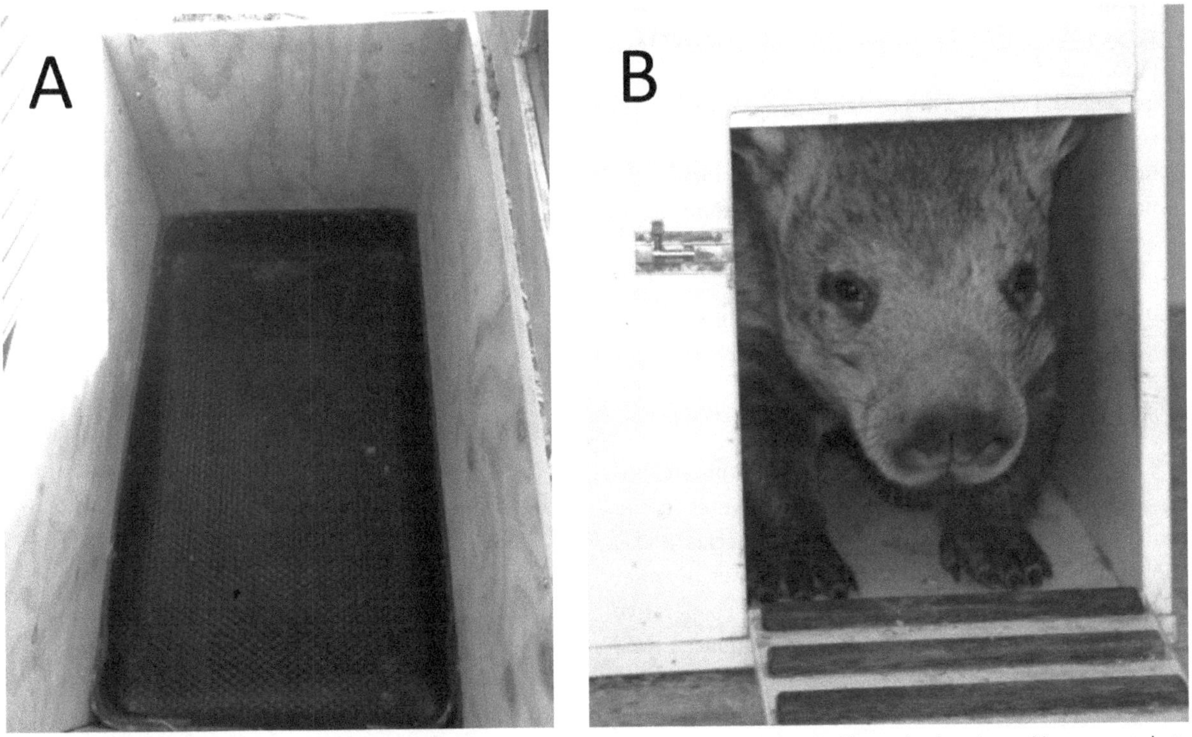


**Fig. 8.4.** Wombat housing boxes. (a) Housing box fitted with rubber floor and (b) lockable box facilitates husbandry and has an opening large enough to allow the wombat to scratch. Photos by Tina Janssen.

often use objects to reach places on their body that they cannot easily reach with their claws to scratch

Attempts can be made to establish plants. Larger shrubs, trees and large tussocks have a better chance of survival than seedlings. To help plants establish, large rocks or logs can be placed around them to make them harder to dig up or electric fencing can be used. If difficulties are encountered growing plants inside the enclosure, they may need to be planted around the enclosure to provide shading.

# 5. GENERAL HUSBANDRY

## 5.1 Hygiene and cleaning

All enclosures should be cleaned daily to remove faeces, urine and uneaten food. Drinking water dishes should be cleaned and refilled daily. When all individuals permanently leave an enclosure, it should be thoroughly cleaned before any new animals are admitted.

## 5.2 Record keeping

It is important to establish a system to monitor the husbandry and health status of captive wombats. The ongoing improvement of animal management is enabled by review of animal records. Records should be kept of:

- identification (all individuals should be identifiable)
- results of veterinary examinations and treatments administered
- behavioural changes or problems
- reproductive behaviour or condition
- bodyweights with matching body measurements, particularly head length and width for a quantitative body condition score
- changes in diet and food consumption
- movements of individuals between enclosures or institutions
- births, with dam and sire if known
- deaths with necropsy results.

The collection of information on physical, physiological, and behavioural patterns of individual wombats can contribute greatly to the husbandry of the species. It also allows individual histories to be transferred to other institutions if required and facilitates a cooperative approach to data collection among institutions. ZIMS, the Zoo Information Management System (Species360 2023), can be used for recording general information on age, sex, parentage, births, identification, bodyweights, transfers and deaths, with separate ZIMS modules for veterinary information and studbooks. These systems have been developed by Species360, which is part of the Conservation Planning Specialist Group (CPSG) software. ZIMS also interfaces with programs, such as PMx, which provides demographic and genetic analysis tools to assist with the management of breeding programs (Lacy *et al.* 2011).

## 5.3 Methods of individual identification

### 5.3.1 Passive Integrated Transponder (PIT) tags

PIT tags (microchips) are implanted between the scapulae of the wombat. PIT tags are an excellent, permanent method of identification. To prevent the tags tracking out along the injection site, the entry point can be sealed with tissue glue (e.g. Vetbond®) or similar fast-setting adhesive. They generally require the animal to be caught to confirm identification with a PIT tag reader. Tags may also move around under the skin to different regions of the body and very occasionally cause secondary complications such as carcinomas (Vogelnest *et al.* 1997).

### 5.3.2 Tattoos

Tattoos have been used successfully on the inside of the ear (Johnson 1991) and the medial aspect of the hindleg (thigh) (Skerratt 2001).

### 5.3.3 Visual identification

As most wombats show variation in pelage and superficial scarring, visual identification via unique physical features can be useful (Triggs 1996). Coat markings, nose colouration patterns, general fur colouration and body size can be useful for identifying southern hairy-nosed wombats (D. Taggart pers. obs.; Swinbourne *et al.* 2018b; O'Brien *et al.* 2021).

### 5.3.4 Ear tags

Although ear tags have the advantage of being read at a distance, they are not recommended for captive animals as they may be lost or torn from the ear when the animals are caught, particularly if being netted, or when they fight with each other (McIlroy 1976).

# 6. FEEDING REQUIREMENTS

## 6.1 Captive diet

Wombats are herbivores that have evolved a dentition and gastrointestinal tract adapted to a high-fibre, low-energy, abrasive natural diet of grasses and other plant material (Kleemann *et al.* 2023; Taggart *et al.* 2024. They are hindgut fermenters (i.e. digestive fermentation occurs in their extensive colon). Captive diets should strive to emulate this feeding ecology whenever possible. Inappropriate rations likely predispose them to obesity, dysbiosis (a reduction in microbial diversity including the loss of beneficial bacteria) and gastrointestinal, kidney and urinary problems. Wild wombats do not eat high starch and high sugar (e.g. grain, fruit, vegetables) or high-protein (e.g. dog food) items and therefore these should be limited in captive wombat diets. Inappropriate supplements (e.g. vitamin D, copper, calcium) are thought to have resulted in nutritional disease in wombats (Booth 1994; Barboza and Vanselow 1990). Historically, captive wombats have been maintained successfully on grass hay-based diets supplemented with fresh grass, browse and smaller quantities of grain and/or pelleted concentrates. The diet recommended for a research colony of breeding southern hairy-nosed wombat is high in fibre but low in protein, in addition to giving wombats access to native grasses, forbes and sedges within their enclosure; native browse and bark may also provide for normal tooth maintenance (T. Janssen pers. obs.).

- Ad lib
  - water
  - meadow hay
  - oaten hay
  - fresh grass is ideal, but time consuming to collect
- Daily diet (per animal)
  - 500 g carrots, large
  - 200 eucalypt or wattle branches
  - lucerne or maize can be given on alternate days
- Supplement
  - none

An alternative diet that has been used with success.

- Daily diet (per animal)
  - 400 g pellets
  - 50 g maize
  - 50 g oats (crushed)
  - 50 g wheat

Diet used during the successful mating of southern hairy-nosed wombats (Hogan *et al.* 2010c):

- Daily diet (per animal)
  - Ad lib water
  - 300 g kangaroo pellets (Riverina Australia Pty Ltd, Brisbane, Qld, Australia)
  - 150 g chopped carrots
  - 50 g oaten hay chaff

Diets used for successful captive health and during the successful mating of southern hairy-nosed wombats.

- Daily diet (per animal)
  - Ad lib water
  - 1 cup rolled oats
  - ½ cup gumnut pellets (Mitavite)
  - ½ cup oaten chaff
  - 200 g sweet potato pieces
  - 3 times per week – ½ corn cob

  (T. Janssen pers. obs.).

- Ad lib water
- Daily mixed ration
  - 120 g rolled oats
  - 120 g gumnut pellets (Mitavite, Australia)
  - 35 g oaten chaff
  - 200 g sliced sweet potato
  - ½ corn cob

  (Swinbourne *et al.* 2015).

The change to high-quality food before the breeding season, associated with increased rainfall, in hairy-nosed wombats, and potentially bare-nosed wombats, is thought to be an important trigger for breeding in the wild (Wells 1978b; Camp *et al.* 2020; Casey *et al.* 2021). To try and replicate this, several studies have used the provision of a bland diet of primarily pellets and hay during most of the year and then a large quantity of fresh green grass from several months before the breeding season until the end of the breeding season to help trigger reproduction. T. Janssen (pers. obs.) has noted that a lack of rain before the breeding at the Australian Animal Care and Education wombat breeding centre is typically associated with reduced breeding behaviours and males showing less aggressive breeding behaviours.

Although artificial diets are largely used, as much grass as possible should be provided fresh daily to supplement the diet. Wombats have a tendency to become obese on high-energy and high-protein diets, so some institutions include a starve day once per week (Gaughwin 1982). However, a better strategy is to provide a consistently low energy, low protein diet.

Due to their low metabolic rate and slow digesta passage rates, wombats should be given a diet based primarily on grass and/or palatable but low-quality food such as hay, ad lib (Booth 1994). High-energy diets, such as pasture replacement pellets, maize, fruit, vegetables and lucerne should not be given, or only occasionally, as these can lead to obesity and other health problems as they are too high in energy for long-term maintenance (Booth 1994). Wombats should also not be fed dry dog food, as it is high in protein, low in fibre content and has a mineral balance designed for carnivores (Booth 1994). There may be a link between feeding inappropriate diets, including excessive vitamin D supplementation, and systemic calcification seen occasionally in captive wombats (Booth 1994; Skerratt *et al.* 1997).

Recent developments in the microbiome research of wild wombats may have significant implications for better understanding and formulating captive wombat diets and species differences (Eisenhofer *et al.* 2021a, 2021b). This work concluded that captivity has a major impact on the gut microbiota of southern hairy-nosed wombats and that different wild populations have unique microbial compositions. These authors speculated on whether such data could be used to manipulate captive diets and improve future management of the species.

### 6.2 Supplements

No specific supplements are needed; however, additional food items such as branches of eucalypts and wattles can be added every few days because wombats occasionally eat bark (Triggs 1996). Take care not to provide diets high in copper, as a diet containing 36 ppm of copper (as $CuSO_4$ in a pig grower supplement) appears to have resulted in death of a captive southern hairy-nosed wombat (Barboza and Vanselow 1990).

### 6.3 Presentation of food

Food is generally provided in stainless steel trays or hoppers placed 20 cm above the ground to prevent the wombats defecating or walking in their food (Gaughwin 1982). One tray is normally provided for each animal. If an animal has just arrived in captivity from the wild, food should be provided on the ground and water as close to the ground as possible, as wombats will initially avoid anything strange/foreign (D. Taggart pers. obs.).

## 7. HANDLING AND TRANSPORT

### 7.1 Timing of capture and handling

Captive wombats are usually best caught during the day when they are less active. Captive wombats are easily contained any time of the day if inside a small den system with a drop door.

### 7.2 Capture equipment

Strong hessian or canvas bags can be used for capture and short-term holding of wombats. When more control is needed, a tapering bag with lace-up inspection ports is useful. If the bag is placed over the head, the wombat will readily climb in until it is firmly wedged (Wells 1971). Wombats will settle very quickly in a hessian sack and often fall asleep if they are maintained in a cool environment (D. Taggart pers. obs.). A more reliable, less stressful approach is to use a transport box placed up against the den door so that the wombat can walk straight into the box and the door is dropped (Fig. 8.5). Transport boxes are best if the walls and door are solid with air holes placed above the animal's head height, so it cannot see out. An access port on the top of the box is ideal for checking on the wombat or for delivering injections for medication or sedation.

### 7.3 Capture and physical restraint

Juvenile wombats less than ~18 months of age can generally be picked up easily under the armpits and carried this way. Adult wombats, however, are large, powerful, muscular animals and can be highly aggressive (particularly common wombats) and, therefore, may be difficult to restrain manually. In captivity, wombats will readily attack and bite legs, arms or hands and can cause significant injury. Some animals will retreat into the den box, pipe or log and present their rear end. In this case, great care needs to be taken when trying to retrieve the wombat as it may endeavour to crush a hand or arm against the side or roof of the hollow. One method used is to grab one of its hindlegs and pull the animal out of the box. Boxes with hinged lids are very useful to gain access to wombats and den areas with multiple sliding doors also work well. The best method is to try and direct the wombat into a hessian sack or net in which it can be handled safely and securely.

Aggressive wombats that charge can often be tricked into charging into a hessian sack or a sturdy net (e.g. steel frame and tennis court netting). Once netted, the wombat can be pushed into a corner of the net and held in place by standing on the net directly behind the wombat. With the net gathered in tightly around the wombat's flanks and across its back, the animal can then be picked up behind the front legs while still within the net and gently dropped into an open hessian sack held by assistants (D. Taggart pers. obs.). If the animal must be picked up without a net, two approaches can be used.

The best and easiest approach is to try and scoop the wombat directly up into an open hessian sack from behind. An alternative and less effective method used by some animal caretakers is to approach the animal from behind and hold it in position by placing a foot against its rump so it cannot reverse, and a hand on each shoulder so that it cannot turn or go forward. Firmly hold its shoulders in place and then move the hands back over the shoulders towards the armpits, sliding one arm under the armpit and across the chest (Fig. 8.5).

Highly aggressive individuals may throw their head backwards and try to bite so be wary of this behaviour. These animals can be sedated with an intramuscular injection of the drug combination tiletamine/zolazepam (Zoletil®) at 3–5 mg/kg to enable them to be more easily handled for simple procedures such as venepuncture. An intramuscular injection of Zoletil® is typically used before further induction and maintenance using gaseous anaesthesia (isoflurane) for more invasive veterinary interventions such as semen collection or surgery.

A study that examined the response of southern hairy-nosed wombats to handing revealed that regular handling lowered the reactivity to and avoidance of the human handler, but did not reduce the stress response (cortisol response), which suggests wombats can enter a state of learned helplessness (Hogan *et al.* 2011a).

## 7.4 Weighing and examination

If the wombat is a juvenile, subadult or adult that has been conditioned to handling it can readily be weighed by holding it, weighing yourself with the wombat and then subtracting your bodyweight; it is not recommended to handle aggressive adult wombats in this manner. Instead, these wombats can be encouraged into a travel box that is weighed and the box weight subtracted to give the wombat's bodyweight. This will help reduce the stress for both the animal and the handler. It is important to collect head measures (length and width) to assess quantitative changes in wombat body condition over time.

## 7.5 Release

Captive wombats need to be released with care and directed away from the handler at release. Aggressive individuals, particularly common wombats, can turn and try to bite. Therefore, the animal caretaker should release animals and quickly exit the enclosure. Ideally, particularly for aggressive animals, they can be released over a small wall (~1 m high) to reduce the risk of biting. If this is not possible, handlers have been known to release a wombat while standing in a plastic or metal garbage bin and then retreating from the enclosure once the animal settles down.

## 7.6 Transport requirements

### 7.6.1 Transport box design

Due to the strength of wombats, boxes must be very sturdily built or the wombat is likely to break out during transit (Fig. 8.6). Further specific details of the box design can be obtained from the International Air Transport Association (IATA 2022). Wombats can be routinely transported by road in hessian sacks within air-conditioned vehicles across short distances (< 8 h) (Fig. 8.7). Care must be taken to tightly tie the sack closed (double ties are recommended) as wombats can sometimes push a single tie directly off the end of the sack. Large, close-weave hessian sacks are recommended.

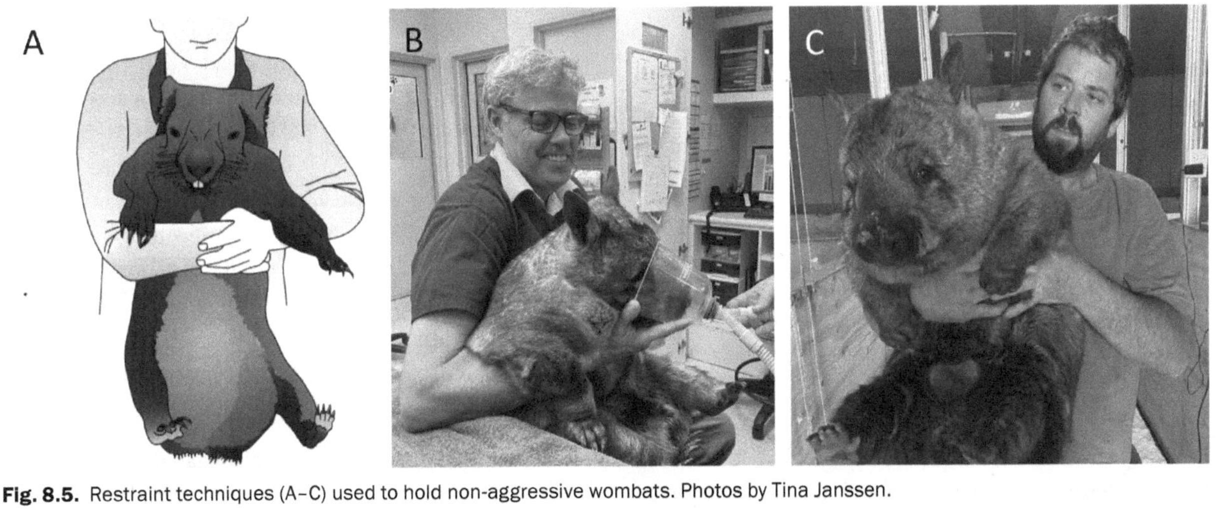


**Fig. 8.5.** Restraint techniques (A–C) used to hold non-aggressive wombats. Photos by Tina Janssen.

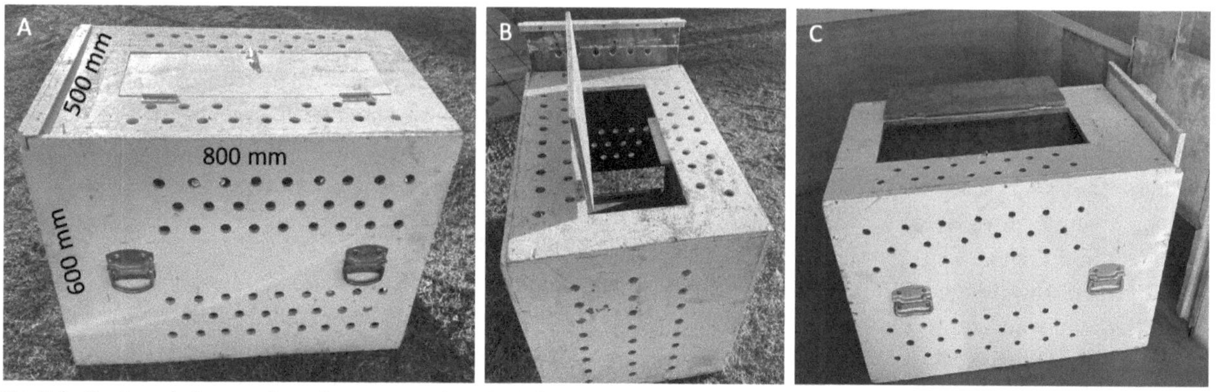


**Fig. 8.6.** Southern hairy-nosed wombat transport box. (a, b) Dimensions and construction of transport box. (c) Transport box up against internal den opening to catch wombat without handling. Photos by Tina Janssen.

**Fig. 8.7.** Green-stripe, close-weave hessian sacks used to secure wild-caught southern hairy-nosed wombats following capture and during air-conditioned road transport across short distances. Photo by David Taggart.

### 7.6.2 Furnishings

Dense rubber matting secured to floor of box for ease of cleaning.

### 7.6.3 Water and food

Due to their low metabolic rate and long digestion times, wombats do not need to be fed or have access to water for trips less than 24 h. Indeed, they can remain underground for days during drier periods when food availability is short (Finlayson *et al.* 2010). For longer journeys, food and water should be provided in a deep dish, although, if the animals have just been removed from the wild, they will likely not touch any food or water for days (D. Taggart and P. Temple-Smith unpubl. data).

### 7.6.4 Animals per transport box

Wombats should be boxed individually for travel. Females with pouch young should not be transferred unless the young are still permanently attached to the teat, although wombats have never been observed to eject young from pouches (D. Taggart pers. obs.).

### 7.6.5 Timing of transportation

Due to the wombat's inability to sweat or tolerate high temperatures, transportation should be overnight or in the morning in cooler weather. Avoid transporting animals in temperatures above 24°C. Wombats should be transported in air-conditioned vehicles and regularly checked to ensure their welfare.

### 7.6.6 Release from the transport box

Generally, the box is opened and the wombat is able to exit in its own time. The box is then removed when the wombat has established another area as its den site.

## 8. HEALTH REQUIREMENTS

### 8.1 Daily health observations

Wombats should be visually inspected at least daily for signs of injury or illness. This should be aligned with the daily keeping routine to enable observation at a minimal distance (e.g. when confined in the night quarters, when feed is offered, during animal movements between enclosure areas etc.). Sometimes wombats need to be encouraged to move to enable meaningful assessment (Bryant and Reiss 2008). Consideration should be given to the animal's recent husbandry context (e.g. look for fight injury if there has been a recent animal introduction, note the pouch profile if there has been breeding opportunity etc.). Visual inspection should include assessment of:

- appetite
- demeanour and behaviour – any changes can be associated with health problems or stress. Depression/

reduced activity, appetite loss, loud vocalisations/screams, teeth grinding, aggression and increased activity such as pacing and repeated escape attempts can be significant. Wombats can be stoic and may not show obvious behavioural signs of stress. Stress responses (e.g. pacing, escape attempts) may only be manifest at night when the animals are unobserved (Bryant and Reiss 2008). Increased scratching can be a sign of pruritis (itchiness), which can be associated with parasitic skin disease (e.g. sarcoptic mange)
- body condition score – visibly obvious declining body condition (see section 8.2 and Fig. 8.8). If there are concerns, a closer assessment may be required as per section 8.2
- gait – wombats walk with a characteristic rolling gait and will also trot and bound. The gait should be symmetrical with their bodyweight borne evenly on the soles of the feet (Bryant and Reiss 2008)
- pelage/skin – the coat should be continuous over the body and head. The skin should be intact and free from inflammation and crusting. Animals in good condition will typically have a glossy pelage
- abdominal profile – bloating of the abdomen is suggestive of gastrointestinal issues
- respiratory movement – movement of the chest wall associated with breathing is normally undetectable at rest. Obvious or laboured breathing can suggest respiratory disease
- eyes/ears/nares/mouth – the eyes should be clear and symmetrically open. Carriage of the ears should be symmetrical. Any discharges or excessive salivation should be noted
- injuries – evidence of physical marks on the rump (via teeth or claw). Note, however, that some bite marks are a consequence of normal mating interactions
- faeces/urine – faeces should be well formed and range in colour from dark brown/black to olive green. Bare-nose wombat scats are cuboidal. Normal faecal output is between 60 and 100 pellets per day
- presence and development of pouch young can be conducted by observation of the bulge in the pouch.

**Fig. 8.8.** Wild southern hairy-nosed wombat in very poor body condition. Note the protruding pelvis, spine, rib cage and shoulders associated with loss of muscle cover in this animal. Alopecia, fur discolouration and increased parasite loads are also common characteristics associated with declining body condition in southern hairy-nosed wombats. Photo by David Taggart.

Daily observations should be documented, and consideration given to reporting signs of ill health to a veterinarian. Investigation into animal deaths should include a veterinary necropsy.

## 8.2 Bodyweight and body condition score

### 8.2.1 Bodyweight

Record and compare with previous weights. Trends in bodyweight give a good general indication of the animal's state of health, provided age, sex and geographical origin are taken into account. Animals in captivity should be weighed monthly to gain an indication of trends. In the wild, wombat bodyweight can vary between late spring and late autumn by up to 30% (D. Taggart pers. obs.)

### 8.2.2 Body condition score

Various body condition indices have been used to assess the condition of wombats. Two have been developed specifically for wombats. A subjective condition index provides a score of 1–5 (D. Taggart pers. obs.; Horsup 1998; Kleemann *et al.* 2023):

(1) ribs, backbone and pelvis visible (very poor condition)
(2) ribs covered but easily felt, backbone still visible and the rump is sunken (poor condition)
(3) pelvis, backbone and ribs covered (average condition)
(4) pelvis, backbone and ribs well covered (very good condition)
(5) no bony processes (pelvis, backbone, ribs) visible / detectable during handling (fat – excellent condition).

However, qualitative indices are open to human error associated with wombat expertise and therefore, unless it is the same person assessing all wombats, generally there is a poor correlation with body fat (Woolnough *et al.* 1997). More accurate methods of determining body condition are quantitative methods that include the use of bioimpedance analysis, which utilises an electric current to predict the amount of body fat and total body water (Woolnough *et al.* 1997) or alternatively, and more simply, an index of body condition derived by dividing bodyweight (grams) by head width (millimetres) (Kleemann *et al.* 2023; Kleemann and Taggart 2024a, 2024b).

## 8.3 Physical examination

The physical examination may include the following:

- temperature - normally 32–36.7°C. Can be taken from the rectum via the cloaca (Gaughwin 1982; Triggs 1996; Booth 1999)
- pulse rate - normally 40–45 beats/min at rest and 55–60 beats/min when active. The systolic pulse rate of the femoral artery can be readily measured (Gaughwin 1982)
- respiratory rate - normally 12–16 breaths/min in deep sleep and 26–32 per minute while dozing (Wünschmann 1966; Triggs 1996; Booth 1999)
- fur - check for alopecia, ectoparasites (e.g. mange), fungal infections or trauma
- eyes
  - should be clear, bright and alert
  - normal bilateral pupillary light response
  - normal corneal reflex
  - should not have any discharges
- teeth and gums - evidence of malocclusion or molar spurs and associated gum ulcers
- presence of lumps over body and auscultation of lungs
- cloaca
  - should be clean
  - check for faeces around the cloaca
- pouch
  - condition of the pouch
  - check whether lactation is occurring by stripping fingers along teats from the base to express any milk
  - if pouch young are present, record the sex, stage of development and bodyweight if detached from the teat and measure head length and width to determine age from growth curves if available
- males
  - check testes - size (length, width, depth) and consistency (firm - not squishy). Note: the size of the testes and epididymides does not change significantly across the year irrespective of whether it is the breeding season or not (Taggart *et al.* 2005)
  - scrotal diameter
  - extrude penis and assess
  - accessory glands, or Cowper's glands, bulge (length and width) is a good indicator of male breeding condition and linked to androgen secretion. These glands are located around the cloaca (Hamilton *et al.* 2000; Taggart *et al.* 2005; Hogan *et al.* 2010d).

## 8.4 Chemical restraint

Pre-anaesthetic fasting is not required for adult wombats as they are not prone to regurgitation (Vogelnest 1999). For detailed discussion of the chemical restraint of wombats, veterinarians are referred to the Australian wildlife medicine texts by Vogelnest and Woods (2008) and Vogelnest and Portas (2025).

Wombats can be sedated if moderate restraint is required and sedation is often used to mitigate the animal's response to a stressful situation. Sedated animals, although less responsive to their environment, will rouse with stimulation. Drugs of the benzodiazepine group (e.g. diazepam, midazolam) have been most commonly used for this in wombats. These drugs are best administered by intramuscular injection and can be given while the animal is restrained in a bag or sack. Drug effects are 'dose dependent' (i.e. higher doses deliver deeper sedation). It is important to monitor sedated animals carefully to ensure they remain safe.

More complete restraint requires general anaesthesia. Anaesthetic drugs have complex effects on the body's physiological systems so anaesthetised animals must be continuously monitored by a veterinarian to ensure patient safety. General anaesthesia in wombats can be induced with injectable intramuscular drugs or by forcing the animal to inhale anaesthetic gas (e.g. isoflurane) in oxygen via a facemask. Mask induction requires the animal to be physically restrained initially, so is only suitable for smaller animals/juveniles. Anaesthesia can be maintained with anaesthetic gas after anaesthesia has been induced.

A combination of sedative and anaesthetic drugs is used for injectable anaesthetic induction. Best results in wombats are with the use of a zolazepam/tiletamine combination (Zoletil®) although medetomidine/ketamine combinations have also been used (Vogelnest 2025).

## 8.5 Common health problems

Numerous infectious agents and diseases have been documented in debilitated, rescued wombats. Table 8.3 lists infectious diseases and Table 8.4 lists non-infectious diseases of wombats.

**Table 8.3.** Common infectious diseases of wombats.

| Disease | Cause | Clinical signs | Diagnosis | Treatment | Prevention | Comment | Reference |
|---|---|---|---|---|---|---|---|
| **Ectoparasites** | | | | | | | |
| Sarcoptic mange | A mite: *Sarcoptes scabiei var. wombatti*. Mange is known to kill many individuals in wild populations. Other mites are known to occur on the skin of wombats, including the genera *Acaroptes* and *Cytostethum*; however, these are apparently harmless. The ear mite (*Raillietia australis*), found in the bare-nosed wombat, does not cause obvious problems | Skin disease. Mild, localised itching, redness and crusting, progressing to fur loss and the presence of severe generalised skin thickening, crusting and fissures | Skin scraping and microscopic demonstration of the mite | Miticide medication (e.g. avermectin group: ivermectin, moxidectin, selamectin) weekly for three treatments Fluralaner | Prevent contact between wild and captive animals | Sarcoptic mange is mostly a disease of wild wombats. Animals acquire the mite indirectly by using burrows previously used by infected animals. Wombats under stress due to for example overcrowding appear at more risk from severe disease. Best treatment results have been with the avermectin group of drugs. A recent study suggests a single topical application of a newer antiparasitic drug, fluralaner, is also safe and effective. Generally, mange treatment is often ineffective in severely affected wombats. People can acquire infestation by contact with infected wombats | 1–12 |
| | Ticks include those of the genera *Ambylomma*, *Aponomma*, and *Ixodes* Several genera of fleas have been identified from each species of wombat including *Lycopsylla* and *Echidnophaga* | None to mild skin irritation. Severe tick infestations can result in significant blood loss/anaemia | Visualise ticks/fleas | Physical removal of ticks. Antiparasite medication (e.g. topical fipronil, topical selamectin) | Prevent contact between wild and captive animals, exclude wombats from vegetation harbouring tick larvae | Ticks and fleas are mostly parasites of wild wombats | 1, 4, 5, 8, 13–21 |
| **Endoparasites** | | | | | | | |
| | Nematodes (roundworms) | None to mild gastrointestinal signs. Lungworm can cause mild respiratory signs | Faecal flotation and microscopic detection of eggs | Antiparasite medication (e.g. moxidection) | Remove faeces from enclosures regularly | Numerous species of roundworm infest the intestines of wombats. One species lives in the lungs. Healthy animals will coexist with low numbers of internal parasites. Treatment is generally unnecessary unless the parasite is causing clinical signs | 5, 17, 22–30 |

**Table 8.3.** continued

| Disease | Cause | Clinical signs | Diagnosis | Treatment | Prevention | Comment | Reference |
|---|---|---|---|---|---|---|---|
| | Cestodes (tapeworms) | Disease in wombats due to tapeworm is rare | Observation of tapeworm segments in faeces. Detection of tissue cysts by ultrasound. Demonstration of liver dysfunction by laboratory testing | Anticestode medication (e.g. praziquantel) | Exclude dogs and foxes from wombat enclosures | Numerous tapeworm species have been described in wombats. Most reside in the intestines and do not cause disease. Some have an extraintestinal stage causing cysts in other tissues. Disease associated with these is also rare. Other tapeworms can cause mild liver disease. Generally, treatment is unnecessary unless the parasite is causing adverse clinical signs | 5, 18, 31 |
| | Trematodes (flukes) | Signs of liver disease (e.g. jaundice, weight loss and fluid retention in the abdomen) | Faecal sedimentation and microscopic detection of fluke eggs | Antifluke medication (e.g. albendazole) | Exclude wombats from swampy habitat occupied by the snail intermediate host | Liver fluke has been described in wild wombats that coexist with the snail intermediate host in marshy areas. It has not been seen in captive wombats | 30, 32 |
| **Protozoal** | | | | | | | |
| | Coccidia<br>*Eimeria* spp. | Diarrhoea, bloating, and ill-thrift | Faecal flotation and microscopic detection of oocysts | Anticoccidia medication (e.g. toltrazuril) | Monitor young wombats during initial access to grazing. Avoid overstocking pasture | Intestinal coccidia are a normal inhabitant of the wombat intestine. They can cause disease in young, hand-reared wombats during the pouch emergence phase when they encounter coccidia oocysts on pasture for the first time | 33–35 |
| | Toxoplasmosis<br>*Toxoplasma gondii* | May be chronic with ill-thrift or acute with respiratory and neurological signs (e.g. seizures, ataxia and blindness) | Serological tests for *Toxoplasma* antibodies in blood | Antitoxoplasma medication (e.g. clindamycin) | Exclude cats from wombats and wombat feed storage areas | Cats are the definitive host for *T. gondii*. Wombats become infected when they consume toxoplasma oocysts passed in cat faeces. Toxoplasmosis is typically a severe and fatal disease in marsupials with poor response to treatment | 3, 5, 35–39 |

| Disease | Cause | Clinical signs | Diagnosis | Treatment | Prevention | Comment | Reference |
|---|---|---|---|---|---|---|---|
| **Bacterial** | | | | | | | |
| Pneumonia | Enterobacteriaceae and others | Variable. There may be obvious respiratory signs (e.g. laboured breathing, nasal discharge). Often the only signs are non-specific depression and loss of appetite | Consideration of history (e.g. orphan being hand-reared), clinical examination including chest radiography etc. | Antibiotic medication, supportive care | Avoid predisposing factors during hand-rearing | Pneumonia is most common in wombats undergoing hand-rearing that aspirate milk formula due to rapid delivery (overlarge hole in teat) or a weak suckle reflex (in hypothermic or compromised joeys). Old, debilitated animals are also at risk of pneumonia | 17, 40–43 |
| Wound infections, especially of toe/foot injuries and fight wounds | Various opportunistic environmental and commensal bacteria (e.g. *Staphylococcus* spp. and *Streptococcus* spp.) | Swelling, heat and purulent discharge. Pain (e.g. lameness associated with infected digits) | Inspection and probing of the injury under anaesthesia. Radiography of infected digits | Local wound care (removal of foreign material and devitalised tissue). Antibiotic medication | Sympathetic husbandry and enclosure design to reduce escape behaviour associated with stress. Separate antagonistic individuals | Digital injuries often result from escape behaviour | 17 |
| Leptospirosis | Several serovars of *Leptospira* spp. | Generally none | Blood testing for antibody evidence of exposure to the organism (serology), urine testing looking for the organism | There are no reports of leptospirosis treatment in a wombat. Antibiotic therapy could be considered should a clinical case present | Exclude animals likely to carry *Leptospira* spp. (e.g. rodents) from wombats | *Leptospira* spp. infections of people and animals are often acquired from rodent carriers. Several wild populations of bare-nosed wombats have shown serological evidence of widespread exposure to *Leptospira* without clinical disease. This suggests that wombats might carry *Leptospira* spp. without getting sick and be a potential source of infection for other animals and people. Leptospirosis has not been reported in captive wombats | 40, 41 |

**Table 8.3.** continued

| Disease | Cause | Clinical signs | Diagnosis | Treatment | Prevention | Comment | Reference |
|---|---|---|---|---|---|---|---|
| Salmonellosis | *Salmonella* spp. | Most animals infected with *Salmonella* have no clinical signs. Infection can become chronic in 'carrier' animals. When present, clinical signs include loose faeces (ranging from soft and semi-formed through to haemorrhagic diarrhoea). Severe cases have lethargy, abdominal pain, anorexia and dehydration. Mild cases usually resolve spontaneously | Culture of the organism from faeces. Shedding can be intermittent in carrier animals, so false-negative results are common | Supportive care (e.g. fluid support, pain relief). Antibiotics can be used but they can increase the risk of animals becoming chronic carriers so the decision to use them needs to be carefully considered | Isolate clinical cases from other animals. Attempt to identify and separate chronic *Salmonella* carriers | Infection is faecal to oral. Salmonellosis is a zoonotic disease. The risk to people is increased by the potential for asymptomatic carrier animals to shed the organism. The risk of zoonotic disease is reduced by rigorous attention to personal hygiene | 35 |
| Tyzzer's disease | *Clostridium piliforme* | Usually an acute, course of < 2 days and few or no clinical signs. Animals are often found dead. Signs, when they occur, include diarrhoea, weakness, depression, altered behaviour fever, dyspnoea, jaundice, anorexia and occasionally convulsions | Because of the rapid course of disease, it is rarely diagnosed before death. A necropsy diagnosis is based on typical lesions containing intracellular bacteria | Antibiotic and supportive therapy | Exclude rodents from captive wombats and wombat food storage areas. Optimise husbandry to minimise stress | Rodents are the natural host for *C. piliforme* and the likely source of infection for other animals. Young and/or stressed animals are more susceptible. Although Tyzzer's disease affects numerous other species, it appears to be uncommon in wombats. There is a single report of Tyzzer's disease in an orphaned, subadult bare-nosed wombat | 43 |

| Disease | Cause | Clinical signs | Diagnosis | Treatment | Prevention | Comment | Reference |
|---|---|---|---|---|---|---|---|
| **Fungal** | | | | | | | |
| Thrush (candidiasis) | *Candida* spp. | Diarrhoea. Occasionally white deposits and inflammation in the mouth | Microscopic detection of organism from faeces or oral lesions | Antifungal medication (e.g. nystatin) | Minimise the use of antibiotics. Minimise stress during hand-rearing | Thrush usually manifests as diarrhoea in joeys being hand-reared that are otherwise bright and hungry. It often occurs after antibiotic treatment | 5, 42, 44 |
| Adiaspiromycosis | *Chrysosporium* spp. | None | Microscopic examination of lung tissue | Unnecessary | Unnecessary | Adiaspiromycosis refers to the presence of a thick-walled fungal spore in tissue. Spores of *Chrysosporium* spp. have been identified coincidentally in the lungs of wild and captive southern hairy-nosed wombats | 5, 42, 44 |

References: (1) Bryant and Reiss (2008); (2) Campbell-Ward (2025); (3) Booth (1999); (4) Doube (1981); (5) Skerratt (1998); (6) Wilkinson *et al.* (2021); (7) Roberts (1964); (8) Roberts (1970); (9) Ruykys *et al.* (2009); (10) Ruykys *et al.* (2013); (11) Death *et al.* (2011); (12) D. Taggart pers. obs.; (13) Green and Munday (1971); (14) McIlroy (1973); (15) Presidente (1982); (16) Smales (1987); (17) Skerratt *et al.* (1998); (18) Gerhardt *et al.* (2000); (19) Skerratt (2001); (20) Gerhardt *et al.* (2000); (21) Skerratt *et al.* (1999); (22) Mawson (1955); (23) Munday and Gregory (1974); 24) Beveridge (1978); (25) Beveridge and Mawson (1978); (26) Spratt (1979); 27) Spratt *et al.* (1991); (28) Smales (1994); (29) Skerratt (1995); (30) Smales (1998); (31) Beveridge (1980); (32) Spratt and Presidente (1981); (33) Barker *et al.* (1979); (34) Hum *et al.* (1991); (35) Rose (1999); (36) Booth (1994); (37) Skerratt *et al.* (1997); (38) Bettiol *et al.* (2000); (39) Miller *et al.* (2000); (40) Durfee and Presidente (1979); (41) Munday and Corbould (1973); (42) Gaughwin (1982); (43) Hum and Best (1988); (44) Munday (1978).

**Table 8.4.** Common non-infectious diseases of wombats.

| Disease/agent | Cause | Clinical signs | Diagnosis | Treatment | Prevention | Comments | Reference |
|---|---|---|---|---|---|---|---|
| Shear mouth (dental malocclusion) | Unknown. Speculation that skull and jaw development is distorted in orphaned joeys that suck on inanimate objects or that are bottle-fed from one side of the mouth | The chewing surfaces of the molar teeth become sloped causing the lower jaw to become offset | Visual inspection of the oral cavity | Regular levelling of the teeth with a dental burr under anaesthesia | Unknown | As the teeth continue to grow, without treatment, the condition worsens rendering eating difficult | 1 |
| Gastrointestinal obstruction | Varies from mild constipation in joeys during hand-rearing to more severe 'gastrointestinal accident' in other animals (e.g. twisted bowel, intussusception) | Speculative. Maybe due to altered gut motility as gut responds to artificial milk formula or captive diet. Maybe secondary to infectious problems (e.g. coccidiosis) | Observation of changed defecation pattern, abdominal distension. Palpation of abdomen. Radiography. Surgical exploration | Enemas and laxatives for constipation. Surgical correction for life-threatening gastrointestinal accident | Feed appropriately | None | 1, 2 |

References: (1) Fagan and Ullrey (2008), (2) Bryant and Reiss (2008).

## 8.6 Preventative medicine

Provision of optimal husbandry and nutrition within sympathetically designed housing will reduce the risk of health issues in captive wombats. Early detection and thorough veterinary investigation of disease problems, including necropsy investigation when deaths occur, will facilitate a good understanding of health trends, enabling the application of preventative strategies targeted to the specific circumstances of the facility. Generally, routine vaccination and anthelminthic treatment are not required for wombats. Although routine, elective veterinary examinations are generally not scheduled for healthy residents, the health status of all new arrivals should be established via clinical examination and quarantine before introduction to an established group.

## 8.7 Zoonoses

Sarcoptic mange is transmissible to people that have direct or indirect contact with affected wombats, causing an often extensive, very itchy, papular rash (D. Taggart pers. obs.). The risk is higher when handling wombats that have severe disease or their bedding material. Manage the risk by avoiding contact with severely affected animals where possible and using PPE and personal insect repellents when contact cannot be avoided. The risk of human bite injury is managed by deployment of appropriate restraint when working with wombats.

# 9. BEHAVIOUR

## 9.1 Activity

Wild wombats are typically crepuscular to nocturnal and display limited activity during daylight hours. They may spend over 16 h each day resting in their burrows in order to conserve energy, which is a behaviour adapted to both their low-energy diet and avoiding high temperatures (Triggs 1996; Johnson 1998; Finlayson *et al.* 2003, 2005, 2010; Taggart and Temple-Smith 2023).

Wombats need burrows due to their inability to regulate their body temperature when temperatures rise above 25°C, whereas they can readily withstand long exposures to air temperatures approaching 0°C in the case of bare-nosed wombats and 5°C for southern hairy-nosed wombats (Brown 1964; Triggs 1996; D. Taggart and S. Kleemann unpubl. data). Burrows also enable wombats to conserve water by avoiding temperature extremes (which may vary from -5°C to 50°C) and low humidity (typically 2–5%) outside the warren, compared with burrow temperatures of 10–27°C (all year) and humidity of 60–70% within tunnels (Wells 1971, 1978a; Shimmin *et al.* 2001, 2002).

Wombats may spend time digging new or extending existing burrows, which can have tunnel lengths that range from only 1.5 m to more than 60 m (Steele and Temple-Smith 1998; Shimmin *et al.* 2001; Swinbourne *et al.* 2015). Burrows typically have 1–2 entrances. Southern hairy-nosed wombats are also known to push up earth plugs within burrows to block the tunnel they occupy, which may be a defensive behaviour (Steele and Temple-Smith 1998) or perhaps act to further increase humidity and reduce moisture loss in a semi-arid environment (Shimmin *et al.* 2001, 2002). Burrows of southern hairy-nosed wombats have a height of 30 cm and a width of 40 cm with terminal sleeping chambers, 80 cm (L) × 50 cm (H) × 60 cm (W), with the roof of the burrow 135 cm below the surface (Steele and Temple-Smith 1998).

Wombats typically emerge from their burrow to feed when the temperature and humidity outside the burrow approach those inside (Wells 1989; Finlayson *et al.* 2005, 2010). In summer, wombats are more active from midnight to early morning (when the temperature is around dewpoint) before the temperature gets too hot, while in winter they are more active in late afternoon to early evening before temperatures have dropped (Wells 1978a; Finlayson *et al.* 2005, 2010). In mild conditions, wombats typically emerge from the burrow after sunset to graze for several hours. The only time in which wombats appear to be diurnal is when they sometimes bask in the sun during the cooler months, especially after cold nights (Wells 1978b; Woolnough 1998; Finlayson *et al.* 2010).

Observations of northern hairy-nosed wombats found no relationship between activity and temperature below ~26°C. However, above this temperature there was a direct relationship, with activity decreasing quickly until no activity above ground was observed at night when the temperature was ~32–34°C (Woolnough 1998). Other observations of the northern hairy-nosed wombat have found a highly significant relationship between temperature and activity, with activity increasing with temperature until 20°C, after which, further increases in temperature result in a decrease in activity (Johnson 1991). Northern hairy-nosed wombats spend only 2–6 h above-ground, with a significant relationship between the time of year (and hence temperature and food availability) and activity, with less activity in the warmer months and more during the cooler months, especially in late winter and early spring (Johnson 1991).

Wild southern hairy-nosed wombats are known to be active for ~6.5 h per 24-h period, but are also known to bask and feed during the day in autumn, winter and spring (usually between 1400 h and dark) (Finlayson *et al.* 2010; D. Taggart and S. Kleemann unpubl. data).

Monitoring of captive bare-nosed wombats found that they spent 53% of their time sleeping, 11% lying resting, 1% exploring, 9% feeding, 9% digging, 4% performing abnormal repetitive behaviour, 6% sitting resting, and 7% in other activities (Hogan 2004). Captive southern hairy-nosed wombats showed a similar time budget to that of the bare-nosed wombats, with animals observed in enclosures that were between 163 and 250 $m^2$. These animals spent 70% of their time sleeping, 9% lying resting, 5% exploring, 5% feeding, 4% performing abnormal repetitive behaviour, 2% digging, 3% sitting resting, and 2% in other activities (Hogan *et al.* 2009; Fig. 8.9). Using video surveillance observations and radio-telemetry, Hogan *et al.* (2009) also showed that southern hairy-nosed wombats exhibit a bimodal nocturnal activity budget, with peaks at the beginning and end of the night between 1800 and 2000 h and 0200 and 0500 h (Hogan *et al.* 2009; Fig. 8.10).

## 9.2 Social behaviour

Wombats are territorial with respect to feeding areas and may have disputes over the use of a warren or burrow, which is often the focus of wombat activity. Among wild bare-nosed wombats, individuals have up to 11 burrows over their home range, although most activity is confined to 3–4 burrows that can be used by more than one animal (McIlroy 1973; 1977; Wells 1989; Taylor 1993).

The focus of social organisation of the southern hairy-nosed wombat is the warren, which can have from 1 to 80 burrows of which many may be interconnected (if there have been collapses over time). There can be 10 or more wombats using a single warren, though not necessarily at the same time (Wells 1989; Steele and Temple-Smith 1998; Taggart *et al.* 2020). Typically, a wombat will have between 1 and 10 warrens within a home range of 2–4 ha. Animals using very large warrens comprising many burrows tend to have fewer warrens within their home range (Finlayson *et al.* 2005).

Colonies consisting of many warrens in a cluster can be spread over kilometres (Löffler and Margules 1980). Females show greater preference for burrows than males, but there appears to be no evidence of burrow ownership among warren occupants (Wells 1973, 1978b; Gaughwin 1982). Although in the wild adult male southern hairy-nosed wombats are dominant to adult females, there appears to be no set rule to dominance in captivity, as in some cases the male is dominant and in other cases

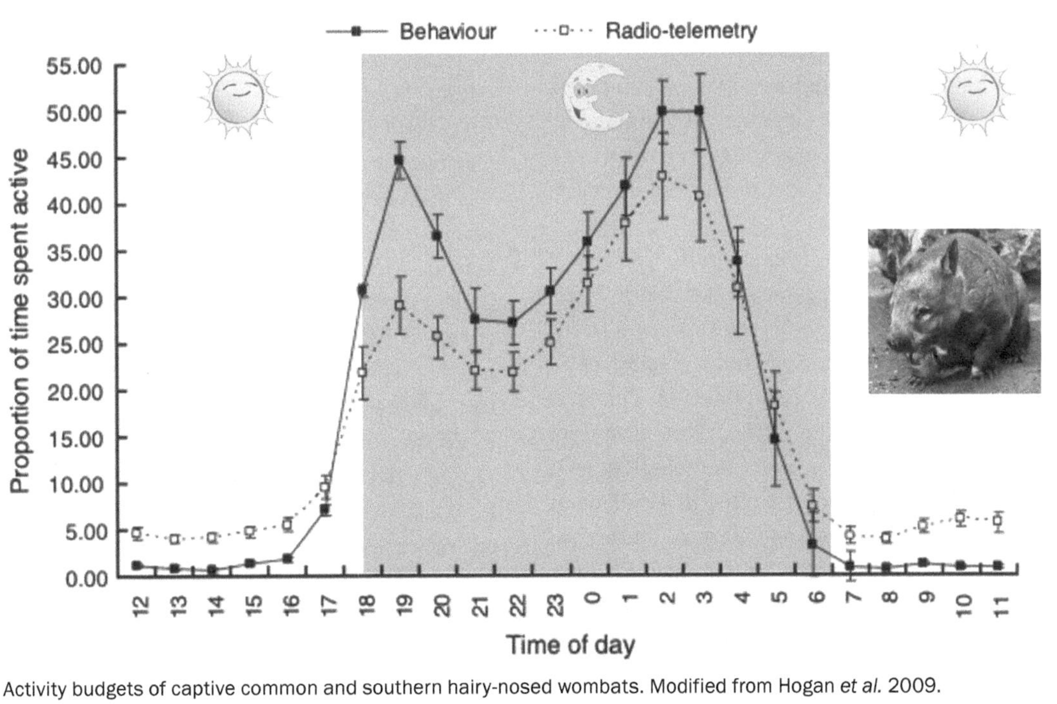


**Fig. 8.9.** Activity budgets of captive common and southern hairy-nosed wombats. Modified from Hogan *et al.* 2009.

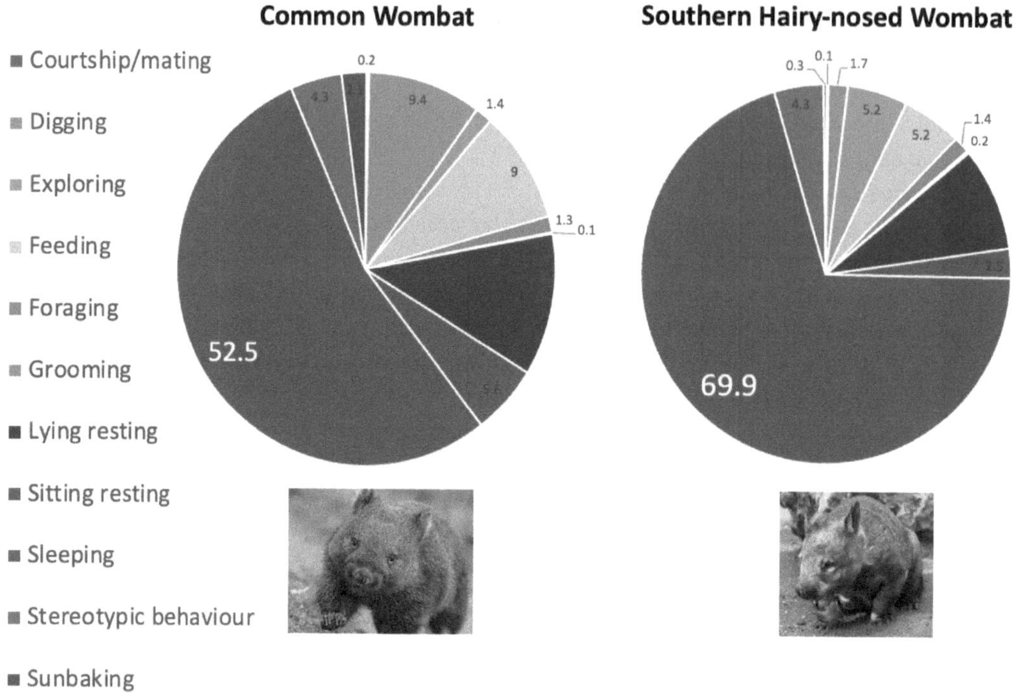


**Fig. 8.10.** Daily activity of captive wombats determined by video surveillance observations and radio-telemetry. Modified from Hogan *et al.* 2009.

the female may be dominant (Gaughwin 1982). If held in pairs or larger groups when in captivity, careful observation needs to be made to ensure that subordinate animals do not suffer poor condition and possible death due to continual harassment (Gaughwin 1982). If housed in groups, it is important to ensure that the animals are compatible. Never house two males together as this will result in aggressive, territorial behaviour and unnecessary injuries.

Fighting in both genera, although rare, consists of bites to the face, ears, rump and flanks (Wells 1989). Aggression is generally begun by a series of vocalisations that include a flat 'chicker chicker' and a rasping 'chur' that can be a prelude to chasing and fighting. This may involve biting the rump of the other animal (the roles can be reversed) with throaty snorts and nasal squeals being produced (Triggs 1996). Gaughwin (1982) notes that captive southern hairy-nosed wombats are territorial with respect to feeding areas and may have disputes over the use of a burrow. Hogan (2010) has reported captive wombats show social head rubbing and reported evidence of animals resting together.

## 9.3 Reproductive behaviour

The reproductive behaviour of wombats comprises six consecutive phases: investigation, attraction, chase, restraint, coitus and recovery (Böer 1998; Hogan *et al.* 2010c). Matings appears to only occur within closed spaces and always accompanied by courtship behaviour. The mean time that female wombats remain receptive is ~13 h; however, the behavioural signs of oestrus are not readily observed, so the interest of the male wombat appears to be the best indicator of oestrus (Hogan *et al.* 2010c). In fact, wombat oestrus detection by humans is currently a primary limitation in the breeding management of wombats. Although successful breeding of captive southern hairy-nosed wombats has been a little more successful in recent years, one study found that reproductive behaviour was only observed in 5 of 12 animals and that despite 49 matings being observed no offspring were produced. Possible causes include anoestrus females, low libido of the males or pairing incompatibilities (Hogan *et al.* 2010c).

Courtship behaviour in wombats is quite extensive, typically occurring over 2–4 days, and begins with the male following the female after he has first inspected her cloaca or urine, often showing flehmen behaviour (Gaughwin 1979), and then a 1–5 day period of guarding by the male after copulation (Taylor 1993; Hogan *et al.* 2010c). Flehmen also occurs in bare-nosed wombats when they encounter the urine of females (Triggs 1996). It usually involves the male standing with his head stretching towards the female's cloaca or urine with his mouth open while he retracts his upper lip, thus baring the gum and wrinkling the nose. As for many species, it appears to be a mechanism to expose the vomeronasal organ to pheromones. When sniffing the male moves his nostrils to and fro in an erratic fashion (Gaughwin 1979; Coulson and Croft 1981) and sometimes associated with rapid licking and mouthing movements during and after showing flehmen. Although this process may appear to be part of how the male detects if the female is in oestrus, a study by Swinbourne *et al.* (2017a) appeared not to support this suggestion, finding no difference in this or any other reproductive behaviour when males were exposed to urine samples recovered from the same female during oestrus and mid-gestation (non-oestrus).

It has been suggested that one of the reasons for the poor breeding success of wombats in captivity is that most of them are hand-reared and thus have been denied the opportunity to acquire learned behaviours integral to courtship, normally gained through exposure to the dam while at heel and during the subadult development phase (Bryant 2000). In the wild, mating probably occurs in the burrow, as the male may have to prevent the female from escaping and also needs to be dominant to her. So, in captivity if the male is not dominant, successful copulation is unlikely to occur (Gaughwin 1982). Janssen and Johnston (unpubl. data) have similarly noted that smaller or subordinate southern hairy-nosed male wombats may have difficulty in manipulating the female into a copulatory hold/posture (Fig. 8.11a–b). Gaughwin (1982) suggested that most successful copulations in southern hairy-nosed wombats occurred in enclosures that simulated the natural burrow environment. A good indication of a successful mating is the presence of a rubbery pale yellow coloured plug of coagulated semen in the enclosure 3–5 days after southern hairy-nosed wombats have mated (Fig. 8.11c; Crowcroft and Soderlund 1977; Brooks *et al.* 1978; Taggart *et al.* 1998a).

Similar to the hairy-nosed wombats, courtship and mating behaviour in wild bare-nosed wombats has been observed to involve the male chasing the female, typically running in circles or figures of eight in an area of 0.5 ha (Marks 1998). The female only stopped if the male delivered a bite to her rump. In this case, mounting occurred outside, with both animals on their sides, which contrasts with the generally held belief that mating normally occurs in the burrow (Triggs 1996; Marks 1998). Further observations by Böer (1998) revealed five major phases to mating behaviour.

(1) The female follows the male, often matching his speed and making nose-to-nose contact with him when they passed one another and presenting her urogenital region to the male by stretching her hindlimbs.
(2) One of the partners utters several grunts that are answered by the other animal and repeated over 2–3 min.
(3) Chasing phase lasting for some 30 min, in which the male chases the female, which bounds away from him while he tries to bite her rump.

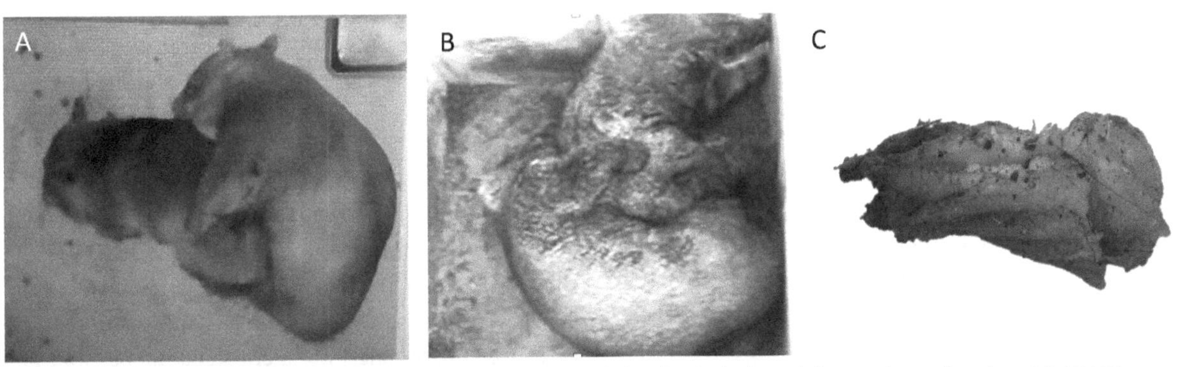


**Fig. 8.11.** (a) Captive southern hairy-nosed wombats (video capture) showing typical copulatory posture – female not in lateral recumbency (Photo by Tina Janssen). (b) Copulation (video capture) with female in lateral recumbency after the male turns the female over into mating position for intromission. Reproduced from Hogan *et al.* (2010c) with permission from CSIRO Publishing. Surveillance footage by Lindsey Hogan. (c) Southern hairy-nosed wombat copulatory plug found in an enclosure post-mating (Photo by Tina Janssen).

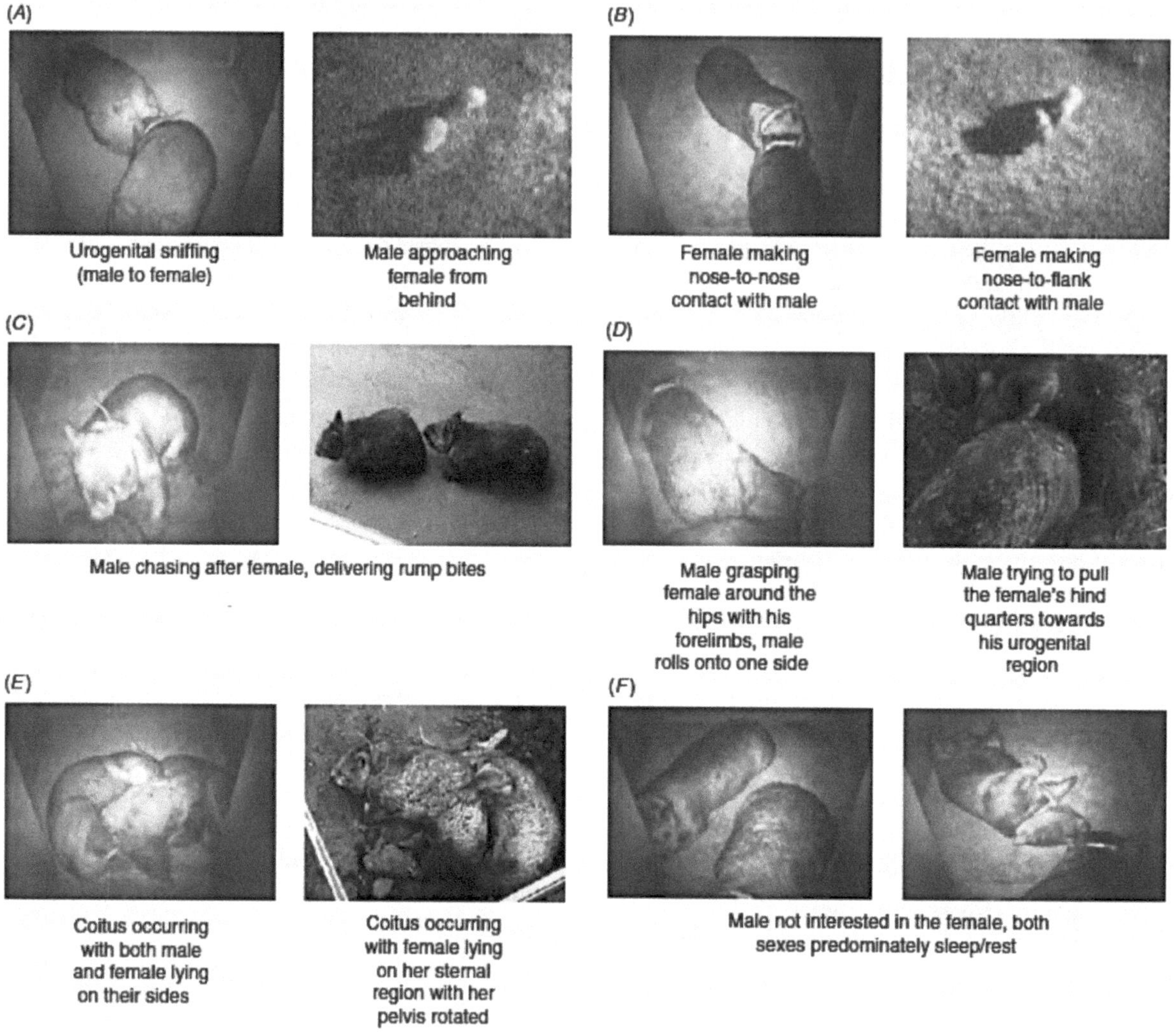


**Fig. 8.12.** Behaviour observed via video surveillance during courtship and mating ritual of captive southern hairy-nosed wombats. Taken from Hogan *et al.* (2010c).

(4) The male approaches from behind and tries to mate the female. After securing a grip he falls on his side and mates for ~30 min.
(5) After mating both individuals roll apart and fall asleep.

Another observation of mating behaviour described how the male followed closely behind the female and put his forepaws on her back several times, but she kept moving. He then grabbed her back leg in his mouth at which time she lay on her belly, resulting in the male laying on his side behind the female and mating her (Taylor 1993).

Hogan *et al.* (2010c) used video surveillance observations to describe the reproductive behaviour of the southern hairy-nosed wombat. Prior to mating the female uses visual and tactile cues to establish and promote sexual interaction with the male. Once the male is aroused a prolonged chase of the female commences, which is terminated when the male catches the female and delivers a powerful bite to her upper rump or grasps one of her hindlegs in his jaws (Hogan *et al.* 2010c). The male then mounts the female's back and forcefully flips her onto her side. Once in this position, the female will often kick at the male and try to right herself. If the male is unable to maintain the female in the mating position she escapes and the chase begins again. Intromission appears to only occur when male is able to successfully turn over the female into lateral recumbency and hold down the female with his teeth and forearms (Fig. 8.11a–b; Fig. 8.12a–f). Copulation has been reported to occur in a range of 18–170 min with repeated

**Table 8.5.** Abnormal repetitive behaviours expressed by captive bare-nosed wombats.

Taken from Hogan and Tribe (2007).

| Form of behaviour | Description of behaviour |
|---|---|
| Boundary pacing | The wombat repeatedly walks a path that corresponds to the perimeter of the enclosure |
| Distinct pattern pacing | The wombat repeatedly walks a path that is a figure eight or circular pattern |
| Fence biting | Chewing and/or licking of the solid bars or wires of the perimeter fence |
| Fence jumping | The wombat is in an excited state and repeatedly runs and jumps against enclosure fencing |
| Gate pawing | The front feet are repeatedly applied to an enclosure gate |
| Paddling | Continuous paddling of the front feet through the air, while the animal is lying on its back |
| Rubbing | The wombat rubs its nose (up and down) against a gate or fencing |
| Screeching | Loud repeating of a high-pitched screech |
| Straight-line pacing | The wombat walks to and fro in a straight-line, usually along a fence |
| Straight-line pacing with fence jumping | Essentially the same as straight-line pacing, but with infrequent fence jumping |
| Wall pawing | The front feet are repeatedly applied to the wall(s) of the enclosure |

copulations over a 13–15 h period (Hogan *et al.* 2010c). Post-mating, the male will typically stay in close proximity to the female and may share the burrow, in what might be interpreted as mate guarding behaviour. Although the female is tolerant of the male initially, ultimately she forces him away. In a video of wombats at the wombat breeding centre at ACCE, Mt Larcom, it is possible to see how the male, despite his attempt, is unable to turn the female over into copulation position for successful intromission; it is possible that female uses this show of strength as a form of mate selection. From a reproductive management perspective, male to female pairing, in addition to genetics, should primarily focus on breeding females with males that are larger in size or more dominant than the female and capable of manipulating the female into a mating position.

## 9.4 Bathing

Although wombats are good swimmers, they usually do not bathe in water. However, they will dust bathe in sand or dusty soil (Triggs 1996).

## 9.5 Behavioural problems

Wombats are typically inactive animals (Finlayson *et al.* 2010), but they will naturally dig a great deal. Stressed animals may vocalise frequently, show escape behaviour by continually trying to climb the walls or bite their way out of the enclosure.

Observations of captive bare-nosed wombats have found them to be highly susceptible to abnormal repetitive behaviour. A total of 11 distinct forms of this behaviour have been identified (Table 8.5), with the average time observed ranging between 1.5 and 5.0 h per day or 20–55% of the day (Hogan and Tribe 2007). Another study found that some hairy-nosed wombats spent 61–129 min undertaking abnormal repetitive behaviour (Hogan *et al.* 2010a). The occurrence of such behaviours may be attributed to several factors, including predictable feeding routines (where food is readily available), lack of space and/or lack of environmental complexity of the enclosure. Previous studies suggest that enclosure sizes up to 400 $m^2$ are necessary to reduce the occurrence of pacing in wombats (Booth, 1994, 1999).

In an effort to reduce abnormal repetitive behaviour in captive hairy-nosed wombats, one institution included the addition of different feeds and olfactory enrichment. Food items included browse (one large eucalyptus branch; grass (one large circular grass sward of *Cenchrus* spp., *Chloris* spp., *Austrostipa* spp., including roots but free of soil, 120–150 cm height × 100 cm in diameter; and buried food, including 290–320 g sweet potato, 290–320 g carrot and 110–250 g corn. The position of these was changed daily (Hogan *et al.* 2010a). Olfactory enrichment included scenting of logs with aniseed, peppermint, vanilla, lemon, coconut and rose. However, despite these efforts they were found to have little effect in reducing the abnormal behaviours (Hogan *et al.* 2010a).

## 9.6 Signs of stress

Stress in wombats can be associated with very loud vocalisations (screechy) and teeth gnashing (D. Taggart and S. Kleemann pers. obs.). When being caught they

**Fig. 8.13.** Southern hairy-nosed wombat showing evidence of emaciation and severe alopecia associated with poor husbandry. Photo by Tina Janssen.

may show a very aggressive defence, biting, scratching or attacking. Chronic stress can result in alopecia (typically starting at the shoulder region) that is usually symmetrical and a result of immunosuppression (especially in cases of severe mange) (Fig. 8.13; Spielman 1994). Other signs of stress may also include a loss of appetite, a corresponding bodyweight loss, pacing and restlessness (abnormal repetitive behaviours) and evacuation of bowels or bladder (T. Janssen pers. obs.).

Faecal and urinary hormone assays have validated and established cortisol metabolites as a means of monitoring stress in the southern hairy-nosed wombat (Hogan *et al.* 2011b; Descovich *et al.* 2012c; Du *et al.* 2017). For example, Du *et al.* (2018) used measurement of urinary cortisol metabolites to explore seasonal changes in this hormone and the effect of shifts in social dynamics; interestingly, these authors describe an elevation in urinary cortisol in winter before spring, but no change when social dynamics were disrupted (changes in enclosure position), although this event did illicit a change in wombat behaviour. Although these techniques might serve as useful tools for assessment of stress in captive wombats for husbandry evaluation, they must also be interpreted in light of the animal's corresponding behaviour and with the knowledge that there is likely to be individual variation in a wombat's response to stress (baselines differ) and that there can be both sample (Descovich *et al.* 2012c) and assay variability (Fanson *et al.* 2017; Fanson 2025).

The significance of stress and its potential to result in reproductive failure have been examined in bare-nosed wombats. Captivity does not appear to suppress progesterone secretion and excretion during the oestrous cycle, whether the animal is held by itself or with another animal (MacCallum 2004). It was suggested that failure to mate might instead be due to a lack of behavioural stimulation (Bryant 2000).

## 9.7 Behavioural enrichment

Although they do not require as much behavioural enrichment as other animals, several things can be done to provide behavioural enrichment to wombats:

- providing browse (see diets)
- providing an adequate depth of soil or regularly maintained soil mounts to allow natural digging behaviours
- planting tussock grasses to allow wombats to chew and dig them up
- walking wombats with the use of lead and harness, but this only suggested if the wombat is quiet. Wombat behaviour can change rapidly and there is always the risk of the animal biting a visitor, particularly if it is a female wombat in oestrus
- providing new scents in the form of urine or faecal pellets from conspecifics
- introducing toys (e.g. bowling ball (T. Janssen pers. obs.))
- providing bedding such as towels; some wombats use them for sleeping, but others just destroy them so need to monitor when introducing to the enclosure.

## 9.8 Training and conditioning

Both bare-nosed and hairy-nosed wombats have been trained to be walk onto a weighing platform (Treby 2005) and to wear a specially designed harness that is attached to a lead so they can be taken for walks (e.g. Moonlit Sanctuary 2023). Southern hairy-nosed wombats have also been conditioned to allow the collection of urine (to facilitate monitoring of the reproductive cycle) by scratching the peri-cloacal region (see section 10.4.1 and Fig. 8.17 for more details).

## 9.9 Introductions and removals

Animals are usually easily removed from each other with few social problems when they return. If introducing animals as a pair, the male should be released first to the enclosure so that he can establish his territory before the female is introduced. If the male is introduced into an enclosure with an established female, the female may be dominant over the male and may not breed. Preferably, introductions should be done with compatible animals on neutral territory and both animals are placed in the den system at the same time to reduce aggression.

## 9.10 Intraspecific compatibility

Most bare-nosed wombats are held singly or as pairs, if compatible, which can be a male and female, two females or two juvenile males (< 18 months of age). Southern hairy-nosed wombats can be held as solitary animals, but preferably should be held as pairs or in small groups. At the AACE wombat breeding facility, southern hairy-nosed wombats are held as solitary animals until such times as breeding is needed, then an introduction is made with the female and male in a neutral enclosure (T. Janssens pers. obs.). Little is known of the northern hairy-nosed wombat's behaviour in captivity, but it is likely that they are best held as solitary animals or as pairs.

## 9.11 Interspecific compatibility

Because of their highly aggressive nature, wombats are not recommended to be housed with other species.

## 9.12 Suitability to captivity

Due to the historically poor breeding success of wombats in captivity, most of those in captivity have been introduced after hand-rearing, as a result of the mother being killed on a road. These animals readily settle into captivity. Of non-hand-reared wombats, juvenile bare-nosed wombats (weighing 10–18 kg) appear to adjust to captivity better than adult animals (≥23 kg) (Presidente 1982). Most eat fresh grass within 2–5 days after being captured and commercial food after 7–14 days (Presidente 1982).

When southern hairy-nosed wombats are caught from the wild some may not eat much for as long as 4 weeks. However, they can generally be encouraged to eat after a couple of weeks if initially provided with fresh grass and left undisturbed. None of the animals in a case reported by Gaughwin (1982) became ill or died. In contrast, Wells (1971) experienced a mortality rate of 40% (thought to be due to starvation) in southern hairy-nosed wombats he caught, but the deaths could have been due to the initial condition of the animals. In some cases, animals brought in have been force-fed a mashed mixture of rolled oats, powdered milk, high-protein cereal and glucose (Gaughwin 1982), though in the case of Wells (1971), doing this did not reduce the mortality rate. The AACE centre has been involved in bringing 30 southern hairy-nosed wombats into captivity from South Australia to the Central Coast of Queensland, ranging from 'pinkie' joeys to subadults with a 100% success rate. If trying to introduce an animal into captivity, it should be kept in quarantine until its suitability for captivity has been assessed. It should be also monitored for bodyweight loss, food intake and general health status.

Similar problems have been observed with the northern hairy-nosed wombat. The only animal brought into captivity in recent times was a juvenile male, 'Solstice', that was trapped and sent to Western Plains Zoo in New South Wales in order to begin a captive breeding program. He refused to eat and had to be force-fed, which required sedation. He lost one-third of his bodyweight and died after 7 months from a twisted bowel (Woodford 2001). The only other northern hairy-nosed wombat to be brought into captivity was a female called 'Joan' that was captured in 1966 when cattle farmers adjacent to the Epping Forest bulldozed a burrow in search of a pet wombat. She was an adult when caught (~2 years old) and survived until 1993 (Woodford 2001), although it is not known if she too found it difficult to adapt initially.

# 10. BREEDING

## 10.1 Mating system

Outside the breeding season, wombats are usually solitary, with most animals being incompatible (e.g. Wünschmann 1966). However, bare-nosed wombats may breed throughout the year depending on the environmental conditions (McIlroy 1973; Triggs 1996; Skerratt 2001). It has been proposed, based on testis to bodyweight ratios and other physiological evidence, that in all three species of wombat the males are polygynous and the females serially monogamous (Taggart *et al.* 1998b). This has been supported by observations at Murrindindi, Victoria, where females were found to have mated with different males in successive years, indicating serial monogamy (S. Banks unpubl. obs.). There appears to be an age-graded ranking system determining access of males to fertile females (R. Stenke unpubl. data). Larger males, weighing ~30 kg, sired most of the young at Murrindindi in 1999 and 2000 (S. Banks unpubl. obs.).

## 10.2 Ease of breeding

Bare-nosed and hairy-nosed wombats have not bred routinely in captivity, despite mating being commonly observed in some captive populations (e.g. Hogan *et al.*

2010c). A possible reason for the lack of successful breeding in captivity may be that most captive wombats have been hand-reared following the death of their mothers in vehicle collisions. These individuals have generally been housed singly. Hence, this source of animals for zoos has precluded the need to focus on improving reproductive management. The few records of breeding include bare-nosed wombats being born at Halle Zoo in Germany in 1914 (Mohr 1942), Whipsnade Zoo in 1931 (Zuckerman 1953), Healesville Sanctuary (Condor 1970), Hannover Zoo (Böer 1998) and Western Plains Zoo in Dubbo New South Wales (C. MacCallum pers. comm.). Southern hairy-nosed wombats bred at Perth Zoo in 1968 (Collins 1973), Melbourne Zoo in 1981 and 1998 (Anon 1982), Taronga Zoo in 1981 (Anon 1982) and Brookfield Zoo (Crowcroft and Soderlund 1977).

The Wombat Breeding Centre as part of the AACE Safe Haven project, in collaboration with the University of Queensland, has focused on research to improve the husbandry and breeding management of its population of over 20 southern hairy-nosed wombats. Initially attention was given to a better understanding the fundamental reproductive physiology and behaviour of the species, which has translated to breeding success; 8 animals have been bred at this facility.

Mating behaviour has not been observed by males or females within their first 3 years (Gaughwin 1982) and even after successful mating the female may not produce young (Crowcroft and Soderlund 1977; Gaughwin 1982). Due to the critically endangered status of the northern hairy-nosed wombat, only several specimens have ever been held in captivity, so there has not been an opportunity to attempt to breed them in captivity to date.

It has been proposed that the primary limitation of successful breeding of wombats, either natural or via techniques such as artificial insemination, has been, that unlike the female koala, which has an overt behavioural oestrus, the female wombats typically shows very little, if any, diagnostic behavioural changes when ready to mate, especially when females are held separately (Johnston (2019); at least for the southern hairy-nosed wombat, this has led to the search for other non-invasive physiological markers of oestrus and ovulation, which will be discussed next.

## 10.3 Oestrous cycle and gestation period

The wombat oestrous cycle physiology has been reviewed by Hogan *et al.* (2013). The southern hairy-nosed wombat has an oestrous cycle of 32–37 days, but can range from 27 to 47 days (Paris *et al.*, 2002; Finlayson *et al.* 2006; Hogan *et al.* 2010a; Swinbourne *et al.* 2018a); the precision of these estimates varies with the sample type examined (plasma, faeces or urine) and individual wombat.

The length of the bare-nosed wombat oestrous cycle ranges from 33 days (range 32–34 days) in Tasmania (Peters and Rose 1979), to ~47 days (range 35–60 days) in Victoria (West *et al.* 2004). In a captive population of bare-nosed wombats, MacCallum (2004) calculated the mean (+ SEM) oestrous cycle length of four females to be 50.7 ± 2.0 (range 47–54) days using plasma progesterone and 52.1 ± 1.9 (range 44 - 69) days using faecal progesterone metabolites.

Unlike its close relative, the koala, in which the luteal phase/ovulation is induced by coitus (physical act of mating and seminal biochemistry), ovulation in both species of wombat appears to be spontaneous. Overt behavioural oestrus is visibly apparent in either species, Hogan *et al.* (2009) have suggested that sexual receptivity in the southern hairy-nosed wombat lasts for ~13 h; other more recent anecdotal observations by Janssen (pers. obs.) suggest that this period could be longer, perhaps 1–2 days, but further confirmation is necessary. Hogan *et al.* (2009) even injected female wombats with a pharmacological dose of oestradiol benzoate in an attempt to induce oestrus, but failed to produce any evidence of sexual receptivity in the female or attraction from the male. Böer (1998) has suggested that oestrus in the bare-nosed wombat lasts 24–81 h.

The gestation period of both species is short: ~20–22 days in the southern hairy-nosed wombat (Hogan *et al.* 2013) and 30–33 days in the bare-nosed wombat; the difference in these estimates seems perplexing and requires further investigation. Janssen (pers. obs.) has indicated that all births at the AACE wombat breeding facility followed a consistent 22-day gestation period. Typically, parturition coincides with decreasing progesterone concentration.

## 10.4 Reproductive status

### 10.4.1 Females

The pouch of adult female wombats opens to the rear, which is particularly apparent when the female has very large, furred pouch young. In southern hairy-nosed wombats, when the pouch young is smaller, the pouch opening is ventral. Wombat pouches contain two teats, which grow in size as the young grows. In almost all cases only one teat is active at any one time (D. Taggart pers. obs.).

Wombats are generally placed in several categories depending on their reproductive status (Fig. 8.14). For females these include:

- non-parous (never bred) - pouch is shallow with no skin folds, clean and dry, teats very small
- parous (bred previously but not presently) - pouch is deep but dry and dirty, the teats are slightly elongated
- oestrus - clitoral swelling present (fleshy/pink) at the time of behavioural oestrus (Taggart and Temple-Smith unpubl. data; Finlayson *et al.* 2006).
- pregnant - pouch is bright red in colour, deep and very moist, skin folds may be observed on the lateral margins
- pouch young present - pouch is deep, very moist and young attached to the teat
- late lactating (young absent from the pouch but still suckling) - pouch area is large, active teat is enlarged and mammary gland skin folds are flaccid, hair sparse and stained, skin smooth and dark pink
- post-lactation - teats are still large, expressing only clear liquid and regressing in size.

The monitoring of the female wombat for oestrus and the introduction of the male for breeding can be challenging, so several different techniques for assessing reproductive status and possible gestation have been investigated. Probably the most significant advance in breeding management at the Wombat Breeding Centre at the AACE has been the incorporation of video surveillance infrastructure in the internal den system and outdoor pens (Fig. 8.15). The recording of this activity provides the baseline behaviour by which physiological (endocrine) parameters of the reproductive cycle can be mapped.

Various studies have mapped the reproductive cycle of the common and southern hairy-nosed wombats in captivity. Some of the early studies investigated changes in basal body temperature, vaginal/urogenital cytology and plasma. However, given the variable tractability of the wombat, and the possibility that handling for cytology and venepuncture could in itself induce

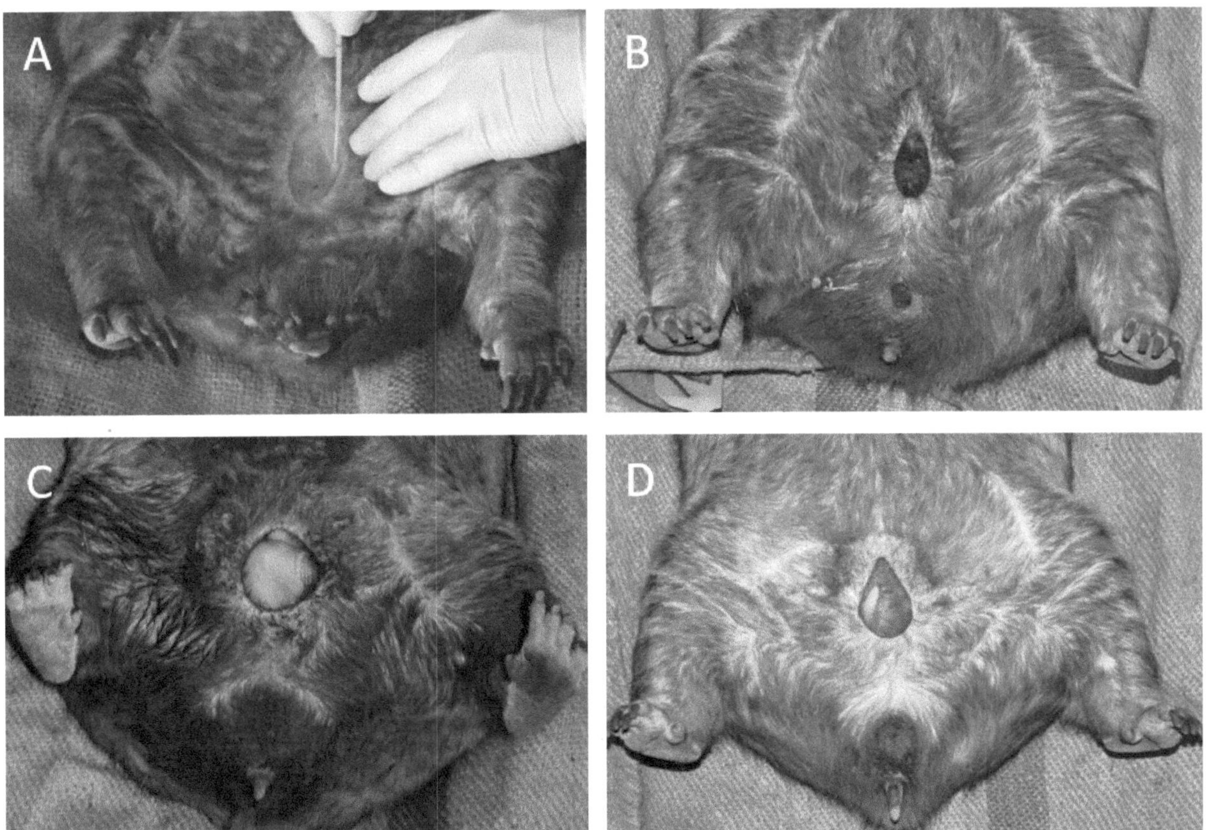


**Fig. 8.14.** Variations in southern hairy-nosed wombat pouch condition, including (a) the shallow, pale pouch of a subadult, (b) the pouch of an adult female in oestrus, (c) the pouch of a female with young present and (d) the pouch of a late-lactation female when the pouch young are no longer present, but the mammary gland is still enlarged and producing milk. Photos by David Taggart.

stress and suppress reproductive activity (Fanson 2025), the focus here is on non-invasive measurements of reproductive assessment. The collection of faecal samples is relatively straight forward in captive wombats, especially when it is possible to use faecal markers (dyes and colour glitter) to monitor group-housed wombats in the same enclosure (Hogan *et al.* 2011b; Fig. 8.16).

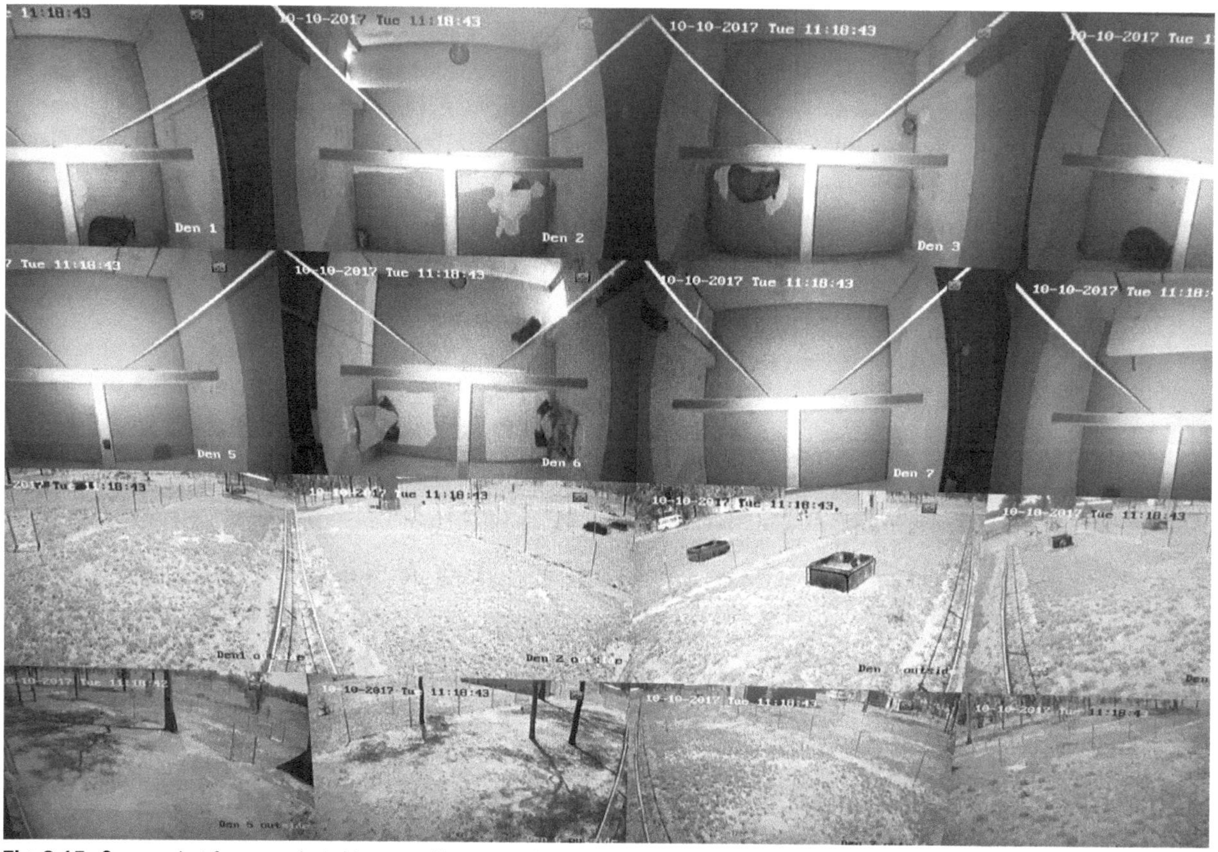

**Fig. 8.15.** Screen shot from wombat video surveillance monitoring system set up at the Wombat Breeding Centre, Australian Animal Care and Education, Mt Larcom, Queensland. Photos taken by Tina Janssen.

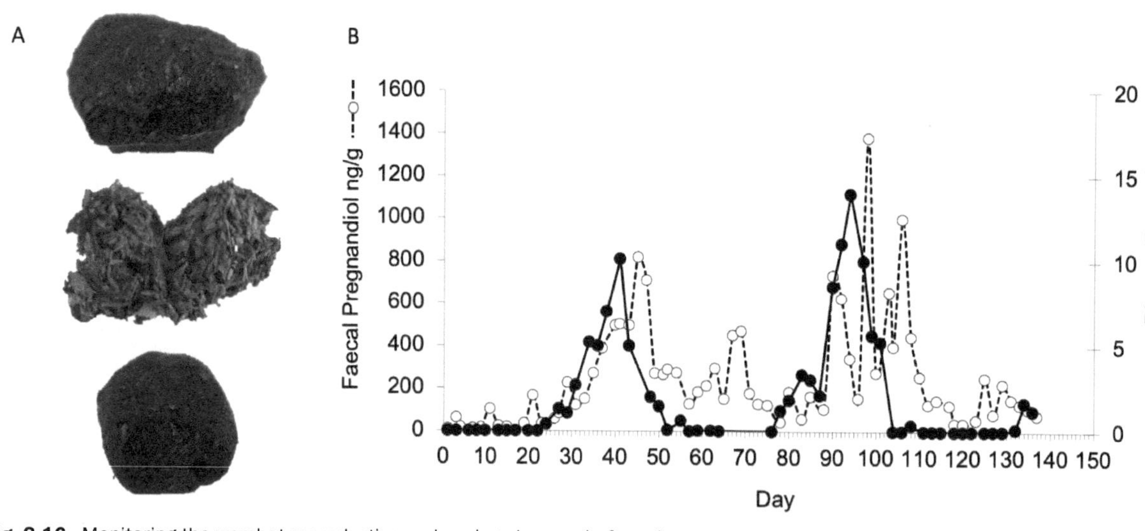


**Fig. 8.16.** Monitoring the wombat reproductive cycle using changes in faecal progesterone metabolites. (a) Faecal pellets of southern hairy-nosed showing glitter in the stool for individual animal identification. (b) Direct comparison between bare nosed wombat plasma progesterone and faecal progesterone metabolites. Note the slight time-lag effect of the hormone profiles and the occurrence of repeated progesterone peaks indicating cyclicity.

The relationship between the progesterone profile obtained using plasma samples and that obtained by extraction of progesterone metabolites from common wombat faeces can be seen in Fig. 8.16. It shows it is possible to observe repeated progesterone peaks indicting that this female is cycling, even though there is a slight time-lag effect associated with the passage of faecal material through the gastrointestinal tract. Although the data from these types of analyses are very informative about reproductive cyclicity, they are less informative about oestrus detection or ovulation, as oestrogens are typically more difficult to measure in faeces and protein hormones, such as luteinising hormone (LH), which is indicative of ovulation, are degraded during gut transit (Swinbourne *et al.* 2017b).

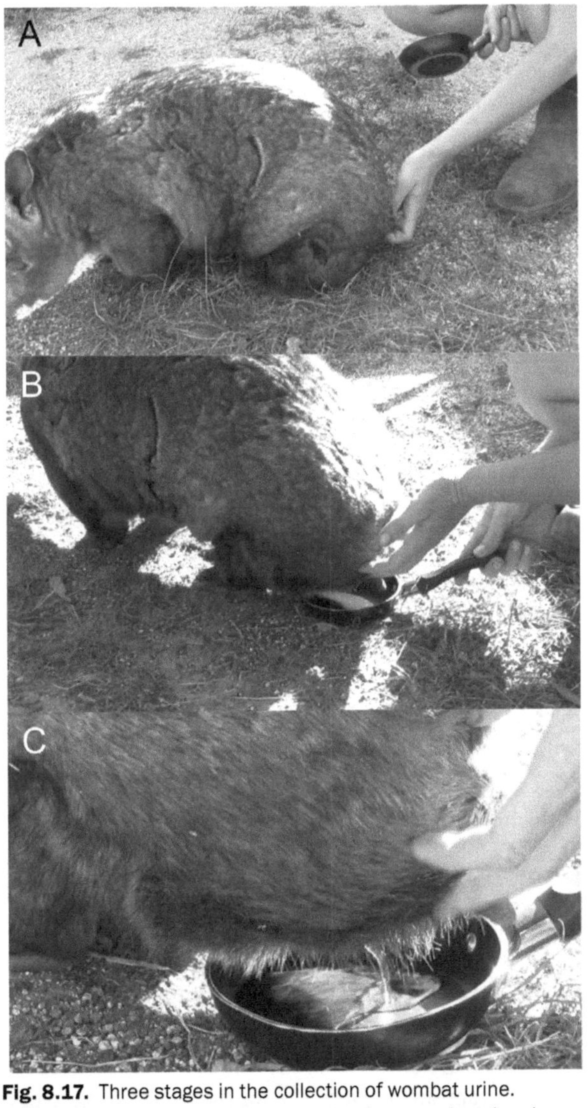


**Fig. 8.17.** Three stages in the collection of wombat urine. (a) Applying tactile stimulation, (b) collection tray placed under urogenital region and (c) sample collected in the tray. Taken from Swinbourne *et al.* (2015).

To facilitate monitoring of the reproductive cycle of southern hairy-nosed wombats, females have been successfully conditioned to allow the collection of urine using a non-invasive technique (Fig. 8.17; Swinbourne *et al.* 2015) that has three steps.

(1) Habituation - the animals are habituated to the collector regularly being in the enclosure to clean the dens and yard enclosures. During the feeding times at ~0700–0900 and 1600 h the animals are watched to observed natural elimination times.
(2) Conditioning of the urination response twice daily (0600 h and 1530 h) just before feeding over a 35-day period. During this process the animal is approached, and woken if asleep, and a small, low-profile collection tray is placed under the urogenital region. The peri-cloacal region is lightly scratched until urination occurs for a maximum of 25 min.
(3) Physical collection of urine sample from each animal per day; if the morning collection is unsuccessful an attempt can be made in the afternoon.

Although Swinbourne *et al.* (2017b) have successfully validated urinary assays under experimental conditions when standardised against creatinine for progesterone, oestrogen and LH (following exogenous gonadotrophin-releasing hormone (GnRH) stimulation), they were unable to use the oestrogen and LH assays to detect oestrus or ovulation, respectively (Fig. 8.18); further studies, using more frequent urine collections, are therefore required in order to map oestrus and ovulation. Hogan *et al.* (2009, 2010b) has also attempted to use activity/pedometry (radio-telemetry collar) to map oestrus activity, but could not establish useful correlations that would allow this technique to be used for oestrus detection, as it has been used in some domestic species.

### 10.4.2 Males

The reproductive condition of male wombats is not easily defined once they have obtained adult body size. Testis and epididymal weight and scrotal diameter do not vary significantly across the year, but the testes must be firm with the tail of the epididymis distinct (Fig. 8.19). In southern hairy-nosed wombats changes in the accessory gland bulge (width), around the base of the penis, provide the best measure of reproductive condition (Fig. 8.19), with all glands, and the prostate, varying seasonally in size with circulating androgen levels (Hamilton *et al.* 2000; Taggart *et al.* 2005; Hogan *et al.* 2010d).

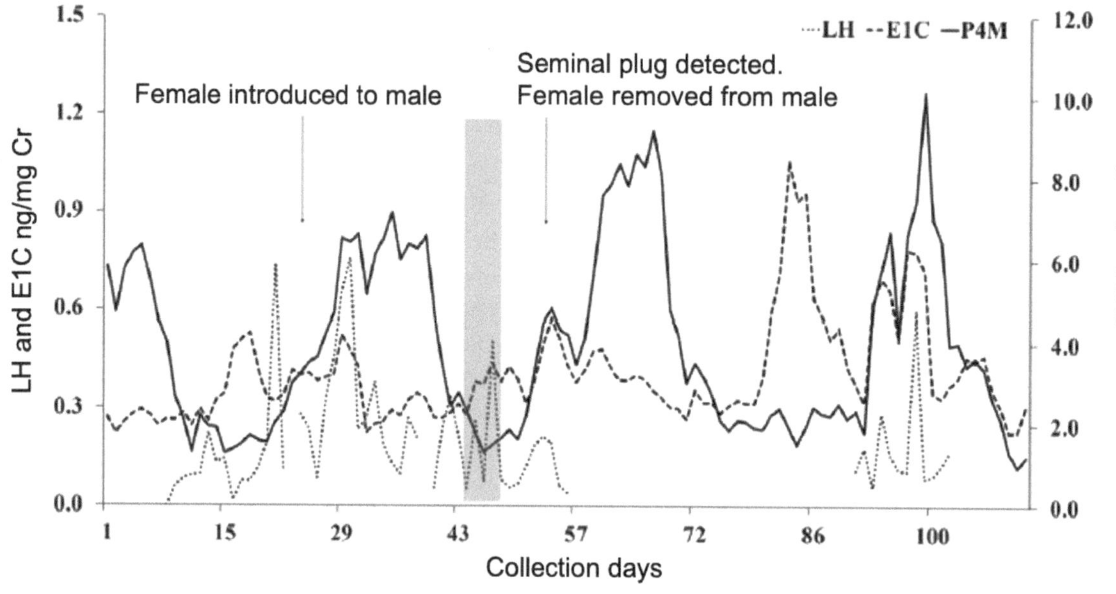


**Fig. 8.18.** Urinary hormone profile of progesterone metabolites (P4M), oestrogen metabolites (E1C), and luteinising hormone (LH) from a female through four cycles. Modified from Swinbourne *et al.* (2017b) with permission from Elsevier.

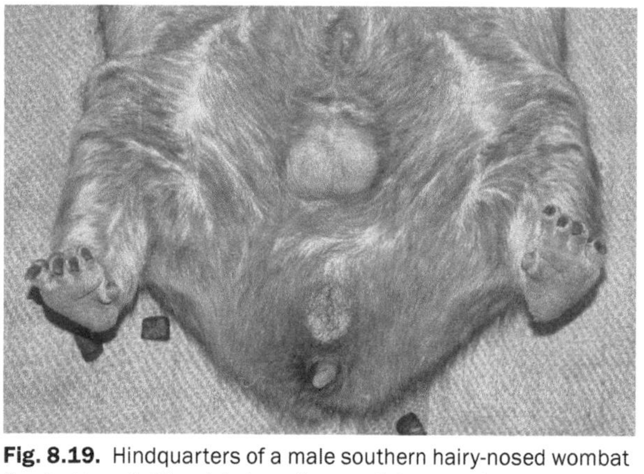

**Fig. 8.19.** Hindquarters of a male southern hairy-nosed wombat (*Lasiorhinus latifrons*) in breeding condition showing scrotum (arrow indicating scrotal diameter) and accessory gland bulge (yellow oval line) surrounding the urogenital sinus and penis. Note: in all marsupials the testes are situated proximal to the urogenital sinus and penis, unlike eutherian mammals. The width of the accessory gland bulge fluctuates seasonally and provides the best external indication of male breeding activity. Photo by David Taggart.

The reproductive condition of a captive male wombat can be monitored by a breeding soundness evaluation that involves a physical evaluation of the external genitalia (see review of male anatomy in Hogan *et al.* 2013), ultrasound evaluation and measurement of the testes, epididymis, prostate and bulbourethral glands and a GnRH stimulation test to assess the steroidogenic capacity (testosterone production) of the testis (Hogan *et al.* 2010d). Spermatogenic activity of the testis can be assessed through the collection of a semen sample via electroejaculation (Fig. 8.20; e.g. bare nosed wombat in MacCallum 2004; southern hairy nosed wombat in Taggart *et al.* 1998a; Johnston 2019) or by means of urine collection in which spermatozoa can be detected (spermatorrhoea; Hogan *et al.* 2010d; Du *et al.* 2017). Although the captive male bare-nosed wombat does not appear to highly seasonal (MacCallum 2004), assessment of reproductive activity in the southern hairy-nosed wombat must take seasonality into account. Recently, Du *et al.* (2017, 2018) have also described their assessment of seasonality in the southern hairy-nosed wombat by measuring testosterone metabolites in the urine.

## 10.5 Reproductive management

Wombats have historically not bred well in captivity, so efforts have been made to breed wombats in captivity with the use of assisted breeding techniques (Hogan *et al.* 2013). Johnston (2019) reported on some of the unique challenges to the successful implementation of these procedures. Johnston (2019) has stated that there are three fundamental components to establishing a successful artificial insemination (AI) program: (1) semen collection and preservation, (2) determining the appropriate timing of insemination and (3) establishing the most appropriate delivery site for the semen sample into the female reproductive tract. All three components are

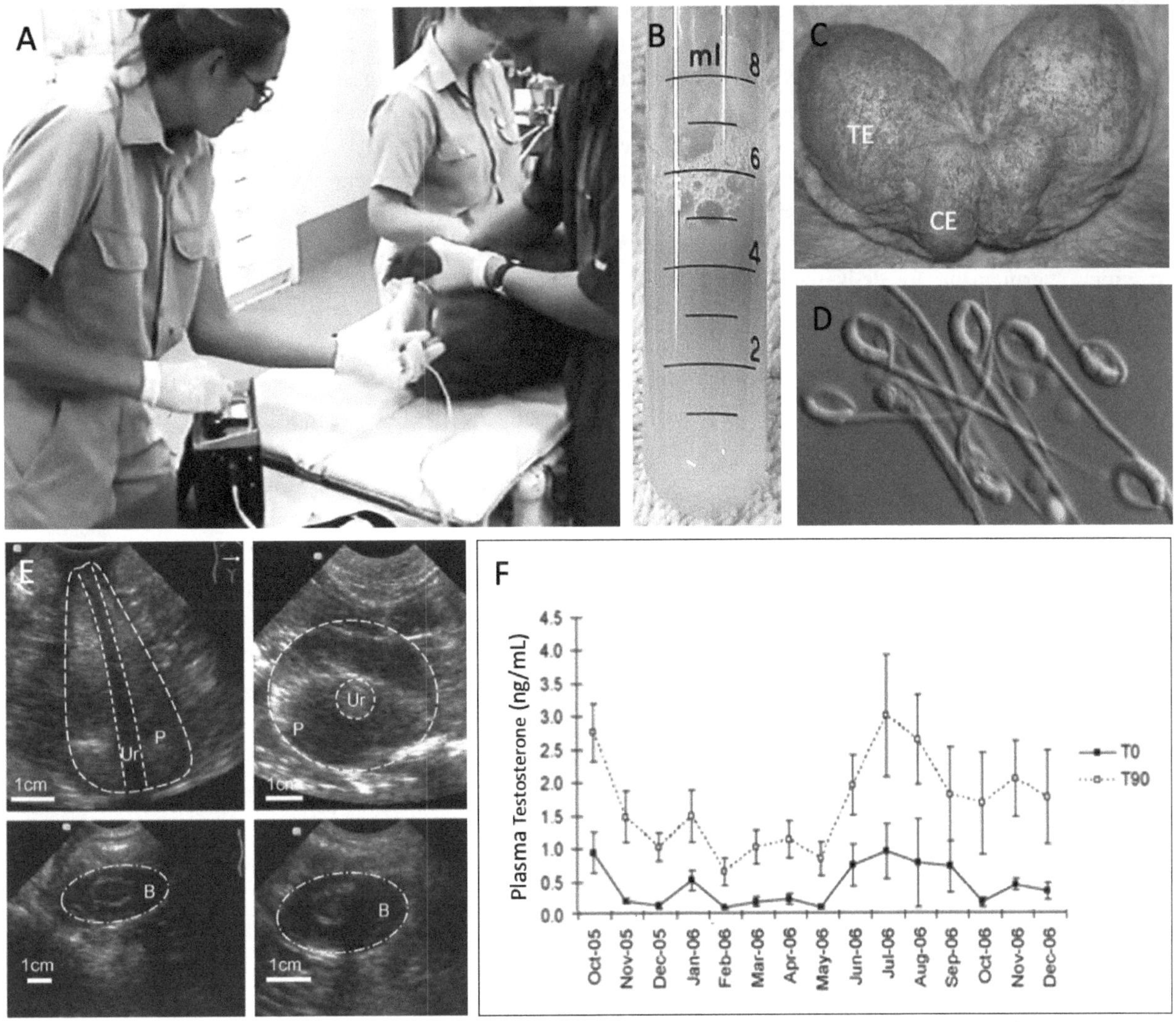


**Fig. 8.20.** Assessment of male wombat reproduction. (a) Electroejaculation under general anaesthesia (BNW). (b) Semen sample showing fractionated sample (BNW). (c) Scrotum for measurement of testicular (TE) volume and distinct presence of cauda epididymis (CE) (BNW). (d) Spermatozoa (BNW). (e) Ultrasonography of prostate (top) and bulbourethral glands (bottom) (SHNW) and (f) Mean ± SEM plasma testosterone concentration of SHNW maintained at Rockhampton Zoo (2005–2006) (T0 - testosterone concentration before GnRH stimulation; T90 - testosterone concentration 90 min after stimulation with GnRH. BNW = bare-nosed wombat, SEM = standard error of the mean, SHNW = southern hairy-nosed wombat. Taken from MacCallum (2004) and Hogan *et al.* (2010d).

reliant on a thorough understanding of reproductive anatomy, physiology and behaviour.

To date, semen collection has been successfully collected in bare-nosed and southern-hairy-nosed wombats by means of electroejaculation from anaesthetised animals and epididymal spermatozoa recovered from necropsy specimens. Semen has also been evaluated and successfully cryopreserved in both species with some impressive results (Taggart *et al.* 1996; Johnston 2019; Johnston and Holt 2019); in fact, although the wombat sperm cell does not appear to survive in a refrigerated or chilled state for as long as the koala spermatozoan, it is far superior to the koala in its tolerance of cryopreservation (Taggart *et al.* 1996). Although there has not been any wombat joey born by AI using frozen-thawed semen, post-thaw recoveries of wombat spermatozoa are typically, in the order of 80% and even up to 90% motility; this post-thaw survival is remarkable compared with other marsupials and even domestic species or humans (Taggart *et al.* 1996). For example, southern-hairy-nosed wombat spermatozoa collected from the cauda epididymis was recently thawed after 10 years storage in liquid nitrogen with some individual straws still showing sperm motility of 90% (S. Johnston pers. obs.).

Given that the gross anatomy of the reproductive tract of the wombat is essentially a scaled-up version of the koala (Temple-Smith and Taggart 1990; Hogan *et al.* 2013), a modified version of the AI pipette that has been successfully developed and used for the koala to produce joeys should also be suitable for the wombat; it may also be possible in the larger-bodied wombat to develop a laparoscopic (keyhole surgery) insemination procedure in which spermatozoa are delivered directly into the uterus.

Nevertheless, the still missing piece of the AI puzzle for the wombat remains the issue of insemination timing. As stated in section 10.3 of this chapter, determining the timing of AI with respect to oestrus and ovulation in wombat requires further research. Apart from observing natural mating, all attempts at visual detection of oestrus in the wombat have so far proven unreliable; including identifying overt diagnostic oestrus behaviour (as is observed in the koala – see Chapter 7), pedometry/activity and non-invasive reproductive hormone monitoring (e.g. LH). Johnston and Janssen (pers. obs.) have even attempted to use a vasectomised male 'teaser' wombat to detect oestrus in the AACE captive southern hairy-nosed wombat population, but unfortunately this particular male lacked sufficient libido and failed to show any subsequent sexual interest. They have also used non-invasive hormone monitoring of progesterone metabolites in urine to map the cycle of five southern hairy-nosed wombats before and after AI to help evaluate the timing of AI (Fig. 8.21). Although none of the AI procedures resulted in females producing pouch young, hormone profiles were used to estimate ovulation; results of these observations suggest that the timing of AI was correct in only 1 of the 5 wombats.

Future studies in the wombat are planned to explore the use of oestrus synchronisation for timed insemination to help overcome the problem of oestrus detection. Janssen (pers. obs.) has also indicated that there is significant individual animal variation in the expression of oestrus behaviour and that close individual monitoring

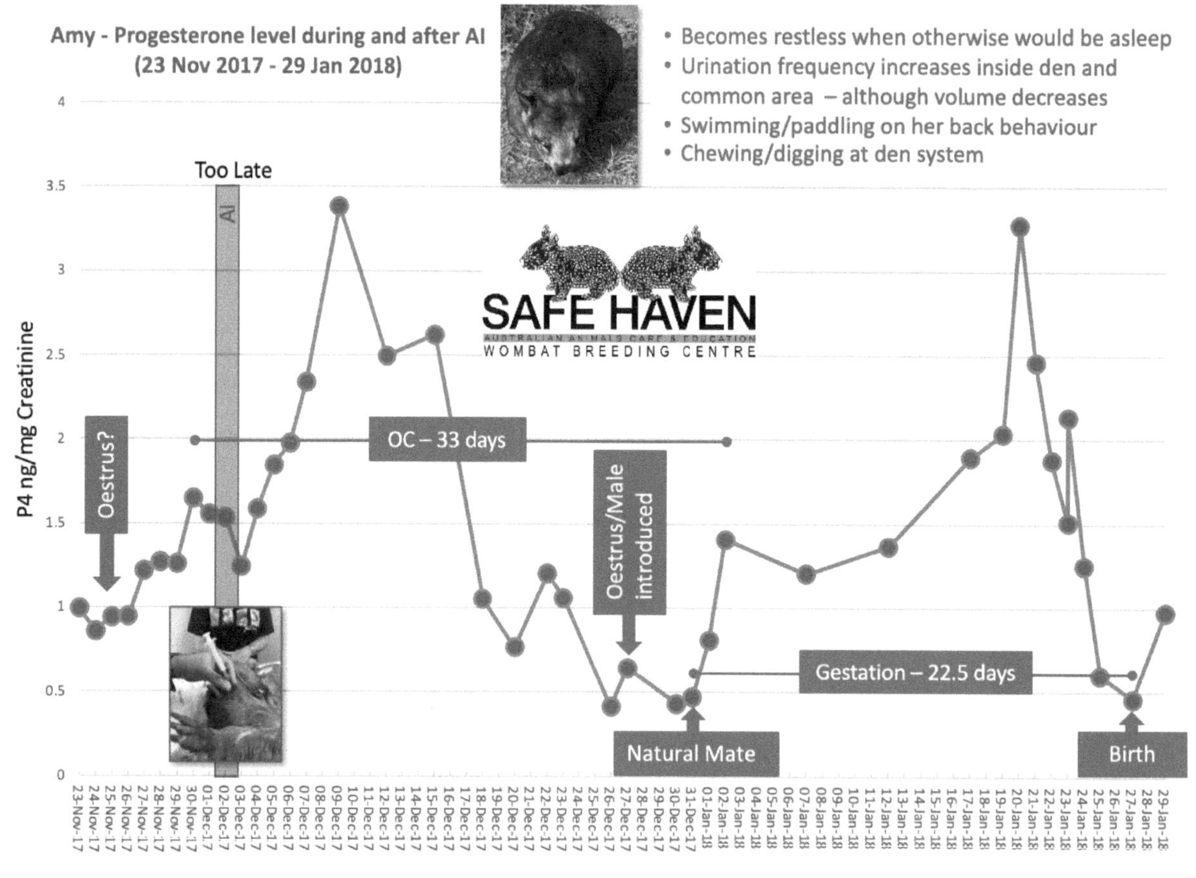


**Fig. 8.21.** Urinary progesterone metabolite analysis mapped through a failed artificial insemination attempt using frozen-thawed semen and a subsequent successful natural mating, pregnancy and parturition. Taken from Johnston *et al.* (2018).

of this behaviour is required to improve the timing of insemination.

Apart from AI, there have been numerous other attempts at developing artificial reproductive techniques such as ovarian hyperstimulation, oocyte collection and maturation and ovarian tissue xenografting, but application of these approaches is currently limited by the lack of fundamental understanding of wombat reproductive physiology and the small size of the reproductive tract. One exciting possibility, which utilises the unique reproductive physiology of marsupials and deserves greater research attention, is the concept of pouch young transfer and intraspecies fostering of wombats (Finlayson *et al.* 2007a), specifically between southern and northern hairy-nosed wombats. Already proven successful in macropod species (Taggart *et al.* 2010; Muhic *et al.* 2012; McLelland *et al.* 2015), the ability to generate northern hairy-nosed wombats from natural mating or AI for transfer to the pouches of synchronised southern hairy-nosed wombats is extremely appealing for captive management. While remaining an exciting possibility, the fact remains that currently there is not even a single captive population (or even pair) of northern hairy-nosed wombats.

## 10.6 Occurrence of hybrids

None are known to occur.

## 10.7 Timing of breeding

### 10.7.1 Bare-nosed wombat

This species of wombat appears to time breeding with both latitude and elevation so that the maximum growth rate of young corresponds with the maximum potential growth period of temperate grasses, which usually occurs in spring between September and January in south-eastern Australia; however, births can occur throughout the year (Mallett and Cooke 1986; Green and Rainbird 1987; Wells 1989). As wombats are polyoestrous, they appear to be able to breed throughout the year in areas of high forage quality and abundance (Peters and Rose 1979; Green and Rainbird 1987). The timing of breeding is different in different locations, with breeding from December to March in the highlands of New South Wales and eastern Victoria (McIlroy 1973; Skerratt 2001) and from March to July in northern Victoria, with most births in June and July (Nicholson 1963; Presidente 1982). Despite these findings MacCallum (2004) noted no detectable female or male reproductive seasonality in a captive population of male bare-nosed wombats housed at Taronga Western Plains Zoo, Dubbo; in fact, oestrous cycles were detected throughout the year and there was no significant change in male reproductive anatomical or seminal parameters.

### 10.7.2 Hairy-nosed wombats

Although there are currently no studies that have monitored reproductive seasonality of female southern hairy-nosed wombats in captivity, anecdotal observations from breeding activity of colonies on the central coast of Queensland (Rockhampton Zoo and AACE Wombat Breeding Centre) indicate a different pattern to that of the wild populations in South Australia from which these animals were originally sourced (T. Janssen and S. Johnston pers. obs.).

Births at the AACE Wombat Breeding Centre have been recorded between September and March. In South Australia, most births have been recorded from late July to September, and dependent upon rainfall, with some births occurring in October–December. This correlates with the growth and germination of native pasture (late September–December) (Wells *et al.* 1986; Wells 1995; Taggart and Temple-Smith 2023). Weaning occurs in spring or early summer and almost 6 months out of phase with the bare-nosed wombat, as these growth periods are associated with the winter rainfall patterns of the arid and semi-arid zones of South Australia (Wells 1989). When rainfall and associated pasture growth declines, bodyweight declines and reproductive activity and breeding may cease completely (Gaughwin *et al.* 1998; Taggart and Temple-Smith 2023). Bodyweight may decline by as much as 25–30% in years of low winter/spring and autumn rainfall (D. Taggart and S. Kleemann pers. obs.).

A study by Hogan *et al.* (2010d) based on four captive males housed at Rockhampton Zoo revealed an increased level of plasma testosterone concentration from late winter and a decrease in spring and an increase in bulbourethral size in late autumn to mid-winter; there was no seasonal pattern detected in faecal testosterone metabolites or other reproductive anatomies and spermatozoa were found in the urine throughout the year. Du *et al.* (2018) have also examined seasonal changes in testosterone urinary metabolites in a captive population at the AACE Mt Larcom, but found only a marginal seasonal effect compared with that reported for wild wombats.

Observations of the northern hairy-nosed wombat show a similar result, with a significant relationship between body condition, breeding rate of females and the amount of summer rainfall. This suggests a nutritional constraint on breeding, although the mechanism is unclear given that the breeding cycle of most females commences before the onset of the summer rainy season (Crossman *et al.* 1994; Woolnough 2000).

## 10.8 Age at first and last breeding

Southern hairy-nosed wombats become sexually mature at ~3years of age and ~23 kg (Taggart *et al.* 2007; Taggart and Temple-Smith 2023). Reproductive behaviour has not been observed until this age (Gaughwin 1982). Male southern hairy-nosed wombats need to be fully developed, which occurs at 4–5 years of age; their bodyweight increase will level out at this stage and they should be large enough to secure a mating. Females can breed as early as 2 years or as late as 4 years (T. Janssen pers. obs.). Bare-nosed wombats breed when they weigh ~22 kg, which is also equivalent to ~3 years of age (McIlroy 1973).

## 10.9 Ability to breed every year

Although bare-nosed wombats appear to breed annually, the two species of hairy-nosed wombats appear to coincide breeding with rainfall and pasture growth, so that in years of low rainfall or drought, few if any females breed (Taggart *et al.* 2020). Northern hairy-nosed wombats appear to breed on average twice per three years (Johnson and Gordon 1995), but southern hairy-nosed wombats are capable of breeding every year if conditions are good enough, but in their semi-arid environment, they typically only breed every 2–3 years (Finlayson *et al.* 2007b). If there is a rain event just before and during the breeding season, captive animals at the AACE facility appear more likely to breed (T. Janssen pers. obs.).

Females wean their young at ~16–18 months and will go into oestrus at the next season (T. Janssen and S. Johnston pers. obs.). One captive female at the AACE facility had a joey at 2 years of age, but it did not survive; this same animal subsequently had two more healthy joeys when she was 3 years of age (T. Janssen pers. obs.). Due to careless mothering the young were taken from her at 7 months of age for hand-rearing. At 4 years of age she carried and weaned a young until it was 16 months of age. Another female was 5 years old when she had a healthy joey and weaned it at 17 months of age (T. Janssen pers. obs.).

## 10.10 Ability to breed more than once per year

All species of wombats can breed only once per year due to the length of time required to raise and wean the young. There may also be significant bodyweight loss associated with lactation that prevents the female from immediately breeding again (Skerratt 2001). Southern hairy-nosed wombats are able to return to oestrus, if a pouch young is lost early in the breeding season. If young are lost between October and December, females do not re-enter oestrus (Finlayson *et al.* 2007a, 2007b).

## 10.11 Nesting requirements

Female wombats should be provided with a well-built den box, large hollow log, artificial burrow or, ideally, an area of earth in which to dig a burrow. Nesting material such as leaf litter and soil should be provided. Wombats bred at the AACE Centre have successfully given birth in their sleeping chambers and raised young to weaning in the same enclosures (T. Janssen pers. obs.).

## 10.12 Breeding diet

Births in wild populations of southern hairy-nosed wombats appear to correlate with rains and associated grass growth after rain when forage quality is maximal (Wells *et al.* 1986; Wells 1989; Woolnough 2000), so the provision of large amounts of fresh grass before the beginning of the breeding season is recommended for these species. It has been suggested that the seasonal variability in the food resources and timing of the reproductive cycle allows the female to invest in fat in order to meet the higher energy demands of reproduction and lactation (Woolnough 2000). T. Janssen (pers. obs.) has noted that southern hairy-nosed wombats at the AACE Wombat Breeding Centre had reduced cycles in years of low rain or drought, even though their body condition remained constant and no amount of increase in their diet made any changes to their reproductive cycle.

## 10.13 Number of young born

Usually only one young is born in all three species of wombat; however, there are records of two being successfully raised in captivity with hand-rearing assistance (Ride 1970; Finlayson *et al.* 2007b; Taggart and Temple-Smith 2023).

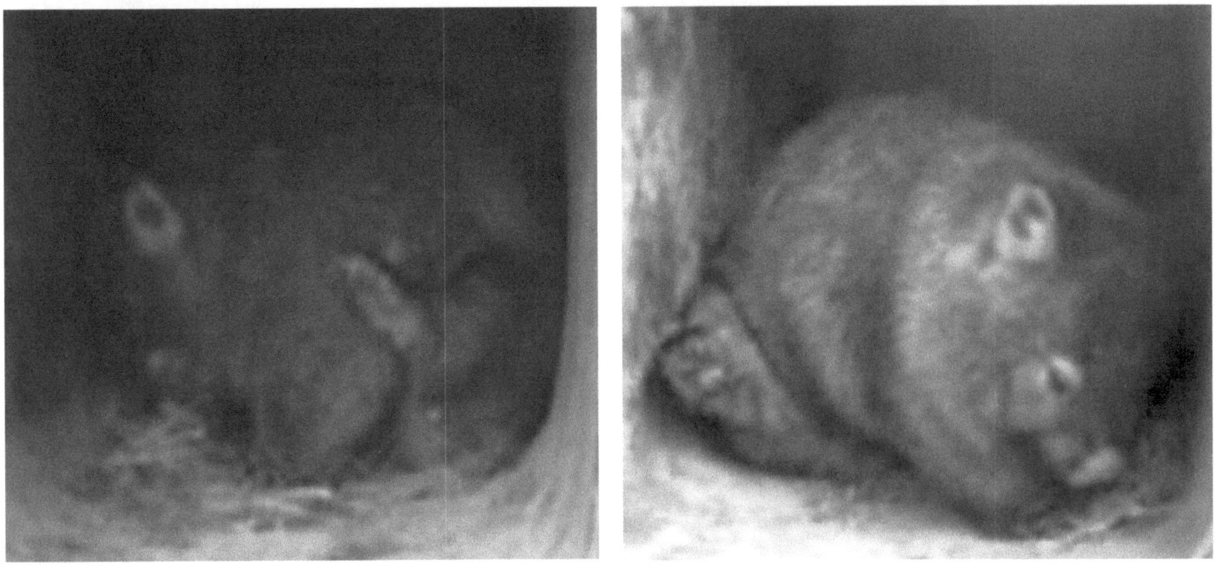

**Fig. 8.22.** Bare-nosed wombat in parturition posture. Taken from Hogan *et al.* (2013).

**Table 8.6.** Reproduction and development of wombats.

| | Permanent pouch exit (days) | Weaning (days) | Sexual maturity Months (F) | Sexual maturity Months (M) | Birth Season | Reference |
|---|---|---|---|---|---|---|
| *Lasiorhinus krefftii* | 300–330 | – | – | – | – | 1 |
| *Lasiorhinus latifrons* | 180–270 | 360–400 | 36 | > 36 | Jul–Dec | 2–5 |
| *Vombatus ursinus* | > 150 | 400 | 24 | – | All year | 6, 7 |

References: (1) Johnson and Gordon (1995); (2) Crowcroft and Soderlund (1977); (3) Gaughwin and Wells (1978); (4) Wells (1995); (5) Taggart *et al.* (2007); (6) Tyndale-Biscoe and Renfree (1987); (7) McIlroy (1995).

Observations of birthing behaviour in the common wombat have been reported by Böer (1998) and Hogan *et al.* (2013) (Fig. 8.22). Both observations indicate the female taking on a squatting or bent forward position, rather than rolling on the side during parturition, presumably associated with reducing the distance for the neonate from the cloaca to the teat in the pouch. The later authors reported the female repeatedly tapping and/or kneading her forepaws on the ground and alternating between squatting and bending forward, while vocalising, grinding her teeth and shaking her body (Hogan *et al.* 2013).

Twins have been observed in wild southern hairy-nosed wombats on only two occasions across > 25 years and > 4000 captured individuals. It is extremely unlikely that both of these young would have survived to term, due to the limitations in pouch space, let alone the ability of the mother to meet the energy requirement of two developing young in a semi-arid environment (D. Taggart and S. Kleemann unpubl. data; Taggart and Temple-Smith 2023). In captivity, once a wombat gives birth, the other wombats in the enclosure should ideally be removed to prevent stress and aggression and to help reduce the risk of pouch young loss (Gaughwin 1982).

## 10.14 Age at weaning

The pouch young begin leaving the pouch and eating solid foods at ~9 months of age and more than double their bodyweight during the next 3–8 months. Southern hairy-nosed wombats generally disperse as large sub-adults at ~2 years of age and at a bodyweight of ~20 kg (Wells 1989). Dispersal of the young is female-biased (Johnson and Crossman 1991; Banks *et al.* 2002; Walker *et al.* 2008). The time in the pouch varies with species and ranges from 10–11 months in the northern hairy-nosed wombat (Johnson and Gordon 1995) to 7–9 months in the southern hairy-nosed and bare-nosed wombats (Table 8.6) (McIlroy 1995; Wells 1995; Taggart *et al.* 2007). Weaning occurs in the bare-nosed wombat at ~12–15 months of age when juvenile females are ~8.65 kg and juvenile males are 8.38 kg. Weaning occurs over ~50 days during which time bodyweight losses between 25 and 420 g have been observed (Triggs 1996; Böer 1998). A male bare-nosed wombat was observed taking solid food when he was ~8.5 months of age and 1330 g while still in the pouch, whereas two females were first observed to eat solids at 1620 g and 1510 g when they were 9 months of age (Böer 1998). This indicates that young wombats feed on a mixture of milk and solids for ~3–6 months

a

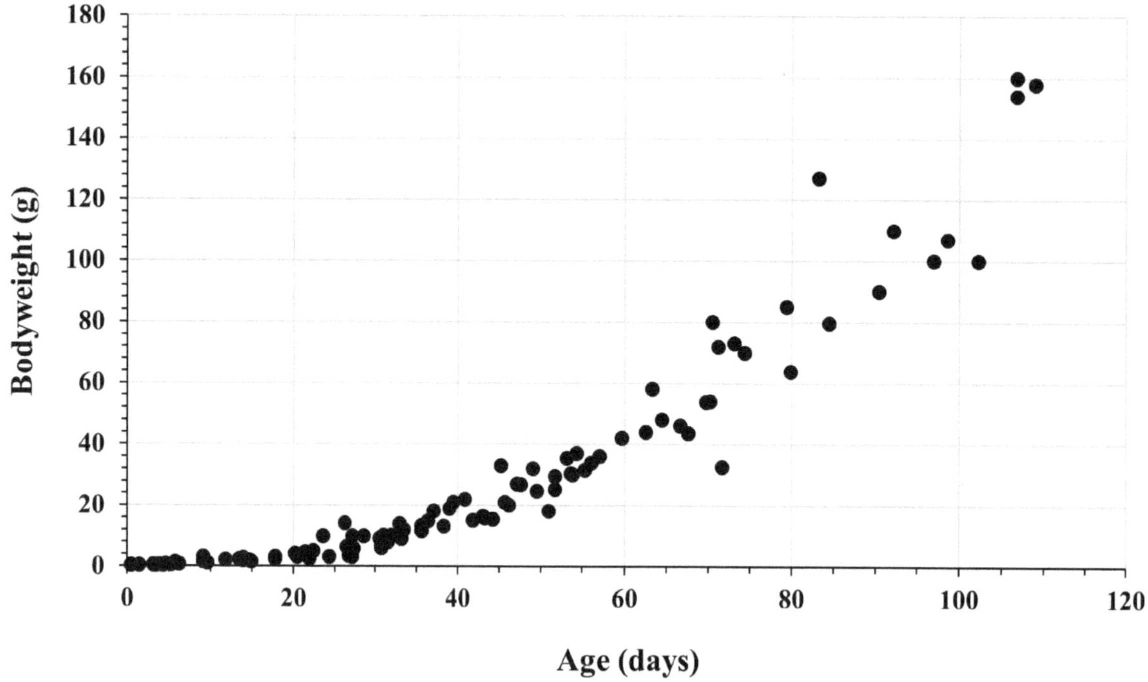


b

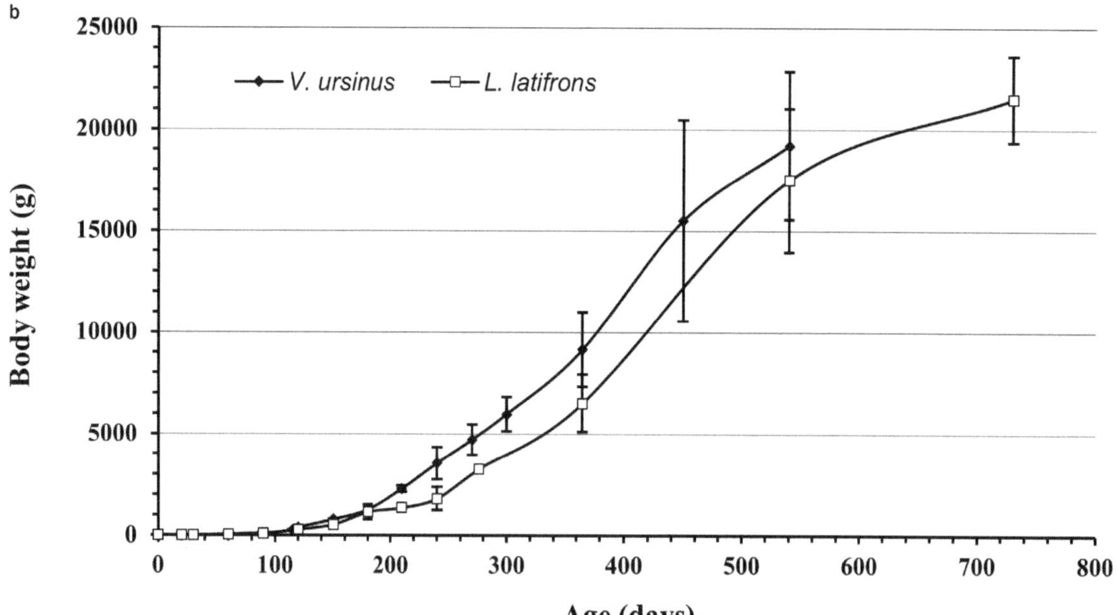


**Fig. 8.23.** (a) Growth in bodyweight and of southern hairy-nosed wombat pouch young between 0 and 100 days. Reproduced from Taggart *et al.* (2007) with permission from CSIRO Publishing. (b) Growth in bodyweight *v.* age of the bare-nosed and southern hairy-nosed wombats. Derived from Gaughwin (1982), Triggs (1996), Austin (1997), Woods (1999) and Taggart *et al.* (2007). Standard deviation error bars are shown on the curve for the bare-nosed wombat.

(Böer 1998). At the AACE Wombat Breeding Centre, southern hairy-nosed wombat young are introduced to solids at ~9 months, although this timing also depends on the individual. Although further research is required, weaning in southern hairy-nosed wombats has been observed to occur at ~16–18 months of age, but the dam and offspring may remain in the burrow for ~2 years.

## 10.15 Age at removal from the mother

In captivity young should be removed from their mothers when they are ~20–28 months old, several months before the female comes into her next oestrus, as she can become increasingly aggressive towards the young (Böer 1998). Severe biting and chasing can occur at this stage, causing the young to run back and forth in the enclosure or hide and cower in a corner (Böer 1998).

## 10.16 Growth and development

If pouch young are present, there are several developmental stages and measurements that can be recorded and compared with existing growth curves or used to establish curves for future reference.

- Developmental stages
  - sex distinguishable
  - tips of ears free
  - papillae of facial vibrissae evident
  - eyelashes visible
  - eyes open
  - fur visible - slight tinge, medium or well-developed
  - tips of first incisors through the gums
  - at foot
  - eating solids
  - self-feeding
  - independent
- Measurements (see Appendix 3)
  - bodyweight (g) - if not on teat
  - head length (mm) - from the occiput to snout tip
  - head width (mm) - maximum width across the zygomatic arches
  - pes length (mm) - from the heel to the distal end of the longest toe, not including the nail
  - tibia length (mm) - from the stifle to the heel.

The growth and development of bare-nosed and southern hairy-nosed wombats is shown in Fig. 8.23. Additional detailed growth and development information references are given in Table 8.7 and Bach (1998) and Taggart *et al.* (2007).

**Table 8.7.** Growth curves measurements that have been developed for the different species of wombats.

HE = head length, LE = leg length, TO = total length, WT = bodyweight.

| | Measurements | Reference |
|---|---|---|
| *Lasiorhinus latifrons* | WT, TO, LE | 1-4 |
| *Vombatus ursinus* | WT, TO, HE | 3, 5–9 |

References: (1) Crowcroft and Soderlund (1977); (2) Gaughwin (1982); (3) Woods (1999); (4) Taggart *et al.* (2007); (5) Young (1980), (6) Presidente (1982); (7) Triggs (1996); (8) Austin (1997); (9) Böer (1998).

# 11. ARTIFICIAL REARING

## 11.1 Housing

As with all native mammals that have been taken into care, minimising stress is a major consideration. Choosing suitable housing can help to create a stress-free environment. To achieve this, several factors should be considered, including:

- securing the area from children and animals
- maintaining the area in a hygienic manner
- escape-proofing the area
- clearing the area of obstacles and hazards
- ensuring the area offers shelter from extreme weather and noise.

Hairy-nosed wombats (e.g. Osborn 1975; Christian 1977; Crowcroft and Soderlund 1977) and particularly bare-nosed wombats (Presidente 1982) are frequently hand-reared. Unfurred or just furred animals should be placed in multiples layers of artificial pouches that simulate, as closely as possible, the security and warmth of the mother's pouch. Natural-fibre products such as cotton are required as they retain warmth better. In unfurred joeys, cotton pouches must be soft to reduce the risk of causing rubbing sores to the hip/pelvic region.

All species of wombat young develop in their mother's pouch, so up to the development stage of first emergence from pouch, wombat joeys need to be suspended in a pouch so they are not weight-bearing as this may harm their normal skeletal development. Wombat joeys at any stage of development should be housed in a climate-controlled room at 24°C. Some rearers do not provide dummies while some do provide one for the joey to suckle on.

The bag should be hung so that it is suspended in a secure box. Unfurred joeys may need extra warm fleecy liners or occasionally an external heat source, such as a warm heat pack, to hold their body temperature, while ensuring they do not become overheated until they are able to thermoregulate themselves.

**Table 8.8.** Concentrations of major constituents of bare-nosed wombat milk.

Derived from Green (1984).

| Total solids (%) | Carbohydrates (%) | Lipids (%) | Protein (%) | Calcium (mg/L) | Iron (mg/L) |
|---|---|---|---|---|---|
| 22.0–51.0 | 4.0–12.0 | 6.0–28.0 | 4.0–9.0 | 4200 | 22 |

Finely furred young that weigh < 2 kg are just beginning to emerge from pouch and should also be housed in a pouch with soft cotton liner that can be lowered in the box to grant the joey safe access in and out of the pouch to start building up muscle. During this stage of development, they should also have supervised time outside the pouch with access to dirt, grasses and filtered sunlight to help build up their immunity and facilitate natural behaviours such as digging and grazing.

As the young begin to fully emerge from the pouch, they can be introduced to their den system with the hanging bag and towels/blankets lining the floor. Once fully emerged from the pouch, the hanging bag can be removed and the joey can be housed in an internal den system similar to the housing for adult wombats, lined with blankets or towels made of natural fibres. Do not use synthetic fibres as these are either too hot or too cold and rub on the joey's skin, producing pressure areas (George *et al.* 1995).

## 11.2 Temperature requirements

Juvenile wombats should be held in a temperature range of ~25–30°C (George *et al.* 1995; Austin 1997). Usually, smaller wombats weighing < 600 g should not be held at a temperature greater than 28°C (George *et al.* 1995). Use a minimum/maximum temperature gauge with a plastic-coated probe that can be placed next to the joey to monitor the temperature. A hot-water bottle (reheated every 4 h) can be used for heating but should be well wrapped inside towels or other fabric and should not be placed too close to the wombat or overheating or dehydration may occur. If the plantar aspect of the joey's feet is red, the joey is too hot; the soles of the feet should be light pink in colour (T. Janssen pers. obs.).

## 11.3 Specific requirements

The skin of unfurred and slightly furred young should be kept moist with the use of Sorbolene cream (without added glycerine) to prevent it becoming dry and cracked (George *et al.* 1995). Baby oil does not appear to be properly absorbed and tends to stay on the skin surface where it rubs off and is absorbed by the liner bag fabric (George *et al.* 1995). Natural vitamin E oil can also be used to keep skin supple.

When first brought in for hand-rearing, the joey may be dehydrated. If so, it can be given plain boiled water, which has been allowed to cool to 36°C, or 1 g of electrolyte replacer if available (Austin 1997). Vytrate® can also be used at a ratio of 20 mL Vytrate® to 250 mL water. It is important to warm the joey before feeding to avoid the risk of inhalation pneumonia. If the joey is cold, place it in a warm water bath and dry it off rather than putting it in a hot box. Organise a veterinary examination.

Stress is a major problem in the successful rearing of native mammals and can be fatal. Therefore, it is important to keep noise to a minimum, not overhandle the animal and maintain high standards of hygiene.

## 11.4 Diet and feeding routine

### 11.4.1 Natural milk

Wombat milk increases in total solids during lactation from 22% in early lactation to 55% in late lactation, while lipids increase from 6% to 28% and proteins from 4% to 9% (Green 1984). In contrast, the concentration of carbohydrates decreases from 12% in early lactation to only 4% in late lactation (Table 8.8) (Green 1984).

### 11.4.2 Milk formulas

The main commercial formulas used for hand-rearing wombats are listed.

- Biolac® – the three formulations are M100 with 2–5 mL of canola oil per 100 mL for furless joeys; M150 as a transitional milk to use when dense fur has developed; and M200, which is used when the animal produces solid dark pellet droppings, as it contains elevated lipid in the form of canola oil. When the joey is nearing weaning, 2–5 mL of canola oil is added per 100 mL of formula. Mixing the formulas assists in the transition from one formula to another. The young animal should be fed 10–15% of its bodyweight per day.
- Di-Vetelact ® –a widely used, low-lactose milk formula. Due to its low energy concentration when prepared as directed, some groups advise the addition of mono- and polyunsaturated fats such as canola oil as with Wombaroo® diets. Adding saturated fats in the

form of cream has been suggested but it is too highly saturated and can lead to malabsorption of calcium. Di-Vetelact ® should be fed at ~20% bodyweight, except in the case of very small joeys (< 100 g).
- Wombaroo ® Wombat Milk - charts are provided to assist in determining the type and volume to be fed. The three formulas range from < 4 for joeys with less than 40% of their pouch life completed; a 0.4 formula for joeys with 40% of their pouch life completed and fine fur, eyes open and erect ears; and a > 0.6 formula for joeys with greater than 60% of their pouch life completed with short, dense fur and spending a lot of time out of the pouch. Evidence suggests that, if given the option, Wombaroo® wombat formula gives better growth rate and hair quality than other milk formulas (Booth 1999).

Folivorous species, such as wombats, need to establish gut flora to break down the vegetable matter in the diet (Eisenhofer *et al.* 2021a, 2021b). This can be achieved by several methods, including offering dry dirt (pica), which they may eat. It can also be assisted by adding half a teaspoon of natural yoghurt to the formula daily (Austin 1997). An alternative method to inoculate with bacteria is by choosing several fresh droppings from a healthy adult wombat, of the same species, grinding them up, adding warm water, straining and adding 5 mL (1 teaspoon) either to the joey's bottle containing milk and mixing it up or giving it directly into the mouth (Austin 1997). The droppings should be checked they are not contaminated with parasites before being used.

At 6 months of age, infant cereal can be added to the Di-Vetelact® formula at ½–1 teaspoon to every 200 mL before feeding (Austin 1997).

### 11.4.3 Feeding apparatus

Very small joeys can be fed using a syringe fitted with a plastic intravenous catheter or 25-mm length of infant gastric feeding tube (Bellamy 1992). Most wombat joeys will, however, be large enough to be fed with a plastic feeder bottle (50 or 100 mL) and a special wombat Type (a) teat (Austin 1997). The teat should be punctured with a hot needle.

### 11.4.4 Feeding routine

During the first few days in captivity wombats usually put up a lot of resistance to taking a bottle and can take several days before feeding willingly. As a result, there is a need to monitor hydration levels and persevere until they willingly take it. When feeding the wombat it should be wrapped in a towel and the hand-rearer should place a hand over its eyes while it is being fed (Austin 1997). The milk should be warmed to ~36°C before feeding. Young wombats should be fed ~15% of their bodyweight daily, although the demand decreases with the introduction of solid food (Booth 1999). Furless joeys should be fed every 2 h and well-furred joeys should be fed every 4 h (Booth 1999). Take care that the milk is not forced at a greater rate than it is sucked, as it can accumulate in the pharynx and be sneezed or coughed out of the nostrils or passed into the lungs where it could result in aspiration pneumonia leading to death (Presidente 1982). Replacing the teat and using a smaller hole in the teat can slow the rate of milk consumption and decrease the incidence of milk aspiration.

Once the young are fully furred and weighing ~2 kg, solid foods can be introduced. This can consist of dry grass with roots and dirt after a few weeks and fresh grass, grated apple, lucerne hay, wallaby pellets and vegetables (initially fed blanched) such as carrot, pumpkin and sweet potato (Presidente 1982; J. Hockley *pers. comm.*).

The number of daily feeds changes as the joey develops (Bellamy 1992). Very young, unfurred joeys should be fed every 2–3 h. When furred, the number of feeds is reduced to five and the volume increased per feed. At full emergence the number of feeds is reduced to 2–3 per day and they should be given access to solid food as described above.

Joeys should be offered fresh water in a sturdy container as they can dehydrate, especially during warm weather. Water is especially important once the joey begins to vacate the pouch and eat solids (George *et al.* 1995).

## 11.5 Data recording

When an animal is first brought in for hand-rearing, its sex, head measurements, bodyweight and approximate age, using growth charts, should be recorded. During the hand-rearing process the following important information should be recorded on a daily basis:

- date when the information is recorded
- time when the information is recorded
- bodyweight to the nearest 1 g if possible
- general activity and demeanour
- characteristics and frequency of defecation and urination

- amount (g) of different food types offered
- food consumption at each feed
- veterinary examinations and results.

This information serves several purposes, including providing important background information such as food consumption, which will assist a veterinarian to reach a diagnosis if the animal becomes sick or fails to grow or gain bodyweight. It also allows a comparison with growth curves to assess progress (section 10.16) and to allow new growth curves to be established for measurements in which they presently do not occur. The developmental stages and measurements outlined in section 10.3.1 should also be recorded on a weekly basis.

## 11.6 Identification methods

Generally, not required, but can be given a PIT tag (see section 5.3.1) once furred.

## 11.7 Hygiene

Maintaining a high standard of hygiene is critical to the survival of the wombat joey.

- Maintain a clean pouch lining at all times. Older joeys may be able to be trained to urinate on newspaper by keeping a piece of newspaper with the smell of urine on it. They have also been trained to urinate into bowl by placing shallow bowl under the cloaca and 'tickling' it until the joey begins urinating (T. Janssen unpubl. obs.). Use a new pouch liner after each feed.
- Personal hygiene – wash and disinfect hands before and after handling the joey. Use antibacterial solution for washing hands with furless joeys, as their immune system is not well developed.
- Wash hands between feeding different joeys.
- Boiled water should be used when making up formulas for joeys of all ages
- Clean up spilt milk formula or paste, faeces and urine from the joey's skin and fur as soon as possible and the fur dried to prevent bacterial or fungal skin infections.
- Wash all feeding equipment in warm water and dishwashing detergent and sterilising it in a microwavable baby steam steriliser or F10® disinfectant.
- Store the teats and bottles in the fridge in an air-tight container after they have been disinfected.
- Only heat the milk once and discarding leftovers.
- Avoid contact with other animals unless certain they pose no health risk. Some carers recommend that joeys, once fully emerged from the pouch, be allowed to socialise with other joeys to avoid human imprinting and encourage normal social behaviour.
- Stimulate the joey to toilet before or after feeding. As with other marsupials, toileting can be done by the application of warm water to the cloaca using cotton wool to stimulate urination and defecation, which allows the animal to keep drier and warmer in its pouch. This practice should be phased out once the animal urinates and defecates regularly without stimulation (Booth 1999).
- Cover the joey's body, if furless, with vitamin E oil/Sorbolene cream as necessary to keep joey's skin supple and prevent drying.

## 11.8 Behavioural considerations

Once a wombat reaches ~18 months of age, it naturally becomes increasingly independent of its hand-rearer and generally becomes quite aggressive. Aggression is a normal behaviour, even in joeys, soon after emergence from the pouch. Mother wombats appear to be quite tolerant of this behaviour, but a joey bite can be very painful. Prior to release they should be completely weaned from human affection, fed a diet of natural grasses and held with other wombats so they can develop their social skills.

## 11.9 Use of foster animals

Fostering within wombat species has been conducted successfully, with 100% success rates being observed in southern hairy-nosed wombats, provided the pouch young is transferred to another female with a young of equivalent or greater size. Young as small as 1.5 g have been transferred successfully to foster mothers, with the growth rates being unaffected (Finlayson *et al.* 2007a). During studies of wild southern hairy-nosed wombats, 15 pouch young ranging in age from 16 to 146 days were transferred to other females. Six of these females were subsequently recaptured between 4 and 11 months later and all were either still carrying the cross-fostered young or showed evidence of late lactation on the same teat (Finlayson *et al.* 2007a).

## 11.10 Weaning

Wombats can be introduced to solid foods by providing freshly cut grass and 'wombat pellets' (Ridley Agriproducts), with weaning usually occurring by 14–15 months of age (12–19 kg for a bare-nosed wombat), and sometimes as early as 12 months (~11 kg) (Austin 1997; Booth

1999). Southern hairy-nosed wombats are weaned at 360–400 days of age (Taggart *et al.* 2007). In the wild, the smallest southern hairy-nosed wombat caught by itself above-ground was ~4 kg in bodyweight, with young usually remaining in the burrow after leaving the pouch until they reach this size (D. Taggart and S. Kleemann unpubl. data).

Fresh clover, leaves from cabbages, or other members of the cabbage family should not be fed as these can cause gut problems and are known to kill young wombats (George *et al.* 1995). The number of bottle feeds can be reduced from four to three over several weeks, but keeping the same volume of milk, and providing small pieces of solid food. After several more weeks when the wombat is at two feeds per day, the amount of formula can be decreased, without watering it down, until the animal is fully weaned after several weeks. A general rule is to decrease the formula by 5% per week as long as the joey continues to gain weight at a minimum of 5–10% of bodyweight per day. Ensure the wombat it is drinking ad lib water and eating solids before fully weaning (Austin 1997). Weaning should be completed by 17–18 months of age. Often wombats wean themselves and refuse to drink any formula, but this should not occur at less than 12 months of age (Austin 1997). At weaning, fresh water should be supplied at all times.

## 11.11 Release procedures

Preparation for release should begin once the wombat begins to leave the pouch (George *et al.* 1995). It should gradually be weaned from the carer into an enclosure with adequate soil depth that allows it to burrow and it should be fed an increasing amount of grass. By 18 months of age, the young wombat is usually driven away by the mother. Ideally the wombat should be soft released so that it can come and go from the carer's home to the wild and then disperse permanently when it is ready. Wombats released as pairs appear to do better than those released on their own, though this bond quickly breaks down after release (George *et al.* 1995).

An understanding of hand-rearing wombats can have a major impact on conservation. For example, in October 2009 caretakers at Epping Forest National Park found a juvenile female northern hairy-nosed wombat that was in extremely poor condition, weighing only 4.2 kg (should have been ~7.5 kg) and with pneumonia. The joey was taken into care for the next 9 months and nursed back to health. She responded well to treatment and hand-rearing, having exceptional bodyweight gains and development despite being initially severely compromised. In July 2010, she weighed ~28 kg and was taken back to the park to be soft-released into the translocated population and was fitted with a tracking collar to monitor its release (Fig. 8.24; T. Janssen pers. obs.). Such a success story provides strong support evidence for the establishment of a northern hairy-nosed captive population and husbandry.

**Fig. 8.24.** 'Hobo' the northern hairy-nosed wombat immediately following transport to the St George second wild population established to reduce the risk of stochastic extinction. Photo by Tina Janssen.

## ACKNOWLEDGEMENTS

Many thanks to Jocelyn Hockley for providing comments on artificial rearing.

## REFERENCES

ABC News (2017) Winnie the wombat: Australia's oldest wombat celebrates 31 years. Posted 16 December 2017. http://www.abc.net.au/news/2017-12-16/winnie-the-wombat-celebrates-31-years/9224232 (accessed 16 December 2020).

Anon (1982) Two zoos breed hairy-nosed wombats. *Thylacinus* 7(2), 44.

Austin MA (1997) *A Practical Guide to the Successful Hand Rearing of Tasmanian Marsupials*. Regal Publications, Melbourne.

Australian Animals Care and Education (2022) Extinction is forever. http://www.aace.org.au (accessed 16 January 2022).

Bach C (1998) *Birth Date Determination in Australasian Marsupials*. Australasian Regional Association of Zoological Parks and Aquaria, Sydney.

Baker AM, Gynther IC (eds) (2023) *Strahan's Mammals of Australia*. 4th edn. Reed New Holland Publishers, Sydney.

Banks SC, Skerratt LF, Taylor AC (2002) Female dispersal relatedness structure in common wombats (*Vombatus ursinus*).

*Journal of Zoology* **256**, 389–399. doi:10.1017/S0952836902000432

Barboza PS, Vanselow BA (1990) Copper toxicity in captive wombats (Marsupialia: Vombatidae). In *Proceedings of the American Association of Zoo Veterinarians*, pp. 204–206.

Barker IK, Munday BL, Presidente PJA (1979) Coccidia of wombats: correction of host parasite relationships. *Eimeria wombati* (Gilruth and Bull, 1912) com. Nov. and *Eimeria ursini* Supperer, 1957 from the hairy-nosed wombat and *Eimeria arundeli* sp. n. from the common wombat. *The Journal of Parasitology* **65**, 451–456. doi:10.2307/3280295

Bellamy T (1992) Marsupial handrearing. In *Urban Wildlife*. Proceedings 204. Post Graduate Committee in Veterinary Science. (Eds L Vogelnest, D Spielman, T Bellamy, M Messer and DM Walker) pp. 49–67. University of Sydney, Sydney.

Bettiol SS, Obendorf DL, Nowarkowski M, Goldsmid JM (2000) Pathology of experimental toxoplasmosis in eastern barred bandicoots in Tasmania. *Journal of Wildlife Diseases* **36**, 141–144. doi:10.7589/0090-3558-36.1.141

Beveridge I (1978) A taxonomic revision of the genera *Phascolostrongylus* Caravan, and *Oesophagostromoides* Schwartz (Nematoda: Trichonematidae) from wombats. *Australian Journal of Zoology* **26**, 585–602. doi:10.1071/ZO9780585

Beveridge I (1980) *Programotaenia* Nybelin (Cestoda: Anoplocephalidae): new species, redescriptions and new host records. *Transactions of the Royal Society of South Australia* **104**, 67–79.

Beveridge I, Mawson PM (1978) A taxonomic revision of the genera *Macropostrongyloides* Tamaguti and *Paramacropostrongylus* Johnston & Mawson (Nematoda: Trichonematidae) from Australian Marsupials. *Australian Journal of Zoology* **26**, 763–787. doi:10.1071/ZO9780763

Böer M (1998) Observations on reproduction in the common wombat *Vombatus ursinus* in captivity. In *Wombats*. (Eds RT Wells and PA Pridmore) pp. 129–146. Surrey Beatty & Sons, Sydney.

Booth R (1994) Medicine and husbandry: Monotremes, wombats and bandicoots. In *Wildlife*. Proceedings 233. Post Graduate Committee in Veterinary Science. (Ed. DI Dryden) pp. 395–420. University of Sydney. Sydney.

Booth R (1999) Wombats: care and treatment of sick, injured and orphaned animals. In *Wildlife in Australia: Healthcare & Management*. Proceedings 327. Post Graduate Foundation in Veterinary Science. (Ed. DI Dryden) pp. 1–11. University of Sydney, Sydney.

Brooks DE, Gaughwin MD, Mann T (1978) Structural and biochemical characteristics of the male accessory organs of reproduction in the hairy-nosed wombat (*Lasiorhinus latifrons*). *Proceedings of the Royal Society of London. Series B, Biological Sciences* **201**, 191–207. doi:10.1098/rspb.1978.0040

Brown GD (1964) Thermoregulation in the common wombat (*Vombatus ursinus*) in an alpine environment. In *Thermal Physiology*. (Ed. JRS Hales) pp. 331–334. Raven Press, New York.

Bryant B (2000) Captivity, stress and reproductive failure in the common wombat (*Vombatus ursinus*) by serial measurements of faecal progesterone metabolites. MSc thesis. University of Sydney, Sydney.

Bryant B, Reiss A (2008) Wombats. In *Medicine of Australian Mammals*. (Eds L Vogelnest and R Woods) pp. 329–358. CSIRO Publishing, Melbourne.

Camp A, Croxford AE, Ford CS, Baumann U, Clements PR, Hiendleder S, Woolford L, Netzel G, Boardman WSJ, Fletcher MT, Wilkinson MJ (2020) Dual-locus DNA metabarcoding reveals southern hairy-nosed wombats (*Lasiorhinus latifrons* Owen) have a summer diet dominated by toxic invasive plants. *PLoS One* **15**(3), e0229390. doi:10.1371/journal.pone.0229390

Campbell-Ward M (2025) Wombats. In *Current Therapy in Medicine of Australian Mammals*. Revised Edition. (Eds L Vogelnest and T Portas) pp. 551–561. CSIRO Publishing, Melbourne.

Casey FF, Stannard HJ, Old JM (2021) A review of wombat diet nutrition. *Australian Mammalogy* **43**, 1–9. doi:10.1071/AM20009

Christian ME (1977) Rearing of marsupials. *Control and Therapy No. 448*. The Post Graduate Committee in Veterinary Science. University of Sydney, Sydney.

Collins LR (1973) *Monotremes and marsupials: A Reference for Zoological Institutions*. Smithsonian Institution Press, Washington DC.

Condor P (1970) Breeding the common wombat in captivity. *Victorian Naturalist* **87**, 322.

Coulson GM, Croft DB (1981) Flehmen in kangaroos. *Australian Mammalogy* **4**, 139–140. doi:10.1071/AM81007

Crandall LS (1964) *The Management of Wild Mammals in Captivity*. University of Chicago Press, Chicago.

Crossman DG, Johnson CN, Horsup AB (1994) Trends in the population of the northern hairy-nosed wombat, *Lasiorhinus krefftii* in Epping Forest National Park, central Queensland. *Pacific Conservation Biology* **1**, 141–149. doi:10.1071/PC940141

Crowcroft P, Soderlund R (1977) Breeding of wombats (*Lasiorhinus latifrons*) in captivity. *Der Zoologische Garten* **47**, 313–322.

Death C, Taggart DA, Williams DB, Milne R, Schultz DJ, Holyoake C, Warren KS (2011) Pharmacokinetics of moxidectin in the southern hairy-nosed wombat (*Lasiorhinus latifrons*). *Journal of Wildlife Diseases* **47**, 643–649. doi:10.7589/0090-3558-47.3.643

Descovich KA, Lisle AT, Johnston S, Nicolson V, Phillips CJC (2012a) Differential responses of captive southern hairy-nosed wombats (*Lasiorhinus latifrons*) to the presence of faeces from different species and male and female conspecifics. *Applied Animal Behaviour Science* **138**, 110–117. doi:10.1016/j.applanim.2012.01.017

Descovich KA, Lisle AT, Johnston S, Phillips CJC (2012b) Space allowance and the behaviour of captive southern hairy-nosed wombats (*Lasiorhinus latifrons*). *Applied Animal Behaviour Science* **140**, 92–98. doi:10.1016/j.applanim.2012.05.009

Descovich KA, Lisle AT, Johnston S, Keeley T, Phillips CJC (2012c) Intrasample variation and the effect of storage delay on faecal metabolite concentrations in the southern hairy-nosed wombat (*Lasiorhinus latifrons*). *Australian Mammalogy* **34**, 217–222. doi:10.1071/AM11047

Doube LJ (1981) Diseases in wombats. In *Wildlife Diseases of the Pacific Basin and Other Countries*. (Ed. ME Fowler) pp. 63–75. Proceedings of the 4th International Conference of the Wildlife Disease Association, Sydney.

Du Z, Keeley T, Janssen T, Nicolson V, Johnston SD (2017) Measurement of testosterone and cortisol metabolites and luteinising hormone in captive southern hairy-nosed wombat (*Lasiorhinus latifrons*) urine. *General and Comparative Endocrinology* **250**, 70–79. doi:10.1016/j.ygcen.2017.06.001

Du Z, Johnston SD, Janssen T, Phillips CJC, Lisle A, Keeley T (2018) Behavioral and endocrine responses to season and social dynamics of captive male southern hairy-nosed wombats

(*Lasiorhinus latifrons*). *General and Comparative Endocrinology* **261**, 40–50. doi:10.1016/j.ygcen.2018.01.021

Durfee PT, Presidente PJA (1979) A seriological survey of Australian wildlife for antibodies to leptospires of the *Hebdomadis* serogroup. *The Australian Journal of Experimental Biology and Medical Science* **57**, 177–189. doi:10.1038/icb.1979.19

Eisenhofer R, Helgen KM, Taggart D (2021*a*) Signatures of landscape and captivity in the gut microbiota of southern hairy-nosed wombats (*Lasiorhinus latifrons*). *Animal Microbiome* **3**, 4. doi:10.1186/s42523-020-00068-y

Eisenhofer R, D'Agnese E, Taggart DA, Carver S, Penrose B (2021*b*) Microbial biogeography of the wombat gastrointestinal tract. *PeerJ* **10**, e12982. doi:10.7717/peerj.12982

Fagan DA, Ullrey DE (2008) Dental malocclusion in a coarse-haired wombat (*Vombatus ursinus*). *Journal of Veterinary Dentistry* **25**, 240–244. doi:10.1177/089875640802500404

Fanson KV (2025) Stress physiology of Australian mammals. In *Current Therapy in Medicine of Australian Mammals*. Revised Edition. (Eds L Vogelnest and T Portas) pp. 112–121. CSIRO Publishing, Melbourne.

Fanson KV, Best EC, Bunce A, Fanson BG, Hogan LA, Keeley T, Narayan EJ, Palme R, Parrott ML, Sharp TM, Skogvold K, Tuthill L, Webster KN, Bashaw M (2017) One size does not fit all: monitoring faecal glucocorticoid metabolites in marsupials. *General and Comparative Endocrinology* **244**, 146–156. doi:10.1016/j.ygcen.2015.10.011

Finlayson GR, Shimmin GA, Temple-Smith PD, Handasyde K, Taggart DA (2003) Monitoring the activity of a southern hairy-nosed wombat, *Lasiorhinus latifrons*, using temperature dataloggers. *Australian Mammalogy* **25**, 205–208. doi:10.1071/AM03205

Finlayson GR, Shimmin GA, Temple-Smith PD, Handasyde KA, Taggart DA (2005) Burrow use and ranging behaviour of the southern hairy-nosed wombat (*Lasiorhinus latifrons*) in the Murraylands, South Australia. *Journal of Zoology* **265**, 189–200 [London]. doi:10.1017/S095283690400620X

Finlayson GR, Shimmin GA, Taggart DA, Skinner JF, Gilmore A, Paris MCJ (2006) Oestrous cycle of captive southern hairy-nosed wombats (*Lasiorhinus latifrons*) in South Australia, Australia. *Animal Reproduction Science* **95**, 295–306. doi:10.1016/j.anireprosci.2005.10.001

Finlayson GR, Taggart DA, Shimmin GA, White CR, Steele VR, Paris MCJ, Temple-Smith PD (2007a) The application of intra-species fostering techniques to the Southern hairy-nosed wombat, *Lasiorhinus latifrons*. *Pacific Conservation Biology* **13**, 259–263. doi:10.1071/PC070259

Finlayson GR, Taggart DA, Shimmin G, Dibben R, White C, Steele V, Paris M, Temple-Smith PD (2007b) Pouch young removal and return to oestrus in wild southern hairy-nosed wombats (*Lasiorhinus latifrons*). *Animal Reproduction Science* **100**, 216–222. doi:10.1016/j.anireprosci.2006.09.013

Finlayson GR, White CR, Dibben R, Shimmin GA, Taggart DA (2010) Activity patterns of the southern hairy-nosed wombat (*Lasiorhinus latifrons*) (Marsupialia: Vombatidae) in the South Australian Murraylands. *Australian Mammalogy* **32**, 39–46. doi:10.1071/AM09024

Fleay R (1957) Growing up with wombats. *Animal Kingdom* **60**, 107–110.

Flower SS (1929) List of vertebrated animals exhibited in the gardens of the Zoological Society of London, 1828–1927. In. 'Mammals'. *Proceedings of the Zoological Society of London* **1929**, 1–419.

Flower SS (1931) Contributions to our knowledge of the duration of life in vertebrate animals. *Proceedings of the Zoological Society of London* **101**, 145–234. doi:10.1111/j.1469-7998.1931.tb06192.x

Gaughwin MD (1979) The occurrence of flehmen in a marsupial: the hairy nosed wombat (*Lasiorhinus latifrons*). *Animal Behaviour* **27**, 1063–1065. doi:10.1016/0003-3472(79)90054-X

Gaughwin MD (1982) Southern hairy-nosed wombat *Lasiorhinus latifrons*: its maintenance, behaviour and reproduction in captivity. In *The Management of Australian Mammals in Captivity*. (Ed. DD Evans) pp. 144-155. Zoological Board of Victoria, Melbourne.

Gaughwin MD, Wells RT (1978) General features of the hairy-nosed wombat (*Lasiorhinus latifrons*) in the Blanche Town region of South Australia. *Australian Mammal Society Bulletin* **5**, 46–47.

Gaughwin MD, Breed WG, Wells RT (1998) Seasonal reproduction in a population of southern hairy-nosed wombats *Lasiorhinus latifrons* in the Blanchtown region of South Australia. In *Wombats*. (Eds RT Wells and PA Pridmore) pp. 109–112. Surrey Beatty & Sons, Sydney.

George H, Parker G, Coote P (1995) Common wombats: rescue rehabilitation release. Unpublished manuscript.

Gerhardt K, Smales LR, McKillup SC (2000) Parasites of the northern hairy-nosed wombat *Lasiorhinus krefftii*: implications for conservation. *Australian Mammalogy* **22**, 17–22. doi:10.1071/AM00017

Gray JE (1863) Notice of three wombats in the Zoological Gardens. *Annals & Magazine of Natural History* **11**(66), 457–459. doi:10.1080/00222936308681458

Green B (1984) Composition of milk and energetics of growth in marsupials. *Symposia of the Zoological Society of London* **51**, 369–387.

Green RH, Munday BL (1971) Parasites of Tasmanian native and feral fauna. Part 1: arthropoda. *Records of the Queen Victoria Museum* **41**, 1–16.

Green RH, Rainbird JL (1987) The common wombat (*Vombatus ursinus*) (Shaw, 1800) in northern Tasmania. Part 1: breeding, growth and development. *Records of the Queen Victoria Museum* **91**, 1–19.

Hamilton RA, Stanton PG, O'Donnell L, Steel VR, Taggart DA, Temple-Smith PD (2000) Determination of seasonality in southern hairy-nosed wombats (*Lasiorhinus latifrons*) by analysis of faecal androgens. *Biology of Reproduction* **63**, 526–531. doi:10.1095/biolreprod63.2.526

Hogan LA (2004) The behaviour of captive common wombats (*Vombatus ursinus*). BSc (Honours) thesis, University of Queensland, Gatton.

Hogan LA (2010) The behaviour and reproductive biology of captive southern hairy-nosed wombats (*Lasiorhinus latifrons*). PhD thesis. The University of Queensland, Gatton.

Hogan LA, Tribe A (2007) Prevalence and cause of stereotypical behaviour in common wombats (*Vombatus ursinus*) residing in Australian zoos. *Applied Animal Behaviour Science* **105**, 180–191. doi:10.1016/j.applanim.2006.06.006

Hogan LA, Phillips C, Lisle A, Horsup A, Janssen T, Johnston S (2009) Remote monitoring of the behaviour and activity of captive southern hairy-nosed wombats (*Lasiorhinus latifrons*). *Australian Mammalogy* **31**, 123–135. doi:10.1071/AM09023

Hogan LA, Johnston SD, Lisle A, Horsup AB, Janssen T, Phillips CJC (2010a) Stereotypies and environmental enrichment in captive southern hairy-nosed wombats, *Lasiorhinus latifrons*. *Applied Animal Behaviour Science* **126**, 85–95. doi:10.1016/j.applanim.2010.05.009

Hogan LA, Phillips CJC, Lisle A, Keeley T, Horsup AB, Janssen T, Johnston SD (2010b) Non-invasive methods of oestrus detection in captive southern hairy-nosed wombats (*Lasiorhinus latifrons*). *Animal Reproduction Science* **119**, 293–304. doi:10.1016/j.anireprosci.2010.01.006

Hogan LA, Phillips C, Lisle A, Horsup A, Janssen T, Johnston S (2010c) Reproductive behaviour of the southern-hairy nosed wombat (*Lasiorhinus latifrons*). *Australian Journal of Zoology* **58**, 350–361. doi:10.1071/ZO10068

Hogan LA, Phillips CJC, Horsup AB, Keeley T, Nicolson V, Janssen T, Lisle A, Johnston SD (2010d) Monitoring male southern hairy-nosed wombat (*Lasiorhinus latifrons*) reproductive function and seasonality in a captive population. *Animal Reproduction Science* **118**, 377–387. doi:10.1016/j.anireprosci.2009.10.001

Hogan LA, Johnston SD, Lisle AT, Keeley T, Wong P, Nicolson V, Horsup AB, Janssen T, Phillips CJC (2011a) Behavioural and physiological responses of captive wombats (*Lasiorhinus latifrons*) to regular handling by humans. *Applied Animal Behaviour Science* **134**, 217–228. doi:10.1016/j.applanim.2011.07.010

Hogan LA, Phillips CJC, Horsup AB, Janssen T, Johnston SD (2011b) Technique for faecal marking in group-housed southern hairy-nosed wombats (*Lasiorhinus latifrons*) (Marsupialia: Vombatidae). *Australian Zoologist* **35**, 649–654. doi:10.7882/AZ.2011.017

Hogan LA, Janssen T, Johnston SD (2013) Wombat reproduction (Marsupialia; Vombatidae): an update and future directions for the development of artificial breeding technology. *Reproduction (Cambridge, England)* **145**, R157–R173. doi:10.1530/REP-13-0012

Home E (1808) An account of some peculiarities in the anatomical structure of the wombat, with observations on the female organs of generation. *Philosophical Transactions of the Royal Society of London* **98**, 304–312. doi:10.1098/rstl.1808.0020

Horsup AB (1998) A trapping survey of the northern hairy-nosed wombat *Lasiorhinus krefftii*. In *Wombats*. (Eds RT Wells and PA Pridmore) pp. 147–155. Surrey Beatty & Sons, Sydney.

Horsup AB (2014) Northern hairy-nosed wombat. Queensland Government. The State of Queensland. Department of Environment and Heritage Protection. www.ehp.qld.gov.au/wildlife/threatened-species/endangered/northern hairynosed _wombat/#how_many_species_of_wombats (accessed 31 August 2014).

Horsup AB, Austin JJ, Fewster RM, Hansen BD, Harper DE, Molyneux JA, White LC, Taylor AC (2021) Demographic trends and reproductive patterns in the northern hairy-nosed wombat (*Lasiorhinus krefftii*) at Epping Forest National Park (Scientific), central Queensland. *Australian Mammalogy* **43**, 72–84. doi:10.1071/AM20030

Hoyle SD, Horsup AB, Johnson CN, Crossman DG, McCallum H (1995) Live-trapping of the northern hairy-nosed wombat (*Lasiorhinus krefftii*): population-size estimates and effects on individuals. *Wildlife Research* **22**, 741–755. doi:10.1071/WR9950741

Hum S, Best FG (1988) Tyzzer's disease in a wombat. *Australian Veterinary Journal* **65**, 89–90. doi:10.1111/j.1751-0813.1988.tb07370.x

Hum S, Barton NJ, Obendorf D, Barker IK (1991) Coccidiosis in common wombats (*Vombatus ursinus*). *Journal of Wildlife Diseases* **27**, 697–700. doi:10.7589/0090-3558-27.4.697

International Air Transport Association (IATA) (2022) *Live Animals Regulations*. 48th edn. International Air Transport Association, Geneva, Switzerland.

Jackson SM, Groves CP (2015) *Taxonomy of Australian Mammals*. CSIRO Publishing, Melbourne.

Jackson SM, Jansen JJFJ, Baglione G, Callou C (2021) Mammals collected and illustrated by the Baudin Expedition to Australia and Timor (1800–1804): a review of the current taxonomy of specimens in the Muséum national d'Histoire naturelle de Paris and the illustrations in the Muséum d'Histoire naturelle du Havre. *Zoosystema* **43**(21), 387–548. doi:10.5252/zoosystema2021v43a21

Johnson CN (1991) Utilisation of habitat by the northern hairy-nosed wombat *Lasiorhinus krefftii*. *Journal of Zoology* **225**, 495–507[London]. doi:10.1111/j.1469-7998.1991.tb03831.x

Johnson CN (1998) The evolutionary ecology of wombats. In *Wombats*. (Eds RT Wells and PA Pridmore) pp. 34–41. Surrey Beatty & Sons, Sydney.

Johnson CN, Crossman D (1991) Dispersal and social organization of the northern hairy-nosed wombat, *Lasiorhinus krefftii*. *Journal of Zoology* **225**, 605–613. doi:10.1111/j.1469-7998.1991.tb04328.x

Johnson CN, Gordon G (1995) Northern hairy-nosed wombat *Lasiorhinus krefftii*. In *The Mammals of Australia*. 2nd edn. (Ed. R Strahan) pp. 200–201. Reed Books, Sydney.

Johnston SD (2019) Challenges associated with the development and transfer of assisted breeding technology in marsupials and monotremes: lessons from the koala, wombat and short-beaked echidna. *Reproduction, Fertility and Development* **31**, 1305–1314. doi:10.1071/RD18113

Johnston SD, Holt WV (2019) Using the koala (*Phascolarctos cinereus*) as a case study to illustrate the development of artificial breeding technology in marsupials: An update. In *Reproductive Sciences in Animal Conservation*. (Eds P Comizzoli, JJL Brown and WV Holt) pp. 327–362. Springer Nature, Cham, Switzerland.

Johnston SD, Janssen T, Brooks P, Hogan L, Swinbourne A, Du Z, Smith M, Nicolson V, Phillips C, McCallum C, Descovich K, Mucci A, Li-Chen H, Shruti D, Keeley T (2018) The development of assisted breeding technology in wombats [Abstract and Oral Presentation]. Wombats Through Time and Space Conference. University of South Australia, 17–19 September 2018, Adelaide.

Kleemann SL, Taggart DA (2024a) Effects of rainfall and drought on body condition and abundance of the southern hairy-nosed wombat (*Lasiorhinus latifrons*). *Journal of Wildlife Management*. In press.

Kleemann SL, Taggart DA (2024b) Age cohort definition and evidence of sexual dimorphism in a large Australian marsupial (*Lasiorhinus latifrons*). *Journal of Mammalogy*. In press.

Kleemann SL, Taggart PL, Hoffman J, Dibben R, Taggart DA (2023) Seasonal changes in body condition and diet of a large burrowing, semi-arid adapted, marsupial herbivore (*Lasiorhinus latifrons*) in degraded and intact landscapes. *The Journal of Wildlife Management*.

Lacy RC, Balou JD, Pollak JP (2011) PMx: software package for demographic and genetic analysis and management of pedigreed populations. *Methods in Ecology and Evolution* **3**, 433–437. doi:10.1111/j.2041-210X.2011.00148.x

Lees C, Johnson K (2002) *Australasian Species Management Program: Regional Census and Plan*. 12th edn. Australasian Regional Association of Zoological Parks and Aquaria, Sydney.

Löffler E, Margules C (1980) Wombats detected from space. *Remote Sensing of Environment* **9**, 47–56. doi:10.1016/0034-4257(80)90046-2

MacCallum C (2004) The reproductive biology of the common wombat, *Vombatus ursinus*: studies towards the development of an artificial insemination protocol. Masters of Philosophy thesis. The University of Queensland, Brisbane.

Mallett KJ, Cooke BD (1986) *The Ecology of the Common Wombat in South Australia*. Nature Conservation Society of South Australia, Adelaide.

Marks CA (1998) Courtship and mating in a pair of free-ranging common wombats *Vombatus ursinus*. In *Wombats*. (Eds RT Wells and PA Pridmore) pp. 125–128. Surrey Beatty & Sons, Sydney.

Marshall VM, Taggart DA, Ostendorf B (2018) Scale-dependent habitat analysis and implications for climate change risk for the southern hairy-nosed wombat. *Australian Mammalogy* **40**, 162–172. doi:10.1071/AM16060

Mawson PM (1955) Some parasites of Australian vertebrates. *Transactions of the Royal Society of South Australia* **78**, 1–7.

McIlroy JC (1973) Aspects on the ecology of the common wombat, *Vombatus ursinus* (Shaw, 1800). PhD thesis. Australian National University, Canberra.

McIlroy JC (1976) Aspects of the ecology of the common wombat, *Vombatus ursinus*. I: capture, handling, marking and radio-tracking techniques. *Australian Wildlife Research* **3**, 105–116. doi:10.1071/WR9760105

McIlroy JC (1977) Aspects of the ecology of the common wombat, *Vombatus ursinus*. II: methods for estimating population numbers. *Australian Wildlife Research* **4**, 223–228. doi:10.1071/WR9770223

McIlroy JC (1995) Common wombat *Vombatus ursinus*. In *The Mammals of Australia*. 2nd edn (Ed. R Strahan) pp. 204–205. Reed Books, Sydney.

McIlroy JC (2008) Common wombat. In *The Mammals of Australia*. 3rd edn. (Eds S Van Dyck and R Strahan) pp. 206–208. Reed Books, Sydney.

McLelland DJ, Fielder K, Males G, Langley N, Schultz D (2015) Successful transfer of a Goodfellow's tree kangaroo (*Dendrolagus goodfellowi*) pouch young to a yellow-footed rock wallaby (*Petrogale xanthopus*) surrogate. *Zoo Biology* **34**, 460–462. doi:10.1002/zoo.21236

Miller DS, Mitchell GF, Biggs B, McCracken H, Myroniuk P, Hewish M (2000) Detection of agglutinating antibodies to *Toxoplasma gondii* in sera from captive mainland Australian eastern barred bandicoot (*Perameles gunnii*). *Journal of Wildlife Diseases* **36**, 213–218. doi:10.7589/0090-3558-36.2.213

Mitchell PC (1911) On longevity and relative viability in mammals and birds: with a note on the theory of longevity. *Proceedings of the Zoological Society of London* **81**, 425–548. doi:10.1111/j.1096-3642.1911.tb01942.x

Mohr E (1942) Einiges uber Wombat-Formen und Marsupialia-Beutel. *Der Zoologische Garten* **14**(1/2), 55–68.

Moonlit Sanctuary (2023) Walk a wombat. https://moonlitsanctuary.com.au/take-a-wombat-for-a-walk/ (accessed 12 March 2023).

Muhic J, Abbott E, Ward MJ (2012) The warru (*Petrogale lateralis* MacDonnell Ranges Race) reintroduction project on the Anangu Pitjantjatjara Yankunytjatjara Lands, South Australia. *Ecological Management & Restoration* **13**, 89–92. doi:10.1111/j.1442-8903.2011.00620.x

Munday BL (1978) Marsupial disease. In *Fauna*. Part B. (Ed. The Post Graduate Committee in Veterinary Science). pp. 335–385. University of Sydney, Sydney.

Munday BL, Corbould A (1973) *Leptospira pomona* infection in wombats. *Journal of Wildlife Diseases* **9**, 72–73. doi:10.7589/0090-3558-9.1.72

Munday BL, Gregory GC (1974) Demonstration of larval forms of *Baylisascaris tasmaniensis* in the wombat (*Vombatus ursinus*). *Journal of Wildlife Diseases* **10**, 241–242. doi:10.7589/0090-3558-10.3.241

Nicholson PJ (1963) Wombats. *Timbertop Magazine* **8**, 32–38.

O'Brien C, Sparrow E, Dibben R, Ostendorf B, Taggart DA (2021) Evaluation of olfactory and visual cues to deter southern hairy-nosed wombats (*Lasiorhinus latifrons*) from their burrows. *Australian Mammalogy* **43**, 110–119. doi:10.1071/AM20007

Osborn DJ (1975) Aussie, Brookfield Zoo's baby wombat. *Brookfield Bison* **10**, 4.

Owen R (1845) On the existence of two species of wombat (*Phascolomys*). *Proceedings of the Zoological Society of London* **13**, 82–83.

Owen R (1872) On the fossil mammals of Australia. Part VI: genus *Phascolomys* Geoffroy. *Philosophical Transactions of the Royal Society of London* **162**, 173–196. doi:10.1098/rstl.1872.0009

Paris MC, White A, Reiss A, West M, Schwarzenberger F (2002) Faecal progesterone metabolites and behavioural observations for the non-invasive assessment of oestrus cycles in the common wombat and Southern hairy-nosed wombat. *Animal Reproduction Science* **72**, 245–257. doi:10.1016/S0378-4320(02)00088-X

Peters DG, Rose RW (1979) The oestrous cycle and basal body temperature in the common wombat *Vombatus ursinus*. *Journal of Reproduction and Fertility* **57**, 453–460. doi:10.1530/jrf.0.0570453

Presidente PJA (1982) Common wombat *Vombatus ursinus*: maintenance in captivity, blood values, infectious and parasitic diseases. In *The Management of Australian Mammals in Captivity*. (Ed. DD Evans) pp. 133–143. The Zoological Board of Victoria, Melbourne.

Queensland Government (2021) Northern hairy-nosed wombat: monitoring and management. https://www.qld.gov.au/environment/plants-animals/conservation/threatened-wildlife/threatened-species/featured-projects/northern-hairy-nosed-wombat/monitoring-management (accessed 27 January 2022).

Queensland Government (2022) About northern hairy-nosed wombats. https://www.qld.gov.au/environment/plants-animals/conservation/threatened-wildlife/threatened-species/featured-projects/northern-hairy-nosed-wombat/about-northern-hairy-nosed-wombats (accessed 27 January 2022).

Ride WDL (1970) *A Guide to the Native Mammals of Australia*. Oxford University Press, London.

Rishworth C, McIlroy JC, Tanton MT (1995) Diet of the common wombat, *Vombatus ursinus*, in plantations of *Pinus radiata*. *Wildlife Research* **22**, 333–339. doi:10.1071/WR9950333

Roberts FHS (1964) The tick fauna of Tasmania. *Records of the Queen Victoria Museum* **17**, 1–8.

Roberts FHS (1970) *Australian Ticks*. CSIRO, Melbourne.

Rose K (1999) Common diseases of urban wildlife. In *Wildlife in Australia: Healthcare & Management*. Proceedings 327. Post Graduate Foundation in Veterinary Science. (Ed. DI Dryden) pp. 365–427. University of Sydney, Sydney.

Ruykys L, Taggart DA, Breed B, Schultz DJ (2009) Sarcoptic mange in southern hairy-nosed wombats (*Lasiorhinus latifrons*): distribution and prevalence in the Murraylands of South Australia. *Australian Journal of Zoology* **57**, 129–138. doi:10.1071/ZO09010

Ruykys L, Breed B, Schultz DJ, Taggart DA (2013) Effects and treatment of sarcoptic mange in southern hairy-nosed wombats (*Lasiorhinus latifrons*). *Journal of Wildlife Diseases* **49**, 312–320. doi:10.7589/2012-10-256

Schmidt M (1880) On the duration of life of the animals in the Zoological Gardens at Frankfurt-on-the-Main. *Proceedings of the Zoological Society* **48**, 299–319. doi:10.1111/j.1469-7998.1880.tb06562.x

Shaw G (1800) *General Zoology or Systematic Natural History*. G. Kearsley, London.

Shimmin G, Skinner J, Baudinette R (2001) Life in a wombat burrow. *Nature Australia* **27**, 62–69.

Shimmin G, Skinner J, Baudinette R (2002) The warren architecture and environment of the southern hairy-nosed wombat (*Lasiorhinus latifrons*). *Journal of Zoology* **258**, 469–477 [London]. doi:10.1017/S0952836902001620

Skerratt LF (1995) *Strongyloides spearei* n. sp. (Nematoda: Strongyloididae) from the common wombat *Vombatus ursinus* (Marsupialia: Vombatidae). *Systematic Parasitology* **32**, 81–89. doi:10.1007/BF00009506

Skerratt LF (1998) Diseases and parasites of the common wombat *Vombatus ursinus* in the Healesville area of Victoria. In *Wombats*. (Eds RT Wells and PA Pridmore) pp. 317–328. Surrey Beatty & Sons, Sydney.

Skerratt LF (2001) Sarcoptic mange in the common wombat, *Vombatus ursinus* (Shaw, 1800). PhD thesis. University of Melbourne, Melbourne.

Skerratt LF, Phelan J, McFarlane R, Speare R (1997) Serodiagnosis of toxoplasmosis in a common wombat. *Journal of Wildlife Diseases* **33**, 346–351. doi:10.7589/0090-3558-33.2.346

Skerratt LF, Martin RW, Handasyde KA (1998) Sarcoptic mange in wombats. *Australian Veterinary Journal* **76**, 408–410. doi:10.1111/j.1751-0813.1998.tb12389.x

Skerratt LF, Middleton D, Beveridge I (1999) Distribution of life cycle stages of *Sarcoptes scabiei* var *wombati* and effects of severe mange on common wombats in Victoria. *Journal of Wildlife Diseases* **35**, 633–646. doi:10.7589/0090-3558-35.4.633

Smales LR (1987) Parasites of the wombat *Vombatus ursinus* from the Gippsland region, Victoria. *Transactions of the Royal Society of South Australia* **111**, 129–130.

Smales LR (1994) A new species of Oesophagostomoides (Nematoda. Cloacinidae) from the northern hairy-nosed wombat *Lasiorhinus krefftii* with a key to species of the genus. *The Journal of Parasitology* **80**, 638–643. doi:10.2307/3283203

Smales LR (1998) Helminth parasites of wombats. In *Wombats*. (Eds RT Wells and PA Pridmore) pp. 312–316. Surrey Beatty & Sons, Sydney.

Species360 (2023) Zoological Information Management System (ZIMS). Minneapolis. www.species360.org.

Spielman D (1994) First aid emergency care for Australian native mammals. In *Wildlife*. Proceedings 233. Post Graduate Committee in Veterinary Science. (Ed. DI Dryden) pp. 339–366. University of Sydney, Sydney.

Spratt DM (1979) A taxonomic revision of the lungworms (Nematoda: Metastrongyloidea) from Australian marsupials. *Australian Journal of Zoology* **67**(Supplement), 1–45.

Spratt DM, Presidente PJA (1981) Prevalence of Fasciola hepatica infection in native mammals in southeastern Australia. *The Australian Journal of Experimental Biology and Medical Science* **59**, 713–721. doi:10.1038/icb.1981.62

Spratt DM, Beveridge I, Walter EL (1991) A catalogue of Australasian monotremes and marsupials and their recorded helminth parasites. *Records of the South Australian Museum. Monograph Series* **1**, 10–105.

Steele VR, Temple-Smith PD (1998) Physical structure of warrens of a small colony of southern hairy-nosed wombats *Lasiorhinus latifrons*. In *Wombats*. (Eds RT Wells and PA Pridmore) pp. 113–124. Surrey Beatty & Sons, Sydney.

Swinbourne AM, Janssen T, Phillips CJC, Johnston SD (2015) Non-invasive urine collection in the female southern hairy-nosed wombat (*Lasiorhinus latifrons*) with the aid of classical conditioning. *Zoo Biology* **34**, 89–93. doi:10.1002/zoo.21171

Swinbourne AM, Johnston SD, Janssen T, Lisle A, Keeley T, Phillips CJC (2017a) Olfactory behavioural responses in captive male southern hairy-nosed wombats (*Lasiorhinus latifrons*) exposed to oestrous and non-oestrous urine samples. *Australian Journal of Zoology* **64**, 395–401. doi:10.1071/ZO16088

Swinbourne AM, Johnston SD, Janssen T, Lisle A, Keeley T (2017b) Validation of techniques to measure reproductive hormones in the urine of female southern hairy-nosed wombats (*Lasiorhinus latifrons*). *General and Comparative Endocrinology* **252**, 130–141. doi:10.1016/j.ygcen.2017.08.003

Swinbourne AM, Johnston SD, Janssen T, Lisle A, Keeley T (2018a) The reproductive biology of captive southern hairy-nosed wombats (*Lasiorhinus latifrons*). Part 1: oestrous cycle characterisation. *Reproduction, Fertility and Development* **30**, 1412–1423. doi:10.1071/RD17497

Swinbourne AM, Phillips CJC, Janssen T, Lisle A, Keeley T, Johnston SD (2018b) Reproductive biology of captive female southern hairy-nosed wombats (*Lasiorhinus latifrons*). Part 2: oestrous behaviour. *Reproduction, Fertility and Development* **30**, 1424–1433. doi:10.1071/RD17539

Swinbourne MJ, Taggart DA, Peacock D, Ostendorf B (2016) Historical changes in the distribution of hairy-nosed wombats (*Lasiorhinus* spp.): a review. *Australian Mammalogy* **39**, 1–16. doi:10.1071/AM15046

Swinbourne M, Taggart DA, Ostendorf B (2021) The population status of southern hairy-nosed wombat (*Lasiorhinus latifrons*). II: landscape factors effecting distribution and abundance. *Australian Mammalogy* **43**, 54–65. doi:10.1071/AM20017

Taggart DA, Temple-Smith P (2023) Southern hairy-nosed wombat *Lasiorhinus latifrons*. In *Strahan's Mammals of Australia*. 4th edn. (Eds AM Baker and IC Gynther) pp. 204–206. Reed New Holland Publishers, Sydney.

Taggart DA, Leigh CM, Steele VR, Breed WG, Temple-Smith PD, Phelan J (1996) The effect of cooling and cryopreservation on sperm motility and morphology of several species of marsupial. *Reproduction, Fertility and Development* **8**, 673–679. doi:10.1071/RD9960673

Taggart DA, Steele VR, Schultz D, Dibben R, Dibben J, Temple-Smith PD (1998a) Semen collection and cryopreservation in the southern hairy-nosed wombat *Lasiorhinus latifrons*: implications for conserving the northern hairy-nosed wombat *Lasiorhinus krefftii*. In *Wombats*. (Eds RT Wells and PA Pridmore) pp. 180–191. Surrey Beatty & Sons, Sydney.

Taggart DA, Breed WG, Temple-Smith PD, Purvis A, Shimmin G (1998b) Reproduction, mating strategies and sperm competition in marsupials and monotremes. In *Sperm Competition and Sexual Selection*. (Eds TR Birkhead and AP Moller) pp. 623–665. Academic Press, London.

Taggart DA, Shimmin GA, Ratcliff JR, Steele VR, Dibben R, Dibben J, White C, Temple-Smith PD (2005) Seasonal changes in the testis, accessory glands and ejaculate characteristics of the southern hairy-nosed wombat, *Lasiorhinus latifrons* (Marsupialia: Vombatidae). *Journal of Zoology* **266**, 95–104. doi:10.1017/S0952836905006722

Taggart DA, Finlayson GR, Shimmin G, Gover C, Dibben R, White C, Steele V, Temple-Smith PD (2007) Growth and development of the southern hairy-nosed wombat, *Lasiorhinus latifrons* (Vombatidae). *Australian Journal of Zoology* **55**, 309–316. doi:10.1071/ZO07056

Taggart DA, Schultz DJ, Fletcher TP, *et al.* (2010) Cross-fostering and short-term pouch young isolation in macropodoid marsupials: Implications for conservation and species management. In *Macropods: The Biology of Kangaroos, Wallabies and Rat Kangaroos*. (Eds G Coulson and M Eldridge) pp. 263–278. CSIRO Publishing, Melbourne.

Taggart DA, Finlayson GR, Sparrow E, Dibben R, Dibben J, Campbell L, Peacock D, Ostendorf B, White C, Temple-Smith PD (2020) Environmental factors influencing hairy-nosed wombat abundance in semi-arid rangelands. *The Journal of Wildlife Management* **84**, 921–929. doi:10.1002/jwmg.21858

Taylor AC, Fewster R (2017) *2016 Northern Hairy-nosed Wombat Hair Census Final Report*. Internal report to the Queensland Department of Environment and Science. Monash University, Melbourne.

Taylor RJ (1993) Observations on the behaviour and ecology of the common wombat *Vombatus ursinus* in northeast Tasmania. *Australian Mammalogy* **16**, 1–7. doi:10.1071/AM93001

Temby ID (1998) The law and wombats in Australia. In *Wombats*. (Eds RT Wells and PA Pridmore) pp. 305–311. Surrey Beatty & Sons, Sydney.

Temple-Smith PD, Taggart DA (1990) On the male generative organs of the koala (*Phascolarctos cinereus*): an update. In *The Biology of the Koala*. (Eds AK Lee and G Sanson) pp. 33–54. Surrey Beatty & Sons, Sydney.

Treby D (2005) *Husbandry Manual for Southern Hairy-nosed Wombat* Lasiorhinus latifrons *(Mammalia; Vombatidae)*. Western Plains Zoo, Dubbo.

Triggs B (1996) *The Wombat: Common Wombats in Australia*. University of New South Wales Press, Sydney.

Tyndale-Biscoe H, Renfree M (1987) *Reproductive Physiology of Marsupials*. Cambridge University Press, Cambridge.

Van Dyck S, Strahan R (Eds) (2008) *The Mammals of Australia*. 3rd edn. New Holland, Sydney.

Vogelnest L (1999) Chemical restraint of Australian fauna. In *Wildlife in Australia: Healthcare & Management*. (Ed. DI Dryden) Post Graduate Foundation in Veterinary Science. pp. 149–186. University of Sydney, Sydney.

Vogelnest L (2025) Capture, physical and chemical restraint. In *Current Therapy in Medicine of Australian Mammals*. Revised Edition. (Eds L Vogelnest and T Portas) pp. 122–151. CSIRO Publishing, Melbourne.

Vogelnest L, Portas T (Eds) (2025) *Current Therapy in Medicine of Australian Mammals*. Revised Edition. CSIRO Publishing, Melbourne.

Vogelnest L, Woods R (2008) *Medicine of Australian Mammals*. CSIRO Publishing, Melbourne.

Vogelnest L, Canfield P, Hartley WJ, Reece RL (1997) Suspected transponder implant associated mesenchymal tumourigenesis in two feathertail gliders (*Acrobates pygmaeus*) and a koala (*Phascolarctos cinereus*). In *Proceedings of the Annual Wildlife Diseases Association (Australasian Section) Conference*, October 1997. Flinders Island. Tasmania. pp. 1–5. Wildlife Diseases Association (Australasian Section).

Walker FM, Taylor A, Sunnocks P (2008) Female dispersal and male kinship-based association in southern hairy-nosed wombats (*Lasiorhinus latifrons*). *Molecular Ecology* **17**, 1361–1374. doi:10.1111/j.1365-294X.2008.03670.x

Walker FM, Upton JR, Sobek CJ, Taggart DA, Gaughwin MD (2020) Genetic monitoring of southern hairy-nosed wombats over two decades reveals that individuals can live for at least 18 years in the same warrens. *Australian Mammalogy* **43**, 22–29. doi:10.1071/AM20012

Wells RT (1971) Maintenance of the hairy-nosed wombat *Lasiorhinus latifrons* in captivity. *International Zoo Yearbook* **11**, 30–31. doi:10.1111/j.1748-1090.1971.tb01837.x

Wells RT (1973) Physiological and behavioural adaptations of the hairy-nosed wombat [(*Lasiorhinus latifrons*)(Owen)] to its arid environment. PhD thesis. University of Adelaide, Adelaide.

Wells RT (1978a) Thermoregulation and activity rhythms in the hairy-nosed wombat, *Lasiorhinus latifrons* (Owen), (Vombatidae). *Australian Journal of Zoology* **26**, 639–651. doi:10.1071/ZO9780639

Wells RT (1978b) Field observations of the hairy-nosed wombat *Lasiorhinus latifrons* (Owen). *Australian Wildlife Research* **5**, 299–303. doi:10.1071/WR9780299

Wells RT (1989) Vombatidae. In *Fauna of Australia. Mammalia*. Vol. 1B. (Eds DW Walton and BJ Richardson) pp. 755–768. Australian Government Publishing Service, Canberra.

Wells RT (1995) Southern hairy-nosed wombat *Lasiorhinus latifrons*. In *The Mammals of Australia*. 2nd edn. (Ed. R Strahan) pp. 202–203. Reed Books, Sydney.

Wells RT, Boreland F, Forward L (1986) The Brookfield Herbivore Study: grazing behaviours and pasture dynamics in a semi-arid zone. *Australian Mammal Society Bulletin* **9**, Abstract.

West M, Galloway D, Shaw J, Tounson A, Paris MCJ (2004) Oestrous cycle of the common wombat, *Vombatus ursinus*, in Victoria, Australia. *Reproduction, Fertility and Development* **16**, 339–346. doi:10.1071/RD03058

Wilkinson V, Tokano K, Nichols D, Martin A, Holme R, Phalen D, Mounsey K, Charleston M, Kreiss A, Pye R, Browne E, Naebord-Nielsen C, Richards SA, Carver S (2021) Fluralaner as a novel treatment for sarcoptic mange in the bare-nosed wombat (*Vombatus ursinus*): safety, pharmacokinetics, efficacy and practicable use. *Parasites & Vectors* **14**(18), 1–18.

Williams R (1990) Wombats. In *Care and Handling of Australian Native Animals*. (Ed. SJ Hand) pp. 91–96. Surrey Beatty & Sons & the Royal Zoological Society of New South Wales, Sydney.

Woinarski JCZ, Burbidge AA, Harrison P (eds.) (2014) *The Action Plan for Australian Mammals 2012*. CSIRO Publishing, Melbourne.

Woodford J (2001) *The Secret Life of Wombats*. Text Publishing, Melbourne.
Woods R (1999) Prevention of disease in hand reared native wildlife. In *Wildlife in Australia: Healthcare & Management*. Proceedings 327. Post Graduate Foundation in Veterinary Science. (Ed. DI Dryden) pp. 455–489. University of Sydney, Sydney.
Woolnough AP (1998) The feeding ecology of the northern hairy-nosed wombat, *Lasiorhinus krefftii* (Marsupialia: Vombatidae). PhD thesis. James Cook University, Townsville.
Woolnough AP (2000) Reproductive strategies of female northern hairy-nosed wombats. *Australian Mammalogy* **21**, 257–258. doi:10.1071/AM00257
Woolnough AP, Johnson CN (2000) Assessment of the potential for competition between two sympatric herbivores: the northern hairy-nosed wombat, *Lasiorhinus krefftii*, and the eastern grey kangaroo, *Macropus giganteus*. *Wildlife Research* **27**, 301–308.
Woolnough AP, Foley WJ, Johnson CN, Evans M (1997) Evaluation of techniques for indirect measurement of body composition in a free-ranging large herbivore, the southern hairy-nosed wombat. *Wildlife Research* **24**, 649–660. doi:10.1071/WR97010
Wünschmann A (1966) Einige Gefangenschaftsbeobachtungen an Breitstirn-wombats (*Lasiorhinus latifrons* Owen, 1845). *Zeitschrift für Tierpsychologie* **23**, 56–71. doi:10.1111/j.1439-0310.1966.tb01591.x
Young GE (1980) Geographic variation in the common wombat, *Vombatus ursinus* (Shaw 1800). *Victorian Naturalist* **97**, 200–204.
Zuckerman S (1953) The breeding seasons of mammals in captivity. *Proceedings of the Zoological Society of London* **122**, 827–950. doi:10.1111/j.1096-3642.1952.tb00251.x

# 9
# Possums and gliders

*Stephen M. Jackson, Ross Goldingay, Marissa L. Parrott, John Winter and Ludovica Valenza*

## 1. INTRODUCTION

The possums and gliders are a highly diverse group of marsupials, comprising six families, 20 genera and 71 species (Baker *et al.* 2023). They occur throughout Australia, New Guinea and numerous islands east of the Wallace Line and range in size from only 6 g for the little pygmy possum (*Cercartetus lepidus*) to > 10 kg for the bear cuscus (*Ailurops ursinus*) that is found on Sulawesi in Indonesia. The possums and gliders fill numerous niches, similar to those of rodents and primates elsewhere in the world, with some being folivores while others consume nectar, pollen and insects. Due to their diverse dietary niches, lifestyles, body size and threatened status, they can make interesting display species and captive management can assist in the conservation of several of the threatened species.

Possums and gliders have long been held in captivity, with records suggesting that considerable numbers of inland sugar gliders (*Petaurus notatus*) were held as pets as early as the 1830–1840s (Gunn 1851). Numerous zoos throughout the world have held possums and gliders, including London Zoo, which had feather-tailed gliders (*Acrobates pygmaeus*) in 1840, common brush-tailed possums (*Trichosurus vulpecula*) since at least 1857, sugar gliders since 1865 and squirrel gliders (*Petaurus norfolcensis*) in 1895 (Collins 1973; Flower 1931; Zuckerman 1953). New York Zoo had feather-tailed gliders in 1920 (Crandall 1964). Healesville Sanctuary, Victoria, and Taronga Zoo, Sydney, have held many species of possums and gliders, including mountain pygmy possum (*Burramys parvus*), Leadbeater's possum (*Gymnobelideus leadbeateri*), yellow-bellied glider (*Petaurus australis*), sugar glider, squirrel glider, eastern ring-tailed possum (*Pseudocheirus peregrinus*), feather-tailed glider and spotted cuscus (*Spilocuscus maculatus*) (Fleay 1949; S. Jackson pers. obs.). Sugar gliders are kept in large numbers as pets in North America and increasingly in Europe (Anon 2018; Campbell *et al.* 2019). London Zoo has held the Torresian striped possum (*Dactylopsila trivirgata*) from animals collected in New Guinea. Greater gliders (*Petauroides volans*) have only been sporadically held in captivity, but have been held privately by David Fleay, by Taronga Zoo and more recently have been successfully maintained and bred by Currumbin Sanctuary, Queensland (Fleay 1933; Taronga Zoo 1968; Fleay-Thomson 2007; Meer 2022).

## 2. TAXONOMY

### 2.1 Nomenclature

Within the Suborder Phalangerida there are six families containing a total of 71 species of possum and gliders. Of these 26 occur only in Australia, 67 occur only in New Guinea and surrounding islands and four occur in both regions (Flannery 1995a, 1995b; Helgen and Jackson 2015; Jackson 2015a, 2015b, 2015c; Jackson and Groves 2015; Baker and Gynther 2023; Baker *et al.* 2023).

The sugar glider has recently been split into three species: savanna glider, inland glider (Krefft's glider) and sugar glider. In the broad sense these species are typically referred to as sugar gliders within this text, as different research may have been completed on each of these under the name of 'sugar glider' (Cremona *et al.* 2021).

The greater glider has also been recognised as three species (*Petauroides volans*, *P. armillatus* and *P. minor*), rather than one, by several authors (Jackson 2015c; Jackson and Groves 2015; McGregor *et al.* 2020). However, the acceptance of the two additional species is not yet widely accepted, pending an upcoming publication that further defines the additional species. The feather-tailed glider has also been split into two species: the narrow-toed feather-tailed glider, *Acrobates pygmaeus*, and broad-toed feather-tailed glider, *A. frontalis*.

**Table 9.1.** Species of possums and gliders in Australia, their bodyweights and conservation status.

*Also occurs in New Guinea and/or surrounding islands. EN = endangered, EX = extinct, LC = least concern, NT = near threatened, UNK = unknown, VU = vulnerable.
Derived from Woinarski *et al.* (2014), Jackson (2015a, 2015b, 2015c); Cremona *et al.* (2021) and Baker and Gynther (2023).

| Common name | Scientific name | Weight (g) | IUCN status |
|---|---|---|---|
| **Superfamily Burramyoidea** | | | |
| **Family Burramyidae** | | | |
| Mountain pygmy possum | *Burramys parvus* | 30–82 | CR |
| Long-tailed pygmy possum (*) | *Cercartetus caudatus* | 25–40 | LC |
| Western pygmy possum | *C. concinnus* | 8–20 | LC |
| Little pygmy possum | *C. lepidus* | 6–10 | LC |
| Eastern pygmy possum | *C. nanus* | 15–43 | LC |
| **Superfamily Petauroidea** | | | |
| **Family Petauridae** | | | |
| Torresian striped possum (*) | *Dactylopsila trivirgata* | 310–545 | LC |
| Leadbeater's possum | *Gymnobelideus leadbeateri* | 100–166 | CR |
| Savanna glider | *Petaurus ariel* | 50–151 | UNK |
| Yellow-bellied glider | *P. australis* | 435–725 | NT |
| Coastal sugar glider (*) | *P. breviceps* | 55–157 | LC |
| Mahogany glider | *P. gracilis* | 310–500 | EN |
| Squirrel glider | *P. norfolcensis* | 190–300 | LC |
| Inland ( = Krefft's) sugar glider | *P. notatus* | 53–205 | LC |
| **Family Pseudocheiridae** | | | |
| Lemuroid ring-tailed possum | *Hemibelideus lemuroides* | 810–1170 | NT |
| Greater glider | *Petauroides volans* | 700–900 | EN |
| Rock ring-tailed possum | *Petropseudes dahlii* | 1280–2000 | LC |
| Western ring-tailed possum | *Pseudocheirus occidentalis* | 750–1300 | CR |
| Eastern ring-tailed possum | *P. peregrinus* | 650–1100 | LC |
| Green ring-tailed possum | *Pseudochirops archeri* | 910–1555 | NT |
| Daintree River ring-tailed possum | *Pseudochirulus cinereus* | 700–1450 | NT |
| Herbert River ring-tailed possum | *P. herbertensis* | 749–1555 | LC |
| **Superfamily Tarsipedoidea** | | | |
| **Family Tarsipedidae** | | | |
| Honey possum | *Tarsipes rostratus* | 7–16 | LC |
| **Family Acrobatidae** | | | |
| Broad-toed feather-tailed glider | *Acrobates frontalis* | 10–19 | LC |
| Narrow-toed feather-tailed glider | *A. pygmaeus* | 8–15 | LC |
| **Superfamily Phalangeroidea** | | | |
| **Family Phalangeridae** | | | |
| Southern common cuscus (*) | *Phalanger mimicus* | 1500–2200 | LC |
| Australian spotted cuscus | *Spilocuscus nudicaudatus* | 1500–4900 | LC |
| Short-eared brush-tailed possum | *Trichosurus caninus* | 2500–4500 | LC |
| Mountain brush-tailed possum | *T. cunninghami* | 2600–4200 | LC |
| Common brush-tailed possum | *T. vulpecula* | 1200–4500 | LC |
| Scaly-tailed possum | *Wyulda squamicaudata* | 900–2000 | NT |

The common brush-tailed possum has several typically recognised subspecies, one of which is sometimes recognised as a distinct species known as the coppery brush-tailed possum (*Trichosurus johnstonii*) (Jackson and Groves 2015; Helgen and Jackson 2015).

**Australian possums**

Class: Mammalia
Supercohort: Marsupialia
Cohort: Australidelphia
Order: Diprotodontia
Suborder: Phalangerida
Superfamily: Burramyoidea
Family: Burramyidae
Genus and species: 5 species in 2 genera
Superfamily: Petauroidea
Family: Petauridae
Genus and species: 8 species in 3 genera
Family: Pseudocheiridae
Genus and species: 8 species in 6 genera
Superfamily: Tarsipedoidea
Family: Tarsipedidae
Genus and species: 1 species
Family: Acrobatidae
Genus and species: 2 species in 1 genus
Superfamily: Phalangeroidea
Family: Phalangeridae
Genus and species: 6 species in 4 genera (see Table 9.1).

### 2.2 Subspecies

See Jackson and Groves (2015).

### 2.3 Recent synonyms

Synonyms for Australian species can be found in Jackson and Groves (2015).

### 2.4 Other common names

See Baker and Gynther (2023).

## 3. NATURAL HISTORY

### 3.1 Bodyweight and morphometrics

The Australian possums and gliders show a very large range in body size from 6 g for the little pygmy possum to 4900 g for the Australian spotted cuscus. The morphometrics for the Australian species can be found in Wilson and Mittermeier (2015) and Baker and Gynther (2023).

### 3.2 Distribution and habitat

Possums and gliders are well known across a wide range and many different habitats in Australia, including the arid regions of central Australia, throughout the east coast, northern Australia, Western Australia and Tasmania. Different species occupy a range of habitat types, from open and closed woodland, rainforest, heathlands, shrublands and even alpine areas (Baker and Gynther (2023).

### 3.3 Conservation status

Of the 32 Australian species, the mountain pygmy possum, Leadbeater's possum and the western ring-tailed possum are critically endangered; the mahogany glider is endangered, as are the southern and central forms of the greater glider (Woinarski *et al.* 2014; Table 9.1). The northern subspecies of yellow-bellied glider is classified as vulnerable. The remaining Australian species of possums and gliders are considered at low risk of becoming extinct, though several are considered close to being threatened due to threats including habitat destruction and fragmentation, predation from introduced species (especially feral cats and red foxes), drought and increasing bushfires (Table 9.1).

### 3.4 Diet in the wild

The dietary niches of possums and gliders limit their body size, with larger species being folivorous and small species, which have higher energy requirements on a mass-specific basis, being limited to energy-rich food items such as nectar, sap and arthropods (Table 9.2; Smith and Lee 1984; Lee and Cockburn 1985).

Of the different food types, leaves and stems are high in structural carbohydrates and low in material that is easily metabolised (Eisenberg 1981). Consumption of plant fibre is more efficient for larger mammals because of their lower energy requirements relative to gut capacity (Van Soest 1982; Freudenberger *et al.* 1989; Justice and Smith 1992). Although plant material is a more ubiquitous food resource than other food types (e.g. exudates or flesh), for arboreal folivores, having eucalypts and other foliage as the major component of their diet, presents several problems, including the presence of toxic secondary compounds (xenobiotics) such as essential oils and tannins, formylated phloroglucinol compounds (FPCs) and the generally low digestible energy and crude protein content of the leaves (Hume *et al.* 1984; Hume 1999; Scriver *et al.* 2004).

**Table 9.2.** Diets of different genera of possums and gliders.

Families are listed in approximately increasing body size to reflect the relationship of diet and body size.

| Genus | Diet | Reference |
|---|---|---|
| **Tarsipedidae** | | |
| *Tarsipes* | Nectar, pollen only - exudivore | 1–3 |
| **Acrobatidae** | | |
| *Acrobates* | Nectar, pollen, manna, honeydew, sap, insects - exudivore | 3–6 |
| **Burramyidae** | | |
| *Burramys* | Invertebrates, seeds, fruits - exudivore | 3, 7, 8 |
| *Cercartetus* | Nectar, pollen, sap, insects, fruit - exudivore | 3, 5, 9–13 |
| **Petauridae** | | |
| *Dactylopsila* | Insects, exudates, fruit, nectar - insectivore | 3, 14–17 |
| *Gymnobelideus* | Exudates, insects, nectar - exudivore | 3, 18 |
| *Petaurus* | Exudates, insects, sap, nectar, pollen, honeydew, manna, lerp, gum, fruit - exudivore | 3, 19–23 |
| **Pseudocheiridae** | | |
| *Hemibelideus* | Almost exclusively leaves - folivore. | 3 |
| *Petauroides* | Almost exclusively leaves of eucalypts - folivore | 3, 24, 25 |
| *Petropseudes* | Flowers, fruits, leaves - folivore | 3 |
| *Pseudocheirus* | Eucalypt leaves, flowers, fruit - folivore | 3, 26 |
| *Pseudochirops* | Almost exclusively leaves, fruit - folivore | 3 |
| *Pseudochirulus* | Leaves, fruits, flowers, fruit - folivore | 3 |
| **Phalangeridae** | | |
| *Phalanger* | Leaves, fruits, flowers, insects, small vertebrates, eggs - folivore | 3 |
| *Spilocuscus* | Leaves, fruits, flowers, some meat - folivore | 3 |
| *Trichosurus* | Eucalypt leaves, fruits, buds, flowers, fungi, occasionally meat such as birds, and bark - folivore | 3, 27 |
| *Wyulda* | Leaves, flowers, fruit, insects - folivore | 3, 28, 29 |

References: (1) Vose (1973); 2) Richardson *et al.* (1986); (3) Van Dyck and Strahan (2008); (4) Turner (1984a); (5) Huang *et al.* (1987); (6) Goldingay and Kavanagh (1995); (7) Mansergh *et al.* (1990); (8) Mansergh and Broome (1994); (9) Hickman and Hickman (1960); (10) Turner (1984b); (11) Arnould (1986); (12) Smith (1986); (13) Hockey *et al.* (2019); (14) Fleay (1942); (15) Smith (1982a); (16) Handasyde and Martin (1996); (17) Rawlins and Handasyde (2002); (18) Smith (1984a); (19) Henry and Craig (1984); (20) Menkhorst and Collier (1988); (21) Howard (1989); (22) Jackson (2001); (23) Goldingay and Jackson (2004); (24) Marples (1973); (25) Kavanagh and Lambert (1990); (26) Pahl (1987a); (27) Anon (1935); (28) Runcie (1999); (29) Hohnen *et al.* (2016).

Among the possums and gliders, the greater gliders (and potentially lemuroid ring-tailed possums) appear to be the only strict folivores, with all other members of the Pseudocheiridae and Phalangeridae supplementing their diet with food that is more easily digested such as blossoms, flowers, fruit, invertebrates and even small vertebrates (Table 9.2). The eastern ring-tailed possum and the northern greater glider are the smallest arboreal marsupials with a predominantly folivorous diet and appear to be at the lower limit in size to be folivorous. To maximise food digestion, the eastern ring-tailed possum is caecotrophic (reingesting soft faeces of high nutritive value derived from caecal contents) in order to obtain access to protein, energy and vitamins that would otherwise be lost because of poor absorption in the caecum and proximal colon (Chilcott 1984; Hume *et al.* 1984; Chilcott and Hume 1985). Cork and Foley (1991) noted that virtually no consumption of tree foliage is seen in primate, marsupial or rodent species smaller than ~700 g. They proposed this as the absolute evolutionary limit for foliage to be a major part of the diet, without supplementation with other more easily digestible matter such as flowers and fruit. Indeed, the smallest pseudocheirid possum, the pygmy ring-tailed possum (*Pseudochirulus mayeri*) of New Guinea, which has a bodyweight of only 105–206 g, appears to eat more digestible food types, such as epiphytic lichens and mosses, and to eat only very small portions of leaves

(Flannery 1995b). Similar observations have been made of the slightly larger (335–380 g) New Guinean lowland ring-tailed possum (*Pseudochirulus canescens*) (Flannery 1995b).

All members of the Petauridae, Burramyidae, Acrobatidae and Tarsipedidae weigh less than the eastern ring-tailed possum and all eat foods that are more easily digested (Table 9.2). Their diets consist of insect and plant exudates such as nectar (and pollen), tree sap, manna, insect honeydew and, in some species, fruit and seeds, in order to obtain their energy requirements (Table 9.2). As exudates are very low in protein, dietary protein requirements are supplied through the consumption of arthropods, pollen and, occasionally, small vertebrates.

Within the Petauridae, Burramyidae, Acrobatidae and Tarsipedidae, member species obtain protein from a variety of sources (Table 9.2). The Torresian striped possum eats large numbers of ants, small stingless bees (*Trigona* spp.), termites, wood-boring larvae and the larvae of several other insects (Troughton 1941; Smith 1982a; Handasyde and Martin 1996). Fleay (1942) found that captive Torresian striped possums caught and ate house mice (*Mus musculus*). Mahogany gliders consume green ants, other insects, spiders, pollen and acacia arils (Van Dyck 1993; Jackson 2001).

Leadbeater's possums consume tree crickets, beetles, moths and spiders (Smith 1984a). The yellow-bellied glider eats a variety of arboreal arthropods, primarily tree crickets, adult and larval beetles, caterpillars, spiders and moths, but is known to feed heavily on nectar and pollen in some locations (Smith and Russell 1982; Henry and Craig 1984; Goldingay 1990; Quin *et al.* 1996).

Wild sugar gliders, which are thought to have been introduced to Tasmania from Victoria, have been observed killing and eating critically endangered hollow-nesting birds including the orange-bellied parrot (*Neophema chrysogaster*) and swift parrot (*Lathamus discolor*) and are a key threat to their conservation (Stojanovic *et al.* 2014). Sugar gliders generally consume nectar, pollen and insects including moths and scarabaeid beetles (Smith 1982b; Howard 1989; J. Winter pers. obs.). Pygmy-possums (*Cercartetus* spp.) consume insects and pollen (Hickman and Hickman 1960).

Mahogany gliders are known to feed on a wide variety of feed types, including nectar and pollen from the genera *Eucalyptus*, *Corymbia* and *Melaleuca*. Other food items include sap from *Acacia mangium* and *Albizia procera*, sap from grass tree flower spikes (*Xanthorrhoea johnsonii*), insects, lerps and honeydew. Fruit from mistletoes and acacia arils are also consumed (Jackson 2001).

Squirrel gliders consume pollen and various insects such as caterpillars and beetles (Menkhorst and Collier 1988; Sharpe and Goldingay 1998; Dobson *et al.* 2005). Squirrel gliders have also been known to kill mice in captivity (Troughton 1941), kill a magpie-lark (*Grallina cyanoleuca*) in the wild and eat the eggs, feed on the egg of a noisy miner (*Manorina melanocephala*) in the wild and hassle a common bronzewing pigeon (*Phaps chalcoptera*) to remove it from its nest so the eggs could be consumed (Winter 1966; Holland 2001; Dobson *et al.* 2005; Ball *et al.* 2009).

Mountain pygmy possums rely on the migratory endangered Bogong moth (*Agrotis infusa*) to gain bodyweight following their annual winter hibernation period and for enough energy to lactate and raise their young. A collapse in Bogong moth numbers (Carruthers *et al.* 2020), especially in the spring-summer of 2017 and 2018, led to wide-spread loss of litters in Victorian populations and the development of Bogong moth-mimicking supplementary food to support possums after food shortage events (including bushfire) (Parrott *et al.* 2020). The possums' diet changes seasonally, with more fruits and seeds being consumed in autumn when animals can often double their bodyweights in preparation for hibernation over winter.

Feather-tailed gliders eat pollen and insects to obtain protein, while the honey possum consumes pollen and has been observed eating mealworms and small moths in captivity (Vose 1973; Turner 1984a, 1984b; Richardson *et al.* 1986).

Pollen is high in protein and although it is protected by a tough exine coat, the nitrogenous cell contents are large components of the diet of several marsupials including the honey possum, eastern pygmy possum and feather-tailed gliders (Stanley and Linskins 1974; Wooller *et al.* 1983; Turner 1984a, 1984b; Richardson *et al.* 1986; van Tets and Whelan 1997; van Tets and Hulbert 1999). Pollen is also a significant protein source for larger possums, including sugar gliders (Goldingay *et al.* 1987; Howard 1989), squirrel gliders (Menkhorst and Collier 1988), mahogany gliders (Jackson 2001) and yellow-bellied gliders (Goldingay 1990; Quin *et al.* 1996).

**Table 9.3.** Maximum longevity (years) of different genera of possums and gliders in the wild and in captivity.

Oldest known longevity is in brackets where known. Families are listed in approximately increasing body size.

| Genus | Wild | Captivity | References |
|---|---|---|---|
| **Tarsipedidae** | | | |
| *Tarsipes* | 1 | 1 (M), 2-3 (F) | 1 |
| **Acrobatidae** | | | |
| *Acrobates* | 3-5 | 2-3 (8) | 2-6 |
| **Burramyidae** | | | |
| *Burramys* | 3 (12) F, 1 (5) M | 5-9 (11) | 7-13 |
| *Cercartetus* | 2-4 | 3-5 (10) | 4, 14-16 |
| **Petauridae** | | | |
| *Dactylopsila* | - | 5-6 | 17, 18 |
| *Gymnobelideus* | - | 5-10 (12) | 19, 20 |
| *Petaurus* | 4-8 | 5-8 (14) | 2, 21-30 |
| **Pseudocheiridae** | | | |
| *Petauroides* | - | 10-12 (15) | 31, 32 |
| *Pseudocheirus* | 4-6+ | 5-6 (8) | 33-38 |
| **Phalangeridae** | | | |
| *Phalanger* | - | 7-9 | 39 |
| *Spilocuscus* | - | 7-9 (11) | 40 |
| *Trichosurus* | 10-11 (13) | 8-12 (17) | 2, 5, 21, 41-47 |

References: (1) F. Bradshaw pers. comm.; (2) Flower (1931); (3) Fleming and Frey (1984); (4) Ward (1990a); (5) Strahan (1995); (6) Dockerill (2013); (7) Mansergh and Scotts (1989); (8) Mansergh and Scotts (1990); (9) Boys (2018); (10) Broome (2001); (11) Heinze *et al.* (2004); (12) Jackson *et al.* (2014); (13) Scheelings and Dobson (2015); (14) Atherton and Haffenden (1982); (15) Ward (1990b); (16) R. Goldingay pers. obs.; (17) F. Wheeler and A. McKenna pers. comm.; (18) T. Carmichael pers. comm.; (19) Smith (1980); (20) Smith (1984b); (21) Mitchell (1911); (22) Henry and Craig (1984); (23) Henry and Suckling (1984); (24) Suckling (1984); (25) Craig (1985); (26) Goldingay and Kavanagh (1990); (27) Slater (1997); (28) Booth (1999); (29) Jackson (2000a); (30) van der Ree *et al.* (2006); (31) Henry (1984); (32) Henry (1985); (33) Thomson and Owen (1964); (34) How *et al.* (1984); (35) Pahl (1987a); (36) Pahl and Lee (1988); (37) Ong (1994); (38) Wayne *et al.* (2005a); (39) S. Jackson pers. obs.; (40) Fleay (1949); (41) MacLean (1967); (42) Crawley (1970); (43) Crawley (1973); (44) How (1981); (45) Barnett *et al.* (1982); (46) Lindenmayer *et al.* (1991); (47) Wayne *et al.* (2005b).

## 3.5 Longevity

### 3.5.1 Wild

There is wide variation in the longevity of the different groups of possums and gliders in the wild. Typical longevity is 1–11 years and generally increases with increasing body size (Table 9.3).

### 3.5.2 Captivity

There is wide variation in the longevity of the different groups of possums and gliders in captivity. The smaller possums in the families Tarsipedidae, Burramyidae and Acrobatidae typically live for 1–12 years while the larger ones such as the Petauridae, Pseudocheiridae and Phalangeridae live for ~4–17 years.

### 3.5.3 Methods used to determine the age of adults

Various parameters, including bodyweight, patagium colour, scent gland development, pouch development and tooth wear, are commonly used in combination to determine the age of possums and gliders.

Three methods of tooth wear assessment can be used to determine the age or relative age of petaurids:

- the degree of flattening (caused by wear) of the upper incisors when viewed from the anterior gives a relative age (Fig. 9.1a)
- the degree of wear and proportion of dentine exposed on the upper incisors when viewing the ventral surface gives an approximate age (Fig. 9.1b; Table 9.4)
- the colour and wear of the lower incisors and the presence of lateral cracks gives an approximate age – young animals have white lower incisors with no lateral cracks, while older animals have teeth that are more discoloured and develop an increasing number of lateral cracks with age. Older animals may have part of the lower incisors chipped off (Table 9.4).

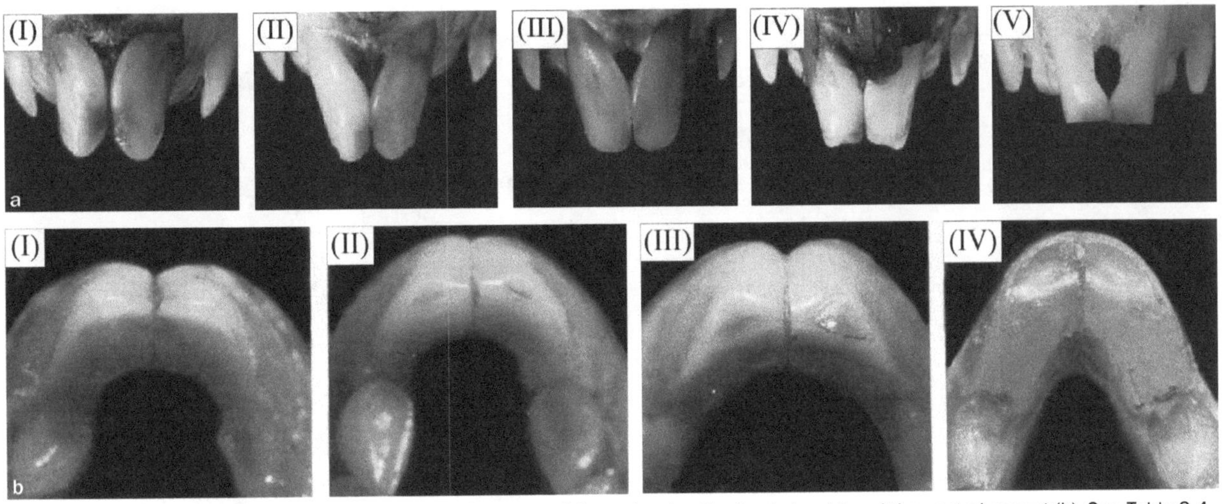


**Fig. 9.1.** Typical wear patterns of upper incisors of *Petaurus* gliders, showing the front view (a) and the ventral aspect (b). See Table 9.4 for information on the age related to each tooth wear stage. Derived from Alexander (1981) with permission from the author and reproduced from Suckling GC (1984) with permission from CSIRO Publishing.

**Table 9.4.** Age-estimation parameters of yellow-bellied, mahogany, squirrel and sugar gliders.

Derived from Alexander (1981), Suckling (1984), Quin (1995), Jackson (2000a) and Goldingay *et al.* (2001).

| **Parameter** | **Estimated age (years)** | | | |
|---|---|---|---|---|
| | **< 1** | **1–2** | **2–3** | **> 3** |
| **Weight of yellow-bellied glider (g)** | | | | |
| Males | < 320 | > 320 | > 450 | > 450 |
| Females | < 250 | > 250 | > 450 | > 430 |
| **Weight of mahogany gliders (g)** | | | | |
| Males | < 300 | > 300 | > 370 | > 370 |
| Females | < 280 | > 280 | > 330 | > 330 |
| **Weight of squirrel gliders (g)** | | | | |
| Males | < 190 | > 190 | > 210 | > 210 |
| Females | < 170 | > 170 | > 190 | > 190 |
| **Weight of sugar gliders in North Queensland (g)** | | | | |
| Males | < 60 | > 60 | > 80 | > 80 |
| Females | < 50 | > 50 | > 70 | > 70 |
| **Weight of sugar gliders in Northern New South Wales (g)** | | | | |
| Males | < 100 | > 100 | > 120 | > 120 |
| Females | < 80 | > 80 | > 100 | > 100 |
| **Yellow-bellied gliders, mahogany gliders, squirrel gliders and sugar gliders** | | | | |
| Wear of upper incisors | None to slight (Fig. 9.1BI) | Slight to Moderate. Sometimes cracked (Fig. 9.1BII) | Moderate to heavy. Often cracked (Fig. 9.1BIII) | Heavy to very heavy. Usually cracked (Fig. 9.1BIV) |
| Wear of lower incisors | White, no cracks | Slight discolouration, lateral cracks slight | Orange discolouration, lateral cracks obvious. Occasionally chipped teeth in old animals | |
| Patagium colour | White | Cream-yellow | Yellow | Yellow-orange |
| Frontal gland (males) | Not developed | Partially to well developed. | | |
| Pouch | Small and shallow with fine white hairs; teats 1 mm long | Larger and deeper than in females that had not bred. Yellow/orange hairs with black scale. Teats > 1 mm | | |

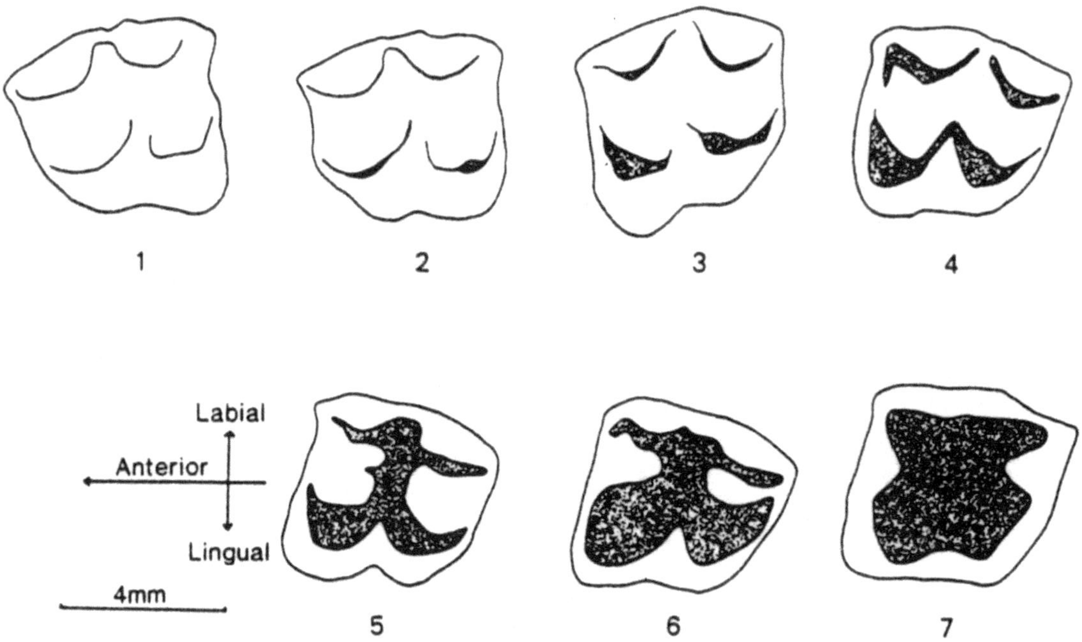


**Fig. 9.2.** Estimates of age in eastern ring-tailed possums using tooth wear. Position and area of dentine (shaded) exposed in seven classes of tooth wear of the first molar (M2) of the upper jaw. Reproduced from Pahl (1987b) with permission from CSIRO Publishing.

Bodyweight is a useful indicator of age until ~18 months for these species (Table 9.4).

Several techniques have been examined to determine age in eastern ring-tailed possums (Fig. 9.2; Table 9.5) and common brush-tailed possums, including tooth wear and growth rings of the teeth; however, only tooth wear can be used for ageing live animals. The tooth wear index for brush-tailed possums uses the upper left first molar to examine wear (generally while the animal is under anaesthetic) (Fig. 9.3), which provides an approximate age, although there is variation in wear between individuals (Winter 1980; Cowan and White 1989).

**Table 9.5.** Age classes of eastern ring-tailed possums responding to the seven tooth wear classes.

Derived from Pahl (1987b).

| Tooth wear class (from Fig. 9.2) | Estimated age class (months) |
|---|---|
| 1 | 0–18 |
| 2 | 7–30 |
| 3 | 7–36 |
| 4 | 13–43 |
| 5 | 19–48 |
| 6 | 25–60 |
| 7 | 31–48 |

Longevity in brush-tailed possums has been determined from dead animals by decalcifying the molars, sectioning with a freezing microtome and staining with haematoxylin (Clout 1982). The sections are examined under a microscope and the numbers of darkly stained bands are counted, with the number of bands equalling the number of years.

## 4. HOUSING REQUIREMENTS

### 4.1 Enclosure design

#### 4.1.1 General principles

All possums and gliders are generally best exhibited in nocturnal houses, due to their nocturnal behaviour. The exception is the Sulawesi bear cuscus (*Ailurops ursinus*) from Sulawesi, which is apparently diurnal (Flannery 1995a; Dwiyahreni *et al.* 1999). Smaller species, such as pygmy possums, feather-tailed gliders and honey possums, can be held in relatively smaller enclosures of ~1 $m^2$ and 1–2 m high, though it is preferable to provide larger and taller enclosures for all species, especially the feather-tailed gliders, due to their highly mobile acrobatic and gliding behaviour. Larger species such as the petaurids, pseudocheirids and phalangerids require a significantly larger area of at least 10 $m^2$ and 3 m high. All possum and glider enclosures need to have good foliage

| Wear Class | Crown M | Age | Description of tooth wear |
|---|---|---|---|
| 1 |  | 0.9 | Cusps high and pointed with no apparent wear |
| 2 |  | 1.2 | Lingual cups with points rounded but with no dentine exposed |
| 3 |  | 1.7 | Small cresents of dentine exposed on ligual cusps, but none on labial cusps |
| 4 |  | 3.7 | Cresents of dentine on lingual cusps larger, but still high and rounded; dentine exposed on at least one labial cusp, but not joined to dentine cresents of lingual cusp |
| 5 |  | 5.6 | Lower limit; dentine of at least one labial cusp joined to dentine cresent of lingual cusp |
| |  | 6.8 | Upper limit: dentine of lingual cusps jined, no longer appearing as cresents; dentine of both labial cusps joined to lingual cusps, but still appear as narrow strips along the cusp ridge |
| 6 |  | 8.9 | Lower limit: lingual cusps flattened, and broad band of exposed dentine between the two; dentin on labial cusps no longer a narrow strip but a broad band |
| |  | 10.7 | Upper limit: both lingual and labial cusps flattened, with large areas of exposed dentine, but still with an enamel indentation between anterior and posterior lingual cusps |
| 7 |  | 11.4 | Cusps completely obliterated and crown of tooth dished; no enamel indentation between anterior and posterior lingual cusps |

**Fig. 9.3.** Estimates of age (years) in brush-tailed possums using tooth wear. Position and area of dentine (shaded) exposed in classes of tooth wear of first molar (M1) of the upper jaw. Adapted from Winter (1980) and Cowan and White (1989) with permission from CSIRO Publishing.

cover, although in enclosures that are viewed by the public the foliage may be thinned out to give better viewing.

### 4.1.2 Enclosure requirements for different groups of possums and gliders

The different species of possums and gliders require different methods of exhibition and maintenance in off-exhibit areas, though structural complexity within the enclosure and appropriate social systems should be provided for each species.

#### 4.1.2.1 Honey possum

Although honey possums have been held in smaller enclosures and indoors, successful breeding has occurred in an outdoor enclosure that measured 4 m (L) × 4 m (W) × 2 m (H) and was constructed of 1 $cm^2$ wire mesh, overlaid with fly screen in order to prevent the escape of any young. It was planted with species of flowering plants such as *Banksia*, *Isopogon*, *Grevillea*, *Eremophila*, *Callistemon* and *Adenanthos*. Refuge areas were provided by planting small saplings of *Corymbia calophylla* and *Kennedia coccinea* (Bradshaw *et al.* 2000). Although nest boxes were provided, they were generally not used, with small rock piles or dense ground cover being used instead. The successful breeding is thought to be the result of the possums having more exercise than in an indoor enclosure, thus preventing obesity, although the synchronicity of births in mid-February also suggests an environmental cue (Bradshaw *et al.* 2000).

#### 4.1.2.2 Feather-tailed gliders

Feather-tailed gliders are easily maintained in small enclosures that contain a network of small (and a few larger) branches, substrate such as leaf litter and a surplus of nest boxes. They are excellent climbers and can easily climb over the smooth surface of glass. Great care is needed to ensure there are no gaps in the walls or they will readily escape. Care also needs to be taken when accessing the enclosure as individuals are often near the doorway, due to their ability to climb walls, and can escape, particularly in the relative darkness of a nocturnal house. The likelihood of this occurring can be reduced by servicing the enclosure during the day (light) cycle when they are generally in their nest boxes or are more easily seen if moving within the enclosure (W. Gleen pers. comm.).

#### 4.1.2.3 Pygmy possums

Pygmy possums should be provided with a network of branches to climb, several rocks to climb over and under (particularly for the mountain pygmy possum) and at least one nest box per animal. Small, open-ended hollow logs are also useful. A substrate of leaf litter, sand or soil works well. Apart from the suggested wooden nest boxes, coconut shells with a large hole cut in the side work well and can be hung up in the branches in the enclosure. Sections of horizontal polyvinyl chloride (PVC) tube or pipe that is 4 × 25 cm and capped at one end are also readily used by pygmy possums (Goldingay 2020a). If groups of animals are being held together, a wooden nest box with two entrance/exit holes will allows animals to enter and exit unhindered.

Mountain pygmy possums should be provided with branches, small logs, rocks, live plants and nest boxes. Nesting materials include moss, sphagnum moss, silver wattle and paperbark (Boys 2018).

Temperature control is highly important for this species as temperatures above 25°C can result in deaths. Cooler temperatures are required to mimic their alpine environment, so temperatures at or below 16°C (and never exceeding 22°C), should be maintained. For mountain pygmy possums to maintain natural behaviours, such as their annual 5–7-month hibernation, cold temperatures down to 2°C in winter are required (Boys 2018). In contrast to mountain pygmy possums, the other species in the family appear to be more heat tolerant. For example, in the wild the eastern pygmy possum is known to use nest boxes and tree hollows in temperatures over 35°C (Goldingay 2020a). However, small mammals in general can overheat quickly, especially in enclosures where an escape to cooler temperatures (such as may be possible in the wild in a deep tree hollow), or cooling air flow, is less accessible. Thus, care must be taken to provide suitable ambient temperatures (generally under 25°C) and a choice of cooler areas within enclosures. Animals should be monitored for signs of heat stress on very warm days.

#### 4.1.2.4 Petaurids

Petaurids generally live in tree hollows, which can be readily substituted by nest boxes. In captivity, Leadbeater's possums need to be provided with a network of branches throughout the enclosure that are vertical and close enough together (e.g. up to 1 m) to allow jumping between them to a network of branches, suspended from the roof, that provides a runway. At least some of these branches should be species of eucalypts with stringy bark, as the possum will tear off strips of this for nesting material. Gliding petaurids require fewer branches to move about, due to their ability to glide. Having larger gaps between branches can encourage these animals to glide.

Torresian striped possums should be provided with branches for climbing and to chew into. Wooden logs, such as eucalypts and acacias, with obvious wood-boring insects should be provided so the possums can feed using their incisors (Carmichael 2000). Care should be taken when housing them in wooden structures, as they are capable of chewing their way through. Hardwood in good condition is strong enough to withstand the occasional nibble but is not recommended for keeping Torresian striped possums over an extended period (Carmichael 2000). Torresian striped possums should be provided with nest boxes that are ~25 cm (L) × 25 cm (W) × 30 cm (H) in several locations, such as on the wall, tree trunk or floor as individuals have their own nesting preferences. The nest box entrance holes should be at least 8–10 cm in diameter (McKenna 2005).

#### 4.1.2.5 Ring-tailed possums and greater gliders

Eastern ring-tailed possums and the greater gliders are canopy dwellers (Davey 1984). However, greater gliders should be provided with opportunities to glide, while ring-tailed possums should be provided with a series of branches arranged in a similar way to those used for the non-gliding petaurids to climb around the enclosure. Green ring-tailed possums need foliage clumps in which to rest, as they do not use hollows (J. Winter pers. obs.), but other ring-tails use tree hollows or fern clumps for resting places and these should be provided. Greater gliders require a nest box located high in the enclosure with access to fresh eucalypt leaves and stringy bark near the nest box entrance to make a nest. They are also reluctant to come to the floor of the enclosure (Fleay 1933).

#### 4.1.2.6 Cuscuses, brush-tailed possums and scaly-tailed possum

These species need several strong branches to allow them to climb around the enclosure. They will also readily forage on the ground so food can be supplied either on the ground or on a feed platform. Due to their tendency to remain inactive for long periods as a result of their diet, it is possible to carefully light the nest box or tree trunk so that they are visible when resting. Lights should be red, and without blue/white light wavelengths, to minimise disruption to the animals and their circadian rhythms. Take care that the nest is not immediately adjacent to the glass partition, as the public will tap it constantly and unnecessarily stress the animals. The lighting may not be beneficial for viewing if the nest box or log is too small, as only the fur on the animal's back will be visible when it curls up to sleep.

For the larger possums that are held in larger enclosures, take care not to fill the enclosure with too much foliage. Ideally the possum should be visible in most positions in the enclosure except the nest box (and even this could potentially be viewed). The Australian spotted cuscus does not use a tree hollow for resting, but sits in leaf clumps to hide itself during daylight hours (J. Winter pers. obs.).

By thinning out some branches it is possible to provide the possums with a feeling of protection while also allowing the public to see them. In the centre (and ideally elsewhere) there should be horizontal branches that can accommodate a food dish and also connect thicker areas of foliage.

### 4.2 Holding area design

Holding areas for possums and gliders can be of relatively simple design but still provide appropriate climbing complexity and have at least a minimum enclosure size, though the larger the enclosure the better to allow natural behaviours and ensure suitable positive welfare (Table 9.6). The small possums, such as the pygmy possums, feather-tailed gliders and honey possums, can be easily held in wooden boxes with one or more panels of wire mesh. Great care needs to be taken with enclosures for small possums such as feather-tailed gliders as they have been observed to escape from enclosures with 1 $cm^2$ mesh (S. Ward pers. comm.).

Eastern pygmy possums, mountain pygmy possums, and honey possums have been bred successfully in enclosures that contained a soil floor and were exposed to the weather so that grass and shrubs could grow. The pygmy possums have been held in large enclosures ~6 m × 3 m (L) that allow ample opportunity for the animals to forage, maintain their social behaviour and experience natural light cycles and weather, which appear to be important in breeding these species. Mountain pygmy possums have been bred with consistent success in indoor alpine-mimicking enclosures, which allow natural winter hibernation and provide suitable temperature (2–16°C) and natural lighting cues (Boys 2018).

Possums such as sugar gliders, Leadbeater's possum and some larger species can be housed in simple, wire-framed structures that can be set up as a series of adjacent enclosures, but Torresian striped possums need to be held in a more sheltered enclosure. In their case, the enclosures can be narrower than recommended but longer (e.g. 2 m (W) × 5+ m (L)). The enclosure should have at least part of the roof area covered (e.g. 1.5–2 m), under which the nest boxes and food should be placed. The floor of these enclosures can be concrete with leaf litter or sand, hollow logs and tussocks. Ideally, an enclosed service area to each enclosure should be provided to allow easier servicing during poor weather and additional security in case an animal escapes from its enclosure.

### 4.3 Spatial requirements

The recommended area requirements of different groups of possums and gliders are given in Table 9.6. An additional 25% floor space is suggested for each extra individual. Some particularly social species, such as honey possums and mountain pygmy possums, should be given larger enclosures for breeding and socialisation.

### 4.4 Position of enclosures

In many cases, possums and gliders will be held in temperature-controlled and enclosed nocturnal houses. If they are held in outdoor enclosures they should be well protected from the prevailing winds and poor weather and with their nest boxes out of full sunlight.

### 4.5 Weather protection

The nest boxes, which are usually hung off one of the walls or on a platform (except for the largely ground- or boulder-dwelling mountain pygmy possum), should be under shelter away from the wind and rain. The remainder of the enclosure can be relatively open to allow airflow.

**Table 9.6.** Minimum area of enclosures recommended for pairs of animals of different genera of Australian possums and gliders.

Families are listed in approximately increasing body size. Small *Petaurus* spp. include the savanna glider, sugar glider and Krefft's glider; medium *Petaurus* spp. include the mahogany glider and squirrel glider; and large *Petaurus* spp. includes the yellow-bellied glider.

| Genus | Area ($m^2$) (m) | Height | Additional floor area for each extra animal (m) |
|---|---|---|---|
| **Burramyidae** | | | |
| *Burramys* | 1–3 (breeding) | 1 | 0.25 |
| *Cercartetus* | 1 | 1 | 0.25 |
| **Petauridae** | | | |
| *Dactylopsila* | 10 | 3 | 4.00 |
| *Gymnobelideus* | 10 | 3 | 4.00 |
| *Petaurus* - small | 10 | 3 | 1.00 |
| *Petaurus* - medium | 12 | 3 | 2.25 |
| *Petaurus* - large | 15 | 3 | 4.00 |
| **Pseudocheiridae** | | | |
| *Hemibelideus* | 10 | 3 | 4.00 |
| *Petauroides* | 10 | 3 | 4.00 |
| *Petropseudes* | 10 | 3 | 4.00 |
| *Pseudocheirus* | 10 | 3 | 4.00 |
| *Pseudochirops* | 10 | 3 | 4.00 |
| *Pseudochirulus* | 10 | 3 | 4.00 |
| **Tarsipedidae** | | | |
| *Tarsipes* | 1–4 (breeding)[1] | 2 | 0.25 |
| **Acrobatidae** | | | |
| *Acrobates* | 1–5 (4 animals) | 2 | 0.25 |
| **Phalangeridae** | | | |
| *Phalanger* | 12 | 3 | 4.00 |
| *Spilocuscus* | 12 | 3 | 4.00 |
| *Trichosurus* | 12 | 3 | 4.00 |
| *Wyulda* | 12 | 3 | 4.00 |

[1]Used by Bradshaw *et al.* (2000) to breed successfully.

## 4.6 Temperature requirements

Heating is generally not required unless there are prolonged periods of low temperatures (e.g. below ~5°C). In most cases the various species are well adapted to low temperatures and go into torpor to conserve energy if required. Indeed, the occurrence of torpor may even be a trigger for synchronised breeding in some species such as the mountain pygmy possum (which live in alpine areas of New South Wales and Victoria above 1200 m). Species that undergo torpor, especially the mountain pygmy possum, should not be overheated and generally do not require additional heat. The mountain pygmy possum should be maintained at temperatures below 22°C (preferably 2–16°C), with colder temperatures inducing winter hibernation (Boys 2018). Temperatures above this are known to cause death (S. Jackson pers. obs.). Similarly, observations by Fleming (1985a) showed that mountain pygmy possums were noticeably stressed by ambient temperatures above 29°C (laying on their flanks, ears fully expanded, tail engorged with blood and saliva spread on their forepaws) and exposure to temperatures of 33°C for less than 1 h was found to be lethal.

Species that typically live in tropical areas in the wild may need additional heat to mimic the wild conditions.

Heating may also increase the activity of the smaller species during cold weather, as they will be less likely to go into torpor; however, the effect this has on reproduction is not known.

## 4.7 Substrate

In most cases the floor substrates are typically sand or leaf litter, though animals have been held successfully on wood shavings. Holding enclosures for small possums may be covered in paper for ease of cleaning.

## 4.8 Nest boxes

In the wild, the different species use various shelter types and locations (Table 9.7). Most of the larger species of possums, such as the ring-tailed possums and brush-tailed possums, often make only rudimentary nests in tree hollows that may be lined with leaves. The rock-tailed possum and scaly-tailed possum find refuge in rocky crevices, while the eastern and western ring-tailed possums may build their own spherical nests of sticks and leaves (known as dreys) independent of tree hollows (Thomson and Owen 1964). Field studies suggest individual eastern ring-tailed possums are either primarily hollow users or primarily drey users (Lindenmayer *et al.* 2008). Smaller species of possums, such as the honey possum and the feather-tailed gliders, nest within hollows, birds' nests, grass tussocks, logs, stumps and even deep in boulder-fields in the case of the mountain pygmy possum. An outline of different known nest types for the various genera of possums and gliders is given in Table 9.7.

Nest boxes should be supplied for all species of possums as they provide security, a place for raising young and, if provided in enclosures that are open to the elements, protection from the weather. Several studies in the wild, and numerous captive observations, have shown that most species of possums, including feather-tailed gliders, pygmy possums, sugar gliders, squirrel gliders, yellow-bellied gliders, Leadbeater's possum, ring-tailed possums, greater gliders, mountain and common brush-tailed possums and cuscus, will readily use nest boxes if they are of adequate size (Menkhorst 1984a, 1984b; Beyer and Goldingay 2006; S. Jackson pers. obs.). Torresian striped possums have been offered hollow logs that provide a comfortable snug fit, without being too roomy (Carmichael 2000). In contrast, some species such as green ring-tailed possums and Australian spotted cuscus do not use nest hollows in the wild, but sit on exposed branches, usually within dense foliage that provides protection (J. Winter pers. obs.). Therefore, good vegetation cover should be provided in their enclosures.

In general, the nest box should allow for several individuals to nest together (as indicated by their social structure) to comfortably inhabit it and the nest opening should be as small as possible while still allowing the possums' access. A large access door (preferably taking up the whole side) should be placed on the side of the nest box to allow easy animal caretaker access to the animals inside. An overhang, of ~10 cm, on the front of the nest box is also recommended, particularly if it is to be placed outdoors. The possums should be able to climb out of the boxes if they are made of timber, but if any individuals (particularly juveniles) have difficulty attach a thin piece of timber, with marks cut into it, to the inside to allow them to climb out of the box. Additional holes can be drilled in the roof or upper sides of the box for nesting females with large young to avoid a build-up of condensation due to multiple animals in the box (especially as young grow larger). Table 9.8 gives an outline of approximate nest box sizes.

Natural nesting materials such as leaves, moss and bark should be provided. Possums and gliders will choose and carry nest material into their nest boxes, thus providing enrichment and increased activity. Increased collection of nest material may also provide an indication of breeding in some species that increase their nest building before the birth of young. For example, the mountain pygmy possum nests throughout the year in bird-nest-shaped nests they construct from soft materials such as mosses and sea grass, but when a female is pregnant and nearing parturition, she will build a large, fully enclosed spherical nest in which to raise her young.

Field studies of nest box preference by eastern pygmy possums for different nest boxes designs in the field suggest there are four main types that are each used relatively frequently (Rueegger *et al.* 2012; Fig. 9.4):

- large-volume plywood nest box. Volume 0.0052 $m^3$. Internal dimensions 24 cm (H) × 15.5 cm (W) × 14 cm (D). Timber 1.5-cm thick (Fig. 9.4a)
- small-volume small plywood nest box. Volume 0.0013 $m^3$. Measurements 13 (H) × 10 (W) × 10 (H) cm; Timber 1.5-cm thick (Fig. 9.4b).
- small-volume PVC pipe capped at both ends. Volume 0.0012 $m^3$. Measurements 10 cm (D) × 15 cm (L);

**Table 9.7.** Nest type and location for various species of possums.

Families are listed in approximately increasing body size.

| Species | Nest type | Location | Reference |
|---|---|---|---|
| **Tarsipedidae** | | | |
| *Tarsipes rostratus* | Not known to build nests | Ground, bird's nests, in grass tussocks or dense vegetation, hollow stems of grass trees | 1, 2 |
| **Acrobatidae** | | | |
| *Acrobates pygmaeus* | Typically make woven egg-shaped nests with leaves such as eucalypt and casuarina, bark and tree fern fibre. Lined with feathers or other soft and flexible material | Tree hollows are used when present, but there are records of other structures being used when hollows are in short supply including deserted ring-tailed possum dreys, telephone interchange boxes, old bird nests, power boxes | 2, 3 |
| **Burramyidae** | | | |
| *Burramys parvus* | Soft grasses and moss | Rocky crevices, under boulders. Nest sites have been associated with rocky areas under dense vegetation, such as mother-shield ferns | 4, 5 |
| *Cercartetus* spp. | Male eastern pygmy possums may not build nests or do so infrequently. Breeding females build nests that can comprise a ball of fibrous bark of eucalypts or grass. A good hollow for breeding nests works well | Tree hollows, forks of tea-trees, leaves of grass tree, on the ground inside logs or hollows, stumps, birds' nests near ground and in the leaf litter at the base of *Banksia, Eucalyptus, Hakea, Leptospermum* and *Xanthorrhoea.* Female eastern pygmy possums appear to be dependent on hollows for breeding | 2, 6–8 |
| **Petauridae** | | | |
| *Dactylopsila* spp. | Leaves | Tree hollow or among clumped epiphytes | 2 |
| *Gymnobelideus leadbeateri* | Large nest of shredded bark | Tree hollows | 2 |
| *Petaurus* spp. | Leaves | Tree hollows. A record of sugar gliders in grass on the ground and even in the roofs of houses | 2, 9–11 |
| **Pseudocheiridae** | | | |
| *Hemibelideus lemuroides* | Leaves? | Tree hollows | 2 |
| *Petauroides* spp. | Leaves? | Tree hollows | 2 |
| *Petropseudes dahlii* | Not known to build a nest | Well-protected rocky crevices. | 2 |
| *Pseudocheirus* spp. | Spherical nests (drey) in densely branching trees or shrubs and made of ferns, leaves, twigs, stringy bark and lined with shredded bark and grass. Hollows may or may not be lined with leaves | Dreys in tree branches or on the ground. Also nest in tree hollows or trees stumps. Primarily use drey if hollows not available | 2, 12, 13 |
| *Pseudochirops* spp. | Does not appear to use nests | Rests on branch in tree canopy | 2 |
| *Pseudochirulus* spp. | Leaves are known to be used | Tree hollows and epiphytic clumps | 2 |
| **Phalangeridae** | | | |
| *Phalanger* spp. | None known | Tree hollow | 2 |
| *Spilocuscus* spp. | Thought to build a sleeping platform of leaves by drawing twigs under it | In thick canopy of a rainforest tree | |
| *Trichosurus* spp. | General leaves of vegetation (e.g. eucalypts), though not always used | Tree hollows, fallen logs, pipes, rock cavities, house roofs and even termite mounds | 2, 11, 14–16 |
| *Wyulda squamicaudata* | Not known | Rock piles, sunken rock piles, large rock slabs, underground rock crevices | 17 |

References: (1) F. Bradshaw pers. comm.; (2) Strahan (1995); (3) Fanning (1980); (4) Kerle (1984a); (5) M. Parrott pers. obs.; (6) Green (1980); 7) Goldingay (2020b); (8) Turner (2020); (9) Morrison (1978); (10) Jackson (2000b); (11) R. Goldingay pers. obs.; (12) Thomson and Owen (1964); (13) Augee *et al.* (1996); (14) Kean (1967); (15) Green and Coleman (1987); (16) S. Jackson pers. obs.; (17) Runcie (1999).

**Table 9.8.** Approximate nest box sizes (cm) for various species of possums.

Families are listed in approximately increasing body size. Measurements are internal sizes.
Derived from Menkhorst (1984b) and Arlidge *et al.* (1993).

| Species | Length × width × height | Entrance size |
|---|---|---|
| **Tarsipedidae** | | |
| *Tarsipes rostratus* | 15 × 20 × 30 | 3 |
| **Acrobatidae** | | |
| *Acrobates* spp. | 15 × 20 × 30 | 3 or 1.5 cm slit |
| **Burramyidae** | | |
| *Burramys parvus* | 14 × 12 × 10 to 25 × 18 × 11 | 5 |
| *Cercartetus* spp. | 14 × 12 × 10 | 2 |
| **Petauridae** | | |
| *Dactylopsila* spp. | 25 × 30 × 45 | 5 |
| *Gymnobelideus leadbeateri* | 20 × 30 × 45 | 4 |
| *Petaurus* spp. | 20 × 30 × 45 | 4–7 |
| **Pseudocheiridae** | | |
| *Hemibelideus lemuroides* | 20 × 30 × 45 | 8 |
| *Petauroides* spp. | 20 × 30 × 45 | 8 |
| *Petropseudes dahlii* | 20 × 30 × 45 | 8 |
| *Pseudocheirus* spp. | 20 × 30 × 45 | 8 |
| *Pseudochirops* spp. | 20 × 30 × 45 | 8 |
| *Pseudochirulus* spp. | 20 × 30 × 45 | 8 |
| **Phalangeridae** | | |
| *Phalanger* spp. | 20 × 30 × 45 | 15 |
| *Spilocuscus* spp. | 30 × 40 × 55 | 20 |
| *Trichosurus* spp. | 20 × 30 × 45 | 15 |
| *Wyulda squamicaudata* | 20 × 30 × 45 | 15 |

Walls 3 mm thick. The PVC pipe nest box was installed horizontally (Fig. 9.4c) and vertically during the study (Fig. 9.4e). The position of the entrance was changed to being on the end in the horizontal position and on the side ~4 cm from the top.

- small wood-cement nest box. Volume 0.0016 $m^3$. Measurements 10.5 cm diameter by internally 17 cm (H); walls 2.5 cm thick (Fig. 9.4d).
- all boxes had similar entrance sizes of ~2.5 cm (H) × ~4.5 cm (W).
- ~1.5 cm thick layer of wood shavings was placed inside each box for nesting.

Other studies of feather-tailed gliders that endeavoured to assess if there was a preference for different shape, volume and entrance found no clear preference (Goldingay *et al.* 2007). In that study nest boxes provided included:

- large (0.025 $m^3$) with an internal dimensions of 48 cm (H) × 28 cm (W) × 18.5 cm (D), slit entrance (1.5 × 15 cm wide) on the side (~30 cm from base)
- medium-sized box (0.008 $m^3$; 40 (H) × 14.5 (W) × 14 (D) cm) had a 45-mm diameter rear-facing circular entrance
- small-sized box (0.005 $m^3$; 23 (H) × 14 (W) × 14 (D) cm) had a 25 mm diameter rear-facing circular entrance
- wedge-shaped bat box (0.003 $m^3$; 19 (H) × 16 (W) × 12.5–5.0 (D) cm) with a slit entrance (2 × 16 cm) at the base.

The boxes were constructed from plywood, with the medium box being a mixture of ply and hardwood.

**Fig. 9.4.** Nest box designs provided to eastern pygmy possums in the field. (a) Large plywood box. (b) Small plywood box. (c) PVC pipe design in a horizontal position. (d) Wood-cement nest box. (e) PVC pipe design in a vertical position. Photos by Niels Rueegger. Reproduced from Rueegger *et al.* (2012) with permission from CSIRO Publishing. Photographs by Ross Goldingay.

The hardwood was ~30 mm thick while the ply was 18–25 mm thick and painted on the outside to enhance longevity.

Strong, wooden nest boxes can also be used to assist release of possums to the wild, by providing a safe transport box and a familiar location and shelter once released. This has been used for species including the mountain pygmy possum and Leadbeater's possum (M. Parrott pers. obs.).

## 4.9 Enclosure furnishings

It is important that the enclosure has a network of climbing branches to allow maximum use of the vertical space available. Fresh branches with leaves are required as they provide both cover to allow the possums and gliders to feel secure and also food for the folivores and behavioural enrichment due to their smell. Exudivores are also likely to explore the leaves of freshly cut branches of foliage for lerps, insects and honeydew; and the flowers for nectar and pollen if present. Gliding possums should ideally be given some open spaces in which to glide.

Smaller species such as mountain pygmy possums may be given rock piles to mimic the scree slopes in which they live; however, great care must be taken stacking them so they do not fall and cause injury or death. It may be an advantage to cover one or more next boxes with rocks.

# 5. GENERAL HUSBANDRY

## 5.1 Hygiene and cleaning

All enclosures should be cleaned daily to remove faecal matter and uneaten food. It is very important to keep the feed area as clean as possible due to the potential for health problems that can result from poor hygiene and bacteria entering the food. Drinking water dishes should be cleaned and refilled daily. When all individuals permanently leave an enclosure, it should be scrubbed out and cleaned as much as possible before new animals enter.

## 5.2 Record keeping

It is important to establish a system whereby the health, condition and reproductive status of captive possums and gliders are routinely monitored and recorded:

- identification numbers (all individuals should be identifiable)
- results of veterinary examinations and treatments
- behavioural changes or problems
- torpor or hibernation
- reproductive behaviour or condition
- bodyweights and measurements
- changes in diet and food consumption
- movements of individuals between enclosures or institutions
- births, with dam and sire if known
- deaths with necropsy results.

Collecting information on the physical, physiological and behavioural patterns of individual possums and gliders can contribute greatly to the husbandry of these species. It also allows individual histories to be transferred to other institutions if required and facilitates a cooperative approach to data collection among institutions. ZIMS, the Zoo Information Management System (Species360 2023) can be used for recording general information on age, sex, parentage, births, identification, bodyweights, transfers and deaths, with separate ZIMS portals for veterinary information and studbooks. These systems have been developed by Species360, which is part of the Conservation Planning Specialist Group (CPSG) software. ZIMS interfaces with programs, such as PMx, which provides demographic and genetic analysis tools to assist with the management of breeding programs (Lacy *et al.* 2011).

## 5.3 Methods of individual identification

### 5.3.1 PIT tags

PIT tags (microchips) are an excellent method of identification because they are permanent. They are implanted between the scapulae of individuals > 10 g in bodyweight and can be used on most species of possums and gliders. Sealing the entry wound with tissue glue (Vetbond®) will ensure that tag stays in place immediately after implantation. The animal generally needs to be caught to confirm identification with a PIT tag reader.

PIT tags are routinely used for the identification of feather-tailed gliders; however, ~11% of animals in the Taronga Zoo collection developed chip-associated tumours. Most of these animals were 6 years of age or older, which is well beyond their normal longevity (Dockerill 2013). Alternative methods that have been used for the identification of feather-tailed gliders include ear notching (Ward 1990a).

### 5.3.2 Tattoos

Tattooing the inside of the ear or inside hindleg has been used, but tattoos tend to fade and they can only be used on the larger species. Tattooing has largely been superseded by PIT tags.

### 5.3.3 Visual identification

Often difficult, especially with smaller species, but visual identification of larger species is often possible using size, colour, sex and markings.

### 5.3.4 Ear tags

Metal ear tags were used with some success in the past, but are not recommended because they can cause sore wounds and are prone to tearing out, particularly in species that have thin ears, such as the petaurids. However, metal ear tags have been used extensively in the field on > 700 petaurids and although some have torn out, most were successful (R. Goldingay pers. obs.; S. Jackson pers. obs.). In mountain pygmy possums, ear tags have caused infection, tearing and growth of scar tissue under the tag and are thus not recommended. If using ear tags, care is needed to avoid veins within the ear when making the hole.

### 5.3.5 Ear notching

Ear notching can be used to individually identify small possums, including wild pygmy possums (Fig. 9.5; Pestell and Petit, 2007). The ears are notched using surgical scissors that are sterilised (dipped in ≥ 70% ethanol and flamed or allowed to dry naturally), and a small piece of tissue ~1–2 $mm^2$ in the appropriate places around the ear margin is removed (Fig. 9.5; Waudby *et al.* 2022). Care needs to be taken to avoid cartilage and any obvious blood vessels. The tissue can be stored in 70% ethanol for later analysis of the DNA if required.

**Fig. 9.5.** Example of an ear notching numbering system to identify individual Australian pygmy possums. Reproduced from Pestell and Petit (2007) with permission from CSIRO Publishing.

## 6. FEEDING REQUIREMENTS

### 6.1 Captive diet

#### 6.1.1 Honey possum

- Ad lib
  - water
- Daily diet (per animal)

(Lactating diet/non-lactating diet)

- warm water 810 mL/810 mL
- honey 350 mL/300 mL
- pollen 70 g/35 g
- balance* 10 g/5 g
- ration: 10 mL/animal/day

*Balance – 100% whey protein powder available from health food shops.

*Diet used by Felicity Bradshaw *et al.* (2000; pers. comm.) & D. Philippe pers. comm.

Blend honey, pollen and approximately half the water in a blender until the pollen is broken down (~3 min). Add remaining water and blend briefly (~30 s). Measure the sugar content using a refractometer (Brix); it should be 20%. Different types of honey will have different sugar content and the volume may need to be adjusted to ensure the sugar percentage is correct.

This mixture is stored frozen in appropriate portions (for no longer than 1 week) to prevent fermentation and is fed at the rate of 10 mL each animal per day in 10 mL syringes fixed upright on a spring clip and inserted through a hole in the enclosure wall. Water is provided in a similar way to the nectar or in bird water holders (Russell and Renfree 1989), though it does not appear to be used, particularly if the enclosure is open to the weather (Bradshaw *et al.* 2000). As males are preferably removed once the female has young, she is then given the lactating diet and the male is given the non-lactating diet. When both males and females are together, they should be given the lactating diet.

Fresh flowers from eucalypts, bloodwoods, melaleucas, banksias and callistemons should be provided daily. After several days the flowers can potentially be sprayed with a fine mist of nectar (see section 6.1.10) to increase possum activity. The enclosure should be planted with species of *Banksia*, *Adenanthos*, *Isopogon*, *Grevillea*, *Dryandra* and *Callistemon*.

#### 6.1.2 Feather-tailed gliders

- Ad lib
  - water
- Daily diet (per animal)*
  - 8 g Wombaroo® Lorikeet & Honeyeater Food
  - 10 g mixed fruit chunks spiked in enclosure (e.g. melon, papaya, kiwifruit, apple, pear, orange, banana)
  - 5 g vegetable chunks spiked in enclosure (e.g. zucchini, pumpkin, spinach, sweet corn, sweet potato)
- Supplement
  - 1 g insects (e.g. mealworms, crickets, worms, maggots, fly pupae, grasshoppers, moths) three times per week
  - 1 blossom/flower as available
  - 0.25 g pollen grain once per week

*Diet used by Healesville Sanctuary.

Fresh flowers from eucalypts, bloodwoods, melaleucas, banksias, grevilleas and callistemons should be provided whenever possible. After several days the flowers can be sprayed with a fine mist of nectar (see section 6.1.10) to increase possum activity. The provision of native flowers has been shown to significantly increase daily foraging activity of gliders (Herrmann *et al.* 2013). Honey and natural peanut butter can also be used in small volumes as an enrichment smear.

#### 6.1.3 Pygmy-possums

Eastern pygmy possum

- Ad lib
  - water
- Daily diet (per animal)
  - 1 tsp Pigeon mix*

- 1 g fruit (apple, banana, orange, pear or fruit in season)
- Supplement
    - 0.5 g nectar mix (see section 6.1.10)
    - 1 g insects (fly pupae, mealworms, crickets, moths, cockroaches) 2–3 times per week
    - 2 g pollen once per week
    - 8 sultanas once per week

Fresh flowers from eucalypts, bloodwoods, melaleucas, banksias and callistemons should be provided wherever possible. After several days the flowers can be sprayed with a fine mist of nectar (see section 6.1.10) to increase possum activity.

The quantity of food offered for pygmy possums should be adjusted according to bodyweight to minimise the potential for over- or underfeeding. In particular, the amount of nectar mix should be monitored as they can become very overweight.

*Diet used by Healesville Sanctuary.

Mountain pygmy possum

- Ad lib
    - water
- Daily diet (per animal)

(Non-breeding animals)

- 4 mealworms
- 6 peas (frozen)
- 6 corn kernels (frozen)
- 4 sprouted mung beans
- 3 Advance™ Puppy kibble
- 2 g baked Wombaroo® Bogong bikkies
- Supplement
    - 1 moth, 3–4 times per week (increase during breeding season)

*Diet used by Healesville Sanctuary.

(Hibernation season)

- Possums > 60 g:
    - 10 sunflower seeds
    - ¼ walnut
- Possums < 60 g:
    - 20 sunflower seeds
    - 3 g walnut pieces

(Lactating females)

- 4 mealworms (gut-loaded with Wombaroo® Insectivore Mix)
- 6 peas (frozen)
- 6 corn kernels (frozen)
- 4 sprouted mung beans
- 3 Advance™ Puppy kibble
- 2 g baked Wombaroo® Bogong bikkies
- 4 to ad lib moths
- 8 almonds
- 8 peanuts
- ¼ walnut
- 1 mL Wombaroo® Lorikeet & Honeyeater Food
- 5 g sunflower seeds

*Diet used by Healesville Sanctuary.

### 6.1.4 *Petaurus* gliders

The quantity of food offered for *Petaurus* gliders should be adjusted according to bodyweight to minimise the potential for over- or underfeeding. In particular, the amount of nectar mix should be monitored as they can become very overweight.

Yellow-bellied glider

- Ad lib
    - water
- Daily diet (per animal)
    - 30 mL nectar mix (see section 6.1.10)
    - 15 g mixed fruit in 10-mm cubes
    - 4 g fly pupae
    - 10 mealworms
- Supplement
    - 3 pieces of Royal Canin Feline Dental kibble
    - 1.5 g Pet Health food once per week
    - 0.5 g egg once per week
    - 0.4 g pollen
    - Insects (including crickets and moths) 3–4 times per week
    - 1 branch acacia/flowering gum per week

*Diet used by Healesville Sanctuary.

Fresh flowers from eucalypts, bloodwoods, melaleucas, banksias and callistemons should be provided wherever possible. After several days the flowers can potentially be sprayed with a fine mist of nectar (see section 6.1.10) to increase possum activity.

Sugar gliders

- Ad lib
    - water
- Daily diet (per animal)
    - 1 Royal Canin Feline Dental Kibble

  - 6 g fruit, chopped
  - 1 teaspoon nectar mix (see section 6.1.10)
  - 1 g insects (fly pupae, crickets, cockroaches)
  - 5 g corn kernels
  - 2 g sprouted seed*
  - 3 mealworms
  - 1 g Wombaroo® Small Carnivore Food (mix with water to soft crumble)
  - 4 g green leafy vegetables (kale, endive, spinach, silverbeet, bok choy)
  - 4 g sweet corn
  - 4 g sweet potato/pumpkin/carrot
  - 10 mL Wombaroo® Lorikeet & Honeyeater Food
- Supplement
  - 10 pollen granules once per week
  - other insects (e.g. mealworms, worms, maggots, moths)
  - acacia, eucalypts, other blossoms as available
  - 3 g fruit
  - peanut butter smear, honey smear once per week each as enrichment

*Diet used by Healesville Sanctuary.

Mahogany glider
- Ad lib
  - water
- Daily diet (per animal)
  - 1 Advance Pet Food Kibble
  - 20 mL nectar mix (see section 6.1.10)
  - 5 g insects (fly pupae, mealworms, crickets, cockroaches)
  - 20 g mixed fruit (e.g. apple, banana, corn, currents, figs, grapes, honey dew melon, mango, paw paw, pear, rock melon, stone fruit (without seeds) and water melon (Whiteford and Booth 2007)
  - 5 g corn kernels
  - 1 g sprouted seed

*Diet used by Jackson *et al.* (2019).

- Supplement
  - 5 g of extra protein should be supplied every second day and can include meal worms, crickets, grass hoppers, fly pupae, moths and egg custard (Whiteford and Booth 2007)
  - acacia, eucalypts, other blossom when available

When feeding insects, it is a good idea to gut-load them with a calcium and phosphate powder such as *Insect Booster* by Passwell® (Gordon Rich, Wombaroo® pers. comm.). The feeding of a diet primarily of insects such as mealworms and crickets without additional calcium can lead to an imbalance of calcium and phosphorous.

Squirrel glider
- Daily diet (per animal)
  - 1 Royal Canin Feline Dental kibble
  - 20 g mixed fruit and vegetables in 10-mm cubes; avoid soft fruits
  - 5 mL nectar mix (see section 6.1.10)
  - 2 g insects (fly pupae, mealworms, crickets, cockroaches)
  - 0.5 g bee pollen
  - 5 g Wombaroo® Small Carnivore Food (mix with water to make soft crumble)
  - 8 g green leafy vegetables (kale, endive, spinach, silverbeet, bok choy)
  - 4 g sweet corn
  - 4 g broccoli
  - 4 g sweet potato/pumpkin/carrot
  - 12 mL Wombaroo® Lorikeet & Honeyeater Food
- Supplement
  - 1 native blossom/flower as available
  - fruit, peanut butter smear, honey smear once per week each as enrichment

*Diet used by Healesville Sanctuary

### 6.1.5 Torresian striped possum

- Ad lib
  - water
- Daily diet (per animal)
  - 10 live crickets or other insects (will readily accept katydid insects and occasionally small beetles and moths)
  - 10–15 mealworms dusted in ½ teaspoon of Wombaroo® Insectivore Mix
  - 10 Mazuri® soaked primate pellets crushed
  - 50 g fruit (e.g. apples, banana, black persimmon, canistel, cherry, custard apple, grape, lychee, mamey sapote, mango, mangosteen, melon, nectarine, orange, paw paw, peach, pear, plum, rambutan, sapodilla; may show little or only occasional interest)

*Diet derived from Carmichael (2000), McKenna (2005) and D. Dickson *pers. comm.*

- Supplement

- nectar mix given occasionally consisting of honey, boiled egg (with shell), high-protein baby cereal and water (F. Wheeler and A. McKenna pers. comm.)
- decomposing rotten logs, such as eucalypts and acacias, with lots of wood-borers, should be supplied whenever possible to encourage natural foraging (Carmichael 2000)
- flowers of trees of the genera *Eucalyptus*, *Corymbia*, *Melaleuca* and *Grevillia* should also be provided

Insectivorous bird mix for striped possums (Carmichael 2000)

- 50% egg cake mix (see below)
- 15% Wombaroo® Insectivore Mix
- 15% egg and biscuit mix
- 10% hard-boiled egg
- 5% grated cheese
- 5% fly pupae

Egg cake recipe

- 2 kg self-raising flour
- 100 g margarine or butter
- 12 eggs
- 750 g sugar

Add water to make up to cake mix consistency.
Bake in flat cake tin in moderate oven.
25–30 mL of Wombaroo® Lorikeet & Honeyeater Food
10–15 mealworms dusted in ½ teaspoon of Wombaroo® insectivore mix
15 g avocado mashed with freshly grated almond (1–2 almonds)
Drizzle of genuine maple syrup and local honey helps with the uptake of the excess insectivore mix.

### 6.1.6 Leadbeater's possum

- Ad lib
  - water
- Daily diet (per animal)
  - 25 mL nectar mix (see section 6.1.10)
  - 4 insects (crickets, fly pupae, mealworms, moths, cockroaches)
- Supplement
  - *Acacia* branch weekly
  - maple syrup smear, fly pupae, pinkie mice, Wombaroo® Lorikeet & Honeyeater Food, native flowers/blossoms, tree sap as enrichment

*Diet used by Healesville Sanctuary.

### 6.1.7 Ring-tailed possums

- Ad lib
  - water
  - new shoots of eucalypt leaves
- Daily diet (per animal)
  - 4 g apple
  - 4 g sweet potato
  - 3 g sprouted seed*
  - 6 g fly pupae

Possums must be offered at least three branches of fresh eucalypt daily so it forms the majority of their diet. If browse is unavailable, the above diet can be supplemented with 4 g apple and 4 g pear/other hard fruits).

- Supplement
  - 2 g Advance® Pet Food Kibble twice per week
  - 8 Almonds, shelled 3–4 times per week
  - native flowers as available
  - fresh branches of foliage to eat (e.g. *Eucalyptus ovata*, *E. dives*, *E. maculata* and *Leptospermum* spp.)

*Diet used by Healesville Sanctuary.

### 6.1.8 Greater glider

- Ad lib
  - water

Fresh branches of young leaves and buds of trees of the genera *Corymbia* and *Eucalyptus*, but may include some other genera including *Acacia*, *Agonis*, *Callistemon* and *Melaleuca* (Table 9.9).
50 mL nectar mix (see section 6.1.10)

### 6.1.9 Cuscuses, brush-tailed possums and scaly-tailed possum

- Ad lib
  - water
- Daily diet (per animal)
  - 4 Advance Pet Food Kibble
  - 20 g apple (¼ medium size)
  - 40 g orange (¼)
  - 10 g banana (¼)
  - 10 g carrot (medium) (¼)
  - 30 g pear (¼)
  - 10 g slice corn

**Table 9.9.** Species of plants known to be eaten by greater gliders throughout their distribution.

***Most preferred species. Initials for the different tree genera include *Ac* = *Acacia*, *Ag* = *Agonis*, *C* = *Corymbia*, *E* = *Eucalyptus* and M = *Melaleuca*.

| Scientific name | Common name | Preference | Reference |
|---|---|---|---|
| *C. citriodora* | Lemon-scented gum | * | 1–3 |
| *C. gummifera* | Red bloodwood | *** | 4 |
| *C. henryi* | Large-leaved spotted gum | *** | 5 |
| *C. intermedia* | Pink bloodwood | * | 2, 3, 6, 7 |
| *C. maculata* | Spotted gum | *** | 5 |
| *C. torelliana* | Cadaghi | *** | 5 |
| *C. trachyphloia* | Brown bloodwood | * | 6 |
| *E. acmenoides* | White mahogany | *** | 2 |
| *E. amplifolia* | Cabbage gum | *** | 5 |
| *E. andrewsi* | New England blackbutt | *** | 8 |
| *E. cameronii* | Diehard stringybark | *** | 8 |
| *E. consideniana* | Yertchuk | *** | 4 |
| *E. creba* | Narrow-leaved red ironbark | *** | 3 |
| *E. cypellocarpa* | Mountain grey gum | * | 9 |
| *E. dalrympleana* | Mountain gum | ** | 4 |
| *E. dives* | Broad-leaved peppermint | *** | 10 |
| *E. drepanophylla* | Grey ironbark | * | 6 |
| *E. dunnii* | Dunn's white gum | *** | 5 |
| *E. eugenoides* | Thin-leaved stringybark | *** | 5 |
| *E. exserta* | Queensland peppermint | * | 6 |
| *E. fastigata* | Brown barrel | *** | 4, 9, 11 |
| *E. fibrosa* | Red ironbark | *** | 1 |
| *E. globoidea* | White stringybark | *** | 11 |
| *E. globulus* | Tasmanian blue gum | * | 10 |
| *E. goniocalyx* | Long-leaved box | *** | 12, 13 |
| *E. haemostoma* | Scribbly gum | *** | 5 |
| *E. laevopinea* | Silver top stringybark | *** | 8 |
| *E. latisinensis* | White mahogany | *** | 3 |
| *E. macrorhyncha* | Red stringybark | *** | 10 |
| *E. microcorys* | Tallowwood | * | 5, 8 |
| *E. moluccana* | Grey box | *** | 1, 7 |
| *E. nicholii* | Narrow-leaved black peppermint | *** | 8 |
| *E. obliqua* | Messmate | * | 8, 9 |
| *E. ovata* | Swamp gum | * | 9, 13 |
| *E. pauciflora* | Snow gum | * | 14 |
| *E. pellita* | Large-fruited red mahogany | ** | 4 |
| *E. piperita* | Sydney peppermint | *** | 4 |
| *E. propinqua* | Small-fruited grey gum | *** | 5 |
| *E. racemosa* | Southern scribbly gum | *** | 5 |
| *E. radiata* | Narrow-leaved peppermint | *** | 4, 8–10,13, 15 |
| *E. robertsoni* | Robertson's peppermint | *** | 14 |
| *E. robusta* | Swamp mahogany | *** | 5 |

| Scientific name | Common name | Preference | Reference |
|---|---|---|---|
| *E. rostrata* | River red gum | * | 13 |
| *E. rudis* | Flooded gum | * | 5 |
| *E. sierberi* | Silvertop ash | *** | 4 |
| *E. signata* | Sydney blue gum | *** | 5,8 |
| *E. stuartiana* | Apple-scented gum | * | 13 |
| *E. tereticornis* | Forest red gum | * | 2, 7 |
| *E. umbra* | Broad-leaved white mahogany | *** | 6 |
| *E. viminalis* | Manna gum | *** | 4, 8–11,13 |
| *Ac. melanoxylon* | Australian blackwood | *** | 10 |
| *Ag. flexuosa* | Western Australian peppermint | *** | 5 |
| *Callistemon* spp. | Bottlebrush | * | 5 |
| *M. leucadendra* | Weeping paperbark | *** | 5 |
| *M. quinquenervia* | Broad-leaved paperbark | * | 3, 6 |

References: (1) Smith *et al.* (2007); (2) Comport *et al.* (1996); (3) Eyre *et al.* (2022); (4) Norton (1988); (5) H. Meer pers. comm.; (6) Foley *et al.* (1990); (7) Starr *et al.* (2021); (8) Foley (1984); (9) Kavanagh and Lambert (1990); (10) O'Donnell (1970); (11) Henry (1995); (12) Smith (1965); (13) Fleay (1947); (14) Marples (1973); (15) Hume *et al.* (1984).

  - 15 g sprouted seed*
  - 13 g green vegetables (e.g. silverbeet, sow thistle, *Tradescantia* sp., spinach)
  - ½ kiwi fruit when available
  - 20 g cantaloupe
  - 20 g honeydew melon
  - 10 g sweet potato
  - 7 g fly pupae
- Supplement
  - 3 g sultanas or sunflower seeds 3–4 times per week
  - 5 currants
  - 1 almond 3–4 times per week
  - other fruits in season (e.g. kiwi fruit)
  - fresh *Acacia* and *Eucalyptus* branches as available

*Diet used by Healesville Sanctuary.

Scaly-tailed possum - although a diet similar to that used for the brush-tailed possum is likely to be adequate, they appear to readily eat eucalypt leaves, casuarina nuts (from which the seeds are prised out) and blossoms of *Callistemon* sp., *Calothamnus* sp., *Eucalyptus* sp., *Grevillea* sp., *Hypocalyma* sp., *Hibbertia* sp., *Melaleuca* sp. and roses. Fruits include apple, banana, cherry, feijoa, grapes, orange, lilly pilly and paw paw. Nuts are readily eaten including brazil, Barcelona, cashew, pistachio, almonds and walnuts, with peanuts being ignored (Fry 1971; Fairfax 1982). Caching of nuts has often been observed (Fry 1971). In the wild they have been observed to feed upon the leaves of *Eucalyptus* sp., *Planchonia careya*, *Xanthostemon eucalyptoides* and *X. paradoxus*, as well as the seeds of the perennial herb *Trachymere didiscordis* (Runcie 1999). So leaves and seeds of these genera should also be provided where possible.

Cuscuses - the brush-tailed possum diet, enlarged or reduced according to body size, works well for cuscuses. Other food items they have been fed include sweet potato, yam, lettuce, leafy vegetation from eucalypts and other plants (Menzies 1972; Shoemaker and Croxton 1982).

### 6.1.10 Nectar mix

- 320 mL water (not boiling)
- 60 g Wombaroo® Lorikeet & Honeyeater Food
- 60 g Wombaroo® High Protein Supplement

Mix to a smooth paste with a little hot water first so the mix forms a slurry with no lumps.

Freeze in suitable volumes for feeding.

Alternative nectar mix

- 900 mL warm water
- 900 mL honey
- 6 shelled hard-boiled eggs
- 150 g Wombaroo® High Protein Supplement
- 6 tsp Sustagen® (vitamin supplement) (Nestlé)

Add the warm water to a 2 L jug and then slowly add the honey and stir until it dissolves.

Blend the eggs (no shells) until mushy.

Add ½ of the honey/water mix, blend. Add remainder of mix. Blend.

Add Sustagen® and ½ of the Wombaroo® High Protein Supplement. Blend.

Add remainder of Wombaroo® High Protein Supplement. Blend for 1.5 min to make lump free.
Can be stored for up to 2 weeks.

## 6.2 Supplements

As per individual diets.

## 6.3 Presentation of food

Most species of possums and gliders are generally fed with a feed dish placed either on the ground or on a platform placed on a side wall or on the trunk or branch of a tree. Observations in captivity suggest that the dishes for greater gliders should be placed in the branches of eucalypts near the top of the enclosure, as they do not like having to descend to feed (Collins 1973). In captivity, both exudivores and folivores have a tendency to eat what they need and then return back to their nest boxes. Therefore, to maximise the time they spend on exhibit (particularly for the exudivores) the food should be spread out in small amounts (see section 9.1). This will also increase their exercise and reduce the incidence of obesity. This strategy can be further enhanced by feeding at irregular times to avoid abnormal repetitive behaviour. It is very difficult to increase the activity time of large folivores due to their slow metabolism and resulting high digestion times; however, replacing foliage in enclosures regularly appears to stimulate activity. If arboreal possums or gliders are to be released to the wild, food should be provided high in the enclosure so as not to habituate possums to foraging on the ground where they may be more likely to fall prey to cats (*Felis catus*) or foxes (*Vulpes vulpes*).

# 7. HANDLING AND TRANSPORT

## 7.1 Timing of capture and handling

Possums and gliders are generally best caught during the day while they are asleep in their nest boxes. If they are held in a nocturnal house, they can be caught early in the morning before the lights go out. Alternatively, they can be netted or trapped within the enclosure.

## 7.2 Capture equipment

Finely woven calico cloth bags are ideal for sugar gliders and pillow cases work well for the petaurids. Larger possums such as ring-tailed possums, greater gliders, brush-tailed possums and cuscus should be placed in larger cotton, calico or hessian bags.

Depending on the species and the enclosure in which they are being captured, the different species of possums and gliders may be captured within Elliott traps or with the use of nets attached to a long handle. Further discussion on the use of different types of capture equipment can found in section 7.3.

## 7.3 Capture and physical restraint

As a general rule when handling possums both large and small do it as gently as possible, with the handler not exhibiting stress. Covering the eyes as soon as possible after capture generally calms the possum. If an individual is handled frequently and gently it will typically become accustomed to being handled. Providing the possum with something to eat while checking the pouch or taking a cloacal swab can help calm the animal.

It is often easy to avoid direct handling or restraint of the possums (unless required for an examination) as quiet individuals can generally be encouraged to move from the nest box into a bag (W. Gleen pers. comm.). If the animal or animals are to be moved to another enclosure, such as during an enclosure renovation, it is often easy to place a catching bag snugly in the nest box entrance and carefully carry the whole nest box to the new enclosure where the catching bag can be removed. Nest boxes may also be constructed with movable door coverings that can be locked in place if catching or movement are likely to be frequent. Care must be taken to ensure the door is opened and latched safely to avoid possums being locked in for longer than necessary. If animals are in large, free-ranging type enclosures, Elliott small live mammal traps baited with favourite diet items can also be used to catch animals, though must be checked first thing in the morning and in the afternoon for any caught animals. Nesting material should be provided and the traps placed in sheltered areas away from wind, rain or sun.

### 7.3.1 Small possums

Small species such as feather-tailed gliders can be easily picked up and held in the hand, as they generally do not bite. The lower incisors of the eastern pygmy possums are very sharp and can inflict a somewhat painful bite so gloves may be used. Although gloves do not have enough feel or dexterity when catching, they may be useful when checking the pouch or taking measurements. These animals can be captured with the use of an everted sock over the hand when catching from a nest box (R. Goldingay pers. obs.).

### 7.3.2 Medium to large possums and gliders

Larger species, such as the petaurids and ring-tailed possums, will bite and scratch with their very sharp claws. These species should be caught and handled with care, which often involves using sturdy gloves. They are best caught in their nest box by plugging up the entrance hole with a cloth bag, taking the nest box off the wall, placing it on the ground in an enclosed well-lit area and then carefully opening the box and placing a catching bag over the animals. An alternative is to use a hand within an everted bag to hold the glider through the bag, remove it from its box, and then evert the bag back over the glider. If there are multiple gliders in a box an extra bag can be placed over the remaining animals as they are captured one by one (R. Goldingay pers. obs.; S Jackson pers. obs.).

If the possum or glider is out in the enclosure, they can generally be caught with a hand-held net that is gently inverted into a catching bag (making sure the claws are not caught). Older, heavier individual brush-tailed possums and possibly cuscuses can be caught by the end of the tail and held aloft, because they cannot climb back up their tail.

Ring-tailed possums will usually try and escape if they are healthy and alert. Although they rarely stand and fight, like brush-tailed possums, they can inflict a bite and have sharp claws that can inflict painful wounds. They can be caught with a catching net, home-made soft catch pole or can be covered with a blanket or bag and placed into a catching bag or cage/box for transport.

Brush-tailed possums and cuscus can be caught in a similar way to petaurids or ring-tailed possums; however, in an urgent or emergency situation they can also be caught using a noose connected to a long pole. Try to gentry place the noose around one armpit and the neck of the animal (like a winner's sash). This lessens the strangulation effect and the animal can be manoeuvred into a waiting carry bag, box or cage with more control. When catching brush-tailed possums or cuscus, use a large hessian bag. A pillowslip is not recommended as they are too lightweight and tear easily; calico bags work well. Always use any bag inside out (i.e. overlock stitching showing on the outside) because it is very easy for animals to catch their claws in the stitching, resulting in the claws being pulled out or the animal becoming entangled.

When using a catchpole, a blanket can be spread out over the animal, giving a wider area in which to control it. When folded, blankets give handlers extra protection against being bitten. With practice and gentle pressure, the head and body of the possum can be determined through the blanket for safe handling of the animal. Animals may also be directly transferred into a waiting handling bag.

In emergency situations, once captured, brush-tailed and ring-tailed possum can be grasped by the base of the tail (only) and held in one of several ways:

- held totally off the ground. Possums tend to climb up themselves and can bite the handler. This can be avoided in brush-tailed possums by slowly twirling the animal
- lift the animal so that its front feet are still touching the ground or a tree trunk, which usually eliminates the desire for the animal to try to climb up its own tail and bite as it is generally trying to get away (should only be done by experienced handlers and for as short a time as possible to avoid damage to the tail, including the potential for broken or dislocated tails).

The scruff of the neck and shoulders may be firmly held or the neck can be held either between the index and middle fingers or between the thumb and index finger for large animals (Fig. 9.6). This is best achieved when the animal is in a bag by placing gentle pressure on the head down to control it with one hand and running the other hand up the back until the grip around the neck can be achieved. Then hold the base of the tail with the other hand. Ensure the weight of the animal is in the lower hand, rather than holding the weight of the animal by its neck or shoulders, which could cause injury and stress (see Fig. 9.6a). The animal's tail can be held securely under the armpit of the handler, freeing up a hand. If an animal is being held by the tail and the neck, it can be placed gently into a bag; this manoeuvre is often easier with two people.

Once the animal is inside a handling bag, unroll the top of the bag and tie it shut about seven-eighths of the way up the bag, making sure not to catch the tail or any other part of the body in the neck of the bag. It is often easier to hold the bag slightly upside down, with the opening held closed by hand just before tying the bag up, as all possums generally climb upwards inside the bag. This should ensure that all parts of the possum are fully inside the bag. It is important not to hold a possum in a hessian (or any other) bag while seated as they can bite through the bag. The darkness of a bag or box generally helps to calm the animal.

**Fig. 9.6.** Handling techniques for handling (a) ring-tailed possums and petaurids and (b) brush-tailed possums. Note that long leather gloves are often used for handling brush-tailed possums. Photo by Stephen Jackson.

If the possum is in a cage for transportation, always cover it with a blanket or place the cage into a hessian bag before transportation. Nose injuries often occur from the possum trying to escape through the wire of the box if they can see out. Usually, the injury consists of a bloody nose, some swelling and scraping on top of the nose.

Many people do not use gloves, as they tend to dramatically reduce the dexterity of the fingers and the manoeuvrability of the hands. Soft rigger's gloves allow better feeling of the animal when handling them.

## 7.4 Weighing and examination

A useful technique for examining the pouches of possums and gliders involves using a transparent plastic tube and an otoscope (Roberts and Kohn 1991). This allows a clear view of the animal through the tube and confines the front limbs for examination. Care must be taken to cover the eyes of the animal to reduce stress. In small possums, the animals can be gently rolled onto its back and the pouch carefully opened. This should not be done with more sensitive species, such as feather-tailed gliders, when they have young as it can cause abortion of the litter.

Weighing is best undertaken by placing the possum in a catching bag as described above and using either hanging or digital scales if available (especially for small species of possums). Conversely, small possums can be placed in covered plastic tubs (with air holes in the lid) for weighing. This is especially useful if urine or faeces is required for reproductive or health analyses as both can be collected directly from the tub (though animals should not be left in the tub for more than a few minutes). Feather-tailed gliders can be weighed in thicker (e.g. snap-lock type) plastic bags that are not sealed or have several small holes at the top to allow plenty of air to enter, using a fine-scale spring balance as the bodyweights will be more accurate and the animals easier to see and handle.

## 7.5 Release

Possums and gliders are generally best released directly into the nest box, if that's where they were first caught, on the ground or onto a branch or tree trunk.

## 7.6 Transport requirements

### 7.6.1 Transport box design

The various species of possums and gliders are relatively easily transported. For short distances (e.g. several hours drive), they are readily transferred in a catching bag, although they should ideally be placed in a nest box with the entrance locked or plugged up, which acts as a secondary barrier if they escape from the bag. The nest box also provides protection, particularly for small species, from other objects that may crush them. Cars or other transportation must be air-conditioned to ensure the animal cannot overheat and is kept as quiet as possible.

Whenever possums and gliders are transported by aeroplane, they should be placed in a recommended wooden box suggested by the International Air Transport Association (IATA 2022). Within this box the possum or glider can be held directly inside a bag (for shorter distances, e.g. several hours) and provided with nesting material so that it does not roll around too much. If the possum or glider is to travel longer distances it should be placed inside the box (without a bag as this can cause overheating) and provided with adequate nesting material. Care must be taken to provide suitable temperatures and air-conditioning, especially for animals such as mountain pygmy possums, which should be transported in cool temperatures (see section 4.6).

### 7.6.2 Furnishings

When placed in a wooden box, possums and gliders should be provided with clean, soft nesting material, such as shredded paper, whether they are placed inside a bag or not, to provide insulation and to stop them rolling around excessively during the trip.

### 7.6.3 Water and food

When transporting animals in wooden boxes secure a water container that has no sharp edges to the side of the box. Water can be provided by placing a piece of clean sponge or rag inside the container and soaking it with fresh water or by using a dripper bottle. A small amount of food that is not likely to be easily spoiled can be provided. Females with pouch young should not be transported unless the young have only recently been born and are still permanently attached to the teat.

### 7.6.4 Animals per transport box

Ideally, all individuals should be housed separately during transport. The very small species such as honey possums and, possibly, feather-tailed gliders can be transported together if they have previously been held together in the same enclosure. Females with pouch young should not be transferred unless the young have only recently been born and are still attached to the teat. However, in an emergency (the 2009 Black Saturday bushfires), mountain pygmy possum females with large young in their nests were evacuated from Healesville Sanctuary to Melbourne Zoo in their familiar nest boxes with no losses of young or obvious issues (M. Parrott pers. obs.).

### 7.6.5 Timing of transportation

Ideally, animals should be transported overnight or in the cooler part of the day, though the temperature should not be too cold (e.g. between 10–20°C).

### 7.6.6 Release from the transport box

Once in the new enclosure, open the bag or box, uncover the animal's head so that it can see outside and then leave it to emerge from the bag or box in its own time. The bag and box are then removed once the animals have fully emerged, if they are not going to be used permanently by the animal as a nest box.

## 8. HEALTH REQUIREMENTS

### 8.1 Daily health observations

Each animal should be observed daily for any signs of injury or illness. It is very important to be familiar with the normal behaviour of the group or individuals, as deviations from this pattern will assist in identifying if there is a potential problem. For example, places where an individual sleeps may change, it may not approach when it normally does or approaches when it normally does not. Providing a small amount of food that the possum or glider prefers on entering the enclosure to encourage the animals to approach will assist in observing their condition, movement and development. The most appropriate time to do this is generally when the enclosure is being cleaned or at feeding. During these times, each animal in the enclosure should be checked for the following:

- coat condition
- fur on the enclosure floor or elsewhere, suggesting fighting or mating, skin diseases
- discharges from the eyes, ears, nose, mouth or cloaca
- appetite
- faeces - number of pellets and consistency
- changes in demeanour
- injuries - including abrasions, swelling around the face, lameness and any asymmetry
- presence and development of pouch young by observation of the bulge in the pouch.

### 8.2 Bodyweight and body condition score

#### 8.2.1 Bodyweight

Record and compare with previous weights. Trends in bodyweight give a good general indication of the animal's state of health, provided age, sex and geographical location are taken into account. Animals in captivity should be weighed monthly to indicate trends, except when females have young and then a hands-off approach is taken.

### 8.2.2 Body condition score

Body condition score can be assessed using several methods, including muscle palpation of the area over the scapula, spine and temporal fossa or by examining the base of the tail and allocating a body condition score (Austin 1997). With experience, condition scores can be allocated (Austin 1997; Viggers *et al.* 1998). Body condition has also been calculated by taking cube root of body mass (g) divided by pes length (mm) (Fleming 1980; Kerle 1998).

## 8.3 Physical examination

The physical examination may include the following:

- temperature - usually 35–36°C, taken via the cloaca
- pulse rate - normally 200–300 beats/min at rest in sugar gliders (Booth 1999). The rate varies greatly with species, decreasing with increasing body size. The rate is taken over the femoral artery or direct heart rate. Note that the heart rate decreases markedly when animals are in torpor or hibernation (to as low as a couple of beats per minute)
- respiratory rate - normally 16–40 breaths/min at rest in sugar gliders (Booth 1999), monitored via auscultation of the lungs. It varies greatly with the species, with the rate decreasing with increasing body size
- fur - check for alopecia, ectoparasites, fungal infections or trauma
- eyes
  - should be clear, bright and alert
  - normal bilateral pupillary light response
  - normal corneal reflex
  - should not have any discharges
- also check for the presence of lumps over body and auscultation of lungs
- cloaca
  - should be clean
  - check for faeces around the cloaca
- pouch
  - condition of the pouch
  - check whether lactation is occurring by milking teats
  - if pouch young are present, record sex, stage of development, bodyweight if detached from the teat and measure to determine age from growth curves, if available. Note that checking the pouch when females have young in some species (such as feather-tailed gliders and some pygmy possums), can cause abortion of the litter and thus should be avoided.
- males
  - check testes - size (length, width, depth) and consistency (firm, not soft)
  - extrude penis and assess
  - check the size and activity of the sternal gland and forehead gland in petaurid gliders.

## 8.4 Chemical restraint

Pre-anaesthetic fasting is not required for adult animals as they are not prone to regurgitation (Vogelnest 1999). If being hand-reared, they should be fasted for 1 h before anaesthesia to prevent regurgitation of the formula. Sedation can be undertaken with diazepam (Valium®) at 0.5–1.0 mg/kg given intramuscularly in the thigh muscle for minor procedures and handling (Vogelnest 1999).

Injectable agents are useful for large or fractious possums. Tiletamine/zolazepam (Zoletil®) is the agent of choice. It can be used at 4–10 mg/kg intramuscularly or 1–3 mg/kg intravenously in the lateral coccygeal vein near the base of the tail (Vogelnest 1999). Tiletamine/zolazepam (Zoletil®) is not recommended in *Petaurus* gliders, as it has been implicated in deaths of three healthy squirrel gliders (Holz 1992; Booth 1999). Inhalation anaesthesia via mask induction is preferred for small possums and gliders (R. Booth pers. comm.; Ludovico pers. obs.). Isoflurane is preferred for inhalation anaesthesia, although halothane in oxygen can also be used. Mask induction is simple, rapid and smooth with maintenance via the mask or intubation in larger species (Vogelnest 1999).

## 8.5 Common health problems

Possums and gliders suffer several problems in captivity. The majority of parasites and diseases that have been recorded are presented below in Tables 9.10 and 9.11.

## 8.6 Preventative medicine

To reduce the risk of infectious diseases in captive populations, several preventative measures should be applied. New animal admissions should be quarantined and routine health checks (including faecal examination) should be undertaken. Hygiene is paramount to reduce the risk of disease, including, but not limited to, daily enclosure cleaning, removal of uneaten food, water changes and bedding changes.

**Table 9.10.** Common infectious and parasitic diseases of zoo-based possums and gliders.

| Disease | Cause | Clinical signs | Diagnosis | Treatment | Prevention | Comments | Reference |
|---|---|---|---|---|---|---|---|
| **Ectoparasites** | | | | | | | |
| Ectoparasitism | Ticks, mites, lice, fleas have been found on possums and gliders, including Leadbeater's possum, *Petaurus* spp., brush-tailed possums and mountain pygmy possum. Brush-tailed possums have also been found with mites, including *Sarcoptes scabei*, resulting in mange | Generally seen on the animal when captured, by excessive grooming, hair loss or inflamed skin | Visual observations or a skin scraping and microscope examination to identify the parasites | Replacement or treatment of nest boxes and bedding to control mites. Injectable or topical ivermectin or moxidectin also controls a range of ectoparasites | Maintaining good hygiene and routine examination of the fur. Quarantine of new arrivals helps prevent the introduction of ectoparasites | | 1–8 |
| **Endoparasites** | | | | | | | |
| Endoparasitism | Various flukes (trematodes), tapeworms (cestodes) and roundworms (nematodes) have been found in possums and gliders. For examples, brush-tailed possum are known to have the lungworm (*Marsupiostrongylus* and *Trichostrongylus* sp.). The trematode (*Athesmia* sp.) has been found in the liver and the nematodes (Parastrongyloides, Paraustrostrongyloides) and potentially Paraustroxyuris have been found in the gastrointestinal tracts of sugar gliders | Not obvious unless diagnosed. May cause diarrhoea, anaemia or ill-thrift | Faecal flotation and a wet preparation to note for the presence of eggs or proglottids (segments that make up the worms) | Anthelmintics can be used without apparent side effects in possums and gliders and include fenbendazole, oxfendazole, and ivermectin | Good hygiene by the removal of faeces and quarantine of new animals | | 2, 7, 9–14 |
| **Fungal** | | | | | | | |
| Dermatophytosis or ringworm | Infection of the skin causing alopecia. It is typically caused by an infection from the genera *Microsporum* sp. or *Trichophyton* sp. Infection generally occurs in immunosuppressed hosts | Can be noted by alopecia, scaling, crusting, itchy skin, papules, hyperpigmentation and erythema | Woods lamp examination of affected areas. Skin biopsy or scrape and a fur pluck can be collected for culture or for examination under the microscope (mineral oil is needed for examination under the microscope) | Itraconazole 5 mg/kg orally once daily on a week on, week off rotation until symptoms resolve. Systemic treatment can be accompanied by topical treatment with antifungal shampoo baths performed twice weekly | Maintaining good hygiene and routine examination of the fur. Quarantine of new arrivals helps prevent the introduction of fungal infections | Dermatophytosis is a zoonotic disease | 15 |

**Table 9.10.** continued

| Disease | Cause | Clinical signs | Diagnosis | Treatment | Prevention | Comments | Reference |
|---|---|---|---|---|---|---|---|
| Cryptococcosis | The yeast *Cryptococcosis neoformans* has been established as the cause of death in Leadbeater's possums and koalas in captivity. The lesions produced were within the nasal sinus. Signs of meningoencephalitis and broncopneumonia. Transmission is via inhalation of contaminated dust and the disease has been reported in a wide range of species. Infection generally occurs in immunosuppressed hosts and usually involves the central nervous system, nasal mucous membranes, lungs or skin. Clinical signs include dilated pupils, head tilt, circling and incoordination. *Cryptococcus gattii* has also been diagnosed in an adult free-ranging koala associated with a pathological fracture from cryptococcal osteomyelitis | Can be sneezing, nasal discharge, sneezing, difficulty breathing, swelling of the nose and face. Central nervous system cryptococcosis can present as seizures, sudden blindness, behavioural changes | Cytology of the lesions and the latex cryptococcal antigen agglutination test of the blood.<br>Lesions within the sinus can be assessed by a computed tomography scan | Fluconazole 15 mg/kg intravenously twice daily × 7 days then 20 mg/kg orally twice daily × 3 weeks | Regular substrate changes and surface treatment of enclosure furniture with antifungal preparations | | 5, 16–18 |
| Candidiasis | *Candida albicans* is often associated with hand-reared animals<br>Infection with *Candida* sp. can result from antibiotic therapy causing candidiasis or thrush. It can also result from less than adequate hygiene or stress in hand-reared joeys | *Candida* can result in diarrhoea that often has a foul yeast-like smell with a yellowish-green and sometimes frothy or curdled appearance. With oral thrush it can result in the mouth becoming sore, ulcers and/or white plaques or crusting around the mouth and a rust-coloured crusty discharge | Diagnosis is made through microscopic examination of scrapings and culture of lesions or faeces for the presence of high numbers of budding or non-budding yeasts. It should be noted that *Candida* is normally present in the gastrointestinal tract of many marsupials in low numbers so the presence of yeasts in faecal smears does not necessarily indicate a problem | Can be given as Nilstat® Oral Drops or Mycostatin® Oral Drops. Persistent infections can also be treated with amphotericin B oral drops | Maintain high hygiene standards by frequently cleaning excess milk formula off the joey. Sick animals should be separated from others. It is also important to minimise stress, which reduces the animal's ability to fight infection | | 19 |

| Disease | Cause | Clinical signs | Diagnosis | Treatment | Prevention | Comments | Reference |
|---|---|---|---|---|---|---|---|
| **Protozoal** | | | | | | | |
| Coccidiosis | *Eimeria* spp., *Isospora* spp. | Diarrhoea, either watery or blood tinged. Dehydration from persistent diarrhoea and ill thrift | Faecal floatation test | Amprolium and sulfonamides | Maintain high hygiene standards, quarantine new animals, routine faecal examinations | | 15, 20 |
| Toxoplasmosis | *Toxoplasma gondii.* From food contaminated with cat faeces | Infection is often not apparent, with clinical illness often occurring in animals that are immunosuppressed or hand reared (Rose 1999). The severity of the illness ranges from mild malaise to peracute death with other signs including depression, weakness, anorexia, pyrexia, dyspnoea, ataxia, hemiplegia, quadriplegia, coma, convulsions, muscle stiffness, diarrhoea, emesis, uveitis, retinitis or cataract formation | Antemortem diagnosis of toxoplasmosis is confirmed by serological testing to detect rising IgG *T. gondii* titres. Direct agglutination test or modified agglutination test using the commercial kit Antigene Toxo-AD | Drugs used to stop the replication of the parasite. Clindamycin is the drug of choice and is administered parenterally or orally at a dose rate of 10–15 mg/kg twice daily for 4 weeks. Sulfadiazine and pyrimethamine are used in combination to treat toxoplasmosis, sulfamethazine at 30–60 mg/kg orally every 12 h and pyrimethamine 0.25–0.5mg/kg orally every 12 h. These drugs act synergistically to inhibit the synthesis of folinic acid, which is required by *T. gondii*. A folinic acid supplement will prevent the development of anaemia or leucopenia by providing folinic acid 5.0 mg/day or brewer's yeast 100 mg/kg/day | Prevention of access to cats and cat faecal matter is required. Cat faeces may contaminate bedding straw, sand or other substrates so cats should be excluded from storage areas | | 14, 15, 21–23 |

**Table 9.10.** continued

| Disease | Cause | Clinical signs | Diagnosis | Treatment | Prevention | Comments | Reference |
|---|---|---|---|---|---|---|---|
| Leishmaniasis | Caused by the protozoan parasite *Leishmania*. Has been reported in a group of captive red kangaroos in the Northern Territory and in other captive *Macropus* species in the same geographical area. Midges are thought to play a role in transmission of the parasite | Lesions of thickened skin or nodules, with crusting and sometimes ulceration | Visualisation of organisms in stained smears of lesion scrapings, tissue biopsies and fine-needle aspirates. Culture of the parasite is gold standard for the diagnosis of leishmania | Treatment safety and efficacy in native Australian wildlife is not known; however, treatment based on dog treatment protocol is suggested | Prevention of fly vector bites and control of sand fly populations | | 24, 25 |
| Bacterial | | | | | | | |
| Leptospirosis | *Leptospira* spp. remains a significant pathogen in possums and gliders species. Brush-tailed possums have been found to be maintenance hosts for this disease. Transmission is via direct or indirect contact with infected urine, placental fluids or milk | Clinical signs can include fever, anorexia, depression, diarrhoea, vomiting, icterus, haemoglobinuria and death | Diagnosis is through the detection of antibodies against the organism and through the detection of the organism in tissue or body fluids. If the animal has renal failure, blood urea nitrogen and creatinine levels may be elevated | Animals should be provided with supportive care and antimicrobial therapy such as β-lactams and tetracyclines | Leptospirosis occurs most commonly during high rainfall periods. As infection can be acquired from contact with the urine of maintenance hosts, contaminated food supplies and bedding. The use of vaccinations should also be considered where appropriate | | 15, 28, 29 |
| Mycobacteriosis | *Mycobacterium* spp. remains a significant pathogen in possums and gliders species | Ulcer or lesion (nose, ear, paws, tail), lethargy, limited movement | Polymerase chain reaction is used for the diagnosis of *M. ulcerans* disease | Some infections have been known to resolve without treatment; however, in diagnosed cases euthanasia should be considered | Avoid contact with affected animals | | 28, 30, 31 |

| Disease | Cause | Clinical signs | Diagnosis | Treatment | Prevention | Comments | Reference |
|---|---|---|---|---|---|---|---|
| Salmonellosis | *Salmonella* spp. have found in possums; however, only in immunosuppressed or stressed individuals. Predisposing factors include overcrowding, transportation, exposure to extended periods of cold conditions and young animals that suffer immunosuppression | Haemorrhagic enteritis, diarrhoea, fever, dehydration and septicaemia Deaths have been recorded in brush-tailed possums and ring-tailed possums | Culture of fresh faecal material with selective growth media. Necropsy findings can include dehydration, enteritis, gastroenteritis, septicaemia | Supportive treatment: Fluid therapy, analgesia. Antimicrobial therapy can be used in immunocompromised and young individuals | Avoid overcrowding animals. Ensure optimal hygiene when handling multiple animals, where possible prevent food and water contamination with salmonella | | 14, 15, 28, 32, 33 |
| Yersiniosis | *Yersinia pseudotuberculosis* has been implicated in the deaths of Leadbeater's possums. Outbreaks of yersiniosis are thought to be associated with stress or immunosuppression factors, including capture, cold weather, diet change, and intraspecific aggression | Although many animals will harbour *Yersinia* without effect, it is capable of causing multisystemic illness. Can result in rapid death from dehydration, diarrhoea, enteritis, or septicaemia | Yersinia is generally diagnosed by isolating organisms from within lesions in the intestine, kidneys, liver, lungs and mesenteric lymph nodes | Prognosis is usually poor, but usually treated with broad-spectrum antibiotic therapy. Once clinical signs are apparent, animals may respond poorly to therapy | High standards of husbandry and hygiene are required and the protection of food and water from wild birds. Minimising stress may also assist in its prevention | | 4, 5, 14, 28 |

References: (1) Presidente (1982a); (2) Presidente *et al.* (1982); (3) Presidente (1984); (4) Munday (1988); (5) Booth (1994); (6) Lindenmayer *et al.* (1994); (7) Booth (1999); (8) Kwak *et al.* (2018); (9) George (1982); (10) De Mendonca (1983); (11) Speare *et al.* (1984); (12) Spratt *et al.* (1990); 12) Viggers and Spratt (1995); (14) Rose (1999); (15) Fowler and Miller (2003); (16) Krockenberger *et al.* (2002); (17) Krockenberger *et al.* (2003); (18) Schmertmann *et al.* (2018); (19) Brust (2013); (20) Yang *et al.* (2012); (21) Bettiol *et al.* (2000); (22) Miller *et al.* (2000); (23) Fowler and Miller (2008); (24) Dougall *et al.* (2009); (25) Dougall *et al.* (2011); 26) Presidente (1982b); (27) Woods (1999); (28) Johnson and Hemsley (2008); (29) Eymann *et al.* (2007a); 30) Buddle and Young (2000); (31) Ban *et al.* (2020); 32) Reddacliff and Spielman (1990); (33) Finnie (1978).

**Table 9.11.** Common non-infectious diseases of zoo-based possums and gliders.

| Disease/agent | Cause | Clinical signs | Diagnosis | Treatment | Prevention | Comments | References |
|---|---|---|---|---|---|---|---|
| Nutritional osteodystrophy or hindlimb paralysis | Commonly reported in pet sugar gliders. Appears to be due to a calcium-deficient diet that often includes only fruit and meat | Sudden onset of hindlimb weakness or paralysis | Radiography of vertebral, pelvic, and long bones demonstrating osteoporosis. Spinal trauma is a differential diagnosis | Cases identified early may respond to a high-calcium, additional vitamin D3 diet and strict enclosure rest | Hindlimb paralysis can be prevented by providing adequate calcium in the diet. Insects fed to gliders or pygmy possums should be supplemented with calcium. Dusting insects with calcium powder is less reliable than feeding (gut loading) a high-calcium diet 48 h before they are fed out | | 1 |
| Bloat | Has been observed in eastern ring-tailed possums due to a build-up of gas in the gastrointestinal tract and has been recorded in both adult and juveniles. The intestines may twist and strangulate resulting in rapid death without surgical intervention | The abdomen is extremely distended, feels tight and makes a drum-like sound when tapped. Low faecal volume and/or small faecal pellets. Increased appetite and thirst | Palpation and radiography | Animals generally succumb. Surgery is potentially an option if cases are caught in time. Early cases may respond to transabdominal removal of gas with a trocar and cannula | Possible causes include poor diet and/or insufficient native browse. Ensuring an appropriate diet and plenty of browse appears to be very important | May be due to high-carbohydrate or fruit-based diets or antibiotic therapy that disturbs the gut flora. Important to provide a mostly folivorous diet when the animals are being weaned to establish proper caecal function | 2–4 |

References: (1) Booth (1999); (2) Woods (1999); (3) Johnson and Hemsley (2008); (4) R. Booth pers. comm.

**Table 9.12.** Zoonotic diseases associated with possums and gliders.

Derived from Hulst (2025).

| Disease | Cause | Clinical signs in humans | Transmission | Prevention |
|---|---|---|---|---|
| Buruli ulcer | *Mycobacteria ulcerans* | Usually a painless, non-tender nodule or papule. The nodule may become fluctuant and erode, forming a characteristic ulcer with undermined edge | Exposure to contaminated water, soil or vegetation | Appropriate personal protective equipment |
| Leptospirosis | *Leptospira* spp. | Ranges from subclinical infection to hepatic and renal failure, severe pulmonary haemorrhage and death | Indirect or direct routes. Urine contamination of water, soil and vegetation providing an avenue for indirect exposure and allowing the organism to enter through cuts, abrasions and mucous membranes. Direct exposure can occur through contact with infected body fluids (including urine) | Workplace health and safety practices |
| Tularaemia | *Francisella tularensis* (bacteria) | Skin ulcer at the site of the bite or skin exposure, and swelling of the nearby lymph glands; can progress to a sore throat and chest pain | Direct and indirect contact. Contact with sick or dead infected animals, including bites/scratches, bites of infected blood-feeding arthropods, drinking contaminated water or eating undercooked meat of an infected animal, contact of broken skin with natural water bodies contaminated by infected animals, inhalation of contaminated dusts or aerosols | Appropriate PPE when handling sick or dead ring-tailed possums |

In the event of a disease outbreak, sick individuals should be isolated immediately from the remainder of the collection animals. PPE should also be worn to prevent zoonoses and to minimise the risk of disease spread.

## 8.7 Zoonoses

Several zoonotic diseases have been associated with possums and gliders (Table 9.12). For many zoonotic diseases, the possum and glider host will show no signs of disease. If zoonotic disease is suspected, seek medical advice.

# 9. BEHAVIOUR

## 9.1 Activity

Observations of captive savanna gliders with different light:dark lengths (L:D of 14:10, 12:12, 9:15) showed that although the amount of time the animal was active increased with increasing night length, the proportion of the night used for activity decreased (Goldingay 1984). Activity within different night lengths increased from 9.7 h at 14:10 to 10.3 h at 12:12 to 10.8 h at 9:15. The mean percentage of the night spent active decreased from 51.7% at 14:10, to 45.7% at 12:12, to 46.3% at 9:15. The peak period of activity was also shortly after the lights went out (Goldingay 1984). These results suggest that night:day lengths of 12:12, or slightly shorter nights, are preferable, although following the natural changes in day length may be important for initiating reproduction. Some species including wild brush-tailed possums and mahogany gliders are known to have a period of inactivity, even sleep, during the middle of the night (Winter 1976; Jackson and Johnson 2002).

The diet can also be important for the activity of possums and gliders, with the *Petaurus* gliders (that are exudivore/omnivores) spending more time foraging (Fig. 9.7; Jackson and Johnson 2002; Sharpe and Goldingay 2018). In contrast, the larger folivores spend less time feeding and this reduces even further as body size increases and becomes increasingly efficient (Fig. 9.7; Jackson and Johnson 2002). The differences in activity observed between exudivorous and folivorous possums is a reflection of the speed that the gut fills and the digestibility of the food they consume. Folivores can fill their gut quickly due to ample availability of foliage, but the digestion of food is slow so they are often inactive while that happens. In contrast, exudivores usually often have to work hard to find and harvest food, but it digests rapidly, so they can continue to forage most of the night. Therefore, there are limits to the activity (and therefore

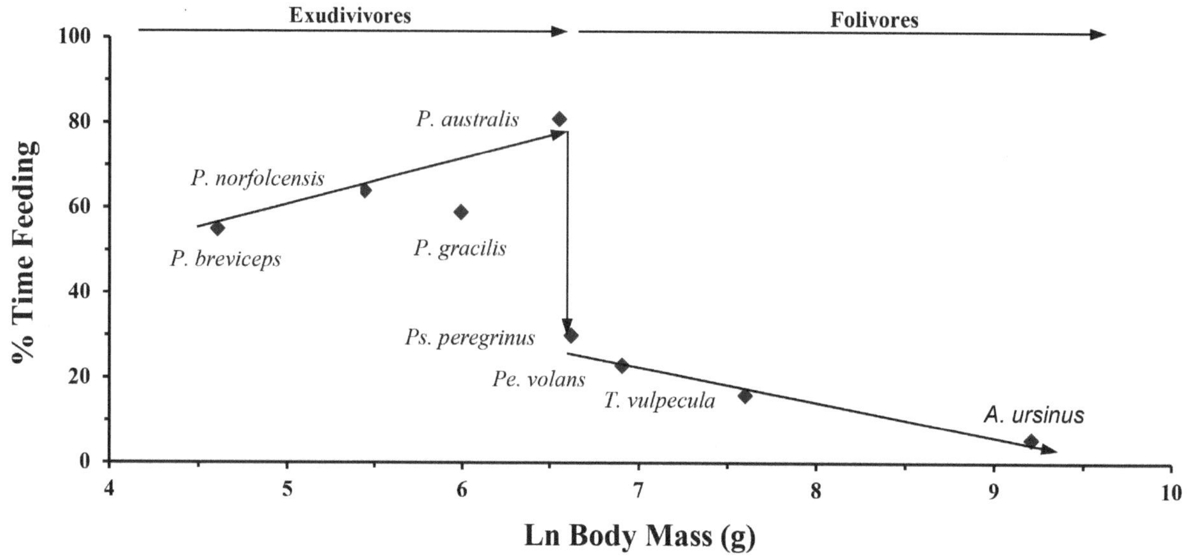


**Fig. 9.7.** Relationship between body size, diet and proportion of night that different species of possums spend feeding. Note that *Ailurops ursinus* is the Sulawesi bear cuscus. Taken from Jackson and Johnson (2002).

visibility to the public) that can be achieved, even with activity feeds, when attempting to exhibit possums and gliders in nocturnal houses.

The thermoregulatory disadvantages of small size (i.e. relative greater heat loss) is significantly reduced in the smaller members of the Petauridae and in all the Burramyidae, Acrobatidae and Tarsipedidae by group huddling, nest construction and daily or seasonal torpor (Smith and Lee 1984; Namekata and Geiser 2009). Several authors have suggested torpor and hibernation as mechanisms to conserve energy and increase fasting endurance during poor weather, lower temperatures and during periods of food shortage (Wakefield 1970; Fleming 1980; Wooller *et al.* 1981; Renfree *et al.* 1984; Jones and Geiser 1992; Geiser 1993; Körtner and Geiser 2000; Geiser 2007; Geiser and Körtner 2010). Though torpor and hibernation patterns can differ between captive and free-ranging animals (Geiser *et al.* 2000), an understanding of it is useful for their captive management and important for species being bred for release.

Wang (1989) described two types of torpor, one involving daily torpor with minimum body temperatures that are metabolically defended during torpor (11–28°C), and a second type that involves deep and prolonged torpor (hibernation) with minimum body temperatures (1–6°C) and torpor bouts lasting between 1 and 3 weeks. Larger species that undergo torpor do so for only short periods while those with lower bodyweights, except the honey possum, undergo periods of hibernation (Table 9.13).

Mountain pygmy possums do not appear to enter hibernation unless their body mass reaches a certain level, which appears to be between 50 and 70 g (Fleming 1985a; Geiser *et al.* 1990). It is not known if seasonal hibernation in mountain pygmy possums is regulated by a circannual rhythm or changes in food availability, temperature or photoperiod, or a mixture of multiple cues (Geiser and Broome 1991). However, in captivity they are known to occasionally wake to feed on cached seeds and nuts (Broome and Geiser 1995). At Healesville Sanctuary, seeds and nuts are provided to hibernating mountain pygmy possums with amounts based on whether they are over or under 60 g, in case they wake during their hibernation cycles (Boys 2018; see section 3.4).

In early studies, mountain pygmy possums showed differences in behaviour and physiology between wild and captive animals. Wild possums fattened extensively during the prehibernation season, which was followed by 7 months of torpor and hibernation, compared with captive-bred animals that neither fattened nor entered torpor in two consecutive winters (Geiser *et al.* 1990). In the more recent mountain pygmy possum breeding program at Healesville Sanctuary, possums have hibernated successfully every year in indoor alpine-mimicking enclosures. A hibernation protocol including seasonal diet changes and autumn weight gain, and induction into hibernation using natural light cycle changes, and a reduction in food and temperature to 2°C, has successfully maintained annual hibernation from May to September with 100% of possums generally hibernating

**Table 9.13.** Bodyweights and the occurrence of torpor or hibernation within the Australian members of the Phalangeroidea.

| Species | Weight (g) | Torpor/hibernation | Reference |
|---|---|---|---|
| *Burramys parvus* | 30–82 | Hibernates | 1–4 |
| *Cercartetus caudatus* | 25–40 | Hibernates? | 1, 5, 6 |
| *Cercartetus concinnus* | 8–20 | Hibernates | 1, 7–10 |
| *Cercartetus lepidus* | 6–9 | Hibernates | 1, 7 |
| *Cercartetus nanus* | 15–43 | Hibernates | 1, 11–15 |
| *Gymnobelideus leadbeateri* | 100–166 | Torpor? | 1, 16 |
| *Petaurus australis* | 450–700 | Torpor? | 6 |
| *Petaurus breviceps* | 85–160 | Torpor | 1, 17–19 |
| *Petaurus notatus* | 60–150 | Torpor | 1, 17–19 |
| *Acrobates pygmaeus* | 8–15 | Hibernates | 1, 6, 20–23 |
| *Tarsipes rostratus* | 7–12 | Torpor | 1, 24, 25 |

References: (1) Strahan (1995); 2) Fleming (1985a); (3) Geiser and Broome (1991); (4) Körtner and Geiser (1998); (5) Atherton and Haffenden (1982); (6) Geiser and Körtner (2004); (7) Geiser (1987); (8) Turner (2020); (9) Turner and Geiser (2017); (10) Turner *et al.* (2012a); (11) Geiser (1993); (12) Bartholomew and Hudson (1962); (13) Geiser (2007); (14) Song *et al.* (1997); (15) Turner *et al.* (2012b); (16) Smith (1980); (17) Fleming (1980); (18) Körtner and Geiser (2000); (19) Geiser *et al.* (2007); (20) Fleming (1985b); (21) Jones and Geiser (1992); (22) Geiser and Ferguson (2001); (23) Frey and Fleming (1984); (24) Withers *et al.* (1990); (25) Collins *et al.* (1987).

each year (see Boys 2018). A decision tree is followed to determine when to provide food to possums during this time (based on weekly bodyweights and a hibernation weight loss regimen; Boys 2018). A protocol to rouse possums from hibernation and increase temperatures in September is followed to wake possums for the spring breeding season (Boys 2018; M. Parrott pers. obs.). Possums following this protocol have successfully been released to the wild and have produced young and survived the following wild hibernation period (M. Parrott unpubl. data).

Observations of western pygmy possums have found they express aseasonal hibernation in captivity, whereas in the field they used highly flexible bouts of torpor that could be either short (< 24 h [6.4 h ± 5.4 h]) or long (> 24 h [89.7 ± 45.9 h]). The initiation of torpor was facilitated by low ambient temperature (8.6°C *v.* 10.6°C or warmer) in the field (Turner *et al.* 2012a). The eastern pygmy possum has also been observed to use short-term hibernation (from 5 to 20 days), with torpor bouts of up to 5.9 days. However, they have also been observed to use torpor for short bouts of < 1 day in summer (Turner *et al.* 2012b).

Within the Petauridae, the sugar glider is known to enter torpor, but anecdotal reports suggest that Leadbeater's possum and the yellow-bellied glider are also occasionally heterothermic (Geiser and Körtner 2010). In captivity the sugar glider appears to be reluctant to enter torpor (Fleming 1980; Dawson and May 1984; Geiser *et al.* 2007), whereas wild sugar gliders from cool-temperate climates displayed more frequent and deeper torpor than their captive counterparts (Körtner and Geiser 2000; Christian and Geiser 2007; Geiser *et al.* 2007).

The honey possum may require torpor rather than hibernation because of the milder temperatures experienced in the areas where they are found in the south-west of Western Australia. Torpor is strongly air temperature-dependent, with higher temperatures resulting in significantly shorter bouts and a higher metabolic rate (Geiser and Broome 1993). During winter when temperatures fall, they can be kept more active or even prevented from entering torpor by maintaining temperatures above ~12°C (Geiser and Broome 1993; L. Andrews pers. comm.).

Captive-bred feather-tailed gliders have been found to differ from wild gliders in behaviour (longer activity periods) and physiology (less frequent torpor, shorter torpor, shallower torpor, higher metabolic rates during rest and torpor and slower rates of rewarming). Captive populations may become hypothermic and are unable to rewarm (Geiser and Ferguson 2001), thus care must be taken to provide adequate temperatures in captivity.

## 9.2 Social behaviour

### 9.2.1 Honey possum

Russell (1986) has made extensive observations of the behaviour of honey possums in captivity. Little is known about their dispersal, mating system or social organisation in the wild, though they are considered to be polyandrous and nest in groups, which is supported by multiple paternity within litters being confirmed by genetic studies (Wooller *et al.* 2000). A hierarchy exists with females, which are larger than males, appearing to show dominance over non-breeding females and males by excluding others from food. Although aggressive encounters are uncommon, they have a range of aggressive agonistic behaviours including pawing, snout jabbing, lunging, jump attacks, chasing and wrestling (Russell 1986).

## 9.2.2 Feather-tailed gliders

Feather-tailed gliders of both species appear to nest communally with multiple males, females and young, in groups of up to 40, though 2–5 individuals is normal. The groups usually consist of adults with the offspring of 1–2 litters (Fleming and Frey 1984; Ward 1990a). They generally show fidelity to 1–2 nests and the use of the same nest by other individuals is tolerated (Fleming and Frey 1984). Their nest sharing, the presence of reproductive females with several males and the little evidence of prolonged associations between males and females suggests that they are promiscuous (Fleming and Frey 1984; Ward 1990a). Genetic studies have now proven this, with female feather-tailed gliders mating with multiple males in a single oestrus and having multiple paternity within litters (Parrott *et al.* 2005). The large testis to bodyweight ratios, male-biased sexual size dimorphism and large sperm suggests there is sperm competition in feather-tailed gliders (Parrott *et al.* 2005). Wild female feather-tailed gliders have also been found with young from other females in 4 out of 10 family groups analysed, which suggests that communal maternal care is a common occurrence (Parrott *et al.* 2005). In captivity, breeding appears to occur best when animals are held in relatively large enclosures that contain several animals rather than in pairs (Woodside 1995).

## 9.2.3 Pygmy possums

In the wild, pygmy possums are normally solitary (Ward 1990b, 1990c), except for the social mountain pygmy possum. In captivity, they generally appear to be socially tolerant, with nest boxes readily shared and males showing low levels of aggression even in the presence of females (e.g. Kerle 1984a). The lack of prolonged associations between the sexes suggests that they are promiscuous (Ward 1990b, 1990c).

Observations of wild long-tailed pygmy possums showed that they build nests beneath the dead fronds of pandanus palms and in tree hollows (Dwyer 1977). The nests, which are spherical in shape, are constructed of dead leaves attached to stems and may be up to 15 cm in diameter and within a few metres of the ground. Adult females with regressing teats have been found together or with males in the same nest. However, there are no other observations of adult or subadult individuals of either sex being found together with a lactating female (Dwyer 1977).

One study of the nest occupancy of eastern pygmy possums found solitary animals on 71% of occasions, lactating female with young on 20% and more than one independent individual on only 8.6% of occasions. None of these groups contained a lactating female or more than one adult female, while four contained more than one adult male (Ward 1990b). In captivity, eastern pygmy possums have been known to routinely share nest boxes with members of the same and opposite sex (Murphy *et al.* 2003).

Female mountain pygmy possums occupy food- and shelter-rich habitats both during and outside the breeding season, whereas males hibernate in habitat, at lower elevations, containing poorer resources (Mansergh and Scotts 1990; Broome 2001). The males visit the female habitats during the breeding season and return to less favourable habitat after the breeding season (Mansergh and Scotts 1989; Mansergh and Broome 1994). A nest discovered in the wild in a granite boulder-field (buried by a layer of soil), was 15–20 cm in diameter and consisted of long, clean strands of moss (*Brachythecium salebrosum*) with a few blades of grass (*Poa* sp.) intertwined, which kept the possums dry while the surrounding soil was damp (Heinze and Olejniczak 2000). A study of captive mountain pygmy possums revealed five non-agonistic social behaviours and one possible agonistic behaviour. The former included walking or running over/pushing under, nose-to-nose sniffing, nose-to-urogenital opening sniffing, huddling and marking that involved a male vigorously rubbing his chest or chin against branches. The only possible agonistic behaviours seemed to be rarely used and involved food stealing (Cash 2015). Social behaviours such as grooming and eating together are observed, even in unrelated animals. Olfactory cues are important for communication in mountain pygmy possums, with males in particular actively scent marking areas using urine, faeces and the paracloacal and sternal glands glands (M. Parrott pers. obs.). Possums sniff marked areas and pick up scent-marked objects or faeces to better examine the scents. Female mate choice is, at least partially, based on scent, with females displaying sniffing and choice behaviours near the bedding and scents of males (see section 9.3.1). Mountain pygmy possums are polygynous, with genetic studies confirming that females mate with multiple males in a single oestrus, both in captivity and in the wild, and can produce multiple paternity within litters (M. Parrott pers. obs.).

In captivity, mountain pygmy possums generally appear to be socially tolerant and commonly share nest boxes, however aggression was observed when a new male was placed with an established pair, resulting in the resident male chasing the outsider (Arlidge *et al.* 1993; Kerle 1984a). The general high level of social tolerance is suggested to be advantageous in allowing huddling, which reduces heat loss (Fleming 1985a). Possums choose to nest together in groups which include both sexes, until females are pregnant and choose to nest individually. Pregnant females will chase other possums away to defend her nest and surrounding areas (Boys 2018). When a large number of males and females were placed together at Healesville Sanctuary (resulting in successful matings) the males inhabited nest boxes on the edge of their large enclosure. This appears to mimic the wild situation with segregation of the sexes during the non-breeding season and the males traversing up the mountain in spring in order to mate (Mansergh and Walsh 1983; Mansergh *et al.* 1988; Broome 2001).

### 9.2.4 Petaurids

Observations of Torresian striped possums at London Zoo found that introductions to each other can lead to high levels of aggression, which can result in fights that can require the separation of animals. Females appear to be the main aggressors, so McKenna (2005) proposed that females should be introduced into the enclosure of the male which was already established in the enclosure (McKenna 2005). When London Zoo successfully bred Torresian striped possums, they observed that the possums often denned separately. However, they do den together during the breeding season when frequent vocalisations were produced. The pair initially slept together after they had mated, which continued and until approximately two weeks after the female had successfully produced pouch young, at which time the female became increasingly aggressive towards the male so he was removed (McKenna 2005).

Observations of wild Torresian striped possums in New Guinea have found den sharing on several occasions, however observations in Australia suggest they rarely share dens and are more solitary than other petaurids (Hide *et al.* 1984; K. Handasyde *pers. comm.*). In one case in the wild, there was intense raucous rivalry between two males over an oestrus female, with all three producing continuous, rolling, guttural shrieks (Van Dyck 1995). They are known to produce several vocalisations (Handasyde and Martin 1996; Carmichael 2000). As with other petaurids, scent marking plays an important role in the social life of the Torresian striped possum. They have an anal gland that exudes a very strong musty odour that is rubbed onto branches by the animal performing a cloacal drag (Biggins 1984; Carmichael 2000). Upon meeting, they raise their tails over and parallel to their backs with the fur on the tail standing on end and they may move their tails from side to side (Carmichael 2000).

Leadbeater's possums nest as socially monogamous pairs in the wild and in captivity with cohesion in the group being maintained by the dominant breeding pair spreading salivary odours during extensive grooming in the nest and mutual licking of the tail-base. When an individual dies the breeding vacancy is quickly filled by immigrants from neighbouring colonies, so introductions of animals are possible (A. Smith 1995; Harley and Lill 2007). Other behaviours include a chatter challenge call, attacking each other, grappling, sniffing and an alarm hiss when held against their will. Colonies in the wild have been observed to contain a single adult female with one to three adult males, suggesting they are monogamous or polyandrous (Smith 1980; 1984a). Females are more socially aggressive than males and readily attack and pursue animals of either sex from another colony (A. Smith 1995). Similar observations in captivity suggest that captive colonies do not like the introduction of females and will fiercely attack them. Juvenile females should also be removed before sexual maturity to avoid being attacked. Scent is important in communication in Leadbeater's possums and individuals can be observed scent marking their areas with paracloacal glands and saliva. Female mate choice trials based on the saliva of males showed differing levels of interest in different males (M. Parrott pers. obs., see section 9.3.1).

Yellow-bellied gliders live in family groups, generally of 2–6 individuals (mean 2.6), comprising a male with 1–2 females and their offspring (Russell 1984; Craig 1985; Goldingay and Kavanagh 1990; Goldingay *et al.* 2001). They can be monogamous, polygynous or interchangeable depending on resource availability (Henry and Craig 1984; Craig 1985; Goldingay and Kavanagh 1990; Goldingay and Kavanagh 1991; Goldingay *et al.* 2001). Although all the petaurids make various calls, the yellow-bellied glider is the most vocal, having an extensive vocal repertoire and making numerous calls throughout the night (Kavanagh and Rohan-Jones 1982; Russell 1984; Goldingay and Kavanagh 1991; Goldingay 1994). As with all members of the genus *Petaurus*, the male has a well-developed scent gland on the top of his head that is used to mark other members of the group. Other glands on the chest and the underside of the tail are also used to mark each other and select parts of the territory (Russell 1984, 1995). When gliders are held in a colony, members are regularly scent-marked by mature males and animals not bearing this scent are often attacked (Dunn 1982). Therefore, the introduction of unfamiliar animals into these groups should be attempted with great caution.

Observations of the coastal sugar glider in coastal New South Wales suggest they are socially monogamous and live in colonies that vary in size from two to seven animals (Quin 1995). Paternal behaviour has also been observed in this species within nest boxes (Goldingay 2010). Observations of captive sugar gliders in New Guinea show they have a complex olfactory communication with scent that is produced by frontal, sternal and cloacal glands. They can distinguish individuals and intra- and intergroup members (Schultze-Westrum 1965, 1969).

Studies of the inland sugar glider (Krefft's glider), which probably has similar social behaviour to the coastal sugar glider, show that it is generally considered to be polygynous in the wild with colonies of 2–7 animals (Henry and Suckling 1984; Suckling 1984; Sadler and Ward 1999). Field-based studies by Sadler and Ward (1999) found clear evidence that these gliders are polygynous and that the associations between adult males and their putative sons reported by Klettenheimer *et al.* (1997) were artefacts of captivity. In captive populations, however, only one adult male in the colony is reproductively active, with the colony consisting of a monogamous pair, offspring and occasionally unrelated adult males (Bradley and Stoddart 1993; Mallick *et al.* 1994; Stoddart *et al.* 1994; Klettenheimer *et al.* 1997).

Observations of the mahogany glider suggest they are territorial and live in pairs with home ranges that almost entirely overlap (86% compared with only ~11% with other males and females). An indication of their territorial behaviour was observed when a male glider viciously attacked a second glider (thought to be a male) and also by their foraging loops where they appear to cover the boundary of the home range every few nights (Jackson 2000b). The frequent den sharing and high degree of home range overlap suggests they are socially monogamous, though males have been observed to mate with the non-paired mate, indicating that extra-pair matings do occur (Van Dyck 1993; Jackson 2000b). Male mahogany gliders have also been observed to show significant paternal behaviour towards their young, including grooming, protecting and play behaviour with the infant. The males also perform most of the nesting behaviour, keeping the young warm and responding to vocalisations (Muller *et al.* 2010).

Squirrel gliders have been observed in both polygynous and monogamous social groups. Groups of between 2 and 9 individuals, with at least one male and several females, have been observed, suggesting a polygynous mating system (Quin 1995). In other studies, a family group typically comprised one mature male (> 2 years old), and one or more adult females and their offspring (Suckling 1995). In another study, up to two adult males were observed in a nest (with a single male aged > 3 years old) (van der Ree 2002). The presence of multiple adults of both sexes within a social group also suggests a polygynous mating system (van der Ree 2002). In contrast to these records, other observations of squirrel gliders have documented monogamous groups (Sharpe and Goldingay 2007). Scent glands on the head are well developed (Suckling 1995).

### 9.2.5 Ring-tailed possums and greater gliders

Adult eastern ring-tailed possums regularly share hollows and/or dreys in a variety of combinations (Thomson and Owen 1964). Observations of eastern ring-tailed possums in the wild found most males (60%) associated with only one female at a time and this bond lasted through the breeding season, although 'covert mating' with other females suggests a socially or facultative monogamous mating system. A few males that were generally older and larger associated with two females throughout the breeding season, suggesting polygyny (Ong 1994). Although ring-tailed possums are largely solitary, males actively initiate and maintain contact with females by constantly visiting their mates, at least once per night, throughout the year and more frequently during the breeding season and actively defend their mates from other males (Ong 1994). Captive observations have shown that a juvenile male viciously attacked a smaller unrelated female (Presidente 1982b) and a castrated male was observed to attack other males and females (L. Andrews pers. comm.).

Observations of captive and wild western ring-tailed possums suggest that adults do not share nest sites (nest boxes or dreys). Captive females do not tolerate adult males, although males rarely initiated agonistic behaviour (Ellis and Jones 1992). Adult females shared their nest boxes with their young until the pouch young were ~90 days old and nearly ready to emerge from the pouch for the first time. At this time the older young were evicted from the nest box (Ellis and Jones 1992).

Rock ring-tailed possums appear to live in family groups of a male, female and any offspring, in mutually exclusive home ranges. Observations suggest that males and females contribute equally to parental care and maintain the pair bond and nest by using scent marking (Runcie 2000). As a result of these observations the mating system in the rock ring-tailed possum is suggested to be obligate social monogamy (Runcie 2000).

Greater gliders are largely solitary, but can be found in pairs or polygynous groups. For example, observations of a high-density population in North Queensland found the home ranges of males and females overlapped, indicating a polygynous mating system (Comport *et al.* 1996). In contrast, observations of low-density populations in southern Australia found most animals in pairs and the remainder were in polygynous groups (Henry 1984; Norton 1988).

Little information is available on the social behaviour of the rainforest ring-tailed possum. Lemuroid ring-tailed possums have been observed to be solitary on 64% of occasions, with green ring-tailed possums (94%) and Herbert River ring-tailed possums generally being found by themselves (Winter and Atherton 1984). When these species were found in groups, which consisted of 2–3 animals, they were thought to comprise an adult male and female and/or a female with young (Winter and Atherton 1984).

## 9.2.6 Cuscuses, brush-tailed possums and scaly-tailed possum

Common brush-tailed possums are highly territorial, polygynous, have a dominance hierarchy and often show aggression towards each other, particularly dominant males towards subordinate males (Biggins and Overstreet 1978). Females and males appear to mate with several mates and are likely to have a serial polygynous mating system (Lee and Cockburn 1985; Taggart *et al.* 1998). When multiple females are held together, they are known to kill and eat each other's young (Anon 1935).

In a field study of brush-tailed possums near Brisbane during the April–May breeding season, males established a consort period with a particular female lasting 28–48 days. They followed the female, giving repetitive soft juvenile-like calls until eventually the female allowed mating without aggressive behaviour. During a secondary breeding season in August–September, consort periods were not established and several males could mate with a single female who responded with overt aggressive behaviour (Winter 1976). Dominant males established exclusive territories with over-lapping subordinate male home ranges (Winter 1976). Males in particular scent-marked their home ranges using chin glands, sternal glands and urine dribbling. They also produced a sticky white glandular substance from the cloaca, which was judged to be indicative of stress or subordination (Winter 1976). Auditory communication was complex, with 19 sounds in the repertoire: 18 vocalisations and a buccal click (Winter 1976).

The mountain brush-tailed possum has a varied social structure, likely depending on resources. Many appear to be socially monogamous, forming strong pair bonds with home ranges that overlap by ~80%, remaining in close proximity to each other and their dens during the night (Martin *et al.* 2001). Genetic analysis of wild populations also suggests that most offspring are fathered by the socially paired mate (Martin *et al.* 2001), though up to 35% may be fathered via extra-pair paternity (Handasyde *et al.* 2007). In other populations, only half the females were found in a socially bonded pair (Blyton *et al.* 2016).

Communication by scent and sound includes well-developed chin, chest and anal glands and a range of vocalisations (How and Kerle 1995). Removing the dominant male for a short time does not appear to produce any significant changes in the dominance hierarchy of the rest of the group and the dominant male will readily reassert his dominance once he returns. When new animals are introduced into the group, they generally occupy a low position in the hierarchy (Biggins and Overstreet 1978). Yearling females with their first pouch young are generally defensive and frequently keep to themselves, while mature females with young in the pouch or on their backs often get on well together (Presidente 1982a).

Very little is known of the social behaviour of scaly-tailed possums, except that they appear to have several dens and be largely solitary (Runcie 1999). Both sexes also appear to lack sternal and paracloacal glands, which are well developed in brush-tailed possums.

The social behaviour of the various species of cuscus is also poorly known; however, they are generally considered to be solitary. Males (e.g. spotted cuscus) are aggressive towards each other in captivity and cannot be housed together (Winter and Leung 1995).

## 9.3 Reproductive behaviour

Courtship behaviour is not well understood for most species. The copulatory behaviour of the mahogany glider has been described (Van Dyck 1993). During this event, the male produced a soft 'chew-chew-chew-chew' sound at which the female made immediate efforts to join him. The female sniffed the male's rump and then followed him up the tree to rest with him in the canopy. The two glided to a nearby tree where they curled up around one another. The male then lunged at the female and they copulated for ~23 min, during which time they both adopted a vertical head-down position on the trunk, with the male thrusting intermittently. The male grasped the female's dorsum in a similar fashion to that adopted by young back-riding gliders newly emerged from the pouch and he bit her neck until they separated (Van Dyck 1993). Mating behaviour has also been observed for wild Torresian striped possums in a rainforest in North Queensland (Van Dyck 1979). The mating process of the striped possums lasted ~10 min on a semi-vertical tree trunk with both sexes making 'gar-gair, gar-gair' vocalisations. The mating behaviour of the brush-tailed possums is also noisy. In the mountain pygmy possum a male and female were observed mating in a nest box, with the male holding the female and massaging her hips with his hindlegs. Following mating some females can have slight fur loss or knotting of fur on their hips from this movement. This can be used as an indication that mating has occurred (M. Parrott pers. obs.).

Mate choice occurs in a wide variety of taxa, including marsupials (Parrott 2019; see also Chapter 4 Carnivorous Marsupials and Chapter 6 Bandicoots), and can significantly affect pairing and breeding success. Studies of mate choice have been conducted on various possum and glider species. In captive mountain pygmy possums at Healesville Sanctuary, mate choice trials were conducted using olfactory cues from males to determine the mate preference of females (following techniques described in Parrott *et al.* 2019). Females showed distinct interest in different males, with the most interest shown in males with high genetic diversity. Females showed 'choice behaviours' similar to those seen in other marsupials such as antechinus (e.g. Parrott *et al.* 2015) and dunnarts (Parrott *et al.* 2019; also see Chapter 4 Carnivorous Marsupials). When placed into groups with multiple males, females mated and produced young with 'preferred males' significantly more times than with untested or non-preferred males. The results show that female mate choice occurs in mountain pygmy possums and allowing a female to choose her mate can increase breeding success in the captive breeding program (M. Parrott pers. obs.).

For Leadbeater's possum, preliminary trials at Healesville Sanctuary using male saliva showed that females were interested in the scents of males when placed in their enclosures. Females had more visits and spent more time with particular male scents, suggesting that there was increased interest in some males over others (M. Parrott pers. obs.).

Wild brush-tailed possums in a Brisbane study established loose pairs between dominant females and dominant males with overlapping home ranges and rare den sharing during consort periods. Subordinate males and females had less well-defined home ranges, often overlapping with the dominant individuals (Winter 1976).

Research into mate choice in wild mountain brush-tailed possums showed two different patterns of mate choice depending on the social context and dispersal patterns of the possums (Blyton *et al.* 2016). Half of the females in the population were in social bonded pairs, which genetic analyses showed were not based on genetic dissimilarity. However, half of the females were not in social pairs and those females chose to mate with males that were genetically dissimilar to themselves (Blyton *et al.* 2016). In different wild populations 10% (Blyton *et al.* 2016) and 35% (Martin *et al.* 2007) of socially bonded females produced young with a male that was not their social partner, which may allow for greater use of mate choice and increased genetic diversity of offspring.

## 9.4 Bathing

Bathing is not normally observed in any of the possums and gliders.

## 9.5 Behavioural problems

The different groups of possums and gliders appear to suffer little from behavioural problems. Greater gliders

appear to be easily stressed. One author has suggested that adult greater gliders do not to do so well in captivity and can lose weight rapidly (Stanvic 1992).
Abnormal repetitive behaviours such as pacing and moving along tracks have been observed in larger species of possums and may be alleviated by providing additional space and environmental enrichment.

## 9.6 Signs of stress

Acute stress can be associated with loud vocalisations, threats and attacks, overgrooming or excessive urination or defecation (Spielman 1994). Ring-tailed possums, for example, may launch themselves at a person, while cuscus and brush-tailed possums can threaten and attack fiercely (Spielman 1994). Under stress, brush-tailed possums may buccal click and in acute stress produce a white secretion from cloacal glands (Winter 1976).

When animals experience stress, the hypothalamic-pituitary-adrenal (HPA) axis is activated, which results in increased production and secretion of glucocorticoid (GCs) hormones (primarily cortisol or corticosterone in mammals) into the bloodstream (Sapolsky *et al.* 2000; Brearley *et al.* 2012). Though blood sampling is the preferred method of measuring circulating GC levels, it is an invasive technique, so some institutions may want to use other techniques, including relying on saliva, faeces, urine and hair measurements (Mallick *et al.* 1994; Brearley *et al.* 2012; Fanson *et al.* 2017). The monitoring of stress hormones via the measurement of faecal glucocorticoid metabolite, using a cortisol enzyme immunoassay, has been successfully used in a captive population of sugar gliders as a tool to non-invasively evaluate their physiological state (Scheun *et al.* 2019). The analysis of hair to determine GC levels has been conducted on squirrel gliders in the wild by collecting hair from their rump using a clean razorblade and cutting the fur at the skin level (Brearley *et al.* 2012).

Fanson *et al.* (2017) showed that GCs can be validated for many marsupial species, with work including species such as mountain pygmy possums and yellow-bellied gliders, though different assays are required to gain the best results in different species. Importantly, it should be remembered that healthy GC levels are important in marsupial reproduction and different levels across oestrus cycles can be positive (Fanson and Parrott 2015).

## 9.7 Behavioural enrichment

Behavioural enrichment activities for possums and gliders can include:

- providing browse such as leaves, flowers or gums (see relevant diets)
- providing live food, such as mealworms or crickets, for insectivorous species, as activity feeds at times throughout the day
- proving food and scent smears to encourage movement and foraging behaviour
- placing food on branches (e.g. fruit spiked on branches) throughout the enclosure rather than in the one location in a feed tray
- providing nesting material such as stringybark or moss to promote nest-building behaviour
- providing opportunities for gliding
- housing with other terrestrial species as appropriate
- providing gums (Gum Arabic Powder food grade by Swift Ltd) in gum feeders for species that feed on gum such as the *Petaurus* gliders (Hawkins 1999); however, feeding gums to Leadbeater's possums has resulted in diarrhoea (Lynch 1995)
- providing mountain pygmy possums with scatter-feeding of live insects, smearing honey on substrates, crushing dog kibble or walnuts and hiding food (Boys 2018). Other opportunities for this species include changing the nesting material; providing new perching, rocks or logs; adding snow to the enclosure; changing the substrate; and adding grass or natural rope
- providing climbing opportunities such as ropes to feather-tailed gliders and other species of pygmy possums to allow them to them to climb around the enclosure.

## 9.8 Training and conditioning

Limited training appears to have been undertaken of most species of possums and gliders, but some individual eastern ring-tailed possums and sugar gliders have been conditioned to be exhibited during educational demonstrations.

## 9.9 Introductions and removals

Most introductions and removals can be undertaken with few problems. Some of the more social species, such as the petaurids, which live in family groups and use scent to maintain the group structure, may need to be carefully observed to assess if there are any problems of aggression. Larger species such as the brush-tailed possums can be very aggressive towards each other, so they also need to be carefully monitored to assess the level of aggression.

## 9.10 Intraspecific compatibility

The social structure and mating system are influenced by several factors, including body size (smaller species often huddle together when it is cold to minimise heat loss), diet and food availability and competition between individuals. In some species of mammals that have more than one mating system, it is generally the result of the males' ability to defend females and/or resources. Polygyny is favoured when:

- food resources are highly abundant or concentrated (Emlen and Oring 1977)
- males can defend access to more than one female
- females do not require males to raise offspring (Emlen and Oring 1977; Kleiman 1977).

Monogamy is favoured when the reverse is true.

Therefore, in a captive environment with plenty of food, species that do at times practise polygyny in the wild should readily be able to practise it in captivity due to the ample food requirements and lack of competition (as they are usually housed in groups of one male and one or more females). Most of the smaller species of possums and gliders can readily be held together; however, as a general rule the larger species are more solitary and aggressive. Therefore, males of species such as brushtailed possums and cuscuses should never be housed together and invariably both sexes are held separately or as pairs to avoid or minimise aggression. The mating system for the different genera of possums and gliders is shown in Table 9.14.

**Table 9.14.** Social structure and mating system of different genera of possums and gliders when held in captivity.

Families are listed in approximately increasing body size.

| Genus | Social structure | Mating system | Suggested sex ratio | Reference |
|---|---|---|---|---|
| **Tarsipedidae** | | | | |
| *Tarsipes* | Colonial | Polyandrous | 3:5 | 1 |
| **Acrobatidae** | | | | |
| *Acrobates* | Colonial | Promiscuous | 1:1-10:10 | 2, 3 |
| **Burramyidae** | | | | |
| *Burramys* | Colonial | Promiscuous | 1:1-5:5 | 4 |
| *Cercartetus* | Solitary | Promiscuous | 1:1-5:5 | 5 |
| **Petauridae** | | | | |
| *Dactylopsila* | Solitary/Pairs | Polygynous? | Solitary or 1:1 | 6 |
| *Gymnobelideus* | Colonial/Pairs | Monogamous | Solitary or 1:1 | 7,8 |
| *Petaurus* | Colonial/Pairs | Monogamous/polygynous | Solitary or 1:1 | 9-13 |
| **Pseudocheiridae** | | | | |
| *Hemibelideus* | Solitary? | Monogamous? | Solitary or 1:1 | |
| *Petauroides* | Solitary | Polygynous/monogamous/ Polygynous | Solitary or 1:1 | 14-16 |
| *Petropseudes* | Solitary/Pairs | Monogamous | Solitary or 1:1 | 17 |
| *Pseudocheirus* | Solitary/Pairs | Monogamous/polygynous | Solitary or 1:1 | 18-20 |
| *Pseudochirops* | Solitary/Pairs | Monogamous? | Solitary or 1:1 | 21 |
| *Pseudochirulus* | Solitary/Pairs | Monogamous? | Solitary or 1:1 | 21 |
| **Phalangeridae** | | | | |
| *Phalanger* | Solitary/Pairs? | Monogamous? | Solitary or 1:1 | |
| *Spilocuscus* | Solitary/Pairs? | Monogamous? | Solitary or 1:1 | |
| *Trichosurus* | Pair/Solitary? | Monogamous/polygynous | Solitary or 1:1 | 22-24 |
| *Wyulda* | Solitary | Monogamous? | Solitary or 1:1 | 25 |

References: (1) Wooller *et al.* (2000); (2) Fleming and Frey (1984); (3) Ward (1990a); (4) Boys (2018); (5) Ward (1988); (6) Carmichael (2000); (7) Smith (1980); 8) Smith (1984b); (9) Henry and Craig (1984); (10) Henry and Suckling (1984); (11) Suckling (1984); (12) Craig (1985); (13) Jackson (2000b); (14) Henry (1984); (15) Henry (1985); (16) Comport *et al.* (1996); (17) Runcie (2000); (18) How *et al.* (1984); (19) Pahl (1987b); (20) Pahl and Lee (1988); (21) Winter and Atherton (1984); (22) How (1976); (23) How (1981); (24) Winter (1976); (25) Runcie (1999).

## 9.11 Interspecific compatibility

The smaller species of possums in the families Tarsipedidae, Acrobatidae and Burramyidae have been held as single-species groups because they are normally kept in smaller enclosures for ease of viewing by the public or for off-exhibit maintenance. There is one record of eastern pygmy possums being introduced to feather-tailed gliders and immediately evicting the gliders from one of their nest boxes; however, no further interspecific fighting was observed (Pepper-Edwards 1988).

Several of the larger possums and gliders in the families Phalangeridae, Petauridae and Pseudocheiridae have been held with other species in captivity. As these species are generally arboreal and require larger enclosures, several terrestrial species can be held with them. These include short-beaked echidna (*Tachyglossus aculeatus*), long-beaked echidna (*Zaglossus bruijnii*), long-nosed bandicoots (*Perameles nasuta*), eastern-barred bandicoots (*P. gunnii*), long-nosed bandicoots (*P. nasuta*), long-footed potoroos (*Potorous longipes*), long-nosed potoroos (*P. tridactylus*) and brush-tailed bettongs (*Bettongia penicillata*). Care should be taken to ensure non-aggressive individuals are used in mixed-species enclosures. Nocturnal birds such as nightjars and tawny frogmouths are generally not recommended to be housed with these species, as there is likely to be aggression between them that may involve the birds being preyed upon. For example, squirrel gliders have been known to attack and kill Australian magpie-larks (*Grallina cyanolueca*) and half-grown guinea fowl (*Numida meleagris*) and sugar gliders have been known to kill mice (*Mus musculus*) in captivity (Troughton 1941; Fleay 1954).

Several species of possums and gliders have been held together, including yellow-bellied gliders, with Leadbeater's possums and greater gliders. Leadbeater's possums have been housed successfully with ring-tailed possums, and brush-tailed possums have lived with tawny frogmouths (*Podargus strigoides*). In the wild, greater gliders and yellow-bellied gliders have been observed to share the same next box (McNabb 2017), so appear to be compatible for captive display. Yellow-bellied gliders have also been held with grey-headed fruit bats (*Pteropus poliocephalus*), which generally worked well, although there was occasional fighting between individuals over food (which frequently occurs in the wild (e.g. S. Jackson pers. obs.; Borsboom 1982)). Some species of possums such as the mountain brush-tailed, common brush-tailed and cuscuses are generally housed by themselves due to their aggressive natures.

# 10. BREEDING

## 10.1 Mating system

The mating system varies between species and is dependent on factors such as social structure, body size and food availability (see section 9.1.3). The mating systems for species where it is known are shown in Table 9.14.

## 10.2 Ease of breeding

### 10.2.1 Honey possum

Until recently, the honey possum has not bred well in captivity, but the use of a high-nitrogen diet and large outdoor, natural-type enclosures appears to have resulted in more consistent breeding. It also appears that holding animals indoors with a fixed artificial photoperiod may inhibit reproductive activity in the female honey possums (Bradshaw *et al.* 2004).

### 10.2.2 Feather-tailed gliders

The feather-tailed glider has bred well at Taronga Zoo in large indoor enclosures and has bred in the nocturnal house at Healesville Sanctuary; however, it has not bred very well anywhere else. The trigger appears to be having large enclosures with a large number of animals (i.e. more than 12–15). They then appear to breed until they reach the carrying capacity of the enclosure and either breeding is reduced or the mortality rate increases. Previous lack of breeding success may be partly due to two different species being held together, before the recent discovery of the speciation. The two species look very similar, but are not able to interbreed and produce young.

### 10.2.3 Pygmy possums

Although most of the pygmy possums held in captivity for any length of time have been bred, there has been little success breeding them with regularity, except for the mountain pygmy possum.

The mountain pygmy possum has bred very well from captive-born males and females and wild-caught males at Healesville Sanctuary. There was initially low breeding success with wild-caught females that came from a genetically depauperate population, with the lack of success potentially due to both inbreeding and husbandry, which had not yet been refined. Wild-caught females collected from a larger, genetically diverse population have now bred well at Healesville Sanctuary. In previous studies, captive-bred mountain pygmy possums lacked the seasonal changes in physiology exhibited by wild animals and generally did not increase body mass to the extent of wild-caught animals (Geiser *et al.* 1990). However, husbandry has now been refined over 15 years, resulting in consistent breeding success in alpine-mimicking facilities that provide appropriate temperature and lighting cues and support annual hibernation (see Boys 2018). Studies of mate choice (Parrott 2019), social system and diet (e.g. Parrott *et al.* 2020) have also been completed to assist breeding success from multiple generations of possums. In the wild, the breeding period has been associated with (1) the spring equinox, (2) the arrival of the Bogong moths, (3) arrival of males in female habitat and final emergence from hibernation and (4) the loss of the snow cover at the end of the winter (Mansergh and Scotts 1990; Körtner and Geiser 1996).

### 10.2.4 Petaurids

The smaller gliders, such as the sugar and squirrel gliders, generally breed well in captivity. Several hand-reared mahogany gliders, one of which was extraordinarily obese, have been held in captivity and have subsequently bred so it is likely they can breed well. The lack of success with breeding yellow-bellied gliders may be due to the small numbers held in captivity, their tendency to obesity and the fact that they have frequently been maintained in captivity after hand-rearing.

Leadbeater's possums generally breed relatively well as part of their international studbook; however, individuals within the long-standing breeding program stopped breeding, so further recruitment from the wild was required. The halt in breeding may be due to increased inbreeding or other unknown factors. More recently, a program to improve breeding success through genetic admixture between wild populations is underway (Zilko *et al.* 2020).

Torresian striped possums have not bred well in captivity over the years, which has contributed to their poor representation in captive institutions. However, in 1999 London Zoo was successful in breeding them and have since bred them several times (McKenna 2005).

### 10.2.5 Ring-tailed possums and greater gliders

The ring-tailed possums have not bred well in captivity, possibly because institutions have not wanted to breed them as orphans are frequently submitted to zoos or shelters and hand-reared.

Greater gliders have bred very poorly in zoos, which may be due to the low numbers brought into captivity.

### 10.2.6 Cuscuses, brush-tailed possums and scaly-tailed possum

Cuscus have bred several times in captivity, but there are few records. Brush-tailed possums can breed well in captivity, but most institutions do not attempt to breed them due to their territorial nature, which means the offspring need to be removed upon maturity, and the difficulty in finding other institutions that will take them. Breeding colonies have been kept for research, such as at the University of Melbourne, Victoria. The scaly-tailed possum has only been held in captivity for a relatively short period and has not had the opportunity to breed.

## 10.3 Oestrous cycle and gestation period

Oestrous cycles typically last 21–30 days for those that are known (Table 9.15). Gestation length is remarkably similar and does not appear to follow any pattern between taxonomic group or body size, generally ranging from 15 to 30 days.

## 10.4 Reproductive status

### 10.4.1 Females

Possums and gliders are generally placed in several categories depending on their reproductive status. The examination of reproductive status in medium to large species can be facilitated by putting them inside a transparent plastic tube and examining the pouch with an otoscope (Roberts and Kohn 1991). Smaller species can be carefully rolled onto their backs against the handler's body and the pouch examined (see Boys 2018). All pouch checks should be completed quickly to minimise stress to the female. For females the categories are:

- nulli-parous (females that have never bred) – pouch is small, clean and dry, teats very small and difficult to distinguish
- non-parous (females that have bred previously but not presently) – pouch is small but distinct, dry and stained, the teats are slightly elongated and obvious
- parous (pregnant) – pouch is pink in colour and glandular in appearance, skin folds may be observed on the lateral margins
- pouch young present – attached to the teat
- lactating (young absent from the pouch but still suckling) – pouch area large, skin folds flaccid, hair sparse and stained, skin smooth and dark pink, teats elongated

**Table 9.15.** Duration of oestrous cycle, gestation of several possums and gliders.

| Species | Oestrous cycle (days) | Gestation (days) | Postpartum oestrus | Embryonic diapause | Reference |
|---|---|---|---|---|---|
| **Burramyidae** | | | | | |
| *Burramys parvus* | ~27 | 13–16 | N | N | 1, 2 |
| *Cercartetus concinnus* | – | < 33 | Y | Y? | 3 |
| *Cercartetus lepidus* | – | – | – | N | 4 |
| **Petauridae** | | | | | |
| *Gymnobelideus leadbeateri* | < 30 | < 20 | N | N | 5 |
| *Petaurus notatus* | 29 | 15–17 | N | N | 6 |
| *Petaurus norfolcensis* | – | 16–17 | – | – | 7 |
| **Pseudocheiridae** | | | | | |
| *Pseudocheirus peregrinus* | 28 | 14–16 | – | – | 8 |
| **Tarsipedidae** | | | | | |
| *Tarsipes rostratus* | 25 | 23 | Y | Y | 9–11 |
| **Acrobatidae** | | | | | |
| *Acrobates pygmaeus* | – | – | Y | Y | 4, 12, 13 |
| **Phalangeridae** | | | | | |
| *Trichosurus caninus* | 26.4 | 15–17 | N | N | 14 |
| *Trichosurus vulpecula* | 21–30 | 16–21 | N | N | 15–17 |

References: (1) Strahan (1995); (2) Scicluna (2017); (3) Clark (1967); (4) Ward (1990c); (5) Smith (1984b); (6) Smith (1971); (7) Woodd *et al.* (2006); (8) Ong (1994); (9) Renfree (1980); (10) Bradshaw *et al.* (2004); (11) Oates *et al.* (2007); (12) Hill (1900); (13) Ward and Renfree (1988); (14) Smith and How (1973); (15) Lyne *et al.* (1959); (16) Pilton and Sharman (1962); (17) Curlewis and Stone (1986).

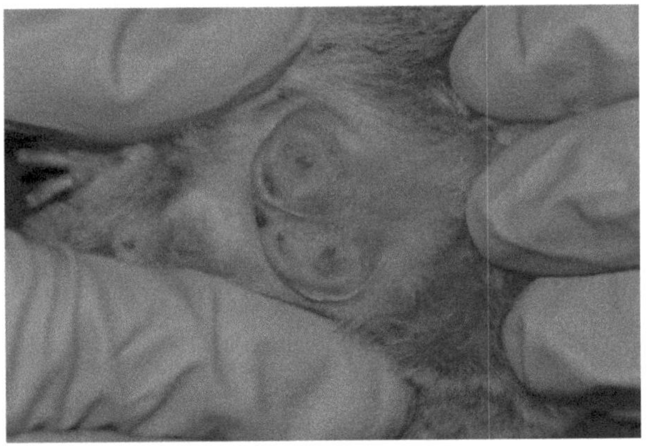

**Fig. 9.8.** Pouch checking a female mountain pygmy possum showing four nipples. Taken from Boys (2018).

- post lactation – teats expressing only clear liquid and/or regressing (Fig. 9.8).

If pouch young are present there are several developmental stages and measurements that can be recorded and compared to existing growth curves (see section 10.16), or used to establish curves for future reference. These include:

- developmental stages
  - sex distinguishable
  - tips of ears free
  - papillae of facial vibrissae evident
  - eyelashes visible
  - eyes open
  - fur visible – slight tinge, medium or well developed
  - tips of first incisors through the gums
  - on back or in nest
  - eating solids
  - self-feeding
  - independent
- measurements (see Appendix 3)
  - bodyweight (g) – if not on teat
  - head length (mm) – from the occiput to snout tip
  - head width (mm) – maximum width across the zygomatic arches
  - crown-rump length (mm) – primarily for very small neonates
  - body length (mm) – from snout tip to cloaca
  - tail length (mm) – from the cloaca to the end of the last vertebra of the tail tip

- total length (mm) – from snout tip to tail tip
- tibia length (mm) – from the hip to the bottom of the pes
- pes length (mm) – from the heel to the base of the longest toe, not including the claw.

### 10.4.2 Males

In some species, such as the *Petaurus* gliders, the males have a scent gland in the middle of their forehead, whereas male brush-tailed possums have a visible cutaneous sternal gland (and not obvious head glands) that becomes increasingly developed with age (see section 9.2). The activity of the gland can be measured from the following scale (Millis and Bradley 2001):

(1) little or no activity – little or no staining of the surrounding hair; little or no hair loss over the gland area; no obvious gland product
(2) medium level activity – some staining of the surrounding hair; some loss of hair over the gland area; waxy glandular products visible
(3) high activity – much staining of the surrounding hair; total loss over gland area; waxy glandular product prominent.

In males of seasonally breeding species, the testes can increase in size during the breeding season. The testes should be measured by measuring the length, width and depth in millimetres. Testis volume can be calculated by using the equation $V = \pi/6 \times (\text{length}) \times (\text{width})^2$ (Spencer 1996). The presence of sperm in the male's urine can also be verified via microscopic analysis to determine if a male is ready to breed (e.g. see Boys 2018).

## 10.5 Reproductive management

The altered day/night lengths that result from maintaining possums and gliders for public display is likely to affect their breeding. In a group of brush-tailed possums moved from an outside enclosure to a room with a 10 h light (L) and 14 h dark (D) cycle, the possums indoors gave birth after 81 days compared with 134 days for the control group outdoors (Gemmell 1990). This suggests that photoperiod plays an important role in the initiation of breeding in the brush-tailed possum (Gemmell 1990). In further experiments on day length, brush-tailed possums were held in enclosures with a short daylength (10L:14D) and long day length (14L:10D) (Gemmell and Sernia 1992), which resulted in the possums in the short day length enclosures breeding earlier in the year (12 Jan to 14 Feb) than the long day length possums (5 May to 8 August), with the possums in natural light giving birth from 3 March to 8 May (Gemmell and Sernia 1992). Over a 2-year period, the possums in both the long and short day length conditions bred three times and those in natural lighting bred twice (Gemmell *et al.* 1993).

To reduce the incidence of breeding in common brush-tailed possums an implant that contains the gonadotrophin-releasing hormone agonist deslorelin has been used successfully, with 80% of females showing no evidence after an average of 381 days (Lohr *et al.* 2009). These results also highlight that it is not completely effective as 20% of female can still produce young.

## 10.6 Occurrence of hybrids

A female Victorian sugar glider and a male Queensland squirrel glider have produced a fertile hybrid (Fleay 1947) in captivity at Healesville Sanctuary, Australia, while Zuckerman (1953) also reported the birth of hybrid between a sugar glider and a squirrel glider at London Zoo. Knipler *et al.* (2021) detected hybridisation of sugar gliders and squirrel gliders in the wild on the central coast of New South Wales.

## 10.7 Timing of breeding

There is a large variation in the start and duration of breeding in the possums and gliders, depending on the season and food availability. Breeding can be seasonal or continuous (Table 9.16). Proving appropriate cues (such as changes in photoperiod, temperature and/or nutrition) is important to promote successful breeding in many captive possum species.

## 10.8 Age at first and last breeding

There is a large variation in the range of first and last breeding in the possums and gliders, with first breeding ranging from less than 5 months in the pygmy possums to ~36 months for the mountain brush-tailed possum. Breeding generally continues until death (Table 9.16).

Oestrus can be determined by examining the urine for the presence of non-keratinised and keratinised epithelial cells, polymorphonuclear leucocytes and in the male, sperm (Duckworth *et al.* 1998). At the time of oestrus, there is a massive increase in the number of epithelial cells and leucocytes in the urine.

In most species it is thought that the females can breed up until their death; however, records for the mountain brush-tailed possum records suggest that females show reproductive senescence before their

**Table 9.16.** Reproduction and development of the possums and gliders.

| Species | Birth season | Litter size (mean) | Litters per year | Permanent pouch exit (days) | Weaning (days) | Sexual maturity (months) | | Reference |
|---|---|---|---|---|---|---|---|---|
| | | | | | | F | M | |
| **Burramyidae** | | | | | | | | |
| *Burramys parvus* | Oct–Nov | 1–4 (3.6) | 1 | 33–37 | 70–75 | 12 | 12 | 1–4 |
| *Cercartetus caudatus* | Aug–Feb | 1–4 | – | 34–45 | 92 | 15 | – | 4, 5 |
| *Cercartetus concinnus* | All year | 3–6 | 2–3 | 25–30 | 50 | 12–15 | – | 4, 6–9 |
| *Cercartetus lepidus* | Sep–Jan | 2–4 | – | – | 90 | – | – | 1, 10 |
| *Cercartetus nanus* | Mar-Jun/Oct-Jan | 2–6 (4) | 2–3 | 30–42 | 50–65 | 4.5–5 | 4.5–5 | 11–14 |
| **Petauridae** | | | | | | | | |
| *Dactylopsila trivirgata* | Mar–Jun | 1–2 | 1 | – | – | – | – | 15, 16 |
| *Gymnobelideus leadbeateri* | Apr–Jun /Oct–Dec All year | 1–2 (1.5) | 2 | 80–93 | 110–120 | 12,24+ | – | 4, 17, 18 |
| *Petaurus australis* | Feb-Dec | 1 | 1 | 90–100 | 180–240 | 24 | 18 | 4, 19–25 |
| *Petaurus notatus* | Apr–Dec | 1.5–2 (1.8) | 1–2 | 70–74 | 110–120 | 8–15 | 12 | 4, 26–30 |
| *Petaurus gracilis* | Apr-Sept | 1–2 (1.6) | 1 | – | – | 12 | 12 | 30 |
| *Petaurus norfolcensis* | Apr–Dec | 1–2 (1.7) | 1 | – | – | 12 | 12 | 31–34 |
| **Pseudocheiridae** | | | | | | | | |
| *Hemibelideus lemuroides* | Aug–Nov | 1 | – | – | – | – | – | 4 |
| *Petauroides volans* | Mar–Jun | 1 | 1 | 90–120 | 180–210 | 12+ | 12+ | 4, 35–37 |
| *Petropseudes dahlii* | Mar–Aug | 1 | – | – | – | – | – | 4 |
| *Pseudochirops archeri* | Aug–Nov | 1 | – | – | – | – | – | 4 |
| *Pseudochirulus cinereus* | All year? | 1 | 1 | – | – | – | – | 4 |
| *Pseudochirulus herbertensis* | All year? | 1 | 1 | 120 | 150–160 | 16 | – | 4, 38 |
| *Pseudocheirus occidentalis* | Apr–Nov | 1–2 | 1 | 104 | | 10–11 | – | 39, 40 |
| *Pseudocheirus peregrinus* | Apr–Nov | 1–4 (2) | 1–2 | 120 | 150–240 | 12 | 10–12 | 41–45 |
| **Tarsipedidae** | | | | | | | | |
| *Tarsipes rostratus* | All year | 2–4 (2.5) | 2+ | 63–70 | 90 | 6 | – | 46–48 |
| **Acrobatidae** | | | | | | | | |
| *Acrobates pygmaeus* | All year/seasonal | 2–4 (2.5) | 2 | 50–65 | 90–100 | 6–8 | 12 | 4, 49–51 |
| **Phalangeridae** | | | | | | | | |
| *Phalanger intercastellanus* | Jun–Sept? | 2 | 1 | – | – | – | – | 4 |
| *Spilocuscus maculatus* | Jun? | 1–3 (1) | 1 | – | – | – | – | 4 |
| *Trichosurus caninus* | Mar–May | 1 | 1 | 150–200 | 275 | 22-36 | 36 | 27, 52, 53 |
| *Trichosurus vulpecula* | Mar–Nov | 1 | 1–2 | 140–150 | 230 | 12–24 | 24 | 35, 52, 54–57, |
| *Wyulda squamicaudata* | Mar–Aug | 1 | 1 | 150–200 | > 240 | 24 | > 18 | 4, 58, 59 |

References: (1) Dimpel and Calaby (1972); (2) Kerle (1984b); (3) Mansergh and Scotts (1990); (4) Strahan (1995); (5) Atherton and Haffenden (1982); (6) Bowley (1939); (7) Casanova (1958); (8) Clark (1967); (9) Ward (1990c); (10) Ward (1992); 11) Turner (1983); (12) Ward (1990b); (13) Goldingay and Rueegger (2018); (14) R. Goldingay unpubl. data; (15) Handasyde and Martin (1996); (16) Handasyde *et al.* (2001); (17) Smith (1984b); (18) Harley and Lill (2007); (19) Russell (1983); (20) Russell (1984); 21) Craig (1985); (22) Goldingay (1992); (23) Goldingay *et al.* (2001); (24) Craig (1986); (25) Goldingay (1992); 26) Smith (1971); (27) Smith (1973); (28) Smith (1979); (29) Suckling (1984); (30) Jackson (2000a); (31) Quin (1995); (32) Millis and Bradley (2001); (33) van der Ree (2002); 34) Sharpe and Goldingay (2010); (35) Smith *et al.* (1969); (36) Tyndale-Biscoe and Smith (1969); (37) Henry (1984); (38) Haffenden (1984); (39) Ellis and Jones (1992); (40) Jones *et al.* (1994); (41) Thomson and Owen (1964); (42) Hughes *et al.* (1965); 4(3) How *et al.* (1984); (44) Pahl and Lee (1988); (45) Ong (1994); 46) Renfree (1980); (47) Wooller *et al.* (1981); (48) Renfree *et al.* (1984); (49) Fleming and Frey (1984); (50) Ward and Renfree (1988); (51) Ward (1990a); (52) How (1976); 53) How (1981); (54) Lyne and Verhagen (1957); (55) Dunnet (1964); (56) Crawley (1973); (57) Smith and How (1973); (58) Humphreys *et al.* (1984); (59) Runcie (1999).

death (Viggers and Lindenmayer 2000). They have been found not to successfully rear young after the age of ~9 years, even though there are records of them living as long as 17 years (Lindenmayer *et al.* 1991; Viggers and Lindenmayer 2000).

## 10.9 Ability to breed every year

All species of possums and gliders appear to be able to breed at least once per year.

## 10.10 Ability to breed more than once per year

Several species of possums have been observed to breed more often in captivity than they do in the wild. Ring-tailed possums have been observed to wean young earlier in captivity than in the wild (160 days *v.* 180–220 days respectively), reach maturity earlier (11 months *v.* 13 months), breed throughout the year and produce up to three litters per year (instead of one or two) (Thomson and Owen 1964; How *et al.* 1984; Roberts *et al.* 1990). Female brush-tailed possums in the wild have been recorded raising two young in a year (Winter 1976).

Similar observations of higher fecundity in captivity have also been observed for the sugar glider, which can also produce three litters per year (MacPherson 1997).

## 10.11 Nesting requirements

Most species require a nest box, even if they do not generally use tree hollows in the wild. Nesting material, which can include fresh branches of eucalypts foliage, casuarina and leptospermum foliage, should also be supplied so the leaves can be used (e.g. pygmy possums, ring-tailed possums, petaurids). Branches of stringybark should be provided for some species (e.g. Leadbeater's possum) so they can use the bark. Other material that has been used includes moss, sea grass, fine dry grasses that are commercially available, hay and shredded paper, though these are generally recommended for off-exhibit enclosures (L. Andrews pers. comm.).

## 10.12 Breeding diet

The only species to have a breeding diet is the honey possum due to the greatly increased nitrogen requirements (see section 6.6.1). However, changes or increases in food availability during lactation are required for females with young of all species.

## 10.13 Number of young born

Litter size ranges from two to six, with 2–3 litters per year in the small possums such as the honey possums, pygmy possums, feather-tailed gliders and sugar gliders to 1–2 in the larger ring-tailed possums and gliders (Atherton and Haffenden 1982; Bradshaw *et al.* 2000). The largest possums such as the yellow-bellied glider, brush-tailed possums and cuscuses usually produce only one young, though they are known to occasionally produce twins (e.g. Craig 1986 for yellow-bellied gliders).

## 10.14 Age at weaning

The age at weaning for a variety of possums and gliders is show in Table 9.16.

## 10.15 Age at removal from the mother

Once weaned, the young of most species, particularly those in the Petauridae and Phalangeridae, should be removed to reflect the natural dispersal and stop aggression between the parents and offspring. Successful rearing of honey possum young in captivity is difficult, but the greatest success is achieved if the lactating females are isolated from one another (Russell and Renfree 1989).

## 10.16 Growth and development

Growth curves using a variety of measurements are known for different species (Table 9.17). Additional growth curves are contained in Bach (1998). Growth curves for several species commonly hand reared are shown in Fig. 9.9.

# 11. ARTIFICIAL REARING

## 11.1 Housing

As with all native mammals that have been taken into care, minimising stress is a major consideration. Choosing suitable housing can help to create a stress-free environment. To achieve this, several factors should be considered:

- securing the area from children and other animals (particularly domestic dogs and cats)
- maintaining high standard of hygiene
- escape-proofing the area
- clearing the area of obstacles and hazards
- offering shelter from extreme weather and noise

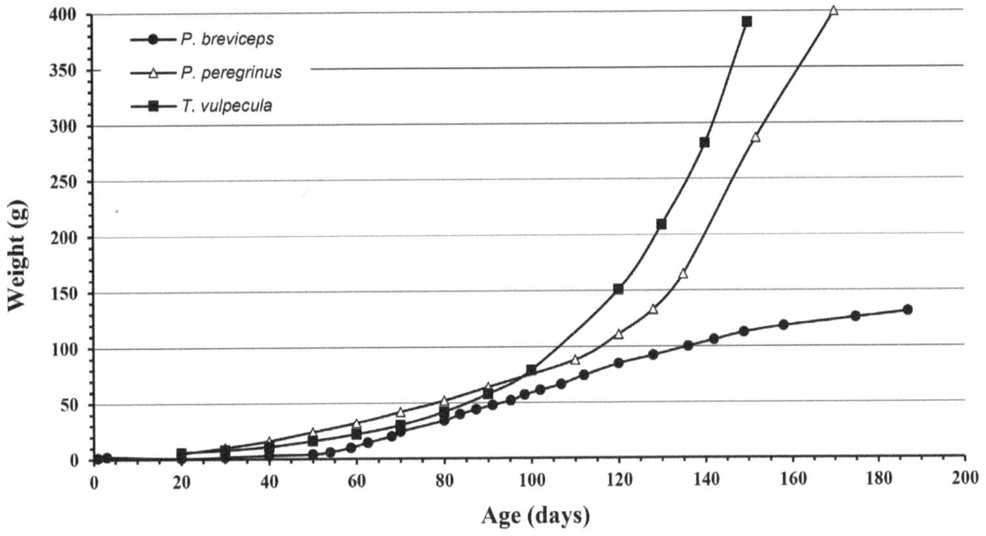


**Fig. 9.9.** Growth in bodyweight of several species of possums and gliders. Derived from Smith (1979), Roberts *et al.* (1990) and Austin (1997).

**Table 9.17.** Growth curve measurements that have been developed for different species of possums and gliders.

AR = arm length, CR = crown-rump, EA = ear length, HB = head body length, HE = head length, HL = head length, HW = head width, LE = leg length, MA = manus length, PE = pes length, TA = tail length, TL = tail length, UL = ulna length, WT = bodyweight.

| Species | Measurements | Reference |
|---|---|---|
| **Burramyidae** | | |
| *Burramys parvus* | WT | 1 |
| *Cercartetus caudatus* | WT, CR, HE, HL, TA | 2 |
| *Cercartetus nanus* | CR, HE | 3, 4 |
| **Petauridae** | | |
| *Gymnobelideus leadbeateri* | WT | 5 |
| *Petaurus notatus* | WT, CR, HL, HW, TA, PE | 6–10 |
| *Petaurus gracilis* | WT, HL, TA, UL | 11 |
| *Petaurus norfolcensis* | WT, HL | 7, 10 |
| **Pseudocheiridae** | | |
| *Petauroides volans* | WT | 12 |
| *Pseudocheirus peregrinus* | WT, CR, HB, HL, PE, TA, TO | 8, 10–17 |
| *Pseudocheirus occidentalis* | WT, HL | 18–20 |
| *Pseudochirulus herbertensis* | WT, TO | 21 |
| **Tarsipedidae** | | |
| *T. rostratus* | WT, HE | 22, 23 |
| **Acrobatidae** | | |
| *Acrobates pygmaeus* | WT, PE, HB, TA | 24–26 |
| **Phalangeridae** | | |
| *Trichosurus caninus* | WT, EA, HE, PE | 14, 27 |
| *Trichosurus vulpecula* | WT, AR, CR, EA, HB, HE, LE, MA, PE, TA | 8, 10, 17, 18, 28–34 |

References: (1) Kerle (1984b); (2) Atherton and Haffenden (1982); (3) Ward (1990b); (4) Westman and Geiser (2004); (5) Smith (1980); (6) Smith (1971); (7) Smith (1979); (8) Austin (1997); (9) Caton (1999); (10) Woods (1999); (11) Booth *et al.* (2019); (12) Collins (1973); (13) Thomson and Owen (1964); (14) How (1976); (15) Presidente (1982b); (16) Roberts *et al.* (1990); (17) Bellamy (1992); (18) Presidente (1982a); (19) Ellis and Jones (1992); (20) Wayne *et al.* (2005a); (21) Haffenden (1984); (22) Bradshaw *et al.* (2000); (23) Wooller *et al.* (1999); (24) Fanning and Watkins (1980); (25) Fleming and Frey (1984); (26) Caton (1995); (27) Viggers and Lindenmayer (2000); (28) Dunnet (1956); (29) Lyne and Verhagen (1957); (30) Kerle and Howe (1992); (31) Gemmell and Hendrikz (1993); (32) Gemmell (1995); (33) B. Smith (1995); (34) Wayne *et al.* (2005b).

Where possible, a humidicrib ICU (e.g. Vetario T30 Intensive Care unit) can be used or otherwise a secure box. The humidicrib or box is lined with a towel and warm pouches are provided for the juvenile to nest in. Unfurred or just furred should be placed in multiples layers of artificial pouches that simulates, as closely as possible, the security and warmth of the mother's pouch. Natural-fibre products such as cotton or wool are required as they retain warmth better and prevent the young from overheating.

Feather-tailed gliders have been successfully hand-reared using a clear Perspex container (315 × 195 × 230 mm (H)) with a lift-up lid containing air holes for ventilation (under which a heat pad, set at 32°C, wrapped in a towel) for animals < 80 days old (Caton 1995). Inside the container with a temperature of ~28°C another towel is folded and placed directly above the heat pad with smaller pieces of fleecy-lined cloth. From 70 days the thermostat for the heat pad is lowered and by 80 days it is the same as the temperature outside. Animals older than 80 days can be placed in a larger box (~700 × 400 × 400 mm (H)) (Caton 1995).

Gliders (e.g. sugar gliders), can be placed in a cotton pouch. The success rate of furless young is lower, though possible; however, once furred the success rate is nearly 100% (Booth 1999). Just furred young require a temperature in the range of 30–34°C, which can be decreased gradually to ambient temperature at ~100 days of age (Booth 1999).

Ring-tailed possums should be placed in a cotton liner inside a woollen pouch. The bag is then placed beside a heat source set at 28–30°C (B. Smith 1995). If the animal is furred the temperature can be kept at 28°C and when it weighs 60 g it should be placed in a small box with a well-wrapped heat pad at 28°C and allowed to leave the pouch and return. At 80–100 g, a larger box with small branches may be required to allow the ring-tailed possum to begin climbing. When it reaches 150 g, the pouch can be placed in an artificial drey and removed, so that only the drey is used when the possum weighs 200 g. Dreys can be made by lining the wire frames of 2 × 30 cm hanging plant baskets with coconut fibre, tying them together to form a sphere and making a small entrance just wide enough for the possum to enter. They appear to have a better survival rate if they are placed with at least one other animal of approximately the same age.

Brush-tailed possums should be placed in a pouch that rests against a heat pad set at 30–32°C. Once the possum reaches 250 g it should be suitably warm in a snug pouch; however, if at any time it feels cold, it should be supplied with artificial warmth (B. Smith 1995). At > 300 g, it should be placed in a small enclosure, such as a cockatoo cage, and given branches to climb. At 400 g, a small nest box should be provided and at 500 g it should be given the opportunity to socialise with another brush-tailed possum, preferably of approximately the same age. They appear to have a better survival rate if they are released in pairs.

## 11.2 Temperature requirements

The temperature of the artificial pouch should be 32–34°C if the joey is furless. As the joey grows fur the temperature can be reduced to 28–30°C (Bellamy 1992). Different temperatures are suggested for different species of possums (e.g. 32°C for feather-tailed gliders (Caton 1995) and 28°C for ring-tailed possums) (B. Smith 1995). Brush-tailed possums should be kept at 30–32°C for individuals less than 250 g and 28°C for new individuals brought in until settled (B. Smith 1995). Use a minimum/maximum temperature gauge with a plastic-coated probe that can be placed in the pouch layers near the joey, as this will ensure that the temperature can be monitored. Do not place the probe in the pouch with the joey as they can chew it. A humidicrib ICU can be very useful for unfurred and just furred joeys as it greatly increases their chance of survival. Furred joeys can be maintained at an appropriate temperature with the use of a heating pad or SnuggleSafe® Microwave Heat Pad.

## 11.3 Specific requirements

The skin of unfurred and slightly furred young should be kept moist by using a humidicrib or Sorbolene cream (not with added glycerine), so that the skin does not become dry and cracked (George *et al.* 1995).

When first brought in for hand-rearing, the young animal may be dehydrated. If so, it can be given plain cooled boiled water, with 5 g (1 teaspoon) of glucose to 250 mL of water or 1 g of electrolyte replacer if available (Austin 1997). Vytrate® can also be used at a ratio of 20 mL Vytrate® to 250 mL water. Alternatively, fluids such as Lectade or Pedialyte, an electrolyte/glucose replacer (if the animal has diarrhoea) or Glucodin® can be given for a period of up to 24–48 h (Barnes 2002). It is important to warm the joey before feeding otherwise there is a greater risk of inhalation pneumonia. If the joey is really cold, place it in a warm water bath and dry it off rather than putting it in a hot box.

**Table 9.18.** Concentrations of the major constituents of milk for different species of possums and gliders.

| Species | Total solids (%) | Carbohydrate (%) | Lipids (%) | Protein (%) | Calcium (mg/L) | Iron (mg/L) | Reference |
|---|---|---|---|---|---|---|---|
| *Petaurus breviceps* | 31.0–38.0 | 11.0 | 22.0 | 8.0 | 2300 | 1.6 | 1, 2 |
| *Pseudocheirus peregrinus* | 12.6–25.0 | 4.0–13.0 | 1.5–4.0 | 4.0–8.4 | 1700–2000 | - | 2 |
| *Trichosurus vulpecula* | 10.0–42.7 | 4.0–12.0 | 1.3–15.0 | 3.0–8.0 | 3000–7700 | 5–9 | 3–6 |

References: (1) Green (1984); (2) Munks *et al.* (1991); (3) Gross and Bolliger (1959); (4) Cowan (1989); (5) Crisp *et al.* (1989); (6) Jolly *et al.* (1996).

Stress is a major problem in successfully rearing native mammals. It can be fatal, especially in ring-tailed possums. Therefore, it is important to keep noise to a minimum, do not overhandle the animals and maintain high standards of hygiene.

## 11.4 Diet and feeding routine

### 11.4.1 Natural milk

Ring-tailed possums' solids increase from 15% in early lactation to 25% at mid-lactation and 13% at late lactation. Hexose concentrations increased from 10% to a peak of 13% and returned to a low of 4–6% in late lactation, while protein and lipids remain relatively constant (Munks *et al.* 1991). Similarly, the concentration of the total solids of the brush-tailed possum's milk varies greatly during lactation, increasing from 10% in early lactation to 24–28% in mid-lactation and 18–43% in late lactation, with lipids increasing from 1.3 to 9% in mid-lactation to 5–15.0% in late lactation. Carbohydrates vary little during lactation while protein can remain relatively constant at ~6% or reach 14%. These changes contrast with other marsupials, except the koala (*Phascolarctos cinereus*), as in most marsupials the solids and lipids increase towards late lactation. Values of different components for several species are shown in Table 9.18, though it should be noted that the different components vary considerably throughout lactation.

### 11.4.2 Milk formula

The three commercial formulas for hand-rearing possums and gliders are described below.

- Biolac ® – several formulas have been developed for marsupials; the M100 formula is recommended for the entire hand-rearing process in ring-tailed possums and brush-tailed possums. They should be fed 10–15% of their bodyweight per day.
- Di-Vetelact ® (Lillelund Pty Ltd) – a widely-used low lactose milk formula. Animals should be fed at ~20% bodyweight except for very small joeys (< 100 g). Sugar gliders, squirrel gliders, greater gliders, ring-tailed possums and brush-tailed possums have been raised on 1:50 mL for possums, 2 drops of Pentavite® and 1 teaspoon of high-protein baby cereal. Many care groups also recommend adding Wombaroo® Impact to Di-Vetelact® at the rate of 1 Impact spoon to 100 mL of made-up milk.
- Wombaroo® Possum Milk Replacer – different formulas are used for the different stages of development to mimic the changes that occur in the female during lactation. The two formulas are the < 0.8 for pouch young before emerging from the pouch and the > 0.8 milk that is used with possums and gliders after pouch emergence. Wombaroo® provide detailed charts.

Feather-tailed gliders have been fed successfully on a mixture of 150 mL warm water, 1 heaped teaspoon of Di-Vetelact®, 1 tablespoon of Complan and 1 tablespoon of pure maple syrup through a 1 mL syringe with a cannula attached or a plastic reusable pipette (Caton 1995; J. Hockley pers. comm.). Biolac® M100 has also been used successfully for rearing feather-tailed gliders. They were fed by wrapping them up in a thin cloth to restrict movement and applying drops to the mouth, which they soon start to lap from a cotton bud or teaspoon (Slater 1985; Caton 1995). When weaning commenced, a small shallow dish with 2 mL of milk was left with the joey and examined each morning to determine if self-feeding. Fruit was introduced by placing ½ teaspoon of soft fruit, covered with a small amount of nectar, in a small shallow bowl until the joey begins to eat it regularly. Mealworms can be introduced by cutting one in half and smearing the insides across the glider's mouth, then leaving the rest with the fruit. Gliders, honey possums and pygmy possums can be fed 1 mL Leadbeater's mix to 70 mL milk formula.

### 11.4.3 Feeding apparatus

Very small joeys can be fed using a silicone or latex syringe teat or plastic intravenous catheter (cannula).

Some possum joeys will, however, accept milk from a small glass bottle (10, 15 or 25 mL) and a special possum teat. There are several Australian suppliers that make a variety of teats suitable for marsupials of various stages/ages. The teat should be punctured with a hot needle or a 1-2 mm slit cut in the tip with a small, sharp pair of scissors. Milk should be fed at ~36°C. Possum joeys will often lap milk from a syringe or syringe and cannula from a very early age, so extensive feeding equipment may not be required or be required for long.

Small possums can be fed by fitting the teat onto a 10 mL syringe, from which it will lap (Austin 1997). Species such as sugar gliders will usually lap readily from the tip of a syringe or a cannula affixed to a 1- or 2.5 mL syringe. Alternatively, they can be taught to lap from a small plastic lid (Booth 1999).

### 11.4.4 Feeding routine

Unfurred young should be fed every 2-3 h, rather than every 1-2 h, as it is very exhausting for the joey. Recently furred young should be fed every 4 h and then approximately once or twice daily before weaning (Booth 1999).

The amount of milk offered per day is 10-20% of the bodyweight, which is then divided up for each feed. The feeding regimen for ring-tailed possums is shown in Table 9.19 and the regimen for brush-tailed possums in Table 9.20. This is then increased as the bodyweight increases by supplying a small amount of fresh, soft, green eucalypt tips and flowers of grevilleas and bottlebrush (B. Smith 1995).

Exudivorous marsupials such as feather-tailed gliders, pygmy possums and gliders (except the greater glider) should be provided with flowers of native species such as eucalypts, banksias, callistemons, acacias, grevilleas and melaleucas.

When feeding, it is important not to feed the milk formula too quickly; the rate at which the milk is squeezed into the mouth should not be faster than the rate at which it is swallowed. Ensuring the hole in the teat is an appropriate size comparable to the animal's size will help. Too much milk results in an accumulation in the pharynx, which is suddenly sneezed or coughed out the nostrils. To avoid this, be very careful of the rate at which milk is released to the joey and use a smaller hole on the teat if required. Encouraging them to lap from a syringe, syringe/cannula or a shallow bowl allows them to feed at their own preferred rate and will prevent aspiration.

The number of daily feeds changes as the joey develops (Bellamy 1992). Very young unfurred joeys should be fed every 2-3 h. When furred, the number of feeds is decreased to five and the volume increased per feed. At full emergence the number of feeds is reduced to 2-3 feeds per day.

**Table 9.19.** Feeding regimen for hand rearing ring-tailed possums using Di-Vetelact®.

Derived from B. Smith (1995).

| Weight (g) | Feeds per day | Example of amount per feed to be given |
|---|---|---|
| 45-60 | 6-5 | 50 g possum = 5-10 mL/day = 1-2 mL/feed |
| 60-80 | 5-4 | 70 g possum = 7-14 mL/day = 2-3 mL/feed + some solids |
| 80-100 | 4-3 | 90 g possum = 9-18 mL/day = 3-4 mL/feed + solids |
| 100 | 3 | 100 g possum = 10-20 mL/day = 5 mL/feed + solids |
| 150 | 1 | 150 g possum = 15-25 mL/day = one feed in a dish/night |
| 200 | 1 | Try to wean at this time onto solids only. |

**Table 9.20.** Feeding regime for hand rearing brush-tailed possums using Di-Vetelact®.

Derived from B. Smith (1995).

| Weight (g) | Feeds per day | Example of amount per feed to be given |
|---|---|---|
| 100-200 | 6-5 | 150 g possum = 15-30 mL/day = 3 mL/feed + solids |
| 200-300 | 5-4 | 200 g possum = 20-40 mL/day = 6 mL/feed + solids |
| 300 | 4 | 300 g possum = 30-60 mL/day = 12 mL/feed + solids |
| 400 | 2 | 400 g possum = 40-80 mL/day = 30 mL/feed + solids |
| 500 | 1 | 500 g possum = 25 mL = one feed in a dish/night + solids |
| | | Should be being weaned by 500 g |

## 11.5 Data recording

When an animal is first brought in for hand-rearing, its sex and approximate age, using growth charts, should be recorded. During the hand-rearing process several important pieces of information should be recorded on a daily basis:

- date when the information is recorded
- time when the information is recorded
- bodyweight to the nearest 1 g if possible
- general activity and demeanour
- characteristics and frequency of defecation and urination
- amount (g) of different food types offered
- food consumption at each feed
- veterinary examinations and results.

This information serves several purposes, including providing important background information such as food consumption, which will assist a veterinarian to reach a diagnosis if the animal becomes sick or fails to grow or gain weight. It also allows a comparison with growth curves to assess progress (section 10.16) and to allow new growth curves to be established for measurements in which they presently do not occur. The developmental stages and measurements outlined in section 10.3.1 should also be recorded on a weekly basis if possible.

## 11.6 Identification methods

Once furred, most species can be identified with an implant chip if required; however, very small species such as feather-tailed gliders, honey possums and pygmy possums should not be implanted until nearly fully grown. A general rule is to not chip individuals that weigh less than 10 g.

## 11.7 Hygiene

Maintaining a high standard of hygiene is critical to the survival of the possum or glider joey. Emphasis needs to be placed on the following:

- maintaining a clean pouch lining at all times. Good hygiene is important; otherwise conditions such as candida (or thrush) due to the fungus *Candida albicans* can occur. Species such as ring-tailed possums and brush-tailed possums appear to be especially prone. Older joeys can be trained to urinate on newspaper by keeping a piece of newspaper with the smell of urine on it
- maintaining personal hygiene by washing and disinfecting hands before and after handling the joey. Use antibacterial solution for washing hands with furless joeys, as their immune system is not well developed
- washing hands between feeding different joeys from different litters
- using boiled water when making up formulas for joeys of all ages
- cleaning spilt milk, faeces and urine from the joeys' skin and fur as soon as possible, drying the animal and replacing their pouch liners
- washing all feeding equipment in warm water and dishwashing detergent and sterilising it in a microwavable baby steam sterilizer or F10® disinfectant
- storing the teats and bottles in the fridge in an air-tight container after they have been disinfected
- only heating up milk once and then discarding leftovers
- making up fresh formula every 24 h
- ensuring the joey is cleaned after feeding, especially the chin and whiskers, as they get messy. If not done properly fungal infections may result.
- avoiding contact with other animals unless certain they pose no health risk
- stimulating the joey to toilet before or after feeding. As with other marsupials, toileting can be done by the application of warm water to the cloaca using a soft tissue or cotton wool cloth to stimulate urination and defecation, which allows the animal to keep drier and warmer in its pouch
- if furless, cover the joey's body with Sorbolene cream (without glycerol) after each feed until fur appears (if not using an ICU humidicrib).

Toileting (wiping the cloaca with a damp cloth) between meals may also be required until the animal defecates by itself (George *et al.* 1995). When toileting possums, be careful with the stimulation of the cloaca as, if done excessively, it can lead to cloacal prolapse and possibly urethral swelling, in which case consult a veterinarian (Bellamy 1992). It can be treated with the use of creams such as Panalog (Squibb), Proctoseyl (Roussel) or Topigol (Squibb) (George *et al.* 1995). Once the joey is ready to start leaving the pouch, it will generally toilet on paper itself rather than in the pouch. Joeys that are wet with urine and faeces should be cleaned with a mild soap solution and dried thoroughly. If the fur becomes soiled, wash it under warm tap water and dry thoroughly (George *et al.* 1995).

## 11.8 Behavioural considerations

Take care that the joey being hand-reared does not become too attached to the raiser, as this will make the weaning process much more difficult. Hand-rearing several individuals together and not overhandling them will help them to socialise properly and reduce these problems.

## 11.9 Use of foster animals

Foster animals have not been readily used to date with any species of possums; however, common brush-tailed possums have been transferred to other mothers of the same species of similar age (Sharman 1962). There is also a record of a wild common brush-tailed possum that adopted a second pouch young that was not her own, so she raised two pouch young concurrently (Eymann *et al.* 2007b).

## 11.10 Weaning

A general rule is to decrease the formula by 5% per week as long as the joey continues to gain bodyweight at a minimum of 5–10% of bodyweight per day. When ready for weaning the possum should be provided with increasing amounts of solid food. This is generally done by hand to begin with and as the possum starts to readily eat solids, the food can be left either in a dish, feeder or in branches firmly wedged in the enclosure in a pot or in the ground. If using a pot, make sure the possum cannot be caught in it, especially if it contains water. In some cases when the possum is slow to wean, it can be encouraged by slightly reducing the amount of milk while providing the solid food. Care needs to be taken that excessive weight loss does not occur during this process. If it does, increase the amount of milk provided.

Depending on the species, different food items should be offered as the young learns to eat solid food. Pygmy possums and petaurids should be given several millilitres of Leadbeater's mix, finely chopped soft fruit, and fresh flowers of eucalypts, banksias and bottlebrush. Brush-tailed possums and ring-tailed possums should be given finely chopped fruit and vegetables, freshly cut leaves and flowers of eucalypts and bottlebrush to eat.

Pygmy possums should be weaned at 2.5–3 months, 4–5 months for petaurids, 6 months for ring-tailed possums and 8 months for brush-tailed possums (Austin 1997; Booth 1999). Fresh water should be supplied at weaning.

## 11.11 Release procedures

If the animal to be hand-reared is to be maintained in captivity there is not as much need to rehabilitate it except in some species where it may affect their effectiveness in breeding. If breeding is not an issue, then the presence of a highly manageable, calm animal is generally of great benefit in captivity and generally results in better displays, as they are more active.

If a possum or glider is to be released into the wild it is important to take every measure to maximise its chances of survival. Measures include:

- minimising the amount of handling by providing milk in a bowl as soon as the animal can lap and leaving solid food for it to consume in its own time. It is very important that the possum does not associate its handler with food; rather, encourage it to explore its enclosure to find the food provided
- minimal petting of the possum or glider to reduce the likelihood of bonding. Placing the food in the enclosure at dusk before the animal emerges for the evening can facilitate this as they will not associate food with humans
- providing as much native food as possible, after dark, so that it becomes used to eating natural food items
- providing lots of climbing opportunities to allow the animals to increase their climbing skills and fitness
- not placing the food on the enclosure floor as this encourages them to come to the ground where they are more likely to be attacked by predators (B. Smith 1995)
- not rearing them near to domestic animals as this habituates them to predators that they need to be wary of in the wild (B. Smith 1995)
- pairing them up, if possible, when they are young with another possum of the same species and approximately the same age.

Prior to release, the possum or glider should be introduced, if it has not been already, to other possums of the same species so that they learn socialisation skills. More nest boxes than individuals should be supplied, as well as natural nesting material to allow them to practise nest building. Generally, soft releases are used in which a nest box or drey is supplied, together with food so that the possum can slowly habituate back into the wild.

The survival of hand-reared possums and gliders released back into the wild is often low, as they invariably fall prey to introduced predators such as cats and foxes. A study that examined 118 hand-reared eastern ring-tailed possums showed that all but eight were killed by predators, which included foxes (52%) and cats (29%), within a mean of 101 days (Augee *et al.* 1996). Similarly, another study of 20 hand-reared brush-tailed possums found that after 40 days only eight animals had survived, nine were killed by foxes or were returned into care and three had unknown fates (Mella *et al.* 2023). The latter authors suggested that personality was a fundamental criterion for the success of rehabilitation program. They also suggested that behavioural traits of hand-reared brush-tailed possums differed significantly from those of wild individuals and that the more time they in care the more likely they were to display highly human-habituated behaviours.

## ACKNOWLEDGEMENTS

Many thanks and Jocelyn Hockley, Daryl Dickson, Karen Scott and Alice Roser for providing valuable comments on the hand-rearing section. Daryl Dickson also provided advice on the diet for the striped possum. Thanks also to Haley Meer for providing important information on the diet of captive greater gliders.

## REFERENCES

Alexander JSA (1981) The status of the squirrel glider, *Petaurus norfolcensis* (Marsupialia: Petauridae) in Victoria. BSc Honours thesis. Latrobe University, Melbourne.

Anon (1935) Carnivorous habits of the brush-tailed opossum (*Trichosurus vulpecula*). *Australian Zoologist* **8**, 149.

Anon (2018) Sugar glider shop. https://sugarglidershop.co.uk (accessed 29 July 2021).

Arlidge J, Halley M, Hunter S (1993) *Manual for the Captive Management of Mountain Pygmy Possum* Burramys parvus. Healesville Sanctuary, Victoria.

Arnould J (1986) Aspects of the diet of the eastern pygmy-possum, *Cercartetus nanus* (Desmarest). *BSc Honours thesis.* Monash University, Victoria.

Atherton RG, Haffenden AT (1982) Observations on the reproduction and growth of the long-tailed pygmy possum, *Cercartetus caudatus* (Marsupialia: Burramyidae), in captivity. *Australian Mammalogy* **5**, 253–259. doi:10.1071/AM82027

Augee ML, Smith B, Rose S (1996) Survival of wild and hand-reared ringtail possums (*Pseudocheirus peregrinus*) in bushland near Sydney. *Wildlife Research* **23**, 99–108. doi:10.1071/WR9960099

Austin MA (1997) *A Practical Guide to the Successful Hand Rearing of Tasmanian Marsupials.* Regal Publications, Melbourne.

Bach C (1998) *Birth Date Determination in Australasian Marsupials.* Australasian Regional Association of Zoological Parks and Aquaria, Sydney.

Baker AM, Gynther IC (eds) (2023) *Strahan's Mammals of Australia.* 4th edn. Reed New Holland Publishers, Sydney.

Baker AM, Eldridge MDB, Fisher DO, Frankham G, Helgen K, Jackson SM, Potter S, Travouillon KJ, Umbrello L (2023) Taxonomy and diversity of living Australasian marsupials. In *American and Australasian Marsupials* (Eds NC Cáceres and CR Dickman). pp. 1–85. Springer, Cham, Switzerland. doi:10.1007/978-3-030-88800-8_32-1

Ball T, Adams E, Goldingay RL (2009) Diet of the squirrel glider in a fragmented landscape near Mackay, central Queensland. *Australian Journal of Zoology* **57**, 295–304. doi:10.1071/ZO08095

Ban S, Cox-Witton K, Grillo T (2020) Mycobacterium ulcerans in a ringtail possum in a new location. *Animal Health Surveillance Quarterly* **25**, 13–17.

Barnes M (2002) Sugar gliders. In *Hand-Rearing Wild and Domestic Mammals.* (Ed. LJ Gage) pp. 55–62. Iowa State Press, Iowa.

Barnett JL, How RA, Humphrey WF (1982) Habitat effects on organ weights, longevity and reproduction in the mountain brushtail possum, *Trichosurus caninus* (Ogilby). *Australian Journal of Zoology* **30**, 23–32. doi:10.1071/ZO9820023

Bartholomew GA, Hudson JW (1962) Hibernation, estivation, temperature regulation, evaporative water loss, and heart rate of the pigmy possum, *Cercaertus [sic] nanus. Physiological Zoology* **35**, 94–107. doi:10.1086/physzool.35.1.30152716

Bellamy T (1992) Marsupial handrearing. In *Urban Wildlife.* Proceedings 204. Post Graduate Committee in Veterinary Science. (Eds L Vogelnest, D Spielman, T Bellamy, M Messer and DM Walker) pp. 49–58. University of Sydney, Sydney.

Bettiol SS, Obendorf DL, Nowarkowski M, Goldsmid JM (2000) Pathology of experimental toxoplasmosis in eastern barred bandicoots in Tasmania. *Journal of Wildlife Diseases* **36**, 141–144. doi:10.7589/0090-3558-36.1.141

Beyer GL, Goldingay RL (2006) The value of nest boxes in the research and management of Australian hollow-using arboreal marsupials. *Wildlife Research* **33**, 161–174. doi:10.1071/WR04109

Biggins JG (1984) Communications in possums: a review. In *Possums and Gliders.* (Eds AP Smith and ID Hume) pp. 35–57. Australian Mammal Society, Sydney.

Biggins JG, Overstreet DH (1978) Aggressive and non-aggressive interactions among captive populations of the brush-tailed possum, *Trichosurus vulpecula* (Marsupialia: Phalangeridae). *Journal of Mammalogy* **59**, 149–159. doi:10.2307/1379884

Blyton MDJ, Shaw RE, Peakall R, Lindenmayer DB, Banks CC (2016) The role of relatedness in mate choice by an arboreal marsupial in the presence of fine-scale genetic structure. *Behavioral Ecology and Sociobiology* **70**, 313–321. doi:10.1007/s00265-015-2049-z

Booth RJ (1994) Medicine and husbandry: Dasyurids, possums and bats. In *Wildlife.* Proceedings 233. Post Graduate Committee in Veterinary Science. (Ed. DI Bryden) pp. 423–441. University of Sydney, Sydney.

Booth RJ (1999) General husbandry and medical care of sugar gliders. In *Current Veterinary Therapy XIII.* (Ed. JD Bonagura) pp. 1157–1163. Saunders & Co., Philadelphia.

Booth RJ, Lack TJ, Jackson SM (2019) Growth and development of the mahogany glider (*Petaurus gracilis*). *Zoo Biology* **38**, 266–271. doi:10.1002/zoo.21479

Borsboom A (1982) Agonistic interactions between bats and arboreal marsupials. *Australian Mammalogy* **5**, 281–282. doi:10.1071/AM82030

Bowley EA (1939) Delayed implantation in *Dromicia. Journal of Mammalogy* **20**, 499. doi:10.2307/1374600

Boys N (2018) *Mountain Pygmy-possum* Burramys parvus *Husbandry Manual.* Healesville Sanctuary, Healesville, Victoria.

Bradley AJ, Stoddart DM (1993) The dorsal paracloacal gland and its relationship with seasonal changes in cutaneous scent gland morphology and plasma androgen in the marsupial sugar glider (*Petaurus breviceps*) (Marsupialia; Petauridae). *Journal of Zoology* **229**, 331–346[London]. doi:10.1111/j.1469-7998.1993.tb02640.x

Bradshaw F, Everett L, Bradshaw SD (2000) On the rearing of honey possums. *Western Australian Naturalist (Perth)* **22**, 281–288.

Bradshaw FJ, Stead-Richardson EJ, Reeder AJ, Oates JE, Bradshaw SD (2004) Reproductive activity in captive female honey possum *Tarsipes rostratus*, assessed by faecal steroid analysis. *General and Comparative Endocrinology* **138**, 20–31. doi:10.1016/j.ygcen.2004.04.008

Brearley G, McAlpine C, Bell S, Bradley A (2012) Influence of urban edges on stress in an arboreal mammal: a case study of squirrel gliders in southeast Queensland, Australia. *Landscape Ecology* **27**, 1407–1419. doi:10.1007/s10980-012-9790-8

Broome LS (2001) Intersite differences in population demography of mountain pygmy-possums *Burramys parvus* Broom (1987–1998): implications for meta-population conservation and ski resorts in Kosciuszko National Park. *Biological Conservation* **102**, 309–323. doi:10.1016/S0006-3207(01)00105-7

Broome LS, Geiser F (1995) Hibernation in free-living mountain pygmy-possums, *Burramys parvus* (Marsupialia: Burramyidae). *Australian Journal of Zoology* **43**, 373–379. doi:10.1071/ZO9950373

Brust DM (2013) Gastrointestinal diseases of marsupials. *Journal of Exotic Pet Medicine* **22**, 132–140. doi:10.1053/j.jepm.2013.05.005

Buddle BM, Young LJ (2000) Immunobiology of mycobacterial infections in marsupials. *Developmental and Comparative Immunology* **24**, 517–529. doi:10.1016/S0145-305X(00)00014-8

Campbell CD, Pecon-Slattery J, Pollak R, Joseph L, Holleley CE (2019) The origin of exotic pet sugar gliders (*Petaurus breviceps*) kept in the United States of America. *PeerJ* **7**, e6180. doi:10.7717/peerj.6180

Carmichael T (2000) Captive husbandry and rehabilitation of the striped possum *Dactylopsila trivirgata picata* in far north Queensland. In *Proceedings of the ARAZPA/ASZK Conference*. pp. 106–119. Sea World. March 2000.

Carruthers D, McLeod E, Costa K, Miller K, Parrott ML (2020) Migrating moths and alpine possums: mobilising community across Eastern Australia. *International Zoo Educators* **2020**, 18–25.

Casanova J (1958) The dormouse or pigmy possum. *Walkabout* **24**, 30–31.

Cash E (2015) A comparison of the behaviours of captive bred and wild caught *Burramys parvus* (Marsupialia: Burramyidae). PhD thesis. La Trobe University, Melbourne.

Caton W (1995) Hand-raising feathertail gliders. *Thylacinus* **20**(2), 9–13.

Caton W (1999) Growth and development of Petaurus breviceps (northern region) in captivity and some comparisons with P. breviceps. *Thylacinus* **23**(2), 6–10.

Chilcott MJ (1984) Coprophagy in the common ringtail possum, *Pseudocheirus peregrinus* (Marsupialia: Petauridae). *Australian Mammalogy* **7**, 107–110. doi:10.1071/AM84011

Chilcott MJ, Hume ID (1985) Coprophagy and selective retention of fluid digesta: their role in the nutrition of the common ringtail possum. *Pseudocheirus peregrinus. Australian Journal of Zoology* **33**, 1–15. doi:10.1071/ZO9850001

Christian N, Geiser F (2007) To use or not to use torpor? Activity and body temperature as predictors. *Naturwissenschaften* **94**, 483–487. doi:10.1007/s00114-007-0215-5

Clark MJ (1967) Pregnancy in the lactating pygmy possum, *Cercartetus concinnus. Australian Journal of Zoology* **15**, 673–683. doi:10.1071/ZO9670673

Clout MN (1982) Determining of age in the brushtail possum using sections from decalcified molar teeth. *New Zealand Journal of Zoology* **9**, 405–408. doi:10.1080/03014223.1982.10423870

Collins BG, Wooller RD, Richardson KC (1987) Torpor by the honey possum, *Tarsipes rostratus* (Marsupialia: Tarsipedidae), in response to food intake and low environmental temperature. *Australian Mammalogy* **11**, 51–57. doi:10.1071/AM88005

Collins LR (1973) *Monotremes and Marsupials*. Smithsonian Institution Press, Washington.

Comport SS, Ward SJ, Foley WJ (1996) Home ranges, time budgets and food-tree use in a high-density tropical population of greater gliders, *Petauroides volans minor* (Pseudocheiridae: Marsupialia). *Wildlife Research* **23**, 401–419. doi:10.1071/WR9960401

Cork SJ, Foley WJ (1991) Digestive and metabolic strategies of arboreal mammalian folivores in relation to chemical defences in temperate and tropical forests. In *Plant Defences Against Mammalian Herbivory*. (Eds RT Palo and CT Robbins) pp. 133–166. CRC Press, New York.

Cowan PE (1989) Changes in milk composition during lactation in the common brushtail possum, *Trichosurus vulpecula* (Marsupialia: Phalangeridae). *Reproduction, Fertility and Development* **1**, 325–335. doi:10.1071/RD9890325

Cowan PE, White AJ (1989) Evaluation of a tooth-wear age index for brushtail possums, *Trichosurus vulpecula. Australian Wildlife Research* **16**, 321–322. doi:10.1071/WR9890321

Craig SA (1985) Social organisation, reproduction and feeding behaviour of a population of yellow-bellied gliders, *Petaurus australis* (Petauridae: Marsupialia). *Australian Wildlife Research* **12**, 1–18. doi:10.1071/WR9850001

Craig SA (1986) A record of twins in the yellow-bellied glider (*Petaurus australis* Shaw) (Marsupialia: Petauridae) with notes on the litter size and reproductive strategy of the species. *Victorian Naturalist* **103**, 72–74.

Crandall LS (1964) *The Management of Wild Mammals in Captivity*. University of Chicago Press, Chicago.

Crawley MC (1970) Longevity of Australian brush-tailed opossums (*Trichosurus vulpecula*) in indigenous forest in New Zealand. *New Zealand Journal of Science* **13**, 348–351.

Crawley MC (1973) A live trapping study of Australian brush-tailed possums, *Trichosurus vulpecula* (Kerr), in the Orongorongo Valley, Wellington, New Zealand. *Australian Journal of Zoology* **21**, 75–90. doi:10.1071/ZO9730075

Cremona T, Baker AM, Cooper SJB, Montague-Drake R, Stobo-Wilson AM, Carthew SM (2021) Integrative taxonomic investigation of *Petaurus breviceps* (Marsupialia: Petauridae) reveals three distinct species. *Zoological Journal of the Linnean Society* **191**, 503–527. doi:10.1093/zoolinnean/zlaa060

Crisp EA, Cowan PE, Messer M (1989) Changes in milk carbohydrates during lactation in the common brushtail possum, *Trichosurus vulpecula* (Marsupialia; Phalangeridae). *Reproduction, Fertility and Development* **1**, 309–314. doi:10.1071/RD9890309

Curlewis JD, Stone GM (1986) Reproduction in captive brushtail possums, *Trichosurus vulpecula. Australian Journal of Zoology* **34**, 47–52. doi:10.1071/ZO9860047

Davey SM (1984) Habitat preferences of arboreal marsupials within a coastal forest in southern new South Wales. In *Possums and Gliders*. (Eds AP Smith and ID Hume) pp. 509–516. Australian Mammal Society, Sydney.

Dawson TJ, May EL (1984) Daily variation in brain temperatures of the sugar glider (*Petaurus breviceps*): some insights into the control of thermoregulation. In *Possums and Gliders* (Eds AP Smith and ID Hume) pp. 375–383. Australian Mammal Society, Sydney.

De Mendonica MM (1983) Bertiella deblocki nouveau cestode Anoplocephalidae D'un marsupial (*Phalanger orientalis* Storr) de Timor Leste. *Annales de Parasitologie Humaine et Comparee* **58**, 203–210. doi:10.1051/parasite/1983583203

Dimpel H, Calaby JH (1972) Further observations on the mountain pygmy possum (*Burramys parvus*). *Victorian Naturalist* **89**, 101–106.

Dobson MDJ, Goldingay RL, Sharpe D (2005) Feeding behaviour of the squirrel glider in remnant habitat in Brisbane. *Australian Mammalogy* **27**, 27–35. doi:10.1071/AM05027

Dockerill R (2013) Problems with identification techniques for feathertail gliders. *Thylacinus* **37**(4), 6–8.

Dougall AM, Shilton C, Choy JL, Alexander B, Walton S (2009) New reports of Australian cutaneous leishmaniasis in Northern Australian macropods. *Epidemiology and Infection* **137**, 1516–1520. doi:10.1017/S0950268809002313

Dougall AM, Alexander B, Holt DC, Harris T, Sultan AH, Bates PA, Rose K, Walton SF (2011) Evidence incriminating midges (Diptera: Ceratopogonidae) as potential vectors of Leishmania in Australia. *International Journal for Parasitology* **41**, 571–579. doi:10.1016/j.ijpara.2010.12.008

Duckworth JA, Scobie S, Jones DE, Selwood L (1998) Determination of oestrous and mating in captive female brushtail possums, *Trichosurus vulpecula* (Marsupialia: Phalangeridae), from urine samples. *Australian Journal of Zoology* **46**, 547–555. doi:10.1071/ZO98042

Dunn RW (1982) Gliders of the genus *Petaurus*: their management in zoos. In *The Management of Australian Mammals in Captivity*. (Ed. DD Evans) pp. 82–84. Zoological Board of Victoria, Melbourne.

Dunnet GM (1956) A live-trapping study of the brush-tailed possum *Trichosurus vulpecula*. *CSIRO Wildlife Research* **1**, 1–18. doi:10.1071/CWR9560001

Dunnet GM (1964) A field study of local populations of the brush-tailed possum *Trichosurus vulpecula* in eastern Australia. *Proceedings of the Zoological Society of London* **142**, 665–695. doi:10.1111/j.1469-7998.1964.tb04634.x

Dwiyahreni AA, Kinnaird MF, O'Brien TG, Supriatna J, Andayani N (1999) Diet and activity of the bear cuscus, *Ailurops ursinus*, in North Sulawesi, Indonesia. *Journal of Mammalogy* **80**, 905–912. doi:10.2307/1383259

Dwyer PD (1977) Notes on *Antechinus* and *Cercartetus* (Marsupialia) in the New Guinea highlands. *Proceedings of the Royal Society of Queensland* **88**, 69–73.

Eisenberg JF (1981) *The Mammalian Radiations: An Analysis of Trends in Evolution, Adaptation, and Behaviour*. Athlone Press, London.

Ellis M, Jones B (1992) Observations of captive and wild western ringtail possums *Pseudocheirus occidentalis*. *Western Australian Naturalist (Perth)* **19**, 1–9.

Emlen ST, Oring LW (1977) Ecology, sexual selection, and the evolution of mating systems. *Science* **197**, 215–223. doi:10.1126/science.327542

Eymann J, Smythe LD, Symonds ML, Dohnt MF, Barnett LJ, Cooper DW, Herbert CA (2007a) Leptospirosis serology in the common brushtail possum (*Trichosurus vulpecula*) from urban Sydney, Australia. *Journal of Wildlife Diseases* **43**, 492–497. doi:10.7589/0090-3558-43.3.492

Eymann J, Neaves L, Eldridge MDB, Cooper DW, Herbert CA (2007b) Surprise in the pouch: a case of adoption in the common brushtail possum (*Trichosurus vulpecular*). *Australian Mammalogy* **29**, 63–68. doi:10.1071/AM07007

Eyre TJ, Smith GC, Venz MF, Mathieson MT, Hogan LD, Starr C, Winter J, McDonald K (2022) *Guide to Greater Glider Habitat in Queensland*. Report prepared for the Department of Agriculture, Water and the Environment, Canberra. Department of Environment and Science, Queensland Government, Brisbane.

Fairfax RA (1982) Notes on the scaly-tailed possum *Wyulda squamicaudata* in captivity. In *The Management of Australian Mammals in Captivity*. (Ed. DD Evans) pp. 73–74. Zoological Board of Victoria, Melbourne.

Fanning FD (1980) Nests of the feathertail glider *Acrobates pygmaeus* (Burramyidae: Marsupialia), from Sydney, New South Wales. *Australian Mammalogy* **3**, 55–56. doi:10.1071/AM80006

Fanning FD, Watkins K (1980) Growth and development in *Acrobates pygmaeus* (Burramyidae: Marsupialia). *Australian Mammalogy* **3**, 57–60. doi:10.1071/AM80007

Fanson KV, Parrott ML (2015) The value of eutherian-marsupial comparisons for understanding the function of glucocorticoids in female mammal reproduction. *Hormones and Behavior* **76**, 41–47. doi:10.1016/j.yhbeh.2015.05.012

Fanson KV, Best EC, Bunce A, Fanson BG, Hogan LA, Keeley T, Narayan EJ, Palme R, Parrott ML, Sharp TM, Skogvold K, Tuthill L, Webster KN, Bashaw M (2017) One size does not fit all: monitoring faecal glucocorticoid metabolites in marsupials. *General and Comparative Endocrinology* **244**, 146–156. doi:10.1016/j.ygcen.2015.10.011

Finnie EP (1978) Marsupial therapy. pp. 619–622, In *Fauna, Proceedings 36*. Post Graduate Committee in Veterinary Science, University of Sydney, Sydney.

Flannery T (1995a) *Mammals of the South-West Pacific & Moluccan Islands*. Reed Books, Sydney.

Flannery T (1995b) *Mammals of New Guinea*. 2nd edn. Australian Museum/Reed Books, Sydney.

Fleay D (1933) The greater flying phalanger. *Victorian Naturalist* **50**, 135–142.

Fleay D (1942) The remarkable striped possum. *Victorian Naturalist* **58**, 151–155.

Fleay D (1947) *Gliders of the Gum Trees*. Bread and Cheese Club, Melbourne.

Fleay D (1949) That curious marsupial, the cuscus. *Animal Kingdom* **52**, 22–25.

Fleay D (1954) The squirrel glider. *Victorian Naturalist* **70**, 208–210.

Fleay-Thomson R (2007) *Animals First: The Story of Pioneer Australian Conservationist and Zoologist Dr David Fleay*. Petaurus Publishing, Nerang, Queensland.

Fleming MR (1980) Thermoregulation and torpor in the sugar glider *Petaurus breviceps* (Marsupialia: Petauridae). *Australian Journal of Zoology* **28**, 521–534. doi:10.1071/ZO9800521

Fleming MR (1985a) The thermal physiology of the mountain pygmy-possum *Burramys parvus* (Marsupialia: Burramyidae). *Australian Mammalogy* **8**, 79–90. doi:10.1071/AM85006

Fleming MR (1985b) The thermal physiology of the feathertail glider *Acrobates pygmaeus* (Marsupialia: Burramyidae).

*Australian Journal of Zoology* **33**, 667–681. doi:10.1071/ZO9850667

Fleming MR, Frey H (1984) Aspects of the natural history of the feathertail glider, *Acrobates pygmaeus* (Marsupialia: Burramyidae) in Victoria. In *Possums and Gliders*. (Eds AP Smith and ID Hume) pp. 403–408. Australian Mammal Society, Sydney.

Flower SS (1931) Contributions to our knowledge of the duration of life in vertebrate animals. *Proceedings of the Zoological Society of London* **101**, 145–234. doi:10.1111/j.1469-7998.1931.tb06192.x

Foley WJ (1984) The utilization of Eucalyptus foliage by the greater glider (*Petauroides volans*) and the brushtail possum (*Trichosurus vulpecula*). PhD thesis. University of New England, Armidale.

Foley WJ, Kehl JC, Nagy KA, Kaplan IR, Borsboom AC (1990) Energy and water metabolism in free-living greater-gliders, *Petauroides volans*. *Australian Journal of Zoology* **38**, 1–9. doi:10.1071/ZO9900001

Fowler ME, Miller RE (Eds) (2003) *Zoo and Wild Animal Medicine*. 5th edn. W.B. Saunders, Philadelphia.

Fowler ME, Miller RE (2008) *Zoo and Wild Animal Medicine: Current Therapy*. Vol. 6. 6th edn. Saunders/Elsevier, St Louis.

Freudenberger DO, Wallis I, Hume ID (1989) Digestive adaptations of kangaroos, wallabies and rat kangaroos. In *Kangaroos, Wallabies and Rat-kangaroos*. Vol. 1. (Eds GC Grigg, PJ Jarman and ID Hume) pp. 179–187. Surrey Beatty & Sons, Sydney.

Frey H, Fleming MR (1984) Torpor and thermoregulatory behaviour in free-ranging feathertail gliders (*Acrobates pygmaeus*) (Marsupialia, Burramyidae) in Victoria. In *Possums and Gliders*. (Eds AP Smith and ID Hume) pp. 393–401. Surrey Beatty & Sons, Sydney.

Fry E (1971) The scaly-tailed possum *Wyulda squamicaudata* in captivity. *International Zoo Yearbook* **11**, 44–45. doi:10.1111/j.1748-1090.1971.tb01842.x

Geiser F (1987) Hibernation and daily torpor in two pygmy possums (*Cercartetus* spp., Marsupialia). *Physiological Zoology* **60**, 93–102. doi:10.1086/physzool.60.1.30158631

Geiser F (1993) Hibernation in the eastern pygmy possum, *Cercartetus nanus* (Marsupialia: Burramyidae). *Australian Journal of Zoology* **41**, 67–75. doi:10.1071/ZO9930067

Geiser F (2007) Yearlong hibernation in a marsupial mammal. *Naturwissenschaften* **94**, 941–944. doi:10.1007/s00114-007-0274-7

Geiser F, Broome LS (1991) Hibernation in the mountain pygmy possum *Burramys parvus* (Marsupialia). *Journal of Zoology* **223**, 593–602 [London]. doi:10.1111/j.1469-7998.1991.tb04390.x

Geiser F, Broome LS (1993) The effect of temperature on the pattern of torpor in a marsupial hibernator. *Journal of Comparative Physiology. B, Biochemical, Systemic, and Environmental Physiology* **163**, 133–137. doi:10.1007/BF00263598

Geiser F, Ferguson C (2001) Intraspecific differences in behaviour and physiology: effects of captive breeding on patterns of torpor in feathertail gliders. *Journal of Comparative Physiology. B, Biochemical, Systemic, and Environmental Physiology* **171**, 569–576. doi:10.1007/s003600100207

Geiser F, Körtner G (2004) Thermal biology, energetics, and torpor in the possums and gliders. In *The Biology of Australian Possums and Gliders*. (Eds RL Goldingay and SM Jackson) pp. 186–198. Surrey Beatty & Sons, Sydney.

Geiser F, Körtner G (2010) Hibernation and daily torpor in Australian mammals. *Australian Zoologist* **35**, 204–215. doi:10.7882/AZ.2010.009

Geiser F, Sink HS, Stahl B, Mansergh IM, Broome LS (1990) Differences in the physiological response to cold in wild and laboratory-bred mountain pygmy possums, *Burramys parvus* (Marsupialia). *Australian Wildlife Research* **17**, 535–539. doi:10.1071/WR9900535

Geiser F, Holloway JC, Körtner G, Maddocks TA, Turbill C, Brigham RM (2000) Do patterns of torpor differ between free-ranging and captive mammals and birds? In *Life in the cold: 11th international hibernation symposium* (Eds G Heldmaier and M Klingenspor) pp. 95–102. Springer-Verlag, Heidelberg.

Geiser F, Holloway JC, Körtner G (2007) Thermal biology, torpor and behaviour in sugar gliders: a laboratory-field comparison. *Journal of Comparative Physiology. B, Biochemical, Systemic, and Environmental Physiology* **177**, 495–501. doi:10.1007/s00360-007-0147-6

Gemmell RT (1990) The influence of day length on the initiation of the breeding season of the marsupial possum *Trichosurus vulpecula*. *Journal of Reproduction and Fertility* **88**, 605–609. doi:10.1530/jrf.0.0880605

Gemmell RT (1995) Breeding biology of brushtail possums *Trichosurus vulpecula* (Marsupialia, Phalangeridae) in captivity. *Australian Mammalogy* **18**, 1–7. doi:10.1071/AM95001

Gemmell RT, Hendrikz JK (1993) Growth rates of the bandicoot *Isoodon macrourus* and the brushtail possum *Trichosurus vulpecula*. *Australian Journal of Zoology* **41**, 141–149. doi:10.1071/ZO9930141

Gemmell RT, Sernia C (1992) The role of photoperiod on the initiation of the breeding season possum *Trichosurus vulpecula*. *Journal of Reproduction and Fertility* **95**, 701–708. doi:10.1530/jrf.0.0950701

Gemmell RT, Cepon G, Sernia C (1993) Effect of photoperiod on the breeding season of the marsupial possum, *Trichosurus vulpecula*. *Journal of Reproduction and Fertility* **98**, 515–520. doi:10.1530/jrf.0.0980515

George GG (1982) Cuscuses Phalanger spp.: their management in captivity. In *The Management of Australian Mammals in Captivity*. (Ed. DD Evans) pp. 67–72. Zoological Board of Victoria, Melbourne.

George H, Parker G, Coote P (1995) *Common Wombats: Rescue Rehabilitation Release*. Unpublished manuscript.

Goldingay RL (1984) Photoperiodic control of diel activity in the Sugar glider (*Petaurus breviceps*). In *Possums and Gliders*. (Eds AP Smith and I Hume) pp. 385–391. Australian Mammal Society, Sydney.

Goldingay RL (1990) The foraging behaviour of a nectar feeding marsupial, *Petaurus australis*. *Oecologia* **85**, 191–199. doi:10.1007/BF00319401

Goldingay RL (1992) Socioecology of the yellow-bellied glider (*Petaurus australis*) in a coastal forest. *Australian Journal of Zoology* **40**, 267–278. doi:10.1071/ZO9920267

Goldingay RL (1994) Loud calls of the yellow-bellied glider, Petaurus australis: territorial behaviour by an arboreal marsupial. *Australian Journal of Zoology* **42**, 279–293. doi:10.1071/ZO9940279

Goldingay RL (2010) Direct male parental care observed in wild sugar gliders. *Australian Mammalogy* **32**, 177–178. doi:10.1071/AM10009

Goldingay RL (2020a) Does temperature variation influence nest box use by the eastern pygmy-possum? *Australian Mammalogy* **42**, 77–84. doi:10.1071/AM19023

Goldingay RL (2020b) Sex and age differences in tree cavity dependence in a small arboreal marsupial. *Journal of Zoology* **312**, 53–62. doi:10.1111/jzo.12793

Goldingay RL, Jackson SM (2004) A review of the ecology of the Australian Petauridae. In *The Biology of Australian Possums and Gliders*. (Eds RL Goldingay and SM Jackson) pp. 376–400. Surrey Beatty, Sydney.

Goldingay RL, Kavanagh RP (1990) Socioecology of the yellow-bellied glider (*Petaurus australis*) at Waratah Creek, NSW. *Australian Journal of Zoology* **38**, 327–341. doi:10.1071/ZO9900327

Goldingay RL, Kavanagh RP (1991) The yellow-bellied glider: a review of its ecology and management considerations. In *Conservation of Australia's Forest Fauna*. (Ed. D Lunney) pp. 365–375. Royal Zoological Society of NSW, Sydney.

Goldingay RL, Kavanagh RP (1995) Foraging behaviour and habitat use of the feathertail glider (*Acrobates pygmaeus*) at Waratah Creek, New South Wales. *Wildlife Research* **22**, 457–470. doi:10.1071/WR9950457

Goldingay RL, Rueegger N (2018) Elevation induced variation in the breeding traits of a nectar-feeding non-flying mammal. *Ecological Research* **33**, 979–988. doi:10.1007/s11284-018-1607-z

Goldingay RL, Carthew SM, Whelan RJ (1987) Transfer of *Banksia spinulosa* pollen by mammals: implications for pollination. *Australian Journal of Zoology* **35**, 319–325. doi:10.1071/ZO9870319

Goldingay RL, Quin DG, Churchill S (2001) Spatial variability in the social organisation of the yellow-bellied glider (*Petaurus australis*) near Ravenshoe, north Queensland. *Australian Journal of Zoology* **49**, 397–409. doi:10.1071/ZO00084

Goldingay RL, Grimson MJ, Smith GC (2007) Do feathertail gliders show a preference for nest box design. *Wildlife Research* **34**, 484–490. doi:10.1071/WR06174

Green B (1984) Composition of milk and energetics of growth in marsupials. *Symposia of the Zoological Society of London* **51**, 369–387.

Green RH (1980) The little pigmy possum, *Cercartetus lepidus* in Tasmania. *Records of the Queen Victoria Museum Launceston* **68**, 1–12.

Green WQ, Coleman JD (1987) Den sites of possums, *Trichosurus vulpecula*, and frequency of use in mixed hardwood forest in Westland, New Zealand. *Australian Wildlife Research* **14**, 285–292. doi:10.1071/WR9870285

Gross R, Bolliger A (1959) Composition of the milk of the marsupial *Trichosurus vulpecular. American Journal of Diseases of Children* **98**, 768–775.

Gunn RC (1851) On the introduction and naturalization of *Petaurus sciueus* in Tasmania. *Papers and Proceedings of the Royal Society of Van Dieman's Land* **1**, 253–255.

Haffenden A (1984) Breeding, growth and development in the Herbert River ringtail possum, *Pseudocheirus herbertensis herbertensis* (Marsupialia: Petauridae). In *Possums and Gliders*. (Eds AP Smith and ID Hume) pp. 277–281. Australian Mammal Society, Sydney.

Handasyde KA, Martin RW (1996) Field observations on the common striped possum (*Dactylopsila trivirgata*) in North Queensland. *Wildlife Research* **23**, 755–766. doi:10.1071/WR9960755

Handasyde KA, Taggart DA, Temple-Smith PD, Martin RW (2001) Data on reproduction, development and growth in the striped possum, *Dactylopsila trivirgata*. Abstract. Possum and Glider Symposium. Queensland Museum, Brisbane.

Handasyde KA, Coulson G, Martin JK, Taylor AC (2007) Long-term pair-bonds without mating fidelity in a mammal. *Behaviour* **144**, 1419–1445. doi:10.1163/156853907782418240

Harley DKP, Lill A (2007) Reproduction in a population of the endangered Leadbeater's possum inhabiting lowland swamp forest. *Journal of Zoology* **272**, 451–457[London]. doi:10.1111/j.1469-7998.2007.00288.x

Hawkins MR (1999) The use of gum feeders: looking at device design and responses from marsupial species. *Thylacinus* **23**(2), 27–31.

Heinze DA, Olejniczak AM (2000) First observations of the mountain pygmy-possum *Burramys parvus* nesting in the wild. *Australian Mammalogy* **22**, 65–67. doi:10.1071/AM00065

Heinze DA, Broome LS, Mansergh IM (2004) A review of the ecology and conservation of the mountain pygmy-possum *Burramys parvus*. In *The Biology of Australian Possums and Gliders*. (Eds RL Goldingay and SM Jackson) pp. 254–267. Surrey Beatty & Sons, Sydney.

Helgen K, Jackson SM (2015) Family Phalangeridae (cuscuses, brush-tailed possums and scaly-tailed possum). In *Handbook of the Mammals of the World*. Vol. 5. *Monotremes and Marsupials*. (Eds DE Wilson and RA Mittermeier) pp. 456–497. Lynx Editions, Barcelona.

Henry SR (1984) Social organisation of the greater glider (*Petauroides volans*) in Victoria. In *Possums and Gliders*. (Eds AP Smith and ID Hume) pp. 221–228. Australian Mammal Society, Sydney.

Henry S (1985) The diet and socioecology of gliding possums in southern Victoria. PhD thesis. Monash University, Victoria.

Henry S (1995) Greater glider Petauroides volans. In *Mammals of Victoria: Distribution, Ecology and Conservation* (Ed. P Menkhorst) pp. 118–120. Oxford University Press, Oxford.

Henry SR, Craig SA (1984) Diet, ranging behaviour and social organisation of the yellow-bellied glider (*Petaurus australis* Shaw) in Victoria. In *Possums and Gliders*. (Eds AP Smith and ID Hume) pp. 331–341. Australian Mammal Society, Sydney.

Henry SR, Suckling G (1984) A review of the ecology of the sugar glider (*Petaurus breviceps*). In *Possums and Gliders*. (Eds AP Smith and I Hume) pp. 355–358. Australian Mammal Society, Sydney.

Herrmann EA, Herrin KV, Gleen W, Davies P, Stapley R, Sebbings V, Wiszniewski J, Spindler R, Faichney GJ, Chaves AV (2013) Partial replacement of an artificial nectar diet with native browse for feather-tail gliders (*Acrobates pygmaeus*) in captivity. *Zoo Biology* **32**, 394–399. doi:10.1002/zoo.21064

Hickman VV, Hickman JL (1960) Notes on the habits of the Tasmanian dormouse *Cercartetus nanus* (Desmarest) and *Eudromica lepida* (Thomas). *Proceedings of the Zoological Society of London* **135**, 365–374. doi:10.1111/j.1469-7998.1960.tb05853.x

Hide RL, Pernetta JC, Senabe T (1984) Exploitation of wild animals. In *The Research Report of the Simbu Land Use Project*, Vol. 4. *South Simbu: studies in Demography, Nutrition, and Subsistence*. pp. 291–380. I.A.S.E.R. Port Moresby.

Hill JP (1900) Contributions to the morphology and development of the female urogenital organs in the Marsupialia.

*Proceedings of the Linnean Society of New South Wales* **25**, 519–532. doi:10.5962/bhl.part.12170

Hockey R, Wilmott L, Hall L, Madani G (2019) Novel field observations of eucalypt sap feeding behaviour in the Eastern Pygmy possum *Cercartetus nanus. Australian Zoologist* **40**, 251–255. doi:10.7882/AZ.2018.039

Hohnen R, Tuft K, Legge S, Walters N, Johanson L, Carver S, Radford IJ, Johnson CN (2016) The significance of topographic complexity in habitat selection and persistence of a declining marsupial in the Kimberley region of Western Australia. *Australian Journal of Zoology* **64**, 198–216. doi:10.1071/ZO16015

Holland GJ (2001) Opportunistic vertebrate predation by the squirrel glider *Petaurus norfolcensis. Victorian Naturalist* **118**, 123–126.

Holz P (1992) Immobilisation of marsupials with tiletamine and zolazepam. *Journal of Zoo and Wildlife Medicine* **23**, 426.

How RA (1976) Reproduction, growth and survival of young in the mountain brushtail possum, *Trichosurus caninus* (Marsupialia). *Australian Journal of Zoology* **24**, 189–199. doi:10.1071/ZO9760189

How RA (1981) Population parameters of two congeneric possums, *Trichosurus* spp. in north-eastern New South Wales. *Australian Journal of Zoology* **29**, 205–215. doi:10.1071/ZO9810205

How RA, Kerle JA (1995) Brushtail possum *Trichosurus vulpecula*. In *The Mammals of Australia*. 2nd edn. (Ed. R Strahan) pp. 273–275. Reed Books, Sydney.

How RA, Barnett JL, Bradley AJ, Humphreys WF, Martin R (1984) The population biology of *Pseudocheirus peregrinus* in a *Leptospermum laevigatum* thicket. In *Possums and Gliders*. (Eds AP Smith and ID Hume) pp. 261–288. Australian Mammal Society, Sydney.

Howard J (1989) Diet of *Petaurus breviceps* (Marsupialia: Petauridae) in a mosaic of coastal woodland and heath. *Australian Mammalogy* **12**, 15–21. doi:10.1071/AM89002

Huang C, Ward S, Lee AK (1987) Comparison of the diets of the feathertail glider, *Acrobates pygmaeus*, and the eastern pygmy-possum, *Cercartetus nanus* (Marsupialia: Burramyidae) in sympatry. *Australian Mammalogy* **10**, 47–50. doi:10.1071/AM87012

Hughes RL, Thomson JA, Owen WH (1965) Reproduction in natural populations of the Australian ringtail possum, *Pseudocheirus peregrinus*, in Victoria. *Australian Journal of Zoology* **13**, 383–406. doi:10.1071/ZO9650383

Hulst F (2025) Zoonoses. In *Current Therapy in Medicine of Australian Mammals*. Revised Edition. (Eds L Vogelnest and T Portas) pp. 273–289. CSIRO Publishing, Melbourne.

Hume ID (1999) *Marsupial Nutrition*. Cambridge University Press, Cambridge.

Hume ID, Foley WJ, Chilcott MJ (1984) Physiological mechanism of foliage digestion in the greater glider and ringtail possum (Marsupialia: Petauridae). In *Possums and Gliders*. (Eds AP Smith and ID Hume) pp. 247–251. Australian Mammal Society, Sydney.

Humphreys WF, How RA, Bradley AK, Kemper CM, Kitchener DJ (1984) The biology of *Wyulda squamicaudata*, Alexander 1919. In *Possums and Gliders*. (Eds AP Smith and ID Hume) pp. 162–169. Australian Mammal Society, Sydney.

International Air Transport Association (IATA) (2022) *Live Animals Regulations*. 48th edn. International Air Transport Association, Geneva Switzerland.

Jackson SM (2000a) Population dynamics and life history of the mahogany glider, *Petaurus gracilis*, and the sugar glider, *Petaurus breviceps*, in North Queensland. *Wildlife Research* **27**, 21–37. doi:10.1071/WR98044

Jackson SM (2000b) Home-range and den use of the mahogany glider, *Petaurus gracilis. Wildlife Research* **27**, 49–60. doi:10.1071/WR98046

Jackson SM (2001) Foraging behaviour and food availability of the mahogany glider *Petaurus gracilis* (Petauridae: Marsupialia). *Journal of Zoology* **253**, 1–13 [London]. doi:10.1017/S0952836901000012

Jackson SM (2015a) Family Burramyidae (pygmy possums). *Handbook of the Mammals of the World*. Vol. 5. *Monotremes and Marsupials*. (Eds DE Wilson and RA Mittermeier) pp. 436–455. Lynx Editions, Barcelona.

Jackson SM (2015b) Family Petauridae (lesser gliders, striped possums and Leadbeater's possum). *Handbook of the Mammals of the World*. Vol. 5. *Monotremes and Marsupials*. (Eds DE Wilson and RA Mittermeier) pp. 532–565. Lynx Editions, Barcelona.

Jackson SM (2015c) Family Pseudocheiridae (ring-tailed possums and greater gliders). *Handbook of the Mammals of the World*. Vol. 5. *Monotremes and Marsupials*. (Eds DE Wilson and RA Mittermeier) pp. 498–530. Lynx Editions, Barcelona.

Jackson SM, Groves CP (2015) *Taxonomy of Australian Mammals*. CSIRO Publishing, Melbourne.

Jackson SM, Johnson CN (2002) Time allocation to foraging in the mahogany glider *Petaurus gracilis* (Marsupialia, Petauridae) and a comparison of activity times in exudivorous and folivorous possums and gliders. *Journal of Zoology* **256**, 271–277[London]. doi:10.1017/S0952836902000304

Jackson SM, McAllen BM, Broome LS, Parrott M (2014) *Mountain Pygmy Possum Captive Breeding Manual for Secret Creek Sanctuary*. Prepared on behalf of the NSW Burramys Captive Breeding Advisory Group, Sydney. 31 pp.

Jackson SM, Dickson D, O'Hara T, Booth R, Nicholson V, Lack TJ (2019) *Mahogany Glider Captive Breeding Manual*. Mahogany Glider Recovery Team, Cardwell, Queensland.

Johnson R, Hemsley S (2008) Gliders and possums. In *Medicine of Australian Mammals*. (Eds L Vogelnest and R Woods) pp. 395–437. CSIRO Publishing, Melbourne.

Jolly SE, Morris GA, Scobie S, Cowan PE (1996) Composition of milk of the common brushtail possum, *Trichosurus vulpecula* (Marsupialia: Phalangeridae): concentrations and elements. *Australian Journal of Zoology* **44**, 479–486. doi:10.1071/ZO9960479

Jones BA, How RA, Kitchener DJ (1994) A field study of *Pseudocheirus occidentalis* (Marsupialia: Petauridae). II: population studies. *Wildlife Research* **21**, 189–201. doi:10.1071/WR9940189

Jones CJ, Geiser F (1992) Prolonged and daily torpor in the feathertail glider, *Acrobates pygmaeus* (Marsupialia: Acrobatidae). *Journal of Zoology* **227**, 101–108[London]. doi:10.1111/j.1469-7998.1992.tb04347.x

Justice KE, Smith FA (1992) A model of dietary fibre utilisation by small mammalian herbivores, with empirical results for *Neotoma. American Naturalist* **139**, 398–416. doi:10.1086/285333

Kavanagh RP, Lambert MJ (1990) Food selection by the greater glider, *Petauroides volans*: is foliar nitrogen a determinant of habitat quality? *Australian Wildlife Research* **17**, 285–299. doi:10.1071/WR9900285

Kavanagh RP, Rohan-Jones WG (1982) Calling behaviour of the yellow-bellied glider, *Petaurus australis* Shaw (Marsupialia:

Petauridae). *Australian Mammalogy* **5**, 95–111. doi:10.1071/AM82011

Kean RI (1967) Behaviour and territorialism in *Trichosurus vulpecula* (Marsupialia). *Proceedings of the New Zealand Ecological Society* **14**, 71–78.

Kerle JA (1984*a*) The behaviour of *Burramys parvus* (Marsupialia) in captivity. *Mammalia* **48**, 317–325. doi:10.1515/mamm.1984.48.3.317

Kerle JA (1984b) Growth and development of *Burramys parvus* in captivity. In *Possums and Gliders*. (Eds AP Smith and ID Hume) pp. 409–412. Australian Mammal Society, Sydney.

Kerle JA (1998) The population dynamics of a tropical possum, *Trichosurus vulpecula arnhemensis* Collett. *Wildlife Research* **25**, 171–181. doi:10.1071/WR96113

Kerle JA, Howe CJ (1992) The breeding biology of a tropical possum, *Trichosurus vulpecula arnhemensis* (Phalangeridae: Marsupialia). *Australian Journal of Zoology* **40**, 653–665. doi:10.1071/ZO9920653

Kleiman DG (1977) Monogamy in mammals. *The Quarterly Review of Biology* **52**, 39–69. doi:10.1086/409721

Klettenheimer BS, Temple-Smith PD, Sofronidis G (1997) Father and son sugar gliders: more than a genetic coalition? *Journal of Zoology* **242**, 741–750[London]. doi:10.1111/j.1469-7998.1997.tb05823.x

Knipler ML, Dowton M, Mikac KM (2021) Genome-Wide SNPs detect hybridisation of marsupial gliders (*Petaurus breviceps breviceps* × *Petaurus norfolcensis*) in the wild. *Genes* **12**, 1327. doi:10.3390/genes12091327

Körtner G, Geiser F (1996) Hibernation in the alpine mountain pygmy possum *Burramys parvus* in the Australian alps. In *Adaptations to the Cold: Tenth International Hibernation Symposium*. (Eds G Körtner and F Geiser) pp. 31–38. University of New England Press, Armidale.

Körtner G, Geiser F (1998) Ecology of natural hibernation in the marsupial mountain pygmy-possum (*Burramys parvus*). *Oecologia* **113**, 170–178. doi:10.1007/s004420050365

Körtner G, Geiser F (2000) Torpor and activity patterns in free-ranging sugar gliders *Petaurus breviceps* (Marsupialia). *Oecologia* **123**, 350–357. doi:10.1007/s004420051021

Krockenberger MB, Canfield PJ, Barnes J, Vogelnest L, Connolly J, Ley C, Malik R (2002) *Cryptococcus neoformans* var. *gattii* in the koala (*Phascolarctos cinereus*): serological evidence for subclinical cryptococcosis. *Medical Mycology* **40**, 273–282. doi:10.1080/mmy.40.3.273.282

Krockenberger MB, Canfield PJ, Malik R (2003) *Cryptococcus neoformans* var. *gattii* in the koala (*Phascolarctos cinereus*): a review of 43 cases of cryptococcosis. *Medical Mycology* **41**, 225–234. doi:10.1080/369378031000137242

Kwak M, Madden C, Wicker LV (2018) *Ixodes heathi* n. sp. (Acari: Ixodidae), a co-endangered tick from the critically endangered mountain pygmy possum (*Burramys parvus*), with notes on its biology and conservation. *Experimental & Applied Acarology* **76**, 413–419. doi:10.1007/s10493-018-0312-5

Lacy RC, Balou JD, Pollak JP (2011) PMx: software package for demographic and genetic analysis and management of pedigreed populations. *Methods in Ecology and Evolution* **3**, 433–437. doi:10.1111/j.2041-210X.2011.00148.x

Lee AK, Cockburn A (1985) *Evolutionary Ecology of Marsupials*. Cambridge University Press, Cambridge.

Lindenmayer DB, Warneke RM, Linga T, Meggs RA, Seebeck JH (1991) A note on the longevity of the mountain brushtail possum, *Trichosurus caninus* in the montane ash forests of the central highlands of Victoria. *Victorian Naturalist* **108**, 4–5.

Lindenmayer DB, Tanton MT, Viggers KL (1994) Fur inhabiting ectoparasites of Leadbeater's possum, *Gymnobelideus leadbeateri* (Marsupialia: Petauridae). *Australian Mammalogy* **17**, 109–111. doi:10.1071/AM94012

Lindenmayer DB, MacGregor C, Welsh A, Donnelly CF, Brown A (2008) The use of hollows and dreys by the common ringtail possum (*Pseudocheirus peregrinus*) in different vegetation types. *Australian Journal of Zoology* **56**, 1–11. doi:10.1071/ZO08054

Lohr CA, Mills H, Robertson H, Bencini R (2009) Deslorelin implants control fertility in urban brushtail possums (*Trichosurus vulpecula*) without impacting negatively on their body condition index. *Wildlife Research* **36**, 324–332. doi:10.1071/WR08050

Lynch M (1995) An investigation of the diet of captive Leadbeater's possum, *Gymnobelideus leadbeateri*. In *International Studbook for Leadbeater's possum*. (Ed. P Myroniuk) pp. 47–54. Zoological Parks Board of Victoria, Parkville.

Lyne AG, Verhagen AMW (1957) Growth of the marsupial *Trichosurus vulpecula* and a comparison with some higher mammals. *Growth* **21**, 167–195.

Lyne AG, Pilton PE, Sharman GB (1959) Oestrous cycle, gestation period and parturition in the marsupial *Trichosurus vulpecula*. *Nature* **183**, 622–623. doi:10.1038/183622b0

MacLean L (1967) A note on the longevity and territoriality of *Trichosurus vulpecula* (Kerr) in the wild. *CSIRO Wildlife Research* **12**, 81–82. doi:10.1071/CWR9670081

MacPherson C (1997) *Sugar Gliders: A Complete Pet Owner's Manual*. Barron's Educational Series, New York.

Mallick J, Stoddart DM, Jones I, Bradley AJ (1994) Behavioural and endocrinological correlates of social status in the male sugar glider (*Petaurus breviceps*, Marsupialia; Petauridae). *Physiology & Behavior* **55**, 1131–1134. doi:10.1016/0031-9384(94)90398-0

Mansergh I, Broome L (1994) *The Mountain Pygmy Possum of the Australian Alps*. University of New South Wales Press, Sydney.

Mansergh I, Scotts D (1989) Habitat continuity and social organisation of the mountain pygmy possum restored by tunnel. *The Journal of Wildlife Management* **53**, 701–707. doi:10.2307/3809200

Mansergh I, Scotts DJ (1990) Aspects of the life history and breeding biology of the mountain pygmy-possum (*Burramys parvus*) (Marsupialia: Burramyidae) in alpine Victoria. *Australian Mammalogy* **13**, 179–192. doi:10.1071/AM90018

Mansergh I, Walsh NG (1983) Observations on the mountain pygmy possum *Burramys parvus* on Mt. Higginbotham, Victoria. *Victorian Naturalist* **100**, 106–115.

Mansergh I, Kelly P, Scotts DJ (1988) Management strategy and guidelines for the conservation of *Burramys parvus* in Victoria. Arthur Rylah Institute. Environmental Research Report Series 66.

Mansergh I, Baxter B, Scotts D, Brady T, Jolley D (1990) Diet of the mountain pygmy-possum, *Burramys parvus* (Marsupialia: Burramyidae) and other small mammals in the alpine environment at Mt. Higginbotham, Victoria. *Australian Mammalogy* **13**, 167–177. doi:10.1071/AM90017

Marples TJ (1973) Studies on the marsupial glider, *Schoinobates volans* (Kerr). IV. Feeding Biology. *Australian Journal of Zoology* **21**, 213–216. doi:10.1071/ZO9730213

Martin JK, Handasyde KA, Taylor AC, Coulson G (2001) The mating system of the bobuck, *Trichosurus caninus* in north-eastern Victoria. Abstract. Possum and Glider Symposium. Queensland Museum, Brisbane.

Martin JK, Handasyde KA, Taylor AC, Coulson G (2007) Long-term pair-bonds without mating fidelity in a mammal. *Behaviour* **144**, 1419–1445. doi:10.1163/156853907782418240

McGregor DC, Padovan A, Georges A, Krockenberger A, Yoon HJ, Youngentob KN (2020) Genetic evidence supports three previously described species of greater glider, *Petauroides volans, P. minor,* and *P. armillatus. Scientific Reports* **10**, 19284. doi:10.1038/s41598-020-76364-z

McKenna A (2005) Husbandry and breeding of the striped possum *Dactylopsila trivirgata* at London ZSL. *International Zoo Yearbook* **39**, 169–176. doi:10.1111/j.1748-1090.2005.tb00016.x

McNabb E (2017) Communal denning by two sympatric gliding possums. *Victorian Naturalist* **134**, 52–56.

Meer H (2022) Greater gliders: an approach to captive populations. *Thylacinus* **46**(3), 6–8.

Mella VSA, Gillies C, McArthur C, Webb E, Herbert C (2023) Factors influencing the success of hand-reared juvenile brushtail possums (*Trichosurus vulpecula*) released into the wild. *Wildlife Research* **50**, 96–107. doi:10.1071/WR22056

Menkhorst PW (1984*a*) The use of nest boxes by forest vertebrates in Gippsland: acceptance, preference and demand. *Australian Wildlife Research* **11**, 255–264. doi:10.1071/WR9840255

Menkhorst PW (1984b) The application of nest boxes in research and management of possums and gliders. In *Possums and Gliders*. (Eds AP Smith and ID Hume) pp. 517–525. Australian Mammal Society, Sydney.

Menkhorst PW, Collier M (1988) Diet of the squirrel glider, *Petaurus norfolcensis* (Marsupialia: Petauridae), in Victoria. *Australian Mammalogy* **11**, 109–116. doi:10.1071/AM88014

Menzies J (1972) Notes on a hand-reared spotted cuscus *Phalanger maculatus. International Zoo Yearbook* **12**, 97–98. doi:10.1111/j.1748-1090.1972.tb02282.x

Miller DS, Mitchell GF, Biggs B, McCracken H, Myroniuk P, Hewish M (2000) Detection of agglutinating antibodies to *Toxoplasma gondii* in sera from captive mainland Australian eastern barred bandicoot (*Perameles gunnii*). *Journal of Wildlife Diseases* **36**, 213–218. doi:10.7589/0090-3558-36.2.213

Millis AL, Bradley AJ (2001) Reproduction in the squirrel glider, *Petaurus norfolcensis* (Petauridae) in south-east Queensland. *Australian Journal of Zoology* **49**, 139–154. doi:10.1071/ZO01003

Mitchell PC (1911) On longevity and relative viability in mammals and birds; with a note on the theory of longevity. *Proceedings of the Zoological Society of London* **81**, 425–548. doi:10.1111/j.1096-3642.1911.tb01942.x

Morrison RGB (1978) An observation of ground nesting in sugar gliders. *South Australian Naturalist* **52**, 54–55.

Muller TL, Ensabella TJ, Booth R, Johnston SD, Phillips CJV (2010) The behavioural and environmental enrichment of captive mahogany gliders (*Petaurus gracilis*). *Australian Mammalogy* **32**, 109–116. doi:10.1071/AM09035

Munday BL (1988) Marsupial diseases. In *Australian Wildlife*. Proceedings 104. Post Graduate Committee in Veterinary Science. (Ed. DI Bryden) pp. 299–365. University of Sydney, Sydney.

Munks SA, Green B, Newgrain K, Messer M (1991) Milk composition in the common ringtail possum, *Pseudocheirus peregrinus* (Petauridae: Marsupialia). *Australian Journal of Zoology* **39**, 403–416. doi:10.1071/ZO9910403

Murphy JA, Phillips BT, Macreadie B (2003) Husbandry and breeding of the eastern pygmy possum *Cercartetus nanus* at Healesville Sanctuary. *International Zoo Yearbook* **38**, 173–178. doi:10.1111/j.1748-1090.2003.tb02077.x

Namekata S, Geiser F (2009) Effects of nest use, huddling, and torpor on thermal energetics of eastern pygmy-possums. *Australian Mammalogy* **31**, 31–34. doi:10.1071/AM08114

Norton TW (1988) Ecology of greater gliders *Petauroides volans* Kerr 1792, in relation to variations in habitat quality in eucalypt forests in south-east New South Wales. PhD thesis. The Australian National University, Canberra.

O'Donnell C (1970) Gliders of the Porepunkah district. *Victorian Naturalist* **87**, 36–39.

Oates JE, Bradshaw FJ, Bradshaw SD, Stead-Richardson EJ, Philippe DL (2007) Reproduction and embryonic diapause in a marsupial: Insights from captive female honey possums, *Tarsipes rostratus* (Tarsipedidae). *General and Comparative Endocrinology* **150**, 445–461. doi:10.1016/j.ygcen.2006.11.004

Ong PS (1994) The social organisation of the common ringtail possum *Pseudocheirus peregrinus* Boddaert 1785. PhD thesis. Monash University, Melbourne.

Pahl LI (1987*a*) Feeding behaviour and diet of the common ringtail possum, *Pseudocheirus peregrinus*, in *Eucalyptus* woodland and *Leptospermum* thickets in southern Victoria. *Australian Journal of Zoology* **35**, 487–506. doi:10.1071/ZO9870487

Pahl LI (1987*b*) Survival, age determination and population age structure of the common ringtail possum, *Pseudocheirus peregrinus*, in *Eucalyptus* woodland and *Leptospermum* thickets in southern Victoria. *Australian Journal of Zoology* **35**, 625–639. doi:10.1071/ZO9870625

Pahl LI, Lee AK (1988) Reproductive traits of two populations of the common ringtail possum, *Pseudocheirus peregrinus*, in Victoria. *Australian Journal of Zoology* **36**, 83–97. doi:10.1071/ZO9880083

Parrott ML (2019) 'Give us a sniff, love': giving marsupials scents from suitors helps breeding programs. *The Conversation*. 18 March. https://theconversation.com/give-us-a-sniff-love-giving-marsupials-scents-from-suitors-helps-breeding-programs-113641

Parrott ML, Ward SJ, Taggart DA (2005) Multiple mating, multiple paternity and sperm competition in the feathertail glider, *Acrobates pygmaeus. Australian Journal of Zoology* **53**, 79–85. doi:10.1071/ZO04025

Parrott ML, Ward SJ, Temple-Smith PD, Selwood L (2015) Females choose mates based on genetic relatedness in a small dasyurid marsupial, the Agile Antechinus (*Antechinus agilis*). *PLoS One* **10**(4), e0122381. doi:10.1371/journal.pone.0122381

Parrott ML, Nation A, Selwood L (2019) Female mate choice significantly increases captive breeding success, and scents can be frozen to determine choice, in the stripe-faced dunnart. *Applied Animal Behaviour Science* **214**, 95–101. doi:10.1016/j.applanim.2019.03.006

Parrott ML, Davis NE, Wicker LV, Broome L, Miller KM (2020) Looks like an ANZAC biscuit, tastes like a protein bar: Bogong Bikkies help mountain pygmy-possums after fire. *The Conversation*. 6 February. https://theconversation.com/looks-like-an-anzac-biscuit-tastes-like-a-protein-bar-bogong-bikkies-help-mountain-pygmy-possums-after-fire-131045

Pepper-Edwards DL (1988) The establishment, maintenance and breeding of feathertail gliders in captivity. Unpublished manuscript.

Pestell AJL, Petit S (2007) Methods and ethical considerations of pitfall trapping for the western pygmy possum (*Cercartetus concinnus* Gould) (Marsupialia: Burramyidae), with observations of capture patterns and nest sites. *Wildlife Research* **34**, 296–305. doi:10.1071/WR06090

Pilton PE, Sharman GB (1962) Reproduction in the marsupial *Trichosurus vulpecula*. *The Journal of Endocrinology* **25**, 119–136. doi:10.1677/joe.0.0250119

Presidente PJA (1982a) Common brushtail possum *Trichosurus vulpecula*: maintenance in captivity, blood values, diseases and parasites. In *The Management of Australian Mammals in Captivity*. (Ed. DD Evans) pp. 55–66. Zoological Board of Victoria, Melbourne.

Presidente PJA (1982b) Common ringtail possum *Pseudocheirus peregrinus*: maintenance in captivity, blood values and diseases. In *The Management of Australian Mammals in Captivity*. (Ed. DD Evans) pp. 75–81. Zoological Board of Victoria, Melbourne.

Presidente PJA (1984) Parasites and diseases of brushtail possums (*Trichosurus* spp.): Occurrence and significance. In *Possums and Gliders*. (Eds AP Smith and ID Hume) pp. 171–190. Australian Mammal Society, Sydney.

Presidente PJA, Barnett JL, How RA, Humphrey WF (1982) Effects of habitat, host sex, and age on the parasites of *Trichosurus caninus* (Marsupialia: Phalangeridae) in north-eastern New South Wales. *Australian Journal of Zoology* **30**, 33–47. doi:10.1071/ZO9820033

Quin DG (1995) Population ecology of the squirrel glider (*Petaurus norfolcensis*) and the sugar glider (*P. breviceps*) (Marsupialia: Petauridae) at Limeburners Creek, on the central north coast of New South Wales. *Wildlife Research* **22**, 471–507. doi:10.1071/WR9950471

Quin D, Goldingay R, Churchill S, Engel D (1996) Feeding behaviour and food availability of the yellow-bellied glider in North Queensland. *Wildlife Research* **23**, 637–646. doi:10.1071/WR9960637

Rawlins DR, Handasyde KA (2002) The feeding ecology of the striped possum *Dactylopsila trivirgata* (Marsupialia: Petauridae) in far north Queensland, Australia. *Journal of Zoology* **257**, 195–206[London]. doi:10.1017/S0952836902000808

Reddacliff G, Spielman D (1990) Diseases and parasites of Australian fauna: a brief introduction. In *Care and Handling of Australian Native Animals: Emergency Care and Captive Management*. (Ed. SJ Hand) pp. 191–197. Royal Zoological Society of New South Wales, Sydney.

Renfree MB (1980) Embryonic diapause in the honey possum, *Tarsipes spencerae*. *Search* **11**, 81.

Renfree MB, Russell EM, Wooller RD (1984) Reproduction and life history of the honey possum *Tarsipes rostratus*. In *Possums and Gliders*. (Eds AP Smith and ID Hume) pp. 427–437. Australian Mammal Society, Sydney.

Richardson KC, Wooller RD, Collins BG (1986) Adaptations to a diet of nectar and pollen in the marsupial *Tarsipes rostratus* (Marsupialia: Tarsipedidae). *Journal of Zoology* **208**, 285–297[London]. doi:10.1111/j.1469-7998.1986.tb01515.x

Roberts M, Kohn F (1991) A technique for obtaining early life history data in pouched marsupials. *Zoo Biology* **10**, 81–86. doi:10.1002/zoo.1430100111

Roberts M, Phillips L, Kohn F (1990) Common ringtail possum (*Pseudocheirus peregrinus*) as a management model for the Pseudocheiridae: reproductive scope, behaviour, and biomedical values on a browse-free diet. *Zoo Biology* **9**, 25–41. doi:10.1002/zoo.1430090104

Rose K (1999) Common diseases of urban wildlife. In *Wildlife in Australia: Healthcare & Management*. Proceedings 327. Post Graduate Foundation in Veterinary Science. (Ed. DI Bryden) pp. 365–427. University of Sydney, Sydney.

Rueegger NN, Goldingay RL, Brookes LO (2012) Does nest box design influence use by the eastern pygmy possum? *Australian Journal of Zoology* **60**, 372–380. doi:10.1071/ZO12117

Runcie MJ (1999) Movements, dens and feeding behaviour of the tropical scaly-tailed possum (*Wyulda squamicaudata*). *Wildlife Research* **26**, 367–373. doi:10.1071/WR98015

Runcie MJ (2000) Biparental obligate monogamy in the rock-haunting possum, *Petropseudes dahli*, from tropical Australia. *Animal Behaviour* **59**, 1001–1008. doi:10.1006/anbe.1999.1392

Russell EM (1986) Observations on the behaviour of the honey possum *Tarsipes rostratus* (Marsupialia: Tarsipedidae) in captivity. *Australian Journal of Zoology Supplementary Series* **34**, 1–63. doi:10.1071/AJZS121

Russell EM, Renfree MB (1989) Tarsipedidae. In *Fauna of Australia. Mammalia*. Vol. 1B. (Eds DW Walton and B Richardson) pp. 769–782. Australian Government Publishing Service, Canberra.

Russell R (1983) Yellow-bellied glider *Petaurus australis*. In *The Complete Book of Australian Mammals*. (Ed. R Strahan) pp. 136. Angus & Robertson, Sydney.

Russell R (1984) Social behaviour of the yellow-bellied glider, *Petaurus australis*, in north Queensland. In *Possums and Gliders*. (Eds AP Smith and ID Hume) pp. 343–353. Australian Mammal Society, Sydney.

Russell R (1995) Yellow-bellied glider *Petaurus australis*. In *The Mammals of Australia*. 2nd edn. (Ed. R Strahan) pp. 226–228. Reed Books, Sydney.

Sadler LM, Ward SJ (1999) Coalitions in male sugar gliders: are they natural? *Journal of Zoology* **248**, 91–96. doi:10.1111/j.1469-7998.1999.tb01026.x

Sapolsky RM, Romero LM, Munck AU (2000) How do glucocorticoids influence stress responses? Integrating permissive, suppressive, stimulatory, and preparative actions. *Endocrine Reviews* **21**, 55–89.

Scheelings TF, Dobson EC (2015) Retrospective analysis of causes of death in mountain pygmy-possums (*Burramys parvus*) at Healesville Sanctuary, Victoria, Australia. *Australian Veterinary Journal* **93**, 424–429. doi:10.1111/avj.12378

Scheun J, Geiser F, Ganswindt A, Nowack J (2019) Non-invasive evaluation of stress hormone responses in a captive population of sugar gliders (*Petaurus breviceps*). *Australian Mammalogy* **42**, 176–184. doi:10.1071/AM18044

Schmertmann LJ, Stalder K, Hudson D, Martin P, Makara M, Meyer W, Malik R, Krockenberger MB (2018) Cryptococcosis in the koala (*Phascolarctos cinereus*): pathogenesis and treatment in the context of two atypical cases. *Medical Mycology* **56**, 926–936. doi:10.1093/mmy/myx146

Schultze-Westrum TG (1965) Innerartliche verständigung durch düfte beim gleitbeutler *Petaurus breviceps papuanus* Thomas (Marsupialia, Phalangeridae). *Zeitschrift für Vergleichende Physiologie* **50**, 151–220. doi:10.1007/BF00302702

Schultze-Westrum TG (1969) Social communication by chemical signals in flying phalangers (*Petaurus breviceps papuanus*). In *Olfaction and Taste*. (Ed. C Pfaffmann) pp. 269–277. (Rockerfeller University Press: New York.).

Scicluna E (2017) The reproductive endocrinology of female mountain pygmy-possums *Burramys parvus* (Marsupialia: Burramyidae); A comparison of successful and unsuccessful breeding females. Honours thesis, Deakin University.

Scriver NJ, Johnson CN, Wallis IR, Takasaki M, Foley WJ, Krockenberger AK (2004) Which trees do wild common brushtail possums (*Trichosurus vulpecula*) prefer? Problems and solutions in scaling laboratory findings to diet selection in the field. *Evolutionary Ecology Research* **6**, 77–87.

Sharman GB (1962) The initiation and maintenance of lactation in the marsupial *Trichosurus vulpecula*. *The Journal of Endocrinology* **25**, 375–385. doi:10.1677/joe.0.0250375

Sharpe DJ, Goldingay RL (1998) Feeding behaviour of the squirrel glider at Bungawalbin Nature Reserve, north-eastern New South Wales. *Wildlife Research* **25**, 243–254. doi:10.1071/WR97037

Sharpe DJ, Goldingay RL (2007) Home range of the Australian squirrel glider, *Petaurus norfolcensis* (Diprotodontia). *Journal of Mammalogy* **88**, 1515–1522. doi:10.1644/06-MAMM-A-404R1.1

Sharpe DJ, Goldingay RL (2010) Population ecology of the nectar-feeding squirrel glider (*Petaurus norfolcensis*) in remnant forest in subtropical Australia. *Wildlife Research* **37**, 77–88. doi:10.1071/WR09051

Sharpe DJ, Goldingay RL (2018) Time budget of the squirrel glider (*Petaurus norfolcensis*) in subtropical Australia. *Australian Journal of Zoology* **66**, 251–260. doi:10.1071/ZO18049

Shoemaker AH, Croxton JM (1982) Husbandry and reproduction of the ground cuscus *Phalanger gymnotis* in captivity. *International Zoo Yearbook* **22**, 176–180. doi:10.1111/j.1748-1090.1982.tb02030.x

Slater G (1985) Handrearing five feathertail gliders. *Thylacinus* **10**(2), 6–8.

Slater G (1997) Longevity of yellow-bellied glider *Petaurus australis*. *Victorian Naturalist* **114**, 241.

Smith AP (1980) The diet and ecology of Leadbeater's possum and the sugar glider. PhD thesis. Monash University, Melbourne.

Smith AP (1982*a*) Is the striped possum (*Dactylopsila trivirgata*; Marsupialia, Petauridae) an arboreal anteater? *Australian Mammalogy* **5**, 229–234. doi:10.1071/AM82023

Smith AP (1982*b*) Diet and feeding strategies of the marsupial sugar glider in temperate Australia. *Journal of Animal Ecology* **51**, 149–166. doi:10.2307/4316

Smith AP (1984*a*) Diet of Leadbeater's possum, *Gymnobelideus leadbeateri* (Marsupialia). *Australian Wildlife Research* **11**, 265–273. doi:10.1071/WR9840265

Smith AP (1984b) Demographic consequences of reproduction, dispersal and social interaction in a population of Leadbeater's possum. In *Possums and Gliders*. (Eds AP Smith and ID Hume) pp. 359–373. Australian Mammal Society, Sydney.

Smith AP (1986) Stomach contents of the long-tailed pygmy possum *Cercartetus caudatus* (Marsupialia: Burramyidae). *Australian Mammalogy* **9**, 135–137. doi:10.1071/AM86019

Smith AP (1995) Leadbeater's possum *Gymnobelideus leadbeateri*. In *The Mammals of Australia*. 2nd edn. (Ed. R Strahan) pp. 224–226. Reed Books, Sydney.

Smith AP, Lee A (1984) The evolution of strategies for survival and reproduction in possums and gliders. In *Possums and Gliders*. (Eds AP Smith and ID Hume) pp. 17–33. Australian Mammal Society, Sydney.

Smith AP, Russell RP (1982) Diet of the yellow-bellied glider *Petaurus australis* (Marsupialia: Petauridae) in north Queensland. *Australian Mammalogy* **5**, 41–45. doi:10.1071/AM82004

Smith B (1995) *Caring for Possums*. Kangaroo Press, Sydney.

Smith GC, Mathieson M, Hogan L (2007) Home range and habitat use of a low-density population of greater glider, *Petauroides volans* (Pseudocheiridae: Marsupialia), in a hollow-limiting environment. *Wildlife Research* **34**, 472–483. doi:10.1071/WR06063

Smith MJ (1971) Breeding the sugar-glider *Petaurus breviceps* in captivity; and growth of pouch young. *International Zoo Yearbook* **11**, 26–28. doi:10.1111/j.1748-1090.1971.tb01835.x

Smith MJ (1973) Petaurus breviceps. *Mammalian Species* **30**, 1–5. doi:10.2307/3503785

Smith MJ (1979) Observations on growth of *Petaurus breviceps* and *P. norfolcensis* (Petauridae: Marsupialia) in captivity. *Australian Wildlife Research* **6**, 141–160. doi:10.1071/WR9790141

Smith MJ, How RA (1973) Reproduction in the mountain possum, *Trichosurus caninus* (Ogilby), in captivity. *Australian Journal of Zoology* **21**, 321–329. doi:10.1071/ZO9730321

Smith MJ, Brown BK, Frith HJ (1969) Breeding of the brushtail possum, *Trichosurus vulpecula* (Kerr), in NSW. *CSIRO Wildlife Research* **14**, 181–193. doi:10.1071/CWR9690181

Smith RFC (1965) Studies on the reproduction and ecology of the greater glider, *Schoinobates volans* (Kerr) (Marsupialia: Phalangeridae). MSc thesis. Australian National University, Canberra.

Smith RFC (1969) Studies on the marsupial glider, *Schoinobates volans* (Kerr). I: reproduction. *Australian Journal of Zoology* **17**, 625–636. doi:10.1071/ZO9690625

Song X, Körtner G, Geiser F (1997) Thermal relations of metabolic rate reduction in a hibernating marsupial. *American Journal of Physiology: Regulatory and Integrative Physiology* **273**, R2097–R2104. doi:10.1152/ajpregu.1997.273.6.R2097

Speare R, Haffenden AT, Daniels PW, Thomas AD, Seawright CD (1984) Disease of the Herbert River ringtail, *Pseudocheirus herbertensis* and other north Queensland rainforest possums. In *Possums and Gliders*. (Eds AP Smith and ID Hume) pp. 283–302. Australian Mammal Society, Sydney.

Species360 (2023) Zoological Information Management System (ZIMS). www.species360.org.

Spencer PBS (1996) Coping with a naturally fragmented environment: a genetic and ecological study of the allied rock wallaby, *Petrogale assimilis*. PhD thesis. James Cook University, Townsville.

Spielman D (1994) First aid emergency care for Australian native mammals. In *Wildlife*. Proceedings 233. Post Graduate Committee in Veterinary Science. (Ed. DI Bryden) pp. 339–366. University of Sydney, Sydney.

Spratt DM, Beveride I, Walter EL (1990) A catalogue of Australasian monotremes and marsupials and their recorded parasites. *Records of the South Australian Museum. Monograph Series* (1), 30.

Stanley RG, Linskins HF (1974) *Pollen: Biology, Biochemistry, and Management*. Springer-Verlag, Berlin.

Stanvic S (1992) *Possums: Rescue – Rearing – Rehabilitation – Release*. The author.

Starr CR, Hughes RT, Hemmings MS, Coase JF, Jess MD (2021) Field studies of a high elevation population of northern greater glider *Petauroides volans minor* in the Bluff State Forest, far north Queensland. *Australian Zoologist* **41**, 743–752. doi:10.7882/AZ.2021.003

Stoddart DM, Bradley AJ, Mallick J (1994) Plasma testosterone concentration, body weight, social dominance and scent marking in male marsupial sugar gliders (*Petaurus breviceps*; Marsupialia: Petauridae). *Journal of Zoology* **232**, 595–601 [London]. doi:10.1111/j.1469-7998.1994.tb04615.x

Stojanovic D, Webb MH, Alderman R, Porfirio LL, Heinsohn R (2014) Discovery of a novel predator reveals extreme but highly variable mortality for an endangered migratory bird. *Diversity & Distributions* **20**, 1200–1207. doi:10.1111/ddi.12214

Strahan R (1995) *Mammals of Australia*. 2nd edn. Reeds Books, Sydney.

Suckling GC (1984) Population ecology of the sugar glider, *Petaurus breviceps*, in a system of fragmented habitats. *Australian Wildlife Research* **11**, 49–75. doi:10.1071/WR9840049

Suckling GC (1995) Sugar glider *Petaurus breviceps*. In *The Mammals of Australia*. 2nd edn. (Ed R Strahan) pp. 234–235. Reed Books, Sydney.

Taggart DA, Breed WG, Temple-Smith PD, Purvis A, Shimmin G (1998) Reproduction, mating strategies and sperm competition in marsupials and monotremes. In *Sperm Competition and Sexual Selection*. (Eds TR Birkhead and AP Moller) pp. 623–665. Academic Press, London.

Taronga Zoo (1968) *Taronga Zoo Guide*. Taronga Zoo, Sydney.

Thomson JA, Owen WH (1964) A field study of the Australian ringtail possum *Pseudocheirus peregrinus* (Marsupialia: Phalangeridae). *Ecological Monographs* **34**, 27–52. doi:10.2307/1948462

Troughton E (1941) *Furred Animals of Australia*. Angus & Robertson, Sydney.

Turner JM (2020) The interrelationship between torpor expression and nest site use of western and eastern pygmy-possums (*Cercartetus* spp.). *Australian Mammalogy* **42**, 85–95. doi:10.1071/AM19005

Turner JM, Geiser F (2017) The influence of natural photoperiod on seasonal torpor expression of two opportunistic marsupial hibernators. *Journal of Comparative Physiology. B, Biochemical, Systemic, and Environmental Physiology* **187**, 375–383. doi:10.1007/s00360-016-1031-z

Turner JM, Warnecke L, Körtner G, Geiser F (2012*a*) Opportunistic hibernation by a free-ranging marsupial. *Journal of Zoology* **286**, 277–284. doi:10.1111/j.1469-7998.2011.00877.x

Turner JM, Körtner G, Warnecke L, Geiser F (2012*b*) Summer and winter torpor use by a free-ranging marsupial. *Comparative Biochemistry and Physiology. Part A, Molecular & Integrative Physiology* **162**, 274–280. doi:10.1016/j.cbpa.2012.03.017

Turner V (1983) Eastern pygmy-possum *Cercartetus nanus*. In *The Complete Book of Australian Mammals*. (Ed. R Strahan) pp. 160. Angus & Robertson, Sydney.

Turner V (1984a) Eucalyptus pollen in the diet of the feathertail glider, *Acrobates pygmaeus* (Marsupialia: Burramyidae). *Australian Wildlife Research* **11**, 77–81. doi:10.1071/WR9840077

Turner V (1984b) *Banksia* pollen as a source of protein in the diet of two Australian marsupials *Cercartetus nanus* and *Tarsipes rostratus*. *Oikos* **43**, 53–61. doi:10.2307/3544245

Tyndale-Biscoe CH, Smith RFC (1969) Studies in the marsupial glider, *Schoinobates volans* (Kerr). II: populatory structure and regulatory mechanisms. *Journal of Animal Ecology* **38**, 637–650. doi:10.2307/3040

van der Ree R (2002) The population ecology of the squirrel glider, *Petaurus norfolcensis*, within a network of remnant linear habitats. *Wildlife Research* **29**, 329–340. doi:10.1071/WR01095

van der Ree R, Harper MJ, Crane M (2006) Longevity in wild populations of the squirrel glider *Petaurus norfolcensis*. *Australian Mammalogy* **28**, 239–242. doi:10.1071/AM06033

Van Dyck S (1979) Mating and other aspects of behaviour in wild striped possums. *Victorian Naturalist* **96**, 84–85.

Van Dyck S (1993) The taxonomy and distribution of *Petaurus gracilis* (Marsupialia: Petauridae), with notes on its ecology and conservation status. *Memoirs of the Queensland Museum* **33**, 77–122.

Van Dyck S (1995) Striped possum *Dactylopsila trivirgata*. In *The Mammals of Australia*. 2nd edn. (Ed. R Strahan) pp. 222–223. Reed Books, Sydney.

Van Dyck SM, Strahan R (Eds) (2008) *The Mammals of Australia*. 3rd edn. Reed New Holland, Sydney.

Van Soest PJ (1982) *Nutritional Ecology of the Ruminant*. Durham and Downey, Portland.

van Tets IG, Hulbert AJ (1999) A comparison of the nitrogen requirements of the eastern pygmy possum, Cercartetus nanus, on a pollen and on a mealworm diet. *Physiological and Biochemical Zoology* **72**, 127–137. doi:10.1086/316649

van Tets IG, Whelan R (1997) Banksia pollen in the diet of Australian mammals. *Ecography* **20**, 499–505. doi:10.1111/j.1600-0587.1997.tb00418.x

Viggers KL, Lindenmayer DB (2000) A population study of the mountain brushtail possum (*Trichosurus caninus*) *in the central highlands of Victoria. Australian Journal of Zoology* **48**, 201–216. doi:10.1071/ZO99040

Viggers KL, Spratt DM (1995) The parasites recorded from *Trichosurus* species (Marsupialia: Phalangeridae). *Wildlife Research* **22**, 311–332. doi:10.1071/WR9950311

Viggers KL, Lindenmayer DB, Cunningham RB, Donnelly CF (1998) Estimating body condition in the mountain brushtail possum, *Trichosurus caninus*. *Wildlife Research* **25**, 499–509. doi:10.1071/WR96110

Vogelnest L (1999) Chemical restraint of Australian fauna. In *Wildlife in Australia: Healthcare & Management*. Proceedings 327. Post Graduate Foundation in Veterinary Science. (Ed DI Bryden) pp. 149–186. University of Sydney, Sydney.

Vose HM (1973) Feeding habits of the Western Australian honey possum, *Tarsipes spenserae*. *Journal of Mammalogy* **54**, 245–247. doi:10.2307/1378883

Wakefield NA (1970) Notes on Australian pigmy possums (*Cercartetus*, Phalangeridae, Marsupialia). *Victorian Naturalist* **7**, 11–18.

Wang LCH (1989) Ecological, physical and biochemical aspects of torpor in mammals and birds. In *Advances in Comparative and Environmental Physiology*. Vol. 4. (Ed. LCH Wand) pp. 361–401. Springer Verlag, Berlin.

Ward S (1988) Life histories of the small diprotodont marsupials. *PhD thesis*. Monash University, Victoria.

Ward SJ (1990a) Life history of the feathertail glider, *Acrobates pygmaeus* (Acrobatidae: Marsupialia), in south-eastern Australia. *Australian Journal of Zoology* **38**, 503–517. doi:10.1071/ZO9900503

Ward SJ (1990b) Life history of the eastern pygmy possum, *Cercartetus nanus* (Burramyidae: Marsupialia), in south-eastern Australia. *Australian Journal of Zoology* **38**, 287–304. doi:10.1071/ZO9900287

Ward SJ (1990c) Reproduction in the western pygmy-possum, *Cercartetus concinnus* (Marsupialia: Burramyidae), with notes on reproduction of some other small possum species. *Australian Journal of Zoology* **38**, 423–438. doi:10.1071/ZO9900423

Ward SJ (1992) Life history of the little pygmy-possum, *Cercartetus lepidus* (Marsupialia: Burramyidae), in the Big Desert, Victoria.

*Australian Journal of Zoology* **40**, 43–55. doi:10.1071/ZO9920043
Ward SJ, Renfree MB (1988) Reproduction in females of the feathertail glider, *Acrobates pygmaeus* (Shaw)(Marsupialia). *Journal of Zoology* **216**, 225–239 [London]. doi:10.1111/j.1469-7998.1988.tb02427.x
Waudby HP, Burns PA, Jensen MA, Hampton JO, Hunter D, McKnight DT, Pestell AJ, Bengsen AJ, Bower D, Coulson G, Heard G, de Laive A, Jolly CJ, Hale S, Fancourt BA, Kelleher SR, Petit S, Behrendorff L, Lumsden LF, McGregor M (2022) Wildlife marking methods. In *Wildlife Research in Australia: Practical and Applied Methods* (Eds BP Smith, HP Waudby, C Alberthsen and JO Hampton). pp. 150–179. CSIRO Publishing: Melbourne.
Wayne AF, Rooney JF, Ward CG, Vellios CV, Lindenmayer DB (2005a) The life history of *Pseudocheirus occidentalis* (Pseudocheiridae) in the jarrah forest of south-western Australia. *Australian Journal of Zoology* **53**, 325–337. doi:10.1071/ZO05026
Wayne AF, Ward CG, Rooney JF, Vellios CV, Lindenmayer DB (2005b) The life history of *Trichosurus vulpecula hypoleucus* (Phalangeridae) in the jarrah forest of south-western Australia. *Australian Journal of Zoology* **53**, 265–278. doi:10.1071/ZO05008
Westman W, Geiser F (2004) Captive breeding, growth, and development of thermoregulation in the eastern pygmy-possum (*Cercartetus nanus*). In *Biology of Australian Possums and Gliders*. (Eds RL Goldingay and SM Jackson) pp. 222–229. Surrey Beatty & Sons, Sydney.
Whiteford J, Booth R (2007) *Mahogany Glider (*Petaurus gracilis*) Husbandry Manual*. David Fleay Wildlife Park, West Burleigh, Gold Coast, Queensland.
Wilson DE, Mittermeier RA (2015) *Handbook of the Mammals of the World*. Vol. 5. *Monotremes and Marsupials*. Lynx Edicions, Barcelona.
Winter JW (1966) Bird predation by the Australian marsupial squirrel glider. *Journal of Mammalogy* **47**, 530. doi:10.2307/1377703
Winter JW (1976) The behaviour and social organisation of the brush-tail possum (*Trichosurus vulpecula*: Kerr). PhD thesis. University of Queensland, Brisbane.
Winter JW (1980) Tooth wear as an age index in a population of the brush-tailed possum, *Trichosurus vulpecula* (Kerr). *Australian Wildlife Research* **7**, 359–363. doi:10.1071/WR9800359
Winter JW, Atherton RG (1984) Social group size in north Queensland ringtail possums of the genera *Pseudocheirus* and *Hemibelideus*. In *Possums and Gliders*. (Eds AP Smith and ID Hume) pp. 311–319. Surrey Beatty & Sons, Sydney.
Winter JW, Leung LKP (1995) Spotted cuscus *Spilocuscus maculatus*. In *The Mammals of Australia*. 2nd edn. (Ed. R Strahan) pp. 266–268. Reed Books, Sydney.
Withers PC, Richardson KC, Wooller RD (1990) Metabolic physiology of euthermic and torpid honey possums, *Tarsipes rostratus*. *Australian Journal of Zoology* **37**, 685–693. doi:10.1071/ZO9890685
Woinarski JCZ, Burbidge AA, Harrison P (eds) (2014) *The Action Plan for Australian Mammals 2012*. CSIRO Publishing, Melbourne.
Woodd C, Czarny NA, Gunn IMG, Sturrock W (2006) Faecal steroid analysis and urinary cytology of the squirrel glider (*Petaurus norfolcensis*). *Australian Journal of Zoology* **54**, 277–285. doi:10.1071/ZO06019
Woods R (1999) Prevention of disease in hand reared native wildlife. In *Wildlife in Australia: Healthcare & Management*. Proceedings 327. Post Graduate Foundation in Veterinary Science. (Ed. DI Bryden) pp. 455–489. University of Sydney, Sydney.
Woodside DP (1995) Feathertail glider *Acrobates pygmaeus*. In *The Mammals of Australia*. 2nd edn. (Ed. R Strahan) pp. 262–264. Reed Books, Sydney.
Wooller RD, Renfree MB, Russell EM, Dunning A, Green SW, Duncan P (1981) Seasonal changes in a population of the nectar-feeding marsupial *Tarsipes spencerae* (Marsupialia: Tarsipedidae). *Journal of Zoology* **195**, 267–279 [London]. doi:10.1111/j.1469-7998.1981.tb03464.x
Wooller RD, Russell EM, Renfree MB, Towers PA (1983) A comparison of seasonal changes in the pollen loads of nectarivorous marsupials and birds. *Australian Wildlife Research* **10**, 311–317. doi:10.1071/WR9830311
Wooller RD, Richardson KC, Bradley GO (1999) Dietary constraints upon reproduction in an obligate pollen and nectar-feeding marsupial, the honey possum (*Tarsipes rostratus*). *Journal of Zoology* **248**, 279–287. doi:10.1111/j.1469-7998.1999.tb01028.x
Wooller RD, Richardson KC, Garavanta CAM, Saffer VM, Bryant KA (2000) Opportunistic breeding in the polyandrous honey possum, *Tarsipes rostratus*. *Australian Journal of Zoology* **48**, 669–680. doi:10.1071/ZO00071
Yang R, Fenwick S, Potter A, Elliot A, Power M, Beveridge I, Ryan U (2012) Molecular characterization of Eimeria species in macropods. *Experimental Parasitology* **132**, 216–221. doi:10.1016/j.exppara.2012.07.003
Zilko JP, Harley D, Hansen B, Pavlova A, Sunnucks P (2020) Accounting for cryptic population substructure enhances detection of inbreeding depression with genomic inbreeding coefficients: an example from a critically endangered marsupial. *Molecular Ecology* **29**, 2978–2993. doi:10.1111/mec.15540
Zuckerman S (1953) The breeding seasons of mammals in captivity. *Proceedings of the Zoological Society of London* **122**, 827–950. doi:10.1111/j.1096-3642.1952.tb00251.x

# 10
# Macropods

*Stephen M. Jackson, Karl Vernes, David Taggart, Karen Coombes and Louise Stockburger*

## 1. INTRODUCTION

Within the superfamily Macropodoidea there are 80 species of kangaroos that occur in various localities and across a broad range of habitats throughout Australia, New Guinea and surrounding islands. In Australia there are 59 species, which includes 1 species within the family Hypsiprymnodontidae, 12 species within the family Potoroidae and 46 species within the family Macropodidae (Baker *et al.* 2023). Many species of macropods have declined in abundance since European settlement, with several having become extinct and many now considered either vulnerable or endangered.

Kangaroos have had a long history in captivity, with records of eastern grey kangaroos (*Macropus giganteus*) going back as early as 1790 when 'The Wonderful Kangaroo from Botany Bay' was held on 'exhibition at the Lyceum in the Strand [London] from 8 o'clock in the Morning, till 8 in the Evening', while others were subsequently held at the Royal Domain at Richmond (England) by 1800, where they bred (Waterhouse 1841; Stanbury and Phipps 1980; Moyal 2001; Jackson and Vernes 2010). Several species of macropods, including the Kangaroo Island western grey kangaroo (*Macropus fuliginosus*) and the red-necked wallaby (*Notamacropus rufogriseus*), were also taken back alive to France as part of the Baudin expedition. Some of these were taken to either the zoo associated with the Paris Museum or the private menagerie of Empress Joséphine (first wife of Napoléon Bonaparte) (Jackson *et al.* 2021).

Kangaroos have also been held at London Zoo since 1830, with records of eastern grey kangaroos, eastern bettongs (*Bettongia gaimardi*) in 1853, Bennett's wallabies in 1867, western grey kangaroos and red kangaroos in 1868 (*Osphranter rufus*), yellow-footed rock-wallabies in 1867 (*Petrogale xanthopus*), bridled nail-tailed wallabies (*Onychogalea frenata*) in 1898, rufous bettongs (*Aepyprymnus rufescens*) in 1882 and brush-tailed bettongs (*Bettongia penicillata*) in 1903 (Zuckerman 1953). Whipsnade Zoo has records of eastern grey kangaroos in 1833 and more recently had a free-ranging population of Bennett's wallabies. Long-nosed potoroos (*Potorous tridactylus*) have been held at the New York Zoo since at least 1901 and tree-kangaroos have been held since 1937 at the National Zoo in Washington. Other species, such as red kangaroos, antilopine wallaroos (*Osphranter antilopinus*) and common wallaroos (*Osphranter robustus*), were known to be in the collection of New York Zoo in 1947 (Crandall 1964). In Australia, zoos have had a long history of holding macropods. Adelaide Zoo, for example, has held brush-tailed rock-wallabies (*Petrogale penicillata*), yellow-footed rock-wallabies (*Petrogale xanthopus*) and various other species since at least 1883 (Hornsby 1979, 1980).

Today, most zoos and fauna parks throughout Australia hold at least a few species of macropods, particularly the larger species such as red kangaroos, eastern grey kangaroos, western grey kangaroos, common wallaroos, red-necked wallabies and swamp wallabies (*Wallabia bicolor*). Some institutions, such as Adelaide Zoo, Monarto Safari Park and Taronga Zoo, have held yellow-footed rock-wallabies and the New Guinean Goodfellow's tree kangaroo; Monarto Safari Park has held Tammar wallabies (*Notamacropus eugenii*); Healesville Sanctuary, Adelaide Zoo and Taronga Zoo have also held brush-tailed rock-wallabies, long-nosed potoroos and brush-tailed bettongs.

Zoos such as Taronga Zoo and Healesville Sanctuary have nocturnal houses where they have exhibited their small macropods such as potoroos and bettongs.

Captive populations serve as educational tools, breeding colonies for threatened species to be released back into the wild and as captive populations for research and exhibition in zoological institutions (Taggart *et al.* 2005; 2010).

# 2. TAXONOMY

## 2.1 Nomenclature

The Superfamily Macropodoidea contains 2 families and 72 species in 17 genera. All 11 species within the Family Potoroidae occur only within Australia, while in the Macropodidae there are 48 species in Australia, 15 in New Guinea or surrounding islands and 3 species that occur in both regions (Flannery 1995a, 1995b; Jackson and Groves 2015; Baker and Gynther 2023).

### Australian macropods

Class: Mammalia
Supercohort: Marsupialia
Cohort: Australidelphia
Order: Diprotodontia
Superfamily: Macropodoidea
Family Hypsiprymnodontidae
Species: 1 species
Family: Potoroidae
Species: 12 species in 4 genera (see Table 10.1).
Family: Macropodidae
Species: 46 species in 11 genera (see Table 10.1).

## 2.2 Subspecies

Synonyms of Australian macropods can be found in Jackson and Groves (2015).

## 2.3 Recent synonyms

See Jackson and Groves (2015).

## 2.4 Other common names

See Baker and Gynther (2023).

**Table 10.1.** Species of macropods in Australia, their bodyweights and conservation status.

*Also occurs in New Guinea and/or surrounding islands. CR = critical, CD = conservation dependent, EN = endangered, EX = extinct, LC = low concern, NT = near threatened, VU = vulnerable.
Derived from Woinarski *et al.* (2014), Eldridge and Frankham (2015), Eldridge and Coulson (2015) and Baker and Gynther (2023).

| Common name | Scientific name | Weight (kg) | IUCN status |
|---|---|---|---|
| **Family Hypsiprymnodontidae** | | | |
| Musky rat-kangaroo | *Hypsiprymnus moschatus* | 0.36–0.68 | LC |
| **Family Potoroidae** | | | |
| Rufous bettong | *Aepyprymnus rufescens* | 1.3–3.0 | LC |
| Desert bettong | *Bettongia anhydra* | Unknown | EX |
| Eastern bettong | *B. gaimardi* | 1.2–2.3 | VU |
| Burrowing bettong | *B. lesueur* | 0.7–1.0 | NT (CD) |
| Brush-tailed bettong | *B. penicillata* | 0.8–1.8 | CR |
| Nullarbor dwarf bettong | *B. pusilla* | Unknown | EX |
| Northern bettong | *B. tropica* | 0.9–1.4 | EN |
| Desert rat-kangaroo | *Caloprymnus campestris* | 0.64–1.06 | EX |
| Gilbert's potoroo | *Potorous gilberti* | 0.7–1.2 | CR |
| Long-footed potoroo | *P. longipes* | 1.6–2.2 | VU |
| Broad-faced potoroo | *P. platyops* | ? | EX |
| Long-nosed potoroo | *P. tridactylus* | 0.7–1.7 | VU |
| **Family Macropodidae** | | | |
| Bennett's tree-kangaroo | *Dendrolagus bennetti* | 6.3–11.5 | NT |
| Lumholtz's tree-kangaroo | *D. lumholtzi* | 4.7–12.5 | NT |
| Allied rock-wallaby | *Petrogale assimilis* | 3.1–5.9 | LC |
| Western short-eared rock-wallaby | *P. brachyotis* | 2.9–6.0 | LC |
| Monjon rock-wallaby | *P. burbidgei* | 0.9–1.4 | LC |

| Common name | Scientific name | Weight (kg) | IUCN status |
|---|---|---|---|
| Cape York rock-wallaby | *P. coenensis* | 3.5–5.0 | EN |
| Nabarlek rock-wallaby | *P. concinna* | 1.1–1.7 | NT |
| Godmans rock-wallaby | *P. godmani* | 3.2–5.9 | NT |
| Herbert's rock-wallaby | *P. herberti* | 3.1–6.7 | LC |
| Unadorned rock-wallaby | *P. inornata* | 3.1–6.6 | LC |
| Black-footed rock-wallaby | *P. lateralis* | 2.9–5.0 | VU |
| Mareeba rock-wallaby | *P. mareeba* | 2.1–6.6 | NT |
| Brush-tailed rock-wallaby | *P. penicillata* | 4.2–10.9 | VU |
| Proserpine rock-wallaby | *P. persephone* | 3.5–10.2 | EN |
| Purple-necked rock-wallaby | *P. purpureicollis* | 2.7–7.1 | NT |
| Rothschild's rock-wallaby | *P. rothschildi* | 3.7–6.6 | LC |
| Mount Claro rock-wallaby | *P. sharmani* | 3.0–4.8 | VU |
| Eastern short-eared rock-wallaby | *P. wilkinsi* | 1.8–5.2 | NT |
| Yellow-footed rock-wallaby | *P. xanthopus* | 6–12 | NT (CD) |
| Rufous-bellied pademelon | *Thylogale billardierii* | 2.4–12.0 | LC |
| Red-legged pademelon (*) | *T. stigmatica* | 2.5–6.8 | LC |
| Red-necked pademelon | *T. thetis* | 1.8–9.1 | LC |
| Central hare-wallaby | *Lagorchestes asomatus* | ? | EX |
| Spectacled hare-wallaby | *L. conspicillatus* | 2.1–4.8 | VU |
| Rufous hare-wallaby | *L. hirsutus* | 0.8–1.9 | EN |
| Eastern hare-wallaby | *L. leporides* | ? | EX |
| Western grey kangaroo | *Macropus fuliginosus* | 17–72 | LC |
| Eastern grey kangaroo | *M. giganteus* | 17–90 | LC |
| Agile wallaby (*) | *Notamacropus agilis* | 9–27 | LC |
| Black-striped wallaby | *N. dorsalis* | 5.2–21 | LC |
| Tammar wallaby | *N. eugenii* | 4–10 | LC |
| Toolache wallaby | *N. greyi* | ? | EX |
| Western brush wallaby | *N. irma* | 7–9 | LC |
| Parma wallaby | *N. parma* | 3.2–5.9 | NT |
| Whip-tailed wallaby | *N. parryi* | 7–26 | LC |
| Red-necked wallaby/Bennett's wallaby | *N. rufogriseus* | 11–27 | LC |
| Bridled nail-tailed wallaby | *Onychogalea frenata* | 1.8–8.3 | VU |
| Crescent nail-tailed wallaby | *O. lunata* | ? | EX |
| Northern nail-tailed wallaby | *O. unguifera* | 3.9–9.0 | LC |
| Antilopine wallaroo | *Osphranter antilopinus* | 14–51 | LC |
| Black wallaroo | *O. bernardus* | 13–22 | NT |
| Common wallaroo/euro | *O. robustus* | 8.6–60 | LC |
| Red kangaroo | *O. rufus* | 17–92 | LC |
| Swamp wallaby | *Wallabia bicolor* | 10.3–20.5 | LC |
| Quokka | *Setonix brachyurus* | 1.6–4.2 | VU |
| Banded hare-wallaby | *Lagostrophus fasciatus* | 1..0–2.3 | VU |

## 3. NATURAL HISTORY

### 3.1 Bodyweight and morphometrics

The body mass of the different Potoroidae and Macropodidae vary greatly, ranging from ~350 g for the musky rat-kangaroo to ~85 kg for the red kangaroo. The morphometrics for the Australian species can be found in Baker and Gynther (2023) and those for New Guinea and surrounding islands are given in Flannery (1995a, 1995b).

### 3.2 Distribution and habitat

The distributions and habitats of the Potoroidae and Macropodidae within Australia are generally well known and are shown in Baker and Gynther (2023), while those in New Guinea are less well known; the details known to date can be found in Flannery (1995a, 1995b).

### 3.3 Conservation status

Australia's macropods have suffered considerably since the arrival of Europeans 200 years ago. Nearly 40% have become extinct or threatened, with 6 species now extinct, 5 endangered and 8 vulnerable (Table 10.1).

### 3.4 Diet in the wild

The diet of macropods is generally related to their body size, with smaller species utilising native rainforest fruits, plant-storage organs or fruiting bodies such as fungi, while the largest species are grazers and the intermediate-sized species mix grazing and browsing (Jarman 1984; Tyndale-Biscoe 2005; Eldridge and Coulson 2015). Within the Potoroidae, musky rat-kangaroos eat fruit and some seeds, while the other potoroids eat seeds, roots and the fruiting bodies of underground fungi (Seebeck and Rose 1989). The amount of leafy vegetation eaten (among the potoroids) increases with increasing body size, culminating in the rufous bettong, which eats mostly vegetation.

Although the larger macropodids are herbivores, they differ greatly in the type of vegetation they consume, with smaller species generally browsing and the larger species grazing (Table 10.2). The medium-sized macropods tend to be more generalist feeders, utilising a wide variety of vegetation that is relatively low in fibre (Dawson 1989). Lumholtz's tree-kangaroos are arboreal folivores that consume mostly leaves, but also some flowers and small fruits. Believed to be a generalist folivores, Coombes (2005) found they are actually quite fussy, similar to koalas, choosing particular leaves from each branch and individuals having favourite native rainforest species available to them in their home range.

The macropods within the Family Macropodidae can be divided into several groups: (1) browsers (> 70% dicotyledonous material that mostly includes stems and leaves of shrubs) and fruit (including that of underground truffles) may also comprise a significant part of the diet of some species; (2) grazers that consume a diet consisting of > 70% monocotyledonous material, mostly grasses; and (3) intermediate browser-grazers that are dietary generalists and defined as consuming < 70% monocotyledons and < 70% dicotyledons (Eldridge and Coulson 2015).

Several species of macropods have also been observed to consume carrion, including a western grey kangaroo seen feeding on a dead seabird (Frick 2014) and a swamp wallaby consuming the wing of a dead short-tailed shearwater (*Ardenna tenuirostris*) (Fitzsimons 2017). Some macropods may also feed upon vertebrates on occasion. For example, Lumholtz's tree-kangaroos have been observed to predate animals that enter their enclosure, including killing and eating a 1.2-m carpet python (*Morelia spilota*), chasing and killing birds to consume the brains, including peaceful doves (*Geopelia striata*) and hatchling Australian brush turkeys (*Alectura lathami*) (Johnson *et al.* 2002; Martin 2005). Lumholtz's tree-kangaroos have been observed grabbing, injuring or killing snakes and birds in their enclosures at Dreamworld and the Tree Roo Rescue and Conservation Centre, though at no time were they observed eating them (K. Coombes pers. obs., A. Legge pers. comm.). It may be related to territoriality or boredom in captivity, so Lumholtz's tree-kangaroos cannot be housed with any other species, or even its own species except for mating (K. Coombes pers. obs.). A male Lumholtz's tree-kangaroo was observed grabbing a bird that continual landed beside him, killing it and throw the body to the ground and on another occasion a female Lumholtz's tree-kangaroo killed a bird in her enclosure (A. Legge pers. obs.). Coombes (2005) investigated the stomach contents of road-killed Lumholtz's and the scats of radio-collared individuals and did not find any evidence of feathers, shell, snake skin or exoskeleton, so concluded that they were not eating insects, snakes, chicks from nests or eggs in the wild, as had been suggested by others.

There is also an observation of a wild euro feeding upon the carcass of a grey kangaroo (Leggett *et al.* 2019), so the degree on carnivory in some species of macropods may be under estimated.

**Table 10.2.** Diet of different genera of macropods.

Each genus of Macropodidae has been designated into one of three feeding grades: browsers, grazers and an intermediate group (Hume 1982; Sanson 1989; Eldridge and Coulson 2015).

| Genus | Food types eaten | Reference |
|---|---|---|
| **Hypsiprymnodontidae** | | |
| *Hypsiprymnus* | Fruit, seeds, invertebrates, fungi | 1–3 |
| **Potoroidae** | | |
| *Aepyprymnus* | Herbs, grasses, roots, tubers, flowers, seeds, leaves, fungi | 1 |
| *Bettongia* | Fungi, grass, seeds, roots and bulbs | 1, 4, 5 |
| *Caloprymnus* | Unknown | |
| *Potorous* | Fungi, invertebrates, fleshy fruit, seeds, flowers, some plant tissue | 1, 6–8 |
| **Macropodidae** | | |
| *Dendrolagus* | Browser - leaves, some fruit and flowers | 1, 9, 10 |
| *Petrogale* | Intermediate - Browser - herbaceous forbs, young grass, leaves of trees and shrubs, ferns, fruit, some fungi | 1, 11–14 |
| *Thylogale* | Browser - succulent short grasses and herbage; fruit, ferns and browse, fallen leaves and fungi | 14–17 |
| *Lagorchestes* | Intermediate - forbs, green grass | 18, 19 |
| *Macropus* | Grazer - grass | 1, 20 |
| *Notamacropus* | Intermediate - short grasses, herbage, some fungi | 1, 11, 14 |
| *Onychogalea.* | Grazer - herbaceous vegetation, green grass, browse, some leaves | 21 |
| *Osphranter* | Grazer - grass | 1 |
| *Wallabia* | Browser - grasses, sedges, forbs, shrubs, vines, ferns, fungus and seeds | 14, 17, 22, 23 |
| *Setonix* | Browser - grasses, shrubs | 24 |
| *Lagostrophus* | Intermediate - grass, shrubs | 1 |

References: (1) Strahan (1995); (2) Johnson and Strahan (1982); (3) Dennis (1997); (4) Johnson and McIlwee (1997); (5) Vernes *et al.* (2001); (6) Bennett and Baxter (1989); (7) Green *et al.* (1999); (8) Nguyen *et al.* (2005); (9) Heise-Pavlov *et al.* (2014); (10) Coombes (2005); (11) Dawson (1989); (12) Dawson and Ellis (1979); (13) Copley and Robinson (1983); (14) Vernes (2010); (15) K. Johnson (1980); (16) Jarman (1984); (17) Vernes (1995); (18) Main and Yadav (1971); (19) Burbidge and Johnson (1995); (20) Sanson (1978); (21) Ellis *et al.* (1977); (22) Harrington (1976); (23) Hollis *et al.* (1986); (24) Hayward (2005).

## 3.5 Longevity

### 3.5.1 Wild

The longevity of macropods in the wild does not appear to be well known for most species, but it is generally accepted as being slightly less than that estimated for the same taxa in captivity (Table 10.3).

### 3.5.2 Captivity

In captivity, wide variation in longevity is reported between the different species of macropods (Table 10.3). For potoroids it is typically 5–8 years, increasing in larger species of wallabies and kangaroos to ~10–15 years for the largest species.

### 3.5.3 Methods used to determine the age of adults

The approximate age of macropods, once they have achieved adult body size (Table 10.1), can often be estimated from examination of the teeth using techniques including molar eruption (i.e. when the molar breaks through the gum line in the jaw), molar progression (i.e. the forward movement of teeth within the mouth) and teeth wear of the premolars and/or molars.

#### 3.5.3.1 Molar eruption

Molar eruption is estimated from the proportion of the molar teeth that have erupted from the gum line. It has the advantage of being able to be conducted with (Kido *et al.* 2018) or without (Driessen and Hocking 1997) X-ray equipment and can potentially be used for genera that do not show molar progression or drift. A limitation of the molar eruption method is that age cannot be determined beyond that at which all molars have erupted (Driessen and Hocking 1997). Five stages are recognised for each molar, each of which is given decimal notations accumulating in fifths (Table 10.4; Fig. 10.1). Each molar is designated a Roman numeral (e.g. III.2 or IV.1).

Tooth eruption can be only used to estimate the age of Lumholtz's tree-kangaroos in the first few years of age (Coombes 2005). Tree-kangaroo dentition has a premolar and four molars in a flat tooth row and their teeth erupt completely early in the animal's maturity and in a relatively short period of time. The deciduous premolars,

**Table 10.3.** Average longevity (years) of different genera of macropods in the wild and in captivity.

Numbers in brackets represent known record longevities.

| Genus | Wild | Captivity | Reference |
|---|---|---|---|
| **Hypsiprymnodontidae** | | | |
| *Hypsiprymnus* | 4+ | 4–7 | 1, 2 |
| **Potoroidae** | | | |
| *Aepyprymnus* | – | 7–12 | 3, 4 |
| *Bettongia* | 5+ | 5–7 (9) | 3–8 |
| *Potorous* | 5–6 (7) | 5–10 (12) | 1, 6, 9–11 |
| **Macropodidae** | | | |
| *Dendrolagus* | 15–17 | 6–11 (20) | 1, 12–15 |
| *Petrogale* | 6–13 | 5–14 | 1, 5, 11, 14, 16–19 |
| *Thylogale* | – | 4–8 | 1 |
| *Lagorchestes* | – | 7–13 | 20 |
| *Macropus* | 19 | 25 | 21, 22 |
| *Notamacropus* | 10–15 | 6–15 | 3, 5, 6, 11, 23–25 |
| *Onychogalea* | – | 6–10 | 1 |
| *Osphranter* | 14–25 | 8–12 (15) | 1, 5, 14, 26, 27 |
| *Wallabia* | – | 5–9+ | 1,14 |
| *Setonix* | – | 6–10 | 13 |
| *Lagostrophus* | – | – | – |

References: (1) P. Johnson pers. comm.; (2) Dennis and Marsh (1997); (3) Healesville Sanctuary; (4) Mitchell (1911); (5) Adelaide Zoo; (6) Flower (1931); (7) Short and Turner (1999); (8) Vernes and Pope (2002); (9) Guiler and Kitchener (1967); (10) Seebeck (1982); (11) Strahan (1995); (12) PNG National Museum and Art Gallery; (13) Crandall (1964); (14) R. Matkovics pers. comm.; (15) Coombes (2005); (16) Kinnear *et al.* (1988a); (17) Robinson *et al.* (1994); (18) Delaney and Marsh (1995); (19) Melbourne Zoo; (20) Western Plains Zoo; (21) Kirkpatrick (1965a); (22) Poole (1982); (23) Auckland Zoo; (24) Inns (1982a); (25) Poole and Brown (1988); (26) Bailey and Best (1992); (27) Oliver (1986).

**Table 10.4.** Eruption stages of partly erupted molars of kangaroos.

Derived from Sharman *et al.* (1964).

| Notation | Position of anterior loph (cusp) | Position of posterior loph (cusp) |
|---|---|---|
| .0 | Below maxilla | Below maxilla |
| .1 | Through maxilla, below gum | Below maxilla |
| .2 | Just through gum | Through maxilla below gum |
| .3 | Part way between gum and full eruption | Just through gum |
| .4 | Fully erupted | Part way between gum and full eruption |

first two incisors and first molars (M1) have erupted by 5–6 months of age, when they are beginning to venture from the pouch; the third molar (M3) erupts before the age of 12 months. The fourth molar (M4) starts to erupt at 12-24 months of age. The eruption of third upper premolar (P3), replacing the deciduous premolars, tends to be delayed until the tooth row is complete in tree-kangaroos when they are approximately 2–3 years old. After this age there are no new teeth, just continuing wear of the biting and grinding surfaces of the existing teeth (Coombes 2005).

The use of molar eruption to determine the age of individuals has been determined for several species, including the Tasmanian bettong (Rose 1989), agile wallaby (Dudzinski *et al.* 1977), Tammar wallaby (Inns 1982b), euro (Ealey 1967), parma wallaby (Maynes 1972), red kangaroo (Newsome 1977; Sharman *et al.* 1964), allied rock-wallaby (Bell *et al.* 1989; Delaney and Marsh 1995), Tasmanian pademelon (Driessen and Hocking 1997) and quokka (Shield 1968; Table 10.5).

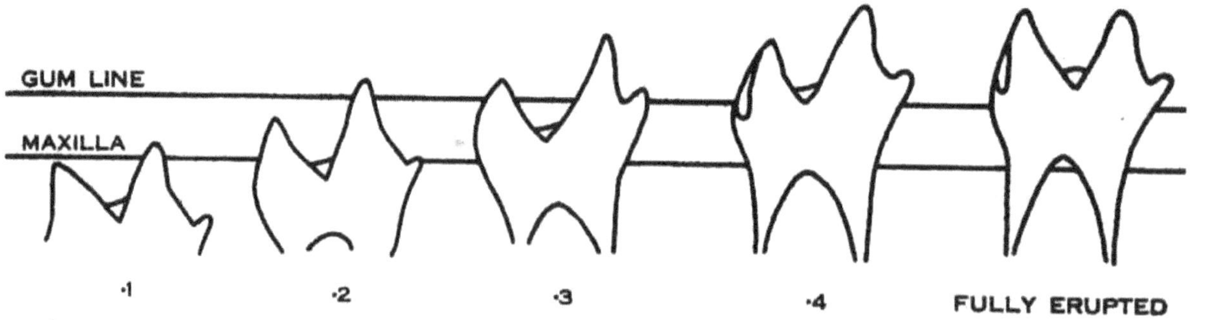


**Fig. 10.1.** Stages of eruption of molar teeth and decimal notation used. Reproduced from Sharman *et al.* (1964) with permission from CSIRO Publishing.

**Table 10.5.** Mean age (months) of the different stages of molar eruption (ME) in various species of macropods.

Derived from Rose (1989), R. Rose *pers. comm.*, Bell *et al.* (1989), Delaney and Marsh (1995), Driessen and Hocking (1997), Dudzinski *et al.* (1977), Sharman *et al.* (1964), Newsome (1977) and Shield (1968).

| ME | *Bettongia gaimardi* | *Petrogale assimilis* | *Thylogale billardierii* | *Notamacropus agilis* (F) | *Notamacropus agilis* (M) | *Osphranter rufus* | *Setonix brachyurus* |
|---|---|---|---|---|---|---|---|
| 0.1 | 3.6 | 6 | 5 | 14 | 16 | 5-7 | - |
| 0.2 | 3.9 | 6-7 | 5 | 15 | 17 | 6-8 | - |
| 0.3 | 4.2 | 7-8 | 6 | 16 | 18 | 6-9 | - |
| 0.4 | 4.4 | 8-9 | 6 | 17 | 19 | 7-11 | - |
| I.1 | 4.6 | 11-12 | 8 | 28 | 29 | 12-14 | - |
| I.2 | 5.5 | 13-14 | 9 | 30 | 31 | 13-17 | - |
| I.3 | 5.8 | 14-16 | 9 | 32 | 33 | 14-20 | - |
| I.4 | 6.1 | 15-19 | 10 | 35 | 35 | 15-24 | - |
| II.1 | 6.5 | 23-25 | 13 | 57 | 53 | 28-29 | 18-19 |
| II.2 | 7.9 | 25-28 | 14 | 61 | 56 | 31-33 | 23 |
| II.3 | 8.2 | 26-32 | 15 | 66 | 60 | 33-40 | 26 |
| II.4 | 9.0 | 28-37 | 17 | 70 | 64 | 36-47 | 29 |
| III.1 | 10.8 | 45-48 | 21 | 115 | 97 | 55-66 | 33-36 |
| III.2 | 14.4 | 48-55 | 23 | 123 | 103 | 63-72 | 39 |
| III.3 | 25.7 | 51-63 | 25 | 132 | 109 | 71-78 | 43 |
| III.4 | 32.2 | 55-72 | 27 | 142 | 116 | 80-85 | 46 |
| IV | 35.9 | 87-93 | 33 | 232 | 177 | 154 | 49 |
| Ref | 1,2 | 3,4 | 5 | 6 | 6 | 7,8 | 9 |

Note: Only figures, either with no equations or were difficult to convert for the table, were provided for *M. eugenii* (Inns 1982b), *M. parma* (Maynes 1972), *M. robustus* (Ealey 1967) and *P. xanthopus* (Poole *et al.* 1985), so accurate individual values were not able to be determined.

### 3.5.3.2 Molar progression

Molar progression involves the forward movement of teeth within the mouth and only occurs in species of macropods with curved tooth rows such as in the genera *Macropus, Osphranter, Petrogale* and some *Notamacropus* (Kirkpatrick 1964).

Although all macropods show molar eruption, molar progression is only seen in some groups. Browsing genera such as *Dendrolagus, Thylogale, Setonix* and *Wallabia* have flat tooth rows and large premolars that block molar progression (Sanson 1989). Some species, such as those in the genus *Petrogale*, generally have a slightly curved tooth row, but also have a large premolar that blocks molar progression, which often results in the first, and sometimes the second, molar being squeezed out from behind the premolar due to molar drift (Sanson 1989). *Petrogale concinna* is unique among marsupials by having unlimited molar replacement, whereas other macropods have only four molars (Sanson 1989).

In intermediate browser-grazers such as *Petrogale, Lagorchestes* and *Notamacropus*, drift occurs because the premolar is smaller and the tooth row slightly curved. In *Lagostrophus* the tooth row is flat, but the premolar is small so drift is possible (Sanson 1989). Heavy grazers, such as *Macropus, Onychogalea*, and *Osphranter*, have well-defined curved tooth rows and small premolars, so all the teeth move forward via molar progression.

Molar progression can only be estimated in living individuals by taking a dorsoventral radiograph through the skull. On each radiograph, the positions of the nearest molar relative to a line drawn across the rostral limits of the orbits are examined (Fig. 10.2a). Using this method, a molar index is assigned to the skull, in which 10 stages of progression in the molars are recognised and given decimal notation that accumulates in tenths (Newsome *et al.* 1977). For example, the 10 stages are designated 0.1–1.0 for M1, then 1.1, 1.2...1.9, 2.0 for M2 and so on (Fig. 10.2b). It should be highlighted that although

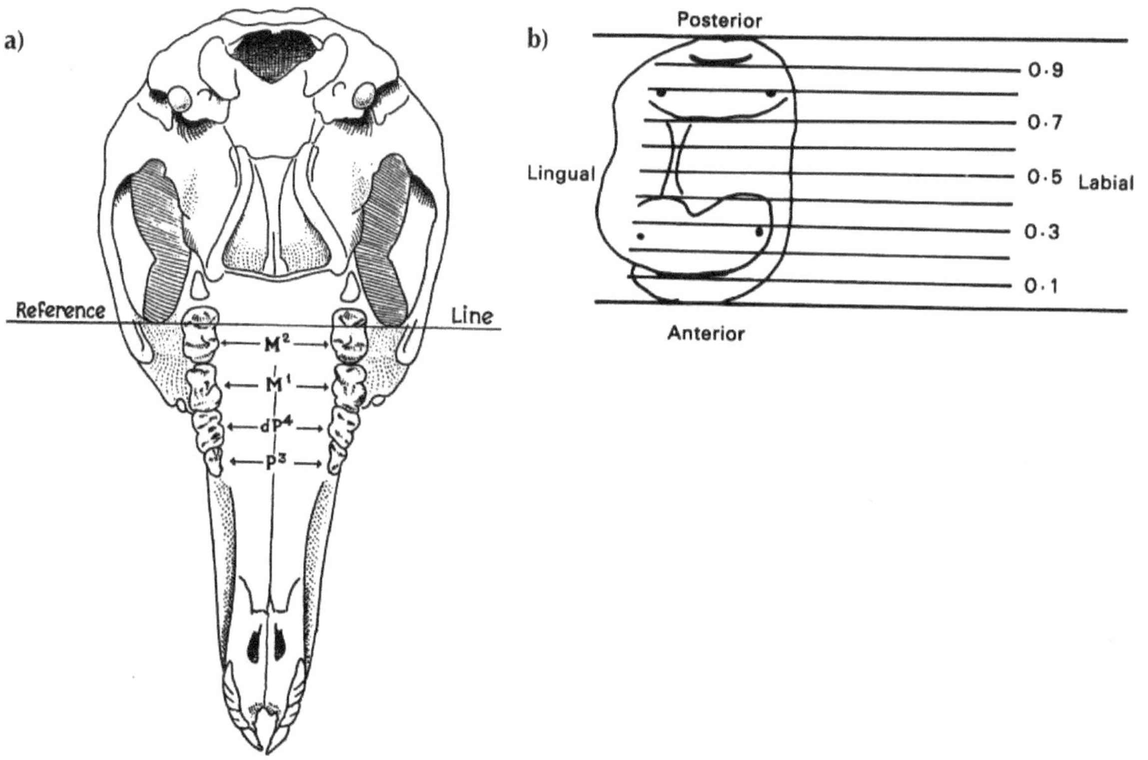


**Fig. 10.2.** Skull showing (a) the reference line used for age determination, via the molar index, using molar progression and (b) the one-tenth divisions of length. Note: this animal has a molar index of 1.7. Taken from Kirkpatrick (1964) and reproduced from Knowlton JE (1984) with permission from CSIRO Publishing.

most species of macropods have only four molars, the occurrence of five molars does occur and so some molar indexes are greater than 4.0 (Kirkpatrick 1965a). To date, species that have been found with five molars include eastern grey kangaroos, red kangaroos and common wallaroos (Kirkpatrick 1965a).

Molar progression has been used to determine the approximate age of several species of macropods (Table 10.6): allied rock-wallaby (Delaney and Marsh 1995), western grey kangaroo (Wilson 1975), eastern grey kangaroo (Kirkpatrick 1964, 1965a, 1967; King and Coulson 2024), agile wallaby (Kirkpatrick and Johnson 1969; Dudzinski *et al.* 1977; Newsome *et al.* 1977), Tammar wallaby (Maynes 1977; Inns 1982b), parma wallaby (Maynes 1972, 1977), red-necked wallaby (Kirkpatrick 1965a), euro (Ealey 1967), wallaroos (Kirkpatrick 1965a; Wilson 1975), red kangaroo (Sharman *et al.* 1964; Kirkpatrick 1965a, 1967, 1970; Wilson 1975; Newsome 1977) and swamp wallaby (Death and Coulson 2016).

Both molar eruption and molar progression are highly correlated with age, although the technique associated with molar progression has narrower confidence limits (Dudzinski *et al.* 1977). Sexual dimorphism has been observed in both molar eruption and molar progression in agile wallabies, as the males grow faster and acquire their teeth earlier than females (Newsome *et al.* 1977). Therefore, it is likely that species that are sexually dimorphic are likely to have similar differences (Newsome *et al.* 1977).

### 3.5.3.3 Molar wear score of the grey kangaroos

Molar wear score (MWS) is a useful indicator in the estimation of age for western grey kangaroos and has a linear relationship to molar progression and therefore leads to similar estimations of age, but requires a complete arcade of molars to estimate the age. Therefore, it is difficult to measure on live animals (McArthur and Sanson 1988; Mayberry *et al.* 2018). This method is used by allocating a score of 1–7, depending on the

**Table 10.6.** Molar indices (MI) of progression for ages (years) of molar teeth in different species of macropods.

Derived from Delaney and Marsh (1995), Wilson (1975), Kirkpatrick (1964, 1965a, 1970), Kirkpatrick and Johnson (1969), Dudzinski *et al.* (1977), Newsome *et al.* (1977), Maynes (1977), Inns (1982b) and Maynes (1972).

| Age | *Petrogale assimilis* | *Macropus fuliginosus* | *Macropus giganteus* | *Notamacropus agilis* (M) | *Notamacropus agilis* (F) | *Notamacropus eugenii* | *Notamacropus parma* | *Notamacropus rufogriseus* | *Osphranter robustus* | *Osphranter rufus* |
|---|---|---|---|---|---|---|---|---|---|---|
| 1 | 1.5 | 0.5–0.8 | 0.4 | 0.8–0.9 | 0.8–1.1 | 0.5–1.2 | 1.0 | 0.9 | 0.8–1.2 | 0.5–1.3 |
| 2 | 2.4 | 0.9–1.6 | 1.4 | 1.8–2.1 | 1.7–2.0 | 1.5–1.9 | 1.7 | 1.7 | 1.3–1.8 | 1.4–2.0 |
| 3 | 2.9 | 1.7–2.1 | 2.0–2.3 | 2.5–2.7 | 2.2–2.5 | 2.2–2.3 | 2.2 | 2.2 | 1.9–2.3 | 2.1–2.3 |
| 4 | 3.3 | 2.2–2.5 | 2.4 | 2.9–3.2 | 2.6–2.9 | 2.4–2.6 | 2.5 | 2.5 | 2.4–2.6 | 2.4–2.6 |
| 5 | 3.5 | 2.6–2.8 | 2.7 | 3.2–3.6 | 2.9–3.2 | 2.6–2.9 | 2.7 | 2.8 | 2.7–2.8 | 2.5–2.9 |
| 6 | 3.8 | 2.9–3.1 | 3.0 | 3.5–3.9 | 3.1–3.4 | 3.1–3.2 | 2.9–3.0 | 3.0 | 2.9–3.0 | 2.8–3.1 |
| 7 | 4.0 | 3.2–3.3 | 3.2 | 3.7–4.0 | 3.3–3.6 | 3.2–3.3 | 3.1 | 3.2 | 3.1–3.2 | 3.0–3.3 |
| 8 | – | 3.4–3.5 | 3.4 | 3.9–4.0 | 3.5–3.8 | 3.3–3.4 | 3.3 | 3.3 | 3.3–3.4 | 3.2–3.5 |
| 9 | – | 3.6–3.7 | 3.6 | – | 3.6–3.9 | 3.5–3.6 | 3.4 | 3.5 | 3.5 | 3.4–3.6 |
| 10 | – | 3.8 | 3.8 | – | 3.8–4.0 | 3.6–3.7 | 3.5 | 3.6 | 3.6 | 3.6–3.7 |
| 11 | – | 3.9 | 3.9 | – | 3.9–4.0 | 3.7–3.8 | 3.6 | 3.7 | 3.7 | 3.7–3.9 |
| 12 | – | 4.0 | 4.0 | – | 4.0 | 3.8–3.9 | 3.7 | 3.8 | 3.8–3.9 | 3.8–4.0 |
| 13 | – | 4.1 | 4.1 | – | – | 3.9–4.0 | 3.8 | 3.9 | 4.0 | 3.9–4.1 |
| 14 | – | 4.2 | 4.2 | – | – | – | 3.9 | 4.0 | 4.0 | 4.0–4.2 |
| 15 | – | 4.3 | 4.3 | – | – | – | 3.9–4.3 | 4.0 | 4.1 | 4.14.2 |
| 16 | – | – | 4.5 | – | – | – | 4.0 | 4.1 | 4.2 | 4.2–4.3 |
| 17 | – | – | 4.6 | – | – | – | – | 4.2 | 4.3 | 4.3–4.4 |
| 18 | – | – | 4.6 | – | – | – | – | 4.3 | 4.4 | 4.4–4.5 |
| 19 | – | – | 4.7 | – | – | – | – | 4.3 | 4.4 | 4.5 |
| 20 | – | – | 4.8 | -- | – | – | – | 4.4 | 4.5 | 4.6 |
| Ref | 1 | 2 | 3,4 | 5,6,7 | 5,6,7 | 8,9 | 8,10 | 4 | 2,4 | 2,4,11 |

Note: Only figures, with no equations, were provided for *M. r. erebescens* (Ealey 1967), so individual values were not able to be determined and were difficult to determine for *P. xanthopus* (Poole *et al.* 1985).

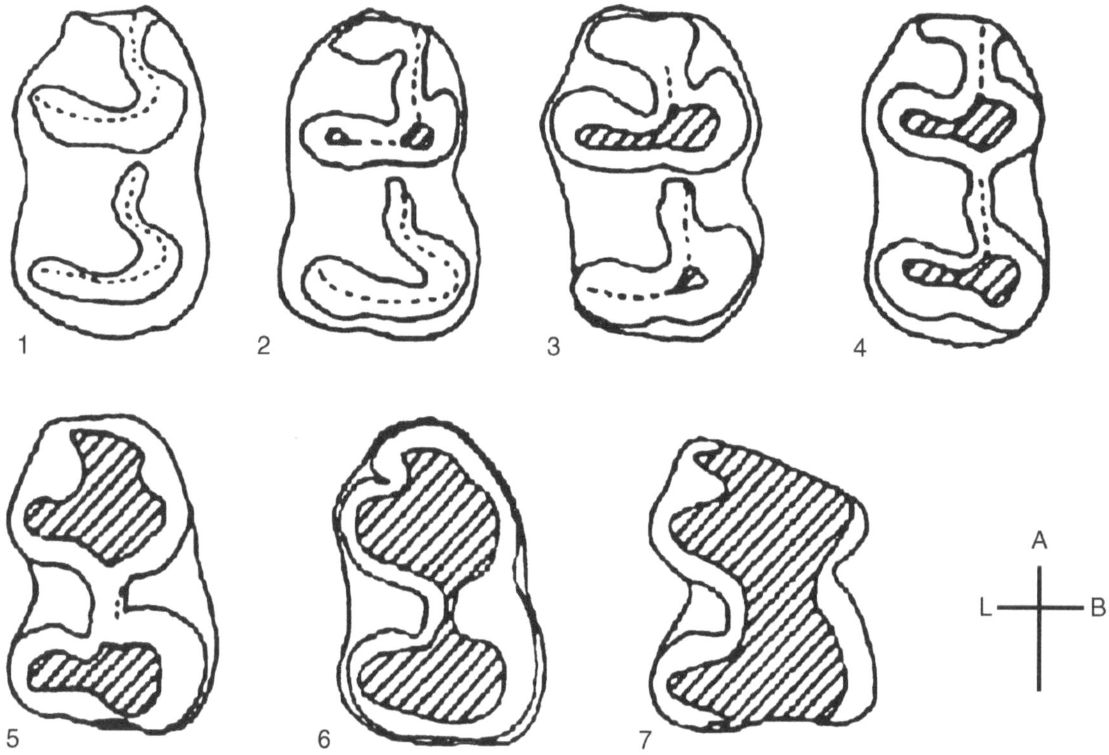


**Fig. 10.3.** Occlusal surface of the lower molar teeth in western grey kangaroos showing teeth wear classes. A = anterior, B = buccal, L = labial. Unerupted teeth are scored as 0 and shed teeth are scored as 6. Taken from McArthur and Sanson (1988) and reproduced from Mayberry *et al.* (2018) with permission from CSIRO Publishing.

degree of wear, and the scores are summed to give a total MWS (Fig. 10.3). Unerupted molars were scored as 0, while shed molars were scored as 7. There are usually up to four molars in a tooth row, but occasionally five.

An example is 4 teeth × a score of 5 to give a MWS score of 20. Some unerupted molars can usually be seen in a defleshed skull behind the last erupted one. If the bone behind the last molar is smooth, then all molars have probably erupted and some may have moved forward and fallen out. The age of the kangaroo can then be estimated using the following equations:

$$\text{Male: age} = 0.411 \times \text{MWS} + 0.33$$

$$\text{(e.g. } 0.411 \times 20 + 0.33 = (0.411 \times 28) + 0.33 = 8.55 \text{ years)}$$

$$\text{Female: age} = 1.229 \times 2.782^{(0.109 \times \text{MWS})}$$

$$\text{(e.g. } 1.229 \times 2.782^{(0.109 \times \text{MWS})} = 1.229 \times 2.782^{(0.109 \times \text{MWS})}$$

$$= 1.229 \times 2.782^{(0.109 \times 20)} = 1.229 \times 2.782^{2.18} = 9.3 \text{ years.}$$

$$\text{Both: age} = 1.301 \times 2.782^{(0.108 \times \text{MWS})}$$

### 3.5.3.4 Tooth wear of tree-kangaroos

Unlike many other macropods of the family Macropodidae, adult tree-kangaroos have a flat tooth row, and their teeth erupt completely early in the animal's maturity and in a relatively short period of time, so can be aged using tooth wear rather than molar progression. An assessment of ageing Lumholtz's tree-kangaroos using tooth wear suggested they can be placed within relative age classes (Coombes 2005). This approach includes assessing the upper right hand side premolar 3: for consistency and because it can be accessed in live animals (Coombes 2005). These relative tooth wear classes were found to correspond with cementum annuli classes, where each cementum layer appears to correspond with 1 year of age. Therefore Class 1 represents animals that are 1 year of age, Class 2 represents animals 2 years of age, and so on, until Class VI represents animals 7–8 years of age, Class VII represents animals 9–10 years of age and Class VIII represents animals 14–17 years of age (Table 10.7; Fig. 10.4).

**Table 10.7.** Key to the eight tooth wear classes used to age Lumholtz's tree-kangaroos.

Table taken from Coombes (2005). See also Fig. 10.4.

| Class | | Description |
|---|---|---|
| I. | Young animals less than 4 molars | Only slight wear of deciduous P1 and P2. Slight wear on molar crests (Fig. 10.4a) |
| II. | Young animals with 2 deciduous premolars and all 4 molars | Showing wear on deciduous premolars and all molars (Fig. 10.4b) |
| III. | Young animals with P3 emerged and all molars | With only slight wear restricted to facets on the crests of P3 with polishing of tips without breaching the enamel (Fig. 10.4c) |
| IV. | 'I shaped wear' on P3 | Moderate wear on P3 cusps. Not quite forming an island (Fig. 10.4d). Rounding of either posterior or both ends of P3, not breaching the enamel. Slight wear on tips of cusps of all molars |
| V. | 'C shaped wear' on P3 | Enamel breached continuously with dentine exposed on P3 leaving an island of enamel in the centre of the tooth surface (Fig. 10.4e). M1 entire longitudinal ridge with exposed dentine. Dentine exposed on most teeth to some extent |
| VI. | 'Basin' on P3 | Dentine exposed on entire P3 with all enamel worn away from occlusal surface forming a basin or cavity (Fig. 10.4f). Sharp ridge may be remaining on buccal side of P3. Dentine exposed to some degree on all molars |
| VII. | 'Flat tooth' P3 and M1 | P3 worn down to almost flat surface (Fig. 10.4 g). Flat occlusal surface on M1, beginning to form a dish |
| VIII. | Extreme wear P3 and M1 | P3 half worn away with wasting between anterior and posterior roots and M1 worn down to half (Fig. 10.4h). Front upper incisors also very worn |

# 4. HOUSING REQUIREMENTS

## 4.1 Enclosure design

### 4.1.1 General principles

Allowing animals to express their natural behaviour, while encouraging choice, should be at the forefront of enclosure design. As a prey species, macropods will require enclosures that are large enough to allow for a flight response if startled. Enclosures should be designed with multiple vegetative areas of varying heights to provide visual barriers as well as protection from the elements. Depending on the species, adding additional height to the enclosure via rock mounds will be of benefit (e.g. rock-dwelling species) and logs, ropes and live trees for tree-kangaroos (K. Coombes pers. obs.). Care should be taken to avoid housing males on shared enclosure fencelines, especially when females are in oestrous as fighting between males can occur through an unsolid fenceline (e.g. consisting only of mesh and shade cloth), leading to significant injury and/or death.

It is important that the fencelines are relatively straight (180°) or slightly curved. Ideally, the corners of the enclosure should not be at right angles (angles ≤ 90°), but be curved at ~45°. The absence of right-angled corners is suggested because macropods generally hop along the fenceline when frightened (e.g. during capture or when startled) and rather than turn at the corners they often hop into the fence at great speed and can cause serious injury to themselves.

The fence should be at least 1.8 m high for larger kangaroos to keep them in and made of wire fencing such as chain mesh (~50 × 50 mm) or corrugated iron or glass panels that will absorb impact from an animal and not allow the head of an animal to go through gaps in the fence. The posts can be made of treated pine or metal tubing and should be located on the outside of the enclosure where possible so that macropods do not collide with them when hopping along the fenceline or climb them (tree-kangaroos). Animals can be further protected from fence-associated injury by covering the inside of the fence to a height of at least 1.8 m with shade cloth, which also reduces their ability to see through the fence. Tree-kangaroos are best kept in enclosures with a small fence gap size (e.g. aviary mesh 12 × 12 mm) that is covered in shade cloth or smooth fencing material such as Perspex/glass or tin (K. Coombes pers. obs.).

Smaller species of macropods that have adapted to captive conditions, such as potoroids, quokkas, Tammar and parma wallabies, may only need a fence height of ~1.2 m to contain them as long as there is an external fence that is predator-proof. Wild-caught animals will require a higher fenceline (e.g. wild-caught quokkas can

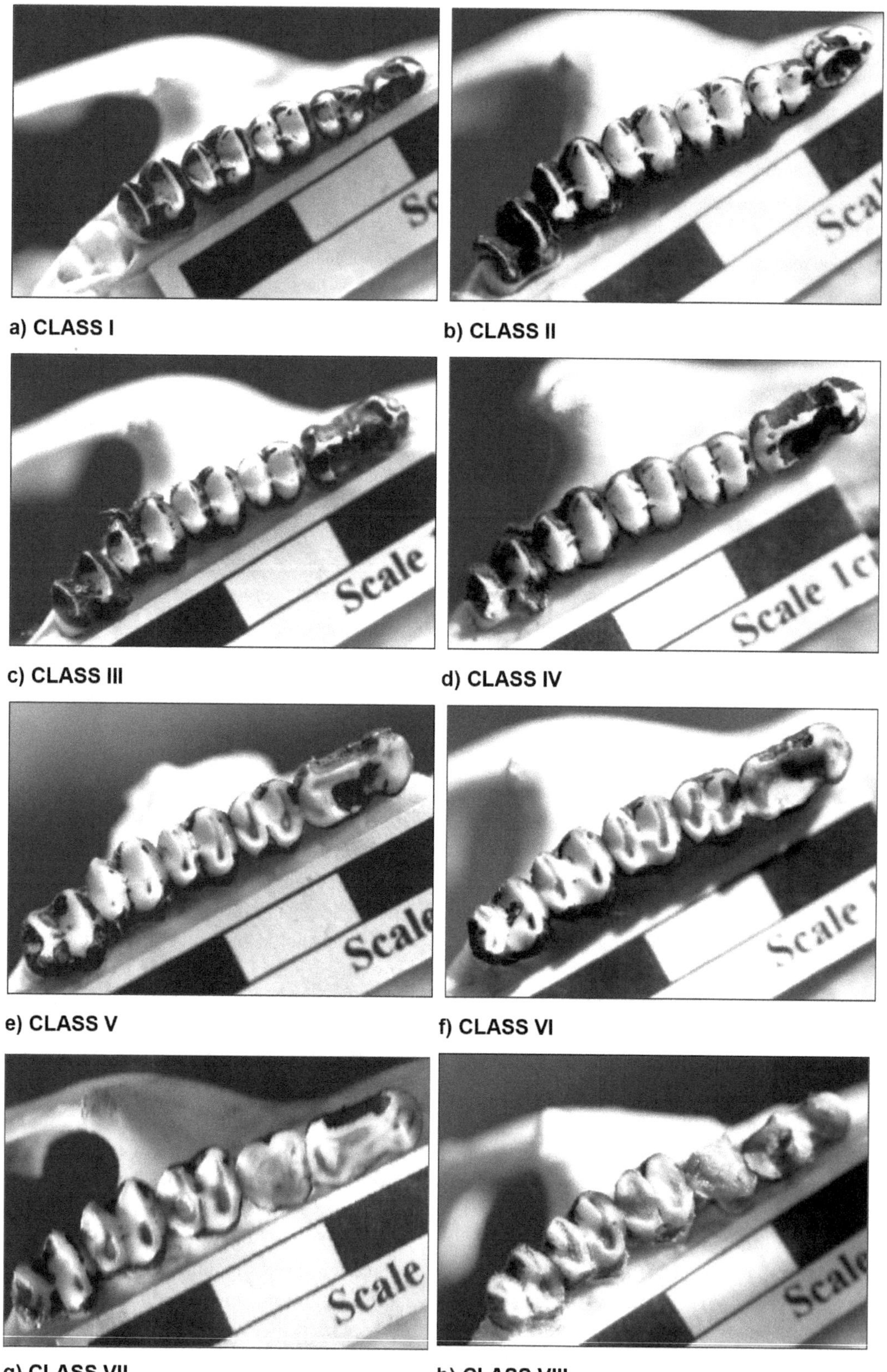

a) CLASS I
b) CLASS II
c) CLASS III
d) CLASS IV
e) CLASS V
f) CLASS VI
g) CLASS VII
h) CLASS VIII

**Fig. 10.4.** Photographs of upper tooth row of *Dendrolagus lumholtzi* showing tooth wear classes. Anterior of tooth row to right. (Upper) Cheek (labial) side and (Lower) tongue (lingual) side of mouth. Photos by Karen Coombes.

scale 1.2-m high fences and wild-born Lumholtz's tree-kangaroos can climb any fence and will jump up and sit on top of fence at 1.5m; K. Coombes pers. obs.; A. Legge pers. obs.). A floppy fence top cannot be used for Lumholtz's kanagroos as they can climb upside down and up steel poles (K. Coombes pers. obs.). The fence used for smaller species will usually consist of a smaller mesh size to ensure the head cannot get through; however, unless the outer area is predator-proof these animals are highly likely to succumb to predation by feral predators, so a standard fence height of at least 1.8 m is required.

As a general rule, species smaller than a swamp wallaby are most likely to be taken by foxes and cats. For example, cats are known to significantly prey upon wild brush-tailed bettongs, allied rock-wallabies, black-footed rock-wallabies, Tasmanian pademelons, rufous hare-wallabies, and banded hare-wallabies (e.g. Spencer 1991; Fancourt 2014, 2015; Hardman *et al.* 2016; Read *et al.* 2019; Taggart pers. obs.).

An outward facing overhang of ~60 cm (set at an angle between 90° and 45°) or electrification of the top of the fence is typically required to keep out predators such as dogs, foxes and cats. If used for tree-kangaroos this needs to be out of their reach when climbing. A second inward-facing overhang may be required to contain some species, particularly rock-wallabies as they can scale 2–3 m fences, and potoroos, bettongs and particularly tree-kangaroos are also good climbers. Dry moats have also been used successfully for a number species, including parma wallabies, quokkas, rock-wallabies, swamp wallabies and western grey kangaroos, although placing hessian at the viewing area of the dry moat until the macropods are settled in is recommended, to minimise any chance of them attempting to escape. Moats have been not recommended for tree-kangaroos.

All macropod facilities should have trees and shrubs to provide shelter and security, though not along the inside of the fenceline as they may collide with solid obstacles when startled and can provide an escape route, especially for Lumholtz's tree-kangaroos and rock-wallabies. All trees should be at least 3 m from the edge of enclosure as tree-kangaroos will climb and jump out (K. Coombes pers. obs.). Tree-kangaroos and rock-wallabies are excellent climbers and have been observed 8–10 m up trees where branches facilitate climbing and/or leaping. Placing tree guards around tall tree trunks will allow the use of the trees, but will help prevent tree-kangaroos and rock-wallabies from climbing too high, necessary to prevent injuries or escape (K. Coombes pers. obs.). Brush-tailed and yellow-footed rock-wallabies have also been observed leaping up to 6–8 m from trees or rock ledges down to the ground in a single bound (Taggart pers. obs.), so large branches hanging over fences must be removed to this height. Ensure that trees on the outside are not too close to the fence as cats and foxes can use them to climb over (e.g. Mella *et al.* 2018). Tammar wallabies will tend to go underneath fences rather than over, so fence boundaries and skirting will need to be checked regularly to ensure they are secure.

A feed shed or hutch should also be provided to macropods held outdoors to keep their food dry and provide shelter. Feed sheds only need to be a simple design, comprising an area of ~4 m (L) × 3 m (D) × 1.8 m (H) for larger species. For tree-kangaroos the hutches should have logs or ropes attached so they can access them easily and have browse pots attached to the sides (K. Coombes pers. obs.). Construction material for the walls can be timber or brick or tin.

If a feed shed is to be also utilised as a catching space, tin walls will provide more 'give' when a rock-wallaby uses the side wall to bounce off. Roofs can be constructed from corrugated tin or shingles and insulated roofing materials can be an efficient way of help reduce the shed temperature in summer. The addition of a second door on the structure will help the animals feel more secure and encourages them to feed there more often, as they have a clear escape route and are less likely to feel trapped. Be mindful that any structures placed within macropod enclosures do not pose a hazard to the animals (all construction materials must be non-toxic and have no exposed sharp edges). Ideally, the hayracks should be removable from the shed before it is to be used for catching to avoid potential injuries from these obstacles.

Shelters should not be positioned close to fencelines as they create an obstacle when the macropods hop along the fenceline. Within tree-kangaroo and rock-wallaby enclosures the animals may use the roof of shelters as basking sites and could potentially use them as launch pads to jump over fences if they are too close to the boundary. Within shelters, the feeding racks should be placed on the walls to keep the food clean and dry.

Rock-wallabies and tree-kangaroos prefer a raised ledge or platform within the shed to perch, typically along the inner wall/s to get them off the ground (D. Taggart pers. obs.). The floor area (which can be sand, hay, sawdust or concrete for easy cleaning) may need to be larger if large numbers of animals are held together or if it is to be used for both feeding and shelter. A thick layer of washed coarse sand is particularly good for Lumholtz's tree-kangaroos, as they will not be hurt when jumping down, it is easy to clean and drains well and appears to reduce the incidence of diseases such as coccidiosis and melioidosis. The sand is also better for wearing down nails (K. Coombes pers. obs.).

Several sheds may be required if there are large numbers of macropods, to minimise aggression and allow sharing of food. The entrance to the shed should ideally face away from the prevailing winds to maximise protection during poor weather (Fig. 10.5). The entrance to the shed should be wide enough to allow multiple animals to enter and exit the space to help reduce aggression and allow subordinate individuals an easy exit if a dominant animal arrives. As discussed above, if the shed is an enclosed space, two doors are better for allowing escape from aggressive encounters.

Some species that live in dry and hot environments (e.g. red kangaroos or euros) may require heating if held in environments that have long periods of cold, wet and windy weather. Heating pads can be buried in the shed floor if it comprises sand or sawdust, to allow the kangaroos to keep warm in cold wet weather, although this should not generally be required. Species from cooler regions or higher elevations, such as Lumholtz's tree-kangaroos, require shade and cooling (including misters on their browse and surrounding areas) during hot weather.

Wherever possible, one or more recovery sheds (at least 3 m (L) × 3m (W) × 1.8 m (H)) that are well ventilated and dark should be attached to the enclosure of larger species of kangaroos and wallabies. This allows macropods to come out of anaesthesia safely or when animals need to be in a confined area in the event of illness/injury to facilitate treatment or monitoring. The shed should have four solid walls that are smooth on the inside, a roof and a concrete floor. A viewing hole to examine the recovering kangaroo is a useful addition.

It is useful to have the recovery shed positioned next to a recovery enclosure (which will vary in size depending upon the size of the species e.g. 3 × 6–12 × 15 m). Shade-clothed fences are a good option. Such sheds can be used for the purposes of conditioning animals to

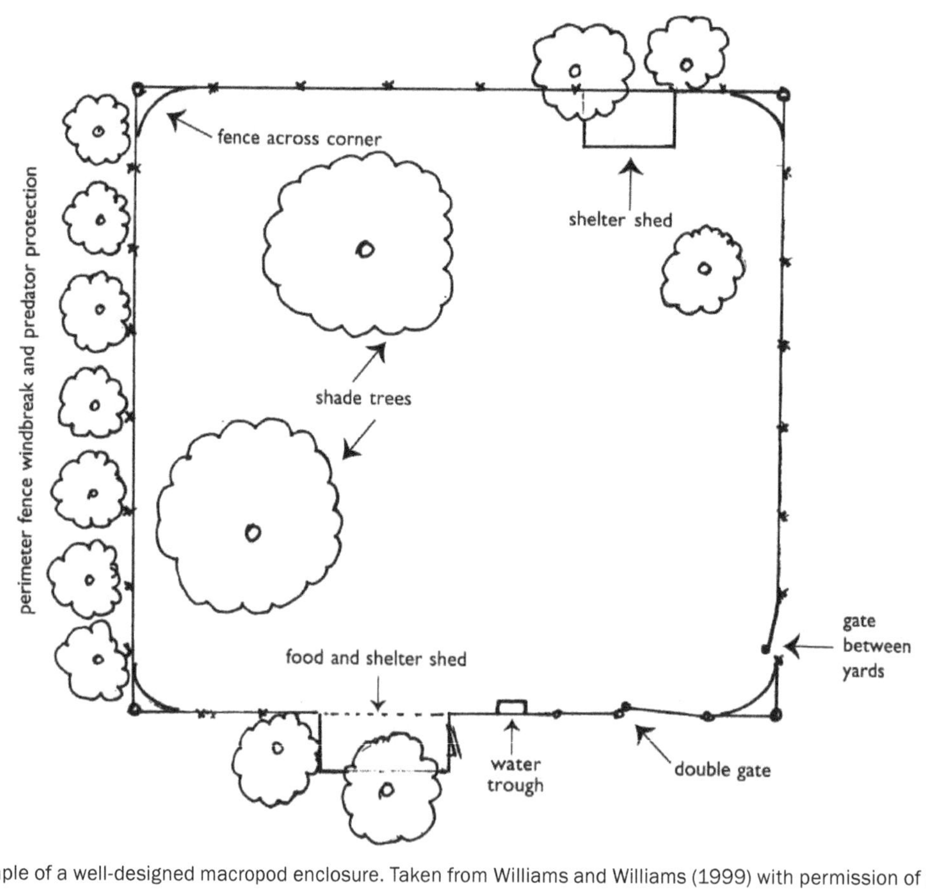


**Fig. 10.5.** Example of a well-designed macropod enclosure. Taken from Williams and Williams (1999) with permission of the publisher.

voluntarily coming in for feed, which can make capturing animals more efficient and less stressful than if they are in a large paddock enclosure. If animals are to be kept in an enclosed environment, it should be a suitable size to allow for a companion animal while the animal is recuperating; this also allows for easier integration back into the group.

## 4.1.2 Enclosure requirements for different groups of macropods

The different groups of kangaroos have their own special requirements within their enclosures in order to keep them healthy. All macropod enclosures should have enough ground cover to reflect their environment in the wild, so that individuals can hide from other kangaroos and the public if required. The larger macropods, such as red kangaroos and grey kangaroos, will require multiple cover such as shrubs and trees to escape the heat while smaller species require a lot of ground cover including grass tussocks, bushes, solid and hollow logs and rock piles. All macropod enclosures should include adequate shelter from wind, rain and extremes of temperature and sunlight, and personal space from other conspecifics but should not have obstacles that could cause injury when they are hopping along the fencelines.

### 4.1.2.1 Musky rat-kangaroos

As musky rat-kangaroos are diurnal they should be kept in natural daylight enclosures. They appear to be susceptible to high and low temperatures and low humidity, particularly when first being acclimatised to their new enclosure (Johnson *et al.* 1983). If the humidity is < 55% or excessive temperatures (> 34°C) are experienced, a misting spray should be provided (Johnson *et al.* 1983).

They are very good climbers, so they should also be provided with angled or horizontal branches to provide runways (Johnson *et al.* 1983). Limit the use of wire mesh or hessian as these can easily be climbed and may result in escaped animals or torn claws. Therefore, it is best to use tin or other smooth surfaces around the enclosure.

Leaf litter should be provided, as coarse sand or concrete can result in sores on the plantar surface of the hindfeet. Fresh leaf litter also provides small arthropod food items. Soft dry grass and tussocks should be made available for nesting material.

### 4.1.2.2 Potoroos and bettongs

Potoroos and bettongs need a lot of areas to shelter for security and nesting. In particular, they should be provided with several tussocks (e.g. *Poa* spp.) or hay and hollow logs in which they can hide. Grass is also used as nesting material. They collect it and carry it to the nest by wrapping their tails around it and hopping back to the nest (Wallis *et al.* 1989; S. Jackson pers. obs.). In the wild, nests are generally located in areas of dense cover such as low, dense clumped vegetation and are constructed over shallow depressions dug in the soil under grass tussocks, the skirts of grass trees or grass close to a log (Christensen and Leftwich 1980; Wallis *et al.* 1989; Vernes and Pope 2001). With the exception of mothers with young at foot, in the wild the nests are generally occupied by single animals (Christensen and Leftwich 1980). However, in captivity, communal nesting is quite common with two animals, usually females or a male and female, sharing the same nest, though males in non-breeding enclosures have also been observed to share nests (Delroy *et al.* 1986). More than two male bettongs within an exhibit generally do not tolerate each other well (L. Stockburger pers. obs.). Adding insulating materials such as sheep wool (free from burrs) to bettongs' enclosures is well received and is quickly woven into existing nests (L. Stockburger pers. obs.).

Burrowing bettongs can have elaborate burrow systems in the wild, with up to 120 entrances and up to 60 individuals (Burbidge 1995). The choice of soil substrate is important for outdoor enclosures with these animals, as it provides structural support for their burrows and shelter from environmental extremes. Loose sandy soil that provides little structural support should be avoided unless it is well vegetated, as it can lead to burrows collapsing and animals dying (Taggart pers. obs.).

If potoroos and bettongs are to be exhibited in outside enclosures, they should be provided with adequate shade so that they do not feel too exposed. Long-nosed potoroos, long-footed potoroos, rufous bettongs and brush-tailed bettongs have been successfully bred and kept in indoor environments (S. Jackson pers. obs.; Bates *et al.* 1972). Bettongs are excellent climbers and can escape from unroofed wire-mesh enclosures, so enclosure design should take this into account.

### 4.1.2.3 Tree kangaroos

Due to their arboreal nature, tree-kangaroos need to be supplied with a framework of interconnecting rough barked branches or trees that are ~10–15 cm in diameter and at least 2.0 m high. When the surfaces of logs start becoming smooth, they should be either replaced or have well-made notches put in them or rope wrapped around the log to provide the tree-kangaroos with good grip. Several shelters or hutches at least 2.0 m off the ground should be provided for them to rest in. Feeding platforms, at least 1.4–1.8 m above the ground, should also be provided for easy keeper access. Wet moats are not recommended as there have been several records of tree-kangaroos drowning in them (Steenberg n.d.). Wire fences used to contain tree-kangaroos should be made of a sufficient gauge to allow the claws plenty of room so they do not get caught, as the nail sheaths easily tear away. At Tree Roo Rescue and Conservation Centre, 12-mm aviary mesh is used and they have not had nails damaged except when animals have been startled and damage was actually inflicted from hitting the side of the enclosure (K. Coombes pers. obs.). As Lumholtz's tree-kangaroos are extremely good climbers, being able to climb upside down along branches and steel rafters in enclosures and up smooth steel fence posts, it is recommended that the outside fence be made of smooth material such as corrugated iron with glass panels or mesh to at least 2 m high with all the posts on the outside. Perimeter fences may also require an overhang around the top to stop them from climbing out (Williams 1990). Concrete floors have been used, particularly in holding facilities, to facilitate cleaning. However, where possible a thick layer of course sand is recommended so they will not be injured when jumping down. It is also easy to clean, drains well and appears to reduce the incidence of diseases such as coccidiosis and melioidosis. The sand is also better for wearing down nails (K. Coombes pers. obs.).

Lumholtz's tree-kangaroos, although from Far North Queensland, prefer cooler temperatures and do not do well in very hot and humid weather as they are usually found in rainforest > 600 m above sea level where it is cooler. On hotter days it is good to have fine misters on the browse and fans to help keep them cool. They do not like direct rain (they hide under the rainforest canopy in the heavy rain in the wild) or mist, but will sit among the browse or in front of fans if hot (K. Coombes pers. obs.). The humidity of the enclosure in which tree-kangaroos are comfortable appears to be species-specific, as there are records of animals developing a dry and scaly tail with humidity > 50%, which may reflect problems due to humidity or diet deficiencies (this condition responds well to hydration with the use of non-toxic agent such as petroleum jelly or vitamin A + D ointment) (Bush and Montali 1999).

#### 4.1.2.4 Hare wallabies, smaller wallabies, nail-tailed wallabies, quokkas and pademelons

As all these species are generally small and live in well-covered habitats, they need to be supplied with lots of grass tussocks, shrubs or spinifex hummocks so they can hide and feel secure (Ingleby and Westoby 1992). They can also be supplied with shelters, such as A-frame wooden shelters, to provide cover from the weather. The importance of shelter has been examined in quokkas in the wild, which showed that they move to bushes with a lower heat load in warm weather and that males defend these sheltered areas (Kitchener 1972).

#### 4.1.2.5 Rock-wallabies and wallaroos

Due to their use of rocks for refuge and security in the wild and their incredible rock-hopping ability, elevation above the ground with the use of rock piles, logs, platforms and branching trees appears to be very important for these species (particularly rock-wallabies). It also appears to be important to provide caves or similar holes small enough for joeys, as dominant female black-footed rock-wallabies have been observed to harass the joeys of lesser ranked females and to protect the best sites for their own young. A study of habitat requirements of the brush-tailed rock-wallaby found that sites occupied by rock-wallabies had twice the number of ledges and three times the number of caves and were invariably oriented so that the cliff received sun for much of the day (northerly aspect) compared with areas that did not have rock-wallabies (Short 1982; Bulinski *et al.* 1997; Mallam 2012). It has also been reported that brush-tailed rock-wallabies prefer sites with high rock coverage and complexity and high shrub density (Bulinski *et al.* 1997; Mallam 2012). Therefore, it is desirable to supply these species with rock piles or elevated platforms and rock shelters, crevices and overhangs so that they feel secure. It is likely that when such features are limited, higher levels of aggression will occur (Bulinski *et al.* 1997). Females can be particularly aggressive to conspecifics (Taggart pers. obs.). It is also preferable to provide animals with plenty of opportunities to sunbathe on these rocky areas and with shelter plants for shade. If animals are to be exhibited as part of an enclosure, the viewing area should be on the northern side where animals are most likely to spend more time.

As rock-wallabies are such agile hoppers and climbers, the fences should be higher than for most other species (~2 m). An inward facing overhang of ~60 cm at 45° is also required and it may also be necessary to electrify the top of the fence, particularly after new animals have been placed in the enclosure and are in a period of exploring for ways to escape. Rock-wallabies will use corners of their enclosure as a spring board to gain more height, so curved corners are also preferred (Taggart pers. obs.). Brush-tailed rock-wallabies have been known to escape over 2.5-m fences without an overhang (Taggart pers. obs.). If possible, the main enclosure and holding areas should be fully enclosed with nylon mesh to prevent animals from injuring their feet on wire mesh fences when bouncing off them. At night these species generally come to the pasture below the rock pile to feed, so pasture can be provided in the same way as for other macropod species.

#### 4.1.2.6 Larger wallabies and kangaroos

Relatively large species will benefit from having shelter, in the form of trees or sheds, that will allow them to escape the direct sun on hot days and provide shelter during poor weather. Additional shelter should be provided in the form of trees. When choosing plants for macropod enclosures it is important to choose species that are non-toxic and which they will not eat as they will readily consume the leaves and bark of many species of plants (S. Jackson pers. obs.). If edible species such as eucalypts and acacias are to be used, they should be well protected with tree guards to above the height of the tallest individual and held in place with the use of several solid wooden stakes.

It is also important to provide dust baths to larger species such as the grey kangaroos and red kangaroos as they readily use these and it may help control some ectoparasites. These can be provided by giving animals access to well-drained soil or sand that allows them to excavate an area in which they can sunbathe (sometimes called hip holes) or dust bathe (S. Jackson pers. obs.).

### 4.1.3 Walk-through enclosures

Several institutions have enclosures that allow members of the public to walk through them. These encounters can work reasonably well as long as the individual animals have a relatively calm demeanour. It is also important to provide fenced-off refuge areas so individual animals can retreat to these areas when they need to be away from the public to feed and rest.

Various species of macropods that have been exhibited in walk-through enclosures are shown in Table 10.8. Some species, such as eastern grey kangaroos, brush-tailed rock-wallabies and parma wallabies, are generally fairly shy unless hand-reared and are not likely to tolerate being approached or touched. Other species, such as red kangaroos, Kangaroo Island western grey kangaroos and red-necked wallabies are often very tolerant and can be approached and patted. It also appears that even species that may be considered to be flighty can be become less so after several generations if they are born within a walk-through enclosure.

It is very important in all walk-through enclosures that feeding the animals by members of the public is discouraged as it can lead to poor health due to inappropriate feed and aggression between macropods trying to gain access to feed and members of the public.

Although hand-reared macropods are generally very approachable and tolerant of the public, larger species (particularly males) can often display aggression towards members of the public in enclosures, so it is not recommended to have them in walk-through enclosures. If these species are present, they should be closely monitored for signs of aggression and removed if they show such behaviour. Macropods will also show aggressive behaviour to visitors in order to acquire food, which is generally the result of previously being fed by the public. Hand-reared female macropods generally do not show the same aggressive behaviour as males. It has generally been found that most problems with public

**Table 10.8.** Species of macropods that have been exhibited in walk through enclosures.

| Species | Temperament |
|---|---|
| *Petrogale penicillata* | Flighty |
| *P. xanthopus* | Flighty? |
| *Thylogale billardierii* | Flighty-Pattable? |
| *T. thetis* | Flighty-Pattable? |
| *Macropus fuliginosus* | Pattable- Flighty? |
| *M. giganteus* | Pattable-Flighty |
| *Notamacropus agilis* | Pattable-Flighty |
| *N. dorsalis* | Flighty |
| *N. eugenii* | Flighty |
| *N. parma* | Flighty |
| *N. parryi* | Pattable-Flighty |
| *N. rufogriseus* | Pattable |
| *Osphranter antilopinus* | Aggressive |
| *O. bernardus* | Pattable-Flighty |
| *O. robustus* | Pattable-Flighty |
| *O. rufus* | Pattable |
| *Wallabia bicolor* | Flighty-Pattable |

interaction in walk-through macropod enclosures, such as scratches or other forms of aggression, are the result of the public harassing the macropods by trying to pat them or interfere with pouch young.

With all walk-through enclosures, there are concerns over the welfare of the macropods held and some species are generally not recommended unless they have been hand-reared. Significant injury, even deaths, has resulted as a consequence of animals being chased and colliding with fences or even having rocks thrown at them. The use of roped off refuge areas, appropriate informative signage and the frequent monitoring of the walk-through enclosure, particularly on busy days, is highly recommended.

## 4.2 Holding area design

The design of holding facilities for macropods can be relatively simple. An area ~8 m (L) × 8 m (W) × 2 m (H) is adequate for holding tree-kangaroos and rock-wallabies. It is important to note that a vertical aspect to enclosures for these species is essential and 'at a bare minimum' the area should contain a well-constructed structure of branches and ropes and a platform at least 2 m off the ground to allow them to feel secure when they rest and feed. Both tree-kangaroos and rock-wallabies will access trees as resting sites if these are incorporated within enclosures. As tree-kangaroos have very tough feet, the floor of hold yards can be concrete, which will facilitate cleaning. However, wild-born Lumholtz's tree-kangaroos are easily stressed when hosing down concrete and some have been known to stop eating and die. As a result, the Tree Roo Rescue and Conservation Centre recommend thick washed coarse sand floors as they are easily cleaned by daily raking, have good drainage and less likely to facilitate diseases (Fig. 10.6) (K. Coombes pers. obs.).

## 4.3 Spatial requirements

The minimum enclosure area recommended for different groups of macropods is shown in Table 10.9. An additional 25% of the recommended area should be added for each additional female and 50% for each additional male. The size of the enclosures provided to the different species of macropods appears to be an important factor in limiting some health issues such as myopathy. For

**Fig. 10.6.** Holding yard design for Lumholtz's tree-kangaroo. Note the sloped branches, ropes and coarse sand substrate. Photo by Karen Coombes.

**Table 10.9.** Minimum area of enclosures recommended for pairs of animals of different genera of Australian macropods.

Derived from Anon (2006).

| Genus | Area ($m^2$) | Area for each additional animal (m) |
|---|---|---|
| **Hypsiprymnodontidae** | | |
| *Hypsiprymnus* | 15 | 5 |
| **Potoroidae** | | |
| *Aepyprymnus* | 15 | 5 |
| *Bettongia* | 15 | 5 |
| *Potorous* | 15 | 5 |
| **Macropodidae** | | |
| *Dendrolagus* | 40 | 10 |
| *Petrogale* | 40 | 10 |
| *Thylogale* | 40 | 10 |
| *Lagorchestes* | 30 | 10 |
| *Macropus* | 250 | 30 |
| *Notamacropus* - small species e.g. Parma and Tammar wallabies | 40 | 10 |
| *Notamacropus* - medium species e.g. Red-necked wallaby | 60 | 20 |
| *Onychogalea* | 40 | 10 |
| *Osphranter* | 250 | 30 |
| *Wallabia* | 60 | 20 |
| *Setonix* | 40 | 10 |
| *Lagostrophus* | 30 | 10 |

example, a study that held quokkas in small (1 × 1 m), medium (3 × 3 m (minus 1 $m^2$)), and large 3 × 10 m enclosures found that the animals held in the two smaller enclosures either developed paralysis, died or showed signs of myopathy when assessed during necropsy. In contrast, the quokkas placed in the large enclosure gained weight and remained healthy (Kakulas 1963).

## 4.4 Position of enclosures

Smaller species of macropods, such as the potoroids, are often held in relatively small, fully enclosed enclosures because of the potential predation risk from feral cats and foxes. Although they are generally nocturnal, it is important that their enclosures are situated so that they can seek both sunlight and shade if required. Larger species of macropods need the enclosure to be situated where it receives good sunlight, particularly during the colder times of the year when they will generally bask. Lumholtz's tree-kangaroos are not nocturnal, but are cathemeral (active night and day with bouts of sleep), so prefer a quiet area with plenty of shade (such as trees and/or shade sails) as in a rainforest.

## 4.5 Weather protection

All species of macropods, particularly the smaller ones, need some form of weather protection, generally in the form of a solid shelter such as an A-frame hut that can hold 1-2 individuals or a larger structure than can potentially allow 6-10 animals to shelter comfortably. The food should also be fully protected so that it is not degraded during poor weather. Wherever possible, numerous feeding sites should be allowed to minimise food dominance and/or aggression between individual animals.

## 4.6 Temperature requirements

Some species are more susceptible to the wet, cold and wind than others, with those that occur in drier areas and in the more northern parts of Australia being most affected. Species that may require additional heating during colder weather include nail-tailed wallabies and hare wallabies. In time they often adjust by growing thicker fur. Rock-wallabies on the other hand, like to bask in an elevated position and will move in and out of the sun as conditions vary and as their body requirements dictate, so capacity for this behaviour should be provided. Lumholtz's tree-kangaroos can cope quite well in cold weather, but stress with hot, humid summer temperatures.

## 4.7 Substrate

A variety of substrates can be used for macropod enclosures. Smaller species that are held indoors are often provided with leaf litter or soil. Coarse wood chips are not recommended as they may cause foot problems. Larger species are typically held on soil as they are normally housed outdoors in a large enclosure. As described before, Lumholtz's tree-kangaroos are best kept on coarse washed sand floors (K. Coombes pers. obs.).

## 4.8 Enclosure furnishings

Few furnishings are required for most species of macropods. Potoroids, however, like numerous tussocks and shrubs to hide under. Medium to larger macropods need shrubs and trees to provide shade and some degree of cover from inclement weather. Tree-kangaroos and rock-wallabies benefit from vertical structures such as living trees, logs, ropes, and branches interconnected with elevated platforms or hutches.

# 5. GENERAL HUSBANDRY

## 5.1 Hygiene and cleaning

Outdoor enclosures should be raked daily over the entire paddock if it is at carrying capacity, being mindful to leave any grass or groundcover intact and not rake the area down to bare soil. The sheds and feeding areas should be thoroughly raked and cleaned daily, cleared of scats and any uneaten supplementary foods such as chick peas as they will attract rodents or go mouldy (K. Coombes pers. obs.). If the enclosures are very large, open-range type enclosures, it may not be necessary to intensively rake them if the natural rate of decomposition of the faecal matter is adequate. It is very important to keep the feed area as clean as possible due to the potential for health problems that can result from poor hygiene. The build-up of faecal contamination may predispose macropods to diseases such as coccidiosis, salmonellosis, lumpy jaw, helminthosis and tetanus (Finnie 1982; Blyde 1994, 1999).

Enclosures and holding areas with asphalt or concrete surfaces used for feeding should be thoroughly swept daily and, ideally, hosed at least every second day and sprayed with F10 disinfectant to provide a clean surface for feeding. When all individuals permanently leave an enclosure, it should be scrubbed out and cleaned as much as possible before new animals enter.

## 5.2 Record keeping

It is important to establish a system whereby the health, welfare, condition and reproductive status of captive macropods are routinely monitored and recorded:

- welfare assessment process where welfare is monitored and enhanced
- exhibit risk assessments are routinely completed
- identification numbers ( all individuals should be identifiable)
- results of veterinary examinations and treatments
- behavioural changes or problems
- reproductive behaviour or condition (pouch status, pouch young emergence, weaning)
- bodyweights and measurements
- changes in diet
- changes in faecal consistency (diarrhoea)
- amount of supplementary food and browse including species recorded daily (tree-kangaroos)
- movements of individuals between enclosures or institutions
- births, with dam and sire if known
- deaths with necropsy results
- enrichment supplied.

The collection of information on physical and behavioural patterns of individual macropods can contribute greatly to the husbandry of these species. It also allows individual histories to be transferred to other institutions if required and facilitates a cooperative approach to data collection and management across institutions. ZIMS, the Zoo Information Management System (Species360 2023), can be used for recording general information on age, sex, parentage, births, identification, bodyweights, transfers and deaths, with separate ZIMS modules for veterinary information and studbooks. These systems have been developed by Species360, which is part of the Conservation Planning Specialist Group (CPSG) software. ZIMS interfaces with programs, such as PMx, which provides demographic and genetic analysis tools to assist with the management of breeding programs (Lacy *et al.* 2011).

## 5.3 Methods of individual identification

Identification techniques must have several attributes in order to be effective (Arlidge and Srb 1992) including:

- being permanent
- positively identifying the animal as an individual
- being inexpensive and easy to apply
- not unreasonably damaging the individual
- being relatively painless to apply
- not interfering with the animal's mobility
- being adaptable to modern data retrieval systems
- being clearly visible if this is required
- being unalterable.

A review of macropod identification (Arlidge and Srb 1992) suggested that not one method could be recommended as the best technique, so it is advisable to use at least two methods to avoid misidentification, which is particularly important when maintaining endangered species (Arlidge and Srb 1992).

### 5.3.1 Passive Integrated Transponder (PIT) tags

PIT tags (microchips) are typically implanted between the scapulae of the individual animal and can be used on all species of macropods. This is an excellent method of animal identification, but can be expensive if many animals are to be implanted. PIT tags are a permanent method of identification, but they can track out along the injection site, which can be avoided by sealing the entry wound with tissue glue (Vetbond®) or similar non-toxic product. They generally require the animal to be caught to confirm identification with a PIT tag reader. Potentially, chip reading 'loops' in raceways or slide openings can be set up to 'remotely' identify the macropods as they move through.

### 5.3.2 Tattoos

Tattoos can be placed on the inside of the ear or inside the hindleg. It is an inexpensive method, but tattoos tend to fade and are restricted to use on light-skinned individuals. Animals will typically need to be restrained to read the tattoo for identification.

### 5.3.3 Visual identification

Using size, sex, colour and markings for physical identification of animals by photos or observing individuals is adequate if people are trained properly. This technique has been used successfully for large numbers of eastern grey and red-necked wallabies (Jarman *et al.* 1989) and for tree-kangaroos (K. Coombes pers. obs.).

### 5.3.4 Ear tags

Metal ear tags are often used on potoroids. They are inexpensive, but not necessarily reliable as they can generally only be read with the animal restrained and also can be ensnared and pulled out. Plastic ear tags have been used in larger macropods to identify individuals from a distance by using different colours and positions in the ear (Fig. 10.7). Ear tags should ideally be placed closer to the centre and the base of the ear, rather than the tip or margins of the pinna where they are more likely to be pulleded out.

### 5.3.5 Ear notching or punching

Various combinations of notches or punches in the ear can be used to represent numbers (see Fig. 10.8). However, either method mutilates the appearance of the animal and often results in torn ears and unfavourable comments

**Fig. 10.7.** Brush-tailed rock-wallaby with an ear tag in right ear and a VHF collar for tracking and monitoring. Photo by David Taggart.

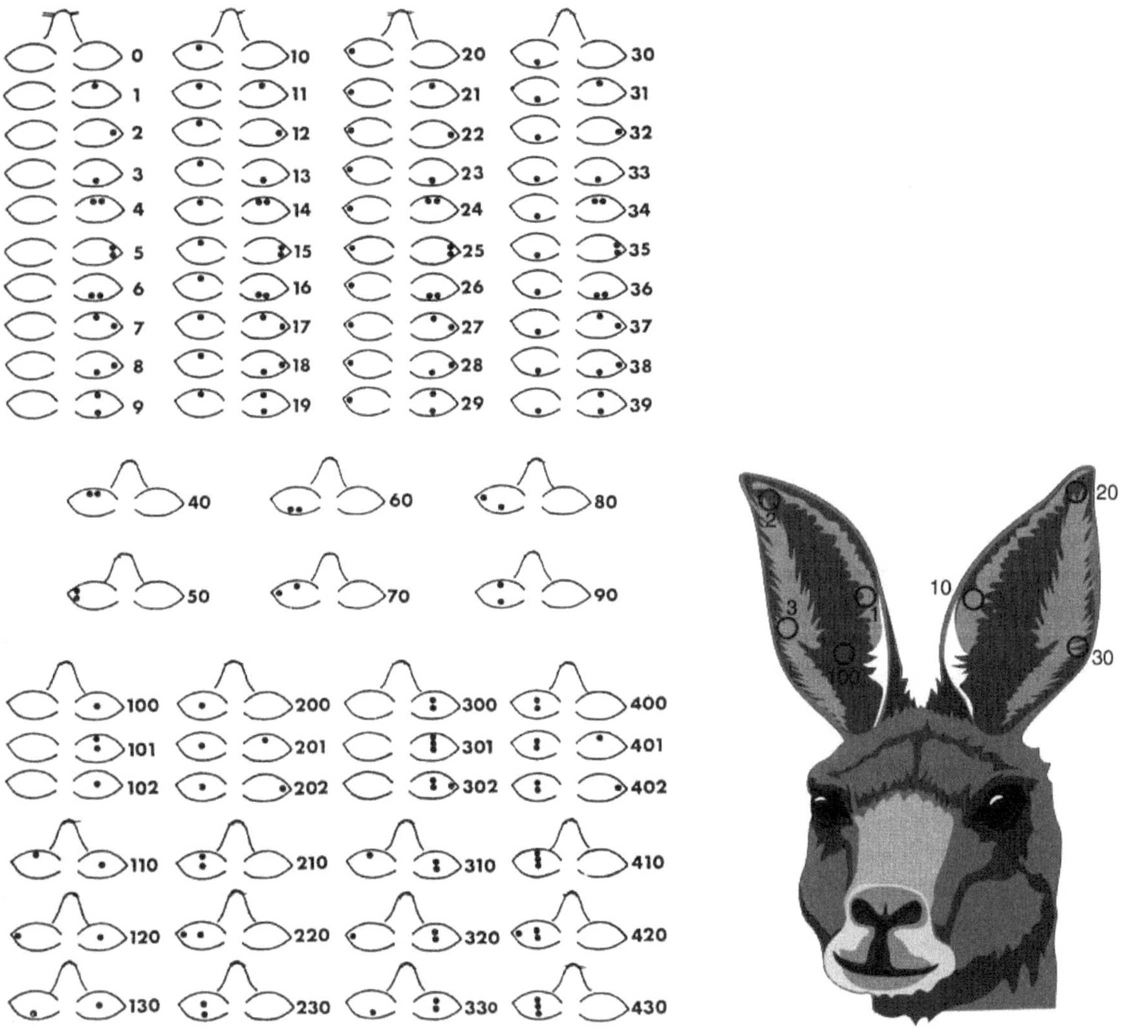


**Fig. 10.8.** Several ear-notching numbering systems that have been used to identify individual macropods in large mobs. Taken from Cooper (1998) and Williams and Williams (1999) with permission from the author and publisher respectively.

from the public. If using this method, take care to avoid veins in the ear, which can be difficult to detect.

### 5.3.6 Freeze banding

This method does not appear to be used anymore and is not recommended. It is expensive because animals need to be captured and anaesthetised for its application to be painless and eliminate the chance of smudging from the animal moving.

### 5.3.7 Collars

One study used collars, ~2.5-cm wide, that were placed round the neck. Each collar included reflective tape of different shapes that presented numbers (Fig. 10.9). The colours and shapes could be seen at a distance (Ealey and Dunnett 1956). Though this method has been used for captive animals it is not recommended for animals on exhibit as it detracts greatly from the natural look of the individuals and may pose a welfare issue, which includes the risk of getting caught on branches or fences, which could incapacitate or potentially hang the animal; the risk of getting a forelimb caught up in the collar; and the risk of animals growing in size or gaining condition, which can lead to constriction and trauma. Therefore, it is important to ensure the collar is properly fitted so it is not too tight or too loose. There are also strong merits to having a 'break point' or elastic component in the collar to allow it to break or stretch if needed.

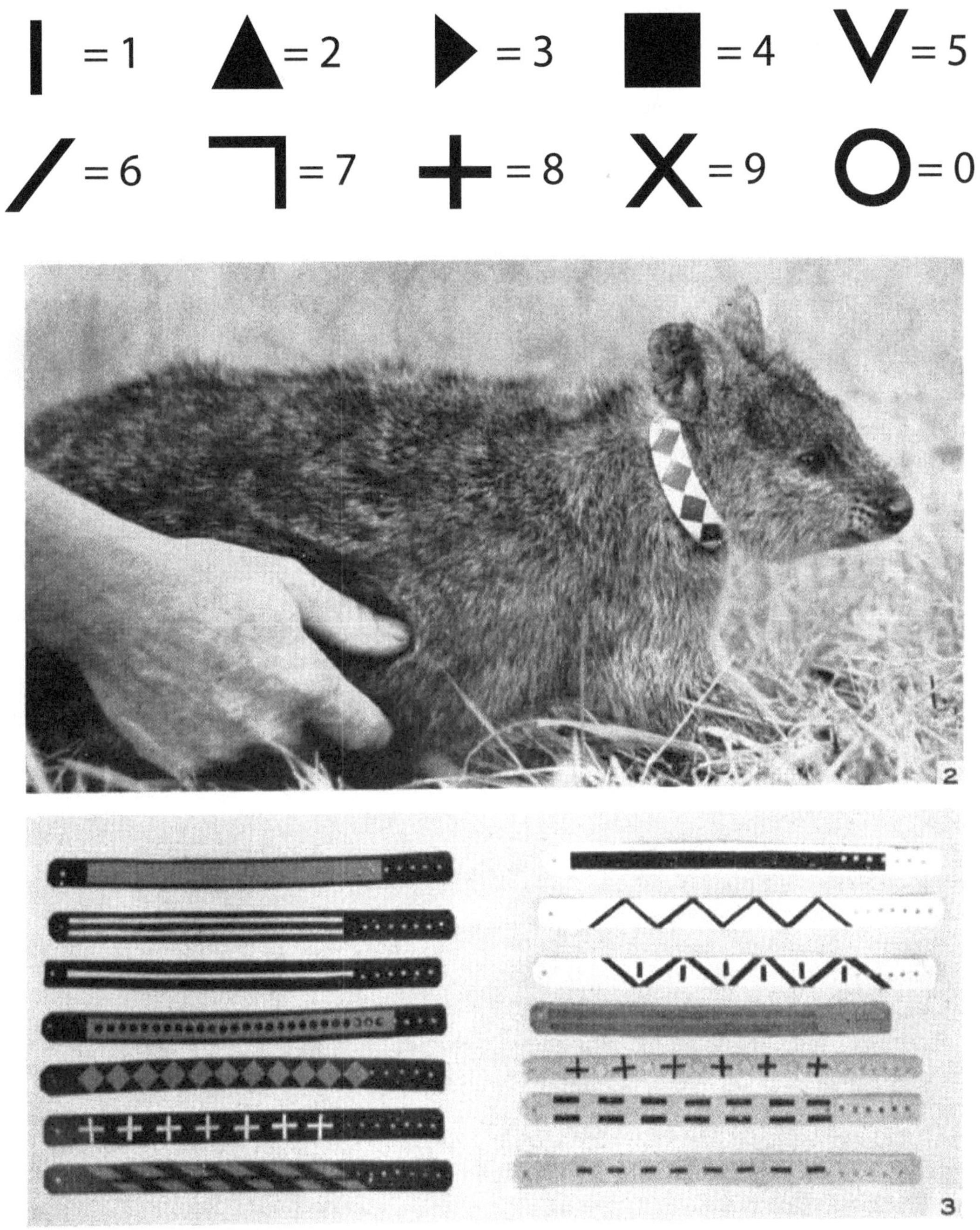


**Fig. 10.9.** (Upper) Examples of shapes to represent numbers on plastic collars. (Middle) Quokka fitted with a plastic collar. (Lower) Plastic collars quokkas showing the patterns of reflective tape. Reproduced from Ealey & Dunnett (1956) with permission from CSIRO Publishing. Photographs by G.M Dunnett.

## 6. FEEDING REQUIREMENTS

### 6.1 Captive diet

The diets of the different groups of macropods differ greatly from each other in the wild and this should be reflected in their diets in captivity. Macropods should be provided with clean fresh water daily and most food can be provided ad lib due to their high nutritional requirements and low incidence of obesity. Food items include hay, hard vegetables such as sweet potato and carrot, apple, dried corn, chick peas and commercially prepared pellets. There has been a general thought that lucerne bales should have a high amount of lucerne to stalk, as the stalk may pierce the gums and allow access to bacteria that cause lumpy jaw; however, there is little evidence to support this. Nonetheless, some institutions have fed lucerne chaff instead, due to concerns with stalks causing gum problems.

Although bread is readily accepted by most macropods, it should be used only when providing treatments (e.g. antibiotics), as it may predispose to lumpy jaw. Instead, treats such as apple or corn are equally effective or even banana balls or macropod biscuits, which are used with tree-kangaroos (K. Coombes pers. obs.).

After feeding, macropods can sometimes be observed regurgitating a bolus of food, which is chewed and reswallowed (known as merycism). This frequently involves a violent heaving motion with vigorous movements of the arms and chest, such that the animal may appear to be violently ill, even projecting food out of its mouth. However, it does not reflect ill health or an unsuitable diet (Owen 1835; Wood Jones 1923; Moir *et al.* 1956; Barker *et al.* 1963; Eldridge and Coulson 2015). Despite the sometimes violent nature of the regurgitation process, merycism plays a similar function to ruminating in bovid ungulates, but differs because it does not involve the transfer of food between different compartments of the stomach. Joeys can become very excited when their mothers regurgitate and will sometimes lick the mother's mouth, which may be important in inoculating the stomach of the joey with the appropriate bacteria to allow proper digestion (S. Jackson pers. obs.; K. Coombes. pers. obs.). Saliva-sharing has been observed in the allied rock-wallaby and Lumholtz tree-kangaroos (Coombes pers obs).. In this species, it has been suggested that the main function of lip-licking is to transfer saliva to the forestomach of the licker to maintain the water content of digesta to assist digestive fermentation and reflects an inadequate water intake from milk, vegetation or drinking water (Davies 2013).

It is also important to note the effect of a sudden change in the quality of the diet. If macropods move from an area of low-quality pasture to one of high quality, there may be a rapid change in the gut flora. Different bacteria, such as *Clostridium* spp. and *Salmonella* spp., can increase greatly in number. In milder cases, the individuals develop diarrhoea that quickly passes after the gut flora have re-adjusted. To overcome the potential for deaths of animals, grass from the new, higher quality enclosure should be cut and added in with the food of the existing lower quality grass so that their stomach flora can slowly adjust before the move. In extreme cases toxins can result in death (S. Jackson pers. obs.).

### 6.1.1 Musky rat-kangaroo

- Ad lib
  - water
  - rolled oats
  - cracked corn
  - poultry layer pellets
- Daily diet (per animal)
  - apple
  - sweet potato
  - orange
  - sultanas
  - peanuts

  (total = 25 g)
  - vitamin E powder added daily
- Supplement
  - insects such as moths, grasshoppers and mealworms
  - hard-boiled egg every 2 weeks

  Diet derived from Johnson *et al.* (1983)

### 6.1.2 Potoroos and rufous bettong

- Ad lib
  - water
- Daily diet (per animal)
  - 3 Eukanuba® Pet Food Kibble
  - 40 g apple in 2-cm cubes
  - 30 g orange
  - 50 g banana
  - 30 g corn
  - 10 g sprouted seed*
  - 30 g carrot sliced
  - 30 g pear
  - 30 g potato/sweet potato
  - 5 g pet health food - 1-cm cubes
  - 20 g pasture replacement pellets
  - 2 g silverbeet
  - 5 mealworms
- Supplement
  - 6 g egg and cheese - 3-4 times per week
  - 2 almonds - 3-4 times per week
  - 5 ml nectar mix - 3-4 times per week
  - 5 sultanas/sunflower seeds - 2-3 times per week
  - Other fruit and vegetables in season

### 6.1.3 Brush-tailed bettong

- Ad lib
  - water
  - lucerne hay
- Daily diet (per animal)
  - ¼ cup apple
  - 1 piece banana
  - 1 piece carrot
  - ½ cup kangaroo cubes
  - 2 pea pods
  - 2 string beans
  - ¼ slice sweet corn
  - ¼ cup sweet potato

### 6.1.4 Tree-kangaroos

- Ad lib
  - water
- Daily diet (per animal)
  - 90% of the diet should be browse cut fresh daily and at least 3 branches per pot per day placed in the tree-kangaroos' favourite spots in enclosure (at least 3–4). Lumholtz's tree-kangaroos should be provided with the leaves of as many species as possible. Tree Roo Rescue and Conservation Centre move older branches to different pots to encourage the tree-kangaroos to find them. Availability of different species of browse will vary depending on the location and season, so flexibility and diversity is important (Table 10.10).

Other browse species that have been provided to the New Guinean Goodfellows tree-kangaroos include poplar (*Populus* spp.), willow (*Salix* spp.), maple (*Acer* spp.) and birch (*Betula* sp.) (Edwards and Ward 2001).

The presentation of browse appears affect the amount consumed by tree-kangaroos (Mullet *et al.* 1990). One study found that when browse was placed on the ground, only 50% was consumed, but when it was 'planted' in the sand, they ate 65–75% and when it was hung in the branches of trees, 90–100% was consumed. Placing browse on the ground can also contaminate it, especially if the ground is muddy. The browse will last longer and be more palatable if placed in PVC pots filled with water and placed at the tree-kangaroos favourite resting places in the enclosure, such as at the side of the hutch, and fresh browse supplied daily. At least three pots should be provided per animal (K. Coombes pers. obs.). To ensure

**Table 10.10.** Browse species for tree-kangaroos.

**Preferred, ***most preferred. Derived from K. Coombes (unpubl. data).

| Common name | Scientific name |
|---|---|
| Red ash | *Alphitonia excelsa* |
| White ash/sarsparilla | *Alphitonia petriei* ** |
| Birds nest fern | *Asplenium australieum* |
| Flame tree | *Brachychiton acerifolius* |
| Swamp she-oak | *Casuarina cooperi* |
| Northern olive | *Chionanthus ramiflora* |
| Five leaf grape vine | *Cissus hypoglauca* *** |
| Three-veined laurel | *Crytocarya triplinervis* |
| Beach tuckerroo | *Cupaniopsis anacardioides* |
| Quandong | *Elaeocarpus grandis* |
| Loquat | *Eriobotrya japonica* |
| Ribbonwood | *Euroshinus facata* *** |
| Native cherry | *Exocarpos latifolius* |
| Weeping fig | *Ficus benjamina* |
| Glycine | *Glycine* spp. *** |
| Grevillea | *Grevillea* spp. |
| Silky tamarin | *Guioa lasioneura* ** |
| Hibiscus | *Hibiscus* spp. |
| Sweet potato vines | *Ipomoea batatas* *** |
| Large leaved privet | *Ligustrum lucidum* *** |
| Small leaved privet | *Ligustrum sinense* *** |
| Bollywood | *Litsea leefeana* ** |
| Mango | *Magifera indica* |
| Woody pear fruit | *Mischocarpus lachnocarpus* ** |
| Mulberry | *Morus rubra* |
| Banana | *Musa acuminata* |
| Native nutmeg | *Myristica insipida* ** |
| Leichhardt tree | *Nauclea orientalis* |
| Grey bollywood | *Neolistsea dealbata* ** |
| Glycine | *Neonotonia wightii* |
| Fishbone fern | *Nephrolepis cordifolia* |
| Celerywood | *Polyscias elegans* *** |
| Plum | *Prunus* spp. |
| Umbrella tree | *Schefflera actinophylla* (leaves and fruit/flowers, they particularly like the new leaves and shoots and will eat the top of the stem) *** |
| Brazilian pepper tree | *Schinus terebinthifolia* |
| Lilly pilly | *Syzgium* spp. |
| Tree lucerne | *Tagasaste* spp. |

that the animals do not prematurely drop the browse, plants may be secured in the canister by large bungee cords or the equivalent (Blessington and Steenberg 2007).

Food bowls should be placed on a platform. Fallen food is rarely consumed by tree-kangaroos and should be removed. Cover the feed station like a hutch to prevent birds form eating food.

In addition to the browse, the captive diet can include the following provided daily in the morning and evening:

- 1 cup chick peas (soaked for 6 h then rinsed well)
- 12 slices sweet potato (sliced into julienne strips ~5 mm thick)
- 12 slices carrot (sliced into julienne strips (easier for them to pick up and eat) 5-mm thick)
- 2 slices sweet corn (25-mm slices)
- 6 slices banana (not too ripe) - 5-mm slices
- 2 slices apple (thin wedges) - 10-mm slices
- green vegetables ad lib (individual animals will like different greens but can include celery and celery tops, carrot tops, pak choy, bok choy, baby spinach, snow peas and beans)
- 1 cup Macropod pellets Or Herbivore Pellets Or Horse Pellets in separate bowl to eat ad lib
- lucerne hay
- no bread

- Supplement
  - Salt blocks or basalt or clay soil should also be supplied in a bowl, as they will occasionally eat soil, potentially for the associated bacteria and possibly to reduce toxins in gut from the leaves (George 1982; Mullet *et al.* 1990; Blessington and Steenberg 2007; Lessiak 2014, S. Jackson pers. obs.; K. Coombes pers. obs.).

### 6.1.5 Kangaroos and wallabies

- Ad lib
  - water
  - lucerne/grass hay
  - pasture/green grass
- Daily diet (per animal)
  - 350 g pasture replacement pellets
  - 60 g maize
  - 1 carrot piece
  - 1 browse: eucalypt/acacia/native mint bush branches every 1–2 days
- Supplement
  - None

### 6.1.6 Rock-wallabies

- Ad lib
  - water
  - lucerne hay
  - pasture/green grass
  - pasture replacement pellets
- Daily diet (per animal)
  - 100 g maize
  - ¾ apple (medium)
  - ¼ carrot (medium)
- Supplement
  - eucalypt branches as available
  - acacia branches as available
  - melaleuca branches as available
  - occasionally 2.5-cm corn on cob
  - occasionally 2.5-cm banana in skin

### 6.1.7 Pademelons and the quokka

- Ad lib
  - Water
  - Gum branches
- Daily diet (per animal)
  - 400 g pasture replacement pellets
  - ½ carrot - chopped
  - ½ apple - chopped
  - 40 g maize
- Supplement
  - eucalypt and acacia branches weekly

Other food items that are frequently given include 'kangaroo cubes' (e.g. speciality feeds, lucerne chaff and horse stud mix).

## 6.2 Supplements

Vitamin E and selenium (in the form of wheat germ or included in kangaroo pellets) are often provided for some species, such as tree-kangaroos and rock-wallabies, as they appear to help reduce the incidence of myopathy. However, the association of vitamin E and myopathy is not well characterised and there is some doubt over the effectiveness of selenium, so further research is needed. Salt blocks may be provided as a source of minerals to larger macropods.

## 6.3 Presentation of food

Food should ideally be provided off the ground, particularly in feeding sheds or in open areas to reduce the risk of contamination. Hay racks can be used to provide food for larger species and can be wall-mounted or double-sided and free-standing. They are best used for holding bales of lucerne. Although hay nets can be used, particularly for smaller species, they are generally more difficult to refill and can result in animals becoming entangled.

Food for the potoroids and most small macropods should be either spread out throughout the enclosure or put in a separate dish for each animal and placed in different parts of the enclosure to minimise fighting and competition for food. It will also reduce the incidence of subordinate individuals being driven away from food sources (Packer 1969; Guiler 1971).

Depending on the species, lucerne chaff, oats, carrots and other food items can be provided in troughs. Water and pellets for tree-kangaroos can be placed in bowls on benches or in the hutch so that it is off the ground. Alternatively, pellet hoppers can be used. The area under the hayracks should be thoroughly raked or hosed every day. Bowls of dirt and salt licks should also be undercover to prevent them getting wet. Some carers provide termite mounds for dirt (K. Coombes pers obs.). Similarly, troughs should be thoroughly cleaned and scrubbed every day. If possible, prevent animals from sitting in the food trays provided, as they may urinate and defecate in them as they eat (S. Jackson pers. obs.).

Water can be provided in several ways, including flow-through or static ponds, troughs or self-filling watering points.

## 7. HANDLING AND TRANSPORT

### 7.1 Timing of capture and handling

The best time to capture macropods for examination is in the morning when the temperature is cooler, particularly important during warmer weather. All macropod captures should be well planned in advance to optimise welfare and minimise risks to the animal and personnel and should be performed just before transport, moving to another enclosure or veterinary examination. Macropods can be stressed easily by capture events and will typically respond to efforts to capture them with flight behaviour that can be frantic and result in significant injuries from colliding with objects or entanglement in fences. Over-exertion, over-heating, and/or excessive stress for sustained periods can cause potentially fatal capture myopathy (muscle breakdown). Some species are more prone to this than others; for example, rock-wallabies, tree-kangaroos and red kangaroos are highly susceptible, whereas Tammar wallabies rarely suffer from capture myopathy. Once captured, inappropriate handling can result in injuries, especially to the hindlimbs.

Female macropods may evict pouch young in response to stressors such as a capture attempt. Strategies should be used to reduce the risk of this and plans for appropriately managing evicted young should be developed.

Once caught macropods will readily kick in defence and can bite. Many species have a large claw on the main fourth digit of the hindlimb that can cause significant injury. Tree-kangaroos have comparatively strong forelimbs with sharp claws.

There are a range of techniques to facilitate the capture and restraint of macropods: behavioural restraint strategies (training and conditioning), physical restraint methods, chemical restraint agents (including sedatives, and anaesthetics, see section 8.4) or a combination thereof. The best technique for each situation should be evaluated on a case-by-case basis to optimise the welfare of the animal and the safety of personnel. A good understanding of species behaviour and the likely response of the macropod to the movements and actions of personnel is important for successful planning and execution of capture attempts. Consideration should be given to: the type of procedure to be carried out; the species, size and demeanour of the animal; the characteristics of the enclosure and associated facilities; equipment available; and the experience of personnel.

Capture procedures should be completed as quickly and efficiently as possible. Measures to cool the animal (such as air-conditioned areas, ice packs, wet towels) should be available to manage hyperthermia if it occurs. Capture teams should ideally include experienced personnel. The plan should be talked through before proceeding to ensure everyone understands their role. Teams should be prepared to withdraw from a capture attempt if it does not go to plan or if the animal becomes noticeably distressed.

Ideally, training/conditioning techniques should be used where possible to facilitate capture and restraint procedures with the cooperation of the animal, thus minimising stress and the risk of injury, and optimising welfare outcomes.

### 7.2 Capture equipment

#### 7.2.1 Capture bags

Selecting the appropriate catching bags for handling macropods is very important. Bags should be:

- checked before every capture to ensure it is safe to use
- the appropriate size for the animal and have a wide opening so it is easy to quickly place the animal inside it and to allow easy access for examination

- made of strong fabric such as heavy cotton, calico, hessian or woolsacks (large species). Large, strongly built bags also help reduce the likelihood of the large 4th toe of the macropod's foot tearing a hole in the bag to ensure the material allows airflow. Capture bags used for tree-kangaroos include large laundry bags, as they have a drawstring tie on the opening, and handmade bags made of 90% shade (K. Coombes pers. obs.)
- ideally made of non-shedding fabrics (hessian (burlap) sacs shed small fibres that can be inhaled)
- made with well-finished seams without loose threads on the inside of the bag, which can cause injury if they wrap around claws or other body parts, especially with smaller macropods
- opaque if possible so the animal cannot see through and is as dark as possible, so they will be calmer and settle faster
- 'breathable' to maintain ventilation and not compromise respiration or thermoregulation. Consideration should be given to the ambient temperature when macropods are held in bags, along with avoiding direct sun where possible, to minimise the risk of hyperthermia.

Two openings (using heavy-duty zippers) can assist with anaesthetic delivery and pouch checking. Ensure bags are fit for purpose before first use.

Some bag designs incorporate short zippers along selected seams to allow flexibility of access to different body parts. Ensure all zippers are closed before placing a macropod in such a bag so as not to provide a potential escape route. The bag also needs to be able to be tied securely with rope to prevent escape and reduce the light penetrating the bag. The tie can be used to suspend the bag for temporary holding, transport or weighing.

### 7.2.2. Nets

Long-handled hoop-nets are typically used to catch macropods. Small species can be caught with a 60–80-cm deep hoop bag with a 60–70 cm opening. Medium-sized macropods (15–40 kg) can be caught in a similar fashion to smaller ones, but with larger hoop bags that have a diameter of 70–80 cm and are 80–100 cm deep.

Open-mesh nets may be advantageous in some situations as the animal may not perceive them as a blind sac and hence be less likely to try and avoid them. However, attention should be paid to the mesh diameter relative to the size of the macropod to avoid injury from limbs getting caught through the net.

Bag nets may be more actively avoided, making capture a little more challenging, but have the advantage that the animal is then contained within a bag without direct visual contact with personnel, thus reducing stress.

### 7.2.3 Personal protective equipment

The capture of tree-kangaroos may be facilitated with the help of welders' gloves that reach to at least the elbow, due to their biting nature (K. Coombes pers. obs.).

## 7.3 Capture and physical restraint

### 7.3.1. Herding macropods to move to adjacent enclosures

Macropods may be moved between adjacent enclosures efficiently using low-stress herding in the desired direction. Whether in a large enclosure or small enclosed shed, it requires a good understanding of macropod behaviour. The terrestrial macropods have strong powerful hindlimbs, capable of fast explosive movements, so great care need to be taken not to trigger erratic behaviour. Unconditioned macropods may perceive herding, or capture attempts, as a threat and respond with flight behaviour, which can have adverse welfare consequences. Therefore, the movements and actions of personnel involved will have a direct influence on the degree of stress experienced by the animals, the risk of physical injury and the success of the manoeuvre.

Getting a macropod to move where needed requires a similar approach whether it involves one person or a large team. By moving towards the animal slowly, encroaching on its flight distance, the animal will be motivated to move away. Wait for the animal to be looking in the direction that you wish it to travel in before physically moving towards the animal again. This response can be utilised to move the animal in the desired direction. One or two people can effectively move animals in a small area, whereas a team of people, appropriately spaced, moving in a line perpendicular to the required direction of movement will be needed in a larger enclosure. It is recommended that one person be responsible for directing the movements of the team according to the behavioural responses of the animal. This will maintain coordination of the team and the likelihood of a successful outcome.

Stress and the resultant sudden frantic flight behaviour can be minimised with calm, slow movements and keeping noise to a minimum. Close observation of behaviour will allow for an appropriate balance between getting the animal(s) to move along steadily and not approaching too close or too quickly and eliciting a frantic flight response. The placement of the team can be adjusted to keep the animal moving in the desired direction, allowing the animal to see that direction as the most likely escape route while closing off alternate escape options in the opposite direction or between members of the capture team. By timing forward movement of the team when the macropod is looking in the desired direction, the animal will most likely move in that direction.

### 7.3.2 Herding macropods for capture

When physical restraint by manual handling is required, consideration of the risk of injury to the animal and the handler is needed. To reduce stress, the period of manual restraint should be kept as short as possible to minimise the risk of capture myopathy.

Whenever possible, positive reinforcement for training and conditioning, and allowing choice and control, are important for optimising animal welfare, regardless of species. Such approaches can readily be applied to macropods to reduce or eliminate stressors associated with capture and restraint and should be adopted where possible and appropriate. Using positive reinforcement, in free-contact or protected-contact situations, macropods can be conditioned readily to the presence of animal care staff, to being touched and handled and to entering small pens, transport boxes or catching bags without the need for physical restraint. Such approaches can be used to perform certain procedures that would otherwise require capture and physical or chemical restraint, including weighing, pouch checking, collection of biological samples and basic clinical examination. Training and conditioning requires adequate staff time and an appropriate training plan to achieve and maintain appropriate behavioural responses.

For a range of reasons it may not be possible or desirable to use training techniques to achieve required outcomes (e.g. the survival success of reintroduced macropods may be reduced by conditioning to human presence in captive settings). In such cases, capture procedures often necessitate a greater degree of negative experience by the animal. It should be the goal of the capture team, throughout the planning and execution of the procedure, to minimise the stress required to achieve the required outcome.

Small macropods (e.g. < ~15 kg), may be captured by slowly herding the animal quietly along a fenceline towards handlers waiting with a hand-held hoop-net or ready to catch the tail of the animal as it hops past. Larger macropods that are being targeted for capture are typically herded along the fenceline by several people with nets (Fig. 10.10). The number of personnel required to capture small or large macropods is influenced by the number of animals being captured and the size and nature of the space the capture is occurring in.

During the capture process it is important that the animal anticipates an escape route ahead. Going too early with the net or attempted tail restraint will block that perceived escaped route and likely cause the animal to change direction and avoid capture. By waiting until the macropod is moving past before going in with the net or tail grab, the animal remains committed to the perceived escape route and be less likely to be able to avoid a well-executed capture. When the animal is just about to hop past the net or bag is quickly swung in front of the animal before it can change direction. Macropods will often make use of narrow pathways, such as behind feed sheds, so these make good points to herd the animal towards, with someone waiting at the other end. If tussocks and other cover are available, the animal may use these for cover (Booth 1994). Lots of vegetation can create various alternate pathways within the enclosure and may hinder the ability to of the catchers to apply appropriate herding pressure and reduce the efficiency of capture.

Giving consideration to the biology and associated behaviour of the species to be captured will increase the likelihood of a successful outcome. For example, Tammar wallabies typically stay close the ground, will generally follow along fencelines and may seek to find an escape route underneath a fenceline. They tend not to leap vertically, so the handler can keep the net close to the ground to facilitate capture. Rock-wallabies on the other hand can be less predictable in following a fenceline and tend to seek to gain elevation to find safety. They can leap high into the air, so handlers need to be prepared when netting rock-wallabies.

Once an animal is caught, the entrance of the hoop is quickly lifted up in the air and the bag twisted, if possible, to prevent the animal from escaping. Alternatively, the opening of the hoop bag is placed on the ground to prevent escape. This can be secured further by standing on the hoop of the net.

Smaller macropods can be transferred from the net or upturned crate to a restraint bag by firmly grasping the base of the animal's tail (not mid-tail or anywhere near the tip) (Fig. 10.11). Holding the base of the tail is preferred for manual restraint as it helps protect the handler

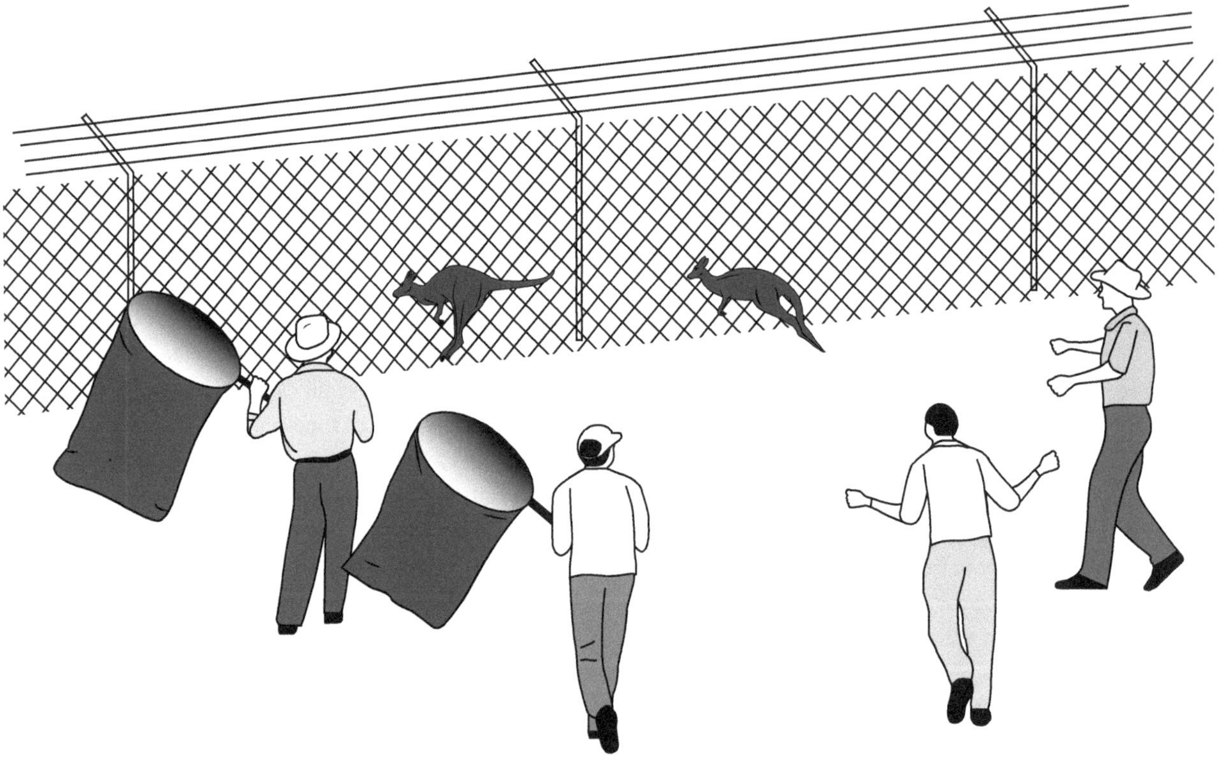

**Fig. 10.10.** Technique used to catch macropods along a fenceline. Note: the catcher with nets and the herders with hands stretched outward (but not waving about) may help create visual barrier and assist with the quiet movement of the animals.

**Fig. 10.11. Technique for handling macropods and transferring them to a cloth bag.**

from being kicked or bitten and reduces the risk of injury to the animal. This restraint should only be relied on for very short periods and have a goal that can be achieved promptly, such as moving the animal into a crate or bag or administering an injection.

A macropod restrained by the tail base can be held up off the ground and moved head-first into the restraint bag that is held open and swiftly lifted up over the macropod by a second operator. Alternatively, the animal may be allowed to move along the ground using its forelimbs, directing it via the tail restraint, into the open bag. For small to medium-sized macropods in a crate, it is possible to reach in, place the bag opening over the animal and 'scoop' them up into the bag, eliminating the need for tail restraint.

Macropods may also be captured by the base of the tail from transport crates, from within a small enclosure or shed, as it is directed through a narrowed section of the enclosure or strategically flushed out from a hiding location. Macropods conditioned to human proximity may be captured by tail base restraint directly from the enclosure if approached without arousing suspicion.

When restrained by the tail base, macropods will typically endeavour to escape by hopping away, kicking out and/or (especially smaller species such as Tammar wallabies and bettongs when suspended from the tail) twisting their body around the axis of the tail. Once caught, the hindlimbs should not be used as a primary mode of restraint; severe injuries such as limb fractures and dislocations can be sustained from the forces created as the macropod kicks out from the point of restraint.

While one person holds the animal's tail, a second person, who is holding a bag of similar diameter to the hoop bag, opens the bag to allow the macropod to be transferred. If the animal is spinning around too much to be bagged when held by the tail, this spinning action can be stopped by moving the animal in a circular motion (like stirring a pot) or allowing its front feet to touch the ground (D. Taggart pers. obs.).

Once the animal is still, it can be bagged. To do this, the still animal is quickly lifted off the ground by the tail, with its head facing towards the bag holder and its hindfeet back towards the handler. The bag holder then scoops the bag up over the head and body of the animal in one quick motion (D. Taggart pers. obs.). The bag is securely tied, making sure the tail does not protrude. The bag can then be placed in a shady position in the middle of the enclosure if there is enough room or, preferably, securely hung from a fence in a quiet area in the shade if more animals are to be caught.

Large animals (e.g. > 40 kg) should ideally be caught by darting them with an anaesthetic agent to induce sedation, unless they have a quiet temperament. The dart should be directed towards the rump due to the large muscle mass, which helps cushion it. Macropods have relatively thin skin (compared with similar-sized eutherians), so darts should not be fired at high speed; rather, they should be 'lobbed' with a high arc to prevent trauma. Once the macropod has been darted, it should be monitored to ensure that it does not injure itself while the anaesthetic takes effect. Once the drug has taken effect, the animal should be transferred to a sturdy capture bag, placed in the shade and monitored to ensure it does not become too hot or cold. Once ready for release, the animal should be placed in the shade and away from any obstacles so that it does not overheat or injure itself while regaining consciousness and co-ordination (as it might hop erratically before regaining full consciousness).

If an animal is to be hand-caught, it can be manoeuvred into a corner of the enclosure by 3–4 people and slowly enclosing it. Once the animal attempts to pass the catchers it should be caught in a large hooped net or, if possible, the tail can be grabbed close to the base and the animal quickly transferred to a catching bag being held by a second person.

To facilitate the capture of both small and, particularly, large macropods, additions can be added to the fenceline to form funnels or other confined spaces where animals can be cornered for capture. When catching animals within large enclosures, a funnel can be set up along one fence to herd the target animal towards. These funnels may comprise shade cloth or hessian mesh ~2 m high and supported on the outside of the funnel by star pickets. The cloth or mesh should be wide enough so there is 1 m in excess at the bottom, which is laid towards the inside of the funnel to reduce the chance of the animal going under the netting. The animals can then be herded into the funnel and, as they go through and out the other side, the individual that is required can be caught up with a net or by the base of the tail. Alternatively, the funnel apex may not have a gap, so the animals will try and make their way back out of the funnel and can be netted one at a time along the fence. To minimise the chance of any problems (including nose injuries from the fence) with multiple animals running past the captors at once, care should be taken not to put too much pressure on the animals within the funnel, just enough to get one animal to try and depart the funnel trap at a time. This can be done by adjusting how close the captor approaches the base of the funnel apex where the animals will be gathered.

Care should be taken to not to hem the animals in too much and, depending on the weather and the animal's temperament, no more than 3–4 attempts should be made at catching an individual to avoid the serious risk of inducing capture myopathy, particularly in the larger species (Booth 1994). Signs of distress in macropods include panting, drooling or licking of the forearms. If any of these occur the catch attempt should be abandoned and an attempt made to catch the animal again later (preferably the next day) or resort to darting it the following day.

Tree-kangaroos and rock-wallabies can often be grabbed by the tail and pulled off their resting platforms or branches and then bagged. Tree-kangaroos can sometimes grab hold of the branch, so it best if they are captured on the ground. Never place your hand under a tree-kangaroo's chest to support it as it will grab the hand with its arms and bite and kick with hindlegs (K. Coombes pers. obs.).

Where possible, do not catch female macropods with known young as they may relax their pouch muscles and drop any large young that are detachable from the teat, from the pouch. When females that are expected, or known to have large pouch young, have to be caught it is very important to maintain good visual contact with them to see if they drop their joey. If a joey is dropped, it should be quickly picked up and put it in a bag and/or under the catcher's shirt or jumper to keep it warm. Once the mother has been caught, the joey can be placed back in the pouch and the pouch should be taped closed to keep the young inside. Once inside the pouch larger joeys will usually re-attach to the teat. The pouch tape will detach over the next day or so, or the female will actively remove it when she settles down.

Once a large macropod is caught it may be necessary for a veterinarian to give diazepam (Valium®) to sedate it to reduce the chance of capture myopathy (Spielman 1994). Great care needs to be taken if this is used as animals cannot control their body temperature effectively after it has been administered (for at least several hours), so there is a risk of overheating or hypothermia unless the animal is being maintained in optimal temperatures (D. Taggart pers. obs.). Kangaroos are generally examined while inside a catching bag, unless they are anaesthetised. They can also be briefly examined outside a bag by grasping the tail at the base and bringing the animal up so the kangaroo's back is resting on the holder's chest.

The tail base of large kangaroos has a large diameter and may require two hands to achieve adequate restraint. Given the strength of these animals, it may be difficult and potentially unsafe for some personnel to attempt. Skilled operators with sufficient strength may be able to then hold a kangaroo up against their body, typically in a squatting position and resting their back against a solid structure (Fig. 10.12). The handler should ensure the hindlegs are always directed away from them. The kangaroo's body is supported with one hand, being careful to avoid the strike zone of the hindlimbs.

**Fig. 10.12.** Manual restraint technique for large macropods.

### 7.3.3 Enclosure design and traps

In large enclosures, use of manual equipment and capture techniques will be difficult, if not impossible, to achieve without considered enclosure design features. The incorporation of one or more such features into macropod enclosures will increase the efficiency of capture procedures and minimise the associated stress and physical exertion of the macropod.

A range of trap designs have been reported for the capture of macropods. Those described for wild populations could be applied in captive settings, especially for more extensively managed animals. Traps may be set and used passively, waiting for the animal to enter the trap voluntarily for a high-value food item, or placed strategically (e.g. along a fenceline or through a hole in the fence at the end of a funnel (Vernes 1993)) to allow animals to be gently herded towards the trap. Leaving traps in a locked-open position permanently, or at least in the lead up to trapping, and placing feed in the traps regularly will condition the animals to readily enter the traps. Whatever system is used, traps should be monitored during operation, or checked at an appropriate frequency, to allow the animals to be removed from the traps promptly, minimising the time spent in a potentially stressful situation or in weather that is too hot or cold. Traps must also be checked frequently to avoid unnecessary injuries.

Treadle-activated traps are useful for small to medium-sized macropods. Cage traps are often used to trap smaller wild macropods (e.g. bettongs). The traps should be positioned carefully to avoid temperature extremes and be covered to provide shelter and to avoid the animal feeling exposed. Macropods may become stressed in a trap and launch themselves explosively in an effort to escape. This can move the cage around in an uncontrolled way, so tether the trap to prevent injuries to the animal. Padding the sides of cage traps or replacing the mesh with a fabric such as shade cloth (Kinnear *et al.* 1988b, 2014) can reduce the risk of injury. A dark material bag suspended on the frame of a treadle-style trap has been used for the capture of swamp wallabies and rock-wallabies. Such traps greatly reduce the risk of injury compared with metal cage traps. The bag can be unhooked from the frame and used as the restraint bag to avoid the need to handle the macropod while removing it from the trap.

Various enclosure and trap designs have been used that may be useful in some circumstances, with some being better suited to smaller or larger species.

- Drive fence for small macropods - originally designed for the captured of small forest-dwelling macropods, including red-legged pademelons. It uses a combination of treadle-activated wire cage traps and a squared U-shaped 150-m long drive fence. The macropods are herded towards the fence which they move along towards traps that are placed in holes cut into the fence; animals attempting to escape through the holes enter the traps and activate the treadle thereby closing the trap (Vernes 1993).
- Darting - an alternative to hand-catching macropods >5 kg, especially larger species. Uses a dart with an immobilising drug such as tiletamine/zolazepam (Zoletil®) using a blowpipe or compressed carbon dioxide or explosive powder charge (which requires a gun licence) (Holz 1992; Coulson 1996). When capturing animal that are habituated to humans King *et al.* (2011) utilised a 'jabstick' that was attached to an extendable aluminium pole. Using this technique, they were able to approach eastern grey kangaroos on foot and inject the hindlimb muscle mass with Zoletil® using the pole syringe that extended up to 4.85 m. Further examination of darting is given by Vogelnest and Portas (2008) and Vogelnest (2025).
- Draw-string trap - originally designed for the capture of larger macropods such as eastern grey kangaroos. Developed to capture macropods moving under fences, using soft netting attached to a steel weldmesh floor and suspended from a weldmesh frame. Draw-string closures at each end of the netting allow selective capture of kangaroos by two operators (Coulson 1996; Coulson *et al.* 2003).
- Drop slide doors at feeding sheds - originally designed to trap animals conditioned to enter feeding sheds. By placing a drop slide door at the entrance to the feed shed and observing which animal enters the feeding shed, the slide door can be remotely dropped and the animal either readily netted inside the shed or netted as it leaves the shed when the slide is raised. This reduces capture distress to an absolute minimum (D. Taggart pers. obs.). Ensure feeding shed is clear of any furnishings and feeding equipment before catching.
- Circular pen trap - circular, small-mesh wire pen traps (~5-6 m in diameter) can be used very effectively to capture semi-wild or wild wallabies. The fence (~1.2 m high depending upon the species) is constructed with a skirt (~0.4 m) on the inside of

the trap to prevent animals pushing under the fence and is supported with droppers on the outside of the trap. Four or five, very light, one-way gates are fitted to the fence and only open to the inside of the trap. Gates are wired open for 7–10 days and the wallabies, potoroos or bandicoots free fed inside the pen trap until it is used by the animals on a regular basis. The one-way gates are then lowered and the animals catch themselves within the pen trap. Staff attend every 1–2 h and capture the wallabies caught within the trap. This type of trap is also very effective in large enclosures and paddocks (D. Taggart pers. obs.). Take care if plastic ear tags have been used because wallabies will 'throw themselves' at a fence when under pressure in an attempt to escape and ear tags can easily be caught in the mesh.

- Enclosure design for netting small wallabies – the most effective mid-sized enclosure design for hand-netting small wallabies is a triangular enclosure with rounded corners. Small groups of wallabies can be pushed towards a corner and then hand-netted with hoop nets on long metal poles as the animals run along the fences past the captors in a bid to leave through a corner of the pen. This type of enclosure is very useful if large numbers of animals need to be caught regularly for pouch checking or other examinations.
- Drop nets – originally designed for the capture of small species such as Tammar wallabies. They utilise a monofilament nylon drop net 60-m long with 117-mm mesh size and 25 m deep, which is set up by tying the upper edge at a height of 2 m at 10-m intervals to tree trunks. Wing fences 50-m long and 2-m high are then attached to funnel the animals towards the drop net (Lentle *et al.* 1997).
- Funnel fencing – originally designed for the capture of larger macropods such as red kangaroos and western grey kangaroos. Two large hessian wings are built (a natural barrier such as an escarpment or lake can be used as a barrier) that are ~50–100 m long and 3 m high to form a funnel. At the apex of the funnel, a 15-$m^2$ yard made of hessian is built to herd the animals into so they can be caught (Keep and Fox 1971).
- Trapping yard with metal funnel trap – originally designed for the capture of larger macropods. It utilises a metal funnel made of parallel rods ~2-m long placed in a hole along a solid fence so that an animal can enter through the funnel into a trapping yard but cannot go back out the same way. These traps were originally used to trap wild kangaroos into enclosures that had water (Gooding and Harrison 1955). Therefore, they could be used to attract animals for capture in large free-range enclosures into smaller enclosures using food and water.
- Baited trap – involves setting up a macropod trap in areas that are most utilised by the target animals. Cage traps work particularly well in very large enclosures and have been used successfully for potoroos, rock-wallabies, swamp wallabies, red-necked wallabies, Tammar wallabies, pademelons and even large macropods such as euros, wallaroos, red kangaroos and western grey kangaroos (Coulson 1996; Di Stefano *et al.* 2005; Department of Parks and Wildlife 2013). An advanced cage has been developed to capture swamp wallabies, which may be useful for other similar-sized species. This involves using strong shade cloth or a modified wool pack, instead of steel mesh walls, to reduce the risk of injury to the trapped wallabies. The trigger mechanism is released when a string is pulled as a result of weight being applied to the shade-cloth floor (Pollock and Montague 1991). Modified treadle traps with padding or shade cloth incorporated into the sides have also been used to provide protection from injury and shelter (D. Taggart pers. obs.).
- Bromilow trap – originally designed for rock-wallabies to reduce the incidence of injury during capture. The trap walls are made of netting or shade cloth that is secured to and suspended from a metal frame. The mesh allows the captured animal to jump about without being injured (Kinnear *et al.* 1988b, 2014).
- Small to medium-sized heavy-duty wire mesh traps –lined with carpet for bettongs, placed in a sheltered area, covered with cloth to darken the space. Do not place in ant-infested areas.
- Drugged bait – baits such as α-chloralose are added to water, chaff or grain. However, the dose is very difficult to control as some animals are virtually unaffected while others can be hand-caught (Arnold *et al.* 1986).
- Stunning – originally designed for the captured of larger macropods such as red kangaroos and western grey kangaroos. It involves shining a (100 W) beam of light into the eyes of a macropod at night usually from the back of a vehicle. The kangaroo is then 'held' in the light while a shot from a 0.22 calibre or 22 calibre magnum rimfire rifle fitted with a

**Table 10.11.** Characteristics and limitations of techniques used to capture macropods.

Derived from Coulson (1996) and Taggart (unpubl. data).

| Method | Selective | Trauma | Myopathy | Personnel required | Limitations |
|---|---|---|---|---|---|
| Hand netting | Yes | Usually minor | Possibly | 1+ | Best suited for smaller species |
| Drive fence | Yes? | Moderate | Possibly | Several | Setting it up and animals need to be herded |
| Darting | Yes | Moderate | Possibly | 1 | Darts or darted animals lost; limited range |
| Draw-string trap | Yes | Minor | No | 1 | Limited to fences with regularly used runways |
| Drop slide doors at feeding sheds | Yes | No | No | 1–2 | Must have animals accustomed to entering sheds to access food |
| Circular pen trap | No | No | No | 2–3 | Animals have to be trained to come into pen traps for food before trapping commences (5 days). Less effective if food is prevalent outside |
| Enclosure design for small wallabies | Yes? | No – minor | Possibly | Several | Animals need to be herded towards corners. Experienced netting and bagging personnel required |
| Drop nets | Yes? | Minor | No | Several | Useful for smaller macropods |
| Funnel fencing | Yes | Moderate | Yes | Several | |
| Trapping yard with metal funnel trap | No | Minor | Possibly | 0 | Animals still need to be caught by hand once inside the smaller trap |
| Baited trap | No | Minor | Yes | 1 | Few species are attracted to baits |
| Bromilow trap | Yes | Minor | No | 1 | Setting it up and waiting time for an animal to move through it |
| Small to medium-sized heavy-duty wire mesh traps | Yes | Usually minor | No | 1 | Limited to small to medium-sized species |
| Drugged bait | No | Usually minor | No | Several | Dose not controlled |
| Stunning | Yes | Usually minor | Possibly | Several | Limited effective range (< 70 m) |

telescopic site is fired between the ears ~10 cm above the top of the skull. Catchers then sprint alongside the margins of the light beam towards the kangaroo and rush it from the sides. Small macropods are caught by the tail, with larger ones caught by netting or a rugby tackle in which the catchers approach from behind (Robertson and Gepp 1982; D. Taggart pers. obs.). The animal is then bagged. Table 10.11 presents an outline of the limitations of the different techniques used to capture macropods.

## 7.4 Weighing and examination

Once a macropod has been caught and bagged various procedures can be performed. During these procedures the handler should remain calm, quiet and use slow, deliberate movements to manipulate the animal while keeping its eyes covered. Procedures may include measuring bodyweight (e.g. with a hanging scale), transportation, examination of select body parts such as the pouch, placement of identification such as ear tags, administration of treatments and placement of a face mask for delivery of gaseous anaesthesia. When placing a face mask, it is preferable to expose the nose of the animal through the bag or zipper opening; placing the mask over the nose through the bag risks compromised delivery of oxygen and anaesthetic gas through the bag material and inadvertent compression of the external nares (nostrils) restricting respiration.

Weighing is relatively easy using hanging scales for smaller species (generally < 15 kg) or by placing the bagged animal on a large walk-on scale for larger species. The pouch can be examined by one or two people for small species such as the potoroids and smaller wallabies or three people for larger species, where one person holds the chest, a second holds the legs and the third examines the pouch.

It is very important, once any macropod is caught and bagged, that its eyes are kept covered during the inspection of its body and pouch, in order to maintain the animal's calm demeanour, to minimise distress and reduce attempts to escape. Macropods can be examined in a bag by placing them on the ground on their side, legs facing away, then kneeling behind with one knee at the animal's shoulder and the other approximately at the hips. Put slight pressure through the bag onto the shoulder region with one hand and with the other hand, keeping the bag against the animal's body to limit light into the sack, open the mouth of the sack proximal to the pouch area. Keep the tail and hindlegs well within the sack while doing this. Next grab the shin of the upper hindleg and pull it back towards the animal's body, which exposes the scrotal area in males and the pouch area, and any pouch young within, in females, for examination. Examination can be done by a single handler if experienced or by a second person if not (D. Taggart pers. obs.). Hindfeet can be measured while the animal is in this position also. Once the scrotal/pouch area has been examined, the hind leg can be released and the tail removed for measurement or blood collection (from the lateral tail vein) and then returned to the sack.

To examine the head and ears, small to medium-sized wallabies can be made to stand within their sack, with their tail between their legs. Kneeling over the animal's body, corralling the animal between the thighs, position the sack opening adjacent to the animal's neck. Be careful not to let unnecessary light into the sack. Move the sack opening up the neck to expose the animal's ears, being careful to keep its eyes covered at all times. To measure or examine the macropod's head, an opaque piece of cloth material can be placed, and held, across the animal's eyes within the sack so it cannot see. The sack can then be removed from the head, exposing it for inspection and/r measurement. It is important to remember that the animal will remain calm if it cannot see and remains in the dark (D. Taggart pers. obs.).

Several techniques have been used to examine macropods within capture bags. Whichever approach is used, the macropod should feel well supported within the bag or it may kick out to stabilise itself or react to a perceived threat. The base of the tail is the primary point of restraint. The hindlimbs should always be contained within the bag to reduce the risk of injury to the handler. Care should be taken to prevent or reduce light entering the bag through openings. The animal's eyes should remain covered. The operator should be always be prepared for sudden movements in response to stimulation and be prepared to quickly adjust the restraint appropriately to prevent escape. Examples of examination techniques include:

- sitting and cradling a smaller to medium-sized macropod (except tree-kangaroos) on its back between the handler's legs. Maintaining gentle pressure over the bag while the required body part is exposed via an opening in the bag and securing the legs so they cannot kick out. This technique is useful for most procedures on small species of macropods, including pouch examinations (Fig. 10.13)
- placing a medium to larger macropod (except tree-kangaroos) in a bag on the ground on its side, with legs directed away from the handler, and kneeling behind it with one knee at the shoulder and the other approximately at the hips. The required body part can be carefully accessed via an opening in the bag (Fig. 10.14). Tree-kangaroos can be examined by the handler kneeling behind the bagged animal with its head pointed away and the opening of the bag to the top of its back for ease of injections. Measurements can be made carefully by keeping the bag over the animal and measuring from outside of the bag.
- holding the base of the tail through the bag, small to medium-sized macropods can be cradled in the crook of the elbow, supporting the animal against the handler's body (Fig. 10.15; D. McLelland pers. comm.). The other hand can be positioned over the front of the body to prevent the hindlimbs kicking up. This technique is useful for a second operator to access the head (e.g. for ear tagging or placement of an anaesthetic mask) or exposing the tail (e.g. for blood collection).

## 7.5 Release

Many macropods have a tendency, when they are released following capture, to immediately hop off quickly and erratically. It is therefore very important that as few people as possible are used to carry the bag

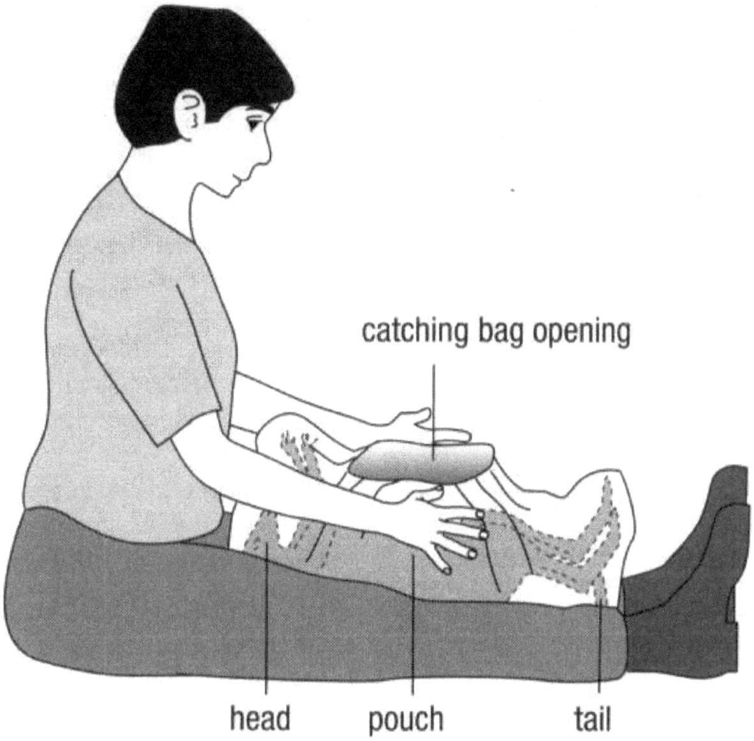


**Fig. 10.13.** Technique for one person to pouch check a small to medium-sized macropod (except tree-kangaroos). Figure derived from D. Taggart (pers. obs.).

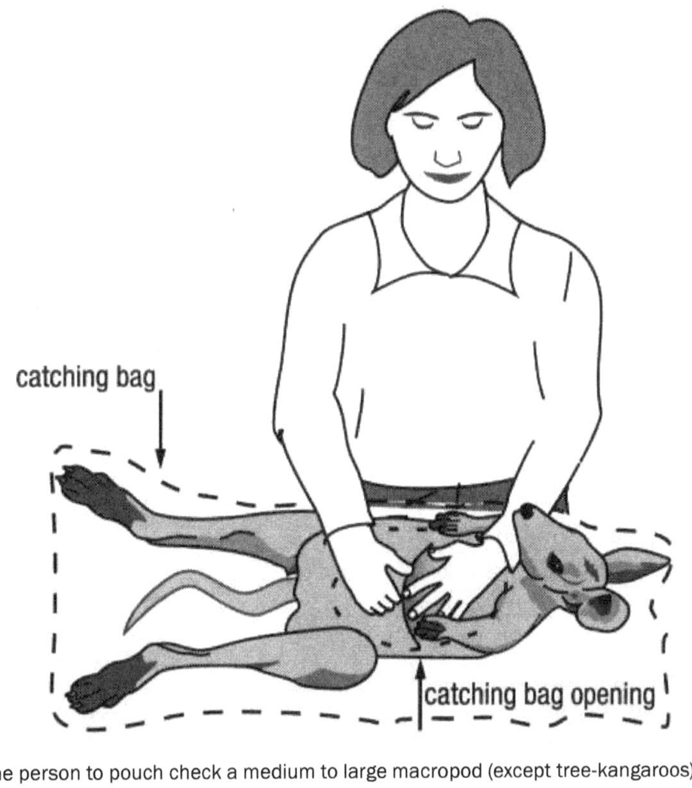


**Fig. 10.14.** Technique for one person to pouch check a medium to large macropod (except tree-kangaroos). Figure derived from D. Taggart (pers. obs.).

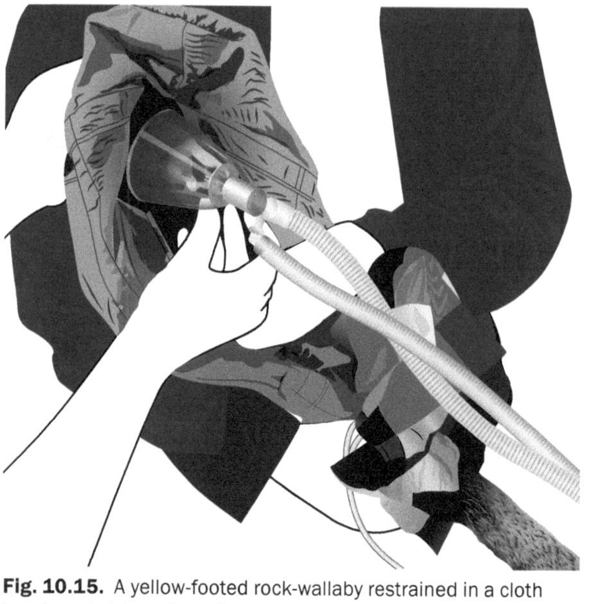

**Fig. 10.15.** A yellow-footed rock-wallaby restrained in a cloth bag for administration of gaseous anaesthesia via facemask. The tail base is held through the bag and the wallaby supported between the arm and body of the handler. The other hand is positioned in front of the hindlimbs to prevent the wallaby from kicking off the facemask. Figure derived from a photo provided by David McLelland.

into the enclosure, to minimise noise and visual stress. The release area should be in as open an area as possible and away from potential sources of collision such as large logs, fences, sheds or trees. This is often a very delicate time as very flighty macropods can break their necks from a collision. Once the animal is released, keep low, slowly retreat towards a fenceline and head towards the exit of the enclosure.

Some macropods, including the potoroids and rock-wallabies, have a tendency to throw their pouch young (especially those of an age that are able to be released from the teat) when released. It is very important that females with young are released with the minimum of stress. If concerns of possible pouch eviction are held the pouch may be taped up (e.g. with Elastoplast). Once the female is released and settles down, she will remove the tape and the young allowed to remain in the pouch. A second option, which is very good for animals that stress such as rock-wallabies, is to secure one of the bottom corners of the sack to the ground with a heavy object (e.g. rock). Then untie the sack and leave it open, so the animal can make its way out in its own, after staff have departed the enclosure. Upon leaving the enclosure, observe the animal to ensure it has exited the bag safely and has not become entangled upon release or any other issues.

## 7.6 Transport requirements

When transporting macropods various factors need to be considered including:

- not removing them from transport boxes or handling them in transit unless it is considered essential by the keeper, research scientist or the veterinary staff
- not exposing macropods to temperatures greater than 28°C or less than 10°C during the transfer
- minimising noise and time from crating to destination
- all available medical and species management documentation accompanying the animal being transferred.

Smaller macropods that are to be transferred over relatively short distances (e.g. 2–3 h) are easily transferred inside a suspended sack. Tammar wallabies have successfully been moved this way with no anaesthetic for up to 12 h at night in cooler weather with lots of ventilation (D. Taggart pers. obs.). Take care to ensure the animal is lying comfortably in the bag, that the bag does not come undone or swing around too much and that the animal does not overheat. Tree-kangaroos have been transported by hanging them in a bag (e.g. made of thick shade cloth) in a large crate, with the bottom of the bag tied to corners of crate to stop the bag swaying. The top of the bag is tied around a dowel across the top of crate. Padding is placed underneath the bag. The animal is lightly sedated with Valium for transport up to 12 h. During transport, animals should be kept in air-conditioning and away from pets while waiting for the flight, as they are required to be at airport 2 h beforehand (K. Coombes pers. obs).

Some authorities never anaesthetise macropods, but place them in dark crates in air-conditioned vehicles (D. Taggart *pers. obs.*). Numerous animals have been transferred successfully this way over 8–24-h periods (D. Taggart pers. obs.).

### 7.6.1 Transport box design

Macropod transport boxes should include:

- no internal frames or edges
- no slatted flooring, but rather a single solid piece

- a substrate that absorbs moisture, is not slippery and is comfortable for the animal to lie on
- padding with foam on the internal walls and ceiling to protect the animal's head, should it become agitated and jump during transfer
- enough space for the animal to turn around, to lie down and to stand comfortably with its paws on the ground. Dimensions should not exceed these criteria as animals may injure themselves by jumping if too much room is provided. When transporting macropods, such as tree-kangaroos, the transport box must be large enough to suspend bag so the animal can curl up and sit on its feet with the tail out front but still touch the floor. It is recommended to never transport tree-kangaroos free in a crate (K. Coombes pers. obs.)
- smooth internal sides of the walls
- appropriate ventilation.

Macropods must be transferred individually inside solid framed boxes that meet IATA (2022) standards when being transferred by plane.

### 7.6.2 Furnishings

No furnishings are required.

### 7.6.3 Water and food

Any container secured inside the box to provide food or water during transfer must be made of plastic with no sharp edges.

### 7.6.4 Animals per transport box

Only one animal should be placed inside a box. Females with pouch young should not be transferred unless the young is recently born and still attached to the teat.

### 7.6.5 Timing of transportation

Due to the potential for overheating, macropods should be transported overnight if on a long journey or in the early morning or evening for shorter journeys.

### 7.6.6 Release from the transport box

The same precautions when releasing an animal from a bag should be used for release from a box. It is often best to open the box, fully remove the door very slowly and immediately exit the enclosure from behind the box. This allows the animal to adjust to the new enclosure and to leave the box when it is ready. The box can then be removed in the next day or two once the animal(s) have had a chance to settle in and calm down.

## 8. HEALTH REQUIREMENTS

*David McLelland*

### 8.1 Daily health observations

Each macropod should be observed daily for signs of injury or illness. Cleaning and feeding activities provide good opportunities for close observation of individuals. Conditioning techniques can be utilised to facilitate close contact with macropods, optimising visual examination. More remote observation of individuals and of groups when they are undisturbed is also beneficial. This can be enhanced through the use of binoculars, camera traps or strategically placed video cameras. Keepers need to be familiar with the normal appearance and behaviour of individuals and with normal group dynamics.

Daily keeper checks should include assessment of the following:

- demeanour and general behaviour.
- mobility and utilisation of the enclosure, including any lameness, stiffness or reluctance to move.
- appetite.
- amount and consistency of faeces; appearance of any urine observed.
- coat and skin condition, including wet fur on the limbs which typically indicates stress-associated licking.
- discharges from the eyes, ears, nose, mouth, pouch or cloaca
- swellings, especially around the face, which may indicate lumpy jaw
- presence and development of pouch young - observation of a bulge and movement in the pouch; conditioning for pouch examination
- semen plugs in the enclosure, suggesting mating
- anything unusual in the appearance or behaviour of animals, or in the enclosure generally.

Any departures from normal should be noted and recorded to assist in identifying potential problems. Any observations of concern should be reported to a veterinarian. If an animal is found dead, the circumstances of when and where the animal was found, how it was positioned and anything unusual about the individual and its surroundings should be recorded; the body should be submitted for necropsy.

## 8.2 Bodyweight and body condition score

### 8.2.1 Bodyweight

Ideally, bodyweight should be recorded regularly and compared with previous weights. The frequency of weight measurements will depend on factors such as the logistics required to weigh particular species/individuals, and the presence of underlying health concerns. Some macropods can be trained to station on scales or be weighed followed physical or chemical capture/restraint (see section 7).

### 8.2.2 Body condition score

Various methods to body condition score (BCS) macropods have been proposed, though there is no accepted standardised approach. Several techniques require handling to allow palpation of the animal or the collection of measurements; some can be applied using observation alone. The validity of BCS methods may be influenced by operator skill/experience and animal-based factors such as species, age and sex. Whichever method is used, it is important to be consistent with the use of that method over time. This will allow for meaningful comparisons of data despite any inherent limitations of the method.

#### 8.2.2.1 Descriptive scale-based methods

These use one or more criteria to define ideal body condition and degrees of departure from that. Detailed descriptions of each category in the scale are required to minimise subjectivity and associated interoperator variability. Descriptive methods developed for use with free-ranging populations typically only document BCS that is less than or equal to ideal body condition (e.g. Johnston *et al.*, 1997). In captive settings, macropods can be subject to both reduced and excessive BCS.

Speare (1988) proposed a 6-point scale for orphaned juvenile macropods from 'very poor' to 'excellent', based on palpation of the tail base, lumbar spine and scapula, and outlined respective prognostic implications. Johnston *et al.* (1997) used a 5-point scale ('poorest' to 'best') based on muscle mass and fat stores over the ribs and lumbar spine to 'subjectively estimate' body condition in free-ranging eastern grey kangaroos. The following BCS system has been used with free-ranging macropod populations (D. Taggart pers. obs.).

(1) Poor – vertebrae at the base of the tail protruding and readily palpated; hips clearly visible with little muscle covering; vertebrae of backbone protruding and readily palpated.
(2) Average – vertebrae at the base of the tail slightly protruding and only just palpable; hips visible but with some muscle covering; vertebrae of backbone barely visible or palpable.
(3) Excellent – vertebrae at the base of the tail well covered by muscle and not protruding; hips barely visible with considerable muscle covering; vertebrae of backbone well covered and not protruding.

A generic 9-point scale for body condition scoring zoo animals is defined by the Zoological Information Management System (ZIMS, Species360 2023), where 1 = emaciated, 5 = ideal condition, and 9 = obese. This scale has been used for the development of a BCS chart for tree-kangaroos with pictorial schematics and detailed descriptions of overall body appearance, forequarters, midsection, hindquarters and tail (Association of Zoos and Aquariums 2017). A similar generic 5-point scale, where 3 = ideal body condition, is also commonly adapted for macropod condition scoring.

#### 8.2.2.2 Mathematical methods based on morphometric data

Various mathematical approaches associating bodyweight with an anatomical length representative of body size have been reported, including leg length (Fisher 1999), pes length (Kerle 1998; Hayward *et al.* 2011, and head length or width (D. Taggart pers. obs.).

Such methods rely on objective measurement, reducing the subjectivity that condition scoring can be prone to; however, there are potential limitations to such approaches (e.g. Green 2001). In macropods, such mathematical methods have not been correlated with qualitative categories of BCS. Thus the results can reveal trends in body condition over time, or between populations, but do not currently allow an individual to be qualitatively designated as having excessive, reduced or ideal body condition.

## 8.3 Physical examination

Detailed physical examinations should be performed by a veterinarian. They may be performed as part of the diagnostic investigation of a clinical problem or as a scheduled health check as part of a preventative health plan. Many components of a physical examination can potentially be performed under physical restraint or through the use of operant conditioning. However, a comprehensive physical examination requires anaesthesia. Additional procedures such as blood collection and diagnostic imaging may be performed concurrently to optimise the information gained on the overall health of the animal.

Standard approaches to performing physical examination in other species can be applied to macropods. Systematic examination of all body parts and evaluation of all body systems should be undertaken. Specific considerations for the physical examination of macropods are outlined in Vogelnest and Woods (2008) and Vogelnest and Portas (2025). Clinical findings associated with common diseases of macropods are outlined in Tables 10.12 and 10.13. Particular attention should be paid to the oral cavity (especially the dentition), gastrointestinal fill, the cloaca and in females the pouch, teats and any pouch young.

The appearance of the pouch varies with reproductive status. In a non-pregnant animal with no pouch young, the pouch may have a dry, dark scale over the inner lining and may be hairy. When preparing for a joey, the pouch will be clean, hairless and moist. Teat and mammary gland development reflects current requirements for young in pouch and/or at foot. Head length of pouch young has been correlated with age for several species.

Body temperature reference ranges for macropod species have not been formally established; an approximate range of 35–36.5°C is generally accepted and commonly cited (D. McLelland unpub. data). Resting heart rate and respiratory rate of macropod species have not been reported.

## 8.4 Chemical restraint

Chemical restraint drugs, including sedatives and anaesthetics, are commonly used to facilitate or otherwise improve outcomes for a range of procedures including disease investigation, assisted reproductive techniques and management needs such as translocation. Some procedures, such as surgery or undertaking a thorough physical examination, necessitate chemical restraint. Many procedures though, such as blood collection and pouch checking, can potentially be performed safely and effectively using physical restraint (see section 7), behavioural restraint (training/conditioning the animal for the procedure), or chemical restraint. The decision of which strategy to use should be evaluated on a case-by-case basis to optimise the welfare of the animal and the safety of personnel. Consideration should be given to: the nature of the procedure to be carried out; the species, size and demeanour of the animal; the characteristics of the enclosure and associated facilities; equipment available; and the experience of personnel. In many cases a combination of restraint methods will be used (e.g. a kangaroo is trained to stand for manual injection with an anaesthetic agent; a wallaby is conditioned to enter a small, darkened pen to improve the efficiency of capture via the tail base, then physically restrained in a catch bag for anaesthetic induction by face mask or manual injection).

Chemical restraint agents are scheduled drugs and many are subject to restrictions because of the potential for their misuse. Accidental or intentional human exposure to these agents can be associated with adverse effects. Measures to prevent exposure, and protocols to effectively manage human-exposure events, should be developed.

Decision making on the type of chemical restraint, the drug protocol and method of delivery should consider the benefits and costs on a case-by-case basis. Sedation and anaesthesia of native mammals should appropriately be performed by, or at least under the direction of, a veterinarian. In limited circumstances, such as some research settings, the possession and use of certain chemical restraint agents for a particular purpose, independent of veterinary involvement, may be permitted by the relevant jurisdiction. General principles of veterinary anaesthesia developed for domestic mammals are broadly applicable to macropods and should be applied. Supplemental oxygen is recommended in all anaesthetic procedures. Appropriate body temperature should be maintained, managing hyperthermia where required and preventing hypothermia. During induction and recovery, macropods should be managed to avoid injury from erratic uncontrolled movements.

Chemical restraint agents are variably administered by injection, inhalation, or orally.

- Injectable agents may be administered, generally by intramuscular injection, by manual injection, pole syringe or a remote delivery system. Remote delivery systems include blow pipes and gas-powered and charge-powered pistols and rifles. Darts are generally plastic or aluminium, with the injection of drug on impact achieved through compressed gas or

**Table 10.12.** Common infectious and parasitic diseases of zoo-based macropods.

PCR = polymerase chain reaction.

| Disease/agent | Cause | Clinical signs | Diagnosis | Treatment | Prevention | Comments | Reference |
|---|---|---|---|---|---|---|---|
| **Ectoparasites** | | | | | | | |
| Ectoparasitism | Ticks, mites, lice, fleas | Often no clinical signs<br>Dermatitis reported occasionally, can be severe with some mites | Clinical and microscopic examination | As appropriate for the parasite involved | Quarantine of animals entering the collection<br>Optimise general health | Typically less common in captive *v.* free-ranging macropods<br>Ectoparasites can increase to large numbers due to an underlying problem | 1 |
| **Endoparasites** | | | | | | | |
| Endoparasitism | Many species of gastrointestinal worms, including hookworms, roundworms, and tapeworms | Typically no clinical signs<br>In some situations, diarrhoea, anorexia, weight loss, and death may occur | Microscopic faecal examination<br>Clinical examination | Recommended only in the face of disease.<br>Appropriate worming medications and supportive care | No evidence that routine worming improves the health of macropods; may be indicated in the face of disease outbreak | Gastrointestinal worms are common in captive and free-ranging macropods<br>Most captive populations are not routinely wormed<br>A small number of worm species have occasionally been associated with disease and deaths | 1–4 |
| **Fungal** | | | | | | | |
| Dermatophytosis (ringworm) | Various species of dermatophyte fungi | Areas of hair loss, usually 1–5 cm diameter, typically on the limbs and tail | Clinical and microscopic examination<br>Fungal culture | Often self-limiting, but no treatment risks further spread<br>Topical and/or oral antifungal treatments as appropriate | Optimise general health and welfare<br>Prophylactic treatment and environmental decontamination may be warranted in outbreak situations | Disease can be associated with periods of stress<br>Zoonotic risk | 5, 6 |
| Candidiasis | *Candida* spp. | Variably, plaques in oral cavity, diarrhoea, occasionally skin lesions or general ill health | Clinical and microscopic examination<br>Fungal culture | Address husbandry issues<br>Topical and/or oral antifungal drugs as appropriate | Optimise general health and welfare, including good hygiene during hand-rearing | Typically only in hand-reared dependent young. Associated with poor hygiene and stress<br>Some antifungal treatments associated with adverse outcomes | 1 |
| **Protozoal** | | | | | | | |

| Disease/agent | Cause | Clinical signs | Diagnosis | Treatment | Prevention | Comments | Reference |
|---|---|---|---|---|---|---|---|
| Coccidiosis | *Eimeria* spp., *Isospora* spp. | Variably, inappetence, abdominal pain, diarrhoea (often bloody), dehydration | Microscopic faecal examination<br>Clinical examination | Anticoccidial medication and supportive care, including fluid therapy and analgesia, as appropriate | Attention to hygiene, optimising nutrition, and minimising stressors such as overcrowding<br>Prophylactic treatments not usually indicated | Juvenile macropods < 12 months, especially hand-reared pouch young, are particularly susceptible<br>Coccidia are commonly identified in macropods without any disease<br>Disease is often associated with underlying factors | 1 |
| Toxoplasmosis | *Toxoplasma gondii* | Wide range of possible presentations, including sudden death, neurological signs, eye disease, diarrhoea, weight loss, respiratory signs | Serological tests for antibodies, in conjunction with clinical signs | Various antimicrobial treatment options with variable efficacy | Preventing cats from accessing enclosures and protecting food and water sources from contamination with cat faeces<br>Research into effective vaccines is ongoing | Toxoplasma replicates in the gastrointestinal tract of cats, with infective oocysts shed in faeces. Oocysts can persist in the environment in food and water for long periods before being ingested by a macropod or other animal | 7 |
| **Bacterial** | | | | | | | |
| Macropod progressive periodontal disease (MPPD, lumpy jaw) | Multifactorial progressive inflammatory and necrotising disease involving anaerobic bacterial infection of the dentition ± surrounding soft tissues and bone | May include drooling, inappetence, dropping food from the mouth, facial swellings, eye problems, weight loss | Dental examination under anaesthesia, diagnostic imaging, ± bacterial culture | Dental cleaning, extraction of affected teeth (or endodontic therapy), debridement of necrotic bone. Regional and/or systemic antibiotics may be indicated Supportive care. Euthanasia in advanced cases | Optimise general health and welfare.<br>Optimise species-appropriate husbandry and nutrition<br>Minimise stressors<br>Attempts at vaccination have not been successful | Often considered the most common cause of disease and death in captive macropods<br>Stressors, including adverse environmental conditions, inappropriate groupings or stocking density, ongoing excessive attention by zoo visitors, repeated capture/restraint events and moves between enclosures/institutions, have been associated with incidence of MPPD | 8–10 |

**Table 10.12.** continued

| Disease/agent | Cause | Clinical signs | Diagnosis | Treatment | Prevention | Comments | Reference |
|---|---|---|---|---|---|---|---|
| Tetanus | *Clostridium tetani* | Muscle rigidity/ spasms, salivation, third eyelid protrusion, death | Clinical signs | Supportive care, tetanus antitoxin | Vaccination | Bacterial spores are inoculated through penetration injuries and other wounds<br>Poor prognosis, even with treatment | 1, 11 |
| Salmonellosis | *Salmonella* spp. | Soft faeces to bloody diarrhoea, inappetence, dehydration, death | Faecal culture | Antimicrobials based on culture and sensitivity testing<br>Supportive care | Optimise general health and welfare<br>Minimise stressors | Typically acquired through exposure to faeces containing pathogenic salmonella (can be carried by healthy mammals, birds, and reptiles)<br>Likelihood of disease varies with the species of *Salmonella* and may be precipitated by underlying stress or other disease<br>Zoonotic risk | 1 |
| Bordetellosis | *Bordetella bronchiseptica* | Nasal discharge, sneezing, laboured breathing | Bacterial culture of nasal swabs or lower respiratory samples | Antimicrobials based on culture and sensitivity testing<br>Supportive care | Vaccination | May be normal respiratory flora, with underlying factors or co-infections involved in disease<br>Spill-over infection from wild rabbits suggested in one outbreak | 1 |
| Pneumonia | Various species of aerobic and anaerobic bacteria | Laboured breathing, coughing, lethargy, inappetence | Clinical examination, radiography, lower respiratory sampling for bacterial culture | Antimicrobials based on culture and sensitivity testing<br>Supportive care | Optimise husbandry<br>Optimise hygiene with hand-reared pouch young, including good bottle-feeding technique to avoid aspiration | Hand-reared pouch young particularly susceptible with inappropriate care | 1 |
| Treponemiasis | *Treponema* spp. | Genital infections with ulceration and inflammatory exudate | Dark-field microscopy, PCR | Appropriate antibiotic treatment | Difficult to prevent in Gilbert's potoroo due to prevalence in the captive population | Venereal disease<br>Common in Gilbert's potoroos<br>Reported in a long-nosed potoroo | 2, 12 |

| Disease/agent | Cause | Clinical signs | Diagnosis | Treatment | Prevention | Comments | Reference |
|---|---|---|---|---|---|---|---|
| Mycobacteriosis | *Mycobacterium* spp. | Variable, non-specific signs (e.g. weight loss, lameness, skin ulcers, abscesses, respiratory or neurological signs, blindness) | Mycobacterial culture and/or PCR<br>Diagnostic imaging may be beneficial, including as a guide for sample collection | Difficult to treat successfully and prognosis often poor<br>Cutaneous mycobacteriosis has been successfully treated with surgery and appropriate antimicrobials | Difficult to prevent as non-tuberculous mycobacteria are ubiquitous in the environment<br>Substrate changes may be considered if repeated cases occur | Most cases involve species of non-tuberculous mycobacteria. However, infection with tuberculous mycobacteria has been documented in macropods<br>Zoonotic risk depends on the species of mycobacteria | 13 |
| **Viral** | | | | | | | |
| Macropod herpesvirus | Various herpesviruses | Variably, conjunctivitis, respiratory signs, ulcers on the cloaca, fever, death | Clinical presentation.<br>Biopsy for histopathology<br>Swabs or biopsy for PCR<br>Serological tests useful to screen for exposure, but do not differentiate active and latent infections | Supportive care<br>No reports of antiviral medications in macropods | Reduce stressors and optimise general health and welfare<br>Targeted quarantine measures<br>Screening of populations to evaluate prior exposure | Stressful situations have been associated with the onset of disease in macropods with subclinical infections | 14 |
| EMC virus | Encephalomyocarditis virus | Sudden death due to heart failure<br>May see wobbly gait, lethargy and breathing difficulty | Necropsy, with histopathology and virus isolation | No treatment reported | Vaccine available<br>Control of rodent populations, which are considered important in virus transmission | Multiple tree-kangaroo cases have been reported; other species likely susceptible also | 1 |
| Tammar sudden death syndrome | Orbiviruses in the Eubanangee serogroup | Sudden death, usually occurs as an outbreak situation over several weeks | Necropsy examination, with histopathology and virus isolation from heart, liver, intestine and brain | No treatment reported | Control of biting insects in high-risk situations | Outbreaks have occurred in early summer in New South Wales, Australian Capital Territory and Queensland, associated with heavy rainfall and increased activity of biting insects | 1, 2, 15 |

References: (1) Vogelnest and Portas (2008); (2) McLelland (2025a); (3) Cripps *et al.* (2014); (4) Northover *et al.* (2017); (5) Boulton *et al.* (2013); (6) Vogelnest (2025); (7) Portas (2025a); (8) McLelland (2025b); 9) Rendle *et al.* (2020); (12) Yip *et al.* (2021); (11) Phillips *et al.* (2012); (12) Vaughan-Higgins *et al.* (2011); (13) Tong Marwood (2025); (14) Devlin and Stalder (2025); (15) Rose *et al.* (2012).

**Table 10.13.** Common non-infectious diseases of zoo-based macropods.

| Disease | Cause | Clinical signs | Diagnosis | Treatment | Prevention | Comments | Reference |
|---|---|---|---|---|---|---|---|
| Capture myopathy | Acute muscle breakdown in response to overexertion, hyperthermia or excessive stress | Weakness; abnormal gait, respiration and mentation; muscle tremors; dark red urine; collapse; death | Clinical signs. Blood and urine tests for muscle injury and renal function | Supportive care, including aggressive fluid therapy, anti-inflammatory drugs, muscle-relaxant sedatives and anxiolytics | Minimise stress. Use appropriate capture/ restraint and transport techniques<br>Vitamin E before stressful events advocated by some | Animals may be acutely affected and can die immediately after the initiating event. Animals that survive this initial period may succumb days to weeks after the event due to heart and/ or kidney failure | 1 |
| Trauma | Collision with fences and other objects<br>Inappropriate handling | Lameness, immobility, death | Clinical signs, physical examination, radiography | Supportive care ± surgery may be possible in some cases. Euthanasia may be indicated | Visual barriers along fencing<br>Minimise stressors to avoid flight responses<br>Use of appropriate capture/restraint techniques | Neck fractures are common with fence collisions at speed<br>Limb and tail injuries can occur with fence entanglement and inappropriate handling | 1 |
| Degenerative joint disease (arthritis) | Often age related | Chronic lameness, stiffness | Clinical signs, physical examination, radiography | Anti-inflammatory drugs, analgesics, certain nutraceuticals, to clinical effect | Provision of species-appropriate diet and substrate, and preventing obesity, may be beneficial | One of several age-related diseases that may be seen in captive macropods | - |
| Mesenteric fat necrosis | Obesity, high-energy/ high-fat diets | Asymptomatic to fatal debility | Radiography – affected fat often mineralised | Not reported. Surgery may be considered | Ensure appropriate diet | Reported cases in quokkas associated with access to meat-based feed in mixed-species enclosure and a walk-through exhibit with public feeding | 2, 3 |
| Xanthomatosis | High-fat diets | Swellings on the skin and internal organs | Biopsy, serum cholesterol | Not reported | Ensure appropriate diet | Reported cases in long-nosed potoroos associated with access to meat mix and sunflower seeds in a mixed-species aviary | 2 |
| Megaoesophagus | Unknown | Regurgitation, weight loss, laboured breathing, death | Radiography, endoscopy | Altered feeding strategy may be palliative<br>No successful treatment described | No preventative measures described | Reported most commonly in parma wallabies | 2, 4, 5 |

| Disease | Cause | Clinical signs | Diagnosis | Treatment | Prevention | Comments | Reference |
|---|---|---|---|---|---|---|---|
| Diverticulosis | Variable, often unknown | Variably, neck swellings, respiratory distress, anorexia, colic, death | Radiography, endoscopy | Surgery in some cases | No preventative measures described | Described in the oesophagus and intestines | 2, 6 |
| Urolithiasis | Urinary stones in the kidney, bladder and/or urethra | Asymptomatic, to difficult urination, abdominal pain, death | Radiography, ultrasonography, urinalysis | Surgery in some cases<br>Urethral anatomy impedes catheterisation | Optimise diet<br>Specific measures depend on the type of stone | Most reported cases involve calcium-based crystals<br>Multiple cases in Tammar wallabies were associated with lucerne hay, with no further cases after change to grass hay | 2, 7 |

References: (1) Vogelnest and Portas (2008); (2) McLelland (2025a); (3) Hulst *et al.* (2010); (4) Woodhouse *et al.* (2010); (5) Burgdorf-Moisuk *et al.* (2012);( 6) Okeson *et al.* (2009); (7) Liptovszky *et al.* (2014).

powder charge, respectively. The safe and effective use of remote delivery systems requires a good level of familiarity and skill with the system being used and an approach that minimises the risk to the individual, other animals in the enclosure and personnel. In many jurisdictions, remote delivery systems require an appropriate firearms licence.

- Inhalation agents, such as isoflurane, are delivered via a face mask or endotracheal tube. A calibrated vaporiser introduces the desired percentage of vaporised anaesthetic to a carrier gas, typically 100% oxygen. The equipment required is routinely maintained in veterinary facilities and can be adapted to be portable for use in the field.
- The use of orally administered chemical restraint agents, typically in food items, can be unrewarding or at least inconsistent in macropods compared with injectable agents. The large sacculated stomach and associated fermenting microbiota likely reduce the availability of certain drugs for absorption in the intestine. However, in some cases, and with certain types of medications, beneficial effects can be achieved.

Sedation may be utilised to improve the ease of performing minor handling procedures and to reduce the stress of certain situations such as hospitalization. Benzodiazepines, such as diazepam, are among the most common sedatives used in macropods. Sedatives may be combined with anaesthetic agents with the goal of improving the quality and safety of anaesthetic induction, maintenance and/or recovery.

Pre-anaesthetic fasting is not required in adult macropods as they are not prone to regurgitation under anaesthesia, though recently chewed food may be present in animals that have recently eaten, which may obscure examination of the oral cavity. Milk-feeding joeys, especially if being hand-reared, should ideally be fasted for at least 1 h to minimise the risk of regurgitation and aspiration.

Anaesthesia of small macropods may be achieved through the use of inhalation agents delivered by face mask, injectable agents or a combination of both, typically with the animal restrained in a catch bag or similar. In large macropods, induction of anaesthesia using injectable agents will be indicated in almost all situations. A range of injectable anaesthetic protocols are commonly used in macropods, including tiletamine-zolazepam (Zoletil®) and medetomidine + ketamine. Response to particular injectable anaesthetic protocols can vary with species, demeanour (tame *v.* wild) and health status. Maintenance of anaesthesia can be achieved with inhalation and/or injectable agents based on the preference and clinical judgement of the anaesthetist. Endotracheal intubation is challenging due the relatively long face and narrow gape and oral cavity, but can be readily achieved with good technique.

Macropods should be monitored closely from induction through recovery to reduce the risk of injury from uncontrolled movements and to identify and manage any anaesthetic complications. Noise should be kept to a minimum during induction and recovery to reduce stress on the animal and avoid flight responses. Macropods should ideally not be released into large enclosures or to the wild until they have sufficiently recovered from the effects of anaesthesia, to minimise the risk of misadventure. Use of restraint bags, transport boxes or small darkened sheds will provide a controlled environment until release.

A range of other psychotropic agents may be used in macropods. Some of these drugs (e.g. trazodone and gabapentin) are known to reduce anxiety in other species and can be expected to have similar effects to reduce stress in macropods. Studies formally evaluating these drugs in macropods have not been performed; however, anecdotal reports suggest they can act to reduce the stress and risk of injury associated with certain procedures and situations with improved welfare outcomes.

Various neuroleptics have been used in macropods to effect quiet behaviour that may be beneficial to reduce the risk of potentially fatal injury in certain situations such as transportation or confinement. These drugs have a general depressant effect, but they do not affect the animal's perception of anxiety or fear. Macropods administered these medications can appear very calm, but this should not be interpreted as a reduction in the level of stress experienced and the welfare implications of this must be considered.

More detailed information on the chemical restraint of macropods is presented in relevant chapters of Vogelnest and Woods (2008) and Vogelnest and Portas (2025).

## 8.5 Common health problems

Macropods are susceptible to a wide range of diseases. Infectious, parasitic and non-infectious diseases commonly documented in macropods managed in captive settings are outlined in Table 10.12 and Table 10.13.

There are a range of other health problems reported less commonly in captive macropods and several diseases that have been predominantly described in free-ranging macropod populations are not included here. For comprehensive information on health problems in captive and free-ranging macropods, see Vogelnest and Woods (2008) and Vogelnest and Portas (2025). The diagnosis and treatment of health problems in macropods should be under the direction of a veterinarian.

The range of health problems likely to occur in populations of captive macropods can be expected to vary between facilities. For many infectious and parasitic diseases, the occurrence of disease is dependent not just on exposure to the infectious agent or parasite, but also a range of other factors that serve to increase the susceptibility of the animal to disease. The likelihood of any particular disease occurring in a captive individual or population can vary, based on factors such as the species, geographical location, climate, the characteristics of the enclosure, the nature and quality of husbandry provided, the appropriateness of the diet and quality of foods provided and other species sharing the enclosure or present in the local environment.

## 8.6 Preventative medicine

The optimisation of species-appropriate husbandry and nutrition, minimisation of stressors and close observation of the captive population as described above are the foundations of a preventative medicine program. Veterinary involvement in the development of a preventative medicine plan for captive macropods is recommended. Specific additional measures that can be considered to aid in the prevention and/or early detection of health problems include:

- veterinary heath checks - performed periodically at a frequency appropriate to the population; optimally performed under anaesthesia to enable thorough physical examination including dental assessment; potentially including blood collection for haematology and biochemistry, and diagnostic imaging
- faecal examination - performed periodically at a frequency appropriate to the population, to document and monitor gastrointestinal parasite populations
- vaccination - macropods may be vaccinated against various diseases including tetanus, bordetellosis and EMC virus. Vaccination programs should be tailored to each facility based on a risk assessment for diseases of concern for the collection
- quarantine - for new animals brought into existing captive populations, a risk assessment should be performed to guide measures to minimise the risk of new diseases entering the population. Such measures should be tailored for each import and the needs of each institution, and often include:
  - health evaluation - ideally under anaesthesia, including physical examination, blood collection for haematology and biochemistry, and diagnostic imaging. May be performed before the shipment or following arrival
  - faecal exams - to screen for gastrointestinal parasites and/or potential pathogens
  - screening tests - for particular diseases of concern based on a case-by-case risk assessment
  - isolation - a period of isolation from the existing population is commonly used to observe for any signs of disease before introducing the new animal(s). This period also allows the animal to adapt to the new environment and diet and provides time for health evaluation and disease screening to be performed
- pathology - all deaths should be investigated by a full necropsy, ideally including histopathology and other ancillary tests as appropriate. Information gained from necropsies can be used to identify health problems of concern in the population. This can lead to more rapid diagnosis and more effective treatment of similar cases in future and the identification of targeted preventative measures to reduce disease incidence in the population.

## 8.7 Zoonoses

Several zoonotic diseases have been associated with macropods (Table 10.14). For many zoonotic diseases, the macropod host will show no signs of disease. For more information on zoonotic diseases associated with macropods, see Hulst (2025). If staff or members of the public suspect they have contracted a zoonotic disease they should seek medical advice.

# 9. BEHAVIOUR

## 9.1 Activity

Almost all species of macropod are considered to be most active at night with bimodal peaks of activity in the first 3 h after sunrise and sunset (usually called crepuscular) (Coulson 1978). In more arid regions, the red kangaroo,

**Table 10.14.** Zoonotic diseases associated with macropods.

Derived from Hulst (2025).

| Disease | Cause | Clinical signs in humans | Transmission | Prevention |
|---|---|---|---|---|
| Ringworm | Various species of dermatophyte fungi | Typically circular rash with reddened, thickened and/or scaly skin | Direct contact with fur/skin of an infected macropod (or many other mammals) or via materials the animals have been in contact with | Good personal hygiene, especially handwashing<br>Use of gloves and other protective measures when handling affected macropods |
| Gastroenteritis | Various pathogens including *E. coli*, salmonella, giardia and cryptosporidia | Diarrhoea, vomiting, abdominal cramping | Contact with macropods or their faeces (also many other animals) | Good personal hygiene, especially handwashing |
| Q fever | *Coxiella burnetti* | No disease through to severe flu-like illness. Potentially hepatitis, pneumonia, chronic heart disease | Bacteria may be shed in birth products, urine and faeces of macropods (and many other mammals) | Vaccination of staff<br>Use of masks if cutting/ slashing grass in enclosures<br>Good personal hygiene, especially handwashing |
| Ross River fever | Mosquito-borne virus | Flu-like illness with muscle/ joint pain, fever, skin rash. Chronic joint pain and fatigue in many patients | Mosquito bites. Macropods and other native mammals are reservoir hosts | Personal mosquito protection<br>Measures to control local mosquito populations |
| Queensland tick typhus | Tick-borne *Rickettsia* bacteria | Fever, skin rash, cough | Tick bites. Macropods and other native mammals are reservoir hosts | Personal tick protection in high-risk areas (e.g. wear repellents, impregnated clothing) |

eastern grey kangaroo and western grey kangaroo appear to be largely nocturnal, showing peak activity between 1800 and 0600 h and being almost totally inactive by 1200 h (McCullough and McCullough 2000). A study of red kangaroos, eastern grey kangaroos and western grey kangaroos in the arid zone found them to vary greatly in their activity throughout the year, depending on temperature and food availability, being active for 40% of the time in the hot summer months and 70% in the coldest months (McCullough and McCullough 2000). The musky rat-kangaroo is considered to be the only truly diurnal macropod.

Tree-kangaroos are cathemeral, active day and night with bouts of sleep (Coombes 2005). They are the only macropods that are able to move their hindlegs independently of each other (others do when they swim) and walk backwards. All members of the Potoroidae build nests in which they rest, while most of the Macropodidae do not have a nest site other than a need for shade. Some, such as the wallaroo and the rock-wallabies, use rock piles, caves and even wombat burrows (Thornett *et al.* 2017) for protection and to help thermoregulate. Tree-kangaroos curl up on a branch (or in captivity a vertical log or bench in hutch in shade) to sleep with their tail underneath and out front from between the hindlegs (Coombes 2005).

A study that assessed the activity of red-necked wallabies in a walk-through enclosure (open enclosure) and two non-walk-though enclosures revealed that there was a significant increase in feeding and interactive behaviours in closed enclosures versus open enclosures. However, behaviours including resting, locomotion and vigilance did not vary between open and closed enclosures, suggesting exhibit type was not a major factor in influencing welfare (Beaudin-Judd *et al.* 2019). Similarly, a study of captive quokkas found there to be a reduction in quokka visibility when visitors were present, which indicates that visitors provoked at least a moderate fear response (Learmonth *et al.* 2018). A further study of the impacts of visitor number on Kangaroo Island western grey kangaroos and red kangaroos found a relationship between increased visitor numbers and increased time spent engaged in visitor-directed vigilance (Sherwen *et al.* 2015). The western grey kangaroos also increased the time spent engaged in locomotion and decreased the time spent resting. However, there was no observed effect of visitor numbers on the distance kangaroos positioned themselves from the visitor pathway or faecal glucocorticoid metabolite concentrations in either species (Sherwen *et al.* 2015).

## 9.2 Social behaviour

The social behaviour of macropods ranges from non-social species (e.g. potoroos and tree-kangaroos), which are generally small and live in more closed forests, to gregarious species (e.g. eastern grey kangaroos and Tammar wallabies) that live in well-organised societies (Hume *et al.* 1989; Paplinska *et al.* 2010). The degree of gregariousness, measured by typical group size, generally increases with body size, openness of habitat and proportion of grasses in the diet (Kaufmann 1974; Russell 1984).

### 9.2.1 Potoroids

Potoroids live in areas of dense vegetation such as forest understorey, grassland or heathland (Russell 1984). They are generally solitary and have areas that they specifically rest in and other areas where they feed, which may or may not overlap. The home ranges of males generally overlap with those of several females, but appear not to overlap as strongly with the home ranges of other males.

In contrast to the other potoroids, which generally nest alone, the burrowing bettong often nests in large communal burrow systems, with as many as 120 entrances and 60 individuals (Burbidge 1995; Sander *et al.* 1997). Fighting often occurs between male potoroids (e.g. brush-tailed bettongs, (S. Jackson pers. obs.; Viola 1977), Tasmanian bettongs (Rose 1982) and long-nosed potoroos), which has led to deaths in captivity (Bates *et al.* 1972). Similar aggression has been observed for burrowing bettongs, with males chasing other males away from females, which are actively defended (Stodart 1966; Sander *et al.* 1997). Female burrowing bettongs show evidence of a hierarchy; they appear to be communal in nature, often using the same nest boxes or burrows (Stodart 1966).

Musky rat-kangaroos have been observed to chase each other in the wild for over 30 s (Dennis and Marsh 1997). In captivity, male musky rat-kangaroos are extremely violent if confined in the same cage in the presence of a female (Dennis and Marsh 1997). Other potoroids, such as rufous bettongs, generally maintain transitory contact with as many females as possible; however, some form male–female pairs even when the female is not in oestrus (Frederick and Johnson 1996). Male rufous bettongs can be extremely aggressive to one another in captivity while females are more tolerant of each other (P. Johnson 1980).

Home ranges of individual long-nosed potoroos overlap with multiple members of each sex, although intrasexual overlap is lower than intersexual overlap (Long 2001). Male home ranges overlap with the squat areas of one or more females, but they do not regularly associate with every female whose squat area they overlap with. It appears that in most cases a male and female pair associate regularly and spend a greater proportion of time in close proximity to each other than would be expected (Long 2001).

### 9.2.2 Tree-kangaroos

Tree-kangaroos live in rainforest and are generally solitary, with only small family groups consisting of the female with joeys from two years, one being a 1-year-old animal and the second from the current year being raised (a visiting male may also be present), or an adult female and male (Coombes 2005). Lumholtz's tree-kangaroo joeys stay with their mother for > 2 years (Coombes 2005). Males are polygynous, with home ranges that overlap with several females (Newell 1999; Coombes 2005). Male home ranges tend to abut those of other males, with antagonistic encounters occurring at the boundaries of the home ranges (Newell 1999). Both male and female Lumholtz's tree-kangaroos hold home ranges of up to 2 ha. Adult males and females are primarily solitary and extremely intolerant of each other; they appear to be one of the few macropods that defends a territory (Martin 1992). Home ranges of Bennett's tree-kangaroos, for example, are 3.8–29.8 ha and overlap with up to three females (Martin 1992). Males appear to be strongly territorial and form dominance hierarchies in captivity (Crook and Skipper 1987). Lumholtz's tree-kangaroos of both sexes cannot be kept together except for a mating pair or young teenagers before they reach sexual maturity (K. Coombes pers. obs.). Female Bennett's tree-kangaroos occupy exclusive home ranges up to 9.8 ha, though it is not known if these are actively defended. The mating system has been described as being facultative polygyny (Martin 1992). The most common agonistic behaviours observed include jumping onto the back of an opponent and biting his neck, ears and tail (Crook and Skipper 1987). Adult male Lumholtz's tree-kangaroos have been observed chasing and fighting other tree-kangaroo in captivity, taking large chunks of fur (K. Coombes pers. obs.). In captivity, coalitions of mothers and daughters have been observed to support each other against males (Ganslosser 1984). Observations of Lumholtz's tree-kangaroos in captivity found males to mostly initiate social behaviour and females tended to respond aggressively towards the male (bite, cuff or swipe) (K. Coombes per. obs.). The aggression of male Lumholtz's tree-kangaroos (Newell 1999; Johnson and Delean 2003) and Bennett's tree kangaroos (Martin 1992) is known to result in fatal wounds as a result of the powerful forearms, powerful claws and biting (Coombes 2005). Allogrooming was extremely rare and non-aggressive contact behaviour consisted primarily of olfactory examination (Hutchins *et al.* 1991). Successful reproduction (two joeys were found dead previously) did not occur until females were isolated, by removing other individuals from the enclosure (Hutchins *et al.* 1991).

### 9.2.3 Hare-wallabies

Hare-wallabies generally live in an environment ranging from open forests to deserts with a well-developed shrub understorey. They are generally solitary, but may occasionally feed together in small groups (Burbidge and Johnson 1995). Observations of rufous hare-wallabies in captivity suggest that females are usually not aggressive to each other, although they do appear to have a dominance hierarchy (Lundie-Jenkins 1993a). Males also show evidence of a hierarchy in relation to access to females and feeding stations (Lundie-Jenkins 1993a). Little aggressive behaviour was observed between mature animals of different sexes, although it has been observed between mature animals of one sex and sub-adult or juveniles of the other sex. In those encounters the mature animal was nearly always the aggressor and the interactions were extremely violent, resulting in the death of several young (Lundie-Jenkins 1993a).

### 9.2.4 Wallabies

Wallabies in the genus *Notamacropus* generally live in open forest and can be gregarious, though a hierarchy exists between males in such species as agile wallabies and black-striped wallabies. Intermale competition for mates can be intense (Rudd 1994), including intermale sperm competition with the presence of mating plugs (Paplinska *et al.* 2010). They can also be quite solitary, particularly during the day while resting.

### 9.2.5 Kangaroos

Kangaroos live in open shrubland, grassland or woodland. Some species, such as wallaroos, are generally found on well-wooded, steep, rocky slopes and terraces of ridge faces (Kaufmann 1974). They generally show strong sexual dimorphism (with males being up to twice the size of females) and live in social groups that can include > 50 individuals and there is a male dominance hierarchy. There is strong sexual selection in the genera *Macropus* and *Osphranter*, which has led to clear sexual dimorphism with males being approximately twice the mass of females. Combat between males occurs when dominant males endeavour to keep other males away from females in oestrus (Frith and Calaby 1969). During combat, individuals approach each other slowly, lick and bite the fur of the upper chest and cluck softly. Upon meeting each other, they stand tall on their hindlegs, lean back on their tail and kick the abdomen of the opponent with the hindlegs, grapple (wrestle) and scratch the head of the opponent with their forearms (Fig. 10.16). Though these fights are usually resolved quickly, mortal combat is known to occur, with animals dying from injuries sustained during fights (Frith and Calaby 1969; Toni 2018).

### 9.2.6 Nail-tailed wallabies

Nail-tailed wallabies generally live in semi-arid shrublands, scrub, dense thickets and grassy woodlands. They are generally solitary, although small groups of 4–5 will often come together to feed. Oestrus females are usually accompanied by a single male (Evans and Gordon 1995; Ingleby and Gordon 1995). Bridled nail-tailed wallabies are sexually promiscuous in both sexes in the wild and aggression between males occurs only around oestrus females (Fisher and Lara 1999). Males do not appear to form a predetermined dominance hierarchy; however, bodyweight (and hence age) strongly influence priority access to females (Fisher and Lara 1999). Females mate with several males within and between oestrous cycles (Fisher and Lara 1999).

### 9.2.7 Rock-wallabies

Rock-wallabies generally live in rocky outcrops or rock piles, are socially monogamous and both males and females exhibit strong territorial behaviour to other members of the same sex. Both males and females appear to show linear dominance hierarchies and adult males and females form relatively stable relationships (though non-partnered breeding does occur in the wild) (Nicholls 1972; Batchelor 1980; Scholz 1980; Sanson *et al.* 1985; Barker 1990; Spencer 1996). In the case of the nabarlek, for example, the dominant female and male (which shared the same nest box) have been observed to

**Fig. 10.16.** Common postures of male red kangaroos threatening and fighting. Redrawn based on a figure from Frith and Calaby (1969).

form a pair, while subordinate females tended to keep apart (Sanson *et al.* 1985). In some species, such as the unadorned rock-wallabies, males are territorial, defending rocky areas, and forming harems comprising a male and one or more females (Russell 1984). Single males are often held with up to four females, resulting in significant agonistic behaviour between the dominant and subordinate females including chasing; two-handed forepaw strikes; severe biting to the back, neck and tail; and kicks to the back of the head, back and shoulders (Sanson *et al.* 1985; Barker 1990; Bulinski *et al.* 1997). Aggression by the dominant female has resulted in the death in captivity of subordinate female yellow-footed rock-wallaby, brush-tailed rock-wallaby (S. Jackson pers. obs.) and nabarlek (Goldstone and Nelson 1986). They appear to breed best in captivity when there are only one or two males in a colony of up to 30 animals (Gasking 1965). Movement of new individuals into a colony must therefore be done with great care, as upsetting established hierarchies can lead to the death of multiple females (Taggart *et al.* 2016; D. Taggart. pers. obs.).

### 9.2.8 Quokka

Kitchener (1973) has observed long-term associations of breeding pairs. He found that males defended shelter sites, which is important during summer when cover by small shrubs is limited. Males are aggressive to one another and form a stable linear dominance hierarchy in which older males are dominant and adult females and juveniles have no social rank (Packer 1969; Kitchener 1972; Uka 1981). Males are generally tolerant of male juveniles until they are sexually mature when, in captivity, they should be removed (Uka 1981). Although they rest together, they are more dispersed when feeding (Kitchener 1972), which allows them to rest during the hottest part of the day (Packer 1965).

### 9.2.9 Pademelons

Pademelons live in dense forest (typically rainforest) and are solitary. They are sexually dimorphic, polygynous and show strong territorial behaviour. They have been observed to generally rest solitarily in more dense forest during the day and come out into more open grassy areas adjacent to the dense forest at night, where several individuals will often feed together. In captivity, aggression has been observed between males and females, with body size important in determining status (Morton and Burton 1973; Paplinska *et al.* 2010).

## 9.3 Reproductive behaviour

Males and females of most macropod species will associate without conflict, but the presence of an oestrus female often induces conflict between males (Tyndale-Biscoe and Renfree 1987; Rudd 1994). In both large and small macropod species, such interactions can lead to severe injuries and even death. As a result of these interactions, a dominance hierarchy is established whereby the dominant male has access to the females when they are in oestrus (Rudd 1994). Care should be taken with larger macropods as they can also become aggressive to humans while females are in oestrus. Testis to body-weight ratio can be used to indicate likelihood of a polygamous or monogamous mating strategy within a species (Paplinska *et al.* 2010).

In some groups of macropods, such as rock-wallabies, a dominance hierarchy is established among the females as well as among males. This can result in the dominant female causing injuries and even death to other females, particularly if there is a male within the enclosure and if the enclosure is not large enough.

The copulatory behaviour is fairly uniform among the different groups of macropods (Rudd 1994), with the male following the female and investigating her cloaca. These behaviours are often associated with the male attempting to grab the female's tail as he emits a clicking sound. If the female is in oestrus the male continues to grab the tail of the female and to follow her. Mating takes place when the female remains stationery and crouches with the male coming up behind. Mating is detectable by examining the female's cloaca, as it will contain a semen plug shortly after the mating, which is later expelled onto the ground.

Flehmen behaviour has been observed in male macropods, which is a type of lip curling when they encounter the urine of females of their own species (generally) (Triggs 1990). Flehmen usually involves the animal standing with his head stretching towards the female's cloaca or urine on the ground with his mouth open while he retracts the upper lip. This bares the gum and wrinkles the nose and appears to be a mechanism to aspirate into the vomeronasal organ (Coulson and Croft 1981; Triggs 1990). Flehmen is sometimes associated with the male making rapid licking and mouthing movements during and after showing the behaviour. It appears to be involved in the male's detecting if the female is in oestrus and therefore ready to mate (Coulson and Croft 1981; Triggs 1990). Lumholtz's tree-kangaroo males will rub their head under the cloaca of females and results in the males head becoming sticky form female paracloacal discharge, which is a good indicator of mating (K. Coombes pers obs).

## 9.4 Bathing

Generally, macropods do not bathe. In the wild, kangaroos have been known to enter water when pursued by dogs; when the dog reaches the animal, the kangaroo will attempt to hold the dog beneath the water to disrupt the attack and sometimes drown the dog. Kangaroos have also been reported on the beach and to enter sea water and 'surf' back to shore. The reason for this rarely observed behaviour is unknown, but may be an attempt to reduce external parasites or perhaps eat exposed marine plant matter (P. Taggart pers. obs.).

## 9.5 Behavioural problems

Some individual male macropods, particularly males that have been hand-reared, can become aggressive and attempt to grab and kick anyone within their enclosure, especially when there is a female in oestrus. It is therefore very important to observe the location of all individuals within the enclosure, particularly troublesome ones. If an animal is showing aggressive behaviours, these can quite often be addressed by crouching down, no longer posing a threat to the animal's dominant position. Alternatively, reach out the head of a rake or shovel over the head of the kangaroo (~30 cm above) to show dominance by height. It is also important not to become cornered as this can allow the animal to grab and kick more easily. A rake or shovel can also be used to keep the animal out of kicking range. If an animal does kick, turn side on and move to the side and away from the kangaroo, as it has to lean back and line up for a kick and by moving away it makes it very difficult for it to kick accurately. If an animal does show consistent aggressive behaviour, it is advisable to leave the enclosure and return later in the day.

Occasionally males will behave aggressively towards each other if a female of a different species is in oestrus. For example, some Lumholtz's tree-kangaroos are also known to be very aggressive towards humans (and kick and bite), so protective clothing (such as welding gloves) needs to be worn (K. Coombes pers. obs.).

## 9.6 Signs of stress

Signs of stress include (Spielman 1994):

- vocalisation
- flinching
- escape attempts
- thumping the ground with the hindfeet
- body trembling
- head shaking (not always a sign of stress in Lumholtz's tree kangaroos, as it appears to be sign of recognition they perform when meeting each other or a familiar animal caretaker (K. Coombes pers. obs.))
- titubation?
- ear flicking
- teeth grinding
- licking the forearms, shoulders, and flanks depending on the degree of stress (resulting in increased thirst)
- dribbling from mouth or dripping mucous from nose (Lumholtz's tree-kangaroos)
- increased breath rate
- reduced food intake (associated with chronic stress) or completely not eating
- diarrhoea.

## 9.7 Behavioural enrichment

The provision of environmental enrichment for kangaroos is increasingly being applied in collections (e.g. Vitória *et al.* 2022). Providing enrichment should be a planned strategy that is introduced to achieve a specific behavioural goal. The idea is to provide species specific challenges, stimulation and opportunities. It is important to evaluate enrichment to ensure that the desired goals are achieved.

When implementing enrichment, it important that it is:

- appropriate for the species
- appropriate for the individual in mind (consider life stage)
- non-hazardous for the animal to use/manipulate/ take part in
- appropriate for animal welfare
- easy to undertake and maintain.

To optimise welfare, consideration of the enrichment value of all aspects of the animal's environment and management, appropriate to the species, should be taken. All components of the diet should be enriching, rather than separating out specific 'enrichment' food items from the regular base diet.

Examples of behavioural enrichment for macropods include:

- providing browse relevant to the species
- providing tussocks and other nesting material for potoroids to promote nest-building behaviour
- increasing enclosure complexity for species such as rock-wallabies and tree-kangaroos so the animals have multiple escape routes from any area, as well as vertical aspects to their enclosure (e.g. rocks or tree branches with cover) and den areas (e.g. shelving off the ground)
- providing truffles (hypogeal fungi) for potoroids
- providing opportunities for age- and species-appropriate social interaction
- providing climbing structures for relevant species
- providing browsing and grazing opportunities
- providing opportunities to dust bathe
- hiding or scattering food for small species such as potoroos and bettongs

- moving feeding locations within the exhibit
- utilising browse holders
- utilising positive reinforcement training or conditioning.

## 9.8 Training and conditioning

Tree-kangaroos (including the New Guinean species *Dendrolagus goodfellowi* and *D. matschiei*) have been successfully trained using operant conditioning (i.e. reinforces or discourages behaviours using a reward as an incentive). Examples of training that has been successful include: standing on scales, moving between enclosures, entering crates, climbing up onto branches on command and remaining stationary for routine examinations (Blessington and Steenberg 2007; K Coombes per. obs.).

## 9.9 Introductions and removals

When animals are introduced for the first time, particularly males, but for some species also females (e.g. rock-wallabies), they can show significant aggression towards each other until they determine their hierarchy (D. Taggart pers. obs.; Schultz *et al.* 2006). Complex enclosure design can help alleviate these issues to some extent. When animals are taken out and returned later there are generally few problems.

Observations of the behaviour of wild-caught western grey kangaroos and red kangaroos brought into captivity suggest they adapt relatively quickly (Munn *et al.* 2017). In their study, Munn *et al.* (2017) observed that on the first day in captivity grooming was higher than by their wild counterparts, which potentially suggested heightened stress or possibly efforts to remove human smells and debris derived from their capture. However by days 7–10 they showed behavioural patterns similar to their free-ranging counterparts (Munn *et al.* 2017).

## 9.10 Intraspecific compatibility

There is a general trend towards larger group sizes in species with larger body size and a link between group size and feeding style (Jarman and Coulson 1989). The use of daytime refuges such as nests or squats is associated with solitary animals, while the use of rocky refuges or burrow systems promotes gregariousness, at least while resting (Jarman and Coulson 1989). An analysis of the social structure of 17 species in the Macropodidae was conducted by Croft (1989) using group size, group structure and home-range overlap patterns for each species during typical active and inactive periods. The type of social organisation designated for each species is shown in Table 10.15. Type 1 species are considered solitary except during reproduction; Type 2 are often solitary but aggregate on favoured resource patches; Type 3 are gregarious. Types 1 and 2 are small to medium-sized with no or little sexual dimorphism, nocturnal, use closed shelter sites and forage in or close to cover. Type 3 species are large and show strong sexual dimorphism, are partially diurnally active, occupy open shelter and foraging habitats and eat predominantly grass. Types 2 and 3 show spacing behaviour where feeding ranges of both sexes overlap and territorial behaviour is absent. Male home ranges are usually larger than

**Table 10.15.** Social organisation of macropods and the suggested sex ratio of different species when held in captivity.

Type 1 are solitary, Type 2 group only when adequate food and Type 3 are gregarious.
Derived from Croft (1989).

| Genus | Social type | Suggested sex ratio (M:F) |
|---|---|---|
| **Hypsiprymnodontidae** | | |
| *Hypsiprymnodon* | 1 | 1:1–2 |
| **Potoroidae** | | |
| *Aepyprymnus* | 1 | 1:1–2 |
| *Bettongia* | 1 | 1:1–2 |
| *Potorous* | 1 | 1:1–2 |
| **Macropodidae** | | |
| *Dendrolagus* | 1 | 1:1 |
| *Petrogale* | 1–2 | 1:1–2 to 2–8 |
| *Thylogale* | 2 | 1:1–2 |
| *Lagorchestes* | 1 | 1:2–3 |
| *Macropus fuliginosus* | 3 | 1:5–6 |
| *Macropus giganteus* | 3 | 1:5–6 |
| *Notamacropus agilis* | 2 | 1:4–5 |
| *Notamacropus dorsalis* | 3 | 1:5–6 |
| *Notamacropus eugenii* | 2 | 1:2–3 |
| *Notamacropus parryi* | 3 | 1:5–6 |
| *Notamacropus rufogriseus* | 2 | 1:2–3 |
| *Osphranter antilopinus* | 3 | 1:5–6 |
| *Osphranter robustus* | 2 | 1:5–6 |
| *Osphranter rufus* | 3 | 1:5–6 |
| *Setonix* | 2 | 1:1–2 |
| *Wallabia* | 1 | 1:2–3 |

females' ranges and more so in Type 2 species where females are not gregarious. Social structure is also very closely linked to the species' mating strategies and intermale competition for mates (Taggart *et al.* 1998; Paplinska *et al.* 2010). It is therefore very important not to overstock enclosures, as this can lead to greater competition for food and resting sites, resulting in higher levels of aggression, stress and disease of, particularly subordinate, animals. The total number of animals held, and even the sex ratio, also depends on the total area available and, in the case of some species such as rock-wallabies, the number of crevices and other shelters. Tree-kangaroos should not be held with any other species.

## 9.11 Interspecific compatibility

Red kangaroos, central euros and yellow-footed rock-wallabies have been held together, but has resulted in fighting between males of the different species, causing minor injuries in open spaces and more serious injuries in confined spaces. In other cases where red kangaroos and brush-tailed rock-wallabies have been held together, no problems were observed. Similar findings have been made for male western grey kangaroos, Kangaroo Island kangaroos and Tammar wallabies. Red kangaroos have been held together as pairs with Tammar wallabies; however, in the absence of the male red kangaroo, Tammar wallabies were observed to be very aggressive to female red kangaroos in oestrus, to the extent that the female red kangaroos have thrown their pouch young due to continual harassment.

Tree-kangaroos have been reported to kill several other species in captivity, including snakes, swamp hens (*Porphyrio porphyrio*), Cape Barren geese (*Cereopsis novaehollandiae*), kookaburras (*Dacelo novaeguineae*), Bourke's parrots (*Neophema bourkii*) and grey goshawks (*Accipiter novaehollandiae*) (Mullet *et al.* 1990; K. Coombes pers. obs.). Tree-kangaroos have also been observed to attack Edward's lories (*Trichoglossus haematodus edwardsi*), water dragons (*Physignathus* sp.), Gouldian finches (*Erythrura gouldiae*) and long-nosed potoroos (Mullet *et al.* 1990). However, Edwards lories have been observed becoming increasingly territorial towards tree-kangaroos after their young had hatched, resulting in the birds sitting on the heads of the tree-kangaroos and biting at their ears (Mullet *et al.* 1990). Lumholtz's tree-kangaroos have been observed to regularly chase and grabbing/attacking red-legged pademelons in captivity (K. Coombes pers. obs.). Bush and Montali (1999) recommend that tree-kangaroos not be housed with avian species because of the danger of avian tuberculosis. As a result of issues with various other species, tree-kangaroos should not be housed with any other species.

Macropods have been held with many species of birds, including Cape Barren geese (*Cereopsis novaehollandiae*), black swans (*Cygnus atratus*), jabirus (*Xenorhynchus asiaticus*), magpie geese (*Anseras semipalmata*), emus (*Dromaius novaehollandiae*), ostriches (*Struthio camelus*), brolgas (*Grus rubicunda*) and numerous species of ducks (family Anatidae), as well as with fallow deer (*Dama dama*). Smaller species of macropods have also been exhibited within koala (*Phascolarctos cinereus*) and echidna (*Tachyglossus aculeatus*) enclosures.

Smaller macropods such as brush-tailed bettongs, rufous bettongs, long-nosed potoroos and short-nosed potoroos have been placed together with echidnas, ring-tailed possums, sugar gliders, squirrel gliders, yellow-bellied gliders and Leadbeater's possums, tuans and tawny frogmouths in a nocturnal house without problems (L. Andrews pers. comm.).

# 10. BREEDING

## 10.1 Mating system

Most macropods are polygynous, the major exception being some rock-wallabies, which appear to be socially monogamous. The degree of intermale sperm competition in females of various species within the Macropodidae has been outlined by Paplinska *et al.* (2010), providing compelling evidence for particular mating strategies and degrees of polygamy and monogamy within some species/groups. Spencer (1996) found that the allied rock-wallaby is socially monogamous, although ~33% of its offspring are fathered by the non-paired mate. For most of the larger macropods within the genera *Macropus* and *Wallabia*, a single male is generally placed with up to 5–6 females. Subadult males are usually permitted within the recommended sex ratios, but should be removed from the enclosure before sexual maturity. Tree-kangaroo and musky rat-kangaroo males should be removed after mating, as they often interfere with the female, causing her to lose her pouch young (Johnson *et al.* 1983). Suggested sex ratios that are often used successfully are shown in Table 10.15.

## 10.2 Ease of breeding

### 10.2.1 Musky rat-kangaroos

Musky rat-kangaroos breed relatively well in captivity (Johnson 1983).

### 10.2.2 Potoroos and bettongs

Most species of potoroos and bettongs breed well in captivity, although several populations, such as those of the long-footed potoroo at Healesville Sanctuary, have greatly reduced breeding, which is thought to be due to inbreeding and avian tuberculosis.

### 10.2.3 Tree-kangaroos

Some institutions have bred tree-kangaroos routinely, while others have not been as successful. The major limiting factor seems to be having at least a few individuals of each sex as research has demonstrated that female choice is important (Kokko *et al.* 2003; Coombes 2005).

### 10.2.4 Hare-wallabies, nail-tailed wallabies, quokkas and pademelons

All species generally breed well in captivity.

### 10.2.5 Rock-wallabies

Rock-wallabies can breed very well in captivity, even in very small enclosures. They often do not breed for an extended period of time and then once they start, they continue to routinely produce young. Upsetting established social structures can be detrimental to breeding success in several rock-wallaby species (Taggart *et al.* 2005, 2016; D. Taggart pers. obs.).

### 10.2.6 Kangaroos and wallabies

Breed well in captivity. The use of artificial insemination of Tammar wallabies has been undertaken successfully (Paris *et al.* 2005).

## 10.3 Oestrous cycle and gestation period

The oestrous cycle of the various species of macropods typically varies between 20 and 40 days. The potoroids tend to have cycles of 20–30 days and the macropodoids have cycles of 30–45 days. The gestation period for most species of macropods is between 25 and 35 days, with the bettongs being around 21 days and the potoroos up to 38 days (Table 10.16).

**Table 10.16.** Duration of oestrous cycle, gestation of several macropods.

Numbers in brackets are mean values.

| Species | Oestrous cycle (days) | Gestation (days) | Postpartum oestrus | Embryonic diapause | Reference |
|---|---|---|---|---|---|
| **Hypsiprymnodontidae** | | | | | |
| *Hypsiprymnodon moschatus* | 25–26 | 19 | N | N? | 1 |
| **Potoroidae** | | | | | |
| *Aepyprymnus rufescens* | 21–36 | 21–30 | Y | Y | 2, 3 |
| *Bettongia gaimardi* | 17-37 (23) | 20–22 (21) | Y | Y | 4 |
| *B. lesueur* | 11–35 (23) | 22–33 (21) | Y | Y | 5 |
| *B. penicillata* | 22–23 | 21 | Y | Y | 6 |
| *B. tropica* | 21–23 (22) | 20–23 (21) | Y | Y | 7–9 |
| *Potorous longipes* | – | 38? | Y? | Y? | 7 |
| *P. tridactylus* | 42 | 38 | Y | Y | 10 |
| **Macropodidae** | | | | | |
| *Dendrolagus lumholtzi* | 47–64 | 42–48 | N | N | 11 |
| *Petrogale assimilis* | – | 29–34 | Y | Y | 12 |
| *P. concinna* | 31–36 | 30–32 | Y | Y | 13 |
| *P. inornata* | 30–32 | 30–32 | Y | Y | 14 |
| *P. penicillata* | – | 31 | Y | Y | 15 |
| *P. persephone* | 33–38 | 30–34 | Y | Y | 16 |
| *P. purpureicollis* | 36–38 | 33–35 | Y | Y | 17 |
| *P. xanthopus* | 32–37 | 31–33 | Y | Y | 18 |

| Species | Oestrous cycle (days) | Gestation (days) | Postpartum oestrus | Embryonic diapause | Reference |
|---|---|---|---|---|---|
| *Thylogale billardierii* | 30 | 30 | Y | Y | 19 |
| *T. stigmatica* | 29-32 | 28-30 | Y | Y | 20 |
| *Lagorchestes conspicillatus* | 30 | 29-31 | Y | Y | 7, 21 |
| *Macropus fuliginosus* | 35 | 31 | N | N | 22, 23 |
| *M. giganteus* | 46 | 34 | N | Y | 22-24 |
| *Notamacropus agilis* | 32 | 29 | Y | Y | 25 |
| *N. dorsalis* | - | 33-35 | - | - | 7 |
| *N. eugenii* | 31 | 29 | Y | Y | 26 |
| *N. parma* | 42 | 35 | N | Y | 7, 27 |
| *N. parryi* | 41-44 | 34-38 | N | Y | 7, 28-30 |
| *N. rufogriseus* | 32-33 | 29-30 | Y | Y | 31 |
| *Onychogalea frenata* | 30-45 | 21-26 | N | Y | 32 |
| *Osphranter antilopinus* | 36-44 | 34 | N? | N? | 7 |
| *O. robustus* | 33-45 | 32 | Y | Y | 4, 7 |
| *O. rufus* | 35 | 33 | Y | Y | 28, 33 |
| *Wallabia bicolor* | 31-33 | 33-38 | N* | Y | 28, 34 |
| *Setonix brachyurus* | 28 | 25-27 | Y | Y | 7, 35-38 |
| *Lagostrophus fasciatus* | - | - | Y | Y | 7 |

References: (1) Lloyd (2001); 2) Moors (1975); (3) Johnson (1978); (4) Rose (1978); (5) Tyndale-Biscoe (1968); (6) Parker (1977); (7) Strahan (1995); (8) Smith (1998); (9) Johnson and Delean (2001); (10) Shaw and Rose (1979); (11) Johnson and Delean (2003); (12) Delaney (1997); (13) Nelson and Goldstone (1986); (14) Johnson, (1979); (15) Taggart *et al.* (2005); 16) Johnson and Delean (1999); (17) Johnson and Delean (2002a); (18) Poole *et al.* (1985); (19) Rose and McCartney (1982a); (20) Johnson and Vernes (1994); (21) Johnson (1993); (22) Poole and Catling (1974); (23) Poole (1975); 24) Kirkpatrick (1965b); (25) Merchant (1976); 26) Merchant (1979); (27) Maynes (1973a); (28) Calaby and Poole (1971); (29) Maynes (1973b); (30) Johnson (1998); (31) Merchant and Calaby (1981); (32) Johnson (1997); (33) Sharman (1963); (34) Paplinska *et al.* (2006); (35) Sharman (1955a); (36) Sharman (1955b); (37) Tyndale-Biscoe (1963); (38) Shield (1968).
*Swamp wallabies are unusual in having a prepartum oestrus with mating up to 8 days before the neonate is born (Strahan 1995; Paplinska *et al.* 2006).

## 10.4 Reproductive status

### 10.4.1 Females

Macropods are generally placed in several categories depending on their reproductive status:

- non-parous (subadult females that have never bred) - pouch shallow with no skin folds, clean and dry, teats very small inverted
- parous (females that have bred previously but not presently) - pouch is small-moderate but distinct, dry and dirty, the teats are slightly elongated, everted
- oestrus/pregnant - pouch is small-moderate, clean, pink in colour, moist and glandular in appearance, skin folds may be observed on the lateral margins of the pouch
- pouch young present - pouch is moderate-deep, moist, pouch young attached to the teat
- lactating (young absent from the pouch but still suckling) - pouch area is large, deep, moist, skin folds flaccid, hair sparse and stained, skin smooth and dark pink, teats elongated
- post lactation - pouch area is moderate-deep, teats expressing only clear liquid and/or regressing.

If pouch young are present, several developmental stages and measurements can be recorded and compared with existing growth curves (see section 10.16) or new curves can be established for future reference. These include:

- developmental stages
  - sex distinguishable
  - tips of ears free
  - papillae of facial vibrissae evident
  - eyelashes visible
  - eyes open
  - fur visible - slight tinge, medium or well developed
  - tips of first incisors through the gums
  - feet soiled (indicating occasional excursions from the pouch)
  - eating solids
  - at foot

    - self-feeding
    - independent
- measurements (see Appendix 3):
    - bodyweight (g) – if not on teat
    - head length (mm) – from the occiput to snout tip
    - head width (mm) – maximum width across the zygomatic arches
    - crown–rump length (mm) – primarily for very small neonates
    - body length (mm) – from snout tip to cloaca
    - tail length (mm) – from the cloaca to the end of the last vertebra of the tail tip
    - total length (mm) – from snout tip to tail tip
    - tibia length (mm) – from the hip to the bottom of the pes
    - pes length (mm) – from the heel to the base of the longest toe, not including the claw.

### 10.4.2 Males

In males of seasonally breeding species such as musky rat-kangaroos, the size of testes can be measured, as they increase in size during the breeding season (Dennis 1997). Measure the length, width and depth in millimetres. Testis volume can be calculated by using the equation $V = \pi/6 \times (\text{length}) \times (\text{width})^2$ and the testes volume can be determined by multiplying this by two (Spencer 1996). Change in scrotal diameter can also be used as a predictor of sexual maturity in macropods (Taggart *et al.* 2005; Paplinska *et al.* 2010).

## 10.5 Reproductive management

Several techniques are used to control breeding in situations where large numbers of macropods are held together (Portas 2025b). These include:

- separation of the sexes
- vasectomy of the dominant male (can include all independent males as well)
- castration of all breeding-age males
- removal of pouch young
- fertility control through hormonal contraceptive implants, including the synthetic progestin levonorgestrel and the gonadotropin-releasing hormone agonist deslorelin (Herbert 2004; Herbert *et al.* 2005)
- tubal ligation in females
- culling.

The advantages and disadvantages of the different methods are listed in Table 10.17 (Booth and Srb 1999). Sometimes a combination of techniques may be used, such as vasectomy of the dominant male and removing any young or non-preferred sex that may be produced from subordinate males.

## 10.6 Occurrence of hybrids

Care needs to be taken that hybridisation does not occur between different species of macropods (Table 10.18). Hybrids have occurred between species that do not have the same number of chromosomes and between genera. Numerous hybrids have been produced to date, but generally they are not fully fertile as they usually have deformed reproductive organs, including empty scrota, small testes, no spermatogonia and small non-everted teats (Close and Lowry 1990).

## 10.7 Timing of breeding

There is a large variation in the initiation of breeding in macropods, depending on the species. Breeding can be seasonal, continuous, dependent on rainfall or the result of summer and winter solstices (Table 10.19).

Some species are continuous breeders and have a postpartum oestrus shortly after they give birth (the swamp wallaby has a prepartum oestrus). Some species are facultative breeders and can enter a drought-induced anoestrus if food and water remain in poor supply for > 6 months (Tyndale-Biscoe and Renfree 1987). These species can have three young in different stages of development: a blastocyst in a state of embryonic diapause that results from the postpartum oestrus, a small neonate that is attached to the teat and a third young that has permanently left the pouch but is still suckling from the mother's teat (and receiving milk that is different to that of the small young) (Tyndale-Biscoe and Renfree 1987).

Seasonal breeders include the quokka, which has a distinct breeding season on Rottnest Island, although it is a continuous breeder on the mainland. The Tammar wallaby and the Bennett's wallaby (the Tasmanian subspecies of the red-necked wallaby) are influenced by the summer and winter solstices, which determine their breeding system (Berger 1966; Berger and Sharman 1969; Sadleir and Tyndale-Biscoe 1977; Tyndale-Biscoe and Renfree 1987). In these species the female has an embryo in diapause, which is reactivated by the summer solstice in late December each year (in the southern hemisphere). That young is then born in late January–March, at which time the female goes into postpartum oestrus, mates and produces another embryo in diapause. If the pouch young is

**Table 10.17.** Advantages and disadvantages of different techniques used to manage captive macropod numbers.

Taken from Booth and Srb (1999).

| Advantages | Disadvantages |
|---|---|
| **Separation of sexes** | |
| Simple and effective | Additional resources needed to hold two populations |
| Resumption of breeding can occur at any time | Animals may pace the fenceline to gain access to the other sex if nearby |
| **Vasectomy of dominant males** | |
| Retention of a functional social group | Surgical procedure |
| Retention of secondary sex characteristics (musculature and dominant behaviour) | Not reversible in practice |
| Ability to retain fertile subordinate males within the social group that can be given mating opportunities later | Reconnection of duct may occur, unless a large piece is removed, resulting in regained fertility |
| | Exhaustion of dominant male due to frequent cycling in females |
| | Sneaky breeding may still occur by subordinate males, though not if independent males are vasectomised as well |
| **Immunocontraception** | |
| Reversible | Still somewhat experimental |
| Can be used on both sexes | Affects social behaviour because of hormonal alteration |
| No surgery required | Duration of effect still unpredictable |
| **Castration of all breeding males** | |
| Simple and effective surgical procedure | Secondary sex characteristics lost (including changed behaviour and appearance) |
| Testicular material can be harvested for genome banking or research purposes | Loss of social structure |
| | Not reversible |
| | Cannot retain subordinate fertile males in the group |
| | Females continue to cycle repeatedly and may behave unpredictably with the public |
| **Tubal ligation of females** | |
| Retention of a functional social group | Invasive surgical procedure<br>Ethically questionable |
| Selection of breeding females is possible | Not reversible in practice |
| | Can maintain a large display as breeding is controlled |
| **Removal of pouch young** | |
| Simple and effective | Ethically questionable |
| Can control the sex of young<br>Can be used to synchronise breeding among females | Generally rapid replacement |
| **Culling** | |
| Simple and effective | Ethically questionable |
| Carcasses can be utilised | Negative public image |

lost before the winter solstice, the embryo in diapause develops and a young is born, whereas if the young survives or dies after the winter solstice, the diapause embryo will not develop until the following summer solstice. In the northern hemisphere, the Bennett's wallaby changes its breeding cycle by exactly 6 months, with births occurring in August and September (Fleming *et al.* 1983).

**Table 10.18.** Hybrid macropods that have been produced in captivity.

2n = number of chromosomes in the genome.

| Male parent species | 2n | Female parent species | 2n | Sex of offspring | Reference |
|---|---|---|---|---|---|
| **Potoroidae** | | | | | |
| *Bettongia penicillata* | 22 | *Bettongia gaimardi* | 22 | – | 1 |
| | | | | | |
| **Macropodidae** | | | | | |
| *Notamacropus agilis* | 16 | *Osphranter rufus* | 20 | F, M | 1 |
| *Notamacropus agilis* | 16 | *Macropus giganteus* | 16 | – | 1 |
| *Notamacropus agilis* | 16 | *Notamacropus rufogriseus* | 16 | F, M | 1 |
| *Osphranter antilopinus* | 16 | *Osphranter robustus robustus* | 16 | – | 1 |
| *Osphranter bernardus* | 18 | *Osphranter antilopinus* | 16 | – | 2 |
| *Notamacropus dorsalis* | 16 | *Notamacropus parryi* | 16 | M | 1 |
| *Notamacropus eugenii* | 16 | *Notamacropus agilis* | 16 | F, M | 1 |
| *Notamacropus eugenii* | 16 | *Notamacropus dorsalis* | 16 | M | 1 |
| *Notamacropus eugenii* | 16 | *Notamacropus rufogriseus* | 16 | M | 1 |
| *Notamacropus eugenii* | 16 | *Setonix brachyurus* | 22 | – | 1 |
| *Macropus fuliginosus melanops?* | 16 | *Macropus fuliginosus fuliginosus?* | 16 | – | 3 |
| *Macropus fuliginosus* | 16 | *Macropus giganteus* | 16 | F, M | 1 |
| *Macropus fuliginosus* | 16 | *Osphranter rufus* | 20 | F, M | 1 |
| *Macropus giganteus* | 16 | *Notamacropus rufogriseus* | 16 | M | 1 |
| *Macropus giganteus* | 16 | *Osphranter rufus* | 20 | F, M | 1 |
| *Notamacropus parma* | 16 | *Notamacropus dorsalis* | 16 | – | 1 |
| *Notamacropus parma* | 16 | *Notamacropus eugenii* | 16 | M | 1 |
| *Notamacropus parryi* | 16 | *Notamacropus rufogriseus* | 16 | ? | 1 |
| *Notamacropus parryi* | 16 | *Macropus giganteus* | 16 | – | 1 |
| *Osphranter robustus robustus* | 16 | *Osphranter robustus erebescens* | 16 | F, M | 1 |
| *Osphranter robustus* | 16 | *Osphranter rufus* | 20 | F, M | 1 |
| *Notamacropus rufogriseus* | 16 | *Notamacropus dorsalis* | 16 | – | 1 |
| *Notamacropus rufogriseus rufogriseus* | 16 | *Notamacropus rufogriseus banksianus* | 16 | M | 1 |
| *Notamacropus rufogriseus* | 16 | *Thylogale thetis* | 22 | – | 1 |
| *Osphranter rufus* | 20 | *Osphranter robustus erebescens* | 16 | M | 3 |
| *Petrogale brachyotis* | 18 | *Petrogale concinna* | 16 | – | 1 |
| *Petrogale godmani* | 20 | *Petrogale assimilis* | 20 | F, M | 1 |
| *Petrogale mareeba* | 18 | *Petrogale assimilis* | 20 | – | 1 |
| *Petrogale rothschildi* | 22 | *Petrogale lateralis hacketii* | 20 | – | 1 |
| *Petrogale xanthopus* | 22 | *Petrogale persephone* | 22 | F | 1 |
| *Petrogale xanthopus* | 22 | *Petrogale persephone* | 22 | F | 4 |
| *Thylogale stigmatica* | 22 | *Thylogale thetis* | 22 | F | 1,5 |
| *Wallabia bicolor* | 11 | *Osphranter robustus robustus* | 16 | M | 1 |
| *Wallabia bicolor* | 11 | *Notamacropus rufogriseus* | 16 | M | 1 |
| *Wallabia bicolor* | 11 | *Notamacropus agilis* | 16 | – | 1 |
| *Wallabia bicolor* | 11 | *Osphranter rufus* | 20 | M? | 6 |

Rferences: (1) Close and Lowry (1990); (2) G. Males pers. comm.; (3) H. Guy pers. comm.; (4) K. Engle pers. comm.; (5) Close and Bell (1997); (6) J. Thomas pers. comm.

**Table 10.19.** Reproduction and development of macropods.

| Species | Birth season | Permanent pouch exit (days) | Weaning (days) | Sexual maturity (months) | | Reference |
|---|---|---|---|---|---|---|
| | | | | Females | Males | |
| **Hypsiprymnodontidae** | | | | | | |
| *Hypsiprymnodon moschatus* | Feb–Apr | 180 | 270 | > 12 | – | 1–3 |
| **Potoroidae** | | | | | | |
| *Aepyprymnus rufescens* | All year | 114 | 155 | 10 | 12 | 4 |
| *Bettongia gaimardi* | All year | 109 | 160 | 8–11 | | 5 |
| *Bettongia lesueur* | All year | 115 | 165 | 7 | 14 | 6 |
| *Bettongia penicillata* | All year | 100 | 130 | 10–12 | 12 | 7, 8 |
| *Potorous longipes* | All year | 140–150 | 106 | 12? | 12? | 9, 10 |
| *Potorous tridactylus* | Jun–Aug | 130 | 147 | 12 | – | 11–13 |
| *Bettongia tropica* | All year | 102–112 | 166–185 | 6–14 | 8–17 | 14–16 |
| **Macropodidae** | | | | | | |
| *Dendrolagus lumholtzi* | All year | 246–275 | 357–511 | 24 | 55 | 17 |
| *Petrogale assimilis* | All year | 180–231 | 253–387 | 18 | 20–24 | 18–20 |
| *P. concinna* | All year | 160–180 | 175 | 12–16 | 12+ | 21, 22 |
| *P. inornata* | All year | 189–227 | 290 | 18 | 20 | 23 |
| *P. penicillata* | All year | – | 230 | 18 | 20–24 | 9, 24, 25 |
| *P. persephone* | All year | 203–215 | 300–350 | 21 | 25 | 26 |
| *P. purpureicollis* | All year | 178–197 | 270–350 | 18 | 22 | 27 |
| *P. xanthopus* | All year | 190–201 | 210–235 | 18 | 18 | 9, 28, 29 |
| *Thylogale billardierii* | Apr–Jun | 200 | 300 | 14–15 | 14–15 | 9, 30–32 |
| *T. stigmatica* | All year* | 174–183 | 220–294 | 11.4 | 15.5 | 33 |
| *T. thetis* | All year | 181 | 210 | 12–17 | 17 | 9, 34 |
| *Lagorchestes conspicillatus* | All year | 152 | – | 11.7 | 15.4 | 35 |
| *L. hirsutus* | All year | 124 | – | 5–18 | 14 | 9 |
| *Macropus fuliginosus* | All year | 300–310 | 540 | 14 | 31 | 36–39 |
| *M. giganteus* | All year | 319 | 540 | 18 | 48 | 36, 37, 39, 40 |
| *Notamacropus agilis* | All year | 219 | 328 | 12 | 14 | 41–43 |
| *N. dorsalis* | All year? | 210 | – | 14 | 20 | 9 |
| *N. eugenii* | Jan–Jun | 250 | 270 | 8 | 24 | 44, 45 |
| *N. irma* | Apr–May? | 180–210? | – | – | – | 9 |
| *N. parma* | All year | 210 | 270–300 | 16 | 22 | 46 |
| *N. parryi* | All year? | 256–267 | 420–450 | 17–22 | 18–24 | 9, 47, 48 |
| *N. rufogriseus* | Jan–June/ All year | 270–280 | 360–510 | 11–21 | 13–19 | 9, 49, 50 |
| *Onychogalea frenata* | All year | 119–126 | 190–245 | 4.5–9 | 9–14 | 51 |
| *Osphranter antilopinus* | All year | 270 | – | – | – | 9 |
| *O. robustus* | All year | 255–260 | 380 | 18–24 | 18–24 | 9, 52 |
| *Wallabia bicolor* | All year | 256 | 450 | 15–18 | 15–18 | 9, 53 |
| *Setonix brachyurus* | All year | 190 | 240 | 9 | 13 | 54 |
| *Lagostrophus fasciatus* | Jan–Feb | 180 | 270 | 12 | 12 | 9 |

*Has been observed to breed between October and June in the wild.

References: (1) Johnson (1983); (2) Dennis (1997); (3) Dennis and Marsh (1997); (4) Johnson (1978); (5) Rose (1978); (6) Tyndale-Biscoe (1968); (7) Sampson (1971); (8) Parker (1977); (9) Strahan (1995); (10) Green and Mitchell (1997); (11) Hughes (1962); (12) Heinsohn (1968); (13) Shaw and Rose (1979); (14) Johnson and McIlwee (1997); (15) Smith (1998); (16) Johnson and Delean (2001); (17) Johnson and Delean (2003); (18) Close and Bell (1990); (19) Eldridge and Close (1995); 20) Delaney (1997); (21) Lee and Cockburn (1985); (22) Nelson and Goldstone (1986); (23) Johnson (1979); (24) Batchelor (1980); (25) Taggart *et al.* (2005); (26) Johnson and Delean (1999); (27) Johnson and Delean (2002b); (28) Poole *et al.* (1985); (29) Robinson *et al.* (1994); (30) Rose and McCartney (1982a); 31) Rose and McCartney (1982b); (32) Rose (1987); (33) Johnson and Vernes (1994); (34) Johnson (1977); (35) Johnson (1993); 36) Poole (1973); 37) Poole (1975); (38) Poole (1976); (39) Poole and Catling (1974); (40) Kirkpatrick (1965b); 41) Kirkpatrick and Johnson (1969); (42) Merchant (1976); (43) Bolton *et al.* (1982); (44) Andrewartha and Barker (1969); 45) Murphy and Smith (1970); (46) Maynes (1973a); (47) Maynes (1973b); 48) Johnson (1998); (49) Catt (1977); 50) Merchant and Calaby (1981); (51) Johnson (1997); (52) Ealey (1967); (53) Calaby and Poole (1971); (54) Shield (1968).

## 10.8 Age at first and last breeding

The age at first breeding varies greatly among the different species of macropods, ranging from ~8 months in the Tammar wallaby to up to 24 months in the whiptail wallaby and the nabarlek (Table 10.19). Most species of macropods will generally breed up until they die.

## 10.9 Ability to breed every year

All macropods have the ability to breed every year, but some species such as the red kangaroo will stop breeding if there is a drought for an extended period of time.

## 10.10 Ability to breed more than once per year

Macropods are not able to fully raise more than one young in a year, although several species are able to mate shortly after birth (postpartum oestrus), which results in an embryo in the state of diapause in which a blastocyst is formed (Table 10.19) (Smith 1981; Selwood 1986). The neonate is then born shortly after the joey in the pouch vacates the pouch permanently. These macropods are able to have three young at different stages of development: one in a state of embryonic diapause, one in the pouch attached to the teat, and one which has vacated the pouch permanently and which puts its head into the pouch to feed. Oestrus can be determined by examining the changes in the proportions of partly cornified epithelial cells in smears taken from the anterior urogenital sinus during the oestrous cycle (Poole *et al.* 1992). In Tammar wallabies a marked and consistent feature of the onset of oestrus is a decrease from ~80% to 20% partly cornified cells (Poole *et al.* 1992).

## 10.11 Nesting requirements

Potoroids need to be supplied with grass tussocks and/or hay to allow them to build their nests. In the wild, some species such as brush-tailed rock-wallabies do not appear to breed until they have several or more refuges or dens available to them (Jarman and Bayne 1997). Therefore, in captivity it appears important to provide at least 2–3 elevated refuges such as small wooden boxes, hollow logs or rocky crevices where they can rest during the day.

## 10.12 Breeding diet

There is no specific diet that is required before, or during, the breeding season.

## 10.13 Number of young born

Only one young is born at a time, although there are rare occurrences of twins. Twinning has been recorded in Tammar wallabies, red kangaroos, red-necked wallabies, eastern grey kangaroos, western grey kangaroos and euros (Troughton 1947; Frith and Sharman 1964; Sharman and Pilton 1964; Newsome 1965; Ealey 1967; Inns 1980; Norbury 1987; van Oorschot and Cooper 1989; Libke and Libke 2000).

There is also one record of triplets in red kangaroos (Troughton 1947). The only species of macropod that routinely produces twins, and even triplets or quadruplets, is the musky rat-kangaroo (Dennis 1997; Dennis and Marsh 1997; Johnson *et al.* 2005).

In tree-kangaroos, the male should be removed ~40 days after copulation, or once pouch young are observed, as there is an extremely high death rate of pouch young as a result of expulsion from the pouch (Bush and Montali 1999). If young are not observed 50 days after copulation, then the male is returned for breeding (Bush and Montali 1999).

## 10.14 Age at weaning

The age at weaning of species for which it is known, is shown in Table 10.19. In several genera, keeping juveniles and parents together after weaning can potentially cause problems:

- in captive populations of brush-tailed bettongs the juveniles were removed at ~550 g, which usually occurred after 120 ± 14 days. When pouch young developed past the 550-g stage, young males were often killed and young females were harassed by the adult male in the pen (Delroy *et al.* 1986). Brush-tailed bettong females can give birth 22 days after the young have left the pouch and can breed continuously (Delroy *et al.* 1986)
- rock-wallaby juveniles should also be removed, unless they are in a large enough enclosure, as juveniles can be attacked by adult animals (Close and Bell 1990).
- tree-kangaroo joeys usually wean around 12 months, once the female gives birth to the next joey, but can be safely kept with mother well after this date. Joeys stay with mother until 2–2.5 years of age. In the wild female can have up to three joeys of varying ages at foot and another in the pouch at one time (Coombes 2005).

## 10.15 Age at removal from the mother

In smaller species, such as the potoroids, the young should generally be removed at weaning as the parents often become intolerant of them. In larger species, they generally remain part of the mob until the males are large enough to begin challenging the dominant male for mating rights.

## 10.16 Growth and development

Growth curves have been developed for several macropod species (Fig. 10.17; Table 10.20). A compilation of some of the growth and development curves that have been prepared can be found in Bach (1998). Growth curves derived from hand-reared animals need to be treated with caution, as studies of Tasmanian bettongs showed wild-reared animals weighed significantly less than those raised in captivity (Taylor and Rose 1987). The age of different growth stages can be found within Table 10.21.

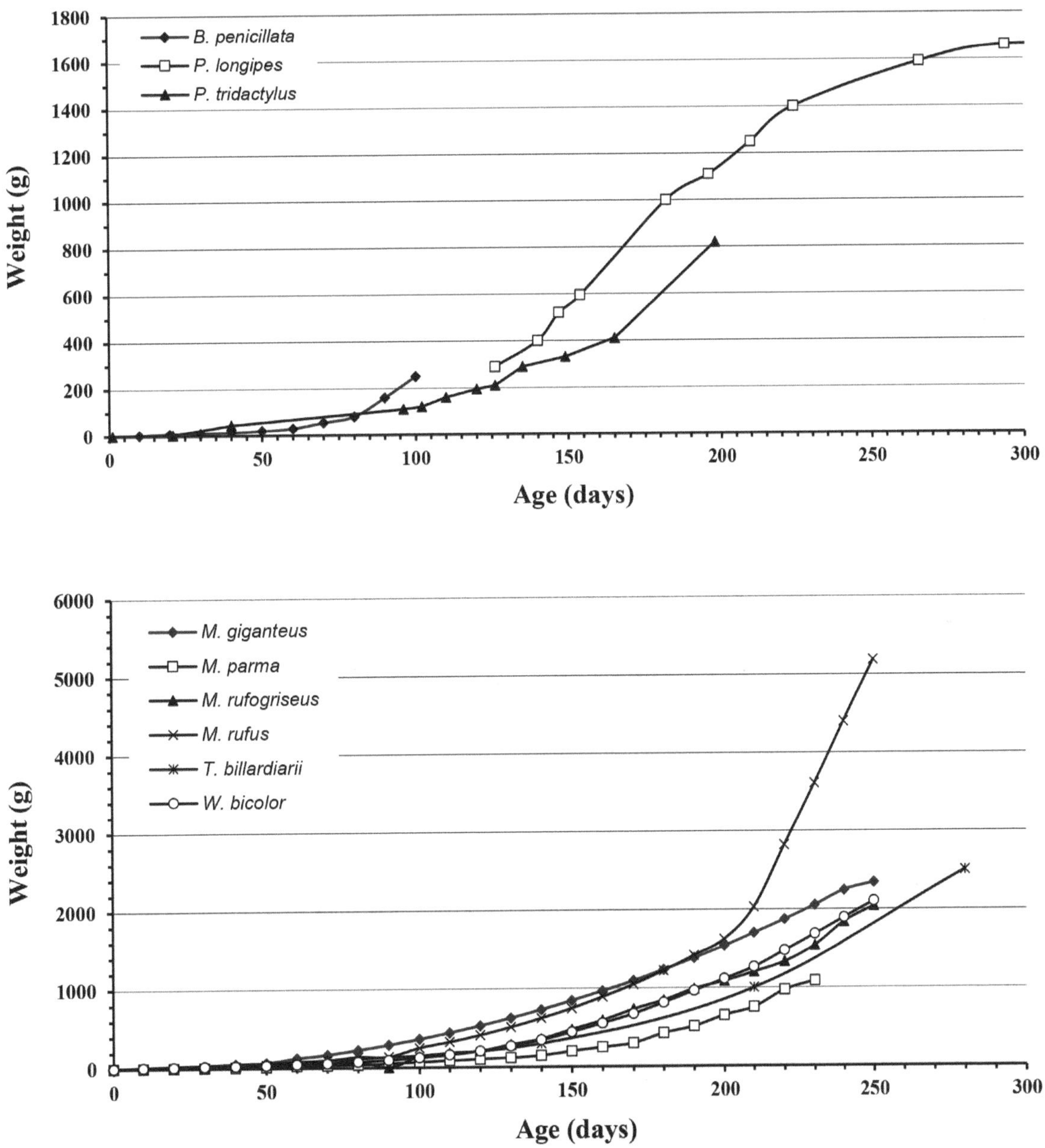


**Fig. 10.17.** Growth in bodyweight of several species of macropods. Derived from Maynes (1976), Rose and McCartney (1982b), Bryant (1989), Bellamy (1992), Seebeck (1992) and Merchant *et al.* (1994).

**Table 10.20.** Growth curve measurements that have been developed for different species of macropods.

CR = crown–rump length, HB = head and body length, HE = head length, PE = pes length, TA = tail length, WT = bodyweight.

| Species | Measurements used | Reference |
|---|---|---|
| **Hypsiprymnodontidae** | | |
| *Hypsiprymnodon moschatus* | HL | 1 |
| **Potoroidae** | | |
| *Aepyprymnus rufescens* | PE, TA | 2 |
| *Bettongia gaimardi* | WT, PE, TA | 3, 4 |
| *B. lesueur* | HE, PE | 5–7 |
| *B. penicillata* | WT, CR, HE, PE | 8, 9 |
| *B. tropica* | HE, PE | 10 |
| *Potorous longipes* | WT, HE, TA, | 8, 11 |
| *P. tridactylus* | WT, HB, HE, PE, TA | 3, 6, 12–15 |
| **Macropodidae** | | |
| *Dendrolagus lumholtzi* | HE, PE | 16 |
| *Petrogale assimilis* | HE, PE | 17, 18 |
| *P. inornata* | PE, TA | 19 |
| *P. lateralis* | HE, PE | 20 |
| *P. penicillata* | PE, TA | 19, 21 |
| *P. persephone* | HE, PE | 22 |
| *P. purpureicollis* | HE, PE | 23 |
| *P. xanthopus* | HE, PE, TA | 24 |
| *Thylogale billardierii* | WT, HE, PE, TA | 3, 4, 25 |
| *T. stigmatica* | HE, PE | 26 |
| *Lagorchestes conspicillatus* | PE, TA, | 27 |
| *L. hirsutus* | HE, PE, TA | 28 |
| *Macropus fuliginosus fuliginosus* | WT, HE, PE, TA | 29 |
| *M. f. melanops* | WT, HE, PE, TA | 29 |
| *M. giganteus* | WT, HE, PE, TA | 3, 4, 30–32 |
| *Notamacropus agilis* | WT, PE, TA | 4, 33–35 |
| *N. dorsalis* | PE, HE | 36 |
| *N. eugenii* | WT, HE, PE | 37, 38 |
| *N. parma* | WT, HB, HE, PE, TA | 39, 40 |
| *N. parryi* | WT, PE, TA | 4, 28, 41 |
| *N. rufogriseus banksianus* | WT, PE, TA | 6, 30, 32, 35, 42 |
| *N. rufogriseus rufogriseus* | WT, PE, TA | 3, 4, 43, 44 |
| *Onychogalea frenata* | HE, PE, TA | 45 |
| *Osphranter robustus erebescens* | WT, PE, TA | 4, 6, 46 |
| *O. r. robustus* | WT, PE, TA | 4, 6, 30, 35 |
| *O. rufus* | WT, HE, PE, TA | 4, 32, 47, 48 |
| *Wallabia bicolor* | WT, HE, PE | 4, 33, 32, 49 |
| *Setonix brachyurus* | WT, HB, PE, TA | 50, 51 |

References: (1) Dennis (1997); (2) Johnson (1978); (3) Austin (1997); (4) Woods (1999); (5) Tyndale-Biscoe (1968); (6) Shephard (1987); (7) Freegard *et al.* (2008); (8) Merchant *et al.* (1994); (9) Thompson *et al.* (2015); (10) Johnson and Delean (2001); (11) Seebeck (1992); (12) Guiler (1960); (13) Hughes (1962); 14) Heinsohn (1968); (15) Bryant (1989); (16) Johnson and Delean (2003); (17) Close and Bell (1990); (18) Delaney and De'ath (1990); (19) Johnson (1979); (20) Jones *et al.* (2004); (21) Taggart *et al.* (2005); (22) Johnson and Delean (1999); (23) Johnson and Delean (2002a); (24) Poole *et al.* (1985); (25) Rose and McCartney (1982b); (26) Johnson and Vernes (1994); (27) Johnson (1993); (28) Lundie-Jenkins (1993b); (29) Poole *et al.* (1982a); (30) Kirkpatrick (1965a); 31) Poole *et al.* (1982b); (32) Bellamy (1992); 33) Kirkpatrick and Johnson (1969); (34) Dudzinski *et al.* (1978); (35) Lavery (1985); (36) Johnson and Delean (2002b); (37) Murphy and Smith (1970); (38) Tyndale-Biscoe and Janssens (1988); (39) Maynes (1972); 40) Maynes (1976); (41) Johnson (1998); 42) Catt (1981); (43) Catt (1979); (44) Fleming *et al.* (1983); (45) Hendrikz and Johnson (1999); (46) Sadleir (1963); (47) Sharman *et al.* (1964); (48) Kirkpatrick (1970); (49) Paplinska *et al.* (2006); 50) Waring *et al.* (1955); (51) Shield and Woolley (1961).

**Table 10.21.** Age (days) of selected growth milestones during pouch life of various macropods.

| Species | Sex distinguishable | Permanently attached to teat | Ears free | Eyes open | Fully furred | In and out of pouch | Reference |
|---|---|---|---|---|---|---|---|
| **Hypsiprymnodontidae** | | | | | | | |
| *Hypsiprymnodon moschatus* | – | – | – | – | – | < 180 | 1 |
| **Potoroidae** | | | | | | | |
| *Aepyprymnus rufescens* | | | | | | | 2 |
| *Bettongia gaimardi* | 10 | 35–50 | 60–85 | 80–85 | 95–100 | 90–110 | 3, 4 |
| *B. lesueur* | | | | | | | 5, 6 |
| *B. penicillata* | | | | | | | 7, 8 |
| *B. tropica* | 14–21 | 42–56 | 84–98 | 84–91 | 77–84 | – | 9 |
| *Potorous longipes* | | | | | | | 7, 10 |
| *P. tridactylus* | 10 | 50–60 | 85–90 | 85–100 | 105–110 | 95–125 | 3, 6, 11–15 |
| **Macropodidae** | | | | | | | |
| *Dendrolagus lumholtzi* | 10–20 | 77–105 | 160–170 | 125–136 | 130–155 | 245–273 | 16 |
| *Petrogale assimilis* | | | | | | | 17, 18 |
| *P. concinna* | – | – | 85 | 110 | – | ?–180 | 11 |
| *P. inornata* | | | | | | | 19 |
| *P. lateralis* | – | – | 75–95 | 98–112 | 95–140 | – | 20 |
| *P. penicillata* | | | | | | 180–200 | 19, 21, 22 |
| *P. persephone* | 10–15 | 88–105 | 115–130 | 120–126 | 125–140 | 200–215 | 23 |
| *P. purpureicollis* | 10–15 | 95–115 | 125–140 | 112–126 | 133–140 | 175–195 | 24 |
| *P. xanthopus* | | | | | | | 25 |
| *Thylogale billardierii* | 15–20 | 100 | 100–140 | 110–125 | 140–160 | 180–210 | 3, 4, 11, 26 |
| *Thylogale stigmatica* | 21–28 | 90–125 | – | 112–126 | 135–150 | 154–180 | 11, 27 |
| *Thylogale thetis* | – | – | 110 | 110 | 160 | –180 | 11 |
| *Lagorchestes conspicillatus* | – | 120 | – | 105 | – | –150 | 11, 28 |
| *Lagorchestes hirsutus* | – | 87 | – | 85 | – | ?–114 | 11, 29 |
| *Macropus fuliginosus fuliginosus* | | | | | | | 30 |
| *M. f. melanops* | | | | | | | 30 |
| *M. giganteus* | – | 130–150 | – | 170–180 | 210 | 250–300 | 3, 4, 11, 31–33 |
| *Notamacropus agilis* | – | – | – | 126 | 175 | 190–220 | 4, 11, 34–36 |
| *N. dorsalis* | 14–18 | 98–122 | 140–155 | 126–135 | 145–155 | 195–224 | 37 |
| *N. eugenii* | – | 100 | – | 140–150 | 170–195 | 180–260 | 11, 38, 39 |
| *N. parma* | – | 110–145 | 130–160 | 110–120 | 200–210 | 160–215 | 4, 11, 28, 40, 41 |
| *N. parryi* | | | | | | | 42 |
| *N. rufogriseus banksianus* | – | – | 135 | 154 | – | 230–275 | 6, 11, 31, 33, 36, 43 |
| *N. r. rufogriseus* | – | – | – | – | – | 185–284 | 3, 4, 44, 45 |
| *Onychogalea frenata* | | | | | | | 46 |

**Table 10.21.** continued

| Species | Sex distinguishable | Permanently attached to teat | Ears free | Eyes open | Fully furred | In and out of pouch | Reference |
|---|---|---|---|---|---|---|---|
| *Osphranter robustus* | - | 120 | 150 | 140 | 210 | 210–245 | 11 |
| *O. r. erebescens* | | | | | | | 4, 6, 47 |
| *O. r. robustus* | | | | | | | 4, 6, 31, 36 |
| *O. rufus* | 10–15 | 90–115 | 80–90 | 100–125 | 160–170 | 180–250 | 4, 11, 33, 48,49 |
| *Wallabia bicolor* | - | - | - | 120–135 | 230 | 160–260 | 4, 11,33, 34, 50 |
| *Setonix brachyurus* | - | 60 | 130–150 | 110 | 200–210 | 185–215 | 11, 51–53 |

References: (1) Dennis (1997); (2) Johnson (1978); (3) Austin (1997); (4) Woods (1999); (5) Tyndale-Biscoe (1968); (6) Shephard (1987); (7) Merchant *et al.* (1994); (8) Thompson *et al.* (2015); (9) Johnson and Delean (2001); (10) Seebeck (1992); (11) Williams and Williams (1999); (12) Guiler (1960); (13) Hughes (1962); (14) Heinsohn (1968); (15) Bryant (1989); (16) Johnson and Delean (2003); (17) Close and Bell (1990); (18) Delaney and De'ath (1990); (19) Johnson (1979); (20) Jones *et al.* (2004); (21) Taggart *et al.* (2005); 22) Kleemann *et al.* (2021); (23) Johnson and Delean (1999); (24) Johnson and Delean (2002a); (25) Poole *et al.* (1985); (26) Rose and McCartney (1982b); (27) Johnson and Vernes (1994); (28) Johnson (1993); (29) Lundie-Jenkins (1993b); (30) Poole *et al.* (1982a); (31) Kirkpatrick (1965a); 32) Poole *et al.* (1982b); (33) Bellamy (1992); (34) Kirkpatrick and Johnson (1969); (35) Dudzinski *et al.* (1978); (36) Lavery (1985); (37) Johnson and Delean (2002b); (38) Murphy and Smith (1970); (39) Tyndale-Biscoe and Janssens (1988); (40) Maynes (1972); 41) Maynes (1976); (42) Johnson (1998); 43) Catt (1981); (44) Catt (1979); (45) Fleming *et al.* (1983); (46) Hendrikz and Johnson (1999); (47) Sadleir (1963); (48) Sharman *et al.* (1964); (49) Kirkpatrick (1970); (50) Paplinska *et al.* (2006); (51) Waring *et al.* (1955); (52) Shield and Woolley (1961); (53) Shield (1968).

# 11. ARTIFICIAL REARING

## 11.1 Housing

As with all native mammals that have been taken into care, minimising stress is a major consideration. With macropods this is even more important because of the risks of the animal developing myopathy. Choosing suitable housing can help to create a stress-free environment. To achieve this, several factors should be considered including:

- securing the area from children and other animals (particularly domestic dogs and cats)
- maintaining high standard of hygiene
- escape-proofing the area
- clearing the area of obstacles and hazards
- ensuring the area offers shelter from extreme weather and noise.

When first presented, joeys are usually hypothermic and need to be warmed in a pouch with a heating unit set at 32–34°C (Booth 2002; Taggart *et al.* 2002). Furless and furred joeys are best kept in an artificial pouch made of non-synthetic fibres such as cotton flannelette that is placed in a woollen or fleece pouch and kept in a warm environment. Pouches need to be washed and disinfected every day as bacteria and yeasts thrive in these warm conditions (Bellamy 1992). The use of an ICU humidicrib increases the success rate of hand-raising joeys, particularly for unfurred young.

Once the joeys become fully furred and ready to start exploring outside the pouch, they can regulate their own body temperature (Taggart *et al.* 2002) and can be placed in hanging pouches that allow them to go in and out as they wish. The pouches should be hung in such a manner that the bottom does not reach the floor. Several examples of pouches are shown in Fig. 10.18. Lumholtz's tree-kangaroo joeys tend to constantly climb out of hanging pouches, so soft cat-style beds, joey sacks or even stand-up shopping bags with an inside pouch are preferred (K. Coombes pers obs.)

## 11.2 Temperature requirements

The temperature of the bag should be 32–36°C if the joey is unfurred. As the joey grows fur the temperature can be reduced to 28–30°C when fully furred (Bellamy 1992; Austin 1997). The correct temperature can be maintained with the use of a heat pad or SnuggleSafe® Microwave Heat Pad. Take care not to overheat. Use a minimum/maximum temperature gauge with a plastic-coated probe that can be placed in the pouch layers near to the joey, as this will ensure that the temperature can be monitored. Do not place the probe in the pouch with the joey as it may chew it. The use of an ICU humidicrib is very useful for unfurred and just furred joeys as it greatly increases their chance of survival by ensuring constant temperature and humidity.

## 11.3 Specific requirements

The skin of unfurred and just furred young should be kept moist with the use of Sorbolene cream (not with added glycerine) so that it does not become dry and cracked (George *et al.* 1995). Other skin creams that have

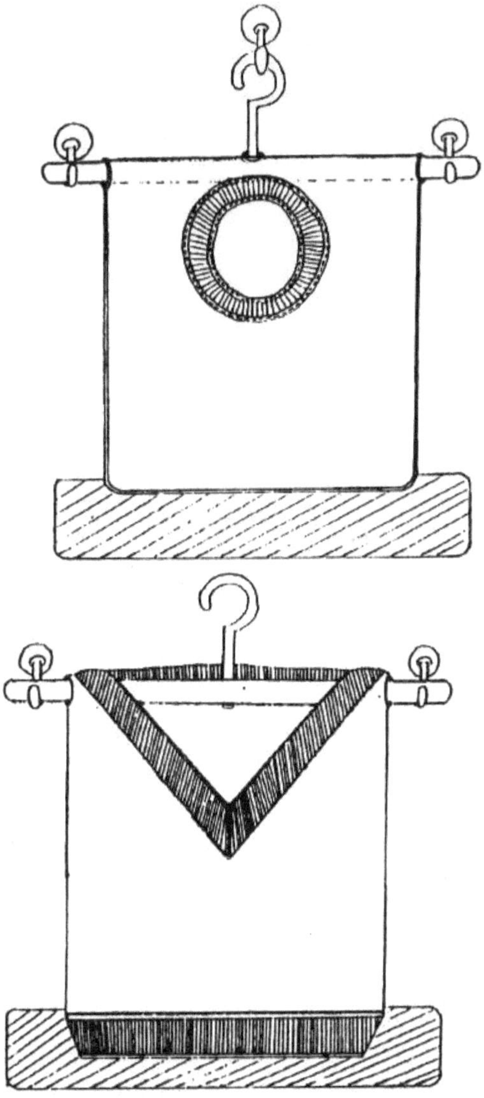

Fig. 10.18. Pouches used for hand-rearing macropods. Taken from Austin (1997) with permission from the author.

been used with success include Wool Fat Bp Standard (pesticides removed), Eucerine ointment and Alpha Keri oil, applied three times per day. Baby oil does not appear to be properly absorbed and tends to stay on the skin surface where it rubs off and is soaked up by the liner bag fabric (George *et al.* 1995). The use of moisturiser is not necessary when using an ICU humidicrib.

When first brought in for hand-rearing, the joey may be dehydrated. If so, it can be given plain boiled water with 5 g (1 teaspoon) of glucose to 250 mL of water or 1 g of electrolyte replacer if available (Austin 1997). They can also be given electrolytes such as Lectade and hydrate. Vytrate® can be made in a dilution ration of 10 mL to 125 mL water. Debilitated neonates can be given an electrolyte such as hydrate in the milk formula for extra energy. It is important to warm the joey before feeding to avoid the risk of inhalation pneumonia. If this takes too long, give warm fluids subcutaneously and bottle-feed later. If the joey is really cold, place it in a warm water bath and dry it off rather than put it in a hot box. Placing joey in a pouch and down your top is a great way to warm and settle joeys. Some arrive stressed and placing them next to your chest provides warmth and a heartbeat (K. Coombes pers. obs).

Stress is a major problem in the successful rearing of native mammals and can be fatal. Therefore it is important to keep noise to a minimum, not overhandle the animal, maintain high standards of hygiene and keep the joey's face covered while bottle feeding.

## 11.4 Diet and feeding routine

### 11.4.1 Natural milk

The concentrations of different milk components generally change throughout lactation in macropods, with the carbohydrates decreasing towards late lactation and proteins remaining relatively steady, but with a small peak in mid-lactation. Lipids, however, show a marked increase in concentration towards late lactation. The ranges in concentrations of the major milk constituents are shown in Table 10.22.

### 11.4.2 Milk formulas

The three commercial formulas for hand-rearing macropods are described below.

- Biolac® - the three formulations are M100 for furless joeys; M150, which is a transitional milk to use when dense fur has developed; and M200 to use when the animal produces solid, dark pellet droppings as it contains elevated lipid in the form of canola oil. To change between formulas use a ratio of 3:1 for the first week, then 2:2 for the second week; 1:3 for the third week and fully onto the next formula the following week. When the joey is nearing weaning, add 2–5 mL of canola oil per 100 mL of formula. Mixing the formulas allows a smooth transition between the two formulas. Joeys should be fed 10–15% of their bodyweight per day.
- Di-Vetelact® - a widely-used low-lactose milk formula. Due to its low energy concentration when prepared as directed, some groups advise the addition of mono- and polyunsaturated fats such as canola oil. ~2–5 mL

**Table 10.22.** Concentrations of the major constituents of milk for different species of macropod milk.

| Species | Total solids (%) | Carbohydrate (%) | Lipids (%) | Protein (%) | Calcium (mg/L) | Iron (mg/L) | Reference |
|---|---|---|---|---|---|---|---|
| **Potoroidae** | | | | | | | |
| *Bettongia gaimardi* | 25–45 | 5.0–13.0 | 4.0–22.5 | 5.0–17.5 | – | – | 1, 2 |
| *B. penicillata* | 16.0–27.0 | 2.0–12.0 | 4.0–12.0 | 7.0–10.0 | – | – | 3 |
| *Potorous tridactylus* | – | 1.0–15.0 | 1.0–27.0 | 2.5–15.0 | – | – | 1, 4 |
| **Macropodidae** | | | | | | | |
| *Petrogale assimilis* | 16.0–22.0 | 2.0–12.0 | 3.0–8.0 | 3.0–6.5 | – | – | 5 |
| *P. xanthopus* | 34.0 | 10.0 | 20.0 | 8.0 | – | – | 6 |
| *Macropus giganteus* | 24.3–32.6 | – | 7.0–16.5 | 6.6–7.0 | 3000–7700 | – | 7 |
| *Notamacropus eugenii* | 12.0–40.0 | 1.0–13.0 | 2.0–23.0 | 4.0–13.0 | 2000–4000 | 7–23 | 6, 8– 12 |
| *N. rufogriseus* | 15.6–27.4 | 1.6–10.9 | 2.5–13.8 | 4.1–9.8 | 1700–5500 | 5–20 | 6, 13 |
| *Osphranter robustus* | 19.3–30.0 | 1.0–7.0 | 2.1–16.2 | 6.3–8.9 | – | – | 14 |
| *O. rufus* | 12.3–26.0 | 1.0–6.0 | 1.0–12.8 | 3.0–8.0 | 4400 | – | 7, 15 |
| *Setonix brachyurus* | – | – | – | – | – | 2–30 | 16–18 |

References: (1) Smolenski and Rose (1988); (2) Rose *et al.* (2003); (3) Merchant *et al.* (1994); (4) Crowley *et al.* (1988); (5) Merchant *et al.* (1996); (6) Green (1984); (7) Poole *et al.* (1982c); (8) Messer and Green (1979); (9) Green *et al.* (1980); 10) Messer *et al.* (1980); (11) Green and Renfree (1982); (12) Janssens and Ternouth (1987); (13) Merchant *et al.* (1989); (14) Bolliger and Pascoe (1953); (15) Lemon and Barker (1967); (16) Bentley and Shield (1962); (17) Kaldor and Ezekiel (1962); (18) Loh and Kaldor (1973).

canola oil can be added per 100 mL milk during the last phase of milk feeding (e.g. 240 days in Tammar wallabies) (Messer and Walker 1992). This should be fed at ~20% bodyweight, except for very small joeys (< 100 g). Some institutions also add 1 tablespoon of high-protein infant cereal per 1 L for furred joeys. Some wildlife groups recommend that Wombaroo® Impact be added at the rate of 1 Impact spoon to 100 mL of made-up milk. Divetelact is preferred for tree-kangaroos and other species, especially in hotter tropical areas (K. Coombes pers. comm.).

- Wombaroo® Kangaroo Milk – different formulas are used for the different stages of development to mimic the changes that occur in the female's milk during lactation: < 0.4 for joeys with less than 40% of pouch life completed, furless with ears closed and downy; 0.4 formula for joeys with 40% of pouch life completed, dark skin (just before fur), eyes open and ears nearly erect; 0.6 formula for joeys with 60% of pouch life completed with fine, short fur and ears erect; > 0.7 formula for joeys that have completed more than 70% of pouch life, have short, dense fur and spend time out of the pouch. Wombaroo® provides detailed charts to help determine the volume to be fed.

Cow's milk should not be fed to marsupials as it has too much fat and lactose, which are poorly digested and result in dehydration and diarrhoea (Stephens 1975). The cataracts that are also observed in macropods that have been reared on cow's milk could be the result of the dehydration, nutritional deficiencies and diarrhoea associated with poor absorption of lactose, excess exposure to sunlight or because some marsupials have a hereditary defect, causing galactosaemia (as in humans) (Stephens *et al.* 1974). Feeding trials have been conducted with red kangaroos and eastern grey kangaroos using cow's milk and milk replacers containing different amounts of glucose or lactose (Walker and Vickery 1988). These showed that there was a decrease in dry matter utilisation and increased diarrhoea at the two highest levels of lactose. No diarrhoea occurred in the joeys given equivalent amounts of glucose. As cow's milk contains ~26% energy as lactose, the intake of milk had to be restricted or the young developed diarrhoea as a result of accumulation of unabsorbed lactose within the intestinal lumen (they cannot absorb lactose at same rate as eutherian mammals) (Messer *et al.* 1989; Walker and Vickery 1988). There appears to be a threshold (~6.5 g lactose per kg or for joeys within the weight range of 1.4–3.0 kg, > 11–12 g lactose per day). Only small weight gains can be expected if cow's milk is used and the energy intake will be well below the amount required to satisfy the normal appetite (Walker and Vickery 1988). These results suggest that cow's milk is a poor substitute for other artificial low-lactose milk formulas and is therefore not recommended. If hand-rearing formulas are not available, products such as soy baby formula can also be used short term (K. Coombes pers. com.).

Gut flora is established by offering dry dirt and fresh grass, as macropods (including adults) are known to eat soil, a habit known as pica. The establishment of gut flora can also be encouraged by including half a teaspoon of natural yoghurt or a pinch of acidophilus powder (unpasteurised) in the formula daily until the faecal consistency appears normal (Austin 1997). An alternative is to add 1 g of ProN8ure Probiotic Soluble® (water-soluble formula containing 180 million colony-forming units/mL of milk for 24 h). Finally, if these are not readily available, bacteria can be added to the digestive system via faeces, though it is advisable to have a veterinarian examine the faeces first to ensure it does not contain parasites. Choose droppings from a healthy adult kangaroo, preferably of the same species, (have them checked for parasites), grind them up, add warm water, strain and adding 5 mL (1 teaspoon) to the joey's milk bottle, mixing it up or feeding directly into the mouth (Austin 1997).

When fine fur appears, some hand-rearers start adding ½–1 teaspoonful of infant cereal to 200 mL of the Di-Vetelact® formula and letting it stand for a few minutes before feeding (Austin 1997).

### 11.4.3 Feeding apparatus

Very small joeys should be fed using a silicone or latex teat specifically designed for the macropod at its current stage of development. Most macropod joeys will, however, be large enough to be fed with a plastic feeder bottle (50 or 100 mL) and a special kangaroo teat. Type (a) teats are used for out of pouch kangaroos and Type (b) for smaller in pouch kangaroos and wallabies (Austin 1997). The teat may be punctured with a hot needle or a tooth pick or a small 1–2 mm slit can be cut in the tip using a small pair of sharp scissors. Milk should be fed at ~36°C.

### 11.4.4 Feeding routine

It is important not to feed the milk formula too quickly; the rate at which the milk is fed should not be faster than the rate at which it is swallowed. Ensuring the hole in the teat is not too large will help (it should only be the size of a pinhole for a small unfurred joey and slightly bigger for a larger furred joey). Too much milk results in an accumulation in the pharynx, which is suddenly sneezed or coughed out the nostrils. To avoid this, be very careful of the rate at which milk is released to the joey and use a smaller hole in the teat if required. Joeys that suckle very strongly on the teat should be fed with a thicker, firmer teat to prevent aspiration.

The number of daily feeds changes as the joey develops (Bellamy 1992). Very young unfurred joeys should be fed every 2–3 h. Once the joey is taking the required volume needed over a 24-h period, the night feeds can be reduced. For furred joeys, the number of feeds is decreased to every 4 h and the volume increased per feed. At this stage they should be given access to grass, foliage and other native vegetation relevant to the species of macropod. Tree-kangaroos should not be offered any fruit, especially when joeys, as it increases the sugar-loving bacteria in their gut and the colic produced causes gas.

At full emergence, when it is fully furred, the number of feeds is reduced to 2–3 per day (the night feeds are not required) and the joey is given continued access to grass and finely chopped carrots, sweet potatoes, kangaroo cubes or pellets and apples. Wallabies should also be given access to leaves, native shrubs such as wattle, eucalypt branches or tree lucerne. Potoroids should also be given access to insects and fungi. If the joey is to be released to the wild, some wildlife carers suggest limiting the amount of commercial fruit/vegetables being offered to prevent the animals becoming dependent on it. This is particularly important in areas where kangaroos may be culled by agricultural producers.

## 11.5 Data recording

When an animal is first brought in for hand-rearing, its sex and approximate age, using growth charts, should be recorded. During the hand-rearing process several important pieces of information should be recorded on a daily basis:

- date when the information is recorded
- time when the information is recorded
- bodyweight to the nearest 1 g if possible
- general activity and demeanour
- characteristics and frequency of defecation and urination
- amount (g) of different food types offered
- food consumption at each feed
- veterinary examinations and results.

This information serves several purposes, including providing important background information such as food consumption which will assist a veterinarian to reach a diagnosis if the animal becomes sick or fails to grow or gain weight. It also allows a comparison with growth curves to assess progress (section 10.16) and to allow new growth curves to be established for measurements in which they presently do not occur. The developmental stages and measurements outlined in section 10.3.1 should also be recorded on a weekly basis if possible, particularly head length and width for accurate aging of young.

## 11.6 Identification methods

Visual identification or implant chips (once the individual is fully furred) can be used until the animal is weaned, when other techniques can be used.

## 11.7 Hygiene

Maintaining a high standard of hygiene is critical to the survival of the macropod joey. Emphasis needs to be placed on the following:

- maintaining a clean pouch lining at all times, using a new pouch liner after each feed.
- stimulating toileting behaviour before or after feeding. As with other marsupials, toileting can be done by the application of warm water to the cloaca using cotton wool to stimulate urination and defecation, which allows the animal to keep drier and warmer in its pouch. Older joeys can be trained to urinate on newspaper by keeping a piece of newspaper with the smell of urine on it
- maintaining personal hygiene by washing and disinfecting hands before and after handling the joey. Use antibacterial solution for washing hands with furless joeys, as their immune system is not well developed
- washing hands between feeding different joeys
- using cooled, boiled water at room temperature when making up formulas for joeys of all ages
- cleaning spilt milk formula, faeces and urine from the joeys' skin and fur as soon as possible and the fur dried to prevent bacterial or fungal skin infections
- washing all feeding equipment in warm water and dishwashing detergent and sterilising it in a microwavable baby steam steriliser or F10® disinfectant
- storing the teats and bottles in the fridge in an air-tight container after they have been disinfected
- only heating up milk once and discarding leftovers
- avoiding contact with other animals unless certain they pose no health risk
- if furless, covering the joey's body with Sorbolene cream after each feed until fur appears.

Toileting between feeds may also be required until good habits are learned (George *et al.* 1995). Be careful with this stimulation because if done excessively it can lead to cloacal prolapse and possibly urethral swelling, in which case consult a veterinarian (Bellamy 1992). It can be treated with the use of creams such as Panalog (Squibb), Proctoseyl (Roussel) and Topigol (Squibb) (George 1990). Once the joey is ready to start leaving the pouch, it should be stimulated in a standing position so that it will eventually learn to defecate when standing (Bellamy 1992). Joeys that are wet with urine and faeces should be cleaned with a mild soap solution and dried thoroughly.

Unfurred joeys have faeces that look like mustard; when just furred it is the consistency of toothpaste; when longer fur appears pellets should be forming; normal pellets should be produced when fully furred.

Several diseases and parasites can occur during hand-rearing, particularly if hygiene is poor (George 1990), including:

- yeast infection - *Candida* and *Torulopsis* can build up and become serious pathogens, particularly when the bacteria in the gut have been reduced by antibiotics (Booth 2002). Infection can also be caused by left-over milk around the mouth (oral thrush). As a result, the animal is reluctant to suck, its faeces become greenish, saliva may appear rusty when the mouth is wiped after feeding and lesions occur in the mouth. This can be treated with Nystatin oral drops.
- bacterial infections -change in faecal colour, acute smelly scouring and lethargy. Bacteria can include: haemolytic *E. coli*, which is strongly associated with acute liquid diarrhoea in newly captured joeys; *Salmonella* spp. strongly associated with chronic pasty diarrhoea and long-term captivity; and *Klebsiella* spp. associated with long-term captivity and acute chronic liquid diarrhoea (Cargill and Frith 1991 in Booth 2002). Treatment is dependent on culturing to identify the bacteria and administering antibiotics if appropriate
- osteoporosis - ack of absorption of calcium resulting in brittle bones that break easily. Can occur due to lack of absorption due to scouring, overheating in the pouch, lack of sunlight (vitamin D), lack of protein in the diet and lack of exercise. Therefore, access to sunlight and exercise are required once the young are well furred and ready to leave the pouch. Diagnosis requires an X-ray

- coccidia - protozoan that appears to be particularly severe in eastern grey kangaroos. There is no satisfactory treatment. Once the disease is recognised the damage to the intestine is often so severe that euthanasia is the only humane solution. Joeys can be treated with a single dose of Baycox® liquid 25 mg/mL: 10 mL/kg orally and check the faeces in 1 week
- worms -overburden can result in scouring, failure to thrive and a dull coat. Worming is required using treatments such as Panacur (15–50 mg/kg depending on severity)
- aspiration pneumonia -infection of the lung resulting from incorrect feeding technique and teats with holes that are too large. It can occur if the joey discharges milk through the nostrils during or after feeding, indicating milk has entered the lungs (Booth 2002). Signs include noisy, laboured breathing, high temperature (fever) and lethargy. Even if fed slowly, if a joey appears to have aspirated milk during or after a feed, seek veterinary assistance promptly. Waiting until the joey shows signs of poor health is often too late. It can be treated with Amoxil PP, Nystatin for thrush and gentle exercise such as physiotherapy or hopping practice, if the joey is old enough.

## 11.8 Behavioural considerations

Take care that the joey being hand-reared does not become too attached to the carer, as this will make the weaning process much more difficult. Some macropods, such as tree-kangaroos, appear to need attachment to their carer until later in rearing. Desensitisation should take place when the joey is older and independent and out of the pouch, not while still on bottle feeds. Tree-kangaroos that are hand-raised and released are very content to stay away from humans once in the wild (K. Coombes pers obs.).

Minimising stress is also very important as stress is the most common cause of illness and death in hand-reared kangaroos (George 1990). Stress includes shock from the initial loss of the mother, inappropriate temperature, constant handling, different people feeding and unusual (especially sharp) sounds such as traffic and dogs (George 1990). Black wallaroo and tree-kangaroo joeys, for example, seem to be extremely sensitive and very prone to stress-related conditions. They have been found to be intolerant of handling by anyone other than the foster carer and any changes to the routine should be managed over a period of time.

Smells of predators such as cats and dogs should be avoided as joeys should not associate with these species while being reared as they will not develop a fear of them and therefore increase the risk of attack from wild cats or dogs once they have been released.

Tree-kangaroos need to be provided with a 'jungle gym' at ~10 months of age when they reach permanent pouch emergence. The enclosure should be at least 3 m × 6 m up to 10 m × 8 m and include climbing opportunities such as logs, thick ropes (~4 cm + in diameter) and branches for them to climb up to ~2–3 m in height. During access to this area the tree-kangaroo should be supervised (Cianelli and Schmidt 2020; K. Coombes pers. obs.).

## 11.9 Use of foster animals

Fostering has been used with success between different species of macropods, even between species of different genera (Sharman and Calaby 1964; Merchant and Sharman 1966; Clark 1968; Johnson 1981; Taggart *et al.* 1997; 2002; 2005; 2010; Taggart 2002; Smith 1998; Schultz *et al.* 2006; Fielder 2015; McLelland *et al.* 2015; Read *et al.* 2018). Cross-fostering involves the transfer of pouch young from a target species into the pouch of a recipient with a pouch young of similar size (Figs 10.19, 10.20). The recipient then raises the young as its own and its young is removed and taken for hand-rearing or is euthanased (Taggart *et al.* 1997, 2002, 2005, 2010; Taggart 2002).

As most macropods have a postpartum oestrus and embryonic diapause, removal of the pouch young allows the blastocyst in diapause to develop. By continually fostering the pouch young of the target species, the breeding frequency can be increased dramatically, up to 8-fold in some species (Taggart *et al.* 2010). For example, rock-wallaby pouch young have been successfully transferred to Tammar wallabies and other species of rock-wallabies. This technique relies on the two species being roughly the same size and the pouch young being approximately the same age. To date it has been used primarily to increase the reproductive output of target species, as it allows them to breed again while the other joey is being raised by the foster parent (Schultz *et al.* 2006; Taggart *et al.* 2002, 2005, 2010; Taggart 2002). The surrogate female, which has been identified to foster the target pouch young, must have a pouch young of its own that is between 10% lighter and 25% heavier in size than the pouch young of the target species whose young it is about to receive. Under no circumstances is a female to be used as a foster female if her natural pouch young is > 10% lighter than the target animal (Taggart *et al.* 2005, 2010).

For example, success in the cross-fostering of brush-tailed rock-wallabies (Fig. 10.21) can be improved if the pouch young are transferred early (between 8 and 20 days) from the brush-tailed rock-wallaby pouch to the pouch of the surrogate mother, the veterinary history

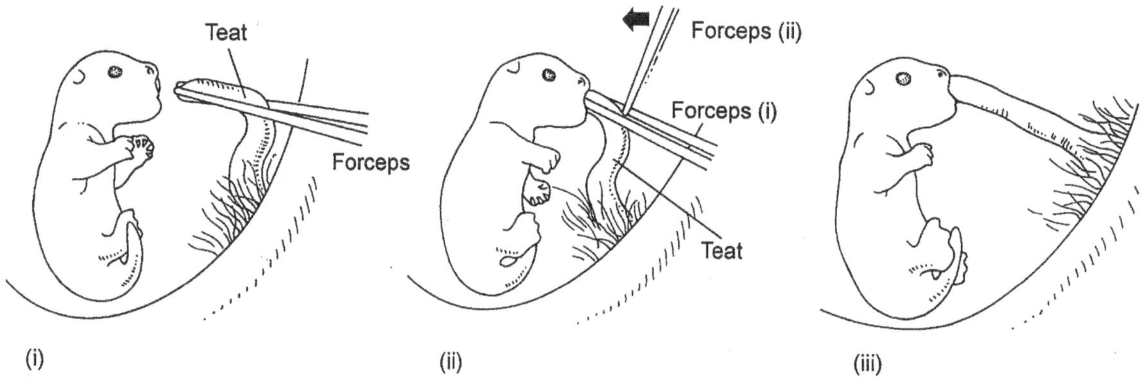


**Fig. 10.19.** Attachment of a wallaby pouch young being cross-fostered to a surrogate mother's teat. (i) Position the tip of the nipple at the end of the fine blunt-ended forceps and hold the head of the pouch young securely within the fingers, (ii) insert the teat into the mouth of the pouch young and (iii) after re-attachment place the young back in the pouch of the foster mother and check 20 min later. Reproduced from Taggart *et al.* (2005) with permission from CSIRO Publishing.

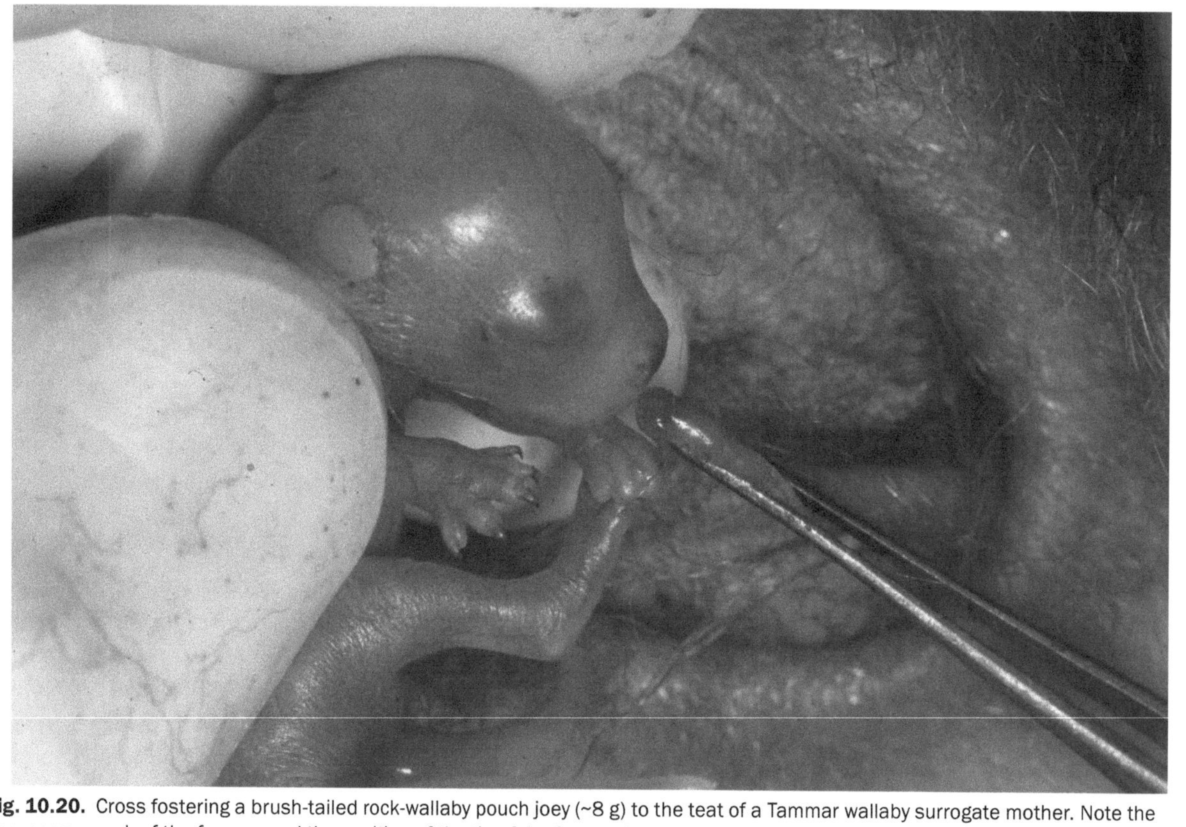

**Fig. 10.20.** Cross fostering a brush-tailed rock-wallaby pouch joey (~8 g) to the teat of a Tammar wallaby surrogate mother. Note the very narrow neck of the forceps and the position of the tip of the forceps in relation to the end of the teat. Photo by David Taggart.

and health of the animals selected to act as surrogate mothers is reviewed prior and the foster mothers have access to increased grazing pasture to improve gum health (Schultz *et al.* 2006).

Tiny unfurred pouch joeys as small as 2 g can also be isolated at a set temperature and high humidity to facilitate their movement across time windows as great as 12 h for the purpose of cross-fostering at another institution (Figs 10.22, 10.23; Taggart *et al.* 2002, 2005, 2010). This technique can enable recruitment of new genetics for captive breeding from wild-based mothers in threatened macropod species without removing valuable wild-based breeding mothers (Taggart *et al.* 2002, 2010).

## 11.10 Weaning

Once the joey is fully furred and starts leaving the pouch, it should be given access to grassy areas, 10–15 min initially, that will allow it to start feeding on grass and begin to hop and so gain muscle tone and coordination in its legs. It is generally obvious when the joey has had enough as it will want to return to its pouch or begin to vocalise. The eruption of teeth is a good indicator of when to begin weaning and offering solid food (Bellamy 1992). This often coincides with the first emergence from the pouch. Initially, the food is only mouthed, but eventually the joey will start eating it as well. Tree-kangaroos can be offered branches and leaves to chew on as early as 4–5 months of age, once they have their head out but sometimes still only velvet fur. They do this in the wild and bite on vegetation while in pouch and mother is climbing around. It also encourages gut bacteria. Tree-kangaroos can still have the normal amount of formula until over 1 year of age. They drink from their mothers until this age and she has a new joey. But is very important they get a lot of browse each day, especially once they are out of pouch (K. Coombes pers. comm.).

Gradual reduction in the amount of milk provided will increase the amount of solid food eaten. A general rule is to decrease the formula by 5% per week as long as the joey continues to gain weight at a minimum of 5–10% of bodyweight per day. Macropods should be given access to fresh grass with new shoots initially, then hard fruits and vegetables (such as apples, carrots and sweet potato) are introduced with lucerne, kangaroo cubes and browse finally being added to the diet.

**Fig. 10.21.** A foster-reared brush-tailed rock-wallaby pouch joey in the pouch of its yellow-footed rock-wallaby surrogate mother late in pouch life. Photo by David Taggart.

**Fig. 10.22.** Two different types of incubators used for transporting pouch joeys. (a) Battery-powered incubator for cooling and with floor heating; (b) simple incubator design relying on the operator to regulate heating and cooling. Long-nosed potoroo in the background. Photos by David Taggart.

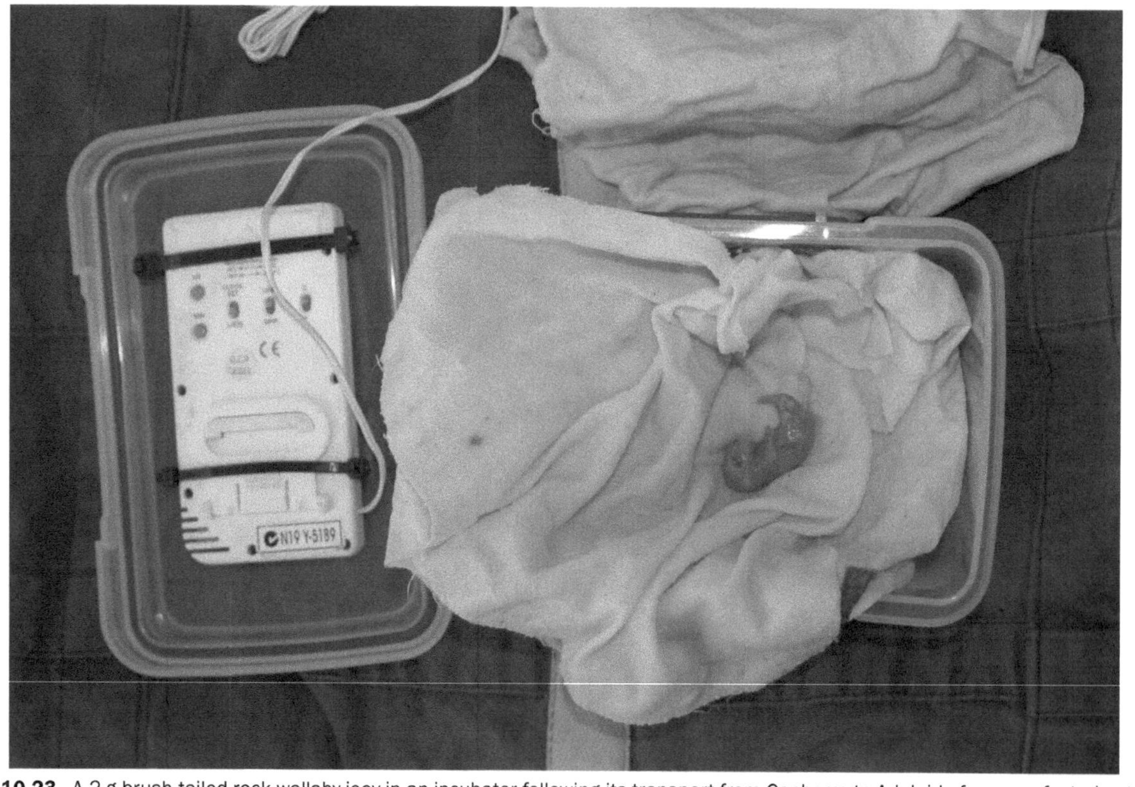

**Fig. 10.23.** A 2 g brush-tailed rock-wallaby joey in an incubator following its transport from Canberra to Adelaide for cross-fostering to a yellow-footed rock-wallaby surrogate mother. Photo by David Taggart.

Before the amount of food is decreased, the macropod should be drinking water from a bowl and consistently eating solids. Initially, start decreasing the number of feeds per day to three and then two and then decrease the volume, without watering down the mixture (Austin 1997). Eastern grey kangaroos can be weaned by 18 months, wallabies by 17 months, 13–14 months for pademelons and 6 months for potoroos and bettongs (Austin 1997).

## 11.11 Release procedures

Macropods that are reintroduced to the wild often fall prey to predators as they have not learnt an appropriate recognition and flight response from potential predators such as dogs and foxes (e.g. Blumstein *et al.* 2002; Taggart *et al.* 2016). In an attempt to overcome this, training has been undertaken for rufous hare-wallabies to condition them to fear model cat and fox predators, which resulted in significantly modified responses to the models (McLean *et al.* 1994). The models used were a taxidermied cat and fox on wheels that burst from a box to surprise the wallabies immediately after a hare-wallaby alarm call was played. The model was pulled through the enclosure while the alarm call continued. The entire sequence lasted ~50 s and occurred at any time of the night. Smell of the predators may be important so placing faecal pellets of these predators in the area immediately after the trial may be of use. A second technique involved the predator models leaping at the wallabies if they approached within 2–3 m and squirting the wallaby with a water pistol (McLean *et al.* 1994). The release of hand-reared eastern grey kangaroos has been shown to work successfully using both hard (released directly into the wild) and soft releases (released after being maintained in acclimatisation yards), with all animals released in a study establishing home ranges, though their reproductive success was not determined (Campbell and Croft 2001). Captive-reared brush-tailed rock-wallabies have also been released into open, unfenced landscapes (Molyneux *et al.* 2011; Taggart *et al.* 2016). Data from that study recommended that captive-reared rock-wallabies are hardened off in a semi-captive facility before release, that they be released in small groups across a large target area to limit disturbance to their social structure and reduce the likelihood of predators detecting all released animals before they have had a chance to learn about their new environment, and that releases occur when yearly fox activity is at a minimum to provide additional opportunities for captive-reared animals to learn how to be wild (Taggart *et al.* 2016; Kleemann *et al.* 2021).

Hand-reared tree-kangaroos have been both soft and hard released. It is difficult procedure as they are highly territorial; if the release area of rainforest has occupants already, the newcomers will be chased out immediately (Coombes 2005). Therefore, releasing into a large area of continuous rainforest, rather than in a fragment of forest, is recommended, where the individual has the opportunity to find an 'empty' area to set up a home range. This also provides safety from human contact and fewer opportunities for dog attacks and vehicle collisions.

In general, before macropods are ready for release, they need to satisfy several criteria (Booth unpubl. data:

- being fit and healthy (physically and mentally)
- maintaining condition on natural foods (natural grasses should be provided)
- being able to recognise its own species
- being familiar with the social behaviour of its species (ideally they should be held with other members of the same species)
- showing appropriate levels of fear of humans and predators
- showing no evidence of being imprinted on humans
- being able to seek shelter under branches that are leant against the fenceline (smaller species such as wallabies and potoroids).

## ACKNOWLEDGEMENTS

Many thanks also to Mark Eldridge for providing information on several macropods. Sincere thanks also to Karen Scott for making numerous edits to the hand-rearing section.

## REFERENCES

Andrewartha HG, Barker S (1969) Introduction to a study of the ecology of the Kangaroo Island wallaby, *Protemnodon eugenii* (Desmarest) within Flinders Chase, Kangaroo Island, South Australia. *Transactions of the Royal Society of South Australia* **93**, 127–132.

Anon (2006) *Standards for Exhibiting Australian Mammals in New South Wales*. NSW Department of Primary Industries, Orange, New South Wales.

Arlidge J, Srb C (1992) Macropod identification. *Thylacinus* **17**, 18–19.

Arnold GW, Steven D, Weeldenburg J, Brown OE (1986) The use of alpha-chloralose for the repeated capture of western grey kangaroos, *Macropus fuliginosus*. *Australian Wildlife Research* **13**, 527–533. doi:10.1071/WR9860527

Association of Zoos and Aquariums (2017). Tree kangaroo body condition score chart. AZA Tree Kangaroo Species Survival Plan. https://nagonline.net/wp-content/uploads/2017/03/Tree-Kangaroo-BCS-Poster.pdf (accessed 1 November 2023)

Austin MA (1997) *A Practical Guide to the Successful Handrearing of Tasmanian Marsupials.* Regal Publications, Melbourne.

Bach C (1998) *Birth Date Determination in Australasian Marsupials.* Australasian Regional Association of Zoological Parks and Aquaria, Sydney.

Bailey P, Best A (1992) A red kangaroo, *Macropus rufus*, rediscovered 25 years after marking in north-western New South Wales. *Australian Mammalogy* **15**, 141. doi:10.1071/AM92022

Baker AM, Gynther IC (eds) (2023) *Strahan's Mammals of Australia.* 4th edn. Reed New Holland Publishers, Sydney.

Baker AM, Eldridge MDB, Fisher DO, Frankham G, Helgen K, Jackson SM, Potter S, Travouillon KJ, Umbrello L (2023) Taxonomy and diversity of living Australasian marsupials. In *American and Australasian Marsupials* (Eds NC Cáceres and CR Dickman). pp. 163–247. Springer, Cham, Switzerland.

Barker SC (1990) Behaviour and social organisation of the allied rock-wallaby *Petrogale assimilis*, Ramsay 1877 (Marsupialia: Macropodidae). *Australian Wildlife Research* **17**, 301–311. doi:10.1071/WR9900301

Barker S, Brown GD, Calaby JH (1963) Food regurgitation in Macropodidae. *Australian Journal of Science* **25**, 430–432.

Batchelor TA (1980) The social organisation of the brush-tailed rock-wallaby (*Petrogale penicillata penicillata*) on Motutapu Island. MSc thesis. University of Auckland, New Zealand.

Bates PC, Bigger TRL, Hulse EV, Palmer A (1972) The management of a small colony of the marsupial *Potorous tridactylus* and a record of its breeding in captivity. *Laboratory Animals* **6**, 301–313. doi:10.1258/002367772781006059

Beaudin-Judd J, Weladji RB, Lazure L, Paré P (2019) Activity budget and spatial distribution of Bennett's wallabies (*Macropus rufogriseus*) in open versus closed exhibit designs. *Zoo Biology* **38**, 258–265. doi:10.1002/zoo.21486

Bell JN, Close RL, Johnson PM (1989) Testicular development in the allied rock-wallaby *Petrogale assimilis*. In *Kangaroos, Wallabies and Rat-kangaroos.* (Eds G Grigg, P Jarman and I Hume) pp. 419–422. Surrey Beatty & Sons, Sydney.

Bellamy T (1992) Marsupial handrearing. In *Urban Wildlife.* Proceedings 204. Post Graduate Committee in Veterinary Science. (Eds L Vogelnest, D Spielman, T Bellamy, M Messer and DM Walker) pp. 49–67. University of Sydney, Sydney.

Bennett AF, Baxter BJ (1989) Diet of the long-nosed potoroo, *Potorous tridactylus* (Marsupialia: Potoroidae), in south-western Victoria. *Australian Wildlife Research* **16**, 263–271. doi:10.1071/WR9890263

Bentley PJ, Shield JW (1962) Metabolism and kidney function in the pouch young of the macropod marsupial *Setonix brachyurus. The Journal of Physiology* **164**, 127–137. doi:10.1113/jphysiol.1962.sp007007

Berger PJ (1966) Eleven-month 'embryonic diapause' in a marsupial. *Nature* **211**(5047), 435–436. doi:10.1038/211435a0

Berger PJ, Sharman GB (1969) Embryonic diapause initiated without the suckling stimulus in the wallaby, *Macropus eugenii. Journal of Mammalogy* **50**, 630–632. doi:10.2307/1378800

Blessington J, Steenberg J (Eds) (2007) *Tree Kangaroo (*Dendrolagus *spp.) Husbandry Manual.* 3rd edn. Minnesota Zoo. Apple Valley, Minnesota.

Blumstein DT, Mari M, Daniel JC, Ardron JG, Griffin AS, Evans CS (2002) Olfactory predator recognition: wallabies may have to learn to be wary. *Animal Conservation* **5**, 87–93. doi:10.1017/S1367943002002123

Blyde D (1994) Management and diseases of macropods. In *Wildlife.* Proceedings 104. Post Graduate Committee in Veterinary Science. (Ed. DI Bryden) pp. 247–251. University of Sydney, Sydney.

Blyde D (1999) Advances of treating diseases of macropods. In *Wildlife in Australia: Healthcare & Management.* Proceedings 327. Post Graduate Foundation in Veterinary Science. (Ed. DI Bryden) pp. 439–452. University of Sydney, Sydney.

Bolliger A, Pascoe JV (1953) Composition of kangaroo milk (wallaroo, *Macropus robustus*). *Australian Journal of Scientific Research* **15**, 215–217.

Bolton BL, Newsome AE, Merchant JC (1982) Reproduction in the agile wallaby, *Macropus agilis* (Gould) in the tropical lowlands of the Northern Territory: opportunism in a seasonal environment. *Australian Journal of Ecology* **7**, 261–277. doi:10.1111/j.1442-9993.1982.tb01505.x

Booth R (1994) Manual and chemical restraint of macropods. In *Wildlife.* Proceedings 233. Post Graduate Committee in Veterinary Science. (Ed. DI Bryden) pp. 443–447. University of Sydney, Sydney.

Booth R (2002) Macropods. In *Hand-Rearing Wild and Domestic Mammals.* (Ed. LJ Gage) pp. 63–74. Iowa State Press, Iowa.

Booth R, Srb C (1999) Population management in captive macropods. Poster ASZK/ARAZPA Conference Alice Springs Desert Park. 21–26 March.

Boulton KA, Vogelnest LJ, Vogelnest L (2013) Dermatophytosis in zoo macropods: a questionnaire study. *Journal of Zoo and Wildlife Medicine* **44**, 555–563. doi:10.1638/2011-0273R2.1

Bryant SL (1989) Growth, development and breeding pattern of the long-nosed potoroo, *Potorous tridactylus* (Kerr, 1792) in Tasmania. In *Kangaroos, Wallabies and Rat-Kangaroos.* (Eds G Grigg, P Jarman and I Hume) pp. 449–456. Surrey Beatty & Sons, Sydney.

Bulinski J, Goldney D, Bauer J (1997) The habitat utilisation and social behaviour of captive rock-wallabies: Implications for management. *Australian Mammalogy* **19**, 191–198. doi:10.1071/AM97191

Burbidge AA (1995) Burrowing Bettong *Bettongia lesueur.* In *The Mammals of Australia.* 2nd edn. (Ed. R Strahan) pp. 289–291. Reed Books, Sydney.

Burbidge AA, Johnson PM (1995) Spectacled hare-wallaby *Lagorchestes conspicillatus.* In *The Mammals of Australia.* (Ed. R Strahan) pp. 313–316. Reed Books, Sydney.

Burgdorf-Moisuk A, Pye GW, Smith JA, Papendick R, Ivy JA, Hamlin-Andrus C (2012) A retrospective and prospective study of megaesophagus in the parma wallaby (*Macropus parma*) at the San Diego Zoo, California, USA. *Journal of Zoo and Wildlife Medicine* **43**, 89–94. doi:10.1638/2011-0202.1

Bush M, Montali R (1999) Medical management of tree kangaroos. In *Zoo and Wild Animal Medicine.* (Eds ME Fowler and EM Miller) pp. 337–343. Saunders, Philadelphia.

Calaby JH, Poole WE (1971) Keeping kangaroos in captivity. *International Zoo Yearbook* **11**, 5–12. doi:10.1111/j.1748-1090.1971.tb01825.x

Campbell L, Croft DB (2001) Comparison of hard and soft release of hand reared eastern grey Kangaroos. In Veterinary conservation biology wildlife health and management in

Australasia. *Proceedings of International Joint Conference*. (Eds A Martin and L Vogelnest) pp. 173–180. Taronga Zoo, Sydney, Australia. 1–6 July 2001.

Cargill C, Frith F (1991) Joey diarrhoea project: preliminary report. *Vetlab Newsletter. Department of Agriculture, South Australia* **36**, 10–13.

Catt DC (1977) The breeding biology of Bennett's wallaby (*Macropus rufogriseus fruticus*) in South Canterbury, New Zealand. *New Zealand Journal of Zoology* **4**, 401–411. doi:10.1080/03014223.1977.9517965

Catt DC (1979) Age determination in Bennett's wallaby, *Macropus rufogriseus fruticus* (Marsupialia) in South Canterbury, New Zealand. *Australian Wildlife Research* **6**, 13–18. doi:10.1071/WR9790013

Catt DC (1981) Growth and condition of Bennett's wallaby (*Macropus rufogriseus fruticus*) in South Canterbury, New Zealand. *New Zealand Journal of Zoology* **8**, 295–300. doi:10.1080/03014223.1981.10427969

Christensen P, Leftwich T (1980) Observations on the nest building habits of the brush-tailed rat-kangaroo or woylie (*Bettongia penicillata*). *Journal of the Royal Society of Western Australia* **63**, 33–38.

Cianelli M, Schmidt K (2020) Rehabilitation of Lumholtz's tree kangaroo joeys. In *Tree Kangaroos: Science and Conservation*. (Eds L Dabek, P Valentine, J Blessington and KR Schwartz) pp. 65–84. Academic Press, London [Note the cover says 2021 but a copy of the book was obtained in December 2020.]

Clark MJ (1968) Growth of pouch-young of the red kangaroo, *Megaleia rufa*, in the pouches of foster mothers of the same and different species. *International Zoo Yearbook* **8**, 102–106. doi:10.1111/j.1748-1090.1968.tb00451.x

Close RL, Bell JN (1990) Age estimation of pouch young of the allied rock-wallaby (*Petrogale assimilis*) in captivity. *Australian Wildlife Research* **17**, 359–367. doi:10.1071/WR9900359

Close RL, Bell JN (1997) Fertile hybrids in two genera of wallabies: *Petrogale* and *Thylogale*. *The Journal of Heredity* **88**, 393–397. doi:10.1093/oxfordjournals.jhered.a023124

Close R, Lowry PS (1990) Hybrids in marsupial research. In *Mammals from Pouches and Eggs: Genetics, Breeding and Evolution of Marsupials and Monotremes*. (Eds JA Marshall Graves, RM Hope and DW Cooper) pp. 117–125. CSIRO Publishing, Melbourne.

Coombes KE (2005) The ecology and habitat utilisation of Lumholtz's tree kangaroos *Dendrolagus lumholtzi* (Marsupialia: Macropodidae), on the Atherton Tablelands, far north Queensland. PhD thesis, James Cook University. Cairns.

Cooper DW (1998) Welfare of kangaroos and wallabies in captivity. Unpublished notes. Macquarie University, Sydney.

Copley PB, Robinson AC (1983) Studies on the yellow-footed rock-wallaby, *Petrogale xanthopus* Gray (Marsupialia: Macropodidae) II: diet. *Australian Wildlife Research* **10**, 63–76. doi:10.1071/WR9830063

Coulson GM (1978) Daily activity patterns of three sympatric macropod species and the effect of a solar eclipse. *Search* **9**, 153–155.

Coulson G (1996) A safe and selective draw-string trap to capture kangaroos moving under fences. *Wildlife Research* **23**, 621–627. doi:10.1071/WR9960621

Coulson GM, Croft DB (1981) Flehmen in kangaroos. *Australian Mammalogy* **4**, 139–140. doi:10.1071/AM81007

Coulson G, Poiani A, Nave CD (2003) A modified draw-string trap for capture of kangaroos at fences. *Australian Mammalogy* **25**, 177–182. doi:10.1071/AM03177

Crandall LS (1964) *The Management of Wild Mammals in Captivity*. University of Chicago Press, Chicago.

Cripps J, Beveridge I, Ploeg R, Coulson G (2014) Experimental manipulation reveals few subclinical impacts of a parasite community in juvenile kangaroos. *International Journal for Parasitology. Parasites and Wildlife* **3**, 88–94. doi:10.1016/j.ijppaw.2014.03.005

Croft DB (1989) Social organisation of the Macropodoidea. In *Kangaroos, Wallabies and Rat-kangaroos*. (Eds G Grigg, P Jarman and I Hume) pp. 505–525. Surrey Beatty & Sons, Sydney.

Crook GA, Skipper G (1987) Husbandry and breeding of Matschie's tree kangaroo *Dendrolagus m. matschiei* at Adelaide Zoological Gardens. *International Zoo Yearbook* **26**, 212–216. doi:10.1111/j.1748-1090.1987.tb03161.x

Crowley H, Woodward D, Rose R (1988) Changes in milk composition during lactation in the potoroo *Potorous tridactylus* (Marsupialia: Potoroidae). *Australian Journal of Biological Sciences* **41**, 289–296. doi:10.1071/BI9880289

Davies WG (2013) Saliva sharing in the allied rock-wallaby, *Petrogale assimilis*. *Australian Mammalogy* **35**, 231–235. doi:10.1071/AM12020

Dawson TJ (1989) Diets of macropodoid marsupials: general patterns and environmental influences. In *Kangaroos, Wallabies and Rat-kangaroos*. (Eds G Grigg, P Jarman and I Hume) pp. 129–142. Surrey Beatty & Sons, Sydney.

Dawson TJ, Ellis BA (1979) Comparison of diets of the yellow-footed rock-wallabies and sympatric herbivores in western New South Wales. *Australian Wildlife Research* **6**, 245–254. doi:10.1071/WR9790245

Death C, Coulson G (2016) A method for age estimation in the swamp wallaby (*Wallabia bicolor*). *Australian Mammalogy* **38**, 246–248. doi:10.1071/AM15021

Delaney RM (1997) Reproductive ecology of the allied rock-wallaby, *Petrogale assimilis*. *Australian Mammalogy* **19**, 209–218. doi:10.1071/AM97209

Delaney R, De'ath G (1990) Age estimation and growth rates of captive and wild pouch young of *Petrogale assimilis*. *Australian Wildlife Research* **17**, 491–499. doi:10.1071/WR9900491

Delaney R, Marsh H (1995) Estimating the age of wild rock-wallabies by dental radiography: a basis for quantifying the age structure of a discrete colony of *Petrogale assimilis*. *Wildlife Research* **22**, 547–559. doi:10.1071/WR9950547

Delroy LB, Earl J, Radbone I, Robinson AC, Hewett M (1986) The breeding and re-establishment of the brush-tailed bettong (*Bettongia penicillata*) in South Australia. *Australian Wildlife Research* **13**, 387–396. doi:10.1071/WR9860387

Dennis AJ (1997) Musky rat-kangaroos, *Hypsiprymnodon moschatus*: cursorial frugivores in Australia's wet-tropical rain forests. PhD thesis. James Cook University, Townsville.

Dennis AJ, Marsh H (1997) Seasonal reproduction in musky rat-kangaroo, *Hypsiprymnodon moschatus*: a response to changes in resource availability. *Wildlife Research* **24**, 561–578. doi:10.1071/WR96038

Department of Parks and Wildlife (2013) *Standard Operating Procedure: Soft Cage Traps for Capture of Macropods*. Department of Parks and Wildlife's Animal Ethics Committee, Perth.

Devlin J, Stalder K (2025) Marsupial herpesviruses. In *Current Therapy in Medicine of Australian Mammals*. Revised Edition. (Eds L Vogelnest and T Portas) pp. 359–364. CSIRO Publishing, Melbourne.

Di Stefano J, Moyle R, Coulson G (2005) A soft-walled double-layered trap for capture of swamp wallabies *Wallabia bicolor*. *Australian Mammalogy* **27**, 235–238. doi:10.1071/AM05235

Driessen MM, Hocking G (1997) Age estimation of the Tasmanian pademelon, *Thylogale billardierii*, by molar progression. *Australian Mammalogy* **20**, 107–110. doi:10.1071/AM97107

Dudzinski ML, Newsome AE, Merchant JC, Bolton BL (1977) Comparing the two usual methods for aging Macropodidae on tooth classes in the agile wallaby. *Australian Wildlife Research* **4**, 219–221. doi:10.1071/WR9770219

Dudzinski ML, Newsome AE, Merchant JC (1978) Growth rhythms in pouch young of the agile wallaby in Australia. *Acta Theriologica* **23**, 401–411. doi:10.4098/AT.arch.78-31

Ealey EHM (1967) Ecology of the euro, *Macropus robustus* (Gould) in north-western Australia. IV. Age and growth. *CSIRO Wildlife Research* **12**, 67–80. doi:10.1071/CWR9670067

Ealey EHM, Dunnett GM (1956) Plastic collars with patterns of reflective tape for marking nocturnal mammals. *CSIRO Wildlife Research* **1**, 59–62. doi:10.1071/CWR9560059

Edwards MS, Ward A (2001) Nutrition, food preparation, and feeding. In *Tree Kangaroo Species Survival Plan* (Eds J Steenberg and J Blessington). Chapter 3. American Association of Zoos and Aquariums, Silver Spring, MD.

Eldridge MDB, Close RL (1995) Allied rock-wallaby *Petrogale assimilis*. In *The Mammals of Australia*. 2nd edn. (Ed. R Strahan) pp. 365–366. Reed Books, Sydney.

Eldridge MDB, Coulson GM (2015) Family Macropodidae (Kanagroos and wallabies). In *Handbook of the Mammals of the World*. Vol. 5. *Monotremes and Marsupials*. (Eds DE Wilson and RA Mittermeier). pp. 630–735. Lynx Edicions, Barcelona.

Eldridge MDB, Frankham GJ (2015) Family Potoroidae (Bettongs and potoroos). In *Handbook of the Mammals of the World*. Vol. 5. *Monotremes and Marsupials*. (Eds DE Wilson and RA Mittermeier). pp. 600–628. Lynx Edicions, Barcelona.

Ellis BA, Russell EM, Dawson TJ, Harrop CJF (1977) Seasonal changes in the diet preferences of free-ranging red kangaroos, euros and sheep in western New South Wales. *Australian Wildlife Research* **4**, 127–144. doi:10.1071/WR9770127

Evans M, Gordon G (1995) Bridled nailtail wallaby *Onychogalea fraenata*. In *The Mammals of Australia*. 2nd edn. (Ed. R Strahan) pp. 356–358. Reed Books, Sydney.

Fancourt BA (2014) Rapid decline in detections of the Tasmanian bettong (*Bettongia gaimardi*) following local incursion of feral cats (*Felis catus*). *Australian Mammalogy* **36**, 247–253. doi:10.1071/AM14004

Fancourt BA (2015) Making a killing: photographic evidence of predation of a Tasmanian pademelon (*Thylogale billardierii*) by a feral cat (*Felis catus*). *Australian Mammalogy* **37**, 120–124. doi:10.1071/AM14044

Fielder K (2015) Makaia's story: the first tree kangaroo to be successfully cross fostered and raised by a wallaby. *Thylacinus* **39**(2), 18–19.

Finnie EP (1982) Husbandry of large macropods at Taronga Zoo, Sydney. In *The Management of Australian Mammals in Captivity*. (Ed. DD Evans) pp. 100–101. Zoological Board of Victoria, Melbourne.

Fisher DO (1999) Offspring sex ratio variation in the bridled nailtail wallaby, *Onychogalea fraenata*. *Behavioral Ecology and Sociobiology* **45**, 411–419. doi:10.1007/s002650050578

Fisher DO, Lara MC (1999) Effects of body size and home range on access to mates and paternity in male bridled nailtail wallabies. *Animal Behaviour* **58**, 121–130. doi:10.1006/anbe.1999.1119

Fitzsimons JA (2017) Carrion consumption by the swamp wallaby (*Wallabia bicolor*). *Australian Mammalogy* **39**, 105–107. doi:10.1071/AM16017

Flannery T (1995a) *Mammals of the South-West Pacific & Moluccan Islands*. Reed Books, Sydney.

Flannery T (1995b) *Mammals of New Guinea*. 2nd edn. Australian Museum/Reed Books, Sydney.

Fleming D, Cinderey RN, Hearn JP (1983) The reproductive biology of Bennett's wallaby (*Macropus rufogriseus rufogriseus*) ranging free at Whipsnade Park. *Journal of Zoology* **201**, 283–291 [London]. doi:10.1111/j.1469-7998.1983.tb04277.x

Flower SS (1931) Contributions to our knowledge of the duration of life in vertebrate animals. *Proceedings of the Zoological Society of London* **101**, 145–234. doi:10.1111/j.1469-7998.1931.tb06192.x

Frederick H, Johnson CN (1996) Social organisation in the rufous bettong, *Aepyprymnus rufescens*. *Australian Journal of Zoology* **44**, 9–17. doi:10.1071/ZO9960009

Freegard CJ, Calver MC, Richard JD, Bradley S (2008) Age estimation of pouch young and growth of the burrowing bettong, *Bettongia lesueur*, (Marsupialia: Potoroidae) on Heirisson Prong, Shark Bay, Western Australia. *Australian Mammalogy* **29**, 157–167. doi:10.1071/AM07020

Frick E (2014). Video: kangaroo eating a bird. Australian Geographic. http://www.australiangeographic.com.au/topics/wildlife/2014/02/video-kangaroo-eats-a-bird (accessed 1 November 2023)

Frith HJ, Calaby JH (1969) *Kangaroos*. Cheshire Publishing, Canberra.

Frith HJ, Sharman GB (1964) Breeding in wild populations of the red kangaroo, *Megaleia rufa*. *CSIRO Wildlife Research* **9**, 86–114. doi:10.1071/CWR9640086

Ganslosser U (1984) On the occurrence of female coalitions in tree kangaroos (Marsupialia, Macropodidae, *Dendrolagus*). *Australian Mammalogy* **7**, 219–221. doi:10.1071/AM84024

Gasking WR (1965) Breeding kangaroos and wallabies in captivity. *International Zoo Yearbook* **5**, 106–109. doi:10.1111/j.1748-1090.1965.tb01589.x

George GG (1982) Tree-kangaroos *Dendrolagus* spp.: their management in captivity. In *The Management of Australian Mammals in Captivity*. (Ed. DD Evans) pp. 102–107. Zoological Board of Victoria, Melbourne.

George H (1990) The care and handling of orphaned kangaroos. In *Care and Handling of Australian Native Animals*. (Ed. SJ Hand) pp. 123–141. Surrey Beatty & Sons & the Royal Zoological Society of New South Wales. Sydney.

George H, Parker G, Coote P (1995) Common wombats: rescue rehabilitation release. Unpublished manuscript.

Goldstone AD, Nelson JE (1986) Aggressive behaviour in two female *Peradorcas concinna* (Macropodidae) and its relation to oestrous. *Australian Wildlife Research* **13**, 375–385. doi:10.1071/WR9860375

Gooding CD, Harrison LA (1955) Trapping yards for kangaroos. *Journal of the Department of Agriculture, Western Australia* **4**(3), 688–694.

Green AJ (2001) Mass/length residuals: measures of body condition or generators of spurious results? *Ecology* **82**, 1473–1483. doi:10.1890/0012-9658(2001)082[1473:MLRMOB]2.0.CO;2

Green B (1984) Composition of milk and energetics of growth in marsupials. *Symposia of the Zoological Society of London* **51**, 369–387.

Green B, Newgrain K, Merchant J (1980) Changes in the milk composition during lactation in the tammar wallaby, *Macropus eugenii*. *Australian Journal of Biological Sciences* **33**, 35–42. doi:10.1071/BI9800035

Green K, Mitchell AT (1997) Breeding of the long-footed potoroo *Potorous longipes* (Marsupialia: Potoroidae) in the wild: behaviour, births and juvenile independence. *Australian Mammalogy* **20**, 1–7. doi:10.1071/AM97001

Green K, Tory MK, Mitchell AT, Tennant P, May TW (1999) The diet of the long-footed potoroo. *Australian Journal of Ecology* **24**, 151–156. doi:10.1046/j.1442-9993.1999.241957.x

Green SW, Renfree MB (1982) Changes in the milk proteins during lactation in the tammar wallaby *Macropus eugenii*. *Australian Journal of Biological Sciences* **35**, 145–152. doi:10.1071/BI9820145

Guiler ER (1960) The pouch young of the potoroo. *Journal of Mammalogy* **41**, 441–451. doi:10.2307/1377531

Guiler ER (1971) Husbandry of the potoroo *Potorous tridactylus*. *International Zoo Yearbook* **11**, 21–22. doi:10.1111/j.1748-1090.1971.tb01831.x

Guiler ER, Kitchener DA (1967) Further observations on longevity in the wild potoroo, *Potorous tridactylus*. *Australian Journal of Science* **30**, 105–106.

Hardman B, Moro D, Calver M (2016) Direct evidence implicates feral cat predation as the primary cause of failure of a mammal reintroduction programme. *Ecological Management & Restoration* **17**, 152–158. doi:10.1111/emr.12210

Harrington J (1976) The diet of the swamp wallaby, *Wallabia bicolor*, at Diamond Flat, New South Wales. Unpublished MNatRes thesis, University of New England, Armidale.

Hayward MW (2005) Diet of the quokka (*Setonix brachyurus*) (Macropodidae: Marsupialia) in the northern jarrah forest of Western Australia. *Wildlife Research* **32**, 15–22. doi:10.1071/WR03051

Hayward MW, Bellchambers K, Herman K, Bentley J, Legge S (2011) Spatial behaviour of yellow-footed rock-wallabies, *Petrogale xanthopus*, changes in response to active conservation management. *Australian Journal of Zoology* **59**, 1–8. doi:10.1071/ZO11007

Heinsohn GE (1968) Habitat requirements and reproductive potential of the macropod marsupial *Potorous tridactylus* in Tasmania. *Mammalia* **32**, 30–43. doi:10.1515/mamm.1968.32.1.30

Heise-Pavlov S, Anderson C, Moshier A (2014) Studying food preferences in captive cryptic folivores can assist in conservation planning: the case of the Lumholtz's tree-kangaroo (*Dendrolagus lumholtzi*). *Australian Mammalogy* **36**, 200–211. doi:10.1071/AM13036

Hendrikz JK, Johnson PM (1999) Development of the bridled nailtail wallaby, *Onychogalea fraenata*, and age estimation of the pouch young. *Wildlife Research* **26**, 239–249. doi:10.1071/WR96002

Herbert CA (2004) Long-acting contraceptives: a new tool to manage overabundant kangaroo populations in nature reserves and urban areas. *Australian Mammalogy* **26**, 67–74. doi:10.1071/AM04067

Herbert CA, Trigg TE, Renfree MB, Shaw G, Eckery DC, Cooper DW (2005) Long-term effects of deslorelin implants on reproduction in the female tammar wallaby (*Macropus eugenii*). *Reproduction Research* **129**, 361–369. doi:10.1530/rep.1.00432

Hollis CJ, Robertson JD, Harden RH (1986) Ecology of the swamp wallaby (*Wallabia bicolor*) in northeastern New South Wales. 1: diet. *Australian Wildlife Research* **13**, 355–365. doi:10.1071/WR9860355

Holz P (1992) Immobilisation of marsupials with tileamine and zolazepam. *Journal of Zoo and Wildlife Medicine* **23**, 426–428.

Hornsby P (1979) The history of the colony of *Petrogale xanthopus* in Adelaide Zoo. *Thylacinus* **4**(1–4), 1–16.

Hornsby P (1980) A history of rock-wallabies *Petrogale* sp. in the Adelaide Zoo. *International Zoo Yearbook* **20**, 254–260. doi:10.1111/j.1748-1090.1980.tb00987.x

Hughes RL (1962) Reproduction of the macropod marsupial *Potorous tridactylus* (Kerr). *Australian Journal of Zoology* **10**, 193–224. doi:10.1071/ZO9620193

Hulst F (2025) Zoonoses. In *Current Therapy in Medicine of Australian Mammals*. Revised Edition. (Eds L Vogelnest and T Portas) pp. 273–289. CSIRO Publishing, Melbourne.

Hulst F, Sangster C, Vogelnest L (2010) Mesenteric fat necrosis in the quokka (*Setonix brachyurus*). In Proceedings of the Wildlife Disease Association, Australasian Section, December 13–17, Dover, Tasmania.

Hume ID (1982) *Digestive Physiology of Marsupials*. Cambridge University Press, Cambridge.

Hume ID, Jarman PJ, Renfree MB, Temple-Smith PD (1989) Macropodidae. In *Fauna of Australia. Mammalia*. (Eds DW Walton and BJ Richardson) pp. 679–715. Australian Government Publishing Service, Canberra.

Hutchins M, Smith GM, Mead DC, Elbin S, Steenberg J (1991) Social behaviour of Matschie's tree kangaroos (*Dendrolagus matschiei*) and its implications for captive management. *Zoo Biology* **10**, 147–164. doi:10.1002/zoo.1430100207

Ingleby S, Gordon G (1995) Northern nailtail wallaby *Onychogalea unguifera*. In *The Mammals of Australia*. 2nd edn. (Ed. R Strahan) pp. 361–362. Reed Books, Sydney.

Ingleby S, Westoby M (1992) Habitat requirements of the spectacled hare-wallaby (*Lagorchestes conspicillatus*) in the Northern Territory and Western Australia. *Wildlife Research* **19**, 721–741. doi:10.1071/WR9920721

Inns RW (1980) Occurrence of twins in macropod marsupials. *Search* **11**, 118–119.

Inns RW (1982a) Life-spans of the tammar wallaby *Macropus eugenii* (Marsupialia: Macropodidae) in wild populations. *Australian Mammalogy* **5**, 283–284. doi:10.1071/AM82031

Inns RW (1982b) Age determination in the Kangaroo Island wallaby, *Macropus eugenii* (Desmarest). *Australian Wildlife Research* **9**, 213–220. doi:10.1071/WR9820213

International Air Transport Association (IATA) (2022) Live Animals Regulations. 48th edn. International Air Transport Association, Geneva, Switzerland.

Jackson SM, Groves CP (2015) *Taxonomy of Australian Mammals*. CSIRO Publishing, Melbourne.

Jackson SM, Vernes K (2010) *Kangaroo: A Portrait of an Extraordinary Marsupial*. Allen and Unwin, Sydney.

Jackson SM, Jansen JJFJ, Callou C, Baglione G (2021) Mammals collected and illustrated by the Baudin Expedition to Australia and Timor (1800–1804): a review of the current taxonomy of specimens in the Muséum National d'Histoire Naturelle de Paris and the illustrations in the Muséum d'Histoire Naturelle du Havre. *Zoosystema* **43**(21), 387–548. doi:10.5252/zoosystema2021v43a21

Janssens PH, Ternouth JH (1987) The transition from milk to forage diets. In *The Nutrition of Herbivores*. (Eds JB Hyacker and JH Ternouth) pp. 281–305. Academic Press, Sydney.

Jarman PJ (1984) The dietary ecology of macropod marsupials. *Proceedings of the Nutrition Society of Australia* **9**, 82–87.

Jarman PJ, Bayne P (1997) Behavioural ecology of *Petrogale penicillata* in relation to conservation. *Australian Mammalogy* **19**, 219–228. doi:10.1071/AM97219

Jarman PJ, Coulson G (1989) Dynamics and adaptiveness of groupings in macropods. In *Kangaroos, Wallabies and Rat Kangaroos*. (Eds G Grigg, P Jarman and I Hume) pp. 527–547. Surrey Beatty & Sons, Sydney.

Jarman PJ, Jones ME, Johnson CN, Southwell CJ, Stuart-Dick RI, Higginbottom KB, Clarke JL (1989) Macropod studies at wallaby creek. VIII. Individual recognition of kangaroos and wallabies. *Australian Wildlife Research* **16**, 179–185. doi:10.1071/WR9890179

Johnson CN, McIlwee AP (1997) Ecology of the northern bettong, *Bettongia tropica*, a tropical mycophagist. *Wildlife Research* **24**, 549–559. doi:10.1071/WR96034

Johnson KA (1977) Ecology and management of the red-necked pademelon, *Thylogale thetis*, on the Dorrigo plateau of northern New South Wales. PhD thesis. University of New England, Armidale.

Johnson KA (1980) Spatial and temporal use of habitat by red-necked pademelons, *Thylogale thetis*. *Australian Wildlife Research* **7**, 157–166. doi:10.1071/WR9800157

Johnson PM (1978) Studies of Macropodidae in Queensland. 9: reproduction of the rufous rat-kangaroo (*Aepyprymnus rufescens* (Gray) in captivity, with age estimation of pouch young. *Queensland Journal of Agriculture and Animal Science* **35**, 69–72.

Johnson PM (1979) Reproduction in the plain rock-wallaby *Petrogale penicillata inornata* Gould, in captivity, with age determination of the pouch young. *Australian Wildlife Research* **6**, 1–4. doi:10.1071/WR9790001

Johnson PM (1980) Observations of the behaviour of the rufous rat-kangaroo, *Aepyprymnus rufescens* (Gray), in captivity. *Australian Wildlife Research* **7**, 347–357. doi:10.1071/WR9800347

Johnson PM (1981) The rearing of marsupial pouch young by foster mothers of different species. *International Zoo Yearbook* **21**, 173–176. doi:10.1111/j.1748-1090.1981.tb01976.x

Johnson PM (1983) Musky rat-kangaroo. In *Complete Book of Australian Mammals*. 1st edn. (Ed. R Strahan) pp. 179–180. Angus and Robertson, Sydney.

Johnson PM (1993) Reproduction of the spectacled hare-wallaby, *Lagorchestes conspicillatus* Gould (Marsupialia: Macropodidae) in captivity with age estimation of the pouch-young. *Wildlife Research* **20**, 97–101. doi:10.1071/WR9930097

Johnson PM (1997) Reproduction in the bridled nailtail wallaby, *Onychogalea fraenata* Gould (Marsupialia: Macropodidae), in captivity. *Wildlife Research* **24**, 411–415. doi:10.1071/WR96028

Johnson PM (1998) Reproduction of the whiptail wallaby, *Macropus parryi* Bennett (Marsupialia: Macropodidae), in captivity with age estimation of the pouch-young. *Wildlife Research* **25**, 635–641. doi:10.1071/WR97128

Johnson PM, Delean JSC (1999) Reproduction in the Proserpine rock-wallaby, *Petrogale persephone* Maynes (Marsupialia: Macropodidae), in captivity, with age estimation and development of pouch young. *Wildlife Research* **26**, 631–639. doi:10.1071/WR98049

Johnson PM, Delean S (2001) Reproduction in the northern bettong, *Bettongia tropica* Wakefield (Marsupialia: Potoroidae), in captivity, with age estimation and development of pouch young. *Wildlife Research* **28**, 79–85. doi:10.1071/WR00007

Johnson PM, Delean S (2002a) Reproduction of the purple-necked rock-wallaby, *Petrogale purpureicollis* Le Souef 1924 (Marsupialia: Macropodidae), in captivity, with age estimation and development of the pouch young. *Wildlife Research* **29**, 463–468. doi:10.1071/WR01096

Johnson PM, Delean S (2002b) Development and age estimation of the pouch young of the black-striped wallaby *Macropus dorsalis*, with notes on reproduction. *Australian Mammalogy* **24**, 193–198. doi:10.1071/AM02193

Johnson PM, Delean S (2003) Reproduction of Lumholtz's tree-kangaroo, *Dendrolagus lumholtzi* (Marsupialia: Macropodidae) in captivity, with age estimation and development of the pouch young. *Wildlife Research* **30**, 505–512. doi:10.1071/WR02090

Johnson PM, Strahan R (1982) A further description of the musky-rat kangaroo, *Hypsiprymnodon moschatus* Ramsay, 1876 (Marsupialia: Potoroidae), with notes on its biology. *Australian Zoologist* **21**, 27–46. doi:10.7882/AZ.1982.002

Johnson PM, Vernes K (1994) Reproduction in the red-legged pademelon, *Thylogale stigmatica* (Gould) (Marsupialia; Macropodidae), and age estimation and development of pouch young. *Wildlife Research* **21**, 553–558. doi:10.1071/WR9940553

Johnson PM, Haffenden AT, Denison J (1983) Husbandry of the musky rat-kangaroo in captivity. *Journal of the Australian Animal Technicians Association* **8**, 1–8.

Johnson P, Hawkes M, Sullivan S (2002) Predation by Lumholtz's tree kangaroos *Dendrolagus lumholtzi* in captivity. *Thylacinus* **26**(3), 6–7.

Johnson PM, Lloyd S, Vallance T, Eldridge MDB (2005) First record of quadruplets in the musky rat-kangaroo *Hypsiprymnodon moschatus*. *Australian Mammalogy* **27**, 95–97. doi:10.1071/AM05095

Johnston SD, Blyde D, Gamble J, Higgins D, Field H, Cooper J (1997) Collection and short-term preservation of semen from free-ranging eastern grey kangaroos (*Macropus giganteus*: Macropodidae). *Australian Veterinary Journal* **75**, 648–651. doi:10.1111/j.1751-0813.1997.tb15362.x

Jones M, Taggart D, Temple-Smith P (2004) Age determination and growth in wild *Petrogale lateralis pearsonii* and captive *Petrogale lateralis* 'MacDonnell Ranges race. *Australian Journal of Zoology* **52**, 447–461. doi:10.1071/ZO03033

Kakulas BA (1963) Influence of the size of enclosure on the development of myopathy in the captive Rottnest quokka. *Nature* **198**, 673–674. doi:10.1038/198673a0

Kaldor I, Ezekiel E (1962) Iron content of mammalian breast milk: measurements in the rat and in a marsupial. *Nature* **196**, 175. doi:10.1038/196175a0

Kaufmann JH (1974) Habitat use and social organisation of nine sympatric species of macropodid marsupials. *Journal of Mammalogy* **55**, 66–80. doi:10.2307/1379257

Keep JM, Fox AM (1971) The capture, restraint and translocation of kangaroos in the wild. *Australian Veterinary Journal* **47**, 141–145. doi:10.1111/j.1751-0813.1971.tb02122.x

Kerle JA (1998) The population dynamics of a tropical possum, *Trichosurus vulpecula arnhemensis* Collett. *Wildlife Research* **25**, 171–181. doi:10.1071/WR96113

Kido N, Tanaka S, Wada Y, Sato S, Omiya T (2018) Molar eruption and identification of the eastern grey kangaroo (*Macropus giganteus*) at different ages. *Journal of Veterinary Medical Science* **80**, 648–652.

King WJ, Coulson G (2024) Estimating age of wild eastern grey kangaroos through molar progression. *Australian Mammalogy* **46**, AM23043.

King WJ, Wilson ME, Allen T, Festa-Bianchet M, Coulson G (2011) A capture technique for free-ranging eastern grey kangaroos (*Macropus giganteus*) habituated to humans. *Australian Mammalogy* **33**, 47–51. doi:10.1071/AM10029

Kinnear JE, Onus ML, Bromilow RN (1988a) Fox control and rock-wallaby population dynamics. *Australian Wildlife Research* **15**, 435–450. doi:10.1071/WR9880435

Kinnear JE, Bromilow RN, Onus RN, Sokolowski RES (1988b) The Bromilow trap: a new risk free soft trap suitable for small to medium-sized macropods. *Australian Wildlife Research* **15**, 235–237. doi:10.1071/WR9880235

Kinnear JE, Bromilow RN, Moore N (2014) The Mark II Bromilow Kinnear humane soft trap: a major upgrade. *Australian Mammalogy* **36**, 121–127. doi:10.1071/AM13023

Kirkpatrick TH (1964) Molar progression and macropod age. *Queensland Journal of Agricultural and Animal Sciences* **21**, 163–165.

Kirkpatrick TH (1965a) Studies of Macropodidae in Queensland. 2: Age estimation in the grey kangaroo, the red kangaroo, the eastern wallaroo, and the red-necked wallaby, with notes on dental abnormalities. *Queensland Journal of Agricultural and Animal Sciences* **22**, 301–317.

Kirkpatrick TH (1965b) Studies of Macropodidae in Queensland. 3: Reproduction in the grey kangaroo (*Macropus major*) in southern Queensland. *Queensland Journal of Agricultural and Animal Sciences* **22**, 319–328.

Kirkpatrick TH (1967) Studies of Macropodidae in Queensland. 6: Sex determination of adult skulls of the grey kangaroo and the red kangaroo. *Queensland Journal of Animal Science* **24**, 131–133.

Kirkpatrick TH (1970) Studies of Macropodidae in Queensland. 8: Age estimation in the red kangaroo (*Megaleia rufa* (Desmarest)). *Queensland Journal of Animal Science* **27**, 461–462.

Kirkpatrick TH, Johnson PM (1969) Studies of Macropodidae in Queensland. 7: Age estimation and reproduction in the agile wallaby (*Wallabia agilis* (Gould)). *Queensland Journal of Agricultural and Animal Sciences* **26**, 691–698.

Kitchener DJ (1972) The importance of shelter to the quokka, *Setonix brachyurus* (Marsupialia) on Rottnest Island. *Australian Journal of Zoology* **20**, 281–299. doi:10.1071/ZO9720281

Kitchener DJ (1973) Notes on home range and movements in two small macropods, the potoroo (*Potorous apicalis*) and the quokka (*Setonix brachyurus*). *Mammalia* **37**, 231–240. doi:10.1515/mamm.1973.37.2.231

Kleemann SL, Sandow D, Stevens M, Schultz DJ, Taggart DA, Croxford A (2021) Non-invasive monitoring and reintroduction biology of the brush-tailed rock-wallaby (*Petrogale penicillata*) in the Grampians National Park, Australia. *Australian Journal of Zoology* **69**, 41–54. doi:10.1071/ZO21009

Knowlton JE (1984) A sighting device for estimating molar index to determine age from macropod skulls. *Australian Wildlife Research* **11**, 451–454. doi:10.1071/WR9840451

Kokko H, Brooks R, Jennions MD, Morley J (2003) The evolution of mate choice and mating biases. *Proceedings of the Royal Society of London. Series B, Biological Sciences* **270**(1515), 653–664. doi:10.1098/rspb.2002.2235

Lacy RC, Balou JD, Pollak JP (2011) PMx: software package for demographic and genetic analysis and management of pedigreed populations. *Methods in Ecology and Evolution* **3**, 433–437. doi:10.1111/j.2041-210X.2011.00148.x

Lavery HJ (1985) *The Kangaroo Keepers*. University of Queensland Press, St Lucia.

Learmonth MJ, Sherwen S, Hemsworth PH (2018) The effects of zoo visitors on Quokka (*Setonix brachyurus*) avoidance behavior in a walk-through exhibit. *Zoo Biology* **37**, 223–228. doi:10.1002/zoo.21433

Lee AK, Cockburn A (1985) *Evolutionary Ecology of Marsupials*. Cambridge University Press, Cambridge.

Leggett KEA, Stubbs-Oliver A, Brun ML, Dawson TJ (2019) Carnivory in euros, inland wallaroos (*Macropus robustus erubescens*). *Australian Mammalogy* **42**, 357–358. doi:10.1071/AM19055

Lemon M, Barker S (1967) Changes in milk composition of the red kangaroo, *Megaleia rufa* (Desmarest), during lactation. *The Australian Journal of Experimental Biology and Medical Science* **45**, 213–219. doi:10.1038/icb.1967.16

Lentle RG, Potter MA, Springgett BP, Stafford KJ (1997) A trapping and immobilisation technique for small macropods. *Wildlife Research* **24**, 373–377. doi:10.1071/WR95052

Lessiak A (2014) Methods for the effective care and rehabilitation of captive Lumholtz's tree-kangaroo. *Independent Study Project (ISP) Collection* **1753**, 1–33.

Libke J, Libke D (2000) Wildlife tales: mumma. *Wildlife Australia* (Summer), 30–32.

Liptovszky M, Sós E, Bende B, Perge E, Molnár V (2014) Urolithiasis in a captive group of tammar wallabies (*Macropus eugenii*). *Zoo Biology* **33**, 455–458. doi:10.1002/zoo.21159

Lloyd S (2001) Oestrous cycle and gestation length in the musky rat-kangaroo, *Hypsiprymnodon moschatus* (Potoroidae: Marsupialia). *Australian Journal of Zoology* **49**, 37–44. doi:10.1071/ZO00010

Loh TT, Kaldor I (1973) Iron in milk and milk fractions of lactating rats, rabbits, and quokkas. *Comparative Biochemistry and Physiology* **44B**, 337–346.

Long KI (2001) Spatio-temporal interactions among male and female long-nosed potoroos, *Potorous tridactylus* (Marsupialia: Macropodidae): mating system implications. *Australian Journal of Zoology* **49**, 17–26. doi:10.1071/ZO00077

Lundie-Jenkins G (1993a) Observations on the behaviour of the rufous hare-wallaby, *Lagorchestes hirsutus* Gould (Marsupialia: Macropodidae) in captivity. *Australian Mammalogy* **16**, 29–34. doi:10.1071/AM93005

Lundie-Jenkins G (1993b) Reproduction and growth to sexual maturity in the rufous hare-wallaby, *Lagorchestes hirsutus* Gould (Marsupialia: Macropodidae) in captivity. *Australian Mammalogy* **16**, 45–49. doi:10.1071/AM93009

Main AR, Yadav M (1971) Conservation of macropods in reserves in Western Australia. *Biological Conservation* **3**, 123–133. doi:10.1016/0006-3207(71)90014-0

Mallam C (2012) Looking for a shadow: predictive habitat suitability modelling for the brush-tailed rock-wallaby (*Petrogale penicillata*) in the Grampians National Park and Black Range State Park, Victoria. Honours thesis, University of Adelaide, South Australia.

Martin R (1992) An ecological study of Bennett's Tree-Kangaroo (*Dendrolagus bennettianus*). Report to the World Wide Fund for Nature. Project 116.

Martin R (2005) *Tree-kangaroos of Australia and New Guinea*. CSIRO Publishing, Melbourne.

Mayberry C, Bencini R, Mawson PR, Maloney SK (2018) Estimating the age of western grey kangaroos (*Macropus fuliginosus ocydromus*). *Australian Mammalogy* **40**, 254–261. doi:10.1071/AM16058

Maynes GM (1972) Age estimation in the parma wallaby, *Macropus parma* (Waterhouse). *Australian Journal of Zoology* **20**, 107–118. doi:10.1071/ZO9720107

Maynes GM (1973a) Reproduction in the parma wallaby, *Macropus parma* Waterhouse. *Australian Journal of Zoology* **21**, 331–351. doi:10.1071/ZO9730331

Maynes GM (1973b) Aspects of reproduction in the whiptail wallaby *Macropus parryi*. *Australian Zoologist* **18**, 43–46.

Maynes GM (1976) Growth of the parma wallaby, *Macropus parma* Waterhouse. *Australian Journal of Zoology* **24**, 217–236. doi:10.1071/ZO9760217

Maynes GM (1977) Breeding and age structure of the populations of *Macropus parma* on Kawau Island, New Zealand. *Australian Journal of Ecology* **2**, 207–214. doi:10.1111/j.1442-9993.1977.tb01138.x

McArthur C, Sanson GD (1988) Tooth wear in eastern grey kangaroos (*Macropus giganteus*) and western grey kangaroos (*Macropus fuliginosus*), and its potential influence on diet selection, digestion and population parameters. *Journal of Zoology* **215**, 491–504. doi:10.1111/j.1469-7998.1988.tb02855.x

McCullough DR, McCullough Y (2000) *Kangaroos in Outback Australia*. Columbia University Press, New York.

McLean IG, Lunde-Jenkins G, Jarman PJ (1994) Training captive rufous hare-wallabies to recognise predators. In *Reintroduction Biology of Australian and New Zealand Fauna*. (Ed. M Serena) pp. 177–182. Surrey Beatty & Sons, Sydney.

McLelland DJ (2025a) Macropods. In *Current Therapy in Medicine of Australian Mammals*. Revised Edition. (Eds L Vogelnest and T Portas) pp. 465–482. CSIRO Publishing, Melbourne.

McLelland DJ (2025b) Macropod progressive periodontal disease ('lumpy jaw'). In *Current Therapy in Medicine of Australian Mammals*. Revised Edition. (Eds L Vogelnest and T Portas) pp. 483–494. CSIRO Publishing, Melbourne.

McLelland DJ, Fielder K, Males G, Langley N, Schultz D (2015) Successful transfer of a Goodfellow's Tree Kangaroo (*Dendrolagus goodfellowi*) pouch young to a yellow-footed rock Wallaby (*Petrogale xanthopus*) surrogate. *Zoo Biology* **34**, 460–462. doi:10.1002/zoo.21236

Mella VSA, McArthur C, Frend R, Crowther MS (2018) Foxes in trees: a threat for Australian arboreal fauna? *Australian Mammalogy* **40**, 103–105. doi:10.1071/AM16049

Merchant JC (1976) Breeding biology of the agile wallaby, *Macropus agilis* (Gould)(Marsupialia: Macropodidae), in captivity. *Australian Wildlife Research* **3**, 93–103. doi:10.1071/WR9760093

Merchant JC (1979) The effect of pregnancy on the interval between one oestrous and the next in the tammar wallaby, *Macropus eugenii*. *Journal of Reproduction and Fertility* **56**, 459–463. doi:10.1530/jrf.0.0560459

Merchant JC, Calaby JH (1981) Reproductive biology of the red-necked wallaby (*Macropus rufogriseus banksianus*) and Bennett's wallaby (*M. r. rufogriseus*) in captivity. *Journal of Zoology* **194**, 203–217. doi:10.1111/j.1469-7998.1981.tb05769.x

Merchant JC, Sharman GB (1966) Observations on the attachment of marsupial pouch young to the teats and on the rearing of pouch young by foster-mothers of the same or different species. *Australian Journal of Zoology* **14**, 593–609. doi:10.1071/ZO9660593

Merchant J, Green B, Messer M, Newgrain K (1989) Milk composition of the red-necked wallaby, *Macropus rufogriseus banksianus* (Marsupialia). *Comparative Biochemistry and Physiology* **93A**, 483–488. doi:10.1016/0300-9629(89)90071-6

Merchant JC, Libke JA, Smith MJ (1994) Lactation and energetics of growth in the brush-tailed bettong *Bettongia penicillata* (Marsupialia: Potoroidae) in captivity. *Australian Journal of Zoology* **42**, 267–277. doi:10.1071/ZO9940267

Merchant JC, Marsh H, Spencer P, De'Ath G (1996) Milk composition and production in free-living allied rock-wallabies, *Petrogale assimilis*. *Australian Journal of Zoology* **44**, 659–674. doi:10.1071/ZO9960659

Messer M, Green B (1979) Milk carbohydrates of marsupials. II: quantitative and qualitative changes in milk carbohydrates during lactation in the tammar wallaby (*Macropus eugenii*). *Australian Journal of Biological Sciences* **32**, 519–531. doi:10.1071/BI9790519

Messer M, Walker DM (1992) Milk substitutes for marsupials: In theory how good (or bad) are they? In *Urban Wildlife*. Proceedings 204. Post Graduate Committee in Veterinary Science. (Eds L Vogelnest, D Spielman, T Bellamy, M Messer and DM Walker) pp. 99–124. University of Sydney, Sydney.

Messer M, Trifonoff E, Stern W, Collins JG, Bradbury JH (1980) Structure of a marsupial-milk trisaccharide. *Carbohydrate Research* **83**, 327–334. doi:10.1016/S0008-6215(00)84545-0

Messer M, Crisp EA, Czolij R (1989) Lactose digestion in suckling macropods. In *Kangaroos, Wallabies and Rat-kangaroos*. (Eds G Grigg, P Jarman and I Hume) pp. 217–221. Surrey Beatty & Sons, Sydney.

Mitchell PC (1911) On longevity and relative viability in mammals and birds; with a note on the theory of longevity. *Proceedings of the Zoological Society of London* **81**, 425–548. doi:10.1111/j.1096-3642.1911.tb01942.x

Moir RJ, Somers M, Waring H (1956) Studies on marsupial nutrition. I: ruminant-like digestion in a herbivorous marsupial (*Setonix brachyurus* Quoy and Gaimard). *Australian Journal of Biological Sciences* **9**, 293–304. doi:10.1071/BI9560293

Molyneaux J, Taggart DA, Corrigan A, Frey S (2011) Home range studies in a reintroduced brush-tailed rock-wallaby (*Petrogale penicillata*) population in the Grampians National Park, Victoria. *Australian Journal of Zoology* **33**, 128–134.

Moors PJ (1975) The urogenital system and notes on the reproductive biology of the female rufous rat-kangaroo, *Aepyprymnus rufescens* (Gray) (Macropodidae). *Australian Journal of Zoology* **23**, 355–361. doi:10.1071/ZO9750355

Morton SR, Burton TC (1973) Observations on the behaviour of the macropodid marsupial *Thylogale billardierii* Desmarest in captivity. *Australian Zoologist* **18**, 1–14.

Moyal A (2001) *Platypus*. Allen & Unwin, Sydney.

Mullet T, Yoshimi D, Steenberg J (1990) *Tree Kangaroo Husbandry Notebook*. Woodland Park Zoological Gardens, Seattle, Washington.

Munn AJ, Phelan S, Rigby M, Roberts JA (2017) Behavioral adjustments of wild-caught kangaroos to captivity. *Australian Mammalogy* **39**, 213–218. doi:10.1071/AM16019

Murphy CR, Smith JR (1970) Age determination of pouch young and juvenile Kangaroo Island wallabies. *Transactions of the Royal Society of South Australia* **94**, 15–20.

Nelson JE, Goldstone A (1986) Reproduction in *Peradorcas concinna* (Marsupialia: Macropodidae). *Australian Wildlife Research* **13**, 501–505. doi:10.1071/WR9860501

Newell GR (1999) Home range and habitat utilisation of Lumholtz's tree-kangaroo (*Dendrolagus lumholtzi*) within a rainforest fragment in north Queensland. *Wildlife Research* **26**, 129–145. doi:10.1071/WR98016

Newsome AE (1965) Reproduction in natural populations of the red kangaroo, *Megaleia rufa* (Desmarest), in central Australia. *Australian Journal of Zoology* **13**, 735–759. doi:10.1071/ZO9650735

Newsome AE (1977) Imbalance in the sex-ratio and age structure of the red kangaroo in central Australia. In *The Biology of Marsupials*. (Eds B Stonehouse and D Gilmore) pp. 221-233. Macmillan, London.

Newsome AE, Merchant JC, Bolton BL, Dudzinski ML (1977) Sexual dimorphism in molar progression and eruption in the agile wallaby. *Australian Wildlife Research* **4**, 1–5. doi:10.1071/WR9770001

Nguyen VP, Needham AD, Friend JA (2005) A quantitative dietary study of the 'critically Endangered' gilbert's potoroo *Potorous gilbertii*. *Australian Mammalogy* **27**, 1–6. doi:10.1071/AM05001

Nicholls C (1972) A study of the behaviour and social organisation of the yellow-footed rock-wallaby *Petrogale xanthopus*. BSc Hons thesis, University of Adelaide, South Australia.

Norbury GL (1987) Twins in the western grey kangaroo, *Macropus fuliginosus* (Marsupialia: Macropodidae), in northwestern Victoria. *Australian Mammalogy* **10**, 33. doi:10.1071/AM87006

Northover AS, Godfrey SS, Lymbery AJ, Morris K, Wayne AF, Thompson RA (2017) Evaluating the effects of ivermectin treatment on communities of gastrointestinal parasites in translocated woylies (*Bettongia penicillata*). *EcoHealth* **14**, 117-127. doi:10.1007/s10393-015-1088-2

Okeson DM, Esterline ML, Coke RL (2009) Esophageal diverticula in parma wallabies (*Macropus parma*). *Journal of Zoo and Wildlife Medicine* **40**, 168–173. doi:10.1638/2008-0039.1

Oliver AJ (1986) Social organisation and dispersal in the Red Kangaroo. PhD thesis. Murdoch University, Perth.

Owen R (1835) Notes on the anatomy of a new species of kangaroo. *Proceedings of the Zoological Society of London* **2**(1834), 152.

Packer WC (1965) Environmental influences on daily and seasonal activity in *Setonix brachyurus* (Quoy & Gaimard) (Marsupialia). *Animal Behaviour* **13**, 270–283. doi:10.1016/0003-3472(65)90046-1

Packer WC (1969) Observations on the behaviour of the marsupial *Setonix brachyurus* (Quoy & Gaimard) in an enclosure. *Journal of Mammalogy* **50**, 8–20. doi:10.2307/1378625

Paplinska JZ, Moyle RLC, Temple-Smith PDM, Renfree MB (2006) Reproduction in female swamp wallabies, *Wallabia bicolor*. *Reproduction, Fertility and Development* **18**, 735–743. doi:10.1071/RD06024

Paplinska JZ, Bencini R, Fisher DO, Newell GR, Goldizen AW, Hazlitt SL, Sigg DP, Finlayson GR, Munn AJ, Chambers BK, Taggart DA (2010) Sperm competition in the Macropodoidea: a review of evidence. In: *Macropods: The Biology of Kangaroos, Wallabies and Rat-kangaroos* (Eds G Coulson and M Eldridge) pp 65–76. CSIRO Publishing, Melbourne.

Paris DB, Taggart DA, Shaw G, Temple-Smith PD, Renfree MB (2005) Birth of pouch young after artificial insemination in the tammar wallaby (*Macropus eugenii*). *Biology of Reproduction* **72**, 451–459. doi:10.1095/biolreprod.104.033282

Parker P (1977) An ecological comparison of marsupial and placental patterns of reproduction. In *The Biology of Marsupials*. (Eds B Stonehouse and D Gilmore) pp. 273–286. Macmillan, London.

Phillips C, Higgins D, Herbert C, Spielman D (2012) Investigating the efficacy of '5 in 1' vaccines in macropods. In: Proceedings of the Australian Wildlife Rehabilitation Conference, Townsville, Queensland.

Pollock DC, Montague TL (1991) A new trap trigger mechanism for the capture of swamp wallabies, *Wallabia bicolor* (Marsupialia: Macropodidae). *Wildlife Research* **18**, 459–461. doi:10.1071/WR9910459

Poole WE (1973) A study of breeding in grey kangaroos, *Macropus giganteus* (Shaw) and *M. fuliginosus* (Desmarest), in central New South Wales. *Australian Journal of Zoology* **21**, 183–212. doi:10.1071/ZO9730183

Poole WE (1975) Reproduction in the two species of grey kangaroos, *Macropus giganteus* (Shaw), and *M. fuliginosus* (Desmarest). II: gestation, parturition and pouch life. *Australian Journal of Zoology* **23**, 333–353. doi:10.1071/ZO9750333

Poole WE (1976) Breeding biology and current status of the grey kangaroos, *Macropus fuliginosus fuliginosus*, of Kangaroo Island, South Australia. *Australian Journal of Zoology* **24**, 169–187. doi:10.1071/ZO9760169

Poole WE (1982) *Macropus giganteus*. *Mammalian Species* **187**, 1–8. doi:10.2307/3504005

Poole WE, Brown GD (1988) Further records of life spans of the tammar-wallaby, *Macropus eugenii* (Marsupialia: Macropodidae), on Kangaroo Island, South Australia. *Australian Mammalogy* **11**, 165–166. doi:10.1071/AM88022

Poole WE, Catling PC (1974) Reproduction in the two species of grey kangaroos, *Macropus giganteus* (Shaw), and *M. fuliginosus* (Desmarest). I: sexual maturity and oestrous. *Australian Journal of Zoology* **22**, 277–302. doi:10.1071/ZO9740277

Poole WE, Carpenter SM, Wood JT (1982a) Growth of grey kangaroos and the reliability of age determination from body measurements. II: the western grey kangaroos, *Macropus fuliginosus fuliginosus, M. f. melanops* and *M. f. ocydromus*. *Australian Wildlife Research* **9**, 203–212. doi:10.1071/WR9820203

Poole WE, Carpenter SM, Wood JT (1982b) Growth of grey kangaroos and the reliability of age determination from body measurements. I. The eastern grey kangaroo, *Macropus giganteus*. *Australian Wildlife Research* **9**, 9–20. doi:10.1071/WR9820009

Poole WE, Sharman GB, Scott KJ, Thompson SY (1982c) Composition of milk from red and grey kangaroos with particular reference to vitamins. *Australian Journal of Biological Sciences* **35**, 607–615. doi:10.1071/BI9820607

Poole WE, Merchant JC, Carpenter SM, Calaby JH (1985) Reproduction, growth and age determination in the yellow-footed rock-wallaby *Petrogale xanthopus* Gray, in captivity. *Australian Wildlife Research* **12**, 127–136. doi:10.1071/WR9850127

Poole WE, Westcott M, Simms NG (1992) Determination of oestrous in the female tammar, *Macropus eugenii*, by analysis of cellular composition of smears from the reproductive tract. *Wildlife Research* **19**, 35–46. doi:10.1071/WR9920035

Portas T (2025a) Toxoplasmosis. In *Current Therapy in Medicine of Australian Mammals*. Revised Edition. (Eds L Vogelnest and T Portas) pp. 343–348. CSIRO Publishing, Melbourne.

Portas T (2025b) Management of overabundant populations. In *Current Therapy in Medicine of Australian Mammals*. Revised Edition. (Eds L Vogelnest and T Portas) pp. 87–98. CSIRO Publishing, Melbourne.

Read J, Copley P, Ward M, Dagg E, Olds L, Taggart DA, West R (2018) Bringing back Warru: return of the black-footed rock wallaby to the APY Lands. In: *Recovery of Australian Threatened Species: A Book of Hope*. (Eds S Garnett, P Latch, D Lindenmayer and J Woinarski) pp. 237–247. CSIRO Publishing, Melbourne.

Read JL, Dagg E, Moseby KE (2019) Prey selectivity by feral cats at central Australian rock-wallaby colonies. *Australian Mammalogy* **41**, 132–141. doi:10.1071/AM17055

Rendle J, Jackson B, Hoorn SV, Yeap L, Warren K, Donaldson R, Ward SJ, Vogelnest L, McLelland D, Lynch M, Vitali S (2020) A retrospective study of macropod progressive periodontal disease ("lumpy jaw") in captive macropods across Australia and Europe: using data from the past to inform future macropod management. *Animals (Basel)* **10**(11), 1954. doi:10.3390/ani10111954

Robertson GG, Gepp B (1982) Capture of kangaroos by 'stunning'. *Australian Wildlife Research* **9**, 393–396. doi:10.1071/WR9820393

Robinson AC, Lim L, Canty PD, Jenkins RB, Macdonald CA (1994) Studies of the yellow-footed rock-wallaby, *Petrogale xanthopus* Gray (Marsupialia: Macropodidae): population studies at middle gorge, South Australia. *Wildlife Research* **21**, 473–481. doi:10.1071/WR9940473

Rose KA, Kirkland PD, Davis RJ, Cooper DW, Blumstein D, Pritchard LI, Newberry KM, Lunt RA (2012) Epizootics of sudden death in tammar wallabies (*Macropus eugenii*) associated with an orbivirus infection. *Australian Veterinary Journal* **90**, 505–509. doi:10.1111/j.1751-0813.2012.00993.x

Rose RW (1978) Reproduction and evolution in female Macropodidae. *Australian Mammalogy* **2**, 65–72. doi:10.1071/AM78008

Rose RW (1982) Tasmanian bettong *Bettongia gaimardi*: maintenance in captivity. In *The Management of Australian Mammals in Captivity*. (Ed. DD Evans) pp. 108–110. Zoological Board of Victoria, Melbourne.

Rose RW (1987) Reproductive biology of the Tasmanian bettong *Bettongia gaimardi* (Macropodidae). *Journal of Zoology* **212**, 59–67. doi:10.1111/j.1469-7998.1987.tb05114.x

Rose RW (1989) Age estimation of the Tasmanian bettong (*Bettongia gaimardi*) (Marsupialia: Potoroidae). *Australian Wildlife Research* **16**, 251–261. doi:10.1071/WR9890251

Rose RW, McCartney DJ (1982a) Reproduction in the red-bellied pademelon, *Thylogale billardierii* (Marsupialia). *Australian Wildlife Research* **9**, 27–32. doi:10.1071/WR9820027

Rose RW, McCartney DJ (1982b) Growth of the red-bellied pademelon, *Thylogale billardierii*, and age estimation of pouch young. *Australian Wildlife Research* **9**, 33–38. doi:10.1071/WR9820033

Rose RW, Morahan TM, Mulchay JE, Ratkowsky DA (2003) Milk composition and growth in wild and captive Tasmanian bettongs, *Bettongia gaimardi* (Marsupialia). *Journal of Comparative Physiology. B, Biochemical, Systemic, and Environmental Physiology* **173**, 125–133. doi:10.1007/s00360-002-0315-7

Rudd C (1994) Sexual behaviour of male and female tammar wallabies (*Macropus eugenii*) at post partum oestrus. *Journal of Zoology* **232**, 151–162. doi:10.1111/j.1469-7998.1994.tb01565.x

Russell EM (1984) Social behaviour and organisation of marsupials. *Mammal Review* **14**, 101–154. doi:10.1111/j.1365-2907.1984.tb00343.x

Sadleir RMFS (1963) Age estimation by measurement of joeys of the euro *Macropus robustus* Gould in Western Australia. *Australian Journal of Zoology* **11**, 241–249. doi:10.1071/ZO9630241

Sadleir RMFS, Tyndale-Biscoe CH (1977) Photoperiod and the termination of embryonic diapause in the marsupial *Macropus eugenii*. *Biology of Reproduction* **16**, 605–608. doi:10.1095/biolreprod16.5.605

Sampson JC (1971) The biology of *Bettongia penicillata* Gray, 1837. PhD thesis. University of Western Australia, Perth.

Sander U, Short J, Turner B (1997) Social organisation and warren use of the burrowing bettong, *Bettongia lesueur* (Macropodoidea: Potoroidae). *Wildlife Research* **24**, 143–157. doi:10.1071/WR96021

Sanson GD (1978) The evolution and significance of mastication in the Macropodidae. *Australian Mammalogy* **2**, 23–28. doi:10.1071/AM78003

Sanson GD (1989) Morphological adaptations of teeth to diets and feeding in the Macropodoidea. In *Kangaroos, Wallabies and Rat-kangaroos*. (Eds G Grigg, P Jarman and I Hume) pp. 151–168. Surrey Beatty & Sons, Sydney.

Sanson GD, Nelson JE, Fell P (1985) Ecology of *Peradorcas concinna* in Arnhem Land in a wet and a dry season. *Proceedings of the Ecological Society of Australia* **13**, 65–72.

Scholz B (1980) The behaviour of the brush-tailed rock-wallaby (*Petrogale penicillata*), in south-east Queensland. Unpublished thesis, Queensland Institute of Technology, Rockhampton.

Schultz DJ, Whitehead PJ, Taggart DA (2006) A review of the surrogacy program for the brush-tailed rock wallaby, *Petrogale penicillata*, with special reference to animal husbandry and veterinary considerations. *Journal of Zoo and Wildlife Medicine* **37**, 33–39. doi:10.1638/04-027.1

Seebeck JH (1982) Long-nosed potoroo *Potorous tridactylus*: husbandry and management of a captive colony. In *The Management of Australian Mammals in Captivity*. (Ed. DD Evans) pp. 111–116. Zoological Board of Victoria, Melbourne.

Seebeck JH (1992) Breeding, growth and development of captive *Potorous longipes* (Marsupialia: Potoroidae); and a comparison with *P. tridactylus*. *Australian Mammalogy* **15**, 37–45. doi:10.1071/AM92005

Seebeck JH, Rose RW (1989) Potoroidae. In *Fauna of Australia*. Vol. 1B. *Mammalia*. (Ed. DW Walton) pp. 716–739. Australian Government Publishing Service, Canberra.

Selwood L (1986) The marsupial blastocyst: a study of the blastocysts in the Hill collection. *Australian Journal of Zoology* **34**, 177–187. doi:10.1071/ZO9860177

Sharman GB (1955*a*) Studies on marsupial *reproduction*. II: the oestrous cycle of *Setonix brachyurus*. *Australian Journal of Zoology* **3**, 44–55. doi:10.1071/ZO9550044

Sharman GB (1955*b*) Studies on marsupial *reproduction*. III: normal and delayed pregnancy in *Setonix brachyurus*. *Australian Journal of Zoology* **3**, 56–70. doi:10.1071/ZO9550056

Sharman GB (1963) Delayed implantation in marsupials. In *Delayed Implantation*. (Ed. AC Enders) pp. 3–14. Chicago University Press, Chicago.

Sharman GB, Calaby JH (1964) Reproductive behaviour in the red kangaroo (*Megaleia rufa*) in captivity. *CSIRO Wildlife Research* **9**, 58–85. doi:10.1071/CWR9640058

Sharman GB, Pilton PE (1964) The life history and reproduction of the red kangaroo (*Megaleia rufa*) in captivity. *Proceedings of the Zoological Society of London* **142**, 29–48. doi:10.1111/j.1469-7998.1964.tb05152.x

Sharman GB, Frith HJ, Calaby JH (1964) Growth of the pouch young, tooth eruption, and age determination in the red kangaroo, *Megaleia rufa*. *CSIRO Wildlife Research* **9**, 20–49. doi:10.1071/CWR9640020

Shaw G, Rose RW (1979) Delayed gestation in the potoroo *Potorous tridactylus* (Kerr). *Australian Journal of Zoology* **27**, 901–912. doi:10.1071/ZO9790901

Shephard NC (1987) *Hand Rearing of Orphan Marsupials*. National Parks and Wildlife Service, Sydney.

Sherwen SL, Hemsworth PH, Butler K, Fanson KV, Magrath MJL (2015) Impacts of visitor number on kangaroos housed in free-range exhibits. *Zoo Biology* **34**, 287–295. doi:10.1002/zoo.21226

Shield J (1968) Reproduction of the quokka, *Setonix brachyurus*, in captivity. *Journal of Zoology* **155**, 427–444. doi:10.1111/j.1469-7998.1968.tb03060.x

Shield JW, Woolley P (1961) Age estimation by measurement of the pouch young of the quokka (*Setonix brachyurus*). *Australian Journal of Zoology* **9**, 14–23. doi:10.1071/ZO9610014

Short J (1982) Habitat requirements of the brush-tailed rock-wallaby, *Petrogale penicillata* in New South Wales. *Australian Wildlife Research* **9**, 239–246. doi:10.1071/WR9820239

Short JC, Turner B (1999) Ecology of burrowing bettongs, *Bettongia lesueur* (Marsupialia: Potoroidae) on Dorre and Bernier Islands, Western Australia. *Wildlife Research* **26**, 651–669. doi:10.1071/WR98039

Smith MJ (1981) Morphological observations on the diapausing blastocyst of some macropodid marsupials. *Journal of Reproduction and Fertility* **61**, 483–486. doi:10.1530/jrf.0.0610483

Smith MJ (1998) Establishment of a captive colony of *Bettongia tropica* (Marsupialia: Potoroidae) by cross-fostering; and observations on reproduction. *Journal of Zoology* **244**, 43–50[London]. doi:10.1111/j.1469-7998.1998.tb00005.x

Smolenski AJ, Rose RW (1988) Comparative lactation in two species of rat-kangaroo (Marsupialia). *Comparative Biochemistry and Physiology* **90A**, 459–463. doi:10.1016/0300-9629(88)90220-4

Speare R (1988) Clinical assessment, disease and management of the orphaned macropod joey. In *Australian Wildlife*. Proceedings 104. Post Graduate Committee in Veterinary Science. (Ed. DI Bryden) pp. 211–296. University of Sydney, Sydney.

Species360 (2023) Zoological Information Management System (ZIMS). www.Species360.org.

Spencer PBS (1991) Evidence of predation by a feral cat, *Felis catus* (Carnivora: Felidae) on an isolated rock-wallaby colony in tropical Queensland. *Australian Mammalogy* **14**, 143–144. doi:10.1071/AM91018

Spencer PBS (1996) Coping with a naturally fragmented environment: a genetic and ecological study of the allied rock wallaby, *Petrogale assimilis*. PhD thesis. James Cook University, Townsville.

Spielman D (1994) First aid and emergency care for Australian native mammals. In *Wildlife*. Proceedings 104. Post Graduate Committee in Veterinary Science. (Ed. DI Bryden) pp. 339–366. University of Sydney, Sydney.

Stanbury P, Phipps G (1980) *Australia's Animals Discovered*. Pergamon Press Sydney.

Steenberg J (n.d.) Tree Kangaroo Husbandry. Part 1. Video. American Zoological and Aquarium Association and Woodland Park Zoological Gardens, Washington.

Stephens T (1975) Nutrition of orphan marsupials. *Australian Veterinary Journal* **51**, 453–458. doi:10.1111/j.1751-0813.1975.tb02377.x

Stephens T, Irvine S, Mutton P, Gupta JD, Harley JD (1974) The case of the cataractous kangaroo. *The Medical Journal of Australia* **2**, 910–911. doi:10.5694/j.1326-5377.1974.tb93759.x

Stodart E (1966) Observations on the behaviour of the marsupial *Bettongia lesueur* (Quoy and Gaimard) in an enclosure. *CSIRO Wildlife Research* **11**, 91–99. doi:10.1071/CWR9660091

Strahan R (1995) *The Mammals of Australia*. 2nd edn. New Holland, Sydney.

Taggart DA (2002) Use of pouch young removal and cross fostering techniques to accelerate breeding and recruitment in the threatened brush-tailed rock wallaby, *Petrogale penicillata*. *ANZCCART News* **15**(3), 7–9.

Taggart DA, Schultz D, Temple-Smith P (1997) Development and application of assisted reproductive technologies in marsupials: their value for the conservation of rock-wallabies. *Australian Mammalogy* **19**, 183–190. doi:10.1071/AM97183

Taggart DA, Breed WG, Temple-Smith PD, Purvis A, Shimmin G (1998) Sperm competition and mating strategies in marsupials and monotremes. In *Sperm Competition and Sexual Selection*. (Eds TR Birkhead and AP Moller) pp. 623–656. Academic Press, London.

Taggart DA, Shimmin GA, Underwood G, Phillips K (2002) Survival of very small macropod pouch young following short-term isolation from the teat at various environmental temperatures. *Animal Conservation* **5**, 275–282. doi:10.1017/S136794300200402X

Taggart DA, Schultz D, White C, Whitehead P, Underwood G, Phillips K (2005) Cross fostering, growth and reproductive studies in the brush-tailed rock-wallaby, *Petrogale penicillata* (Marsupialia: Macropodidae): efforts to accelerate breeding in a threatened marsupial species. *Australian Journal of Zoology* **53**, 313-323. doi:10.1071/ZO05002

Taggart DA, Schultz DS, Fletcher TP, Friend T, Smith I, Breed WG, Temple-Smith PD (2010) Cross-fostering and short-term pouch young isolation in Macropodoid marsupials: implications for conservation and species management. In *Macropods: The biology of Kangaroos, Wallabies and Rat Kangaroos*. (Eds G Coulson and M Eldridge). CSIRO Publishing, Melbourne. pp 263–278.

Taggart DA, Schultz DJ, Corrigan TC, Schultz TJ, Stevens M, Panther D, White CR (2016) Reintroduction methods and a review of mortality in the brush-tailed rock-wallaby, Grampians National Park, Australia. *Australian Journal of Zoology* **63**, 383–397. doi:10.1071/ZO15029

Taylor RJ, Rose RW (1987) Comparison of growth of pouch young of the Tasmanian Bettong, *Bettongia gaimardi*, in captivity and in the wild. *Australian Wildlife Research* **14**, 257–262. doi:10.1071/WR9870257

Thompson CK, Wayne AF, Godfrey SS, Thompson RCA (2015) Survival, age estimation and sexual maturity of pouch young of the brush-tailed bettong (*Bettongia penicillata*) in captivity. *Australian Mammalogy* **37**, 29–38. doi:10.1071/AM14025

Thornett E, Ostendorf B, Taggart DA (2017) Interspecies co-use of southern hairy-nosed wombat (*Lasiorhinus latifrons*) burrows: Implications for wildlife conservation and management. *Australian Mammalogy* **39**, 205–212. doi:10.1071/AM15052

Tong Marwood L (2025) Mycobacteriosis. In *Current Therapy in Medicine of Australian Mammals*. Revised Edition. (Eds L Vogelnest and T Portas) pp. 349–358. CSIRO Publishing, Melbourne.

Toni P (2018) Combat leads to intraspecific killing in eastern grey kangaroos. *Australian Mammalogy* **40**, 109–111. doi:10.1071/AM16057

Triggs B (1990) *The Wombat: Common Wombats in Australia*. University of New South Wales Press, Sydney.

Troughton E (1947) Kangaroo twins and triplets. *The Australian Museum Magazine* **9**, 160–164.

Tyndale-Biscoe CH (1963) The role of the corpus luteum in the delayed implantation in marsupials. In *Delayed Implantation*. (Ed. AC Enders) pp 15–32. Chicago University Press, Chicago.

Tyndale-Biscoe CH (1968) Reproduction and post-natal development in the marsupial *Bettongia lesueur* (Quoy and Gaimard). *Australian Journal of Zoology* **16**, 577–602. doi:10.1071/ZO9680577

Tyndale-Biscoe H (2005) *Life of Marsupials*. CSIRO Publishing, Melbourne.

Tyndale-Biscoe CH, Janssens PA (1988) Introduction. *The Developing Marsupial, Models for Biomedical Research*. Springer-Verlag, Berlin.

Tyndale-Biscoe CH, Renfree M (1987) *Reproductive Physiology of Marsupials*. Cambridge University Press, Cambridge.

Uka D (1981) Notes on aggression in male quokkas (*Setonix brachyurus*) at the Royal Melbourne Zoo. *Thylacinus* **6**(2), 27–29.

van Oorschot RAH, Cooper DW (1989) Twinning in the genus *Macropus*, especially *M. eugenii* (Marsupialia: Macropodidae). *Australian Mammalogy* **12**, 83–84. doi:10.1071/AM89015

Vaughan-Higgins R, Buller N, Friend JA, Robertson I, Monaghan CL, Fenwick S, Warren K (2011) Balanoposthitis, dyspareunia, and Treponema in the critically endangered Gilbert's potoroo (*Potorous gilbertii*). *Journal of Wildlife Diseases* **47**, 1019–1025. doi:10.7589/0090-3558-47.4.1019

Vernes K (1993) A drive fence for capturing small forest-dwelling macropods. *Wildlife Research* **20**, 189–191. doi:10.1071/WR9930189

Vernes K (1995) The diet of the red-legged pademelon *Thylogale stigmatica* (Gould) (Marsupialia: Macropodidae) in fragmented tropical rainforest, north Queensland, Australia. *Mammalia* **59**, 517–525. doi:10.1515/mamm.1995.59.4.517

Vernes K (2010) Mycophagy in a community of macropodoid species. In *Macropods: The Biology of Kangaroos, Wallabies and Rat-kangaroos*. (Eds GM Coulson and MDB Eldridge). pp 155–169. CSIRO Publishing, Melbourne.

Vernes K, Pope LC (2001) Stability of nest range, home range and movement of the northern bettong (*Bettongia tropica*) following moderate-intensity fire in a tropical woodland, north-eastern Queensland. *Wildlife Research* **28**, 141–150. doi:10.1071/WR00054

Vernes K, Pope LC (2002) Fecundity, pouch young survivorship and breeding season of the northern bettong (*Bettongia tropica*) in the wild. *Australian Mammalogy* **23**, 95–100. doi:10.1071/AM01095

Vernes K, Castellano M, Johnson CN (2001) Effects of season and fire on the diversity of hypogeous fungi consumed by a tropical mycophagous marsupial. *Journal of Animal Ecology* **70**, 945–954. doi:10.1046/j.0021-8790.2001.00564.x

Viola S (1977) Observations on the brush-tailed bettong *Bettongia penicillata* at the New York Zoological Park. *International Zoo Yearbook* **17**, 156–157. doi:10.1111/j.1748-1090.1977.tb00893.x

Vitória da BA, Padilha FLA, Colbachini H, Duarte RM, Costa FR (2022) Effects of application of environmental enrichment in kangaroos (*Osphranter rufus*) at Aquário de São Paulo. *Pesquisa e Ensino em Ciências Exatas e da Natureza* **6**, e1945.

Vogelnest L (2025a) Capture, physical and chemical restraint. In *Current Therapy in Medicine of Australian Mammals*. Revised Edition. (Eds L Vogelnest and T Portas) pp. 122–151. CSIRO Publishing, Melbourne.

Vogelnest L (2025b) Dermatology. In *Current Therapy in Medicine of Australian Mammals*. Revised Edition. (Eds L Vogelnest and T Portas) pp. 186–211. CSIRO Publishing, Melbourne.

Vogelnest L, Portas T (2008) Macropods. In *Medicine of Australian Mammals*. (Eds L Vogelnest and R Woods). pp 133–225. CSIRO Publishing, Melbourne.

Vogelnest L, Portas T (Eds) (2025) *Current Therapy in Medicine of Australian Mammals*. Revised Edition. CSIRO Publishing, Melbourne.

Vogelnest L, Woods R (2008) *Medicine of Australian Mammals*. CSIRO Publishing, Melbourne.

Walker DM, Vickery K (1988) Tolerance of pouch young kangaroos (Macropodidae) for cows' milk replacers containing different amounts of glucose or lactose. *Australian Mammalogy* **11**, 125–133. doi:10.1071/AM88016

Wallis IR, Jarman PJ, Johnson BE, Liddle RW (1989) Nest sites and use of nests by rufous bettongs *Aepyprymnus rufescens*. In *Kangaroos, Wallabies and Rat-kangaroos*. Vol. 2. (Eds G Grigg, P Jarman and I Hume) pp. 619–623. Surrey Beatty & Sons, Sydney.

Waring H, Sharman GB, Lovat D, Kahan M (1955) Studies on marsupial reproduction. I: general features and techniques. *Australian Journal of Zoology* **3**, 34–43. doi:10.1071/ZO9550034

Waterhouse GR (1841) The Naturalists Library. Mammalia. Vol. XI. *Marsupialia or Pouched Animals*. W.H. Lizars, London.

Williams A, Williams R (1999) *Caring for Kangaroos and Wallabies*. Kangaroo Press, Sydney.

Williams R (1990) Kangaroos in captivity. In *Care and Handling of Australian Native Animals*. (Ed. SJ Hand) pp. 109–121. Surrey Beatty & Sons & the Royal Zoological Society of New South Wales. Sydney.

Wilson GR (1975) Age structures of populations of kangaroos (Macropodidae) taken by professional shooters in New South Wales. *Australian Wildlife Research* **2**, 1–9. doi:10.1071/WR9750001

Woinarski JCZ, Burbidge AA, Harrison P (eds.) (2014) *The Action Plan for Australian Mammals 2012*. CSIRO Publishing, Melbourne.

Wood Jones F (1923) *The Mammals of Southern Australia: Part 1.* Government Printer, Adelaide.

Woodhouse SJ, Duncan AE, Shellabarger W (2010) Idiopathic megaesophagus in a red kangaroo (*Macropus rufus*). In: *Proceedings of the American Association of Zoo Veterinarians*, October 23–39, South Padre Island, Texas.

Woods R (1999) Prevention of disease in hand reared native wildlife. In *Wildlife in Australia: Healthcare & Management*. Post Graduate Foundation in Veterinary Science. (Ed. DI Bryden) pp. 455–489. University of Sydney, Sydney.

Yip S, Dehcheshmeh MM, McLelland DJ, Boardman WS, Saputra S, Ebrahimie E, Weyrich LS, Bird PS, Trott DJ (2021) *Porphyromonas* spp., *Fusobacterium* spp., and *Bacteroides* spp. dominate microbiota in the course of macropod progressive periodontal disease. *Scientific Reports* **11**, 17775. doi:10.1038/s41598-021-97057-1

Zuckerman S (1953) The breeding seasons of mammals in captivity. *Proceedings of the Zoological Society of London* **122**, 827–950. doi:10.1111/j.1096-3642.1952.tb00251.x

# 11
# Dugong

*Kerrie Dixon, David Blyde, Helene Marsh, Paola Unda Marron and Stephen M. Jackson*

## 1. INTRODUCTION

The dugong (*Dugong dugon*) is the only extant species within the family Dugongidae and one of only four living species in the order Sirenia (sea cows) (Marsh *et al.* 2011). The two families and three recent genera of sirenians include three species of manatees in the family Trichechidae; the extinct Steller's sea cow (*Hydrodamalis gigas*) and the dugong are in the family Dugongidae. Sirenians are more closely related to elephants and hyraxes than to other marine mammals (Marsh *et al.* 2011).

Dugongs have more specialist diets and feeding behaviours than manatees (Marsh *et al.* 2011, 2018; Keith-Diagne *et al.* 2022) and holding them under human care for either rehabilitation or exhibition presents many challenges for the facility and the wellbeing of the animals. The first recorded attempt to keep a dugong in captivity was in 1955, when a male dugong was transported from Palau to San Francisco but survived for only 45 days (Goto *et al.* 2004). Since the 1960s, dugongs have been sporadically maintained in zoos, oceanaria and other facilities such as fish-traps and sea pens in at least 10 countries in their range: Australia, China, India, Indonesia, Japan, Malaysia, Myanmar, Qatar, Singapore, Thailand and the United Arab Emirates (UAE) (Kataoka and Asano 1981; H. Marsh unpubl. data). Nonetheless, many of these attempts to keep dugongs in captivity have been short-lived. Marsh (1991) reported on the fate of 30 dugongs held in 13 institutions and found that 9 died within 1 month and only 11 lived longer than 6 months in captivity.

Nonetheless, some animals have lived in captivity for many years. A female was maintained at Underwater World, Singapore from 1998 to 2014, after she was rescued as a suckling calf when her lactating mother drowned in a fishing net (Chua *et al.* 2001). This female grew from 59 kg to 150 kg by 2013 when she was aged 15 years (Underwater World Singapore 2013), much lighter than the animals maintained at Toba, Japan (Goto *et al.* 2004). For example, a female at that facility weighed more than 300 kg by age 12 years. That animal was found as an orphaned calf in the Philippines in October 1986 and estimated to be less than 1 year of age when she was transferred to Toba Aquarium in April 1987 where she was reared by staff (Goto *et al.* 2004).

Dugongs have never bred in captivity. Jones (1959, 1967) described the behaviours of two dugongs held in in a small tank at Mandapam Camp in India, including mating behaviour. Toba Aquarium attempted a breeding program after the first observed captive mating behaviour in 1984. During that program a female was introduced into the breeding program in 1995; however, the program was unsuccessful and the male dugong died in 2011 (Society for Marine Mammalogy 2023).

Despite several dugong calves being brought into captivity for hand-rearing, there have been no recorded successful releases of hand-reared animals back to the wild and several facilities have been required to maintain the animals over the long-term. Dugongs are far from ideal captive animals as they are difficult and expensive to maintain. Providing the environment and diet required to maintain appropriate animal welfare is challenging and costly. Therefore, consideration of the long-term welfare of a dependent calf needs to be considered before it is taken into captivity. Marsh (2022) outlines several options for orphaned calves.

Dugongs are culturally significant to traditional and contemporary human communities throughout their range (Ponnampalam *et al.* 2022). Some captive dugongs have received considerable publicity. The dugong is the symbol for Toba Aquarium and features prominently in marketing for the SEA LIFE Sydney Aquarium in Australia, which has exhibited a male dugong since 2008 as explained below. In 2019, an orphaned dugong calf in Thailand was the centre of global attention (Ponnampalam *et al.* 2022).

# 2. TAXONOMY

## 2.1. Nomenclature

Class: Mammalia
Order: Sirenia
Family Dugongidae
Genus and species: *Dugong dugon*

## 2.2 Subspecies

None. See Jackson and Groves (2015).

## 2.3 Recent synonyms

Synonyms for dugongs can be found in Jackson and Groves (2015).

## 2.4 Other common names

See Baker and Gynther (2023).

# 3. NATURAL HISTORY

## 3.1 Bodyweight and morphometrics

The bodyweight of an adult dugong is up to 570 kg (Jefferson *et al.* 2008; Lanyon *et al.* 2010) and they have a total length of up to 3.3 m (Jefferson *et al.* 2008). Dugongs display very little sexual dimorphism. The easiest method to sex individuals is to observe the genital openings on the abdomen. The males have two relatively small openings as well as an umbilical scar. The anus is located low on the ventral side/abdomen near the middle of the peduncle, with the genital opening half-way between the anus and the umbilical scar (Fig. 11.1). Females have one longer urogenital opening, as well as the umbilical scar (Fig. 11.1). Both male and female dugongs have tusks (upper posterior incisors). In males, the tusks erupt around puberty. The erupted tusk is short (Fig. 11.2) as it is worn to a blade shape while the animal is bottom feeding. In females, tusks erupt only in very old females as the tusks grow posteriorly up the premaxilla (Marsh 1980). Adult females have large axillary teats, which are easily visible.

**Fig. 11.1.** How to determine the sex of a dugong. The female (a) has a contiguous genital slit and anus. The male (b) has a genital slit between the umbilicus and the anus. Photo (a) by Brooke Witherby, SEA LIFE Sydney Aquarium and photo (b) by Karen Wilshaw. The picture of the male is reproduced from Marsh *et al.* (2011) under the contractual copyright arrangements that Marsh has with Cambridge University Press.

## 3.2 Distribution and habitat

Dugongs inhabit coastal and island waters between East Africa and Vanuatu between the latitudes of ~27° north and south of the Equator. Their range spans ~128 000 km of coastline (860 000 $km^2$) across at least 37, and possibly up to 44, states and territories (Marsh 1989; Marsh *et al.* 2011; Marsh and Sobtzick 2015). Dugongs are generally associated with seagrass communities in waters to 35 m deep, although most individuals are sighted in shallower waters (Marsh *et al.* 2011).

## 3.3 Conservation status

Dugongs are listed as Vulnerable to extinction by the IUCN (Marsh and Sobtzick 2015). Threats vary among the different populations as detailed in Marsh *et al.* (2011) and Hines *et al.* (2012) and summarised in Marsh and Sobtzick (2015). Major threats include: incidental capture in fishing gear (e.g. gill-nets); shark nets for bather

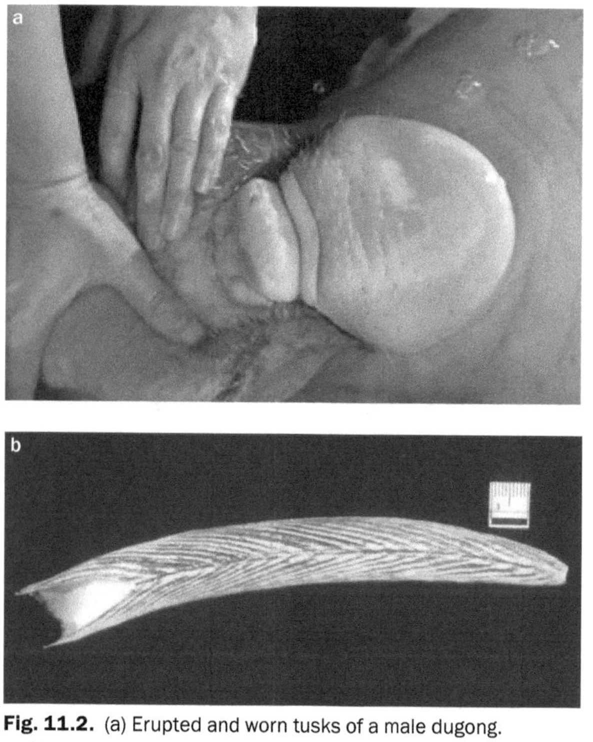

**Fig. 11.2.** (a) Erupted and worn tusks of a male dugong. (b) Unerupted female dugong tusk prepared for age determination. One dark line plus one light line represents one year's growth. Photo (a) is from SEA LIFE Sydney Aquarium and photo (b) is reproduced from Marsh *et al.* (2011) under the contractual copyright arrangements that Marsh has with Cambridge University Press.

protection; illegal, unreported and unregulated (IUU) fishing, especially if the meat is subsequently sold (Pilcher *et al.* 2017); hunting both legal (i.e. culturally sanctioned) and illegal; vessel strikes and boating activities (e.g. acoustic pollution); damage/modification/loss of habitat caused by human settlement on coasts from shipping and trawling and destructive fishing practices; natural processes (e.g. cyclones and tsunamis); threats to seagrass (including untreated sewage disposal, coastal dredging and reclamation, inshore commercial trawling, agricultural pollution); chemical pollution (e.g. oil spills); and climate change (sea level rise, extreme weather events and high temperatures) (Marsh *et al.* 2022).

## 3.4 Diet in the wild

Dugongs are 'seagrass community specialists' (Marsh *et al.* 2018; Keith-Diagne *et al.* 2022) exploiting both plants and animals in seagrass meadows, including macro-invertebrates and algae, as well as seagrasses. They feed on most species of seagrass that occur within their range, except for some robust, fibrous, temperate seagrasses in the genus *Posidonia*. The relative importance of seagrasses to the diet of the dugong differs among locations and changes at the same location during times of seagrass loss from plant dieback and extreme weather events (Marsh *et al.* 2011). Dugongs apparently subsist largely on algae at some locations (Whiting 2002a, 2002b, 2008) and when seagrass communities have been damaged by extreme weather events (Spain and Heinsohn 1973; Marsh *et al.* 1982). At least at the high latitude limits of their range in winter, some dugongs also deliberately eat infaunal invertebrates, including thin-shelled burrowing mussels (*Botula vagina*) and possibly sea pens (*Virgularia*), ascidians (especially *Sycozoa pulchra*; Anderson 1989) and unidentified chaetopterid worms (Preen 1995). Preen (1995) suggested that this behaviour enables dugongs to meet their nitrogen requirement in times of seasonal shortage in the subtropics. Some researchers have inferred that dugongs selectively target pioneer, low biomass seagrass genera, especially *Halodule* and *Halophila*, in preference to more fibrous climax genera such as *Zostera* or *Thalassia*. Most of these claims about diet selectivity are not supported by confirmatory data on the biomass or nutrient concentrations of the available seagrasses and no food choice experiments have been conducted. It appears that despite their specialist feeding apparatus and behaviour (see section 6.1) dugongs are capable of consuming a relatively diverse diet, although not nearly as diverse as that of any manatee species (Keith-Diagne *et al.* 2022).

Dugongs live entirely in marine systems in the wild and apparently meet their water requirements from their food (Smoll *et al.* 2020).

## 3.5 Longevity

### 3.5.1 Wild

The maximum recorded life span for a dugong is 73 years. This specimen was a wild female that was aged by dental analysis (Marsh 1995).

### 3.5.2 Captivity

The oldest captive dugongs are a 35-year-old female in Toba Aquarium in Japan, a male at SEA LIFE Sydney Aquarium that is 25 years old, and a 4-year-old male in the UAE (P. Marron pers. obs.) (as of 2023).

### 3.5.3 Methods used to determine the age of adults

It is not possible to age a live dugong unless it has been in captivity since a neonate. The age of a deceased dugong can be determined by counting the incisor dentinal growth layer groups (Fig. 11.2 right). Dugongs deposit one growth layer group per year. Once the incisors erupt and become worn (Fig. 11.2 left), it is no longer possible to accurately determine age from teeth (Marsh 1980).

## 4. HOUSING REQUIREMENTS

### 4.1 Enclosure design

Enclosures can be either natural or artificial. If artificial, the enclosure must have walls that are durable, watertight, non-porous, non-abrasive, non-toxic and easily cleaned and disinfected. Sharp projections and loose fittings must be avoided. Artificial pools have the advantage of being able to maintain water parameters better than natural pools. Natural pools have the advantage that they can be constructed to be much larger for less cost than artificial pools. The pool surface area should be large enough to accommodate the natural behaviours of each animal and deep enough to allow it to dive (at least 4 m) and avoid the potential effects of ultraviolet radiation. Structures that allow the dugong to swim through them and enclosure furniture should be provided for enrichment opportunities to stimulate their most important senses including touch and their hydrodynamic receptors (Marshall *et al.* 2022). Animals should be protected from hazards and loud and continuous noise. Shade should be supplied as well as shelter from any strong winds or rain.

Dugongs should be housed in a tropical marine environment that allows them to swim freely and express their natural behaviours. The exhibit should include an area where an animal can be restrained easily. This may involve constructing a submerged platform that can be raised to lift the animal out of the water.

### 4.2 Holding area design

A holding pool should be included in the design of the enclosure. This area can serve to hold animals for veterinary purposes, before and after procedures, training and conditioning sessions and husbandry requirements. If possible, the pool should be able to be drained when needed or include a false bottom that can be raised when needed. The pool size should be at least 4 m (L) × 3 m (W) × 3 m (D). It is important to note that this 'medical' pool is only for temporary use (von Fersen and Walb 2018). Ideally, the holding pool should be connected to the main enclosure and separated by a gate. The dugong can be conditioned to swim through the gate into the medical pool; otherwise, hard panels can be used to guide the animal into the medical pool.

### 4.3 Spatial requirements

The minimum dimensions for one animal in long-term care are 11 m (L) × 9 m (W) × 4 m (D). If more than one animal is to be housed in the enclosure then the length and the width should be increased by 50%. The dugong held at SEA LIFE Sydney Aquarium is housed in a 1.5 million litre enclosure.

*The Guidelines for Exhibiting Antillean Manatees* (Trichechus manatus manatus) suggest a minimum water surface area for two animals is 150 $m^2$ and a volume of 270 $m^3$ and a depth that varies from 0.7 m to at least 3.0 m, preferably at least 4.0 m. For each additional animal, an increase of 25 $m^2$ in water surface is required, with an equivalent volume varying between 50 and 75 $m^3$ (von Fersen and Walb 2018). However, given that dugongs dive deeper than manatees (Keith-Diagne *et al.* 2022), a maximum pool depth of at least 4 m is suggested for dugongs.

### 4.4 Position of enclosures

There are no guidelines or stipulations around the position of where a dugong enclosure should be within a facility. If it is an outdoor enclosure, it is important to take into consideration the movement of the sun, the direction of prevailing winds and where the shade will fall throughout the day.

### 4.5 Weather protection

The animals should be provided with appropriate shaded and sheltered areas to protect them from adverse ambient conditions caused by weather (such as strong winds and rain), sunlight or artificial lighting, glare or other environmental factors. Dugongs should also be provided with access to natural vitamin D via access to sunlight.

### 4.6 Temperature requirements and water quality

#### 4.6.1 Temperature

Water samples should be obtained daily to determine the temperature and water quality. The temperature of the water in the exhibit should reflect that of a tropical/temperate region. Water temperature in the wild fluctuates with time of year as well as location. However, the temperature in a facility under human care should be maintained between 23°C and 29°C. The water temperature used to hold dugongs should not be below 20°C for extended periods due to the potential risk of cold stress syndrome (Owen *et al.* 2013).

The coastal water temperatures across the dugong's range in Australia vary between ~16°C and 30°C. A recommended water temperature for dugong is ~24.5°C in winter and 27°C during summer. Changes in temperature should be made gradually to allow acclimatisation for all the animals in the enclosure (not just the dugong). It is recommended that water temperature is increased or decreased by 0.5°C per day until the desired temperature is reached.

When setting temperature parameters, it is important to consider the other animals in the dugong's enclosure. Although the habitat range of these animals may overlap with that of the dugong, in the wild they can move if the temperature becomes too cold or too hot, an option unavailable in a captive environment.

### 4.6.2 Coliform bacteria count

Bacterial counts should be undertaken at least weekly. The coliform bacteria count of the water in the enclosure must not exceed 500 MPN (most probable number) per 100 mL. If the count exceeds this reading it must be rectified immediately (Crane 1994).

### 4.6.3 Bromine

Many disinfectants will produce a residual bromine level in the water body. The bromine residual in the water should not exceed 1.0 mg/L and should be tested at least daily. If chlorine rather than bromine is measured it should not exceed 0.5 mg/L.

### 4.6.4 Salinity

Salinity should be within the range of 34–36 ppt.

### 4.6.5 Dissolved oxygen

Dissolved oxygen is not an issue for dugongs as they breathe air. However, it may be very important for any other tank animals such as fish, sharks or rays. Dissolved oxygen levels should be as close to 100% as possible. Any reading below 85% should be followed up and rectified. A reading of 103% and above should also be followed up as it may be a symptom of a super-saturation event.

### 4.6.6 pH

The pH of a solution refers to the level of acidity or basicity of a solution. A low pH is acidic and a high pH is alkaline or basic. The scale runs from 1 to 14, with 7 considered neutral pH. The pH has an effect on the toxicity of ammonia. As the pH increases, the percentage of toxic ammonia in the water increases. The pH of the water should be maintained between 7.8 and 8.5 and tested on a daily basis.

### 4.6.7 Ammonium (NH4)/ammonia (NH3)

Ammonia levels for mammals should remain below 0.7 mg/L and be tested daily. Ammonia is produced by marine animals as the end product of protein (nitrogen) metabolism. In fish, most of the ammonia is excreted via the gills. Some other sources of ammonia in the water are from fish faeces, dead fish, decaying plant matter and uneaten food. Ammonia is very toxic to fish, whereas ammonium is relatively non-toxic for mammals. Ammonia is broken down into a safer compound by nitrifying bacteria. This process is called the nitrogen cycle. Ammonia is broken down into nitrite, which is broken down into nitrate, which is broken down to nitrogen gas that evaporates into the air. Ammonia does not appear to cause systemic disease in mammals, but high levels can cause irritation to mucous membranes and eyes.

### 4.6.8 Nitrite ($NO_2$)

Nitrite is very toxic to fish. It enters via the gills and then causes haemoglobin changes by altering the iron molecule. Due to this damaging effect, a chromic low-grade exposure will cause anaemia.

### 4.6.9 Nitrate ($NO_3$)

Nitrate is the final product in the nitrifying process. It is much less harmful to fish than nitrite. If there are fish in the exhibit the nitrate level should be below 100 ppm (Jones 2012). von Fersen and Walb (2018) advise that levels should be kept below 20 ppm in manatee pools. If the nitrate levels are allowed to rise, algal blooms may occur because nitrate is a good plant fertiliser and will encourage algal growth.

### 4.6.10 Phosphorus (P)

Increasing phosphate levels in water are not usually toxic to fish, but will increase algal growth.

### 4.6.11 Carbonate hardness

Carbonate hardness is a measure of the water hardness caused by the presence of calcium carbonate ($CaCO_3$) and calcium bicarbonate ($CaHCO_3$). As a general rule, soft water fish, with few exceptions, can tolerate transfer to hard water conditions. They may need soft water conditions for breeding but not for general living.

### 4.6.12 Oxidation reduction potential/redox

Redox potential is a measure of the tendency of a body of water to acquire electrons from or lose electrons to an electrode and thereby be reduced or oxidised respectively. ORP is essentially a measurement of sanitiser effectiveness in water bodies. The higher the ORP the more effective is the disinfecting agent.

## 4.7 Life support systems

A guiding principle for life support systems is to over-engineer the required component. As a long-term investment, having equipment that runs at 80% of its maximum capacity will be cheaper from an ongoing power cost and will last longer than equipment that is under-engineered and running constantly at 100% capacity (Jones 2014).

Adequate life support is essential for any dugong enclosure. Aquatic animals are at the mercy of the water around them. When designing exhibits for dugongs, it is important not to follow the life support used with other marine mammals such as dolphins or seals; the waste produced by a herbivore is much greater than by a carnivore. Consulting with experts who deal with hippopotamus and elephant wastewater treatment is recommended because these systems deal with water containing large volumes of organic material (von Fersen and Walb 2018).

### 4.7.1 Filtration

Several options can be used to provide filtration of the water to maintain its quality.

- Sand filters are commonly used in large-scale oceanariums. They are filled with a medium, usually sand. The water from the tank is pumped under pressure through this medium and as the water passes through the media, particulate matter is retained. A small amount of biological filtration will also happen once the filter has been in operation for some time. While the filter is working the internal pressure will increase because of the particulate matter building up. Every filter has a different ideal working pressure (refer to the manufacturing details of individual models). Once the pressure reaches a specific level the filter will need to be backwashed by changing the valves to push water through the filter in the reverse direction, disturbing and churning up the sand and sending the dirty water to waste.
- Biological filtration is the action of bacteria on nitrogenous compounds. It does not remove solids or affect other water contaminants. It facilitates the nitrogen cycle, converting ammonia to nitrite then nitrate. Biological filtration is usually achieved by passing water through a bio-chamber containing some sort of bio-ball or bio-media. These balls or media have a large surface area to which the nitrifying bacteria can attach.
- Membrane filtration is the removal of substances by absorption into filter material such as activated charcoal, zeolites, ionising resin or peat moss.
- A protein skimmer is a mechanical system used to remove organic matter and reduce nitrogen compounds. The exhibit water is mixed with air and pumped into the reaction chamber. Proteins and solids become attached to the air bubbles and flow upward and concentrate in foam that is being eliminated from the system. The foam generators on these systems can also incorporate ozone, which increases the oxidation of organic material (von Fersen and Walb 2018).

### 4.7.2 Disinfection

If chemical treatment is used to maintain water quality and cleanliness, it should not cause harm to the animals.

- Chlorine – the most common form of disinfection. It is a strong oxidant that rapidly kills many harmful microorganisms. However, chlorine is a toxic gas and has significant risks associated with its use. This problem can be avoided by the use of hypochlorite solution, which is a relatively inexpensive solution that releases free chlorine when it is dissolved in water. Calcium hypochlorite is a solid form of chlorine; however, greater human contact is needed to use this form of the disinfectant and therefore it is more dangerous to use. The disadvantage of chlorine is that it reacts with natural organic compounds in the water to form potentially harmful byproducts such as trihalomethanes and haloacetic acid, both of which are carcinogenic in large quantities. Chlorine is also limited in its effectiveness against protozoal cysts and giardia.
- Chlorine dioxide – a faster acting disinfectant than elemental chlorine but works in a similar fashion.
- Chloramine disinfection – not as strong an oxidant as chlorine, but provides a longer lasting residual. In

addition, it does not form potentially harmful byproducts such as trihalomethanes and haloacetic acid. Chlorine can be converted to chloramines by the addition of ammonia.

- Ozone – has been applied to marine mammal life support systems for decades and its success as an oxidant and disinfectant is well documented. In many marine mammal facilities, ozone has superseded chlorine as the primary chemical treatment. An ozone-generator charges oxygen to create the ozone, which is an oxidiser. Once in the water, it attacks and neutralises the contaminants and then reverts back to oxygen. It also enhances the efficiency of waste removal systems such as protein skimmers by coagulating the suspended solids, which are then more easily removed via the air bubbles. The size of the enclosure, the number of animals within the enclosure, the salinity as well as the temperature all determine the amount of ozone that will be needed to effectively treat the water. Oxidation reduction potential probes are used to measure the water level, but have been known to be unreliable. If inhaled by humans, ozone gas can cause damage to the lungs, so this system requires the installation of ambient ozone gas monitors to detect any leaks (Aiken 1995; LaBonne 1998).
- Ultraviolet (UV) light irradiation – UV light is used to clean water and works because the light energy penetrates organisms' cell walls and disrupts their genetic material and inhibits reproduction. Similar to ozone, the efficacy of UV light depends on the time of contact, so the speed of water flow is crucial. The effectiveness of the treatment is also affected by the turbidity of the water. Advantages of UV light compared with chemical disinfection include: no toxic byproducts; no volatile compound emissions or toxic air emissions; not requiring storage of dangerous material; and not changing the chemical quality of the water. As a result, UV light is the safest among the current techniques for cleaning water (von Fersen and Walb 2018).

### 4.7.3 Closed versus open systems

A closed system is one that has a finite amount of water that constantly recirculates around the tank. Water changes, backwashes or dive cleans require a capacity to replace lost water with either artificial sea water or filtered sea water. An open system requires a constant flow of filtered salt water drawn into the system from the ocean and returned back to the ocean via an overflow set up. Open systems have advantages over closed systems because there is a constant change of water, which makes it easier to maintain good water quality. One downside to an open system is that the temperature of the ocean may drop below the desired temperature of the tank, especially during winter, which can be avoided by heating the new water before it goes into the tank or by reducing the volume of new water. The second method reduces the capacity to do large backwashes or water changes because there is less water coming into the tank and the water level does not return to normal as quickly as before. The efficacy of this system depends on the quality of the new water, including its salinity, turbidity and level of contamination. It is also more difficult to maintain levels of disinfection in open systems.

## 4.8 Substrate

Wild dugongs are generally found in shallow sandy and muddy coastal areas that support seagrass communities (Marsh *et al.* 2011). Sand, however, is not an ideal substrate medium for captivity because it is easily disturbed by the movements of animals in the tank and transferred into the water column, which creates water clarity issues that reduce visitor experience. Soft coral gravel can be used as a substitute, provided it is soft enough to reduce the risk of pressure sores on animals eating and resting on the floor and large enough to prevent disturbance.

## 4.9 Enclosure furnishings

Dugongs are considered to largely perceive their environment through somatosensation (touch and hydrodynamic reception) and hearing; aversion and taste (chemoreception) are also important to some degree (Marshall *et al.* 2022). All body hairs are sensory hairs, which mediate hydrodynamic reception analogous to the function of the lateral line systems in fish and amphibians. The importance of somatosensation to dugongs should inform enclosure fittings (Marshall *et al.* 2022). Dugongs need to be able to stimulate their sense of touch by having places on which to rub their bodies and stimulate their hydrodynamic senses with moving water and bubbles. Swim throughs and non-hazardous furniture should be included in the exhibit design to add environmental enrichment. Furnishings that do not significantly reduce the floor space can provide cover for other animals and create places where the dugong can scratch and stimulate their sensory hairs.

The male dugong at SEA LIFE Sydney Aquarium does not interact with exhibit furnishings very often. He spends most of the time feeding on the exhibit floor. When not feeding, he will occasionally put his head into the air and play in the bubbles from the pool filtration system. A second animal was fond of the surface recirculation that creates a powerful stream of water along the surface of part of the exhibit and would swim into it almost like a 'dugong treadmill'. Thus, creating different swimming patterns through the use of recirculation pumps and movement of the feeding trays around the exhibit is more beneficial than filling the tank with furnishings.

## 5. GENERAL HUSBANDRY

### 5.1 Hygiene and cleaning

As dugongs urinate and defecate in the water, some form of disinfection is imperative. Options include chlorine, ozone or UV light as outlined above. If chemical treatment is used to maintain water quality and cleanliness, it should not cause harm to any of the other animals in the tank. The bromine residual in the water should not exceed 1.0 mg/L. The pH of the water should be maintained between 7.8 and 8.5. Ammonia levels should remain below 0.7 mg/L. Each of these values should be measured daily. Faecal coliform counts should remain < 500 MPN per 100 mL water and be measured weekly. If the coliform count exceeds this level, action should be taken to decrease bacterial contamination, including a water change or increasing disinfection.

### 5.2 Record keeping

It is important to establish a system whereby the health, condition and reproductive status of captive dugongs are routinely monitored and recorded:

- identification numbers (all individuals should be individually identifiable; see section 5.3)
- results of veterinary examinations and treatments
- behavioural changes or problems
- reproductive behaviour or condition
- bodyweights and measurements
- amount of food provided (including components) and consumed. It should be recognised that a lot of food supplied to a dugong is lost into the water and needs to be accounted for
- changes in diet and food consumption
- movements of individuals between enclosures or institutions
- births, with dam and sire recorded if this occurs
- deaths with necropsy results.

The collection of information on the physical, physiological, and behavioural patterns of individual dugongs can contribute greatly to the husbandry of this species. It also allows individual histories to be shared with other institutions and facilitates a cooperative approach to data collection among institutions. ZIMS, the Zoo Information Management System (Species360 2023) can be used for recording general information on age, sex, parentage, births, identification, bodyweights, transfers and deaths, with separate ZIMS modules for veterinary information and studbooks. These systems have been developed by Species360, which is part of the Conservation Planning Specialist Group (CPSG) software.

### 5.3 Methods of individual identification

As the number of dugongs held in a facility is likely to be low, individuals should easily be identified by their size and physical markings. Individuals in wild populations of dugongs have been identified using passive integrated transponder (PIT) tags (microchips) (Lanyon *et al.* 2002) inserted at a 90° angle into the musculature of the neck, turtle tags inserted on the edge of the tail fluke or genetic identification (Cope *et al.* 2015).

### 5.4 Routine data collection

#### 5.4.1 Water data collection

See section 4.6 for the correct ranges for coliform bacteria count, salinity, dissolved oxygen, temperature and water chemistry.

Routine timings of environmental (water) data collection:

- coliform bacteria count - weekly
- salinity - daily
- temperature - daily
- dissolved oxygen - daily
- water chemistry (bromine/chlorine levels) - daily

#### 5.4.2 Individual animal information

Routine timings of physical data collection:

- distant visual inspection from the tunnel and water surface - daily
- complete physical evaluation - weekly

- biological samples collection (blood, gastric, faecal and respiratory samples) - every 6 months
- bodyweight - every 3 months
- morphometrics - every 3 months
- abdominal ultrasound - once per year

# 6 FEEDING REQUIREMENTS

## 6.1 Captive diet

Ingredients of captive dugong diets have included seagrass, cos (romaine) lettuce, snow pea sprouts, cabbage, endive, wheat grass and spinach. Cos lettuce has been the predominant diet of some captive animals. A healthy adult dugong may consume up to 30 kg of cos lettuce per day, making the feeding of these animals very time consuming and very expensive. Alternatively, other captive dugongs have been fed seagrass obtained from local sources. Diets based on artificial ingredients such as cos lettuce may be deficient in essential vitamins and minerals and supplementation may be necessary. Attempts to supplement captive diets have included darting the animals with 10 mL Hemoplex (Troy Laboratories, Glendenning, NSW, Australia). A healthy adult dugong may need to consume at least 20–25 kg seagrass per day to maintain bodyweight in captivity. For the purpose of training husbandry behaviours, fruits (banana, apple and papaya) can be used as reinforcement.

Sourcing seagrass from the wild is very time consuming and in certain jurisdictions permits are required from the local conservation agency. The shape of the jaw and the position of the mouth (Fig. 11.3) make a dugong an obligate bottom feeder and in a captivity efforts should be made to feed the animals from the bottom of their enclosures. This requirement may necessitate stacking the food into trays and placing the trays on the bottom of the exhibit. Supplemental fresh water is not necessary for captive animals fed an adequate diet.

Another complementary diet that has been used successfully is Mazuri Herbivorous Sea Turtle Gel Diet (for herbivorous sea turtles), which is a complete, balanced diet plus a good extra source of water. This diet is useful during rehabilitation and for dehydrated animals (P. Marron pers. obs.).

## 6.2 Supplements

Attempts to supplement the diets of captive dugongs have included darting the animals with 10 mL Hemoplex (Troy Laboratories) intramuscularly monthly and adding

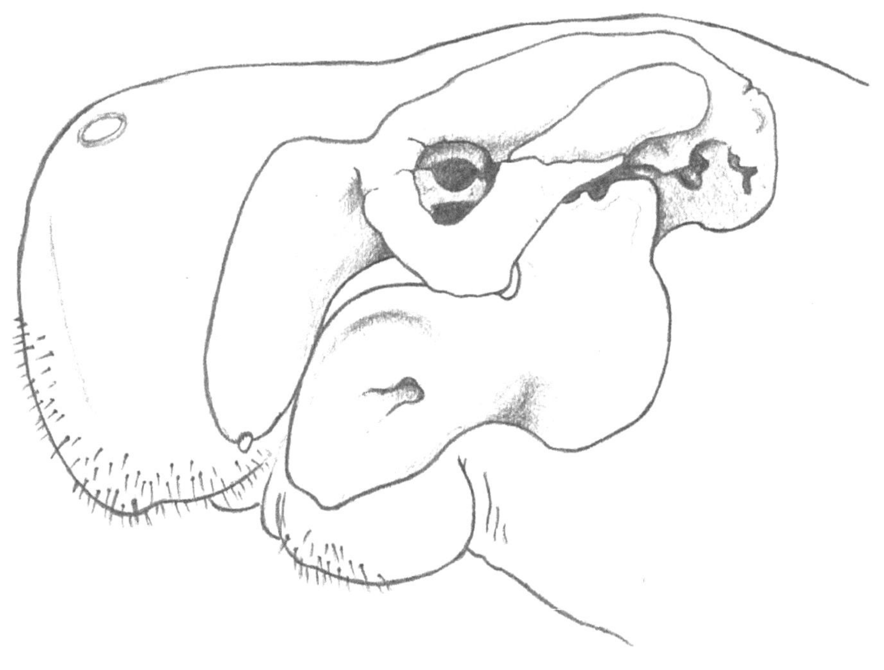

**Fig. 11.3.** Dugong skull showing the distinctive downward deflection of the premaxilla and mandible. This figure has been reproduced from Marsh *et al.* (2011) under the contractual copyright arrangements that Marsh has with Cambridge University Press. The figure was drawn by Gareth Wild.

vitamins/minerals to the diet by injecting them into the stems of cos lettuce before feeding.

Replicating the dugong's natural diet in captivity is very difficult. It is therefore important that some supplements are provided to make up for any dietary imbalances. At SEA LIFE Sydney Aquarium, the following supplements have been used.

- Probiotics - Protexin paste (5 g). An equine probiotic that is provided once per day to help counteract any colic issues.
- Prebiotics - help to feed the gut bacteria within the animal's digestive system and encourage the healthy growth of the indigenous bacteria. As captive dugongs eat different food types to wild dugongs, they obviously require different bacteria to help break their food down.
- Spirulina - hand-feeding 24 g of spirulina per day to replace the lack of iodine in lettuce.
- Feroglobin - oral supplemental source of iron, zinc and B vitamins.
- Hemoplex - 10 mL of Hemoplex is hand-fed twice per week or 10 mL intramuscular injection every 3 months administered in the dorsal epaxial muscles.

## 6.3 Presentation of food

Dugongs are obligate bottom feeders because of the downward deflection of their premaxillla and mandible (Fig. 11.3; Domning 2022; Keith-Diagne *et al.* 2022). Providing food on the floor of the enclosure will be important for the dugong to feed efficiently, but requires a method of anchoring the food to the bottom. At SEA LIFE Sydney Aquarium, this is achieved using large rectangular metal trays wrapped in airline tubing (thin stretch tubing that is used in small aquariums that attach to an air stone and an air pump; Fig. 11.4). The lettuce is tucked between the tubing causing it to sit upright like grass growing from the floor. These trays are heavy enough to sink to the floor when thrown in. A long PVC pipe with a hook attachment is used to retrieve the cleared trays.

# 7. HANDLING AND TRANSPORT

## 7.1 Timing of capture and handling

The best time to catch, or restrain, a captive animal is first thing in the morning to provide maximum time to undertake any movements or medical procedures before the public arrive and to enable monitoring of the animal

**Fig. 11.4.** Washing the cos lettuce leaves (left) and placement in trays (right). Photo by Stephen Jackson.

for the remainder of the day. Wild dugongs are held out of the water for a maximum of 60 min for research (Lanyon *et al.* 2010) and this time is also considered the maximum for captive animals, except during transit.

## 7.2 Capture equipment

Capture equipment can include:

- davit arm – a large, heavy strong metal piece of equipment that is purpose-built to allow the hoist and sling to be manoeuvred over the water in the dugong enclosure. It slots onto a holder that is fixed in place on the deck above the enclosure
- engine hoist
- large mesh stretcher strong enough to hold the weight of an adult dugong (Fig. 11.5) or a purpose-built sling may be used
- rated straps for attaching the sling to the winch
- D-rings, hooks and metal rings
- chain block winch – an electrical winch can be used, but there must be a hand block and chain winch onsite as a backup
- extension cords
- large padded mattresses – large enough for the dugong to rest on after being removed from the water as well as a small mattress to cover its fluke
- bucket – a small bucket of water may be needed to induce respirations while the animal is on the deck if it is holding its breath for too long
- minimum of two team members in wetsuits, who will be in the water during the catch. Other staff are required on the dry side to assist. Usually a minimum of 6–8 staff is required on the dry side
- mung bean sprouts for positive reinforcement.

## 7.3 Capture and physical restraint

A captive environment enables training and conditioning to help capture and restrain the dugong. The steps involved are:

(1) test all equipment needed and ensure adequate staff and other resources are available
(2) encourage the animal into the stretcher or onto a submerged platform with a stretcher in place
(3) restrain the animal in the stretcher or on a mattress on the platform (Lanyon *et al.* 2006)
(4) lift stretcher/platform out of the water
(5) once the animal is out of the water, ensure the tail, head and pectoral fins are not covered
(6) ensure staff maintain the animal in ventral recumbency such that it is not struggling, rolling or moving laterally
(7) carry out the procedure(s)
(8) when the procedures are concluded, lower the stretcher or platform back into the water and release the animal.

## 7.4 Weighing and examination

- Heart rate – the heart rate of 30 wild dugongs ranged from 40 to 96 beats/min (Lanyon *et al.* 2010) when restrained out of the water, but varied markedly between individuals, with most having a mean heart rate between 60 and 85 beats/min. Heart rates for wild dugongs appeared to be slightly elevated compared with wild-caught West Indian manatees (40–70 beats/min) (Wong *et al.* 2012). The biggest variability in heart rates was recorded in dugongs less than 2.6 m body length (40–96 beats/min)

**Fig. 11.5.** Stretcher used to hold a dugong. Photo by SEA LIFE Sydney Aquarium.

(Lanyon *et al.* 2010). Heart rate during restraint can be monitored by electrocardiography, Doppler or via a stethoscope. Heart rate will increase during capture and restraint.

- Respiration - Dugongs should be monitored while any restraint is attempted. The breathing rate should be monitored, especially when the animal is out of the water. Time between breaths should be around 90–120 s when on land. Pouring water over the nostrils of dugongs directed caudally can stimulate the animal to take a breath. Dousing the head or snout region with water or sometimes running a slow trickle over the snout is usually enough to elicit a breath. Manual stimulation of the nares does not work. End-tidal carbon dioxide can be monitored via a face mask placed loosely over the nostrils with a capnometer attached. As soon as a dugong is brought out of water, the priority should be to facilitate regular voluntary respirations because some dugongs have a tendency for slow breathing when first removed from the water. Before a full medical assessment commences, the focus should be on attaining at least one unassisted breath and any number of assisted breaths during the first 7 min out of water and establishing regular unassisted breathing within 15 min. Those few dugongs that do not meet these criteria should be returned to the water.

Pulse oximetry readings to measure oxygen saturation in the blood can occasionally be collected from the nasal septum, buccal mucosa or the rostral gingival ridge of the lower jaw.

- Temperature - body temperature should also be monitored because dugongs have a tendency to lose heat in cool, windy conditions. Flexible rectal thermometers that are well-lubricated can be used. Placing an oral thermometer alongside the caudal buccal mucosa gives fairly accurate results, but thermometers can be damaged by the dugong's molars (Chittick and Walsh 2007).
- Weighing - weigh every 3 months (Fig. 11.6). Animals can be trained to haul out onto a weigh platform voluntarily, but otherwise it involves encouraging the animal into a stretcher, or onto a platform, and lifting it out of the water by a chain and block winch. Attached to the winch is a hanging scale that allows visualisation of the weight. After releasing the animal back into the water, reattach the stretcher and all the straps to get a weight of the equipment once it is wet.

## 7.5 Release

When releasing a dugong back into the water, the number of people in the water should be at a minimum. Slowly lower the stretcher into the water until the water is taking the animal's bodyweight. Detach all straps on

**Fig. 11.6.** Weighing a dugong. Photo by Josephine Chong, SEA LIFE Sydney Aquarium.

the far side of the animal keeping the straps closest to the animal caretaker still attached. Only one person is needed in the water to release an animal. The person in the water should be able to stand on a platform and guide the unattached side of the stretcher down and away from the animal, ensuring the metal pole does not come down on top of it. As soon as the stretcher is clear and dropped down, the person in the water exits. The winch is then used to lift the stretcher from the water.

## 7.6 Transport requirements

### 7.6.1 Transport box design

Dugongs should be transported on individual stretchers with thick foam padding (Geraci and Lounsbury 1993). Care should be taken regarding exhaust fumes when traveling in a motor vehicle. Transport time should be kept to a minimum. Sedatives may be useful when transporting an anxious animal.

### 7.6.2 Furnishings

None.

### 7.6.3 Water and food

During transportation the dugong needs to be kept cool, moist and calm. It should be shaded from the sun and the amount of noise should be limited. If possible, the animal should be kept at 23–28°C during the transport. Care should be taken not to overheat or overcool it.

### 7.6.4 Animals per transport box

Only one animal should be held in a given box.

### 7.6.5 Timing of transportation

Transportation is best undertaken overnight or early in the day and later in the evening to avoid high temperatures and direct sunlight.

## 7.7 Release from the transport box

This should be done in a similar fashion to the release after weighing in a stretcher. See section 7.5.

# 8. HEALTH REQUIREMENTS

## 8.1 Daily health observations

Each dugong should be observed daily for any signs of injury or illness, especially after they have been introduced to conspecifics, as they often show aggression towards each other. The most appropriate time to do this check is first thing in the morning, generally when the dugong is being fed for the first time in the day. Particular areas that should be checked include:

- skin condition – in particular, rake marks if males are housed with other individuals
- discharges from the eyes, ears, nose, mouth, genital opening or anus
- appetite
- faeces – amount and consistency
- eyes – cloudiness, opacities and swelling
- changes in demeanour, activity level and behaviour
- injuries.

## 8.2 Bodyweight and body condition score

### 8.2.1 Bodyweight

Ideally bodyweight and condition should be recorded regularly. The frequency of weight measurements will depend on factors such as the logistics required to weigh particular individuals and the presence of underlying health concerns.

### 8.2.2 Body condition score

A method of estimating the body condition score of dugong ranges from 5 (excellent) to 1 (poor) using the information below.

- Body condition 5 – Excellent
  - Large dorsal area, no pronounced visible backbone
  - No concaving/indentation around the head
  - No concaving/indentation around the pectoral flippers
  - Large ventral surface
  - Peduncle/tailstock full
  - Cannot feel ribs or backbone
- Body condition 4 – Very good
  - Large dorsal area, backbone slightly pronounced
  - Slightly convex between spine and ribs, can be difficult to feel backbone, but cannot feel ribs
  - Slight concaving/indentation around pectoral flippers
  - Slight decrease in condition around the head, facial features still quite full
  - Large ventral surface
  - Peduncle/tailstock muscular with good condition around the tail
- Body condition 3 – Good
  - Neither convex or concave between spine and ribs; can feel the backbone, although it does not stand out. The ribs can be felt

  - Head more pronounced
  - Indentation around the pectoral flippers more pronounced
  - Ventral surface slightly concave
  - Facial features becoming slightly sunken
- Body condition 2 - Fair
  - Slightly concave between spine and ribs
  - Backbone palpable
  - Ribs noticeable
  - Decrease in condition of the head and shoulders; sunken facial features around the head leading into the shoulders
  - Ventral surface slightly concave
  - Peduncle/tailstock thinner and muscle is more pronounced
  - Condition decreased in and around the pectoral flippers
- Body condition 1 - Poor
  - Severely concave between spine and ribs
  - Backbone is very noticeable, ribs clearly palpable
  - Elongated/pronounced head, 'peanut head'
  - Ventral surface highly concave
  - More concave from behind the ear leading to pectoral flippers
  - Facial features sunken

## 8.3 Physical examination

A dugong can be manually restrained, but catching it in the water is difficult. If possible, each dugong should be conditioned to swim up onto a platform, or onto a stretcher, which can be raised out of the water. The platform should be smooth and devoid of any protruding objects. When an animal is out of the water it may thrash about on the deck, so the tail fluke should be covered to protect it from hitting anything and causing injury. If the platform is open to the sun, the animal should be covered with a wet towel or sheet, avoiding the head and tail. The animal should be maintained in ventral recumbency. A weigh platform should be incorporated into the design of the main platform to allow animals to be readily weighed.

The physical examination should be systematic and may include the following:

- body length and girth (Fig. 11.7; Eros *et al.* 2007)
- pulse rate (40–60 beats/min)
- respiratory rate (4–6 breaths/5 min)
- temperature
- eyes (clear, bright and alert)
- normal bilateral pupillary light response
- normal palpebral reflex
- no discharges
- check for the presence of lumps over the body
- integument.

## 8.4 Chemical restraint

General anaesthesia should not be necessary for a dugong, but may be considered following a boat strike to

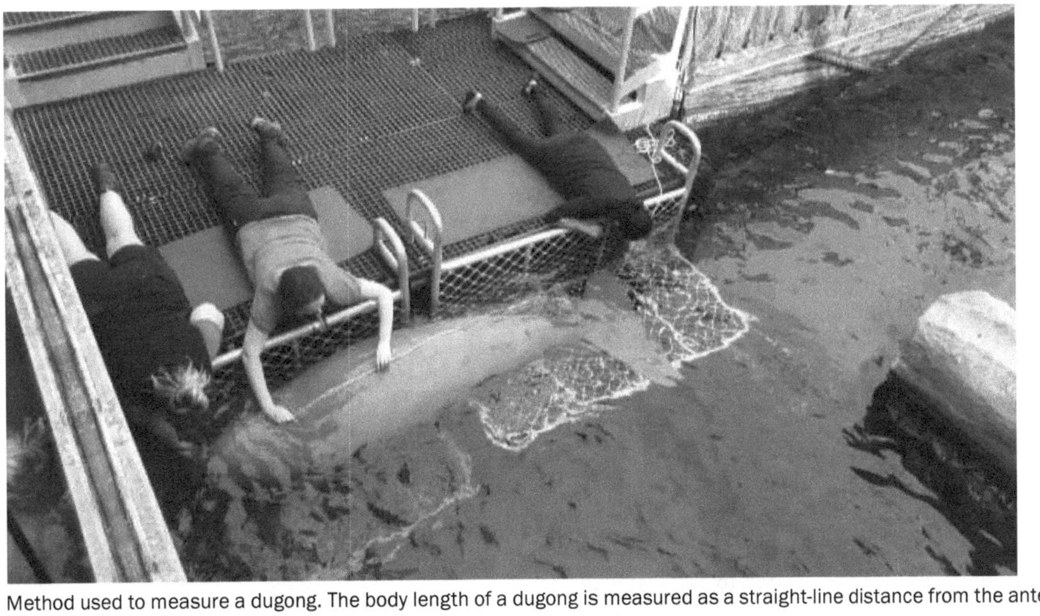

**Fig. 11.7.** Method used to measure a dugong. The body length of a dugong is measured as a straight-line distance from the anterior tip of the snout to the tail notch. Photo by SEA LIFE Sydney Aquarium.

repair a fractured limb or a major wound. Even then, anaesthesia should be considered very carefully as the postoperative care required is substantial and the facilities necessary to house and rehabilitate the animal could be problematic. If anaesthesia is necessary, the procedure for a manatee outlined by Chittick and Walsh (2007) should be followed as outlined below. Chemical restraint or sedation using benzodiazepines (midazolam and diazepam) may be used for minor procedures that do not necessary requires general anaesthesia (P. Marron pers. obs.).

The animal should be fasted for 24 h before the procedure. A sedation dose of midazolam should be given at a dose of 0.05 mg/kg intramuscularly using spinal needles in the peduncle area 30 min before the induction of anaesthesia. A mask can then be placed over the animal's nostrils and the animal can be induced with oxygen and either isoflurane or sevoflurane. Once the animal is in a deep enough plane of anaesthesia, an endotracheal tube can be placed via the nasal passages using a flexible endoscope to guide it. Intubation is more difficult in dugongs than in manatees as the anatomy of the nasal bones is somewhat different (Blyde 2023). The premaxilla and mandible are much more deflected in dugongs compared with manatees (Fig. 11.3; see also fig. 4.4 in Marsh *et al.* 2011). Once intubated the animal should be given intermittent positive-pressure ventilation. Monitoring should include end-tidal $CO_2$, pulse rate, $PO_2$ and temperature. Antagonism of butorphanol or meperidine with naltrexone and diazepam or midazolam with flumazenil (intramuscular) early during the anaesthetic procedure is advised to improve ventilation and perfusion to vital organs and facilitate a speedy recovery. The animal should remain intubated as long as possible, until it shows signs of consciousness and reflexes return, as apnoea is common following extubation. Doxopram has been used to stimulate respiration in marine mammals, including manatees.

Placement of an oral thermometer alongside the caudal buccal mucosa is the most appropriate site for temperature monitoring. It should be expected that body temperature will decrease under prolonged anaesthesia and so a means of warming the patient during the procedure should be available. Normal core body temperature is 35.5–36°C; however, this temperature can rapidly decrease in colder water or air temperatures during restraint. Dermal temperatures can be monitored with laser thermometers, but are related to the water or air temperature and do not necessarily correlate with body temperature. In contrast, thermography can be used to show all the variations in skin temperature present and for discriminating true skin temperature ranges across the body from those areas arising from handler contact. Thermography of eyes and nasal exhalation has also been considered as a method of monitoring core body temperature, but is still undergoing evaluation.

Metabolism and heat production during cold weather are lower in newly weaned juveniles and this may affect their body temperature. Electrocardiography leads can be placed on the animal to monitor heart rate and rhythm. Because both dolphins and manatees can have exhaled $CO_2$ levels over 100 mmHg, the capnometer used must be able to read up to 150 mmHg. Routes for fluid administration if needed include the intraperitoneal and intra-osseous routes. Vessels of the distal pectoral fluke may be accessible using ultrasound guidance and provide a route for fluid administration.

If intubation is too daunting, the animal could be left on a mask and maintained on oxygen and isoflurane or sevoflurane, but ventilation would not be possible.

The animal should not be placed back into water until it is clear that it is breathing voluntarily and responsive to stimuli. It should be placed in shallow water initially until demonstrably able to swim unaided.

## 8.5 Common health problems

There are few infectious diseases recorded in dugongs. Table 11.1 summarises the most common infectious diseases and Table 11.2 includes the most common non-infectious diseases.

The most common problems encountered in captive dugongs include:

- Gastrointestinal disease: pain, constipation, diarrhoea
- Conspecific trauma
- Nutritional deficiencies: iron, copper because of an inadequate captive diet.

## 8.6 Preventative medicine

No vaccinations or routine anthelminthic treatments are warranted. Vitamin/mineral supplements may be provided on a regular basis. Vaccination against *Clostridium perfringens* using a 5-in-1 vaccine could be considered.

## 8.7 Zoonoses

None.

**Table 11.1.** Infectious diseases or agents of captive, free-range and rescued dugongs of clinical significance.

Adapted from Blyde (unpubl. data).
b.i.d. = twice daily; IM = intramuscular; PO = per os (oral); s.i.d. = once daily.

| Disease/ infectious agent | Cause | Clinical signs | Prevention/ comments | Diagnosis | Treatment | Reference |
|---|---|---|---|---|---|---|
| **Endoparasites** | | | | | | |
| Intestinal worms | *Paradujardinia halicoris* | Impacted stomach and upper small intestine. Ruptured small intestinal abscess | Not seen in captive dugongs | Observation at necropsy of stranded animals | Ivermectin (200 ug/kg) PO | 1, 2 |
| Respiratory worms | *Cochleotrema indicum* | Tracheitis, bronchitis | Not seen in captive dugongs | Observation at necropsy of stranded animals | Ivermectin (200 ug/kg) PO | 1, 3 |
| Liver parasites | *Folitrema. Jecoris* | Liver abscess | Not seen in captive dugongs | Observation at necropsy of stranded animals | Ivermectin (200 ug/kg) PO | 3 |
| **Protozoa** | | | | | | |
| Toxoplasmosis | *Toxoplasma gondii* | Toxoplasma was associated with death of a free-range adult male dugong. Clinical signs included cachexia and multifocal hepatitis | Limit cat access to food sources | Necropsy. Increasing antibody titres | Clindamycin (5 mg/kg b.i.d.) PO or Atovaquone (100 mg/kg s.i.d.) PO or sulfadiazine 40 mg/kg and trimethoprim 8 mg/kg IM b.i.d. | 1 |
| Cryptosporidiosis | *Cryptosporidium parvum* | Gastrointestinal disease of juvenile animals being hand-reared | Appropriate hygiene when hand-rearing juveniles | Faecal examination, histopathology | Only warranted if considered clinically significant. Treatment is basically supportive (i.e. oral fluids) | 4, 5 |
| Blood parasites | Fish blood fluke | Fish blood fluke found in the heart of a dead stranded animal | Not seen in captive dugongs | Blood smear | Praziquantel (5–20 mg/kg) PO, IM | 6 |
| **Fungal** | | | | | | |
| Systemic fungal infections | *Madurella mycetomi* | Bronchiolitis and myocardial necrosis | Not seen in captive dugongs | Observation at necropsy of stranded animals | None reported but antifungals could be attempted if diagnosed antemortem | 7 |
| **Bacterial** | | | | | | |
| Respiratory disease | *Bordetella bronchiseptica* | Bronchiolitis and myocardial necrosis | Not seen in captive dugongs | Observation at necropsy of stranded animals | Enrofloxacin (5 mg/kg s.i.d. IM) | 8 |
| Intestinal disease | *Clostridium* sp. | Peritonitis, septicaemia, enteritis, colic, volvulus, intussusception | Vaccination with 5-in-1 vaccine | Abdominal cramping. Clinical signs. Necropsy | Long-acting penicillin (22 000 IU/kg IM) | 7 |

**Table 11.1.** continued

| Disease/ infectious agent | Cause | Clinical signs | Prevention/ comments | Diagnosis | Treatment | Reference |
|---|---|---|---|---|---|---|
| Respiratory disease | *Pasteurella multocida* | Pneumonia and pleuritis | Not seen in captive dugongs | Observation at necropsy of stranded animals | Antibacterial drugs as selected by culture and sensitivity | 7 |
| Urogenital disease | *Pseudomonas* spp. | Pyometra | Not seen in captive dugongs | Observation at necropsy of stranded animals | Gentamicin (6.6 mg/kg s.i.d. IM) Amikacin (7 mg/kg b.i.d. IM) | 9 |
| Intestinal disease | *Salmonella lohbruegge* | Lethargy, weight loss, dehydration, diarrhoea | Hygiene, correct treatment and disinfection of water body | Culture of faeces | Antibacterial drugs as selected by culture and sensitivity | 10 |

References: (1) Owen *et al.* (2012); 2) Marsh *et al.* (2000); (3) Woolford *et al.* (2015); (4) Hill *et al.* (1997); (5) Morgan *et al.* (2015); (6) Hutson *et al.* (2019); (7) Nielsen *et al.* (2013); (8) Norima *et al.* (2018); (9) Chansue *et al.* (2006); (10) Elliott *et al.* (1981).

**Table 11.2.** Common non-infectious diseases and causes of mortality of captive and rescued dugongs.

Adapted from Blyde (unpubl. data).

| Disease/ agent | Cause | Clinical signs | Diagnosis | Treatment | Prevention/ comments | Reference |
|---|---|---|---|---|---|---|
| Starvation | Occurs when animals stray outside of their normal geographic ranges. It may also occur when seagrass beds are damaged after floods, large tropical storms or cyclones as this is their only source of food | Emaciation, death | Signalment, clinical signs | Euthanasia, supportive care | Not relevant to captive dugongs | 1 |
| Predation | Sharks | Wounds, death | Observation/ history, physical examination, necropsy | Medical and surgical treatment if not fatal | Not relevant to captive dugongs | 1 |
| Cold stress syndrome | Dugongs exposed to long periods of water temperatures < 20°C | Weight loss, dermatitis over dorsum and dorsal and ventral tail surfaces | Signallment, clinical signs, climatic conditions | Relocate to more favourable conditions. Antibiotics. Supportive care | Ensure water temperature is > 20°C | 2 |
| Anthropogenic threats | Trauma from boats, propellers. Drowning. Entanglement. | Emaciation, trauma, dehydration, entanglement, entrapment, drowning, death | Observation/ history, physical examination, necropsy | Medical and surgical treatment if not fatal | Increased conservation action, climate change action, policy changes. Not relevant to captive dugongs | 1 |

| Disease/ agent | Cause | Clinical signs | Diagnosis | Treatment | Prevention/ comments | Reference |
|---|---|---|---|---|---|---|
| Oil spills | Container leaks | Signs of oil contamination | History. Signs of oil contamination | Removal of oil from external and internal surfaces | Policy. Not relevant to captive dugongs | 3 |
| Nutritional deficiencies and toxicities | Artificial diets deficient in iodine, copper, iron | Poor body condition, lethargic, decreased appetite, goitre, metastatic calcification | Blood results, including mineral and vitamin levels | Supplementing captive diets with multivitamin products | Ensuring a balanced diet in captivity | 1 |

References: (1) Blyde (2023); (2) Owen *et al.* (2013); (3) Preen (2004).

## 9. BEHAVIOUR

### 9.1 Activity

A daylight time-budget for dugongs using the shallow seagrass beds in eastern Moreton Bay at high tide was developed by Hodgson (2004). Those animals spent most of their time feeding (41%), travelling (32%), moving between the surface and bottom (18%) and little time resting, socialising or in other behaviours. It is not known whether this time-budget applies in other wild settings, stages of the tidal cycle, seasons, times of day or whether it is relevant to captive behaviour. It is problematic to infer time-budgets of dugongs from data collected by time depth recorders mounted on wild dugongs (Hagihara *et al.* 2011).

### 9.2 Social behaviour

Dugong behaviour appears to lack much social complexity and persistent social groupings are unknown (O'Shea *et al.* 2022). The strongest social bonds are between females and nursing calves and vocal communication appears to be most prevalent between mothers and calves. Socially transmitted knowledge (tradition) is likely to be important to dugongs. For example, dugongs may rely on traditional knowledge to find food in response to regional loss of food supply from extreme weather events. Thus captive-reared dugongs may lack the knowledge to succeed in the wild as evidenced by the histories of 'Pig' and 'Mariam' (see section 11.11).

### 9.3 Reproductive behaviour

Several mating systems have been described in dugongs (O'Shea *et al.* 2022), including scramble promiscuity (mating herds) (Preen 1989) and lone mating pairs in areas of low dugong density (Adulyanukosol *et al.* 2007; Infantes *et al.* 2020; O'Shea *et al.* 2022). Further research into the hypothesised leks reported by Anderson (1997) is needed, because they have only been reported at a single site, a cove in Shark Bay, Western Australia, and mating herds have been observed in the same region (Holley 2006).

During the breeding season, the focus of reproductively active males shifts from feeding to fighting other males, which results in tusk scarring (Preen 1989; Anderson 1995; Burgess *et al.* 2013a; Lanyon *et al.* 2021). By the end of winter and the beginning of spring, some males may be starving (J. Lanyon pers. comm.). In captivity, mature males may intermittently show a change in behaviour, including chasing equipment such as the pool hook, bashing toys around, increased playing with feed trays, reduced focus during training sessions and increased intensity of rough, high-energy play. After a few days to a week this behaviour suddenly reduces and the focus returns to food and food intake will increase to make up for the recent lack of feeding. A captive male at SEA LIFE Sydney Aquarium has been known to display these behaviours throughout the year, rather than just winter.

### 9.4 Behavioural problems

Both male and female dugongs have been successfully kept in captivity for long periods. For example, Jones (1967) describes a male and female dugong that were housed together in a small tank at Mandapam Camp in India for 5.5 years during which they were so attached to each other that it was almost impossible to separate them. The male became sexually mature by 1966 and mating behaviour was recorded without major incident, although the male was recorded biting the female.

Nonetheless, the adult male dugong at SEA LIFE Sydney Aquarium goes through periods where his interest in food drops off, after which he begins to show undesirable behaviours as described above.

## 9.5 Signs of stress

Signs of stress in captive dugongs are similar to other mammalian species – decreased appetite, weight loss and changes in behaviour including looking for escape routes. Animals being harassed by other individuals may have rake marks or other skin trauma from interactions within the enclosure.

## 9.6 Behavioural enrichment

Dugongs need enrichment to keep them mentally stimulated in captivity. They can be provided with many different enrichment opportunities including various toys (Fig. 11.8), water stimulations such as bubble streams or training sessions with marine mammal specialists. Some individuals appear to enjoy interaction with their animal caretakers and stay engaged for longer if it involves a game or interaction. Toys can vary in their size and texture, which allows the dugong to use the very sensitive bristles around the face to explore different textures. Dugongs also appear to enjoy being able to manipulate items with their pectoral flippers and use their strength to grab and sink floating items. Adult males also appear to enjoy tusking the play items.

Care must be taken to ensure that the toys are safe for interaction. Dugongs are very strong and capable of breaking even the toughest of toys. Entanglement can also be an issue if using anything with rope or firehose and they may require constant observation.

There are differences in the play behaviour of individual dugongs. The adult female at Toba Aquarium plays more gently than the male at SEA LIFE Sydney Aquarium. Dugongs are also known to interact with other species, including green sea turtles (*Chelonia mydas*), which creates further opportunities for interaction and enrichment.

**Fig. 11.8.** (Top) Examples of toys that may be provided. (Bottom) Dugong playing with one of the toys. Photos by SEA LIFE Sydney Aquarium.

## 9.7 Training and conditioning

Dugongs can be trained for specific husbandry purposes, including stretcher training and presentation of various body areas, including pectoral flippers, eyes and tail flukes. They can also be gate-trained and trained for ventral body examinations including ultrasound examinations. A male dugong named 'Malqout' was successfully trained by Worldwide Zoo in Abu Dhabi to allow the collection of blood, respiratory chuff and voluntary body measurements (P. Marron pers. obs.).

## 9.8 Introductions and removals

Introductions of animals should proceed cautiously ideally via the use of a 'howdy' door. Animals should ideally get used to each other in adjoining pens before being introduced. Adult males may be problematic and may attack and injure other males, females and infants.

## 9.9 Intraspecific compatibility

It may be possible to house multiple females together as long as there is sufficient space and food available for them to feed separately. A male and a female have been kept together successfully for long periods (Jones 1967); however, holding multiple males, or a male with a female, may not be appropriate as mature males are likely to fight with each other. Once a male reaches sexual maturity, it may also harass female(s) and unless there is capacity to separate them, the female(s) may be tusked and harassed. The possibility of reproduction should be considered and, if males are housed with females, reproductive control should be considered.

## 9.10 Interspecific compatibility

Dugongs have been held successfully with a variety of other marine life, including fish, rays and turtles. Careful observations should be made to determine that no adverse interactions are occurring between the different species.

Dugongs in captivity have been known to kill other animals kept in the same enclosure (e.g. marine turtles by lying on them and drowning or crushing them). Fish that eat the dugong's food could be a problem. On the other hand, they could be helpful in cleaning up the waste food in the enclosure.

## 9.11 Suitability for captivity

Dugongs are charismatic megafauna. However, because of the resources and cost involved in keeping dugongs in captivity they are rarely kept in zoos or aquaria. The costs of maintaining and controlling temperature and water quality in a large water body in which there are large sources of organic material (i.e. waste food and faeces) is also a factor.

# 10. BREEDING

## 10.1 Mating system

Wild dugongs appear to have a promiscuous mating system; multiple males endeavour to simultaneously mate with a receptive female (Preen 1989). In locations where dugong density is low, lone mating pairs have been observed (Adulyanukosol *et al.* 2007; Infantes *et al.* 2020; O'Shea *et al.* 2022).

## 10.2 Ease of breeding

No dugongs have been bred in captivity, despite mating being observed between animals kept in India (Jones 1967) and Japan (Wakai *et al.* 2002). At this time, there is a low success rate of hand-rearing and releasing rescued calves, as well as the proven difficulty of meeting the nutritional needs of adults. There is also a lack of space to house these very long-lived intelligent animals.

## 10.3 Oestrous cycle and gestation period

Females are polyovular and polyoestrus and may undergo several sterile cycles before becoming pregnant (Marsh *et al.* 2011). Records of urinary progesterone and oestradiol-17β in a captive 10–12-year-old female named 'Serena' at Toba Aquarium over 2 years revealed 14 oestrous cycles with an estimated ovulatory cycle of ~50 days (Wakai *et al.* 2002). Gestation is estimated to last ~14 months (Marsh *et* al. 2011).

## 10.4 Reproductive status

### 10.4.1 Females

Pregnancy can be confirmed via three methods (Burgess *et al.* 2012a).

(1) Enzyme-immunoassay of faecal progesterone metabolites (fP), which differs between sexes and reproductive size classes ($p < 0.001$), and 30-fold higher in confirmed pregnant dugongs (2017–7760 ng/g) compared with presumed non-pregnant females (30–221 ng/g), juvenile females (29–195 ng/g), and males (24–261 ng/g) ($p < 0.001$).
(2) External body morphometrics. In Moreton Bay, pregnant dugongs had body lengths ≥ 2.60 m, maximum girth ≥ 2.15 m, anal girth ≥ 1.26 m and teat length ≥ 5 cm
(3) Ultrasonography.

It is possible that captive dugongs will reach sexual maturity at smaller sizes than reported by Burgess *et al.* (2012a). The smallest wild female with a placental scar (evidence of a previous pregnancy) had a body length of 2.27 m (Kwan 2002).

### 10.4.2 Males

Using carcass analysis of dugongs from several locations in the tropics and histological examination of their testes, Marsh *et al.* (1984) and Kwan (2002) concluded that: (1) puberty in male dugongs is reached at a range of body lengths between 1.9 m and 2.5 m; and (2) testicular maturity preceded tusk eruption. Burgess *et al.* (2012b) examined the relationships between body length, tusk eruption, seasonality and group association on faecal testosterone metabolite concentrations in 159 free-ranging males, in subtropical Moreton Bay, Australia. Puberty apparently occurred at body lengths between 2.4 m and 2.6 m when faecal testosterone levels increase 4-fold (> 500ng/g) over juvenile levels and was associated with tusk eruption. However, social maturity may be delayed until male dugongs were more than 2.6 long and had well-developed, erupted tusks.

## 10.5 Reproductive management

There is currently no research around or methods of contraception for dugongs as there are only a few individuals worldwide in human care, including SEA LIFE Sydney Aquarium (Australia), Toba Aquarium (Japan) and the SeaWorld Abu Dhabi (UAE). Deslorelin implants (GnRH agonist) have been surgically inserted in a male dugong in an attempt to decrease testosterone levels and aggression. Dosages of 9.4 mg and 15.6 mg given via an intramuscular slow-release implant positioned just caudal to the neck were ineffective in decreasing testosterone levels, semen production and aggression (Burgess *et al.* 2013b). Contraceptives have not been used in females to date.

## 10.6 Occurrence of hybrids

None.

## 10.7 Timing of breeding

There is some reproductive seasonality in dugongs, even though the timing of births varies within the different geographic ranges. In tropical Australia, most dugong births are in the second half of the year, coinciding with the season of seagrass growth (Marsh *et al.* 2011). In subtropical Moreton Bay, mating occurs in spring (Burgess *et al.* 2012b).

## 10.8 Age at first and last breeding

Males generally reach maturity at around 10–15 years of age and 2.4 m and 2.6 m in length, which coincides with the eruption of their tusks (Marsh *et al.* 2011). Females generally reach sexual maturity at ~10–17 years (Marsh *et al.* 2011). The age at first breeding varies, apparently with nutritional condition. The youngest parous female observed in the wild was aged 6 years old, the oldest nulliparous female had an estimated age of 17.5 years (Marsh *et al.* 2011). There is no evidence of reproductive senescence.

Age of sexual maturity in captivity is likely to be similar to the wild; however, it may occur earlier due to improved nutrition.

## 10.9 Ability to breed every year

Considering that pregnancy is estimated to last ~14 months (Marsh *et al.* 2011) and the calf suckles from its mother for at least 18 months before weaning, a dugong cannot breed every year. After the birth of a single calf there is an extended period of maternal investment and a prolonged intercalving interval (3–7 years) (Marsh *et al.* 2011).

## 10.10 Ability to breed more than once per year

An individual female dugong gives birth at irregular intervals every few years (Marsh *et al.* 2011).

## 10.11 Number of young born

Dugongs typically bear a single young, although there are occasional records of twins (Marsh *et al.* 2011).

## 10.12 Age at weaning

Size at birth is variable in the wild with neonates ranging from ~1.1 to 1.25 m long and ~27–35 kg in bodyweight. Calves begin to consume some seagrass soon after birth, but they continue to suckle from their mothers for at least 18 months (Marsh *et al.* 2011). Data on the duration of lactation are limited.

## 10.13 Age at removal from the mother

Unknown as dugongs have not bred in captivity.

## 10.14 Growth and development

Growth and development vary with nutritional plane and the rate of growth in captive animals is highly variable as explained above. Marsh (1980) produced a growth curve for wild dugongs using the number of tusk growth layers and body length. The growth curve indicated rapid growth until around puberty and slow growth after that. There was considerable variation in the relationship between body length and estimated age. The growth of dugongs from shortly after birth to 500 days has been recorded for the bodyweight (Fig. 11.9) and length (Fig. 11.10) of one animal (Blanshard 2000).

# 11. ARTIFICIAL REARING

Although wild dugongs start to eat solid food soon after birth, animals less than 1.5 m in length will be suckling and totally milk dependent, while those less than 1.8 m long may be still suckling (Marsh *et al.* 2011). Attempts

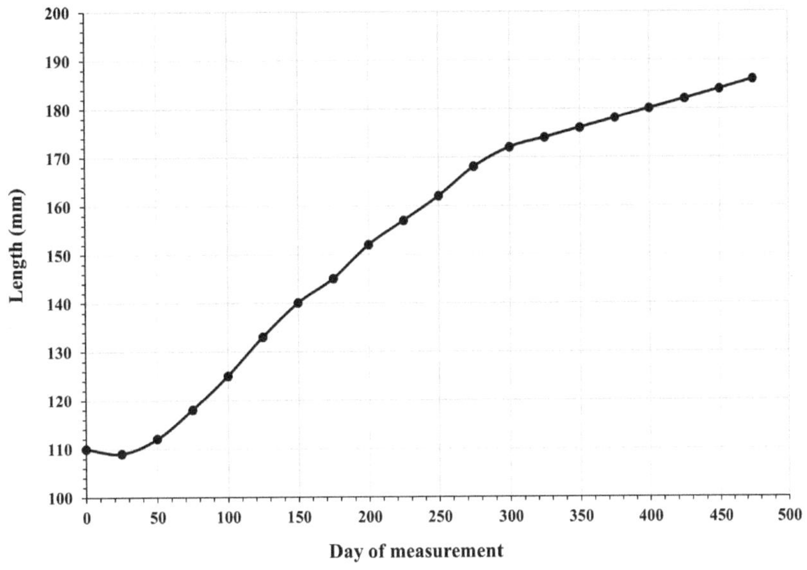


**Fig. 11.9.** Growth in bodyweight of a hand-reared dugong. Derived from Blanshard (2000).

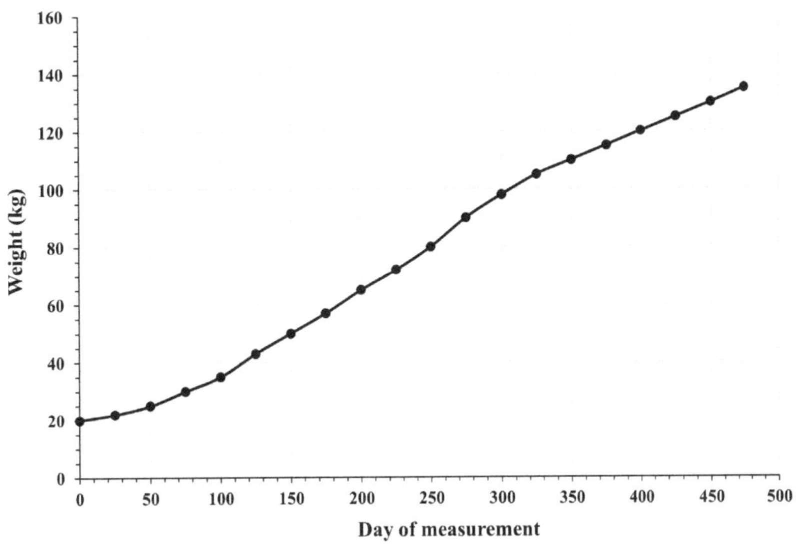


**Fig. 11.10.** Growth in length of a hand-reared dugong. Derived from Blanshard (2000).

have been made to hand-rear neonatal dugongs; however, most attempts have been unsuccessful with notable exceptions discussed here. Therefore, a decision of what to do with an orphaned dugong calf is complex. The final decision must be made on a case-by-case basis and will vary by jurisdiction according to the relevant laws on keeping wild animals in captivity, cultural norms and specific circumstances. The most pragmatic option may be to euthanase the calf. However, this approach is culturally unacceptable in many countries in the dugong's range (Marsh 2022). A second option, which has been adopted by some managers, is to return the orphaned dugong to the sea in an area where there are known to be dugongs, in the hope it is adopted by a lactating female dugong. There are no data on the success of such initiatives and the animal welfare issues have not been addressed (H. Marsh pers. obs.). Cases of adoption have never been documented in dugongs, but persistent allomaternal care has been observed in Florida manatees at aggregation sites. Allomaternal care has also been recorded in captive Florida, Antillean and Amazonian manatees (O'Shea *et al.* 2022). The advantages and disadvantages of various options for an orphaned dependent calf can be found in Table 11.3 (Marsh 2022).

## 11.1 Housing

A male new-born dugong calf that was received by Sea World on the Gold Coast (Queensland, Australia) on 28 November 1998 was 109 cm in length and emaciated, weighing 19.7 kg on arrival. The animal appeared to be a new-born because there were shreds of umbilical cord attached to the umbilicus, the bristles were very short on his facial disc, there was an absence of calluses on the undersides of the tips of the pectoral flippers and the horny plates in his top and bottom jaws were unstained (Blanshard 2000).

The dugong was held in a 2.1-m diameter, circular polycarbonate trough with a wall height of 60 cm, which had a water depth of ~55 cm. On 31 December 1998, the animal was moved to a 3.1-m diameter pool with a 78-cm wall height and a water depth of ~72–73 cm. Each pool was heated to a temperature of ~25°C by several aquarium heaters that were protected inside a fenestrated PVC pipe hung from the side of the pool (Blanshard 2000). The water was pumped through a sand filter. Faeces were manually removed from the pool by skimming the surface with a hand net.

The dugong was placed in an aboveground domestic swimming pool (~11.5 m (L) × 4.5 m (W) × 1 m (D)) on 7 October 1999. The water was warmed using four 1500 W heaters that were covered by a PVC pipe barrier at one end of the pool. The temperature of the water was ~25°C. One-third of the pool was covered by a vinyl roof, while the remainder was covered by 90% shade cloth. The water was cleaned using a separate filtration system that contained a prefilter strainer, two pumps, two large sand filters and a sodium hypochlorite dosing pump (Blanshard 2000).

## 11.2 Temperature requirements

Water temperature for a facility holding a dependent calf should be maintained at ~25°C (Blanshard 2000).

## 11.3 Specific requirements

When a juvenile, wild dugong is first discovered and it is decided to take it into captivity, it is important to provide good supportive care as this will not only improve the animal's immediate welfare but also help prevent injuries and reduce the likelihood of shock due to physiological stress. Therefore:

- ensure no water, sand or objects obstruct the nares
- protect the skin and keep it moist
- cover the animal in wet sheets or towels
- keep the animal floating in shallow water, if possible, to facilitate keeping it moist and less stressed
- perform the first initial physical evaluation as soon as possible
- check for wounds or signs of trauma such haematomas or fresh blood from any orifice
- record the respiratory rate (average between 5 and 9 breaths/5 min) and observe for any respiratory effort, which can indicate respiratory disease. An elevated respiratory rate can also be indicative of pain or stress.

**Table 11.3.** Advantages and disadvantages of managing stranded dugong calves.

Derived from Marsh (2022).

| # | Option | Advantages | Disadvantages |
|---|---|---|---|
| 1 | Release back into the sea in a known dugong area | Expedient | Fate of animal will likely be unknown. Welfare issues likely due to starvation or predation |
| 2 | Hand-rear in captivity in the home country | Available for public education | Likely requires life-long commitment to captive care that will be expensive; animal welfare challenges; high risk of death |
| 3 | Hand-rear in an unrestricted wild setting in the home country | No need for expensive facility; potential for dugong to contribute to wild gene pool if it survives to maturity | Exposed to local threats; likelihood of escape and/or survival uncertain; costs of hand-rearing are substantial |
| 4 | Transfer to an aquarium with the capacity for long-term care | Higher likelihood of long-term survival than option 2 or 3 | Transfer bureaucracy and costs may be prohibitive; costs of hand-rearing are substantial; dugong will not contribute to wild gene pool |
| 5 | Euthanasia | Rapid; opportunity for posthumous learning by scientists | Unacceptable in some countries; no opportunity for public education; dugong will not contribute to wild gene pool |

Some animals may keep the nostrils open for a more prolonged period; this also could be related to stress
- collect a blood sample for complete blood count, serum biochemistry panel, erythrocyte sedimentation rate, electrolytes, venous blood gas and lactate levels. Blood glucose should be measured immediately with a portable glucometer
- collect samples, including gastric, faecal, and respiratory chuff, as they also provide valuable medical information on the animal's current status. When feasible, collect the mentioned samples as well for culture and sensitivity.

Rehydration is critical in the initial part of the rehabilitation. The skin elasticity and the blood results will indicate the hydration status. Most animals will be at least a 5–7% dehydrated. Oral rehydration is preferred via gastric intubation. Oral electrolytes or water at room temperature can be provided using a 12–16F feeding tube. It is critical to ensure the feeding tube is in the stomach; aspiration pneumonia is a real risk during the procedure. It is recommended to measure the length of the feeding tube from the mouth to the end of the pectoral fin as this will give an approximation of where the stomach is located. When introducing the gastric tube, be alert for any respiratory signs such as coughing, sneezing or abnormal respiratory sounds, which could indicate the tube is in the trachea instead of the oesophagus/ stomach. Performing rehydration with the animal in the water will facilitate the procedure. The size of the animal will determine amount of water or electrolytes that can be delivered. For a 20-kg calf, ~300–400 mL of fluids can be delivered at a time.

Most stranded calves will have some degree of hypoglycaemia. Initially, blood glucose should be measured every 2–4 h until normoglycaemic. Glucose can be supplemented orally at 0.5–1 mL/kg of 50% dextrose; intravenous supplementation is challenging.

## 11.4 Diet and feeding routine

### 11.4.1 Natural milk

Unknown. Bachman and Irvine (1979) provide details of the composition of Florida Manatee milk.

### 11.4.2 Milk formulas

The main formulas for hand-rearing dugongs include the following products.

- Di-Vetelact® (Lillelund Pty Ltd) – a low-lactose powdered milk replacement. The contents of acidophilus capsules can be added to each batch to assist in the breakdown of the residual lactose (Blanshard 2000). Other ingredients include canned coconut extract (Creamy Coconut®; Chef's Choice Food Manufacturer Co. Ltd, Thailand) to increase the fat content of the milk. Calogen (N.V. Nutricia, Zoelemeer, The Netherlands) can also be added to the Di-Vetelact® to increase nutrient density.
- Esbilac® Puppy Milk Replacer and Zoologic Milk Matrix milk (P. Marron pers. obs.).
  - 300 mL Esbilac® Puppy Milk Replacer (1 part powder to 2 parts water)
  - 400 mL Zoologic® Milk Matrix 33/40 (1 part powder to 3 parts water)
  - 300 mL coconut milk
  - 32 mL canola oil
  - 250 mg taurine
  - 1 capsule vitamin B complex
  - 40 mg simethicone (anti-bloat agent), per feed

Note: lactobacillus and vitamin E can also be added to the formula.

Once made up, the milk should not be kept more than 24 h.

Hand-rearing is an extremely labour-intensive and prolonged commitment and should not be commenced lightly. Orphaned dugongs can be bottle-fed and generally nurse better while being cradled upside down in water.

Faeces collected from wild dugongs in Moreton Bay were administered orally to two calves in an attempt to trans-faunate and establish a healthy gastrointestinal microbial community. A female was inoculated once post-weaning, and a male both before and after weaning and twice (December 2002 and September 2005) after recapture from the wild in an attempt to alleviate perceived digestive problems (Eigeland *et al.* 2012).

### 11.4.3 Feeding apparatus

Blanshard (2000) successfully fed an orphaned dugong ('Pig' see below) with formula from a baby's bottle fitted with a rubber calf teat. It may be necessary to experiment with different types of rubber nipples (Fig. 11.11). Some dugongs have difficulties consuming enough milk and supplementation via an orogastric tube may be necessary. The temperature of the milk should always be checked with a digital thermometer before use.

### 11.4.4 Feeding routine

The introduction of milk formula should occur as soon as the animal has been rehydrated. Initially, the milk

**Fig. 11.11.** Dugong being bottle-fed. Photo by Paola Marron.

formula should be introduced in a diluted form to allow the gastrointestinal tract to adapt to it. Start with 10% milk formula and 90% water and increase the concentration daily. Calves should be fed every 3–5 h. A 20-kg calf initially needs between 800 and 1100 mL (= 50–80 kcal/kg) of milk formula daily. The calf should be weighed every day. Ideally there should be an increase of bodyweight of 300–500 g/day (P. Marron pers. obs.).

It is recommended to measure gastric residuals before each feeding via gastric tube. If the gastric residuals are significant, the feeding should be postponed for a couple of hours. The calf should defecate daily; defecation will indicate how well the milk formula is being digested.

The introduction to small amounts of solid food such as seagrass and lettuce can be done as soon as the dugong is fully rehydrated. Solid food will also help to stimulate gastrointestinal movements (P. Marron pers. obs.).

At Sea World, a male dugong ('Pig') was bottle-fed on a mixture of Di-Vetelact® supplemented with canned coconut extract. The animal was given 11 feeds (2 h apart) per day until 12 months (Blanshard 2000). It was inoculated with wild dugong faeces on two occasions to establish the gut flora required to digest plant foods. After 16 months in captivity, it was 1.9 m long and weighed 130 kg (Figs 11.9, 11.10; Blanshard 2000).

In April, 2019, an orphaned dugong calf ('Mariam') was found in the coastal waters of Krabi, Thailand, by local villagers (Ponnampalam *et al.* 2022). They alerted government personnel, who relocated her to a sheltered area around Koh Libong in Trang Province, the province that supports the largest population of dugongs in Thailand (Hines *et al.* 2005). Although allowed to roam freely, 'Mariam' was bottle-fed by her caretakers, who entered the water to feed her using a custom-made nipple made from a rubber glove finger stuffed with gauze and with an intravenous extension line going through it, through which she sucked. She was fed seagrass daily and apparently grew fond of her caretakers, playfully interacting with them and imprinting on their kayak and swimming under it as it moved. Mariam beached nightly during the lowest tide, suggesting she lacked the experience to know when to swim into deeper waters, and eventually died (Ponnampalam *et al.* 2022).

Issues associated with feeding include the following.

- Pneumonia: as a result of being stranded in shallow water and weak animals may aspirate sea water and later develop pneumonia. Others develop aspiration pneumonia due to poor feeding technique during rehabilitation. Daily monitoring of respiratory rate and effort is crucial. The clinical signs to be alert for are an increase in respiratory rate and effort. Some animals may also develop a nasal discharge. Broad-spectrum antibiotics can be used prophylactically (P. Marron pers. obs.).
- Constipation: can be associated with the poor assimilation of the milk formula and dehydration. If the calf is not defecating on a daily basis and

constipation is suspected, oral lactulose and mineral oil can be used. Mineral oil can also be used as an enema. Adequate maintenance of hydration is essential when using lactulose (P. Marron pers. obs.).
- Intestinal bloat: clinical signs include an inability to dive, body arching and buoyancy. Medications such as simethicone and antispasmodics such as Buscopan® (hyoscine butylbromide) may help, but pain management should be also considered. Medications for pain and inflammation that can be used are non-steroidal anti-inflammatory drugs such as flunixin meglumine and meloxicam (P. Marron pers. obs.).
- Pneumatosis cystoides intestinalis secondary to intestinal obstruction, gastrointestinal poor perfusion and necrotic enterocolitis are common. Successful treatment can be achieved using antibiotics such as metronidazole and gentamicin. After antibiotic treatment, diarrhoea and intestinal dysbiosis are common. Seeding with the faeces of a healthy dugong is beneficial if access is available.

## 11.5 Data recording

When an animal is first brought in for hand-rearing, its sex and approximate age, using growth charts, should be recorded. During the hand-rearing process information should be recorded on a daily basis:

- date and time when the information is recorded
- gastric residuals before each feeding via gastric tube. If the gastric residuals are significant, feeding should be postponed for a couple of hours
- bodyweight - should increase 300–500 g per day
- general activity and demeanour
- characteristics and frequency of defecation and urination
- amount of food offered and when
- food consumption at each feed
- defecation - calf should defecate daily. Defecation will indicate how well the milk formula is being digested. Constipation is one of the most common medical conditions
- veterinary examinations and results.

This information serves several purposes, including providing important background information such as food consumption, that will assist a veterinarian to reach a diagnosis if the animal becomes sick or fails to grow or gain weight. It also allows a comparison with growth curves to assess progress (see section 10.16).

## 11.6 Identification methods

Not required.

## 11.7 Hygiene

Pool water needs to be maintained in a similar manner to that for captive adults. Faeces can be manually removed from small pools as it appears. If there is not a filtration/disinfection plant attached to the hand-rearing pool, regular water changes (at least every 24 h) will be required.

## 11.8 Behavioural considerations

Hand-reared dugong have typically struggled to survive post-release due to a failure to adapt to their natural environment. As a result, they typically die as a result of starvation and predation that may have been facilitated by malnutrition (Blanshard 2000). More research is required on how to successfully reintegrate a hand-reared animal back into the wild.

## 11.9 Use of foster animals

Using foster species is not an option.

## 11.10 Weaning

A male dugong at SeaWorld on the Gold Coast ('Pig') was gradually weaned onto plant food by 19 months of age. His staple diet comprised cos lettuce, hydroponic butter crunch and coral lettuces, endive, choy sum, pak choy and English spinach (Blanshard 2000).

## 11.11 Release procedures

'Pig' was released into the wild at ~3 years of age, when he weighed 150 kg. He was first released into a closed coastal lagoon on Moreton Island (latitude 27.5°S), which supported a seagrass community but no other dugongs, before being released into the wild on the nearby Eastern Banks, Moreton Bay, a region that supports some 1000 dugongs (Lanyon 2019). The animal was fitted with a satellite tracking device (H. Marsh pers. obs.) before release, but the device detached and was found washed up covered with shark bites. After 8 months, the dugong was taken back into captivity when it was found to be severely malnourished and covered in wounds inflicted by the tusks of adult male dugongs. This animal's health was gradually restored after further faecal transplants. 'Pig' was moved to SEA LIFE Sydney Aquarium in 2008, along with a young female dugong that was also rescued by Sea World. Both animals were deemed unreleasable.

The two animals had to be separated after the male reached sexual maturity because he often behaved violently towards the female.

## ACKNOWLEDGEMENTS

We thank the dugong animal caretakers at SEA LIFE Sydney Aquarium and Toba Aquarium for their insights. They have worked hard to further the dugongs' care, as well as improving information on dugong husbandry.

## REFERENCES

Adulyanukosol K, Thongsukdee S, Hara T, Arai N, Tsuchiya M (2007) Observations of dugong reproductive behavior in Trang Province, Thailand: further evidence of intraspecies variation in dugong behavior. *Marine Biology* **151**, 1887–1891. doi:10.1007/s00227-007-0619-y

Aiken A (1995) Use of ozone to improve water quality in aquatic exhibits. *International Zoo Yearbook* **34**, 106–114. doi:10.1111/j.1748-1090.1995.tb00668.x

Anderson PK (1989) Deliberate foraging on macroinvertebrates by dugongs. *National Geographic Research* **5**, 4–6.

Anderson PK (1995) Scarring and photoidentification of dugongs (*Dugong dugon*) in Shark Bay, Western Australia. *Aquatic Mammals* **21**, 205–211.

Anderson P (1997) Shark Bay dugongs in summer. I: Lek mating. *Behaviour* **134**, 433–462. doi:10.1163/156853997X00629

Bachman KC, Irvine AB (1979) Composition of milk from the Florida manatee, *Trichechus manatus latirostris*. *Comparative Biochemistry and Physiology. Part A, Physiology* **62**, 873–878. doi:10.1016/0300-9629(79)90020-3

Baker AM, Gynther IC (eds) (2023) *Strahan's Mammals of Australia*. 4th edn. Reed New Holland Publishers, Sydney.

Blanshard W (2000) Current achievements and future challenges in dugong rehabilitation. In Achievements and Challenges. Australasian Regional Association of Zoological Parks and Aquaria & Australasian Society of Zoo Keeping 2000 Conference Proceedings, Sea World, Gold Coast.

Blyde D (2023) Dugong medicine. In *Fowler's Zoo and Wild Animal Medicine Current Therapy*. Vol. 10 (Eds E Miller, N Lamberski and P Calle) pp. 547–556. Elsevier, Amsterdam.

Burgess EA, Lanyon JM, Brown JL, Blyde DJ, Keeley T (2012a) Diagnosing pregnancy in free-ranging dugongs using fecal progesterone metabolite concentrations and body morphometrics: A population application. *General and Comparative Endocrinology* **177**, 82–92. doi:10.1016/j.ygcen.2012.02.008

Burgess E, Keely T, Lanyon JL (2012b) Testosterone and tusks: maturation and seasonal reproductive patterns of live, free-ranging male dugongs (*Dugong dugon*). *Reproduction (Cambridge, England)* **143**, 683–697. doi:10.1530/REP-11-0434

Burgess E, Brown J, Lanyon J (2013a) Sex, scarring, and stress: understanding seasonal costs in a cryptic marine mammal. *Conservation Physiology* **1**, cot014. doi:10.1093/conphys/cot014

Burgess EA, Blanshard WH, Barnes AD, Gilchrist S, Keeley T, Chua J, Lanyon JM (2013b) Reproductive hormone monitoring of dugongs in captivity: detecting the onset of sexual maturity in a cryptic marine mammal. *Animal Reproduction Science* **140**, 255–267. doi:10.1016/j.anireprosci.2013.06.005

Chansue N, Monanunsap S, Sailasuta A (2006) Pyometra with diffuse fibrinopurulent peritonitis in a dugong (*Dugong dugon*). In Proceedings of Asian Zoo and Wildlife Medicine and Pathology. pp. 26–29. Chulalongkorn University, Faculty of Veterinary Science, Bangkok, Thailand.

Chittick EJ, Walsh MT (2007) Sirenians (manatees and dugongs). In *Zoo and Wildlife Immobilisation and Anaesthesia*. (Eds G West, D Herd and N Caulkett) pp 497–505. Blackwell Publishing: Iowa.

Chua FHC, Mackay B, Whitfield M, Oi WL (2001) Rehabilitation and display of a stranded dugong (*Dugong dugon*) calf in Singapore. *Bulletin de l'Institut océanographique*, Monaco, n° spécial 20, fascicule 1.

Cope RC, Pollett PK, Lanyon JM, Seddon JM (2015) Indirect detection of genetic dispersal (movement and breeding events) through pedigree analysis of dugong populations in southern Queensland, Australia. *Biological Conservation* **181**, 91–101. doi:10.1016/j.biocon.2014.11.011

Crane M (1994) Standards for Exhibiting Bottle-nosed Dolphins (*Tursiops truncatus*) in New South Wales. New South Wales Agriculture, Orange, New South Wales.

Domning DP (2022) What can we infer about the behaviour of extinct sirenians? In *Ethology and Behavioral Ecology of Sirenians*. (Ed. H Marsh) pp. 1–17. Springer, Berlin.

Eigeland KA, Lanyon JM, Trott DJ, Ouwerkerk D, Blanshard W, Milinovich GJ, Gulino LM, Martinez E, Merson S, Klieve AV (2012) Bacterial community structure in the hindgut of wild and captive dugongs (*Dugong dugon*). *Aquatic Mammals* **38**, 402–411. doi:10.1578/AM.38.4.2012.402

Elliott H, Thomas A, Ladds PW, Heinsohn GE (1981) A fatal case of salmonellosis in a dugong. *Journal of Wildlife Diseases* **17**, 203–208. doi:10.7589/0090-3558-17.2.203

Eros C, Marsh H, Bonde R, O'Shea T, Beck C, Recchia C, Dobbs K, Turner M, Lemm S, Pears R, Bowater R (2007) *Procedures for the Salvage and Necropsy of the Dugong (*Dugong dugon*)*, 2nd edn. Great Barrier Reef Marine Park Authority Research Publication 85, 1–98.

Geraci JR, Lounsbury VJ (1993) *Marine Mammals Ashore: A Field Guide for Strandings*. pp 145–158 Texas A&M Sea Grant Publications, Galvaston TX.

Goto M, Ito C, Sani Yahaya M, Wakamura K, Asano S, Wakai Y, Furuta M, Kataoka T (2004) Effects of age, body size and season on food consumption and digestion of captive dugongs (*Dugong dugon*). *Marine and Freshwater Behaviour and Physiology* **37**, 89–97. doi:10.1080/1023624042000199935

Hagihara R, Jones R, Sheppard J, Hodgson AJ, Marsh H (2011) Minimizing errors in the analysis of dive recordings from shallow-diving animals. *Journal of Experimental Marine Biology and Ecology* **399**, 173–181. doi:10.1016/j.jembe.2011.01.001

Hill BD, Fraser IR, Prior HC (1997) *Cryptosporidium* infection in a dugong (*Dugong dugon*). *Australian Veterinary Journal* **75**, 670–671. doi:10.1111/j.1751-0813.1997.tb15369.x

Hines EM, Adulyanukosol K, Duffus DA (2005) Dugong (*Dugong dugon*) abundance along the Andaman coast of Thailand. *Marine Mammal Science* **21**, 536–549. doi:10.1111/j.1748-7692.2005.tb01247.x

Hines EM, Adulyanukosol K, Poochaviranon S, Somany P, Ath SL, Cox N, Symington K, Tun T, Ilangakoon A, Delongh HH, Aragones LV, Lu S, Jiang X, Jing X, D'Souza E, Patankar V, Sutaria D, Jethva B, Solanki P (2012) Dugongs in Asia. In *Sirenian Conservation: Issues and Strategies in Developing Countries*. (Eds E Hines, J Reynolds, A Mignucci-Giannoni and A Marmontel) pp. 58–76. University Press of Florida, Gainesville.

Hodgson AJ (2004) Dugong behaviour and responses to human influences. PhD thesis, James Cook University, Townsville.

Holley DK (2006) Movement patterns and habitat usage of Shark Bay dugongs. MS thesis, Edith Cowan University, Perth.

Hutson KS, Vaughan DB, Blair D (2019) First record of a 'fish' blood fluke (Digenea: Aporocotylidae) from a marine mammal: *Cardicola dhangali* n. sp. *International Journal for Parasitology. Parasites and Wildlife* **10**, 23–28. doi:10.1016/j.ijppaw.2019.06.009

Infantes E, Cossa D, Stankovik M, Panyawai J, Tuntiprapas P, Daochai C, Prathep A (2020) Dugong reproductive behaviour in Koh Libong, Thailand: observations using drones. *Aquatic Mammals* **46**, 603–608. doi:10.1578/AM.46.6.2020.603

Jackson SM, Groves CP (2015) *Taxonomy of Australian Mammals*. CSIRO Publishing, Melbourne.

Jefferson T, Webber M, Pitman R (2008) *Marine Mammals of the World: A Comprehensive Guide to Their Identification*. Academic Press, Cambridge, MA.

Jones R (2012) Module Three: water chemistry and quality. The Aquarium Vet. https://www.theaquariumvet.com/store-2/module-three-water-chemistry-and-quality (accessed 1 November 2023).

Jones R (2014) Module Four: life support systems. The Aquarium Vet. https://www.theaquariumvet.com/store-2/module-four-life-support-systems (accessed 1 November 2023).

Jones S (1959) On a pair of captive dugongs (*Dugong dugon* (Erxleben). *Journal of the Marine Biological Association of India* **1**, 198–202.

Jones S (1967) The dugong *Dugong dugon* (Müller) its present status in the seas around India with observations on its behaviour in captivity. *International Zoo Yearbook* **7**, 215–220. doi:10.1111/j.1748-1090.1967.tb00398.x

Kataoka T, Asano S (1981) On the keeping of dugong (*Dugong dugon*) in Toba Aquarium. In *The Dugong: Proceedings of a Seminar/Workshop*. 8–13 May 1979. (Ed. H Marsh) pp. 115–117. Department of Zoology, James Cook University, Townsville.

Keith-Diagne LW, Barlas ME, Reid JP, Hodgson AJ, Marsh H (2022) Diving and foraging behaviors. In *Ethology and Behavioral Ecology of Sirenians*. (Ed. H Marsh) pp. 67–100. Springer, Berlin.

Kwan D (2002) Towards a sustainable Indigenous fishery for dugongs in Torres Strait: a contribution of empirical data and process. PhD thesis, James Cook University, Townsville.

LaBonne DL (1998) Ozonation of marine mammal pool waters. In *Marine Mammal Water Quality: Proceedings of a Symposium*. (Eds J Coakley and RL Crawford) pp. 9–20. *United States Department of Agriculture. Technical Bulletin* 1868, 1–79.

Lanyon JM (2019) Management of megafauna in estuaries and coastal waters: Moreton Bay as a case study. In *Coasts and Estuaries: The Future*. (Eds E Wolanski, JW Day, M Elliott and R Ramachandran) pp. 87–101. Elsevier, Amsterdam.

Lanyon JM, Sneath HL, Kirkwood JM, Slade RW (2002) Establishing a mark-recapture program for dugongs in Moreton Bay, south-east Queensland. *Australian Mammalogy* **24**, 51–56. doi:10.1071/AM02051

Lanyon JM, Slade RW, Sneath HL, Broderick D, Kirkwood JM, Limpus D, Limpus CJ, Jessop T (2006) A method for capturing dugongs (*Dugong dugon*) in open water. *Aquatic Mammals* **32**, 196–201. doi:10.1578/AM.32.2.2006.196

Lanyon JM, Sneath HL, Long T, Bonde RK (2010) Physiological response of wild dugongs (*Dugong dugon*) to out-of-water sampling for health assessment. *Aquatic Mammals* **36**, 46–58. doi:10.1578/AM.36.1.2010.46

Lanyon JM, Athousis C, Sneath HL, Burgess EA (2021) Body scarring as an indicator of social function of dugong (*Dugong dugon*) tusks. *Marine Mammal Science* **37**, 962–981. doi:10.1111/mms.12788

Marsh H (1980) Age determination of the dugong (*Dugong dugon* (Müller) in Northern Australia and its biological implications. *Report of the International Whaling Commission. Special Issue* **3**, 181–201.

Marsh H (1989) Dugongidae. In *Fauna of Australia. Mammalia* Vol 1B. (Eds DW Walton and BJ Richardson) pp 1030–1038, Australian Government Publishing Service, Canberra.

Marsh H (1991) Our tropical siren. *The Australian Geographer* **21**, 42–57.

Marsh H (1995) The life history, pattern of breeding, and population dynamics of the dugong. In *Population Biology of the Florida Manatee*. (Eds TJ O'Shea, BB Ackerman and HF Percival). *US Department of the Interior, National Biological Service, Information and Technology Report* (Washington, DC) 1, 75–83.

Marsh H (2022) *Options for Handling a Stranded Orphaned Dugong Calf: Advice to Policy Makers and Managers*. CMS Technical Series No. 44. Coordinating Unit of the CMS Dugong MOU, Abu Dhabi, United Arab Emirates. https://www.cms.int/en/publication/options-handling-stranded-orphaned-dugong-calf-advice-policy-makers-and-managers (accessed 26 February 2023).

Marsh H, Sobtzick S (2015) *Dugong dugon*. The IUCN Red List of Threatened Species 2015: e.T6909A43792211. doi:10.2305/IUCN.UK.2015-4.RLTS.T6909A43792211.en

Marsh H, Channells PW, Heinsohn GE, Morrissey J (1982) Analysis of stomach contents of dugongs from Queensland. *Australian Wildlife Research* **9**, 55–68. doi:10.1071/WR9820055

Marsh H, Heinsohn GE, Marsh LM (1984) Breeding cycle, life history and population dynamics of the dugong, *Dugong dugon* (Sirenia: Dugongidae). *Australian Journal of Zoology* **32**, 767–788. doi:10.1071/ZO9840767

Marsh H, Eros C, Webb R (2000) Dugongs in health and disease. In *Marine Wildlife: the Fabian Fay Course for Veterinarians, Proceedings 335*. pp. 301–317. Post Graduate Foundation in Veterinary Science, Sydney.

Marsh H, O'Shea TJ, Reynolds JE (2011) *Ecology and Conservation of Sirenia: Dugongs and Manatees*. Cambridge University Press, Cambridge.

Marsh H, Grech A, McMahon K (2018) Dugongs: Seagrass community specialists. In *Seagrasses of Australia*. (Eds A Larkum, G Kendrick and P Ralph) pp 629–661. Springer, Cham.

Marsh H, Albouy C, Arraut E, Castelblanco-Martínez DN, Collier C, Edwards H, James C, Keith-Diagne L (2022) How might climate change affect the ethology and behavioral ecology of dugongs and manatees? In *Ethology and Behavioral Ecology of Sirenians*. (Ed H Marsh) pp. 351–406. Springer, Berlin.

Marshall CD, Sarko DK, Reep RL (2022) Morphological and sensory innovations for an aquatic lifestyle. In *Ethology and Behavioral Ecology of Sirenians*. (Ed H Marsh) pp. 19–65. Springer, Berlin.

Morgan UM, Xiao L, Hill BD, O'Donoghue P, Limor J, Lal A, Andrew Thompson RC (2015) Detection of the *Cryptosporidium parvum* "Human" genotype in a dugon *Dugong dugon*). *The Journal of Parasitology* **86**, 1352–1354.

Nielsen KA, Owen HC, Mills PC, Flint M, Gibson JS (2013) Bacteria isolated from dugongs (*Dugong dugon*) submitted for postmortem examination in Queensland, Australia, 2000–2011. *Journal of Zoo and Wildlife Medicine* **44**, 35–41. doi:10.1638/1042-7260-44.1.35

Norima L, Aida M, Zaidin A, Syed Abdullah SAK, Tamimi MAA, Sarol K, Hanafi H, Noordin MM, Mazlima M (2018) First report of *Madurella mycetomi* isolate that induced systemic pulmonary and myocardium mycosis with verminious enteritis in *Dugong dugon*, of Mersing, Johore. *Malaysian Journal of Veterinary Research* **9**, 30–39.

O'Shea TJ, Beck CA, Hodgson AJ, Keith-Diagne L, Marmontel M (2022) Social and reproductive behaviors. In *Ethology and Behavioral Ecology of Sirenians*. (Ed. H Marsh) pp. 101–154. Springer, Berlin.

Owen H, Gillespie A, Wilkie I (2012) Postmortem findings from dugong (*Dugong dugon*) submissions to The University of Queensland: 1997–2010. *Journal of Wildlife Diseases* **48**, 962–970. doi:10.7589/2011-03-080

Owen HC, Flint M, Limpus CL, Palmieri C, Mills PC (2013) Evidence of sirenian cold stress syndrome in dugongs *Dugong dugon* from southeast Queensland, Australia. *Diseases of Aquatic Organisms* **103**, 1–7. doi:10.3354/dao02568

Pilcher NJ, Adulyanukosol K, Das H, Davis P, Hines E, Marsh H, Ponnampalam L, Reynold J (2017) A low-cost solution for documenting distribution and abundance of endangered marine fauna and impacts from fisheries. *PLoS One* **12**(12), e0190021. doi:10.1371/journal.pone.0190021

Ponnampalam LS, Keith-Diagne LW, Marmontel M, Marshall CD, Reep RL, Powell J, Marsh H (2022) Historical and current interactions with humans. In *Ethology and Behavioral Ecology of Sirenians*. (Ed. H Marsh) pp. 299–349. Springer, Berlin.

Preen A (1989) Observations of mating behaviour in dugongs (*Dugong dugon*). *Marine Mammal Science* **5**, 382–387. doi:10.1111/j.1748-7692.1989.tb00350.x

Preen A (1995) Diet of dugongs: are they omnivores? *Journal of Mammalogy* **76**, 163–171. doi:10.2307/1382325

Preen A (2004) Distribution, abundance and conservation status of dugongs and dolphins in the dolphins in the southern and western Arabian Gulf. *Biological Conservation* **118**, 205–218. doi:10.1016/j.biocon.2003.08.014

Smoll L, Beard L, Lanyon J (2020) Osmoregulation and electrolyte balance in a fully marine mammal, the dugong (*Dugong dugon*). *Journal of Comparative Physiology. B, Biochemical, Systemic, and Environmental Physiology* **190**, 139–148. doi:10.1007/s00360-019-01250-8

Society for Marine Mammalogy (2023) *Dugong dugon* (Dugong). https://marinemammalscience.org/facts/dugong-dugon (accessed 24 September 2023).

Spain AV, Heinsohn GE (1973) Cyclone associated feeding changes in the dugong (Mammalia: Sirenia). *Mammalia* **37**, 678–680.

Species360 (2023) Zoological Information Management System (ZIMS). Minneapolis, USA. www.species360.org.

Underwater World Singapore (2013) Facebook, 29 April 2013 (accessed 26 September 2021).

von Fersen L, Walb R (Eds) (2018) *EAZA Best Practice Guidelines: Antillean Manatee* (Trichechus manatus manatus). European Association of Zoos and Aquaria, Tiergarten Nürnberg, Germany.

Wakai Y, Hasegawa K, Sakamoto S, Asano S, Watanabe G, Taya K (2002) Annual changes of urinary progesterone and estradiol-17β of the dugong (*Dugong dugon*) in captivity. *Zoological Science* **19**, 679–682. doi:10.2108/zsj.19.679

Whiting SD (2002a) Dive times for foraging dugongs in the Northern Territory. *Australian Mammalogy* **23**, 167–168. doi:10.1071/AM01167

Whiting SD (2002b) Rocky reefs provide foraging habitat for dugongs in the Darwin region of Northern Australia. *Australian Mammalogy* **24**, 147–150. doi:10.1071/AM02147

Whiting SD (2008) Movements and distribution of dugongs (*Dugong dugon*) in a macro-tidal environment in northern Australia. *Australian Journal of Zoology* **56**, 215–222. doi:10.1071/ZO08033

Wong AW, Bonde RK, Siegal-Willott J, Stamper MA, Colee J, Powell JA, Reid JP, Deutsch CJ, Harr KE (2012) Monitoring oral temperature, heart rate, and respiration rate of West Indian manatees (*Trichechus manatus*) during capture and handling in the field. *Aquatic Mammals* **38**, 1–16. doi:10.1578/AM.38.1.2012.1

Woolford L, Franklin C, Whap T, Loban F, Lanyon JM (2015) Pathological findings in wild harvested dugongs *Dugong dugon* of central Torres Strait, Australia. *Diseases of Aquatic Organisms* **113**, 89–102. doi:10.3354/dao02825

# 12
# Rodents

*Stephen M. Jackson, Chris R. Dickman, Julie Old, Frances Hulst and Wendy Gleen*

## 1. INTRODUCTION

The rodents (Order Rodentia) comprise a highly diverse group of mammals comprising over 2475 species in 34 families. The group includes animals such as beavers, porcupines, squirrels, flying squirrels, gophers, agoutis, chinchillas, coypu, mole-rats, 'true' rats, mice and the capybara, which is the largest, with new species being described frequently (Wilson *et al.* 2016, 2017). All Australian native rodents are in the family Muridae and are only relatively recent arrivals to the continent (Godthelp 1999; Aplin 2006; Baker and Gynther 2023). Ancestors of the 'old endemics' (members of the tribe Hydromyini) arrived from Asia between 6 and 8 million years ago, while the 'new endemics' (members of tribe Rattini that contains species of the genus *Rattus*) arrived approximately 1 million years ago. Despite only one family of rodents occurring in Australia, there is a large range in their size, physical appearance and ecological niches (Table 12.1; Baker and Gynther 2023).

Rodents, especially smaller species such as the murids, have not been held widely in zoos. This may reflect their perceived lack of charisma and their negative image as carriers of disease. For example, the introduced black rat *Rattus rattus* played a key role in the bubonic plague or Black Death in which the fleas it carried contained the bacterium *Yersinia pestis* that killed ~20 million people between 1346 and 1352 (McEvedy 1988; Barnett 2001). A further major plague in 1665 killed some 100 000 people. Since then, other plague outbreaks have occurred, including a small outbreak in January 1900 in Sydney (Banks *et al.* 2011), with others subsequently in Africa, South America and in India as recently as 1994 (Ashton 1986; Barnett 2001). The image of rodents in Australia is further tarnished by localised or widespread population explosions of the introduced brown or Norway rat (*Rattus norvegicus*) and house mouse (*Mus musculus*) and occasionally of native species, including the long-haired rat (*Rattus villosissimus*) and canefield rat (*Rattus sordidus*), that cause crop damage and economic and social upheavals (Le Souef and Burrell 1926; Plomley 1972). Despite the stigma that surrounds murid rodents, they are a diverse and highly interesting group of animals that are often very appealing (e.g. stick-nest rats, rock-rats, water-rats, tree-rats and hopping-mice). In recent years attempts have been made to introduce common names that avoid use of the words 'mouse' and 'rat' to improve public perception of native rodents. Names derived from Indigenous languages, such as rakali (water-rat: *Hydromys chrysogaster*), yirrkoo (water-mouse: *Xeromys myoides*) and mootit (bush rat: *Rattus fuscipes*), have been proposed, but uptake is slow.

Native Australian rodents have often been held in research institutions with bush rats being flown to the USA where they successfully bred as early as 1955 (Horner and Taylor 1958; Finlayson 1960) and (*Notomys alexis*) and plains mice (*Pseudomys australis*) have been held as laboratory animals since the mid-1960s (Smith *et al.* 1972; Happold 2021). Subsequently, rodents have been held in various Australian zoos (including Adelaide Zoo, Alice Springs Desert Park, Taronga Zoo and the Australian Wildlife Park) where they have proven to be popular exhibits. Species held include water-rats, stick-nest rats (*Leporillus conditor*), bush rats, swamp rats (*Rattus lutreola*), several species of *Pseudomys*, hopping-mice, tree-rats, rock-rats and several species of *Rattus* (Lees and Johnson 2002). Despite their often-poor representation in zoological collections, there is a great need to increase public education about Australian native rodents, which have suffered the highest rate of extinction and decline of any group of native mammals over the past 230 years (Burbidge and McKenzie 1989; Dickman *et al.* 2000).

Australian native rodents are beautiful mammals that fulfil essential functions across many different ecosystems and need to be protected (Dickman 1999; Woinarski *et al.* 2014; Roycroft *et al.* 2021).

## 2. TAXONOMY

### 2.1. Nomenclature

The family Muridae has ~816 species in 155 genera worldwide (Wilson *et al.* 2017). Within Australia, all 67 described species of native rodents are placed in the subfamily Murinae, which is further divided into two tribes: the Hydromyini (58 species) and Rattini (9 species) (Jackson and Groves 2015) (Table 12.1). Several new species await description.

**Australian rodents**

Class: Mammalia
Order: Rodentia
Family: Muridae
Subfamily: Murinae
Tribe: Hydromyini
Genus and species: 58 species in 13 genera
Tribe: Rattini
Genus and species: 9 species in 1 genus

### 2.2 Subspecies

See Jackson and Groves (2015).

### 2.3 Recent synonyms

Synonyms can be found in Jackson and Groves (2015).

**Table 12.1.** Australian native rodents including their bodyweight and conservation status.

*Also occurs in New Guinea and/or surrounding islands. CR = Critically Endangered, EX = Extinct, EN = Endangered, LC = Least Concern, NT = Near Threatened, NT (CD) = Near Threatened–Conservation Dependent, VU = Vulnerable.
Derived from Woinarski *et al.* (2014) and Baker and Gynther (2023).

| Common name | Scientific name | Weight (g) | IUCN status |
|---|---|---|---|
| **Family Muridae** | | | |
| **Subfamily Murinae** | | | |
| **Tribe Hydromyini** | | | |
| White-footed tree-rat | *Conilurus albipes* | c. 200 | EX |
| Capricorn rabbit-rat | *C. capricornensis* | Unknown | EX |
| Brush-tailed tree-rat (*) | *C. penicillatus* | 102–242 | VU |
| Water-rat (*) | *Hydromys chrysogaster* | 340–1275 | LC |
| Central short-tailed mouse | *Leggadina forresti* | 13–30 | LC |
| Northern short-tailed mouse | *L. lakedownensis* | 15–20 | LC |
| Lesser stick-nest rat | *Leporillus apicalis* | c. 150 | EX |
| Greater stick-nest rat | *L. conditor* | 180–450 | NT (CD) |
| Broad-toothed rat | *Mastacomys fuscus* | 97–145 | NT |
| Grassland melomys | *Melomys burtoni* | 26–124 | LC |
| Cape York melomys | *M. capensis* | 45–116 | LC |
| Fawn-footed melomys | *M. cervinipes* | 45–126 | LC |
| Bramble Cay melomys | *M. rubicola* | 78–164 | EX |
| Black-footed tree-rat | *Mesembriomys gouldii* | 580–882 | VU |
| Golden-backed tree-rat | *M. macrurus* | 207–330 | NT |
| Spinifex hopping-mouse | *Notomys alexis* | 27–45 | LC |
| Short-tailed hopping-mouse | *N. amplus* | c. 100 | EX |
| Northern hopping-mouse | *N. aquilo* | 31–47 | VU |
| Fawn hopping-mouse | *N. cervinus* | 30–50 | NT |
| Dusky hopping-mouse | *N. fuscus* | 26–52 | VU |
| Long-tailed hopping-mouse | *N. longicaudatus* | c. 100 | EX |
| Big-eared hopping-mouse | *N. macrotis* | c. 45 | EX |

| Common name | Scientific name | Weight (g) | IUCN status |
|---|---|---|---|
| Mitchell's hopping-mouse | *N. mitchellii* | 40–60 | LC |
| Darling Downs hopping-mouse | *N. mordax* | Unknown | EX |
| Broad-cheeked hopping-mouse | *N. robustus* | Unknown | EX |
| Tree mouse (*) | *Pogonomys* sp. | 40–85 | LC |
| Ash-grey mouse | *Pseudomys albocinereus* | 14–40 | LC |
| Silky mouse | *P. apodemoides* | 16–22 | LC |
| Long-eared mouse | *P. auritus* | Unknown | EX |
| Plains mouse | *P. australis* | 30–65 | VU |
| Bolam's mouse | *P. bolami* | 9–16 | LC |
| Kakadu pebble-mouse | *P. calabyi* | 12–21 | NT |
| Western pebble-mouse | *P. chapmani* | 10–15 | LC |
| Northern delicate mouse (*) | *P. delicatulus* | 6–15 | LC |
| Desert mouse | *P. desertor* | 11–35 | LC |
| Smoky mouse | *P. fumeus* | 38–90 | VU |
| Blue-grey mouse | *P. glaucus* | 25–30 | EX |
| Gould's mouse | *P. gouldii* | 27–76 | VU |
| Eastern chestnut mouse | *P. gracilicaudatus* | 45–118 | LC |
| Sandy inland mouse | *P. hermannsburgensis* | 9–15 | LC |
| Long-tailed mouse | *P. higginsi* | 50–90 | LC |
| Central pebble-mouse | *P. johnsoni* | 7–20 | LC |
| Eastern delicate mouse | *P. mimulus* | 6–10 | LC? |
| Western chestnut mouse | *P. nanus* | 18–56 | LC |
| New Holland mouse | *P. novaehollandiae* | 20–26 | VU |
| Western mouse | *P. occidentalis* | 33–53 | NT |
| Hastings River mouse | *P. oralis* | 70–110 | VU |
| Eastern pebble-mouse | *P. patrius* | 10–19 | LC |
| Western delicate mouse | *P. pilbarensis* | 6–10 | LC? |
| Heath mouse | *P. shortridgei* | 55–90 | NT |
| Giant white-tailed rat (*) | *Uromys caudimaculatus* | 500–820 | LC |
| Pygmy white-tailed rat | *Uromys hadrourus* | 172–230 | NT |
| Water-mouse | *Xeromys myoides* | 32–56 | VU |
| Common rock-rat | *Zyzomys argurus* | 26–55 | LC |
| Arnhem Land rock-rat | *Z. maini* | 65–140 | VU |
| Carpentarian rock-rat | *Z. palatalis* | 65–161 | CR |
| Central rock-rat | *Z. pedunculatus* | 70–150 | CR |
| Kimberley rock-rat | *Z. woodwardi* | 80–190 | LC |
| **Tribe Rattini** | | | |
| Dusky rat | *Rattus colletti* | 50–213 | LC |
| Bush rat | *R. fuscipes* | 40–225 | LC |
| Cape York rat (*) | *R. leucopus* | 95–207 | LC |
| Swamp rat | *R. lutreola* | 50–200 | LC |
| Maclear's rat | *R. macleari* | c. 200 | EX |
| Bulldog rat | *R. nativitatis* | 250–300 | EX |
| Canefield rat (*) | *R. sordidus* | 50–260 | LC |
| Pale field rat | *R. tunneyi* | 42–85 | LC |
| Long-haired rat | *R. villosissimus* | 54–280 | LC |

## 2.4 Other common names

See Baker and Gynther (2023).

# 3. NATURAL HISTORY

## 3.1 Bodyweight and morphometrics

The Australian native rodents range in size from less than 10 g for some *Pseudomys* species to 1275 g for the water-rat (Baker and Gynther 2023). The morphometrics of individual Australian rodents can be found in Watts and Aslin (1981) and Baker and Gynther (2023).

## 3.2 Distribution and habitat

Australian native rodents occupy all habitats, including deserts, rainforest, woodland, grassland and alpine areas. They also occupy nearly every available niche, including aquatic, terrestrial or arboreal. Further details on the distributions and habitats of individual species are given in Baker and Gynther (2023).

## 3.3 Conservation status

Many species of Australian rodents have become extinct or threatened since European settlement 230 years ago. Of the 67 species, over 30% are extinct or threatened: 13 are extinct, 2 critically endangered and 11 vulnerable to extinction (Table 12.1). The Bramble Cay melomys, which was last recorded in 2009, has been declared the first mammal to be driven to extinction by the effects of climate change (Woinarski *et al.* 2014; Gynther *et al.* 2016). Recently, the Shark Bay mouse (*Pseudomys fieldi*) has been proposed to be a junior synonym of the extinct Gould's mouse (*Pseudomys gouldii*). Based on this, Gould's mouse may no longer be considered extinct (Roycroft *et al.* 2021).

## 3.4 Diet in the wild

Most Australian rodents are omnivorous, consuming plant matter such as seeds, leaves, roots, fungi, pollen and fruits with invertebrates also an important part of their diet, particularly during spring and summer when most species reach a peak in reproduction (Watts 1970; Murray and Dickman 1994; Murray *et al.* 1999). Some species, such as the water-rat and water-mouse, consume mostly freshwater and saltwater (respectively) invertebrates and the water-rat also eats fish, frogs and birds (Table 12.2).

**Table 12.2.** Wild diets of the different genera of Australian rodents.

| Genus | Diet | Reference |
|---|---|---|
| **Murinae** | | |
| **Hydromyini** | | |
| *Conilurus* | Grasses, seeds, insects - herbivorous/omnivorous | 1, 2 |
| *Hydromys* | Fish, aquatic insects, birds, crustaceans, bivalves - carnivorous | 3 |
| *Leggadina* | Seeds, vegetation, arthropods - omnivorous | 4–6 |
| *Leporillus* | Leaves, stems, seeds - herbivorous | 7–12 |
| *Mastacomys* | Sedges, fungus, grass, seeds, sphagnum moss - herbivorous | 13, 14 |
| *Melomys* | Fruits, seeds, grass, insects - omnivorous | 1, 15 |
| *Mesembriomys* | Fruits, seeds, grass, flowers, invertebrates, molluscs - omnivorous | 1 |
| *Notomys* | Seeds, arthropods, fungi, grass, roots, flowers - omnivorous | 16, 17 |
| *Pogonomys* | Fruits, leaves - herbivorous | 1 |
| *Pseudomys* | Fungi, arthropods, seeds, grass, roots, nectar and pollen, ferns - omnivorous | 17–19 |
| *Uromys* | Fruits, nuts, fungi, insects, small reptiles, amphibians, crustaceans, bird eggs - omnivorous | 1 |
| *Xeromys* | Small marine crustaceans - crabs, polyclads, pulmonates, bivalves - carnivorous | 20, 21 |
| *Zyzomys* | Seeds, some fruits, fungus, flowers, stems, leaves, insects - mostly herbivorous | 22–25 |
| **Rattini** | | |
| *Rattus* | Insects, seeds, fruit, grass, roots, fungus - generalist fungivores | 14, 22–28 |

References: (1) Strahan (1995); (2) Firth *et al.* (2005); (3) Woollard *et al.* (1978); (4) Watts (1972); (5) Read (1984); (6) Morton *et al.* (1994); (7) Finlayson (1941); (8) Robinson (1975); 9) Watts and Eves (1976); (10) Watts and Aslin (1981); (11) Copley (1988); (12) Copley (1999); (13) Green (1968); 14) Carron *et al.* (1990); 15) Leung (1999a); (16) Murray and Dickman (1994); (17) Murray *et al.* (1999); (18) Norton (1987a); (19) Luo *et al.* (1994); (20) Magnusson *et al.* (1976); (21) Van Dyck (1996); (22) Watts (1977); (23) Begg and Dunlop (1980); (24) Begg and Dunlop (1985); (25 Nano *et al.* (2003); (26) Leung (1999b); (27) Norton (1987b); (28) Dickman and Happold (2022).

## 3.5 Longevity

### 3.5.1 Wild

The limited information on the longevity of rodents in the wild suggests they typically live 1-2 years (Table 12.3).

### 3.5.2 Captivity

In captivity, rodents generally have short life spans, although longer than in the wild, and typically live more than 24 months, with some species living to over 60 months (Table 12.3).

### 3.5.3 Methods used to determine the age of adults

It is difficult to determine the age of live rodents once they reach adulthood. Reasonably accurate estimates have been determined using eye-lens weight (e.g. Berry and Truslove 1968; Myers *et al.* 1977); however, this is unsuitable for use in live animals. A technique using X-rays to determine age using bone fusion and linear measurements of bone dimensions for small mammals may be useful (Myers 1978). Measuring the incisor width of live animals has been used, with studies of bush rats utilising both the upper incisors to determine age and size (Freeland 1972; Press 1987). Tooth wear has been used to assign age classes to different stages of wear of the molar teeth (Breakey 1963; Lidicker 1966; Bellamy *et al.* 1973).

Most of these methods are time-consuming and limited in the range over which age can be estimated, particularly as most rodent species only live for 2-3 years in captivity. In most cases, it is adequate to use growth curves based on body mass to determine the age of juveniles and then divide the population into adult reproducing animals and subadult non-reproducing animals.

**Table 12.3.** Longevity (months) of different genera of rodents in the wild and in captivity.

| Genus | Wild | Captivity | Reference |
|---|---|---|---|
| **Hydromyini** | | | |
| *Conilurus* | - | 45-60 | 1, 2 |
| *Hydromys* | - | 24-70 | 1, 3 |
| *Leggadina* | - | 12-18 | 1 |
| *Leporillus* | - | 24-55 | 1, 3 |
| *Mastacomys* | 48 | 15-24 | 1, 4, 5 |
| *Melomys* | 24 | 36+ | 1, 6 |
| *Mesembriomys* | - | 24-68 | 1, 3 |
| *Notomys* | 24 | 36-66 | 1, 7 |
| *Pseudomys* | 18-24 | 36-60 | 1, 2, 8, 9 |
| *Uromys* | - | 60+ | 1 |
| *Zyzomys* | 12-24 | 60 | 1, 2, 10 |
| **Rattini** | | | |
| *Rattus* | 12-24 | 24-36 | 11-14 |

References: (1) Watts (1982a); (2) Strahan (1995); (3) W. Gleen pers. obs.; (4) Calaby and Wimbush (1964); (5) Happold (2015); (6) Smith (1985); (7) C. Dickman pers. obs.; (8) Brandle and Moseby (1999); (9) Cockburn (1981a); 10) Lloyd (1999); (11) Lunney (1978) in Taylor and Calaby (1988a); (12) Watts (1982b); (13) Robinson (1987); (14) Happold (1989).

## 4. HOUSING REQUIREMENTS

## 4.1 Enclosure design

Most rodents can be exhibited within enclosures that mimic their natural habitat by using the appropriate substrate and furnishings. The enclosure surfaces need to be lined with resistant material such as tin, steel or glass (or other non-toxic chewing resistant material) to stop animals chewing through the walls and escaping. As with other small mammal species, it is important to avoid crevices or small holes or hollows in enclosure walls where they might trap themselves. The enclosures also need to be predator-proof.

Enclosures with a floor space of ~100-200 × 100-200 × 40-60 cm high are adequate for smaller terrestrial animals. Arboreal species such as tree-rats, fawn-footed melomys and prehensile-tailed rats should be given additional vertical space (1.5-2.0 m) to allow a variety of climbing structures to be provided and regularly changed. These species should be given several vertical and cross branches to allow them to move freely above ground and the feed platform should also be elevated so they do not have to come to the ground (Table 12.4).

Greater stick-nest rats should be given larger areas (e.g. at least 400 (L) × 500 (W) cm) so that branches can be supplied for the building of stick nests, which provide significant behaviour enrichment. Watching the construction of these impressive nests and the finished product is entertaining, with the animals' engineering prowess making a very good nocturnal exhibit if the lighting is adequate.

The desert rodents, such as most of the hopping-mice, some *Pseudomys* and some rock-rats, can readily be held in enclosures with a substrate of sand and rocks with grass tussocks. Ensure that rocks are placed on the enclosure floor so they do not collapse into burrows dug underneath. Rock-rats prefer to move quietly and will therefore avoid dried leaf litter in favour of rocks, so this should be

**Table 12.4.** Minimum areas of enclosures recommended for breeding pairs of animals of different genera of Australian rodents.

Derived from Watts (1979a, 1980, 1982a), Calver *et al.* (1991), Brisbane (1998) and Kerle and Fleming (2024).

| Genus | Area (L × W × H) (cm) | Additional floor area for each extra animal (cm) |
|---|---|---|
| **Murinae** | | |
| **Hydromyini** | | |
| *Conilurus* | 150 × 150 × 180 | 75 × 75 |
| *Hydromys* | 300 × 300 × 100 water area = 200 × 100 × 50 | 200 × 200 |
| *Leggadina* | 100 × 100 × 40 | 25 × 25 |
| *Leporillus* | 400 × 500 × 100 | 100 × 100 |
| *Mastacomys* | 100 × 100 × 40 | 40 × 40 |
| *Melomys* | 100 × 100 × 100 | 50 × 50 |
| *Mesembriomys* | 245 × 245 × 180 | 50 × 50 |
| *Notomys* | 100 × 100 × 40 | 20 × 20 |
| *Pogonomys* | 100 × 100 × 180 | 50 × 50 |
| *Pseudomys* | 100 × 100 × 40 | 20 × 20 |
| *Uromys* | 245 × 245 × 180 | 100 × 100 |
| *Xeromys* | 100 × 100 × 40 | 50 × 50 |
| *Zyzomys* | 100 × 100 × 50 | 40 × 40 |
| **Rattini** | | |
| *Rattus* | 100 × 100 × 40 | 50 × 50 |

considered during enclosure design (Lloyd 1999). Some species, such as long-haired rats, hopping-mice and *Pseudomys*, have well-developed underground tunnels in the wild (Pye 1991; Predavec and Dickman 1994; Diete *et al.* 2014) so the enclosure could make a feature by representing a cut away version of a tunnel system. Small rocks should be provided for pebble-mice so that these can be used as part of the animals' burrow designs.

Water-rats should be exhibited within an enclosure with a land area of at least 300 (L) × 200 (W) cm plus a pool of water at least 200 × 100 × 50 cm deep, so they can swim, forage and play. As they spend a lot of time in the nest box, a viewing window into the nest box with appropriate illumination (not too bright) may also be possible.

## 4.2 Holding area design

Holding areas for Australian rodents can be very simple in design, in most cases consisting of a simple enclosure with the walls and floor made of chew-resistant and non-toxic material such as glass, tin or hard wood (ideally lined with tin or other suitable chew-resistant and non-toxic metal) rather than soft ply. Ideally, the roof should be made of mesh and be fully removable and the front should be made of glass to allow easy viewing. Plastic tubs are typically not suitable because rodents can often chew through them. Larger climbing species, such as tree-rats, can be held in enclosures constructed from wire mesh. It is not advisable to use wire mesh enclosure sides in tall enclosures for smaller species such as *Pseudomys* and *Notomys* as they risk injury if they fall from a height.

Water-rats should have access to water for swimming at all times and in holding areas this may consist of a dry area and a pond ~100 cm (L) × 100 cm (W) × 50 cm (H) in which their food can be provided.

Nest areas suitable for most species can be provided by the addition of nest boxes ~10 × 10 × 10 cm – 20 × 20 × 15 cm, with a suitably sized entry hole or tunnel and a sliding lid for easy access by animal caretakers. A dividing wall is useful to allow introductions to occur with fewer problems, particularly in highly aggressive species such as rock-rats. When providing nesting areas, ensure the height is enough for the animals to sit up and ensure there are perforations to allow airflow (and escape) to avoid overcrowding and suffocation.

Nesting materials that can be provided include coconuts, cardboard tubes, shredded paper, perforated plastic cylinders and rolls of decorticated eucalypt bark (e.g. ribbon gum) with the diameter depending on the species.

## 4.3 Spatial requirements

Australian native rodents have been held in a wide variety of enclosure sizes. Examples are given below.

- Short-tailed mice, *Pseudomys*, rock-rats and hopping-mice have been bred successfully in enclosures 38 cm (L) × 25 cm (W) × 15 cm (H) to 55 × 38 × 20 cm; with tree-rats and white-tailed rats being held successfully in areas 90 × 65 × 40 cm (Watts 1979a).
- 34 species of native rodents, including tree-rats, short-tailed mice, stick-nest rats, broad-toothed rats, hopping-mice, *Pseudomys*, rock-rats, water-rats, melomys, white-tailed rats and true rats (Watts 1980), have been placed in enclosures 40 cm (L) × 25 cm (W) × 20 cm (H) made of clear plastic to 91 × 61 × 33 cm wooden and glass enclosures. Animals

held by this author were previously placed in larger enclosures with a floor area of 1.2 × 2.4 m; however, these were reduced due to a lack of evidence that the larger enclosure sizes resulted in better breeding success (as the smaller sizes still allowed pairs to live together with no problems).

- Rodents of the genera *Leggadina* and *Pseudomys* have been held in enclosures 20 cm (L) × 40 cm (W) × 40 cm (H); while rock-rats, hopping-mice and short-tailed mice have been held in enclosures 23 × 30 × 53 cm (Calver *et al.* 1991).
- Rock-rats have been held in enclosures 90 cm (L) × 40 cm (W) × 45 cm (H) with nest boxes, native grasses, red sand rocks and branches (Brisbane 1998).

Despite the success of these early studies in maintaining rodents in small enclosures, larger enclosures are encouraged because the animals can display a fuller range of natural behaviours and should be used wherever possible. For example, stick-nest rats can readily be held in smaller enclosures, but are much more active if held in larger enclosures (20–25 $m^2$) where they can be provided with nest-building opportunities through the provision of branches (W. Gleen pers. obs.). A summary of minimum recommended enclosure sizes can be found in Table 12.4.

## 4.4 Position of enclosures

Native rodents are often housed indoors and the enclosures should be positioned out of direct sunlight to prevent overheating. Good ventilation should be promoted to prevent ammonia build-up from urine-soiled substrate, although draughts should be avoided to prevent fluctuating temperatures. Some facilities use outdoor enclosures with natural light cycles (potentially to trigger breeding).

## 4.5 Weather protection

Rodents are usually housed inside weatherproof buildings. Outdoor enclosures should be provided with adequate weather protection and refuges from direct sunlight, excess heat, cold, rain and wind and have good drainage.

## 4.6 Temperature requirements

The recommended temperature range is 15–25°C for most species; however, the broad-toothed rat, which lives in subalpine and alpine environments, should ideally be kept at lower temperatures of ~10–20°C. Ample bedding material should be provided as additional thermal insulation. In the case of social species, such as some hopping-mice and *Pseudomys*, co-housing several animals will allow huddling to keep warm.

## 4.7 Substrate

Different substrates can be used, including soil, paper, sawdust, sand, or leaf litter and paper-based kitty litter. For exhibit purposes, the species should be given a substrate most appropriate to its wild habitat, such as sand or leaf litter. Fine sand is recommended because bricklayers sand is coarse and can be abrasive to animals' feet (Williams 1990). A further advantage of using fine sand is that it can be sieved daily for easy cleaning. In off-exhibit holding enclosures, substrates that are easily changed, such as sawdust, paper, or kitty litter, should be used.

Burrowing species can be provided with substrate that allows this behaviour in a confined section of the enclosure so that there are safe areas for staff to stand on and to clean and provide food and water stations.

## 4.8 Nest boxes

Nesting areas should be provided to allow rodents somewhere to retreat to sleep, raise young or feed. Nest boxes can be provided; however, other less elaborate apparatus can be used, especially for the smaller species, including plastic or thick cardboard tubes, hollow logs, pieces of bark, cardboard boxes and shredded paper. If plastic or cardboard or plastic tubes are used, ensure they have a large enough diameter or are perforated with small holes to allow airflow. Plastic tubes should also be regularly checked to ensure the rodents are not chewing and ingesting plastic.

## 4.9 Enclosure furnishings

Enclosure furnishings should be provided that are consistent with the ecology of the species being housed and which maximise the expression of natural behaviours. Desert species, such as rock-rats, should be provided with several rocks that the animals can climb over and potentially nest between (though great care needs to be taken that the rocks do not become dislodged and injure animals). In order to minimise injuries, rocks and other materials should be placed directly on the enclosure floor rather than on the substrate. For pebble-mice, smaller rocks that the animals can comfortably carry should be provided. Arboreal species such as tree-rats, prehensile-tailed rats and fawn-footed melomys should be given numerous vertical and cross branches that allow them to climb vertically throughout the enclosure. Water-rats should be provided with a pond where they can swim, demonstrate natural behaviour and keep their coats clean.

## 5. GENERAL HUSBANDRY

### 5.1 Hygiene and cleaning

Each enclosure should be cleaned every 1–2 days to remove faecal matter and uneaten food. Small enclosures can be spot-cleaned daily and given a full substrate clean weekly or more often if required. Cleaning includes inspecting the nest boxes for urine and faeces and cleaning them when required. Drinking water dishes should be cleaned daily and water bottles should be checked daily to make sure the nozzle is working properly and that the bottle is at least two-thirds full. When all individuals leave an enclosure, it should be scrubbed out using deodorising disinfectant and ideally rested before new animals enter. Large indoor enclosures (> 4 $m^2$) should have a substrate change every 3 months and the furniture should be inspected and cleaned or replaced with new structures when needed. Enclosure surfaces and nest boxes should be scrubbed and disinfected.

### 5.2 Record keeping

It is important to establish a system whereby the health, condition and reproductive status of the captive rodent population are routinely monitored and recorded:

- identification (all individuals should be identifiable)
- results of veterinary examinations and treatments administered
- behavioural changes or problems
- reproductive behaviour or condition
- bodyweights and measurements
- changes in diet and food consumption
- movements of individuals between enclosures or institutions
- births, with dam and sire if known
- deaths with necropsy results.

The collection of information on physical, physiological, and behavioural patterns of individual rodents can contribute greatly to the husbandry of most or all species. It also allows individual histories to be transferred to other institutions if required and facilitates a cooperative approach to data collection among institutions. ZIMS, the Zoo Information Management System (Species360 2023) can be used for recording general information on age, sex, parentage, births, identification, bodyweights, transfers and deaths, with separate ZIMS modules for veterinary/medical information and studbooks. These systems have been developed by Species360, which is part of the Conservation Planning Specialist Group (CPSG) software. ZIMS interfaces with programs, such as PMx, which provides demographic and genetic analysis tools to assist with the management of breeding programs (Lacy *et al.* 2011).

### 5.3 Methods of individual identification

#### 5.3.1 Passive Integrated Transponder (PIT) tags

PIT tags (microchips) can be placed subcutaneously between the scapulae in rodents over 10–12 g bodyweight. This is an excellent and permanent method of identification, but it can be expensive if there are many animals to be identified. PIT tags can track out along the injection site, so the entry wound can be sealed with tissue glue (Vetbond®) or similar fast-setting adhesive. Animals may need to be caught to confirm identification with a PIT tag reader, but in some cases the tag can be read through the wall of a hide or nest box.

#### 5.3.2 Ear tattooing

Ear-tattoos have been used successfully on small rodents (e.g. Lindner and Fuelling 2002) and, because they are minimally invasive, can be a useful way to identify individuals. The disadvantage of this technique is the animals need to be caught to be identified and the tattoos may become unclear over time.

#### 5.3.3 Ear tags

Small metal rodent ear tags have been used in medium to large-sized rodents, including stick-nest rats, but they can be pulled out and the resulting ear wound can become infected. When placing the tags, take care to avoid veins in the ear. A modified ear tag has been developed and used successfully on small mammals (Salamon and Klettenheimer 1994). However, ear tags are not recommended.

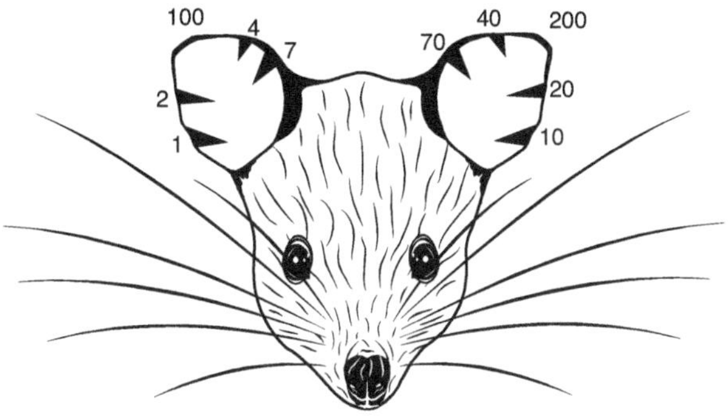


**Fig. 12.1.** Example of an ear notch numbering system to identify individual Australian rodents. Figure redrawn from C. Dickman.

### 5.3.4 Ear notching

Ear notching can be used to individually identify small rodents (Fig. 12.1). Care should be taken to clip small notches only in the margin of the ears rather than punching holes within the pinna because holes can get snagged on sticks or thin leaves such as those of the hummock grasses. This technique has been used routinely in field studies of several species (C. Dickman pers. obs.).

The ears are notched using surgical scissors that are sterilised (dipped in ≥ 70% ethanol and flamed or allowed to dry naturally) and a small piece of tissue ~1–2 $mm^2$ from around the ear margin is removed (Fig. 12.1; Waudby *et al.* 2022; C. Dickman pers. obs.). Care needs to be taken to avoid cartilage and any obvious blood vessels. The tissue can be stored in 100% ethanol for later DNA analysis if required. The clipping process takes ~30 s (quicker for a single clip) and rarely results in bleeding or obvious distress in the animals being processed (C. Dickman pers. obs.).

### 5.3.5 Hair bleaching

Hair bleaching could potentially be used as it has been trialled successfully on other small mammals such as antechinuses (see Chapter 4).

## 6. FEEDING REQUIREMENT

### 6.1 Captive diet

The wild diets of Australian native rodents suggest there are few strictly herbivorous species and that most are omnivorous, with a diet of seeds, grass and fungi that is supplemented with insects (Table 12.2). Some species, such as the water-rat and water-mouse, are almost entirely carnivorous. Captive diets should reflect these dietary adaptations. Wherever possible, diets should be developed with input from an animal nutritionist or veterinarian.

Diets should meet the behavioural and nutritional needs of the individuals and a wide variety of food items should be offered. A balanced commercial herbivore pellet (e.g. kangaroo/guinea-pig/rabbit pellet) may form the basis of an omnivorous diet with additional fresh vegetables, seeds, insects or egg depending on the level of omnivory. Compared with herbivore pellets, commercial rodent (mouse and rat) pellets may be too high in easily digested starch and sugar for native species. High-sugar items (e.g. dried fruit) should not be provided to reduce the risk of obesity and other health problems. High-fat items are also often favoured; for example, spinifex hopping-mice increase their body mass significantly when provided with ad lib sunflower seeds. Therefore, care should be taken with the volume and nutritional content of all dietary items provided (Stannard *et al.* 2018). The following are examples of possible diets that could be adapted for individual institution's use. The body condition of the animals should be closely monitored and the amount of food offered increased or decreased to ensure good body condition is maintained and all animals in an enclosure have access to sufficient food.

Although the provision of clean water is advisable for all species (e.g. water bottles), species such as the spinifex hopping-mouse and the sandy inland mouse have been shown to survive and rear young with no free water, instead obtaining it from their food, including dry seed (C. Dickman pers. obs.). If water is provided, it should be provided in very shallow dishes to prevent accidental drowning.

### 6.1.1 *Conilurus* (tree-rats)

Daily diet unknown. May be similar to that used for *Notomys*.

### 6.1.2 *Hydromys* (water-rat)

- Daily diet (per animal): ~100–120 g
  - 4 yabbies
  - or 1–2 medium-sized fish (e.g. pilchards or whiting)
  - or 2 day-old chicks
  - 4–5 mussels

Alternatively, a mix of whitebait, prawns, hard-boiled egg, mussels, fly pupae, meat mix (based on kangaroo or beef mince with hard-boiled egg and dog kibble) and yabbies.

There may be some seasonal variation in food consumption (e.g. an increase when animals are in reproductive condition and possibly a similar increase in colder periods).

- Enrichment (select from)
  - 10 locusts/crickets when available
  - 1 raw egg weekly
  - apple, corn, coarse grasses, rushes and grass seeds, mixed sprouted seed
  - 4–5 dog kibble pellets
  - 10 g cheese, peanuts or raw coconut

Water-rats have been held in captivity where they favoured insects, crabs, fish and lizards.

### 6.1.3 *Leggadina* (short-tailed mice)

Daily diet: unknown. May be similar to that used for *Notomys*.

### 6.1.4 *Leporillus* (stick-nest rats)

- Daily diet (per animal): 12–16 g
  - 75% plain canary seed
  - 20% white French millet
  - 5% hulled oats
- Supplement
  - 30 g mixed greens (e.g. endive, spinach, beans)
  - 10 g herbivore pellet
  - 3 g enrichment (e.g. peanut, parrot mix)

Changing the types of greens offered each day and changing around the enrichment items helps to ensure that the diet is balanced.

The diet for *Leporillus* needs to be low in sugar, so supplements need to be carefully managed.

  - sow thistle
  - saltbush (*Rhagodia candolleana*)
  - pigface (*Carpobrotus glaucescens*) and other coastal succulents
  - 3 cm carrot
  - 1 rat and mouse cube/herbivore pellet
  - soaked or sprouted seed

### 6.1.5 *Mastacomys* (broad-toothed rat)

- Daily diet (per animal): 40–50 g
  - birdseed or herbivore pellet
  - leafy vegetables
  - grasses and sedges
- Enrichment
  - soaked or sprouted seed

### 6.1.6 *Melomys* (melomys)

- Daily diet (per animal): 20–30 g
  - finch birdseed
  - fruit or vegetable
- Enrichment
  - soaked or sprouted seed
  - occasional insects

### 6.1.7 *Mesembriomys* (black-footed tree-rats)

- Daily diet (per animal):
  - ¼ apple
  - 10 carrots
  - 20 g lettuce
  - ¾ cup mixed seed, with low or no sunflower seed content
  - 20 g spinach or endive
- Enrichment
  - occasional insects
  - soaked or sprouted seed
  - hard-shelled nuts (e.g. walnuts)

### 6.1.8 *Notomys* (hopping-mice) and *Pseudomys* (mice)

- Daily diet (per animal):
  (Breeding season)
  - endive, spinach, broccoli, kale, silverbeet
  - 5 g millet or canary mix
  - 5 g fresh vegetation (grass, browse)
  - 5 g fruit (e.g. apple)
  - 5 g invertebrates (e.g. mealworms, moths, cockroaches)
  - fresh cut flowers (e.g. *Callistemon*, *Leptospermum*, *Acacia* sp.) as available
  - fungi (mushroom)

  (Non-breeding season)
  - 5–10 g high-fibre vegetables (broccoli, sweet potato, carrot)
  - 10 g coarse seed mix
  - 10 g fresh vegetation (grass, browse)
- Enrichment
  - seaberry saltbush (*Rhagodia candolleana*), *Acacia* sp. and *Banksia* sp.
  - fungi (mushrooms)
  - 1–2 rat or mouse cubes
  - 1 piece apple or corn on the cob
  - soaked or sprouted seed
  - sweet potato and carrot (small amounts)

### 6.1.9 *Uromys* (white-tailed rats)

Daily diet unknown, but should be similar to that used for *Rattus*, with large seeds and fruits included.

### 6.1.10 *Zyzomys* (rock-rats)

- Daily diet (per animal)
  - seed mix (50% finch, 25% budgerigar and 25% pigeon)
  - seeding grasses with stems attached (e.g. winter grass (*Poa* sp.), wild oats (*Avena fatua*), elephant grass (*Pennisetum purpureum*), fountain grass (*Pennisetum setaceom*), Rhodes grass (*Chloris guyana*), coastal spinifex (*Spinifex longifolius*), and native tussock grasses (*Damondia* sp.))
- Enrichment
  - seeds of *Acacia* sp. when available
  - insects, including beetles, butterflies, moths, large ants or termites, but also mealworms and cockroaches
  - small pieces of sweet potato, apple or carrot
  - green vegetation including salt-bush (*Atriplex* sp.), sea spinach (*Tetragonia decumbens*) and pig-face (*Carpobrotus virescens*)

Central rock-rats are prone to obesity, so the amount of food provided should be adjusted to maintain healthy body condition (Gaikhorst and Lambert 2009).

### 6.1.11 *Rattus* (rats)

- Daily diet (per animal)
  - 20 g seed
  - 10 g fruit (1 cm cube) (may include carrot)
  - 10 g greens (e.g. silverbeet, sow thistle, small-leaf spiderwort (*Tradescantia fluminensis*), spinach)
  - 2 mealworms
  - dog kibble or rat or mouse cube
- Enrichment
  - 1 almond (2–3 times per week)
  - woody plants to chew on (weekly)
  - *Banksia* flowers as available
  - soaked or sprouted seed

## 6.2 Supplements

If a balanced diet is offered with plenty of variety, no dietary supplements are needed.

## 6.3 Presentation of food

Food is easily provided in small dishes for most rodent species. Food for the water-rat can be placed next to or in the water body. Food, supplementary activity feeds and enrichment can be provided during the day to increase activity levels for optimal day-time display.

Because rodent teeth grow continuously, material suitable for gnawing should always be provided such as non-toxic branches or nuts with hard shells. However, avoid overfeeding nuts as their high nutrient content can lead to obesity.

# 7. HANDLING AND TRANSPORT

## 7.1 Timing of capture and handling

Captive rodents are most easily caught while they are asleep in their nest boxes. If held in a reverse light cycle nocturnal house, this is usually in the morning before the lights go off. Most species of rodents will retreat to a shelter when disturbed and can be accessed from there if needed at other times.

## 7.2 Capture equipment

Both small and large species are readily held in calico cloth bags. Ensure that the hems are on the outside of the bag to prevent the animal's toes becoming entangled in the cotton string. Rodents should not be left in bags on benches where they may roll off. Rodents should not be transported or held for any length of time in calico bags without a secondary container as they can chew through the bag in several minutes (W. Gleen pers. obs.).

## 7.3 Capture and physical restraint

Small rodents can be tipped directly from a nest box into a catching bag or, for the smallest species weighing < 20 g, caught with a bare hand or using a calico catching bag turned inside out. Once caught, tie the top of the bag with a small rope or draw string. Many rodents, especially rock-rats, hopping-mice and *Pseudomys*, should never be caught by the tail, as the tail skin is likely to deglove, necessitating surgical amputation to reduce the risk of infection and prevent the exposed section of the tail drying out and dropping off.

As a general rule it is better to not pick up any native rodents by the tail. It is safest to catch larger rodents in their nest box and tip them into a bag or use a small but strong calico net (to avoid toes being injured) or trap them in a large Elliott or cage trap). Mid-sized and smaller species, such as the stick-nest rat, may need to be trapped with an appropriately sized Elliott trap, as they can be difficult to catch if they have built a stick-nest with nest chambers that are inaccessible to human handlers.

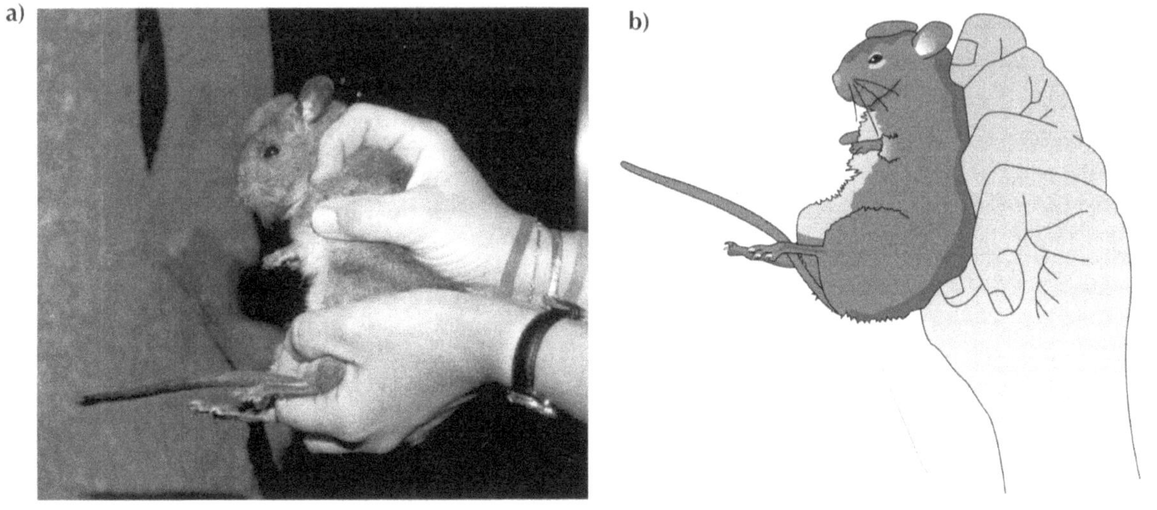


**Fig. 12.2.** Techniques for holding rodents between the index and middle fingers. Photo by Stephen Jackson.

The principles for holding all species of rodents, once they are caught, are similar to those used for holding small to medium-sized dasyurids (see Chapter 4). Once in the catching bag, the animal can be handled by holding its head down with one hand and using the other hand to gently grasp it over the shoulders so that the head is held securely between the index and middle fingers (Fig. 12.2a). An alternative method for smaller rodents (i.e. < 50 g) is to hold them by the scruff of the neck with the thumb and forefinger (Fig. 12.2b) ensuring they are not held too tightly, which may suffocate the animal. Many smaller species do not readily bite; stick-nest rats are renowned for their placid temperament. Larger species, such as tree-rats, white-tailed rats and the water-rat need to be handled with care as they are very strong and can inflict nasty bites. Tree-rats and white-tailed rats also have very sharp claws that can inflict painful lacerations if they are not handled carefully. Gloves may prevent lacerations from these larger species.

## 7.4 Weighing and examination

Once they have been caught, many rodents can be readily handled. The smallest species (weighing < ~10 g) can be weighed in plastic bags (that are not sealed or have several small holes at the top to allow air flow), using a fine-scale spring balance or electric scales to obtain more accurate bodyweights (1 g or ideally 0.1 g increments). Ideally, the bag can be weighed before capture to minimise handling and the weight of the animal calculated by subtracting the bag and animal weight. To reduce handling and stress, most examinations can be achieved by keeping the animal in the bag and only exposing the relevant parts of the body to be examined.

## 7.5 Release

Rodents should be released either directly into their nest box or on the ground within their enclosure in order to observe them moving.

## 7.6 Transport requirements

Most rodents are relatively easily transported. It is important to consider adequate containment against chewing, protection from extremes of temperature and physical trauma and to allow adequate ventilation. For short periods (1–2 h) individual rodents can be contained in a catching bag that is then placed inside a pet carrier or nest box with air holes and the entrance properly plugged. The outer container acts as a secondary barrier to escape if the rodent chews out of the bag and also provides physical protection from other objects that may crush them.

### 7.6.1 Transport box design

For smaller species, such as hopping-mice, *Pseudomys* and short-tailed mice, the transport box can be divided into two or more compartments for convenience. The box needs to be strong enough to ensure the rodents do not chew out, have secure doors, adequate ventilation and an option for providing a water bottle. Whenever rodents are transported by plane, they should be placed in a recommended wooden box using specifications outlined by the International Air Transport Association (IATA 2022).

### 7.6.2 Furnishings

Wood shavings or shredded paper should be provided to minimise animal movements during transport and to provide insulation against heat and cold.

### 7.6.3 Water and food

Provide water on all except short trips (i.e. 1–2 h) in cool weather. Water can be given in a small water bottle, but the animals should be familiar with their use before being transported. A small dish of food should be provided. Alternatively, high-water content food such as greens can be provided if water provision is difficult.

### 7.6.4 Animals per transport box

Only one animal of the larger species per box, but smaller/social species may be transported together in small groups or in separate compartments of the same box.

### 7.6.5 Timing of transportation

Whenever possible, transport should be done either in the early morning or overnight or in an air-conditioned vehicle so that animals do not become overheated.

### 7.6.6 Release from the transport box

Smaller species are generally released by picking them up and placing them in the enclosure or nest box, while larger species can be released by opening the box and allowing them to emerge from the box in their own time.

## 8. HEALTH REQUIREMENTS

### 8.1 Daily health observations

Each rodent should be observed daily for signs of injury or illness, with extra time needed for large groups of small rodents. Daily individual observation is particularly important after introductions to new animals or enclosures. The most appropriate time to do this is generally when the enclosure is being cleaned or when animals are being fed, as many of the larger rodent species will learn to approach for food; however, distant observation of the undisturbed animal is also beneficial.

Each animal should be checked for the following:

- coat/skin appearance
- discharges from eyes, ears, nose, mouth
- appetite, and eating behaviour (e.g. not coming readily for food, difficulty chewing)
- faeces: number, colour and consistency of pellets
- changes in demeanour, activity level and behaviour (e.g. separating from the group, huddled appearance, not coming out for food)
- injuries, swellings or lameness.

Any signs of injury or illness or unusual changes in behaviour should be documented and reported to a veterinarian. If an animal is found dead, make note of the circumstances and anything unusual or abnormal on the body or surrounding area. Carefully collect the body (using PPE where appropriate) and submit it for necropsy by a veterinarian.

### 8.2 Bodyweight and body condition score

#### 8.2.1 Bodyweight

Ideally bodyweight is recorded regularly (e.g. monthly or ideally weekly) and results compared with previous weights. The frequency of weight measurements will depend on factors such as the logistics required to weigh particular species/individuals and the presence of underlying health concerns. Trends in bodyweight may reflect seasonal changes in some species, but can also give an early indication of the animal's state of health.

#### 8.2.2 Body condition score

Body condition is best assessed by palpation of the muscle and subcutaneous cover over the shoulders, hips and spine.

### 8.3 Physical examination

Because close visual examination of each animal is difficult when in a group, all rodents should be closely examined each time they are caught. A physical examination can be conducted with an unrestrained rodent that is accustomed to handling or one that is restrained; however, most rodents seldom remain still for long and the procedure may be stressful. If any abnormalities are found or the animal is very stressed, it should be examined under anaesthesia.

The physical examination may include the following:

- fur/skin – check for alopecia, ectoparasites, fungal infections or wounds
- eyes
  - should be symmetrical, clear, bright with a smooth glistening surface and no discharges

- check for lumps or swelling over body and limbs
  - anus should be clean
- males
  - check testes for symmetry and consistency.

Testes are inguinal in juveniles and subadults of many species and therefore may not be palpable until adult size (C. Dickman pers. obs.).

A more detailed physical examination should be conducted by a veterinarian, with the animal under anaesthesia. Veterinarians should refer to Breed and Eden (2008) and Scheelings (2025) for detailed information on examination and medicine for Australian native rodents. All body areas and systems should be evaluated: cardiovascular; respiratory; oral cavity (including dentition), abdomen; musculoskeletal; genitourinary; ocular; skin and coat; ears; teats and mammary glands; and lymph nodes. Bodyweight and condition should be recorded and, if possible, heart rate; respiratory rate; and body temperature.

## 8.4 Chemical restraint

Rodents are usually anaesthetised using gaseous anaesthetics delivered to the unrestrained animal via an induction chamber or by facemask with the rodent manually restrained. Pre-anaesthetic fasting is not required and intubation is difficult and seldom undertaken (Breed and Eden 2008). Thermal support is required during anaesthesia.

## 8.5 Common health problems

A summary of the most common infectious diseases affecting captive native rodents is given in Table 12.5 and for common non-infectious diseases in Table 12.6. A detailed discussion of diseases in wild and zoo-based rodents can be found in Vogelnest and Woods (2008) and Vogelnest and Portas (2025). Some haematology and blood biochemistry values for native rodents are available in Old *et al.* (2005, 2007a) and Tulk *et al.* (2016).

**Table 12.5.** Common infectious diseases of captive rodents.

| Disease/ agent | Cause | Clinical signs | Diagnosis | Treatment | Prevention | Reference |
|---|---|---|---|---|---|---|
| **Ectoparasites** | | | | | | |
| Fur mites | Multiple genera | Itching, fur loss, self-trauma or may not cause any signs | Identification of mites on microscopy together with clinical signs | Topical miticide (e.g. ivermectin, selamectin, fipronil). For example, an appropriate dose of ivermectin diluted in propylene glycol has been effective in treating mites on spinifex hopping-mice | Quarantine new arrivals, manage stocking density. Change nest material and wash the nest boxes frequently. Straw used in native rodent enclosures should be stored in rodent-proof conditions to prevent cross-contamination | 1–6 |
| **Endoparasites** | | | | | | |
| | Trematodes (water- rats) | Usually none | Faecal flotation for eggs | Praziquantel | Faecal monitoring. Molluscs, insects, crustaceans, fish or amphibians act as intermediate hosts | 5, 7–9 |
| | Cestodes *Rodentolepsis fraterna* | Enteritis and death | Identification of eggs on faecal flotation and worms at necropsy | Praziquantel | Have been observed in captive spinifex hopping-mice | 10 |
| | Nematodes *Capillaria hepatica* | Ill-thrift, weight loss, death | Identification of worms and eggs in liver at necropsy | Anthelmintics may be tried, but destocking may be required | Indirect life cycle. Adult worms deposit eggs in the rodent liver, which can be consumed by predator or during cannibalism or released to substrate through decomposition after death | 11 |

| Disease/ agent | Cause | Clinical signs | Diagnosis | Treatment | Prevention | Reference |
|---|---|---|---|---|---|---|
| **Bacterial, viral and fungal** | | | | | | |
| Respiratory infections | Bacteria, mycoplasma, fungi including *Cryptococcus* spp. (yeasts) | Sneezing, nasal discharge, difficulty breathing, inappetence, weight loss | Clinical examination, X-ray, microscopy, necropsy | Supportive care, antimicrobial drugs according to cause | Quarantine new arrivals, ensure good husbandry especially temperature, humidity and ventilation | 12, 13 |
| Tyzzer's disease | *Clostridium piliformes* has been reported in spinifex hopping-mice | Short incubation period and is difficult to diagnose antemortem. Some animals may show diarrhoea and lethargy before death | Bacterial identification using special stains on histopathology | Supportive care, antibiotics | Prevent overcrowding, improve husbandry. *C. piliformes* is highly resistant in the environment, so euthanasia of infected or potentially asymptomatic individuals is recommended to prevent contamination of the enclosure. Enclosures should be thoroughly disinfected with 0.3% hypochlorite or other suitable disinfectant. All enclosure furniture should be destroyed | 4, 5, 14 |

References: (1) Robinson *et al.* (1978); (2) Smales *et al.* (1990); (3) Old *et al.* (2007b); (5) Breed and Eden (2008); (6) Scheelings (2025); (7) Smales (1997); (8) Obendorf and Smales (1985); (9) Smales and Cribb (1997); (10) Hill *et al.* (2007); (11) Taronga Zoo records; (12) Mackie *et al.* (2001); (13) Johnson (2008); (14) Stannard *et al.* (2017).

**Table 12.6.** Common non-infectious diseases of captive rodents.

Adapted from Breed and Eden (2008) and Scheelings (2025).

| Disease | Cause | Clinical signs | Diagnosis | Treatment | Prevention | Reference |
|---|---|---|---|---|---|---|
| Fight wounds | Inter-animal aggression | Wounds, bruising, fur loss | Clinical signs | Wound care under anaesthesia; antimicrobial drugs may be indicated | Attention to social grouping and stocking density. Successful reintegration of a rodent into its group after separation may be difficult | 1, 2 |
| Dental malocclusion | Trauma, nutrition, lack of gnawing opportunity | Overgrown or abnormally worn teeth, weight loss | Dental examination | Trimming of teeth may be possible | Provide appropriate diet and gnawing material | 1 |
| Cataracts | Unknown. Some may be age-related or occasionally associated with diabetes mellitus | Opacity visible through the pupil | Ophthalmic examination | None | Optimal diet and weight management | 1–3 |
| Obesity | Overfeeding, inappropriate diet | Obesity, diabetes (excessive drinking) | Physical examination; urine or blood test (for diabetes) | Reduce feed or alter dietary components | Monitor body condition, optimise diet | 2, 3 |
| Cannibalism | Stress | Usually affects neonates | | | Avoid overstocking, minimise stress and handling | 1, 2 |

References: (1) Breed and Eden (2008); (2) Taronga Zoo records; (3) Lendon *et al.* (2015).

## 8.6 Preventative medicine

No routine treatments are recommended for Australian rodents, but regular faecal parasitology and anthelmintic treatments may be considered. Wherever possible, all animals that die should undergo necropsy to inform animal health decisions.

## 8.7 Zoonoses

Australian rodents, especially those that have been wild-caught, may harbour pathogens capable of causing disease in people without apparent signs in the animal. Some zoonotic diseases acquired from rodents may be limited to particular geographical areas and include leptospirosis (*Leptospira* spp.), rat-bite fever (*Streptobacillus moniliformis*), scrub-typhus (*Orientia tsutsugamushi*) and spotted fever (*Rickettsia australis*) (Hulst 2025). Carers should avoid bites from rodents and practice good personal hygiene.

# 9. BEHAVIOUR

## 9.1 Activity

Most rodents are nocturnal; however, some species can spend a considerable amount of time active during the day. Broad-toothed rats are mostly active during the night, but can be found active during the day if dense cover is available (Green 1968; Wallis *et al.* 1982). Maximum activity in captivity is generally between 1900 and 2300 h and individuals are least active during the afternoon (1400-1700 h) (Wallis *et al.* 1982). In general, broad-toothed rats are more active throughout the 24-h period in January and February than in March or April (Bubela *et al.* 1991). In winter they are more likely to be active in the late afternoon and early evening, although overall activity is lower (Bubela *et al.* 1991).

Most *Pseudomys* are nocturnal (Stanley 1971; Woods and Kennedy 1997); however, some, such as the desert mouse and long-tailed mouse, are sometimes diurnal, possibly due to the time-consuming and regular foraging requirements that stem from their mostly folivorous diet (Green 1995; Read *et al.* 1999). Activity in the heath mouse is bimodally distributed, with the highest levels occurring just before dawn and a second peak after dusk (Woods and Kennedy 1997). Similarly, water-rats are most active at sunset, but they can forage in full daylight (Olsen 1995).

*Rattus* spp. are generally nocturnal (Wood 1971; Warden and Wallis 1979); however, the swamp rat can often be seen during the day, especially during cooler overcast weather, and can spend up to 50% of its active time during the day (Braithwaite 1977; Braithwaite and Lee 1979). Long-haired rats can also spend ~15% of their active time during daylight hours, but usually do so under cloudy conditions and towards the end of a population eruption when food resources have become scarce (C. Dickman pers. obs.). Stick-nest rats have been observed basking by day on top of their nests in the wild (Watts and Aslin 1981).

## 9.2 Social behaviour

Social organisation of Australian rodents is diverse and reflects both the nature and fluctuations in the carrying capacity of resources in the species' different wild habitats (or enclosures in captive populations) and the severity of environmental conditions such as inhospitable temperatures and aridity (Bubela and Happold 1993).

### 9.2.1 *Conilurus* (tree-rats)

Little is known of the social behaviour of these species of rodents. Animals in captivity produce soft, high-pitched vocalisations and males nest with the mother and her young, suggesting a high level of social bonding and tolerance (Kemper 1995). The brush-tailed rabbit-rat has been observed to nest primarily within fallen logs and hollows of eucalypts and bloodwoods (*Corymbia* spp.) (Firth *et al.* 2006). Although the exact details are unknown, Knut Dahl (1897) kept several brush-tailed rabbit-rats and wrote 'Whenever I kept a number of them together in captivity they would always fight, and very often kill each other.' These observations are supported by Watts (1982a) who suggested that these rodents should be held in small groups or be kept singly.

### 9.2.2 *Hydromys* (water-rat)

Water-rats build their nests in tunnels in the banks on the side of freshwater streams or occasionally in logs. Not much is known about their social behaviour, but they appear to be territorial and, where populations are dense, there can be considerable fighting, particularly among males, which may have damaged tails (Olsen 1980). They should be held in small groups, which generally comprise a pair and their offspring (Watts 1982a; S. Jackson pers. obs.).

### 9.2.3 *Leggadina* (short-tailed mice)

Not much is known of the social behaviour of these mice. Central short-tailed mice build underground burrows that have a constant temperature (28°C) and high humidity (72–97%) in the hot environments where they live (Moro and Morris 2000a). These burrows have been excavated and found to be inhabited by solitary animals or by females with young (Reid and Morton 1995). Socially intolerant behaviour can also be inferred from the animals' core home ranges, which show a low level of overlap (Moro and Morris 2000a). From the little that is known of these rodents, they should be held in single-sex groups (Watts 1982a).

### 9.2.4 *Leporillus* (stick-nest rats)

Stick-nest rats build large aboveground nests that can be approximately 1 m high and 1.5–3 m wide, usually made of sticks and stones and are around a tree, bush, cave or over rabbit burrows (Le Souef 1922; Aslin 1972; Robinson 1975; Watts and Eves 1976). In areas where rocky outcrops occur, they will often build the nests within these structures (Robinson 1975). Nests built in open areas comprise a central chamber from which a series of tunnels radiate (Robinson 1975). If not many sticks are available, they will place stones 2–5 cm in diameter over and through the sticks to give the nest more weight and stability (LeSouef 1922). In captivity, up to 10 rats (consisting of an adult male, female and two or three successive litters of male and female young) may occupy one nest, whereas in the wild very large nests may contain communities of 10–20 animals (Copley 1999). Adult females are frequently aggressive towards adult males, which often seek shelter away from the main family group (Copley 1999). Only established pairs should be held together (Watts 1982a; S. Jackson pers. obs.).

### 9.2.5 *Mastacomys* (broad-toothed rat)

Wild observations indicate that females have exclusive non-overlapping home ranges with their young, mostly in their preferred habitat of heath bordering streams, while males have either separate or overlapping home ranges, but not necessarily in preferred habitat and mostly where females do not occur (Happold 1989). The spacing of home ranges and the intolerance of females towards males except at mating suggests a promiscuous mating system (Happold 1989, 2015). Other observations have shown broad-toothed rats to nest solitarily in summer and communally in winter, occurring in groups of up to five individuals of both sexes (Carron 1985 in Bubela *et al.* 1991; Happold 1989). This behaviour is most likely to reduce heat loss due to the cold climate in their alpine and subalpine habitat (Bubela and Happold 1993). However, outside the nest, spacing of individuals and agonistic behaviour are more pronounced in winter than in other seasons, perhaps because animal activity is restricted to the subnivean space (Happold 2015). In captivity, pairs will live together amicably and share the same nest when the female does not have young; however, once the young are born, the female becomes dominant and will savagely attack the male if he comes near or tries to enter the nest box (Calaby and Wimbush 1964). In captivity, these rodents appear to be best held in single-sex groups (Watts 1982a).

### 9.2.6 *Melomys* (melomys)

Melomys are semi-arboreal, especially the fawn-footed melomys (Leung 1999a). Cape York melomys nest in tree hollows lined with dried leaves (Leung 1999a). The grassland melomys build nests in *Pandanus* sp., grass tussocks and sugarcane (McDougall 1944). In captivity, males will use nests built and abandoned by females but, if available, they generally use wood shavings as cover (McDougall 1944). They are recommended to be held singly in captivity (Watts 1982a).

### 9.2.7 *Mesembriomys* (tree-rats)

Black-footed tree-rats shelter in hollows in standing or fallen trees (Griffiths *et al.* 2002). Not much is known of their social behaviour except that females outnumber males throughout most of the year and it appears that most animals lead a relatively solitary existence (Friend and Calaby 1995). In captivity, pairs have been kept together successfully and although they have bred, have not done so routinely. Golden-backed tree-rats appear to have home ranges that are occupied by one pair of adults, and probably their young, and although they are highly arboreal they spend a lot of time on the ground (Baker and Gynther 2023). They have been kept and bred successfully in captivity, where observations suggest that females may be dominant to males, especially when mating is about to occur (Kerle and Fleming 2024). It also appears that movement between enclosures and nest boxes by pairs is typically initiated by the female. One captive facility that held them as pairs gave each animal their own enclosure with an access hole cut between them. Given they can be aggressive to each other, they are held singly or in pairs in captivity (Watts 1982a; Kerle and Fleming 2024).

### 9.2.8 *Notomys* (hopping-mice)

These species show a high degree of social behaviour and will socially groom, huddle and communally nest regardless of sex, reproductive state, age or ambient temperature (Stanley 1971; Happold 1976a). In observations of the northern hopping-mouse, for example, Donald Thomson commented that this species was strongly gregarious and that captive animals would pile into a 'great heap' when put together (Dixon and Huxley 1985). Agonistic behaviour does occur (especially when an animal is first introduced) and includes stalking, rushing, attacks, chasing and aggressive mounting, sparring and fighting. Aggression occurs between males, but it is not as serious as fights between females, which often result in wounds (Stanley 1971). Despite the aggression towards some individuals, the social system is thought to be communal, with strong bonds between females and their mates (Stanley 1971; Happold 2021). Group members are tolerant of neonates and juveniles and some assist in rearing the young (Stanley 1971; Happold 2021). Although males have small testes and are smaller than females, which suggests a monogamous mating system, pairs do not form long-term monogamous relationships and both sexes may mate with different partners in subsequent years (Breed 1997).

Observations of spinifex hopping-mice in the field confirm that animals generally are tolerant of each other, but sociality changes with environmental conditions. During extended dry periods animals are dispersed and pairs spend short periods (4–5 days) near small stands of shrubs or low trees before moving 2–3 km to other groups of shrubs (Dickman *et al.* 2011; Pavey *et al.* 2017). Breeding occurs during spring under these conditions and young accompany adults on their dispersal movements between shrubs, but survival is low. In contrast, after heavy summer rains, when abundant food resources appear, animals become more sedentary and occupy the same burrow system for weeks at a time. The number of animals using each burrow system doubles at these times, with up to seven individuals observed to cohabit (Dickman *et al.* 2010). Direct observations suggest that these are family units, with one or two adult animals shepherding juveniles during brief forays near the burrow entrance (Dickman *et al.* 2010). Breeding can extend over several seasons and continue for as long as resources last; juvenile survival also increases when conditions are good, so that populations can increase by two orders of magnitude within a year (Dickman *et al.* 1999; Southgate and Masters 1996).

In captivity, Mitchell's hopping-mouse lives in small groups of up to four that are composed of one or more adult females with one or more adult males (Happold 1976a). The spinifex hopping-mouse lives in stable groups with two or more adult females and two or more adult males, nesting communally in a burrow and rearing their young together. The mating system in the wild is unknown, but it has been suggested that males are polygynous and females polyandrous despite the males having very small testes (0.14% body mass), which suggests monogamy (Happold 1976a; Breed 1997). If polyandry occurs, this allows the possibility of sperm competition (Richards and Burke 2007). However, captive studies to examine multiple paternity showed that it did not occur in the spinifex hopping-mouse but did in the plains mouse, which may instead indicate serial monogamy (Breed and Adams 1992). Further experiments have revealed that mating involves locking or tying between individuals (Breed 1990a) and that more than one male may inseminate oestrous females, suggesting that sperm competition may, at times, take place, although it is not known if this occurs in wild populations (Dewsbury and Hodges 1987; Breed 1990a; Breed and Washington 1991).

Limited information suggests the fawn hopping-mouse, northern hopping-mouse and the dusky hopping-mouse live in large groups and have a similar organisation to the spinifex hopping-mouse (Bubela and Happold 1993; Dixon and Huxley 1985). However, the fawn hopping-mouse has comparatively larger testes than other hopping-mice (0.65% *v.* 0.14–0.17% of body mass) and is the only species of hopping-mouse in which a copulatory plug has been recorded (Crichton 1974). They associate in both small and large groups (Watts 1982a).

### 9.2.9 *Pogonomys* (tree mouse)

The tree mouse is highly arboreal, but nests in burrow complexes during the day (Winter and Whitford 1995). Little is known of its social behaviour. Observations in New Guinea found them to live in small groups (Dwyer 1975), but groups of up to 15 have also been recorded (Winter and Whitford 1995).

### 9.2.10 *Pseudomys* (mice; pseudomys)

*Pseudomys* spp. often nest underground or in other covered locations such as rotten fallen logs, with nests including materials such as eucalypt leaves for the New Holland mouse and Hastings River mouse (Kemper 1981; Meek 2002), shredded bark for the silky mouse (Watts and Aslin 1981) and dry grass for the desert mouse (Read *et al.* 1999). Other species, including the delicate mouse,

Gould's mouse, eastern chestnut mouse, and long-tailed mouse, construct their nests of plant material (Watts and Aslin 1981; Green 1993; Fox 2008).

The *Pseudomys* typically have either a dispersed social organisation or variations of communal social organisation (e.g. Happold 1976a, 2021; Meek 2002). The different species of pebble-mice build mounds of pebbles around the entrance of their burrows by carrying the stones in their mouths and moving them in place with their forelimbs. These can reach over 3 m in diameter and 10 $m^2$ in area (Anstee 1996; Anstee *et al.* 1997; Breed and Ford 2007). The western pebble-mouse appears to be social, living in groups of up to 12, with individuals showing a high degree of fidelity with home ranges that overlap (Anstee *et al.* 1997).

Adult female ash-grey mice live communally in stable groups of two or three in one burrow and rear their young together. One or more adult males visit these females, apparently one at a time. The mating system is unknown (Happold 1976a). In captivity the strongest repulsion has been between females of different groups and the weakest between males, which appear to intermingle (Happold 1976a).

Silky mice live in burrow systems in which two or more males may be present; females are visited by one or more males during oestrus, suggesting a promiscuous mating system (Cockburn 1981b). However, their small testicular size of only 0.6%t of body mass suggests low levels of intermale competition (Breed 1997).

Plains mice appear to live in large groups of males and females, which can number up to 20 during the non-breeding season. During the breeding season the burrow system seldom has more than one adult male or three females (Breed and Adams 1992; Watts 1995). Plains mice make nesting chambers comprised of the dominant vegetation in the immediate area of the burrow, shaped into a fist-sized cup or ball, usually 30–50 cm below the surface, which appears to be maintained only when females are breeding (Brandle and Moseby 1999). They have been seen in captivity carrying their young around the enclosure attached to their teats (J. Old pers. obs.).

Delicate mice are thought to have a promiscuous mating system (Braithwaite and Brady 1993). No pair bonding appears to occur and multiple paternity within litters has been observed (see section 10.1.2).

Desert mice are largely solitary, forming only brief male–female pairs for mating and when females associate with their unweaned young (Happold 1976a, 2021). In captivity they nest singly and defend their nest-boxes, showing territorial behaviour and/or temporal avoidance when two adults are enclosed together (Happold 1976a). Observations on western chestnut mice have indicated a similar pattern of social behaviour (Breed 1989). Although the desert mouse has been thought to have a monogamous mating system in the wild (Bubela and Happold 1993), the relatively large testes (1.4–2.5% of body mass) suggest a polyandrous mating system, despite little being known of its social behaviour (Breed 1997).

Observations of smoky mice have shown them to nest communally during the breeding season, behaviour not observed in other *Pseudomys* from similar habitats (Woods and Ford 2000). Their nests were discovered to be of three distinct types: covered scrapes, uncovered surface nests and in burrows (Woods and Ford 2000). Captive observations found males to show high levels of aggression when placed in the same enclosure, with the resident male being the aggressor and biting the back, rump and base of tail of the other animal. When a female was placed in the same enclosure as another female, they showed mutual avoidance; however, the resident female showed aggression to the introduced animal if it went near her nest. Male and female introductions were amicable except when additional animals were introduced. Males were always introduced to female enclosures and after some chasing, they generally nested together during daylight hours (Woods and Ford 2000).

The sandy inland mouse appears to be social, but also shows considerable flexibility in sociality depending on the environmental conditions and perhaps habitat. In one study, animals were found living in small groups of reproductively active individuals and groups of up to 22 animals that were reproductively inactive, with individuals showing a high degree of fidelity and having home ranges that largely overlapped (Breed 1995). In another study, individuals were relatively sedentary, lived in large groups (up to 11 animals) and were reproductively active during highly productive periods after heavy summer rains, but became more mobile and less social during dry periods (Dickman *et al.* 2010). During the dry times animals retreated to stands of gidgee trees within a landscape that was dominated by extensive areas of hummock grass, but dispersed out into the hummock grass after heavy rains (Dickman *et al.* 2011). These combined observations suggest that sandy inland mice are gregarious and highly social in the wild when conditions are favourable, but much less so when stressed by adverse environmental conditions. This species is potentially polyandrous, but in captivity the reproductive output of females is the same irrespective of whether they have mated with one or with two males (Firman 2014).

Heath mice and long-tailed mice live in pairs that nest together and there is agonistic and territorial behaviour between adult males, between adult females and between males and females that are not pair-bonded (Green 1968; Happold 1976a). The small testes of 0.4% body mass in heath mice and the pair bonds that form before the breeding season, throughout pregnancy and while raising the young all suggest monogamy, but polygamy may occur occasionally (Happold 1976a; Breed 1997). Most *Pseudomys* can be held in small groups; however, plains mice, sandy inland mice, western chestnut mice and New Holland mice are best kept in single-sex groups and desert mice are best kept as pairs or singly (Watts 1982a).

### 9.2.11 *Uromys* (white-tailed rats)

Giant white-tailed rats nest in burrows or tree hollows. Although they have large overlapping home ranges, they also show a high level of site fidelity and probably defence of core areas, indicating a strong degree of territoriality (Moore 1995). Territories may be held by structured family groups, with adults showing a high degree of parental care of their young (Eiris 2011). In captivity, they are best kept singly or as pairs (Watts 1982a).

### 9.2.12 *Xeromys* (water-mouse)

The water-mouse lives in nests in mounds that are composed of peat, sedge, mud and sand (Breed and Ford 2007). The mounds are up to 60 cm in height, 48 cm in circumference, with the entrance located just above the high tide level of well-watered habitats such as mangroves, freshwater lagoons and swamps. The nest is lined with dried melaleuca leaves. It appears to be a social species with nests containing up to eight individuals of all age groups and either sex, with one adult male and more than one adult female being resident in the same mound (Van Dyck 1995, 1996; Breed and Ford 2007). Home ranges of groups overlap slightly with other groups, but core ranges do not overlap (Van Dyck 1996).

### 9.2.13 *Zyzomys* (rock-rats)

Rock-rats are generally nervous and flighty, although some individuals are calmer than others (Lloyd 1999). As with most rodents, they move their tails in a wiggling or swirling motion when stressed. In the wild they are generally solitary, with no evidence of a colonial lifestyle or family groups. In captivity, aggression and fighting are common, especially between females, and can result in the tail stripping very easily, so they should be housed separately or as pairs (with great caution) (Watts 1982a; Lloyd 1999). Similarly, other observations have found breeding pairs to rarely nest together, with animals rarely seen together, apart from males seeking out females for mating and aggressive behaviours of the females (Gaikhorst and Lambert 2009).

Various behaviours include nose to nose sniffing, pawing at each other and paw wrestling where animals grasp paws, using high-pitched vocalisations (Lloyd 1999). Males have been observed rubbing their chin on a female's nest site in captivity, which suggests they use scent to mark their territories and attract mates (Lloyd 1999). They do not construct burrows and they show no inclination to dig; their nests are sometimes lined with twigs and leaves, which may be used by both males and females to block the entrance of the nest area. Nesting behaviour greatly increases in females in the late stages of pregnancy and can be used as an indicator of impending birth (Lloyd 1999).

### 9.2.14 *Rattus* (rats)

The Cape York rat nests communally in burrows (Leung 1999b). Urine marking has been recorded in male and female swamp rats to mark their territory and attract mates (Mallick 1992). Peak activity is generally bimodal, with animals being most active after sunset and just before sunrise (Warneke 1971). Several aggressive behaviours are seen, including chasing, vocalising, fighting, wrestling, lunging and submissiveness (Begg and Nelson 1977). As swamp rats often nest alone and males are widely dispersed with largely overlapping home ranges, they are likely to be promiscuous (Braithwaite and Lee 1979).

The level of aggression increases from August to November at the onset of the breeding season in bush rats (Robinson 1987). Low levels of agonistic behaviour occur in December and January in female bush rats (Robinson 1987). Females appear to be tolerant of the male's presence around parturition and while the young are present (Woodside 1983). Paternal behaviour has been observed, including nest attendance, washing the young and guarding (Horner and Taylor 1969; Watts 1982b; Taylor and Calaby 1988b). This species has been observed to preferentially shelter within grass trees

(*Xanthorrhoea* sp.) with a dense canopy cover (Frazer and Petit 2007).

Captive long-haired rats have been held together successfully in groups of three males, within enclosures 30 cm (L) × 36 cm (W) × 25 cm (H); however, when animals were placed in larger enclosures (100 × 50 × 50 cm) and unfamiliar individuals were introduced, conflicts were frequent and deaths often resulted (Barnett *et al.* 1975). Social behaviour in the wild is flexible, with one or two animals occupying simple burrows in dry season refuge sites, and groups of adults and young occupying deep (1 m) and complex burrow systems with multiple entrance and exit holes when populations are expanding after rainfall (Predavec and Dickman 1994).

Groups of juvenile *Rattus* or single sexes can be housed together with little risk, but it is not recommended to have groups of adults of both sexes unless the enclosure is large enough for the animals to readily hide from each other (Watts 1982b).

## 9.3 Reproductive behaviour

The female Carpentarian rock rat has been observed to sometimes lift its tail to allow a male to smell its anus (Lloyd 1999), but mating behaviour has not been observed yet. Hopping-mice show sexual behaviour by the male following or chasing the female and attempting to mate. The male often rides the female while trying to mount and pats her rump during mounting and copulation (Stanley 1971). When mounting, hopping-mice appear to lock or tie together for generally 12–51 s but up to 553 s (average 106 s) (Dewsbury and Hodges 1987; Breed 1990a). During the lock, the pair generally fall on their side with the female often struggling violently, sometimes dragging the male around the enclosure and even biting him (Breed 1990a). In *Rattus*, copulation occurs at night regardless of the timing of pairing. During copulation the male clasps the lower abdomen of the female and rests his chin on her mid-dorsum (Taylor 1961).

Anecdotally, it appears that if breeding of small rodents is interrupted by separating sexes for too long (maybe more than 1–2 months), it can be difficult to recommence breeding.

## 9.4 Bathing

With the exception of the water-rat there are few records of rodents utilising water for bathing; indeed, many species of desert rodents are capable of sustaining themselves without access to free water for considerable periods as their natural diet contains adequate water.

## 9.5 Behavioural problems

Rodents generally suffer from few behavioural problems. Bush rats have, however, been observed to carry out abnormal repetitive behaviour such as boxing, threat postures, biting, scratching and rising up on their hindlegs (Barnett and Stewart 1975; Woodside 1983). In overcrowded enclosures there is a greater likelihood of increased aggression, where dominant animals can exclude others from resources such as nest boxes and food.

## 9.6 Signs of stress

Signs of stress in rodents include increased vocalisation and fighting. In some species, such as the water-rat, stress can result in a decrease in bodyweight, escape behaviour or a general lack of activity.

## 9.7 Behavioural enrichment

Although rodents generally do not suffer from obvious behavioural problems, there are several behavioural enrichment stimuli that can be provided to optimise psychological and physiological wellbeing. These include:

- varying the enclosure surface by changing the soil profile and by the addition of furniture such as rocks, sticks, wood shavings and plastic or cardboard pipes
- scattering food, such as insects, throughout the enclosure at different times of the day
- providing arboreal species with a network of branches to climb throughout the enclosure
- providing running wheels (sparingly, to ensure they do not develop abnormal repetitive behaviour)
- frequently changing rocks, grass tussocks and sticks and climbing branches
- providing stick-nest rats with lots of sticks and small branches to allow them to build a nest
- providing water-rats with soft grasses
- providing plains rats and hopping-mice with leaf litter.

## 9.8 Training and conditioning

There are limited reports of training Australian rodents. A hand-raised male black-footed tree-rat at Taronga Zoo was conditioned as a visitor-encounter animal (W. Gleen pers. obs.). Training sessions for this animal were initiated in the enclosure within the nocturnal exhibit, during the animal's active phase. A 'finger tap' was used to target the station (stay still), and a high value food reward (such as a piece of nut) was used to reinforce the desired response. The animal was trained to station anywhere within the enclosure and on a person's shoulder. The animal would typically come out of its nest box when the animal caretaker entered the enclosure; however, if it did not, the session was cancelled (W. Gleen pers. obs.).

Once the tree-rat consistently stationed on cue, he was brought in his nest box into an enclosed animal encounter room with dimmed lighting and the same training was repeated in the new environment. Though the animal quickly learnt the concept of stationing at different points around his enclosure, it took patience to overcome his inherent timidity when outside his enclosure. Training progressed at his pace and was entirely voluntary on his behalf. Eventually small groups of zoo visitors were able to meet him.

Over time the distance he was asked to travel from the nest box was increased as was the visitor group size. Eventually the animal would walk across visitors' shoulders as they linked arms, with up to three people forming what could be described as a 'human tree' to demonstrate the animal's natural climbing behaviour. The tree-rat would climb across the human tree using 'tap and reward' with the visitors, if comfortable to do so, handing him the piece of nut. Visitors were asked to volunteer for the close encounter and were chosen carefully to ensure they were comfortable with the level of interaction; they were briefed about the potential for very minor scratches from his claws and the fact that, being a wild animal, he was not toilet trained. An animal caretaker stood close by with the nest box so that he could return to it quickly if he wanted to. Occasionally, having completed his traverse he would wander a little further, apparently exploring, before returning to his nest box, which may have indicated his increasing confidence.

## 9.9 Introductions and removals

Introductions of new animals should be conducted with caution, even in social species, as any animal can react unfavourably to unfamiliar individuals. Although hopping-mice are highly amicable to members of their group there is an initial period of aggression towards a new animal until the stranger becomes accepted as one of the members of the group and will huddle with the rest (Happold 1976a).

Introductions of rock-rats should be made under strict supervision due to their highly aggressive nature (Lloyd 1999). This is best achieved by using enclosures with removable dividers (see Chapter 4). Rock-rats have been successfully introduced using enclosures with a removable mesh divider so that animals have visual, auditory and olfactory communication for 9 days before their introduction (Brisbane 1998). There was aggression, with one male chasing the female away, so the female was removed and placed with another, more tolerant male, and they were nesting together after 14 days (Brisbane 1998).

The ease with which different species can be paired was reviewed by Watts (1982a) who suggested that most species of rodents can be introduced relatively easily; however, stick-nest rats and grassland melomys are difficult to pair. Stick-nest rats are generally best introduced as juveniles rather than as adults (as they are very likely to kill each other) and once the pairing is established, they form a strong bond. Other species, such as short-tailed mice, broad-toothed rats, fawn-footed melomys, fawn hopping-mice, desert and delicate mice, white-tailed rats and rock-rats, should be introduced with care (Watts 1982a).

When species of *Rattus* are first introduced they can fight, sometimes quite viciously, but this seldom results in serious injury as they quickly establish a dominant-subordinate hierarchy (Watts 1982b).

The application of talcum powder to some species, such as spinifex hopping-mice and plains mice, can help reduce aggression when placing animals together because it stimulates mutual grooming. As a general rule, if a rodent is separated from others in its group for ≥ 1 day, it may be more difficult to place it back later (J. Old pers. obs.).

Other options that may be used to improve the chances of a successful introduction include the following. (1) If introducing one animal into a group, pair it up with another animal first so that two animals are introduced. (2) Introduce animals into neutral territory (just after a substrate change for instance). (3) Let the new individual have some time in the enclosure by itself so it is familiar with the layout before returning the other animals. (4) For relevant species, apply talcum powder, as discussed above.

## 9.10 Intraspecific compatibility

Rodents are highly variable in their social behaviour, ranging from solitary to social species that form pairs and have complex family groups (Table 12.7; section 9.2). As a result of the broad range of social structures, rodents

**Table 12.7.** Social behaviour of rodents and the suggested sex ratio of different species when held in captivity.

Presented as male:female ratio.

| Genus | Social behaviour | Suggested sex ratio |
|---|---|---|
| **Hydromyini** | | |
| *Conilurus* | Social | 1:1; 0:2 |
| *Hydromys* | Solitary? | 1:1; 0:2 |
| *Leggadina* | Solitary | 1:1; 0:2 |
| *Leporillus* | Pairs | 1:1; 0:2 |
| *Mastacomys* | Social | 1:1; 0:2 |
| *Melomys* | Solitary | Solitary |
| *Mesembriomys* | Solitary | 1:1; 0:2 |
| *Notomys* | Social | 1:1; 0:2; groups |
| *Pogonomys* | Social | 1:1; 0:2; groups? |
| *Pseudomys* | Social | 1:1; 0:2; groups |
| *Uromys* | Solitary | 1:1, 0:2; solitary |
| *Xeromys* | Social | 1:1; 0:2; groups |
| *Zyzomys* | Solitary | 1:1; solitary |
| **Rattini** | | |
| *Rattus* | Solitary/social | 1:1; single-sex groups |

should be kept as solitary animals, pairs or groups depending on the species. Some species, such as many hopping-mice and *Pseudomys*, are typically more social, while others such as rock-rats are comparatively more difficult to hold together. Regardless of the suggested sex ratio listed in Table 12.7, it is advisable that continued observation is undertaken after animals are introduced, as well as during the breeding season.

## 9.11 Interspecific compatibility

Due to their small size, rodents are generally not housed with other species as they may be hard to see in the enclosure or they hide or may be preyed upon. Food items for water-rats, such as yabbies, can be held with them as part of the exhibit until they are eaten.

# 10. BREEDING

## 10.1 Mating system

### 10.1.1 Reproductive strategies

Most species of rodents, particularly those living in unpredictable environments such as deserts, can breed at most times of the year, but maximise their reproductive output after rain when there is a surge in the quality and quantity of available food. These species tend to have short oestrous cycles and gestation periods and quickly reach sexual maturity. Hydromyine rodents such as hopping-mice and *Pseudomys* spp. breed soon after rain, with the young recruiting into the population some 4–6 months later, compared with 'new endemics' such as house mice and *Rattus* that breed much more quickly after rain and produce more offspring (Dickman *et al.* 1999; Southgate and Masters 1996; Greenville *et al.* 2012, 2016). In contrast to these rodents, those that occur in more stable environments tend to live longer, breed seasonally and have a lower rate of reproduction and include species such as black-footed tree-rats and the long-tailed mouse (Watts and Aslin 1981; Strahan 1995).

### 10.1.2 Multiple paternity

At least some species show sperm competition, resulting in multiple paternity within litters. This is usually related to relative testis size and has been observed in plains mice and captive sandy inland mice, which have large testes and produce large numbers of sperm. In contrast, spinifex hopping-mice have very small testes, produce fewer sperm and appear not to have sperm competition or multiple paternity within litters (Breed and Adams 1992; Richards and Burke 2007). Despite this, female hopping-mice will mate with more than one male when in oestrus in captivity; however, the first male to mate leaves behind a coagulated vaginal plug that may stop further sperm entering. So, although there is the potential for sperm competition to occur it may usually be stopped by this plug (Breed 1990a; Breed and Washington 1991). In species such as *Rattus*, in which the males have very large testes, multiple paternity is highly likely.

## 10.2 Ease of breeding

Although most species of rodent breed well initially, the medium to longer-term breeding success of captive-bred offspring is often low. Watts (1980) found that ~35% of wild-caught rodents bred in captivity, so it is important to obtain an adequate founding stock in order to develop a healthy and genetically diverse population. Watts (1980) suggested that founding with at least eight individuals is important to maintain population size. In his attempts to breed rodents, Watts (1980) found that of the 34 species he tried to establish, all colonies became extinct within several generations, with the exception of Mitchell's hopping-mouse, spinifex hopping-mouse and the plains mouse, which continued to breed well with no decline in population size after 4 years.

Similar to various other species of native rodents, the central rock-rat has been observed to breed very well when taken in from the wild in full reproductive condition. However, if breeding is stopped for management reasons it may not be possible to recommence breeding activity (Gaikhorst and Lambert 2009).

Bush rats have been found to produce as many as 14 litters in 15 months, but such output dwindles rapidly by the third generation (Taylor 1961; Watts 1982b). For example, Watts (1982b) noted that breeding in 24 separate colonies of *Rattus* led to rapid initial increases in numbers due to high output by wild-caught animals and their offspring, but in all cases the grand offspring of the wild-caught animals bred poorly and the colonies eventually died out because of the lack of fecundity of second- and third-generation animals.

In other results from captive colonies, approximately one-third of female native rodents brought into captivity bred and, of the 27 species maintained in captivity at various times, only the Mitchell's hopping-mouse, spinifex hopping-mouse, plains mouse and the western chestnut mouse have continued to breed past the second generation in captivity (Watts 1973, 1980). In a more recent study, Ricci (2003) successfully bred sandy inland mice for more than three generations, although most animals showed poor reproductive success after the first generation.

## 10.3 Oestrous cycle and gestation period

Oestrous cycles for most species are very short, generally ranging from 4 to 12 days, with several notable exceptions, including the black-footed tree-rat, which has an oestrous cycle of ~26 days, and the Kimberley rock-rat, which has an oestrous cycle of 16–18 days (Table 12.8). Gestation lengths for most species are typically 30–40 days for the Hydromyini and 20–30 days for the Murinae, with the notable exception of the western chestnut mouse, which has a shorter gestation of only 23 days (Table 12.8). There is a positive correlation between bodyweight and length of pregnancy in Australian rodents (Yom-Tov 1985).

## 10.4 Reproductive status

The sex and reproductive status of Australian rodents can sometimes be difficult to determine, especially if they are juveniles or in non-breeding condition (Gleen 2009; Berris *et al.* 2020a). The best way to determine the sex is to look at the area anterior to the anus. Males can be distinguished from females by males lacking teats and the presence of two urogenital openings in females (i.e. the vagina and urethra opening of the genital papilla), except in juvenile or non-breeding females when the vagina is closed (Fig. 12.3). The observation of a furless stripe between the anus and genital papilla in females has been used by some authors (Watts and Aslin 1981), but this is not always characteristic of the sexes. The presence of teats in females can also be used but these are usually difficult to find in fully furred nulliparous individuals. In some larger species the penis can be gently extruded from the genital papilla.

### 10.4.1 Reproductive anatomy

- Males
  - Testes – paired and may be scrotal or non-scrotal, in which case the testes are not descended and are withheld in the body cavity or the inguinal canal
  - Epididymes – sometimes visible as small dark pigmented sacs on the end of the testes
  - Genital papilla – small protuberance through which the penis can sometimes be extruded during handling in larger species. This papilla appears similar in males and females.
- Females
  - Vagina – may be open and visible in some cases, especially in adult sexually mature animals. In juveniles or animals out of breeding condition, it may be closed and/or more difficult to see

**Table 12.8.** Oestrous cycle (days) and gestation (days) for Australian native rodents.

Numbers in brackets are mean values.

| Species | Oestrus (days) | Gestation (days) | Postpartum oestrus | Litters per year | Reference |
|---|---|---|---|---|---|
| **Hydromyini** | | | | | |
| *Conilurus penicillatus* | 9–12 (9) | 36 | Yes | 2 | 1–3 |
| *Hydromys chrysogaster* | 7–17 (10) | 33–41 (34) | Yes | 1–5 (2) | 3, 4 |
| *Leggadina forresti* | 7–8 | 35 | - | 2+ | 2, 3 |

| Species | Oestrus (days) | Gestation (days) | Postpartum oestrus | Litters per year | Reference |
|---|---|---|---|---|---|
| *L. lakedownensis* | 7 | 30–33 | – | – | 2, 3 |
| *Leporillus conditor* | 14 | 44–45 | – | – | 3, 5–7 |
| *Mastacomys fuscus* | – | 38–40 | – | 2–3 | 3, 8 |
| *Melomys cervinipes* | – | 38–40 | Yes | 2+ | 3, 9, 10 |
| *Mesembriomys gouldii* | 21–35 (26) | 42–44 | Yes | – | 3, 11 |
| *M. macrurus* | 9–10 | 44–49 | Yes | – | 12, 13 |
| *Notomys alexis* | 4–8 (6) | 32–47 (34) | Yes | 2+ | 3, 9, 14–20 |
| *N. aquilo* | – | 49? | – | 2+ | 12, 21, 22 |
| *N. cervinus* | 8 | 32–51 | Yes | – | 3, 9, 16 |
| *N. fuscus* | 6–9 (9) | 33–39 (33) | Yes | – | 2, 3, 16 |
| *N. mitchellii* | 7–8 (8) | 34–42 | Yes | – | 3, 9, 16 |
| *Pseudomys albocinereus* | 6–10 (7) | 37–39 | Yes | – | 2, 3, 17 |
| *P. apodemoides* | – | 34–36 | – | – | 2, 3 |
| *P. australis* | 5–21 (7) | 30–35 (32) | Yes | 2+ | 3, 9, 15, 20 |
| *P. delicatulus* | 6 | 29–37 (31) | Yes | 2–6 | 2, 3, 12, 23 |
| *P. desertor* | 7–9 | 27–34 (28) | Yes | – | 3, 9, 17 |
| *P. gouldii* | – | 18–30 | – | – | 3, 12 |
| *P. gracilicaudatus* | – | 27 | Yes | 3 | 3, 12 |
| *P. hermannsburgensis* | 7–13 | 30–34 | – | 3 | 2, 3 |
| *P. higginsi* | 12–19 | 31–32 | – | 1–2 | 2, 3 |
| *P. nanus* | 5–7 (6) | 22–27 (24) | – | – | 2, 3, 20, 24 |
| *P. novaehollandiae* | 4–10 (6) | 29–33 (32) | – | 3–4 | 3, 25–27 |
| *P. shortridgei* | 10–11 | 28–34 | No? | 1–2 | 2, 17 |
| *Uromys caudimaculatus* | 6–7 (7) | 36–41 | – | 2? | 2, 3, 28 |
| *Zyzomys argurus* | 5–7 (7) | 25–35 | Yes | – | 2, 3, 29 |
| *Z. pedunculatus* | – | 30–40 | Yes | – | 30 |
| *Z. woodwardi* | 16–18 (18) | 35 | Yes | – | 2, 3, 24, 29 |
| | | | | | |
| **Rattini** | | | | | |
| *Rattus colletti* | 5.5 | 22 | – | – | 3, 31 |
| *R. exulans* | – | 22–23 | Yes | 1–13 (5) | 3, 32 |
| *R. fuscipes* | 4.5–5.5 | 22–33 | Yes | 3 | 3, 31, 33–38 |
| *R. leucopus* | 4.3–5.3 | 23 | Yes | 3 | 3, 31, 36, 37 |
| *R. lutreolus* | 4–5.4 | 21–25 | Yes | 2–7 | 3, 32, 35, 38–41 |
| *R. sordidus* | 4–6.8 | 21–23 | Yes | – | 3, 31, 36, 42 |
| *R. tunneyi* | 4.3–4.9 | 21–22 | Yes | – | 3, 31, 36 |
| *R. villosissimus* | 5.0–5.4 | 22–23 | – | 3+ | 3, 31, 36, 37 |

References: (1) Taylor and Horner (1971a); (2) Watts (1979a); (3) Yom-Tov (1985); (4) Olsen (1982); (5) Robinson (1975); (6) Copley (1988); (7) Copley (1999); (8) Calaby and Wimbush (1964); (9) Watts (1982a); (10) Taylor and Horner (1970a); (11) Crichton (1969); (12) Strahan (1995); (13) Kerle and Fleming (2024); (14) Stanley (1971); (15) Smith *et al.* (1972); (16) Crichton (1974); (17) Happold (1976b); (18) Telfer and Breed (1976); (19) Breed (1979); (20) Breed (1989); (21) Dixon and Huxley (1985); (22) Diete *et al.* (2016); (23) Taylor and Horner (1970b); (24) Watts (1979b); (25) Kemper (1976a); 26) Kemper (1980); (27) Pye (1991); (28) Eiris (2011); (29) Calaby and Taylor (1983); (30) Gaikhorst and Lambert (2009); (31) Breed (1978); (32) Egoscue (1970); (33) Taylor (1961); (34) Warneke (1971); (35) Taylor and Horner (1972a); (36) Taylor and Horner (1973); (37) Breed (1976a); (38) Robinson (1987); (39) Green (1967); (40) Braithwaite and Lee (1979); (41) Fox (1985); (42) McDougall (1946).

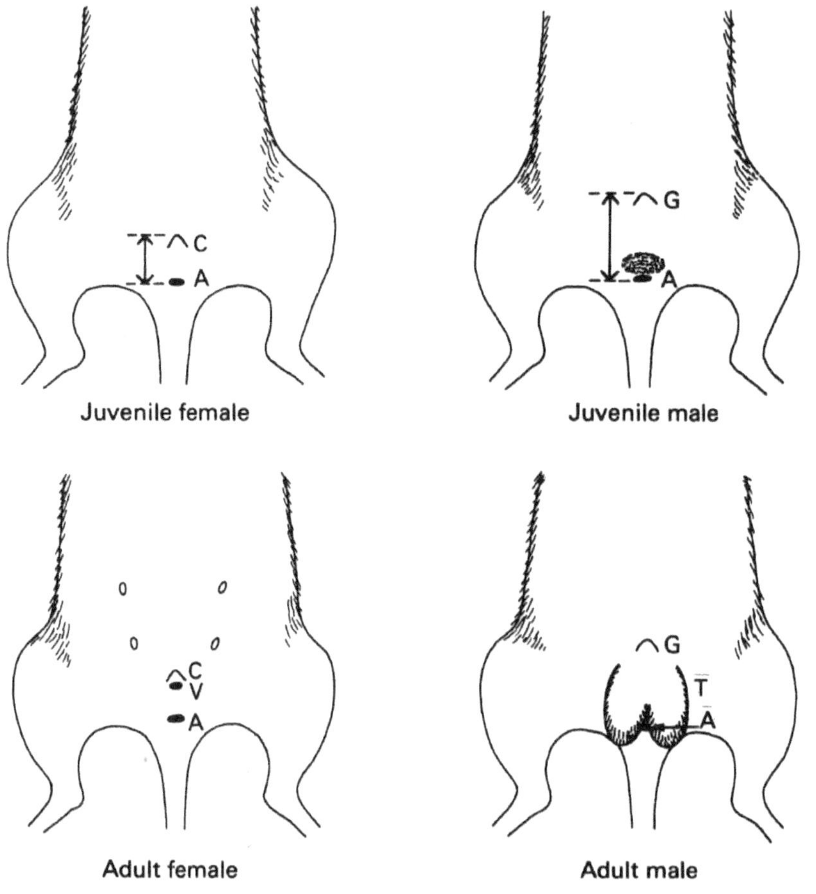


**Fig. 12.3.** External sex differences of Australian rodents. A = anus; C = clitoris; G = genital papilla; T = testes; V = vaginal opening. Taken from Watts and Aslin (1981).

- Teats – may be visible or palpable in adults and/or animals in breeding condition. They can be hard to find, especially in juveniles of small species
- Genital papilla – the urethra opens through the genital papilla. This small protuberance looks similar in males and females.

## 10.4.2 Reproductive condition

Reproductive condition of males of most species can be divided into several different categories (Smith 1985):

- testes not descended into the scrotal sac
- testes scrotal but small
- medium to large scrotal testes with an unenlarged cauda epididymal sac
- large scrotal testes with enlarged cauda epididymal sac
- testes that had an enlarged cauda epididymal sac but which has collapsed.

These categories provide a good indication of male reproductive state, but care must be taken to ensure correct assessment in species with very small testes, such as the spinifex hopping-mouse (Berris *et al.* 2020a).

The reproductive condition of females can be divided into several categories (Begg 1981; Smith 1985):

- juvenile – imperforate vagina; nipples not clearly visible and bodyweight is below that of adult
- non-breeding adult – adult bodyweight with imperforate vagina, no sperm plug (Berris *et al.* 2020a)
- breeding adult – adult bodyweight with perforate vagina, with or without sperm plug (may represent just mated or early pregnancy)
- pregnant – can be diagnosed by palpation in the more advanced stages of pregnancy; the nipples are usually enlarged
- lactating – teats are large and elongated and surrounded by rings of bare skin. It is often difficult to express milk from the teats (e.g. *Notomys*, *Pseudomys*, *Rattus*).

The reproductive stage of female rodents can be determined with some precision by examining a vaginal smear: a cotton bud dipped in saline or distilled water is gently inserted into the vagina and then rolled onto a glass slide that is allowed to air dry and then stained with methylene blue or toluidine blue (Crichton 1974; Olsen 1982). The vaginal smears are assigned to one of four categories according to the cell type present (Long and Evans 1922):

(1) proestrus - predominantly nucleated epithelial cells
(2) oestrus - almost solely cornified epithelial cells
(3) metoestrus - mostly a mixture of cornified cells and leucocytes
(4) anoestrus - predominantly leucocytes.

Although assessment of vaginal smears is the most frequently used technique for determining the stage of the oestrous cycle of female rodents, a technique that assesses the visual appearance of the vagina of the house mouse (*Mus musculus*) can be used (Champlin *et al.* 1973). The stages, which have different lengths, are:

(1) dioestrus - vagina has a small opening, the tissues are bluish-purple in colour and very moist
(2) proestrus - vagina is gaping and the tissues are reddish-pink and moist. Numerous longitudinal folds or striations are visible on both the dorsal and ventral lips
(3) oestrus - vaginal signs are similar to those of proestrus but the tissues are lighter pink and less moist, and the striations are more pronounced
(4) metoestrus-1 - vaginal tissues are pale and dry. Dorsal lip is not as swollen as in oestrus
(5) metoestrus-2 - vaginal signs are similar to those of metoestrus-1, but the lip is less swollen and has receded. Whitish cellular debris lines the inner walls or fills the vagina.

Changes in the vagina throughout oestrus have also been observed in the New Holland mouse, but with considerable variation among individuals (Kemper 1976a).

If young are present, several developmental stages and measurements can be recorded and compared to existing growth curves (see section 10.16) or new curves can be established for future reference. These include:

- developmental stage
  - sex distinguishable
  - tips of ears free
  - papillae of facial vibrissae evident
  - eyes open
  - fur visible - slight tinge, medium or well developed
  - tips of first incisors observable through the gums
  - eating solids
  - self-feeding
  - independent.
- measurements (see Appendix 3)
  - bodyweight (g)
  - head length (mm) - from the occiput to snout tip
  - head width (mm) - maximum width across the zygomatic arches
  - crown-rump length (mm) - primarily for very small individuals
  - body length (mm) - from snout to anus
  - tail length (mm) - from the anus to the end of the last vertebrate of the tail tip
  - total length (mm) - from snout tip to tail tip
  - tibia length (mm) - from the hip to the bottom of the pes
  - pes length (mm) - from the heel to the base of the longest toe, not including the claw.

## 10.5 Reproductive management

Previous studies attempted to reduce the age at which female hopping-mice reach sexual maturity by placing them with unrelated adult males or females; however, it was ineffective (Thompson and Breed 1982).

Individuals placed in pairs that do not breed after several months during the breeding season should be separated and repaired with another mate, as this can often stimulate breeding. For example, Watts (1982b) interchanged *Rattus* spp. after 3-4 months.

Studies of wild populations of Cape York melomys and the Cape York rat found their reproduction to be related to the availability of food, which includes rainforest fruits that peak in summer (Leung 1999a, 1999b). In some species, an ad lib diet has resulted in some normally seasonally breeding species breeding throughout the year (e.g. bush rats; White *et al.* 1996). However, diet is not the only factor affecting breeding, as captive bush rats show seasonal changes in both bodyweight and activity (Stewart and Barnett 1983; White *et al.* 1996). Decreasing temperature and photoperiod may also be important (Robinson 1987; Taylor and Horner 1973).

Studies of captive bush rats showed that reproduction was not greatly decreased by constant or changing light cycles or by temperature, only by a decrease in food (Irby *et al.* 1984). Day length can influence the growth rate of rats, e.g. weanling black rats *R. rattus*, which were found to gain weight 10% more slowly over a period of 5–6 weeks when the day length of artificial lighting was decreased from a cycle of 12 h light (L):12 h dark (D) by daily increments of 5 min than when kept on a constant day length of 12 h (Young 1965).

Care needs to be taken when maintaining populations of rodents to prevent overbreeding. As many species have the ability to breed rapidly, captive populations can become highly inbred, resulting in inbreeding depression. For example, a population of long-haired rats was found to show significant inbreeding depression in the number of young born, with a 3.4% decrease for each 10% increase in the inbreeding coefficient (Lacy and Horner 1997). This inbreeding depression showed no evidence of amelioration and the population became highly inbred over 25 generations. Intriguingly, however, inbreeding had no effect on survival or sex ratio of litters, with larger litters containing more females.

Environmental factors influence reproduction in many rodent species. For example, the age at which first oestrus occurs in female spinifex hopping-mice is influenced by a short photoperiod (6L:18D), and the 7–8 day oestrous cycle is prolonged by short photoperiods or water deprivation (Breed 1975). Further studies have found high population density to decrease reproduction in spinifex hopping-mice (Breed 1976b), principally through suppression of female fecundity (Berris *et al.* 2020b). It has also been found that in a controlled environment with decreased lighting (shortened photoperiod; 6L:18D), temperature (12°C) and food quality (Kellogg's Rice Bubbles, cod liver oil, glucose and dicalcium phosphate), the testes of sexually active adult male bush rats regressed at any time of the year to be the same size as in the non-breeding season (Irby *et al.* 1984). Full testicular recovery was achieved only when the photoperiod (14L:10D), temperature (20°C) and food quality (rat cubes) were changed. Experiments in which only one or two of these factors were altered failed to reverse the sexual regression (Irby *et al.* 1984).

## 10.6 Occurrence of hybrids

Although many species within the rodent genera are likely to be able to hybridise, few instances have been recorded. There is a record of a bush rat hybridising with a swamp rat, but all the offspring except one died before 50 days of age. The one surviving offspring was fertile (Fox and Murray 1979). The different subspecies of the bush rat (*Rattus fuscipes assimilis*, *R. f. greyi* and *R. f. fuscipes*) breed together readily in captivity, with no reduction in either the parent or hybrid's reproductive ability (Horner and Taylor 1965).

## 10.7 Timing of breeding

In species that live in temperate environments, where abundant food generally occurs predictably at about the same time each year, seasonal reproduction usually occurs, whereas in desert environments, where resources are unpredictably available, breeding is usually aseasonal (i.e. occurs immediately after rain so that the young are born when food is available) (Breed 1990b). Such breeding events can occur locally or regionally (e.g. over thousands of square kilometres), depending on the scale of the rainfall event (Greenville *et al.* 2012, 2016; Bennison *et al.* 2018). In southern Australia, breeding is generally associated with food availability and the winter and spring rains that bring an influx of food, while in tropical northern Australia, where the climate is monsoonal, the rains occur primarily over the summer. As a result, most species breed throughout the spring and summer months or in some cases year-round. In captivity, with continuous food and water availability, many species will breed more regularly or continuously (Table 12.9).

## 10.8 Age at first breeding and last breeding

Due to their short life spans, rodents quickly reach sexual maturity, as early as 36 days of age in *Rattus* spp. and typically 60–90 days of age in most of the Hydromyine species. Once sexually mature, most species of rodent are sexually active for only 1–2 breeding seasons.

## 10.9 Ability to breed every year

All species of rodents that live for more than 1 year can breed every year of their lives.

## 10.10 Ability to breed more than once per year

Most species of Australian rodents have a postpartum oestrus and can breed at least several times per year.

## 10.11 Nesting requirements

During the breeding season a nest box or other refuge should be provided. For small species this can include perforated plastic tubes, cardboard tubes or shredded paper.

## 10.12 Breeding diet

In all the desert species, breeding occurs from spring to autumn (i.e. over the summer wet season) irrespective of

**Table 12.9.** Reproduction and development of Australian native rodents.

The numbers in brackets in the litter size column are the mean and the numbers in brackets in the sexual maturity column are the bodyweight in grams (g). Numbers listed are derived from the references listed and Geffen *et al.* (2011).

| Species | Nipples | Litter size (mean) | Weaning (days) | Sexual maturity (days) M/F (BW g) | Birth season | Reproductive senescence (months) | Reference |
|---|---|---|---|---|---|---|---|
| **Hydromyini** | | | | | | | |
| *Conilurus albipes* | 4 | (3) | – | – | – | – | 1 |
| *C. penicillatus* | 4 | 1-4 (2) | 20–49 | 42–77 | Mar–Oct | – | 2–4 |
| *Hydromys chrysogaster* | 4 | 1-7 (4) | 28–32 | 124 (400)/ 130 (425) | Sept–Mar | 42 | 1, 3, 5–7 |
| *Leggadina forresti* | 4 | 3-4 (3) | 28–30 | 90 | – | – | 3, 8 |
| *L. lakedownensis* | 4 | 2-3 (3) | 30 | 90 | Apr–Aug | – | 3, 8, 9 |
| *Leporillus conditor* | 4 | 1-4 (2) | 30–40 | 240 | All year | – | 1, 3, 10, 11 |
| *Mastacomys fuscus* | 4 | 1-4 (2) | 35–40 | – | Oct–Mar | – | 1, 3, 12, 13 |
| *Melomys burtoni* | 4 | 1-5 (3) | 21 | – | All year | – | 4, 14, 15 |
| *M. capensis* | 4 | 1-2 (2) | 14 | 43 / 80 | All year | 24 | 1, 16 |
| *M. cervinipes* | 4 | 1-5 (2) | 20–40 | – | Aug–Mar | – | 1, 3, 15, 17, 18 |
| *Mesembriomys gouldii* | 4 | 1-3 (2) | 40–42 | 80–90 | All year | – | 1–3, 19, 20 |
| *M. macrurus* | 4 | 1-2 (2) | 42–49 | 60–72/76+ | All year? | – | 1, 4, 21 |
| *Notomys alexis* | 4 | 1-6 (4) | 28–33 | 60–84 (?/27) | All year | 24 | 22–27 |
| *N. aquilo* | 4 | 1-5 (3) | ~30 | – | All year | – | 1, 28, 29 |
| *N. cervinus* | 4 | 1-5 (3) | 28–35 | 90/180 | All year | – | 1, 3, 4 |
| *N. fuscus* | 4 | 1-5 (3) | 30 | 70–90 | All year | F 26/M 38 | 1, 3, 4, 8, 30 |
| *N. mitchellii* | 4 | 1-5 (4) | 30–35 | 90 | Jul–Feb | – | 27 |
| *Pogonomys* sp. | 6 | 2-3 (3) | – | – | Oct–Jan | – | 1, 31 |
| *Pseudomys albocinereus* | 4 | 1-6 (4) | 25–35 | 75 | Aug–Mar | – | 3, 8, 24, 27 |
| *P. apodemoides* | 4 | 1-4 (2) | 35–40 | – | – | – | 1, 8 |
| *P. australis* | 4 | 1-7 (4) | 28–30 | 84–180 | All year | – | 1, 3, 23, 26, 32 |
| *P. chapmani* | 4 | 1-4 (4) | – | – | – | – | 1 |
| *P. delicatulus* | 4 | 1-4 (3) | 30 | – | All year? | – | 1, 4, 8, 27, 33, 34 |
| *P. desertor* | 4 | 1-4 (3) | 20–30 | 70–80 | All year | – | 1, 3, 24 |
| *P. gouldii* | 4 | 1-4 (3) | 30 | – | Jun–Oct | – | 4 |
| *P. fumeus* | 4 | 1-4 (3) | – | – | Sept–May | – | 1, 35 |
| *P. gracilicaudatus* | 4 | 1-5 (3) | 28 | – | Sep–Mar | 24 | 1, 4, 36 |
| *P. hermannsburgensis* | 4 | 1-7 (3) | 30 | 90 | All year | – | 1, 8, 27, 33 |
| *P. higginsi* | 4 | 1-4 (3) | 25–30 | 270 | Nov–Apr | – | 3, 8, 13 |
| *P. nanus* | 4 | 1-5 (3) | 21–25 | 50–100 | Dec–Oct | – | 1, 3, 4, 8, 26 |
| *P. novaehollandiae* | 4 | 1-6 (4) | – | 49–120 (14/15) | Sep–Mar | – | 1, 36–39 |
| *P. occidentalis* | 4 | 1-4 (3) | – | – | Jul–Dec | – | 1, 4 |
| *P. oralis* | 4 | 2-3 | – | – | Jun–Feb | – | 4 |
| *P. shortridgei* | 4 | 1-3 (3) | – | 300–330 | Oct–Dec | – | 1, 24 |
| *Uromys caudimaculatus* | 4 | 1-4 (2) | 40 | 180 | All year | – | 3, 4, 40 |
| *Xeromys myoides* | 4 | 1-4 | – | – | All year? | – | 4, 41 |

**Table 12.9.** continued

| Species | Nipples | Litter size (mean) | Weaning (days) | Sexual maturity (days) M/F (BW g) | Birth season | Reproductive senescence (months) | Reference |
|---|---|---|---|---|---|---|---|
| *Zyzomys argurus* | 4 | 1–4 | 28–35 | 150–180 (29) | All year | – | 1, 3, 8, 42–44 |
| *Z. maini* | 4 | 1–4 (3) | 28 | 150–180 | All year | – | 4 |
| *Z. palatalis* | 4 | 1–4 | – | – | – | 48 | 45 |
| *Z. pedunculatus* | 4 | 1–4 | 35 | 229–243 | – | – | 46 |
| *Z. woodwardi* | 4 | 1–4 (2) | 28–35 | 150–180 (70) | All year | – | 1, 3, 8, 42, 43 |
| **Rattini** | | | | | | | |
| *Rattus colletti* | 12 | 1–9 (7) | 20 | 30–76 | All year | 24–35 | 1, 47 |
| *R. exulans* | 10 | 1–10 (4) | 20 | 75 | Dec–Aug | 12–24 | 1, 47–49 |
| *R. fuscipes* | 10 | 3–8 (5) | 20–31 | 68–120 | Aug–Mar | 12–30 | 1, 18, 31, 47, 50–57 |
| *R. leucopus* | 6 | 1–5 (3) | 22–30 | 64/120 | All year? | 12–24 | 1, 47, 52, 53, 58 |
| *R. lutreolus* | 10 | 1–10 (5) | 21–25 | 63 (100/) (60) | Aug–May | 12–24 | 1, 47, 52, 54, 59–63 |
| *R. sordidus* | 12 | 1–12 (6) | 20 | 63–70 | Nov–May | 12–24 | 1, 47, 52, 64 |
| *R. tunneyi* | 10 | 2–10 (4) | 21 | 35–56 | Oct–Apr | 16–33 | 1, 47, 52 |
| *R. villosissimus* | 12 | 5–9 (8) | 21 | 40–77 | All year | 12–24 | 1, 48, 53, 65, 66 |

References: (1) Yom-Tov (1985); (2) Taylor and Horner (1971a); (3) Watts (1982a); (4) Strahan (1995); (5) Troughton (1941); (6) McNally (1960); (7) Olsen (1982); (8) Watts (1979b); (9) Moro and Morris (2000b); (10) Robinson (1975); (11) Copley (1988); (12) Calaby and Wimbush (1964); (13) Green (1968); (14) Begg *et al.* (1983); (15) Smith (1985); (16) Leung (1999a); (17) Taylor and Horner (1970a); (18) Wood (1971); 19) Crichton (1969); (20) Friend (1987); (21) Kerle and Fleming (2024); (22) Stanley (1971); 23) Smith *et al.* (1972); (24) Happold (1976b); (25) Thompson and Breed (1982); (26) Breed (1989); (27) Breed (1990b); (28) Dixon and Huxley (1985); (29) Diete *et al.* (2016); (30) Aslin and Watts (1980); (31) Dwyer (1975); (32) Young (2018); (33) Taylor and Horner (1970b); (34) Braithwaite and Brady (1993); (35) Cockburn (1981a); (36) Taylor and Horner (1972b); (37) Kemper (1976a); (38) Kemper (1980); (39) Pye (1991); (40) Eiris (2011); (41) Van Dyck (1996); 42) Begg (1981); (43) Calaby and Taylor (1983); (44) Bradley *et al.* (1988); (45) Lloyd (1999); (46) Gaikhorst and Lambert (2009); (47) Watts (1982b); (48) Egoscue (1970); (49) Dwyer (1978); (50) Taylor (1961); (51) Taylor and Horner (1971b); (52) Taylor and Horner (1973); (53) Breed (1976a); (54) Braithwaite (1980); (55) Press (1987); (56) Robinson (1987); 57) White *et al.* (1996); 58) Leung (1999b); (59) Green (1967); (60) Monamy (1995); (61) Braithwaite and Lee (1979); (62) Fox (1979); (63) Norton (1987b); (64) Redhead (1979); (65) Carstairs (1972); (66) Predavec and Dickman (1994).

rainfall and therefore the availability of food. This is probably because, if it does not rain for 3–4 years, some breeding is needed or populations will become extinct. Nonetheless, a small increase in food during the breeding season will help to increase survival of the young.

## 10.13 Number of young born

Litter sizes range from one to six and are typically three or four in Hydromyine rodents, often more in *Rattus* spp. (Table 12.9). When considering all Australian rodents, Yom-Tov (1985) found no significant correlation between female bodyweight and litter size.

## 10.14 Age at weaning

Weaning occurs at a very young age in rodents and in Australian species typically occurs after only 20–30 days (Table 12.9).

## 10.15 Age at removal from the mother

Once weaned, juvenile rodents can be removed from their mothers.

## 10.16 Growth and development

The growth and development of several species of rodent have been determined and several are shown in Fig. 12.4. References for further growth and development information are given in Table 12.10.

# 11. ARTIFICIAL REARING

## 11.1 Housing

Choosing suitable housing can help to create an environment with minimal stress. To achieve this, several factors should be considered including:

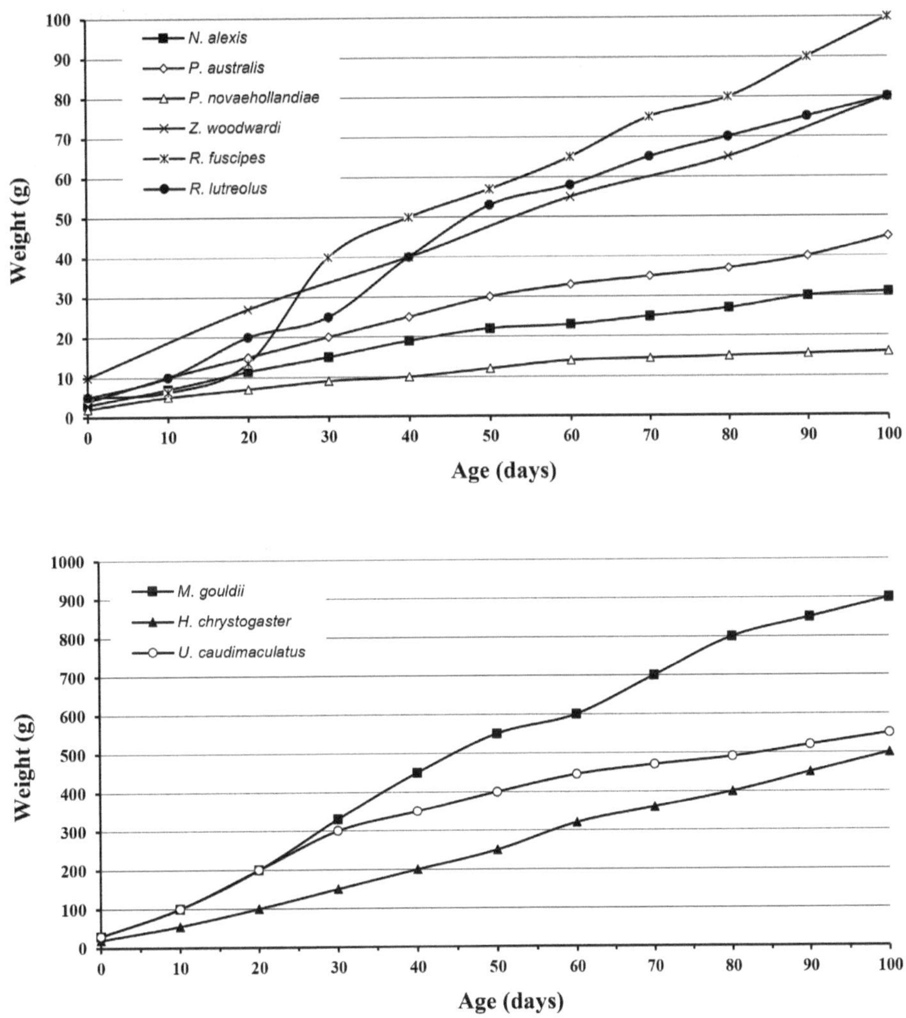


**Fig. 12.4.** Growth in bodyweight of several species of rodents. Derived from Taylor (1961), Crichton (1969, 1974), Smith *et al.* (1972), Kemper (1976b), Green (1967), Watts (1979a) and Olsen (1982).

- securing the area from children and other animals (particularly domestic dogs and cats)
- maintaining a high standard of hygiene
- escape-proofing the area
- ensuring the area offers shelter from the weather and noise.

During the 'dependency' stage, rodents can be kept in a small, well-ventilated, chew-proof container. The container must be large enough to allow the animal to move away from a heat source (see below), but not so large that it can get too far away from the warmth (e.g. a PVC box used for a stick-nest rat was 70 cm (L) × 40 cm (W) × 50 cm (H) and the roof was perforated to allow good ventilation. A warm 'nest' is required: a soft cotton bag inside a woollen one has worked well. As the animal becomes active, it must be given ample opportunity for exercise in a larger enclosure.

## 11.2 Temperature requirements

Heat should be supplied (but not enforced) until the animal is fully furred and active. A heat pad set at 28–32°C can be used, with the animal encouraged to self-regulate temperature by moving closer to or further away from the source (a hot water bottle wrapped in a soft cloth can also be used, but these cool down relatively quickly, making it difficult to maintain a constant temperature). Note: this heat source could potentially be chewed, so unless it is rendered chew-proof, it should be removed as the animal becomes furred, toothed and more active. It is vital that temperature is carefully monitored with a thermometer

**Table 12.10.** Growth curve measurements that have been developed for different species of rodents.

EA = ear length, HB = head body length, HE = head length, PE = pes length, TA = tail length, WT= bodyweight.

| Species | Growth measurements | References |
|---|---|---|
| **Hydromyini** | | |
| *Conilurus penicillatus* | WT, HB, HE, PE, TA | 1, 2 |
| *Hydromys chrysogaster* | WT, EA | 3 |
| *Melomys cervinipes* | WT, HB, PE, TA | 1, 4 |
| *M. gouldii* | WT, HB, PE, TA | 5 |
| *Notomys alexis* | WT | 6, 7 |
| *N. cervinus* | WT | 7 |
| *N. mitchellii* | WT | 7 |
| *Pseudomys apodemoides* | WT, HB, PE, TA, | 1 |
| *P. australis* | WT | 6 |
| *P. delicatulus* | WT, HB, PE, TA | 1 |
| *P. gracilicaudatus* | WT, HB, HE, PE, TA | 9 |
| *P. higginsi* | WT, HB, PE, TA | 1, 10 |
| *P. nanus* | WT, HB, PE, TA | 1 |
| *P. novaehollandiae* | WT, HE, PE, | 11, 12 |
| *Uromys caudimaculatus* | WT, HB, PE, TA | 1 |
| *Zyzomys argurus* | WT, HB, PE, TA | 1 |
| *Z. woodwardi* | WT, HB, PE, TA | 1 |
| **Rattini** | | |
| *Rattus colletti* | WT, HB, PE, TA | 1 |
| *R. exulans* | WT, EA, HB, PE, TA | 13 |
| *R. fuscipes* | WT, HB, PE, TA | 1, 14, 15 |
| *R. leucopus* | WT, HB, PE, TA | 1 |
| *R. lutreolus* | WT, HB, HE, PE, TA | 16, 17 |
| *R. sordidus* | WT, HB | 18 |
| *R. tunneyi* | WT, HB, PE, TA | 1, 17 |

References: (1) Watts (1979a); (2) Kemper (2020); (3) Olsen (1982); (4) Rowell and Rymer (2020); (5) Crichton (1969); (6) Kerle and Fleming (2024); (7) Smith *et al.* (1972); (8) Crichton (1974); (9) Fox and Kemper (1982); (10) Green (1968); (11) Kemper (1976b); (12) Kemper (1979); (13) Wirtz (1973); (14) Taylor (1961); (15) Taylor and Horner (1971a); (16) Green (1967); (17) Fox (1979); (18) McDougall (1946).

next to the heat source at all times. If the rodent is unfurred, it can be assumed that heat requirements are similar to those of unfurred marsupials (i.e. 32°C). Use a minimum/maximum thermometer with a plastic-coated probe that can be placed next to the pup as this will ensure correct temperature tracking and regulation.

## 11.3 Specific requirements

When first brought in for hand-rearing, young rodents may be cold and dehydrated. Before attempting to feed, the pup should be warmed so that it is more responsive and can swallow properly. The pup can be placed in an incubator or in a small pouch with an external heat source such as hot water bottle or heat pad, taking care to prevent direct contact with the heat source and avoid any risk of burning the animal (see section 11.2).

If the pup is severely dehydrated, under the direction of a veterinarian warmed fluids for injection may be given subcutaneously while the pup is warming up. Once warmed, an electrolyte replacer such as Vytrate® can be offered orally. Formula feeding is commenced when the pup is at normal body temperature and hydration is improved.

Stress is a major problem in the successful rearing of native mammals, as it can be fatal. Therefore, it is important to keep noise to a minimum, not to overhandle the animals and to maintain high standards of hygiene.

While some species of rodents are more precocious than others, activity levels may increase almost overnight, so a larger enclosure needs to be readied in time. Once the animals start to leave the nest to explore and play it signals that weaning is imminent. These young animals can be readily imprinted. Most rodents are inquisitive, but easily startled and incredibly quick. Activity sessions take place as the animals emerge from the nest, usually shortly after dusk, at which time they should be given access to a secure enclosure set up for the species' specific needs. For example, for the stick-nest rat, use a large pet carrier (~700 × 500 × 500 mm), which is fortified where necessary by wire mesh on the outside. A leaf litter substrate can be used and sturdy climbing branches set up. Security is a vital factor, as young rodents can chew through anything but the hardest of materials.

### 11.3.1 *Hydromys* (water-rat): case study 1

This animal was found clinging to sticks during floods at Grafton, New South Wales, and was rescued and hand-reared by a local family. At day 1, bodyweight was 55 g and head-body length ~15 cm. Eyes were open, but not fully and fur was downy but not waterproof. The rat was lethargic and not eating well for the first 2 days. It was fed on warm, low-lactose kitty milk, by eyedropper and it soon learned to lap. By day 6, the animal weighed ~75 g and was much more active, swimming for short periods and lapping milk. Favoured food items were bait prawns and boiled egg.

On day 15 of hand-rearing it was surrendered to a zoo where it was deemed unsuitable for eventual release due to human imprinting. The water-rat was taken home overnight for 1 week by an animal caretaker, to ensure the weaning process was not made too abruptly (although eating solids and lapping milk well, the animal was still small and used to human contact for reassurance, and to encourage feeding). During this time, it was offered and ate the following items: prawns, egg, yabby, fly pupae, mussels, meat mix (a mix of mince (beef or kangaroo), egg and dog kibble), avocado, peanuts and crickets. Di-Vetelact® milk was also offered until day 20 of hand-rearing. Most food was eaten overnight.

During the day, the animal was kept in a similar container to that described for the stick-nest rat (see case study 2). At night it was given access to an aquarium (glass tank) with swimming water ~30 cm deep and a dry bank constructed of dirt and rocks. Easy access into and out of the water was ensured. The water-rat was also given exercise in the bathtub, again with easy access in and out, and under close supervision: 5–7 min was the maximum amount of time spent in the water before seeking new areas to explore.

### 11.3.2 *Leporillus* (stick-nest rat): case study 2

Bodyweight at the commencement of hand-rearing was 42.6 g. The head-body length was ~85 mm. Eyes were beginning to open, ears still flat against the head, fur soft and downy and the animal was very firmly attached to the teat.

Generally, six feeds per day were given, with the animal taking an average of 7.2 mL of milk per day. On day 6 of hand-rearing, feeds were reduced to four daily and the animal weaned abruptly on day 9, at which time stimulating to make it toilet after feeding was no longer necessary. Hard food had been offered from day 5: sprouted seed, a variety of greens, fruit and vegetables, nuts and dry seed. Favourites were the sprouted seed, apple and sweet corn (adults particularly favour greens, e.g. spinach, endive, and especially beans and the succulent plant pigface).

Milk was offered at body temperature from a teat attached to a bottle. A ringtail possum-sized teat was preferred, with a tiny needle hole.

Stick-nest rats are placid by nature, so hand-rearing can readily result in imprinting. This was deliberately encouraged in this instance because the animal was required as a contact animal. Hand-reared stick-nest rats have been introduced into a social group, so hand-rearing does not preclude this. Hand-reared stick-nest rats have been used to educate visitors about native rodents to try to counteract the negative image visitors often have of them. Due to the species' attractive appearance, soft fur and friendly behaviour they make an ideal ambassador in public settings.

### 11.3.3 *Mesembriomys* (black-footed tree-rat): case study 3

Bodyweight at commencement of hand-rearing was 175 g, head-body length was ~130 mm. The animal was furred, eyes open and detached from the teat. Di-Vetelact® was fed using an eyedropper for five feeds per day, with 7–16 mL taken each time. The animal was stimulated to toilet after each feed, until day 4 when it became no longer necessary. Hard food was offered from day 1, but the animal did not begin to wean until day 6 of hand-rearing when the amount of milk offered was decreased to a maximum of 10 mL per feed. Weaning was complete by day 11. Hard food offered included a variety of fruit, vegetables, greens, seed and nuts. Favourites were cucumber, avocado and nuts.

This animal was by nature much more timid than the stick-nest rat, but remained extremely gentle in human contact. Because of her impressive size and appearance, she was a great representative of this group of mammals and has made several appearances on television. She was introduced to a male at age 4.5 years old with no resulting problems; however, breeding did not occur (possibly due to her age).

## 11.4 Diet and feeding routine

### 11.4.1 Natural milk

The milk of native rodents changes throughout lactation, with solids typically ranging from 20–23% in early lactation to 30–35% in late lactation and lipids increasing from 10–5% to 20–31% in late lactation. In contrast, the content of proteins and carbohydrates remains relatively constant throughout lactation (Table 12.11). Iron content decreases steadily during lactation.

**Table 12.11.** Concentrations of the major constituents of milk for different species of rodents.

*Exotic species in Australia that are used here as analogues for native rodents of the *Rattus* genus.

| Species | Total solids (%) | Carbohydrates (%) | Lipids (%) | Protein (%) | Calcium (mg/L) | Iron (mg/L) | Reference |
|---|---|---|---|---|---|---|---|
| *Notomys alexis* | 20.6–32.7 | 2.3–2.6 | 14.2–21.0 | 5.3–6.0 | – | – | 1 |
| *N. cervinus* | 24.0–34.2 | 2.3–2.8 | 10.3–19.7 | 5.6–6.0 | – | – | 1 |
| *N. mitchellii* | 23.9–33.3 | 2.6–2.7 | 7.5–11.6 | 6.5 | – | – | 1 |
| *Pseudomys australis* | 25.4–29.7 | 2.9–3.6 | 9.2–22.6 | 5.3–6.4 | – | – | 1 |
| *Rattus rattus** | 22.2–32.6 | 2.0–3.8 | 8.7–14.8 | 8.1–9.2 | 750–1000 | 3–14 | 2–5 |
| *R. norvegicus** | 21.0 | 2.4–3.6 | 9.3–10.3 | 8.7 | 2680 | – | 6, 7 |

References: (1) Baverstock *et al.* (1976); (2) Kaldor and Ezekiel (1962); (3) Venkatachalam and Ramanathan (1964); (4) Oftedal and Iverson (1995); (5) Cox and Mueller (1936); (6) Luckey *et al.* (1955); (7) Jenness and Sloan (1970).

**Table 12.12.** Volume of Wombaroo® Dog Milk Replacer (mL) for native rodents of different bodyweights (g).

| Bodyweight | Feed volume | Bodyweight | Feed volume | Bodyweight | Feed volume |
|---|---|---|---|---|---|
| 0.1 | 0.15 | 1 | 0.8 | 10 | 4.4 |
| 0.2 | 0.25 | 2 | 1.3 | 20 | 7.8 |
| 0.3 | 0.35 | 3 | 1.8 | 30 | 10.0 |
| 0.4 | 0.40 | 4 | 2.2 | 40 | 12.4 |
| 0.5 | 0.45 | 5 | 2.6 | 50 | 14.7 |

### 11.4.2 Milk formulas

The main commercial formulas for hand-rearing rodents include:

- Biolac® – can potentially be used for rodents
- Di-Vetelact® (Lillelund Pty Ltd) – widely used low-lactose milk formula. Due to its low energy concentration when prepared as directed, some groups advise the addition of mono- and polyunsaturated fats such as canola oil-.
- Wombaroo® dog milk replacer has been used successfully to raise spinifex and Mitchell's hopping-mice.

### 11.4.3 Feeding apparatus

The type of equipment and teat used may be decided by preference and trial and error. For smaller species (e.g. spinifex hopping-mouse and plains mouse) an eyedropper, 1 ml syringe or catheter attached to a syringe can be used (remove the needle from the plastic coating and cut the plastic down to size). For medium to large-sized rodents (e.g. stick-nest rat and black-footed tree-rat) a ring-tail possum-sized teat attached to a small bottle has been used effectively. The teat should be punctured with a hot needle.

### 11.4.4 Feeding routine

Initially the manufacturer's instructions should be followed, but as they are only guidelines it may be necessary to adjust for specific circumstances (i.e. for the first few days more frequent feeding may be required if the animal only takes small amounts per feed). Young rodents should be fed at least 10–20% of their bodyweight in milk per day. Approximate volumes of milk that can be supplied to rodents of different body size are shown in Table 12.12. The animal must be weighed daily and feeds adjusted accordingly if necessary. The animal should be toileted after each feed, as per normal hand-rearing practices.

## 11.5 Data recording

When an animal is first brought in for hand-rearing, its sex and approximate age, using growth charts, should be recorded. During the hand-rearing process the following important information should be recorded on a daily basis:

- date when the information is recorded
- time when the information is recorded
- bodyweight to the nearest 1 g

- general activity and demeanour
- characteristics and frequency of defecation and urination
- amount (g) of different food types offered
- food consumption at each feed
- veterinary examinations and results.

These records provide important background information, such as food consumption, which will assist a veterinarian to reach a diagnosis if the animal becomes sick or fails to grow or gain weight. It also allows a comparison with growth curves to assess progress (section 10.16) and to allow new growth curves to be established for measurements in which they presently do not occur. The developmental stages and measurements outlined in section 10.3.1 should also be recorded on a weekly basis if possible.

## 11.6 Identification methods

Generally not required, though when furred a small patch of the fur could be painted with non-toxic paint.

## 11.7 Hygiene

Maintaining a high standard of hygiene is critical to the survival of the rodent. Emphasis needs to be placed on:

- maintaining clean bedding at all times
- maintaining personal hygiene by washing and disinfecting hands before and after handling the rodent
- washing hands between feeding different rodents
- using boiled water when making up formulas for rodents of all ages
- cleaning up spilt milk formula or paste, faeces and urine from the rodent's skin and fur as soon as possible and the fur dried to prevent bacterial or fungal skin infections
- washing all feeding equipment in warm water and dishwashing detergent and sterilising it in a microwavable baby steam steriliser or F10 ® disinfectant
- storing the teats and bottles in the fridge in an air-tight container after they have been disinfected
- heating up milk once and discarding leftovers
- avoiding contact with other animals unless certain they pose no health risk
- stimulating each animal to toilet before or after feeding. As with other mammals, toileting can be done by the application of warm water to the cloaca using cotton wool to stimulate urination and defecation, which allows the animal to keep drier and warmer in its bedding.

## 11.8 Behavioural considerations

None.

## 11.9 Use of foster animals

Cross-fostering has been undertaken with *Rattus* spp. and lactating female rats of all species will rear fostered young of their own or different species (Taylor 1961; Watts 1982b; Lloyd 1998). The older the young the greater the risk of rejection and, if possible, young should not be fostered when they are more than 7 days old (Watts 1982b). There is also the potential of cross-fostering native species under domestic mice as the hopping mice have been cross-fostered to domestic mice (M. Parrott pers. comm.).

## 11.10 Weaning

At weaning, fresh water should be supplied. A variety of fresh branches from native shrubs suitable for chewing should be made available at this time (non-toxic) as rodent incisor teeth need constant gnawing to ensure they do not overgrow. Native grasses and other nesting material should also be supplied. Regular weighing during weaning is particularly important to ensure body-weight is maintained as milk is reduced.

## 11.11 Release procedures

Hand-reared rodents are invariably maintained in captivity. If the animal is to be released, a wide variety of adult food items and a water dish (ensure the animal cannot drown in it) should be offered as the animal approaches weaning stage. Also, a variety of fresh branches from non-toxic native shrubs suitable for gnawing should be made available at this time, as rodent incisor teeth need constant maintenance in this manner. Native grasses or other material for nesting should also be supplied.

## ACKNOWLEDGEMENTS

Sincere thanks to Emeritus Professor Bill Breed for numerous suggestions and valuable references provided that contributed greatly to this chapter.

## REFERENCES

Anstee SD (1996) Use of external mound structures as indicators of the presence of the pebble mouse, *Pseudomys chapmani*, in mound systems. *Wildlife Research* **23**, 429–434. doi:10.1071/WR9960429

Anstee SD, Roberts JD, O'Shea JEO (1997) Social structure and patterns of movement of the western pebble-mouse,

*Pseudomys chapmani*, at Marandoo, Western Australia. *Wildlife Research* **24**, 295–305. doi:10.1071/WR96093

Aplin KP (2006) Ten million years of rodent evolution in Australasia: phylogenetic evidence and a speculative historical biogeography. In *Evolution and Biogeography of Australasian Vertebrates*. (Eds JR Merrick, M Archer, GM Hickey and MSY Lee) pp. 707–744. Auscipub, Sydney.

Ashton P (1986) Rats: public health in Sydney, 1900. In *Case Studies in Australian History*. (Ed. D Stewart) pp. 94–109. Heinemann Educational Australia, Richmond, Victoria.

Aslin HJ (1972) Nest-building of *Leporillus conditor* in captivity. *South Australian Naturalist* **47**, 43–46.

Aslin HJ, Watts CHS (1980) Breeding of a captive colony of *Notomys fuscus* Wood Jones (Rodentia: Muridae). *Australian Wildlife Research* **7**, 379–383. doi:10.1071/WR9800379

Baker AM, Gynther IC (eds) (2023) *Strahan's Mammals of Australia*. 4th edn. Reed New Holland Publishers, Sydney.

Banks PB, Cleary GP, Dickman CR (2011) Sydney's bubonic plague outbreak 1900–1910: a disaster for foreshore wildlife? *Australian Zoologist* **35**, 1033–1039. doi:10.7882/AZ.2011.058

Barnett SA (2001) *The Story of Rats: Their Impact on Us, and Our Impact on Them*. Allen & Unwin, Sydney.

Barnett SA, Stewart AP (1975) Audible signals during intolerant behaviour of *Rattus fuscipes*. *Australian Journal of Zoology* **23**, 103–112. doi:10.1071/ZO9750103

Barnett SA, Hocking WE, Munro KMH, Walker KZ (1975) Socially induced renal pathology of captive wild rats (*Rattus villosissimus*). *Aggressive Behavior* **1**, 123–133. doi:10.1002/1098-2337(1975)1:2<123::AID-AB2480010203>3.0.CO;2-U

Baverstock PR, Spencer L, Pollard C (1976) Water balance of small lactating rodents. II: concentration and composition of milk of females on *ad libitum* and restricted water intakes. *Comparative Biochemistry and Physiology. A. Comparative Physiology* **53**, 47–52. doi:10.1016/S0300-9629(76)80008-4

Begg RJ (1981) The small mammals of Little Nourlangie Rock, N.T. IV. Ecology of *Zyzomys woodwardi*, the large rock-rat, and *Z. argurus*, the common rock-rat. *Australian Wildlife Research* **8**, 307–320. doi:10.1071/WR9810307

Begg RJ, Dunlop CR (1980) Security eating, and diet in the large rock-rat, *Zyzomys woodwardi* (Rodentia: Muridae). *Australian Wildlife Research* **7**, 63–70. doi:10.1071/WR9800063

Begg RJ, Dunlop CR (1985) The diets of the large rock-rat, *Zyzomys woodwardi*, and the common rock-rat, *Z. argurus* (Rodentia: Muridae). *Australian Wildlife Research* **12**, 19–24. doi:10.1071/WR9850019

Begg RJ, Nelson JE (1977) The agonistic behaviour of *Rattus villosissimus*. *Australian Journal of Zoology* **25**, 291–327. doi:10.1071/ZO9770291

Begg R, Walsh B, Woerle F, King S (1983) Ecology of *Melomys burtoni*, the grassland melomys (Rodentia: Muridae) at Coburg Peninsula, N.T. *Australian Wildlife Research* **10**, 259–267. doi:10.1071/WR9830259

Bellamy D, Berry RJ, Jakobson ME, Lidicker WZ, Morgan J, Murphy HM (1973) Ageing in an island population of the house mouse. *Age and Ageing* **2**, 235–250. doi:10.1093/ageing/2.4.235

Bennison K, Godfree R, Dickman CR (2018) Synchronous boom-bust cycles in central Australian rodents and marsupials in response to rainfall and fire. *Journal of Mammalogy* **99**, 1137–1148. doi:10.1093/jmammal/gyy105

Berris KK, Breed WG, Carthew SM (2020*a*) What can we deduce about the reproductive condition of spinifex hopping mice (*Notomys alexis*) from external examination? *Australian Mammalogy* **42**, 11–15. doi:10.1071/AM18038

Berris KK, Breed WG, Moseby KE, Carthew SM (2020*b*) Female reproductive suppression in an Australian arid zone rodent, the spinifex hopping mouse. *Journal of Zoology* **312**, 163–173. doi:10.1111/jzo.12813

Berry RJ, Truslove GM (1968) Age and eye lens weight in the house mouse. *Journal of Zoology* **155**, 247–252. doi:10.1111/j.1469-7998.1968.tb03051.x

Bradley AJ, Kemper CM, Kitchener DJ, Humphreys WF, How RA, Schmitt LH (1988) Population ecology and physiology of the common rock-rat, *Zyzomys argurus* (Rodentia: Muridae) in tropical northwestern Australia. *Journal of Mammalogy* **69**, 749–764. doi:10.2307/1381630

Braithwaite RW (1977) Preliminary observations on the activity patterns of *Rattus lutreolus* and other Victorian small mammals. *Victorian Naturalist* **94**, 216–219.

Braithwaite RW (1980) The ecology of *Rattus lutreolus*. II: reproductive tactics. *Australian Wildlife Research* **7**, 53–62. doi:10.1071/WR9800053

Braithwaite RW, Brady P (1993) The delicate mouse, *Pseudomys delicatulus*: a continuous breeder waiting for the good times. *Australian Mammalogy* **16**, 93–98. doi:10.1071/AM93022

Braithwaite RW, Lee AK (1979) The ecology of *Rattus lutreolus*. I: a Victorian heathland population. *Australian Wildlife Research* **6**, 173–189. doi:10.1071/WR9790173

Brandle R, Moseby KE (1999) Comparative ecology of two populations of *Pseudomys australis* in northern South Australia. *Wildlife Research* **26**, 541–564. doi:10.1071/WR97049

Breakey DR (1963) The breeding season and age structure of feral house mouse populations near San Francisco Bay, California. *Journal of Mammalogy* **44**, 153–168. doi:10.2307/1377448

Breed A, Eden P (2008) Rodents. In *Medicine of Australian Mammals*. (Eds L Vogelnest and R Woods) pp. 503–526. CSIRO Publishing, Melbourne.

Breed B, Ford F (2007) *Native Mice and Rats*. CSIRO Publishing, Melbourne.

Breed WG (1975) Environmental factors and reproduction in the female hopping mouse *Notomys alexis*. *Journal of Reproduction and Fertility* **45**, 273–281. doi:10.1530/jrf.0.0450273

Breed WG (1976a) Effect of differing environments on the oestrous cycle and ovulation rates in several species of native Australian rats. *Journal of Reproduction and Fertility* **46**, 513–514. doi:10.1530/jrf.0.0460513

Breed WG (1976b) Effects of environment on ovarian activity of wild hopping mice (*Notomys alexis*). *Journal of Reproduction and Fertility* **47**, 395–397. doi:10.1530/jrf.0.0470395

Breed WG (1978) Ovulation rates oestrous cycle lengths in several species of Australian native rats (*Rattus* sp.) from various habitats. *Australian Journal of Zoology* **26**, 475–480. doi:10.1071/ZO9780475

Breed WG (1979) The reproductive rate of the hopping-mouse *Notomys alexis* and its ecological significance. *Australian Journal of Zoology* **27**, 177–194. doi:10.1071/ZO9790177

Breed WG (1989) Comparative studies on the reproductive biology of three species of laboratory bred Australian conilurine rodent (Muridae: Hydromyidae). *Journal of Zoology* **217**, 683–699[London]. doi:10.1111/j.1469-7998.1989.tb02517.x

Breed WG (1990*a*) Copulatory behaviour and coagulum formation in the female reproductive tract of the Australian hopping mouse, *Notomys alexis*. *Journal of Reproduction and Fertility* **88**, 17–24. doi:10.1530/jrf.0.0880017

Breed WG (1990*b*) Comparative studies on the timing of reproduction and foetal number in six species of Australian conilurine rodents (Muridae: Hydromyinae). *Journal of Zoology* **221**, 1–10[London]. doi:10.1111/j.1469-7998.1990.tb03770.x

Breed WG (1995) Sandy inland mouse *Pseudomys hermannsburgensis*. In *The Mammals of Australia*. 2nd edn. (Ed. R Strahan) pp. 604–605. Reed Books, Sydney.

Breed WG (1997) Interspecific variation of testis size and epididymal sperm numbers in Australasian rodents with special reference to the genus *Notomys*. *Australian Journal of Zoology* **45**, 651–669. doi:10.1071/ZO97010

Breed WG, Adams M (1992) Breeding systems of spinifex hopping mice (*Notomys alexis*) and plains rats (*Pseudomys australis*): a test for multiple paternity within the laboratory. *Australian Journal of Zoology* **40**, 13–20. doi:10.1071/ZO9920013

Breed WG, Washington JM (1991) Mating behaviour and insemination in the hopping mouse, *Notomys alexis*. *Journal of Reproduction and Fertility* **93**, 187–194. doi:10.1530/jrf.0.0930187

Brisbane K (1998) A rare occasion! The rediscovery of the central rock rat *Zyzomys pedunculatus* in the Western MacDonnell National Park. ARAZPA/ASZK Conference Proceedings. Taronga Zoo 22–27 March.

Bubela TM, Happold DCD (1993) The social organisation and mating system of an Australian subalpine rodent, the broad-toothed rat, *Mastacomys fuscus* Thomas. *Wildlife Research* **20**, 405–417. doi:10.1071/WR9930405

Bubela TM, Happold DCD, Broome LS (1991) Home range and activity of the broad-toothed rat, *Mastacomys fuscus*, in subalpine heathland. *Wildlife Research* **18**, 39–48. doi:10.1071/WR9910039

Burbidge AA, McKenzie NL (1989) Patterns in the modern decline of Western Australia's vertebrate fauna: causes and conservation implications. *Biological Conservation* **50**, 143–198. doi:10.1016/0006-3207(89)90009-8

Calaby JH, Taylor JM (1983) Breeding in wild populations of the Australian rock-rats, *Zyzomys argurus* and *Z. woodwardi*. *Journal of Mammalogy* **64**, 610–616. doi:10.2307/1380516

Calaby JH, Wimbush DJ (1964) Observations on the broad-toothed rat *Mastacomys fuscus* Thomas. *CSIRO Wildlife Research* **9**, 123–133. doi:10.1071/CWR9640123

Calver MC, King DR, Gardner JL, Martin GR (1991) Total food consumption of some native Australian small mammals in the laboratory. *Australian Mammalogy* **14**, 139–142. doi:10.1071/AM91017

Carron PL (1985) The ecology of three species of small mammals in subalpine habitat. PhD thesis. Australian National University, Canberra.

Carron PL, Happold CD, Bubela TM (1990) Diet of two sympatric Australian subalpine rodents, *Mastacomys fuscus* and *Rattus fuscipes*. *Australian Wildlife Research* **17**, 479–489. doi:10.1071/WR9900479

Carstairs JL (1972) Reproduction and development of *Rattus villosissimus* in captivity. *Australian Mammalogy* **1**, 403[Abstract].

Champlin AK, Dorr DL, Gates AH (1973) Determining the stage of estrous cycle in the mouse by the appearance of the vagina. *Biology of Reproduction* **8**, 491–494. doi:10.1093/biolreprod/8.4.491

Cockburn A (1981*a*) Population regulation and dispersion of the smoky mouse *Pseudomys fumeus*. II: spring decline, breeding success and habitat heterogeneity. *Australian Journal of Ecology* **6**, 255–266. doi:10.1111/j.1442-9993.1981.tb01575.x

Cockburn A (1981*b*) Population processes of the silky desert mouse *Pseudomys apodemoides* (Rodentia) in mature heathlands. *Australian Wildlife Research* **8**, 499–514. doi:10.1071/WR9810499

Copley P (1988) *The Stick-nest Rats of Australia*. Department of Environment and Planning, South Australia.

Copley P (1999) Natural histories of Australia's stick-nest rats, genus *Leporillus* (Rodentia: Muridae). *Wildlife Research* **26**, 513–539. doi:10.1071/WR97056

Cox WM, Mueller AJ (1936) The composition of milk from stock rats and an apparatus for milking small laboratory animals. *The Journal of Nutrition* **13**, 249–261. doi:10.1093/jn/13.3.249

Crichton EG (1969) Reproduction in the pseudomyine rodent *Mesembriomys gouldii* (Gray)(Muridae). *Australian Journal of Zoology* **17**, 785–797. doi:10.1071/ZO9690785

Crichton EG (1974) Aspects of reproduction in the genus *Notomys* (Muridae). *Australian Journal of Zoology* **22**, 439–447. doi:10.1071/ZO9740439

Dahl K (1897) Biological notes on North-Australian mammals. *The Zoologist* **1**(4), 189–216.

Dewsbury DA, Hodges AW (1987) Copulatory behavior and related phenomena in spiny mice (*Acomys caharinus*) and hopping mice (*Notomys alexis*). *Journal of Mammalogy* **68**, 49–57. doi:10.2307/1381044

Dickman CR (1999) Rodent-ecosystem relationships: a review. In *Ecologically-based Management of Rodent Pests*. (Eds GR Singleton, LA Hinds, H Leirs and Z Zhang). pp. 113–133. Australian Centre for International Agricultural Research, Canberra.

Dickman CR, Happold DCD (2022) Demographic and dietary responses of small mammals to high severity fire. *Australian Zoologist* **42**, 256–277. doi:10.7882/AZ.2022.018

Dickman CR, Mahon PS, Masters P, Gibson DF (1999) Long-term dynamics of rodent populations in arid Australia: the influence of rainfall. *Wildlife Research* **26**, 389–403. doi:10.1071/WR97057

Dickman CR, Leung LKP, Van Dyck SM (2000) Status, ecological attributes and conservation of native rodents in Queensland. *Wildlife Research* **27**, 333–346. doi:10.1071/WR97130

Dickman CR, Greenville AC, Beh CL, Tamayo B, Wardle GM (2010) Social organization and movements of desert rodents during population "booms" and "busts" in central Australia. *Journal of Mammalogy* **91**, 798–810. doi:10.1644/09-MAMM-S-205.1

Dickman CR, Greenville AC, Tamayo B, Wardle GM (2011) Spatial dynamics of small mammals in central Australian desert habitats: the role of drought refugia. *Journal of Mammalogy* **92**, 1193–1209. doi:10.1644/10-MAMM-S-329.1

Diete RL, Meek PD, Dickman CR, Leung LKP (2014) Burrowing behaviour of the northern hopping-mouse (*Notomys aquilo*): field observations. *Australian Mammalogy* **36**, 242–246. doi:10.1071/AM13039

Diete RL, Meek PD, Dickman CR, Leung LKP (2016) Ecology and conservation of the northern hopping-mouse (*Notomys aquilo*). *Australian Journal of Zoology* **64**, 21–32. doi:10.1071/ZO15082

Dixon J, Huxley L (1985) *Donald Thomson's Mammals and Fishes of Northern Australia*. Thomas Nelson Australia, Melbourne.

Dwyer PD (1975) Observations on the breeding biology of some New Guinea murid rodents. *Australian Wildlife Research* **2**, 33–45. doi:10.1071/WR9750033

Dwyer PD (1978) A study of *Rattus exulans* (Peale) (Rodentia: Muridae) in the New Guinea highlands. *Australian Wildlife Research* **5**, 221–248. doi:10.1071/WR9780221

Egoscue HJ (1970) A laboratory colony of the Polynesian rat, *Rattus exulans*. *Journal of Mammalogy* **51**, 261–266. doi:10.2307/1378476

Eiris GC (2011) Factors that limit the distribution and abundance of the giant white-tailed rat *Uromys caudimaculatus*. PhD thesis, University of Sydney, Sydney.

Finlayson HH (1941) On central Australian mammals. Pt 2: the Muridae. *Transactions of the Royal Society of South Australia* **65**, 215–233.

Finlayson HH (1960) *Rattus greyi* Gray and its derivatives. *Transactions of the Royal Society of South Australia* **83**, 123–147.

Firman RC (2014) Female fitness, sperm traits and patterns of paternity in an Australian polyandrous mouse. *Behavioral Ecology and Sociobiology* **68**, 283–290. doi:10.1007/s00265-013-1643-1

Firth RSC, Jefferys E, Woinarski JCZ, Noske RA (2005) The diet of the brush-tailed rabbit-rat (*Conilurus penicillatus*) from the monsoonal tropics of the Northern Territory, Australia. *Wildlife Research* **32**, 517–523. doi:10.1071/WR04127

Firth RSC, Woinarski JCZ, Noske RA (2006) Home range and den characteristics of the brush-tailed rabbit-rat (*Conilurus penicillatus*) in the monsoonal tropics of the Northern Territory, Australia. *Wildlife Research* **33**, 397–407. doi:10.1071/WR05057

Fox BJ (1979) Growth and development of *Rattus lutreolus* (Rodentia: Muridae) in the laboratory. *Australian Journal of Zoology* **27**, 945–957. doi:10.1071/ZO9790945

Fox BJ (1985) A graphical method for estimating length of gestation and estrous cycle length from birth intervals in rodents. *Journal of Mammalogy* **66**, 168–173. doi:10.2307/1380977

Fox BJ (2008) Eastern chestnut mouse. *Pseudomys gracilicaudatus* (Gould, 1845). In *The Mammals of Australia*. 2nd edn. (Eds S Van Dyck and R Strahan) pp. 634–635. Reed Books, Sydney.

Fox BJ, Kemper CM (1982) Growth and development of *Pseudomys gracilicaudatus* (Rodentia: Muridae) in the laboratory. *Australian Journal of Zoology* **30**, 175–185. doi:10.1071/ZO9820175

Fox BJ, Murray JD (1979) Laboratory hybridisation of Australian *Rattus fuscipes* and *Rattus lutreolus* and its karyotypic confirmation. *Australian Journal of Zoology* **27**, 691–698. doi:10.1071/ZO9790691

Frazer DS, Petit S (2007) Use of *Xanthorrhoea semiplana* (grass-trees) for refuge by *Rattus fuscipes* (southern bush rat). *Wildlife Research* **34**, 379–386. doi:10.1071/WR06123

Freeland WJ (1972) A rainforest and its rodents. MSc thesis. University of Queensland, Brisbane.

Friend GR (1987) Population ecology of *Mesembriomys gouldii* (Rodentia: Muridae) in the wet-dry tropics of the Northern Territory. *Australian Wildlife Research* **14**, 293–303. doi:10.1071/WR9870293

Friend GR, Calaby JH (1995) Black-footed tree-rat *Mesembriomys gouldii*. In *The Mammals of Australia*. 2nd edn. (Ed. R Strahan) pp. 564–566. Reed Books, Sydney.

Gaikhorst G, Lambert C (2009) Breeding and maintenance of the central rock-rat *Zyzomys pedunculatus* at Perth Zoo. *International Zoo Yearbook* **43**, 212–221. doi:10.1111/j.1748-1090.2008.00075.x

Geffen E, Rowe KC, Yom-Tov Y (2011) Reproductive rates in Australian rodents are related to phylogeny. *PLoS One* **6**(4), e19199. doi:10.1371/journal.pone.0019199

Gleen W (2009) Gender determination, by examination of external features, of some species of Australian rodents. Unpublished report. Taronga Zoo, Sydney.

Godthelp H (1999) Diversity, relationships and origins of the tertiary and quaternary rodents of Australia. *Australian Mammalogy* **21**, 32–34.

Green RH (1967) The murids and small dasyurids of in Tasmania: parts 1 and 2. *Records of the Queen Victoria Museum Launceston* **28**, 1–19.

Green RH (1968) The murids and small dasyurids in Tasmania: parts 3 and 4. *Records of the Queen Victoria Museum* **32**, 1–19.

Green RH (1993) *The Fauna of Tasmania: Mammals*. Potoroo Publishing, Launceston.

Green RH (1995) Long-tailed mouse *Pseudomys higginsi*. In *The Mammals of Australia*. 2nd edn. (Ed. R Strahan) pp. 605–606. Reed Books, Sydney.

Greenville AC, Wardle GM, Dickman CR (2012) Extreme climatic events drive mammal irruptions: regression analysis of 100-year trends in desert rainfall and temperature. *Ecology and Evolution* **2**, 2645–2658. doi:10.1002/ece3.377

Greenville AC, Wardle GM, Nguyen V, Dickman CR (2016) Population dynamics of desert mammals: similarities and contrasts within a multispecies assemblage. *Ecosphere* **7**(5), e01343. doi:10.1002/ecs2.1343

Griffiths AD, Koenig J, Carrol F, Price O (2002) Activity area and day-time tree use of the black-footed tree-rat *Mesembriomys gouldii*. *Australian Mammalogy* **23**, 181–183. doi:10.1071/AM01181

Gynther I, Waller N, Leung LKP (2016) Confirmation of the extinction of the Bramble Cay Melomys *Melomys rubicola* on Bramble Cay, Torres Strait: results and conclusions from a comprehensive survey in August–September 2014. Unpublished report to the Department of Environment and Heritage Protection, Queensland Government, Brisbane.

Happold DCD (1989) Small mammals of the Australian Alps. In *The Scientific Significance of the Australian Alps*. (Ed. R Good) pp. 221–239. Australian Academy of Science, Canberra.

Happold DCD (2015) A 10-year demographic study of a small mammal community in the Australian Alps. *Australian Journal of Zoology* **63**, 338–349. doi:10.1071/ZO15033

Happold M (1976*a*) Social behaviour of the conilurine rodents (Muridae) of Australia. *Zeitschrift für Tierpsychologie* **40**, 113–182. doi:10.1111/j.1439-0310.1976.tb00930.x

Happold M (1976*b*) Reproductive biology and development in the conilurine rodents (Muridae) of Australia. *Australian Journal of Zoology* **24**, 19–26. doi:10.1071/ZO9760019

Happold M (2021) *Journey Among Animals*. The Book Guild, Kibworth, UK.

Hill NJ, Rose K, Deane EM, Old JM (2007) *Rodentolepis fraterna*: the cause of mortality in a new host, the spinifex hopping-mouse (*Notomys alexis*)? *Australian Veterinary Journal* **85**, 62–64. doi:10.1111/j.1751-0813.2006.00095.x

Horner BE, Taylor JM (1958) Breeding of *Rattus assimilis* in captivity. *Journal of Mammalogy* **39**, 301–302. doi:10.2307/1376215

Horner BE, Taylor JM (1965) Systematic relationships among *Rattus* in southern Australia: evidence from cross-breeding experiments. *CSIRO Wildlife Research* **10**, 101–109. doi:10.1071/CWR9650101

Horner BE, Taylor JM (1969) Paternal behavior in *Rattus fuscipes*. *Journal of Mammalogy* **50**, 803–805. doi:10.2307/1378259

Hulst F (2025) Zoonoses. In *Current Therapy in Medicine of Australian Mammals*. Revised Edition. (Eds. L Vogelnest and T Portas) pp. 273–289. CSIRO Publishing, Melbourne.

International Air Transport Association (IATA) (2022) Live Animals Regulations. 48th edn. International Air Transport Association, Geneva Switzerland.

Irby DC, Kerr JB, Risbridger GP, De Kretser DM (1984) Seasonally and experimentally induced changes in testicular function of the Australian bush rat (*Rattus fuscipes*). *Journal of Reproduction and Fertility* **70**, 657–666. doi:10.1530/jrf.0.0700657

Jackson SM, Groves C (2015) *Taxonomy of Australian Mammals*. CSIRO Publishing, Melbourne.

Jenness R, Sloan RE (1970) The composition of milks of various species: a review. *Dairy Science Abstracts* **32**, 599–612.

Johnson R (2008) Cryptococcal infections in Australian mammals in a zoo. In Proceedings of the AAZV ARAV Joint Conference. Los Angeles, CA. pp. 27–28.

Kaldor I, Ezekiel E (1962) Iron content of mammalian breast milk: measurements in the rat and in a marsupial. *Nature* **196**, 175. doi:10.1038/196175a0

Kemper CM (1976*a*) Reproduction of *Pseudomys novaehollandiae* (Muridae) in the laboratory. *Australian Journal of Zoology* **24**, 159–167. doi:10.1071/ZO9760159

Kemper CM (1976*b*) Growth and development of the Australian murid *Pseudomys novaehollandiae*. *Australian Journal of Zoology* **24**, 27–37. doi:10.1071/ZO9760027

Kemper CM (1979) Growth of an Australian murid (*Pseudomys novaehollandiae*) in the wild. *Acta Theriologica* **24**, 257–266. doi:10.4098/AT.arch.79-26

Kemper CM (1980) Reproduction of *Pseudomys novaehollandiae* (Muridae) in the wild. *Australian Wildlife Research* **7**, 385–402. doi:10.1071/WR9800385

Kemper CM (1981) Description of *Pseudomys novaehollandiae* burrows located with radio-isotopes. *Australian Mammalogy* **4**, 141–143. doi:10.1071/AM81008

Kemper CM (1995) Brush-tailed tree-rat *Conilurus penicillatus*. In *The Mammals of Australia*. 2nd edn. (Ed. R Strahan) pp. 553–554. Reed Books, Sydney.

Kemper CM (2020) Growth and development of the brush-tailed rabbit-rat (*Conilurus penicillatus*), a threatened tree-rat from northern Australia. *Australian Mammalogy* **43**, 151–159. doi:10.1071/AM19027

Kerle JA, Fleming MR (2024) The golden-backed tree-rat (*Mesembriomys macrurus*) (Peters 1876): reproduction, growth, development and behaviour in captivity. *Australian Mammalogy* **46**, AM24007. doi:10.1071/AM24007

Lacy RC, Horner BE (1997) Effects of inbreeding on reproduction and sex ratio of *Rattus villosissimus*. *Journal of Mammalogy* **78**, 877–887. doi:10.2307/1382946

Lacy RC, Balou JD, Pollak JP (2011) PMx: software package for demographic and genetic analysis and management of pedigreed populations. *Methods in Ecology and Evolution* **3**, 433–437. doi:10.1111/j.2041-210X.2011.00148.x

Le Souef AS (1922) Notes on the nesting and breeding habits of the house building rat (*Conilurus conditor*) and Banfield's rat (*Uromys banfieldi*). *Australian Zoologist* **3**, 15–16.

Le Souef AS, Burrell H (1926) *The Wild Animals of Australasia*. Harrap, London.

Lees C, Johnson K (2002) *Australasian Species Management Program: Regional Census and Plan*. 12th edn. Australasian Regional Association of Zoological Parks and Aquaria, Sydney.

Lendon C, Hulst F, Shaw M (2015) Managing diabetes through diet in black-footed tree rats (*Mesembriomys gouldii*) and greater stick-nest rats (*Leporillus conditor*). In Proceedings of the AAZ/NAG Joint Conference. Portland, Oregon. p. 90.

Leung LK-P (1999*a*) Ecology of Australian tropical rainforest mammals. II: the Cape York melomys, *Melomys capensis* (Muridae: Rodentia). *Wildlife Research* **26**, 307–316. doi:10.1071/WR96043

Leung LK-P (1999*b*) Ecology of Australian tropical rainforest mammals. III: the Cape York rat, *Rattus leucopus* (Muridae: Rodentia). *Wildlife Research* **26**, 317–328. doi:10.1071/WR96044

Lidicker WZ (1966) Ecological observations on a feral house mouse population declining to extinction. *Ecological Monographs* **36**, 27–50. doi:10.2307/1948487

Lindner E, Fuelling O (2002) Marking methods in small mammals: ear-tattoo as an alternative to toe-clipping. *Journal of Zoology* **256**, 159–163[London]. doi:10.1017/S0952836902000195

Lloyd K (1998) The laboratory rat as an inter-species foster-mother for an Australian rodent. *Thylacinus* **22**, 6–7.

Lloyd K (1999) The behaviour of the endangered Carpentarian rock rat (*Zyzomys palatalis*) in captivity. ARAZPA/ASZK Conference, Alice Springs Desert Park, Alice Springs.

Long JA, Evans HM (1922) The oestrous cycle of the rat and its associated phenomena. *Memoirs of the University of California* **6**, 1–148.

Luckey TD, Mende TJ, Pleasants J (1955) The physical and chemical characterisation of rat's milk. *The Journal of Nutrition* **54**, 345–359. doi:10.1093/jn/54.3.345

Lunney D (1978) Ecology of *Rattus lutreolus*. MSc thesis. University of Sydney, Sydney.

Luo J, Fox BJ, Jeffreys E (1994) Diet of the eastern chestnut mouse (*Pseudomys gracilicaudatus*). I. Composition, diversity and individual variation. *Wildlife Research* **21**, 401–417. doi:10.1071/WR9940401

Mackie JT, Booth R, Caton W, Stevenson R (2001) Concurrent infections with cilia-associated respiratory bacillus and mycoplasmas in spinifex hopping mice (*Notomys alexis*) with pneumonia. *Australian Veterinary Journal* **79**, 502–504. doi:10.1111/j.1751-0813.2001.tb13026.x

Magnusson WE, Webb GJW, Taylor JA (1976) Two new locality records, a new habitat and a nest description for *Xeromys myoides* Thomas (Rodentia: Muridae). *Australian Wildlife Research* **3**, 153–157. doi:10.1071/WR9760153

Mallick SA (1992) Urine-marking in three species of *Rattus*. *Wildlife Research* **19**, 89–93. doi:10.1071/WR9920089

McDougall WA (1944) An investigation of the rat pest problem in Queensland canefields: 2. Species and general habits. *Queensland Journal of Agricultural Science* **1**, 48–78.

McDougall WA (1946) An investigation of the rat pest problem in Queensland canefields. Part 4: breeding and life histories. *Queensland Journal of Agricultural Science* **3**, 1–43.

McEvedy C (1988) The bubonic plague. *Scientific American* **258**, 74–79. doi:10.1038/scientificamerican0288-118

McKenzie NL, Kerle JA (1995) Golden-backed tree-rat *Mesembriomys macrurus*. In *The Mammals of Australia*. 2nd edn. (Ed. R Strahan) pp. 566–568. Reed Books, Sydney.

McNally J (1960) The biology of the water rat *Hydromys chrysogaster* Geoffroy (Muridae: Hydromyinae) in Victoria. *Australian Journal of Zoology* **8**, 170–180. doi:10.1071/ZO9600170

Meek PD (2002) The nest of the Hastings River mouse *Pseudomys oralis*. *Australian Mammalogy* **24**, 225–227. doi:10.1071/AM02225

Monamy V (1995) Population dynamics of, and habitat use by, Australian native rodents in wet sclerophyll forest, Tasmania. I: *Rattus lutreolus velutinus* (Rodentia: Muridae). *Wildlife Research* **22**, 647–660. doi:10.1071/WR9950647

Moore LA (1995) Giant white-tailed rat *Uromys caudimaculatus*. In *The Mammals of Australia*. 2nd edn. (Ed. R Strahan) pp. 638–640. Reed Books, Sydney.

Moro D, Morris K (2000*a*) Movements and refugia of Lakeland Downs short-tailed mice, *Leggadina lakedownensis*, and house mice, *Mus domesticus*, on Thevenard Island, Western Australia. *Wildlife Research* **27**, 11–20. doi:10.1071/WR99016

Moro D, Morris K (2000*b*) Population structure and dynamics of sympatric house mice, *Mus domesticus*, and Lakeland Downs short-tailed mice, *Leggadina lakedownensis*, on Thevenard Island, Western Australia. *Wildlife Research* **27**, 257–268. doi:10.1071/WR99019

Morton SR, Brown JH, Kelt DA, Reid JRW (1994) Comparisons of community structure among small mammals of North American and Australian deserts. *Australian Journal of Zoology* **42**, 501–525. doi:10.1071/ZO9940501

Murray BR, Dickman CR (1994) Granivory and microhabitat use in Australian desert rodents: are seeds important? *Oecologia* **99**, 216–225. doi:10.1007/BF00627733

Murray BR, Dickman CR, Watts CHS, Morton SR (1999) The dietary ecology of Australian desert rodents. *Wildlife Research* **26**, 421–437. doi:10.1071/WR97046

Myers P (1978) A method for determining the age of living small mammals. *Journal of Zoology* **186**, 551–556.

Myers K, Carstairs J, Gilbert N (1977) Determination of age of indigenous rats in Australia. *The Journal of Wildlife Management* **41**, 322–326. doi:10.2307/3800616

Nano TJ, Smith CM, Jefferys E (2003) Investigation into the diet of the central rock-rat (*Zyzomys pedunculatus*). *Wildlife Research* **30**, 513–518. doi:10.1071/WR01084

Norton TW (1987*a*) The ecology of small mammals in north-eastern Tasmania. II. *Pseudomys novaehollandiae* and the introduced *Mus musculus*. *Australian Wildlife Research* **14**, 435–441. doi:10.1071/WR9870435

Norton TW (1987*b*) The ecology of small mammals in north-eastern Tasmania. I. *Rattus lutreolus velutinus*. *Australian Wildlife Research* **14**, 415–433. doi:10.1071/WR9870415

Obendorf DL, Smales LR (1985) The internal parasites and pathological findings in *Hydromys chrysogaster* (Muridae: Hydromyinae) from Tasmania. *Australian Journal of Zoology* **33**, 33–38. doi:10.1071/ZO9850033

Oftedal IT, Iverson SJ (1995) Comparative analysis of nonhuman milks. In *Handbook of Milk Composition*. (Ed. RG Jensen) pp. 749–789. Academic Press, New York.

Old JM, Connelly L, Francis J, Branch K, Fry G, Deane EM (2005) Haematology and serum biochemistry of three Australian desert murids: the plains rat (*Pseudomys australis*), the spinifex hopping-mouse (*Notomys alexis*) and the central rock-rat (*Zyzomys pedunculatus*). *Comparative Clinical Pathology* **14**, 130–137. doi:10.1007/s00580-005-0586-y

Old JM, Connelly L, Francis J, Gogler J (2007*a*) Haematology and serum biochemistry of the Carpentarian rock-rat (*Zyzomys palatalis*). *Comparative Clinical Pathology* **16**, 249–252. doi:10.1007/s00580-007-0694-y

Old JM, Hill NJ, Deane EM (2007*b*) Isolation of the mite *Mycoptes musculinus* Koch from the spinifex hopping-mouse (*Notomys alexis*). *Laboratory Animals* **41**, 292–295. doi:10.1258/002367707780378186

Olsen P (1980) Seasonal and maturational pelage changes, and injuries, in the eastern water rat, *Hydromys chrysogaster*, at Griffith, N.S.W. *Australian Wildlife Research* **7**, 217–233. doi:10.1071/WR9800217

Olsen P (1982) Reproductive biology and development of the water rat, *Hydromys chrysogaster*, in captivity. *Australian Wildlife Research* **9**, 39–53. doi:10.1071/WR9820039

Olsen PD (1995) Water rat *Hydromys chrysogaster*. In *The Mammals of Australia*. 2nd edn. (Ed. R Strahan) pp. 628–629. Reed Books, Sydney.

Pavey CR, Addison J, Brandle R, Dickman CR, McDonald PJ, Moseby KE, Young LI (2017) The role of refuges in the persistence of Australian dryland mammals. *Biological Reviews of the Cambridge Philosophical Society* **92**, 647–664. doi:10.1111/brv.12247

Plomley NJB (1972) Some notes on plagues of small mammals in Australia. *Journal of Natural History* **6**, 363–384. doi:10.1080/00222937200770351

Predavec M, Dickman CR (1994) Population dynamics and habitat use of the long-haired rat (*Rattus villosissimus*) in south-western Queensland. *Wildlife Research* **21**, 1–10. doi:10.1071/WR9940001

Press AJ (1987) Comparison of the demography of populations of *Rattus fuscipes* living in cool temperate rainforests and dry sclerophyll forests. *Australian Wildlife Research* **14**, 45–63. doi:10.1071/WR9870045

Pye T (1991) The New Holland mouse (*Pseudomys novaehollandiae*) (Rodentia: Muridae) in Tasmania: a field study. *Wildlife Research* **18**, 521–531. doi:10.1071/WR9910521

Read DG (1984) Diet and habitat preferences of *Leggadina forresti* (Rodentia: Muridae) in western New South Wales. *Australian Mammalogy* **7**, 215–217. doi:10.1071/AM84023

Read JL, Copley P, Bird P (1999) The distribution, ecology and current status of *Pseudomys desertor* in South Australia. *Wildlife Research* **26**, 453–462. doi:10.1071/WR97051

Redhead TD (1979) On the demography of *Rattus sordidus colletti* in monsoonal Australia. *Australian Journal of Ecology* **4**, 115–136. doi:10.1111/j.1442-9993.1979.tb01202.x

Reid JWR, Morton SR (1995) Forrest's mouse *Leggadina forresti*. In *The Mammals of Australia*. 2nd edn. (Ed. R Strahan) pp. 555–556. Reed Books, Sydney.

Ricci S (2003) Population dynamics and trophic ecology of two species of Australian desert rodents. PhD thesis. University of Sydney, Sydney.

Richards N, Burke D (2007) Physiological and behavioural indicators of social organisation in *Notomys alexis*. p. 10, in Abstract from Australian Society for the Study of Animal Behaviour Conference (33rd: 2006), Sydney.

Robinson AC (1975) The stick-nest rat, *Leporillus conditor* on Franklin Island, Nuyts Archipelago, South Australia. *Australian Mammalogy* **1**, 319–327. doi:10.1071/AM75002

Robinson AC (1987) The ecology of the bush rat, *Rattus fuscipes* (Rodentia: Muridae) in Sherbrooke Forest, Victoria. *Australian Mammalogy* **11**, 35–49. doi:10.1071/AM88004

Robinson JF, Robinson AC, Watts CHS, Baverstock PR (1978) Notes on rodents and marsupials and their ectoparasites collected in Australia 1974–1975. *Transactions of the Royal Society of South Australia* **102**, 59–70.

Rowell MK, Rymer TL (2020) Growth and behavioural development of the fawn-footed mosaic-tailed rat (*Melomys cervinipes*). *Australian Mammalogy* **43**, 330–334. doi:10.1071/AM20037

Roycroft E, MacDonald AJ, Moritz C, Moussalli A, Miguez RP, Rowe KC (2021) Museum genomics reveals the rapid decline and extinction of Australian rodents since European settlement. *Proceedings of the National Academy of Sciences of the United States of America* **118**(27), e2021390118. doi:10.1073/pnas.2021390118

Salamon M, Klettenheimer B (1994) A new technique for marking and later recognising small mammals in the field. *Journal of Zoology* **233**, 314–317. doi:10.1111/j.1469-7998.1994.tb08592.x

Scheelings TF (2025) Rodents. In *Current Therapy in Medicine of Australian Mammals*. Revised Edition. (Eds L Vogelnest and T Portas) pp. 618–620. CSIRO Publishing, Melbourne.

Smales LR (1997) A review of the helminth parasites of Australian rodents. *Australian Journal of Zoology* **45**, 505–521. doi:10.1071/ZO97013

Smales LR, Cribb TH (1997) Helminth parasite communities of the water-rat, *Hydromys chrysogaster*, from Queensland. *Wildlife Research* **24**, 445–457. doi:10.1071/WR96074

Smales LR, Obendorf DL, Miller AK (1990) Parasites of the water-rat *Hydromys chrysogaster* from Victoria and South Australia. *Australian Journal of Zoology* **37**, 657–663. doi:10.1071/ZO9890657

Smith GC (1985) Biology and habitat usage of sympatric populations of fawn-footed melomys (*Melomys cervinipes*) and the grassland melomys (*M. burtoni*) (Rodentia: Muridae). *Australian Zoologist* **21**, 551–563.

Smith JR, Watts CHS, Crichton EG (1972) Reproduction in the Australian desert rodents *Notomys alexis* and *Pseudomys australis* (Muridae). *Australian Mammalogy* **1**, 1–7. doi:10.1071/AM72001

Southgate R, Masters P (1996) Fluctuations of rodent populations in response to rainfall and fire in a central Australian hummock grassland dominated by *Plectrachne schinzii*. *Wildlife Research* **23**, 289–303. doi:10.1071/WR9960289

Species360 (2023) Zoological Information Management System (ZIMS). Minneapolis, USA. www.species360.org.

Stanley M (1971) An ethogram of the hopping mouse *Notomys alexis*. *Zeitschrift für Tierpsychologie* **29**, 225–258. doi:10.1111/j.1439-0310.1971.tb01735.x

Stannard HJ, Tulk ML, Old JM (2017) Dead mouse hopping: Tyzzer's disease in spinifex hopping-mice (*Notomys alexis*). *Veterinary Microbiology* **201**, 201–207. doi:10.1016/j.vetmic.2017.01.018

Stannard HJ, Tulk ML, Bortolazzo MJ, Old JM (2018) Go with your gut: Digestibility and digestive function of two arid-zone Australian murids, *Pseudomys australis* and *Notomys alexis*. *Zoology (Jena, Germany)* **128**, 55–61. doi:10.1016/j.zool.2018.03.003

Stewart AP, Barnett SA (1983) Seasonal changes in body weight and composition of Australian bush rats, *Rattus fuscipes*, and adaptation to winter. *Australian Journal of Zoology* **31**, 29–37. doi:10.1071/ZO9830029

Strahan R (1995) *The Mammals of Australia*. 2nd edn. New Holland, Sydney.

Taylor JM (1961) Reproductive biology of the Australian bush rat, *Rattus assimilis*. *University of California Publications in Zoology* **60**, 1–66.

Taylor JM, Calaby JH (1988a) *Rattus lutreolus*. *Mammalian Species* **299**, 1–7.

Taylor JM, Calaby JH (1988b) *Rattus fuscipes*. *Mammalian Species* **298**, 1–8.

Taylor JM, Horner BE (1970a) Reproduction in the mosaic tailed rat *Melomys cervinipes* (Rodentia: Muridae). *Australian Journal of Zoology* **18**, 171–184. doi:10.1071/ZO9700171

Taylor JM, Horner BE (1970b) Observations on reproduction in *Leggadina* (Rodentia: Muridae). *Journal of Mammalogy* **51**, 10–17. doi:10.2307/1378526

Taylor JM, Horner BE (1971a) Reproduction in the Australian tree-rat *Conilurus penicillatus* (Rodentia: Muridae). *CSIRO Wildlife Research* **16**, 1–9. doi:10.1071/CWR9710001

Taylor JM, Horner BE (1971b) Sexual maturation in the Australian rodent *Rattus fuscipes assimilis*. *Australian Journal of Zoology* **19**, 1–17. doi:10.1071/ZO9710001

Taylor JM, Horner BE (1972a) Breeding biology of three subspecies of the native Australian rat *Rattus fuscipes* in the laboratory. *Australian Mammalogy* **1**, 8–13. doi:10.1071/AM72002

Taylor JM, Horner BE (1972b) Observations on the reproductive biology of *Pseudomys* (Rodentia: Muridae). *Journal of Mammalogy* **53**, 318–328. doi:10.2307/1379168

Taylor JM, Horner BE (1973) Reproductive characteristics of wild native Australian *Rattus* (Rodentia: Muridae). *Australian Journal of Zoology* **21**, 437–475. doi:10.1071/ZO9730437

Telfer S, Breed WG (1976) The effect of age on the female reproductive tract of the hopping mouse *Notomys alexis*. *Australian Journal of Zoology* **24**, 533–540. doi:10.1071/ZO9760533

Thompson JGE, Breed WG (1982) The effect of the social environment on vaginal perforation and oestrous in the hopping-mouse *Notomys alexis*. *Australian Journal of Zoology* **30**, 169–173.

Troughton E (1941) Australian water-rats: their origins and habits. *Australian Museum Magazine* **7**, 377–381.

Tulk ML, Stannard HJ, Old JM (2016) Haematology and blood biochemistry in captive Australian native murids: black-footed tree-rat (*Mesembriomys gouldii*) and greater stick-nest rat (*Leporillus conditor*). *SpringerPlus* **5**, 1479. doi:10.1186/s40064-016-3111-7

Van Dyck S (1995) False water rat *Xeromys myoides*. In *The Mammals of Australia*. 2nd edn. (Ed. R Strahan) pp. 630–631. Reed Books, Sydney.

Van Dyck S (1996) *Xeromys myoides* Thomas, 1889 (Rodentia: Muridae) in mangrove communities of North Stradbroke Island, southeast Queensland. *Memoirs of the Queensland Museum* **42**, 337–366.

Venkatachalam PA, Ramanathan KS (1964) Effect of protein deficiency during gestation and lactation on body weight and composition of offspring. *The Journal of Nutrition* **84**, 38–42. doi:10.1093/jn/84.1.38

Vogelnest L, Portas T (Eds) (2025) *Current Therapy in Medicine of Australian Mammals*. Revised Edition. CSIRO Publishing, Melbourne.

Vogelnest L, Woods R (2008) *Medicine of Australian Mammals*. CSIRO Publishing, Melbourne.

Wallis RL, Brunner H, Menkhorst PW (1982) Victorian field studies on the broad-toothed rat (*Mastacomys fuscus* Thomas). *Victorian Naturalist* **99**, 12–21.

Warden JC, Wallis RL (1979) Further data on activity patterns of small mammals. *Victorian Naturalist* **96**, 54–55.

Warneke RM (1971) Field study of the bush rat (*Rattus fuscipes*). *Wildlife Contributions, Victoria* **14**, 1–115.

Watts CHS (1970) The foods eaten by some Australian desert rodents. *South Australian Naturalist* **44**, 71–74.

Watts CHS (1972) *Handbook of South Australian Rodents and Small Marsupials*. Field Naturalists Society of South Australia, Adelaide.

Watts CHS (1973) The Australian rodents *Notomys alexis* and *Pseudomys australis* as laboratory animals. *Experimental Animals* **22**(suppl), 179–185.

Watts CHS (1977) The foods eaten by some Australian rodents (Muridae). *Australian Wildlife Research* **4**, 151–157. doi:10.1071/WR9770151

Watts CHS (1979*a*) Body growth in some Australian rodents. *Australian Zoologist* **20**, 297–303.

Watts CHS (1979*b*) Reproductive parameters of some Australian rodents. *Australian Zoologist* **20**, 305–310.

Watts CHS (1980) Success rates in founding captive colonies of Australian rodents and marsupials. *International Zoo Yearbook* **20**, 245–253. doi:10.1111/j.1748-1090.1980.tb00985.x

Watts CHS (1982a) Australian hydromyine rodents: maintenance of captive colonies. In *The Management of Australian Mammals in Captivity*. (Ed. DD Evans) pp. 180–184. The Zoological Board of Victoria, Melbourne.

Watts CHS (1982b) The husbandry of Australian *Rattus*. In *The Management of Australian Mammals in Captivity*. (Ed. DD Evans) pp. 177–179. The Zoological Board of Victoria, Melbourne.

Watts CHS (1995) Plains Rat *Pseudomys australis*. In *The Mammals of Australia*. 2nd edn. (Ed. R Strahan) pp. 586–587. Reed Books, Sydney.

Watts CHS, Aslin HJ (1981) *The Rodents of Australia*. Angus and Robertson, Sydney.

Watts CHS, Eves BM (1976) Notes on the nests and diet of the white-tailed sticknest rat *Leporillus apicalis* in northern South Australia. *South Australian Naturalist* **51**, 9–12.

Waudby HP, Burns PA, Jensen MA, Hampton JO, Hunter D, McKnight DT, Pestell AJ, Bengsen AJ, Bower D, Coulson G, Heard G, de Laive A, Jolly CJ, Hale S, Fancourt BA, Kelleher SR, Petit S, Behrendorff L, Lumsden LF, McGregor M (2022) Wildlife marking methods. In *Wildlife Research in Australia: Practical and Applied Methods*. (Eds BP Smith, HP Waudby, C Alberthsen and JO Hampton). pp. 150–179. CSIRO Publishing, Melbourne.

White RM, Kennaway DJ, Seamark RF (1996) Reproductive seasonality of the bush rat (*Rattus fuscipes greyi*) in South Australia. *Wildlife Research* **23**, 317–336. doi:10.1071/WR9960317

Williams R (1990) Rodents. In *Care and Handling of Australian Native Animals*. (Ed. SJ Hand) pp. 185–190. Surrey Beatty & Sons & the Royal Zoological Society of New South Wales. Sydney.

Wilson DE, Lacher TE, Jr, Mittermeier RA (eds.) (2016) *Handbook of the Mammals of the World*. Vol. 6. *Lagomorphs and Rodents I*. Lynx Editions, Barcelona.

Wilson DE, Lacher TE, Jr, Mittermeier RA (eds.) (2017) *Handbook of the Mammals of the World*. Vol. 7. *Rodents II*. Lynx Editions, Barcelona.

Winter JW, Whitford D (1995) Prehensile-tailed rat *Pogonomys mollipilosus*. In *The Mammals of Australia*. 2nd edn. (Ed. R Strahan) pp. 643–645. Reed Books, Sydney.

Wirtz WO, II (1973) Growth and development of *Rattus exulans*. *Journal of Mammalogy* **54**, 189–202. doi:10.2307/1378879

Woinarski JCZ, Burbidge AA, Harrison P (eds) (2014) *The Action Plan for Australian Mammals 2012*. CSIRO Publishing, Melbourne.

Wood DH (1971) The ecology of *Rattus fuscipes* and *Melomys cervinipes* (Rodentia: Muridae) in a south-east Queensland rain forest. *Australian Journal of Zoology* **19**, 371–392. doi:10.1071/ZO9710371

Woods RE, Ford FD (2000) Observations on the behaviour of the smoky mouse *Pseudomys fumeus* (Rodentia: Muridae). *Australian Mammalogy* **22**, 35–42. doi:10.1071/AM00035

Woods R, Kennedy G (1997) Circadian activity rhythms of captive heath rats (*Pseudomys shortridgei*). *Wildlife Research* **24**, 459–468. doi:10.1071/WR96046

Woodside DP (1983) The role of social behaviour and spacing in populations of the bush rat, *Rattus fuscipes*. PhD thesis, Australian National University, Canberra.

Woollard P, Vestjens WJM, Maclean L (1978) The ecology of the eastern water rat *Hydromys chrysogaster* at Griffiths, NSW: food and feeding habits. *Australian Wildlife Research* **5**, 59–73. doi:10.1071/WR9780059

Yom-Tov Y (1985) The reproductive rates of Australian rodents. *Oecologia* **66**, 250–255. doi:10.1007/BF00379862

Young DAB (1965) A photoperiod influence on the growth rates in rats. *The Journal of Experimental Zoology* **160**, 241–246. doi:10.1002/jez.1401600212

Young LI (2018) Understanding the characteristics and role of refuges in the persistence of the plains mouse (*Pseudomys australis*) in arid landscapes. PhD thesis, University of Sydney, Sydney.

# 13
# Bats

*Stephen M. Jackson, Peter Holz, Jennefer Mclean, Rachel Lyons and Lindy Lumsden*

## 1. INTRODUCTION

The bats (Order Chiroptera) are a highly diverse group of mammals that occur on all continents except Antarctica. The group includes 20% of all mammals, with over 1400 species and many new species continually being recognised (Wilson and Mittermeier 2019). Within Australia 80 species of bats are currently recognised, including the flying-foxes and fruit bats of the family Pteropodidae, with other bat species being typically insectivorous (Baker and Gynther 2023). Bats are the only mammals able to sustain flight, which they achieve with wings that consist of a thin skin membrane that covers elongated fingers, except for the thumb. Most species also have a membrane that extends between the hindlegs and partially or totally covers the tail if one is present.

Despite the large number of bat species within Australia and their unique ability to fly, they have been poorly represented within Australian zoo collections and are rarely held in collections outside Australia. Bats from other regions of the world are, however, frequently kept in collections overseas, especially in the USA. Within Australian zoos the maintenance of bats has generally been limited to flying-foxes (primarily grey-headed flying-foxes (*Pteropus poliocephalus*) and spectacled flying-foxes (*Pteropus conspicillatus*)), though several institutions, including Perth Zoo, Taronga Zoo and Territory Wildlife Park, have also held ghost bats (*Macroderma gigas*), with Taronga Zoo being the only institution to have held the eastern blossom-bat (*Syconycteris australis*) (Lees and Johnson 2002).

The lack of Australian bats, other than flying-foxes and ghost bats, on public exhibition is probably due to the small size of some species, difficulties in keeping most species in large numbers, their lack of diurnal activity, difficulties in keeping many species for long periods, difficulties in breeding them and the need to be vaccinated against the Australian bat lyssavirus to work with them. Therefore, there is a great need to advance the husbandry techniques of this interesting group of mammals.

## 2. TAXONOMY

### 2.1 Nomenclature

All bats belong to the Order Chiroptera, which is divided into 231 genera and ~1400 species (Wilson and Mittermeier 2019). The bats were previously divided into the suborders Megachiroptera, which contained the Old World fruit bats and flying-foxes, and Microchiroptera, which contained the insectivorous and carnivorous bats. However, the bats are now separated within the suborders Yinpterochiroptera and the Yangochiroptera. The suborder Yinpterochiroptera includes the Old World fruit bats and flying-foxes, as well as the ghost bat, horseshoe-bats, leaf-nosed bats and orange diamond-faced bats, while the suborder Yangochiroptera includes the sheath-tailed bats, free-tailed bats, bent-winged bats and the vespertilionid bats (Jackson and Groves 2015) (Table 13.1). As there are closer similarities in the captive care of the ghost bat, horseshoe-bats, leaf-nosed bats and orange leaf-nosed bats to the yangochiropteran families than to the fruit bats and flying-foxes, these are treated together in this chapter and are collectively referred to as 'insectivorous bats'.

**Australian bats**

Class: Mammalia
Order: Chiroptera
Suborder: Yinpterochiroptera
Family: Pteropodidae – Old World fruit bats
Genus and species: 11 species within 5 genera
Family: Megadermatidae – ghost bat
Genus and species: 1 species within 1 genus
Family: Rhinolophidae – horseshoe-bats

Genus and species: 2 species within 1 genus
Family: Hipposideridae - leaf-nosed bats
Genus and species: 6 species within 1 genus
Family Rhinonycteridae - diamond-faced bats
Genus and species: 1 species within 1 genus
Suborder: Yangochiroptera
Family: Emballonuridae - sheath-tailed bats
Genus and species: 8 species within 2 genera
Family: Molossidae - free-tailed bats
Genus and species: 11 species within 5 genera
Family: Miniopteridae - bent-winged bats
Genus and species: 2 species within 1 genus
Family: Vespertilionidae - vespertilionid bats
Genus and species: 38 species within 10 genera

## 2.2 Subspecies

Synonyms can be found in Jackson and Groves (2015).

## 2.3 Recent synonyms

See Jackson and Groves (2015).

## 2.4 Other common names

See Baker and Gynther (2023).

# 3. NATURAL HISTORY

## 3.1 Bodyweight and morphometrics

The morphometrics of all Australian species of bat are generally well known, with adult bats varying between < 5 g to > 1000 g (Table 13.1). The morphometrics for the Australian species can be found in Baker and Gynther (2023).

## 3.2 Distribution and habitat

Broad distribution patterns and habitat requirements of the bats within Australia are generally well known and can be found in Churchill (2008), Wilson and Mittermeier (2019), Australasian Bat Society (2021) and Baker and Gynther (2023).

## 3.3 Conservation status

Of the 11 species of Pteropodidae, the Percy Island flying-fox is extinct, the Christmas Island flying-fox is critically endangered, the spectacled flying-fox is endangered and the grey-headed flying-fox is vulnerable, with the other species considered to have a low risk of extinction. In general, the insectivorous bats are doing better from a conservation perspective, except for the Christmas Island pipistrelle and the Lord Howe long-eared bat, which are extinct, and the ghost bat, Arnhem leaf-nosed bat, northern leaf-nosed bat, golden-tipped bat, eastern false pipistrelle, large-eared pied bat, Corben's long-eared bat and Tasmanian long-eared bat, which are all listed as vulnerable (Table 13.1). In addition, at the subspecies level the southern bent-wing bat (*Miniopterus orianae bassanii*) is listed as critically endangered under Australian legislation.

**Table 13.1.** Species of bats in Australia, their bodyweights and conservation status.

CR = critically endangered, EN = endangered, EX = extinct, LC = least concern, NT = near threatened, VU = vulnerable.
Derived from Churchill (2008), Woinarski *et al.* (2014), Armstrong *et al.* (2020) and Baker and Gynther (2023).

| Common name | Scientific name | Weight (g) | IUCN status |
|---|---|---|---|
| **Suborder Yinpterochiroptera** | | | |
| **Family Pteropodidae** | | | |
| Northern blossom-bat | *Macroglossus minimus* | 10–17 | LC |
| Eastern blossom-bat | *Syconycteris australis* | 16–26 | LC |
| Eastern tube-nosed bat | *Nyctimene robinsoni* | 42–62 | LC |
| Bare-backed fruit bat | *Dobsonia magna* | 350–550 | LC |
| Black flying-fox | *Pteropus alecto* | 443–986 | LC |
| Percy Island flying-fox | *P. brunneus* | c. 200 | EX |
| Spectacled flying-fox | *P. conspicillatus* | 500–1100 | EN |
| Large-eared flying-fox | *P. macrotis* | 280–450 | LC |
| Christmas Island flying-fox | *P. natalis* | 163–556 | CR |
| Grey-headed flying-fox | *P. poliocephalus* | 522–1106 | VU |

| Common name | Scientific name | Weight (g) | IUCN status |
|---|---|---|---|
| Little red flying-fox | *P. scapulatus* | 310–604 | LC |
| **Family Megadermatidae** | | | |
| Ghost bat | *Macroderma gigas* | 140–172 | VU |
| **Family Rhinolophidae** | | | |
| Eastern horseshoe-bat | *Rhinolophus megaphyllus* | 7–13 | LC |
| Large-eared horseshoe-bat | *R. robertsi* | 8–16 | LC |
| **Family Hipposideridae** | | | |
| Dusky leaf-nosed bat | *Hipposideros ater* | 4.5–10 | LC |
| Fawn leaf-nosed bat | *H. cervinus* | 6–9 | LC |
| Diadem leaf-nosed bat | *H. diadema* | 30–57 | LC |
| Arnhem leaf-nosed bat | *H. inornatus* | 22–35 | VU |
| Semon's leaf-nosed bat | *H. semoni* | 6–10 | LC |
| Northern leaf-nosed bat | *H. stenotis* | 4.5–8 | VU |
| **Family Rhinonycteridae** | | | |
| Orange diamond-faced bat | *Rhinonicteris aurantia* | 7–11 | LC |
| **Suborder Yangochiroptera** | | | |
| **Family Emballonuridae** | | | |
| Yellow-bellied sheath-tailed bat | *Saccolaimus flaviventris* | 30–60 | LC |
| Cape York sheath-tailed bat | *S. mixtus* | 21–27 | NT |
| Bare-rumped sheath-tailed bat | *S. saccolaimus* | 38–69 | LC |
| Coastal sheath-tailed bat | *Taphozous australis* | 25–50 | NT |
| Common sheath-tailed bat | *T. georgianus* | 19–51 | LC |
| Hill's sheath-tailed bat | *T. hilli* | 20–25 | LC |
| Arnhem sheath-tailed bat | *T. kapalgensis* | 26–30 | LC |
| Troughton's sheath-tailed bat | *T. troughtoni* | 30–35 | LC |
| **Family Molossidae** | | | |
| White-striped free-tailed bat | *Austronomus australis* | 26–48 | LC |
| Greater northern free-tailed bat | *Chaerephon jobensis* | 20–30 | LC |
| Eastern coastal free-tailed bat | *Micronomus norfolkensis* | 6–12 | NT |
| Northern coastal free-tailed bat | *Ozimops cobourgianus* | 7–10.5 | LC |
| Cape York free-tailed bat | *O. halli* | 5–7 | DD |
| Western free-tailed bat | *O. kitcheneri* | 7.5–10.5 | LC |
| Northern free-tailed bat | *O. lumsdenae* | 11.0–19.5 | LC |
| Inland free-tailed bat | *O. petersi* | 7–12 | LC |
| Southern free-tailed bat | *O. planiceps* | 7–13 | LC |
| Ride's free-tailed bat | *O. ridei* | 8–13 | LC |
| Bristle-faced free-tailed bat | *Setirostris eleryi* | 4.5–7.5 | NT |
| **Family Miniopteridae** | | | |
| Little bent-winged bat | *Miniopterus australis* | 5–10 | LC |
| Large bent-winged bat | *M. orianae* | 9–19 | LC |
| **Family Vespertilionidae** | | | |
| Large-footed myotis | *Myotis macropus* | 7–12 | LC |
| Golden-tipped bat | *Phoniscus papuensis* | 5–9.3 | VU |

**Table 13.1.** continued

| Common name | Scientific name | Weight (g) | IUCN status |
|---|---|---|---|
| Flute-nosed bat | *Murina florium* | 6–11 | LC |
| Large-eared pied bat | *Chalinolobus dwyeri* | 6.8–12.0 | VU |
| Gould's wattled bat | *C. gouldii* | 7–20 | LC |
| Chocolate wattled bat | *C. morio* | 8–11 | LC |
| Hoary wattled bat | *C. nigrogriseus* | 7.5–10.0 | LC |
| Little pied bat | *C. picatus* | 3–8 | NT |
| Western false pipistrelle | *Falsistrellus mackenziei* | 17–26 | NT |
| Eastern false pipistrelle | *F. tasmaniensis* | 17–28 | VU |
| Arnhem long-eared bat | *Nyctophilus arnhemensis* | 5–8 | LC |
| Eastern long-eared bat | *N. bifax* | 5–12 | LC |
| Corben's long-eared bat | *N. corbeni* | 11–26 | VU |
| Pallid long-eared bat | *N. daedalus* | 6.5–14.0 | LC |
| Lesser long-eared bat | *N. geoffroyi* | 6–12 | LC |
| Gould's long-eared bat | *N. gouldi* | 9–13 | LC |
| Holt's long-eared bat | *N. holtorum* | 9 | LC |
| Lord Howe long-eared bat | *N. howensis* | Unknown | EX |
| Greater long-eared bat | *N. major* | 12.2–15.4 | LC |
| Tasmanian long-eared bat | *N. sherrini* | 13–17.5 | VU |
| Pygmy long-eared bat | *N. walkeri* | 3.3–7.0 | LC |
| Inland broad-nosed bat | *Scotorepens balstoni* | 7–14 | LC |
| Little broad-nosed bat | *S. greyii* | 4.0–8.5 | LC |
| Eastern broad-nosed bat | *S. orion* | 9–15 | LC |
| Northern broad-nosed bat | *S. sanborni* | 6–9 | LC |
| Inland forest-bat | *Vespadelus baverstocki* | 3–6 | LC |
| Northern cave-bat | *V. caurinus* | 2.3–4.5 | LC |
| Large forest-bat | *V. darlingtoni* | 4–10 | LC |
| Yellow-lipped cave-bat | *V. douglasorum* | 4.5–6.1 | LC |
| Finlayson's cave-bat | *V. finlaysoni* | 2.8–6.3 | LC |
| Eastern forest-bat | *V. pumilus* | 3.5–6 | LC |
| Southern forest-bat | *V. regulus* | 3.6–7.0 | LC |
| Eastern cave-bat | *V. troughtoni* | 4–7 | LC |
| Little forest-bat | *V. vulturnus* | 2.6–5.5 | LC |
| Forest pipistrelle | *Pipistrellus adamsi* | 3–6 | LC |
| Christmas Island pipistrelle | *P. murrayi* | 2.6–4.6 | EX |
| Northern pipistrelle | *P. westralis* | 2.7–3.3 | LC |
| Greater broad-nosed bat | *Scoteanax rueppellii* | 25–35 | LC |

## 3.4 Diet in the wild

The diet of bats is related to body size and taxonomic grouping. The bats of the family Pteropodidae are typically fruit and/or nectar feeders, while the other bats are primarily insectivorous. Exceptions to this are the ghost bat, which is carnivorous, feeding on rodents, birds, frogs and other bats, in addition to invertebrates, and the large-footed myotis, which includes small fish in its diet. The diets of different genera of Australian bats are shown in Table 13.2.

**Table 13.2.** Wild diet of different genera of Australian bats.

Derived from Flannery (1995a, 1995b), Churchill (2008), Clulow and Blundell (2011), Baker and Gynther (2023), and Smith *et al.* (2020).

| Genus | Diet |
|---|---|
| **Yinpterochiroptera** | |
| **Pteropodidae** | |
| *Macroglossus* | Nectar and pollen (e.g. *Melaleuca, Syzygium, Sonneratia* and *Musa* (banana))<br>Fruits (e.g. *Ficus* and *Timonius*) |
| *Syconycteris* | Nectar and pollen (e.g. *Banksia, Eucalyptus, Grevillea, Melaleuca, Syzygium*)<br>Fruits (e.g. *Ficus*) |
| *Nyctimene* | Fruit (e.g. *Eugenia, Ficus, Syzygium*)<br>Nectar (e.g. *Banksia*) |
| *Dobsonia* | Fruit (e.g. *Carica* (pawpaw), *Ficus, Musa* (banana))<br>Nectar and pollen (e.g. *Corymbia*) |
| *Pteropus* | Nectar and pollen (e.g. *Corymbia, Eucalyptus, Melaleuca, Syncarpia* and various others species)<br>Fruit from a variety of species (e.g. *Ficus, Terminalia, Syzygium* and *Eugenia*)<br>*Albizia* and *Ficus* leaves and large insects (e.g. cicadas) are sometimes eaten |
| **Megadermatidae** | |
| *Macroderma* | Vertebrates and invertebrates (e.g. birds, bats, rodents, frogs, geckos, locusts, millipedes, spiders, cockroaches, termites, crickets, moths, beetles, caterpillars and ants) |
| **Rhinolophidae** | |
| *Rhinolophus* | Invertebrates (e.g. moths, bugs, grasshoppers, beetles, flies and wasps)<br>Forage in shrubs and subcanopy |
| **Hipposideridae** | |
| *Hipposideros* | Invertebrates (e.g. moths, mosquitoes, beetles, bugs, cockroaches, grasshoppers, weevils and flies)<br>Forage in dense vegetation low to the ground |
| **Rhinonycteridae** | |
| *Rhinonicteris* | Invertebrates (e.g. moths and beetles, termites, ants, wasps, bugs, flies, cockroaches)<br>Forage low to the ground |
| **Yangochiroptera** | |
| **Emballonuridae** | |
| *Saccolaimus* | Invertebrates (e.g. beetles, grasshoppers, flying ants, bugs, moths)<br>Forage above tree canopy |
| *Taphozous* | Invertebrates (e.g. beetles, flies, cockroaches, moths, bugs, grasshoppers and ants)<br>Forage above tree canopy |
| **Molossidae** | |
| *Austronomus* | Invertebrates (e.g. moths, beetles, bugs and grasshoppers)<br>Forage above tree canopy |
| *Chaerephon* | Invertebrates (e.g. bugs, moths, ants, ground beetles, flies, crickets, sawflies and earwigs)<br>Forage in subcanopy |
| *Micronomus* | Invertebrates (specific details unknown) |
| *Ozimops* | Invertebrates (e.g. bugs, moths, beetles, bugs, flies, ants, grasshoppers)<br>Forage above the tree canopy and in the subcanopy |
| *Setirostris* | Invertebrates (specific details unknown) |
| **Miniopteridae** | |
| *Miniopterus* | Invertebrates (e.g. moths, grasshoppers, flies, ants and wasps)<br>Forage in tree canopy and understory |
| **Vespertilionidae** | |
| *Myotis* | Invertebrates (e.g. midges, mosquitoes, moths, beetles, crickets, cockroaches, flies, water boatmen, water striders and mayflies) and small fish<br>Forage over water surface |

**Table 13.2.** continued

| | |
|---|---|
| *Phoniscus* | Spiders (e.g. orb weaving spiders)<br>Forage in dense understory |
| *Murina* | Invertebrates (e.g. beetles and spiders)<br>Forage in subcanopy. |
| *Chalinolobus* | Invertebrates (e.g. moths, beetles, wingless ants, cockroaches, stoneflies, katydids, crickets, cicadas, flies, spiders, grasshoppers and termites)<br>Forage in tree canopy and understory |
| *Falsistrellus* | Invertebrates (e.g. moths, beetles, weevils, bugs, flies and ants)<br>Forage in tree canopy and understory |
| *Nyctophilus* | Invertebrates (e.g. moths, beetles, ants, spiders, crickets, bugs, lacewings, caterpillars and flies)<br>Forage in tree canopy, subcanopy and understory |
| *Scotorepens* | Invertebrates (e.g. cockroaches, termites, crickets, katydids, cicadas, bugs, beetles, moths, flies, mosquitoes, grasshoppers and ants)<br>Forage below tree canopy |
| *Vespadelus* | Invertebrates (e.g. moths, beetles, termites, spiders, flies, bugs, caddisflies)<br>Forage between tree canopy and understory |
| *Pipistrellus* | Invertebrates (e.g. moths, beetles, ants, bugs and flies)<br>Forage above and below the tree canopy |
| *Scoteanax* | Invertebrates (e.g. moths, beetles). Other bats<br>Forage in the tree canopy and subcanopy |

## 3.5 Longevity

### 3.5.1 Wild

Little is known of the longevity of most species of bats; however, information is available for some species which suggests that bats can live between 10 and 15 years, and sometimes can live over 20 years in the wild and in captivity (though the average life span is probably a lot less than this). A long-term study of the big brown bat (*Eptesicus fuscus*) in the USA revealed that animals lived as long as 17–18 years (Goehring 1972); another study found two male little brown bats (*Myotis lucifugus*) in the USA to be recaptured after 29 and 30 years respectively (Keen and Hitchcock 1980). A Brandt's bat (*Myotis brandtii*) was recaught 41 years after banding in Russia, with many others recorded living between 20 and 40 years (Podlutsky *et al.* 2005). Within Australia longevity records are typically less than these extremes from the northern hemisphere, with 5–12 years being more typical; for example, 12 years for Gould's wattled bat (L. Lumsden pers. obs.), 7–11 years for three species of forest bats (*Vespadelus* spp.) (Law *et al.* 2018), 8.5 years for the large-footed myotis (Law *et al.* 2020), 7 years for Gould's long-eared bats and golden-tipped bats and 5 years for the eastern false pipistrelle (Law *et al.* 2023). The exception are the bent-wing bats, with one southern bent-wing bat recorded living in the wild for 22 years (Lumsden and Gray 2001). Longevities that are known for different genera of bats can be found in Table 13.3.

### 3.5.2 Captivity

The Pteropodidae generally do well in captivity, with flying-foxes typically living 10–25 years and the smaller blossom-bats, which are more difficult to keep, typically living 2–6 years of age. The ghost bat appears to do the best, with longevities from 8 to 16 years. Typical longevity for captive yangochiropteran bats has often been much less than that observed for their wild counterparts, with some species being difficult to keep and others that have been readily kept for many years in captivity (Table 13.3). For example, two Ride's free-tailed bats were kept for 23 years (Lumsden 2012).

### 3.5.3 Methods used to determine the age of adults

Determining absolute age in bats is difficult unless the time of birth is known; however, there are generally three age classes or life stages that are more easily recognised. These are (Reardon and Flavel 1987):

(1) juvenile – not fully weaned and usually less than 40 days old (for insectivorous bats)
(2) subadult – from juvenile to adult
(3) adult – fully grown and sexually mature.

The appearance of epiphyseal cartilage of the finger joints is a useful indicator of general age for insectivorous bats, with this best done by shining a light from behind the wing. In very young bats, the transparent

**Table 13.3.** Longevity of different genera of bats in captivity or in the wild, including for overseas species of genera that occur in Australia.

| Genus | Known longevity Wild (years) | Known longevity Captive (years) | Reference |
|---|---|---|---|
| **Yinpterochiroptera** | | | |
| **Pteropodidae** | | | |
| *Pteropus* | 5–20+ | 10–30 | 1–8 |
| *Syconycteris* | > 5 | 2–6 | 9, 10 |
| **Megadermatidae** | | | |
| *Macroderma* | – | 8–16 | 9, 11 |
| **Rhinolophidae** | | | |
| *Rhinolophus* | 18–26 | – | 12–14 |
| **Hipposideridae** | | | |
| *Hipposideros* | 10 | – | 15 |
| **Yangochiroptera** | | | |
| **Emballonuridae** | | | |
| *Saccolaimus* | – | 13 | 16 |
| **Molossidae** | | | |
| *Austronomus* | – | 1 | 15 |
| *Ozimops* | – | 23 | 17 |
| **Miniopteridae** | | | |
| *Miniopterus* | 14–23 | 2 | 14, 16, 18–20 |
| **Vespertilionidae** | | | |
| *Myotis* | 8–41 | – | 2, 14, 23–29 |
| *Phoniscus* | 7 | – | 21 |
| *Chalinolobus* | 8–12 | 7 | 16, 22 |
| *Falsistrellus* | 5 | – | 21 |
| *Nyctophilus* | 7–8 | 6 | 16, 21, 23 |
| *Scotorepens* | – | 3+ | 16 |
| *Vespadelus* | 7–11 | 5 | 16, 22 |
| *Pipistrellus* | 6–15 | – | 3, 13, 14, 23 |

References: (1) Healesville Sanctuary ARKS records; (2) Ratcliffe (1932); (3) Martin *et al.* (1987); (4) Fascione (1995); (5) Vardon and Tidemann (2000); (6) Tidemann and Nelson (2011); (7) Fox *et al.* (2008); (8) Divljan *et al.* (2006); (9) C. Bach pers. comm.; (10) Law (1996); (11) Nelson (1989); (12) Sluiter *et al.* (1971); (13) Stebbings (1977); (14) Tuttle and Stevenson (1982); (15) Power *et al.* (2022); (16) L. Lumsden pers. obs.; 17) Lumsden (2012); (18) Reardon and Flavel (1987); (19) Churchill (2008); (20) Lumsden and Gray (2001); (21) Law *et al.* (2023); (22) Law *et al.* (2018); 23) Paradiso and Greenhall (1967); (24) Hall *et al.* (1957); (25) Griffin and Hitchcock (1965); (26) Keen and Hitchcock (1980); (27) Sommers *et al.* (1993); (28) Podlutsky *et al.* (2005); (29) Law *et al.* (2020).

cartilaginous bands in the wing joint are quite obvious, becoming more difficult to see with increasing age (Parnaby 1999). Juveniles have flat unfused wing joints and adults have fused knobby wing joints (Fig. 13.1). A general guide to ageing bats is given in Table 13.4. Prior to the young flying, there are three bands of cartilage. For approximately the first month of flight there are two bands. This transitions to just one band for 1–2 months after the bat starts flying, while in animals that have been flying for more than ~2 months, the joint is circular as in adults (L. Lumsden pers. obs.; Reardon and Flavel 1987; Parnaby 1999).

Several techniques have been used with variable success to determine the age of adult bats, ranging from tooth wear that can be used on live animals to analysing incremental lines of dentin, cementum and bone tissue that can only be examined on dead animals (e.g. Andersen 1917; Laws 1952; Schowalter *et al.* 1978; Phillips *et al.* 1982; Cool *et al.* 1994; Divljan *et al.* 2006; Fox *et al.* 2008).

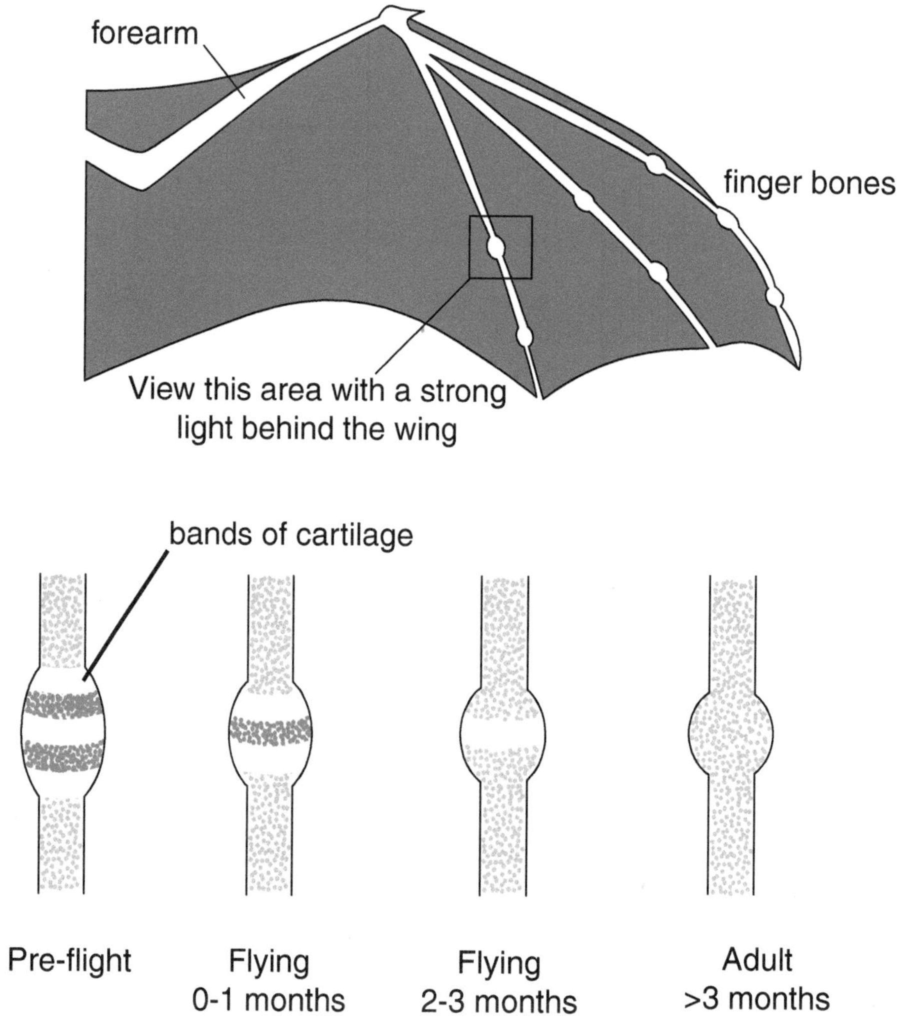


**Fig. 13.1.** Aging bats using fused and unfused wing joints of juvenile and adult bats. Redrawn based on a figure from Lumsden (1995).

**Table 13.4.** General characteristics of different life stages in insectivorous bats.

Derived from Reardon and Flavel (1987) and Parnaby (1999).

| Character | Juvenile | Subadult | Adult |
|---|---|---|---|
| Size (weight and forearm length) | < 80% adult | 80–100% adult | Adult |
| Teeth | Milk teeth may still be present, others needle sharp | Sharp and unworn | May be showing wear |
| Teats (females only) | Almost invisible | Almost invisible | Clearly seen |
| Finger joints | Unfused large and obvious cartilaginous bands (see Fig. 13.1) | Not fully fused Cartilaginous band and blood vessels distinct | Fully fused, knobby Cartilaginous gap not visible |

One technique that has been used on living insectivorous bats uses the extent of wear on the canine teeth:

(1) tip unworn and pointed
(2) tip slightly worn
(3) worn nearly half way to the gum
(4) worn half way to the gum
(5) projecting only slightly above gum
(6) worn completely to the gum (Twente 1955).

Similarly, good relationships between age and canine length, upper canine width and upper incisor length in the American big brown bat have been observed (Christian 1956). Although these techniques are useful in determining the relative age, or an approximate age, it is difficult to assign absolute ages as teeth of known-aged animals for several North American species of *Myotis*, recovered 18–19 years after tagging, showed little wear (Hall *et al.* 1957). The slow rate of change in tooth wear suggests that this technique is best used by those familiar with the variation found between individuals (Anthony 1990).

## 4. HOUSING REQUIREMENTS

### 4.1 Enclosure design

#### 4.1.1 Pteropodids

In the wild, flying-fox camps are generally associated with water: either salt water or fresh water and in riparian or mangrove vegetation (Ratcliffe 1931, 1932; Tidemann *et al.* 1999). It is therefore recommended to have a shallow water body available to provide humidity under the flying-foxes and a network of branches or ropes to mimic trees to climb on. Water in bottles that hang near the food or in dishes need to be provided.

The enclosure should be roofed as high as possible because the bats feel more secure if they are high up and will always climb to the high point (Fig. 13.2). Some people suggest having enclosures no more than ~2 m high so that animals can be caught easily if sick or injured; however, bats have been held in enclosures with a roof ~4–6 m high and have been readily caught using long-handled poles (S. Jackson pers. obs.) or through providing a feeding station trapping system at a lower height. The enclosure used by the Ku-ring-gai Bat Conservation Society in Sydney uses both high- and low-roofed areas. The bats can roost high to feel secure and sun themselves, while an adjoining lower roofed area is used for feeding and can be closed off to make capture easier. The enclosure should be double-wired with ~10–15 cm between the two layers to minimise any chance of the captive bats acquiring disease from wild bats and to keep members of the public away from the captive bats to reduce the possible transfer of zoonotic diseases.

It is important that enclosure surfaces are non-abrasive, ideally sealed and able to withstand regular hosing (Fascione 1995). If wire is used on the enclosure, it should ideally be vinyl coated, Teflon sprayed or non-galvanised because the wire often causes abrasions to the wings and particularly the wrist joints. Polyethylene mesh is an excellent material (Barnard 1995). Some authors have suggested that galvanised wire should be avoided because bat urine corrodes tinned surfaces and may cause zinc toxicity if ingested (Wilson DE 1988); however, others have used it for over 30 years without issues (J. Mclean pers. obs.). The size of the mesh opening should be small enough to prevent animals from pushing a wing or foot through it, but not so fine that animals get stuck in the mesh (Wilson DE 1988), as bats are known to get hyperextension thumb nail injuries if the wire mesh is too small (J. Mclean pers. obs.).

Although flying-foxes have been successfully held indoors in nocturnal houses, they appear to be more prone to fungal disease in such enclosures. They also have excellent sight and enjoy basking in the sun and cleaning themselves in the rain, so they should be held outdoors where they have access to the sun, natural light cycles and weather conditions (George 1990; Barnard 1995; S. Jackson pers. obs.). When held outdoors, sheltered and shaded areas should always be provided in the form of thick vegetation or a roofed area so that they can retreat from the wind, rain and direct sun as required (Fascione 1995). The shelter should ideally not be made of a material that makes the area dark, but preferably allows some light to pass through it. When held indoors, light cycles of 12:12 h light/dark have been used successfully; however, lighting cycles that more closely mimic those found naturally are likely to provide better breeding success (see section 10.2).

In the wild, flying-foxes generally establish their congregations, or camps, in areas that have lots of branches to hang onto and allow them to access the sun. Shade from the sun also must be provided when needed, especially if heat-stress events are likely (Ratcliffe 1931, 1932; J. Mclean pers. obs.). In captivity plenty of climbing opportunities should be provided for the bats, in the form of the wire enclosure itself, if held outdoors, and/or the provision of an extensive network of branches, natural or artificial vines or vinyl-coated wire (2.5 cm diameter) hung horizontally from the ceiling, or thick rope (though this is difficult to clean) (Fascione 1995; S. Jackson pers. obs.). Take care with branches and other climbing apparatus to ensure that the animals still have room to fly and that the branches will not cause wing tears. At least some of the climbing apparatus should extend to the ground as bats fall to the ground on occasion (or after release when captured). This allows them to easily climb back up to the roof or higher areas of the enclosure.

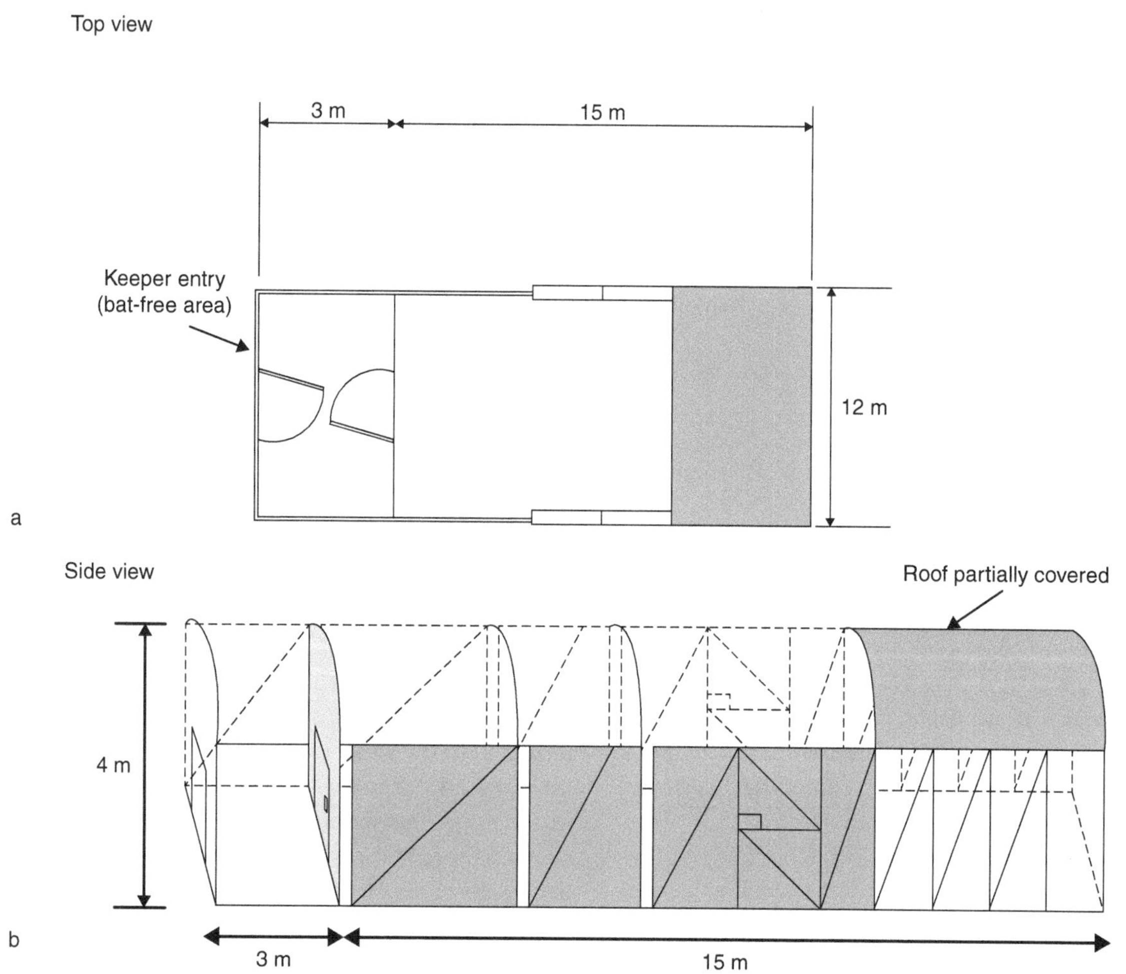


**Fig. 13.2.** Example of a flight enclosure for flying-foxes. (a) Plan view and (b) side view (not to scale).

## 4.1.2 Insectivorous bats

For long-term housing or pre-release flight training, tree-hole dwelling bats can be held in outside enclosures with high-quality heavy gauge polyester insect screening mesh walls instead of wire mesh, as wire can be abrasive and is corroded by urine. Mesh size is important and must be less than 6 mm because larger mesh sizes allow escape or entanglement (Hopkins 1990; Barnard 1995). Natural hollow logs or nest boxes can be added to provide roosting sites. All species should have a smaller roosting retreat zone comprising a smaller meshed enclosure placed as high as possible on a side wall shelf. This is typically where they are raised within or previously occupied before transfer to the larger enclosure (Lyons 2020). Feeding stations initially need to be located within the smaller roost retreat enclosures, but as the animals become accustomed to the larger enclosure can be moved slowly to other locations.

The larger enclosure should be at least 8 m in diameter and 2.5 m high and ideally round or hexagonal in shape for small species, and larger if housing larger species. The enclosure should have galvanised wire on the outside (to keep out predators) and lined with plastic mesh inside so the bats do not damage themselves (J. Mclean pers. obs.). The enclosure should also have several feeding bays on ledges on the walls so that the bats can hang from the wire walls while feeding (Fig. 13.3). Several shallow dishes or bottles of water should be placed near the food for easy access.

Insect-attracting light sources, although not intended or able to replace a full natural diet in scope or scale, are useful for enrichment purposes and in the initiation of

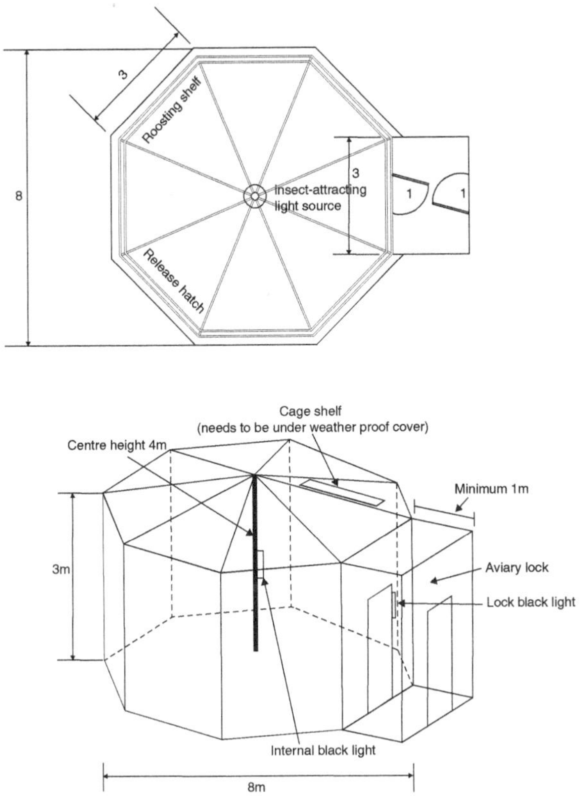


**Fig. 13.3.** Examples of a flight enclosure for insectivorous bats.

hunting training for young bats (Lyons 2020). Due to the small aperture mesh used for enclosures, lighting sources are best when actively managed and when double lighting systems are used. Lights are placed both within the enclosure and the lock area. Insects are attracted initially into an open lock area and then actively transferred into the main enclosure by turning the main light on, the lock door being opened and the lock light switched off. Although enclosures should be provided with an insect-attracting light system, their effect can be limited especially during cooler months as the abundance of insects depends on temperature, atmospheric pressure, atmospheric ions, humidity, moonlight, season, rainfall and wind (e.g. Williams *et al.* 1956; Humphrey-Smith 1982). Lighting systems should not be relied on at any time of the year to sustain the bats but only to supplement their diet and activity.

Due to their ability to escape through small openings, the maximum allowable gap in construction is only 5 mm (Hall 1982a). If using wire enclosures, particularly galvanised ones, care needs to be taken to ensure there are no sharp projections and that the bats can climb around in them (Hall 1982a; Department of Planning, Industry and Environment 2021). Metal-framed enclosures are more durable than timber frames and polyethylene plastic mesh is preferable to nylon insect screen, which deteriorates quickly and appears to be harsher on bat thumbs (J. Mclean pers. obs.). Wall materials such as glass, metals, plastic-coated vinyl, plexiglass or other hard, slick materials may be problematic because they can prevent the bats from gripping the walls properly and cause friction burns as the bats attempt to climb them unsuccessfully (Lollar and Schmidt-French 1998; Lyons 2020).

Outdoor flight enclosures should be double wired. The inner enclosure can be constructed of high-quality heavy gauge polyester insect screening mesh. The outside enclosure should be covered in a sturdy metal screening such as hail wire (6 mm) in order to prevent other animals from breaking into the enclosure (Lollar and Schmidt-French 1998). However, double-wiring reduces visibility, so the provision of glassed viewing areas may be required for exhibit purposes, but is not needed for rehabilitation purposes.

Bats held in nocturnal houses should be also provided with natural light cycles and a roosting retreat zone comprising a smaller enclosure with roosting pouches to allow them to retreat from the light as required (Lollar and Schmidt-French 1998).

## 4.2 Short-term holding area design for insectivorous bats

The short-term containment of insectivorous bats can be either indoors or in smaller enclosures, while teaching them to feed or if required for quarantine or veterinary treatment. A small-mesh enclosure or pet carrier can be used, with a roost retreat zone similar to the larger enclosures (Fig. 13.4). Feed dishes are placed on the floor of the enclosure at the base of the walls and the floor is covered with absorbent paper for ease of cleaning. The enclosure should have multiple roosting options, such as suspended pouches, several layers of hanging closely woven material (i.e. not towelling) or a small nest box for hollow-dwelling species. For cave-dwelling species (e.g. bent-wing bats) provide a reasonable-sized, stable rock for the bats to hang on, as these species prefer rocks to cloth-based materials. The mesh of the enclosure should be able to be dismantled for high-pressure and machine washing and sterilisation when required. Cover the sides, top and rear of the enclosure to simulate a dark roost, but position it to allow some light (twilight intensity) to enter for exposure to natural light cycles.

**Fig. 13.4.** Short-term holding enclosure for insectivorous bats held indoors. Photo from Rachel Lyons.

## 4.3 Spatial requirements

### 4.3.1 Pteropodids

Flying-foxes should be provided with enough room to fully stretch both wings and be able to make (at least) short flights. Bats that have been deprived of flight for 1 month or more have been found to lose the ability to fly due to a loss of muscle condition; though they can generally be exercised to regain their fitness for flying (Wilson DE 1988).

Recommended minimum sizes for enclosures for the long-term holding of Australian bats are given in Table 13.5 (DES 2020; Department of Planning, Industry and Environment 2021). The bats also need to be given adequate vegetation or other hanging material, such as ropes, at either end of the enclosure and the area between should allow free unhindered flight.

### 4.3.2 Insectivorous bats

For long-term housing or prerelease flight practice, the area of the flight enclosure depends on the species and their natural foraging style. Species that forage in more cluttered environments, such as within the tree canopy and understory vegetation (e.g. the slower flying eastern horseshoe-bat and species of long-eared bats (*Nyctophilus* spp.)), find it easy to fly in smaller aviaries in captivity, but require at least the dimensions given in Table 13.5 for full muscle conditioning and long-term fitness (Lyons 2020). In contrast, other species that are open or above-canopy fliers and foragers (typically fast fliers with relatively poor manoeuvrability), such as free-tailed bats (e.g. *Ozimops* spp.), require a much larger enclosure (Hall 1982a; Hopkins 1990; Lyons 2020) that ideally is circular or octagonal in shape to encourage both fast flight and manoeuvrable flight that works different muscle groups (Lyons 2020). Further details of flight enclosure requirements for Australian insectivorous bats are contained in Lyons (2020).

As insectivorous bats need to be able to hunt food on the wing immediately on release, flight muscle conditioning and cardio-fitness is imperative for survival in the wild (Lyons 2020). Survival studies post-release show high or total mortality rates when flight practice of long duration (> 4 weeks) in a sufficient sized area for the specific species is not provided (Kelly *et al.* 2008).

## 4.4 Position of enclosures

Flying-foxes are usually held outdoors with the enclosure sited so that it has good sun exposure because the bats will readily utilise the sun, particularly in cool weather, though shaded areas should also be provided for protection during hot weather. Black flying-foxes when housed outdoors in winter (including in subtropical zones) often

**Table 13.5.** Minimum enclosure areas recommended for the long-term holding of each genus of Australian bats.

Note: for cave-roosting species the nesting box should be constructed of material that simulates rock.

| Species | No. of bats per enclosure of shown dimensions | Area (L × W × H) (m) | Additional floor area for each extra animal (m) | Mesh size (mm) | Roost box dimensions (cm) |
|---|---|---|---|---|---|
| **Yinpterochiroptera** | | | | | |
| **Pteropodidae** | | | | | |
| *Macroglossus* | 10 | 5 × 5 × 2 | 1.0 × 1.0 | 10 | Not required |
| *Syconycteris* | 10 | 5 × 5 × 2 | 1.0 × 1.0 | 10 | Not required |
| *Nyctimene* | 10 | 5 × 5 × 2 | 1.5 × 1.5 | 10 | Not required |
| *Dobsonia* | 30 | 10 × 6 × 3 | 2.0 × 2.0 | 20 | Not required |
| *Pteropus* | 30 | 15 × 12 × 4 | 2.0 × 2.0 | 20 | Not required |
| **Megadermatidae** | | | | | |
| *Macroderma* | 10 | 6 × 6 × 2 | 1.5 × 1.5 | 10 | Not required |
| **Rhinolophidae** | | | | | |
| *Rhinolophus* | 10 | 6 × 6 × 2 | 0.5 × 0.5 | 6 | 20 × 20 × 40–75 |
| **Hipposideridae** | | | | | |
| *Hipposideros* | 10 | 6 × 6 × 2 | 0.5 × 0.5 | 6 | 20 × 20 × 40–75 |
| **Rhinonycteridae** | | | | | |
| *Rhinonicteris* | 10 | 6 × 6 × 2 | 0.5 × 0.5 | 6 | 20 × 20 × 40–75 |
| **Yangochiroptera** | | | | | |
| **Emballonuridae** | | | | | |
| *Saccolaimus* | 10 | 12 × 12 × 2 | 0.5 × 0.5 | 6 | 20 × 20 × 40–75 |
| *Taphozous* | 10 | 12 × 12 × 3 | 0.5 × 0.5 | 6 | 20 × 20 × 40–75 |
| **Molossidae** | | | | | |
| *Austronomus* | 10 | 8 × 8 × 4 | 0.5 × 0.5 | 6 | 20 × 20 × 40–75 |
| *Chaerephon* | 10 | 8 × 8 × 3 | 0.5 × 0.5 | 6 | 20 × 20 × 40–75 |
| *Micronomus* | 10 | 8 × 8 × 3 | 0.5 × 0.5 | 6 | 20 × 20 × 40–75 |
| *Ozimops* | 10 | 8 × 8 × 4 | 0.5 × 0.5 | 6 | 20 × 20 × 40–75 |
| *Setirostris* | 10 | 5 × 5 × 3 | 0.5 × 0.5 | 6 | 20 × 20 × 40–75 |
| **Miniopteridae** | | | | | |
| *Miniopterus* | 10 | 8 × 8 × 3 | 0.5 × 0.5 | 6 | 20 × 20 × 40–75 |
| **Vespertilionidae** | | | | | |
| *Myotis* | 10 | 5 × 5 × 3 | 0.5 × 0.5 | 6 | 20 × 20 × 40–75 |
| *Phoniscus* | 10 | 3–5 × 3–5 × 2 | 0.5 × 0.5 | 6 | 20 × 20 × 40–75 |
| *Murina* | 10 | 8 × 8 × 3 | 0.5 × 0.5 | 6 | 20 × 20 × 40–75 |
| *Chalinolobus* | 10 | 8 × 8 × 3 | 0.5 × 0.5 | 6 | 20 × 20 × 40–75 |
| *Falsistrellus* | 10 | 8 × 8 × 3 | 0.5 × 0.5 | 6 | 20 × 20 × 40–75 |
| *Nyctophilus* | 10 | 3–5 × 3–5 × 2 | 0.5 × 0.5 | 6 | 20 × 20 × 40–75 |
| *Scotorepens* | 10 | 8 × 8 × 3 | 0.5 × 0.5 | 6 | 20 × 20 × 40–75 |
| *Vespadelus* | 10 | 8 × 8 × 3 | 0.5 × 0.5 | 6 | 20 × 20 × 40–75 |
| *Pipistrellus* | 10 | 5 × 5 × 3 | 0.5 × 0.5 | 6 | 20 × 20 × 40–75 |
| *Scoteanax* | 10 | 8 × 8 × 3 | 0.5 × 0.5 | 6 | 20 × 2 0 × 40–75 |

suffer necrosis of the ear tips from exposure to the cold (frost bite). Warmth supplied through hanging blanket nooks and artificial heating is required for this species. Small pteropodid bats, such as blossom-bats, and insectivorous bats are more often held indoors, which better allows the climate to be controlled, except where captive-breeding and release programs are in operation and acclimatisation is required.

## 4.5 Weather protection

If held outdoors, protective roofing and side-walls for a section of the enclosure should be provided so the bats can retreat from the prevailing rain, direct sunlight and wind, with the remaining enclosure being relatively open.

## 4.6 Temperature requirements

Most species of flying-foxes are highly susceptible to temperatures above 40°C and relative humidity below 30%, with high mortality rates known to occur in wild populations (Welbergen *et al.* 2008; Kim and Stephen 2018; Parry-Jones 2018; Mounter 2019; Ratnayake *et al.* 2019). Nursing females and juvenile bats are particularly prone to heat stress (Snoyman *et al.* 2012; Parry-Jones 2018). Therefore, these species need to be provided with shade and sprinklers that can be turned on to mist the enclosure to reduce the ambient temperature when temperatures exceed ~35°C (Mo and Roache 2020; Department of Planning, Industry and Environment 2021). Misting also increases humidity, which reduces fluid loss and prevents dehydration.

Insectivorous and small pteropodid bats often have very specific temperature and humidity requirements, particularly when raising offspring. It is important that the enclosure mimics as best as possible the roosting conditions in the wild. Roost sites are chosen in the wild for several reasons, including protection from the weather and predators, optimal microclimate, reduced commuting costs, improved mating opportunities and improved maternal care (Altringham 1999). Bats roost in a variety of locations from caves, mines and houses (with relatively constant temperatures and humidity) to within tree hollows, under bark or sometimes in dead logs, as well as just free hanging on branches, situations that have much greater variability in temperature and humidity (Churchill 2008). Some species of bats roost in unusual places such as in bird nests (e.g. golden-tipped bats) or within epiphytic ferns (e.g. flute-nosed bats) (Churchill 2008). Other human structures have provided roosts for bats, including drainpipes, chimneys, under floors, water tanks, cars, rolled up swags, old stoves, old tractor exhausts, under stones, in cracks in wooden fence posts and tin cans (Reardon and Flavel 1987; Churchill 2008).

The location in which cave-roosting bats roost varies greatly, ranging from near the entrance in the twilight zone to deeper in the cave. Roosts include a range of surface types, including open surfaces to very small cracks and from over water to under boulders (Fig. 13.5). Caves that offer a wide thermal range combined with structural and elevational complexity provide the greatest diversity of roosting sites (Kunz 1982). The distance that bats roost apart also varies between individuals and, more importantly, between species, from huddling close together to ~30 cm apart. This, in turn, has implications on the number of animals that can be adequately housed together without overcrowding. An outline of roost locations, environments and colony sizes of those known for different genera of bats is shown in Table 13.6. Care needs to be taken, however, to reflect the seasonal life cycle colony dynamics for each species. Expert knowledge should be sought for each species to avoid territorial fighting, breeding season injury and unwanted pregnancies due to incorrect housing density and arrangements for the time of year (Lyons 2020).

Several species of insectivorous bats are exceptionally prone to low humidity conditions and should be housed in humidity-controlled settings, particularly during climatic heat events. Cave and/or aquatic habitat-dependent species are most susceptible, including diamond-faced bats, leaf-nosed bats, horseshoe-bats and the large-footed myotis (Lyons 2020).

Most species, and particularly those that live in tropical climates such as ghost bats, require heating provided by a wall-mounted covered heat source such as infrared light or a heating mat so that a thermal gradient can be established. The bats can be provided with several roosting sites by hanging pieces of polar fleece (~30 $cm^2$) or roosting pouches located at different distances from the heat source. Polar fleece is preferable to towelling, in which thumbs and toes can get caught and cause injuries.

Various species of insectivorous bats have been observed to follow the same activity and bodyweight patterns as wild individuals (L. Lumsden pers. obs.), when provided with natural conditions. Many Australian species of bats use torpor extensively to reduce energy expenditure, especially during the cooler months

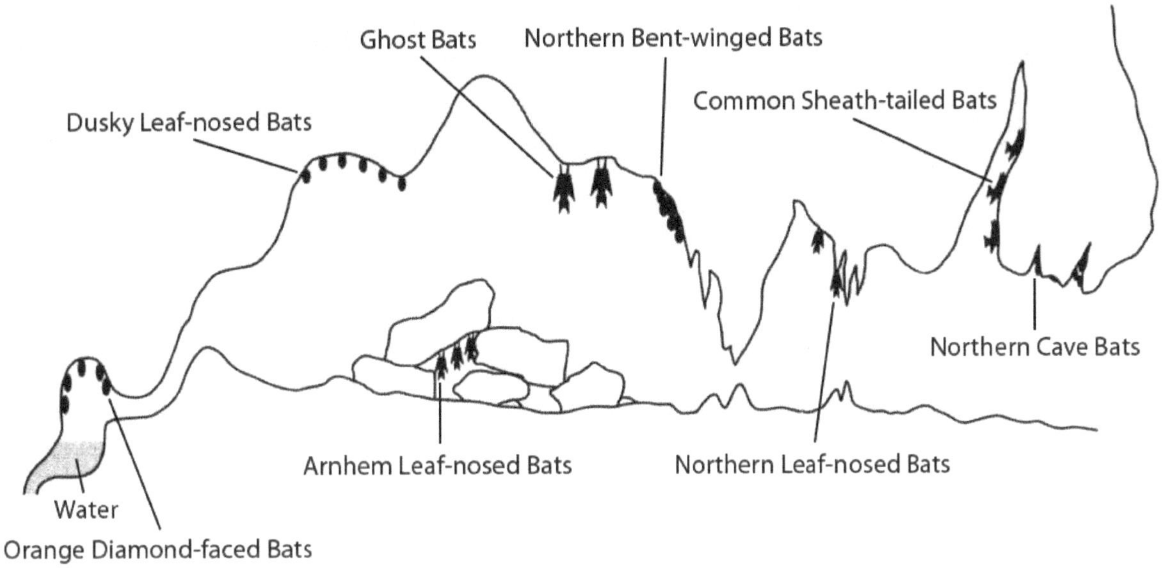


**Fig. 13.5.** Profile of a cave in the Northern Territory, showing how bats select different areas of the cave for roosting. Derived from Churchill (2008).

(Geiser 2006; Geiser and Stawski 2011; Bondarenco *et al.* 2013). In captivity, eastern free-tailed bats, for example, can range from 8 g in spring to 13 g in autumn, which provides ample fat reserves to go into torpor over winter where they need only one feed every 5–6 days (L. Lumsden pers. obs.). Bent-wing bats have been induced into torpor by reducing the temperature to 9–10°C in a small container in which the humidity was 75–85% (Hall 1982b). Species from temperate regions are likely to tolerate a wider range of temperatures and humidity than those from more tropical regions (L. Lumsden pers. obs.). More recently, grey-headed flying-foxes have been recorded to enter torpor while day-roosting in southern Australia during winter. During torpor their minimum body temperature lowered to 27°C (Turbill *et al.* 2024). Torpor has not previously been recorded in bats weighing > 50 g and occurred after a period of wet, windy and cold weather, suggesting it is an adaptation to reduce energy expenditure when they have depleted body energy stores and increased thermoregulatory costs (Turbill *et al.* 2024).

The number of bats held together depends greatly on the species (Table 13.6). Within the Pteropodidae, species such as the tube-nosed bat appear to be largely solitary, blossom-bats appear to be solitary or to live in small groups whereas flying-foxes can live in groups of > 100 000 individuals. Similarly, among the insectivorous bats many species are solitary, some roost in small groups and others roost in large groups of up to 100 000. Roost size does, however, vary for most species of insectivorous bat according to the time of year and season. Therefore, the species should be held according to its social behaviour at the particular time of year in order to fulfil thermoregulatory requirements (though temperature can be artificially controlled in captivity) or to prevent aggression due to mating stress and overcrowding, which can sometimes result in fatal injuries (Lyons 2020).

It is important to regulate the temperature with a thermostat to ensure the bats are exposed to temperatures similar to those they experience in the wild, especially for bats from tropical regions of Australia being held in more temperate regions. When required, they can be held indoors in mesh enclosures with heat pads mounted externally on one end (Lyons 2020) or in enclosures with solid walls constructed of wood (e.g. 2-cm thick plywood with horizontally grooved sides (0.6–1.3 cm apart) which help the bats grip the wall while roosting) (Barnard 1995).

Although heating is not always required to maintain bats, it can be used in enclosures to make them more active and therefore more visible throughout the year. It may also be required to facilitate more rapid recovery from illness or injury, which can be slowed if the bats enter torpor. Heating can be provided by clamp-on lights with 25 W red light globes (with care to ensure there is no ability for the bats to come in contact with

**Table 13.6.** Roost location, environment and colony sizes of the different genera of Australian bats.

Derived from McKean and Hamilton-Smith (1967), Churchill (2008), Law (1993), Law and Chidel (2004), Drury and Geiser (2014), Lumsden *et al.* (2020) and L. Lumsden (pers. obs.).

| Genus | Location | Colony size |
|---|---|---|
| **Yinpterochiroptera** | | |
| **Pteropodidae** | | |
| *Macroglossus* | Thick foliage in shady vegetation, bamboo thickets, palm fronds, rolled up leaves of bananas, eaves of huts | 1–3 |
| *Syconycteris* | Littoral rainforest subcanopy and sometimes canopy or wet sclerophyll forest. Use a new roost each night but in the same vicinity | Solitary |
| *Nyctimene* | Canopy and understory of rainforest trees | Solitary |
| *Dobsonia* | Caves in the twilight zone, boulder piles, disused mines, concrete bunkers, dark rainforest thickets and large tree hollows | < 100 in Australia<br>1000s in New Guinea |
| *Pteropus* | Roost in the canopy of trees in woodlands, gullies, mangroves or rainforest. One record from within a cave. Camps can often contain multiple species of flying-foxes | 10s to > 100 000 |
| **Megadermatidae** | | |
| *Macroderma* | Sandstone caves, boulder piles, abandoned mines. Preferences for roost microclimates of 27.5°C and 80% humidity | 1–1500<br>25 cm apart |
| **Rhinolophidae** | | |
| *Rhinolophus* | Caves, abandoned mines and buildings, rock piles, road culverts. Usually in complete darkness but can be in the twilight zone. Generally, with high temperature and humidity. *R. robertsi* is likely to also roost in dense vegetation and tree hollows | Typically, < 20<br>Up to 2000<br>15–20 cm apart |
| **Hipposideridae** | | |
| *Hipposideros* | Caves, boulder piles, disused mines. Generally, with high temperature and humidity. Some have been recorded in tree hollows or in the tree canopy. *H. semoni* has been found in an oven, clothes closet and door handle of a car | Usually 10–30<br>From 1–3000<br>10–25 cm apart |
| **Rhinonycteridae** | | |
| *Rhinonicteris* | Caves and abandoned mines that are hot (28–32°C) and humid (96–100%). Thought to be forest dwellers during the wet season | 5–20 000<br>10–15 cm apart |
| **Yangochiroptera** | | |
| **Emballonuridae** | | |
| *Saccolaimus* | Primarily tree hollows, but also recorded in abandoned sugar glider nest, external walls of buildings and in caves | 1–6 |
| *Taphozous* | Caves in twilight zone, disused mines, boulder piles, rock fissures and concrete bunkers. *T. kapalgensis* is a tree dweller | 2–25 but up to 260 |
| **Molossidae** | | |
| *Austronomus* | Usually in tree hollows | 1–25, maternity colonies up to 300 |
| *Chaerephon* | Usually in tree hollows, but also in coconut palms, caves, buildings and wooden jetties | Usually 1–15, but can be up to 300 |
| *Micronomus* | Usually in tree hollows, but also in caves and roofs or cavities of houses | 1–10? |
| *Ozimops* | Usually in tree hollows and under bark | Up to 100 |
| *Setirostris* | Usually in tree hollows | 1–15 |
| **Miniopteridae** | | |
| *Miniopterus* | Caves, mines and road culverts | Up to 100 000 |
| **Vespertilionidae** | | |
| *Myotis* | Caves, tree hollows, dense vegetation, under bridges, mines, tunnels, storm water drains. Usually roost close to water | Usually 10–15<br>Up to 300 |

| Genus | Location | Colony size |
|---|---|---|
| *Phoniscus* | Abandoned nests of gerygones (*Gerygone* sp.) and scrub wrens (*Sericornis* sp.), beneath hanging moss and in foliage. There is an also a record in a cave | Solitary or small groups |
| *Murina* | Rainforest trees (e.g. base of fronds of epiphytic ferns or tree ferns and bird nests) | 1–12 |
| *Chalinolobus* | *C. dwyeri* roosts in caves and mines in twilight zone. *C. nigrogriseus* and *C. gouldii* roosts in tree hollows (e.g. *Eucalyptus camaldulensis*) or buildings. *C. morio* roosts in tree hollows, houses, exfoliating bark of trees, fairy martin nests, culverts, bridges and caves. *C. picatus* roosts in trees, caves and abandoned mines | Usually 1–40<br>Up to 400 |
| *Falsistrellus* | Tree hollows, branches and stumps. Sexes segregate during roosting, at least during much of spring and summer | 3–36 |
| *Nyctophilus* | Dead tree hollows and loose bark, low to the ground under bark, and rainforest foliage, Pandanus leaves, in houses | 1–3<br>Maternity colonies of 10–40 |
| *Scotorepens* | Tree hollows, roofs of buildings | 2–20 |
| *Vespadelus* | Tree hollows, buildings such as tin sheds, or caves in cracks and crevices near the entrance to caves and mines | Usually < 80<br>Can be > 500 |
| *Pipistrellus* | Tree hollows and among vegetation (e.g. in dry bamboo, coconut palms and Pandanus). Several records in caves | ~6 |
| *Scoteanax* | Tree hollows, branches, roofs of old buildings | Unknown |

the globe) or heating pads attached to the outside of roosting enclosures to create a temperature gradient so that the bats can choose the preferred temperature (Barnard 1995). Insectivorous bats can survive in a wide temperature range, but to increase activity the species from temperate regions can be held at 22–23°C and 50–65% humidity during winter and 25–28°C and 50–80% humidity during summer (Barnard 1995). Tropical bats should be held all year round at 25–28°C with a relative humidity of 55–90% (Barnard 1995). The temperatures and humidity of roosts in the wild for species for which they are known are shown in Table 13.7. The effect of manipulating the temperature on reproduction and fertility is poorly known; however, an experiment that placed Gould's long-eared bats in warmer than usual conditions (22°C), with natural lighting cycles, after the winter solstice resulted in the advancement of the onset of the spring phase of the reproductive cycle in both males and females (Phillips and Inwards 1985). Holding impregnated wild female bats over winter with heating and ample food will bring forward ovulation and fertilisation by stored sperm in their oviducts. Captive bats of some species, including the greater broad-nosed bat, eastern broad-nosed bat, little broad-nosed bat, Gould's long-eared bat, eastern long-eared bat, northern free-tailed bat, Ride's free-tailed bat and white-striped free-tailed bat, are generally born 2–3 months earlier than normal wild pup births for the species. It is possible that this phenomenon also occurs in other species of insectivorous bats if given the opportunity (R. Lyons pers. obs.).

## 4.7 Substrate

In outdoor enclosures the substrate should be non-abrasive, easy to clean (ideally by hosing) and have good drainage, such as smooth concrete, linoleum, rubber lining or weedmat covered in shade cloth. Other substrates such as soil, grass, mulch or sand are not recommended due to the potential for bats to ingest them (Barnard 1995; Fascione 1995), difficulties in hygiene management and difficulties in seeing grounded smaller bats, thus endangering them when entering the enclosure. In indoor enclosures, padded puppy training liners are ideal to soak up spills and urine and easy to replace.

## 4.8 Roost boxes

Given the variety of locations and microhabitats in which insectivorous bats choose to roost (Tables 13.6, 13.7), the size and type of roost box provided needs to vary accordingly. Roost boxes that are used for wild bats have also been used with success for captive populations. Roost boxes should be made from untreated wood and can be hung with closely woven cloth such as denim pleated and stapled on to the internal wall through which the passageway to the flight area protrudes.

The number of compartments depends on the number of bats that are to use it and the species. Several commonly utilised roost boxes preferred over others in

**Table 13.7.** Temperatures, humidity and spacing in roosts of Australian bats known from the wild.

| Species | Temperature (°C) | Humidity (%) | Spacing (cm) | Reference |
|---|---|---|---|---|
| **Yinpterochiroptera** | | | | |
| *Syconycteris australis* | 16–24 | 40–80 | Solitary | 1 |
| *Macroderma gigas* | 23–26 | – | – | 2 |
| *Rhinolophus megaphyllus* | 12–33 | 85–100 | 2–3 | 3 |
| *Rhinonicteris aurantia* | 27–32 | 85–100 | 12 | 4–6 |
| **Yangochiroptera** | | | | |
| *Myotis macropus* | 14–24 | | Cluster | 7 |
| *Miniopterus orianae* | 10–30 | 80–100 | Cluster | 8–11 |
| *Chalinolobus gouldii* | 12–40 | – | Cluster | 11–12 |
| *Nyctophilus corbeni* | 15–45 | | Solitary | 13 |
| *Nyctophilus geoffroyi* | 12–46 | – | Solitary, Cluster | 11, 14 |
| *Nyctophilus major* | 12–18 | – | Solitary | 14 |
| *Vespadelus regulus* | 20–45 | – | Cluster | 15 |

References: (1) Law (1993); (2) Toop (1985); 3) Hall *et al.* (1975); (4) Jolly (1988a); (5) Churchill (1991); (6) Armstrong (2000); (7) Campbell *et al.* (2010); (8) Hall (1982c); (9) Baudinette *et al.* (1994); (10) Hall *et al.* (1997); (11) L. Lumsden pers. obs.; (12) Dixon and Huxley (1989); (13) Lumsden *et al.* (2008); (14) Hosken (1996); (15) Menkhorst (1995).

captive situations include wedge-shaped designs with bottom slot entries and box-shaped designs with one or more compartments, again with bottom slot entries. Both styles have landing boards below the entry slot with grooved markings for grip and top-opening hinged lids for inspection.

One design uses untreated 25-mm, or slightly thicker, timber that is rough sawn or roughened on all surfaces so that the bats are able to land and investigate by crawling all over the box (Fig. 13.6; Stebbings and Walsh 1991). If the wood is relatively smooth, horizontal saw cuts should be made ~1 mm deep on the inside and around the entrance to allow the bats' sharp claws to grip the crevices. The lid fits into a groove that is cut into the top of the back plate or, alternatively, a hinge can be made and a strip of tyre inner tube tacked over the joint (a well-fitted lid, which should have a hook at the front to secure it down, increases the likelihood of the box being used). The entrance should be at the bottom of the box and either in the front or in the base, with a rough landing area, and the slit should be the full length of the box, a minimum of 15 mm (W) × 40 mm (L), although the entrance dimensions will be dependent on the species. Dimensions will need to increase for larger species. Joints are nailed and glued with waterproof glue to ensure the box is draught-proof. Larger boxes can be divided by a vertical partition of rough timber, 10-mm thick, with a 40-mm space above the base (Stebbings and Walsh 1991).

The colour of roost boxes also appears to be important and will depend on the likely conditions at the site. A roost selection study of Gould's long-eared bats (conducted in winter) found that when given the choice they chose the warmer black boxes rather than the white. The black boxes were recorded to be an average of 5.1°C (maximum 7.5°C) warmer than the white boxes at the maximum daytime temperature (Doty *et al.* 2016). This preference assists in minimising energy use in arousal from torpor/diurnal lethargy in winter. Similarly, west-facing roost walls with solar access have the same effect. However, the situation may be different in summer and where extreme temperatures may occur, boxes painted pale colours are recommended to reflect the heat (Griffiths *et al.* 2017), as temperatures up to 45°C have been recorded in nest boxes (Griffiths *et al.* 2022). An examination of the parameters selected by tree-hole roosting bats, including the little forest bat, chocolate wattled bat, lesser long-eared bat, Gould's long-eared bat and the southern free-tailed bat, showed that all selected roosts had cavities with one dimension not much larger than the bat. None were far from water and colony size was variable, but tended to be small and segregated by sex and species (Tidemann and Flavel 1987). Similarly, the large-footed myotis is known to roost in tree hollows in riparian vegetation, crevices of an old aqueduct tunnel near water (Campbell 2009) and most commonly under bridges and in road culverts (Gorecki 2020).

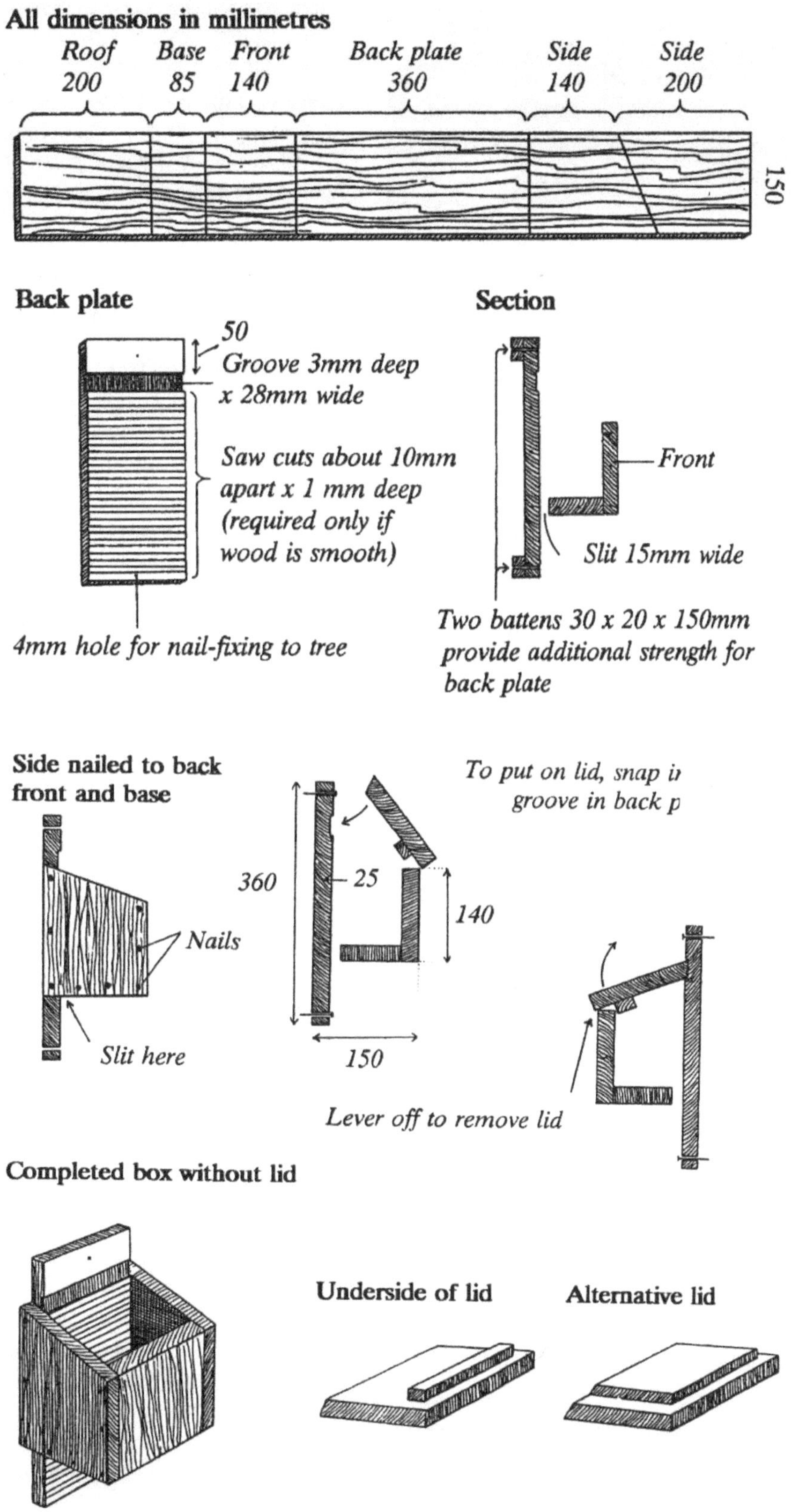


**Fig. 13.6.** Roost box for tree-hole bats. Reprinted with permission from the Bat Conservation Trust (n.d.)/www.bats.org.uk.

## 4.9 Enclosure furnishings

Enclosures should be relatively free from obstacles to maximise the ability of bats to fly. Large pteropodids such as flying-foxes do not need additional fittings, such as branches to roost on, if they are given adequate wire or ropes to hang on, though they can be supplied as part of behavioural enrichment and to provide shade. Smaller pteropodids, such as blossom-bats, should be given foliage in which they can hide. Tube-nosed bats appear to prefer hanging from the roof and do not take advantage of foliage when it is provided (J. Mclean pers. obs.). Natural vegetation can be used; however, artificial rainforest plants have also been used in conjunction with natural branches to minimise the requirement for branch turnover, which can be disruptive to roosting (S. Jackson pers. obs.). The only furnishings required for insectivorous bats are those that are required for roosting and feeding.

# 5. GENERAL HUSBANDRY

## 5.1 Hygiene and cleaning

Enclosures for large species such as flying-foxes and ghost bats should be cleaned daily to minimise the build-up of uneaten food and faecal matter. The enclosure should be cleaned routinely with a 1% bleach solution and then rinsed to help reduce unpleasant odours from accumulated urine (Barnard 1995). Apart from concrete floors, heat-welded polyethylene floors may be used, which can be cleaned with just water without bleach. Many Australian bat carers use sand on the floor of aviaries and rake the solids out in the morning (J. Mclean pers. obs.). If cleaning products such as bleach are used, the bats should generally be removed from the feeding area into an adjacent part of the enclosure until the enclosure is dry.

Wire screening of the enclosure can be cleaned by high-pressure water followed by disinfectant treatment. The bats can be returned once the enclosure is completely dry (Lollar and Schmidt-French 1998). Pressurised steam cleaning with a wire scrubber can also be used to cleaning enclosures as can boiling water (Barnard 1995). Small enclosures can be dried under natural sunlight for several days (Barnard 1995). Mesh enclosures with plastic tubing can be taken apart and put through washing machines or soaked before being left in the shade to dry (Lyons 2020); however, full sun exposure for extended periods can distort the mesh (R. Lyons pers. obs.).

While cleaning the enclosure of flying-foxes, endeavour to check the bats by looking for any injuries or asymmetry in the wings that may indicate fractures or dislocations (Wilson P 1988). Pteropod bats generally wrap their wings around themselves, but larger species such as flying-foxes may stretch their wings in a threat display which assists in assessing their condition. Pteropod bats generally have their head bent towards the chest at 90° to the trunk, while insectivorous bats hold the head in alignment with the body (Wilson P 1988).

In large populations of bats, histoplasmosis is a zoonotic risk if there is a build-up of faecal material (see also section 8.7). Histoplasmosis is caused by inhalation of the fungus *Histoplasma capsulatum*, which is found worldwide in soils, especially if enriched by the excreta of bats or birds, and is generally transferred by spores in the environment rather than from host to host (Constantine 1985; Hine 1988; Booth 1994). This disease can occur as a rare, consistently fatal form or as a more prevalent acute non-fatal form in humans (Booth 1994). The risk of histoplasmosis is highest in densely populated caves, though even in caves in southern Australia the risk is low because few caves have *Histoplasma* in them (Holz *et al.* 2018a). In captivity, the risks are much lower as long as enclosures are routinely cleaned to prevent build-up of fungus within guano piles.

## 5.2 Record keeping

It is important to establish a system whereby the health, condition and reproductive status of the captive populations of bats are routinely monitored and recorded:

- identification (all individuals should be identifiable)
- results of veterinary examinations and treatments administered
- behavioural changes or problems
- reproductive behaviour or condition
- bodyweights and measurements
- changes in diet and food consumption
- movements of individuals between enclosures or institutions
- births, with parentage if known
- deaths with necropsy results.

The collection of information on the physical, physiological, and behavioural patterns of individual bats can contribute greatly to the husbandry of these species. It also allows individual histories to be transferred to other institutions if required and facilitates a cooperative approach to data collection among institutions. ZIMS, the Zoo Information Management System (Species360 2023), can be used for recording general information on age, sex, parentage, births, identification, bodyweights, transfers and deaths, with separate ZIMS modules for veterinary information and studbooks. These systems have been developed by Species360, which is part of the Conservation Planning Specialist Group (CPSG) software. ZIMS interfaces with programs, such as PMx, which provides demographic and genetic analysis tools to assist with the management of breeding programs (Lacy *et al.* 2011).

## 5.3 Methods of individual identification

Various techniques have been used to mark bats, each having its advantages and disadvantages, with some methods being better than others (Lumsden *et al.* 2022).

### 5.3.1 Passive Integrated Transponder (PIT) tags

PIT tags (microchips) are the most commonly used form of identification by bat rehabilitators, due to the very low rate of injury for both groups of bats. For smaller species, some carers recommend PIT tags should be inserted by a veterinarian under anaesthetic (Lyons 2020). In contrast, field researchers have successfully PIT-tagged 4000 southern bent-wing bats without anaesthesia (van Harten *et al.* 2020; L. Lumsden pers. obs.), as well as large numbers of Gould's wattled bats (van Harten *et al.* 2021) and they believe the procedure can be undertaken safely and quickly by trained individuals. Some authors do not recommend anaesthetic for wild insectivorous bats due to the associated risks and the longer time the bats need to be held (Lumsden *et al.* 2022).

PIT tags are inserted between the scapulae and have been used with great success on several species of bats from large ones, such as flying-foxes, to very small species including the eastern blossom-bat. Tracking of the tag along the injection site can be prevented by sealing the entry wound with a single drop of tissue glue (e.g. Vetbond®), which has been found successful in wild populations of southern bent-wing bats and Gould's wattled bats (van Harten *et al.* 2020, 2021). Insectivorous bats can, however, be irritated by the use of adhesive and will scratch the glue and fur off, leaving a bald patch that will eventually regrow (R. Lyons pers. obs.). There are records of these implants migrating under the skin and lodging elsewhere. Microchip use generally requires the animal to be caught to confirm identification with a microchip scanner, but readers can be set up in nest boxes, caves or other roosting sites to record animals as they move past the scanner (Kerth and König 1996; Kunz 2001; van Harten *et al.* 2019).

### 5.3.2 Tattoos

In the past, tattooing the patagium was done with limited success, as there was often a need for further tattooing. However, more recently it has been found to be a safe, quick and effective long-term marking technique for insectivorous bats, with tattoos on the propatagium lasting up to 10 years (Lumsden *et al.* 2022). Ghost bats have been tattooed successfully on the ears using a series of lines that allows visual recognition of each individual for more accurate recording of behaviours (Fig. 13.7). Due to the curvature of the ear in positions one and two, the lines are drawn on both sides of the ear in these positions in ghost bats (Jackson and Worthington-Wilmer 1994).

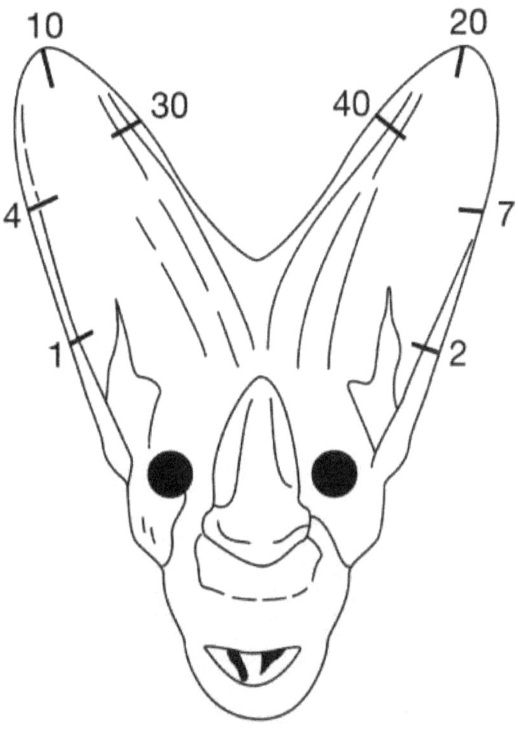


**Fig. 13.7.** Ear tattoo numbering used to identify the ghost bat. Taken from Jackson and Worthington-Wilmer (1994).

### 5.3.3 Ear notching

Ear notching has been used successfully with eastern blossom-bats. The notches could be cut in the ears using the same system as for the tattooing, but this technique should not be used on any species of insectivorous bat as it is likely to interfere with their echolocation (Lumsden *et al.* 2022).

### 5.3.4 Forearm bands

All banding of bats in Australia is regulated by the Australian Bird and Bat Banding Scheme (ABBBS), so in order to band bats personnel involved must be licenced under that scheme. The ABBBS recommends the type and size of bands for each species and can supply the bands.

In the past, the use of forearm bands on wild bats resulted in injuries and a significant decrease in the survival of several species of bats (eastern horseshoe-bats, common sheath-tailed bats, coastal sheath-tailed bats and southern free-tailed bats), depending on the band type, band size and metal type (Baker *et al.* 2001). Recent research suggests that each species needs to be considered individually when assessing the suitability of band types, rather than assuming that all will react the same way. As a result, the ABBBS has adopted a precautionary principle, including a moratorium on the banding of several families of insectivorous bats and not approving new banding permits (DCCEEW 2021), and is working with researchers to determine the most appropriate options (L. Lumsden pers. obs.). Baker *et al.* (2001) suggested there was a need to develop alternative methods for marking insectivorous bats for ecological studies and the most frequently used techniques are listed below.

The main type of bands used on insectivorous bats as part of the ABBBS are specifically constructed metal alloy bat bands with a rolled flange or lipped wing at each end (Baker *et al.* 2001). In the past, split metal bird bands were also used but found to cause unacceptable injury rates (Baker *et al.* 2001) and are no longer recommended. It is important to determine the appropriate size of bands for a species, with tighter fitting (smaller) sizes invariably resulting in injuries (Baker *et al.* 2001). Flanged bat bands should not be used on the forearms of pteropodids (Rudran and Kunz 1996).

For bats with a small propatagium, bat bands are attached by holding the wing extended, placing an open hand over the shaft of the forearm near the wrist and closing it with pressure from the thumb and forefinger (Fig. 13.8). In some countries, males are banded on the right forearm and females on the left (Lollar and Schmidt-French 1998); however, in Australia this convention is not used when banding wild bats (L. Lumsden pers. obs.), though could be useful in captivity. When the lip-end is closed around a bat's forearm, it should be nearly round and have a small space between the ends and the wing membrane to minimise the possibility of the wing membrane becoming irritated and leading to an overgrowth of tissue (Rudran and Kunz 1996). The inside diameter should also be large enough to prevent pressure on the underlying tissue (Rudran and Kunz 1996).

**Fig. 13.8.** Band used on the forearm of a greater broad-nosed bat. Photo courtesy of George Madani.

Four main injuries have been observed as a result of bands being too small (Zambelli *et al.* 2009) or otherwise causing irritation: (1) ingrown forearm bands partially embedded in the flesh; (2) swelling of the forearm as a result of constriction, possibly the result of the bats biting the band and surrounding area, resulting in reduced blood flow; (3) perforation of the wing membrane from the band rubbing; and (4) scraping and rubbing of the breast skin where the forearms fold during rest (Fig. 13.9).

### 5.3.5 Thumb bands

The ABBBS recommends the use of stainless steel or incoloy bands to mark the thumbs of larger pteropodids. Butt-end bands are attached by placing the band into the closing pliers and then placing them around the thumb and carefully closing the band (Fig. 13.10). Metal bands are better than plastic ones because they are harder and more resistant to chewing and therefore less likely to be damaged or cause injury to the underlying tissue. Before attaching a thumb band, determine the best fit by trying different sized bands, keeping in mind that the band should be loose enough to move freely over the shaft of the thumb without causing irritation, but not so loose as to slip over the thumb claw, and the ends should never overlap (Rudran and Kunz 1996). They should never be used on animals in the wild that are still growing but can be used on captive juveniles; 5-mm thumb bands have been used extensively for flying-fox orphans, but need to be removed from the thumb by 12 weeks of age and, if required, replaced with a larger band size (J. Mclean pers. obs.).

### 5.3.6 Ball-chain necklaces

Necklaces made from stainless steel ball-chains onto which butt-end bands or split-rings are strung for identification purposes have been useful for long-term marking of both pteropodids and insectivorous bats with few injuries or deaths (Rudran and Kunz 1996). They are durable, lightweight, flexible and rarely lost, but should only be used on adult animals (Rudran and Kunz 1996). Ball-chain necklaces can be made or bought (see Appendix 2).

Fitting of ball-chain necklaces usually requires two people, although it can be done by one person with practice and a restraining device (Handley *et al.* 1991). The necklace should be small enough to prevent removal by the bat and large enough so that it does not abrade the underlying skin or choke the bat (Rudran

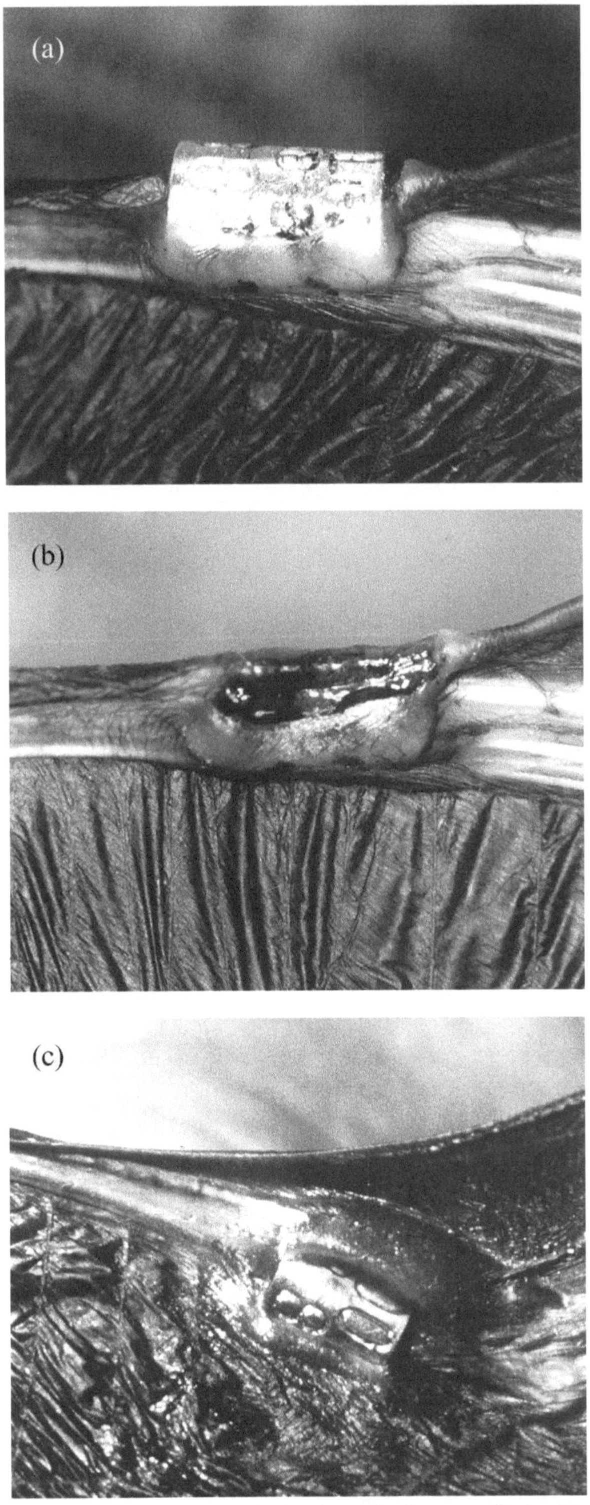


**Fig. 13.9.** Injuries caused by bands on the forearm of bats: (a) band embedded with skin growing over the edge and causing swelling, (b) raw and swollen area after the band was removed and (c) band lodged and embedded near the elbow. Reproduced from Baker *et al.* (2001) with permission from CSIRO Publishing.

**Fig. 13.10.** A butt-end band attached to the thumb of a pteropodid. Redrawn based on a figure by Jennefer Mclean.

and Kunz 1996). If skin irritation occurs, necklaces should be enlarged or replaced with another marking device (Rudran and Kunz 1996). Necklaces have not been used extensively in wild studies in Australia, however (L. Lumsden pers. obs.; Temby 1995). Although they have been used previously for species of bats in captivity, it has been suggested that they may not be appropriate for marking bats maintained in captivity as skin irritation has been observed in some bats, especially if they become obese or if food accumulates under the necklace when bats feed from dishes (Rudran and Kunz 1996).

### 5.3.7 Velcro collars

Velcro collars have been commonly used on all flying-fox species and different ages (e.g. PAA Puppy ID Collars). They can be purchased online in different colours and should be cut longitudinally with the edges burned to prevent fraying. Check the collars weekly for fraying as the ends could be annoying for the bats to get in their mouths (J. Mclean pers. obs.).

### 5.3.8 Visual identification and temporary marking

Captive bats in small groups can often be identified by individual markings such as white patches of fur, scars on ears and the colour of fur (L. Lumsden pers. obs.). For short-term identification small patches of fur in different positions on the dorsal surface can be trimmed with fine scissors to reveal the lower section of the fur sheaths, which are often a different colour to the tips, and remains visible for several weeks to several months (Lumsden *et al.* 2022). Non-toxic paint pens used for marking cattle can be used, which last longer if placed on areas that cannot be licked off such as the back of the ears (Lyons 2020). Other forms of temporary marking such as nail polish or correction fluid are not recommended due to potential toxicity if ingested through grooming (Lumsden *et al.* 2022).

# 6. FEEDING REQUIREMENTS

## 6.1 Captive diet

### 6.1.1 Flying-foxes

- Ad lib
  - water (salt and fresh water) preferably in dripper bottles to keep water clean, different coloured lids for salt and fresh.
- Daily diet (per animal)
  - 350 g chopped fruit (e.g. apple, banana, pawpaw, pear, melon, mangoes, mandarins, some oranges, pears or any soft fruit in season)

  Where available:
  - fresh flowers of eucalypts, melaleucas, banksias and other blossoms
  - fresh mulberry (*Morus* sp.), fiddlewood (*Citharexylum spinosum*) and tree lettuce (*Lactuca indica*) leaves.
- Supplement
  - Little red flying-foxes may lose the quality of their fur unless a formula supplement is provided. Mix 1 × 237-mL tetrapack of Resource® Plus into a 5-L banana smoothie made with full-cream milk powder and Wombaroo® High Protein Supplement. Provide in a bowl hung on the side of the enclosure. It is also important to provide all flying-foxes, but especially little red flying-foxes, lots of licking food as much of their diet is nectar (e.g. smoothie and ripe bananas in a suet feeder cage) (J. Mclean pers. obs.).

Flying-foxes readily eat 25–35% of their bodyweight in fruit daily. They should generally be provided with at least three varieties depending on what fruit is in season. Flying-foxes squeeze the pieces of fruit in their mouth, swallowing the juice and spitting out the seeds and pulp (Van Dyck 1982). Be careful when feeding homogenised fruit to adult flying-foxes as this may cause them to ingest large quantities of fibrous material, which may impair the absorption of nutrients (George 1990).

The Tolga Bat Hospital near Atherton, Queensland, hang whole or halved fruit on galvanised wire necklaces (J. Mclean pers. obs.) and soft fruits such as bananas are provided within a suet feeder for licking (also see section 6.3).

Alternatively, fruit can be cut up into bite-sized pieces (~2 $cm^2$ cubes) to reduce wastage as bats may take a bite from larger pieces of fruit and drop the rest (S. Jackson pers. obs.). Food dishes, such as 1-2-L small plastic buckets, can be hung from branches, ropes or from the roof to allow the bats easy access. One container should be provided for every 1–2 animals and they should be spread throughout the enclosure to minimise fighting over food.

Water containers, either a dish, pond or a water bottle (also called drippers), should be supplied and the bats will readily learn to use them. Water bottles must be checked daily to make sure the ball in the nozzle has not jammed with saliva or other matter. Spectacled flying-foxes are known to drink more salt water than fresh water, so a mixture of 1 teaspoon of salt to 2 L of water can be provided in some of the water bottles (differentiated by different coloured lids) to give them a choice (J. Mclean pers. obs.).

### 6.1.2 Blossom-bats

- Ad lib
  - water
- Daily diet (per animal)
  - 10 mL blossom-bat mix (see details below)
  - 10 mL water

  Mixed together and poured into a nectar feeder or bowl attached to the side of the enclosure.

  Or

  - 3 scoops Wombaroo® Lorikeet & Honeyeater Food
  - ½ teaspoon Wombaroo® High Protein Powder
  - ¼ teaspoon Wombaroo® Small Carnivore mix or The Missing Link® Skin & Coat

  And

  - 50:50 mix of maple syrup and water mix in a separate container
  - fresh flowers of banksias, eucalypts, melaleucas, or other blossoms – as much as possible.
- Blossom-bat mix
  - 2 bananas
  - 500 mL apple juice
  - 150 g raw sugar
  - 150 g Glucodin®
  - 120 g Infasoy

(1) Break up the bananas in a jug and add apple juice to make up to 500 mL total; blend
(2) Add sugar and Glucodin®; blend
(3) Add Infasoy; blend 1.5 min
(4) Pour equal amounts into three containers
(5) Freeze and thaw as required

Can add additional protein powder, swapped out occasionally with 50:50 water and pure maple syrup.

Note: keep fibre low in diet as fruit is limited within the natural diet.

### 6.1.3 Ghost bat

- Ad lib: water
- Daily diet (per animal)
  - 1 day-old chick
  - 1 adult mouse (may be reduced if does not eat everything supplied)
  - ad lib mealworms – twice daily
- Supplement
  - Crickets, grasshoppers, cockroaches, moths, other large invertebrates

Carnivorous bats eat 10–20% of bodyweight in food per night (Hopkins 1990).

### 6.1.4 Insectivorous bats

- Ad lib
  - Water. Shallow water dishes are preferable, but if the water is placed in a deep dish, marbles (or round stones or pebbles) should be added to prevent the bats from potentially drowning.
- Daily diet (per animal)
  - ad lib live mealworms sprinkled immediately before being provided with preferably The Missing Link® Skin & Coat Supplement Powder or Wombaroo® Small Carnivore mix (avoid Wombaroo® Insectivore mix as it can cause scouring). Rate of 100 g mealworms to 1 teaspoon of The Missing Link® Skin & Coat Supplement Powder.
- Supplement
  - ad lib moths, flies, crickets and cockroaches or other flying insects.

Insects should not be frozen, as bats may refuse to eat them, and because they will begin to decay immediately after being defrosted the potentially spoiled food could cause illness if ingested (Barnard 1995).

### 6.1.5 Mealworms

The standard diet for most insectivorous bats in captivity consists of live mealworms (*Tenebrio molitor*), which are not worms but larvae of the darkling beetle. Bats can be fed for long periods of time purely on mealworms that are supplemented as described above. Individuals have been successfully kept for 23 years on supplemented mealworms (L. Lumsden pers. obs.). As mealworms are low in calcium, phosphorous, vitamins $D_3$, A, E and B complex, bats fed solely on unsupplemented mealworms can suffer deficiencies such as metabolic bone disorder and alopecia (this usually disappears after the diet is corrected) (Rasweiler 1977; Lyons 2020). The first place the hair loss is apparent is usually under the chin, so any bats showing these signs should have the amount of supplement increased (L. Lumsden pers. obs.).

Immediately before feeding the mealworms should be coated with The Missing Link® Skin & Coat Supplement Powder (preferred) or Wombaroo® Small Carnivore Food (Lyons 2020). Note that coating the insects with Wombaroo® Insectivore Rearing Mix is not recommended (Lyons 2020).

Note that insectivorous bats should not be fed dried mealworms or live giant mealworms (*Zophobas atratus*) as they can bite and injure small bats (Lyons and Wimberley 2014). Further details on mealworms can be found in Chapter 4.

### 6.1.6 Mealworm media

Various mealworm media have been used to maintain the insects, including a wheat bran/pollard medium that can include vegetables such as cabbage, carrots, lettuce, potato, and pumpkin, and as well as wholegrain bread. Ideally, the vegetables and bread should be on a plastic lid, not on the bedding material, to reduce the potential of rotting. Fruits are typically avoided because they become mouldy too quickly (Lyons 2020).

The nutritional value of the mealworms should be improved by either gut loading them for 1–2 days before they are fed to the bats or coating them immediately before feeding. Gut loading or coating media can include:

- 50:50 mix of Insect Booster (not the Insectivore Rearing Mix) (Wombaroo®/Passwell ®) and wheat bran in addition to vegetables such as leafy greens, sweet potato, and carrots (Lyons 2020)
- blended chick-starter (60%), wheat bran (38.5%) and calcium carbonate powder (Balanced Calcium Powder® Mavlab™ Animal Health; 1.5%) (Lyons 2020)
- 50:50 mix of Wombaroo® Small Carnivore Mix and bran or generously coat the mealworms with Wombaroo® Small Carnivore Mix immediately before feeding to the bats. Mealworms can be supplemented with banana peel when in the bran mix (L. Lumsden pers. obs.)

- The Missing Link® (Skin and Coat Supplement Powder) can be used as a coating of the mealworms immediately before feeding them to bats. Sprinkle 1 heaped teaspoon over ~100 g mealworms or pro-rata (R. Lyon pers. obs.).

  Alternatively, in preparation for each feed, more mealworms than are needed can be placed in a container and The Missing Link® is added at the rate of ~1 teaspoon per 100 g of mealworms. Between the morning and afternoon feeds the mealworms are in the powder for ~7 h and those placed in it overnight for ~17 h during which they consume most of the powder (J. Mclean pers. obs.).

Mealworms are easily cared for in a shallow container with a fine mesh floor. The waste falls through the mesh, keeping the media clean. A large-sized bran media can be used so that it does not readily go through the floor sieve (e.g. McKenzie's Natural Process Bran) and then a sieve can be used to collect the mealworms to be fed to the bats.

Other mealworm media, prepared for up to 5000 mealworms, that have also been used successfully include (from Lollar and Schmidt-French 1998; Barnard 1995):

- Option 1
  - 240 mL vitamin/mineral supplement containing vitamin D3
  - 240 mL sterilised bonemeal powder
  - 454 g oat bran
  - 340 g wheat germ
  - 240 mL green leafy vegetables
  - apple thinly sliced
  - sweet potato (small) thinly sliced
  - carrots peeled
  - corn cob
- Option 2
  - 480 mL wheat germ
  - 480 mL oat bran
  - 120 mL sterilised bone meal powder
  - 240 mL canine vitamin/mineral supplement
  - 120 mL avian vitamin/mineral supplement
  - 60 mL green leafy vegetables
  - sliced apple
  - sliced sweet potato

The ingredients for the mealworm powders should be finely ground and well mixed to allow the mealworms to be easily sifted out (Barnard 1995; Lollar and Schmidt-French 1998). The mealworms can be maintained in a plastic storage container. They should have a lid with a large section cut out and replaced with nylon fly screen, if made from solid plastic, to allow plenty of ventilation so that the fruit and vegetables do not go mouldy (Lollar and Schmidt-French 1998). The mealworms also like heavy breads and will accumulate under them (J. Mclean pers. obs.).

The mealworms should be sieved or hand-picked to remove them from the fine matter. They can also be separated from coarse debris by placing them onto a clean, loosely woven cloth laid flat inside a kitty-litter tray that has a 60-W lamp placed 30–45 cm away. The live mealworms will then migrate under the cloth to avoid the light, leaving the debris on the surface of the cloth where it can be discarded easily (Barnard 1995).

Once on the medium, they should be placed in a cool dry place and preferably refrigerated to prevent contamination of the medium by parasites such as grain mites (*Tyrophagus* spp.) and to slow down the rate of maturation into adult beetles (Barnard 1995; Lollar and Schmidt-French 1998). The mealworm media should be discarded and replaced every 2–4 weeks (Lollar and Schmidt-French 1998); however, if the aim is to develop a self-sustaining population of mealworms, the adults can be removed and placed in a separate container. Adding layers of paper or hessian and leaving the bottom powdery mix where the eggs are laid undisturbed appears to help the colony through the full lifecycle (L. Lumsden pers. obs.).

Insectivorous bats can eat 25–50% of bodyweight in food per night (Hopkins 1990), so the smaller species need 6–10 mealworms and larger species require up to 30. These proportions are highly dependent on the temperature where the bats are being housed and the size of the enclosure and will be less for small enclosures where the bats are not doing the full range of activities and therefore do not require the same amount of food. Obesity may result if they are overfed, so it is important to regularly weigh the bats, keeping in mind that some species will put on weight before winter. In the wild they put on up to 20–30% of their bodyweight in autumn and then lose it in spring and can follow the same pattern in captivity (L. Lumsden pers. obs.). Many bat species need only be fed every second or third night during winter in subtropical and temperate climates due to increased incidence of torpor. Many species become overweight over winter if fed daily (R. Lyons pers. obs.).

### 6.1.7 Artificial diet: blended food diet

For weaning, injured or dietary compromised Australian insectivorous bats, rehabilitators regularly use a blended food diet, which is essentially a blended version of a mealworm diet with additional supplements (Lyons 2020).

The diet must only be handfed by syringe or lapping from a small dish and care taken to clean residue from around the bats mouth after feeding (R. Lyons pers. obs.).

Blended food diet (Lyons 2020):

- 1½ cups frozen mealworms
- ½ cup cold water
- 2 teaspoons of The Missing Link® or Wombaroo® Small Carnivore Mix
- ⅛ teaspoon Soluvet® vitamin powder
- ½ teaspoon of Liquid Oral Care® or similar equivalent (when feeding to Molossidae species in particular)
- ½ teaspoon of Megaderm® Supplement* or similar equivalent
- 1 scoop S26 Soy Powder or equivalent

*If Megaderm® is not available, then add 3 teaspoons of The Missing Link® or Wombaroo® Small Carnivore Mix instead of 2 teaspoons.

Using a blender, add the water, then gradually add the frozen mealworms and blend to the consistency of honey. Mix in the remaining ingredients and transfer the mix into either flattened zip-lock bags or ice cube trays that are then placed within a zip-lock bag in the freezer. Keep the mixture cool throughout the process to prevent spoilage. The frozen blended food can be kept up to 30 days and when needed can be defrosted within the fridge for up to 2 days. Do not refreeze the mixture. The ice cube blocks can feed approximately six small bats per cube per night. If the mixture is added directly into a zip-lock bag, the amount required can be snapped off and left to defrost.

### 6.1.8 Water

Water dishes should be provided at all times. They should be very shallow, as bats have drowned in as little as 1.25 cm water (Lollar and Schmidt-French 1998), or contain marbles or pebbles. Different containers can be used for fresh and salt water depending on the species.

For pteropodids, water containers need to be hung high up and on the sides of enclosure walls. They can be either of the following:

- drippers with metal ball in the nozzle, ensuring that the nozzle does not get clogged with saliva or food
- bird cage (coop) cups (these should be filled with marbles so that small bats cannot fall in) (Lollar and Schmidt-French 1998); they need to be shallow but heavy to avoid tipping.

Water dishes can include:

- small, plastic pet water bowls
- small jar lids or glass Petri dishes
- small plates
- tealight holders.

Vitamin and mineral supplements such as Aviation (1 drop per 25 mL) and Aiming (2 drops per 25 mL) should be added to the water if not added to the food (Barnard 1995).

*Ozimops* species of bats are highly prone to tooth decay in medium to longer term care and in captive situations when fed captive diets. Untreated severe tooth decay can cause tooth and gum abscesses and subsequent eye protrusion and rupture. Liquid oral care products (e.g. Petkin's Liquid Oral Care) have been used successfully in reducing the incidence of tooth decay (Lyons 2020) when added daily to the water dishes.

## 6.2 Supplements

It is important to add calcium, vitamins and minerals to the diet.

## 6.3 Presentation of food

Food for flying-foxes can be presented in various ways, including small plastic buckets (~2 L), long stainless-steel trays or by hanging the fruit by wires. It is very important to make sure there are plenty of climbing and hanging structures to ensure each bat gets access to the food. As flying-foxes compete over their food, several feed stations should be provided and adequately spaced to reduce aggression during feeding (Fig. 13.11). If needed, cloth can be hung from the roof around the enclosure to provide visual barriers between feeding locations.

Ghost bats are readily fed from plastic trays on the ground or from shelves that have wire behind them so they can hang on the wire to feed. Ghost bats are excellent flyers and will readily land on the floor with wings spread out to take their food and will even eat on the floor. From this position they can readily take flight directly from the floor. Other species of Australian insectivorous bats can be given their food from shallow dishes on platforms next to the walls on which they can land and climb on to gain access to their food.

**Fig. 13.11.** Feeding spectacled flying-foxes. Note the hanging fruit held by wires (with both ends of the wire attached to the roof), tubs of banana smoothie, bananas placed within mesh suet feeders and water containers. Photo courtesy of Tolga Bat Hospital.

## 6.4 Teaching insectivorous bats how to feed in captivity

When insectivorous bats first come into captivity they usually need to be taught how to feed on a captive diet. Force-feeding is undertaken by wrapping the bat in a cloth or a hand with a loose fist so the bat's head protrudes between the thumb and first finger (Fig. 13.12; Hopkins 1990). The bat is offered the viscera of a decapitated mealworm directly into its mouth if it is open. If the mouth is tightly closed, the viscera can be smeared onto its lips and when the bat licks its lips, the viscera can be squeezed into its mouth, which usually results in the bat starting to chew. If the bat clamps down on the mealworm, but does not chew, the mealworm can be lightly pulled several times, which often stimulates chewing (L. Lumsden pers. obs.). After a few tries, the bat usually bites, then eats the mealworms. It can sometimes take multiple feeding attempts before whole mealworms are taken. The bat should be encouraged to eat whole mealworms as soon as possible, as the viscera alone does not provide sufficient fibre (found in the exoskeleton), resulting in loose faeces (L. Lumsden pers. obs.).

Once the bats are readily taking mealworms while being hand-fed, they can then be trained to self-feed mealworms from a bowl. Training is undertaken by holding the bat's head close to the mealworms in a bowl while hand-feeding, and progressively encouraging it to grab a mealworm from the bowl rather than from the hand. The speed at which bats learn to self-feed typically varies according to the feeding habitats of the species, with species that glean prey from surfaces learning faster than species that catch all their prey in flight (L. Lumsden pers. obs.). Learning can range from a single feed (e.g. long-eared bats *Nyctophilus*) to several nights (e.g. ghost bats) or up to 2 weeks. Repeat the process over several feeds to ensure the bat has sufficiently learnt the technique before allowing it to solely self-feed. Some species may require retraining if they lose bodyweight while others may not learn how to feed independently (e.g. some free-tailed bats *Ozimops* spp.) (Wilson P 1988; Hopkins 1990; L. Lumsden pers. obs.). The condition of the bat can be checked by checking the distention of the abdominal area periodically as it feeds, by examining the area on the underside of the bat just between the legs (which should not be sunken

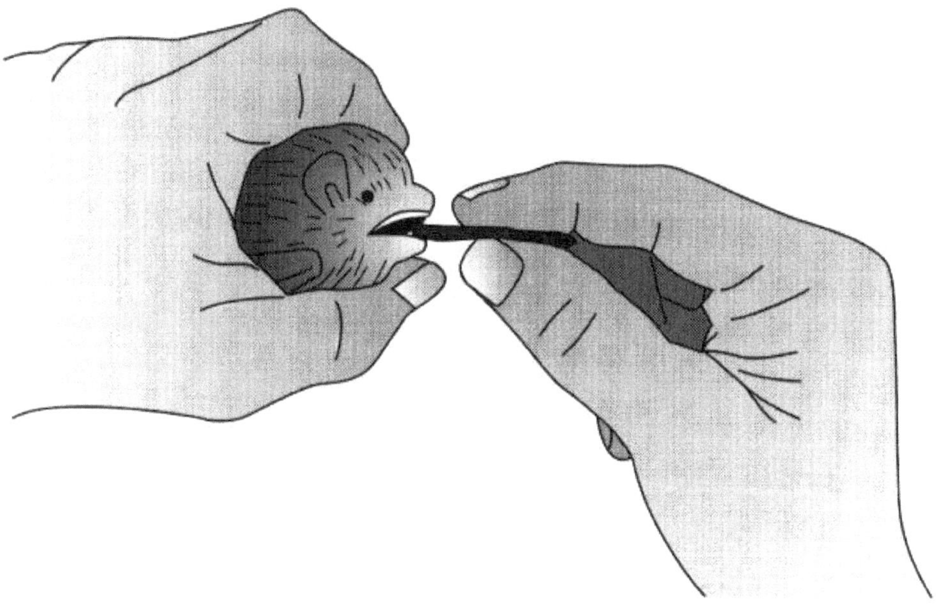

**Fig. 13.12.** Holding position when hand-feeding insectivorous bats. Note that gloves should be worn during feeding. Derived from Hall (1982a).

in or over distended) and by weighing regularly (Lollar and Schmidt-French 1998). All species of *Ozimops* and the chocolate wattled bat can overeat and become obese. If this occurs, factoring in the seasonal changes in bodyweight, they should be encouraged to fly regularly and their intake of food reduced (Hall 1982a). Bats being rehabilitated are preferably hand-fed, at least initially, so that their condition can be monitored (L. Lumsden pers. obs.).

Any bowls used for mealworms need to be deep enough to prevent the mealworms from climbing out, but not so deep that the bats cannot get out; some species will hang above the container and just put their head in, others will sit in the bowl to feed (L. Lumsden pers. obs.).

## 6.5 Conditioning insectivorous bats to enclosures in captivity

While an individual is being acclimatised to captivity, it is useful to place it in a small holding enclosure to discourage flying and increase the frequency of encounters with food (Hopkins 1990). A small enclosure also allows the bat to be more easily caught and therefore more likely to be handled (Hopkins 1990).

Ghost bats may be placed in preflight enclosures (e.g. 100 (L) × 100 (W) × 100 (H) cm) that allow them to stretch out their wings while they learn to feed from a bowl. The enclosures should have at least one wall of wire mesh (~1 cm aperture) and the whole enclosure should be covered to block out light, draughts and visual access to prevent them being startled (Hopkins 1990).

Insectivorous bats from the wild can be held in mesh pet carry cases or fly breeding enclosures (40 (L) × 40 (W) × 40 (H) cm). At least two walls should be lined with cloth (e.g. finely woven cotton) that extends the full height. As most insectivorous bats prefer to roost in the highest, most sheltered position, a 10–15 cm flap of cloth should be added at the top of the enclosure for the bats to crawl under.

## 6.6 Observational learning of insectivorous bats in captivity

Ghost bats and several other species have been observed to learn to feed themselves if placed with bats that already know how to feed (Gaudet and Fenton 1984; Hopkins 1990). Observations have shown that conditioning a bat to feed by hand can take longer than if it learns from observing other bats (Gaudet and Fenton 1984).

# 7. HANDLING AND TRANSPORT

## 7.1 Timing of capture and handling

Bats are generally most easily caught when roosting, which is usually either while the lights are on if held in a nocturnal house or during daytime if outdoors. However, if insectivorous bats are being caught to be hand-fed this should occur in the evening when they are most active (L. Lumsden, pers. obs.).

## 7.2 Capture equipment

### 7.2.1 Capture bags

Most insectivorous bats can be held easily within a calico bag or sewn cotton pouch (30 cm (W) × 45 cm (D)) and kept in a cool place away from direct sunlight. Depending on the size of the bag, several individuals of a given species can be temporarily held together (Jones *et al.* 1996). Particular care is required when holding horseshoe-bats, which are highly prone to desiccation in low humidity, where the wing and tail membranes can become completely inflexible, like dried leaves (Parnaby 1999). This can be avoided by moistening the membranes (Parnaby 1999). In some species, individuals, especially males, may bite or claw one another if they are held in the same bag. For example, Hill's sheath-tailed bat should be held individually in bags as they claw one another, sometimes inflicting fatal injuries (Reardon and Flavel 1987). Individuals of different species should be bagged separately. Each bag should be tightly secured to prevent the bats from escaping. Cloth bags should be double seamed, or turned inside out if only single-seamed, to avoid any risk of bats becoming tangled in loose threads. Bags containing bats should always be hung up (e.g. using hangers used for ties and scarfs; L. Lumsden pers. obs.) and not laid on surfaces (e.g. tables or chairs) to prevent inadvertent crushing.

Bats should be processed as quickly as possible, as prolonged restraint may cause unnecessary stress or depletion of energy reserves and body water (Jones *et al.* 1996). Allowing small groups of the same species (most species) to cluster together in the same bag may minimise stress and reduce energy and water loss (Jones *et al.* 1996). Bats should not be left in potentially stressful environments (e.g. hot, cold or noisy) (Kunz and Kurta 1988).

### 7.2.2 Personal protection

All handlers of bats must be vaccinated with the rabies vaccine to provide protection against the Australian bat lyssavirus and potentially other diseases. When handling bats PPE includes long-sleeved shirt and pants, covered boots and gloves. Different types of gloves include heavy duty nitrile gloves with or without additional fingerless gloves for additional protection over the nitrile gloves; leather gardening gloves; gloves made of Kevlar; leather gauntlet gloves. Oversized cotton work gloves can be used as they allow the animal to be manipulated and the excess material protects the hands from bites (Barnard 1995); however, there is a risk that claws can get caught. Specific details of PPE recommended for handing bats can be found within Wildlife Health Australia (2020a).

## 7.3 Capture and physical restraint

### 7.3.1 Pteropodids

All bats have delicate wing bones that can easily break during capture, so take great care to prevent any harm.

In large, free-flight enclosures, it may be useful to erect a food trap. This allows the animal to be fed in a small enclosed area on a regular basis so that when capture is necessary, the trap can be closed when the bat(s) enter to feed (Fascione 1995).

Pteropodids can generally be caught in captivity with the use of a hoop net, made with strong netting and with an adequately long pole for the height of the enclosure. Alternatively, a large-mouthed bucket (~36 cm diameter) on the end of a telescopic pole has been used to catch flying-foxes from enclosures up to 6 m high (J. Mclean pers. obs.). When approached, the bats will generally panic and often crash into walls, each other and even land on the captors, so execute the capture as quickly as possible and in the coolest time of the day to minimise stress and overheating. In addition to hoop netting, a smaller portable version of the pteropodid harp trap developed by Tidemann and Loughland (1993) could assist trapping of flying-foxes within large enclosures, although potentially they could learn to avoid it.

Pteropodids generally screech loudly, wriggle, try to bite and move their wings and feet wildly so the handler uses both hands to restrain the bat, one to hold the wings against the body and the other to securely hold the head (Fig. 13.13).

Bats can also be securely held with a folded bath towel wrapped around the bat. Its feet are then unhooked from the enclosure, bag or net and held firmly through the towel. The feet can be allowed to grip a piece of cloth or towelling, which generally helps to reassure the bat (Wilson P 1988; Bishop *et al.* 2020). The scruff of the neck can be held to control the head or the bat can be grasped around the neck to stabilise it and prevent it from biting while being examined.

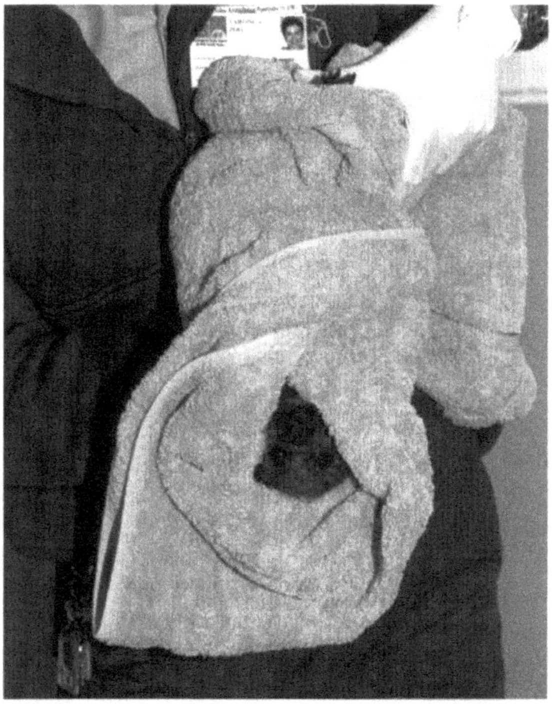

**Fig. 13.13.** Technique used to hold flying-foxes. Photo by Stephen Jackson.

## 7.3.2 Insectivorous bats

If the insectivorous bats are roosting (e.g. within artificial houses or small enclosures), they can generally be easily hand-caught and placed directly into bags. Wherever possible, avoid attempting to capture flying insectivorous bats using a net, as not only is it difficult and stressful, but their delicate wings can be easily injured (Lyons 2020).

If capture of flying bats is required, a lightweight hoop net (with mosquito netting) with a handle of appropriate length to allow quick movement can be used (Kunz and Kurta 1988). The inside of the lip of the net can be lined with a flap of plastic so that if the bat climbs up the net it will climb under the flap or the net can be twisted immediately after the bat is caught so that that the top of the net is closed. Bats are less likely to escape if the net is relatively deep (~0.5 m) (Jones *et al.* 1996). Handles can be made of various materials, including bamboo, aluminium, plastic or wood, as long as they are lightweight and easy to move in the air. Caution should be taken with hoop nets, as there is the risk of damage to wings if the edge of the hoop hits the bat (L. Lumsden pers. obs.). Apart from using a hoop net, insectivorous bats could also be caught within enclosures with the use of either a mist net or harp trap (Tidemann and Woodside 1978) or, most easily, by waiting until they land and then picking them up.

Great care needs to be taken in handling bats, particularly the smaller species, as they are very fragile, particularly the wings. It is best to hold bats by lightly holding the wings in the folded position and applying only enough pressure to restrain the bat's attempts to escape. Strips of polar fleece cloth are used very successfully in insectivorous bat rehabilitation to cover, hold and assess bats. They feel protected and cushioned and are less stressed; furthermore the polar fleece does not get caught in the claws and is also useful to prevent being bitten (R. Lyons pers. obs.). However, remain calm if bitten and do not quickly pull away as this can easily injure the bat. Blowing on the bat's face generally makes it release its hold (Fascione 1995; S. Jackson pers. obs.). The wearing of nitrile gloves when handling insectivorous bats reduces this risk.

Insectivorous bats can generally be held loosely in the palm of the hand with fingers gently but firmly wrapped around the body (Finnemore and Richardson 1999). The bat can be orientated within the palm so that the head protrudes between the thumb and forefinger (Jones *et al.* 1996; Fig. 13.14a.) or with the thumb encircling it (Fig. 13.14b).

To examine the wing, cup the bat in the palm of the hand to extend the wing (Fig. 13.14a). When holding a bat for banding one wing should be extended and stabilised while the band is placed over the forearm (Jones *et al.* 1996). When taking a forearm measurement, hold the bat loosely in one hand with the thumb resting gently along the body and covering the head, with the forearm exposed underneath the thumb and the feet gripping onto a finger (Fig. 13.14a). In this position the bat remains calm and it is easy to measure the forearm (L. Lumsden pers. obs.). To sex the bat, hold it on its back with the thumb gently under the chin and use the other hand to uncurl the tail membrane and expose the lower part of the abdomen (Fig. 13.14b).

Do not hold bats by the scruff of the neck and take great care if holding the wings out, as the bat will usually struggle, and do not hold them by the wing tips (Reardon and Flavel 1987).

Although generally placid and non-aggressive, care needs to be taken when handling bats, as even very small species can inflict painful bites and there is also the potential for the transmission of zoonotic diseases such as the Australian bat lyssavirus (see section 7.3.3). It is recommended that gloves should always be worn when handling bats, regardless of the handler's vaccination status.

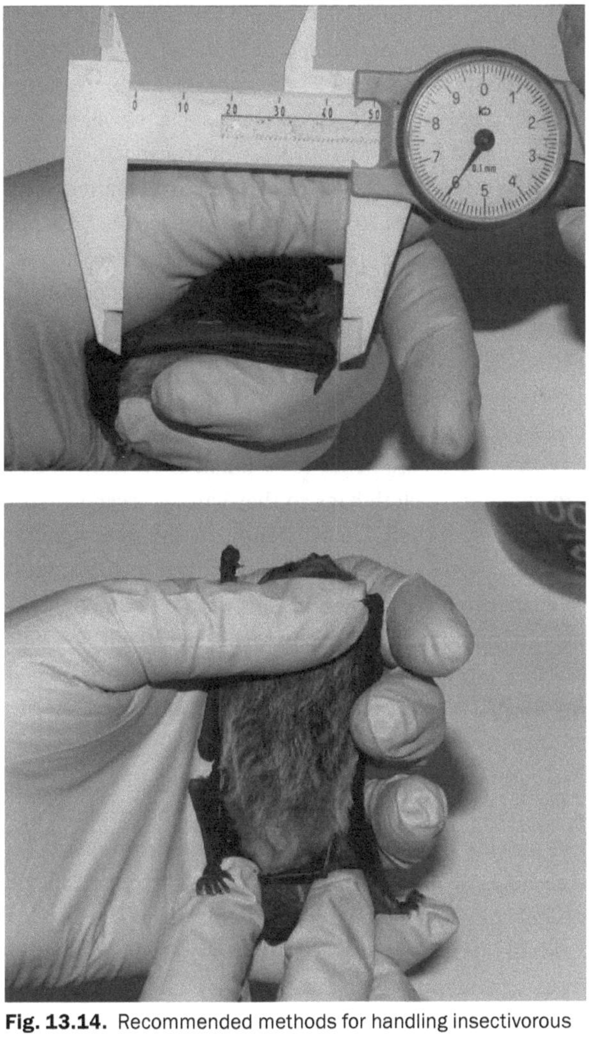

**Fig. 13.14.** Recommended methods for handling insectivorous bats. Photos by Lindy Lumsden.

## 7.3.3 Health precautions when handling bats

The Australian bat lyssavirus (family Rhabdoviridae), which belongs to the same genus as rabies, presents a direct health risk to humans bitten or scratched by an infected bat, as it causes acute fatal encephalomyelitis with death from cardiac or respiratory failure (Allworth *et al.* 1996; Fraser *et al.* 1996; Field and Ross 1999; Rupprecht 1999; Wildlife Health Australia 2020b). It was first reported in mid-1996 from black flying-foxes, but may have been in Australia for some time. The virus has now been isolated from black flying-foxes, little red flying-foxes, grey-headed flying-foxes, spectacled flying-foxes and the yellow-bellied sheath-tailed bat.

Australian bat lyssavirus has been isolated in five states of Australia (Allworth *et al.* 1996; Tidemann *et al.* 1997; Field and Ross 1999; Thompson 1999; Field 2005). In addition to the flying-foxes, and the yellow-bellied sheath-tailed bat, a range of insectivorous bats has since been tested and serological evidence of infection was found in 7 of the 14 genera assessed (Field 2005; Prada *et al.* 2019); however, active virus has only been found in the yellow-bellied sheath-tailed bat to date. Rabies and the Australian bat lyssavirus can only be transmitted via the saliva of an infected bat coming into contact with an open wound or mucous membrane via a bite or scratch (Brass 1994). As the early clinical signs of the infection (such as fever, physical discomfort and inappetence in bats) are subtle and non-specific there is a risk associated with handling any bats of unknown health status as they could be carrying the virus. There is no accurate ante-mortem test. Individuals handling bats should be given pre-exposure vaccination against the disease and, if bitten or scratched, must wash the wound immediately with soap or detergent for 5–10 min and seek urgent medical advice. Once a person shows symptoms (which can take up to 9 years in rabies and may be similar for the Australian bat lyssavirus) it is too late and death is inevitable within a few weeks (Lyssavirus Expert Group 1996).

Despite its occurrence, the prevalence of the Hendra virus in flying-foxes (see section 8.3.3.2) should be put into context, as 128 bat carers were tested for the virus shortly after its discovery and none was found with detectable antibodies even though they were known to have been bitten and scratched by numerous bats. It was proposed that these results suggest that neither prolonged close conduct nor casual contact with flying-foxes presents a risk of this disease in humans (Selvey *et al.* 1996). It is now known that the virus is transmitted in bat urine and not via bites or scratches. To date, no people have been infected directly by bats. All human cases developed after contact with infected horses.

## 7.4 Weighing and examination

Large species of bats, such as flying-foxes and ghost bats, can be readily weighed in cloth bags using hanging scales or electronic kitchen scales. Flying foxes can also be weighed by wrapping them in a towel and placing them in a shallow tub on a kitchen scale (J. Mclean pers. obs.). Smaller species (some weighing as little as 5 g) can be weighed using electric scales with 1-g (ideally to 0.1 g) increments, by placing the bat in a light-weight secure container on the scales. Alternatively, they can be weighed using a spring balance with intervals of 1 g or less, with the bat placed in a small plastic or cloth bag.

Physical examination of bats should include a detailed examination of the eyes, ears, teeth, body, wings and legs (see sections 8.1 and 8.2; Wilson P 1988).

## 7.5 Release

Pteropodids can be released back into the enclosure directly from the bag or container onto a climbing surface or by placing the bag in which they are being held on the ground near a surface which they can climb up to fly off or find a secure roosting position. If the animal is a female with young attached, then it is important to allow the young time in the bag to gain a firm grip on the mother before release. Captive insectivorous bats are readily released back into the roost box. If being released back to the wild they should be able to fly well in captivity first, for at least several weeks in a flight enclosure of suitable size for the species. Insectivorous bats should be released at dusk on a warm and calm night. They should be returned to the area they were found, as bats are social animals and need to be returned to their social group, and because they are familiar with that area and know the location of the roost sites. Insectivorous bats often have a defined roosting area within which they move roost sites frequently, sometimes on a daily basis, regularly returning to previously used roosts (Lumsden *et al.* 2020).

Release the bat directly from the hand in a position where you can see where it flies to. Ensure it is very warm before being placed on an outstretched hand with the head of the bat angled downwards. If it does not fly after a couple of minutes, encourage it to fly by moving it around on, and between, hands, to get it more active. If it is reluctant to fly, or falls to the ground, it should be reassessed as to its suitability for release and rewarmed if necessary or returned to captivity and given more flight training before trying again. It is preferable to release bats from the hand rather than placing them on a tree trunk as the bat may be taken by a predator, fall on the ground or climb higher up the tree or under bark, and not be retrievable if unable to fly (L. Lumsden pers. obs.).

## 7.6 Transport requirements

### 7.6.1 Transport box design

Pteropodids should be provided with a sturdy box for transport. It should be ~30 cm (W) × 45 cm (H) × 30 cm (D). The roof should be ventilated with ~1 $cm^2$ wire that also provides roosting access for the bats during the trip. Further specific details of box design are provided by the International Air Transport Association (IATA 2022).

Insectivorous bats can be readily transported by securing them inside a cloth bag that is placed inside a sturdy box or plastic pet carrier. They should not be transported in a container with any loose objects (e.g. wood, bark or water dishes) that could move during transport.

Some authors have suggested that Styrofoam coolers should not be used to transport or house bats with the bats loose inside them, as the particles adhere to the bats' claws and fur and may subsequently be ingested during grooming, resulting in illness or death (Lollar and Schmidt-French 1998). However, other authors (Barnard 1995) have used them successfully by placing the bats in cloth bags that are put inside Styrofoam containers. In any container, care needs to be taken to ensure the bats are enclosed in cloth bags so they cannot escape during transport and there is sufficient air volume or ventilation. Take care also when opening pouches with zippers to ensure that wings, legs or tails do not become entangled in them (Lollar and Schmidt-French 1998).

For longer distances, and whenever a plane trip is involved, a sturdy box should be provided. Most species of insectivorous bats, with the exception of the ghost bat, can be held successfully in a wooden box ~20 (H) × 20 (W) × 20 (D) cm. Further specific details of box design are provided by the IATA (2022). American species of tree-hole dwelling bats have been transported on numerous occasions in the USA in plastic tool boxes that have an interior box constructed of wood veneer that fits tightly inside an outer box. One bag per bat, which is securely tied, is then placed inside the wooden box so that they do not slide around. Additional empty bags can be used as padding but should not be placed on top of the bats (Barnard 1995). Care must be taken to monitor the ambient temperature inside the box or container to ensure overheating does not occur during transport and ensure there is adequate ventilation.

### 7.6.2 Furnishings

Pteropodids should be provided with wire at the top of the box, ideally as part of the ventilation. Potentially branches could be used, but must not have any sharp points that could harm the bat.

Insectivorous bats are generally readily transported in a cloth bag inside a sturdy box with no furnishings required (L. Lumsden pers. obs.). Any furnishings, aromatic bedding such as pine bark or cedar chips or loose material should not be placed with the bat due to the risk of these causing injury or respiratory distress during transport (Barnard 1995; L. Lumsden pers. obs.).

### 7.6.3 Water and food

Depending on the bat species, and the mode and duration of transport, provide the necessary requirements such as food and water, though on short journeys, where the animals are moved during the day, food may not be required (Barnard 1995). During longer trips, or trips that occur when the bats are usually active, the bats should be intermittently removed from the container and provided with food and water. Insectivorous bats should not be transported with containers of water because the water can spill and wet the bats or the containers can move and injure the bats (Barnard 1995).

### 7.6.4 Animals per transport box

Adult flying-foxes should generally be housed separately to minimise any problems; however, young animals can be housed together. Small numbers of insectivorous bats that usually roost together can be transported together, which will provide them with additional heating and reduce stress. Different species should never be transported in the same container (Barnard 1995). Care should be taken with the transportation of dominant males of some insectivorous species. Ideally, they should be housed in separate bags but placed adjoining each other.

### 7.6.5 Timing of transportation

Ideally, bats should be transported during the coolest part of the day or when they are less active such as in the early morning. It is also important to plan transport around the bats' feeding times, being sure to feed just before shipment (Barnard 1995). Ensure the bats are not transported or left in hot cars or placed in the sun (L. Lumsden pers. obs.).

### 7.6.6 Release from the transport box

Pteropodids can be allowed to leave the box in their own time or removed and either placed on the floor next to a surface they can readily climb or, preferably, placed directly onto a climbing surface where they can climb to their preferred position.

Insectivorous bats should generally be removed from the transport box and placed in their roosting box. The large ghost bat can be placed on a wall or even on the open floor, as these bats are able to take off from the floor, unlike pteropodids, and fly to their preferred roosting site.

## 8. HEALTH REQUIREMENTS

### 8.1 Daily health observations

Larger bat species, such as the ghost bat and the flying-foxes, should be individually checked daily as they roost by examining the wings and seeing how freely and evenly the wings move (which could suggest a break).

Due to their small size, insectivorous bats and small pteropodids are difficult to observe for the subtle signs of illness from a distance, so these should be checked whenever the bats are being handled.

The most appropriate time to observe flying-foxes and insectivorous bats is generally when the enclosure is being cleaned or food is provided. During these times, each animal within the enclosure should be checked for the following:

- general condition
- coat condition - fur, ectoparasites
- discharges from the eyes, ears, nose, mouth or anus
- eyes - bright and alert (if the bat is not in torpor), check for conjunctivitis, injury, corneal ulceration
- dental condition - tooth discolouration, decay and gum infections. *Ozimops* species are highly susceptible to teeth decay in medium- to long-term care
- appetite and eating behaviour (Note: when insectivorous bats go into periods of torpor over winter, they will not want to eat each day (L. Lumsden pers. obs.))
- faeces - number and consistency
- changes in demeanour, activity level and behaviour (e.g. hanging alone when it is normally with the other bats)
- injuries - including lacerations and injuries to the wings (including holes, tears, swelling of the fingers and, if possible, checking for broken bones and asymmetry of the wings) and climbing ability. Legs are rarely injured but check the toes for breaks, burns and other injuries and gripping ability. Toenails should not be clipped.
- presence and development of young

If the animals are held in a larger colony, then the enclosure floor and the roost area floor should be checked carefully for any animals that have died or are unwell.

As far as possible, all individuals should be accounted for each day. Any signs of injury or illness or unusual changes in behaviour should be documented and reported to a veterinarian. If an animal is found dead, make note of the circumstances, anything unusual or abnormal on the body or surrounding area. Carefully collect the body (using PPE where appropriate) and submit it for necropsy by a veterinarian.

## 8.2 Bodyweight and body condition score

### 8.2.1 Bodyweight

Record regularly (ideally, monthly) and compare with previous weights. Trends in bodyweight of bats give a good general indication of the animal's state of health, provided age, sex, season and geographical location are taken into account. If there are concerns, or an animal is newly transferred, weighing every few days or weekly is necessary.

### 8.2.2 Body condition score

A body condition score is best determined by palpation of muscle in the area over the scapula, spine and temporal fossa.

## 8.3 Physical examination

The physical examination may include the following:

- condition of the wings, wing membranes, and digits (Fig. 13.15)
- temperature - normally 37-39°C for pteropodids; temperature can be taken through the anus in flying-foxes. Remember that insectivorous bats and small pteropodids can go into torpor several times a day, so their temperature can vary considerably. It can also be difficult to take the temperature of insectivorous bats rectally due to their small size
- pulse rate - ~100-400 beats/min in active pteropodids (e.g. 360 beats/min in flying-foxes; J. Mclean pers. obs.) and ~250-450 beats/min in insectivorous bats (though heart rates as high as 900-1000 beats/min have frequently been recorded). The heart rate can decrease during diurnal torpor and hibernation to only 10-60 beats/min (Hill and Smith 1985; Altringham 1999). The heart rate also varies greatly with the species, temperature and body size

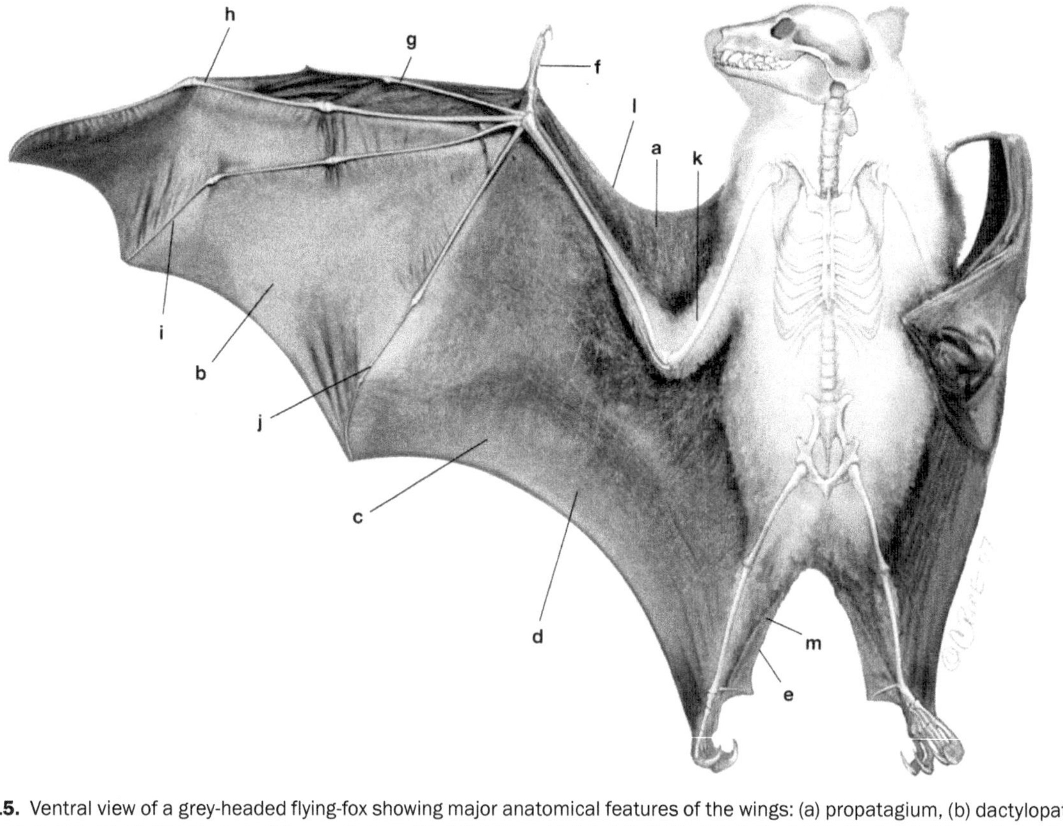


**Fig. 13.15.** Ventral view of a grey-headed flying-fox showing major anatomical features of the wings: (a) propatagium, (b) dactylopatagium, (c) bands of elastin, (d) plagiopatagium, (e) uropatagium, (f) digit 1 (thumb), (g) digit 2, (h) digit 3, (i) digit 4, (j) digit 5, (k) median vein, (l) wing or cephalic vein and (m) uropatagial vein. Reproduced from Olsson and Woods (2008) with permission from CSIRO Publishing.

- respiratory rate - varies greatly with the species, temperature, if in torpor and body size, with rate decreasing with increasing body size; auscultation of lungs
- fur/skin - check for alopecia, ectoparasites, fungal infections or wounds
- eyes
  - should be clear, bright and alert
  - normal bilateral pupillary light response
  - normal corneal reflex
  - should not have any discharges
- check for the presence of lumps over body
- anus
  - should be clean
  - check for faeces around the anus, which may indicate diarrhoea or improper toileting that may indicate a problem, although faeces often remain in the furled tail membrane of healthy, but inactive, insectivorous bats
- males
  - check testes - size (length, width, depth) and consistency (firm - not squishy). In insectivorous bats this varies seasonally and between species
  - extrude penis and assess (flying-foxes).

## 8.4 Chemical restraint

Pre-anaesthetic fasting is not required as bats are not prone to regurgitation, except immediately after eating or drinking. Inhalation anaesthesia using isoflurane (5% for induction and 2–2.5% for maintenance) with oxygen is the anaesthetic technique of choice. Induction and recoveries are rapid and smooth. Bats can be induced using a mask. For prolonged procedures, bats > 150 g can be intubated with a > 2-mm internal diameter endotracheal tube. If required, injectable agents can also be used: ketamine (10 mg/kg) plus xylazine (2 mg/kg) or ketamine (5 mg/kg) plus medetomidine (50 µg/kg). The latter combination can be partially reversed with atipamezole given at five times the medetomidine dose (Heard 2014; Vogelnest 2025).

## 8.5 Common health problems

Bats suffer few problems in captivity; most of the parasites and diseases that have been recorded are presented in Table 13.8 and the most common non-infectious diseases of captive bats are listed in Table 13.9. Injured or sick bats should be assessed by a veterinarian to obtain an accurate diagnosis and receive the appropriate treatment. Indiscriminate use of antibiotics can delay recovery and increase the likelihood of antibiotic resistance.

## 8.6 Preventative medicine

All bats should be quarantined for at least 30 days before entering a collection. Preventative medicine focusses on daily monitoring and, ideally, an annual physical examination. No routine vaccinations are required. Enclosures should be designed to deny flying-foxes access to cat faeces, wild invertebrates and wild flying-foxes to prevent transmission of diseases such as toxoplasmosis, angiostrongyliasis and Australian bat lyssavirus. Necropsies should be performed on any bats that die (Heard 2003).

## 8.7 Zoonoses

Many new and emerging human diseases have potential bat origins (Dobson 2005). The most significant Australian bat related zoonoses are listed in Table 13.10.

## 8.8 Housing small bats under treatment or rehabilitation

### 8.8.1 Housing

Small bats that are being held for treatment or rehabilitation should be placed in a small enclosure to allow ready observation. An enclosure that is ~40 × 40 × 40 cm is adequate for most species of small bats.

These enclosures are ideally made of mesh with a heat pad at one end and several layers of cloth against it so that the bat can roost between these layers (Lyons 2020). The roost should also have a cover over it to so the bat can retreat into the darkness.

Due to their ability to enter torpor and reduce their metabolic rate, during any treatment requiring medication insectivorous bats and smaller species of pteropodids (e.g. blossom-bats) should be housed in a temperature-controlled setting consistent with their thermoneutral zone (30–35°C). Adequate temperature, ample food and hydration will reduce torpor incidence and will allow for more consistent metabolising of medication treatments (Lyons 2020).

### 8.8.2 Water

A shallow dish of water that is available at all times must be provided to all species.

# 9. BEHAVIOUR

## 9.1 Activity

Bats are nocturnal, but flying-foxes typically commence flight at dusk and return to the roost at dawn. The pteropodids use sight and smell when travelling to find their

**Table 13.8.** Known health problems in bats.

PCR = polymerase chain reaction.

| Disease/ infectious agent | Cause | Clinical signs | Diagnosis | Treatment | Prevention/comments | Reference |
|---|---|---|---|---|---|---|
| **Ectoparasites** | | | | | | |
| Bat flies | The most common ectoparasites are nycteriibid and streblid flies, which are generally considered to be harmless but are potential vectors of disease including the protozoan *Trypanosoma pteropi* | None. Possibly increased grooming | Observed on bat | None. Healthy bats have low numbers that do not seem to cause any problems. The only time numbers seem to build up is if the bat is debilitated or otherwise unwell | Pest strips located on the outside of the enclosure have been known to deter ectoparasites confined to the host if the bats are held indoors; otherwise, repeated treatment with ivermectin can be used. Pest strips should not be used within roosting boxes | 1–5 |
| Mites | Bats are known to carry several species of mites | Mites are known to cause irritation, scurfy skin and hair loss (alopecia). Possibly increased grooming | Observed on bat | Mites have been successfully treated with fipronil at 1 drop per bat (for flying-foxes) applied to the back and neck | See above | 5, 6 |
| Tick paralysis: *Ixodes holocyclus* | Ticks (Ixodidae) have been found on bats | Paralysis has been observed in spectacled flying-foxes. Signs include laboured breathing and cardiac abnormalities | Ticks observed on the bat. Clinical signs | Tick removal combined with tick antiserum | Check wild flying-foxes for ticks and manually remove any that are found | 7–9 |
| Nematode round worm: *Riouxgolvania beveridgei* | Muspiceoid nematode | Skin nodules on the fore and hindlimbs of bent-winged bats. No obvious effect on health | Observed on bat | None | Nematode found in the skin | 5, 10 |
| **Endoparasites** | | | | | | |
| *Toxocara pteropodis* (nematode round worm) | A large ascaridoid nematode round worm that infects the suckling young of all Australian *Pteropus* species | May cause ill thrift (failure to grow) but usually no clinical signs | Faecal float to identify eggs in suckling young | Wormed with pyrantel pamoate (115 mg/g,). The paste (1 mL/kg) is smeared on the fur and the bat licks it off during grooming | Nematode found in flying fox intestines | 2, 11–13 |
| *Angiostrongylus cantonensis* | A nematode round worm that is transmitted by slugs and snails carrying the intermediate stage and inadvertently ingested with fruit | Hindlimb paresis or tetraparesis, depression and anorexia. Has caused fatal encephalitis in wild and captive flying-foxes | Clinical signs. Nematode found in the brain of flying-foxes | Panacur 100® (fenbendazole 100mg/mL) at 10mg/kg can be used | Prevent bats ingesting the snail or slug intermediate host | 14, 15 |

| Disease/ infectious agent | Cause | Clinical signs | Diagnosis | Treatment | Prevention/comments | Reference |
|---|---|---|---|---|---|---|
| **Protozoa** | | | | | | |
| *Toxoplasma gondii* | Apicomplexan protozoan | Paresis, paralysis, lethargy, anorexia | Clinical signs. Histology | Prevent bats having access to cat faeces | Cats are the definitive host | 16 |
| *Polychromophilus* spp. | Apicomplexan protozoan closely related to the malaria parasite | None. Possible weight loss | Blood film examination or PCR on blood | None | Transmitted by bat flies | 17 |
| *Trypanosoma* spp. | Flagellate protozoan | Usually asymptomatic but can cause lethargy, icterus, anaemia | Blood film examination or PCR on blood | None | Symptoms described in little red flying-foxes | 18, 19 |
| **Viruses** | | | | | | |
| Australian bat lyssavirus (ABLV) | First seen in a black flying-fox on 24 May 1996, ABLV is from the family Rhabdovirus, which includes classic rabies and five other rabies-like serotypes. Has been isolated from *Pteropus* bats and the yellow-bellied sheath-tailed bat | Clinical signs include an inability to swallow, an inability to fly, hindquarter paresis (slight or incomplete paralysis), apparent general weakness and progressive neurological signs including sudden aggression, vocalisation, dysphagia (difficulty in swallowing) and paresis | PCR, fluorescent antibody testing, virus isolation or immunohistochemistry of brain tissue are most often used to establish a diagnosis of ABLV infection. Any dead bats or bats that show clinical signs of ABLV infection should be euthanased and tested | Untreatable. Affected animals should be euthanased | Prevalence in healthy flying-foxes is ~1%. Prevalence in flying-foxes submitted for testing is ~5%. There is no reliable antemortem test. Enclosures in which bats have been found with this virus should be cleaned with strong acids, bases or detergents. All handlers should be vaccinated against rabies and take precautions when handling bats to minimise the risk of being bitten | 3, 20–32 |
| Hendra virus | All species of Australian flying-foxes appear to be the probable host of Hendra virus and they appear to carry it throughout their natural distribution | There appears to be an endemic pattern of subclinical infection of Hendra virus in flying-fox populations throughout Australia | PCR of urine and tissues | No treatment. Horse vaccine is commercially available | Found in flying-foxes and can be transmitted to horses. Keep horses away from areas where flying-foxes congregate | 26, 30, 33–41 |
| Menangle virus | The Menangle virus (paramyxovirus) was first observed in stillborn piglets during an outbreak of reproductive disease in 1997 in pigs at Menangle in New South Wales | None | Testing bat blood for antibodies to the virus | None | The virus is believed to be carried asymptomatically by flying-foxes. No cases have been detected since the initial outbreak | 42, 43 |

**Table 13.8.** continued

| Disease/ infectious agent | Cause | Clinical signs | Diagnosis | Treatment | Prevention/comments | Reference |
|---|---|---|---|---|---|---|
| **Bacterial** | | | | | | |
| Leptospirosis | *Leptospira* spp. | None | PCR of urine and kidney | None | Found in flying-foxes | 44–46 |
| Miscellaneous | *Salmonella, Shigella, Yersinia, Pasteurella* | Septicaemia, abscesses, pneumonia | Necropsy and culture | Antibiotics, supportive care | Appropriate husbandry | |
| **Fungal** | | | | | | |
| White nose syndrome | *Pseudogymnoascus destructans*. Not currently known within Australia. Appears to have originated in Europe, but has caused widespread mortality in North America | Affected bats are emaciated and dehydrated, frequently with white fungal material on their muzzle and wings | PCR and histology | No treatment. Decontaminate all gear between caves | Exotic to Australia. Affects cave dwelling insectivorous bats. Does not grow above 20°C | 32, 47–51 |
| Histoplasmosis | *Histoplasma capsulatum* | Usually asymptomatic | PCR of bats and cave surfaces | None | Commonly found associated with large quantities of bat guano, such as in caves | 51–52 |

References: (1) Wilson P (1988); (2) Booth (1994); 3) Olsson and Woods (2008); (4) Fascione (1995); (5) Holz *et al.* (2018b); (6) Domrow (1967); (7) Buettner *et al.* (2013); (8) Campbell *et al.* (2003); (9) Durden and Wilson (1985); (10) McLelland *et al.* (2013); (11) Constantine (1993); (12) Heard *et al.* (1995); (13) Prociv (1990); (14) Barrett *et al.* (2002); (15) Reddacliff *et al.* (1999); (16) Sangster *et al.* (2012); (17) Holz *et al.* (2019); (18) Austen *et al.* (2015); (19) Thompson *et al.* (2014); (20) Crerar *et al.* (1996); (21) Barrett (2004); (22) Allworth *et al.* (1996); (23) Tidemann *et al.* (1997); (24) Field *et al.* (1999); (25) Thompson (1999); (26) Field and Ross (1999); (27) Skerratt *et al.* (1998); (28) Hooper *et al.* (1997); (29) McCall *et al.* (2000); (30) Field (2005); (31) Rose (1999); (32) Cox-Witton and Bishop (2025); (33) Murray *et al.* (1995); (34) Young (1996); (35) Young *et al.* (1996); (36) Philbey *et al.* (1998); (37) Halpin *et al.* (1996); (38) Williamson *et al.* (1998); (39) Hooper *et al.* (1996); (40) Field *et al.* (2000); (41) Field *et al.* (2007); (42) Ross *et al.* (2001); (43) Hall and Richards (2000); (44) Constantine (1970); (45) Smythe *et al.* (2002); (46) Cox *et al.* (2005); (47) Blehert *et al.* (2009); (48) Turbill and Welbergen (2020); (49) Frick *et al.* (2015); (50) Drees *et al.* (2017); (51) Holz *et al.* (2018a); (52) Heard (2003).

**Table 13.9.** Common non-infectious health issues of captive bats.

| Disease/agent | Cause | Clinical signs | Diagnosis | Treatment | Prevention/comments | Reference |
|---|---|---|---|---|---|---|
| Abrasions | Usually caused by crawling on rough surfaces, particularly untreated cement and galvanised wire enclosures | Abrasions to their wrists | Clinical signs | Topical antibiotic cream can be applied and adhesive dressing such as DuoDERM® Dressings | Ensuring that surfaces the bats climb on are smooth and non-abrasive | 1 |
| Alopecia | May be due to ectoparasites, fungi, bacteria or a deficiency in vitamins and minerals | Fur loss around the eyes or elsewhere, including under the chin and throat of insectivorous bats | Visual signs, microscopic examination of skin scrapings, culture and/or skin biopsy | Dependent on cause: antibiotics, antifungals, parasiticides and examine the diet | Provide an adequate diet including nutritional supplements such as Wombaroo® High Protein Supplement for flying-foxes and Wombaroo® Insect Booster for mealworms fed to insectivorous bats | 2 |
| Fractures | Generally rare in flying-foxes but can occur in insectivorous bats, due to impact while flying in aviaries, fighting and mishandling | One of the wings or legs not moving as freely or evenly as the other | Radiography or visual signs, such as protruding bones through the skin in the case of compound fractures | Fractures involving joints rarely heal well enough for flight, although midshaft fractures of long bones can be more readily treated | Providing adequate climbing substrate and properly handling the bats | 3 |
| Heat stress | Flying-fox deaths are known to occur when temperatures > 40°C and humidity is < 30%. Insectivorous bats have been reported to die at temperatures > 44°C | Panting, wing fanning, salivating, shade seeking | Based on ambient temperature and clinical signs | Treatment options for flying-foxes include wrapping the bat in a wet towel or immersing the bat in tepid water. Ensure they do not become hypothermic from the body temperature dropping too quickly. The rectal temperature is recorded every minute initially and every 2–5 min after the body temperatures falls below 40°C. Hydration can be achieved by providing: (1) moisture-rich food (e.g. watermelon), (2) mix of 1 teaspoon of Glucodin® powder to 250 mL preboiled water. Provide a maximum of 10% bodyweight over a 24-h period or (3) rehydration with subcutaneous injections of fluids. Animals should be held in small enclosures with wet towels attached to the roof and allowed to hang down | Prevention includes: moving bats to cooler locations, and for flying-foxes providing shade and sprinklers that can be turned on when ambient temperatures exceed ~35°C. See section 4.6 | 4–8 |

**Table 13.9.** continued

| Disease/agent | Cause | Clinical signs | Diagnosis | Treatment | Prevention/comments | Reference |
|---|---|---|---|---|---|---|
| Holes in wing membranes | May be due to fighting or rough surfaces, such as loose wire within the enclosure. Wing tears can be very large and, in at least some cases, the bat can still fly without any apparent problems | Obvious holes in the wing membranes | Visual inspection | Small, and even some large, holes usually heal well by themselves with no treatment; however, a weak disinfectant such as diluted Betadine or an antibiotic solution can be applied twice daily to fresh wounds using an eyedropper. Holes and tears in wing membranes usually take between several weeks and several months to heal properly, though holes generally heal quicker than tears. Large tears, particularly if they extend all the way through the leading or trailing edge of the wing membrane, are unlikely to heal unless the tissue on the edges of the tear is reattached in a manner consistent with new tissue growth, such as with a tissue adhesive (e.g. VetBond®) every 1 cm (not in a continuous line) along the tear. It is not recommended that such tears be sutured as it causes further trauma to the tissue, including muscle fibres | Ensure there are no sharp objects in the enclosures that can tear wing membranes | 3, 9 |
| Neoplasia | Usually unknown. More common in older animals | Variable. May see skin masses but depends on organs involved | Histology | Surgical removal, euthanasia | Uncommon. Some neoplasms have been associated with microchips | 10 |
| Wing membrane infections | Usually occurs when flying-foxes are unable to hang in the sun, lack access to fresh air or are unable to flap their wings properly. These infections are commonly fungal (e.g. candidiasis) or bacterial. Often caused by stress in hand-reared animals | The condition is often called 'slimy wing' and manifests as cream-white patches on the wings with a slimy feel and a bad odour. It can cause permanent and serious wing damage and ultimately may result in death | Visual signs and microscopic examination of the fungus/yeast | The infections can be treated with topical lovone®, Conofite® or Panalog® or one-part Malaseb® to 30 parts water and then thoroughly drying the wings with soft towels. In serious cases, oral antimicrobial drugs may be required | Allowing access to sun, fresh air and adequate room to fully flap the wings | 11 |

References: (1) Hall (1982a); (2) Wilson P (1988); (3) Lollar and Schmidt-French (1998); (4) Parry-Jones (2018); (5) Lyons (2020); (6) Welbergen *et al.* (2008); (7) Pinson (2020); (8) Mo and Roache (2020); (9) Richards (2000); (10) Olsson and Woods (2008); (11) Hall and Richards (2000).

**Table 13.10.** Zoonotic diseases observed in bats.

| Disease | Cause | Clinical signs in humans | Transmission | Prevention | Reference |
|---|---|---|---|---|---|
| Australian bat lyssavirus | *Lyssavirus* | Neurological signs, death | Present in bat saliva. Usually transmitted by bites | Anyone handling bats should wear appropriate PPE and have current rabies immunity. If bitten, wash wound with soap and water for 5–10 min and seek immediate medical advice | 1 |
| Hendra virus | *Henipavirus* | Neurological signs, death | Flying-foxes infect horses through contact with urine or other bodily fluids. All human cases have been contracted from infected horses | Wear appropriate PPE when handling bats and potentially infected horses. Wash hands with soap and water | 2 |
| Leptospirosis | *Leptospira* spp. | Fever, headache, chills, vomiting, jaundice, kidney failure | Found in flying-foxes | Wear appropriate PPE | 3 |
| Histoplasmosis | *Histoplasma capsulatum* | Commonly asymptomatic, but can include gingivitis, pneumonia septicaemia and death | Associated with bat caves and guano | Wear appropriate PPE. Seek medical advice | 4, 5 |

References: (1) Animal Health Australia (2021a); (2) Animal Health Australia (2021b); (3) Olsson and Woods (2008); (4) O'Sullivan *et al.* (2004); (5) McLeod *et al.* (2011).

food. The insectivorous bats use a very high-pitched sound called sonar or ultrasonic echolocation in addition to sight. Echolocation is produced by emitting short pulses of only a few thousandths of a second, typically at a rate of 10–15 pulses/s, but can be up to 100–200 pulses/s when the bat is homing in on an insect. The type of sound varies from species to species but is in the range of 11–160 kHz (human hearing is typically < 15 kHz). The echoes returning from the objects such as trees, cave walls, buildings or food are collected by the forward-facing ears, converted to nerve impulses by the highly specialised inner ear and transmitted to the brain for processing, which allows them to rapidly judge distance, shape and texture (Reardon and Flavel 1987).

Insectivorous bats often travel large distances foraging, with many species having two main feed periods during the night: one just after sunset and the other before dawn, each usually lasting ~2.5 h (Reardon and Flavel 1987; Taylor and O'Neill 1988). Some bats will feed continuously throughout the night with short resting periods where they hang up and digest their food (L. Lumsden pers. obs.). Light levels appear to influence activity, as some species are thought to fly in shadows to avoid moonlight and appear more often on moonless nights (Reith 1982). Some species emerge from their roosts early in the dusk period while there is still a reasonable amount of ambient light (e.g. Gould's wattled bat and little forest bat) while others do not emerge until it is quite dark (e.g. lesser long-eared bat) (L. Lumsden pers. obs.). Therefore, the level of light in the enclosure may be very important and may need to be targeted to the specific species, to encourage them to fly more often and therefore provide a more active exhibit.

Little information is known about the time budgets of Australian bats; however, studies of the little brown bat (*Myotis lucifiugus*) in the USA provide information that may be of use. The proportion of the night the little brown bat spends foraging varies both daily and seasonally in relation to reproductive condition, prey density and ambient temperature (Anthony *et al.* 1981). Although the effect of light has been suggested as important in determining activity, another study of little brown bats showed that light intensity did not affect activity, but rather activity was significantly related to ambient temperature (Negraeff and Brigham 1995). A single continuous night roosting period was observed during pregnancy, then during lactation females returned to maternity roosts between foraging bouts and in late summer after the young become volant, they used night roosts. When there were cool nights and low prey density the bats spent longer periods at night roosts with shorter foraging periods. Other observations of pipistelle bats (*Pipistrellus pipistrellus*) showed that they left hibernation to feed in all months of the year, with activity being most likely on warmer nights (Avery 1985).

In Victoria, female lesser long-eared bats with dependent young have been recorded leaving the roost on dusk, foraging for ~2.5 h and then returning to the roost to suckle their young for 30 min. They then undertake another 2-h foraging bout, followed by another 30-min period back at the roost suckling young, before a final 2-h foraging period before dawn (Lumsden *et al.* 2020). Outside the breeding season individuals are less likely to return to the day roost, but periodically use night roosts to hang up between foraging (L. Lumsden pers. obs.).

Most Australian species of insectivorous bats, with the exception of the ghost bat, can enter periods of torpor to save energy (Ellis *et al.* 1991), although some potentially overcome food shortages by migrating to warmer areas where food is still available (Kitchener and Hudson 1982; Allison 1989). To enable extended periods of torpor or hibernation, the insectivorous bats deposit fat in late summer and autumn when food availability is high, which is then used to sustain themselves until the onset of warm weather when insects again become plentiful (Tidemann 1982; Reardon and Flavel 1987). The length of time that bats remain in torpor is dependent upon temperature and humidity. For example, Gould's long-eared bats are known to enter torpor for < 2 h at an ambient temperature of 25°C. However, at an ambient temperature lower than this they can be torpid for > 24 h and hence not be foraging on some nights (Geiser & Brigham 2000). This can occur at all times of the year, but more often and for longer periods over winter. However, Australian species typically do not enter as prolonged periods of hibernation as bats in the northern hemisphere, with some activity occurring over the cooler months. For example, in a study in New South Wales over winter, long-eared bats remained torpid for up to 15 days, although shorter bouts were also common, before the bats aroused to forage and drink, which typically occurred on warmer nights (Turbill and Geiser 2008). Tropical species of bats are more active throughout the year as there is adequate food availability, however, bats in subtropical climates can also enter periods of torpor during the day to conserve energy (Turbill *et al.* 2003a) or even undertake torpor bouts of up to 5 days in length (Stawski *et al.* 2009).

Among the pteropodids, several species are known to enter torpor. The eastern blossom-bat, for example, has been recorded to undergo torpor-during daylight at temperatures of 18°C (Geiser *et al.* 1996). The known temperature at which the different bat species enter torpor, and the length of time they remain torpid, can be found in Table 13.11. As discussed in section 4.6, the grey-headed flying-fox has also been observed to enter torpor for short periods (Turbill *et al.* 2024).

When holding insectivorous bats from tropical regions in captivity in temperate climates, it is important to provide additional heating, in order to better mimic their natural environment, by providing a 25-W heat lamp with a red-light bulb or placing a heat pad over a section of the top or side of the roost box to allow some warmth (Lollar and Schmidt-French 1998). If heating is required it is also important to provide cloth drapes or roosting pouches in several locations to allow the bats to choose the most appropriate microclimate. In most cases, however, where the bats are being held within their natural distribution, heating is not required as long as well-insulated roost boxes, which mimic tree hollows are provided, allowing bats to enter torpor. However, it is important not to induce torpor in non-hibernating species or bats that are sick or injured as this will slow down the healing process (Lollar and Schmidt-French 1998).

Trapping studies in north Queensland have revealed that several species of bat genera, such as *Taphozous*, *Rhinolophus*, *Hipposideros*, *Miniopterus*, *Myotis*, *Scoteanax*, *Vespadelus*, *Nyctophilus*, *Chaerephon* and *Ozimops*, are more active when the temperature range is 15.5–26°C. When the temperature range is lower, 13.8–17.3°C, they are much less active, presumably as they are in torpor due to the lack of insect prey (Richards 1989). In Victoria in autumn, 11°C appears to be a threshold below which activity levels of *Nyctophilus* and *Chalinolobus* markedly decline (L. Lumsden pers. obs.).

## 9.2 Social behaviour

### 9.2.1 Pteropodids

Among the pteropodids such as the flying-foxes, the aggregations or camps they establish form the social unit of these highly social species. These camps may comprise more than one species of flying-fox and vary throughout the year, with summer camps, at which the young are born and raised, being occupied from September to April or June. The winter camps are occupied the rest of the time and the sexes are usually segregated and contain a large number of immature animals. The social structure of flying-fox camps is complex, with several subgroups as seen in the section below (Nelson 1965a). Flying-foxes are often seen and heard fighting with a variety of vocalisations and behaviours that are

**Table 13.11.** Air temperatures at which some species of bats are known to enter torpor.

*Was published as *Miniopterus schreibersii*.

| Species | Temperature (°C) | Duration (h) | Reference |
|---|---|---|---|
| **Yinpterochiroptera** | | | |
| *Macroglossus minimus* | 14–25 | ~8 | 1 |
| *Syconycteris australis* | < 26 | 5.5–7.3 | 2, 3 |
| *Nyctimene robinsoni* | | < 1 | 4 |
| *Pteropus poliocephalus* | ~6 | 3–5 | 5 |
| **Yangochiroptera** | | | |
| *Chalinolobus gouldii* | 9–16 | - | 6, 7 |
| *Miniopterus orianae* | 9.5–11 | | 8 |
| *Nyctophilus bifax* | 9–28 | < 1–64 (3.5) | 9–12, |
| *Nyctophilus geoffroyi* | < 25 | < 1– < 15 | 13–15 |
| *Nyctophilus gouldi* | < 25 | < 1– > 24 | 13 |
| *Nyctophilus major* | < 23 | - | 16 |
| *Vespadelus pumilus* | | 4 | 17 |

References: (1) Bartels *et al.* (1998); (2) Geiser *et al.* (1996); (3) Coburn and Geiser (1998); (4) Hall and Pettigrew (1995); (5) Turbill *et al.* (2024); (6) Dixon and Huxley (1989); (7) Hosken and Withers (1997); (8) Hall (1982c); (9) Stawski and Geiser (2009); (10) Stawski (2010); (11) Stawski *et al.* (2009); (12) Stawski *et al.* (2008); (13) Geiser and Brigham (2000); (14) Turbill *et al.* (2003b); (15) Geiser *et al.* (2011); (16) Hosken (1997); (17) Turbill *et al.* (2003a).

generally ritualised (Nelson 1964, 1965a). Little is known of the social behaviour of other species of pteropodids; however, most of the other Australian species appear to be either solitary or form small groups (Table 13.6). In captivity, flying-foxes are generally best kept with more females than males, with a ratio of one male to three females being used successfully (S. Jackson pers. obs.).

## 9.2.2 Insectivorous bats

Little is known of the social behaviour of insectivorous bats, other than estimates of colony sizes that vary from single animals to many thousands (Table 13.6). One of the few observations is of the large-footed myotis, which showed that males are strongly attached to a particular site and apparently defend it (Dwyer 1970a). Females are attracted to these sites and form harems during the mating season. The bond between the mother and her young extends beyond lactation (Dwyer 1970a).

Several species of tree hole roosting insectivorous bats have been found to have a fusion-fission social structure whereby individuals form close associations but the total population may be spread over multiple roosts on any one day, regularly separating and reforming as a group. For example, a group of white-striped free-tailed bats in Queensland regularly moved between a large communal roost used by up to 300 individuals, and multiple smaller satellite roosts (Rhodes 2007). Gould's wattled bats roosting in bat boxes in Victoria formed into cryptic subgroupings within a larger roosting population, with females showing stronger associations with roost-mates than males (Godinho *et al.* 2015).

A study of captive ghost bats developed an ethogram of their behaviour to determine the relationship between social vocalisations and behaviours including eating, flying, fighting, grooming, huddling and mating (Hanrahan *et al.* 2022). This study found a strong association between flight behaviour and the 'chirp-trill' vocalisation by the male member of the colony, which was associated with a territorial or mate attraction function. The study also found a strong association between the 'squabble', 'rasp' and 'grumble' vocalisations and fighting behaviour. The squabble and rasp were thought to likely to represent levels of agonistic vocalisations produced by aggressive bats during altercations. The 'grumble' vocalisation appears to be an appeasement call, as it was produced by the target of the aggressor (Hanrahan *et al.* 2022).

## 9.3 Reproductive behaviour

The reproductive behaviour of several species of flying-foxes is well known, but for many others it is very poorly known (Hall and Richards 2000). Observations in the wild found that the sexes of grey-headed flying-foxes and little red flying-foxes segregate, although the extent varied. The separation can range from only one sex occurring in an individual tree to both sexes occurring in the same tree, but one sex occupying a higher part of the tree than the other (Nelson 1965a). During the breeding season, males gain bodyweight, show an increase in odour and marking behaviour and become increasingly aggressive (although fighting is generally ritualised and rarely leads to injury or death) as they begin to establish their territories (Nelson 1965a; Martin *et al.* 1995). During the breeding season, several different social groupings are present within a camp of flying-foxes, including:

- guard groups that extend around the perimeter of the camp
- family groups, comprising a male, a female and young conceived by the female in the previous breeding season
- adult groups differing from the family groups in that there is no young present and the male may be either monogamous or polygamous
- juvenile packs (Nelson 1965a).

Other observations have shown that males form short-term unstable harems during the breeding season, with the size of the harem being related to the size of the roost population, and that females join different harems at different roosts (Eby and Jones 2002). After mating, the females generally leave their territories and form groups of pregnant females (Nelson 1965a). Generally, the pregnant females and those with newborn young are nervous and fly away at the least disturbance (Nelson 1965a).

Reproductive behaviour in the wild is less well known for insectivorous bats. For most species mating occurs in autumn with either sperm storage or delayed implantation until pregnancy commences in spring (see section 10). For bats in captivity, unless breeding is desired, males and females should be separated during the mating period to avoid both unwanted pregnancies and harassment and injury by mating males. Density reductions and ample roosting site provision is also necessary in male enclosures as males will simulate mating behaviour with other males, which can lead to injury, infections from bites and death (Lyons 2020). Bald patches on the rear of the neck in both females and less dominant males are the result of repeated mating harassment.

## 9.4 Bathing

Sprinklers should be provided over outdoor flying-fox enclosures, particularly during warm weather, to allow them to cool down and bathe. Male flying-foxes often urine bathe whereby they produce urine and then spread it over themselves, which may be associated with scenting themselves rather than bathing (J. Mclean pers. obs.).

## 9.5 Behavioural problems

There is little information recorded on behavioural problems with captive bats. One behavioural problem that has been observed in some yellow-bellied sheath-tailed bats in southern Australia is where individuals have all started scratching their belly fur out and end up bald over most of the belly (L. Lumsden pers. obs.). Veterinary examination of these animals found no parasites or obvious reasons so it was proposed that it was likely due to stress (L. Lumsden pers. obs.). These bats were picked up in unusual situations, such as found on exposed walls during the middle of the day and all were grossly underweight. Initially, they put on weight rapidly but after 1–2 months started to lose weight with some individuals dying after 3–4 months (L. Lumsden pers. obs.). This problem was addressed by providing a heat pad at all times for yellow-bellied sheath-tailed bats in captivity in southern Australia to reduce stress, as it appears this species is more adapted to warmer climates (L. Lumsden pers. obs.).

## 9.6 Signs of stress

Signs of stress in pteropodids include excessive vocalisations, inappetence and loss of bodyweight. Signs of stress in insectivorous bats include agitation, aggression, inappetence, weight loss, fur loss, reluctance to exit roosts and death through myopathy due to excessive stress flying and extended elevated heart rate (Lyons 2020). A sign of stress that has been observed in yellow-bellied sheath-tailed bats when first brought into captivity is a wide-eyed stare, showing a blue ring around the eye and the body extremely still, which can be mistaken for the animal being calm and placid, rather than in a severe state of stress (L. Lumsden pers. obs.).

## 9.7 Behavioural enrichment

Enrichment can be provided in various ways, including providing foliage for cover, scented plants for olfactory enrichment and branches, ropes and logs to climb on and over (LeBlanc 1998). Browse in the form of eucalypt, bloodwood and melaleuca branches containing lerps and/or honeydew or fresh flowers can also be added whenever possible in order to stimulate natural foraging behaviours. Other techniques that assist flying-foxes enrichment include the addition or changing or ropes, mirrors and allowing adequate flying space (Fascione 1995).

Insectivorous bats can be stimulated by releasing live moths, crickets or other invertebrates for them to hunt, but it depends on the species and if they are in a flight enclosure if they will recognise these insects as food (L. Lumsden pers. obs.).

Behavioural enrichment can include:

- holding flying-foxes and gregarious insectivorous bats in a group so that they can show normal social behaviour
- providing a diversity of fruit items to flying-foxes
- providing pteropodids with browse (see section 6). This can be provided by placing large branches with numerous fresh flowers or with the use of browse tubes. Browse tubes consist of a piece of PVC pipe (~5 cm diameter and 30 cm long) with two end caps, in which one has an eyebolt and chain attached (LeBlanc 2001). The tube has numerous 7-9 mm holes into which stems of leaves and flowers of various food trees can be inserted. Alternatively, wooden dowel can be placed through the holes or various scents can be added
- providing flying-foxes with a variety of climbing opportunities such as rope (at least 2.5 cm in diameter. Some carers have also used pool noodles enclosed in materials and hung vertically from the roof (J. Mclean pers. obs.). The position of these should be changed every 4-5 months to add variety
- providing adequate area to make at least short flights
- providing hessian bags (tightly woven and ~100 × 60 cm) to flying-foxes by hanging them over clothes hangers in the bats' roosting areas (with the entrance facing downward). These are used to decrease aggression and are often used in cool weather for them to keep warm. Some bats will even climb into them in cold weather. When the bags are replaced, the old bag should be placed over the new bag for several days to allow the scent to be transferred (M. Beck pers. comm.)
- providing flying-foxes with sprinklers to bathe under, particularly on hot days
- providing insectivorous bats with flying insects, as available
- providing insectivorous bats adequate climbing opportunities and multiple roosting sites so they can roost together or apart
- providing insectivorous bats with various climbing opportunities and crevices
- providing insectivorous bats with natural foliage and bark features while ensuring these do not reduce the flight space.

## 9.8 Training and conditioning

Flying-foxes have been trained to be held during public exhibits and educational talks. Individuals could potentially be trained to hang off a perch that could be attached to a scale so they could be weighed.

## 9.9 Introductions and removals

When introducing flying-foxes, it may help in some cases for the bat or bats to be introduced initially by holding them in a small enclosure near where the other bats hang so that they can all become familiarised (Fascione 1995). Any introductions should also be undertaken at the beginning of the day to allow the whole day for observation and attending to any problems that may arise. Aggression can be minimised by offering unscheduled food and adding additional branches to confuse the colony slightly (Fascione 1995). Despite these changes, flying-foxes, particularly males, will still bicker, as they re-establish their territories (Fascione 1995; S. Jackson pers. obs.).

Similarly, introductions of insectivorous bats into established colonies are best done through housing in small enclosures within the larger colony enclosure for several days. Careful assessment of males both within existing colony and new males should be made for several days, particularly during breeding season (R. Lyons pers. obs.).

## 9.10 Intraspecific compatibility

The fruit bats are highly social species so they can readily be held in groups and although other species, such as blossom-bats, appear to be largely solitary, they also can readily be held in groups (Table 13.12). Among the insectivorous bats, most species are known to roost either in small numbers or in very large numbers, which probably reflects thermal requirements and roost type. In maintaining any species in captivity, it is important to ensure adequate roosting opportunities to minimise conflicts.

Greater broad-nosed bats are known to attack and eat other insectivorous bats in captivity and should not be housed with any other species. Secure enclosures are required to ensure they are unable to gain access to nearby housed insectivorous bats.

## 9.11 Interspecific compatibility

Flying-foxes have been held with yellow-bellied gliders (*Petaurus australis*), although aggressive encounters, mainly over food, are known to occur. Flying-foxes are

**Table 13.12.** Roosting population size of bats and the suggested group size of different species when held in captivity.

Roost size derived from Table 13.6.

| Genus | Roost population size | Suggested captive number |
|---|---|---|
| **Yinpterochiroptera** | | |
| **Pteropodidae** | | |
| *Macroglossus* | 1–3 | Solitary or groups |
| *Syconycteris* | Solitary | Solitary or groups |
| *Nyctimene* | Solitary | Solitary or groups |
| *Dobsonia* | < 100 in Australia. 1000s in New Guinea | Groups |
| *Pteropus* | 10s to > 100 000 | Groups |
| **Megadermatidae** | | |
| *Macroderma* | 1-1500 | Groups |
| **Rhinolophidae** | | |
| *Rhinolophus* | Typically < 20. Up to 2000 | Groups |
| **Hipposideridae** | | |
| *Hipposideros* | Usually 10–30. From 1–3000 | Groups |
| **Rhinonycteridae** | | |
| *Rhinonicteris* | 5–20 000 | Groups |
| **Yangochiroptera** | | |
| **Emballonuridae** | | |
| *Saccolaimus* | 1–6 | Solitary or groups |
| *Taphozous* | 2–25 but up to 260 | Groups |
| **Molossidae** | | |
| *Austronomus* | 1–25, maternity colonies up to 300 | Solitary or groups |
| *Chaerephon* | 1–7 | Solitary or groups |
| *Micronomus* | Up to 100 | Groups |
| *Ozimops* | Up to 100 | Groups |
| *Setirostris* | 1–15 | Groups |
| **Miniopteridae** | | |
| *Miniopterus* | Up to 100 000 | Groups |
| **Vespertilionidae** | | |
| *Myotis* | Usually 10–15. Up to 300 | Groups |
| *Phoniscus* | Solitary or small groups | Groups |
| *Murina* | 1–12 | Solitary or groups |
| *Chalinolobus* | Usually 3–40. Up to 400 | Groups |
| *Falsistrellus* | 3–36 | Groups |
| *Nyctophilus* | 2–3. Maternity colonies of 10–40 | Groups |
| *Scotorepens* | 2–20. Maternity colonies up to 300 | Groups |
| *Vespadelus* | Usually < 80. Can be > 500 | Groups |
| *Pipistrellus* | ~ 6 individuals | Groups |
| *Scoteanax* | Solitary or small groups | Solitary or groups? |

known to chase and vocalise at several possum species in the wild, including sugar gliders (*Petaurus breviceps*), mahogany gliders (*Petaurus gracilis*), yellow-bellied gliders and greater gliders (*Petauroides volans*) (Russell 1981; Borsboom 1982; Jackson 2001). Flying-foxes have been kept successfully with terrestrial animals such as bettongs and potoroos; however, care needs to be taken to ensure the substrate is frequently cleaned as the waste from the bats can quickly build up, potentially causing health problems to animals on the ground.

## 9.12 Suitability to captivity

There are large differences between different species of bats in their ability to adapt to captivity. Some species, such as the flying-foxes and ghost bats, are highly robust and breed readily while others can be maintained relatively easily, but are difficult to breed, and some do not do well in captivity at all. An outline of species that have been kept in captivity and their relative success is outlined in Table 13.13.

A review in 1982 (Hall 1982a) found the longest period that a bat was held in captivity was only 26 months for the common bent-winged bat, with an eastern horseshoe-bat lasting only 4 days and most species dying within 12 months. In more recent times, bats have been more successfully held in captivity, with some *Ozimops* being held for 23 years (Lumsden 2012), *Saccolaimus* for 13 years, *Scotorepens* for 8 years, *Chalinolobus* for 7 years, *Nyctophilus* for 6 years, and *Miniopterus* and *Vespadelus* at least several years (L. Lumsden pers. obs.). Female diadem leaf-nosed bats have been kept in captivity for over 9 years (J. Mclean pers. obs.).

Numerous species of bats were maintained in captivity by Hall (1982a), including the genera *Macroderma*, *Rhinolophus*, *Pipistrellus*, *Myotis*, *Chalinolobus*, *Vespadelus*, *Miniopterus* and *Nyctophilus*, with *Nyctophilus* adapting the easiest to captivity while the *Austronomus* quickly learned to eat but were rather inactive. Due to the ease with which it can be taught to feed, the large bent-winged bat is also recommended (Hall 1982a). Ghost bats are also easy to feed, breed well, are spectacular to watch and live long in captivity, so are highly recommended for captive populations (S. Jackson pers. obs.).

Pteropodids, especially flying-foxes, generally take to captivity readily, whereas insectivorous bats are sometimes difficult to establish and some species do

**Table 13.13.** Suitability to captivity of various species of bats that are known.

| Genus | Notes | Reference |
|---|---|---|
| **Yinpterochiroptera** | | |
| **Pteropodidae** | | |
| *Macroglossus* | Unknown | - |
| *Syconycteris* | Maintains and breeds relatively well in captivity | 1 |
| *Nyctimene* | Fussy eaters and not easy to maintain in captivity | 2 |
| *Dobsonia* | Readily maintained in captivity, but breeding unknown | 1 |
| *Pteropus* | Maintains and breeds well in captivity | 1 |
| **Megadermatidae** | | |
| *Macroderma* | Maintains and breeds well in captivity | 1 |
| **Rhinolophidae** | | |
| *Rhinolophidae* | Difficult to maintain in captivity medium to long term | 3 |
| **Hipposideridae** | | |
| *Hipposideros* | Keep well in captivity, but breeding maybe problematic | 2 |
| **Rhinonycteridae** | | |
| *Rhinonicteris* | Unknown | - |
| **Yangochiroptera** | | |
| **Emballonuridae** | | |
| *Saccolaimus* | Survive well in captivity where attention to diet is paid (mealworms only a portion - alternate items offered). Usually require hand-feeding | 3 |

**Table 13.13.** continued

| Genus | Notes | Reference |
|---|---|---|
| *Taphozous australis* | Survive well in captivity, though may need to hand-feed > 40 mealworms per night | 4 |
| **Molossidae** | | |
| *Austronomus* | Hardy animals that are easily kept in captivity, though some may require continued hand-feeding. Prone to obesity | 3 |
| *Chaerephon* | Hardy animals that are easily kept in captivity, though some may require continued hand-feeding. Prone to obesity | 3 |
| *Micronomus* | Hardy animals that are easily kept and bred in captivity, though some may require continued hand-feeding. Prone to obesity | 3 |
| *Ozimops* | Hardy animals that are easily kept and bred in captivity, though some may require continued hand-feeding. Prone to teeth decay using captive diets and obesity | 3 |
| *Setirostris* | No known observations | – |
| **Miniopteridae** | | |
| *Miniopterus australis* | Difficult to maintain in captivity as they are prone to stress | 3, 5 |
| *Miniopterus orianae* | Can be kept in captivity | 3 |
| **Vespertilionidae** | | |
| *Myotis* | Difficult to maintain for even short duration – avoid attempts | 3 |
| *Phoniscus* | Survive well in captivity; however, access to specific spider species may prove difficult to maintain | 3 |
| *Murina* | No known observations | 3 |
| *Chalinolobus gouldii* | Survives well in captivity | 4, 6 |
| *Chalinolobus picatus* | Proved to be fragile species in captivity | 4 |
| *Falsistrellus* | Hardy animals and survive well in captivity, but prone to obesity | 3 |
| *Nyctophilus geoffroyi* | Survives well in captivity and quickly adapt to feeding themselves | 4, 6 |
| *Nyctophilus gouldi* | Survives and breeds well in captivity | 7 |
| *Scotorepens balstoni* | Hardy animals and survive well in captivity | 4 |
| *Scotorepens greyii* | Hardy animals and survive and breed well in captivity | 3, 4 |
| *Scotorepens orion* | Hardy animals and survive and breed well in captivity | 3, 6 |
| *Vespadelus* | Can be kept in captivity for extended periods, but at times can be difficult to maintain as they are prone to stress | 3, 6 |
| *Pipistrellus* | Unknown for Australian species, but overseas species in this genus survive and breed well in captivity | 7, 9 |
| *Scoteanax* | Hardy animals and survive well in captivity, but prone to obesity | 3 |

References: (1) S. Jackson pers. obs.; (2) J. Mclean pers. obs.; (3) R. Lyons pers. obs.; (4) Reardon and Flavel (1987); (5) Hall (1982a); (6) L. Lumsden pers. obs.; (7) Phillips and Inwards (1985); (8) Hughes *et al.* (1995); (9) Racey and Kleiman (1970).

not do well. The success of introducing insectivorous bats into captivity can be increased by choosing the correct time of year. Greater success generally occurs if bats are caught at the end of summer or early autumn as they are in the best condition in preparation for the upcoming winter (Humphrey-Smith 1982; Hall 1982b). Alternately hand-reared unreleasable bats or bats that are the result of unsuccessful rehabilitation attempts are ideal candidates for colonies as they are well adjusted to captivity.

## 10. BREEDING

### 10.1 Mating system

Most Australian bats appear to be polygynous; however, some flying-foxes will pair up during the mating season (Nelson 1965a). Most insectivorous bats are polygynous.

### 10.2 Ease of breeding

Flying-foxes generally breed well when they are housed outside, but indoor enclosures with constant light cycles appear to affect them. The little red flying-fox has a robust testis cycle and is resistant to modifications of photoperiod, while the grey-headed flying-fox can be manipulated by photoperiod but responds slowly and incompletely (O'Brien 1993). Species such as the eastern blossom-bat and ghost bats have, however, bred regularly in the artificial light cycles at Taronga Zoo (S. Jackson pers. obs.).

For several species of overseas bats, the ability to breed in captivity is affected by changes in light cycle and resulting changes in melatonin (Beasley *et al.* 1984). A study of pallid bats (*Antrozous pallidus*) in the USA found that bats held in short day conditions (10 h light : 14 h dark) had regressed testes, epididymal spermatozoa and fully developed sex glands corresponding with autumn reproductive condition in the wild (Beasley and Zucker 1984). In contrast, those bats held in long day conditions (14 h light : 10 h dark) had testes undergoing spermatogenesis, few epididymal spermatozoa and undeveloped sex glands, which corresponded with the summer reproductive condition (Beasley and Zucker 1984). Therefore, it appears very important in captivity to provide light cycles that are as close as possible to natural light cycles.

## 10.3 Oestrous cycle and gestation period

The oestrous cycle of most species of Australian bats is poorly understood. The gestation periods are known for a few species and typically last 90–180 days (Table 13.14).

## 10.4 Reproductive status

The identification of reproductive condition requires familiarity with the various conditions of teat size and shape and the relative size of the testes and

**Table 13.14.** Timing of reproduction and number of young born in bats for which it is known.

#Species that can breed twice in the same year.

| Species | Age at first birth (months) | No. of young | Mating | Birth season | Gestation (~ days) | Weaning (months) | Reference |
|---|---|---|---|---|---|---|---|
| **Yinpterochiroptera** | | | | | | | |
| **Pteropodidae** | | | | | | | |
| *Macroglossus minimus* | – | 1 | – | Dec–Feb | – | – | 1, 2 |
| *Syconycteris australis* | – | 1 | – | Oct–Nov<br>Feb–Apr | – | – | 2, 3 |
| *Dobsonia magna* | > 24 | 1 | May–Jun | Aug–Nov | – | – | 2–4 |
| *Pteropus alecto* | – | 1 | Mar–Apr | Oct–Dec | – | – | 2, 5 |
| *P. conspicillatus* | – | 1 | Apr–May | Oct–Dec | – | – | 2, 6 |
| *P. poliocephalus* | 16(f),30(m) | 1 | Mar–Apr | Sep–Nov | 180 | 5–6 | 2, 3, 7–9 |
| *P. scapulatus* | 16 | 1 | Oct–Jan | Apr–June | – | – | 2, 3, 6, 8–10 |
| **Megadermatidae** | | | | | | | |
| *Macroderma gigas* | – | 1 | Jul–Aug | Sep–Nov | – | 2 | 2, 4 |
| **Rhinolophidae** | | | | | | | |
| *Rhinolophus megaphyllus* | 12–24 | 1 | Jun–Jul | Nov–Dec | 120–135 | (Jan–Feb) | 2, 3, 11, 12 |
| *R. robertsi* | – | – | – | Oct–Nov | – | – | 2 |
| **Hipposideridae** | | | | | | | |
| *Hipposideros ater* | 6(f) | 1–2 | – | Oct–Dec | 190–200 | 2.5 (Dec–Jan) | 2, 4, 12, 13 |
| *H. cervinus* | – | – | – | Nov–Dec | – | (Dec–Jan) | 2, 12 |
| *H. diadema* | – | 1 | – | Sep–Dec | – | (Jan–Feb) | 1, 2 |
| *H. stenotis* | – | 1–2 | Jun–Jul | Dec–Jan | – | – | 1, 2 |
| **Rhinonycteridae** | | | | | | | |

**Table 13.14.** continued

| Species | Age at first birth (months) | No. of young | Mating | Birth season | Gestation (~ days) | Weaning (months) | Reference |
|---|---|---|---|---|---|---|---|
| *Rhinonicteris aurantia* | 16(m), 7(f) | 1 | | Dec–Jan | 150 | 2–2.5 (Feb) | 1, 14 |
| **Yangochiroptera** | | | | | | | |
| **Emballonuridae** | | | | | | | |
| *Saccolaimus flaviventris* | – | 1 | – | Dec–Mar | – | – | 1, 15 |
| *S. saccolaimus* | | 1 | – | Nov–Mar | – | – | 1 |
| *Taphozous georgianus* | 9 (f), 21 (m) | 1 | Aug–Sep | Oct–Dec | 90 | 1 (Dec–Apr) | 1–3, 12, 16 |
| *T. hilli* | – | 1 | – | Oct–Feb | 120 | – | 1 |
| **Molossidae** | | | | | | | |
| *Austronomus australis* | 12(f), 24(m) | 1 | – | Nov–Feb | 98 | (Dec–Jan) | 1, 12, 19 |
| *Chaerephon jobensis* | – | 1 | – | Nov–Dec | – | (Dec–Jan) | 1 |
| *M. loriae* | – | – | – | Oct–Nov | – | (Dec–Jan) | 12 |
| *Ozimops lumsdenae* | – | 1 | – | Nov–Dec? | – | (Jan) | 1 |
| *O. planiceps* | 12(f), 24(m) | 1 | Jun–Jul | Nov–Jan | – | (Feb–Mar) | 1, 9, 18, 19 |
| **Miniopteridae** | | | | | | | |
| *Miniopterus australis* | 16 | 1 | Jul–Aug | Nov–Dec | – | (Feb) | 3, 12, 20 |
| *M. orianae* | 16 | 1 | May–Jun | Oct–Jan | 195–225 | (Jan–Feb) | 3, 9, 12, 21, 22 |
| **Vespertilionidae** | | | | | | | |
| *Myotis macropus* | – | 1–2 | – | Oct–Feb# | – | (Jan–Mar) | 1, 3, 9, 22, 23 |
| *Phoniscus papuensis* | – | – | – | Oct–Feb# | 195–225 | (Nov–Apr) | 24 |
| *Chalinolobus dwyeri* | – | 1 | – | Nov–Dec | – | – | 2, 23 |
| *C. gouldii* | – | 2 | May–Aug | Sep–Jan# | 90 | 1.5 (Dec–Jan) | 9, 23, 25, 26 |
| *C. morio* | – | 1–2 | – | Sep–Nov | – | (Jan–Mar) | 12, 22, 27, 28 |
| *C. nigrogriseus* | – | ? | – | Oct–Dec | – | (Jan) | 1, 12 |
| *C. picatus* | – | 2 | – | Dec–Jan | – | – | 29 |
| *Falsistrellus tasmaniensis* | – | | – | Nov | – | (Jan–Feb) | 9 |
| *Nyctophilus arnhemensis* | – | 1–2 | – | Oct–Dec | – | – | 1 |
| *N. bifax* | – | 1 | – | Oct–Nov | – | – | 1 |
| *N. corbeni* | – | 2 | – | Nov | – | (Dec–Jan) | 30 |
| *N. geoffroyi* | – | 2 | – | Oct–Nov | – | (Dec–Jan) | 1, 23 |
| *N. gouldi* | 12–15(m), 7–9(f) | 1–2 | – | Oct–Dec | – | (Dec–Jan) | 9, 29 |
| *N. walkeri* | – | 1–2 | – | Oct–Dec | – | – | 1 |
| *Scotorepens balstoni* | – | 1–2 | Apr–May | Oct–Dec | – | (Jan) | 1, 12, 23 |
| *S. greyii* | – | 1–2 | Apr–May | Oct–Dec | – | – | 23, 31 |

| Species | Age at first birth (months) | No. of young | Mating | Birth season | Gestation (~ days) | Weaning (months) | Reference |
|---|---|---|---|---|---|---|---|
| *S. sanborni* | – | | – | Nov | – | (Jan) | 12 |
| *Vespadelus darlingtoni* | < 12 (f), < 24(m) | 1 | – | Nov–Dec | – | (Jan–Feb) | 12, 32 |
| *V. baverstocki* | – | 1 | – | Oct–Dec | – | – | 1 |
| *V. caurinus* | – | 1–2 | – | Oct–Nov | – | – | 2 |
| *V. finlaysoni* | – | 1–2 | – | Aug–Mar[#] | – | – | 1, 23 |
| *V. pumilus* | – | 2 | – | Nov–Dec | – | – | 3 |
| *V. regulus* | – | 1 | – | Nov–Dec | – | – | 3, 33 |
| *V. vulturnus* | 12(f), 24(m) | 1 | Feb–Mar | Nov–Dec | – | (Dec–Jan) | 9, 32, 34, 35 |
| *Pipistrellus adamsi* | – | 1 | – | Oct–Nov | – | – | 2 |
| *S. rueppellii* | – | 1 | – | Jan | – | – | 2 |

References: (1) Thomson (1991); (2) Strahan (1995); (3) Hall and Richards (1979); (4) Dwyer (1975); (5) Vardon and Tidemann (1998); (6) Martin *et al.* (1987); (7) Nelson (1965a); (8) Nelson (1965b); (9) Menkhorst (1995); (10) Ratcliffe (1931); (11) Young (1975); (12) Richards (1989); (13) Hall (1983); (14) Churchill (1995); (15) Chimimba and Kitchener (1987); (16) Jolly (1990); (17) Kitchener and Hudson (1982); (18) Crichton and Krutzsch (1987); (19) Krutzsch and Crichton (1987); (20) Dwyer (1968); (21) Dwyer (1963a); (22) Reardon and Flavel (1987); (23) Dwyer (1970b); (24) Madani *et al.* (2023); (25) Kitchener (1975); (26) Dixon and Huxley (1989); (27) Young (1979); 28) Kitchener and Coster (1981); (29) Phillips and Inwards (1985); (30) Baker and Gynther (2023); (31) R. Lyons pers. obs.; (32) Tidemann (1993); (33) Kitchener and Halse (1978); (34) Tidemann (1982); (35) L. Lumsden pers. obs.

epididymides for each species (Parnaby 1999). The sexes are generally relatively easy to determine, particularly during the breeding season (Fig. 13.16).

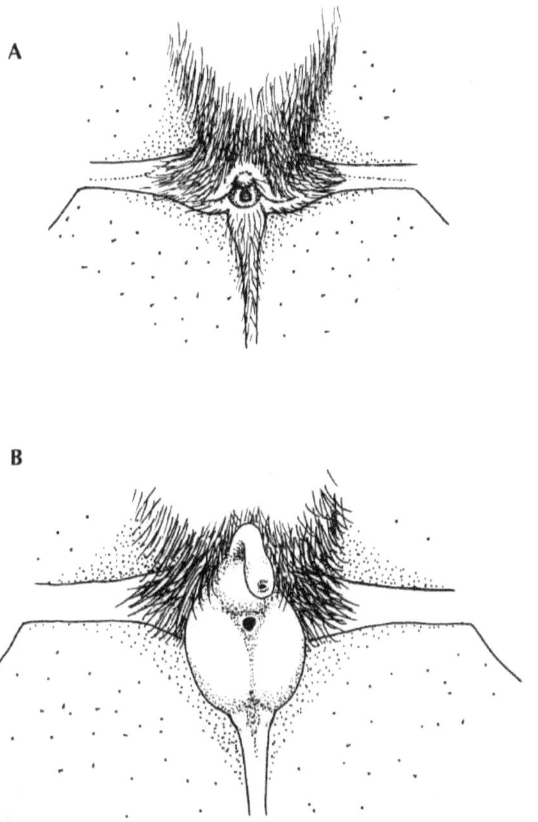


**Fig. 13.16.** Bat genitalia: (a) female and (b) male. Taken from Parnaby (1999) with permission from the publisher.

## 10.4.1 Females

Reproductive status of female bats can be allocated into one of the following:

- nulliparous: the nipple is very small and dome-like and often surrounded by dense fur (Fig. 13.17a). The surrounding skin is neither pigmented, raised nor wrinkled (Parnaby 1999)
- pregnant: diagnosed by gentle abdominal palpation or veterinary ultrasound machine (Bail *et al.* 2022). To notice the difference, it may be worth palpating males as well to distinguish a full stomach from pregnancy (Parnaby 1999). In captive animals it is suggested to do this before feeding (L. Lumsden pers. obs.)
- lactating: nipple large and pendulous, surrounding skin raised by conspicuous, distended, subcutaneous white milk glands. Milk can often be expressed by gently squeezing the nipple (Fig. 13.17b) (Parnaby 1999)
- post-lactating: nipple pendulous and surrounded by a circular patch of naked wrinkled skin (Parnaby 1999)
- regressed: nipple small, often sub-triangular rather than a small dome, surrounding skin often of darker pigmentation (Fig. 13.17c) (Parnaby 1999).

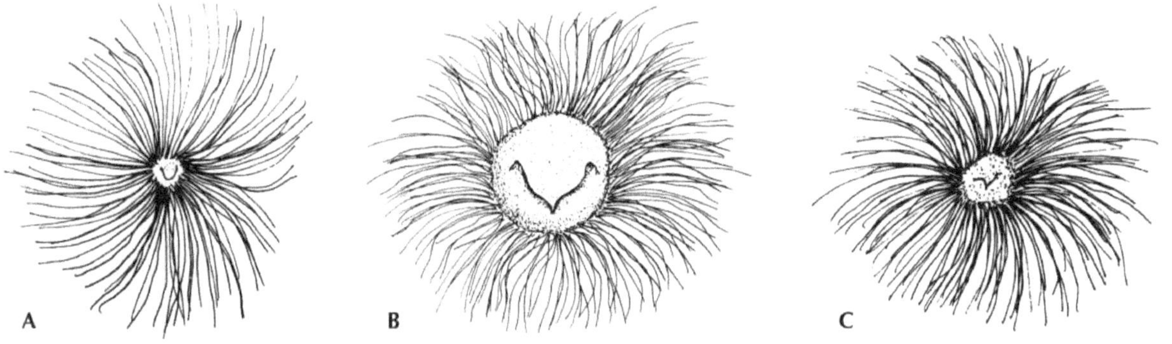


**Fig. 13.17.** Mammary area at different stages of bat reproduction: (a) nulliparous, (b) lactation and (c) teats regressed. Taken from Parnaby (1999) with permission from the publisher.

If young are present, there are several developmental stages and measurements that can be recorded and compared with existing growth curves (see section 10.16), or new curves established for future reference.

- Developmental stages
  - skin starting to get pigmented (changing from pink to blackish)
  - eyes open
  - fur visible on the back
  - fur visible on stomach
  - appearance of teeth through the gums
  - self-hanging
  - stretching and flapping wings
  - feeding on solids
  - self-feeding
  - able to fly
  - independent
- Measurements (see Fig. 13.18; Appendix 3)
  - bodyweight (g)
  - head length (mm) - from the occiput to snout tip
  - head width (mm) - the maximum width across the zygomatic arches
  - body length (mm) - from the snout to anus (mm)
  - hindleg length (mm), see Fig. 13.18
  - forearm length (mm), see Fig. 13.18
  - pes length (mm) - from the heel to the base of the longest toe, not including the claw.

### 10.4.2 Males

The size of the testes and the epididymal sac, which is attached to and extends from, the ventral surface of the testes along the tail membrane, can be used to assess the reproductive condition of males. The epididymal sac, which is used to store sperm, increases in size as the testes decrease in size (Parnaby 1999). Some experience is needed to detect differences. Care should be taken when looking for the testes, as in some species they ascend inguinally or abdominally outside of the breeding season (e.g. Kitchener 1976; Nelson 1965b; Racey 1988). In some species such as molossid bats the testes stay abdominally even in the breeding season (L. Lumsden pers. obs.).

## 10.5 Reproductive management

Separating sexes is generally the most used option, although new techniques are presently being developed. Melengestrol acetate contraceptive implants have been used successfully on female Rodrigues flying-foxes (*Pteropus rodricensis*) and appear to have comparatively little effect on their social behaviour (Hayes *et al.* 1996). Although these implants worked well, obvious physical effects were observed, including weight gain and retarded hair growth above the incision site. There was also a 22% rate of rejection/loss of the implants.

## 10.6 Occurrence of hybrids

Hybridisation is suggested to have occurred between the black and grey-headed flying-foxes and potentially between the black flying-foxes and the spectacled flying-foxes in the wild. Care needs to be taken with these results, as they may be at least partly the result of the genetic tests undertaken at the time, which may not have been as discriminating as tests available now (Webb and Tidemann 1995).

## 10.7 Timing of breeding

Most species of insectivorous bats in Australia usually have only have one breeding season each year, with young being born from late spring to early summer, though some species in subtropical and tropical

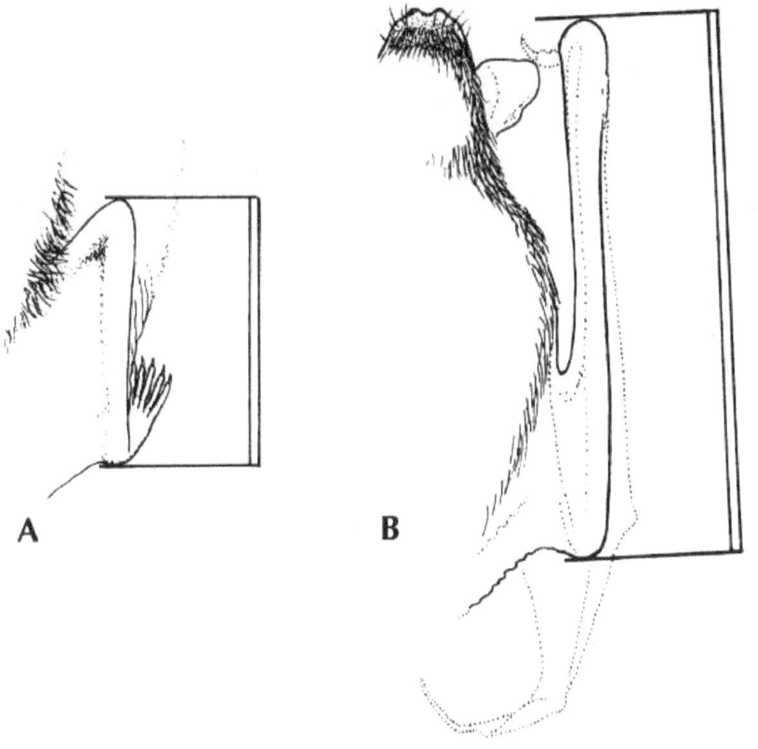


**Fig. 13.18.** Position for measuring hindleg (a) and forearm (b). Taken from Parnaby (1999) with permission from the publisher.

climates, such as *Myotis*, can have 2–3 breeding seasons per year (Dwyer 1970b) (Table 13.14).

Insectivorous bats that hibernate use several strategies to breed. The males produce sperm during summer, with mating typically occurring in autumn, but any resulting pregnancy would be during the period when the females slow their metabolism during hibernation, which would inhibit embryonic development. Different mechanisms have been developed by bats to manage the timing of breeding including (Altringham 2011):

- storage of sperm by males: mating may occur while the female is awake or in torpor or hibernation
- delayed ovulation and fertilisation by storing the sperm in the oviduct and uterus of the female
- delayed implantation of the fertilised egg by storing in the oviduct
- embryonic diapause
- delaying birth until good weather and adequate insect supply
- synchronising births to maximise heating opportunities.

As a result of this system, mating and pregnancy can be separated by many months, with the main phase of pregnancy occurring in spring. Therefore, any female bat brought into captivity from the wild between autumn and early summer could be pregnant, or have the ability to become pregnant, without males being present, and this should be factored into to assessments of body-weight variation.

The timing of reproduction can differ greatly between location (and food availability) and particularly in different latitudes. Black flying-foxes generally give birth to most young in January–March in the northern part of their range (12° S) and October–November in the southern part of their range (27° S) (Vardon and Tidemann 1998). Changes in day length appear to play a major role in determining breeding season in flying-foxes, as male grey-headed flying-foxes that were progressively moved over a 120-day period from natural short days (16 h light : 8 h dark) to longer days (9 h light : 15 h dark) resulted in testicular volume peaking during decreasing photoperiod. In contrast, those animals that were held in long day lengths did not show as much variation in testes volume, suggesting they do not rely on an endogenous rhythm (McGuckin and Blackshaw 1992). Therefore, flying-foxes held in nocturnal houses should have light–dark phases that change throughout the year to reflect as close as possible the natural light cycles.

Similar observations of changes in the timing of breeding have been observed for several species of insectivorous bats. The little bent-winged bat has been observed to copulate in August at latitude 15° 15' S (Vanuatu) and in June at 30° S (New South Wales) (Dwyer 1963b). The common bent-wing bat was observed to copulate in May at latitude 30° S (New South Wales) and the equivalent of March (September) at 45° N (France); however, in all four locations the time of birth was the same, December. These observations show the increased need for sperm storage with increasing latitude and associated length and severity of winter (Dwyer 1963b).

Vaginal smears have been used successfully to determine oestrous cycles in the common sheath-tailed bat (Jolly 1988b). Cottonwool swabs, moistened with saline, are used to collect the smear, which is then spread over a glass slide and allowed to air dry before being fixed with 10% buffered formalin for 20 min. The slides are then stained with acid fuchsin and toluidine blue, which differentiates the cornified cells as red and the non-cornified cells as blue (Dix and Billings 1969). The cells are then classified as negative (no cornified cells), light cornification (small numbers of cornified cells) or heavy cornification (large numbers of cornified cells consistent with oestrus) (Jolly 1988b).

## 10.8 Age at first and last breeding

The age of first breeding is poorly known for many species. Some species of insectivorous bats give birth in the first year of their life when the female is 12 months old, with mating occurring when the females are ~6 months of age; however, some males might not mate in their first year (see Table 13.14). In other groups (e.g. bent-winged bats), females do not breed until the second year of their life and flying-foxes are also typically older at first breeding.

## 10.9 Ability to breed every year

Although most species of bats appear to be able to breed every year, not all individuals actually do breed each year, such as the Gould's wattled bat (e.g. Martin *et al.* 1987; Lumsden and Bennett 1995). Pteropodids, and particularly flying-foxes, breed very well in captivity (S. Jackson pers. obs.). The lack of breeding success for insectivorous bats has been attributed by some authors (at least partly) to a lack of exercise, as captive bats have a tendency to become very lethargic, unwilling to fly and slow to arouse as a result of reduced exercise and overeating (Baer and Holgiun 1971; Humphrey-Smith 1982). Stress and environmental setup are the predominant issues.

## 10.10 Ability to breed more than once per year

Most species appear to be able to breed only once per year; however, there are some, especially those that occur in more arid or tropical regions where there is greater year-round food availability, that are able to breed more than once. These include the eastern blossom-bat and large-footed myotis (can breed twice per year in Sydney and Brisbane, and three times per year in North Queensland (R. Lyons pers. obs.). Finlayson's cave bat and the eastern forest bat can breed twice or more per year (Strahan 1995). In temperate areas, although a single breeding season is the norm, in years when the conditions are good, such as when there is higher-than-average rainfall associated with La Nina events, some individuals can breed twice, such as the Gould's wattled bat (Eastick *et al.* 2022) and golden-tipped bat (Madani *et al.* 2023).

## 10.11 Roosting requirements

There appears to be no specific roosting requirements for pteropodids in captivity. Though there appears to be specific roosting requirements for insectivorous bats, particularly for cave roosting species, in the wild (e.g. Baudinette *et al.* 1994), little specific information is known of the exact requirements for most species. Therefore, the roost requirements of each individual species should be researched. Details of roosting requirements that are known can be found in section 4.3.2. In captivity, cave-roosting species need to be given access to rock or a similar substrate, for roosting on (L. Lumsden pers. obs.).

Many species develop maternity colonies, which are dominated by females and young, so in captivity, males may need to be moved into a bachelor group during the lactation period. Separating males and females and decreasing the male densities is advisable. Cave-dwelling bats that are known to form maternity colonies include ghost bats (Pettigrew *et al.* 1986), eastern horseshoe-bats (Dwyer 1966) and bent-winged bats (Dwyer and Hamilton-Smith 1965). Most species of tree-hole roosting bats appear to form maternity colonies (e.g. Gould's wattled bats and lesser long-eared bats), with the maternity colonies larger than groupings at other times of the year (Lumsden *et al.* 2020).

## 10.12 Breeding diet

No specific diet for breeding exists. Pregnant and lactating female insectivorous bats require a more substantial diet than supplemented mealworms. Blended food diet can be offered by hand to such bats to increase nutritional intake.

## 10.13 Number of young born

Some species of insectivorous bats have one young and others usually have two (Table 13.14). Most species leave their young in the roost from birth, or within a few days of birth, as females are rarely caught in flight carrying young (Reardon and Flavel 1987; L. Lumsden pers. obs.). When females are caught with young, they are likely to be moving them between roosts rather than carrying them while foraging. Lactating female tree-hole roosting bats (e.g. Gould's wattled bat and lesser long-eared bat) are known to regularly shift young between roosts, often daily (Lumsden *et al.* 2020). Female lesser long-eared bats have been observed shifting their young between roosts when they return to the roost to suckle their young after a 2-h foraging bout after dusk (Lumsden *et al.* 2020). In contrast, flying-foxes carry their single young while foraging for ~3–6 weeks before leaving them in a crèche at the camp during the night (J. Mclean pers. obs.).

## 10.14 Age at weaning

The approximate age or month of weaning for various species of bats is shown in Table 13.14.

## 10.15 Age at removal from the mother

In captivity, once weaned, they can be removed from the parent if required; however, they generally can be left in the group as part of the colony. For insectivorous bats in the wild, there appears to be a period of at least a few weeks when the young have started flying but are still suckling, with some individuals observed to have both insects and milk in their stomach at this stage (P. Holz pers. obs.). Therefore, in captivity it should not be assumed they are weaned as soon as they can fly and milk should be continued but gradually phased out (L. Lumsden pers. obs.).

## 10.16 Growth and development

The growth and development of grey-headed flying-fox is shown in Fig. 13.19 with a curve for the inland

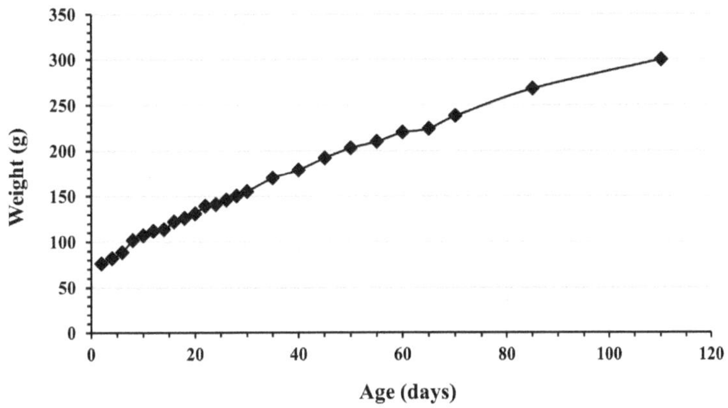


**Fig. 13.19.** Growth in bodyweight of the grey-headed flying-fox. Derived from George (1990) and Parry-Jones (2000).

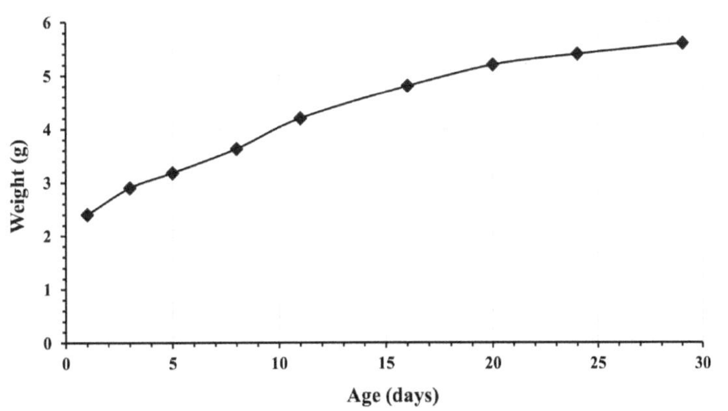


**Fig. 13.20.** Growth in bodyweight with age for the inland broad-nosed bat. Derived from Ryan (1966).

broad-nosed bat in Fig. 13.20. Additional growth and development references can be found in Table 13.15. The significant variation of bodyweights and sizes of bats of the same species has prevented comprehensive relatable growth chart development for the majority of insectivorous bats.

When grey-headed flying-foxes are born they are fully furred, except for the belly, their eyes are open and they usually weigh ~75 g, though can weigh as little as 50 g if there has been little food available to the female during pregnancy (George 1990) or if they are premature. The growth and development of grey-headed flying-foxes should be checked at least weekly and the weight gains should parallel those of mother-reared young (Table 13.16). Developmental stages observed in Gould's wattled bats in the wild can be found in Table 13.17.

**Table 13.15.** Growth curves measurements that have been developed for different species of bats.

CR = crown–rump length, FA = forearm length, RL = rump length, TL = tail length, WT = bodyweight.
*African subspecies; however, the curves could be used as an analogue of the Australian bent-winged bats.

| Species | Parameters | Reference |
|---|---|---|
| *Pteropus conspicillatus* | FA | 1 |
| *Pteropus poliocephalus* | WT, FA | 2–5 |
| *Taphozous georgianus* | WT, RL, FA, TL | 6 |
| *Chalinolobus gouldii* | WT, FA | 7 |
| *Scotorepens balstoni* | WT, FA | 8 |
| *Miniopterus schreibersii natalensis** | WT, CR | 9 |
| *Myotis adversus* | FA | 10 |

References: (1) Mclean *et al.* (2018); (2) George (1990); (3) Parry-Jones (2000); (4) Divljan *et al.* (2011); (5) Welbergen (2010); (6) Jolly (1990); (7) Eastick *et al.* (2022); (8) Ryan (1966); (9) Van der Merwe (1979); (10) Dwyer (1970b).

**Table 13.16.** Growth and development stages in captive mother-reared grey-headed flying-foxes.

Derived from George (1990) and Pinson (2020).

| Time | Forearm (mm) | Hindleg length (mm) | Head–body (mm) | Bodyweight (g) |
|---|---|---|---|---|
| Day 1 | 59 | 33 | 116 | 75 |
| | | Young fully furred except for belly. Milk teeth, four incisors just showing. Two canines upper and lower. Two premolars upper and lower | | |
| Day 2 | | Young trying to flap wings. Stretching head and neck muscles. Arching and pushing on mother's chest | | |
| Week 1 | 65 | 33 | 114 | 88 |
| Week 2 | 74 | 38 | 120 | 133 |
| Week 3 | 85 | 43 | 130 | 140 |
| | | Young left mother, hung on and explored branch. Belly well furred over | | |
| Week 4 | 92 | 46 | 142.4 | 149 |
| | | Mother uses mouth to retrieve young. Young supporting its own weight on branch while suckling. Well-furred guard hairs and adult colour. Lower first molars erupted. Premolars top and bottom erupted | | |
| Week 5 | 104 | 50.5 | 150 | 170 |
| Week 8 | 114 | 54.7 | 160 | 189 |
| | | Young licks juice from around mother's mouth when feeding and takes chewed food from her. Two pairs of upper and lower molars and premolars. Two upper and lower incisors replace milk teeth | | |
| Week 9 | 119 | 55.7 | 175 | 237 |
| Week 10 | 123 | 56.3 | 180 | 250 |
| | | Free of mother in daylight hours and associating with other bats. Mother no longer very protective. All incisors erupted and have replaced milk teeth | | |
| Week 11 | | Young taking food from container independently | | |
| Week 12 | 130 | 61 | 195 | 265 |
| | | Young socialising and eating well. Separated from mother for long periods. Three pairs of upper and lower incisors. Lower milk canines lost | | |
| Week 13 | | All milk teeth gone. Fourth lower molars about to erupt | | |
| Week 16 | 138 | 66.5 | 210 | 360 |

| Time | Forearm (mm) | Hindleg length (mm) | Head–body (mm) | Bodyweight (g) |
|---|---|---|---|---|
| | | Eating fruit well. Four pairs of lower molars and three pairs of upper molars | | |
| Week 19 | 146 | 69 | 230 | 370 |
| | | Fur luxurious and glossy | | |
| Week 22 | 153 | 71.4 | 235 | 450 |
| | | Hangs with other juveniles. Mother still lactating | | |
| Week 30 | 155 | 73.6 | 240 | 480 |
| Week 39 | 157 | 75 | 257 | 552 |
| Week 47 | 160 | 75 | 260 | 580 |
| Week 52 | 168 | 78 | 260 | 610 |

**Table 13.17.** Developmental stages in Gould's wattled bats in the wild.

Derived from Eastick *et al.* (2022).

| Time | Appearance |
|---|---|
| Day 1 | Pink, furless |
| Days 4–5 | Light brown pigmentation on head |
| Days 5–8 | Medium pigmentation on head and shoulders |
| Days 7–9 | Dark pigmentation on head and shoulders, eyes open |
| Day 8 | Sparse fur emerging |
| Day 11 | Velvet fur |
| Days 15–17 | No longer attached to mother in roost |
| Days 30–36 | Starting to fly |

## 11. ARTIFICIAL REARING

### 11.1 Housing

As with all native mammals that have been taken into care, minimising stress is a major consideration. Choosing suitable housing can help to create a stress-free environment. To achieve this, several factors should be considered including:

- securing the area from children and other animals
- maintaining a high standard of hygiene
- escape-proofing the area
- clearing the area of obstacles and hazards
- ensuring the area offers shelter from the weather and noise.

#### 11.1.1 Pteropodids

Most orphaned young grey-headed flying-foxes and spectacled flying-foxes are received between October and November, whereas little red flying-foxes are born between April and May (George 1990; J. Mclean pers. obs.). A common method used to assist rearing young flying-foxes is to use a wrap (Figs 13.21, 13.22). These can be made of flannelette or 100% cotton material and in different sizes for different ages, with the largest being ~22 × 45 cm and the smallest being 15 × 25 cm. One end of the wrap includes a 'mumma roll' that is filled with polyester (similar to that found in cushions) as it is light and easy to wash. The 'mumma roll' for the large wrap is ~11 cm wide (22 cm in circumference) and 9 cm wide (18 cm circumference) for the small wrap. The bat is placed belly down on the 'mumma roll' and then wrapped around with the rest of the material.

**Fig. 13.21.** Flying-fox wrap with 'mumma roll' that can be placed around flying-fox pups. Photo courtesy of Tolga Bat Hospital.

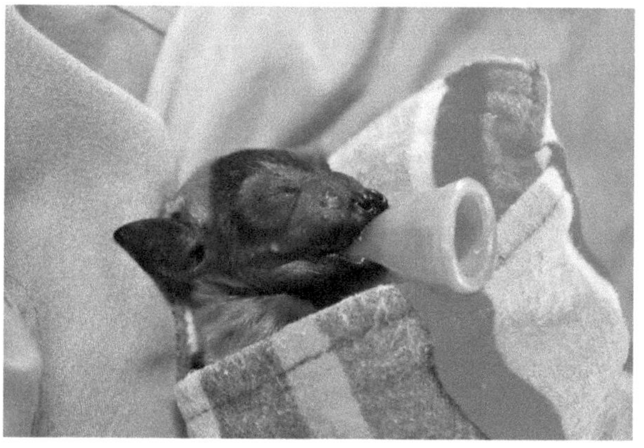

**Fig. 13.22.** Spectacled flying-fox pup in a wrap. Photo courtesy of Tolga Bat Hospital.

Due to the large number of bats received by the Tolga Bat Hospital they have developed specialised facilities (Tolga Bat Hospital n.d.). A synthesis of the procedure they follow for flying-foxes (mostly spectacled flying-foxes) in each of the age groups is given below.

### 0–4 weeks

- Flying-fox pups are placed within large plastic boxes (~40 (W) × 60 (L) × 30 (D) cm, but use a size that suits size of the heat pad) with a mesh top to allow the bats to climb around within the box. The depth of the box depends on the size of the pup (J. Mclean pers. obs.).
- Heating pads (65 × 40 cm) are placed on the bottom and up the sides of the box, because the pups will hang against the sides. Temperature of the mat is set manually and depends on thickness of material over the heat pad. Any pup requiring a critical temperature is in an intensive care unit.
- Pups are given a 'sock mumma' to hang onto and may like to suck on dummies or pacifiers.
- Each day the mesh tops are pegged onto a railing outside to allow the bats to stretch and sun themselves for ~20 min (depending on ambient temperature).
- Pups spend time out of the box on a modified clothes airer, especially at 3–4 weeks old.
- Some pups commence feeding through a teat via a syringe, while others commence straight onto a teat and bottle (Fig. 13.23).
- When bottle-feeding the hole in the teat must not be too large or the pup cannot control the milk flow and may get pneumonia.

### 4 weeks

- Transition period when bats are housed mainly in hanging boxes and become confident with self-feeding from dripper bottles. Pups vary quite a lot in their readiness for hanging longer periods of time and some require more opportunities to lie down. This is the age most wild pups are left in the trees at night while mum goes feeding.
- Bats stay wrapped after hand-feeding, laid down for ~20 mins, toileted, and put back in box hanging on a sock mumma.

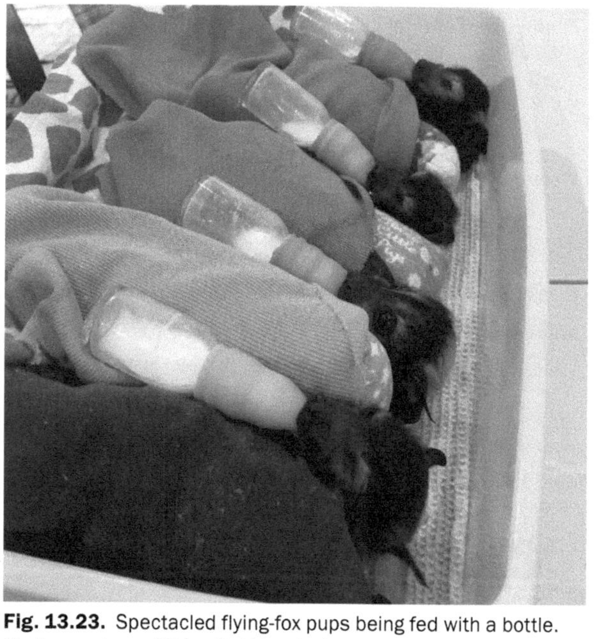

**Fig. 13.23.** Spectacled flying-fox pups being fed with a bottle. Photo courtesy of Tolga Bat Hospital.

### 4–8 weeks

- Pups are moved to a climbing frame that allows them to move about at night (Fig. 13.24).
    - The climbing frame has a 25 × 25 mm mesh top that is 50 (W) × 1800 (L) cm, solid adjustable-height floor and no solid sides to allow wing flapping.
    - The frames can be easily carried to allow large numbers of bats to be moved outside, inside or to the lie-down position.
- The jungle gym room has polyethylene netting in different locations to allow the pups to climb about at night between the roof wire and the moveable frames.
- The frames allow the floor to be raised so the pups can lie down while still hanging with their feet for 2–3 h in the morning and again at ~1700 h. (Fig. 13.25). The floor is raised to ~20 cm below the mesh they hang from. Pups at this age would normally be on their mothers all day, so they need a break from hanging and taking all their weight.

**Fig. 13.24.** Climbing frame for juvenile flying-foxes. Photo courtesy of Tolga Bat Hospital.

**Fig. 13.25.** Spectacled flying-foxes lying down under the frame on a light sheet. Photo courtesy of Tolga Bat Hospital.

- During the day the frames are carried outside on the covered verandah (weather permitting) to allow the bats to sun themselves (Fig. 13.26). The bats are outside for at least 7 h, but the time spent in the sun depends on how hot the day is, aiming for at least 20 min of sun each day at this age. They are moved out of the sun if they start fanning themselves.

**Fig. 13.26.** Spectacled flying-foxes placed outside to sun themselves. Photo courtesy of Tolga Bat Hospital.

#### 8–12 weeks

- Pups are moved to an outside enclosure.
- Soft fruits (including watermelon, rock melon and pears) are introduced into the diet, but milk is still important because wild pups are breast-fed for up to 6 months. Some carers replace milk with high-protein formula on fruit or in a smoothie.
- Pups need lots of opportunities for climbing and flapping their wings.

#### 12–16 weeks

- Pups are moved to a large flight aviary, but for most carers with a small number of pups in care the pups are moved to a crèche in which the pups from several carers come together for socialisation and breaking bonds with humans.

#### Release

- At ~450 g and a minimum forearm length of 145 mm, the bats are moved to a soft-release enclosure in the flying fox colony. They are moved in small groups of 10–30 at a time, kept in the cage for ~ 1 week, then fed on the outside of the cage. The next group is brought out, until all the bats are out. The advantage of this system is that the older pups do not have to wait for the younger ones to be a good size. The duration of support feeding varies with weather and food availability in the forest.

### 11.1.2 Insectivorous bats

Young insectivorous bats are best reared in temperature- and humidity-controlled conditions within a humidicrib. Specific humidity and heat are required to reduce evapotranspiration and dehydration and to also maintain the pup in its thermoneutral zone to enable digestion of milk.

If newborn, or finely furred, they can be housed within a small cotton or flannelette pouch within a small mesh enclosure placed within a humidicrib; however, as they develop, they are more comfortable hanging independently from polar fleece or cotton wraps draped on the walls of the mesh enclosure.

When a pup has reached the stage of having fluffy fur its mesh container can be removed from the humidicrib and placed inside an enclosure at normal ambient temperature with a heat pad hung from the side of the enclosure.

Pouches should be made of finely woven natural cotton, polar fleece or flannelette material that has no loose strings. Material such as towels are not recommended as the claws can get caught and cause injuries. These pouches should be changed at each feed to remove any faeces and urine (Lyons 2020).

## 11.2 Temperature requirements

### 11.2.1 Pteropodids

Pteropodids should be kept at ~28°C using a hot water bottle, heat lamp or heat pad and lining the enclosure with cloths for the bat to hang between. The temperature should be monitored closely with a thermometer or touch (George 1990; J. Mclean pers. obs.). Use a minimum/maximum thermometer with a plastic-coated probe that can be placed next to the juvenile. Incubators are also used regularly with young pteropodids. When toileting flying-fox pups measure the temperature of urine to check if it is warm.

### 11.2.2 Insectivorous bats

The recommended temperature at which insectivorous bats should be kept depends on the developmental age of the pup. Newborns and finely furred bats require temperatures of 32–38°C (Lollar and Schmidt-French 1998, Lyons 2020) and high humidity of ~55–80% (Barnard 2002). These conditions are best gained through the use of modern humidicribs with both temperature and humidity controls. Maintaining these temperatures and humidity levels is critical as digestion of food is dependent on the thermoneutral zone being maintained and torpor avoided where possible (Lyons 2020). A minimum/maximum thermometer with a plastic-coated probe can be placed next to the pup to monitor the temperature. Older fully to fluffy furred bats have the ability to thermoregulate and can be housed in variable temperatures but must have access to a heat source (heat lamp hung on one wall of the enclosure).

## 11.3 Specific requirements

Each newly arrived flying-fox must be isolated in a separate area until its disease status can be determined by a veterinarian or experienced wildlife rehabilitator, except in environmental events with mass casualties (e.g. heat stress event or major hailstorm) (Department of Planning, Industry and Environment 2021). If a bat is suspected to have a notifiable disease, such as Australian bat lyssavirus, the relevant government health department should be contacted.

When first brought in for hand-rearing, the bat may be dehydrated. If so, it can be given cooled boiled water, with 5 g (1 teaspoon) of glucose to 100 mL of water or 1 g of electrolyte replacer if available (Austin 1997). Vytrate® can also be used at a ratio of 20 mL Vytrate® to 250 mL water. It is important to warm the pup before feeding to avoid the risk of inhalation pneumonia. If this takes too long, give fluids subcutaneously and feed only when hydrated. Maintaining adequate hydration is very important when rearing pups and may involve subcutaneous injections of fluids even for adequately feeding pups due to their large fluid evaporation from their naked surface areas (Barnard 2002; Lyons 2020). A pup should not be fed until it is hydrated and warmed to a body temperature of ~35°C (Barnard 2002).

After each feed, the flying-fox pup's wings should be extended and preferably given a little exercise. At ~3 weeks of age the young bat becomes more active and may be housed so that it can climb out of its wrapping and stretch its wings. Take care that they do not become chilled overnight (George 1990). The bats should be provided with more activity time each day and should be allowed to hang onto slim rods (e.g. plastic-coated clothes airer), with the wrapping cloth pegged close by for security, where they can practise flapping their wings (George 1990). A second towel can be secured underneath the bat, which should hopefully catch it if it loses its grip.

At ~12 weeks of age the young flying-fox generally starts learning to fly in the wild. They can be encouraged to fly by placing them in a large enclosure and calling them (George 1990). Hang towels at each end of the enclosure for the bat to fly onto. Leafy branches can be placed around the enclosure to soften any falls. Pool noodles covered in cloth or 220-mm netting hung vertically can also be used for the bats to fly onto.

Juvenile insectivorous bats should naturally start flying between 5 and 12 weeks of age, provided their housing is large enough for them to fly (species specific) (R. Lyons pers. obs.). While usually not needed if providing the right housing environment and housing with other bats, they can be trained to fly near the time of weaning by placing them flat on the hand and giving them lift by raising and lowering the hand (L. Lumsden pers. obs.). It is best to have a soft landing in front of them (e.g. a piece of foam mattress) and lots of soft hanging material for them to land, or crash land, on. Housing the pups with adult bats when they are approaching weaning age encourages them to partake in flight and enhances normal social behaviour (Lyons 2020).

## 11.4 Diet and feeding routine

### 11.4.1. Natural milk

The natural milk has only been determined for the grey-headed flying-fox among the Australian pteropodids, with no information available on the milk of Australian insectivorous bats. Milk compositions of several species of insectivorous bats from the USA with genera represented in Australia are shown in Table 13.18.

### 11.4.2 Milk formulas

#### 11.4.2.1 Pteropodids

The main commercial formulas for hand-rearing pteropodid bats include:

- Biolac® Flying Fox Milk Powder - contains fully digestible nutrients including carbohydrates, proteins, lipids and elevated levels of many vitamins and minerals that are essential for flying-foxes
- Di-Vetelact® - not the preferred choice but can be used to hand-rear flying-foxes
- Wombaroo® Flying-foxes' Milk - can be fed from birth to weaning. Neonates can also be given artificial colostrum supplied by Wombaroo®
- Nan® 1 and Nan® 2 (Nestle Australia) - the higher protein levels of Nan 2 (or any human infant formula for babies >6 months) are closer to that of flying-fox milk than Nan 1 (Messer and Parry-Jones 1997).

Made-up formula can be refrigerated and the amount required removed and warmed as required.

#### 11.4.2.2 Insectivorous bats

Learnings from several rehabilitation groups around the world have found consistent issues with the use of milk substitutes. These include a higher incidence of metabolic bone disorder and bloat when bovine-/whey-based milk-substitutes are used in the smaller species of bat, even when homogenised (Lollar 2018; Lyons 2020). Alternatives to bovine-based formulas are used more commonly in insectivorous bat rehabilitation in Australia with success.

The three main formulas used for hand-rearing insectivorous bats globally are:

- Ghost Bat milk - developed by the Territory Wildlife Park and successfully used. It consists of 100 mL cow's milk, 1 egg yolk and 1 teaspoon of glucose powder (Glucodin®), heated in the microwave until custard consistency, stirred again and heated for another 20 s. This is made up daily
- Microbat Milk Recipe 1 (most species of insectivorous bats except molossids) - 1.5 scoops of S26 Infasoy® Powder, 100 mL goat's milk, 2 mL Megaderm® Supplement, 2 level teaspoons (4.3 g) dried egg white or one medium egg, and ¼ teaspoon of Human Paediatric Pro-biotic powder (Lyons 2020)

**Table 13.18.** Concentrations of the major constituents of milk for different species of bats.

*Overseas species that have genera represented in Australia.

| Species | Total solids (%) | Carbohydrate (%) | Lipids (%) | Protein (%) | Reference |
|---|---|---|---|---|---|
| **Yinpterochiroptera** | | | | | |
| *Pteropus poliocephalus* | 11.1-12.7 | 5.9-6.4 | 1.9-2.2 | 2.6-3.6 | 1 |
| **Yangochiroptera** | | | | | |
| *Myotis lucifugus** | 25.1-31.9 | 3.8-4.0 | 10.9-15.8 | 8.5-9.7 | 2, 3 |
| *M. thysanodes** | 40.5 | 3.4 | 17.9 | 12.1 | 4 |
| *M. velifer** | 22.0-32.4 | 3.9-4.5 | 7.6-19.9 | 9.0-10.7 | 2 |

References: (1) Messer and Parry-Jones (1997); (2) Kunz *et al.* (1995); (3) Kunz *et al.* (1983); (4) Jenness and Sloan (1970).

- Microbat Milk Recipe 2 (molossid bats) – 2 scoops of S26 Infasoy® Powder, 100 mL goat's milk, 3 mL Megaderm® Supplement and ¼ teaspoon of Human Paediatric Pro-biotic powder (Lyons 2020).

Other milk formulas include:

- Biolac® Puppy Milk – has fully digestible carbohydrates with a proven balance of proteins and increased levels of solids. It also has elevated levels of vitamins and minerals with the correct balance of saturated, mono-unsaturated and polyunsaturated fats. To make up, add 20 mL of milk powder to 100 mL sterile water
- Wombaroo® Insectivorous Bat Milk – can be fed from birth to weaning. Neonates can also be given Wombaroo's® artificial colostrum
- Blended diet (for slow introduction once fine-furred) – 1½ cups frozen mealworms, ½ cup cold water, 2 teaspoons of The Missing Link®, ¼ teaspoon of vitamin powder (e.g. Soluvet®), ½ teaspoon of Megaderm® Supplement, and 1 scoop Infasoy® Powder or equivalent brand. Mix and place in ice-cube trays for portions (Lyons 2020).

## 11.4.3 Feeding apparatus

### 11.4.3.1 Pteropodids

Flying-fox FF teats are available through Wombaroo® or use possum-sized marsupial teats as these are most easily accepted; small dog teats can be used. The teat should be punctured with a hot needle. Use a 21-gauge needle and check the size of the hole by letting water drip through the teat; it should not run through the teat, just drip through. The milk is supplied using a 10-mL syringe. When the flying-fox drinks 12 mL at a feed and is hanging most of the time, it should be able to self-feed from a dripper bottle, such as those used for guinea-pigs, with the milk being replaced every 4 h (Mclean 2000).

### 11.4.3.2 Insectivorous bats

Feeding milk to such small animals is difficult; small eyedroppers, 1-mL syringes, surgical tubing, eye makeup sponges, rubber catheters or specially designed artificial feeding devices can be used (Adams and Baer 1988; Hopkins 1990; Taylor *et al.* 1974; Barnard 2002). Some species, such as the vespertilionids, will lap milk so droppers work well, while others, such as the molossid bats, do not lap milk and it should be offered through a soft high-density foam sponge that they can suck (Lollar and Schmidt-French 1998). Some species will also readily lap from a spoon (e.g. lesser long-eared bats) (L. Lumsden pers. obs.). Whichever method is used, it is important to hold the bat upside down, at least initially, so that fluid is not aspirated into the lungs when it is learning to feed (Lollar and Schmidt-French 1998).

## 11.4.4 Feeding routine

### 11.4.4.1 Pteropodids

Pups should be at the right temperature all the time not just before feeding. So ensure the young bat is warm to touch and that it struggles when it is unwrapped, otherwise the milk may be inhaled, causing pneumonia.

During feeding, young are best wrapped in a small towel or blanket to restrain the wings and feet and fed on its side with the head slightly lower than the feet to reduce the potential of the milk formula entering the lungs (Wilson P 1988; George 1990). A flying-fox's teat over the end of a 10-mL bottle controls the milk flow perfectly and allows close monitoring of the volume of milk consumed. Hold the bottle parallel so only a trickle of milk is always in the teat. Feeds should be in accordance with growth charts appropriate for the pup's age and stage of development.

At ~ 8 weeks of age, flying-foxes can be offered small quantities of baby tinned puree fruit (apple is preferred) or soft fresh fruit at night (e.g. melons, pears). It is important that the pups form spats when they consume fruit and really just swallow the juice. They will still be drinking milk from dripper bottles, remembering that wild mothers will breast feed for up to 6 months. In care, pups will gradually wean themselves at ~3-4 months of age, but need Wombaroo® High Protein powder on cut fruit or put into smoothies. After 3 months of age, they should be consuming ~200 g of mixed chopped fruit with 10 g of Wombaroo® High Protein powder.

For about the first month of life, the pup should be toileted both before and 20 min after each feed. The pup is inverted and the bladder area tickled with a tissue to stimulate the bat to urinate and defecate (George 1990). The faeces should be golden-yellow in colour and of thin porridge consistency. The bat soon learns to invert to go to the toilet.

### 11.4.4.2 Insectivorous bats

Infant bats will typically accept milk readily, provided they are warm and not dehydrated. To determine the feeding schedule for insectivorous bats, check the pup's abdomen before and after the first feed (Lollar and Schmidt-French 1998). Digestion of milk is generally visible through the skin of the abdomen for most small bats, so time the feeding intervals to ensure the stomach is never completely empty (Hopkins 1990; Lollar and Schmidt-French 1998). Note the appearance of the milk in the stomach and how the abdomen feels when it is gently palpated with the fingertip. Check the pup each hour after the first feed to determine changes and ensure the abdomen remains slightly rounded; feed again when there is only a small amount of milk visible in the stomach (Lollar and Schmidt-French 1998). If the pup is furred the milk in the stomach will not be visible, so use gentle palpation. Take care not to overfeed as ruptures and bloat can result, which commonly leads to death (Lollar 2018). Even slight overfeeding can cause rupture and death. The width of the stomach when full should never be greater than the width of the ribcage (Lyons 2020). Pups should only be offered food once the prior food is digested. If digestion has not occurred and a fair amount of milk is still present in the abdomen after 4 h, and it was not initially overfed, the pup is either dehydrated or too cold. Immediate rectification of these issues is imperative before any further feeds are provided (Lyons 2020).

After each feed, a cotton swab wet with lukewarm water can be used to massage the pup's genitals to stimulate urination and defecation, although they may not defecate after every meal (Barnard 1995). Normal stools are soft and black, but it is not unusual for them to be cream-coloured for 1–2 days until they adjust to the milk replacer (Barnard 1995). Insectivorous bat pups are able to go to the toilet themselves, but it often remains on their body. When pouched in groups, cleaning the pups may be required at each feed.

Ghost bats have been fed half-strength Vytrate®/milk formula and 2.5 mL of ghost bat milk plus mealworms and mouse pieces at weaning when they are ~2 months old. A bowl of milk is also left in the enclosure overnight directly below the hanging towels. Ghost bats begin self-feeding within a very short period of time from the milk bowls and consume up to 20 mL of milk overnight plus three feeds of insects during the day depending on their age.

## 11.5 Data recording

When an animal is first brought in for hand-rearing, its sex and approximate age, using growth charts, should be recorded. During the hand-rearing process the following important information should be recorded on a daily basis:

- date when the information is recorded
- time when the information is recorded
- bodyweight to the nearest 1 g for pteropodids and 0.1 g for insectivorous bats
- general activity and demeanour
- characteristics and frequency of defecation and urination
- amount (g or mL) of different food types offered
- food consumption at each feed
- veterinary examinations and results.

This information provides important background data, such as food consumption, which will assist a veterinarian to reach a diagnosis if the animal becomes sick or fails to grow or gain weight. It also allows a comparison with growth curves to assess progress (section 10.16) and to allow new growth curves to be established for measurements and species in which they presently do not occur. The developmental stages and measurements outlined in section 10.3.1 should also be recorded on a weekly basis if possible.

## 11.6 Identification methods

If hand-rearing several individuals, they can be identified by either PIT tags, bird bands (flying-foxes), bodyweight, appearance, non-toxic cattle crayon markings or potentially the claws could be coloured on different toes and/or using different colours to distinguish individuals. Insectivorous bat pups can be PIT tagged from a young age; for example, Gould's wattled bats can be PIT tagged from birth (Eastick *et al.* 2022).

## 11.7 Hygiene

Maintaining a high standard of hygiene is critical to the survival of the juvenile bat. Emphasis needs to be placed on the following:

- maintaining clean bedding or roost area at all times
- maintaining personal hygiene by washing and disinfecting hands before and after handling the bat
- washing hands between feeding different bats
- using cooled boiled water when making up formulas for very young pups
- cleaning up spilt milk formula, faeces and urine from the bat's skin and fur as soon as possible, and then drying the animal
- washing all feeding equipment in cold or warm soapy water or sterilising in a suitable agent such as F10® or Milton® Anti-bacterial Solution or boiling it for 10 min. Need suitable bottle brushes for the bottle and the spouts of the dripper bottles. After sterilising, rinse the equipment in preboiled cold water

- storing the teats and bottles in the fridge in an air-tight container after they have been disinfected
- heating up milk once and discarding leftovers
- avoiding contact with other animals unless certain they pose no health risk
- stimulating the animal to toilet before, during or after feeding, to keep bedding clean
- keeping the pup clean from food, urine and faeces. Their bodies should be wiped over every day, including the wings and fur. Every few days they can be washed with a cloth and warm water and carefully dried (Wilson P 1988)
- feeding sick or debilitated bats last when feeding multiple bats and cleaning hands and equipment well to avoid cross-contamination.

## 11.8 Behavioural considerations

Socialisation is important when semi-weaned so that they develop normal bat behaviour and are not too imprinted. They should therefore be placed with other hand-reared bats, and adult bats, if possible, to allow normal bat socialisation to occur (Wilson P 1988; Lyons 2020).

## 11.9 Use of foster animals

Bats are not known to have been fostered to other bats of the same or other species. However, juveniles do form social roosting bonds with other adults if raised in the company of adults.

## 11.10 Weaning

### 11.10.1 Pteropodids

Flying-foxes are generally weaned in the wild at ~6 months of age (Wilson P 1988), but in captivity fruit is introduced at ~8 weeks of age and milk ceases usually by 4 months. Pups in the wild are crèched as soon as they are too big to carry, so young can be socialised from ~4 weeks or so. Flying-foxes should have learnt to fly by ~3 months of age and have gradually been supplementing their diet with fruit/nectar, whether captive or free living.

### 11.10.2 Insectivorous bats

Depending on the species, insectivorous bats can generally start to be weaned at ~6 weeks of age. However, even before this time, the bat blended food diet can be gradually added to their milk feeds, eventually replacing it when approaching full weaning age. Intermediate level formulas can include the following mixture:

- 480 mL frozen mealworms
- 142 g veal baby food
- 71 g banana baby food (no sugar or preservatives)
- 1 raw egg.

The mixture is poured into plastic zip-lock bags and laid flat to freeze or in ice-cube containers kept within zip-lock bags and frozen. Appropriately sized amounts can then be portioned and defrosted. The blended diet can be added to the milk at varying strengths depending on the age of the pup.

Some rearers do not use this intermediate formula; instead they wean bats directly onto mealworms by cutting off the insect's head and squeezing the viscera into the pup's mouth like a tube of tooth paste. Whenever the bats are willing, also allowing them to chew on the chitinous exoskeletons of the mealworms strengthens their jaws (Barnard 1995). To teach a juvenile to associate food with a dish, feed it over a dish containing mealworms using blunt forceps, to reduce the risk of being bitten. Once the bat has eaten all it desires, offer it a few drops of fresh water. Once the bat feeds regularly on whole mealworms (~10–40 or more, depending on the species), it should be housed and fed as an adult (Barnard 1995).

## 11.11 Release procedures

### 11.11.1 Pteropodids

Orphaned pups should be placed with other similar-aged pups of the same species after 4 weeks of age, as this is the age at which the mothers leave them in the trees at night. Contact with humans should be kept to a minimum after 3 months of age to increase the time that the pups socialise with other bats.

When flying-foxes are ready for release (350 g in grey-headed flying-foxes), several are placed together in a large enclosure within hearing distance and sight of the colony they are to join (George 1990).

When placed in the release enclosure, they are kept together for varying periods of time (7–30 days) to allow them to socialise, become accustomed to the wild surroundings and to give them an opportunity for flight practice (Augee and Ford 1999). During this time it is preferable for the young to be fed by people who were not involved with their rearing, as the young would naturally be breaking the bond with their mother at this stage (George 1990). When the young bats weigh ~450 g and forearms measure 145 mm they are ready for release, normally during mid-February. The top of the enclosure is opened so the young bats can come and go as they please, with food provided for up to another 8 weeks until they are flying well, at which time the food provided is slowly reduced (George 1990; Augee and Ford 1999). Observations of released grey-headed flying-foxes have found them to successfully establish in the wild, with various local movements being recorded and other roost sites being used. They have been known to fly up to 310 km within several months of their release (Augee and Ford 1999). One hand-reared grey-headed flying-fox that was released in 2004 and fitted with a metal identification band was recaptured alive in 2022, making the survival of the animal over 18 years of age. The animal was also captured ~297 km from where it was released (Mo *et al.* 2023).

## 11.11.2 Insectivorous bats

When releasing insectivorous bats, it is important to make sure the animal is in good condition, socialised and with good bodyweight for the species. It must also have had access to a large flight enclosure for at least 4 weeks of flying in a sustained fashion, not just building up to flying (Lyons 2020). This is to ensure that it obtains the necessary fitness and muscle condition to catch food on the wing immediately after release. Unlike young starting to fly in the wild, which have several weeks to adjust to catching their own food while they continue to receive milk from their mother, hand-reared young need to completely feed themselves from the day of their release. During the time in the flight enclosure the bat needs access to flying insects to develop its prey catching abilities. Before it is released, it should be assessed for sufficient development of these skills to survive in the wild.

Different species require different sized aviaries to undertaken flight practice within, with the faster above-canopy species requiring aviary sizes > 8 m (W) × 8 m (L) × 4 m (H) in size. Cluttered vegetation forager bats can obtain necessary flight fitness in smaller enclosures. Post-release survival has shown that it is imperative to provide flight practice in a large enclosure for an extended period before release to increase survival rates (Kelly *et al.* 2008).

Hand-reared bats are best released in the area where they were found to integrate back into their original colony. Depending on their age when brought into care, they may or may not recognise the surrounding area and know the location of roosting sites. A soft or hard release can be undertaken. For a soft release the box the bat was using as a roost in the flight enclosure is transferred to the release site, attached to a tree with its entrance blocked and the bat placed back inside. On dusk, the entrance is opened from the ground (e.g. using a string attached to the cloth temporarily closing the entrance) and the bat is allowed to leave when it is ready. This provides a known roost that the bat can return to while learning where the natural roosts are; however, this does not always occur (L. Lumsden pers. obs.). Unlike flying-foxes, which are known to return to soft-release enclosures to feed, providing food for insectivorous bats at the release site has not proved successful (L. Lumsden pers. obs.). For a hard release, the bat should be released after dark to avoid dusk predators such as kookaburras (Lyons 2020). Make sure there are no predators such as cats and dogs and no obstacles such as branches that may interfere with a clean take-off. In preparation for release the bat should be fed about half a normal feed and not be overfed (otherwise it may want to go to sleep) and it should be very warm and active (L. Lumsden pers. obs.). It can then be released in a similar way to releasing adults (section 7.5) by holding it on an outstretched hand with the head angled downwards and allowed to fly off. If it does not fly within several minutes, it can be moved around on the hand to encourage it to be more active and take off. Individuals should not be placed on a tree or other structure as there is a greater risk of predation or the bat moving higher and being unable to be retrieved if required. If the bat does not fly after several attempts it should be reassessed and consideration given to further flight training. Using the hard release approach the bat needs to quickly learn where the natural roost sites are. Little is known of the success of hand-reared young, but some at least have been known to survive for at least several months (L. Lumsden pers. obs.).

## REFERENCES


Adams DB, Baer GM (1988) Caesarean section and artificial feeding device for suckling bats. *Journal of Mammalogy* **47**, 524. doi:10.2307/1377698

Allison FR (1989) Molossidae. In *Fauna of Australia. Mammalia.* Vol. 1B. (Eds DW Walton and BJ Richardson) pp. 892–909. Australian Government Publishing Service, Canberra.

Allworth A, Murray K, Morgan J (1996) A human case of encephalitis due to a Lyssavirus recently identified in flying-foxes. *Communicable Diseases Intelligence* **20**, 504.

Altringham JD (1999) *Bats: Biology and Behaviour.* Oxford University Press, Oxford.

Altringham JD (2011) *Bats from Evolution to Conservation.* Oxford University Press, Oxford.

Andersen K (1917) On the determination of age in bats. *Journal of the Bombay Natural History Society* **25**, 249–259.

Animal Health Australia (2021a) Response strategy: Lyssaviruses (version 5.0). Australian Veterinary Emergency Plan (AUSVETPLAN), edition 5, Canberra, ACT.

Animal Health Australia (2021b). Response strategy: Hendra virus (version 5.0). Australian Veterinary Emergency Plan (AUSVETPLAN), edition 5, Canberra, ACT.

Anthony ELP (1990) Age determination in bats. In *Ecological and Behavioural Methods for the Study of Bats*. (Ed. TH Kunz) pp. 47–58. Smithsonian Institution Press, Washington DC.

Anthony ELP, Stack MH, Kunz TH (1981) Night roosting and nocturnal time budgets of the little brown bat, *Myotis lucifugus*: effects of reproductive status, prey density and environmental conditions. *Oecologia* **51**, 151–156. doi:10.1007/BF00540593

Armstrong KN (2000) Roost microclimates of the bat *Rhinonicteris aurantius* in a limestone cave in Geike Gorge, Western Australia. *Australian Mammalogy* **22**, 69–70. doi:10.1071/AM00069

Armstrong KN, Reardon TB, Jackson SM (2020) A current taxonomic list of Australian Chiroptera. Australasian Bat Society website. Version 2020-06-09. http://ausbats.org.au/species-list/4593775065

Augee ML, Ford D (1999) Radio-tracking studies of grey-headed flying-foxes, *Pteropus poliocephalus*, from the Gordon Colony, Sydney. *Proceedings of the Linnean Society of New South Wales* **121**, 61–70.

Austen JM, O'Dea M, Jackson B, Ryan U (2015) High prevalence of *Trypanosoma vegrandis* in bats from Western Australia. *Veterinary Parasitology* **214**, 342–347. doi:10.1016/j.vetpar.2015.10.016

Austin MA (1997) *A Practical Guide to the Successful Hand Rearing of Tasmanian Marsupials*. Regal Publications, Melbourne.

Australasian Bat Society (2021) BatMap. http://ausbats.org.au/batmap (accessed 21 April 2021).

Avery MI (1985) Winter activity of pipistrelle bats. *Journal of Animal Ecology* **54**, 721–738. doi:10.2307/4374

Baer GM, Holgiun GM (1971) Breeding Mexican freetail bats in captivity. *American Midland Naturalist* **85**, 515–517. doi:10.2307/2423774

Bail N, Ma M, Boardman W, Lumsden L, Reardon T, Prowse T (2022) Pregnancy testing in bats: a case study on Southern Bent-wing Bats (*Miniopterus orianae bassanii*). 20th Australasian Bat Society Conference, Brisbane, Queensland, 11–13 April 2022. *Australasian Bat Society Newsletter* **58**, 28–29.

Baker AM, Gynther IC (eds) (2023) *Strahan's Mammals of Australia*. 4th edn. Reed New Holland Publishers, Sydney.

Baker GB, Lumsden LF, Dettman EB, Schedvin NK, Schulz M, Watkins D, Jansen L (2001) The effect of forearm bands on insectivorous bats (Microchiroptera) in Australia. *Wildlife Research* **28**, 229–237. doi:10.1071/WR99068

Barnard S (1995) *Bats in Captivity*. Wild Ones Animal Books, Springville, CA.

Barnard SM (2002) Insectivorous bats. In *Hand-rearing Wild and Domestic Mammals*. (Ed. LJ Gage) pp. 96–103. Iowa State Press, Iowa.

Barrett J (2004) Australian Bat Lyssavirus. PhD thesis. University of Queensland, Brisbane.

Barrett JL, Carlisle MS, Prociv P (2002) Neuro-angiostrongylosis in wild black and grey-headed flying foxes (*Pteropus* spp). *Australian Veterinary Journal* **80**, 554–558. doi:10.1111/j.1751-0813.2002.tb11039.x

Bartels W, Law BS, Geiser F (1998) Daily torpor and energetics in a tropical mammal, the northern blossom-bat *Macroglossus minimus* (Megachiroptera). *Journal of Comparative Physiology. B, Biochemical, Systemic, and Environmental Physiology* **168**, 233–239. doi:10.1007/s003600050141

Bat Conservation Trust (n.d.) Bat box information pack. https://cdn.bats.org.uk/uploads/pdf/Bat-Box-Information-Pack-Sept-2020-JF.pdf (accessed 11 November 2023).

Baudinette RV, Wells RT, Sanderson KJ, Clark B (1994) Microclimatic conditions in maternity caves of the bent-wing bat, *Miniopterus schreibersii*: an attempted restoration of a former maternity site. *Wildlife Research* **21**, 607–619. doi:10.1071/WR9940607

Beasley LJ, Zucker I (1984) Photoperiod influences the annual reproductive cycle of the male pallid bat (*Antrozous pallidus*). *Journal of Reproduction and Fertility* **70**, 567–573. doi:10.1530/jrf.0.0700567

Beasley LJ, Smale L, Smith ER (1984) Melatonin influences the reproductive physiology of male pallid bats. *Biology of Reproduction* **30**, 300–305. doi:10.1095/biolreprod30.2.300

Bishop T, Mclean J, Peel A (2020) Personal protective equipment (PPE) information for bat handlers. Wildlife Health Australia, Sydney. https://wildlifehealthaustralia.com.au/Portals/0/Documents/ProgramProjects/PPE_Info_for_Bat_Handlers.pdf (accessed 23 July 2021).

Blehert DS, Hicks AC, Behr M, *et al.* (2009) Bat white-nose syndrome: an emerging fungal pathogen? *Science* **323**, 227. doi:10.1126/science.1163874

Bondarenco A, Körtner G, Geiser F (2013) Some like it cold: summer torpor by freetail bats in the Australian arid zone. *Journal of Comparative Physiology. B, Biochemical, Systemic, and Environmental Physiology* **183**, 1113–1122. doi:10.1007/s00360-013-0779-7

Booth R (1994) Medicine and husbandry: Dasyurids, possums and bats. In *Wildlife*. Proceedings 233. Post Graduate Committee in Veterinary Science. (Ed. DI Bryden) pp. 423–441. University of Sydney, Sydney.

Borsboom A (1982) Agonistic interactions between bats and arboreal marsupials. *Australian Mammalogy* **5**, 281–282. doi:10.1071/AM82030

Brass DA (1994) *Rabies in Bats: Natural History and Public Health Implications*. Livia Press, Ridgefield, CT.

Buettner P, Westcott DA, Mclean J, Brown L, McKeown A, Johnson A, Wilson K, Blair D, Luly J, Skerratt L, Muller R, Speare R (2013) Tick paralysis in spectacled flying-foxes (*Pteropus conspicillatus*) in north Queensland, Australia: impact of a ground-dwelling ectoparasite finding an arboreal host. *PLoS One* **8**(9), e73078. doi:10.1371/journal.pone.0073078

Campbell FE, Atwell RB, Smart L (2003) Effects of the paralysis tick, *Ixodes holocyclus*, on the electrocardiogram of the spectacled flying-fox, *Pteropus conspicillatus*. *Australian Veterinary Journal* **81**, 328–331. doi:10.1111/j.1751-0813.2003.tb11505.x

Campbell S (2009) So long as it's near water: variable roosting behaviour of the large-footed myotis (*Myotis macropus*). *Australian Journal of Zoology* **57**, 89–98. doi:10.1071/ZO09006

Campbell S, Coulson G, Lumsden LF (2010) Divergent microclimates in artificial and natural roosts of the large-footed myotis (*Myotis macropus*). *Acta Chiropterologica* **12**, 173–185. doi:10.3161/150811010X504671

Chimimba CT, Kitchener DJ (1987) Breeding in the yellow-bellied sheath-tailed bat *Saccolaimus flaviventris* (Peters, 1867) (Chiroptera: Emballonuridae). *Records of the Western Australian Museum* **13**, 241–248.

Christian JJ (1956) The natural history of a summer aggregation of the big brown bat, *Eptesicus fuscus*. *American Midland Naturalist* **55**, 66–95. doi:10.2307/2422322

Churchill SK (1991) Distribution, abundance and roost selection of the orange horseshoe-bat, *Rhinonycteris aurantius*, a tropical cave-dweller. *Wildlife Research* **18**, 343–353. doi:10.1071/WR9910343

Churchill SK (1995) Reproductive ecology of the orange horseshoe bat, *Rhinonycteris aurantius* (Hipposideridae: Chiroptera), a tropical cave-dweller. *Wildlife Research* **22**, 687–698. doi:10.1071/WR9950687

Churchill S (2008) *Australian Bats*. 2nd edn Allen & Unwin, Sydney.

Clulow S, Blundell AT (2011) Deliberate insectivory by the fruit bat *Pteropus poliocephalus* by aerial hunting. *Acta Chiropterologica* **13**, 201–205. doi:10.3161/150811011X578750

Coburn DK, Geiser F (1998) Seasonal changes in energetics and torpor patterns in the subtropical blossom-bat *Syconycteris australis* (Megachiroptera). *Oecologia* **113**, 467–473. doi:10.1007/s004420050399

Constantine DG (1970) Bats in relation to health, welfare, and economy of man. In *Biology of Bats*. (Ed. WA Wimsatt) pp. 319–449. Academic Press, New York.

Constantine DG (1985) Disease exchange between bats and researchers: problems and precautions. *Australian Mammalogy* **8**, 325–329. doi:10.1071/AM85033

Constantine DG (1993) Chiroptera: bat medicine, management and conservation. In *Zoo and Wild Animal Medicine. Current Therapy 3*. (Ed. ME Fowler) pp. 310–321. Saunders, Philadelphia.

Cool SM, Bennett MB, Romaniuk K (1994) Age estimation of pteropodid bats (Megachiroptera) from hard tissue parameters. *Wildlife Research* **21**, 353–364. doi:10.1071/WR9940353

Cox TE, Smythe LD, Leung LKP (2005) Flying foxes of pathogenic Leptospira species. *Journal of Wildlife Diseases* **41**, 753–757. doi:10.7589/0090-3558-41.4.753

Cox-Witton K, Bishop T (2025) Bats. In *Current Therapy in Medicine of Australian Mammals*. Revised Edition. (Eds L Vogelnest and T Portas) pp. 594–609. CSIRO Publishing, Melbourne.

Crerar S, Longbottom H, Rooney J, Thornber P (1996) Human health aspects of a possible Lyssavirus in a black flying fox. *Communicable Diseases Intelligence* **20**, 325.

Crichton EG, Krutzsch PH (1987) Reproductive biology of the little mastiff bat, *Mormopterus planiceps* (Chiroptera: Molossidae), in southeast Australia. *The American Journal of Anatomy* **178**, 369–386. doi:10.1002/aja.1001780408

DCCEEW (Department of Climate Change, Energy, the Environment and Water) (2021) About banding. Australian Government, Canberra. www.dcceew.gov.au/science-research/bird-bat-banding/about-banding (accessed 4 May 2023).

Department of Planning, Industry and Environment (2021) *Code of Practice for Injured, Sick and Orphaned Flying-foxes*. New South Wales Government, Sydney.

DES (Department of Environment and Science, Wildlife and Threatened Species Operations Branch) (2020) *Code of Practice Care of Sick, Injured or Orphaned Protected Animals in Queensland*. Queensland Government, Brisbane.

Divljan A, Parry-Jones K, Wardle GM (2006) Age determination in the grey-headed flying fox. *The Journal of Wildlife Management* **70**, 607–611. doi:10.2193/0022-541X(2006)70[607:ADITGF]2.0.CO;2

Divljan A, Parry-Jones K, Wardle GM (2011) One hundred and forty days in the life of a flying fox tooth-fairy: estimating the age of pups using tooth eruption and replacement. In *The Biology and Conservation of Australasian Bats*. (Eds B Law, P Eby, D Lunney and L Lumsden) pp. 97–105. Royal Zoological Society of New South Wales, Sydney.

Dix WM, Billings SM (1969) Technique for permanent vaginal smear preparations from rodents and other mammals. *Georgia Academy of Science* **27**, 122–126.

Dixon JM, Huxley L (1989) Observations on a maternity colony of Gould's wattled bat *Chalinolobus gouldii* (Chiroptera: Vespertilionidae). *Mammalia* **53**, 395–414. doi:10.1515/mamm.1989.53.3.395

Dobson AP (2005) What links bats to emerging infectious diseases? *Science* **310**(5748), 628–629. doi:10.1126/science.1120872

Domrow R (1967) Mite parasites of small mammals from scrub typhus foci in Australia. *Australian Journal of Zoology* **15**, 759–798. doi:10.1071/ZO9670759

Doty AC, Stawski C, Currie SE, Geiser F (2016) Black or white? Physiological implications of roost colour and choice in a microbat. *Journal of Thermal Biology* **60**, 162–170. doi:10.1016/j.jtherbio.2016.07.015

Drees KP, Lorch JM, Puechmaille SJ, *et al.* (2017) Phylogenetics of a fungal invasion: origins and widespread dispersal of white-nose syndrome. *mBio* **8**, e01941-17. doi:10.1128/mBio.01941-17

Drury RL, Geiser F (2014) Activity patterns and roosting of the eastern blossom-bat (*Syconycteris australis*). *Australian Mammalogy* **36**, 29–34. doi:10.1071/AM13025

Durden LA, Wilson N (1985) Ectoparasites from the grey-headed flying fox, *Pteropus poliocephalus* and the red flying fox, *P. scapulatus* (Chiroptera: Pteropodidae) from southeastern Queensland, Australia. *Macroderma* **1**, 51–53.

Dwyer PD (1963a) The breeding biology of *Miniopterus schreibersii blepotis* (Temminck) (Chiroptera) in north-eastern New South Wales. *Australian Journal of Zoology* **11**, 219–240. doi:10.1071/ZO9630219

Dwyer PD (1963b) Reproduction and distribution of *Miniopterus* (Chiroptera). *Australian Journal of Science* **25**, 435–436.

Dwyer PD (1966) Observations on the eastern horseshoe bat in north-eastern New South Wales. *Helictite* **4**, 73–82.

Dwyer PD (1968) The biology, origin and adaptation of *Miniopterus australis* (Chiroptera) in New South Wales. *Australian Journal of Zoology* **16**, 49–68. doi:10.1071/ZO9680049

Dwyer PD (1970a) Social organization in the bat *Myotis adversus*. *Science* **168**, 1006–1008. doi:10.1126/science.168.3934.1006

Dwyer PD (1970b) Latitude and breeding season in a polyoestrous species of *Myotis*. *Journal of Mammalogy* **51**, 405–410. doi:10.2307/1378506

Dwyer PD (1975) Notes on *Dobsonia moluccense* (Chiroptera) in the New Guinea highlands. *Mammalia* **39**, 113–118. doi:10.1515/mamm.1975.39.1.113

Dwyer PD, Hamilton-Smith E (1965) Breeding caves and maternity colonies of the bent-winged bat in south eastern Australia. *Helictite* **4**, 3–21.

Eastick DL, Griffiths SR, Yen JDL, Robert KA (2022) Size at birth, postnatal growth, and reproductive timing in an Australian microbat. *Integrative Organismal Biology (Oxford, England)* **4**(1), obac030. doi:10.1093/iob/obac030

Eby P, Jones V (2002) A bloke in every port: group composition and gender bias in the migration patterns of grey-headed flying-foxes *Pteropus poliocephalus*. 10th Australasian Bat Society Conference, Cairns. 2–5 April.

Ellis WAH, Marples TG, Phillips WR (1991) The effects of a temperature-determined food supply on the annual activity cycle of the lesser long-eared bat, *Nyctophilus geoffroyi* Leach, 1821 (Microchiroptera: Vespertilionidae). *Australian Journal of Zoology* **39**, 263–271. doi:10.1071/ZO9910263

Fascione N (1995) *Flying-fox Husbandry Manual.* American Zoo and Aquarium Association, Chiroptera Advisory Group.

Field HE (2005) The ecology of Hendra virus and Australian bat lyssavirus. PhD thesis. School of Veterinary Science, University of Queensland, Brisbane.

Field H, Ross T (1999) Emerging viral diseases. In *Wildlife in Australia: Healthcare & Management.* Proceedings 327. Post Graduate Foundation in Veterinary Science. (Ed. DI Bryden) pp. 501–513. University of Sydney, Sydney.

Field HE, McCall B, Barrett J (1999) Australian bat lyssavirus infection in a captive juvenile black flying fox. *Emerging Infectious Diseases* **5**, 438–440. doi:10.3201/eid0503.990316

Field HE, Barrett PC, Hughes RC, Shield J, Sullivan ND (2000) A fatal case of Hendra virus infection in a horse in north Queensland: clinical and epidemiological features. *Australian Veterinary Journal* **78**, 279–280. doi:10.1111/j.1751-0813.2000.tb11758.x

Field HE, Breed AC, Shield J, Hedlefs RM, Pittard K, Pott B, Summers PM (2007) Epidemiological perspectives on Hendra virus infection in horses and flying foxes. *Australian Veterinary Journal* **85**, 268–270. doi:10.1111/j.1751-0813.2007.00170.x

Finnemore M, Richardson PW (1999) Catching Bats. In *The Bat Workers Manual.* (Eds AJ Mitchell-Jones and AP McLeish) pp. 33–38. Joint Nature Conservation Council, Peterborough, UK.

Flannery TF (1995a) *Mammals of New Guinea.* Reed Books, Sydney.

Flannery TF (1995b) *Mammals of the South-West Pacific & Moluccan Islands.* Reed Books, Sydney.

Fox S, Luly J, Mitchell C, Mclean J, Westcott DA (2008) Demographic indications of decline in the spectacled flying fox (*Pteropus conspicillatus*) on the Atherton Tablelands of northern Queensland. *Wildlife Research* **35**, 417–424. doi:10.1071/WR07127

Fraser GC, Hooper PT, Lunt RA, Gould AR, Gleeson LJ, Hyatt AD, Russell GM, Kattenbelt JA (1996) Encephalitis caused by a lyssavirus in flying-foxes in Australia. *Emerging Infectious Diseases* **2**, 327–331. doi:10.3201/eid0204.960408

Frick WF, Puechmaille SJ, Hoyt JR, *et al.* (2015) Disease alters macroecological patterns of North American bats. *Global Ecology and Biogeography* **24**, 741–749. doi:10.1111/geb.12290

Gaudet CL, Fenton MB (1984) Observational learning in three species of insectivorous bats (Chiroptera). *Animal Behaviour* **32**, 385–388. doi:10.1016/S0003-3472(84)80273-0

Geiser F (2006) Energetics, thermal biology, and torpor in Australian bats. In *Functional and Evolutionary Ecology of Bats.* (Eds TH Kunz, A Zubaid and GF McCracken) pp. 5–22. Oxford University Press, Oxford.

Geiser F, Brigham RM (2000) Torpor, thermal biology, and energetics in Australian long-eared bats (*Nyctophilus*). *Journal of Comparative Physiology. B, Biochemical, Systemic, and Environmental Physiology* **170**, 153–162. doi:10.1007/s003600050270

Geiser F, Stawski C (2011) Hibernation and torpor in tropical and subtropical bats in relation to energetics, extinctions and the evolution of endothermy. *Integrative and Comparative Biology* **51**, 337–348. doi:10.1093/icb/icr042

Geiser F, Cockburn DK, Körtner G, Law BS (1996) Thermoregulation, energy metabolism, and torpor in blossom bats, *Syconycteris australis* (Megachiroptera). *Journal of Zoology* **239**, 583–590. doi:10.1111/j.1469-7998.1996.tb05944.x

Geiser F, Stawski C, Bonarenco A, Pavey CR (2011) Torpor and activity in a free-ranging tropical bat: implications for the distribution and conservation of mammals? *Naturwissenschaften* **98**, 447–452. doi:10.1007/s00114-011-0779-y

George H (1990) Grey-headed flying-foxes. In *Care and Handling of Australian Native Animals.* (Ed. SJ Hand) pp. 159–169. Surrey Beatty & Sons & the Royal Zoological Society of New South Wales. Sydney.

Godinho LN, Lumsden LF, Coulson G, Griffiths SR (2015) Network analysis reveals cryptic seasonal patterns of association in Gould's wattled bats (*Chalinolobus gouldii*) roosting in bat-boxes. *Behaviour* **152**, 2079–2105. doi:10.1163/1568539X-00003315

Goehring HH (1972) Twenty-year study of *Eptesicus fuscus* in Minnesota. *Journal of Mammalogy* **53**, 201–207. doi:10.2307/1378850

Gorecki V (2020) The ecology and conservation of the large-footed myotis (*Myotis macropus*) in an urban environment. PhD thesis, Queensland University of Technology, Brisbane.

Griffin DR, Hitchcock HB (1965) Probable 24-year longevity records for *Myotis lucifugus. Journal of Mammalogy* **46**, 332. doi:10.2307/1377857

Griffiths SR, Rowland JA, Briscoe NJ, Lentini PE, Handasyde KA, Lumsden LF, Robert KA (2017) Surface reflectance drives nest box temperature profiles and thermal suitability for target wildlife. *PLoS One* **12**, e0176951. doi:10.1371/journal.pone.0176951

Griffiths SR, Robert KA, Jones CS (2022) Chainsaw hollows carved into live trees provide well insulated supplementary shelters for wildlife during extreme heat. *Wildlife Research* **49**, 596–609. doi:10.1071/WR21112

Hall JS, Cloutier RJ, Griffin D (1957) Longevity records and notes on tooth wear of bats. *Journal of Mammalogy* **38**, 407–409. doi:10.2307/1376240

Hall LS (1982a) Management of microchiropterans in captivity. In *The Management of Australian Mammals in Captivity.* (Ed. DD Evans) pp. 157–160. Zoological Board of Victoria, Melbourne.

Hall LS (1982b) Common bent-winged bat *Miniopterus schreibersii*: maintenance of large colony. In *The Management of Australian Mammals in Captivity.* (Ed. DD Evans) pp. 172–174. Zoological Board of Victoria, Melbourne.

Hall LS (1982c) The effect of cave microclimate on winter roosting behaviour in the bat, *Miniopterus schreibersii blepotis. Australian Journal of Ecology* **7**, 129–136. doi:10.1111/j.1442-9993.1982.tb01586.x

Hall LS (1983) *Hipposideros ater.* In *The Complete Book of Australian Mammals.* (Ed. R Strahan) pp. 298–299. Angus and Robertson, Sydney.

Hall LS, Pettigrew J (1995) The bat with the stereo nose. *Australian Natural History* **24**, 26–28.

Hall LS, Richards GC (1979) *Bats of Eastern Australia.* Queensland Museum, Brisbane.

Hall LS, Richards G (2000) *Flying Foxes: Fruit and Blossom Bats of Australia.* University of New South Wales Press, Sydney.

Hall LS, Young RA, Spate AP (1975) Roost selection of the eastern horseshoe bat *Rhinolophus megaphyllus.* In Proceedings of the 10th Biennial Conference of the Australian Speleological Federation. pp. 47–56. Australian Speleological Federation, Sydney.

Hall LS, Richards G, McKenzie N, Dunlop N (1997) The importance of abandoned mines as habitat for bats. In *Conservation Outside Nature Reserves*. (Eds P Hale and D Lamb) pp. 326–333. Centre for Conservation Biology, University of Queensland, Brisbane.

Halpin K, Young P, Field H (1996) Identification of likely natural hosts for equine morbilivirus. *Communicable Diseases Intelligence* **20**, 476.

Handley CO, Jr, Wilson DE, Gardner AL (Eds) (1991) *Demography and Natural History of the Common Flying-foxes,* Artibeus jamaicensis *on Barro Colorado Island, Panama*. Smithsonian Institution Press, Washington DC.

Hanrahan N, Dalziell AH, Turbill C, Armstrong KN, Welbergen JA (2022) Ethogram of ghost bat (*Macroderma gigas*) behaviours and associated social vocalisations. *Acta Chiropterologica* **24**, 195–208. doi:10.3161/15081109ACC2022.24.1.016

Hayes KT, Feistner ATC, Halliwell EC (1996) The effect of contraceptive implants on the behaviour of female Rodrigues flying-foxes, *Pteropus rodricensis*. *Zoo Biology* **15**, 21–36. doi:10.1002/(SICI)1098-2361(1996)15:1<21::AID-ZOO3>3.0.CO;2-E

Heard DJ (2003) Chiroptera. In *Zoo and Wildlife Animal Medicine*. (Eds ME Fowler and RE Miller) pp. 315–333. WB Saunders, St Louis, MO.

Heard D (2014) Chiropterans (Bats). In *Zoo Animal and Wildlife Immobilization and Anesthesia*. 2nd edn. (Eds G West, D Heard and N Caulkett) pp. 543–550. John Wiley and Sons, Ames.

Heard DJ, Garner J, Greiner E (1995) Toxocariasis and intestinal volvulus in an island flying fox (*Pteropus hypomelanus*). *Journal of Zoo and Wildlife Medicine* **26**, 550–552.

Hill JE, Smith JD (1985) *Bats: A Natural History*. British Museum, London.

Hine RS (Ed.) (1988) *Concise Veterinary Dictionary*. Oxford University Press, Oxford.

Holz PH, Lumsden LF, Marenda MS, Browning GF, Hufschmid J (2018a) Two subspecies of bent-winged bats (*Miniopterus orianae bassanii* and *oceanensis*) in southern Australia have diverse fungal skin flora but not *Pseudogymnoascus destructans*. *PLoS One* **13**(10), e0204282. doi:10.1371/journal.pone.0204282

Holz PH, Lumsden LF, Hufschmid J (2018b) Ectoparasites are unlikely to be a primary cause of population declines of bent-winged bats in south-eastern Australia. *Parasites and Wildlife* **7**, 423–428. doi:10.1016/j.ijppaw.2018.10.006

Holz PH, Lumsden LF, Legione AR, Hufschmid J (2019) *Polychromophilus melanipherus* and haemoplasma infections not associated with clinical signs in southern bent-winged bats (*Miniopterus orianae bassanii*) and eastern bent-winged bats (*Miniopterus orianae oceanensis*). *Parasites and Wildlife* **8**, 10–18. doi:10.1016/j.ijppaw.2018.11.008

Hooper PT, Gould AR, Russell GM, Kattenbelt JA, Mitchell G (1996) The retrospective diagnosis of a second outbreak of equine morbillivirus infection. *Australian Veterinary Journal* **74**, 244–245. doi:10.1111/j.1751-0813.1996.tb15414.x

Hooper PT, Lunt RA, Gould AR, Samaratunga H, Hyatt AD, Gleeson LJ, Rodwell BJ, Rupprecht CE, Smith JS, Murray PK (1997) A new Lyssavirus: the first endemic rabies related virus recognised in Australia. *Bulletin de l'Institut Pasteur* **95**, 209–218. doi:10.1016/S0020-2452(97)83529-5

Hopkins CS (1990) Carnivorous and insectivorous bats. In *Care and Handling of Australian Native Animals*. (Ed. SJ Hand) pp. 171–184. Surrey Beatty & Sons, Sydney.

Hosken DJ (1996) Roost selection by the lesser long-eared bat, *Nyctophilus geoffroyi*, and the greater long-eared bat, *N. major* (Chiroptera: Vespertilionidae) in Banksia woodlands. *Journal of the Royal Society of Western Australia* **79**, 211–216.

Hosken DJ (1997) Thermal biology and metabolism of the greater long-eared bat, *Nyctophilus major* (Chiroptera: Vespertilionidae). *Australian Journal of Zoology* **45**, 145–156. doi:10.1071/ZO96043

Hosken DJ, Withers PC (1997) Temperature regulation and metabolism of an Australian bat, *Chalinolobus gouldii* (Chiroptera: Vespertilionidae) when euthermic and torpid. *Journal of Comparative Physiology. B, Biochemical, Systemic, and Environmental Physiology* **167**, 71–80. doi:10.1007/s003600050049

Hughes PM, Rayner JMV, Jones G (1995) Ontogeny of 'true' flight and other aspects of growth in the bat *Pipistrellus pipistrellus*. *Journal of Zoology* **236**, 291–318. doi:10.1111/j.1469-7998.1995.tb04494.x

Humphrey-Smith I (1982) Survival of captive Microchiroptera feeding on prey attracted to artificial lights. In *The Management of Australian Mammals in Captivity*. (Ed. DD Evans) pp. 164–171. Zoological Board of Victoria, Melbourne.

International Air Transport Association (IATA) (2022) *Live Animals Regulations*. 48th edn. International Air Transport Association, Geneva, Switzerland.

Jackson SM (2001) Foraging behaviour and food availability of the mahogany glider *Petaurus gracilis* (Petauridae: Marsupialia). *Journal of Zoology* **253**, 1–13[London]. doi:10.1017/S0952836901000012

Jackson SM, Groves C (2015) *Taxonomy of Australian Mammals*. CSIRO Publishing, Melbourne.

Jackson SM, Worthington-Wilmer J (1994) New methods in the management of the ghost bat in captivity. ARAZPA/ASZK Conference Proceedings: Notogea, Our World. pp. 188–195. Darwin, Northern Territory.

Jenness R, Sloan RE (1970) The composition of milks of various species: a review. *Dairy Science Abstracts* **32**, 599–612.

Jolly S (1988a) Five colonies of the orange horseshoe bat, *Rhinonycteris aurantius* (Chiroptera: Hipposideridae) in the Northern Territory. *Australian Wildlife Research* **15**, 41–49. doi:10.1071/WR9880041

Jolly S (1988b) Vaginal smears and the reproductive cycle of the common sheath-tail bat, *Taphozous georgianus* (Chiroptera: Emballonuridae). *Australian Mammalogy* **11**, 75–76. doi:10.1071/AM88010

Jolly S (1990) The biology of the common sheath-tail bat, *Taphozous georgianus* (Chiroptera: Emballonuridae), in central Queensland. *Australian Journal of Zoology* **38**, 65–77. doi:10.1071/ZO9900065

Jones C, McShea WJ, Conroy MJ, Kunz TH (1996) Capturing mammals. In *Measuring and Monitoring Biological Diversity: Standard Methods for Mammals*. (Eds DE Wilson, ER Cole, JD Nichols, R Rudran and MS Foster) pp. 115–155. Smithsonian Institution Press, Washington, DC.

Keen R, Hitchcock HB (1980) Survival and longevity of the little brown bat (*Myotis lucifugus*) in southeastern Ontario. *Journal of Mammalogy* **61**, 1–7. doi:10.2307/1379951

Kelly A, Goodwin S, Grogan A, Mathews F (2008) Post-release survival of hand-reared pipistrelle bats (*Pipistrellus* spp). *Animal Welfare (South Mimms, England)* **17**, 375–382. doi:10.1017/S0962728600027871

Kerth B, König B (1996) Transponder and an infrared-video camera as methods used in a field study on the special behaviour of Bechstein's bats (*Myotis bechsteini*). *Myotis* **34**, 27–34.

Kim S, Stephen A (2018) Extreme heat wipes out almost one third of Australia's spectacled flying fox population. ABC News 19 December 2018. https://www.abc.net.au/news/2018-12-19/heat-wipes-out-one-third-of-flying-fox-species/10632940 (accessed 15 December 2019).

Kitchener DJ (1975) Reproduction in female Gould's wattled bat, *Chalinolobus gouldii* (Gray) (Vespertilionidae), in Western Australian. *Australian Journal of Zoology* **23**, 29–42. doi:10.1071/ZO9750029

Kitchener DJ (1976) Further observations on reproduction in the common sheath-tail bat, *Taphozous georgianus* Thomas, 1915 in Western Australian, with notes on the gular pouch. *Records of the Western Australian Museum* **4**, 335–347.

Kitchener DJ, Coster P (1981) Reproduction in female *Chalinolobus morio* (Gray) (Vespertilionidae), in south-western Australian. *Australian Journal of Zoology* **29**, 305–320. doi:10.1071/ZO9810305

Kitchener DJ, Halse SA (1978) Reproduction in female *Eptesicus regulus* (Thomas) (Vespertilionidae), in south-western Australian. *Australian Journal of Zoology* **26**, 257–267. doi:10.1071/ZO9780257

Kitchener DJ, Hudson CJ (1982) Reproduction in the female white-striped mastiff bat, *Tadarida australis* (Gray) (Molossidae). *Australian Journal of Zoology* **30**, 1–14. doi:10.1071/ZO9820001

Krutzsch PH, Crichton EG (1987) Reproductive biology of the male little mastiff bat *Mormopterus planiceps* (Chiroptera: Molossidae) in southeast Australia. *The American Journal of Anatomy* **178**, 352–368. doi:10.1002/aja.1001780407

Kunz TH (1982) Roosting ecology of bats. In *Ecology of Bats*. (Ed. TH Kunz) pp. 1–55. Plenum Press, New York.

Kunz TH (2001) Seeing in the dark: recent technological advances for the study of free-ranging bats. Abstract. In 12th International Bat Research Conference. Hotel Equatorial, Bangi, Selangor, Malaysia. (Eds BMM Zain and N Ahmad) Faculty of Science and Technology Universiti Kebangsaan Malaysia, Bangi, Selangor.

Kunz TH, Kurta A (1988) Capture methods and holding devices. In *Ecological and Behavioural Methods for the Study of Bats*. (Ed. TH Kunz) pp. 1–30. Smithsonian Institution Press, Washington, DC.

Kunz TH, Stack MH, Jenness R (1983) A comparison of milk composition in *Myotis lucifugus* and *Eptesicus fuscus* (Chiroptera: Vespertilionidae). *Biology of Reproduction* **28**, 229–234. doi:10.1095/biolreprod28.1.229

Kunz TH, Oftedal OT, Robson SK, Kretzmann MB, Kirk C (1995) Changes in milk composition during lactation in three species of insectivorous bats. *Journal of Comparative Physiology. B, Biochemical, Systemic, and Environmental Physiology* **164**, 543–551. doi:10.1007/BF00261395

Lacy RC, Balou JD, Pollak JP (2011) PMx: software package for demographic and genetic analysis and management of pedigreed populations. *Methods in Ecology and Evolution* **3**, 433–437. doi:10.1111/j.2041-210X.2011.00148.x

Law BS (1993) Roosting and foraging ecology of the Queensland blossom-bat (*Syconycteris australis*) in north-eastern New South Wales: flexibility in response to seasonal variation. *Wildlife Research* **20**, 419–431. doi:10.1071/WR9930419

Law BS (1996) Residency and site fidelity of marked populations of the common blossom bat *Syconycteris australis* in relation to the availability of banksia inflorescences in New South Wales, Australia. *Oikos* **77**, 447–458. doi:10.2307/3545934

Law BS, Chidel M (2004) Roosting and foraging ecology of the golden-tipped bat (*Kerivoula papuensis*) on the south coast of New South Wales. *Wildlife Research* **31**, 73–82. doi:10.1071/WR03001

Law BS, Chidel M, Law PR (2018) Forest bat population dynamics over 14 years at a climate refuge: effects of timber harvesting and weather extremes. *PLoS One* **13**(2), e0191471. doi:10.1371/journal.pone.0191471

Law B, Chidel M, Law PR (2020) Multi-year population dynamics of a specialist trawling bat at streams with contrasting disturbance. *Journal of Mammalogy* **101**(2), 433–447. doi:10.1093/jmammal/gyz210

Law B, Brassil T, Chidel M (2023) Site fidelity and other attributes of infrequently trapped bats over two decades in a montane wet sclerophyll forest. *Australian Mammalogy* **45**, 91–97. doi:10.1071/AM22019

Laws RM (1952) A new method of age determination for mammals. *Nature* **169**, 972–973. doi:10.1038/169972b0

LeBlanc D (1998) Horticultural options for flying-foxes enrichment. In Proceedings of the AZH National Conference. pp. 205–211. Indianapolis, Indiana.

LeBlanc D (2001) A browse tube for plant-visiting bats. *The Shape of Enrichment* **10**(3), 10.

Lees C, Johnson K (2002) *Australasian Species Management Program: Regional Census and Plan*. 12th edn. Australasian Regional Association of Zoological Parks and Aquaria, Sydney.

Lollar A (2018) *The Rehabilitation and Captive Care of Insectivorous Bats*. A Bat World Sanctuary Publication, Weatherford, TX.

Lollar A, Schmidt-French B (1998) *Captive Care and Medical Reference for the Rehabilitation of Insectivorous Bats*. Bat World, TX.

Lumsden L (1995) Insectivorous bats. *Wildcry: Newsletter of the Wildlife Care Network* Autumn/Winter **1**, 3–4.

Lumsden L (2012) Farewell to 'George': a 23 year old eastern freetail bat. *The Australasian Bat Society Newsletter* **39**, 38.

Lumsden L, Bennett AF (1995) Bats of a semi-arid environment in south-eastern Australia: biogeography, ecology and conservation. *Wildlife Research* **22**, 217–240. doi:10.1071/WR9950217

Lumsden L, Gray P (2001) Longevity record for a southern bent-wing bat *Miniopterus schreibersii bassanii*. *The Australasian Bat Society Newsletter* **16**, 43–44.

Lumsden L, Nelson J, Lindeman M (2008) Ecological research on the Eastern Long-eared Bat *Nyctophilus timoriensis* (south-eastern form): a report to the Mallee Catchment Management Authority. Arthur Rylah Institute for Environmental Research, Department of Sustainability and Environment, Heidelberg, Vic.

Lumsden LF, Griffiths SR, Silins JE, Bennett AF (2020) Roosting behaviour and the tree-hole requirements of bats: insights from the lesser long-eared bat *Nyctophilus geoffroyi* and Gould's wattled bat *Chalinolobus gouldii* in south-eastern Australia. *Australian Journal of Zoology* **68**, 296–306. doi:10.1071/ZO20072

Lumsden LF, Morrant DS, Broken-Brow J, Holz PH, McKeown A, Gill SA, Vanderduys E (2022) Bats. In *Wildlife Research in Australia: Practical and Applied Methods*. (Eds BP Smith, H

Waudby, CA Alberthsen and JO Hampton) pp. 437–449. CSIRO Publishing, Melbourne.

Lyons R (2020) *Introduction to the Care and Rehabilitation of Microbats (Focussing on Species of South East Queensland).* Version 6.0. Wildcare Australia, Nerang, Queensland.

Lyons R, Wimberley T (2014) *Introduction to the Care and Rehabilitation of Microbats: Focussing on Species of South East Queensland.* Wildcare Australia, Nerang, Queensland.

Lyssavirus Expert Group (1996) Prevention of human Lyssavirus infection. *Communicable Diseases Intelligence* **20**, 505–507.

Madani G, Turbill C, Law B (2023) The golden tip for better breeding: evidence for polyoestry in the golden-tipped bat (*Phoniscus papuensis*). *Australian Mammalogy* doi:10.1071/AM22032.

Martin L, Towers PA, McGuckin MA, Little L, Luckhoff H, Blackshaw AW (1987) Reproductive biology of flying-foxes (Chiroptera: Pteropodidae). *Australian Mammalogy* **10**, 115–118. doi:10.1071/AM87024

Martin L, Kennedy JH, Little L, Luckhoff HC, O'Brien GM, Pow CST, Towers PA, Waldon AK, Wang DY (1995) The reproductive biology of Australian flying-foxes (genus *Pteropus*). *Symposia of the Zoological Society of London* **67**, 167–184. doi:10.1093/oso/9780198549451.003.0011

McCall BJ, Epstein JH, Neill AS, Heel K, Field HE, Barrett J, Smith GA, Selvey LA, Rodwell B, Lunt R (2000) Potential human exposure to Australian bat lyssavirus: Brisbane South and South Coast, Queensland, 1996–999. *Emerging Infectious Diseases* **6**, 259–264. doi:10.3201/eid0603.000305

McGuckin MA, Blackshaw AW (1992) Effects of photoperiod on the reproductive physiology of male flying foxes, *Pteropus poliocephalus. Reproduction, Fertility and Development* **4**, 43–53. doi:10.1071/RD9920043

McKean JL, Hamilton-Smith E (1967) Litter size and maternity sites in Australian bats (Chiroptera). *Victorian Naturalist* **84**, 203–206.

Mclean J (2000) Rearing and release of spectacled flying-foxes at Tolga in far north Queensland, Australia. *The Australasian Bat Society Newsletter* **15**, 43–46.

Mclean J, Johnson A, Woods D, Muller R, Blair D, Buettner PG (2018) Growth rates of, and milk feeding schedules for, juvenile spectacled flying-foxes (*Pteropus conspicillatus*) reared for release at a rehabilitation centre in north Queensland, Australia. *Australian Journal of Zoology* **66**, 201–213. doi:10.1071/ZO18076

McLelland DJ, Reardon T, Bourne S, Dickason C, Kessell A, Boardman W (2013) Outbreak of skin nodules associated with *Riouxgolvania beveridgei* (Nematoda: Muspiceida) in the southern bentwing bat (*Miniopterus schreibersii bassanii*), South Australia. *Journal of Wildlife Diseases* **49**, 1009–1013. doi:10.7589/2012-11-288

McLeod DSA, Mortimer RH, Perry-Keene DA, Allworth A, Woods ML, Perry-Keene J, McBride WJH, Coulter C, Robson JMB (2011) Histoplasmosis in Australia: report of 16 cases and literature review. *Medicine* **90**, 61–68. doi:10.1097/MD.0b013e318206e499

Menkhorst PW (Ed.) (1995) *Mammals of Victoria: Distribution, Ecology and Conservation.* Oxford University Press, Sydney.

Messer M, Parry-Jones K (1997) Milk composition in the grey headed flying fox, *Pteropus poliocephalus* (Pteropodidae: Chiroptera). *Australian Journal of Zoology* **45**, 65–73. doi:10.1071/ZO96052

Mo M, Roache M (2020) A review of intervention methods used to reduce flying-fox mortalities in heat stress events. *Australian Mammalogy* **43**, 137–150. doi:10.1071/AM20038

Mo M, Cross S, Boyd K (2023) Post-release survivorship of 18 years in a hand-reared grey-headed flying-fox (*Pteropus poliocephalus*) revealed by a metal identification band. *Australian Mammalogy* **45**, 241–245. doi:10.1071/AM22021

Mounter B (2019) Amended threatened species list set to grow as Wet Tropics suffers through extreme heat events. ABC News. 23 February 2019. https://www.abc.net.au/news/2019-02-23/flying-fox-decline-signals-dire-warning-for-health-of-tropics/10838580 (accessed 26 February 2019).

Murray K, Selleck P, Hooper P, Hyatt A, Gould A, Gleeson L, Westerbury H, Hiley L, Selvey L, Rodwell B, Ketterer P (1995) A morbilivirus that caused fatal disease in horses and humans. *Science* **268**, 94–97. doi:10.1126/science.7701348

Negraeff OE, Brigham RM (1995) The influences of moonlight on the activity of little brown bats (*Myotis lucifugus*). *Zeitschrift für Säugetierkunde* **60**, 330–336.

Nelson JE (1964) Vocal communications in Australian flying-foxes (Pteropodidae; Megachiroptera). *Zeitschrift für Tierpsychologie* **21**, 857–870. doi:10.1111/j.1439-0310.1964.tb01224.x

Nelson JE (1965a) Behaviour of Australian Pteropodidae (Megachiroptera). *Animal Behaviour* **13**, 544–557. doi:10.1016/0003-3472(65)90118-1

Nelson JE (1965b) Movements of Australian flying-foxes (Pteropodidae: Megachiroptera). *Australian Journal of Zoology* **13**, 53–73. doi:10.1071/ZO9650053

Nelson JE (1989) Megadermatidae. In *Fauna of Australia.* Vol. 1B. (Eds D Walton and BJ Richardson) pp. 852–856. Australian Government Publishing Service, Canberra.

O'Brien GM (1993) Seasonal reproduction in flying-foxes, reviewed in the context of other tropical mammals. *Reproduction, Fertility and Development* **5**, 499–521. doi:10.1071/RD9930499

O'Sullivan MVN, Whitby M, Chahoud C, Miller SM (2004) Histoplasmosis in Australia: a report of a case with a review of the literature. *Australian Dental Journal* **49**(2), 94–97. doi:10.1111/j.1834-7819.2004.tb00057.x

Olsson A, Woods R (2008) Bats. In *Medicine of Australian Mammals.* (Eds L Vogelnest and R Woods) pp. 465–502. CSIRO Publishing, Melbourne.

Paradiso JL, Greenhall AM (1967) Longevity records for American bats. *American Midland Naturalist* **78**, 251–252. doi:10.2307/2423387

Parnaby HE (1999) An interim guide to identification of insectivorous bats of southeastern Australia. *Technical Reports of the Australian Museum* **8**, 1–33[2nd reprint 1999]. doi:10.3853/j.1031-8062.8.1999.1528

Parry-Jones G (2000) *The Grey-headed Flying Fox: Care & Rescue.* Wambina Flying Fox Research Centre, Gosford, NSW.

Parry-Jones K (2018) Report on the heat proofing of colony sites: The Central Coast/Singleton experience. Wildlife Animal Rescue and Care Society Inc., Gosford, NSW.

Pettigrew J, Baker GB, Baker-Gabb D, Baverstock G, Coles R, Conole L, Churchill S, Fitzherbert K, Guppy A, Hall L, Helman P, Nelson J, Priddel D, Pulsford I, Richards G, Schulz M, Tidemann CR (1986) The Australian ghost bat, *Macroderma gigas*, at Pine Creek, Northern Territory. *Macroderma* **2**, 8–19.

Philbey AW, Kirkland PD, Ross AD, Davis RJ, Gleeson AB, Love RJ, Daniels PW, Gould AR, Hyatt AD (1998) An apparently new virus (Family Paramyxoviridae) infectious for pigs, humans and fruit bats. *Emerging Infectious Diseases* **4**, 269–271. doi:10.3201/eid0402.980214

Phillips CJ, Steinberg B, Kunz TH (1982) Dentin, cementum, and age determination in bats: a critical evaluation. *Journal of Mammalogy* **63**, 197–207. doi:10.2307/1380628

Phillips WR, Inwards SJ (1985) The annual activity and breeding cycles of Gould's long-eared bat *Nyctophilus gouldi* (Microchiroptera: Vespertilionidae). *Australian Journal of Zoology* **33**, 111–126. doi:10.1071/ZO9850111

Pinson D (2020) *The Flying-Fox Manual: A New Handbook for Wildlife Carers Involved in the Care of Orphaned Baby, and the Rescue, Rehabilitation & Release of Adult Flying-Foxes in Australia*. StickeeBatz Publishing, Cow Bay, Queensland.

Podlutsky AJ, Khritankov AM, Ovodov ND, Austad SN (2005) A new field record for bat longevity. *Journal of Gerontology* **60A**, 1366–1368.

Power ML, Foley NM, Jones G, Teeling EC (2022) Taking flight: an ecological, evolutionary and genomic perspective on bat telomeres. *Molecular Ecology* **31**, 6053–6068. doi:10.1111/mec.16117

Prada D, Boyd V, Baker M, Jackson B, O'Dea M (2019) Insights into Australian Bat Lyssavirus in insectivorous bats of Western Australia. *Tropical Medicine and Infectious Disease* **4**, 46. doi:10.3390/tropicalmed4010046

Prociv P (1990) Aberrant migration by *Toxocara pteropodis* in flying foxes: two case reports. *Journal of Wildlife Diseases* **26**, 532–534. doi:10.7589/0090-3558-26.4.532

Racey PA (1988) Reproductive assessment in bats. In *Ecological and Behavioural Methods for the Study of Bats*. (Ed. TH Kunz) pp. 31–45. Smithsonian Institution Press, Washington, DC.

Racey PA, Kleiman DG (1970) Maintenance and breeding in captivity of some vespertilionid bats, with special reference to the noctule. *International Zoo Yearbook* **10**, 65–70. doi:10.1111/j.1748-1090.1970.tb01280.x

Rasweiler JJ (1977) The care and management of bats as laboratory animals. In *Biology of Bats*. (Ed. WA Wimsatt) pp. 519–617. Academic Press, New York.

Ratcliffe FN (1931) The flying-fox (*Pteropus*) in Australia. *Bulletin of the Council of Science Industry and Research in Australia* **53**, 1–80.

Ratcliffe FN (1932) Notes on the flying-foxes (*Pteropus* spp.) of Australia. *Journal of Animal Ecology* **1**, 32–57. doi:10.2307/993

Ratnayake HU, Kearney MR, Govekar P, Karoly D, Welbergen JA (2019) Forecasting wildlife die-offs from extreme heat events. *Animal Conservation* **22**, 386–395. doi:10.1111/acv.12476

Reardon T, Flavel S (1987) *A Guide to the Bats of South Australia*. South Australian Museum and Field Naturalists Society, Adelaide.

Reddacliff LA, Bellamy TA, Hartley WJ (1999) *Angiostrongylus cantonensis* infection in grey-headed fruit bats (*Pteropus poliocephalus*). *Australian Veterinary Journal* **77**, 466–468. doi:10.1111/j.1751-0813.1999.tb12095.x

Reith CC (1982) Insectivorous bats fly in shadows to avoid moonlight. *Journal of Mammalogy* **63**, 685–688. doi:10.2307/1380284

Rhodes M (2007) Roost fidelity and fission-fusion dynamics of white-striped free-tailed bat (*Tadarida australis*). *Journal of Mammalogy* **88**, 1252–1260. doi:10.1644/06-MAMM-A-374R1.1

Richards GC (1989) Nocturnal activity of insectivorous bats relative to temperature and prey availability in tropical Queensland. *Australian Wildlife Research* **16**, 151–158. doi:10.1071/WR9890151

Richards G (2000) Natural wing tears in large bentwing bats (*Miniopterus schreibersii*). *The Australasian Bat Society Newsletter* **15**, 48.

Rose K (1999) Common diseases of urban wildlife. In *Wildlife in Australia: Healthcare & Management*. (Ed. DI Bryden) pp. 365–427. Post Graduate Foundation in Veterinary Science. University of Sydney, Sydney.

Ross T, Philbey A, Kirklan P, Field H (2001) Megachiroptera and Menagle Virus in Australia. In Proceedings of International Joint Conference. Veterinary Conservation Biology Wildlife Health and Management in AUSTRALASIA. (Eds A Martin and L Vogelnest) pp. 241–244. Taronga Zoo, Sydney. 1–6 July 2001.

Rudran R, Kunz TH (1996) Methods for marking mammals. In *Measuring and Monitoring Biological Diversity: Standard Methods for Mammals*. (Eds DE Wilson, ER Cole, JD Nichols, R Rudran and MS Foster) pp. 299–310. Smithsonian Institution Press, Washington, DC.

Rupprecht CE (1999) Rabies: Global problems, zoonotic threat and preventative management. In *Zoo & Wild Animal Medicine*. Current Therapy 4. (Ed. ME Fowler) pp. 136–146. Saunders, Philadelphia.

Russell R (1981) How fluffy gliders led me to the tree of life. *Habitat* **9**, 7–8.

Ryan MR (1966) Observations on the broad-nosed bat, *Scoteinus balstoni*, in Victoria. *Journal of Zoology* **148**, 162–166[London]. doi:10.1111/j.1469-7998.1966.tb02945.x

Sangster CR, Gordon AN, Hayes D (2012) Systemic toxoplasmosis in captive flying-foxes. *Australian Veterinary Journal* **90**, 140–142. doi:10.1111/j.1751-0813.2011.00868.x

Schowalter DB, Harder LD, Treichel BH (1978) Age composition of some vespertilionid bats as determined by dental annuli. *Canadian Journal of Zoology* **56**, 355–358. doi:10.1139/z78-049

Selvey L, Taylor R, Arklay A, Gerrard J (1996) Screening of bat carers for antibodies to equine morbilivirus. *Communicable Diseases Intelligence* **20**, 477–478.

Skerratt LF, Speare R, Berger L, Winsor H (1998) Lyssaviral infection and lead poisoning in black flying foxes from Queensland. *Journal of Wildlife Diseases* **34**, 355–361. doi:10.7589/0090-3558-34.2.355

Sluiter JW, Van Heerdt PF, Bezem JJ (1971) Parametres de population chez le grand Rhinolophe fer-a-cheval (*Rhinolophus ferrumequinum* Schreber), estimes par la methode de reprises apres baguages. *Mammalia* **35**, 254–272. doi:10.1515/mamm.1971.35.2.254

Smith HM, Neaves LE, Divljan A (2020) Predation on cicadas by an Australian flying-fox *Pteropus poliocephalus* based on DNA evidence. *Australian Zoologist* **40**, 515–528. doi:10.7882/AZ.2018.029

Smythe LD, Field HE, Barnett LJ, Smith CS, Dohnt MF, Symonds ML, Moore MR, Rolfe PF (2002) Leptospiral antibodies in flying-foxes in Australia. *Journal of Wildlife Diseases* **38**, 182–186. doi:10.7589/0090-3558-38.1.182

Snoyman S, Munich J, Brown C (2012) Nursing females are more prone to heat stress: demography matters when managing flying-foxes for climate change. *Applied Animal Behaviour Science* **142**, 90–97. doi:10.1016/j.applanim.2012.09.003

Sommers LA, Davis WH, Hitchcock HB (1993) Longevity records for *Myotis lucifugus*. *Bat Research News* **34**, 3.

Species360 (2023) Zoological Information Management System (ZIMS). Minneapolis, USA. www.species360.org.

Stawski C (2010) Torpor during the reproductive season in a free-ranging subtropical bat, *Nyctophilus bifax*. *Journal of Thermal Biology* **35**, 245–249. doi:10.1016/j.jtherbio.2010.05.009

Stawski C, Geiser F (2009) Fat and fed: frequent use of summer torpor in a subtropical bat. *Naturwissenschaften* **97**, 29–35. doi:10.1007/s00114-009-0606-x

Stawski C, Turbill C, Geiser F (2008) Prolonged torpor use during winter by a free-ranging bat in subtropical Australia. In *Hypermetabolism in Animals: Hibernation, Torpor and Cryobiology*. (Eds BG Lovegrove and AE McKechnie) pp. 353–360. University of KwaZulu-Natal, Pietermaritzburg, South Africa.

Stawski C, Turbill C, Geiser F (2009) Hibernation by a free-ranging subtropical bat (*Nyctophilus bifax*). *Journal of Comparative Physiology. B, Biochemical, Systemic, and Environmental Physiology* **179**, 433–441. doi:10.1007/s00360-008-0328-y

Stebbings RE (1977) Order Chiroptera. In *Handbook of British Mammals*. (Eds GB Corbet and HN Southern) pp. 68–128. Blackwell Scientific, Oxford.

Stebbings RE, Walsh ST (1991) *Bat Boxes: A Guide to the History, Function, Construction and Use in the Conservation of Bats*. The Bat Conservation Trust, London.

Strahan R (1995) *The Mammals of Australia*. 2nd edn. Reed Books, Sydney.

Taylor H, Gould E, Frank A, Woolf N (1974) Successful hand-raising of one week old bats, *Eptesicus* and *Antrozous*, by stomach catheter. *Journal of Mammalogy* **55**, 228–231. doi:10.2307/1379279

Taylor RJ, O'Neill MG (1988) Summer activity patterns of insectivorous bats and their prey in Tasmania. *Australian Wildlife Research* **15**, 533–539. doi:10.1071/WR9880533

Temby I (1995) Bats in chains come in to roost. *Australasian Bat Society Newsletter* **4**, 35–39.

Thompson CK, Godfrey SS, Thompson RCA (2014) Trypanosomes of Australian mammals: a review. *International Journal for Parasitology. Parasites and Wildlife* **3**, 57–66. doi:10.1016/j.ijppaw.2014.02.002

Thompson GK (1999) Veterinary surgeon's guide to Australian Bat Lyssavirus. *Australian Veterinary Journal* **77**, 710–712. doi:10.1111/j.1751-0813.1999.tb12904.x

Thomson BG (1991) *A Field Guide to the Bats of the Northern Territory*. Conservation Commission of the Northern Territory, Darwin.

Tidemann CR (1982) Sex differences in seasonal changes of brown adipose tissue and activity of the Australian vespertilionid bat *Eptesicus vulturnus*. *Australian Journal of Zoology* **30**, 15–22. doi:10.1071/ZO9820015

Tidemann CR (1993) Reproduction in the bats, *V. vulturnus*, *V. regulus* and *V. darlingtoni* (Microchiroptera: Vespertilionidae) in coastal south-eastern Australia. *Australian Journal of Zoology* **41**, 21–35. doi:10.1071/ZO9930021

Tidemann CR, Flavel SC (1987) Factors affecting choice of diurnal roost site by tree hole bats (Microchiroptera) in south-eastern Australia. *Australian Wildlife Research* **14**, 459–473. doi:10.1071/WR9870459

Tidemann CR, Loughland RA (1993) A harp trap for large megachiropterans. *Wildlife Research* **20**, 607–611. doi:10.1071/WR9930607

Tidemann CR, Nelson JE (2011) Life expectancy, causes of death and movements of the grey-headed flying-fox (*Pteropus poliocephalus*) inferred from banding. *Acta Chiropterologica* **13**, 419–429. doi:10.3161/150811011X624901

Tidemann CR, Woodside DP (1978) A collapsible bat-trap and a comparison of results obtained with the trap and with mist nets. *Australian Wildlife Research* **5**, 355–362. doi:10.1071/WR9780355

Tidemann CR, Vardon MJ, Nelson JE, Speare R, Gleeson LJ (1997) Health and conservation implications of Australian bat *Lyssavirus*. *Australian Zoologist* **30**, 369–376. doi:10.7882/AZ.1997.015

Tidemann CR, Vardon MJ, Loughland RA, Brocklehurst PJ (1999) Dry season camps of flying-foxes (*Pteropus* spp.) in Kakadu World Heritage Area, north Australia. *Journal of Zoology* **247**, 155–163[London]. doi:10.1111/j.1469-7998.1999.tb00979.x

Tolga Bat Hospital (n.d.) Rearing. https://tolgabathospital.org/rearing/ (accessed 25 July 2021).

Toop J (1985) Habitat requirements, survival strategies and ecology of the ghost bat *Macroderma gigas* Dobson, (Microchiroptera, Megadermatidae) in central coastal Queensland. *Macroderma* **1**, 37–41.

Turbill C, Geiser F (2008) Hibernation by tree-roosting bats. *Journal of Comparative Physiology. B, Biochemical, Systemic, and Environmental Physiology* **178**, 597–605. doi:10.1007/s00360-007-0249-1

Turbill C, Welbergen J (2020) Australia's threatened bats need protection from a silent killer: white-nose syndrome. *The Conversation* 20 January 2020. https://theconversation.com/australias-threatened-bats-need-protection-from-a-silent-killer-white-nose-syndrome-129186 (accessed 21 January 2020).

Turbill C, Law BS, Geiser F (2003a) Summer torpor in a free-ranging bat from subtropical Australia. *Journal of Thermal Biology* **28**, 223–226. doi:10.1016/S0306-4565(02)00067-0

Turbill C, Körtner G, Geiser F (2003b) Natural use of torpor by a small, tree-roosting bat during summer. *Physiological and Biochemical Zoology* **76**, 868–876. doi:10.1086/378915

Turbill C, Walker M, Boardman W, Martin JM, McKeown A, Meade J, Welbergen JA (2024) Torpor use in the wild by one of the world's largest bats. *Proceedings of the Royal Society B. Biological Sciences* **291**, 20241137. doi:10.1098/rspb.2024.1137

Tuttle MD, Stevenson D (1982) Growth and survival in bats. In *Ecology of Bats*. (Ed. TH Kunz) pp. 105–150. Plenum Press, New York.

Twente JW, Jr (1955) Aspects of a population of cavern-dwelling bats. *Journal of Mammalogy* **36**, 379–390. doi:10.2307/1375680

Van der Merwe M (1979) Foetal growth curves and seasonal breeding in Natal clinging bat *Miniopterus schreibersii natalensis*. *South African Journal of Zoology* **14**, 17–21. doi:10.1080/02541858.1979.11447643

Van Dyck S (1982) Management of captive Megachiroptera. In *The Management of Australian Mammals in Captivity*. (Ed. DD Evans) pp. 161–163. Zoological Board of Victoria, Melbourne.

van Harten E, Reardon T, Lumsden LF, Meyers N, Prowse TAA, Weyland J, Lawrence R (2019) High detectability with low impact: optimizing large PIT tracking systems for cave-dwelling bats. *Ecology and Evolution* **9**, 10916–10928. doi:10.1002/ece3.5482 van Harten E, Reardon T, Holz PH, Lawrence R, Prowse TAA, Lumsden LF (2020) Recovery of southern bent-winged bats (*Miniopterus orianae bassanii*) after PIT-tagging and the use of surgical adhesive. *Australian Mammalogy* **42**, 216–219. doi:10.1071/AM19024

van Harten E, Lentini PE, Eastick DL, Bender R, Lumsden LF, Visintin C, Griffiths SR (2021) Low rates of PIT-tag loss in an insectivorous bat species. *The Journal of Wildlife Management* **85**, 1739–1743. doi:10.1002/jwmg.22114

Vardon MJ, Tidemann CR (1998) Reproduction, growth and maturity in the black flying-fox, *Pteropus alecto* (Megachiroptera: Pteropodidae). *Australian Journal of Zoology* **46**, 329–344. doi:10.1071/ZO98023

Vardon MJ, Tidemann CR (2000) The black flying-fox (*Pteropus alecto*) in north Australia: juvenile mortality and longevity. *Australian Journal of Zoology* **48**, 91–97. doi:10.1071/ZO99060

Vogelnest L (2025) Capture, physical and chemical restraint. In *Current Therapy in Medicine of Australian Mammals*. Revised Edition. (Eds L Vogelnest and T Portas) pp. 122–151. CSIRO Publishing, Melbourne.

Webb N, Tidemann C (1995) Hybridisation between black (*Pteropus alecto*) and grey-headed (*P. poliocephalus*) flying-foxes (Megachiroptera: Pteropodidae). *Australian Mammalogy* **18**, 19–26. doi:10.1071/AM95019

Welbergen JA (2010) Growth, bimaturation and sexual size dimorphism in wild gray-headed flying foxes (*Pteropus poliocephalus*). *Journal of Mammalogy* **91**, 38–47. doi:10.1644/09-MAMM-A-157R.1.

Welbergen JA, Klose SM, Markus N, Eby P (2008) Climate change and the effects of temperature extremes on Australian flying-foxes. *Proceedings. Biological Sciences* **275**, 419–425. doi:10.1098/rspb.2007.1385

Wildlife Health Australia (2020a) Personal protective equipment (PPE) information for bat handlers. https://wildlifehealthaustralia.com.au/Portals/0/Documents/ProgramProjects/PPE_Info_for_Bat_Handlers.pdf (accessed 11 April 2023).

Wildlife Health Australia (2020b) Australian Bat Lyssavirus. https://wildlifehealthaustralia.com.au/Portals/0/Documents/FactSheets/mammals/Australian_Bat_Lyssavirus.pdf (accessed 11 April 2023).

Williams CB, Singh BP, El Ziady S (1956) An investigation into the possible effects of moonlight on the activity of insects in the field. *Proceedings of the Royal Entomological Society of London A* **31**, 135–144. doi:10.1111/j.1365-3032.1956.tb00197.x

Williamson MM, Hooper PT, Selleck PW, Gleeson LJ, Daniels PW, Westbury HA, Murray PK (1998) Transmission studies of Hendra virus (equine morbillivirus) in flying-foxes, horses and cats. *Australian Veterinary Journal* **76**, 813–818. doi:10.1111/j.1751-0813.1998.tb12335.x

Wilson DE (1988) Maintaining bats for captive studies. In *Ecological and Behavioural Methods for the Study of Bats*. (Ed. TH Kunz) pp. 247–264. Smithsonian Institution Press, Washington, DC.

Wilson DE, Mittermeier RA (2019) *Handbook of the Mammals of the World*. Vol. 9. *Bats*. Lynx Edicions, Barcelona.

Wilson P (1988) Veterinary treatment of bats. In *Australian Wildlife*. Post Graduate Committee in Veterinary Science. (Ed. DI Bryden) pp. 517–529. University of Sydney, Sydney.

Woinarski JCZ, Burbidge AA, Harrison P (eds) (2014) *The Action Plan for Australian Mammals 2012*. CSIRO Publishing, Melbourne.

Young RA (1975) Aging criteria, pelage colour polymorphism and moulting in *Rhinolophus megaphyllus* (Chiroptera) from south-eastern Queensland, Australia. *Mammalia* **39**, 75–111. doi:10.1515/mamm.1975.39.1.75

Young RA (1979) Observations on parturition, litter size and foetal development at birth in the chocolate wattled bat, *Chalinolobus morio* (Vespertilionidae). *Victorian Naturalist* **96**, 90–91.

Young PL (1996) Possible reservoir host of equine morbilivirus identified. *Communicable Diseases Intelligence* **20**, 262.

Young PL, Halpin K, Selleck PW, Field H, Gravel JL, *et al.* (1996) Seriological evidence for the presence in *Pteropus* bats of a paramyxovirus related to equine morbilivirus. *Emerging Infectious Diseases* **2**, 239–240. doi:10.3201/eid0203.960315

Zambelli N, Moretti M, Mattei-Roesli M, Bontadina F (2009) Negative consequences of forearm bands that are too small for bats. *Acta Chiropterologica* **11**, 216–219. doi:10.3161/150811009X465857

# 14
# Dingo

*Stephen M. Jackson, Bradley P. Smith, Peter J. S. Fleming, Frances Hulst and Kylie M. Cairns*

## 1. INTRODUCTION

The dingo (*Canis familiaris*) arrived in Australia from south-east Asia at least 3500 years ago, likely via seafarers who may have used them as companions, for trade or as a source of food (Corbett 2001a; Cairns and Wilton 2016; Balme *et al.* 2018). Due to their recent arrival and shared evolutionary lineage with ancient dogs from Asia, some controversy exists as to whether they should be considered an ancient breed of dog or a native distinct species (Jackson *et al.* 2017, 2019, 2021; Smith *et al.* 2019; Cairns 2021). Advances in genetic study of dingoes have identified multiple ecotypes, or evolutionarily significant units (Cairns *et al.* 2017, 2018, 2020; Koungoulos 2020), and confirmed hybridisation of dingoes with modern dogs (Wilton *et al.* 1999; Stephens *et al.* 2015; Cairns *et al.* 2020, 2021a).

Dingoes continue to live in the wild without direct association with people. However, First Nations Australians have had a long practical, cultural and spiritual relationship with dingoes, as tame individuals were used as companions, bedwarmers, guards, food and to assist in hunting (Tench, 1789; Lee 1925; Corbett 2001a; Smith and Litchfield 2009; Koungoulos and Fillios 2020). Such was the status of dingoes with First Nations Australian cultures, they were sometimes given individual names, groomed of parasites and treated with great affection (Lumholtz 1889; Berndt and Berndt 1942; Hamilton 1972). Although First Nations Australians raised and kept tamed dingoes as companions, their relationship with them might not have constituted intensive captivity.

Dingoes were kept in captivity by Europeans soon after the arrival of the First Fleet in Sydney in 1788 when Captain Phillip (1789) sent a female to the Marchioness of Salisbury at Hatfield-House in London. Early observations by Hunter (1793: pp. 56–57) suggested that 'although they are very good-natured animals when domesticated, they are impossible to cure of their 'savageness', which all I have seen seem to possess'. London Zoo has bred dingoes as early as 1830 and the Bronx Zoo has held several animals since 1933 (Zuckerman 1953; Crandall 1964). As early as 1932, Adelaide Zoo experimented with dingo–wolf hybrids (Lockyer 1932). In Australia, dingoes are widely held in zoos, sanctuaries and fauna parks across the country.

Captive dingoes can also be found across America, with several founding breeding populations exported from Australia in 2010. Dingoes are generally considered a popular exhibit, with many institutions routinely walking them around the paths for the public to interact with, and offering 'behind the scenes' encounters, as well as associated educational talks. Despite the success of dingoes as exhibits within zoos and fauna parks, many aspects of their behaviour and the relationship between humans and dingoes does not resemble that of a typical human–dog relationship (Smith 2014; Smith *et al.* 2017) and many dingoes struggle to adjust to being kept in private residences. Therefore, keeping dingoes as pets is not recommended (Oakman 2001; Smith 2015a).

There has been strong support to retain the purity of dingoes, with organisations such as the Australian Native Dog Conservation Society in Bargo, New South Wales (Walters 1995), and the Dingo Discovery Sanctuary and Research Centre/Australian Dingo Foundation in Victoria (Smith and Watson 2015; Smith *et al.* 2021) initiating careful captive-breeding programs with DNA-tested animals. Hybridisation with domestic dogs is considered a serious threat to the purity of dingoes in some wild populations (Stephens *et al.* 2015; Allen *et al.* 2017; Cairns *et al.* 2020, 2021a) and hybrids are disfavoured for keeping in captivity. Critical breakthroughs in DNA testing, beginning with Wilton (2001), now allow captive and wild dingo populations to be monitored for the presence and extent of modern dog ancestry so that informed captive breeding and management plans can be developed.

## 2. TAXONOMY

### 2.1. Nomenclature

There is debate about the taxonomic status and nomenclature of dingoes due to uncertainty about the distinctiveness of their ancestry and close relationship to domestic dogs. Some authors have argued that morphological and genetic data clearly show the dingo is an ancient breed of domestic dog (i.e. *Canis familiaris*) derived from grey wolves (*Canis lupus*) by a process of active or passive domestication or their combination (see Jackson *et al.* 2017, 2019, 2021). Others have argued that dingoes are a distinct species of proto-dog and not a domestic dog or breed of domestic dog (Smith *et al.* 2019; Cairns *et al.* 2020). However, most taxonomists agree that the scientific name of the dingo should be *Canis familiaris* (Jackson *et al.* 2017, 2019, 2021; AMTC 2024).

Class: Mammalia
Order: Carnivora
Family: Canidae
Genus and species: *Canis familiaris*

### 2.2 Subspecies

Although there are no recognised subspecies, there is evidence of multiple ecotypes, races or varieties of dingo based on phenotypes, genetic differences and skull morphology. The currently differentiated ecotypes include south-eastern (alpine) dingoes, north-western (desert) dingoes and K'Gari (Fraser Island) dingoes (Cairns *et al.* 2017, 2018; Ballard and Wilson 2019; Koungoulos 2020). Ancient DNA data confirms that this south-eastern versus north-western population structure in dingoes is prehistoric and has persisted for several thousand years (Souilmi *et al.* 2024).

### 2.3 Recent synonyms

The dingo has had many synonyms since its first description in 1792 as *Canis antarticus* by Kerr. It was subsequently described as *C. dingo* (Meyer 1793) and a series of other names (See Jackson and Groves 2015) until the earliest name *C. antarticus* was suppressed in 1956 and the more popular *C. dingo* was made available for usage (Jackson *et al.* 2017; Smith *et al.* 2019). In 1982, the name *C. lupus dingo* was often used to reflect its ancestry (Corbett 2006). Since then, the dingo has been referred to as *C. familiaris*, *C. familiaris dingo*, *C. lupus familiaris*, or *C. lupus dingo* (Kreplins *et al.* 2019) until the name *C. dingo* was revived in 2014 (Crowther *et al.* 2014) and used by some authors.

### 2.4 Other common names

The names 'Warrigal' and 'Australian native dog' occur commonly in books and geographically (Corbett 2001a; Breckwoldt 1988; Fleming *et al.* 2001). There are many regional names for the dingo in First Nations languages, such as Wongari on K'gari (Miller *et al.* 2024) and Willkurr in north-western Victoria. Including local names approved by local First Nations people could benefit visitors' experience of exhibited dingoes.

## 3. NATURAL HISTORY

### 3.1 Bodyweight and morphometrics

The dingo is a typical medium-sized canid in size, weighing 10–24 kg, measuring 44–63 cm at the shoulder, with a head and body length of 86–122 cm and a tail length of 26–38 cm (Corbett 1995; 2001a). Dingoes hit their mean bodyweight at 12 months of age, with males generally being larger than females (Corbett 1995; Smith 2015b; Behrendorff *et al.* 2016). Captive dingoes tend to be heavier and larger than their free-ranging counterparts (Smith 2015b).

The dingo's coat can vary from a wiry single coat in the tropical regions, to a thick double coat in the cooler regions. Although the typical colouration of the dingo is a light tan/ginger, various other colours exist including white, sandy yellow, red-ginger, sable, black and tan or black (Corbett 1995, 2001a; Elledge *et al.* 2008; Cairns *et al.* 2021b). Dingoes also typically (but do not always) have white markings on the feet, tail tip and chest, some have black muzzles and all have pricked ears and bushy tails (Corbett 1995). It is not uncommon in the wild to see dingoes with more extensive white markings such as a collar, Irish marking or spotting (white on chest, muzzle, feet and tail tip) or even piebald. Coat colour is not a reliable indicator of purity and atypical coat colourations are occasionally observed in pure dingoes (Cairns *et al.* 2021b).

Dingoes have many morphological differences that can be used to help distinguish them from modern dogs of similar size (Smith 2015b). The dingo's body is longer than its height; the ribcage is long and extends to the rear; the head is the widest part of its body (i.e. it has a narrow chest and does not extend below the elbow); the skull is high behind the ears, and the ears are positioned forward of the occiput; the eyes are slanted, almond-shaped and obliquely placed; and the tail is typically carried low, usually does not curl over the back, and a precaudal scent gland is positioned on the tail. There are several other morphological traits that show limited evidence of recent domestication. For example, dingoes have a longer muzzle, larger canine and carnassial teeth, and a flatter cranium with larger nuchal and sagittal crest than modern dogs (see Crowther *et al.* 2014; Smith *et al.* 2019 for reviews). Dingoes also have a larger brain (relative to body size) than modern dogs (Smith *et al.* 2018).

## 3.2 Distribution and habitat

Being generalists, dingoes are found in all habitat types, ranging from alpine, woodland, grassland, arid and tropical regions throughout most of the mainland of Australia (Allen and West 2013; Corbett 2001a; Fleming *et al.* 2014). They naturally occur on some large northern islands (e.g. K'Gari, Groote Eylandt and Melville and Bathurst islands), but are notably absent from Tasmania and the large southern islands (Flinders Island, King Island and Kangaroo Island).

## 3.3 Conservation status

Dingoes have some legislative protection in the Northern Territory and northern South Australia and the pure dingo is regarded as a threatened species in Victoria (Fleming *et al.* 2014). Elsewhere, they are either a declared pest or unprotected (e.g. New South Wales Fleming *et al.* 2021).

Although their range has decreased in southern Australia since 1788, dingoes are present throughout most of their current range on mainland Australia and several islands (Allen and West 2013; Allen *et al.* 2017). Their numbers have increased in some areas because of anthropogenic changes such as establishment of permanent water and reliable food associated with livestock production and mining activities (e.g. Newsome *et al.* 2013). Dingoes have long been considered a pest by many livestock producers because they can kill and injure cattle (*Bos taurus*), goats (*Capra hircus*), and sheep (*Ovis aries*) (e.g. Thomson 1984; Fleming *et al.* 2001) and most Australian legislation related to dingoes reflects this agricultural impact (e.g. Fleming *et al.* 2021). As a result, many are shot, trapped and poisoned and they have been extirpated from most major urban vicinities and their hinterlands, much of the sheep-wheat zone in south-east Australia and the agricultural zone of south-west Western Australia (Fleming *et al.* 2014). There are also private and government-built dingo-proof exclusion fences, such as the dingo barrier fence that runs from the Great Australian Bight in South Australia to south-eastern Queensland (a length of 5635 km, Breckwoldt 1988), and the vermin fences in Western Australia (Pacioni *et al.* 2020), which are designed to keep dingo numbers low in the areas protected by the fences.

In addition to population suppression by humans, another issue facing dingo conservation is hybridisation, either through cross-breeding with modern dogs or interbreeding among hybrids (Corbett 2001b). The extent of hybridisation, however, is currently being revised, with genome-wide DNA testing identifying less domestic dog ancestry in wild dingoes than suggested by earlier genetic surveys (Corbett 2001b; Stephens *et al.* 2015; Cairns *et al.* 2021a, 2023).

When the dingo's conservation status was recently reviewed against the Australian Governments *Environment Protection and Biodiversity Conservation Act 1999* and the *Environment Protection and Biodiversity Conservation Regulations 2000*, as well as criteria of the Union for the Conservation of Nature (IUCN), the dingo was not recognised as threatened in any Australian jurisdiction (Allen *et al.* 2017). However, if regional ecotypes and or isolated populations are considered separately and are at conservation risk, then they could meet the conditions of endangered, threatened or vulnerable populations (Cairns *et al.* 2018, 2021a).

## 3.4 Diet in the wild

Dingoes have a flexible, generalist diet that varies with prey availability and between different ecosystems across Australia, which is useful when selecting captive diets. Doherty *et al.* (2019) reviewed 73 published and unpublished datasets, revealing that at least 229 species of vertebrates were consumed by dingoes, including 66% mammals, 22% birds, 11% reptiles and 1% other taxa. Kangaroos and wallabies were the most frequently consumed native mammals and rabbits (*Oryctolagus cuniculus*) and hares (*Lepus capensis*) were commonly eaten introduced species (Marsack and Campbell 1990; Doherty *et al.* 2019). Dingoes also eat vegetation, carrion, poultry and small livestock and calves (Gould 1845–1863; Thomson 1984; Allen and Fleming 2004; Fleming *et al.*

2014; Corbett 2001b). Rarely, cannibalism has been recorded and usually demonstrates opportunism on finding dead dingoes rather than food shortage (e.g. Meek and Brown 2017; Smith *et al.* 2020) or occurs when captive bitches consume dead pups (Corbett 1988).

## 3.5 Longevity

### 3.5.1 Wild

Little is known about the lifespan of wild dingoes. Typically they live to 3–5 years, with not many living past 7–8 years. However, ongoing research on K'gari has documented dingoes living up to 13 years of age and producing young for at least 10 years (Behrendorff and Allen 2016; G. Ballard pers. comm.).

### 3.5.2 Captivity

In captivity, dingoes generally live 12–14 years (S. Jackson pers. obs.) but can live up to 20 years (Smith 2015b).

### 3.5.3 Methods used to determine the age of adults

Several methods have been tested to reliably determine the age of adult canids such as dingoes, coyotes, wolves and foxes. These include looking at the cementum annuli (Linhart and Knowlton 1967; Fancy 1980; Jones 1990), wear of the incisors (Gier 1968), wear on the conules on the first upper molar (Rogers 1965), socket tightness (Nellis *et al.* 1978), eye-lens weight (Catling *et al.* 1991) and basic skull measurements. All the methods have proved unreliable for distinguishing cohorts of known age dingoes (Thomson and Rose 1992) and have the added disadvantage that they can only be conducted on deceased animals.

A dental examination method has been developed for ageing live dingoes that uses aspects such as the presence of milk teeth, wear of the canine teeth and incisors (Table 14.1; Allen and O'Bryan 2011). This method is most accurate for animals under 4 years of age.

**Table 14.1.** Estimating the age of dingoes from teeth wear.

Derived from Allen and O'Bryan (2011).

| Approximate age (years) | Description |
|---|---|
| < 0.5 | Milk teeth present<br>Adult canines not yet emerged |
| 1 | Canines full in place with rounded tips<br>Pointed cusps and notches on incisors<br>Large pulp cavity (identified by dental radiography) |
| 2 | Canines slightly flattened tips<br>Cusps flatten and notches on incisors begin to wear off<br>Pulp cavity slowly closes |
| 4 | Canines with flattened tips<br>Discolorations and show evidence of wear marks from opposing teeth<br>Notch on incisors worn off |
| 10 | Canine teeth have wear marks from the opposing canine<br>Some incisors worn to the gum-line<br>Teeth flattened to pegs<br>Erosion and discoloration are common<br>Pulp cavity longer visible or just a thin line |

# 4. HOUSING REQUIREMENTS

## 4.1 Enclosure design

Dingoes are social and should be housed in compatible pairs or family groups (see section 9.10 for intraspecific compatibility). Dingoes are highly agile climbers and jumpers, as well as being efficient diggers, so the enclosure needs to be very secure to contain them. Fences should be made of heavy-gauge chain mesh as they are able to chew through some types of fencing wire (B. Smith and P. Fleming pers. obs.). The mesh should have small apertures so that dingoes and unwary visitors cannot put any part of their body through the fence. Dingoes can climb vertical wire fences 2 m high (Fig. 14.1), so fences should be at least 2.5 m high with an additional overhang of ~0.5 m that is curved inwards at the top (at ~35–45 degrees) or enclosures should have a mesh roof. Care must also be taken to prevent dingoes using the corner stays as a step or ramp when leaping to escape (P. Fleming pers. obs.). Dingoes habitually exploit right angles in fencing to jump or climb out, so escapes can be prevented by either using obtuse angles around the top of the perimeter, or protecting the top corner/s where right angles occur with a triangle of mesh. An electric wire or two around the inside of the perimeter is effective in preventing jumping or climbing escapes.

Digging under fences can be prevented by fitting an inward return skirt or apron of mesh that is at least 900 mm wide (note: this is the standard width of most commercially available mesh). Mesh should be pinned to the ground and buried. Alternatively, the vertical material can be sunk into the ground ~1 m or imbedded into a 600-mm deep strip of perimeter concrete, if flexible, or

**Fig. 14.1.** (Left) Adult dingo climbing a 2-m vertical fence. (Right) Juvenile dingo getting through a small gap in the fencing. Wherever its head can fit, so can the rest of its body. Photos by Lyn Watson/Australian Dingo Foundation.

butting directly onto the concrete strip if rigid. Holes dug near the perimeter should be repaired and/or filled in with large rocks.

The front of enclosures can contain a wet moat that is ~2 m (D) × 2.0–2.5 m (W) with the height of the wall above the water at least 1 m. Dry moats can also be used, where the ground at the bottom slopes backward up towards visitor height, but require the addition of electric fencing. Great care should be taken to ensure there are no footholds in the front wall, as the dingoes will use them to jump out of the enclosure.

Gates should fit tightly in frames to prevent feet, human fingers or pup parts being injured by neighbouring dingoes. Gate latches should be designed so dingoes cannot open them with their paw or snout (see Smith *et al.* 2012) and fitted high enough (at least 1.5 m high) to prevent unauthorised entry by young children. Gates should open inwards. A double-gate system (airlock) will also help prevent escapes through open gates and a concrete or stout timber bed log should be placed under the gates to prevent dingoes from digging out.

Dingoes can navigate through extremely small openings or slits in wire netting (Fig. 14.1). Consequently, all fences need to be closely monitored daily and repaired as required. Upon escape, dingoes are strongly territorial and will rarely travel far from bonded territory and so searches to recapture them should commence near the holding facility.

In conjunction with the main enclosure, there should be an area that allows the dingoes to be locked off-exhibit while the main enclosure is being cleaned and to separate dingoes for feeding individually if required or to hold an individual while it is being introduced to its mate or group. In the case of a family group of dingoes, a series of compartments ~1.5 m (W) × 5 m (L) × 2 m (H) is suggested. The total off-exhibit area available to a pair of dingoes should be at least 25 $m^2$ (see section 4.3). The holding area should have a den in each compartment and sliding gates between each compartment for easy movement of the dingoes. It can be arranged with separate entrances for each compartment from the enclosure or with the compartments arranged so that the dingoes run through one entrance and are locked off as they reach their respective compartments before they run back out. Though this second method usually works well, a disadvantage is getting the appropriate individual or individuals into the correct compartment. The presence of these compartments, which can each contain 1–2 dingoes, allows them to be brought in and appropriately fed, groomed, checked and given medication if required. This area can be locked off during the day and the animals can be given access at night.

## 4.2 Holding area design

Holding area design is very similar to that required for the main enclosure, except high fences are used on all sides. If there are a series of dens near each other in which different pairs or family groups are held next to each other the materials require consideration to prevent damage from the inevitable fence fighting, unless each enclosure is free-standing. The lower 1 m should be constructed of solid material to preclude constant stress from eye contact and mesh should be 25 × 25 mm welded type to prevent damage to ears and feet from dingoes in the adjacent enclosures (Fig. 14.2).

It is important that animals from different groups or pairs, particularly males, do not have access to each other; even when being led past each other with only cyclone meshing between them, they may try and bite each other through the fence and potentially injure themselves or staff. It is highly recommended that a solid material (e.g. tin sheeting) is used to stop visual and physical contact so that the teeth cannot be hooked on any part of the fence (Fig. 14.2). Also keep in mind that they will try to get to each other under doors and through holes in the door, such as where the bolt slides through if these openings are large enough.

**Fig. 14.2.** An example of a holding enclosure with gravel substrate floors, covered area with artificial den, raised platform and solid material separating the compartments. Photo by Bradley Smith.

All outside gates should be securely locked to prevent unfamiliar people from gaining access and all internal latches to which dingoes have access should be securely fastened with either a padlock or clip to prevent dingoes from using their paws or snouts to unlock the gate.

## 4.3 Spatial requirements

Enclosures must be of a size that allows all individuals to exercise properly and for subordinate dingoes to retreat from aggressive encounters. An area of at least 300 $m^2$ with an adjacent holding area of 25 $m^2$ for each pair is recommended, although an area of 220 $m^2$ has also been proposed (NSW DPI 2006). The enclosure area calculations should not include the space provided for a wet or dry moat, as dingoes do not normally swim and may be less likely to use the space within the dry moat, especially during visitor hours. An additional area of 40 $m^2$ should be added for each additional animal, although this should be assessed per individual context (number of dingoes, enclosure design). Smaller areas than suggested here are possible, but should be assessed on usability of the space rather than size alone.

## 4.4 Position of enclosures

All enclosures should be placed to allow the dingoes to bask in the sun or to seek shade, as required.

## 4.5 Weather protection

Protection from weather in display enclosures can be provided by naturalistic caves and in the holding area by roofed dens that allow the dingoes to retreat from rain and wind and keep dry and warm. Trees and bushes should be planted to provide shade on hot days.

## 4.6 Temperature requirements

Shade is required for cooling in hot weather but heating is not required. Dingoes are double-coated, which provides them with insulation in the winter months. The coat is seasonal and varies from a wiry single coat to a thick double coat and tail in the colder southern mountain regions. When desert dingoes are moved to colder climates, they produce an undercoat (Smith 2015b).

## 4.7 Substrate

The main enclosure, and/or holding area, should preferably have a mixture of surfaces such as gravel, earth and sand to allow digging and rolling. The holding area can be made with a concrete floor for easy cleaning, but a crushed rubble/gravel floor also works well because it is absorbent and easily cleaned (Fig. 14.2).

### 4.8 Shelter

A solid shelter must be provided that allows each dingo to retreat from poor weather. It need only be large enough to allow individual animals to curl up and whelp in if required. An area of at least 1.0 m (W) × 0.8 m (D) × 1.0 m (H) with the entrance facing away from the prevailing winds is suitable. Dingoes may choose not to sleep in the shelter, preferring to sleep on top of it, even on cold nights. Shelters are also useful for providing shade in summer. Details relating to whelping dens (which can also be used all year round), are detailed in section 10.11.

### 4.9 Enclosure furnishings

The enclosure should provide structural complexity to allow dingoes to express natural behaviours such as climbing, scent marking and foraging and to stimulate their senses. Large rocks and logs, undulations in soil topography and vegetation such as trees, grass and shrubs all provide structural and sensory complexity and a sandpit located in a sunny position can be played in or slept in.

## 5. GENERAL HUSBANDRY

### 5.1 Hygiene and cleaning

The enclosure should be cleaned daily of faeces and leftover food such as uneaten bones and the remains of carcasses. Dingoes will kill free-ranging animals that enter their enclosure, including birds, possums and rodents, so these remains should also be removed when possible. If used, bedding substrate should be replaced weekly, or as needed. For details related to den bedding when pups are present, see section 11.7.

### 5.2 Record keeping

It is important to establish a system whereby the health, condition and reproductive status of each captive dingo is routinely monitored and recorded:

- identification numbers, sex, age and reproductive status of all individuals
- results of veterinary examinations and treatments administered
- changes in defecation and urination
- behavioural changes or problems (e.g. pacing, excessive grooming, aggression)
- interactions between conspecifics
- reproductive behaviour or condition
- bodyweights and measurements
- changes in diet and food consumption
- movements of individuals between enclosures or institutions
- births (including dam and sire if known and details such as litter size, sex, bodyweight of pups)
- deaths (age, context and cause of death, with necropsy results if available).

The collection of information on the physical, physiological, and behavioural patterns of dingoes can contribute greatly to the husbandry of these species. It also allows individual histories to be transferred to other institutions if required and facilitates a cooperative approach to data collection among institutions. ZIMS, the Zoo Information Management System (Species360 2023), can be used for recording general information on age, sex, parentage, births, identification, bodyweights, transfers and deaths, with separate ZIMS modules for veterinary information and studbooks. Systems have been developed by Species360, which is part of the Conservation Planning Specialist Group (CPSG) software. ZIMS interfaces with programs, such as PMx, which provides demographic and genetic analysis tools to assist with the management of breeding programs (Lacy *et al.* 2011).

### 5.3 Methods of individual identification

#### 5.3.1 Passive Integrated Transponder (PIT) tags

PIT tags (microchips) can be implanted subcutaneously between the scapulae of individuals and should be used on all dingoes because microchipping is an excellent permanent method of individual identification. To prevent the tags tracking out along the injection site, the entry point can be sealed with tissue glue (e.g. Vetbond®) or similar fast-setting adhesive. Animals usually need to be caught for scanning with a PIT tag reader to confirm identification.

#### 5.3.2 Visual identification

Most adult dingoes can be identified by their physical appearance (e.g. size, genitals, coat colour, white markings on feet, legs, chest and neck) and this is generally the technique used by regular carers. A visual record of photographs of both sides, dorsal views and front-on head shots, and with descriptions of points of differentiation, is a useful aid for identification. Pups are born with distinguishable markings that stay with them into adulthood, which can make them easy to identify despite maturational changes in size and shape. See section 11.6 for options available for marking of pups. Ensuring name and information tags are placed on the pens and updated when animals are moved between enclosures is important for assisting identification.

### 5.3.3 Tattoos

Tattoos can be placed inside the ears or inner thigh/flank of the dingoes to allow identification if there are many dingoes and PIT tags/scanner are not readily available, although it does mean that they may need to be caught to be identified. There are welfare implications for tattooing small pups and adults without sedation and if not done correctly can cause drooping of the ears in rare cases. The use of tattoos is generally impractical for easy identification and they fade over time, making them difficult to read (Fancourt *et al.* 2022). By puberty, the ear and inner thigh of dingoes is completely covered by hair, making tattoos almost impossible to discern without shaving the site.

### 5.3.4 Collars

Collars can be used for identification purposes, but are not recommended because: (1) if worn for long periods they can rub the fur off the animal's necks;( 2) they may be caught on enclosure fencing,( 3) they may not remain intact if they are chewed by other dingoes; and (4) they do not promote dingoes as wildlife to visitors. Collars and harnesses are generally only suitable for leading dingoes and are removed once the walk is over.

### 5.3.5 Ear tags

Ear tags can provide a method of identification and are currently used with some success as permanent individual markers on free-ranging dingoes on K'gari and on research sites throughout mainland Australia. Combinations of tag colour, size, shape and number/letter can be used to identify individuals. Ear location can be used to signify sex (e.g. left ear – male; right ear – female). One major benefit of ear tags is that they offer a way to identify individuals from a distance. However, as with collars, they may not be visually appropriate for an exhibit animal. They can also be damaged or snagged and poor-quality tags can fade and become unclear to read over time (Fancourt *et al.* 2022). Ear tags should be placed centrally in the ear to maintain the structural integrity and shape of the ear and to reduce the chance of snagging or ripping out during altercations with other dingoes. Incorrect positioning on the ear or tagging an animal too young can result in infection or permanent damage to the ear. See DBCA (2017) for standard operating procedure.

## 6. FEEDING REQUIREMENTS

### 6.1 Captive diet

Dingo dietary requirements are similar to those of modern domestic dogs. A suitable balanced diet can be made up of high-quality dry dog kibble with fresh meat/gutted carcasses. The diet may need to be modified at times to ensure it meets changing physiological demands such as pregnancy, lactation, growth, old age, exercise and season. Enrichment items may be included as part of the daily diet. A diet sheet should be developed for individual animals and situation rather than be prescriptive.

Free-ranging dingoes feed on a wide variety of animals (see section 3.4), so the diets of captive dingoes should also be primarily meat-based. It is essential that dingoes are regularly provided with raw meaty bones for good dental health. Fresh beef or lamb bones from a butcher, or chicken necks/wings/carcasses will give dingoes plenty to chew, which, in addition to being good for their health, will also relieve boredom. Cooked bones must not be fed, as they can splinter and lodge in a dingo's throat or intestine, causing serious health problems.

A high standard of food hygiene and meat quality is necessary to prevent potential disease transmission and only offal from butchers (that have undergone a meat inspection process) should be fed to dingoes. Dingoes can act as definitive host for the hydatid tapeworm *Echinococcus granulosis* after infection from parasitic cysts in the organs of macropods or sheep, which act as intermediate hosts (Jenkins and MacPherson 2003). Similarly, dingoes can host the tapeworms *Taenia pisiformis* and *T. serialis* through eating cysts in rabbit carcasses. Meat may also carry disease-causing bacteria (e.g. salmonella). The risk of food-borne disease can be minimised by not feeding offal and the freezing of meat or carcasses for at least 3 weeks may kill some parasites.

Various diets are fed to captive dingoes and often contain dried or tinned dog food as the staple, with fresh meat provided for variation. Captive dingoes can be conveniently maintained on a commercial dried food ration of meat-based pellets as the major ingredient. Dry dog food is a complete nutritional diet for dingoes, is easily kept without refrigeration and does not attract insects. The quantities provided on the packet are for domestic dogs and the amount fed to dingoes should be adjusted as needed to maintain good lean body condition. Dingoes may not have evolved with the same ability as most dog breeds to process commercial cereal grains (Arendt *et al.* 2016) and may benefit from dog food that contains alternative carbohydrate sources, which is often labelled 'grain free'.

The commercial dog foods selected should also be low in fat because many wild foods of dingoes are usually not high in fat (e.g. rabbit and kangaroo). High-fat diets can lead to obesity in captive dingoes and may predispose them to pancreatitis.

For variation and enrichment, meaty mutton and beef bones, gutted rabbits, venison, raw fish, adult raw chickens (minced or frames with fat removed), mice/rats and day-old chicks can be used. Lamb and pork meat should be given sparingly due to their potentially high fat content. Other food items may include boiled eggs, whole eggs in shells, apples, carrots and whole coconut. Dingoes should also be given access to fresh grass. Captive dingoes cope well with a predictable and similar diet, with variation used for enrichment (L. Watson pers. comm.). If a raw diet is preferred, this should be managed under the supervision of a veterinarian.

A suggested diet for an adult dingo based primarily on commercial dog food and fresh meat that is used by the Dingo Discovery Sanctuary and Research Centre (Melbourne) is presented below.

- Ad lib
  - water
- Daily diet (per adult dingo)

| | |
|---|---|
| Sunday | AM: 2/3 dry dog* food plus 1/3 raw lean meat** |
| | PM: enrichment/activity foods# |
| Monday | AM: 2/3 dry food plus 1/3 raw lean meat |
| | PM: Enrichment |
| Tuesday | AM: 2/3 dry food plus 1/3 raw lean meat |
| | PM: Enrichment |
| Wednesday | AM: Starve (fasting) day† |
| | PM: 1 large bone (e.g. beef knuckle bone) |
| Thursday | AM: 2/3 dry food plus 1/3 raw lean meat |
| | PM: Enrichment |
| Friday | AM: 2/3 dry food plus 1/3 raw lean meat |
| | PM: Enrichment |
| Saturday | AM: 2/3 dry food plus 1/3 raw lean meat |
| | PM: Enrichment |

*Portion according to guidelines. Recommended to be meat-based, low-fat and low in commercial cereal grains.

**Lean meat (with fat removed) can include chicken, kangaroo, horse, beef, or rabbit.

#Enrichment food items can be used during presentations, training, to stimulate foraging behaviour, and to alleviate boredom. Options can include pigs' ears, apples, carrots, whole eggs in shell, meaty bones, chicken frames, whole coconut, mice, and day-old chicks.

†Starve day can be altered each week so that it is unpredictable.

## 6.2 Supplements

When high-quality commercial dog food is used there should be no requirement for any additives. A veterinarian may recommend nutritional supplements in the case of dietary imbalance or in cases where a specific nutritional requirement is higher.

## 6.3 Presentation of food

Adult dingoes need only be fed once per day, which will help reduce overfeeding and obesity. Food quantities may need to be increased in winter when a higher energy intake is needed for maintaining body coat, warmth and communal provisioning. There are several techniques that can be used when feeding that will encourage activity and help alleviate boredom (see section 9.7 for food-related enrichment ideas). If feeding multiple times per day, ensure the daily food requirement is not exceeded to avoid obesity. For feeding pregnant and lactating bitches and pups refer to section 10.12. Starve days are not necessary, but can be used infrequently to reduce predictability of daily routines. Starve days should be avoided during pup rearing so that adults can regurgitate food to pups (note: regurgitation occurs soon after feeding).

It is important that dingoes are fed separately and are monitored to ensure that each individual gets the allocated portion. Otherwise larger dingoes, males or the more dominant animals will either hoard the food or consume that of others as well as their own. The method of achieving individual feeding will depend on the number of dingoes in the enclosure. For a small number of dingoes, use separate food bowls and either feed them in different parts of the enclosure or hand feed as a group. For larger groups, it may be necessary to physically separate certain individuals or if the dingoes are fed together, provide a large amount of food that is spread out and ensure there are large rocks and other refugia so the subordinates can quickly grab some food and retreat to cover and eat. When feeding groups, be mindful of a dominance hierarchy. For example, if you feed the more subordinate animals first, dominant animals will notice and may become aggressive towards these subordinate animals once the group is put together again. A better approach is to feed animals in different parts of the enclosure, with the dominant fed first and the subordinate animals shortly after.

Water should always be available, preferably via an automatic system, either bowl type or 'lick operated valve' on a metal tap. Poly piping will soon be destroyed or excavated, so if used it must be protected. Otherwise, galvanized metal buckets should be clipped to the fence and stabilised with a bracket and fixed high enough to prevent small pups from climbing in (L. Watson pers. comm.).

## 7. HANDLING AND TRANSPORT

### 7.1 Timing of capture and handling

Capture and handling can be carried out any time of the day or night, but in the morning when they are in their den, or after feeding, might increase success and reduce distress. Animals that are to undergo anaesthesia should not be fed before capture.

### 7.2 Capture equipment

Most captive dingoes can be readily handled by familiar people using a collar and lead without the need for muzzling. When being handled by strangers, a properly fitted muzzle is advisable (depending on the dingo and rules of the facility). Dingoes can be conditioned to enter a crate or box for transport. Catching bags or nets are not used, because the process is stressful and may lead to injury. If the dingo cannot be approached, then the use of a jab-stick to remotely deliver chemical restraint drugs is preferable.

### 7.3 Capture and physical restraint

The capture and restraint method will depend on the level of tameness of the individual dingo and should be done in a way that does not cause physical harm and minimises stress. This may differ according to individual dingoes and the person attempting the capture. Dingoes are easily frightened by unexpected noises, novel equipment and unfamiliar handlers, even when well socialised. Therefore, quiet handling techniques are recommended.

It is vital that dingoes are trained from a young age to allow capture and lifting so that animal caretakers can put on a muzzle, collar and/or harness. If some level of handling and training is not achieved at an early age, handling might not be possible in adulthood when the dingo is not familiar with the handler. Dingoes should not be chased around the yard, as they are very fast and highly agile and it can make them very wary of people in the future, through loss of trust, and greatly hinder the development of working relationships with them. If the dingo is extremely distressed and backed into a corner, it may react aggressively.

Tame dingoes should be relatively easy to catch by hand. The use of a food lure/treat and a second or third person helping to herd the dingo may aid capture. Capture can be a stressful event, so should be carried out by a person/s familiar to the dingo and in a manner that causes the least amount of distress.

Once captured, a dog collar or harness can typically be placed on the animal as most dingoes can be trained to walk on a lead. A properly fitting commercially available soft/material-based dog muzzle suitable for a medium-sized dog should always be available for any physical capture event, even if it is not used. Dingoes can be briefly examined and given injections while restrained and muzzled .

Fig. 14.3 illustrates the best way to firmly hold and carry a dingo. This method should not be used on highly agitated or aggressive dingoes.

### 7.4 Weighing and examination

The easiest way to weigh an adult dingo is by using an animal scale (with a weighing capacity to at least 30 kg). A model that has a low profile and rubber or non-slip surface covering the platform pan is preferable. The dingo can be easily trained or coaxed to stand stationary on the platform. This is preferable because it is a quick and accurate method that does not require handling the animal. If such scales are not available, and the dingo is comfortable being held, an alternative is to weigh the handler and the animal, and then subtracting the handler's weight. Pups can be weighed using standard kitchen scales and then transitioned to animal scales.

**Fig. 14.3.** (Left) Technique used to hold dingoes during routine examinations. (Right) Dingo being held upside down by its hips. Photos by Lyn Watson/Australian Dingo Foundation.

## 7.5 Release

Dingoes can be released by removing the lead, harness or muzzle. If an individual has been removed from its mate or family group for more than 1 day, especially if a subordinate animal, it may need to be placed in a holding area before being let into the enclosure to minimise aggression upon its return (see section 9.6).

## 7.6 Transport requirements

### 7.6.1 Transport box design

The box needs to be large enough to allow the dingo to lie down comfortably and tall enough for it to stand comfortably in a natural position with its head extended and to turn around. A sturdy box or other transport container suited for a large dog breed will suffice. Adequate inside dimensions are: 100 cm (H) × 150 cm (D) × 80 cm (W). The box should be well ventilated with handles on each side for easy lifting by two people. Containers for air travel need to meet the standards set by the International Air Transport Association (IATA 2022). Some dingoes find being in a box/crate highly stressful and take a long time to adjust. Providing some crate training before transport will likely reduce this.

### 7.6.2 Furnishings

Make sure the floor of the box is non-slip. The addition of a towel or blanket may improve comfort, add insulation and help to soak up any urine, saliva or vomit. The inclusion of a blanket or items with familiar scent is recommended to provide comfort and aid transition to a new environment.

### 7.6.3 Food and water

Food is not generally required for travel times less than 24 h because it may induce vomiting during travel. Water is not generally required for short journeys less than 4 h in cool weather. For travel times greater than 4 h or in hot weather water should be made available at regular intervals. When water is provided, ensure it is in an anchored deep bowl to reduce the amount of splashing.

### 7.6.4 Animals per transport box

Adult animals should be transported individually and when separate boxes are used, they should be placed side-by-side to prevent fretting. Dingo pups should always travel two to a crate because if they are separated from siblings before 12 weeks of age they may become panicked and distressed, which can result in long-term psychological and behavioural issues (L. Watson pers. comm.).

### 7.6.5 Timing of transportation

Dingoes will adapt more successfully to new surroundings and conditions if transportation is carried out before they are 14 weeks of age. If considering international travel, pups will travel more successfully if they are over 8 weeks of age, but under 3 months of age. Translocation becomes less successful with each advancing week (L. Watson pers. comm.).

It is preferable that dingoes arrive at their destination during daylight, so they can visually explore their new surroundings. For short (< 4 h) intra- and interstate travel by road or air, any time of day is adequate. For longer journeys, and in hot weather, it is best to organise an early morning departure or travel overnight. Travel by road should be considered when possible because it is less stressful than travel by air.

### 7.6.6 Release from the transport box

At release, a dingo is likely to be stressed and nervous, so the process of release should not be rushed and care taken to minimise additional stress. Place the box in the off-exhibit or quarantine area and wait a short period (~30 min) so the dingo can get familiar with the sounds and smells of the new environment. Release the door of the box and let each dingo come out in its own time. Try to keep the box in the holding area for several days to offer a safe space (especially if blankets or items with familiar scent are present). During this adjustment period it is beneficial for animal caretakers who will work with the dingoes to start interacting with them. Short training sessions can also commence. Ensure that each dingo is comfortable before releasing into the main enclosure.

## 8. HEALTH REQUIREMENTS

### 8.1 Daily health observations

All dingoes should be observed daily for any signs of illness, injury or distress, ensuring prompt veterinary care if necessary. Any concerns should be recorded so that information can be shared among staff, as well as providing long-term retrospective health and behaviour information. The most convenient time to do these daily checks is when the enclosure is being cleaned or when the dingoes are fed or groomed. Regular animal caretakers can identify any changes in behaviour or appearance. Daily health observations/checks should include the following:

- coat condition – not oily nor have an odour. Dingoes will normally groom and maintain a clean coat without the need for human intervention
- external parasites – fleas, lice, ticks, and mites. Ectoparasites may be more difficult to see on darker coated animals. Their presence may be suggested by the dingo scratching or biting affected areas
- discharges – blood (dried and fresh), pus or mucus from the eyes, ears, nose, mouth, anal region or genitals
- appetite – lack of appetite may indicate an underlying medical condition. A pregnant bitch will not eat the day before giving birth
- faeces – number and consistency. Faeces should be firm and moist, without evidence of blood or visible worms. Constipation or diarrhoea should be investigated if recurrent or persistent.
- eyes – cloudiness, squinting, colour change, uneven pupil shape
- changes in demeanour, activity level and behaviour – withdrawn, submissive and nervous demeanour might indicate health problems or distress (e.g. aggression from a conspecific). Signs of distress may also include frequent crying and whining, chewing of the body or damage to enclosure, escape behaviour and aggression towards other dingoes and humans.
- injuries – wounds (e.g. lacerations or punctures), bleeding, swelling or lameness.

Any signs of injury or illness or unusual changes in behaviour should be documented and reported to a veterinarian and a suitable course of action determined between animal caretakers and veterinary staff. This may include a change in diet, separation from conspecifics or providing medical attention. If an animal is found dead, make note of the circumstances, anything unusual or abnormal on the body or surrounding area. Carefully collect the body (using PPE such as gloves) and submit it to the attending veterinarian for necropsy.

### 8.2 Bodyweight and body condition score

#### 8.2.1 Bodyweight

All animals should be weighed regularly, at least monthly, or more frequently for very young or aged animals or those with health concerns. Bodyweight should be compared with previous weights to assess trends over time.

### 8.2.2 Body condition score

Dingoes are naturally lean in appearance. Over-conditioned animals will require diet adjustment to maintain optimal lean condition. Reduced body condition may result if animals are outcompeted for food or have an underlying disease. Obesity or emaciation needs to be managed.

Body condition is best determined by using palpation to assess the soft tissue cover over the spine, shoulders and pelvic bones. Because dingoes are naturally lean and usually carry little excess fat (Fig. 14.4; Behrendorff *et al.* 2016), it is common to see the ribs of healthy free-ranging dingoes. Body condition scoring for dingoes is based on a score out of 5.

- 1/5: emaciated - spine, ribs and pelvic bones are obvious bony protrusions as a result of a loss of muscle mass and body fat. Bones of the skull may also appear more prominent. .
- 2/5: lean - spine, ribs and pelvic bone may be seen and are easily felt and there is limited muscle cover or subcutaneous fat. The animal has an obvious waist when viewed from above. Many wild dingoes have this appearance, especially during reduced prey times.
- 3/5: ideal - spine, ribs and pelvic bone are easily felt but are not necessarily seen. The animal has a robust muscular appearance with no obvious accumulation of fat.
- 4/5: over-conditioned - spine, ribs and pelvic bone cannot be seen and ribs and spine may be more difficult to feel. The animal has a rounded contour when viewed from above. There is some fat cover over the base of the tail, hips and chest. .

**Fig. 14.4.** Side profile of a healthy captive female adult dingo nursing one of her pups. Photo by Bradley Smith.

- 5/5: obese - obvious fat deposits on tail, pelvic bone, chest and hindquarters. The animal has no waist when viewed from above and bony prominences of the hips are difficult to palpate.

### 8.3 Physical examination

In addition to daily health inspections, all dingoes should receive an annual health check by a veterinarian. More frequent veterinary examinations may be required for older dingoes or those with health or welfare concerns. Animals trained for examination or those that are readily handled can be examined without the need for chemical restraint, though timid or apprehensive animals or those requiring more detailed health assessments will require sedation or anaesthesia.

A veterinary examination should include all body systems as for a domestic dog. Resting heart and respiratory rates and normal rectal temperature do not differ from similar sized domestic dogs. Dingoes have naturally flexible joints. They show greater flexibility in hip rotation and abduction and can support a wider range of movement compared to domestic dogs (B. Smith, unpubl. data). This should be considered when assessing dingo hips according to domestic dog criteria for hip dysplasia.

### 8.4 Chemical restraint

Chemical restraint is indicated to facilitate examination of unapproachable animals and when invasive or painful procedures are undertaken. Even confident and readily handled dingoes may become apprehensive and distressed in unfamiliar surroundings or with unfamiliar people, making chemical restraint for examination a less stressful option than struggling with manual restraint. Dingoes should be fasted for 12 h before general anaesthesia (Vogelnest 2025).

Sedatives, premedications and anaesthetic agents used in domestic dogs can be used in the dingo at domestic dog dose rates. However, because dingoes tend to be highly anxious and unpredictable in unusual settings, use similar precautions as with similar-natured domestic dogs. Dingoes are naturally lean-bodied and hence may appear more sensitive to the effects of lipophilic anaesthetic drugs in a manner similar to lean-muscled domestic dog breeds (e.g. sighthounds).

The anaesthetic regimen will depend on the procedure being undertaken, the health status of the dingo, whether the dingo can be restrained for injection, the preference and experience of the veterinarian and whether a more rapid recovery through drug reversal is required (e.g. to expedite return to a group).

For dingoes that can be handled, the use of premedication with which the veterinarian is familiar will allow a controlled and non-stressful anaesthetic induction. Typically, subcutaneous or intramuscular premedication, followed by carefully titrated intravenous induction and maintenance with an inhalation agent is suitable.

Unapproachable dingoes will need to be darted (or possibly pole syringed with a jab-stick if confined to a trap or enclosure). Slightly higher drug doses are generally used for wild animals and accurate estimates of body mass are needed to prevent overdose. The use of reversible combinations, usually based on alpha-2 agonist use, allows rapid recovery. The larger muscles of the rump are the preferred site for intramuscular injection or dart placement.

## 8.5 Common health problems

High standards of husbandry and hygiene, veterinary input into the development of a locally appropriate preventative health program and prompt veterinary consultation when signs of disease are noted will minimise health problems. Dingoes maintained in human care are susceptible to the same disease processes as domestic dogs. Diagnostic tests and treatment protocols follow those recommended for domestic dogs. Route and frequency of drug administration and in some cases, drug choice may be influenced by the ability to handle the animal and by its social situation (Hulst 2008). Choice of ecto- and endoparasite treatments should be done in consultation with a veterinarian to ensure appropriate coverage.

More detailed descriptions of infectious and non-infectious diseases reported in wild and captive dingoes are presented in the Tables 14.2 and 14.3 and in Hulst (2008) and Vogelnest and Portas (2025).

## 8.6 Preventative medicine

Preventative medicine programs for dingoes follow those recommended for domestic dogs and should be developed in consultation with a veterinarian familiar with local canine infectious diseases. Together with an annual veterinary examination, the program may include regular intestinal worming, faecal parasite checks, ectoparasite control and heartworm prevention, vaccination for canine distemper, hepatitis, parvovirus and for kennel cough and/or leptospirosis if indicated. Many combined treatments for dogs are marketed that cover different selections of intestinal worms, heartworm and ectoparasites. A veterinarian should help select the appropriate products so that likely diseases are prevented using the most effective drugs without doubling up or using unnecessary drugs.

## 8.7 Zoonoses

Captive dingoes that are maintained in a clean environment with preventative health care are an unlikely source of zoonotic infection. The most important zoonotic diseases associated with dingoes are covered in Table 14.2.

- Hydatid cysts can develop in humans when eggs of the hydatid tapeworm *Echinococcus granulosus* are accidentally ingested from handling the faeces or fur of an infected dingo. Wild dingoes are commonly infected by eating cysts present in intermediate hosts (Jenkins and Morris 2003; Jenkins *et al.* 2008). The tapeworm can easily be eliminated from captive dingoes by regular worming and by not allowing the transmission cycle to be completed by never feeding offal from intermediate hosts (usually macropods or sheep).
- Ringworm, a fungal skin infection, can be transmitted to people who handle infected dingoes (Hulst 2008). It can result in circular red itchy or scaly patches on contact skin.
- *Toxocara canis* can cause visceral larva migrans as round worm larvae migrate through body tissues after accidental ingestion of eggs. *Ancylostoma caninum* can cause cutaneous larva migrans as hookworm larvae penetrate the skin. *A. ceylanicum* has been reported in wild dingoes in northern Queensland. It can cause abdominal discomfort and diarrhoea in people who can also develop patent infections (Smout *et al.* 2013).

Attention to personal hygiene practices including regular handwashing, use of gloves when handling faeces and not eating, drinking or smoking in animal areas will reduce the risk of zoonotic transmission.

# 9. BEHAVIOUR

## 9.1 Activity

In general, dingo behaviour and activity follow a seasonal pattern, which revolve around their annual breeding cycle. The focus of dingo social behaviour is on

**Table 14.2.** Common or important infectious diseases of captive dingoes.

PCR = polymerase chain reaction.

| Disease | Cause | Clinical signs | Diagnosis | Treatment | Prevention/comment | Reference |
|---|---|---|---|---|---|---|
| **Ectoparasites** | | | | | | |
| Flea infestation; flea allergy dermatitis | *Ctenocephalides felis; C. canis* | Itching, fur loss, skin irritation, especially around the rump and tail base. Flea allergy dermatitis with more severe signs may develop in some dingoes. Fleas can occur year-round but seasonal flea activity usually peaks in late summer and early autumn | Visualisation of fleas/flea dirt in fur<br>Flea allergy dermatitis may result from hypersensitivity to flea saliva from a single flea bite and the flea may not be seen. Diagnosis is based on typical clinical signs and response to flea elimination | Strategic use of oral or topical (spray/spot-on) dog flea control products designed to kill adult and larval fleas and interrupt the life cycle (e.g. fipronil, imidacloprid, nitempyram or isoxazolines)<br>Flea allergy dermatitis may require anti-inflammatory treatment | Regular use of flea control products. Wash/change any bedding weekly. Consider use of environmental flea killer in yard or shelter. Fleas can transmit tapeworms (*Dipilydium canis*) | 1, 9 |
| Stickfast fleas | *Echidnophaga myrmecobii* | Small brown fleas attached around the edge of ears or face; may result in rubbing or itching | Visualisation of fleas | Flea control products (see above) | Common in dingoes that feed on rabbits, which harbour stickfast fleas | 4 |
| Lice | *Trichodectes canis* | Biting lice that cause scratching, irritation, poor hair coat | Visualisation of lice or eggs attached to the base of hairs | Topical/oral ectoparasiticides for dogs (e.g. fipronil, imidacloprid, selamectin). Collars and rinses may also be used | Reported in wild dingoes. Bedding should be changed regularly to prevent reinfestation | 2 |
| Demodectic mange | *Demodex canis* (a mite) | Localised facial hair loss, often around the eyes | Deep skin scrapings to identify mites burrowed into the skin | Oral/topical miticides (e.g. isoxazolines) | Uncommon. Rarely may be generalised, affecting whole body | 13 |
| Sarcoptic mange | *Sarcoptes scabiei* (a mite) | Alopecia, scaling, crusting and thickening of the skin with intense itching | Several skin scrapings to identify the mites | Oral/topical miticides (e.g. isoxazolines) | Seen in wild dingoes. Rare in captivity. Zoonotic | 2, 15 |
| Tick paralysis | *Toxin of adult Ixodes holocyclus* tick. The toxin is injected through tick mouthparts as it feeds on host blood. The toxin affects the nervous system causing paralysis | Early signs of toxicity include hindlimb wobbliness or weakness, reluctance to move, voice change, respiratory noise, vomiting; signs progress rapidly over 12–24 h resulting in collapse and death | Clinical signs together with identification of a paralysis tick or previous attachment site (tick crater). Ticks are mostly found around the head and neck, but may be anywhere on the body. Signs usually develop 3–4 days after tick attachment, so can still occur once a tick has dropped off or been removed | Tick removal using a tick removal hook or tweezers applied at skin level without squeezing the body of the tick. A thorough search for additional ticks should always be done. Topical fipronil spray will kill any ticks that may have been missed. Tick antiserum to counteract the toxin is administered intravenously. Hospitalisation may be required for intensive supportive care. Animals may die despite treatment if signs are advanced | Paralysis tick range is restricted to eastern seaboard of Australia. Wild dingoes in endemic tick areas generally not affected by tick paralysis; can occur in captive animals. Paralysis tick prevention products for dogs can be used, including oral and topical (spray or spot-on) preparations. Daily tick searches should be undertaken if tolerated and any ticks removed. Animal then monitored closely for signs and veterinary advice sought early if seen | 1, 9–11 |

**Table 14.2.** continued

| Disease | Cause | Clinical signs | Diagnosis | Treatment | Prevention/comment | Reference |
|---|---|---|---|---|---|---|
| **Endoparasites** | | | | | | |
| Intestinal worms | *Toxocara canis* (round worm); *Trichuris vulpis* (whipworm); *Ancylostoma caninum, A. ceylanicum, Uncinaria stenocephalica* (hook worms) | Ill-thrift, pot belly in pups (round worms); ill-thrift, diarrhoea, dark tarry faeces and anaemia (whipworm and hookworm) | Faecal examination for typical eggs. Round worms may be seen in faeces | Anthelmintic products for dogs. Combined oral or topical products for worms, fleas and ticks may be used | Regular disposal of faeces from environment to prevent egg build-up; regular faecal examination and worming<br>Heavy infection in pups may be fatal<br>Hookworm can cause significant disease in wild dingoes | 2, 4, 16, 17 |
| Tapeworm | *Dipylidium caninum, Taenia pisiformis, T. serialis; Spirometra erinacei* | Usually none.<br>Anal itching or scooting may be seen. Weight loss may occur with heavy taenia or spirometra infections | Tapeworm segments seen as small white rice grain-like specks on perineal fur (*D. caninum*); faecal examination for segments (proglottids) or eggs | Praziquantel | *D. caninum* is carried by fleas, and less commonly lice, so their control will prevent infection. *T. serialis, T. pisformis* infection occurs through feeding infected rabbits, which act as intermediate hosts. *S. erinacei* is common in wild dingoes.<br>Infection through consuming intermediate host (usually reptile or amphibian). Higher treatment dosage required to eliminate spirometra | 2,4, 16, 17 |
| Hydatid tapeworm | *Echinococcus granulosus.* Dogs/dingoes are definitive host for adult worms, with tissue cysts forming in intermediate host. Wild dingoes form a sylvatic cycle with macropods | None. Dingoes may have thousands of worms without signs | Faecal examination for eggs | Praziquantel | Regular worming. Do not feed offal. Intermediate hosts, primarily sheep or macropods, are infected by ingesting eggs from canid faeces. Cycle is completed when definitive host consumes cysts from intermediate host. An important zoonosis, see section 8.6 | 5,7 |

| Disease | Cause | Clinical signs | Diagnosis | Treatment | Prevention/comment | Reference |
|---|---|---|---|---|---|---|
| Heartworm | *Dirofilaria immitis*, spread by mosquito vector. Mosquito injects immature larvae into the bloodstream, which mature into adult worms in the pulmonary arteries/heart. Adult female worms produce microfilaria, which circulate in the blood to infect a feeding mosquito, completing the cycle | Signs depend on the number of worms and their position in the cardiovascular system<br>Cough, exercise intolerance, weight loss; eventual heart failure may ensue | Antigen test on blood; detection of circulating microfilaria | Adulticide (e.g. melarsomine) followed by microfilaricide (e.g. ivermectin, moxidectin) | Annual injection, oral or topical spot-on heartworm prevention for dogs containing macrocyclic lactones (avermectins and milbemycins)<br>Endemic in many parts of Australia with variable prevalence. Wild dingoes in northern Australia infected | 2, 3, 17–19 |
| Canine lungworm | *Oslerus osleri* | Cough, difficulty breathing, weight loss. Infection may be unapparent. Signs mainly in young animals and occur due to nodules containing worms in trachea and bronchi | Identification of larvae in tracheal washes or faeces | Nematode worming products for dogs (e.g. ivermectin, fenbendazole) | Common in wild dingoes in south-east Australia. Direct transmission to weaning pups when bitch regurgitates food<br>Not seen in captivity | 2, 12 |
| **Bacterial** | | | | | | |
| Canine ehrlichiosis | *Ehrlichia canis*, a rickettsial blood parasite transmitted by the brown dog tick *Rhipicephalus sanguineus* | Dingoes are assumed to show similar signs as dogs: acute phase - fever, lethargy, anorexia, enlarged lymph nodes, bleeding disorders (e.g. nose bleeds). A subclinical phase with no signs may ensue and may be followed by a chronic, more severe form with wasting and eventual death, which may follow months to years after an acute phase | PCR, serology, blood smear (insensitive) | Doxycycline daily for 4 weeks is usually effective in acute phase. Chronic form is usually fatal | Exotic to Australia until 2020 and is a notifiable disease in all states. Dingoes are assumed susceptible if exposed to infected brown dog ticks, which are mainly found in the north of Australia. Prevention relies on strict environmental tick control conducted by a licenced pest controller, regular body searches for ticks and use of topical spot-on or oral products that repel ticks on dogs as well as kill them (e.g. flumethrin collar, permethrin spot-on). Products based on isoxazolines do not kill attached ticks fast enough to prevent infection but may be used in conjunction with repellent products | 8, 14 |

**Table 14.2.** continued

| Disease | Cause | Clinical signs | Diagnosis | Treatment | Prevention/comment | Reference |
|---|---|---|---|---|---|---|
| **Viral** | | | | | | |
| Parvovirus | Canine parvovirus | Fever, lethargy, bloody diarrhoea. Usually in unvaccinated pups/young dogs. May be fatal | Clinical signs, faecal antigen testing or faecal PCR | Follows protocols for dogs, primarily intensive symptomatic and supportive treatment | Vaccination following recommended protocols for dogs. Highly contagious and spread by direct or indirect contact with faeces. The virus is stable in the environment. Strict hygiene and isolation of infected animals important | 1, 6 |
| Distemper; infectious hepatitis | Canine distemper virus; canine adenovirus 1 (infectious hepatitis) | Wide range of signs including fever, depression, anorexia | Definitive diagnosis difficult in live animal | Supportive treatment – both diseases may be fatal | Uncommon since widespread canine vaccination. Included in routine vaccination protocols. Recorded historically in wild dingoes | 2 |
| Kennel cough (canine infectious tracheobronchitis) | Canine parainfluenza virus and *Bordetella bronchiseptica* (a bacteria) | Persistent harsh cough, bouts of coughing may cause gagging or vomiting. Animals usually remain otherwise well | Typical clinical signs | Symptomatic. Disease is self-limiting | Dingoes presumed susceptible. Vaccination for at risk animals follows recommended protocols for dogs<br>Highly contagious between in-contact dogs | |
| **Fungal** | | | | | | |
| Ringworm | *Microsporum canis, M. gypseum* | Circular, hairless, slightly flaky grey or sometimes red and inflamed skin lesions, usually on face or limbs. Not usually itchy | Microscopy, fungal culture | Topical or systemic antifungal drugs | May be self-limiting. Can spread between animals either directly or via bedding or brushes. Disinfect items (e.g. with F10®) and maintain separation from uninfected animals. Zoonosis | 1 |

(1) Hulst (2008); (2) Corbett (2001a); (3) Starr and Mulley (1988); (4) Brown and Copeman (2003); (5) Jenkins and Morris (2003); (6) Hooper *et al.* (1983); (7) Jenkins *et al.* (2008); (8) Wildlife Health Australia (2020); (9) Boden (1998); (10) Turner (1990); (11) Barnes *et al.* (2001); (12) Dunsmore and Spratt (1979); (13) Newsome and Coman (1989); (14) Northern Territory Government Department of Industry, Tourism and Trade (2021); (15) Thomson (1992a); (16) Smout *et al.* (2013); (17) Smout *et al.* (2018); (18) Smout *et al.* (2016); (19) Nguyen *et al.* (2016)

**Table 14.3.** Common non-infectious diseases and causes of death of captive dingoes.

Derived from Hulst (2008)

| Disease | Cause | Clinical signs | Diagnosis | Treatment | Prevention/comments | Reference |
|---|---|---|---|---|---|---|
| Trauma | Conspecific aggression most common; collision with obstacles, capture injuries | Varies from minor puncture wounds/lacerations to bone or teeth fractures, severe soft tissue trauma, life-threatening injuries or death | Observation of wounds, surgical exploration, diagnostic imaging | Dependent on severity. May require antibiotics, wound cleaning, suturing, fracture repair | Conspecific aggression is the most common cause of injury/disease in captive dingoes | 1 |
| Stress | Stressors can be divided into: (1) institutional stressors (e.g. inadequate enclosure size, loud noises, receiving medical treatment); (2) human stressors (e.g. lack of socialisation to humans, high frequency of visitors, changes to keepers/staff); and (3) social stressors (e.g. introduction of unfamiliar animals or removal of familiar ones, maintaining too many animals in the space. Also see section 9.2.1. | Behavioural signs of stress include frequent vocalisations (crying, whining, howling), chewing own body or the enclosure, hypervigilance, freezing, exaggerated startles, pacing, avoidance and aggression towards other dingoes and humans. Body language signs include tail tucked under body, ears lowered, yawning, lip licking, panting and raised hackles. The presence of stress can be clinically assessed by examining salivary corticosteroid concentrations or the presence of inflammatory cytokines in saliva, blood or faeces | Review of enclosure size and design; the number, sex and age of occupants; and dingo capture and handling procedures. Observation of nearby stressors (permanent and temporary) | Treatment actions may depend on urgency and severity. Remove the cause of the stressor where possible and/or adjust management procedures (according to those provided in this chapter) | Breeding season represents the most stressful period (mostly relating to rank disputes). Most stress-related problems can be prevented with early socialisation, good management (e.g. separation) and an understanding of each particular dingo's propensities and personality | 2 |
| Obesity | Overfeeding | Excessive body condition, obvious fat deposits under skin | Observation | Dietary adjustment | Appropriate nutrition (see section 6.1)<br>Dingoes are naturally lean and obesity should be avoided | |
| Hot spot (acute moist dermatitis) | Localised skin irritation, possibly due to insect bite, with subsequent licking and chewing followed by secondary bacterial infection | Acute onset of red oozing areas of inflamed skin, with matted or missing fur. Painful and itchy which perpetuates licking and further inflammation | Typical appearance | Clip surrounding fur, cleaning and flushing of skin with saline or dilute chlorhexidine solution.<br>Topical anti-inflammatory/antibacterial treatment may be indicated | Dingoes have dense fur, which creates an ideal environment for moist dermatitis to develop.<br>Close observation and regular ectoparasite control will reduce the incidence | |
| Hip dysplasia | Multifactorial, usually with a genetic component | Hindlimb lameness, abnormal hindlimb gait | Radiography | Dependent on severity, surgery may be indicated | Not reported in wild dingoes, rare in captivity. Dingoes have naturally lax hips compared with domestic dogs, so parameters used for diagnosis in dogs may not be appropriate | 1, 3 |

References: (1) Hulst 2008; (2) Smith *et al.* 2016; (3) Smith and Watson (2015).

**Table 14.4.** Summary of the reproductive activities of the dingo.

Timing may vary slightly across Australia. Reproduced from Smith (2015c).

| Month | Season | Activity |
| --- | --- | --- |
| March<br>April<br>May | Autumn | ▪ Breeding season<br>▪ The frequency of howling, scent marking, ground scratching and aggression increases and signifies pre-breeding<br>▪ Before mating, dingoes spend this time to either find a mate and/or strengthen existing pair bonds<br>▪ Pre-oestrus lasts between 10–12 days and oestrus between 7–16 days in captivity and 30–60 days in the wild<br>▪ Gestation period is ~63 days |
| June<br>July<br>August | Winter | ▪ Whelping and weaning pups<br>▪ Female selects suitable den for whelping<br>▪ Maternal care occurs in two stages: (1) intensive maternal care of pups, which are confined to the den, and (2) when pups are weaned and moved between locations (total food requirements of the pups is much greater) |
| September<br>October<br>November | Spring | ▪ Post-weaning<br>▪ Pups start to become independent around 3–4 months<br>▪ Parents teach and train pups |
| December<br>January<br>February | Summer | ▪ Non-breeding period<br>▪ Pups either remain with the family and share activities, are abandoned or leave territory |

reproduction during Autumn and shifts to the rearing of pups in winter and spring. The calmest period occurs in during summer. Table 14.4 provides a breakdown of the major periods of activity across the year.

Wild dingoes are typically crepuscular, although this is seasonal with daytime activity more prominent during the winter months (Tatler *et al.* 2021). Observations on a population of dingoes in north-eastern New South Wales showed that they travelled the same distance each hour at night but there were peaks at dusk and dawn (Harden 1985). Their movements were characterised by two different patterns: the intense use of a small area with many changes in direction, believed to be associated with hunting and a more directed movement, where they frequently traversed a large part of the home range. However, south-eastern Australian dingoes have also been recorded making large dispersal or exploratory forays (Robley *et al.* 2010). The second pattern may serve to maintain communication between individuals by regular visits to scent posts. In both cases, the activity periods were generally short with 65% being less than an hour and interspersed with shorter rest periods, of which 70% were less than 30 min. They also spent 65% of the day active and 35% of the day resting (Harden 1985).

In contrast to this study, observations in the hotter, less vegetated environment of north-west Western Australia and the Simpson Desert, showed that dingoes tended to be inactive during the middle of the day, with major peaks of activity and movement occurring around dawn and dusk (Thomson 1992b; Corbett 2001a). In north-west Western Australia, dingoes did not regularly travel large distances across their home range, which had a mean width of 10.5 km (Thomson 1992c). There, the movements were localised and groups tended to focus their activity in a particular section of their range for a time, with occasional forays into other parts of their range. Dingoes also showed strong site fidelity and seldom left their home range, but a few were recorded making long-distance (> 150 km) movements (Thomson 1992c).

In captivity dingoes are generally diurnal. Activity occurs in short bursts, notably at cooler times of the day (at dawn and dusk) and around scheduled mealtimes, routine maintenance and animal caretaker talks. Dingoes are generally inactive, but have intermittent active periods lasting up to 30 min. Energy is conserved during most of the day and night where animals are well provided for (L. Watson pers. comm.).

## 9.2 Social behaviour

### 9.2.1 Group formation

Dingoes, like other wild canids, are social animals and where conditions are favourable, they maintain stable family groups with distinct territories that overlap little with neighbouring groups (Thomson 1992c).

Nonetheless, regional variations are seen, reflecting the flexible nature of dingo social structure and space use (e.g. Newsome *et al.* 2013). Observations in the wild have found dingoes living in situations ranging from solitary to family groups of up to 12 individuals (Corbett and Newsome 1975; Corbett 1995). One study found that of over 1000 dingoes seen, 73% were alone, 16.2% were in pairs, 5.1% in trios, 2.8% in quartets and isolated sightings were observed of groups of up to seven animals (Corbett and Newsome 1975).

When dingoes live in groups, the size seems largely the result of food availability, with solitary animals being more common when there is a primarily small to medium-sized prey available and groups more common when there is a greater abundance of larger prey such as wallabies and kangaroos (Corbett and Newsome 1987; Corbett 2001a; Robertshaw and Harden 1985, 1986; Thomson 1992c).

The dingo pack generally comprises an extended family, which similar to other canids includes a mated pair, their offspring of the year and sometimes offspring of previous years (Corbett 2001a). In captivity, dingoes display a dominance hierarchy between both males and females and within males and females, with males generally being more dominant than females and older animals being more dominant than younger ones. Some captive males have been observed to be subordinate to females. Most rank changes occur during the breeding season with plasma cortisol values suggesting that the highest or more dominant and the lowest ranking males show the highest stress levels (Corbett 1988).

Captive dingoes should be kept in pairs. The establishment of non-family groups has been tried, but is problematic. For example, a group established at Taronga Zoo comprised 6-month-old pups from three different litters. This group contained a large male dingo, two females from a second litter and a male and two females from a third litter. Despite the early age at which these animals were placed together and the considerably larger size of the single male dingo, he eventually had to be removed as the three smaller dingoes from the same litter, with the male leading them, overpowered this larger individual.

If breeding is permitted in a captive setting, restricting the number of generations residing in the same enclosure to one or two (size of enclosure depending) is also essential to avoid conflict, as offspring that would otherwise disperse cannot (see section 9.9).

### 9.2.2 Aggression

When dingoes live in a pair or family group, aggressive behaviour can occur over food, attention from other animals or people, as part of play (especially between adults and pups), when an adult is disciplining a pup or to establish and maintain dominance hierarchies. Dominance displays are a natural and important part of dingo social behaviour and are generally greater as the group establishes, as pups are learning the rules of dingo society, and during the breeding season in both sexes (see section 9.2.1; Corbett 2001a; Smith *et al.* 2016). Typical threat displays include an upright stance, the tail being raised in the air, raising the lips to show the teeth, snarling and raising the hair on the back (Fig. 14.5; Corbett 1988). As a result of this aggression the subordinate animal may squeal, rear back and flee or keep its tail down between its legs against the stomach, hold its ears flat against the head and avoid eye contact (Corbett 1988). The lips are generally stretched backward without baring the teeth (submissive grin) and the tongue often protrudes (Corbett 1988). The subordinate animal may also show ritualised passive submission, which is generally slow and deliberate and includes retreating or rolling over on its back to expose its belly. These responses are often associated with the dominant animal standing over the subordinate and licking its genital regions (Fig. 14.6; Corbett 1988).

On occasion, contact between two or more individuals can occur involving snapping only or fierce biting. Physical attacks leading to serious fighting and wounding of combatants either occurs spontaneously or results from less intensive threat behaviour or from returned threat (Corbett 1988).

If two unrelated adult dominant males come together in a captive environment they can cause significant injury to each other, which may result in the death of one animal, so they should never be placed together or have close access. Generally, when a subordinate animal is being attacked it will vocalise a lot but, invariably, few if any injuries result.

Subordinate animals may show ritualised, active submissive behaviour to attain friendly or harmonic social integration towards more dominant animals. During this they approach more dominant animals in a slightly crouched posture with ears flattened, the tail down and wagging and often push the muzzle of the superior animal with their nose and lick its muzzle (Fig. 14.7; Schenkel 1947, 1967; Corbett 1988).

**Fig. 14.5.** An adult female dingo showing a typical threat display. Photo by Bradley Smith.

**Fig. 14.6.** Passive submissive behaviour of a juvenile dingo towards an adult. Photo by Bradley Smith.

## 9.2.3 Friendly behaviour

Dingoes display various types of friendly behaviours, including play, play fighting (with inhibited biting), chasing, greeting behaviour, grooming other individuals and resting close together (Corbett 1988, 2001a; Byosiere *et al.* 2018).

**Fig. 14.7.** Active submission shown by juvenile and subordinate dingoes. Photo by Bradley Smith.

## 9.2.4 Vocalisations

Auditory communication is important for dingoes and other wild dogs because they are often spatially separated (Fleming *et al.* 2001; Déaux and Clarke 2013). Dingoes have a broad vocal repertoire and make several different types of sounds and vocalisations, including howls for long-range communication, bark-howls and snuffs, each with several variations. The types of vocalisations and their context are listed in Table 14.5.

The howl can be made either by an individual or as a chorus where several dingoes howl together and is able to convey individual information (e.g. sex) and group information (e.g. size) (Déaux *et al.* 2016a). Howling serves several functions (Corbett and Newsome 1975; Breckwoldt 1988; Corbett 2001a; Déaux *et al.* 2016a):

- locating members of the same group for group behaviours such as cooperative hunting
- locating mates for breeding
- defining territory and avoiding strangers
- calling pups in
- expressing alarm
- signalling members of the group that they are moving.

In the wild, howls are mostly made during the night, with peaks at dawn and dusk (Corbett 2001a). This is also generally true in captivity, but other sounds, such as sirens, loud noises and even particular people (e.g. veterinarians) walking past, can trigger dingoes to howl during the day (S. Jackson pers. obs.).

Although it is generally considered that dingoes cannot bark, they can, as can all wild canids. It is the context in which barking occurs that differs between wild and

**Table 14.5.** Common dingo sounds and vocalisations and the primary context for their production.

Derived from Déaux and Clarke (2013).

| Sound category | Primary context of production |
|---|---|
| Snuff | Frustration and anxiousness |
| Snort | Frustration and anxiousness |
| Woof | Warning and defence |
| Bark | Warning and defence |
| Growl | Threat and attack |
| Snarl | Threat and attack |
| Bark-howl | Warning and defence |
| Howl | Long-range social communication |
| Whine | Frustration and anxiousness |
| Whimper | Frustration and anxiousness |
| Yelp | Fear and pain |

domestic canids. Dingo barks are generally sharper, more abrupt and throaty than the barking of domestic dogs and usually comprise only a single or low number of barks, as opposed to the frequently prolonged barking of modern domestic dogs. The bark-howl generally starts with one or several barks and then follows with a plateau howl or sometimes an inflection howl and is generally used in cases of extreme alarm in the wild (Corbett 2001a; Déaux and Clarke 2013; Déaux *et al.* 2016b).

Dingoes sometimes moan, a soft-pitched plateau howl, which is used for medium range (100 m) communication (Corbett 2001a). They snuff by repeatedly expelling air through the nose and generally do this when an individual is feeling threatened, unsure or startled by a human (Corbett 2001a; Déaux and Clarke 2013).

In captivity, vocalisation remains an important part of the dingo's social behaviour. Lone or separated dingoes will often vocalise by howling. Dingoes will also howl socially upon certain cues (e.g. at sunrise and sunset), at feeding time or upon the arrival of a familiar person (B. Smith pers. obs.).

### 9.2.5 Pheromones

Dingoes have a highly developed sense of smell that they use in social communication (Corbett 2001a; Fleming *et al.* 2001). They possess scent glands, including anal glands, interdigital glands and caudal glands, and spread their scent via urine, faeces and directly from their scent glands (Corbett 2001a; Fleming *et al.* 2001). Scent posts are often used to deposit scent and include grass tussocks, small bushes, fallen logs, fence posts, rocks and the faeces of other species. Dingoes will also scratch the ground, particularly after urinating or defecating. Scent posts are most common around shared resources such as water sources, hunting grounds and frequently used tracks (Corbett 2001a). The pheromones released by different individuals no doubt serve to convey their sex, age, social position and reproductive status. Observing scent marking can reveal insights into the hierarchy of dingoes, as only more dominant individuals urinate over the site of a subordinate (B. Smith pers. obs.).

## 9.3 Reproductive behaviour

The onset of breeding season varies according to age (dingoes can reproduce in their first year but most do not bear pups until they are 2 years old); social status (often the oldest or most dominant females come into oestrus before others); latitude (with those in northern Australia whelping earlier than the south); and seasonal conditions (can vary in response to environmental conditions such as drought) (Smith 2015c). During the breeding season, dingoes reach a peak in activities such as raised leg urination and increased howling behaviour, which correlates with increased testosterone and oestrogen levels (Thomson 1992b). See also section 10 on breeding.

Reproductive behaviour in subordinate males is generally suppressed by the dominant male with threats, intense attacks and imposition behaviour to drive them away from females (Corbett 2001a). The dominant female also suppresses subordinate females' courtship activities, but less frequently and less intensely (Corbett 2001a). It is suggested that this behaviour also helps to reduce the incidence of litter-mate breeding (Corbett 2001a). Although many mating attempts are usually suppressed, subordinate females have been observed to copulate and produce young in favourable conditions (e.g. at mine sites (P. Fleming pers. obs.)) and in captivity (B. Smith pers. obs.). However, in captivity, these young may be killed by the dominant female (Corbett 1988), highlighting the importance of carefully managing breeding and adults' access to other dingoes' pups (see section 9.10 for further details).

## 9.4 Bathing

Dingoes usually avoid large bodies of water, but in hot weather they may stand, lay or play in water to cool off. Providing a pond or moat with walk-in access or even a large bathing pool can be a beneficial addition to an enclosure.

## 9.5 Behavioural problems

Behavioural problems include escape behaviour, high levels of aggression towards other dingoes and humans and destruction of the enclosure. Captivity can be a highly stressful environment for dingoes (see Table 14.3), particularly during the breeding season (see section 8.3).

## 9.6 Signs of stress

Behavioural signs of stress include frequent vocalisations (whining, howling), chewing parts of their own body or the enclosure, hypervigilance, freezing, exaggerated startles, pacing, avoidance and aggression towards other dingoes and humans. Body language signs include tail tucked under body, ears lowered, yawning, lip licking, panting and raised hackles. If the cause of stress is not managed, it can lead to chronic stress and post-traumatic stress disorder (as has been seen in captive wolves) (Mallonee and Joslin 2004). The presence of stress in dingoes can also be assessed by measuring corticosteroid concentrations in saliva, blood or faeces. Studies suggest that dingo pups have higher basal corticosteroid concentrations than adults and that adult female and male concentrations do not significantly differ, except during the breeding season when male concentrations are elevated (Smith *et al.* 2016).

## 9.7 Behavioural enrichment

Behavioural enrichment can reduce boredom, increase physical fitness and promote natural behaviours in captivity. Enrichment can be divided into the following categories:

- social (human and conspecific)
  - maintaining dingoes as either pairs or family groups rather than as solitary individuals
  - animal caretaker interactions (patting, feeding, playing games)
  - training (including lead training and basic targeting behaviours)
  - grooming individuals using a dog brush, which also allows a close physical inspection and increases the bonds between the animal caretaker and the individual dingoes
- food-related
  - varying the diet throughout the week (see section 6). Note that enrichment food items should be considered when formulating the daily diet to avoid overfeeding
  - varying feed times so that the dingoes are not anticipating and waiting to be fed
  - feeding more than once per day
  - scattering a portion of their favourite dry food into bushes or hard to get at places can create feeding stimulation and hide and seek games
  - hiding food in containers (including boxes and bags)
  - providing food during visitor experiences
  - providing raw bones to chew on – this can provide significant enrichment as the dingoes gain access to bone marrow and chewing helps to maintain dental health
  - providing insects, which promote curiosity and hunting behaviour
  - putting blood and/or meat or bones in ice blocks so that it takes longer to gain access to the food
  - providing treat balls or puzzle feeders that require manipulation to obtain treats
  - wrapping food in an old sugar sack or horse hide and throwing it into the enclosure; this will promote tugs of war, group feeding and demonstrate the group hierarchy (Allsopp 1998)
- olfactory and auditory
  - spraying blood of food items in different parts of the enclosure to stimulate smell and hunting behaviour
  - walking dingoes outside the enclosure (around the facility as well as in unoccupied enclosures of other species) so that they can smell the various scents in their surroundings
  - scattering spices (and other strong odours) to stimulate the sense of smell
  - dragging food items around the enclosure and hiding them to stimulate smell and hunting behaviour
  - playing audio of other carnivores or prey animals to stimulate their hearing (Allsopp 1998)
- physical
  - daily exercise, such as walking outside the enclosure
  - providing room to run within the enclosure and encourage activity by providing lures and scent-finding games
  - providing pieces of leather or hide for the dingoes to chew or pull on
  - providing platforms to allow dingoes to survey their enclosure and beyond from a height
  - providing dirt and large hollow logs to allow dingoes to dig dens and hide

- tunnels, climbing frames, logs, rocks
- rotation of dingoes through different yards/parts of the enclosure
- planting various species of plants such as grass tussocks and bushes.

## 9.8 Training and conditioning

Dingoes have a reputation for being difficult to train because they are considered highly independent, neophobic and strong-willed. This claim is supported by many anecdotal accounts of dingo behaviour in captivity, high rates of relinquishment and studies comparing owner-rated behaviour of dogs and dingoes (see Smith 2014; Smith *et al.* 2017). However, dingoes are not difficult to train once their needs are understood and responded to appropriately. Dingoes are less predisposed than modern domestic dogs to monitoring people (and considering humans as dispensers of 'good things'), which is a compelling and unique aspect of their evolutionary pathway (N. Bishop pers. comm.). Dingoes are also generally more neophobic as they age (from 12–18 months onwards) and unless well socialised from an early age become less able to respond to new or changing places, people, training and experiences in adulthood.

Although training dingoes has its challenges, it is highly encouraged in captive contexts. Some degree of training and cooperative ability will improve animal welfare and management (e.g. crate training to reduce stress during transport and training to allow veterinary examination). Training can offer an enrichment opportunity, enhance visitor experiences (Fig. 14.8) and improve the relationship between the animal caretaker and the dingo. Dingoes have been trained as scent detector animals; as certified animal-assistance animals; and to howl on command (Smith and Watson 2015).

Recommended key behaviours for training dingoes in captive settings include:

- habituation to novel stimuli
- sit and stay
- lead, harness and muzzle training from an early age. Front attaching harnesses are preferred so the dingo cannot pull with the full force of its weight from the shoulders
- stationing – to go to a designated area and stay there until cued otherwise. For example, standing on a scale to be weighed (see Fig. 14.8)
- targeting – to touch a target (e.g. hand, stick, board) with nose or paw.

The importance of exposing each dingo to a wide variety of stimuli at the appropriate time in the young dingo's development cannot be understated (N. Bishop pers. comm.). It should begin at approximately 8 weeks of age or as soon as possible. It will require animal caretakers who are experienced in positive reinforcement techniques, have knowledge of canine and other carnivore's behaviour, are patient and observant and have the time to dedicate to regular training sessions. Dingoes will form strong attachments with familiar people (particularly as pups and juveniles, less so as adults), so

**Fig. 14.8.** The dingoes at Cleland Wildlife Park have been trained to station on logs during the daily animal caretaker presentation. Photo by Bradley Smith.

ensuring several personnel form reliable relationships with the dingoes early is important for long-term success.

All training should utilise positive reinforcement methods. That is, any consequence for a behaviour should aim to increase the likelihood of that behaviour occurring again in the future. Initial training sessions should involve play, feeding and grooming. When playing with a pup make sure that it is taught to be gentle when taking food from the hand. Natural behaviours can be easily trained by calling 'come' when it comes towards you or 'sit' when it sits and offering a food reinforcer. The food reinforcer is the most meaningful aspect of this, with the verbal 'praise' paired with it as a secondary reinforcer (N. Bishop pers. comm.).

When training, adhere to the modern principles of animal training for training dingoes. A training program should be developed in consultation with a person who is a qualified/experienced animal trainer. The following principles and considerations for training dingoes are based on the dingo training guides developed by Nicholas Bishop and the staff at Zoos South Australia. Importantly, develop a training schedule and plan with the following aspects:

- establishes the *purpose* of the training (e.g. goal of the behaviour and reason for the behaviour)
- considers the *practical aspects* of the training (e.g. the individual's learning history, frequency and duration of training sessions, safety considerations and how sessions and progress will be documented)
- determines the *training methods* to be used (e.g. a definition of the behaviour, the cues, bridge and reinforcers to be used, the learning environment and learning steps).

Other considerations when working with dingoes include:

- keeping training sessions brief and being willing to cease a session early if required. Reasons for not commencing a session or for ending a session early include when the dingo is distracted, agitated, unwell, satiated or appears disinterested in the task
- limiting distracting stimuli (e.g. enrichment objects, food, conspecifics) before commencing training sessions
- always being patient and altering expectations as needed during training sessions
- ensuring consistency of training methods when multiple trainers are involved. The more trainers entering the program, the harder it can be to keep track of inevitable variations in technical approach - with subsequent confusion and behaviour breakdown for the dingoes
- ensuring trainer roles are clear and assigned. For example, a primary trainer (develops the training plan, consults relevant partners, coordinates, records and updates plans) *and* support trainers (assists, collaborates with and supports the primary trainer)
- utilises positive reinforcement training methods
- establishing a reinforcer that the dingo finds most rewarding. This will likely come down to the individual and its learning history. When using food, do not discount food that they receive as part of their daily diet as being highly rewarding because some dingoes avoid novelty such as commercial dog treats (B. Smith pers. obs.).
- avoiding aversive responses such as harsh verbal or physical correction. Behaviour orients to reinforcement and dingoes set up their approaches to minimise punishers and maximise reinforcers (N. Bishop pers. comm.). As such, make it clear what is expected in the first place and make it worth their while. For example, using a firm 'NO' is not meaningful unless it has been attached to a punisher - which should be avoided.
- with food reinforcers, use the smallest unit of food that is still reinforcing. Attention can be improved or maintained by varying the reinforcer occasionally in the stream of the training session
- the trainer does not need to be dominant over the dingo to successfully train or achieve obedience (e.g. never grab the dingo by the neck or roll it on its back in an act of dominance).

## 9.9 Introductions and removals

Unrelated pups can be introduced to existing family groups at 6–10 weeks of age, under close supervision. It is essential that such introductions are closely monitored to ensure the pups are not harmed. Introduction of new dingoes into existing groups can be successful during the natural nurturing period, from late winter to autumn when young are highly dependent and adults naturally inclined towards altruistic behaviour (L. Watson pers. comm.).

Dingoes that are part of a social group become stressed when separated. Any animals that are removed from either a pair or group should be returned as soon as possible, especially in the case of more subordinate individuals. If a subordinate individual has to be removed for more than 1 day (e.g. for treatment), it may be worthwhile to hold it in an adjacent area to the social group area for a further day or two to allow the other dingoes to re-establish auditory, olfactory and visual contact with it before its return. The release of an animal to its mate or group in this situation is best undertaken in the morning to maximise the time that it can be closely monitored to assess how well it reintegrates into the group. Upon return, some aggression is expected as the dingoes re-establish their position within the group; however, they should socially re-establish within several hours. If, by the end of the day, the dingo has not settled into the group, it may be worthwhile removing the dingo overnight into the holding area and trying again the next morning.

## 9.10 Intraspecific compatibility

Dingoes can be held as solitary individuals, but it is preferable that they be held in male–female pairs or in family groups (i.e. a pair and their offspring or siblings). Keeping three or more unrelated dingoes together can often end in aggression and fighting between individuals (even when neutered), so they should be regularly monitored. Dingoes, particularly if unrelated, display extreme same-sex-directed aggression, which will likely result in serious injury and/or death. Family groups should only be maintained until the offspring are ~18 months old for females and 12 months for males, as this is when the natural dispersal occurs. If they are not separated (and rehoused) at this time, animals become highly stressed and fearful and serious fights and injuries can result as there is nowhere for these dingoes to go because they are restricted by their enclosure boundaries (see section 9.2.1). Subadults can be successfully kept in groups of up to six but when they reach 12–18 months will require close monitoring and housing in smaller groups and ultimately in pairs as dominant animals and hierarchies become manifest (L. Watson pers. comm.).

Whether actively breeding, or an incidental litter is born, restricting the number of generations residing in the same enclosure is essential because social and spatial restriction is one of the major causes of stress in dingoes. Corbett (1988), for instance, housed multiple generations of dingoes (reaching a maximum of 16) in a small enclosure (45 × 25 m). Individuals that would otherwise have hidden or naturally dispersed in the wild were prevented from doing so, resulting in high levels of aggression and subsequently death. Other aberrant behaviours reported in Corbett's (1988) captive groups included infanticide by dominant females, which has yet to be observed outside of captivity (in fact the opposite is true, see Smith and Vague 2017).

## 9.11 Interspecific compatibility

Due to their large size and territorial and carnivorous nature, dingoes are unsuitable for housing with most species. Any young child or animal that is likely to panic, run, squeal in alarm or act like prey should be carefully protected. Pet dingo pups while young will bond with already resident household pets and will readily accept a domestic dog as a companion; however, given their highly territorial nature, it is not recommended to attempt to introduce a new animal to an adult dingo. Potential options for mixed exhibits include fish or tortoises (although they may eat them if given access) if a large water body such as a moat is used to keep the dingoes in the enclosure.

## 9.12 Socialisation with humans

In captivity, dingoes need to be socialised with humans from as early an age as possible, with the period between 2 and 8 weeks being critical. Socialisation and training after this is possible, but becomes more limited as the weeks advance (see also section 7 for timings relating to relocating dingoes). Socialisation includes getting the dingoes to experience as many sights, sounds, settings and situations as they will likely encounter in their life. It is likely that timid dingoes will always be timid, and more outgoing and dominant dingoes will generally remain outgoing and dominant. A key change in behaviour and temperament occurs around 12–18 months of age when even the most outgoing pups can become highly neophobic, timid and aloof (B. Smith pers. obs.). During this transition stage into adulthood is another key period when negative experiences may lead to long-term psychological impacts.

Despite the best attempts to socialise some individuals they can still be easily frightened by the unexpected, particularly when outside their enclosure. When scared, they can react unpredictably so great care needs to be taken. Even a person they know well who looks different by wearing a hat, a coat that rustles, different clothes or sunglasses may be treated warily until the dingo establishes who they are. Interactions and relationships between dingoes and humans are on their terms – they can highly affectionate (especially as pups), but also very aloof and independent. See section 9.8 for suggestions relating to interacting and training dingoes.

At different times, dingoes can change in behaviour in response to seasonal changes (see also section 9.1). At the onset of the second breeding season their behaviour undergoes a subtle change, whereby outgoing adolescents may become more cautious and aloof (ADCA 1996). Other changes in behaviour include during oestrus when the female and nearby males can become more aggressive. Males can also become aggressive towards their animal caretakers if they get between them and their mate at this time (ADCA 1996).

## 10. BREEDING

### 10.1 Mating system

Dingoes have an annual breeding season, which commences with pre-oestrus approximately late January and can continue through to late June. Adults may become aggressive towards subadults of the same sex from December onwards, so arrangements should be in place to hold animals not meant for breeding separately at that time to avoid stress and potential injury.

Dingoes are primarily polygynous, with the dominant male mating with all the females in the group. However, subordinate males may also mate with subordinate females. Females have only one oestrous cycle each year and in the wild they normally do not breed until their second year. Males are fertile all year, except in hot arid climates where spermatogenesis is diminished during the hotter months (Catling *et al.* 1992).

Matings tend to occur over 7–10 days. During mating, as with all canids, the male mounts the female from behind, which results in a copulatory tie whereby the two dingoes become 'locked' together and cannot break away, lasting anywhere between 5 and 30 min (Fig. 14.9). Pups are always born during winter months, so appropriate shelter must be provided.

### 10.2 Ease of breeding

Dingoes breed readily in captivity and females are able to breed in their first year (although this is not recommended – see section 10.7). When planning to breed from a male and female, it is important to introduce them as early as possible before the breeding season commences to give them the opportunity to bond, which will increase the likelihood of conception (Smith 2015c).

### 10.3 Oestrous cycle and gestation period

Oestrous typically lasts 7–16 days and the gestation period for dingoes is typically between 61 and 69 (typically ~63) days for captive dingoes so that births occur from May to August each year (Jones and Stevens 1988; Catling *et al.* 1992; Corbett 1995; Smith 2015c).

### 10.4 Reproductive status

#### 10.4.1 Females

The different reproductive stages of dingoes are:

- anoestrus – the non-breeding phase in which the female has a small vulva with no vaginal discharge. On vaginal cytology, epithelial cells are non-cornified, small and rounded (Corbett 2001a)
- pro-oestrus – typically lasts 10–12 days. Vaginal bleeding occurs, the vulva begins to swell, there is a marked increase in the frequency of urination (often with evidence of territorial marking) and an increase

**Fig. 14.9.** (Left) A pair of dingoes mating. (Right) Copulatory tie. Photos by Bradley Smith.

in interest from males (Jones and Joshua 1982; Corbett 2001a). On vaginal cytology most epithelial cells are non-cornified, but becoming large, angular and cornified (Corbett 2001a)
- oestrus – typically lasts 7–16 days. The vulvovaginal swelling is maximal, vulval discharge becomes less haemorrhagic, even colourless, and the female becomes more accepting of the male to mate (Jones and Joshua 1982; Corbett 2001a). On vaginal cytology most epithelial cells are large, angular and cornified (Corbett 2001a). During oestrus, females may go days without eating and can lose several kilograms (ADCA 1996). If not mated, females can prolong oestrus by going off and coming back into oestrus 30 days later (especially in dominant females) (B. Oakman pers. comm.). Therefore, if not required for breeding, pairs should be separated for at least 3 months or contraception is used.
- metoestrus – follows oestrus when females do not conceive. The vulval size is variable but decreasing in size and the discharge is variable in volume and composition. On vaginal cytology, most epithelial cells are non-cornified, becoming small and rounded (Corbett 2001a).
- pregnancy – follows successful mating with conception
- lactation – follows birth of a litter. The teats are enlarged for milk production
- post lactation – follows weaning of pup with teats regressing and no milk or only a clear liquid being produced.

If young are present there are several developmental stages and measurements that can be recorded and compared with existing growth curves (see section 10.16) or used to establish curves for future reference (see Smith 2015c). These include:

- developmental stages
  - physical
    - teeth eruption (partial, full)
    - eyes open (partial, full)
    - crawling
    - walking
    - leave den independently
    - move freely outside of den
    - running
    - urinate/defecate outside of den
  - responses
    - respond to handler's voice
    - response to loud noise
    - approaches fence to greet humans
    - exploratory behaviour
    - fear of objects
  - feeding (parental)
    - suckling (mother laying down)
    - suckling (mother standing up)
    - regurgitation by adult to pups
    - soliciting behaviour for food
    - weaning
  - feeding (independent)
    - eating solids
    - self-feeding
  - vocalisation
    - bark, howl, growl, whine, yelp/cry
  - social/play
    - wags tail
    - plays with littermates
    - group play with objects
  - independence
- measurements (see Appendix 3)
  - body mass (g)
  - head length (mm) – from the occiput to snout tip
  - head width (mm) – maximum width across the zygomatic arches
  - head/skull circumference – taken just behind the eyes. Measure all the way around the head and across the top of the head between the ears
  - ear length (mm)
  - shoulder height (mm) – from the ground to withers (highest point of the shoulder blades)
  - body length (mm) – from snout tip to anus/base of tail
  - tail length (mm) – from the anus to the end of the last vertebra of the tail tip
  - total length (mm) – from snout tip to tail tip
  - chest/girth (mm) – measured at the widest part of the chest just behind the front legs. Measure around the entire chest
  - tibia length (mm) – from the hip to the bottom of the pes.

### 10.4.2 Males

Reproductive stages are usually externally visible, as the males' testes swell to twice the normal size and the urine excreted is darker in colour and stronger in odour (B. Oakman pers. comm.). There is also a large increase in prostate weight, testes weight, the number of tubules in the testes and the number of sperm in the ejaculate (L. Corbett pers. comm.). When females are in oestrus, males may go for days without eating and can lose up to 5 kg, which is a third of their bodyweight (ADCA 1996).

## 10.5 Reproductive management

Dingoes breed readily in captivity so reproductive management usually focusses on prevention of reproduction. Techniques used for prevention of breeding may be temporary or permanent. Temporary methods include separation of the sexes before the start of the breeding season or the use of hormones (oral, injectable or implants). Gonadotropin releasing hormone (GNRH) agonists are considered the safest reversible contraceptives and progestins should only be used short-term due to undesirable side effects (AZA Reproductive Management Centre 2021).

The Dingo Discovery Sanctuary and Research Centre recommends temporary fertility control of males using a 4.7-mg GNRH agonist implant (Suprelorin®) placed annually by the end of January in advance of the breeding season (or 6 weeks before the autumn equinox if in northern hemisphere) (L. Watson pers. comm.).

Dingoes that are no longer required for breeding can be permanently sterilised through vasectomy or surgical removal of testes in males and ovariohysterectomy in females. Removal of gonads is best done after 12 months of age to allow full physical development.

## 10.6 Occurrence of hybrids

Over the past 230 years, some dingoes have mated with modern breeds of dogs, resulting in hybrids (Newsome and Corbett 1985; Stephens *et al.* 2015; Cairns *et al.* 2021a). Although wild dingoes usually maintain predominately dingo traits according to morphology (Parr *et al.* 2016) and genetic testing (Stephens *et al.* 2015; Cairns *et al.* 2021a), hybridisation constitutes an existential threat to the purity of the dingo population in some parts of Australia.

Genetic testing provides an accurate and reliable tool for wildlife managers and conservation organisations to assess the ancestry of a dingo population or individual to inform conservation or management policy and captive breeding. Ideally, any dingoes kept in captivity that might be used for breeding should be DNA tested. In general, animals that are identified to have a significant portion of dog ancestry should be excluded from breeding programs and surgically sterilised.

Although skull morphometrics were traditionally used to assess the purity of dingoes (Corbett 2001a), genetic testing is now considered to be the only method of reliably distinguishing pure dingoes, mixed dingoes and domestic dogs (Stephens *et al.* 2015; Cairns *et al.* 2021a). Genetic testing can be performed on a range of biological samples, including saliva, buccal swabs, blood, tissue biopsies and sometimes scats or hair. Where possible, non-invasive samples such as saliva and buccal swabs are preferred. Sedation might be required to collect blood, saliva or buccal swabs from some dingoes. There are two main methods of genetic testing that are used to assess the ancestry of dingoes. The first genetic test was developed in the late 1990s by Dr Alan Wilton and now uses 23 microsatellite markers to estimate whether an animal is a dingo, dingo hybrid or domestic dog (Wilton *et al.* 1999; Wilton 2001). Due to statistical uncertainty, geographical variation and the relatively low number of genetic markers, microsatellite DNA testing results should be considered estimates rather than a specific percentage of dingo ancestry. Despite this limitation, microsatellite genetic testing has been the basis of extensive genetic surveys of dingo populations across Australia (Stephens *et al.* 2015; Cairns *et al.* 2021a) and has been used to verify the ancestry of captive dingoes kept by conservation organisations before breeding (Smith 2015b). It is important to understand that no genetic test is without limitations. Users should carefully evaluate this microsatellite method, as it is restricted by the small number of markers, gives only a broad estimate of dingo ancestry percentage, cannot identify specific dog breed ancestry and may be biased by regional differences in dingoes (Cairns *et al.* 2020, 2021a). In practice, any animal with a microsatellite genetic test result over 90% dingo should be classified as a 'pure' dingo and samples with a result between 80–90% dingo as a 'likely pure dingo' (Cairns *et al.* 2021a; D. Stephens pers. comm.).

The second genetic testing method uses thousands of single-nucleotide polymorphism (SNP) markers and ancestry modelling (Cairns *et al.* 2011; Cairns 2021), similar to the human genealogy tests now widely available. The SNP method harnesses large datasets of wild or captive dingoes, domestic dog samples and commercially available SNP testing platforms to provide a much higher degree of accuracy and precision in identifying the percentage dingo versus dog ancestry. SNP genetic testing can also be used to identify the specific breeds contributing to dog ancestry components, as well as providing information about the animal's dingo regional ecotype and level of genetic inbreeding. Future advances may be able to assess the presence of common disease or phenotype markers, drawing upon the wealth of knowledge about domestic dogs and inherited diseases. Additionally, SNP genetic testing uses dingo samples from across Australia so is less likely to be biased by geographical variation than other methods. One drawback of SNP genetic testing is a higher cost per sample and longer turnaround time. Either genetic testing method can be used to inform and critically assess the identity of an individual dingo for captive breeding or the monitoring of wild populations, but accuracy may be preferable in a captive-breeding context.

## 10.7 Timing of breeding

Dingoes are typically monoestrus, with mating occurring in the wild between March and June, though in captivity mating has been observed as early as late January (Catling *et al.* 1992; Corbett 1995; Smith 2015c). Births in the winter months are timed to coincide with conditions optimal for raising young (e.g. beginning of seasonal blooms of food) (Smith 2015c). The breeding season for dingoes is triggered by changes in daylight/sunlight (photoperiodism) (see Fig. 14.10). Dingo breeding season begins when the daylight hours are decreasing. It is rare for matings that result in pregnancy to occur after the winter solstice (shortest day of the year) and after the summer solstice (longest day of the year) (Lord *et al.* 2013; Smith 2015c). Outside of this timing, it is not required to try and prevent mating. By contrast, modern dogs can conceive at any time of the year and have multiple litters per year.

If the females are living in a stable family unit, the most dominant female (which is usually the oldest or mother) generally comes into oestrus before other females in the group, with some subordinate females going through a pseudopregnancy (Corbett 1995). If no mating occurs the oestrus period is prolonged and some females have been observed to come into oestrus a second time at 1 month after finishing oestrus (ADCA 1996; Lord *et al.* 2013).

## 10.8 Age at first and last breeding

Young female dingoes become sexually mature in their first season when they are 9–12 months old; however, few of them breed in their first year in free-ranging populations (Green and Catling 1977). In the wild, females generally commence breeding when they are 2 years of age and males reach full sexual maturity at 1–3 years of age (Corbett 1995). Observations of wild animals found females to have their first oestrus between 1 and 4 years of age, with only 36% of those less than 2 years old being sexually mature (Jones and Stevens 1988). However, captive females can breed in their first year/season as early as 9–10 months of age (Thomson *et al.* 1992; Corbett 1995; ADCA 1996), but it is recommended that they do not breed at this age as they are still in adolescence. Males reach sexual maturity at 1–3 years of age and observations in the wild found 63% of males less than 2 years of age were sexually mature and 97% of those aged 3 years or more were sexually mature (Jones and Stevens 1988; Corbett 2001a).

## 10.9 Ability to breed every year

Dingoes are able to breed every year (Corbett 1995; Smith 2015c), but adhere to a strict annual breeding season (see section 10.7).

## 10.10 Ability to breed more than once per year

As a true wild canid, female dingoes only produce one litter per year during the breeding season (winter), unlike domestic dogs that can breed twice per year at any time of year (Lord *et al.* 2013; Section 10.6; Fig. 14.10). As noted in section 10.4.1 female dingoes can experience a second oestrus if not successfully mated. Dingoes carrying some modern dog ancestry are more likely to breed once per year like pure dingoes (Cursino *et al.* 2017), but some may be capable of breeding twice per year.

## 10.11 Whelping dens

In the wild, whelping dens are often made in enlarged rabbit-holes, caves in rocky hills, hollow logs, under debris in dry creek beds, under large tussocks of grass, among protruding tree roots and under rock ledges along water courses (Corbett and Newsome 1975; Corbett 1995; Smith and Vague 2017). Dens in the wild mostly occur very close to water sources, including small perennial springs and soaks, and are elevated to provide the dingoes with an overview of the immediate area (Corbett 1995; Smith and Vague 2017). Smith and Vague (2017) measured 24 dens in the Great Sandy Desert, with the openings of the dens, on average, measuring 456 mm wide and 421 mm in height and were typically domed. The average depth of the chamber measured 1045 mm. Dens typically had only one opening and faced south-west to avoid the harsh midday sun (Smith and Vague 2017). In captivity, dingoes should be provided with a secure den that allows them to shelter from the

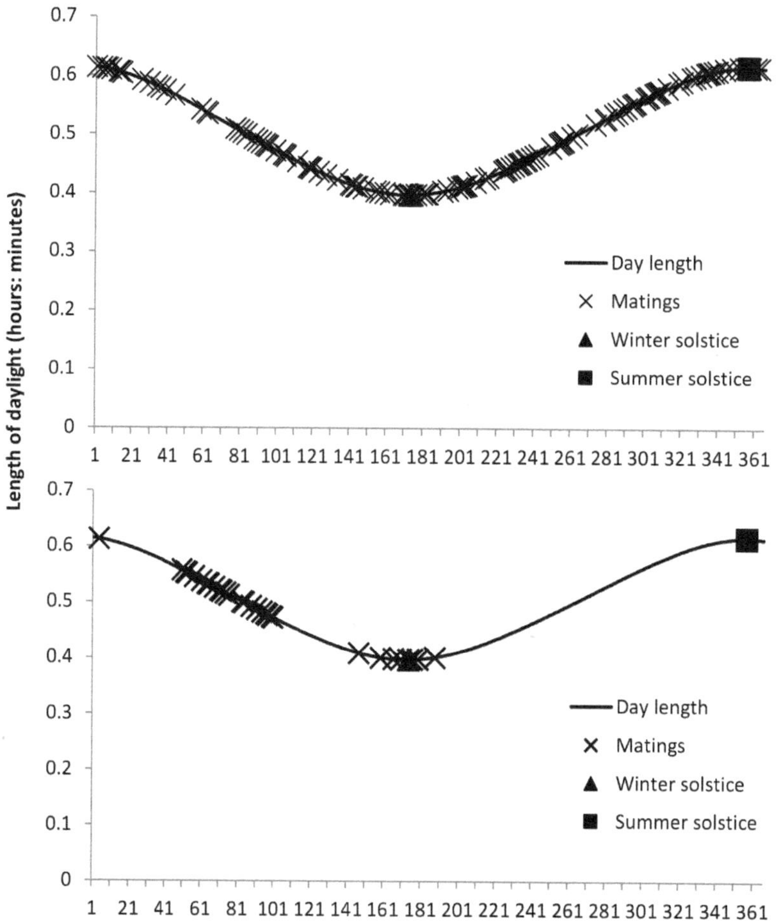


**Fig. 14.10.** Number of conceptions of (top) domestic dogs (*n* = 141, from Guide Dogs Victoria) and (bottom) dingoes (*n* = 36 from Dingo Discovery Sanctuary and Research Centre) plotted against the length of daylight of the 2008 calendar year. Reprinted from Lord *et al.* (2013), with permission from Elsevier.

weather and faces away from the predominant winter winds. These dens should be large enough to allow the female to stand and lie down comfortably and ideally have a piece of wood at the entrance that is high enough to keep the young inside. Nesting material can include hay, straw or shredded paper (Fig. 14.11). It is also important to make sure the dens are accessible by animal caretakers and for ease of cleaning once the pups are weaned.

Shelter boxes should be roofed with only a slight fall angle to allow dingoes to use it as a vantage point and resting place. Openings should be small to give the dingo privacy and a refuge that can be defended during rough play, with a lip across the base to keep bedding material and young pups contained (Fig. 14.12). They should be raised off the ground sufficient for the dingo to get underneath and not to give shelter to vermin, but not too high to preclude any pups from entering from outside. Natural log ramps and clumps of native grass can be planted to give the opening further cover (L. Watson pers. comm.).

Observations of captive dingoes found that females with pups occupied the den almost continuously during the night, but significantly less so during the day. When in the den females spent most of the time nursing their pups. In contrast, the father of the pups did not typically spend time inside the den despite space within the den, low outside temperatures and lack of apparent aggression from females (Hudson *et al.* 2019).

## 10.12 Breeding diet

Pregnant and lactating females have a higher caloric requirement. During the last 2 weeks of pregnancy and during lactation the mother should be fed additional rations that are split between two feeds (e.g. up to twice

**Fig. 14.11.** (Top) Dam and pups inside a captive den at the Dingo Discovery Sanctuary and Research Centre. Either straw (left) or shredded paper (right) can be used inside the den. (Bottom) Pups outside the den. Photos by Lyn Watson/Australian Dingo Foundation (left), Gianluca Caretta Photography (right) and Bradley Smith (bottom).

the maintenance diet), increasing as required to maintain her physical condition.

Pups do not need to be interfered with while they are being provisioned by their mother. Weaning can commence with the eruption of milk teeth or from 3 weeks as the pups show inclination and interest. Pups can be weaned onto raw meat, such as minced chicken, beef or kangaroo that is mixed with a high-quality commercial (puppy formula) ration twice daily for total nutrition as they grow. Large meaty bones and water should be available ad libitum. Cow's milk is not recommended as a substitute for natural mother's milk (see section 11.3). Water and a ration of high-quality commercial dog food will result in healthy bone and muscle development (for additional details, see section 11.10). It is normal behaviour for adult dingoes to regurgitate food for pups during the nurturing phase of each year (see section 10.14).

**Fig. 14.12.** Custom-made wooden whelping box used at the Dingo Discovery Sanctuary and Research Centre (860 mm (H) × 940 mm (W) × 635 mm (D) with 360 × 460 mm opening). When the pups develop the box can be turned upside down to allow the pups to exit. Photo by Kevin Newman.

## 10.13 Number of young born

The average litter size is ~5 (range 1–10) and is slightly female-biased (Corbett 1995; Smith 2015c). Over 36 litters born in captivity over 9 years, Smith (2015c) reported a sex split of 2.2 males to 3.4 females. When dingoes are in a family unit, usually only one litter is successfully raised, usually that of the oldest or parental breeding pair, indicating reproductive suppression of the other females in the group. However, when conditions are favourable, multiple females will produce litters and communal denning occurs (Smith and Vague 2017). Dingoes encounter few problems during whelping, but unrelated male and female dingoes should be separated from the breeding pair and pups as they are likely to attempt to kill them (B. Smith pers. obs.).

## 10.14 Age at weaning

Alloparenting (helping) behaviour occurs in dingoes, with both male and female pack members helping the dominant female bring food to the pups (Lord *et al.* 2013; Corbett 1995). Allonursing has also been observed in captive dingoes when a subordinate female continued to nurse after her pups were killed (Corbett 1988) and in the wild where three litters were observed in the same den (Smith and Vague 2017). The helpers also teach the pups hunting, fighting and social behaviours such as submission, begging and greeting (ADCA 1996). In the wild, dingo pups are normally suckle for 7–10 weeks and have been observed to follow the pack to kills from 9 weeks, with the pups being fed by adults up to 20 weeks of age (Corbett and Newsome 1975; Breckwoldt 1988; Thomson 1992a). Pups are typically weaned by 12–24 weeks and are either abandoned by the mother and left to fend for themselves, or remain in the company of adults and share in pack activities (Corbett and Newsome 1975; Corbett 1995; ADCA 1996; Lord *et al.* 2013).

## 10.15 Age at removal from the mother

Family bonding is immensely important to dingoes, so the sire and dam should be left together to care for the young for as long as possible. Fathers play an important role in pup socialisation and should be allowed to interact with pups. As the mother begins to reject the pups, the father will gradually take over the socialisation. In some cases, the father or other adults may show aggression to young pups. In this situation, those adults should be separated from the pups. In free-ranging dingoes, dispersal of offspring usually does not occur until around 18 months, but can occur at any time.

If pups are intended to live in close proximity to humans, then early human socialisation is recommended, with handling of pups commencing during the first 48 h after birth and continuing regularly thereafter.

## 10.16 Growth and development

The growth of dingoes has been recorded for several parameters, including head length (Fig. 14.13) and body mass and measurements for pups and adults (Fig. 14.14). Head length is a useful indicator of age up until ~120 days, but after this a reliable estimate can only be obtained by determining the eye-lens mass, which is reliable up to 500 days (Catling *et al.* 1991). Front paw print size can be used as a simple way to distinguish

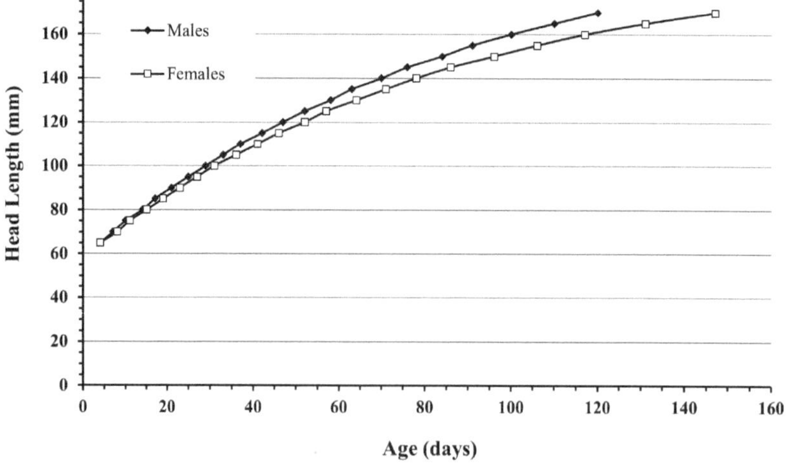


**Fig. 14.13.** Growth in head length of the dingo. Adapted from Catling *et al.* (1991) with permission from CSIRO Publishing.

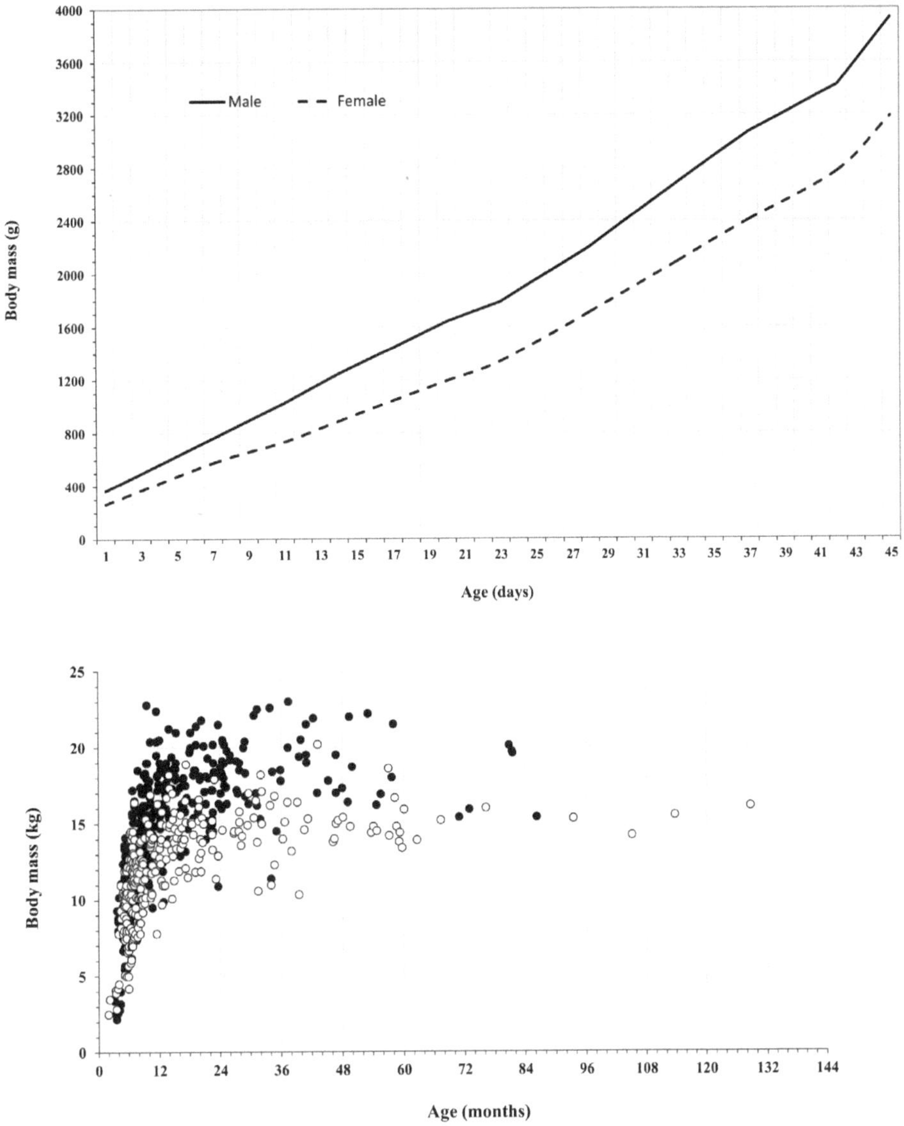


**Fig. 14.14.** (Top) Changes in body mass of a litter of captive dingo pups (3 males, 2 females) during the first 50 days (derived from Smith (2015b). (Bottom) Body mass of male (solid marks, *n* = 392) and female (hollow marks, *n* = 292) dingoes captured, euthanased or found dead on K'gari between August 2001 and December 2021. Updated from Smith (2015b) and Behrendorff *et al.* (2016).

pups and juveniles (4–6 cm footprints) from adults (Allen 2015). Bodyweight from skulls and sedated dingoes has been determined from two simple measurements of the hard palate with high accuracy (Lucas *et al.* 2019).

## 11. ARTIFICIAL REARING

### 11.1 Housing

Natural rearing of dingo pups by the mother in the den is preferred, but there may be occasions (such as mismothering or a health issue) that may preclude this. When hand-rearing is required, minimising stress is a major consideration (see sections 8.3 and 9.6 for causes and signs of stress and trauma). Choosing suitable housing can help to create a stress-free environment. To achieve this, several factors should be considered including:

- is the area secure from children and animals?
- is the area easy to maintain in a hygienic manner?
- is the area escape-proof?
- is the area clear of obstacles and hazards?
- does the area offer shelter from the weather and noise?

## 11.2 Temperature requirements

Orphaned pups will require alternative methods of warmth such as incubators, heat lamps, heating pads or hot water bottles. During the first 4 days of its life, the orphaned pup should be maintained in an environmental temperature of 29.4–32.2°C. The temperature may gradually be reduced to 26.7°C by days 7–10 and to 22.2°C by the end of the 4th week.

## 11.3 Specific requirements

No attempt should be made to feed milk replacement to a pup that is hypothermic. If possible, the pup should be kept with the mother or other dingoes, as this is critical for the development of its social behaviour.

## 11.4 Diet and feeding routine

### 11.4.1 Natural milk

Canid (dog and wolf) milk is much richer in fat and protein than cow's milk and also contains far more calories (Turner 1990). The basic constituents are given in Table 14.6.

### 11.4.2 Milk formulas

The main commercial formulas for hand-rearing dingoes include:

- Biolac® Puppy Milk
- Di-Vetelact® Dog Milk Replacer (Lillelund Pty Ltd)
- Wombaroo® Dog Milk Replacer.

### 11.4.3 Feeding apparatus

Pups can be fed with a puppy feeding bottle, syringe or dropper if very young (e.g. < 1–2 weeks old) and with a bowl when they are old enough to lap at about 3 weeks of age.

### 11.4.4 Feeding routine

The milk formula should be made as per the recommendations and fed at 37°C and the amount fed is normally ~10–15% of the animal's bodyweight per day. Shortly after birth the pup should be fed every 2–3 h and feeding is subsequently decreased to intervals of 3–4 h after the first week (Iben and Leibetseder 1994). It may be necessary to feed weak or compromised pups once every hour or two around the clock until they are stabilised (Barrette 2002). The best time to feed them is when they wake (Hodgman 1976). As described above, solid food can be introduced at approximately 2–3 weeks (in conjunction with regular milk feeds) and can include dried dog food that has been well soaked in hot water to soften. Once this food is readily eaten and the teeth are beginning to develop, puppy food or soft dried food can be provided.

## 11.5 Data recording

When an animal is first brought in for hand-rearing, its sex and approximate age, using growth charts, should be recorded. During the hand-rearing process the following important information should be recorded on a daily basis:

- date and time when the information is recorded
- body mass to the nearest 1 g if possible
- general activity and demeanour
- characteristics and frequency of defecation and urination
- amount (g) of different food types offered
- food consumption at each feed and daily water consumption
- veterinary examinations and results.

This information serves several purposes, including providing important background information, such as food consumption, which will assist a veterinarian to reach a diagnosis if the animal becomes sick or fails to grow or gain bodyweight. It also allows a comparison with growth curves to assess progress (section 10.16) and to allow new growth curves to be established for measurements in which they presently do not occur. The developmental stages and measurements outlined in section 10.3.1 should also be recorded on a weekly basis if possible.

**Table 14.6.** Concentrations of the major constituents of milk for the wolf and dog.

| Species | Total solids (%) | Carbohydrate (%) | Lipids (%) | Protein (%) | Calcium (mg/L) | Iron (µg/L) | Reference |
|---|---|---|---|---|---|---|---|
| *Canis lupus* | 23.5 | – | 6.6 | 12.4 | – | – | 1,2 |
| *C. familiaris* | 22.1–28.3 | 2.6–4.1 | 8.3–14.3 | 6.6–17.3 | 1800–5300 | 4–14 | 3–7 |

References: (1) Ben Shaul (1963); (2) Lauer *et al.* (1969); (3) Jenness and Sloan (1970); (4) Anderson *et al.* (1940); (5) Anderson *et al.* (1991); (6) Oftedal (1984); (7) Zhang *et al.* (2022).

## 11.6 Identification methods

Pups can usually be identified by distinct colour or markings on their body, legs or face (Fig. 14.11). In the event that it is difficult to distinguish pups in a litter, marking their body (head, back of neck, paws) with a non-toxic paint, dye or marker pen, using nail polish to mark a nail or by clipping or shaving hair from different areas are temporary, effective methods until they are old enough to be microchipped. Paint must be non-toxic and will need to be reapplied as the mother is likely to lick it. Coloured bands or collars do not tend to be effective because the mother or littermates will remove or destroy them. See section 5.3 for further details regarding marking and identification.

## 11.7 Hygiene

Maintaining a high standard of hygiene is critical to neonatal health. Emphasis needs to be placed on the following:

- the box or basket and bedding must be kept clean and fresh at all times, especially because bottle-fed pups are more prone to diarrhoea. Maintain personal hygiene by washing and disinfecting hands before and after handling the dingo
- washing hands between feeding different dingoes
- using boiled water when making up formulas for very young dingoes
- cleaning spilt milk formula from the face, clean faeces and urine from the skin and fur as soon as possible and then drying the animal
- washing all feeding equipment in warm water and dishwashing detergent and sterilising it in a microwavable baby steam sterilizer or F10® disinfectant
- storing the teats and bottles in the fridge in an airtight container after they have been disinfected
- only heating up milk once and discarding leftovers
- avoiding contact with other animals unless certain they pose no health risk
- toileting for the first 1–2 weeks after birth - the pups may not be able to urinate or defecate in the absence of the mother as she generally stimulates this by licking the pup. Gently massage the abdomen with a small wad of cotton wool or gauze until the pup urinates and defecates and do this regularly until the pup eliminates without assistance, which is at approximately 3 weeks of age (Hodgman 1976)

## 11.8 Behavioural considerations

Take care to ensure pups are socialised with other people and dingoes. Pups raised in isolation may have difficulty integrating properly with other pups, as they have not learnt submissive behaviours. They may be more aggressive because this is instinctive (Corbett 2001a).

Being raised by the parent is highly important in the social behaviour of dingoes. In a study of pups that were not raised by their parents (isolated), they performed only aggressive behaviours, whereas those pups raised by their parents (normal) performed both aggressive and submissive behaviours, depending on their experiences and social status. For example, even when an isolated pup was being mauled by a more dominant animal it persisted with aggressive behaviours. These observations suggest that dominant behaviour is innate (inherited), whereas submissive behaviour is learned from other animals in the pack (Corbett 2001a).

## 11.9 Use of foster animals

There is no record of fostering by other species, but domestic dogs could potentially be used to foster pups from birth to weaning. Fostering should not be required post weaning, when contact with dingoes is important and recommended.

## 11.10 Weaning

There are many prepared weaning formulas available that can be mixed with the milk replacement. As the pup begins to lap, the milk replacer can be reduced until the pup is eating solid food (Barrette 2002). Solids can start to be introduced with the eruption of milk teeth from 3 weeks as the pups show interest in sold food. Pups should be weaned onto a balanced diet that may include shredded lean meat (chicken, beef, kangaroo). A high-quality commercial puppy food ration can also be given for total nutrition as they grow. Large meaty bones can be given. At 6 weeks of age the pups should have two solid meals per day and three milk meals, with fresh drinking water always available (Hodgman 1976). As with parent-weaned pups, weaning can begin at 3–4 months. At weaning, fresh water should be supplied ad libitum.

## 11.11 Release procedures

Wherever possible, rehabilitated pups and adults should be removed for only as long as necessary because problems (e.g. distress at being separated and conflict upon return) increase with time. Long-term isolation of a pup from its mother and litter during critical social developmental stages is also likely to lead to social problems over its lifespan (e.g. pup does not learn sufficient social skills and thus does not respond to another dingo using context-appropriate behaviours such as submission, which can lead to fighting). It is preferable to provide care and rehabilitation within the enclosure or in an adjacent enclosure. Short-term separation and veterinary care over several days should not create problems. Reintroductions should not be rushed and animals well monitored. Refer to sections 7.6.6 and 9.9 for guidance on staging reintroductions.

If a dingo is not compatible in an existing social group, and can no longer be housed at the facility (either independently or in a new pair), there are several zoos, wildlife parks, and state-based dingo rescue organisations that may be willing to take the animal and should be contacted in circumstances where rehabilitated dingoes require rehousing. Some dingoes that are extremely distressed in captive settings (see Table 14.3 for potential causes of stress, and section 9.6 for signs of stress) may not cope with ongoing or intensive veterinary care or a change in social structure or relocation. In such cases, euthanasia may be a more humane option.

Hand-reared animals should not be released into the wild. This is not only illegal in many jurisdictions, but chances of survival without ongoing human intervention are relatively low.

## ACKNOWLEDGEMENTS

Sincere thanks to Lyn Watson and Kevin Newman from the Dingo Discovery Sanctuary and Research Centre/ Australian Dingo Foundation for value comments and discussion relating to all aspects of dingo husbandry, particularly diet and enclosure design; Nicholas Bishop, the animal behaviour and creative programs manager at Adelaide Zoo (Zoos SA) for guidance with the training section; and Linda Behrendorff from QPWS for discussions relating to dingo body condition. We thank Dr Danielle Stephens (Zoologenetics) and the late Dr Alan Wilton (UNSW) for commentary on hybridisation and genetic testing of dingoes.

## REFERENCES

Allen BJ, Fleming P (2004) Review of canid management in Australia for the protection of livestock and wildlife: potential application to coyote management. *Sheep & Goat Research Journal* **19**, 97–104.

Allen BJ, Allen LR, Ballard G, Jackson SM, Fleming PJS (2017) A roadmap to meaningful dingo conservation. *Canid Biology and Conservation* **20**(11), 45–56.

Allen LR (2015) Demographic and functional responses of wild dogs to poison baiting. *Ecological Management & Restoration* **16**, 58–66. doi:10.1111/emr.12138

Allen LR, O'Bryan L (2011) *Guide to Field Estimating the Age of Wild Dogs*. Biosecurity Queensland, Toowoomba.

Allen L, West P (2013) Influence of dingoes on sheep distribution in Australia. *Australian Veterinary Journal* **91**, 261–267. doi:10.1111/avj.12075

Allsopp NB (1998) Canine enrichment program. *Thylacinus* **22**(1), 39–43.

AMTC (2024) The AMTC Australian Mammal Species List. Version 4.0. https://australianmammals.org.au/publications/amtc-species-list (accessed 4 August 2024).

Anderson HD, Johnson BC, Arnold A (1940) The composition of dog's milk. *The American Journal of Physiology* **129**, 631–634. doi:10.1152/ajplegacy.1940.129.3.631

Anderson RS, Carlos GM, Robinson IP, Booles D, Burger IH, Whyte AL (1991) Zinc, copper, iron and calcium concentrations in bitch milk. *The Journal of Nutrition* **121**, S81–S82. doi:10.1093/jn/121.suppl_11.S81

Arendt M, Cairns KM, Ballard JWO, Savolainen P, Axelsson E (2016) Diet adaptation in dog reflects spread of prehistoric agriculture. *Heredity* **117**, 301–306. doi:10.1038/hdy.2016.48

Australian Dingo Conservation Association (ADCA) (1996) *The Australian Dingo Handbook*. Australian Dingo Conservation Association Incorporated, Australian Capital Territory.

AZA Reproductive Management Centre (2021) Contraception recommendations, Canidae. American Zoo and Aquarium Reproductive Management Centre. https://www.stlzoo.org/animals/scienceresearch/reproductivemanagementcenter/contraceptionrecommendatio/contraceptionmethods/canidae (accessed 25 October 2021).

Ballard JWO, Wilson LAB (2019) The Australian dingo: untamed or feral? *Frontiers in Zoology* **16**(2), 1–19.

Balme J, O'Connor S, Fallon S (2018) New dates on dingo bones from Madura Cave provide oldest firm evidence for arrival of the species in Australia. *Scientific Reports* **8**, 9933. doi:10.1038/s41598-018-28324-x

Barnes J, Vogelnest L, Hulst F, Rose K, Bryant B (2001) Tick toxicity in zoo animals in eastern Australia. In Proceedings of the American Association of Zoo Veterinarians, American association of Wildlife Veterinarians, Association of Reptile and Amphibian Veterinarians and National Association of Zoo and Wildlife Veterinarians Joint Conference. Orlando, Florida.

Barrette VT (2002) Puppies. In *Hand-Rearing Wild and Domestic Mammals*. (Ed. LJ Gage) pp. 13–18. Iowa State Press, Iowa.

Behrendorff L, Allen BL (2016) From den to dust: longevity of three dingoes (*Canis lupus dingo*) on Fraser Island (K'gari). *Australian Mammalogy* **38**, 256–260. doi:10.1071/AM16005

Behrendorff L, Leung LKP, McKinnon A, Hanger J, Belonje G, Tapply J, Jones D, Allen BL (2016) Insects for breakfast and whales for dinner: the diet and body condition of dingoes on Fraser Island (K'gari). *Scientific Reports* **6**, 23469. doi:10.1038/srep23469

Ben Shaul DM (1963) The composition of the milk of wild animals. *International Zoo Yearbook* **4**, 333–342. doi:10.1111/j.1748-1090.1963.tb03691.x

Berndt RM, Berndt CH (1942) A preliminary report of fieldwork in the Ooldea region, western South Australia. *Oceania* **13**, 143–169. doi:10.1002/j.1834-4461.1942.tb00375.x

Boden E (1998) *Black's Veterinary Dictionary*. A & C Black, London.

Breckwoldt R (1988) *The Dingo: A Very Elegant Animal*. Angus & Robertson, Sydney.

Brown B, Copeman DB (2003) Zoonotic importance of parasites in wild dogs caught in the vicinity of Townsville. *Australian Veterinary Journal* **81**, 700–702. doi:10.1111/j.1751-0813.2003.tb12546.x

Byosiere SE, Espinosa J, Smith B (2018) The function of play bows in *Canis lupus* and its variants: a comparison of dingo (*Canis lupus dingo*), dog (*Canis lupus familiaris*) and wolf puppies (*Canis lupus*). *Behaviour* **155**, 369–388. doi:10.1163/1568539X-00003495

Cairns KM (2021) What is a dingo: origins, hybridisation and identity. *Australian Zoologist* **41**, 322–337. doi:10.7882/AZ.2021.004

Cairns KM, Wilton AN (2016) New insights on the history of canids in Oceania based on mitochondrial and nuclear data. *Genetica* **144**, 553–565. doi:10.1007/s10709-016-9924-z

Cairns KM, Wilton AN, Ballard JWO (2011) The identification of dingoes in a background of hybrids. In *Advances in Genetics Research*. (Ed. KV Urbano) pp. 309–327. Nova Science Publishers, New York.

Cairns KM, Brown SK, Sacks BN, Ballard JWO (2017) Conservation implications for dingoes from the maternal and paternal genome: multiple populations, dog introgression and demography. *Ecology and Evolution* **7**, 9787–9807. doi:10.1002/ece3.3487

Cairns KM, Shannon LM, Koler-Matznick J, Ballard JWO, Boyko AR (2018) Elucidating biogeographical patterns in Australian native canids using genome wide SNPs. *PLoS One* **13**, e0198754. doi:10.1371/journal.pone.0198754

Cairns KM, Nesbitt BJ, Laffan SW, Letnic M, Crowther MS (2020) Geographic hot spots of dingo genetic ancestry in southeastern Australia despite hybridisation with domestic dogs. *Conservation Genetics* **21**, 77–90. doi:10.1007/s10592-019-01230-z

Cairns KM, Crowther MS, Nesbitt B, Letnic M (2021a) The myth of wild dogs in Australia: are there any out there? *Australian Mammalogy* **44**, 67–75. doi:10.1071/AM20055

Cairns KM, Newman KD, Crowther MS, Letnic M (2021b) Pelage variation in dingoes across southeastern Australia: implications for conservation and management. *Journal of Zoology* **314**, 104–115. doi:10.1111/jzo.12875

Cairns KM, Crowther MS, Parker HG, Ostrander EA, Letnic M (2023) Genome-wide variant analyses reveal new patterns of admixture and population structure in Australian dingoes. *Molecular Ecology* **32**, 4133–4150. doi:10.1111/mec.16998

Catling PC, Corbett LK, Westcott M (1991) Age determination in the dingo and crossbreeds. *Wildlife Research* **18**, 75–83. doi:10.1071/WR9910075

Catling PC, Corbett LK, Newsome AE (1992) Reproduction in captive and wild dingoes (*Canis familiaris dingo*) in temperate and arid environments of Australia. *Wildlife Research* **19**, 195–209. doi:10.1071/WR9920195

Corbett LK (1988) Social dynamics of a captive dingo pack: population regulation by dominant female infanticide. *Ethology* **78**, 177–198. doi:10.1111/j.1439-0310.1988.tb00229.x

Corbett LK (1995) Dingo *Canis lupus dingo*. In *The Mammals of Australia*. 2nd edn. (Ed. R Strahan) pp. 696–698. Reed Books, Sydney.

Corbett L (2001a) *The Dingo in Australia and Asia*. 2nd edn. JB Books Australia, Marleston.

Corbett L (2001b) The conservation status of dingoes (*Canis lupus dingo*) in Australia, with particular reference to New South Wales: threats to pure dingoes and potential solution. In *A Symposium on the Dingo*. (Eds CR Dickman and D Lunney) pp. 10–19. Royal Zoological Society of New South Wales, Sydney.

Corbett L (2006) The Australian dingo. In *Evolution and Zoogeography of Australasian Fauna*. (Eds JR Merrick, M Archer, G Hickey and M Lee) pp. 745–753. Australian Scientific Publishing, Sydney.

Corbett LK, Newsome AE (1975) Dingo society and its maintenance: A preliminary analysis. In *The Wild Canids: Their Systematics, Behavioural Ecology and Evolution*. (Ed. MW Fox) pp. 369–379. Krieger, Florida.

Corbett LK, Newsome AE (1987) The feeding ecology of the dingo. III: Dietary relationships with widely fluctuating prey populations in arid Australia: an hypothesis of alteration of predation. *Oecologia* **74**, 215–227. doi:10.1007/BF00379362

Crandall LS (1964) *The Management of Wild Mammals in Captivity*. University of Chicago Press, Chicago.

Crowther MS, Fillios M, Colman N, Letnic M (2014) An updated description of the Australian dingo (*Canis dingo* Meyer, 1793). *Journal of Zoology* **293**, 192–203. doi:10.1111/jzo.12134

Cursino MS, Harriott L, Allen BL, Gentle M, Leung LKP (2017) Do female dingo-dog hybrids breed like dingoes or dogs? *Australian Journal of Zoology* **65**, 112–119. doi:10.1071/ZO17005

Déaux ÉC, Clarke JA (2013) Dingo (*Canis lupus dingo*) acoustic repertoire: form and contexts. *Behaviour* **150**, 75–101. doi:10.1163/1568539X-00003038

Déaux EC, Clarke JA, Charrier I (2016*a*) Dingo howls: the content and efficacy of a long-range vocal signal. *Ethology* **122**, 649–659. doi:10.1111/eth.12510

Déaux EC, Allen AP, Clarke JA, Charrier I (2016*b*) Concatenation of 'alert' and 'identity' segments in dingoes' alarm calls. *Scientific Reports* **6**, 30556. doi:10.1038/srep30556

Department of Biodiversity, Conservation and Attractions (DBCA) (2017) *Standard Operating Procedure: Semi-permanent Marking of Mammals using Ear Tags*. Department of Biodiversity, Conservation and Attractions. Perth, WA.

Doherty TS, Davis NE, Dickman CR, Forsyth DM, Letnic M, Nimmo DG, Palmer R, Ritchie EG, Benshemesh J, Edwards G, Lawrence J, Lumsden L, Pascoe C, Sharp A, Stokeld D, Myers C, Story G, Triggs B, Venosta M, Wysong M, Newsome T (2019) Continental patterns in the diet of a top predator: Australia's dingo. *Mammal Review* **49**, 31–44. doi:10.1111/mam.12139

Dunsmore JD, Spratt DM (1979) The life history of *Filaroides osleri* in wild and domestic canids in Australia. *Veterinary Parasitology* **5**, 275–286. doi:10.1016/0304-4017(79)90017-7

Elledge AE, Allen LR, Carlsson B, Wilton AN, Leung LK (2008) An evaluation of genetic analyses, skull morphology and visual appearance for assessing dingo purity: implications for dingo conservation. *Wildlife Research* **35**, 812–820. doi:10.1071/WR07056

Fancourt BA, Allen BL, Jackson SM Meek Pd, Zewe F, Ballard G, Behrendorff L, Claridge AW, Fleming PJS (2022) Eutherian carnivores. In *Wildlife Research in Australia: Practical and Applied Methods*. (Eds BP Smith, HP Waudby, C Alberthsen and JO Hampton). pp. 428–436. CSIRO Publishing, Melbourne.

Fancy SG (1980) Preparation of mammalian teeth for age determination by cementum layers: a review. *Wildlife Society Bulletin* **8**, 242–248.

Fleming PJS, Corbett L, Harden B, Thomson P (2001) *Managing the Impacts of Dingoes and other Wild Dogs*. Bureau of Rural Sciences, Canberra.

Fleming PJS, Allen BL, Allen LR, Ballard G, Bengsen AJ, Gentle MN, McLeod LJ, Meek PD, Saunders GR (2014) Management of wild canids in Australia: free-ranging dogs and red foxes. In *Carnivores of Australia: Past, Present and Future*. (Eds AS Glen and CR Dickman) pp. 105–149. CSIRO Publishing, Melbourne.

Fleming PJS, Ballard G, Cutter N (2021) There is no dingo dilemma: legislation facilitates culling, containment and conservation of Dingoes in New South Wales. *Australian Zoologist* **41**, 408–416. doi:10.7882/AZ.2020.035

Gier HT (1968) Coyotes in Kansas. Kansas Agricultural Experiment Station Bulletin 393.

Gould J (1845–1863) *The Mammals of Australia*. The Author, London.

Green B, Catling P (1977) The biology of the dingo. In *Australian Animals and their Environment*. (Eds H Messel and ST Butler) pp. 51–60. Shakespeare Head Press, Sydney.

Hamilton A (1972) Aboriginal man's best friend? *Mankind* **8**, 287–295.

Harden RH (1985) The ecology of the dingo in north-eastern new South Wales. I: movements and home range. *Australian Wildlife Research* **12**, 25–37. doi:10.1071/WR9850025

Hodgman SFJ (1976) Obstetrical and paediatric nursing and the principles of breeding. (a) obstetrical nursing and the principles of breeding in the dog. In *Jones's Animal Nursing*. (Ed. RS Pinniger) pp. 427–439. Pergamon Press, Oxford.

Hooper PT, Reece RL, Grayton RJ (1983) Pathology of selected cases from two Melbourne zoos. Proceedings of the Wildlife Diseases Association (Australasian Section) Annual Conference. Melbourne Zoo, Melbourne.

Hudson R, Rodel HG, Elizalde MT, Kennedy GA, Smith BP (2019) Diurnal pattern of pre-weaning den visits and nursing in breeding pairs of captive dingoes (*Canis dingo*). *Mammalian Biology* **94**, 86–91. doi:10.1016/j.mambio.2018.07.002

Hulst F (2008) Dingoes. In *Medicine of Australian Mammals*. (Eds L Vogelnest and R Woods) pp. 527–540. CSIRO Publishing, Melbourne.

Hunter J (1793) *An Historical Journal of the Transactions at Port Jackson and Norfolk Island, with the Discoveries which have been made in New South Wales and in the Southern Ocean*. J. Stockdale, London.

Iben C, Leibetseder J (1994) Handrearing of orphaned puppies and kittens. *The Journal of Nutrition* **124**, 2630S–2632S. doi:10.1093/jn/124.suppl_12.2630S

International Air Transport Association (IATA) (2022) *Live Animals Regulations*. 48th edn. International Air Transport Association, Geneva Switzerland.

Jackson SM, Groves C (2015) *Taxonomy of Australian Mammals*. CSIRO Publishing, Melbourne.

Jackson SM, Groves CP, Fleming P, Aplin K, Eldridge MDB, Gonzalez A, Helgen K (2017) The wayward dog: an assessment of the taxonomy of the dingo. *Zootaxa* **4317**, 201–224.

Jackson SM, Fleming PJS, Eldridge MDB, Ingleby S, Flannery TF, Johnson RN, Cooper SJB, Mitchell KJ, Souili T, Cooper A, Wilson DE, Helgen KM (2019) The dogma of dingoes: taxonomic status of the dingo. A reply to Smith *et al. Zootaxa* **4564**, 198–212. doi:10.11646/zootaxa.4564.1.7

Jackson SM, Fleming PJS, Eldridge MDB, Archer M, Ingleby S, Johnson RN, Helgen KM (2021) Taxonomy of the dingo: it's an ancient dog. *Australian Zoologist* **41**, 347–357. doi:10.7882/AZ.2020.049

Jenkins DJ, MacPherson CNL (2003) Transmission ecology of *Echinococcus* in wild-life in Australia and Africa. *Parasitology* **127**, S63–S72. doi:10.1017/S0031182003003871

Jenkins DJ, Morris B (2003) *Echinococcus granulosus* in wildlife in and around the Kosciusko National Park, southeastern Australia. *Australian Veterinary Journal* **81**, 81–85. doi:10.1111/j.1751-0813.2003.tb11440.x

Jenkins DJ, Allen L, Goullet M (2008) Encroachment of *Echinococcus granulosus* into urban areas in eastern Queensland, Australia. *Australian Veterinary Journal* **86**, 294–300. doi:10.1111/j.1751-0813.2008.00327.x

Jenness R, Sloan RE (1970) The composition of milks of various species: a review. *Dairy Science Abstracts* **32**, 599–612.

Jones DE, Joshua JO (1982) *Reproductive Clinical Problems in the Dog*. Wright, Bristol.

Jones E (1990) Physical characteristics and taxonomic status of wild canids, *Canis familiaris*, from the eastern highlands of Victoria. *Australian Wildlife Research* **17**, 69–81. doi:10.1071/WR9900069

Jones E, Stevens PL (1988) Reproduction in wild canids, *Canis familiaris*, from the eastern highlands of Victoria. *Australian Wildlife Research* **15**, 385–394. doi:10.1071/WR9880385

Kerr R (1792) *The Animal Kingdom. Class 1: Mammalia*. Murray & Faulder, London.

Koungoulos L (2020) Old dogs, new tricks: 3D geometric analysis of cranial morphology supports ancient population substructure in the Australian dingo. *Zoomorphology* **139**, 263–275. doi:10.1007/s00435-019-00475-z

Koungoulos L, Fillios M (2020) Hunting dogs down under? On the Aboriginal use of tame dingoes in dietary game acquisition and its relevance to Australian prehistory. *Journal of Anthropological Archaeology* **58**, 101146. doi:10.1016/j.jaa.2020.101146

Kreplins TL, Gaynor A, Kennedy MS, Baudains CM, Adams P, Bateman PW, Fleming PA (2019) What to call a dog? A review of the common names for Australian free-ranging dogs. *Pacific Conservation Biology* **25**, 124–134. doi:10.1071/PC18018

Lacy RC, Balou JD, Pollak JP (2011) PMx: software package for demographic and genetic analysis and management of pedigreed populations. *Methods in Ecology and Evolution* **3**, 433–437. doi:10.1111/j.2041-210X.2011.00148.x

Lauer BH, Kuyt E, Baker BE (1969) Wolf milk. I: Arctic wolf (*Canis lupus arctos*) and husky milk: gross composition and fatty acid constitution. *Canadian Journal of Zoology* **47**, 99–102. doi:10.1139/z69-018

Lee I (1925) *Early Explorers in Australia: From the Log-books and Journals, with Maps and Illustrations*. Methuen & Co. Ltd, London.

Linhart SB, Knowlton RFF (1967) Determination of age of coyotes by tooth cementum layers. *The Journal of Wildlife Management* **31**, 362–368. doi:10.2307/3798334

Lockyer N (1932) Hybrid between wolf and dingo. *Nature* **130**, 280.

Lord K, Feinstein M, Smith B, Coppinger R (2013) Variation in reproductive traits of members of the genus *Canis* with special attention to the domestic dog (*Canis familiaris*). *Behavioural Processes* **92**, 131–142. doi:10.1016/j.beproc.2012.10.009

Lucas T, Smith B, Norris R, Henneberg M (2019) Reconstructing body weight of the Australian dingo (*Canis dingo*) from two simple measurements of the hard palate. *Journal of*

*Archaeological Science: Reports* **23**, 534–539. doi:10.1016/j.jasrep.2018.11.018

Lumholtz C (1889) *Among Cannibals: An Account of Four Years' Travels in Australia and of Camp Life with the Aborigines of Queensland*. J. Murray, London.

Mallonee J, Joslin P (2004) Traumatic stress disorder observed in an adult wild captive wolf (*Canis lupus*). *Journal of Applied Animal Welfare Science* **7**, 107–126. doi:10.1207/s15327604jaws0702_3

Marsack P, Campbell G (1990) Feeding behaviour and diet of dingoes in the Nullabor region, Western Australia. *Australian Wildlife Research* **17**, 349–357. doi:10.1071/WR9900349

Meek PD, Brown SC (2017) It's a dog eat dog world: observations of dingo (*Canis familiaris*) cannibalism. *Australian Mammalogy* **39**, 92–94. doi:10.1071/AM16018

Meyer FAA (1793) *Systematisch-Summarische Uebersicht der neuesten Zoologischen Entdeckungen in Neuholland und Afrika. Nebst zwey andern Zoologischen Abhandlunge*. Dykischen, Leipzig.

Miller SM, Behrendorff L, Allen BL, Andrew RL, Ballard G, Ballard JWO, Cairns KM, Conroy GC, Fleming PJ, Grueber CE, Oakey J, Smith DA, Stephens D, Wade C, Bishop JM (2024) Isolation, small population size, and management influence inbreeding and reduced genetic variation in k'gari dingoes. *Conservation Genetics* **25**, 955–971.

Nellis CH, Wetmore SP, Keith LB (1978) Age-related characteristics of coyote canines. *The Journal of Wildlife Management* **42**, 680–683. doi:10.2307/3800844

New South Wales Department of Primary Industries (2006) *Standards for Exhibiting Australian Mammals in New South Wales*. New South Wales Department of Primary Industries, Orange, NSW.

Newsome AE, Coman BJ (1989) Canidae. In *Fauna of Australia. Mammalia*. Vol. 1B. (Eds DW Walton and BJ Richardson) pp. 993–1005. Australian Government Publishing Service, Canberra.

Newsome AE, Corbett LK (1985) The identity of the dingo. III: the incidence of dingoes, dogs and hybrids and their coat colours in remote and settled regions of Australia. *Australian Journal of Zoology* **33**, 363–375. doi:10.1071/ZO9850363

Newsome TM, Ballard GA, Dickman CR, Fleming PJS, Howden C (2013) Anthropogenic resource subsidies determine space use by Australian arid zone dingoes: an improved resource selection modelling approach. *PLoS One* **8**, e63931. doi:10.1371/journal.pone.0063931

Nguyen C, Koh WL, Casteriano A, Beijerink N, Godfrey C, Brown G, Emery D, Slapeta J (2016) Mosquito-borne heartworm *Dirofilaria immitis* in dogs from Australia. *Parasites & Vectors* **9**, 535. doi:10.1186/s13071-016-1821-x

Northern Territory Government Department of Industry, Tourism and Trade (2021) Infection with *Ehrlichia canis* (Ehrlichiosis) Guidelines for Veterinarians. Northern Territory Government. https://nt.gov.au/data/assets/pdf_file/0007/979720/e-canis-guidelines-for-vets.pdf (accessed 31 Oct 2021).

Oakman B (2001) The problems with keeping dingoes as pets and dingo conservation. In *A Symposium on the Dingo*. (Eds CR Dickman and D Lunney) pp. 34–38. Royal Zoological Society of New South Wales, Sydney.

Oftedal OT (1984) Lactation in the dog: milk composition and intake by puppies. *The Journal of Nutrition* **114**, 803–812. doi:10.1093/jn/114.5.803

Pacioni C, Kennedy MS, Ramsey DSL (2020) When do predator exclusion fences work best? A spatially explicit modelling approach. *Wildlife Research* **48**, 209–217. doi:10.1071/WR19192

Parr WC, Wilson LA, Wroe S, Colman NJ, Crowther MS, Letnic M (2016) Cranial shape and the modularity of hybridization in dingoes and dogs: hybridization does not spell the end for native morphology. *Evolutionary Biology* **43**, 171–187. doi:10.1007/s11692-016-9371-x

Phillip A (1789) *The Voyage of Governor Phillip to Botany Bay, With an Account of the Establishment of the Colonies of Port Jackson & Norfolk Island Compiled from Authentic Papers...To which are added the Journals of Lieuts Shortland, Watts, Ball & Capt. Marshall, With an Account of their Discoveries*. John Stockdale, London.

Robertshaw JD, Harden RH (1985) The ecology of the dingo in north-eastern New South Wales. II: diet. *Australian Wildlife Research* **12**, 39–50. doi:10.1071/WR9850039

Robertshaw JD, Harden RH (1986) The ecology of the dingo in north-eastern New South Wales. IV: prey selection by dingoes, and its effect on the major prey species, the swamp wallaby, *Wallabia bicolor* (Desmarest). *Australian Wildlife Research* **13**, 141–163. doi:10.1071/WR9860141

Robley A, Gormley A, Forsyth DM, Wilton AN, Stephens D (2010) Movements and habitat selection by wild dogs in eastern Victoria. *Australian Mammalogy* **32**, 23–32. doi:10.1071/AM09030

Rogers JG (1965) Analysis of the coyote population of Dona Ana County, New Mexico. MSc thesis. New Mexico State University, University Park.

Schenkel R (1947) Expression studies of wolves. *Behaviour* **1**, 81–129. doi:10.1163/156853948X00065

Schenkel R (1967) Submission: its features and function in the wolf and dog. *American Zoologist* **7**, 319–329. doi:10.1093/icb/7.2.319

Smith BP (2014) Living with wild dogs: personality dimensions in captive dingoes (*Canis dingo*) and implications for ownership. *Anthrozoos* **27**, 423–433. doi:10.2752/1753037 14X14023922797869

Smith BP (2015a) The personality, behaviour and suitability of dingoes as companion animals. In *The Dingo Debate: Origins, Behaviour and Conservation*. (Ed. B Smith) pp. 251–275. CSIRO Publishing, Melbourne.

Smith BP (2015b) Characteristics of the Australian dingo (*Canis dingo* Meyer, 1793). In *The Dingo Debate: Origins, Behaviour and Conservation*. (Ed. B Smith) pp. 1–23. CSIRO Publishing, Melbourne.

Smith BP (2015c) Biology and behaviour of the dingo. In *The Dingo Debate: Origins, Behaviour and Conservation*. (Ed. B Smith) pp. 25–53. CSIRO Publishing, Melbourne.

Smith BP, Litchfield CA (2009) A review of the relationship between Indigenous Australians, dingoes (*Canis dingo*) and domestic dogs (*Canis familiaris*). *Anthrozoos* **22**, 111–128. doi:10.2752/175303709X434149

Smith BP, Vague AL (2017) The denning behaviour of dingoes (*Canis dingo*) living in a human-modified environment. *Australian Mammalogy* **39**, 161–168. doi:10.1071/AM16027

Smith BP, Watson L (2015) The role of private sanctuaries in dingo conservation and the management of dingoes in captivity. In *The Dingo Debate: Origins, Behaviour and Conservation*. (Ed. B Smith) pp. 277–299. CSIRO Publishing, Melbourne.

Smith BP, Appleby R, Litchfield C (2012) Spontaneous tool-use: an observation of a dingo (*Canis dingo*) using a table to access an out-of-reach food reward. *Behavioural Processes* **89**, 219–224. doi:10.1016/j.beproc.2011.11.004

Smith BP, Flavel M, Simpson B (2016) Quantification of salivary cortisol from captive dingoes (*Canis dingo*) in relation to age, gender, and breeding season: implications for captive management. *Australian Mammalogy* **38**, 21–28. doi:10.1071/AM15017

Smith BP, Browne M, Serpell J (2017) Owner-reported behavioural characteristics of dingoes (*Canis dingo*) living as companion animals: a comparison to 'modern' and 'ancient' dog breeds. *Applied Animal Behaviour Science* **187**, 77–84. doi:10.1016/j.applanim.2016.11.010

Smith BP, Lucas T, Norris R, Henneberg M (2018) Brain size/body weight in the dingo (*Canis dingo*): Comparisons with domestic and wild canids. *Australian Journal of Zoology* **65**, 292–301. doi:10.1071/ZO17040

Smith BP, Cairns KM, Adams JW, Newsome TM, Fillios M, Déaux EC, Parr WCH, Letnic M, Van Eeden LM, Appleby RG, Bradshaw CJA, Savolainen P, Ritchie EG, Nimmo DG, Archer-lean C, Greenville AC, Dickman CR, Watson L, Moseby KE, Doherty TS, Wallach AD, Morrant DS, Crowther MS (2019) Taxonomic status of the Australian dingo: the case for *Canis dingo* Meyer, 1793? *Zootaxa* **4564**, 173–197. doi:10.11646/zootaxa.4564.1.6

Smith BP, Morrant DS, Vague AL, Doherty TS (2020) High rates of cannibalism and food waste consumption by dingoes living at a remote mining operation in the Great Sandy Desert, Western Australia. *Australian Mammalogy* **42**, 230–234. doi:10.1071/AM19033

Smith BP, Palermo S, Watson L (2021) The important role of private wildlife organisations in facilitating scientific research: a case study of the Australian Dingo Foundation. *Australian Zoologist* **41**, 568–579. doi:10.7882/AZ.2020.023

Smout FA, Thompson RCA, Skerratt LF (2013) First report of *Ancylostoma ceylanicum* in wild canids. *International Journal for Parasitology. Parasites and Wildlife* **2**, 173–177. doi:10.1016/j.ijppaw.2013.04.003

Smout FA, Skerratt LF, Butler JR, Johnson CN, Congdon BC (2016) Dingoes (*Canis dingo* Meyer, 1793) continue to be an important reservoir host of *Dirofilaria immitis* in low density housing areas in Australia. *Veterinary Parasitology* **215**, 6–10. doi:10.1016/j.vetpar.2015.10.020

Smout FA, Skerratt LF, Johnson CN, Butler J, Congdon BC (2018) Zoonotic helminth diseases in dogs and dingoes utilising shared resources in an Australian Aboriginal community. *Tropical Medicine and Infectious Disease* **3**(4), 110. doi:10.3390/tropicalmed3040110

Souilmi Y, Wasef S, Williams MP, Conroy G, Bar I, Bover P, Mitchell KJ (2024) Ancient genomes reveal over two thousand years of dingo population structure. *Proceedings of the National Academy of Sciences* **121**, e2407584121.

Species360 (2023) Zoological Information Management System (ZIMS). Minneapolis, USA. www.species360.org.

Starr TW, Mulley RC (1988) *Dinofilaria immitis* in the dingo (*Canis familiaris dingo*) in a tropical region of the Northern Territory, Australia. *Journal of Wildlife Diseases* **24**, 164–165. doi:10.7589/0090-3558-24.1.164

Stephens D, Wilton AN, Fleming PJS, Berry O (2015) Death by sex in an Australian icon: a continent-wide survey reveals extensive hybridisation between dingoes and domestic dogs. *Molecular Ecology* **24**, 5643–5656. doi:10.1111/mec.13416

Tatler J, Currie SE, Cassey P, Scharf AK, Roshier DA, Prowse TA (2021) Accelerometer informed time-energy budgets reveal the importance of temperature to the activity of a wild, arid zone canid. *Movement Ecology* **9**, 11. doi:10.1186/s40462-021-00246-w

Tench W (1789) *A Narrative of the Expedition to Botany Bay; with an Account of New South Wales, its Productions, Inhabitants, &c, to which is subjoined a List of the Civil and Military Establishments at Port Jackson.* J. Debrett, London.

Thomson PC (1984) Dingoes and sheep in pastoral areas. *Journal of Agriculture of Western Australia* **25**, 27–31.

Thomson PC (1992a) The behavioural ecology of dingoes in northwestern Australia. 1: study areas and animals. *Wildlife Research* **19**, 509–518. doi:10.1071/WR9920509

Thomson PC (1992b) The behavioural ecology of dingoes in north-western Australia. II: activity patterns, breeding season and pup rearing. *Wildlife Research* **19**, 519–530. doi:10.1071/WR9920519

Thomson PC (1992c) The behavioural ecology of dingoes in north-western Australia. IV: social and spatial organisation, and movements. *Wildlife Research* **19**, 543–563. doi:10.1071/WR9920543

Thomson PC, Rose K (1992) Age determination of dingoes from characteristics of canine teeth. *Wildlife Research* **19**, 597–599. doi:10.1071/WR9920597

Thomson PC, Rose K, Kok NE (1992) The behavioural ecology of dingoes in north-western Australia. V: population dynamics and variation in the social system. *Wildlife Research* **19**, 565–584. doi:10.1071/WR9920565

Turner T (1990) *Veterinary Notes for Dog Owners*. Popular Dogs, London.

Vogelnest L (2025) Capture, physical and chemical restraint. In *Current Therapy in Medicine of Australian Mammals*. Revised Edition. (Eds L Vogelnest and T Portas) pp. 122–151. CSIRO Publishing, Melbourne.

Vogelnest L, Portas T (Eds) (2025) *Current Therapy in Medicine of Australian Mammals*. Revised Edition. CSIRO Publishing, Melbourne.

Walters B (1995) *The Company of Dingoes: Two Decades with our Native Dog*. Australian Native Dog Conservation Society, Bargo, New South Wales.

Wildlife Health Australia (2020) Canine ehrlichiosis in Australia. Fact Sheet. www.wildlifehealthaustralia.com.au/Portals/0/Documents/FactSheets/Mammals/Canine_ehrlichiosis_in_Australia.pdf (accessed 19 November 2021).

Wilton AN (2001) DNA methods of assessing dingo purity. In *A Symposium on the Dingo*. (Eds CR Dickman and D Lunney) pp. 49–56. Royal Zoological Society on New South Wales, Sydney.

Wilton AN, Steward DJ, Zafiris K (1999) Microsatellite variation in the Australian dingo. *The Journal of Heredity* **90**, 108–111. doi:10.1093/jhered/90.1.108

Zhang M, Sun X, Cheng J, Guo M (2022) Analysis and comparison of nutrition profiles of canine milk with bovine and caprine milk. *Foods* **11**, 472. doi:10.3390/foods11030472

Zuckerman S (1953) The breeding seasons of mammals in captivity. *Proceedings of the Zoological Society of London* **122**, 827–950. doi:10.1111/j.1096-3642.1952.tb00251.x

# 15
# Seals

*Danielle Fox, Benjamin J. Pitcher and Joy S. Tripovich*

## 1. INTRODUCTION

There are extant 34 species of seals in the world, divided into three families: the Otariidae, Phocidae and Odobenidae (Wilson and Mittermeier 2014). The Otariidae, which includes 15 species of fur seals and sea lions, possess external pinnae (i.e. ear flaps) and are typically more upright when on land. The Phocidae includes 18 species of earless seals (also known as 'true seals') that have no external ear appendages, only a hole, and includes species such as the leopard seal and the southern elephant seal that are known to frequent Australian waters. On land the phocids move in a lunging caterpillar-like motion. The third family, Odobenidae, contains only the walrus, which is restricted to the Northern Hemisphere.

There are 11 species of seals that occur within Australian waters. Of these, three are commonly found in waters along southern Australia, while the remainder occur in Australia's Antarctic Territory including Macquarie Island (Webber 2014; Baker and Gynther 2023). The three species that frequent the southern waters of Australia are the long-nosed fur seal (*Arctocephalus forsteri*), Australian fur seal (*Arctocephalus pusillus doriferus*) and the Australian sea-lion (*Neophoca cinerea*).

Zoos and aquariums in Australia have been housing pinnipeds for over 100 years. Taronga Zoo has housed fur seals since its opening in 1916, listing two 'Australian Seals *Arctocephalus forsteri*' (Taronga Zoological Park 1916). Taronga's Seal Pool was completed in 1926, enabling the housing of larger groups of seals (Strahan 1991). This chapter focuses on the long-nosed fur seal, Australian fur seal, Australian sea lion and leopard seal, as these are the most commonly held (or previously held in the case of leopard seals) pinnipeds in the captive setting.

## 2. TAXONOMY

### 2.1. Nomenclature

The seals that occur within Australia comprise two families and 11 species (Table 15.1; Jackson and Groves 2015; Baker and Gynther 2023). Seals have historically been recognised as the order Pinnipedia; however, more recent evidence places them within the order Carnivora (Jackson and Groves 2015; Baker and Gynther 2023). The species *Arctocephalus pusillus* has two subspecies: *A. p. pusillus*, the South African (Cape) fur seal, and *A. p. dorifeus*, the Australian fur seal (Wickens and York 1997). These two species have been recognised as subspecies

**Table 15.1.** Species of seals recorded within Australian waters and their conservation status.

Derived from Woinarski *et al.* (2014), Chilvers and Harcourt (2021) and Baker and Gynther (2023).
EN = Endangered; LC = Least Concern.

| Common name | Scientific name | IUCN status |
|---|---|---|
| **Family Otariidae** | | |
| Long-nosed fur seal | *Arctocephalus forsteri* | LC |
| Antarctic fur seal | *Arctocephalus gazella* | LC |
| Australian fur seal | *Arctocephalus pusillus doriferus* | LC |
| Subantarctic fur seal | *Arctocephalus tropicalis* | LC |
| Australian sea lion | *Neophoca cinerea* | EN |
| New Zealand sea lion | *Phocarctos hookeri* | EN |
| **Family Phocidae** | | |
| Leopard seal | *Hydrurga leptonyx* | LC |
| Weddell seal | *Leptonychotes weddellii* | LC |
| Crabeater seal | *Lobodon carcinophaga* | LC |
| Southern elephant seal | *Mirounga leonina* | LC |
| Ross seal | *Ommatophoca rossii* | LC |

on the basis of slight cranial differences (Warneke and Shaughnessy 1985).

**Australian seals**
Class: Mammalia
Superorder: Fereuungulata
Order: Carnivora
Suborder: Caniformia
Infraorder: Arctoidea
Superfamily: Phocoidea
Family: Otariidae
Genus and species: 6 species in 3 genera
Family: Phocidae
Genus and species: 5 species in 5 genera

## 2.2 Subspecies

The fur seal, *Arctocephalus pusillus* is represented by two subspecies with disjunct distributions: *A. p. pusillus* in South African waters and *A. p. doriferus* in Australian waters. The Australian subspecies has been reported from South Australia, Victoria, Tasmania and New South Wales.

## 2.3 Recent synonyms

Historically, the naming of the Australian sea lion, *Neophoca cinerea*, has led to some confusion. In 1816, the French zoologist François Péron erroneously gave the species two names, *Otaria cinerea* and *O. albicolis*, one applying to females and young males and the other adult males (Péron 1816; Wood Jones 1925). The Australian sea lion and New Zealand sea lion (*Phocarctos hookeri*) were variously regarded as conspecific, congeneric or heterogeneric until 1960 when King (1960) suggested generic separation based on skull morphology. This was later confirmed through a detailed behavioural comparison by Marlow (1975). A review of recent synonyms can be found within Jackson and Groves (2015).

## 2.4 Other common names

The common name 'long-nosed fur seal' was widely adopted for *Arctocephalus forsteri* in Australia after a proposal by Shaughnessy and Goldsworthy (2015). It was suggested that the common name 'New Zealand fur seal' was inaccurate and misleading with regards to the distribution of the species and contributed to the incorrect assumption that it was not native to Australia and subsequent persecution of the species for fisheries interactions.

The seal *A. pusillus* (i.e. Australian and South African Cape fur seals) was also known as brown seals, but this term is rarely used today. The Australian sea lion is the dominant common name for this species, but references may be found to the white-necked hair seal, cowled seal, counsellor seal and South Australian sea lion (Wood Jones 1925). Other common names can be found within Baker and Gynther (2023).

# 3. NATURAL HISTORY

## 3.1 Bodyweight and morphometrics

Seals range in size from only 30 kg to over 3800 kg. The morphometrics of each species found in Australian waters can be found in Table 15.2 and Baker and Gynther (2023).

**Table 15.2.** Body lengths and weights of different species of Australian seals.

Derived from Stewart (2014), Webber (2014), and Chilvers and Harcourt (2021).

| Species | Body length (males) (m) | Bodyweight (males) (kg) | Body length (females) (m) | Bodyweight (females) (kg) |
|---|---|---|---|---|
| **Family Otariidae** | | | | |
| *Arctocephalus forsteri* | 1.40–1.80 | 80–160 | 1.10–1.50 | 30–50 |
| *A. gazella* | 1.60–2.00 | 90–204 | 1.05–1.40 | 22–50 |
| *A pusillus doriferus* | 1.90–2.30 | 135–289 | 1.30–1.76 | 41–120 |
| *A tropicalis* | 1.50–1.90 | 70–165 | 1.00–1.50 | 25–67 |
| *Neophoca cinerea* | 1.80–2.50 | 180–250 | 1.30–1.85 | 60–105 |
| *Phocarctos hookeri* | 2.40–3.50 | 320–450 | 1.80–2.00 | 85–165 |
| **Family Phocidae** | | | | |
| *Hydrurga leptonyx* | 2.50–3.20 | 200–400 | 2.40–3.40 | 225–600 |
| *Leptonychotes weddellii* | 2.50–3.00 | 350–450 | 2.50–3.30 | 350–450 |
| *Lobodon carcinophaga* | 2.30–2.60 | 180–225 | 2.35–2.60 | 185–225 |
| *Mirounga leonina* | 3.50–4.20 | 1500–3000 | 2.00–2.60 | 250–800 |
| *Ommatophoca rossii* | 1.70–2.50 | 130–250 | 1.90–3.00 | 160–250 |

## 3.2 Distribution and habitat

Three species of otariids (Australian sea lion, and Australian and long-nosed fur seals) are resident to the southern coasts of mainland Australia. The Antarctic and subantarctic fur seals breed on islands in the Southern Ocean and are occasional visitors to mainland Australia. Similarly, the phocids breed on islands (e.g. elephant seal) or the Antarctic continent and associated ice and are infrequent visitors to the mainland.

Australian fur seals are a temperate latitude species, breeding at 20 sites in southern Australia. These sites include the 10 previously recognised breeding colonies (Lady Julia Percy Island, Seal Rocks, The Skerries, Judgement Rocks, Kanowna Island, Moriarty Rocks, Reid Rocks, West Moncoeur Island, Tenth Island, and Rag Island (in the Cliffy Group)) (Kirkwood *et al.* 2005), as well as the new breeding sites of Walker Island and Baudin Rocks in South Australia and Sloop Rocks in Tasmania (McIntosh *et al.* 2018) and seven haul-out sites where pups have been occasionally born (Kirkwood *et al.* 2010). These seals prefer to breed on rocky parts of islands with flat and open parts (Shaughnessy 1999). Outside of the breeding season Australian fur seals are found in South Australia, southern Tasmania and Jervis Bay Territory (Brothers and Pemberton 1990; Fortescue in Shaughnessy 1999; Shaughnessy 2000).

Long-nosed fur seals breed in southern Australia from Western Australia to New South Wales. The species also breeds around New Zealand and its subantarctic islands. The majority of the Australian population occurs in South Australia, between Kangaroo Island and the southern tip of the Eyre Peninsula (Shaughnessy *et al.* 2015). Individuals will spend extended periods at sea while foraging and cover large distances. Adult males typically forage at the edge of the continental shelf. Adult female foraging patterns vary seasonally as females and juveniles forage closer to their colonies while moulting and when rearing young pups. Females forage in pelagic waters associated with the subtropical front during winter and continental shelf or near-shelf waters in summer and autumn (Shaughnessy and Goldsworthy 2019).

As both the Australian and long-nosed fur seal populations recover from their depletion during sealing and persecution, they are returning to their historic ranges and expanding along the coast of New South Wales. Regular haul-outs are establishing around Sydney and the population is expected to continue to recover. The first record of a long-nosed fur seal pup on Montague Island (Barunguba) was documented in 1999 (Shaughnessy *et al.* 2001) and recent pup surveys between 2014 and 2017 indicate the haul-out site has transitioned from occasional pupping (1–14 pups) into a small breeding colony (≥ 15 pups) (S. Reinhold pers. comm). Similarly, McIntosh *et al.* (2018) reported an estimated 19 Australian fur seal pups on Montague Island in 2013, up from 1 in 2003 and 2 in 2007.

The Australian sea lions extant range extends from the east of Kangaroo Island, South Australia, west to the Houtman Abrolhos in Western Australia (McIntosh and Pitcher 2021). Breeding occurs at ~28 locations in Western Australia and 48 in South Australia (Shaughnessy *et al.* 2011; DSEWPAC 2013). Only nine of these colonies in South Australia have produced more than 100 pups in a breeding season, while 51 produced less than 30 pups per season (DEWHA 2010). Historic records indicate that the Australian sea lion was extirpated from the islands of Bass Straight during the sealing era of the early 1800s; however, the population was likely already lower than that of other seals in the region and it has not recovered to its former range (Ling 1999; McIntosh and Pitcher 2021). Infrequent sightings of Australian sea lions in waters of New South Wales have been reported (Shaughnessy *et al.* 2001).

Leopard seals are primarily found within the circumpolar region of the Antarctic pack ice and are regularly seen frequenting the subantarctic islands of the southern oceans and southern continents (Erickson *et al.* 1971; Gilbert and Erickson 1977). Beyond this range, they have been found in South America (Rodríguez *et al.* 2003; Aguayo-Lobo *et al.* 2011), South Africa (Berry 1960), Australia (Mawson and Coughran 1999) and as far north as the Cook Islands (Berry 1960). As juvenile seals are not required to return to the pack ice to breed, they are seen to move further north during winter (Rogers 2009).

Leopard seals, on occasion, will visit the Australian coastline and are protected under various conventions including the Scientific Committee on Antarctic Research agreement. This agreement states that any Antarctic species found requiring rehabilitation runs the risk of introducing pathogens into the Antarctica and is discouraged (see XXIV-3 of the recommendations (SCAR 1996)).

## 3.3 Conservation status

Pinnipeds in Australia have historically been heavily exploited. Sealing of elephant seals, fur seals and to a lesser extent sea lions was one of colonial Australia's first export industries (Gill 1967; Harcourt *et al.* 2015). Species were extirpated from large parts of their range due to unsustainable harvesting. Industrial sealing commenced with European colonisation and was most intense by the early 1800s. However, the population quickly became overexploited and by 1807 was notably in decline and sealers were voyaging to more distant islands. By the late 1830s the industry was no longer profitable and largely ceased, apart from some brief attempts to revive it (Gill 1967).

In Australia all marine mammals that occur within the Exclusive Economic Zone are protected under the *Environment Protection and Biodiversity Conservation Act 1999* (Harcourt *et al.* 2015). This includes areas under Australian claim around the Antarctic continent, but this is not recognised by most countries. Additional state legislation applies within each state and surrounding waters to three nautical miles offshore.

Populations of long-nosed fur seal appear to be increasing and expanding their range following protection from hunting and persecution. It is estimated that between 1989–90 and 2013–14 the pup abundance in South Australia increased 3.6-fold; 83% of the population occurs in South Australia, 14% in Western Australia and the remainder spread across Tasmania, Victoria and New South Wales (Shaughnessy *et al.* 2015). Although the species is increasing, it is still susceptible to many of the threats faced by pinnipeds, including collision with marine traffic, entanglement in fishing gear and debris, disturbance at breeding sites and disease. A further issue facing long-nosed fur seals is growing human-wildlife conflict. As numbers increase, interactions with commercial and recreational fisheries are becoming more common. Fur seals have the potential to impact on prey resources or compete directly with fisheries. The species is facing calls to manage the population, which must be considered carefully to ensure its conservation.

Australian fur seals are classified as Vulnerable in New South Wales under the *Threatened Species Conservation Act*, as past commercial harvesting had significantly affected the original population size and distribution (Warneke and Shaughnessy 1985). In general, numbers of Australian fur seals are increasing, with estimates the population has doubled in size between 1986 and 2002. However, in 2013–14 counts of live pups on colonies reported a decrease in pup production, with the suspected cause potentially indicating that the population has approached regional carrying capacity at established colonies (Kirkwood *et al.* 2010) or that there were shortages in food, although the reason remains unclear (McIntosh *et al.* 2018).

Like other demersal foraging species, the Australian sea lion is vulnerable to direct and indirect interactions with humans, climate change and extreme weather events (e.g. El Niño) (Kovacs *et al.* 2012; Goldsworthy *et al.* 2021). The pup abundance of the species has declined 64% over 3 generations (Goldsworthy *et al.* 2021) and the species is listed as Endangered in both the IUCN Red List and the *Environment Protection and Biodiversity Conservation Act 1999*. Current threats to the species include by-catch in fisheries, interaction with aquaculture, marine debris entanglement and disease (Goldsworthy *et al.* 2021).

In contrast to species inhabiting the Australian continent and subantarctic islands, the leopard seal was not a major target of commercial sealing. The species is not listed as endangered or threatened and it is estimated to have a population of between 220 000 and 440 000 individuals (Jefferson *et al.* 2015). The species may face some interactions with commercial Antarctic krill fisheries, as well as effects of climate change on ice formation and prey distribution, but the extent of these potential impacts are largely unknown.

## 3.4 Diet in the wild

The diets of pinnipeds are largely dependent on their foraging strategy. The Australian sea lion and Australian fur seal are benthic foragers, whereas the long-nosed fur seal and leopard seal are pelagic foragers consuming prey throughout the water column. Diets of all species range from crustacea, cephalopods, fish, birds, including penguins, diving petrels and shearwaters, and even other seals (Table 15.3).

The long-nosed fur seal diet is likely to vary seasonally as individuals access different foraging areas during and outside the breeding season (Harcourt 2005; Hardy *et al.* 2017). The species feeds mainly on cephalopods, fish and occasional seabirds (including penguins and diving petrels). The Australian fur seal feeds on a variety of fish species (mainly redbait *Emmelichthys nitidus*), jack mackerel (*Trachurus declivis*) and leatherjackets (Monacanthidae) and also consumes cephalopods (Gales and Pemberton 1994).

Although fresh water may not be essential for fur seal survival, it is utilised if present. Males have been observed drinking from freshwater pools, particularly on hot days during the breeding season, while females will drink directly from water running off cliffs above colonies (Harcourt 2005).

Australian sea lions appear to be opportunistic predators that consume a wide range of benthopelagic and

**Table 15.3.** Wild diet of different species of Australian seals.

Derived from Chilvers and Harcourt (2021), and Baker and Gynther (2023).

| Species | Diet |
|---|---|
| **Family Otariidae** | |
| *Arctocephalus forsteri* | Fish, cephalopods, birds (penguins and short-tailed shearwater) |
| *A. pusillus doriferus* | Schooling fish, squid |
| *A. gazella* | Lanternfish, mackerel ice-fish, myctophid fish, krill |
| *A. tropicalis* | Myctophid fish (lanternfish), cephalopods |
| *Neophoca cinerea* | Fish, cephalopods, crustaceans |
| *Phocarctos hookeri* | Fish, cephalopods, crustaceans |
| **Family Phocidae** | |
| *Mirounga leonina* | Fish, crustaceans, octopus, squid |
| *Hydrurga leptonyx* | Penguins, seals, cephalopods, fish, krill |
| *Leptonychotes weddellii* | Fish, cephalopods, small crustaceans |
| *Lobodon carcinophaga* | Krill |
| *Ommatophoca rossii* | Squid, fish |

**Table 15.4.** Longevity of seals in the wild and in captivity.

| Species | Wild males (years) | Wild females (years) | Captivity (years) | Reference |
|---|---|---|---|---|
| *Arctocephalus forsteri* | 19 | 25 | 28 | 1, 2 |
| *A. pusillus doriferus)* | 16.9 | 20.9 | 29 | 1, 3 |
| *A. tropicalis* | > 18 | > 23 | 21 | 1, 4 |
| *Neophoca cinerea* | 21 | 24 | 30 | 1, 5, 6 |
| *Hydrurga leptonyx* | c. > 26 | c. > 26 | 10 | 1, 7 |

Reference: (1) ZIMS 36 records extraction date (2021); (2) McKenzie *et al.* (2007); (3) Arnould and Warneke (2002); (4) Bester (1987); (5) McIntosh (2007); (6) McIntosh and Pitcher (2021); (7) Kooyman (1981).

demersal species (Peters *et al.* 2015; Berry *et al.* 2017). DNA, regurgitant and stomach analyses have identified several fish, cephalopod and crustacean species present in the sea lion diet, and less commonly sea birds such as little penguins (Gales and Cheal 1992; McIntosh *et al.* 2006; Peters *et al.* 2015; Berry *et al.* 2017).

Leopard seals have a wide range of prey (Hall-Aspland and Rogers 2004) and will consume penguins, fish, sea birds, cephalopods and seals. Leopard seal teeth are capable of gripping and tearing prey as well as sieving krill. Krill forms the majority of the leopard seals diet, especially in the winter months (Rogers 2009).

## 3.5 Longevity

In general, increases in longevity are recorded across seal species in the captive setting because these environments lack predators and there are no shortages of food.

### 3.5.1 Wild

The maximum recorded longevity of wild Australian fur seals is 16.9 years for males and 20.9 years for females (Arnould and Warneke 2002). The oldest recorded long-nosed fur seals in the wild were a 19-year-old male and 25-year-old female (McKenzie *et al.* 2007). The oldest known female Australian sea lion reached 26 years of age, gave birth to a pup at the age of 24, but died while still suckling the pup (McIntosh 2007).

### 3.5.2 Captivity

In captivity Australian and long-nosed fur seals have similar life spans with longevity estimated at 25 and 26 years respectively (Table 15.4). The oldest recorded Australian sea lion lived to 30 years and 9 months at Kamogawa Sea World, Japan. The female was born in September 1988 and was the offspring of a pair that were part of a group of sea lions sent to the 1975 World's Fair in Okinawa, Japan, by the South Australian Government (McIntosh and Pitcher 2021).

### 3.5.3 Methods to determine the age of adults

Determining the age of individuals in pinnipeds can be difficult. Body length can be used to estimate age from birth to approximately sexual maturity, but most individuals slow or cease growing at maturity. Growth curves exist for several species (e.g. *H. leptonyx* McLaren 1993; *N cinerea* McIntosh 2007). Ageing of adults can be done more reliably by examining dental growth of post-canine teeth, including the pulp cavity and growth layer groups (GLG) in the acellular cementum (Cebuhar *et al.* 2021). One GLG of alternating translucent and opaque layers typically represents 1 year of growth and can be counted from the neonatal line (McIntosh 2007). A significant disadvantage to ageing by dental growth is the need to remove a tooth from the individual. Although studies of wild pinnipeds have shown little long-term effects of tooth removal (Arnbom *et al.* 1992), it is unlikely that such accurate age estimation of a living captive individual would be needed to warrant tooth removal. Genetic techniques, including telomere length estimation, have been trialled and can classify individuals into broad age groups (pup/juvenile *v.* adult), but require further refinement to allow more accurate age determination (Izzo *et al.* 2011).

## 4. HOUSING REQUIREMENTS

### 4.1 Enclosure design

Any enclosure housing seals must accommodate their terrestrial and aquatic needs. All enclosures require a pool (preferably inbuilt) with adequate filtration or the ability to change pool water readily with a reliable salt water supply. The enclosure must also have sufficient land area to allow seals to haul out of the water and lie with adequate spacing between them (i.e. ability to rollover or extend flippers without touching another seal). There should be an off-exhibit holding area to allow animals to spend time off-exhibit for management activities, including veterinary procedures, to recover from illness or injury and to allow mother and pups to bond, and for enclosure maintenance.

Some species, such as fur seals, are agile climbers and prefer to haul out above water level, so appropriate mock-rock or climbing areas to allow them to be at a raised height (~1–2 m) is desirable. Therefore, walls need to be a minimum of 1.8 m high to provide adequate containment (E. Bombonato pers. comm.).

Access to and from the water needs to be on a gradual slant/gradient, as phocids seals (e.g. leopard seals) are unable to climb. Providing a slope also aids seal pups that are learning to enter and exit the water, as they can be cumbersome when they first start to swim.

Any furnishings within the enclosure need to be fastened securely to the enclosure and cannot have any sharp or hanging edges that a seal can bite. Australian sea lions (in particular) are very inquisitive and will chew on any loose objects and can even dig their teeth into silicone strips and remove them from underwater window frames. For this reason, silicone edges are to be covered with metal strips that are adhered to the pool surface and/or window surface. Any part of an enclosure or furniture that can become loose or free is at risk of being ingested by the seals.

Drains and water inlet pipe holes must be covered by a metal grate or metal bars (welded into place) so that a seal pup cannot put its head into the pipe and potentially become stuck.

There needs to be access to water for cleaning/hosing purposes; however, the hose must be kept out of the animals' reach or removed from the enclosure altogether when not in use.

As many seals develop ocular problems that may be caused or exacerbated by high levels of solar radiation, seal enclosures should have shade cover or characteristics that minimise the amount of reflected light (e.g. dull, non-reflective surfaces on walls and above water structures that provide shade) (New South Wales Government 2008).

Haul-out areas are required to be smooth to avoid abrasions, particularly for phocids that bear most of their bodyweight on their abdomens as they move. Drains are also required on haul-out spaces so that run-off from hosed surfaces do not directly enter pools.

Water quality and chemistry is to be monitored regularly and maintained via appropriate filtration systems and/or chemical treatment.

### 4.2 Holding area design

The off-exhibit holding enclosure must include a pool if seals are to be held in this space for more than a few hours (especially in warmer conditions) as the seal may overheat. Although it is not a physical requirement for seals to swim, a pool provides them with an area where they feel safe and where they can exercise. Holding areas must have a haul-out area where the seals can rest.

Dry pens (i.e. pens without access to a pool) can be used for unwell seals where there is a requirement for them to be kept dry, for seals recovering from an anaesthetic where they risk drowning if given access to a pool or for pregnant females that are due to give birth. Dry pens can also be used to hold or separate seals for a short period of time (up to 2 h) if a holding area with pool is not available.

Pinnipeds undergoing veterinary care can be held in pens without access to water for swimming if indicated (e.g. after a surgical procedure). However, they are much more content and surgical wounds heal better (particular if on the underside of the body) if they have free access to water (ideally, salt water as it is better for wounds). When animals have no access to water, fresh drinking water must be provided and the animals will enjoy being hosed. Dry pens should have non-abrasive washable floors, be well ventilated and large enough for the animal to move around freely. A slightly raised, preferably slatted, platform should be provided as a dry resting area. Pups and juveniles like to hide or sleep in a roomy PetPak or crate within the pen (Barnes *et al.* 2008).

## 4.3 Spatial requirements

Enclosures must provide enough room both horizontally and vertically so that the seals within it have a sufficient area to exercise, evade one another to avoid conflict and provide for social and breeding needs and wellbeing (New South Wales Government 2008). Minimum recommended pool sizes and haul-out areas for long-nosed fur seals, Australian fur seal, Australian sea lions and leopard seals can be found in Tables 15.5 and 15.6.

## 4.4 Position of enclosures

In some institutions pinnipeds are housed in indoor facilities. In these cases, appropriate climate control that is sufficiently regulated by heating or cooling to protect captive seals from extremes of temperature, and to prevent discomfort must be provided (New South Wales Government 2008). Outdoor facilities must facilitate a natural range of air and water temperatures and have some protection from extreme wind, rain and heat.

## 4.5 Weather protection

A sheltered area for protection from rain and wind is desirable, though not essential as seals naturally live along coastlines where water, rain and wind are prevalent. Shade structures are essential because seals can develop ocular conditions related to ultraviolet radiation. Shade will also allow seals to shelter from heat.

**Table 15.5.** Pool dimensions required for different species of seals.

Derived from the *Standards for Exhibiting Seals in New South Wales* (New South Wales Government 2008).

| Species | Species length | Surface area | Area per additional animal | Minimum dimension |
|---|---|---|---|---|
| *Arctocephalus forsteri*<br>*A. pusillus*<br>*Neophoca cinerea*<br>**1–4 animals** | ≤ 2.5 m | 20 $m^2$ | 5 $m^2$ | 1.5 m |
| *Hydrurga leptonyx*<br>**1–2 animals** | > 2.5 m in length | 32 $m^2$ | 16 $m^2$ | 2.0 m |

**Table 15.6.** Haul-out area dimensions.

Derived from the *Standards for Exhibiting Seals in New South Wales* (New South Wales Government 2008).

| Species | Species length | Surface area | Volume | Average depth | Volume per additional animal | Surface area per additional animal |
|---|---|---|---|---|---|---|
| *Arctocephalus forsteri*<br>*A. pusillus*<br>*Neophoca cinerea*<br>**1–4 animals** | ≤ 2.5 m | 60 $m^2$ | 200 $m^3$ | 2.0 m<br>Must incorporate a vertical column of water that is 2.5 m deep and 5 m in diameter | 25 $m^3$ | Surface area per additional seal: 15 $m^2$ |
| *Hydrurga leptonyx*<br>**1–2 animals** | > 2.5 m | 180 $m^2$ | 600 $m^3$ | 3.0 m<br>Must incorporate a vertical column of water that is 3.4 m deep and 7 m diameter | 200 $m^3$ | 90 $m^2$ |

## 4.6 Temperature requirements and water quality

### 4.6.1 Temperature

Daily water samples should be obtained daily to determine the temperature and water quality. Many captive seals in Australia are housed in outdoor facilities and can adapt to the weather and seasonal changes. However, leopard seals are negatively affected by heat because they are naturally found in Antarctic waters; therefore, they have adaptations that enable them to insulate heat. It is advised that leopard seals be housed in an environment where the water can be chilled and land areas can remain cool during warmer months. Cooling misters and or cold/chilled platforms are recommended for the haul-out areas for leopard seals.

As a guide, the temperature range should be within the natural temperature range of the species; however, if this is not achievable, the maximum temperature range is 5–28°C (New South Wales Government 2008).

Although salt water is preferred, pinnipeds can be successfully housed in fresh water. Oral sodium chloride supplementation is required for all pinnipeds held in fresh water to ensure general wellbeing. Some institutions use artificial seawater salts when natural seawater in not accessible. Good water quality is essential for captive pinnipeds to thrive (Veterian Key 2017).

Animal and food waste materials must be regularly removed from the water, either by scooping or vacuuming it or flushing it down an open drain. The smaller the water body or higher volume of animals within it, the more frequently this will need to be done. Walls and other surfaces need to be cleaned of algae on a regular basis, more often in the warmer months if the pool has no temperature control. Filtration and chemical treatment should also be used in order to maintain the health of the animals (New South Wales Government 2008).

### 4.6.2 Coliform bacteria count

Faecal coliform testing is mandatory for all water systems housing seals and pH and chemical testing is required for all systems apart from open-water systems (Table 15.7). It is recommended that a minimum of five water collection samples at 6-day intervals are obtained to analyse water quality. The suggested intervals are proposed to ensure variation in the days of collection to correlate with variation in animal usage of the pools (New South Wales Government 2008).

### 4.6.3 Bromine

The residual chlorine in the water should not exceed 1.0 mg/L and should be tested at least daily.

### 4.6.4 Salinity

Salinity should be within the range of 25–35 ppt.

### 4.6.5 pH

The pH of the water should be maintained in the range of 7.2–8.4, tested daily and adjusted when needed.

### 4.6.6 Ammonium (NH4)/ammonia (NH3)

Ammonia levels for mammals should remain below 0.7 mg/L and be tested daily. Ammonia is produced by marine animals as the end product of protein (nitrogen) metabolism. Though ammonia is relatively non-toxic for mammals, high levels can cause irritation to mucous membranes and eyes.

### 4.6.7 Phosphorus (P)

Increasing phosphate levels in water are not usually toxic to seals, but will increase algal growth. Phosphorus should be kept below 0.07 ppm.

### 4.6.8 Oxidation reduction potential (ORP)/redox

Redox potential is a measure of the tendency of a chemical species to acquire electrons from or lose electrons to

**Table 15.7.** Water quality standards for keeping seals.

cfu = colony-forming units.

| Water quality factor | Minimum requirements |
|---|---|
| Faecal coliform bacteria | The median bacterial content in samples taken from each water body should not exceed 150 cfu per 100 mL of water. The second highest sample taken must be < 600 cfu/100 mL |
| Enterococci bacteria | The median bacterial content in samples taken from each water body should not exceed 35 cfu/100 mL of water. The second highest sample taken must be < 100 cfu/100 mL |
| pH | The pH of any pool water must be maintained within the range 7.2–8.4 |
| Salinity | The salinity of any pool water must be maintained within a range of 26–36 PPT (g/L) |

an electrode and thereby be reduced or oxidised respectively. ORP is basically a measurement of sanitiser effectiveness in water bodies. The higher the ORP the more effective is the disinfecting agent.
ORP can effectively remove chlorine and heavy metals (including iron, lead, mercury, magnesium, hydrogen sulfide, calcium carbonate, and chromium) from the water and can also help control microorganisms (such as bacteria, algae, and fungi) and other inorganic contaminants.

### 4.7 Life support systems

It is best practice to house pinnipeds in water bodies that have regularly maintained and monitored life support systems. When functioning at full capacity, these systems will maintain water quality and minimise the need for manual pool cleaning. Life support systems for seal enclosures should include the following elements (B. McKenzie pers. comm.):

- sand filters - to extract particulate matter from the water and aid in improving water clarity by trapping and extracting solid matter
- protein skimmers - to extract proteins from water, aid in water clarity and extract protein waste
- chlorine dosing - a sterilising agent that treats and sterilises water (usually before it goes into the main water body)
- ozone dosing - a sterilising agent that usually treats water before it goes into the main water body (will kill off other organisms)
- balance tanks - filtered sea water from a harbour (or other sea water source) into balance tank
- vacuum system - large pools often have built-in vacuum lines for waste removal.

Open water systems do not require any mechanical filtration devices and are often used to increase water clarity. The water supply to an open water system enters from either a natural source or tap, flows through the pool and then exits through a waste point. There is a continuous exchange of water (Spotte 1991).

Semi-closed water systems also have an on-demand supply of pool water, but at a lower rate than for an open system and therefore requires a granular medium filter (usually sand) to improve water clarity (Spotte 1991).

Closed water systems use 100% cent recycled pool water through a granular medium filter (Spotte 1991).

### 4.8 Substrate

Enclosures can have sand, pebble-crete, polished concrete or other smooth surfaces as flooring (the latter preferred especially for phocids as rough surfaces can cause abrasions when they manoeuvre on land). The pools should be painted in dark colours to prevent glare and reduce the incidence of eye problems.

For variety, enclosures can be predominately smooth rock/concrete flooring and then a sand area and/or mock rock structures for investigation and play. Vegetation that is safe if ingested can be planted within the enclosures; however, these are easily destructed by inquisitive seals (i.e. Australian sea lions).

### 4.9 Enclosure furnishings

For otariids (particularly fur seals), climbing rock or mock-rock structures are appropriate. They need to be set back away from enclosure walls so that the seals cannot climb out of their enclosure. Climbing structures should be tiered so that the otariids can easily climb up and down. Seal enclosures may be fitted with a 'bed'. This is a slightly elevated wooden platform that has small gaps between the slats for water to drain through. It provides a dry space for seals to haul out when land areas receive splash or water overflow from the pool.

Any furnishings within the enclosure need to be fastened securely to the enclosure and cannot have any sharp or hanging edges that a seal can bite. Australian sea lions are very inquisitive and will chew on any loose objects and can even dig their teeth into silicone strips and remove them from underwater window frames. For this reason, silicone edges are to be covered with metal strips that are adhered to the pool surface and/or window surface. Any part of an enclosure that can become loose or free is at risk of being ingested by the seals.

## 5. GENERAL HUSBANDRY

### 5.1 Hygiene and cleaning

Haul-out areas must be hosed down daily, and all faeces removed and discarded appropriately. Soiled areas may be soaked and then scrubbed with a deck scrubber, to remove any faecal matter that cannot be shifted with hosing alone. Faecal matter should be removed from pools (depending on their size) every day or up to every week for large pools. Faeces can be removed from pools by using a pool scoop or underwater vacuums, operated by in larger pools. Algae growth can be scrubbed off pool surfaces or pools can be emptied of water and hosed out.

Disinfection of contaminated surfaces can be done, but if chemicals are used, they need to be rinsed off thoroughly before seals can have access to the area.

Pools should be emptied and refilled with fresh salt water daily unless they have filtration and sterilisation. Depending on the size of the pool, the water may not need to be drained from the pool regularly if algae growth is managed and faeces removed from water.

## 5.2 Record keeping

It is important to establish a system whereby the health, condition and reproductive status of captive seals are routinely monitored to record all animal health and welfare observations, including:

- identification (all individuals should be identifiable)
- results of veterinary examinations and treatments provided
- behavioural changes or problems
- reproductive behaviour or condition
- bodyweights and measurements
- changes in diet and daily food intake
- movements of individuals between enclosures or institutions
- advancements in training and conditioning
- participation in encounters or presentations
- introductions with other animals
- aggression between animals or towards animal caretakers
- births, with dam and sire recorded if known
- deaths with necropsy results.

The collection of information on the physical, physiological, and behavioural patterns of individual seals can contribute greatly to their husbandry. It also allows individual histories to be transferred to other institutions if required and facilitates a cooperative approach to data collection amongst institutions. ZIMS, the Zoo Information Management System (Species360 2023) can be used for recording general information on age, sex, parentage, births, identification, bodyweights, transfers and deaths, with separate ZIMS modules for veterinary/medical information and studbooks. These systems have been developed by Species360, which is part of the Conservation Planning Specialist Group (CPSG) software. ZIMS interfaces with programs, such as PMx, which provides demographic and genetic analysis tools to assist with the management of breeding programs (Lacy *et al.* 2011).

## 5.3 Methods of individual identification

### 5.3.1 Passive Integrated Transponder (PIT) tags

PIT tags (microchips) are the most common method used for permanent identification (Barnes *et al.* 2008). The PIT tag is implanted below the base of the back of the neck or between the shoulder blades of the seal. Because some seals can grow beyond 250 kg, the implant may shift as the animal grows and can become difficult to locate.

### 5.3.2 Livestock tags

Livestock tags are ~15 × 40 mm and are placed in the interdigital webbing ~25 mm from the edge of the webbing between the upper two digits of the hind-flipper of phocids or into the loose skin on the trailing edge of the upper aspect of the fore-flipper of otariids. It is not uncommon for plastic flipper tags to be lost due to damage to the interdigital webbing or from being torn out or by damage or failure of the tag itself (Barnes *et al.* 2008).

### 5.3.3 Paint markings

Individuals can be identified temporarily and before moulting with oil-based paint markings, by shaving or by bleaching the fur with a mixture of potassium persulfate and hydrogen peroxide (Barnes *et al.* 2008).

### 5.3.4 Visual identification

Visual identification of physical features is the most common method of seal identification used in zoos and aquariums. Animal caretakers work closely with the seals in their care and can recognise individual characteristics that can identify them. Many of these characteristics are visual (e.g. scars, size or shape of features, fur colour), but some unique characteristics may be behavioural as well. For example, if there are two seals of similar size and appearance, they may show different behavioural traits or may be trained to come to or feed at a particular station. This is a less reliable means of identification, but can be used in some instances.

Branding has been used previously as a method to permanently identify individuals, particularly for long-term conservation studies, such as of wild southern elephant seals. Hot branding was found to be effective and did not cause increased deaths when compared with flipper tagging. By contrast, cryobranding was less effective because the marks did not persist (McMahon *et al.* 2006). However, due to public concerns about welfare, hot iron branding of pinnipeds was banned on Australian subantarctic islands in 2000 and is no longer being used in many jurisdictions (Monamy 2007).

## 6. FEEDING REQUIREMENTS

### 6.1 Captive diet

The food provided to captive seals must be high quality, have appropriate nutritional value, be free from contamination and must be of sufficient quantity to maintain all seals held in a state of good health. (New South Wales Government 2008).

Seals should be provided with a variety of fish species, preferably those that they would naturally feed upon in the wild. Some common species of fish fed to captive seals in Australia are red spot whiting, slimy mackerel, red bait, herring, pilchards and squid. Substitutions can be made with fish that they would not naturally find within their range if availability of natural food species is limited.

The fat and protein content of fish fed to captive seals should be made known (through testing samples of the fish) and this knowledge can used to provide the seal with a balanced diet of appropriate quantity, seasonally or as required for lifestyle stage (e.g. for growing animals, for maintaining or losing bodyweight, for pregnant or lactating females).

All food provided must be fit for human consumption. Fish being fed out needs to be undamaged and it should not be freezer burnt. Fish is to be fed out within 12–24 h of being defrosted. It must be stored and frozen in a food-safe environment so as to prevent contamination from other substances and prevent freezer burn. Freezer temperatures must not exceed -20°C. Stored fish needs to be clearly labelled and have the catch date clearly displayed. Frozen fish needs to be fed out within 4–6 months (New South Wales Government 2008).

### 6.2 Supplements

Appropriate food supplements must be routinely added to the seals diet, as should medication directed by a veterinarian (New South Wales Government 2008). A multivitamin can be fed out by inserting it behind the operculum of a fish before feeding it to the seal. Taronga Zoo uses Aquatic Mammal Chewable™ (Vetafarm) tablets, which contain vitamin A, vitamin E, thiamine, riboflavin, pyridoxine, pantothenic, folic acid and biotin. These supplements replace vital minerals that may be lost when fish are frozen and stored for a length of time.

### 6.3 Presentation of food

All food should be thawed in salt water or a mixture of fresh and salt water with a ratio of 1:2 (fresh water to salt water) to assist in maintaining osmotic pressure and reducing nutrient leaching. Food is weighed daily according to each seal's daily intake. Once prepared it can be stored within buckets (preferably stainless steel) and kept in a fridge between feeds.

Seals should be fed at least 2–3 times per day unless otherwise directed by a veterinarian. Food can be presented in the form of an enrichment feed or can be hand fed by animal caretakers. Between feeds, any unused fish should be kept in the fridge and any uneaten fish should be removed from the seals' enclosure and discarded.

Sick or recently rescued seals may need assist-feeding (Barnes *et al.* 2008), which requires several people and the animal to be restrained (see section 7). The mouth is opened using soft rope on the upper and lower jaws, then whole fish are placed head first into the back of the mouth using tongs or forceps. The animal will usually swallow voluntarily if the fish is placed correctly. Massaging the throat can encourage and assist swallowing. Once swallowed, introduce the next fish. Start with only a few fish at each feed and gradually increase the number of fish. With successive feeds the seal will take fish more readily and the level of restraint can be reduced. Generally, pinnipeds will feed voluntarily within a few days. Force-feeding of adult pinnipeds can be difficult and dangerous. Fluids, milk formulas and fish gruel can be administered by stomach tube. The animal is restrained and the mouth held open as described above. A gag with a central hole can be used, with the stomach tube passed through the central hole into the pharynx. Allowing the animal to assist in swallowing the tube reduces the risk of passing it into the trachea. If the tube is in the oesophagus, it can usually be palpated in the neck of young animals, but is frequently not palpable in adults. Volumes of fluid that can typically be safely given by stomach tube are 100–200 mL for young pups (< 10 kg), 200–400 mL for older pups (10–20 kg), 1–2 L for sub-adults/adults less than 100 kg and 2–4 L for adults over 100 kg. Pinnipeds undergoing rehabilitation that are in poor body condition are likely to have compromised gastrointestinal function and should be slowly reintroduced to food. For the first 24–48 h they should be stomach tubed and fed fluids only (water or electrolyte solution), then semi-solid food such as blended fish with electrolytes or EmerAid Piscivore® can be introduced, starting with a small quantity mixed with the fluids. The volume of semi-solid food per feed can be gradually increased over 48–72 h. If no problems occur, solid food can be gradually introduced.

## 7. HANDLING AND TRANSPORT

### 7.1 Timing of capture and handling

Ideally, seals to be captured would be already hauled out and away from the water. Captive seals can be dry penned before capture for the safety of animal caretakers and for ease of restraint. Wild seals would only need to be captured if they are unwell or injured. In most cases, this would result in the seal being hauled out.

When planning a catch, select the time of day when visibility is high (daylight hours) and if the catch is on or near the beach, personnel would need to be aware of the tide and what effect this may have on the intended catch. Consideration needs to be given to access and public visibility. In some cases, where access to the animal is difficult or dangerous (e.g. on a rock ledge by the water), experts suggest to wait and see if the seal chooses to haul out at a more suitable location before attempting capture.

### 7.2 Capture equipment

Cargo nets can assist in the restraint of larger pinnipeds. When handlers are holding the animal down, having a cargo net on top of the seal can assist with restricting its movement, particularly during a procedure.

Some institutions will use squeeze cages (Fig. 15.1) to assist in the restraint of larger pinnipeds. Seal can be trained to walk into a squeeze cage and to tolerate a level of pressure from the squeeze. This can assist with fast procedures (e.g. manual injections) or the seal can be trained at this point to put its muzzle into a gas mask for sedation.

**Fig. 15.1.** Squeeze cage used to restrain a seal. Photo by Danielle Fox.

Restraint boards can also be used. The animal lays down on a board and is restrained by straps across the body. These can also be incorporated into the bottom of squeeze cages and once the seal is strapped to the board, the squeeze cage can be opened and the seal is removed on the board. These are particularly used in field situations (Vogelnest 2025).

### 7.3 Capture and physical restraint

Capture and restraint techniques for pinnipeds may vary depending on the age, size, species, location and demeanour of the seal (Vogelnest 2025). Experienced handlers may choose to physically restrain the animal if it is small enough to do so, while also ensuring that safety of handlers is not compromised

Capture techniques used to secure an otariid seal can include the use of a herding board (or several) to encourage the animal into an area or space that is more conducive to restraining the seal or a sausage net to capture leopard seals (and other phocid seals) (Fig. 15.2). A head bag can be used to place over the head of the seal to restrict its vision before the catch (Fig. 15.3). Once the head bag (or towel for a small seal) is covering the face of the seal the handlers can then proceed to restrain it by swiftly reaching for the back of the seal's neck and pinning it to the floor. At this point any extra handlers needed can come in and assist with holding the seal flat

**Fig. 15.2.** Using a sausage net to capture a leopard seal. Photo by Ady D'Ettorre.

**Fig. 15.3.** Using a head bag on a leopard seal. Photo by Ady D'Ettorre.

on the floor, which will include tucking its pectoral flippers underneath its body to restrict movement. If handlers cannot get close enough to the seal to place a head bag over its head, they can instead use a hoop net to cover/scoop the animal. Once in the net handlers can quickly approach and using the same techniques as with a head bag to physically restrain the seal.

It is important not to underestimate the speed and flexibility of pinnipeds, especially otariids, when attempting a catch and to always have a clear escape route in mind (Vogelnest 2025). For larger otariids (≥ 50 kg), animals can be captured using both hoop net and head bag. Approaching quietly and slowly from behind, throwing the large net over the seal as quickly as possible to encase the seal. An additional head bag can be placed, if needed. Once the net is in place handlers can approach and physically restrain the seal (Vogelnest 2025). Capturing large seals can be dangerous, so if the animal's demeanour seems aggressive, handlers and veterinarians may choose to chemically restrain the animal for the safety of all involved.

When capturing a phocid, most handlers will use a sausage net for a ≥ 50-kg seal. The sausage net has a very wide opening so that two handlers can operate it, each holding one side of the opening. The handlers should approach the seal so that the opening of the net is directly in front and as the seal moves forward it bounces into the length of the net. Once the net is down far enough and partially or fully encasing the seal, a head bag can be placed and the handlers can physically restrain the seal.

## 7.4 Weighing and examination

Weighing seals in a captive environment is commonly a trained behaviour whereby the seals are conditioned to station on to a weigh board sitting on top of the scales. Seals that are not conditioned or have only recently come into human care can be weighed in a pet carrier or transport cage (seals ≤ 30–50 kg). The seal is encouraged into the container, which then can be lifted onto the scales. Larger seals (≥ 50 kg), if they are hand feeding, can be lured onto a weigh board during a feeding session. Ideally, all seals should be weighed before anaesthesia so that drug doses can be accurately calculated. If this is not possible, an estimated weight can be used and an exact weight obtained once the seal is anaesthetised.

## 7.5 Release

Releasing of seals needs to be done with caution, as a seal in close proximity to humans may bite if it feels threatened. If a seal is being manually restrained, then clear communication among the handlers is necessary. The handlers restraining the rear flippers may release and clear out of the way first, while the handlers closer to the head need to release and quickly move away from the seal, in the opposite direction to the way the seal is facing.

If the seal is in a container, opening the door of the container either in the release space or at entrance to the release space and letting the seal to walk out of its own accord would be most appropriate. Once the door is opened, the handler should quickly move away from the container and remain at a safe distance until the seal has moved away.

## 7.6 Transport requirements

### 7.6.1 Transport box design

Transport box design will vary depending on the age, size and species of the seal (Fig. 15.4). Leopard seals require long and low boxes to enable them to lie flat and

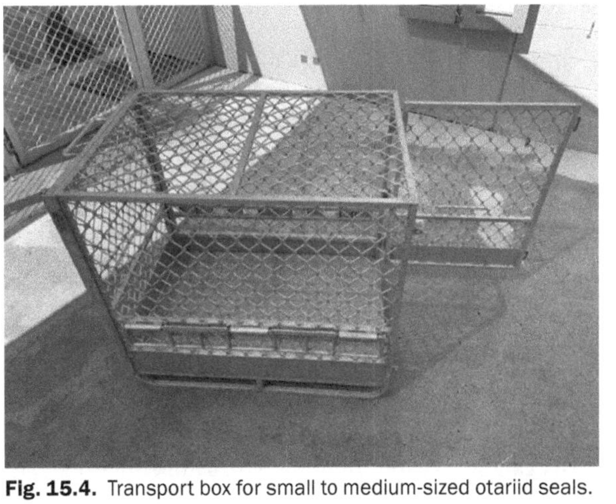

**Fig. 15.4.** Transport box for small to medium-sized otariid seals. Photo by Danielle Fox.

straight. The box should not allow enough room for the leopard seal to arch and bend backwards or sideways because it can get stuck in this position and injure itself.

Otariid seals are comfortable in a box that allows them the space to turn around in and to lie flat in. Boxes should be constructed with a strong steel or wooden frame. For long-haul trips during which the seal will be unsupervised for some time (e.g. flights) it is preferable that the inside of the box is to be covered with smooth boards so that a seal cannot grab onto anything with its teeth. All boxes need to provide adequate ventilation and any open wire/mesh areas can be covered (on the outside so the seal cannot chew/swallow) with shade mesh to reduce exposure to visual stimuli.

### 7.6.2 Furnishings

No furnishings are required.

### 7.6.3 Water and food

It is not common for seals to be offered food (or water) during transportation. If the seal will be travelling for a long period (i.e. over 8–12 h) then food can be offered at intervals if the option is available (e.g. if the animal is travelling by vehicle); however, if this is not possible, the seal is offered food when departing and then on arrival at the destination.

### 7.6.4 Animals per transport box

Only one pinniped should be transported in a given box. Aggression between individuals can occur if transported together in the same box for a length of time. An exception to this would be a mother and pup travelling together, or possibly two young pups; however, this should be assessed on a case-by-case basis by expert staff.

### 7.6.5 Timing of transportation

Careful consideration should be taken as to the time of day that the seal will be travelling. In summer months, if the seal will be spending time in the box without access to air conditioning, then this should be done during the coolest part of the day to minimise the possibility of heat stress and dehydration. Seals can be sprayed with a cool mist of water on their body for extra comfort on warmer days. Ideally, seals should travel in a temperature-controlled cabin.

### 7.6.6 Release from the transport box

Seals can be released from the box by opening the door while inside the enclosure or at the entrance to the enclosure, making sure that there are no gaps between the box and the doorway itself. Herding boards can be used to close any gaps between the box and the doorway. Once the door is opened, the handler should quickly move away from the box and remain at a safe distance until the seal has moved on.

## 8. HEALTH REQUIREMENTS

*Larry Vogelnest*

### 8.1 Daily health observations

Captive pinnipeds are readily trained for a wide range of behaviours. Performing daily health observations and examinations should therefore be relatively easy. Each animal should be checked for signs of injury or illness, particularly after events such as introductions with other animals (conspecific or other species) or new enclosures and breeding. Distant observation of the undisturbed animal, either swimming or when hauled out is important.

Each animal should be assessed and checked for the following:

- appetite
- faeces – amount and consistency (if available)
- changes in demeanour, activity level and behaviour
- injuries
- gait, particularly lameness and posture when in water and on land
- fur and skin – check for alopecia, wounds, inflammation, blisters (on bare flipper skin)
- discharges from the eyes, ears, nose or mouth
- eyes – squinting (blepharospasm), corneal opacity, redness.

Any signs of injury or illness or unusual changes in behaviour should be documented and reported to a veterinarian. If an animal is found dead, make note of the circumstances and anything unusual or abnormal on the body or surrounding area. A veterinarian should be informed immediately so that arrangements can be made for a necropsy. Appropriate PPE (gloves, face mask) must be worn when handling the body.

### 8.2 Bodyweight and body condition score

All animals should be weighed and have body condition assessed on a regular basis.

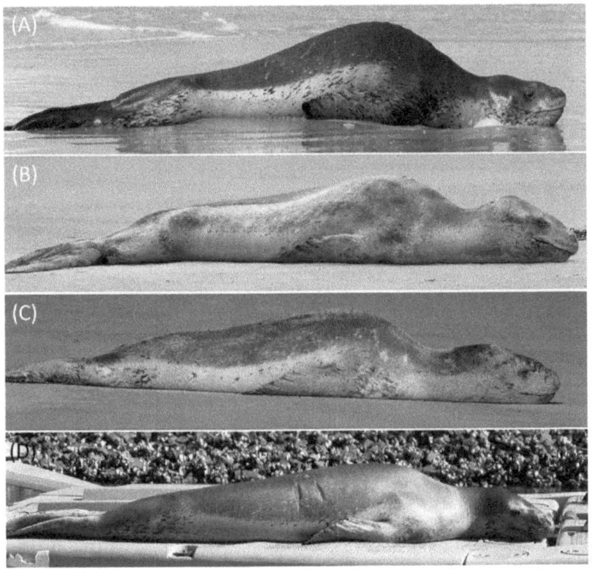


**Fig. 15.5.** Body condition of leopard seals classified as: (a) severe, (b) poor, (c) good or (d) excellent. Modified from Gray *et al.* (2009) and Hupman *et al.* (2020). Photographs from LeopardSeals.org.

### 8.2.1 Bodyweight

Bodyweight should be taken weekly. Pinnipeds show significant seasonal bodyweight variation, with the fur seals being heaviest in December–January and Australian sea lions in June–September. A persistent decline in bodyweight outside of the normal seasonal variation is an indication for more detailed examination and should be reported to a veterinarian.

### 8.2.2 Body condition score

A specific body condition scoring system has been developed for pinnipeds. Similar to bodyweight, there is seasonal variation in body condition. However, in rescued or sick pinnipeds, body condition may be an important clinical consideration (Fig. 15.5). A specific body condition score has been developed for leopard seals by Gray *et al.* (2009) and Hupman *et al.* (2020) (Fig. 15.5). The main things to look for in an under-condition animals are obvious projections of the: (1) neural spines along the back, (2) sagittal crest on the head, (3) ribs, (4) zygomatic arch (or cheek bone) and (5) pelvic bones. These features are not visible in animals with good to excellent body condition scores.

## 8.3 Physical examination

A comprehensive physical examination of a trained pinniped can be performed by animal caretakers/trainers without the need for sedation. Depending on the temperament of the individual animal, and with the animal under behavioural control by its trainer, much of the examination can be performed by a veterinarian if required. This includes the eyes, oral cavity and palpation of the fore- and hind-flippers. In addition to the list above (section 8.1) the following can be assessed:

- abnormal swellings, masses or deformities of the body and limbs
- anus, scrotum and vulva – should be clean and free of swelling (normal vulvar swelling can be seen in late pregnancy), discharge or faeces.

A detailed physical examination of sick pinnipeds should be conducted by a veterinarian. Anaesthesia may be indicated if more detailed examination is required, keeping in mind the challenges of anaesthesia in pinnipeds and the increased risk associated with anaesthesia of sick pinnipeds. Veterinarians should refer to Vogelnest (2025) and Vogelnest and Portas (2025) for detailed information on examination and medicine for Australian pinnipeds. All body areas and systems should be evaluated: cardiovascular; respiratory (thoracic auscultation is less indicative than in terrestrial mammals and pinnipeds with significant pulmonary disease can have apparently normal lung sounds (Van Bonn 2015); oral cavity (teeth, tongue, oral mucosa), abdomen (due to the small 'window' [the rib cage extends far caudally], size of the animals and the subcutaneous fat layer, abdominal palpation is generally unrewarding (Bodley 2025)); musculoskeletal; genitourinary; ocular; and integument. Important parameters and vital signs should be measured and recorded: bodyweight and condition; heart rate; respiratory rate; temperature. Recording the body temperature of a conscious pinniped can be challenging; aural thermometers can be used in otariids and high-precision infrared thermometers directed at the back of the mouth may be of use.

Medical training of pinnipeds facilitates a cooperative physical examination, administration of medication, collection of routine samples and non-invasive diagnostic procedures (Ramirez 2012), avoiding the need for sedation or anaesthesia. Under behavioural control, a well-trained pinniped should allow close proximity of a veterinarian and palpation of the body and flippers. The challenge, however, is that even the most well-trained pinnipeds no longer respond to behavioural cues when they become sick and inappetent.

Desensitisation is key to preparing animals for exposure to syringes and needles, new smells, the presence of veterinarians and specialists, and other unique stimuli associated with various medical procedures. This involves habituation (exposure to new stimuli repeatedly over time so the animal becomes accustomed and accepts it) and counterconditioning (controlled exposure to a new stimulus, paired with positive reinforcement). Using these techniques, pinnipeds can be trained for various healthcare assessments and drug administrations, including:

- venepuncture (taking blood), with the most commonly used veins being the caudal gluteal vein, interdigital veins of the hind-flippers or brachial vein
- injection of medications, sedative or anaesthetic drugs and contraceptive implants (deslorelin)
- intradermal tuberculosis testing
- mask induction for inhalation anaesthesia
- administration of eye and ear drops (Fig. 15.6)
- ultrasound examination (Fig. 15.7)
- ocular examination (Fig. 15.7)
- radiography (Figs 15.8, 15.9).

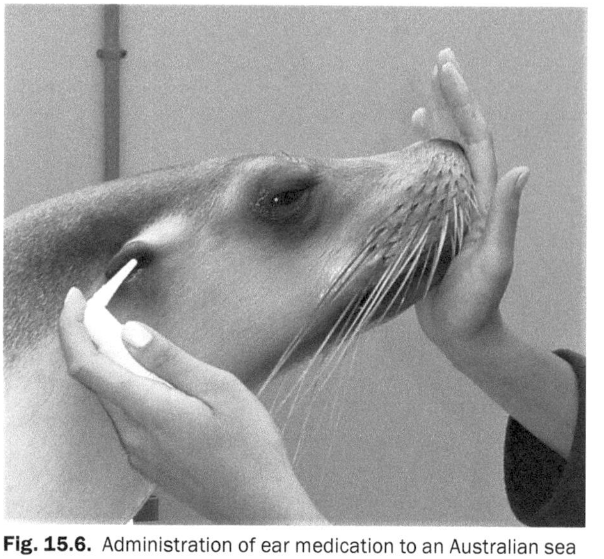

**Fig. 15.6.** Administration of ear medication to an Australian sea lion. Photo Larry Vogelnest.

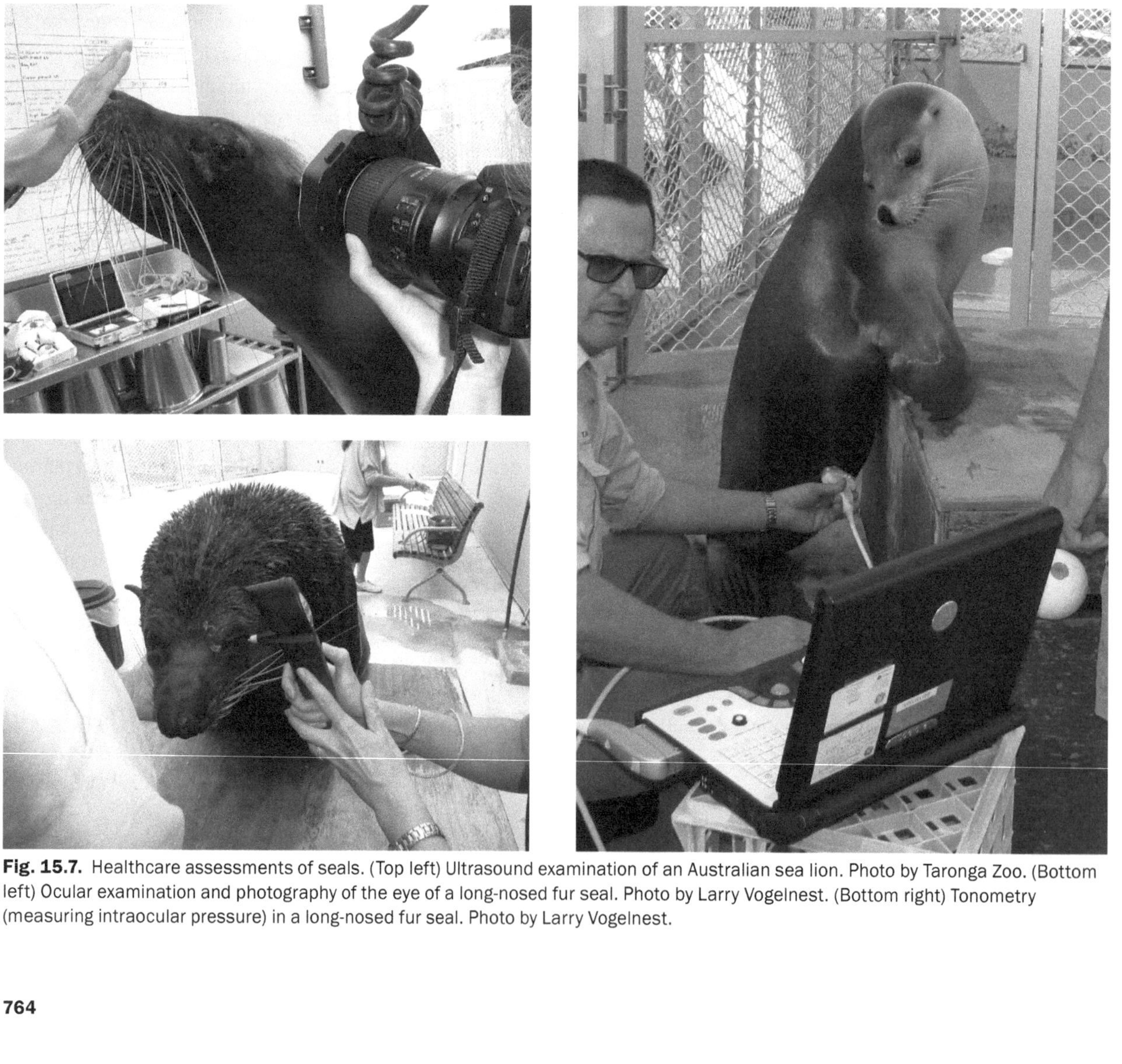

**Fig. 15.7.** Healthcare assessments of seals. (Top left) Ultrasound examination of an Australian sea lion. Photo by Taronga Zoo. (Bottom left) Ocular examination and photography of the eye of a long-nosed fur seal. Photo by Larry Vogelnest. (Bottom right) Tonometry (measuring intraocular pressure) in a long-nosed fur seal. Photo by Larry Vogelnest.

**Fig. 15.8.** Radiography of seals. (Top) Radiographic sensor plate (Photo by Larry Vogelnest). (Bottom) Seal being X-rayed. Photo by Elizabeth Arthur.

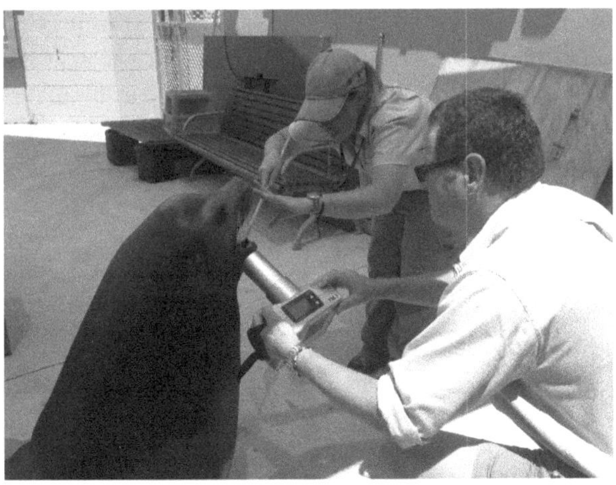

**Fig. 15.9.** X-ray of seal using a small dental plate held within the mouth of the seal while keeping its mouth open. The X-ray generator is held close to the outside of the jaw. Photo by Taronga Zoo.

Clinical procedures that can conducted under behavioural control include the following.

- Ultrasound examination – can be used in pinnipeds for pregnancy diagnosis, investigation of superficial masses or swellings, or examination of internal structures and organs. Training the seal to accept an ultrasound transducer and machine is relatively straightforward because it is often an extension of other tactile behaviours that may already be taught for other reasons. The animal is only required to remain still for short periods of time (3–4 min). However, one of the challenges of ultrasound examination in furred animals is that air can be trapped within or under the fur, blocking the penetration of ultrasound waves into the tissue beneath. This is particularly a problem in fur seals whose dense underfur traps air against the skin. Wetting the fur with medical grade alcohol and ruffling the fur to remove the air may help (but will require additional training). Clipping the fur may help, but still requires significant wetting with alcohol. Application of acoustic coupling gel may also help. Sea lions on the other hand have a single layer of fur that when wet with water allows perfect acoustic coupling.
- Radiography – can be used in pinnipeds for investigating injuries or illness as well as pregnancy diagnosis. It is particularly useful for detection of bone injuries or degenerative changes (e.g. arthritis) and ingested foreign bodies, particularly those that are radiopaque. Any part of the body can be radiographed. Dental radiography is useful for investigation of orodental trauma and disease. The use of a board into which the radiographic sensor plate can be recessed is useful (Fig. 15.8). It can be challenging to train the animal to feel comfortable with the radiographic equipment (stand and X-ray generator), people wearing lead gowns and ensuring the animal remains motionless for long enough to obtain the image (Ramirez 2012). Obtaining intra-oral dental radiographs requires the animal to have a small dental plate held within the mouth while keeping its mouth open and the X-ray generator held close to the outside of the jaw (Fig. 15.9). As most captive pinnipeds are trained for mouth open behaviours for routine inspection of the oral cavity and teeth, training for dental radiography is a relatively easy extension of this behaviour.
- Ocular examination and treatments. Ocular disease is relatively common in captive pinnipeds. Allowing close examination of the eye using equipment such as an ophthalmoscope, slit lamp or tonometer and application of drops or ointment to dilate the pupil, anaesthetise the cornea or for treatment are useful.

Training a pinniped to keep its eye open, particularly when diseased or injured, can be challenging, because medications and equipment need to be held close to the eye. Tonometry (measuring pressure in the eye) is even more challenging because it requires a brief targeted puff of air onto the cornea (pneumotonometry) or bouncing a small plastic tipped metal probe against the cornea (rebound tonometry) (Fig. 15.7).

## 8.4 Chemical restraint

Chemical restraint (sedation and anaesthesia) of pinnipeds is complex and may pose considerable risk to the animal. Planning and preparation for a procedure are important. It is recommended that a veterinarian experienced in chemical restraint of pinnipeds performs the procedure and either the veterinarian or an experienced veterinary nurse is dedicated to monitoring the sedation or anaesthesia and take no part in any other aspects of the procedure. Other risk factors in pinniped anaesthesia include (Barnes *et al.* 2008; Stringer *et al.* 2012; Vogelnest 2025):

- debility and poor health status
- administration of atropine before or during anaesthesia
- overestimation of bodyweight in calculating the drug dose.

There are considerable differences in response to chemical restraint, particularly between otariids and phocids and even within these groups there are differences between species (e.g. leopard seals) (Vogelnest *et al.* 2010; Pussini and Goebel 2015). Before performing chemical restraint, it is important that the veterinarian is familiar with the idiosyncrasies of the species. Anaesthetic-related deaths in pinnipeds, particularly phocids, are higher than for other mammals (Barnes *et al.* 2008; Stringer *et al.* 2012). A resuscitation plan must always be developed before a procedure. The effect of the dive response in pinnipeds and other aquatic mammals in relation to anaesthesia is poorly understood. The dive response is a group of physiological changes that occur during submersion in water and has been described in a wide variety of aquatic and terrestrial vertebrate species. The changes consist of breath holding, low heart rate (bradycardia) and redistribution of blood to the heart and central nervous system (Panneton 2013). It is highly likely that anaesthetic-related deaths of pinnipeds may be at least in part associated with other anaesthetic complications (see Vogelnest 2025), rather than the influence of the dive response alone (Macgregor *et al.* 2014). Key considerations and risk factors for the chemical restraint of pinnipeds are discussed further in Vogelnest (2025).

### 8.4.1 Sedation

Sedation to facilitate minor procedures (venepuncture, other sampling, insertion of contraceptive implants, intradermal tuberculosis testing, mask induction of inhalation anaesthesia), examination and before physical restraint is often beneficial (Vogelnest 2025). Sedative drugs can be given orally or in trained captive pinnipeds by hand injection. If the seal is untrained, injections can be given by hand using physical restraint or by pole syringe or dart.

### 8.4.2 Anaesthesia

Anaesthesia is required for more complex and invasive procedures. Most well-trained pinnipeds will not respond to behavioural cues when they become sick and inappetent. Anaesthesia (or sedation) is generally required to investigate and administer treatment to these animals, albeit high risk. Fasting for 12 h before anaesthesia is recommended. Anaesthetic agents are usually administered by injection as per sedation described above or using gaseous agents administered via mask in pinnipeds that have been trained for this or can be safely physically restrained, or sedated and then physically restrained. Anaesthesia can be maintained via a mask; however, intubation is strongly recommended. Continuous-rate intravenous infusion of anaesthetic agents may also be used (Vogelnest 2025).

## 8.5 Common health problems

Pinnipeds are susceptible to a range of infectious and non-infectious diseases. Infectious disease is uncommon in captive pinnipeds. Detailed information on infectious and non-infectious diseases of Australian pinnipeds is covered in Barnes *et al.* (2008) and Bodley (2025). Several serological and faecal screening surveys of free-ranging Australian pinniped species have indicated exposure to a range of pathogens without any evidence of clinical disease. Examples include *Toxoplasma gondii*, *Giardia duodenalis* and *Brucella* spp. Important diseases are summarised in Tables 15.8 and 15.9.

**Table 15.8.** Selected infectious diseases or agents of captive and rescued Australian pinnipeds.

CT = computed tomography; PCR = polymerase chain reaction.

| Disease/ infectious agent | Cause | Clinical signs | Diagnosis | Treatment | Prevention/comments | Reference |
|---|---|---|---|---|---|---|
| **Mites** | | | | | | |
| Nasal mite | *Orthohalarachne* spp. | None described. Mites found in nasal cavities, trachea and bronchi. Lesions associated with the mites have been described | Mites are very large and are clearly visible to the naked eye | None reported | Seen in free-ranging Australian sea lions and fur seals | 1–4 |
| **Endoparasites** | | | | | | |
| Hookworm | Hookworm (*Uncinaria* sp.). *U. sanguinis* in Australian sea lions | Anaemia, debility and death in free-ranging Australian fur seal, long-nosed fur seals and Australian sea lion pups | Faecal analysis, necropsy | Ivermectin is highly effective at eliminating *U. sanguinis* in Australian sea lion pups | Pups acquire hookworms during their first days of life through their mothers' colostrum. Not seen in captive pinnipeds | 1, 5–7 |
| Heartworm | *Dirofilaria immitis* | No clinical cases have been reported in Australian pinnipeds. Mild persistent cough, exercise intolerance, decreased appetite, weight loss. May be asymptomatic | Various blood tests are available | None reported in Australian pinnipeds | As Australian pinnipeds are likely to be susceptible, prophylaxis is recommended for captive pinnipeds in endemic regions. Oral ivermectin and other products available for domestic dogs can be used | 8 |
| Others | A wide range of other endoparasites have been reported in both captive and free-ranging Australian pinnipeds, including but not limited to lungworms (*Otostrongylus circumlitus*, *Parafilaroides* spp.); roundworms (*Contracaecum osculatum, Anisakis* spp.); tapeworms (*Diphyllobothrium* spp.); acanthocephalans (*Corynosoma australe*); *Anisakis simplex* and *Contracaecum* spp.; *Diphyllobothrium* spp. *Diphyllobothrium delphini* | | | | Endoparasites are rarely if ever seen in captive pinnipeds as most parasites are killed when fish is frozen | 1 |
| **Bacterial** | | | | | | |
| Tuberculosis | *Mycobacterium pinnipedii* | Asymptomatic, inappetence, weight loss, exercise intolerance, coughing, difficulty or abnormal respiration (dyspnoea) | Serology (DPP® VetTB Assay, Chembio Diagnostic Systems); comparative intradermal tuberculin test; examination of sputum or respiratory samples using special stains, PCR and culture; radiography, CT scanning; necropsy | Not usually attempted. Euthanasia if diagnosis confirmed. Valuable animals may be treated using specific drugs | Diagnosed in captive and free-ranging Australian sea lions, long-nosed fur seals, Australian fur seals, Antarctic fur seals and a Sub-Antarctic fur seal. Non-releasable free-ranging pinnipeds should undergo testing and quarantine before joining a captive population. An important zoonotic disease | 1, 4, 9, 10 |

**Table 15.8.** continued

| Disease/ infectious agent | Cause | Clinical signs | Diagnosis | Treatment | Prevention/comments | Reference |
|---|---|---|---|---|---|---|
| Gastroenteritis and peritonitis (see Table 15.9 for gastrointestinal ulceration) | A range of bacteria, including but not limited to *Clostridium* spp., *Proteus mirabilis, Escherichia coli, Helicobacter* sp. Peritonitis may be secondary to gastrointestinal perforation | Inappetence, vomiting, weight loss, diarrhoea, abdominal guarding, adoption of unusual postures, vague behavioural changes, death. Signs are often non-specific and confirmation of diagnosis in the live animal can be difficult | Clinical signs, faecal culture, necropsy | Affected animals can be challenging to medicate as they are inappetent and will not accept oral medication or tolerate administration of fluids or gastrointestinal protectants parenterally or by injection. Antibiotics, anti-emetics, gastrointestinal motility-modifying agents can be administered by injection (hand or pole syringe, dart) | Prognosis is often poor. Early recognition and treatment are important. Use of clostridial vaccines is considered by some marine mammal veterinarians to be beneficial in preventing clostridia-associated disease | 1 |

References: (1) Barnes *et al.* (2008); (2) Spratt *et al.* (2008); (3) Ladds (2009); (4) Bodley (2025); (5) Norman (1995); (6) Morgan *et al.* (2000); (7) Marcus *et al.* (2015); (8) Duignan (2000); (9) Forshaw and Phelps (1991); (10) Woods *et al.* (1995).

**Table 15.9.** Non-infectious diseases of captive and rescued Australian pinnipeds.

CT = computed tomography; MRI = magnetic resonance imaging; NSW = New South Wales; Qld = Queensland; UV = ultraviolet.

| Disease/agent | Cause | Clinical signs | Diagnosis | Treatment | Prevention/comments | Reference |
|---|---|---|---|---|---|---|
| Ophthalmic disease | In captivity: multifactorial, including water quality, chlorine, ozone and bromine concentrations, exposure to UV light, trauma, fresh water pools and opportunistic pathogens.<br>Free-ranging: most commonly associated with trauma.<br>The cause of cataracts may be due to trauma, metabolic or infectious disease, nutritional or congenital factors, advanced age and possibly UV light; however, the cause is often undetermined | Corneal opacities (focal or diffuse), ulceration, scarring; chronic keratitis (inflammation of the cornea; descemetoceles; corneal perforation.<br>Blepharospasm (involuntary tight closure of the eyelids) is common and may persist for days to months. This may be a response to pain and/or photophobia.<br>Cataracts (opaque lens seen behind the pupil).<br>Diffuse corneal opacity, conjunctivitis, ocular swelling and pain are seen with generalised inflammation (uveitis) | Ocular examination, preferably by a specialist ophthalmologist. Examination can be complicated by blepharospasm, which precludes visualisation of the eye.<br>Application of topical local anaesthetic drops along the eyelids may relieve pain and facilitate opening of the eye for examination | Topical preparations (antibiotics, anti-inflammatories): can be difficult to administer due to blepharospasm; animal must remain out of water for 15–20 min. Subconjunctival immunosuppressive drug implants have been used successfully in pinnipeds.<br>Oral anti-inflammatory and analgesic drugs (see precautions in relation to gastrointestinal ulceration).<br>Surgery (superficial keratectomy, corneal flap, conjunctival pedicle grafts, cataract removal, removal of the eye) | More common in captive than free-ranging pinnipeds. Seen in all species. In captivity prevention includes provision of shade, non-reflective dark painted pools and surrounds to minimise glare; avoiding excessive amounts of oxidative disinfectant in water, such as chlorine, ozone and bromine; optimise water quality; natural seawater is preferable; avoid 'sky fish feeding' (do not feed from above the animal or throw fish high) | 1–4 |
| Gastrointestinal foreign bodies | Ingestions of foreign bodies (rocks/stones, coins, silicone, rubber, metal objects, items dropped into pools by visitors such as hats, plastic toys, sunglasses) | Often asymptomatic. Inappetence, vomiting, weight loss, abdominal guarding, adoption of unusual postures and vague behavioural changes resembling neurological signs | Clinical signs, abdominal palpation, radiography | Gorge feeding may result in regurgitation of foreign bodies; surgical removal | Ingestion and regurgitation of small rocks is a normal behaviour in free-ranging Australian sea lions and is thought to assist fish breakdown in the stomach. This species is particularly prone to ingestion of foreign bodies in captivity | 1 |

**Table 15.9.** continued

| Disease/agent | Cause | Clinical signs | Diagnosis | Treatment | Prevention/comments | Reference |
|---|---|---|---|---|---|---|
| Gastric ulcers | Stress, inappetence, drugs (especially long-term use) such as non-steroidal anti-inflammatory drugs and corticosteroids; other diseases (liver, kidney, parasitism, bacterial infections) | See gastroenteritis Table 15.8. | History, clinical signs, gastroscopy, biopsy, necropsy | As per gastroenteritis (Table 15.8) plus specific therapy to reduce gastric acidity and gastric acid production and protect the stomach lining. This prevents further destruction of the stomach lining and promotes ulcer healing | Minimising stress and the judicious use of drugs such as non-steroidal anti-inflammatory drugs and corticosteroids are important preventative measures | 1 |
| Starvation/ exhaustion | Free-ranging 'weanling' long-nosed and Australian fur seals haul out on land starving and exhausted. Injuries and/or parasitism may be a consequence of or further contribute to debility. Failure to thrive post weaning, reduced fish stocks.<br>Thin, weak, juvenile leopard seals haul out on mainland beaches. Dispersal into temperate regions is thought to be influenced by resource availability, particularly krill, which is thought to be more heavily utilised by juvenile leopard seals | Poor body condition; lethargy; unconcerned about presence of humans and dogs; 'cookie cutter' shark bites or other wounds and injuries | Observation, examination, time of year | Prognosis is poor for milk-dependent pups and emaciated or severely injured seals older than 2.5 years. Prognosis for weanlings is better with or without intervention. Intervention to euthanase weak, poorly responsive, injured or emaciated pinnipeds is appropriate on welfare grounds. A decision to rescue and rehabilitate must be considered carefully, in consultation with the relevant government wildlife agency | Australian and long-nosed fur seal dependant pups (< 12 months old), weanlings (~18 months old), juveniles (~32 months old). Leopard seals 1–3 year old.<br>Victorian coast – mostly between November and January; NSW and south-east Qld coast – mostly between June and August.<br>Leopard seals seen from July to October, with a peak in August – September | 3 |
| Entanglement, marine debris | Fishing lines, hooks, nets, plastic packing straps, other plastics | Strangulation, lacerations, amputation, gastrointestinal foreign bodies, exhaustion, starvation, drowning | Observation, examination | Capture and removal | Free-ranging pinnipeds only. Responsible disposal of plastics and fishing gear | 3 |

| Disease/agent | Cause | Clinical signs | Diagnosis | Treatment | Prevention/comments | Reference |
|---|---|---|---|---|---|---|
| Trauma | Bite wounds due to conspecific aggression (captive and free-ranging) or predators; boat strike; gunshot; spears; embedded stingray barbs; lightning strike | Wounds, fractures, amputation, swelling, abscesses, lameness | Observation, examination | Open bite wounds (even if severe) heal remarkably well in the saltwater environment. Antibiotics are rarely indicated; analgesia may be indicated in some cases; surgical intervention and repair may be required in some cases | Major trauma more common in free-ranging pinnipeds | 1, 3 |
| Neoplasia | A range of benign and malignant neoplasms have been diagnosed in Australian pinnipeds | Variable depending on site and nature of the tumour | Clinical signs, biopsies, radiography, CT scanning, necropsy | None reported | Neoplasia is relatively rare in Australian pinnipeds and mostly seen in aged captive animals | 1, 5 |
| Visceral lepidopterism | Ingestion of caterpillars, the hairs (setae) of which migrate into and through the gastrointestinal wall and into other organs inciting a foreign body inflammatory reaction. The setae are morphologically consistent with those shed by the bag-shelter moth (*Ochrogaster lunifer*) larvae, which are known to cause visceral lepidopterism in Australian horses | Often non-specific. Inappetence, weight loss, abdominal guarding, adoption of unusual postures, vague behavioural changes, death | Antemortem diagnosis is not possible. Necropsy, histopathology, presence or caterpillars in area of pinniped enclosures | None reported | Has been diagnosed in captive California sea lions, Australian sea lions and long-nosed fur seals in Australia | 6 |
| Idiopathic epilepsy | Unknown | Seizures | By exclusion based on examination, blood work, advanced imaging (CT, MRI) | Oral phenobarbitone | Has been reported in captive Australian fur seals and Australian sea lions | 1 |

References: (1) Barnes *et al.* (2008); (2) Ladds (2009); (3) Bodley (2025); (4) Miller *et al.* (2013); (5) Tong Marwood (2025); (6) Tong *et al.* (2017).

## 8.6 Preventative medicine

The cornerstones of preventative medicine are: good biosecurity; optimal husbandry, housing and environment; good nutrition; disease screening; vaccination; routine preventative treatments; and conducting a necropsy on all animals that die. All new pinnipeds, particularly if from the wild, entering a captive population should undergo a period of quarantine and relevant disease screening (e.g. tuberculosis testing, screening for parasites and treatment for parasites if indicated). The health of captive pinnipeds, particularly ocular health, is reliant on optimal enclosure and environmental conditions (see Tables 15.5 and 15.6). As thawed frozen fish is almost exclusively fed to captive pinnipeds, optimal freezing and thawing protocols and supplementation with vitamins (particularly thiamine and vitamin E), and variety are important for maintaining good health (see section 6). Routine screening for parasites in long-term captive pinnipeds is usually unnecessary, as once eliminated reinfection is unlikely. Vaccination and routine preventative treatments are listed in Table 15.8. Necropsy facilitates a diagnosis, and if relevant informs preventative action to protect other pinnipeds in a zoo population.

## 8.7 Zoonoses

There are few zoonoses associated with pinnipeds. Important zoonoses are discussed in Table 15.10.

Pinnipeds can be dangerous, with the potential to inflict serious bites. Appropriate restraint, handling and training are important to prevent injuries. Safe manual handling techniques must be used when lifting or carrying heavy pinnipeds when anaesthetised or dead. Consideration must also be given to zoonotic risks associated with handling fish during food preparation and feeding sessions. Cuts, scrapes or fish spine-penetrating wounds can become infected with bacteria. Hygiene and use of appropriate PPE are important preventative measures.

# 9. BEHAVIOUR

## 9.1 Activity

Pinnipeds are top-order predators with similar social and reproductive behaviours (Gentry and Kooyman, 1986). Seals will thermoregulate in the ocean by raising their flippers in the air to either warm up or cool down. On land, during cold days fur seals will tuck their flippers under them and slightly curl their body to retain

**Table 15.10.** Zoonoses associated with Australian pinnipeds.

| Disease agent | Animal host | Transmission method | Signs and symptoms in people | Prevention | Reference |
|---|---|---|---|---|---|
| Tuberculosis (*Mycobacterium pinnipedii*) | Australian sea lion), long-nosed fur seal, Australian fur seal | Aerosol | Weight loss, night sweats, cough, difficulty breathing | Regular screening of people who work with pinnipeds (captive or free-ranging). Use of PPE (N95 masks) when conducting a necropsy | 1–3 |
| 'Seal finger' *Mycoplasma phocicerebrale* | Potentially any species | Direct through abraded skin from bites or during handling infected tissues | Severe cellulitis, mainly on fingers or hand. If untreated may progress to chronic inflammation of tendons and joints | Appropriate restraint and handling to prevent bites | 3 |
| Brucellosis *Brucella pinnipedialis* | Potentially Australian fur seal, Australian sea lion, leopard seal | Direct contact of abraded skin or mucosa with fluids or tissue from infected animals. Possibly inhalation | Mild to chronic influenza-like illness, weight loss, chronic joint pain. More severe disease possible and pregnant women are at risk of abortion, fetal defects or fetal death | Use of appropriate PPE, personal hygiene | 3 |

References: (1) Forshaw and Phelps (1991); (2) Cousins *et al.* (1993); (3) Hulst (2025).

heat. During hotter days they will extend their body and flippers to maximise heat loss (Beentjes 2006).

## 9.2 Social behaviour

### 9.2.1 Long-nosed fur seal and Australian fur seal

Long-nosed and Australian fur seals spend a large proportion of their time hauled out on rocky shores. During the non-breeding season, fur seals will move beyond the breeding colonies and then return to the breeding colonies in late spring.

Australian fur seals are positively thigmotactic, tolerating a high degree of bodily contact, a trait that sets them apart from all other *Arctocephalus* species (Warneke and Shaughnessy 1985).

During the breeding season, male Australian fur seals compete aggressively to establish and maintain territories in which the females give birth and subsequently mate shortly thereafter (Riedman 1990). At this time, females then alternate between feeding out at sea and suckling their young on land (Tripovich *et al.* 2008a). The use of vocalisations, olfaction and postural displays are intrinsic to the breeding behaviour of Australian fur seals (Tripovich *et al.* 2009).

Vocalisation in pinniped communication has varying functions, being used in species recognition, territorial defence and for mate and kin identification (Insley *et al.* 2003). In Australian fur seals, mothers and pups use a vocal recognition system to reunite after the separation of foraging trips and this recognition is required throughout the maternal dependency period, especially as the pups move around the colony as they grow older (Tripovich *et al.* 2006, 2008a, 2009). Olfaction is used at a closer range to confirm identity.

Most male fur seals typically fight most aggressively during the period of territory establishment (Roux and Jouventin 1987). Once boundary areas are recognised, combats between neighbouring males are less frequent (Roux and Jouventin 1987), with stereotypical postures and vocalisations used to maintain boundary areas (Warneke and Shaughnessy 1985; Fernández-Juricic *et al.* 1999). Territorial male otariids fast throughout the breeding season, adopting territory maintenance techniques that are energetically conservative and may be important for extending territory tenure (Arnould and Duck 1997). For this reason, vocal recognition or the ability to distinguish neighbours from intruding challengers may be important for conserving energy expenditure and reducing the number of physical conflicts. Neighbour-stranger recognition has been demonstrated in Australian fur seals (Tripovich *et al.* 2008b) and is assumed to be the same for long-nosed fur seals.

### 9.2.2 Australian sea lion

In the wild Australian sea lions inhabit both sandy beaches and rocky shorelines throughout their range. Because of the extended maternal care in this species mothers and dependant offspring are often found in close association in colonies. Pups typically stay close to the location of their birth until around 2 months of age, when they begin moving about the colony and associating with other pups, particularly while the mother is at sea foraging (Pitcher *et al.* 2009). Outside of the perinatal attendance period, females and juveniles are often thigmotactic, particularly in response to colder weather. However, mothers will not tolerate close approaches from other adult females, and often pups, during the perinatal attendance period and can be very aggressive. Males are more social as juveniles and subadults, but become more solitary when they reach maturity.

Mothers and offspring have a well-developed recognition system and used vocal, olfactory and visual cues to recognise each other (Charrier *et al.* 2009; Pitcher *et al.* 2012; Wierucka *et al.* 2017, 2018; Charrier *et al.* 2022). Vocal recognition of the pup by the mother is established within 48 h of birth (Pitcher *et al.* 2010), but pups can take several weeks to learn their mother's voice (Pitcher *et al.* 2009). It is likely that olfactory recognition of the pup is learnt rapidly after birth (Pitcher *et al.* 2011a; Wierucka *et al.* 2018, 2019).

### 9.2.3 Leopard seal

Leopard seals are solitary and will breed in summer on the outer pack ice. The pack ice coverage varies with the season, with maximum pack ice reported between August and October and a minimum between February and March (Rogers 2009). Vocal communication is important to the mating system of leopard seals, with both males and females actively vocalising in the breeding season (Rogers 2009).

## 9.3 Reproductive behaviour

### 9.3.1 Long-nosed fur seal and Australian fur seal

Fur seals and sea lions are polygynous and gregarious during the breeding season with pronounced sexual dimorphism in all species (Riedman 1990; Bartholomew

1970). Males reach sexual maturity at 3–6 years of age (Atkinson 1997), but do not attain social maturity until 9–14 years. Females are both sexually and socially mature by 3 years of age, usually producing one young per season until approaching 20 years of age (Gentry and Kooyman 1986). Among the otariids, the length of lactation ranges from 4 to 24 months (Bowen 1991). A long neonatal dependency period in some otariid species requires the female to adopt a nursing-foraging cycle, whereby females alternate between foraging at sea and nursing their pups. When returning from sea the mothers must relocate their offspring among the large number of other pups in the breeding colony. Therefore, it is imperative that there is some means whereby individuals are able to reunite so that maternal investment is not misplaced (Hanggi 1992).

### 9.3.2 Australian sea lion

Male Australian sea lions mate-guard females in anticipation of mating. Females typically come ashore and select a suitable location to give birth. They are then guarded by a mature male that will actively repel other males and attempt to prevent the female from moving away. The perinatal attendance period typically lasts for 6–12 days (Higgins and Gass 1993; Pitcher *et al.* 2010). Typically, females come into oestrus and mate 1–2 days before the end of the perinatal attendance period (Higgins and Gass 1993; Pitcher *et al.* 2010).

### 9.3.3 Leopard seal

Vocalisations play an important role in the breeding behaviour of leopard seals as evidenced by the increase in vocal activity of leopard seals before and during their breeding season (Rogers 2009). Lone male will vocalise for long periods each day (early November through to January), while the females will vocalise for a shorter period of time, presumably to indicate sexual receptivity (Rogers 2009).

## 9.4 Bathing

Pinnipeds display grooming behaviours both on land and in the water. Otariids have nails on their rear flippers and can reach most parts of their body when scratching. Their foreflippers do not have nails but are used to rub their hind areas and face. They will also use nibble at their hind areas.

Leopard seals will groom by rubbing against a solid surface. They are not as flexible as the fur seals and sea lions and rarely use their flippers to groom; however, they have been seen scratching one rear flipper with the other rear flipper (A. D'Ettorre pers. comm.). Time spent grooming greatly increases when the seal is moulting.

## 9.5 Behavioural problems

Behavioural problems in captive pinnipeds can include abnormal repetitive behaviours (e.g. 'pattern swimming'), ongoing aggression, regular ingestion of foreign materials and anticipatory behaviours.

The most common repetitive behaviour in captive pinnipeds is pattern swimming, which may not necessarily reflect negative welfare, but rather may represent a selective advantage on the individuals performing the behaviour because it may be analogous to the long periods of swimming behaviour seals undertake in the wild to forage and socialise (Newberry 1995; Arnould and Kirkwood 2008).

Aggression between individuals is normal, particularly among males during the breeding season. Regular states of heightened aggression can be due to overcrowding, changes in hierarchy or limited space to establish territories. When males reach maturity, they can become extremely territorial and will rarely tolerate other adult males within the same vicinity as them.

Ingestion of foreign materials is common in otariids, particularly Australian sea lions. This behaviour is observed both in the wild and in captive environments. Otariids have been observed swallowing stones and can retain these in their stomach for a period. These may help the seal to sink, help stability, aid in the breakdown of food or potentially reduce hunger pangs (Kirkwood and Goldsworthy 2013).

Anticipatory behaviours have been observed in captive pinnipeds that are conditioned or have become accustomed to the routines of their carers. Certain behaviours may be elicited in anticipation of an event (e.g. scheduled feeds or training sessions) and can sometimes be related to frustration. Seals have been observed in a heightened state of excitement/arousal, increased vocalisation, rapidly swimming or pacing at the entrance of their enclosure in anticipation of a desired event.

## 9.6 Signs of stress

Signs of stress in pinnipeds can include increased aggression (towards carers or companion seals), bodyweight loss/loss of appetite, avoidance of other seals or people, decreased activity or hyperactivity. Some less obvious signs that can be indicators of immediate stress are wide eyes, flared nostrils, yawning, vocalising, huffing and an overall change in demeanour.

## 9.7 Behavioural enrichment

Goal-based enrichment programs are beneficial for captive pinnipeds. Providing stimuli that promote behavioural responses that are within their natural repertoire can assist with ensuring captive pinnipeds remain mentally and physically stimulated. Such enrichment activities can include:

- use of food-based enrichment items to promote investigative behaviours
- providing a variety of sinking items and floating items to promote diving or surface swimming
- providing olfactory stimulants
- providing auditory stimulants
- changes in environment
- providing opportunities to socialise with different pinnipeds
- providing a variety of food in different shapes and sizes.

Another common form of enrichment for captive pinnipeds is positive reinforcement training and conditioning, which provide the seals with choice and variety through participation.

Efforts to stimulate varied foraging behaviour in Australian fur seals has also included trials that presented food as a scatter in open water, in a free-floating ball device or in a static box device (Hocking *et al.* 2015). The study found that: animals spent more time interacting with the ball and static box devices than the scatter feed; the total time spent pattern swimming was reduced in the enrichment treatments compared with controls; and the time spent performing random swimming behaviours increased. The authors suggested that the best way to increase the diversity of foraging behaviours is to provide seals with a range of foraging scenarios where food is encountered in different ways (Hocking *et al.* 2015).

## 9.8 Training and conditioning

Many institutions that house pinnipeds will train them for mental and physical stimulation, for cooperative care (husbandry and veterinary) and for presentation purposes.

Pinnipeds are usually highly food motivated, so using food as their primary reinforcer is likely to achieve results. Some pinnipeds (e.g. Australian sea lions) are motivated by toys and play, so using these as secondary reinforcers also works well.

### 9.8.1 Classical and operant conditioning

Before any training and conditioning takes place, the trainer should take the time to get to know the animal's history and build a relationship with the individual animal. Training can take place through classical conditioning or operant conditioning.

- Classical conditioning is a form of passive learning that pairs a novel stimulus with something that is reinforcing, therefore giving reinforcing value (over time) to the novel stimuli itself. This method is often used when conditioning a bridge (i.e. a signal the trainer makes that marks the instant, in time, when an animal has given a correct response).
- Operant conditioning is a form of active learning that is consequence based, either positive or negative:
  - positive reinforcement increases the likelihood of a behaviour re-occurring because something desirable has been added to the scenario after the behaviour has occurred
  - negative reinforcement increases the likelihood of a behaviour re-occurring because something undesirable has been removed from the scenario after the behaviour has occurred
  - positive punishment decreases the likelihood of a behaviour reoccurring because something undesirable has been added to the scenario after the behaviour has occurred
  - negative punishment decreases the likelihood of a behaviour reoccurring because something desired has been removed from the scenario after the behaviour has occurred.

### 9.8.2 Conditioning a bridge

A bridge is a training tool that bridges the time between the correct response from the animal to when the reinforcement is presented. Often, a whistle or word (e.g. using the word 'good') will be chosen as the bridge signal. In order to give the bridge reinforcing value the trainer will sound the bridge and immediately present a primary reinforcer (most often food); this will need to be repeated many times over several days or more until the desired response has been successfully learnt. Over time the animal will hear the bridge, stop what they are doing and go to collect their reinforcement. Once this stage is reached, the bridge has been conditioned.

### 9.8.3 Cooperative behaviours

Cooperative behaviours are behaviours that are trained to assist animal caretakers in the day-to-day maintenance of the animals. Some examples are:

- stationing - conditioning an animal to go to a station (a place designated by the animal caretaker) can be useful when wanting to separate an animal from a group or for when a animal caretaker needs to enter or exit an enclosure (Fig. 15.10)
- recall -generally used to send or receive an animal. The animal is conditioned to come or go when the recall is sounded. This behaviour is useful in an emergency to send a seal away from danger and back to safety
- shifting -training the animal/s to go through an opening (e.g. doorway, gate or slide). This behaviour is needed when an animal has to be moved from one area to another and they can be trained so that behaviour is performed without the need for the animal caretaker to enter into the animal's enclosure
- control walking - training the animal walk beside the handler in a calm manner.

**Fig. 15.10.** A young male Australian sea lion stationing on floor station (also known as a 'bed'. Photo by Danielle Fox.

### 9.8.4 Shaping techniques

There are many different methods of training or shaping a behaviour. Several common techniques used in various animal conditioning environments include:

- target training - training the animal to touch a part of its body to animal caretaker's hand or an object (Figs 15.11, 15.12). Common targets with pinnipeds

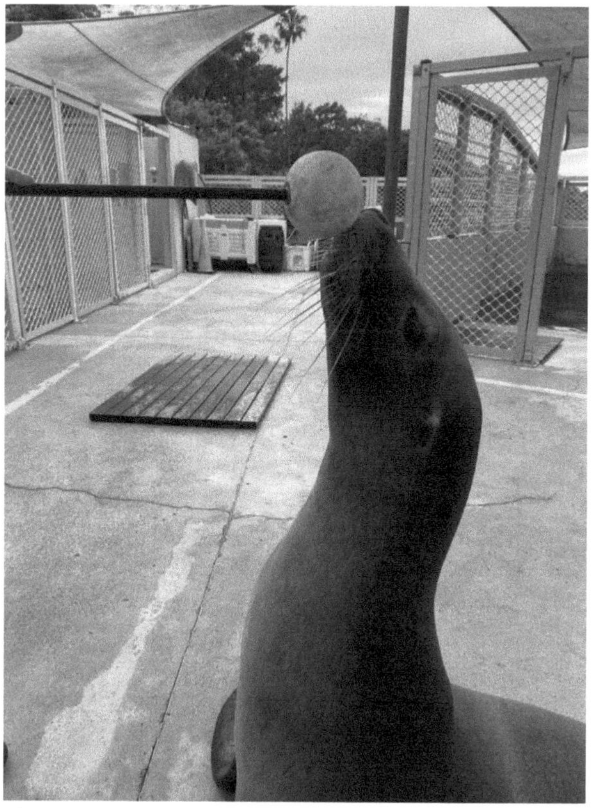

**Fig. 15.11.** A young male Australian sea lion targeting his nose to a target pole. Photo by Danielle Fox.

**Fig. 15.12.** A young male Australian sea lion targeting his nose to a trainer's hand. Photo by Danielle Fox.

are nose targets, body targets, targeting on a target pole or stick etc. It is through the use of target training that more complex behaviours can be shaped (e.g. waving, bowing or porpoising)
- scanning/capturing - this technique focusses on a behaviour that the animal does naturally and then capturing it by reinforcing the animal whenever it displays the behaviour. Once the animal is offering the behaviour regularly, the animal caretaker can pair the behaviour with the chosen cue
- moulding/modelling - this technique focusses on physically manipulating the animal into the desired posture or position.

#### 9.8.5 Training plans and record keeping

Preplanning and record keeping is an essential part of any animal training program to keep track of which techniques were or were not successful.

### 9.9 Introductions and removals

Most introductions need to be done slowly and under close observation. Many pinniped species when introduced to another individual can display aggressive behaviours and potentially injure each other; therefore, it is advisable for carers to have a clear plan on how to separate the seals should they need to intervene.

Adult male Australian sea lions generally do not interact well with other adult males and can be dangerous towards juveniles; however, they are likely to accept the company of other females. In other species, adult or adolescent males may socialise well for most of the year, but can become territorial during the breeding season. For this reason, it is recommended not to introduce adult or adolescent males during their breeding seasons.

### 9.10 Intraspecific compatibility

Long-nosed fur seals and Australian fur seals can be housed together without issues. During the breeding season there may be some aggression between males, so carers should have a plan to separate them if needed.

Australian sea lions will live peacefully in harems, but it is not recommended that the male be housed with small pups due to the risk of injury to the pups. Therefore, heavily pregnant females should be moved from the harem to a pupping facility.

Large groups of leopard seals have not been held in captivity, mainly due to the lack of individuals in care at one time. Male/female pairs have been successfully housed together for years at a time.

### 9.11 Interspecific compatibility

Australian fur seals have been successfully housed with other pinniped species, including long-nosed fur seals, Australian sea lions (females and juveniles) and leopard seals. Long-nosed fur seals can also be housed with female or juvenile Australian sea lions and, on rare occasions, adult male Australian sea lions, however, the aggressive temperament of bull Australian sea lions make them less suited to socialising with other species. Long-nosed furs seals have not been known to socialise with leopard seals in captive environments.

## 10. BREEDING

### 10.1 Mating system

#### 10.1.1 Long-nosed fur seal

The long-nosed fur seal breeding season commences in late October when males arrive at breeding colonies, before the females, and acquire and defend territories with ritualised behavioural posturing, vocal displays and physical combats. On average a male will have 5-8 females on his territory, but this varies between male and location (Chilvers and Goldsworthy 2015). Long-nosed fur seal pups are born from mid-November to January, (with peak pupping occurring in December) (Chilvers and Goldsworthy 2015). Long-nosed fur seals are in oestrus for 7-8 days postpartum. After a short period of time ashore with their pups (~1-2 days), the mothers will depart to sea and commence a nursing-feeding cycle that continues until the pups are ~9-10 months old (Goldsworthy 2006).

#### 10.1.2 Australian fur seal

The Australian fur seal breeding season commences in late October when territorial males begin to arrive at the breeding islands. Males use vocalisations, behavioural posturing and physical contact in an aggressive manner to secure, establish and protect a territory frequented by females in oestrus (Stirling and Warneke 1971; Warneke and Shaughnessy 1985). Territory size and location depends on the male's fighting ability, but the average size of a territory is 62 $m^2$ (Warneke and Shaughnessy 1985). From late October to late December, females will come to the breeding rookeries, where they will give birth (median pupping date of 1 December (Warneke and Shaughnessy 1985) and later mate. Females are sexually receptive for 5-6 days postpartum (Shaughnessy 1999). They will then alternate between nursing their young and foraging at sea, continuing this nursing-feeding cycle until the pups are ~10–11 months old (Arnould and Hindell 2001). A small proportion of females suckle their pups for a second or even third year (Warneke 1982; Hume *et al.* 2001).

### 10.1.3 Australian sea lion

In the wild Australian sea lions can show serial polygyny, with individual mate-guarding occurring where females are more dispersed and harem-like behaviour where females are denser and males can guard several at a time. Potential lek-like behaviour may also be seen in some colonies (McIntosh and Pitcher 2021).

### 10.1.4 Leopard seal

Leopard seals are suspected to give birth from October to mid-November and then mating occurs from December to early January, once pups have been weaned. Females will typically give birth to their pups and wean them on the ice floes of the Antarctic pack ice, where pups will undergo rapid growth during the first 6 months postpartum. Lactation is suspected to last for up to 4 weeks (Rogers 2009)

## 10.2 Ease of breeding

### 10.2.1 Long-nosed fur seal

Long nosed fur seals breed well in captive environments. They have an annual breeding cycle, which will allow females to produce up to one pup each year if she has access to a reproductive male during the breeding season.

### 10.2.2 Australian fur seal

Australian fur seals breed well in captive environments. They have an annual breeding cycle, which will allow females to produce up to one pup each year if she has access to a reproductive male during the breeding season.

### 10.2.3 Australian sea lion

Australian sea lions generally breed well if the oestrous cycle of the female is known. If oestrogen levels are monitored and females are given access to a reproductive male when they are in oestrus, then it is likely that the female will become pregnant. Oestrogen levels can be monitored through faecal analysis to determine the cycle of the individual. Alternatively, if females and males are displaying an increased amount of interest towards one another (e.g. difficult to separate, increased vocalisation towards one another) it is likely that the female is in oestrus and that mating will take place. In the wild, most females (83%) will pup every breeding season (Goldsworthy *et al.* 2019).

### 10.2.4 Leopard seal

There has been no known successful breeding of leopard seals in captive environments, which may relate to the limited number of leopard seals that have been held in captivity, particularly a male and female paired together. At Taronga Zoo, a male and female pair were housed together for several years and although courting behaviour was observed, neither mating nor pregnancies occurred.

## 10.3 Oestrous cycle and gestation period

### 10.3.1 Long-nosed fur seal

Long-nosed fur seals have an annual synchronous breeding season, with most pups being born between late November and early January, peaking around 25–26 December. They have an embryonic diapause of 4 months, with an active gestation period of 8 months. Females will come into oestrous ~8 days after giving birth (Crawley and Wilson 1976; Goldsworthy 2023a).

### 10.3.2 Australian fur seal

Australian fur seals have a synchronous annual breeding cycle, giving birth to their young during the early Austral summer months. Females will then come into oestrus 5–7 days after giving birth and will mate with the territory holder. Females have a 3–4-month embryonic diapause, with an 8-month active gestation period to have their next pup 12 months after mating (Kirkwood and Arnould 2008).

### 10.3.3 Australian sea lion

In Australian sea lions oestrus occurs ~10 days after giving birth (Higgins and Gass 1993). Gales *et al.* (1997) found a similar duration of embryonic diapause to that of other pinnipeds, identifying that the blastocyst reactivates and implants at between 3.5 and 5 months of gestation. They also found that the species has a prolonged post-implantation period of up to 14 months, the longest post-implantation period of pregnancy recorded in any pinniped.

Determining the timing of oestrus in females without a recent birth, and when individuals are housed separately, is more difficult than when natural mating is observed. Techniques include looking for a swollen vulva in females although this is not always evident when in oestrus. Often there will be increased interest from the male towards the female (and sometimes the female towards male) when a female comes into oestrus. Signs of increased interest will include vocalisations directed at the seal of interest, following or chasing and mating.

### 10.3.4 Leopard seal

The female leopard seal will come into oestrus at or shortly after weaning her pup (Jefferson *et al.* 2015). Mating then occurs in the water and there are reports of a short, delayed implantation (Riedman 1990) by some authors, but others report that there is no delayed implantation period (Maxwell 1967; Harrison 1968). Gestation for leopard seals is estimated to be 8–9 months (Gwynn 1953; Tikhomirov 1975; Ledingham 1979), but there are reports that it can extend to 12 months (Atkinson 1997).

## 10.4 Reproductive status

### 10.4.1 Long-nosed fur seal

Male long-nosed fur seals will hold territories from 8 to 10 years of age, with the majority of males having held a territory for at least one breeding season by 12 years of age (McKenzie *et al.* 2007). Females give birth between 4 and 8 years of age, with an average age at reproductive maturity of 5 years (McKenzie *et al.* 2007).

### 10.4.2 Australian fur seal

Australian fur seal females are both sexually and socially mature by 3 years of age, usually producing one young per season until approaching 20 years of age (Gentry and Kooyman 1986).

Australian fur seals males reach sexual maturity at 3–6 years of age (Atkinson 1997), but do not attain social maturity until 9–14 years. Similarly, Long-nosed fur seal males do not hold territories until they are 8–10 years of age (McKenzie *et al.* 2007).

### 10.4.3 Australian sea lion

Male Australian sea lions begin to show physical signs of maturity with the development of a characteristic white mane at between 7.6 and 9.7 years of age (mean 8.7 ± 0.73 years). The mane is fully developed at ~10 years (range 8.7-12 years). In the wild males are typically able to successfully mate-guard females between the ages of 13 and 16 years (McIntosh 2007).

Age at first parturition for wild female Australian sea lions has been observed between 3.8 years and up to 6.1 years with a mean of 5 ± 0.28 years (McIntosh 2007). Females aged 8 and 9 years had the highest pupping frequency (100% of individuals studied). The maximum number of offspring observed from one individual was 8 pups in 8 consecutive seasons. The mean reproductive output is 5.4 ± 0.39. The oldest recorded female to give birth in the wild was 24 years (McIntosh 2007).

### 10.4.4 Leopard seal

Leopard seals are sexually mature by 4.5 years in males and 4 years in females (Kooyman 1981; Bonner 1994) and females are considerably larger than males in size (Rogers 2009).

## 10.5 Reproductive management

Physical separation during the breeding season is the most common form of breeding control. Some institutions have been known to castrate males or use contraceptive implants (e.g. deslorelin). Due to the small number of reproductive Australian sea lions in captivity, and the rarity of obtaining founders from the wild (typically only through rescue of individuals unfit for release), chemical or surgical fertility control is not typically used. Physical separation of sexually mature individuals is the most suitable method to control breeding.

## 10.6 Occurrence of hybrids

Hybridisation across fur seals has been reported between Antarctic fur seal (*Arctocephalus gazella*), subantarctic fur seals (*Arctocephalus tropicalis*) and long-nosed fur seals (*Arctocephalus forsteri*) (Page *et al.* 2001; Lancaster *et al.* 2006).

There have been at least two known hybridisations between Australian sea lions and long-nosed fur seals in captive environments. In both of these instances the pups did not thrive and died at very young ages (D. Fox unpubl. data).

## 10.7 Timing of breeding

The timing of breeding for most species of Australian seal occurs from approximately August to January (Table 15.11; Stewart 2014; Webber 2014). The exception to this is the Australian sea lion, which has a non-synchronised, non-annual breeding cycle. Due to the extended gestation period (see section 10.3.3), mature females breed at ~18-month cycles. Further, breeding is asynchronous between colonies and often within colonies. Breeding seasons of individual colonies occur at various times of the year (Gales *et al.* 1992, Goldsworthy *et al.* 2019) and at

**Table 15.11.** Reproduction and development for different species of seals.

| Species | Timing of breeding | Gestation (months) | Weaning | Age sexual maturity (years M: F)* |
|---|---|---|---|---|
| **Family Otariidae** | | | | |
| *Arctocephalus forsteri* | Nov–Jan | 11–12 | c. 10 months | 8–10: 4–8 |
| *A. pusillus doriferus* | Oct–Jan | 11–12 | 10–12, but up to 36, months | 9–12: 3–6 |
| *A. gazella* | Nov–Dec | 11–12 | c. 4 months | c 7: 3 |
| *A. tropicalis* | Oct–Jan | 12 | 10–11 months | 8–10 (puberty 4): 4–6 |
| *Neophoca cinerea* | Asynchronous | 17.5 | 15–18 months | c. 6: 3.8–6 |
| *Phocarctos hookeri* | Oct–Dec | 11–12 | c. 10–12 months | c. 7: 3 |
| **Family Phocidae** | | | | |
| *Hydrurga leptonyx* | Oct–Dec | 10–12 | 21–30 days | 3–6: 2–7 |
| *Leptonychotes weddellii* | Sep–Nov | 10 | 42–49 days | 7–8: 3–6 |
| *Lobodon carcinophaga* | Sept–Oct | 9–10 | 14–21 days | 2–7: 1.3–6 |
| *Mirounga leonina* | Aug–Oct | 11 | 24 days | 4–6: 2–4 |
| *Ommatophoca rossii* | Oct–Jan | ~11–12 | c. 42 days | 2–7: 3–4 |

Derived from: Atkinson (1997); Bester (1995); Jefferson *et al.* (2015); Laws (1984); McKenzie *et al.* (2007); McMahon *et al.* (1997); Pistorius *et al.* (2001); Reijnders *et al.* (1990); Shaughnessy and Kerry (1989); Skinner (1984); Stewart (2014); Webber (2014).

the Seal Bay colony 90% of births occur over 124 days (McIntosh *et al.* 2012).

## 10.8 Age at first breeding and last breeding

Most species of seals typically first breed when ~2–12 years of age depending on the species, with males typically mating when older than females (Table 15.11). In pinnipeds, male sexual maturity observed in the wild is often 'social maturity', when an individual is capable of holding a territory or guarding a mate; as such it may be later than physiological maturity or puberty when reproduction first becomes possible. It is likely that age of physiological maturity is similar between females and males and genetic evidence suggests that social maturity may not reliably indicate mating success (e.g. sneakier males that obtain mating success without territory holding or mate guarding) (Boyd 2009). Most individuals continue to breed throughout their lifespan, although the ability to compete for mates is likely to affect male breeding success in senescent individuals, particularly in the wild.

## 10.9 Ability to breed every year

### 10.9.1 Long-nosed fur seal

Long-nosed fur seals breed annually, with peak pupping occurring in mid-December. Seals will generally produce one pup after a gestation period of ~10 months (Crawley and Wilson 1976).

### 10.9.2 Australian fur seal

Australian fur seals have a synchronous annual breeding cycle and the majority of pups are born in a 3–4-week period with a peak in late November to early December (Warneke and Shaughnessy 1985; Gibbens and Arnould 2009; McIntosh *et al.* 2018).

### 10.9.3 Australian sea lion

Wild Australian sea lion females have been observed breeding in successive breeding seasons for up to at least eight seasons (McIntosh 2007). However, the breeding cycle is ~17–18 months. In zoos mating postpartum is often attempted but not always successful, possibly because of mistiming the re-introduction of the female and male. Introductions take place when the female shows signs of interest towards the male and usually occur on day 5 or 6 after pupping.

### 10.9.4 Leopard seal

There have no instances of leopard seals breeding in captivity.

## 10.10 Ability to breed more than once per year

### 10.10.1 Long-nose fur seal

No, due to > 12-month gestation period.

### 10.10.2 Australia fur seal

No, due to > 12-month gestation period.

### 10.10.3 Australian sea lion

Due to the extended gestation of 17–18 months, including 3–4 months of embryonic diapause (Gales *et al.* 1997), Australian sea lions cannot breed more than once per year.

### 10.10.4 Leopard seal

No, due to > 12-month gestation period.

## 10.11 Haul-out requirements

Pinnipeds commonly give birth on land, so need to have adequate haul-out space for them to give birth comfortably. The pregnant female should have enough space to lie flat with flippers spread and enough room to roll over and walk in a circle. Birthing mothers can often move around a lot while seeking a comfortable position, so giving them the space to do this on land is important.

## 10.12 Breeding diet

No specific changes are made during pregnancy. Seals moderate their own intake due to seasonal fluctuations and body condition.

## 10.13 Number of young born

Single pups are typical for all species of seals, with the occurrence of twins being rare.

## 10.14 Age at weaning

### 10.14.1 Long-nosed fur seal

Long-nosed fur seals will nurse their young for ~10 months and pups have a mean weaning bodyweight of 13–16 kg (Kirkwood and Goldsworthy 2013).

### 10.14.2 Australian fur seal

The lactation period lasts 11–12 months, but most pups are capable of foraging by 7 months of age and there is considerable variation in the weaning period, with a small number of pups suckling into their second and even third year (Stirling and Warneke 1971).

### 10.14.3 Australian sea lion

Australian sea lion pups are typically weaned after 12 months of age, over a 3–6-month period before the birth of the next pup (Lowther and Goldsworthy 2016; Higgins and Gass 1993). However, nearly one-third of mothers will continue to provision juveniles after the birth of the new pup or if the birth is unsuccessful (Higgins and Gass 1993; McIntosh and Kennedy 2013).

In a zoo setting pups are usually weaned between 12 and 18 months. When the pup starts to show interest in solid foods (sometimes as early as 9 months old), carers will supplement their milk diet with a small amount of solids and separate the pup from the dam for increased periods of time.

### 10.14.4 Leopard seal

Leopard seals will give birth (between October and December) to a single pup that will be nursed for ~30 days (Kirkwood and Goldsworthy 2013).

## 10.15 Age at removal from the mother

### 10.15.1 Long-nosed fur seal

Long nosed fur seals will nurse their young for ~10 months and pups have a mean weaning bodyweight of 13–16 kg (Kirkwood and Goldsworthy 2013).

### 10.15.2 Australian fur seal

The lactation period lasts 11–12 months, but most pups are capable of foraging by 7 months and there is considerable variation in the weaning period, with a small number of pups suckling into their second and even third year (Stirling and Warneke 1971).

### 10.15.3 Australian sea lion

Australian sea lion pups are typically weaned after 12 months of age, over a 3–6-month period before the birth of the next pup (Lowther and Goldsworthy 2016; Higgins and Gass 1993). However, nearly one-third of mothers will continue to provision juveniles after the birth of the new pup or if the birth is unsuccessful (McIntosh and Kennedy 2013; Higgins and Gass 1993).

In a zoo setting pups are usually weaned between 12–18 months. When the pup starts to show and interest in solid foods (sometimes as early as 9 months old), carers will supplement their milk diet with a small amount of solids and separate the pup from the dam for increased periods of time.

### 10.15.4 Leopard seal

Leopard seals have not bred in captivity. However, in the wild the mother would typically leave the pup at ~1 month after birth (Kirkwood and Goldsworthy 2013).

## 10.16 Growth and development

The growth of seals has been recorded for several species as either bodyweight and standard body length. Standard length is the straight-line length from the nose to tip of the tail on the ventral surface of the seal, or the dorsal surface (American Society of Mammalogists 1967; McLaren 1993). Species of Australian seal that have growth curves developed to date are given in Table 15.12.

### 10.16.1 Long-nosed fur seal

Separation between mother and pup would begin when the pup is ~6–8 months old. Initially, these separations would only be for 1 h and slowly increase to several hours. Depending on the intake of solid foods from the pup, separations continue to increase in duration until overnight and then progress to periods in excess of 24 h until the pup is fully weaned. Growth curves for this species can be found in Fig. 15.13.

### 10.16.2 Australian fur seal

The Australian fur seal is weaned at ~10 months. Though weaning is typically completed before the commencement of the next breeding season, some young may continue to be suckled into their second or even third year (Kirkwood and Arnould 2023). Growth curves for this species can be found in Fig. 15.14.

### 10.16.3 Australian sea lion

In captivity Australian sea lions are typically weaned at ~12–18 months, or as soon as they are eating enough solid foods to maintain their bodyweight. However, the weaning period may extend to ≥ 3 years if the female does not reproduce in subsequent breeding season or the new pup dies (Goldsworthy 2023b). Growth curves for this species in captivity and in the wild can be found in Figs 15.15, 15.16.

### 10.16.4 Leopard seal

This species has not been bred in a captive environment, but given their very short nursing period, separations would take place at ~3–4 weeks of age. Growth curves for this species are shown in Fig. 15.17.

**Table 15.12.** Growth curve measurements that have been developed for different species of seals.

Length assessed using standard body length

| Species | Measurements | Reference |
|---|---|---|
| *Arctocephalus forsteri* | Body length | 1 |
| *A. gazella* | Body length, weight | 2 |
| *A. pusillus doriferus* | Body length, weight | 3 |
| *A. tropicalis* | Bodyweight | 4 |
| *Hydrurga leptonyx* | Body length | 5, 6 |
| *Neophoca cinerea* | Body length, weight | 7 |
| *Leptonychotes weddellii* | Bodyweight | 6, 8, 9 |
| *Lobodon carcinophaga* | Body length | 10 |
| *Mirounga leonina* | Body length, weight | 6, 11, 12 |

References: (1) McKenzie *et al.* (2007); (2) Payne (1979); (3) Arnould and Warneke (2002); (4) Guinet and Georges (2000); (5) Laws (1957); (6) McLaren (1993); (7) McIntosh (2007); (8) Mansfield (1958); (9) Bryden *et al.* (1984); (10) Laws (1958); (11) Laws (1953); (12) Carrick *et al.* (1962).

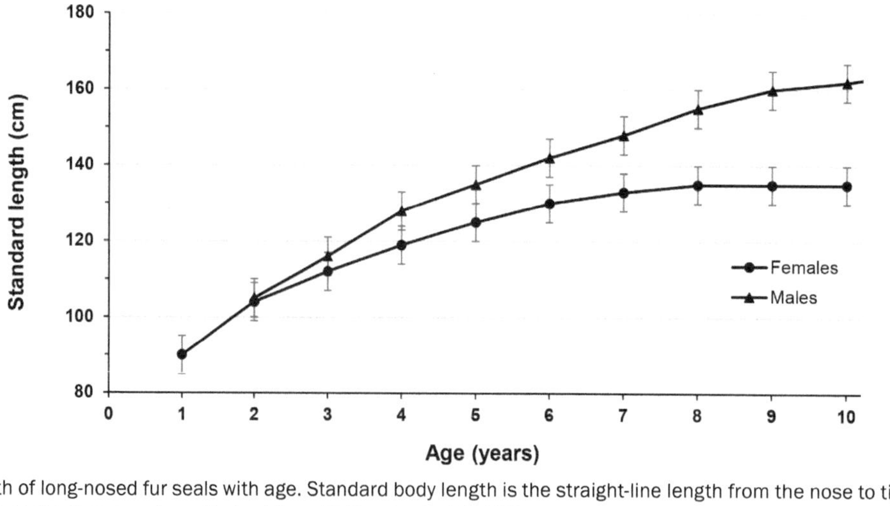


**Fig. 15.13.** Growth of long-nosed fur seals with age. Standard body length is the straight-line length from the nose to tip of the tail on the ventral surface or the dorsal surface. Derived from McKenzie *et al.* (2007).

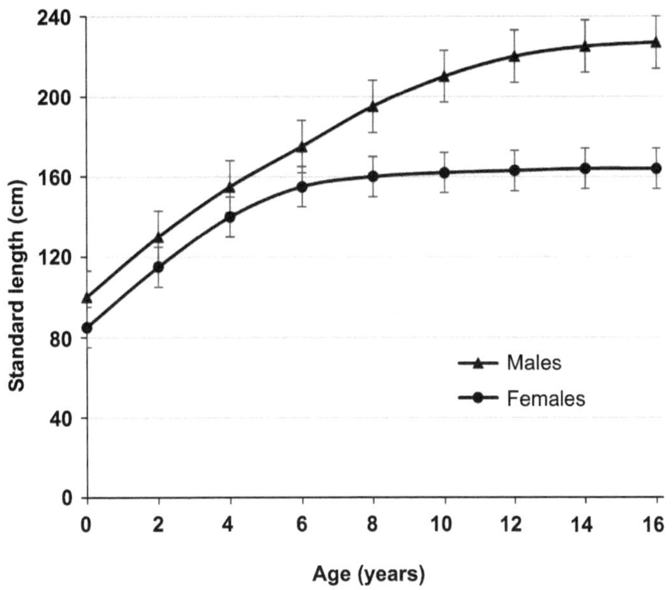


**Fig. 15.14.** Growth of Australian fur seals with age. Data derived from Arnould and Warneke (2002).

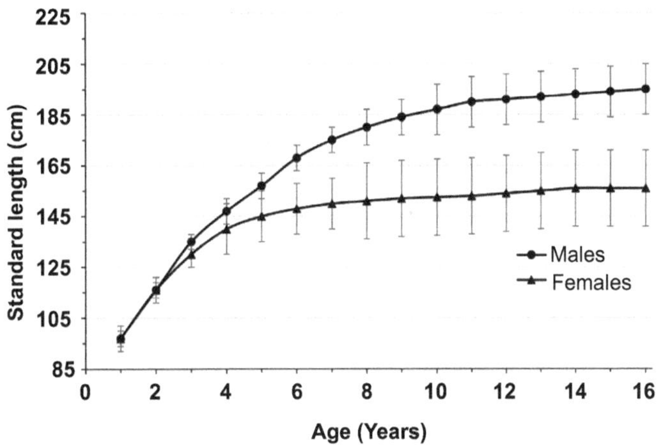


**Fig. 15.15.** Growth of wild Australian sea lions with age. Data derived from McIntosh (2007).

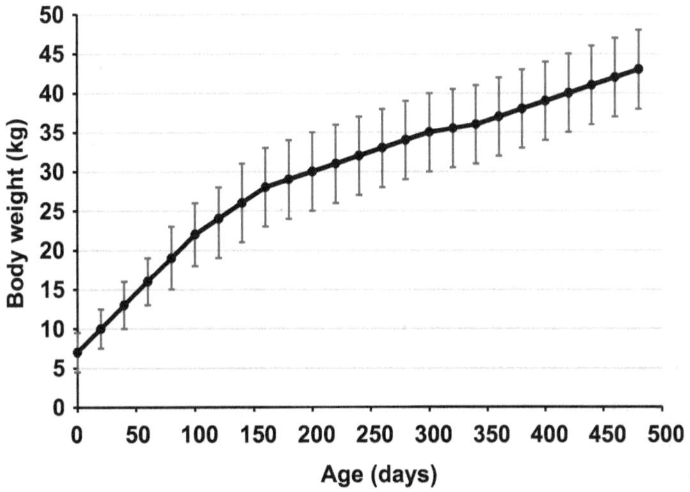


**Fig. 15.16.** Growth of captive Australian sea lions with age. Data from Taronga Zoo.

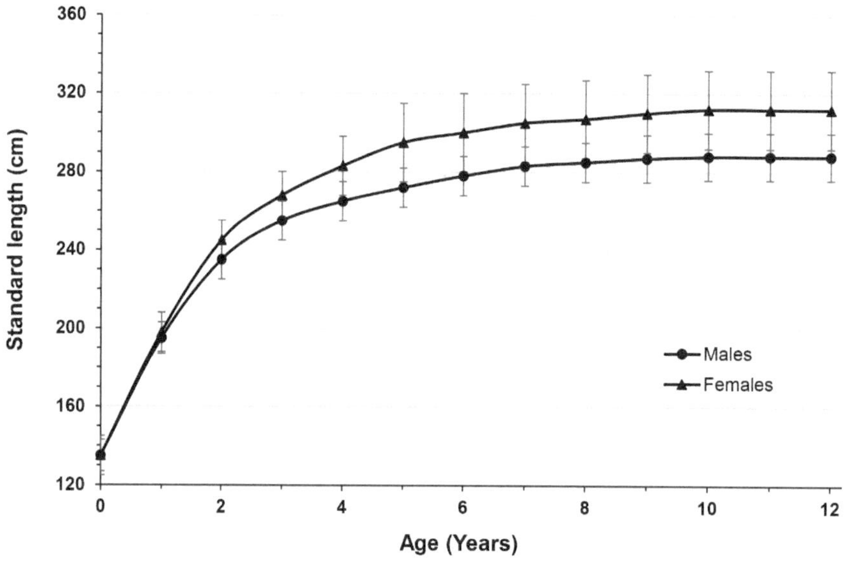


**Fig. 15.17.** Growth of wild leopard seals with age. Figure derived from McLaren (1993) via Laws (1957).

# 11. ARTIFICIAL REARING

## 11.1 Housing

As with all native mammals that have been taken into care, minimising stress is a major consideration. Choosing suitable housing can help to create a stress-free environment, taking into consideration the following:

- securing the area from children and animals
- maintaining a high standard of hygiene
- escape-proofing the area
- clearing the area of obstacles and hazards
- ensuring the area offers shelter from extreme weather and noise.

For otariid species a 'bed' area can be provided (i.e. boards or wooden slats) that is slightly elevated from floor level (1–3 cm) to provide a dry haul-out place that is not on concrete and/or a PetPak or small shelter with towels for extra warmth (Fig. 15.18). A small children's pool (made from hard plastic) may be introduced into the environment when the pup is ~3–4 weeks old. The introduction of the water body will allow natural swimming and exploration behaviours.

Phocids (e.g. leopard seal pups) require a sloped entry to the pool for ease of entering and exiting the water. Alternatively, a pool with an adjustable floor can assist with the introduction of the pup to water, allowing the depth of the pool to be controlled and to increase with the pup's swimming ability (Fig. 15.19).

If the mother of the pup is accessible and not a danger to the pup then all efforts should be made to keep the two housed together in an off-exhibit pupping facility. The pup can be removed for feeds and placed back with its mother for all other aspects of mothering and care. Access to swimming facilities should be withheld from the pup initially and then gradually introduced at increasing depths from 3–4 weeks of age.

**Fig. 15.18.** Holding-yard seal bed. Photo by Danielle Fox.

**Fig. 15.19.** Pool with adjustable floor. (a) Floor has been raised to the water surface and (b) the floor has been lowered below the water. Photos by Danielle Fox.

## 11.2 Temperature requirements

Seal pups should be kept in temperatures similar to that of their natural habitat. Long-nosed fur seals, Australian fur seals and subantarctic fur seals are all typically born in summer, so an indoor facility where temperatures can be maintained between 22°C and 25°C would be suitable. If the temperature drops during the night, access to a heat lamp is recommended. Australian sea lions can be born at varied times throughout the year as they have a non-annual breeding cycle, so access to a heat lamp in colder months is necessary.

Leopard seals have never been artificially reared so there is no information regarding their specific requirements.

## 11.3 Specific requirements

Where possible, the pup should be kept with the mother to assist in the development of social behaviours.

## 11.4 Diet and feeding routine

### 11.4.1 Natural milk

Australian fur seal milk has an average of 42% lipid and 10% protein (Arnould and Hindell 1999). The lipid levels begin at an average of 30% in early lactation and increase through to late lactation, peaking at an average of 50% before decreasing to 45% prior to weaning (Arnould and Hindell 1999). Protein content increases slightly from an average of 10–12% throughout lactation (Arnould and Hindell 1999). Australian fur seals consume an average of 557 g of milk per day ranging from 400 g/day shortly after birth to 675 g/day at 210 days of age (Kirkwood and Goldsworthy 2013). Long-nosed fur seal milk lipid content ranges from 22% to 57% (mean 37%) and protein levels are between 15% and 16% (Kirkwood and Goldsworthy 2013).

Australian sea lion milk lipid content increases throughout lactation. In the first 250 days of lactation it is on average 26.1% (± 8.7%), but after 250 days it is on average 39.1% (± 5.5%). Protein content is on average 9.97% (± 2.52%) (Gales *et al.* 1996).

Both Australian fur seals and Australian sea lions show a peak in milk lipid content at approximately 70% of the duration of lactation. This late-lactation decrease in lipid content may reflect the metabolic demands on the mother of concurrent gestation or changes associated with the weaning process (Arnould and Hindell 1999).

### 11.4.2 Milk formulas

At Taronga Zoo, Wombaroo ® Sea Lion Milk Replacer has long-life thickened cream added to make up the fat content and Lacteeze drops are used to reduce the lactose levels. There is some evidence to suggest that sea lions have an intolerance to lactose and develop problems in digestion that can lead to diarrhoea, hence the use of Lacteeze.

### 11.4.3 Feeding apparatus

Small teats connected to bottles have been successfully used for artificially rearing pinniped species, but it is not guaranteed that the pup will be receptive to this form of feeding. If not, tube feeding the milk formula is the next best option (Fig. 15.20).

Equipment needed includes:

- 9-mm diameter sterile tubes for tube feeding
- large syringe (50–70 mL) to inject the formula into the tube
- mouth ropes to hold the upper and lower jaws open during the insertion of the tube
- mouth chocks to keep mouth open (optional)
- sterile jug/ container for making up and storing milk formula
- measuring cups
- mixing spoons.

### 11.4.4 Feeding routine

Formula is divided and fed over four feeds initially and slowly reduced to two feeds per day. A new born seal pup should be fed milk warmed to 32–34°C monitored by thermometer. Recommended feeds of Wombaroo® Sea lion Milk formula are given in Table 15.13.

When using Wombaroo® formula (Tables 15.14, 15.15) for seal pups, the milk must be introduced slowly. Alternatively, the milk can be diluted with water before feeding whether by bottle or tube feed for the first 4 days of hand-rearing the pup.

**Fig. 15.20.** Tube-feeding an Australian sea lion pup. Photo by Danielle Fox.

## 11.5 Data recording

When an animal is first brought in for artificial rearing, its sex and approximate age, using growth charts, should be recorded. During the hand-rearing process several important pieces of information should be recorded on a daily basis:

- date when the information is recorded
- time when the information is recorded
- bodyweight to the nearest 1 g if possible (very important for small species)
- general activity and demeanour
- characteristics and frequency of defecation and urination
- amount (mLs) of formula consumed
- veterinary examinations and results.

This information serves several purposes, including providing important background information such as food consumption, that will assist a veterinarian to reach a diagnosis if the animal becomes sick or fails to grow or gain weight. It also allows a comparison with growth curves to assess progress (section 10.16).

## 11.6 Identification methods

Bleaching, dying or colouring sections of the animal's fur can be used as a means to identify individuals if they are not easily identifiable from physical characteristics. Implanted PIT tags are used very successfully both in captivity and in the wild. Livestock ear tags can be used in the flippers. Typically, two tags are attached to each animal, both containing the same identification information. On phocids, tags are attached to the posterior edge of the interdigital web, usually between the third and fourth digits of each hindflipper. On otariids the tags are attached through the connective tissue on the posterior edge of each foreflipper, midway between the body of the seal and the distal extension of fur on the dorsal surface of the flipper (Condy and Bester 1975). In wild Australian sea lions, embossed tags tend to remain legible for longer than printed tags due to sand wearing off the text printed on the surface of the tag (B Pitcher pers. obs.).

**Table 15.13.** Recommended intakes of Wombaroo® Sea Lion Milk Replacer.

| Bodyweight (kg) | Feed volume (mL/day) | Bodyweight (kg) | Feed volume (mL/day) | Bodyweight (kg) | Feed volume (mL/day) |
|---|---|---|---|---|---|
| 2 | 150 | 12 | 670 | 25 | 1225 |
| 4 | 270 | 14 | 760 | 35 | 1600 |
| 6 | 380 | 16 | 850 | 40 | 1800 |
| 8 | 480 | 18 | 930 | 45 | 2000 |
| 10 | 570 | 20 | 1000 | 50 | 2200 |

**Table 15.14.** Composition of Wombaroo® formula provided to seals.

| Day | Formula (%) | Water (%) |
|---|---|---|
| 1 | 25 | 75 |
| 2 | 50 | 50 |
| 3 | 75 | 25 |
| 4 | 100 | 0 |

**Table 15.15.** Recommended intake of Wombaroo® Sea Lion Milk (as per manufacturer's instructions).

| Bodyweight (kg) | Feed volume (mL/day) | Bodyweight (kg) | Feed volume (mL/day) | Bodyweight (kg) | Feed volume (mL/day) |
|---|---|---|---|---|---|
| 2 | 150 | 12 | 670 | 25 | 1225 |
| 4 | 270 | 14 | 760 | 35 | 1600 |
| 6 | 380 | 16 | 850 | 40 | 1800 |
| 8 | 480 | 18 | 930 | 45 | 2000 |
| 10 | 570 | 20 | 1000 | 50 | 2200 |

## 11.7 Hygiene

Maintaining a high level of hygiene is critical when hand-rearing seal pups. All equipment used for feeding must be sterilised after each use. Handlers must wash and disinfect hands before handling or feeding pups. Milk must only be heated up once and any leftovers are to be discarded.

## 11.8 Behavioural considerations

The decision to artificially rear seal pups should be made in consultation with pinniped specialists and veterinary experts, as many species of pinniped can develop behavioural issues when artificially reared. They are likely to be more aggressive towards handlers in a captive environment. Long-nosed fur seals, Australian fur seals and Australian sea lions that have been artificially reared would likely be deemed unsuitable for release due to the high possibility of imprinting occurring from the lengthy process and time in human care. Leopard seals, although they have never been artificially reared, would make better candidates for release as the artificial rearing process would only be ~4 weeks. This species is fully weaned at 4 weeks old and therefore would have less interaction with humans. Experienced handlers that have knowledge of artificially rearing pinnipeds need to be available to consult or undertake the process.

Many institutions will recommend to only intervene through assisted feeding, particularly in circumstances where the pup's dam is not producing sufficient milk, but all other aspects of mothering is sufficient. If possible, the pup should be kept with the mother to assist in the development of social behaviour.

## 11.9 Use of foster animals

Dams have been known to allow pups other than their own offspring to suckle and this is quite common in some species of pinniped. Some reasons for this occurring are a female losing her pup and adopting another, a female nursing her own pup and allowing an orphan to suckle as well or cases of mistaken identity (Riedman 1990; Pitcher *et al.* 2011b; McIntosh and Pitcher 2021).

## 11.10 Weaning

Weaning is abrupt for most phocid species and pups usually eat no solid food during lactation (Riedman 1990). Leopard seals are weaned at ~4weeks of age when they will start to feed on krill and other solid foods.

Otariids have a much slower weaning process, with the milk diet often supplemented with solid food later in the lactation process (Riedman 1990). Weaning of captive otariids will begin at ~6–8 months, with fish being introduced from time to time to initiate interest. The goal will be to achieve initial weaning as soon as possible after this time. If the pup is being artificially reared or supplement-fed, weaning can begin as early as 4 months of age, but this should be assessed on an individual basis. Live fish can be used for weaning if a consistent supply of the appropriate size fish is available. If not, this method is not recommended. One should be certain about the fish supply before beginning live fish weaning. If live fish is used the veterinarian must be informed so that a worming prophylaxis can be used. Live fish should be used for a short time only and all efforts should be made to approximate onto dead fish.

## 11.11 Release procedures

A seal that is being rehabilitated and released must be in human care for the least amount of time as possible to prevent imprinting. Feeding should take place in a way that the animal dos not know that the food has come from humans. Ideally, the fish fed to a rehabilitated seal would be alive so that the animal does not lose its hunting skills. Seals that are released after rehabilitation need to not to have any physical impairment that would prevent thriving in the wild.

Long-nosed fur seal, Australian fur seal and Australian sea lion pups that are artificially reared may not be good release candidates due to habituation to humans during the hand-rearing process. Further, unlike many other otariids, Australian sea lions begin learning where, when and how to forage at least 8 months before weaning, by conducting foraging trips into adult female foraging habitat (Lowther and Goldsworthy 2012). This makes successful release of captive-bred individuals highly unlikely because they will lack vital foraging experience. Only juvenile (weaned) and adults that require short-term veterinary treatment should be considered for rehabilitation and release.

Leopard seals have been successfully rehabilitated and released from Australian coastlines in the past, although this is not practised anymore due to the Scientific Committee of Antarctic Research recommendations that state that rehabilitation and release of native species serve no conservation purpose and pose risks of introducing pathogens to wild populations (SCAR 1996).

## REFERENCES

Aguayo-Lobo A, Acevedo J, Brito JL, Acuña P, Bassoi M, Secchi ER, Rosa LD (2011) Presence of the leopard seal, *Hydrurga leptonyx* (De Blainville, 1820), on the coast of Chile: an example of the Antarctica–South America connection in the marine environment. *Oecologia Australis* **15**, 69–85. doi:10.4257/oeco.2011.1501.07

American Society of Mammalogists (1967) Standard measurements of seals. *Journal of Mammalogy* **48**, 459–462. doi:10.2307/1377778

Arnbom TA, Lunn NJ, Boyd IL, Barton T (1992) Aging live Antarctic fur seals and southern elephant seals. *Marine Mammal Science* **8**, 37–43. doi:10.1111/j.1748-7692.1992.tb00123.x

Arnould JPY, Duck CD (1997) The costs and benefits of territorial tenure, and factors affecting mating success, in male Antarctic fur seals. *Journal of Zoology* **241**, 649–664. doi:10.1111/j.1469-7998.1997.tb05739.x

Arnould JP, Hindell MA (1999) The composition of Australian fur seal (*Arctocephalus pusillus doriferus*) milk throughout lactation. *Physiological and Biochemical Zoology* **72**, 605–612.

Arnould JPY, Hindell MA (2001) Dive behavior, foraging locations, and maternal attendance patterns of Australian fur seals (*Arctocephalus pusillus doriferus*). *Canadian Journal of Zoology* **79**, 35–48. doi:10.1139/z00-178

Arnould JPY, Kirkwood R (2008) Habitat selection by female Australian fur seals *Arctocephalus pusillus doriferus*. *Aquatic Conservation* **17**(Suppl 1), S53–S67. doi:10.1002/aqc.908

Arnould JPY, Warneke RM (2002) Growth and condition in Australian fur seals (*Arctocephalus pusillus doriferus*) (Carnivora: Pinnipedia). *Australian Journal of Zoology* **50**, 53–66. doi:10.1071/ZO01077

Atkinson S (1997) Reproductive biology of seals. *Reviews of Reproduction* **2**, 175–194. doi:10.1530/ror.0.0020175

Baker AM, Gynther IC (Eds) (2023) *Strahan's Mammals of Australia.* 4th edn. Reed New Holland Publishers, Sydney.

Barnes J, Higgins D, Gray R (2008) Pinnipeds. In *Medicine of Australian Mammals.* (Eds L Vogelnest and R Woods) pp. 541–589. CSIRO Publishing, Melbourne.

Bartholomew GA (1970) A model for the evolution of pinniped polygyny. *Evolution* **24**, 546–559. doi:10.2307/2406835

Beentjes MP (2006) Behavioral thermoregulation of the New Zealand sea lion (*Phocarctos hookeri*). *Marine Mammal Science* **22**, 311–325. doi:10.1111/j.1748-7692.2006.00022.x

Berry JA (1960) The occurrence of a leopard seal (*Hydrurga leptonyx*) in the tropics. *Annals & Magazine of Natural History* **3**, 591. doi:10.1080/00222936008651062

Berry TE, Osterrieder SK, Murray DC, Coghlan ML, Richardson AJ, Grealy AK, Stat M, Bejder L, Bunce M (2017) DNA metabarcoding for diet analysis and biodiversity: A case study using the endangered Australian sea lion (*Neophoca cinerea*). *Ecology and Evolution* **7**, 5435–5453. doi:10.1002/ece3.3123

Bester MN (1987) Subantarctic fur seal, *Arctocephalus tropicalis*, at Gough Island (Tristan da Cunha Group). *NOAA Technical Report NMFS* **51**, 57–60.

Bester MN (1995) Reproduction in the female Subantarctic fur seal, *Arctocephalus tropicalis*. *Marine Mammal Science* **11**, 362–375. doi:10.1111/j.1748-7692.1995.tb00291.x

Bodley K (2025) Pinnipeds. In *Current Therapy in Medicine of Australian Mammals*. Revised Edition. (Eds L Vogelnest and T Portas) pp. 621–634. CSIRO Publishing, Melbourne.

Bonner WN (1994) *Seals and Sea Lions of the World*. Blandford, London.

Bowen WD 1991. Behavioral ecology of pinniped neonates. In *The Behavior of Pinnipeds*. (Ed. D Renouf) pp. 66–127. Chapman and Hall, London.

Boyd IL (2009) Pinniped Life History. In *Encyclopaedia of Marine Mammals* (2nd edn) (Eds WF Perrin, B Würsig and JGM Thewissen) pp. 868–873. Academic Press, Burlington, MA.

Brothers N, Pemberton D (1990) Status of Australian and New Zealand fur seals at Maatsuyker Island, southwestern Tasmania. *Australian Wildlife Research* **17**, 563–569. doi:10.1071/WR9900563

Bryden MM, Smith MSR, Tedman RA, Featherston DW (1984) Growth of the Weddell seal *Leptonychotes weddellii* (Pinnipedia). *Australian Journal of Zoology* **32**, 33–41. doi:10.1071/ZO9840033

Carrick R, Csordas S, Ingham SE (1962) Studies on the southern elephant seal, *Mirounga leonina* (L.). IV: breeding and development. *CSIRO Wildlife Research* **7**, 161–197. doi:10.1071/CWR9620161

Cebuhar J, Negrete J, Botta S (2021) Age estimation and growth layer patterns in teeth of crabeater seals: a comparison of techniques. *Marine Mammal Science* **37**, 385–398. doi:10.1111/mms.12749

Charrier I, Pitcher BJ, Harcourt RG (2009) Vocal recognition of mothers by Australian sea lion pups: individual signature and environmental constraints. *Animal Behaviour* **78**, 1127–1134. doi:10.1016/j.anbehav.2009.07.032

Charrier I, Pitcher BJ, Harcourt RG (2022) Mother-pup recognition mechanisms in Australian sea lion (*Neophoca cinerea*) using uni- and multi-modal approaches. *Animal Cognition* **25**, 1019–1028. doi:10.1007/s10071-022-01641-5

Chilvers BL, Goldsworthy SD (2015) *Arctocephalus forsteri*, New Zealand fur seal. The IUCN Red List of Threatened Species 2015: e.T41664A45230026. doi:10.2305/IUCN.UK.2015-2.RLTS.T41664A45230026.en

Chilvers BL, Harcourt RG (2021) Families Otariidae and Phocidae. In *The Handbook of New Zealand Mammals*. (Eds CM King and DM Forsyth) pp. 241–277. CSIRO Publishing, Melbourne.

Condy PR, Bester MN (1975) Notes on the tagging of seals on Marion and Gough Islands. *South African Journal of Antarctic Research* **5**, 45–47.

Cousins DV, Williams SN, Reuter R, Forshaw D, Chadwick B, Coughran D, Collins P, Gales N (1993) Tuberculosis in wild seals and characterisation of the seal bacillus. *Australian Veterinary Journal* **70**, 92–97. doi:10.1111/j.1751-0813.1993.tb03284.x

Crawley M, Wilson G (1976) The natural history and behaviour of the New Zealand fur seal (*Arctocephalus forsteri*). *Tuatara: Journal of the Biological Society* **22**, 1–28.

Department of Environment, Water, Heritage and the Arts (DEWHA) (2010) Recovery Plan for the Australian Sea Lion (*Neophoca cinerea*). Technical Issues Paper. Australian Government, Canberra.

Department of Sustainability, Environment, Water, Population and Communities (DSEWPAC) (2013) Recovery Plan for the Australian Sea Lion (*Neophoca cinerea*). Commonwealth of Australia, Canberra.

Duignan PJ (2000) Diseases of cetaceans and pinnipeds. In *Marine Wildlife, Proceedings 335*. pp. 431–462. Post Graduate Foundation in Veterinary Science, Sydney.

Erickson AW, Siniff D, Cline D, Hofman R (1971) Distributional ecology of Antarctic seals. Symposium on Antarctic ice and water masses. Scientific Committee on Antarctic Research Cambridge, UK, 55–76.

Fernández-Juricic E, Campagna C, Enriquez V, Ortiz CL (1999) Vocal communication and individual variation in breeding South American sea lions. *Behaviour* **136**, 495–517. doi:10.1163/156853999501441

Forshaw D, Phelps GR (1991) Tuberculosis in a captive colony of pinnipeds. *Journal of Wildlife Diseases* **27**, 288–295. doi:10.7589/0090-3558-27.2.288

Gales N, Cheal AJ (1992) Estimating diet composition of the Australian sea lion (*Neophoca cinerea*) from scat analysis: an unreliable technique. *Wildlife Research* **19**, 447–455. doi:10.1071/WR9920447

Gales N, Cheal AJ, Pobar GJ, Williamson P (1992) Breeding biology and movements of Australian sea-lions, *Neophoca cinerea*, off the west coast of Western Australia. *Wildlife Research* **19**, 405–416. doi:10.1071/WR9920405

Gales NJ, Costa DP, Kretzmann M (1996) Proximate composition of Australian sea lion milk throughout the entire Supra-annual lactation period. *Australian Journal of Zoology* **44**, 651–657.

Gales NJ, Williamson P, Higgins LV, Blackberry MA, James I (1997) Evidence for a prolonged postimplantation period in the Australian sea lion (*Neophoca cinerea*). *Journal of Reproduction and Fertility* **111**, 159–163. doi:10.1530/jrf.0.1110159

Gales R, Pemberton D (1994) Diet of the Australian fur seal in Tasmania. *Marine and Freshwater Research* **45**, 653–664. doi:10.1071/MF9940653

Gentry RL, Kooyman GL (1986) *Fur Seals: Maternal Strategies on Land and at Sea*. Princeton University Press, New Jersey.

Gibbens J, Arnould JPY (2009) Interannual variation in pup production and the timing of breeding in benthic foraging Australian fur seals. *Marine Mammal Science* **25**, 573–587. doi:10.1111/j.1748-7692.2008.00270.x

Gilbert JR, Erickson AW (1977) Distribution and abundance of seals in the pack ice of the Pacific sector of the Southern Ocean. In *Adaptations within Antarctic Ecosystems*. (Ed. L Llano) pp. 703–740. Smithsonian Institute: Washington, DC.

Gill JCH (1967) Notes on the sealing industry of early Australia. *Journal of the Royal Historical Society of Queensland* **8**, 218–245.

Goldsworthy SD (2006) Maternal strategies of the New Zealand fur seal: evidence for interannual variability in provisioning and pup growth strategies. *Australian Journal of Zoology* **54**, 31–44. doi:10.1071/ZO05041

Goldsworthy SD (2023a) Long-nosed fur seal *Arctocephalus forsteri*. In *Strahan's Mammals of Australia*. 4th edn. (Eds AM Baker and IC Gynther). pp. 683–685. New Holland Publishers, Sydney.

Goldsworthy SD (2023b) Australia sea lion *Neophoca cinerea*. In *Strahan's Mammals of Australia*. 4th edn. (Eds AM Baker and IC Gynther). pp. 691–693. New Holland Publishers, Sydney.

Goldsworthy SD, Shaughnessy PD, Smart J, Bailleul F, Stonnill M, Lashmar K (2019) Monitoring of Seal Bay and other pinniped populations on Kangaroo Island: 2017/2018. Report to the Department of Environment and Water. South Australian Research and Development Institute (Aquatic Sciences), Adelaide. SARDI Publication No. F2014/000322-5. SARDI Research Report Series No. 1018.

Goldsworthy SD, Shaughnessy PD, Mackay AI, Bailleul F, Holman D, Lowther AD, Page B, Waples K, Raudino H, Bryars S, Anderson T (2021) Assessment of the status and trends in abundance of a coastal pinniped, the Australian sea lion *Neophoca cinerea*. *Endangered Species Research* **44**, 421–437. doi:10.3354/esr01118

Gray RB, Rogers TL, Canfield PJ (2009) Health assessment of the leopard seal, *Hydrurga leptonyx*, in Prydz Bay, Eastern Antarctica and NSW, Australia. In *Health of Antarctic Wildlife*. (Eds KR Kerry and MJ Riddle) pp. 167–192. Springer, Berlin.

Guinet C, Georges JY (2000) Growth in pups of the Subantarctic fur seal (*Arctocephalus tropicalis*) on Amsterdam Island. *Journal of Zoology* **251**, 289–296. doi:10.1111/j.1469-7998.2000.tb01079.x

Gwynn AM (1953) The status of the leopard seal at Heard Island and Macquarie Island, 1948–1950. ANARE Interim Reports 3, 1–33. Antarctic Division, Melbourne.

Hall-Aspland SA, Rogers TL (2004) Summer diet of leopard seals (*Hydrurga leptonyx*) in Prydz Bay, Eastern Antarctica. *Polar Biology* **27**, 729–734. doi:10.1007/s00300-004-0662-9

Hanggi EB (1992) The importance of vocal cues in mother-pup recognition in a California sea lion. *Marine Mammal Science* **8**, 430–432. doi:10.1111/j.1748-7692.1992.tb00061.x

Harcourt R (2005) New Zealand Fur Seal. In *The Handbook of New Zealand Mammals* 2nd edn. (Ed. CM King). pp. 225–235. Oxford University Press, Auckland.

Harcourt R, Marsh H, Slip D, Chilvers L, Noad M, Dunlop R (2015) Marine mammals, back from the brink? Contemporary conservation issues. In *Austral Ark: The State of Wildlife in Australia and New Zealand*. (Eds A Stow, N Maclean and GI Holwell) pp. 322–353. Cambridge University Press, Cambridge.

Hardy N, Berry T, Kelaher BP, Goldsworthy SD, Bunce M, Coleman MA, Gillanders BM, Connell SD, Blewitt M, Figueira W (2017) Assessing the trophic ecology of top predators across a recolonisation frontier using DNA metabarcoding of diets. *Marine Ecology Progress Series* **573**, 237–254. doi:10.3354/meps12165

Harrison RJ (1968) *The Behavior and Physiology of Pinnipeds*. Appleton-Century-Crofts, New York.

Higgins LV, Gass L (1993) Birth to weaning: parturition, duration of lactation, and attendance cycles of Australian sea lions (*Neophoca cinerea*). *Canadian Journal of Zoology* **71**, 2047–2055. doi:10.1139/z93-290

Hocking DP, Salverson M, Evans AR (2015) Foraging-based enrichment promotes more varied behaviour in captive Australian fur seals (*Arctocephalus pusillus doriferus*). *PLoS One* **10**(5), e0124615. doi:10.1371/journal.pone.0124615

Hulst F (2025) Zoonoses. In *Current Therapy in Medicine of Australian Mammals*. Revised Edition. (Eds L Vogelnest and T Portas) pp. 273–289. CSIRO Publishing, Melbourne.

Hume F, Arnould J, Kirkwood R, Davis P (2001) Extended maternal dependence by juvenile Australian fur seals (*Arctocephalus pusillus doriferus*). *Australian Mammalogy* **23**, 67–70. doi:10.1071/AM01067

Hupman K, Visser IN, Fyfe J, Cawthorn M, Forbes G, Grabham AA, Bout R, Mathias B, Benninghaus E, Matucci K, Cooper T, Fletcher L, Godoy D (2020) From vagrant to resident: occurrence, residency and births of leopard seals (*Hydrurga leptonyx*) in New Zealand waters. *New Zealand Journal of Marine and Freshwater Research* **54**, 1–23. doi:10.1080/00288330.2019.1619598

Insley SJ, Phillips AV, Charrier I (2003) A review of social recognition in pinnipeds. *Aquatic Mammals* **29**, 181–201. doi:10.1578/016754203101024149

Izzo C, Hamer DJ, Bertozzi T, Donnellan SC, Gillanders BM (2011) Telomere length and age in pinnipeds: the endangered Australian sea lion as a case study. *Marine Mammal Science* **27**, 841–851. doi:10.1111/j.1748-7692.2010.00450.x

Jackson SM, Groves CP (2015) *Taxonomy of Australian Mammals*. CSIRO Publishing, Melbourne.

Jefferson TA, Webber MA, Pitman RL (2015) *Marine Mammals of the World*. 2nd edn. Academic Press, Cambridge, MA.

King JE (1960) Sea lions of the genera *Neophoca* and *Phocarctos*. *Mammalia* **24**, 445–456.

Kirkwood R, Arnould JPY (2008) Australian fur seal *Arctocephalus pusillus*. In *The Mammals of Australia*. 3rd edn. (Ed S Van Dyck and R Strahan). pp. 723–725. Reed Books, Sydney.

Kirkwood R, Arnould JPY (2023) Australian fur seal *Arctocephalus pusillus doriferus*. In *Strahan's Mammals of Australia*. 4th edn. (Eds AM Baker and IC Gynther). pp. 687–690. New Holland Publishing, Sydney.

Kirkwood R, Goldsworthy S (2013) *Fur Seals and Sea Lions*. CSIRO Publishing, Melbourne.

Kirkwood R, Gales R, Terauds A, Arnould JPY, Pemberton D, Shaughnessy PD, Mitchell AT, Gibbens J (2005) Pup production and population trends of the Australian fur seal (*Arctocephalus pusillus doriferus*). *Marine Mammal Science* **21**, 260–282. doi:10.1111/j.1748-7692.2005.tb01227.x

Kirkwood R, Pemberton D, Gales R, Hoskins AJ, Mitchell T, Shaughnessy PD, Arnould JPY (2010) Continued population recovery by Australian fur seals. *Marine and Freshwater Research* **61**, 695–701. doi:10.1071/MF09213

Kooyman GL (1981) Leopard seal *Hydrurga leptonyx* Blainville, 1820. Academic Press, London, UK.

Kovacs KM, Aguilar A, Aurioles D, Burkanov V, Campagna C, Gales N, Gelatt T, Goldsworthy SD, Goodman SJ, Hofmeyr GJG, Härkönen T, Lowry L, Lydersen C, Schipper J, Sipilä T, Southwell C, Stuart S, Thompson D, Trillmich F (2012) Global threats to pinnipeds. *Marine Mammal Science* **28**, 414–436. doi:10.1111/j.1748-7692.2011.00479.x

Lacy RC, Balou JD, Pollak JP (2011) PMx: software package for demographic and genetic analysis and management of pedigreed populations. *Methods in Ecology and Evolution* **3**, 433–437. doi:10.1111/j.2041-210X.2011.00148.x

Ladds P (2009) *Pathology of Australian Native Wildlife*. CSIRO Publishing, Melbourne.

Lancaster ML, Gemmell NJ, Negro S, Goldsworthy S, Sunnucks P (2006) Ménage à trois on Macquarie Island: hybridization among three species of fur seal (*Arctocephalus* spp.) following historical population extinction. *Molecular Ecology* **15**, 3681–3692. doi:10.1111/j.1365-294X.2006.03041.x

Laws M (1958) Growth rates and ages of crabeater seals, *Lobodon carcinophaga* Jacquinot and Pucheran. *Proceedings of the Zoological Society of London* **130**, 275–288. doi:10.1111/j.1096-3642.1958.tb00574.x

Laws RM (1953) The elephant seal (*Mirounga leonina* Linn.). I: growth and age. *Falkland Island Dependencies Scientific Survey Report* **8**, 1–62.

Laws RM (1957) On the growth rates of the leopard seal, *Hydrurga leptonyx* (De Blainville, 1820). *Säugetierkundliche Mitteilungen* **5**, 49–55.

Laws RM (1984) Seals. In *Antarctic Ecology* Vol. 2. (Ed. RM Laws) pp. 621–715. Academic Press, London.

Ledingham RB (1979) 1978 expedition to renovate the 1912–13 Australasian Antarctic Expedition base hut. *The Polar Record* **19**(122), 485–488. doi:10.1017/S0032247400002424

Ling JK (1999) Exploitation of fur seals and sea lions from Australian, New Zealand and adjacent subantarctic islands during the eighteenth, nineteenth and twentieth centuries. *Australian Zoologist* **31**, 323–350. doi:10.7882/AZ.1999.036

Lowther AD, Goldsworthy SD (2012) Head start: Australian sea lion pups gain experience of adult foraging grounds before weaning. *Marine Biology* **159**, 2687–2696. doi:10.1007/s00227-012-2026-2

Lowther AD, Goldsworthy SD (2016) When were the weaners weaned? Identifying the onset of Australian sea lion nutritional independence. *Journal of Mammalogy* **97**, 1304–1311. doi:10.1093/jmammal/gyw106

Macgregor JW, Holyoake C, Fleming PA, Robertson ID, Connolly JH, Warren KS (2014) Investigation into the characteristics, triggers and mechanism of apnoea and bradycardia in the anaesthetized platypus (*Ornithorhynchus anatinus*). *Conservation Physiology* **2**, cou053. doi:10.1093/conphys/cou053

Mansfield AW (1958) The breeding behaviour and reproductive cycle of the Weddell seal (*Leptonychotes weddelli* Lesson). Falkland Island Dependencies Scientific Survey. *Scientific Reports* **18**, 1–41.

Marcus AD, Higgins DP, Gray R (2015) Ivermectin treatment of free-ranging Australian sea lion (*Neophoca cinerea*) pups: effect on hookworm and lice infection status, haematological parameters, growth, and survival. *Parasitology Research* **114**, 2743–2755. doi:10.1007/s00436-015-4481-4

Marlow BJ (1975) The comparative behaviour of the Australasian sea lions *Neophoca cinerea* and *Phocarctos hookeri* (Pinnipedia: Otariidae). *Mammalia* **39**, 159–230. doi:10.1515/mamm.1975.39.2.159

Mawson PR, Coughran D (1999) Records of sick, injured and dead pinnipeds in Western Australia 1980–1996. *Journal of the Royal Society of Western Australia* **82**, 121–128.

Maxwell G (1967) *Seals of the World*. Constable, London.

McIntosh RR (2007) The life history and population demographics of the Australian sea lion, *Neophoca cinerea*. PhD thesis, La Trobe University, Melbourne.

McIntosh RR, Kennedy CW (2013) Morphology, sex ratio and cause of death in Australian sea lion (*Neophoca cinerea*) pups. *Australian Mammalogy* **35**, 93–100. doi:10.1071/AM12037

McIntosh RR, Pitcher BJ (2021) The enigmatic life history of the Australian sea lion. In *Ethology and Behavioral Ecology of Otariids and the Odobenid*. (Eds C Campagna and R Harcourt). pp 557–585. Springer Nature, Switzerland.

McIntosh RR, Page B, Goldsworthy SD (2006) Dietary analysis of regurgitates and stomach samples from free-living Australian sea lions. *Wildlife Research* **33**, 661–669. doi:10.1071/WR06025

McIntosh RR, Goldsworthy SD, Shaughnessy PD, Kennedy CW, Burch P (2012) Estimating pup production in a mammal with an extended aseasonal breeding season, the Australian sea lion (*Neophoca cinerea*). *Wildlife Research* **39**, 137–148. doi:10.1071/WR10222

McIntosh RR, Kirkman SP, Thalmann S, Sutherland DR, Mitchell A, Arnould JPY, Salton M, Slip DJ, Dann P, Kirkwood R (2018) Understanding meta-population trends of the Australian fur seal, with insights for adaptive monitoring. *PLoS One* **13**(9), e0200253. doi:10.1371/journal.pone.0200253

McKenzie J, Page B, Shaughnessy PD, Hindell MA (2007) Age and reproductive maturity of New Zealand fur seals (*Arctocephalus forsteri*) in southern Australia. *Journal of Mammalogy* **88**, 639–648. doi:10.1644/06-MAMM-A-150R1.1

McLaren IA (1993) Growth in pinnipeds. *Biological Reviews of the Cambridge Philosophical Society* **68**, 1–79. doi:10.1111/j.1469-185X.1993.tb00731.x

McMahon CR, van den Hoff J, Burton HR, Davis PD (1997) Evidence for precocious development in female pups of the southern elephant seal, *Mirounga leonina*, at Macquarie Island. *Marine Mammal Research in the Southern Hemisphere* **1**, 92–96.

McMahon CR, Burton HR, van den Hoff J, Woods R, Bradshaw CJA (2006) Assessing hot-iron and cryo-branding for permanently marking southern elephant seals. *The Journal of Wildlife Management* **70**, 1484–1489. doi:10.2193/0022-541X(2006)70[1484:AHACFP]2.0.CO;2

Miller S, Colitz CMH, St Leger J, Dubelzeig R (2013) A retrospective survey of the ocular histopathology of the pinniped eye with emphasis on corneal disease. *Veterinary Ophthalmology* **16**, 119–129. doi:10.1111/j.1463-5224.2012.01040.x

Monamy V (2007) Hot iron branding of seals and sea lions: why the ban will remain. *Australian Veterinary Journal* **85**, 485–486. doi:10.1111/j.1751-0813.2007.00239.x

Morgan K, Norman TJdeB, Duignan PJ, Gibbs N, Best H (2000) Studies on hookworm in New Zealand otariid seals. In *Proceedings of Wildlife Disease Association (Australasian Section) and NZVA Wildlife Society Joint Conference*. Warkworth, New Zealand.

New South Wales Government (2008) *Standards for Exhibiting Seals in New South Wales*. Department of Primary Industries, Orange New South Wales.

Newberry RC (1995) Environmental enrichment: increasing the biological relevance of captive environments. *Applied Animal Behaviour Science* **44**, 229–243. doi:10.1016/0168-1591(95)00616-Z

Norman RJdeB (1995) Disease in free-living Australian seals. In *Proceedings of the Australian Association of Veterinary Conservation Biologists*. 23 May 1995. pp. 41–50. Marine Mammal Symposium, Melbourne.

Page B, Goldsworthy SD, Hindell MA (2001) Vocal traits of hybrid fur seals: intermediate to their parental species. *Animal Behaviour* **61**, 959–967. doi:10.1006/anbe.2000.1663

Panneton WM (2013) The mammalian diving response: an enigmatic reflex to preserve life? *Physiology (Bethesda, MD)* **28**, 284–297. doi:10.1152/physiol.00020.2013

Payne MR (1979) Growth in the Antarctic fur seal *Arctocephalus gazella*. *Journal of Zoology* **187**, 1–20. doi:10.1111/j.1469-7998.1979.tb07709.x

Péron F (1816) *Voyage de Découvertes aux Terres Australes, Exécuté par Ordre de Sa Majesté l'Empereur et roi, sur les Corvettes le Géographe, le Naturaliste, et la Goelette le Casuarina, Pendant les Années 1800, 1801, 1802 et 1804*. Vol. 2. Historique. l'Imprimerie Royale, Paris.

Peters KJ, Ophelkeller K, Bott NJ, Goldsworthy SD (2015) PCR-based techniques to determine diet of the Australian sea lion (*Neophoca cinerea*): a comparison with morphological analysis. *Marine Ecology (Berlin)* **36**, 1428–1439. doi:10.1111/maec.12242

Pistorius PA, Bester MN, Kirkman SP, Taylor FE (2001) Temporal changes in fecundity and age at sexual maturity of southern elephant seals at Marion Island. *Polar Biology* **24**, 343–348. doi:10.1007/s003000000217

Pitcher BJ, Ahonen H, Harcourt RG, Charrier I (2009) Delayed onset of vocal recognition in Australian sea lion pups (*Neophoca cinerea*). *Naturwissenschaften* **96**, 901–909. doi:10.1007/s00114-009-0546-5

Pitcher BJ, Harcourt RG, Charrier I (2010) Rapid onset of maternal vocal recognition in a colonially breeding mammal, the Australian sea lion. *PLoS One* **5**(8), e12195. doi:10.1371/journal.pone.0012195

Pitcher BJ, Harcourt RG, Schaal B, Charrier I (2011a) Social olfaction in marine mammals: wild female Australian sea lions can identify their pup's scent. *Biology Letters* **7**, 60–62. doi:10.1098/rsbl.2010.0569

Pitcher BJ, Ahonen H, Charrier I, Harcourt RG (2011b) Allosuckling behavior in the Australian sea lion (*Neophoca cinerea*): an updated understanding. *Marine Mammal Science* **27**, 881–888.

Pitcher BJ, Harcourt RG, Charrier I (2012) Individual identity encoding and environmental constraints in vocal recognition of pups by Australian sea lion mothers. *Animal Behaviour* **83**, 681–690. doi:10.1016/j.anbehav.2011.12.012

Pussini N, Goebel ME (2015) A safer protocol for field immobilization of leopard seals (*Hydrurga leptonyx*). *Marine Mammal Science* **31**, 1549–1558. doi:10.1111/mms.12232

Ramirez K (2012) Marine mammal training: the history of training animals for medical behaviors and keys to their success. *The Veterinary Clinics of North America. Exotic Animal Practice* **15**, 413–423. doi:10.1016/j.cvex.2012.06.005

Reijnders PJH, Plötz J, Zegers J, Gräfe M (1990) Breeding biology of Weddell seals (*Leptonychotes weddellii*) at Drescher Inlet, Riiser Larsen Ice Shelf, Antarctica. *Polar Biology* **10**, 301–306. doi:10.1007/BF00238429

Riedman M (1990) *The Pinnipeds: Seals, Sea Lions and Walruses.* University of California Press, Berkeley and Los Angeles.

Rodríguez D, Bastida R, Morón S, Heredia S, Loureiro J (2003) Occurrence of leopard seals in northern Argentina. *The Latin American Journal of Aquatic Mammals* **2**, 51–54. doi:10.5597/lajam00031

Rogers TL (2009) Leopard Seal, *Hydrurga leptonyx*. In *Encyclopedia of Marine Mammals* 2nd edn. (Eds WF Perrin, B Würsig and JGM Thewissen) pp. 673–674. Academic Press, London.

Roux JP, Jouventin P (1987) Behavioural cues to individual recognition in the Subantarctic fur seal, *Arctocephalus tropicalis*. In: *Status, Biology, and Ecology of Fur Seals.* Proceedings of an International Symposium and Workshop Cambridge, England, 23–27 April 1984. (Eds JP Croxall and RL Gentry) pp. 95–103. NOAA Technical Report NMFS 51. Cambridge, UK.

Scientific Committee on Antarctic Research (SCAR) (1996) Recommendation SCAR XXIV-3. Concerning the re-introduction of indigenous species. *SCAR Bulletin* **125**, 9.

Shaughnessy PD (1999) *Action Plan for Australian Seals.* Environment Australia, Canberra.

Shaughnessy P (2000) Seal research in South Australia, 1999/2000: abundance of New Zealand fur seal pups on Kangaroo Island and the Neptune Islands. Report to National Parks and Wildlife, South Australia, Department of Environment and Heritage.

Shaughnessy PD, Goldsworthy SD (2015) Long-nosed fur seal: a new vernacular name for the fur seal, *Arctocephalus forsteri*, in Australia. *Marine Mammal Science* **31**, 830–832. doi:10.1111/mms.12203

Shaughnessy PD, Goldsworthy SD (2019) Dispersion of long-nosed fur seals (*Arctocephalus forsteri*) determined by tagging. *Australian Journal of Zoology* **67**, 173–179. doi:10.1071/ZO20032

Shaughnessy PD, Kerry KR (1989) Crabeater seals *Lobodon carcinophagus* during the breeding season: observations on five groups near Enderby Land, Antarctica. *Marine Mammal Science* **5**, 68–77. doi:10.1111/j.1748-7692.1989.tb00214.x

Shaughnessy PD, Briggs SV, Constable R (2001) Observations on seals at Montague Island, New South Wales. *Australian Mammalogy* **23**, 1–7. doi:10.1071/AM01001

Shaughnessy PD, Goldsworthy SD, Hamer DJ, Page B, McIntosh RR (2011) Australian sea lions *Neophoca cinerea* at colonies in South Australia: distribution and abundance, 2004 to 2008. *Endangered Species Research* **13**, 87–98. doi:10.3354/esr00317

Shaughnessy PD, Goldsworthy SD, Mackay AI (2015) The long-nosed fur seal (*Arctocephalus forsteri*) in South Australia in 2013–14: abundance, status and trends. *Australian Journal of Zoology* **63**, 101–110. doi:10.1071/ZO14103

Skinner J (1984) Research on the Ross Seal *Ommatophoca rossii*, in the King Haakon VII Sea, Antarctica. *South African Journal of Science* **80**, 30–31.

Species360 (2023) Zoological Information Management System (ZIMS). Minneapolis, USA. www.species360.org.

Spotte S (1991) *Sterilisation of Marine Mammal Pool Waters: Theoretical and Health Considerations.* United States Department of Agriculture. Animal and Plant Health Inspection Service. Bulletin No. 1797.

Spratt D, Beveridge I, Skerratt L, Spear R (2008) Appendix 2: guide to the identification of common parasites of Australian mammals. In *Medicine of Australian Mammals.* (Eds L Vogelnest and R Woods) pp. 639–658. CSIRO Publishing, Melbourne.

Stewart BS (2014) Family Phocidae (Earless Seals). In *Handbook of the Mammals of the World.* Vol. 4. *Sea Mammals.* (Eds DE Wilson, RA Mittermeier, A Martinez-Vilalta and D Leslie) pp. 120–183. Lynx Edicions, Barcelona.

Stirling I, Warneke RM (1971) Implications of a comparison of the airborne vocalisations and some aspects of the behaviour of the two Australian fur seals. *Arctocephalus* spp, on the evolution and present taxonomy of the genus. *Australian Journal of Zoology* **19**, 227–241. doi:10.1071/ZO9710227

Strahan R (1991) *Beauty and the Beasts: A History of Taronga Zoo, Western Plains Zoo and their Antecedents.* Surrey Beatty and Sons, Sydney.

Stringer EM, Van Bonn W, Chinnadurai SK, Gulland FMD (2012) Risk factors associated with perianesthetic mortality of stranded free-ranging California sea lions (*Zalophus californianus*) undergoing rehabilitation. *Journal of Zoo and Wildlife Medicine* **43**, 233–239. doi:10.1638/2010-0148.1

Taronga Zoological Park (1916) *Annual Report*. Trustees of Taronga Zoological Park, Sydney.

Tikhomirov EA (1975) Biology of the ice forms of seals in the Pacific section of the Antarctic. *Rapports Et Proces-Verbaux Des Reunions* **169**, 409–412.

Tong Marwood LJ (2025) Neoplasia. In *Current Therapy in Medicine of Australian Mammals*. Revised Edition. (Eds L Vogelnest and T Portas) pp. 299–320. CSIRO Publishing, Melbourne.

Tong LJ, Hulst F, Tobias G, Vinette Herrin K, Meagher P, Vogelnest L (2017) Visceral lepidopterism in captive otariids due to migration of caterpillar setae associated with systemic vasculopathy and death. In 49th American Association of Zoo Veterinarians Annual Conference. (Ed K Baer) pp. 113–114. Friscoe, Texas, United States of America.

Tripovich JS, Rogers TL, Canfield R, Arnould JPY (2006) Individual variation in the pup attraction call produced by female Australian fur seals during early lactation. *The Journal of the Acoustical Society of America* **120**, 502–509. doi:10.1121/1.2202864

Tripovich JS, Canfield R, Rogers TL, Arnould JPY (2008*a*) Characterization of Australian fur seal vocalizations during the breeding season. *Marine Mammal Science* **24**, 913–928. doi:10.1111/j.1748-7692.2008.00229.x

Tripovich JS, Charrier I, Rogers TL, Canfield R, Arnould JPY (2008*b*) Who goes there? Differential responses to neighbor and stranger vocalizations in male Australian fur seals. *Marine Mammal Science* **24**, 941–948. doi:10.1111/j.1748-7692.2008.00222.x

Tripovich JS, Canfield R, Rogers TL, Arnould JPY (2009) Individual variation of the female attraction call produced by Australian fur seal pups throughout the maternal dependence period. *Bioacoustics* **18**, 259–276. doi:10.1080/09524622.2009.9753605

Van Bonn W (2015) Pinnipedia. In *Fowler's Zoo and Wild Animal Medicine*. (Eds RE Miller and ME Fowler) pp. 436–450. Elsevier Saunders, St Louis.

Veterian Key (2017) Pinnipeds. https://veteriankey.com/pinnipeds (accessed 30 August 2022).

Vogelnest L (2025) Capture, physical and chemical restraint. In *Current Therapy in Medicine of Australian Mammals*. Revised Edition. (Eds L Vogelnest and T Portas) pp. 122–151. CSIRO Publishing, Melbourne.

Vogelnest L, Portas T (Eds) (2025) *Current Therapy in Medicine of Australian Mammals*. Revised Edition. CSIRO Publishing, Melbourne.

Vogelnest L, Edwards N, Ciaglia M, Carlini A, Slip D, Rogers T (2010) Anaesthesia of leopard seals (*Hydrurga leptonyx*) on the western Antarctic Peninsula. *Kokako* **17**, 31–33.

Warneke RM (1982) The distribution and abundance of seals in the Australasian region, with summaries of biology and current research. In *Mammals in the Seas*, FAO Fisheries Series No. 5. pp. 431–474. FAO, Rome.

Warneke RM, Shaughnessy PD (1985) *Arctocephalus pusillus*, the South African and Australian Fur Seal: taxonomy, biogeography and life history. In *Studies of Sea Mammals in South Latitudes*. (Eds JK Ling and MM Bryden) pp. 53–77. South Australian Museum, Adelaide.

Webber MS (2014) Family Otariidae (eared seals). In *Handbook of the Mammals of the World*. Vol. 4. *Sea Mammals* (Eds DE Wilson and RA Mittermeier). pp 34–101. Lynx Edicions, Barcelona.

Wickens P, York AE (1997) Comparative population dynamics of fur seals. *Marine Mammal Science* **13**, 241–292. doi:10.1111/j.1748-7692.1997.tb00631.x

Wierucka K, Pitcher BJ, Harcourt R, Charrier I (2017) The role of visual cues in mother-pup reunions in a colonially breeding mammal. *Biology Letters* **13**, 20170444. doi:10.1098/rsbl.2017.0444

Wierucka K, Pitcher BJ, Harcourt R, Charrier I (2018) Multimodal mother-offspring recognition: the relative importance of sensory cues in a colonial mammal. *Animal Behaviour* **146**, 135–142. doi:10.1016/j.anbehav.2018.10.019

Wierucka K, Barthes N, Harcourt R, Schaal B, Charrier I, Pitcher BJ (2019) Chemical fingerprints suggest direct familiarisation rather than phenotype matching during olfactory recognition in Australian sea lions (*Neophoca cinerea*). *Journal of Experimental Marine Biology and Ecology* **517**, 49–53. doi:10.1016/j.jembe.2019.06.001

Wilson DE, Mittermeier RA (2014) *Handbook of the Mammals of the World*. Vol. 4. *Sea Mammals*. Lynx Edicions, Barcelona.

Woinarski JCZ, Burbidge AA, Harrison P (eds) (2014) *The Action Plan for Australian Mammals 2012*. CSIRO Publishing, Melbourne.

Wood Jones F (1925) *The Mammals of South Australia*. Part III. R.E.E. Rogers, Adelaide.

Woods R, Cousins DV, Kirkwood R, Obendorf DL (1995) Tuberculosis in a wild Australian fur seal (*Arctocephalus pusillus doriferus*) from Tasmania. *Journal of Wildlife Diseases* **31**, 83–86. doi:10.7589/0090-3558-31.1.83

# 16
# Dolphins

*David Blyde, Guy Bedford, Justin Meager and Stephen M. Jackson*

## 1. INTRODUCTION

Cetaceans are a large group of specialised marine mammals that are divided into two subgroups: the Mysticeti (baleen whales, containing 15 species) and the Odontoceti (toothed whales, containing 79 species) (Wilson and Mittermeier 2014; Rosel *et al.* 2021). The dolphins of the family Delphinidae are a diverse group of odontocete whales that includes 38 species, which occur throughout the world.

Within Australian waters 20 species of delphinids are recognised, of which only a few have been kept in captivity for exhibit. These include the false killer whale (*Pseudorca crassidens*), Australian humpback dolphin (*Sousa sahulensis*), Australian snubfin dolphin (*Orcaella heinsohni*), Indo-Pacific bottlenose dolphin (*Tursiops aduncus*) and common bottlenose dolphin (*Tursiops truncatus*) (G. Bedford unpubl. data). Other species such as Risso's dolphin (*Grampus griseus*) and rough-toothed dolphin (*Steno bredanensis*) have been rescued and kept for short periods of time for treatment before being released again (D. Blyde pers. obs.). There are many examples of successful rescues, rehabilitation and releases of wild delphinids in Australia, but some of the deep-water species such as the Fraser's dolphin (*Lagenodelphis hosei*), striped dolphin (*Stenella coeruleoalba*) and Pantropical spotted dolphins (*Stenella attenuate*) are poorly suited to even short-term captivity and are therefore unsuitable for rescue and rehabilitation (G. Bedford pers. obs.). The popularity of exhibiting dolphins in Australia reached its peak from the 1950s to the 1970s (G. Bedford unpubl. data).

There have been changing public opinions on the appropriateness of keeping dolphins in captivity as a result of concerns that the very high intellect of dolphins makes them prone to boredom and that this is compounded by inadequate social group structures, severe enclosure space limitations and simple enclosure complexity that further contribute to boredom and chronic stress (e.g. Marino *et al.* 2007; Corkeron 2009; Grimm 2011; RSPCA 2019). As a result, the wild capture of dolphins for marine parks and oceanariums in Australia was banned in the 1990s and only Indo-Pacific and common bottlenose dolphins are held in captivity at two locations in Australia: Sea World, Queensland and Coffs Coast Wildlife Sanctuary, New South Wales.

Of the dolphin species that have been kept in captivity in Australia, the most commonly kept are the Indo-Pacific bottlenose and the common bottlenose dolphin. Survival and breeding rates of these species in captivity are comparable if not higher than in their wild counterparts (Wells and Scott 2002). Knowledge of bottlenose dolphin biology and husbandry stems not just from captive dolphins, but also from long-term research on wild populations. For example, there is a long history of research on provisioned (hand-fed) bottlenose dolphins in Australia, notably from over 30 years of research on Indo-Pacific bottlenose dolphins at Monkey Mia, Western Australia (e.g. Mann and Kemps 2003; Smith *et al.* 2008). This chapter focuses on the bottlenose dolphins (*Tursiops* spp.), which are arguably the most familiar of the smaller dolphins because they have been the most common species in captivity in Australia and abroad.

## 2. TAXONOMY

Of the 94 known cetacean species, 45 are found in waters around Australia. These include 10 species in the suborder Mysticeti (baleen whales) and 35 species in the suborder Odontoceti (toothed whales) (Bryden *et al.* 1998; Menkhorst 2001; Baker and Gynther 2023). The 20 species of delphinids that occur in Australian waters range in size from only 25 kg for an adult dwarf spinner dolphin to up to 10 tonnes for an adult killer whale – a 400-fold difference (Table 16.1). All bottlenose dolphins were

**Table 16.1.** Species of dolphin within the family Delphinidae found in Australia waters, their bodyweights and conservation status. Derived from Jefferson *et al.* (2008), Jackson and Groves (2015), Rosel *et al.* (2021) and Baker and Gynther (2023).

| Common name | Scientific name | Weight (kg) | IUCN status |
|---|---|---|---|
| Common dolphin | *Delphinus delphis* | 46–110 | Least concern |
| Pygmy killer whale | *Feresa attenuata* | 225 | Least concern |
| Short-finned pilot whale | *Globicephala macrorhynchus* | 1500–2000 | Least concern |
| Long-finned pilot whale | *Globicephala melas* | 1800–3000 | Least concern |
| Risso's dolphin | *Grampus griseus* | 400 | Least concern |
| Fraser's dolphin | *Lagenodelphis hosei* | Up to 210 | Least concern |
| Hourglass dolphin | *Lagenorhynchus cruciger* | 88–100 | Least concern |
| Dusky dolphin | *Lagenorhynchus obscurus* | 140 | Least concern |
| Southern right whale dolphin | *Lissodelphis peronii* | 120 | Least concern |
| Australian snubfin dolphin | *Orcaella heinsohni* | Up to 130 | Vulnerable |
| Killer whale | *Orcinus orca* | 6000–8000 | Data deficient |
| Melon-headed whale | *Peponocephala electra* | 190–228 | Least concern |
| False killer whale | *Pseudorca crassidens* | 2000 | Near threatened |
| Australian humpback dolphin | *Sousa sahulensis* | 150–280 | Vulnerable |
| Pantropical spotted dolphin | *Stenella attenuata* | 90–115 | Least concern |
| Striped dolphin | *Stenella coeruleoalba* | 156 | Least concern |
| Spinner dolphin[1] | *Stenella longirostris* | 23–78 | Least concern |
| Rough-toothed dolphin | *Steno bredanensis* | 155 | Least concern |
| Indo-Pacific bottlenose dolphin | *Tursiops aduncus* | 90–180 | Near threatened |
| Common bottlenose dolphin | *Tursiops truncatus* | 160–250 | Least concern |

[1]Includes Gray's spinner dolphin (*S. l. longirostris*) and the dwarf spinner dolphin (*S. l. roseiventris*).

classified as *Tursiops truncatus* until the Indo-Pacific bottlenose dolphin was recognised as a separate species in 1998. There has been some contention as to whether a third species exists in Australian waters (Charlton-Robb *et al.* 2011; Jedensjö *et al.* 2020), but the Society for Marine Mammalogy's Committee on Taxonomy currently only recognises the Indo-Pacific bottlenose dolphin and the common bottlenose dolphin (Rosel *et al.* 2021). Both species are known to hybridise in captivity (D. Blyde pers. obs.).

Bottlenose dolphins are distinguishable from other dolphins by the absence of obvious colour stripes and a mid-grey to brownish grey colour. They also have a relatively short, stout beak and a large, falcate dorsal fin. Common bottlenose dolphins are much larger than Indo-Pacific bottlenose dolphins, reaching up to 3.8 m in body length, compared with only 2.7 m for Indo-Pacific bottlenose dolphins. A dolphin reliably measured as greater than 3 m from the fluke notch to the tip of the rostrum (in a straight line) can therefore be identified as a common bottlenose dolphin. Indo-Pacific bottlenose dolphins also have proportionally larger flippers (15–23% of body length) than common bottlenose dolphins (14–18% of body length) (Best and Folkens 2007) and a proportionally larger rostrum. Teeth counts range from 21 to 29 for Indo-Pacific bottlenose dolphins and from 18 to 27 for common bottlenose dolphins. Common bottlenose dolphins do not have the extensive dark speckling on their ventrum that develops with age in Indo-Pacific bottlenose dolphins. The presence of urogenital and throat patches in common bottlenose dolphins, and a spinal blaze in Indo-Pacific bottlenose dolphins, can also be used to distinguish them (Jefferson *et al.* 2008; J. Meager unpubl. data).

In many areas of the world, such as the western North Atlantic, eastern Pacific and Peru, there are coastal and offshore ecotypes of common bottlenose dolphins (Hoelzel *et al.* 1998; Fruet *et al.* 2017). The coastal ecotype is generally smaller and lighter grey, whereas the offshore ecotype is larger and darker and occurs in larger populations (Wells and Scott 2002). In Australia and elsewhere across the Western Pacific and Indian oceans, it is thought that the Indian-Pacific bottlenose dolphin occupies the inshore ecological niche and the common bottlenose dolphin occupies the offshore ecological niche (Tezanos-Pinto *et al.* 2008).

Features that are important to identify the other delphinids in Australian waters include body colouration, body length, dorsal fin shape, flipper shape and rostrum. Taxonomists also use the number of pairs of teeth to help identify dolphins.

## 2.1. Nomenclature

**Australian dolphins**
Class: Mammalia
Order: Cetacea
Suborder Odontoceti
Family Delphinidae
Genus and species: *Tursiops aduncus* and *T. truncatus*

## 2.2 Subspecies

See Jackson and Groves (2015).

## 2.3 Recent synonyms

Synonyms for Australian dolphins can be found in Jackson and Groves (2015).

## 2.4 Other common names

See Baker and Gynther (2023).

# 3. NATURAL HISTORY

## 3.1 Bodyweight and morphometrics

The Indo-Pacific bottlenose dolphin and the common bottlenose dolphin have considerable geographical and clinal variations in morphology (Wells and Scott 2002). Common bottlenose dolphins are born at lengths of ~1–1.3 m, compared with ~0.85–1.12 m for Indo-Pacific bottlenose dolphins, and as discussed above, reach much larger sizes (see also Table 16.1). Important morphometric measurements for dolphins that can help to identify species include total length, fin height, length of the dorsal fin base, length of the flipper and girth on a transverse plane that intersects the axilla (Norris 1961). The latter measure of body girth, together with body length, can also inform on body condition (Best and Folkens 2007; Hart *et al.* 2013, 2017). Additionally, total body length can be used to estimate bodyweight (Ridgway and Fenner 1982; Bedford 2000). Recently, the distance from the blowhole to the dorsal fin has also been used to estimate body length of free-ranging bottlenose dolphins using stereo photogrammetry (van Aswegen *et al.* 2019).

There is no pronounced sexual dimorphism for either species, although male common bottlenose dolphins are often significantly larger (Tolley *et al.* 1995) and have proportionally taller dorsal fins than females (Rowe and Dawson 2008). Of the delphinids, sexual dimorphism in body size is the most striking in the short and long-finned pilot whales and the killer whale. Male pilot whales also develop a more pronounced melon and have a proportionally much larger dorsal fin (Perrin *et al.* 2002).

Adult male bottlenose dolphins sometimes have more scars than adult females, which has been used as an indicator of the sex of free-ranging dolphins (Rowe and Dawson 2009). Females can be distinguished at sea when they are in close association of dependant calves. Stranded or captive dolphins can be sexed if there is a clear view of the ventral genital slit, as females have a long external genital slit in which both the vagina and anus are located while males have separate genital and anal slits.

## 3.2 Distribution and habitat

The common bottlenose dolphin is globally widespread in temperate to tropical waters, whereas the Indo-Pacific bottlenose dolphin is limited to waters from eastern Africa, Australasia and northward to southern Japan (Natoli *et al.* 2004; Wells and Scott 2018). In the Australian region, the common bottlenose dolphin tends to inhabit deeper offshore waters around Australia (Allen *et al.* 2016).

In southern and eastern Australia Indo-Pacific bottlenose dolphins occur in coastal waters and often enter estuaries (Fury 2009), but they rarely enter estuaries in tropical Australia where they are mostly restricted to nearshore waters (J. Meager unpubl. data). Seasonal variation in the distribution of Indo-Pacific bottlenose dolphins has been reported, which has been attributed to variation in sea surface temperature (e.g. Hanf *et al.* 2022). Globally, common bottlenose dolphins demonstrate a variety of movement patterns from local residency to seasonal migrations (e.g. Wells and Scott 2002), but no resident populations have been described in Australian waters. Habitat use is also variable between individuals and communities of Indo-Pacific bottlenose dolphins (Ansmann *et al.* 2013, 2015; Kopps *et al.* 2014).

Many of the delphinid species around Australia are wide-ranging over neritic or open-ocean areas, but are rarely encountered near the coast. Other than the Indo-Pacific bottlenose dolphin, only the Australian humpback and the Australian snubfin have resident populations in coastal areas, although resident populations of common dolphins have been described (Ansmann *et al.* 2015).

## 3.3 Conservation status

Both species of bottlenose dolphins are listed as 'Least Concern' by Commonwealth (*Environment Protection and Biodiversity Conservation Act 1999*) and state legislation around Australia. Populations of Indo-Pacific dolphins in the Arafura and Timor Sea are listed on Appendix II of the Convention of Migratory Species. Under the non-statutory IUCN Red List, the Indo-Pacific bottlenose dolphin is listed as Near Threatened and the common bottlenose is listed as Least Concern. Both the Australian snubfin dolphin and the Australian humpback dolphin are listed as Vulnerable by the *Nature Conservation Act 1992* (Queensland) and by the IUCN.

No reliable population estimates are available for common bottlenose dolphins in Australian waters. Population estimates for Indo-Pacific bottlenose dolphins vary between study sites. Most studies report several hundreds of individuals in resident populations (e.g. Lukoschek and Chilvers 2008; Haughey *et al.* 2020), but population sizes are variable. For example, in Moreton Bay, Queensland, a small subpopulation of ~174 dolphins resides in the southern part of the Bay (Ansmann *et al.* 2013), whereas a larger population of ~446 is found in the northern Bay and a much larger population occurs in the adjacent oceanic area off Point Lookout (estimated to be 700–1000 individuals by Chilvers and Corkeron 2003). In most studies to date in northern Australia re-sighting rates have been too low for reliable estimates (e.g. Palmer *et al.* 2014), suggesting that home ranges are larger than study sites. A decline in reproductive rates and survival in the bottlenose population in western Shark Bay, Western Australia, was described after a marine heatwave (Wild *et al.* 2019a).

## 3.4 Diet in the wild

The diet of free-range bottlenose dolphins is well understood from studies of the stomachs of stranded animals and observations of their foraging behaviour in the wild. The primary prey of both species are fishes and cephalopods (Amir *et al.* 2005). Common bottlenose dolphins demonstrate consistent preferences for scianids, scobrids and mugilids (Wells and Scott 2018). Indo-Pacific bottlenose dolphins consume a variety of prey, including demersal, benthic and pelagic fishes, cuttlefish and squid (Lear and Bryden 1980; Cockcroft and Ross 1990; Smith and Sprogis 2016; Sprogis *et al.* 2017; J. Meager unpubl. data). Schooling fishes such as mullet and tailor are also often eaten (Lear and Bryden 1980). Studies overseas have reported a high proportion of soniferous fishes in the stomachs of common bottlenose dolphins (Gannon and Waples 2004), suggesting that passive listening could be used to locate prey.

Several studies have suggested that the diets of bottlenose dolphins vary between location (Gannon and Waples 2004) and life-history stage. Lactating females typically have the highest energy demands (Cheal and Gales 1992), which manifests in a greater diversity of prey species (Cockcroft and Ross 1990) and a tendency to feed in the nearshore zone (Wells and Scott 2018). In contrast, mature males tend to feed on generally larger prey further offshore (Cockcroft and Ross 1990).

Bottlenose dolphins can develop foraging specialisations that are socially transmitted (Krützen *et al.* 2005; Wild *et al.* 2019b; Wild *et al.* 2020) and they are renowned for complex foraging behaviours such as 'mud ringing' whereby dolphins create a ring around fish by disturbing the mud with their flukes (Torres and Read 2009), foraging on trawler discards (Chilvers and Corkeron 2001), 'kerpunking' where tail slapping is used to startle fishes (Connor *et al.* 2000a) and strand feeding where dolphins deliberately strand themselves to hunt fish at the water's edge and use their powerful flukes to return to the water (Jiménez and Alava 2015). In Shark Bay, Western Australia, bottlenose dolphins are thought to use marine sponges as foraging tools (Smolker *et al.* 1997). Foraging specialisations and home ranges are often associated with family groups.

## 3.5 Longevity

### 3.5.1 Wild

Wild common bottlenose dolphins can live up to 58 years of age, with females typically living up to 50 years of age and males 40–45 years of age. The Indo-Pacific bottlenose dolphin typically lives ~40 years, but some records suggest they may live up to 50 years (Wang *et al.* 2014).

### 3.5.2 Captivity

The maximum age for a common bottlenose dolphin in the US Navy Marine Mammal Program was reported to be 52 years old (Venn-Watson *et al.* 2015). A similar longevity of Indo-Pacific bottlenose dolphins in captivity has been reported (47–50 years; Koga *et al.* 2019). An Australian humpback dolphin was caught in Moreton Bay in 1966 at an estimated 3–4 years of age and was kept in captivity for 54 years at Sea World until she died in 2019, meaning that she was likely to be have been 57–58 years old at death (D. Blyde unpubl. data).

### 3.5.3 Methods used to determine the age of adults

The most accurate ageing of wild bottlenose dolphins occurs when known calves are monitored from birth in long-term sight-re-sight studies that use distinguishing features on dorsal fins to identify individuals (e.g. Karniski *et al.* 2018), but most studies are relatively short-term. The longevity of bottlenose dolphins in Australia has also been recorded in the long-running provisioning programs at Monkey Mia, Western Australia, and Tangalooma, Queensland, where dolphins have been observed from birth. Morphometrics may also give some indication of the age of individual animals. Animals increase in length until maturity. Teeth wear as animals age, although diet can affect the wear of teeth. Many Indo-Pacific bottlenose dolphins tend to develop more ventral spots as they age.

Several other techniques have been developed to estimate the age of dolphins, including: (1) extracting a tooth and examining the dental growth layers (Hohn *et al.* 1989); (2) dental radiography and assessing the pulp:tooth area ratio (Herrman *et al.* 2020); (3) radiography of the pectoral flipper to assess the growth rate closures of individual bones within the flipper (Barratclough *et al.* 2019); and (4) epigenetic ageing (i.e. using a biopsy dart) (Barratclough *et al.* 2021), which can allow dolphins to be aged in the wild without the need for capture.

## 4. HOUSING REQUIREMENTS

### 4.1 Enclosure design

Most dolphins adapt well to a captive environment. They are active, intelligent and inquisitive animals and their housing and husbandry must meet their behavioural and physical requirements. Housing standards have been developed for captive cetaceans in Queensland (Department of Agriculture, Fisheries and Forestry, Queensland 1992) and New South Wales (Crane 1994). These standards cover such things as minimum space, containment structures, water and climatic conditions.

Successfully housing bottlenose dolphins requires carefully designed enclosures that include specialised areas for feeding purposes, veterinary/husbandry equipment and the ability to individually separate animals from conspecifics if required. Bottlenose dolphins require saltwater enclosures (concrete pools or natural tidal basins/lagoons) that incorporate the following features.

(1) Sufficient water depth to promote vigorous physical activity, while also allowing individuals to be separated during times of aggression, breeding and health issues.
(2) Large surface area to ensure the ability of nursing mothers to swim while suckling calves.
(3) A medical/health pool that includes in its design:
   - the ability to be rapidly drained of water or
   - feature a platform that can be raised above the water surface to allow animal care and veterinary staff the ability to physically restrain an animal if required.
(4) Natural or artificial sea water with a salinity range of 1.026–1.036 ppt and a pH range of 7.5–8.4.
(5) Protection against adverse weather events.
(6) Vehicular access.
(7) Horizontal land area for fish kitchen, staff amenities and other associated infrastructure.

An increasing number of studies are being undertaken on the requirements of dolphins being held in captivity. Several studies have been undertaken on enclosure construction and open and closed pools. In the general the findings have revealed that:

- a significant positive relationship exists between behavioural diversity and the number of habitats to which common and Indo-Pacific dolphins had access (Miller *et al.* 2021a)
- enclosure depth is important; a study found that common dolphins utilised moderate depth locations (5.5 m deep; 68%) in the enclosure rather than shallow (0.25–3.5 m; 29%) or deeper areas (> 8 m; 3%) (Shyan *et al.* 2002)
- common bottlenose dolphins also display higher activity levels in larger enclosures than in smaller enclosures (Bassos-Hull and Wells 1997)
- dolphins in closed facilities spent less time swimming and more time floating compared with animals in open facilities. Similarly, dolphins in closed facilities spent more time swimming in circles and interacting socially, while those in open facilities swam in linear patterns and spent less time

socially interacting (Ruiz *et al.* 2009; Blasio *et al.* 2012; Ugaz *et al.* 2013).

Ideally, dolphins should be kept in natural sea water. Artificially produced salt water can be used, but is expensive and depends on the availability of the constituents. Dolphins do better in natural, outdoor enclosures. Indoor enclosures are associated with a higher incidence of disease, particularly respiratory disease. Dolphin enclosures should be designed to ensure that there are no areas of inadequate water circulation. Additionally, filtration and disinfection of the body of water may be required to limit algal growth and bacterial loads. The filtration system should remove animal waste matter, prevent the growth of microorganisms and maintain water clarity. If chemicals such as chlorine, bromide or ozone are used to ensure water quality, daily water analysis is required to ensure appropriate levels are maintained. Attention to the ambient acoustic levels must be considered (both above ground and underwater) when housing bottlenose dolphins, particularly noise generated by mechanical systems, which may be transmitted underwater.

Pool construction can be either natural or artificial. If artificial (concrete or fibreglass), the enclosure should be durable, watertight, non-porous, non-abrasive, non-toxic and easily cleaned and disinfected. Sharp projections and loose fittings should be avoided. Artificial pools have the advantage of allowing easier maintenance and better control of water quality than natural pools. They also offer less opportunity for animals to encounter foreign objects, such as stones, that may be ingested. Natural pools can be constructed much larger for less cost than artificial pools. The pools should be large enough to accommodate the natural behaviours of the animals and deep enough for the animals to dive and avoid the effects of ultraviolet radiation. Bottlenose dolphins should have access to natural sunlight; however, areas of shade should also be available.

Shade should be supplied and the temperature of the water should be maintained between 19.5°C and 28.5°C. If chemical treatment is used to maintain water quality and cleanliness, it should not cause harm to the animals. Many different systems can be used to disinfect or sterilise the water. Chlorine dosing and ozone are the most common. Both systems are suitable if water quality is monitored carefully and regularly. As dolphins live entirely in water, water quality needs to be adequate in order to maintain healthy animals. Table 16.2 below shows important water parameters and their acceptable levels.

**Table 16.2.** Ideal water parameters for cetacean pools.

MPN = most probable number

| Water parameter | Acceptable range |
|---|---|
| Temperature | 19.5–28.5°C |
| pH | 7.5–8.4 |
| Salinity | 1.026–1.036 ppt |
| Alkalinity | 80–100 ppm |
| Coliform bacteria count | < 500 MPN/100 mL |
| Chlorine level | < 0.5 ppm |

### 4.1.2 Segregation and/or isolation

Bottlenose dolphins are highly social (Connor *et al.* 2000b; Tsai and Mann 2013), but at times animals will need to be separated from their conspecifics for health, husbandry, breeding or social reasons. If pregnancies are to be avoided and segregation between the sexes is required, the design of the enclosure should incorporate enough separate areas whereby the sexes can be moved from one section to another without either group occupying the same body of water at the same time. To achieve such segregation, a minimum of three enclosures with interconnecting gates or channels between areas is required.

### 4.1.3 Medical/health enclosure

A medical or health pool should be incorporated into the overall design of the facility (Fig. 16.1). It should be a smaller pool that is shallow with a flat bottom to facilitate it being easily netted off to assist with the capture of an individual animal. Additionally, this pool would ideally contain a lifting platform that allows animal care and veterinary staff to elevate an animal out of the water for medical examinations and procedures (Fig. 16.2). Additionally, the medical/isolation pool should incorporate some form of shade cover. This is of particular importance to avoid sunburn if an animal becomes lethargic and remains stationary at the surface of the pool for extended periods of time.

### 4.1.4 Design and position of feeding stations

Bottlenose dolphin enclosures require feeding stations that facilitate animal care staff to feed animals, monitor fluctuations in appetite and observe changes in behaviour and train the animals if required. Feeding stations should be large enough to accommodate the feeding of several animals at the same time. The feeding stations should be positioned around the enclosure in areas where there is sufficient depth for the animals to be able to assemble vertically in the water column. Stations should not be located where the animals are looking directly into the sun while feeding. Ideally, feeding stations should be protected from sunlight and rain.

**Fig. 16.1.** Medical pool used for dolphin management with lifting platform. Photo from the collection of Guy Bedford.

**Fig. 16.2.** Dolphin being stretchered on a lifting platform. Photo from the collection of Guy Bedford.

## 4.2 Holding area design

A holding area should be incorporated into the design of the pools so that animals can be separated off for medical procedures or to alleviate intraspecific aggression.

## 4.3 Spatial requirements

Spatial requirements proposed for keeping *Tursiops* dolphins can be found within Table 16.3.

## 4.4 Position of enclosures

Ideally, enclosures should be positioned in quiet, well-ventilated areas. Training and feeding areas should be positioned so that animals are not looking into the sun during training/feeding sessions.

## 4.5 Weather protection

If animals are housed in sea pens, it is recommended that an emergency accommodation and evacuation plan be developed for use during extremely adverse weather events such as tropical storms and cyclones. Such emergency accommodation may utilise domestic or public swimming pools or inflatable pools (Figs 16.3-16.5). Evacuation plans must be commenced with sufficient time for transporting animals before the weather event occurring.

## 4.6 Temperature requirements and water quality

### 4.6.1 Temperature

Water samples should be obtained daily to determine the temperature and water quality. The natural water temperature for the species in the wild should be reproduced as closely as possible. Free-ranging dolphins in south-east Queensland experience annual sea surface

**Table 16.3.** Minimum areas of enclosures recommended for Tursiops for up to five animals.

Derived from Crane (1994).

| Pool type | Pool volume ($m^3$) | Minimum dimensions (m) | Minimum surface area ($m^2$) | Volume per additional animal ($m^3$) |
|---|---|---|---|---|
| Primary pool | 1400 | Lengths not narrower than 14 m or shallower than 2.5 m. At least half the pool must be at > 3 m deep | 245<br>Additional 49 $m^2$ for each additional animal | 300 |
| Secondary pool | 600 | Lengths not narrower than 12 m or shallower than 1.8 m. At least half the pool must be at > 2.5 m deep | 245<br>Additional 49 $m^2$ for each additional animal | 200 |
| Isolation pool | | Not narrower than 7 m or shallower than 2 m | 49 | |

**Fig. 16.3.** Domestic swimming pool used as emergency dolphin enclosure. Photo from the collection of Guy Bedford.

**Fig. 16.4.** Inflatable pool used as emergency dolphin enclosure. Photo from the collection of Guy Bedford.

**Fig. 16.5.** Public swimming pool used as emergency dolphin enclosure. Photo from the collection of Guy Bedford.

temperatures ranging from 19.5°C to 28.5°C and therefore this is considered to be the optimum water temperature range for captive animals from this region. Bottlenose dolphins can adapt their thermal neutral zone (the water temperature at which the metabolic rate is lowest) by varying blubber thickness. In this adaptation the food intake can increase or decrease depending on the direction of the temperature change. It is likely that this response is to maintain body temperature stability (G. Bedford unpubl. data).

Unlike their free-ranging counterparts, captive bottlenose dolphins cannot migrate to avoid extreme seasonal variations in water temperatures and the relatively small volume of water in captive facilities can be subject to large temperature variations. Captive animal experience in south-east Queensland has demonstrated that bottlenose dolphins can become heat stressed at water temperatures above 30°C. At this temperature animals become lethargic, demonstrate a significantly decreased interest in feeding and lose bodyweight in response to a reduced thermogenic demand. As such, when water temperature decreases as winter approaches, dolphins housed in warmer water temperatures during the summer months may be underweight at the time when thermogenic demand increases. This may have health implications because some animals may not experience sufficient weight gain during the winter months, particularly if the temperature of a small body of water falls below the optimum range. The most critical component of water temperature within the enclosures therefore is not just the absolute temperature but also the temperature range.

### 4.6.2 Coliform bacteria count

Bacterial counts should be performed at least weekly. The coliform bacteria count of the enclosure must not exceed 500 MPN (most probable number) per 100 mL. If the count exceeds this reading it must be rectified immediately. If chemical treatment is used to maintain water quality and cleanliness, it must not cause harm to the animals.

### 4.6.3 Bromine

The residual bromine in the water should not exceed 1.0 mg/L and should be tested at least daily. If chlorine rather than bromine is measured it should not exceed 0.5 mg/L.

### 4.6.4 Salinity

Salinity should be within the range of 26–36 ppt. Low salinity for prolonged periods of time can lead to skin and eye disease and mild electrolyte abnormalities (Deming *et al.* 2020).

### 4.6.5 pH

The pH of the water should be maintained between 7.5 and 8.4 and should be tested daily.

### 4.6.6 Ammonium (NH4)/ammonia (NH3)

Ammonia levels for mammals should remain below 0.7 mg/L and be tested daily. Ammonia is produced when faeces and urine are broken down during decomposition. Some other sources of ammonia in the water are from fish faeces, dead fish, decaying plant matter and uneaten food. Ammonia is very toxic to fish, but relatively non-toxic for mammals. Ammonia is broken down into a safer compound by nitrifying bacteria. This process is called the nitrogen cycle. Ammonia is broken down into nitrite, which is broken down into nitrate, which is broken down to nitrogen gas that evaporates into the air. Ammonia does not appear to cause systemic disease in mammals, but high levels can cause irritation to mucous membranes and eyes.

### 4.6.7 Phosphorus (P)

Increasing phosphate levels in water are not usually toxic to dolphins, but will increase algal growth. Phosphorus should be kept below 0.07 ppm.

### 4.6.8 Oxidation reduction potential (ORP)/ redox

Redox potential is a measure of the tendency of a chemical species to acquire electrons from or lose electrons to an electrode and thereby be reduced or oxidised respectively. ORP is essentially a measurement of sanitiser effectiveness in water bodies. The higher the ORP the more effective is the disinfecting agent.

ORP can effectively remove chlorine and heavy metals (including iron, lead, mercury, magnesium, hydrogen sulfide, calcium carbonate and chromium) from the water and can also help control microorganisms (e.g. bacteria, algae and fungi) and other inorganic contaminants.

## 4.7 Life support systems

A guiding principle for life support systems is to over-engineer the required component. As a long-term investment, having equipment that runs at 80% of its maximum capacity will be cheaper from an ongoing power cost and will last longer than equipment that is under-engineered and running constantly at 100% capacity (Jones 2014).

Adequate life support is essential for any dolphin enclosure. Aquatic animals are at the mercy of the water around them, even more so than terrestrial animals are influenced by the air they live in. The waste produced by an individual dolphin is quite small as they are very efficient in converting food to energy.

### 4.7.1 Filtration

Several options can be used to provide filtration of the water to maintain its quality.

- Sand filters are commonly used in large-scale oceanariums. They are filled with a filtration medium, usually sand. The water from the tank is pumped under pressure through this medium and as the water passes through the media, particulate matter is retained. A small amount of biological filtration will also happen once the filter has been in operation for some time due to the particulate matter building up (see below). Every filter has a different ideal working pressure (refer to the manufacturing details of individual models). Once the pressure reaches a specific level the filter will need to be backwashed by changing the valves to push water through the filter in the reverse direction, disturbing and churning up the sand and sending the dirty water to waste.
- A protein skimmer is a mechanical system to remove organic matter and to reduce nitrogen compounds. The water is mixed with air and pumped into a reaction chamber. Proteins and solids become attached to the air bubbles and flow upward and concentrate in foam that is then eliminated from the system. The foam generators on these systems can also incorporate ozone that increases the oxidation of organic material (von Fersen and Walb 2018).
- Biological filtration is the action of bacteria on nitrogenous compounds. It does not remove solids or affect other water contaminants. It facilitates the nitrogen cycle, converting ammonia to nitrite then nitrate. Biological filtration is usually achieved by passing water through a bio-chamber that has some sort of bio-ball or bio-media. These balls or media have a large surface area to which the nitrifying bacteria can attach.

### 4.7.2 Disinfection

If chemical treatment is used to maintain water quality and cleanliness, it should not cause harm to the animals.

- Chlorine - the most common form of disinfection. Chlorine is a strong oxidant that rapidly kills many harmful microorganisms. However, chlorine is a toxic gas and has significant risks associated with its use. This problem is avoided by using hypochlorite solution, which is a relatively inexpensive solution that releases free chlorine when it is dissolved in water. Calcium hypochlorite is a solid form of chlorine that is available however, greater human contact is needed to use this form of the disinfectant and therefore it is more dangerous to use. The disadvantage of chlorine is that it reacts with natural organic compounds in the water to form potentially harmful byproducts such as trihalomethanes and haloacetic acid, both of which are carcinogenic in large quantities. Chlorine is also limited in its effectiveness against protozoal cysts and giardia.
- Chlorine dioxide - a faster acting disinfectant than elemental chlorine but works in a similar fashion.
- Chloramine disinfection - not as strong an oxidant as chlorine, but provides a longer lasting residual. In addition, it does not form potentially harmful byproducts such as trihalomethanes and haloacetic acid. Chlorine can be converted to chloramines by the addition of ammonia.
- Ozone - has been applied to marine mammal life support systems for decades. Its success as an oxidant and disinfectant is well documented. In many marine mammal facilities ozone has superseded chlorine as the primary chemical treatment. An ozone-generator charges oxygen to create the ozone, which is an oxidiser. Once in the water, it attacks and neutralises the contaminants and then reverts back to oxygen. It also enhances the efficiency of waste removal systems such as protein skimmers by coagulating the suspended solids, which are then more easily removed via the air bubbles. The size of the enclosure, the number of animals within the enclosure, the salinity as well as the temperature all determine the amount of ozone that will be needed to effectively treat the water. If inhaled by humans, ozone gas can cause damage to the lungs, so this system requires the installation of ambient ozone gas monitors to detect any leaks (Coakley and Crawford 1998).
- Ultraviolet (UV) light irradiation - UV light is used to clean water and works because the light penetrates organisms' cell walls and disrupts their genetic material and inhibits reproduction. Similar to ozone, the efficacy of UV light depends on the time of contact, so the speed of water flow is crucial. The effectiveness of the treatment is also affected by the turbidity of the water. Advantages of UV light compared with chemical disinfection include: no toxic byproducts; no volatile compound emissions or toxic air emissions; no requirement for storage of dangerous material; and not changing the chemical quality of the water. As a result, it is the safest among the current techniques for cleaning water (von Fersen and Walb 2018), but is expensive to install.

## 4.8 Substrate

Natural substrates such as sand are acceptable, but captive dolphins have a tendency to ingest foreign objects and as such small rocks should be avoided in the substrate. Concrete bottoms with no substrate are also acceptable, but not as aesthetically pleasing.

## 4.9 Enclosure furnishings

Dolphin pools typically have little variety or complexity in the underwater environment due to the need for the tank to filter and circulate water and to maintain water hygiene (RSPCA 2019). There has, however, been an increased recognition of the importance of environmental enrichment within dolphin enclosures. As a result, some facilities have more natural substrates, including sand substrates, and obstacles such as boulders, islands and coves (RSPCA 2019). It should be noted that captive dolphins have a tendency to ingest foreign objects, including rocks, sand and vegetation, and as such small rocks should be avoided in the substrate.

Bottlenose dolphin enclosures should have pools that interconnect via channels and/or gates. Nets should not

**Fig. 16.6.** Dolphin enclosure gate between pools. Photo from the collection of Guy Bedford.

be used to partition adjoining pools as they represent a serious risk of entanglement. Gates need to be constructed robustly because dolphins have been known to destroy gates in order to gain access to another group of animals (Fig. 16.6). Swim throughs and enrichment items such as artificial kelp can be incorporated into the enclosure. Care should be taken that the furnishings cannot be ingested or act as entanglement threats.

## 5. GENERAL HUSBANDRY

### 5.1 Hygiene and cleaning

All food preparation areas, utensils and food containers must be cleaned with an appropriate detergent immediately after use. All such items should be sanitised with an appropriate disinfectant daily. Provision for the safe removal of excess food, waste, trash and debris must also be made. Disposal facilities must be operated in such a manner that will minimise odours, disease hazards and pest infestation. The design of the enclosures should facilitate easy cleaning of the pools to remove excess algal growth, uneaten food or detritus. Drainage from the surrounding pool areas should ensure that water from these zones does not flow into the enclosure.

### 5.2 Record keeping

It is important to establish a system whereby the health, condition and reproductive status of captive dolphin population are routinely monitored and recorded:

- identification ( all individuals should be identifiable)
- results of veterinary examinations and treatments provided, including contraception medications
- behavioural changes or problems
- reproductive behaviour or condition
- bodyweights and measurements
- changes in diet and daily food intake (DFI)
- any supplements, vitamins and minerals given
- movements of individuals between enclosures or institutions
- births, with dam and sire recorded if known
- deaths with necropsy results.

The collection of information on physical and behavioural patterns of individual dolphins can contribute greatly to the husbandry of these species. It also allows individual histories to be transferred to other institutions if required and facilitates a cooperative approach to data collection and management across institutions. ZIMS, the Zoo Information Management System (Species360 2023) can be used for recording general information on age, sex, parentage, births, identification, bodyweights, transfers and deaths, with separate ZIMS modules for veterinary information and studbooks. These systems have been developed by Species360, which is part of the Conservation Planning Specialist Group (CPSG) software. Records pertaining to the animal's physical environment should include: water temperatures, pH, salinity tests, levels of disinfectants, life support system maintenance and cleaning regimens.

### 5.3 Methods of individual identification

In the Australian captive environment, bottlenose dolphins are usually identified through a combination of individual body markings (including colour, pigmentation, and speckling) and unique dorsal or tail fin profiles. Additionally, an experienced handler can also recognise an animal by the way it swims, or by its unique facial appearance.

PIT tags (microchips) have been successfully utilised in overseas captive facilities as a means of permanently identifying individuals. PIT tags are usually placed intramuscularly, anteriorly below the dorsal fin on the right side of the animal to identify males and the left side of the animal to identify females.

## 6. FEEDING REQUIREMENTS

### 6.1 Captive diet

A variety of fish species should be fed in captivity, but poisonous or spiny fish (such as puffer fish) should be avoided. Each animal should be fed at least three times a day with a variety of species, ideally the normal prey species of the dolphins in the particular area from which they originate. Each fish species has a different nutritional content that will vary with sex, season and reproductive stage.

Captive dolphins should be fed thawed frozen whole fish rather than fillets or cleaned, gutted fish. The calcium in the bones, vitamin A in the liver and the fats present in the gonads of fish are all essential in providing a balanced diet. As a guide, captive animals should be fed 5–7% percent of their bodyweight daily. This should be increased in pregnant, lactating and growing animals. Captive dolphins, like other species in captivity, have a tendency to become obese and therefore bodyweight should be monitored and controlled.

Ideally, fish should be frozen for at least 6 weeks to kill parasites. However, appropriate storage and thawing of frozen fish are important to maintain nutrient content and quality (Geraci and St Aubin 1980; Geraci 1986). The ideal thawing conditions to maintain hydration and nutrients is in air at 4–8°C. Thawing fish in water will leach essential vitamins and minerals out of the fish.

Some fish contain thiaminases that destroy thiamine. Fat-soluble vitamins are also readily lost. All fish fed should be supplemented daily with vitamin E (50 IU/kg of fish fed) and thiamine (20 mg/kg of fish fed). These supplements can be given as tablets or capsules in the fish. Because of the short transit time of food in the gastrointestinal tract, supplements should be easily digestible. Human products are superior to the commercially available veterinary products.

An artificial diet has been developed in the USA. It is formulated using traditional fish feed industry ingredients. It has been trialled on dolphins and was found to be palatable. However, the animals needed 25% more of the artificial diet on a daily weight basis in order to maintain bodyweight (Wright *et al.* 2005). It could be useful if storage space becomes an issue or where supplies of fish are disrupted. Care should be taken when captive dolphins are fed unusual non-prey items. The feeding of some North Atlantic fish such as herring (*Clupeidae* spp.), capelin (*Osmeridae* spp.) and mackerel (*Scombridae* spp.) has been associated with nephroliths and liver disease in captive animals (Le-Bert *et al.* 2018).

It should be noted that when free-ranging cetaceans are brought into captivity, they may not eat dead fish for several days, which may necessitate force feeding if the animal is in poor body condition. Force-feeding is easily accomplished by restraining the animal, opening the jaws and introducing the fish deep into the oesophagus. Because of the nature of their anatomy there is no risk of entering the respiratory tract.

### 6.2 Supplements

Vitamin B and vitamin E are necessary supplements for captive-based diets. These two vitamins are destroyed in the freezing process and all captive diets are based on thawed frozen whole fish.

### 6.3 Presentation of food

Bottlenose dolphins should be hand fed a minimum of three times during a period of 24 h. It may be necessary to increase the frequency of feeds during times of illness, lactation and parturition. Food should be kept refrigerated or iced before feeding in order to reduce microbial build up and/or nutrient loss. Fish should ideally be 15–20 cm in body length and fed head first, one fish at a time to the animals to avoid the dorsal and caudal spines being lodged in the animal's oesophagus. As a general guide, bottlenose dolphins require between 5–7% of their bodyweight in daily food intake. Nutritional intake will change throughout the year in response to changes in water temperature. Nutritional demand will be increased in growing animals, pregnancy and lactation. Animals should be conditioned to eat in an attentive, yet non-aggressive manner. Aggression with conspecifics over food is considered inappropriate behaviour and should be addressed through conditioning and/or a change in daily food target. Satiation should not be allowed to occur so as to provide animal care staff with baseline information of a particular animal's eating habits. Satiation can provide misleading information on the initial health assessment of an animal.

## 7. HANDLING AND TRANSPORT

### 7.1 Timing of capture and handling

Elective captures and handling should be scheduled for early morning to give staff the opportunity to proceed carefully and monitor the animal after the procedure.

This is also the time when there are generally less visitors. It is also generally cooler in the mornings than later in the day particularly in temperate and tropical regions.

Jewellery and watches should be removed as they present an entanglement risk and may cause damage to animals.

### 7.2 Capture equipment

Depending on the design of the enclosure the following equipment may be needed:

- nets (can be an entanglement hazard). If nets are being utilised, knives should be available to facilitate disentanglement if required
- stretchers
- foam mattresses
- towels or sheets (wet and placed on the dorsum of the animal to prevent sunburn)
- shade of some description (e.g. marquees if permanent shade is not available)
- buckets and/or hoses/sprayers/sponges to keep the animals wet (this should be salt water as fresh water may cause skin and eye issues)

- food fish if reinforcement is considered appropriate
- sufficient number of experienced staff.

## 7.3 Capture and physical restraint

Handling and restraint of captive cetaceans is frequently required to facilitate moving them between facilities or conducting routine husbandry or veterinary procedures. There is always potential for injury to both the animal and staff so it is important that these procedures are carried out by appropriately trained personnel. Many procedures can be performed using operant conditioning, making them safer and less stressful for both the animals and personnel. A training program to facilitate transport and handling should be incorporated into the husbandry program of all captive dolphins.

In general, the capture and restraint of bottlenose dolphins requires coercing the animals into smaller and shallow isolation/medical areas within an enclosure. This can be achieved through regularly feeding the animals within the medical area so there is a willingness by the dolphins to enter such a location. Oftentimes however, animals displaying poor eating because of illness will be reluctant to enter an isolation pool and will need to be coerced by the use of a haul net. Haul netting operations can be hazardous for animals and personnel and should only be performed with an appropriate number of experienced staff available to assist an animal out of the net should it become entangled. A haul net should be of suitable length and depth so that the entire width and depth of the pool is able to be netted without excess net being available for an animal to become entangled. Once suspended, the net is slowly drawn towards the isolation pool, coercing the animal/s to swim into it.

After the dolphin has entered the smaller isolation/medical pool, personnel can similarly coerce the animal to swim onto a submerged section of a lifting platform and then raise it above the surface of the water for restraint purposes. Alternatively, two or three staff can enter the shallow netted section of the pool and physically restrain an animal (Fig. 16.7). Typically, one staff member will hold their arms around the body with their hands holding the pectoral flippers close to the torso while the additional staff will hold the stock area behind the dorsal fin in a similar fashion.

Once restrained, dolphins can be placed on a flat smooth surface. If the restraint procedure will be longer than 10–15 min, they should be positioned on foam pads, ~7.5–10 cm thick, or mattresses to alleviate pressure on the stomach and lungs. Even wild dolphins tolerate

**Fig. 16.7.** Restraining a dolphin. Photo from the collection of Guy Bedford.

physical restraint for up to 3 h without undue stress occurring. However, most procedures can be completed within 90 min.

Care should always be taken with the tail because it can cause significant damage to handlers if they are hit by it. Dolphins can bite and their teeth are sharp. They can also inflict damage on themselves if not appropriately restrained. They should be kept cool and wet during the restraint procedure. Care should be taken not to get water into the blow hole if they are being sprayed or hosed with water. They should not be subjected to long periods of direct sunlight as they can be damaged by solar radiation and suffer severe solar dermatitis. Even 30 min of direct sunlight can cause solar damage in Australian conditions. If the procedure is likely to take longer than this, a shade shelter should be erected.

A restrained dolphin should be placed on a foam mattress with additional mattresses placed over the body of the animal. Animal care staff can also place additional weight on the animal if required. Care should be exercised to make sure the animal can expand its thorax to breathe and there is not too much pressure on the abdomen. The blowhole should remain unobstructed throughout the procedure.

## 7.4 Weighing and examination

The bodyweight of a bottlenose dolphin can be an indicator of health or illness. Dolphins that are ill can lose as much as 10 kg in 1 week and it is prudent to incorporate regular recording of bodyweights into the husbandry routine – ideally fortnightly or monthly. There are various methods used in the captive environment to obtain such bodyweights. One method involves collecting the animal and placing it in a purpose-designed sling and suspending it under a set of hanging scales (Fig. 16.8). Bottlenose dolphins can be conditioned to positively accept such procedures.

In recent years, the most common method to collect bodyweights is to train a dolphin to slide on to a purpose-designed set of platform scales (Fig. 16.9). This procedure can be performed with a minimum of staff and on a daily basis if required.

Animals can be trained to haul out onto a weigh platform voluntarily. Ideally, weighing should occur every month. Pregnant animals in the last two trimesters of pregnancy should not be weighed.

If weigh scales are not available an equation can be used to estimate normal bodyweight (ENBW) from measurements of the length of an animal (Bedford 2000). For common bottlenose dolphins ENBW = (1.616 × body length in cm) - 236.42 and for Indo-Pacific bottlenose dolphins Log ENBW = (0.1389 × Log body length$^{3.2019}$).

**Fig. 16.8.** Weighing a dolphin utilising suspended scales. Photo from the collection of Rodney S Able. Used by permission of Grant Able.

**Fig. 16.9.** Weighing a dolphin utilising platform scales. Photo from the collection of Guy Bedford.

Weight relationships for common bottlenose dolphin can also be calculated using (Best and Folkens 2007):

- $W = 11.32L^{2.9869}$, where W = weight in kilograms and L = total length in metres.
- This implies that a 2.0 m animal weighs ~90 kg and a 3.0 m animal 301 kg.

Weight relationships for Indo-Pacific bottlenose dolphin can also be calculated using (Best and Folkens 2007):

- $W = 12.365L^{2.9495}$, where W = weight in kilograms and L = total length in metres.
- This implies that a 1.5 m animal weighs ~41 kg and a 2.5 m animal 184 kg.

Normal adult bodyweights for the Indo-Pacific bottlenose dolphin are 120–180 kg for males and 90–150 kg for females; for the common bottlenose dolphin 180–250 kg for males and 160–220 kg for females; bodyweight does fluctuate seasonally, with animals tending to be at their heaviest over winter and lightest over summer. This generally corresponds with daily food intake. Body condition can be assessed visually by experienced observers. Severe and pathological weight loss is evident when the ribs are easily visualised and when there is muscle wastage along the dorsal thoracic musculature and the musculature just posterior to the cranium, giving a concave shape to the animal lateral to the midline and a 'peanut' shape when viewed from above (see section 8.2). Any sudden or unexpected loss of weight in the captive situation should be investigated.

## 7.5 Release

Animals are generally released into the holding or medical pool in the reverse of the capture procedure. Animals should always be monitored post-capture.

## 7.6 Transport requirements

### 7.6.1 Transport box design

Purpose-made padded stretchers with openings through which the flippers can protrude and openings in the genital region to prevent urine scalding are ideal. Stretchers without these openings should only be used for short periods (20 min) to avoid crushing the flippers and overheating. The fabric of the stretcher should be smooth and easily disinfected. Cetaceans should always be transported in ventral recumbency and never in lateral recumbency. The stretcher can be suspended in a water-filled container for long travel periods with the dorsal surface exposed to the air so the animal can breathe. Foam padding can be used to help protect and support the animal. Cetaceans should never be transported free in a water-filled container as there is a risk of drowning and impact with the side of the container, which may result in serious injury. The transport time should be kept to a minimum. The animal should be kept wet by lying moist towels or sheets over it or by spraying with water continuously. The blowhole and the tail fluke should remain free of coverings so that the animal can breathe unobstructed. The tail fluke assists in temperature exchange with the environment. Temperatures during transport should not exceed 30°C or be less than 15°C. Overheating is more of a problem than cold. It may be useful to have a supply of ice on hand. At all times the animal must be protected from direct sunlight, wind and exhaust fumes. Cetaceans should never be rolled to move them as this is very distressing for the animal and may result in damage to the flippers and dorsal fin. A small crane either fixed to the transport vehicle or separate is useful for lifting heavier animals. The use of sedatives in transporting cetaceans is debatable. Unless the animal is highly agitated, it may be better to forego the use of these agents, as they can cause respiratory depression and alter thermoregulation. In recent years the use of football socks to cover the pectoral flippers and tail flukes has been utilised during transport operations to minimise accidental trauma to these body areas.

### 7.6.2 Furnishings

None.

### 7.6.3 Water and food

Dolphins acquire all their necessary water through the oxidation of fats in the fish that are fed to them. No supplemental fresh water is necessary for captive animals fed an adequate diet.

Bottlenose dolphins should not be fed 24 h before being subjected to an extended transport period. This is to limit the amount of faecal matter contaminating water in the transport container. Animals are usually transported in a transport container individually, although there have been some attempts overseas to transport animals in pairs within a transport container.

### 7.6.4 Animals per transport box

Only one animal should be held in a given box.

### 7.6.5 Timing of transportation

Transportation is best undertaken overnight or early in the day and later in the evening to avoid high temperatures and direct sunlight.

### 7.6.6 Release from the transport box

Animals should be lowered into the water in the transport stretcher and then gently extricated from the stretcher by experienced handlers present in the water. Care should be taken not to allow the blowhole to be immersed in water for an extended period of time (> 2 min) as animals can potentially drown. Animals can be taken to the edge of a deep enclosure with a sheer drop off in the stretcher and gently rolled in to the water ensuring no part of the animal comes in contact with the enclosure wall.

# 8. HEALTH REQUIREMENTS

## 8.1 Daily health observations

Each individual dolphin should be observed daily for any signs of injury or illness, especially after they have been introduced to other animals, as they often show aggression towards each other. The most appropriate time to do this check is first thing in the morning, generally when they are being fed for the first time in the day. Particular areas that should be checked include:

- skin condition – in particular, rake marks from play/aggression behaviour
- discharges – from the eyes, ears, blowhole, mouth, genital openings or anus
- appetite
- faeces – amount and consistency; normal faeces should dissipate quickly in the water.
- eyes – cloudiness, opacities and swelling
- changes in demeanour, activity level and behaviour
- injuries.

## 8.2 Bodyweight and body condition score

### 8.2.1 Bodyweight

Ideally, bodyweight should be recorded regularly and compared with previous weights. The frequency of weight measurements will depend on factors such as the logistics required to weigh particular species/individuals and the presence of underlying health concerns.

### 8.2.2. Body condition score

A body condition scoring system has been developed by Clegg *et al.* (2015; Fig. 16.10) that includes an index of 1-5 including:

Body condition 1 - Emaciated

- Head - concave depression behind skull, scapula bones visible
- Flukes - transverse processes angular and well defined
- Lateral view - ribs clearly visible, hollow area under dorsal fin above spine
- Dorsal view - concave indent behind skull (i.e. peanut-shaped head), minimal fat cover

Body condition 3: Ideal

- Head - palpable fat cover around the skull, scapula bones only visible when the pectoral fin is manipulated
- Flukes - transverse processes only faintly visible and not clearly defined
- Lateral view - ribs not visible, notable layer of fat parallel and adjacent to spine

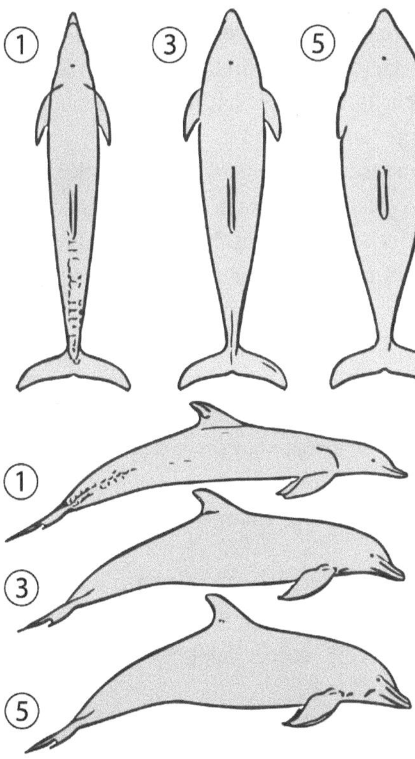


**Fig. 16.10.** Dolphin body condition scoring. (1) Emaciated, (3) Ideal and (5) Obese. (See details in section 8.2.) Derived from Clegg *et al.* 2015. Drawing by Mark Blyde.

- Dorsal view - smooth transition from head to neck, rounded fat deposition along spine

Body condition 5: Obese

- Head - convex fat deposition behind the skull, palpable fat layer above the scapula
- Flukes - transverse processes are not visible
- Lateral view - ribs are not visible, ample fat deposition in the area under the dorsal fin and mid-abdomen, folds of skin caused by excess fat deposition, folds may be visible under chin when head is not bent
- Dorsal view - no definition from the head to neck, body is barrel-shaped with excessive fat deposition, particularly behind the dorsal fin

## 8.3 Physical examination

Physical restraint in captive dolphins is often the type of restraint chosen for many minor procedures in dolphins. A physical examination may be undertaken when an animal is restrained. Particular areas that should be checked include:

- skin condition - in particular, rake marks from play/aggression behaviour
- discharges - from the eyes, ears, blowhole, mouth, genital openings or anus
- faeces - amount and consistency
- eyes - cloudiness, opacities and swelling
- injuries
- mouth and teeth examinations by carefully holding the jaws open with rolled up towels.

Additional examinations that can be undertaken while a dolphin is restrained include:

- ultrasound examinations and radiography.
- biopsies
- collection of blood, urine and any other samples when needed.

## 8.4 Chemical restraint

Chemical restraint is rarely used with cetaceans because it is difficult and associated with many problems. Procedures such as dentistry, endoscopy and minor surgery can be accomplished with a combination of sedation and local anaesthesia.

General anaesthesia in cetaceans should not be undertaken lightly. Critical to the success of the procedure is the ability to catheterise the lateral subcutaneous caudal vein. This is generally done via ultrasound-guided cannulation using a micropuncture venous kit. The technique for reliably placing an ultrasound-guided superficial intravenous catheter in the lateral subcutaneous caudal vein in *T. truncatus* is described by Ivancic *et al.* (2015).

Cetaceans should be fasted 24 h before a general anaesthesia. In preparation for general anaesthesia animals are given 0.08–0.1 mg/kg midazolam intramuscularly (IM) as a premedication injection. Other drugs have been used for premedication, both on their own or in conjunction with midazolam. Pethidine IM can be given just before induction and after the midazolam has taken effect in order to reduce the possibility of excitement in the animal. Dosages of 0.2–0.5 mg/kg IM have been used. Once sedated, a catheter is placed in the lateral subcutaneous caudal vein and propofol is administered at a dose rate of 2–4 mg/kg by slow intravenous injection. The dose rate of propofol can be reduced by a second IM injection of midazolam at the same dose as the first or 0.05–0.08 mg/kg IV in combination with the propofol. Once the animal is in a deep enough plane of anaesthesia, the mouth can be opened and the elongated epiglottal and cricoarytenoid cartilages or 'goosebeak' can be rostrally dislocated, allowing access to the trachea. An endotracheal tube of suitable diameter can be introduced into the trachea. Once the animal is intubated it can be maintained on a combination of oxygen and sevoflurane or isoflurane. Intermittent positive pressure ventilation should be instituted in order to keep the end-tidal carbon dioxide level less than 40 mmHg. Intensive monitoring of the anaesthesia will provide the clinician with better success. Pulse rate and oxygen saturation should be monitored.

Electrocardiographic monitoring should be undertaken. In-line end-tidal carbon dioxide levels should also be measured and the minimum inhibitory concentration of the volatile anaesthetic agent should be monitored if possible. Direct arterial blood pressure can be measured by cannulation of the median artery of the pectoral flipper using ultrasound guided catheterisation (Ivancic *et al.* 2014). The median blood pressure in 10 general anaesthetic procedures in cetaceans was 44 mmHg, but support improved values to 75 mmHg. Support can include increasing the intravenous fluid rates, decreasing the inhalational anaesthetic levels and dobutamine (0.2–2.0 ug/kg/min) intravenously (IV).

Extubation and replacement of the 'goosebeak' remains critical for the success of the procedure and re-intubation should be considered if the patient does not begin to breathe spontaneously. Bailey (2016) describes a series of 10 general anaesthetic procedures in cetaceans. Midazolam can be reversed with flumazenil (0.005 mg/kg IM/IV). Pethidine can be reversed by naloxone 0.01 mg/kg IV. Naloxone only lasts ~20 min so this dose can be repeated if necessary. If there are signs of respiratory depression or general excitement, naltrexone can be given at a dose rate of 0.05–0.1 mg/kg IM/IV to obtain 24 h of opioid reversal. Again, this can be repeated if necessary.

Sedation of animals for painful procedures such as bone marrow biopsy has been accomplished with a mixture of midazolam (0.1 mg/kg) and butorphanol (0.1 mg/kg) IM. Less painful procedures can be undertaken using midazolam alone or using diazepam (0.1–0.2 mg/kg) IM. The clinician may choose to reverse the midazolam or diazepam with flumazenil, but this is generally not necessary.

Free-ranging animals can be difficult to catch and restrain. If an animal in the wild can be conditioned to hand feed, sedatives can be disguised in the fish. A combination of 0.6 mg/kg diazepam and 0.4 mg/kg butorphanol given orally can give mild sedation to an animal in 2–3 h (Martelli *et al.* 2016). This level of sedation can be sufficient to enable a boat to get close enough to an animal to get a net around it.

Sedation at sea of entangled north Atlantic right whales (*Eubalaena glacialis*) has been used to enhance disentanglement efforts (Moore *et al.* 2010). Midazolam (0.1 mg/kg IM) in combination with butorphanol (0.1 mg/kg IM) was delivered via a purpose-designed pole syringe. This combination of drugs produced an increased respiratory rate, an increased swimming speed and a marked reduction in boat avoidance, allowing improved disentanglement. This combination of drugs may have applications for free-range cetaceans that need to be restrained but where it is not possible to get close enough to net them.

## 8.5 Common health problems

Table 16.4 lists infectious disease and Table 16.5 lists non-infectious diseases of dolphins.

## 8.6 Preventative medicine

Yearly vaccinations for erysipelas (*Erysipelothrix rhusiopathiae*) and with Ultravac 5-in-1 are undertaken in some captive facilities and should be considered.

**Table 16.4.** Infectious diseases or agents of captive, free-range and rescued dolphins of clinical significance.

b.i.d. = twice daily; IM = intramuscular; PO = per os (oral); s.i.d. = once daily.

| Disease/ infectious agent | Cause | Clinical signs | Diagnosis | Treatment | Prevention/comments | Reference |
|---|---|---|---|---|---|---|
| **Endoparasites** | | | | | | |
| Respiratory worms | *Nasitrema* spp. | Foamy discharge, irritation of the blowhole. Can travel to the brain causing neurological signs such as seizures and abnormal swimming patterns | Ova can be seen in wet mounts of the blowhole secretions | Ivermectin (200 ug/kg) PO | Good biosecurity procedures by not bringing infected animals into the collection | 1 |
| **Protozoal** | | | | | | |
| Toxoplasmosis | *Toxoplasma gondii* | Generally produces non-specific signs such as lethargy and decreased appetite. Can be fatal | Increasing antibody titres. Extremely high erythrocyte sediment rates > 50 mm/h are indicative of toxoplasmosis | Clindamycin (5 mg/kg) b.i.d. PO<br>Prednisolone should also be given 0.25–0.4 mg/kg PO s.i.d. as an adjunct therapy to decrease inflammation | Limit cat access to food sources. Some fish can act as paratenic hosts and this organism can survive freezing for many months | 2 |
| **Fungal** | | | | | | |
| Systemic fungal infections | *Aspergillosis fumigatus* | Non -specific signs depending on what organ is affected. Often neurological signs when the brain is affected | Blood parameters and fungal antibodies. Observation at necropsy of stranded animal | Anti-fungal drugs depending on cultures and sensitivities | Maintain ideal water quality parameters | 1 |
| **Bacteria** | | | | | | |
| Erysipelas | *Erysipelothrix rhusiopathiae* | Clinical signs can vary from rhomboid or diamond-shaped skin lesions to anorexia and depression to sudden death depending on the severity and the acuteness of the infection. Non-specific signs depending on what organ is affected | Diagnosis is made by culturing the skin lesions or blood cultures from those animals showing systemic disease. Animals that die acutely can be diagnosed by culturing organs from necropsy material | Amoxycillin (12.5 mg/kg) b.i.d. PO | Food fish are often the source of infection. Freezing and good food hygiene may assist in prevention. Yearly vaccinations are advised | 3 |
| Intestinal disease | *Clostridium* spp. | Peritonitis, septicaemia, enteritis, colic, volvulus, intussusception. Sudden death | Abdominal cramping. Clinical signs. Necropsy | Long-acting penicillin (22 000 IU/kg IM) | Vaccination with 5-in-1 vaccine | 4 |

**Table 16.4.** continued

| Disease/ infectious agent | Cause | Clinical signs | Diagnosis | Treatment | Prevention/comments | Reference |
|---|---|---|---|---|---|---|
| Brucellosis | *Brucella ceti* | Abortion, pneumonia and neurological signs | Culturing the organism from aborted fetuses and associated afterbirth. Transtracheal washes can sometimes be useful in identifying the organism in animals with pneumonia. Cerebrospinal taps could potentially identify the organism in animals with neurological signs. Histopathology and bacteriology from organs and tissues at necropsy is often the only way the disease is diagnosed. Positive serology for brucella can be a useful indicator the animal is infected or has been in contact with the organism | Antibacterial drugs as selected by culture and sensitivity | Good biosecurity procedures by not bringing infected animals into the collection. It is spread via blowhole secretions and is also spread venereally | 5 |
| Pasteurellosis | *Pastuerella multocida* | Non-specific. Signs can range from lethargy and anorexia to sudden death | Diagnosis is made by culturing blowhole secretions or blood cultures from those animals showing systemic disease. Animals that die acutely can be diagnosed by culturing organs from necropsy | Appropriate antibiotics based on culture and sensitivity results<br>Vaccination with autogenous vaccines | | |
| **Viruses** | | | | | | |
| Cetacean pox or tattoo disease | *Cetacean pox virus* | Causes small to moderate 'tattoo'-like lesions on the skin of the infected animals | Diagnosis can be made by visualising the specific lesions. Histopathology of biopsied tissues will confirm the diagnosis | Lesions are generally self-limiting | The disease is spread from animal to animal by biting insects. Limiting breeding areas for mosquitoes and midges may help limit the spread of the disease | 7 |
| Cetacean morbillivirus | *Cetacean morbillivirus* | Generally causes neurological and respiratory signs. Affected animals appear dull and have decreased mentation. They may have some increased respiratory effort and some discharge from the blowhole. Death usually ensues 3–5 days after showing clinical signs | Can be made by clinical signs in association with high titres to the virus. Diagnosis is most often made by histopathology on necropsy specimens | There are no specific treatments and therefore treatment is generalised and supportive and includes antibiotics, anti-inflammatories, fluid therapy and nutritional support | Biosecurity is the key to preventing animals from becoming infected. The virus is shed in respiratory vapour and is passed from animal to animal via blowhole secretions | 7 |

References: (1) Blyde (2019); (2) Blyde (2016); (3) Sweeney and Ridgway (1975); (4) Begeman *et al.* (2013); (5) Mackie *et al.* (2020); (6) Van Bressem *et al.* (2009); (7) Stone *et al.* (2011).

**Table 16.5.** Common non-infectious diseases and causes of mortality of captive and rescued dolphins.

Adapted from Blyde unpublished.

| Disease/agent | Cause | Clinical signs | Diagnosis | Treatment | Prevention/comments | Reference |
|---|---|---|---|---|---|---|
| Predation | Sharks | Wounds, death | Observation/history, physical examination, necropsy | Medical and surgical treatment if not fatal | Not relevant to captive dolphins | 1 |
| Anthropogenic threats | Trauma from boats and propellors<br>Drowning<br>Entanglement | Emaciation, trauma, dehydration, entanglement, entrapment, drowning, death | Observation/history, physical examination, necropsy | Medical and surgical treatment if not fatal | Increased conservation action, climate change action, policy changes. Not relevant to captive dolphins | 1 |
| Gastrointestinal volvulus | Multifactorial causes including foreign bodies in the stomach leading to delayed gastric emptying, feeding large fish that sit for longer than normal in the stomach and overgrowth of clostridial bacteria producing enterotoxins | Severe depression, reluctance to move, complete anorexia | Ultrasonography, radiology, haematology and biochemistry | None | Bi-annual vaccination with Ultravac 5-in-1 vaccine. Limiting foreign body ingestion. Avoid feeding large fish when gastric foreign bodies are present | 2 |
| Foreign body ingestion | Dolphins are very curious animals and will often ingest foreign objects that enter the pool, including rocks, nails, screws and vegetative material | Decreased appetite, lethargy | Radiology, endoscopy and haematology and biochemistry | Remove the foreign objects manually or endoscopically. Often animals will regurgitate these objects if they are maintained in concrete pools for a period of time | Limit animals' access to foreign bodies | 1 |
| Keratopathy | Solar radiation impinging on the corneas | Corneal opacities particularly at the 10 o'clock and 2 o'clock positions | Clinical signs are diagnostic | Topical anti-inflammatory eye drops and topical antibiotic drops | Provision of shade for captive animals. Do not put animals in positions where they are looking directly into the sun | 3 |

References: (1) Blyde (2019); (2) Begeman *et al.* (2013); (3) Colitz *et al.* (2016).

## 8.7 Zoonoses

No common zoonotic diseases have been recorded. Good hygiene, such as washing hands after handling animals and their food and avoiding the exhaled breath from the blowhole of these animals, should be practised.

Dolphins in Australia may be infected with brucellosis, but it is a cetacean strain and not particularly infectious to humans. If animals are known to be infected with *Brucella ceti*, then wearing gloves and masks when handling them or performing necropsies may be wise. This disease causes abortion in these animals and care should be taken when handling aborted fetuses.

# 9. BEHAVIOUR

## 9.1 Activity

Bottlenose dolphins in the wild are active day and night, dividing their time between foraging, resting, travelling and socialising, with foraging typically taking up most time (Steiner 2011; Foroughirad and Mann 2013). Notwithstanding the difficulties in comparing studies that have used different approaches (Steiner 2011), activity budgets vary between age and sex classes because of different energetic or sociosexual requirements. For example, populations with a larger home range, such as offshore common bottlenose dolphins, are likely to spend more time travelling between food patches than resident inshore Indo-Pacific bottlenose dolphins. In captivity, behaviour is a powerful indicator of welfare.

## 9.2 Social behaviour

Bottlenose dolphins are highly social and live in dynamic groups called fission-fusion societies (Connor *et al.* 2000b; Tsai and Mann 2013). In both species of bottlenose dolphins, associations are typically stronger between adult males than between adult females and male–female associations are typically short lived (Connor *et al.* 2019; Wells and Scott 2002). Males are often found in pairs and trios, whereas females often form larger pods with their offspring present. Male bottlenose dolphins in some populations form multilevel male alliances to gain access to oestrous females and exclude competing males (Gerber *et al.* 2021). Sexual segregation is common in bottlenose dolphins, where females avoid male harassment outside of courtship periods (Galezo *et al.* 2017). Dominance hierarchies have been reported in both species in captivity, with dominant large adult males maintaining the hierarchy through aggressive behaviours and posturing (Shane *et al.* 1986).

## 9.3 Reproductive behaviour

Males will aggressively pursue females in oestrus. Males will fight over females in oestrus (D. Blyde pers. obs.).

## 9.4 Behavioural problems

Being rather intelligent animals, this species requires a great deal of mental stimulation in the captive setting. Behavioural issues could include aggression to conspecifics, self-harm and continuous regurgitating and re-ingestion and ingestion of foreign objects (D. Blyde unpubl. data). A formal enrichment program for all animals in captivity should be developed and implemented.

## 9.5 Signs of stress

Signs of stress include decreased appetite and associated weight loss. In addition, animals may self-harm by knocking their bodies and tail flukes against enclosure perimeters. Increased amounts of rake marks may indicate the animal is unwell and/or is being aggressed by conspecifics. If not rectified, stress can lead to serious disease and potentially deaths (Waples and Gales 2002)

## 9.6 Behavioural enrichment

Any facility that houses dolphins must have a formal enrichment program to provide a stimulating environment to these highly intelligent animals. Enrichment can be both environmental (social structure, enclosure furniture, substrate) or behavioural (toys, training programs, interactive public programs). Environmental enrichment can include changing the social structures of animals, providing interesting enclosure furniture and changing substrates. Behavioural enrichment tools can include training and conditioning periods or interactions with staff and visitors. The use of novel food items, fish in frozen ice blocks, novel food locations or feeding regimens (including grovelling food items) and the safe temporary introduction of other species (e.g. fish and seals) can provide behavioural enrichment. Management practices including social management and training programs may help to influence dolphin welfare and reduce the impacts of limited pool sizes in captivity (Miller *et al.* 2021a). Research suggests that training needs to be implemented, providing cognitive enrichment with challenges that are controllable, motivating, relevant and possible to master (Clark 2013). An example of cognitive enrichment is an underwater maze device that was developed by Clark *et al.* (2013), which requires a dolphin to navigate a rubber ball through a maze of pipes to an exit pipe that may provide reinforcement once the puzzle is solved. They found that some animals used a variety of problem-solving strategies without prior training, whereas others did not use the maze at all. Nonetheless, the maze did not have a significant effect on repetitive (circular) swimming behaviour, but males did spend significantly more time underwater when the device was present. They also suggested that further research was needed to identify appropriate environmental enrichment (Clark *et al.* 2013).

Providing enrichment can increase the welfare of animals in captivity and can decrease destructive behaviours such as abnormal repetitive swim patterns, ingestion of foreign objects and trauma (both self-trauma and conspecific trauma). The effectiveness of environmental enrichment objects varies between individuals. For example, one study investigated the use of 21 familiar objects with a group of six common bottlenose dolphins at Asterix Park in France and found that only half of them elicited manipulative behaviours (Delfour and Beyer 2012). The objects that elicited behavioural responses included (1) two outdoor water hoses, (2) a square foam mattress, (3) a plastic hoop with a piece of foam attached; (4) a plastic hoop with a piece of foam attached by tape with additional hanging pieces of plastic; (5) a square foam mattress with holes in it; (6) a fish-shaped foam mattress; (7) a collection of several objects; (8) a suspended ball (target used during dolphin shows); (9) a buoy and funnel; and (10) pieces of fire hose. Therefore, behavioural changes observed after objects have been provided do not necessarily indicate enrichment has been provided, so behaviours observed towards the objects must be investigated in relation to their social context and individual behaviour patterns to ensure the behavioural needs of the animals are met (Delfour and Beyer 2012).

A large study of 47 animals at 25 facilities around the world found that frequency of tracing behaviour (a form of abnormal repetitive behaviour that involves an individual animal swimming in an identical path in a fixed repetitive pattern) was inversely related to the amount of behavioural diversity present. The study also found that animals that were trained on a predictable schedule had higher behavioural diversity than those trained on a semi-predictable schedule. There was also a significant inverse relationship between behavioural diversity and the ratio of faecal cortisol to dehydroepiandrosterone metabolites (Miller *et al.* 2021b). Animals that were split into groups and reunited or rotated between subgroups had higher behavioural diversity than animals managed in the same group (Miller *et al.* 2021a). Care should be taken to ensure that the enrichment items are not harmful to the animals ( i.e. able to be swallowed, provide entanglement risks or have sharp projections on them).

Observations of the behaviour of dolphins in interactive swim programs were equivocal in a study by Brensing *et al.* (2005), which found that at one location dolphins showed signs of stress with adult swimmers such as avoidance, speed increase, higher metabolism and intensification of a subgroup. In contrast at a second location in the same study dolphins seemed to be attracted to adult swimmers. Therefore, the behaviour and welfare of animals participating in presentations and interactive programs should be closely monitored. Included in this is the prevention of diseases potentially acquired by these animals from close association with people.

## 9.7 Training and conditioning

All training and conditioning of captive animals should follow the principles of positive reinforcement, utilising an approximation shaping process to achieve training goals. Additionally, the use of a bridging stimulus augments the training process by precisely pinpointing appropriate behaviour during training sessions. Punishment of animals including hitting, yelling, isolation and withholding of food is not acceptable. Training and conditioning are important for all animals held in captivity, but can be more complex in dolphins because of their relatively high intelligence. Techniques can include a combination of approximation and scanning, targeting, shaping and modelling. Techniques need to be consistent within the facility to avoid animal confusion when being trained by multiple trainers. Dolphins are also highly food motivated and can be fed individual fish to reward them for displaying correct behaviours. Training sessions should be highly variable in content and avoid a routine approach.

Husbandry training should include, but not be limited to:

- slide-outs for weighing, body morphometrics and general health assessments
- physical handling of all parts of the animal's body
- eye examinations and the application of eye drops
- faecal collection
- blowhole examination and collection of 'blow'
- mouth opening and examination of the teeth and tongue

- blood collection from the tail fluke (both dorsally and ventrally)
- maintaining the animals in dorsal and lateral recumbency to facilitate ultrasound examinations particularly pregnancy determinations
- assisted reproductive behaviours such as semen collection, urine collection and artificial insemination.

More intricate behaviours can be conditioned in dolphins and exciting educational displays can be provided for visitors as well as interactive swimming programs. All training with dolphins is predicated on the relationship between the animal and the trainer being a positive experience.

## 9.8 Introductions and removals

Generally, dolphins are not aggressive to other animals that are introduced from a new group. However, males will show dominance behaviour to other males and adult males caught from different groups have been observed to fight viciously, especially during the breeding season, whereas adult males caught from the same groups have been observed to maintain predominantly size-based social hierarchies with comparatively little aggression (Shane *et al.* 1986). Young calves have been known to be killed by adult males and females in captivity and the introduction of young calves to other animals should be done slowly and cautiously. Often it is better to introduce young calves to the group one animal at a time.

## 9.9 Intraspecific compatibility

Dolphins generally tolerate other animals of the same species reasonably well.

## 9.10 Interspecific compatibility

In the wild, mixed species groups of dolphins are often observed, especially when associated with a common food source. Interactions between the Australian humpback and Australian snubfin dolphins in the wild have been reported to be predominantly aggressive and sexual (Parra 2005) and a hybrid between the species has been documented (Brown *et al.* 2014). Historically, different species of dolphins have been housed together in captivity without problems. Common bottlenose dolphins and Indo-Pacific bottlenose dolphins have been routinely housed together. Melon-headed whales have been known to be aggressive to other cetaceans in a captive setting. Australian humpback dolphins and bottlenose dolphins have also been kept in the same pool (D. Blyde pers. obs.) and mixed species groups of these species have been reported frequently in the field (Corkeron 1990; J. Meager, unpubl. data).

Seal species have occasionally been housed with cetacean species for short periods of time in captivity. However, seals are more manoeuvrable and have been known to inflict bites on dolphins in captivity. Housing fish with dolphins is common, though care needs to be taken that the fish species are not toxic (e.g. puffer fish) and that the fish have some sanctuary to retreat to so as to avoid predation.

## 9.11 Suitability for captivity

The keeping of cetaceans in captivity is a controversial issue. Although they have been kept in captivity for decades, there are those who argue that this group of animals is not suitable for captivity. Clegg *et al.* (2015) have developed an objective method of measuring welfare in cetaceans in captivity, which can help to address this issue (The C-Well® assessment). This assessment comprises 36 measures, which constitute 11 welfare criteria and 4 welfare principles. It allows for measurements of the welfare of individual animals and entire facilities, which would allow for the development of 'best practices' in the keeping of cetaceans in captivity. These criteria can be recorded over time to assess whether the individual and overall welfare of cetaceans in captivity is improving or not. It can also measure animal welfare between facilities, allowing benchmarking to occur. Lauderdale *et al.* (2021a) undertook a study to identify factors that were related to the welfare of bottlenose dolphins in captivity and to develop reference intervals and values for common and novel indicators of health and welfare for various dolphin species. They found that environmental enrichment programs and social management factors were more strongly related to positive welfare than captive enclosure characteristics. Animal management techniques such as mixing groups up and changing physical surroundings also contributed to positive animal welfare (Miller *et al.* 2021b). Providing enclosures with a variety of characteristics and implementing formal environmental and behavioural enrichment and training programs are key components to maintaining positive welfare in captive dolphin populations (Lauderdale *at al.* 2021b).

# 10. BREEDING

## 10.1 Mating system

In the wild, bottlenose dolphins often form groups of females with their offspring, whereas males often form pair bonds, trios or more complex alliances. When females are in oestrus they seek out males, mate with them and return to the pod to give birth. The captive environment should try and mimic this system as much as possible.

## 10.2 Ease of breeding

Given the right conditions and environment the breeding of dolphins in captivity is not difficult. Assisted reproductive techniques are well advanced in this species, including semen collection, semen preservation and artificial insemination.

Although breeding of these animals in captivity is common, there have been a few issues with dystocia, maternal rejection of calves and early medical problems with the dam and offspring. Baumgartner *et al.* (2018) have developed a birth protocol that addresses many of these issues. It includes the facilities necessary, personnel requirements and procedures that can be undertaken to minimise problems during parturition and early calf rearing.

It should be noted here that the breeding of captive dolphins has recently been made illegal in New South Wales.

## 10.3 Oestrous cycle and gestation period

Dolphins are polyoestrous with a seasonal peak in spring. The oestrous cycle is 21–42 days with a mean of 30 days (Schroeder 1990). The gestion period of Indo-Pacific bottlenose dolphins and common bottle-nosed dolphins is ~12 months (Perrin and Reilly 1984; Baker and Gynther 2023).

## 10.4 Reproductive status

### 10.4.1 Females

Female dolphins in the wild are generally not mature enough to breed until 9–10 years of age; however, in captivity they can be sexually mature as young as 5 years and potentially as young as 3 years (D. Blyde pers. obs.). In a study of wild Indo-Pacific bottlenose dolphins, sexual maturity was documented at 6–14 years of age in females (Kemper *et al.* 2019).

### 10.4.2 Males

Male dolphins in the wild are generally not mature enough to breed until 7–8 years of age; however, in captivity they can be sexually mature as young as 5 years and potentially as young as 3 years (Blyde and Vogelnest 2008). In a study of wild Indo-Pacific bottlenose dolphins, male sexual maturity was estimated to be at 12–15 years of age (Kemper *et al.* 2014).

## 10.5 Reproductive management

There have been few indications for contraception in cetaceans. If facilities are available, separation of sexes may be used. Mature females can be contracepted using oral altrenogest at a dose rate of 0.044 mg/kg orally to control oestrus and breeding in females. This can be given continuously for many years without causing reproductive disease or permanent contraception. Ovulation generally occurs 15–27 days after the cessation of altrenogest therapy. Care must be taken when administering the medication. Gloves should be worn and male trainers should preferably administer the treatment (David Blyde pers. obs.).

Deslorelin implants can decrease testosterone levels in mature males and reduce reproductive drive and sperm concentration. Although deslorelin may be used as a male contraceptive, there is no guarantee of complete sterility. Theoretically, deslorelin could be a useful contraceptive in female dolphins, but its use in this regard has yet to be evaluated. In males, two 4.7-mg implants or one 9.5-mg implant are usually inserted IM, just lateral to the ventral mid-line. One 9.5-mg implant administered IM will last ~2.5 years. Implants can be removed surgically. If not removed, male testosterone levels and sperm production return to normal 2.5–3 years after insertion (David Blyde pers. obs.).

## 10.6 Occurrence of hybrids

Common bottlenose and Indo-Pacific bottlenose dolphins hybridise in captivity and in the wild. In a captive setting, common bottlenose dolphins have been documented to hybridise with species from at least six other genera (Gridley *et al.* 2018). Hybrids between false killer whales and bottlenose dolphins have often occurred in captivity (G. Bedford pers. obs.). The hybridisation rate of both species of bottlenose dolphins in the wild has almost certainly been underestimated because it is difficult to confirm without a genetic sample.

## 10.7 Timing of breeding

The oestrous cycle is ~21 days, but long periods of anoestrous are common in females in captivity. No effective chemical method has been found to induce oestrus in dolphins; however, changing the social structure or introducing a new breeding male may bring females into oestrus.

Detection of oestrus can be accomplished by behavioural observations, ultrasonography, hormonal levels in serum and/or urine or vaginal cytology. Oestrus can be monitored in captive female cetaceans by ultrasound examination of the ovaries, which are located caudal to the kidneys. A 3.5-MHz probe is probably the most useful for this procedure. A follicle can be followed by ultrasound examination until it is ~30 mm in diameter. It can then be examined on a daily basis until it is seen to ovulate. Mating generally occurs at ovulation time. Monitoring the oestrous cycle of captive female cetaceans allows more precise timing of male–female introductions, as males and females are generally kept separately outside of breeding attempts.

Pregnancy in captive females can be diagnosed in two ways. Three successive serum progesterone levels, taken 2 weeks apart exceeding 9.5 nmol/L (3 ng/mL) are diagnostic for pregnancy. Alternatively, pregnancy can be detected using ultrasound from 14 weeks of gestation. The probe is best positioned over the lateral or ventral abdomen, depending on which is easiest. There are many advantages to ultrasound diagnosis, including the ongoing monitoring of pregnancy. Measurement of the diameter of the fetal cranium and the diameter of the fetal thoracic cavity can give a reasonable prediction of time of parturition (Lacave *et al.* 2004). In addition, routine ultrasound examinations of pregnancy assist in monitoring fetal health.

If animals are conditioned for this procedure, they can be positioned in right or left lateral recumbency or dorsal recumbency while in the water. For animals that have to be physically restrained, they should be in either right or left lateral recumbency. The exact position of the probe depends on the size of the fetus, but the operator should begin at the genital slit and move cranially and dorsally.

Assisted reproduction techniques are being used in captive cetaceans. Males can be conditioned to produce semen samples, which can be collected and cryopreserved. Artificial insemination of females has been successful, with many live calves being produced. Female dolphins can be synchronised by treating them with oral altrenogest at ~1 mL/50 kg s.i.d. or 0.044 mg/kg. When the altrenogest is stopped, female dolphins should come into oestrus between day 15 and 27 following cessation of therapy. If not inseminated at this particular oestrus, the female will cycle again in ~21 days. Sperm sex selection has been successful in dolphins using flow cytometric sorting using a high-speed cell sorter (O'Brien and Robeck 2006). This allows the production of known sex offspring, which has significant implications for captive cetacean management.

## 10.8 Age at first breeding and last breeding

Both males and females in captivity are potentially reproductively mature at 5 years of age, which contrasts with the situation in the wild where males reach sexually maturity at 7–8 years of age and females at 9–10 years of age. There is no evidence to suggest there is any breeding senescence in this species.

## 10.9 Ability to breed every year

Gestation is 12 months and the calf suckles from the dam for at least 12–18 months. The intercalving interval in the wild is 3 years.

## 10.10 Ability to breed more than once per year

No, due to 12-month gestation period and subsequent suckling period.

## 10.11 Number of young born

Generally, only one calf is born at a time. Twins are rare and are often aborted before term.

Depending on the nature of the animals it may be useful to separate the dam just before calving to allow for uninterrupted parturition and initial suckling of the calf. In other situations it may be advantageous to have the dam calve with one or two other females if there is seen to be little risk of the other animals causing harm to the calf. This may be of particular benefit to first-time mothers.

## 10.12 Breeding diet

Care should be taken not to allow the pregnant female to get too fat prior to parturition as it can lead to dystocia. The diet should be balanced with the attitude and body score of the individual animal. In the last trimester of pregnancy daily food intake usually rises. After parturition the daily food intake should increase gradually and animals fed ad lib.

## 10.13 Age at weaning

Calves are totally dependent on their dam for probably the first 6 months of life during which time milk is their only form of nutrition. Calves begin to eat fish at ~6 months of age but continue to suckle for ~18–24 months. Close associations between mothers and calves after weaning can persist for up to 6 years in the wild and sometimes longer (Shane *et al.* 1986).

## 10.14 Age of removal from the mother

Calves are usually with their dam in the wild until 3-6 years of age and there is no need to alter this structure in captivity.

## 10.16 Growth and development

The growth of wild Indo-Pacific bottlenose dolphins has been recorded from south-western Australia and Shark Bay, Western Australia, and animals from the south-west reach a significantly larger body size than those from Shark Bay (Fig. 16.11; van Aswegen *et al.* 2019). These results indicate growth curves may be highly variable and depend on the region where the dolphins occur.

Growth and development of calves in captivity should be monitored routinely. Growth curves should be developed for each species within the facility (Fig. 16.12). Calf growth in captivity can be compared with that in the wild to determine optimum values (Wells 2009).

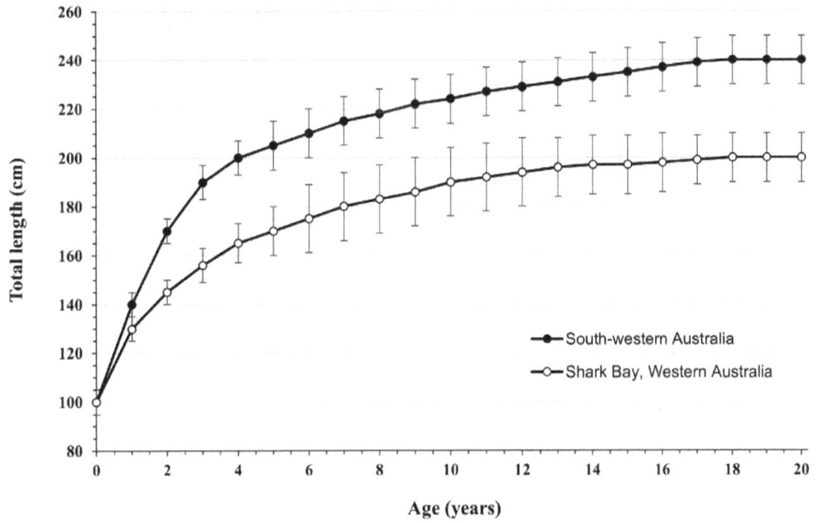


**Fig. 16.11.** Growth curves of free-ranging Indo-Pacific bottlenose dolphins from south-western Australia and Shark Bay, Western Australia. Data derived from van Aswegen *et al.* (2019).

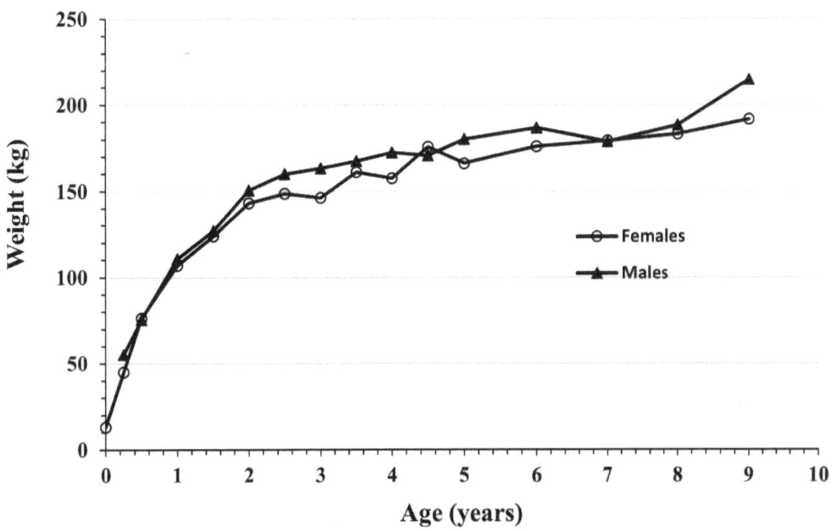


**Fig. 16.12.** Average growth curve for male and female captive *Tursiops truncatus*. Data from Sea World on the Gold Coast, Queensland, Australia.

## 11. ARTIFICIAL REARING

### 11.1 Housing

Young hand-reared neonates are best kept in aboveground pools. Ideal water heights are around 1–1.2 m so that carers can walk around the pool assisting the animal to swim and stay afloat if necessary. Smooth sides are ideal to prevent animals from damaging themselves on sharp projections. Round pools are preferred so that animals cannot wedge themselves into corners.

### 11.2 Temperature requirements

Water temperatures for neonates should be in the range of 25–28°C.

### 11.3 Specific requirements

Water parameters for captive animals should be as described in sections 4.6 and 4.7. Lighting above the pool is useful as feeding at night is often undertaken. Shelter from wind and rain is necessary; however, some natural light should be provided during the day.

### 11.4 Diet and feeding routine

Orphaned wild or mismothered captive cetaceans may need to be hand-reared. The milk of dolphins has been analysed (Martini *et al.* 2022) and various milk formulas have been used (Table 16.6). Note: the two formulas developed in the USA use some ingredients that are not available in Australia. Divetelact® is readily available in Australia and has proven safe and useful for hand-rearing dolphins. It can be started at the normal dilution of 1 scoop per 60 mL and gradually increased in concentration to 1 scoop per 30 mL. Other ingredients such as Calogen (N.V. Nutricia, Zoelemeer, The Netherlands) can be added to the Divetelact® to increase nutrient density. The milk can be given by stomach tube initially and the calf can be gradually introduced to a bottle and teat. In dolphins the best teat to try initially is a lamb teat, but some experimentation may be required to find the teat that the neonate responds to best. The volume given per day should be 20% of bodyweight, divided into 10–12 feeds because the dolphin calf's stomach is quite small. Blended fish fillets can be added to the formula at ~8 weeks of age and small whole fish can be introduced to the orphan at ~12–14 weeks of age. Orphan animals should be vaccinated for *Erysipelothrix* sp. and *Clostridium perfringens* before introducing fish products. Goodall (1980) describes the hand-rearing of a small dolphin calf from ~5 months of age. Animals should be weighed on a regular basis (at least weekly) to ensure they are gaining weight. Antibiotics can be given either PO or IM if indicated for the first 5–7 days. Ceftiofur (2 mg/kg s.i.d. IM for 5–7 days) or amoxicillin-clavulinic acid (12.5 mg/kg b.i.d. IM for 5–7 days) are the drugs of choice. Flower *et al.* (2018) described the interventions that were undertaken to assist an ill neonatal dolphin (*T. truncatus*) in captivity. They describe some equipment and procedures that were necessary in this particular case. They also describe dolphin milk formulas, medications used and growth rate charts for this individual animal (Fig. 16.13).

**Table 16.6.** Milk formulas used for dolphins.

Reproduced from Townsend (1999).

| Sea World formula | 'April' formula |
|---|---|
| Herring fillets with viscera – 750 mL | Herring fillets with viscera – 750 mL |
| Zoological 30/55 – 225 g | Multimilk powder |
| Zoological 33/40 – 450 g | Safflower oil – 50 mL |
| Salmon oil – 50 mL | Lecithin – 15 mL |
| Taurine – 500 mg | Taurine – 250 mg |
| Lactobacillus – 1.5 tablets | Lactobacillus – 3 tablets |
| Osteoform – 1 tablespoon | Osteoform – 1 tablespoon |
| Sea World marine vitamin tablets | Bottled water – 1100 mL |
| Whipped cream – 100 mL | Multivitamins with zinc |
| 50% dextrose – 60 mL | |
| Bottled water – 1100 mL | |

Osteoform: vitamin/mineral mix available in the USA. Manufactured by Vetamix.

Zoological: milk substitute available in the USA. Manufactured by Pet-Ag Inc. 30/55 contains 30% protein and 55% cent fat whereas 33/40 contains 33% protein and 40% fat.

Young orphaned animals can be introduced to experienced mothers in a captive situation for fostering. Allonursing has been accomplished on several occasions. The female cetacean does not need to be lactating at the time. Generally, the experienced female will begin producing milk within 24 h of the introduction of the calf and will allow the orphan to suckle. Close observations and regular weights of the orphan will determine the success or failure of this technique. It is recommended to vaccinate calves against *Clostridium perfringens* toxin. The first dose can be given on day 2 or 3 and a booster should be given 4 weeks later.

Calves need to suckle within the first 24 h in order to obtain the antibody-rich colostrum. Cetaceans have a diffuse epitheliochorial placenta and it is therefore most likely that the neonate receives its passive immunity through the colostrum rather than transplacentally. If the calf does not suckle in the first 24 h it may not absorb sufficient antibodies to provide adequate immunity. There are, however, reports of calves not suckling for the first 24 h and surviving. If there is concern about adequate passive transfer from direct observations that the

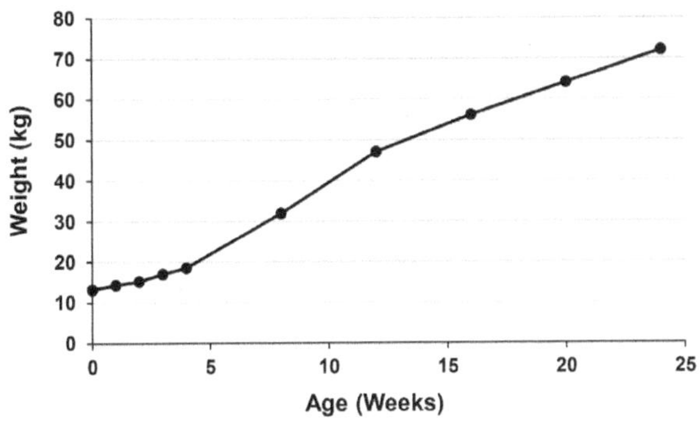


**Fig. 16.13.** Growth curve for a hand-reared male *Tursiops truncatus*. Data from Flower *et al.* (2018).

calf has not suckled, the dam can be manually restrained and milk can be collected from the mammary gland by applying negative pressure to the gland by using a cut-off 60-mL syringe. This milk can then be given to the calf by stomach tube. Care should always be taken when restraining newborn calves as the mothers are usually very protective and may attack. Alternatively, plasma can be collected from a healthy adult animal and given to the calf IV, IP or PO at a rate of 10 mL/kg. Lactation lasts ~7–18 months and in captivity calves will start eating whole fish at 3–6 months of age.

### 11.4.1 Natural milk

The milk of dolphins has been analysed (Martini *et al.* 2022). High percentages of dry matter, fat and protein were found in the colostrum and the percentages of these components decreased over time. The milk was found to be rich in unsaturated fatty acids, which were more than double compared with the saturated acids.

### 11.4.2 Milk formulas

Milk formulas can be based on Divetelact® with added caloric rich supplements such as Calogen.

### 11.4.3 Feeding apparatus

Teats, bottles and feeding tubes.

### 11.4.4 Feeding routine

Neonate calves may need to be fed every 2 h.

## 11.5 Data recording

Records should be kept daily and include: the amount and composition of the milk fed; how it was delivered to the animal; daily weights; times of urination and defecation; water quality parameters and any health concerns and medical treatments.

## 11.6 Identification methods

As for adults.

## 11.7 Hygiene

As with all young hand-reared animals, hygiene is vitally important. All utensils including bottles and teats should be washed after every feed and left soaking in a disinfection solution (e.g. 2% chlorine, Milton Australia Pty Ltd, Rozelle, NSW, Australia) that is changed daily. Hands should be washed before and after feeding the animal. The water should remain well disinfected throughout.

## 11.8 Behavioural considerations

As dolphins are social animals, young hand-reared animals should be introduced to other animals as soon as possible. Young calves and quiet females are the ideal first contacts.

## 11.9 Use of foster animals

Female dolphins and other species of cetaceans have been known to foster orphan calves on occasions. Young, orphaned animals can be introduced to experienced mothers in a captive situation, which has been done successfully on several occasions. The female cetacean does not need to be lactating at the time. Generally, the experienced female will begin producing milk within 24 h of the introduction of the calf and will allow the orphan to suckle (Blyde and Vogelnest 2008). This has been observed within the same species and also between species. Some examples include a killer whale female fostering a pilot whale calf, an Indo-Pacific bottlenose dolphin fostering a common dolphin calf and a common dolphin female fostering a melon-headed whale calf (Carzon *et al.* 2019).

## 11.10 Weaning

Dolphins in captivity are usually weaned around 2–3 years of age, either gradually (by gradually increasing the time spent away from the dam) or abruptly (by moving the calf to another pool). The strategy will be determined by the resources available at the facility, which will include staff numbers, pool array and the ability to easily separate and re-unite the calf with its dam. If an abrupt weaning is considered, it is beneficial to introduce the calf to other dolphins of similar age.

## 11.11 Release procedures

Many dolphins are rescued from the wild and can be treated, rehabilitated and released. The criteria for non-release of rescued animals are:

- calf less than six months of age and had been totally dependent on its dam. These animals do not have the skills to find food and avoid predators in the wild
- captive-born animals
- animals with severe injuries that will prevent them finding food or avoiding predators in the wild (e.g. missing a limb, tail or portion of jaw, sight in only one eye or deaf).

## REFERENCES

Allen SJ, Bryant KA, Kraus RH, Loneragan NR, Kopps AM, Brown AM, Gerber L, Krutzen M (2016) Genetic isolation between coastal and fishery-impacted, offshore bottlenose dolphin (*Tursiops* spp.) populations. *Molecular Ecology* **25**, 2735–2753. doi:10.1111/mec.13622

Amir OA, Berggren P, Ndaro SGM, Jiddawi NS (2005) Feeding ecology of the Indo-Pacific bottlenose dolphin (Tursiops aduncus) incidentally caught in the gillnet fisheries off Zanzibar, Tanzania. *Estuarine, Coastal and Shelf Science* **63**, 429–437. doi:10.1016/j.ecss.2004.12.006

Ansmann IC, Lanyon JM, Seddon JM, Parra GJ (2013) Monitoring dolphins in an urban marine system: total and effective population size estimates of Indo-Pacific bottlenose dolphins in Moreton Bay, Australia. *PLoS One* **8**, e65239. doi:10.1371/journal.pone.0065239

Ansmann IC, Lanyon JM, Seddon JM, Parra GJ (2015) Habitat and resource partitioning among Indo-Pacific bottlenose dolphins in Moreton Bay, Australia. *Marine Mammal Science* **31**, 211–230. doi:10.1111/mms.12153

Bailey JE (2016) Cetacean anesthesia: A review of 10 clinical anesthesia events, lessons learned and future plans. In *International Association for Aquatic Animal Medicine*. 47th Annual IAAAM Meeting & Conference. 21–26 May 2016. Virginia Beach, Virginia, USA. www.vin.com/apputil/content/defaultadv1.aspx?pId=14818&meta=Generic&catId=75118&id=7312395 (accessed 19 November 2022).

Baker AM, Gynther IC (eds) (2023) *Strahan's Mammals of Australia*. 4th edn. Reed New Holland Publishers, Sydney.

Barratclough A, Sanz-Requena R, Marti-Bonmati L, Schmitt TL, Jensen E, García-Párraga D (2019) Radiographic assessment of pectoral flipper bone maturation in bottlenose dolphins (*Tursiops truncatus*), as a novel technique to accurately estimate chronological age. *PLoS One* **14**(9), e0222722. doi:10.1371/journal.pone.0222722

Barratclough A, Smith CR, Gomez FM, Photopoulou T, Takeshita R, Pirotta E, Thomas L, McClain AM, Parry C, Zoller JA (2021) Accurate epigenetic aging in bottlenose dolphins (*Tursiops truncatus*), an essential step in the conservation of at-risk dolphins. *Journal of Zoological and Botanical Gardens* **2**, 416–420. doi:10.3390/jzbg2030030

Bassos-Hull K, Wells RS (1997) Effect of pool features on the behavior of captive dolphins. *Marine Mammal Science* **13**, 533–534. doi:10.1111/j.1748-7692.1997.tb00665.x

Baumgartner K, Lacave G, Sweeney JC, Will H (2018) A suggested birth protocol for bottlenose dolphins (*Tursiops truncatus*) – Updated 2015, Zoo Nuremberg. *Aquatic Mammals* **44**, 100–109. doi:10.1578/AM.44.1.2018.100

Bedford G (2000) The use of body length, girth and weight relationships as a husbandry tool for captive bottlenose dolphins. In *Proceedings of 10th ARAZPA/ASZK conference.* pp 264–277. ARAZPA/ASZK, Gold Coast, Queensland.

Begeman L, St Leger JA, Blyde DJ, Jauniaux TP, Lair S, Lovewell G, Raverty S, Seibel H, Siebert U, Staggs SL, Martelli P, Keesler RI (2013) Intestinal volvulus in cetaceans. *Veterinary Pathology* **50**, 590–596. doi:10.1177/0300985812465327

Best PB, Folkens PA (2007) *Whales and Dolphins of the Southern African Subregion*. Cambridge University Press, Cambridge.

Blasio AL, Pérez RV, Pardo MR, Maldonado FG (2012) Maintenance behaviour and cortisol levels in bottlenose dolphins (*Tursiops truncatus*) in closed and open facilities. *Veterinaria (México)* **43**, 103–112.

Blyde D (2016) Toxoplasmosis: An emerging disease of marine mammals in Australasia. In *Proceedings of 47th Annual Conference of International Association of Aquatic Animal Medicine*. Virginia Beach, USA.

Blyde D (2025) Cetaceans. In *Current Therapy in Medicine of Australian Mammals*. Revised Edition. (Eds L Vogelnest L and T Portas) pp. 635–647. CSIRO Publishing, Melbourne.

Blyde D, Vogelnest L (2008) Cetaceans. In *Medicine of Australian Mammals*. (Eds L Vogelnest and R Woods) pp. 591–614. CSIRO Publishing, Melbourne.

Brensing K, Linke K, Busch M, Matthes I, van der Woude SE (2005) Impact of different groups of swimmers on dolphins in swim-with-the-dolphin programs in two settings. *Anthrozoos* **18**, 409–429. doi:10.2752/089279305785593956

Brown AM, Kopps AM, Allen SJ, Bejder L, Littleford-Colquhoun B, Parra GJ, Cagnazzi D, Thiele D, Palmer C, Frere CH (2014) Population differentiation and hybridisation of Australian snubfin (*Orcaella heinsohni*) and Indo-Pacific humpback (*Sousa chinensis*) dolphins in north-western Australia. *PLoS One* **9**, e101427. doi:10.1371/journal.pone.0101427

Bryden MM, Marsh H, Shaughnessy P (1998) *Dugongs, Whales, Dolphins and Seals: A Guide to the Sea Mammals of Australasia.* Allen & Unwin: Sydney.

Carzon P, Delfour F, Dudzinski K, Oremus M, Clua É (2019) Cross-genus adoptions in delphinids: one example with taxonomic discussion. *Ethology* **125**(9), 669–676. doi:10.1111/eth.12916

Charlton-Robb K, Gershwin LA, Thompson R, Austin J, Owen K, McKechnie S (2011) A new dolphin species, the Burrunan Dolphin *Tursiops australis* sp. nov., endemic to southern Australian coastal waters. *PLoS One* **6**, e24047. doi:10.1371/journal.pone.0024047

Cheal AJ, Gales NJ (1992) Growth, sexual maturity and food intake of Australian Indian Ocean bottlenose dolphins, *Tursiops truncatus*, in captivity. *Australian Journal of Zoology* **40**, 215–223. doi:10.1071/ZO9920215

Chilvers BL, Corkeron PJ (2001) Trawling and bottlenose dolphins' social structure. *Proceedings of the Royal Society of London. Series B, Containing Papers of a Biological Character* **268**, 1901–1905. doi:10.1098/rspb.2001.1732

Chilvers BL, Corkeron PJ (2003) Abundance of Indo-Pacific bottlenose dolphins, *Tursiops aduncus*, off Point Lookout, Queensland, Australia. *Marine Mammal Science* **19**, 85–95. doi:10.1111/j.1748-7692.2003.tb01094.x

Clark FE (2013) Marine mammal cognition and captive care: a proposal for cognitive enrichment in zoos and aquariums. *The Journal of Zoo and Aquarium Research* **1**(11), 1–6.

Clark FE, Davies SL, Madigan AW, *et al.* (2013) Cognitive enrichment for bottlenose dolphins (*Tursiops truncatus*): evaluation of a novel underwater maze device. *Zoo Biology* **32**, 608–619. doi:10.1002/zoo.21096

Clegg ILK, Borger-Turner JL, Eskelinen HC (2015) C-Well: the development of a welfare assessment index for captive bottlenose dolphins (*Tursiops truncatus*). *Animal Welfare (South Mimms, England)* **24**, 267–282. doi:10.7120/09627286.24.3.267

Coakley J, Crawford RL (1998) *Marine Mammal Water Quality: Proceedings of a Symposium.* US Department of Agriculture, Animal and Plant Health Inspection Service. pp. 79.

Cockcroft VG, Ross GJB (1990) Food and feeding of the Indian Ocean bottlenose dolphin off southern Natal, South Africa. In *The Bottlenose Dolphin.* (Ed SR Leatherwood) pp. 295–328. Academic Press, San Diego.

Colitz CMH, Walsh MT, McCulloch SD (2016) Characterization of anterior segment ophthalmologic lesions identified in free-ranging dolphins and those under human care. *Journal of Zoo and Wildlife Medicine* **47**, 56–75. doi:10.1638/2014-0157.1

Connor RC, Heithaus MR, Berggren P, Miksis JL (2000*a*) Kerplunking: Surface fluke-splashes during shallow-water bottom foraging by bottlenose dolphins. *Marine Mammal Science* **16**, 646–653. doi:10.1111/j.1748-7692.2000.tb00959.x

Connor RC, Wells RS, Mann J, Read AJ (2000b) The bottlenose dolphin: social relationships in a fission-fusion society. In *Cetacean Societies: Field Studies of Dolphins and Whales.* (Eds J Mann, R O'Connor, P Tyack and H Whitehead) pp. 91–126. University of Chicago Press, Chicago.

Connor RC, Sakai M, Morisaka T, Allen SJ (2019) The Indo-Pacific bottlenose dolphin (*Tursiops aduncus*). In *Ethology and Behavioral Ecology of Odontocetes* (Ed. B Würsig) pp. 345–368. Springer, Cham.

Corkeron PJ (1990) Aspects of the behavioral ecology of inshore dolphins *Tursiops truncatus* and *Sousa chinensis* in Moreton Bay, Australia. In *The Bottlenose Dolphin.* (Eds S Leatherwood and RR Reeves) pp. 285–293. Academic Press, San Diego.

Corkeron P (2009) Captivity. In *Encyclopedia of Marine Mammals.* 2nd edn (Eds WF Perrin, B Würsig and JGM Thewissen). pp. 183–188.

Crane M (1994) *Standards for Exhibiting Bottle-nosed Dolphins (*Tursiops truncatus*) in New South Wales.* NSW Agriculture, Orange, New South Wales.

Delfour F, Beyer H (2012) Assessing the effectiveness of environmental enrichment in bottlenose dolphins (*Tursiops truncatus*). *Zoo Biology* **31**, 137–150. doi:10.1002/zoo.20383

Deming AC, Wingers NL, Moore DP, Rotstein D, Wells RS, Ewing R, Hodanbosi MR, Carmichael RH (2020) Health impacts and recovery from prolonged freshwater exposure in a common bottlenose dolphin (Tursiops truncatus). *Frontiers in Veterinary Science* **7**, 235. doi:10.3389/fvets.2020.00235

Department of Agriculture, Fisheries and Forestry, Queensland (1992) *Code of Practice of the Australasian Regional Association of Zoological Parks and Aquaria: Minimum Standards for Exhibiting Wildlife in Queensland.* pp. 95–97.

Flower JE, Langan JN, Nevitt BN, Chinnadurai SK, Stacey R, Ivančić M, Adkesson MJ (2018) Neonatal critical care and hand-rearing of a bottlenose dolphin (*Tursiops truncatus*). *Aquatic Mammals* **44**, 482–490. doi:10.1578/AM.44.5.2018.482

Foroughirad V, Mann J (2013) Human fish provisioning has long-term impacts on the behavior and survival of bottlenose dolphins. *Biological Conservation* **160**, 242–249. doi:10.1016/j.biocon.2013.01.001

Fruet PF, Secchi ER, Di Tullio JC, Simões-Lopes PC, Daura-Jorge F, Costa APB, Vermeulen E, Flores PAC, Genoves RC, Laporta P, Beheregaray LB, Möller LM (2017) Genetic divergence between two phenotypically distinct bottlenose dolphin ecotypes suggests separate evolutionary trajectories. *Ecology and Evolution* **7**(21), 9131–9143. doi:10.1002/ece3.3335

Fury CA (2009) Abundance, behaviour and habitat use patterns of Indo-Pacific bottlenose dolphins (*Tursiops aduncus*) in the Clarence and Richmond River estuaries in northern New South Wales, Australia. PhD thesis. Southern Cross University, Lismore.

Galezo AA, Krzyszczyk E, Mann J (2017) Sexual segregation in Indo-Pacific bottlenose dolphins is driven by female avoidance of males. *Behavioral Ecology* **29**, 377–386. doi:10.1093/beheco/arx177

Gannon DP, Waples DM (2004) Diets of coastal bottlenose dolphins from the US mid-Atlantic coast differ by habitat. *Marine Mammal Science* **20**, 527–545. doi:10.1111/j.1748-7692.2004.tb01177.x

Geraci JR (1986) Marine mammals: nutrition and nutritional disorders. In *Zoo and Wild Animal Medicine.* (Ed. ME Fowler) pp. 760–764. Saunders, Philadelphia.

Geraci JR, St Aubin DL (1980) Nutritional disorders of captive fish eating animals. In *The Comparative Pathology of Zoo Animals.* (Eds RJ Montali and G Migaki) pp. 41–49. Smithsonian Institution Press, Washington, DC.

Gerber L, Wittwer S, Allen SJ, Holmes KG, King SL, Sherwin WB, Wild S, Willems EP, Connor RC, Krützen M (2021) Cooperative partner choice in multi-level male dolphin alliances. *Scientific Reports* **11**, 6901. doi:10.1038/s41598-021-85583-x

Goodall SH (1980) Rearing of an orphaned bottle-nosed dolphin (*Tursiops truncatus*). Unpublished notes.

Gridley T, Elwen SH, Harris G, Moore D, Hoelzel A, Lampen F (2018) Hybridization in bottlenose dolphins: a case study of *Tursiops aduncus* × *T. truncatus* hybrids and successful backcross hybridization events. *PLoS One* **13**, e0201722. doi:10.1371/journal.pone.0201722

Grimm D (2011) Are dolphins too smart for captivity? *Science* **332**, 526–529. doi:10.1126/science.332.6029.526

Hanf D, Hodgson AJ, Kobryn H, Bejder L, Smith JN (2022) Dolphin distribution and habitat suitability in north western Australia: applications and implications of a broad-scale, non-targeted dataset. *Frontiers in Marine Science* **8**, 733841. doi:10.3389/fmars.2021.733841

Hart LB, Wells RS, Schwacke LH (2013) Reference ranges for body condition in wild bottlenose dolphins *Tursiops truncatus*. *Aquatic Biology* **18**, 63–68. doi:10.3354/ab00491

Hart LB, Wischusen K, Wells RS (2017) Rapid assessment of bottlenose dolphin (*Tursiops truncatus*) body condition: there's an app for that. *Aquatic Mammals* **43**, 635–643. doi:10.1578/AM.43.6.2017.635

Haughey R, Hunt T, Hanf D, Rankin RW, Parra GJ (2020) Photographic capture-recapture analysis reveals a large population of Indo-Pacific bottlenose dolphins (*Tursiops aduncus*) with low site fidelity off the North West Cape, Western Australia. *Frontiers in Marine Science* **6**, 781. doi:10.3389/fmars.2019.00781

Herrman JM, Morey JS, Takeshita R, De Guise S, Wells RS, McFee W, Speakman T, Townsend F, Smith CR, Rowles T, Schwacke L

(2020) Age determination of common bottlenose dolphins (*Tursiops truncatus*) using dental radiography pulp: tooth area ratio measurements. *PLoS One* **15**(11), e0242273. doi:10.1371/journal.pone.0242273

Hoelzel AR, Potter CW, Best PB (1998) Genetic differentiation between parapatric 'nearshore' and 'offshore' populations of the bottlenose dolphin. *Proceedings. Biological Sciences* **265**, 1177–1183. doi:10.1098/rspb.1998.0416

Hohn AA, Scott MD, Wells RS, Sweeney JC, Irvine AB (1989) Growth layers in teeth from known-age, free-ranging bottlenose dolphins. *Marine Mammal Science* **5**(4), 315–342. doi:10.1111/j.1748-7692.1989.tb00346.x

Ivancic M, Bailey JE, Costidis AM (2014) Technique for ultrasound-guided arterial catheter placement in the pectoral flipper of the bottlenose dolphin (*Tursiops truncatus*). In *Proceedings of 45th Annual Conference of International Association of Aquatic Animal Medicine* Gold Coast, Queensland.

Ivancic M, Johnson S, Costidis AM, Renner MS (2015) A technique for ultrasound-guided catheterization of a peripheral vein in the bottlenose dolphin (*Tursiops truncatus*). In *Proceedings of 46th Annual Conference of International Association of Aquatic Animal Medicine* Chicago, Illinois.

Jackson SM, Groves CP (2015) *Taxonomy of Australian Mammals.* CSIRO Publishing, Melbourne.

Jedensjö M, Kemper CM, Milella M, Willems EP, Krützen M (2020) Taxonomy and distribution of bottlenose dolphins (genus *Tursiops*) in Australian waters: an osteological clarification. *Canadian Journal of Zoology* **98**, 461–479. doi:10.1139/cjz-2018-0270

Jefferson TA, Webber MA, Pitman RL (2008) *Marine Mammals of the World: A Comprehensive Guide to their Identification.* Academic Press, London.

Jiménez PJ, Alava JJ (2015) Strand-feeding by coastal bottlenose dolphins (*Tursiops truncatus*) in the Gulf of Guayaquil, Ecuador. *The Latin American Journal of Aquatic Mammals* **10**, 33–37. doi:10.5597/lajam00191

Jones R (2014) Module Four: life support systems. The Aquarium Vet. www.theaquariumvet.com/store-2/module-four-life-support-systems (accessed 19 November 2022).

Karniski C, Krzyszczyk E, Mann J (2018) Senescence impacts reproduction and maternal investment in bottlenose dolphins. *Proceedings. Biological Sciences* **285**, 20181123. doi:10.1098/rspb.2018.1123

Kemper CM, Trentin E, Tomo I (2014) Sexual maturity in male Indo-Pacific bottlenose dolphins (*Tursiops aduncus*): evidence for regressed/pathological adults. *Journal of Mammalogy* **95**, 357–368. doi:10.1644/13-MAMM-A-007.1

Kemper CM, Talamonti M, Bossley M, Steiner A (2019) Sexual maturity and estimated fecundity in female Indo-Pacific bottlenose dolphins (*Tursiops aduncus*) from South Australia: combining field observations and postmortem results. *Marine Mammal Science* **35**, 40–57. doi:10.1111/mms.12509

Koga S, Yanagisawa M, Koga H, Ueda K, Kawazu I, Tokutake K, Funasaka N, Yoshioka M, Miyahara H (2019) Reproductive ability of elderly male Indo-Pacific bottlenose dolphins (*Tursiops aduncus*) in captivity. *Mammal Study* **44**, 135–139. doi:10.3106/ms2018-0044

Kopps AM, Ackermann CY, Sherwin WB, Allen SJ, Bejder L, Krützen M (2014) Cultural transmission of tool use combined with habitat specializations leads to fine-scale genetic structure in bottlenose dolphins. *Proceedings of the Royal Society B* **281**(1782), 20133245.

Krützen M, Mann J, Heithaus MR, Connor RC, Bejder L, Sherwin WB (2005) Cultural transmission of tool use in bottlenose dolphins. *Proceedings of the National Academy of Sciences of the United States of America* **102**, 8939–8943. doi:10.1073/pnas.0500232102

Lacave G, Eggermont M, Verslycke T, Brook F, Salbany A, Roque L, Kinoshita R (2004) Prediction from ultrasonographic measurements of the expected delivery date in two species of bottlenosed dolphin (*Tursiops truncatus* and *Tursiops aduncus*). *The Veterinary Record* **154**(8), 228–233. doi:10.1136/vr.154.8.228

Lauderdale LK, Mellen JD, Walsh MT, Granger DA, Miller LJ (2021*a*) Towards understanding the welfare of cetaceans in accredited zoos and aquariums. *PLoS One* **16**(8), e0255506. doi:10.1371/journal.pone.0255506

Lauderdale LK, Walsh MT, Mellen JD, Granger DA, Miller LJ (2021*b*) Environmental enrichment, training, and habitat characteristics of common bottlenose dolphins (*Tursiops truncatus*) and Indo-Pacific bottlenose dolphins (*Tursiops aduncus*). *PLoS One* **16**(8), e0253688. doi:10.1371/journal.pone.0253688

Le-Bert CR, Smith CR, Poindexter J, Ardente A, Meegan J, Wells RS, Venn-Watson S, Jensen ED, Sakhaee K (2018) Comparison of potential dietary and urinary risk factors for ammonium urate nephrolithiasis in two bottlenose dolphin (Tursiops truncatus) populations. *American Journal of Physiology. Renal Physiology* **315**, 231–237. doi:10.1152/ajprenal.00606.2017

Lear RJ, Bryden M (1980) A study of the bottlenose dolphin, *Tursiops truncatus,* in eastern Australian waters, Occasional Paper No. 4. Australian National Parks and Wildlife Service, Canberra.

Lukoschek V, Chilvers BL (2008) A robust baseline for bottlenose dolphin abundance in coastal Moreton Bay: a large carnivore living in a region of escalating anthropogenic impacts. *Wildlife Research* **35**, 593–605. doi:10.1071/WR07021

Mackie JT, Blyde D, Harris L, Roe WD, Keyburn AL (2020) Brucellosis associated with stillbirth in a bottlenose dolphin in Australia. *Australian Veterinary Journal* **98**, 92–95. doi:10.1111/avj.12903

Mann J, Kemps C (2003) The effects of provisioning on maternal care in wild bottlenose dolphins, Shark Bay, Australia. In *Marine Mammals: Fisheries, Tourism and Management* (Eds N Gales, M Hindell and R Kirkwood). pp. 304–317. CSIRO Publishing, Melbourne.

Marino L, Connor RC, Fordyce RE, Herman LM, Hof PR, Lefebvre L, Lusseau D, McCowan B, Nimchinsky EA, Pack AA, Rendell L, Reidenberg JS, Reiss D, Uhen MD, Van Der Gucht E, Whitehead H (2007) Cetaceans have complex brains for complex cognition. *PLoS Biology* **5**(5), e139. doi:10.1371/journal.pbio.0050139

Martelli PR, Churgin SM, Fernando N, Abel G, Wong G, Lee FK (2016) Oral sedation in open water to facilitate recovery and care of a critically injured Indo-Pacific humpback dolphin (*Sousa chinensis*). In *Proceedings of 47th Annual Conference of International Association of Aquatic Animal Medicine.* Virginia Beach, USA.

Martini M, Altomonte I, Federico Sommer M, Gill C, Biancani B, Licitra R, Salari F (2022) Milk composition, fatty acids profile and fat globule size of bottlenose dolphin (*Tursiops truncatus*, Montagu 1821) milk at early lactation. *Veterinary Research Communications* **46**, 577–583. doi:10.1007/s11259-021-09845-9

Menkhorst PW (2001) *A Field Guide to the Mammals of Australia*. Oxford University Press, Melbourne.

Miller LJ, Lauderdale LK, Mellen JD, Walsh MT, Granger DA (2021a) Relationships between animal management and habitat characteristics with two potential indicators of welfare for bottlenose dolphins under professional care. *PLoS One* **16**(8), e0252861. doi:10.1371/journal.pone.0252861

Miller LJ, Lauderdale LK, Bryant JL, Mellen JD, Walsh MT, Granger DA (2021b) Behavioral diversity as a potential positive indicator of animal welfare in bottlenose dolphins. *PLoS One* **16**(8), e0253113. doi:10.1371/journal.pone.0253113

Moore M, Walsh M, Bailey J, Brunson D, Gulland F, Landry S, Mattila D, Mayo C, Slay C, Smith J, Rowles T (2010) Sedation at sea of entangled North Atlantic Right Whales (*Eubalaena glacialis*) to enhance disentanglement. *PLoS One* **5**(3), e9597. doi:10.1371/journal.pone.0009597

Natoli A, Peddemors VM, Rus Hoelzel A (2004) Population structure and speciation in the genus *Tursiops* based on microsatellite and mitochondrial DNA analyses. *Journal of Evolutionary Biology* **17**, 363–375. doi:10.1046/j.1420-9101.2003.00672.x

Norris KS (1961) Standardized methods for measuring and recording data on the smaller cetaceans. *Journal of Mammalogy* **42**, 471–476. doi:10.2307/1377364

O'Brien JK, Robeck TR (2006) Development of sperm sexing and associated reproductive technology for sex preselection of captive bottlenose dolphins (*Tursiops truncatus*). *Reproduction, Fertility and Development* **18**, 319–329. doi:10.1071/RD05108

Palmer C, Brooks L, Parra GJ, Rogers T, Glasgow D, Woinarski JCZ (2014) Estimates of abundance and apparent survival of coastal dolphins in Port Essington harbour, Northern Territory, Australia. *Wildlife Research* **41**, 35–45. doi:10.1071/WR14031

Parra GJ (2005) Behavioural ecology of Irrawaddy, *Orcaella brevirostris* (Owen in Gray, 1866), and Indo-Pacific humpback dolphins, *Sousa chinensis* (Osbeck, 1765), in northeast Queensland, Australia: a comparative study. PhD thesis, James Cook University, Townsville.

Perrin WF, Reilly SB (1984) Reproductive parameters of dolphins and small whales of the Family Delphinidae. *Report of the International Whaling Commission. Special Issue* **6**, 97–133.

Perrin WF, Würsig BG, Thewissen JGM (2002) *Encyclopedia of Marine Mammals*. Academic Press, San Diego.

Ridgway S, Fenner C (1982) Weight-length relationships of wild-caught and captive Atlantic bottlenose dolphins. *Journal of the American Veterinary Medical Association* **181**, 1310–1315.

Rosel PE, Archer FI, Baker CS, Boness DJ, Brownell RL, Jr, Churchill M, Costa AP, Domning DP, Fordyce RE, Jefferson TA, Kinze C, McGowen M, Oliveira LR, Wang JY, Yamada TK (2021) List of marine mammal species and subspecies. https://marinemammalscience.org/science-and-publications/list-marine-mammal-species-subspecies/ (accessed 22 May 2022).

Rowe LE, Dawson SM (2008) Laser photogrammetry to determine dorsal fin size in a population of bottlenose dolphins from Doubtful Sound, New Zealand. *Australian Journal of Zoology* **56**, 239–248. doi:10.1071/ZO08051

Rowe LE, Dawson SM (2009) Determining the sex of bottlenose dolphins from Doubtful Sound using dorsal fin photographs. *Marine Mammal Science* **25**, 19–34. doi:10.1111/j.1748-7692.2008.00235.x

RSPCA (2019) *Research Report: The Welfare of Dolphins in Captivity*. RSPCA Australia, Canberra.

Ruiz CU, Sánchez A, Maldonado FG (2009) Social and individual behavior of a group of bottlenose dolphins (*Tursiops truncatus*) in open and closed facilities. *Veterinaria (México)* **40**, 381–387.

Schroeder JP (1990) Reproductive aspects of marine mammals. In *Handbook of Marine Mammal Medicine: Health, Disease, and Rehabilitation*. (Ed. LA Dierauf) pp 353–369. CRC Press, Boca Raton.

Shane SH, Wells RS, Würsig B (1986) Ecology, behavior and social organization of the bottlenose dolphin: a review. *Marine Mammal Science* **2**, 34–63. doi:10.1111/j.1748-7692.1986.tb00026.x

Shyan MR, Merritt D, Kohlmeier NM, Barton K, Tenge J (2002) Effects of pool size on free-choice selections by Atlantic bottlenosed dolphins at one zoo facility. *Journal of Applied Animal Welfare Science* **5**, 215–225. doi:10.1207/S15327604JAWS0503_05

Smith HC, Sprogis KR (2016) Seasonal feeding on giant cuttlefish (*Sepia apama*) by Indo-Pacific bottlenose dolphins (*Tursiops aduncus*) in south-western Australia. *Australian Journal of Zoology* **64**, 8–13. doi:10.1071/ZO15075

Smith H, Samuels A, Bradley S (2008) Reducing risky interactions between tourists and free-ranging dolphins (*Tursiops* sp.) in an artificial feeding program at Monkey Mia, Western Australia. *Tourism Management* **29**, 994–1001. doi:10.1016/j.tourman.2008.01.001

Smolker R, Richards A, Connor R, Mann J, Berggren P (1997) Sponge carrying by dolphins (Delphinidae, *Tursiops* sp.): A foraging specialization involving tool use? *Ethology* **103**, 454–465. doi:10.1111/j.1439-0310.1997.tb00160.x

Species360 (2023) Zoological Information Management System (ZIMS). Minneapolis, USA. www.species360.org.

Sprogis KR, Raudino HC, Hocking D, Bejder L (2017) Complex prey handling of octopus by bottlenose dolphins (*Tursiops aduncus*). *Marine Mammal Science* **33**, 934–945. doi:10.1111/mms.12405

Steiner A (2011) Activity budget of inshore Indo-Pacific bottlenose dolphins (*Tursiops aduncus*): a critical evaluation of methods and comparison among other populations. *Marine Mammal Science* **27**, 20–38. doi:10.1111/j.1748-7692.2010.00388.x

Stone BM, Blyde DJ, Saliki JT, Blas-Machado U, Bingham J, Hyatt A, Wang J, Payne J, Crameri S (2011) Fatal cetacean morbillivirus infection in an Australian offshore bottlenose dolphin (*Tursiops truncatus*). *Australian Veterinary Journal* **89**, 452–457. doi:10.1111/j.1751-0813.2011.00849.x

Sweeney JC, Ridgway SH (1975) Common diseases of small cetaceans. *Journal of the American Veterinary Medical Association* **167**, 533.

Tezanos-Pinto G, Baker CS, Russell K, Martien K, Baird RW, Hutt A, Stone G, Mignucci-Giannoni AA, Caballero S, Endo T, Lavery S, Oremus M, Olavarría C, Garrigue C (2008) A worldwide perspective on the population structure and genetic diversity of bottlenose dolphins (*Tursiops truncatus*) in New Zealand. *The Journal of Heredity* **100**, 11–24. doi:10.1093/jhered/esn039

Tolley KA, Read AJ, Wells RS, Urian KW, Scott MD, Irvine AB, Hohn AA (1995) Sexual dimorphism in wild bottlenose

dolphins (*Tursiops truncatus*) from Sarasota, Florida. *Journal of Mammalogy* **76**, 1190–1198. doi:10.2307/1382611
Torres LG, Read AJ (2009) Where to catch a fish? The influence of foraging tactics on the ecology of bottlenose dolphins (*Tursiops truncatus*) in Florida Bay, Florida. *Marine Mammal Science* **25**, 797–815. doi:10.1111/j.1748-7692.2009.00297.x
Townsend F (1999) Hand-rearing techniques for neonate cetaceans. In *Zoo and Wild Animal Medicine Current Therapy 4.* (Eds ME Fowler and RE Miller) pp. 493–497. W.B. Saunders, Philadelphia.
Tsai YJJ, Mann J (2013) Dispersal, philopatry, and the role of fission-fusion dynamics in bottlenose dolphins. *Marine Mammal Science* **29**, 261–279. doi:10.1111/j.1748-7692.2011.00559.x
Ugaz C, Valdez RA, Romano MC, Galindo F (2013) Behavior and salivary cortisol of captive dolphins (*Tursiops truncatus*) kept in open and closed facilities. *Journal of Veterinary Behavior* **8**, 285–290. doi:10.1016/j.jveb.2012.10.006
van Aswegen M, Christiansen F, Symons J, Mann J, Nicholson K, Sprogis K, Bejder L (2019) Morphological differences between coastal bottlenose dolphin (*Tursiops aduncus*) populations identified using non-invasive stereo-laser photogrammetry. *Scientific Reports* **9**, 12235. doi:10.1038/s41598-019-48419-3
Van Bressem MF, Van Waerebeek K, Aznar FJ, Raga JA, Jepson PD, Duignan P, Deaville R, Flach L, Viddi F, Baker JR, Di Beneditto AP, Echegaray M, Genov T, Reyes J, Felix F, Gaspar R, Ramos R, Peddemors V, Sanino GP, Siebert U (2009) Epidemiological pattern of tattoo skin disease: a potential general health indicator for cetaceans. *Diseases of Aquatic Organisms* **85**, 225–237. doi:10.3354/dao02080
Venn-Watson SK, Jensen ED, Smith CR, Xitco M, Ridgway SH (2015) Evaluation of annual survival and mortality rates and longevity of bottlenose dolphins (*Tursiops truncatus*) at the United States Navy Marine Mammal Program from 2004 through 2013. *Journal of the American Veterinary Medical Association* **246**, 893–898. doi:10.2460/javma.246.8.893
von Fersen L, Walb R (Eds) (2018) EAZA Best Practice Guidelines: Antillean manatee (*Trichechus manatus manatus*). European Association of Zoos and Aquaria. Tiergarten Nürnberg, Germany.
Wang JY, Riehl KN, Dungan SZ (2014) Family Delphinidae (ocean dolphins) In *Handbook of the Mammals of the World.* Vol. 4. *Sea Mammals.* (Eds DE Wilson and RA Mittermeier) pp. 410–527. Lynx Edicions, Barcelona.
Waples KA, Gales NJ (2002) Evaluating and minimising social stress in the care of captive bottlenose dolphins (Tursiops aduncus). *Zoo Biology* **21**, 5–26. doi:10.1002/zoo.10004
Wells RS (2009) Learning from nature: bottlenose dolphin care and husbandry. *Zoo Biology* **28**, 635–651. doi:10.1002/zoo.20252
Wells RS, Scott MD (2002) Bottlenose dolphins. In *Encyclopedia of Marine Mammals.* (Eds WF Perrin, B Würsig and JGM Thewissen) pp. 122–129. Academic Press, San Diego.
Wells RS, Scott MD (2018) Bottlenose Dolphin, *Tursiops Truncatus.* In *Encyclopedia of Marine Mammals.* 3rd edn. (Eds B Würsig, JGM Thewissen and KM Kovacs) pp. 118–125. Academic Press, San Diego.
Wild S, Krützen M, Rankin RW, Hoppitt WJE, Gerber L, Allen SJ (2019*a*) Long-term decline in survival and reproduction of dolphins following a marine heatwave. *Current Biology* **29**, R239–R240. doi:10.1016/j.cub.2019.02.047
Wild S, Allen SJ, Krützen M, King SL, Gerber L, Hoppitt WJE (2019*b*) Multi-network-based diffusion analysis reveals vertical cultural transmission of sponge tool use within dolphin matrilines. *Biology Letters* **15**(7), 20190227. doi:10.1098/rsbl.2019.0227
Wild S, Hoppitt WJE, Allen SJ, Krutzen M (2020) Integrating genetic, environmental, and social networks to reveal transmission pathways of a dolphin foraging innovation. *Current Biology* **30**, 3024–3030. doi:10.1016/j.cub.2020.05.069
Wilson DE, Mittermeier RA (Eds) (2014) *Handbook of the Mammals of the World.* Vol. 4. *Sea Mammals.* Lynx Editions, Barcelona.
Wright KC, Griffin M, Lanter K, Salmon D (2005) The development and research of a commercially prepared fish substitute for captive marine birds and mammals. In *Proceedings of the Dolphin Neonatal and Reproduction Symposium.* (Ed. D Olson) p. 46. Indianapolis Zoo, Indianapolis.

# Appendix 1: Glossary

**Alpha** - Dominant or top-ranking animal.

**Ambient temperature** - Air temperature.

**Anaemia** - A reduction below normal in the number or volume of erythrocytes (red blood cells) or in the quantity of haemoglobin (blood protein) in the blood. Clinically, it manifests as weakness, exercise intolerance, hyperpnoea (abnormal increase in depth and rate of respiration), which is only moderate, paleness of the mucous membranes, abnormal rapid heart rate and a large increase in the intensity of the heart sounds.

**Anaesthesia** - Loss of feeling or sensation. Usually the result of artificial means.

**Anorexia** - Complete loss of appetite.

**Anterior** - Towards the front.

**Anxiolytic** - A mild sedative, such as diazepam, used for the relief of anxiety.

**Arboreal** - Tree climbing.

**Ascites** - Abnormal accumulation of serous (oedematous) fluid within the peritoneal cavity. Characterised by distension of the abdomen, a fluid thrill on percussion, typical ground glass appearance on radiography and a positive result on paracentesis.

**Ataxia** - Failure of muscular coordination; irregularity of muscular action.

**Auscultation** - Listening for sounds produced within the body, chiefly to ascertain the condition of the thoracic or abdominal viscera. May be performed with the unaided ear or with a stethoscope.

**Bacula** - Penis bone that occurs in many species of mammals, including dogs and bats.

**Behavioural enrichment** - Food or devices provided to captive animals to enhance natural behaviours and activity patterns.

**Blastocyst** - An early stage of embryonic development that consists of a hollow ball of cells ≤ 70–100 cells and is only 0.25 mm in diameter) with a localised thickening (embryonic disk) that will develop into the actual embryo.

**Bronchitis** - Inflammation of one or more bronchi (any of the larger passages conveying air within the lungs). Signs of acute bronchitis include fever and an irritating cough.

**Caecotrophy** - The reingestion of faeces of high nutritive value derived from the caecal contents. This material is excreted as soft faeces (caecotrophes) at regular intervals during daylight hours (e.g. when common ringtails are resting in tree hollows or dreys). The caecotrophs are taken directly from the cloaca and are chewed before swallowing.

**Carnivorous** - Feeds on other animals.

**Cholangitis** - inflammation of the bile ducts.

**Chorioallantoic placenta** - More developed placenta formed of trophoblast and allantois and is vascularised by the allantoic or umbilical vessels. Compare with yolk sac placenta.

**Cloaca** - A common passage for faecal, urinary and reproductive discharges.

**Conjunctivitis** - Inflammation of the delicate membrane lining the eyelids and covering parts of the eyeball.

**Convulsions** - A series of involuntary contractions of the voluntary muscles.

**Coprophagy** - Ingestion of faeces.

**Cornified epithelial cells** - Keratinised tissue cells that are shed during oestrus.

**Corticosteroids** - Any of the hormones produced by the outer part of the adrenal gland.

**Crepuscular** - Of the twilight. Applied to animals that are active at dusk and dawn.

**Cystitis** - Inflammation of the urinary bladder.

**Dehydration** - An excessive loss of water from the body. It can be tested by pinching the skin and letting it go. If it does not fall back in a few seconds the animal may be severely dehydrated.

**Dentine** - One of the hard tissues of the teeth that constitutes most of its bulk. It lies between the pulp cavity and the enamel and where it is not covered by enamel is covered by cementum, the third hard substance of the tooth.

**Dermatomycosis** - A fungal infection of the skin.

**Distribution** - The area or range that is occupied by a species.

**Diurnal** - Active during the day.

**Drey** - A nest constructed by ring-tailed possums. A typical drey consists of a leaf, twig, grass or fern frond structure situated in the branches or trunk forks of a shrub or tree. Live branches may be bent around and incorporated together with their leaves into the structure.

**Dysentery** - Disease of the intestinal tract causing severe diarrhoea with blood and mucus.

**Dyspnoea** - Laboured or difficult breathing.

**Dysphagia** - Difficulty in swallowing.

**Echolocation** - The location of objects by means of reflected sound.

**Ectoparasite** - A parasite that lives on the surface of the host's body.

**Embryonic diapause** - During diapause a viable embryo is carried in the uterus for long periods (many months) with its development arrested at the stage of a blastocyst. In all macropods (except the Tammar wallaby and the Bennett's wallaby) this is lactational anoestrous where the suckling stimuli stops the development of the blastocyst. However, in both of these species the development is determined by the summer solstice.

**Emesis** - The act of vomiting.

**Endoparasite** - A parasite that lives within the host's body.

**Enteritis** - Inflammation of the intestinal mucosa resulting in the clinical signs of diarrhoea, sometimes dysentery, abdominal pain and dehydration and electrolyte loss and imbalance.

**Eosophilic** - Staining readily with eosin. Pertaining to eosinophils or eosinophilia (i.e. a granular leukocyte with a nucleus that has two lobes connected by a thread of chromatin and cytoplasm containing course, round or rod-shaped eosinophilic granules (lysosomes) of uniform size).

**Epiphyseal fusion** - Fusion of the long bones.

**Etymology** - The formation or development of words and their meanings.

**Fibrosis** - Formation of fibrous tissue.

**Flaccidity** - Lack of tone of muscular vascular organ or tissue.

**Flehmen** - A type of lip curling exhibited by males when encountering the urine of females - baring the gum and wrinkling the nose- which appears to be a mechanism to aspirate the vomeronasal organ. It is sometimes associated with the animal making rapid licking and mouthing movements during and after showing flehmen. This process appears to be involved in males detecting if the female is in oestrus and is therefore ready to mate.

**Fossorial** - An animal that burrows or digs.

**Foster species** - When a lactating female of one species is used to rear a juvenile of another species.

**Gastritis** - Inflammation of the lining of the stomach.

**Gestation period** - The duration of pregnancy.

**Granuloma** - A tumour-like mass or nodule of granulation tissue.

**Growth chart/curve** - A graph that shows the relationship between age and a body measurement such as bodyweight, head length, head width or total length.

**Habitat** - The natural environment occupied by a particular organism.

**Harem** - A form of social organisation in which the male controls more than one female for mating.

**Hemiplegic** - Paralysis of one side of the body. Usually caused by a brain lesion such as a tumour.

**Hepatic** - Emanating from or pertaining to the liver.

**Herbivorous** - Animals that subsist in their natural state entirely by eating plants and plant products.

**Hibernacula** - The location in which an individual hibernates.

**Hibernation** - A type of deep and prolonged torpor in which the minimum body temperature decreases to within several degrees of the ambient temperature and can decrease to 1-6°C and last several weeks or more.

**Hierarchy** - A form of social organisation in which there is an order and each individual has a rank or status from the top-ranked most dominant animal to the least-ranked most subordinate individual. The rank is turn reflects the individual's priority to feeding and mating success.

**Home range** - The area in which an individual travels in order to fulfil feeding requirements, social behaviour, reproduction and nests.

**Honeydew** - The sugary waste excreted by psyllids on the leaves and other parts of the plant on which they feed. The presence of honeydew can be recognised by the growth of a black fungus called 'sooty mould'.

**Hybrid** - An offspring of parents of different strains, variations, species or subspecies.

**IATA** - International Air Transport Association supports many areas of aviation activity and has developed standards for the transport of live animals within aircraft.

**Illthrift** - Failure to grow, increase in weight or maintain weight in the presence of apparently adequate food supplies and the absence of recognisable disease.

**Immunosuppression** - Diminished immune responsiveness.

**Inanition** - The exhausted state due to prolonged under nutrition. Starvation.

**Incubation period** - The period from when the egg is laid to the individual inside hatching from the egg in oviparous species.

**Interspecific** - Between species.

**Intraspecific** - Within species.

**Iteroparity** - At least some individuals breed more than once (e.g. most species of dasyurids and most mammals. Compare with semelparity.

**IUCN Status** - International Union for the Conservation of Natural Resources Status.

- Extinct: A taxon in which there is no reasonable doubt that the last individual has died.
- Critically Endangered: A taxon facing an extremely high risk of extinction in the wild in the immediate future.
- Endangered: A taxon that is not endangered but is facing a very high risk of extinction in the wild in the near future.
- Vulnerable: A taxon that is not critically endangered or endangered but is facing a high risk of extinction in the wild in the medium-term future.
- Unknown
- Lower Risk: A taxon that does not fulfil any of the threatened categories. Taxa in this category can be separated into three subcategories:
  1) Conservation Dependent (cd). Taxa that are the focus of a continuing taxon-specific or habitat-specific conservation program, the cessation of which would result in the taxon qualifying for one of the three threatened categories listed above within a period of 5 years.
  2) Near Threatened (nt). Taxa that do not qualify for Conservation Dependent, but which are close to qualifying for Vulnerable.
  3) Least Concern (lc). Taxa that do not qualify for Conservation Dependent or Near Threatened.
- Unknown (Data Deficient): When there is inadequate information to make a direct, or indirect, assessment of a species risk of extinction based on its distribution and/or population status.

**Jaundice** - Yellowness of the skin, sclerae (outer coat of the eyeball), mucous membranes and excretions due to hyperbilirubinaemia and deposition of bile pigments.

**Labial** - Edge of the tooth closest to the lips (i.e. the outer edge of the tooth).

**Laryngitis** - Inflammation of the mucous membrane of the larynx, characterised by cough, pain on palpation over the larynx, dysphagia and possibly regurgitation through the nose.

**Lerps** - The coverings or testa excreted by the nymphs of psyllids, under which they shelter and feed and which enlarge as they develop. The adult then emerges from the lerp and lives a free existence.

**Lethargy** - A condition of drowsiness or indifference.

**Lingual** - Edge of the tooth closest to the tongue (i.e. the inner edge of the tooth).

**Longevity** - Duration of an individual animal's life.

**Lymphopenia** - Decrease in the number of lymphocytes of the blood.

**Lymphadenopathy** - Disease of the lymph nodes.

**Mating system**

- Monogamous - Mating of a male with a female involving no extra individuals of either sex. Usually the bond operates through the breeding season and in some cases may extend through the adult life of two individuals.
- Polygynous -A male has more than one female partner during a single breeding season.
- Polyandrous - A female has more than one male at a time (usually during a single breeding season).
- Polygamous - A pattern of mating in which both males and females have more than one sexual partner during a single breeding season.

**Meningomyelitis** - Inflammation of the spinal cord and its meninges. Meninges are the three membranes covering the brain and spinal cord.

**Mono-oestrus** -Only one oestrous cycle per year.

**Morphometrics** - The measurement of body parts.

**MPN** - most probable number

**Mucocutaneous** - Pertaining to mucous membranes and skin.

**Mucopurulent** - Marked by an exudate containing both mucous and pus.

**Multiple paternity** - A litter in which the offspring have more than one father. Is the result of the female mating with more than one male and the occurrence of sperm competition after mating.

**Myoglobinuria** - The presence of myoglobin in the urine. The urine appears dark red, to the naked eye, is indistinguishable from urine containing haemoglobin, although the two conditions can be differentiated by laboratory tests. Serious kidney damage is a possible sequel to myoglobinuria. It is seen in conditions where there is sudden, extensive damage to skeletal muscle in mature animals.

**Myopathy** - Disease of the muscles. Usually the result of increased lactic acid in the muscles.

**Nasopharyngeal** - Pertaining to the nasal and pharyngeal cavities.

**Necrobacillosis** - Tissue damage that manifests as areas of necrosis and the pus has a characteristic rotting odour.

**Necropsy** - A post mortem or autopsy undertaken on an animal to learn more about its disease status and/ or cause of death.

**Neonate** - A newly born animal.

**Neoplasia** - The formation of a neoplasm (tumour or abnormal growth).

**Nephritis** - Inflammation of the kidney; a focal or diffuse proliferation or destructive disease that may involve the glomerulus, tubule or interstitial renal tissue.

**Neuroleptic agent** - A drug that immobilises; however, the animal is not completely unconscious but is insensitive to painful stimuli. This helps to relax the animal (e.g. kangaroo) so that it does not become stressed or struggle to escape.

**Neutrophilia** - An increase in the number of neutrophils (one of three granular leukocytes) in the blood.

**Nocturnal** - Active during the night.

**Nomenclature** - The naming of species.

**Nulliparous** - Never given birth.

**Nystagmus** - A periodic, rhythmic, involuntary movement of both eyeballs in unison. There is a slow component in one direction and a quick return. The movement may be vertical, horizontal or rotary.

**Oedematous** - An abnormal accumulation of fluid.

**Oestrus cycle** - The regular interval between periods during which the female is sexually active. See also Mono-oestrus and Polyoestrus.

**Omnivorous** - Eating both plant and animal food.

**Ophthalmia** - Severe inflammation of the eye or of the conjunctiva (delicate membrane lining the eyelids and covering parts of the eyeball) or deeper structures of the eye.

**Osteosarcoma** - Bone-producing malignant tumour.

**Otoscope** - An instrument for inspecting the ear.

**Ovulation number** - The number of ova discharged from the Graafian follicle (a small sac embedded in the ovary that encloses the ovary).

**Papillae** - A small, nipple-shaped projection or elevation. Can be the precurser to hair growth.

**Paracentesis** - Surgical puncture of a cavity for the aspiration of fluid.

**Parakeratotic** - Referring to parakeratosis, which is the persistence of the nuclei of keratinocytes (cells of the epidermis that produce keratin) as they rise into the horny layer of the skin causing a lesion.

**Paratenic host** - animal acting as a transport host in a parasite's lifecycle. The parasite persists but does not multiply in the animal. The main host is infected after ingesting the paratenic host.

**Parenteral** - Not through the alimentary canal (by subcutaneous, intramuscular, intrasternal or intravenous injection; e.g. parenteral fluid therapy).

**Paresis** - Slight or incomplete paralysis.

**PCR** - polymerase chain reaction

**Pica** - Licking and eating unnatural articles of food (e.g. kangaroos eating soil).

**Photoperiod** - The length of time between consecutive light phases or dark phases.

**Pleroceroids** - The second larval stage of a pseudophyllidean cestodes; follows the proceroid.

**Polyoestrus** -More than one oestrous cycle per year.

**Postpartum oestrus** - When a female is ready for mating shortly after giving birth, generally 1-3 days postpartum.

**Posterior** - Towards the back or rear.

**PPE** - personal protective equipment

**ppt** - parts per thousand (g/L)

**Proglottids** - Segments that make up the body of a tapeworm.

**Protozoan** - A phylum comprising the unicellular eukaryotic organisms.

**Pruritus** - Symptom of itching, which is the prominent feature of most parasitic skin diseases.

**Psyllids** - Insects of the Family Psyllidae, Order Homoptera. They feed by sucking sap and the nymphs live in galls or free on the plant and often produce honeydew or protective wax.

**Puggle** - Juvenile echidna.

**Purulent** - Containing or forming pus.

**Pyrexic** - Pertaining to a fever.

**Quarantine** - Restrictions placed on entering or leaving premises where a case of communicable disease exists or is suspected.

**Retinitis** - Inflammation of the retina.

**Reverse cycle lighting** - The use of artificial lighting such as white, blue or red lighting during the day to simulate night time so that nocturnal animals are more active. During the night, bright lights are lit to simulate day time and the nocturnal species retreat to their nests.

**Reproductive status** - Condition of males and females with respect to reproduction (e.g. non-breeding, pregnant or lactating).

**Rhinitis** - Inflammation of the mucous membranes of the nose.

**Salmonellosis** - A highly contagious disease of all animal species caused by *Salmonella.*

**Sedation** - The allaying of irritability or excitement, especially by administration of a sedative.

**Sedentary** - Of inactive habits.

**Semelparity** - Reproduce only once in a lifetime. The males die after their first mating season and it is usually associated with testicular failure (e.g. *Antechinus* spp. and *Phascogale* sp.) Compare with iteroparity.

**Septicaemia** - Systemic disease associated with the presence and persistence of pathogenic micro-organisms and their toxins in the blood.

**Serous** - Pertaining to serum that is thin and watery.

**Sexual dimorphism** -A difference in body size between the two sexes.

**Social behaviour** - Behaviour between individuals within a species.

**Solitary** - Individuals that avoid the company of others.

**Sperm competition** - Process where the sperm from more than one male competes in the female reproductive tract after mating to fertilise the female's ova. Can result in multiple paternity within litters in some species.

**Stereotypic behaviour** - Constant and repetitive actions, such as vocalisations, grooming, walking or weaving, which would otherwise not be seen normally in the species.

**Stress** - An individual under pressure or tension.

**Subspecies** - The rank below the species level.

**Synonym** - Each of two or more scientific names of the same rank used to denote the same taxon.

**Tachycardia** - Abnormally rapid heart rate.

**Tachypnoea** - Very rapid respiration that is shallow in depth.

**Temporal** - Lasting or existing only for a time.

**Torpor** - A state of physical and physiological inactivity, especially in excessive heat or cold. Usually involves daily torpor with minimum body temperatures that are metabolically defended (body temperature generally ranges from 11°C to 28°C). Compare with hibernation.

**Toxoplasmosis** - A contagious disease of all species caused by the sporozoan parasite *Toxoplasma gondii.* It can cause pneumonia and central nervous system disease.

**Tracheitis** - Inflammation of the trachea, characterised by cough, pain and coughing on compression of the trachea, and in severe cases with obstruction of the airway.

**Tremor** - A continuous repetitive twitching of skeletal muscle, usually palpable and visible. Tremor is also a sign in many other diseases of the nervous system.

**Trocar** - A sharp-pointed rod-like instrument used with a cannula to puncture the wall of a body cavity and withdraw fluid or gas.

**Tumour** - A growth of tissue in which cell multiplication is uncontrolled and progressive. A cancer.

**Tyzzer's disease** - A fatal necrotising hepatitis caused by the bacteria *Bacillis piliformis.* Signs are severe diarrhoea, jaundice and high serum levels of liver enzymes.

**Uveitis** - Inflammation of the uvea (the iris, ciliary and choroid together).

**Vibrissae** -Hairs growing on the skin that is reflected into nostrils and on the skin about the nose (muzzle) of an animal. Whiskers.

**Weaning** - The act of separating the young from the dam that it has been suckling, or receiving a milk diet provided by the dam or from artificial sources.

**Yolk sac placenta** - Less invasive placenta that does not establish contact with the maternal tissues, but the allantois and fetus are more completely enclosed in the folds of the yolk sac than in the chorioallantoic placenta..

**Zeitgeber** - A rhythmically occurring event, especially in the environment, which acts as a cue in the regulation of an organism's circadian rhythm.

# Appendix 2: Suppliers and wildlife agencies

## COMPUTER SOFTWARE

**Conservation Planning Specialist Group**
Species Survival Commission, IUCN – The World Conservation Union
12101 Johnny Cake Ridge Road, Apple Valley, MN 55124, USA
Email: office@cpsg.org
Website: www.cpsg.org
- Vortex – Population Viability Analysis program

**Zoological Information Management System (ZIMS)**
7900 International Drive, Suite 300, Bloomington, MN 55425, USA
Ph: +1 952 223 8030
Email: support@species360.org
Website: https://www.species360.org
- Zoological Information Management System (ZIMS) software

## IDENTIFICATION EQUIPMENT

**4Tags**
PO Box 2002, New Farm, Qld 4005, Australia
12/76 Doggett Street, Newstead, Qld 4006, Australia
3/55 Currumbin Cree Road, Currumbin Waters, Qld 4223, Australia
Ph: 1300 813 500
Email: sales@4tags.com.au
Website: www.4tags.com.au
- Leader tags and applicators (for koalas)

**Avian ID**
Unit 3, Tescan Units, Pool Business Park, Wilson Way, Redruth, TR15 3RX, UK
Ph: +44 (0) 1209 212775
Email: info@avianid.co.uk
Website: https://www.avianid.co.uk/
- Plastic split-rings
- Aluminium rings

**Animal Electronics I.D. Systems**
PO Box 6804, Baulkham Hills, NSW 2153, Australia
Ph: +61 2 9899 7101
Email: lifechip@digivet.com.au
Website: www.animal-id.com.au
- Implant chips and scanners

**Australian Bird and Bat Banding Database**
GPO Box 8, Canberra, ACT 2601, Australia
Ph: +61 2 6274 2407
Email: abbbs@environment.gov.au
Website: www.environment.gov.au/science/bird-and-bat-banding
- Bat bands and information on the scheme

**Ball-Chain Manufacturing Co. Inc.**
741 South Fulton Avenue, Mount Vernon, NY 10550, USA
Ph: +1 914 664 7500
Email: contact@ballchain.com
Website: www.ballchain.com
- Ball-chains and connectors (stainless steel)

**Central Animal Records**
22 Fiveways Boulevard, Keysborough, Vic. 3173, Australia
Ph: +61 3 9706 3187
Email: info@car.com.au
Website: www.car.com.au
- PIT tags and scanners

**Destron Fearing**
2805 East 14th Street, Dallas/Fort Worth Airport, TX 75261, USA
Ph: +1 800 328 0118
Email: customerservice@destronfearing.com
Website: www.destronfearing.com
- PIT tags and scanners

**DLC Aust Pty Ltd**
17–19 Horne Street, Hoppers Crossing, Vic. 3029, Australia
Ph: +61 3 9360 9700
Email: support@dlc.com.au
Website: www.dlc.com.au
- Tattoo equipment

**Gey Band and Tag Co.**
2940 Felton Road, Norristown, PA 19401, USA
Ph: +1 610 277 3280

Website: www.industrynet.com/listing/1146346/gey-band-tag-co-inc
- Lipped bat bands (aluminium)
- Butt-end bird bands (monel, aluminium and stainless steel)
- Closing pliers
- Ball-chains and connectors (stainless steel)
- Ear tags (monel)

**Microchips Australia Pty Ltd**
22 Fiveways Boulevard, Keysborough, Vic. 3173, Australia
Phone: +61 3 9706 3165
Email: info@microchips.com.au
Website: https://microchips.com.au/
- Trovan Microchip details

**MS Schippers**
Smaragdweg 60, 5527 LB HAPERT, The Netherlands
Ph: +31 (0) 497 339 774
Email: export@msschippers.com
Website: www.msschippers.com/twintag-ear-tag-printed-both-sides-0403134RANGE.html
- TwinTag ear tags (for koalas)

**National Band and Tag Company**
PO Box 72430 or 721 York St
Newport, KY 41072-0430, USA
Ph: +1 859 261 2035
Email: tags@nationalband.com
Website: www.nationalband.com
- Spiral leg bands
- Butt-end bird bands and applicators
- Ear tags and applicators

**PAA Puppy ID Collars**
Available at: All Things Canine
Email: info@allthingscanine.com.au
Website: www.allthingscanine.com.au/products/ppa-puppy-id-colours
- Velcro collars for bats

**Porzana**
Elms Farm, Pett Lane, Icklesham, East Sussex, TN36 4AH, UK
Ph: +44 179 722 9591
Email: porzana@bto.org
Website: http://porzana.co.uk/bat-rings
- Aluminium bat rings
- Incoloy bat rings

**Sulyn Industries Incorporated**
Advantus Corp., 12276 San Jose Blvd, Building 618, Jacksonville, FL 32223, USA
Phone: 904-482-0091
Email: crafts@advantus.com
Website: https://sulyn.com/contact.html
- Plastic glitter to identify individual wombat faeces

## FOOD PRODUCTS

**Aristopet Animal Health**
Lot 2/31 Topham Road, Smeaton Grange, NSW 2567, Australia
Ph: 1300 651 111
Website: https://aristopet.com.au
- Tick and flea spray
- Flea powder

**Australia's Own**
50–52 The Entrance Road, The Entrance, NSW 2261, Australia
Ph: 1300 888 012
Email: info@australiasown.com.au
Website: https://australiasown.com.au
- Bee pollen.

**Biolac**
PO Box 4026, Burill Lake, NSW 2539, Australia
Ph: +61 (2) 4637 6626
Ph: +61 (0) 406 501 595
Email: biolac@optusnet.com.au
Website: www.biolac.com.au
- Biolac milk for marsupials
- Flying fox formula
- Marsupial feeding teats, bottles and syringes
- Pouches and baskets
- Heat pads

**Darling Bee Products**
325 Anthony Place, Sawyers Valley, WA 6074, Australia
Ph: +61 8 9295 1534
Ph: +61 (0) 439 939 012
Email: darlingbeehoney@bigpond.com
Website: www.darlingbeeproducts.com
- Banksia honey

**Di-Vetelact**
PO Box 496, Vaucluse, NSW 2030, Australia
Ph: +61 1800 336 547

Website: www.divetelact.com
- Di-Vetelact®: low lactose animal milk formula

**Eukanuba**
Ph: +61 1300 657 021
Website: www.eukanuba.com/au
- Eukanuba® dog food

**Koala Browse Inc.**
9528 87th Place South, Boynton Beach, FL 33437, USA
Ph: +1 (561) 436 0577
Email: kevin@kbrowse.com
Website: www.kbrowse.com/
- Providers of eucalyptus foliage

**Mac Foliage**
Knapmore Hill, Lounston, Bickington, Newton Abbott, Devon TQ12 6LB, UK
Ph: +44 (0)77 8529 3518
Website: www.macfoliage.co.uk
- Providers of eucalyptus foliage

**Mavlab Animal Health**
PO Box 3186, Logan City DC, Qld 4114, Australia
Phone: +61 7 3808 1399
Website: www.mavlab.com.au/
- Calcium carbonate powder

**Mazuri Zoo Foods**
PO Box 705, Witham Essex CM8 3AD, UK
Ph: +44 (0) 1376 511260
Email: info@mazurizoofoods.com
Website: www.mazurizoofoods.com
- Mazuri food

**Mazuri Exotic Animal Nutrition**
PO Box 66812, St Louis, MO 63166, USA
Ph: +1 833 462 9874
Email: info@mazuri.com
Website: www.mazuri.com
- Mazuri high-calcium diet
- Mazuri primate pellets

**Mead-Johnson Nutrition**
2400 West Lloyd Expressway, Evansville, IN 47721, USA
Ph: +1 812 429 5000
Email: orders@mjn.com
Website: www.meadjohnson.com/company
- Portagen®

**The Missing Link Products®**
Ph: +1 (800) 628-9653
Email: customerservice@missinglinkproducts.com
Website: https://missinglinkproducts.com/
- The Missing Link® Skin & Coat Supplement Powder for Dogs

**Nestle®Australia Ltd**
Building D, 1 Homebush Bay Drive, Rhodes, NSW 2138, Australia
Ph: +61 2 8756 2000
Website: www.nestle.com.au/en
Nestle® USA Inc.
800 North Brand Blvd, Glendale, CA 91203, USA
Ph: +1 800 225 2270
Website: www.nestleusa.com
- Nan Supreme

**Pentavite**
74–86 Garden Road, Clayton, Vic. 3168, Australia
Ph: +61 3 9828 9400
Website: https://www.pentavite.com/
- Multivitamin + iron liquid

**Riverina Australia Pty Ltd**
PO Box 1146, Milton, Qld 4064, Australia
Ph: +61 7 3007 6888
Website: www.riverina.com.au
- Macropod pellets

**Specialty Feeds**
3150 Great Eastern Hwy, Glen Forrest, WA 6071, Australia
Ph: +61 8 9298 8111
Email: info@specialtyfeeds.com
Website: www.specialtyfeeds.com
- Kangaroo cubes
- Quokka cubes

**The Gold Leaf Factory International**
26 Moresby Avenue, Seaford, Vic. 3198, Australia
Ph: +61 3 9786 2247
Website: https://www.goldleaf.com.au
- Gum Arabic powder (food grade)

**Vetafarm**
3 Bye Street, Wagga Wagga, NSW 2650, Australia
Ph: +61 2 6933 0400

Email: sales@vetafarm.com.au
Website: www.vetafarm.com.au
- Poly-aid Plus: energy drink for sick/injured wildlife.
- Soluvet: water-soluble vitamin supplement.

**Veterinary Nutrition Solutions**
Ph: +61 2 8011 4291 or +61 466 941 637
Email: contact@vetnutritionsolutions.com.au
Website: https://www.vetnutritionsolutions.com.au/
Website: https://emeraid.com/vet/emeraid-herbivore
- EmerAid Intensive Care Herbivore®

**Wombaroo Passwell Food Products**
PO Box 151, Glen Osmond, SA 5064, Australia
Email: wombaroo@adelaide.on.net
Website: www.wombaroo.com.au
- Echidna Milk Replacer - Early Lactation and Late Lactation
- Echidna Food
- Small Carnivore Food: for insectivorous and carnivorous mammals and bandicoots
- Koala Milk Replacer - Early Lactation and Late Lactation
- Wombat Milk Replacer - <0.4, 0.4 and >0.6
- Possum Milk Replacer - <0.8 and >0.8
- Kangaroo Milk Replacer - <0.4, 0.4, 0.6, >0.7
- Kangaroo Pellets
- Flying Fox Milk Replacer
- Bat Milk Replacer: for insectivorous bats
- High Protein Supplement
- Impact Colostrum Supplement
- Insect Booster: for insects such as crickets, woodies & mealworms
- Teats
- Feeding bottles
- Cosy heat pads
- Latex teats:

C - Carnivorous marsupials
F - Flying foxes
FM - Out of pouch kangaroos, wombats and koalas
LD - Possums, wombats and koalas
MTM - In pouch kangaroos, wallabies and koalas
P - Small, difficult to feed animals
SD - Possums and gliders
STM - Small in pouch kangaroos, wallabies and possums
TM - Out of pouch kangaroos and wallabies

## CATCHING AND HANDLING EQUIPMENT

**ACES Animal Care Equipment and Services**
Unit 9, 59-65 Keys Road
PO Box 591, Moorabbin, Vic. 3189, Australia
Ph: +61 3 9532 6069
Email: aces@animal-care.com.au
Website: https://animal-care.com.au/
- Ketch poles, gloves, nets, traps

**Fuhrman Diversified Inc.**
2912 Bayport Boulevard, Seabrook, TX 77586, USA
Ph: +1 281 470 8397
Email: fdi@flash.net
Website: www.fieldcam.com
- Flexi Nets and other equipment for catching animals

**Wiretainers Pty Ltd**
(Mascott Wireworks was bought out by Wiretainers Pty Ltd)
5-9 Fink Street, Preston, Vic. 3072, Australia
Ph: +61 3 9480 1200
Email: sales@wiretainers.com.au
Website: http://www.wiretainers.com.au
- Cage traps

## SUBSTRATE

**Fibrecycle**
Unit 1/18 Millennium Circuit, Helensvale, Qld 4212, Australia
Ph: +61 7 3382 0711
Email: info@fibrecycle.com
Website: https://fibrecycle.com/contact-us
- Kitty litter

**Solite Corporation**
962 Kings Highway, Saugerties, NY 12477, USA
Ph: +1 (845) 246-2177
Email: solitenes@nesolite.com
Website: https://northeastsolite.com/
- Lightweight aggregate concrete

## CLEANING AGENTS

**Chemical Essentials**
PO Box 1369, Mitcham North, Vic. 3132, Australia
Ph: +61 3 9761 6382
Email: chemicalessentials@bigpond.com
Website: www.chemicalessentials.com.au

**F10 cleaner**
Milton
Ph: +61 3 8586 0500

Email: info@nicepak.com.au
Website: www.miltonbaby.com.au
Milton®: sterilising milk feeding apparatus

## STATE AND OTHER WILDLIFE AGENCIES

**Department of Climate Change, Energy, the Environment and Water**
John Gorton Building, corner King Edward Terrace and Parkes Place
Parkes, ACT 2600, Australia
GPO Box 3090, Canberra, ACT 2601, Australia
Ph: +61 1800 920 528
Website: www.dcceew.gov.au/

**New South Wales**
Department of Planning and the Environment
43 Bridge St
PO Box 1967, Hurstville, NSW 2220, Australia
Ph: +61 2 9585 6444
Ph: +61 2 9585 6406
Email: info@environment.nsw.gov.au
Email: wildlife.licensing@environment.nsw.gov.au
Website: www.environment.nsw.gov.au/

**Northern Territory**
Department of Environment, Parks and Water Security
25 Chung Wah Terrace, Palmerston, NT 0830, Australia
Ph: +61 8 8999 5511
Website: https://depws.nt.gov.au/

**Queensland**
Queensland Department of Environment and Service
GPO Box 2454, Brisbane, Qld 4001, Australia,
Ph: 13 74 68
Website: https://parks.des.qld.gov.au/

**South Australia**
Department for Environment and Water
GPO Box 1047, Adelaide, SA 5001, Australia
Ph: +61 8 8204 1910
Email: DEWHeritage@sa.gov.au
Website: www.environment.sa.gov.au

**Tasmania**
Parks and Wildlife Service, Tasmania
GPO Box 1751, Hobart, Tas. 7001, Australia.
Ph: 1300 827 727
Email: webenquiries@parks.tas.gov.au
Website: https://parks.tas.gov.au/

**Victoria**
Department of Energy, Environment and Climate Action
8 Nicholson Street, East Melbourne, Vic. 3002
PO Box 500, Melbourne, Vic. 8002, Australia
Ph: 136 186
Email: customer.service@delwp.vic.gov.au
Website: www.deeca.vic.gov.au/

**Western Australia**
Department of Biodiversity, Conservation and Attractions
Locked Bag 104, Bentley Delivery Centre, WA 6983, Australia
Ph: +61 8 9442 0300
Email: info@conservation.wa.gov.au
Website: www.dbca.wa.gov.au

**Dingo Sanctuary Bargo**
3105A Remembrance Drive, Bargo, NSW 2574, Australia
Email: enquiries@dingosanctuary.org.au
Website: https://dingosanctuary.org.au/

**Ku-ring-gai Bat Conservation Society Inc.**
PO Box 607, Gordon, NSW 2072, Australia
Email: web@sydneybats.org.au
Website: https://sydneybats.org.au/

**Marsupial Society of Australia**
Email: marsupialsocietyinc@gmail.com
Website: www.marsupialsaustralia.org/

# Appendix 3: Taking body measurements

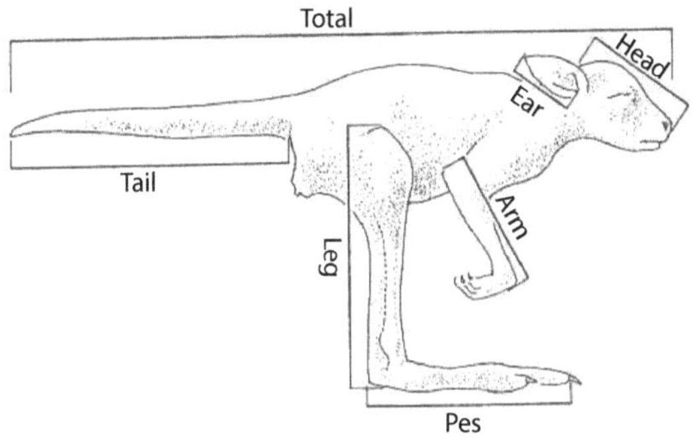


Recommended measurements for young mammals. Taken from Poole *et al.* (1982) with permission from the publisher.

Crown-rump and head length for very small marsupials. Derived from Taplin (1980) with permission from the editor.

## REFERENCES

Poole WE, Carpenter SM, Wood JT (1982) Growth of grey kangaroos and the reliability of age determination from body measurements. I. The eastern grey kangaroo, Macropus giganteus. *Australian Wildlife Research* **9**, 9-20. doi:10.1071/WR9820009

Taplin LE (1980) Some observations on the reproductive biology of Sminthopsis virginiae (Tarragon) (Marsupialia: Dasyuridae). *Australian Zoologist* **20**, 407-418.

# Index

Kolibri 360 GmbH
Ferdinand-Jühlke-Straße 7
99095 Erfurt, Deutschland
produktsicherheit@kolibri360.de